able of Problem-Solving Strategies

ote for users of the five-volume edition:

hapters 37–42 are not in the Standard Edition.

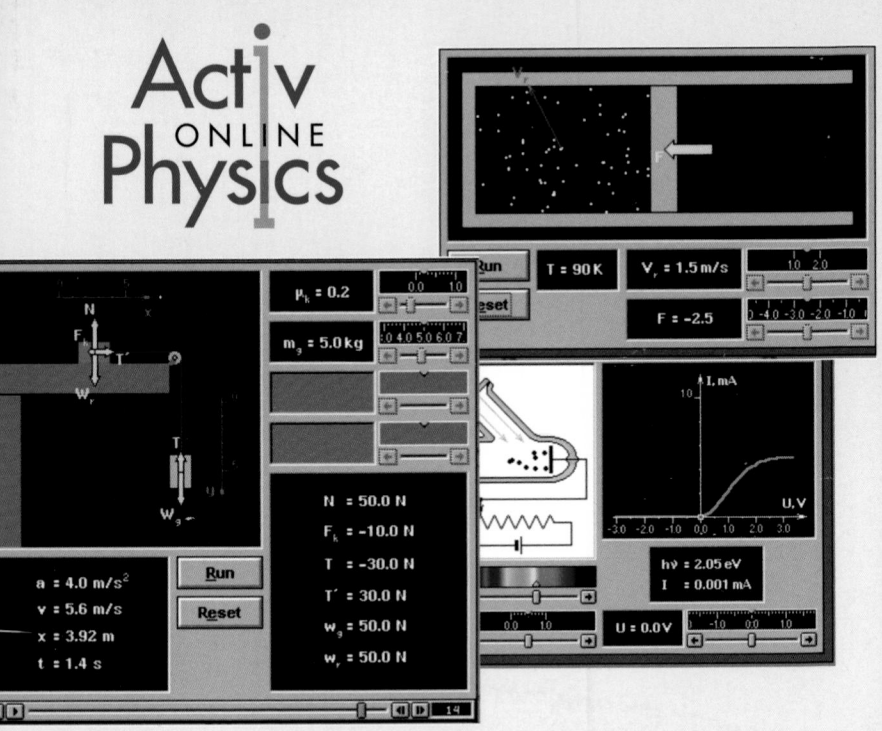

ActivPhysics™ OnLine utilizes visualization, simulation and multiple representations to help you better understand key physical processes, experiment quantitatively, and develop your critical-thinking skills. This library of online interactive simulations are coupled with thought-provoking questions and activities to guide your understanding of physics.

Website: www.aw-bc.com/knight

Minimum System Requirements:
Windows: 250 MHz; OS 98, NT, ME, 2000, XP
Macintosh: 233 MHz; OS 9.2, 10
Both:
• 64 RAM installed
• 1024 x 768 screen resolution
• Browsers: Internet Explorer 5.5, 6.0; Netscape 6.2.3, 7.0
• Plug Ins: Macromedia's Flash 6.2.3

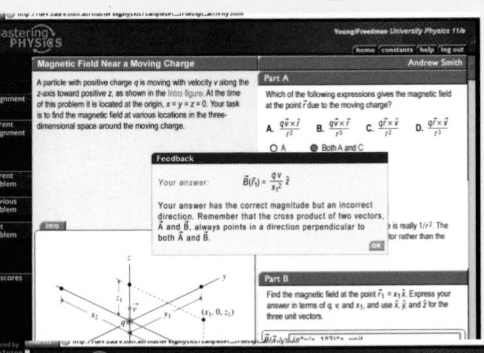

MasteringPhysics™ is the first Socratic tutoring system developed specifically for physics students like you. It is the result of years of detailed studies of how students work physics problems, and where they get stuck and need help. Studies show students who used MasteringPhysics significantly improved their scores on traditional final exams and the Force Concept Inventory (a conceptual test) when compared with traditional hand-graded homework.

With your purchase of a new copy of Knight's *Physics for Scientists and Engineers*, you should have received a Student Access Kit for **MasteringPhysics™** if your professor required it as a component of your course. The kit contains instructions and a code for you to access MasteringPhysics.

If you did not purchase a new textbook and your professor requires you to enroll in the **MasteringPhysics** online homework and tutorial program, you may purchase an online subscription with a major credit card. Go to www.masteringphysics.com and follow the links to purchasing online.

Minimum System Requirements:
Windows: 250 MHz; OS 98, NT, ME, 2000, XP
Macintosh: 233 MHz; OS 9.2, 10
RedHat Linux 8.0
All:
• 64 RAM installed
• 1024 x 768 screen resolution
• Browsers: Internet Explorer 5.0, 5.5, 6.0; Netscape 6.2.3, 7.0; Mozilla 1.2, 1.3

MasteringPhysics™ is powered by MyCyberTutor by Effective Educational Technologies

STUDENT ACCESS KIT

Physics for Scientists and Engineers

A Strategic Approach

ActivPhysics™ OnLine Activities

Activ ONLINE Physics   www.aw-bc.com/knight

Physics for Scientists and Engineers

with Modern Physics

A Strategic Approach

Randall D. Knight

California Polytechnic State University, San Luis Obispo

PEARSON

Addison
Wesley

San Francisco   Boston   New York
Cape Town   Hong Kong   London   Madrid   Mexico City
Montreal   Munich   Paris   Singapore   Sydney   Tokyo   Toronto

Executive Editor:	Adam Black, Ph.D.
Development Editor:	Alice Houston, Ph.D.
Project Manager:	Laura Kenney Editorial & Production Services
Associate Editor:	Liana Allday
Media Producer:	Claire Masson
Marketing Manager:	Christy Lawrence
Market Development:	Susan Winslow
Manufacturing Supervisor:	Vivian McDougal
Art Director:	Blakely Kim
Production Service:	Thompson Steele, Inc.
Text Design:	Mark Ong, Side by Side Studios
Cover Design:	Yvo Riezebos Design
Illustrations:	Precision Graphics
Photo Research:	Cypress Integrated Systems
Cover Printer:	Phoenix Color Corporation
Printer and Binder:	R. R. Donnelley & Sons
Cover Image:	Rainbow/PictureQuest
Credits:	see page C–1

Library of Congress Cataloging-in-Publication Data
Knight, Randall Dewey.
    Physics for scientists and engineers : a strategic approach / Randall D. Knight.
        p. cm.
    Includes index.
    ISBN 0-8053-8960-1 (extended ed. with MasteringPhysics)
      1. Physics   I. Title.

    QC23.2.K65 2004
    530--dc22

2003062809

ISBN 0-8053-8960-1 Extended Edition with MasteringPhysics
ISBN 0-8053-9006-5 Extended Edition without MasteringPhysics

5 6 7 8 9 10—DOW—06 05
www.aw-bc.com

Brief Contents

About the Author

Randy Knight has taught introductory physics for over 20 years at Ohio State University and California Polytechnic University, where he is currently Professor of Physics. Professor Knight received a bachelor's degree in physics from Washington University in St. Louis and a Ph.D. in physics from the University of California, Berkeley. He was a post-doctoral fellow at the Harvard-Smithsonian Center for Astrophysics before joining the faculty at Ohio State University. It was at Ohio State that he began to learn about the research in physics education that, many years later, led to this book.

Professor Knight's research interests are in the field of lasers and spectroscopy, and he has published over 25 research papers. He recently led the effort to establish an environmental studies program at Cal Poly, where, in addition to teaching introductory physics, he also teaches classes on energy, oceanography, and environmental issues. When he's not in the classroom or in front of a computer, you can find Randy hiking, sea kayaking, playing the piano, or spending time with his wife Sally and their seven cats.

Preface to the Instructor

In 1997 we published *Physics: A Contemporary Perspective.* This was the first comprehensive, calculus-based textbook to make extensive use of results from physics education research. The development and testing that led to this book had been partially funded by the National Science Foundation. In the preface we noted that it was a "work in progress" and that we very much wanted to hear from users—both instructors and students—to help us shape the book into a final form.

And hear from you we did! We received feedback and reviews from roughly 150 professors and, especially important, 4500 of their students. This textbook, the newly titled *Physics for Scientists and Engineers: A Strategic Approach*, is the result of synthesizing that feedback and using it to produce a book that we hope is uniquely tuned to helping today's students succeed. It is the first introductory textbook built from the ground up on research into how students can more effectively learn physics.

Objectives

My primary goals in writing *Physics for Scientists and Engineers: A Strategic Approach* have been:

- To produce a textbook that is more focused and coherent, less encyclopedic.
- To move key results from physics education research into the classroom in a way that allows instructors to use a range of teaching styles.
- To provide a balance of quantitative reasoning and conceptual understanding, with special attention to concepts known to cause student difficulties.
- To develop students' problem-solving skills in a systematic manner.
- To support an active-learning environment.

These goals and the rationale behind them are discussed at length in my small paperback book, *Five Easy Lessons: Strategies for Successful Physics Teaching* (Addison Wesley, 2002). Please request a copy from your local Addison Wesley sales representative if it would be of interest to you (ISBN 0-8053-8702-1).

FIVE EASY LESSONS

Strategies for Successful
Physics Teaching

RANDALL D. KNIGHT

Textbook Organization

The 42-chapter extended edition (ISBN 0-8053-8685-8) of *Physics for Scientists and Engineers* is intended for use in a three-semester course. Most of the 36-chapter standard edition (ISBN 0-8053-8982-2), ending with relativity, can be covered in two semesters, but the judicious omission of a few chapters will avoid rushing through the material and give students more time to develop their knowledge and skills.

There's a growing sentiment that quantum physics is quickly becoming the province of engineers, not just scientists, and that even a two–semester course should include a reasonable introduction to quantum ideas. The *Instructor's Guide* outlines a couple of routes through the book that allow most of the quantum physics chapters to be reached in two semesters. I've written the book with the hope that an increasing number of instructors will choose one of these routes.

- **Extended edition,** with modern physics (ISBN 0-8053-8685-8): chapters 1–42.
- **Standard edition** (ISBN 0-8053-8982-2): chapters 1–36.
- **Volume 1** (ISBN 0-8053-8963-6) covers mechanics: chapters 1–15.
- **Volume 2** (ISBN 0-8053-8966-0) covers thermodynamics: chapters 16–19.
- **Volume 3** (ISBN 0-8053-8969-5) covers waves and optics: chapters 20–24.
- **Volume 4** (ISBN 0-8053-8972-5) covers electricity and magnetism, plus relativity: chapters 25–36.
- **Volume 5** (ISBN 0-8053-8975-X) covers relativity and quantum physics: chapters 36–42.
- **Volumes 1–5** boxed set (ISBN 0-8053-8978-4).

The full textbook is divided into seven parts: Part I: *Newton's Laws*, Part II: *Conservation Laws*, Part III: *Applications of Newtonian Mechanics*, Part IV: *Thermodynamics*, Part V: *Waves and Optics*, Part VI: *Electricity and Magnetism*, and Part VII: *Relativity and Quantum Mechanics*. Although I recommend covering the parts in this order (see below), doing so is by no means essential. Each topic is self-contained, and Parts III–VI can be rearranged to suit an instructor's needs. To facilitate a reordering of topics, the full text is available in the five individual volumes listed in the margin.

Organization Rationale: Thermodynamics is placed before waves because it is a continuation of ideas from mechanics. The key idea in thermodynamics is energy, and moving from mechanics into thermodynamics allows the uninterrupted development of this important idea. Further, waves introduce students to functions of two variables, and the mathematics of waves is more akin to electricity and magnetism than to mechanics. Thus moving from waves to fields to quantum physics provides a gradual transition of ideas and skills.

The purpose of placing optics with waves is to provide a coherent presentation of wave physics, one of the two pillars of classical physics. Optics as it is presented in introductory physics makes no use of the properties of electromagnetic fields. There's little reason other than historical tradition to delay optics until after E&M. The documented difficulties that students have with optics are difficulties with waves, not difficulties with electricity and magnetism. However, the optics chapters are easily deferred until the end of Part VI for instructors who prefer that ordering of topics.

More Effective Problem-Solving Instruction

Careful and systematic instruction is provided on all aspects of problem solving. Some of the features that support this approach are described here, and more details are provided in the *Instructor's Guide*.

- An emphasis on using *multiple representations*—descriptions in words, pictures, graphs, and mathematics—to look at a problem from many perspectives.
- The explicit use of *models*, such as the particle model, the wave model, and the field model, to help students recognize and isolate the essential features of a physical process.
- TACTICS BOXES for the development of particular skills, such as drawing a free-body diagram or using Lenz's law. Tactics Box steps are explicitly illustrated in subsequent worked examples, and these are often the starting point of a full problem-solving strategy.

TACTICS BOX 4.3   **Drawing a free-body diagram**

➊ **Identify all forces acting on the object.** This step was described in Tactics Box 4.2.

➋ **Draw a coordinate system.** Use the axes defined in your pictorial representation. If those axes are tilted, for motion along an incline, then the axes of the free-body diagram should be similarly tilted.

➌ **Represent the object as a dot at the origin of the coordinate axes.** This is the particle model.

➍ **Draw vectors representing each of the identified forces.** This was described in Tactics Box 4.1. Be sure to label each force vector.

➎ **Draw and label the *net force* vector $\vec{F}_{net}$.** Draw this vector beside the diagram, not on the particle. Or, if appropriate, write $\vec{F}_{net} = \vec{0}$. Then check that $\vec{F}_{net}$ points in the same direction as the acceleration vector $\vec{a}$ on your motion diagram.

TACTICS BOX 32.2   **Evaluating line integrals**

➊ If $\vec{B}$ is everywhere perpendicular to a line, the line integral of $\vec{B}$ is

$$\int_i^f \vec{B} \cdot d\vec{s} = 0$$

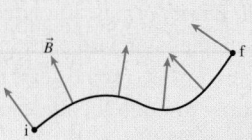

➋ If $\vec{B}$ is everywhere tangent to a line of length L *and* has the same magnitude B at every point, the line integral of $\vec{B}$ is

$$\int_i^f \vec{B} \cdot d\vec{s} = BL$$

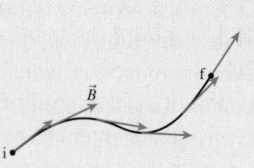

■ PROBLEM-SOLVING STRATEGIES that help students develop confidence and more proficient problem-solving skills through the use of a consistent four-step approach: MODEL, VISUALIZE, SOLVE, ASSESS. Strategies are provided for each broad class of problems, such as dynamics problems or problems involving electromagnetic induction. The icon directs students to the specially developed *Skill Builder* tutorial problems in MasteringPhysics™ (see page xi), where they can interactively work through each of these strategies online.

■ Worked EXAMPLES that illustrate good problem-solving practices through the consistent use of the four-step problem-solving approach and, where appropriate, the Tactics Box steps. The worked examples are often very detailed and carefully lead the student step by step through the *reasoning* behind the solution, not just through the numerical calculations. Steps that are often implicit or omitted in other textbooks, because they seem so obvious to experts, are explicitly discussed since research has shown these are often the points where students become confused.

■ NOTE ► Paragraphs within worked examples caution against common mistakes and point out useful tips for tackling problems.

■ The *Student Workbook* (see page xi), a unique component of this text, bridges the gap between worked examples and end-of-chapter problems. It provides qualitative problems and exercises that focus on developing the skills and conceptual understanding necessary to solve problems with confidence.

■ Approximately 3000 original and diverse *end-of-chapter problems* have been carefully crafted to exercise and test the full range of qualitative and quantitative problem-solving skills. *Exercises*, which are keyed to specific sections, allow students to practice basic skills and computations. *Problems* require a better understanding of the material and often draw upon multiple representations of knowledge. *Challenge Problems* are more likely to use calculus, utilize ideas from more than one chapter, and sometimes lead students to explore topics that weren't explicitly covered in the chapter.

PROBLEM-SOLVING STRATEGY 5.2 **Dynamics problems**

MODEL Make simplifying assumptions.

VISUALIZE

Pictorial representation. Show important points in the motion with a sketch, establish a coordinate system, define symbols, and identify what the problem is trying to find. This is the process of translating words to symbols.

Physical representation. Use a motion diagram to determine the object's acceleration vector $\vec{a}$. Then identify all forces acting on the object and show them on a free-body diagram.

It's OK to go back and forth between these two steps as you visualize the situation.

SOLVE The mathematical representation is based on Newton's second law

$$\vec{F}_{net} = \sum_i \vec{F}_i = m\vec{a}$$

The vector sum of the forces is found directly from the free-body diagram. Depending on the problem, either

■ Solve for the acceleration, then use kinematics to find velocities and positions, or
■ Use kinematics to determine the acceleration, then solve for unknown forces.

ASSESS Check that your result has the correct units, is reasonable, and answers the question.

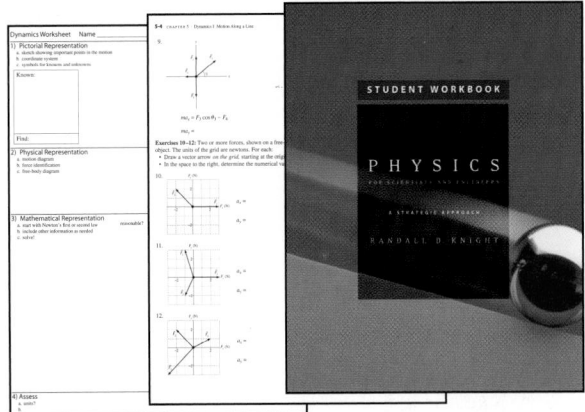

Proven Features to Promote Deeper Understanding

Research has shown that many students taking calculus-based physics arrive with a wealth of misconceptions and subsequently struggle to develop a coherent understanding of the subject. Using a number of unique, reinforcing techniques, this book tackles these issues head-on to enable students to build a solid foundation of understanding.

■ A *concrete-to-abstract* approach introduces new concepts through observations about the real world and everyday experience. Step by step, the text then builds up the concepts and principles needed by a theory that will make sense of the observations and make new, testable predictions. This inductive approach better matches how students learn, and it reinforces how physics—and science in general—operates.

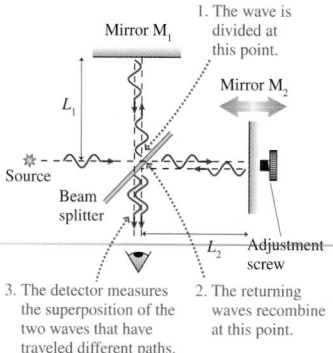

1. The wave is divided at this point.

Mirror M₁

Mirror M₂

L_1

Source

Beam splitter

L_2 Adjustment screw

3. The detector measures the superposition of the two waves that have traveled different paths.

2. The returning waves recombine at this point.

Annotated **FIGURE** showing the operation of the Michelson interferometer.

- **STOP TO THINK** questions embedded in each chapter allow students to assess whether they've understood the main idea of a section. The *Stop to Think* questions, which include concept questions, ratio reasoning, and ranking tasks, are primarily derived from physics education research.

- **NOTE ▶** paragraphs draw attention to common misconceptions, clarify possible confusions in terminology and notation, and provide important links to previous topics.

- Unique *annotated figures*, based on research into visual learning modes, make the artwork a teaching tool on a par with the written text. Commentary in blue—the "instructor's voice"—helps students "read" the figure. Students "learn by viewing" how to interpret a graph, how to translate between multiple representations, how to grasp a difficult concept through a visual analogy, and many other important skills.

- The learning goals and links that begin each chapter outline what the student needs to remember from previous chapters and what to focus on in the chapter ahead.

  - **▶ Looking Ahead** lists key concepts and skills the student will learn in the coming chapter.

  - **◀ Looking Back** suggests important topics students should review from previous chapters.

- Unique schematic *Chapter Summaries* help students organize their knowledge in an expert-like hierarchy, from general principles (top) to applications (bottom). Side-by-side pictorial, graphical, textual, and mathematical representations are used to help students with different learning styles and enable them to better translate between these key representations.

- *Part Overviews and Summaries* provide a global framework for the student's learning. Each part begins with an overview of the chapters ahead. It then concludes with a broad summary to help students draw connections between the concepts presented in that set of chapters. **KNOWLEDGE STRUCTURE** tables in the part summaries, similar to the chapter summaries, help students see a forest rather than dozens of individual trees.

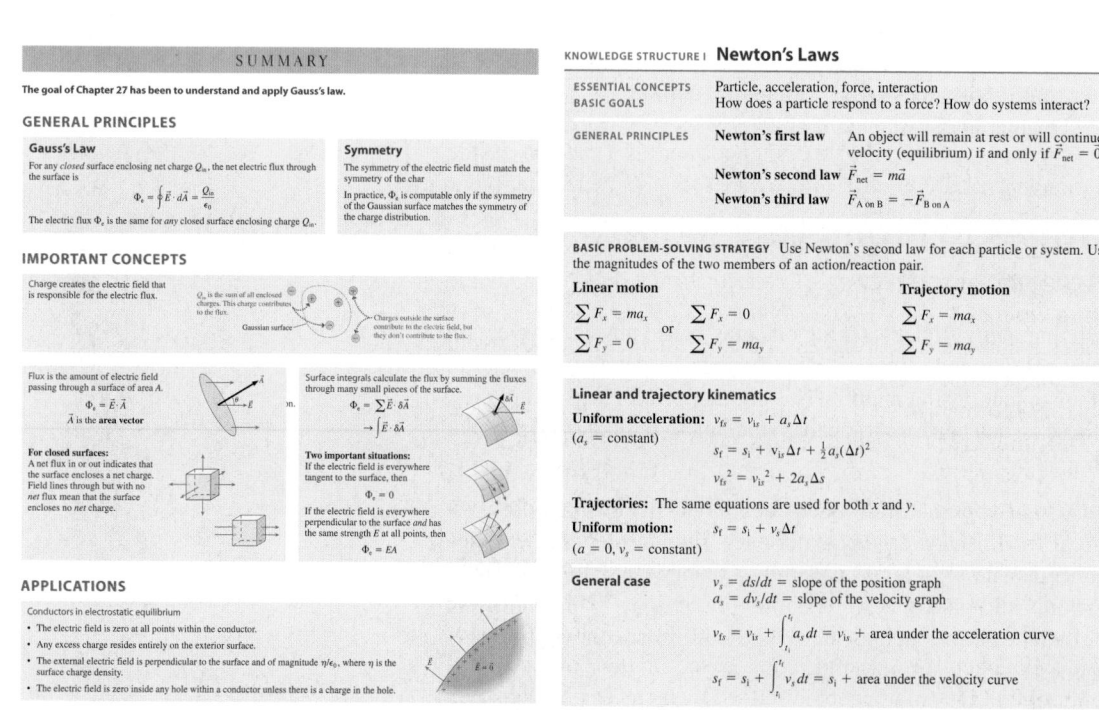

SUMMARY

The goal of Chapter 27 has been to understand and apply Gauss's law.

GENERAL PRINCIPLES

Gauss's Law

For any *closed* surface enclosing net charge Q_{in}, the net electric flux through the surface is

$$\Phi_e = \oint \vec{E} \cdot d\vec{A} = \frac{Q_{in}}{\epsilon_0}$$

The electric flux Φ_e is the same for *any* closed surface enclosing charge Q_{in}.

Symmetry

The symmetry of the electric field must match the symmetry of the charge.

In practice, Φ_e is computable only if the symmetry of the Gaussian surface matches the symmetry of the charge distribution.

IMPORTANT CONCEPTS

Charge creates the electric field that is responsible for the electric flux.

Q_{in} is the sum of all enclosed charges. This charge contributes to the flux.

Charges outside the surface contribute to the electric field, but they don't contribute to the flux.

Gaussian surface

Flux is the amount of electric field passing through a surface of area A.

$$\Phi_e = \vec{E} \cdot \vec{A}$$

$\vec{A}$ is the **area vector**

Surface integrals calculate the flux by summing the fluxes through many small pieces of the surface.

$$\Phi_e = \sum \vec{E} \cdot \delta \vec{A}$$
$$\rightarrow \int \vec{E} \cdot \delta \vec{A}$$

For closed surfaces:
A net flux in or out indicates that the surface encloses a net charge. Field lines through but with no *net* flux mean that the surface encloses no *net* charge.

Two important situations:
If the electric field is everywhere tangent to the surface, then

$$\Phi_e = 0$$

If the electric field is everywhere perpendicular to the surface *and* has the same strength E at all points, then

$$\Phi_e = EA$$

APPLICATIONS

Conductors in electrostatic equilibrium

- The electric field is zero at all points within the conductor.
- Any excess charge resides entirely on the exterior surface.
- The external electric field is perpendicular to the surface and of magnitude η/ϵ_0, where η is the surface charge density.
- The electric field is zero inside any hole within a conductor unless there is a charge in the hole.

KNOWLEDGE STRUCTURE I **Newton's Laws**

ESSENTIAL CONCEPTS	Particle, acceleration, force, interaction
BASIC GOALS	How does a particle respond to a force? How do systems interact?

GENERAL PRINCIPLES	**Newton's first law**	An object will remain at rest or will continue to move with constant velocity (equilibrium) if and only if $\vec{F}_{net} = \vec{0}$.
	Newton's second law	$\vec{F}_{net} = m\vec{a}$
	Newton's third law	$\vec{F}_{A \text{ on } B} = -\vec{F}_{B \text{ on } A}$

BASIC PROBLEM-SOLVING STRATEGY Use Newton's second law for each particle or system. Use Newton's third law to equate the magnitudes of the two members of an action/reaction pair.

Linear motion		**Trajectory motion**	**Circular motion**
$\sum F_x = ma_x$	$\sum F_x = 0$	$\sum F_x = ma_x$	$\sum F_r = mv^2/r = m\omega^2 r$
$\sum F_y = 0$ or	$\sum F_y = ma_y$	$\sum F_y = ma_y$	$\sum F_z = 0$

Linear and trajectory kinematics	**Circular kinematics**
Uniform acceleration: $v_{fs} = v_{is} + a_s \Delta t$	**Uniform circular motion:**
(a_s = constant) $\quad s_f = s_i + v_{is} \Delta t + \frac{1}{2}a_s(\Delta t)^2$	$\theta_f = \theta_i + \omega \Delta t$
$\quad v_{fs}^2 = v_{is}^2 + 2a_s \Delta s$	$a_r = v^2/r = \omega^2 r$
Trajectories: The same equations are used for both x and y.	$v = \omega r$
Uniform motion: $s_f = s_i + v_s \Delta t$	$T = 2\pi r/v = 2\pi/\omega$
($a = 0$, v_s = constant)	

General case	$v_s = ds/dt$ = slope of the position graph
	$a_s = dv_s/dt$ = slope of the velocity graph
	$v_{fs} = v_{is} + \int_{t_i}^{t_f} a_s \, dt = v_{is}$ + area under the acceleration curve
	$s_f = s_i + \int_{t_i}^{t_f} v_s \, dt = s_i$ + area under the velocity curve

The Student Workbook

A key component of *Physics for Scientists and Engineers: A Strategic Approach* is the accompanying *Student Workbook*. The workbook bridges the gap between textbook and homework problems by providing students the opportunity to learn and practice skills prior to using those skills in quantitative end-of-chapter problems, much as a musician practices technique separately from performance pieces. The workbook exercises, which are keyed to each section of the textbook, focus on developing specific skills, ranging from identifying forces and drawing free-body diagrams to interpreting wave functions.

The workbook exercises, which are generally qualitative and/or graphical, draw heavily upon the physics education research literature. The exercises deal with issues known to cause student difficulties and employ techniques that have proven to be effective at overcoming those difficulties. The workbook exercises can be used in-class as part of an active-learning teaching strategy, in recitation sections, or as assigned homework. More information about effective use of the *Student Workbook* can be found in the *Instructor's Guide*.

Available versions: Extended (ISBN 0-8053-8961-X), Standard (ISBN 0-8053-8984-9), Volume 1 (ISBN 0-8053-8965-2), Volume 2 (ISBN 0-8053-8968-7), Volume 3 (ISBN 0-8053-8971-7), Volume 4 (ISBN 0-8053-8974-1), and Volume 5 (ISBN 0-8053-8977-6).

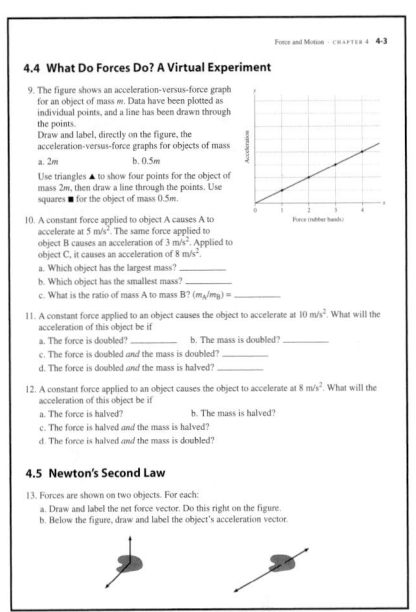

Instructor Supplements

- The **Instructor's Guide for Physics for Scientists and Engineers** (ISBN 0-8053-8985-7) offers detailed comments and suggested teaching ideas for every chapter, an extensive review of what has been learned from physics education research, and guidelines for using active-learning techniques in your classroom.

- The **Instructor's Solutions Manuals**, **Chapters 1–19** (ISBN 0-8053-8986-5), and **Chapters 20–42** (ISBN 0-8053-8989-X), written by Professors Pawan Kahol and Donald Foster, at Wichita State University, provide *complete* solutions to all the end-of-chapter problems. The solutions follow the four-step Model/Visualize/Solve/Assess procedure used in the *Problem-Solving Strategies* and all worked examples. Emphasis is placed on the reasoning behind the solution, rather than just the numerical manipulations. The full text of each solution is available as an editable Word document and as a pdf file on the *Instructor's Supplement CD-ROM* for your own use or for posting on your course website.

- The cross-platform **Instructor's Resource CD-ROMs** (ISBN 0-8053-8996-2) consists of the **Simulation and Image Presentation CD-ROM** and the **Instructor's Supplement CD-ROM**. The *Simulation and Image Presentation CD-ROM* provides a comprehensive library of more than 220 applets from *ActivPhysics OnLine*, as well as all the figures from the textbook (excluding photographs) in JPEG format. In addition, all the tables, chapter summaries, and knowledge structures are provided as JPEGs, and the Tactics Boxes, Problem-Solving Strategies, and key (boxed) equations are provided in editable Word format. The *Instructor's Supplement CD-ROM* provides editable Word versions and pdf files of the *Instructor's Guide* and the *Instructor's Solutions Manuals*. Complete *Student Workbook* solutions are also provided as pdf files.

- **MasteringPhysics**™ (www.masteringphysics.com) is a sophisticated, research-proven online tutorial and homework assignment system that provides students with individualized feedback and hints based on their input. It provides a comprehensive library of conceptual tutorials (including one for each

Problem-Solving Strategy in this textbook), multistep self-tutoring problems, and end-of-chapter problems from *Physics for Scientists and Engineers*. *MasteringPhysics*™ provides instructors with a fast and effective way to assign online homework assignments that comprise a range of problem types. The powerful post-assignment diagnostics allow instructors to assess the progress of their class as a whole or to quickly identify individual students' areas of difficulty.

- **ActivPhysics**™ **OnLine** (www.aw-bc.com/knight) provides a comprehensive library of more than 420 tried and tested *ActivPhysics* applets updated for web delivery using the latest online technologies. In addition, it provides a suite of highly regarded applet-based tutorials developed by education pioneers Professors Alan Van Heuvelen and Paul D'Alessandris. The *ActivPhysics* margin icon directs students to specific exercises that complement the textbook discussion.

  The online exercises are designed to encourage students to confront misconceptions, reason qualitatively about physical processes, experiment quantitatively, and learn to think critically. They cover all topics from mechanics to electricity and magnetism and from optics to modern physics. The highly acclaimed *ActivPhysics OnLine* companion workbooks help students work through complex concepts and understand them more clearly. More than 220 applets from the *ActivPhysics OnLine* library are also available on the *Simulation and Image Presentation CD-ROM*.

- The **Printed Test Bank** (ISBN 0-8053-8994-6) and cross-platform **Computerized Test Bank** (ISBN 0-8053-8995-4), prepared by Professor Benjamin Grinstein, at the University of California, San Diego, contain more than 1500 high-quality problems, with a range of multiple-choice, true/false, short-answer, and regular homework-type questions. In the computerized version, more than half of the questions have numerical values that can be randomly assigned for each student.

- The **Transparency Acetates** (ISBN 0-8053-8993-8) provide more than 200 key figures from *Physics for Scientists and Engineers* for classroom presentation.

Student Supplements

- The **Student Solutions Manuals Chapters 1–19** (ISBN 0-8053-8708-0) and **Chapters 20–42** (ISBN 0-8053-8998-9), written by Professors Pawan Kahol and Donald Foster at Wichita State University, provides *detailed* solutions to more than half of the odd-numbered end-of-chapter problems. The solutions follow the four-step Model/Visualize/Solve/Assess procedure used in the *Problem-Solving Strategies* and all worked examples.

- **MasteringPhysics**™ (www.masteringphysics.com) provides students with individualized online tutoring by responding to their wrong answers and providing hints for solving multistep problems. It gives them immediate and up-to-date assessment of their progress, and shows where they need to practice more.

- **ActivPhysics**™ **OnLine** (www.aw-bc.com/knight) provides students with a suite of highly regarded applet-based tutorials (see above). The accompanying workbooks help students work though complex concepts and understand them more clearly. The *ActivPhysics* margin icon directs students to specific exercises that complement the textbook discussion.

- **ActivPhysics OnLine Workbook Volume 1: Mechanics • Thermal Physics • Oscillations & Waves** (ISBN 0-8053-9060-X)

- **ActivPhysics OnLine Workbook Volume 2: Electricity & Magnetism • Optics • Modern Physics** (ISBN 0-8053-9061-8)

■ The **Addison-Wesley Tutor Center** (www.aw.com/tutorcenter) provides one-on-one tutoring via telephone, fax, email, or interactive website during evening hours and on weekends. Qualified college instructors answer questions and provide instruction for *Mastering Physics*™ and for the examples, exercises, and problems in *Physics for Scientists and Engineers*.

Acknowledgments

I have relied upon conversations with and, especially, the written publications of many members of the physics education community. Those who may recognize their influence include Arnold Arons, Uri Ganiel, Ibrahim Halloun, Richard Hake, David Hestenes, Leonard Jossem, Jill Larkin, Priscilla Laws, John Mallinckrodt, Lillian McDermott, Edward "Joe" Redish, Fred Reif, Rachel Scherr, Bruce Sherwood, David Sokoloff, Ronald Thornton, Sheila Tobias, and Alan Van Heuleven. John Rigden, founder and director of the Introductory University Physics Project, provided the impetus that got me started down this path. Early development of the materials was supported by the National Science Foundation as the *Physics for the Year 2000* project; their support is gratefully acknowledged.

I am grateful to Pawan Kahol and Don Foster for the difficult task of writing the *Instructor's Solutions Manuals*; to Jim Andrews and Susan Cable for writing the workbook answers; to Wayne Anderson, Jim Andrews, Dave Ettestad, Stuart Field, Robert Glosser, and Charlie Hibbard for their contributions to the end-of-chapter problems; and to my colleague Matt Moelter for many valuable contributions and suggestions.

I especially want to thank my editor Adam Black, development editor Alice Houston, editorial assistant Liana Allday, and all the other staff at Addison Wesley for their enthusiasm and hard work on this project. Project manager Laura Kenney, Carolyn Field and the team at Thompson Steele, Inc., copy editor Kevin Gleason, photo researcher Brian Donnelly, and page-layout artist Judy Maenle get much of the credit for making this complex project all come together. In addition to the reviewers and classroom testers listed below, who gave invaluable feedback, I am particularly grateful to Wendell Potter and Susan Cable for their close scrutiny of every word and figure.

Finally, I am endlessly grateful to my wife Sally for her love, encouragement, and patience, and to our many cats for their innate abilities to hold down piles of papers and to type qqqqqqqq whenever it was needed.

Randy Knight, September 2003
rknight@calpoly.edu

Reviewers and Classroom Testers

Gary B. Adams, *Arizona State University*
Wayne R. Anderson, *Sacramento City College*
James H. Andrews, *Youngstown State University*
David Balogh, *Fresno City College*
Dewayne Beery, *Buffalo State College*
Joseph Bellina, *Saint Mary's College*
James R. Benbrook, *University of Houston*
David Besson, *University of Kansas*

Randy Bohn, *University of Toledo*
Art Braundmeier, *University of Southern Illinois, Edwardsville*
Carl Bromberg, *Michigan State University*
Douglas Brown, *Cabrillo College*
Ronald Brown, *California Polytechnic State University, San Luis Obispo*
Mike Broyles, *Collin County Community College*

James Carolan, *University of British Columbia*
Michael Crescimanno, *Youngstown State University*
Wei Cui, *Purdue University*
Robert J. Culbertson, *Arizona State University*
Purna C. Das, *Purdue University North Central*
Dwain Desbien, *Estrella Mountain Community College*
John F. Devlin, *University of Michigan, Dearborn*
Alex Dickison, *Seminole Community College*
Chaden Djalali, *University of South Carolina*
Sandra Doty, *Denison University*
Miles J. Dresser, *Washington State University*
Charlotte Elster, *Ohio University*
Robert J. Endorf, *University of Cincinnati*
Tilahun Eneyew, *Embry-Riddle Aeronautical University*
F. Paul Esposito, *University of Cincinnati*
John Evans, *Lee University*
Michael R. Falvo, *University of North Carolina*
Abbas Faridi, *Orange Coast College*
Stuart Field, *Colorado State University*
Daniel Finley, *University of New Mexico*
Jane D. Flood, *Muhlenberg College*
Thomas Furtak, *Colorado School of Mines*
Richard Gass, *University of Cincinnati*
J. David Gavenda, *University of Texas, Austin*
Stuart Gazes, *University of Chicago*
Katherine M. Gietzen, *Southwest Missouri State University*
Robert Glosser, *University of Texas, Dallas*
William Golightly, *University of California, Berkeley*
Paul Gresser, *University of Maryland*
C. Frank Griffin, *University of Akron*
John B. Gruber, *San Jose State University*
Randy Harris, *University of California, Davis*
Stephen Haas, *University of Southern California*
Nicole Herbots, *Arizona State University*
Scott Hildreth, *Chabot College*
David Hobbs, *South Plains College*
Laurent Hodges, *Iowa State University*
John L. Hubisz, *North Carolina State University*
George Igo, *University of California, Los Angeles*
Bob Jacobsen, *University of California, Berkeley*
Rong-Sheng Jin, *Florida Institute of Technology*
Marty Johnston, *University of St. Thomas*
Stanley T. Jones, *University of Alabama*
Darrell Judge, *University of Southern California*
Pawan Kahol, *Wichita State University*
Teruki Kamon, *Texas A&M University*
Richard Karas, *California State University, San Marcos*
Deborah Katz, *U.S. Naval Academy*
Miron Kaufman, *Cleveland State University*
M. Kotlarchyk, *Rochester Institute of Technology*
Cagliyan Kurdak, *University of Michigan*
Fred Krauss, *Delta College*
H. Sarma Lakkaraju, *San Jose State University*

Darrell R. Lamm, *Georgia Institute of Technology*
Robert LaMontagne, *Providence College*
Alessandra Lanzara, *University of California, Berkeley*
Sen-Ben Liao, *Massachusetts Institute of Technology*
Dean Livelybrooks, *University of Oregon*
Chun-Min Lo, *University of South Florida*
Richard McCorkle, *University of Rhode Island*
James McGuire, *Tulane University*
Theresa Moreau, *Amherst College*
Gary Morris, *Rice University*
Michael A. Morrison, *University of Oklahoma*
Richard Mowat, *North Carolina State University*
Taha Mzoughi, *Mississippi State University*
Vaman M. Naik, *University of Michigan, Dearborn*
Craig Ogilvie, *Iowa State University*
Martin Okafor, *Georgia Perimeter College*
Benedict Y. Oh, *University of Wisconsin*
Georgia Papaefthymiou, *Villanova University*
Peggy Perozzo, *Mary Baldwin College*
Brian K. Pickett, *Purdue University, Calumet*
Joe Pifer, *Rutgers University*
Dale Pleticha, *Gordon College*
Robert Pompi, *SUNY-Binghamton*
David Potter, *Austin Community College*
Chandra Prayaga, *University of West Florida*
Didarul Qadir, *Central Michigan University*
Michael Read, *College of the Siskiyous*
Michael Rodman, *Spokane Falls Community College*
Sharon Rosell, *Central Washington University*
Anthony Russo, *Okaloosa-Walton Community College*
Otto F. Sankey, *Arizona State University*
Rachel E. Scherr, *University of Maryland*
Bruce Schumm, *University of California, Santa Cruz*
Douglas Sherman, *San Jose State University*
Elizabeth H. Simmons, *Boston University*
Alan Slavin, *Trent College*
William Smith, *Boise State University*
Paul Sokol, *Pennsylvania State University*
Chris Sorensen, *Kansas State University*
Anna and Ivan Stern, *AW Tutor Center*
Michael Strauss, *University of Oklahoma*
Arthur Viescas, *Pennsylvania State University*
Chris Vuille, *Embry-Riddle Aeronautical University*
Ernst D. Von Meerwall, *University of Akron*
Robert Webb, *Texas A&M University*
Zodiac Webster, *California State University, San Bernardino*
Robert Weidman, *Michigan Technical University*
Jeff Allen Winger, *Mississippi State University*
Ronald Zammit, *California Polytechnic State University, San Luis Obispo*
Darin T. Zimmerman, *Pennsylvania State University, Altoona*

Preface to the Student

From Me to You

The most incomprehensible thing about the universe is that it is comprehensible.
   —Albert Einstein

The day I went into physics class it was death.
   —Sylvia Plath, *The Bell Jar*

Let's have a little chat before we start. A rather one-sided chat, admittedly, because you can't respond, but that's OK. I've heard from many of your fellow students over the years, so I have a pretty good idea of what's on your mind.

What's your reaction to taking physics? Fear and loathing? Uncertainty? Excitement? All of the above? Let's face it, physics has a bit of an image problem on campus. You've probably heard that it's difficult, maybe downright impossible unless you're an Einstein. Things that you've heard, your experiences in other science courses, and many other factors all color your *expectations* about what this course is going to be like.

It's true that there are many new ideas to be learned in physics and that the course, like college courses in general, is going to be much faster paced than science courses you had in high school. I think it's fair to say that it will be an *intense* course. But we can avoid many potential problems and difficulties if we can establish, here at the beginning, what this course is about and what is expected of you—and of me!

Just what is physics, anyway? Physics is a way of thinking about the physical aspects of nature. Physics is not better than art or biology or poetry or religion, which are also ways to think about nature; it's simply different. One of the things this course will emphasize is that physics is a human endeavor. The information content of this book was not found in a cave or conveyed to us by aliens; it was discovered by real people engaged in a struggle with real issues. I hope to convey to you something of the history and the process by which we have come to accept the principles that form the foundation of today's science and engineering.

You might be surprised to hear that physics is not about "facts." Oh, not that facts are unimportant, but physics is far more focused on discovering *relationships* that exist between facts and *patterns* that exist in nature than on learning facts for their own sake. As a consequence, there's not a lot of memorization when you study physics. Some—there are still definitions and equations to learn—but less than in many other courses. Our emphasis, instead, will be on thinking and reasoning. This is important to factor into your expectations for the course.

Perhaps most important of all, *physics is not math!* Physics is much broader. We're going to look for patterns and relationships in nature, develop the logic that relates different ideas, and search for the reasons *why* things happen as they do. In doing so, we're going to stress qualitative reasoning, pictorial and graphical reasoning, and reasoning by analogy. And yes, we will use math, but it's just one tool among many.

It will save you much frustration if you're aware of this physics–math distinction up front. Many of you, I know, want to find a formula and plug numbers into it—that is, to do a math problem. Maybe that's what you learned in high school science courses, but it is *not* what this course expects of you. We'll certainly do

(a) X-ray diffraction pattern

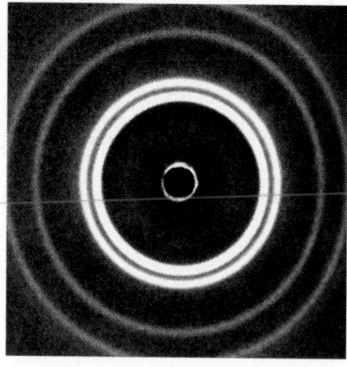

(b) Electron diffraction pattern

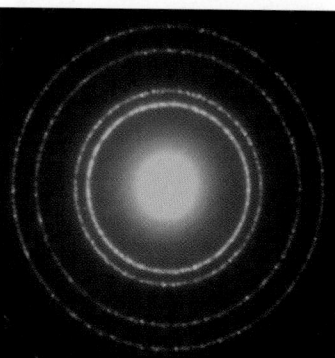

many calculations, but the specific numbers are usually the last and least important step in the analysis.

Physics is about recognizing patterns. The top photograph is an x-ray diffraction pattern that shows how a collimated beam of x rays spreads out after passing through a crystal. The bottom photograph shows what happens when a collimated beam of electrons is shot through the same crystal. What does the obvious similarity in these two photographs tell us about the nature of light and about the nature of matter?

As you study, you'll sometimes be baffled, puzzled, and confused. That's perfectly normal and to be expected. Making mistakes is OK too *if* you're willing to learn from the experience. No one is born knowing how to do physics any more than he or she is born knowing how to play the piano or shoot basketballs. The ability to do physics comes from practice, repetition, and struggling with the ideas until you "own" them and can apply them yourself in new situations. There's no way to make learning effortless, at least for anything worth learning, so expect to have some difficult moments ahead.

But also expect to have some moments of excitement at the joy of discovery. There will be instants at which the pieces suddenly click into place and you *know* that you understand a difficult idea. There will be times when you'll surprise yourself by successfully working a difficult problem that you didn't think you could solve. My hope, as an author, is that the excitement and sense of adventure will far outweigh the difficulties and frustrations.

Many of you, I suspect, would like to know the "best" way to study for this course. There is no best way. People are too different, and what works for one student works less effectively for another. But I do want to stress that *reading the text* is vitally important. Class time will be used to clarify difficulties and to develop tools for using the knowledge, but your instructor will *not* use class time simply to repeat information in the text. The basic knowledge for this course is written down within these pages, and the *number one expectation* is that you will read carefully and thoroughly to find and learn that knowledge.

Despite there being no best way to study, I will suggest *one* way that is successful for many students. It consists of the following four steps:

1. **Read each chapter *before* it is discussed in class.** I cannot stress too highly how important this step is. Class attendance is largely ineffective if you have not prepared. When you first read a chapter, focus on learning new vocabulary, definitions, and notation. There's a list of terms and notations at the end of each chapter. Learn them! You won't understand what's being discussed or how the ideas are being used if you don't know what the terms and symbols mean.

2. **Participate actively in class.** Take notes, ask and answer questions, take part in discussion groups. There is ample scientific evidence that *active participation* is far more effective for learning science than is passive listening.

3. **After class, go back for a *careful* rereading of the chapter.** In your second reading, pay closer attention to the details and the worked examples. Look for the *logic* behind each example (and I've tried to help make this clear), not just at what formula is being used. Do the *Student Workbook* exercises for each section as you finish your reading of it.

4. **Finally, apply what you have learned to the homework problems at the end of each chapter.** I strongly encourage you to form a study group with two or three classmates. There's good evidence that students who study regularly with a group do better than the rugged individualists who try to go it alone.

Did someone mention a workbook? The companion *Student Workbook* is a vital part of this course. It contains questions and exercises that ask you to reason *qualitatively*, to use graphical information, and to give explanations. It is through these exercises that you will learn what the concepts mean and will practice the reasoning skills appropriate to the chapter. You will then have acquired the baseline knowledge that you need *before* turning to the end-of-chapter homework problems. In sports or in music, you would never think of performing before you practice, so why would you want to do so in physics? The workbook is where you practice and work on basic skills.

Many of you, I know, would like to go straight to the homework problems and then thumb through the text looking for a formula that seems like it will work. That approach will not succeed in this course, and it's guaranteed to make you frustrated and discouraged. Very few homework problems are "plug and chug" problems where you simply put numbers into a formula. To work the homework problems successfully, you need a better study strategy—either that outlined above or your own—that helps you learn the concepts and the relationships between the ideas. Many of the chapters in this book have Problem-Solving Strategies to help you develop effective problem-solving skills.

A traditional guideline in college is to study two hours outside of class for every hour spent in class, and this text is designed with that expectation. Of course, two hours is an average. Some chapters are fairly straightforward and will go quickly. Others likely will require much more than two study hours per class hour.

Now that you know more about what is expected of you, what can you expect of me? That's a little trickier, because the book is already written! Nonetheless, it was prepared on the basis of what I think my students throughout the years have expected—and wanted—from their physics textbook.

You should know that these course materials—the text and the workbook—are based upon extensive research about how students learn physics and the challenges they face. The effectiveness of many of the exercises has been demonstrated through extensive class testing. I've written the book in an informal style that I hope you will find appealing and that will encourage you to do the reading. And finally, I have endeavored to make clear not only that physics, as a technical body of knowledge, is relevant to your profession but also that physics is an exciting adventure of the human mind.

I hope you'll enjoy the time we're going to spend together.

Detailed Contents

Part II Conservation Laws

Part III Applications of Newtonian Mechanics

Part IV Thermodynamics

Part V   Waves and Optics

Part VII   Relativity and Quantum
                Physics

Introduction

Journey into Physics

Said Alice to the Cheshire cat,

"Cheshire-Puss, would you tell me, please, which way I ought to go from here?"
"That depends a good deal on where you want to go," said the Cat.
"I don't much care where—" said Alice.
"Then it doesn't matter which way you go," said the Cat.
   —Lewis Carroll, *Alice in Wonderland*

Have you ever wondered about questions such as

> Why is the sky blue?
>
> Why is glass an insulator but metal a conductor?
>
> What, really, is an atom?

These are the questions of which physics is made. Physicists try to understand the universe in which we live by observing the phenomena of nature—such as the sky being blue—and by looking for patterns and principles to explain these phenomena. Many of the discoveries made by physicists, from electromagnetic waves to nuclear energy, have forever altered the ways in which we live and think.

You are about to embark on a journey into the realm of physics. It is a journey in which you will learn about many physical phenomena and find the answers to questions such as the ones posed above. Along the way, you will also learn how to use physics to analyze and solve many practical problems.

As you proceed, you are going to see the methods by which physicists have come to understand the laws of nature. The ideas and theories of physics are not arbitrary; they are firmly grounded in experiments and measurements. By the time you finish this text, you will be able to recognize the *evidence* upon which our present knowledge of the universe is based.

Which Way Should We Go?

We are rather like Alice in Wonderland, here at the start of the journey, in that we must decide which way to go. Physics is an immense body of knowledge, and without specific goals it would not much matter which topics we study. But unlike Alice, we *do* have some particular destinations that we would like to visit.

The physics that provides the foundation for all of modern science and engineering can be divided into three broad categories:

- Particles and energy.
- Fields and waves.
- The atomic structure of matter.

A particle, in the sense that we'll use the term, is an idealization of a physical object. We will use particles to understand how objects move and how they interact with each other. One of the most important properties of a particle or a collection of particles is *energy*. We will study energy both for its value in understanding physical processes and because of its practical importance in a technological society.

A scanning tunneling microscope allows us to "see" the individual atoms on a surface. One of our goals is to understand how an image such as this is made.

Particles are discrete, localized objects. Although many phenomena can be understood in terms of particles and their interactions, the long-range interactions of gravity, electricity, and magnetism are best understood in terms of *fields*, such as the gravitational field and the electric field. Rather than being discrete, fields spread continuously through space. Much of the second half of this book will be focused on understanding fields and the interactions between fields and particles.

Certainly one of the most significant discoveries of the past 500 years is that matter consists of atoms. Atoms and their properties are described by quantum physics, but we cannot leap directly into that subject and expect that it would make any sense. To reach our destination, we are going to have to study many other topics along the way—rather like having to visit the Rocky Mountains if you want to drive from New York to San Francisco. All our knowledge of particles and fields will come into play as we end our journey by studying the atomic structure of matter.

The Route Ahead

Here at the beginning, we can survey the route ahead. Where will our journey take us? What scenic vistas will we view along the way?

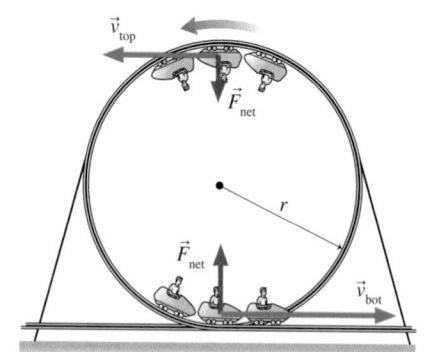

Parts I and II, *Newton's Laws* and *Conservation Laws*, form the basis of what is called *classical mechanics*. Classical mechanics is the study of motion. (It is called *classical* to distinguish it from the modern theory of motion at the atomic level, which is called *quantum mechanics*.) The first two parts of this textbook establish the basic language and concepts of motion. Part I will look at motion in terms of *particles* and *forces*. We will use these concepts to study the motion of everything from accelerating sprinters to orbiting satellites. Then, in Part II, we will introduce the ideas of *momentum* and *energy*. These concepts—especially energy—will give us a new perspective on motion and extend our ability to analyze motion.

Part III, *Applications of Newtonian Mechanics*, will pause to look at four important applications of classical mechanics: Newton's theory of gravity, rotational motion, oscillatory motion, and the motion of fluids. Only oscillatory motion is a prerequisite for later chapters. Your instructor may choose to cover some or all of the other chapters, depending upon the time available, but your study of Parts IV–VII will not be hampered if these chapters are omitted.

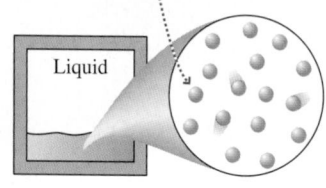

Atoms are held close together by weak molecular bonds, but they can slide around each other.

Part IV, *Thermodynamics*, extends the ideas of particles and energy to systems such as liquids and gases that contain vast numbers of particles. Here we will look for connections between the *microscopic* behavior of large numbers of atoms and the *macroscopic* properties of bulk matter. You will find that some of the properties of gases that you know from chemistry, such as the ideal gas law, turn out to be direct consequences of the underlying atomic structure of the gas. We will also expand the concept of energy and study how energy is transferred and utilized.

Waves are ubiquitous in nature, whether they be large-scale oscillations like ocean waves, the less obvious motions of sound waves, or the subtle undulations of light waves and matter waves that go to the heart of the atomic structure of matter. In **Part V,** *Waves and Optics,* we will emphasize the unity of wave physics and find that many diverse wave phenomena can be analyzed with the same concepts and mathematical language. It is here we will begin to accumulate evidence that the theory of classical mechanics is inadequate to explain the observed behavior of atoms, and we will end this section with some atomic puzzles that seem to defy understanding.

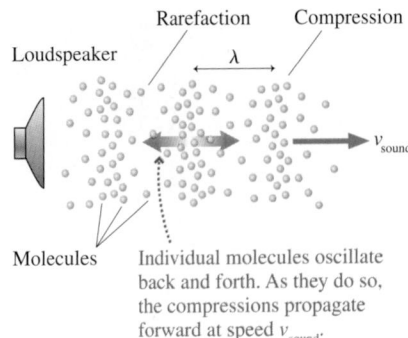

Individual molecules oscillate back and forth. As they do so, the compressions propagate forward at speed v_{sound}.

Part VI, *Electricity and Magnetism,* is devoted to the *electromagnetic force,* one of the most important forces in nature. In essence, the electromagnetic force is the "glue" that holds atoms together. It is also the force that makes this the "electronic age." We'll begin this part of the journey with simple observations of static electricity. Bit by bit, we'll be led to the basic ideas behind electrical circuits, to magnetism, and eventually to the discovery of electromagnetic waves.

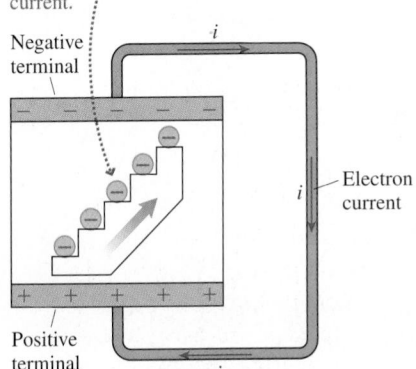

The "charge escalator" inside a battery continuously "lifts" electrons from the positive to the negative terminal. This renewal of charge sustains the electron current.

Part VII is *Relativity and Quantum Physics.* We'll start by exploring the strange world of Einstein's theory of *relativity,* a world in which space and time aren't quite what they appear to be. Then we will enter the microscopic domain of *atoms,* where the behaviors of light and matter are at complete odds with what our common sense tells us is possible. Although the mathematics of quantum theory quickly gets beyond the level of this text, and time will be running out, you will see that the quantum theory of atoms and nuclei explains many of the things that you learned simply as rules in chemistry.

We will not have visited all of physics on our travels. There just isn't time. Many exciting topics, ranging from quarks to black holes, will have to remain unexplored. But this particular journey need not be the last. As you finish this text, you will have the background and the experience to explore new topics further in more advanced courses or for yourself.

With that said, let us take the first step.

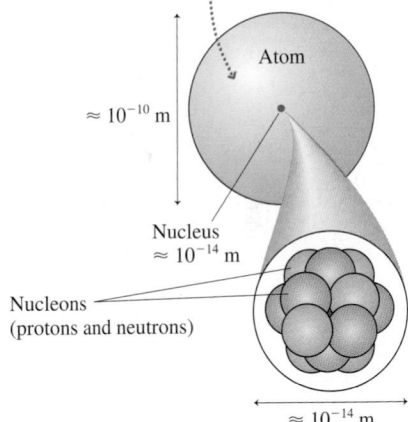

This picture of an atom would need to be 10 m in diameter if it were drawn to the same scale as the dot representing the nucleus.

If the acceleration of a Podracer can reach 50 m/s², what are the maximum tensions in the two large cables? To find out, what quantities do you need to estimate?

Newton's Laws

Why Things Change

Each of the seven parts of this book opens with an overview. The overview gives you a look ahead, a glimpse of where your journey will take you in the next few chapters. It's easy to lose sight of the big picture while you're busy negotiating the terrain of each chapter. You may find it helpful to look back at the overviews several times as your travels take you through the different parts of the book.

Change

Simple observations of the world around you show that most things change, few things remain the same. Some changes, such as aging, are biological. Others, such as sugar dissolving in your coffee, are chemical. We're going to be concerned about changes that involve *motion* of one form or another. For example, you were standing, now you're sitting. The leaf that was on the tree has fallen. The air molecules in the room have moved to new positions.

Part I of this textbook, *Newton's Laws,* is about motion. The "laws of motion" were discovered by Isaac Newton roughly 350 years ago, so our study of motion is hardly cutting edge science. Nonetheless, it is still extremely important. Mechanics—the science of motion—is the basis for much of engineering and applied science. And many of the ideas introduced in mechanics will be needed later to understand things like the motion of waves or the motion of electrons in a semiconductor. Newton's mechanics is the foundation of much of contemporary science, so it is important that we start at the beginning.

Part I is going to focus on the motion of "things" such as balls, cars, rockets, and satellites. These are *macroscopic* objects, as opposed to microscopic atoms and molecules. They are also objects with well-defined boundaries, in contrast to the spread-out motion of a wave. We'll get to the motion of waves and atoms at later stages of our journey.

There are two big questions we must tackle to study how things change by moving:

- **How do we describe motion?** It is easy to say that an object moves, but it's not obvious how we should measure or characterize the motion if we want to analyze it mathematically. The mathematical description of motion is called *kinematics,* and it is the subject matter of Chapters 1 and 2.
- **How do we explain motion?** Why do objects have the particular motion they do? Why, when you toss a ball upward, does it go up and then come back down rather than keep going up? Are there "laws of nature" that allow us to predict

an object's motion? The explanation of motion in terms of its causes is called *dynamics,* and it is the topic of Chapters 4 through 8.

Chapter 3, wedged between kinematics and dynamics, is about *vectors.* Our universe is three dimensional, and vectors are a mathematical tool for describing motion in three dimensions. Much of physics and engineering is written in the language of vectors, hence it is important that you become fluent in this language at an early stage of your journey.

Two key ideas that will help answer these questions are *force* (the "cause") and *acceleration* (the "effect"). One of our goals is to develop a "Newtonian intuition" for the connection between force and acceleration. Much of Chapters 1 through 4 is devoted to helping you develop your intuition about the nature of motion. Pictures and graphs will play a big role in this development. You'll then put this knowledge to use in Chapters 5 through 8 as you analyze motion of increasing complexity.

Models: A Vehicle for Understanding

Reality is extremely complicated. We would never be able to develop a science if we had to keep track of every little detail of every situation. Consider a simple example: throwing a ball. To understand the motion of the ball, is it necessary to keep track of every atom inside? Of every quark inside every proton inside every nucleus inside every atom inside the ball? Do we need to analyze what you ate for breakfast and the biochemistry of how that was translated into muscle power? In principle, the answer to all these questions is "Yes." But in practice these are all

We'll make simplifying assumptions when we analyze the motion of a ball in Chapter 5. This simplified description of reality is a *model* of the situation.

details that have no influence at all on the measurements you might make on the ball.

We can do a perfectly fine analysis if we treat the ball as a round solid and your hand as another solid that exerts a force on the ball. This is now what we call a *model* of the situation. A model is a simplified description of reality— much as a model airplane is a simplified version of a real airplane—that is used to reduce the complexity of a problem to the point where it can be analyzed and understood.

Much of physics and engineering is a matter of model building—simplifying the situation, isolating the essential features, and developing a set of equations that provides an adequate, although not perfect, description of reality. Physics, in particular, attempts to strip a phenomenon down to its barest essentials in order to illustrate the physical principles involved. Many of the features neglected by the physicist as "unnecessary details" would be considered absolutely essential by an engineer or a chemist. Both are right. The model each investigator uses has to match the needs each of them is trying to meet.

In building a model, you need to isolate and keep just those features that are essential to the problem. In the case of the ball, for example, you can ignore the ball's atomic structure because it doesn't affect the motion of the ball as a whole. But you cannot ignore gravity. It is an essential feature of the model because it directly influences the motion of the ball. What about air resistance? If you are throwing a very hard, dense ball a very short distance, then ignoring air resistance is probably acceptable. But if you are throwing a ping-pong ball from the Empire State Building, then certainly air resistance is an essential feature of the model. The more details you omit, the simpler the model and the easier it will be to solve the model's equations—but at the expense of less accuracy and poorer agreement with reality.

Model building is a major part of the strategy that we will develop for solving problems. It is, however, a skill that takes some practice and experience to acquire. We will pay close attention, especially in the earlier chapters, to where simplifying assumptions are being made, and why. Learning *how* to simplify a situation is the essence of successful modeling. As you begin to apply these ideas in your own homework, you will be gaining experience with model building and learning how to be a sophisticated problem solver.

1 Concepts of Motion

This snowboarder is demonstrating a fairly extreme form of motion.

▶ **Looking Ahead**

The goal of Chapter 1 is to introduce the fundamental concepts of motion. In this chapter you will learn to:

- Draw and interpret motion diagrams.
- Describe motion with vectors.
- Use the concepts of position, velocity, and acceleration.
- Use multiple representations of motion.
- Analyze and interpret motion problems.

Socrates: *The nature of motion appears to be the question with which we begin.*

Plato, 375 BCE

The universe in which we live is one of change and motion. This snowboarder was clearly in motion when the photograph was taken. In the course of a day you probably walk, run, bicycle, or drive your car, all forms of motion. The clock hands are moving inexorably forward as you read this text. The pages of this book may look quite still, but a microscopic view would reveal jostling atoms and whirling electrons. The stars look as permanent as anything, yet the astronomer's telescope reveals them to be ceaselessly moving within galaxies that rotate and orbit yet other galaxies.

Motion is a theme that will appear in one form or another throughout this entire book. Although we all have intuition about motion, based on our experiences, some of the important aspects of motion turn out to be rather subtle. So rather than jumping immediately into a lot of mathematics and calculations, this first chapter focuses on *visualizing* motion and becoming familiar with the *concepts* needed to describe a moving object. We will use mathematical ideas when needed, because they increase the precision of our thoughts, but we will defer actual calculations until Chapter 2. Our goal is to lay the foundations for understanding motion.

1.1 Motion Diagrams

The quest to understand motion dates to antiquity. The ancient Babylonians, Chinese, and Greeks were especially interested in the celestial motions of the night sky. The Greek philosopher and scientist Aristotle wrote extensively about the nature of moving objects. However, our modern understanding of motion did not begin until Galileo (1564–1642) first formulated the concepts of motion in

Translational motion

Circular motion

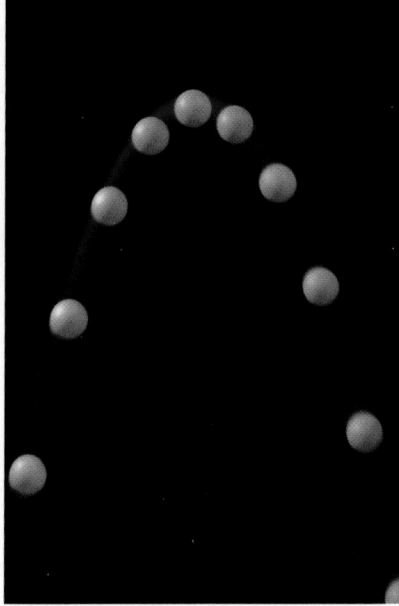
Projectile motion

Rotational motion

FIGURE 1.1 Four basic types of motion.

FIGURE 1.2 Several frames from the movie of a car.

mathematical terms. And it took Newton (1642–1727) and the invention of calculus to put the concepts of motion on a firm and rigorous footing.

As a starting point, let's define **motion** as the change of an object's position with time. Examples of motion are easy to list. Bicycles, baseballs, cars, airplanes, and rockets are all objects that move. The path along which an object moves, which might be a straight line or might be curved, is called the object's **trajectory.**

Figure 1.1 shows four basic types of motion that we will study in this book. Rotational motion is somewhat different from the other three in that rotation is a change of the object's *angular* position. We'll defer rotational motion until later and, for now, focus on motion along a line, circular motion, and projectile motion.

The fundamental question we want to ask is: What *concepts* are needed to give a full and accurate description of motion?

Making a Motion Diagram

An easy way to study motion is to make a movie of a moving object. A movie camera, as you probably know, takes photographs at a fixed rate, typically 30 photographs every second. Each separate photo is called a *frame,* and the frames are all lined up one after the other in a *filmstrip.* As an example, Figure 1.2 shows a few frames from the movie of a car going past. Not surprisingly, the car is in a somewhat different position in each frame.

Suppose we now cut the individual frames of the film strip apart, stack them on top of each other, and then project the entire stack at once onto a screen for viewing. The result is shown in Figure 1.3. This composite photo, showing an object's position at several *equally spaced instants of time,* is called a **motion diagram.** As simple as motion diagrams seem, they will turn out to be a powerful tool for analyzing motion.

NOTE ▶ It's important to keep the camera in a *fixed position* as the object moves by. Don't "pan" it to track the moving object. ◀

Now let's take our camera out into the world and make a few motion diagrams. The following table shows how we can see important aspects of the motion in a motion diagram.

The same amount of time elapses between each image and the next.

FIGURE 1.3 A motion diagram of the car shows all the frames simultaneously.

Examples of motion diagrams

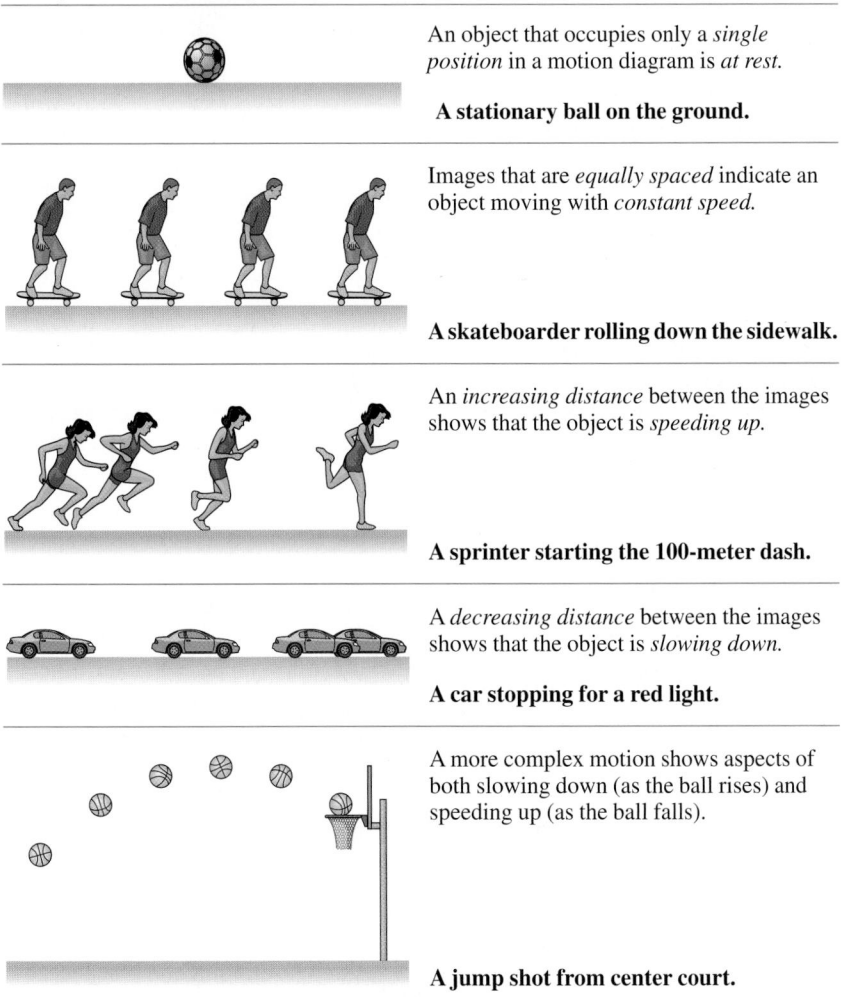

An object that occupies only a *single position* in a motion diagram is *at rest.*

A stationary ball on the ground.

Images that are *equally spaced* indicate an object moving with *constant speed.*

A skateboarder rolling down the sidewalk.

An *increasing distance* between the images shows that the object is *speeding up.*

A sprinter starting the 100-meter dash.

A *decreasing distance* between the images shows that the object is *slowing down.*

A car stopping for a red light.

A more complex motion shows aspects of both slowing down (as the ball rises) and speeding up (as the ball falls).

A jump shot from center court.

We have defined several concepts (at rest, constant speed, speeding up, and slowing down) in terms of how the moving object appears in a motion diagram. These are called **operational definitions,** meaning that the concepts are defined in terms of a particular procedure or operation performed by the investigator. For example, we could answer the question "Is the airplane speeding up?" by checking whether or not the images in the plane's motion diagram are getting farther apart. Many of the concepts in physics will be introduced as operational definitions. This reminds us that physics is an experimental science.

STOP TO THINK 1.1   Which car is going faster, A or B? Assume there are equal intervals of time between the frames of both movies.

Car A   Car B

NOTE ▶ Each chapter in this textbook will have several *Stop to Think* questions. These questions are designed to see if you've understood the basic ideas that have been presented. The answers are given at the end of the chapter, but you should make a serious effort to think about these questions before turning to the answers. If you answer correctly, and are sure of your answer rather than just guessing, you can proceed to the next section with confidence. But if you answer incorrectly, it would be wise to reread the preceding sections carefully before proceeding onward. ◄

1.2 The Particle Model

For many objects, such as cars and rockets, the motion of the object *as a whole* is not influenced by the "details" of the object's size and shape. To describe the object's motion, all we really need to keep track of is the motion of a single point, such as a white dot painted on the side of the object.

If we restrict our attention to objects undergoing **translational motion,** which is the motion of an object along a trajectory, we can consider the object *as if* it were just a single point, without size or shape. We can also treat the object *as if* all of its mass were concentrated into this single point. An object that can be represented as a mass at a single point in space is called a **particle.** A particle has no size, no shape, and no distinction between top and bottom or between front and back.

If we treat an object as a particle, we can represent the object in each frame of a motion diagram as a simple dot rather than having to draw a full picture. Figure 1.4 shows how much simpler motion diagrams appear when the object is represented as a particle. Note that the dots have been numbered 0, 1, 2, . . . to tell the sequence in which the frames were exposed. These diagrams are more abstract than the pictures, but they are easier to draw and they still convey our full understanding of the object's motion.

Using the Particle Model

Treating an object as a particle is, of course, a simplification of reality. As we noted in the overview, such a simplification is called a *model.* Models allow us to focus on the important aspects of a phenomenon by excluding those aspects that play only a minor role. The **particle model** of motion is a simplification in which we treat a moving object as if all of its mass were concentrated at a single point.

The particle model is an excellent approximation of reality for the motion of cars, planes, rockets, and similar objects. People are somewhat more complex, because of moving arms and legs, but the motion of a person's body as a whole is still described reasonably well within the particle model. A ball rolling on a surface is a combination of translational and rotational motion, but we can treat the ball as a particle if its translational motion is all we care about. In fact, so successful

(a) Motion diagram of a rocket launch

4 ●

Numbers show
order in which
frames were   3 ●
exposed.

2 ●

1 ●
0 ●

(b) Motion diagram of a car stopping

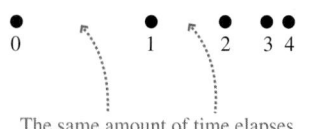

The same amount of time elapses
between each image and the next.

FIGURE 1.4 Motion diagrams in which the object is represented as a particle.

has the particle model been that the motion of more complex objects, which cannot be treated as a single particle, is often analyzed as if the object were a collection of particles.

Not all motions can be reduced to the motion of a single point. Consider a rotating gear. The center of the gear doesn't move at all, and each tooth on the gear is moving in a different direction. Rotational motion is qualitatively different than translational motion, and we'll need to go beyond the particle model later when we study rotational motion.

STOP TO THINK 1.2 Three motion diagrams are shown. Which is a dust particle settling to the floor at constant speed, which is a ball dropped from the roof of a building, and which is a descending rocket slowing to make a soft landing on Mars?

(a)	(b)	(c)
0 ●	0 ●	0 ●
1 ●		
2 ●	1 ●	
		1 ●
3 ●	2 ●	
		2 ●
4 ●	3 ●	
		3 ●
	4 ●	4 ●
5 ●	5 ●	5 ●

1.3 Position and Time

To develop motion diagrams further, we need to be able to make measurements. As we look at a motion diagram, it would be useful to know *where* the object is (i.e., its *position*) and *when* the object was at that position (i.e., the *time*). These are easy measurements to make.

Position measurements can be made by laying a coordinate system grid over a motion diagram. You can then measure the (x, y) coordinates of each point in the motion diagram. Of course, the world does not come with a coordinate system attached. A coordinate system is an artificial grid that *you* place over a problem in order to analyze the motion. You place the origin of your coordinate system wherever you wish, and different observers of a moving object might all choose to use different origins. Likewise, you can choose the orientation of the *x*-axis and *y*-axis to be helpful for that particular problem. The conventional choice is for the *x*-axis to point to the right and the *y*-axis to point upward, but there is nothing sacred about this choice. We will soon have many occasions to tilt the axes at an angle.

Time, in a sense, is also a coordinate system, although you may never have thought of time this way. You can pick an arbitrary point in the motion and label it "$t = 0$ seconds." This is simply the instant you decide to start your clock or stopwatch, so it is the origin of your time coordinate. Different observers might choose to start their clocks at different moments. A movie frame labeled "$t = 4$ seconds" means it was taken 4 seconds after you started your clock.

We typically choose $t = 0$ to represent the "beginning" of a problem, but the object may have been moving before then. Those earlier instants would be measured as negative times, just as objects on the *x*-axis to the left of the origin have negative values of position. Negative numbers are not to be avoided; they simply locate an event in space or time *relative to an origin.*

To illustrate, Figure 1.5a on the next page shows an *xy*-coordinate system and time information superimposed over the motion diagram of a basketball. You can see that ball's position is $(x_4, y_4) = (4 \text{ m}, 3 \text{ m})$ at time $t_4 = 2.0$ s. Notice how we've used subscripts to indicate the time and the object's position in a specific frame of the motion diagram.

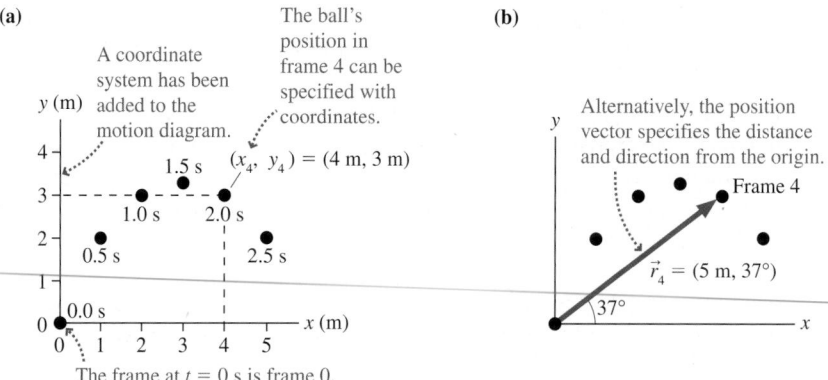

FIGURE 1.5 Position and time measurements made on the motion diagram of a basketball.

NOTE ▶ The first frame is labeled 0 to correspond with time $t = 0$. That is why the fifth frame is labeled 4. ◀

Another way to locate the ball is to draw an arrow from the origin to the point representing the ball. You can then specify the length and direction of the arrow. An arrow drawn from the origin to an object's position is called the **position vector** of the object, and it is given the symbol $\vec{r}$. Figure 1.5b shows the position vector $\vec{r}_4 = (5 \text{ m}, 37°)$.

The position vector $\vec{r}$ does not tell us anything different than the coordinates (x, y). It simply provides the information in an alternative form. Although you're probably more familiar with (x, y) coordinates than with vectors, you will find that vectors are a useful and powerful way to describe many concepts in physics.

A Word About Vectors and Notation

Before we continue our discussion, we should take a closer look at what a vector is. Vectors will be studied thoroughly in Chapter 3, so all we need for now is a little basic information. Some physical quantities, such as time, mass, and temperature, can be described completely by a single number with a unit. For example, the mass of an object is 6 kg and its temperature is 30°C. When a physical quantity is described by a single number (with a unit), we call it a **scalar quantity.** A scalar can be positive, negative, or zero.

Many other quantities, however, have a directional quality and cannot be described by a single number. To describe the motion of a car, for example, you must specify not only how fast it is moving, but also the *direction* in which it is moving. A **vector quantity** is a quantity that has both a *size* (the "How far?" or "How fast?") and a *direction* (the "Which way?"). The size or length of a vector is called its *magnitude*. The magnitude of a vector can be positive or zero, but it cannot be negative.

When we want to represent a vector quantity with a symbol, we need somehow to indicate that the symbol is for a vector rather than for a scalar. We do this by drawing an arrow over the letter that represents the quantity. Thus $\vec{r}$ and $\vec{A}$ are symbols for vectors, whereas r and A, without the arrows, are symbols for scalars. In handwritten work you must draw arrows over all symbols that represent vectors. This may seem strange until you get used to it, but it is very important because we will often use both r and $\vec{r}$, or both A and $\vec{A}$, in the same problem, and they mean different things! Without the arrow, you will be using the same symbol with two different meanings and will likely end up making a mistake. Note that the arrow over the symbol always points to the right, regardless of which direction the actual vector points. Thus we write $\vec{r}$ or $\vec{A}$, never $\overleftarrow{r}$ or $\overleftarrow{A}$.

NOTE ▶ Some textbooks represent vectors with boldface type, such as **r** or **A**. This book will consistently display the vector arrow over vector symbols, just as you should do in handwritten work. ◄

Change in Position

Now that you've seen how to measure position and time, let's return to the problem of motion, where we need to measure *changes* in position that occur over time. Consider the following:

Sam is standing 50 feet (ft) east of the corner of 12th Street and Vine. He then walks northeast for 100 ft to a second point. What is Sam's change of position?

Figure 1.6 shows Sam's motion in terms of position vectors. Because we're free to place the origin of our coordinate system wherever we wish, we've placed it at the intersection. Sam's initial position is the vector $\vec{r}_0$ drawn from the origin to the point where he starts walking. Vector $\vec{r}_1$ is his position after he finishes walking. You can see that Sam has changed position, and a *change* of position is called a *displacement*. His displacement is the vector labeled $\Delta\vec{r}$. The Greek letter delta (Δ) is used in math and science to indicate the *change* in a quantity. Here it indicates a change in the position $\vec{r}$.

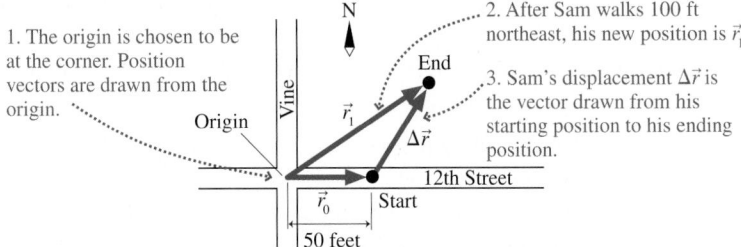

1. The origin is chosen to be at the corner. Position vectors are drawn from the origin.

2. After Sam walks 100 ft northeast, his new position is $\vec{r}_1$.

3. Sam's displacement $\Delta\vec{r}$ is the vector drawn from his starting position to his ending position.

FIGURE 1.6 Sam undergoes a displacement $\Delta\vec{r}$ from position $\vec{r}_0$ to position $\vec{r}_1$.

NOTE ▶ $\Delta\vec{r}$ is a *single* symbol. You cannot cancel out or remove the Δ in algebraic operations. ◄

Displacement is a vector quantity; it requires both a length and a direction to describe it. Specifically, the displacement $\Delta\vec{r}$ is a vector drawn *from* a starting position *to* an ending position. Sam's displacement vector is written

$$\Delta\vec{r} = (100 \text{ ft, northeast})$$

where we've given both the length and the direction. The length, or magnitude, of a displacement vector is simply the straight-line distance between the starting and ending positions.

Suppose you start 10 ft from a door and walk directly away from the door for 5 ft. Although it's clear that you end up 15 ft from the door, the *procedure* by which you learn this is to *add* your change in position (5 ft) to your initial position (10 ft).

Similarly, we can answer the question "Where does Sam end up?" if we *add* his change in position (his displacement $\Delta\vec{r}$) to his initial position, the vector $\vec{r}_0$. Sam's final position in Figure 1.6, vector $\vec{r}_1$, can be seen as a combination of vector $\vec{r}_0$ *plus* vector $\Delta\vec{r}$. In fact, $\vec{r}_1$ is the *vector sum* of vectors $\vec{r}_0$ and $\Delta\vec{r}$. This is written

$$\vec{r}_1 = \vec{r}_0 + \Delta\vec{r} \tag{1.1}$$

Notice, however, that we are now adding vector quantities, not scalar quantities. Vector addition is a different process from "regular" addition. You can add two vectors $\vec{A}$ and $\vec{B}$ with the following three-step procedure.

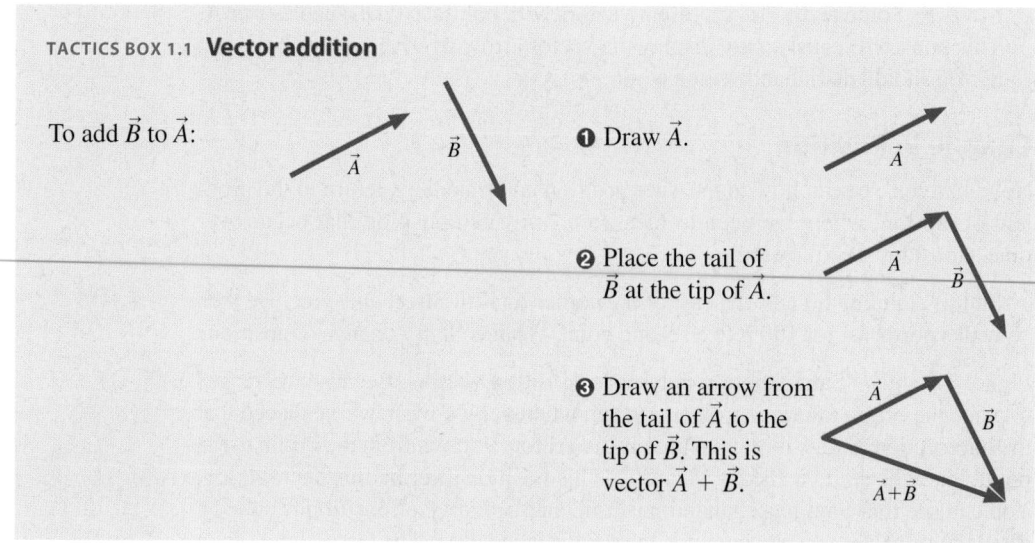

TACTICS BOX 1.1 **Vector addition**

To add $\vec{B}$ to $\vec{A}$:

❶ Draw $\vec{A}$.

❷ Place the tail of $\vec{B}$ at the tip of $\vec{A}$.

❸ Draw an arrow from the tail of $\vec{A}$ to the tip of $\vec{B}$. This is vector $\vec{A} + \vec{B}$.

This is exactly how $\vec{r}_0$ and $\Delta\vec{r}$ are added in Figure 1.6 to give $\vec{r}_1$.

NOTE ▶ A vector is not tied to a particular location on the page. You can move a vector around on a figure as long as you don't change its length or the direction it points. Vector $\vec{B}$ is not changed by sliding it over to where its tail is at the tip of $\vec{A}$. ◀

In Figure 1.6, we chose *arbitrarily* to put the origin of the coordinate system at the corner. While this might be convenient, it certainly is not mandatory. Figure 1.7 shows a different choice of where to place the origin. Notice something interesting. Vectors $\vec{r}_0$ and $\vec{r}_1$ have become new vectors $\vec{r}_2$ and $\vec{r}_3$, but the displacement vector $\Delta\vec{r}$ has not changed! **The displacement is a quantity that is independent of the coordinate system.** In other words, the arrow drawn from the one position of an object to the next is the same no matter what coordinate system you choose. This independence gives the displacement $\Delta\vec{r}$ more *physical significance* than the position vectors themselves have.

This observation suggests that the displacement, rather than the actual position, is what we want to focus on as we analyze the motion of an object. Equation 1.1 told us that $\vec{r}_1 = \vec{r}_0 + \Delta\vec{r}$. This is easily rearranged to give a more precise definition of displacement: **The displacement $\Delta\vec{r}$ of an object as it moves from an initial position $\vec{r}_i$ to a final position $\vec{r}_f$ is**

$$\Delta\vec{r} = \vec{r}_f - \vec{r}_i \qquad (1.2)$$

Graphically, $\Delta\vec{r}$ is a vector arrow drawn *from* position $\vec{r}_i$ *to* position $\vec{r}_f$. The displacement vector is independent of the coordinate system.

NOTE ▶ To be more general, we've written Equation 1.2 in terms of an *initial position* and a *final position*, indicated by subscripts i and f. We'll frequently use i and f when writing general equations, then use specific numbers or values, such as 0 and 1, when working a problem. ◀

This definition of $\Delta\vec{r}$ involves *vector subtraction*. With numbers, subtraction is the same as the addition of a negative number. That is, $5 - 3$ is the same as $5 + (-3)$. Similarly, we can use the rules for vector addition to find $\vec{A} - \vec{B} = \vec{A} + (-\vec{B})$ if we first define what we mean by $-\vec{B}$. As Figure 1.8 shows, the negative of vector $\vec{B}$ is a vector with the same length but pointing in the opposite direction. This makes sense because $\vec{B} - \vec{B} = \vec{B} + (-\vec{B}) = \vec{0}$, where $\vec{0}$, a vector with zero length, is called the **zero vector.**

The displacement vector is not affected by the choice of origin.

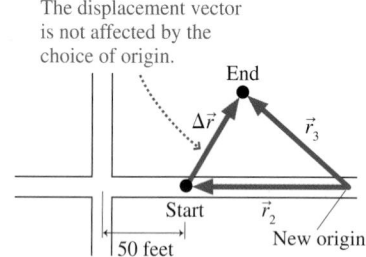

FIGURE 1.7 Sam's displacement $\Delta\vec{r}$ is unchanged by using a different coordinate system.

Vector $-\vec{B}$ has the same length as $\vec{B}$ but points in the opposite direction.

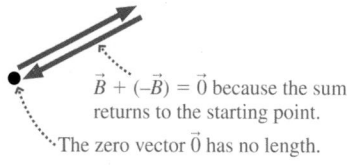

$\vec{B} + (-\vec{B}) = \vec{0}$ because the sum returns to the starting point.

The zero vector $\vec{0}$ has no length.

FIGURE 1.8 The negative of a vector.

TACTICS BOX 1.2 **Vector subtraction**

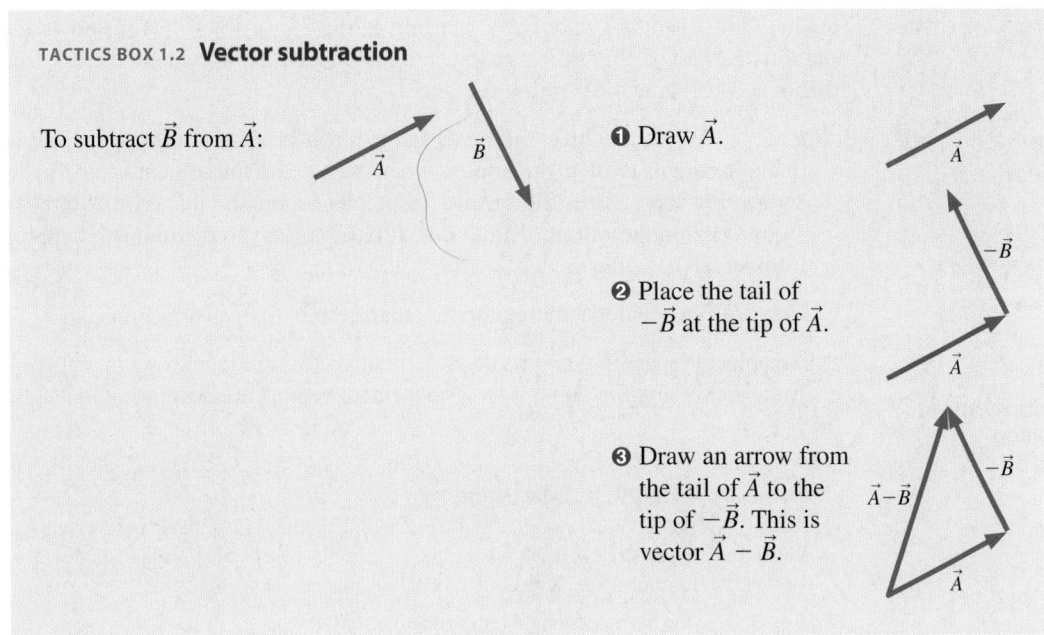

To subtract $\vec{B}$ from $\vec{A}$:

❶ Draw $\vec{A}$.

❷ Place the tail of $-\vec{B}$ at the tip of $\vec{A}$.

❸ Draw an arrow from the tail of $\vec{A}$ to the tip of $-\vec{B}$. This is vector $\vec{A} - \vec{B}$.

Figure 1.9 shows how to use the vector subtraction rules to find the displacement $\Delta\vec{r}$ in going from position $\vec{r}_i$ to position $\vec{r}_f$.

(a)

(b)

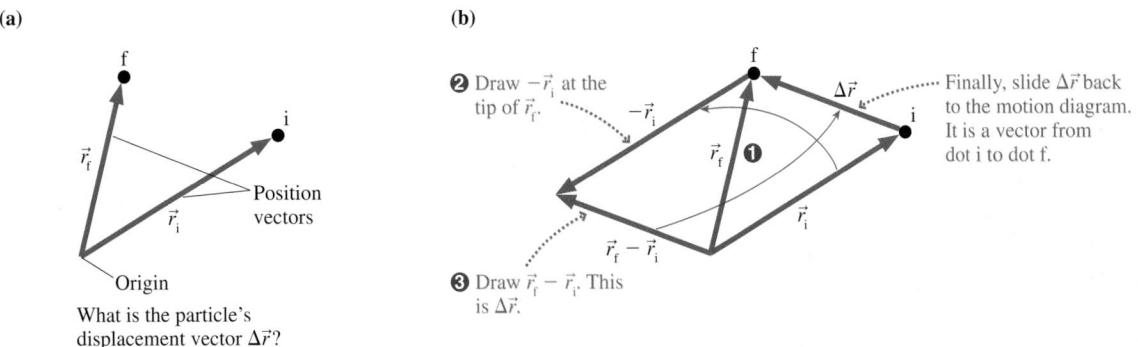

❷ Draw $-\vec{r}_i$ at the tip of $\vec{r}_f$.

Finally, slide $\Delta\vec{r}$ back to the motion diagram. It is a vector from dot i to dot f.

❸ Draw $\vec{r}_f - \vec{r}_i$. This is $\Delta\vec{r}$.

Position vectors

Origin

What is the particle's displacement vector $\Delta\vec{r}$?

FIGURE 1.9 Using vector subtraction to find $\Delta\vec{r}$.

Change in Time

It's also useful to consider a *change* in time. For example, the clock readings of two frames of film might be t_1 and t_2. The specific values are arbitrary because they are timed relative to an arbitrary instant that you chose to call $t = 0$. But the **time interval** $\Delta t = t_2 - t_1$ is *not* arbitrary. It represents the elapsed time for the object to move from one position to the next. All observers will measure the same value for Δt, regardless of when they choose to start their clocks.

The time interval $\Delta t = t_f - t_i$ measures the elapsed time as an object moves from an initial position $\vec{r}_i$ at time t_i to a final position $\vec{r}_f$ at time t_f. The value of Δt is independent of the specific clock used to measure the times.

Application to Motion Diagrams

A motion diagram is a series of position measurements. Suppose an object moves from position $\vec{r}_n$ in frame n of a motion diagram to position $\vec{r}_{n+1}$ in frame $n + 1$. As Figure 1.10 on the next page shows, the displacement vector $\Delta\vec{r}$ is simply the vector drawn from dot n to dot $n + 1$.

The first step in analyzing a motion diagram is to determine all of the displacement vectors by drawing arrows connecting each dot to the next. Label each

A stopwatch is used to measure a time interval.

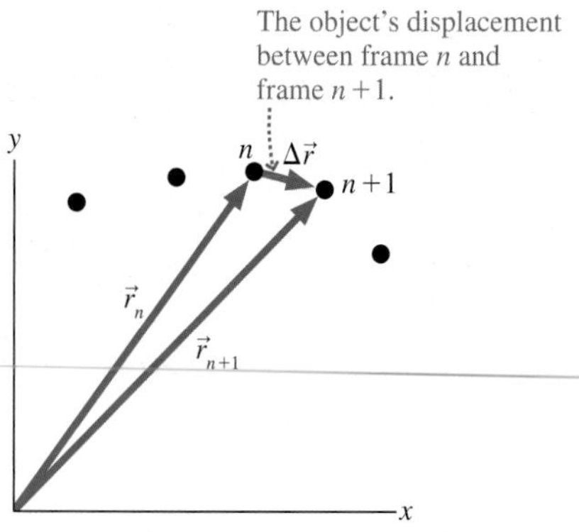

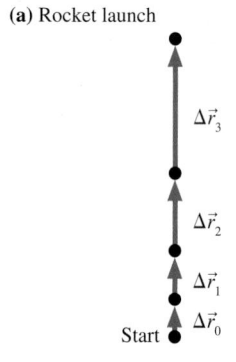

The object's displacement between frame n and frame $n+1$.

FIGURE 1.10 Drawing a displacement vector on a motion diagram.

(a) Rocket launch

$\Delta \vec{r}_3$

$\Delta \vec{r}_2$

$\Delta \vec{r}_1$

Start $\Delta \vec{r}_0$

(b) Car stopping

$\Delta \vec{r}_0 \quad \Delta \vec{r}_1 \quad \Delta \vec{r}_2 \quad \Delta \vec{r}_3$

Stop

FIGURE 1.11 Motion diagrams with the displacement vectors.

arrow with a *vector* symbol $\Delta \vec{r}_n$, starting with $n = 0$. Figure 1.11 shows the motion diagrams of Figure 1.4 redrawn to include the displacement vectors. You do not need to show the position vectors.

NOTE ▶ When an object either starts from rest or ends at rest, the initial or final dots are *as close together* as you can draw the displacement vector arrow connecting them. In addition, just to be clear, you should write "Start" or "Stop" beside the initial or final dot. It is important to distinguish stopping from merely slowing down. ◀

Now we can conclude, more precisely than before, that, as time proceeds:

- An object is speeding up if its displacement vectors are increasing in length.
- An object is slowing down if its displacement vectors are decreasing in length.

EXAMPLE 1.1 Headfirst into the snow
Alice is sliding along a smooth, icy road on her sled when she suddenly runs head-first into a large, very soft snowbank that gradually brings her to a halt. Draw a motion diagram for Alice. Show and label all displacement vectors.

MODEL Use the particle model to represent Alice as a dot.

VISUALIZE Figure 1.12 shows Alice's motion diagram. The problem statement suggests that Alice's speed is very nearly constant until she hits the snowbank. Thus her displacement vectors are of equal length as she slides along the icy road. She begins slowing when she hits the snowbank, so the displacement vectors then get shorter until she stops. It is reasonable to assume that her stopping distance in the snow is less than the distance she had slid along the road, but we do not want to make her stop *too* quickly.

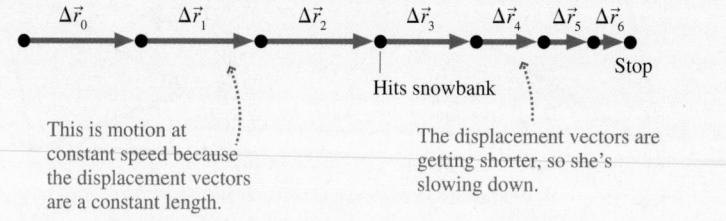

$\Delta \vec{r}_0 \qquad \Delta \vec{r}_1 \qquad \Delta \vec{r}_2 \qquad \Delta \vec{r}_3 \quad \Delta \vec{r}_4 \quad \Delta \vec{r}_5 \ \Delta \vec{r}_6$

Stop

Hits snowbank

This is motion at constant speed because the displacement vectors are a constant length.

The displacement vectors are getting shorter, so she's slowing down.

FIGURE 1.12 Alice's motion diagram.

To summarize the main idea of this section, we have added coordinate systems and clocks to our motion diagrams in order to measure *when* each frame was exposed and *where* the object was located at that time. Different observers of the motion may choose different coordinate systems and different clocks. Thus a particular value of the position $\vec{r}$ or the time t is arbitrary because each is measured relative to an arbitrarily chosen origin. However, all observers find the *same* values for the displacements $\Delta \vec{r}$ and the time intervals Δt because these are independent of the specific coordinate system used to measure them.

1.4 Velocity

It's no surprise that, during a given time interval, a speeding bullet travels farther than a speeding snail. To extend our study of motion so that we can compare the bullet to the snail, we need a way to measure how fast or how slowly an object moves.

One quantity that measures an object's fastness or slowness is its **average speed,** defined as the ratio

$$\text{average speed} = \frac{\text{distance traveled}}{\text{time interval spent traveling}} \qquad (1.3)$$

If you drive 15 miles (mi) in 30 minutes ($\frac{1}{2}$ hour), your average speed is

$$\text{average speed} = \frac{15 \text{ mi}}{\frac{1}{2} \text{ hour}} = 30 \text{ mph} \qquad (1.4)$$

Although the concept of speed is widely used in our day-to-day lives, it is not a sufficient basis for a science of motion. To see why, imagine you're trying to land a jet plane on an aircraft carrier. It matters a great deal to you whether the aircraft carrier is moving at 20 mph (miles per hour) to the north or 20 mph to the east. Simply knowing that the boat's speed is 20 mph is not enough information! The difficulty with speed is that it tells us nothing about the *direction* in which an object is moving.

A more useful question than "How fast does an object move?" is the question "How quickly (or slowly) does an object move from one position to another?" These may seem to be the same question, but they're not. An object's change of position is measured by its displacement $\Delta \vec{r}$, a vector quantity. The displacement tells us not only the distance traveled by an object, but also the *direction* of motion.

Consider the ratio $\Delta \vec{r}/\Delta t$ for an object that undergoes a displacement $\Delta \vec{r}$ during the time interval Δt. This ratio is a vector, because $\Delta \vec{r}$ is a vector, so it has both a magnitude and a direction. The size, or magnitude, of this ratio is very similar to the definition of speed: The ratio will be larger for a fast object than for a slow object. But in addition to measuring how fast an object moves, this ratio is a vector that points in the same direction as $\Delta \vec{r}$. That is, it points in the direction of motion.

It is convenient to give this ratio a name. We call it the **average velocity,** and it has the symbol $\vec{v}_{\text{avg}}$. **The average velocity of an object during the time interval Δt, in which the object undergoes a displacement $\Delta \vec{r}$, is the vector**

$$\vec{v}_{\text{avg}} = \frac{\Delta \vec{r}}{\Delta t} \qquad (1.5)$$

An object's average-velocity vector points in the same direction as the displacement vector $\Delta \vec{r}$. This is the direction of motion.

> NOTE ▶ In everyday language we do not make a distinction between speed and velocity, but in physics *the distinction is very important*. In particular, speed is simply "How fast," whereas velocity is "How fast, and in which direction." As we go along we will be giving other words more precise meaning in physics than they have in everyday language. ◀

As an example, Figure 1.13a shows two ships that start from the same position and move 5 miles in 15 minutes. Both ships have a speed of 20 mph, but their velocities are different. Because their displacements during Δt are $\Delta \vec{r}_A = $ (5 mi, north) and $\Delta \vec{r}_B = $ (5 mi, east), we can write their velocities as

$$\vec{v}_{\text{avg A}} = (20 \text{ mph, north})$$
$$\vec{v}_{\text{avg B}} = (20 \text{ mph, east})$$
$$(1.6)$$

Notice how the velocity *vectors* in Figure 1.13b point in the direction of motion.

> NOTE ▶ Our goal in this chapter is to *visualize* motion with motion diagrams. Strictly speaking, the vector we have defined in Equation 1.5, and the vector we will show on motion diagrams, is the *average* velocity $\vec{v}_{\text{avg}}$. But to allow the motion diagram to be a useful tool, we will drop the subscript and refer to the average velocity as simply $\vec{v}$. Our definitions and symbols, which somewhat blur the distinction between average and instantaneous quantities, are adequate for visualization purposes, but they're not the final word on the subject. We will refine these definitions in Chapter 2, where our goal will be to develop the mathematics of motion. ◀

The victory goes to the runner with the highest average speed.

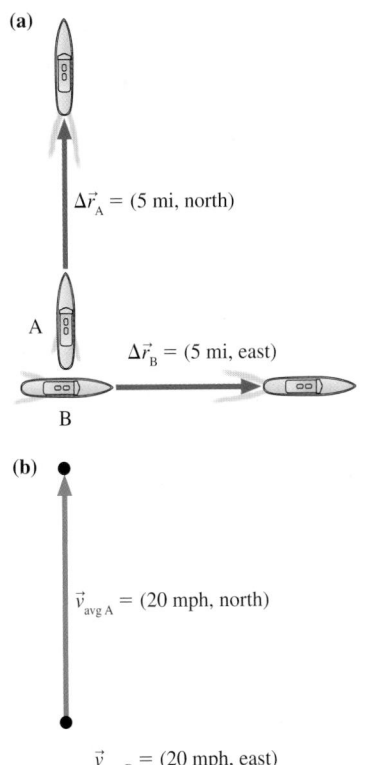

FIGURE 1.13 The displacement vectors and velocities of ships A and B.

Motion Diagrams with Velocity Vectors

The velocity vector, as we've defined it, points in the same direction as the displacement $\Delta \vec{r}$, and the length of $\vec{v}$ is directly proportional to the length of $\Delta \vec{r}$. Consequently, the vectors connecting each dot of a motion diagram to the next, which we previously labeled as displacement vectors, could equally well be identified as velocity vectors.

This idea is illustrated in Figure 1.14, which shows four frames from the motion diagram of a tortoise racing a hare. The vectors connecting the dots are now labeled as velocity vectors $\vec{v}$. The length of each velocity vector is no longer a distance. Instead, **the length of a velocity vector represents the average speed with which the object moves between the two points.** Longer velocity vectors indicate faster motion. You can see from the diagram that the hare moves faster than the tortoise.

Notice that the hare's velocity vectors do not change; each has the same length and direction. We say the hare is moving with *constant velocity*. The tortoise is also moving with its own constant velocity.

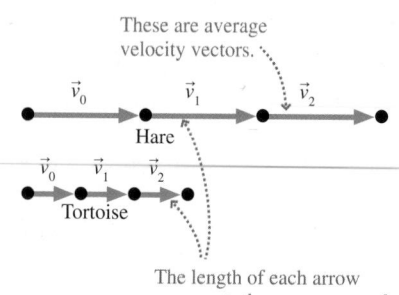

These are average velocity vectors.

The length of each arrow represents the average speed. The hare moves faster than the tortoise.

FIGURE 1.14 Motion diagram of the tortoise racing the hare.

EXAMPLE 1.2 Accelerating up a hill

The light turns green and a car accelerates, starting from rest, up a 20° hill. Draw a motion diagram that shows the car's velocity.

MODEL Use the particle model to represent the car as a dot.

VISUALIZE The car's motion takes place along a straight line, but the line is neither horizontal nor vertical. Because a motion diagram is made from frames of a movie, it will show the object moving with the correct orientation—in this case, at an angle of 20°. Figure 1.15 shows several frames of the motion diagram, where we see the car speeding up. The car starts from rest, so the first arrow is drawn as short as possible and the first dot is labeled "Start." The displacement vectors have been drawn from each dot to the next, but then they have been identified and labeled as average velocity vectors $\vec{v}$.

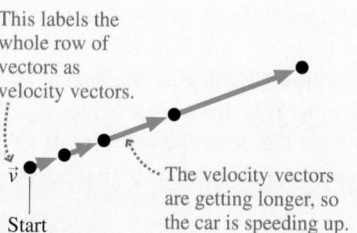

This labels the whole row of vectors as velocity vectors.

Start

The velocity vectors are getting longer, so the car is speeding up.

FIGURE 1.15 Motion diagram of a car accelerating up a hill.

NOTE ▶ Rather than label every single vector, it's easier to give one label to the entire row of velocity vectors. You can see this in Figure 1.15. ◀

EXAMPLE 1.3 It's a hit!

Jake hits a ball at a 60° angle. It is caught by Jim. Draw a motion diagram of the ball.

MODEL This example is typical of how many problems in science and engineering are worded. The problem does not give a clear statement of where the motion begins or ends. Are we interested in the motion of the ball just during the time it is in the air between Jake and Jim? What about the motion *as* Jake hits it (ball rapidly speeding up) or *as* Jim catches it (ball rapidly slowing down)? Should we include Jim dropping the ball after he catches it? The point is that *you* will often be called on to make a *reasonable interpretation* of a problem statement. In this problem, the details of hitting and catching the ball are complex. The motion of the ball through the air is easier to describe, and it's a motion you might expect to learn about in a physics class. So our *interpretation* is that the motion diagram should start as the ball leaves Jake's bat (ball already moving) and should end the instant it touches Jim's hand (ball still moving). We will model the ball as a particle.

VISUALIZE With this interpretation in mind, Figure 1.16 shows the motion diagram of the ball. Notice how, in contrast to the car of Figure 1.15, the ball is already moving as the motion diagram movie begins. As before, the average velocity vectors are found by connecting the dots with arrows. You can see that the average velocity vectors get shorter (ball slowing down), get longer (ball speeding up), and change direction. Each $\vec{v}$ is different, so this is *not* constant velocity motion.

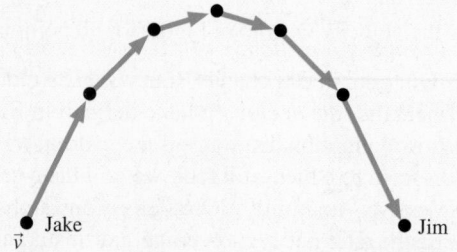

Jake

Jim

FIGURE 1.16 Motion diagram of a ball traveling from Jake to Jim.

Relating Position to Velocity

We defined the average velocity $\vec{v}$ in terms of the displacement $\Delta\vec{r}$, but let's turn that around. Suppose a ball at position $\vec{r}_1$ has velocity $\vec{v}$. As you can see in Figure 1.17, the ball is displaced from its starting position $\vec{r}_1$ to a new position $\vec{r}_2$ during the time interval Δt. We can combine $\vec{v} = \Delta\vec{r}/\Delta t$ and $\Delta\vec{r} = \vec{r}_2 - \vec{r}_1$ to write

$$\vec{r}_2 = \vec{r}_1 + \vec{v}\,\Delta t \tag{1.7}$$

Knowing an object's velocity is the key to finding its position at later instants of time.

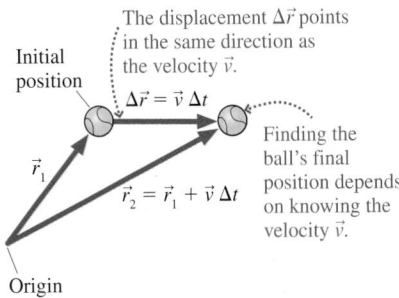

FIGURE 1.17 A ball with average velocity $\vec{v}$ moves from position $\vec{r}_1$ to position $\vec{r}_2$.

STOP TO THINK 1.3 A particle moves from position 1 to position 2 during the interval Δt. Which vector shows the particle's average velocity?

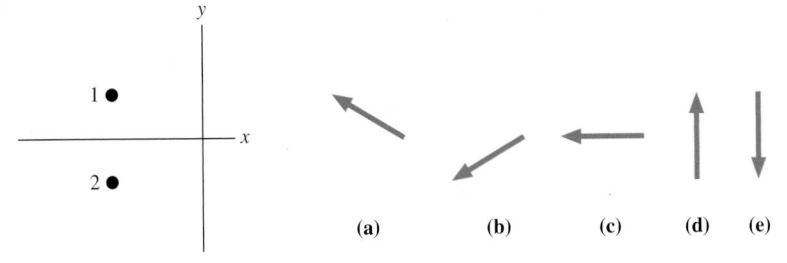

1.5 Acceleration

The goal of this chapter is to find a set of concepts with which to describe motion. Position, time, and velocity are important concepts, and at first glance they might appear to be sufficient. But that is not the case. Sometimes an object's velocity is constant, as it was in Figure 1.14. More often, an object's velocity changes as it moves, as in Figures 1.15 and 1.16. We need one more motion concept, one that will describe a *change* in the velocity.

Because velocity is a vector, it can change in two possible ways:

1. The magnitude can change, indicating a change in speed, or
2. The direction can change, indicating that the object has changed direction.

Figure 1.15 showed the motion diagram of a car speeding up. That was an example in which the magnitude of the velocity vector changed but not the direction.

As an example where only the direction changes, but not the magnitude, Figure 1.18 is the motion diagram of a runner going at constant speed around a circular track. The lengths of all the average velocity vectors are the same, showing that the runner's speed is constant, but the direction of each velocity vector is different. Thus Figure 1.18 also represents a changing velocity.

How can we measure the change of velocity in a meaningful way? When we wanted to measure changes in position, the ratio $\Delta\vec{r}/\Delta t$ was useful. This ratio is the *rate of change of position*. By analogy, consider an object whose velocity changes from $\vec{v}_1$ to $\vec{v}_2$ during the time interval Δt. The ratio $\Delta\vec{v}/\Delta t$, where $\Delta\vec{v} = \vec{v}_2 - \vec{v}_1$, is the *rate of change of velocity*. But what does it measure?

Consider two cars, a Volkswagen Beetle and a fancy Porsche. Let them start from rest, and measure their velocities after an elapsed time of 10 seconds. The Porsche, we can assume, will have a larger $\Delta\vec{v}$. Consequently, it will have the larger value of the ratio $\Delta\vec{v}/\Delta t$.

Restart the two cars, but now measure the time interval Δt it takes for each of them to reach a speed of 60 mph. The Porsche will have a smaller value of Δt, which again gives it a larger value of the ratio $\Delta\vec{v}/\Delta t$. Thus this ratio appears to measure how quickly the car speeds up. It has a large magnitude for objects that

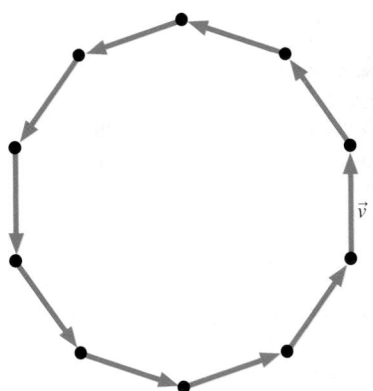

The lengths of the velocity vectors are the same, indicating constant speed, but the direction of each vector is different. This is a changing velocity.

FIGURE 1.18 Motion diagram of a runner on a circular track.

speed up quickly and a small magnitude for objects that speed up slowly. The ratio $\Delta\vec{v}/\Delta t$ is called the **average acceleration,** and its symbol is $\vec{a}_{avg}$. The average acceleration of an object during the time interval Δt, in which the object's velocity changes by $\Delta\vec{v}$, is the vector

$$\vec{a}_{avg} = \frac{\Delta\vec{v}}{\Delta t} \tag{1.8}$$

An object's average-acceleration vector points in the same direction as the vector $\Delta\vec{v}$. Note that acceleration, like position and velocity, is a vector. Both its magnitude and its direction are important pieces of information.

Acceleration is a fairly abstract concept. Position and time are our real hands-on measurements of an object, and they are easy to understand. You can "see" where the object is located and the time on the clock. Velocity is a bit more abstract, being a relationship between the change of position and the change of time. Motion diagrams help us visualize velocity as the vector arrows connecting one position of the object to the next. Acceleration is an even more abstract idea about changes in the velocity. Yet it is essential to develop a good intuition about acceleration because it will be a key concept for understanding why objects move as they do.

NOTE ▶ As we did with velocity, we will drop the subscript and refer to the average acceleration as simply $\vec{a}$. This is adequate for visualization purposes, but not the final word on the subject. We will refine the definition of acceleration in Chapter 2. ◀

Finding the Acceleration Vectors on a Motion Diagram

Let's look at how we can determine the average acceleration vector $\vec{a}$ from a motion diagram. From its definition, we see that $\vec{a}$ points in the same direction as $\Delta\vec{v}$, the change of velocity. This critically important idea is the basis for a technique to find $\vec{a}$.

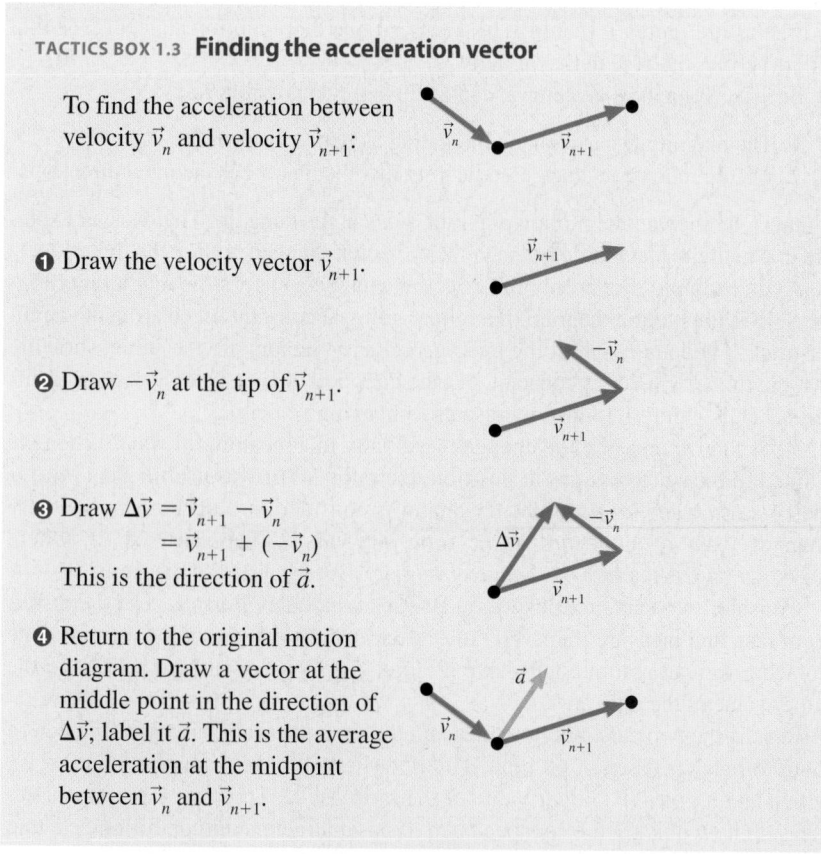

TACTICS BOX 1.3 Finding the acceleration vector

To find the acceleration between velocity $\vec{v}_n$ and velocity $\vec{v}_{n+1}$:

❶ Draw the velocity vector $\vec{v}_{n+1}$.

❷ Draw $-\vec{v}_n$ at the tip of $\vec{v}_{n+1}$.

❸ Draw $\Delta\vec{v} = \vec{v}_{n+1} - \vec{v}_n$
$\quad\quad = \vec{v}_{n+1} + (-\vec{v}_n)$
This is the direction of $\vec{a}$.

❹ Return to the original motion diagram. Draw a vector at the middle point in the direction of $\Delta\vec{v}$; label it $\vec{a}$. This is the average acceleration at the midpoint between $\vec{v}_n$ and $\vec{v}_{n+1}$.

Notice that the acceleration vector goes beside the dot, not beside the velocity vectors. This is because each acceleration vector is determined as the *difference* between the two velocity vectors on either side of a dot. The length of $\vec{a}$ does not have to be the exact length of $\Delta\vec{v}$; it is the direction of $\vec{a}$ that is most important.

The procedure of Tactics Box 1.3 can be repeated to find $\vec{a}$ at each point in the motion diagram. Note that we cannot determine $\vec{a}$ at the first and last points because we have only one velocity vector and can't find $\Delta\vec{v}$.

We can turn the Equation 1.8 definition of average acceleration around. Suppose a ball moving with velocity $\vec{v}_1$ has acceleration $\vec{a}$. Because of the acceleration, the ball's velocity changes during the time interval Δt to velocity $\vec{v}_2$. We can combine $\vec{a} = \Delta\vec{v}/\Delta t$ and $\Delta\vec{v} = \vec{v}_2 - \vec{v}_1$ to write

$$\vec{v}_2 = \vec{v}_1 + \vec{a}\,\Delta t \qquad (1.9)$$

Knowing an object's acceleration is the key to finding its velocity at later instants of time.

The Complete Motion Diagram

You've now seen several *Tactics Boxes* that help you achieve specific tasks. Tactics Boxes will appear in nearly every chapter in this book. We'll also, where appropriate, provide *Problem-Solving Strategies*. Problem solving will be discussed in more detail later in the chapter, but this is a good place for the first problem-solving strategy.

(MP) PROBLEM-SOLVING STRATEGY 1.1 **Motion diagrams**

MODEL Represent the moving object as a particle. Make simplifying assumptions when interpreting the problem statement.

VISUALIZE A complete motion diagram consists of:

- The position of the object in each frame of the film, shown as a dot. Use five or six dots to make the motion clear but without overcrowding the picture. More complex motions may need more dots.
- The average velocity vectors, found by connecting each dot in the motion diagram to the next with a vector arrow. There is *one* velocity vector linking each *two* position dots. Label the row of velocity vectors $\vec{v}$.
- The average acceleration vectors, found using Tactics Box 1.3. There is *one* acceleration vector linking each *two* velocity vectors. Each acceleration vector is drawn at the dot between the two velocity vectors it links. Use $\vec{0}$ to indicate a point at which the acceleration is zero. Label the row of acceleration vectors $\vec{a}$.

STOP TO THINK 1.4   A particle undergoes acceleration $\vec{a}$ while moving from point 1 to point 2. Which of the choices shows the velocity vector $\vec{v}_2$ as the particle moves away from point 2?

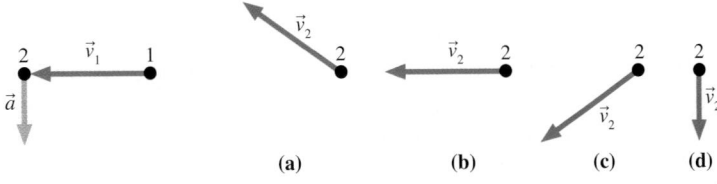

(a)   (b)   (c)   (d)

1.6 Examples of Motion Diagrams

This section will look at examples of the full strategy for drawing motion diagrams.

EXAMPLE 1.4 Skiing through the woods

A skier glides along smooth, horizontal snow at constant speed, then speeds up going down a hill. Draw the skier's motion diagram.

MODEL Represent the skier as a particle. It's reasonable to assume that the downhill slope is a straight line.

VISUALIZE Figure 1.19 shows a complete motion diagram of the skier. The dots are equally spaced for the horizontal motion, indicating constant speed, then the dots get further apart as the skier speeds up down the hill. The insets show how the average acceleration vector $\vec{a}$ is determined. All the other acceleration vectors along the slope will be similar to the one shown because each velocity vector is longer than the preceding one. The acceleration at the point where the slope changes is a little tricky because the velocity changes in both length and direction, but the procedure in Tactics Box 1.3 is designed to handle situations exactly like this. Notice that we've explicitly written $\vec{0}$ for the acceleration beside the dots where the velocity is constant.

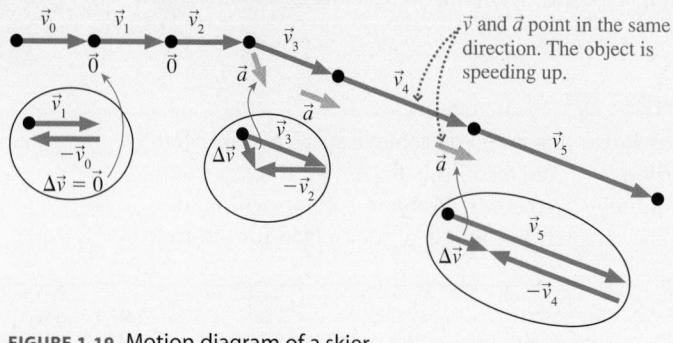

FIGURE 1.19 Motion diagram of a skier.

EXAMPLE 1.5 The first astronauts land on Mars

A spaceship carrying the first astronauts to Mars descends safely to the surface. Draw a motion diagram for the last few seconds of the descent.

MODEL Represent the spaceship as a particle. It's reasonable to assume that its motion in the last few seconds is straight down. The problem ends as the spacecraft touches the surface.

VISUALIZE Figure 1.20 shows a complete motion diagram as the spaceship descends and slows, using its rockets, until it comes to rest on the surface. Notice how the dots get closer together as it slows. The inset shows how the acceleration vector $\vec{a}$ is determined at one point. All the other acceleration vectors will be similar, because for each pair of velocity vectors the earlier one is longer than the later one.

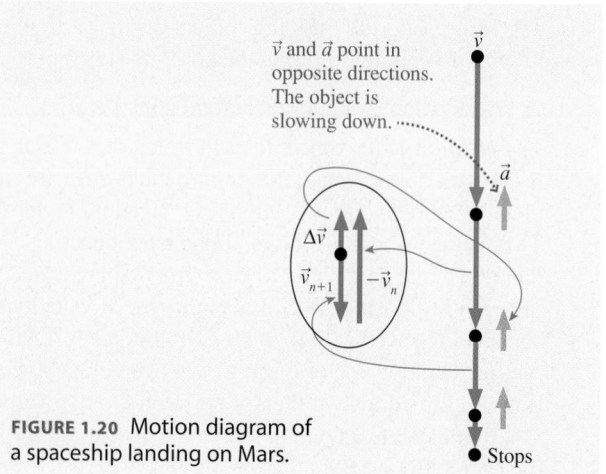

FIGURE 1.20 Motion diagram of a spaceship landing on Mars.

Speeding Up and Slowing Down

Notice something interesting in Figures 1.19 and 1.20. Where the object is speeding up, the acceleration and velocity vectors point in the *same direction*. Where the object is slowing down, the acceleration and velocity vectors point in *opposite directions*. These results, which are consistent with Equation 1.9, are always true for motion in a straight line. **For motion along a line:**

- An object is speeding up if and only if $\vec{v}$ and $\vec{a}$ point in the same direction.
- An object is slowing down if and only if $\vec{v}$ and $\vec{a}$ point in opposite directions.
- An object's velocity is constant if and only if $\vec{a} = \vec{0}$.

NOTE ▶ In everyday language, we use the word *accelerate* to mean "speed up" and the word *decelerate* to mean "slow down." But speeding up and slowing down are both changes in the velocity and consequently, by our definition, *both* are accelerations. In physics, *acceleration* refers to changing the velocity, no matter what the change is, and not just to speeding up. ◄

More Examples of Motion Diagrams

EXAMPLE 1.6 At the amusement park

Anne rides the Ferris wheel at an amusement park. Draw Anne's motion diagram.

MODEL Represent Anne as a particle. Assume that the Ferris wheel turns at constant speed.

VISUALIZE The motion diagram of Figure 1.21 uses 10 frames of film to show one complete revolution of the Ferris wheel. The seat on the Ferris wheel moves in a circle at a constant speed, so we've shown equal distances between each dot and the next. As before, the velocity vectors are found by connecting each dot to the next. Note that the velocity vectors are *straight lines,* not curves. This is because the displacement vectors $\Delta\vec{r}$ are *not* the same as the actual path followed.

Although Anne moves at constant speed, she does *not* move with constant velocity. All the velocity vectors have the same length, but each has a different *direction,* and that means Anne is accelerating. This is not a "speeding up" or "slowing down" acceleration, but it is still a change of the velocity with time. The inset to Figure 1.21 shows how to find the acceleration at the bottom of the circle. Vector $\vec{v}_n$ is the velocity vector that leads into this dot, while $\vec{v}_{n+1}$ moves away from it. From the circular geometry of the main figure, the two angles marked α are equal. Thus we see that $\vec{v}_{n+1}$ and $-\vec{v}_n$ form an isosceles triangle and vector $\Delta\vec{v}$ is exactly vertical. When $\vec{a}$ is drawn on the

motion diagram, in the same direction as $\Delta\vec{v}$, we see that the acceleration vector points directly to the center of the circle.

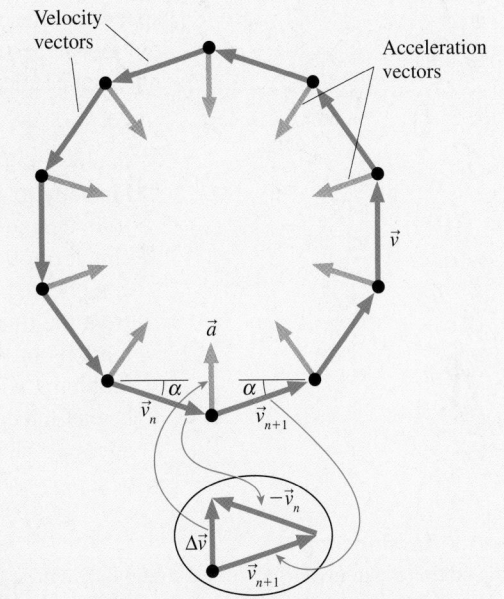

FIGURE 1.21 Anne's motion diagram on the Ferris wheel.

No matter which dot you select on the motion diagram in Figure 1.21, the velocities leading to and away from that dot change in such a way as to cause the acceleration to point directly to the center of the circle. You should convince yourself of this by finding $\vec{a}$ at several other points around the circle. An acceleration that always points directly toward the center of a circle is called a *centripetal acceleration*. The word "centripetal" comes from a Greek root meaning "center seeking." We will have a lot to say about centripetal acceleration in later chapters.

EXAMPLE 1.7 Throwing the shot

The Hulk is an Olympic shot putter. Draw a motion diagram for the shot from the moment it leaves the Hulk's hand until it hits the ground.

MODEL Represent the shot as a particle.

VISUALIZE The shot is an example of projectile motion. You probably know, from watching baseballs, basketballs, rocks, and so on, that projectiles move in some sort of an arc. Projectiles also slow down as they rise and speed up again as they fall. The top of the arc is where the object has the slowest motion, which is why a ball seems to hesitate there before falling, but it is not at rest. Figure 1.22a shows a photo of a projectile moving through the air. The motion diagram of the shot is shown in Figure 1.22b. Notice that the shot does *not* start from rest; it leaves the Hulk's hand with an initial velocity. The velocity vectors along the ascent get

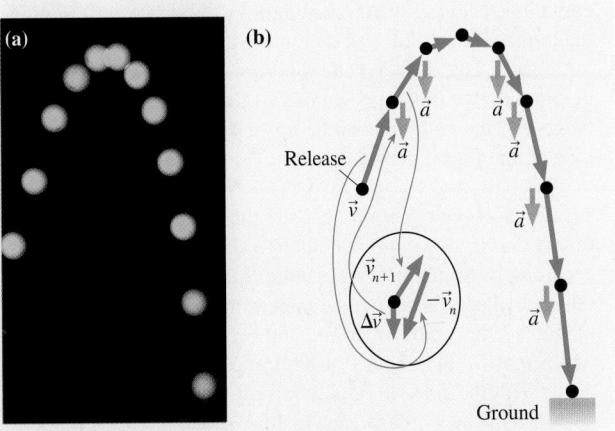

FIGURE 1.22 Photograph and motion diagram of a projectile.

shorter, because the shot slows while rising, then lengthen again as the shot falls. The last position is at the instant of impact.

Both the length and the direction of the velocity vector are changing, so the shot must have an acceleration. The inset shows the determination of $\vec{a}$ at one point along the trajectory. The acceleration must be pointed more or less downward, but we cannot determine the exact direction of $\vec{a}$ without further information about the exact shape of the arc. It turns out, for projectile motion, that the acceleration vector at every point is *straight down.* This is a conclusion that we will justify in Chapter 6 but will simply assert, without proof, for now. This very specific acceleration is called the *acceleration due to gravity.*

It is a good idea to compare the Example 1.7 projectile motion to the constant-speed circular motion in Example 1.6. Both objects follow curved trajectories, but the shapes of the curves are different. Consequently, the motions have very different acceleration vectors.

NOTE ▶ Example 1.7 revisits the important issue of how we interpret problems in physics. The Hulk clearly had to flex his muscles and move various parts of his body to get the shot moving. That part of the motion is complicated. But from the moment the shot leaves his hand until it hits the ground, it has a fairly simple motion. The impact and any subsequent motion again become very complicated. In this case, you were explicitly asked to consider the motion only from the time the Hulk releases the shot until it hits the ground; the complex motions of the throw and the impact are not part of the problem. But many problems will not be this specific. *You* have to interpret when the problem begins and ends. The key is to focus on isolating those parts of the problem that are essential while disregarding the nonessential aspects. ◀

EXAMPLE 1.8 **Tossing a ball**

Draw the motion diagram of a ball tossed straight up in the air.

MODEL This problem calls for some interpretation. Should we include the toss itself, or only the motion after the tosser releases the ball? Should we include the ball hitting the ground? It appears that this problem is really concerned with the ball's motion through the air. Consequently, we begin the motion diagram at the moment that the tosser releases the ball and end the diagram at the moment the ball hits the ground. We will consider neither the toss nor the impact. And, of course, we will represent the ball as a particle.

VISUALIZE We have a slight difficulty here because the ball retraces its route as it falls. A literal motion diagram would show the upward motion and downward motion on top of each other, leading to confusion. We can avoid this difficulty by horizontally separating the upward motion and downward motion diagrams. This will not affect our conclusions because it does not change any of the vectors. Figure 1.23 shows the motion diagram drawn this way. Notice that the very top dot is shown twice—as the end point of the upward motion and the beginning point of the downward motion.

The ball slows down as it rises. You've learned that the acceleration vectors point opposite the velocity vectors for an object that is slowing down along a line, and they are shown accordingly. Similarly, $\vec{a}$ and $\vec{v}$ point in the same direction as the falling ball speeds up. Notice something interesting: the acceleration vectors point downward both while the ball is rising *and* while it is falling. Both "speeding up" and "slowing down" occur with the *same* acceleration vector. This is an important conclusion, one worth pausing to think about.

Now let's look at the top point on the ball's trajectory. The velocity vectors are pointing upward but getting shorter as the ball approaches the top. As it starts to fall, the velocity vectors are pointing downward and getting longer. There must be a moment—just an instant as $\vec{v}$ switches from pointing up to pointing down—when the velocity is zero. Indeed, the ball's velocity *is* zero for an instant at the precise top of the motion!

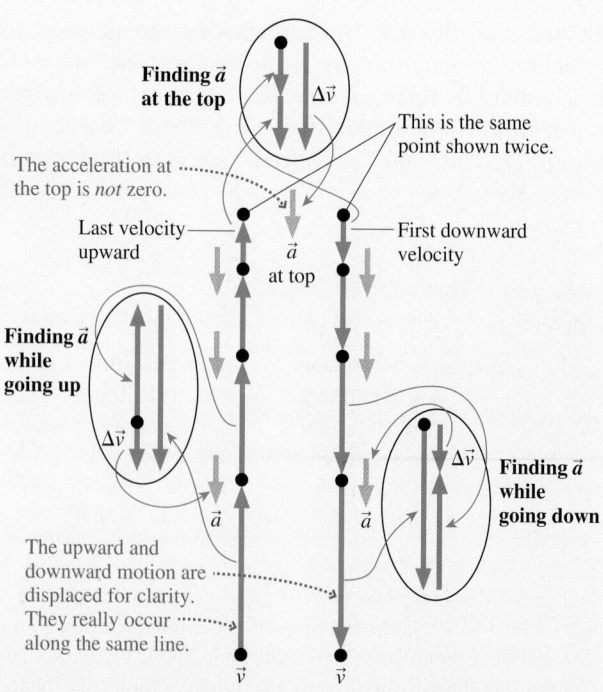

FIGURE 1.23 Motion diagram of a ball tossed straight up in the air.

But what about the acceleration at the top? The inset shows how the average acceleration is determined from the last upward velocity before the top point and the first downward velocity. We find that the acceleration at the top is pointing downward, just as it does elsewhere in the motion.

Many people expect the acceleration to be zero at the highest point. But recall that the velocity at the top point *is* changing—from up to down. If the velocity is changing, there *must* be an acceleration. A downward-pointing acceleration vector is needed to turn the velocity vector from up to down. Another way to think about this is to note that zero acceleration would mean no change of velocity. When the ball reached zero velocity at the top, it would hang there and not fall if the acceleration were also zero!

The motion of the ball in Example 1.8 is really a special case of projectile motion. The difference between the tossed ball and the shot thrown by the Hulk is that the ball happens to start out moving exactly vertically. We saw in Example 1.7 that the acceleration is a constant downward vector for projectile motion. That is exactly what we have found again for the ball.

1.7 From Words to Symbols

Physics is not mathematics. Math problems are clearly stated, such as "What is 2 + 2?" Physics is about the world around us, and to describe that world we must use language. Now, language is wonderful—we couldn't communicate without it—but it can sometimes be imprecise or ambiguous.

The challenge when reading a physics problem is to translate the words into symbols that can be manipulated, calculated, and graphed. This translation from words to symbols is the heart of problem solving in physics. This is the point where ambiguous words and phrases must be clarified, where the imprecise must be made precise, and where you arrive at an understanding of exactly what the question is asking.

You may have been told that the first step in solving a physics problem is to "draw a picture," but perhaps you didn't know why or what to draw. The purpose of drawing a picture is to aid you in the words-to-symbols translation. And there really is a *method* for drawing pictures, one that will help you be a better problem solver. It is called the **pictorial representation** of the problem.

TACTICS BOX 1.4 Drawing a pictorial representation

❶ **Sketch the situation.** Not just any sketch. Show the object at the *beginning* of the motion, at the *end,* and at any point where the character of the motion changes. Very simple drawings are adequate.

❷ **Establish a coordinate system.** Select your axes and origin to match the motion.

❸ **Define symbols.** Use the sketch to define symbols representing quantities such as position, velocity, acceleration, and time. *Every* variable used later in the mathematical solution should be defined on the sketch. Some will have known values, others are initially unknown, but all should be given symbolic names.

❹ **List known information.** Make a table of the quantities whose values you can determine from the problem statement or that can be found quickly with simple geometry or unit conversions. Some quantities are implied by the problem, rather than explicitly given. Others are determined by your choice of coordinate system.

❺ **Identify the desired unknowns.** What quantity or quantities will allow you to answer the question? These should have been defined as symbols in step 3. Don't list every unknown; only the one or two needed to answer the question.

It's not an overstatement to say that a well-done pictorial representation of the problem will take you halfway to the solution. The following example illustrates how to construct a pictorial representation for a problem that is typical of problems you will see in the next few chapters.

EXAMPLE 1.9 Drawing a pictorial representation

Draw a pictorial representation for the following problem:

A rocket sled accelerates at 50 m/s² for 5 seconds, then coasts for 3 seconds. What is the total distance traveled?

VISUALIZE The motion has a beginning, an end, and a point where the nature of the motion changes from accelerating to coasting. These are the three points sketched in Figure 1.24. A coordinate system has been chosen with the origin at the starting point. The quantities x, v, and t are needed at each of three *points,* so these have been defined on the sketch and distinguished by subscripts. Accelerations are associated with *intervals* between the points, so only two accelerations are defined. Values for three quantities are given in the problem statement. Others, such as $x_0 = 0$ m and $t_0 = 0$ s, are inferred from our choice of coordinate system. The value $v_{0x} = 0$ m/s is part of our *interpretation* of the problem. Finally, we identify x_2 as the quantity that will answer the question. We now understand quite a bit about the problem and would be ready to start a quantitative analysis.

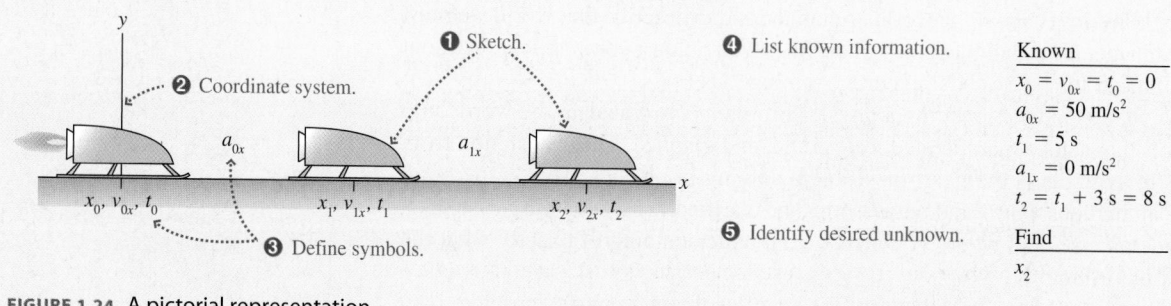

FIGURE 1.24 A pictorial representation.

We didn't *solve* the problem; that is not the purpose of the pictorial representation. The pictorial representation is a systematic way to go about interpreting a problem and getting ready for a mathematical solution. Although this is a simple problem, and you probably know how to solve it if you've taken physics before, you will soon be faced with much more challenging problems. Learning good problem-solving skills at the beginning, while the problems are easy, will make them second nature later when you really need them.

Using Symbols

Symbols are a language that allows us to talk with precision about the relationships in a problem. As with any language, we all need to agree to use words or symbols in the same way if we want to be able to communicate with each other. Many of the ways we use symbols in science and engineering are somewhat arbitrary, often reflecting historical roots. Nonetheless, practicing scientists and engineers have come to agree on how to use the language of symbols. Learning this language is part of learning physics.

We use subscripts in physics to designate a particular point in the problem. Scientists usually label the starting point of the problem with the subscript "0," not the subscript "1" that you might expect. When using subscripts, make sure that all symbols referring to the same point in the problem have the *same numerical subscript.* To have the one point in a problem characterized by position x_1 but velocity v_{2x} is guaranteed to lead to confusion!

Notation for Motion Along a Line

Our goal in this section is simply to learn how to *analyze* a problem statement, not to solve it. We will limit ourselves to motion along a straight line. Although the position, average velocity, and average acceleration of an object are vectors, the full vector notation is not needed for motion along a line. Instead, we can measure the one-dimensional position, velocity, and acceleration of an object with the quantities x, v_x, and a_x (or y, v_y, and a_y if the motion is vertical). The vector nature of these quantities appears through their *signs:*

■ v_x (or v_y) is positive if the velocity vector $\vec{v}$ points to the right (or up). It is negative if the velocity vector $\vec{v}$ points to the left (or down).
■ a_x (or a_y) is positive if the acceleration vector $\vec{a}$ points to the right (or up). It is negative if the acceleration vector $\vec{a}$ points to the left (or down).

The appropriate sign for v is usually clear. Determining the sign of a is more difficult, and this is where a motion diagram can help.

We will examine one-dimensional motion in detail in Chapter 2, but these simple rules are enough to understand the examples and homework problems of this chapter.

Representations

A picture is one way to *represent* your knowledge of a situation. You could also represent your knowledge using words, graphs, or equations. Each **representation of knowledge** gives us a different perspective on the problem. The more tools you have for thinking about a complex problem, the more likely you are to solve it.

There are five representations of knowledge that we will use over and over:

1. The *verbal* representation. A problem statement, in words, is a verbal representation of knowledge. So is an explanation that you write.
2. The *pictorial* representation. The pictorial representation, which we've just presented, is the most literal depiction of the situation.
3. The *physical* representation. We will develop many special techniques and diagrams that allow us to analyze aspects of the physics. The motion diagram, which allows us to find the acceleration vectors, is part of the physical representation.
4. The *graphical* representation. We will make extensive use of graphs to portray information.
5. The *mathematical* representation. Equations that can be used to find the numerical values of specific quantities are the mathematical representation.

NOTE ▶ The mathematical representation is only one of many. Much of physics is more about thinking and reasoning than it is about solving equations. ◀

1.8 A Problem-Solving Strategy

One of the goals of this textbook is to help you learn a *strategy* for solving physics problems. The purpose of a strategy is to guide you in the right direction with minimal wasted effort. The problem-solving strategy shown on the next page is based on using different representations of knowledge.

Throughout this textbook we will emphasize the first two steps. They are the *physics* of the problem, as opposed to the mathematics of solving the resulting equations. This is not to say that those mathematical operations are always easy— in many cases they are not. But our primary goal is to understand the physics.

Building a house requires careful planning. The architect's visualization and drawings have to be complete before the detailed procedures of construction get underway. The same is true for solving problems in physics.

Problem-Solving Strategy

MODEL It's impossible to treat every detail of a situation. Simplify the situation with a model that captures the essential features. For example, the object in a mechanics problem is usually represented as a particle.

VISUALIZE This is where expert problem solvers put most of their effort.

- Draw a *pictorial representation*. This helps you assess the information you are given and starts the process of translating the problem into symbols.
- Draw a *physical representation*. This helps you visualize important aspects of the physics. Motion diagrams are part of the physical representation. Chapter 4 will introduce free-body diagrams to display information about forces.
- Use a *graphical representation* if it is appropriate for the problem.
- Go back and forth between these three representations; they need not be done in any particular order.

SOLVE Only after modeling and visualizing are complete is it time to develop a *mathematical representation* with specific equations that must be solved. All symbols used here should have been defined in the pictorial representation.

ASSESS Is your result believable? Does it have proper units? Does it make sense?

Using the Problem-Solving Strategy

A couple of final examples will summarize the ideas of this chapter. Our task, at this point, is not to *solve* the problem, but to focus on what is happening in the problem—in other words, to make the translation from words to symbols in preparation for subsequent mathematical analysis. We will do so, in each case, by developing the pictorial representation and drawing the motion diagram.

EXAMPLE 1.10 Launching a weather rocket

Use the first two steps of the problem-solving strategy to analyze the following problem:

A small rocket, such as those used for meteorological measurements of the atmosphere, is launched vertically with an acceleration of 30 m/s². It runs out of fuel after 30 seconds. What is its maximum altitude?

MODEL We need to do some interpretation. Common sense tells us that the rocket does not stop the instant it runs out of fuel. Instead, it continues upward, while slowing, until reaching its maximum altitude. This second half of the motion, after running out of fuel, is like the ball that was tossed upward in the first half of Example 1.8. Because the problem does not ask about the rocket's descent, we conclude that the problem ends at the point of maximum altitude. We'll represent the rocket as a particle.

VISUALIZE Figure 1.25 shows the physical represention (i.e., the motion diagram) and the pictorial representation. The rocket is speeding up during the first half of the motion, so $\vec{a}_0$ is parallel to $\vec{v}$. The initial acceleration $a_{0y} = 30$ m/s² is given in the problem. During the second half, as the rocket slows, $\vec{a}_1$ points downward. Thus a_{1y} is a negative number.

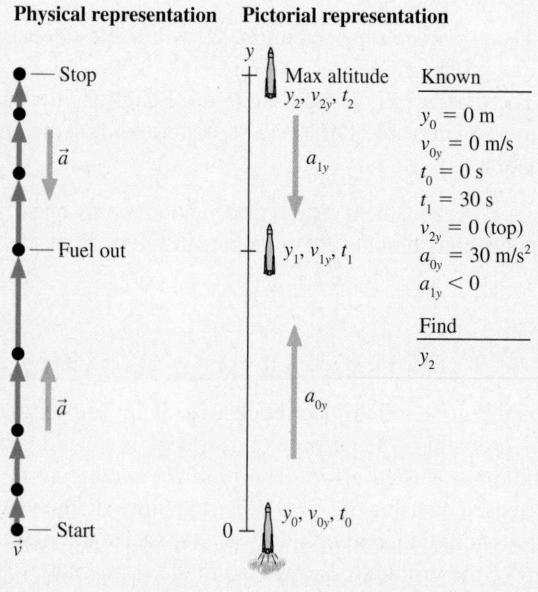

Physical representation   **Pictorial representation**

Stop

$\vec{a}$

Fuel out

$\vec{a}$

Start

$\vec{v}$

y_2, v_{2y}, t_2 — Max altitude

a_{1y}

y_1, v_{1y}, t_1

a_{0y}

y_0, v_{0y}, t_0

Known
$y_0 = 0$ m
$v_{0y} = 0$ m/s
$t_0 = 0$ s
$t_1 = 30$ s
$v_{2y} = 0$ (top)
$a_{0y} = 30$ m/s²
$a_{1y} < 0$

Find
y_2

FIGURE 1.25 Physical and pictorial representations for the rocket.

This information is then transferred to the pictorial representation. Although the velocity v_{2y} wasn't given in the problem statement, we know it must be zero at the very top of the trajectory. Lastly, we have identified y_2 as the desired unknown. This, of course, is not the only unknown in the problem, but it is the one we are specifically asked to find.

ASSESS If you've had a previous physics class, you may be tempted to assign a_{1y} the value -9.8 m/s^2, the acceleration due to gravity. However, that would be true only if there is no air resistance on the rocket. We will need to consider the *forces* acting on the rocket during the second half of its motion before we can determine a value for a_{1y}. For now, all that we can safely conclude is that a_{1y} is negative.

EXAMPLE 1.11 Making the tackle
Use the first two steps of the problem-solving strategy to analyze the following problem:

> Fred catches the football while standing directly on the goal line. He immediately starts running forward with an acceleration of 6 ft/s². At the moment the catch is made, Tommy is 20 yards away and heading directly toward Fred with a steady speed of 15 ft/s. If neither deviates from a straight-ahead path, where will Tommy tackle Fred?

MODEL Both Fred and Tommy will be represented as particles.

VISUALIZE Here is a problem with two moving objects. Figure 1.26 first shows the physical representation, with Fred accelerating toward the right and Tommy moving to the left. This is our choice—the problem didn't say which way Fred runs. Our choice gives positive numbers for Fred's velocity and

acceleration, but you could just as well set up the problem with Fred running toward the left. Tommy moves toward the left, so he has a *negative* value for $(v_{0x})_T$. The ending point of the problem is fairly clear: when Fred and Tommy collide (ouch!).

Notice how we have used subscripts F and T to distinguish symbols for Fred's motion from similar symbols describing Tommy's motion. Using the same symbol to mean more than one thing is guaranteed to lead to confusion and errors. The time, however, does not have an extra subscript because only one clock is needed to time both Fred and Tommy. Thus *both* starting positions are described with the same time symbol t_0.

We know intuitively how the problem ends—they collide—but how do we state that more precisely? To collide, they must both reach the same position at the same time (t_1), and so the relation we are looking for is $(x_1)_F = (x_1)_T$. Keep in mind that $(x_1)_T$ is Tommy's *position*, measured with respect to the coordinate system, not the distance that Tommy has traveled.

Physical representation

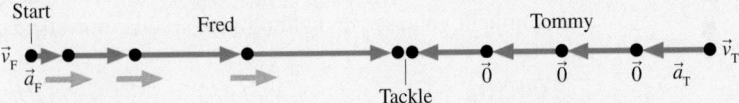

Pictorial representation

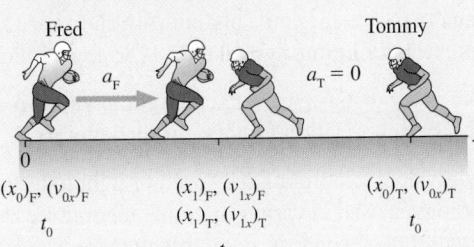

Known
$(x_0)_F = (v_{0x})_F = t_0 = 0$
$(x_0)_T = 20 \text{ yards} = 60 \text{ ft}$
$(v_{0x})_T = -15 \text{ ft/s}$
$a_F = 6 \text{ ft/s}^2 \quad a_T = 0 \text{ ft/s}^2$

Find
$(x_1)_F$ at t_1 when $(x_1)_F = (x_1)_T$

FIGURE 1.26 Physical and pictorial representations.

As you finish this chapter, you now have some powerful tools for thinking about and analyzing problems of motion. Continued practice will make these second nature to you, and coming chapters will give you plenty of opportunity for that practice.

1.9 Units and Significant Figures

Science is based upon experimental measurements, and measurements require *units*. The system of units currently used in science is called *le Système Internationale d'Unités*. These are commonly referred to as **SI units.** Older books often referred to *mks units,* which stands for "meter-kilogram-second," or *cgs units,* which is "centimeter-gram-second." For practical purposes, SI units are the same as mks units. In casual speaking we often refer to *metric units,* although this could mean either mks or cgs units. To be precise, we will follow the internationally accepted custom of calling our system of units *SI units.*

All of the quantities needed to understand motion can be expressed in terms of the three basic SI units shown in Table 1.1. Other quantities can be expressed as a combination of these basic units. Velocity, expressed in meters per second or m/s, is a ratio of the length unit to the time unit.

TABLE 1.1 Some basic SI units

Quantity	Unit	Abbreviation
time	second	s
length	meter	m
mass	kilogram	kg

Time

The standard of time prior to 1960 was based on the *mean solar day.* As time-keeping accuracy and astronomical observations improved, it became apparent that the earth's rotation is not perfectly steady. Meanwhile, physicists had been developing a device called an *atomic clock.* This instrument is able to measure, with incredibly high precision, the frequency of radio waves absorbed by atoms as they move between two closely spaced energy levels. This frequency can be reproduced with great accuracy at many laboratories around the world. Consequently, the SI unit of time—the second—was redefined in 1967 as follows:

One *second* is the time required for 9,192,631,770 oscillations of the radio wave absorbed by the cesium-133 atom. The abbreviation for second is the letter s.

A radio station operated (in the United States) by the National Institute of Standards and Technology broadcasts a signal whose frequency is linked directly to the atomic clocks. This signal is the time standard, and any time-measuring equipment you use was calibrated from this time standard.

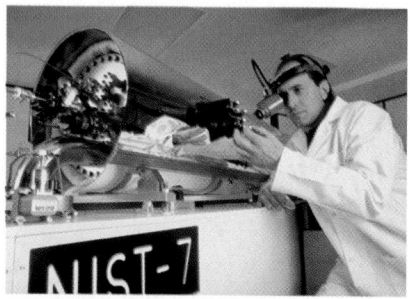

An atomic clock at the National Institute of Standards and Technology is the primary standard of time.

Length

The SI unit of length—the meter—also has a long and interesting history. It was originally defined as one ten-millionth of the distance from the North Pole to the equator along a line passing through Paris. There are obvious practical difficulties with implementing this definition, and it was later abandoned in favor of the distance between two scratches on a platinum-iridium bar stored in a special vault in Paris. The present definition, agreed to in 1983, is as follows:

One *meter* is the distance traveled by light in vacuum during 1/299,792,458 of a second. The abbreviation for meter is the letter m.

This is equivalent to defining the speed of light to be exactly 299,792,458 m/s. Laser technology is used in various national laboratories to implement this definition and to calibrate secondary standards that are easier to use. These standards ultimately make their way to your ruler or to a meter stick. It is worth keeping in mind that any measuring device you use is only as accurate as the care with which it was calibrated.

Mass

The original unit of mass, the gram, was defined as the mass of 1 cubic centimeter of water. That is why you know the density of water as 1 g/cm³. This definition proved to be impractical when scientists needed to make very accurate measurements. The SI unit of mass—the kilogram—was redefined in 1889 as:

By international agreement, this metal cylinder, stored in Paris, is the definition of the kilogram.

One *kilogram* is the mass of the international standard kilogram, a polished platinum-iridium cylinder stored in Paris. The abbreviation for kilogram is the symbol kg.

The kilogram is the only SI unit still defined by a manufactured object. Despite the prefix *kilo,* it is the kilogram, not the gram, that is the proper SI unit.

Using Prefixes

We will have many occasions to use lengths, times, and masses that are either much less or much greater than the standards of 1 meter, 1 second, and 1 kilogram. We will do so by using *prefixes* to denote various powers of ten. Table 1.2 lists the common prefixes that will be used frequently throughout this book. Memorize it! Few things in science are learned by rote memory, but this list is one of them. A more extensive list of prefixes is shown inside the cover of the book.

Although prefixes make it easier to talk about quantities, the proper SI units are meters, seconds, and kilograms. Quantities given with prefixed units must be converted to SI units before any calculations are done. Unit conversions are best done at the very beginning of a problem, as part of the pictorial representation.

TABLE 1.2 Common prefixes

Prefix	Power of 10	Abbreviation
mega-	10^6	M
kilo-	10^3	k
centi-	10^{-2}	c
milli-	10^{-3}	m
micro-	10^{-6}	μ
nano-	10^{-9}	n

Unit Conversions

Although SI units are our standard, we cannot entirely forget that the United States still uses English units. Many engineering calculations are done in English units. And even after repeated exposure to metric units in classes, most of us "think" in the English units we grew up with. Thus it remains important to be able to convert back and forth between SI units and English units. Table 1.3 shows a few frequently used conversions, and these are worth memorizing if you do not already know them. While the English system was originally based on the length of the king's foot, it is interesting to note that today the conversion 1 in = 2.54 cm is the *definition* of the inch. In other words, the English system for lengths is now based on the meter!

TABLE 1.3 Useful unit conversions

1 in = 2.54 cm
1 mi = 1.609 km
1 mph = 0.447 m/s
1 m = 39.37 in
1 km = 0.621 mi
1 m/s = 2.24 mph

There are various techniques for doing unit conversions. One effective method is to write the conversion factor as a ratio equal to one. For example, using information in Tables 1.2 and 1.3,

$$\frac{10^{-6} \text{ m}}{1 \text{ } \mu\text{m}} = 1 \quad \text{and} \quad \frac{2.54 \text{ cm}}{1 \text{ in}} = 1$$

Because multiplying any expression by 1 does not change its value, these ratios are easily used for conversions. To convert 3.5 μm to meters we would compute

$$3.5 \text{ } \mu\text{m} \times \frac{10^{-6} \text{ m}}{1 \text{ } \mu\text{m}} = 3.5 \times 10^{-6} \text{ m}.$$

Similarly, the conversion of 2 feet to meters would be

$$2 \text{ ft} \times \frac{12 \text{ in}}{1 \text{ ft}} \times \frac{2.54 \text{ cm}}{1 \text{ in}} \times \frac{10^{-2} \text{ m}}{1 \text{ cm}} = 0.610 \text{ m}.$$

Notice how units in the numerator and in the denominator cancel until just the desired units remain at the end. You can continue this process of multiplying by 1 as many times as necessary to complete all the conversions.

Assessment

As we get further into problem solving, we will need to decide whether or not the answer to a problem "makes sense." To determine this, at least until you have more experience with SI units, you may need to convert from SI units back to the English units in which you think. But this conversion does not need to be very accurate. For example, if you are working a problem about automobile speeds and reach an answer of 35 m/s, all you really want to know is whether or not this is a realistic speed for a car. That requires a "quick and dirty" conversion, not a conversion of great accuracy.

Table 1.4 shows a number of approximate conversion factors that can be used to assess the answer to a problem. Using 1 m/s ≈ 2 mph, you find that 35 m/s is roughly 70 mph, a reasonable speed for a car. But an answer of 350 m/s, which you might get after making a calculation error, would be an unreasonable 700 mph. Practice with these will allow you to develop intuition for metric units.

NOTE ▶ These approximate conversion factors are accurate only to one significant figure. This is sufficient to assess the answer to a problem, but do *not* use the conversion factors from Table 1.4 for converting English units to SI units at the start of a problem. Use Table 1.3. ◀

TABLE 1.4 Approximate conversion factors

1 cm ≈ $\frac{1}{2}$ in
10 cm ≈ 4 in
1 m ≈ 1 yard
1 m ≈ 3 feet
1 km ≈ 0.6 mile
1 m/s ≈ 2 mph

Significant Figures

It is necessary to say a few words about a perennial source of difficulty: significant figures. Mathematics is a subject where numbers and relationships can be as precise as desired, but physics deals with a real world of ambiguity and imprecision. It is important in all areas of science and engineering to state clearly what you know about a situation—no less and, especially, no more. Numbers provide one way to specify your knowledge.

If you report that a length has a value of 6.2 m, the implication is that the actual value falls between 6.15 m and 6.25 m and thus rounds to 6.2 m. If that is the case, then reporting a value of simply 6 m is saying less than you know; you are withholding information. On the other hand, to report the number as 6.213 m is wrong. Any person reviewing your work—perhaps a client who hired you—would interpret the number 6.213 m as meaning that the actual length falls between 6.2125 m and 6.2135 m, thus rounding to 6.213 m. In this case, you are claiming to have knowledge and information that you do not really possess.

The way to state your knowledge precisely is through the proper use of **significant figures.** You can think of a significant figure as being a digit that is reliably known. A number such as 6.2 m has *two* significant figures because the next decimal place—the one-hundredths—is not reliably known. As Figure 1.27 shows, the best way to determine how many significant figures a number has is to write it in scientific notation.

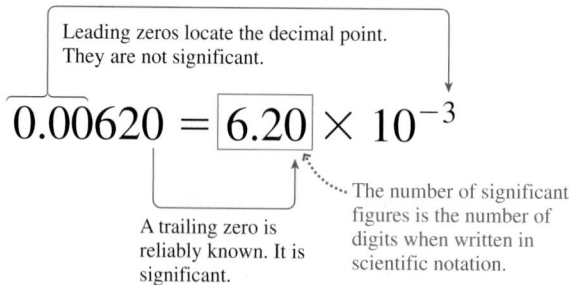

Leading zeros locate the decimal point. They are not significant.

$$0.00620 = 6.20 \times 10^{-3}$$

A trailing zero is reliably known. It is significant.

The number of significant figures is the number of digits when written in scientific notation.

■ The number of significant figures ≠ the number of decimal places.

■ Changing units shifts the decimal point but does not change the number of significant figures.

FIGURE 1.27 Determining significant figures.

Calculations with numbers follow the "weakest link" rule. The saying, which you probably know, is that "a chain is only as strong as its weakest link." If nine out of ten links in a chain can support a 1000 pound weight, that strength is meaningless if the tenth link can support only 200 pounds. Nine out of the ten numbers used in a calculation might be known with a precision of 0.01%; but if the tenth number is poorly known, with a precision of only 10%, then the result of the calculation cannot possibly be more precise than 10%. The weak link rules!

TACTICS BOX 1.5 Using significant figures

❶ When multiplying or dividing several numbers, or taking roots, the number of significant figures in the answer should match the number of significant figures of the *least* precisely known number used in the calculation.

❷ When adding or subtracting several numbers, the number of decimal places in the answer should match the *smallest* number of decimal places of any number used in the calculation.

EXAMPLE 1.12 Using significant figures

An object consists of two pieces. The mass of one piece has been measured to be 6.47 kg. The volume of the second piece, which is made of aluminum, has been measured to be 4.44×10^{-4} m^3. A handbook lists the density of aluminum as 2.7×10^3 kg/m^3. What is the total mass of the object?

SOLVE First, calculate the mass of the second piece:

$$m = (4.44 \times 10^{-4} \text{ m}^3)(2.7 \times 10^3 \text{ kg/m}^3)$$
$$= 1.199 \text{ kg} = 1.2 \text{ kg}$$

The number of significant figures of a product must match that of the *least* precisely known number, which is the two-significant-figure density of aluminum. Now add the two masses:

$$
\begin{array}{r}
6.47 \text{ kg} \\
+ \ 1.2 \ \text{ kg} \\
\hline
7.7 \ \text{ kg}
\end{array}
$$

The sum is 7.67 kg, but the hundredths place is not reliable because the second mass has no reliable information about this digit. Thus we must round to the one decimal place of the 1.2 kg. The best we can say, with reliability, is that the total mass is 7.7 kg.

There are two notable exceptions to these rules:

1. It is customary to keep one extra significant figure if (and only if) the number starts with a 1. For example, 10.43 could be used in a calculation with 8.91. The rationale for this exception is that four significant figures for numbers starting with 1 has roughly the same percentage accuracy as three significant figures for numbers starting with 2–9.
2. It is acceptable to keep one or two extra digits during intermediate steps of a calculation, as long as the final answer is reported with the proper number of significant figures. The goal is to minimize round-off errors in the calculation. But only one or two extra digits, not the seven or eight shown in your calculator display.

In laboratory work, the proper number of significant figures is determined by the experiment. The least precisely measured quantity sets the proper number. Textbook problems in science and engineering use an accepted standard of *three* significant figures for nearly all calculations. Two significant figures is too imprecise for most problems, while four is unnecessary *unless* the problem happens to give very accurate data with which to work. **Three significant figures will be the standard in this textbook, unless a problem provides data with more than three significant figures.** All data supplied with a problem can be assumed to have three-significant-figure accuracy even when, to avoid being unduly pedantic, integer values are given as "20 kg" rather than "2.00×10^1 kg."

NOTE ▶ Be careful! Many calculators have a default setting that shows two decimal places, such as 5.23. This is dangerous. If you need to calculate

TABLE 1.5 Some approximate lengths

	Length (m)
Circumference of the earth	4×10^7
New York to Los Angeles	5×10^6
Distance you can drive in 1 hour	1×10^5
Altitude of jet planes	1×10^4
Distance across a college campus	1000
Length of a football field	100
Length of a classroom	10
Length of your arm	1
Width of a textbook	0.1
Length of your little fingernail	0.01
Diameter of a pencil lead	1×10^{-3}
Thickness of a sheet of paper	1×10^{-4}
Diameter of a dust particle	1×10^{-5}

TABLE 1.6 Some approximate masses

	Mass (kg)
Large airliner	1×10^5
Small car	1000
Large human	100
Medium-size dog	10
Science textbook	1
Apple	0.1
Pencil	0.01
Raisin	1×10^{-3}
Fly	1×10^{-4}

5.23/58.5, your calculator will show a result of 0.09 and it is all too easy to write that down as an answer. But by doing so, you have reduced a calculation of two numbers having three significant figures to an answer with only one significant figure. The proper result of this division is 0.0894 or 8.94×10^{-2}. You will avoid this error if you keep your calculator set to display numbers in *scientific notation* with two decimal places. ◄

Proper use of significant figures is part of the "culture" of science and engineering. We will frequently emphasize these "cultural issues" because you must learn to speak the same language as the natives if you wish to communicate effectively. Most students "know" the rules of significant figures, having learned them in high school, but many fail to apply them. It is important that you understand the reasons for significant figures and that you get in the habit of using them properly.

Orders of Magnitude and Estimating

Precise calculations are appropriate when we have precise data, but there are many times when a very rough estimate is sufficient. Suppose you see a rock fall off a cliff and would like to know how fast it was going when it hit the ground. By doing a mental comparison with the speeds of familiar objects, such as cars and bicycles, you might judge that the rock was traveling at "about" 20 mph.

This is a one-significant-figure estimate. With some luck, you can probably distinguish 20 mph from either 10 mph or 30 mph, but you certainly, just from a visual appearance, cannot distinguish 20 mph from 21 mph. A one-significant-figure estimate or calculation, such as this, is called an **order-of-magnitude estimate.** An order-of-magnitude estimate is indicated by the symbol $\sim$, which indicates even less precision than the "approximately equal" symbol $\approx$. You would say that the speed of the falling rock is $v \sim 20$ mph.

A useful skill is to make reliable order-of-magnitude estimates on the basis of known information, simple reasoning, and common sense. This is a skill that is acquired by practice. Most chapters in this book will have homework problems that ask you to make order-of-magnitude estimates. The following example is typical of an estimation problem.

Tables 1.5 and 1.6 have information that will be useful for doing estimates.

EXAMPLE 1.13 **Estimating a sprinter's speed**
Estimate the speed with which an Olympic sprinter crosses the finish line of the 100 m dash.

SOLVE We do need one piece of information, but it is a widely known piece of sports trivia. That is, world-class sprinters run the 100 m dash in about 10 s. Their *average* speed is

$v_{avg} \approx (100 \text{ m})/(10 \text{ s}) \approx 10 \text{ m/s}$. But that's only average. They go slower than average at the beginning, and they cross the finish line at a speed faster than average. How much faster? Twice as fast, 20 m/s, would be ≈ 40 mph. Sprinters don't seem like they're running as fast as a 40 mph car, so this probably is too fast. Let's *estimate* that their final speed is 50% faster than the average. Thus they cross the finish line at $v \sim 15$ m/s.

STOP TO THINK 1.5 Rank in order, from the most to the least, the number of significant figures in the following numbers. For example, if b has more than c, c has the same number as a, and a has more than d, you could give your answer as b > c = a > d.

a. 8200   b. 0.0052   c. 0.430   d. 4.321×10^{-10}

SUMMARY

The goal of Chapter 1 has been to introduce the fundamental concepts of motion.

GENERAL STRATEGY

Motion Diagrams

- Help visualize motion.
- Provide a tool for finding acceleration vectors.

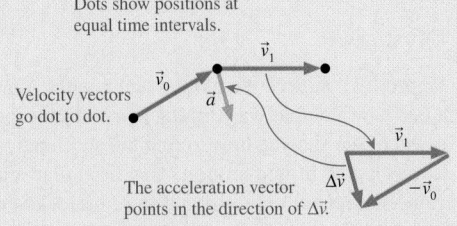

Dots show positions at equal time intervals.

Velocity vectors go dot to dot.

The acceleration vector points in the direction of $\Delta \vec{v}$.

▶ These are the average velocity and the average acceleration vectors.

Problem Solving

MODEL Make simplifying assumptions.

VISUALIZE Use:

- **Pictorial representation**
- **Physical representation**
- **Graphical representation**

SOLVE Use a **mathematical representation** to find numerical answers.

ASSESS Does the answer have the proper units? Does it make sense?

IMPORTANT CONCEPTS

The particle model represents a moving object as if all its mass were concentrated at a single point.

Position locates an object with respect to a chosen coordinate system. Change in position is called displacement.

Velocity is the rate of change of the position vector $\vec{r}$.

Acceleration is the rate of change of the velocity vector $\vec{v}$. An object has an acceleration if it

- Changes speed and/or
- Changes direction.

Pictorial Representation

❶ Sketch the situation.

❷ Establish coordinates.

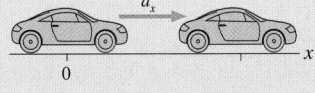

❸ Define symbols.

x_0, v_{0x}, t_0 x_1, v_{1x}, t_1

❹ List knowns.

Known
$x_0 = v_{0x} = t_0 = 0$
$a_x = 2 \text{ m/s}^2$ $t_1 = 2 \text{ s}$

❺ Identify desired unknown.

Find
x_1

APPLICATIONS

For **motion along a line:**

- Speeding up: $\vec{v}$ and $\vec{a}$ point in the same direction.
- Slowing down: $\vec{v}$ and $\vec{a}$ point in opposite directions.
- Constant speed: $\vec{a} = \vec{0}$.

Significant figures are reliably known digits. Three significant figures is the standard for this book. The number of significant figures for:

- **Multiplication, division, powers** is set by the value with the fewest significant figures.
- **Addition, subtraction** is set by the value with the smallest number of decimal places.

TERMS AND NOTATION

motion	particle model	time interval, Δt	SI units
trajectory	position vector, $\vec{r}$	average speed	significant figures
motion diagram	scalar quantity	average velocity, $\vec{v}$	order-of-magnitude estimate
operational definition	vector quantity	average acceleration, $\vec{a}$	
translational motion	displacement, $\Delta \vec{r}$	pictorial representation	
particle	zero vector, $\vec{0}$	representation of knowledge	

EXERCISES AND PROBLEMS

Exercises

Section 1.1 Motion Diagrams

1. A car skids to a halt to avoid hitting an object in the road. Draw a basic motion diagram, using the images from the movie, from the time the skid begins until the car is stopped.
2. You drop a soccer ball from your third-story balcony. Draw a basic motion diagram, using the images from the movie, from the time you release the ball until it touches the ground.

Section 1.2 The Particle Model

3. a. Write a paragraph describing the *particle model*. What is it, and why is it important?
   b. Give two examples of situations, different from those described in the text, for which the particle model is appropriate.
   c. Give an example of a situation, different from those described in the text, for which it would be inappropriate.

Section 1.3 Position and Time

4. Write a sentence or two describing the difference between position and displacement. Give one example of each.

Section 1.4 Velocity

5. a. What is an *operational definition?*
   b. Give operational definitions of displacement and velocity. Your definition should be given mostly in words and pictures, with a minimum of symbols or mathematics.
6. A softball player hits the ball and starts running toward first base. Draw a motion diagram, using the particle model, showing her position and her average velocity vectors during the first few seconds of her run.
7. A softball player slides into second base. Draw a basic motion diagram, using the particle model, showing his position and his average velocity vectors from the time he begins to slide until he reaches the base.

Section 1.5 Acceleration

8. Give an operational definition of acceleration. Your definition should be given mostly in words and pictures, with a minimum of symbols or mathematics.
9. a. Find the average acceleration vector at point 1 of this three-point motion diagram.
   b. Is the object's average speed between points 1 and 2 greater than, less than, or equal to its average speed between points 0 and 1? Explain how you can tell.

FIGURE EX1.9

10. a. Find the average acceleration vector at point 1 of this three-point motion diagram.
    b. Is the object's average speed between points 1 and 2 greater than, less than, or equal to its average speed between points 0 and 1? Explain how you can tell.

FIGURE EX1.10

11. Figure 1.21 showed the motion diagram for Anne as she rode a Ferris wheel that was turning at a constant speed. The inset to the figure showed how to find the acceleration vector at the lowest point in her motion. Use a similar analysis to find Anne's acceleration vector at the 12 o'clock, 4 o'clock, and 8 o'clock positions of the motion diagram. Use a ruler so that your analysis is accurate.
12. Figure 1.18 showed the motion diagram of a runner on a circular track. Find the acceleration vector when the runner is at the top of the diagram and when the runner is at the bottom of the diagram.

Section 1.6 Examples of Motion Diagrams

13. A car travels to the left at a steady speed for a few seconds, then brakes for a stop sign. Draw a complete motion diagram of the car.
14. A child is sledding on a smooth, level patch of snow. She encounters a rocky patch and slows to a stop. Draw a complete motion diagram of the child and her sled.
15. A roof tile falls straight down from a two-story building. It lands in a swimming pool and settles gently to the bottom. Draw a complete motion diagram of the tile.
16. Your roommate drops a tennis ball from a third story balcony. It hits the sidewalk and bounces as high as the second story. Draw a complete motion diagram of the tennis ball from the time it is released until it reaches the maximum height on its bounce. Be sure to determine and show the acceleration at the lowest point.
17. A car is driving north at steady speed. It makes a gradual 90° left turn without losing speed, then continues driving to the west. Draw a complete motion diagram as seen from a helicopter hovering over the highway.
18. A toy car rolls down a ramp, then across a smooth, horizontal floor. Draw a complete motion diagram of the toy car.

Section 1.7 From Words to Symbols
Section 1.8 A Problem-Solving Strategy

19. Draw a pictorial representation for the following problem. Do *not* solve the problem. The light turns green, and a bicyclist starts forward with an acceleration of 1.5 m/s². How far must she travel to reach a speed of 7.5 m/s?
20. Draw a pictorial representation for the following problem. Do *not* solve the problem. What acceleration does a rocket need to reach a speed of 200 m/s at a height of 1.0 km?

Section 1.9 Units and Significant Figures

21. Convert the following to SI units:
    a. 9.12 μs
    b. 3.42 km
    c. 44 cm/ms
    d. 80 km/hour
22. Convert the following to SI units:
    a. 8 in
    b. 66 ft/s
    c. 60 mph
    d. 14 in^2
23. Convert the following to SI units:
    a. 1 hour
    b. 1 day
    c. 1 year
    d. 32 ft/s^2
24. Using the approximate conversion factors in Table 1.4, convert the following to SI units *without* using your calculator.
    a. 20 ft
    b. 60 mi
    c. 60 mph
    d. 8 in
25. A regulation soccer field for international play is a rectangle with a length between 100 m and 110 m and a width between 64 m and 75 m. What are the smallest and largest areas that the field could be?
26. The quantity called *mass density* is the mass per unit volume of a substance. Express the following mass densities in SI units.
    a. Aluminum, 2.7×10^{-3} kg/cm^3
    b. Alcohol, 0.81 g/cm^3
27. How many significant figures does each of the following numbers have?
    a. 6.21
    b. 62.1
    c. 0.620
    d. 0.062
28. How many significant figures does each of the following numbers have?
    a. 6200
    b. 0.006200
    c. 1.0621
    d. 6.21×10^3
29. Compute the following numbers, applying the significant figure rule adopted in this textbook.
    a. 33.3×25.4
    b. $33.3 - 25.4$
    c. $\sqrt{33.3}$
    d. $333.3 \div 25.4$
30. Compute the following numbers, applying the significant figure rule adopted in this textbook.
    a. 33.3^2
    b. 33.3×45.1
    c. $\sqrt{22.2} - 1.2$
    d. 44.4^{-1}
31. Estimate (don't measure!) the length of a typical car. Give your answer in both feet and meters. Briefly describe how you arrived at this estimate.
32. Estimate the height of a telephone pole. Give your answer in both feet and meters. Briefly describe how you arrived at this estimate.
33. Estimate the average speed with which you go from home to campus via whatever mode of transportation you use most commonly. Give your answer in both mph and m/s. Briefly describe how you arrived at this estimate.
34. Estimate the average speed with which the hair on your head grows. Give your answer in both m/s and μm/hour. Briefly describe how you arrived at this estimate.

Problems

For Problems 35 through 44, draw a complete motion diagram *and* a pictorial representation. Do *not* solve these problems or do any mathematics.

35. A Porsche accelerates from a stoplight at 5.0 m/s^2 for five seconds, then coasts for three more seconds. How far has it traveled?
36. Billy drops a watermelon from the top of a three-story building, 10 m above the sidewalk. How fast is the watermelon going when it hits?
37. Sam is recklessly driving 60 mph in a 30 mph speed zone when he suddenly sees the police. He steps on the brakes and slows to 30 mph in three seconds, looking nonchalant as he passes the officer. How far does he travel while braking?
38. A speed skater moving across frictionless ice at 8.0 m/s hits a 5.0-m-wide patch of rough ice. She slows steadily, then continues on at 6.0 m/s. What is her acceleration on the rough ice?
39. You would like to stick a wet spit wad on the ceiling, so you toss it straight up with a speed of 10 m/s. How long does it take to reach the ceiling, 3.0 m above?
40. A student standing on the ground throws a ball straight up. The ball leaves the student's hand with a speed of 15 m/s when the hand is 2.0 m above the ground. How long is the ball in the air before it hits the ground? (The student moves her hand out of the way.)
41. A ball rolls along a smooth horizontal floor at 10 m/s, then starts up a 20° ramp. How high does it go before rolling back down?
42. A motorist is traveling at 20 m/s. He is 60 m from a stop light when he sees it turn yellow. His reaction time, before stepping on the brake, is 0.50 s. What steady deceleration while braking will bring him to a stop right at the light?
43. Ice hockey star Bruce Blades is 5.0 m from the blue line and gliding toward it at a speed of 4.0 m/s. You are 20 m from the blue line, directly behind Bruce. You want to pass the puck to Bruce. With what speed should you shoot the puck down the ice so that it reaches Bruce exactly as he crosses the blue line?
44. You are standing still as Fred runs past you with the football at a speed of 6.0 yards per second. He has only 30 yards left to go before reaching the goal line to score the winning touchdown. If you begin running at the exact instant he passes you, what acceleration must you maintain to catch him 5.0 yards in front of the goal line?

Problems 45 through 50 show a motion diagram. For each of these problems, write a one or two sentence "story" about a *real object* that has this motion diagram. Your stories should talk about people or objects by name and say what they are doing. Problems 35–44 are examples of motion short stories.

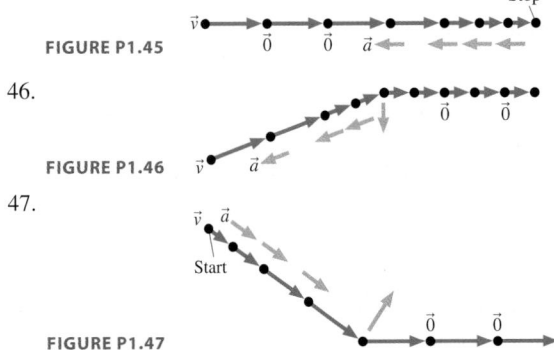

45.

FIGURE P1.45

46.

FIGURE P1.46

47.

FIGURE P1.47

48.

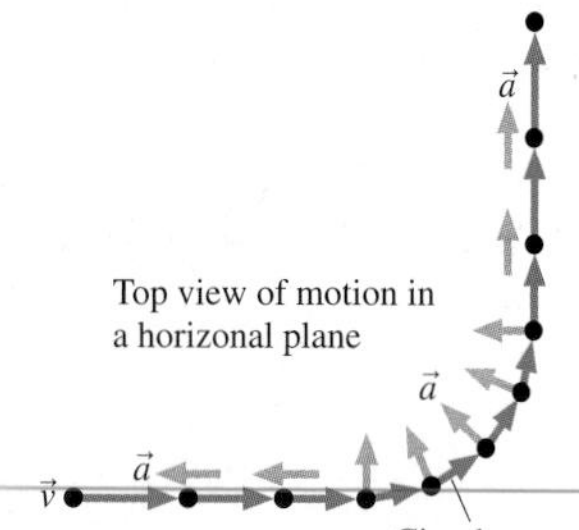

FIGURE P1.48

Top view of motion in a horizonal plane

Circular arc

49.

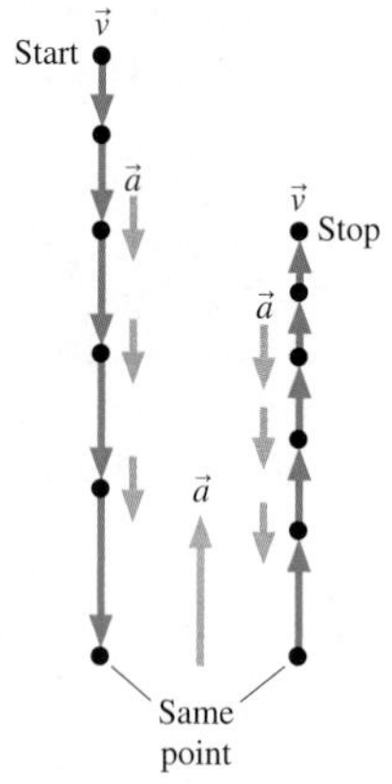

Start

Stop

$\vec{a}$

Same point

FIGURE P1.49

50.

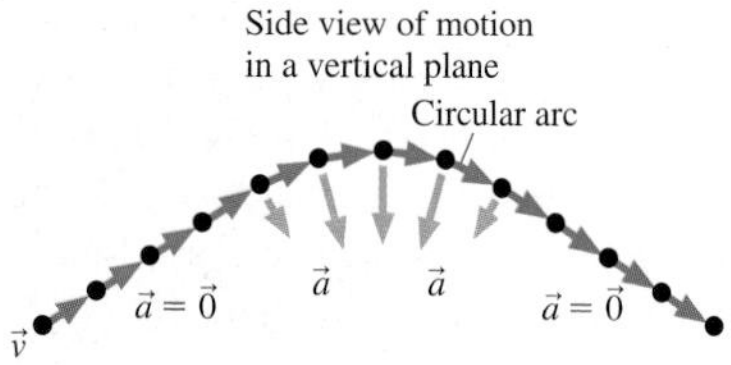

Side view of motion in a vertical plane

Circular arc

$\vec{a} = \vec{0}$ $\vec{a}$ $\vec{a}$ $\vec{a} = \vec{0}$

FIGURE P1.50

Problems 51 through 56 show a partial motion diagram. For each:
  a. Complete the motion diagram by adding acceleration vectors.
  b. Write a physics *problem* for which this is the correct motion diagram. Be imaginative! Don't forget to include enough information to make the problem complete and to state clearly what is to be found.
  c. Draw a full pictorial representation for your problem.

51.

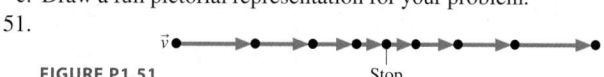

FIGURE P1.51   Stop

52.

FIGURE P1.52 $\vec{v}$

53.

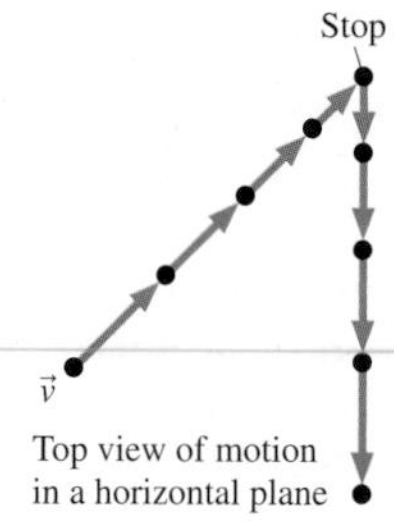

Stop

$\vec{v}$

Top view of motion in a horizontal plane

FIGURE P1.53

54.   Start

$\vec{v}_A$

$\vec{v}_B$

Start

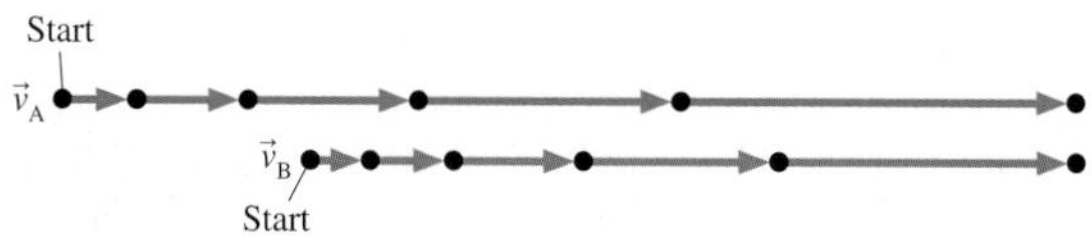

FIGURE P1.54

55.

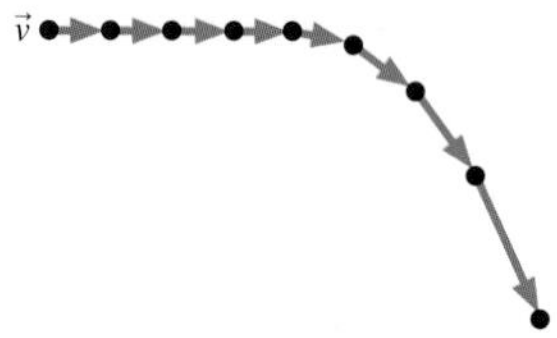

$\vec{v}$

FIGURE P1.55

56.

$\vec{v}$

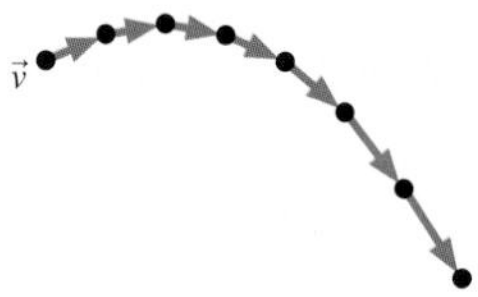

FIGURE P1.56

57. Consider a pendulum swinging back and forth on a string. Use a motion diagram analysis and a written explanation to answer the following questions.
  a. At the lowest point in the motion, is the velocity zero or nonzero? Is the acceleration zero or nonzero? If these vectors aren't zero, which way do they point?
  b. At the end of its arc, when the pendulum is at the highest point on the right or left side, is the velocity zero or nonzero? Is the acceleration zero or nonzero? If these vectors aren't zero, which way do they point?

STOP TO THINK ANSWERS

Stop to Think 1.1: B. The images of B are farther apart, so it travels a larger distance than does A during the same intervals of time.

Stop to Think 1.2: a. Dropped ball. **b.** Dust particle. **c.** Descending rocket.

Stop to Think 1.3: e. The average velocity vector is found by connecting one dot in the motion diagram to the next.

Stop to Think 1.4: c. The velocity leading away from this point is $\vec{v}_2 = \vec{v}_1 + \vec{a}\Delta t$.

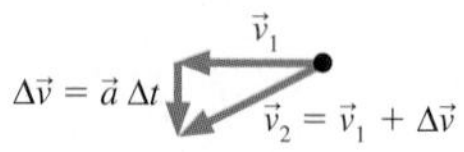

$\Delta\vec{v} = \vec{a}\,\Delta t$ $\vec{v}_1$ $\vec{v}_2 = \vec{v}_1 + \Delta\vec{v}$

Stop to Think 1.5: d > c > b = a.

2 Kinematics: The Mathematics of Motion

World-class sprinters have a tremendous acceleration at the start of a race.

▶ Looking Ahead

The goal of Chapter 2 is to learn how to solve problems about motion in a straight line. In this chapter you will learn to:

- Understand the mathematics of position, velocity, and acceleration for motion along a straight line.
- Use a graphical representation of motion.
- Use an explicit problem-solving strategy for kinematics problems.
- Understand free-fall motion and motion along inclined planes.

◀ Looking Back

Each chapter in this textbook builds on ideas and techniques from previous chapters. The Looking Back feature calls your attention to specific sections that are of major significance to the present chapter. A brief review of these sections will improve your study of this chapter. Please review:

- Sections 1.4–1.5 Velocity and acceleration.
- Sections 1.7–1.8 Problem solving in physics.

A race, whether between runners, bicyclists, or drag racers, exemplifies the idea of motion. Today, we use electronic stopwatches, video recorders, and other sophisticated instruments to analyze motion, but it hasn't always been so. Galileo, who in the early 1600s was the first scientist to study motion experimentally, used his pulse to measure time!

Galileo made a useful distinction between the *cause* of motion and the *description* of motion. **Kinematics** is the modern name for the mathematical description of motion without regard to causes. The term comes from the Greek word *kinema,* meaning "movement." You know this word through its English variation *cinema*—motion pictures! In this chapter on kinematics we'll develop the mathematical tools for describing motion. Then, in Chapter 4, we'll turn our attention to the *cause* of motion.

We will begin our study of kinematics with motion in one dimension; that is, motion along a straight line. Runners, drag racers, and skiers are just a few examples of motion in one dimension. The kinematics of two-dimensional motion—projectile motion and circular motion—will be considered in later chapters.

2.1 Motion in One Dimension

You learned in Chapter 1 that an object's motion can be described in terms of three fundamental quantities: its position $\vec{r}$, velocity $\vec{v}$, and acceleration $\vec{a}$. These quantities are vectors, having a direction as well as a magnitude. But for motion in one dimension, the vectors are restricted to point only "forward" or "backward." Consequently, we can describe one-dimensional motion with the simpler quantities x, v_x, and a_x (or y, v_y, and a_y). However, we need to give each of these quantities an explicit *sign,* positive or negative, to indicate whether the position, velocity, or acceleration vector points forward or backward. Learning to use the signs correctly is an important goal of this chapter.

Determining the Signs of Position, Velocity, and Acceleration

 1.1

Position, velocity, and acceleration are measured with respect to a coordinate system. You will recall that a coordinate system is a grid or axis that *you* impose on a problem to analyze the motion. We will find it convenient to use an x-axis to describe both horizontal motion and motion along an inclined plane. A y-axis will be used for vertical motion. A coordinate axis has two essential features:

1. An origin, to define zero, and
2. An x or y label to indicate the positive end of the axis.

We will adopt the convention that the positive end of an x-axis is to the right and the positive end of a y-axis is up. The signs of position, velocity, and acceleration are based on this convention.

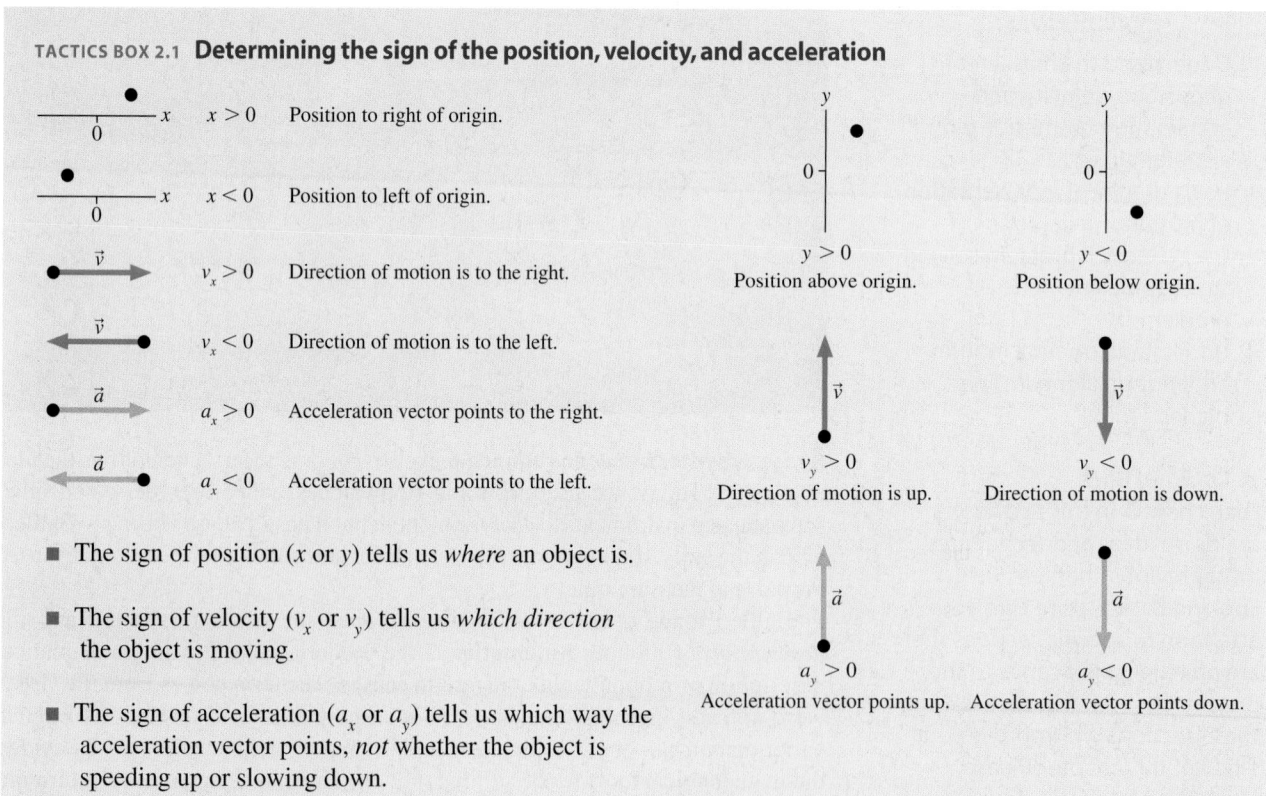

TACTICS BOX 2.1 **Determining the sign of the position, velocity, and acceleration**

$x > 0$ Position to right of origin.

$x < 0$ Position to left of origin.

$v_x > 0$ Direction of motion is to the right.

$v_x < 0$ Direction of motion is to the left.

$a_x > 0$ Acceleration vector points to the right.

$a_x < 0$ Acceleration vector points to the left.

$y > 0$ Position above origin.

$y < 0$ Position below origin.

$v_y > 0$ Direction of motion is up.

$v_y < 0$ Direction of motion is down.

$a_y > 0$ Acceleration vector points up.

$a_y < 0$ Acceleration vector points down.

■ The sign of position (x or y) tells us *where* an object is.

■ The sign of velocity (v_x or v_y) tells us *which direction* the object is moving.

■ The sign of acceleration (a_x or a_y) tells us which way the acceleration vector points, *not* whether the object is speeding up or slowing down.

Acceleration is where things get a bit tricky. A natural tendency is to think that a positive value of a_x or a_y describes an object that is speeding up while a negative value describes an object that is slowing down (decelerating). However, this interpretation *does not work.*

Acceleration was defined as $\vec{a}_{avg} = \Delta\vec{v}/\Delta t$. The direction of $\vec{a}$ can be determined by using a motion diagram to find the direction of $\Delta\vec{v}$. The one-dimensional acceleration a_x (or a_y) is then positive if the vector $\vec{a}$ points to the right (or up), negative if $\vec{a}$ points to the left (or down).

Figure 2.1 shows that this method for determining the sign of a does not conform to the simple idea of speeding up and slowing down. The object in Figure 2.1a has a positive acceleration ($a_x > 0$) not because it is speeding up but because the vector $\vec{a}$ points to the right. Compare this with the motion diagram of Figure 2.1b. Here the object is slowing down, but it still has a positive acceleration ($a_x > 0$) because $\vec{a}$ points to the right.

In Chapter 1 we found that an object is speeding up if $\vec{v}$ and $\vec{a}$ point in the same direction, slowing down if they point in opposite directions. For one-dimensional motion this rule becomes:

- **An object is speeding up if and only if v_x and a_x have the same sign.**
- **An object is slowing down if and only if v_x and a_x have opposite signs.**
- **An object's velocity is constant if and only if $a_x = 0$.**

Notice how the first two of these rules are at work in Figure 2.1

Position-versus-Time Graphs

Figure 2.2 is a motion diagram, made at 1 frame per minute, of a student walking to school. You can see that she leaves home at a time we choose to call $t = 0$ min and makes steady progress for a while. Beginning at $t = 3$ min there is a period where the distance traveled during each time interval becomes less—perhaps she slowed down to speak with a friend. Then she picks up the pace and the distances within each interval are longer.

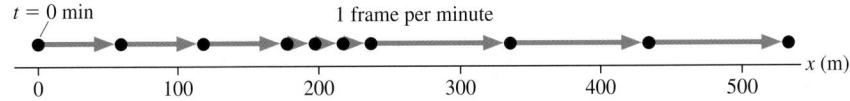

FIGURE 2.2 The motion diagram of a student walking to school and a coordinate axis for making measurements.

Figure 2.2 includes a coordinate axis, and you can see that every dot in a motion diagram occurs at a specific position. Table 2.1 shows the student's positions at different times as measured along this axis. For example, she is at position $x = 120$ m at $t = 2$ min.

The motion diagram is one way to represent the student's motion. Another is to make a graph of the measurements in Table 2.1. Figure 2.3a is a graph of x versus t for the student. The motion diagram tells us only where the student is at a few discrete points of time, so this graph of the data shows only points, no lines.

> **NOTE** ▶ A graph of "a versus b" means that a is graphed on the vertical axis and b on the horizontal axis. Saying "graph a versus b" is really a shorthand way of saying "graph a as a function of b." ◀

However, common sense tells us the following. First, the student was *somewhere specific* at all times. That is, there was never a time when she failed to have a well-defined position, nor could she occupy two positions at one time. (As reasonable as this belief appears to be, it will be severely questioned and found not entirely accurate when we get to quantum physics!) Second, the student moved *continuously* through all intervening points of space. She could not go from $x = 100$ m to $x = 200$ m without passing through every point in between. It is thus quite reasonable to believe that her motion can be shown as a continuous

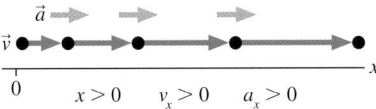

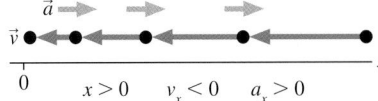

FIGURE 2.1 One of these objects is speeding up, the other slowing down, but they both have a positive acceleration a_x.

TABLE 2.1 Measured positions of a student walking to school

Time t (min)	Position x (m)
0	0
1	60
2	120
3	180
4	200
5	220
6	240
7	340
8	440
9	540

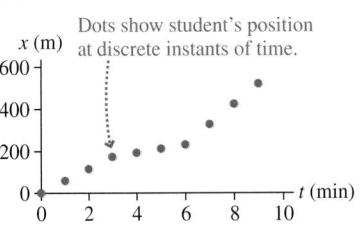

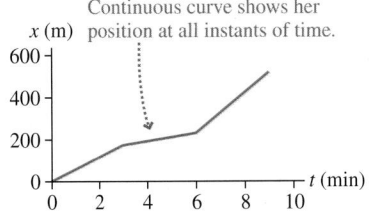

FIGURE 2.3 Position graphs of the student's motion.

curve passing through the measured points, as shown in Figure 2.3b. A continuous curve that shows an object's position as a function of time is called a **position-versus-time graph** or, sometimes, just a *position graph*.

NOTE ▶ A graph is *not* a "picture" of the motion. The student is walking along a straight line, but the graph itself is not a straight line. Further, we've graphed her position on the vertical axis even though her motion is horizontal. Graphs are *abstract representations* of motion. We will place significant emphasis on the process of interpreting graphs, and many of the exercises and problems will give you a chance to practice these skills. ◀

EXAMPLE 2.1 Interpreting a position graph

The graph in Figure 2.4a represents the motion of a car along a straight road. Describe the motion of the car.

MODEL Consider the car to be a particle that occupies a single point in space.

VISUALIZE As Figure 2.4b shows, the graph represents a car that travels to the left for 30 minutes, stops for 10 minutes, then travels back to the right for 40 minutes.

(a)

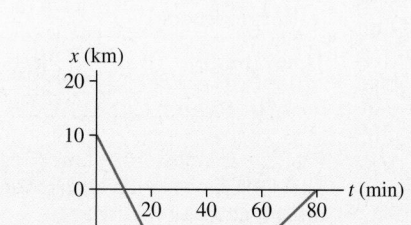

FIGURE 2.4 Position-versus-time graph of a car.

(b)

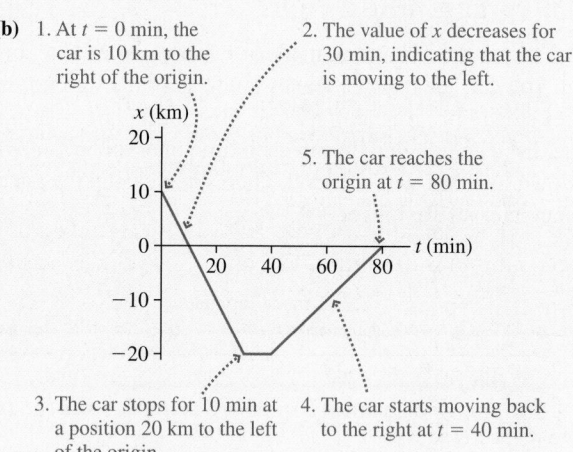

1. At $t = 0$ min, the car is 10 km to the right of the origin.

2. The value of x decreases for 30 min, indicating that the car is moving to the left.

5. The car reaches the origin at $t = 80$ min.

3. The car stops for 10 min at a position 20 km to the left of the origin.

4. The car starts moving back to the right at $t = 40$ min.

2.2 Uniform Motion

If you drive your car at a perfectly steady 60 miles per hour (mph), you will cover 60 mi during the first hour, another 60 mi during the second hour, yet another 60 mi during the third hour, and so on. This is an example of what we call *uniform motion*. In this case, 60 mi is not your position, but rather the *change* in your position during each hour; that is, your displacement Δx. Similarly, 1 hour is a time interval Δt rather than a specific instant of time. This suggests the following definition: **Straight-line motion in which equal displacements occur during *any* successive equal-time intervals is called uniform motion.**

The qualifier "any" is important. If during each hour you drive 120 mph for 30 minutes and stop for 30 minutes, you will cover 60 mi during each successive 1-hour interval. But you would *not* have equal displacements during successive 30-minute intervals, so this motion is not uniform. Your constant 60 mph driving is uniform motion because you will find equal displacements no matter how you choose your successive time intervals.

Figure 2.5 shows how uniform and nonuniform motion appear in motion diagrams and position-versus-time graphs. Notice that the position-versus-time graph for uniform motion is a straight line. This follows from the requirement that all Δx corresponding to the same Δt be equal. In fact, an alternative definition of uniform motion is: **An object's motion is uniform if and only if its position-versus-time graph is a straight line.**

The slope of a straight-line graph is defined as "rise over run." Because position is graphed on the vertical axis, the "rise" of a position-versus-time graph is the object's displacement Δx. The "run" is the time interval Δt. Consequently, the slope is $\Delta x/\Delta t$. The slope of a straight-line graph is constant, so an object in uniform motion has the *same* value of $\Delta x/\Delta t$ during *any* time interval Δt.

Chapter 1 defined the *average velocity* as $\Delta \vec{r}/\Delta t$. For one-dimensional motion this is simply

$$v_{avg} \equiv \frac{\Delta x}{\Delta t} \text{ or } \frac{\Delta y}{\Delta t} = \text{slope of the position-versus-time graph} \qquad (2.1)$$

That is, **the average velocity is the slope of the position-versus-time graph.** Velocity has units of "length per time," such as "miles per hour." The SI units of velocity are meters per second, abbreviated m/s.

NOTE ▶ The symbol $\equiv$ in Equation 2.1 stands for "is defined as" or "is equivalent to." This is a stronger statement than the two sides simply being equal. ◀

Equation 2.1 allows us to associate the slope of the position-versus-time graph, a *geometrical* quantity, with the *physical* quantity that we call the average velocity v_{avg}. This is an extremely important idea. In the case of uniform motion, where the slope $\Delta x/\Delta t$ is the same at all times, it appears that the average velocity is constant and unchanging. Consequently, a final definition of uniform motion is: **An object's motion is uniform if and only if its velocity v_x or v_y is constant and unchanging.** There's no real need to specify "average" for a velocity that doesn't change, so we will drop the subscript and refer to the average velocity as v_x or v_y.

Uniform motion

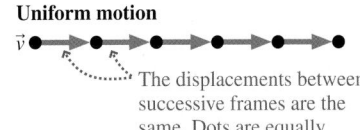

The displacements between successive frames are the same. Dots are equally spaced. v_x is constant.

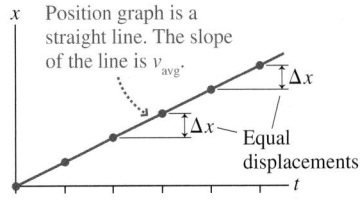

x Position graph is a straight line. The slope of the line is v_{avg}.

Δx

Δx Equal displacements

Nonuniform motion

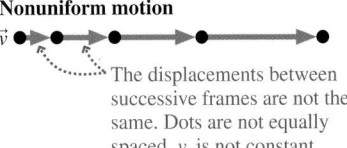

The displacements between successive frames are not the same. Dots are not equally spaced. v_x is not constant.

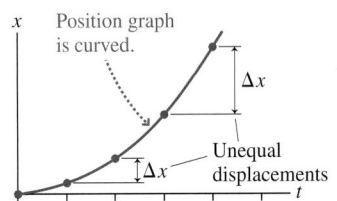

x Position graph is curved.

Δx

Δx Unequal displacements

FIGURE 2.5 Motion diagrams and position graphs for uniform motion and nonuniform motion.

EXAMPLE 2.2 Skating with constant velocity
The position-versus-time graph of Figure 2.6a represents the motion of two students on roller blades. Determine their velocities and describe their motion.

MODEL Represent the two students as particles.

VISUALIZE Figure 2.6a is a graphical representation of the students' motion. Both graphs are straight lines, telling us that both skaters are moving uniformly with constant velocities.

(a) x (m)

$(v_x)_A = \text{slope} = \frac{\Delta x}{\Delta t} = 5.0$ m/s

$\Delta t_A = 0.4$ s

$\Delta x_A = 2.0$ m

$\Delta x_B = -1.0$ m

$\Delta t_B = 0.5$ s

$(v_x)_B = \text{slope} = \frac{\Delta x}{\Delta t} = -2.0$ m/s

(b)

FIGURE 2.6 Graphical and photographic representations of two students on roller blades.

SOLVE We can determine the students' velocities by measuring the slopes of the graphs. Skater A undergoes a displacement $\Delta x_A = 2.0$ m during the time interval $\Delta t_A = 0.40$ s. Thus his velocity is

$$(v_x)_A = \frac{\Delta x_A}{\Delta t_A} = \frac{2.0 \text{ m}}{0.40 \text{ s}} = 5.0 \text{ m/s}$$

We need to be more careful with skater B. Although he moves a distance of 1.0 m in 0.50 s, his *displacement* Δx has a very precise definition:

$$\Delta x_B = x_{\text{at } 0.5 \text{ s}} - x_{\text{at } 0.0 \text{ s}} = 0.0 \text{ m} - 1.0 \text{ m} = -1.0 \text{ m}$$

Careful attention to the signs is very important! This leads to

$$(v_x)_B = \frac{\Delta x_B}{\Delta t_B} = \frac{-1.0 \text{ m}}{0.50 \text{ s}} = -2.0 \text{ m/s}$$

ASSESS The minus sign indicates that skater B is moving to the left. Our interpretation of this graph is that two students on roller blades are moving with constant velocities in opposite directions, as the photograph of Figure 2.6b on page 39 shows. Skater A starts at $x = 2.0$ m and moves to the right with a velocity of 5.0 m/s. Skater B starts at $x = 1.0$ m and moves to the left with a velocity of -2.0 m/s. Their speeds, of ≈ 10 mph and ≈ 4 mph, are reasonable for skaters on roller blades.

Example 2.2 brought out several points that are worth emphasizing.

TACTICS BOX 2.2   Interpreting position-versus-time graphs

❶ Steeper slopes correspond to faster speeds.

❷ Negative slopes correspond to negative velocities and, hence, to motion to the left (or down).

❸ The slope is a ratio of intervals, $\Delta x/\Delta t$, not a ratio of coordinates. That is, the slope is *not* simply x/t.

❹ We are distinguishing between the *actual* slope and the *physically meaningful* slope. If you were to use a ruler to measure the rise and the run of the graph, you could compute the actual slope of the line as drawn on the page. That is not the slope to which we are referring when we equate the velocity with the slope of the line. Instead, we find the *physically meaningful* slope by measuring the rise and run using the scales along the axes. The "rise" Δx is some number of meters; the "run" Δt is some number of seconds. The physically meaningful rise and run include units, and the ratio of these units gives the units of the slope.

An object's **speed** v is how fast it's going, independent of direction. This is simply $v = |v_x|$ or $v = |v_y|$, the magnitude or absolute value of its velocity. In Example 2.2, for example, skater B's *velocity* is -2.0 m/s but his *speed* is 2.0 m/s. Speed is a scalar quantity, not a vector.

NOTE ▶ Our mathematical analysis of motion is based on velocity, not speed. The subscript in v_x or v_y is an essential part of the notation, reminding us that, even in one dimension, the velocity is a vector. ◀

The Mathematics of Uniform Motion

We need a mathematical analysis of motion that will be valid regardless of whether an object moves along the x-axis, the y-axis, or any other straight line. Consequently, it will be convenient to write equations for a "generic axis" that we will call the s-axis. The position of an object will be represented by the symbol s and its velocity by v_s.

NOTE ▶ Equations written in terms of s are valid for any one-dimensional motion. In a specific problem, however, you should use either x or y, whichever is appropriate, rather than s. ◀

Consider an object in uniform motion along the *s*-axis with the linear position-versus-time graph shown in Figure 2.7. The object's **initial position** is s_i at time t_i. The term *initial position* refers to the starting point of our analysis or the starting point in a problem; the object may or may not have been in motion prior to t_i. At a later time t_f, the ending point of our analysis or the ending point of a problem, the object's **final position** is s_f.

The object's velocity v_s along the *s*-axis can be determined by finding the slope of the graph:

$$v_s = \frac{\text{rise}}{\text{run}} = \frac{\Delta s}{\Delta t} = \frac{s_f - s_i}{t_f - t_i} \qquad (2.2)$$

Equation 2.2 is easily rearranged to give

$$s_f = s_i + v_s \Delta t \text{ (uniform motion)} \qquad (2.3)$$

Equation 2.3 applies to any time interval Δt during which the velocity is constant.

The velocity of a uniformly moving object tells us the amount by which its position changes during each second. A particle with a velocity of 20 m/s *changes* its position by 20 m during every second of motion: by 20 m during the first second of its motion, by another 20 m during the next second, and so on. If the object starts at $s_i = 10$ m, it will be at $s = 30$ m after 1 second of motion and at $s = 50$ m after 2 seconds of motion. Thinking of velocity like this will help you develop an intuitive understanding of the connection between velocity and position.

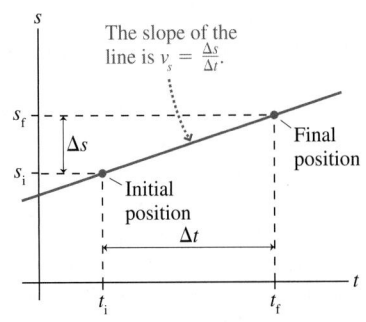

FIGURE 2.7 The velocity is found from the slope of the position-versus-time graph.

EXAMPLE 2.3 Lunch in Cleveland?

Bob leaves home in Chicago at 9:00 A.M. and travels east at a steady 60 mph. Susan, 400 miles to the east in Pittsburgh, leaves at the same time and travels west at a steady 40 mph. Where will they meet for lunch?

MODEL Here is a problem where, for the first time, we can really put all four aspects of our problem-solving strategy into play. To begin, represent Bob and Susan as particles.

VISUALIZE Figure 2.8 shows the physical representation (the motion diagram) and the pictorial representation. The equal spacings of the dots in the motion diagram indicate that the motion is uniform. In evaluating the given information, we recognize that the starting time of 9:00 A.M. is not relevant to the problem. Consequently, the initial time is chosen as simply $t_0 = 0$ hr. Bob and Susan are traveling in opposite directions, hence one of the velocities must be a negative number. We have

Physical representation

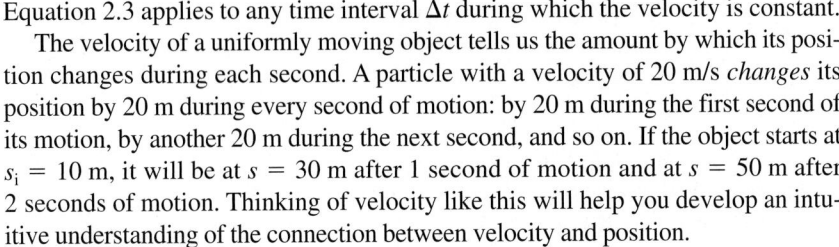

Pictorial representation

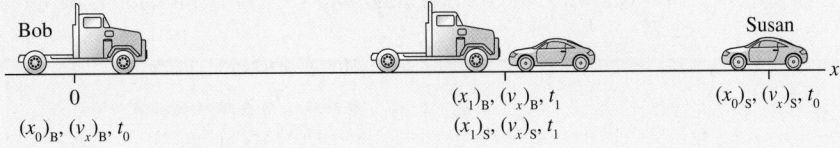

Known
$(x_0)_B = 0$ mi $(v_x)_B = 60$ mph $t_0 = 0$ hr
$(x_0)_S = 400$ mi $(v_x)_S = -40$ mph
t_1 is when $(x_1)_B = (x_1)_S$

Find
$(x_1)_B$

FIGURE 2.8 Physical representation and pictorial representation for Example 2.3.

chosen a coordinate system in which Bob starts at the origin and moves to the right (east) while Susan is moving to the left (west). Thus Susan has the negative velocity. Notice how we've assigned position, velocity, and time symbols to each point in the motion. Pay special attention to how subscripts are used to distinguish different points in the problem and to distinguish Bob's symbols from Susan's.

One purpose of the pictorial representation is to establish what we need to find. Bob and Susan meet when they have the same position at the same time t_1. Thus we want to find $(x_1)_B$ at the time when $(x_1)_B = (x_1)_S$. Notice that $(x_1)_B$ and $(x_1)_S$ are their *positions*, which are equal when they meet, not the distances they have traveled.

SOLVE The goal of the mathematical representation is to proceed from the pictorial representation to a mathematical solution of the problem. We can begin by using Equation 2.3 to find Bob's and Susan's positions at time t_1 when they meet:

$$(x_1)_B = (x_0)_B + (v_x)_B(t_1 - t_0) = (v_x)_B t_1$$
$$(x_1)_S = (x_0)_S + (v_x)_S(t_1 - t_0) = (x_0)_S + (v_x)_S t_1$$

Notice two things. First, we started by writing the *full* statement of Equation 2.3. Only then did we simplify by dropping those terms known to be zero. You're less likely to make accidental errors if you follow this procedure. Second, we replaced the generic symbol s with the specific horizontal-position symbol x, and we replaced the generic subscripts i and f with the specific symbols 0 and 1 that we defined in the pictorial representation. This is also good problem-solving technique.

The condition that Bob and Susan meet is

$$(x_1)_B = (x_1)_S$$

By equating the right-hand sides of the above equations, we get

$$(v_x)_B t_1 = (x_0)_S + (v_x)_S t_1$$

Solving for t_1, we find that they meet at time

$$t_1 = \frac{(x_0)_S}{(v_x)_B - (v_x)_S} = \frac{400 \text{ miles}}{60 \text{ mph} - (-40) \text{ mph}} = 4.0 \text{ hours}$$

Finally, inserting this time back into the equation for x_{B1} gives

$$(x_1)_B = \left(60 \frac{\text{miles}}{\text{hour}}\right) \times (4.0 \text{ hours}) = 240 \text{ miles}$$

While this is a number, it is not yet the answer to the question. The phrase "240 miles" by itself does not say anything meaningful. Because this is the value of Bob's *position*, and Bob was driving east, the answer to the question is, "They meet 240 miles east of Chicago."

ASSESS Before stopping, we should check whether or not this answer seems reasonable. We certainly expected an answer between 0 miles and 400 miles. We also know that Bob is driving faster than Susan, so we expect that their meeting point will be *more* than halfway from Chicago to Pittsburgh. Our assessment tells us that 240 miles is a reasonable answer.

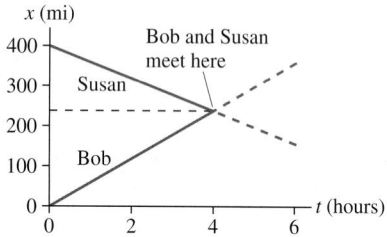

FIGURE 2.9 Position-versus-time graphs for Bob and Susan.

It is instructive to look at this example from a graphical perspective. Figure 2.9 shows position-versus-time graphs for Bob and Susan. Notice the negative slope for Susan's graph, indicating her negative velocity. The point of interest is the intersection of the two lines; this is where Bob and Susan have the same position at the same time. Our method of solution, in which we equated $(x_1)_B$ and $(x_1)_S$, is really just solving the mathematical problem of finding the intersection of two lines.

STOP TO THINK 2.1 Which position-versus-time graph represents the motion shown in the motion diagram?

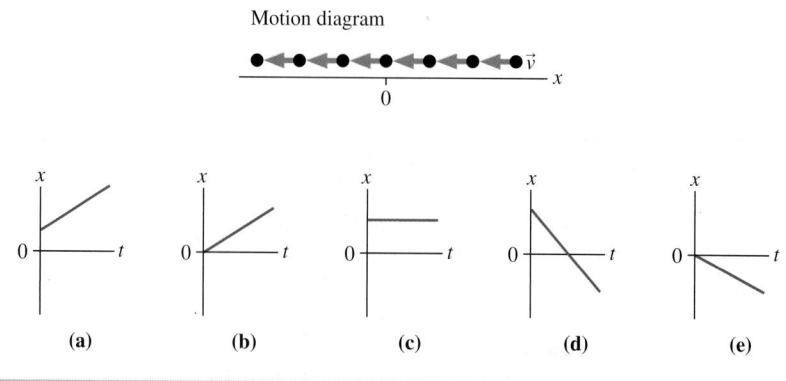

2.3 Instantaneous Velocity

Not many objects in the universe move with constant velocity. Consider, for example, what happens when you drive a car. You start from rest, accelerate, perhaps drive at steady velocity for a while (constant v), decelerate, and ultimately stop. This is clearly a more complex problem than we considered in the last section. If we were to graph your car's position at various times, the graph would not be a straight line.

Figure 2.10a shows the motion diagram of a jet as it takes off. Figure 2.10b is the corresponding position-versus-time graph. The graph curves upward as the spacing between the motion-diagram dots increases. We can determine the jet's average speed v_{avg} between any two times t_i and t_f by selecting those two points on the graph, drawing the straight-line connection between them, measuring Δx and Δt, and finally using these to compute $v_{avg} = \Delta x/\Delta t$. Graphically, v_{avg} is simply the slope of the straight-line connection between the two points.

Average velocity has only limited usefulness for an object whose velocity isn't constant. Suppose, for example, that you and Frank and Karen all leave at precisely 8:00 A.M. in your cars. At 9:00 A.M. you have each traveled exactly 60 mi. All you discern from this information is that each of you had the same average velocity $v_{avg} = 60$ mph. Suppose, however, that Frank started out going faster than 60 mph and later slowed down; Karen got a slow start, but later sped up; and you drove at a steady but boring 60 mph the entire hour. What would each of you see if you read your car's speedometer at 8:10? You would see 60 mph, but Frank would see a speed greater than 60 mph while Karen would see a speed less than 60 mph. Later, at 8:50, Frank's speedometer reads less than 60 mph, Karen's reads more than 60 mph, while yours has not changed.

The speedometer reading tells you how fast you're going *at that instant*, rather than averaged over the entire hour. We can define an object's **instantaneous velocity** to be its velocity—a speed *and* a direction—at a single *instant* of time t.

Such a definition, though, raises some difficult issues. Just what does it mean to have a velocity "at an instant"? Suppose a police officer pulls you over and says, "I just clocked you going 80 miles per hour." You might respond, "But that's impossible. I've only been driving for 20 minutes, so I can't possibly have gone 80 miles." Unfortunately for you, the police officer was a physics major. He replies, "I mean that at the instant I measured your velocity, you were moving at a rate such that you *would* cover a distance of 80 miles *if* you were to continue at that velocity without change for 1 hour. That will be a $200 fine."

Here, again, is the idea that velocity is the *rate* at which an object changes its position. Rates tell us how quickly or how slowly things change, and that idea is conveyed by the word "per." An instantaneous velocity of 80 mph means that the rate at which your car's position is changing—at that exact instant—is such that it would travel a distance of 80 miles in 1 hour *if* it continued at that rate without change. Whether or not it actually does travel at that velocity for another hour, or even for another millisecond, is not relevant.

Using Motion Diagrams and Graphs

Let's use motion diagrams to analyze an accelerating jet plane. Figure 2.11a on the next page shows a motion diagram made using a normal 30-frames-per-second camera. The third velocity vector is the *average* velocity during the time interval Δt it took the object to move the distance Δs from point 2 to point 3. We would like to determine the *instantaneous* velocity v_s at the position marked with the red $\times$, slightly before point 3. Because the jet is accelerating, its velocity at this point—like Karen's at 8:50—is larger than the average velocity between 2 and 3. How can we measure it?

Suppose we use a high-speed camera, one that takes 100 frames per second, to film just the segment of motion between points 2 and 3. This "magnified" motion

This jet plane most definitely does not move with a constant velocity.

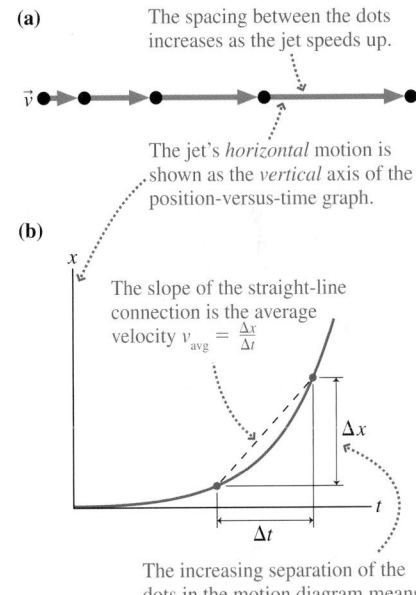

(a) The spacing between the dots increases as the jet speeds up.

$\vec{v}$

The jet's *horizontal* motion is shown as the *vertical* axis of the position-versus-time graph.

(b)

x

The slope of the straight-line connection is the average velocity $v_{avg} = \dfrac{\Delta x}{\Delta t}$

Δx

Δt

t

The increasing separation of the dots in the motion diagram means that Δx increases and the graph curves upward.

FIGURE 2.10 Motion diagram and position graph of a jet during take off.

The speedometer reading tells you how fast you're going *at that instant*.

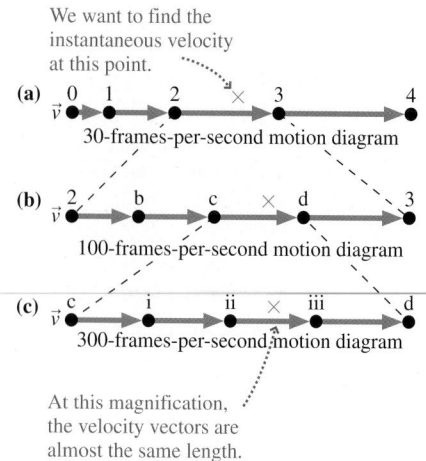

We want to find the instantaneous velocity at this point.

(a) 30-frames-per-second motion diagram

(b) 100-frames-per-second motion diagram

(c) 300-frames-per-second motion diagram

At this magnification, the velocity vectors are almost the same length.

FIGURE 2.11 Motion diagram of an accelerating jet made with increasingly fast movie cameras.

diagram is shown in Figure 2.11b. While the velocity changed a lot between points 0 and 4, causing the length of the arrows in Figure 2.11a to change greatly, the velocity change between 2 and 3 is much less. The velocity arrow from c to d is the *average* velocity between these two points. This average is a much closer approximation to the instantaneous velocity at × because the lengths of the velocity arrows on either side are nearly the same. Even so, the velocity is still changing. The instantaneous velocity we seek is not the same as the average velocity.

Finally we get out our really-high-speed 300-frames-per-second camera and film just the interval between points c and d. The result is shown in Figure 2.11c. Now, at this level of magnification, each velocity vector is *almost* the same length. On this time scale, the motion near × appears very nearly uniform! *If the motion were to continue at a constant velocity after ×, then the velocity it would have is the velocity of Figure 2.11c.*

The point of Figure 2.11 is that the average velocity $v_{avg} = \Delta s/\Delta t$ becomes a better and better approximation to the instantaneous velocity v_s as the time interval Δt over which the displacement is measured gets smaller and smaller. By magnifying the motion diagram, we are using smaller and smaller time intervals Δt. But even 300 frames per second isn't fast enough. We need to let $\Delta t \to 0$.

We can state this idea mathematically in terms of a limit:

$$v_s \equiv \lim_{\Delta t \to 0} \frac{\Delta s}{\Delta t} = \frac{ds}{dt} \text{ (instantaneous velocity)} \qquad (2.4)$$

As Δt gets smaller and smaller, the average velocity $v_{avg} = \Delta s/\Delta t$ reaches a constant value—the limit—and no longer changes. This limit is called *the derivative of s with respect to t,* and it is denoted ds/dt. We'll look at derivatives in the next section.

Now let's analyze the same accelerating jet with position-versus-time graphs. Figure 2.12a shows two points that are separated by $\Delta t = 1/30$ s. These are points 2 and 3 in the 30-frames-per-second motion diagram of Figure 2.11a. The slope of the straight line connecting these points is the average velocity over this time interval. Figures 2.12b and 2.12c show the effect of decreasing the time interval Δt.

As Δt gets smaller, the straight line becomes a better and better approximation of the curve between the two points and the average velocity becomes a better and better approximation of the instantaneous velocity at point ×. Notice that the slope in Figure 2.12c is a little larger than the slope in Figure 2.12a. This is because, as we had noted, the instantaneous velocity at × is larger than the average velocity between points 2 and 3.

Finally, in Figure 2.12d, we reach the limit $\Delta t \to 0$. In this limit, the straight line is tangent to the curve at point ×. **The instantaneous velocity at time t is the slope of the line that is tangent to the position-versus-time graph at time t.** The practical issue will be how to determine the slope of the tangent line.

(a) The slope between 2 and 3 is v_{avg}.

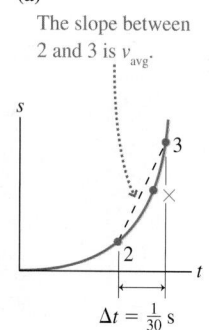

$\Delta t = \frac{1}{30}$ s

(b) The slope between c and d is a better approximation to the velocity at ×.

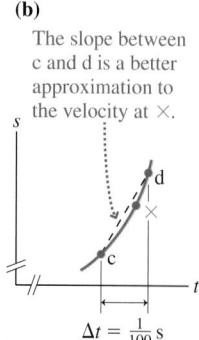

$\Delta t = \frac{1}{100}$ s

(c) The velocity at × is the slope as $\Delta t \to 0$.

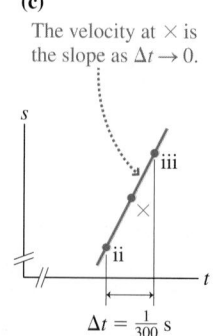

$\Delta t = \frac{1}{300}$ s

(d) The velocity at × is the slope of the line tangent to the curve at ×.

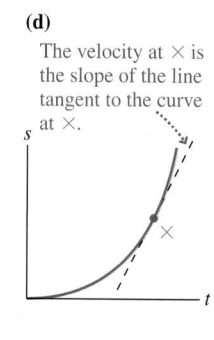

FIGURE 2.12 Position-versus-time graphs that correspond to the motion diagrams of Figure 2.11.

EXAMPLE 2.4 Relating a velocity graph to a position graph

Figure 2.13a is the position-versus-time graph of a car.

a. Draw the car's velocity-versus-time graph.
b. Describe the car's motion.

MODEL Represent the car as a particle, with a well-defined position at each instant of time.

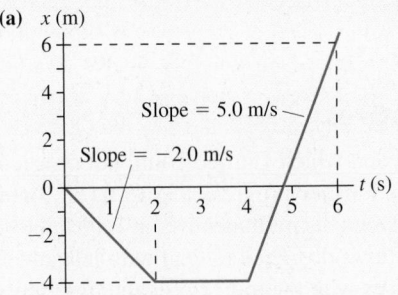

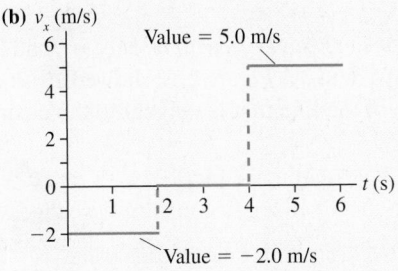

FIGURE 2.13 Position-versus-time graph and the corresponding velocity-versus-time graph.

VISUALIZE Figure 2.13a is a graphical representation of the motion.

SOLVE

a. The car's position-versus-time graph is a sequence of three straight lines. Each of these straight lines represents uniform motion at a constant velocity. We can determine the car's velocity during each interval of time by measuring the slope of the line. From $t = 0$ s to $t = 2$ s ($\Delta t = 2$ s) the car's displacement is $\Delta x = -4$ m $- 0$ m $= -4$ m. The velocity during this interval is

$$v_x = \frac{\Delta x}{\Delta t} = \frac{-4.0 \text{ m}}{2.0 \text{ s}} = -2.0 \text{ m/s}$$

The car's position does not change from $t = 2$ s to $t = 4$ s ($\Delta x = 0$), so $v_x = 0$. Finally, the displacement between $t = 4$ s and $t = 6$ s ($\Delta t = 2$ s) is $\Delta x = 10$ m. Thus the velocity during this interval is

$$v_x = \frac{10 \text{ m}}{2.0 \text{ s}} = 5.0 \text{ m/s}$$

These velocities are shown on the velocity-versus-time graph of Figure 2.13b.

b. The car backs up for 2 s at 2 m/s, sits at rest for 2 s, then drives forward at 5 m/s for at least 2 s. We can't tell from the graph what happens for $t > 6$ s.

ASSESS The velocity graph and the position graph look completely different. The *value* of the velocity graph at any instant of time equals the *slope* of the position graph.

EXAMPLE 2.5 Finding velocity from position graphically

Figure 2.14 shows the position-versus-time graph of a particle that moves along the y-axis.

a. At which labeled point or points is the particle moving the slowest?
b. At which point or points is the particle moving the fastest?
c. Sketch an approximate velocity-versus-time graph for the particle.

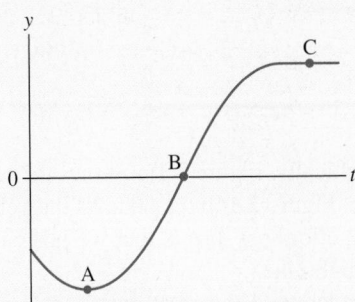

FIGURE 2.14 Position-versus-time graph.

MODEL The problem statement tells us the object is a particle.

VISUALIZE Figure 2.15 is a graphical representation of the motion.

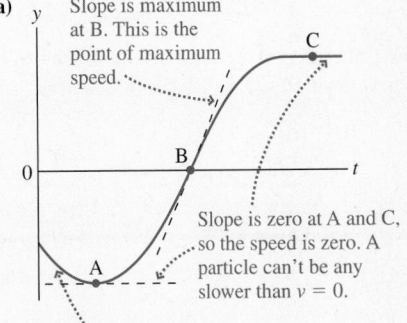

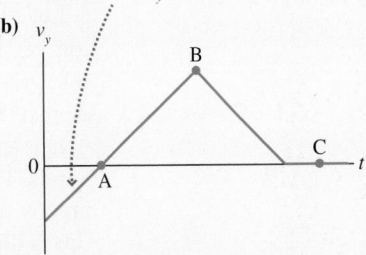

FIGURE 2.15 The velocity-versus-time graph is found from the position graph.

SOLVE

a. Figure 2.15a shows that the particle is slowest—no speed at all!—at points A and C. At point A, the speed is only instantaneously zero. At point C, the particle has actually stopped and remains at rest.

b. The particle moves the fastest at point B.

c. Although we cannot find an exact velocity-versus-time graph, we can see that the slope, and hence v_y, is initially negative, becomes zero at point A, rises to a maximum value at point B, decreases back to zero a little before point C, then remains at zero thereafter. Thus Figure 2.15b shows, at least approximately, the particle's velocity-versus-time graph.

ASSESS Once again, the shape of the velocity graph bears no resemblance to the shape of the position graph. You must transfer *slope* information from the position graph to *value* information on the velocity graph.

A Little Calculus: Derivatives

We have reached the point beyond which Galileo could not proceed because he lacked the mathematical tools. Further progress had to await a new branch of mathematics called *calculus,* invented simultaneously in England by Newton and in Germany by Leibniz. Calculus is designed to deal with instantaneous quantities. In other words, it provides us with the tools for evaluating limits such as the one in Equation 2.4.

The notation ds/dt is called *the derivative of s with respect to t*, and Equation 2.4 defines it as the limiting value of a ratio. As Figure 2.12 showed, ds/dt can be interpreted graphically as the slope of the line that is tangent to the position-versus-time graph at time t.

EXAMPLE 2.6 Finding velocity from position as a derivative

The position of a particle as a function of time is $s = 2t^2$ m, where t is in s. What is the velocity v_s as a function of time?

SOLVE To solve this problem we need to "take the derivative" of s.

$$v_s = \frac{ds}{dt} = \lim_{\Delta t \to 0} \frac{\Delta s}{\Delta t}$$

During the time interval Δt, the particle moves from position $s_{\text{at } t}$ to the new position $s_{\text{at } t+\Delta t}$. Its displacement is

$$\Delta s = s_{\text{at } t+\Delta t} - s_{\text{at } t}$$

$$= 2(t + \Delta t)^2 - 2t^2$$

$$= 2(t^2 + 2t\Delta t + (\Delta t)^2) - 2t^2$$

$$= 4t\Delta t + 2(\Delta t)^2$$

The average velocity during the time interval Δt is

$$v_{\text{avg}} = \frac{\Delta s}{\Delta t} = \frac{4t\Delta t + 2(\Delta t)^2}{\Delta t} = 4t + 2\Delta t$$

We can finish by taking the limit $\Delta t \to 0$ to find

$$v_s = \frac{ds}{dt} = \lim_{\Delta t \to 0}(4t + 2\Delta t) = 4t \text{ m/s}$$

In other words, the function for calculating the velocity at any instant of time is $v_s = 4t$ m/s, where t is in s. At $t = 3$ s, for example, the particle is located at position $s = 18$ m and its instantaneous velocity, at just that instant, is $v_s = 12$ m/s.

Let's look at this example graphically. Figure 2.16a shows the particle's position-versus-time graph $s = 2t^2$ m. Figure 2.16b then shows the velocity-versus-time graph, using the velocity function $v = 4t$ m/s that we just calculated. You see that the velocity graph is a straight line.

It is critically important to understand the relationship between these two graphs. The *value* of the velocity graph at any instant of time, which we can read directly off the vertical axis, is the *slope* of the position graph at that same time. This is illustrated at $t = 1$ s and $t = 3$ s.

Example 2.6 showed how the limit of $\Delta s/\Delta t$ can be evaluated to find a derivative, but the procedure is clearly rather tedious. It would hinder us significantly if

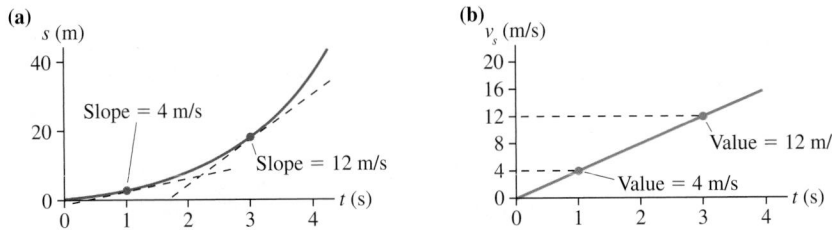

FIGURE 2.16 Position-versus-time and velocity-versus-time graphs for Example 2.6.

we had to do this for every new situation. Fortunately, we need only a few basic derivatives in this text. Learn these, and you do not have to go all the way back to the definition in terms of limits.

The only functions we will use in Parts I and II of this book are powers and polynomials. Consider the function $u = ct^n$, where c and n are constants. The following result is proven in calculus:

$$\text{The derivative of } u = ct^n \text{ is } \frac{du}{dt} = nct^{n-1} \qquad (2.5)$$

NOTE ▶ The symbol u is a "dummy name." Equation 2.5 can be used to take the derivative of *any* function of the form ct^n. ◀

Example 2.6 needed to find the derivative of the function $s = 2t^2$. Using Equation 2.5 with $c = 2$ and $n = 2$, the derivative of $s = 2t^2$ with respect to t is

$$v_s = \frac{ds}{dt} = 2 \cdot 2t^{2-1} = 4t$$

Similarly, the derivative of the function $x = 3/t^2 = 3t^{-2}$ is

$$\frac{dx}{dt} = (-2) \cdot 3t^{-2-1} = -6t^{-3} = -\frac{6}{t^3}$$

A value that doesn't change with time, such as the position of an object at rest, can be represented by the function $u = c = $ constant. That is, the exponent of t^n is $n = 0$. You can see from Equation 2.5 that the derivative of a constant is zero. That is,

$$\frac{du}{dt} = 0 \text{ if } u = c = \text{constant} \qquad (2.6)$$

This makes sense. The graph of the function $u = c$ is simply a horizontal line at height c. The slope of a horizontal line—which is what the derivative du/dt measures—is zero.

The only other information we need about derivatives for now is how to evaluate the derivative of the sum of two or more functions. Let u and w be two separate functions of time. You will learn in calculus that

$$\frac{d}{dt}(u + w) = \frac{du}{dt} + \frac{dw}{dt} \qquad (2.7)$$

That is, the derivative of a sum is the sum of the derivatives.

NOTE ▶ You may have learned in calculus to take the derivative dy/dx, where y is a function of x. The derivatives we use in physics are the same; only the notation is different. We're interested in how quantities change with time, so our derivatives are with respect to t instead of x. ◀

EXAMPLE 2.7 Using calculus to find the velocity

A particle's position is given by the function $x = (-t^3 + 3t)$ m, where t is in s.

a. What is the particle's position and velocity at $t = 2$ s?
b. Draw graphs of x and v_x during the interval -3 s $\leq t \leq 3$ s.
c. Draw a motion diagram to illustrate this motion.

SOLVE

a. We can compute the position at $t = 2$ s directly from the function x:

$$x(\text{at } t = 2 \text{ s}) = -(2)^3 + (3)(2) = -8 + 6 = -2 \text{ m}$$

The velocity is then $v_x = dx/dt$. The function for x is the sum of two polynomials, so

$$v_x = \frac{dx}{dt} = \frac{d}{dt}(-t^3 + 3t) = \frac{d}{dt}(-t^3) + \frac{d}{dt}(3t)$$

The first derivative is a power with $c = -1$ and $n = 3$; the second has $c = 3$ and $n = 1$. Using Equation 2.5,

$$v_x = (-3t^2 + 3) \text{ m/s}$$

where t is in s. Evaluating the velocity at $t = 2$ s gives

$$v_x(\text{at } t = 2 \text{ s}) = -3(2)^2 + 3 = -9 \text{ m/s}$$

The negative sign indicates that the particle, at this instant of time, is moving to the *left* at a speed of 9 m/s.

b. Figure 2.17 shows the position graph and the velocity graph. These were created by computing, and then graphing, the values of x and v_x at several points between -3 s and 3 s. The slope of the position-versus-time graph at $t = 2$ s is -9 m/s; this becomes the *value* that is graphed for the velocity at $t = 2$ s. Similar measurements are shown at $t = -1$ s, where the velocity is instantaneously zero.

c. Finally, we can interpret the graphs of part b to draw the motion diagram shown in Figure 2.18.

- The particle is initially to the right of the origin ($x > 0$ at $t = -3$ s) but moving to the left ($v_x < 0$). Its *speed* is slowing ($v = |v_x|$ is decreasing), so the velocity vector arrows are getting shorter.
- The particle passes the origin at $t \approx -1.5$ s, but it is still moving to the left.
- The position reaches a minimum at $t = -1$ s; the particle is as far left as it is going. The velocity is *instantaneously* $v_x = 0$ m/s as the particle reverses direction.
- The particle moves back to the right between $t = -1$ s and $t = 1$ s ($v_x > 0$).
- The particle turns around again at $t = 1$ s and begins moving back to the left ($v_x < 0$). It keeps speeding up, then disappears off to the left.

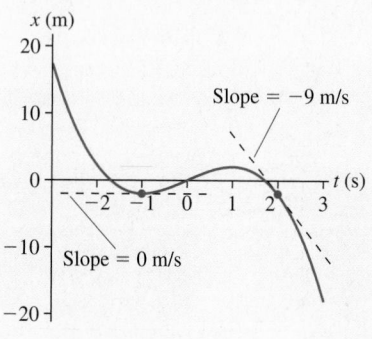

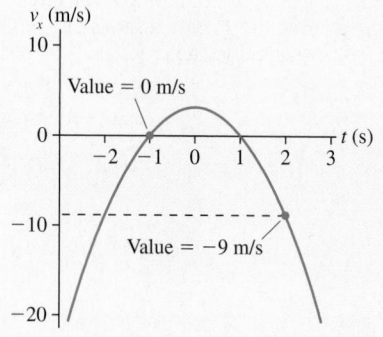

FIGURE 2.17 Position and velocity graphs.

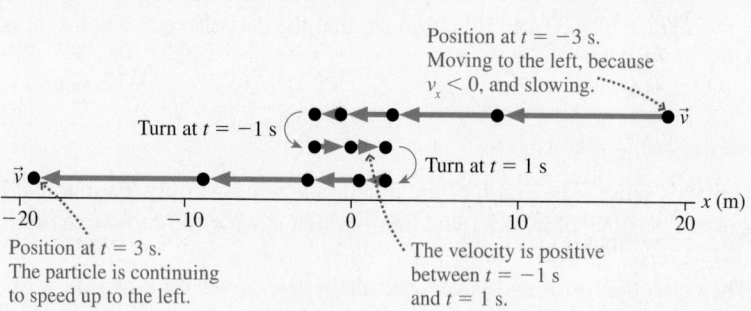

FIGURE 2.18 The motion diagram for Example 2.7.

The particle in this example moved out to $x = -2$ m at $t = -1$ s, then returned. The point in its motion where it reversed direction is called a *turning point*. Because the velocity was negative just before reaching the turning point and positive just after, it had to pass through $v_x = 0$ m/s. Thus, a **turning point** is a point where the velocity is instantaneously zero as the particle reverses direction. A second turning point occurs at $t = 1$ s as the particle reaches $x = 2$ m. We will see many future examples of turning points.

STOP TO THINK 2.2   Which velocity-versus-time graph goes with the position-versus-time graph on the left?

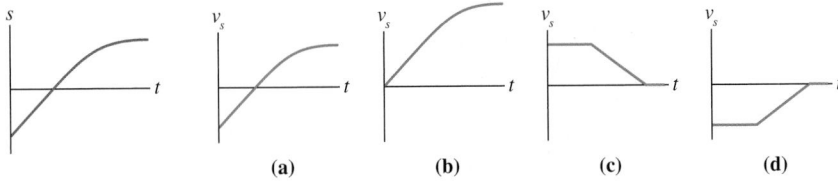

(a)   **(b)**   **(c)**   **(d)**

2.4 Finding Position from Velocity

Equation 2.4 provides a means of finding the instantaneous velocity v_s if we know the position s as a function of time. In mathematical terms, the velocity is the derivative of the position function. Graphically, the velocity is the slope of the position-versus-time graph.

But what about the reverse problem? Can we use the object's velocity to predict its position at some future time t? Equation 2.3, $s_f = s_i + v_s\Delta t$, does this for the case of uniform motion with a constant velocity. We need to find a more general expression that is valid when v_s is not constant.

Figure 2.19a is a velocity-versus-time graph for a particle whose velocity varies with time. Suppose we know the object's position to be s_i at an initial time t_i. Our goal is to find its position s_f at a later time t_f.

Because we know how to handle constant velocities, using Equation 2.3, let's *approximate* the velocity function of Figure 2.19a as a series of constant-velocity steps of width Δt. This is illustrated in Figure 2.19b. During the first step, from time t_i to time $t_i + \Delta t$, the velocity has the constant value $(v_s)_1$. The velocity is a constant $(v_s)_2$ during the second step from $t_i + \Delta t$ to $t_i + 2\Delta t$, and so on. The velocity during step k has the constant value $(v_s)_k$. Altogether the velocity-versus-time curve has been divided into N constant-velocity steps of equal width Δt. Although the approximation shown in the figure is rather rough, with only nine steps, we can easily imagine that it could be made as accurate as desired by having more and more ever-narrower steps .

The velocity during each step is constant (uniform motion), so we can apply Equation 2.3 to each step. The object's displacement Δs_1 during the first step is simply $\Delta s_1 = (v_s)_1\Delta t$. The displacement during the second step $\Delta s_2 = (v_s)_2\Delta t$, and during step k the displacement is $\Delta s_k = (v_s)_k\Delta t$.

The total displacement of the object between t_i and t_f can be approximated as the sum of the all the individual displacements during each of the N constant-velocity steps. That is,

$$\Delta s = s_f - s_i \approx \Delta s_1 + \Delta s_2 + \cdots + \Delta s_N = \sum_{k=1}^{N}(v_s)_k\Delta t \qquad (2.8)$$

where Σ (Greek sigma) is the symbol for summation. With a simple rearrangement, the particle's final position is

$$s_f \approx s_i + \sum_{k=1}^{N}(v_s)_k\Delta t \qquad (2.9)$$

Our goal was to use the velocity to find the final position s_f. Equation 2.9 nearly reaches that goal, but Equation 2.9 is only approximate because the constant-velocity steps are only an approximation of the true velocity graph. But if we now let $\Delta t \to 0$, each step's width approaches zero while the total number of

(a) 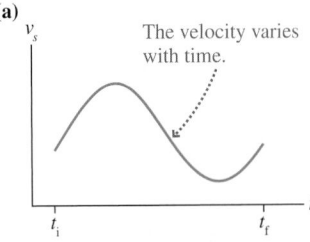 The velocity varies with time.

(b) 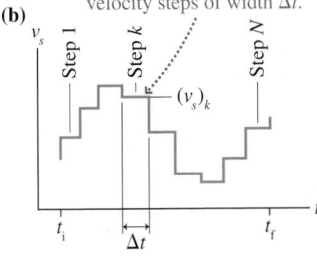 The velocity curve is approximated by constant-velocity steps of width Δt.

FIGURE 2.19 Approximating a velocity-versus-time graph with a series of constant-velocity steps.

steps N approaches infinity. In this limit, the series of steps becomes a perfect replica of the velocity-versus-time graph and Equation 2.9 becomes exact. Thus

$$s_{\text{f}} = s_{\text{i}} + \lim_{\Delta t \to 0} \sum_{k=1}^{N} (v_s)_k \Delta t = s_{\text{i}} + \int_{t_{\text{i}}}^{t_{\text{f}}} v_s \, dt \tag{2.10}$$

The curlicue symbol is called an *integral*. The expression on the right is read, "the integral of $v_s \, dt$ from t_{i} to t_{f}." Equation 2.10 is the result that we were seeking. It allows us to predict an object's position s_{f} at a future time t_{f}.

We can give Equation 2.10 an important geometric interpretation. Figure 2.20 shows step k in the approximation of the velocity graph as a long, thin rectangle of height $(v_s)_k$ and width Δt. The product $\Delta s_k = (v_s)_k \Delta t$ is the area (base × height) of this small rectangle. The sum in Equation 2.10 adds up all of these rectangular areas to give the total area enclosed between the t-axis and the tops of the steps. The limit of this sum as $\Delta t \to 0$ is the total area enclosed between the t-axis and the velocity curve. This is called the "area under the curve." Thus a graphical interpretation of Equation 2.10 is:

$$s_{\text{f}} = s_{\text{i}} + \text{area under the velocity curve } v_s \text{ between } t_{\text{i}} \text{ and } t_{\text{f}} \tag{2.11}$$

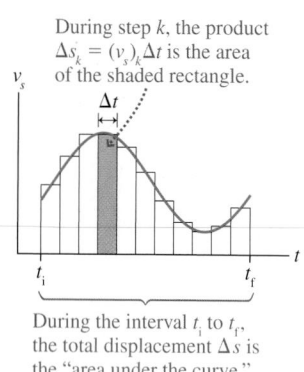

During step k, the product $\Delta s_k = (v_s)_k \Delta t$ is the area of the shaded rectangle.

During the interval t_{i} to t_{f}, the total displacement Δs is the "area under the curve."

FIGURE 2.20 The total displacement Δs is the "area under the curve."

NOTE ▶ Wait a minute! The displacement $\Delta s = s_{\text{f}} - s_{\text{i}}$ is a length. How can a length equal an area? Recall earlier, when we found that the velocity is the slope of the position graph, we made a distinction between the *actual* slope and the *physically meaningful* slope? The same distinction applies here. The velocity graph does indeed bound a certain area on the page. That is the actual area, but it is *not* the area to which we are referring. Once again, we need to measure the quantities we are using, v_s and Δt, by referring to the scales on the axes. Δt is some number of seconds while v_s is some number of meters per second. When these are multiplied together, the *physically meaningful* area has units of meters, appropriate for a displacement. The following examples will help make this clear. ◀

EXAMPLE 2.8 The displacement during a drag race
Figure 2.21 shows the velocity-versus-time graph of a drag racer. How far does the racer move during the first 3.0 s?

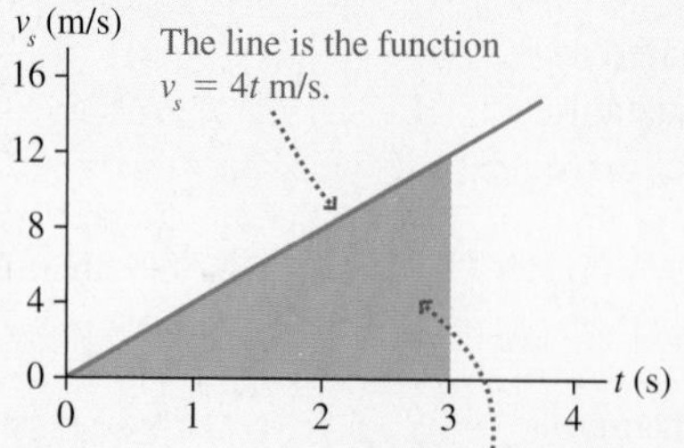

v_s (m/s)

The line is the function $v_s = 4t$ m/s.

The displacement Δs is the area of the shaded triangle.

FIGURE 2.21 Velocity-versus-time graph for the drag racer of Example 2.8.

MODEL Represent the drag racer as a particle with a well-defined position at all times.

VISUALIZE Figure 2.21 is a graphical representation of the motion.

SOLVE The question "how far" indicates that we need to find a displacement Δs rather than a position s. According to Equation 2.11, the car's displacement $\Delta s = s_{\text{f}} - s_{\text{i}}$ between $t = 0$ s and $t = 3$ s is the area under the curve from $t = 0$ s to $t = 3$ s. The curve in this case is an angled line, so the area is that of a triangle:

$$\Delta s = \text{area of triangle between } t = 0 \text{ s and } t = 3 \text{ s}$$
$$= \tfrac{1}{2} \times \text{base} \times \text{height}$$
$$= \tfrac{1}{2} \times 3 \text{ s} \times 12 \text{ m/s}$$
$$= 18 \text{ m}$$

The drag racer moves 18 m during the first 3 seconds.

ASSESS The "area" is a product of s with m/s, so Δs has the proper units of m.

EXAMPLE 2.9 Finding an expression for the racer's position

a. Find an algebraic expression for the position s as a function of time t for the drag racer whose velocity-versus-time graph was shown in Figure 2.21. Assume the car's initial position is $s_i = 0$ m at $t_i = 0$ s.
b. Draw the car's position-versus-time graph.

SOLVE

a. Let $s_i = 0$ at $t_i = 0$ and let s be the position at later time t. The straight line for v_s in Figure 2.21 is described by the linear function $v_s = 4t$ m/s, where t is in s. Then

$$s = s_i + \int_0^t v_s dt = 0 + \text{area under the triangle between 0 and } t$$

$$= 0 + \tfrac{1}{2}(t - 0)(4t - 0)$$

$$= 2t^2 \text{ m, where } t \text{ is in s}$$

b. Figure 2.22 shows the drag racer's position-versus-time graph. It's simply a graph of the function $s = 2t^2$ m, where t is in s. Notice that the *linear* velocity graph of Figure 2.21 is

associated with a *parabolic* position graph. This is a general result that we will see again.

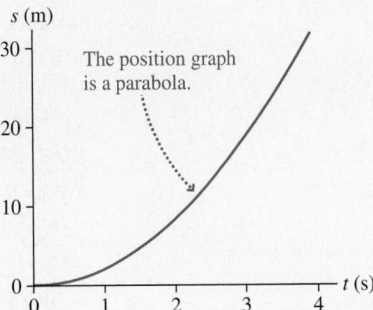

FIGURE 2.22 The position-versus-time graph for the drag racer whose velocity graph was shown in Figure 2.21.

ASSESS This is exactly Example 2.6 in reverse! There we found, by taking the derivative, that a particle whose position is $s = 2t^2$ m has a velocity described by $v_s = 4t$ m/s. Here we have found, by integration, that a drag racer whose velocity is $v_s = 4t$ m/s has a position described by $s = 2t^2$ m.

EXAMPLE 2.10 Finding the turning point

Figure 2.23 is the velocity graph for a particle that starts at $x_i = 30$ m at time $t_i = 0$ s.

a. Draw a motion diagram for the particle.
b. Where is the particle's turning point?
c. At what time does the particle reach the origin?

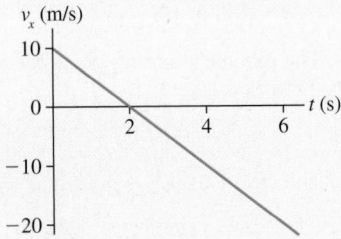

FIGURE 2.23 Velocity-versus-time graph for the particle of Example 2.10.

VISUALIZE The particle is initially 30 m to the right of the origin and moving *to the right* ($v_x > 0$) with a speed of 10 m/s. But v_x is decreasing, so the particle is slowing down. At $t = 2$ s the velocity, just for an instant, is zero before becoming negative. This is the turning point. The velocity is negative for $t > 2$ s, so the particle has reversed direction and moves back toward the origin. At some later time, which we want to find, the particle will pass $x = 0$ m.

SOLVE

a. Figure 2.24 shows the motion diagram. The distance scale will be established in parts b and c but is shown here for convenience.

b. The particle reaches the turning point at $t = 2$ s. To learn *where* it is at that time we need to find the displacement during the first two seconds. We can do this by finding the area under the curve between $t = 0$ s and $t = 2$ s:

$$x(\text{at } t = 2\text{ s}) = x_i + \int_{0\text{ s}}^{2\text{ s}} v_x dt$$

$$= x_i + \text{ area under the curve between 0 s and 2 s}$$

$$= 30 \text{ m} + \tfrac{1}{2}(2\text{ s} - 0\text{ s})(10\text{ m/s} - 0\text{ m/s})$$

$$= 40 \text{ m}$$

The turning point is at $x = 40$ m.

c. The particle needs to move $\Delta x = -40$ m to get from the turning point to the origin. That is, the area under the curve from $t = 2$ s to the desired time t needs to be -40 m. Because the curve is below the axis, with negative values of v_x, the area to the right of $t = 2$ s is a *negative* area. With a bit of geometry, you will find that the triangle with a base extending from $t = 2$ s to $t = 6$ s has an area of -40 m. Thus the particle reaches the origin at $t = 6$ s.

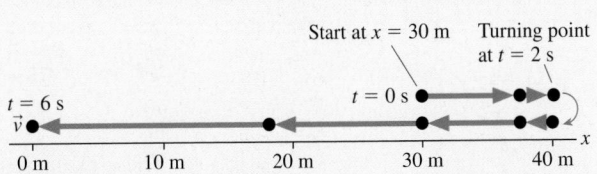

FIGURE 2.24 Motion diagram for the particle whose velocity graph was shown in Figure 2.23.

A Little More Calculus: Integrals

Taking the derivative of a function is equivalent to finding the slope of a graph of the function. Similarly, evaluating an integral is equivalent to finding the area under a graph of the function. The graphical method is very important for building intuition about motion but is limited in its practical application. Just as derivatives of standard functions can be evaluated and tabulated, so can integrals.

The integral in Equation 2.10 is called a *definite integral* because there are two definite boundaries to the area we want to find. These boundaries are called the lower (t_i) and upper (t_f) *limits of integration*. For the important function $u = ct^n$, the essential result from calculus is that

$$\int_{t_i}^{t_f} u\, dt = \int_{t_i}^{t_f} ct^n\, dt = \left. \frac{ct^{n+1}}{n+1} \right|_{t_i}^{t_f} = \frac{ct_f^{n+1}}{n+1} - \frac{ct_i^{n+1}}{n+1} \qquad (n \neq -1) \quad (2.12)$$

The vertical bar in the third step with subscript t_i and superscript t_f is a shorthand notation from calculus that means—as seen in the last step—the integral evaluated at the upper limit t_f *minus* the integral evaluated at the lower limit t_i. You also need to know that for two functions u and w,

$$\int_{t_i}^{t_f} (u + w)\, dt = \int_{t_i}^{t_f} u\, dt + \int_{t_i}^{t_f} w\, dt \qquad (2.13)$$

That is, the integral of a sum is equal to the sum of the integrals.

EXAMPLE 2.11 Using calculus to find the position
Use calculus to solve Example 2.10.

SOLVE Figure 2.23 is a linear graph. Its "y-intercept" is seen to be 10 m/s and its slope is -5 (m/s)/s. Thus the velocity graphed here can be described by the equation

$$v_x = (10 - 5t) \text{ m/s}$$

where t is in s. We can find the position x at time t by using Equation 2.10:

$$x = x_i + \int_0^t v_x\, dt = 30 \text{ m} + \int_0^t (10 - 5t)\, dt$$

$$= 30 \text{ m} + \int_0^t 10\, dt - \int_0^t 5t\, dt$$

We used Equation 2.13 for the integral of a sum to get the final expression. The first integral is a function of the form $u = ct^n$ with $c = 10$ and $n = 0$; the second is of the form $u = ct^n$ with $c = 5$ and $n = 1$. Using Equation 2.12,

$$\int_0^t 10\, dt = 10t \Big|_0^t = 10 \cdot t - 10 \cdot 0 = 10t \text{ m}$$

and

$$\int_0^t 5t\, dt = \tfrac{5}{2}t^2 \Big|_0^t = \tfrac{5}{2} \cdot t^2 - \tfrac{5}{2} \cdot 0^2 = \tfrac{5}{2}t^2 \text{ m}$$

Combining the pieces gives

$$x = (30 + 10t - \tfrac{5}{2}t^2) \text{ m}$$

where t is in s. The particle's turning point occurs at $t = 2$ s, and its position at that time is

$$x(\text{at } t = 2 \text{ s}) = 30 + (10)(2) - \tfrac{5}{2}(2)^2 = 40 \text{ m}$$

The time at which the particle reaches the origin is found by setting $x = 0$ m:

$$30 + 10t - \tfrac{5}{2}t^2 = 0$$

This quadratic equation has two solutions: $t = -2$ s or $t = 6$ s.

When we solve a quadratic equation, we cannot just arbitrarily select the root we want. Instead, we must decide which is the *meaningful* root. Here the negative root refers to a time before the problem began, so the meaningful one is the positive root, $t = 6$ s.

ASSESS The results agree with the answers we found previously from a graphical solution.

These examples make the point that there are often many ways to solve a problem. The graphical procedures for finding derivatives and integrals are simple, but they work only for a limited range of problems—those where the geometry is simple. The techniques of calculus are more demanding, but these techniques allow us to deal with functions whose graphs are quite complex.

Summing Up

As you work on building intuition about motion, you need to be able to move back and forth between four different representations of the motion:

- The motion diagram;
- The position-versus-time graph;
- The velocity-versus-time graph;
- The description in words.

Given a description of a certain motion, you should be able to sketch the motion diagram and the position and velocity graphs. Given one graph, you should be able to generate the other. And given position and velocity graphs, you should be able to "interpret" them by describing the motion in words or in a motion diagram.

STOP TO THINK 2.3 Which position-versus-time graph goes with the velocity-versus-time graph on the left? The particle's position at $t_i = 0$ s is $x_i = -10$ m.

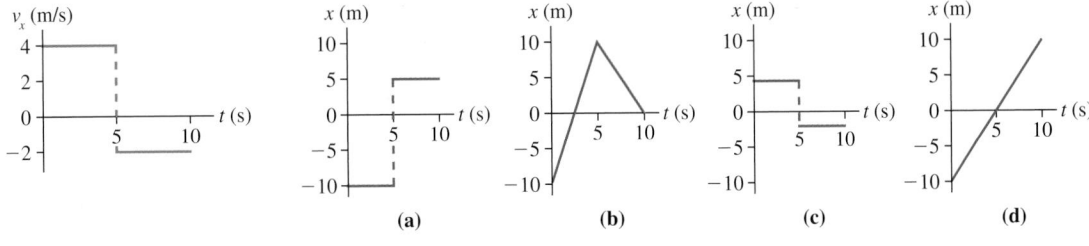

(a)   (b)   (c)   (d)

2.5 Motion with Constant Acceleration

We need one more major concept to describe one-dimensional motion: acceleration. Acceleration, as we noted in Chapter 1, is a rather abstract concept. You cannot "see" the value of acceleration, as you can that of position, nor can you judge it by looking to see if an object is moving quickly or slowly. Nonetheless, acceleration is the linchpin of mechanics. We will see very shortly that Newton's laws relate the acceleration of an object to the forces that are exerted on it.

Let's conduct a race between a Volkswagen Beetle and a Porsche to see which can achieve a velocity of 30 m/s (≈ 60 mph) in the shortest time. Both cars are equipped with computers that will record the speedometer reading 10 times each second. This gives a nearly continuous record of the *instantaneous* velocity of each car. Table 2.2 shows some of the data. The velocity-versus-time graphs, based on these data, are shown in Figure 2.25.

How can we describe the difference in performance of the two cars? It is not that one has a different velocity from the other; both achieve every velocity between 0 and 30 m/s. The distinction is how long it took each to *change* its velocity from 0 to 30 m/s. The Porsche changed velocity quickly, in 6 s, while the VW needed 15 s to make the same velocity change. This suggests that the distinction is, once again, a *rate*.

In this case, as we compare the two cars, we are looking at the rate at which their velocities change. Because the Porsche had a velocity change $\Delta v_s = 30$ m/s during a time interval $\Delta t = 6$ s, the *rate* at which its velocity changed was

$$\text{rate of velocity change} = \frac{\Delta v_s}{\Delta t} = \frac{30 \text{ m/s}}{6.0 \text{ s}} = 5.0 \text{ (m/s)/s} \qquad (2.14)$$

Notice the units. They are units of "velocity per second." A rate of velocity change of 5.0 "meters per second per second" means that the velocity increases

TABLE 2.2 Velocities of a Porsche and a Volkswagen Beetle

$t(s)$	v_{Porsche} (m/s)	v_{VW} (m/s)
0.0	0.0	0.0
0.1	0.5	0.2
0.2	1.0	0.4
0.3	1.5	0.6
0.4	2.0	0.8
$\vdots$	$\vdots$	$\vdots$

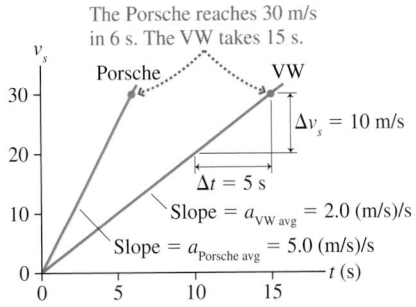

FIGURE 2.25 Velocity-versus-time graphs for the Porsche and the VW Beetle.

by 5.0 m/s during the first second, by another 5.0 m/s during the next second, and so on. In fact, the velocity will increase by 5.0 m/s during any second in which it is changing at the rate of 5.0 (m/s)/s.

Chapter 1 introduced *acceleration* as "the rate of change of velocity." That is, acceleration measures how quickly or slowly an object's velocity changes. The Porsche's velocity changed quickly, so it had a large acceleration. The VW's velocity changed more slowly, so its acceleration was less. In parallel with our treatment of velocity, let's call this the **average acceleration** a_{avg} during the time interval Δt:

$$a_{avg} \equiv \frac{\Delta v_s}{\Delta t} \text{ (average acceleration)} \tag{2.15}$$

Because Δv_s and Δt are the "rise" and "run" of a velocity-versus-time graph, we see that a_{avg} can be interpreted graphically as the *slope* of a straight-line velocity-versus-time graph. Figure 2.25 uses this idea to show that the VW's average acceleration is

$$a_{VW\,avg} = \frac{\Delta v_s}{\Delta t} = \frac{10 \text{ m/s}}{5.0 \text{ s}} = 2.0 \text{ (m/s)/s} \tag{2.16}$$

This is less than the acceleration of the Porsche, as expected.

1.2, 1.3   Activ Physics ONLINE

An object whose velocity-versus-time graph is a straight-line graph has a steady and unchanging acceleration. Such a graph represents motion with *constant acceleration*, which we call **uniformly accelerated motion: An object has uniformly accelerated motion if and only if its acceleration a_s is constant and unchanging. The object's velocity-versus-time graph is a straight line, and a_s is the slope of the line.** There's no need to specify "average" if the acceleration is constant, so we'll use the symbol a_s as we discuss motion along the s-axis with constant acceleration.

NOTE ▶ An important aspect of acceleration is its *sign*. Acceleration $\vec{a}$, like position $\vec{r}$ and velocity $\vec{v}$, is a vector. For motion in one dimension the sign of a_x (or a_y) is positive if the vector $\vec{a}$ points to the right (or up), negative if it points to the left (or down). This was illustrated in Figure 2.1, which you may wish to review. It's particularly important to emphasize that positive and negative values of a_s do *not* correspond to "speeding up" and "slowing down." ◀

EXAMPLE 2.12 Relating acceleration to velocity

a. A particle has a velocity of 10 m/s and a constant acceleration of 2 (m/s)/s. What is its velocity 1 s later? 2 s later?

b. A particle has a velocity of −10 m/s and a constant acceleration of 2 (m/s)/s. What is its velocity 1 s later? 2 s later?

SOLVE

a. An acceleration of 2 (m/s)/s *means* that the velocity increases by 2 m/s every 1 s. If the particle's initial velocity is 10 m/s, then 1 s later its velocity will be 12 m/s. After 2 s,

which is 1 additional second later, it will increase by another 2 m/s to 14 m/s. After 3 s it will be 16 m/s. Here a positive a_s is causing the particle to speed up.

b. If the particle's initial velocity is a *negative* −10 m/s but the acceleration is a positive +2 (m/s)/s, then 1 s later the velocity will be −8 m/s. After 2 s it will be −6 m/s, and so on. In this case, a positive a_s is causing the object to *slow down* (decreasing speed v). This agrees with our rule from Section 2.1: An object is slowing down if and only if v_s and a_s have opposite signs.

NOTE ▶ It is customary to abbreviate the acceleration units (m/s)/s as m/s². For example, the particles in Example 2.12 had an acceleration of 2 m/s². We will use this notation, but keep in mind the *meaning* of the notation as "(meters per second) per second." ◀

EXAMPLE 2.13 Running the court

A basketball player starts at the left end of the court and moves with the velocity shown in Figure 2.26. Draw a motion diagram and an acceleration-versus-time graph for the basketball player.

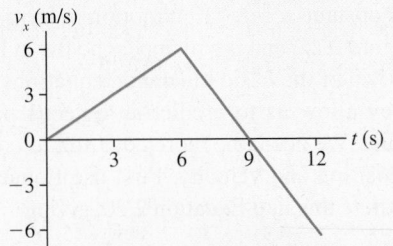

FIGURE 2.26 Velocity-versus-time graph for the basketball player of Example 2.13.

VISUALIZE The velocity is positive (motion to the right) and increasing for the first 6 seconds, so the velocity arrows in the motion diagram are to the right and getting longer. From $t = 6$ s to 9 s the motion is still to the right (v_x is still positive), but the arrows are getting shorter because v_x is decreasing. There's a turning point at $t = 9$ s, when $v_x = 0$, and after that the motion is to the left (v_x is negative) and getting faster. The motion diagram of Figure 2.27a shows the velocity vectors and the acceleration vectors.

SOLVE Acceleration is the slope of the velocity graph. For the first 6 s, the slope has the constant value

$$a_x = \frac{\Delta v_x}{\Delta t} = \frac{6.0 \text{ m/s}}{6.0 \text{ s}} = 1.0 \text{ m/s}^2$$

The velocity decreases by 12 m/s during the 6-s interval from $t = 6$ s to $t = 12$ s, so

$$a_x = \frac{\Delta v_x}{\Delta t} = \frac{-12 \text{ m/s}}{6.0 \text{ s}} = -2.0 \text{ m/s}^2$$

The acceleration graph for these 12 s is shown in Figure 2.27b. Although there are two segments of the motion, each segment is uniformly accelerated motion with constant acceleration. Notice that there is no change in the acceleration at $t = 9$ s, the turning point.

(a)

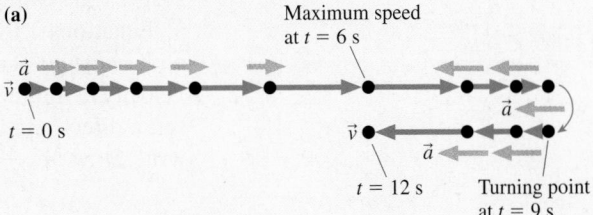

(b)

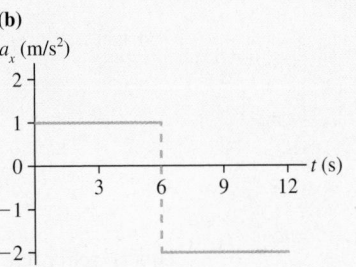

FIGURE 2.27 Motion diagram and acceleration graph for Example 2.13.

ASSESS The *sign* of a_x does *not* tell us whether or not the object is speeding up or slowing down. The basketball player is slowing down from $t = 6$ s to $t = 9$ s, then speeding up from $t = 9$ s to $t = 12$ s. Nonetheless, his acceleration is negative during this entire interval because his acceleration vector, as seen in the motion diagram, always points to the left.

The Kinematic Equations of Constant Acceleration

Consider an object whose acceleration a_s remains constant during the time interval $\Delta t = t_f - t_i$. At the beginning of this interval, at time t_i, the object has initial velocity v_{is} and initial position s_i. Note that t_i is often zero, but it does not have to be. Figure 2.28a shows the acceleration-versus-time graph. It is a horizontal line between t_i and t_f, indicating a *constant* acceleration.

The object's velocity is changing because the object is accelerating. It is not hard to find the object's velocity v_{fs} at a later time t_f. By definition,

$$a_s = \frac{\Delta v_s}{\Delta t} = \frac{v_{fs} - v_{is}}{\Delta t} \tag{2.17}$$

which is easily rearranged to give

$$v_{fs} = v_{is} + a_s \Delta t \tag{2.18}$$

The velocity-versus-time graph, shown in Figure 2.28b, is a straight line that starts at v_{is} and has slope a_s.

We would also like to know the object's position s_f at time t_f. As you learned in the last section,

$$s_f = s_i + \text{area under the velocity curve } v_s \text{ between } t_i \text{ and } t_f \tag{2.19}$$

(a) Acceleration

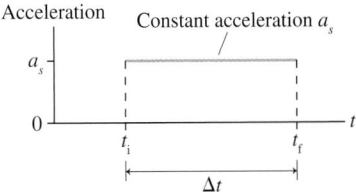

Displacement Δs is the area under the curve. The area can be divided into a rectangle of height v_{is} and a triangle of height $a_s \Delta t$.

(b) Velocity

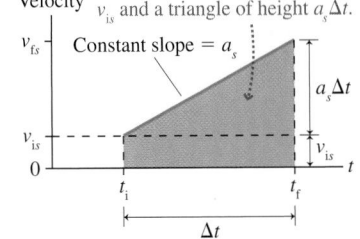

FIGURE 2.28 Acceleration and velocity graphs for motion with constant acceleration.

The shaded area in Figure 2.28b can be subdivided into a rectangle of area $v_{is}\Delta t$ and a triangle of area $\frac{1}{2}(a_s\Delta t)(\Delta t) = \frac{1}{2}a_s(\Delta t)^2$. Adding these gives

$$s_f = s_i + v_{is}\Delta t + \tfrac{1}{2}a_s(\Delta t)^2 \qquad (2.20)$$

where $\Delta t = t_f - t_i$ is the elapsed time. The quadratic dependence on Δt causes the position-versus-time graph for constant-acceleration motion to have a parabolic shape. You saw this earlier in Figure 2.22, and it will appear below in Figure 2.29.

Equations 2.18 and 2.20 are two of the basic kinematic equations for motion with *constant* acceleration. They allow us to predict an object's position and velocity at a future instant of time. We need one more equation to complete our set, a direct relation between position and velocity. First use Equation 2.18 to write $\Delta t = (v_{fs} - v_{is})/a_s$. Substitute this into Equation 2.20, giving

$$
\begin{aligned}
s_f &= s_i + v_{is}\left(\frac{v_{fs} - v_{is}}{a_s}\right) + \tfrac{1}{2}a_s\left(\frac{v_{fs} - v_{is}}{a_s}\right)^2 \\[2mm]
&= s_i + \left(\frac{v_{is}v_{fs}}{a_s} - \frac{v_{is}^2}{a_s}\right) + \left(\frac{v_{fs}^2}{2a_s} - \frac{v_{is}v_{fs}}{a_s} + \frac{v_{is}^2}{2a_s}\right) \qquad (2.21) \\[2mm]
&= s_i + \frac{v_{fs}^2 - v_{is}^2}{2a_s}
\end{aligned}
$$

This is easily rearranged to read

$$v_{fs}^2 = v_{is}^2 + 2a_s\,\Delta s \qquad (2.22)$$

where $\Delta s = s_f - s_i$ is the *displacement* (not the distance!).

Equations 2.18, 2.20, and 2.22, which are summarized in Table 2.3, are the key results for motion with constant acceleration.

Figure 2.29 is a comparison of motion with constant velocity (uniform motion) and motion with constant acceleration (uniformly accelerated motion). Notice that uniform motion is really a special case of uniformly accelerated motion in which the constant acceleration happens to be zero. The graphs for a negative acceleration are left as an exercise.

TABLE 2.3 The kinematic equations for motion with constant acceleration

$v_{fs} = v_{is} + a_s\Delta t$

$s_f = s_i + v_{is}\Delta t + \tfrac{1}{2}a_s(\Delta t)^2$

$v_{fs}^2 = v_{is}^2 + 2a_s\,\Delta s$

(a) Motion at constant velocity

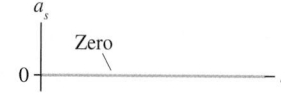

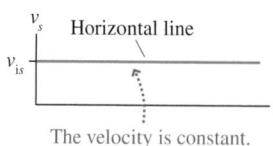

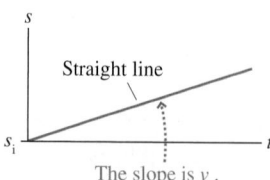

(b) Motion at constant acceleration

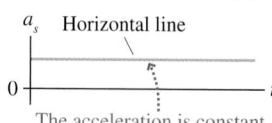

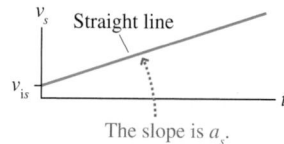

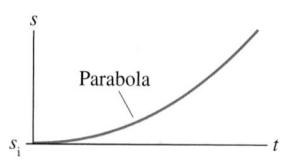

FIGURE 2.29 Motion with constant velocity and constant acceleration. These graphs assume $s_i = 0$, $v_{is} > 0$, and (for constant acceleration) $a_s > 0$.

A Problem-Solving Strategy

This information can be assembled into a problem-solving strategy for kinematics with constant acceleration.

PROBLEM-SOLVING STRATEGY 2.1 Kinematics with constant acceleration

MODEL Use the particle model. Make simplifying assumptions.

VISUALIZE Use different representations of the information in the problem.

- Draw a *motion diagram*. Motion diagrams are part of the physical representation.
- Draw a *pictorial representation*. This helps you assess the information you are given and starts the process of translating the problem into symbols.
- Use a *graphical representation* if it is appropriate for the problem.
- Go back and forth between these three representations as needed.

SOLVE The mathematical representation is based on the three kinematic equations

$$v_{fs} = v_{is} + a_s \Delta t$$

$$s_f = s_i + v_{is} \Delta t + \tfrac{1}{2} a_s (\Delta t)^2$$

$$v_{fs}^2 = v_{is}^2 + 2 a_s \Delta s$$

- Use x or y, as appropriate to the problem, rather than the generic s.
- Replace i and f with numerical subscripts defined in the pictorial representation.
- Uniform motion with constant velocity has $a_s = 0$.

ASSESS Is your result believable? Does it have proper units? Does it make sense?

EXAMPLE 2.14 The motion of a rocket sled

A rocket sled accelerates at 50 m/s^2 for 5.0 s, coasts for 3.0 s, then deploys a braking parachute and decelerates at 3.0 m/s^2 until coming to a halt.

a. What is the maximum velocity of the rocket sled?
b. What is the total distance traveled?

MODEL Represent the rocket sled as a particle.

VISUALIZE Figure 2.30 shows the physical and pictorial representations. Recall that we discussed the first two-thirds of this problem as Example 1.9 in Chapter 1.

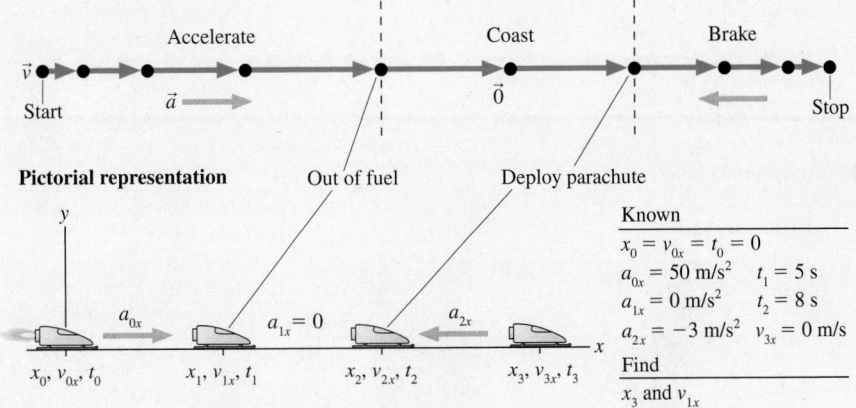

FIGURE 2.30 The physical and pictorial representations of the rocket sled.

SOLVE

a. The maximum velocity is identified in the pictorial representation as v_{1x}, the velocity at time t_1 when the acceleration phase ends. The first kinematic equation in Table 2.3 gives

$$v_{1x} = v_{0x} + a_{0x}(t_1 - t_0) = a_{0x}t_1$$
$$= (50 \text{ m/s}^2)(5.0 \text{ s}) = 250 \text{ m/s}$$

We started with the complete equation, then simplified by noting which terms were zero.

b. Finding the total distance requires several steps. First, the sled's position when the acceleration ends at t_1 is found from the second equation in Table 2.3:

$$x_1 = x_0 + v_{0x}(t_1 - t_0) + \tfrac{1}{2}a_{0x}(t_1 - t_0)^2 = \tfrac{1}{2}a_{0x}t_1^2$$
$$= \tfrac{1}{2}(50 \text{ m/s}^2)(5.0 \text{ s})^2 = 625 \text{ m}$$

During the coasting phase, which is uniform motion with no acceleration ($a_{1x} = 0$),

$$x_2 = x_1 + v_{1x}\Delta t = x_1 + v_{1x}(t_2 - t_1)$$
$$= 625 \text{ m} + (250 \text{ m/s})(3.0 \text{ s}) = 1375 \text{ m}$$

Notice that, in this case, Δt is not simply t. The braking phase is a little different because we don't know how long it lasts. But we do know that the sled ends with $v_{3x} = 0$ m/s, so we can use the third equation in Table 2.3:

$$v_{3x}^2 = v_{2x}^2 + 2a_{2x}\Delta x = v_{2x}^2 + 2a_{2x}(x_3 - x_2)$$

This can be solved for x_3:

$$x_3 = x_2 + \frac{v_{3x}^2 - v_{2x}^2}{2a_{2x}}$$
$$= 1375 \text{ m} + \frac{0 - (250 \text{ m/s})^2}{2(-3.0 \text{ m/s}^2)} = 11{,}800 \text{ m}$$

ASSESS Using the approximate conversion factor 1 m/s ≈ 2 mph from Table 1.4, we see that the top speed is ≈ 500 mph. The total distance traveled is ≈ 12 km ≈ 7 mi. This is reasonable because it takes a very long distance to stop from a top speed of 500 mph!

NOTE ▶ We used explicit numerical subscripts throughout the mathematical representation, each referring to a symbol that was defined in the pictorial representation. The subscripts i and f in the Table 2.3 equations are just generic "place holders" and don't have unique values. During the acceleration phase we had i = 0 and f = 1. Later, during the coasting phase, these became i = 1 and f = 2. The numerical subscripts have a clear meaning and are less likely to lead to confusion. ◀

EXAMPLE 2.15 **Friday night football**

Fred catches the football while standing directly on the goal line. He immediately starts running forward with an acceleration of 6 ft/s^2. At the moment the catch is made, Tommy is 20 yards away and heading directly toward Fred with a steady speed of 15 ft/s. If neither deviates from a straight-ahead path, where will Tommy tackle Fred?

MODEL Represent Fred and Tommy as particles.

VISUALIZE This problem statement was analyzed in Example 1.11. The physical and pictorial representations are shown again in Figure 2.31.

SOLVE We want to find *where* Fred and Tommy have the same position. The pictorial representation designates time t_1 as *when* they meet. The axes have been chosen so that Fred starts at

Physical representation

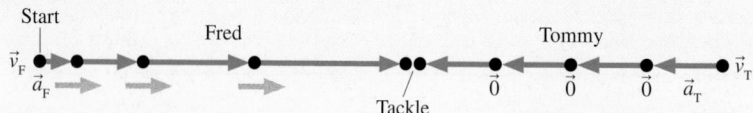

Pictorial representation

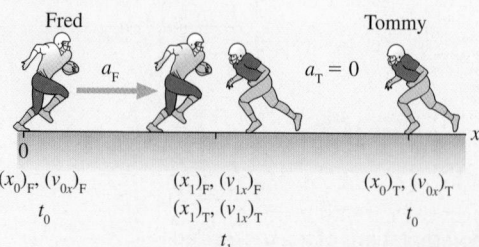

Known
$(x_0)_F = 0$ yards $(v_{0x})_F = 0$ ft/s $t_0 = 0$ s
$(x_0)_T = 20$ yards = 60 ft
$(v_{0x})_T = -15$ ft/s
$a_F = 6$ ft/s^2 $a_T = 0$ ft/s^2

Find
$(x_1)_F$ at t_1 when $(x_1)_F = (x_1)_T$

FIGURE 2.31 The physical and pictorial representations for Example 2.15.

$(x_0)_F = 0$ ft and moves to the right while Tommy starts at $(x_0)_T = 60$ ft and runs to the left with a *negative* velocity. The first equation of Table 2.3 allows us to find their positions at time t_1. These are:

$$(x_1)_F = (x_0)_F + (v_{0x})_F(t_1 - t_0) + \tfrac{1}{2}(a_x)_F(t_1 - t_0)^2$$

$$= \tfrac{1}{2}(a_x)_F t_1^2$$

$$(x_1)_T = (x_0)_T + (v_{0x})_T(t_1 - t_0) + \tfrac{1}{2}(a_x)_T(t_1 - t_0)^2$$

$$= (x_0)_T + (v_{0x})_T t_1$$

Notice that Tommy's position equation contains the term $(v_{0x})_T t_1$, not $-(v_{0x})_T t_1$. The fact that he is moving to the left has already been considered in assigning a *negative value* to $(v_{0x})_T$, hence we don't want to add any additional negative signs in the equation. If we now set $(x_1)_F$ and $(x_1)_T$ equal to each other, indicating the point of the tackle, we can solve for t_1:

$$\tfrac{1}{2}(a_x)_F t_1^2 = (x_0)_T + (v_{0x})_T t_1$$

$$\tfrac{1}{2}(a_x)_F t_1^2 - (v_{0x})_T t_1 - (x_0)_T = 0$$

$$3t_1^2 + 15t_1 - 60 = 0$$

The solutions of this quadratic equation for t_1 are $t_1 = (-7.62 \text{ s}, +2.62 \text{ s})$. The negative time is not meaningful in this problem, so the time of the tackle is $t_1 = 2.62$ s. Using this to compute $(x_1)_F$ gives

$$(x_1)_F = \tfrac{1}{2}(a_x)_F t_1^2 = 20.6 \text{ feet} = 6.9 \text{ yards}$$

Tommy makes the tackle at just about the 7-yard line!

ASSESS The answer had to be between 0 yards and 20 yards. Because Tommy was already running, whereas Fred started from rest, it is reasonable that Fred will cover less than half the 20-yard separation before meeting Tommy. Thus 6.9 yards is a reasonable answer.

NOTE ▶ The purpose of the assessment step is not to prove that an answer must be right but to rule out answers that, with a little thought, are clearly wrong. ◀

It is worth exploring Example 2.15 graphically. Figure 2.32 shows position-versus-time graphs for Fred and Tommy. The curves intersect at $t = 2.62$ s, and that is where the tackle occurs. You should compare this problem to Example 2.3 and Figure 2.9 for Bob and Susan to notice the similarities and the differences.

STOP TO THINK 2.4 Which velocity-versus-time graph or graphs goes with this acceleration-versus-time graph? The particle is initially moving to the right.

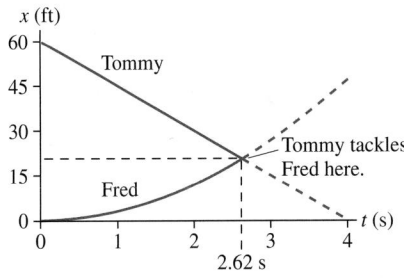

FIGURE 2.32 Position-versus-time graphs for Fred and Tommy.

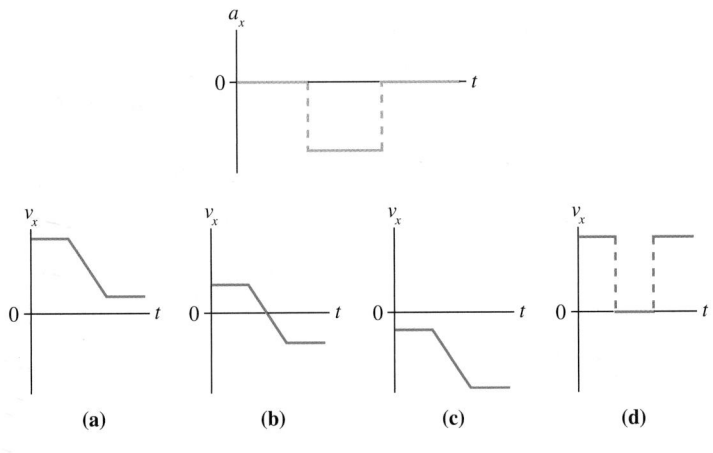

(a)   (b)   (c)   (d)

Act**i**v
ONLINE
Physics

1.4, 1.5, 1.6, 1.8, 1.9,
1.11, 1.12, 1.13, 1.14

2.6 Free Fall

The motion of an object moving under the influence of gravity only, and no other forces, is called **free fall.** Strictly speaking, free fall occurs only in a vacuum, where there is no air resistance. Fortunately, the effect of air resistance is small for "heavy objects," so we'll make only a very slight error in treating these objects *as if* they were in free fall. For very light objects, such as a feather, or for objects that fall through very large distances and gain very high speeds, the effect

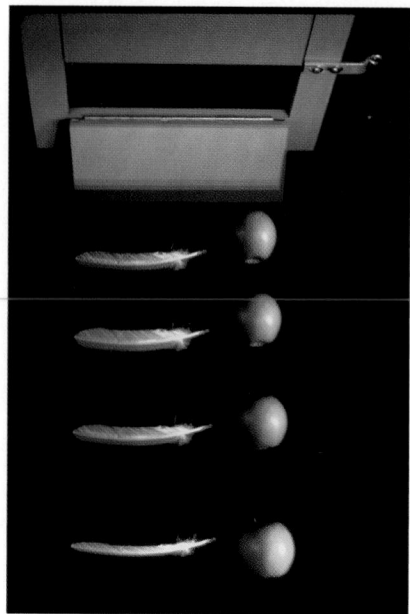

In the absence of air resistance, any two objects fall at the same rate and hit the ground at the same time. The apple and feather in this photograph are falling in a vacuum.

of air resistance is *not* negligible. Motion with air resistance is a problem we will study in Chapter 5. Until then, we will restrict our attention to "heavy objects" and will make the reasonable assumption that falling objects are in free fall.

The motion of falling objects has interested scientists since antiquity. The ancient Greek scientist and philosopher Aristotle asserted that heavier objects "fall faster" than do light objects. After all, a rock falls to the ground much more quickly than a feather. Aristotle's claim was based on casual observations and on what he thought "should" happen rather than on actual experiments.

Galileo, in the 17th century, was the first to challenge Aristotle and to put Aristotle's assertion to a rigorous experimental test. The story of Galileo dropping different weights from the leaning bell tower at the cathedral in Pisa is well known, although historians cannot confirm its truth. But bell towers were common in the Italy of Galileo's day, so he had ample opportunity to make the measurements and observations that he describes in his writings.

Careful observations show that falling objects *don't* "hit the ground" at the same time. There are slight differences in the arrival times, but Galileo correctly identified these differences as due to air resistance. He then formulated a general conclusion for an idealized situation of motion in a vacuum. In doing so, Galileo developed a *model* of motion—motion in the absence of air resistance—that could only be approximated by any real object. It was Galileo's innovative use of experiments, models, and mathematics that made him the first "modern" scientist.

Galileo's discovery can be summarized as follows:

- Two objects dropped from the same height will, if air resistance can be neglected, hit the ground at the same time and with the same speed.
- Consequently, **any two objects in free fall, regardless of their mass, have the same acceleration** $\vec{a}_{\text{free fall}}$. This is an especially important conclusion.

Figure 2.33a shows the motion diagram of an object that was released from rest and falls freely. Figure 2.33b shows the object's velocity graph. The motion diagram and graph are identical for a falling pea and a falling boulder. The acceleration $a_{\text{free fall}}$ is easily found from the slope of the velocity graph. Careful measurements show that the value of the free-fall acceleration varies ever-so-slightly at different places on the earth, due to the slightly nonspherical shape of the earth and to the fact that the earth is rotating. A global average, at sea level, is

$$\vec{a}_{\text{free fall}} = (9.80 \text{ m/s}^2, \text{ vertically downward}) \tag{2.23}$$

where *vertically downward* means along a line toward the center of the earth.

The length, or magnitude, of the free-fall acceleration is known as the **acceleration due to gravity,** and has the special symbol g:

$$g = 9.80 \text{ m/s}^2 \quad \text{(acceleration due to gravity)}$$

Several points about free fall are worthy of note:

- g, by definition, is *always* positive. **There will never be a problem that will use a negative value for g.** But, you say, objects fall when you release them rather than rise, so how can g be positive?
- g is *not* the acceleration $a_{\text{free fall}}$, but simply its magnitude. Because we've chosen the y-axis to point vertically up, the downward acceleration vector $\vec{a}_{\text{free fall}}$ has the one-dimensional acceleration

$$a_y = a_{\text{free fall}} = -g \tag{2.24}$$

It is a_y that is negative, not g.

- Because free fall is motion with constant acceleration, we can use the kinematic equations of Table 2.3 with the acceleration being due to gravity, $a_y = -g$.

(a)

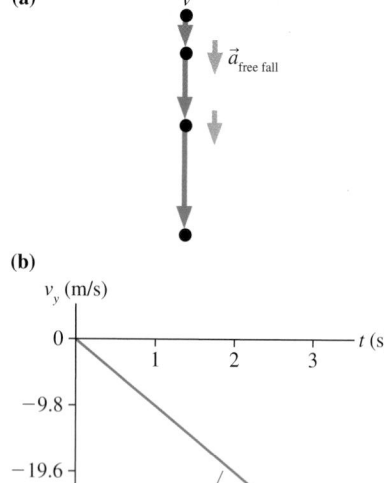

(b)

FIGURE 2.33 Motion of an object in free fall.

- g is not called "gravity." Gravity is a force, not an acceleration. g is *the acceleration due to gravity*.
- $g = 9.80 \text{ m/s}^2$ only on earth. Other planets have different values of g. You will learn in Chapter 12 how to determine g for other planets.

NOTE ▶ Despite the name, free fall is not restricted to objects that are literally falling. Any object moving under the influence of gravity only, and no other forces, is in free fall. This includes objects falling straight down, objects that have been tossed or shot straight up, and projectile motion. This chapter considers only objects that move up and down along a vertical line; projectile motion will be studied in Chapter 6. ◀

ActivPhysics 1.7, 1.10

EXAMPLE 2.16 A falling rock

A rock is released from rest at the top of a 100-m-tall building. How long does the rock take to fall to the ground, and what is its impact velocity?

MODEL Represent the rock as a particle. Assume air resistance is negligible.

VISUALIZE Figure 2.34 shows the physical and pictorial representations. We have placed the origin at the ground, which makes $y_0 = 100$ m. Although the rock falls 100 m, it is important to notice that the *displacement* is $\Delta y = y_1 - y_0 = -100$ m.

SOLVE Free fall is motion with the specific constant acceleration $a_y = -g$. The first question involves a relation between time and distance, so only the second equation in Table 2.3 is relevant. Using $v_{0y} = 0$ m/s and $t_0 = 0$ s, we find

$$y_1 = y_0 + v_{0y}\Delta t + \tfrac{1}{2}a_y\Delta t^2 = y_0 + v_{0y}\Delta t - \tfrac{1}{2}g\Delta t^2 = y_0 - \tfrac{1}{2}gt_1^2$$

We can now solve for t_1, finding:

$$t_1 = \sqrt{\frac{2(y_0 - y_1)}{g}} = \sqrt{\frac{2(100 \text{ m} - 0 \text{ m})}{9.80 \text{ m/s}^2}} = \pm 4.52 \text{ s}$$

The $\pm$ sign indicates that there are two mathematical solutions; therefore we have to use physical reasoning to choose between them. A negative t_1 would refer to a time before we dropped the rock, so we select the positive root: $t_1 = 4.52$ s.

Now that we know the fall time, we can use the first kinematic equation to find v_{1y}:

$$v_{1y} = v_{0y} - g\Delta t = -gt_1 = -(9.80 \text{ m/s}^2)(4.52 \text{ s})$$

$$= -44.3 \text{ m/s}.$$

Alternatively, we could work directly from the third kinematic equation:

$$v_{1y} = \sqrt{v_{0y}^2 - 2g\Delta y} = \sqrt{-2g(y_1 - y_0)}$$

$$= \sqrt{-2(9.80 \text{ m/s}^2)(0 \text{ m} - 100 \text{ m})} = \pm 44.3 \text{ m/s}$$

This method is useful if you don't know Δt. However, we must again choose the correct sign of the square root. Because the velocity vector points downward, the sign of v_y has to be negative. Thus $v_{1y} = -44.3$ m/s. The importance of careful attention to the signs cannot be overemphasized!

A common error would be to say "The rock fell 100 m, so $\Delta y = 100$ m." This would have you trying to take the square root of a negative number. As noted above, Δy is not a distance. It is a *displacement,* with a carefully defined meaning of $y_f - y_i$. In this case, $\Delta y = y_1 - y_0 = -100$ m.

ASSESS Are the answers reasonable? Well, 100 m is about 300 feet, which is about the height of a 30-floor building. How long does it take something to fall 30 floors? Four or five seconds seems pretty reasonable. How fast would it be going at the bottom? Using 1 m/s ≈ 2 mph, we find that 44.3 m/s ≈ 90 mph. That also seems pretty reasonable after falling 30 floors. Had we misplaced a decimal point, though, and found 443 m/s, we would be suspicious when we converted this to ≈ 900 mph! The answers all seem reasonable.

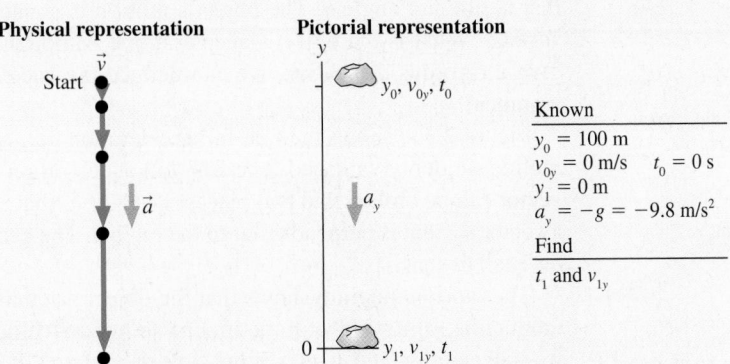

FIGURE 2.34 Physical and pictorial representations of a falling rock.

EXAMPLE 2.17 A vertical cannonball
A cannonball is shot straight up with an initial speed of 100 m/s. How high does it go?

MODEL Represent the cannonball as a particle. Assume air resistance is negligible.

VISUALIZE Figure 2.35 shows the physical and pictorial representations for the cannonball's motion. Even though the ball was shot upward, this is a free-fall problem because the ball (after being launched) is moving under the influence of gravity *only*. A critical aspect of the problem is knowing where it ends. How do we put "how high" into symbols? The clue is that the very top point of the trajectory is a *turning point*. Recall that the instantaneous velocity at a turning point is $v = 0$. Thus we can characterize the "top" of the trajectory as the point where $v_{1y} = 0$ m/s. This was not explicitly stated but is part of our interpretation of the problem.

SOLVE We are looking for a relationship between distance and velocity, without knowing the time interval. This relationship is described mathematically by the third kinematic equation in Table 2.3. Using $y_0 = 0$ m and $v_{1y} = 0$ m/s, we have

$$v_{1y}^2 = 0 = v_{0y}^2 - 2g\Delta y = v_{0y}^2 - 2gy_1$$

Solving for y_1, we find that the cannonball reaches a height

$$y_1 = \frac{v_{0y}^2}{2g} = \frac{(100 \text{ m/s})^2}{2(9.80 \text{ m/s}^2)} = 510 \text{ m}$$

ASSESS Is this answer reasonable? A speed of 100 m/s is ≈ 200 mph—that's pretty fast! The calculated height is 510 m ≈ 1500 ft. In Example 2.16 we found that an object dropped from 100 m is going 44 m/s when it hits the ground, so it seems reasonable that an object shot upward at 100 m/s will go significantly higher than 100 m. While we cannot say that 510 m is necessarily better than 400 m or 600 m, we can say that it is not unreasonable. The point of the assessment is not to prove that the answer *has* to be right, but to find answers that are obviously wrong.

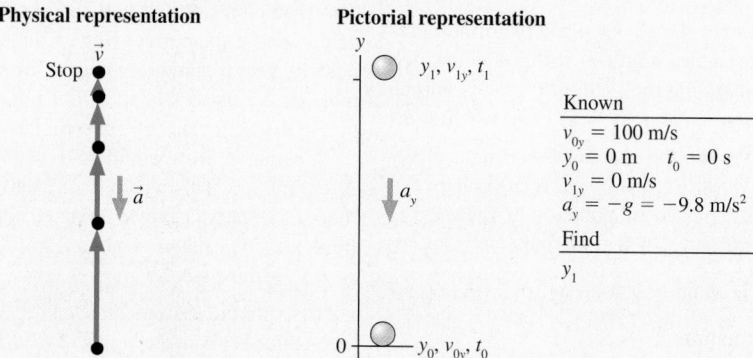

Physical representation

Stop

$\vec{v}$

$\vec{a}$

Pictorial representation

y

y_1, v_{1y}, t_1

a_y

y_0, v_{0y}, t_0

0

Known
$v_{0y} = 100$ m/s
$y_0 = 0$ m $t_0 = 0$ s
$v_{1y} = 0$ m/s
$a_y = -g = -9.8$ m/s²

Find
y_1

FIGURE 2.35 Physical and pictorial representations for Example 2.17.

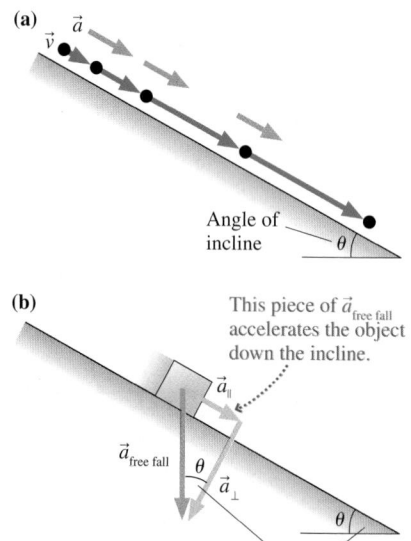

(a)

$\vec{a}$

$\vec{v}$

Angle of incline θ

(b)

This piece of $\vec{a}_{\text{free fall}}$ accelerates the object down the incline.

$\vec{a}_{\parallel}$

$\vec{a}_{\text{free fall}}$ θ

$\vec{a}_{\perp}$

θ

Same angle

FIGURE 2.36 Acceleration on an inclined plane.

2.7 Motion on an Inclined Plane

A problem closely related to free fall is that of an object moving down a straight, but frictionless, inclined plane, such as a skier going down a slope on frictionless snow. In practice, we can come very close to this ideal by having a ball rolling on a very smooth surface.

Figure 2.36a shows an object accelerating down a frictionless, inclined plane that is tilted at angle θ. The object's motion is constrained to be parallel to the surface. What is the object's acceleration? Although we're not yet prepared to give a rigorous derivation, we can deduce the acceleration with a plausibility argument.

Figure 2.36b shows the free-fall acceleration $\vec{a}_{\text{free fall}}$ the ball would have if the incline suddenly vanished. The free-fall acceleration points straight down. This vector can be broken into two pieces: a vector $\vec{a}_{\parallel}$ that is parallel to the incline and a vector $\vec{a}_{\perp}$ that is perpendicular to the incline. The vector addition rules of Chapter 1 tell us that $\vec{a}_{\text{free fall}} = \vec{a}_{\parallel} + \vec{a}_{\perp}$.

The motion diagram shows that the object's actual acceleration is parallel to the incline. The surface of the incline somehow "blocks" $\vec{a}_{\perp}$, through a process we will examine in Chapter 5, but $\vec{a}_{\parallel}$ is unhindered. It is this piece of $\vec{a}_{\text{free fall}}$, parallel to the incline, that accelerates the object.

Consider an inclined plane tilted at angle θ. Figure 2.36b shows that the three vectors form a right triangle with angle θ at the bottom. By definition, the length, or magnitude, of $\vec{a}_{\text{free fall}}$ is g. Vector $\vec{a}_{\parallel}$ is opposite angle θ, so the length, or magnitude, of $\vec{a}_{\parallel}$ must be $g\sin\theta$. Consequently, the one-dimensional acceleration along the incline is

$$a_s = \pm g\sin\theta \qquad (2.25)$$

The correct sign depends on the direction in which the ramp is tilted, as the following examples will illustrate.

Equation 2.25 makes sense. Suppose the plane is perfectly horizontal. If you place an object on a horizontal surface, you expect it to stay at rest with no acceleration. Equation 2.25 gives $a_s = 0$ when $\theta = 0°$, in agreement with our expectations. Now suppose you tilt the plane until it becomes vertical, at $\theta = 90°$. Without friction, an object would simply fall, in free fall, parallel to the vertical surface. Equation 2.25 gives $a_s = -g = a_{\text{free fall}}$ when $\theta = 90°$, again in agreement with our expectations. We'll use Newton's laws of motion in Chapter 5 to verify Equation 2.25, but we can have confidence in it now because we see that it gives the correct result in these *limiting cases*.

Skiing is an example of motion on an inclined plane.

EXAMPLE 2.18 Skiing down an incline

A skier's speed at the bottom of a 100-m-long, frictionless, snow-covered slope is 20 m/s. What is the angle of the slope?

MODEL Represent the skier as a particle. Assume that air resistance is negligible. Assume that the slope is a straight line.

VISUALIZE Figure 2.37 shows the physical and pictorial representations of the skier.

SOLVE The motion diagram shows that the acceleration vector $\vec{a}_{\parallel}$ points in the positive s-direction. Thus the one-dimensional acceleration is $a_s = +g\sin\theta$. This is constant-acceleration motion. The third kinematic equation from Table 2.3 is

$$v_{1s}^2 = v_{0s}^2 + 2a_s\Delta x = 2g\sin\theta\,\Delta x$$

where we used $v_{0s} = 0$ m/s. Solving for $\sin\theta$, we find

$$\sin\theta = \frac{v_{1s}^2}{2g\Delta x} = \frac{(20 \text{ m/s})^2}{2(9.80 \text{ m/s}^2)(100 \text{ m})} = 0.204$$

Thus

$$\theta = \sin^{-1}(0.204) = 11.8°$$

ASSESS A 100-m-long slope and a speed of 20 m/s $\approx$ 40 mph are fairly typical parameters for skiing. A 1° angle or an 80° angle would be unrealistic, but 12° seems plausible.

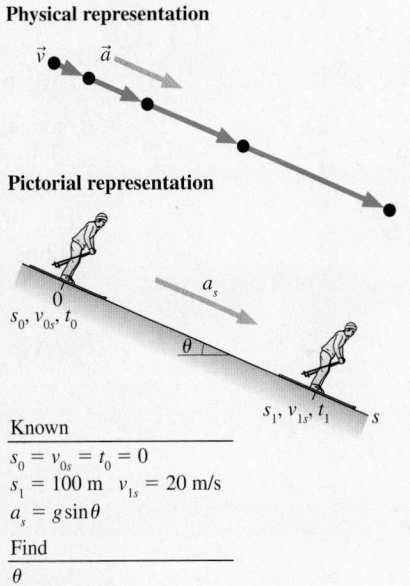

Physical representation

Pictorial representation

Known
$s_0 = v_{0s} = t_0 = 0$
$s_1 = 100$ m $v_{1s} = 20$ m/s
$a_s = g\sin\theta$

Find
θ

FIGURE 2.37 Physical and pictorial representations for the skier of Example 2.18.

EXAMPLE 2.19 At the amusement park

An amusement park ride shoots a car up a frictionless track inclined at 30°. The car rolls up, then rolls back down. If the height of the track is 20 m, what is the maximum allowable speed with which the car can start?

MODEL Represent the car as a particle. Assume air resistance is negligible.

VISUALIZE Figure 2.38 on the next page shows the physical and pictorial representations of the car. The problem starts as the car is shot up the incline, and it ends when the car reaches its highest point. The highest point is a turning point, so $v_{1s} = 0$ m/s. The motion diagram shows that the acceleration vector $\vec{a}_{\parallel}$ points in the negative s-direction, so $a_s = -g\sin\theta$. The *maximum* starting speed is that at which the car goes to the very top of the ramp, a height of 20 m.

Physical representation

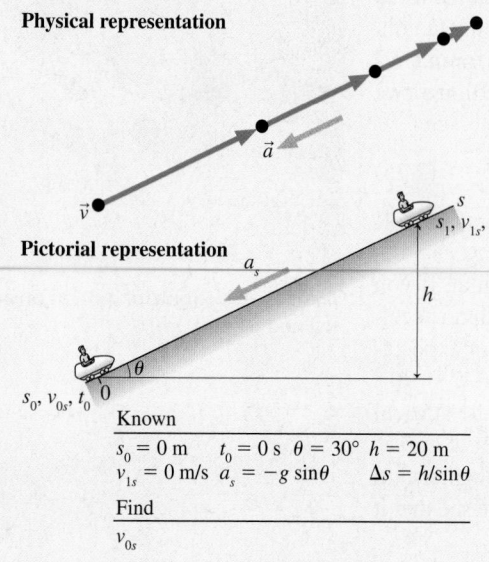

Pictorial representation

Known		
$s_0 = 0$ m	$t_0 = 0$ s $\theta = 30°$	$h = 20$ m
$v_{1s} = 0$ m/s	$a_s = -g\sin\theta$	$\Delta s = h/\sin\theta$

Find

v_{0s}

FIGURE 2.38 Physical and pictorial representations for the car of Example 2.19.

SOLVE The maximum possible displacement Δs_{max} is related to the height h by

$$\Delta s_{max} = s_1 - s_0 = \frac{h}{\sin 30°} = \frac{20\text{ m}}{\sin 30°} = 40\text{ m}$$

The initial speed v_{0s} that allows the car to travel this distance is found from

$$v_{1s}{}^2 = 0 = v_{0s}{}^2 + 2a_s\Delta x = v_{0s}{}^2 - 2g\sin\theta\Delta x$$

$$v_{0s} = \sqrt{2g\sin 30°\Delta x} = \sqrt{2(9.8\text{ m/s}^2)(0.500)(40\text{ m})}$$

$$= 19.8\text{ m/s}$$

This is the maximum speed, because a car starting any faster will run off the top.

ASSESS 20 m $\approx$ 60 feet and 19.8 m/s $\approx$ 40 mph. It seems plausible that a car would need to be going this fast to gain 60 feet of elevation rolling up a ramp. Be sure you understand why the sign of a_s is negative here but positive in Example 2.18.

Thinking Graphically

Kinematics is the language of motion. We will spend the entire rest of this course studying moving objects, from baseballs to electrons, and the concepts we have developed in this chapter will be used extensively. One of the most important ideas, summarized in Tactics Box 2.3, has been that the relationships between position, velocity, and acceleration can be expressed graphically.

TACTICS BOX 2.3 **Interpreting graphical representations of motion**

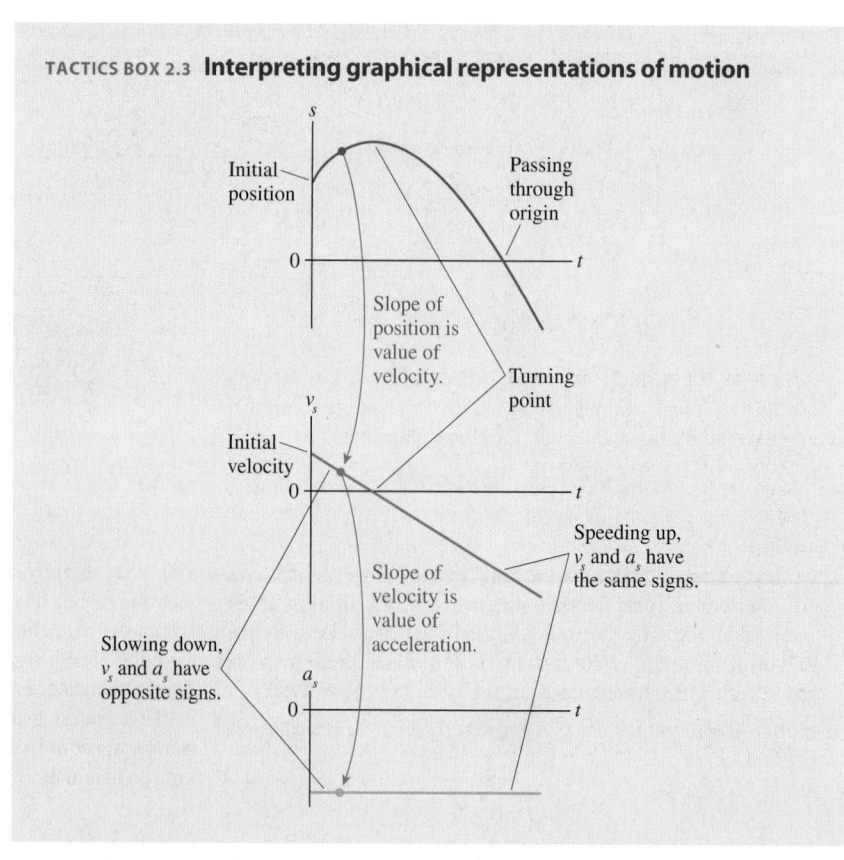

A good way to solidify your understanding of motion graphs is to consider the problem of a hard, smooth ball rolling on a smooth (i.e., frictionless) track. The track is made up of several straight segments connected together. Each segment may be either horizontal or inclined. Your task will be to analyze the ball's motion graphically. This will require you to reason about, rather than calculate, the relationships between s, v_s, and a_s.

There are two variations to this type of problem. In the first, you are given a picture of a track and the initial condition of the ball. The problem is then to draw graphs of s, v_s, and a_s. In the second, you are given the graphs, and the problem is to deduce the shape of the track on which the ball is rolling.

There are a small number of rules to follow in each of these problems:

1. Assume that the ball passes smoothly from one segment of the track to the next, with no loss of speed and without ever leaving the track.
2. The position, velocity, and acceleration graphs should be stacked vertically. They should each have the same horizontal scale so that a vertical line drawn through all three connects points describing the same instant of time.
3. Although the graphs have no numbers, they should show the correct *relationships*. For example, if the velocity is greater during the first part of the motion than during the second part, then the position graph should be steeper in the first part than in the second. Similarly, longer-lasting motions should span a greater horizontal range than shorter-lasting motions.
4. The position s is the position measured *along* the track. Similarly, v_s and a_s are the velocity and acceleration parallel to the track.

EXAMPLE 2.20 From track to graphs I
Draw position, velocity, and acceleration graphs for the ball on the track of Figure 2.39.

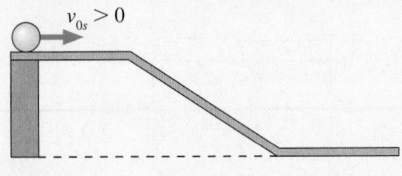

FIGURE 2.39 A ball rolling along a frictionless track.

VISUALIZE It is often easiest to begin with the velocity. Here the ball starts with an initial velocity v_{0s}. There is no acceleration on the horizontal surface ($a_s = 0$ if $\theta = 0°$), so the velocity remains constant until the ball reaches the slope. The slope is an inclined plane that, as we have learned, has constant acceleration. The velocity increases linearly with time during constant-acceleration motion. The ball returns to constant-velocity motion after reaching the bottom horizontal segment. The middle graph of Figure 2.40 shows the velocity.

We have enough information to draw the acceleration graph. We noted that the acceleration is zero while the ball is on the horizontal segments, and a_s has a constant positive value on the slope. These accelerations are consistent with the slope of the velocity graph: zero slope, then positive slope, then a return to zero slope. The bottom part of Figure 2.40 shows the acceleration graph. The acceleration cannot *really* change instantly from zero to a nonzero value, but the change can be so quick

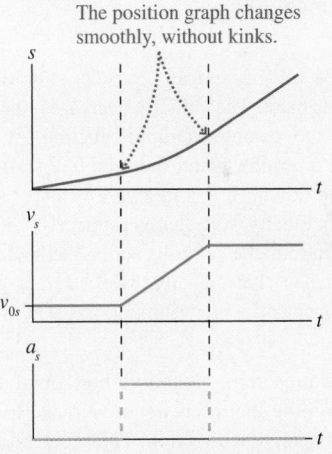

The position graph changes smoothly, without kinks.

FIGURE 2.40 Motion graphs for the ball in Example 2.20.

that we do not see it on the time scale of the graph. That is what the vertical dotted lines imply.

Finally, we need to find the position-versus-time graph. You might want to refer back to Figure 2.29 to review how the position graph looks for constant-velocity and constant-acceleration motion. The position increases linearly with time during the first segment at constant velocity. It also does so during the third segment of motion, but with a steeper slope to indicate a faster velocity. In between, while the acceleration is nonzero but constant, the position graph has a *parabolic* shape. The top part of Figure 2.40 shows the position-versus-time graph.

Two points are worth noting:

1. The dotted vertical lines through the graphs show the instants when the ball moves from one segment of the track to the next. Because of Rule 1, the speed does not change abruptly at these points; it changes gradually.
2. The parabolic section of the position-versus-time graph blends *smoothly* into the straight lines on either side. This is a consequence of Rule 1. An abrupt change of slope (a "kink") would indicate an abrupt change in velocity and would violate Rule 1.

EXAMPLE 2.21 From track to graphs II

Figure 2.41 shows a track with a "switch." A ball moving left-to-right passes through and heads up the incline, but a ball rolling down the incline goes straight through and continues downhill. Draw position, velocity, and acceleration graphs of the ball's motion.

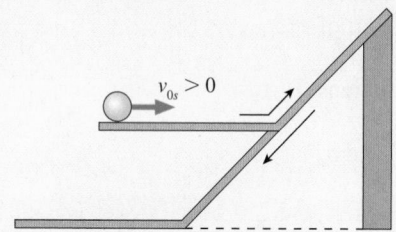

FIGURE 2.41 The ball and track for Example 2.21.

VISUALIZE The velocity remains constant at v_{0s} while the ball is on the level segment. The velocity decreases linearly with time after the ball moves onto the uphill incline. At some point v_s reaches zero, a turning point, then the ball starts rolling back down. Rolling downhill is a *negative* velocity (motion to the left), but the velocity still changes linearly with time. Upon reaching the bottom, the ball rolls across the horizontal segment with constant negative velocity. The *speed* is greater than it was on the upper horizontal segment, so $|v_s|$ at the end is larger than v_{0s}.

The acceleration is zero on the two horizontal segments. The acceleration on the incline is *negative* because the vector $\vec{a}_\parallel$ points in the negative s-direction. The acceleration is constant the whole time that the ball is on the incline, regardless of whether it is moving up or down. "Moving uphill" or "moving downhill" is determined by the sign of the velocity, not by the acceleration. This constant negative acceleration is consistent with the constant negative slope of the velocity graph. The velocity changes sign at the turning point, where $v = 0$, but **the acceleration does not change at this point.**

The position changes linearly while the velocity is constant. The position changes parabolically while the acceleration is constant and reaches a *maximum value* at the turning point, which is the top of the parabola. The constant negative velocity during the last segment implies a straight-line position graph with negative slope. All three motion graphs are shown in Figure 2.42.

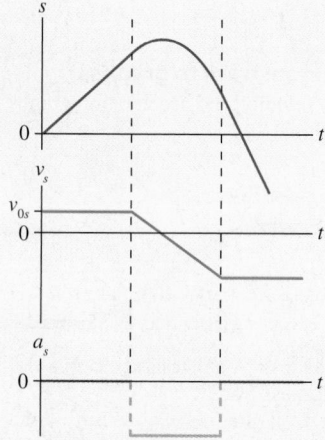

FIGURE 2.42 Motion graphs for the ball in Example 2.21.

EXAMPLE 2.22 From graphs to track

Figure 2.43 shows a set of motion graphs for a ball moving on a track. Draw a picture of the track and describe the ball's initial condition. Each segment of the track is *straight,* but the segments may be tilted.

VISUALIZE As with the last two examples, let's start by examining the velocity graph. The ball starts with initial velocity $v_{0s} = 0$, then the velocity increases linearly with time. This indicates that the ball is released from rest while on a slope, then starts rolling downhill in the positive s-direction. The acceleration graph confirms this: The initial acceleration is positive. That the motion is to the right (positive s-direction) is also seen from the position graph, where s becomes increasingly positive. After a while, the acceleration drops to zero and the velocity holds constant—a horizontal segment! Then the acceleration becomes negative while the velocity is *positive* but decreasing. This is motion to the right while slowing down, implying that the ball is going uphill. The acceleration has a smaller magnitude in the third segment than it had during the first segment, and the velocity graph is less steep. This indicates that the uphill tilt is less than the initial downhill tilt.

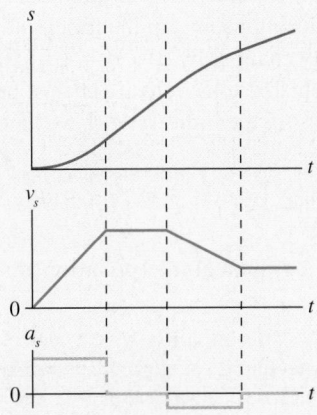

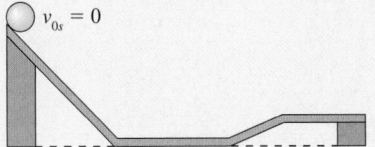

Lastly, the acceleration is again zero and the velocity is constant, so the final segment is horizontal. The position has increased throughout and v_s is never negative, so the motion is purely left-to-right with no turning points. Figure 2.44 shows the track and the initial conditions that are responsible for the graphs of Figure 2.43.

FIGURE 2.43 Motion graphs of a ball rolling on a track of unknown shape.

FIGURE 2.44 Track responsible for the motion graphs of Figure 2.43.

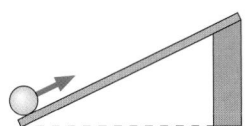

STOP TO THINK 2.5 The ball rolls up the ramp, then back down. Which is the correct acceleration graph?

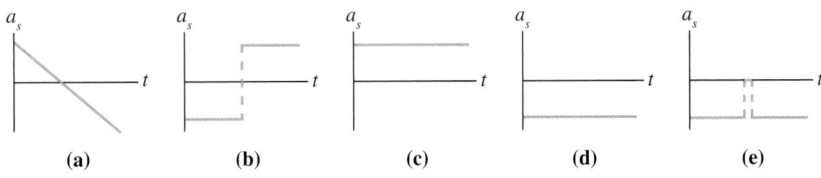

2.8 Instantaneous Acceleration

Figure 2.45 shows a velocity that increases with time, reaches a maximum, then decreases. This is *not* uniformly accelerated motion. We can still define an average acceleration, as we did in Section 2.5, but the average acceleration does not give a complete description of motion with nonuniform acceleration. Instead, we need the acceleration at each *instant* of time.

We can define an instantaneous acceleration in much the same way that we defined the instantaneous velocity. The instantaneous velocity was found to be the limit of the average velocity as the time interval $\Delta t \rightarrow 0$. Graphically, the instantaneous velocity at time t is the slope of the position-versus-time graph at that time. By analogy: **The instantaneous acceleration a_s at a specific instant of time t is the slope of the line that is tangent to the velocity-versus-time curve at time t.** Mathematically, this is

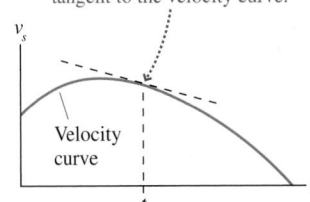

FIGURE 2.45 Motion with nonuniform acceleration.

$$a_s \equiv \lim_{\Delta t \to 0} \frac{\Delta v_s}{\Delta t} = \frac{dv_s}{dt} \text{ (instantaneous acceleration)} \qquad (2.26)$$

The instantaneous acceleration is the derivative (i.e., the rate of change) of the velocity.

The reverse problem—to find the velocity v_s if we know the acceleration a_s at all instants of time—is also important. When we wanted to find the position from the velocity, we took a velocity curve, divided it into N steps, found that the displacement Δs_k during step k was the area $(v_s)_k \Delta t$ of a small rectangle, then we added all the steps (i.e., integrated) to find s_f.

We can do the same with acceleration. An acceleration curve can be divided into N very narrow steps so that during each step the acceleration is essentially constant. During step k, the velocity changes by $\Delta(v_s)_k = (a_s)_k \Delta t$. This is the area of the small rectangle under the step. The total velocity change between t_i and t_f is found by adding all the small $\Delta(v_s)_k$. In the limit $\Delta t \to 0$, we have

$$v_{fs} = v_{is} + \lim_{\Delta t \to 0} \sum_{k=1}^{N} (a_s)_k \Delta t = v_{is} + \int_{t_i}^{t_f} a_s \, dt \qquad (2.27)$$

This mathematical statement has a graphical interpretation analogous to Equation 2.11. In this case:

> $v_{fs} = v_{is}$ + area under the acceleration curve a_s between t_i and t_f (2.28)

The constant acceleration equation $v_{fs} = v_{is} + a_s \Delta t$ is a special example of Equation 2.28. If you look back at Figure 2.28a you will see that the quantity $a_s \Delta t$ is the rectangular area under the horizontal acceleration curve.

EXAMPLE 2.23 Finding velocity from acceleration

Figure 2.46 shows the acceleration graph for a particle with an initial velocity of 10 m/s. What is the particle's velocity at $t = 8$ s?

MODEL We're told this is the motion of a particle.

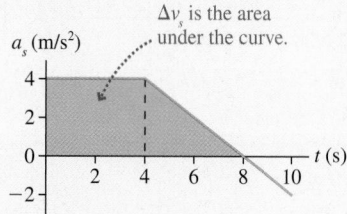

FIGURE 2.46 Acceleration graph for Example 2.23.

VISUALIZE Figure 2.46 is a graphical representation of the motion.

SOLVE The change in velocity is found as the area under the acceleration curve:

$$v_{fs} = v_{is} + \text{area under the acceleration curve } a_s$$
$$\text{between } t_i \text{ and } t_f$$

The area under the curve between $t_i = 0$ s and $t_f = 8$ s can be subdivided into a rectangle ($0 \text{ s} \le t \le 4 \text{ s}$) and a triangle ($4 \text{ s} \le t \le 8 \text{ s}$). These areas are easily computed. Thus

$$v_s(\text{at } t = 8 \text{ s}) = 10 \text{ m/s} + (4 \text{ (m/s)/s})(4 \text{ s})$$
$$+ \tfrac{1}{2}(4 \text{ (m/s)/s})(4 \text{ s})$$
$$= 34 \text{ m/s}$$

EXAMPLE 2.24 A nonuniform acceleration

Figure 2.47a shows the velocity-versus-time graph for a particle whose velocity is given by $v_s = \left[10 - (t - 5)^2 \right]$ m/s, where t is in s.

a. Find an expression for the particle's acceleration a_s and draw the acceleration-versus-time graph.
b. Describe the motion.

MODEL We're told that this is a particle.

VISUALIZE The figure shows the velocity graph. It is a parabola centered at $t = 5$ s with an apex $v_{max} = 10$ m/s. The slope of v_s is positive but decreasing in magnitude for $t < 5$ s. The slope is zero at $t = 5$ s, and it is negative and increasing in magnitude for $t > 5$ s. Thus the acceleration graph should start positive, decrease steadily, pass through zero at $t = 5$ s, then become increasingly negative.

SOLVE

a. We can find an expression for a_s by taking the derivative of v_s. First, expand the square to give

$$v_s = (-t^2 + 10t - 15) \text{ m/s}$$

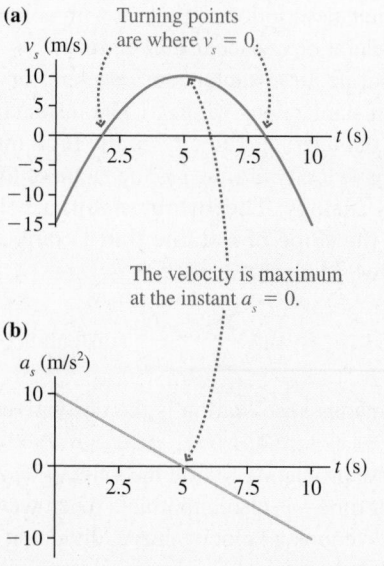

FIGURE 2.47 Velocity and acceleration graphs for Example 2.24.

Then use the derivative rule (Equation 2.5) to find

$$a_s = \frac{dv_s}{dt} = (-2t + 10) \text{ m/s}^2$$

where t is in s. This is a linear equation that is graphed in Figure 2.47b. The graph meets our expectations.

b. This is a complex motion. The particle starts out moving to the left ($v_s < 0$) at 15 m/s. The positive acceleration causes the speed to decrease (slowing down because v_s and a_s have opposite signs) until the particle reaches a turning point ($v_s = 0$) just before $t = 2$ s. The particle then moves to the right ($v_s > 0$) and speeds up until reaching maximum speed at $t = 5$ s. From $t = 5$ s to just after $t = 8$ s, the particle is still moving to the right ($v_s > 0$) but slowing down. Another turning point occurs just after $t = 8$ s. Then the particle moves back to the left and gains speed as the negative a_s makes the velocity ever more negative.

STOP TO THINK 2.6   Rank in order, from largest to smallest, the accelerations at points A to C.

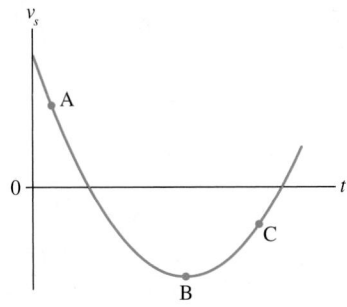

a. $a_A > a_B > a_C$.

c. $a_C > a_B > a_A$.

b. $a_C > a_A > a_B$.

d. $a_B > a_A > a_C$.

SUMMARY

The goal of Chapter 2 has been to learn how to solve problems about motion in a straight line.

GENERAL PRINCIPLES

Kinematics describes motion in terms of position, velocity, and acceleration. General kinematic relationships are given **mathematically** by:

Instantaneous velocity $v_s = ds/dt = $ slope of position graph

Instantaneous acceleration $a_s = dv_s/dt = $ slope of velocity graph

Final position $s_f = s_i + \int_{t_i}^{t_f} v_s \, dt = s_i + \begin{cases} \text{area under the velocity curve} \\ \text{from } t_i \text{ to } t_f \end{cases}$

Final velocity $v_{fs} = v_{is} + \int_{t_i}^{t_f} a_s \, dt = v_{is} + \begin{cases} \text{area under the acceleration} \\ \text{curve from } t_i \text{ to } t_f \end{cases}$

The kinematic equations for **motion with constant acceleration:**

$$v_{fs} = v_{is} + a_s \Delta t$$

$$s_f = s_i + v_{is}\Delta t + \tfrac{1}{2}a_s(\Delta t)^2$$

$$v_{fs}^2 = v_{is}^2 + 2a_s \Delta s$$

IMPORTANT CONCEPTS

Position, velocity, and acceleration are related **graphically.**

- The slope of the position-versus-time graph is the value on the velocity graph.

- The slope of the velocity graph is the value on the acceleration graph.

- s is a maximum or minimum at a turning point, and $v_s = 0$.

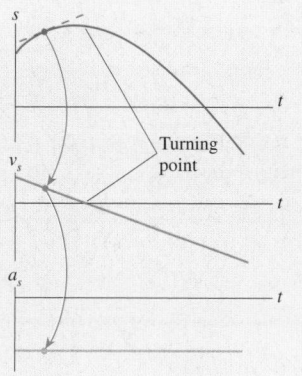

Motion with constant acceleration is uniformly accelerated motion.

Uniform motion is motion with constant velocity and zero acceleration.

$$s_f = s_i + v_s \Delta t$$

APPLICATIONS

The **sign of** v_s indicates the direction of motion.

- $v_s > 0$ is motion to the right or up.

- $v_s < 0$ is motion to the left or down.

The **sign of** a_s indicates which way $\vec{a}$ points, *not* whether the object is speeding up or slowing down.

- $a_s > 0$ if $\vec{a}$ points to the right or up.

- $a_s < 0$ if $\vec{a}$ points to the left or down.

- The direction of $\vec{a}$ is found with a motion diagram.

An object is **speeding up** if and only if v_s and a_s have the same sign. An object is **slowing down** if and only if v_s and a_s have opposite signs.

Free fall is constant-acceleration motion with
$$a_y = -g = -9.80 \text{ m/s}^2.$$

Motion on an inclined plane has $a_s = \pm g\sin\theta$. The sign depends on the direction of the tilt.

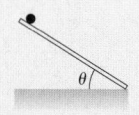

TERMS AND NOTATION

kinematics	final position, s_f	uniformly accelerated motion
position-versus-time graph	instantaneous velocity, v_s	free fall
uniform motion	turning point	acceleration due to gravity, g
speed, v	average acceleration, a_{avg}	instantaneous acceleration, a_s
initial position, s_i		

EXERCISES AND PROBLEMS

The ✐ icon in front of a problem indicates that the problem can be done on a Dynamics Worksheet. Dynamics Worksheets are found at the back of the *Student Workbook*. If you use a worksheet, draw a motion diagram in the Physical Representation section, establish your coordinate system and symbols in the Pictorial Representation section, then solve the problem in the Mathematical Representation section.

Exercises

Section 2.1 Motion in One Dimension

1. Figure Ex2.1 shows a motion diagram of a car traveling down a street. The camera took one frame every second. A distance scale is provided.
   a. Measure the x-value of the car at each dot. Place your data in a table, similar to Table 2.1, showing each position and the instant of time at which it occurred.
   b. Make a position-versus-time graph for the ball. Because you have data only at certain instants of time, your graph should consist of dots that are not connected together.

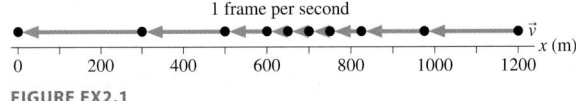

FIGURE EX2.1

2. For each motion diagram, determine the sign (positive or negative) of the position, the velocity, and the acceleration.

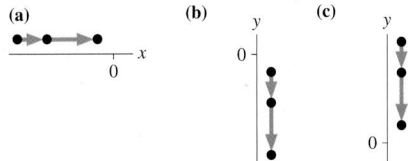

FIGURE EX2.2

3. Write a short description of the motion of a real object for which this would be a realistic position-versus-time graph.

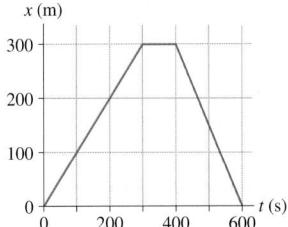

FIGURE EX2.3

4. Write a short description of the motion of a real object for which this would be a realistic position-versus-time graph.

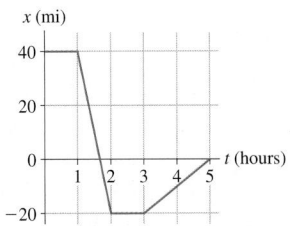

FIGURE EX2.4

Section 2.2 Uniform Motion

5. A car starts at the origin and moves with velocity $\vec{v} = (10 \text{ m/s, northeast})$. How far from the origin will the car be after traveling for 45 s?

6. Larry leaves home at 9:05 and runs at constant speed to the lamppost. He reaches the lamppost at 9:07, immediately turns, and runs to the tree. Larry arrives at the tree at 9:10.
   a. What is Larry's average velocity during each of these two intervals?
   b. What is the average velocity for Larry's entire run?

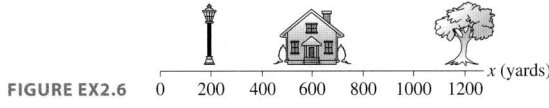

FIGURE EX2.6

7. Alan leaves Los Angeles at 8:00 A.M. to drive to San Francisco, 400 mi away. He travels at a steady 50 mph. Beth leaves Los Angeles at 9:00 A.M. and drives a steady 60 mph.
   a. Who gets to San Francisco first?
   b. How long does the first to arrive have to wait for the second?

8. A bicyclist has the position-versus-time graph shown. What is the bicyclist's velocity at $t = 10$ s, at $t = 25$ s, and at $t = 35$ s?

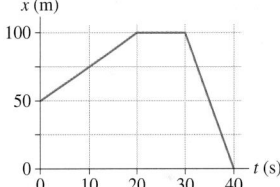

FIGURE EX2.8

9. Julie drives 100 mi to Grandmother's house. On the way to Grandmother's, Julie drives half the *distance* at 40 mph and half the distance at 60 mph. On her return trip, she drives half the *time* at 40 mph and half the time at 60 mph.
   a. What is Julie's average speed on the way to Grandmother's house?
   b. What is her average speed on the return trip?

Section 2.3 Instantaneous Velocity

Section 2.4 Finding Position from Velocity

10. Figure Ex2.10 shows the position graph of a particle.
    a. Draw the particle's velocity graph for the interval $0 \text{ s} \le t \le 4 \text{ s}$.
    b. Does this particle have a turning point or points? If so, at what time or times?

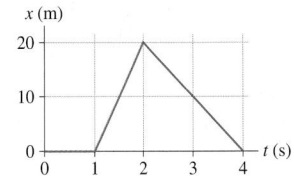

FIGURE EX2.10

11. A particle starts from $x_0 = 10$ m at $t_0 = 0$ and moves with the velocity graph shown in Figure Ex2.11.
    a. What is the object's position at $t = 2$ s, 3 s, and 4 s?
    b. Does this particle have a turning point? If so, at what time?

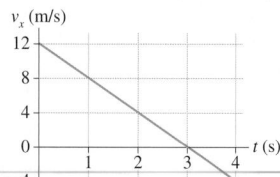

FIGURE EX2.11

Section 2.5 Motion with Constant Acceleration

12. Figure Ex2.12 shows the velocity graph of a particle. Draw the particle's acceleration graph for the interval 0 s $\le t \le 4$ s. Give both axes an appropriate numerical scale.

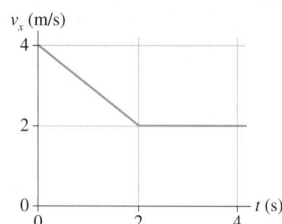

FIGURE EX2.12

13. Figure Ex2.13 shows the velocity graph of a train that starts from the origin at $t = 0$ s.
    a. Draw position and acceleration graphs for the train.
    b. Find the acceleration of the train at $t = 3.0$ s.

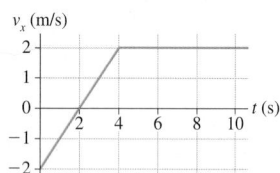

FIGURE EX2.13

14. Figure Ex2.14 shows the velocity graph of a particle moving along the x-axis. Its initial position is $x_0 = 2$ m at $t_0 = 0$ s. At $t = 2$ s, what are the particle's (a) position, (b) velocity, and (c) acceleration?

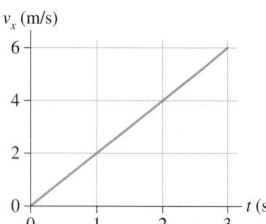

FIGURE EX2.14

15. a. What constant acceleration, in SI units, must a car have to go from zero to 60 mph in 10 s?
    b. What fraction of g is this?
    c. How far has the car traveled when it reaches 60 mph? Give your answer both in SI units and in feet.
16. A jet plane is cruising at 300 m/s when suddenly the pilot turns the engines up to full throttle. After traveling 4.0 km, the jet is moving with a speed of 400 m/s.
    a. What is the jet's acceleration, assuming it to be a constant acceleration?
    b. Is your answer reasonable? Explain.

17. A speed skater moving across frictionless ice at 8.0 m/s hits a 5.0-m-wide patch of rough ice. She slows steadily, then continues on at 6.0 m/s. What is her acceleration on the rough ice?
18. a. How many days will it take a spaceship to accelerate to the speed of light (3.0×10^8 m/s) with the acceleration g?
    b. How far will it travel during this interval?
    c. What fraction of a light year is your answer to part b? A *light year* is the distance light travels in one year.

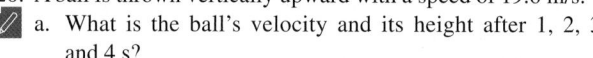

NOTE ▶ We know, from Einstein's theory of relativity, that no object can travel at the speed of light. So this problem, while interesting and instructive, is not realistic. ◀

Section 2.6 Free Fall

19. Ball bearings are made by letting spherical drops of molten metal fall inside a tall tower—called a *shot tower*—and solidify as they fall.
    a. If a bearing needs 4.0 s to solidify enough for impact, how high must the tower be?
    b. What is the bearing's impact velocity?
20. A ball is thrown vertically upward with a speed of 19.6 m/s.
    a. What is the ball's velocity and its height after 1, 2, 3, and 4 s?
    b. Draw the ball's velocity-versus-time graph. Give both axes an appropriate numerical scale.
21. A student standing on the ground throws a ball straight up. The ball leaves the student's hand with a speed of 15 m/s when the hand is 2.0 m above the ground. How long is the ball in the air before it hits the ground? (The student moves her hand out of the way.)
22. A rock is tossed straight up with a speed of 20 m/s. When it returns, it falls into a hole 10 m deep.
    a. What is the rock's velocity as it hits the bottom of the hole?
    b. How long is the rock in the air, from the instant it is released until it hits the bottom of the hole?

Section 2.7 Motion on an Inclined Plane

23. A car traveling at 30 m/s runs out of gas while traveling up a 20° slope. How far up the hill will it coast before starting to roll back down?
24. A skier is gliding along at 3.0 m/s on horizontal, frictionless snow. He suddenly starts down a 10° incline. His speed at the bottom is 15 m/s.
    a. What is the length of the incline?
    b. How long does it take him to reach the bottom?

Section 2.8 Instantaneous Acceleration

25. A particle moving along the x-axis has its position described by the function $x = (2t^3 - t + 1)$ m, where t is in s. At $t = 2$ s, what are the particle's (a) position, (b) velocity, and (c) acceleration?
26. A particle moving along the x-axis has its velocity described by the function $v_x = 2t^2$ m/s, where t is in s. Its initial position is $x_0 = 1$ m at $t_0 = 0$ s. At $t = 1$ s, what are the particle's (a) position, (b) velocity, and (c) acceleration?

27. Figure Ex2.27 shows the acceleration-versus-time graph of a particle moving along the x-axis. Its initial velocity is $v_{0x} = 8.0$ m/s at $t_0 = 0$ s. What is the particle's velocity at $t = 4.0$ s?

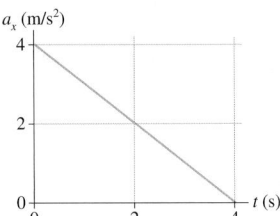

FIGURE EX2.27

Problems

28. Figure P2.28 shows the motion diagram, made at two frames of film per second, of a ball rolling along a track. The track has a 3.0-m-long sticky section.
    a. Use the meter stick to measure the positions of the center of the ball. Place your data in a table, similar to Table 2.1, showing each position and the instant of time at which it occurred.
    b. Make a position-versus-time graph for the ball. Because you have data only at certain instants of time, your graph should consist of dots that are not connected together.
    c. What is the *change* in the ball's position from $t = 0$ s to $t = 1.0$ s?
    d. What is the *change* in the ball's position from $t = 2.0$ s to $t = 4.0$ s?
    e. What is the ball's velocity before reaching the sticky section?
    f. What is the ball's velocity after passing the sticky section?
    g. Determine the ball's acceleration on the sticky section of the track.

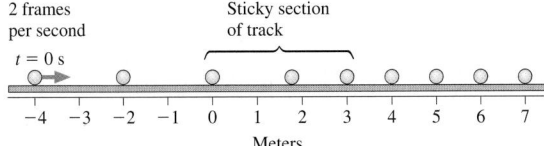

FIGURE P2.28

29. A particle's position on the x-axis is given by the function $x = (t^2 - 4t + 2)$ m, where t is in s.
    a. Make a position-versus-time graph for the interval 0 s $\leq t \leq 5$ s. Do this by calculating and plotting x every 0.5 s from 0 s to 5 s, then drawing a smooth curve through the points.
    b. Determine the particle's velocity at $t = 1.0$ s by drawing the tangent line on your graph and measuring its slope.
    c. Determine the particle's velocity at $t = 1.0$ s by evaluating the derivative at that instant. Compare this to your result from part b.
    d. Are there any turning points in the particle's motion? If so, at what position or positions?
    e. Where is the particle when $v_x = 4.0$ m/s?
    f. Draw a motion diagram for the particle.

30. The velocity-versus-time graph is shown for a particle moving along the x-axis. Its initial position is $x_0 = 2.0$ m at $t_0 = 0$ s.
    a. What are the particle's position, velocity, and acceleration at $t = 1.0$ s?
    b. What are the particle's position, velocity, and acceleration at $t = 3.0$ s?

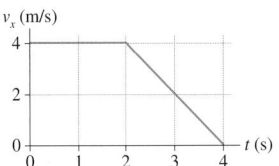

FIGURE P2.30

31. Three particles move along the x-axis, each starting with $v_{0x} = 10$ m/s at $t_0 = 0$ s. The graph for A is a position-versus-time graph; the graph for B is a velocity-versus-time graph; the graph for C is an acceleration-versus-time graph. Find each particle's velocity at $t = 7.0$ s. Work with the geometry of the graphs, not with kinematic equations.

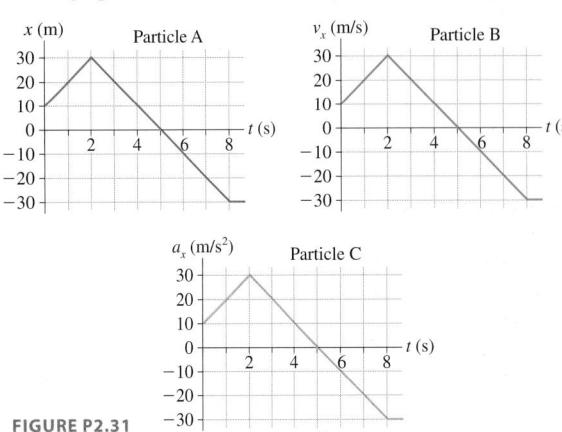

FIGURE P2.31

32. The velocity graph is shown for a particle having initial position $x_0 = 0$ m at $t_0 = 0$ s.
    a. At what time or times is the particle found at $x = 35$ m? Work with the geometry of the graph, not with kinematic equations.
    b. Draw a motion diagram for the particle.

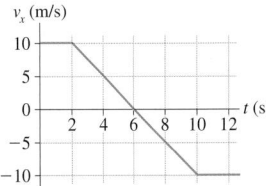

FIGURE P2.32

33. The acceleration graph is shown for a particle that starts from rest at $t = 0$ s. Determine the object's velocity at times $t = 0$ s, 2 s, 4 s, 6 s, and 8 s.

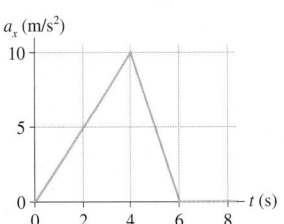

FIGURE P2.33

34. A block is suspended from a spring, pulled down, and released. The block's position-versus-time graph is shown in Figure P2.34.
    a. At what times is the velocity zero? At what times is the velocity most positive? Most negative?
    b. Draw a reasonable velocity-versus-time graph.

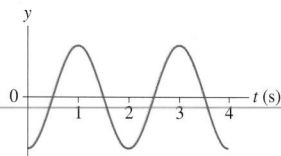

FIGURE P2.34

35. The acceleration graph is shown for a particle that starts from rest at $t = 0$ s.
    a. Draw the particle's velocity graph over the interval 0 s $\le t \le 10$ s. Include an appropriate numerical scale on both axes.
    b. Describe, in words, how the velocity graph would differ if the particle had an initial velocity of 2.0 m/s.

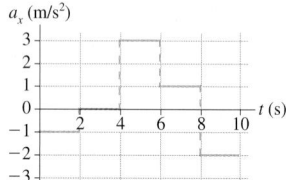

FIGURE P2.35

36. Drop a rubber ball or a tennis ball from a height of about 25 cm (≈ 1 ft) and watch carefully as it bounces. Then draw a position graph, a velocity graph, and an acceleration graph from the instant you drop the ball until it returns to its maximum height. Stack your three graphs vertically so that the time axes are aligned with each other. Pay particular attention to the time when the ball is in contact with the ground. This is a short interval of time, but it's not zero.

37. The position of a particle is given by the function $x = (2t^3 - 9t^2 + 12)$ m, where t is in s.
    a. At what time or times is $v_x = 0$ m/s?
    b. What are the particle's position and its acceleration at this time(s)?

38. An object starts from rest at $x = 0$ m at time $t = 0$ s. Five seconds later, at $t = 5.0$ s, the object is observed to be at $x = 40.0$ m and to have velocity $v_x = 11$ m/s.
    a. Was the object's acceleration uniform or nonuniform? Explain your reasoning.
    b. Sketch the velocity-versus-time graph implied by these data. Is the graph a straight line or curved? If curved, is it concave upward or downward?

39. A particle's velocity is described by the function $v_x = kt^2$ m/s, where k is a constant and t is in s. The particle's position at $t_0 = 0$ s is $x_0 = -9.0$ m. At $t_1 = 3.0$ s, the particle is at $x_1 = 9.0$ m. Determine the value of the constant k. Be sure to include the proper units.

40. A particle's acceleration is described by the function $a_x = (10 - t)$ m/s^2, where t is in s. Its initial conditions are $x_0 = 0$ m and $v_{0x} = 0$ m/s at $t = 0$ s.
    a. At what time is the velocity again zero?
    b. What is the particle's position at that time?

41. A ball rolls along the frictionless track shown. Each segment of the track is straight, and the ball passes smoothly from one segment to the next without changing speed or leaving the track. Draw three vertically stacked graphs showing position, velocity, and acceleration versus time. Each graph should have the same time axis, and the proportions of the graph should be qualitatively correct. Assume that the ball has enough speed to reach the top.

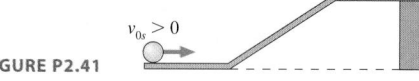

FIGURE P2.41

42. Draw position, velocity, and acceleration graphs for the ball shown here. See Problem 41 for more information.

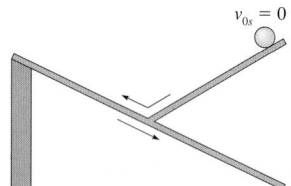

FIGURE P2.42

43. Draw position, velocity, and acceleration graphs for the ball shown here. See Problem 41 for more information. The ball changes direction but not speed as it bounces from the reflecting wall.

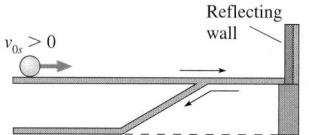

FIGURE P2.43

44. Figure P2.44 shows a set of kinematic graphs for a ball rolling on a track. All segments of the track are straight lines, but some may be tilted. Draw a picture of the track and also indicate the ball's initial condition.

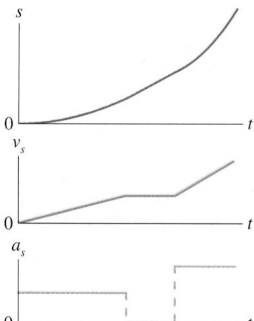

FIGURE P2.44

45. Figure P2.45 shows a set of kinematic graphs for a ball rolling on a track. All segments of the track are straight lines, but some may be tilted. Draw a picture of the track and also indicate the ball's initial condition.

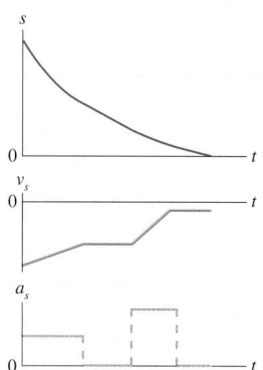

FIGURE P2.45

46. Figure P2.46 shows a set of kinematic graphs for a ball rolling on a track. All segments of the track are straight lines, but some may be tilted. Draw a picture of the track and also indicate the ball's initial condition.

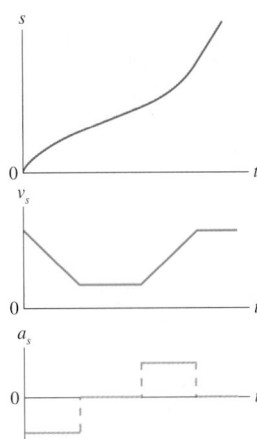

FIGURE P2.46

47. The takeoff speed for an Airbus A320 jetliner is 80 m/s. Velocity data measured during takeoff are as follows:

$t(s)$	v_s (m/s)
0	0
10	23
20	46
30	69

a. What is the takeoff speed in miles per hour?
b. Is the jetliner's acceleration constant during takeoff? Explain.
c. At what time do the wheels leave the ground?
d. For safety reasons, in case of an aborted takeoff, the runway must be three times the takeoff distance. Can an A320 take off safely on a 2.5-mi-long runway?

48. Does a real automobile have constant acceleration? Measured data for a Porsche 944 Turbo at maximum acceleration is as follows:

$t(s)$	v_s (mph)
0	0
2	28
4	46
6	60
8	70
10	78

a. Make a graph of velocity versus time. Based on your graph, is the acceleration constant? Explain.
b. Draw a smooth curve through the points on your graph, then use your graph to estimate the car's acceleration at 2.0 s and 8.0 s. Give your answer in SI units.
c. Use your graph to estimate the distance traveled in the first 10 s.
Hint: Approximate the curve with five steps, each of width $\Delta t = 2$ s. Let the height of each step be the average of the velocities at the beginning and end of the step.

49. A driver has a reaction time of 0.50 s, and the maximum deceleration of her car is 6.0 m/s^2. She is driving at 20 m/s when suddenly she sees an obstacle in the road 50 m in front of her. Can she stop the car in time to avoid a collision?

50. You are driving to the grocery store at 20 m/s. You are 110 m from an intersection when the traffic light turns red. Assume that your reaction time is 0.50 s and that your car brakes with constant acceleration.
a. How far are you from the intersection when you begin to apply the brakes?
b. What acceleration will bring you to rest right at the intersection?
c. How long does it take you to stop?

51. The minimum stopping distance for a car traveling at a speed of 30 m/s is 60 m, including the distance traveled during the driver's reaction time of 0.50 s.
a. What is the minimum stopping distance for the same car traveling at a speed of 40 m/s?
b. Draw a position-versus-time graph for the motion of the car in part a. Assume the car is at $x_0 = 0$ m when the driver first sees the emergency situation ahead that calls for a rapid halt.

52. You're driving down the highway late one night at 20 m/s when a deer steps onto the road 35 m in front of you. Your reaction time before stepping on the brakes is 0.5 s, and the maximum deceleration of your car is 10 m/s^2.
a. How much distance is between you and the deer when you come to a stop?
b. What is the maximum speed you could have and still not hit the deer?

53. A 200kg weather rocket is loaded with 100 kg of fuel and fired straight up. It accelerates upward at 30 m/s^2 for 30 s, then runs out of fuel. Ignore any air resistance effects.
a. What is the rocket's maximum altitude?
b. How long is the rocket in the air?
c. Draw a velocity-versus-time graph for the rocket from liftoff until it hits the ground.

54. A 1000kg weather rocket is launched straight up. The rocket motor provides a constant acceleration for 16 s, then the motor stops. The rocket altitude 20 s after launch is 5100 m. You can ignore any effects of air resistance.
a. What was the rocket's acceleration during the first 16 s?
b. What is the rocket's speed as it passes through a cloud 5100 m above the ground?

55. A lead ball is dropped into a lake from a diving board 5.0 m above the water. After entering the water, it sinks to the bottom with a constant velocity equal to the velocity with which it hit the water. The ball reaches the bottom 3.0 s after it is released. How deep is the lake?

56. A hotel elevator ascends 200 m with a maximum speed of 5.0 m/s. Its acceleration and deceleration both have a magnitude of 1.0 m/s^2.
a. How far does the elevator move while accelerating to full speed from rest?
b. How long does it take to make the complete trip from bottom to top?

57. A car starts from rest at a stop sign. It accelerates at 4.0 m/s^2 for 6 seconds, coasts for 2 s, and then slows down at a rate of 3.0 m/s^2 for the next stop sign. How far apart are the stop signs?

58. A car accelerates at 2.0 m/s² along a straight road. It passes two marks that are 30 m apart at times $t = 4.0$ s and $t = 5.0$ s. What was the car's initial velocity?

59. Santa loses his footing and slides down a frictionless, snowy roof that is tilted at an angle of 30°. If Santa slides 10 m before reaching the edge, what is his speed as he leaves the roof?

60. Ann and Carol are driving their cars along the same straight road. Carol is located at $x = 2.4$ mi at $t = 0$ hours and drives at a steady 36 mph. Ann, who is traveling in the same direction, is located at $x = 0.0$ mi at $t = 0.50$ hours and drives at a steady 50 mph.
    a. At what time does Ann overtake Carol?
    b. What is their position at this instant?
    c. Draw a position-versus-time graph showing the motion of both Ann and Carol.

61. A ball rolls along the smooth track shown in the figure with an initial speed of 5.0 m/s. Assume that the ball turns all the corners smoothly, with no loss of speed.
    a. What is the ball's speed as it goes over the top?
    b. What is its speed when it reaches the level track on the right side?
    c. By what percentage does the ball's final speed differ from its initial speed? Is this surprising?

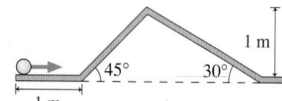

FIGURE P2.61

62. A toy train is pushed forward and released at $x_0 = 2.0$ m with a speed of 2.0 m/s. It rolls at a steady speed for 2.0 s, then one wheel begins to stick. The train comes to a stop 6.0 m from the point at which it was released. What is the train's acceleration after its wheel begins to stick?

63. Bob is driving the getaway car after the big bank robbery. He's going 50 m/s when his headlights suddenly reveal a nail strip that the cops have placed across the road 150 m in front of him. If Bob can stop in time, he can throw the car into reverse and escape. But if he crosses the nail strip, all his tires will go flat and he will be caught. Bob's reaction time before he can hit the brakes is 0.60 s, and his car's maximum deceleration is 10 m/s². Is Bob in jail?

64. One game at the amusement park has you push a puck up a long, frictionless ramp. You win a stuffed animal if the puck, at its highest point, comes to within 10 cm of the end of the ramp without going off. You give the puck a push, releasing it with a speed of 5.0 m/s when it is 8.50 m from the end of the ramp. The puck's speed after traveling 3.0 m is 4.0 m/s. Are you a winner?

65. A professional skier's *initial* acceleration on fresh snow is 90% of the acceleration expected on a frictionless, inclined plane, the loss being due to friction. Due to air resistance, his acceleration slowly decreases as he picks up speed. The speed record on a mountain in Oregon is 180 kilometers per hour at the bottom of a 25° slope that drops 200 m.
    a. What exit speed could a skier reach in the absence of air resistance?
    b. What percentage of this ideal speed is lost to air resistance?

66. Heather and Jerry are standing on a bridge 50 m above a river. Heather throws a rock straight down with a speed of 20 m/s. Jerry, at exactly the same instant of time, throws a rock straight up with the same speed. Ignore air resistance.
    a. How much time elapses between the first splash and the second splash?
    b. Which rock has the faster speed as it hits the water?

67. Nicole throws a ball straight up. Chad watches the ball from a window 5.0 m above the point where Nicole released it. The ball passes Chad on the way up, and it has a speed of 10 m/s as it passes him on the way back down. How fast did Nicole throw the ball?

68. A motorist is driving at 20 m/s when she sees that a traffic light 200 m ahead has just turned red. She knows that this light stays red for 15 s, and she wants to reach the light just as it turns green again. It takes her 1.0 s to step on the brakes and begin slowing. What is her speed as she reaches the light at the instant it turns green?

69. David is driving a steady 30 m/s when he passes Tina, who is sitting in her car at rest. Tina begins to accelerate at a steady 2.0 m/s² at the instant when David passes.
    a. How far does Tina drive before passing David?
    b. What is her speed as she passes him?

70. A Porsche challenges a Honda to a 400-m race. Because the Porsche's acceleration of 3.5 m/s² is larger than the Honda's 3.0 m/s², the Honda gets a 50-m head start. Both cars start accelerating at the same instant. Who wins?

71. A cat is sleeping on the floor in the middle of a 3.0-m-wide room when a barking dog enters with a speed of 1.50 m/s. As the dog enters, the cat (as only cats can do) immediately accelerates at 0.85 m/s² toward an open window on the opposite side of the room. The dog (all bark and no bite) is a bit startled by the cat and begins to slow down at 0.10 m/s² as soon as it enters the room. Does the dog catch the cat before the cat is able to leap through the window?

72. You want to visit your friend in Seattle during spring break. To save money, you decide to travel there by train. Unfortunately, your physics final exam took the full 3 hours, so you are late in arriving at the train station. You run as fast as you can, but just as you reach the platform you see your train, 30 m ahead of you down the platform, begin to accelerate at 1.0 m/s². You chase after the train at your maximum speed of 8.0 m/s, but there's a barrier 50 m ahead. Will you be able to leap onto the back step of the train before you crash into the barrier?

73. A rocket is launched straight up with constant acceleration. Four seconds after liftoff, a bolt falls off the side of the rocket. The bolt hits the ground 6.0 s later. What was the rocket's acceleration?

In Problems 74 through 77 you are given the kinematic equation or equations that are used to solve a problem. For each of these, you are to
    a. Write a *realistic* problem for which this is the correct equation(s). Be sure that the answer your problem requests is consistent with the equation(s) given.
    b. Draw the motion diagram and the pictorial representation for your problem.
    c. Finish the solution of the problem.

74. $64 \text{ m} = 0 \text{ m} + (32 \text{ m/s})(4 \text{ s} - 0 \text{ s}) + \frac{1}{2}a_x(4 \text{ s} - 0 \text{ s})^2$

75. $(10 \text{ m/s})^2 = v_{0y}^2 - 2(9.8 \text{ m/s}^2)(10 \text{ m} - 0 \text{ m})$

76. $(0 \text{ m/s})^2 = (5 \text{ m/s})^2 - 2(9.8 \text{ m/s}^2)(\sin 10°)(x_1 - 0 \text{ m})$

77. $v_{1x} = 0 \text{ m/s} + (20 \text{ m/s}^2)(5 \text{ s} - 0 \text{ s})$
    $x_1 = 0 \text{ m} + (0 \text{ m/s})(5 \text{ s} - 0 \text{ s}) + \frac{1}{2}(20 \text{ m/s}^2)(5 \text{ s} - 0 \text{ s})^2$
    $x_2 = x_1 + v_{1x}(10 \text{ s} - 5 \text{ s})$

Challenge Problems

78. Jill has just gotten out of her car in the grocery store parking lot. The parking lot is on a hill and is tilted 3°. Fifty meters downhill from Jill, a little old lady lets go of a fully loaded shopping cart. The cart, with frictionless wheels, starts to roll straight downhill. Jill immediately starts to sprint after the cart with her top acceleration of 2.0 m/s². How far has the cart rolled before Jill catches it?

79. As a science project, you drop a watermelon off the top of the Empire State Building, 320 m above the sidewalk. It so happens that Superman flies by at the instant you release the watermelon. Superman is headed straight down with a speed of 35 m/s. How fast is the watermelon going when it passes Superman?

80. Your school science club has devised a special event for homecoming. You've attached a rocket to the rear of a small car that has been decorated in the blue-and-gold school colors. The rocket provides a constant acceleration for 9.0 s. As the rocket shuts off, a parachute opens and slows the car at a rate of 5.0 m/s². The car passes the judges' box in the center of the grandstand, 990 m from the starting line, exactly 12 s after you fire the rocket. What is the car's speed as it passes the judges?

81. Careful measurements have been made of Olympic sprinters in the 100-meter dash. A simple but reasonably accurate model is that a sprinter accelerates at 3.6 m/s² for $3\frac{1}{3}$ s, then runs at constant velocity to the finish line.
    a. What is the race time for a sprinter who follows this model?
    b. A sprinter could run a faster race by accelerating faster at the beginning, thus reaching top speed sooner. If a sprinter's top speed is the same as in part a, what acceleration would he need to run the 100-meter dash in 9.9 s?
    c. By what percent did the sprinter need to increase his acceleration in order to decrease his time by 1%?

82. A sprinter can accelerate with constant acceleration for 4.0 s before reaching top speed. He can run the 100-meter dash in 10 s. What is his speed as he crosses the finish line?

83. The Starship Enterprise returns from warp drive to ordinary space with a forward speed of 50 km/s. To the crew's great surprise, a Klingon ship is 100 km directly ahead, traveling in the same direction at a mere 20 km/s. Without evasive action, the Enterprise will overtake and collide with the Klingons in just slightly over 3.0 s. The Enterprise's computers react instantly to brake the ship. What acceleration does the Enterprise need to just barely avoid a collision with the Klingon ship? Assume the acceleration is constant.
    Hint: Draw a position-versus-time graph showing the motions of both the Enterprise and the Klingon ship. Let $x_0 = 0$ km be the location of the Enterprise as it returns from warp drive. How do you show graphically the situation in which the collision is "barely avoided"? Once you decide what it looks like graphically, express that situation mathematically.

STOP TO THINK ANSWERS

Stop to Think 2.1: d. The particle starts with positive x and moves to negative x.

Stop to Think 2.2: c. The velocity is the slope of the position graph. The slope is positive and constant until the position graph crosses the axis, then positive but decreasing, and finally zero when the position graph is horizontal.

Stop to Think 2.3: b. A constant positive v_x corresponds to a linearly increasing x, starting from $x_i = -10$ m. The constant negative v_x then corresponds to a linearly decreasing x.

Stop to Think 2.4: a or b. The velocity is constant while $a = 0$, it decreases linearly while a is negative. Graphs a, b, and c all have the same acceleration, but only graphs a and b have a positive initial velocity that represents a particle moving to the right.

Stop to Think 2.5: d. The acceleration vector $\vec{a}_{\parallel}$ points downhill (negative s-direction) and has the constant value $-g\sin\theta$ throughout the motion.

Stop to Think 2.6: c. Acceleration is the slope of the graph. The slope is zero at B. Although the graph is steepest at A, the slope at that point is negative, and so $a_A < a_B$. Only C has a positive slope, so $a_C > a_B$.

3 Vectors and Coordinate Systems

Wind has both a speed and a direction, hence the motion of the wind is described by a vector.

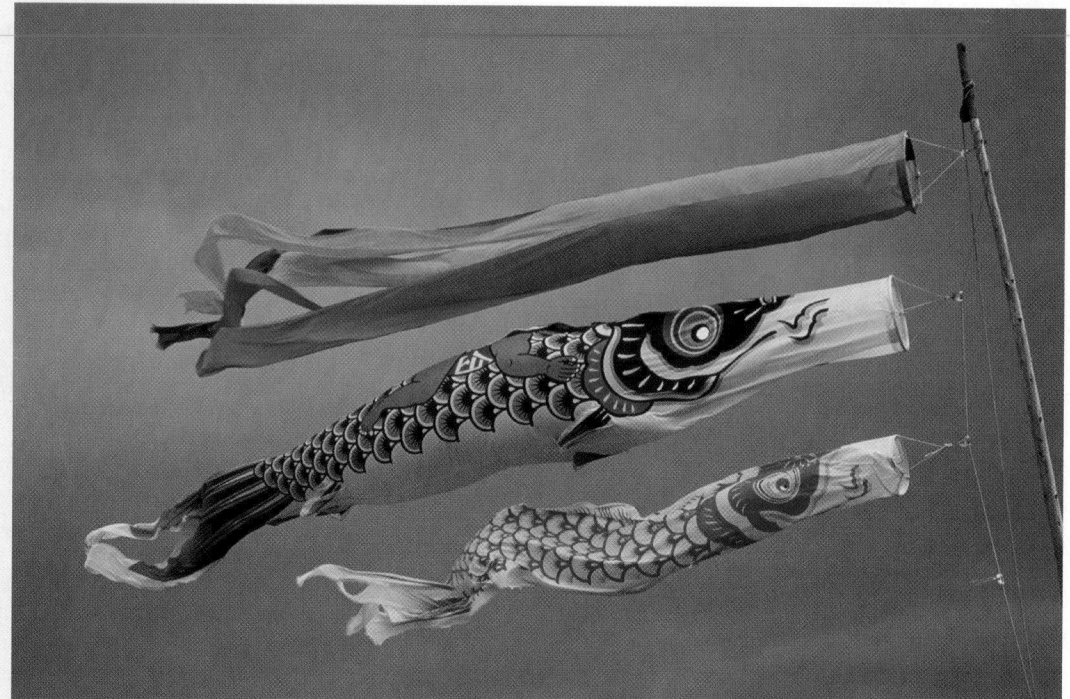

► Looking Ahead
The goals of Chapter 3 are to learn how vectors are represented and used. In this chapter you will learn to:

- Understand and use the basic properties of vectors.
- Decompose a vector into its components and reassemble vector components into a magnitude and direction.
- Add and subtract vectors both graphically and using components.

◄ Looking Back
This chapter continues the development of vectors that was begun in Chapter 1. Please review:

- Section 1.3 Vector addition and subtraction.

Many of the quantities that we use to describe the physical world are simply numbers. For example, the mass of an object is 2 kg, its temperature is 21°C, and it occupies a volume of 250 cm³. A quantity that is fully described by a single number (with units) is called a **scalar quantity.** Mass, temperature, and volume are all scalars. Other scalar quantities include pressure, density, energy, charge, and voltage. We will often use an algebraic symbol to represent a scalar quantity. Thus m will represent mass, T temperature, V volume, E energy, and so on. Notice that scalars, in printed text, are shown in italics.

Our universe has three dimensions, so some quantities also need a direction for a full description. If you ask someone for directions to the post office, the reply "Go three blocks" will not be very helpful. A full description might be, "Go three blocks south." A quantity having both a size and a direction is called a **vector quantity.**

You met examples of vector quantities in Chapter 1: position, displacement, velocity, and acceleration. You will soon make the acquaintance of others, such as force, momentum, and the electric field. Now, before we begin a study of forces, it's worth spending a little time to look more closely at vectors.

3.1 Scalars and Vectors

Suppose you are assigned the task of measuring the temperature at various points throughout a building and then showing the information on a building floor plan. To do this, you could put little dots on the floor plan, to show the points at which you made measurements, then write the temperature at that point beside the dot.

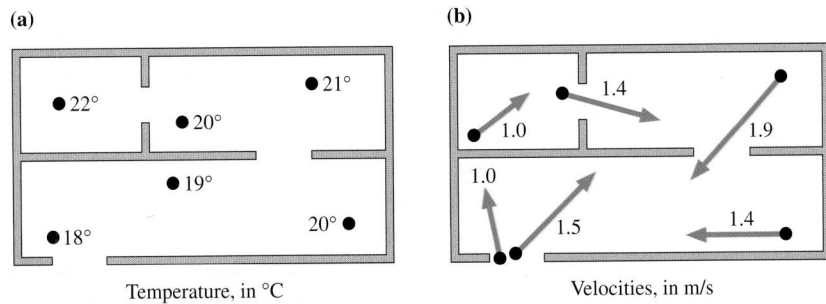

(a)   **(b)**

Temperature, in °C   Velocities, in m/s

FIGURE 3.1 Measurements of scalar and vector quantities.

In other words, as Figure 3.1a shows, you can represent the temperature at each point with a simple number (with units). Temperature is a scalar quantity.

Having done such a good job on your first assignment, you are next assigned the task of measuring the velocities of several employees as they move about in their work. Recall from Chapter 1 that velocity is a vector; it has both a size and a direction. Simply writing each employee's speed is not sufficient because speed doesn't take into account the direction in which the person moved. After some thought, you conclude that a good way to represent the velocity is by drawing an arrow whose length is proportional to the speed and that points in the direction of motion. Further, as Figure 3.1b shows, you decide to place the *tail* of an arrow at the point where you measured the velocity.

As this example illustrates, the *geometric representation* of a vector is an arrow, with the tail of the arrow (not its tip!) placed at the point where the measurement is made. The vector then seems to radiate outward from the point to which it is attached. An arrow makes a natural representation of a vector because it inherently has both a length and a direction.

The mathematical term for the length, or size, of a vector is **magnitude,** so we can say that **a vector is a quantity having a magnitude and a direction.** As an example, Figure 3.2 shows the geometric representation of a particle's velocity vector $\vec{v}$. The particle's speed at this point is 5 m/s, *and* it is moving in the direction indicated by the arrow. The arrow is drawn with its tail at the point where the velocity was measured.

NOTE ► Although the vector arrow is drawn across the page, from its tail to its tip, this does *not* indicate that the vector "stretches" across this distance. Instead, the vector arrow tells us the value of the vector quantity only at the one point where the tail of the vector is placed. ◄

Arrows are good for pictures, but we also need an *algebraic representation* of vectors to use in labels and in equations. We do this by drawing a small arrow over the letter that represents the vector: $\vec{r}$ for position, $\vec{v}$ for velocity, $\vec{a}$ for acceleration, and so on.

The *magnitude* of a vector is indicated by the letter without the arrow. For example, the magnitude of the velocity vector in Figure 3.2 is $v = 5$ m/s. This is the object's *speed*. The magnitude of the acceleration vector $\vec{a}$ is written a. **The magnitude of a vector is a scalar quantity.**

NOTE ► The magnitude of a vector cannot be a negative number; it must be positive or zero, with appropriate units. ◄

It is important to get in the habit of using the arrow symbol for vectors. If you omit the vector arrow from the velocity vector $\vec{v}$ and write only v, then you're referring only to the object's speed, not its velocity. The symbols $\vec{r}$ and r, or $\vec{v}$ and v, do *not* represent the same thing, so if you omit the vector arrow from vector symbols you will soon have confusion and mistakes.

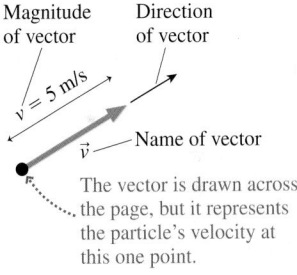

Magnitude of vector   Direction of vector

$v = 5$ m/s

$\vec{v}$ — Name of vector

The vector is drawn across the page, but it represents the particle's velocity at this one point.

FIGURE 3.2 The velocity vector $\vec{v}$ has both a magnitude and a direction.

The boat's displacement is the straight-line connection from its initial to its final position.

3.2 Properties of Vectors

Recall from Chapter 1 that the *displacement* is a vector drawn from an object's initial position to its position at some later time. Because displacement is an easy concept to think about, we can use it to introduce some of the properties of vectors. However, these properties apply to *all* vectors, not just to displacement.

Suppose that Sam starts from his front door, walks across the street, and ends up 200 ft to the northeast of where he started. Sam's displacement, which we will label $\vec{S}$, is shown in Figure 3.3a. The displacement vector is a *straight-line connection* from his initial to his final position, not necessarily his actual path. The dotted line indicates a possible route Sam might have taken, but his displacement is the vector $\vec{S}$.

To describe a vector we must specify both its magnitude and its direction. We can write Sam's displacement as

$$\vec{S} = (200 \text{ ft, northeast})$$

where the first number specifies the magnitude and the second number is the direction. The magnitude of Sam's displacement is $S = 200$ ft, the distance between his initial and final points.

Sam's next-door neighbor Bill also walks 200 ft to the northeast, starting from his own front door. Bill's displacement $\vec{B} = (200 \text{ ft, northeast})$ has the same magnitude and direction as Sam's displacement $\vec{S}$. Because vectors are defined by their magnitude and direction, **two vectors are equal if they have the same magnitude and direction.** This is true regardless of the starting points of the vectors. Thus the two displacements in Figure 3.3b are equal to each other, and we can write $\vec{B} = \vec{S}$.

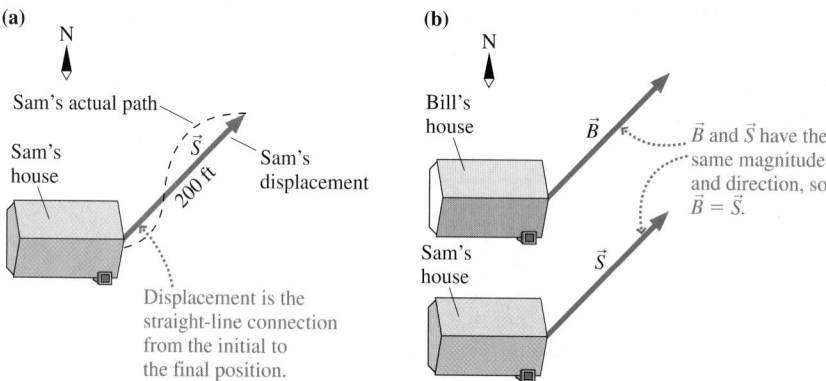

FIGURE 3.3 Displacement vectors.

NOTE ▶ A vector is unchanged if you move it to a different point on the page as long you don't change its length or the direction it points. We used this idea in Chapter 1 when we moved velocity vectors around in order to find the average acceleration vector $\vec{a}$. ◀

Vector Addition

Figure 3.4 shows the displacement of a hiker who starts at point P and ends at point S. She first hikes 4 miles to the east, then 3 miles to the north. The first leg of the hike is described by the displacement $\vec{A} = (4 \text{ mi, east})$. The second leg of the hike has displacement $\vec{B} = (3 \text{ mi, north})$. Now, by definition, a vector from the initial position P to the final position S is also a displacement. This is vector $\vec{C}$ on the figure. $\vec{C}$ is the *net displacement* because it describes the net result of the hiker's first having displacement $\vec{A}$, then displacement $\vec{B}$.

If you earn $50 on Saturday and $60 on Sunday, your *net* income for the weekend is the sum of $50 and $60. With scalars, the word *net* implies addition. The

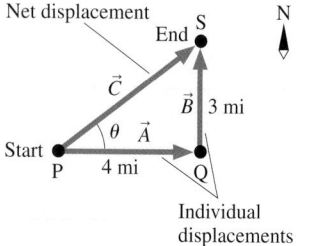

FIGURE 3.4 The net displacement $\vec{C}$ resulting from two displacements $\vec{A}$ and $\vec{B}$.

same is true with vectors. The net displacement $\vec{C}$ is an initial displacement $\vec{A}$ *plus* a second displacement $\vec{B}$, or

$$\vec{C} = \vec{A} + \vec{B} \tag{3.1}$$

The sum of two vectors is called the **resultant vector.** It's not hard to show that vector addition is commutative: $\vec{A} + \vec{B} = \vec{B} + \vec{A}$. That is, you can add vectors in any order you wish.

Look back at Tactics Box 1.1 on page 10 to see the three-step procedure for adding two vectors. This tip-to-tail method for adding vectors, which is used to find $\vec{C} = \vec{A} + \vec{B}$ in Figure 3.4, is called **graphical addition.** Any two vectors of the same type—two velocity vectors or two force vectors—can be added in exactly the same way.

The graphical method for adding vectors is straightforward, but we need to do a little geometry to come up with a complete description of the resultant vector $\vec{C}$. Vector $\vec{C}$ of Figure 3.4 is defined by its magnitude C and by its direction. Because the three vectors $\vec{A}$, $\vec{B}$, and $\vec{C}$ form a right triangle, the magnitude, or length, of $\vec{C}$ is given by the Pythagorean theorem:

$$C = \sqrt{A^2 + B^2} = \sqrt{(4 \text{ mi})^2 + (3 \text{ mi})^2} = 5 \text{ mi} \tag{3.2}$$

Notice that Equation 3.2 uses the magnitudes A and B of the vectors $\vec{A}$ and $\vec{B}$. The angle θ, which is used in Figure 3.4 to describe the direction of $\vec{C}$, is easily found for a right triangle:

$$\theta = \tan^{-1}\left(\frac{B}{A}\right) = \tan^{-1}\left(\frac{3 \text{ mi}}{4 \text{ mi}}\right) = 37° \tag{3.3}$$

Altogether, the hiker's net displacement is

$$\vec{C} = \vec{A} + \vec{B} = (5 \text{ mi}, 37° \text{ north of east}) \tag{3.4}$$

NOTE ▶ Vector mathematics makes extensive use of geometry and trigonometry. Appendix A, at the end of this book, contains a brief review of these topics. ◀

EXAMPLE 3.1 Using graphical addition to find a displacement

A bird flies 100 m due east from a tree, then 200 m northwest (that is, 45° north of west). What is the bird's net displacement?

VISUALIZE Figure 3.5 shows the two individual displacements, which we've called $\vec{A}$ and $\vec{B}$. The net displacement is the vector sum $\vec{C} = \vec{A} + \vec{B}$, which is found graphically.

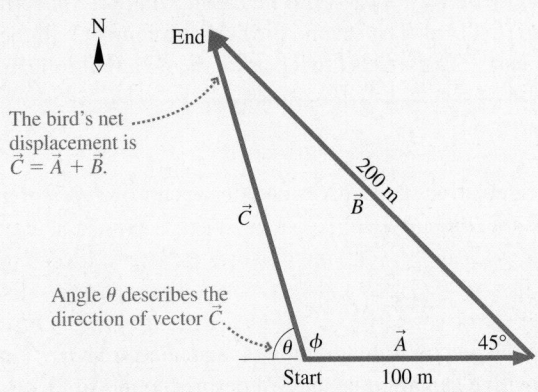

FIGURE 3.5 The bird's net displacement is $\vec{C} = \vec{A} + \vec{B}$.

SOLVE The two displacements are $\vec{A} = (100 \text{ m}, \text{east})$ and $\vec{B} = (200 \text{ m}, \text{northwest})$. The net displacement $\vec{C} = \vec{A} + \vec{B}$ is found by drawing a vector from the initial to the final position. But describing $\vec{C}$ is a bit trickier than the example of the hiker because $\vec{A}$ and $\vec{B}$ are not at right angles. First, we can find the magnitude of $\vec{C}$ by using the law of cosines from trigonometry:

$$C^2 = A^2 + B^2 - 2AB\cos(45°)$$
$$= (100 \text{ m})^2 + (200 \text{ m})^2 - 2(100 \text{ m})(200 \text{ m})\cos(45°)$$
$$= 21{,}720 \text{ m}^2$$

Thus $C = \sqrt{21{,}720 \text{ m}^2} = 147 \text{ m}$. Then a second use of the law of cosines can determine angle ϕ (the Greek letter phi):

$$B^2 = A^2 + C^2 - 2AC\cos\phi$$

$$\phi = \cos^{-1}\left[\frac{A^2 + C^2 - B^2}{2AC}\right] = 106°$$

It is easier to describe $\vec{C}$ with the angle $\theta = 180° - \phi = 74°$. The bird's net displacement is

$$\vec{C} = (147 \text{ m}, 74° \text{ north of west})$$

When two vectors are to be added, it is often convenient to draw them with their tails together, as shown in Figure 3.6a. To evaluate $\vec{D} + \vec{E}$, you could move vector $\vec{E}$ over to where its tail is on the tip of $\vec{D}$, then use the tip-to-tail rule of graphical addition. The gives vector $\vec{F} = \vec{D} + \vec{E}$ in Figure 3.6b. Alternatively, Figure 3.6c shows that the vector sum $\vec{D} + \vec{E}$ can be found as the diagonal of the parallelogram defined by $\vec{D}$ and $\vec{E}$. This method for vector addition, which some of you may have learned, is called the *parallelogram rule* of vector addition.

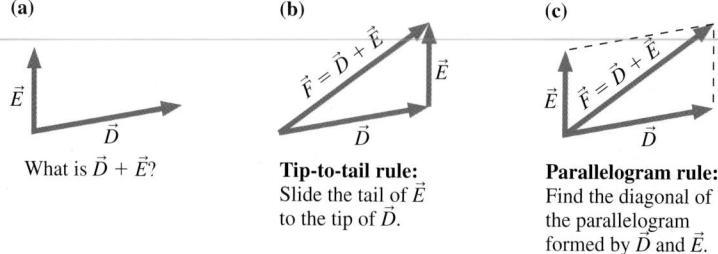

FIGURE 3.6 Two vectors can be added using the tip-to-tail rule or the parallelogram rule.

Vector addition is easily extended to more than two vectors. Figure 3.7 shows a hiker moving from initial position 0 to position 1, then position 2, then position 3, and finally arriving at position 4. These four segments are described by displacement vectors $\vec{D}_1$, $\vec{D}_2$, $\vec{D}_3$, and $\vec{D}_4$. The hiker's *net* displacement, an arrow from position 0 to position 4, is the vector $\vec{D}_{net}$. In this case,

$$\vec{D}_{net} = \vec{D}_1 + \vec{D}_2 + \vec{D}_3 + \vec{D}_4 \tag{3.5}$$

The vector sum is found by using the tip-to-tail method three times in succession.

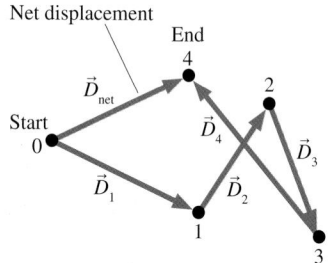

FIGURE 3.7 The net displacement after four individual displacements.

STOP TO THINK 3.1 Which figure shows $\vec{A}_1 + \vec{A}_2 + \vec{A}_3$?

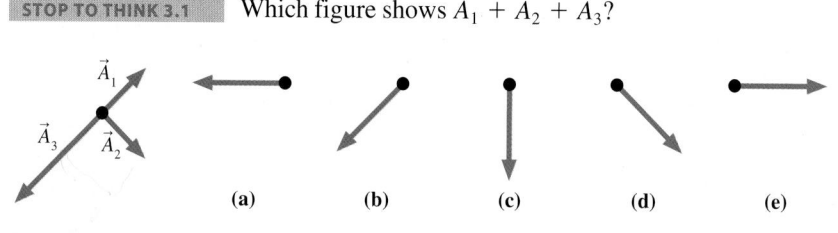

Multiplication by a Scalar

Suppose a second bird flies twice as far to the east as the bird in Example 3.1. The first bird's displacement was $\vec{A}_1 = (100 \text{ m, east})$, where a subscript has been added to denote the first bird. The second bird's displacement will then certainly be $\vec{A}_2 = (200 \text{ m, east})$. The words "twice as" indicate a multiplication, so we can say

$$\vec{A}_2 = 2\vec{A}_1$$

Multiplying a vector by a positive scalar gives another vector of *different magnitude* but pointing in the *same direction*.

Let the vector $\vec{A}$ be

$$\vec{A} = (A, \theta_A) \tag{3.6}$$

where we've specified the vector's magnitude A and direction θ_A. Now let $\vec{B} = c\vec{A}$, where c is a positive scalar constant. We define the multiplication of a vector by a scalar such that

$$\vec{B} = c\vec{A} \text{ means that } (B, \theta_B) = (cA, \theta_A) \tag{3.7}$$

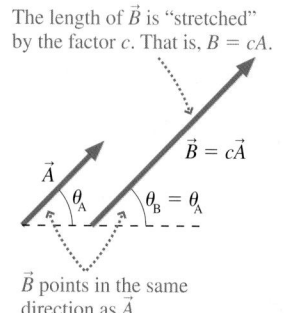

The length of $\vec{B}$ is "stretched" by the factor c. That is, $B = cA$.

$\vec{B}$ points in the same direction as $\vec{A}$.

FIGURE 3.8 Multiplication of a vector by a scalar.

In other words, the vector is stretched or compressed by the factor c (i.e., vector $\vec{B}$ has magnitude $B = cA$), but $\vec{B}$ points in the same direction as $\vec{A}$. This is illustrated in Figure 3.8 on the previous page.

We used this property of vectors in Chapter 1 when we asserted that vector $\vec{a}$ points in the same direction as $\Delta\vec{v}$. From the definition

$$\vec{a} = \frac{\Delta\vec{v}}{\Delta t} = \left(\frac{1}{\Delta t}\right)\Delta\vec{v} \tag{3.8}$$

where $(1/\Delta t)$ is a scalar constant, we see that $\vec{a}$ points in the same direction as $\Delta\vec{v}$ but differs in length by the factor $(1/\Delta t)$.

Suppose we multiply $\vec{A}$ by zero. Using Equation 3.7,

$$0 \cdot \vec{A} = \vec{0} = (0 \text{ m, direction undefined}) \tag{3.9}$$

The product is a vector having zero length or magnitude. This vector is known as the **zero vector,** denoted $\vec{0}$. The direction of the zero vector is irrelevant; you cannot describe the direction of an arrow of zero length!

What happens if we multiply a vector by a negative number? Equation 3.7 does not apply if $c < 0$ because vector $\vec{B}$ cannot have a negative magnitude. Consider the vector $-\vec{A}$, which is equivalent to multiplying $\vec{A}$ by -1. Because

$$\vec{A} + (-\vec{A}) = \vec{0} \tag{3.10}$$

the vector $-\vec{A}$ must be such that, when it is added to $\vec{A}$, the resultant is the zero vector $\vec{0}$. In other words, the *tip* of $-\vec{A}$ must return to the *tail* of $\vec{A}$, as shown in Figure 3.9. This will be true only if $-\vec{A}$ is equal in magnitude to $\vec{A}$, but opposite in direction. Thus we can conclude that

$$-\vec{A} = (A, \text{ direction opposite } \vec{A}) \tag{3.11}$$

That is, **multiplying a vector by -1 reverses its direction without changing its length.**

As an example, Figure 3.10 shows vectors $\vec{A}$, $2\vec{A}$, and $-3\vec{A}$. Multiplication by 2 doubles the length of the vector but does not change its direction. Multiplication by -3 stretches the length by a factor of 3 *and* reverses the direction.

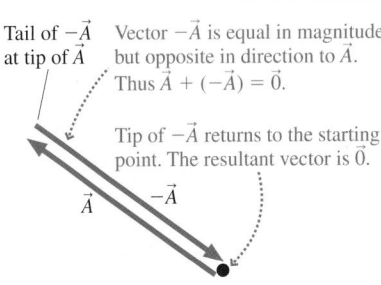

Tail of $-\vec{A}$ at tip of $\vec{A}$

Vector $-\vec{A}$ is equal in magnitude but opposite in direction to $\vec{A}$. Thus $\vec{A} + (-\vec{A}) = \vec{0}$.

Tip of $-\vec{A}$ returns to the starting point. The resultant vector is $\vec{0}$.

FIGURE 3.9 Vector $-\vec{A}$.

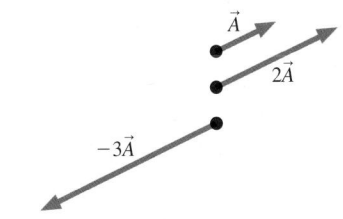

FIGURE 3.10 Vectors $\vec{A}$, $2\vec{A}$, and $-3\vec{A}$.

EXAMPLE 3.2 Velocity and displacement
Carolyn drives her car north at 30 km/hr for 1 hour, east at 60 km/hr for 2 hours, then north at 50 km/hr for 1 hour. What is Carolyn's net displacement?

SOLVE Chapter 1 defined velocity as

$$\vec{v} = \frac{\Delta\vec{r}}{\Delta t}$$

so the displacement $\Delta\vec{r}$ during the time interval Δt is $\Delta\vec{r} = (\Delta t)\vec{v}$. This is multiplication of the vector $\vec{v}$ by the scalar Δt. Carolyn's velocity during the first hour is $\vec{v}_1 = (30 \text{ km/hr, north})$, so her displacement during this interval is

$$\Delta\vec{r}_1 = (1 \text{ hour})(30 \text{ km/hr, north}) = (30 \text{ km, north})$$

Similarly,

$$\Delta\vec{r}_2 = (2 \text{ hours})(60 \text{ km/hr, east}) = (120 \text{ km, east})$$

$$\Delta\vec{r}_3 = (1 \text{ hour})(50 \text{ km/hr, north}) = (50 \text{ km, north})$$

In this case, multiplication by a scalar changes not only the length of the vector but also its units, from km/hr to km. The direction, however, is unchanged. Carolyn's net displacement is

$$\Delta\vec{r}_{net} = \Delta\vec{r}_1 + \Delta\vec{r}_2 + \Delta\vec{r}_3$$

This addition of the three vectors is shown in Figure 3.11, using the tip-to-tail method. $\Delta\vec{r}_{net}$ stretches from Carolyn's initial position to her final position. The magnitude of her net displacement is found using the Pythagorean theorem:

$$r_{net} = \sqrt{(120 \text{ km})^2 + (80 \text{ km})^2} = 144 \text{ km}$$

The direction of $\Delta\vec{r}_{net}$ is described by angle θ, which is

$$\theta = \tan^{-1}\left(\frac{80 \text{ km}}{120 \text{ km}}\right) = 33.7°$$

Thus Carolyn's net displacement is $\Delta\vec{r}_{net} = (144 \text{ km}, 33.7°$ north of east).

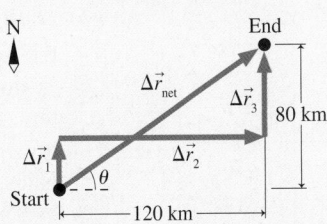

FIGURE 3.11 The net displacement is the vector sum $\Delta\vec{r}_{net} = \Delta\vec{r}_1 + \Delta\vec{r}_2 + \Delta\vec{r}_3$.

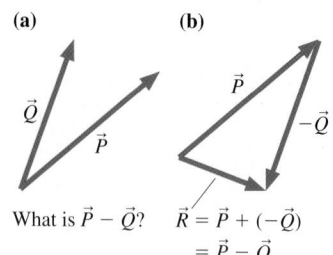

(a)   **(b)**

$\vec{Q}$

$\vec{P}$

$\vec{P}$

$-\vec{Q}$

What is $\vec{P} - \vec{Q}$? $\vec{R} = \vec{P} + (-\vec{Q})$
$= \vec{P} - \vec{Q}$

FIGURE 3.12 Vector subtraction.

Vector Subtraction

Figure 3.12a shows two vectors, $\vec{P}$ and $\vec{Q}$. What is $\vec{R} = \vec{P} - \vec{Q}$? Look back at Tactics Box 1.2 on page 11, which showed how to perform vector subtraction graphically. Figure 3.12b finds $\vec{P} - \vec{Q}$ by writing $\vec{R} = \vec{P} + (-\vec{Q})$, then using the rules of vector addition.

STOP TO THINK 3.2   Which figure shows $2\vec{A} - \vec{B}$?

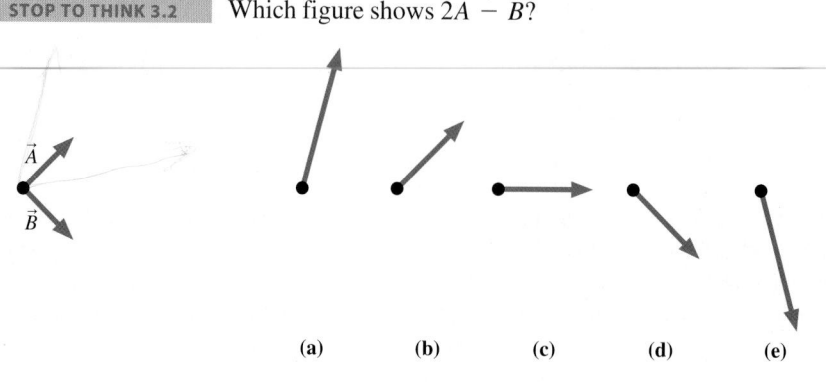

$\vec{A}$

$\vec{B}$

(a)   **(b)**   **(c)**   **(d)**   **(e)**

3.3 Coordinate Systems and Vector Components

Thus far, our discussion of vectors and their properties has not used a coordinate system at all. Vectors do not require a coordinate system. We can add and subtract vectors graphically, and we will do so frequently to clarify our understanding of a situation. But the graphical addition of vectors is not an especially good way to find quantitative results. In this section we will introduce a *coordinate description* of vectors that will be the basis of an easier method for doing vector calculations.

Coordinate Systems

As we noted in the first chapter, the world does not come with a coordinate system attached to it. A coordinate system is an artificially imposed grid that you place on a problem in order to make quantitative measurements. It may be helpful to think of drawing a grid on a piece of transparent plastic that you can then overlay on top of the problem. This conveys the idea that *you* choose:

- Where to place the origin, and
- How to orient the axes.

Different problem solvers may choose to use different coordinate systems; that is perfectly acceptable. However, some coordinate systems will make a problem easier to solve. Part of our goal is to learn how to choose an appropriate coordinate system for each problem.

We will generally use **Cartesian coordinates.** This is a coordinate system with the axes perpendicular to each other, forming a rectangular grid. The standard *xy*-coordinate system with which you are familiar is a Cartesian coordinate system. An *xyz*-coordinate system would be a Cartesian coordinate system in three dimensions. There are other possible coordinate systems, such as polar coordinates, but we will not be concerned with those for now.

The placement of the axes is not entirely arbitrary. By convention, the positive *y*-axis is located 90° *counterclockwise* (ccw) from the positive *x*-axis, as illustrated in Figure 3.13. Figure 3.13 also identifies the four **quadrants** of the coordinate system, I through IV. Notice that the quadrants are counted ccw from the positive *x*-axis.

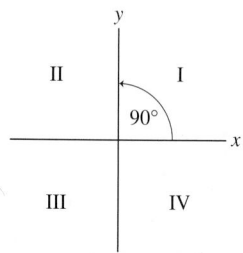

FIGURE 3.13 A conventional Cartesian coordinate system and the quadrants of the *xy*-plane.

Coordinate axes have a positive end and a negative end, separated by zero at the origin where the two axes cross. When you draw a coordinate system, it is important to label the axes. This is done by placing x and y labels at the *positive* ends of the axes, as in Figure 3.13. The purpose of the labels is twofold:

- To identify which axis is which, and
- To identify the positive ends of the axes.

This will be important when you need to determine whether the quantities in a problem should be assigned positive or negative values.

Component Vectors

Let's see how we can use a coordinate system to describe a vector. Figure 3.14 shows a vector $\vec{A}$ and an xy-coordinate system that we've chosen. Once the directions of the axes are known, we can define two new vectors *parallel to the axes* that we call the **component vectors** of $\vec{A}$. Vector $\vec{A}_x$, called the *x-component vector,* is the projection of $\vec{A}$ along the x-axis. Vector $\vec{A}_y$, the *y-component vector*, is the projection of $\vec{A}$ along the y-axis. Notice that the component vectors are perpendicular to each other.

You can see, using the parallelogram rule, that $\vec{A}$ is the vector sum of the two component vectors:

$$\vec{A} = \vec{A}_x + \vec{A}_y \tag{3.12}$$

In essence, we have broken vector $\vec{A}$ into two perpendicular vectors that are parallel to the coordinate axes. This process is called the **decomposition** of vector $\vec{A}$ into its component vectors.

> NOTE ▶ It is not necessary for the tail of $\vec{A}$ to be at the origin. All we need to know is the *orientation* of the coordinate system so that we can draw $\vec{A}_x$ and $\vec{A}_y$ parallel to the axes. ◄

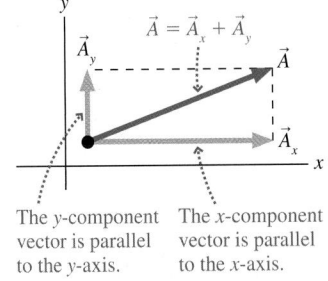

The *y*-component vector is parallel to the *y*-axis. The *x*-component vector is parallel to the *x*-axis.

FIGURE 3.14 Component vectors $\vec{A}_x$ and $\vec{A}_y$ are drawn parallel to the coordinate axes such that $\vec{A} = \vec{A}_x + \vec{A}_y$.

Components

You learned in Chapter 2 to give the one-dimensional kinematic variable v_x a positive sign if the velocity vector $\vec{v}$ points toward the positive end of the x-axis, a negative sign if $\vec{v}$ points in the negative x-direction. The basis of that rule is that v_x is what we call the *x-component* of the velocity vector. We need to extend this idea to vectors in general.

Suppose vector $\vec{A}$ has been decomposed into component vectors $\vec{A}_x$ and $\vec{A}_y$ parallel to the coordinate axes. We can describe each component vector with a single number (a scalar) called the **component.** The *x-component* and *y-component* of vector $\vec{A}$, denoted A_x and A_y, are determined as follows:

TACTICS BOX 3.1 **Determining the components of a vector**

❶ The absolute value $|A_x|$ of the x-component A_x is the magnitude of the component vector $\vec{A}_x$.
❷ The *sign* of A_x is positive if $\vec{A}_x$ points in the positive x-direction, negative if $\vec{A}_x$ points in the negative x-direction.
❸ The *y*-component A_y is determined similarly.

In other words, the component A_x tells us two things: how big $\vec{A}_x$ is and, with its sign, which end of the axis $\vec{A}_x$ points toward. Figure 3.15 on the next page shows three examples of determining the components of a vector.

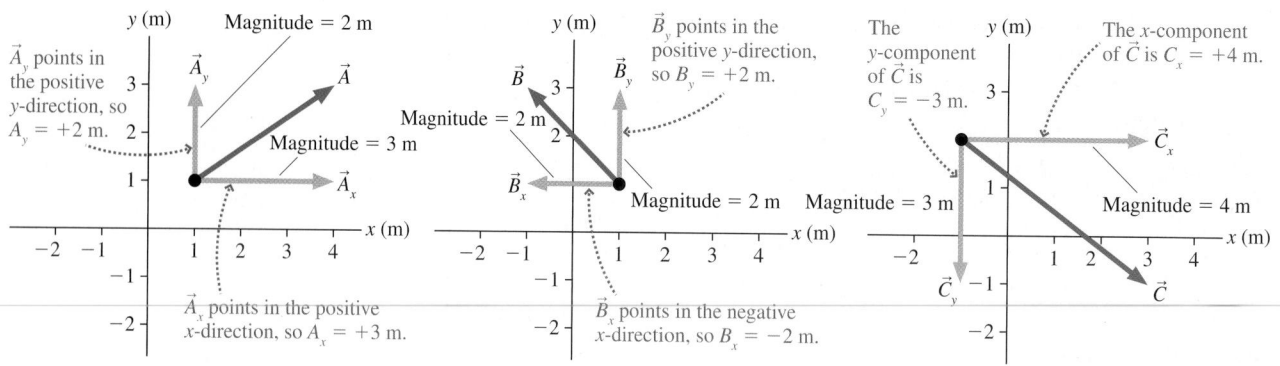

FIGURE 3.15 Determining the components of a vector.

NOTE ▶ Beware of the somewhat confusing terminology. $\vec{A}_x$ and $\vec{A}_y$ are called *component vectors,* whereas A_x and A_y are simply called *components.* The components A_x and A_y are scalars—just numbers (with units)—so make sure you do *not* put arrow symbols over the components. ◀

Much of physics is expressed in the language of vectors. We will frequently need to decompose a vector into its components. We will also need to "reassemble" a vector from its components. In other words, we need to move back and forth between the graphical and the component representations of a vector. To do so we apply geometry and trigonometry.

Consider first the problem of decomposing a vector into its x- and y-components. Figure 3.16a shows a vector $\vec{A}$ at angle θ from the x-axis. It is *essential* to use a picture or diagram such as this to define the angle you are using to describe the vector's direction.

$\vec{A}$ points to the right and up, so Tactics Box 3.1 tells us that the components A_x and A_y are both positive. We can use trigonometry to find

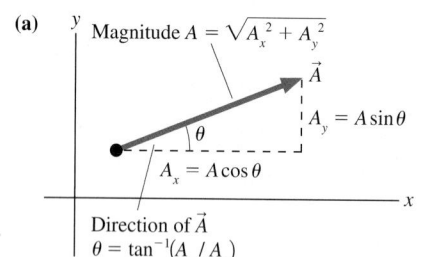

$$A_x = A\cos\theta$$
$$A_y = A\sin\theta \tag{3.13}$$

where A is the magnitude, or length, of $\vec{A}$. These equations convert the length and angle description of vector $\vec{A}$ into the vector's components, but they are correct *only* if $\vec{A}$ is in the first quadrant.

Figure 3.16b shows vector $\vec{C}$ in the fourth quadrant. In this case, where the component vector $\vec{A}_y$ is pointing *down,* in the negative y-direction, the y-component C_y is a *negative* number. The angle ϕ is measured from the y-axis, so the components of $\vec{C}$ are

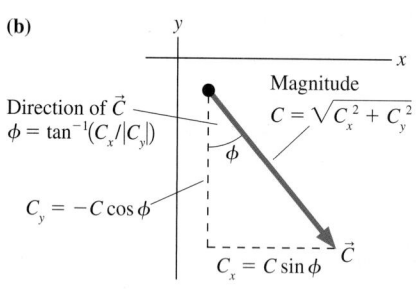

$$C_x = C\sin\phi$$
$$C_y = -C\cos\phi \tag{3.14}$$

FIGURE 3.16 Moving between the graphical representation and the component representation.

The role of sine and cosine is reversed from that in Equations 3.13 because we are using a different angle.

NOTE ▶ Each decomposition requires that you pay close attention to the direction in which the vector points and the angles that are defined. The minus sign, when needed, must be inserted manually. ◀

We can also go in the opposite direction and determine the length and angle of a vector from its x- and y-components. Because A in Figure 3.16a is the hypotenuse of a right triangle, its length is given by the Pythagorean theorem:

$$A = \sqrt{A_x^2 + A_y^2} \tag{3.15}$$

Similarly, the tangent of angle θ is the ratio of the far side to the adjacent side, so

$$\theta = \tan^{-1}\left(\frac{A_y}{A_x}\right) \tag{3.16}$$

where $\tan^{-1}$ is the inverse tangent function. Equations 3.15 and 3.16 can be thought of as the "reverse" of Equations 3.13.

Equation 3.15 always works for finding the length or magnitude of a vector because the squares eliminate any concerns over the signs of the components. But finding the angle, just like finding the components, requires close attention to how the angle is defined and to the signs of the components. For example, finding the angle of vector $\vec{C}$ in Figure 3.16b requires the length of C_y *without* the minus sign. Thus vector $\vec{C}$ has magnitude and direction

$$C = \sqrt{C_x^2 + C_y^2}$$

$$\phi = \tan^{-1}\left(\frac{C_x}{|C_y|}\right) \tag{3.17}$$

Notice that the roles of x and y differ from those in Equation 3.16.

EXAMPLE 3.3 Finding the components of an acceleration vector

Find the x- and y-components of the acceleration vector $\vec{a}$ shown in Figure 3.17a.

VISUALIZE It's important to *draw* vectors. Figure 3.17b shows the original vector $\vec{a}$ decomposed into components parallel to the axes.

SOLVE The acceleration vector $\vec{a} = (6\ \text{m/s}^2,\ 30°$ below the negative x-axis) points to the left (negative x-direction) and down (negative y-direction), so the components a_x and a_y are both negative:

$$a_x = -a\cos 30° = -(6\ \text{m/s}^2)\cos 30° = -5.2\ \text{m/s}^2$$

$$a_y = -a\sin 30° = -(6\ \text{m/s}^2)\sin 30° = -3.0\ \text{m/s}^2$$

ASSESS The units of a_x and a_y are the same as the units of vector $\vec{a}$. Notice that we had to insert the minus signs manually by observing that the vector is in the third quadrant.

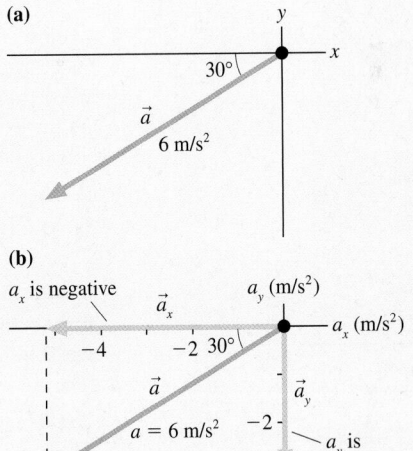

FIGURE 3.17 The acceleration vector $\vec{a}$ of Example 3.3.

EXAMPLE 3.4 Finding the direction of motion

Figure 3.18a shows a particle's velocity vector $\vec{v}$. Determine the particle's speed and direction of motion.

VISUALIZE Figure 3.18b shows the components v_x and v_y and defines an angle θ with which we can specify the direction of motion.

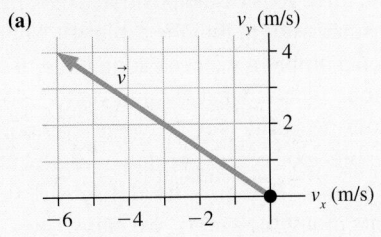

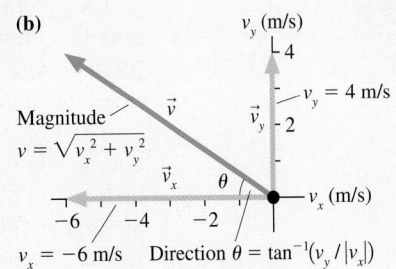

FIGURE 3.18 The velocity vector $\vec{v}$ of Example 3.4.

SOLVE We can read the components of $\vec{v}$ directly from the axes: $v_x = -6$ m/s and $v_y = 4$ m/s. Notice that v_x is negative. This is enough information to find the particle's speed v, which is the magnitude of $\vec{v}$:

$$v = \sqrt{v_x^2 + v_y^2} = \sqrt{(-6 \text{ m/s})^2 + (4 \text{ m/s})^2} = 7.2 \text{ m/s}$$

From trigonometry, angle θ is

$$\theta = \tan^{-1}\left(\frac{v_y}{|v_x|}\right) = \tan^{-1}\left(\frac{4 \text{ m/s}}{6 \text{ m/s}}\right) = 33.7°$$

The absolute value signs are necessary because v_x is a negative number. The velocity vector $\vec{v}$ can be written in terms of the speed and the direction of motion as

$$\vec{v} = (7.2 \text{ m/s}, 33.7° \text{ above the negative } x\text{-axis})$$

STOP TO THINK 3.3   What are the x- and y-components C_x and C_y of vector $\vec{C}$?

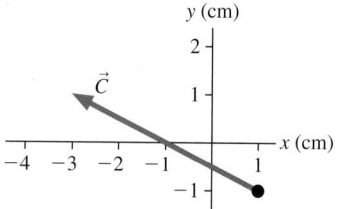

3.4 Vector Algebra

Vector components are a powerful tool for doing mathematics with vectors. In this section you'll learn how to use components to add and subtract vectors. First, we'll introduce an efficient way to write a vector in terms of its components.

Unit Vectors

The vectors (1, $+x$-direction) and (1, $+y$-direction), shown in Figure 3.19, have some interesting and useful properties. Each has a magnitude of 1, no units, and is parallel to a coordinate axis. A vector with these properties is called a **unit vector.** These unit vectors have the special symbols

$$\hat{\imath} \equiv (1, +x\text{-direction})$$

$$\hat{\jmath} \equiv (1, +y\text{-direction})$$

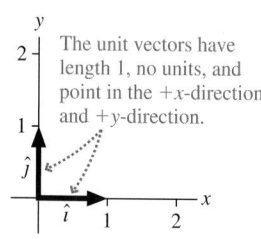

The unit vectors have length 1, no units, and point in the +x-direction and +y-direction.

FIGURE 3.19 The unit vectors $\hat{\imath}$ and $\hat{\jmath}$.

The notation $\hat{\imath}$ (read "i hat") and $\hat{\jmath}$ (read "j hat") indicates a unit vector with a magnitude of 1.

Unit vectors establish the directions of the positive axes of the coordinate system. Our choice of a coordinate system may be arbitrary, but once we decide to place a coordinate system on a problem we need something to tell us "That direction is the positive x-direction." This is what the unit vectors do.

The unit vectors provide a useful way to write component vectors. The component vector $\vec{A}_x$ is the piece of vector $\vec{A}$ that is parallel to the x-axis. Similarly, $\vec{A}_y$ is parallel to the y-axis. Because, by definition, the vector $\hat{\imath}$ points along the x-axis and $\hat{\jmath}$ points along the y-axis, we can write

$$\vec{A}_x = A_x\hat{\imath}$$

$$\vec{A}_y = A_y\hat{\jmath}$$

(3.18)

Equations 3.18 separate each component vector into a scalar piece of length A_x (or A_y) and a directional piece $\hat{\imath}$ (or $\hat{\jmath}$). The full decomposition of vector $\vec{A}$ can then be written

$$\vec{A} = \vec{A}_x + \vec{A}_y = A_x\hat{\imath} + A_y\hat{\jmath} \tag{3.19}$$

Figure 3.20 shows how the unit vectors and the components fit together to form vector $\vec{A}$.

NOTE ▶ In three dimensions, the unit vector along the $+z$-direction is called $\hat{k}$, and to describe vector $\vec{A}$ we would include an additional component vector $\vec{A}_z = A_z\hat{k}$. ◀

You may have learned in a math class to think of vectors as pairs or triplets of numbers, such as $(4, -2, 5)$. This is another, and completely equivalent, way to write the components of a vector. Thus we could write, for a vector in three dimensions,

$$\vec{B} = 4\hat{\imath} - 2\hat{\jmath} + 5\hat{k} = (4,-2,5)$$

You will find the notation using unit vectors to be more convenient for the equations we will use in physics, but rest assured that you already know a lot about vectors if you learned about them as pairs or triplets of numbers.

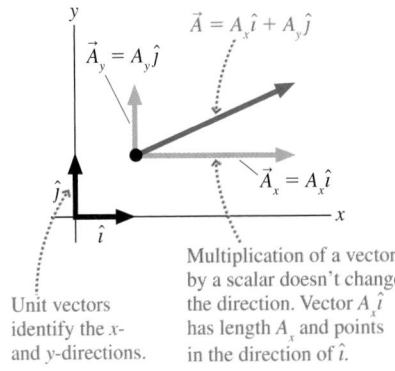

Multiplication of a vector by a scalar doesn't change the direction. Vector $A_x\hat{\imath}$ has length A_x and points in the direction of $\hat{\imath}$.

Unit vectors identify the x- and y-directions.

FIGURE 3.20 The decomposition of vector $\vec{A}$ is $A_x\hat{\imath} + A_y\hat{\jmath}$.

EXAMPLE 3.5 Run rabbit run!

A rabbit, escaping a fox, runs 40° north of west at 10 m/s. A coordinate system is established with the positive x-axis to the east and the positive y-axis to the north. Write the rabbit's velocity in terms of components and unit vectors.

VISUALIZE Figure 3.21 shows the rabbit's velocity vector and the coordinate axes. We're showing a velocity vector, so the axes are labeled v_x and v_y rather than x and y.

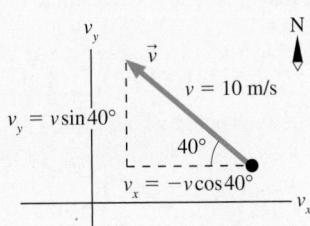

FIGURE 3.21 The velocity vector $\vec{v}$ is decomposed into components v_x and v_y.

SOLVE 10 m/s is the rabbit's *speed*, not its velocity. The velocity, which includes directional information, is

$$\vec{v} = (10 \text{ m/s}, 40° \text{ north of west})$$

Vector $\vec{v}$ points to the left and up, so the components v_x and v_y are negative and positive, respectively. The components are

$$v_x = -(10 \text{ m/s})\cos 40° = -7.66 \text{ m/s}$$

$$v_y = +(10 \text{ m/s})\sin 40° = 6.43 \text{ m/s}$$

With v_x and v_y now known, the rabbit's velocity vector is

$$\vec{v} = v_x\hat{\imath} + v_y\hat{\jmath} = (-7.66\hat{\imath} + 6.43\hat{\jmath}) \text{ m/s}$$

Notice that we've pulled the units to the end, rather than writing them with each component.

ASSESS Notice that the minus sign for v_x was inserted manually. Signs don't occur automatically; you have to set them after checking the vector's direction.

Working with Vectors

You learned in Section 3.2 how to add vectors graphically, but it is a tedious problem in geometry and trigonometry to find precise values for the magnitude and direction of the resultant. The addition and subtraction of vectors becomes much easier if we use components and unit vectors.

To see this, let's evaluate the vector sum $\vec{D} = \vec{A} + \vec{B} + \vec{C}$. To begin, write this sum in terms of the components of each vector:

$$\vec{D} = D_x\hat{\imath} + D_y\hat{\jmath} = \vec{A} + \vec{B} + \vec{C} = (A_x\hat{\imath} + A_y\hat{\jmath}) + (B_x\hat{\imath} + B_y\hat{\jmath}) + (C_x\hat{\imath} + C_y\hat{\jmath})$$

$$\tag{3.20}$$

We can group together all the x-components and all the y-components on the right side, in which case Equation 3.20 is

$$(D_x)\hat{\imath} + (D_y)\hat{\jmath} = (A_x + B_x + C_x)\hat{\imath} + (A_y + B_y + C_y)\hat{\jmath} \qquad (3.21)$$

Comparing the x- and y-components on the left and right sides of Equation 3.21, we find:

$$D_x = A_x + B_x + C_x$$
$$D_y = A_y + B_y + C_y \qquad (3.22)$$

Stated in words, Equation 3.22 says that we can perform vector addition by adding the x-components of the individual vectors to give the x-component of the resultant and by adding the y-components of the individual vectors to give the y-component of the resultant. This method of vector addition is called **algebraic addition.**

EXAMPLE 3.6 **Using algebraic addition to find a displacement**

Example 3.1 was about a bird that flew 100 m to the east, then 200 m to the northwest. Use the algebraic addition of vectors to find the bird's net displacement. Compare the result to Example 3.1.

VISUALIZE Figure 3.22 shows displacement vectors $\vec{A}$ = (100 m, east) and $\vec{B}$ = (200 m, northwest). We draw vectors tip-to-tail if we are going to add them graphically, but it's usually easier to draw them all from the origin if we are going to use algebraic addition.

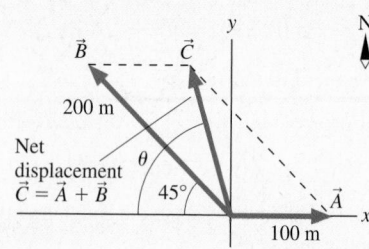

FIGURE 3.22 The net displacement is $\vec{C} = \vec{A} + \vec{B}$.

SOLVE To add the vectors algebraically we must know their components. From the figure these are seen to be

$$\vec{A} = 100\,\hat{\imath}\,\text{m}$$
$$\vec{B} = (-200 \cos 45°\,\hat{\imath} + 200 \sin 45°\,\hat{\jmath})\,\text{m}$$
$$= (-141\hat{\imath} + 141\hat{\jmath})\,\text{m}$$

Notice that vector quantities must include units. Also notice, as you would expect from the figure, that $\vec{B}$ has a negative x-component. Adding $\vec{A}$ and $\vec{B}$ by components gives

$$\vec{C} = \vec{A} + \vec{B} = 100\hat{\imath}\,\text{m} + (-141\hat{\imath} + 141\hat{\jmath})\,\text{m}$$
$$= (100\,\text{m} - 141\,\text{m})\hat{\imath} + (141\,\text{m})\hat{\jmath}$$
$$= (-41\hat{\imath} + 141\hat{\jmath})\,\text{m}$$

This would be a perfectly acceptable answer for many purposes. However, we need to calculate the magnitude and direction of $\vec{C}$ if we want to compare this result to our earlier answer. The magnitude of $\vec{C}$ is

$$C = \sqrt{C_x^2 + C_y^2} = \sqrt{(-41\,\text{m})^2 + (141\,\text{m})^2} = 147\,\text{m}$$

The angle θ, as defined in Figure 3.22, is

$$\theta = \tan^{-1}\left(\frac{C_y}{|C_x|}\right) = \tan^{-1}\left(\frac{141\,\text{m}}{41\,\text{m}}\right) = 74°$$

Thus $\vec{C}$ = (147 m, 74° north of west), in perfect agreement with Example 3.1.

Vector subtraction and the multiplication of a vector by a scalar, using components, are very much like vector addition. To find $\vec{R} = \vec{P} - \vec{Q}$ we would compute

$$R_x = P_x - Q_x$$
$$R_y = P_y - Q_y \qquad (3.23)$$

Similarly, $\vec{T} = c\vec{S}$ would be

$$T_x = cS_x$$
$$T_y = cS_y \qquad (3.24)$$

The next few chapters will make frequent use of *vector equations*. For example, you will learn that the equation to calculate the force on a car skidding to a stop is

$$\vec{F} = \vec{n} + \vec{w} + \mu\vec{f} \tag{3.25}$$

The following general rule is used to evaluate such an equation:

The *x*-component of the left-hand side of a vector equation is found by doing scalar calculations (addition, subtraction, multiplication) with just the *x*-components of all the vectors on the right-hand side. A separate set of calculations uses just the *y*-components and, if needed, the *z*-components.

Thus Equation 3.25 is really just a shorthand way of writing three simultaneous equations:

$$\begin{aligned}
F_x &= n_x + w_x + \mu f_x \\
F_y &= n_y + w_y + \mu f_y \\
F_z &= n_z + w_z + \mu f_z
\end{aligned} \tag{3.26}$$

In other words, a vector equation is interpreted as meaning: Equate the *x*-components on both sides of the equals sign, then equate the *y*-components, and then the *z*-components. Vector notation allows us to write these three equations in a much more compact form.

Tilted Axes and Arbitrary Directions

As we've noted, the coordinate system is entirely your choice. It is a grid that you impose on the problem in a manner that will make the problem easiest to solve. We will soon meet problems where it will be convenient to tilt the axes of the coordinate system, such as those shown in Figure 3.23. Although you may not have seen such a coordinate system before, it is perfectly legitimate. The axes are perpendicular, and the *y*-axis is oriented correctly with respect to the *x*-axis. While we are used to having the *x*-axis horizontal, there is no requirement that it has to be that way.

Finding components with tilted axes is no harder than what we have done so far. Vector $\vec{C}$ in Figure 3.23 can be decomposed $\vec{C} = C_x\hat{i} + C_y\hat{j}$, where $C_x = C\cos\theta$ and $C_y = C\sin\theta$. Note that the unit vectors $\hat{i}$ and $\hat{j}$ correspond to the *axes*, not to "horizontal" and "vertical," so they are also tilted.

Tilted axes are useful if you need to determine component vectors "parallel to" and "perpendicular to" an arbitrary line or surface. For example, we will soon need to decompose a force vector into component vectors parallel to and perpendicular to a surface.

Figure 3.24a shows a vector $\vec{A}$ and a tilted line. Suppose we would like to find the component vectors of $\vec{A}$ parallel and perpendicular to the line. To do so, establish a tilted coordinate system with the *x*-axis parallel to the line and the *y*-axis perpendicular to the line, as shown in Figure 3.24b. Then $\vec{A}_x$ is equivalent to vector $\vec{A}_{\parallel}$, the component of $\vec{A}$ parallel to the line, and $\vec{A}_y$ is equivalent to the perpendicular component vector $\vec{A}_{\perp}$. Notice that $\vec{A} = \vec{A}_{\parallel} + \vec{A}_{\perp}$.

If ϕ is the angle between $\vec{A}$ and the line, we can easily calculate the parallel and perpendicular components of $\vec{A}$:

$$\begin{aligned}
A_{\parallel} &= A_x = A\cos\phi \\
A_{\perp} &= A_y = A\sin\phi
\end{aligned} \tag{3.27}$$

It was not necessary to have the tail of $\vec{A}$ on the line in order to find a component of $\vec{A}$ parallel to the line. The line simply indicates a direction, and the component vector $\vec{A}_{\parallel}$ points in that direction.

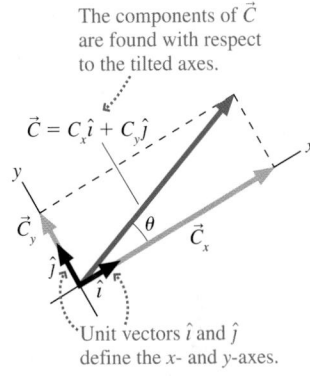

The components of $\vec{C}$ are found with respect to the tilted axes.

$\vec{C} = C_x\hat{i} + C_y\hat{j}$

Unit vectors $\hat{i}$ and $\hat{j}$ define the *x*- and *y*-axes.

FIGURE 3.23 A coordinate system with tilted axes.

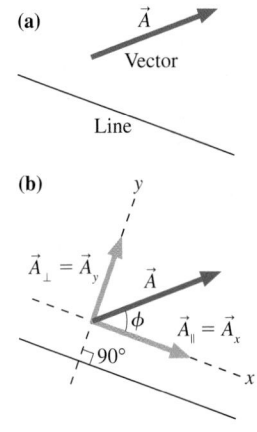

(a) $\vec{A}$ — Vector — Line

(b) $\vec{A}_{\perp} = \vec{A}_y$ $\vec{A}$ ϕ $\vec{A}_{\parallel} = \vec{A}_x$ 90°

FIGURE 3.24 Finding the components of $\vec{A}$ parallel to and perpendicular to the line.

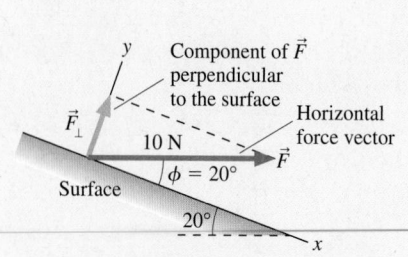

FIGURE 3.25 Finding the component of a force vector perpendicular to a surface.

EXAMPLE 3.7 **Finding the force perpendicular to a surface**

A horizontal force $\vec{F}$ with a strength of 10 N is applied to a surface. (You'll learn in Chapter 4 that force is a vector quantity measured in units of *newtons,* abbreviated N.) The surface is tilted at a 20° angle. Find the component of the force vector perpendicular to the surface.

VISUALIZE Figure 3.25 shows a horizontal force $\vec{F}$ applied to the surface. A tilted coordinate system has its y-axis perpendicular to the surface, so the perpendicular component is $F_\perp = F_y$.

SOLVE From geometry, the force vector $\vec{F}$ makes an angle $\phi = 20°$ with the tilted x-axis. The perpendicular component of $\vec{F}$ is thus

$$F_\perp = F\sin 20° = (10\text{ N})\sin 20° = 3.42\text{ N}$$

STOP TO THINK 3.4   Angle ϕ that specifies the direction of $\vec{C}$ is given by

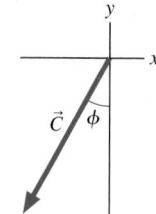

a. $\tan^{-1}(C_x/C_y)$.   b. $\tan^{-1}(C_x/|C_y|)$.

c. $\tan^{-1}(|C_x|/|C_y|)$.   d. $\tan^{-1}(C_y/C_x)$.

e. $\tan^{-1}(C_y/|C_x|)$.   f. $\tan^{-1}(|C_y|/|C_x|)$.

SUMMARY

The goal of Chapter 3 has been to learn how vectors are represented and used.

GENERAL PRINCIPLES

A vector is a quantity described by both a magnitude and a direction.

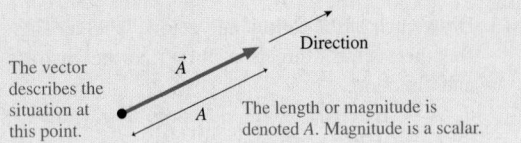

The vector describes the situation at this point.

$\vec{A}$

Direction

The length or magnitude is denoted A. Magnitude is a scalar.

Unit Vectors

Unit vectors have magnitude 1 and no units. Unit vectors $\hat{\imath}$ and $\hat{\jmath}$ define the directions of the x- and y-axes.

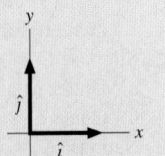

USING VECTORS

Components

The component vectors are parallel to the x- and y-axes.

$$\vec{A} = \vec{A}_x + \vec{A}_y = A_x\hat{\imath} + A_y\hat{\jmath}$$

In the figure at the right, for example:

$$A_x = A\cos\theta \quad A = \sqrt{A_x^2 + A_y^2}$$

$$A_y = A\sin\theta \quad \theta = \tan^{-1}(A_y/A_x)$$

▶ Minus signs need to be included if the vector points down or left.

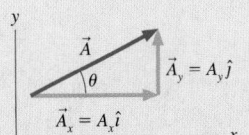

$\vec{A}_y = A_y\hat{\jmath}$

$\vec{A}_x = A_x\hat{\imath}$

$A_x < 0$	$A_x > 0$
$A_y > 0$	$A_y > 0$
$A_x < 0$	$A_x > 0$
$A_y < 0$	$A_y < 0$

The components A_x and A_y are the magnitudes of the component vectors $\vec{A}_x$ and $\vec{A}_y$ *and* a plus or minus sign to show whether the component vector points toward the positive end or the negative end of the axis.

Working Graphically

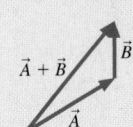

Addition $\quad$ $\vec{A} + \vec{B}$ $\quad$ $\vec{A} + \vec{B}$

Negative $\quad$ Subtraction $\quad$ Multiplication

$\vec{A} - \vec{B}$

Working Algebraically

Vector calculations are done component by component.

$$\vec{C} = 2\vec{A} + \vec{B} \quad \text{means} \quad \begin{cases} C_x = 2A_x + B_x \\ C_y = 2A_y + B_y \end{cases}$$

The magnitude of $\vec{C}$ is then $C = \sqrt{C_x^2 + C_y^2}$ and its direction is found using $\tan^{-1}$.

TERMS AND NOTATION

scalar quantity
vector quantity
magnitude
resultant vector
graphical addition

zero vector, $\vec{0}$
Cartesian coordinates
quadrants
component vector

decomposition
component
unit vector, $\hat{\imath}$ or $\hat{\jmath}$
algebraic addition

EXERCISES AND PROBLEMS

Exercises

Section 3.2 Properties of Vectors

1. a. Can a vector have nonzero magnitude if a component is zero? If no, why not? If yes, give an example.
   b. Can a vector have zero magnitude and a nonzero component? If no, why not? If yes, give an example.
2. Suppose $\vec{C} = \vec{A} + \vec{B}$.
   a. Under what circumstances does $C = A + B$?
   b. Could $C = A - B$? If so, how? If not, why not?
3. Suppose $\vec{C} = \vec{A} - \vec{B}$.
   a. Under what circumstances does $C = A - B$?
   b. Could $C = A + B$? If so, how? If not, why not?
4. Trace the vectors in Figure Ex3.4 onto your paper. Then find (a) $\vec{A} + \vec{B}$ and (b) $\vec{A} - \vec{B}$.

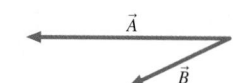

FIGURE EX3.4

5. Trace the vectors in Figure Ex3.5 onto your paper. Then find (a) $\vec{A} + \vec{B}$ and (b) $\vec{A} - \vec{B}$.

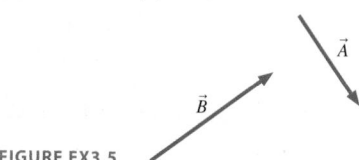

FIGURE EX3.5

Section 3.3 Coordinate Systems and Vector Components

6. A position vector in the first quadrant has an x-component of 6 m and a magnitude of 10 m. What is the value of its y-component?
7. A velocity vector 40° below the positive x-axis has a y-component of 10 m/s. What is the value of its x-component?
8. a. What are the x- and y-components of vector $\vec{E}$ in terms of the angle θ and the magnitude E shown in Figure Ex3.8?
   b. For the same vector, what are the x- and y-components in terms of the angle ϕ and the magnitude E?

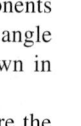

FIGURE EX3.8

9. Draw each of the following vectors, then find its x- and y-components.
   a. $\vec{r} = (100 \text{ m}, 45° \text{ below } +x\text{-axis})$
   b. $\vec{v} = (300 \text{ m/s}, 20° \text{ above } +x\text{-axis})$
   c. $\vec{a} = (5.0 \text{ m/s}^2, -y\text{-direction})$
   d. $\vec{F} = (50 \text{ N}, 36.9° \text{ above } -x\text{-axis})$
10. Draw each of the following vectors, then find its x- and y-components.
    a. $\vec{r} = (2 \text{ km}, 30° \text{ left of } +y\text{-axis})$
    b. $\vec{v} = (5 \text{ cm/s}, -x\text{-direction})$
    c. $\vec{a} = (10 \text{ m/s}^2, 40° \text{ left of } -y\text{-axis})$
    d. $\vec{F} = (50 \text{ N}, 36.9° \text{ right of } +y\text{-axis})$
11. Let $\vec{C} = (3.15 \text{ m}, 15° \text{ above the negative } x\text{-axis})$ and $\vec{D} = (25.67, 30° \text{ to the right of the negative } y\text{-axis})$. Find the magnitude, the x-component, and the y-component of each vector.

12. The quantity called the *electric field* is a vector. The electric field inside a scientific instrument is $\vec{E} = (125\hat{\imath} - 250\hat{\jmath})$ V/m, where V/m stands for volts per meter. What are the magnitude and direction of the electric field?

Section 3.4 Vector Algebra

13. Draw each of the following vectors, label an angle that specifies the vector's direction, then find the vector's magnitude and direction.
    a. $\vec{A} = 4\hat{\imath} - 6\hat{\jmath}$
    b. $\vec{r} = (50\hat{\imath} + 80\hat{\jmath})$ m
    c. $\vec{v} = (-20\hat{\imath} + 40\hat{\jmath})$ m/s
    d. $\vec{a} = (2\hat{\imath} - 6\hat{\jmath})$ m/s^2
14. Draw each of the following vectors, label an angle that specifies the vector's direction, then find its magnitude and direction.
    a. $\vec{B} = -4\hat{\imath} + 4\hat{\jmath}$
    b. $\vec{r} = (-2\hat{\imath} - \hat{\jmath})$ cm
    c. $\vec{v} = (-10\hat{\imath} - 100\hat{\jmath})$ mph
    d. $\vec{a} = (20\hat{\imath} + 10\hat{\jmath})$ m/s^2
15. Let $\vec{A} = 2\hat{\imath} + 3\hat{\jmath}$ and $\vec{B} = 4\hat{\imath} - 2\hat{\jmath}$.
    a. Draw a coordinate system and on it show vectors $\vec{A}$ and $\vec{B}$.
    b. Use graphical vector subtraction to find $\vec{C} = \vec{A} - \vec{B}$.
16. Let $\vec{A} = 5\hat{\imath} + 2\hat{\jmath}$, $\vec{B} = -3\hat{\imath} - 5\hat{\jmath}$, and $\vec{C} = \vec{A} + \vec{B}$.
    a. Write vector $\vec{C}$ in component form.
    b. Draw a coordinate system and on it show vectors $\vec{A}$, $\vec{B}$, and $\vec{C}$.
    c. What are the magnitude and direction of vector $\vec{C}$?
17. Let $\vec{A} = 5\hat{\imath} + 2\hat{\jmath}$, $\vec{B} = -3\hat{\imath} - 5\hat{\jmath}$, and $\vec{D} = \vec{A} - \vec{B}$.
    a. Write vector $\vec{D}$ in component form.
    b. Draw a coordinate system and on it show vectors $\vec{A}$, $\vec{B}$, and $\vec{D}$.
    c. What are the magnitude and direction of vector $\vec{D}$?
18. Let $\vec{A} = 5\hat{\imath} + 2\hat{\jmath}$, $\vec{B} = -3\hat{\imath} - 5\hat{\jmath}$, and $\vec{E} = 2\vec{A} + 3\vec{B}$.
    a. Write vector $\vec{E}$ in component form.
    b. Draw a coordinate system and on it show vectors $\vec{A}$, $\vec{B}$, and $\vec{E}$.
    c. What are the magnitude and direction of vector $\vec{E}$?
19. Let $\vec{A} = 5\hat{\imath} + 2\hat{\jmath}$, $\vec{B} = -3\hat{\imath} - 5\hat{\jmath}$, and $\vec{F} = \vec{A} - 4\vec{B}$.
    a. Write vector $\vec{F}$ in component form.
    b. Draw a coordinate system and on it show vectors $\vec{A}$, $\vec{B}$, and $\vec{F}$.
    c. What are the magnitude and direction of vector $\vec{F}$?
20. Are the following statements true or false? Explain your answer.
    a. The magnitude of a vector can be different in different coordinate systems.
    b. The direction of a vector can be different in different coordinate systems.
    c. The components of a vector can be different in different coordinate systems.
21. Let $\vec{A} = (4.0 \text{ m, vertically downward})$ and $\vec{B} = (5.0 \text{ m}, 120°$ clockwise from $\vec{A})$. Find the x- and y-components of $\vec{A}$ and $\vec{B}$ in each of the two coordinate systems shown in Figure Ex3.21.

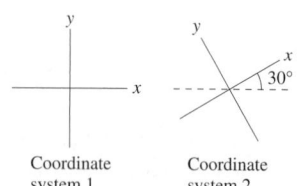

FIGURE EX3.21   Coordinate system 1   Coordinate system 2

22. What are the *x*- and *y*-components of the velocity vector shown in Figure Ex3.22?

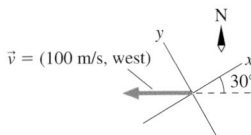

FIGURE EX3.22

Problems

23. Figure P3.23 shows vectors $\vec{A}$ and $\vec{B}$. Let $\vec{C} = \vec{A} + \vec{B}$.
    a. Reproduce the figure on your page as accurately as possible, using a ruler and protractor. Draw vector $\vec{C}$ on your figure, using the graphical addition of $\vec{A}$ and $\vec{B}$. Then determine the magnitude and direction of $\vec{C}$ by *measuring* it with a ruler and protractor.
    b. Based on your figure of part a, use geometry and trigonometry to *calculate* the magnitude and direction of $\vec{C}$.
    c. Decompose vectors $\vec{A}$ and $\vec{B}$ into components, then use these to calculate algebraically the magnitude and direction of $\vec{C}$.

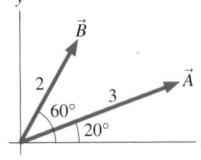

FIGURE P3.23

24. a. What is the angle ϕ between vectors $\vec{E}$ and $\vec{F}$ in Figure P3.24?
    b. Use geometry and trigonometry to determine the magnitude and direction of $\vec{G} = \vec{E} + \vec{F}$.
    c. Use components to determine the magnitude and direction of $\vec{G} = \vec{E} + \vec{F}$.

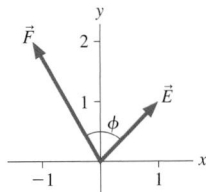

FIGURE P3.24   **FIGURE P3.25**

25. For the three vectors shown above in Figure P3.25, $\vec{A} + \vec{B} + \vec{C} = -2\hat{\imath}$. What is vector $\vec{B}$?
    a. Write $\vec{B}$ in component form.
    b. Write $\vec{B}$ as a magnitude and a direction.
26. Figure P3.26 shows vectors $\vec{A}$ and $\vec{B}$. Find vector $\vec{C}$ such that $\vec{A} + \vec{B} + \vec{C} = \vec{0}$. Write your answer in component form.

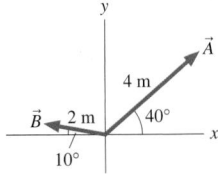

FIGURE P3.26   **FIGURE P3.27**

27. Figure P3.27 shows vectors $\vec{A}$ and $\vec{B}$. Find $\vec{D} = 2\vec{A} + \vec{B}$. Write your answer in component form.

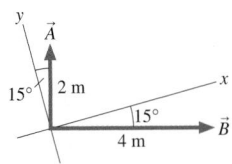

28. Let $\vec{A} = (3.0\text{ m}, 20° \text{ south of east})$, $\vec{B} = (2.0\text{ m}, \text{north})$, and $\vec{C} = (5.0\text{ m}, 70° \text{ south of west})$.
    a. Draw and label $\vec{A}$, $\vec{B}$, and $\vec{C}$ with their tails at the origin. Use a coordinate system with the *x*-axis to the east.
    b. Write $\vec{A}$, $\vec{B}$, and $\vec{C}$ in component form, using unit vectors.
    c. Find the magnitude and the direction of $\vec{D} = \vec{A} + \vec{B} + \vec{C}$.
29. Trace the vectors in Figure P3.29 onto your paper. Use the graphical method of vector addition and subtraction to find the following.
    a. $\vec{D} + \vec{E} + \vec{F}$
    b. $\vec{D} + 2\vec{E}$
    c. $\vec{D} - 2\vec{E} + \vec{F}$

FIGURE P3.29

30. Let $\vec{E} = 2\hat{\imath} + 3\hat{\jmath}$ and $\vec{F} = 2\hat{\imath} - 2\hat{\jmath}$. Find the magnitude of
    a. $\vec{E}$ and $\vec{F}$ b. $\vec{E} + \vec{F}$ c. $-\vec{E} - 2\vec{F}$
31. Find a vector that points in the same direction as the vector $(\hat{\imath} + \hat{\jmath})$ and whose magnitude is 1.
32. The position of a particle as a function of time is given by $\vec{r} = (5\hat{\imath} + 4\hat{\jmath})t^2$ m, where *t* is in seconds.
    a. What is the particle's distance from the origin at $t = 0, 2$, and 5 s?
    b. Find an expression for the particle's velocity $\vec{v}$ as a function of time.
    c. What is the particle's speed at $t = 0, 2$, and 5 s?
33. While vacationing in the mountains you do some hiking. In the morning, your displacement is $\vec{S}_{\text{morning}} = (2000\text{ m, east}) + (3000\text{ m, north}) + (200\text{ m, vertical})$. After lunch, your displacement is $\vec{S}_{\text{afternoon}} = (1500\text{ m, west}) + (2000\text{ m, north}) - (300\text{ m, vertical})$.
    a. At the end of the hike, how much higher or lower are you compared to your starting point?
    b. What is your total displacement?
34. The minute hand on a watch is 2.0 cm in length. What is the displacement vector of the tip of the minute hand
    a. From 8:00 to 8:20 A.M.?
    b. From 8:00 to 9:00 A.M.?
35. Bob walks 200 m south, then jogs 400 m southwest, then walks 200 m in a direction 30° east of north.
    a. Draw an accurate graphical representation of Bob's motion. Use a ruler and a protractor!
    b. Use either trigonometry or components to find the displacement that will return Bob to his starting point by the most direct route. Give your answer as a distance and a direction.
    c. Does your answer to part b agree with what you can measure on your diagram of part a?
36. Jim's dog Sparky runs 50 m northeast to a tree, then 70 m west to a second tree, and finally 20 m south to a third tree.
    a. Draw a picture and establish a coordinate system.
    b. Calculate Sparky's net displacement in component form.
    c. Calculate Sparky's net displacement as a magnitude and an angle.
37. A field mouse trying to escape a hawk runs east for 5.0 m, darts southeast for 3.0 m, then drops 1.0 m down a hole into its burrow. What is the magnitude of the net displacement of the mouse?
38. Carlos runs with velocity $\vec{v} = (5\text{ m/s}, 25° \text{ north of east})$ for 10 minutes. How far to the north of his starting position does Carlos end up?
39. A cannon tilted upward at 30° fires a cannonball with a speed of 100 m/s. What is the component of the cannonball's velocity parallel to the ground?

40. Jack and Jill ran up the hill at 3.0 m/s. The horizontal component of Jill's velocity vector was 2.5 m/s.
    a. What was the angle of the hill?
    b. What was the vertical component of Jill's velocity?

41. The treasure map in Figure P3.41 gives the following directions to the buried treasure: "Start at the old oak tree, walk due north for 500 paces, then due east for 100 paces. Dig." But when you arrive, you find an angry dragon just north of the tree. To avoid the dragon, you set off along the yellow brick road at an angle 60° east of

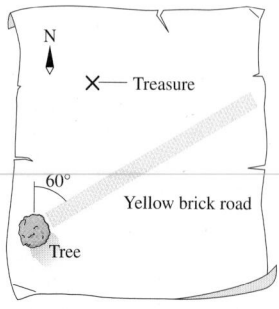

FIGURE P3.41

north. After walking 300 paces you see an opening through the woods. Which direction should you go, and how far, to reach the treasure?

42. Mary needs to row her boat across a 100-m-wide river that is flowing to the east at a speed of 1.0 m/s. Mary can row the boat with a speed of 2.0 m/s relative to the water.
    a. If Mary rows straight north, how far downstream will she land?
    b. Draw a picture showing Mary's displacement due to rowing, her displacement due to the river's motion, and her net displacement.

43. A jet plane is flying horizontally with a speed of 500 m/s over a hill that slopes upward with a 3% grade (i.e., the "rise" is 3% of the "run"). What is the component of the plane's velocity perpendicular to the ground?

44. A flock of ducks is trying to migrate south for the winter, but they keep being blown off course by a wind blowing from the west at 6.0 m/s. A wise elder duck finally realizes that the solution is to fly at an angle to the wind. If the ducks can fly at 8.0 m/s relative to the air, what direction should they head in order to move directly south?

45. A pine cone falls straight down from a pine tree growing on a 20° slope. The pine cone hits the ground with a speed of 10 m/s. What is the component of the pine cone's impact velocity (a) parallel to the ground and (b) perpendicular to the ground?

46. The car in Figure P3.46 speeds up as it turns a quarter-circle curve from north to east. When exactly halfway around the curve, the car's acceleration is $\vec{a} = (2 \text{ m/s}^2, 15°$ south of east). At this point, what is the component of $\vec{a}$ (a) tangent to the circle and (b) perpendicular to the circle?

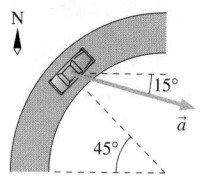

FIGURE P3.46

47. Figure P3.47 shows three ropes tied together in a knot. One of your friends pulls on a rope with 3 units of force and another pulls on a second rope with 5 units of force. How hard and in what direction must you pull on the third rope to keep the knot from moving?

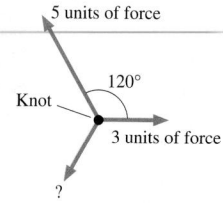

FIGURE P3.47

48. Three forces are exerted on an object placed on a tilted floor in Figure P3.48. The forces are measured in newtons (N). Assuming that forces are vectors,
    a. What is the component of the *net force* $\vec{F}_{net} = \vec{F}_1 + \vec{F}_2 + \vec{F}_3$ parallel to the floor?
    b. What is the component of $\vec{F}_{net}$ perpendicular to the floor?
    c. What are the magnitude and direction of $\vec{F}_{net}$?

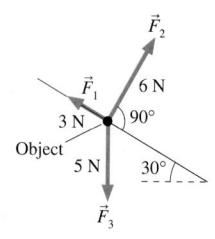

FIGURE P3.48

49. Figure P3.49 shows four electrical charges located at the corners of a rectangle. Like charges, you will recall, repel each other while opposite charges attract. Charge B exerts a repulsive force (directly *away from* B) on charge A of 3 N. Charge C exerts an attractive force (directly *toward* C) on charge A of 6 N. Finally, charge D exerts an attractive force of 2 N on charge A. Assuming that forces are vectors, what is the magnitude and direction of the net force $\vec{F}_{net}$ exerted on charge A?

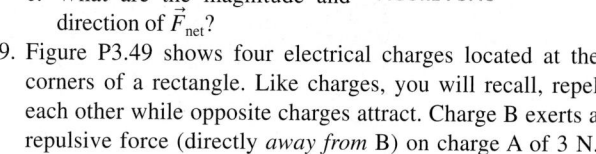

FIGURE P3.49

STOP TO THINK ANSWERS

Stop to Think 3.1: c. The graphical construction of $\vec{A}_1 + \vec{A}_2 + \vec{A}_3$ is shown below.

Stop to Think 3.2: a. The graphical construction of $2\vec{A} - \vec{B}$ is shown below.

Stop to Think 3.3: $C_x = -4$ cm, $C_y = 2$ cm.

Stop to Think 3.4: c. Vector $\vec{C}$ points to the left and down, so both C_x and C_y are negative. C_x is in the numerator because it is the side opposite ϕ.

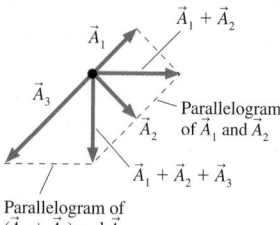

STOP TO THINK 3.1

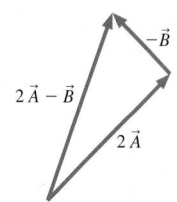

STOP TO THINK 3.2

4 Force and Motion

A drag race is a memorable example of the connection between force and motion.

▶ **Looking Ahead**
The goal of Chapter 4 is to establish a connection between force and motion. In this chapter you will learn to:

■ Recognize what a force is and is not.
■ Identify the specific forces acting on an object.
■ Draw free-body diagrams.
■ Understand the connection between force and motion.

◀ **Looking Back**
To master the material introduced in this chapter, you must understand how acceleration is determined and how vectors are used. Please review:

■ Section 1.5 Acceleration.
■ Section 3.2 Properties of vectors.

This drag racer can cover a quarter mile from a standing start in less than 5 seconds. Not bad! We could use kinematics to describe the car's motion with pictures, graphs, and equations. By defining position, velocity, and acceleration and dressing them in mathematical clothing, kinematics provides a language to describe *how* something moves. But kinematics would tell us nothing about *why* the car accelerates so quickly. For the more fundamental task of understanding the *cause* of motion, we turn our attention to **dynamics.** Dynamics joins with kinematics to form **mechanics,** the general science of motion. We study dynamics qualitatively in this chapter, then develop it quantitatively in the next four chapters.

The theory of mechanics originated in the mid-1600s when Sir Isaac Newton formulated his laws of motion. These fundamental principles of mechanics explain how motion occurs as a consequence of forces. Newton's laws are more than 300 years old, but they still form the basis for our contemporary understanding of motion.

A challenge in learning physics is that a textbook is not an experiment. The book can assert that an experiment will have a certain outcome, but you may not be convinced unless you see or do the experiment yourself. Newton's laws are frequently contrary to our intuition, and a lack of familiarity with the evidence for Newton's laws is a source of difficulty for many people. You will have an opportunity through lecture demonstrations and in the laboratory to see for yourself the evidence supporting Newton's laws. Physics is not an arbitrary collection of definitions and formulas, but a consistent theory as to how the universe really works. It is only with experience and evidence that we learn to separate physical fact from fantasy.

4.1 Force

If you kick a ball, it rolls across the floor. If you pull on a door handle, the door opens. You know, from many years of experience, that some sort of *force* is required to move these objects. Our goal is to understand *why* motion occurs, and the observation that force and motion are related is a good place to start.

The two major issues that this chapter will examine are:

- What is a force?
- What is the connection between force and motion?

We begin with the first of these questions in the table below.

What is a force?

A force is a push or a pull.

> Our commonsense idea of a **force** is that it is a *push* or a *pull*. We will refine this idea as we go along, but it is an adequate starting point. Notice our careful choice of words: We refer to "*a* force," rather than simply "force." We want to think of a force as a very specific *action,* so that we can talk about a single force or perhaps about two or three individual forces that we can clearly distinguish. Hence the concrete idea of "a force" acting on an object.

A force acts on an object.

> Implicit in our concept of force is that **a force acts on an object.** In other words, pushes and pulls are applied *to* something—an object. From the object's perspective, it has a force *exerted* on it. Forces do not exist in isolation from the object that experiences them.

A force requires an agent.

> Every force has an **agent,** something that acts or exerts power. That is, a force has a specific, identifiable *cause.* As you throw a ball, it is your hand, while in contact with the ball, that is the agent or the cause of the force exerted on the ball. *If* a force is being exerted on an object, you must be able to identify a specific cause (i.e., the agent) of that force. Conversely, a force is not exerted on an object *unless* you can identify a specific cause or agent. Although this idea may seem to be stating the obvious, you will find it to be a powerful tool for avoiding some common misconceptions about what is and is not a force.

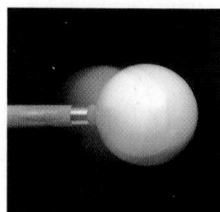

A force is a vector.

> If you push an object, you can push either gently or very hard. Similarly, you can push either left or right, up or down. To quantify a push, we need to specify both a magnitude *and* a direction. It should thus come as no surprise that a force is a vector quantity. The symbol for a force is the vector symbol $\vec{F}$. The size or strength of a force is its magnitude F.

A force can be either a contact force . . .

> There are two basic classes of forces, depending on whether the agent touches the object or not. **Contact forces** are forces that act on an object by touching it at a point of contact. The bat must touch the ball to hit it. A string must be tied to an object to pull it. The majority of forces that we will examine are contact forces.

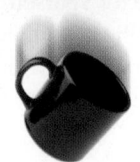

. . . or a long-range force.

> **Long-range forces** are forces that act on an object without physical contact. Magnetism is an example of a long-range force. You have undoubtedly held a magnet over a paper clip and seen the paper clip leap up to the magnet. A coffee cup released from your hand is pulled to the earth by the long-range force of gravity.

Let's summarize these ideas as our definition of force:

- A force is a push or a pull on an object.
- A force is a vector. It has both a magnitude and a direction.
- A force requires an agent. Something does the pushing or pulling.
- A force is either a contact force or a long-range force. Gravity is the only long-range force we will deal with until much later in the book.

There's one more important aspect of forces. If you push against a door (the object) to close it, the door pushes back against your hand (the agent). If a tow rope pulls on a car (the object), the car pulls back on the rope (the agent). Thus, in general, if an agent exerts a force on an object, the object exerts a force on the agent. We really need to think of a force as an *interaction* between two objects. Although the interaction perspective is a more exact way to view forces, it adds complications that we would like to avoid for now. Our approach will be to start by focusing on how a single object responds to forces exerted on it. Then, in Chapter 8, we'll return to the larger issue of how two or more objects interact with each other.

NOTE ▶ In the particle model, objects cannot exert forces on themselves. A force on an object will always have an agent or cause external to the object. Now, there are certainly objects that have internal forces (think of all the forces inside the engine of your car!), but the particle model is not valid if you need to consider those internal forces. If you are going to treat your car as a particle and look only at the overall motion of the car as a whole, that motion will be a consequence of external forces acting on the car. ◀

Force Vectors

We can use a simple diagram to visualize how forces are exerted on objects. Because we are using the particle model, in which objects are treated as points, the process of drawing a force vector is straightforward. Here is how it goes:

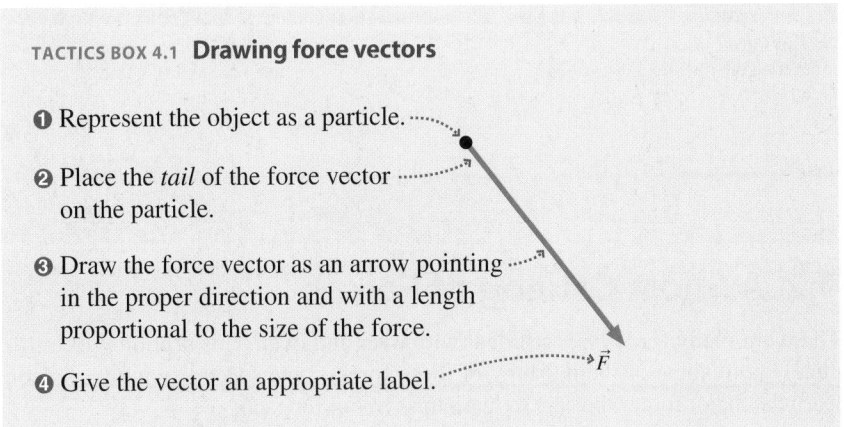

TACTICS BOX 4.1 **Drawing force vectors**

❶ Represent the object as a particle.

❷ Place the *tail* of the force vector on the particle.

❸ Draw the force vector as an arrow pointing in the proper direction and with a length proportional to the size of the force.

❹ Give the vector an appropriate label.

$\vec{F}$

Step 2 may seem contrary to what a "push" should do, but recall that moving a vector does not change it as long as the length and angle do not change. The vector $\vec{F}$ is the same regardless of whether the tail or the tip is placed on the particle. Our reason for using the tail will become clear when we consider how to combine several forces.

Figure 4.1 on the next page shows three examples of force vectors. One is a push, one a pull, and one a long-range force, but in all three the *tail* of the force vector is placed on the particle representing the object.

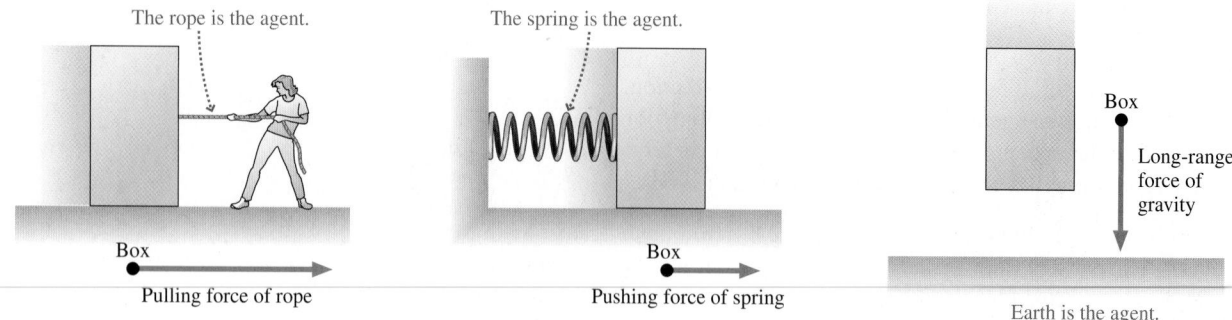

FIGURE 4.1 Three force vectors.

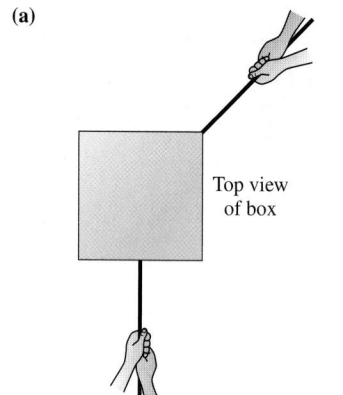

(a)

Top view of box

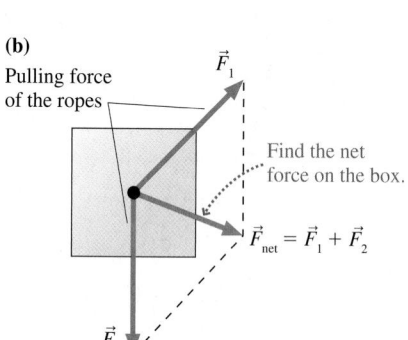

(b)

Pulling force of the ropes

Find the net force on the box.

$\vec{F}_{net} = \vec{F}_1 + \vec{F}_2$

FIGURE 4.2 Two forces applied to a box.

Combining Forces

Figure 4.2a shows a box being pulled by two ropes, each exerting a force on the box. How will the box respond? Experimentally, we find that when several individual forces $\vec{F}_1$, $\vec{F}_2$, $\vec{F}_3$, . . . are exerted on an object, they combine to form a **net force** given by the *vector* sum of the individual forces:

$$\vec{F}_{net} \equiv \sum_{i=1}^{N} \vec{F}_i = \vec{F}_1 + \vec{F}_2 + \cdots + \vec{F}_N \tag{4.1}$$

Recall that $\equiv$ is the symbol meaning "is defined as." Mathematically, this summation is called a **superposition of forces.** The net force is sometimes called the *resultant force.* Figure 4.2b shows the net force on the box.

STOP TO THINK 4.1 Two forces are exerted on an object. What third force would make the net force point to the left?

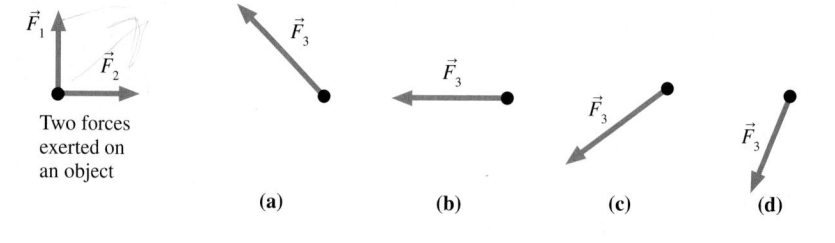

Two forces exerted on an object

(a)   (b)   (c)   (d)

4.2 A Short Catalog of Forces

There are many forces we will deal with over and over. This section will introduce you to some of them. Many of these forces have special symbols. As you learn the major forces, be sure to learn the symbol for each.

Weight

A falling rock is pulled toward the earth by the long-range force of gravity. Gravity is what keeps you in your chair, keeps the planets in their orbits around the sun, and shapes the large-scale structure of the universe. We'll have a thorough look at gravity in Chapter 12. For now we'll concentrate on objects on or near the surface of the earth (or other planet).

The gravitational pull of the earth on an object on or near the surface of the earth is called **weight.** The symbol for weight is $\vec{w}$. Weight is the only long-range force we will encounter in the next few chapters. The agent for the weight force is

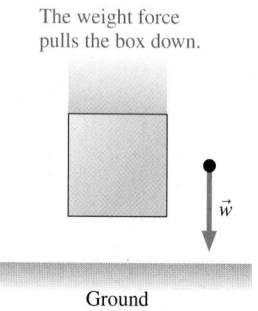

The weight force pulls the box down.

Ground

FIGURE 4.3 Weight.

the *entire earth* pulling on an object. Weight acts on an object whether the object is moving or at rest. The weight vector always points vertically downward, as shown in Figure 4.3 on the previous page.

NOTE ▶ We often refer to "the weight" of an object. This is an informal expression for *w*, the magnitude of the weight force exerted on the object. Note that **weight is not the same thing as mass.** We will briefly examine mass later in the chapter and explore the connection between weight and mass in Chapter 5. ◀

Spring Force

Springs exert one of the most common contact forces. A spring can either push (when compressed) or pull (when stretched). Figure 4.4 shows the spring force. In both cases, pushing and pulling, the tail of the force vector is placed on the particle in the force diagram. There is no special symbol for a spring force, so we simply use a subscript label: $\vec{F}_{sp}$.

A stretched spring exerts a force on an object.

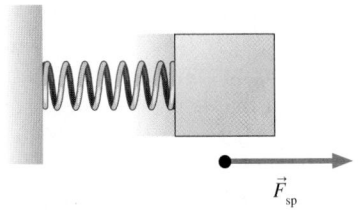

A compressed spring exerts a pushing force on an object.

$\vec{F}_{sp}$

(a)

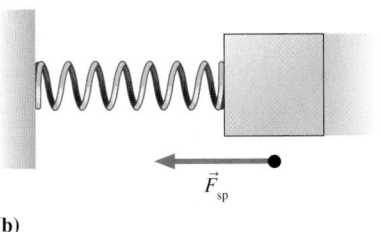

A stretched spring exerts a pulling force on an object.

$\vec{F}_{sp}$

(b)

FIGURE 4.4 The spring force.

Although you may think of a spring as a metal coil that can be stretched or compressed, this is only one type of spring. Hold a ruler, or any other thin piece of wood or metal, by the ends and bend it slightly. It flexes. When you let go, it "springs" back to its original shape. This is just as much a spring as is a metal coil.

Tension Force

When a string or rope or wire pulls on an object, it exerts a contact force that we call the **tension force,** represented by a capital $\vec{T}$. The direction of the tension force is always in the direction of the string or rope, as you can see in Figure 4.5. The commonplace reference to "the tension" in a string is an informal expression for *T*, the size or magnitude of the tension force.

If you were to use a very powerful microscope to look inside a rope, you would "see" that it is made of *atoms* joined together by *molecular bonds*. Molecular bonds are not rigid connections between the atoms. They are more accurately thought of as tiny *springs* holding the atoms together, as in Figure 4.6. These are very stiff springs, to be sure, but pulling on the ends of a string or rope stretches the molecular springs ever so slightly. The tension within a rope and the tension experienced by an object at the end of the rope are really the net spring force being exerted by billions and billions of microscopic springs.

This atomic-level view of tension introduces a new idea: a microscopic **atomic model** for understanding the behavior and properties of macroscopic objects. We will frequently use an atomic model to obtain a deeper understanding of our observations.

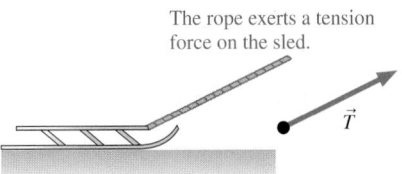

The rope exerts a tension force on the sled.

$\vec{T}$

FIGURE 4.5 Tension.

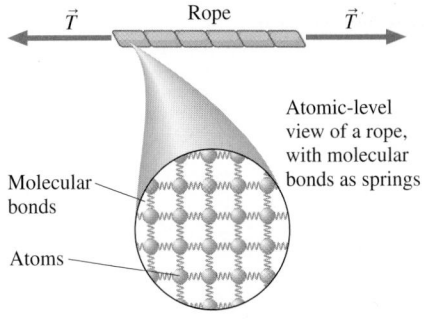

$\vec{T}$ Rope $\vec{T}$

Atomic-level view of a rope, with molecular bonds as springs

Molecular bonds

Atoms

FIGURE 4.6 An atomic-level view of tension.

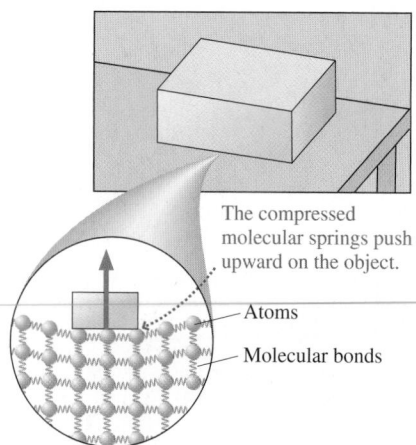

FIGURE 4.7 Atomic-level view of the force exerted by a table.

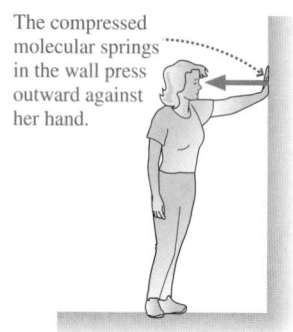

FIGURE 4.8 The wall pushes outward against your hand.

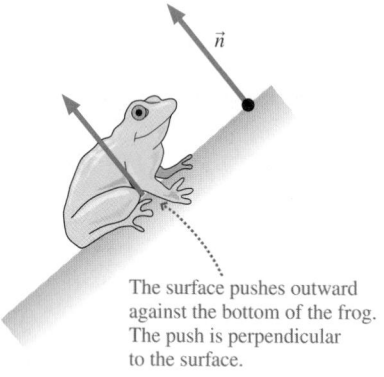

FIGURE 4.9 The normal force.

Normal Force

If you sit on a bed, the springs in the mattress compress and, as a consequence of the compression, exert an upward force on you. Stiffer springs would show less compression but still exert an upward force. The compression of extremely stiff springs might be measurable only by sensitive instruments. Nonetheless, the springs would compress ever so slightly and exert an upward spring force on you.

Figure 4.7 shows an object resting on top of a sturdy table. The table may not visibly flex or sag, but—just as you do to the bed—the object compresses the molecular springs in the table. The size of the compression is very small because molecular springs are so stiff, but it is not zero. As a consequence, the compressed molecular springs *push upward* on the object. We say that "the table" exerts the upward force, but it is important to understand that the pushing is *really* done by molecular springs. Similarly, an object resting on the ground compresses the molecular springs holding the ground together and, as a consequence, the ground pushes up on the object.

We can extend this idea. Suppose you place your hand on a wall and lean against it, as shown in Figure 4.8. Does the wall exert a force on your hand? As you lean, you compress the molecular springs in the wall and, as a consequence, they push outward against your hand. So the answer is "yes," the wall does exert a force on you.

The force the table surface exerts is vertical, the force the wall exerts is horizontal. But in all cases, the force exerted on an object that is pressing against a surface is in a direction *perpendicular* to the surface. Mathematicians refer to a line that is perpendicular to a surface as being *normal* to the surface. In keeping with this terminology, we define the **normal force** as the force exerted by a surface (the agent) against an object that is pressing against the surface. The symbol for the normal force is $\vec{n}$.

We're not using the word *normal* to imply that the force is an "ordinary" force or to distinguish it from an "abnormal force." A surface exerts a force *perpendicular* (i.e., normal) to itself as the molecular springs press *outward*. Figure 4.9 shows an object on an inclined surface, a common situation. Notice how the normal force $\vec{n}$ is perpendicular to the surface.

We have spent a lot of time describing the normal force because many people have a difficult time understanding it. The normal force is a very real force arising from the very real compression of molecular bonds. It is in essence just a spring force, but one exerted by a vast number of microscopic springs acting at once. The normal force is responsible for the "solidness" of solids. It is what prevents you from passing right through the chair you are sitting in and what causes the pain and the lump if you bang your head into a door. Your head can then tell you that the force exerted on it by the door was very real!

Friction

You've certainly observed that a rolling or sliding object, if not pushed or propelled, slows down and eventually stops. You've probably discovered that you can slide better across a sheet of ice than across asphalt. And you also know that most objects stay in place on a table without sliding off even if the table isn't absolutely level. The force responsible for these sorts of behavior is **friction.** The symbol for friction is a lower case $\vec{f}$.

Friction, like the normal force, is exerted by a surface. On a microscopic level, friction arises as atoms from the object and atoms on the surface run into each other. The rougher the surface is, the more these atoms are forced into close proximity and, as a result, the larger the friction force. We will develop a simple model of friction in the next chapter that will be sufficient for our needs. For now, it is useful to distinguish between two kinds of friction:

- *Kinetic friction,* denoted $\vec{f}_k$, appears as an object slides across a surface. This is a force that "opposes the motion," meaning that the friction force vector $\vec{f}_k$ points in a direction opposite the velocity vector $\vec{v}$ (i.e., "the motion").
- *Static friction,* denoted $\vec{f}_s$, is the force that keeps an object "stuck" on a surface and prevents its motion. Finding the direction of $\vec{f}_s$ is a little trickier than finding it for $\vec{f}_k$. Static friction points opposite the direction in which the object *would* move if there were no friction. That is, it points in the direction necessary to *prevent* motion.

Figure 4.10 shows examples of kinetic and static friction.

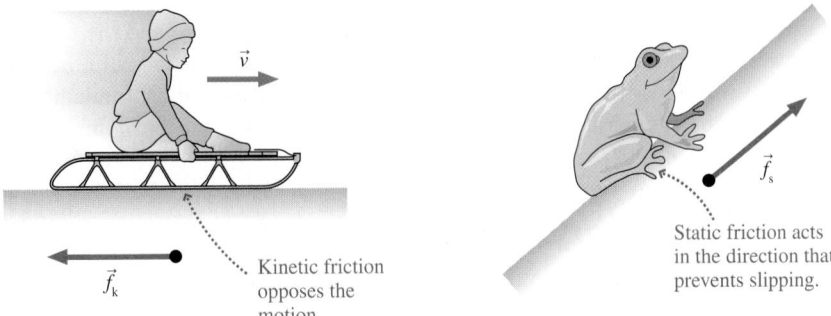

Kinetic friction opposes the motion.

Static friction acts in the direction that prevents slipping.

FIGURE 4.10 Kinetic and static friction.

NOTE ▶ A surface exerts a kinetic friction force when an object moves *relative to* the surface. A package on a conveyor belt is in motion, but it does not experience a kinetic friction force because it is not moving relative to the belt. So to be precise, we should say that the kinetic friction force points opposite to an object's motion *relative to* a surface. ◀

Drag

Friction at a surface is one example of a *resistive force,* a force that opposes or resists motion. Resistive forces are also experienced by objects moving through fluids—gases and liquids. The resistive force of a fluid is called **drag** and is symbolized as $\vec{D}$. Drag, like kinetic friction, points opposite the direction of motion. Figure 4.11 shows an example of drag.

Drag can be a large force for objects moving at high speeds or in dense fluids. Hold your arm out the window as you ride in a car and feel how the air resistance against it increases rapidly as the car's speed increases. Drop a lightweight object into a beaker of water and watch how slowly it settles to the bottom. In both cases the drag force is very significant.

For objects that are heavy and compact, that move in air, and whose speed is not too great, the drag force of air resistance is fairly small. To keep things as simple as possible, **you can neglect air resistance in all problems unless a problem explicitly asks you to include it.** The error introduced into calculations by this approximation is generally pretty small. This textbook will not consider objects moving in liquids.

Thrust

A jet airplane obviously has a force that propels it forward during takeoff. Likewise for the rocket being launched in Figure 4.12. This force, called **thrust,** occurs when a jet or rocket engine expels gas molecules at high speed. Thrust is a

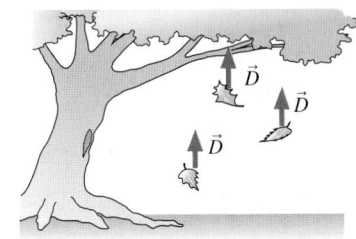

Air resistance is a significant force on falling leaves. It points opposite the direction of motion.

FIGURE 4.11 Air resistance is an example of drag.

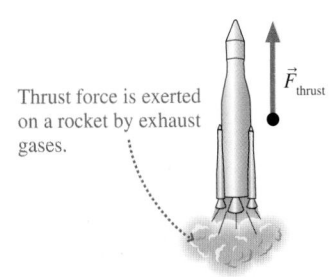

Thrust force is exerted on a rocket by exhaust gases.

FIGURE 4.12 Thrust force on a rocket.

contact force, with the exhaust gas being the agent that pushes on the engine. The process by which thrust is generated is rather subtle, and we will postpone a full discussion until we introduce Newton's third law in Chapter 8. For now, we will treat thrust as a force opposite the direction in which the exhaust gas is expelled. There's no special symbol for thrust, so we will call it $\vec{F}_{\text{thrust}}$.

Force	Notation
General force	$\vec{F}$
Weight	$\vec{w}$
Spring force	$\vec{F}_{\text{sp}}$
Tension	$\vec{T}$
Normal force	$\vec{n}$
Static friction	$\vec{f}_{\text{s}}$
Kinetic friction	$\vec{f}_{\text{k}}$
Drag	$\vec{D}$
Thrust	$\vec{F}_{\text{thrust}}$

Electric and Magnetic Forces

Electricity and magnetism, like gravity, exert long-range forces. The forces of electricity and magnetism act on charged particles. We will study electric and magnetic forces in detail in Part VI of this textbook. For now, it is worth noting that the forces holding molecules together—the molecular bonds—are not actually tiny springs. Atoms and molecules are made of charged particles—electrons and protons—and what we call a molecular bond is really an attractive electric force between these particles. So when we say that the normal force and the tension force are due to "molecular springs," or that friction is due to atoms running into each other, what we're really saying is that these forces, at the most fundamental level, are actually electric forces between the charged particles in the atoms.

4.3 Identifying Forces

Force and motion problems generally have two basic steps:

1. Identify all of the forces acting on an object.
2. Use Newton's laws and kinematics to determine the motion.

Understanding the first step is the primary goal of this chapter. We'll turn our attention to step 2 in the next chapter.

A typical physics problem describes an object that is being pushed and pulled in various directions. Some forces are given explicitly, others are only implied. In order to proceed, it is necessary to determine all the forces that act on the object. It is also necessary to avoid including forces that do not really exist. Now that you have learned the properties of forces and seen a catalog of typical forces, we can develop a step-by-step method for identifying each force in a problem. This procedure for identifying forces is part of the *physical representation* of the problem.

TACTICS BOX 4.2 **Identifying forces**

❶ **Identify "the system" and "the environment."** The system is the object whose motion you wish to study; the environment is everything else.

❷ **Draw a picture of the situation.** Show the object—the system—and everything in the environment that touches the system. Ropes, springs, and surfaces are all parts of the environment.

❸ **Draw a closed curve around the system.** Only the object is inside the curve; everything else is outside.

❹ **Locate every point on the boundary of this curve where the environment touches the system.** These are the points where the environment exerts *contact forces* on the object.

❺ **Name and label each contact force acting on the object.** There is at least one force at each point of contact; there may be more than one. When necessary, use subscripts to distinguish forces of the same type.

❻ **Name and label each long-range force acting on the object.** For now, the only long-range force is weight.

EXAMPLE 4.1 **Forces on a bungee jumper**

A bungee jumper has leapt off a bridge and is nearing the bottom of her fall. What forces are being exerted on the bungee jumper?

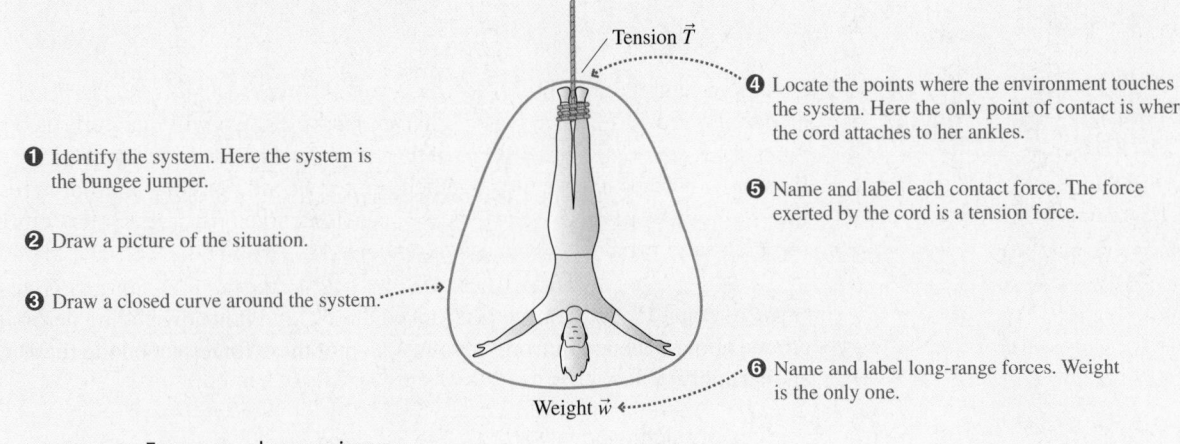

❶ Identify the system. Here the system is the bungee jumper.

❷ Draw a picture of the situation.

❸ Draw a closed curve around the system.

Tension $\vec{T}$

❹ Locate the points where the environment touches the system. Here the only point of contact is where the cord attaches to her ankles.

❺ Name and label each contact force. The force exerted by the cord is a tension force.

❻ Name and label long-range forces. Weight is the only one.

Weight $\vec{w}$

FIGURE 4.13 Forces on a bungee jumper.

EXAMPLE 4.2 **Forces on a skier**

A skier is being towed up a snow-covered hill by a tow rope. What forces are being exerted on the skier?

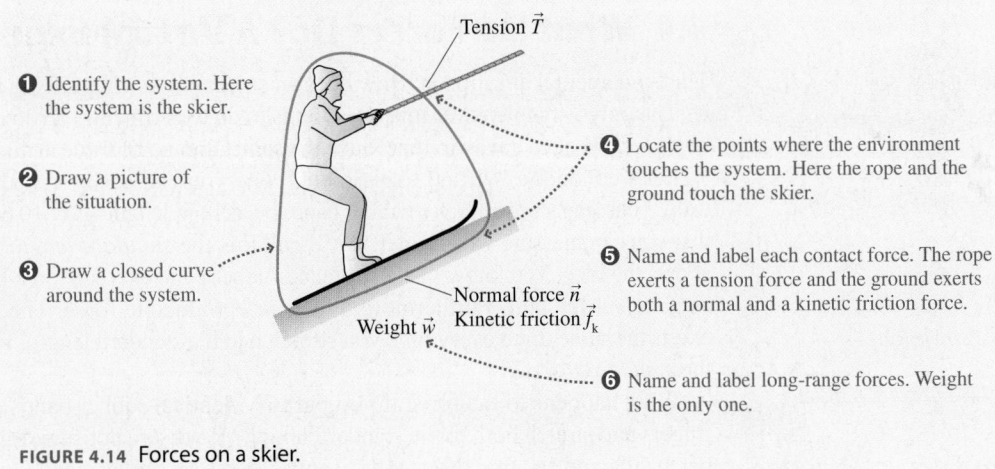

Tension $\vec{T}$

❶ Identify the system. Here the system is the skier.

❷ Draw a picture of the situation.

❸ Draw a closed curve around the system.

❹ Locate the points where the environment touches the system. Here the rope and the ground touch the skier.

❺ Name and label each contact force. The rope exerts a tension force and the ground exerts both a normal and a kinetic friction force.

Normal force $\vec{n}$
Kinetic friction $\vec{f}_k$
Weight $\vec{w}$

❻ Name and label long-range forces. Weight is the only one.

FIGURE 4.14 Forces on a skier.

NOTE ▶ You might have expected two friction forces and two normal forces in Example 4.2, one on each ski. Keep in mind, however, that we're working within the particle model, which represents the skier by a single point. A particle has only one contact with the ground, so there is a single normal force and a single friction force. The particle model is valid if we want to analyze the translational motion of the skier as a whole, but we would have to go beyond the particle model to find out what happens to each ski. ◀

Now that you're getting the hang of this, the next example is meant to look much more like a sketch you should make when asked to identify forces in a homework problem.

EXAMPLE 4.3 Forces on a rocket

A rocket is being launched to place a new satellite in orbit. Air resistance is not negligible. What forces are being exerted on the rocket?

Air — Drag $\vec{D}$

Weight $\vec{w}$

Thrust $\vec{F}_{\text{thrust}}$

Exhaust

FIGURE 4.15 Forces on a rocket.

STOP TO THINK 4.2 You've just kicked a rock, and it is now sliding across the ground about 2 meters in front of you. Which of these forces act on the rock? List all that apply.

    a. Gravity, acting downward.
    b. The normal force, acting upward.
    c. The force of the kick, acting in the direction of motion.
    d. Friction, acting opposite the direction of motion.
    e. Air resistance, acting opposite the direction of motion.

4.4 What Do Forces Do? A Virtual Experiment

The fundamental question is: How does an object move when a force is exerted on it? The only way to answer this question is to do experiments. To do experiments, however, we need a way to reproduce the same amount of force again and again.

Let's conduct a "virtual experiment," one you can easily visualize. Imagine using your fingers to stretch a rubber band to a certain length—say 10 centimeters—that you can measure with a ruler. We'll call this the *standard length*. Figure 4.16 shows the idea. You know that a stretched rubber band exerts a force because your fingers *feel* the pull. Furthermore, this is a reproducible force. The rubber band exerts the same force every time you stretch it to the standard length. We'll call this the *standard force F*.

What happens to the force if you put *two* identical rubber bands around your fingers and stretch both to the standard length? If you are not sure, find a few rubber bands and try this. You will discover that two rubber bands exert a larger pulling force than one rubber band. This is not surprising. If two rubber bands are each pulling equally hard, the net pull is twice that of one rubber band: $F_{\text{net}} = 2F$. N side-by-side rubber bands, each pulled to the standard length, will exert N times the standard force: $F_{\text{net}} = NF$.

Now we're ready to start the virtual experiment. Imagine an object to which you can attach rubber bands, such as a block of wood with a hook. If you attach a rubber band and stretch it to the standard length, the object experiences the same force F as did your finger. N rubber bands attached to the object will exert N times the force of one rubber band. The rubber bands give us a way of applying a known and reproducible force to an object.

Our next task is to measure the object's motion in response to these forces. Imagine using the rubber bands to pull the object across a horizontal table. Friction between the object and the surface might affect our results, so let's just eliminate friction. This is, after all, a virtual experiment! (In practice you could nearly

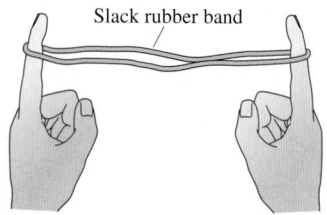
Slack rubber band

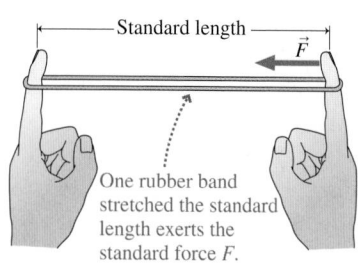
Standard length
$\vec{F}$
One rubber band stretched the standard length exerts the standard force F.

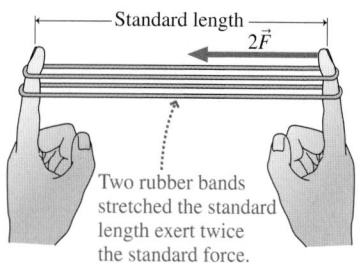
Standard length
$2\vec{F}$
Two rubber bands stretched the standard length exert twice the standard force.

FIGURE 4.16 A reproducible force.

eliminate friction by pulling a smooth block over a smooth sheet of ice or by supporting the object on a cushion of air.) To make measurements, lay a meter stick along the edge of the table and hang a movie camera over the table to record the motion.

Our experiment will be easiest to interpret if the force is *constant* throughout the object's motion. If you stretch the rubber band and then release the object, it moves toward your hand. But as it does so, the rubber band gets shorter and the pulling force decreases. To keep the pulling force constant, you must *move your hand* at just the right speed to keep the length of the rubber band from changing! Figure 4.17 shows the experiment being carried out. Once the motion is complete, you can use motion diagrams (made from the movie frames) and kinematics to analyze the object's motion.

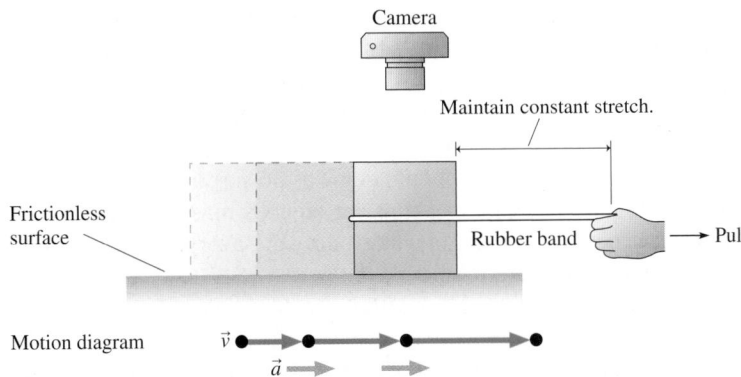

FIGURE 4.17 Measuring the motion of an object that is pulled with a constant force.

The first important finding of this experiment is that **an object pulled with a constant force moves with a constant acceleration.** This finding could not have been anticipated in advance. It's conceivable that the object would speed up for a while, then move with a steady speed. Or that it would continue to speed up, but that the *rate* of increase, the acceleration, would steadily decline. These are conceivable motions, but they're not what happens. Instead, the object continues to accelerate *with a constant acceleration* for as long as you pull it with a constant force.

The next question is: What happens if you increase the force by using several rubber bands? To find out, use 2 rubber bands. Stretch both to the standard length to double the force, then measure the acceleration. Then measure the acceleration due to 3 rubber bands, then 4, and so on. Table 4.1 shows the results of this experiment. You can see that doubling the force causes twice the acceleration, tripling the force causes three times the acceleration, and so on.

Figure 4.18 is a graph of the data. Force is the independent variable, the one you can control, so we've placed force on the horizontal axis to make an acceleration-versus-force graph. The graph shows that **the acceleration is directly proportional to the force.** This is our second important finding. Recall that proportionality indicates a linear relationship whose graph passes through the origin (*y*-intercept of zero). This result can be written

$$a = cF \tag{4.2}$$

where *c* is called the *proportionality constant*. The proportionality constant *c* is the slope of the graph.

The final question for our virtual experiment is: How does the acceleration depend on the size of the object? (The "size" of an object is somewhat ambiguous. We'll be more precise below.) To find out, glue the original object and an identical copy together, and then, applying the *same force* as you applied to the

TABLE 4.1 Acceleration due to an increasing force

Rubber bands	Force	Acceleration
1	F	a_1
2	$2F$	$a_2 = 2a_1$
3	$3F$	$a_3 = 3a_1$
$\vdots$	$\vdots$	$\vdots$
N	NF	$a_N = Na_1$

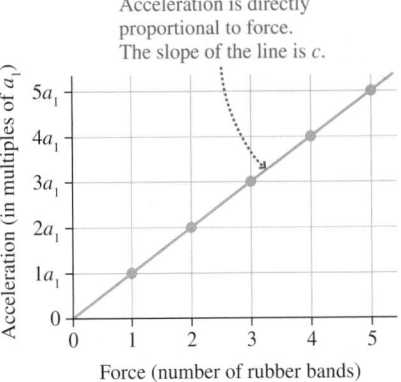

FIGURE 4.18 Graph of acceleration versus force.

TABLE 4.2 Acceleration with different numbers of objects

Number of objects	Acceleration
1	a_1
2	$a_2 = \frac{1}{2}a_1$
3	$a_3 = \frac{1}{3}a_1$
⋮	⋮
N	$a_N = \frac{1}{N}a_1$

original, single object, measure the acceleration of this new object. Three objects glued together make an object three times the size of the original. Doing several such experiments, applying the same force to each object, would give you the results shown in Table 4.2. An object twice the size of the original has only half the acceleration of the original object when both are subjected to the same force. An object three times the size of the original has one-third the acceleration.

Figure 4.19 shows these results added to the graph of Figure 4.18. You can see that the proportionality constant c between acceleration and force—the slope of the line—changes with the size of the object. The graph for an object twice the size of the original is a line with half the slope. It may seem surprising that larger objects have smaller slopes, so you'll want to think about this carefully.

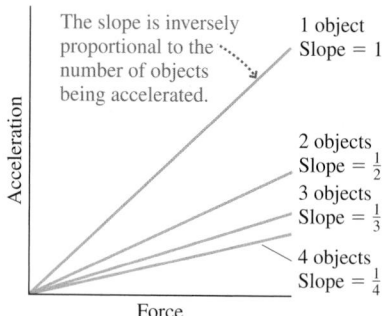

FIGURE 4.19 Acceleration-versus-force graphs for objects of different size.

Mass

Now, "twice the size" is a little vague; we could mean the object's external dimensions or some other measure. Although *mass* is a common word, we've avoided the term so far because we first need to define what mass is. Because we made the larger objects in our experiment from the same material as the original object, an object twice the size has twice as many atoms—twice the amount of matter—as the original. Thus it should come as no surprise that it has twice the mass as the original. Loosely speaking, **an object's mass is a measure of the amount of matter it contains.** This is certainly our everyday meaning of *mass*, but it is not yet a precise definition.

Figure 4.19 showed that an object with twice the amount of matter as the original accelerates only half as quickly if both experience the same force. An object with N times as much matter has only $\frac{1}{N}$ of the original acceleration. The more matter an object has, the more it *resists* accelerating in response to a force. You're familiar with this idea: Your car is much harder to push than your bicycle. The tendency of an object to resist a *change* in its velocity (i.e., to resist acceleration) is called **inertia.** Figure 4.19 tells us that larger objects have more inertia than smaller objects of the same material.

We can make this idea precise by defining the **inertial mass** m of an object to be

$$m \equiv \frac{1}{\text{slope of the acceleration-versus-force graph}} = \frac{F}{a}$$

We usually refer to the inertial mass as simply "the mass." Mass is an *intrinsic* property of an object. It is the property that determines how an object accelerates in response to an applied force.

> **STOP TO THINK 4.3** Two rubber bands stretched to the standard length cause an object to accelerate at 2 m/s^2. Suppose another object with twice the mass is pulled by four rubber bands stretched to the standard length. The acceleration of this second object is
>
> a. 1 m/s^2.   b. 2 m/s^2.   c. 4 m/s^2.   d. 8 m/s^2.   e. 16 m/s^2.

4.5 Newton's Second Law

We can now summarize the results of our experiment. Figure 4.18 showed that the acceleration is directly proportional to the force, a conclusion that we wrote in Equation 4.2 with the unspecified proportionality constant c. Now we see that c, the slope of the acceleration-versus-force graph, is the inverse of the inertial mass m.

Thus we've found that a force of magnitude F causes an object of mass m to accelerate with

$$a = \frac{F}{m}$$

This simple equation answers the question with which we started: How does an object move when a force is exerted on it? A force causes an object to *accelerate!* Furthermore, the size of the acceleration is directly proportional to the size of the force and inversely proportional to the object's mass.

This is an important finding, but our experiment was limited to looking at an object's response to a single applied force. Realistically, an object is likely to be subjected to several distinct forces $\vec{F}_1$, $\vec{F}_2$, $\vec{F}_3$, ... that may point in different directions. What happens then? In that case, it is found experimentally that the acceleration is determined by the *net* force.

Newton was the first to recognize the connection between force and motion. This relationship is known today as Newton's second law.

> **Newton's second law** An object of mass m subjected to forces $\vec{F}_1$, $\vec{F}_2$, $\vec{F}_3$, ... will undergo an acceleration $\vec{a}$ given by
>
> $$\vec{a} = \frac{\vec{F}_{net}}{m} \qquad (4.3)$$
>
> where the net force $\vec{F}_{net} = \vec{F}_1 + \vec{F}_2 + \vec{F}_3 + \cdots$ is the vector sum of the individual forces. The acceleration vector $\vec{a}$ points in the same direction as the net force vector $\vec{F}_{net}$.

It may seem puzzling that we've skipped over Newton's first law. The reasons for this will become clear as we continue our discussion of dynamics. For now, the critical idea is that **an object's acceleration vector $\vec{a}$ points in the same direction as the net force vector $\vec{F}_{net}$.**

The significance of Newton's second law cannot be overstated. There was no reason to suspect that there should be any simple relationship between force and acceleration. Yet there it is, a simple but exceedingly powerful equation relating the two. Newton's work, preceded to some extent by Galileo's, marks the beginning of a highly successful period in the history of science during which it was learned that the behavior of physical objects can often be described and predicted by mathematical relationships. While some relationships are found to apply only in special circumstances, others seem to have universal applicability. Those equations that appear to apply at all times and under all conditions have come to be called "laws of nature." Newton's second law is a law of nature; you will meet others as we go through this book.

We can rewrite Newton's second law in the form

$$\vec{F}_{net} = m\vec{a} \qquad (4.4)$$

which is how you'll see it presented in many textbooks. Equations 4.3 and 4.4 are mathematically equivalent, but Equation 4.3 better describes the central idea of Newtonian mechanics: A force applied to an object causes the object to accelerate.

Be careful not to think that one force "overcomes" the others to determine the motion. Forces are not in competition with each other! It is $\vec{F}_{net}$, the sum of *all* the forces, that determines the acceleration $\vec{a}$.

As an example, Figure 4.20a shows a box being pulled by two ropes. The ropes exert tension forces $\vec{T}_1$ and $\vec{T}_2$ on the box. Figure 4.20b represents the box as a particle, shows the forces acting on the box, and adds them graphically to

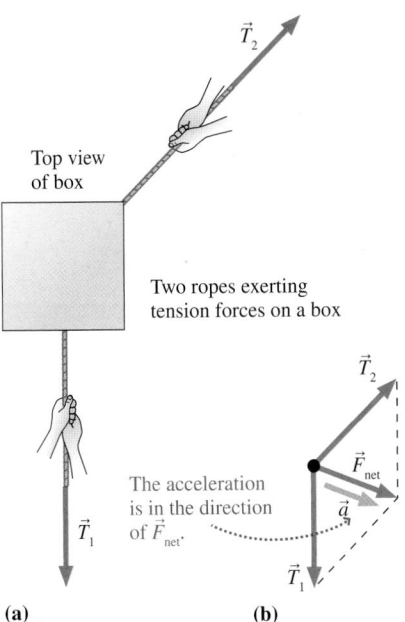

Top view of box

Two ropes exerting tension forces on a box

The acceleration is in the direction of $\vec{F}_{net}$.

(a)   **(b)**

FIGURE 4.20 Acceleration of a pulled box.

find the net force $\vec{F}_{\text{net}}$. The box will accelerate in the direction of $\vec{F}_{\text{net}}$ with an acceleration of magnitude

$$\vec{a} = \frac{\vec{F}_{\text{net}}}{m} = \frac{\vec{T}_1 + \vec{T}_2}{m}$$

NOTE ▶ The acceleration is *not* $(T_1 + T_2)/m$. You must add the forces as *vectors,* not merely add their magnitudes as scalars. ◀

Units of Force

Because $\vec{F}_{\text{net}} = m\vec{a}$, the units of force must be mass units multiplied by acceleration units. We've previously specified the SI unit of mass as the kilogram. We can now define the basic unit of force as "the force that causes a 1 kg mass to accelerate at 1 m/s^2." From the second law, this force is

$$1 \text{ basic unit of force} \equiv 1 \text{ kg} \times 1\frac{\text{m}}{\text{s}^2} = 1\frac{\text{kg m}}{\text{s}^2}$$

This basic unit of force is called a newton:

> One **newton** is the force that causes a 1 kg mass to accelerate at 1 m/s^2. The abbreviation for newton is N. Mathematically, 1 N = 1 kg m/s^2.

The newton is a *secondary unit,* meaning that it is defined in terms of the *primary units* of kilograms, meters, and seconds. We will introduce other secondary units as needed.

It is important to develop a feeling for what the size of forces should be. Table 4.3 shows some typical forces. As you can see, "typical" forces on "typical" objects are likely to be in the range 0.01–10,000 N. Forces less than 0.01 N are too small to consider unless you are dealing with very small objects. Forces greater than 10,000 N would make sense only if applied to very massive objects.

The unit of force in the English system is the *pound* (abbreviated lb). Although the definition of the pound has varied throughout history, it is now defined in terms of the newton:

$$1 \text{ pound} = 1 \text{ lb} \equiv 4.45 \text{ N}$$

You very likely associate pounds with kilograms rather than with newtons. Everyday language often confuses the ideas of mass and weight, but we're going to need to make a clear distinction between them. More on this in the next chapter.

TABLE 4.3 Approximate magnitude of some typical forces

Force	Approximate magnitude (newtons)
Weight of a U.S. quarter	0.05
Weight of a 1-pound object	5
Weight of a 110-pound person	500
Propulsion force of a car	5,000
Thrust force of a rocket motor	5,000,000

STOP TO THINK 4.4 Three forces act on an object. In which direction does the object accelerate?

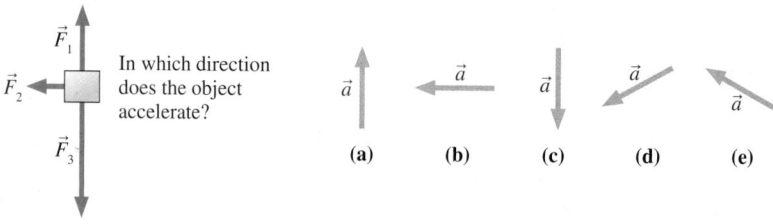

4.6 Newton's First Law

As we remarked earlier, Aristotle and his contemporaries in the world of ancient Greece were very interested in motion. One question they asked was: What is the "natural state" of an object if left to itself? It does not take an expensive research program to see that every moving object on earth, if left to itself, eventually comes to rest. Aristotle concluded that the natural state of an earthly object is to be at rest. An object at rest requires no explanation; it is doing precisely what comes naturally to it. A moving object, though, is not in its natural state and thus requires an explanation: Why is this object moving? What keeps it going and prevents it from being in its natural state?

Galileo reopened the question of the "natural state" of objects. He suggested focusing on the *limiting case* in which resistance to the motion (e.g., friction or air resistance) is zero. This is an idealization that may not be realizable in practice, but Galileo had asserted previously, with great success, that the idealized case can establish a *general principle*. Many careful experiments in which he minimized the influence of friction led Galileo to a conclusion that was in sharp contrast to Aristotle's belief that rest is an object's natural state.

Galileo found that an external influence (i.e., a force) is needed to make an object accelerate—to *change* its velocity. In particular, a force is needed to put an object in motion. But, in the absence of friction or air resistance, a moving object continues to move along a straight line forever with no loss of speed. In other words, the natural state of an object—its behavior if free of external influences—is *uniform motion* with constant velocity! This does not happen in practice because friction or air resistance prevents the object from being left alone. "At rest" has no special significance in Galileo's view of motion; it is simply uniform motion that happens to have $\vec{v} = \vec{0}$.

Galileo's experiments were limited to motion along horizontal surfaces. It was left to Newton to generalize this result, and today we call it Newton's first law of motion.

> **Newton's first law** An object that is at rest will remain at rest, or an object that is moving will continue to move in a straight line with constant velocity, if and only if the net force acting on the object is zero.

Newton's first law is also known as the *law of inertia*. If an object is at rest, it has a tendency to stay at rest. If it is moving, it has a tendency to continue moving with the *same velocity*.

> **NOTE** ▶ The first law refers to *net* force. An object can remain at rest, or can move in a straight line with constant velocity, even though forces are exerted on it as long as the *net* force is zero. ◀

Notice the "if and only if" aspect of Newton's first law. If an object is at rest or moves with constant velocity, then we can conclude that there is no net force acting on it. Conversely, if no net force is acting on it, we can conclude that the object will have constant velocity, not just constant speed. The direction remains constant, too!

An object on which the net force is zero, $\vec{F}_{net} = \vec{0}$ is said to be in **mechanical equilibrium.** According to Newton's first law, there are two distinct forms of mechanical equilibrium:

1. The object is at rest. This is **static equilibrium.**
2. The object is moving in a straight line with constant velocity. This is **dynamic equilibrium.**

Two examples of mechanical equilibrium are shown in Figure 4.21. Both share the common feature that the acceleration is zero: $\vec{a} = \vec{0}$.

What Good Is Newton's First Law?

The first law completes our definition of force. It answers the question: What is a force? If an "influence" on an object causes the object's velocity to change, the influence is a force.

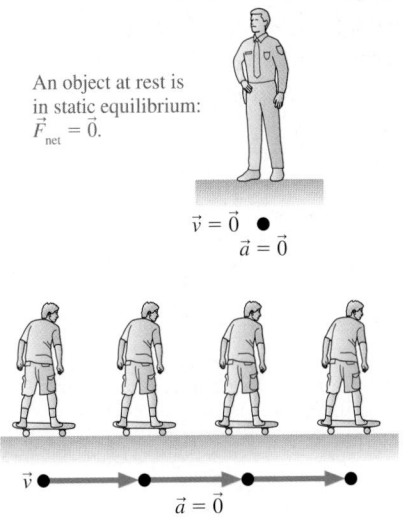

An object at rest is in static equilibrium: $\vec{F}_{net} = \vec{0}$.

$\vec{v} = \vec{0}$
$\vec{a} = \vec{0}$

$\vec{v}$
$\vec{a} = \vec{0}$

An object moving in a straight line at constant velocity is in dynamic equilibrium: $\vec{F}_{net} = \vec{0}$.

FIGURE 4.21 Two examples of mechanical equilibrium.

Newton's first law changes the question the ancient Greeks were trying to answer: What causes an object to move? Newton's first law says **no cause is needed for an object to move!** Uniform motion is the object's natural state. Nothing at all is required for it to remain in that state. The proper question, according to Newton, is: What causes an object to *change* its velocity? Newton, with Galileo's help, also gave us the answer. **A *force* is what causes an object to change its velocity.**

The preceding paragraph contains the essence of Newtonian mechanics. This new perspective on motion, however, is often contrary to our common experience. We all know perfectly well that you must keep pushing an object—exerting a force on it—to keep it moving. Newton is asking us to change our point of view and to consider motion *from the object's perspective* rather than from our personal perspective. As far as the object is concerned, our push is just one of several forces acting on it. Others might include friction, air resistance, or gravity. Only by knowing the *net* force can we determine the object's motion.

Newton's first law may seem to be merely a special case of Newton's second law. After all, the equation $\vec{F}_{net} = m\vec{a}$ tells us that an object moving with constant velocity ($\vec{a} = \vec{0}$) has $\vec{F}_{net} = \vec{0}$. The difficulty is that the second law assumes that we already know what force is. The purpose of the first law is to *identify* a force as something that disturbs a state of equilibrium. The second law then describes how the object responds to this force. Thus from a *logical* perspective, the first law really is a separate statement that must precede the second law. But this is a rather formal distinction. From a pedagogical perspective it is better—as we have done—to use a commonsense understanding of force and start with Newton's second law.

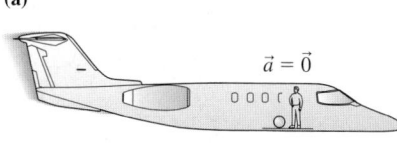

This guy thinks there's a force hurling him into the windshield. What a dummy!

(a)

$\vec{a} = \vec{0}$

The ball stays in place.

A ball with no horizontal forces stays at rest in an airplane cruising at constant velocity. The airplane is an inertial reference frame.

(b)

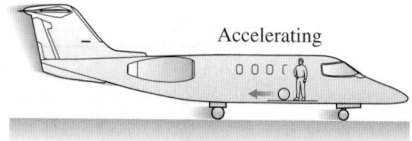

Accelerating

The ball rolls to the back.

The ball rolls to the back of the plane during takeoff. An accelerating plane is not an inertial reference frame.

FIGURE 4.22 Reference frames.

Inertial Reference Frames

If a car stops suddenly, you may be "thrown" into the windshield if you're not wearing your seat belt. You have a very real forward acceleration *relative to the car,* but is there a force pushing you forward? A force is a push or a pull caused by an identifiable agent in contact with the object. Although you *seem* to be pushed forward, there's no agent to do the pushing.

The difficulty—an acceleration without an apparent force—comes from using an inappropriate coordinate system. Your acceleration measured in a coordinate system attached to the car is not the same as your acceleration measured in a coordinate system attached to the ground. Newton's second law says $\vec{F}_{net} = m\vec{a}$. But which $\vec{a}$? Measured in which coordinate system?

We define an **inertial reference frame** as a coordinate system in which Newton's laws are valid. The first law provides a convenient way to test whether a coordinate system is an inertial reference frame. If $\vec{a} = \vec{0}$ (an object is at rest or moving with constant velocity) only when $\vec{F}_{net} = \vec{0}$, then the coordinate system in which $\vec{a}$ is measured is an inertial reference frame.

Not all coordinate systems are inertial reference frames. Figure 4.22a shows a physics student cruising at constant velocity in an airplane. If the student places a ball on the floor, it stays there. There are no horizontal forces, and the ball remains at rest relative to the airplane. That is, $\vec{a} = \vec{0}$ in the airplane's coordinate system when $\vec{F}_{net} = \vec{0}$. Newton's first law is satisfied, so this airplane is an inertial reference frame.

The physics student in Figure 4.22b conducts the same experiment during takeoff. She carefully places the ball on the floor just as the airplane starts to accelerate down the runway. You can imagine what happens. The ball rolls to the back of the plane as the passengers are being pressed back into their seats. Nothing exerts a horizontal contact force on the ball, yet the ball accelerates *in the plane's coordinate system.* This violates Newton's first law, so the plane is *not* an inertial reference frame during takeoff.

In the first example, the plane is traveling with constant velocity. In the second, the plane is accelerating. **Accelerating reference frames are not inertial reference frames.** Consequently, Newton's laws are not valid in a coordinate system attached to an accelerating object.

But accelerating with respect to what? The plane accelerated with respect to the earth. But the earth is accelerating as it rotates on its axis and revolves around the sun. The entire solar system is accelerating as our Milky Way galaxy rotates.

This is a subtle and difficult question, one that Einstein grappled with as he developed his theory of relativity. We cannot give a complete answer here, but suffice it to say that an inertial reference frame is a reference frame that is not accelerating *with respect to the distant stars*. Thus the earth is not exactly an inertial reference frame. However, the earth's acceleration with respect to the distant stars is so small that violations of Newton's laws can be measured only in extremely high-precision experiments. We will treat the earth and laboratories attached to the earth as inertial reference frames, an approximation that is exceedingly well justified.

We will prove in Chapter 6 that a coordinate system moving with constant velocity relative to an inertial reference frame is also an inertial reference frame, a reference frame in which Newton's laws are valid. Because the earth is an inertial reference frame, the airplane of Figure 4.22a is also an inertial reference frame. But a car braking to a stop is not, so you *cannot* use Newton's laws in the car's reference frame.

To understand the motion of objects in the car, such as the passengers, you need to measure velocities and accelerations *relative to the ground*. From the perspective of an observer on the ground, the body of a passenger in a braking car tries to continue moving forward with constant velocity, exactly as we would expect on the basis of Newton's first law, while his immediate surroundings are decelerating. The passenger is not "thrown" into the windshield. Instead, the windshield runs into the passenger!

Common Misconceptions About Force

It is important to identify correctly all the forces acting on an object. It is equally important not to include forces that do not really exist. We have established a number of criteria for identifying forces; the two critical ones are:

- A force has an agent. Something tangible and identifiable causes the force.
- Forces exist at the point of contact between the agent and the object experiencing the force (except for the few special cases of long-range forces).

We all have had many experiences suggesting that a force is necessary to keep something moving. Consider a bowling ball rolling along on a smooth floor. It is very tempting to think that a horizontal "force of motion" keeps it moving in the forward direction. But if we draw a closed curve around the ball, *nothing contacts it* except the floor. No agent is giving the ball a forward push. According to our definition, then, there is *no* forward "force of motion" acting on the ball. So what keeps it going? Recall our discussion of the first law: *no* cause is needed to keep an object moving at constant velocity. It continues to move forward simply because of its inertia.

One reason for wanting to include a "force of motion" is that we tend to view the problem from our perspective as one of the agents of force. You certainly have to keep pushing to shove a box across the floor at constant velocity. If you stop, it stops. Newton's laws, though, require that we adopt the object's perspective. The box experiences your pushing force in one direction *and* a friction force in the opposite direction. The box moves at constant velocity if the *net* force is zero. This will be true as long as your pushing force exactly balances the friction force.

When you stop pushing, the friction force causes an acceleration that slows and stops the box.

A related problem occurs if you throw a ball. A pushing force was indeed required to accelerate the ball *as it was thrown*. But that force disappears the instant the ball loses contact with your hand. The force does not stick with the ball as the ball travels through the air. Once the ball has acquired a velocity, *nothing* is needed to keep it moving with that velocity.

A final difficulty worth noting is the force due to air pressure. You may have learned in an earlier science class that air, like any fluid, exerts forces on objects. Perhaps you learned this idea as "the air presses down with a weight of 15 pounds on every square inch." There is only one error here, but it is a serious one: the word *down*. Air pressure, at sea level, does indeed exert a force of 15 pounds per square inch, but in *all* directions. It presses down on the top of an object, but also inward on the sides and upward on the bottom. For most purposes, the *net* force due to air pressure is zero! The only way to experience an air pressure force is to form a seal around one side of the object and then remove the air, creating a *vacuum*. When you press a suction cup against the wall, you press the air out and the rubber forms a seal that prevents the air from returning. Now the air pressure does hold the suction cup in place! We do not need to be concerned with air pressure until Part III of this book.

4.7 Free-Body Diagrams

Having discussed at length what is and is not a force, we are ready to assemble our knowledge about force and motion into a single diagram called a *free-body diagram.* You will learn in the next chapter how to write the equations of motion directly from the free-body diagram. Solution of the equations is a mathematical exercise—possibly a difficult one, but nonetheless an exercise that could be done by a computer. The *physics* of the problem, as distinct from the purely calculational aspects, are the steps that lead to the free-body diagram.

A **free-body diagram** represents the object as a particle and shows *all* of the forces acting on the object. The free-body diagram joins with the motion diagram and force identification to form the full *physical representation* of a problem.

TACTICS BOX 4.3 **Drawing a free-body diagram**

❶ **Identify all forces acting on the object.** This step was described in Tactics Box 4.2.

❷ **Draw a coordinate system.** Use the axes defined in your pictorial representation. If those axes are tilted, for motion along an incline, then the axes of the free-body diagram should be similarly tilted.

❸ **Represent the object as a dot at the origin of the coordinate axes.** This is the particle model.

❹ **Draw vectors representing each of the identified forces.** This was described in Tactics Box 4.1. Be sure to label each force vector.

❺ **Draw and label the *net force* vector $\vec{F}_{net}$.** Draw this vector beside the diagram, not on the particle. Or, if appropriate, write $\vec{F}_{net} = \vec{0}$. Then check that $\vec{F}_{net}$ points in the same direction as the acceleration vector $\vec{a}$ on your motion diagram.

EXAMPLE 4.4 **An elevator accelerates upward**

An elevator, suspended by a cable, speeds up as it moves upward from the ground floor. Draw a free-body diagram of the elevator.

MODEL Treat the elevator as a particle.

VISUALIZE

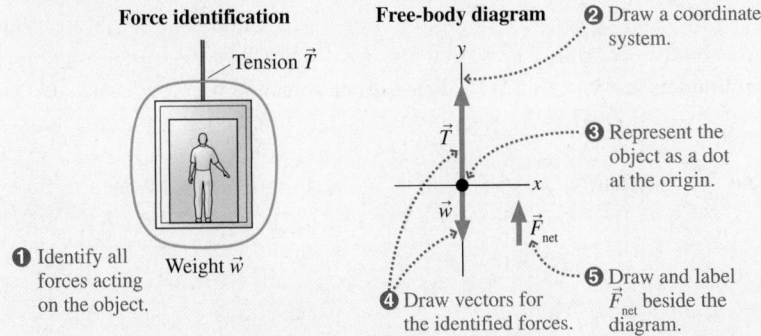

FIGURE 4.23 Free-body diagram of an elevator accelerating upward.

ASSESS The coordinate axes, with a vertical y-axis, are the ones we would use in a pictorial representation of the motion. The elevator is accelerating upward, so $\vec{F}_{net}$ must point upward. For this to be true, the magnitude of $\vec{T}$ must be larger than the magnitude of $\vec{w}$. The diagram has been drawn accordingly.

EXAMPLE 4.5 **An ice block shoots across a frozen lake**

Bobby straps a small model rocket to a block of ice and shoots it across the smooth surface of a frozen lake. Friction is negligible. Draw a full physical representation of the block of ice.

MODEL Treat the block of ice as a particle. The full physical representation consists of a motion diagram to determine $\vec{a}$, a force identification picture, and a free-body diagram. The statement of the situation implies that friction is negligible.

VISUALIZE

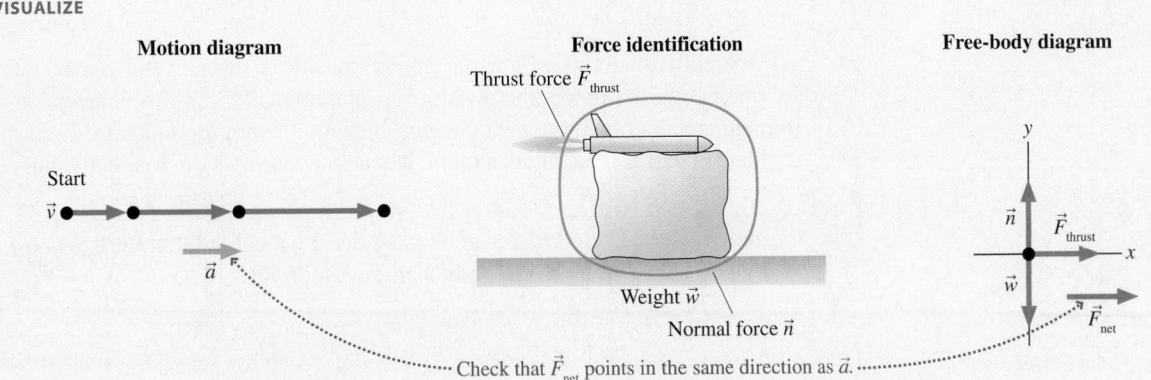

FIGURE 4.24 Physical representation for a block of ice shooting across a frictionless frozen lake.

ASSESS The motion diagram tells us that the acceleration is in the $+x$-direction. According to the rules of vector addition, this can be true only if the upward-pointing $\vec{n}$ and the downward-pointing $\vec{w}$ are equal in magnitude and thus cancel each other $(w_y = -n_y)$. The vectors have been drawn accordingly, and this leaves the net force vector pointing toward the right, in agreement with $\vec{a}$ from the motion diagram.

EXAMPLE 4.6 A skier is pulled up a hill

A tow rope pulls a skier up a snow-covered hill at a constant speed. Draw a full physical representation of the skier.

MODEL This is Example 4.2 again with the additional information that the skier is moving at constant speed. The skier will be treated as a particle in *dynamic equilibrium*. If we were doing a kinematics problem, the pictorial representation would use a tilted coordinate system with the *x*-axis parallel to the slope, so we use these same tilted coordinate axes for the free-body diagram.

VISUALIZE

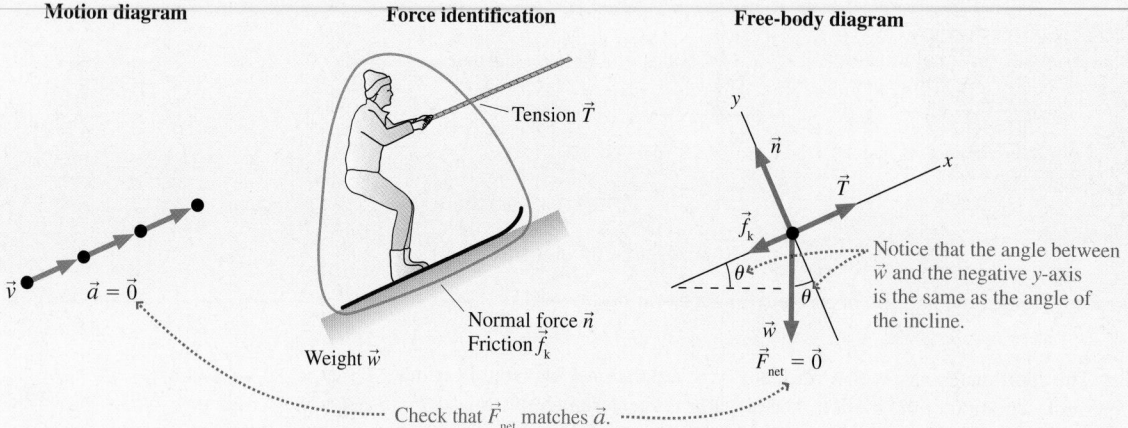

FIGURE 4.25 Physical representation for a skier being towed at a constant speed.

ASSESS We have shown $\vec{T}$ pulling parallel to the slope and $\vec{f}_k$, which opposes the direction of motion, pointing down the slope. $\vec{n}$ is perpendicular to the surface and thus along the *y*-axis. Finally, and this is important, the weight $\vec{w}$ is *vertically* downward, *not* along the negative *y*-axis. In fact, you should convince yourself from the geometry that the angle θ between the $\vec{w}$ vector and the negative *y*-axis is the same as the angle θ of the incline above the horizontal. The skier moves in a straight line with constant speed, so $\vec{a} = \vec{0}$ and, from Newton's first law, $\vec{F}_{net} = \vec{0}$. Thus we have drawn the vectors such that the *y*-component of $\vec{w}$ is equal in magnitude to $\vec{n}$. Similarly, $\vec{T}$ must be large enough to match the negative *x*-components of both $\vec{f}_k$ and $\vec{w}$.

Free-body diagrams will be our major tool for the next several chapters. Careful practice with the workbook exercises and homework in this chapter will pay immediate benefits in the next chapter. Indeed, it is not too much to assert that a problem is half solved, or even more, when you complete the free-body diagram.

STOP TO THINK 4.5 An elevator suspended by a cable is moving upward and slowing to a stop. Which free-body diagram is correct?

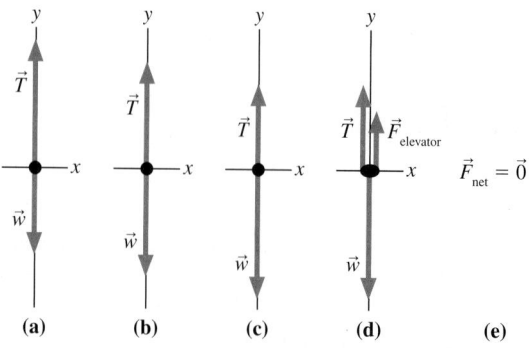

SUMMARY

The goal of Chapter 4 has been to learn how force and motion are connected.

GENERAL PRINCIPLES

Newton's First Law

An object at rest will remain at rest, or an object that is moving will continue to move in a straight line with constant velocity, if and only if the net force on the object is zero.

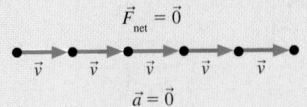

The first law tells us that no "cause" is needed for motion. Uniform motion is the "natural state" of an object.

Newton's laws are valid only in inertial reference frames.

Newton's Second Law

An object with mass m will undergo acceleration

$$\vec{a} = \frac{1}{m}\vec{F}_{\text{net}}$$

where $\vec{F}_{\text{net}} = \vec{F}_1 + \vec{F}_2 + \vec{F}_3 + \cdots$ is the vector sum of all the individual forces acting on the object.

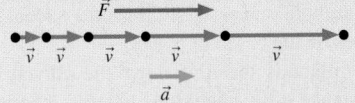

The second law tells us that a net force causes an object to accelerate. This is the connection between force and motion that we are seeking.

IMPORTANT CONCEPTS

Acceleration is the link to kinematics.

From a, find v and x.
From v and x, find a.

$\vec{a} = \vec{0}$ is the condition for equilibrium.

Static equilibrium if $\vec{v} = \vec{0}$.
Dynamic equilibrium if $\vec{v} = $ constant.

Equilibrium occurs if and only if $\vec{F}_{\text{net}} = \vec{0}$.

Mass is the resistance of an object to acceleration. It is an intrinsic property of an object.

Force is a push or a pull on an object.

- Force is a vector, with a magnitude and a direction.

- Force requires an agent.

- Force is either a contact force or a long-range force.

KEY SKILLS

Identifying Forces

Forces are identified by locating the points where the environment touches the system. These are points where contact forces are exerted. In addition, objects with mass feel a long-range weight force.

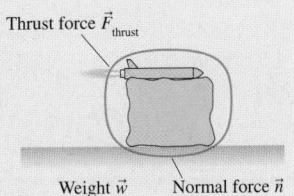

Free-Body Diagrams

A free-body diagram represents the object as a particle at the origin of a coordinate system. Force vectors are drawn with their tails on the particle. The net force vector is drawn beside the diagram.

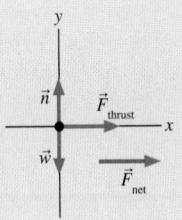

TERMS AND NOTATION

dynamics	superposition of forces	thrust, $\vec{F}_{\text{thrust}}$	static equilibrium
mechanics	weight, $\vec{w}$	inertia	dynamic equilibrium
force, $\vec{F}$	tension force, $\vec{T}$	inertial mass, m	inertial reference frame
agent	atomic model	Newton's second law	free-body diagram
contact force	normal force, $\vec{n}$	newton, N	
long-range force	friction, $\vec{f}_k$ or $\vec{f}_s$	Newton's first law	
net force, $\vec{F}_{\text{net}}$	drag, $\vec{D}$	mechanical equilibrium	

EXERCISES AND PROBLEMS

Exercises

Section 4.1 Force

1. Write a one-paragraph essay on the topic "What Is a Force?" Explain in your own words how you recognize a force and what the properties of forces are.

Section 4.3 Identifying Forces

2. A mountain climber is hanging from a rope in the middle of a crevasse. The rope is vertical. Identify the forces on the mountain climber.
3. A baseball player is sliding into second base. Identify the forces on the baseball player.
4. A jet plane is speeding down the runway during takeoff. Air resistance is not negligible. Identify the forces on the jet.

Section 4.4 What Do Forces Do?

5. Figure Ex4.5 shows an acceleration-versus-force graph for three objects pulled by rubber bands. The mass of object 2 is 0.20 kg. What are the masses of objects 1 and 3? Explain your reasoning.

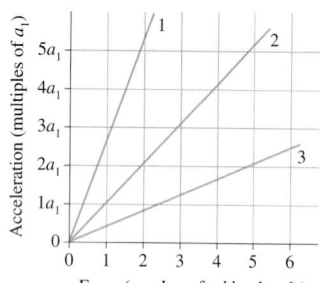

FIGURE EX4.5

6. Two rubber bands pulling on an object cause it to accelerate at $1.2 \, \text{m/s}^2$.
   a. What will be the object's acceleration if it is pulled by four rubber bands?
   b. What will be the acceleration of two of these objects glued together if they are pulled by two rubber bands?

Section 4.5 Newton's Second Law

7. Write a one-paragraph essay on the topic "Force and Motion." Explain in your own words the connection between force and motion. Where possible, cite *evidence* supporting your statements.
8. Figure Ex4.8 shows an acceleration-versus-force graph for a 500 g object. Redraw this graph and add appropriate acceleration values on the vertical scale.

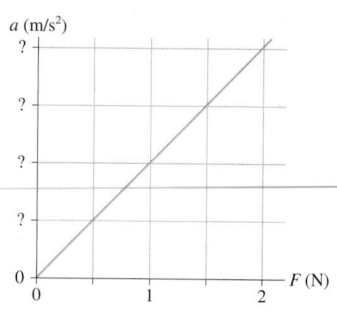

FIGURE EX4.8

9. Figure Ex4.9 shows an object's acceleration-versus-force graph. What is the object's mass?

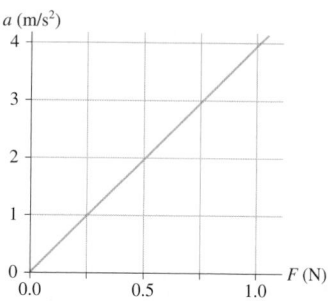

FIGURE EX4.9

10. Based on the information in Table 4.3, estimate
    a. The weight of this textbook.
    b. The propulsion force of a bicycle.
11. Based on the information in Table 4.3, estimate
    a. The weight of a pencil.
    b. The propulsion force of a sprinter.

Section 4.6 Newton's First Law

Exercises 12 through 14 show two forces acting on an object. Redraw the diagram, then add a third force that will cause the object to be in equilibrium. Label the new force $\vec{F}_3$.

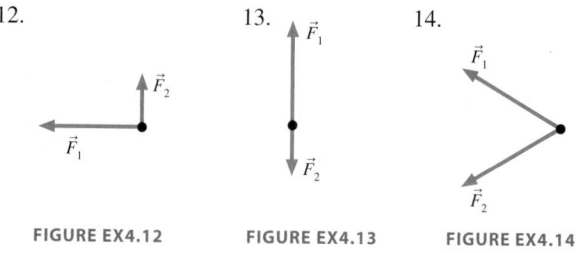

FIGURE EX4.12   FIGURE EX4.13   FIGURE EX4.14

Section 4.7 Free-Body Diagrams

Exercises 15 through 17 show a free-body diagram. For each:
   a. Redraw the free-body diagram.
   b. Write a short description of a real object for which this is the correct free-body diagram. Use Examples 4.4, 4.5, and 4.6 as models of what a description should be like.

15.

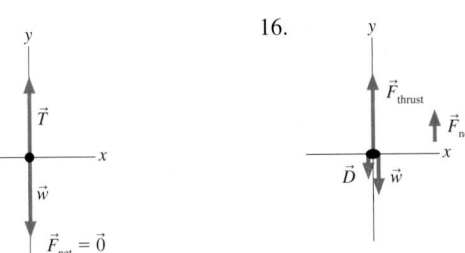

FIGURE EX4.15

16.

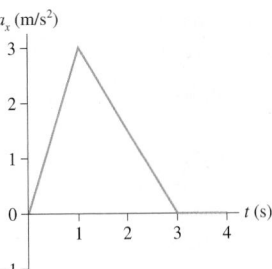

FIGURE EX4.16

17.

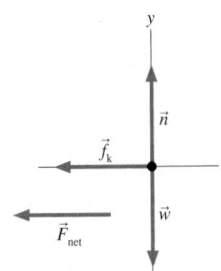

FIGURE EX4.17

Exercises 18 through 21 describe a situation. For each, identify all forces acting on the object and draw a free-body diagram of the object.

18. Your car is sitting in the parking lot.
19. An ice hockey puck glides across frictionless ice.
20. An elevator, hanging from a cable, descends at steady speed.
21. Your physics textbook is sliding across the table.

Problems

22. Redraw the two motion diagrams shown in Figure P4.22, then draw a vector beside each one to show the direction of the net force acting on the object. Explain your reasoning.

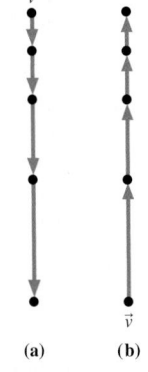

FIGURE P4.22   **(a)**   **(b)**

23. Redraw the two motion diagrams shown in Figure P4.23, then draw a vector beside each one to show the direction of the net force acting on the object. Explain your reasoning.

(a)   **(b)**

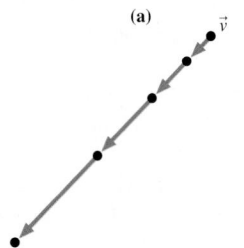   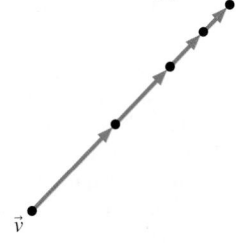

FIGURE P4.23

24. A force with x-component F_x acts on a 2.0 kg object as it moves along the x-axis. The object's acceleration graph (a_x versus t) is shown in Figure P4.24. Draw a graph of F_x versus t.

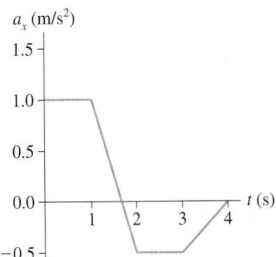

FIGURE P4.24

25. A force with x-component F_x acts on a 500 g object as it moves along the x-axis. The object's acceleration graph (a_x versus t) is shown in Figure P4.25. Draw a graph of F_x versus t.

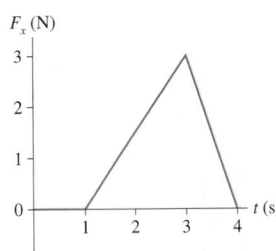

FIGURE P4.25

26. A force with x-component F_x acts on a 2.0 kg object as it moves along the x-axis. A graph of F_x versus t is shown in Figure P4.26. Draw an acceleration graph (a_x versus t) for this object.

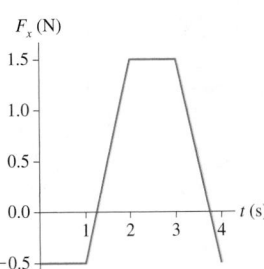

FIGURE P4.26

27. A force with x-component F_x acts on a 500 g object as it moves along the x-axis. A graph of F_x versus t is shown in Figure P4.27. Draw an acceleration graph (a_x versus t) for this object.

FIGURE P4.27

28. A constant force is applied to an object, causing the object to accelerate at 10 m/s². What will the acceleration be if
   a. The force is halved?
   b. The object's mass is halved?
   c. The force and the object's mass are both halved?
   d. The force is halved and the object's mass is doubled?

29. A constant force is applied to an object, causing the object to accelerate at 8.0 m/s². What will the acceleration be if
    a. The force is doubled?
    b. The object's mass is doubled?
    c. The force and the object's mass are both doubled?
    d. The force is doubled and the object's mass is halved?

Problems 30 through 36 show a free-body diagram. For each:
   a. Redraw the diagram.
   b. Identify the direction of the acceleration vector $\vec{a}$ and show it as a vector next to your diagram. Or, if appropriate, write $\vec{a} = \vec{0}$.
   c. If possible, identify the direction of the velocity vector $\vec{v}$ and show it as a labeled vector.
   d. Write a short description of a real object for which this is the correct free-body diagram. Use Examples 4.4, 4.5, and 4.6 as models of what a description should be like.

30.

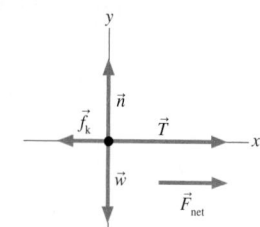

FIGURE P4.30

31.

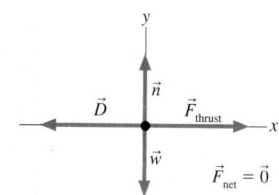

FIGURE P4.31

32.

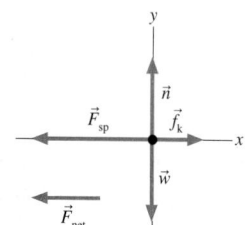

FIGURE P4.32

33.

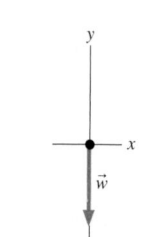

FIGURE P4.33

34.

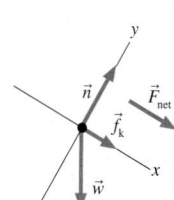

FIGURE P4.34

35.

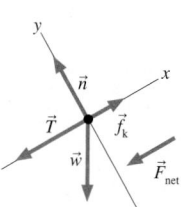

FIGURE P4.35

36.

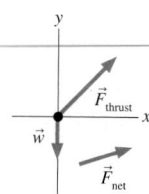

FIGURE P4.36

Problems 37 through 46 describe a situation. For each, draw a full physical representation. A full physical representation is a motion diagram, force identification, and a free-body diagram.

37. An elevator, suspended by a single cable, has just left the tenth floor and is speeding up as it descends toward the ground floor.
38. A rocket is being launched straight up. Air resistance is not negligible.
39. A jet plane is speeding down the runway during takeoff. Air resistance is not negligible.
40. You've slammed on the brakes and your car is skidding to a stop while going down a 20° hill.
41. A skier is going down a 20° slope. A *horizontal* headwind is blowing in the skier's face. Friction is small, but not zero.
42. You've just kicked a soccer ball and it is now rolling across the grass.
43. A styrofoam ball has just been shot straight up. Air resistance is not negligible.
44. A spring-loaded gun shoots a plastic ball. The trigger has just been pulled and the ball is starting to move down the barrel. The barrel is horizontal.
45. A person on a bridge throws a rock straight down toward the water. The rock has just been released.
46. A gymnast has just landed on a trampoline. She's still moving downward as the trampoline stretches.

Challenge Problems

47. A heavy box is in the back of a truck. The truck is accelerating to the right. Draw a full physical representation of the box.
48. A bag of groceries is on the back seat of your car as you stop for a stop light. The bag does not slide. Draw a full physical representation of the bag.
49. A rubber ball bounces. We'd like to understand *how* the ball bounces.
    a. A rubber ball has been dropped and is bouncing off the floor. Draw a motion diagram of the ball during the brief time interval that it is in contact with the floor. Show 4 or 5 frames as the ball compresses, then another 4 or 5 frames as it expands. What is the direction of $\vec{a}$ during each of these parts of the motion?
    b. Draw a picture of the ball in contact with the floor and identify all forces acting on the ball.

c. Draw a free-body diagram of the ball during its contact with the ground. Is there a net force acting on the ball? If so, in which direction?

d. Write a paragraph in which you describe what you learned from parts a to c and in which you answer the question: How does a ball bounce?

50. If a car stops suddenly, you feel "thrown forward." We'd like to understand what happens to the passengers as a car stops. Imagine yourself sitting on a *very* slippery bench inside a car. This bench has no friction, no seat back, and there's nothing for you to hold to.

a. Draw a picture and identify all of the forces acting on you as the car travels at a perfectly steady speed on level ground.

b. Draw your free-body diagram. Is there a net force on you? If so, in which direction?

c. Repeat parts a and b with the car slowing down.

d. Describe what happens to you as the car slows down.

e. Use Newton's laws to explain why you seem to be "thrown forward" as the car stops. Is there really a force pushing you forward?

f. Suppose now that the bench is not slippery. As the car slows down, you stay on the bench and don't slide off. What force is responsible for your deceleration? In which direction does this force point? Include a free-body diagram as part of your answer.

Stop to Think 4.1: c.

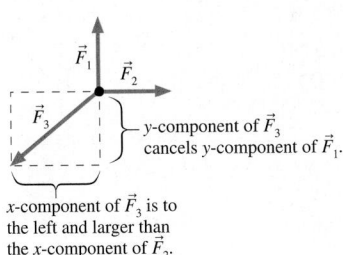

Stop to Think 4.4: d.

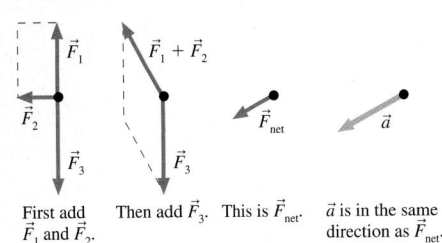

Stop to Think 4.2: a, b, and d. Friction and the normal force are the only contact forces. Nothing is touching the rock to provide a "force of the kick." We've agreed to ignore air resistance unless a problem specifically calls for it.

Stop to Think 4.3: b. Acceleration is proportional to force, so doubling the number of rubber bands doubles the acceleration of the original object from 2 m/s^2 to 4 m/s^2. But acceleration is also inversely proportional to mass. Doubling the mass cuts the acceleration in half, back to 2 m/s^2.

Stop to Think 4.5: c. The acceleration vector points downward as the elevator slows. $\vec{F}_{net}$ points in the same direction as $\vec{a}$, so $\vec{F}_{net}$ also points down. This will be true if the tension is less than the weight: $T < w$.

5 Dynamics I: Motion Along a Line

This skydiver may not know it, but he is testing Newton's second law as he plunges toward the ground below.

▶ **Looking Ahead**

The goal of Chapter 5 is to learn how to solve problems about motion in a straight line. In this chapter you will learn to:

- Solve static and dynamic equilibrium problems by applying a Newton's-first-law strategy.
- Solve dynamics problems by applying a Newton's-second-law strategy.
- Understand how mass, weight, and apparent weight differ.
- Use simple models of friction and drag.

◀ **Looking Back**

This chapter pulls together many strands of thought from Chapters 1–4. Please review:

- Sections 2.5–2.7 Constant acceleration kinematics, including free fall.
- Sections 3.3–3.4 Working with vectors and vector components.
- Sections 4.2, 4.3, and 4.7 Identifying forces and drawing free-body diagrams.

A skydiver accelerates until reaching a *terminal speed* of about 140 mph. To understand the skydiver's motion, we need to look closely at the forces exerted on him. We also need to understand how those forces determine his motion.

In Chapter 4 we learned what a force is and is not. We also discovered the fundamental relationship between force and motion: Newton's second law. Chapter 5 begins to develop a *strategy* for solving force and motion problems. Our strategy is to learn a set of *procedures,* not to memorize a set of equations.

This chapter focuses on objects that move in a straight line, such as runners, bicycles, cars, planes, and rockets. Weight, tension, thrust, friction, and drag forces will be essential to our understanding. Projectile motion and motion in a circle are the topics of the next two chapters.

122

5.1 Equilibrium

An object on which the net force is zero is said to be in *equilibrium*. The object might be at rest in *static equilibrium*, or it might be moving along a straight line with constant velocity in *dynamic equilibrium*. Both are identical from a Newtonian perspective because $\vec{F}_{net} = \vec{0}$ and $\vec{a} = \vec{0}$.

Newton's first law is the basis for a four-step *strategy* for solving equilibrium problems.

 PROBLEM-SOLVING STRATEGY 5.1 Equilibrium problems

MODEL Make simplifying assumptions.

VISUALIZE

Physical representation. Identify all forces acting on the object and show them on a free-body diagram.

Pictorial representation. The free-body diagram is usually sufficient as a picture for equilibrium problems, but you still must translate words to symbols and identify what the problem is trying to find.

It's OK to go back and forth between these two steps as you visualize the situation.

SOLVE The mathematical representation is based on Newton's first law

$$\vec{F}_{net} = \sum_i \vec{F}_i = \vec{0}$$

The vector sum of the forces is found directly from the free-body diagram.

ASSESS Check that your result has the correct units, is reasonable, and answers the question.

The concept of equilibrium is essential for the engineering analysis of stationary objects such as bridges.

Newton's laws are *vector equations*. Recall from Chapter 3 that the vector equation in the step labeled Solve is a shorthand way of writing two simultaneous equations:

$$(F_{net})_x = \sum_i (F_i)_x = 0$$

$$(F_{net})_y = \sum_i (F_i)_y = 0$$

(5.1)

In other words, each component of $\vec{F}_{net}$ must simultaneously be zero. Although real-world situations often have forces pointing in three dimensions, thus requiring a third equation for the *z*-component of $\vec{F}_{net}$, we will restrict ourselves for now to problems that can be analyzed in two dimensions.

Equilibrium problems occur frequently, especially in engineering applications. Let's look at a couple of examples.

Static Equilibrium

EXAMPLE 5.1 Three-way tug-of-war
You and two friends find three ropes tied together with a single knot and decide to have a three-way tug-of-war. Alice pulls to the west with 100 N of force while Bob pulls to the south with

200 N. How hard, and in which direction, should you pull to keep the knot from moving?

MODEL We'll treat the *knot* in the rope as a particle in static equilibrium.

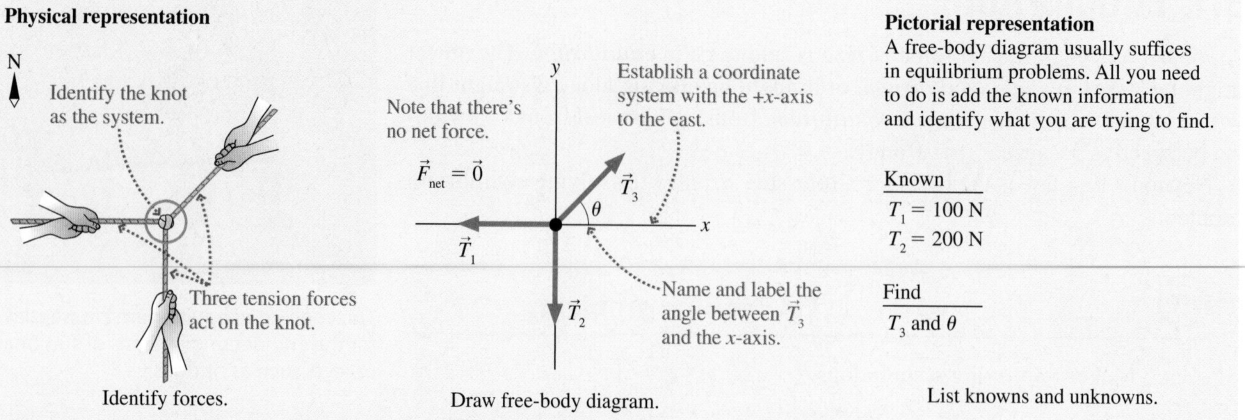

Physical representation

Identify the knot as the system.

N

Three tension forces act on the knot.

Identify forces.

Note that there's no net force.

$$\vec{F}_{net} = \vec{0}$$

Establish a coordinate system with the +x-axis to the east.

$\vec{T}_3$

θ

$\vec{T}_1$

$\vec{T}_2$

Name and label the angle between $\vec{T}_3$ and the x-axis.

Draw free-body diagram.

Pictorial representation
A free-body diagram usually suffices in equilibrium problems. All you need to do is add the known information and identify what you are trying to find.

Known
$T_1 = 100$ N
$T_2 = 200$ N

Find
T_3 and θ

List knowns and unknowns.

FIGURE 5.1 Physical and pictorial representations for a knot in static equilibrium.

VISUALIZE Figure 5.1 shows how to carry out the physical and pictorial representations. Notice that we've *defined* angle θ to indicate the direction of your pull.

SOLVE The free-body diagram shows tension forces $\vec{T}_1$, $\vec{T}_2$, and $\vec{T}_3$ acting on the knot. Newton's first law, written in component form, is

$$(F_{net})_x = \sum_i (F_i)_x = T_{1x} + T_{2x} + T_{3x} = 0$$

$$(F_{net})_y = \sum_i (F_i)_y = T_{1y} + T_{2y} + T_{3y} = 0$$

NOTE ▶ You might have been tempted to write $-T_{1x}$ in the first equation because $\vec{T}_1$ points in the negative x-direction. But the net force, by definition, is the *sum* of all the individual forces. The fact that $\vec{T}_1$ points to the left will be taken into account when we *evaluate* the components. ◀

The components of the force vectors can be evaluated directly from the free-body diagram:

$$T_{1x} = -T_1 \qquad T_{1y} = 0$$
$$T_{2x} = 0 \qquad T_{2y} = -T_2$$
$$T_{3x} = +T_3 \cos\theta \qquad T_{3y} = +T_3 \sin\theta$$

This is where the signs enter, with T_{1x} being assigned a negative value because $\vec{T}_1$ points to the left. Similarly, $T_{2y} = -T_2$. With these components, Newton's first law becomes

$$-T_1 + T_3 \cos\theta = 0$$
$$-T_2 + T_3 \sin\theta = 0$$

These are two simultaneous equations for the two unknowns T_3 and θ. We will encounter equations of this form on many occasions, so make a note of the method of solution. First, rewrite the two equations as

$$T_1 = T_3 \cos\theta$$
$$T_2 = T_3 \sin\theta$$

Next, divide the second equation by the first to eliminate T_3:

$$\frac{T_2}{T_1} = \frac{T_3 \sin\theta}{T_3 \cos\theta} = \tan\theta$$

Then solve for θ:

$$\theta = \tan^{-1}\left(\frac{T_2}{T_1}\right) = \tan^{-1}\left(\frac{200 \text{ N}}{100 \text{ N}}\right) = 63.4°$$

Finally, use θ to find T_3:

$$T_3 = \frac{T_1}{\cos\theta} = \frac{100 \text{ N}}{\cos 63.4°} = 224 \text{ N}$$

The force that maintains equilibrium and prevents the knot from moving is thus

$$\vec{T}_3 = (224 \text{ N}, 63.4° \text{ north of east})$$

ASSESS Is this result reasonable? Because your friends pulled west and south, you expected to pull in a generally northeast direction. You also expected to pull harder than either of them but, because they didn't pull in the same direction, less than the sum of their pulls. The result for $\vec{T}_3$ meets these expectations.

Dynamic Equilibrium

EXAMPLE 5.2 Towing a car up a hill
A car with a weight of 15,000 N is being towed up a 20° slope at constant velocity. Friction is negligible. The tow rope is rated at 6000 N maximum tension. Will it break?

MODEL We'll treat the car as a particle in dynamic equilibrium. We'll ignore friction.

VISUALIZE This problem asks for a yes or no answer, not a number, but we still need a quantitative analysis. Part of our analysis

Physical representation

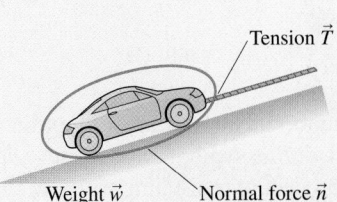

Tension $\vec{T}$

Weight $\vec{w}$ Normal force $\vec{n}$

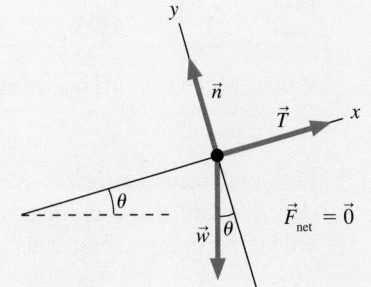

$\vec{F}_{net} = \vec{0}$

Pictorial representation

Known

$\theta = 20°$
$w = 15{,}000 \text{ N}$

Find

T

FIGURE 5.2 The physical and pictorial representations of a car being towed up a hill.

of the problem statement is to determine which quantity or quantities allow us to answer the question. In this case the answer is clear: We need to calculate the tension in the rope. Figure 5.2 shows the physical and pictorial representations. Note the similarities to Examples 4.2 and 4.6 in Chapter 4, which you may want to review.

SOLVE The free-body diagram shows forces $\vec{T}$, $\vec{n}$, and $\vec{w}$ acting on the car. Newton's first law is

$$(F_{net})_x = \sum F_x = T_x + n_x + w_x = 0$$
$$(F_{net})_y = \sum F_y = T_y + n_y + w_y = 0$$

Notice that we dropped the label i from the sum. From here on, we'll use $\sum F_x$ and $\sum F_y$ as a simple shorthand notation to indicate that we're adding all the x-components and all the y-components of the forces.

We can deduce the components directly from the free-body diagram:

$T_x = T$	$T_y = 0$
$n_x = 0$	$n_y = n$
$w_x = -w\sin\theta$	$w_y = -w\cos\theta$

NOTE ▶ The weight has both x- and y-components in this coordinate system, both of which are negative due to the direction of the vector $\vec{w}$. You'll see this situation often, so be sure you understand where w_x and w_y come from. ◀

With these components, the first law becomes

$$T - w\sin\theta = 0$$
$$n - w\cos\theta = 0$$

The first of these can be rewritten as

$$T = w\sin\theta$$
$$= (15{,}000 \text{ N})\sin 20° = 5130 \text{ N}$$

Because $T < 6000$ N, we conclude that the rope will *not* break. It turned out that we did not need the y-component equation in this problem.

ASSESS Because there's no friction, it would not take *any* tension force to keep the car rolling along a horizontal surface ($\theta = 0°$). At the other extreme, $\theta = 90°$, the tension force would need to equal the car's weight ($T = w = 15{,}000$ N) to lift the car straight up at constant velocity. The tension force for a 20° slope should be somewhere in between, and 5130 N is a little less than half the weight of the car. That our result is reasonable doesn't prove it's right, but we have at least ruled out careless errors that give unreasonable results.

5.2 Using Newton's Second Law

Equilibrium is important, but it is a special case of motion. Newton's second law is a more general link between force and motion. We now need a strategy for using Newton's second law to solve dynamics problems.

The essence of Newtonian mechanics can be expressed in two steps:

- The forces on an object determine its acceleration $\vec{a} = \vec{F}_{net}/m$.
- The object's trajectory can be determined by using $\vec{a}$ in the equations of kinematics.

These two ideas are the basis of a strategy for solving dynamics problems.

 2.1, 2.2, 2.3, 2.4

PROBLEM-SOLVING STRATEGY 5.2 Dynamics problems

MODEL Make simplifying assumptions.

VISUALIZE

Pictorial representation. Show important points in the motion with a sketch, establish a coordinate system, define symbols, and identify what the problem is trying to find. This is the process of translating words to symbols.

Physical representation. Use a motion diagram to determine the object's acceleration vector $\vec{a}$. Then identify all forces acting on the object and show them on a free-body diagram.

It's OK to go back and forth between these two steps as you visualize the situation.

SOLVE The mathematical representation is based on Newton's second law

$$\vec{F}_{net} = \sum_i \vec{F}_i = m\vec{a}$$

The vector sum of the forces is found directly from the free-body diagram. Depending on the problem, either

- Solve for the acceleration, then use kinematics to find velocities and positions, or
- Use kinematics to determine the acceleration, then solve for unknown forces.

ASSESS Check that your result has the correct units, is reasonable, and answers the question.

Newton's second law is a vector equation. To apply the step labeled Solve, you must write the second law as two simultaneous equations:

$$(F_{net})_x = \sum F_x = ma_x$$
$$(F_{net})_y = \sum F_y = ma_y$$

(5.2)

The primary goal of this chapter is to illustrate the use of this strategy. Let's start with some examples.

EXAMPLE 5.3 Speed of a towed car

A 1500 kg car is pulled by a tow truck. The tension in the tow rope is 2500 N, and a 200 N friction force opposes the motion. If the car starts from rest, what is its speed after 5.0 seconds?

MODEL We'll treat the car as an accelerating particle. We'll assume, as part of our *interpretation* of the problem, that the road is horizontal and that the direction of motion is to the right.

VISUALIZE Figure 5.3 shows the pictorial and physical representations. We've established a coordinate system and defined symbols to represent kinematic quantities. We've identified the speed v_1, rather than the velocity v_{1x}, as what we're trying to find.

SOLVE We begin the mathematical representation with Newton's second law:

$$(F_{net})_x = \sum F_x = T_x + f_x + n_x + w_x = ma_x$$
$$(F_{net})_y = \sum F_y = T_y + f_y + n_y + w_y = ma_y$$

All four forces acting on the car have been included in the vector sum. The equations are perfectly general, with $+$ signs everywhere, because the four vectors are *added* to give $\vec{F}_{net}$. We can now "read" the vector components directly from the free-body diagram:

$$T_x = +T \qquad T_y = 0$$
$$n_x = 0 \qquad n_y = +n$$
$$f_x = -f \qquad f_y = 0$$
$$w_x = 0 \qquad w_y = -w$$

Pictorial representation

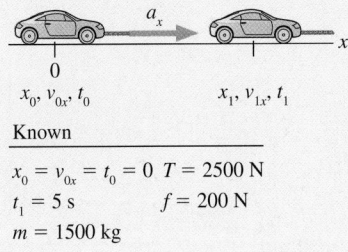

Known

$x_0 = v_{0x} = t_0 = 0$ $T = 2500$ N
$t_1 = 5$ s $f = 200$ N
$m = 1500$ kg

Find

v_1

Physical representation

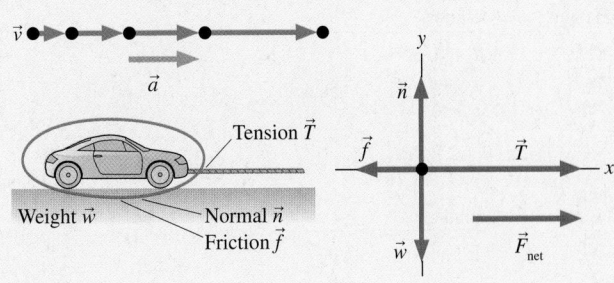

FIGURE 5.3 Pictorial and physical representations of a car being towed.

The signs of the components depend on which way the vectors point. Substituting these into the second-law equations and dividing by m gives

$$a_x = \frac{1}{m}(T - f)$$
$$= \frac{1}{1500 \text{ kg}}(2500 \text{ N} - 200 \text{ N}) = 1.53 \text{ m/s}^2$$
$$a_y = \frac{1}{m}(n - w)$$

NOTE ▶ Newton's second law has allowed us to determine a_x exactly but has given only an algebraic expression for a_y. However, we know *from the motion diagram* that $a_y = 0$! That is, the motion is purely along the x-axis, so there is *no* acceleration

along the y-axis. The requirement $a_y = 0$ allows us to conclude that $n = w$. Although we do not need n for this problem, it will be important in many future problems. ◀

We can finish by using constant-acceleration kinematics to find the velocity:

$$v_{1x} = v_{0x} + a_x \Delta t$$
$$= 0 + (1.53 \text{ m/s}^2)(5.0 \text{ s})$$
$$= 7.65 \text{ m/s}$$

The problem asked for the *speed* after 5.0 s, which is $v_1 = 7.65$ m/s.

ASSESS 7.65 m/s ≈ 15 mph, a reasonable speed after 5 s of acceleration.

EXAMPLE 5.4 Altitude of a rocket

A 500 g model rocket with a weight of 4.9 N is launched straight up. The small rocket motor burns for 5.0 s and has a steady thrust of 20 N. What maximum altitude does the rocket reach? Assume that the mass loss of the burned fuel is negligible.

MODEL We'll treat the rocket as an accelerating particle. Air resistance will be neglected.

VISUALIZE The pictorial representation of Figure 5.4 on the next page finds that this is a two-part problem. First, the rocket accelerates straight up. Second, the rocket continues going up as it slows down, a free-fall situation. The maximum altitude is at the end of the second part of the motion.

SOLVE We now know what the problem is asking, have established relevant symbols and coordinates, and know what the forces are. We begin the mathematical representation by writing Newton's second law, in component form, as the rocket accelerates upward. The free-body diagram shows two forces, so

$$(F_{\text{net}})_x = \sum F_x = (F_{\text{thrust}})_x + w_x = ma_{0x}$$
$$(F_{\text{net}})_y = \sum F_y = (F_{\text{thrust}})_y + w_y = ma_{0y}$$

The fact that vector $\vec{w}$ points downward—and which might have tempted you to use a minus sign in the y-equation—will be taken into account when we *evaluate* the components. None of the vectors in this problem has an x-component, so only the y-component of the second law is needed. We can use the free-body diagram to see that

$$(F_{\text{thrust}})_y = +F_{\text{thrust}}$$
$$w_y = -w$$

This is the point at which the directional information about the force vectors enters. The y-component of the second law is then

$$a_{0y} = \frac{1}{m}(F_{\text{thrust}} - w)$$
$$= \frac{20.0 \text{ N} - 4.9 \text{ N}}{0.500 \text{ kg}} = 30.2 \text{ m/s}^2$$

Notice that we converted the mass to SI units of kilograms before doing any calculations and that, because of the definition of the newton, the division of newtons by kilograms automatically gives the correct SI units of acceleration.

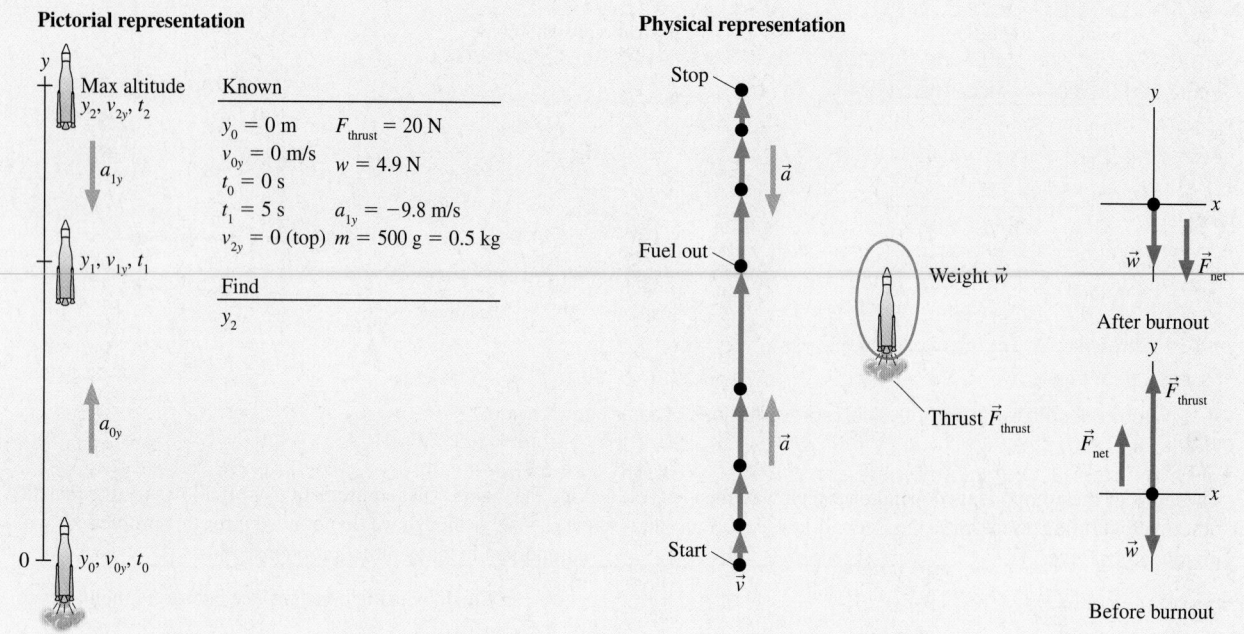

FIGURE 5.4 Pictorial and physical representations of a rocket launch.

The acceleration of the rocket is constant until it runs out of fuel, so we can use constant-acceleration kinematics to find the altitude and velocity at burnout ($\Delta t = t_1 = 5.0$ s):

$$y_1 = y_0 + v_{0y}\,\Delta t + \tfrac{1}{2}a_{0y}(\Delta t)^2$$

$$= \tfrac{1}{2}a_{0y}(\Delta t)^2 = 377 \text{ m}$$

$$v_{1y} = v_{0y} + a_{0y}\,\Delta t = a_{0y}\,\Delta t = 151 \text{ m/s}$$

The only force on the rocket after burnout is gravity, so the second part of the motion is free-fall with $a_{1y} = -g$. We do not know how long it takes to reach the top, but we do know that the final velocity is $v_{2y} = 0$.

We can use free-fall kinematics to find the maximum altitude:

$$v_{2y}{}^2 = 0 = v_{1y}{}^2 - 2g\,\Delta y = v_{1y}{}^2 - 2g(y_2 - y_1)$$

which we can solve to find

$$y_2 = y_1 + \frac{v_{1y}{}^2}{2g} = 377 \text{ m} + \frac{(151 \text{ m/s})^2}{2(9.80 \text{ m/s}^2)}$$

$$= 1540 \text{ m} = 1.54 \text{ km}$$

ASSESS The maximum altitude reached by this rocket is 1.54 km, or just slightly under one mile. While this is fairly high, it does not seem unreasonable for a high-acceleration rocket.

These first examples have shown all the details. Our purpose has been to show how the problem-solving strategy is put into practice. Future examples will be briefer, but the basic *procedure* will remain the same.

STOP TO THINK 5.1   A Martian lander is approaching the surface. It is slowing its descent by firing its rocket motor. Which is the correct free-body diagram for the lander?

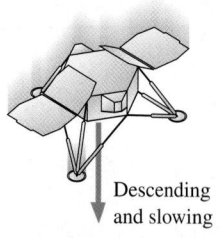

Descending and slowing

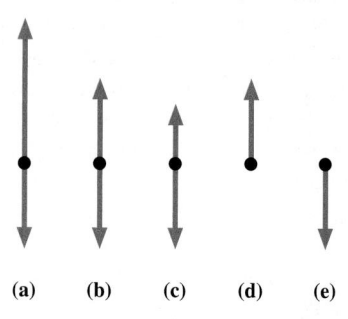

(a)   (b)   (c)   (d)   (e)

5.3 Mass and Weight

When the doctor asks what you weigh, what does she really mean? We do not make a large distinction in our ordinary use of language between the terms *weight* and *mass*, but in physics their distinction is of critical importance.

Mass, you'll recall from Chapter 4, is a scalar quantity that describes an object's inertia. Loosely speaking, it also describes the amount of matter in an object. Mass, measured in kilograms, is an intrinsic property of an object; it has the same value wherever the object may be and whatever forces might be acting on it.

Weight, on the other hand, is a *force*. Specifically, it is the gravitational force exerted on an object by a planet. Weight is a vector, not a scalar, and the vector's direction is always straight down. Weight is measured in newtons.

Mass and weight are not the same thing, but they are related. We can use Galileo's discovery about free fall to make the connection. Figure 5.5 shows the free-body diagram of an object in free fall. The *only* force acting on this object is its weight, the downward pull of gravity, so $\vec{F}_{net} = \vec{w}$. Newton's second law is thus

$$\vec{F}_{net} = \vec{w} = m\vec{a} \tag{5.3}$$

Recall Galileo's discovery that *any* object in free fall, regardless of its mass, has the same acceleration:

$$\vec{a}_{free\ fall} = (9.80\ \text{m/s}^2, \text{downward}) = (g, \text{downward}) \tag{5.4}$$

where $g = 9.80\ \text{m/s}^2$ is the acceleration due to gravity at the Earth's surface. If we use the free-fall acceleration in Equation 5.3, the object's weight is

$$\vec{w} = (mg, \text{downward}) \tag{5.5}$$

The magnitude of the weight force, which we call simply "the weight," is directly proportional to the mass, with g as the constant of proportionality:

$$w = mg \tag{5.6}$$

Because an object's weight depends on g, and the value of g varies from planet to planet, weight is not a fixed, constant property of an object. The value of g at the surface of the moon is about one-sixth its earthly value, so an object on the moon would have only one-sixth its weight on Earth. The object's weight on Jupiter would be larger than its weight on Earth. Its mass, however, would be the same. The amount of matter has not changed, only the gravitational force exerted on that matter.

So when the doctor asks what you weigh, she really wants to know your *mass*. That's the amount of matter in your body. You can't really "lose weight" by going to the moon, even though you would weigh less there!

Measuring Mass and Weight

A *pan balance*, shown in Figure 5.6, is a device for measuring *mass*. You may have used a pan balance to "weigh" chemicals in a chemistry lab. An unknown mass is placed in one pan, then known masses are added to the other until the pans balance. Gravity pulls down on both sides, effectively *comparing* the masses, and the unknown mass equals the sum of the known masses that balance it. Although a pan balance requires gravity in order to function, it does not depend on the value of g. Consequently, the pan balance would give the same result on another planet.

Spring scales, such as the two shown in Figure 5.7 on the next page, measure weight, not mass. Hanging an item on the scale in Figure 5.7a, which might be used to weigh items in the grocery store, stretches the spring. The spring in the "bathroom scale" in Figure 5.7b is compressed when you stand on it.

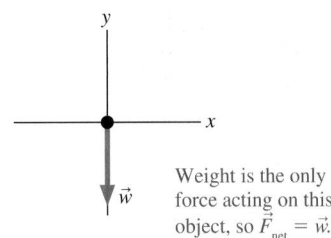

FIGURE 5.5 The free-body diagram of an object in free fall.

Weight is the only force acting on this object, so $\vec{F}_{net} = \vec{w}$.

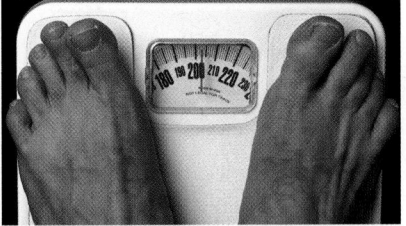

A spring scale, such as the familiar bathroom scale, measures weight, not mass.

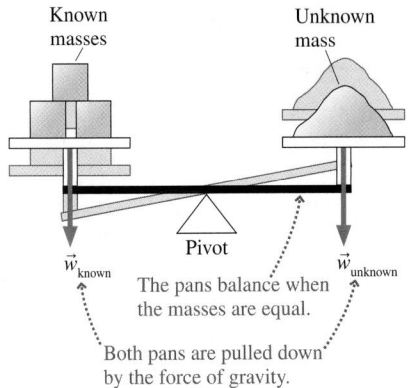

If the unknown mass differs from the known masses, the beam will rotate about the pivot.

Known masses

Unknown mass

$\vec{w}_{known}$

Pivot

$\vec{w}_{unknown}$

The pans balance when the masses are equal.

Both pans are pulled down by the force of gravity.

FIGURE 5.6 A pan balance measures mass.

(a)

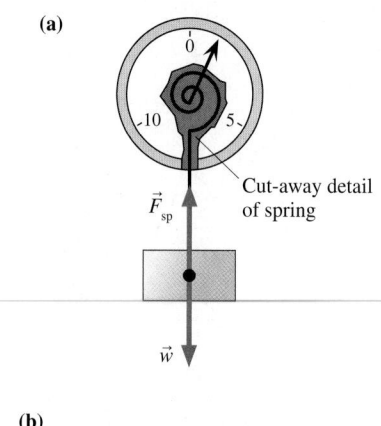

$\vec{F}_{sp}$ · Cut-away detail of spring

$\vec{w}$

(b)

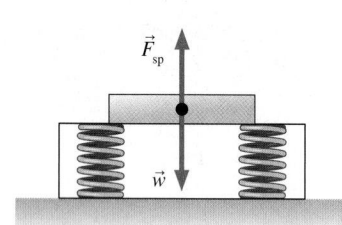

$\vec{F}_{sp}$

$\vec{w}$

FIGURE 5.7 A spring scale measures weight.

A spring scale can be understood on the basis of Newton's first law. The object being weighed is at rest, in static equilibrium, so the net force on it must be zero. The stretched spring in Figure 5.7a *pulls* up; the compressed spring in Figure 5.7b *pushes* up. But in both cases, in order to have $\vec{F}_{net} = \vec{0}$, the upward spring force must exactly balance the downward weight force:

$$F_{sp} = w = mg \qquad (5.7)$$

The *reading* of a spring scale is F_{sp}, the magnitude of the force that the spring is exerting. If the object is in equilibrium, then F_{sp} is exactly equal to the object's weight w. The scale does not "know" the weight of the object. All it can do is to measure how much the spring is stretched or compressed. On a different planet, with a different value for g, the expansion or compression of the spring would be different and the scale's reading would be different.

The unit of force in the English system is the *pound*. We noted in Chapter 4 that the pound is defined as 1 lb ≡ 4.45 N. An object whose weight $w = mg$ is 4.45 N has a mass

$$m = \frac{w}{g} = \frac{4.45 \text{ N}}{9.80 \text{ m/s}^2} = 0.454 \text{ kg} = 454 \text{ g}$$

You may have learned in previous science classes that "1 pound = 454 grams" or, equivalently, that "1 kg = 2.2 lb." Strictly speaking, these well-known "conversion factors" are not true. They are comparing a weight (pounds) to a mass (kilograms). The correct statement would be, "A mass of 1 kg has a weight on *earth* of 2.2 pounds." On another planet, the weight of a 1 kg mass would be something other than 2.2 pounds.

EXAMPLE 5.5 Mass and weight on Jupiter
What is the kilograms-to-pounds conversion factor on Jupiter, where the acceleration due to gravity is 25.9 m/s²?

SOLVE Consider an object with a mass of 1 kg. Its weight on Jupiter is

$$w_{Jupiter} = mg_{Jupiter} = (1 \text{ kg})(25.9 \text{ m/s}^2)$$
$$= 25.9 \text{ N} \times \frac{1 \text{ lb}}{4.45 \text{ N}} = 5.82 \text{ lb}$$

If you had gone to school on Jupiter, you would have learned that 1 kg = 5.82 lb.

Apparent Weight

The weight of an object is the force of gravity on that object. You may never have thought about it, but gravity is not a force that you can feel or sense directly. Your *sensation* of weight—how heavy you feel—is due to *contact forces* pressing against you. Surfaces touch you and activate nerve endings in your skin. As you read this, your sensation of weight is due to the normal force exerted on you by the chair in which you are sitting. When you stand, you feel the contact force of the floor pushing against your feet. If you hang from a rope, your sensation of weight is due to the tension force pulling up on you.

If you stand at rest, with $\vec{a} = \vec{0}$, then the force you *feel*, the normal force pressing against your feet, is exactly equal in magnitude to your weight. This would be a trivial conclusion if objects were always at rest, but what happens if $\vec{a} \neq \vec{0}$?

Recall the sensations you feel while being accelerated. You feel "heavy" when an elevator suddenly accelerates upward or when an airplane accelerates for take-off. This sensation vanishes as soon as the elevator or airplane reaches a steady cruising speed. Your stomach seems to rise a little and you feel lighter than normal as the upward-moving elevator brakes to a halt or a roller coaster goes over the top. Your true weight $w = mg$ has not changed during these events, but your *apparent weight* has.

To investigate this, imagine a man weighing himself by standing on a spring scale in an elevator as it accelerates upward. What does the scale read? How does the scale reading correspond to the man's *sensation* of weight?

As Figure 5.8 shows, the only forces acting on the man are the upward spring force of the scale and the downward weight force. This seems to be the same situation as Figure 5.7b, but there's one big difference. The man is accelerating; he's not in equilibrium. Thus, according to Newton's laws, there must be a net force acting on the man in the direction of $\vec{a}$.

For the net force $\vec{F}_{net}$ to point upward, the magnitude of the spring force must be *greater* than the magnitude of the weight force. That is, $F_{sp} > w$. This conclusion has major implications. Looking at the free-body diagram in Figure 5.8, we see that the y-component of Newton's second law is

$$(F_{net})_y = (F_{sp})_y + w_y = F_{sp} - w = F_{sp} - mg = ma_y \qquad (5.8)$$

where m is the man's mass.

The scale reading is the value of F_{sp}, the magnitude of the force that the scale exerts on the man. Solving Equation 5.8 for F_{sp} gives

$$F_{sp} = \text{scale reading} = mg + ma_y = mg\left(1 + \frac{a_y}{g}\right) = w\left(1 + \frac{a_y}{g}\right) \qquad (5.9)$$

If the elevator is either at rest or moving with constant velocity, then $a_y = 0$ and the man is in equilibrium. In that case, $F_{sp} = w$ and the scale correctly reads his weight. But if $a_y \neq 0$, the scale's reading is *not* the man's true weight.

Let's define an object's **apparent weight** w_{app} as the magnitude of the contact force that supports the object. From Equation 5.9, this is

$$w_{app} = w\left(1 + \frac{a_y}{g}\right) \qquad (5.10)$$

We've found w_{app} for a situation where a contact force supports the object from below. A homework problem will let you show that the expression is also valid when the object is supported from above by the tension in a rope or cable.

An object *appears* to weigh whatever the scale reads, although, as the saying goes, appearances can be deceiving. As the elevator accelerates upward, $a_y > 0$ and $w_{app} > w$. You *feel* heavier than normal, and a scale would read more than your true weight. The acceleration vector $\vec{a}$ points downward and $a_y < 0$ when the elevator brakes, so $w_{app} < w$. You feel lighter and, if you were standing on a scale, it would read less than your true weight.

An object doesn't have to be on a scale for its apparent weight to differ from its true weight. An object's apparent weight is the magnitude of the contact force supporting it. It makes no difference whether this is the spring force of the scale or simply the normal force of the floor.

The idea of apparent weight has important applications. Astronauts are nearly crushed by their apparent weight during a rocket launch when $a_y \gg g$. Much of the thrill of amusement park rides, such as roller coasters, comes from rapid changes in your apparent weight. In Chapter 7, the concept of apparent weight will help us understand how you can swing a bucket of water over your head without the water falling out!

Weightlessness

One last issue before leaving this topic: Suppose the elevator cable breaks and the elevator, along with the man and his scale, plunges straight down in free fall! What will the scale read? The acceleration in free fall is $a_y = -g$. When this

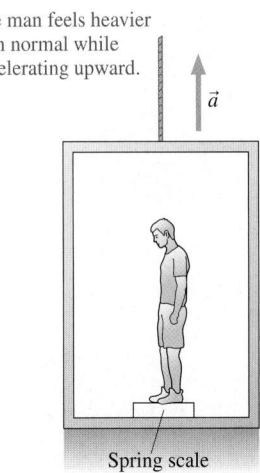

The man feels heavier than normal while accelerating upward.

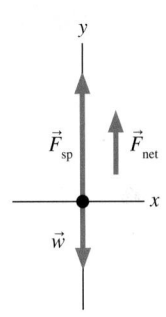

FIGURE 5.8 A man weighing himself in an accelerating elevator.

Astronauts are weightless as they orbit the earth.

acceleration is used in Equation 5.10, we find that $w_{app} = 0$! In other words, the man has *no sensation* of weight.

Think about this carefully. Suppose, as the elevator falls, the man inside releases a ball from his hand. In the absence of air resistance, as Galileo discovered, both the man and the ball would fall at the same rate. From the man's perspective, the ball would appear to "float" beside him. Similarly, the scale would float beneath him and not press against his feet. He is what we call *weightless*.

Surprisingly, "weightless" does *not* mean "no weight." An object that is **weightless** has no *apparent* weight. The distinction is significant. The man's weight is still *mg*, because gravity is still pulling down on him, but he has no *sensation* of weight as he free falls. The term "weightless" is a very poor one, likely to cause confusion because it implies that objects have no weight. As we see, that is not the case.

But isn't this exactly what happens to astronauts orbiting the earth? You've seen films of astronauts and various objects floating inside the Space Shuttle. If an astronaut tries to stand on a scale, it does not exert any force against her feet and reads zero. She is said to be weightless. But if the criterion to be weightless is to be in free fall, and if astronauts orbiting the earth are weightless, does this mean that they are in free fall? This is a very interesting question to which we shall return in Chapter 7.

STOP TO THINK 5.2 An elevator that has descended from the 50th floor is coming to a halt at the 1st floor. As it does, your apparent weight is

a. More than your true weight. b. Less than your true weight.
c. Equal to your true weight.   d. Zero.

5.4 Friction

In everyday life, friction is everywhere. Friction is absolutely essential for many things we do. Without friction you could not walk, drive, or even sit down (you would slide right off the chair!). It is sometimes useful to think about idealized frictionless situations, but it is equally necessary to understand a real world where friction is present. Although friction is a complicated force, many aspects of friction can be described with a simple model.

Static Friction

Chapter 4 defined *static friction* $\vec{f}_s$ as the force on an object that keeps it from slipping. Figure 5.9 shows a person pushing on a box with horizontal force $\vec{F}_{push}$. If the box remains at rest, "stuck" to the floor, it must be because of a static friction force pushing back to the left. The box is in static equilibrium, so the static friction must exactly balance the pushing force:

$$f_s = F_{push} \tag{5.11}$$

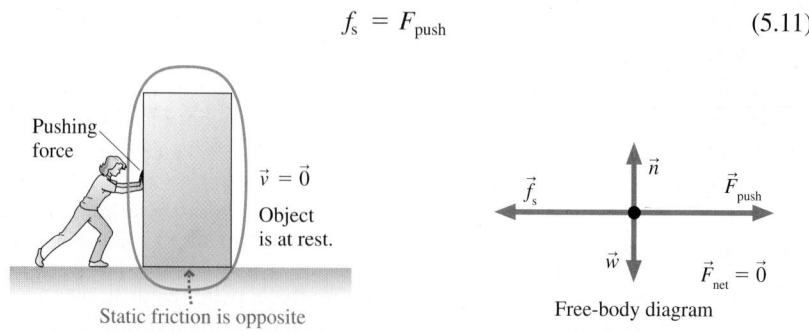

FIGURE 5.9 Static friction keeps an object from slipping.

To determine the direction of $\vec{f}_s$, decide which way the object would move if there were no friction. The static friction force $\vec{f}_s$ points in the *opposite* direction to prevent the motion.

Unlike weight, which has the precise and unambiguous magnitude $w = mg$, the size of the static friction force depends on how hard you push. The harder the person in Figure 5.9 pushes, the harder the floor pushes back. Reduce the pushing force, and the static friction force will automatically be reduced to match. Static friction acts in *response* to an applied force. Figure 5.10 illustrates this idea.

But there's clearly a limit to how big f_s can get. If you push hard enough, the object slips and starts to move. In other words, the static friction force has a *maximum* possible size $f_{s\,max}$.

- An object remains at rest as long as $f_s < f_{s\,max}$.
- The object slips when $f_s = f_{s\,max}$.
- A static friction force $f_s > f_{s\,max}$ is not physically possible.

Experiments with friction (first done by Leonardo da Vinci) show that $f_{s\,max}$ is proportional to the magnitude of the normal force. That is,

$$f_{s\,max} = \mu_s n \qquad (5.12)$$

where μ_s is called the **coefficient of static friction.** The coefficient is a dimensionless number that depends on the materials of which the object and the surface are made. Table 5.1 shows some typical values of coefficients of friction. It is to be emphasized that these are only approximate. The exact value of the coefficient depends on the roughness, cleanliness, and dryness of the surfaces.

> NOTE ▶ Equation 5.12 does *not* say $f_s = \mu_s n$. The value of f_s depends on the force or forces that static friction has to balance to keep the object from moving. It can have any value from 0 up to, but not exceeding, $\mu_s n$. ◀

Kinetic Friction

Once the box starts to slide, in Figure 5.11, the static friction force is replaced by a kinetic friction force $\vec{f}_k$. Experiments show that kinetic friction, unlike static friction, has a nearly *constant* magnitude. Furthermore, the size of the kinetic friction force is *less* than the maximum static friction, $f_k < f_{s\,max}$, which explains why it is easier to keep the box moving than it was to start it moving. The direction of $\vec{f}_k$ is always opposite to the direction in which an object slides across the surface.

The kinetic friction force is also proportional to the magnitude of the normal force:

$$f_k = \mu_k n \qquad (5.13)$$

where μ_k is called the **coefficient of kinetic friction.** Table 5.1 includes typical values of μ_k. You can see that $\mu_k < \mu_s$, causing the kinetic friction to be less than the maximum static friction.

Rolling Friction

If you slam on the brakes hard enough, your car tires slide against the road surface and leave skid marks. This is kinetic friction. A wheel *rolling* on a surface also experiences friction, but not kinetic friction. The portion of the wheel that contacts the surface is stationary with respect to the surface, not sliding. To see this, roll a wheel slowly and watch how it touches the ground.

Textbooks draw wheels as circles, but no wheel is perfectly round. The weight of the wheel, and of any object supported by the wheel, causes the bottom of the wheel to flatten where it touches the surface, as Figure 5.12 on the next page

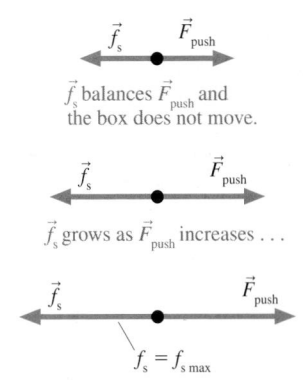

$\vec{f}_s$ balances $\vec{F}_{push}$ and the box does not move.

$\vec{f}_s$ grows as $\vec{F}_{push}$ increases . . .

$f_s = f_{s\,max}$

. . . until f_s reaches $f_{s\,max}$. Now, if $\vec{F}_{push}$ gets any bigger, the object will start to move.

FIGURE 5.10 Static friction acts in *response* to an applied force.

TABLE 5.1 Coefficients of friction

Materials	Static μ_s	Kinetic μ_k	Rolling μ_r
Rubber on concrete	1.00	0.80	0.02
Steel on steel (dry)	0.80	0.60	0.002
Steel on steel (lubricated)	0.10	0.05	
Wood on wood	0.50	0.20	
Wood on snow	0.12	0.06	
Ice on ice	0.10	0.03	

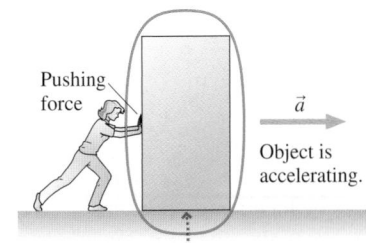

Kinetic friction is opposite the motion.

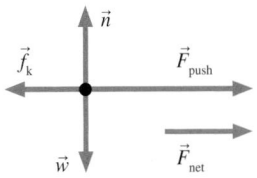

FIGURE 5.11 The kinetic friction force is opposite the direction of motion.

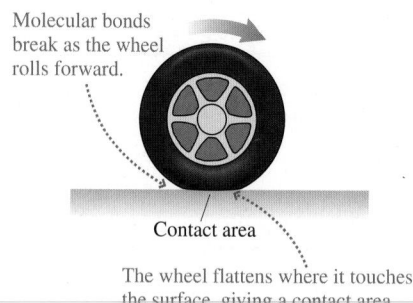

Molecular bonds break as the wheel rolls forward.

Contact area

The wheel flattens where it touches the surface, giving a contact area rather than a point of contact.

FIGURE 5.12 Rolling friction is due to the contact area between a wheel and the surface.

2.5, 2.6

shows. The contact area between a car tire and the road is fairly large. The contact area between a steel locomotive wheel and a steel rail is much less, but it's not zero.

Molecular bonds are quickly established where the wheel presses against the surface. These bonds have to be broken as the wheel rolls forward, and the effort needed to break them causes **rolling friction.** (Think how it is to walk with a wad of chewing gum stuck to the sole of your shoe!) The force of rolling friction can be calculated in terms of a **coefficient of rolling friction** μ_r:

$$\vec{f}_r = (\mu_r n, \text{ direction opposite the motion}) \tag{5.14}$$

Rolling friction acts very much like kinetic friction, but values of μ_r (see Table 5.1) are much less than values of μ_k. This is why it is easier to roll an object on wheels than to slide it.

A Model of Friction

These ideas can be summarized in a *model* of friction:

> Static: $\vec{f}_s \leq (\mu_s n, \text{ direction as necessary to prevent motion})$
>
> Kinetic: $\vec{f}_k = (\mu_k n, \text{ direction opposite the motion})$ (5.15)
>
> Rolling: $\vec{f}_r = (\mu_r n, \text{ direction opposite the motion})$

Here "motion" means "motion relative to the surface." The maximum value of static friction $f_{s\,max} = \mu_s n$ occurs at the point where the object slips and begins to move.

NOTE ▶ Equations 5.15 are a "model" of friction, not a "law" of friction. These equations provide a reasonably accurate, but not perfect, description of how friction forces act. For example, we've ignored the surface area of the object because surface area has little effect. Likewise, our model assumes that the kinetic friction force is independent of the object's speed. This is a fairly good, but not perfect, approximation. Equations 5.15 are a simplification of reality that works reasonably well, which is what we mean by a "model." They are not a "law of nature" on a level with Newton's laws. ◀

Figure 5.13 summarizes these ideas graphically by showing how the friction force changes as the magnitude of an applied force $\vec{F}_{push}$ increases.

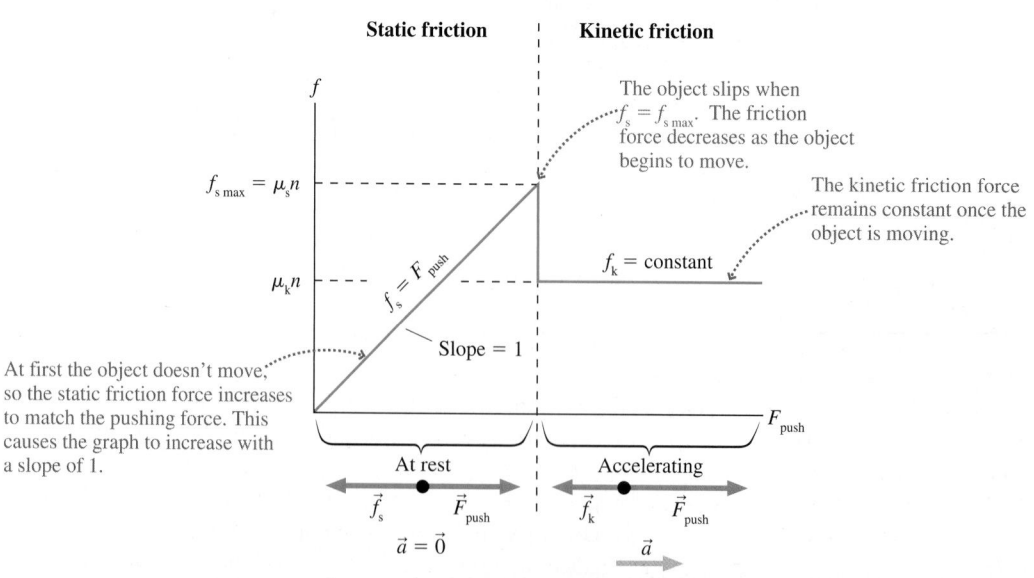

FIGURE 5.13 The friction force response to an increasing applied force.

STOP TO THINK 5.3   Rank in order, from largest to smallest, the size of the friction forces $\vec{f}_a$ to $\vec{f}_e$ in these 5 different situations. The box and the floor are made of the same materials in all situations.

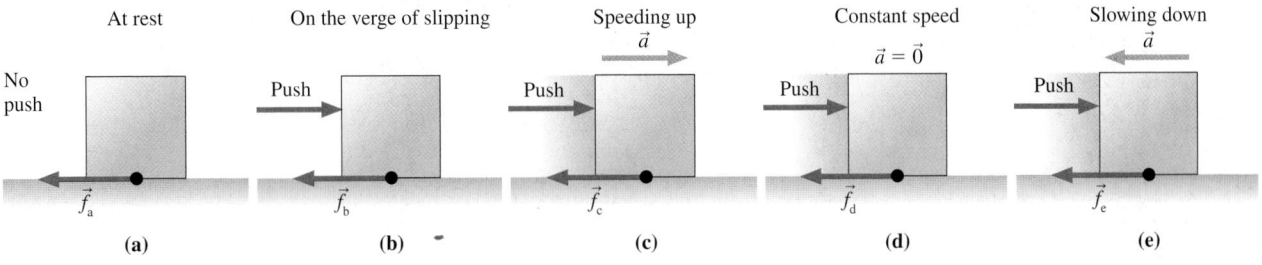

(a)   (b)   (c)   (d)   (e)

EXAMPLE 5.6 **How far does a box slide?**

Carol pushes a 50.0 kg wood box across a wood floor at a steady speed of 2.0 m/s. How much force does Carol exert on the box? If she stops pushing, how far will the box slide before coming to rest?

MODEL We model the box as a particle and we describe the friction forces with the model of static and kinetic friction. This is a two-part problem: first while Carol is pushing the box, then as it slides after she releases it.

VISUALIZE This is a fairly complex situation, one that calls for careful visualization. Figure 5.14 shows the pictorial and physical representations both while Carol pushes, when $\vec{a} = 0$, and after she stops. We've placed $x = 0$ at the point where she stops pushing because this is the point where the kinematics calculation for "How far?" will begin. Notice that each part of the motion needs its own free-body diagram. The box is moving until the very instant that the problem ends, so only kinetic friction is relevant.

SOLVE We'll start by finding how hard Carol has to push to keep the box moving at a steady speed. The box is in dynamic equilibrium ($\vec{a} = 0$), and Newton's first law is

$$\sum F_x = F_{push} - f_k = 0$$
$$\sum F_y = n - w = n - mg = 0$$

where we've used $w = mg$ for the weight. The negative sign occurs in the first equation because $\vec{f}_k$ points to the left and thus the *component* is negative: $(f_k)_x = -f_k$. Similarly, $w_y = -w$ because the weight vector points down. In addition to Newton's laws, we also have our model of kinetic friction:

$$f_k = \mu_k n$$

Altogether we have three simultaneous equations in the three unknowns F_{push}, f_k, and n. Fortunately, these equations are easy to solve. The y-component of Newton's law tells us that $n = mg$. We can then find the friction force to be

$$f_k = \mu_k mg$$

Substitute this into the x-component of the first law, giving

$$F_{push} = f_k = \mu_k mg$$
$$= (0.20)(50.0 \text{ kg})(9.80 \text{ m/s}^2) = 98.0 \text{ N}$$

where μ_k for wood on wood was taken from Table 5.1. This is how hard Carol pushes to keep the box moving at a steady speed.

The box is not in equilibrium after Carol stops pushing it. Our strategy for the second half of the problem is to use Newton's second law to find the acceleration, then use kinematics to find how far the box moves before stopping. We see from the

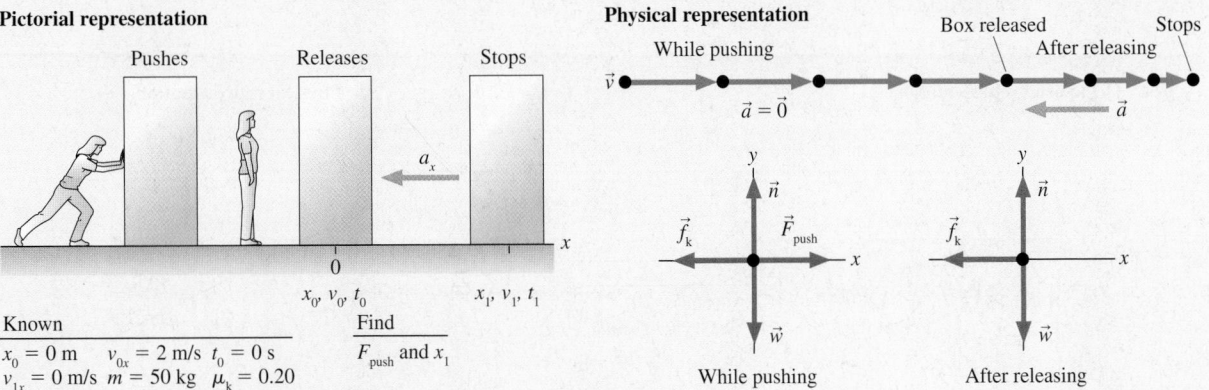

FIGURE 5.14 Pictorial and physical representations of a box sliding across a floor.

motion diagram that $a_y = 0$. Newton's second law, applied to the second free-body diagram of Figure 5.14, is

$$\sum F_x = -f_k = ma_x$$

$$\sum F_y = n - mg = ma_y = 0$$

We also have our model of friction,

$$f_k = \mu_k n$$

We see from the y-component equation that $n = mg$, and thus $f_k = \mu_k mg$. Using this in the x-component equation gives

$$ma_x = -f_k = -\mu_k mg$$

This is easily solved to find the box's acceleration:

$$a_x = -\mu_k g = -(0.20)(9.80 \text{ m/s}^2) = -1.96 \text{ m/s}^2$$

The acceleration component a_x is negative because the acceleration vector $\vec{a}$ points to the left, as we see from the motion diagram.

Now we are left with a problem of constant-acceleration kinematics. We are interested in a distance, rather than a time interval, so the easiest way to proceed is

$$v_{1x}^2 = 0 = v_{0x}^2 + 2a_x \Delta x = v_{0x}^2 + 2a_x x_1$$

from which the distance that the box slides is

$$x_1 = \frac{-v_{0x}^2}{2a_x} = \frac{-(2.0 \text{ m/s})^2}{2(-1.96 \text{ m/s}^2)} = 1.02 \text{ m}$$

We get a positive answer because the two negative signs cancel.

ASSESS Carol was pushing at 2 m/s $\approx$ 4 mph, which is fairly fast. The box slides 1.02 m, which is slightly over 3 feet. That sounds reasonable.

NOTE ▶ We needed both the horizontal and the vertical components of the second law even though the motion was entirely horizontal. This need is typical when friction is involved because we must find the normal force before we can evaluate the friction force. ◀

EXAMPLE 5.7 Dumping a file cabinet

A 50.0 kg steel file cabinet is in the back of a dump truck. The truck's bed, also made of steel, is slowly tilted. What is the size of the static friction force on the cabinet when the bed is tilted 20°? At what angle will the file cabinet begin to slide?

MODEL We'll model the file cabinet as a particle. We'll also use the model of static friction. The file cabinet will slip when the static friction force reaches its maximum value $f_{s\,max}$.

VISUALIZE Figure 5.15 shows the pictorial and physical representations when the truck bed is tilted at angle θ. This is a static equilibrium problem, so the free-body diagram suffices for a pictorial representation. We can make the analysis easier if we tilt the coordinate system to match the bed of the truck. To prevent the file cabinet from slipping, the static friction force must point *up* the slope.

SOLVE The file cabinet is in static equilibrium. Newton's first law is

$$(F_{net})_x = \sum F_x = n_x + w_x + (f_s)_x = 0$$

$$(F_{net})_y = \sum F_y = n_y + w_y + (f_s)_y = 0$$

From the free-body diagram we see that f_s has only a *negative* x-component and that n has only a positive y-component. The weight vector can be written $\vec{w} = +w\sin\theta\,\hat{i} - w\cos\theta\,\hat{j}$, so $\vec{w}$ has both x- and y-components in this coordinate system. Thus the first law becomes

$$\sum F_x = w\sin\theta - f_s = mg\sin\theta - f_s = 0$$

$$\sum F_y = n - w\cos\theta = n - mg\cos\theta = 0$$

where we've used $w = mg$. The x-component equation allows us to determine the size of the static friction force when $\theta = 20°$:

$$f_s = mg\sin\theta = (50.0 \text{ kg})(9.80 \text{ m/s}^2)\sin 20°$$

$$= 168 \text{ N}$$

This value does not require knowing μ_s. We simply have to find the size of the friction force that will balance the component of $\vec{w}$ that points down the slope. The coefficient of static friction only enters when we want to find the angle at which the file

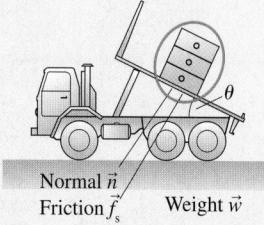

Pictorial representation

Normal $\vec{n}$
Friction $\vec{f}_s$ Weight $\vec{w}$

Known
$\mu_s = 0.80$ $m = 50$ kg
$\mu_k = 0.60$

Find
f_s where $\theta = 20°$
θ where cabinet slips

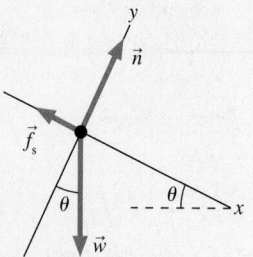

Physical representation

FIGURE 5.15 The pictorial and physical representations of a file cabinet in a tilted dump truck.

cabinet slips. Slipping occurs when the static friction reaches its maximum value

$$f_s = f_{s\,max} = \mu_s n$$

From the y-component of Newton's law we see that $n = mg\cos\theta$. Consequently,

$$f_{s\,max} = \mu_s mg\cos\theta$$

Substituting this into the x-component of the first law gives

$$mg\sin\theta - \mu_s mg\cos\theta = 0$$

The mg in both terms cancels, and we find

$$\frac{\sin\theta}{\cos\theta} = \tan\theta = \mu_s$$

$$\theta = \tan^{-1}\mu_s = \tan^{-1}(0.80) = 38.7°$$

ASSESS Steel doesn't slide all that well on unlubricated steel, so a fairly large angle is not surprising. The answer seems reasonable. It is worth noting that $n = mg\cos\theta$ in this example. A common error is to use simply $n = mg$. Be sure to evaluate the normal force within the context of each specific problem.

The angle at which slipping begins is called the *angle of repose.* Figure 5.16 shows that knowing the angle of repose can be very important because it is the angle at which loose materials (gravel, sand, snow, etc.) begin to slide on a mountainside, leading to landslides and avalanches.

Causes of Friction

It is worth a brief pause to look at the *causes* of friction. All surfaces, even those quite smooth to the touch, are very rough on a microscopic scale. When two objects are placed in contact, they do not make a smooth fit. Instead, as Figure 5.17 shows, the high points on one surface become jammed against the high points on the other surface while the low points are not in contact at all. Only a very small fraction (typically 10^{-4}) of the surface area is in actual contact. The amount of contact depends on how hard the surfaces are pushed together, which is why friction forces are proportional to n.

At the points of actual contact, the atoms in the two materials are pressed closely together and molecular bonds are established between them. These bonds are the "cause" of the static friction force. For an object to slip, you must push it hard enough to break these molecular bonds between the surfaces. Once they are broken, and the two surfaces are sliding against each other, there are still attractive forces between the atoms on the opposing surfaces as the high points of the materials push past each other. However, the atoms move past each other so quickly that they do not have time to establish the tight bonds of static friction. That is why the kinetic friction force is smaller.

Occasionally, in the course of sliding, two high points will be forced together so closely that they do form a tight bond. As the motion continues, it is not this surface bond that breaks but weaker bonds at the *base* of one of the high points. When this happens, a small piece of the object is left behind "embedded" in the surface. This is what we call *abrasion.* Abrasion causes materials to wear out as a result of friction, be they the piston rings in your car or the seat of your pants. In machines, abrasion is minimized with lubrication, a very thin film of liquid between the surfaces that allows them to "float" past each other with many fewer points in actual contact.

Friction, as seen at the atomic level, is a very complex phenomenon. A detailed understanding of friction is at the forefront of engineering research today, where it is especially important for designing highly miniaturized machines and nanostructures.

5.5 Drag

The air exerts a drag force on objects as they move through the air. You experience drag forces every day as you jog, bicycle, ski, or drive your car. The drag force is especially important for the skydiver at the beginning of the chapter.

FIGURE 5.16 The angle of repose is the angle at which loose materials, such as gravel or snow, begin to slide.

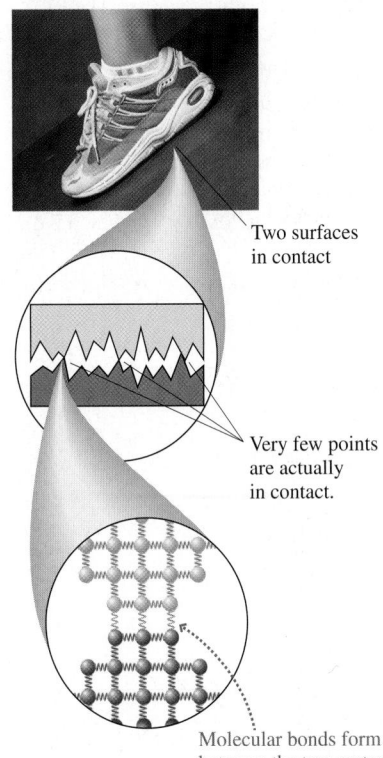

Two surfaces in contact

Very few points are actually in contact.

Molecular bonds form between the two materials. These bonds have to be broken as the object slides.

FIGURE 5.17 An atomic-level view of friction.

FIGURE 5.18 The drag force on a high-speed motorcyclist is significant.

The drag force $\vec{D}$

- Is opposite in direction to $\vec{v}$.
- Increases in magnitude as the object's speed increases.

Figure 5.18 illustrates the drag force.

Drag is a more complex force than ordinary friction because drag depends on the object's speed. At relatively low speeds the drag force is small and can usually be neglected, but drag plays an important role as speeds increase.

Experimental studies have found that the drag force depends on an object's speed in a complicated way. Fortunately, we can use a fairly simple *model* of drag if the following three conditions are met:

- The object's size (diameter) is between a few millimeters and a few meters.
- The object's speed is less than a few hundred meters per second.
- The object is moving through the air near the earth's surface.

These conditions are usually satisfied for balls, people, cars, and many other objects of the everyday world. Under these conditions, the drag force can be written

$$\vec{D} \approx (\tfrac{1}{4}Av^2, \text{ direction opposite the motion})\qquad(5.16)$$

where A is the cross-section area of the object. The size of the drag force is proportional to the *square* of the object's speed. This model of drag fails for objects that are very small (such as dust particles), very fast (such as jet planes), or that move in other media (such as water). We'll leave those situations to more advanced textbooks.

NOTE ▶ Let's look at this model more closely. You may have noticed that an area multiplied by a speed squared does not give units of force. Unlike the $\frac{1}{2}$ in $\Delta x = \frac{1}{2}a(\Delta t)^2$, which is a "pure" number, the $\frac{1}{4}$ in the expression for $\vec{D}$ has units. This number depends on the air's density and viscosity, and it's actually $\frac{1}{4}$ kg/m³. We've suppressed the units in Equation 5.16, but doing so gives us an expression that works *only* if A is in m² and v is in m/s. Equation 5.16 cannot be converted to other units. And the number is not exactly $\frac{1}{4}$, which is why Equation 5.16 has an $\approx$ sign rather than an $=$ sign, but it's close enough to allow Equation 5.16 to be a reasonable yet simple model of drag. ◀

The area in Equation 5.16 is the cross section of the object as it "faces into the wind." Figure 5.19 shows how to calculate the cross-section area for objects of different shape. It's interesting to note that the magnitude of the drag force, $\frac{1}{4}Av^2$, depends on the object's *size and shape* but not on its *mass*. We will see shortly that the irrelevance of the mass has important consequences.

A falling sphere

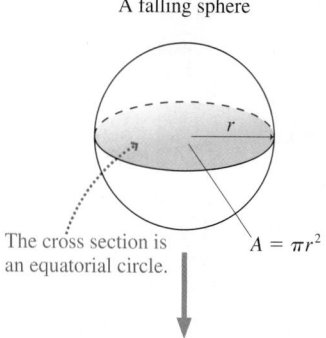

The cross section is an equatorial circle. $A = \pi r^2$

A cylinder falling end down

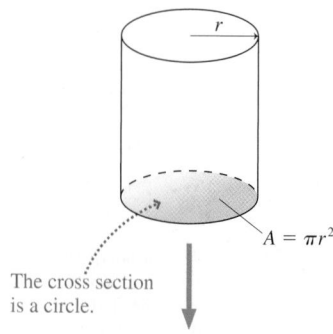

The cross section is a circle. $A = \pi r^2$

A cylinder falling side down

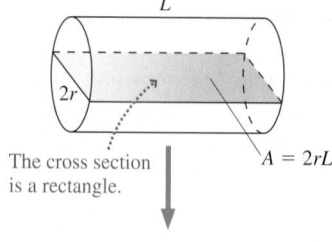

The cross section is a rectangle. $A = 2rL$

A box falling end down

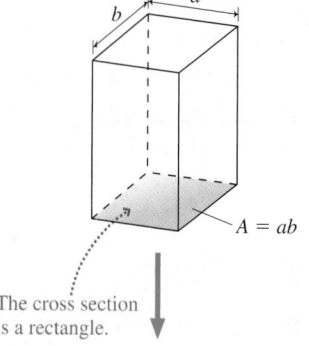

The cross section is a rectangle. $A = ab$

FIGURE 5.19 Cross-section areas for objects of different shape.

EXAMPLE 5.8 Air resistance compared to rolling friction

The profile of a typical 1500 kg passenger car, as seen from the front, is 1.6 m wide and 1.4 m high. At what speed does the magnitude of the drag equal the magnitude of the rolling friction?

MODEL Treat the car as a particle. Use the models of rolling friction and drag.

VISUALIZE Figure 5.20 shows the car and a free-body diagram. A full pictorial representation is not needed because we won't be doing any kinematics calculations.

SOLVE Drag is less than friction at low speeds, where air resistance is negligible. But drag increases as v increases, so there will be a speed at which the two forces are equal in size. Above this speed, drag is more important than rolling friction.

The magnitudes of the forces are $D \approx \frac{1}{4}Av^2$ and $f_r = \mu_r n$. There's no motion and no acceleration in the vertical direction, so we can see from the free-body diagram that $n = w = mg$. Thus $f_r = \mu_r mg$. Equating friction and drag, we have

$$\frac{1}{4}Av^2 = \mu_r mg$$

Solving for v, we find

$$v = \sqrt{\frac{4\mu_r mg}{A}} = \sqrt{\frac{4(0.02)(1500 \text{ kg})(9.8 \text{ m/s}^2)}{(1.4 \text{ m})(1.6 \text{ m})}} = 23 \text{ m/s}$$

where the value of μ_r for rubber on concrete was taken from Table 5.1.

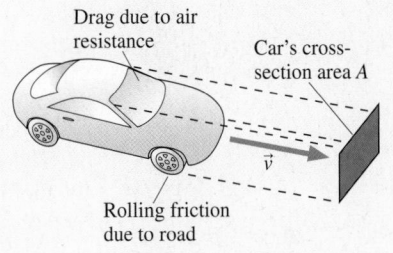

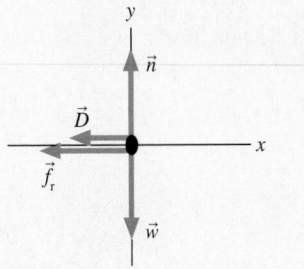

FIGURE 5.20 A car experiences both rolling friction and drag.

ASSESS 23 m/s is approximately 50 mph, a reasonable result. This calculation shows that our assumption that we can ignore air resistance is really quite good for car speeds less than 30 or 40 mph. Calculations that neglect drag will be increasingly inaccurate as speeds go above 50 mph.

Figure 5.21 shows a ball moving up and down vertically. If there were no air resistance, the ball would be in free fall with $a_{\text{free fall}} = -g$ throughout its flight. Let's see how drag changes this.

Referring to Figure 5.21:

1. The drag force $\vec{D}$ points down as the ball rises. This *increases* the net force on the ball and causes the ball to slow down *more quickly* than it would in a vacuum. The magnitude of the acceleration, which we'll calculate below, is $a > g$.
2. The drag force decreases as the ball slows.
3. $\vec{v} = \vec{0}$ at the highest point in the ball's motion, so there's no drag and the acceleration is simply $a_{\text{free fall}} = -g$.
4. The drag force increases as the ball speeds up.
5. The drag force $\vec{D}$ points up as the ball falls. This *decreases* the net force on the ball and causes the ball to speed up *less quickly* than it would in a vacuum. The magnitude of the acceleration is $a < g$.

We can use Newton's second law to find the ball's acceleration $a_\uparrow$ as it rises. You can see from the forces in Figure 5.21 that

$$a_\uparrow = \frac{(F_{\text{net}})_y}{m} = \frac{-mg - D}{m} = -\left(g + \frac{D}{m}\right) \tag{5.17}$$

The magnitude of $a_\uparrow$, which is the ball's deceleration as it rises, is $g + D/m$. Air resistance causes the ball to slow down *more quickly* than it would in a vacuum. But Equation 5.17 tells us more. Because D depends on the object's size but not on its mass, drag has a larger *effect* (larger acceleration) on a less massive ball than on a more massive ball of the same size.

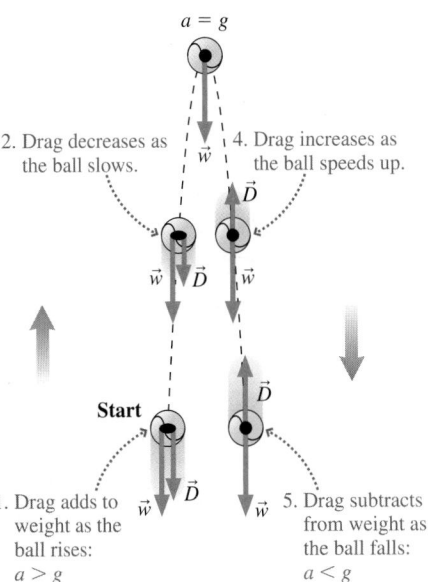

FIGURE 5.21 Drag force on a ball moving vertically.

A Ping-Pong ball and a golf ball are about the same size, but it's harder to throw the Ping-Pong ball than the golf ball. We can now give an *explanation:*

- The drag force has the same magnitude for two objects of equal size.
- According to Newton's second law, the acceleration (the *effect* of the force) depends inversely on the mass.
- Therefore, the effect of the drag force is larger on a less massive ball than on a more massive ball of equal size.

As the ball in Figure 5.21 falls, its acceleration $a_\downarrow$ is

$$a_\downarrow = \frac{(F_{\text{net}})_y}{m} = \frac{-mg + D}{m} = -\left(g - \frac{D}{m}\right) \qquad (5.18)$$

The magnitude of $a_\downarrow$ is $g - D/m$, so the ball speeds up *less quickly* than it would in a vacuum. Once again, the effect is larger for a less massive ball than for a more massive ball of equal size.

Terminal Speed

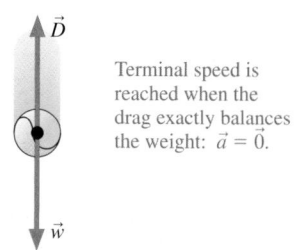

FIGURE 5.22 An object falling at terminal speed.

Terminal speed is reached when the drag exactly balances the weight: $\vec{a} = \vec{0}$.

The drag force increases as an object falls and gains speed. If the object falls far enough, it will eventually reach a speed, shown in Figure 5.22, at which $D = w$. That is, the drag force will be equal and opposite to the weight force. The net force at this speed is $\vec{F}_{\text{net}} = \vec{0}$, so there is no further acceleration and the object falls with a *constant* speed. The speed at which the exact balance between the upward drag force and the downward weight force causes an object to fall without acceleration is called the **terminal speed** v_{term}. Once an object has reached terminal speed, it will continue falling at that speed until it hits the ground.

It's not hard to compute the terminal speed. It is the speed, by definition, at which $D = w$ or, equivalently, $\frac{1}{4}Av^2 \approx mg$. This speed is

$$v_{\text{term}} \approx \sqrt{\frac{4mg}{A}} \qquad (5.19)$$

A more massive object has a larger terminal speed than a less massive object of equal size. A 10-cm-diameter lead ball, with a mass of 6 kg, has a terminal speed of 170 m/s while a 10-cm-diameter Styrofoam ball, with a mass of 50 g, has a terminal speed of only 15 m/s.

A popular use of Equation 5.19 is to find the terminal speed of a skydiver. A skydiver is rather like the cylinder of Figure 5.19 falling "side down." A typical skydiver is 1.8 m long and 0.40 m wide ($A = 0.72$ m^2) and has a mass of 75 kg. His terminal speed is

$$v_{\text{term}} \approx \sqrt{\frac{4mg}{A}} = \sqrt{\frac{4(75 \text{ kg})(9.8 \text{ m/s}^2)}{0.72 \text{ m}^2}} = 64 \text{ m/s}$$

This is roughly 140 mph. A higher speed can be reached by falling feet first or head first, which reduces the area A.

Figure 5.23 shows the results of a more detailed calculation for a falling object. Without drag, the velocity graph is a straight line with slope $= a_y = -g$. When drag is included, the slope steadily decreases in magnitude and approaches zero (no further acceleration) as the object reaches terminal speed.

Although we've focused our analysis on objects moving vertically, the same ideas apply to objects moving horizontally. If an object is thrown or shot horizontally, $\vec{D}$ causes the object to slow down. An airplane reaches its maximum speed, which is analogous to the terminal speed, when the drag is equal and opposite to the thrust: $D = F_{\text{thrust}}$. The net force is then zero and the plane cannot go any faster. The maximum speed of a passenger jet is about 550 mph.

We will continue to neglect drag unless a problem specifically calls for drag to be considered.

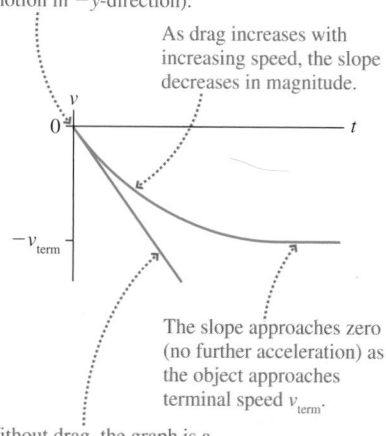

The velocity starts at zero, then becomes increasingly negative (motion in $-y$-direction).

As drag increases with increasing speed, the slope decreases in magnitude.

The slope approaches zero (no further acceleration) as the object approaches terminal speed v_{term}.

Without drag, the graph is a straight line with slope $a_y = -g$.

FIGURE 5.23 The velocity-versus-time graph of a falling object with and without drag.

STOP TO THINK 5.4 The terminal speed of a Styrofoam ball is 15 m/s. Suppose a Styrofoam ball is shot straight down with an initial speed of 30 m/s. Which velocity graph is correct?

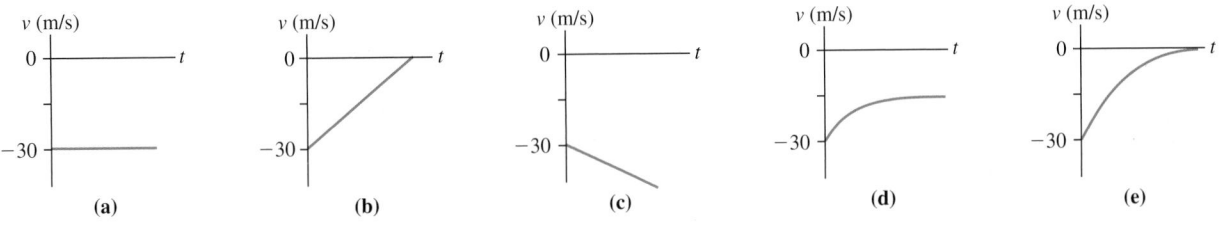

5.6 More Examples of Newton's Second Law

We will finish this chapter with several additional examples in which we use the problem-solving strategy in more complex scenarios.

2.7, 2.8, 2.9

EXAMPLE 5.9 Stopping distances

A 1500 kg car is traveling at a speed of 30 m/s when the driver slams on the brakes and skids to a halt. Determine the stopping distance if the car is traveling up a 10° slope, down a 10° slope, or on a level road.

MODEL We'll represent the car as a particle and we'll use the model of kinetic friction. We want to solve the problem only once, not three separate times, so we'll leave the slope angle θ unspecified until the end.

VISUALIZE Figure 5.24 shows pictorial and physical representations. We've shown the car sliding uphill, but these representations work equally well for a level or downhill slide if we let θ be zero or negative, respectively. We've used a tilted coordinate system so that the motion is along one of the axes. We've *assumed* that the car is traveling to the right, although the problem didn't state this. You could equally well make the opposite assumption, but you would have to be careful with negative values of x. The car *skids* to a halt, so we've taken the coefficient of *kinetic* friction for rubber on concrete from Table 5.1.

Pictorial representation

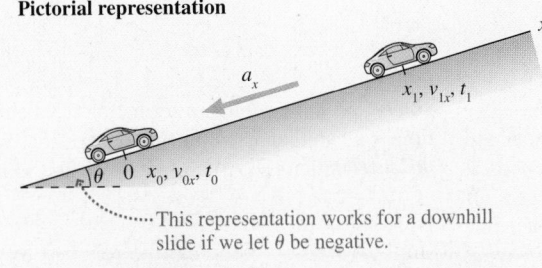

········This representation works for a downhill slide if we let θ be negative.

Known	
$x_0 = t_0 = 0$	$v_{0x} = 30$ m/s
$m = 1500$ kg	$v_{1x} = 0$ m/s
$\mu_k = 0.80$	
$\theta = -10°, 0°, 10°$	

Find

$\Delta x = x_1 - x_0 = x_1$

Physical representation

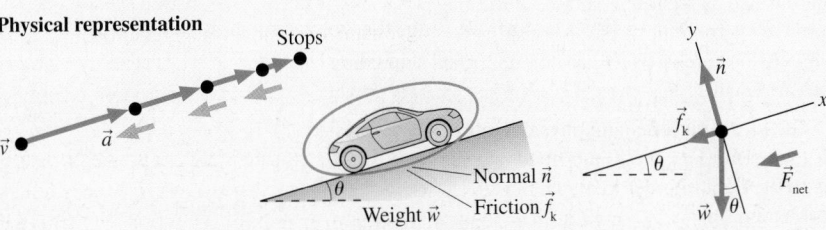

FIGURE 5.24 Pictorial and physical representations of a skidding car.

SOLVE Newton's second law and the model of kinetic friction are

$$\sum F_x = n_x + w_x + (f_k)_x$$
$$= -mg\sin\theta - f_k = ma_x$$
$$\sum F_y = n_y + w_y + (f_k)_y$$
$$= n - mg\cos\theta = ma_y = 0$$
$$f_k = \mu_k n$$

We've written these equations by "reading" the motion diagram and the free-body diagram. Notice that both components of the weight vector $\vec{w}$ are negative. $a_y = 0$, because the motion is entirely along the x-axis.

The second equation gives $n = mg\cos\theta$. Using this in the friction model, we find $f_k = \mu_k mg\cos\theta$. Inserting this result back into the first equation then gives

$$ma_x = -mg\sin\theta - \mu_k mg\cos\theta$$
$$= -mg(\sin\theta + \mu_k\cos\theta)$$
$$a_x = -g(\sin\theta + \mu_k\cos\theta)$$

This is a constant acceleration. Constant-acceleration kinematics gives

$$v_{1x}^2 = 0 = v_{0x}^2 + 2a_x(x_1 - x_0) = v_{0x}^2 + 2a_x x_1$$

which we can solve for the stopping distance x_1:

$$x_1 = -\frac{v_{0x}^2}{2a_x} = \frac{v_{0x}^2}{2g(\sin\theta + \mu_k\cos\theta)}$$

Notice how the minus sign in the expression for a canceled the minus sign in the expression for x_1. Evaluating our result at the three different angles gives the stopping distances:

$$x_1 = \begin{cases} 48\text{ m} & \theta = 10° & \text{uphill} \\ 57\text{ m} & \theta = 0° & \text{level} \\ 75\text{ m} & \theta = -10° & \text{downhill} \end{cases}$$

The implications are clear about the danger of driving downhill too fast!

ASSESS 30 m/s $\approx$ 60 mph and 57 m $\approx$ 180 feet on a level surface. This is similar to the stopping distances you learned when you got your driver's license, so the results seem reasonable. Additional confirmation comes from noting that the expression for a_x becomes $-g\sin\theta$ if $\mu_k = 0$. This is what you learned in Chapter 3 for the acceleration on a frictionless inclined plane.

This is a good example for pointing out the advantages of working problems *algebraically*. If you had started plugging in numbers early, you would not have found that the mass eventually cancels out and you would have done several needless calculations. In addition, it is now easy to calculate the stopping distance for different angles. Had you been computing numbers, rather than algebraic expressions, you would have had to go all the way back to the beginning for each angle.

EXAMPLE 5.10 A dog sled race

It's dog sled race day in Alaska! A wooden sled, with rider and supplies, has a mass of 200 kg. When the starting gun sounds, it takes the dogs 15 meters to reach their "cruising speed" of 5.0 m/s across the snow. Two ropes are attached to the sled, one on each side of the dogs. The ropes pull upward at 10°. What are the tensions in the ropes at the start of the race?

MODEL We'll represent the sled as a particle and we'll use the model of kinetic friction. We interpret the question as asking for the *magnitude* T of the tension forces. We'll assume that the tensions in the two ropes are equal.

VISUALIZE Figure 5.25 shows the pictorial and physical representations. Notice that the tensions $\vec{T}_1$ and $\vec{T}_2$ are tilted up, but the net force is directly to the right in order to match the acceleration $\vec{a}$ of the motion diagram.

SOLVE We have enough information to calculate the acceleration. We can then use $\vec{a}$ to find the tension. We'll assume that the acceleration is constant during the first 15 m. Then

$$v_{1x}^2 = v_{0x}^2 + 2a_x(x_1 - x_0) = 2a_x x_1$$
$$a_x = \frac{v_{1x}^2}{2x_1} = \frac{(5.0\text{ m/s})^2}{2(15\text{ m})} = 0.833\text{ m/s}^2$$

Newton's second law can be written by "reading" the free-body diagram:

$$\sum F_x = n_x + T_{1x} + T_{2x} + w_x + (f_k)_x$$
$$= 2T\cos\theta - f_k = ma_x$$
$$\sum F_y = n_y + T_{1y} + T_{2y} + w_y + (f_k)_y$$
$$= n + 2T\sin\theta - mg = ma_y = 0$$

Pictorial representation

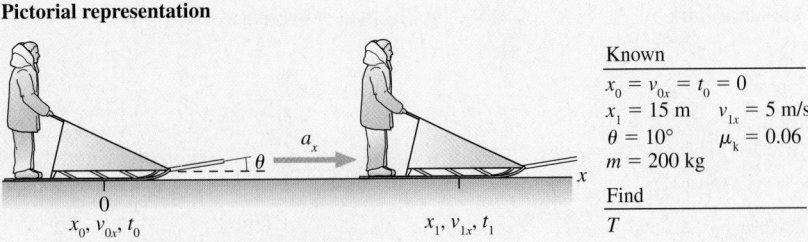

Known
$x_0 = v_{0x} = t_0 = 0$
$x_1 = 15 \text{ m}$ $v_{1x} = 5 \text{ m/s}$
$\theta = 10°$ $\mu_k = 0.06$
$m = 200 \text{ kg}$

Find
T

Physical representation

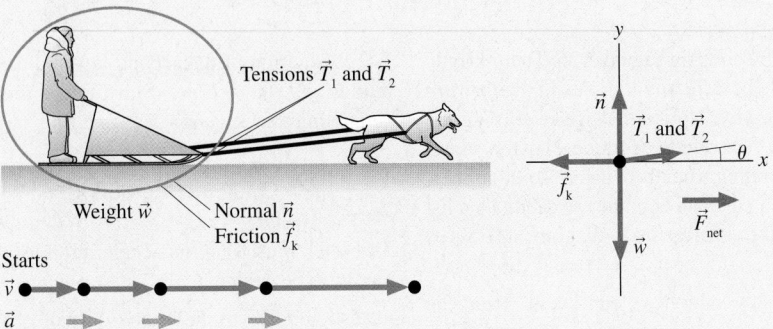

FIGURE 5.25 Pictorial and physical representations of an accelerating dog sled.

Make sure you understand where all the terms come from, including their signs. We've used $w = mg$ and our knowledge that $\vec{a}$ has only an x-component. The tensions $\vec{T}_1$ and $\vec{T}_2$ have both x- and y-components. The assumption of equal tensions allows us to write $T_1 = T_2 = T$, and this introduces the factors of 2.

In addition, we have the model of kinetic friction

$$f_k = \mu_k n$$

From the y-equation and the friction equation,

$$n = mg - 2T \sin\theta$$

$$f_k = \mu_k n = \mu_k mg - 2\mu_k T \sin\theta$$

Notice that n is *not* equal to mg. The y-components of the tension forces support part of the weight, so the ground does not press against the bottom of the sled as hard as it would otherwise.

Substituting the friction back into the x-equation gives

$$2T \cos\theta - (\mu_k mg - 2\mu_k T \sin\theta)$$
$$= 2T(\cos\theta + \mu_k \sin\theta) - \mu_k mg = ma_x$$

$$T = \frac{1}{2} \frac{m(a_x + \mu_k g)}{\cos\theta + \mu_k \sin\theta}$$

Using $a_x = 0.833$ from above with $m = 200$ kg and $\theta = 10°$, we find that the tension is

$$T = 143 \text{ N}$$

ASSESS It's a bit hard to assess this result. We do know that the weight of the sled is $mg \approx 2000$ N. We also know that the dogs can drag a sled over snow (small μ_k) but probably can't lift the sled straight up, so we anticipate that $T \ll 2000$ N. Our calculation agrees.

EXAMPLE 5.11 Make sure the cargo doesn't slide
A 100 kg box of dimensions 50 cm × 50 cm × 50 cm is in the back of a flatbed truck. The coefficients of friction between the box and the bed of the truck are $\mu_s = 0.4$ and $\mu_k = 0.2$. What is the maximum acceleration the truck can have without the box slipping?

MODEL This is a somewhat different problem from any we have looked at thus far. Let the box, which we'll model as a particle, be the system. It contacts its environment only where it touches the truck bed, so only the truck can exert contact forces on the

box. If the box does *not* slip, then there is no motion of the box *relative to the truck* and the box must accelerate *with the truck*: $a_{box} = a_{truck}$. As the box accelerates, it must, according to Newton's second law, have a net force acting on it. But from what?

Imagine, for a moment, that the truck bed is frictionless. The box would slide backwards (as seen in the truck's reference frame) as the truck accelerates. The force that prevents sliding is *static friction*, so the truck must exert a static friction force on the box to "pull" the box along with it and prevent the box from sliding *relative to the truck*.

Pictorial representation

Known

$m = 100$ kg
Box dimensions 50 cm × 50 cm × 50 cm
$\mu_s = 0.4$ $\mu_k = 0.2$

Find

Acceleration at which box slips

Physical representation

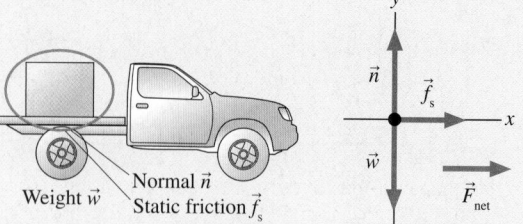

Weight $\vec{w}$ Normal $\vec{n}$ Static friction $\vec{f_s}$

FIGURE 5.26 Pictorial and physical representations for the box in a flatbed truck.

VISUALIZE This situation is shown in Figure 5.26. There is only one horizontal force on the box, $\vec{f_s}$, and it points in the *forward* direction to accelerate the box. Notice that we're solving the problem with the ground as our reference frame. Newton's laws are not valid in the accelerating truck because it is not an inertial reference. There are no kinematics in this problem, so a list of known information next to the free-body diagram suffices as a pictorial representation.

SOLVE Newton's second law, which we can "read" from the free-body diagram, is

$$\sum F_x = f_s = ma_x$$

$$\sum F_y = n - w = n - mg = ma_y = 0$$

Now, static friction, you will recall, can be *any* value between 0 and $f_{s\,max}$. If the truck accelerates slowly, so that the box doesn't slip, then $f_s < f_{s\,max}$. However, we're interested in the acceleration a_{max} at which the box begins to slip. This is the acceleration at which f_s reaches its maximum possible value

$$f_s = f_{s\,max} = \mu_s n$$

The y-equation of the second law and the friction model combine to give $f_s = \mu_s mg$. Substituting this into the x-equation, and noting that a_x is now a_{max}, we find

$$a_{max} = \frac{f_s}{m} = \mu_s g = 3.9 \text{ m/s}^2$$

The truck must keep its acceleration less than 3.9 m/s² if slipping is to be avoided.

ASSESS 3.9 m/s² is about one-third of g. You may have noticed that items in a car or truck are likely to *tip over* when you start or stop, but they slide only if you really floor it and accelerate very quickly. So this answer seems reasonable. Notice that the dimensions of the crate were not needed. Real-world situations rarely have exactly the information you need, no more and no less. Many problems in this textbook will require you to assess the information in the problem statement in order to learn which is relevant to the solution.

The mathematical representation of this last example was quite straightforward. The challenge was in the analysis that preceded the mathematics—that is, in the *physics* of the problem rather than the mathematics. It is here that our analysis tools—motion diagrams, force identification, and free-body diagrams—prove their value.

SUMMARY

The goal of Chapter 5 has been to learn how to solve problems about motion in a straight line.

GENERAL STRATEGY

All examples in this chapter follow a four-part strategy. You'll become a better problem solver if you adhere to it as you do the homework problems. The *Dynamics Worksheets* will help you structure your work in this way.

Equilibrium Problems

Object at rest or moving with constant velocity.

MODEL Make simplifying assumptions.

VISUALIZE
  Physical representation:
    Forces and free-body diagram
  Pictorial representation:
    Translate words to symbols.

SOLVE Use Newton's first law

$$\vec{F}_{net} = \sum_i \vec{F}_i = \vec{0}$$

"Read" the vectors from the free-body diagram.

ASSESS Is the result reasonable?

Go back and forth between represen-tations as needed.

Dynamics Problems

Object accelerating.

MODEL Make simplifying assumptions.

VISUALIZE
  Pictorial representation:
    Sketch to define situation.
    Translate words to symbols.
  Physical representation:
    Forces and free-body diagram

SOLVE Use Newton's second law

$$\vec{F}_{net} = \sum_i \vec{F}_i = m\vec{a}$$

"Read" the vectors from the free-body diagram.
Use kinematics to find velocities and positions.

ASSESS Is the result reasonable?

IMPORTANT CONCEPTS

Specific information about three important forces:

Weight $\vec{w} = (mg, \text{downwards})$

Friction $\vec{f}_s = (0 \text{ to } \mu_s n, \text{ direction as necessary to prevent motion})$

$\vec{f}_k = (\mu_k n, \text{ direction opposite the motion})$

$\vec{f}_r = (\mu_r n, \text{ direction opposite the motion})$

Drag $\vec{D} \approx (\frac{1}{4}Av^2, \text{ direction opposite the motion})$

Newton's laws are vector expressions. You must write them out by components:

$$(F_{net})_x = \sum F_x = ma_x \text{ or } 0$$

$$(F_{net})_y = \sum F_y = ma_y \text{ or } 0$$

APPLICATIONS

Apparent weight is the magnitude of the contact force supporting an object. It is what a scale would read, and it is your sensation of weight. It equals your true weight $w = mg$ only when $a = 0$.

$$w_{app} = w\left(1 + \frac{a_y}{g}\right)$$

Terminal speed is $v_{term} \approx \sqrt{\dfrac{4mg}{A}}$

TERMS AND NOTATION

apparent weight, w_{app}
weightless
coefficient of static friction, μ_s

coefficient of kinetic friction, μ_k
rolling friction

coefficient of rolling friction, μ_r
terminal speed, v_{term}

EXERCISES AND PROBLEMS

The ✎ icon indicates that the problem can be done on a Dynamics Worksheet.

Exercises

Section 5.1 Equilibrium

1. The three ropes in the figure are tied to a small, very light ring. Two of the ropes are anchored to walls at right angles, and the third rope pulls as shown. What are T_1 and T_2, the magnitudes of the tension forces in the first two ropes?

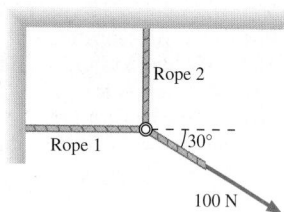

FIGURE EX5.1

2. The three ropes in the figure are tied to a small, very light ring. Two of these ropes are anchored to walls at right angles with the tensions shown in the figure. What are the magnitude and direction of the tension $\vec{T}_3$ in the third rope?

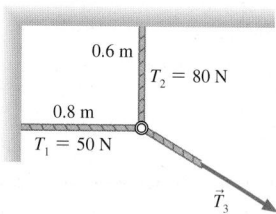

FIGURE EX5.2

3. A 20 kg loudspeaker is suspended 2.0 m below the ceiling by two 3.0-m-long cables that angle outward at equal angles. What is the tension in the cables?

4. A football coach sits on a sled while two of his players build their strength by dragging the sled across the field with ropes. The friction force on the sled is 1000 N and the angle between the two ropes is 20°. How hard must each player pull to drag the coach at a steady 2.0 m/s?

Section 5.2 Using Newton's Second Law

5. In each of the two free-body diagrams, the forces are acting on a 2.0 kg object. For each diagram, find the values of a_x and a_y, the x- and y-components of the acceleration.

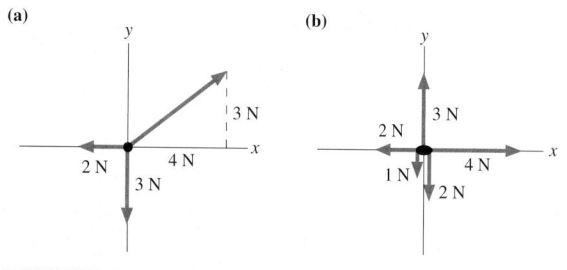

FIGURE EX5.5

6. In each of the two free-body diagrams, the forces are acting on a 2.0 kg object. For each diagram, find the values of a_x and a_y, the x- and y-components of the acceleration.

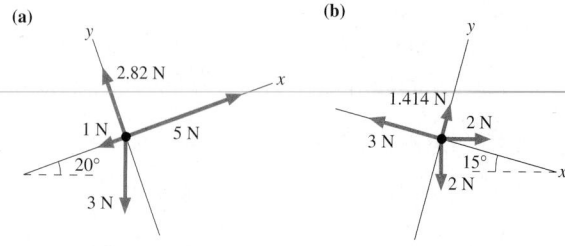

FIGURE EX5.6

7. Figure Ex5.7 shows the velocity graph of a 2.0 kg object as it moves along the x-axis. What is the net force acting on this object at $t = 1$ s? At 4 s? At 7 s?

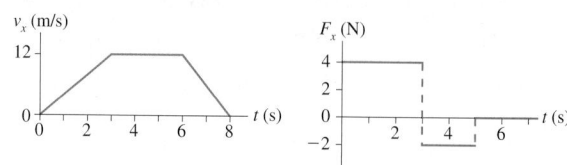

FIGURE EX5.7   **FIGURE EX. 5.8**

8. Figure Ex5.8 shows the force acting on a 2.0 kg object as it moves along the x-axis. The object is at rest at the origin at $t = 0$ s. What are its acceleration and velocity at $t = 6$ s?

9. A horizontal rope is tied to a 50 kg box on frictionless ice. What is the tension in the rope if:
   a. The box is at rest?
   b. The box moves at a steady 5.0 m/s?
   c. The box has $v_x = 5.0$ m/s and $a_x = 5.0$ m/s²?

10. A 50 kg box hangs from a rope. What is the tension in the rope if:
   a. The box is at rest?
   b. The box moves up a steady 5.0 m/s?
   c. The box has $v_y = 5.0$ m/s and is speeding up at 5.0 m/s²?
   d. The box has $v_y = 5.0$ m/s and is slowing down at 5.0 m/s²?

Section 5.3 Mass and Weight

11. An astronaut's weight on earth is 800 N. What is his weight on Mars, where $g = 3.76$ m/s²?

12. A woman has a mass of 55 kg.
   a. What is her weight on earth?
   b. What are her mass and her weight on the moon, where $g = 1.62$ m/s²?

13. It takes the elevator in a skyscraper 4.0 s to reach its cruising speed of 10 m/s. A 60 kg passenger gets aboard on the ground floor. What is the passenger's apparent weight
   a. Before the elevator starts moving?
   b. While the elevator is speeding up?
   c. After the elevator reaches its cruising speed?

14. Figure Ex5.14 shows the velocity graph of a 75 kg passenger in an elevator. What is the passenger's apparent weight at $t = 1$ s? At 5 s? At 9 s?

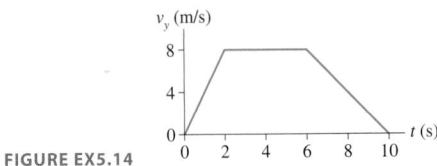

FIGURE EX5.14

Section 5.4 Friction

15. Bonnie and Clyde are sliding a 300 kg bank safe across the floor to their getaway car. The safe slides with a constant speed if Clyde pushes from behind with 385 N of force while Bonnie pulls forward on a rope with 350 N of force. What is the safe's coefficient of kinetic friction on the bank floor?

16. A 4000 kg truck is parked on a 15° slope. How big is the friction force on the truck?

17. A 1000 kg car traveling at a speed of 40 m/s skids to a halt on wet concrete where $\mu_k = 0.6$. How long are the skid marks?

18. An Airbus A320 jetliner has a takeoff mass of 75,000 kg. It reaches its takeoff speed of 82 m/s (180 mph) in 35 s. What is the thrust of the engines? You can neglect air resistance but not rolling friction.

19. A 50,000 kg locomotive is traveling at 10 m/s when its engine and brakes both fail. How far will the locomotive roll before it comes to a stop?

20. A stubborn, 120 kg mule sits down and refuses to move. To drag the mule to the barn, the exasperated farmer ties a rope around the mule and pulls with his maximum force of 800 N. The coefficients of friction between the mule and the ground are $\mu_s = 0.8$ and $\mu_k = 0.5$. Is the farmer able to move the mule?

Section 5.5 Drag

21. A 75 kg skydiver can be modeled as a rectangular "box" with dimensions 20 cm × 40 cm × 180 cm. What is his terminal speed if he falls feet first?

22. A 22-cm-diameter bowling ball has a terminal speed of 85 m/s. What is the ball's mass?

Problems

23. Use information that you can find in Chapters 4 and 5 to *estimate* the acceleration of a car.

24. Use information that you can find in Chapters 4 and 5 to *estimate* the size of the friction force on a baseball player sliding into second base.

25. A 5.0 kg object initially at rest at the origin is subjected to the time-varying force shown in Figure P5.25. What is the object's velocity at $t = 6$ s?

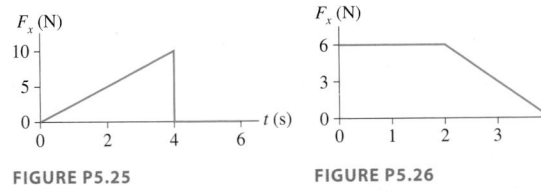

FIGURE P5.25   **FIGURE P5.26**

26. A 2.0 kg object initially at rest at the origin is subjected to the time-varying force shown in Figure P5.26. What is the object's velocity at $t = 4$ s?

27. A 1000 kg steel beam is supported by two ropes. Each rope has a maximum sustained tension of 6000 N. Does either rope break? If so, which one(s)?

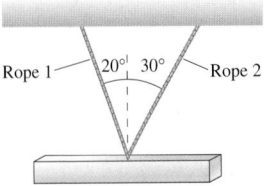

FIGURE P5.27

28. A 500 kg piano is being lowered into position by a crane while two people steady it with ropes pulling to the sides. Bob's rope pulls to the left, 15° below horizontal, with 500 N of tension. Ellen's rope pulls toward the right, 25° below horizontal.
    a. What tension must Ellen maintain in her rope to keep the piano descending at a steady speed?
    b. What is the tension in the main cable supporting the piano?

29. In an electricity experiment, a 1.0 g plastic ball is suspended on a 60-cm-long string and given an electric charge. A charged rod brought near the ball exerts a horizontal electrical force $\vec{F}_{elec}$ on it, causing the ball to swing out to a 20° angle and remain there.
    a. What is the magnitude of $\vec{F}_{elec}$?
    b. What is the tension in the string?

30. Henry gets into an elevator on the 50th floor of a building and it begins moving at $t = 0$ s. The figure shows his apparent weight over the next 12 s.
    a. Is the elevator's initial direction up or down? Explain how you can tell.
    b. What is Henry's mass?
    c. How far has Henry traveled at $t = 12$ s?

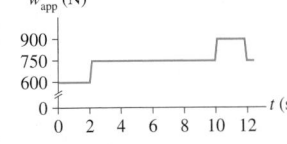

FIGURE P5.30

31. Zach, whose mass is 80 kg, is in an elevator descending at 10 m/s. The elevator takes 3.0 s to brake to a stop at the first floor.
    a. What is Zach's apparent weight before the elevator starts braking?
    b. What is Zach's apparent weight while the elevator is braking?

32. You've always wondered about the acceleration of the elevators in the 101-story-tall Empire State Building. One day, while visiting New York, you take your bathroom scale into the elevator and stand on them. The scales read 150 lb as the door closes. The reading varies between 120 lb and 170 lb as the elevator travels 101 floors. What conclusions can you draw?

33. An accident victim with a broken leg is being placed in traction. The patient wears a special boot with a pulley attached to the sole. The foot and boot together have a mass of 4.0 kg, and the doctor has decided to hang a 6.0 kg mass from the rope. The boot is held suspended by the ropes and does not touch the bed.
    a. Determine the amount of tension in the rope by using Newton's laws to analyze the hanging mass.

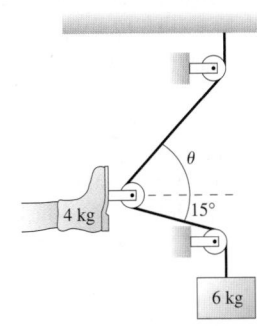

FIGURE P5.33

b. The net traction force needs to pull straight out on the leg. What is the proper angle θ for the upper rope?

c. What is the net traction force pulling on the leg?

Hint: If the pulleys are frictionless, which we will assume, the tension in the rope is constant from one end to the other.

34. Seat belts and air bags save lives by reducing the forces exerted on the driver and passengers in an automobile collision. Cars are designed with a "crumple zone" in the front of the car. In the event of an impact, the passenger compartment decelerates over a distance of about 1 m as the front of the car crumples. An occupant restrained by seat belts and air bags decelerates with the car. By contrast, an unrestrained occupant keeps moving forward with no loss of speed (Newton's first law!) until hitting the dashboard or windshield. These are unyielding surfaces, and the unfortunate occupant then decelerates over a distance of only about 5 mm.

a. A 60 kg person is in a head-on collision. The car's speed at impact is 15 m/s. Estimate the net force on the person if he or she is wearing a seat belt and if the air bag deploys.

b. Estimate the net force that ultimately stops the person if he or she is not restrained by a seat belt or air bag.

c. How do these two forces compare to the person's weight?

35. A rifle with a barrel length of 60 cm fires a 10 g bullet with a horizontal speed of 400 m/s. The bullet strikes a block of wood and penetrates to a depth of 12 cm.

a. What frictional force (assumed to be constant) does the wood exert on the bullet?

b. How long does it take the bullet to come to rest?

c. Draw a velocity-versus-time graph for the bullet in the wood.

36. Compressed air is used to fire a 50 g ball vertically upward from a 1.0-m-tall tube. The air exerts an upward force of 2.0 N on the ball as long as it is in the tube. How high does the ball go above the top of the tube?

37. What thrust does a 200 g model rocket need in order to have a vertical acceleration of 10 m/s²?

a. On Earth?

b. On the moon, where $g = 1.62$ m/s²?

38. A 20,000 kg rocket has a rocket motor that generates 3.0×10^5 N of thrust.

a. What is the rocket's initial upward acceleration?

b. At an altitude of 5000 m the rocket's acceleration has increased to 6.0 m/s². What mass of fuel has it burned?

39. A 2.0 kg steel block is at rest on a steel table. A horizontal string pulls on the block.

a. What is the minimum string tension needed to move the block?

b. If the string tension is 20 N, what is the block's speed after moving 1.0 m?

c. If the string tension is 20 N and the table is coated with oil, what is the block's speed after moving 1.0 m?

40. Sam, whose mass is 75 kg, takes off across level snow on his jet-powered skis. The skis have a thrust of 200 N and a coefficient of kinetic friction on snow of 0.1. Unfortunately, the skis run out of fuel after only 10 s.

a. What is Sam's top speed?

b. How far has Sam traveled when he finally coasts to a stop?

41. Sam, whose mass is 75 kg, takes off down a 50-m-high, 10° slope on his jet-powered skis. The skis have a thrust of 200 N. Sam's speed at the bottom is 40 m/s. What is the coefficient of kinetic friction of his skis on snow?

42. A 10 kg crate is placed on a horizontal conveyor belt. The materials are such that $\mu_s = 0.5$ and $\mu_k = 0.3$.

a. Draw a free-body diagram showing all the forces on the crate if the conveyer belt runs at constant speed.

b. Draw a free-body diagram showing all the forces on the crate if the conveyer belt is speeding up.

c. What is the maximum acceleration the belt can have without the crate slipping?

43. A baggage handler drops your 10 kg suitcase onto a conveyor belt running at 2.0 m/s. The materials are such that $\mu_s = 0.5$ and $\mu_k = 0.3$. How far is your suitcase dragged before it is riding smoothly on the belt?

44. Johnny jumps off a swing, lands sitting down on a grassy 20° slope, and slides 3.5 m down the slope before stopping. The coefficient of kinetic friction between grass and the seat of Johnny's pants is 0.5. What was his initial speed on the grass?

45. A 2.0 kg wood block is launched up a wooden ramp that is inclined at a 30° angle. The block's initial speed is 10 m/s.

a. What vertical height does the block reach above its starting point?

b. What speed does it have when it slides back down to its starting point?

46. It's moving day, and you need to push a 100 kg box up a 20° ramp into the truck. The coefficients of friction for the box on the ramp are $\mu_s = 0.9$ and $\mu_k = 0.6$. Your largest pushing force is 1000 N. Can you get the box into the truck without assistance if you get a running start at the ramp? If you stop on the ramp, will you be able to get the box moving again?

47. It's a snowy day and you're pulling a friend along a level road on a sled. You've both been taking physics, so she asks what you think the coefficient of friction between the sled and the snow is. You've been walking at a steady 1.5 m/s, and the rope pulls up on the sled at a 30° angle. You estimate that the mass of the sled, with your friend on it, is 60 kg and that you're pulling with a force of 75 N. What answer will you give?

48. A horizontal rope pulls a 10 kg wood sled across frictionless snow. A 5.0 kg wood box rides on the sled. What is the largest tension force for which the box doesn't slip?

49. A pickup truck with a steel bed is carrying a steel file cabinet. If the truck's speed is 15 m/s, what is shortest distance in which it can stop without the file cabinet sliding?

50. You're driving along at 25 m/s with your aunt's valuable antiques in the back of your pickup truck when suddenly you see a giant hole in the road 55 m ahead of you. Fortunately, your foot is right beside the brake and your reaction time is zero! Will the antiques be as fortunate?

a. Can you stop the truck before it falls into the hole?

b. If your answer to part a is yes, can you stop without the antiques sliding and being damaged? Their coefficients of friction are $\mu_s = 0.6$ and $\mu_k = 0.3$.

Hint: You're not trying to stop in the shortest possible distance. What's your best strategy for avoiding damage to the antiques?

51. A 2.0 kg wood box slides down a vertical wood wall while you push on it at a 45° angle. What magnitude of force should you apply to cause the box to slide down at a constant speed?

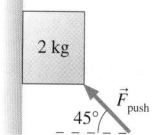

FIGURE P5.51

52. A 1.0 kg wood block is pressed against a vertical wood wall by the 12 N force shown. If the block is initially at rest, will it move upward, move downward, or stay at rest?

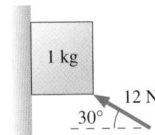

FIGURE P5.52

53. What is the terminal speed for an 80 kg skier going down a 40° snow-covered slope on wooden skis? Assume that the skier is 1.8 m tall and 0.40 m wide.

54. A 10 g Ping-Pong ball has a diameter of 3.5 cm.
    a. The ball is shot straight up at twice its terminal speed. What is its initial acceleration?
    b. The ball is shot straight down at twice its terminal speed. What is its initial acceleration?
    c. The ball is shot straight down at twice its terminal speed. Draw a plausible velocity-versus-time graph.

55. Try this! Hold your right hand out with your palm perpendicular to the ground, as if you were getting ready to shake hands. You can't hold anything in your palm this way because it would fall straight down. Use your left hand to hold a small object, such as a ball or a coin, against your outstretched palm, then let go as you quickly swing your hand to the left across your body, parallel to the ground. You'll find that the object stays against your palm; it doesn't slip or fall.
    a. Is the condition for keeping the object against your palm one of maintaining a certain minimum velocity v_{min}? Or one of maintaining a certain minimum acceleration a_{min}? Explain.
    b. Suppose the object's mass is 50 g, with $\mu_s = 0.8$ and $\mu_k = 0.4$. Determine either v_{min} or a_{min}, whichever you answered in part a.

56. Suppose you use your hand to *push* a ball straight down toward the floor. The ball's weight is 1.0 N.
    a. Draw the ball's free-body diagram while you're pushing it.
    b. Is F_{net} larger than, smaller than, or equal to the ball's weight w? Explain.
    c. Find an expression for the ball's apparent weight when its acceleration is a_y. Evaluate w_{app} for a_y equal to $-g$, $-1.5g$, and $-2g$.

57. You've been called in to investigate a construction accident in which the cable broke while a crane was lifting a 4500 kg container. The steel cable is 2.0 cm in diameter and has a safety rating of 50,000 N. The crane is designed not to exceed speeds of 3.0 m/s or accelerations of 1.0 m/s², and your tests find that the crane is not defective. What is your conclusion? Did the crane operator recklessly lift too heavy a load? Or was the cable defective?

58. An artist friend of yours needs help hanging a 500 lb sculpture from the ceiling. For artistic reasons, she wants to use just two ropes. One will be 30° from vertical, the other 60°. She needs you to determine the smallest diameter rope that can safely support this expensive piece of art. On a visit to the hardware store you find that rope is sold in increments of $\frac{1}{8}$-inch diameter and that the safety rating is 4000 pounds per square inch of cross section. What diameter rope should you buy?

59. A machine has an 800 g steel shuttle that is pulled along a square steel rail by an elastic cord. The shuttle is released when the elastic cord has 20 N tension at a 45° angle. What is the initial acceleration of the shuttle?

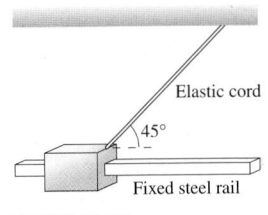

FIGURE P5.59

60. A 1.0 kg ball hangs from the ceiling of a truck by a 1.0-m-long string. The back of the truck, where you are riding with the ball, has no windows and has been completely soundproofed. The truck travels along an exceedingly smooth test track, and you feel no bumps or bounces as it moves. Your only instruments are a meter stick, a protractor, and a stopwatch.
    a. The driver tells you, over a loudspeaker, that the truck is either at rest, or it is moving forward at a steady speed of 5 m/s. Can you determine which it is? If so, how? If not, why not?
    b. Next, the driver tells you that the truck is either moving forward with a steady speed of 5 m/s, or it is accelerating at 5 m/s². Can you determine which it is? If so, how? If not, why not?
    c. Suppose the truck has been accelerating forward at 5 m/s² long enough for the ball to achieve a steady position. Does the ball have an acceleration? If so, what are the magnitude and direction of the ball's acceleration?
    d. Draw a free-body diagram that shows all forces acting on the ball as the truck accelerates.
    e. Suppose the ball makes a 10° angle with the vertical. If possible, determine the truck's velocity. If possible, determine the truck's acceleration.

61. Imagine *hanging* from a big spring scale as it moves vertically with acceleration a. Show that Equation 5.10 is the correct expression for your apparent weight.

Problems 62 through 65 show a free-body diagram. For each:
    a. Write a realistic dynamics problem for which this is the correct free-body diagram. Your problem should ask a question that can be answered with a value of position or velocity (such as "How far?" or "How fast?"), and should give sufficient information to allow a solution.
    b. Solve your problem!

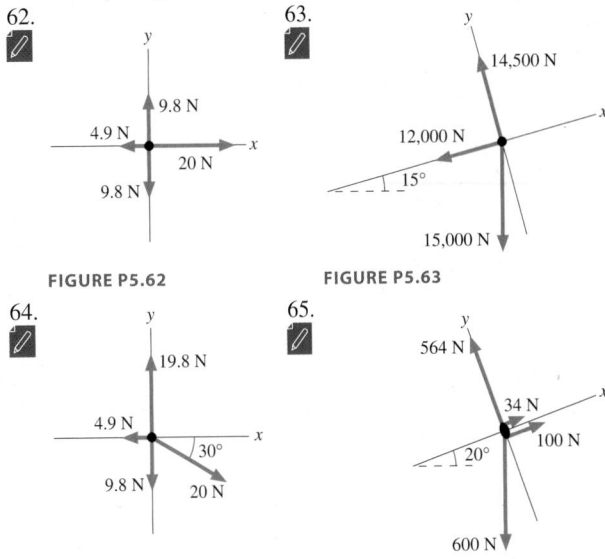

62.

FIGURE P5.62

63.

FIGURE P5.63

64.

FIGURE P5.64

65.

FIGURE P5.65

In Problems 66 through 69 you are given the dynamics equations that are used to solve a problem. For each of these, you are to

a. Write a realistic problem for which these are the correct equations.

b. Draw the free-body diagram and the pictorial representation for your problem.

c. Finish the solution of the problem.

66. $-0.8n = (1500 \text{ kg})a_x$

$n - (1500 \text{ kg})(9.80 \text{ m/s}^2) = 0$

67. $T - 0.2n - (20 \text{ kg})(9.80 \text{ m/s}^2) \sin 20°$

$= (20 \text{ kg})(2 \text{ m/s}^2)$

$n - (20 \text{ kg})(9.80 \text{ m/s}^2) \cos 20° = 0$

68. $(100 \text{ N}) \cos 30° - f_k = (20 \text{ kg})a_x$

$n + (100 \text{ N}) \sin 30° - (20 \text{ kg})(9.80 \text{ m/s}^2) = 0$

$f_k = 0.2n$

69. $-f_k + (20 \text{ kg})(9.80 \text{ m/s}^2) \sin \theta = 0$

$n - (20 \text{ kg})(9.80 \text{ m/s}^2) \cos \theta = 0$

$f_k = 0.2n$

Challenge Problems

70. The figure shows an *accelerometer,* a device for measuring the horizontal acceleration of cars and airplanes. A ball is free to roll on a parabolic track described by the equation $y = x^2$, where both x and y are in meters. A scale along the bottom is used to measure the ball's horizontal position x.

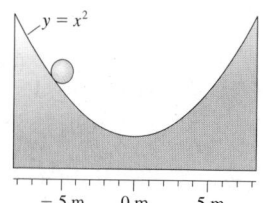

a. Find an expression that allows you to use a measured position x (in m) to compute the acceleration a_x (in m/s²). (For example, $a_x = 3x$ is a possible expression.)

FIGURE CP5.70

b. What is the acceleration if $x = 20$ cm?

71. You've entered a "slow ski race" where the winner is the skier who takes the *longest* time to go down a 15° slope without ever stopping. You need to choose the best wax to apply to your skis. Red wax has a coefficient of kinetic friction 0.25, yellow is 0.20, green is 0.15, and blue is 0.10. Having just finished taking physics, you realize that a wax too slippery will cause you to accelerate down the slope and lose the race. But

a wax that's too sticky will cause you to stop and be disqualified. You know that a strong headwind will apply a 50 N horizontal force against you as you ski, and you know that your mass is 75 kg. Which wax do you choose?

72. A testing laboratory wants to determine if a new widget can withstand large accelerations and decelerations. To find out, they glue a 5.0 kg widget to a test stand that will drive it vertically up and down. The graph shows its acceleration during the first second, starting from rest.

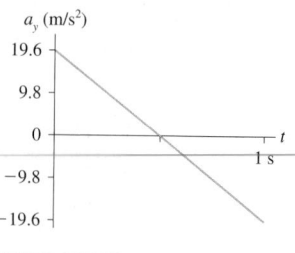

FIGURE CP5.72

a. Identify the forces acting on the widget and draw a free-body diagram.

b. Determine the value of n_y, the y-component of the normal force acting on the widget, during the first second of motion. Give your answer as a graph of n_y-versus-t.

c. Your answer to part b should show an interval of time during which n_y is negative. How can this be? Explain what it means physically for n_y to be negative.

d. At what time is the apparent weight of the widget a maximum? What is the acceleration at this time?

e. Is the apparent weight of the widget ever zero? If so, at what instant of time does this happen? What is the acceleration at that time?

f. Suppose the technician forgets to glue the widget to the test stand. Will the widget remain on the test stand throughout the first second, or will it fly off the stand at some instant of time? If so, at what time will this occur?

73. An object with cross section A is shot horizontally across frictionless ice. Its initial velocity is v_{0x} at $t_0 = 0$ s. Air resistance is not negligible.

a. Show that the velocity at time t is given by the expression

$$v_x = \frac{v_{0x}}{1 + Av_{0x}t/4m}$$

b. A 1.6 m wide, 1.4 m high, 1500 kg car hits a very slick patch of ice while going 20 m/s. If friction is neglected, how long will it take until the car's speed drops to 10 m/s? To 5 m/s?

c. Assess whether or not it is reasonable to neglect kinetic friction.

STOP TO THINK ANSWERS

Stop to Think 5.1: a. The lander is descending and slowing. The acceleration vector points upward, and so $\vec{F}_{net}$ points upward. This can be true only if the thrust has a larger magnitude than the weight.

Stop to Think 5.2: a. You are descending and slowing, so your acceleration vector points upward and there is a net upward force on you. The floor pushes up against your feet harder than gravity pulls down.

Stop to Think 5.3: $f_b > f_c = f_d = f_e > f_a$. Situations c, d, and e are all kinetic friction, which does not depend on either velocity or

acceleration. Kinetic friction is smaller than the maximum static friction that is exerted in b. $f_a = 0$ because no friction is needed to keep the object at rest.

Stop to Think 5.4: d. The ball is shot *down* at 30 m/s, so $v_{0y} = -30$ m/s. This exceeds the terminal speed, so the upward drag force is *larger* than the downward weight force. Thus the ball *slows down* even though it is "falling." It will slow until $v_y = -15$ m/s, the terminal velocity, then maintain that velocity.

6 Dynamics II: Motion in a Plane

This diver is a spinning projectile following a parabolic trajectory.

▶ Looking Ahead

The goal of Chapter 6 is to learn to solve problems about motion in a plane. In this chapter you will learn to:

- Understand kinematics and dynamics in two dimensions.
- Understand projectile motion.
- Explore the issues of relative motion.

◀ Looking Back

This chapter is an extension of several ideas introduced in Chapters 1–5. Please review:

- Section 1.5 Finding acceleration vectors on a motion diagram.
- Sections 2.5–2.6 Constant-acceleration kinematics and free fall.
- Section 4.6 Inertial reference frames.
- Section 5.2 Solving problems with Newton's second law.

We have limited ourselves thus far to motion along a straight line. One-dimensional motion includes a lot of interesting physics and applications, but motion in the real world is often in two or more dimensions. A car turning a corner, a planet orbiting the sun, and the diver in the photograph are examples of two-dimensional motion. Restricting ourselves to one dimension has allowed us to concentrate on basic physics principles, but the time has come to broaden our horizons and consider a wider variety of motion.

Newton's laws are "laws of nature," meaning that they describe all motion, not just motion along a straight line. This chapter and the next will extend the application of Newton's laws to new situations. Chapter 6 will focus on motion in which we can treat the x- and y-components of the acceleration independently of each other. Projectile motion is an important example. Chapter 7 will cover circular motion, where the components are *not* independent.

Motion in a plane will provide us with the tools to look at an important question. Suppose you and I are moving relative to each other. Perhaps I'm standing still while you drive past in your car. How do physical measurements that I make in my coordinate system compare to measurements you make in your coordinate system? Are "my physics" and "your physics" the same? Questions such as these led Einstein to his theory of relativity. We will begin to answer these questions in the context of what is called Galilean relativity.

6.1 Kinematics in Two Dimensions

To begin, let's look at how motion in two dimensions differs from motion in one dimension. As an example, Figure 6.1 shows the motion diagram of a roller coaster car. Here the object moves in a vertical plane, but we'll also look at situations where the plane of motion is horizontal. We'll call it the *xy*-plane regardless of its orientation.

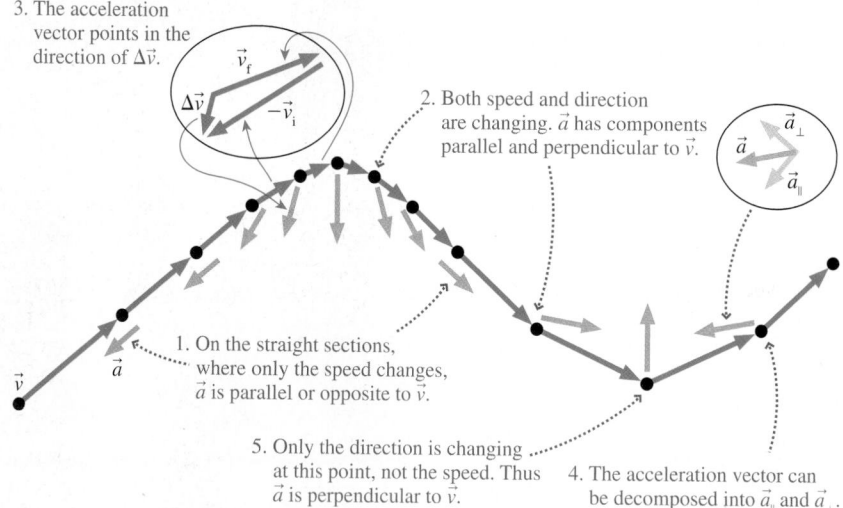

FIGURE 6.1 The motion diagram of a roller coaster car.

What we most need to understand is how velocity and acceleration are related in two dimensions. The average acceleration was defined in Chapter 1 as $\vec{a}_{avg} = \Delta\vec{v}/\Delta t$. That is, the acceleration vector $\vec{a}$ points in the direction of $\Delta\vec{v}$, the change in velocity. Because velocity is a *vector*, a change can be either a change in length (i.e., a change of speed) or a change in direction.

On the straight sections of the roller coaster, you can see that a particle moving along a *straight line* speeds up if $\vec{a}$ and $\vec{v}$ point in the same direction and slows down if $\vec{a}$ and $\vec{v}$ point in opposite directions. This idea was the basis for the one-dimensional kinematics we developed in Chapter 2. For linear motion, acceleration is a change of speed.

When the direction of $\vec{v}$ changes, as it does when the roller coaster car goes over the hill or through the valley, the acceleration gets a bit tricky. You learned in Section 1.5 how to use vector subtraction to find the direction of $\vec{a}$, and a review is well worthwhile. The procedure is shown at one point in the motion diagram.

Chapter 3 showed how to decompose a vector into two perpendicular components. At point 4 in Figure 6.1 the acceleration vector $\vec{a}$ has been decomposed into a piece $\vec{a}_{\parallel}$ that is parallel to $\vec{v}$ and a piece $\vec{a}_{\perp}$ that is perpendicular to $\vec{v}$. $\vec{a}_{\parallel}$ **is the piece of the acceleration vector that changes the speed.** In this case it is slowing the car because $\vec{a}_{\parallel}$ is opposite the motion. **The component $\vec{a}_{\perp}$ is the piece of the acceleration that causes the velocity to change direction.**

Notice that $\vec{a}$ *always* has a perpendicular component at points where the direction of $\vec{v}$ is changing. At point 5, where only the direction is changing, not the speed, the parallel component vanishes and $\vec{a}$ is perpendicular to $\vec{v}$.

Position and Velocity

Motion diagrams are an important tool for visualizing motion, but we also need to develop a mathematical description of motion in two dimensions. It will be easiest to use x- and y-components of vectors, rather than components parallel and perpendicular to the motion. We'll point out, as we go along, the connection between these two points of view.

Figure 6.2 shows a particle moving along a curved path—its *trajectory*—in the xy-plane. At time t_1, the particle is at point 1. We can locate the particle in terms of its position vector

$$\vec{r}_1 = r_{1x}\hat{i} + r_{1y}\hat{j}$$

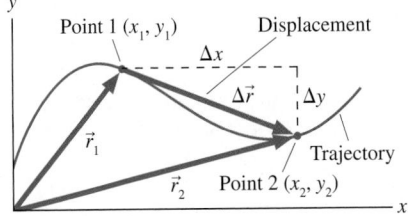

FIGURE 6.2 A particle moving along a trajectory in the xy-plane.

But r_{1x}, the x-component of $\vec{r}_1$, is simply x_1, the x-coordinate of the point. Similarly, r_{1y} is the y-coordinate y_1. Hence the position vector is

$$\vec{r}_1 = x_1\hat{i} + y_1\hat{j} \tag{6.1}$$

A short time later, at t_2, the position vector is $\vec{r}_2 = x_2\hat{i} + y_2\hat{j}$.

NOTE ▶ In Chapter 2 we made extensive use of position-versus-time graphs, either x-versus-t or y-versus-t. Figure 6.2, like many of the graphs we'll use in this chapter, is a graph of y-versus-x. In other words, it's an actual *picture* of the trajectory, not an abstract representation of the motion. ◀

The vector connecting these two points on the trajectory is the *displacement vector*

$$\Delta\vec{r} = \vec{r}_2 - \vec{r}_1$$

We can write the displacement vector in component form as

$$\Delta\vec{r} = \Delta x\hat{i} + \Delta y\hat{j} \tag{6.2}$$

where $\Delta x = x_2 - x_1$ and $\Delta y = y_2 - y_1$ are the horizontal and vertical changes of position.

Chapter 1 defined the *average velocity* of a particle moving through a displacement $\Delta\vec{r}$ in a time interval Δt as

$$\vec{v}_{avg} = \frac{\Delta\vec{r}}{\Delta t} = \frac{\Delta x}{\Delta t}\hat{i} + \frac{\Delta y}{\Delta t}\hat{j} \tag{6.3}$$

You learned in Chapter 2 that the *instantaneous velocity* is the limit of $\vec{v}_{avg}$ as $\Delta t \to 0$. Taking the limit of Equation 6.3 gives the instantaneous velocity in two dimensions:

$$\vec{v} = \lim_{\Delta t \to 0} \frac{\Delta\vec{r}}{\Delta t} = \frac{d\vec{r}}{dt} = \frac{dx}{dt}\hat{i} + \frac{dy}{dt}\hat{j} \tag{6.4}$$

But we can also write the velocity vector in terms of its x- and y-components as

$$\vec{v} = v_x\hat{i} + v_y\hat{j} \tag{6.5}$$

Comparing Equations 6.4 and 6.5, you can see that the velocity vector $\vec{v}$ has x- and y-components

$$v_x = \frac{dx}{dt} \quad \text{and} \quad v_y = \frac{dy}{dt} \tag{6.6}$$

That is, the x-component v_x of the velocity vector is the rate dx/dt at which the particle's x-coordinate is changing. The y-component is similar.

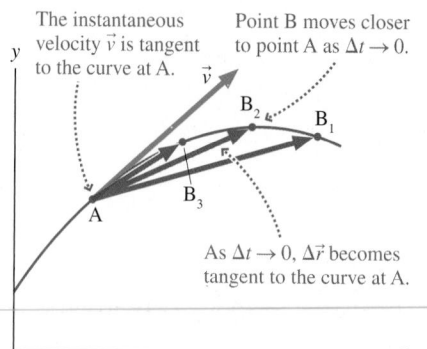

FIGURE 6.3 The instantaneous velocity vector $\vec{v}$ is tangent to the trajectory.

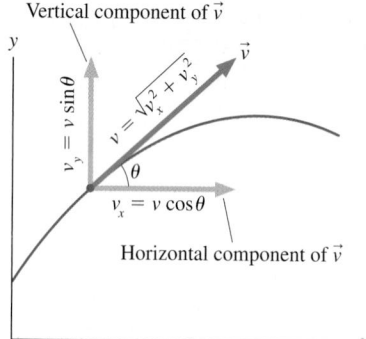

FIGURE 6.4 Relating the components of $\vec{v}$ to the speed and direction.

The average velocity $\vec{v}_{avg}$ points in the direction of $\Delta\vec{r}$, a fact we used in Chapter 1 to draw the velocity vectors on motion diagrams. Figure 6.3 shows that $\Delta\vec{r}$ becomes tangent to the trajectory as $\Delta t \to 0$. Consequently, **the instantaneous velocity vector $\vec{v}$ is tangent to the trajectory.**

Figure 6.4 illustrates another important feature of the velocity vector. If the vector's angle θ is measured from the positive x-axis, the velocity vector components are

$$v_x = \frac{dx}{dt} = v\cos\theta$$

$$v_y = \frac{dy}{dt} = v\sin\theta \qquad (6.7)$$

where

$$v = \sqrt{v_x^2 + v_y^2} \qquad (6.8)$$

is the particle's *speed* at that point. Speed is always a positive number (or zero), whereas the components are *signed* quantities to convey information about the direction of the velocity vector. Conversely, we can use the two velocity components to determine the direction of motion:

$$\tan\theta = \frac{v_y}{v_x} \qquad (6.9)$$

NOTE ▶ In Chapter 2, you learned that the *value* of the velocity component v_s at time t is given by the *slope* of the position-versus-time graph at time t. Now we see that the *direction* of the velocity vector $\vec{v}$ is given by the *tangent* to the y-versus-x graph of the trajectory. Figure 6.5 reminds you that these two graphs use different interpretations of the tangent lines. The tangent to the trajectory does not tell us anything about how fast the particle is moving, only its direction. ◀

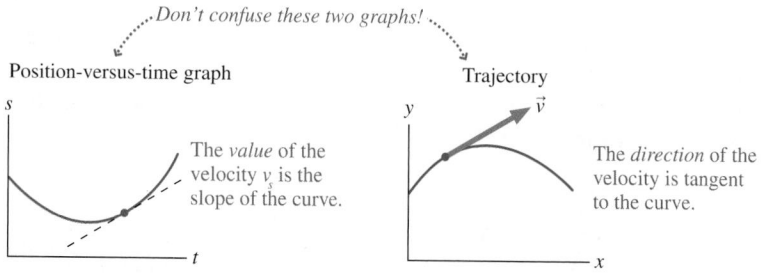

FIGURE 6.5 Two different uses of tangent lines.

EXAMPLE 6.1 Describing the motion with graphs

A particle's motion is described by the two equations

$$x = 2t^2 \text{ m}$$
$$y = (5t + 5) \text{ m}$$

where the time t is in s.

a. Draw a graph of the particle's trajectory.
b. Draw a graph of the particle's speed as a function of time.

MODEL These are *parametric equations* that give the particle's coordinates x and y separately in terms of the parameter t.

SOLVE

a. The trajectory is a curve in the xy-plane. The easiest way to proceed is to calculate x and y at several instants of time.

t (s)	x (m)	y (m)	v (m/s)
0	0	5	5.0
1	2	10	6.4
2	8	15	9.4
3	18	20	13.0
4	32	25	16.8

These points are plotted in Figure 6.6a, then a smooth curve is drawn through them to show the trajectory.

b. The particle's speed is given by Equation 6.8. We first need to use Equation 6.6 to find the components of the velocity vector:

$$v_x = \frac{dx}{dt} = 4t \text{ m/s} \quad \text{and} \quad v_y = \frac{dy}{dt} = 5 \text{ m/s}$$

Using these gives the particle's speed at time t:

$$v = \sqrt{v_x^2 + v_y^2} = \sqrt{16t^2 + 25} \text{ m/s}$$

The speed was computed in the table above and is graphed in Figure 6.6b.

ASSESS The y-versus-x graph of Figure 6.6a is a trajectory, not a position-versus-time graph. Thus the slope is *not* the particle's speed. The particle is speeding up, as you can see in the second graph, even though the slope of the trajectory is decreasing.

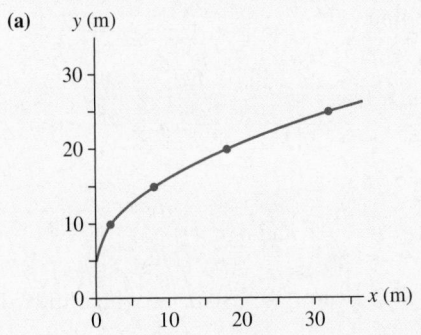

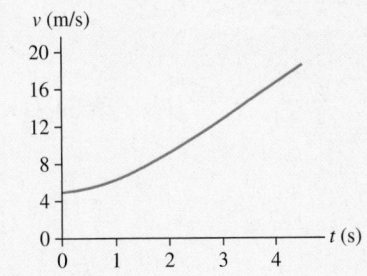

FIGURE 6.6 Two motion graphs for the particle of Example 6.1.

Acceleration

Let's return to the particle moving along a trajectory in the xy-plane. Figure 6.7 shows the instantaneous velocity $\vec{v}_1$ at point 1 and, a short time later, velocity $\vec{v}_2$ at point 2. These two vectors are tangent to the trajectory.

In Chapter 1 we defined the particle's *average acceleration* as

$$\vec{a}_{\text{avg}} = \frac{\Delta \vec{v}}{\Delta t} \tag{6.10}$$

where $\Delta \vec{v} = \vec{v}_2 - \vec{v}_1$ is the change in velocity during the interval Δt. We can use the vector-subtraction technique of Chapter 1 to find $\vec{a}_{\text{avg}}$ on this segment of the trajectory. This is shown in the insert to Figure 6.7.

If we now take the limit $\Delta t \rightarrow 0$, the *instantaneous acceleration* is

$$\vec{a} = \lim_{\Delta t \rightarrow 0} \frac{\Delta \vec{v}}{\Delta t} = \frac{d\vec{v}}{dt} \tag{6.11}$$

As $\Delta t \rightarrow 0$, points 1 and 2 in Figure 6.7 merge, and the instantaneous acceleration $\vec{a}$ is found at the same point on the trajectory (and the same instant of time) as the instantaneous velocity $\vec{v}$. This is shown in Figure 6.8.

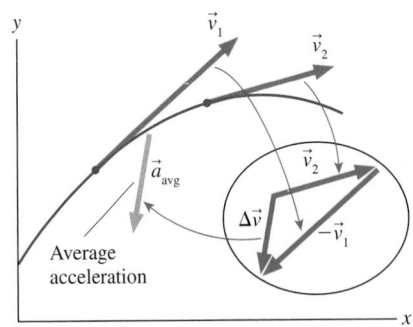

FIGURE 6.7 The average acceleration vector at a point on a curved trajectory.

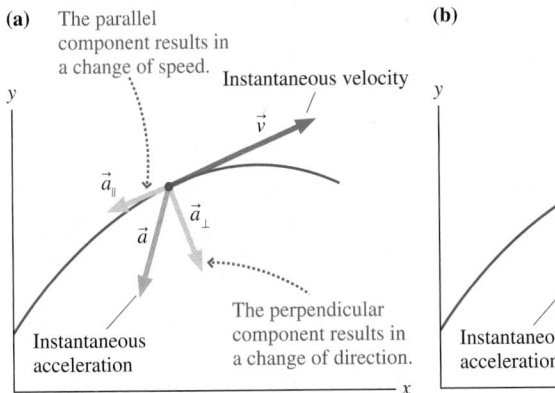

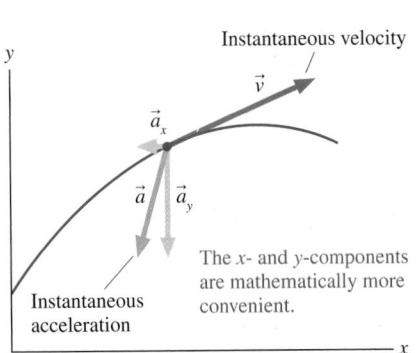

FIGURE 6.8 The instantaneous acceleration $\vec{a}$ can be decomposed into parallel and perpendicular components or into x- and y-components.

The acceleration vector $\vec{a}$ is the rate at which $\vec{v}$ is changing at that instant. To show this, Figure 6.8a decomposes $\vec{a}$ into components $\vec{a}_\parallel$ and $\vec{a}_\perp$ that are parallel and perpendicular to the trajectory. $\vec{a}_\parallel$ is associated with a change of speed and $\vec{a}_\perp$ is associated with a change in direction. Both kinds of changes are accelerations. Notice that $\vec{a}_\perp$ always points toward the "inside" of the curve because that is the direction in which $\vec{v}$ is changing.

The parallel and perpendicular components of $\vec{a}$ convey important ideas about acceleration, but the directions of $\vec{a}_\parallel$ and $\vec{a}_\perp$ keep changing. It's usually more practical to write $\vec{a}$ in terms of the x- and y-components shown in Figure 6.8b. Because $\vec{v} = v_x\hat{\imath} + v_y\hat{\jmath}$, we find

$$\vec{a} = a_x\hat{\imath} + a_y\hat{\jmath} = \frac{d\vec{v}}{dt} = \frac{dv_x}{dt}\hat{\imath} + \frac{dv_y}{dt}\hat{\jmath} \tag{6.12}$$

from which we see that

$$a_x = \frac{dv_x}{dt} \quad \text{and} \quad a_y = \frac{dv_y}{dt} \tag{6.13}$$

That is, the x-component of $\vec{a}$ is the rate dv_x/dt at which the x-component of velocity is changing.

Constant Acceleration

We're going to restrict our study of motion in a plane to situations where the acceleration $\vec{a} = a_x\hat{\imath} + a_y\hat{\jmath}$ is constant. This implies that the two components a_x and a_y are both constant (including, perhaps, zero). In this case, everything you learned about constant-acceleration kinematics in Chapter 2 carries over to the x- and y-components of two-dimensional motion.

Consider a particle that moves with constant acceleration from an initial position $\vec{r}_i = x_i\hat{\imath} + y_i\hat{\jmath}$, starting with initial velocity $\vec{v}_i = v_{ix}\hat{\imath} + v_{iy}\hat{\jmath}$. Its position and velocity at a final point f are

$$\begin{aligned} x_f &= x_i + v_{ix}\Delta t + \tfrac{1}{2}a_x(\Delta t)^2 & y_f &= y_i + v_{iy}\Delta t + \tfrac{1}{2}a_y(\Delta t)^2 \\ v_{fx} &= v_{ix} + a_x\Delta t & v_{fy} &= v_{iy} + a_y\Delta t \end{aligned} \tag{6.14}$$

There are *many* quantities to keep track of in two-dimensional kinematics, making the pictorial representation all the more important.

NOTE ▶ For constant acceleration, the x-component of the motion and the y-component of the motion are independent of each other. However, they remain connected through the fact that Δt must be the same for both. ◀

EXAMPLE 6.2 Plotting the trajectory of the shuttlecraft

The up thrusters on the shuttlecraft of the starship *Enterprise* give it an upward acceleration of 5.0 m/s². Its forward thrusters provide a forward acceleration of 20 m/s². As it leaves the *Enterprise*, the shuttlecraft turns on only the upthrusters. After clearing the flight deck, 3.0 s later, it adds the forward thrusters. Plot a trajectory of the shuttlecraft for its first 6 s.

MODEL Represent the shuttlecraft as a particle. There are two segments of constant-acceleration motion.

VISUALIZE Figure 6.9 shows a pictorial representation. The coordinate system has been chosen so that the shuttlecraft starts at the origin and initially moves along the y-axis. The craft moves vertically for 3 s, then begins to acquire a forward motion. There are three points in the motion: the beginning, the end, and the point at which forward thrusters are turned on. These points are labeled (x_0, y_0), (x_1, y_1), and (x_2, y_2). The velocities are (v_{0x}, v_{0y}), (v_{1x}, v_{1y}), and (v_{2x}, v_{2y}). This will be our standard labeling scheme for trajectories, where it is essential to keep the x-components and y-components separate.

Pictorial representation

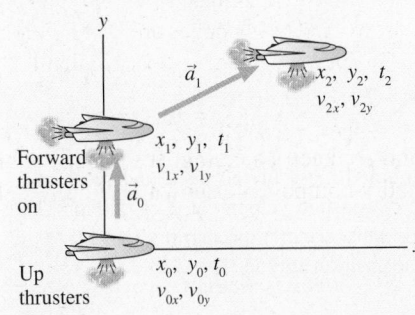

FIGURE 6.9 Pictorial representation of the motion of the shuttlecraft.

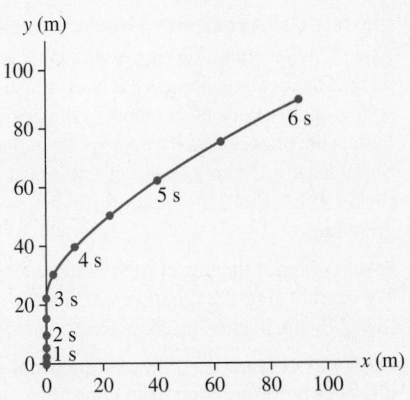

FIGURE 6.10 The shuttlecraft trajectory.

Known

$x_0 = y_0 = v_{0x} = v_{0y} = t_0 = 0$

$a_{0x} = 0 \text{ m/s}^2 \quad a_{0y} = 5.0 \text{ m/s}^2 \quad t_1 = 3 \text{ s}$

$a_{1x} = 20 \text{ m/s}^2 \quad a_{1y} = 5.0 \text{ m/s}^2 \quad t_2 = 6 \text{ s}$

Find

x and y at time t

SOLVE During the first phase of the acceleration, when $a_{0x} = 0 \text{ m/s}^2$ and $a_{0y} = 5.0 \text{ m/s}^2$, the motion is described by

$$y = y_0 + v_{0y}(t - t_0) + \frac{1}{2}a_{0y}(t - t_0)^2 = 2.5t^2 \text{ m}$$

$$v_y = v_{0y} + a_{0y}(t - t_0) = 5.0t \text{ m/s}$$

where the time t is in s. These equations allow us to calculate the position and velocity at any time t. At $t_1 = 3.0$ s, when the first phase of the motion ends, we find that

$$x_1 = 0 \text{ m} \qquad v_{1x} = 0 \text{ m/s}$$

$$y_1 = 22.5 \text{ m} \qquad v_{1y} = 15 \text{ m/s}$$

During the next 3 s, when $a_{1x} = 20 \text{ m/s}^2$ and $a_{1y} = 5.0 \text{ m/s}^2$, the x- and y-coordinates are

$$x = x_1 + v_{1x}(t - t_1) + \frac{1}{2}a_{1x}(t - t_1)^2$$

$$= 10(t - 3.0)^2 \text{ m}$$

$$y = y_1 + v_{1y}(t - t_1) + \frac{1}{2}a_{1y}(t - t_1)^2$$

$$= (22.5 + 15(t - 3.0) + 2.5(t - 3.0)^2) \text{ m}$$

where, again, t is in s. To show the trajectory, we've calculated x and y every 0.5 s, plotted the points in Figure 6.10, and drawn a smooth curve through the points. You can see the shuttlecraft "lift off" during the first 3 s, then begin to accelerate forward.

STOP TO THINK 6.1   This acceleration will cause the particle to

a. Speed up and curve upward.
b. Speed up and curve downward.
c. Slow down and curve upward.
d. Slow down and curve downward.
e. Move to the right and down.
f. Reverse direction.

6.2 Dynamics in Two Dimensions

Newton's second law $\vec{a} = \vec{F}_{net}/m$ determines an object's acceleration. It makes no distinction between linear motion and nonlinear motion. The x- and y-components of the acceleration vector are given by

$$a_x = \frac{(F_{net})_x}{m} \quad \text{and} \quad a_y = \frac{(F_{net})_y}{m} \tag{6.15}$$

Problem-Solving Strategy 5.2 for dynamics problems, on page 126, is still valid. As a quick review, you should

1. Draw a pictorial representation and a physical representation (motion diagram and free-body diagram).
2. Use Newton's second law in component form:

$$(F_{net})_x = \sum F_x = ma_x \quad \text{and} \quad (F_{net})_y = \sum F_y = ma_y$$

The force components (including proper signs) are found from the free-body diagram. Solve for the acceleration, then use the Equation 6.14 kinematic equations to find velocities and positions.

EXAMPLE 6.3 A rocketing hockey puck

Alice tapes a small 200 g model rocket to a 400 g ice hockey puck. The rocket generates 8.0 N of thrust. She orients the puck so that the rocket's nose points in the positive y-direction, then pushes the puck across frictionless ice in the positive x-direction. She releases it with a speed of 2.0 m/s at the exact instant the rocket fires. Find an equation for the puck's trajectory, then graph it.

MODEL Model the puck with the attached rocket as a particle. We need to find the *function* $y(x)$ that describes the curve followed by the puck in the xy-plane.

VISUALIZE Figure 6.11 shows a pictorial representation and a physical representation. The coordinate axes were defined in the problem statement, and we'll specify that the motion starts at the origin. We can identify three forces acting on the puck, but only $\vec{F}_{thrust}$ acts in the plane of motion. The normal force $\vec{n}$ and the weight $\vec{w}$ are perpendicular to the plane of motion (along the z-axis) and cancel each other to give $(F_{net})_z = 0$.

SOLVE The net force in the xy-plane is $\vec{F}_{net} = F_{thrust}\,\hat{j}$. The x-component of $\vec{F}_{net}$ is zero, so $a_x = 0$ and the motion along the x-axis is uniform motion with constant velocity. That is, the rocket thrust along the y-axis causes no change in the x-component of the puck's velocity. It will remain at $v_x = 2.0$ m/s. Simultaneously, $\vec{F}_{thrust}$ will cause the puck to accelerate in the positive y-direction. Hence y and v_y will steadily increase from

their initial values of zero. We can use the free-body diagram of Figure 6.11 and Newton's second law to find the acceleration:

$$a_x = \frac{(F_{net})_x}{m} = 0$$

$$a_y = \frac{(F_{net})_y}{m} = \frac{F_{thrust}}{m} = \frac{8.0\ \text{N}}{0.60\ \text{kg}} = 13.33\ \text{m/s}^2$$

where m is the combined mass of the puck and the rocket. The x-motion is one of constant velocity at $v_{0y} = 2.0$ m/s. The y-motion is one of constant acceleration, starting from $y_0 = 0$ m with $v_{0y} = 0$ m/s. The position at time t is

$$x = x_0 + v_{0x}(t - t_0) + \frac{1}{2}a_x(t - t_0)^2$$

$$= v_{0x}t = 2.0t\ \text{m}$$

$$y = y_0 + v_{0y}(t - t_0) + \frac{1}{2}a_y(t - t_0)^2$$

$$= \frac{1}{2}a_y t^2 = 6.67t^2\ \text{m}$$

where t is in s. This solution gives x and y as explicit functions of time, but it is not the $y(x)$ solution we were looking for. To express y in terms of x, we need to eliminate the time variable from these two equations. Using the x-equation, we find that $t = x/2.0$. Substituting this into the y-equation, the function that describes the trajectory is

$$y(x) = 6.67\left(\frac{x}{2.0}\right)^2 = 1.67x^2\ \text{m}$$

where x is in m. The trajectory equation is of the form $y(x) = cx^2$, which is the equation of a parabola. Figure 6.12 shows the trajectory calculated from this equation.

Pictorial representation

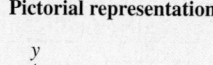
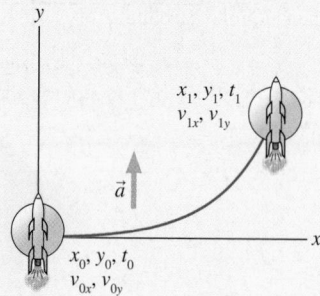

Known
$x_0 = y_0 = t_0 = 0$
$v_{0x} = 2.0$ m/s $v_{0y} = 0$ m/s
$(F_{net})_x = 0$ N $(F_{net})_y = 8.0$ N
$m = 0.60$ kg

Find
Trajectory $y(x)$

Physical representation

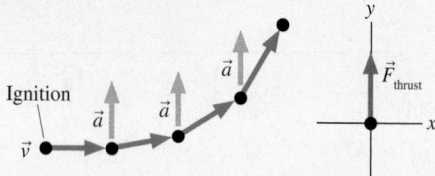

FIGURE 6.11 Pictorial and physical representations of the hockey puck.

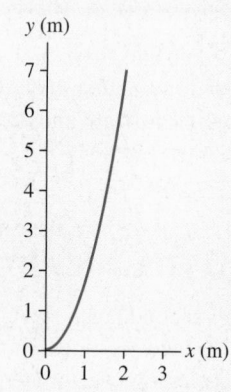

FIGURE 6.12 Parabolic trajectory of the rocketing hockey puck.

ASSESS The solution depended on the fact that the time parameter t is the *same* for both components of the motion.

Although Example 6.3 was about a rocketing hockey puck, our conclusion is quite general. **Any object for which one component of the acceleration is zero while the other has a constant, nonzero value follows a parabolic trajectory.** Important examples of motion along a parabolic trajectory include projectiles, which we will study in the next section, and electrons moving through the "deflection plates" that steer them to the face of your computer display terminal.

STOP TO THINK 6.2   The components of this particle's acceleration are

a. $a_x > 0, a_y > 0$.   b. $a_x = 0, a_y > 0$.   c. $a_x < 0, a_y > 0$.

d. $a_x > 0, a_y < 0$.   e. $a_x = 0, a_y < 0$.   f. $a_x < 0, a_y < 0$.

6.3 Projectile Motion

Baseballs and tennis balls flying through the air, Olympic divers, daredevils shot from cannons, and bullets shot from guns all exhibit what we call *projectile motion*. **A projectile is an object that moves in two dimensions under the influence of only the gravitational force.** Projectile motion is an extension of the free-fall motion we studied in Chapter 2. We will continue to neglect the influence of air resistance, leading to results that are a good approximation of reality for objects moving relatively slowly over relatively short distances.

If the only force acting on an object is its weight, pointing in the negative *y*-direction, then $a_x = 0$ while a_y has a constant, nonzero value. As we just noted, these are the conditions for a *parabolic trajectory* and, indeed, projectiles do move along parabolic paths. Figure 6.13 shows the parabolic trajectory of a bouncing ball. You should also look back at Example 1.7, where we examined the motion diagram of a projectile.

The start of a projectile's motion, be it thrown by hand or shot from a gun, is called the *launch*, and the angle θ of the initial velocity $\vec{v}_i$ above the horizontal (i.e., above the *x*-axis) is called the **launch angle.** Figure 6.14 illustrates the relationship between the initial velocity vector $\vec{v}_i$ and the initial values of the components v_{ix} and v_{iy}. You can see that

$$v_{ix} = v_i \cos\theta$$
$$v_{iy} = v_i \sin\theta \tag{6.16}$$

where v_i is the initial speed.

> **NOTE** ▶ The components v_{ix} and v_{iy} are not always positive. In particular, a projectile launched at an angle *below* the horizontal (such as a ball thrown downward from the roof of a building) has *negative* values for θ and v_{iy}. However, the *speed* v_i is always positive. ◀

The net force on a projectile is simply its weight: $\vec{F}_{net} = \vec{w} = -mg\hat{j}$, as shown in Figure 6.14. We can use Newton's second law to find the acceleration:

$$a_x = \frac{(F_{net})_x}{m} = 0$$
$$a_y = \frac{(F_{net})_y}{m} = \frac{-mg}{m} = -g \tag{6.17}$$

In other words, **the vertical component of acceleration a_y is just the familiar $-g$ of free fall while the horizontal component a_x is zero.**

To see how these conditions influence the motion, Figure 6.15 shows a projectile launched from $(x_i, y_i) = (0\,\text{m}, 0\,\text{m})$ with an initial velocity $\vec{v}_i = (9.8\hat{i} + 19.6\hat{j})$ m/s. The velocity and acceleration vectors are then shown every 1.0 s. The value of v_x never changes, but v_y decreases by 9.8 m/s² every second. This is what it *means* to accelerate at $a_y = -9.8$ m/s² $= (-9.8$ m/s) per second. Be sure to notice that nothing *pushes* the projectile along the curve. Instead, the downward weight force causes a downward acceleration that changes the velocity vector as shown.

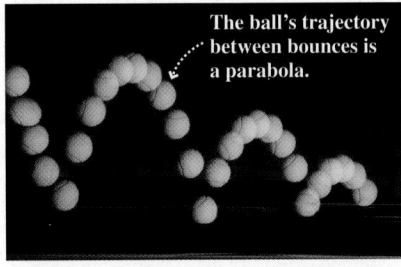

FIGURE 6.13 The parabolic trajectory of a bouncing ball.

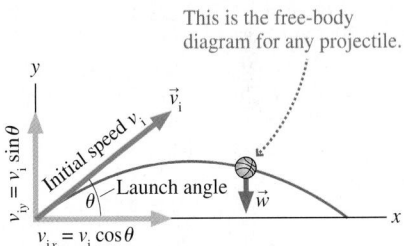

FIGURE 6.14 A projectile that is launched with initial velocity $\vec{v}_i$ follows a parabolic trajectory.

3.1–3.7

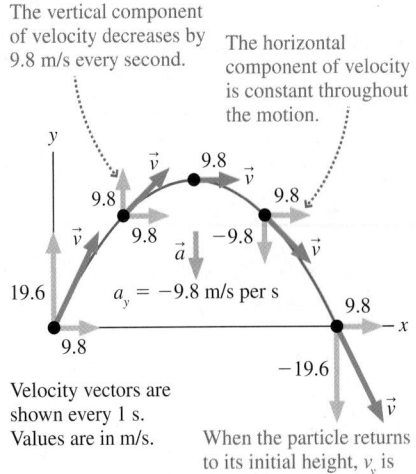

The vertical component of velocity decreases by 9.8 m/s every second.

The horizontal component of velocity is constant throughout the motion.

Velocity vectors are shown every 1 s. Values are in m/s.

When the particle returns to its initial height, v_y is opposite its initial value.

FIGURE 6.15 The velocity and acceleration vectors of a projectile moving along a parabolic trajectory.

You can see from Figure 6.15 that **projectile motion is made up of two independent motions:** uniform motion at constant velocity in the horizontal direction and free-fall motion in the vertical direction. The kinematic equations that describe these two motions are

$$x_f = x_i + v_{ix}\Delta t \qquad\qquad y_f = y_i + v_{iy}\Delta t - \tfrac{1}{2}g(\Delta t)^2$$

$$v_{fx} = v_{ix} = \text{constant} \qquad\qquad v_{fy} = v_{iy} - g\Delta t$$

(6.18)

These are parametric equations for the parabolic trajectory of a projectile.

EXAMPLE 6.4 **Don't try this at home!**

A stunt man drives a car off a 10-m-high cliff at a speed of 20 m/s. How far does the car land from the base of the cliff?

MODEL Represent the car as a particle in free fall. Assume that the car is moving horizontally as it leaves the cliff.

VISUALIZE The pictorial representation, shown in Figure 6.16, is *very* important because the number of symbols in projectile motion problems can be quite large. We have chosen to put the origin at the base of the cliff. The assumption that the car is

Pictorial representation

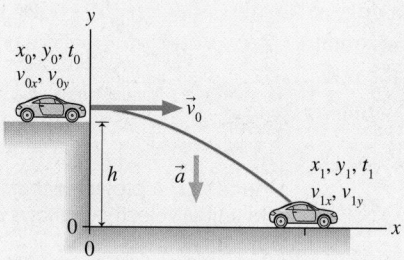

Known
$x_0 = v_{0y} = t_0 = 0$
$y_0 = 10 \text{ m} \quad v_{0x} = v_0 = 20 \text{ m/s}$
$a_x = 0 \text{ m/s}^2 \quad a_y = -g \quad y_1 = 0 \text{ m}$

Find
x_1

FIGURE 6.16 Pictorial representation for the car of Example 6.4.

moving horizontally as it leaves the cliff leads to $v_{0x} = v_0$ and $v_{0y} = 0$ m/s. A physical representation is not needed in projectile motion problems because we already know that a free-body diagram consists of a single force vector—the downward weight force—and that $\vec{a} = -g\hat{j}$.

SOLVE Each point on the trajectory has x- and y-components of position, velocity, and acceleration but only *one* value of time. The time needed to move horizontally to x_1 is the *same* time needed to fall vertically through distance h. **Although the horizontal and vertical motions are independent, they are connected through the time t.** This is a critical observation for solving projectile motion problems. The kinematics equations are

$$x_1 = x_0 + v_{0x}(t_1 - t_0) = v_0 t_1$$

$$y_1 = 0 = y_0 + v_{0y}(t_1 - t_0) - \frac{1}{2}g(t_1 - t_0)^2 = h - \frac{1}{2}g t_1^2$$

We can use the vertical equation to determine the time t_1 needed to fall distance h:

$$t_1 = \sqrt{\frac{2h}{g}} = \sqrt{\frac{2(10 \text{ m})}{9.80 \text{ m/s}^2}} = 1.43 \text{ s}$$

We then insert this expression for t into the horizontal equation to find the distance traveled:

$$x_1 = v_0 t_1 = (20 \text{ m/s})(1.43 \text{ s}) = 28.6 \text{ m}$$

ASSESS The cliff height is $h \approx 33$ ft and the initial speed is $v_0 \approx 40$ mph. Traveling $x_1 = 29$ m ≈ 95 ft before hitting the ground seems reasonable.

Reasoning About Projectile Motion

Think about the following question:

A rifle fires a bullet exactly horizontally at height h above a horizontal field. At the exact instant that the bullet is fired, a second bullet is simply dropped from height h. Which bullet hits the ground first?

It may seem hard to believe, but they hit the ground *simultaneously*. They do so because the horizontal and vertical components of projectile motion are independent of each other. The initial horizontal velocity of the first bullet has *no* influence over its vertical motion. Neither bullet has any initial motion in the vertical direction, so both fall distance h in the same amount of time. You can see this in Figure 6.17, where one ball is shot horizontally and the other released from rest at the same instant. The *vertical* motions of the two balls are identical, and they hit the floor simultaneously.

Figure 6.18a shows a useful way to think about the trajectory of a projectile. Without gravity, a projectile would follow a straight line. Because of gravity, the particle at time t has "fallen" a distance $\frac{1}{2}gt^2$ below this line. The separation grows as $\frac{1}{2}gt^2$, giving the trajectory its parabolic shape.

Figure 6.18a can help you understand a "classic" problem in physics.

A hungry hunter in the jungle wants to shoot down a coconut that is hanging from the branch of a tree. He aims the gun directly at the coconut, but as luck would have it the coconut falls from the branch at the *exact* instant the hunter pulls the trigger. Does the bullet hit the coconut?

You might think that the bullet will miss, but it doesn't. Although the bullet travels very fast, it follows a slightly curved trajectory, not a straight line. Had the coconut stayed on the tree, the bullet would have curved under its target as gravity causes it to fall a distance $\frac{1}{2}gt^2$ below the straight line. But $\frac{1}{2}gt^2$ is also the distance the coconut falls while the bullet is in flight. Thus, as Figure 6.18b shows, the bullet and the coconut fall the same distance and meet at the same point!

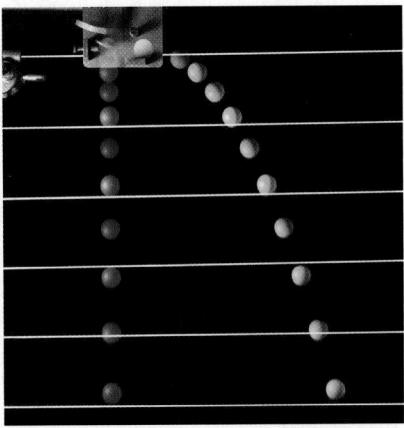

FIGURE 6.17 A projectile launched horizontally falls in the same time as a projectile that is released from rest.

(a)

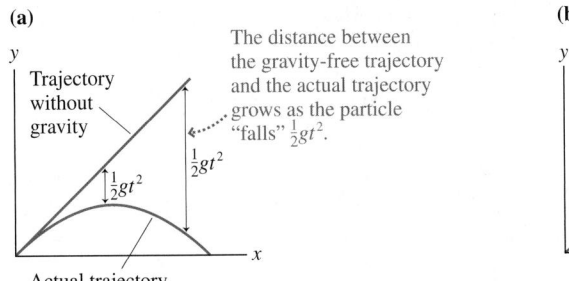

(b)

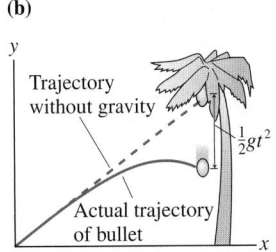

FIGURE 6.18 A projectile follows a parabolic trajectory because it "falls" a distance $\frac{1}{2}gt^2$ below a straight-line trajectory.

Solving Projectile Motion Problems

Let's summarize this information with a problem-solving strategy.

 PROBLEM-SOLVING STRATEGY 6.1 **Projectile motion problems**

MODEL Make simplifying assumptions.

VISUALIZE Use a pictorial representation. Establish a coordinate system with the x-axis horizontal and the y-axis vertical. Show important points in the motion on a sketch. Define symbols and identify what the problem is trying to find.

SOLVE The acceleration is known: $a_x = 0$ and $a_y = -g$. Thus the problem becomes one of kinematics. The kinematic equations are

$$x_f = x_i + v_{ix}\Delta t \qquad y_f = y_i + v_{iy}\Delta t - \frac{1}{2}g(\Delta t)^2$$

$$v_{fx} = v_{ix} = \text{constant} \qquad v_{fy} = v_{iy} - g\Delta t$$

Δt is the same for the horizontal and vertical components of the motion. Find Δt from one component, then use that value for the other component.

ASSESS Check that your result has the correct units, is reasonable, and answers the question.

EXAMPLE 6.5 The distance of a fly ball

A baseball is hit at angle θ and is caught at the height from which it was hit.

a. If the ball is hit at a 30° angle, with what speed must it leave the bat to travel 100 m?

b. What angle causes the ball to go the maximum distance?

MODEL Represent the ball as a particle. Ignore air resistance.

VISUALIZE Figure 6.19 shows the pictorial representation. The height above the ground at which the ball was hit and caught is not relevant, because they are the same, so we have placed the origin at the point where the ball is hit. The ball travels distance x_1.

Pictorial representation

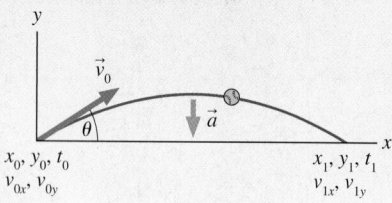

Known
$x_0 = y_0 = 0 \quad \theta = 30° \quad t_0 = 0 \text{ s}$
$x_1 = 100 \text{ m} \quad y_1 = 0 \text{ m}$

Find
v_0

FIGURE 6.19 Pictorial representation for the baseball of Example 6.5.

SOLVE

a. The initial x- and y-components of the ball's velocity are

$$v_{0x} = v_0 \cos\theta$$

$$v_{0y} = v_0 \sin\theta$$

where v_0 is the initial speed that we need to find. The kinematic equations of projectile motion are

$$x_1 = x_0 + v_{0x}(t_1 - t_0)$$
$$= (v_0 \cos\theta)t_1$$

$$y_1 = 0 = y_0 + v_{0y}(t_1 - t_0) - \frac{1}{2}g(t_1 - t_0)^2$$
$$= (v_0 \sin\theta)t_1 - \frac{1}{2}gt_1^2$$

We can use the vertical equation to find the time of flight:

$$0 = (v_0 \sin\theta)t_1 - \frac{1}{2}gt_1^2 = \left(v_0 \sin\theta - \frac{1}{2}gt_1\right)t_1$$

$$t_1 = 0 \quad \text{or} \quad \frac{2v_0 \sin\theta}{g}$$

Both values are legitimate solutions. The first corresponds to the instant when $y = 0$ at the beginning of the trajectory and the second to when $y = 0$ at the end. Clearly, though, we want the second solution. Substituting this expression for t_1 into the equation for x_1 gives

$$x_1 = (v_0 \cos\theta)\frac{2v_0 \sin\theta}{g} = \frac{2v_0^2 \sin\theta \cos\theta}{g}$$

We can simplify this result by using the trigonometric identity $2\sin\theta\cos\theta = \sin(2\theta)$. The distance traveled by the ball when hit at angle θ is

$$x_1 = \frac{v_0^2 \sin(2\theta)}{g}$$

Setting $x_1 = 100$ m and solving for the speed v_0 gives

$$v_0 = \sqrt{\frac{gx_1}{\sin(2\theta)}} = \sqrt{\frac{(9.80 \text{ m/s}^2)(100 \text{ m})}{\sin 60°}} = 33.6 \text{ m/s}$$

b. What value of θ will maximize x_1? As you know, the sine function has a maximum value of 1 at an angle of 90°. Because the equation for x_1 contains the expression $\sin(2\theta)$, it will reach a maximum for $\theta_{\max} = 45°$. If the batter hitting the ball with a speed of 33.6 m/s had hit it at a 45° angle, the ball would have traveled 115 m and just cleared the left field wall (at 110 m)—winning the World Series and bringing fame and fortune to our hero. Instead, he hit at a mere 30° angle, the ball was caught, and history has forgotten his name.

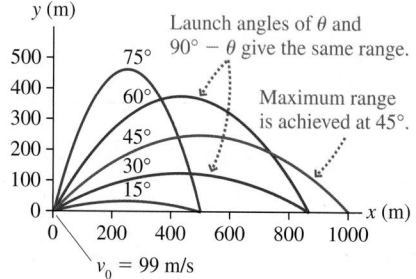

FIGURE 6.20 Trajectories of a projectile launched at different angles with a speed of 99 m/s.

As Example 6.5 found, a projectile that lands at the same elevation from which it was fired travels distance

$$\text{distance} = \frac{v_0^2 \sin(2\theta)}{g} \tag{6.19}$$

The maximum distance occurs for $\theta = 45°$, where $\sin(2\theta) = 1$. But there's more that we can learn from this equation. Because $\sin(180° - x) = \sin x$, it follows that $\sin(2(90° - \theta)) = \sin(2\theta)$. Consequently, a projectile launched either at angle θ or at angle $(90° - \theta)$ will travel the same distance. Figure 6.20 shows the trajectories of projectiles launched with the same initial speed in 15° increments of angle.

NOTE ▶ Equation 6.19 is *not* a general result. It applies *only* in situations where the projectile lands at the same elevation from which it was fired. ◀

EXAMPLE 6.6 Santa's sleigh ride

Santa parked his sleigh 5.0 m from the edge of a 20° roof. Unfortunately, the parking brake failed. Santa's sleigh slid down the roof and landed on the ground 6.0 m from the wall of the house. The sleigh's coefficient of kinetic friction on the snowy roof is 0.08. How high was the wall of the house?

MODEL Represent the sleigh as a particle. This is a two-part problem. First, the sleigh slides down the roof—motion in a straight line. The sleigh then becomes a projectile until it hits the ground. These two problems are connected through the fact that the *final* speed of the linear motion problem is the *initial* speed of the projectile problem.

VISUALIZE Figure 6.21 shows separate pictorial representations for the two parts of the motion. Notice that we're using different coordinate systems for the two parts of the problem; each is chosen to serve a specific need. Point 2 is the same as point 1, but we've given it a new number in order to write its coordinates in the new coordinate system without causing confusion. We've explicitly noted that $v_2 = v_1$, which connects the linear motion problem to the projectile motion problem. The linear motion includes a free-body diagram, but none is needed for the projectile motion.

SOLVE The first part of the motion is similar to problems you solved in Chapter 5. Newton's second law and the model of kinetic friction are

$$\sum F_x = w_x + (f_k)_x = mg\sin\theta - f_k = ma_x$$
$$\sum F_y = w_y + n_y = n - mg\cos\theta = ma_y = 0$$
$$f_k = \mu_k n$$

We've written these equations, as we did in Chapter 5, by "reading" the free-body diagram. We know that $a_y = 0$ because the motion is entirely along the x-axis. The y-equation gives $n = mg\cos\theta$, from which we find that the friction is $f_k = \mu_k mg\cos\theta$. Substituting this into the x-equation gives

$$a = \frac{mg\sin\theta - \mu_k mg\cos\theta}{m}$$
$$= g(\sin\theta - \mu_k\cos\theta) = 2.62 \text{ m/s}^2$$

Notice that the mass canceled out, which is fortunate because we weren't given a value. The velocity after sliding 5.0 m is found from one-dimensional kinematics:

$$v_{1x}^2 = v_{0x}^2 + 2a_x\Delta x = 2a_x x_1$$
$$v_{1x} = \sqrt{2a_x x_1} = 5.11 \text{ m/s}$$

Now we have to solve a projectile problem. Notice that in the pictorial model we've restarted the clock: $t_2 = 0$ s. The initial velocity $\vec{v}_2$ of the projectile motion is the *same* as the final velocity $\vec{v}_1$ of the linear motion, which we can write

$$\vec{v}_1 = (5.11 \text{ m/s}, 20° \text{ below horizontal})$$

The initial velocity components are

$$v_{2x} = v_2\cos\theta = (5.11 \text{ m/s})\cos(-20°) = 4.80 \text{ m/s}$$
$$v_{2y} = v_2\sin\theta = (5.11 \text{ m/s})\sin(-20°) = -1.75 \text{ m/s}$$

This is an example where the initial value of v_y is negative. The kinematic equations are

$$x_3 = x_2 + v_{2x}(t_3 - t_2) = v_{2x}t_3$$
$$y_3 = 0 = y_2 + v_{2y}(t_3 - t_2) - \frac{1}{2}g(t_3 - t_2)^2$$
$$= h + v_{2y}t_3 - \frac{1}{2}gt_3^2$$

We can use the horizontal motion to find that the sleigh hits the ground at time

$$t_3 = \frac{x_3}{v_{2x}} = \frac{6.0 \text{ m}}{4.80 \text{ m/s}} = 1.25 \text{ s}$$

We can now use t_3 in the vertical-motion equation to solve for the wall height:

$$h = \frac{1}{2}gt_3^2 - v_{2y}t_3$$
$$= \frac{1}{2}(9.80 \text{ m/s}^2)(1.25 \text{ s})^2 - (-1.75 \text{ m/s})(1.25 \text{ s})$$
$$= 9.84 \text{ m}$$

The wall of the house is 9.84 m high, or roughly 30 ft.

Part 1
Pictorial representation

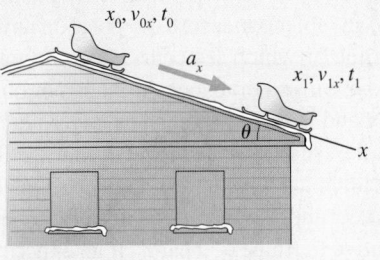

Physical representation

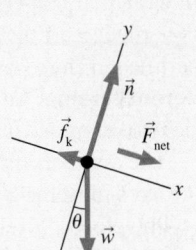

Part 2
Pictorial representation

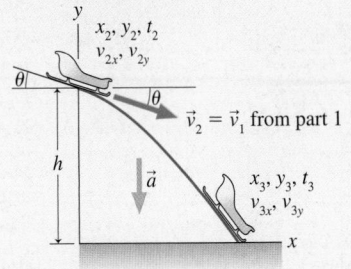

FIGURE 6.21 The pictorial and physical representations of Santa's sleigh as it slides down and off the roof.

A 100 g ball rolls off a table and lands 2 m from the base of the table. A 200 g ball rolls off the same table with the same speed. It lands at distance

a. < 1 m.   b. 1 m.   c. Between 1 m and 2 m.
d. 2 m.   e. Between 2 m and 4 m.   f. 4 m.

6.4 Relative Motion

You've now dealt many times with problems that say something like "A car travels at 30 m/s" or "A plane travels at 300 m/s." But just what do these statements really mean?

In Figure 6.22, Amy, Bill, and Carlos are watching a runner. According to Amy, the runner's velocity is $v_x = 5$ m/s. But to Bill, who's riding alongside, the runner is lifting his legs up and down but going neither forward nor backward relative to Bill. As far as Bill is concerned, the runner's velocity is $v_x = 0$ m/s. Carlos sees the runner receding in his rearview mirror, in the *negative x*-direction, getting 10 m further away from him every second. According to Carlos, the runner's velocity is $v_x = -10$ m/s. Which is the runner's *true* velocity?

Velocity is not a concept that can be true or false. The runner's velocity *relative to Amy* is 5 m/s. That is, his velocity is 5 m/s in a coordinate system attached to Amy and in which Amy is at rest. The runner's velocity relative to Bill is 0 m/s, and the velocity relative to Carlos is -10 m/s. These are all valid descriptions of the runner's motion.

What about the jet plane, which is speeding up? Suppose Amy, Bill, and Carlos each uses his or her coordinate system to measure the plane's acceleration. Do their values agree or disagree? Acceleration is important because the question we ultimately want to address in this section is "In which coordinate systems are Newton's laws valid?" If Newton's laws are to be the foundation of mechanics, we certainly should know the coordinate systems in which we can use these laws. And because Newton's second law is about acceleration, not velocity, we need to explore how acceleration is measured in different coordinate systems. First, however, we need to look at relative position and relative velocity.

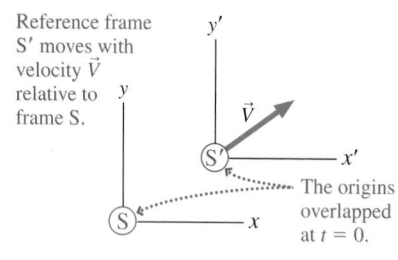

FIGURE 6.22 Amy, Bill, and Carlos each measure the velocity of the runner and the acceleration of the jet plane. The velocities are shown in Amy's reference frame.

Relative Position

Suppose that Amy and Bill each have a coordinate system attached to their bodies. As Bill bicycles past Amy, he carries his coordinate system with him. Each is at rest in his or her coordinate system. Further, let's imagine that Amy and Bill each have helpers in their coordinate systems with meter sticks and stopwatches. Amy and Bill, with their helpers, are able to measure the position at which a physical event takes place and the time at which it occurs. A coordinate system in which an experimenter makes position and time measurements of physical events is called a **reference frame.** Amy and Bill each have their own reference frame.

Let's define two reference frames, shown in Figure 6.23, that we'll call frame S and frame S'. (The symbol ' is called a *prime*, and S' is pronounced "S prime.") The coordinate axes in frame S are x and y, while those in S' are x' and y'. Frame S' is moving with velocity $\vec{V}$ relative to frame S. That is, if an experimenter at rest in S measures the motion of the origin of S' as it goes past, she finds that the origin of S' has velocity $\vec{V}$. Of course, an experimenter at rest in S' would say that frame S has velocity $-\vec{V}$. We'll use an uppercase V for the velocity of reference frames, reserving lowercase v for the velocity of objects that move in the reference frames.

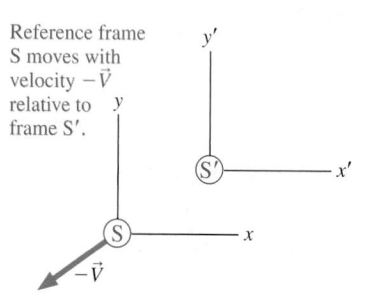

FIGURE 6.23 Reference frame S' is moving with velocity $\vec{V}$ relative to reference frame S.

NOTE ▶ There's no implication that either frame is "at rest." All we know is that the two frames are moving *relative* to each other with velocity $\vec{V}$. ◀

We will stipulate four conditions for reference frames:

1. The frames are oriented the same, with the x- and x'-axes parallel to each other.
2. The origins of frame S and frame S$'$ coincide at $t = 0$.
3. All motion is in the xy-plane, so we don't need to consider the z-axis.
4. The relative velocity $\vec{V}$ is *constant*.

The first three are a matter of how we define the coordinate systems. Item 4, by contrast, is a choice with consequences. It says that we will consider only reference frames that move with constant speed in a straight line. These are called **inertial reference frames,** a term introduced in Chapter 4. We'll see that these are the reference frames in which Newton's laws are valid.

Suppose a light bulb flashes at time t. Experimenters in both reference frames see the flash and measure its position. Observers in S place the flash at position $\vec{r}$, as measured with respect to the coordinate system of frame S. Similarly, experimenters in S$'$ determine that the flash occurred at position $\vec{r}'$, relative to the origin of S$'$. (We'll use primes to indicate positions and velocities measured in frame S$'$.)

What is the relationship between the position vectors $\vec{r}$ and $\vec{r}'$? It's not hard to see, from Figure 6.24, that

$$\vec{r} = \vec{r}' + \vec{R} \tag{6.20}$$

where $\vec{R}$ is the position vector of the origin of frame S$'$ as measured in frame S.

Frame S$'$ is traveling with velocity $\vec{V}$ relative to frame S, and their origins coincide at $t = 0$. At time t, when the light flashes, the origin of S$'$ has moved to position $\vec{R} = t\vec{V}$. (We've written $t\vec{V}$ rather than $\vec{V}t$ because it is customary to write the scalar first.) Thus

$$\vec{r} = \vec{r}' + t\vec{V} \quad \text{or} \quad \vec{r}' = \vec{r} - t\vec{V} \tag{6.21}$$

Equation 6.21 is called the **Galilean transformation of position.** It will be easiest for most purposes to write this in terms of components:

$$\begin{aligned} x = x' + V_x t \qquad x' = x - V_x t \\ \text{or} \\ y = y' + V_y t \qquad y' = y - V_y t \end{aligned} \tag{6.22}$$

If we know *where* and *when* an event occurred in one reference frame, we can *transform* that position into any other reference frame that moves relative to the first with constant velocity $\vec{V}$.

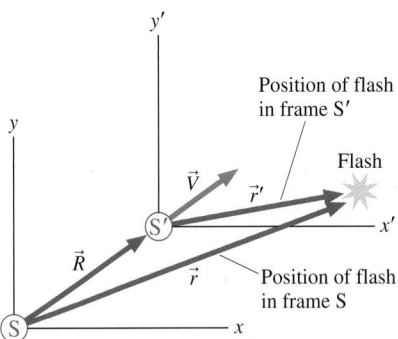

FIGURE 6.24 Measurements in frame S find that a light flash occurred at position $\vec{r}$. The same flash occurred at position $\vec{r}'$ in frame S$'$.

EXAMPLE 6.7 Watching a ball toss
Mike throws a ball upward at a 63° angle with a speed of 22 m/s. Nancy rides past Mike on her bicycle at 10 m/s at the instant he releases the ball.

a. Find and graph the ball's trajectory as seen by Mike.
b. Find and graph the ball's trajectory as seen by Nancy.

SOLVE

a. For Mike, the ball is a projectile that follows a parabolic trajectory. This problem is almost exactly the same as Example 6.5. The components of the initial velocity are

$$v_{0x} = v_0 \cos\theta = (22 \text{ m/s}) \cos 63° = 10.0 \text{ m/s}$$

$$v_{0y} = v_0 \sin\theta = (22 \text{ m/s}) \sin 63° = 19.6 \text{ m/s}$$

The x- and y-equations of motion for the ball's position at time t are

$$x = x_0 + v_{0x}(t - t_0) = 10.0t \text{ m}$$

$$y = y_0 + v_{0y}(t - t_0) - \frac{1}{2}g(t - t_0)^2 = (19.6t - 4.9t^2) \text{ m}$$

where t is in s. It's not hard to show that the ball reaches height $y_{\max} = 19.6$ m at $t = 2$ s and hits the ground at $t = 4$ s. The trajectory is shown in Figure 6.25a on the next page.

b. We can determine the trajectory Nancy sees by using Equations 6.22 to transform the ball's position from Mike's reference frame into Nancy's reference frame. Let Mike be in frame S and Nancy in frame S$'$. Nancy moves with velocity $\vec{V} = 10.0\hat{\imath}$ m/s relative to S. In terms of components,

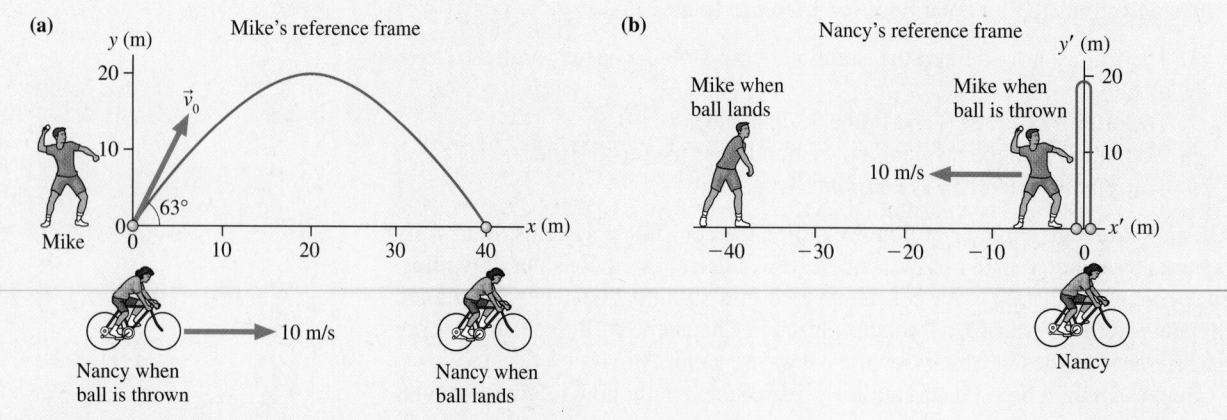

FIGURE 6.25 A ball's trajectory as seen by Mike and Nancy.

$V_x = 10$ m/s and $V_y = 0$ m/s. When Mike, at time t, measures the ball at position (x, y) in frame S, Nancy finds the ball at

$$x' = x - V_x t = 10.0t - 10.0t = 0$$

$$y' = y - V_y t = y$$

Because Nancy's horizontal motion is the same as the ball's ($V_x = v_x = 10$ m/s in Mike's frame), she doesn't see the ball moving either right or left. Nancy's experience is like that of Bill riding beside the runner in Figure 6.22. The ball moves *vertically* up and down in frame S'. Further, the vertical position y' in S' is the same as the vertical position y in S. According to Nancy, the ball goes straight up, reaches a height of 19.6 m, and falls straight back down. It hits the ground right beside her bicycle at $t = 4$ s. This is seen in Figure 6.25b.

In Chapter 2 we studied *free fall*, vertical motion straight up and down. In Section 6.3 we studied the parabolic trajectories of *projectile motion*. Now, from Example 6.7 we see that **free-fall motion and projectile motion are really the same motion, simply seen from two different reference frames.** The motion is vertical in the *one* reference frame whose horizontal motion is the same as the ball's. The trajectory is a parabola in any other reference frame.

Relative Velocity

Let's think a bit more about Example 6.7. According to an observer in Mike's reference frame, Mike throws the ball with velocity $\vec{v}_0 = (10.0\hat{i} + 19.6\hat{j})$ m/s. The ball's initial speed is $v_0 = 22.0$ m/s. But in frame S', where Nancy sees the ball go straight up and down, Mike throws the ball with velocity $\vec{v}_0' = 19.6\hat{j}$ m/s. An object's velocity measured in frame S is *not* the same as its velocity measured in frame S'. Our goal is to find a general relationship between an object's velocity $\vec{v}$ as measured in reference frame S and its velocity $\vec{v}'$ as measured in a different frame S'.

Figure 6.26 shows a *moving object* that is observed from reference frames S and S'. Experimenters in frame S locate the object at position $\vec{r}$ and measure its velocity to be $\vec{v}$. Simultaneously, experimenters in S' measure position $\vec{r}'$ and velocity $\vec{v}'$. The position vectors, which are related by $\vec{r} = \vec{r}' + \vec{R}$, change as the object moves. In addition, $\vec{R}$ changes as the reference frames move relative to each other. The *rate* of change is

$$\frac{d\vec{r}}{dt} = \frac{d\vec{r}'}{dt} + \frac{d\vec{R}}{dt} \tag{6.23}$$

The derivative $d\vec{r}/dt$, by definition, is the object's velocity $\vec{v}$ measured in frame S. Similarly, $d\vec{r}'/dt$ is the object's velocity $\vec{v}'$ measured in frame S'. And

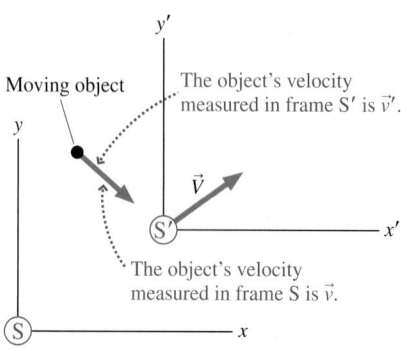

FIGURE 6.26 A velocity of a moving object is measured by experimenters in two different reference frames.

$d\vec{R}/dt$ is the velocity $\vec{V}$ of frame S′ relative to frame S. Consequently, Equation 6.23 tells us that

$$\vec{v} = \vec{v}' + \vec{V} \quad \text{or} \quad \vec{v}' = \vec{v} - \vec{V} \tag{6.24}$$

Equation 6.24 is the **Galilean transformation of velocity.** If we know an object's velocity measured in one reference frame, we can transform it into the velocity that would be measured by an experimenter in a different reference frame. As Figure 6.27 shows, doing so is an exercise in vector addition.

We will often find it convenient, as we did with position, to write Equation 6.24 in terms of components:

$$\begin{array}{ll} v_x = v_x' + V_x & v_x' = v_x - V_x \\ & \text{or} \\ v_y = v_y' + V_y & v_y' = v_y - V_y \end{array} \tag{6.25}$$

This relationship between velocities measured by experimenters in different frames of reference was recognized by Galileo in his pioneering studies of motion, hence its name.

Let's apply Equation 6.25 to Mike and Nancy. We've already noted that the ball's initial velocity in Mike's frame, frame S, is $\vec{v}_0 = (10.0\hat{\imath} + 19.6\hat{\jmath})$ m/s. Nancy was moving relative to Mike at velocity $\vec{V} = 10.0\hat{\imath}$ m/s. We can use Equation 6.25 to transform the velocity to Nancy's frame, frame S′, finding

$$v_x' = v_x - V_x = 10.0 \text{ m/s} - 10.0 \text{ m/s} = 0 \text{ m/s}$$

$$v_y' = v_y - V_y = 19.6 \text{ m/s} - 0 \text{ m/s} = 19.6 \text{ m/s}$$

Thus $\vec{v}_0' = 19.6\,\hat{\jmath}$ m/s. This agrees with our conclusion from Example 6.7.

It's important to understand the distinction between the three velocities $\vec{v}$, $\vec{v}'$, and $\vec{V}$. $\vec{v}$ and $\vec{v}'$ are the velocities of an *object* that is observed from both reference frames. Experimenters in S use their meter sticks and stopwatches to measure the object's velocity $\vec{v}$ in their reference frame. At the same time, experimenters in S′ measure the velocity of the same object to be $\vec{v}'$. $\vec{V}$ is the relative velocity between two *reference frames*; the velocity of S′ as measured by an experimenter in S. $\vec{V}$ has nothing to do with the object. It may happen that either $\vec{v}$ or $\vec{v}'$ is zero, meaning that the object is at rest in one reference frame, but we still must distinguish between the object and the reference frame.

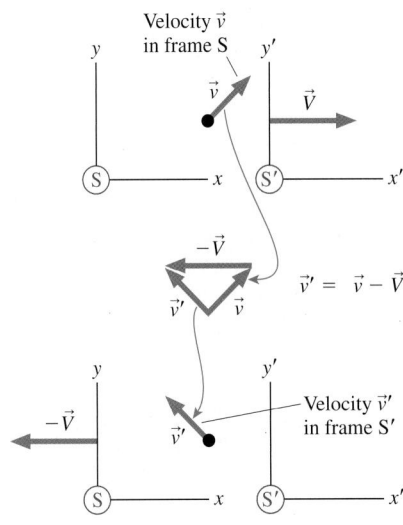

FIGURE 6.27 Velocities $\vec{v}$ and $\vec{v}'$, as measured in frames S and S′, are related by vector addition.

EXAMPLE 6.8 A speeding bullet
The police are chasing a bank robber. While driving at 50 m/s, they fire a bullet to shoot out a tire of his car. The police gun shoots bullets at 300 m/s. What is the bullet's speed as measured by a TV camera crew parked beside the road?

MODEL Assume that all motion is along the x-axis. Let the earth be frame S and a frame attached to the police car be S′. Frame S′ moves relative to frame S with $V_x = 50$ m/s.

SOLVE The bullet is the moving object that will be observed from both frames. The gun is in frame S′, so the bullet travels in this frame with $v_x' = 300$ m/s. We can use Equation 6.25 to transform the bullet's velocity into the earth reference frame:

$$v_x = v_x' + V_x = 300 \text{ m/s} + 50 \text{ m/s} = 350 \text{ m/s}$$

The Galilean velocity transformations are pretty much common sense for one-dimensional motion. Their real usefulness appears when an object travels in a *medium* that moves with respect to the earth. For example, a boat moves relative to the water. What is the boat's net motion if the water is a flowing river? Airplanes fly relative to the air, but the air at high altitudes often flows at high speed. Navigation of boats and planes requires knowing both the motion of the vessel in the medium and the motion of the medium relative to the earth.

EXAMPLE 6.9 Flying to Cleveland I

Cleveland is 300 miles east of Chicago. A plane leaves Chicago flying due east at 500 mph. The pilot forgot to check the weather and doesn't know that the wind is blowing to the south at 50 mph. What is the plane's ground speed? Where is the plane 0.60 hours later, when the pilot expects to land in Cleveland?

MODEL Let the earth be reference frame S. Chicago and Cleveland are at rest in the earth's frame. Let the air be frame S'. If the x-axis points east and the y-axis north, then the air is moving with respect to the earth at $\vec{V} = -50\hat{j}$ mph. The plane flies in the air, so its velocity in frame S' is $\vec{v}' = 500\hat{i}$ mph.

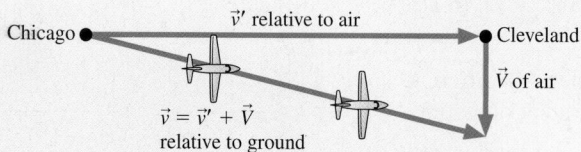

FIGURE 6.28 The wind causes a plane flying due east in the air to move to the southeast relative to the earth.

SOLVE The velocity transformation equation $\vec{v} = \vec{v}' + \vec{V}$ is a vector addition equation. Figure 6.28 shows graphically what happens. Although the nose of the plane points east, the wind carries the plane in a direction somewhat south of east. The plane's velocity relative to the ground is

$$\vec{v} = \vec{v}' + \vec{V} = (500\hat{i} - 50\hat{j}) \text{ mph}$$

The plane's ground speed, its speed in frame S, is

$$v = \sqrt{v_x^2 + v_y^2} = 502 \text{ mph}$$

After flying for 0.6 hours at this velocity, the plane's location (relative to Chicago) is

$$x = v_x t = (500 \text{ mph})(0.6 \text{ hr}) = 300 \text{ mi}$$
$$y = v_y t = (-50 \text{ mph})(0.6 \text{ hr}) = -30 \text{ mi}$$

The plane is 30 mi due south of Cleveland! Although the pilot thought he was flying to the east, his actual heading has been $\tan^{-1}(V/v) = \tan^{-1}(0.10) = 5.71°$ south of east.

EXAMPLE 6.10 Flying to Cleveland II

A wiser pilot flying from Chicago to Cleveland on the same day plots a course that will take her directly to Cleveland. In which direction does she fly the plane? How long does it take to reach Cleveland?

MODEL Let the earth be reference frame S. Let the air be frame S'. If the x-axis points east and the y-axis north, then the air is moving with respect to the earth at $\vec{V} = -50\hat{j}$ mph.

SOLVE The objective of navigation is to move between two points on the earth's surface, in frame S. The wiser pilot, who knows that the wind will affect her plane, draws the vector picture of Figure 6.29. The plane's velocity in frame S is

$$v_x = v_x' + V_x = (500 \text{ mph}) \cos\theta$$
$$v_y = v_y' + V_y = (500 \text{ mph}) \sin\theta - 50 \text{ mph}$$

In plotting her course, the pilot knows that she wants $v_y = 0$ in order to fly due east to Cleveland in the earth's frame. To achieve this, she'll actually have to point the nose of the plane somewhat north of east. The proper heading is

$$\theta = \sin^{-1}\left(\frac{50 \text{ mph}}{500 \text{ mph}}\right) = 5.74°$$

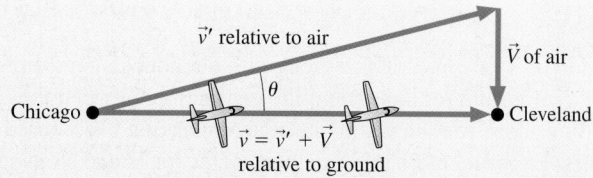

FIGURE 6.29 To travel due east in a south wind, a pilot has to point the plane somewhat to the northeast.

The plane's velocity in frame S is then $\vec{v} = (500 \text{ mph}) \cos 5.74° \hat{i} = 497\hat{i}$ mph. You can see from Figure 6.29 that the plane's speed v in the earth's frame is slower than its speed v' in the air's reference frame. The time needed to fly to Cleveland at this speed is

$$t = \frac{300 \text{ mi}}{497 \text{ mph}} = 0.604 \text{ hr}$$

It takes 0.004 hr = 14 s more time to reach Cleveland than it would on a day without wind.

ASSESS A boat crossing a river or an ocean current faces the same difficulties. These are exactly the kinds of calculations performed by pilots of boats and planes as part of navigation.

The Galilean Principle of Relativity

The most important question we raised at the beginning of this section was "In which reference frames are Newton's laws valid?" That is, in which reference frames does $\vec{F}_{net} = m\vec{a}$?

Suppose a net force $\vec{F}_{net}$ acts on an object in frame S. Further, suppose that Newton's laws have been tested and found valid in frame S. Then experimenters in S will find that the object has acceleration $\vec{a}$ such that $\vec{F}_{net} = m\vec{a}$.

What is the situation in frame S′ that moves relative to frame S with velocity $\vec{V}$? Does $\vec{F}'_{net} = m\vec{a}'$? To answer this question we must transform the force and acceleration measurements of frame S to frame S′.

Recall that a force is a push or pull, the strength of which can be measured with a spring scale. The strength doesn't depend on a coordinate system. If experimenters in frame S see the scale reading 5 N, experimenters in frame S′ will see the same reading. In other words, the size and direction of a force are the same in all reference frames. Thus $\vec{F}'_{net} = \vec{F}_{net}$.

So what about acceleration? The velocity transformation equation, Equation 6.24, is $\vec{v}' = \vec{v} - \vec{V}$. If we take the time derivative of this equation, we find

$$\frac{d\vec{v}'}{dt} = \frac{d\vec{v}}{dt} - \frac{d\vec{V}}{dt}$$

$$\vec{a}' = \vec{a} - \frac{d\vec{V}}{dt}$$

(6.26)

where we've used the definitions of $\vec{a}$ and $\vec{a}'$. Now, one of the four conditions we placed on reference frames S and S′ was the requirement that $\vec{V}$ be a *constant* velocity. Thus $d\vec{V}/dt = 0$, and the **Galilean transformation of acceleration** is

$$\vec{a}' = \vec{a}$$

(6.27)

Observers in frames S and S′ may measure different positions and velocities for an object, but they *agree* on its acceleration.

Reference frames that move with constant velocity are inertial reference frames. Equation 6.27 tells us that experimenters in two inertial reference frames measure the *same acceleration* for an object. Those experimenters also measure the *same force* exerted on the object. Thus if Newton's laws are tested and found to be valid in any one particular inertial reference—and they have been—we can conclude that Newton's laws are valid in *all* inertial reference frames. This idea is known as the Galilean principle of relativity:

> **Galilean principle of relativity** Newton's laws of motion are valid in all inertial reference frames.

A reference frame that is speeding up, slowing down, or turning is not an inertial reference frame. Equation 6.27 is *not* true if S′ is a noninertial frame because $d\vec{V}/dt \neq 0$. Thus $\vec{F}_{net} \neq m\vec{a}$ in a noninertial reference frame and Newton's laws cannot be used.

While the Galilean transformations and the Galilean principle of relativity seem to be almost obvious, they're actually based on quite fundamental assumptions about the nature of space and time. Those assumptions were unquestioned—in fact, not even recognized—until the beginning of the 20th century. To see where problems arose, consider the situation in Figure 6.30. Tom is shooting his laser pointer in the positive x-direction. The laser beam is moving away from Tom at the speed of light, $v_x = 3.0 \times 10^8$ m/s.

Sue flies by in her spaceship, traveling in the positive x-direction with velocity $V_x = 2.0 \times 10^8$ m/s. This is the velocity of her reference frame relative to Tom's. Sue sees the laser beam traveling past the window of her spaceship and uses instruments on board her spaceship to measure the velocity of the light. According to Equation 6.25, the Galilean transformation of velocity, Sue *should* find that the light travels with velocity

$$v'_x = v_x - V_x = 1.0 \times 10^8 \text{ m/s}$$

After all, this is really no different from the police and the bullet of Example 6.8.

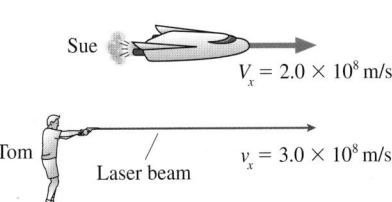

FIGURE 6.30 The laser beam moves away from Tom at the speed of light, $v_x = 3.0 \times 10^8$ m/s. How fast does the laser beam go past Sue's window?

But when the experiment is done, Sue finds that the laser beam travels in her reference frame with $v'_x = 3.0 \times 10^8$ m/s, exactly the same velocity that Tom measured. That is, $v'_x = v_x$! This is the same as if the TV crew and the police in Example 6.8 *both* observed the bullet to have $v_x = 300$ m/s.

Now, it's true that we can't do the experiment exactly as described here, but physicists did figure out over 100 years ago how to measure the speed of light in moving reference frames. Many experiments that are the equivalent of Tom and Sue's have been done, and they all find the same result: **The speed of light is the same in all inertial reference frames, no matter how fast the reference frames are moving with respect to each other.**

This finding is totally at odds with the Galilean transformation of velocity. It is also totally at odds with common sense. It turns out that the universal constancy of the speed of light can be understood only if *time* is different for experimenters moving relative to each other. It was Albert Einstein who, in 1905, first suggested that space and time are more complex and more subtle than is assumed in Newtonian mechanics. This is the basis of his theory of relativity, a topic that we'll return to in Chapter 36.

Fortunately for us, Galilean relativity works extremely well for objects whose speed is much less than the speed of light. Einstein's relativity becomes important only when speeds exceed about 10% of the speed of light. The fastest human spacecraft doesn't come anywhere close to this, so relativity is rarely an issue in engineering or everyday life. But atomic particles such as electrons and protons routinely travel at more than 99% of the speed of light in particle accelerators. Relativity, which is one of the cornerstones of modern physics, is essential for understanding their behavior.

STOP TO THINK 6.4   A plane traveling horizontally to the right at 100 m/s flies past a helicopter that is going straight up at 20 m/s. From the helicopter's perspective, the plane's direction and speed are

a. Right and up, less than 100 m/s.
b. Right and up, 100 m/s.
c. Right and up, more than 100 m/s.
d. Right and down, less than 100 m/s.
e. Right and down, 100 m/s.
f. Right and down, more than 100 m/s.

SUMMARY

The goal of Chapter 6 has been to learn to solve problems about motion in a plane.

GENERAL PRINCIPLES

Galilean Principle of Relativity

Newton's laws of motion are valid in all inertial reference frames.

Newton's Second Law

Expressed in x- and y-component form:

$$(F_{net})_x = \sum F_x = ma_x$$

$$(F_{net})_y = \sum F_y = ma_y$$

IMPORTANT CONCEPTS

Relative motion

Inertial reference frames move relative to each other with constant velocity $\vec{V}$. Measurements of position and velocity measured in frame S are related to measurements in frame S′ by the Galilean transformations

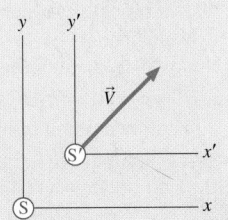

$$x' = x - V_x t \quad v'_x = v_x - V_x$$

$$y' = y - V_y t \quad v'_y = v_y - V_y$$

The instantaneous velocity

$$\vec{v} = d\vec{r}/dt,$$

is a vector tangent to the trajectory.

The instantaneous acceleration is

$$\vec{a} = d\vec{v}/dt$$

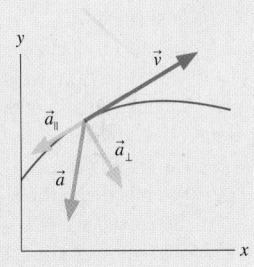

$\vec{a}_{\parallel}$, the component of $\vec{a}$ parallel to $\vec{v}$, is responsible for change of *speed*. $\vec{a}_{\perp}$, the component of $\vec{a}$ perpendicular to $\vec{v}$, is responsible for change of *direction*.

APPLICATIONS

Kinematics in two dimensions

If $\vec{a}$ is constant, then the x- and y-components of motion are independent of each other. For a particle that starts from initial position $\vec{r}_i$ and velocity $\vec{v}_i$, its position and velocity at a final point f are

$$x_f = x_i + v_{ix}\Delta t + \tfrac{1}{2}a_x(\Delta t)^2$$

$$y_f = y_i + v_{iy}\Delta t + \tfrac{1}{2}a_y(\Delta t)^2$$

$$v_{fx} = v_{ix} + a_x\Delta t$$

$$v_{fy} = v_{iy} + a_y\Delta t$$

Projectile motion occurs if the only force on the object is its weight.

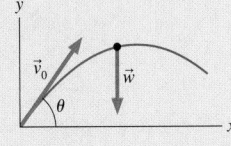

- Uniform motion in the horizontal direction with $v_{0x} = v_0\cos\theta$.
- Free-fall motion in the vertical direction with $a_y = -g$ and $v_{0y} = v_0\sin\theta$.
- The combined motion is a parabola.
- The x and y kinematic equations have the *same* value for Δt.

TERMS AND NOTATION

projectile
launch angle, θ
reference frame
inertial reference frame

Galilean transformation of position
Galilean transformation of velocity
Galilean transformation of acceleration
Galilean principle of relativity

EXERCISES AND PROBLEMS

The 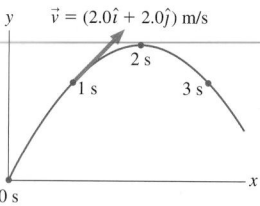 icon in front of a problem indicates that the problem can be done on a Dynamics Worksheet.

Exercises

Section 6.1 Kinematics in Two Dimensions

1. A particle moves in the xy-plane with constant acceleration. The particle is located at $\vec{r} = (2\hat{i} + 4\hat{j})$ m at $t = 0$ s. At $t = 3$ s it is at $\vec{r} = (8\hat{i} - 2\hat{j})$ m and has velocity $\vec{v} = (5\hat{i} - 5\hat{j})$ m/s.
   a. What is the particle's acceleration vector $\vec{a}$?
   b. What are its position, velocity, and speed at $t = 5$ s?

2. A sailboat is traveling east at 5.0 m/s. A sudden gust of wind gives the boat an acceleration $\vec{a} = (0.80 \text{ m/s}^2, 40° \text{ north of east})$. What are the boat's speed and direction 6.0 s later when the gust subsides?

3. A particle's trajectory is described by $x = (\frac{1}{2}t^3 - 2t^2)$ m and $y = (\frac{1}{2}t^2 - 2t)$ m, where t is in s.
   a. Calculate and plot the trajectory from $t = -2$ s to $t = 5$ s.
   b. What are the particle's position and speed at $t = 0$ s and $t = 4$ s?
   c. What is the particle's direction of motion, measured from the x-axis, at $t = 0$ s and $t = 4$ s?

4. A rocket-powered hockey puck moves on a horizontal frictionless table. Figure Ex6.4 shows graphs of v_x and v_y, the x- and y-components of the puck's velocity. The puck starts from the origin.
   a. In which direction is the puck moving at $t = 2.0$ s?
   b. How far from the origin is the puck at $t = 5.0$ s?

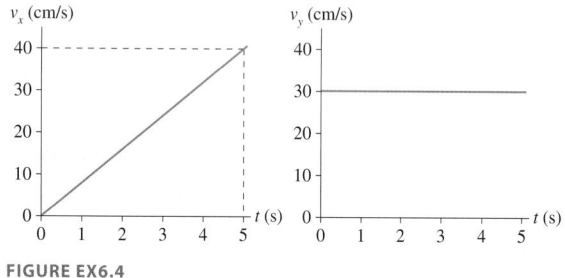

FIGURE EX6.4

Section 6.2 Dynamics in Two Dimensions

5. As a science fair project, you want to launch an 800 g model rocket straight up and hit a horizontally moving target as it passes 30 m above the launch point. The rocket engine provides a constant thrust of 15.0 N. The target is approaching at a speed of 15 m/s. At what horizontal distance between the target and the rocket should you launch?

6. A 500 g model rocket is on a cart that is rolling to the right at a speed of 3.0 m/s. The rocket engine, when it is fired, exerts an 8.0 N thrust on the rocket. Your goal is to have the rocket pass through a small horizontal hoop that is 20 m above the launch point. At what horizontal distance left of the hoop should you launch?

Section 6.3 Projectile Motion

7. For a projectile, which of the following quantities are constant during the flight: x, y, r, v_x, v_y, v, a_x, a_y, F_x, F_y? Which of the quantities are zero throughout the flight?

8. A physics student on Planet Exidor throws a ball, and it follows the parabolic trajectory shown in Figure Ex6.8. The ball's position is shown at 1 s intervals until $t = 3$ s. At $t = 1$ s, the ball's velocity is $\vec{v} = (2.0\hat{i} + 2.0\hat{j})$ m/s.
   a. Determine the ball's velocity at $t = 0$ s, 2 s, and 3 s.
   b. What is the value of g on Planet Exidor?
   c. What was the ball's launch angle?

FIGURE EX6.8

9. An object is launched with an initial velocity of 50 m/s at a launch angle of 36.9° above the horizontal.
   a. Make a table showing values of x, y, v_x, v_y, and the speed v every 1 s from $t = 0$ s to $t = 6$ s.
   b. Plot a graph of the object's trajectory during the first 6 s of motion.

10. Two spheres are launched horizontally from a 1.0-m-high table. Sphere A has a mass of 1.0 kg and is launched with an initial speed of 5.0 m/s. Sphere B has a mass of 0.4 kg and is launched with an initial speed of 2.5 m/s.
   a. Compare the times for each sphere to hit the floor.
   b. Compare the distances that each travels from the edge of the table.

11. A rifle is aimed horizontally at a target 50 m away. The bullet hits the target 2.0 cm below the aim point.
   a. What was the bullet's flight time?
   b. What was the bullet's speed as it left the barrel?

12. A ball thrown horizontally at 25 m/s travels a horizontal distance of 50 m before hitting the ground. From what height was the ball thrown?

Section 6.4 Relative Motion

13. A boat takes 3.0 hours to travel 30 km down a river, then 5.0 hours to return. How fast is the river flowing?

14. When the moving sidewalk at the airport is broken, as it often seems to be, it takes you 50 s to walk from your gate to baggage claim. When it is working and you stand on the moving sidewalk the entire way, without walking, it takes 75 s to travel the same distance. How long will it take you to travel from the gate to baggage claim if you walk while riding on the moving sidewalk?

15. An assembly line has a staple gun that rolls to the left at 1.0 m/s while parts to be stapled roll past it to the right at 3.0 m/s. The staple gun fires 10 staples per second. How far apart are the staples in the finished part?

16. Ted is sitting in his lawn chair when Stella flies directly overhead, going southeast at 100 m/s. Five seconds later, a firecracker explodes 200 m east of Ted. What are the coordinates of the explosion in Stella's reference frame? Let Stella be at the origin, with her x-axis pointing to the east.

17. Ships A and B leave port together. For the next two hours, ship A travels at 20 mph in a direction 30° west of north while the ship B travels 20° east of north at 25 mph.
    a. What is the distance between the two ships two hours after they depart?
    b. What is the speed of ship A as seen by ship B?

Problems

18. A particle starts from rest at $\vec{r}_0 = 9.0\hat{j}$ m and moves in the xy-plane with the velocity shown in Figure P6.18. The particle passes through a wire hoop located at $\vec{r}_1 = 20\hat{i}$ m, then continues onward.
    a. At what time does the particle pass through the hoop?
    b. What is the value of v_{4y}, the y-component of the particle's velocity at $t = 4$ s?
    c. Calculate and plot the particle's trajectory from $t = 0$ s to $t = 4$ s.

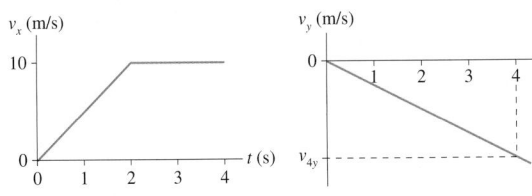

FIGURE P6.18

19. A 4.0×10^{10} kg asteroid is heading directly toward the center of the earth at a steady 20 km/s. To save the planet, astronauts strap a giant rocket to the asteroid perpendicular to its direction of travel. The rocket generates 5.0×10^9 N of thrust. The rocket is fired when the asteroid is 4.0×10^6 km away from earth. You can ignore the rotational motion of the earth and asteroid around the sun.
    a. If the mission fails, how many hours is it until the asteroid impacts the earth?
    b. The radius of the earth is 6400 km. By what minimum angle must the asteroid be deflected to just miss the earth?
    c. The rocket fires at full thrust for 300 s before running out of fuel. Is the earth saved?

20. A rocket-powered hockey puck on frictionless ice, such as in Example 6.3, slides along the y-axis with speed v_0. The front of the rocket is tilted at angle θ from the x-axis. The rocket motor ignites as the puck crosses the origin, exerting force $\vec{F}_{\text{thrust}}$ on the puck.
    a. Find an algebraic expression $y(x)$ for the puck's trajectory.
    b. Suppose the thrust is 2.0 N, the combined mass of the puck and rocket is 1.0 kg, and the initial speed is 2.0 m/s. Make a graph of your function $y(x)$ from $x = 0$ to $x = 20$ m for the two cases $\theta = 45°$ and $\theta = -45°$.

21. You are asked to consult for the city's research hospital, where a group of doctors is investigating the bombardment of cancer tumors with high-energy ions. The ions are fired directly toward the center of the tumor at speeds of 5.0×10^6 m/s. To cover the entire tumor area, the ions are deflected sideways by passing them between two charged metal

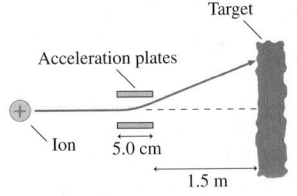

FIGURE P6.21

plates that accelerate the ions perpendicular to the direction of their initial motion. The acceleration region is 5.0 cm long, and the ends of the acceleration plates are 1.5 m from the patient. What acceleration is required to move an ion 2.0 cm across the tumor?

22. A projectile's horizontal range on level ground is $R = v_0^2 \sin 2\theta/g$. At what launch angle or angles will the projectile land at half of its maximum possible range?

23. a. A projectile is launched with speed v_0 and angle θ. Derive an expression for the projectile's maximum height h.
    b. A baseball is hit with a speed of 33.6 m/s. Calculate its height and the distance traveled if it is hit at angles of 30°, 45°, and 60°.

24. A sailor climbs to the top of the mast, 15 m above the deck, to look for land while his ship moves steadily forward through calm waters at 4.0 m/s. Unfortunately, he drops his spyglass to the deck below.
    a. Where does it land with respect to the base of the mast below him?
    b. Where does it land with respect to a fisherman sitting at rest in his dinghy as the ship goes past? Assume that the fisherman is even with the mast at the instant the spyglass is dropped.

25. A projectile is fired with an initial speed of 30 m/s at an angle of 60° above the horizontal. The object hits the ground 7.5 s later.
    a. How much higher or lower is the launch point relative to the point where the projectile hits the ground?
    b. To what maximum height above the launch point does the projectile rise?
    c. What are the magnitude and direction of the projectile's velocity at the instant it hits the ground?

26. In the Olympic shotput event, an athlete throws the shot with an initial speed of 12 m/s at a 40.0° angle from the horizontal. The shot leaves her hand at a height of 1.8 m above the ground.
    a. How far does the shot travel?
    b. Repeat the calculation of part (a) for angles 42.5°, 45.0°, and 47.5°. Put all your results, including 40.0°, in a table. At what angle of release does she throw the farthest?

27. On the Apollo 14 mission to the moon, astronaut Alan Shepard hit a golf ball with a 6 iron. The acceleration due to gravity on the moon is 1/6 of its value on earth. Suppose he hits the ball with a speed of 25 m/s at an angle 30° above the horizontal.
    a. How much farther did the ball travel on the moon than it would have on earth?
    b. For how much more time was the ball in flight?

28. A ball is thrown toward a cliff of height h with a speed of 30 m/s and an angle of 60° above horizontal. It lands on the edge of the cliff 4.0 s later.
    a. How high is the cliff?
    b. What was the maximum height of the ball?
    c. What is the ball's impact speed?

29. A tennis player hits a ball 2.0 m above the ground. The ball leaves his racquet with a speed of 20.0 m/s at an angle 5° above the horizontal. The horizontal distance to the net is 7.0 m, and the net is 1.0 m high. Does the ball clear the net? If so, by how much? If not, by how much does it miss?

30. A baseball player friend of yours wants to determine his pitching speed. You have him stand on a ledge and throw the ball

horizontally from an elevation 4.0 m above the ground. The ball lands 25 m away.

   a. What is his pitching speed?

   b. As you think about it, you're not sure he threw the ball exactly horizontally. As you watch him throw, the pitches seem to vary from 5° below horizontal to 5° above horizontal. What is the *range* of speeds with which the ball might have left his hand?

31. You are playing right field for the baseball team. Your team is up by one run in the bottom of the last inning of the game when a ground ball slips through the infield and comes straight toward you. As you pick up the ball 65 m from home plate, you see a runner rounding third base and heading for home with the tying run. You throw the ball at an angle of 30° above the horizontal with just the right speed so that the ball is caught by the catcher, standing on home plate, at the same height as you threw it. As you release the ball, the runner is 20.0 m from home plate and running full speed at 8.0 m/s. Will the ball arrive in time for your team's catcher to make the tag and win the game?

32. A stunt man drives a car at a speed of 20 m/s off a 30-m-high cliff. The road leading to the cliff is inclined upward at an angle of 20°.

   a. How far from the base of the cliff does the car land?

   b. What is the car's impact speed?

33. In one contest at the county fair, a spring-loaded plunger launches a ball at a speed of 3.0 m/s from one corner of a smooth, flat board that is tilted up at a 20° angle. To win, you must make the ball hit a small target at the adjacent corner, 2.50 m away. At what angle θ should you tilt the ball launcher?

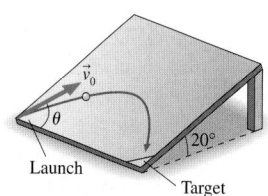

FIGURE P6.33

34. You're 6.0 m from one wall of a house. You want to toss a ball to your friend who is 6.0 m from the opposite wall. The throw and catch each occur 1.0 m above the ground.

   a. What minimum speed will allow the ball to clear the roof?

   b. At what angle should you toss the ball?

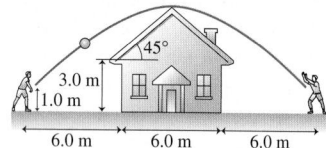

FIGURE P6.34

35. A supply plane needs to drop a package of food to scientists working on a glacier in Greenland. The plane flies 100 m above the glacier at a speed of 150 m/s. How far short of the target should it drop the package?

36. An antiaircraft gun fires shells at 200 m/s at a 60° angle. An enemy plane flies directly toward the gun at 300 m/s, 500 m off the ground. How far away (horizontally) must the plane be when the gun fires for the shell to hit the plane? Explain why you get two answers to this problem.

37. King Arthur's knights fire a cannon from the top of the castle wall. The cannonball is fired at a speed of 50 m/s and an angle of 30°. A cannonball that was accidentally dropped hits the moat below in 1.5 s.

   a. How far from the castle wall does the cannonball hit the ground?

   b. What is the ball's maximum height above the ground?

38. Quarterback Fred is going to throw a pass to tight end Doug. Doug is 20 m in front of Fred and running straight away at 6.0 m/s when Fred throws the 500 g football at a 40° angle. Doug catches the ball without having to alter his speed and runs for the game-winning touchdown. How fast did Fred throw the ball?

39. You are watching an archery tournament when you start wondering how fast an arrow is shot from the bow. Remembering your physics, you ask one of the archers to shoot an arrow parallel to the ground. You find the arrow stuck in the ground 60 m away, making a 3° angle with the ground. How fast was the arrow shot?

40. An archer standing on a 15° slope shoots an arrow 20° above the horizontal, as shown in Figure P6.40. How far down the slope does the arrow hit if it is shot with a speed of 50 m/s from 1.75 m above the ground?

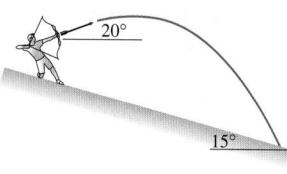

FIGURE P6.40

41. A popular pastime is to see who can push an object closest to the edge of a table without its going off. You push the 100 g object and release it 2.0 m from the table edge. Unfortunately, you push a little too hard. The object slides across, sails off the edge, falls 1.0 m to the floor, and lands 30 cm from the edge of the table. If the coefficient of kinetic friction is 0.5, what was the object's speed as you released it?

42. Sand moves without slipping at 6.0 m/s down a conveyer that is tilted at 15°. The sand enters a pipe 3.0 m below the end of the conveyer belt, as shown in Figure P6.42. What is the horizontal distance *d* between the conveyer belt and the pipe?

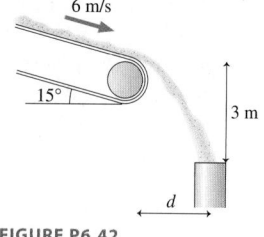

FIGURE P6.42

43. Sam (75 kg) takes off up a 50-m-high, 10° frictionless slope on his jet-powered skis. The skis have a thrust of 200 N. He keeps his skis tilted at 10° after becoming airborne, as shown in Figure P6.43. How far does Sam land from the base of the cliff?

FIGURE P6.43

44. A skateboarder starts up a 1.0-m-high, 30° ramp at a speed of 7.0 m/s. The skateboard wheels roll without friction. How far from the end of the ramp does the skateboarder touch down?

45. A motorcycle daredevil plans to ride up a 2.0-m-high, 20° ramp, sail across a 10-m-wide pool filled with hungry crocodiles, and land at ground level on the other side. He has done this stunt many times and approaches it with confidence. Unfortunately, the motorcycle engine dies just as he starts up the ramp. He is going 11 m/s at that instant, and the rolling

friction of his rubber tires is not negligible. Does he survive, or does he become crocodile food?

46. A 5000 kg interceptor rocket is launched at an angle of 44.7°. The thrust of the rocket motor is 140,700 N.
    a. Find an equation $y(x)$ that describes the rocket's trajectory.
    b. What is the shape of the trajectory?
    c. At what elevation does the rocket reach the speed of sound, 330 m/s?

47. A rocket-powered hockey puck has a thrust of 2.0 N and a total mass of 1.0 kg. It is released from rest on a frictionless table, 4.0 m from the edge of a 2.0 m drop. The front of the rocket is pointed directly toward the edge. How far does the puck land from the base of the table?

48. A 500 g model rocket is resting horizontally at the top edge of a 40-m-high wall when it is accidentally bumped. The bump pushes it off the edge with a horizontal speed of 0.5 m/s and at the same time causes the engine to ignite. When the engine fires, it exerts a constant 20 N horizontal thrust away from the wall.
    a. How far from the base of the wall does the rocket land?
    b. Describe the trajectory of the rocket while it travels to the ground.

In Problems 49 through 51 you are given the equations that are used to solve a problem. For each of these, you are to

a. Write a realistic problem for which these are the correct equations. Be sure that the answer your problem requests is consistent with the equations given.
b. Finish the solution of the problem.

49. $100 \text{ m} = 0 \text{ m} + (50\cos\theta \text{ m/s})t_1$
    $0 \text{ m} = 0 \text{ m} + (50\sin\theta \text{ m/s})t_1 - \frac{1}{2}(9.80 \text{ m/s}^2)t_1^2$

50. $x_1 = 0 \text{ m} + (30 \text{ m/s})t_1$
    $0 \text{ m} = 300 \text{ m} - \frac{1}{2}(9.80 \text{ m/s}^2)t_1^2$

51. $v_x = -(6.0\cos 45°) \text{ m/s} + 3.0 \text{ m/s}$
    $v_y = (6.0\sin 45°) \text{ m/s} + 0 \text{ m/s}$
    $100 \text{ m} = v_y t_1$
    $x_1 = v_x t_1$

52. Write a realistic problem for which the x-versus-t and y-versus-t graphs shown in Figure P6.52 represent the motion of an object. Be sure the answer your problem requests is consistent with the graphs. Then finish the solution of the problem.

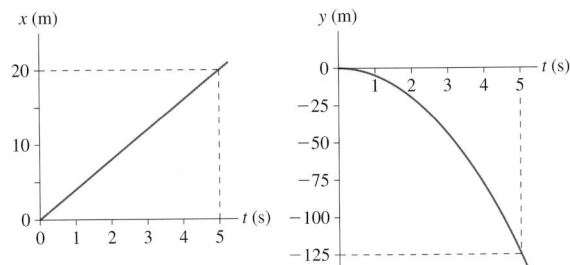

FIGURE P6.52

53. Mary needs to row her boat across a 100-m-wide river that is flowing to the east at a speed of 3.0 m/s. Mary can row with a speed of 2.0 m/s.
    a. If Mary rows straight north, where will she land?
    b. Draw a picture showing her displacement due to rowing, her displacement due to the river's motion, and her net displacement.

54. A kayaker needs to paddle north across a 100-m-wide harbor. The tide is going out, creating a tidal current that flows to the east at 2.0 m/s. The kayaker can paddle with a speed of 3.0 m/s.
    a. In which direction should he paddle in order to travel straight across the harbor?
    b. How long will it take him to cross?

55. Mike throws a ball upward at a 63° angle with a speed of 22 m/s. Nancy drives past Mike at 30 m/s at the instant he releases the ball.
    a. What is the ball's initial angle in Nancy's reference frame?
    b. Find and graph the ball's trajectory as seen by Nancy.

56. A sailboat is sailing due east at 8.0 mph. The wind appears to blow from the southwest at 12.0 mph.
    a. What are the true wind speed and direction?
    b. What are the true wind speed and direction if the wind appears to blow from the northeast at 12.0 mph?

57. A child in danger of drowning in a river is being carried downstream by a current that flows uniformly with a speed of 2.0 m/s. The child is 200 m from the shore and 1500 m upstream of the boat dock from which the rescue team sets out. If their boat speed is 8.0 m/s with respect to the water, at what angle should the pilot leave the shore to go directly to the child?

58. Quarterback Fred is going to throw a pass to tight end Alberto. Fred is being chased, however, and he throws the ball at 20 m/s while running directly toward the nearest sideline at 4.0 m/s. Alberto is standing still directly upfield from Fred at the moment the ball is released. In what direction should Fred throw so that Alberto can catch the ball without moving?

59. The paper delivery boy tries to throw the paper into your narrow driveway without slowing down. His pickup truck travels at 10 mph, and he throws the paper at 20 mph just as the truck passes the driveway.
    a. In what direction should he throw the paper in order for it to land in the driveway?
    b. What is the paper's speed relative to the ground?

60. While driving north at 25 m/s during a rainstorm you notice that the rain makes an angle of 38° with the vertical. While driving back home moments later at the same speed but in the opposite direction, you see that the rain is falling straight down. From these observations, determine the speed and angle of the raindrops relative to the ground.

61. A plane has an airspeed of 200 mph. The pilot wishes to reach a destination 600 mi due east, but a wind is blowing at 50 mph in the direction 30° north of east.
    a. In what direction must the pilot head the plane in order to reach her destination?
    b. How long will the trip take?

62. Susan, driving north at 60 mph, and Shawn, driving east at 45 mph, are approaching an intersection. What is Shawn's speed relative to Susan's reference frame?

63. As is discussed more fully in Chapter 42, one of the processes by which a radioactive nucleus can decay to a more stable state is by the emission of a very energetic particle called a gamma-ray photon. Like the laser beam discussed in this chapter, the gamma-ray photon is a form of light and travels at the speed of light, 3.0×10^8 m/s. Suppose a radioactive nucleus emits a gamma-ray photon while moving through the laboratory toward the east at 1.5×10^8 m/s. If the photon is emitted toward the east, parallel to the motion of the nucleus, what is the speed of the gamma-ray photon as determined by a scientist in the laboratory reference frame?

Challenge Problems

64. In the absence of air resistance, a projectile that lands at the elevation from which it was launched achieves maximum range when launched at a 45° angle. Suppose a projectile of mass m is launched with speed v_0 into a headwind that exerts a constant, horizontal retarding force $\vec{F}_{wind} = -F_{wind}\hat{\imath}$.
    a. Find an expression for the angle at which the range is maximum.
    b. By what percentage is the maximum range of a 0.50 kg ball reduced if $F_{wind} = 0.60$ N?

65. You have been hired to assist with stunts for a new action movie. In one scene, the writers want to drop a package from a small plane into a moving convertible sports car. The car will drive along a horizontal road at 30 m/s. The plane will approach the car from behind at an altitude of 60 m and a speed of 50 m/s relative to the ground. The copilot will view the car through a sighting tube that measures the angle below horizontal. At what angle should the package be released?

66. A cat is chasing a mouse. The mouse runs in a straight line at a speed of 1.5 m/s. If the cat leaps off the floor at a 30° angle and a speed of 4.0 m/s, at what distance behind the mouse should the cat leap in order to land on the poor mouse?

67. A rubber ball is dropped onto a ramp that is tilted at 20°, as shown in Figure CP6.67. A bouncing ball obeys the "law of reflection," which says that the ball leaves the surface at the same angle it approached the surface. The ball's next bounce is 3 m to the right of its first bounce. What is the ball's rebound speed on its first bounce?

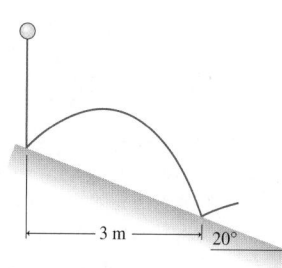

FIGURE CP6.67

68. A motorcycle daredevil wants to set a record for jumping over burning school buses. He has hired you to help with the design. He intends to ride off a horizontal platform at 40 m/s, cross the burning buses in a pit below him, then land on a ramp sloping down at 20°. It's very important that he not bounce when he hits the landing ramp because that could cause him to lose control and crash. You immediately recognize that he won't bounce if his velocity is parallel to the ramp as he touches down. This can be accomplished if the ramp is tangent to his trajectory *and* if he lands right on the front edge of the ramp. There's no room for error! Your task is to determine where to place the landing ramp. That is, how far from the edge of the launching platform should the front edge of the landing ramp be horizontally and how far below it? There's a clause in your contract that requires you to test your design before the hero goes on national television to set the record.

69. Driving a spaceship isn't as easy as it looks in the movies. Imagine you're a physics student in the 31st century. You live in a remote space colony where the gravitational force from any stars or planets is negligible. You're on your way home from school, coasting along in your 20,000 kg personal spacecraft at 2.0 km/s, when the computer alerts you to the fact that the entrance to your pod is 500 km away along a line 30° from your present heading, as shown in Figure CP6.69. You need to make a left turn so that you can enter the pod going straight ahead at 1.0 km/s. You could do this with a series of small rocket burns, but you want to impress the girls in the spacecraft behind you by getting through the entrance with a single rocket burn. You can use small thrusters to quickly rotate your spacecraft to a different orientation before and after the main rocket burn.

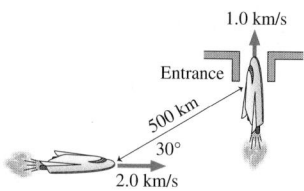

FIGURE CP6.69

    a. You need to determine three things: How to orient your spacecraft for the main rocket burn, the magnitude F_{thrust} of the rocket burn, and the length of the burn. Use a coordinate system in which you start at the origin and are initially moving along the x-axis. Measure the orientation of your spacecraft by the angle it makes with the positive x-axis. Your initial orientation is 0°. You can end the burn before you reach the entrance, but you're not allowed to have the engine on as you pass through the entrance. Mass loss during the burn is negligible.
    b. Calculate your position coordinates every 50 s until you reach the entrance, then plot a graph of your trajectory. Be sure to label the position of the entrance.

70. Uri is on a flight from Boston to Los Angeles. His plane is traveling 20° south of west at 500 mph. Val is on a flight from Miami to Seattle. Her plane is traveling 30° north of west at 500 mph. Somewhere over Kansas, Uri's plane passes 1000 ft directly over Val's plane. Uri is sitting on the right side and can see Val's plane below him after they pass. Uri notices that the fuselage of Val's plane doesn't point in the direction that her plane is moving. What is the angle between the fuselage and the direction of motion?

STOP TO THINK ANSWERS

Stop to Think 6.1: d. The parallel component of $\vec{a}$ is opposite $\vec{v}$ and will cause the particle to slow down. The perpendicular component of $\vec{a}$ will cause the particle to change directions in a downward direction.

Stop to Think 6.2: b. A parabola requires acceleration in one direction but not the other. The horizontal motion is constant velocity, with $a_x = 0$.

Stop to Think 6.3: d. A projectile's acceleration $\vec{a} = -g\hat{\jmath}$ does not depend on its mass. The second ball has the same initial velocity and same acceleration, so it follows the same trajectory and lands at the same position.

Stop to Think 6.4: f. The helicopter frame S′ moves with $\vec{V} = 20\hat{\jmath}$ m/s relative to the earth frame S. The plane moves with $\vec{v} = 100\hat{\imath}$ m/s in the earth's frame. The vector addition in the figure shows that $\vec{v}'$ is longer than $\vec{v}$.

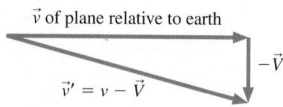

7 Dynamics III: Motion in a Circle

Why doesn't the roller coaster fall off the track at the top of the loop?

◀ **Looking Ahead**

The goal of Chapter 7 is to learn to solve problems about motion in a circle. In this chapter you will learn to:

- Understand the mathematics of circular kinematics.
- Use Newton's laws to analyze the dynamics of circular motion.
- Understand circular orbits of satellites and planets.
- Think about apparent weight and fictitious forces for objects in circular motion.

◀ **Looking Back**

This chapter, like Chapter 6, extends ideas of one-dimensional kinematics and dynamics into two dimensions. Please review:

- Sections 1.5–1.6 Finding acceleration vectors on a motion diagram.
- Sections 2.2 and 2.5 Uniform motion and constant-acceleration kinematics.
- Sections 5.2 and 5.3 Newton's second law and apparent weight.
- Section 6.1 Kinematics in two dimensions.

A roller coaster doing a loop-the-loop is a dramatic example of circular motion. But why doesn't it fall off the track when it's upside-down at the top of the loop? How is it that you can swing a bucket of water over your head without the water falling out? Why does your car "spin out" of a curve on an wet or icy road but not when the pavement is dry?

To answer these questions, we must study how objects move in circles. We can understand circular motion in terms of forces and Newton's laws, but first we'll need to develop some new concepts and mathematical tools for describing motion in a circle. We'll be able to apply these new ideas to a wide variety of interesting and important problems, from a car rounding a curve to the orbit of the earth around the sun.

Although this chapter will continue to focus on the motion of particles, circular motion is a prelude to studying the rotational motion of solid objects. The ideas introduced in this chapter will be the basis for Chapter 13, "Rotation of a Rigid Body."

7.1 Uniform Circular Motion

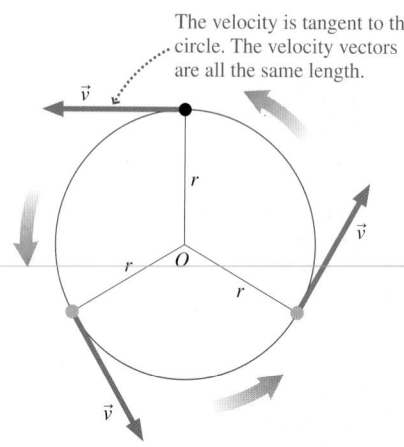

The velocity is tangent to the circle. The velocity vectors are all the same length.

FIGURE 7.1 A particle in uniform circular motion.

To begin, consider a particle that moves at *constant speed* around a circle of radius r. This is called **uniform circular motion.** Our first task is to *describe* the motion, and we will do so in this section and the next by developing the kinematics of uniform circular motion. We'll need some new mathematical ideas because the xy-coordinate system we've been using is not the best coordinate system for circular motion. Our second task, in Section 7.3, will be to *explain* the motion in terms of forces.

Figure 7.1 shows a particle moving around a circle of radius r. The particle might be a satellite moving in an orbit, a ball on the end of a string, or even just a dot painted on the side of a wheel. Regardless of what the particle represents, its velocity vector $\vec{v}$ is always tangent to the circle. The particle's speed v is constant, so the vector $\vec{v}$ stays the same length as the particle moves around the circle.

Period

The time interval it takes the particle to go around the circle once, completing one revolution (abbreviated rev), is called the **period** of the motion. Period is represented by the symbol T. This is a logical symbol, because period is an interval of time, but there's a risk of confusing the period T with the symbol $\vec{T}$ for tension or, even worse, the identical symbol T for the magnitude of the tension force.

> **NOTE** ▶ The number of symbols used in science and engineering far exceeds the number of letters in the English alphabet. Even after we've borrowed from the Greek alphabet, it's inevitable that some letters are used several times to represent entirely different quantities. The use of T is the first time we've run into this problem, but it won't be the last. You must be alert to the *context* of a symbol's use in order to deduce its meaning. Always remembering to use vector arrows over vector symbols, such as $\vec{T}$, helps to clarify the meaning of symbols. ◀

It's easy to relate the particle's period T to its speed v. For a particle moving with constant speed, speed is simply distance/time. In one period, the particle moves once around a circle of radius r and travels the circumference $2\pi r$. Thus

$$v = \frac{1 \text{ circumference}}{1 \text{ period}} = \frac{2\pi r}{T} \tag{7.1}$$

EXAMPLE 7.1 A rotating crankshaft
A 4.0-cm-diameter crankshaft turns at 2400 rpm (revolutions per minute). What is the speed of a point on the surface of the crankshaft?

SOLVE We need to determine the time it takes the crankshaft to make 1 rev. First, convert 2400 rpm to revolutions per second:

$$\frac{2400 \text{ rev}}{1 \text{ min}} \times \frac{1 \text{ min}}{60 \text{ s}} = 40 \text{ rev/s}$$

If the crankshaft turns 40 times in 1 s, the time for 1 rev is

$$T = \frac{1}{40} \text{ s} = 0.025 \text{ s}$$

Thus the speed of a point on the surface, where $r = 2.0$ cm $= 0.020$ m, is

$$v = \frac{2\pi r}{T} = \frac{2\pi(0.020 \text{ m})}{0.025 \text{ s}} = 5.03 \text{ m/s}$$

Angular Position

Rather than using xy-coordinates, it will be more convenient to describe the position of the particle by its distance r from the center of the circle (labeled O) and its angle θ from the positive x-axis. This is shown in Figure 7.2. The angle θ is the **angular position** of the particle.

We can distinguish a position above the *x*-axis from a position that is an equal angle below the *x*-axis by *defining* θ to be positive when measured *counterclockwise* (ccw) from the positive *x*-axis. An angle measured clockwise (cw) from the positive *x*-axis has a negative value. "Clockwise" and "counterclockwise" in circular motion are analogous, respectively, to "left of the origin" and "right of the origin" in linear motion, which we associated with negative and positive values of *x*. A particle 30° below the positive *x*-axis is equally well described by either $\theta = -30°$ or $\theta = +330°$. We could also describe this particle by $\theta = \frac{11}{12}$ rev, where *revolutions* are another way to measure the angle.

Although degrees and revolutions are widely used measures of angle, mathematicians and scientists usually find it more convenient to measure the angle θ in Figure 7.2 by using the **arc length** *s* that the particle travels along the edge of a circle of radius *r*. We define the angular unit of **radians** such that

$$\theta \text{ (radians)} \equiv \frac{s}{r} \tag{7.2}$$

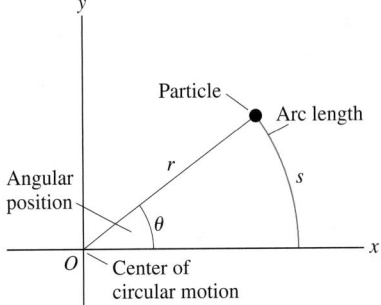

FIGURE 7.2 A particle's position is described by distance *r* and angle θ.

The radian, which is abbreviated rad, is the SI unit of an angle. An angle of 1 rad has an arc length *s* exactly equal to the radius *r*.

The arc length completely around a circle is the circle's circumference $2\pi r$. Thus the angle of a full circle is

$$\theta_{\text{full circle}} = \frac{2\pi r}{r} = 2\pi \text{ rad}$$

This relationship is the basis for the well-known conversion factors

$$1 \text{ rev} = 360° = 2\pi \text{ rad}$$

As a simple example of converting between radians and degrees, let's convert an angle of 1 rad to degrees:

$$1 \text{ rad} = 1 \text{ rad} \times \frac{360°}{2\pi \text{ rad}} = 57.3°$$

Thus a reasonable approximation is 1 rad $\approx$ 60°. We will often specify angles in degrees, but keep in mind that the SI unit is the radian.

An important consequence of Equation 7.2 is that the arc length spanning angle θ is

$$s = r\theta \tag{7.3}$$

This is a result that we will use often, but it is valid *only* if θ is measured in radians and not in degrees. This very simple relationship between angle and arc length is one of the primary motivations for using radians.

NOTE ▶ While the concept of measuring an angle is simple, units of angle are often troublesome. Unlike the kilogram or the second, for which we have standards, the radian is a *defined* unit. Further, its definition as a ratio of two lengths makes it a *pure number* without dimensions. Thus the unit of angle, be it radians or degrees or revolutions, is really just a *name* to remind us that we're dealing with an angle. The practical implication is that the radian unit sometimes appears or disappears without warning. This seems rather mysterious until you get used to it. This textbook will call your attention to such behavior the first few times it occurs. With a little practice, you'll soon learn when the rad unit is needed and when it's not. ◀

Angular Velocity

Figure 7.3 shows a particle moving in a circle from an initial angular position θ_i at time t_i to a final angular position θ_f at a later time t_f. The change $\Delta\theta = \theta_f - \theta_i$ is called the **angular displacement.** We can measure the particle's circular motion

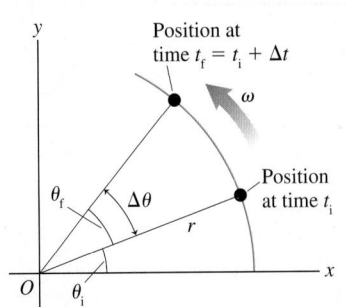

FIGURE 7.3 A particle moves from θ_i to θ_f with angular velocity ω.

in terms of the rate of change of θ, just as we measured the particle's linear motion in terms of the rate of change of its position x.

In analogy with linear motion, let's define the *average angular velocity* to be

$$\text{average angular velocity} \equiv \frac{\Delta\theta}{\Delta t} \qquad (7.4)$$

As the time interval Δt becomes very small, $\Delta t \rightarrow 0$, we arrive at the definition of the instantaneous **angular velocity**

$$\omega \equiv \lim_{\Delta t \rightarrow 0} \frac{\Delta\theta}{\Delta t} = \frac{d\theta}{dt} \qquad \text{(angular velocity)} \qquad (7.5)$$

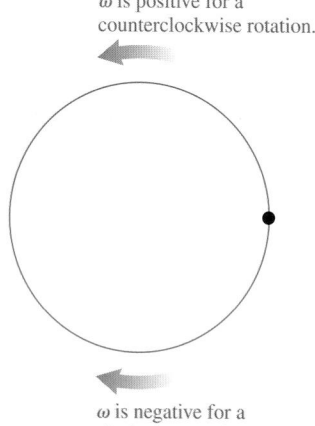

ω is positive for a counterclockwise rotation.

ω is negative for a clockwise rotation.

FIGURE 7.4 Positive and negative angular velocities.

The symbol ω is a lowercase Greek omega, *not* an ordinary w. The SI unit of angular velocity is rad/s, but °/s, rev/s, and rev/min are also common units. Revolutions per minute is often abbreviated rpm.

Angular velocity is the *rate* at which a particle's angular position is changing. A particle that starts from $\theta = 0$ rad with an angular velocity of 0.5 rad/s will be at angle $\theta = 0.5$ rad after 1 s, at $\theta = 1.0$ rad after 2 s, at $\theta = 1.5$ rad after 3 s, and so on. Its angular position is increasing at the *rate* of 0.5 radians per second. In analogy with uniform linear motion, which you studied in Chapter 2, uniform circular motion is motion in which the angle increases at a *constant* rate: **A particle moves with uniform circular motion if and only if its angular velocity ω is constant and unchanging.**

Angular velocity, like the velocity v_s of one-dimensional motion, can be positive or negative. The signs shown in Figure 7.4 are based on the fact that θ was defined to be positive for a counterclockwise rotation. Because the definition $\omega = d\theta/dt$ for circular motion parallels the definition $v_s = ds/dt$ for linear motion, the graphical relationships we found between v_s and s in Chapter 2 apply equally well to ω and θ:

- $\omega = $ slope of the θ-versus-t graph at time t
- $\theta_f = \theta_i + $ area under the ω-versus-t graph between t_i and t_f

---

EXAMPLE 7.2 A graphical representation of circular motion

Figure 7.5 shows the angular position of a particle moving around a circle of radius r. Describe the particle's motion and draw an ω-versus-t graph.

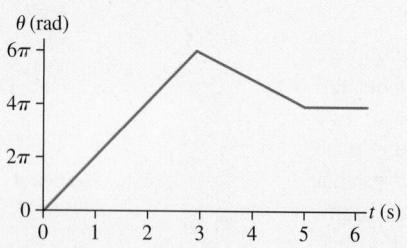

FIGURE 7.5 Angular position graph for the particle of Example 7.2.

SOLVE Although circular motion seems to "start over" every revolution (every 2π rad), the angular position θ continues to increase. $\theta = 6\pi$ rad corresponds to three revolutions. This particle makes 3 ccw (because θ is getting more positive) rev in

3 s, immediately reverses direction and makes 1 cw rev in 2 s, then stops at $t = 5$ s and holds the position $\theta = 4\pi$ rad. The angular velocity is found by measuring the slope of the graph:

$t = 0 - 3$ s   slope $= \Delta\theta/\Delta t = 6\pi$ rad/3 s $= 2\pi$ rad/s

$t = 3 - 5$ s   slope $= \Delta\theta/\Delta t = -2\pi$ rad/2 s $= -\pi$ rad/s

$t > 5$ s   slope $= \Delta\theta/\Delta t = 0$ rad/s

These results are shown as an ω-versus-t graph in Figure 7.6. For the first 3 s, the motion is uniform circular motion with $\omega = 2\pi$ rad/s. The particle then changes to a different uniform circular motion with $\omega = -\pi$ rad/s for 2 s, then stops.

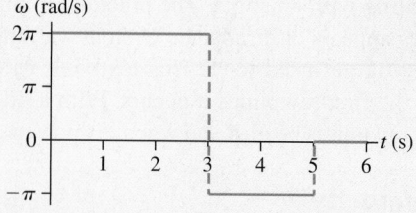

FIGURE 7.6 ω-versus-t graph for the particle of Example 7.2.

NOTE ▶ In physics, we nearly always want to give results as numerical values. Example 7.1 had a π in the equation, but we used its numerical value to compute $v = 5.03$ m/s. However, angles in radians are an exception to this rule. It's okay to leave a π in the value of θ or ω, and we have done so in Example 7.2. ◀

The angular velocity is constant during uniform circular motion, so the ω-versus-t graph is a horizontal line. It's easy to see from Figure 7.7 that the area under the curve from t_i to t_f is simply $\omega \Delta t$. Consequently,

$$\theta_f = \theta_i + \omega \Delta t \qquad \text{(uniform circular motion)} \qquad (7.6)$$

Equation 7.6 is equivalent, with different variables, to the result $s_f = s_i + v_s \Delta t$ for uniform linear motion. You will see many more instances where circular motion is analogous to linear motion with angular variables replacing linear variables. Thus much of what you learned about linear kinematics and dynamics carries over to circular motion.

Not surprisingly, the angular velocity ω is closely related to the *period T* of the motion. As a particle goes around a circle one time, its angular displacement is $\Delta \theta = 2\pi$ rad during the interval $\Delta t = T$. Thus, using the definition of angular velocity, we find

$$|\omega| = \frac{2\pi \text{ rad}}{T} \quad \text{or} \quad T = \frac{2\pi \text{ rad}}{|\omega|} \qquad (7.7)$$

The period alone gives only the absolute value of $|\omega|$. To determine the sign of ω you need to know the direction of motion.

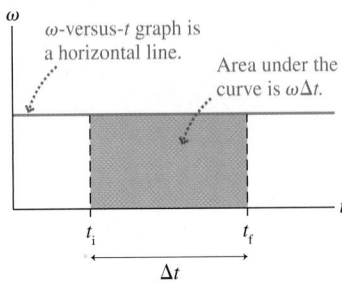

FIGURE 7.7 The ω-versus-t graph for uniform circular motion is a horizontal line.

EXAMPLE 7.3 At the roulette wheel

A small, steel roulette ball rolls around the inside of a 30-cm-diameter roulette wheel. The ball completes 2 rev in 1.20 s.

a. What is the ball's angular velocity?
b. What is the ball's position at $t = 2.0$ s? Assume $\theta_i = 0$.

MODEL Model the ball as a particle in uniform circular motion.

SOLVE

a. The period of the ball's motion, the time for 1 rev, is $T = 0.60$ s. Thus

$$\omega = \frac{2\pi \text{ rad}}{T} = \frac{2\pi \text{ rad}}{0.60 \text{ s}} = 10.47 \text{ rad/s}$$

b. The ball starts at $\theta_i = 0$ rad. After $\Delta t = 2.0$ s, its position is given by Equation 7.6:

$$\theta_f = 0 \text{ rad} + (10.47 \text{ rad/s})(2.0 \text{ s}) = 20.94 \text{ rad}$$

Although this is a mathematically acceptable answer, an observer would say that the ball is always located somewhere between 0° and 360°. Thus it is common practice to subtract off an integer number of 2π, representing the completed revolutions. Because $20.94/2\pi = 3.333$, we can write

$$\theta_f = 20.94 \text{ rad} = 3.333 \times 2\pi \text{ rad}$$
$$= 3 \times 2\pi \text{ rad} + 0.333 \times 2\pi \text{ rad}$$
$$= 3 \times 2\pi \text{ rad} + 2.09 \text{ rad}$$

In other words, at $t = 2.0$ s the ball has completed 3 rev and is 2.09 rad = 120° into its fourth revolution. An observer would say that the ball's position is $\theta = 120°$.

STOP TO THINK 7.1 A particle moves cw around a circle at constant speed for 2.0 s. It then reverses direction and moves ccw at half the original speed until it has traveled through the same angle. Which is the particle's angle-versus-time graph?

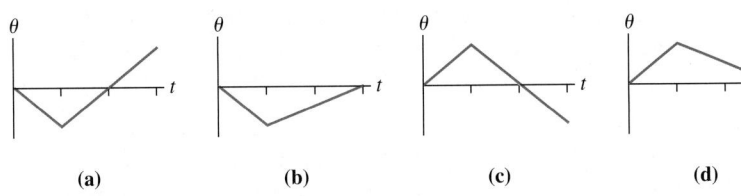

(a)   (b)   (c)   (d)

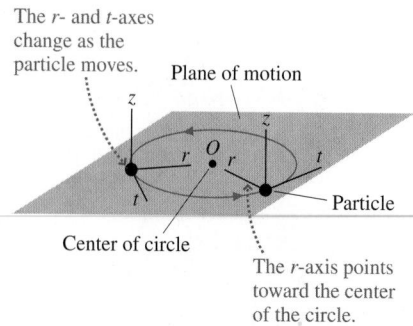

The *r*- and *t*-axes change as the particle moves.

Plane of motion

Center of circle

Particle

The *r*-axis points toward the center of the circle.

FIGURE 7.8 The axes of the *rtz*-coordinate system.

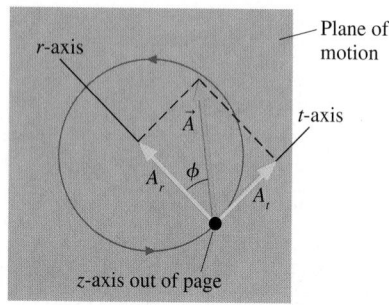

r-axis

Plane of motion

t-axis

z-axis out of page

FIGURE 7.9 Vector $\vec{A}$ can be decomposed into radial and tangential components.

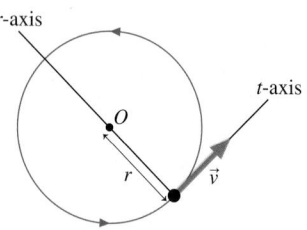

r-axis

t-axis

FIGURE 7.10 The velocity vector $\vec{v}$ has only a tangential component v_t.

7.2 Velocity and Acceleration in Uniform Circular Motion

The *xy*-coordinate system we've been using for linear motion and projectile motion is not the best coordinate system for circular motion. Figure 7.8 shows a circular trajectory and the plane in which the circle lies. Let's establish a coordinate system with its origin at the point where the particle is located. The axes are defined as follows:

- The *r*-axis (radial axis) points *from* the particle *toward* the center of the circle.
- The *t*-axis (tangential axis) is tangent to the circle, pointing in the ccw direction.
- The *z*-axis is perpendicular to the plane of motion.

The three axes of this *rtz*-coordinate system are mutually perpendicular, just like the axes of the familiar *xyz*-coordinate system. Notice how the axes move with the particle so that the *r*-axis always points to the center of the circle. It will take a little getting used to, but you will soon see that circular motion problems are most easily described in these coordinates.

Figure 7.9 shows a vector $\vec{A}$ in the plane of motion. We can decompose $\vec{A}$ into its radial and tangential components:

$$A_r = A\cos\phi$$
$$A_t = A\sin\phi$$

where ϕ is the angle with the *r*-axis. The positive *r*-direction, by definition, is toward the center of the circle, so the radial component A_r has a positive value. $\vec{A}$ lies in the plane of motion, so its perpendicular component is $A_z = 0$.

NOTE ▶ In Chapter 6, we noted that the acceleration vector $\vec{a}$ can be decomposed into a component $\vec{a}_\parallel$ parallel to the motion and a component $\vec{a}_\perp$ perpendicular to the motion. That idea is the basis for the *rtz*-coordinate system. Because the velocity vector $\vec{v}$ is tangent to the circle, the tangential component A_t of vector $\vec{A}$ is the component of $\vec{A}$ parallel to the motion. The radial component A_r is the component perpendicular to the motion. ◀

Velocity

For a particle in circular motion, such as the one seen in Figure 7.10, the velocity vector $\vec{v}$ is tangent to the circle. In other words, the velocity vector has only a tangential component v_t. The radial and perpendicular components of $\vec{v}$ are always zero.

The tangential velocity component v_t is the rate ds/dt at which the particle moves *around* the circle, where *s* is the arc length measured from the positive *x*-axis. From Equation 7.3, the arc length is $s = r\theta$. Taking the derivative, we find

$$v_t = \frac{ds}{dt} = r\frac{d\theta}{dt}$$

But $d\theta/dt$ is the angular velocity ω. Thus the tangential velocity and the angular velocity are related by

$$v_t = \omega r \qquad \text{(with } \omega \text{ in rad/s)} \tag{7.8}$$

NOTE ▶ ω is restricted to rad/s because the relationship $s = r\theta$ is the definition of radians. While it may be convenient in some problems to measure ω in rev/s or rpm, you must convert to SI units of rad/s before using Equation 7.8. ◀

The tangential velocity v_t is positive for ccw motion, negative for cw motion. Because v_t is the only nonzero component of $\vec{v}$, the particle's speed is $v = |v_t| = |\omega|r$. We'll sometimes write this as $v = \omega r$ if there's no ambiguity about the sign of ω.

As a simple example, a particle moving cw at 2.0 m/s in a circle of radius 40 cm has an angular velocity

$$\omega = \frac{v_t}{r} = \frac{-2.0 \text{ m/s}}{0.40 \text{ m}} = -5.0 \text{ rad/s}$$

where v_t and ω are negative because the motion is clockwise. Notice the units. Velocity divided by distance has units of s^{-1}. But because the division, in this case, gives us an angular quantity, we've inserted the *dimensionless* unit rad to give ω the appropriate units of rad/s.

To summarize, the velocity in the *rtz*-coordinates is

$$v_r = 0$$
$$v_t = \omega r \qquad (7.9)$$
$$v_z = 0$$

Acceleration

Example 1.6 in Chapter 1 looked at the uniform circular motion of a Ferris wheel. The motion diagram from that example is shown again as Figure 7.11a. Although the particle moves with constant speed, it has an acceleration because the *direction* of the velocity vector $\vec{v}$ is always changing. A motion diagram analysis shows that the **acceleration $\vec{a}$ points toward the center of the circle.** The instantaneous velocity is tangent to the circle, so $\vec{v}$ and $\vec{a}$ are perpendicular to each other at all points on the circle, as Figure 7.11b shows.

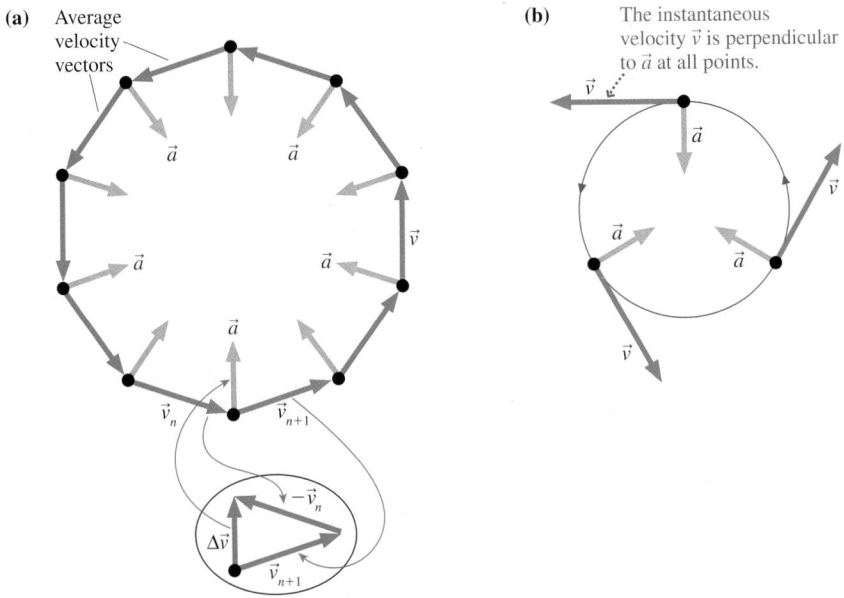

FIGURE 7.11 Motion diagram for uniform circular motion. The acceleration $\vec{a}$ always points to the center.

The acceleration of uniform circular motion is called **centripetal acceleration,** a term from a Greek root meaning "center seeking." Centripetal acceleration is not a new type of acceleration; all we are doing is *naming* an acceleration that corresponds to a particular type of motion. The magnitude of the centripetal acceleration is constant because each successive $\Delta\vec{v}$ in the motion diagram has the same length.

The motion diagram tells us the direction that $\vec{a}$ points in, but it doesn't give us a value for a. To complete our description of circular motion, we need to find a quantitative relationship between a and the tangential velocity v_t.

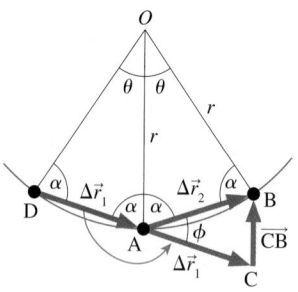

FIGURE 7.12 Finding the relationship between velocity and acceleration for circular motion.

Figure 7.12 shows in more detail the three points at the bottom of the motion diagram of Figure 7.11. Vectors $\Delta\vec{r}_1$ and $\Delta\vec{r}_2$ have the same length, namely $v\Delta t$. Vector $\Delta\vec{r}_1$ has been redrawn with the tails of the two vectors together at point A. The vector that points straight up from C to B is

$$\overrightarrow{CB} = \Delta\vec{r}_2 - \Delta\vec{r}_1 = \vec{v}_2\Delta t - \vec{v}_1\Delta t = (\vec{v}_2 - \vec{v}_1)\Delta t = \Delta\vec{v}\Delta t$$

Triangle ABC is an isosceles triangle because the motion is at constant speed. Triangle ABO is also isosceles because sides AO and BO are both of length r. Furthermore, because $\theta + \alpha + \alpha = 180°$ in triangle ABO and $\phi + \alpha + \alpha = 180°$ along line DAC, you can see that $\phi = \theta$ and hence ABC and ABO are *similar triangles* with side AB in common.

Recall, from geometry, that the ratios of the sides of similar triangles are equal. Thus

$$\frac{CB}{AB} = \frac{AB}{AO} \quad \text{or} \quad \frac{|\Delta\vec{v}|\Delta t}{v\Delta t} = \frac{v\Delta t}{r} \tag{7.10}$$

where $|\Delta\vec{v}|$ indicates the magnitude of vector $\Delta\vec{v}$. Rearranging this equation gives

$$\frac{|\Delta\vec{v}|}{\Delta t} = \frac{v^2}{r} \tag{7.11}$$

But $|\Delta\vec{v}|/\Delta t$ is the magnitude of the average acceleration vector: $\vec{a}_{avg} = \Delta\vec{v}/\Delta t$. Thus

$$a_{avg} = \frac{v^2}{r} \tag{7.12}$$

This analysis has been in terms of the *average* velocities and the *average* acceleration. If we now let $\Delta t \to 0$, the three points in Figure 7.12 move together but the geometrical relationships do not change. Thus Equation 7.12 remains valid when we take the limit, and it applies equally well to instantaneous velocities and accelerations.

4.1   Actıv
        Physıcs

Thus the acceleration of uniform circular motion points to the center of the circle, because $\Delta\vec{v}$ does, and has magnitude v^2/r. In vector notation,

$$\vec{a} = \left(\frac{v^2}{r}, \text{ toward center of circle}\right) \quad \text{(centripetal acceleration)} \tag{7.13}$$

This acceleration is conveniently written in the *rtz*-coordinate system as

$$a_r = \frac{v^2}{r} = \omega^2 r$$
$$a_t = 0 \tag{7.14}$$
$$a_z = 0$$

We used Equation 7.8, $v = \omega r$, to write a_r in terms of the angular velocity ω. For convenience, we'll often refer to the component a_r as "the centripetal acceleration."

Figure 7.13 shows the acceleration vector in the *rtz*-coordinate system. Compare this to the velocity vector in Figure 7.10. You can begin to see the advantages of the *rtz*-coordinate system. Only one component of $\vec{v}$ and one component of $\vec{a}$ are nonzero.

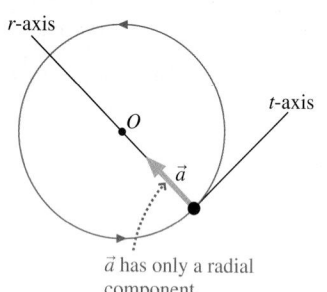

$\vec{a}$ has only a radial component.

FIGURE 7.13 The acceleration vector in the *rtz*-coordinate system.

EXAMPLE 7.4 The acceleration of an atomic electron
We will later study the Bohr atom. This is a simple model of the hydrogen atom in which an electron orbits a proton at a radius of 5.29×10^{-11} m with a period of 1.52×10^{-16} s. What is the electron's centripetal acceleration?

SOLVE From Equation 7.1, the electron's speed is

$$v = \frac{2\pi r}{T} = \frac{2\pi(5.29 \times 10^{-11} \text{ m})}{1.52 \times 10^{-16} \text{ s}} = 2.19 \times 10^6 \text{ m/s}$$

Then from Equation 7.14,

$$a_r = \frac{v^2}{r} = \frac{(2.19 \times 10^6 \text{ m/s})^2}{5.29 \times 10^{-11} \text{ m}} = 9.07 \times 10^{22} \text{ m/s}^2$$

ASSESS This was not intended as a profound problem, merely to illustrate how a centripetal acceleration is computed. In addition, it demonstrates the unbelievably enormous accelerations that take place at the atomic level. It should then come as no surprise that atomic particles may behave in ways that our intuition, trained by accelerations of only a few m/s², cannot easily grasp.

STOP TO THINK 7.2   Rank in order, from largest to smallest, the centripetal accelerations $(a_r)_a$ to $(a_r)_e$ of particles a to e.

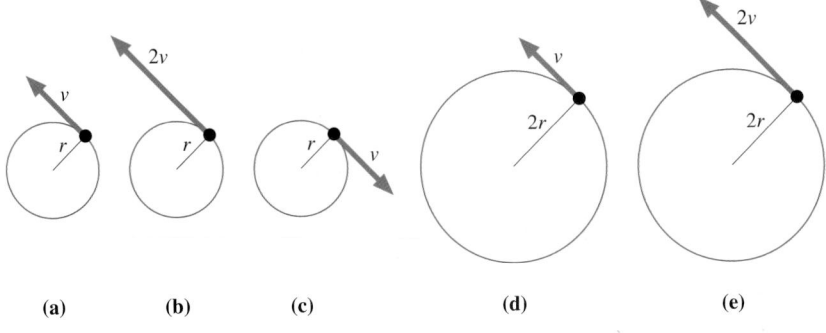

(a)   (b)   (c)   (d)   (e)

7.3 Dynamics of Uniform Circular Motion

A particle in uniform circular motion is clearly not traveling at constant velocity in a straight line. Consequently, according to Newton's first law, the particle *must* have a net force acting on it. We've already determined the acceleration of a particle in uniform circular motion—the centripetal acceleration of Equation 7.13. Newton's second law tells us exactly how much net force is needed to cause this acceleration:

$$\vec{F}_{\text{net}} = m\vec{a} = \left(\frac{mv^2}{r}, \text{ toward center of circle}\right) \tag{7.15}$$

In other words, a particle of mass m moving at constant speed v around a circle of radius r must have a net force of magnitude mv^2/r pointing toward the center of the circle. Without such a force, the particle would move off in a straight line tangent to the circle.

Figure 7.14 shows the net force $\vec{F}_{\text{net}}$ acting on a particle as it undergoes uniform circular motion. You can see that $\vec{F}_{\text{net}}$ **points along the radial axis of the rtz-coordinate system, toward the center of the circle**. The tangential and perpendicular components of $\vec{F}_{\text{net}}$ are zero.

NOTE ▶ The force described by Equation 7.15 is not a *new* force. Our rules for identifying forces have not changed. What we are saying is that a particle moves with uniform circular motion *if and only if* a net force always points toward the center of the circle. The force itself must have an identifiable agent and will be one of our familiar forces, such as tension, friction, or the normal force. Equation 7.15 simply tells us how the force needs to act—how strong and in which direction—to cause the particle to move with speed v in a circle of radius r. ◀

Highway and racetrack curves are banked to allow the normal force of the road to provide the centripetal acceleration of the turn.

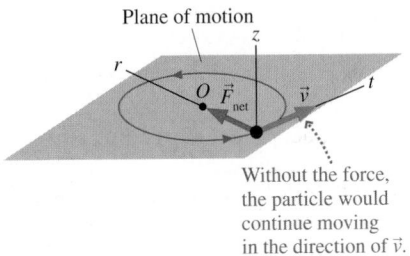

FIGURE 7.14 The net force points in the radial direction, toward the center of the circle.

The usefulness of the *rtz*-coordinate system becomes apparent when we write Newton's second law, Equation 7.15, in terms of the *r*-, *t*-, and *z*-components:

$$(F_{net})_r = \sum F_r = ma_r = \frac{mv^2}{r} = m\omega^2 r$$

$$(F_{net})_t = \sum F_t = ma_t = 0 \qquad (7.16)$$

$$(F_{net})_z = \sum F_z = ma_z = 0$$

Notice that we've used our explicit knowledge of the acceleration, as given in Equation 7.14, to write the right-hand side of these equations. **For uniform circular motion, the sum of the forces along the *t*-axis and along the *z*-axis *must* equal zero, and the sum of the forces along the *r*-axis *must* equal ma_r where a_r is the centripetal acceleration.**

It is time for some examples to clarify these ideas and to see how some of the forces you've come to know can be involved in circular motion.

EXAMPLE 7.5 Spinning in a circle

An energetic father places his 20 kg child on a 5.0 kg cart to which a 2.0-m-long rope is attached. He then holds the end of the rope and spins the cart and child around in a circle, keeping the rope parallel to the ground. If the tension in the rope is 100 N, how many revolutions per minute (rpm) does the cart make? Rolling friction between the cart's wheels and the ground is negligible.

MODEL Model the child in the cart as a particle in uniform circular motion.

VISUALIZE Figure 7.15 shows the pictorial and physical representations. The main idea of the pictorial representation is to illustrate the relevant geometry and to define the symbols that will be used. A circular-motion problem usually does not have starting and ending points like a projectile problem, so numerical subscripts such as x_1 or y_2 are usually not needed. Here we need to define the cart's speed v and the radius r of the circle.

A motion diagram is not needed for uniform circular motion because we know the acceleration $\vec{a}$ points to the center of the circle. Motion diagrams have been very helpful, but they should be second nature to you now. We'll continue to use motion diagrams when they help clarify a situation, but we won't include them in every solution. The essential part of the physical representation is the free-body diagram. **For uniform circular motion we'll draw the free-body diagram in the *rz*-plane, looking at the edge of the circle, because this is the plane of the forces.** The contact forces acting on the cart are the normal force of the ground and the tension force of the rope. The normal force is perpendicular to the plane of the motion and thus in the *z*-direction. The direction of $\vec{T}$ is determined by the statement that the rope is parallel to the ground. In addition, there is the long-range weight force $\vec{w}$.

SOLVE We defined the *r*-axis to point toward the center of the circle, so $\vec{T}$ points in the positive *r*-direction and has *r*-component $T_r = T$. Newton's second law, using the *rtz*-components of Equation 7.16, is

$$\sum F_r = T = \frac{mv^2}{r}$$

$$\sum F_z = n - w = 0$$

We've taken the *r*- and *z*-components of the forces directly from the free-body diagram, as you learned to do in Chapter 5. Then we've *explicitly* equated the sums to $a_r = v^2/r$ and $a_z = 0$. This is the basic strategy for all uniform circular-motion problems. From the *z*-equation we can find that $n = w$. This would be useful if we needed to determine a friction force, but it's not

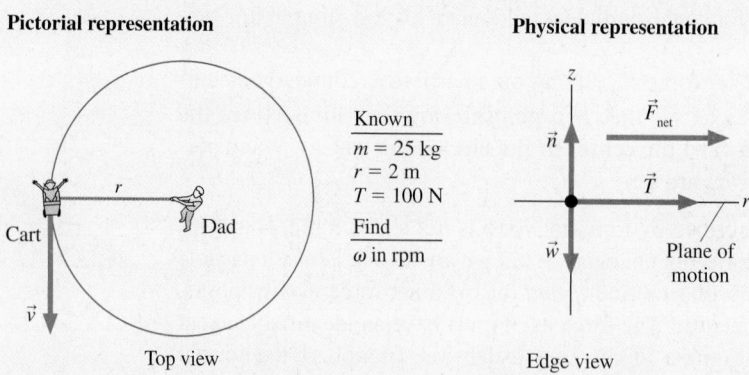

Pictorial representation

Cart

r

Dad

$\vec{v}$

Top view

Known
$m = 25$ kg
$r = 2$ m
$T = 100$ N

Find
ω in rpm

Physical representation

z

$\vec{n}$

$\vec{F}_{net}$

$\vec{T}$

r

$\vec{w}$

Plane of motion

Edge view

FIGURE 7.15 Pictorial and physical representations of a cart spinning in a circle.

needed in this problem. From the *r*-equation, the speed of the cart is

$$v = \sqrt{\frac{rT}{m}} = \sqrt{\frac{(2.0 \text{ m})(100 \text{ N})}{25 \text{ kg}}} = 2.83 \text{ m/s}$$

The cart's angular velocity is then found from Equation 7.8:

$$\omega = \frac{v_t}{r} = \frac{v}{r} = \frac{2.83 \text{ m/s}}{2.0 \text{ m}} = 1.41 \text{ rad/s}$$

This is another case where we inserted the radian unit because ω is specifically an *angular* velocity. Finally, we need to convert ω to rpm:

$$\omega = \frac{1.41 \text{ rad}}{1 \text{ s}} \times \frac{1 \text{ rev}}{2\pi \text{ rad}} \times \frac{60 \text{ s}}{1 \text{ min}} = 13.5 \text{ rpm}$$

ASSESS 13.5 rpm corresponds to a period $T \approx 4.5$ s. This result is reasonable.

This has been a fairly typical circular-motion problem. You might want to think about how the solution would change if the rope is *not* parallel to the ground.

EXAMPLE 7.6 **Turning the corner I**

What is the maximum speed with which a 1500 kg car can make a left turn around a curve of radius 50 m on a level (unbanked) road without sliding?

MODEL Although the car turns only a quarter of a circle, we can model the car as a particle in uniform circular motion as it goes around the turn. Assume that rolling friction is negligible.

VISUALIZE Figure 7.16 shows the pictorial and physical representations. The car moves along a circular arc at constant speed for the quarter-circle necessary to complete the turn. The motion before and after the turn is not relevant to the problem. The more interesting issue is *how* a car turns a corner. What force or forces can we identify that cause the direction of the velocity vector to change? Imagine you are driving a car on a completely frictionless road, such as a very icy road. You would not be able to turn a corner. Turning the steering wheel would be of no use; the car would slide straight ahead, in accordance with both Newton's first law and the experience of anyone who has ever driven on ice! So it must be *friction* that somehow allows the car to turn.

The force-identification section of Figure 7.16 shows the top view of a tire as it turns a corner. If the road surface were frictionless, the tire would slide straight ahead. The force that prevents an object from sliding across a surface is *static friction*.

Static friction $\vec{f_s}$ pushes *sideways* on the tire, toward the center of the circle. How do we know the direction is sideways? If $\vec{f_s}$ had a component either parallel to $\vec{v}$ or opposite to $\vec{v}$, it would cause the car to speed up or slow down. Because the car changes direction but not speed, static friction must be perpendicular to $\vec{v}$. Thus $\vec{f_s}$ causes the centripetal acceleration of circular motion around the curve. With this in mind, the free-body diagram, drawn from behind the car, shows the static friction force pointing toward the center of the circle.

SOLVE Because the static friction force has a maximum value, there will be a maximum speed with which a car can turn without sliding. The maximum speed is reached when the static friction force reaches its maximum $f_{s\,max} = \mu_s n$. If the car enters the curve at a speed higher than the maximum, static friction will not be large enough to provide the necessary centripetal acceleration and the car will slide.

The static friction force points in the positive *r*-direction, so its radial component is simply the magnitude of the vector: $(f_s)_r = f_s$. Newton's second law in the *rtz*-coordinate system is

$$\sum F_r = f_s = \frac{mv^2}{r}$$

$$\sum F_z = n - w = 0$$

Pictorial representation

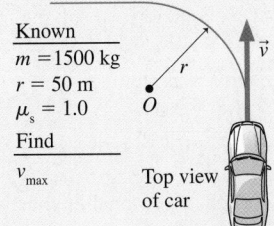

Known
$m = 1500$ kg
$r = 50$ m
$\mu_s = 1.0$

Find

v_{max}

Top view of car

Physical representation

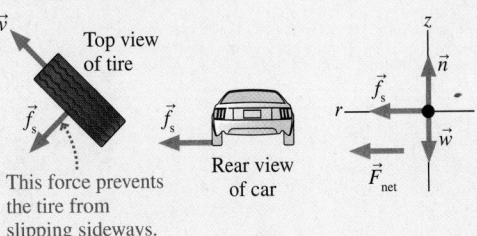

Top view of tire

This force prevents the tire from slipping sideways.

Rear view of car

FIGURE 7.16 Pictorial and physical representations of a car turning a corner.

The only difference from Example 7.5 is that the tension force toward the center has been replaced by a static friction force toward the center. From the radial equation, the speed is

$$v = \sqrt{\frac{rf_s}{m}}$$

The speed will be a maximum when f_s reaches its maximum value

$$f_s = f_{s\,max} = \mu_s n = \mu_s w = \mu_s mg$$

where we used $n = w$ from the z-equation. At that point,

$$v_{max} = \sqrt{\frac{rf_{s\,max}}{m}} = \sqrt{\mu_s rg}$$

$$= \sqrt{(1.00)(50\text{ m})(9.80\text{ m/s}^2)} = 22.1\text{ m/s}$$

where we found μ_s in Table 5.1.

ASSESS 22.1 m/s $\approx$ 45 mph, a reasonable answer for how fast a car can take an unbanked curve. Notice that the car's mass canceled out and that the final equation for v_{max} is quite simple. This is another example of why it pays to work algebraically until the very end.

4.5

Because μ_s depends on road conditions, the maximum safe speed through turns can vary dramatically. Wet roads, in particular, lower the value of μ_s and thus lower the speed of turns. A car that handles normally while driving straight ahead on a wet road can suddenly slide out of control when turning a corner. Icy conditions are even worse. The corner you turn every day at 45 mph will require a speed of no more than 15 mph if the coefficient of static friction drops to 0.1.

EXAMPLE 7.7 **Turning the corner II**

A highway curve of radius 70 m is banked at a 15° angle. At what speed v_0 can a car take this curve without assistance from friction?

MODEL The car is a particle in uniform circular motion.

VISUALIZE Having just discussed the role of friction in turning corners, it is perhaps surprising to suggest that the same turn can also be accomplished without friction. Example 7.6 considered a level roadway, but real highway curves are *banked* by being tilted up at the outside edge of the curve. The angle is modest on ordinary highways, but it can be quite large on high-speed racetracks. The purpose of banking becomes clear if you look at the free-body diagram in Figure 7.17. The normal force $\vec{n}$ is perpendicular to the road, so tilting the road causes $\vec{n}$ to have a component toward the center of the circle. The radial

component n_r is the inward force that causes the centripetal acceleration needed to turn the car. Notice that we are *not* using a tilted coordinate system, although this looks rather like an inclined-plane problem. The center of the circle is in the same horizontal plane as the car, and for circular motion problems we need the r-axis to pass through the center. Tilted axes are for *linear* motion along an incline.

SOLVE Without friction, $n_r = n\sin\theta$ is the only component of force in the radial direction. It is this inward component of the normal force on the car that causes it to turn the corner. Newton's second law is

$$\sum F_r = n\sin\theta = \frac{mv_0^2}{r}$$

$$\sum F_z = n\cos\theta - w = 0$$

where θ is the angle at which the road is banked and we've assumed that the car is traveling at the correct speed v_0. From the z-equation,

$$n = \frac{w}{\cos\theta} = \frac{mg}{\cos\theta}$$

Substituting this into the r-equation and solving for v_0 gives

$$\frac{mg}{\cos\theta}\sin\theta = mg\tan\theta = \frac{mv_0^2}{r}$$

$$v_0 = \sqrt{rg\tan\theta} = 13.6\text{ m/s}$$

ASSESS This is $\approx$ 27 mph, a reasonable speed. Only at this very specific speed can the turn be negotiated without reliance on friction forces.

Pictorial representation

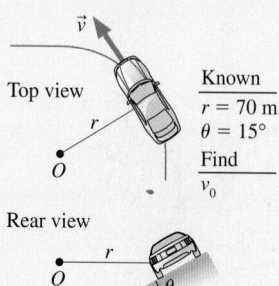

Top view

Known
$r = 70$ m
$\theta = 15°$

Find
v_0

Rear view

Physical representation

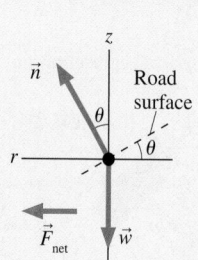

Road surface

FIGURE 7.17 Pictorial and physical representations of a car on a banked curve.

It's interesting to explore what happens at other speeds. The car will need to rely on both the banking *and* friction if it takes the curve at a speed higher or lower than v_0. Figure 7.18a has modified the free-body diagram to include a static friction force. Remember that $\vec{f}_s$ must be parallel to the surface, so it is tilted downward at angle θ. Because $\vec{f}_s$ has a component in the positive r-direction, the *net* radial force is larger than that provided by $\vec{n}$ alone. This will allow the car to take the curve at $v > v_0$. We could use a quantitative analysis similar to Example 7.6 to determine the maximum speed on a banked curve by analyzing Figure 7.18a when $f_s = f_{s\,max}$.

But what about taking the curve at a speed $v < v_0$? In this situation, the r-component of the normal force is too big; not that much center-directed force is needed. As Figure 7.18b shows, the net force can be reduced by having $\vec{f}_s$ point *up* the slope! This seems very strange at first, but consider the limiting case in which the car is parked on the banked curve, with $v = 0$. Were it not for a static friction force pointing *up* the slope, the car would slide sideways down the incline. In fact, for any speed less than v_0 the car will slip to the inside of the curve unless it is prevented from doing so by a static friction force pointing up the slope.

Our analysis thus finds three divisions of speed. At v_0, the car turns the corner with no assistance from friction. At greater speeds, the car will slide out of the curve unless an inward-directed friction force increases the size of the net force. And lastly, at lesser speeds the car will slip down the incline unless an outward-directed friction force prevents it from doing so.

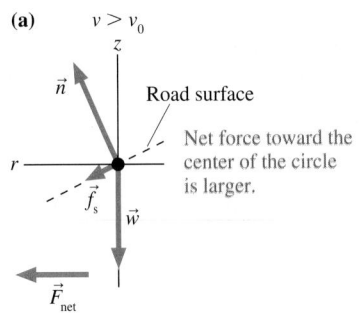

(a) $v > v_0$

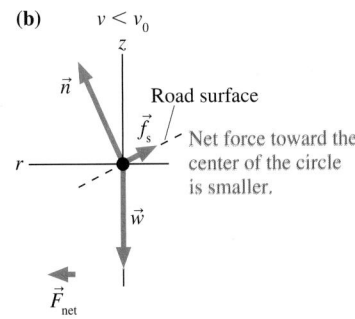

(b) $v < v_0$

FIGURE 7.18 Free-body diagrams showing the static friction force when $v > v_0$ and when $v < v_0$.

EXAMPLE 7.8 **A rock in a sling**

A Stone Age hunter places a 1.0 kg rock in a sling and swings it in a horizontal circle around his head on a 1.0-m-long vine. If the vine breaks at a tension of 200 N, what is the maximum angular velocity, in rpm, with which he can swing the rock?

MODEL Model the rock as a particle in uniform circular motion.

VISUALIZE This problem appears, at first, to be essentially the same as Example 7.5, where the father spun his child around on a rope. However, the lack of a normal force from a supporting surface makes a *big* difference. In this case, the *only* contact force on the rock is the tension in the vine. Because the rock moves in a horizontal circle, you may be tempted to draw a free-body diagram like Figure 7.19a where $\vec{T}$ is directed along the r-axis. You will quickly run into trouble, however, because this diagram has a net force in the z-direction and it is impossible to satisfy $\sum F_z = 0$. The weight $\vec{w}$ certainly points vertically downward, so the difficulty must be with $\vec{T}$.

As an experiment, tie a small weight to a string, swing it over your head, and check the *angle* of the string. You will quickly discover that the string is *not* horizontal but, instead, is angled downward. The pictorial model of Figure 7.19b labels the angle θ. Notice that the rock moves in a *horizontal* circle, so the center of the circle is *not* at his hand. The r-axis point to the center of the circle, but the tension force is directed along the vine. Thus the correct free-body diagram is the one in Figure 7.19b.

SOLVE The free-body diagram shows that the downward weight force is balanced by an upward component of the tension, leaving the radial component of the tension to cause the centripetal acceleration. Newton's second law is

$$\sum F_r = T\cos\theta = \frac{mv^2}{r}$$

$$\sum F_z = T\sin\theta - w = T\sin\theta - mg = 0$$

(a)

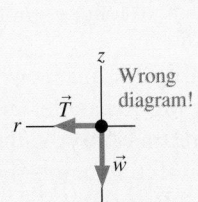

Wrong diagram!

(b)

Pictorial representation

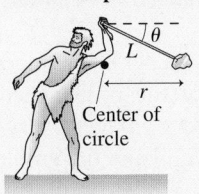

Known
$m = 1.0$ kg
$L = 1.0$ m
$T_{max} = 200$ N
Find
ω_{max}

Physical representation

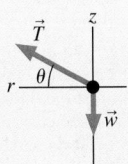

FIGURE 7.19 Pictorial and physical representations of a rock in a sling.

where θ is the angle of the vine below horizontal. From the z-equation we find

$$\sin\theta = \frac{mg}{T}$$

$$\theta = \sin^{-1}\left(\frac{(1.0\ \text{kg})(9.8\ \text{m/s}^2)}{200\ \text{N}}\right) = 2.81°$$

where we've evaluated the angle at the maximum tension of 200 N. The vine's angle of inclination is small but not zero. Turning now to the r-equation, we find the rock's speed is

$$v = \sqrt{\frac{rT\cos\theta}{m}}$$

Careful! The radius r of the circle is *not* the length L of the vine. You can see in Figure 7.19b that $r = L\cos\theta$. Thus

$$v = \sqrt{\frac{LT\cos^2\theta}{m}} = \sqrt{\frac{(1.0\ \text{m})(200\ \text{N})(\cos 2.81°)^2}{1.0\ \text{kg}}} = 14.1\ \text{m/s}$$

We can now find the maximum angular velocity, the value of ω that brings the tension to the breaking point, to be

$$\omega_{max} = \frac{v}{r} = \frac{v}{L\cos\theta} = \frac{14.1\ \text{rad}}{1\ \text{s}} \times \frac{1\ \text{rev}}{2\pi\ \text{rad}} \times \frac{60\ \text{s}}{1\ \text{min}} = 135\ \text{rpm}$$

STOP TO THINK 7.3   A block on a string spins in a horizontal circle on a frictionless table. Rank order, from largest to smallest, the tensions T_a to T_e acting on blocks a to e.

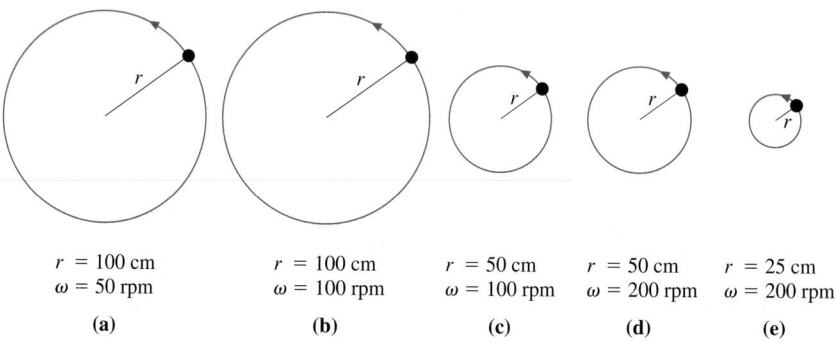

$r = 100$ cm	$r = 100$ cm	$r = 50$ cm	$r = 50$ cm	$r = 25$ cm
$\omega = 50$ rpm	$\omega = 100$ rpm	$\omega = 100$ rpm	$\omega = 200$ rpm	$\omega = 200$ rpm
(a)	**(b)**	**(c)**	**(d)**	**(e)**

7.4 Circular Orbits

Satellites orbit the earth, the earth orbits the sun, and our entire solar system orbits the center of the Milky Way galaxy. Not all orbits are circular, but in this section we'll limit our analysis to circular orbits. We'll look at the elliptical orbits of satellites and planets in Chapter 12.

How does a satellite orbit the earth? What forces act on it? Why does it move in a circle? To answer these important questions, let's return, for a moment, to projectile motion. Projectile motion occurs when the only force on an object is gravity. Our analysis of projectiles made an implicit assumption that the earth is flat and that the acceleration due to gravity is everywhere straight down. This is an acceptable approximation for projectiles of limited range, such as baseballs or cannon balls, but there comes a point where we can no longer ignore the curvature of the earth.

Figure 7.20 shows a perfectly smooth, spherical, airless planet with one tower of height h. A projectile is launched from this tower parallel to the ground ($\theta = 0°$) with speed v_0. If v_0 is very small, as in trajectory A, the "flat-earth approximation" is valid and the problem is identical to Example 6.4 in which a car drove off a cliff. The projectile simply falls to the ground along a parabolic trajectory.

As the initial speed v_0 is increased, the projectile begins to notice that the ground is curving out from beneath it. It is falling the entire time, always getting

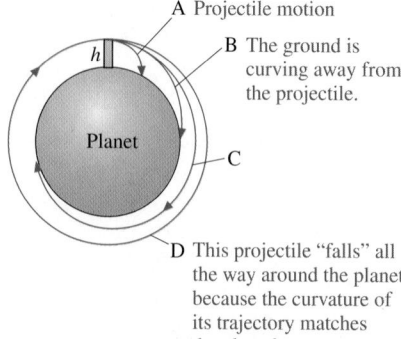

A Projectile motion

B The ground is curving away from the projectile.

C

D This projectile "falls" all the way around the planet because the curvature of its trajectory matches the planet's.

FIGURE 7.20 Projectiles being launched at increasing speeds from height h on a smooth, airless planet.

closer to the ground, but the distance that the projectile travels before finally reaching the ground—that is, its range—increases because the projectile must "catch up" with the ground that is curving away from it. Trajectories B and C are of this type. The actual calculation of these trajectories is beyond the scope of this textbook, but you should be able to understand the factors that influence the trajectory.

If the launch speed v_0 is sufficiently large, there comes a point where the curve of the trajectory and the curve of the earth are parallel. In this case, the projectile "falls" but it never gets any closer to the ground! This is the situation for trajectory D. A closed trajectory around a planet or star, such as trajectory D, is called an **orbit.**

The most important point of this qualitative analysis is that **an orbiting projectile is in free fall.** This is, admittedly, a strange idea, but one worth careful thought. An orbiting projectile is really no different from a thrown baseball or a car driving off a cliff. The only force acting on it is gravity, but its tangential velocity is so large that the curvature of its trajectory matches the curvature of the earth. When this happens, the projectile "falls" under the influence of gravity but never gets any closer to the surface, which curves away beneath it.

In the flat-earth approximation, shown in Figure 7.21a, the weight force acting on an object of mass m is

$$\vec{w} = (mg, \text{vertically downward}) \qquad \text{(flat-earth approximation)} \qquad (7.17)$$

But since stars and planets are actually spherical (or very close to it), the "real" force of gravity acting on an object is directed toward the *center* of the planet, as shown in Figure 7.21b. In this case the weight is

$$\vec{w} = (mg, \text{toward center}) \qquad \text{(spherical planet)} \qquad (7.18)$$

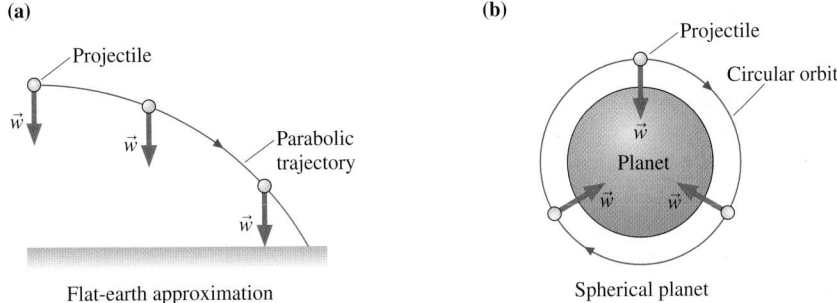

(a)

Projectile

$\vec{w}$

$\vec{w}$

Parabolic trajectory

$\vec{w}$

Flat-earth approximation

(b)

Projectile

Circular orbit

$\vec{w}$

Planet

$\vec{w}$

$\vec{w}$

Spherical planet

FIGURE 7.21 The "real" weight force is always directed toward the center of the planet.

As you have learned, a force of constant magnitude that always points toward the center of a circle causes the centripetal acceleration of uniform circular motion. Thus the weight force of Equation 7.18 on the object in Figure 7.21b causes it to have acceleration

$$\vec{a} = \frac{\vec{F}_{net}}{m} = (g, \text{toward center}) \qquad (7.19)$$

An object moving in a circle of radius r at speed v_{orbit} will have this centripetal acceleration if

$$a_r = \frac{(v_{orbit})^2}{r} = g \qquad (7.20)$$

That is, if an object moves parallel to the surface with the speed

$$v_{orbit} = \sqrt{rg} \qquad (7.21)$$

then the acceleration due to gravity provides exactly the centripetal acceleration needed for a circular orbit of radius r. An object with any other speed will not follow a circular orbit.

The earth's radius is $r = R_e = 6.37 \times 10^6$ m. (A table of useful astronomical data is inside the cover of this book.) The orbital speed of a projectile just skimming the surface of an airless, bald earth is

$$v_{orbit} = \sqrt{rg} = \sqrt{(6.37 \times 10^6 \text{ m})(9.80 \text{ m/s}^2)} = 7900 \text{ m/s} \approx 16,000 \text{ mph}$$

Even if there were no trees and mountains, a real projectile moving at this speed would burn up from the friction of air resistance.

Suppose, however, that we launched the projectile from a tower of height $h = 200$ mi $\approx 3.2 \times 10^5$ m, just above the earth's atmosphere. This is approximately the height of low-earth-orbit satellites, such as the Space Shuttle. Note that $h \ll R_e$, so the radius of the orbit $r = R_e + h = 6.69 \times 10^6$ m is only 5% greater than the earth's radius. Many people have a mental image that satellites orbit far above the earth, but in fact many satellites come pretty close to skimming the surface. Our calculation of v_{orbit} thus turns out to be quite a good estimate of the speed of a satellite in low earth orbit. (You will see later that the value of g differs slightly at a satellite's altitude, but not by much. Our calculation remains a good estimate.)

We can use v_{orbit} to calculate the period of a satellite orbit:

$$T = \frac{2\pi r}{v_{orbit}} = 2\pi \sqrt{\frac{r}{g}} \tag{7.22}$$

For a low earth orbit, with $r = R_e + 200$ miles, we find $T = 5192$ s $= 87$ min. The period of the space shuttle at an altitude of 200 mi is, indeed, just about 87 minutes.

When we discussed *weightlessness* in Chapter 5, we discovered that it occurs during free fall. We asked the question, at the end of Section 5.3, whether astronauts and their spacecraft were in free fall. We can now give an affirmative answer: They are, indeed, in free fall. They are falling continuously around the earth, under the influence of only the gravitational force, but never getting any closer to the ground because the earth's surface curves beneath them. Weightlessness in space is no different from the weightlessness in a free-falling elevator. It does *not* occur from an absence of weight or an absence of gravity. Instead, the astronaut, the spacecraft, and everything in it are "weightless" because they are all falling together.

Gravity

We can leave this section with a glance ahead, where we will look at the gravitational force more closely. If a satellite is simply "falling" around the earth, with the gravitational force causing a centripetal acceleration, then what about the moon? Is it obeying the same laws of physics? Or do celestial objects obey laws that we cannot discover by experiments here on earth?

The radius of the moon's orbit around the earth is $r = R_m = 3.84 \times 10^8$ m. If we use Equation 7.22 to calculate the period of the moon's orbit, the time it takes the moon to circle the earth once, we get

$$T = 2\pi \sqrt{\frac{r}{g}} = 2\pi \sqrt{\frac{3.84 \times 10^8 \text{ m}}{9.80 \text{ m/s}^2}} = 655 \text{ min} \approx 11 \text{ hours}$$

This is clearly wrong. As you probably know, the full moon occurs roughly once a month. More exactly, we know from astronomical measurements that the period of the moon's orbit is $T = 27.3$ days $= 2.36 \times 10^6$ s, a factor of 60 longer than we calculated it to be.

Saturn's beautiful rings consist of dust particles and small rocks orbiting the planet.

Newton believed that the laws of motion he had discovered were *universal.* That is, they should apply to the motion of the moon as well as to the motion of objects in the laboratory. But why should we assume that the acceleration due to gravity g is the same at the distance of the moon as it is on or near the earth's surface? If gravity is the force of the earth pulling on an object, it seems plausible that the size of that force, and thus the size of g, should diminish with increasing distance from the earth.

If the moon orbits the earth because of the earth's gravitational pull, what value of g would be needed to explain the moon's period? We can calculate $g_{\text{at moon}}$ from Equation 7.22 and the observed value of the moon's period:

$$g_{\text{at moon}} = \frac{4\pi^2 R_{\text{m}}}{T_{\text{moon}}^2} = 0.00272 \text{ m/s}^2$$

This is much less than the earth-bound value of 9.8 m/s^2.

Newton proposed the idea that the earth's force of gravity decreases inversely with the square of the distance from the earth. This is the basis of *Newton's law of gravity*, a topic you will study in Chapter 12. There we will be able to use the mass of the earth and the distance to the moon to *predict* that $g_{\text{at moon}} = 0.00272$ m/s^2, exactly as expected. The moon, just like the space shuttle, is simply "falling" around the earth!

7.5 Fictitious Forces and Apparent Weight

If you are riding in a car that makes a sudden stop, you may feel as if a force "throws" you forward toward the windshield. But there really is no such force. You can not identify any agent that does the throwing. An observer watching from beside the road would simply see you continuing forward as the car stops.

The decelerating car is not an inertial reference frame. You learned in Chapter 4, and in more detail in Chapter 6, that Newton's laws are valid only in inertial reference frames. The roadside observer is in the earth's inertial reference frame. His observations of the car decelerating relative to the earth while you continue forward with constant velocity are in accord with Newton's laws.

Nonetheless, the fact that you *seem* to be hurled forward relative to the car is a very real experience. You can describe your experience in terms of what are called **fictitious forces**. These are not real forces, because no agent is exerting them, but they describe your motion *relative to a noninertial reference frame.* Figure 7.22 shows the situation from both reference frames.

Centrifugal Force?

If the car turns a corner quickly, you feel "thrown" against the door. But is there really such a force? Figure 7.23 shows a bird's-eye view of you riding in a car as it makes a left turn. You try to continue moving in a straight line, obeying Newton's first law, when—without having been provoked—the door suddenly turns in front of you and runs into you! You do, indeed, then feel the force of the door because it is now the normal force of the door, pointing *inward* toward the center of the curve, that is causing you to turn the corner. But you were not "thrown" into the door; the door ran into you. The bird's-eye view, from an inertial reference frame, gives the proper perspective of what happens.

The "force" that seems to push an object to the outside of a circle is called the *centrifugal force.* Despite having a name, the centrifugal force is a fictitious force. It describes your experience *relative to a noninertial reference frame*, but there really is no such force. You must always use Newton's laws in an inertial reference frame, such as the reference frame of the ground. **A centrifugal force will never appear on a free-body diagram and never be included in Newton's laws.**

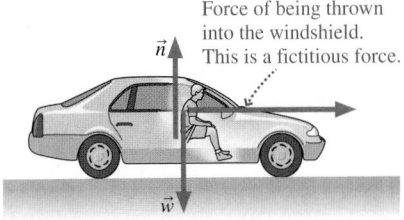

Noninertial reference frame of passenger

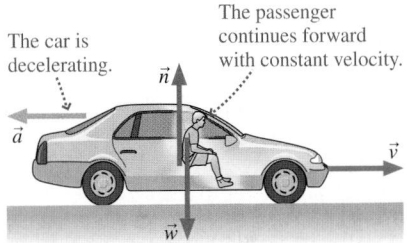

Inertial reference frame of the ground

FIGURE 7.22 The forces are properly identified only in an inertial reference frame.

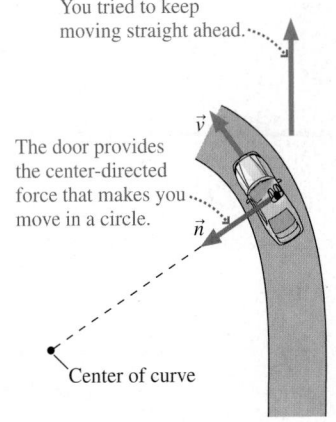

FIGURE 7.23 Bird's-eye view of a passenger as a car turns a corner.

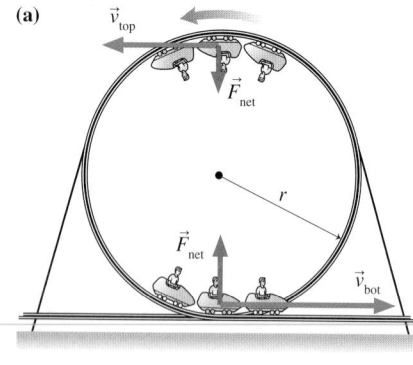

(a)

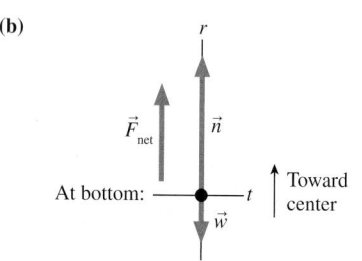

(b)

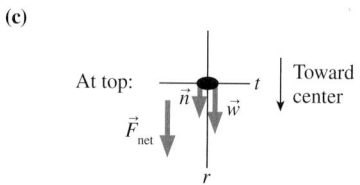

(c)

FIGURE 7.24 A roller coaster car going around a loop-the-loop.

Many popular amusement park rides are based on circular motion.

NOTE ▶ You might wonder if the *rtz*-coordinate system is an inertial reference frame. It is, and Newton's laws apply, although the reason is rather subtle. We're using the *rtz*-coordinates to establish directions for decomposing vectors, but we're not making measurements in the *rtz*-system. That is, velocities and accelerations are measured in the laboratory reference frame. The particle would always be at rest ($\vec{v} = \vec{0}$) if we measured velocities in a reference frame attached to the particle. Thus the analysis of this chapter really is in the laboratory's inertial reference frame. ◀

Why Does the Water Stay in the Bucket?

Imagine swinging a bucket of water over your head. If you swing the bucket quickly, the water stays in. But you'll get a shower if you swing too slowly. Why does the water stay in the bucket? You might have thought there was a centrifugal force holding the water in, but we see that there isn't a centrifugal force. Analyzing this question will tell us a lot about forces in general and circular motion in particular. We'll begin by looking at a similar situation—a roller coaster—then return to the water in the bucket.

Figure 7.24a shows a roller coaster car going around a vertical loop-the-loop of radius *r*. We'll assume that the motion makes a complete circle and not worry about the entrance to and exit from the loop. Now, motion in a vertical circle is *not* uniform circular motion. The car slows down as it goes up one side and speeds up as it comes back down the other, so there is a component of the acceleration $\vec{a}$ that is *tangent* to the circle. But when the car is at the very top and very bottom points, it is changing only direction, not speed, so at those points the acceleration is purely centripetal and the circular-motion version of Newton's second law from Section 7.3 applies.

If you've ever ridden a roller coaster, you know that your sensation of weight changes as you go over the crests and through the dips. To understand why, Figure 7.24b shows a passenger's free-body diagram at the *bottom* of the loop. The only forces acting on the passenger are her weight $\vec{w}$ and the normal force $\vec{n}$ of the seat pushing up on her. Recall, from Chapter 5, that the passenger's apparent weight, her sensation of weight, is the magnitude of the force supporting her. Here the seat is supporting her with the normal force $\vec{n}$, so her apparent weight is $w_{app} = n$.

From this information we can deduce the following:

- She's moving in a circle, so there *must* be a net force toward the center of the circle—above her head—to provide the centripetal acceleration.
- The net force points *upward*, so it must be the case that $n > w$.
- Her apparent weight is $w_{app} = n$, so her apparent weight is larger than her true weight ($w_{app} > w$). Thus she "feels heavy" at the bottom of the circle.

In short, the normal force has to *exceed* the weight force to provide the net force she needs to "turn the corner" at the bottom of the circle. The logic of this analysis is especially important.

To analyze the situation quantitatively, notice that the *r*-axis, which must point toward the center of the circle, points *upward*. Thus the *r*-component of Newton's second law for circular motion is

$$\sum F_r = n_r + w_r = n - w = ma_r = \frac{m(v_{bot})^2}{r} \qquad (7.23)$$

From Equation 7.23 we find

$$w_{app} = n = w + \frac{m(v_{bot})^2}{r} \qquad (7.24)$$

The passenger's apparent weight at the bottom is *larger* than her true weight w, which agrees with your experience when you go through a dip or a valley.

Now let's look at the roller coaster car as it crosses the top of the loop. This looks more like the water in the bucket, but things are a little trickier here. Whereas the normal force of the track pushes up when the car is at the bottom of the circle, it *presses down* when the car is at the top and the track is above the car. Figure 7.24c shows the car's free-body diagram at the top of the loop. Think about this diagram carefully to make sure you agree.

The car is still moving in a circle, so there *must* be a net force toward the center of the circle to provide the centripetal acceleration. The *r*-axis, which must point toward the center of the circle, now points *downward*. Consequently, both forces have *positive* components. Newton's second law at the top of the circle is

$$\sum F_r = n_r + w_r = n + w = \frac{m(v_{top})^2}{r} \tag{7.25}$$

Be sure you understand why this equation differs from Equation 7.23, describing what happens at the bottom.

From Equation 7.25, the normal force that the track exerts on the car is

$$n = \frac{m(v_{top})^2}{r} - w \tag{7.26}$$

If v_{top} is sufficiently large, the apparent weight $w_{app} = n$ of the car (and of the passengers in the car) can be larger than the true weight.

Our interest, however, is in what happens as the car gets slower and slower. Notice from Equation 7.26 that, as v_{top} decreases, there comes a point when n reaches zero. At that point, the track is *not* pushing against the car. Instead, the car is able to complete the circle because the weight force alone provides sufficient centripetal acceleration.

The speed at which $n = 0$ is called the *critical speed* v_c:

$$v_c = \sqrt{\frac{rw}{m}} = \sqrt{\frac{rmg}{m}} = \sqrt{rg} \tag{7.27}$$

The critical speed is the slowest speed at which the car can complete the circle. To understand why, notice that Equation 7.26 gives a negative value for n if $v < v_c$. But that is physically impossible. The track can push against the wheels of the car ($n > 0$), but it can't pull on them. When a solution becomes physically impossible, it usually indicates that we've made an incorrect assumption about the situation. In this case, we *assumed* that the motion was circular. But if we find that $n < 0$, our assumption is no longer valid. If $v < v_c$, the car cannot turn the full loop but, instead, comes off the track and becomes a projectile!

If you look back at the free-body diagram, **the critical speed v_c is the speed at which the weight force alone is sufficient to cause circular motion at the top.** The normal force has shrunk to zero. Circular motion with a speed less than v_c isn't possible because there's *too much* downward force. If the car attempts to go around at a lower speed, the normal force drops to zero *before* the car reaches the top. "No normal force" means "no contact." The car leaves the track when n reaches zero, becoming a projectile moving under the influence of only the weight force. Figure 7.25 summarizes this reasoning for the car on the loop-the-loop.

Returning now to the water in the bucket, Figure 7.26 on the next page shows a water-filled bucket at the top of a circle of radius r. Notice that the water has a tangential velocity. If the bucket suddenly disappeared, the water wouldn't fall straight down. Instead, it would fall along a parabolic trajectory like a ball that is thrown horizontally. This is the motion of an object acted on only by gravity.

If the water is to follow a circular trajectory with *more curvature* than the parabola, it needs *more force* than just its weight. The extra force is provided by

The normal force adds to the weight to make a large enough force for the car to turn the circle.

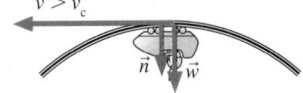

At v_c, the weight force alone is enough force for the car to turn the circle. $\vec{n} = \vec{0}$ at the top point.

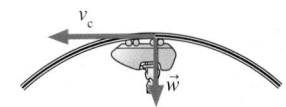

The weight force is too large for the car to stay in the circle!   Normal force became zero here.

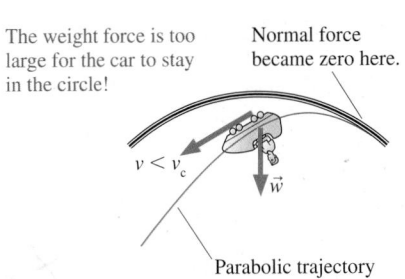

Parabolic trajectory

FIGURE 7.25 The motion of a roller coaster car for speeds above and below the critical speed.

The normal force of the bucket pushing on the water keeps the water in the bucket by forcing it to move in a circle.

If there were no bucket, the water would follow this parabolic trajectory.

$\omega > \omega_c$

$\vec{v}$

$\vec{n}$ $\vec{w}$

O r

At $\omega = \omega_c$, the weight force alone is enough force to keep the water moving around the circle. $\vec{n} = \vec{0}$ at the top point.

$\omega = \omega_c$

$\vec{v}$

$\vec{w}$

If $\omega < \omega_c$, the weight force is too large. It pulls the water out of the circle and into a tighter parabolic trajectory.

$\omega < \omega_c$

$\vec{v}$

$\vec{w}$

Parabolic trajectory

Normal force became zero here.

FIGURE 7.26 The forces on the water in a bucket.

the bottom of the bucket pushing on the water. As long as the bucket is pushing on the water, the water and the bucket are in contact and thus the water is "in" the bucket. (Water is a deformable substance, so the sides of the bucket are needed to keep the water together but otherwise aren't relevant to the motion.)

Notice the similarity to the car making the left turn in Figure 7.23. The passenger feels like he's being "hurled" into the door by a centrifugal force, but it's actually the pushing force from the door, pushing inward toward the center of the circle, that causes the passenger to turn the corner instead of moving straight ahead. Here it seems like the water is being "pinned" against the bottom of the bucket by a centrifugal force, but it's really the pushing force from the bottom of the bucket that causes the water to move in a circle instead of following a free-fall parabola.

As you gradually slow the speed of the bucket, the normal force of the bucket on the water gets smaller and smaller. There comes a point, as the angular velocity ω decreases, when n reaches zero. At that point, the bucket is *not* pushing against the water. Instead, the water is able to complete the circle because the weight force alone provides sufficient centripetal acceleration.

The critical angular velocity ω_c is the angular velocity at which the weight force alone is sufficient to cause circular motion at the top. Circular motion with an angular velocity less than ω_c isn't possible because there's *too much* downward force. If you attempt to swing the bucket with a smaller angular velocity, the normal force drops to zero *before* the water reaches the top. "No normal force" means "no contact." The water leaves the bucket when n becomes zero, becoming a projectile moving under the influence of only the weight force. That's when you get wet!

The analysis is exactly the same as for the roller coaster car. The only difference is that you're likely to swing a bucket with constant angular velocity (the roller coaster car does *not* have constant angular velocity), so it's more useful to calculate the critical angular velocity rather than the critical speed. We can find the critical angular velocity from Equation 7.27 by using $v_c = \omega_c r$. This gives

$$\omega_c = \sqrt{\frac{g}{r}} \qquad (7.28)$$

It's not easy to understand why the water stays in the bucket. Careful thought about the *reasoning* presented in this section will greatly increase your understanding of forces and circular motion.

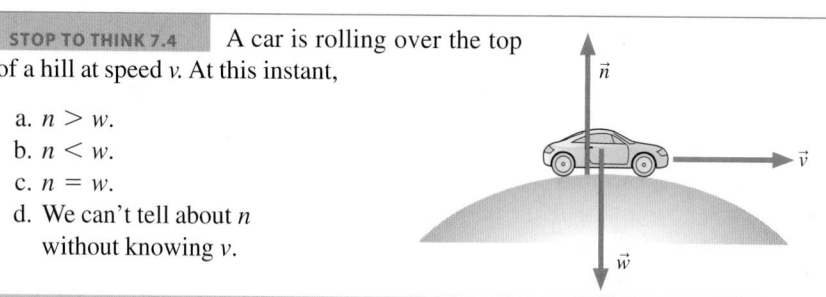

STOP TO THINK 7.4   A car is rolling over the top of a hill at speed v. At this instant,

a. $n > w$.
b. $n < w$.
c. $n = w$.
d. We can't tell about n without knowing v.

$\vec{n}$

$\vec{v}$

$\vec{w}$

7.6 Nonuniform Circular Motion

Many interesting examples of circular motion involve objects whose speed changes. A roller coaster car doing a loop-the-loop slows down as it goes up one side, speeds up as it comes back down the other. The same is true if a ball on a string is swung in a vertical circle. The ball in a roulette wheel gradually slows until it stops. Circular motion with a changing speed is called **nonuniform circular motion.**

Figure 7.27a reminds you of uniform circular motion. The nonuniform circular motion of Figure 7.27b still has $v_r = 0$, because the particle is constrained to follow the circle, but v_t is no longer constant. The tangential component of velocity v_t changes if the particle speeds up or slows down. A changing tangential velocity implies that the particle has a **tangential acceleration**

$$a_t = \frac{dv_t}{dt} \tag{7.29}$$

The tangential acceleration is what causes the particle to change the speed with which it goes *around* the circle.

The tangential acceleration that changes the particle's speed is in addition to the radial, or centripetal, acceleration that changes the particle's direction of motion. The full acceleration vector $\vec{a}$, shown in Figure 7.28, can be written in terms of its radial and tangential component vectors as

$$\vec{a} = \vec{a}_r + \vec{a}_t \tag{7.30}$$

The magnitude of the acceleration is

$$a = \sqrt{a_r^2 + a_t^2} \tag{7.31}$$

and its angle from the r-axis, which we'll call ϕ, is

$$\phi = \tan^{-1}\left(\frac{a_t}{a_r}\right) \tag{7.32}$$

The acceleration vector is a pure centripetal acceleration, pointing toward the center of the circle ($\phi = 0$), only if $a_t = 0$.

If a_t is constant, then the arc length s traveled by the particle around the circle and the tangential velocity v_t are found from constant-acceleration kinematics:

$$s_f = s_i + v_{it}\Delta t + \frac{1}{2}a_t(\Delta t)^2$$
$$v_{ft} = v_{it} + a_t\Delta t \tag{7.33}$$

We can express these kinematic equations in terms of angular quantities by dividing both sides of Equation 7.33 by the radius r:

$$\frac{s_f}{r} = \frac{s_i}{r} + \frac{v_{it}}{r}\Delta t + \frac{a_t}{2r}(\Delta t)^2$$
$$\frac{v_{ft}}{r} = \frac{v_{it}}{r} + \frac{a_t}{r}\Delta t \tag{7.34}$$

You'll recognize that s/r is the angular position θ and v_t/r is the angular velocity ω. Thus the angular position and velocity after undergoing tangential acceleration a_t are

$$\theta_f = \theta_i + \omega_i\Delta t + \frac{a_t}{2r}(\Delta t)^2$$
$$\text{(nonuniform circular motion)} \tag{7.35}$$
$$\omega_f = \omega_i + \frac{a_t}{r}(\Delta t)$$

Equations 7.35 are an extension of Equation 7.6 to the case of nonuniform circular motion. In addition, the centripetal acceleration equation $a_r = v^2/r = \omega^2 r$ is still valid.

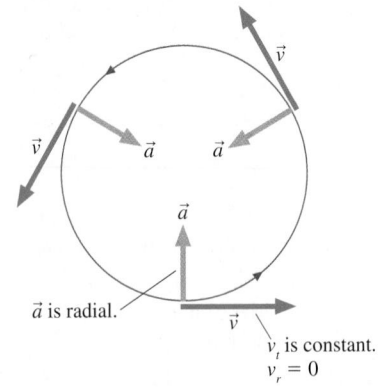

(a) Uniform circular motion

$\vec{a}$ is radial.

v_t is constant.
$v_r = 0$

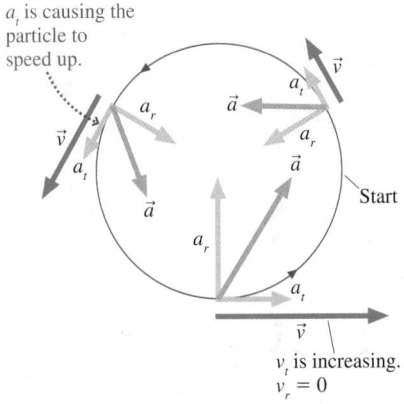

(b) Nonuniform circular motion

a_t is causing the particle to speed up.

Start

v_t is increasing.
$v_r = 0$

FIGURE 7.27 Uniform and nonuniform circular motion.

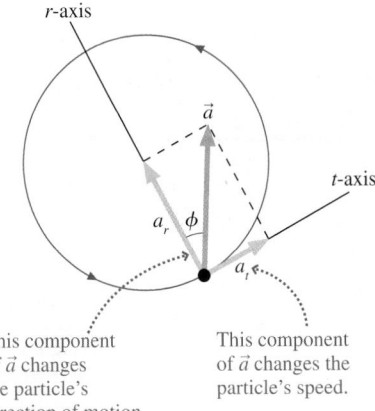

r-axis

t-axis

This component of $\vec{a}$ changes the particle's direction of motion.

This component of $\vec{a}$ changes the particle's speed.

FIGURE 7.28 The acceleration vector has a radial and tangential component.

EXAMPLE 7.9 Circular rocket motion

A model rocket is attached to the end of a 2.0-m-long massless, rigid rod. The other end of the rod rotates on a frictionless pivot, causing the rocket to move in a horizontal circle. The rocket accelerates at 1.0 m/s^2 for 10 s, then runs out of fuel.

a. What is the rocket's angular velocity, in rpm, when it runs out of fuel?
b. How many revolutions has the rocket made at that time?
c. What is the magnitude of $\vec{a}$ and its angle from the r-axis at $t = 2$ s?

MODEL Model the rocket as a particle in nonuniform circular motion. Assume that the rocket starts from rest.

VISUALIZE Figure 7.29 is a pictorial representation of the situation. The acceleration caused by the rocket motor is the tangential acceleration, $a_t = 1.0 \text{ m/s}^2$.

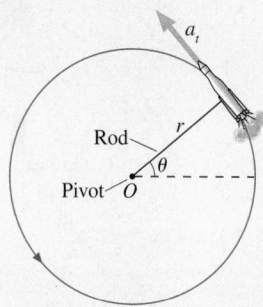

Known
$$\theta_i = \omega_i = t_i = 0$$
$$a_t = 1.0 \text{ m/s}^2$$
$$r = 2.0 \text{ m}$$
$$t_f = 10 \text{ s}$$

Find
$$\omega_f \text{ and } \theta_f$$
$$\vec{a} \text{ at } t = 2 \text{ s}$$

FIGURE 7.29 The nonuniform circular motion of a model rocket.

SOLVE

a. The angular velocity after 10 s is

$$\omega_f = \omega_i + \frac{a_t}{r}\Delta t = 0 + \frac{1.0 \text{ m/s}^2}{2.0 \text{ m}}10 \text{ s} = 5.0 \text{ rad/s}$$

This is another situation in which we explicitly inserted the rad unit. Converting to rpm:

$$\omega_f = \frac{5.0 \text{ rad}}{1 \text{ s}} \times \frac{1 \text{ rev}}{2\pi \text{ rad}} \times \frac{60 \text{ s}}{1 \text{ min}} = 48 \text{ rpm}$$

b. The angular position at 10 s is

$$\theta_f = \theta_i + \omega_i\Delta t + \frac{a_t}{2r}(\Delta t)^2$$

$$= 0 + 0 + \frac{1.0 \text{ m/s}^2}{2(2.0 \text{ m})}(10 \text{ s})^2 = 25 \text{ rad}$$

Converting to revolutions:

$$\theta_f = 25 \text{ rad} \times \frac{1 \text{ rev}}{2\pi \text{ rad}} = 3.98 \text{ rev} \approx 4 \text{ rev}$$

c. The rocket motor creates the tangential acceleration $a_t = 1.0 \text{ m/s}^2$. As the rocket speeds up, tension in the rod causes a radial acceleration $a_r = \omega^2 r$. At $t = 2$ s,

$$\omega_{2\text{ s}} = \omega_i + \frac{a_t}{r}\Delta t = 0 + \frac{1.0 \text{ m/s}^2}{2.0 \text{ m}}2.0 \text{ s} = 1.0 \text{ rad/s}$$

$$a_r = (\omega_{2\text{ s}})^2 r = (1.0 \text{ rad/s})^2(2.0 \text{ m}) = 2.0 \text{ m/s}^2$$

This is a situation where we explicitly *dropped* rad² from the units of the answer because the final result is a linear quantity, not an angular quantity. Now we can use Equations 7.31 and 7.32 to find

$$a = \sqrt{a_r^2 + a_t^2} = \sqrt{(2.0 \text{ m/s}^2)^2 + (1.0 \text{ m/s}^2)^2} = 2.24 \text{ m/s}^2$$

$$\phi = \tan^{-1}\left(\frac{a_t}{a_r}\right) = \tan^{-1}\left(\frac{1.0 \text{ m/s}^2}{2.00 \text{ m/s}^2}\right) = 27°$$

Dynamics of Nonuniform Circular Motion

Figure 7.30 shows a net force $\vec{F}_{net}$ acting on a particle as it moves around a circle of radius r. $\vec{F}_{net}$ is likely to be a superposition of several forces, such as a tension force in a string, a thrust force, a friction force, and so on.

We can decompose the force vector $\vec{F}_{net}$ into a *tangential* component $(F_{net})_t$ and a *radial* component $(F_{net})_r$. The component $(F_{net})_t$ is positive for a tangential force in the ccw direction, negative for a tangential force in the cw direction. Because of our definition of the r-axis, the component $(F_{net})_r$ is positive for a radial force *toward* the center, negative for a radial force away from the center. For example, the particular force illustrated in Figure 7.30 has positive values for both $(F_{net})_t$ and $(F_{net})_r$.

The force component $(F_{net})_r$ perpendicular to the trajectory creates a centripetal acceleration and causes the particle to change directions. It is the component $(F_{net})_t$ parallel to the trajectory that creates a tangential acceleration and

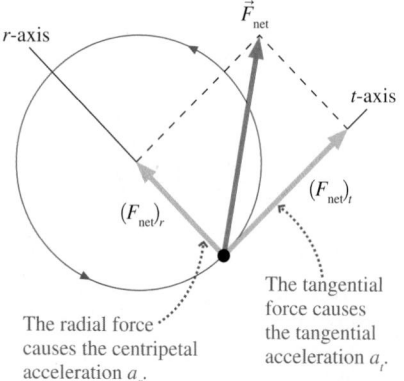

The radial force causes the centripetal acceleration a_r.

The tangential force causes the tangential acceleration a_t.

FIGURE 7.30 Net force $\vec{F}_{net}$ is applied to a particle moving in a circle.

causes the particle to change speed. Force and acceleration are related to each other through Newton's second law:

$$(F_{\text{net}})_r = \sum F_r = ma_r = \frac{mv^2}{r} = m\omega^2 r$$

$$(F_{\text{net}})_t = \sum F_t = ma_t \qquad\qquad (7.36)$$

$$(F_{\text{net}})_z = \sum F_z = 0$$

NOTE ▶ Equations 7.36 differ from Equations 7.16 for uniform circular motion only in the fact that a_t is no longer constrained to be zero. **◀**

EXAMPLE 7.10 Slowing circular motion
A motor spins a 2.0 kg steel block around on an 80-cm-long arm at 200 rpm. The block is supported by a steel table. After the motor stops, how long does the block take to come to rest? How many revolutions does the block make during this time? Assume that the axle is frictionless.

MODEL Model the steel block as a particle in nonuniform circular motion.

VISUALIZE Figure 7.31 shows the pictorial and physical representations. Notice that, for the first time, we need a free-body diagram showing forces in three dimensions.

SOLVE If the table were frictionless, the block would spin around forever because of the frictionless axle. However, friction between the block and table exerts a retarding force $\vec{f}_k$ on the block. Kinetic friction is always opposite the direction of motion $\vec{v}$, so $\vec{f}_k$ is *tangent* to the circle.

The magnitude of the friction force is $f_k = \mu_k n$. The vertical forces, perpendicular to the plane of the motion, are the normal force $\vec{n}$ and the weight $\vec{w}$. There's no net force in the vertical direction, so the z-component of the second law is

$$\sum F_z = n - w = 0$$

from which we can conclude that $n = w = mg$ and thus $f_k = \mu_k mg$. The friction force is the only tangential component of force, so the t-component of Newton's second law is

$$\sum F_t = (f_k)_t = -f_k = ma_t$$

$$a_t = \frac{-f_k}{m} = \frac{-\mu_k mg}{m} = -\mu_k g = -5.88 \text{ m/s}^2$$

The coefficient of friction for steel on steel was taken from Table 5.1. The component $(f_k)_t$ is negative because the friction force vector points in the clockwise direction. The initial angular velocity needs to be converted to rad/s:

$$\omega_i = \frac{200 \text{ rev}}{1 \text{ min}} \times \frac{2\pi \text{ rad}}{1 \text{ rev}} \times \frac{1 \text{ min}}{60 \text{ s}} = 20.9 \text{ rad/s}$$

We can now use Equation 7.35 for circular kinematics to find the time it takes the block to come to rest:

$$\omega_f = 0 \text{ rad/s} = \omega_i + \frac{a_t}{r}(\Delta t) = \omega_i + \frac{a_t}{r}t_f$$

$$t_f = -\frac{r\omega_i}{a_t} = -\frac{(0.80 \text{ m})(20.9 \text{ rad/s})}{-5.88 \text{ m/s}^2} = 2.84 \text{ s}$$

The angular displacement while the block slows to a stop is then

$$\Delta\theta = \theta_f - \theta_i = \omega_i t_f + \frac{a_t}{2r}t_f^2$$

$$= (20.9 \text{ rad/s})(2.84 \text{ s}) + \frac{(-5.88 \text{ m/s})}{2(0.80 \text{ m})}(2.84 \text{ s})^2$$

$$= 29.7 \text{ rad} \times \frac{1 \text{ rev}}{2\pi \text{ rad}} = 4.73 \text{ rev}$$

ASSESS Is this answer reasonable? The block was moving pretty fast—200 rpm on an arm about 30 in long. Even though friction of steel on steel is fairly large, it's reasonable that the block would make several revolutions before stopping. The purpose of the assessment, as always, is not to prove that the answer is right but to rule out obviously unreasonable answers that have been reached by mistake.

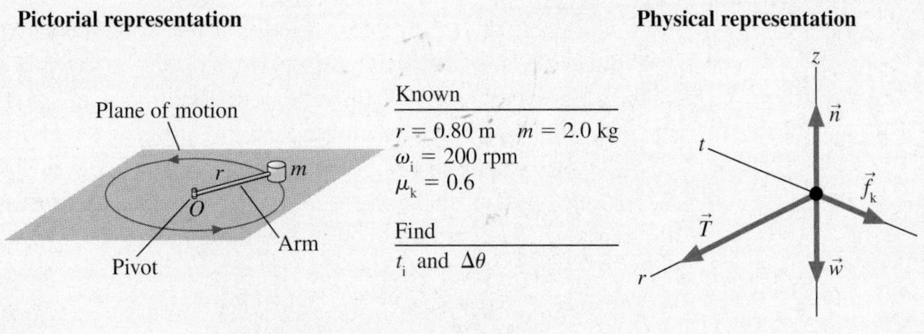

Pictorial representation

Physical representation

Known	
$r = 0.80$ m	$m = 2.0$ kg
$\omega_i = 200$ rpm	
$\mu_k = 0.6$	

Find
t_i and $\Delta\theta$

FIGURE 7.31 Pictorial and physical representations of the block of Example 7.10.

4.2, 4.3, 4.4   Activ ONLINE Physics

We've come a long way since our first dynamics problems in Chapter 5, but our basic strategy has not changed.

(MP) PROBLEM-SOLVING STRATEGY 7.1 **Circular motion problems**

MODEL Make simplifying assumptions.

VISUALIZE **Pictorial representation.** Establish a coordinate system with the r-axis pointing toward the center of the circle. Show important points in the motion on a sketch. Define symbols and identify what the problem is trying to find.

Physical representation. Identify the forces and show them on a free-body diagram.

SOLVE Newton's second law is

$$(F_{\text{net}})_r = \sum F_r = ma_r = \frac{mv^2}{r} = m\omega^2 r$$

$$(F_{\text{net}})_t = \sum F_t = ma_t$$

$$(F_{\text{net}})_z = \sum F_z = 0$$

- Determine the force components from the free-body diagram. Be careful with signs.
- Solve for the acceleration, then use kinematics to find velocities and positions.

ASSESS Check that your result has the correct units, is reasonable, and answers the question.

STOP TO THINK 7.5   A ball on a string is swung in a vertical circle. The string happens to break when it is parallel to the ground and the ball is moving up. Which trajectory does the ball follow?

String breaks.

SUMMARY

The goal of Chapter 7 has been to learn to solve problems about motion in a circle.

GENERAL PRINCIPLES

Newton's Second Law

Expressed in *rtz*-component form:

$$(F_{net})_r = \sum F_r = ma_r = \frac{mv^2}{r} = m\omega^2 r \qquad (F_{net})_t = \sum F_t = \begin{cases} 0 & \text{uniform motion} \\ ma_t & \text{nonuniform motion} \end{cases} \qquad (F_{net})_z = \sum F_z = 0$$

Uniform Circular Motion

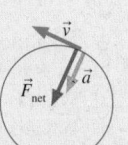

- v is constant.
- $\vec{F}_{net}$ points toward the center of the circle.
- The **centripetal acceleration** $\vec{a}$ points toward the center of the circle. It changes the particle's direction but not its speed.

Nonuniform Circular Motion

- v changes.
- $\vec{a}$ is parallel to $\vec{F}_{net}$.
- The radial component a_r changes the particle's direction.
- The tangential component a_t changes the particle's speed.

IMPORTANT CONCEPTS

rtz-coordinates

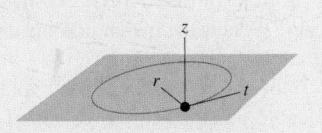

Angular position
$$\theta = s/r$$
Angular velocity
$$\omega = d\theta/dt$$
$$v_t = \omega r$$

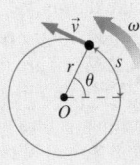

APPLICATIONS

Circular motion kinematics

Period $T = \dfrac{2\pi r}{v} = \dfrac{2\pi}{\omega}$

Uniform circular motion

$v_t = $ constant $\qquad \omega = $ constant

$\theta_f = \theta_i + \omega\Delta t$

Nonuniform circular motion

$\theta_f = \theta_i + \omega_i\Delta t + \dfrac{a_t}{2r}(\Delta t)^2$

$\omega_f = \omega_i + \dfrac{a_t}{r}\Delta t$

Orbits

A circular orbit has radius r if

$$v = \sqrt{rg}$$

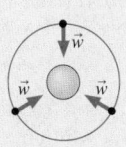

Apparent weight

Circular motion requires a net force pointing to the center. The apparent weight $w_{app} = n$ is usually not the same as the true weight w. n must be > 0 for the object to be in contact with a surface.

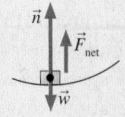

TERMS AND NOTATION

uniform circular motion	radians	orbit
period, T	angular displacement, $\Delta\theta$	fictitious force
angular position, θ	angular velocity, ω	nonuniform circular motion
arc length, s	centripetal acceleration, a_r	tangential acceleration, a_t

EXERCISES AND PROBLEMS

The  icon indicates that the problem can be done on a Dynamics Worksheet.

Exercises

Section 7.1 Uniform Circular Motion

1. Figure Ex7.1 shows the angular-position-versus-time graph for a particle moving in a circle.
   a. Write a description of the particle's motion.
   b. Draw the angular-velocity-versus-time graph.

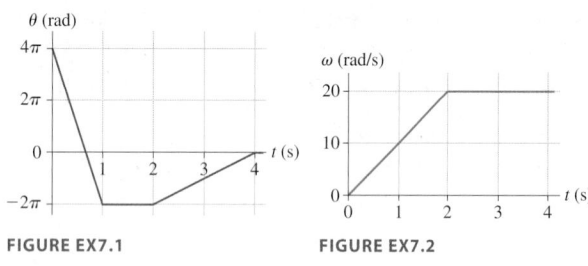

FIGURE EX7.1   **FIGURE EX7.2**

2. Figure Ex7.2 shows the angular-velocity-versus-time graph for a particle moving in a circle. How many revolutions does the object make during the first 4 s?
3. An old-fashioned single-play vinyl record rotates on a turntable at 45 rpm. What are (a) the angular velocity in rad/s and (b) the period of the motion?
4. The earth's radius is about 4000 miles. Kampala, the capital of Uganda, and Singapore are both nearly on the equator. The distance between them is 5000 miles.
   a. Through what angle do you turn, relative to the earth, if you fly from Kampala to Singapore? Give your answer in both radians and degrees.
   b. The flight from Kampala to Singapore takes 9 hours. What is the plane's angular velocity?

Section 7.2 Velocity and Acceleration in Uniform Circular Motion

5. The radius of the earth's very nearly circular orbit around the sun is 1.5×10^{11} m. Find the magnitude of the earth's (a) velocity, (b) angular velocity, and (c) centripetal acceleration as it travels around the sun. Assume a year of 365 days.
6. In uniform circular motion, which of the following quantities are constant: speed, instantaneous velocity, radial velocity, radial acceleration, tangential acceleration, the magnitude of the net force? Which of the quantities are zero throughout the motion?
7. Your roommate is working on his bicycle and has the bike upside down. He spins the 60-cm-diameter wheel, and you notice that a pebble stuck in the tread goes by three times every second. What are the pebble's speed and acceleration?
8. A 300-m-tall tower is built on the equator. How much faster does a point at the top of the tower move than a point at the bottom?

9. To withstand "g-forces" of up to 10 g's, caused by suddenly pulling out of a steep dive, fighter jet pilots train on a "human centrifuge." 10 g's is an acceleration of 98 m/s^2. If the length of the centrifuge arm is 12 m, at what speed is the rider moving when she experiences 10 g's?
10. How fast must a plane fly along the earth's equator so that the sun stands still relative to the passengers? In which direction must the plane fly, east to west or west to east? Give your answer in both km/hr and mph. The radius of the earth is 6400 km.

Section 7.3 Dynamics of Uniform Circular Motion

11. A 200 g block on a 50-cm-long string swings in a circle on a horizontal, frictionless table at 75 rpm.
    a. What is the speed of the block?
    b. What is the tension in the string?
12. In the Bohr model of the hydrogen atom, an electron (mass $m = 9.1 \times 10^{-31}$ kg) orbits a proton at a distance of 5.3×10^{-11} m. The proton pulls on the electron with an electric force of 9.2×10^{-8} N. How many revolutions per second does the electron make?
13. A highway curve of radius 500 m is designed for traffic moving at a speed of 90 km/hr. What is the correct banking angle of the road?
14. A 1500 kg car drives around a flat 200-m-diameter circular track at 25 m/s. What is the magnitude and direction of the net force on the car? What causes this force?
15. Suppose the moon were held in its orbit not by gravity but by a massless cable attached to the center of the earth. What would be the tension in the cable? Use the table of astronomical data inside the cover of the book.
16. A 30 g ball rolls around a 40-cm-diameter L-shaped track, shown in Figure Ex7.16, at 60 rpm. What is the magnitude of the net force that the track exerts on the ball? Rolling friction can be neglected.

FIGURE EX7.16

Section 7.4 Circular Orbits

17. A satellite orbiting the moon very near the surface has a period of 110 min. What is the moon's acceleration due to gravity?
18. What is the acceleration due to gravity of the sun at the distance of the earth's orbit?

Section 7.5 Fictitious Forces and Apparent Weight

19. The passengers in a roller coaster car feel 50% heavier than their true weight as the car goes through a dip with a 30 m radius of curvature. What is the car's speed at the bottom of the dip?
20. A roller coaster car crosses the top of a circular loop-the-loop at twice the critical speed. What is the ratio of the car's apparent weight to its true weight?

21. As a roller coaster car crosses the top of a 40-m-diameter loop-the-loop, its apparent weight is the same as its true weight. What is the car's speed at the top?

22. Measure the length of your arms. Then estimate the minimum angular velocity (in rpm) for swinging a bucket of water in a vertical circle without spilling any.

Section 7.6 Nonuniform Circular Motion

23. A car speeds up as it turns from traveling due south to heading due east. When exactly halfway around the curve, the car's acceleration is 3.0 m/s^2, 20° north of east. What are the radial and tangential components of the acceleration at that point?

24. A 5.0-m-diameter merry-go-round is initially turning with a 4.0 s period. It slows down and stops in 20 s.
    a. Before slowing, what is the speed of a child on the rim?
    b. How many revolutions does the merry-go-round make as it stops?

25. A 3.0-cm-diameter crankshaft that is rotating at 2500 rpm comes to a halt in 1.5 s.
    a. What is the tangential acceleration of a point on the surface of the crankshaft?
    b. How many revolutions does the crankshaft make as it stops?

26. A computer disk is 8.0 cm in diameter. A reference dot on the edge of the disk is initially located at $\theta = 45°$. The disk accelerates steadily for $\frac{1}{2}$ second, reaching 2000 rpm, then coasts at steady angular velocity for another $\frac{1}{2}$ second. What are the location and speed of the reference dot at $t = 1$ s?

Problems

27. A car starts from rest on a curve with a radius of 120 m and accelerates at 1.0 m/s^2. Through what angle will the car have traveled when the magnitude of its total acceleration is 2.0 m/s^2?

28. A typical laboratory centrifuge rotates at 4000 rpm. Test tubes have to be placed into a centrifuge very carefully because of the very large accelerations.
    a. What is the acceleration at the end of a test tube that is 10 cm from the axis of rotation?
    b. For comparison, what is the magnitude of the acceleration a test tube would experience if dropped from a height of 1.0 m and stopped in a 1.0-ms-long encounter with a hard floor?

29. Astronauts use a centrifuge to simulate the acceleration of a rocket launch. The centrifuge takes 30 s to speed up from rest to its top speed of 1 rotation every 1.3 s. The astronaut is strapped into a seat 6.0 m from the axis.
    a. What is the astronaut's tangential acceleration during the first 30 s?
    b. How many g's of acceleration does the astronaut experience when the device is rotating at top speed? Each 9.8 m/s^2 of acceleration is 1 g.

30. Communications satellites are placed in a circular orbit where they stay directly over a fixed point on the equator as the earth rotates. These are called *geosynchronous orbits*. The altitude of a geosynchronous orbit is 3.58×10^7 m ($\approx$22,000 miles).
    a. What is the period of a satellite in a geosynchronous orbit?
    b. Find the value of g at this altitude.

c. What is the apparent weight of a 2000 kg satellite in a geosynchronous orbit?

31. A 75 kg man weighs himself at the north pole and at the equator. Which scale reading is higher? By how much?

32. A 1500 kg car takes a 50-m-radius unbanked curve at 15 m/s. What is the size of the friction force on the car?

33. The father of Example 7.5 stands at the summit of a conical hill as he spins his 20 kg child around on a 5.0 kg cart with a 2.0 m long rope. The sides of the hill are inclined at 20°. He again keeps the rope parallel to the ground, and friction is negligible. What rope tension will allow the cart to spin with the same 13.5 rpm it had in the example?

34. A 500 g ball swings in a vertical circle at the end of a 1.5-m-long string. When the ball is at the bottom of the circle, the tension in the string is 15 N. What is the speed of the ball at that point?

35. A concrete highway curve of radius 70 m is banked at a 15° angle. What is the maximum speed with which a 1500 kg rubber-tired car can take this curve without sliding?

36. A student ties a 500 g rock to a 1.0-m-long string and swings it around her head in a horizontal circle. At what angular velocity, in rpm, does the string tilt down at a 10° angle?

37. A 5.0 g coin is placed 15 cm from the center of a turntable. The coin has static and kinetic coefficients of friction with the turntable surface of $\mu_s = 0.80$ and $\mu_k = 0.50$. The turntable very slowly speeds up to 60 rpm. Does the coin slide off?

38. You've taken your neighbor's young child to the carnival to ride the rides. She wants to ride The Rocket. Eight rocket-shaped cars hang by chains from the outside edge of a large steel disk. A vertical axle through the center of the ride turns the disk, causing the cars to revolve in a circle. You've just finished taking physics, so you decide to figure out the speed of the cars while you wait. You estimate that the disk is 5 m in diameter and the chains are 6 m long. The ride takes 10 s to reach full speed, then the cars swing out until the chains are 20° from vertical. What is the car's speed?

39. A *conical pendulum* is formed by attaching a 500 g ball to a 1.0-m-long string, then allowing the mass to move in a horizontal circle of radius 20 cm. Figure P7.39 shows that the string traces out the surface of a cone, hence the name.
    a. What is the tension in the string?
    b. What is the ball's angular velocity, in rpm?

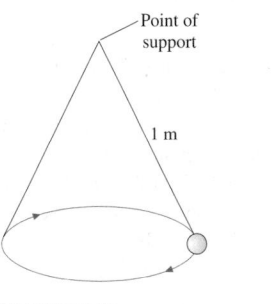

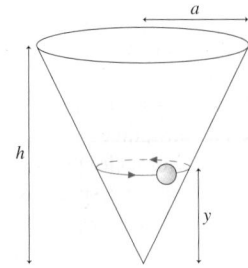

FIGURE P7.39   **FIGURE P7.40**

40. A small ball rolls around a horizontal circle at height y inside the cone shown in Figure P7.40. Find an expression of the ball's speed in terms of a, h, y, and g.

41. In an old-fashioned amusement park ride, passengers stand inside a 5.0-m-diameter hollow steel cylinder with their backs against the wall. The cylinder begins to rotate about a vertical axis. Then the floor on which the passengers are standing

suddenly drops away! If all goes well, the passengers will "stick" to the wall and not slide. Clothing has a static coefficient of friction against steel in the range 0.6 to 1.0 and a kinetic coefficient in the range 0.4 to 0.7. A sign next to the entrance says "No children under 30 kg allowed." What is the minimum angular velocity, in rpm, for which the ride is safe?

42. A 10 g steel marble is spun so that it rolls at 150 rpm around the *inside* of a vertically oriented steel tube. The tube, shown in Figure P7.42, is 12 cm in diameter. Assume that the rolling resistance is small enough for the marble to maintain 150 rpm for several seconds. During this time, will the marble spin in a horizontal circle, at constant height, or will it spiral down the inside of the tube?

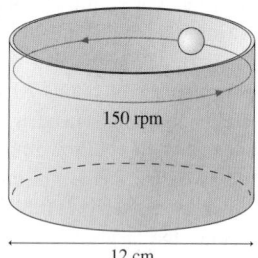

FIGURE P7.42

43. Three cars are driving at 25 m/s along the road shown in Figure P7.43. Car B is at the bottom of the hill and car C is at the top. Suppose each car suddenly brakes hard and starts to skid. What is the tangential acceleration (i.e., the acceleration parallel to the road) of each car? Assume $\mu_k = 1.0$.

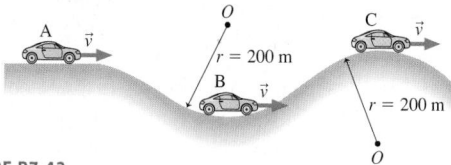

FIGURE P7.43

44. A car drives over the top of a hill that has a radius of 50 m. What maximum speed can the car have without flying off the road at the top of the hill?

45. A 500 g ball moves in a vertical circle on a 102-cm-long string. If the speed at the top is 4.0 m/s, then the speed at the bottom will be 7.5 m/s. (You'll learn how to show this in Chapter 10.)
    a. What is the ball's weight?
    b. What is the tension in the string when the ball is at the top?
    c. What is the tension in the string when the ball is at the bottom?

46. While at the county fair, you decide to ride the Ferris wheel. Having eaten too many candy apples and elephant ears, you find the motion somewhat unpleasant. To take your mind off your stomach, you wonder about the motion of the ride. You estimate the radius of the big wheel to be 15 m, and you use your watch to find that each loop around takes 25 s.
    a. What are your speed and magnitude of your acceleration?
    b. What is the ratio of your apparent weight to your true weight at the top of the ride?
    c. What is the ratio of your apparent weight to your true weight at the bottom?

47. In an amusement park ride called The Roundup, passengers stand inside a 16-m-diameter rotating ring. After the ring has acquired sufficient speed, it tilts into a vertical plane, as shown in Figure P7.47.

a. Suppose the ring rotates once every 4.5 s. If a rider's mass is 55 kg, with how much force does the ring push on her at the top of the ride? At the bottom?

b. What is the longest rotation period of the wheel that will prevent the riders from falling off at the top?

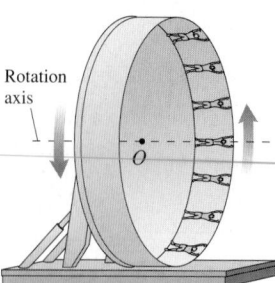

FIGURE P7.47

48. You have a new job designing rides for an amusement park. In one ride, the rider's chair is attached by a 9.0-m-long chain to the top of a tall rotating tower. The tower spins the chair and rider around at the rate of 1 rev every 4.0 s. In your design, you've assumed that the maximum possible combined weight of the chair and rider is 150 kg. You've found a great price for chain at the local discount store, but your supervisor wonders if the chain is strong enough. You contact the manufacturer and learn that the chain is rated to withstand a tension of 3000 N. Will this chain be strong enough for the ride?

49. Suppose you swing a ball in a vertical circle on a 1.0-m-long string. As you probably know from experience, there is a *minimum* angular velocity ω_{min} you must maintain if you want the ball to complete the full circle. If you swing the ball at $\omega < \omega_{min}$, then the string goes slack before the ball reaches the top of the circle. What is ω_{min}? Give your answer in rpm.

50. It is proposed that future space stations create an artificial gravity by rotating. Suppose a space station is constructed as a 1000-m-diameter cylinder that rotates about its axis. The inside surface is the deck of the space station. What rotation period will provide "normal" gravity?

51. A 100 g ball on a 60-cm-long string is swung in a vertical circle about a point 200 cm above the floor. The tension in the string when the ball is at the very bottom of the circle is 5.0 N. A very sharp knife is suddenly inserted, as shown in Figure P7.51, to cut the string directly below the point of support. Where does the ball hit the floor?

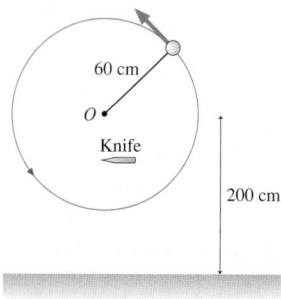

FIGURE P7.51

52. A 100 g ball on a 60-cm-long string is swung in a vertical circle about a point 200 cm above the floor. The string suddenly breaks when it is parallel to the ground and the ball is moving upward. The ball reaches a height 600 cm above the floor. What was the tension in the string an instant before it broke?

53. A 1500 kg car starts from rest and drives around a flat 50-m-diameter circular track. The forward force provided by the car's drive wheels is a constant 1000 N.
    a. What are the magnitude and direction of the car's acceleration at $t = 10$ s?
    b. If the car has rubber tires and the track is concrete, at what time does the car begin to slide out of the circle?

54. A 500 g steel block rotates on a steel table while attached to a 2.0-m-long massless rod. Compressed air fed through the rod is ejected from a nozzle on the back of the block, exerting a thrust force of 3.5 N. The nozzle is 70° from the radial line, as shown in Figure P7.54. The block starts from rest.
    a. What is the block's angular velocity after 10 rev?
    b. What is the tension in the rod after 10 rev?

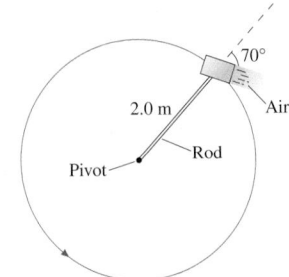

FIGURE P7.54

55. A 500 g steel block rotates on a steel table while attached to a 1.2-m-long hollow tube. Compressed air fed through the tube and ejected from a nozzle on the back of the block exerts a thrust force of 4.0 N perpendicular to the tube. The maximum tension the tube can withstand without breaking is 50 N. If the block starts from rest, how many revolutions does it make before the tube breaks?

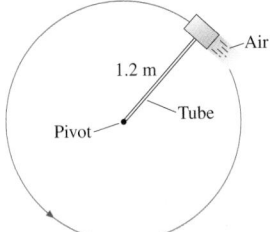

FIGURE P7.55

56. A 2.0 kg ball swings in a vertical circle on the end of an 80-cm-long string. The tension in the string is 20 N when its angle from the highest point on the circle is $\theta = 30°$.
    a. What is the ball's speed when $\theta = 30°$?
    b. What are the magnitude and direction of the ball's acceleration when $\theta = 30°$?

In Problems 57 through 59 you are given the equation (or equations) used to solve a problem. For each of these, you are to
   a. Write a realistic problem for which this is the correct equation. Be sure that the answer your problem requests is consistent with the equation given.
   b. Finish the solution of the problem.

57. $60 \text{ N} = (0.30 \text{ kg})\omega^2(0.50 \text{ m})$

58. $(1500 \text{ kg})(9.8 \text{ m/s}^2) - 11,760 \text{ N} = (1500 \text{ kg}) v^2/(200 \text{ m})$

59. $2.5 \text{ rad} = 0 \text{ rad} + \omega_i(10 \text{ s}) + ((1.5 \text{ m/s}^2)/2(50 \text{ m}))(10 \text{ s})^2$

    $\omega_f = \omega_i + ((1.5 \text{ m/s}^2)/(50 \text{ m}))(10 \text{ s})$

Challenge Problems

60. Two wires are tied to the 2.0 kg sphere shown in Figure CP7.60. The sphere revolves in a horizontal circle at constant speed.
    a. For what speed is the tension the same in both wires?
    b. What is the tension?

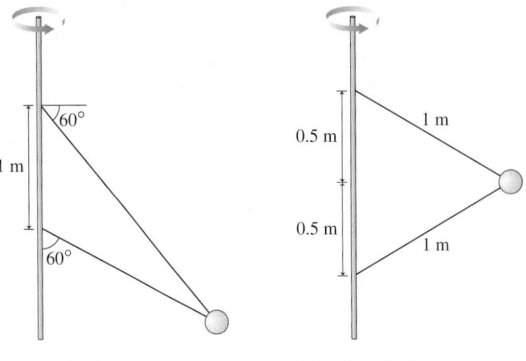

FIGURE CP7.60   **FIGURE CP7.61**

61. Two wires are tied to the 300 g sphere shown in Figure CP7.61. The sphere revolves in a horizontal circle at a constant speed of 7.5 m/s. What is the tension in each of the wires?

62. A 60 g ball is tied to the end of a 50-cm-long string and swung in a vertical circle. The center of the circle, as shown in Figure CP7.62, is 150 cm above the floor. The ball is swung at the minimum speed necessary to make it over the top without the string going slack. If the string is released at the instant the ball is at the top of the loop, where does the ball hit the ground?

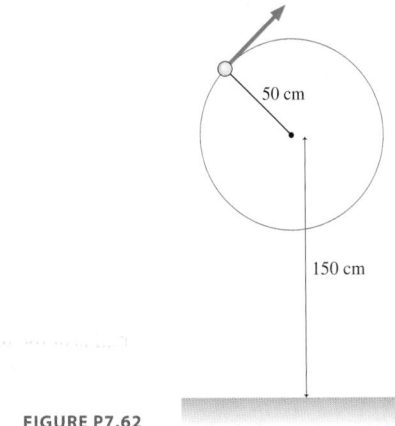

FIGURE P7.62

63. A small ball rolls around a horizontal circle at height y inside a frictionless hemispherical bowl of radius R, as shown in Figure CP7.63.
    a. Find an expression for the ball's angular velocity in terms of R, y, and g.
    b. What is the minimum value of ω for which the ball can move in a circle?
    c. What is ω in rpm if $R = 20$ cm and the ball is halfway up?

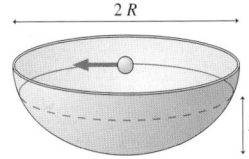

FIGURE CP7.63

64. You are flying to New York. You've been reading the in-flight magazine, which has an article about the physics of flying. You learned that the airflow over the wings creates a *lift force* that is always perpendicular to the wings. In level flight, the upward lift force exactly balances the downward weight force. The pilot comes on to say that, because of heavy traffic, the plane is going to circle the airport for a while. She says that you'll maintain a speed of 400 mph at an altitude of 20,000 ft. You start to wonder what the diameter of the plane's circle around the airport is. You notice that the pilot has banked the plane so that the wings are 10° from horizontal. The safety card in the seatback pocket informs you that the plane's wing span is 250 ft. What can you learn about the diameter?

65. If a vertical cylinder of water (or any other liquid) rotates about its axis, as shown in Figure CP7.65, the surface forms a smooth curve. Assuming that the water rotates as a unit (i.e., all the water rotates with the same angular velocity), show that the shape of the surface is a parabola described by the equation $z = (\omega^2/2\,g)r^2$.

Hint: Each particle of water on the surface is subject to only two forces: gravity and the normal force due to the water underneath it. The normal force, as always, acts perpendicular to the surface.

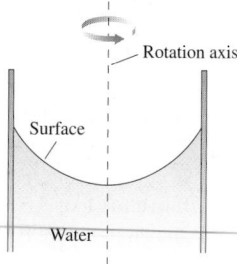

FIGURE CP7.65

STOP TO THINK ANSWERS

Stop to Think 7.1: b. An initial cw rotation causes the particle's angular position to become increasingly negative. The speed drops to half after reversing direction, so the slope becomes positive and is half as steep as the initial slope. Turning through the same angle returns the particle to $\theta = 0°$.

Stop to Think 7.2: $(a_r)_b > (a_r)_e > (a_r)_a = (a_r)_c > (a_r)_d$. Centripetal acceleration is v^2/r. Doubling r decreases a_r by a factor of 2. Doubling v increases a_r by a factor of 4. Reversing direction doesn't change a_r.

Stop to Think 7.3: $T_d > T_b = T_e > T_c > T_a$. The center-directed force is $m\omega^2 r$. Changing r by a factor of 2 changes the tension by a factor of 2, but changing ω by a factor of 2 changes the tension by a factor of 4.

Stop to Think 7.4: b. The car is moving in a circle, so there must be a net force toward the center of the circle. The circle is below the car, so the net force must point downward. This can be true only if $w > n$.

Stop to Think 7.5: c. The ball does not have a "memory" of its previous motion. The velocity $\vec{v}$ is straight up at the instant the string breaks. The only force on the ball after the string breaks is the weight force, straight down. This is just like tossing a ball straight up.

8 Newton's Third Law

These two sumo wrestlers are *interacting* with each other.

▶ **Looking Ahead**
The goal of Chapter 8 is to use Newton's third law to understand interacting systems. In this chapter you will learn to:

- Identify action/reaction pairs of forces in interacting systems.
- Understand and use Newton's third law.
- Use an expanded problem-solving strategy for dynamics problems.
- Understand the role of strings, ropes, and pulleys.

◀ **Looking Back**
This chapter further develops the concept of force. Please review:

- Sections 4.1–4.3 The basic concept of force and the atomic-level view of tension.
- Section 5.2 The basic problem-solving strategy for dynamics.

Rather than a single particle responding to a well-defined force, such as you've learned to deal with in the last few chapters, these sumo wrestlers are two systems *interacting* with each other. The harder one sumo wrestler pushes, the harder the other pushes back. A hammer and a nail, your foot and a soccer ball, and the earth-moon system are other examples of interacting systems.

Newton's second law is not sufficient to explain what happens when two or more objects interact. Newton's second law, the essence of single-particle dynamics, treats a system as an isolated entity being acting upon by external forces. Chapter 8 will introduce a new law of physics, Newton's *third* law, that describes how two systems interact with each other. We will then expand our problem-solving strategy to make simultaneous use of Newton's second and third laws when solving problems of interacting systems.

Newton's third law brings us to the pinnacle of Newton's theory of forces and motion. The tools you will have learned when you finish this chapter can be used to solve complex but realistic dynamics problems.

8.1 Interacting Systems

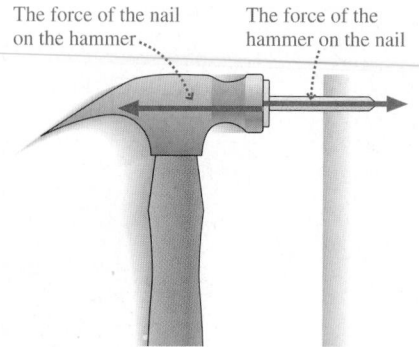

The force of the nail on the hammer

The force of the hammer on the nail

FIGURE 8.1 The hammer and nail are interacting with each other.

Our goal is to understand how two systems interact. Think about the hammer and nail in Figure 8.1. The hammer certainly exerts a force on the nail as it drives the nail forward. At the same time, the nail exerts a force on the hammer. If you are not sure that it does, imagine hitting the nail with a glass hammer. It's the force of the nail on the hammer that causes the glass to shatter.

If you stop to think about it, any time that object A pushes or pulls on object B, object B pushes or pulls back on object A. As sumo wrestler A pushes on sumo wrestler B, B pushes back on A. (If A pushed forward without B pushing back, A would fall over in the same way you do if someone suddenly opens a door you're leaning against.) Your chair pushes upward on you (a normal force) while, at the same time, you push down on the chair. These are examples of what we call an *interaction*. An **interaction** is the mutual influence of two systems on each other.

To be more specific, if object A exerts a force $\vec{F}_{\text{A on B}}$ on object B, then object B exerts a force $\vec{F}_{\text{B on A}}$ on object A. This pair of forces, shown in Figure 8.2, is called an **action/reaction pair.** Two systems interact by exerting an action/reaction pair of forces on each other. Notice the very explicit subscripts on the force vectors. The first letter is the *agent*, the second letter is the object on which the force acts. $\vec{F}_{\text{A on B}}$ is a force exerted *by* A *on* B. The distinction is important, and we will use this explicit notation for much of this chapter.

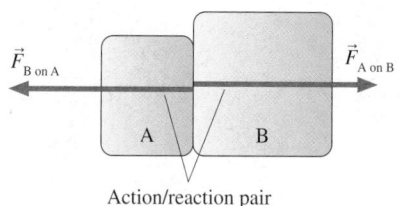

$\vec{F}_{\text{B on A}}$ $\vec{F}_{\text{A on B}}$

A   B

Action/reaction pair

FIGURE 8.2 An action/reaction pair of forces.

NOTE ▶ The name "action/reaction pair" is somewhat misleading. The forces occur simultaneously, and we cannot say which is the "action" and which the "reaction." Neither is there any implication about cause and effect; the action does not *cause* the reaction. **An action/reaction pair of forces exists as a pair, or not at all.** In identifying action/reaction pairs, the labels are the key. Force $\vec{F}_{\text{A on B}}$ is paired with force $\vec{F}_{\text{B on A}}$. ◀

The sumo wrestlers and the hammer and nail interact through contact forces. The same idea holds true for long-range forces. You probably have played with kitchen magnets or bar magnets. As you hold two magnets, you can feel with your fingertips that *both* have forces pulling on them.

But what about gravity? If you release a ball, it falls because the earth's gravity exerts a downward force $\vec{F}_{\text{earth on ball}}$ on it. This force is what we've been calling *weight*. But does the ball also pull upward on the earth? That is, is there a force $\vec{F}_{\text{ball on earth}}$?

Newton was the first to recognize that, indeed, the ball *does* pull upward on the earth. Likewise, the moon pulls on the earth in response to the earth's gravity pulling on the moon. Newton's evidence was the tides. Scientists and astronomers have studied and timed the ocean's tides since antiquity. It was known that the tides depend on the phase of the moon, but Newton was the first to understand that the tides are the ocean's response to the gravitational pull of the moon on the earth. As Figure 8.3 shows, the flexible water bulges toward the moon while the relatively inflexible crust of the earth remains stationary.

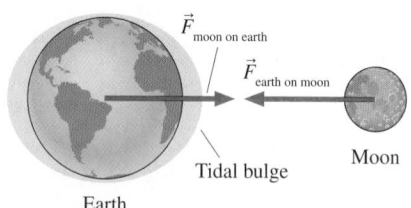

$\vec{F}_{\text{moon on earth}}$

$\vec{F}_{\text{earth on moon}}$

Moon

Tidal bulge

Earth

FIGURE 8.3 The ocean tides are an indication of the long-range gravitational interaction of the earth and the moon.

Systems and the Environment

In earlier chapters we considered forces acting on a single object that we called the *system*. The forces originated from agents in the *environment*. Figure 8.4a shows a diagrammatic representation of single-particle dynamics. If all the forces acting on the particle are known, we can use Newton's second law $\vec{F}_{\text{net}} = m\vec{a}$ to determine the particle's acceleration.

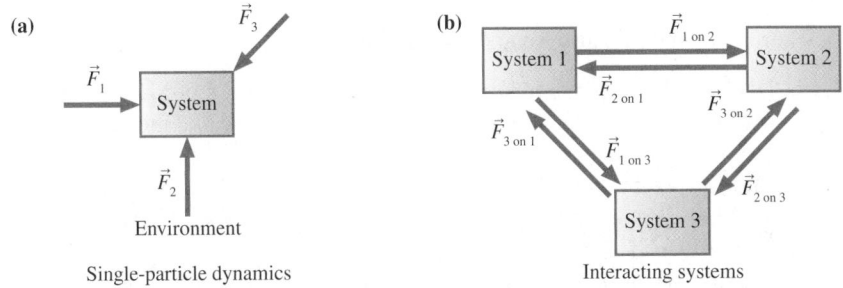

FIGURE 8.4 Single-particle dynamics and a model of interacting systems.

We now want to extend the particle model to situations in which two or more objects, each represented as a particle, interact with each other. For example, Figure 8.4b shows three systems interacting via action/reaction pairs of forces. The forces can be given labels such as $\vec{F}_{1 \text{ on } 2}$ and $\vec{F}_{2 \text{ on } 1}$.

We will often be interested in the motion of Systems 1 and 2 but not of System 3. For example, Systems 1 and 2 might be the hammer and the nail while System 3 is the earth. The earth interacts with both the hammer and the nail, but in a practical sense the earth remains "at rest" while the hammer and nail move. It will be convenient to separate "systems of interest" from "systems in the environment," as shown in Figure 8.5a. Forces originating in the environment are called **external forces.**

> NOTE ▶ This is a practical distinction, not a fundamental distinction. If object B pushes or pulls on object A whenever A pushes or pulls on B, then *every* force is one member of an action/reaction pair. There is no such thing as a true "external force." What we call an external force is an interaction between a system of interest and a system whose motion is not of interest. ◀

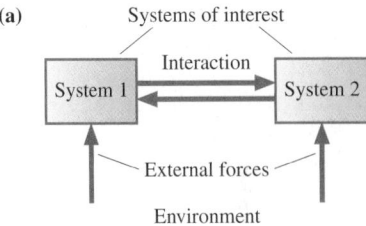

Figure 8.5b illustrates this idea for the hammer and the nail. $\vec{F}_{\text{hammer on nail}}$ and $\vec{F}_{\text{nail on hammer}}$ are an action/reaction pair that describe the interaction between the hammer and the nail. The force labeled $\vec{w}_{\text{hammer}}$ is really force $\vec{F}_{\text{earth on hammer}}$. That is, the hammer interacts with the earth through the action/reaction pair of forces $\vec{F}_{\text{earth on hammer}}$ and $\vec{F}_{\text{hammer on earth}}$. But we only want to know how the hammer moves, not how the earth moves, so the force $\vec{F}_{\text{earth on hammer}}$ can be treated as an external force.

Newton's second law $\vec{a} = \vec{F}_{\text{net}}/m$ applies *separately* to Systems 1 and 2 in Figure 8.5a:

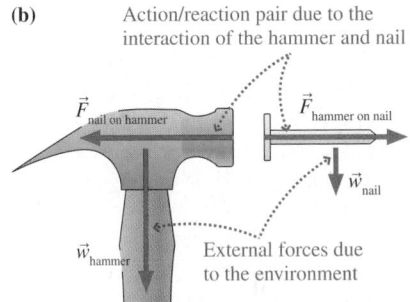

FIGURE 8.5 As a practical matter, an interaction with a system whose motion is not of interest can be called an external force.

$$\text{System 1:} \quad \vec{a}_1 = \frac{\vec{F}_{1 \text{ net}}}{m_1} = \frac{1}{m_1} \sum \vec{F}_{\text{on } 1}$$

$$\text{System 2:} \quad \vec{a}_2 = \frac{\vec{F}_{2 \text{ net}}}{m_2} = \frac{1}{m_2} \sum \vec{F}_{\text{on } 2}$$

(8.1)

The net force on System 1, denoted $\sum \vec{F}_{\text{on } 1}$, is the sum of *all* forces acting *on* System 1. The sum includes both forces due to System 2 ($\vec{F}_{2 \text{ on } 1}$) and any external forces originating in the environment.

> NOTE ▶ Forces exerted *by* System 1, such as $\vec{F}_{1 \text{ on } 2}$, do *not* appear in the equation for System 1. Objects change their motion in response to forces exerted *on* them, not by forces exerted *by* them. ◀

8.2 Identifying Action/Reaction Pairs

The key step for analyzing interacting systems is the identification of all the action/reaction pairs of forces.

The bat and the ball are interacting with each other.

We'll illustrate these ideas with two concrete examples. The first example will
be much longer than usual because we'll go carefully through all the steps in the
reasoning.

EXAMPLE 8.1   The forces involved in pushing a crate

Figure 8.6a shows a person pushing a large crate across a rough
surface. Identify all action/reaction pairs, show them on a figure, then draw free-body diagrams of the person and the crate.

VISUALIZE Figure 8.6b redraws the figure with every object in
the correct position but separated from all other objects. The
person and the crate are obvious objects. The earth is also an
object that both exerts and experiences forces. We've represented the earth rather abstractly as a long rectangle so that it
can be "under" both the person and the crate. It will be useful to
distinguish between the surface, which exerts contact forces,
and the earth as a whole, which exerts the long-range force of
gravity. The letters P (person), C (crate), S (surface), and E (the
earth as a whole) will be used for labels.

Figure 8.6b also identifies the various interactions. Some,
like the pushing interaction between the person and the crate,

are fairly obvious. The interactions with the earth are a little
trickier. Gravity, which is a long-range force, is an interaction
between each object and the earth as a whole. Friction forces
and normal forces are a contact interaction between each object
and the earth's surface. Altogether, there are seven interactions.

NOTE ▶ Interactions are between two *different* objects. None
of the interactions are between an object and itself. ◀

We can now start identifying and labeling forces, as you
learned to do in Chapter 4. These are shown in Figure 8.7. We'll
begin with the crate: It is pushed with force $\vec{F}_{P\,on\,C}$, it experiences an upward normal force $\vec{n}_{S\,on\,C}$, and it has a kinetic friction force $\vec{f}_{S\,on\,C}$ in the direction opposite the motion. The crate
also has a downward weight force $\vec{w}_{E\,on\,C}$ exerted on it by the
earth as a whole. These are all forces $\vec{F}_{something\,on\,C}$, and all the
force vectors are drawn *on* crate C.

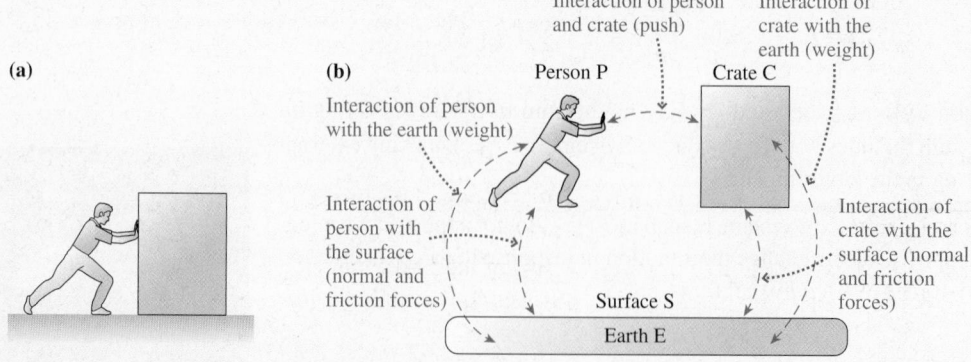

(a)   **(b)**

Interaction of person
and crate (push)

Interaction of
crate with the
earth (weight)

Person P   Crate C

Interaction of person
with the earth (weight)

Interaction of
person with
the surface
(normal and
friction forces)

Interaction of
crate with the
surface (normal
and friction
forces)

Surface S

Earth E

FIGURE 8.6 A person pushes a crate across a rough floor.

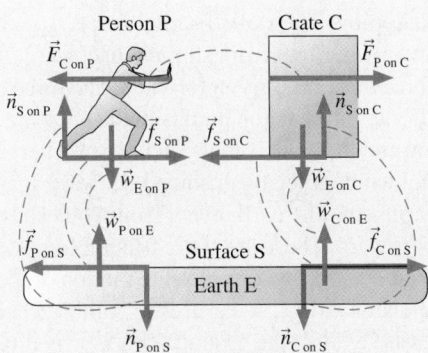

FIGURE 8.7 The forces involved in pushing a crate.

If A pushes or pulls on B, then B pushes or pulls back on A. The reaction to force $\vec{F}_{A \text{ on } B}$ is $\vec{F}_{B \text{ on } A}$. We can use this reasoning to deduce the reactions to the four forces exerted on crate C. The person pushes the crate, so the crate pushes against the person with force $\vec{F}_{C \text{ on } P}$. This is a force exerted *on the person*, so it is drawn on person P. Forces $\vec{F}_{P \text{ on } C}$ and $\vec{F}_{C \text{ on } P}$ are then connected with a dotted line to show that they are an action/reaction pair. Similarly, force $\vec{n}_{C \text{ on } S}$ is the force of the crate pushing down on the surface. This is a contact force, so it is drawn on the earth's surface. The weight force $\vec{w}_{E \text{ on } C}$ is the gravitational force *by* the earth *on* the crate. By "earth" we mean the *entire* earth, not just the surface. To identify the reaction force, simply reverse the letters in the labels. The reaction to $\vec{w}_{E \text{ on } C}$ is $\vec{w}_{C \text{ on } E}$. In other words, the crate pulls up on the entire earth.

We have to be careful with friction. Force $\vec{f}_{S \text{ on } C}$ is a kinetic friction force retarding the crate's motion, so it points to the left. The crate exerts force $\vec{f}_{C \text{ on } S}$ on the surface. But in which direction? Imagine the floor is covered with sand. As the crate slides, it tries to push the sand to the right. Thus force $\vec{f}_{C \text{ on } S}$, which forms an action/reaction pair with $\vec{f}_{S \text{ on } C}$, points to the right.

The person experiences similar forces. Force $\vec{F}_{C \text{ on } P}$ pushes against the person's hands while the normal force $\vec{n}_{S \text{ on } P}$ pushes up and the weight force $\vec{w}_{E \text{ on } P}$ pulls down. These are forces exerted *on* the person, so the force vectors are drawn on P.

The hardest interaction to understand is the friction between the person and the surface. It is tempting to draw force $\vec{f}_{S \text{ on } P}$ pointing to the left. After all, friction forces are supposed to be in the direction opposite the motion. But if we did so, the person would have two forces to the left ($\vec{F}_{C \text{ on } P}$ and $\vec{f}_{S \text{ on } P}$) and none to

the right, causing the person to accelerate *backward*! That is clearly not what happens, so what is wrong?

Imagine pushing a crate to the right across loose sand. Each time you take a step, you tend to kick the sand to the *left*, behind you. Thus friction force $\vec{f}_{P \text{ on } S}$, the force of the person pushing against the earth, is to the *left*. In reaction, the force of the earth's surface against the person is a friction force to the *right*. This force, $\vec{f}_{S \text{ on } P}$, causes the person to accelerate in the forward direction. Forces $\vec{f}_{S \text{ on } P}$ to the right and $\vec{f}_{P \text{ on } S}$ to the left form another action/reaction pair.

Notice that *every* force is one member of an action/reaction pair. Further, each member of a pair is attached to a different object. **The two forces of an action/reaction pair never occur on the same object.**

But surely 14 forces is an excessive number for such a simple situation! While these forces all exist, we don't need all of them to understand how the crate moves. This is the point where we can distinguish between systems of interest and the environment. The *crate* and the *person* are systems of interest because they move. The surface and the earth do not move, so we can locate them in the environment.

Figure 8.8 shows a free body diagram of just the two systems of interest. Forces $\vec{F}_{P \text{ on } C}$ and $\vec{F}_{C \text{ on } P}$ are an action/reaction pair between the systems of interest, so we've kept their full labels and connected them with a dotted line. We can now treat all other forces as external forces, and can simplify their labels. Even so, it is important to use subscript labels such as $\vec{w}_P$ and $\vec{w}_C$, or $\vec{n}_P$ and $\vec{n}_C$, to distinguish external forces on the crate from external forces on the person.

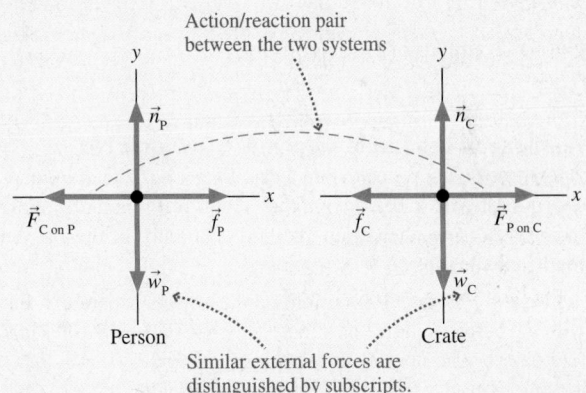

FIGURE 8.8 Free-body diagrams of the person and the crate.

Propulsion

The friction force $\vec{f}_{S \text{ on } P}$ is an example of **propulsion.** It is the force that a system with an internal source of energy uses to drive itself forward. Propulsion is an important feature not only of walking or running but also of the forward motion of cars, jets, and rockets. Propulsion is somewhat counterintuitive, so it is worth a closer look.

If you try to walk across a frictionless floor, your foot slips and slides *backward*. In order for you to walk, the floor needs to have friction so that your foot *sticks* to the floor as you straighten your leg, moving your body forward. The

What force causes this sprinter to accelerate?

friction that prevents slipping is *static* friction. Static friction, you will recall, acts in the direction that prevents slipping. The static friction force $\vec{f}_{S \text{ on } P}$ has to point in the *forward* direction to prevent your foot from slipping backward. It is this forward-directed static friction force that propels you forward! The force of your foot on the floor, the other half of the action/reaction pair, is in the opposite direction.

The distinction between you and the crate is that you have an *internal source of energy* that allows you to straighten your leg by pushing backward against the surface. In essence, you walk by pushing the earth away from you. The earth's surface responds by pushing you forward. These are static friction forces. In contrast, all the crate can do is slide, so *kinetic* friction opposes the motion of the crate.

Figure 8.9 shows how propulsion works. A car uses its motor to spin the tires, causing the tires to push backward against the ground. This is why dirt and gravel are kicked backward, not forward. The earth's surface responds by pushing the car forward. These are also *static* friction forces. The tire is rolling, but the bottom of the tire, where it contacts the road, is instantaneously at rest. If it weren't, you would leave one giant skid mark as you drove and would burn off the tread within a few miles.

Rocket motors are somewhat different because they are not pushing *against* anything. That's why rocket propulsion works in the vacuum of space. Instead, the rocket engine pushes hot, expanding gases out of the back of the rocket. In response, the exhaust gases push the rocket forward with the force we've called *thrust*.

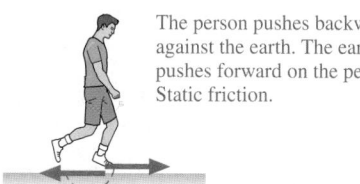

The person pushes backward against the earth. The earth pushes forward on the person. Static friction.

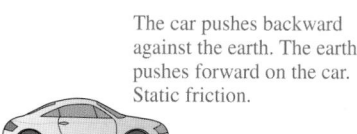

The car pushes backward against the earth. The earth pushes forward on the car. Static friction.

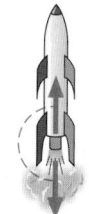

The rocket pushes the hot gases backward. The gases push the rocket forward. Thrust force.

FIGURE 8.9 Examples of propulsion.

EXAMPLE 8.2 **The forces involved in towing a car**

A tow truck uses a rope to pull a car along a horizontal road, as shown in Figure 8.10a. Identify all action/reaction pairs, show them on a figure, then draw free-body diagrams of the car, the truck, and the rope.

VISUALIZE Figure 8.10b has drawn the objects separately, but with the correct relative positions. The rope is shown as a separate object. The normal force and weight force action/reaction pairs are identical to those in Example 8.1. We've assumed that the weight of the rope is negligible in comparison with all the other forces.

NOTE ▶ Make sure you avoid the common error of considering $\vec{n}$ and $\vec{w}$ to be an action/reaction pair. These are both forces on the *same* object, whereas the two forces of an action/reaction pair are always on two *different* objects that are interacting with each other. ◀

What about the friction forces? The car is an inert object rolling along. It would slow and stop if the rope were cut, so the

(a)

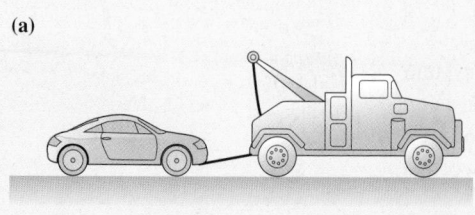

(b)

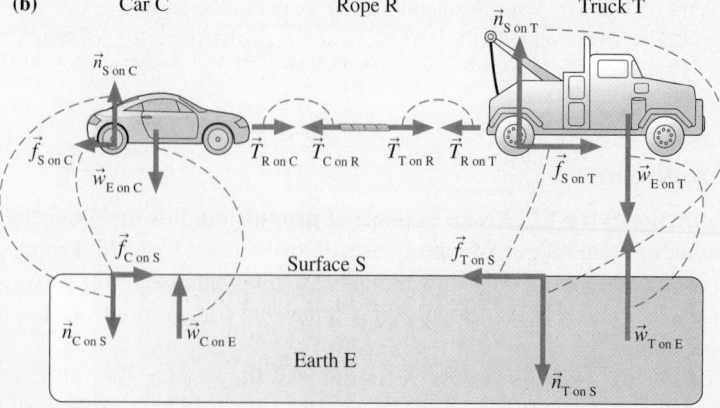

FIGURE 8.10 Identifying the forces as a truck tows a car.

surface must exert a rolling friction force $\vec{f}_{S\,on\,C}$ to the left. The reaction of the car trying to drag the ground along with it is the force $\vec{f}_{C\,on\,S}$ to the right. The truck, however, has an internal source of energy. The truck's drive wheels push the ground to the left with force $\vec{f}_{T\,on\,S}$. In reaction, the ground propels the truck forward, to the right, with force $\vec{f}_{S\,on\,T}$.

Finally, we need to identify the forces between the car, the truck, and the rope. What pulls on what in the horizontal direction? The rope pulls on the car with a tension force $\vec{T}_{R\,on\,C}$. You might be tempted to put the reaction force on the truck, because we say that "the truck pulls the car," but the truck is not in contact with the car. The truck pulls on the rope, then the rope pulls on the car. Thus the reaction to $\vec{T}_{R\,on\,C}$ is a force on the *rope*: $\vec{T}_{C\,on\,R}$. At the other end, $\vec{T}_{T\,on\,R}$ and $\vec{T}_{R\,on\,T}$ are an action/reaction pair.

We have *three* systems of interest: the truck, the rope, and the car. The surface and the earth are in the environment, so normal forces, weight forces, and friction forces can be treated as external forces. This information is shown on the free-body diagrams of Figure 8.11.

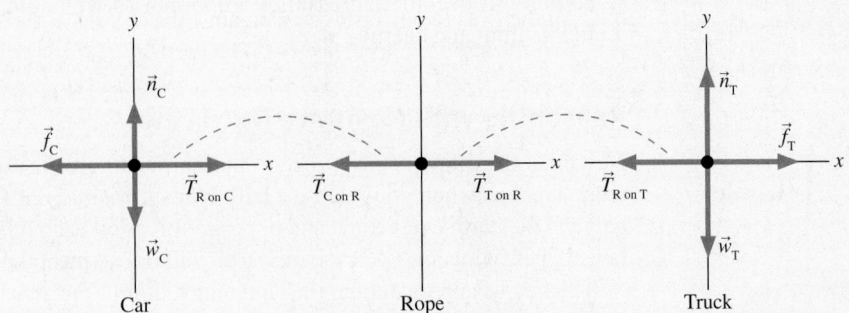

FIGURE 8.11 Free-body diagrams of the three systems of interest in Example 8.2.

STOP TO THINK 8.1 What, if anything, is wrong with this force diagram for a bicycle that is accelerating toward the right?

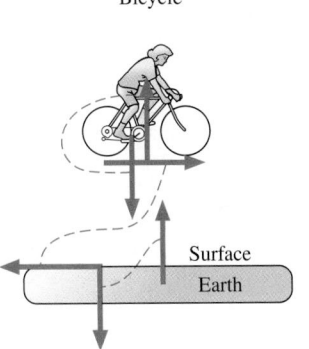

8.3 Newton's Third Law

Newton was the first to recognize how the two members of an action/reaction pair of forces are related to each other. Today we know this as Newton's third law:

> **Newton's third law** Every force occurs as one member of an action/reaction pair of forces.
>
> - The two members of an action/reaction pair act on two *different* objects.
> - The two members of an action/reaction pair are equal in magnitude but opposite in direction: $\vec{F}_{A\,on\,B} = -\vec{F}_{B\,on\,A}$.

We deduced most of the third law in Section 8.2. There we found that the two members of an action/reaction pair are always opposite in direction (see Figures 8.7 and 8.10). According to the third law, this will always be true. But the most significant portion of the third law, which is by no means obvious, is that the two members of an action/reaction pair have *equal* magnitudes. That is, $F_{\text{A on B}} = F_{\text{B on A}}$. This is the quantitative relationship that will allow you to solve problems of interacting systems.

Newton's third law is frequently stated as "For every action there is an equal but opposite reaction." While this is indeed a catchy phrase, it lacks the preciseness of our preferred version. In particular, it fails to capture an essential feature of action/reaction pairs—that they each act on a *different* object.

> **NOTE** ▶ Newton's third law extends and completes our concept of *force*. We can now recognize force as an *interaction* between objects rather than as some "thing" with an independent existence of its own. The concept of an interaction will become increasingly important as we begin to study the laws of momentum and energy. ◀

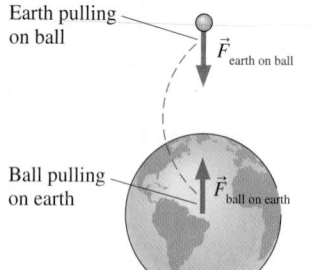

FIGURE 8.12 The action/reaction forces of a ball and the earth are equal in magnitude.

Reasoning with Newton's Third Law

Newton's third law is easy to state but harder to grasp. For example, consider what happens when you release a ball. Not surprisingly, it falls down. But if the ball and the earth exert equal and opposite forces on each other, as Newton's third law alleges, why don't you see the earth "fall up" to meet the ball?

The key to understanding this and many similar puzzles is that **the forces are equal but the accelerations are not.** Equal causes can produce very unequal effects. Figure 8.12 shows equal-magnitude forces on the ball and the earth. The force on ball B is simply the weight force of Chapter 5:

$$\vec{F}_{\text{earth on ball}} = \vec{w}_{\text{B}} = -m_{\text{B}} g \hat{j} \tag{8.2}$$

where m_{B} is the mass of the ball. According to Newton's second law, this force gives the ball an acceleration

$$\vec{a}_{\text{B}} = \frac{\vec{w}_{\text{B}}}{m_{\text{B}}} = -g \hat{j} \tag{8.3}$$

This is just the familiar free-fall acceleration due to gravity.

According to Newton's third law, the ball pulls up on the earth with force $\vec{F}_{\text{ball on earth}}$. As the ball accelerates down, the earth as a whole has an upward acceleration

$$\vec{a}_{\text{E}} = \frac{\vec{F}_{\text{ball on earth}}}{m_{\text{E}}} \tag{8.4}$$

where m_{E} is the mass of the earth. Because $\vec{F}_{\text{earth on ball}}$ and $\vec{F}_{\text{ball on earth}}$ are an action/reaction pair, $\vec{F}_{\text{ball on earth}}$ must be equal in magnitude and opposite in direction to $\vec{F}_{\text{earth on ball}}$. That is,

$$\vec{F}_{\text{ball on earth}} = -\vec{F}_{\text{earth on ball}} = -\vec{w}_{\text{B}} = +m_{\text{B}} g \hat{j} \tag{8.5}$$

Using this result in Equation 8.4, the upward acceleration of the earth as a whole is

$$\vec{a}_{\text{E}} = \frac{\vec{F}_{\text{ball on earth}}}{m_{\text{E}}} = \frac{m_{\text{B}} g \hat{j}}{m_{\text{E}}} = \left(\frac{m_{\text{B}}}{m_{\text{E}}}\right) g \hat{j} \tag{8.6}$$

The upward acceleration of the earth is less than the downward acceleration of the ball by the factor $m_{\text{B}}/m_{\text{E}}$. If we assume a 1 kg ball, we can estimate the magnitude of $\vec{a}_{\text{E}}$:

$$a_{\text{E}} = \frac{1 \text{ kg}}{6 \times 10^{24} \text{ kg}} g \approx 2 \times 10^{-24} \text{ m/s}^2$$

With this incredibly small acceleration, it would take the earth 8×10^{15} years, approximately 500,000 times the age of the universe, to reach a speed of 1 mph! So we certainly would not expect to see or feel the earth "fall up" after dropping a ball.

NOTE ▶ Newton's third law equates the size of two forces, not two accelerations. The acceleration continues to depend on the mass, as Newton's second law states. **In an interaction between two objects of different mass, the lighter mass will do essentially all of the accelerating even though the forces exerted on the two objects are equal.** ◀

EXAMPLE 8.3 **The forces on accelerating boxes**

The hand shown in Figure 8.13 pushes boxes A and B to the right across a frictionless table. The mass of B is larger than the mass of A.

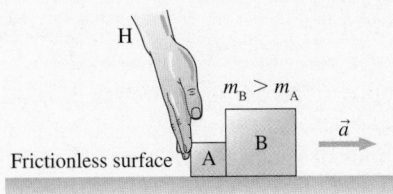

FIGURE 8.13 Hand H pushes boxes A and B across a frictionless table.

a. Draw free-body diagrams of A, B, and the hand H, showing only the *horizontal* forces. Connect action/reaction pairs with dotted lines.
b. Rank in order, from largest to smallest, the horizontal forces shown on your free-body diagrams.

VISUALIZE

a. The hand H pushes on box A, and A pushes back on H. Thus $\vec{F}_{\text{H on A}}$ and $\vec{F}_{\text{A on H}}$ are an action/reaction pair. Similarly, A pushes on B and B pushes back on A. The hand H does not touch box B, so there is no interaction between them. There is no friction. Figure 8.14 shows the four horizontal forces and identifies two action/reaction pairs. (We've chosen to ignore forces of the wrist or arm on the hand because our systems of interest are the boxes A and B.) Notice that each force is shown on the free-body diagram of the object that it acts *on*.

b. According to Newton's third law, $F_{\text{A on H}} = F_{\text{H on A}}$ and $F_{\text{A on B}} = F_{\text{B on A}}$. But the third law is not our only tool. Because the boxes are accelerating to the right, Newton's *second* law tells us that box A must have a net force to the right. Consequently, $F_{\text{H on A}} > F_{\text{B on A}}$. Thus

$$F_{\text{A on H}} = F_{\text{H on A}} > F_{\text{A on B}} = F_{\text{B on A}}$$

ASSESS You might have expected $F_{\text{A on B}}$ to be larger than $F_{\text{H on A}}$ because $m_{\text{B}} > m_{\text{A}}$. It's true that the *net* force on B is larger than the *net* force on A, but we have to reason more closely to judge the individual forces. Notice how we used both the second and the third laws to answer this question.

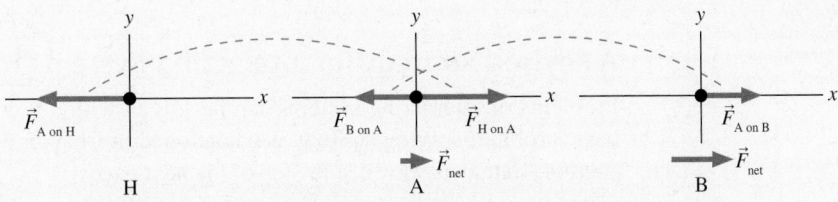

FIGURE 8.14 The free-body diagrams, showing only the horizontal forces.

STOP TO THINK 8.2   Car B is stopped for a red light. Car A, which has the same mass as car B, doesn't see the red light and runs into the back of B. Which of the following statements is true?

a. B exerts a force on A but A doesn't exert a force on B.
b. B exerts a larger force on A than A exerts on B.
c. B exerts the same amount of force on A as A exerts on B.
d. A exerts a larger force on B than B exerts on A.
e. A exerts a force on B but B doesn't exert a force on A.

Acceleration Constraints

Newton's third law is one quantitative relationship you can use to solve problems of interacting systems. In addition, we frequently have other information about the motion in a problem. For example, think about the two boxes in Example 8.3. As long as they're touching, box A *has* to have exactly the same acceleration as box B. If they were to accelerate differently, either box B would take off on its own or it would suddenly slow down and box A would run over it! Our problem implicitly assumes that neither of these is happening. Thus the two accelerations are *constrained* to be equal: $\vec{a}_A = \vec{a}_B$. A well-defined relationship between the accelerations of two or more systems is called an **acceleration constraint.** It is an independent piece of information that can help solve a problem.

In practice, we'll express acceleration constraints in terms of the x- and y-components of $\vec{a}$. Consider the car being towed in Figure 8.15. As long as the rope is under tension, the accelerations are constrained to be equal: $\vec{a}_C = \vec{a}_T$. This is one-dimensional motion, so for problem solving we would use just the x-components a_{Cx} and a_{Tx}. In terms of these components, the acceleration constraint is

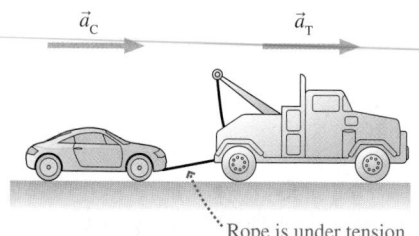

FIGURE 8.15 The car and the truck have the same acceleration.

$$a_{Cx} = a_{Tx} = a_x$$

Because the accelerations of both systems are equal, we can drop the subscripts C and T and call both of them a_x.

Don't assume the accelerations of A and B will always have the same sign. Consider blocks A and B in Figure 8.16. The blocks are connected by a string, so they are constrained to move together and their accelerations have equal magnitudes. But A has a positive acceleration (to the right) in the x-direction while B has a negative acceleration (downward) in the y-direction. Thus the acceleration constraint is

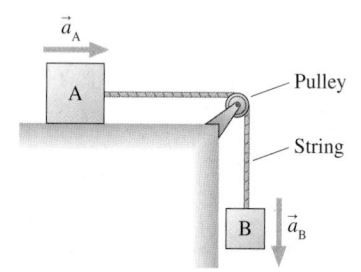

FIGURE 8.16 The string constrains the two systems to accelerate together.

$$a_{Ax} = -a_{By}$$

This relationship does *not* say that a_{Ax} is a negative number. It is simply a relational statement, saying that a_{Ax} is (-1) times whatever a_{By} happens to be. The acceleration a_{By} in Figure 8.16 is a negative number, so a_{Ax} is positive. In some problems, the signs of a_{Ax} and a_{By} may not be known until the problem is solved, but the *relationship* is known from the beginning.

A Revised Strategy for Interacting-System Problems

Problems of interacting systems can be solved with a few modifications to the basic problem-solving strategy we developed in Chapter 5. A revised problem-solving strategy is shown at the top of the next page.

NOTE ▶ We have dropped the motion diagram from the physical representation. Motion diagrams served a useful function in the early chapters, but by now you should be able to determine the directions of the acceleration vectors without the need for an explicit diagram. But if you are uncertain—use one! ◀

You might be puzzled that the Solve step calls for the use of the third law to equate just the *magnitudes* of action/reaction forces. What about the "opposite in direction" part of the third law? You have already used it! Your free-body diagrams should show the two members of an action/reaction pair to be opposite in direction, and that information will have been utilized in writing the second-law equations. Because the directional information has already been used, all that is left is the magnitude information.

PROBLEM-SOLVING STRATEGY 8.1 Interacting-system problems

MODEL Identify which objects are systems and which are part of the environment. Make simplifying assumptions.

VISUALIZE Pictorial representation. Show important points in the motion with a sketch. You may want to give each system a separate coordinate system. Define symbols and identify what the problem is trying to find. Include acceleration constraints as part of the pictorial model.
Physical representation. Identify all forces acting on each system and all action/reaction pairs. Draw a *separate* free-body diagram for each system. Connect the force vectors of action/reaction pairs with dotted lines. Use subscript labels to distinguish forces, such as $\vec{n}$ and $\vec{w}$, that act independently on more than one system.

SOLVE Use Newton's second and third laws:

- Write the equations of Newton's second law for each system, using the force information from the free-body diagrams.
- Equate the magnitudes of action/reaction pairs.
- Include the acceleration constraints, the friction model, and other quantitative information relevant to the problem.
- Solve for the acceleration, then use kinematics to find velocities and positions.

ASSESS Check that your result has the correct units, is reasonable, and answers the question.

NOTE ▶ Two steps are especially important when drawing the free-body diagrams. First, draw a *separate* diagram for each system. They need not have the same coordinate system. Second, show only the forces acting *on* that system. The force $\vec{F}_{\text{A on B}}$ goes on the free-body diagram of System B, but $\vec{F}_{\text{B on A}}$ goes on the diagram of System A. The two members of an action/reaction pair *always* appear on two different free-body diagrams—*never* on the same diagram. ◀

EXAMPLE 8.4 Keep the crate from sliding
You and a friend have just loaded a 200 kg crate filled with priceless art objects into the back of a 2000 kg truck. As you press down on the accelerator, force $\vec{F}_{\text{surface on truck}}$ propels the truck forward. To keep things simple, call this just $\vec{F}_T$. What is the maximum magnitude $\vec{F}_T$ can have without the crate sliding? The static and kinetic coefficients of friction between the crate and the bed of the truck are 0.8 and 0.3. Rolling friction of the truck is negligible.

MODEL The crate and the truck are separate systems that we'll call C and T. We'll model them as particles. The earth and the road surface are part of the environment.

VISUALIZE We're not doing any kinematics in this problem, so the pictorial representation of Figure 8.17 on the next page is minimal. We need a coordinate system, the known information, and—new to problems of interacting systems—the acceleration constraint. As long as the crate doesn't slip, it must accelerate *with* the truck. Both accelerations are in the positive *x*-direction, so the acceleration constraint in this problem is

$$a_{Cx} = a_{Tx} = a_x$$

Figure 8.17 also shows the free-body diagrams. The crate's diagram is drawn above the truck's diagram to reflect their relative positions. Force $\vec{n}_{\text{T on C}}$ is the normal force of the *truck* pushing up on the crate. The crate does not contact the ground, so the ground cannot exert forces on the crate. Although the crate moves, there's no motion of the crate *relative to* the truck. The force that prevents slipping is static friction. To prevent the crate from sliding out the back of the truck, a *static* friction force $\vec{f}_{\text{T on C}}$ points in the forward direction. The value of μ_k isn't relevant, so only μ_s is included as known information.

The truck has a weight force $\vec{w}_T$ and a normal force $\vec{n}_T$ of the *ground* pushing up on the truck. The truck tires try to push the ground backward as they spin, so the ground reacts with the forward force $\vec{F}_T$ that propels the truck forward. The truck experiences two other forces. Because $\vec{n}_{\text{T on C}}$ is a force of the truck on the crate, the truck experiences the reaction force $\vec{n}_{\text{C on T}}$ of the crate pressing down on the truck. Similarly, $\vec{f}_{\text{C on T}}$ is the reaction to friction force $\vec{f}_{\text{T on C}}$ of the truck on the crate. The truck tries to drag the crate forward; the crate tries to hold the truck back.

Notice that $\vec{n}_T$ and $\vec{w}_T$ are *not* an action/reaction pair. They are really the forces $\vec{n}_{\text{surface on T}}$ and $\vec{w}_{\text{earth on T}}$, so their reactions

Pictorial representation   **Physical representation**

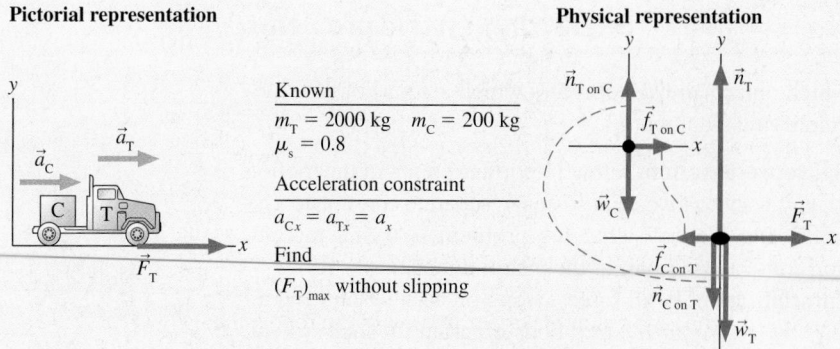

FIGURE 8.17 Pictorial and physical representations of the crate and truck in Example 8.4.

act on the surface and the earth as a whole. The two members of an action/reaction pair *never* appear on the same free-body diagram.

SOLVE Now we're ready to write Newton's second law. For the crate:

$$\sum (F_{\text{on crate}})_x = f_{\text{T on C}} = m_C a_{Cx} = m_C a_x$$

$$\sum (F_{\text{on crate}})_y = n_{\text{T on C}} - w_C = n_{\text{T on C}} - m_C g = 0$$

For the truck:

$$\sum (F_{\text{on truck}})_x = F_T - f_{\text{C on T}} = m_T a_{Tx} = m_T a_x$$

$$\sum (F_{\text{on truck}})_y = n_T - w_T - n_{\text{C on T}} = n_T - m_T g - n_{\text{C on T}} = 0$$

Be sure you agree with all the signs, which are based on the free-body diagrams. The net force in the *y*-direction is zero because there's no motion in the *y*-direction. It may seem like a lot of effort to write all the subscripts, but it is very important in problems with more than one system.

Notice that we've already used the acceleration constraint $a_{Cx} = a_{Tx} = a_x$. Another important piece of information is Newton's third law, which tells us that $f_{\text{C on T}} = f_{\text{T on C}}$ and $n_{\text{C on T}} = n_{\text{T on C}}$. Finally, we know that the maximum value of F_T will occur when the static friction on the crate reaches its maximum value

$$f_{\text{T on C}} = f_{s\,\text{max}} = \mu_s n_{\text{T on C}}$$

The friction depends on the normal force on the crate, not the normal force on the truck.

Now we can assemble all the pieces. From the *y*-equation of the crate, $n_{\text{T on C}} = m_C g$. Thus

$$f_{\text{T on C}} = \mu_s n_{\text{T on C}} = \mu_s m_C g$$

Using this in the *x*-equation of the crate, we find that the acceleration is

$$a_x = \frac{f_{\text{T on C}}}{m_C} = \mu_s g$$

This is the crate's maximum acceleration without slipping. Now use this acceleration *and* the fact that $f_{\text{C on T}} = f_{\text{T on C}} = \mu_s m_C g$ in the *x*-equation of the truck to find

$$F_T - f_{\text{C on T}} = F_T - \mu_s m_C g = m_T a_x = m_T \mu_s g$$

Solving for F_T, the maximum propulsion without the crate sliding is

$$(F_T)_{\text{max}} = \mu_s(m_T + m_C)g$$

$$= 0.8(2200 \text{ kg})(9.80 \text{ m/s}^2) = 17,200 \text{ N}$$

ASSESS This is a hard result to assess. Few of us have any intuition about the size of forces that propel cars and trucks. Even so, the fact that the forward force on the truck is a significant fraction (80%) of the combined weight of the truck and the crate seems plausible. We might have been suspicious if F_T had been only a tiny fraction of the weight or much greater than the weight.

As you can see, there are many equations and many pieces of information to keep track of when solving a problem of interacting systems. These problems are not inherently harder than the problems you learned to solve in Chapters 5–7, but they do require a high level of organization. Using the systematic approach of the problem-solving strategy will help you solve similar problems successfully.

STOP TO THINK 8.3 Boxes A and B are sliding to the right across a frictionless table. The hand H is slowing them down. The mass of A is larger than the mass of B. Rank in order, from largest to smallest, the *horizontal* forces on A, B, and H.

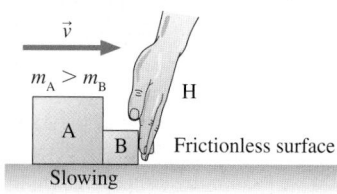

a. $F_{\text{B on H}} = F_{\text{H on B}} = F_{\text{A on B}} = F_{\text{B on A}}$ b. $F_{\text{B on H}} = F_{\text{H on B}} > F_{\text{A on B}} = F_{\text{B on A}}$

c. $F_{\text{B on H}} = F_{\text{H on B}} < F_{\text{A on B}} = F_{\text{B on A}}$ d. $F_{\text{H on B}} = F_{\text{H on A}} > F_{\text{A on B}}$

8.4 Ropes and Pulleys

Many systems are connected by strings, ropes, cables, and so on. In single-particle dynamics, we defined *tension* as the force exerted on an object by a rope or string. Now we need to think more carefully about the string itself. Just what do we mean when we talk about the tension "in" a string?

Tension Revisited

Figure 8.18a shows a heavy safe hanging from a rope, placing the rope under tension. If you cut the rope, the safe and the lower portion of the rope will fall. Thus there must be a force *within* the rope by which the upper portion of the rope pulls upward on the lower portion to prevent it from falling.

Chapter 4 introduced an atomic-level model in which tension is due to the stretching of spring-like molecular bonds within the rope. Stretched springs exert pulling forces, and the combined pulling force of billions of stretched molecular springs in a string or rope is what we call *tension*.

An important aspect of tension is that it pulls equally *in both directions*. Figure 8.18b is a very thin cross section through the rope. This small piece of rope is in equilibrium, so it must be pulled equally from both sides. To gain a mental picture, imagine holding your arms outstretched and having two friends pull on them. You'll remain at rest—but "in tension"—as long as they pull with equal strength in opposite directions. But if one lets go, analogous to the breaking of molecular bonds if a rope breaks or is cut, you'll fly off in the other direction!

The bottom layer of molecules in the rope is in contact with the safe. Here the combined pulling force of the stretched molecular springs pulls up on the safe with force $\vec{T}_{\text{R on S}}$. This is the force that we've been calling simply $\vec{T}$, the tension force of the rope pulling on the safe. Because the tension pulls equally in both directions, the tension *on the safe* at the end of the rope is the same strength as the tension *in the rope* near the end of the rope.

Figure 8.19 shows forces on the safe, the rope, and the earth. Newton's third law tells us that $T_{\text{R on S}} = T_{\text{S on R}}$ and $w_{\text{E on S}} = w_{\text{S on E}}$. These relationships are true whether the safe is at rest or accelerating. We need Newton's second law to compare $T_{\text{R on S}}$ to $w_{\text{E on S}}$.

- If the safe is in equilibrium, either at rest or moving with constant velocity, then $\vec{F}_{\text{net}} = \vec{0}$. Thus $T_{\text{R on S}} = w_{\text{E on S}}$. The tension in the rope is equal to the weight of the safe.
- If the safe is accelerating, it must have a net force acting on it. Thus $T_{\text{R on S}} \neq w_{\text{E on S}}$. The tension in the rope is *not* equal to the weight of the safe.

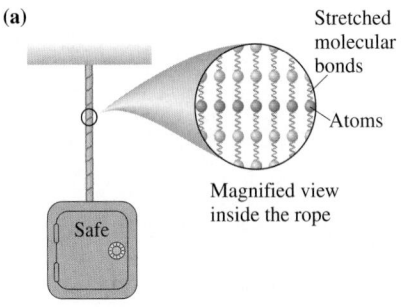

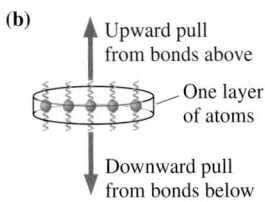

FIGURE 8.18 Tension forces within the rope are due to stretching the spring-like molecular bonds.

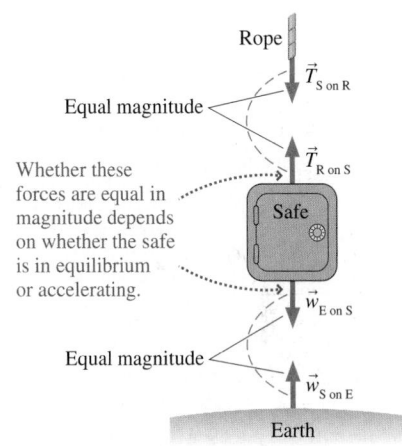

FIGURE 8.19 The forces between the rope, the safe, and the earth.

EXAMPLE 8.5 Pulling a rope

Figure 8.20a shows a student pulling horizontally with a 100 N force on a rope that is attached to a wall. In Figure 8.20b, two students in a tug-of-war pull on opposite ends of a rope with 100 N each. Is the tension in the second rope larger, smaller, or the same as that in the first?

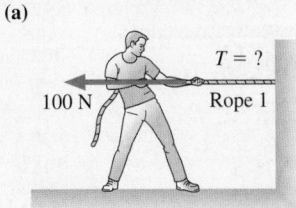

FIGURE 8.20 Pulling on a rope. Which produces a larger tension?

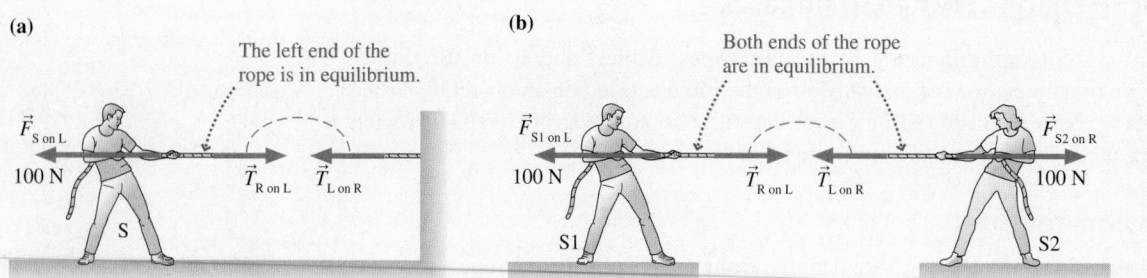

FIGURE 8.21 Analysis of tension forces.

SOLVE Surely pulling on a rope from both ends causes more tension than pulling on one end. Right? Before jumping to conclusions, let's analyze the situation carefully.

Suppose we make an imaginary slice through the rope, as shown in Figure 8.21. The right half of the rope pulls on the left half with $\vec{T}_{\text{R on L}}$ while the left half pulls back on the right half with $\vec{T}_{\text{L on R}}$. These two forces are an action/reaction pair, and their magnitude is what we *mean* by "the tension in the rope." The left half of the rope is in equilibrium, so force $\vec{T}_{\text{R on L}}$ has to balance exactly the 100 N force with which the student is pulling. That is, $T_{\text{R on L}} = F_{\text{S on L}}$. Thus

$$T_{\text{L on R}} = T_{\text{R on L}} = F_{\text{S on L}} = 100 \text{ N}$$

The first equality is based on Newton's third law (action/reaction pair). The second equality follows from Newton's second law (left half is in equilibrium). This reasoning leads us to the conclusion that the tension in the first rope is 100 N.

Now make an imaginary slice through the second rope. The left half of the rope is pulled by forces $\vec{T}_{\text{R on L}}$ and $\vec{F}_{\text{S1 on L}}$. This half of the rope is again in equilibrium, because the rope is at rest, so from Newton's second law

$$T_{\text{R on L}} = F_{\text{S1 on L}} = 100 \text{ N}$$

Similarly, the right half of the rope is pulled by forces $\vec{T}_{\text{L on R}}$ and $\vec{F}_{\text{S2 on R}}$. This piece of the rope is also in equilibrium, so

$$T_{\text{L on R}} = F_{\text{S2 on R}} = 100 \text{ N}$$

And $\vec{T}_{\text{L on R}}$ and $\vec{T}_{\text{R on L}}$ are again an action/reaction pair, so from Newton's third law

$$T_{\text{L on R}} = T_{\text{R on L}} = 100 \text{ N}$$

The tension in the rope has not changed! It is still 100 N.

You may have *assumed* that the student on the right in Figure 8.20b is doing something to the rope that the wall in Figure 8.20a does not do. But let's look more closely. Figure 8.22 shows a detailed view of the point at which the rope of Figure 8.20a is tied to the wall. Because the rope pulls on the wall with force $\vec{F}_{\text{R on W}}$, the wall must pull back on the rope (action/reaction pair) with force $\vec{F}_{\text{W on R}}$. And because the rope as a whole is in equilibrium, the wall's pull to the right must balance the student's pull to the left: $F_{\text{W on R}} = F_{\text{S on R}} = 100 \text{ N}$.

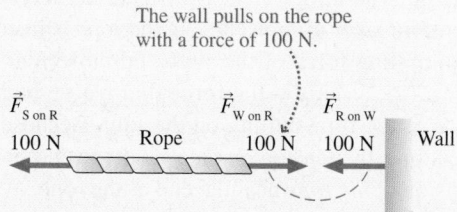

FIGURE 8.22 A closer look at the forces on the rope.

In other words, the wall in Figure 8.20a pulls the right end of the rope with a force of 100 N. The student in Figure 8.20b pulls the right end of the rope with a force of 100 N. The forces are the same in both situations! The rope does not care whether it is pulled by a wall or by a hand. It experiences the same forces in both cases, so the rope's tension is the same 100 N in both.

STOP TO THINK 8.4   All three 50 kg blocks are at rest. Is the tension in rope 2 greater than, less than, or equal to the tension in rope 1?

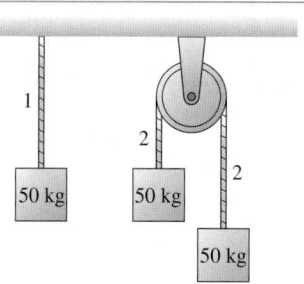

The Massless String Approximation

Example 8.5 showed that the tension is constant throughout a rope that is in equilibrium. But what happens if the rope is accelerating? For example, Figure 8.23a shows two blocks connected by a string and being pulled by force $\vec{F}$. Is the string's tension at the right end, where it pulls back on B, the same as the tension at the left end, where it pulls on A?

(a)

(b)

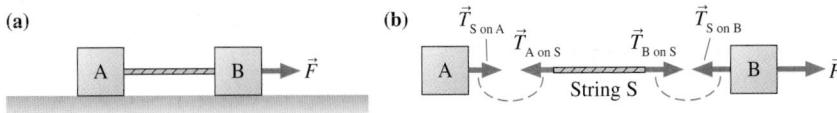

FIGURE 8.23 The string's tension pulls forward on block A, backward on block B.

Figure 8.23b shows the horizontal forces acting on the blocks and the string. If the string is accelerating, then it must have a net force applied to it. The only forces acting on the string are $\vec{T}_{\text{A on S}}$ and $\vec{T}_{\text{B on S}}$, so Newton's second law *for the string* is

$$(F_{\text{net}})_x = T_{\text{B on S}} - T_{\text{A on S}} = m_s a_x \tag{8.7}$$

where m_s is the mass of the string.

If the string is accelerating, then the tensions at the two ends can *not* be the same. In fact, you can see that

$$T_{\text{B on S}} = T_{\text{A on S}} + m_s a_x \tag{8.8}$$

The tension at the "front" of the string is higher than the tension at the "back." This difference in the tensions is necessary to accelerate the string! On the other hand, the tension is constant throughout a string in equilibrium ($a_x = 0$). This was the situation in Example 8.5.

Often in physics and engineering problems the mass of the string or rope is much less than the masses of the objects that it connects. In such cases, we can adopt the **massless string approximation.** In the limit $m_s \to 0$, Equation 8.8 becomes

$$T_{\text{B on S}} = T_{\text{A on S}} \quad \text{(massless string approximation)} \tag{8.9}$$

In other words, **the tension in a massless string is constant.** This is nice, but it isn't the primary justification for the massless string approximation.

Look again at Figure 8.23b. If $T_{\text{B on S}} = T_{\text{A on S}}$, then

$$\vec{T}_{\text{S on A}} = -\vec{T}_{\text{S on B}} \tag{8.10}$$

That is, the force on block A is equal and opposite to the force on block B. Forces $\vec{T}_{\text{S on A}}$ and $\vec{T}_{\text{S on B}}$ act *as if* they are an action/reaction pair of forces. Thus we can draw the simplified diagram of Figure 8.24 in which the string is missing and blocks A and B interact directly with each other through forces that we can call $\vec{T}_{\text{A on B}}$ and $\vec{T}_{\text{B on A}}$.

In other words, **if systems A and B interact with each other through a massless string, we can omit the string and treat forces $\vec{F}_{\text{A on B}}$ and $\vec{F}_{\text{B on A}}$ as if they are an action/reaction pair.** This is not literally true, because A and B are not in contact. Nonetheless, all a massless string does is transmit a force from A to B without changing the magnitude of that force. This is the real significance of the massless string approximation.

NOTE ▶ For problems in this book, you can assume that any strings or ropes are massless unless the problem explicitly states otherwise. The simplified view of Figure 8.24 is appropriate under these conditions. But if the string has a mass, it must be treated as a separate system. ◀

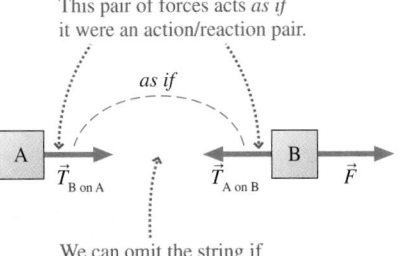

This pair of forces acts *as if* it were an action/reaction pair.

We can omit the string if we assume it is massless.

FIGURE 8.24 The massless string approximation allows systems A and B to act *as if* they are directly interacting.

EXAMPLE 8.6 **Comparing two tensions**

Blocks A and B in Figure 8.25 are connected by massless string 2 and pulled across a frictionless table by massless string 1. B has a larger mass than A. Is the tension in string 2 larger, smaller, or equal to the tension in string 1?

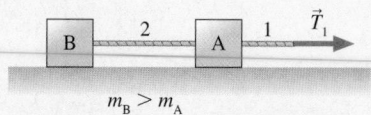

$m_B > m_A$

FIGURE 8.25 Blocks A and B are pulled across a frictionless table by massless strings.

MODEL The massless string approximation allows us to treat A and B *as if* they interact directly with each other. The blocks and strings must be accelerating because there's a force to the right and no friction.

SOLVE Both blocks have the same acceleration ($a_{Ax} = a_{Bx}$). B has a larger mass, so it may be tempting to conclude that string 2, which pulls B, has a greater tension than string 1, which pulls A. The flaw in this reasoning is that Newton's second law tells us only about the *net* force. The net force on B *is* larger than the net force on A, but the net force on A is *not* just the tension $\vec{T}_1$ in the forward direction. The tension in string 2 also pulls *backward* on A!

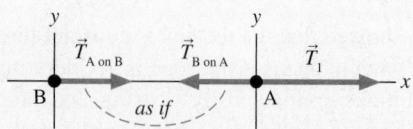

FIGURE 8.26 The horizontal forces on blocks A and B.

Figure 8.26 shows the horizontal forces in this frictionless situation. Forces $\vec{T}_{A \text{ on } B}$ and $\vec{T}_{B \text{ on } A}$ act *as if* they are an action/reaction pair. From Newton's third law,

$$T_{A \text{ on } B} = T_{B \text{ on } A} = T_2$$

where T_2 is the tension in string 2. From Newton's second law, the net force on A is

$$(F_{A \text{ net}})_x = T_1 - T_{B \text{ on } A} = T_1 - T_2 = m_A a_{Ax}$$

The net force on A is the *difference* in tension between string 1 pulling forward and string 2 pulling backward. The blocks are accelerating to the right, making $a_{Ax} > 0$, so

$$T_1 > T_2$$

The tension in string 2 is *smaller* than the tension in string 1.

ASSESS This is not an intuitively obvious result. A careful study of the reasoning in this example is worthwhile. An alternative analysis would note that $\vec{T}_1$ pulls *both* blocks, of combined mass $(m_A + m_B)$, whereas $\vec{T}_2$ pulls only block B. Thus string 1 must have the larger tension.

Pulleys

Strings and ropes often pass over pulleys. The application might be as simple as lifting a heavy weight or as complex as the internal cable-and-pulley arrangement that precisely moves a robot arm.

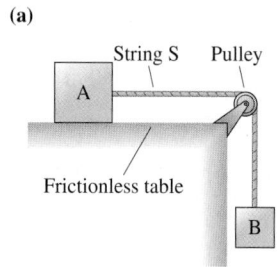

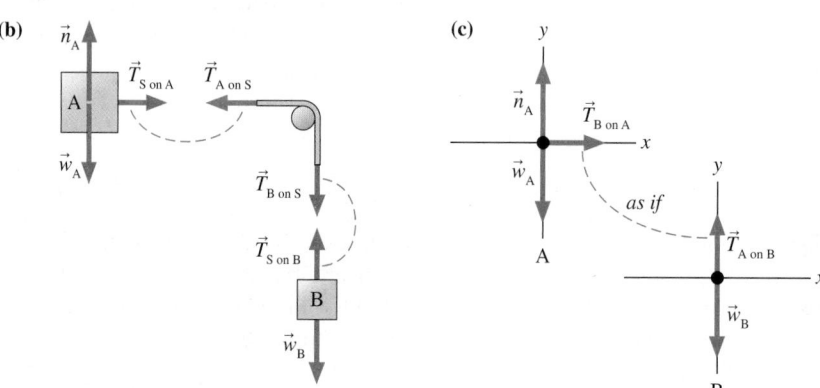

FIGURE 8.27 Blocks A and B are connected by a string that passes over a pulley.

Figure 8.27a shows a simple situation in which block B drags block A across a frictionless table as it falls. Figure 8.27b has drawn the objects separately and shown the forces. As the string moves, static friction between the string and pulley causes the pulley to turn. If we assume that

- The string *and* the pulley are both massless, and
- There is no friction where the pulley turns on its axle,

then no net force is needed to accelerate the string or turn the pulley. In this case,

$$T_{\text{A on S}} = T_{\text{B on S}}$$

In other words, the **tension in a massless string remains constant as it passes over a massless, frictionless pulley.**

Tension forces $\vec{T}_{\text{A on S}}$ and $\vec{T}_{\text{S on A}}$ are a true action/reaction pair, as are $\vec{T}_{\text{B on S}}$ and $T_{\text{S on B}}$. If we have a massless string and a massless, frictionless pulley, so that $T_{\text{A on S}} = T_{\text{B on S}}$, then it must be the case that

$$T_{\text{S on A}} = T_{\text{S on B}}$$

That is, the force of the string pulling on A equals the force of the string pulling on B.

Because of this, we can draw the simplified free-body diagram of Figure 8.27c in which the string and pulley are omitted. Forces $\vec{T}_{\text{A on B}}$ and $\vec{T}_{\text{B on A}}$ act *as if* they are an action/reaction pair, even though they are not opposite in direction. We can again say that A and B are systems that interact with each other *through the string*, and thus the force of A on B is paired with the force of B on A. The tension force gets "turned" by the pulley, which is why the two forces are not opposite each other but we can still equate their magnitudes.

STOP TO THINK 8.5 In Figure 8.27, is the tension in the string greater than, less than, or equal to the weight of block B?

8.5 Examples of Interacting-System Problems

We will conclude this chapter with several extended examples. Although the mathematics will be more involved than in any of our work up to this point, we will continue to emphasize the *reasoning* one uses in approaching problems such as these. The solutions will be based on Problem-Solving Strategy 8.1. In fact, these problems are now reaching a level of complexity that, for all practical purposes, it becomes impossible to work them unless you are following a well-planned strategy. Our earlier emphasis upon identifying forces and using free-body diagrams will now really begin to pay off!

Activ Physics 2.10, 2.11

EXAMPLE 8.7 Mountain climbing

A 90 kg mountain climber is suspended from the ropes shown in Figure 8.28a. The maximum tension that rope 3 can withstand before breaking is 1500 N. What is the smallest that angle θ can become before the rope breaks and the climber falls into the gorge?

MODEL Climber C, who can be modeled as a particle, is one system. The other point where forces are exerted is the knot, where the three ropes are tied together. We'll consider knot K to be a second system. Both systems are in static equilibrium. We'll assume massless ropes.

VISUALIZE Figure 8.28b shows two free-body diagrams. Forces $\vec{T}_{\text{C on K}}$ and $\vec{T}_{\text{K on C}}$ are not, strictly speaking, an action/reaction pair because the climber is not in contact with the knot. But if the ropes are massless, $\vec{T}_{\text{C on K}}$ and $\vec{T}_{\text{K on C}}$ act *as if* they are an action/reaction pair.

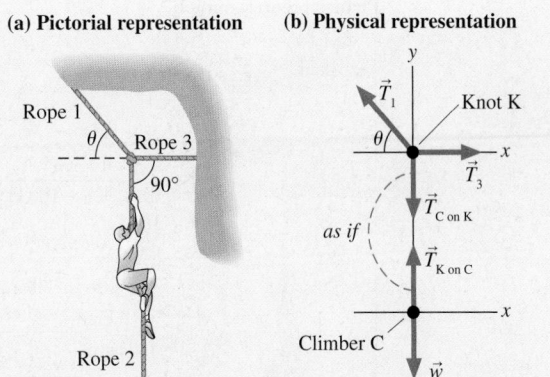

(a) Pictorial representation   **(b) Physical representation**

FIGURE 8.28 A mountain climber hanging from ropes and the free-body diagrams of the knot and of the climber.

SOLVE This is static equilibrium, so the net force on the climber and on the knot are zero. For the climber:

$$\sum (F_{\text{on C}})_y = T_{\text{K on C}} - w = T_{\text{K on C}} - mg = 0$$

And for the knot:

$$\sum (F_{\text{on K}})_x = T_3 - T_1 \cos\theta = 0$$
$$\sum (F_{\text{on K}})_y = T_1 \sin\theta - T_{\text{C on K}} = 0$$

From Newton's third law,

$$T_{\text{C on K}} = T_{\text{K on C}}$$

But $T_{\text{K on C}} = mg$, from the climber's equation, so $T_{\text{C on K}} = mg$. Using this gives us the knot's equations:

$$T_1 \cos\theta = T_3$$
$$T_1 \sin\theta = mg$$

Dividing the second of these by the first gives

$$\frac{T_1 \sin\theta}{T_1 \cos\theta} = \tan\theta = \frac{mg}{T_3}$$

If angle θ is too small, tension T_3 will exceed 1500 N. The smallest possible θ, at which T_3 reaches its maximum value of 1500 N, is

$$\theta_{\min} = \tan^{-1}\left(\frac{mg}{T_{3\,\max}}\right) = \tan^{-1}\left(\frac{(90\text{ kg})(9.80\text{ m/s}^2)}{1500\text{ N}}\right) = 30.5°$$

EXAMPLE 8.8 **The show must go on!**

A 200 kg set used in a play is stored in the loft above the stage. The rope holding the set passes up and over a pulley, then is tied backstage. The director tells a 100 kg stagehand to lower the set. When he unties the rope, the set falls and the unfortunate man is hoisted into the loft. What is the stagehand's acceleration?

MODEL The systems of interest are the stagehand M and the set S, which we will model as particles. Assume a massless rope and a massless, frictionless pulley.

VISUALIZE Figure 8.29 shows the pictorial and physical representations. The man's acceleration a_{My} is positive while the set's acceleration a_{Sy} is negative. These two accelerations have the same magnitude, because the two systems are connected by a rope, but they have opposite signs. Thus the acceleration constraint is $a_{\text{Sy}} = -a_{\text{My}}$. Forces $\vec{T}_{\text{M on S}}$ and $\vec{T}_{\text{S on M}}$ are not literally an action/reaction pair, but they act *as if* they are because the rope is massless and the pulley is massless and frictionless. Notice that the pulley has "turned" the tension force so that

$\vec{T}_{\text{M on S}}$ and $\vec{T}_{\text{S on M}}$ are *parallel* to each other rather than opposite, as members of a true action/reaction pair would have to be.

SOLVE Newton's second law for the man and the set are

$$\sum (F_{\text{on M}})_y = T_{\text{S on M}} - w_{\text{M}} = T_{\text{S on M}} - m_{\text{M}}g = m_{\text{M}}a_{\text{My}}$$
$$\sum (F_{\text{on S}})_y = T_{\text{M on S}} - w_{\text{S}}$$
$$= T_{\text{M on S}} - m_{\text{S}}g = m_{\text{S}}a_{\text{Sy}} = -m_{\text{S}}a_{\text{My}}$$

Only the y-equations are needed. Notice that we used the acceleration constraint in the last step. Newton's third law is

$$T_{\text{M on S}} = T_{\text{S on M}} = T$$

where we can drop the subscripts and call the tension simply T. With this substitution, the two second-law equations can be written

$$T - m_{\text{M}}g = m_{\text{M}}a_{\text{My}}$$
$$T - m_{\text{S}}g = -m_{\text{S}}a_{\text{My}}$$

Pictorial representation

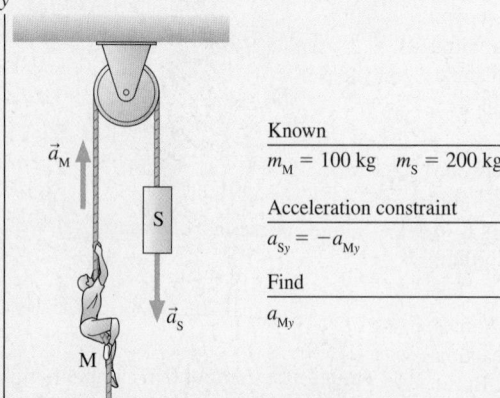

Known
$m_{\text{M}} = 100\text{ kg}$ $m_{\text{S}} = 200\text{ kg}$

Acceleration constraint
$a_{\text{Sy}} = -a_{\text{My}}$

Find
a_{My}

Physical representation

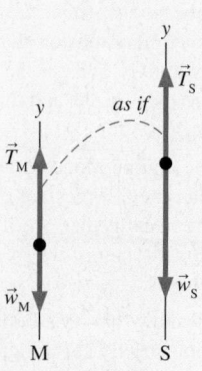

FIGURE 8.29 The pictorial and physical representations for Example 8.8.

These are simultaneous equations in the two unknowns T and a_{My}. We can eliminate T by subtracting the second equation from the first to give

$$(m_S - m_M)g = (m_S + m_M)a_{My}$$

Finally, we can solve for the hapless stagehand's acceleration:

$$a_{My} = \frac{m_S - m_M}{m_S + m_M}g = \frac{100 \text{ kg}}{300 \text{ kg}}9.80 \text{ m/s}^2 = 3.27 \text{ m/s}^2$$

This is also the acceleration with which the set falls. If the rope's tension was needed, we could now find it from $T = m_M a_{My} + m_M g$.

ASSESS If the stagehand weren't holding on, the set would fall with free-fall acceleration g. The stagehand acts as a *counterweight* to reduce the acceleration.

EXAMPLE 8.9 **A not-so-clever bank robbery**

Bank robbers have pushed a 1000 kg safe to a second-story floor-to-ceiling window. They plan to break the window, then lower the safe 3.0 m to their truck. Not being too clever, they stack up 500 kg of furniture, tie a rope between the safe and the furniture, and place the rope over a pulley. Then they push the safe out the window. What is the safe's speed when it hits the truck? The coefficient of kinetic friction between the furniture and the floor is 0.5.

MODEL This is a continuation of the situation that we analyzed in Figures 8.16 and 8.27. The systems of interest are the safe S and the furniture F, which we will model as particles. We will assume a massless rope and a massless, frictionless pulley.

VISUALIZE The pictorial representation in Figure 8.30 establishes a coordinate system and defines the symbols that will be needed to calculate the safe's motion. The safe and the furniture are tied together, so their accelerations have the same magnitude. The safe has a y-component of acceleration a_{Sy} that is negative because the safe accelerates in the $-y$-direction. The furniture has an x-component a_F that is positive. Thus the acceleration constraint is

$$a_{Fx} = -a_{Sy}$$

The free-body diagrams of Figure 8.30 are modeled after Figure 8.27 but now include a kinetic friction force on the furniture.

Forces $\vec{T}_{F \text{ on } S}$ and $\vec{T}_{S \text{ on } F}$ act *as if* they are an action/reaction pair, so they have been connected with a dotted line.

SOLVE We can write Newton's second law directly from the free-body diagrams. For the furniture,

$$\sum (F_{\text{on } F})_x = T_{S \text{ on } F} - f_k = T - f_k = m_F a_{Fx} = -m_F a_{Sy}$$
$$\sum (F_{\text{on } F})_y = n - w_F = n - m_F g = 0$$

And for the safe,

$$\sum (F_{\text{on } S})_y = T_{F \text{ on } S} - w_S = T - m_S g = m_S a_{Sy}$$

Notice how we used the acceleration constraint in the first equation. We also went ahead and made use of Newton's third law: $T_{F \text{ on } S} = T_{S \text{ on } F} = T$. We have one additional piece of information, the model of kinetic friction:

$$f_k = \mu_k n = \mu_k m_F g$$

where we used the y-equation of the furniture to deduce that $n = m_F g$. Substitute this result for f_k into the x-equation of the furniture, then rewrite the furniture's x-equation and the safe's y-equation:

$$T - \mu_k m_F g = -m_F a_{Sy}$$
$$T - m_S g = m_S a_{Sy}$$

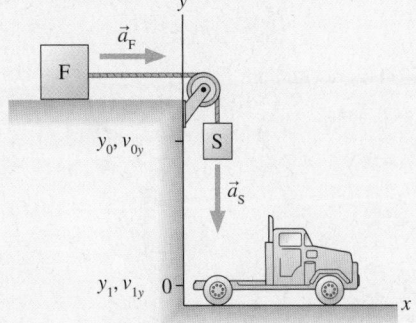

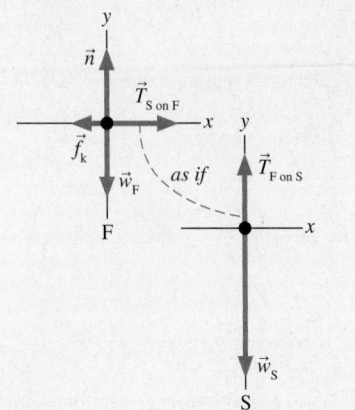

Known

$y_0 = 3$ m	$v_{0y} = 0$ m/s
$y_1 = 0$ m	
$m_F = 500$ kg	$m_S = 1000$ kg

Acceleration constraint

$a_{Fx} = -a_{Sy}$

Find

v_1

FIGURE 8.30 The pictorial and physical representations for Example 8.9.

We have succeeded in reducing our knowledge to two simultaneous equations in the two unknowns a_{Sy} and T. Subtract the second equation from the first to eliminate T:

$$(m_S - \mu_k m_F)g = -(m_S + m_F)a_{Sy}$$

Finally, solve for the safe's acceleration:

$$a_{Sy} = -\left(\frac{m_S - \mu_k m_F}{m_S + m_F}\right)g$$

$$= -\frac{1000 \text{ kg} - 0.5(500 \text{ kg})}{1000 \text{ kg} + 500 \text{ kg}}9.80 \text{ m/s}^2 = -4.9 \text{ m/s}^2$$

Now we need to calculate the kinematics of the falling safe. Because the time of the fall is not known or needed, we can use

$$v_{1y}^2 = v_{0y}^2 + 2a_{Sy}\Delta y = 0 + 2a_{Sy}(y_1 - y_0) = -2a_{Sy}y_0$$

$$v_1 = \sqrt{-2a_{Sy}y_0} = \sqrt{-2(-4.9 \text{ m/s}^2)(3.0 \text{ m})} = 5.42 \text{ m/s}$$

The value of v_{1y} is negative, but we only needed to find the speed so we took the absolute value. It seems unlikely that the truck will survive the impact of the 1000 kg safe!

EXAMPLE 8.10 Pushing a package

A 40 kg boy works at his dad's hardware store. One of the boy's jobs is to unload the delivery truck. He places each package on a 30° ramp and shoves it up the ramp into the storeroom. He needs to shove the package with an acceleration of at least 1.0 m/s² in order for the package to make it to the top of the ramp. One day the ground is wet with rain and he's wearing slick leather-soled shoes. The coefficient of static friction between his shoes and the ground is only 0.25. The largest package of the day is 15 kg, and its coefficient of kinetic friction on the ramp is 0.40. Can he give the package a big enough shove to reach the top of the ramp without his feet slipping?

MODEL The systems of interest are the boy B and the package P, which we will model as particles.

VISUALIZE There's a lot of information in this problem, so the pictorial representation of Figure 8.31 is essential. The package is moving up an incline whereas the boy, if he slips, will move horizontally. Consequently, it is useful to give them different coordinate systems. The free-body diagrams show that the boy pushes the package with force $\vec{F}_{B \text{ on } P}$ and the package pushes back with force $\vec{F}_{P \text{ on } B}$. If static friction does its job, it must point *forward* to prevent the boy's feet from slipping backward. To answer the question, we'll first calculate how much static friction is needed for the boy to push the package with an acceleration of 1.0 m/s². Then we'll compare that to the maximum possible static friction $f_{s \text{ max}}$.

SOLVE Now we're ready to write Newton's second law. The boy is in static equilibrium, with $\vec{F}_{net} = \vec{0}$, so his equations are

$$\sum (F_{\text{on } B})_x = f_s - F_{P \text{ on } B}\cos\theta = f_s - F\cos\theta = 0$$

$$\sum (F_{\text{on } B})_y = n_B - w_B - F_{P \text{ on } B}\sin\theta$$
$$= n_B - m_B g - F\sin\theta = 0$$

The package is accelerating up the ramp as he pushes it, so the package's equations are

$$\sum (F_{\text{on } P})_x = F_{B \text{ on } P} - f_k - w_P\sin\theta$$
$$= F - f_k - m_P g\sin\theta = m_P a_x$$

$$\sum (F_{\text{on } P})_y = n_P - w_P\cos\theta = n_P - m_P g\cos\theta = 0$$

We went ahead and made use of Newton's third law: $F_{P \text{ on } B} = F_{B \text{ on } P} = F$. The package's y-equation tells us that $n_P = m_P g\cos\theta$, so the kinetic friction on the package is

$$f_k = \mu_k n_P = \mu_k m_P g\cos\theta$$

If we substitute this into the package's x-equation, we can solve for force F:

$$F - \mu_k m_P g\cos\theta - m_P g\sin\theta = m_P a_x$$
$$F = m_P(a_x + g\sin\theta + \mu_k g\cos\theta) = 139 \text{ N}$$

Pictorial representation

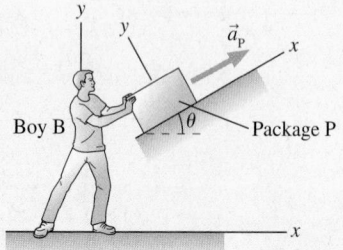

Known	
$m_B = 40$ kg	$m_P = 15$ kg
$\theta = 30°$	$a_x = 1.0$ m/s²
$\mu_s = 0.25$	$\mu_k = 0.40$

Find

Compare f_s to $f_{s \text{ max}}$

Physical representation

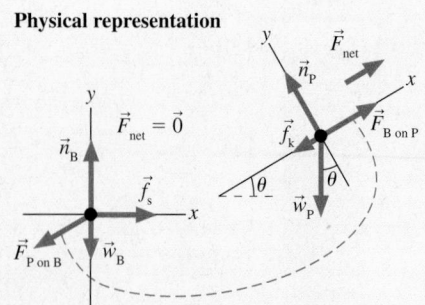

FIGURE 8.31 The pictorial and physical representations for Example 8.10.

This is the size of the force that will accelerate the package up the ramp at 1.0 m/s². If we now use this in the boy's x-equation, we find

$$f_s = F\cos\theta = 120 \text{ N}$$

The boy *needs* this much static friction to push the package without slipping. But needing 120 N of friction doesn't mean that 120 N is available. The maximum possible static friction is $f_{s\,max} = \mu_s n_B$. The normal force acting on the boy is not simply his weight. The normal force is affected by the vertical component of $\vec{F}_{P \text{ on } B}$. From his y-equation we find

$$n_B = m_B g + F\sin\theta = 462 \text{ N}$$

Thus the maximum static friction without slipping is

$$f_{s\,max} = \mu_s n_B = 115 \text{ N}$$

Consequently, the boy *cannot* shove hard enough without slipping.

ASSESS This is an excellent illustration of how crucial it is to focus on clarifying information, identifying forces, and drawing free-body diagrams. The rest of the problem was not trivial, but we could work our way through it with confidence after having found the free-body diagrams. It would be hopeless, even for an experienced physicist, to try to go directly to Newton's laws without this analysis.

Newton's third law brings us to the end of Part I on Newtonian mechanics. Rather than pursue ever more complex dynamics problems, our road will turn in a new direction as we begin to look at the new and important concepts of momentum and energy. Forces will remain with us, always forming the most basic level of our understanding of dynamics, but they need to begin to share the stage with newer ideas.

STOP TO THINK 8.6 A small car is pushing a larger truck that has a dead battery. The mass of the truck is larger than the mass of the car. Which of the following statements is true?

a. The car exerts a force on the truck, but the truck doesn't exert a force on the car.
b. The car exerts a larger force on the truck than the truck exerts on the car.
c. The car exerts the same amount of force on the truck as the truck exerts on the car.
d. The truck exerts a larger force on the car than the car exerts on the truck.
e. The truck exerts a force on the car, but the car doesn't exert a force on the truck.

SUMMARY

The goal of Chapter 8 has been to learn to use Newton's third law to understand interacting systems.

GENERAL PRINCIPLES

Newton's Third Law

Every force occurs as one member of an **action/reaction pair** of forces. The two members of an action/reaction pair:

- Act on two *different* objects.

- Are equal in magnitude but opposite in direction:

$$\vec{F}_{\text{A on B}} = -\vec{F}_{\text{B on A}}$$

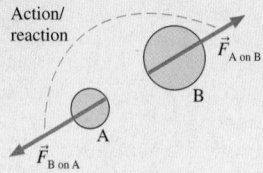

Solving Interacting-System Problems

MODEL   Choose the systems of interest.

VISUALIZE
   Pictorial representation:
      Sketch and define coordinates.
      Identify acceleration constants.
   Physical representation:
      Draw a separate free-body diagram for each system.
      Connect action/reaction pairs with dotted lines.

SOLVE   Write Newton's second law for each system.
      Include *all* forces acting *on* each system.
      Use Newton's third law to equate the magnitudes
         of action/reaction pairs.
      Include acceleration constraints and friction.

ASSESS   Is the result reasonable?

IMPORTANT CONCEPTS

Interacting systems and the environment

Two systems interact by exerting forces on each other.
Systems whose motion is not of interest form the environment.
The systems of interest interact with the environment, but
those interactions can be considered external forces.

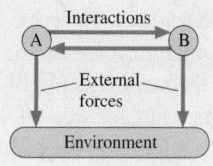

APPLICATIONS

Acceleration constraints

Objects that are constrained
to move together must have
accelerations of equal
magnitude: $a_{\text{A}} = a_{\text{B}}$.
This must be expressed in
terms of components, such
as $a_{\text{Ax}} = -a_{\text{By}}$.

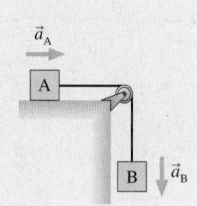

Strings and pulleys

The tension in a string or rope pulls in
both directions. The tension is constant
in a string if the string is:

- Massless, or

- In equilibrium

Systems connected by massless strings
passing over massless, frictionless
pulleys act *as if* they interact via an
action/reaction pair of forces.

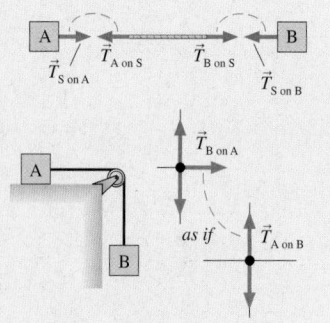

TERMS AND NOTATION

interaction	propulsion	acceleration constraint
action/reaction pair	Newton's third law	massless string approximation
external force		

EXERCISES AND PROBLEMS

The icon indicates that the problem can be done on a Dynamics Worksheet.

Exercises

Section 8.2 Identifying Action/Reaction Pairs

1. A weight lifter stands up from a squatting position while holding a heavy barbell across his shoulders. Identify all action/reaction pairs, show them on a figure, then draw free-body diagrams for the weight lifter and the barbell. Use dotted lines to connect the members of an action/reaction pair.

2. A softball player is throwing the ball. Her arm has come forward to where it is beside her head, but she hasn't yet released the ball. Identify all action/reaction pairs, show them on a figure, then draw free-body diagrams for the ball player and the ball. Use dotted lines to connect the members of an action/reaction pair.

3. A soccer ball and a bowling ball roll across a hard floor and collide head on. Identify all action/reaction pairs during the time that the balls are in contact, show them on a figure, then draw free-body diagrams for each ball. Use dotted lines to connect the members of an action/reaction pair. Rolling friction is negligible.

4. A mountain climber is using a massless rope to pull a bag of supplies up a 45° slope. Identify all action/reaction pairs, show them on a figure, then draw free-body diagrams for the mountain climber, the rope, and the bag. Use dotted lines to connect the members of an action/reaction pair.

5. Block A in Figure Ex8.5 is heavier than block B and is sliding down the incline. All surfaces have kinetic friction. Identify all action/reaction pairs, show them on a figure, then draw free-body diagrams for both blocks, the rope, and the pulley. The pulley holder can be considered part of the earth. Use dotted lines to connect the members of an action/reaction pair.

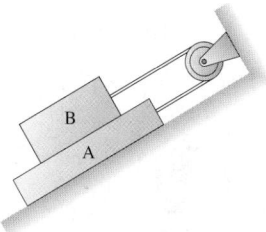

FIGURE EX8.5

Section 8.3 Newton's Third Law

6. Figure Ex8.6 shows two strong magnets on opposite sides of a small table. The long-range attractive force between the magnets keeps the lower magnet in place.
   a. Identify all action/reaction pairs, show them on a figure, then draw free-body diagrams for both magnets and the table. Use dotted lines to connect the members of an action/reaction pair.

   b. Suppose the weight of the table is 20 N, the weight of each magnet is 2.0 N, and the magnetic force on the lower magnet is three times its weight. Find the magnitude of each of the forces shown on your free-body diagrams.

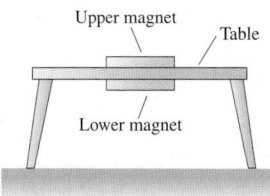

FIGURE EX8.6

7. a. How much force does an 80 kg astronaut exert on his chair while sitting at rest on the launch pad?
   b. How much force does the astronaut exert on his chair while accelerating straight up at 10 m/s^2?

8. A 1000 kg car pushes a 2000 kg truck that has a dead battery. When the driver steps on the accelerator, the drive wheels of the car push against the ground with a force of 4500 N.
   a. What is the magnitude of the force of the car on the truck?
   b. What is the magnitude of the force of the truck on the car?

9. Blocks with masses of 1 kg, 2 kg, and 3 kg are lined up in a row on a frictionless table. All three are pushed forward by a 12 N force applied to the 1 kg block. How much force does the 2 kg block exert on the 3 kg block? How much force does the 2 kg block exert on the 1 kg block?

10. An 80 kg spacewalking astronaut pushes off a 640 kg satellite, exerting a 100 N force for the 0.50 s it takes him to straighten his arms. How far apart are the astronaut and the satellite after 1.0 min?

Section 8.4 Ropes and Pulleys

11. What is the tension in the rope of Figure Ex8.11?

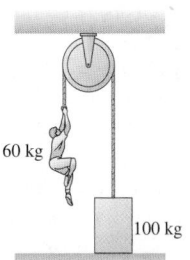

FIGURE EX8.11

12. Jimmy has caught two fish in Yellow Creek. He has tied the line holding the 3.0 kg steelhead trout to the tail of the 1.5 kg carp. To show the fish to a friend, he lifts upward on the carp with a force of 60 N.
    a. Draw separate free-body diagrams for the trout and the carp. Label all forces, then use dotted lines to connect action/reaction pairs or forces that act as if they are a pair.
    b. Rank in order, from largest to smallest, the magnitudes of all the forces shown on your free-body diagrams. Explain your reasoning.

13. A 2-m-long, 500 g rope pulls a 10 kg block of ice across a horizontal, frictionless surface. The block accelerates at 2.0 m/s². How much force pulls forward on (a) the ice, (b) the rope?

14. The cable cars in San Francisco are pulled along their tracks by an underground steel cable that moves along at 9.5 mph. The cable is driven by large motors at a central power station and extends, via an intricate pulley arrangement, for several miles beneath the city streets. The length of a cable stretches by up to 100 ft during its lifetime. To keep the tension constant, the cable passes around a 1.5-m-diameter "tensioning pulley" that rolls back and forth on rails, as shown in Figure Ex8.14. A 2000 kg block is attached to the tensioning pulley's cart,

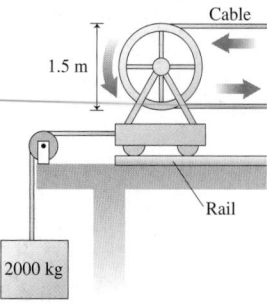

FIGURE EX8.14

via a rope and pulley, and is suspended in a deep hole. What is the tension in the cable car's cable?

15. A mobile at the art museum has a 2.0 kg steel cat and 4.0 kg steel dog suspended from a lightweight cable, as shown in Figure Ex8.15. It is found that $\theta_1 = 20°$ when the center rope is adjusted to be perfectly horizontal. What are the tension and the angle of rope 3?

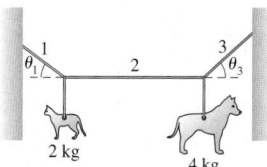

FIGURE EX8.15

Problems

16. A massive steel cable drags a 20 kg block across a horizontal, frictionless surface. A 100 N force applied to the cable causes the block to reach a speed of 4.0 m/s in a distance of 2.0 m. What is the mass of the cable?

17. A massive steel cable drags a 20 kg block across a horizontal, frictionless surface. A 100 N force applied to the cable causes the block to reach a speed of 4.0 m/s in 2.0 s. What is the difference in tension between the two ends of the cable?

18. A 1.0-m-long massive steel cable drags a 20 kg block across a horizontal, frictionless surface. A 100 N force applied to the cable causes the block to travel 4.0 m in 2.0 s. Graph the tension in the cable as a function of position along the cable, starting at the point where the cable is attached to the block.

19. A 3.0-m-long, 2.2 kg rope is suspended from the ceiling. Graph the tension in the rope as a function of position along the rope, starting from the bottom.

20. The sled dog in Figure P8.20 drags sleds A and B across the snow. The coefficient of friction between the sleds and the snow is 0.10. If the tension in rope 1 is 150 N, what is the tension in rope 2?

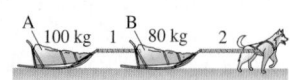

FIGURE P8.20

21. While driving to work last year, I was holding my coffee mug in my left hand while changing the CD with my right hand. Then the cell phone rang, so I placed the mug on the flat part of my dashboard. Then, believe it or not, a deer ran out of the woods and on to the road right in front of me. Fortunately, my reaction time was zero, and I was able to stop from a speed of 20 m/s in a mere 50 m, just barely avoiding the deer. Later tests revealed that the static and kinetic coefficients of friction of the coffee mug on the dash are 0.50 and 0.30, respectively; the coffee and mug had a mass of 0.50 kg; and the mass of the deer was 120 kg. Did my coffee mug slide?

22. a. Describe how a car accelerates from rest. Your explanation should be in terms of forces and physical laws. You should include and use a free-body diagram.
    b. Why can a car accelerate but a house cannot? Again, your explanation should be in terms of forces and their properties.
    c. Two-thirds of the weight of a 1500 kg car rests on the drive wheels. What is the maximum acceleration of this car on a concrete surface?

23. A Federation starship (2.0×10^6 kg) uses its tractor beam to pull a shuttlecraft (2.0×10^4 kg) aboard from a distance of 10 km away. The tractor beam exerts a constant force of 4.0×10^4 N on the shuttlecraft. Both spacecraft are initially at rest. How far does the starship move as it pulls the shuttlecraft aboard?

24. Bob, who has a mass of 75 kg, can throw a 500 g rock with a speed of 30 m/s. The distance through which his hand moves as he accelerates the rock forward from rest until he releases it is 1.0 m.
    a. What constant force must Bob exert on the rock to throw it with this speed?
    b. If Bob is standing on frictionless ice, what is his recoil speed after releasing the rock?

25. Two packages at UPS start sliding down the 20° ramp shown in Figure P8.25. Package A has a mass of 5.0 kg and a coefficient of friction of 0.20. Package B has a mass of 10 kg and a coefficient of friction of 0.15. How long does it take package A to reach the bottom?

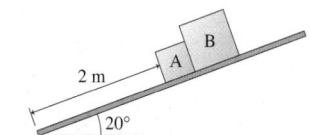

FIGURE P8.25

26. Figure P8.26 shows two 1.0 kg blocks connected by a rope. A second rope hangs beneath the lower block. Both ropes have a mass of 250 g. The entire assembly is accelerated upward at 3.0 m/s² by force $\vec{F}$.
    a. What is F?
    b. What is the tension at the top end of rope 1?
    c. What is the tension at the bottom end of rope 1?
    d. What is the tension at the top end of rope 2?

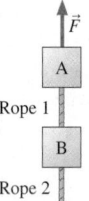

FIGURE P8.26

27. The 1.0 kg block in Figure P8.27 is tied to the wall with a rope. It sits on top of the 2.0 kg block. The lower block is pulled to the right with a tension force of 20 N. The coefficient of kinetic friction at both the lower and upper surfaces of the 2.0 kg block is $\mu_k = 0.40$.
    a. What is the tension in the rope holding the 1.0 kg block to the wall?
    b. What is the acceleration of the 2.0 kg block?

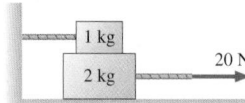

FIGURE P8.27

28. The lower block in Figure P8.28 is pulled on by a rope with a tension force of 20 N. The coefficient of kinetic friction between the lower block and the surface is 0.30. The coefficient of kinetic friction between the lower block and the upper block is also 0.30. What is the acceleration of the 2.0 kg block?

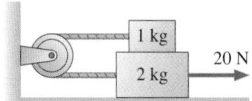

FIGURE P8.28

29. A rope attached to a 20 kg wood sled pulls the sled up a 20° snow-covered hill. A 10 kg wood box rides on top of the sled. If the tension in the rope steadily increases, at what value of the tension does the box slip?

30. Mass m_1 on the frictionless table of Figure P8.30 is connected by a string through a hole in the table to a hanging mass m_2. With what speed must m_1 rotate in a circle of radius r if m_2 is to remain hanging at rest?

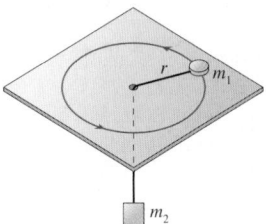

FIGURE P8.30

31. You see the boy next door trying to push a crate down the sidewalk. He can barely keep it moving, and his feet occasionally slip. You start to wonder how heavy the crate is. You call to ask the boy his mass, and he replies "50 kg." From your recent physics class you estimate that the static and kinetic coefficients of friction are 0.8 and 0.4 for the boy's shoes, and 0.5 and 0.2 for the crate. Estimate the mass of the crate.

32. The coefficient of static friction is 0.60 between the two blocks in Figure P8.32. The coefficient of kinetic friction between the lower block and the floor is 0.20. Force $\vec{F}$ causes both blocks to cross a distance of 5.0 m, starting from rest. What is the least amount of time in which this motion can be completed without the top block sliding on the lower block?

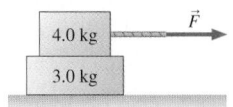

FIGURE P8.32

33. The 100 kg block in Figure P8.33 takes 6.0 s to reach the floor after being released from rest. What is the mass of the block on the left?

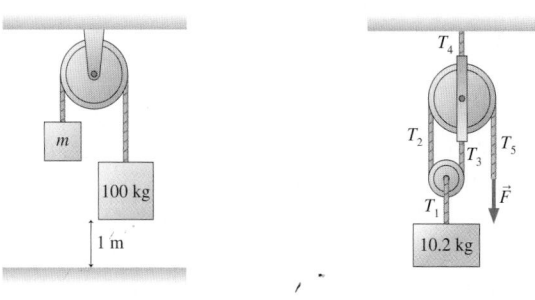

FIGURE P8.33   **FIGURE P8.34**

34. The 10.2 kg block in Figure P8.34 is held in place by the massless rope passing over two massless, frictionless pulleys. Find the tensions T_1 to T_5 and the magnitude of force $\vec{F}$.

35. The coefficient of kinetic friction between the 2.0 kg block in Figure P8.35 and the table is 0.30. What is the acceleration of the 2.0 kg block?

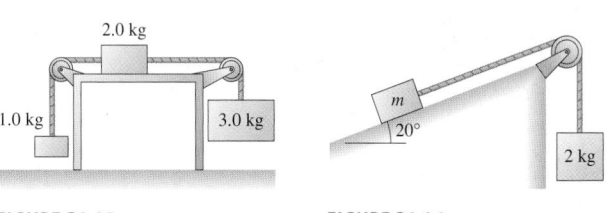

FIGURE P8.35   **FIGURE P8.36**

36. Figure P8.36 shows a block of mass m resting on a 20° slope. The block has coefficients of friction $\mu_s = 0.80$ and $\mu_k = 0.50$ with the surface. It is connected via a massless string over a massless, frictionless pulley to a hanging block of mass 2.0 kg.
    a. What is the minimum mass m that will stick and not slip?
    b. If this minimum mass is nudged ever so slightly, it will start being pulled up the incline. What acceleration will it have?

37. A 4.0 kg box is on a frictionless 35° slope and is connected via a massless string over a massless, frictionless pulley to a hanging 2.0 kg weight. The picture for this situation is similar to Figure P8.36.
    a. What is the tension in the string if the 4.0 kg box is *held* in place, so that it cannot move?
    b. If the box is then released, which way will it move on the slope?
    c. What is the tension in the string once the box begins to move?

38. The 1.0 kg physics book in Figure P8.38 is connected by a string to a 500 g coffee cup. The book is given a push up the slope and released with a speed of 3.0 m/s. The coefficients of friction are $\mu_s = 0.50$ and $\mu_k = 0.20$.
    a. How far does the book slide?
    b. At the highest point, does the book stick to the slope, or does it slide back down?

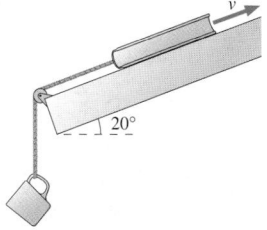

FIGURE P8.38

39. The 2000 kg cable car shown in Figure P8.39 descends a 200-m-high hill. In addition to its brakes, the cable car controls its speed by pulling an 1800 kg counterweight up the other side of the hill. The rolling friction of both the cable car and the counterweight are negligible.
    a. How much braking force does the cable car need to descend at constant speed?
    b. One day the brakes fail just as the cable car leaves the top on its downward journey. What is the runaway car's speed at the bottom of the hill?

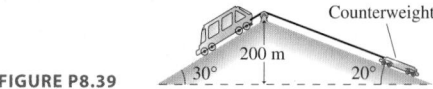

FIGURE P8.39

40. In Figure P8.40, find an expression for the acceleration of m_1.
    Hint: Think carefully about the acceleration constraint.

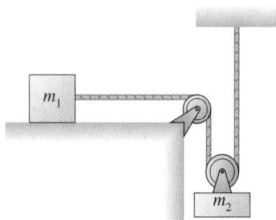

FIGURE P8.40

41. What is the acceleration of the 2.0 kg block in Figure P8.41 across the frictionless table?
    Hint: Think carefully about the acceleration constraint.

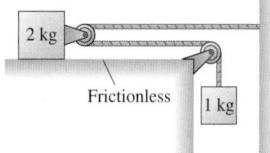

FIGURE P8.41

42. A house painter uses the chair and pulley arrangement of Figure P8.42 to lift himself up the side of a house. The painter's mass is 70 kg and the chair's mass is 10 kg. With what force must he pull down on the rope in order to accelerate upward at 0.20 m/s²?

FIGURE P8.42

43. A 70 kg tightrope walker stands at the center of a rope. The rope supports are 10 m apart and the rope sags 10° at each end. The tightrope walker crouches down, then leaps straight up with an acceleration of 8.0 m/s² to catch a passing trapeze. What is the tension in the rope as he jumps?

44. Find an expression for the magnitude of the horizontal force F in Figure P8.44 for which m_1 does not slip either up or down along the wedge. All surfaces are frictionless.

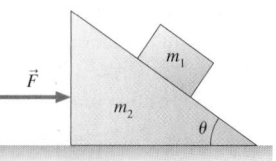

FIGURE P8.44

Problems 45 and 46 show the free-body diagrams of two interacting systems. For each of these, you are to
   a. Write a realistic problem for which these are the correct free-body diagrams. Be sure that the answer your problem requests is consistent with the diagrams shown.
   b. Finish the solution of the problem.

45.   46.

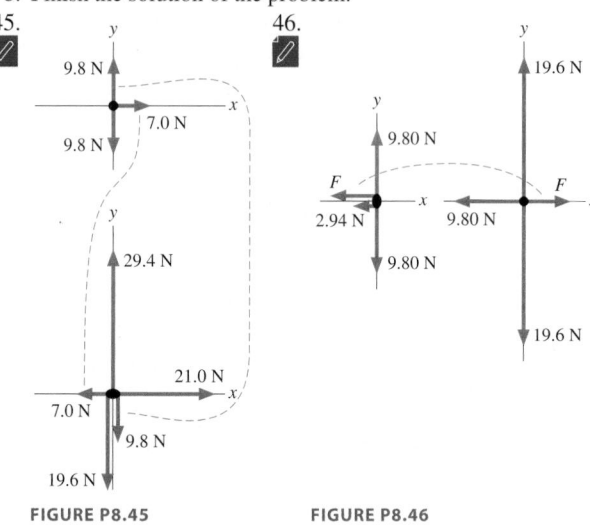

FIGURE P8.45   **FIGURE P8.46**

Challenge Problems

47. A 100 g ball of clay is thrown horizontally with a speed of 10 m/s toward a 900 g block resting on a frictionless surface. It hits the block and sticks. The clay exerts a constant force on the block during the 10 ms it takes the clay to come to rest relative to the block. After 10 ms, the block and the clay are sliding along the surface as a single system.
    a. What is their speed after the collision?
    b. What is the force of the clay on the block during the collision?
    c. What is the force of the block on the clay?

    NOTE ► This problem can be worked using the conservation laws you will be learning in the next few chapters. However, here you're asked to solve the problem using Newton's laws. ◄

48. A 100 kg basketball player can leap straight up in the air to a height of 80 cm, as shown in Figure CP8.48. You can understand how by analyzing the situation as follows:
    a. The player bends his legs until the upper part of his body has dropped by 60 cm, then he begins his jump. Draw separate free-body diagrams for the player and for the floor *as* he is jumping, but before his feet leave the ground.

b. Is there a net force on the player as he jumps (before his feet leave the ground)? How can that be? Explain.

c. With what speed must the player leave the ground to reach a height of 80 cm?

d. What was his acceleration, assumed to be constant, as he jumped?

e. Suppose the player jumps while standing on a bathroom scale that reads in newtons. What does the scale read before he jumps, as he is jumping, and after his feet leave the ground?

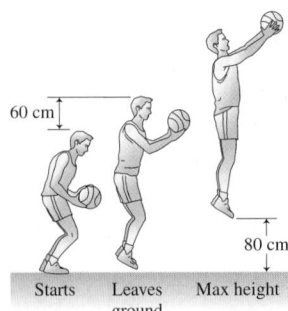

60 cm

80 cm

Starts Leaves Max height
        ground

FIGURE CP8.48

49. Figure CP8.49 shows a 200 g hamster sitting on an 800 g wedge-shaped block. The block, in turn, rests on a spring scale.

a. Initially, static friction is sufficient to keep the hamster from moving. In this case, the hamster and the block are effectively a single 1000 g mass and the scale should read 9.8 N. Show that this is the case by treating the hamster and the block as *separate* systems and analyzing the forces.

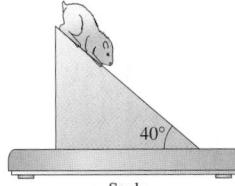

40°

Scale

FIGURE CP8.49

b. An extra-fine lubricating oil having $\mu_s = \mu_k = 0$ is sprayed on the top surface of the block, causing the hamster to slide down. Friction between the block and the scale is large enough that the block does *not* slip on the scale. What does the scale read as the hamster slides down?

50. Figure CP8.50 shows three hanging masses connected by massless strings over two massless, frictionless pulleys.

a. Find the acceleration constraint for this system. It is a single equation relating a_{1y}, a_{2y}, and a_{3y}.

Hint: y_A isn't constant.

b. Find an expression for the tension in string A.

Hint: You should be able to write four second-law equations. These, plus the acceleration constraint, are five equations in five unknowns.

c. Suppose: $m_1 = 2.5$ kg, $m_2 = 1.5$ kg, and $m_3 = 4.0$ kg. Find the acceleration of each.

d. The 4.0 kg mass would appear to be in equilibrium. Explain why it accelerates.

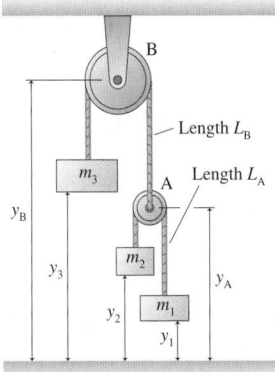

B

Length L_B

m_3

A Length L_A

y_B

y_3

m_2

y_A

y_2

m_1

y_1

FIGURE CP8.50

STOP TO THINK ANSWERS

Stop to Think 8.1: The weight force and the normal force are incorrectly identified as an action/reaction pair. The normal force is $\vec{n}_{S \text{ on } B}$, so it is paired with force $\vec{n}_{B \text{ on } S}$ that acts *on the surface*. The weight force is $\vec{w}_{E \text{ on } B}$, so it is paired with force $\vec{w}_{B \text{ on } E}$ that acts *on the earth*.

Stop to Think 8.2: c. Newton's third law says that the force of A on B is *equal* and opposite to the force of B on A. This is always true. The speed of the objects isn't relevant.

Stop to Think 8.3: b. $F_{B \text{ on } H} = F_{H \text{ on } B}$ and $F_{A \text{ on } B} = F_{B \text{ on } A}$ because these are action/reaction pairs. Box B is slowing down and therefore must have a net force to the left. So from Newton's second law we also know that $F_{H \text{ on } B} > F_{A \text{ on } B}$.

Stop to Think 8.4: Equal to. Each block is hanging in equilibrium, with no net force, so the upward tension force is *mg*.

Stop to Think 8.5: Less. Block B is *accelerating* downward, so the net force on B must point down. The only forces acting on B are the tension and the weight, so $T_{S \text{ on } B} < w_{E \text{ on } B}$.

Stop to Think 8.6: c. Newton's third law says that the force of A on B is *equal* and opposite to the force of B on A. This is always true. The mass of the objects isn't relevant.

Newton's Laws

The goal of Part I has been to discover the connection between force and motion. We started with *kinematics*, which is the mathematical description of motion, then we proceeded to *dynamics*, which is the explanation of motion in terms of forces. Newton's three laws of motion form the basis of our explanation. All of the examples we have studied so far are applications of Newton's laws.

The table below is called a *knowledge structure* for Newton's laws. A knowledge structure summarizes the essential concepts, the general principles, and the primary applications of a theory. The first section of the table tells us that Newtonian mechanics is concerned with how *particles* respond to *forces*. The second section indicates that we have introduced only three general principles, Newton's three laws of motion.

You use this knowledge structure by working your way through it, from top to bottom. Once you recognize a prob-lem as a dynamics problem, you immediately know to start with Newton's laws. You can then determine the category of motion and apply Newton's second law in the appro-priate form. Newton's third law will help you identify the forces acting on particles as they interact. Finally, the kine-matic equations for that category of motion allow you to reach the solution you seek.

The knowledge structure provides the *procedural knowledge* for solving dynamics problems, but it does not represent the total knowledge required. You must add to it knowledge about what position and velocity are, about how forces are identified, about action/reaction pairs, about drawing and using free body diagrams, and so on. These are specific *tools* for problem solving. The problem-solving strategies of Chapters 5 and 8 combine the proce-dures and the tools into a powerful method for thinking about and solving problems.

KNOWLEDGE STRUCTURE I **Newton's Laws**

ESSENTIAL CONCEPTS	Particle, acceleration, force, interaction
BASIC GOALS	How does a particle respond to a force? How do systems interact?

GENERAL PRINCIPLES	**Newton's first law**	An object will remain at rest or will continue to move with constant velocity (equilibrium) if and only if $\vec{F}_{net} = \vec{0}$.
	Newton's second law	$\vec{F}_{net} = m\vec{a}$
	Newton's third law	$\vec{F}_{A \, on \, B} = -\vec{F}_{B \, on \, A}$

BASIC PROBLEM-SOLVING STRATEGY Use Newton's second law for each particle or system. Use Newton's third law to equate the magnitudes of the two members of an action/reaction pair.

Linear motion

$$\sum F_x = ma_x \quad \text{or} \quad \sum F_x = 0$$
$$\sum F_y = 0 \quad\quad\quad \sum F_y = ma_y$$

Trajectory motion

$$\sum F_x = ma_x$$
$$\sum F_y = ma_y$$

Circular motion

$$\sum F_r = mv^2/r = m\omega^2 r$$
$$\sum F_z = 0$$

Linear and trajectory kinematics

Uniform acceleration: $v_{fs} = v_{is} + a_s \Delta t$
(a_s = constant)
$$s_f = s_i + v_{is}\Delta t + \tfrac{1}{2}a_s(\Delta t)^2$$
$$v_{fs}^2 = v_{is}^2 + 2a_s\Delta s$$

Trajectories: The same equations are used for both x and y.

Uniform motion: $\quad s_f = s_i + v_s\Delta t$
($a = 0$, v_s = constant)

General case $\quad v_s = ds/dt$ = slope of the position graph
$\quad\quad\quad a_s = dv_s/dt$ = slope of the velocity graph

$$v_{fs} = v_{is} + \int_{t_i}^{t_f} a_s\, dt = v_{is} + \text{area under the acceleration curve}$$

$$s_f = s_i + \int_{t_i}^{t_f} v_s\, dt = s_i + \text{area under the velocity curve}$$

Circular kinematics

Uniform circular motion:

$$\theta_f = \theta_i + \omega\Delta t$$
$$a_r = v^2/r = \omega^2 r$$
$$v = \omega r$$
$$T = 2\pi r/v = 2\pi/\omega$$

The Forces of Nature

What are the fundamental forces of nature? That is, what set of distinct, irreducible forces can explain everything we know about nature? This is a question that has long intrigued physicists. For example, friction is not a fundamental force because it can be reduced to electric forces between atoms. What about other forces?

Physicists have long recognized three basic forces: the gravitational force, the electric force, and the magnetic force. The gravitational force is an inherent attraction between two masses. The electric force is a force between charges. The magnetic force, which is a bit more mysterious, causes compass needles to point north and holds your shopping list on the refrigerator door.

In the 1860s, the Scottish physicist James Clerk Maxwell developed a theory that *unified* the electric and magnetic forces into a single *electromagnetic force*. Where there had appeared to be two separate forces, Maxwell found there to be a single force that, under appropriate conditions, exhibits "electric behavior" or "magnetic behavior." Maxwell used his theory to predict the existence of *electromagnetic waves*, including light. Our entire telecommunications industry is testimony to Maxwell's genius.

Maxwell's electromagnetic force was soon found to be the "glue" holding atoms, molecules, and solids together. With the exception of gravity, *every* force we have considered so far can be traced to electromagnetic forces between atoms.

The discovery of the atomic nucleus, about 1910, presented difficulties that could not be explained by either gravitational or electromagnetic forces. The atomic nucleus is an unimaginably dense ball of protons and neutrons. But what holds it together against the repulsive electric forces between the protons? There must be an attractive force inside the nucleus that is stronger than the repulsive electric force. This force, called the *strong force*, is the force that holds atomic nuclei together. The strong force is a *short-range* force, extending only about about 10^{-14} m. It is completely negligible outside the nucleus. The subatomic particles called *quarks*, of which you have likely heard, are part of our understanding of how the strong force works.

In the 1930s, physicists found that the nuclear radioactivity called *beta decay* could not be explained by either the electromagnetic or the strong force. Careful experiments established that the decay is due to a previously undiscovered force within the nucleus. The strength of this force is less than either the strong force or the electromagnetic force, so this new force was named the *weak force*. Although discovered in conjunction with radioactivity, it is now known to play an important role in the fusion reactions that power the stars.

By 1940, the recognized forces of nature were four: the gravitational force, the electromagnetic force, the strong force, and the weak force. Physicists were understandably curious whether all four of these were truly fundamental or if some of them could be further unified. Indeed, innovative work in the 1960s and 1970s produced a theory that unified the electromagnetic force and the weak force.

Predictions of this new theory were confirmed during the 1980s at some of the world's largest particle accelerators, and we now speak of the *electroweak force*. Under appropriate conditions, the electroweak force exhibits "electromagnetic behavior" or "weak behavior." But under other conditions, new phenomena appear that are consequences of the full electroweak force. These conditions appear on earth only in the largest and most energetic particle accelerators, which is why we were not previously aware of the unified nature of the these two forces. However, the earliest moments of the Big Bang provided the right conditions for the electroweak force to play a significant role. Thus a theory developed to help us understand the workings of nature on the smallest subatomic scale has unexpectedly given us powerful new insights into the origin of the universe.

The success of the electroweak theory has prompted efforts to unify the electroweak force and the strong force into a *grand unified theory*. Only time will tell if the strong force and the electroweak force are really just two different aspects of a single force, or if they are truly distinct. Some physicists even envision a day when all the forces of nature will be unified in a single theory, the so-called *Theory of Everything*! For today, however, our understanding of the forces of nature is in terms of three fundamental forces: the gravitational force, the electroweak force, and the strong force.

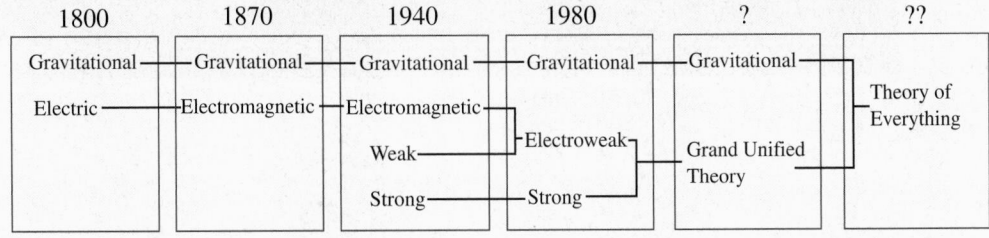

FIGURE I.1 An historical progression of our understanding of the fundamental forces of nature.

The exploding Death Star releases 10^{33} J of energy. If the expanding shock wave exerts an impulse of 2.5×10^6 N s on the *Millennium Falcon*, by how much does the starship's velocity change? To find out, what property of the starship do you need to estimate?

Conservation Laws

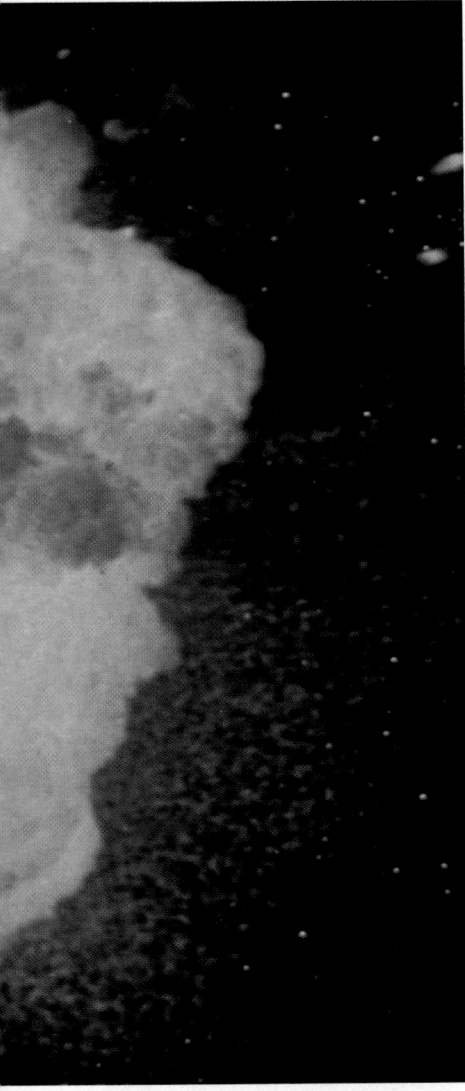

Why Some Things Stay the Same

Part I of this textbook was about *change*. Simple observations show us that most things in the world around us are changing. One particular type of change, motion, is governed by Newton's second law: $\vec{a} = \vec{F}_{net}/m$. If you know the net force acting on an object, then you can predict its motion. Although Newton's second law is a very powerful statement, it isn't the whole story. In Part II, we will be interested in things that stay the same while other things around them change.

Some Things Don't Change

Suppose you have a strong, sealed box from which you have removed all the air. First put 2×10^{24} molecules of hydrogen gas into the box (weighing 6.64 g), then add 1×10^{24} molecules of oxygen gas (weighing 53.16 g). The box, you'll surely agree, now contains 59.80 g of gas. You could confirm this by weighing the box and subtracting the weight of the empty box. No matter how much the gases inside swirl around, with molecules bouncing off each other and the walls, the mass stays the same. No surprise.

Two wires extend into the box. If you connect a battery, a spark jumps between the wires and causes an explosion inside the box! Fortunately the box withstands the explosion without breaking or leaking. If you now peek inside the box, the gases are gone and there's a puddle of water in the bottom. How much does the water weigh? You guessed it—59.80 g! Despite the complex changes that occurred during the explosion, the mass inside the box remained the same.

Now you may be thinking, "Of course. There were two parts hydrogen to one part oxygen, so the chemical reaction $2H_2 + O_2 \rightarrow 2H_2O$ says that all the gas will react to produce water." But set aside what you know from chemistry and simply think about the process. Isn't it quite amazing that the mass is completely unchanged by the explosion? And if the mass doesn't change, might the box have other properties that don't change?

Conservation of Mass

Newton's third law says that a force is really an *interaction* between two objects, each exerting a force on the other via an action/reaction pair. Our primary goal in Part II will be to discover certain properties of a system of particles that *don't change* as the particles interact. A quantity that stays the same throughout an interaction is said to be *conserved*. The existence of conserved quantities is both a profound statement about nature and, at the same time, a very practical tool for problem solving.

In the case of chemical reactions, the final mass M_f after the reaction is complete is the same as the initial mass M_i *if the system is closed*. That is, once the box is closed you're not allowed to add or remove any matter. Our knowledge about mass can be stated as a *conservation law*.

> **Law of conservation of mass** The total mass in a closed system is constant. Mathematically, $M_f = M_i$.

The qualification "in a closed system" is important. The final mass certainly won't equal the initial mass if you open the box halfway through and remove some of the matter. Other conservation laws that we discover will also have qualifications stating the circumstances under which they apply. Knowing when a law can be used is just as important as knowing the law! The fact that some quantity Q is conserved under some circumstances doesn't mean that it will always be conserved.

The law of conservation of mass is a critical underpinning of all of chemistry. In fact, conservation of mass is tacitly assumed in writing a reaction equation such as $2H_2 + O_2 \rightarrow 2H_2O$. Although our common sense and everyday experience suggests that mass is conserved, it was only after many long and precise experiments that conservation of mass was accepted as a fundamental law of nature.

Surprisingly, just as the laws of chemistry were becoming well established at the beginning of the twentieth century, Einstein's 1905 theory of relativity showed that there are circumstances in which mass actually is *not* conserved but can be converted to energy in accordance with his famous formula $E = mc^2$. Nonetheless, conservation of mass is an exceedingly good approximation in nearly all applications of science and engineering.

Conservation Laws

A system of interacting particles has another curious property. Each system is characterized by a certain number, and no matter how complex the interactions, the value of this number never changes. This number is called the *energy* of the system, and the fact that it never changes is called the *law of conservation of energy*. It is, perhaps, the single most important physical law ever discovered.

The law of conservation of energy is more fundamental than Newton's laws of motion. Newton's laws are a very good description of motion under many circumstances, but they fail for objects moving at extremely high speeds or for objects that are the size of atoms. But as far as we know, the law of conservation of energy is valid under all circumstances. No violation of this law has ever been observed. Energy will be *the* most important concept throughout the remainder of this textbook.

But what is energy? How do you determine the energy number for a system? These are not easy questions. Energy is an abstract idea, not as tangible or easy to picture as mass or force. Our modern concept of energy wasn't fully formulated until the middle of the nineteenth century, nearly two hundred years after Newton. Conservation of energy was not recognized as a law until the relationship between *energy* and *heat* was understood. That is a topic we will take up in Part IV, where the full concept of energy will be found to be the basis of thermodynamics. But all that in due time. In Part II we will be content to introduce the concept of energy and show how energy gives us an important new perspective on interacting particles.

Another important quantity we will study is called *momentum*. The law of conservation of mass is perhaps not surprising because of our everyday experience with mass. But we don't have everyday experience with momentum, so we have no expectation that it should be conserved. But under the proper circumstances, analogous to having the box tightly sealed, only those interactions can occur that do not change the momentum of a system.

Conservation laws tell us that nature is not free to act in arbitrary ways. Many conceivable interactions simply don't happen. We could *imagine* a situation in which 10 g of A and 10 g of B react to give 19 g of product, but such reactions simply don't happen. The only possible reactions are ones that conserve mass. In other situations, the only possible outcomes are ones that conserve momentum or energy. This is a very powerful statement about nature.

Conservation laws will give us a new and different *perspective* on motion. This is not insignificant. You probably have seen optical illusions where a figure appears first one way, then another, even though the basic information in the figure has not changed. Likewise with motion. We will soon see that there are some situations most easily analyzed from the perspective of Newton's laws, but others that make much more sense when analyzed from a conservation-law perspective. An important goal of Part II is to learn which perspective is best for a given problem.

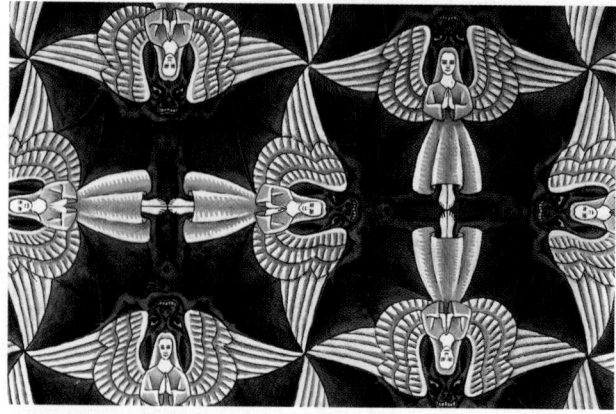

You can see angels or bats, depending on your perspective, but not both at once.

9 Impulse and Momentum

A tennis ball collides with a racket. Notice that the right side of the ball is flattened.

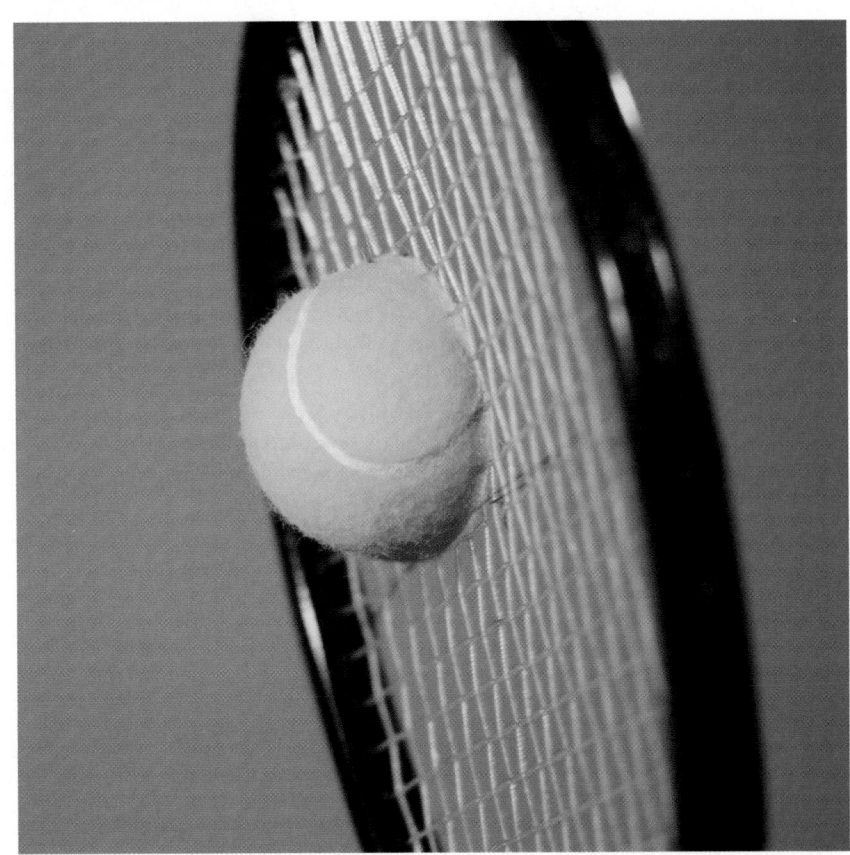

▶ Looking Ahead

The goal of Chapter 9 is to introduce the ideas of impulse, momentum, and angular momentum and to learn a new problem-solving strategy based on conservation laws. In this chapter you will learn to:

- Understand and use the concepts of impulse and momentum.
- Use a new before-and-after pictorial representation.
- Solve problems using the law of conservation of momentum.
- Apply these ideas to explosions and collisions.
- Use the law of conservation of angular momentum in simple situations.

◀ Looking Back

The law of conservation of momentum is based on Newton's third law. Please review:

- Sections 8.2–8.3 Action/reaction force pairs and Newton's third law.

A racket hitting a tennis ball is an example of what we'll call a *collision*. A collision is a complex interaction between two objects, and using Newton's second law to predict the outcome of a collision would be a daunting challenge. Nevertheless, some collisions have very simple outcomes. For example, consider a train car rolling along the tracks toward an identical car at rest. The two cars couple together upon impact and then roll down the tracks together. The forces between the train cars during the collision are unimaginably complex. Yet if you were to measure their speeds, you would find that the two coupled cars have exactly half the speed of the single car before impact. How can such a complex interaction give rise to such a simple outcome?

The opposite of a collision is an interaction that forces two objects apart. These interactions are called *explosions*, even though they may lack a flash or a pop. As an example, imagine a 75 kg archer on ice skates. If the archer shoots a 75 g arrow forward, the archer recoils backward. The interaction between the archer, the bow, and the arrow is very complex, yet the archer's recoil speed is always 1/1000 of the speed of the arrow. Another simple outcome.

Our goal in this chapter is to learn how to predict these simple outcomes without having to know all the details of the interaction forces. The new idea that will make this possible is *momentum*, a concept we will use to relate the situation "before" an interaction to the situation "after" the interaction. This before-and-after perspective will be a powerful new problem-solving tool.

9.1 Momentum and Impulse

Suppose that two or more objects have an intense and perhaps complex interaction, such as a collision or an explosion. Our goal is to find a relationship between the velocities of the objects before the interaction and their velocities after the interaction. We'll start by looking at collisions.

A **collision** is a short-duration interaction between two objects. The collision between a tennis ball and a racket, or a baseball and a bat, may seem instantaneous to your eye, but that is a limitation of your perception. A careful look at the photograph that opens this chapter reveals that the right side of the ball is flattened and pressed up against the strings of the racket. It takes time to compress the ball, and more time for the ball to re-expand as it leaves the racket.

The duration of a collision depends on the materials from which the objects are made, but 1 to 10 ms (0.001 to 0.010 s) is typical. This is the time during which the two objects are in contact with each other. The harder the objects, the shorter the contact time. A collision between two steel balls lasts less than 1 ms.

Figure 9.1 shows a microscopic view of a collision in which object A bounces off object B. The spring-like molecular bonds—the same bonds that cause normal forces and tension forces—compress during the collision, then re-expand as A bounces back. The molecular springs of a hard material, such as steel, are stiffer than the molecular springs of a rubber ball. Thus the deformation of steel is less than rubber and the duration of the collision is shorter, but neither is zero.

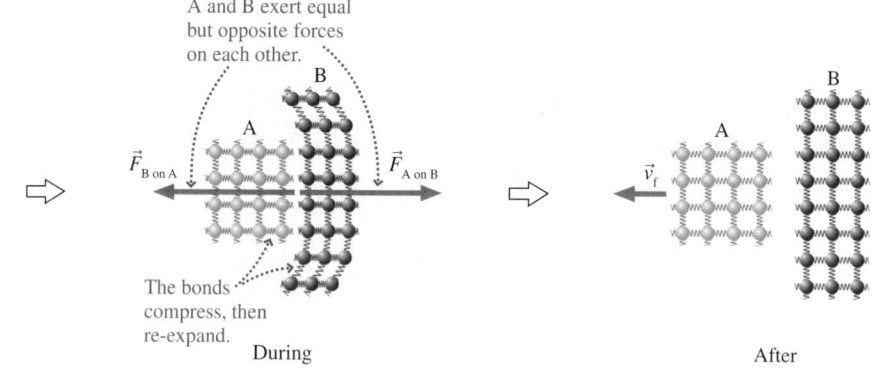

FIGURE 9.1 A microscopic view of a collision.

As A and B come into contact, the molecular springs begin to compress. The forces $\vec{F}_{A \text{ on } B}$ and $\vec{F}_{B \text{ on } A}$ are an action/reaction pair and, according to Newton's third law, have equal magnitudes: $F_{A \text{ on } B} = F_{B \text{ on } A}$. The force increases rapidly as the bonds compress, reaches a maximum at the instant A is at rest (point of maximum compression), then decreases as the bonds re-expand. This idea is shown graphically in Figure 9.2.

A large force exerted during a small interval of time is called an **impulsive force.** The force of a tennis racket on a ball, which would look much like Figure 9.2, is a good example of an impulsive force. Notice that an impulsive force has a well-defined duration.

NOTE ▶ Until now, we have not dealt with forces that change with time. Because an impulsive force is a function of time, we will write it as $F(t)$. ◀

Consider a particle traveling in a straight line along the x-axis with initial velocity v_{ix}. The particle suddenly collides with another object and experiences an impulsive force $F_x(t)$ that begins at time t_i and ends at time t_f. After the collision, the particle has final velocity v_{fx}.

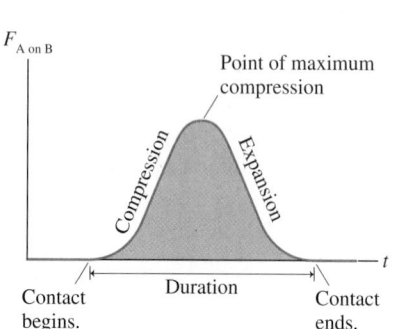

FIGURE 9.2 The rapidly changing magnitude of the force during a collision.

NOTE ▶ Both v_x and F_x are components of vectors and thus have *signs* indicating which way the vectors point. ◀

We can analyze the collision with Newton's second law to find the final velocity. Acceleration in one dimension is $a_x = dv_x/dt$, so the second law is

$$ma_x = m\frac{dv_x}{dt} = F_x(t)$$

After multiplying both sides by dt, we can write the second law as

$$m\,dv_x = F_x(t)\,dt \tag{9.1}$$

The force is nonzero only during the interval of time from t_i to t_f, so let's integrate Equation 9.1 over this interval. The velocity changes from v_{ix} to v_{fx} during the collision, thus

$$m\int_{v_i}^{v_f} dv_x = mv_{fx} - mv_{ix} = \int_{t_i}^{t_f} F_x(t)\,dt \tag{9.2}$$

We need some new tools to help us make sense of Equation 9.2.

Momentum

The product of the particle's mass and velocity is called the *momentum* of the particle:

$$\textbf{momentum} = \vec{p} = m\vec{v} \tag{9.3}$$

Momentum, like velocity, is a vector. The units of momentum are kg m/s.

Figure 9.3 shows that the momentum vector $\vec{p}$ is parallel to the velocity vector $\vec{v}$. Like any vector, $\vec{p}$ can be decomposed into x- and y-components. Equation 9.3, which is a vector equation, is a shorthand way to write the simultaneous equations

$$p_x = mv_x$$
$$p_y = mv_y$$

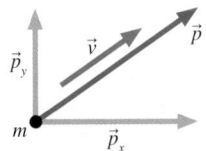

FIGURE 9.3 A particle's momentum vector $\vec{p}$ can be decomposed into x- and y-components.

NOTE ▶ One of the most common errors in momentum problems is a failure to use the appropriate signs. The momentum component p_x has the same sign as v_x. Momentum is *negative* for a particle moving to the left (on the x-axis) or down (on the y-axis). ◀

Momentum is another term that we use in everyday speech without a precise definition. In physics and engineering, momentum is a technical term whose meaning is defined in Equation 9.3. An object can have a large momentum either by having a small mass but a large velocity (a bullet fired from a rifle) or a small velocity but a large mass (a large truck rolling at a slow 1 mph).

Newton actually formulated his second law in terms of momentum rather than acceleration:

$$\vec{F} = m\vec{a} = m\frac{d\vec{v}}{dt} = \frac{d(m\vec{v})}{dt} = \frac{d\vec{p}}{dt} \tag{9.4}$$

This statement of the second law, saying that **force is the rate of change of momentum,** is more general than our earlier version $\vec{F} = m\vec{a}$. It allows for the possibility that the mass of the object might change, such as a rocket that is losing mass as it burns fuel.

NOTE ▶ The plural of *momentum* is *momenta*, from its Latin origin. ◀

Returning to Equation 9.2, you can see that mv_{ix} and mv_{fx} are p_{ix} and p_{fx}, the x-component of the particle's momentum before and after the collision. In terms of momentum, Equation 9.2 is

$$\Delta p_x = p_{fx} - p_{ix} = \int_{t_i}^{t_f} F_x(t)\, dt \qquad (9.5)$$

Now we need to examine the right-hand side of Equation 9.5.

Impulse

Equation 9.5 tells us that the particle's change in momentum is related to the time integral of the force. Let's define a quantity J_x called the *impulse* to be

$$\textbf{impulse} = J_x = \int_{t_i}^{t_f} F_x(t)\, dt \qquad (9.6)$$

$$= \text{area under the } F_x(t) \text{ curve between } t_i \text{ and } t_f$$

Strictly speaking, impulse has units of N s, but you should be able to show that N s are equivalent to kg m/s, the units of momentum.

Figure 9.4a portrays the impulse graphically as the area under the force curve. Because the force changes in a complicated way during a collision, it is often useful to describe the collision in terms of an *average* force F_{avg}. As Figure 9.4b shows, F_{avg} is the height of a rectangle that has the same area, and thus the same impulse, as the real force curve. The impulse exerted during the collision is

$$J_x = F_{avg}\Delta t \qquad (9.7)$$

Equation 9.2, which we found by integrating Newton's second law, can now be rewritten in terms of impulse and momentum as

$$\Delta p_x = J_x \quad \text{(impulse-momentum theorem)} \qquad (9.8)$$

This result, called the **impulse-momentum theorem,** tells us that **an impulse delivered to a particle changes the particle's momentum.** The momentum p_{fx} "after" an interaction, such as a collision or an explosion, is equal to the momentum p_{ix} "before" the interaction *plus* the impulse that arises from the interaction:

$$p_f = p_i + J_x \qquad (9.9)$$

The impulse-momentum theorem tells us that we do *not* need to know all the details of the force function $F_x(t)$. No matter how complicated the force, only the integral of the force—the area under the force curve—is needed to find p_{fx}. We opened this chapter with examples in which complex interactions led to very simple outcomes. The impulse-momentum theorem is an important step toward understanding these outcomes.

Figure 9.5 illustrates the impulse-momentum theorem for a rubber ball bouncing off a wall. Notice the signs; they are very important. The ball is initially traveling toward the right, so v_{ix} and p_{ix} are positive. After the bounce, v_{fx} and p_{fx} are negative. The force *on the ball* is toward the left, so F_x is also negative. The graphs show how the force and the velocity change with time.

Although the interaction is very complex, the impulse—the area under the force graph—is all we need to know to find the ball's velocity as it rebounds from the wall. The final momentum is

$$p_{fx} = p_{ix} + J_x = p_{ix} + \text{area under the force curve}$$

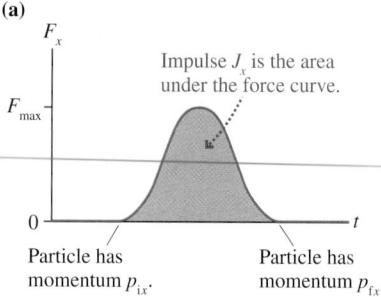

(a)

Impulse J_x is the area under the force curve.

Particle has momentum p_{ix}.

Particle has momentum p_{fx}.

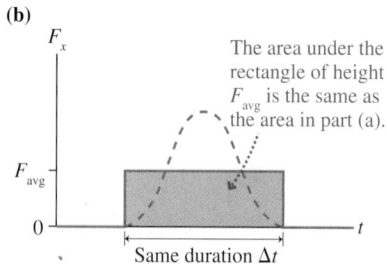

(b)

The area under the rectangle of height F_{avg} is the same as the area in part (a).

Same duration Δt

FIGURE 9.4 Looking at the impulse graphically.

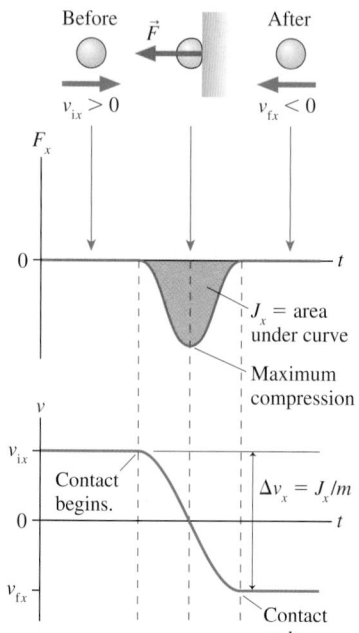

FIGURE 9.5 The impulse-momentum theorem helps us understand a rubber ball bouncing off a wall.

Thus the final velocity is

$$v_{fx} = \frac{p_{fx}}{m} = v_{1x} + \frac{\text{area under the force curve}}{m}$$

In this example, the area has a negative value.

STOP TO THINK 9.1 The cart's change of momentum is

a. -30 kg m/s.
b. -20 kg m/s.
c. 0 kg m/s.
d. 10 kg m/s.
e. 20 kg m/s.
f. 30 kg m/s.

10 kg

2 m/s

1 m/s

9.2 Solving Impulse and Momentum Problems

Pictorial representations have become an important problem-solving tool. The pictorial representations and free-body diagrams that you learned to draw in Part I were oriented toward the use of Newton's laws and a subsequent kinematical analysis. Now we are interested in making a connection between "before" and "after."

TACTICS BOX 9.1 **Drawing a before-and-after pictorial representation**

❶ **Sketch the situation.** Use two drawings, labeled "Before" and "After," to show the objects *before* they interact and again *after* they interact.
❷ **Establish a coordinate system.** Select your axes to match the motion.
❸ **Define symbols.** Define symbols for the masses and for the velocities before and after the interaction. Position and time are not needed.
❹ **List known information.** Give the values of quantities known from the problem statement or that can be found quickly with simple geometry or unit conversions. Before-and-after pictures are usually simpler than the pictures you used for dynamics problems, so listing known information on the sketch is adequate.
❺ **Identify the desired unknowns.** What quantity or quantities will allow you to answer the question? These should have been defined as symbols in step 3.

NOTE ▶ The generic subscripts i and f, for *initial* and *final* are adequate in equations for a simple problem, but in more complex problems using numerical subscripts, such as v_{1x} and v_{2x}, will help keep all the symbols straight. ◀

EXAMPLE 9.1 **Hitting a baseball**

A 150 g baseball is thrown with a speed of 20 m/s. It is hit straight back toward the pitcher at a speed of 40 m/s. The interaction force between the ball and the bat has the shape shown in Figure 9.6. What is the *maximum* force F_{max} that the bat exerts on the ball? What is the *average* force that the bat exerts on the ball?

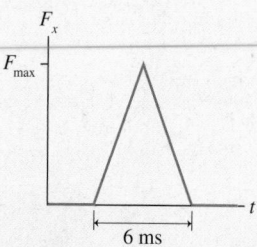

FIGURE 9.6 The interaction force between the baseball and the bat.

MODEL Model the baseball as a particle and the interaction as a collision.

VISUALIZE Figure 9.7 is a before-and-after pictorial representation. The steps from Tactics Box 9.1 are explicitly noted. Because F_x is positive (a force to the right), we know the ball was initially moving toward the left and is hit back toward the right. Thus we converted the statements about *speeds* into information about *velocities*, with v_{ix} negative.

SOLVE So far we've consistently started the mathematical representation with Newton's second law. Now we want to use the impulse-momentum theorem:

$$\Delta p_x = J_x = \text{area under the force curve}$$

We know the velocities before and after the collision, so we can find the change in the ball's momentum:

$$\Delta p_x = mv_{fx} - mv_{ix} = (0.15 \text{ kg})(40 \text{ m/s} - (-20 \text{ m/s}))$$
$$= 9.0 \text{ kg m/s}$$

The force curve is a triangle with height F_{max} and width 6.0 ms. The area under the curve is

$$J_x = \text{area} = \frac{1}{2} \times F_{max} \times (0.0060 \text{ s}) = (F_{max})(0.0030 \text{ s})$$

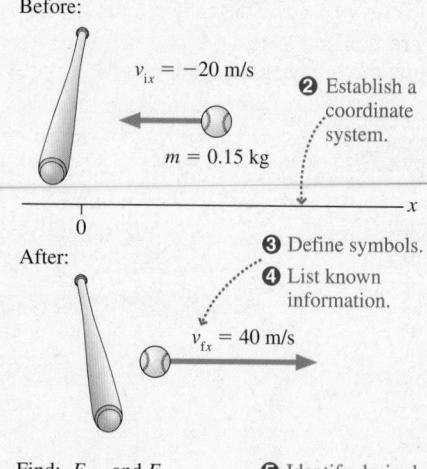

❶ Draw the before-and-after pictures.

Before:

$v_{ix} = -20$ m/s

❷ Establish a coordinate system.

$m = 0.15$ kg

After:

❸ Define symbols.
❹ List known information.

$v_{fx} = 40$ m/s

Find: F_{max} and F_{avg}

❺ Identify desired unknowns.

FIGURE 9.7 A before-and-after pictorial representation.

According to the impulse-momentum theorem,

$$9.0 \text{ kg m/s} = (F_{max})(0.0030 \text{ s})$$

Thus the *maximum* force is

$$F_{max} = \frac{9.0 \text{ kg m/s}}{0.0030 \text{ s}} = 3000 \text{ N}$$

The *average* force, which depends on the collision duration $\Delta t = 0.0060$ s, has the smaller value

$$F_{avg} = \frac{J_x}{\Delta t} = \frac{\Delta p_x}{\Delta t} = \frac{9.0 \text{ kg m/s}}{0.0060 \text{ s}} = 1500 \text{ N}$$

ASSESS F_{max} is a large force, but quite typical of the impulsive forces during collisions. The main thing to focus on is our new perspective: an impulse changes the momentum of an object.

Other forces often act on an object during a collision or other brief interaction. In Example 9.1, for instance, the baseball also has a weight force acting on it. Usually these other forces are *much* smaller than the interaction forces. The 1.5 N weight of the ball is vastly less than the 3000 N force of the bat on the ball. We can reasonably neglect these small forces *during* the brief time of the impulsive force by using what is called the **impulse approximation.**

When we use the impulse approximation, p_{ix} and p_{fx} (and v_{ix} and v_{fx}) are then the momenta (and velocities) *immediately* before and *immediately* after the collision. For example, the velocities in Example 9.1 are those of the ball just before and after it collides with the bat. We could then do a follow-up problem, including weight and drag, to find the ball's speed a second later as the second baseman catches it. We'll look at some two-part examples later in the chapter.

Momentum Bar Charts

The impulse-momentum theorem tells us that **impulse transfers momentum to an object.** If an object has 2 kg m/s of momentum, a 1 kg m/s impulse exerted on the object increases its momentum to 3 kg m/s. That is, $p_{fx} = p_{ix} + J_x$.

We can represent this "momentum accounting" with a **momentum bar chart.** Figure 9.8a shows a bar chart in which one unit of impulse adds to an initial two units of momentum to give three units of momentum. The bar chart of Figure 9.8b represents the baseball of Example 9.1, which started with negative momentum because it was moving to the left. Momentum bar charts, like before-and-after pictorial representations, are a tool for visualizing an interaction.

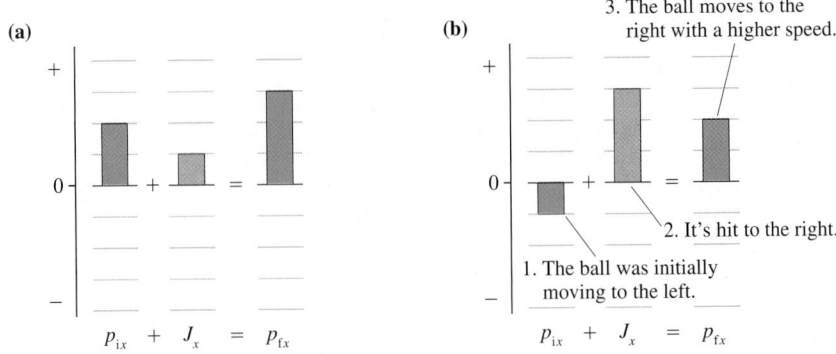

FIGURE 9.8 Two examples of momentum bar charts.

NOTE ▶ The vertical scale of a momentum bar chart has no numbers; it can be adjusted to match any problem. However, be sure that all bars in a given problem use a consistent scale. ◀

EXAMPLE 9.2 A bouncing ball
A 100 g rubber ball is dropped from a height of 2.0 m onto a hard floor. Figure 9.9 shows the force that the floor exerts on the ball. How high does the ball bounce?

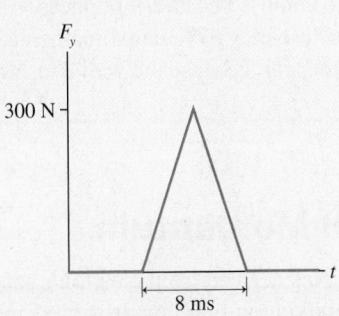

FIGURE 9.9 The force of the floor on a bouncing rubber ball.

MODEL Model the ball as a particle that is subjected to an impulsive force while in contact with the floor. Using the impulse approximation, we'll neglect the ball's weight during these 8 ms. The fall and subsequent rise are free-fall motion.

VISUALIZE Figure 9.10a is a pictorial representation. Here we have a three-part problem (downward free fall, impulsive collision, upward free fall), so the pictorial motion includes both the before and after of the collision (v_{1y} changing to v_{2y}) and the beginning and end of the free-fall motion.

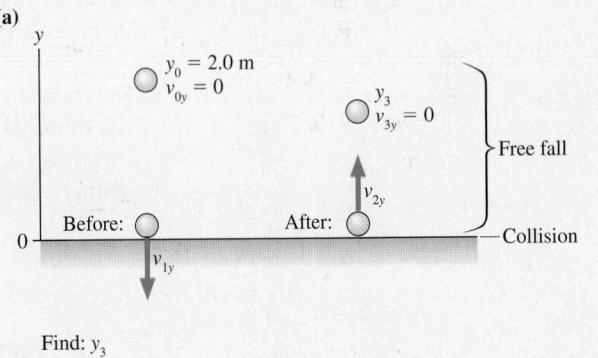

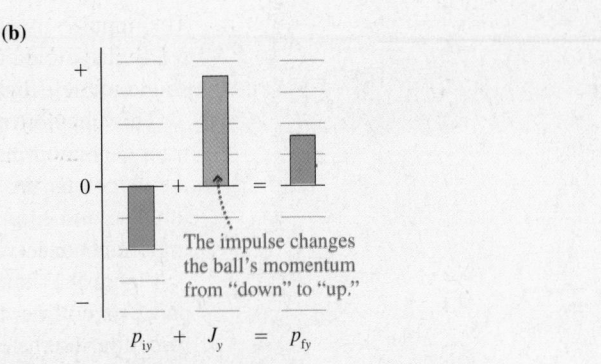

FIGURE 9.10 Pictorial representation of the ball and a momentum bar chart of the collision with the floor.

SOLVE Velocity v_{1y}, the ball's velocity *immediately* before the collision, is found using free-fall kinematics with $\Delta y = -2.0$ m:

$$v_{1y}^2 = v_{0y}^2 - 2g\Delta y = 0 - 2g\Delta y$$

$$v_{1y} = \sqrt{-2g\Delta y} = \sqrt{-2(9.80 \text{ m/s}^2)(-2.0 \text{ m})} = -6.26 \text{ m/s}$$

We've chosen the negative root because the ball is moving in the negative y-direction.

The impulse-momentum theorem is $p_{2y} = p_{1y} + J_y$. The initial momentum, just before the collision, is $p_{1y} = mv_{1y} = -0.626$ kg m/s. The force of the floor is upward, so J_y is positive. The final momentum p_{2y}, as the ball leaves the floor, is positive but probably smaller in magnitude than p_{1y} because we know that rubber balls don't bounce back to their initial height. This relationship between p_{1y}, p_{2y}, and J_y is shown in the momentum bar chart of Figure 9.10b.

From Figure 9.8, the impulse J_y is

$$J_y = \text{area under the force curve} = \frac{1}{2} \times (300 \text{ N}) \times (0.0080 \text{ s})$$

$$= 1.200 \text{ N s}$$

Thus

$$p_{2y} = p_{1y} + J_y = (-0.626 \text{ kg m/s}) + 1.200 \text{ N s} = 0.574 \text{ kg m/s}$$

and the post-collision velocity is

$$v_{2y} = \frac{p_{2y}}{m} = \frac{0.574 \text{ kg m/s}}{0.10 \text{ kg}} = 5.74 \text{ m/s}$$

The rebound speed is less than the impact speed, as expected. Finally a second use of free-fall kinematics yields

$$v_{3y}^2 = 0 = v_{2y}^2 - 2g\Delta y = v_{2y}^2 - 2gy_3$$

$$y_3 = \frac{v_{2y}^2}{2g} = \frac{(5.74 \text{ m/s})^2}{2(9.80 \text{ m/s}^2)} = 1.68 \text{ m}$$

The ball bounces back to a height of 1.68 m.

ASSESS The ball bounces back to less than its initial height, which is realistic.

NOTE ▶ Example 9.2 illustrates an important point: The impulse-momentum theorem applies *only* during the brief interval in which an impulsive force is applied. Many problems will have segments of the motion that must be analyzed with kinematics or Newton's laws. The impulse-momentum theorem is a new and useful tool, but it doesn't replace all that you've learned up until now. **◀**

STOP TO THINK 9.2   A 10 g rubber ball and a 10 g clay ball are thrown at a wall with equal speeds. The rubber ball bounces, the clay ball sticks. Which ball exerts a larger impulse on the wall?

a. The clay ball exerts a larger impulse because it sticks.
b. The rubber ball exerts a larger impulse because it bounces.
c. They exert equal impulses because they have equal momenta.
d. Neither exerts an impulse on the wall because the wall doesn't move.

9.3 Conservation of Momentum

The impulse-momentum theorem was derived from Newton's second law and is really just an alternative way of looking at that law. It is used in the context of single-particle dynamics, much as we used Newton's law in Chapters 4–7.

This chapter opened by noting that very complex interactions, such as two train cars coupling together, sometimes have very simple outcomes. To predict the outcomes, we need to see how Newton's *third* law looks in the language of impulse and momentum. Newton's third law will lead us to one of the most important conservation laws in physics.

Figure 9.11 shows two particles with initial velocities $(v_{ix})_1$ and $(v_{ix})_2$. The particles collide, then bounce apart with final velocities $(v_{fx})_1$ and $(v_{fx})_2$. The forces during the collision, as the particles are interacting, are the action/reaction pair $\vec{F}_{1 \text{ on } 2}$ and $\vec{F}_{2 \text{ on } 1}$. For now, we'll continue to assume that the motion is one dimensional along the x-axis.

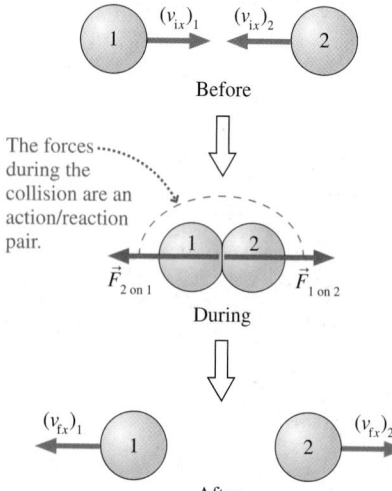

FIGURE 9.11 A collision between two particles.

NOTE ▶ The notation, with all the subscripts, may seem excessive. But there are two particles, and each has an initial and a final velocity, so we need to distinguish among four different velocities. ◀

Newton's second law for each particle *during* the collision is

$$\frac{d(p_x)_1}{dt} = (F_x)_{2 \text{ on } 1}$$

$$\frac{d(p_x)_2}{dt} = (F_x)_{1 \text{ on } 2} = -(F_x)_{2 \text{ on } 1}$$ (9.10)

We made explicit use of Newton's third law in the second equation.

Although Equations 9.10 are for two different particles, suppose—just to see what happens—we were to *add* these two equations. If we do, we find that

$$\frac{d(p_x)_1}{dt} + \frac{d(p_x)_2}{dt} = \frac{d}{dt}((p_x)_1 + (p_x)_2) = (F_x)_{2 \text{ on } 1} + (-(F_x)_{2 \text{ on } 1}) = 0 \quad (9.11)$$

If the time derivative of the quantity $(p_x)_1 + (p_x)_2$ is zero, it must be the case that

$$(p_x)_1 + (p_x)_2 = \text{constant}$$ (9.12)

Equation 9.12 is a conservation law! If $(p_x)_1 + (p_x)_2$ is a constant, then the sum of the momenta *after* the collision equals the sum of the momenta *before* the collision. That is,

$$(p_{fx})_1 + (p_{fx})_2 = (p_{ix})_1 + (p_{ix})_2$$ (9.13)

Furthermore, this equality is independent of the interaction force. We don't need to know anything about $\vec{F}_{1 \text{ on } 2}$ and $\vec{F}_{2 \text{ on } 1}$ to make use of Equation 9.13.

As an example, Figure 9.12 is a before-and-after pictorial representation of two equal-mass train cars colliding and coupling. Equation 9.13 relates the momenta of the cars after the collision to their momenta before the collision:

$$m_1(v_{fx})_1 + m_2(v_{fx})_2 = m_1(v_{ix})_1 + m_2(v_{ix})_2$$

Initially, car 1 is moving with velocity $(v_{ix})_1 = v_i$ while car 2 is at rest. Afterward, they roll together with the common final velocity v_f. Furthermore, $m_1 = m_2 = m$. With this information, the sum of the momenta is

$$mv_f + mv_f = 2mv_f = mv_i + 0$$

The mass cancels, and we find that the train cars' final velocity is $v_f = \frac{1}{2}v_i$. We were able to make this prediction of a simple outcome without knowing anything at all about the very complex interaction between the two cars as they collide.

Law of Conservation of Momentum

Equation 9.13 illustrates the idea of a conservation law for momentum, but it was derived for the specific case of two particles colliding in one dimension. Our goal is to develop a more general law of conservation of momentum, a law that will be valid in three dimensions and that will work for any type of interaction. The next few paragraphs are fairly mathematical, so you might want to begin by looking ahead to Equation 9.20 and the statement of the law of conservation of momentum to see where we're heading.

Consider a *system* consisting of N particles. Figure 9.13 shows a simple case where $N = 3$. The particles might be large entities (cars, baseballs, etc.), or they might be the microscopic atoms in a gas. We can identify each particle by an identification number k. Every particle in the system *interacts* with every other particle via action/reaction pairs of forces $\vec{F}_{j \text{ on } k}$ and $\vec{F}_{k \text{ on } j}$. In addition, every particle is subjected to possible *external forces* $\vec{F}_{\text{ext on } k}$ from agents outside the system.

Before:

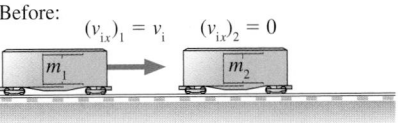

After:

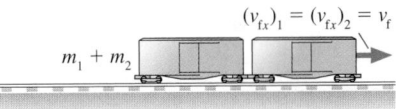

FIGURE 9.12 Two colliding train cars.

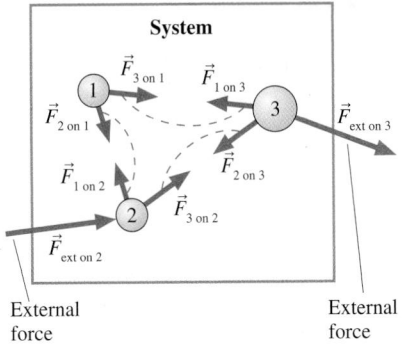

FIGURE 9.13 A system of three particles.

NOTE ▶ This definition of "the system" differs from the one we used in Chapter 8. There, where we were interested in how objects interact with each other, we identified each particle-like object as a separate system. When we use conservation laws, it is more useful to think of all the interacting objects as a single entity that we'll call "the system." ◀

If particle k has velocity $\vec{v}_k$, its momentum is $\vec{p}_k = m_k \vec{v}_k$. Define the **total momentum** $\vec{P}$ of the system as the vector sum

$$\vec{P} = \text{total momentum} = \vec{p}_1 + \vec{p}_2 + \vec{p}_3 + \cdots + \vec{p}_N = \sum_{k=1}^{N} \vec{p}_k \qquad (9.14)$$

In other words, the total momentum *of the system* is the sum of all the individual momenta.

The time derivative of $\vec{P}$ tells us how the total momentum of the system changes with time:

$$\frac{d\vec{P}}{dt} = \sum_k \frac{d\vec{p}_k}{dt} = \sum_k \vec{F}_k \qquad (9.15)$$

where we used Newton's second law from Equation 9.4 for each particle in the form $\vec{F}_k = d\vec{p}_k/dt$.

The net force acting on particle k can be divided into *external forces*, from outside the system, and *interaction forces* due to the other particles in the system:

$$\vec{F}_k = \sum_{j \neq k} \vec{F}_{j \text{ on } k} + \vec{F}_{\text{ext on } k} \qquad (9.16)$$

The restriction $j \neq k$ expresses the fact that particle k does not exert a force on itself. Using this in Equation 9.15 gives the rate of change of the total momentum P of the system:

$$\frac{d\vec{P}}{dt} = \sum_k \sum_{j \neq k} \vec{F}_{j \text{ on } k} + \sum_k \vec{F}_{\text{ext on } k} \qquad (9.17)$$

The double sum on $\vec{F}_{j \text{ on } k}$ adds *every* interaction force within the system. But the interaction forces come in action/reaction pairs, with $\vec{F}_{k \text{ on } j} = -\vec{F}_{j \text{ on } k}$, so $\vec{F}_{k \text{ on } j} + \vec{F}_{j \text{ on } k} = \vec{0}$. Consequently, **the sum of all the interaction forces is zero.** As a result, Equation 9.17 becomes

$$\frac{d\vec{P}}{dt} = \sum_k \vec{F}_{\text{ext on } k} = \vec{F}_{\text{net}} \qquad (9.18)$$

where $\vec{F}_{\text{net}}$ is the net force exerted on the system by agents outside the system. But this is just Newton's second law written for the system as a whole! That is, the rate of change of the total momentum of the whole system is equal to the net force applied to the whole system.

Equation 9.18 has two very important implications. First, it tells us that we can analyze the motion of the system as a whole without needing to consider interaction forces between the particles that make up the system. In fact, we have been using this idea all along as an *assumption* of the particle model. When we treat cars and rocks and baseballs as particles, we assume that the internal forces between the atoms—the forces that hold the object together—do not affect the motion of the object as a whole. Now we have *justified* that assumption.

The second implication of Equation 9.18, and the more important one from the perspective of this chapter, applies to what we call an *isolated system*. An **isolated system** is a system for which the *net* external force is zero: $\vec{F}_{\text{net}} = \vec{0}$.

That is, an isolated system is one on which there are *no* external forces or for which the external forces are balanced and add to zero.

For an isolated system, Equation 9.18 is simply

$$\frac{d\vec{P}}{dt} = \vec{0} \quad \text{(isolated system)} \qquad (9.19)$$

In other words, **the *total* momentum of an isolated system does not change.** The total momentum $\vec{P}$ remains constant, *regardless* of whatever interactions are going on *inside* the system. The importance of this result is sufficient to elevate it to a law of nature, alongside Newton's laws.

Law of conservation of momentum The total momentum $\vec{P}$ of an isolated system is a constant. Interactions within the system do not change the system's total momentum.

NOTE ▶ It is worth emphasizing the critical role of Newton's third law in the derivation of Equation 9.19. The law of conservation of momentum is a direct consequence of the fact that interactions within an isolated system are action/reaction pairs. ◀

Mathematically, the law of conservation of momentum for an isolated system is

$$\vec{P}_{\text{f}} = \vec{P}_{\text{i}} \qquad (9.20)$$

The total momentum after an interaction is equal to the total momentum before the interaction. Because Equation 9.20 is a vector equation, the equality is true for each of the components of the momentum vector. That is,

$$(p_{\text{f}x})_1 + (p_{\text{f}x})_2 + (p_{\text{f}x})_3 + \cdots = (p_{\text{i}x})_1 + (p_{\text{i}x})_2 + (p_{\text{i}x})_3 + \cdots$$
$$(p_{\text{f}y})_1 + (p_{\text{f}y})_2 + (p_{\text{f}y})_3 + \cdots = (p_{\text{i}y})_1 + (p_{\text{i}y})_2 + (p_{\text{i}y})_3 + \cdots \qquad (9.21)$$

The *x*-equation is an extension of Equation 9.13 to *N* interacting particles.

EXAMPLE 9.3 Two balls shot from a tube
A 10 g ball and a 30 g ball are placed in a tube with a massless compressed spring between them. When the spring is released, the 10 g ball flies out of the tube at a speed of 6.0 m/s. With what speed does the 30 g ball emerge from the other end?

MODEL The two balls are the system. The balls interact with each other, but they form an isolated system because, for each ball, the upward normal force of the tube balances the downward weight force to make $\vec{F}_{\text{net}} = \vec{0}$. Thus the total momentum of the system is conserved.

VISUALIZE Figure 9.14 shows a before-and-after pictorial representation for the two balls. The total momentum before the spring is released is $\vec{P}_{\text{i}} = \vec{0}$ because both balls are at rest. Consequently, the *total* momentum will be $\vec{0}$ after the spring is released. The mathematical statement of momentum conservation, Equation 9.21, is

$$m_1(v_{\text{f}x})_1 + m_2(v_{\text{f}x})_2 = m_1(v_{\text{i}x})_1 + m_2(v_{\text{i}x})_2 = 0$$

where we've written the *x*-component of the momenta in terms of v_x and made use of the fact that the initial velocities are both zero.

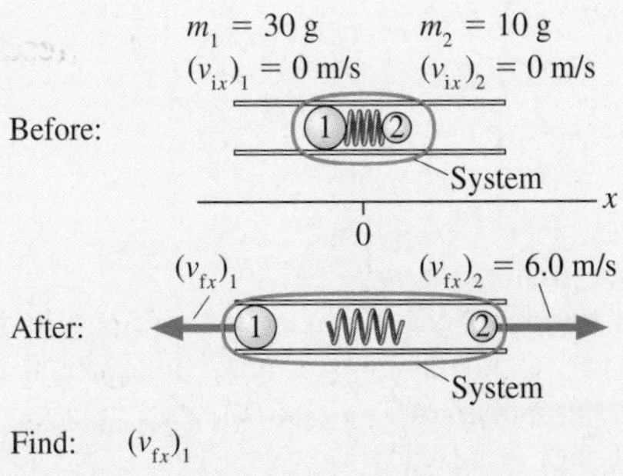

FIGURE 9.14 Before-and-after pictorial representation for two balls shot out of a tube.

Solving for $(v_{\text{f}x})_1$, we find

$$(v_{\text{f}x})_1 = -\frac{m_2}{m_1}(v_{\text{f}x})_2 = -\frac{1}{3}(v_{\text{f}x})_2 = -2.0 \text{ m/s}$$

The 30 g ball emerges with a *speed* of 2.0 m/s, one-third the speed of the 10 g ball.

ASSESS The *total* momentum of the system is zero, but the individual momenta are not. Because $(p_{fx})_2$ is positive (the 10 g ball moves to the right), $(p_{fx})_1$ must have the same magnitude but the opposite sign (the 30 g ball moves to the left). Notice that we didn't need to know any details about the spring to conclude that the 30 g ball has one-third the speed of the 10 g ball. Conservation of momentum *mandates* this result.

A Strategy for Conservation of Momentum Problems

6.3, 6.4, 6.6, 6.7, 6.10

Our derivation of the law of conservation of momentum, and the conditions under which it holds, suggests a problem-solving strategy.

 PROBLEM-SOLVING STRATEGY 9.1 Conservation of momentum

MODEL Clearly define *the system*.

- If possible, choose a system that is isolated ($\vec{F}_{net} = \vec{0}$) or within which the interactions are sufficiently short and intense that you can ignore external forces for the duration of the interaction (the impulse approximation). Momentum is conserved.
- If it's not possible to choose an isolated system, try to divide the problem into parts such that momentum is conserved during one segment of the motion. Other segments of the motion can be analyzed using Newton's laws or, as you'll learn in Chapters 10 and 11, conservation of energy.

VISUALIZE Draw a before-and-after pictorial representation. Define symbols that will be used in the problem, list known values, and identify what you're trying to find.

SOLVE The mathematical representation is based on the law of conservation of momentum: $\vec{P}_f = \vec{P}_i$. In component form, this is

$$(p_{fx})_1 + (p_{fx})_2 + (p_{fx})_3 + \cdots = (p_{ix})_1 + (p_{ix})_2 + (p_{ix})_3 + \cdots$$

$$(p_{fy})_1 + (p_{fy})_2 + (p_{fy})_3 + \cdots = (p_{iy})_1 + (p_{iy})_2 + (p_{iy})_3 + \cdots$$

ASSESS Check that your result has the correct units, is reasonable, and answers the question.

EXAMPLE 9.4 **Rolling away**

Bob sees a stationary cart 8.0 m in front of him. He decides to run to the cart as fast as he can, jump on, and roll down the street. Bob has a mass of 75 kg and the cart's mass is 25 kg. If Bob accelerates at a steady 1.0 m/s^2, what is the cart's speed just after Bob jumps on?

MODEL This is a two-part problem. First Bob accelerates across the ground. Then Bob lands on and sticks to the cart, a "collision" between Bob and the cart. The interaction forces between Bob and the cart (i.e., friction) act only over the fraction of a second it takes Bob's feet to become stuck to the cart. Using the impulse approximation allows the system Bob + cart to be treated as an isolated system during the brief interval of the "collision," and thus the total momentum of Bob + cart is conserved during this interaction. But the system Bob + cart is *not* an isolated system for the entire problem because Bob's initial acceleration has nothing to do with the cart.

VISUALIZE Our strategy is to divide the problem into an *acceleration* part, which we can analyze using kinematics, and a *collision* part that we can analyze with momentum conservation. The pictorial representation of Figure 9.15 includes information about both parts. Notice two important points. First, Bob's velocity $(v_{1x})_B$ at the end of his run is his "before" velocity for the collision. Second, Bob and the cart move together at the end, so v_{2x} is their common final velocity.

SOLVE The first part of the mathematical representation is kinematics. We don't know how long Bob accelerates, but we do know his acceleration and the distance. Thus

$$(v_{1x})_B^2 = (v_{0x})_B^2 + 2a_x(x_1 - x_0) = 2a_x x_1$$

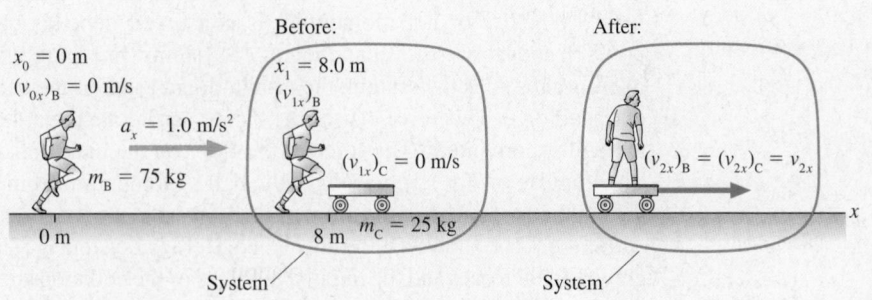

FIGURE 9.15 Pictorial representation of Bob and the cart.

His velocity after accelerating for 8.0 m is

$$(v_{1x})_B = \sqrt{2a_x x_1} = 4.0 \text{ m/s}$$

The second part of the problem, the collision, uses conservation of momentum: $P_{2x} = P_{1x}$. Written in terms of the individual momenta, this is

$$m_B(v_{2x})_B + m_C(v_{2x})_C = (m_B + m_C)v_{2x}$$
$$= m_B(v_{1x})_B + m_C(v_{1x})_C$$
$$= m_B(v_{1x})_B$$

where we've used $(v_{1x})_C = 0$ m/s because the cart starts at rest. Solving for v_{2x}, we find

$$v_{2x} = \frac{m_B}{m_B + m_C}(v_{1x})_B = \frac{75 \text{ kg}}{100 \text{ kg}} \times 4.0 \text{ m/s} = 3.0 \text{ m/s}$$

The cart's speed is 3.0 m/s immediately after Bob jumps on.

Notice how easy this was! No forces, no acceleration constraints, no simultaneous equations. Why didn't we think of this before? While conservation laws are indeed powerful, they can only answer certain questions. Had we wanted to know how far Bob slid across the cart before sticking to it, how long the slide took, or what the cart's acceleration was during the collision, we would not have been able to answer such questions on the basis of the conservation law. There is a price to pay for finding a simple connection between before and after, and that price is the loss of information about the details of the interaction. If we are satisfied with knowing only about before and after, then conservation laws are a simple and straightforward way to proceed. But many problems *do* require us to understand the interaction, and for these there is no avoiding Newton's laws and all they entail.

It Depends on the System

The first step in the problem-solving strategy asks you to clearly define *the system*. This is worth emphasizing, because many problem-solving errors arise from trying to apply momentum conservation to an inappropriate system. **The goal is to choose a system whose momentum will be conserved.** Even then, it is the *total* momentum of the system that is conserved, not the momenta of the individual particles within the system.

As an example, consider what happens if you drop a rubber ball and let it bounce off a hard floor. Is momentum conserved during the collision of the ball with the floor? You might be tempted to answer yes because the ball's rebound speed is very nearly equal to its impact speed. But there are two errors in this reasoning.

First, momentum depends on *velocity*, not speed. The ball's velocity and momentum just before the collision are negative. They are positive after the collision. Even if their magnitudes are equal, the ball's momentum after the collision is *not* equal to its momentum before the collision.

(a)

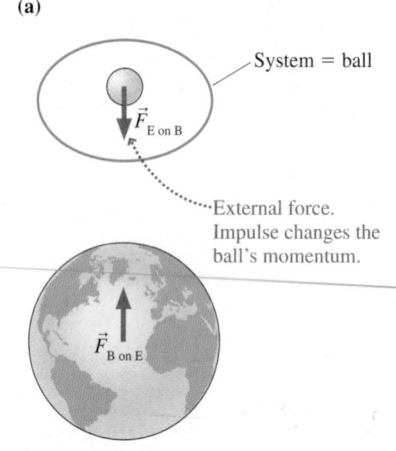

System = ball

External force. Impulse changes the ball's momentum.

(b)

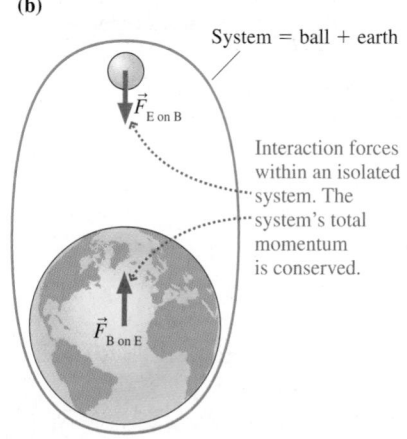

System = ball + earth

Interaction forces within an isolated system. The system's total momentum is conserved.

FIGURE 9.16 Whether or not momentum is conserved as a ball falls to earth depends on your choice of the system.

But more importantly, we haven't defined the system. The momentum of what? Whether or not momentum is conserved depends on the system. Figure 9.16 shows two different choices of systems. In Figure 9.16a, where the ball itself is chosen as the system, the gravitational force of the earth on the ball is an external force. This force causes the ball to accelerate toward the earth, changing the ball's momentum. The force of the floor on the ball is also an external force. The impulse of $\vec{F}_{\text{floor on ball}}$ changes the ball's momentum from "down" to "up" as the ball bounces. The momentum of this system is most definitely *not* conserved.

Figure 9.16b shows a different choice. Here the system is ball + earth. Now the gravitational forces and the impulsive forces of the collision are interactions *within* the system. This is an isolated system, so the *total* momentum $\vec{P} = \vec{p}_{\text{ball}} + \vec{p}_{\text{earth}}$ is conserved.

In fact, the total momentum is $\vec{P} = \vec{0}$. Before you release the ball, both the ball and the earth are at rest (in the earth's reference frame). The total momentum is zero before you release the ball, so it will *always* be zero. Consider the situation just before the ball hits the floor. If the ball's velocity is v_{By}, it must be the case that

$$m_{\text{B}} v_{\text{By}} + m_{\text{E}} v_{\text{Ey}} = 0$$

and thus

$$v_{\text{Ey}} = -\frac{m_{\text{B}}}{m_{\text{E}}} v_{\text{By}}$$

In other words, as the ball is pulled down toward the earth, the ball pulls up on the earth (action/reaction pair of forces) until the entire earth reaches velocity v_{Ey}. The earth's momentum is equal and opposite to the ball's momentum.

Why don't we notice the earth "leaping up" toward us each time we drop something? Because of the earth's enormous mass relative to everyday objects. A typical rubber ball has a mass of 60 g and hits the ground with a velocity of about -5 m/s. The earth's upward velocity is thus

$$v_{\text{Ey}} \approx -\frac{6 \times 10^{-2} \text{ kg}}{6 \times 10^{24} \text{ kg}} (-5 \text{ m/s}) = 5 \times 10^{-26} \text{ m/s}$$

The earth does, indeed, have a momentum equal and opposite to that of the ball, but the earth is so massive that it needs only an infinitesimal velocity to match the ball's momentum. At this speed, it would take the earth 300 million years to move the diameter of an atom!

STOP TO THINK 9.3   Objects A and C are made of different materials, with different "springiness," but they have the same mass and are initially at rest. When ball B collides with object A, the ball ends up at rest. When ball B is thrown with the same speed and collides with object C, the ball rebounds to the left. Compare the velocities of A and C after the collisions. Is v_A greater than, equal to, or less than v_C?

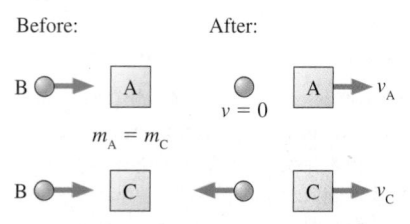

9.4 Explosions

An **explosion,** where the particles of the system move apart from each other after a brief, intense interaction, is the opposite of a collision. The explosive forces, which could be from an expanding spring or from expanding hot gases, are *internal* forces. If the system is isolated, its total momentum during the explosion will be conserved.

EXAMPLE 9.5 **Recoil**

A 10 g bullet is fired from a 3.0 kg rifle with a speed of 500 m/s. What is the recoil speed of the rifle?

MODEL A simple analysis would say that the rifle exerts a force on the bullet and the bullet, by Newton's third law, exerts a force on the rifle, causing the rifle to recoil. However, this is a little *too* simple. After all, the rifle has no means by which to exert a force on the bullet. Instead, the rifle causes a small mass of gunpowder to explode, releasing energy stored in the gunpowder. The expanding gas exerts forces on *both* the bullet and the rifle.

Let's define the system to be bullet + gas + rifle. The forces due to the expanding gas during the explosion are internal forces, within the system. Any friction forces between the bullet and the rifle as the bullet travels down the barrel are also internal forces. The only external forces, the weights, are balanced by the normal forces of the barrel on the bullet and the person holding the rifle, so $\vec{F}_{net} = \vec{0}$. This is an isolated system and the law of conservation of momentum applies.

VISUALIZE Figure 9.17 shows a pictorial representation before and after the bullet is fired. We'll assume the bullet is fired in the positive x-direction.

SOLVE The x-component of the total momentum is $P_x = (p_x)_{bullet} + (p_x)_{rifle} + (p_x)_{gas}$. Everything is at rest before the trigger is pulled, so the initial momentum is zero. After the trigger is pulled, the momentum of the expanding gas is actually the sum of the momenta of all the molecules in the gas. For every molecule moving in the forward direction with velocity v and momentum mv there is, on average, another molecule mov-

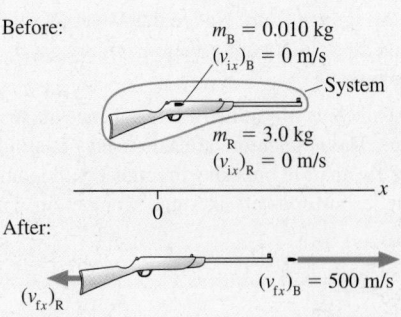

Before:
$m_B = 0.010$ kg
$(v_{ix})_B = 0$ m/s
System
$m_R = 3.0$ kg
$(v_{ix})_R = 0$ m/s

After:
$(v_{fx})_R$
$(v_{fx})_B = 500$ m/s

Find: $(v_{fx})_R$

FIGURE 9.17 Before-and-after pictorial representation of a rifle firing a bullet.

ing in the opposite direction with velocity $-v$ and thus momentum $-mv$. When summed over the enormous number of molecules in the gas, we will be left with $p_{gas} \approx 0$. Thus the final momentum is that of the rifle and bullet. The law of conservation of momentum is

$$P_{fx} = m_B(v_{fx})_B + m_R(v_{fx})_R = P_{ix} = 0$$

Solving for the rifle's velocity, we find

$$(v_{fx})_R = -\frac{m_B}{m_R}(v_{fx})_B = -\frac{0.010 \text{ kg}}{3.0 \text{ kg}} \times 500 \text{ m/s} = -1.67 \text{ m/s}$$

The minus sign indicates that the rifle's recoil is to the left. The recoil *speed* is 1.67 m/s.

We would not know where to begin to solve a problem such as this using Newton's laws. But Example 9.5 is a simple problem when approached from the before-and-after perspective of a conservation law. The selection of bullet + gas + rifle as "the system" was the critical step. For momentum conservation to be a useful principle, we had to select a system in which the complicated forces due to expanding gas and friction were all internal forces. The rifle by itself is *not* an isolated system, so its momentum is *not* conserved.

EXAMPLE 9.6 **Radioactivity**

A ^{238}U uranium nucleus is radioactive. It spontaneously disintegrates into a small fragment that is ejected with a measured speed of 1.50×10^7 m/s and a "daughter nucleus" that recoils with a measured speed of 2.56×10^5 m/s. What are the atomic masses of the ejected fragment and the daughter nucleus?

MODEL The notation ^{238}U indicates the isotope of uranium with an atomic mass of 238 u, where u is the abbreviation for the *unified atomic mass unit*. The nucleus contains 92 protons (uranium is atomic number 92) and 146 neutrons. The disintegration of a nucleus is, in essence, an explosion. Only *internal* nuclear forces are involved, so the total momentum is conserved in the radioactive decay.

VISUALIZE Figure 9.18 shows the before-and-after pictorial representation. The mass of the daughter nucleus is m_1 and that of

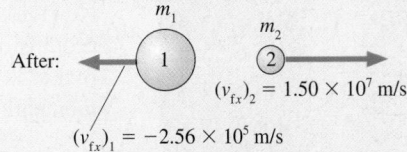

Before:
^{238}U
$m = 238$ u
$v_{ix} = 0$ m/s

After:
m_1
m_2
1
2
$(v_{fx})_2 = 1.50 \times 10^7$ m/s
$(v_{fx})_1 = -2.56 \times 10^5$ m/s

Find: m_1 and m_2

FIGURE 9.18 Before-and-after pictorial representation of the decay of a ^{238}U nucleus.

the ejected fragment is m_2. Notice that we converted the speed information to velocity information, giving $(v_{fx})_1$ and $(v_{fx})_2$ opposite signs.

SOLVE The nucleus was initially at rest, hence the total momentum is zero. The momentum after the decay is still zero if the two pieces fly apart in opposite directions with momenta equal in magnitude but opposite in sign. Momentum conservation requires

$$P_{fx} = m_1(v_{fx})_1 + m_2(v_{fx})_2 = P_{ix} = 0$$

Although we know both final velocities, this is not enough information to find the two unknown masses. However, we also have another conservation law, conservation of mass, that requires

$$m_1 + m_2 = 238 \text{ u}$$

Combining these two conservation laws gives

$$m_1(v_{fx})_1 + (238 \text{ u} - m_1)(v_{fx})_2 = 0$$

The mass of the daughter nucleus is

$$m_1 = \frac{(v_{fx})_2}{(v_{fx})_2 - (v_{fx})_1} \times 238 \text{ u}$$

$$= \frac{1.50 \times 10^7 \text{ m/s}}{(1.50 \times 10^7 - (-2.56 \times 10^5)) \text{ m/s}} \times 238 \text{ u} = 234 \text{ u}$$

With m_1 known, the mass of the ejected fragment is $m_2 = 238 - m_1 = 4 \text{ u}$.

ASSESS All we learn from a momentum analysis is the masses. Chemical analysis of the daughter nucleus shows that it is the element thorium, atomic number 90, with two fewer protons than the uranium nucleus. The ejected fragment carried away two protons as part of its mass of 4 u, so it must be a particle consisting of two protons and two neutrons. This is the nucleus of a helium atom, ^{4}He, which in nuclear physics is called an *alpha particle* α. Thus the radioactive decay of ^{238}U can be written as ^{238}U $\rightarrow$ ^{234}Th $+ \alpha$.

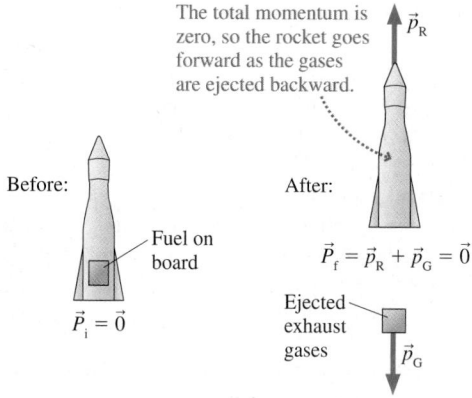

FIGURE 9.19 Rocket propulsion is an example of conservation of momentum.

The total momentum of the rocket + gases system must be conserved, so the rocket accelerates forward as the gases are expelled backward.

Much the same reasoning explains how a rocket or jet aircraft accelerates. Figure 9.19 shows a rocket with a parcel of fuel on board. Burning converts the fuel to hot gases that are expelled from the rocket motor. If we choose rocket + gases to be the system, the burning and expulsion are all internal forces. There are no other forces, so the total momentum of the rocket + gases system must be conserved. The rocket gains forward velocity and momentum as the exhaust gases are shot out the back, but the *total* momentum of the system remains zero.

Many people find it hard to understand how a rocket can accelerate in the vacuum of space because there is nothing to "push against." Thinking in terms of momentum, you can see that the rocket does not push against anything *external*, but only against the gases that it pushes out the back. In return, in accordance with Newton's third law, the gases push forward on the rocket. The details of rocket propulsion are more complex than we want to handle, because the mass of the rocket is changing, but you should be able to use the law of conservation of momentum to understand the basic principle by which rocket propulsion occurs.

STOP TO THINK 9.4 An explosion in a rigid pipe shoots out three pieces. A 6 g piece comes out the right end. A 4 g piece comes out the left end with twice the speed of the 6 g piece. From which end, left or right, does the third piece emerge?

9.5 Inelastic Collisions

Collisions can have different possible outcomes. A rubber ball dropped on the floor bounces, but a ball of clay sticks to the floor without bouncing. A golf club hitting a golf ball causes the ball to rebound away from the club, but a bullet striking a block of wood embeds itself in the block.

A collision in which the two objects stick together and move with a common final velocity is called a **perfectly inelastic collision.** The clay hitting the floor and the bullet embedding itself in the wood are examples of perfectly inelastic collisions. Other examples include railroad cars coupling together upon impact and darts hitting a dart board. Figure 9.20 emphasizes the fact that the two objects have a common final velocity after they collide.

In an *elastic collision*, by contrast, the two objects bounce apart. We've looked at some examples of elastic collisions, but a full analysis requires some ideas about energy. We will return to elastic collisions in Chapter 10.

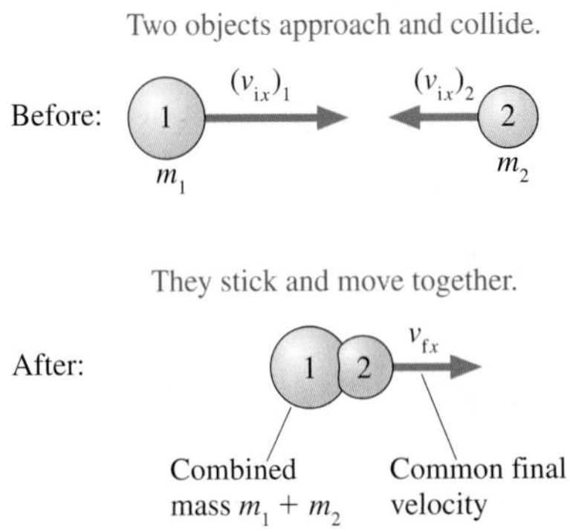

FIGURE 9.20 An inelastic collision.

EXAMPLE 9.7 An inelastic glider collision

In a laboratory experiment, a 200 g air-track glider and a 400 g air-track glider are pushed toward each other from opposite ends of the track. The gliders have Velcro tabs on the front so that they will stick together when they collide. The 200 g glider is pushed with an initial speed of 3.0 m/s. The collision causes it to reverse direction at 0.50 m/s. What was the initial speed of the 400 g glider?

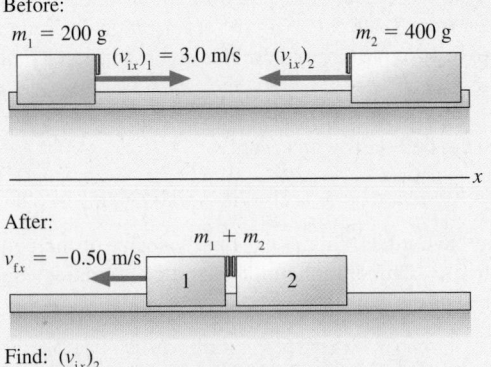

FIGURE 9.21 The before-and-after pictorial representation of two gliders colliding on an air track.

MODEL Model the gliders as particles. Define the two gliders together as the system. This is an isolated system, so its total momentum is conserved in the collision. The gliders stick together, so this is a perfectly inelastic collision.

VISUALIZE Figure 9.21 shows a pictorial representation. We've chosen to let the 200 g glider (glider 1) start out moving to the right, so $(v_{ix})_1$ is a positive 3.0 m/s. The gliders move to the left after the collision, so their common final velocity is $v_{fx} = -0.50$ m/s. Velocity $(v_{ix})_2$ will be negative.

SOLVE The law of conservation of momentum, $P_{fx} = P_{ix}$, is

$$(m_1 + m_2)v_{fx} = m_1(v_{ix})_1 + m_2(v_{ix})_2$$

where we made use of the fact that the combined mass $m_1 + m_2$ moves together after the collision. We can easily solve for the initial velocity of the 400 g glider:

$$(v_{ix})_2 = \frac{(m_1 + m_2)v_{fx} - m_1(v_{ix})_1}{m_2}$$

$$= \frac{(0.60\ \text{kg})(-0.50\ \text{m/s}) - (0.20\ \text{kg})(3.0\ \text{m/s})}{0.40\ \text{kg}}$$

$$= -2.25\ \text{m/s}$$

The negative sign, which we anticipated, indicates that the 400 g glider started out moving to the left. The initial *speed* of the glider, which we were asked to find, is 2.25 m/s.

EXAMPLE 9.8 Momentum in a car crash

A 2000 kg Cadillac had just started forward from a stop sign when it was struck from behind by a 1000 kg Volkswagen. The bumpers became entangled, and the two cars skidded forward together until they come to rest. Fortunately, both cars were equipped with airbags and the drivers were using seat belts, so no one was injured. Officer Tom, responding to the accident, measured the skid marks to be 3.0 m long. He also took testimony from the driver that the Cadillac's speed just before the impact was 5.0 m/s. Officer Tom charged the Volkswagen driver with reckless driving. Should the Volkswagen driver also be charged with exceeding the 50 km/hr speed limit? The judge

calls you as an "expert witness" to analyze the evidence. What is your conclusion?

MODEL This is really *two* problems. First, there is an inelastic collision. The two cars are not an isolated system, because of external friction forces, but friction is not going to be significant during the brief collision. Within the impulse approximation, the momentum of the Volkswagen + Cadillac system will be conserved in the collision. Then we have a second problem, a dynamics problem of the two cars sliding.

VISUALIZE Figure 9.22a on the next page is a pictorial representation showing both the before and after of the collision and the

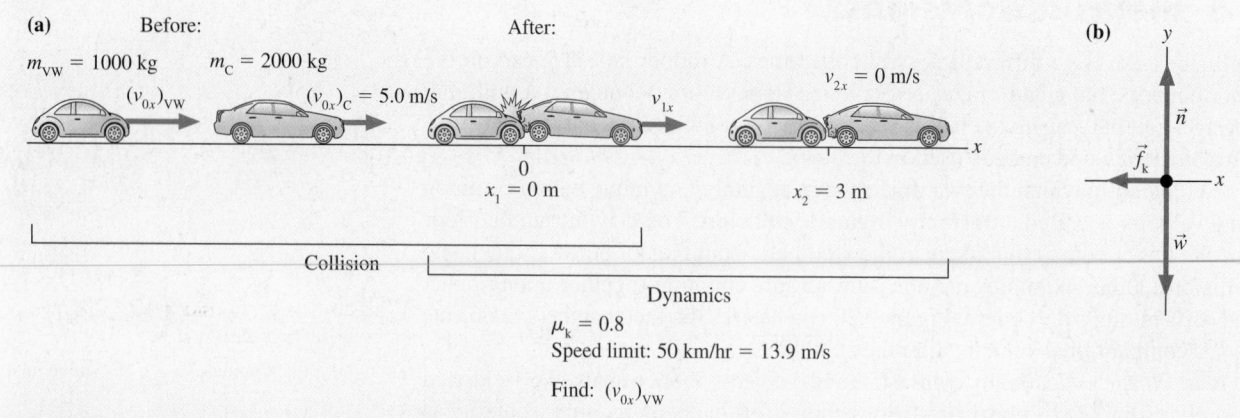

FIGURE 9.22 Pictorial representation and a free-body diagram of the cars as they skid.

more familiar picture for the dynamics of the skidding. We do not need to consider forces during the collision, because we will use the law of conservation of momentum, but we do need a free-body diagram of the cars during the subsequent skid. This is shown in Figure 9.22b.

The cars have a common velocity v_{1x} just after the collision. This is the *initial* velocity for the dynamics problem. Our goal is to find $(v_{0x})_{VW}$, the Volkswagen's velocity at the moment of impact. The 50 km/hr speed limit has been converted to 13.9 m/s.

SOLVE First, the inelastic collision. The law of conservation of momentum is

$$(m_{VW} + m_C)v_{1x} = m_{VW}(v_{0x})_{VW} + m_C(v_{0x})_C$$

Solving for the initial velocity of the Volkswagen, we find

$$(v_{0x})_{VW} = \frac{(m_{VW} + m_C)v_{1x} - m_C(v_{0x})_C}{m_{VW}}$$

To evaluate $(v_{0x})_{VW}$, we need to know v_{1x}, the velocity *immediately* after the collision as the cars begin to skid. This information will come out of the dynamics of the skid. Newton's second law, based on the free-body diagram, and the model of kinetic friction are

$$\sum F_x = -f_k = (m_{VW} + m_C)a_x$$
$$\sum F_y = n - (m_{VW} + m_C)g = 0$$
$$f_k = \mu_k n$$

where we have noted that $\vec{f}_k$ points to the left (negative x-component) and that the total mass is $m_{VW} + m_C$. From the y-equation and the friction equation,

$$f_k = \mu_k(m_{VW} + m_C)g$$

Using this in the x-equation gives us the acceleration during the skid,

$$a_x = \frac{-f_k}{m_{VW} + m_C} = -\mu_k g = -7.84 \text{ m/s}^2$$

where the coefficient of kinetic friction for rubber on concrete is taken from Table 5.1. With the acceleration determined, we can move on to the kinematics. This is constant acceleration, so

$$v_{2x}^2 = 0 = v_{1x}^2 + 2a_x(x_2 - x_1) = v_{1x}^2 + 2a_x x_2$$

Hence the skid starts with velocity

$$v_{1x} = \sqrt{-2a_x x_1} = \sqrt{-2(-7.84 \text{ m/s}^2)(3.0 \text{ m})} = 6.86 \text{ m/s}$$

As we have noted, this is the final velocity of the collision. Inserting v_{1x} back into the momentum conservation equation, we finally determine that

$$(v_{0x})_{VW} = \frac{(3000 \text{ kg})(6.86 \text{ m/s}) - (2000 \text{ kg})(5.0 \text{ m/s})}{1000 \text{ kg}}$$
$$= 10.6 \text{ m/s}$$

On the basis of your testimony, the Volkswagen driver is *not* charged with speeding!

NOTE ▶ Momentum is *not always* conserved. In this example, momentum was conserved during the collision but *not* during the skid. Momentum is conserved only for an isolated system. In practice, it is not unusual for momentum to be conserved in one part or one aspect of a problem but not in others. ◀

STOP TO THINK 9.5   The two particles are both moving to the right. Particle 1 catches up with particle 2 and collides with it. The particles stick together and continue on with velocity v_f. Which of these statements is true?

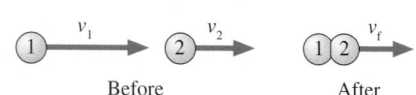

a. v_f is greater than v_1.   b. $v_f = v_1$.   c. v_f is greater than v_2 but less than v_1.
d. $v_f = v_2$.   e. v_f is less than v_2.   f. Can't tell without knowing the masses.

9.6 Momentum in Two Dimensions

Our examples thus far have been confined to motion along a one-dimensional axis. Many practical examples of momentum conservation involve motion in a plane. The total momentum $\vec{P}$ is a *vector* sum of the momenta $\vec{p} = m\vec{v}$ of the individual particles. Consequently, as we found in Section 9.3, momentum is conserved only if each component of $\vec{P}$ is conserved:

$$(p_{fx})_1 + (p_{fx})_2 + (p_{fx})_3 + \cdots = (p_{ix})_1 + (p_{ix})_2 + (p_{ix})_3 + \cdots$$
$$(p_{fy})_1 + (p_{fy})_2 + (p_{fy})_3 + \cdots = (p_{iy})_1 + (p_{iy})_2 + (p_{iy})_3 + \cdots \quad (9.22)$$

In this section we'll apply momentum conservation to motion in two dimensions.

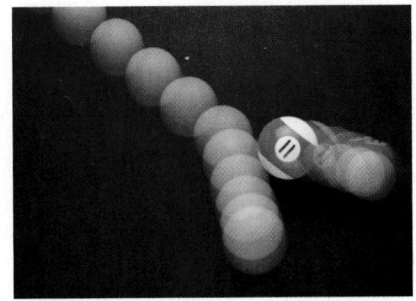

Collisions and explosions often involve motion in two dimensions.

EXAMPLE 9.9 Momentum in a 2D car crash

The 2000 kg Cadillac and the 1000 kg Volkswagen of Example 9.8 meet again the following week, just after leaving the auto body shop where they had been repaired. The stoplight has just turned green, and the Cadillac, heading north, drives forward into the intersection. The Volkswagen, traveling east, fails to stop. The Volkswagen crashes into the left front fender of the Cadillac, then the cars stick together and slide to a halt. Officer Tom, responding to the accident, sees that the skid marks go 35° northeast from the point of impact. The Cadillac driver, who keeps a close eye on the speedometer, reports that he was traveling at 3.0 m/s when the accident occurred. How fast was the Volkswagen going just before the impact?

MODEL This is another inelastic collision. The total momentum of the Volkswagen + Cadillac system is conserved.

VISUALIZE Figure 9.23 is a before-and-after pictorial representation. The Volkswagen travels on the x-axis and the Cadillac on the y-axis, hence $(v_{0y})_{VW} = 0$ and $(v_{0x})_C = 0$.

SOLVE After the collision, the two cars move with the common velocity $\vec{v}_1$. The velocity components, as in a projectile motion problem, are $v_{1x} = v_1 \cos\theta$ and $v_{1y} = v_1 \sin\theta$. Thus the simultaneous x- and y-momentum equations are

$$(m_C + m_{VW})v_{1x} = (m_C + m_{VW})v_1\cos\theta$$
$$= m_C(v_{0x})_C + m_{VW}(v_{0x})_{VW} = m_{VW}(v_{0x})_{VW}$$
$$(m_C + m_{VW})v_{1y} = (m_C + m_{VW})v_1\sin\theta$$
$$= m_C(v_{0y})_C + m_{VW}(v_{0y})_{VW} = m_C(v_{0y})_C$$

We can use the y-equation to find:

$$v_1 = \frac{m_C(v_{0y})_C}{(m_C + m_{VW})\sin\theta} = \frac{(2000\ \text{kg})(3.0\ \text{m/s})}{(3000\ \text{kg})\sin 35°} = 3.49\ \text{m/s}$$

Using this value for v_1 in the x-equation, we find that the Volkswagen's velocity was

$$(v_{0x})_{VW} = \frac{(m_C + m_{VW})v_1\cos\theta}{m_{VW}} = 8.6\ \text{m/s}$$

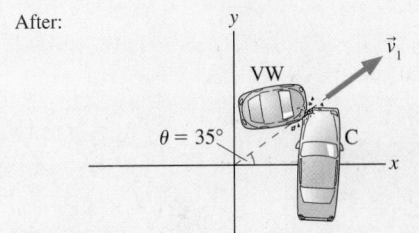

Before:

$(v_{0x})_{VW}$

VW
$m_{VW} = 1000$ kg

$(v_{0y})_C = 3.0$ m/s

C $m_C = 2000$ kg

After:

$\vec{v}_1$

VW

$\theta = 35°$ C

Find: $(v_{0x})_{VW}$

FIGURE 9.23 Pictorial representation of the collision between the Cadillac and the Volkswagen.

It's instructive to examine this collision with a picture of the momentum vectors. Before the collision, $\vec{p}_{VW} = (1000\ \text{kg})(8.6\ \text{m/s})\hat{\imath} = 8600\hat{\imath}\ \text{kg}\,\text{m/s}$ and $\vec{p}_C = (2000\ \text{kg})(3.0\ \text{m/s})\hat{\jmath} = 6000\hat{\jmath}\ \text{kg}\,\text{m/s}$. These vectors, and their sum $\vec{P} = \vec{p}_{VW} + \vec{p}_C$ are shown in Figure 9.24. You can see that the total momentum vector makes a 35° angle with the x-axis. The individual momenta change in the collision, *but the total momentum does not*. That is why the skid marks are 35° north of east.

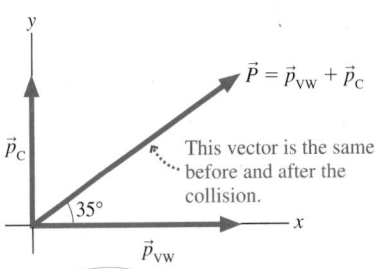

$\vec{P} = \vec{p}_{VW} + \vec{p}_C$

$\vec{p}_C$

This vector is the same before and after the collision.

35°

$\vec{p}_{VW}$

FIGURE 9.24 The momentum vectors of the car crash.

EXAMPLE 9.10 **A three-piece explosion**

A 10 g projectile is traveling east at 2.0 m/s when it suddenly explodes into three pieces. A 3.0 g fragment is shot due west at 10 m/s while another 3.0 g fragment travels 40° north of east at 12 m/s. What are the speed and direction of the third fragment?

MODEL Although many complex forces are involved in the explosion, they are all internal to the system. There are no external forces, so this is an isolated system and its total momentum is conserved.

VISUALIZE Figure 9.25 shows a before-and-after pictorial representation. We'll use uppercase M and V to distinguish the initial object from the three pieces into which it explodes.

Before:

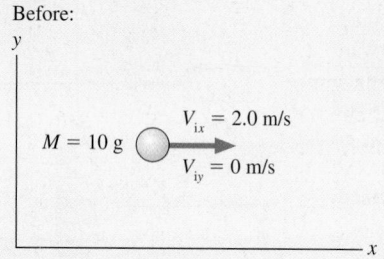

After:

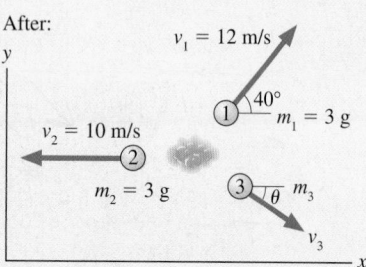

Find: v_3 and θ

FIGURE 9.25 Before-and-after pictorial representation of the three-piece explosion.

SOLVE The system is the initial object and the subsequent three pieces. Conservation of momentum requires

$$m_1(v_{fx})_1 + m_2(v_{fx})_2 + m_3(v_{fx})_3 = MV_{ix}$$

$$m_1(v_{fy})_1 + m_2(v_{fy})_2 + m_3(v_{fy})_3 = MV_{iy}$$

Conservation of mass implies that

$$m_3 = M - m_1 - m_2 = 4.0 \text{ g}$$

Neither the original object nor m_2 have any momentum along the y-axis. We can use Figure 9.24 to write out the x- and y-components of $\vec{v}_1$ and $\vec{v}_3$, leading to

$$m_1 v_1 \cos 40° - m_2 v_2 + m_3 v_3 \cos\theta = MV$$

$$m_1 v_1 \sin 40° - m_3 v_3 \sin\theta = 0$$

where we used $(v_{fx})_2 = -v_2$ because m_2 is moving in the negative x-direction. Inserting known values in these equations gives us

$$-2.42 + 4v_3 \cos\theta = 20$$

$$23.14 - 4v_3 \sin\theta = 0$$

We can leave the masses in grams in this situation because the conversion factor to kilograms appears on both sides of the equation and thus cancels out. To solve, first use the second equation to write $v_3 = 5.79/\sin\theta$. Substitute this result into the first equation, noting that $\cos\theta/\sin\theta = 1/\tan\theta$, to get

$$-2.42 + 4\left(\frac{5.79}{\sin\theta}\right)\cos\theta = -2.42 + \frac{23.14}{\tan\theta} = 20$$

Now solve for θ:

$$\tan\theta = \frac{23.14}{20 + 2.42} = 1.032$$

$$\theta = \tan^{-1}(1.032) = 45.9°$$

Finally, use this result in the earlier expression for v_3 to find

$$v_3 = \frac{5.79}{\sin 45.9°} = 8.06 \text{ m/s.}$$

The third fragment, with a mass of 4.0 g, is shot 45.9° south of east at a speed of 8.06 m/s.

9.7 Angular Momentum

The momentum $\vec{p}$ describes linear motion, such as motion along a line or motion in a plane. For a single particle, the law of conservation of momentum is an alternative way of stating Newton's first law. Rather than saying that a particle will continue to move in a straight line at constant velocity unless acted on by a net force, we can say that the momentum of an isolated particle is conserved. Both express the idea that a particle in linear motion tends to "keep going" unless something acts on it to change its motion.

Another important class of motion that you've studied is motion in a circle. The momentum $\vec{p}$ is *not* a conserved quantity for a particle in circular motion. Momentum is a vector, and the momentum of a particle in circular motion changes as the direction of motion changes.

Nonetheless, a spinning bicycle wheel would keep turning if it were not for friction. A ball moving in a circle at the end of string tends to "keep going," like the bicycle wheel, as long as the conditions for circular motion haven't changed. The quantity that expresses this idea for circular motion is called *angular momentum*.

This section is a first introduction to the concept of angular momentum. We will restrict our attention to particles in circular motion, and we will not present any proofs or derivations. A more rigorous discussion is deferred to Chapter 13, where we will find that angular momentum is an important idea for understanding the rotational motion of solid objects.

Figure 9.26a shows a particle moving in a circle of radius r with velocity $\vec{v}$. We define the particle's **angular momentum** as

$$L = mv_t \qquad (9.23)$$

where v_t is the tangential component of $\vec{v}$. The angular momentum L has the same sign as v_t, which you will recall from Chapter 7 is positive for counterclockwise rotation, negative for clockwise rotation. The units of angular momentum are $\text{kg}\,\text{m}^2/\text{s}$.

Figure 9.26b shows a net force $\vec{F}_{\text{net}}$ acting on the particle. As in Chapter 7, we can decompose the vector $\vec{F}_{\text{net}}$ into a tangential component F_t and a radial component F_r. The radial component of the force causes the centripetal acceleration of circular motion. The component F_t parallel to the trajectory causes the particle to speed up or slow down.

If there's a tangential force, the particle's angular momentum $L = mv_t$ changes as v_t changes. If there is no tangential force, $F_t = 0$, the angular momentum remains constant. Without formal proof, we'll state this idea as the *law of conservation of angular momentum*:

> **Law of conservation of angular momentum** The angular momentum of a particle (or system of particles) in circular motion does not change unless there is a net tangential force on the particle. That is, $L_f = L_i$ if $F_t = 0$.

As stated, the law of conservation of angular momentum seems rather limited in scope. The significance of angular momentum will become more apparent in later chapters, where you'll see how it can be applied to noncircular orbits and to the rotation of solid objects. But we can give the law of conservation of momentum one interesting application in its current form: circular trajectories that change diameter.

Suppose you tie a ball to a string and swing it in a horizontal circle over your head. The tension in the rope is the radial force F_r that causes the ball's centripetal acceleration. There is no tangential force (assuming air resistance is negligible) and, once you get it going, the ball will continue to turn in a circle with constant tangential velocity v_t and constant angular momentum L. What if you now begin to pull the rope in, decreasing the radius of the ball's trajectory? Because there's no tangential force, *the ball's angular momentum is conserved* as it slowly spirals inward into smaller and smaller circles. Using the definition $L = mv_t$, you can see that the ball's speed has to increase as r decreases in order to keep L constant.

(a)
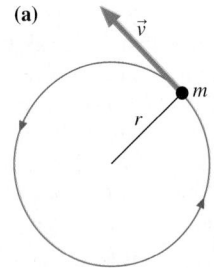

(b) The tangential component of $\vec{F}$ causes the particle's speed and angular momentum to change.

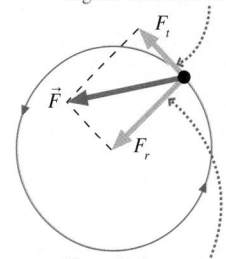

The radial component of $\vec{F}$ causes the centripetal acceleration of circular motion.

FIGURE 9.26 A particle in circular motion has angular momentum.

The angular momentum of a spinning skater is conserved. This allows the skater to use her arms to control her speed.

EXAMPLE 9.11 A spinning ice skater

An ice skater spins around on the tips of his blades while holding a 5.0 kg weight in each hand. He begins with his arms straight out from his body and his hands 140 cm apart. While spinning at 2.0 rev/s, he pulls the weights in and holds them 50 cm apart against his shoulders. If we neglect the mass of the skater, how fast is he spinning after pulling the weights in?

MODEL Although the mass of the skater is larger than the mass of the weights, neglecting the skater's mass is not a bad approximation. Angular momentum depends on the radius of the circle. The skater's mass is concentrated in his torso, which has an effective radius (i.e., where most of the mass is concentrated) of only 9 or 10 cm. The weights move in much larger circles and have a disproportionate influence on his motion. The skater's arms exert radial forces on the weights just to keep them moving in circles, and even larger radial forces as he pulls them in. But there is no tangential force on the weights, so their total angular momentum is conserved.

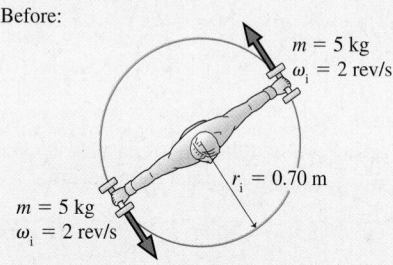

Before:

$m = 5$ kg
$\omega_i = 2$ rev/s

$r_i = 0.70$ m

$m = 5$ kg
$\omega_i = 2$ rev/s

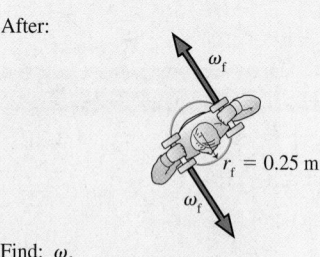

After:

ω_f

$r_f = 0.25$ m

ω_f

Find: ω_f

FIGURE 9.27 Top view pictorial representation of the spinning ice skater.

VISUALIZE Figure 9.27 shows a before-and-after pictorial representation, as seen from above.

SOLVE The two weights have the same mass, move in circles with the same radius, and have the same tangential velocity. Thus the total angular momentum is twice that of one weight. The mathematical statement of angular momentum conservation, $L_f = L_i$, is

$$2mr_f v_{ft} = 2mr_i v_{it}$$

You learned in Chapter 7 that a particle's tangential velocity is related to its angular velocity by $v_t = r\omega$. Using this result, and canceling the $2m$, we find

$$r_f^2 \omega_f = r_i^2 \omega_i$$

$$\omega_f = \left(\frac{r_i}{r_f}\right)^2 \omega_i$$

Although we would need ω in rad/s to calculate v_t from $v_t = r\omega$, we don't need to convert ω because the conversion factor would be the same on both sides of the equation and would cancel. We can leave the skater's initial angular velocity as $\omega_i = 2.0$ rev/s. When he pulls the weights in, his angular velocity increases to

$$\omega_f = \left(\frac{0.70 \text{ m}}{0.25 \text{ m}}\right)^2 \times 2.0 \text{ rev/s} = 15.7 \text{ rev/s}$$

ASSESS Pulling in the weights increases the skater's spin from 2 rev/s to nearly 16 rev/s. This is actually somewhat high, because we neglected the mass of the skater. Nonetheless, you've probably seen how figure skaters start a slow spin with arms outstretched, then dramatically speed up by raising their arms over their heads. They use the mass of their arms, not extra weights, to control their speed. While this example has been over simplified, it does illustrate that a skater's spin is governed by conservation of angular momentum.

STOP TO THINK 9.6 A dry ice (solid carbon dioxide) puck revolves in a circle on the end of a lightweight rigid rod that turns on frictionless bearings. A cushion of CO_2 gas allows the puck to glide across the surface without friction. As the puck sublimates (changes from a solid to a gas), does its speed increase, decrease, or stay the same?

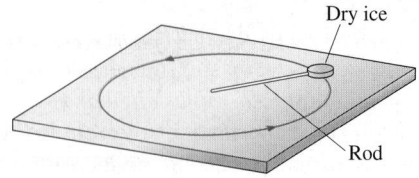

Dry ice

Rod

SUMMARY

The goal of Chapter 9 has been to introduce the ideas of impulse, momentum, and angular momentum and to learn a new problem-solving strategy based on conservation laws.

GENERAL PRINCIPLES

Law of Conservation of Momentum

The total momentum $\vec{P} = \vec{p}_1 + \vec{p}_2 + \cdots$ of an isolated system is a constant. Thus

$$\vec{P}_f = \vec{P}_i$$

Law of Conservation of Angular Momentum

The angular momentum L of a particle or system of particles in circular motion does not change unless there is a net tangential force. Thus

$$L_f = L_i$$

Solving Momentum Conservation Problems

MODEL Choose an isolated system or a system that is isolated during at least part of the problem.

VISUALIZE Draw a pictorial representation of the system before and after the interaction.

SOLVE Write the law of conservation of momentum in terms of vector components

$$(p_{fx})_1 + (p_{fx})_2 + \cdots = (p_{ix})_1 + (p_{ix})_2 + \cdots$$

$$(p_{fy})_1 + (p_{fy})_2 + \cdots = (p_{iy})_1 + (p_{iy})_2 + \cdots$$

ASSESS Is the result reasonable?

IMPORTANT CONCEPTS

Momentum $\vec{p} = m\vec{v}$

Impulse $J_x = \int_{t_i}^{t_f} F_x(t)\, dt$ = area under force curve

Impulse and momentum are related by the impulse-momentum theorem

$$\Delta p_x = J_x$$

This is an alternative statement of Newton's second law.

Angular momentum $L = mrv_t$

System A group of interacting particles.

Isolated system A system on which there are no external forces or the net external force is zero.

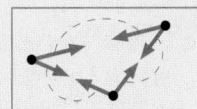

Before-and-after pictorial representation

- Define the system.
- Use two drawings to show the system *before* and *after* the interaction.
- List known information and identify what you are trying to find.

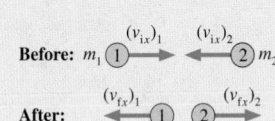

APPLICATIONS

Collisions Two or more particles come together. In a perfectly inelastic collision, they stick together and move with a common final velocity.

Explosions Two or more particles move away from each other.

Two dimensions No new ideas, but both the x- and y-components of P must be conserved, giving two simultaneous equations.

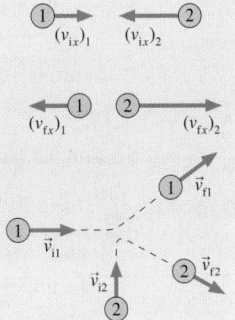

Momentum bar charts display the impulse-momentum theorem $p_{fx} = p_{ix} + J_x$ in graphical form.

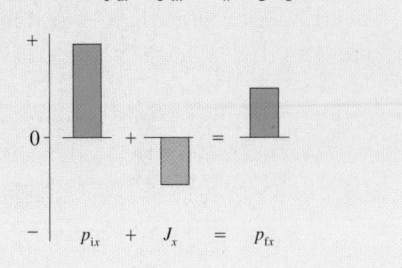

TERMS AND NOTATION

collision	impulse-momentum theorem	isolated system	perfectly inelastic collision
impulsive force	impulse approximation	law of conservation	angular momentum, L
momentum, $\vec{p}$	momentum bar chart	of momentum	law of conservation of
impulse, J	total momentum, $\vec{P}$	explosion	angular momentum

EXERCISES AND PROBLEMS

The 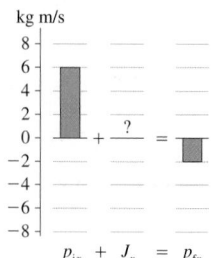 icon indicates that the problem can be done on a Momentum Worksheet.

Exercises

Section 9.1 Momentum and Impulse

1. What is the magnitude of the momentum of
   a. A 1500 kg car traveling at 10 m/s?
   b. A 200 g baseball thrown at 40 m/s?
2. At what speed do a bicycle and its rider, with a combined mass of 100 kg, have the same momentum as a 1500 kg car traveling at 5.0 m/s?
3. What value of F_{max} gives an impulse of 6.0 N s?

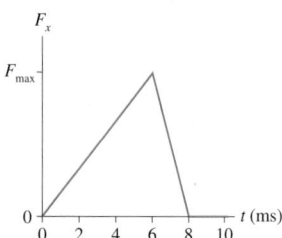

FIGURE EX9.3

4. Suppose a rubber ball and a steel ball collide. Which, if either, receives the larger impulse? Explain.
5. What is the impulse on a 3.0 kg particle that experiences this force?

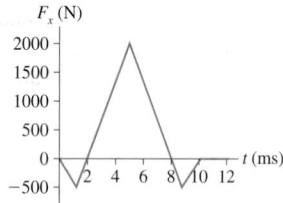

FIGURE EX9.5

Section 9.2 Solving Impulse and Momentum Problems

6. Figure Ex9.6 is an incomplete momentum bar chart for a collision that lasts 10 ms. What are the magnitude and direction of the average collision force exerted on the object?

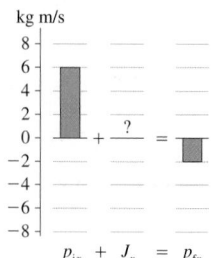

FIGURE EX9.6 $p_{ix} + J_x = p_{fx}$

7. a. A 2.0 kg object is moving to the right with a speed of 1.0 m/s when it experiences the force shown in Figure Ex9.7a. What are the object's speed and direction after the force ends?

b. Answer this question for the force shown in Figure Ex9.7b.

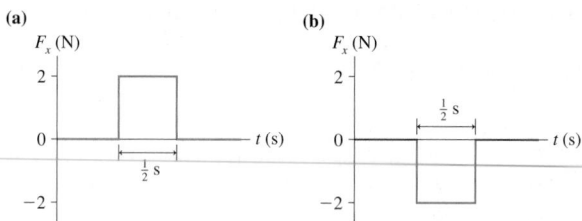

FIGURE EX9.7

8. A sled slides along a horizontal surface on which the coefficient of kinetic friction is 0.25. Its velocity at point A is 8.0 m/s and at point B is 5.0 m/s. Use the impulse-momentum theorem to find how long the sled takes to travel from A to B.
9. Use the impulse-momentum theorem to find how long a falling object takes to increase its speed from 5.5 m/s to 10.4 m/s.
10. A 60 g tennis ball with an initial speed of 32 m/s hits a wall and rebounds with the same speed. Figure Ex9.10 shows the force of the wall on the ball during the collision. What is the value of F_{max}, the maximum value of the contact force during the collision?

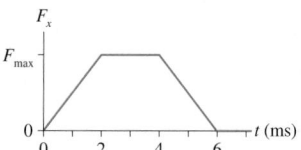

FIGURE EX9.10

11. A 600 g air-track glider collides with a spring at one end of the track. Figure Ex9.11 shows the glider's velocity and the force exerted on the glider by the spring. How long is the glider in contact with the spring?

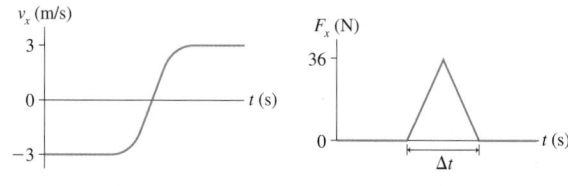

FIGURE EX9.11

Section 9.3 Conservation of Momentum

12. A 10-m-long glider with a mass of 680 kg (including the passengers) is gliding horizontally through the air at 30 m/s when a 60 kg skydiver drops out by releasing his grip on the glider. What is the glider's velocity just after the skydiver lets go?
13. A 10,000 kg railroad car is rolling at 2.0 m/s when a 4000 kg load of gravel is suddenly dropped in. What is the car's speed just after the gravel is loaded?
14. A 5000 kg open train car is rolling on frictionless rails at 22 m/s when it starts pouring rain. A few minutes later, the car's speed is 20 m/s. What mass of water has collected in the car?

Section 9.4 Explosions

15. A 50 kg archer, standing on frictionless ice, shoots a 100 g arrow at a speed of 100 m/s. What is the recoil speed of the archer?

16. In Problem 24 of Chapter 8 you found the recoil speed of Bob as he throws a rock while standing on frictionless ice. Bob has a mass of 75 kg and can throw a 500 g rock with a speed of 30 m/s. Find Bob's recoil speed again, this time using momentum.

17. Dan is gliding on his skateboard at 4.0 m/s. He suddenly jumps backward off the skateboard, kicking the skateboard forward at 8.0 m/s. How fast is Dan going as his feet hit the ground? Dan's mass is 50 kg and the skateboard's mass is 5.0 kg.

Section 9.5 Inelastic Collisions

18. A 300 g bird flying along at 6.0 m/s sees a 10 g insect heading straight toward it with a speed of 30 m/s. The bird opens its mouth wide and enjoys a nice lunch. What is the bird's speed immediately after swallowing?

19. The parking brake on a 2000 kg Cadillac has failed, and it is rolling slowly, at 1 mph, toward a group of small children. Seeing the situation, you realize you have just enough time to drive your 1000 kg Volkswagen head-on into the Cadillac and save the children. With what speed should you impact the Cadillac to bring it to a halt?

20. A 1500 kg car is rolling at 2.0 m/s. You would like to stop the car by firing a 10 kg blob of sticky clay at it. How fast should you fire the clay?

Section 9.6 Momentum in Two Dimensions

21. A 20 g ball of clay traveling east at 3.0 m/s collides with a 30 g ball of clay traveling north at 2.0 m/s. What are the speed and the direction of the resulting 50 g ball of clay?

22. Two particles collide and bounce apart. Figure Ex9.22 shows the initial momenta of both and the final momentum of particle 2. What is the final momentum of particle 1? Show the momentum vector on the diagram *and* write it in component form.

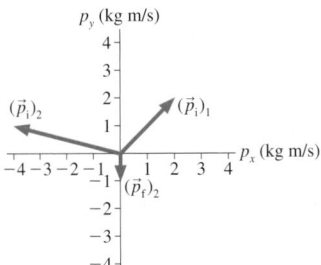

FIGURE EX9.22

Section 9.7 Angular Momentum

23. What is the angular momentum of the moon around the earth? Relevant astronomical data are found inside the cover of the book.

24. A 200 g puck revolves in a circle on a frictionless table at the end of a 50-cm-long string. The puck's angular momentum is 3.0 kg m^2/s. What is the tension in the string?

Problems

25. A 50 g ball is launched from ground level at an angle 30° above the horizon. Its initial speed is 25 m/s.
    a. What are the values of p_x and p_y an instant after the ball is thrown, at the point of maximum altitude, and an instant before the ball hits the ground?
    b. Why is one component of $\vec{p}$ constant? Explain.
    c. For the component of $\vec{p}$ that changes, show that the change in momentum is equal to the weight of the ball multiplied by the time of flight. Explain why this is so.

26. Far in space, where gravity is negligible, a 425 kg rocket traveling at 75 m/s fires its engines. Figure P9.26 shows the thrust force as a function of time. The mass lost by the rocket during these 30 s is negligible.
    a. What impulse does the engine impart to the rocket?
    b. At what time does the rocket reach its maximum speed? What is the maximum speed?

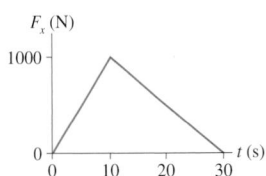

FIGURE P9.26

27. A tennis player swings her 1000 g racket with a speed of 10 m/s. She hits a 60 g tennis ball that was approaching her at a speed of 20 m/s. The ball rebounds at 40 m/s.
    a. How fast is her racket moving immediately after the impact? You can ignore the interaction of the racket with her hand for the brief duration of the collision.
    b. If the tennis ball and racket are in contact for 10 ms, what is the average force that the racket exerts on the ball? How does this compare to the ball's weight?

28. A 200 g ball is dropped from a height of 2.0 m, bounces on a hard floor, and rebounds to a height of 1.5 m. Figure P9.28 shows the impulse received from the floor. What maximum force does the floor exert on the ball?

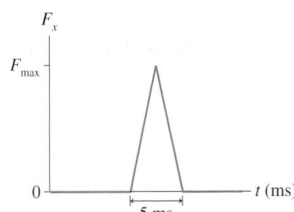

FIGURE P9.28

29. A 40 g rubber ball is dropped from a height of 1.8 m and rebounds to two-thirds of its initial height.
    a. Find the magnitude and direction of the impulse that the floor exerts on the ball.
    b. Using simple observations of an ordinary rubber ball, sketch a physically plausible graph of the force of the floor on the ball as a function of time.
    c. Make a plausible estimate of how long the ball is in contact with the floor, then use this quantity to calculate the approximate average force of the floor on the ball.

30. A 500 g cart is released from rest 1.0 m from the bottom of a frictionless, 30° ramp. The cart rolls down the ramp and bounces off a rubber block at the bottom. Figure P9.30 shows the force during the collision. After the cart bounces, how far does it roll back up the ramp?

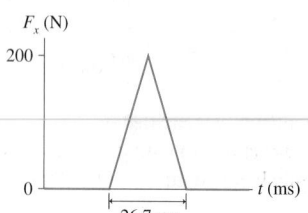

FIGURE P9.30

31. A 20 g ball of clay is thrown horizontally at 30 m/s toward a 1.0 kg block sitting at rest on a frictionless surface. The clay hits and sticks to the block.
    a. What impulse does the clay exert on the block?
    b. What impulse does the block exert on the clay?
    c. Does $J_{\text{block on clay}} = -J_{\text{clay on block}}$?

32. Three identical train cars, coupled together, are rolling east at 2.0 m/s. A fourth car traveling east at 4.0 m/s catches up with the three and couples to make a four-car train. A moment later, the train cars hit a fifth car that was at rest on the tracks, and it couples to make a five-car train. What is the speed of the five-car train?

33. Most geologists believe that the dinosaurs became extinct 65 million years ago when a large comet or asteroid struck the earth, throwing up so much dust that the sun was blocked out for a period of many months. Suppose an asteroid with a diameter of 2.0 km and a mass of 1.0×10^{13} kg hits the earth with an impact speed of 4.0×10^4 m/s.
    a. What is the earth's recoil speed after such a collision? (Use a reference frame in which the earth was initially at rest.)
    b. What percentage is this of the earth's speed around the sun? (Use the astronomical data inside the cover.)

34. At the center of a 50-m-diameter circular ice rink, a 75 kg skater traveling north at 2.5 m/s collides with and holds onto a 60 kg skater who had been heading west at 3.5 m/s.
    a. How long will it take them to glide to the edge of the rink?
    b. Where will they reach it? Give your answer as an angle north of west.

35. Two ice skaters, with masses of 50 kg and 75 kg, are at the center of a 60-m-diameter circular rink. The skaters push off against each other and glide to opposite edges of the rink. If the heavier skater reaches the edge in 20 s, how long does the lighter skater take to reach the edge?

36. A firecracker in a coconut blows the coconut into three pieces. Two pieces of equal mass fly off south and west, perpendicular to each other, at 20 m/s. The third piece has twice the mass as the other two. What are the speed and direction of the third piece?

37. One billiard ball is shot east at 2.0 m/s. A second, identical billiard ball is shot west at 1.0 m/s. The balls have a glancing collision, not a head-on collision, deflecting the second ball by 90° and sending it north at 1.41 m/s. What are the speed and direction of the first ball after the collision?

38. A 10 g bullet is fired into a 10 kg wood block that is at rest on a wood table. The block, with the bullet embedded, slides 5.0 cm across the table. What was the speed of the bullet?

39. Fred (mass 60 kg) is running with the football at a speed of 6.0 m/s when he is met head-on by Brutus (mass 120 kg), who is moving at 4.0 m/s. Brutus grabs Fred in a tight grip, and they fall to the ground. Which way do they slide, and how far? The coefficient of kinetic friction between football uniforms and Astroturf is 0.30.

40. You are part of a search-and-rescue mission that has been called out to look for a lost explorer. You've found the missing explorer, but you're separated from him by a 200-m-high cliff and a 30-m-wide raging river. To save his life, you need to get a 5.0 kg package of emergency supplies across the river. Unfortunately, you can't throw the package hard enough to make it across. Fortunately, you happen to have a 1.0 kg rocket intended for launching flares. Improvising quickly, you attach a sharpened stick to the front of the rocket, so that it will impale itself into the package of supplies, then fire the rocket at ground level toward the supplies. What minimum speed must the rocket have just before impact in order to save the explorer's life?

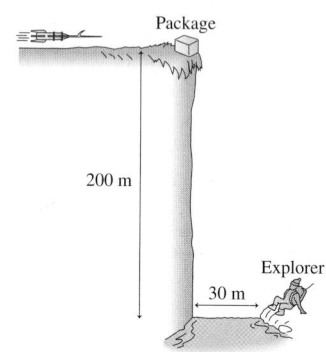

FIGURE P9.40

41. An object at rest on a flat, horizontal surface explodes into two fragments, one seven times as massive as the other. The heavier fragment slides 8.2 m before stopping. How far does the lighter fragment slide? Assume that both fragments have the same coefficient of kinetic friction.

42. A 1500 kg weather rocket accelerates upward at 10 m/s². It explodes 2.0 s after liftoff and breaks into two fragments, one twice as massive as the other. Photos reveal that the lighter fragment traveled straight up and reached a maximum height of 530 m. What were the speed and direction of the heavier fragment just after the explosion?

43. In a ballistics test, a 25 g bullet traveling horizontally at 1200 m/s goes through a 30-cm-thick 350 kg stationary target and emerges with a speed of 900 m/s. The target is free to slide on a smooth horizontal surface.
    a. How long is the bullet in the target? What average force does it exert on the target?
    b. What is the target's speed just after the bullet emerges?

44. Two 500 g blocks of wood are 2.0 m apart on a frictionless table. A 10 g bullet is fired at 400 m/s toward the blocks. It passes all the way through the first block, then embeds itself in the second block. The speed of the first block immediately afterward is 6.0 m/s. What is the speed of the second block after the bullet stops?

45. The skiing duo of Brian (80 kg) and Ashley (50 kg) is always a crowd pleaser. In one routine, Brian, wearing wood skis, starts at the top of a 200-m-long, 20° slope. Ashley waits for

him halfway down. As he skis past, she leaps into his arms and he carries her the rest of the way down. What is their speed at the bottom of the slope?

46. In a military test, a 575 kg unmanned spy plane is traveling north at an altitude of 2700 m and a speed of 450 m/s. It is intercepted by a 1280 kg rocket traveling east at 725 m/s. If the rocket and the spy plane become enmeshed in a tangled mess, where, relative to the point of impact, do they hit the ground?

47. The Army of the Nation of Whynot has a plan to propel small vehicles across the battlefield by shooting them, from behind, with a machine gun. They've hired you as a consultant to help with an upcoming test. A 100 kg test vehicle will roll along frictionless rails. The vehicle has a tall "sail" made of ultra-hard steel. Previous tests have shown that a 20 g bullet traveling at 400 m/s rebounds from the sail at 200 m/s. The design objective is for the cart to reach a speed of 12 m/s in 20 s. You need to tell them how many bullets to fire per second from the machine gun. A five-star general will watch the test, so your ability to get future consulting jobs will depend on the outcome.

48. A 500 kg cannon fires a 10 kg cannonball with a speed of 200 m/s relative to the muzzle. The cannon is on wheels that roll without friction. When the cannon fires, what is the speed of the cannonball relative to the earth?

49. A spaceship of mass 2.0×10^6 kg is cruising at a speed of 5.0×10^6 m/s when the antimatter reactor fails, blowing the ship into three pieces. One section, having a mass of 5.0×10^5 kg, is blown straight backward with a speed of 2.0×10^6 m/s. A second piece, with mass 8.0×10^5 kg, continues forward at 1.0×10^6 m/s. What are the direction and speed of the third piece?

50. A proton (mass 1 u) is shot at a speed of 5.0×10^7 m/s toward a gold target. The nucleus of a gold atom (mass 197 u) repels the proton and deflects it straight back toward the source with 90% of its initial speed. What is the recoil velocity of the gold nucleus?

51. A proton (mass 1 u) is shot toward an unknown target nucleus at a speed of 2.5×10^6 m/s. The proton rebounds with its speed reduced by 25% while the target nucleus acquires a speed of 3.12×10^5 m/s. What is the mass, in atomic mass units, of the target nucleus?

52. The nucleus of the polonium isotope ^{214}Po (mass 214 u) is radioactive and decays by emitting an alpha particle (a helium nucleus with mass 4 u). Laboratory experiments measure the speed of the alpha particle to be 1.92×10^7 m/s. Assuming the polonium nucleus was initially at rest, what is the recoil velocity of the nucleus that remains after the decay?

53. A neutron is an electrically neutral subatomic particle with a mass just slightly greater than that of a proton. A free neutron is radioactive and decays after a few minutes into other subatomic particles. In one experiment, a neutron at rest was observed to decay into a proton (mass 1.67×10^{-27} kg) and an electron (mass 9.11×10^{-31} kg). The proton and electron were shot out back-to-back. The proton speed was measured to be 1.0×10^5 m/s and the electron speed was 3.0×10^7 m/s. No other decay products were detected.
   a. Was momentum conserved in the decay of this neutron?

   NOTE ▶ Experiments such as this were first performed in the 1930s and seemed to indicate a failure of the law of conser-

vation of momentum. However, physicists were reluctant to give up a conservation law that had worked so well in every other circumstance. In 1933, Wolfgang Pauli postulated that the neutron might have a *third* decay product that is virtually impossible to detect. Even so, it can carry away just enough momentum to keep the total momentum conserved. This proposed particle was named the *neutrino*, meaning "little neutral one." Neutrinos were, indeed, discovered nearly 20 years later. ◀

   b. If a neutrino was emitted in the above neutron decay, in which direction did it travel? Explain your reasoning.
   c. How much momentum did this neutrino "carry away" with it?

54. A 20 g ball of clay traveling east at 2.0 m/s collides with a 30 g ball of clay traveling 30° south of west at 1.0 m/s. What are the speed and direction of the resulting 50 g blob of clay?

55. Figure P9.55 shows a collision between three balls of clay. The three hit simultaneously and stick together. What are the speed and direction of the resulting blob of clay?

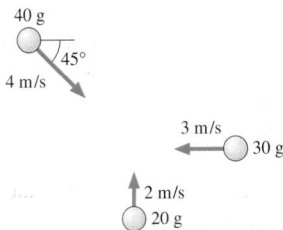

FIGURE P9.55

56. A 2100 kg truck is traveling east through an intersection at 2.0 m/s when it is hit simultaneously from the side and the rear. (Some people have all the luck!) One car is a 1200 kg compact traveling north at 5.0 m/s. The other is a 1500 kg midsize traveling east at 10 m/s. The three vehicles become entangled and slide as one body. What are their speed and direction just after the collision?

57. The carbon isotope ^{14}C is used for carbon dating of archeological artifacts. ^{14}C (mass 2.34×10^{-26} kg) decays by the process known as *beta decay* in which the nucleus emits an electron (the beta particle) and a subatomic particle called a neutrino. In one such decay, the electron and the neutrino are emitted at right angles to each other. The electron (mass 9.11×10^{-31} kg) has a speed of 5.0×10^7 m/s and the neutrino has a momentum of 8.0×10^{-24} kg m/s. What is the recoil speed of the nucleus?

58. A block of dry ice (solid carbon dioxide) tied to a 50-cm-long string swings in a circle with an initial speed of 2.0 m/s. As the CO_2 sublimates (that is, goes from a solid directly to a gas without melting), the gas vapor provides a "cushion" on which the block slides without friction. What is the speed of the block after half its mass has sublimated?

59. A 1.0-m-long massless rod is pivoted at one end and swings around in a circle on a frictionless table. A block with a hole through the center can slide in and out along the rod. Initially, a small piece of wax holds the block 30 cm from the pivot. The block is spun at 50 rpm, then the temperature of the rod is slowly increased. When the wax melts, the block slides out to the end of the rod. What is the final angular velocity? Give your answer in rpm.

60. Two people are standing on a very light board that is balanced on a fulcrum. The lighter person suddenly jumps straight up at 1.5 m/s. Just after he jumps, how fast, and in which direction, will the heavier person be moving?

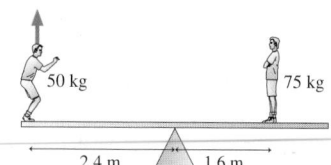

FIGURE P9.60   2.4 m   1.6 m

61. Figure P9.61 shows a 100 g puck revolving in a 20-cm-radius circle on a frictionless table. The string passes through a hole in the center of the table and is tied to two 200 g weights.
    a. What speed does the puck need to support the two weights?
    b. Suppose a flame burns through the string and causes the lower weight to fall off while the puck is revolving. What will be the puck's speed and the radius of its trajectory after the weight drops?

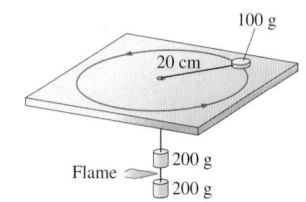

FIGURE P9.61

In Problems 62 through 65 you are given the equation used to solve a problem. For each of these, you are to
   a. Write a realistic problem for which this is the correct equation.
   b. Draw the before-and-after pictorial representation.
   c. Finish the solution of the problem.

62. $(0.10 \text{ kg})(40 \text{ m/s}) - (0.10 \text{ kg})(-30 \text{ m/s}) = \frac{1}{2}(1400 \text{ N})\Delta t$

63. $(600 \text{ g})(4.0 \text{ m/s}) = (400 \text{ g})(3.0 \text{ m/s}) + (200 \text{ g})(v_{ix})_2$

64. $(3000 \text{ kg})v_{fx} = (2000 \text{ kg})(5.0 \text{ m/s}) + (1000 \text{ kg})(-4.0 \text{ m/s})$

65. $(50 \text{ g})(v_{fx})_1 + (100 \text{ g})(7.5 \text{ m/s}) = (150 \text{ g})(1.0 \text{ m/s})$

Challenge Problems

66. A 75 kg shell is fired with an initial speed of 125 m/s at an angle 55° above horizontal. Air resistance is negligible. At its highest point, the shell explodes into two fragments, one four times more massive than the other. The heavier fragment lands directly below the point of the explosion. If the explosion exerts forces only in the horizontal direction, how far from the launch point does the lighter fragment land?

67. A 1000 kg cart is rolling to the right at 5.0 m/s. A 70 kg man is standing on the right end of the cart. What is the speed of the cart if the man suddenly starts running to the left with a speed of 10 m/s relative to the cart?

68. Ann (mass 50 kg) is standing at the left end of a 15-m-long, 500 kg cart that has frictionless wheels and rolls on a frictionless track. Initially both Ann and the cart are at rest. Suddenly, Ann starts running along the cart at a speed of 5.0 m/s relative to the cart. How far will Ann have run *relative to the ground* when she reaches the right end of the cart?

69. A 20 kg wood ball hangs from a 2.0-m-long wire. The maximum tension the wire can withstand without breaking is 400 N. A 1.0 kg projectile traveling horizontally hits and embeds itself in the wood ball. What is the largest speed this projectile can have without causing the cable to break?

70. A two-stage rocket is traveling at 1200 m/s with respect to the earth when the first stage runs out of fuel. Explosive bolts release the first stage and push it backward with a speed of 35 m/s relative to the second stage. The first stage is three times as massive as the second stage. What is the speed of the second stage after the separation?

71. You are the ground-control commander of a 2000 kg scientific rocket that is approaching Mars at a speed of 25,000 km/hr. It needs to quickly slow to 15,000 km/hr to begin a controlled descent to the surface. If the rocket enters the Martian atmosphere too fast it will burn up, and if it enters too slowly, it will use up its maneuvering fuel before reaching the surface and will crash. The rocket has a new braking system: Several 5.0 kg "bullets" on the front of the rocket can be fired straight ahead. Each has a high-explosive charge that fires it at a speed of 139,000 m/s relative to the rocket. You need to send the rocket an instruction to tell it how many bullets to fire. Success will bring you fame and glory, but failure of this $500,000,000 mission will ruin your career.

72. You are a world-famous physicist-lawyer defending a client who has been charged with murder. It is alleged that your client, Mr. Smith, shot the victim, Mr. Wesson. The detective who investigated the scene of the crime found a second bullet, from a shot that missed Mr. Wesson, that had embedded itself into a chair. You arise to cross-examine the detective.

You: In what type of chair did you find the bullet?
Det: A wooden chair.
You: How massive was this chair?
Det: It had a mass of 20 kg.
You: How did the chair respond to being struck with a bullet?
Det: It slid across the floor.
You: How far?
Det: Three centimeters. The slide marks on the dusty floor are quite distinct.
You: What kind of floor was it?
Det: A wood floor, very nice oak planks.
You: What was the mass of the bullet you retrieved from the chair?
Det: Its mass was 10 g.
You: And how far had it penetrated into the chair?
Det: A distance of 4 cm.
You: Have you tested the gun you found in Mr. Smith's possession?
Det: I have.
You: What is the muzzle velocity of bullets fired from that gun?
Det: The muzzle velocity is 450 m/s.
You: And the barrel length?
Det: The gun has a barrel length of 62 cm.

With only a slight hesitation, you turn confidently to the jury and proclaim, "My client's gun did not fire these shots!" How are you going to convince the jury and the judge?

Stop to Think 9.1: f. The cart is initially moving in the negative x-direction, so $p_{ix} = -20 \text{ kg m/s}$. After it bounces, $p_{fx} = 10 \text{ kg m/s}$. Thus $\Delta p = (10 \text{ kg m/s}) - (-20 \text{ kg m/s}) = 30 \text{ kg m/s}$.

Stop to Think 9.2: b. The clay ball goes from $v_{ix} = v$ to $v_{fx} = 0$, so $J_{clay} = \Delta p_x = -mv$. The rubber ball rebounds, going from $v_{ix} = v$ to $v_{fx} = -v$ (same speed, opposite direction). Thus $J_{rubber} = \Delta p_x = -2mv$. The rubber ball has a larger momentum change, and this requires a larger impulse.

Stop to Think 9.3: Less than. The ball's momentum $m_B v_B$ is the same in both cases. Momentum is conserved, so the *total* momentum is the same after both collisions. The ball that rebounds from C has *negative* momentum, so C must have a larger momentum than A.

Stop to Think 9.4: Right end. The pieces started at rest, so the total momentum of the system is zero. It's an isolated system, so the total momentum after the explosion is still zero. The 6 g piece has momentum $6v$. The 4 g piece, with velocity $-2v$, has momentum $-8v$. The combined momentum of these two pieces is $-2v$. In order for P to be zero, the third piece must have a *positive* momentum ($+2v$) and thus a positive velocity.

Stop to Think 9.5: c. Momentum conservation requires $(m_1 + m_2) \times v_f = m_1 v_1 + m_2 v_2$. Because $v_1 > v_2$, it must be that $(m_1 + m_2) \times v_f = m_1 v_1 + m_2 v_2 > m_1 v_2 + m_2 v_2 = (m_1 + m_2)v_2$. Thus $v_f > v_2$. Similarly, $v_2 < v_1$ so $(m_1 + m_2)v_f = m_1 v_1 + m_2 v_2 < m_1 v_1 + m_2 v_1 = (m_1 + m_2)v_1$. Thus $v_f < v_1$. The collision causes m_1 to slow down and m_2 to speed up.

Stop to Think 9.6: Increase. There are no tangential forces as the puck sublimates, so its angular momentum $L = mrv_t$ is conserved. The tangential velocity has to increase as m decreases to keep L constant.

10 Energy

A pole vaulter can lift himself nearly 6 m (20 ft) off the ground by transforming the kinetic energy of his run into gravitational potential energy.

▶ **Looking Ahead**

The goal of Chapter 10 is to introduce the ideas of kinetic and potential energy and to learn a new problem-solving strategy based on conservation of energy. In this chapter you will learn to:

- Understand and use the concepts of kinetic and potential energy.
- Use energy bar graphs.
- Use and interpret energy diagrams.
- Solve problems using the law of conservation of mechanical energy.
- Apply these ideas to elastic collisions.

◀ **Looking Back**

Our introduction to energy will be based on free fall. We will also use the before-and-after pictorial representation developed for impulse and momentum problems. Please review

- Section 2.6 Free-fall kinematics.
- Sections 9.2–9.3 Before-and-after pictorial representations and conservation of momentum.

Energy. It's a word you hear all the time. We use chemical energy to heat our homes and bodies, electrical energy to power our lights and computers, and solar energy to grow our crops and forests. We're told to use energy wisely and not to waste it. Athletes and weary students consume "energy bars" and "energy drinks" to gain quick energy.

But just what is energy? The concept of energy has grown and changed with time, and it is not easy to define in a general way just what energy is. Rather than starting with a formal definition, we're going to let the concept of energy expand slowly over the course of several chapters. The purpose of this chapter is to introduce the two most fundamental forms of energy, kinetic energy and potential energy. Our goal is to understand the characteristics of energy, how energy is used, and, especially important, how energy is transformed from one form to another. For example, this pole vaulter, after years of training, has become extraordinarily proficient at transforming kinetic energy into gravitational potential energy.

Ultimately we will discover a very powerful conservation law for energy. Some scientists consider the law of conservation of energy to be the most important of all the laws of nature. But all that in due time; first we have to start with the basic ideas.

10.1 A "Natural Money" Called Energy

We will start by discussing what seems to be a completely unrelated topic: money. As you will discover, monetary systems have much in common with energy. Let's begin with a short story.

The Parable of the Lost Penny

John was a hard worker. His only source of income was the paycheck he received each month. Even though most of each paycheck had to be spent on basic necessities, John managed to keep a respectable balance in his checking account. He even saved enough to occasionally buy a few stocks and bonds, his investment in the future.

John never cared much for pennies, so he kept a jar by the door and dropped all his pennies into it at the end of each day. Eventually, he reasoned, his saved pennies would be worth taking to the bank and converting into crisp new dollar bills.

John found it fascinating to keep track of these various forms of money. He noticed, to his dismay, that the amount of money in his checking account did not spontaneously increase overnight. Furthermore, there seemed to be a definite correlation between the size of his paycheck and the amount of money he had in the bank. So John decided to embark on a systematic study of money.

He began, as would any good scientist, by using his initial observations to formulate a hypothesis. John called this hypothesis a *model* of the monetary system. He found that he could represent his monetary model with the flowchart shown in Figure 10.1.

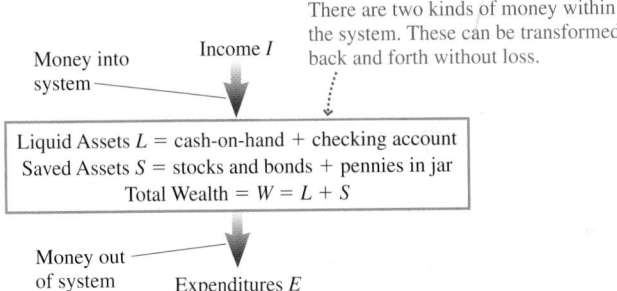

There are two kinds of money within the system. These can be transformed back and forth without loss.

Money into system — Income I

Liquid Assets L = cash-on-hand + checking account
Saved Assets S = stocks and bonds + pennies in jar
Total Wealth = $W = L + S$

Money out of system — Expenditures E

FIGURE 10.1 John's model of the monetary system.

As the chart shows, John divided his money into two basic types, liquid assets and saved assets. The *liquid assets L*, which included his checking account and the cash in his pockets, were moneys available for immediate use. His *saved assets S*, which included his stocks and bonds as well as the jar of pennies, had the *potential* to be converted into liquid assets, but they were not available for immediate use. John decided to call the sum total of assets his *wealth*: $W = L + S$.

John's assets were, more or less, simply definitions. The more interesting question, he thought, was how his wealth depended on his *income I* and *expenditures E*. These represented money transfers *to* him by his employer and money transferred *by* him to stores and bill collectors. After painstakingly collecting and analyzing his data, John finally determined that the relationship between monetary transfers and wealth is

$$\Delta W = I - E$$

John interpreted this equation to mean that the *change* in his wealth, ΔW, was numerically equal to the *net* monetary transfer $I - E$.

During a week-long period when John stayed home sick, isolated from the rest of the world, he had neither income nor expenses. In grand confirmation of his hypothesis, he found that his wealth W_f at the end of the week was identical to his wealth W_i at the week's beginning. That is, $W_f = W_i$. This occurred despite the fact that he had moved pennies from his pocket to the jar and also, by telephone, had sold some stocks and transferred the money to his checking account. In other words, John found that he could make all of the *internal* conversions of assets from one form to another that he wanted, but his total wealth remained constant (W = constant) as long as he was isolated from the world. This seemed such a remarkable rule that John named it the *law of conservation of wealth*.

One day, however, John added up his income and expenditures for the week, and the changes in his various assets, and he was 1¢ off! Inexplicably, some money seemed to have vanished. He was devastated. All those years of careful research, and now it seemed that his monetary hypothesis might not be true. Under some circumstances, yet to be discovered, it looked like $\Delta W \neq I - E$. Off by a measly penny. A wasted scientific life. . . .

But wait! In a flash of inspiration, John realized that perhaps there were other types of assets, yet to be discovered, and that his monetary hypothesis would still be valid if *all* assets were included. Weeks went by as John, in frantic activity, searched fruitlessly for previously *hidden* assets. Then one day, as John lifted the cushion off the sofa to vacuum out the potato chip crumbs—lo and behold, there it was!—the missing penny!

John raced to complete his theory, now including the sofa (as well as the washing machine) as a previously unknown form of saved assets that needed to be included in *S*. Other researchers soon discovered other types of assets, such as the remarkable find of the "cash in the mattress." To this day, when *all* known assets are included, monetary scientists have never found a violation of John's simple hypothesis that $\Delta W = I - E$. John was last seen sailing for Stockholm to collect the Nobel Prize for his Theory of Wealth.

This photovoltaic panel is transforming solar energy into electrical energy.

There are two kinds of energy within the system. These can be transformed back and forth without loss.

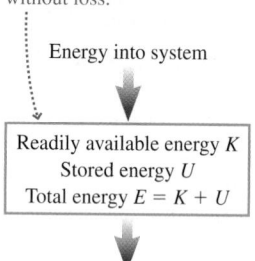

Energy into system

Readily available energy K
Stored energy U
Total energy $E = K + U$

Energy out of system

FIGURE 10.2 An initial model of energy.

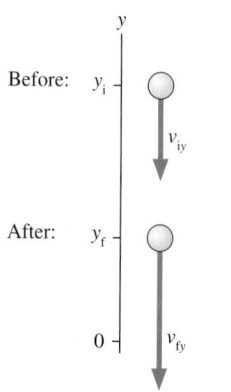

FIGURE 10.3 The before-and-after representation of an object in free fall.

Energy

John, despite his diligent efforts, did not discover a law of nature. The monetary system is a human construction that, by design, obeys John's "laws." Monetary system laws, such that you cannot print money in your basement, are enforced by society, not by nature. But suppose that physical objects possessed a "natural money" that was governed by a theory, or model, similar to John's. An object might have several different forms of natural money that could be converted back and forth, but the total amount of an object's natural money would *change* only if natural money were *transferred* to or from the object. Two key words here, as in John's model, are *transfer* and *change*.

One of the greatest and most significant discoveries of science is that there is such a "natural money" called **energy.** You have heard of some of the many forms of energy, such as solar energy or nuclear energy, but others may be new to you. These forms of energy can differ as much as a checking account differs from loose change in the sofa. Much of our study is going to be focused on the *transformation* of energy from one form to another. Much of modern technology is concerned with transforming energy, such as changing the chemical energy of oil molecules to electrical energy or to the kinetic energy of your car.

As we use energy concepts, we will be "accounting" for energy that is transferred in or out of a system or that is transformed from one form to another within a system. Figure 10.2 shows a simple model of energy that is based on John's model of the monetary system. There are many details that must be added to this model, but it's a good starting point. The fact that nature "balances the books" for energy is one of the most profound discoveries of science.

A major goal is to discover the conditions under which energy is conserved. Surprisingly, the *law of conservation of energy* was not recognized until the mid-nineteenth century, long after Newton. The reason, similar to John's lost penny, was that it took scientists a long time to realize how many types of energy there are and the various ways that energy can be converted from one form to another. As you'll soon learn, energy ideas go well beyond Newtonian mechanics to include new concepts about heat, about chemical energy, and about the energy of the individual atoms and molecules that comprise a system. All of these forms of energy will ultimately have to be included in our accounting scheme for energy.

There's a lot to say about energy, and energy is an abstract idea, so we'll take it one step at a time. Much of the "theory" will be postponed until Chapter 11, after you've had some practice using the basic concepts of energy introduced in this chapter. We will extend these ideas in Part IV when we reach the study of thermodynamics.

10.2 Kinetic Energy and Gravitational Potential Energy

To begin, consider an object in vertical free fall. It can only move up or down, and the only force acting on it is the gravitational weight force $\vec{w}$. The object's position and velocity are given by the free-fall kinematics of Chapter 2, with $a_y = -g$.

Figure 10.3 is a before-and-after pictorial representation of an object in free fall, as you learned to draw in Chapter 9. We didn't call attention to it in Chapter 2, but one of the free-fall equations also relates "before" and "after." In particular, the kinematic equation

$$v_{fy}^2 = v_{iy}^2 + 2a_y\Delta y = v_{iy}^2 - 2g(y_f - y_i) \tag{10.1}$$

can easily be rewritten as

$$v_{fy}^2 + 2gy_f = v_{iy}^2 + 2gy_i \tag{10.2}$$

Equation 10.2 is a conservation law for free-fall motion. It tells us that the quantity $v_y^2 + 2gy$ has the same value *after* free fall (regardless of whether the motion is upward or downward) that it had *before* free fall. But free fall is a very specific type of motion, so it's not clear if this "law" has any wider validity. Let's introduce a more general technique to arrive at the same result, but a technique that can be extended to other types of motion.

Newton's second law for one-dimensional motion along the y-axis is

$$(F_{net})_y = ma_y = m\frac{dv_y}{dt} \qquad (10.3)$$

The net force on an object in free fall is $(F_{net})_y = -mg$, so Equation 10.3 is

$$m\frac{dv_y}{dt} = -mg \qquad (10.4)$$

Recall, from calculus, that we can use the chain rule to write

$$\frac{dv_y}{dt} = \frac{dv_y}{dy}\frac{dy}{dt} = v_y\frac{dv_y}{dy} \qquad (10.5)$$

where we used $v_y = dy/dt$. Substituting this into Equation 10.4 gives

$$mv_y\frac{dv_y}{dy} = -mg \qquad (10.6)$$

The chain rule has allowed us to change from a derivative of v_y with respect to time to a derivative of v_y with respect to position. This simple change will be the key step on the road to energy.

Rewrite Equation 10.6 as

$$mv_y\,dv_y = -mg\,dy \qquad (10.7)$$

Now we can integrate both sides of the equation. However, we have to be careful to make sure the limits of integration match. We want to integrate from "before," when the object is at position y_i and has velocity v_{iy}, to "after," when the object is at position y_f and has velocity v_{fy}. Figure 10.3 shows these points in the motion. With these limits, the integrals are

$$\int_{v_{iy}}^{v_{fy}} mv_y\,dv_y = -\int_{y_i}^{y_f} mg\,dy \qquad (10.8)$$

Carrying out the integrations, with m and g as constants, we find

$$\frac{1}{2}mv_y^2\Big|_{v_{iy}}^{v_{fy}} = \frac{1}{2}mv_{fy}^2 - \frac{1}{2}mv_{iy}^2 = -mgy\Big|_{y_i}^{y_f} = -mgy_f + mgy_i \qquad (10.9)$$

The quantity v_y is the y-component of the vector $\vec{v}$. The sign of v_y indicates the direction of motion. But because v_y is squared wherever it appears in Equation 10.9, the sign is not relevant. All we need to know are the initial and final *speeds* v_i and v_f. With this, Equation 10.9 can be written

$$\frac{1}{2}mv_f^2 + mgy_f = \frac{1}{2}mv_i^2 + mgy_i \qquad (10.10)$$

You should recognize that Equation 10.10, other than a constant factor of $\frac{1}{2}m$, is the same as Equation 10.2. This seems like a lot of effort to get to a result we already knew. However, our purpose was not to get the answer but to introduce a *procedure* that will turn out to have other valuable applications.

Kinetic and Potential Energy

6.1

The quantity

$$K = \frac{1}{2}mv^2 \quad \text{(kinetic energy)} \tag{10.11}$$

is called the **kinetic energy** of the object. The quantity

$$U_g = mgy \quad \text{(gravitational potential energy)} \tag{10.12}$$

is the object's **gravitational potential energy.** These are the two basic forms of energy. **Kinetic energy is an energy of motion.** It depends on the object's speed but not its location. **Potential energy is an energy of position.** It depends on the object's position but not its speed.

NOTE ▶ Gravitational potential energy is called simply *potential energy* if there's no ambiguity. ◄

One of the most important characteristics of energy is that it is a scalar, not a vector. Kinetic energy depends on an object's speed v but *not* on the direction of motion. The kinetic energy of a particle will be the same regardless of whether it is moving up or down or left or right. Consequently, the mathematics of using energy is often much easier than the vector mathematics required by force and acceleration.

NOTE ▶ By its definition, kinetic energy can never be a negative number. If you find, in the course of solving a problem, that K is negative—stop! You have made an error somewhere. Don't just "lose" the minus sign and hope that everything turns out OK. ◄

The unit of kinetic energy is that of mass multiplied by velocity squared. In the SI system of units, this is $\text{kg}\,\text{m}^2/\text{s}^2$. The unit of energy is so important that is has been given its own name, the **joule.** We define:

$$1 \text{ joule} = 1 \text{ J} \equiv 1 \text{ kg}\,\text{m}^2/\text{s}^2$$

The unit of potential energy, $\text{kg} \times \text{m/s}^2 \times \text{m} = \text{kg}\,\text{m}^2/\text{s}^2$, is also the joule.

To give you an idea about the size of a joule, consider a 0.5 kg mass (weight on earth $\approx$ 1 lb) moving at 4 m/s ($\approx$10 mph). Its kinetic energy is

$$K = \frac{1}{2}mv^2 = \frac{1}{2}(0.5 \text{ kg})(4 \text{ m/s})^2 = 4 \text{ J}$$

Its gravitational potential energy at a height of 1 m ($\approx$3 ft) is

$$U_g = mgy = (0.5 \text{ kg})(9.8 \text{ m/s}^2)(1 \text{ m}) \approx 5 \text{ J}$$

This suggests that ordinary-sized objects moving at ordinary speeds will have energies of a fraction of a joule up to, perhaps, a few thousand joules (a running person has $K \approx 1000$ J). A high-speed truck might have $K \approx 10^6$ J.

NOTE ▶ You *must* have masses in kg and velocities in m/s before doing energy calculations. ◄

In terms of energy, Equation 10.10 says that for an object in free fall,

$$K_f + U_{gf} = K_i + U_{gi} \tag{10.13}$$

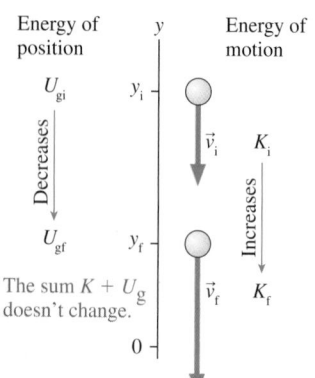

FIGURE 10.4 Kinetic energy and gravitational potential energy.

In other words, the sum $K + U_g$ of kinetic energy and gravitational potential energy is not changed by free fall. Its value *after* free fall (regardless of whether the motion is upward or downward) is the same as *before* free fall. Figure 10.4 illustrates this important idea.

EXAMPLE 10.1 **Launching a pebble**

Bob uses a slingshot to shoot a 20 g pebble straight up with a speed of 25 m/s. How high does the pebble go?

MODEL This is free-fall motion, so the sum of the kinetic and gravitational potential energy does not change as the pebble rises.

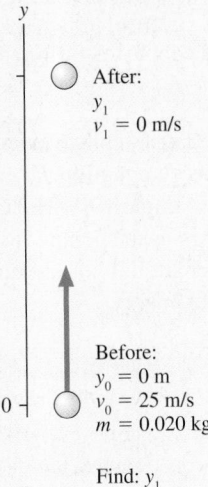

Before:
$y_0 = 0$ m
$v_0 = 25$ m/s
$m = 0.020$ kg

After:
y_1
$v_1 = 0$ m/s

Find: y_1

FIGURE 10.5 The before-and-after pictorial representation of a pebble shot upward from a slingshot.

VISUALIZE Figure 10.5 shows a before-and-after pictorial representation. The pictorial representation for energy problems is essentially the same as the pictorial representation you learned in Chapter 9 for momentum problems. We'll use numerical subscripts 0 and 1 for the initial and final points.

SOLVE Equation 10.13,

$$K_1 + U_{g1} = K_0 + U_{g0}$$

tells us that the sum $K + U_g$ is not changed by the motion. Using the definitions of K and U_g gives us this statement about energy:

$$\frac{1}{2}mv_1^2 + mgy_1 = \frac{1}{2}mv_0^2 + mgy_0$$

Here $y_0 = 0$ m and $v_1 = 0$ m/s, so the energy equation simplifies to

$$mgy_1 = \frac{1}{2}mv_0^2$$

This is easily solved for the height y_1:

$$y_1 = \frac{v_0^2}{2g} = \frac{(25 \text{ m/s})^2}{2(9.8 \text{ m/s}^2)} = 31.9 \text{ m}$$

ASSESS Notice that the mass canceled and wasn't needed, a fact about free fall that you should remember from Chapter 2.

STOP TO THINK 10.1 Rank in order, from largest to smallest, the gravitational potential energies of balls 1 to 4.

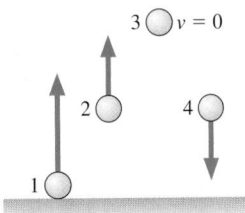

Energy Bar Charts

The pebble of Example 10.1 starts out with all kinetic energy, an energy of motion. As the pebble ascends, kinetic energy is converted into gravitational potential energy, *but the sum of the two doesn't change*. At the top, the pebble's energy is entirely potential energy. The simple bar chart in Figure 10.6 on the next page shows graphically how kinetic energy is transformed into gravitational potential energy as a pebble rises. The potential energy is then transformed back into kinetic energy as the pebble falls. The sum $K + U_g$ remains constant throughout the motion.

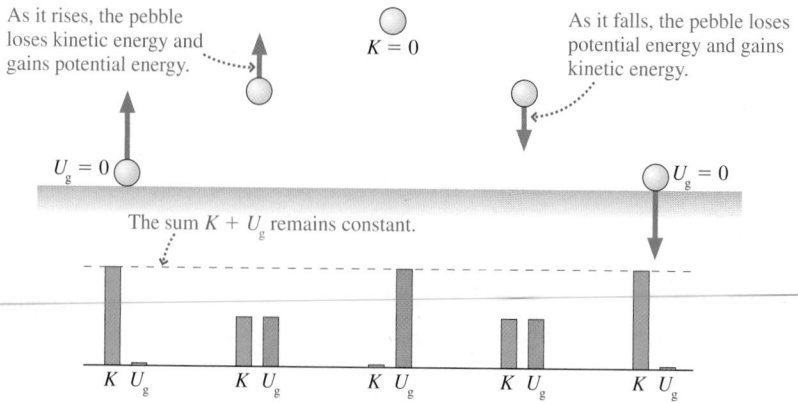

As it rises, the pebble loses kinetic energy and gains potential energy.

$K = 0$

As it falls, the pebble loses potential energy and gains kinetic energy.

$U_g = 0$ $U_g = 0$

The sum $K + U_g$ remains constant.

$K \ U_g$ $K \ U_g$ $K \ U_g$ $K \ U_g$ $K \ U_g$

FIGURE 10.6 Simple energy bar charts for a pebble tossed into the air.

Figure 10.7a is an energy bar chart more suitable to problem solving. The chart is a graphical representation of the energy equation $K_f + U_{gf} = K_i + U_{gi}$. Figure 10.7b applies this to the pebble of Example 10.1. The initial kinetic energy is transformed entirely into potential energy as the pebble reaches its highest point. There are no numerical scales on a bar chart, but you should draw the bar heights proportional to the amount of each type of energy.

The initial kinetic energy is transformed entirely into potential energy.

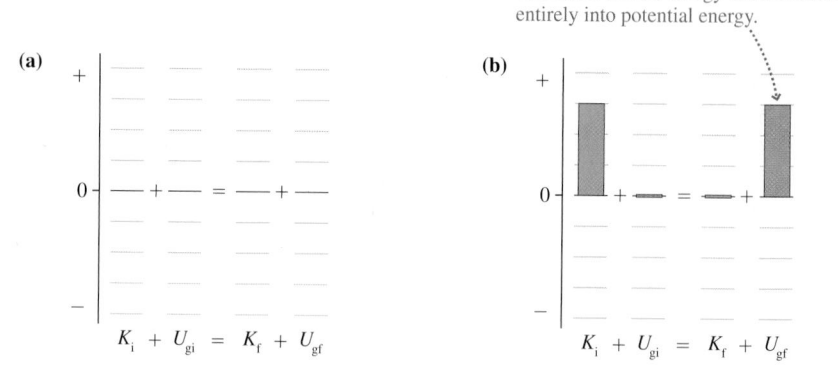

(a)

$K_i \ + \ U_{gi} \ = \ K_f \ + \ U_{gf}$

(b)

$K_i \ + \ U_{gi} \ = \ K_f \ + \ U_{gf}$

FIGURE 10.7 An energy bar chart suitable for problem solving.

The Zero of Potential Energy

Our expression for the gravitational potential energy $U_g = mgy$ seems straightforward. But you might notice, on further reflection, that the value of U_g depends on where you choose to put the origin of your coordinate system. Consider Figure 10.8, where Betty and Bill are attempting to determine the potential energy of a 1 kg rock that is 1 m above the ground. Betty chooses to put the origin of her coordinate system on the ground, measures $y_{rock} = 1$ m, and quickly computes $U_g = mgy = 9.8$ J. Bill, on the other hand, read Chapter 1 very carefully and recalls that it is entirely up to him where to locate the origin of his coordinate system. So he places his origin next to the rock, measures $y_{rock} = 0$ m, and declares that $U_g = mgy = 0$ J!

How can the potential energy of one rock at one position in space have two different values? The source of this apparent difficulty comes from our interpretation of Equation 10.9. The integral of $-mgdy$ resulted in the expression $-mg(y_f - y_i)$, and this led us to propose that $U_g = mgy$. But all we are *really* justified in concluding is that the potential energy *changes* by $\Delta U = -mg(y_f - y_i)$. To go

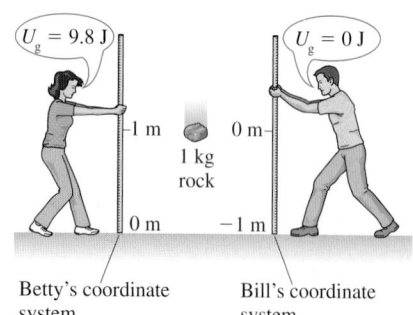

$U_g = 9.8$ J $U_g = 0$ J

-1 m   0 m

1 kg rock

0 m   -1 m

Betty's coordinate system   Bill's coordinate system

FIGURE 10.8 Betty and Bill use coordinate systems with different origins to determine the potential energy of a rock.

beyond this and claim $U_g = mgy$ is consistent with $\Delta U = -mg(y_f - y_i)$, but so also would be a claim that $U_g = mgy + C$ where C is any constant.

No matter where the rock is located, Betty's value of y will always equal Bill's value plus 1 m. Consequently, her value of the potential energy will always equal Bill's value plus 9.8 J. That is, their values of U_g differ by a constant. Nonetheless, both will calculate exactly the *same* value for ΔU if the rock changes position.

EXAMPLE 10.2 The speed of a falling rock

The rock shown in Figure 10.8 is released from rest. Use both Betty's and Bill's perspectives to calculate its speed just before it hits the ground.

MODEL This is free-fall motion, so the sum of the kinetic and gravitational potential energy does not change as the rock falls.

VISUALIZE Figure 10.9 shows a before-and-after pictorial representation using both Betty's and Bill's coordinate systems.

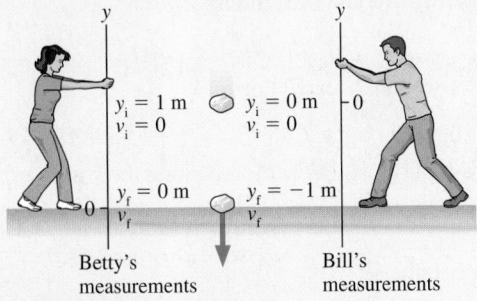

Betty's measurements   Bill's measurements

FIGURE 10.9 The before-and-after pictorial representation of a falling rock.

SOLVE The energy equation is $K_f + U_{gf} = K_i + U_{gi}$. Bill and Betty both agree that $K_i = 0$, because the rock was released from rest, so we have

$$K_f = \frac{1}{2}mv_f^2 = -(U_{gf} - U_{gi}) = -\Delta U$$

According to Betty, $U_{gi} = mgy_i = 9.8$ J and $U_{gf} = mgy_f = 0$ J. Thus

$$\Delta U_{Betty} = U_{gf} - U_{gi} = -9.8 \text{ J}$$

The rock *loses* potential energy as it falls. According to Bill, $U_{gi} = mgy_i = 0$ J and $U_{gf} = mgy_f = -9.8$ J. Thus

$$\Delta U_{Bill} = U_{gf} - U_{gi} = -9.8 \text{ J}$$

Bill has different values for U_{gi} and U_{gf} but the *same* value for ΔU. Thus they both agree that the rock hits the ground with speed

$$v_f = \sqrt{\frac{-2\Delta U}{m}} = \sqrt{\frac{-2(-9.8 \text{ J})}{1.0 \text{ kg}}} = 4.43 \text{ m/s}$$

Figure 10.10 shows energy bar charts for Betty and Bill. Despite their disagreement over the value of U_g, Betty and Bill arrive at the same value for v_f and their K_f bars are the same height. The reason is that only ΔU has physical significance, not U_g itself, and Betty and Bill found the same value for ΔU. You can place the origin of your coordinate system, and thus the "zero of potential energy," wherever you choose and be assured of getting the correct answer to a problem.

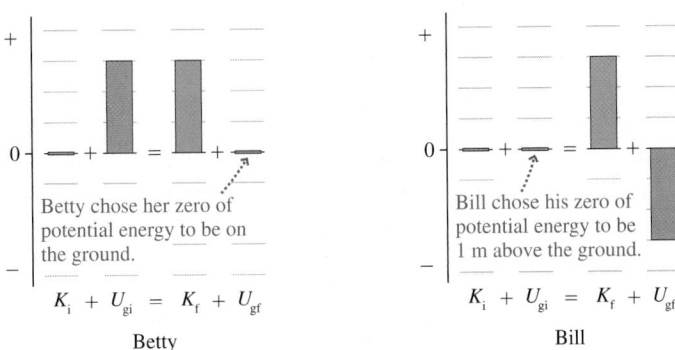

FIGURE 10.10 Betty's and Bill's energy bar charts for the falling rock.

NOTE ▶ Gravitational potential energy can be negative, as U_{gf} is for Bill. **A negative value for U_g means that the particle has *less* potential for motion at that point than it does at $y = 0$.** But there's nothing wrong with that. Contrast this with kinetic energy, which *cannot* be negative. ◀

10.3 A Closer Look at Gravitational Potential Energy

(a)

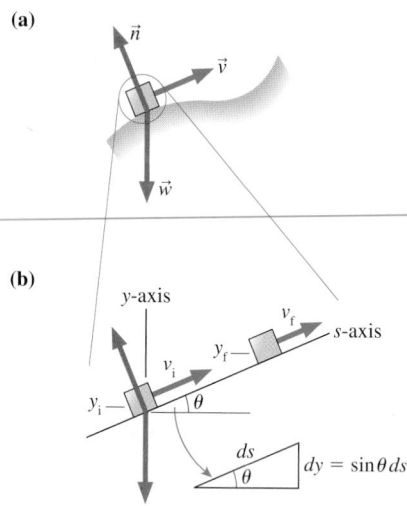

(b)

FIGURE 10.11 A particle moving along a frictionless surface of arbitrary shape.

The concept of energy would be of little interest or use if it applied only to free fall. Let's begin to expand the idea. Figure 10.11a shows an object of mass m sliding along a frictionless surface. The only forces acting on the object are its weight $\vec{w}$ and the normal force $\vec{n}$ from the surface. If the surface is curved, you know from calculus that we can subdivide the surface into many small (perhaps infinitesimal) straight-line segments. Figure 10.11b shows a magnified segment of the surface that, over some small distance, is a straight line at angle θ.

We can analyze the motion along this small segment using the procedure of Equations 10.3 through 10.10. Define an s-axis parallel to the direction of motion. Newton's second law along this axis is

$$(F_{net})_s = ma_s = m\frac{dv_s}{dt} \tag{10.14}$$

Using the chain rule, Equation 10.14 can be written

$$(F_{net})_s = m\frac{dv_s}{dt} = m\frac{dv_s}{ds}\frac{ds}{dt} = mv_s\frac{dv_s}{ds} \tag{10.15}$$

where, in the last step, we used $ds/dt = v_s$.

You can see from Figure 10.11b that the net force along the s-axis is

$$(F_{net})_s = -w\sin\theta = -mg\sin\theta \tag{10.16}$$

Thus Newton's second law becomes

$$-mg\sin\theta = mv_s\frac{dv_s}{ds} \tag{10.17}$$

Multiplying both sides by ds gives

$$mv_s\,dv_s = -mg\sin\theta\,ds \tag{10.18}$$

But you can see from the figure that $\sin\theta\,ds$ is simply dy, so Equation 10.18 becomes

$$mv_s\,dv_s = -mg\,dy \tag{10.19}$$

This is *identical* to Equation 10.7, which we found for free fall. Consequently, integrating this equation from "before" to "after" leads again to Equation 10.10:

$$\frac{1}{2}mv_f^2 + mgy_f = \frac{1}{2}mv_i^2 + mgy_i \tag{10.20}$$

where v_i^2 and v_f^2 are the squares of the *speeds* at the beginning and end of this segment of the motion.

We previously defined the kinetic energy $K = \frac{1}{2}mv^2$ and the gravitational potential energy $U_g = mgy$. Equation 10.20 shows that

$$K_f + U_{gf} = K_i + U_{gi} \tag{10.21}$$

for a particle moving along *any* frictionless surface, regardless of the shape.

NOTE ▶ For energy calculations, the y-axis is specifically a *vertical* axis. Gravitational potential energy depends on the *height* above the earth's surface. A tilted coordinate system, such as we often used in dynamics problems, doesn't work for problems with gravitational potential energy. ◀

The roller coaster's potential energy is converted to kinetic energy as it speeds toward the bottom of the hill.

A small child slides down the four frictionless slides A–D. Each has the same height. Rank in order, from largest to smallest, her speeds v_A to v_D at the bottom.

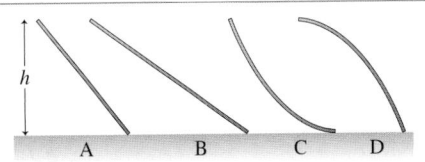

EXAMPLE 10.3 The speed of a sled

Christine runs forward with her sled at 2.0 m/s. She hops onto the sled at the top of a 5.0-m-high, very slippery slope. What is her speed at the bottom?

MODEL Model Christine and the sled as a particle. Assume the slope is frictionless. In that case, the sum of her kinetic and gravitational potential energy does not change as she slides down.

VISUALIZE Figure 10.12a shows a before-and-after pictorial representation. We are not told the angle of the slope, or even if it is a straight slope, but the *change* in potential energy depends only on the height Christine descends and *not* on the shape of the hill. Figure 10.12b shows an energy bar chart in which we see an initial kinetic *and* potential energy being transformed entirely into kinetic energy as she goes down the slope. The

purpose of the bar chart is to visualize how the energy changes, and we can show that without yet knowing any numerical values of the energy.

SOLVE The quantity $K + U_g$ is the same at the bottom of the hill as it was at the top. Thus

$$\frac{1}{2}mv_1^2 + mgy_1 = \frac{1}{2}mv_0^2 + mgy_0$$

This is easily solved for Christine's speed at the bottom:

$$v_1 = \sqrt{v_0^2 + 2g(y_0 - y_1)} = \sqrt{v_0^2 + 2gh} = 10.1 \text{ m/s}$$

ASSESS We did not need to know the mass of either Christine or the sled.

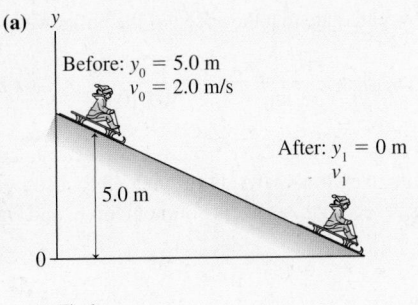

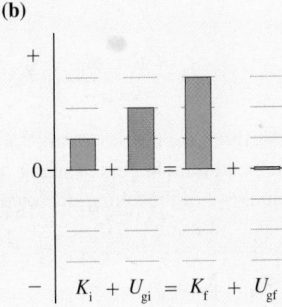

FIGURE 10.12 Pictorial representation and energy bar chart of Christine sliding down the hill.

Notice that the normal force $\vec{n}$ in Figure 10.12 didn't enter our analysis. The equation $K_f + U_{gf} = K_i + U_{gi}$ is a statement about how the particle's speed changes as it changes position. $\vec{n}$ does not have a component in the direction of motion, so it cannot change the particle's speed. Thus the normal force doesn't play a role in energy calculations.

The same is true for an object tied to a string and moving in a circle. The tension in the string causes the direction to change, but $\vec{T}$ does not have a component in the direction of motion and does not change the speed of the object. Hence Equation 10.21 also applies to a *pendulum*.

EXAMPLE 10.4 A ballistic pendulum

A 10 g bullet is fired into a 1200 g wood block that hangs from a 150-cm-long string. The bullet embeds itself into the block, and the block then swings out to an angle of 40°. What was the speed of the bullet? (This is called a *ballistic pendulum*.)

MODEL This is a two-part problem. The impact of the bullet with the block is an inelastic collision. We haven't done any analysis to let us know what happens to energy during a colli-

sion, but you learned in Chapter 9 that *momentum* is conserved in an inelastic collision. After the collision is over, the block swings out as a pendulum. The sum of the kinetic and gravitational potential energy does not change as the block swings to its largest angle.

VISUALIZE Figure 10.13 on the next page is a pictorial representation in which we've identified before-and-after quantities for both the collision and the swing.

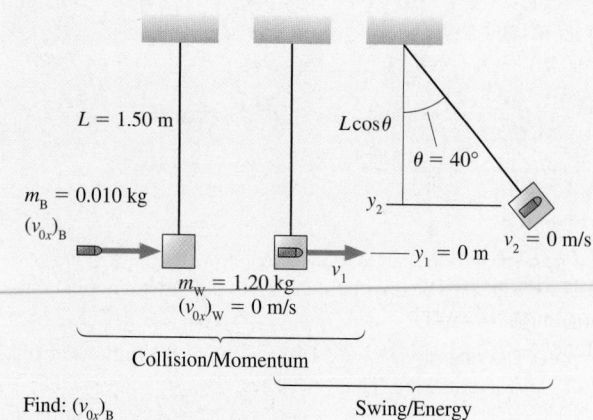

FIGURE 10.13 A ballistic pendulum is used to measure the speed of a bullet.

SOLVE The momentum conservation equation $P_f = P_i$ applied to the inelastic collision gives

$$(m_W + m_B)v_{1x} = m_W(v_{0x})_W + m_B(v_{0x})_B$$

The wood block is initially at rest, with $(v_{0x})_W = 0$, so the bullet's velocity is

$$(v_{0x})_B = \frac{m_W + m_B}{m_B} v_{1x}$$

where v_{1x} is the velocity of the block + bullet *immediately* after the collision, as the pendulum begins to swing. If we can determine v_{1x} from an analysis of the swing, then we will be able to

calculate the speed of the bullet. Turning our attention to the swing, the energy equation $K_f + U_{gf} = K_i + U_{gi}$ is

$$\frac{1}{2}(m_W + m_B)v_2^2 + (m_W + m_B)gy_2$$
$$= \frac{1}{2}(m_W + m_B)v_1^2 + (m_W + m_B)gy_1$$

We used the *total* mass $(m_W + m_B)$ of the block and embedded bullet, but notice that it cancels out. We also dropped the x-subscript on v_1 because for energy calculations we need only speed, not velocity. The speed is zero at the top of the swing ($v_2 = 0$), and we've defined the y-axis such that $y_1 = 0$ m thus

$$v_1 = \sqrt{2gy_2}$$

The initial speed is found simply from the maximum height of the swing. You can see from the geometry of Figure 10.13 that

$$y_2 = L - L\cos\theta = L(1 - \cos\theta) = 0.351 \text{ m}$$

With this, the initial velocity of the pendulum, immediately after the collision, is

$$v_{1x} = v_1 = \sqrt{2gy_2} = \sqrt{2(9.80 \text{ m/s}^2)(0.351 \text{ m})} = 2.62 \text{ m/s}$$

Having found v_{1x} from an energy analysis of the swing, we can now calculate that the speed of the bullet was

$$(v_{0x})_B = \frac{m_W + m_B}{m_B}v_{1x} = \frac{1.210 \text{ kg}}{0.010 \text{ kg}} \times 2.62 \text{ m/s} = 317 \text{ m/s}$$

ASSESS It would have been very difficult to solve this problem using Newton's laws, but it yielded to a straightforward analysis based on the concepts of momentum and energy.

Conservation of Mechanical Energy

The sum of the kinetic energy and the potential energy of a system is called the **mechanical energy**

$$E_{mech} = K + U \tag{10.22}$$

Here K is the total kinetic energy of all the particles in the system and U is the potential energy stored in the system. Our examples thus far suggest that a particle's mechanical energy does not change as it moves under the influence of gravity. The kinetic energy and the potential energy can change, as they are transformed back and forth into each other, but their sum remains constant. We can express the unchanging value of E_{mech} as

$$\Delta E_{mech} = \Delta K + \Delta U = 0 \tag{10.23}$$

This statement is called the **law of conservation of mechanical energy.**

> NOTE ▶ The law of conservation of mechanical energy does *not* say $\Delta K = \Delta U$. One of the changes has to be positive while the other is negative if energy is to be conserved. ◀

But is this really a law of nature? Consider a box that is given a shove and then slides along the floor until it stops. The box loses kinetic energy as it slows down, but $\Delta U_g = 0$ because y doesn't change. Thus ΔE_{mech} is *not* zero for the box; its mechanical energy is not conserved.

In Chapter 9 you learned that momentum is conserved only for an isolated system. One of the important goals of this chapter and the next is to learn the conditions under which mechanical energy is conserved. We've seen thus far that mechanical energy *is* conserved for a particle that moves along a frictionless trajectory under the influence of gravity, but mechanical energy is *not* conserved when there is friction.

You know, of course, that after the box slides across the floor, both the box and the floor are slightly warmer than before. The kinetic energy has not been transformed into potential energy, but *something* has happened. We've already noted that there are different kinds of energy. Perhaps friction causes the kinetic energy to be transformed into a form of energy other than potential energy. This is a very important issue, one that we'll begin to explore in the next chapter.

The Basic Energy Model

We're beginning to develop what we'll call the *basic energy model*. This is a model of energy as a form of "natural money." It is based on three hypotheses:

1. There are two kinds of energy, a kinetic energy associated with the motion of a particle and a potential energy associated with the position of a particle.
2. Kinetic energy can be transformed into potential energy, and potential energy can be transformed into kinetic energy.
3. Under some circumstances the mechanical energy $E_{mech} = K + U$ is conserved. Its value at the end of a process equals its value at the beginning.

Our task, if the basic energy model is to be useful, is to answer three crucial questions:

1. Under what conditions is E_{mech} conserved?
2. What happens to the energy when E_{mech} isn't conserved?
3. How do you calculate the potential energy U for forces other than gravity?

There are many parts to the energy puzzle, and we must put them together piece by piece. This chapter is focused on how to use energy in situations were E_{mech} is conserved. We will turn our attention to answering these three important questions in the next chapter. For now, we can begin to develop a strategy for using energy to solve problems.

 PROBLEM-SOLVING STRATEGY 10.1 Conservation of mechanical energy

MODEL Choose a system without friction or other losses of mechanical energy.

VISUALIZE Draw a before-and-after pictorial representation. Define symbols that will be used in the problem, list known values, and identify what you're trying to find.

SOLVE The mathematical representation is based on the law of conservation of mechanical energy:

$$K_f + U_f = K_i + U_i$$

ASSESS Check that your result has the correct units, is reasonable, and answers the question.

A box slides along the frictionless surface shown in the figure. It is released from rest at the position shown. Is the highest point the box reaches on the other side at level a, level b, or level c?

10.4 Restoring Forces and Hooke's Law

Springs and rubber bands store energy—potential energy—that can be transformed into kinetic energy.

If you stretch a rubber band, a force appears that tries to pull the rubber band back to its equilibrium, or unstretched, length. A force that restores a system to an equilibrium position is called a **restoring force.** Systems that exhibit restoring forces are called **elastic.** The most basic examples of elasticity are things like springs and rubber bands. If you stretch a spring, a tension-like force pulls back. Similarly, a compressed spring tries to re-expand to its equilibrium length. Other examples of elasticity and restoring forces abound. The steel beams bend slightly as you drive your car over a bridge, but they are restored to equilibrium after your car passes by. Nearly everything that stretches, compresses, flexes, bends, or twists exhibits a restoring force and can be called elastic.

Elasticity has many important applications, but we have another motive for introducing elasticity at this point. One of the goals of this textbook is to understand the atomic structure of matter. We have already introduced a simple model of a solid in which particle-like atoms are held together by spring-like molecular bonds. Figure 10.14 reminds you of this model. We devoted the first nine chapters to understanding the dynamics of particles. Now it's time to look more closely at the elastic behavior of the bonds, the "glue" that holds matter together.

We're going to use a simple spring as a prototype of elasticity. Suppose you have a spring whose **equilibrium length** is L_0. This is the length of the spring when it is neither pushing nor pulling. If you now stretch the spring to length L, how hard does it pull back? One way to find out is to attach the spring to a bar, as shown in Figure 10.15, then to hang a mass m from the spring. The mass stretches the spring to length L. Lengths L_0 and L are easily measured with a meter stick.

The mass hangs in static equilibrium, so the upward spring force $\vec{F}_{sp}$ exactly balances the downward weight force $\vec{w}$ to give $\vec{F}_{net} = \vec{0}$. That is,

$$F_{sp} = w = mg \tag{10.24}$$

By using different masses to stretch the spring to different lengths, we can determine how F_{sp}, the magnitude of the spring's restoring force, depends on the length L.

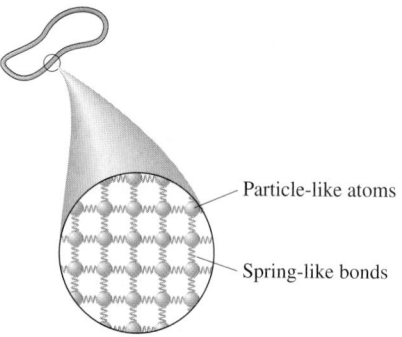

FIGURE 10.14 An elastic solid can be modeled as particle-like atoms held together by spring-like molecular bonds.

— Particle-like atoms

— Spring-like bonds

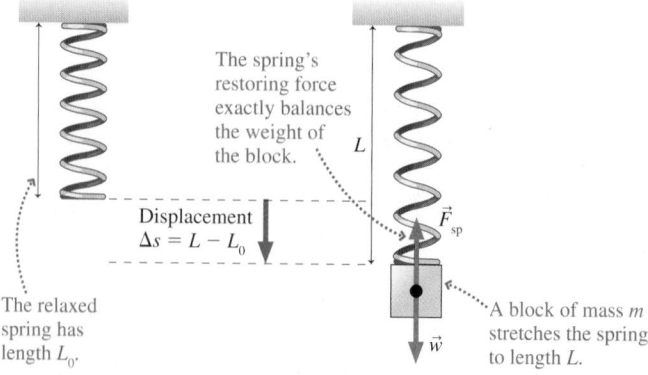

The spring's restoring force exactly balances the weight of the block.

L

$\vec{F}_{sp}$

Displacement $\Delta s = L - L_0$

The relaxed spring has length L_0.

A block of mass m stretches the spring to length L.

$\vec{w}$

FIGURE 10.15 A spring of equilibrium length L_0 is stretched to length L by a hanging mass.

Figure 10.16 shows measured data for the restoring force of a real spring. Notice that the quantity graphed along the horizontal axis is $\Delta s = L - L_0$. This is the distance that the end of the spring has moved, which we call the **displacement from equilibrium.** The graph shows that the restoring force is proportional to the displacement. That is, the data fall along the straight line

$$F_{sp} = k\Delta s \qquad (10.25)$$

The proportionality constant k, which is the slope of the force-versus-displacement graph, is called the **spring constant.** The units of the spring constant are N/m.

NOTE ▶ The force does not depend on the spring's physical length L but, instead, on the *displacement* Δs of the end of the spring. ◀

The spring constant k is a property that characterizes a spring, just as mass m characterizes a particle. If k is large, it takes a large pull to cause a significant stretch, and we call the spring a "stiff" spring. If k is small, the spring can be stretched with very little force, and we call it a "soft" spring. Every spring has its own, unique value of k that remains constant for that spring. The spring constant for the spring in Figure 10.16 can be determined from the slope of the straight line to be $k = 3.5$ N/m.

NOTE ▶ Just as we used massless strings, we will adopt the idealization of a *massless spring.* While not a perfect description, it is a good approximation if the mass attached to a spring is much larger than the mass of the spring itself. ◀

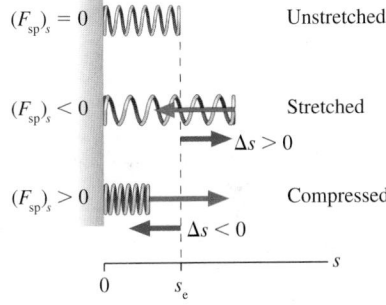

FIGURE 10.16 Measured data for the restoring force of a real spring.

Hooke's Law

Experiments show that Equation 10.25 is valid whether the spring is stretched or compressed. The only difference is the *direction* of the force—pulling in one case, pushing in the other. But because the spring force really is a vector, we do need to write Equation 10.25 in a form that gives the correct sign to the vector component of $\vec{F}_{sp}$.

Figure 10.17 shows a spring along a generic s-axis. The equilibrium position of the end of the spring is denoted s_e. This is the *position,* or coordinate, of the free end of the spring, *not* the spring's equilibrium length L_0.

When the spring is stretched, the displacement from equilibrium $\Delta s = s - s_e$ is *positive* while $(F_{sp})_s$, the s-component of the restoring force pointing to the left, is *negative*. If the spring is compressed, the displacement from equilibrium Δs is negative while the s-component of $\vec{F}_{sp}$, which now points to the right, is positive. No matter which way the end of the spring is displaced from equilibrium, the sign of the force component $(F_{sp})_s$ is always opposite to the sign of the displacement Δs. We can write this mathematically as

$$(F_{sp})_s = -k\Delta s \quad \text{(Hooke's law)} \qquad (10.26)$$

where $\Delta s = s - s_e$ is the displacement of the end of the spring from equilibrium. The minus sign is the mathematical indication of a *restoring* force.

Equation 10.26 for the restoring force of a spring is called **Hooke's law.** This "law" was first suggested by Robert Hooke, a contemporary (and sometimes bitter rival) of Newton. Hooke's law is not a true "law of nature," in the sense that Newton's laws are, but is actually just a *model* of a restoring force. It works extremely well for some springs, as in Figure 10.16, but less well for others. Hooke's law will fail for any spring if it is compressed or stretched too far.

NOTE ▶ Some of you, in an earlier physics course, may have learned to write Hooke's law as $F_{sp} = -kx$ (assuming the spring is along the x-axis), rather than as $-k\Delta x$. This can be misleading, and it is a common source of errors.

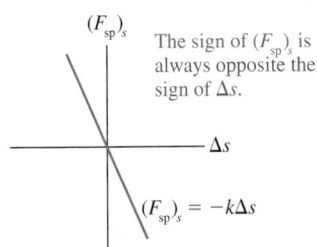

FIGURE 10.17 The direction of $\vec{F}_{sp}$ is always opposite the displacement $\Delta\vec{s}$.

The restoring force will be $-kx$ *only* if the coordinate system in the problem is chosen such that the origin is at the equilibrium position of the free end of the spring. That is, $\Delta x = x$ only if $x_e = 0$. This is often done, but in complex problems it will sometimes be more convenient to locate the origin of the coordinate system elsewhere. If you try to use $F_{sp} = -kx$ after the origin has moved elsewhere—big trouble! So make sure you learn Hooke's law as $(F_{sp})_s = -k\Delta s$. ◄

EXAMPLE 10.5 Pull until it slips

Figure 10.18a shows a spring attached to a 2.0 kg block. The other end of the spring is pulled by a motorized toy train that moves forward at 5.0 cm/s. The spring constant is 50 N/m and the coefficient of static friction between the block and the surface is 0.60. The spring is at its equilibrium length at $t = 0$ s when the train starts to move. When does the block slip?

(a)

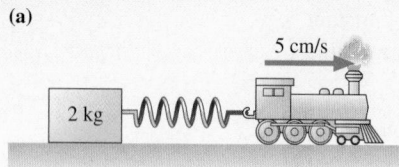

(b)

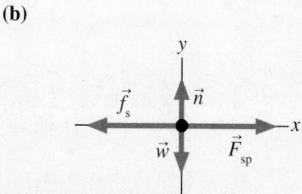

FIGURE 10.18 A toy train stretches the spring until the block slips.

MODEL Model the block as a particle and the spring as an ideal spring obeying Hooke's law.

VISUALIZE Figure 10.18b is a free-body diagram for the block.

SOLVE Recall that the tension in a massless string pulls equally at *both* ends of the string. The same is true for the spring force: It pulls (or pushes) equally at *both* ends. Imagine holding a rubber band with your left hand and stretching it with your right hand. Your left hand feels the pulling force, even though it was the right end of the rubber band that moved.

This is the key to solving the problem. As the right end of the spring moves, stretching the spring, the spring pulls backward on the train *and* forward on the block with equal strength. As the spring stretches, the static friction force on the block increases in magnitude to keep the block at rest. The block is in static equilibrium, so

$$\sum (F_{net})_x = (F_{sp})_x + (f_s)_x = F_{sp} - f_s = 0$$

where F_{sp} is the *magnitude* of the spring force. The magnitude is $F_{sp} = k\Delta x$, where $\Delta x = v_x t$ is the distance the train has moved. Thus

$$f_s = F_{sp} = k\Delta x$$

The block slips when the static friction force reaches its maximum value $f_{s\,max} = \mu_s n = \mu_s mg$. This occurs when the train has moved

$$\Delta x = \frac{f_{s\,max}}{k} = \frac{\mu_s mg}{k} = \frac{(0.60)(2.0 \text{ kg})(9.8 \text{ m/s}^2)}{50 \text{ N/m}}$$
$$= 0.235 \text{ m} = 23.5 \text{ cm}$$

The time at which the block slips is

$$t = \frac{\Delta x}{v} = \frac{23.5 \text{ cm}}{5.0 \text{ cm/s}} = 4.7 \text{ s}$$

The slip can range from a few centimeters in a relatively small earthquake to several meters in a very large earthquake.

This example illustrates an important class of motion called *stick-slip motion*. Once the block slips, it will shoot forward some distance, then stop and stick again. If the train continues to move forward, there will be a recurring sequence of stick, slip, stick, slip, stick. . . . Calculating the period of this stick-slip motion is not hard, although it's a bit beyond where we are right now.

Earthquakes are an important example of stick-slip motion. The large tectonic plates that make up the earth's crust are attempting to slide past each other, but friction forces cause the edges of the plates to stick together. The continued motion of the plates bends and deforms the rocks along the boundary. You may think of rocks as rigid and brittle, but large masses of rock, especially under the immense pressures within the earth, are somewhat elastic and can be "stretched." Eventually the elastic force of the deformed rocks exceeds the static friction force between the plates. An earthquake occurs as the plates slip and lurch forward. Once the tension is released, the plates stick together again and the process starts all over.

The graph shows force versus displacement for three springs. Rank in order, from largest to smallest, the spring constants k_1, k_2, and k_3.

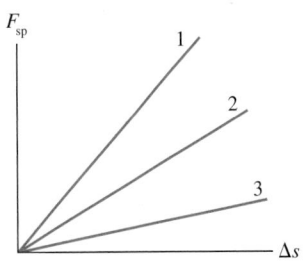

10.5 Elastic Potential Energy

The forces we have worked with thus far—gravity, friction, tension—have been constant forces. That is, their magnitudes do not change as an object moves. That feature has been important, because the kinematic equations we developed in Chapter 2 are for motion with constant acceleration. But a spring force exerts a *variable* force. The force is zero if $L = L_0$ (no displacement), and it steadily increases as the stretching length increases. The "natural motion" of a mass on a spring—think of pulling down on a spring and then releasing it—is an *oscillation*. This is *not* constant-acceleration motion, and we haven't yet developed the kinematics to handle oscillatory motion. That is why we haven't considered spring forces before now.

But suppose we're interested not in the time dependence of motion but only in before-and-after situations. For example, Figure 10.19 shows a before-and-after situation in which a spring launches a ball. Asking how the compression of the spring (the "before") affects the speed of the ball (the "after") is very different from wanting to know the ball's position as a function of time as the spring expands.

You certainly have a sense that a compressed spring has "stored energy," and Figure 10.19 shows clearly that the stored energy is transferred to the kinetic energy of the ball. Let's analyze this process with the same method we developed for motion under the influence of gravity. Newton's second law for the ball is

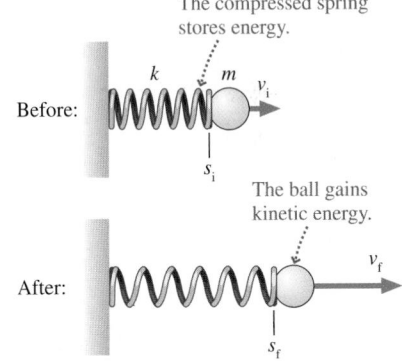

The compressed spring stores energy.

Before:

The ball gains kinetic energy.

After:

FIGURE 10.19 Before and after a spring launches a ball.

$$(F_{net})_s = ma_s = m\frac{dv_s}{dt} \qquad (10.27)$$

The net force on the ball is given by Hooke's law, $(F_{net})_s = -k(s - s_e)$. Thus

$$m\frac{dv_s}{dt} = -k(s - s_e) \qquad (10.28)$$

We'll use a generic s-axis, although it is better in actual problem solving to use x or y, depending on whether the motion is horizontal or vertical.

As we did before, use the chain rule to write

$$\frac{dv_s}{dt} = \frac{dv_s}{ds}\frac{ds}{dt} = v_s\frac{dv_s}{ds} \qquad (10.29)$$

Substitute this into Equation 10.28, then multiply both sides by ds to get

$$mv_s\,dv_s = -k(s - s_e)\,ds \qquad (10.30)$$

We can integrate both sides of the equation from the initial conditions i to the final conditions f—that is, integrate "from before to after," giving

$$\int_{v_i}^{v_f} mv_s\,dv_s = \frac{1}{2}mv_f^2 - \frac{1}{2}mv_i^2 = -k\int_{s_i}^{s_f}(s - s_e)\,ds \qquad (10.31)$$

The integral on the right is not difficult, but many of you are new to calculus so we'll proceed step by step. The easiest way to get the answer in the most useful

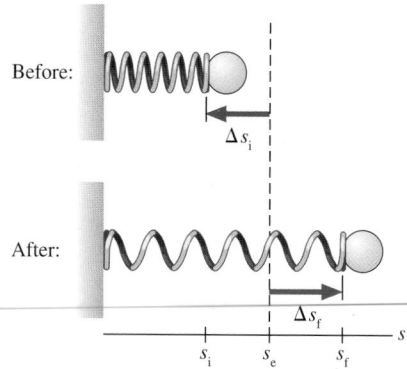

Before:

Δs_i

After:

Δs_f

s

s_i s_e s_f

FIGURE 10.20 The initial and final displacements of the spring.

form is to make a change of variables. Define $u = (s - s_e)$, in which case $ds = du$. This changes the integrand from $(s - s_e)ds$ to $u\,du$.

When we change variables, we also must change the limits of integration. In particular, $s = s_i$ at the lower integration limit makes $u = s_i - s_e = \Delta s_i$, where Δs_i is the initial displacement of the spring from equilibrium. Likewise, $s = s_f$ makes $u = s_f - s_e = \Delta s_f$ at the upper limit. Figure 10.20 clarifies the meaning of Δs_i and Δs_f.

With this change of variables, the integral is

$$-k\int_{s_i}^{s_f} (s - s_e)\,ds = -k\int_{\Delta s_i}^{\Delta s_f} u\,du = -\frac{1}{2}ku^2\Big|_{\Delta s_i}^{\Delta s_f}$$

$$= -\frac{1}{2}k(\Delta s_f)^2 + \frac{1}{2}k(\Delta s_i)^2 \tag{10.32}$$

Using this result makes Equation 10.31 become

$$\frac{1}{2}mv_f^2 - \frac{1}{2}mv_i^2 = -\frac{1}{2}k(\Delta s_f)^2 + \frac{1}{2}k(\Delta s_i)^2 \tag{10.33}$$

which can be rewritten as

$$\frac{1}{2}mv_f^2 + \frac{1}{2}k(\Delta s_f)^2 = \frac{1}{2}mv_i^2 + \frac{1}{2}k(\Delta s_i)^2 \tag{10.34}$$

We've succeeded in our goal of relating before and after. In particular, the quantity

$$\frac{1}{2}mv^2 + \frac{1}{2}k(\Delta s)^2 \tag{10.35}$$

does not change as the spring compresses or expands. You recognize $\frac{1}{2}mv^2$ as the kinetic energy K. Let's define the **elastic potential energy** U_s of a spring to be

$$U_s = \frac{1}{2}k(\Delta s)^2 \quad \text{(elastic potential energy)} \tag{10.36}$$

Then Equation 10.34 tells us that an object moving on a spring obeys

$$K_f + U_{sf} = K_i + U_{si} \tag{10.37}$$

In other words, the mechanical energy $E_{mech} = K + U_s$ is conserved for an object moving *without friction* on an ideal spring.

NOTE ▶ Because Δs is squared, the elastic potential energy is positive for a spring that is either stretched or compressed. U_s is zero when the spring is at its equilibrium length L_0 and $\Delta s = 0$. ◀

EXAMPLE 10.6 **A spring-launched plastic ball**

A spring-loaded toy gun is used to launch a 10 g plastic ball. The spring, which has a spring constant of 10 N/m, is compressed by 10 cm as the ball is pushed into the barrel. When the trigger is pulled, the spring is released and shoots the ball back out. What is the ball's speed as it leaves the barrel? Assume that friction is negligible.

MODEL Assume an ideal spring that obeys Hooke's law. Also assume that the gun is held firmly enough to prevent recoil. There's no friction, hence the mechanical energy $K + U_s$ is conserved.

VISUALIZE Figure 10.21a shows a before-and-after pictorial representation. The compressed spring will push on the ball until the spring has returned to its equilibrium length. We have chosen to put the origin of the coordinate system at the equilibrium position of the free end of the spring, making $x_1 = -10$ cm and $x_2 = x_e = 0$ cm. It's also useful to look at an energy bar chart. The bar chart of Figure 10.21b shows the potential energy stored in the compressed spring being entirely transformed into the kinetic energy of the ball.

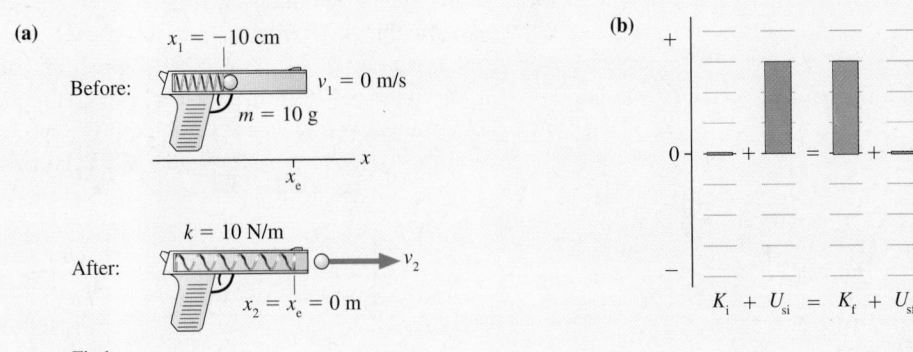

FIGURE 10.21 The before-and-after pictorial representation and energy bar chart of a ball being shot out of a spring-loaded toy gun. The gun is held and does not recoil.

SOLVE The energy conservation equation is $K_2 + U_{s2} = K_1 + U_{s1}$. We can use the elastic potential energy of the spring, Equation 10.36, to write this as

$$\frac{1}{2}mv_2^2 + \frac{1}{2}k(x_2 - x_e)^2 = \frac{1}{2}mv_1^2 + \frac{1}{2}k(x_1 - x_e)^2$$

Notice that we used x, rather than the generic s, and that we explicitly wrote out the meaning of Δx_1 and Δx_2. Using $x_2 = x_e = 0$ m and $v_1 = 0$ m/s simplifies this to

$$\frac{1}{2}mv_2^2 = \frac{1}{2}kx_1^2$$

It is now straightforward to solve for the ball's speed:

$$v_2 = \sqrt{\frac{kx_1^2}{m}} = \sqrt{\frac{(10 \text{ N/m})(-0.10 \text{ m})^2}{0.010 \text{ kg}}} = 3.16 \text{ m/s}$$

ASSESS This is a problem that we could *not* have solved with Newton's laws. The acceleration is not constant, and we have not learned how to handle the kinematics of nonconstant acceleration. But with conservation of energy—it's easy!

If an object attached to a spring moves vertically, the system has both elastic *and* gravitational potential energy. The mechanical energy then contains *two* potential energy terms:

$$E_{\text{mech}} = K + U_g + U_s \tag{10.38}$$

In other words, there are now two distinct ways of storing energy inside the system. The following example demonstrates this idea.

EXAMPLE 10.7 A spring-launched satellite

Prince Harry the Horrible wanted to be the first to launch a satellite. He placed a 2.0 kg payload on top of a very stiff 2.0-m-long spring with a spring constant of 50,000 N/m. Then the prince had his strongest men use a winch to crank the spring down to a length of 80 cm. When released, the spring shot the payload straight up. How high did it go?

MODEL Assume an ideal spring that obeys Hooke's law. There's no friction, hence the mechanical energy $K + U_g + U_s$ is conserved.

VISUALIZE Figure 10.22a on the next page shows the satellite ready for launch. We have chosen to place the origin of the coordinate system on the ground, which means that the equilibrium position of the end of the unstretched spring is *not* $y_e = 0$ m but, instead, $y_e = 2.0$ m. The payload reaches height y_2, where $v_2 = 0$ m/s.

SOLVE The energy conservation equation $K_2 + U_{s2} + U_{g2} = K_1 + U_{s1} + U_{g1}$ is

$$\frac{1}{2}mv_2^2 + \frac{1}{2}k(y_e - y_e)^2 + mgy_2$$
$$= \frac{1}{2}mv_1^2 + \frac{1}{2}k(y_1 - y_e)^2 + mgy_1$$

Notice that the elastic potential energy term on the left has $(y_e - y_e)^2$, not $(y_2 - y_e)^2$. The payload moves to position y_2, *but the spring does not!* The end of the spring stops at y_e. Both the initial and final speeds v_1 and v_2 are zero. Solving for the height:

$$y_2 = y_1 + \frac{k(y_1 - y_e)^2}{2mg} = 1840 \text{ m}$$

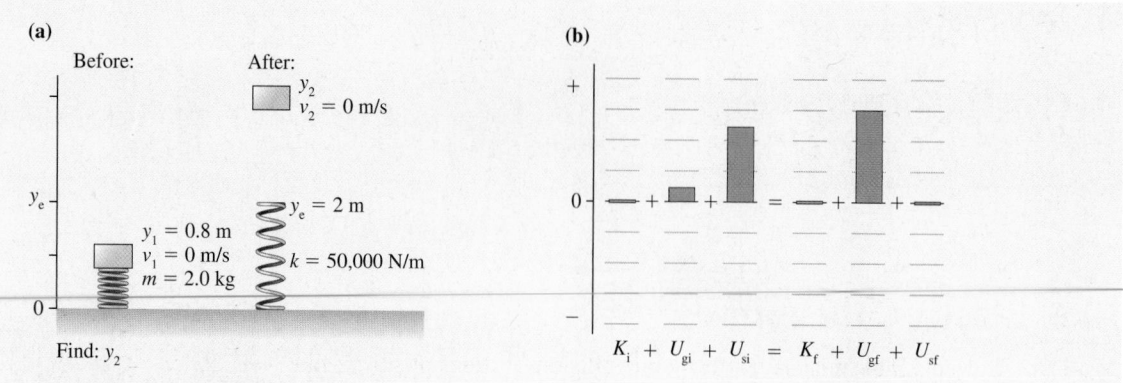

FIGURE 10.22 Before-and-after pictorial representation and energy bar chart of Prince Harry's spring-launched payload.

ASSESS Figure 10.22b shows an energy bar chart. The net effect of the launch is to transform the potential energy stored in the spring entirely into gravitational potential energy. The kinetic energy is zero at the beginning and zero again at the highest point. The payload does have kinetic energy as it comes off the spring; we did not need to know this energy to solve the problem.

EXAMPLE 10.8 Pushing apart

A spring with spring constant 2000 N/m is sandwiched between a 1.0 kg block and a 2.0 kg block on a frictionless table. The blocks are pushed together to compress the spring by 10 cm, then released. What are the velocities of the blocks as they fly apart?

MODEL Assume an ideal spring that obeys Hooke's law. There's no friction, hence the mechanical energy $K + U_s$ is conserved. In addition, because the blocks and spring form an isolated system, their total momentum is conserved.

VISUALIZE Figure 10.23 is a before-and-after pictorial representation.

Before:

$m_1 = 1.0$ kg
$(v_{ix})_1 = 0$ ⬜ 1 〰〰〰 2 $(v_{ix})_2 = 0$
$m_2 = 2.0$ kg
$k = 2000$ N/m
$\Delta x_i = 10$ cm

After:

$(v_{fx})_1$ ⬅ 1 〰〰〰〰 2 ➡ $(v_{fx})_2$

Find: $(v_{fx})_1$ and $(v_{fx})_2$

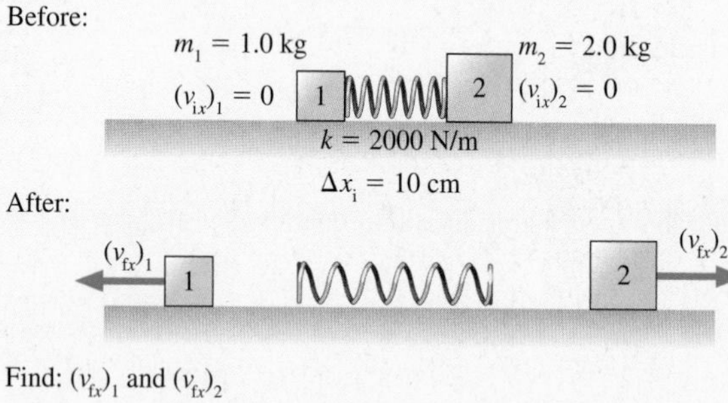

FIGURE 10.23 Before-and-after pictorial representation of the blocks and spring.

SOLVE The initial energy, with the spring compressed, is entirely potential. The final energy is entirely kinetic. The energy conservation equation $K_f + U_{sf} = K_i + U_{si}$ is

$$\frac{1}{2}m_1(v_f)_1^2 + \frac{1}{2}m_2(v_f)_2^2 + 0 = 0 + 0 + \frac{1}{2}k(\Delta x_i)^2$$

Notice that *both* blocks contribute to the kinetic energy. The energy equation has two unknowns, $(v_f)_1$ and $(v_f)_2$, and one

equation is not enough to solve the problem. Fortunately, momentum is also conserved. The initial momentum is zero, because both blocks are at rest, so the momentum equation is

$$m_1(v_{fx})_1 + m_2(v_{fx})_2 = 0$$

which can be solved to give

$$(v_{fx})_1 = -\frac{m_2}{m_1}(v_{fx})_2$$

The minus sign indicates that the blocks move in opposite directions. The speed $(v_f)_1 = (m_2/m_1)(v_f)_2$ is all we need to calculate the kinetic energy. Substituting $(v_f)_1$ into the energy equation gives

$$\frac{1}{2}m_1\left(\frac{m_2}{m_1}(v_f)_2\right)^2 + \frac{1}{2}m_2(v_f)_2^2 = \frac{1}{2}k(\Delta x_i)^2$$

which simplifies to

$$m_2\left(1 + \frac{m_2}{m_1}\right)(v_f)_2^2 = k(\Delta x_i)^2$$

Solving for $(v_f)_2$, we find

$$(v_f)_2 = \sqrt{\frac{k(\Delta x_i)^2}{m_2(1 + m_2/m_1)}} = 1.826 \text{ m/s}$$

Finally, we can go back to find

$$(v_{fx})_1 = -\frac{m_2}{m_1}(v_{fx})_2 = -3.65 \text{ m/s}$$

The 2.0 kg block moves to the right at 1.826 m/s while the 1.0 kg block goes left at 3.65 m/s.

ASSESS This example shows just how powerful a problem-solving tool the conservation laws are.

10.6 Elastic Collisions

Figure 9.1 showed a molecular-level view of a collision. Billions of spring-like molecular bonds are compressed as two objects collide, then the bonds expand and push the objects apart. In the language of energy, the kinetic energy of the objects is transformed into the elastic potential energy of molecular bonds, then back into kinetic energy as the two objects spring apart.

In some cases, such as the inelastic collisions of Chapter 9, some of the mechanical energy is dissipated inside the objects and not all of the kinetic energy is recovered. That is, $K_f < K_i$. (A homework problem will let you show this explicitly.) We're now interested in collisions in which *all* of the kinetic energy is stored as elastic potential energy in the bonds, and then *all* of the stored energy is transformed back into the post-collision kinetic energy of the objects. Such a collision, in which the mechanical energy is conserved, is called a **perfectly elastic collision.**

Needless to say, most real collisions fall somewhere between perfectly elastic and perfectly inelastic. A rubber ball bouncing on the floor might "lose" 20% of its kinetic energy on each bounce and return to only 80% of the height of the previous bounce. Perfectly elastic and perfectly inelastic collisions are limiting cases rarely seen in the real world, but they are nonetheless instructive for demonstrating the major ideas without making the mathematics too complex. Collisions between two very hard objects, such as two billiard balls or two steel balls, come close to being perfectly elastic.

Figure 10.24 shows a head-on, perfectly elastic collision of a ball of mass m_1, having initial velocity $(v_{ix})_1$, with a ball of mass m_2 that is initially at rest. The balls' velocities after the collision are $(v_{fx})_1$ and $(v_{fx})_2$. These are velocities, not speeds, and have signs. Ball 1, in particular, might bounce backward and have a negative value for $(v_{fx})_1$.

The collision must obey two conservation laws: conservation of momentum (obeyed in any collision) and conservation of mechanical energy (because the collision is perfectly elastic). Although the energy is transformed into potential energy during the collision, the mechanical energy before and after the collision is purely kinetic energy. Thus

momentum conservation: $m_1(v_{fx})_1 + m_2(v_{fx})_2 = m_1(v_{ix})_1$ (10.39)

energy conservation: $\frac{1}{2}m_1(v_{fx})_1^2 + \frac{1}{2}m_2(v_{fx})_2^2 = \frac{1}{2}m_1(v_{ix})_1^2$ (10.40)

Momentum conservation alone is not sufficient to analyze the collision because there are two unknowns: the two final velocities. That is why we did not consider perfectly elastic collisions in Chapter 9. Energy conservation gives us another condition. Isolating $(v_{fx})_1$ in Equation 10.39 gives

$$(v_{fx})_1 = (v_{ix})_1 - \frac{m_2}{m_1}(v_{fx})_2$$ (10.41)

A perfectly elastic collision conserves both momentum and mechanical energy.

Act**iv**
Phys**i**cs   6.2

Before: ① $\xrightarrow{(v_{ix})_1}$ ② K_i

During: ①②   Energy is stored in compressed molecular bonds, then released as the bonds re-expand.

After: ① ⟶ ② ⟶ $K_f = K_i$
            $(v_{fx})_1$ $(v_{fx})_2$

FIGURE 10.24 A perfectly elastic collision.

Substitute this into Equation 10.40:

$$\frac{1}{2}m_1\left((v_{ix})_1 - \frac{m_2}{m_1}(v_{fx})_2\right)^2 + \frac{1}{2}m_2(v_{fx})_2^2$$

$$= \frac{1}{2}\left(m_1(v_{ix})_1^2 - 2m_2(v_{ix})_1(v_{fx})_2 + \frac{m_2^2}{m_1}(v_{fx})_2^2 + m_2(v_{fx})_2^2\right)$$

$$= \frac{1}{2}m_1(v_{ix})_1^2$$

This looks rather gruesome, but the first two terms on each side cancel and the resulting equation can be rearranged to give

$$(v_{fx})_2\left[\left(1 + \frac{m_2}{m_1}\right)(v_{fx})_2 - 2(v_{ix})_1\right] = 0 \qquad (10.42)$$

One possible solution to this equation is seen to be $(v_{fx})_2 = 0$. However, this solution is of no interest; it is the case where ball 1 misses ball 2. The other solution is

$$(v_{fx})_2 = \frac{2}{1 + m_2/m_1}(v_{ix})_1 = \frac{2m_1}{m_1 + m_2}(v_{ix})_1$$

which, finally, can be substituted back into to Equation 10.41 to yield $(v_{fx})_1$. The complete solution is

$$\begin{aligned}(v_{fx})_1 &= \frac{m_1 - m_2}{m_1 + m_2}(v_{ix})_1 \qquad \text{(perfectly elastic collision}\\(v_{fx})_2 &= \frac{2m_1}{m_1 + m_2}(v_{ix})_1 \qquad \text{with ball 2 initially at rest)}\end{aligned} \qquad (10.43)$$

Equations 10.43 allow us to compute the final velocity of each ball. These equations are a little difficult to interpret, so let us look at the three special cases shown in Figure 10.25.

Case 1: $m_1 = m_2$. This is the case of one billiard ball striking another of equal mass. For this case, Equations 10.43 give

$$v_{1f} = 0$$
$$v_{2f} = v_{1i}$$

Case 2: $m_1 \gg m_2$. This is the case of a bowling ball running into a Ping-Pong ball. We do not want an exact solution here, but an approximate solution for the limiting case that $m_1 \to \infty$. Equations 10.43 in this limit give

$$v_{1f} \approx v_{1i}$$
$$v_{2f} \approx 2v_{1i}$$

Case 3: $m_1 \ll m_2$. Now we have the reverse case of a Ping-Pong ball colliding with a bowling ball. Here we are interested in the limit $m_1 \to 0$, in which case Equations 10.43 become

$$v_{1f} \approx -v_{1i}$$
$$v_{2f} \approx 0$$

These cases agree well with our expectations and give us confidence that Equations 10.43 accurately describe a perfectly elastic collision.

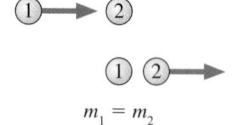

$m_1 = m_2$

Ball 1 stops. Ball 2 goes forward with $v_2 = v_1$.

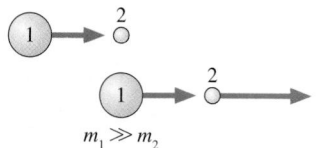

$m_1 \gg m_2$

Ball 1 hardly slows down. Ball 2 is knocked forward at $v_2 \approx 2v_1$.

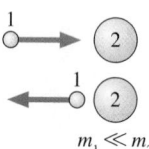

$m_1 \ll m_2$

Ball 1 bounces off ball 2 with almost no loss of speed. Ball 2 hardly moves.

FIGURE 10.25 Three special elastic collisions.

EXAMPLE 10.9 **A rebounding pendulum**

A 200 g steel ball hangs on a 1.0-m-long string. The ball is pulled sideways so that the string is at a 45° angle, then released. At the very bottom of its swing the ball strikes a 500 g steel block that is resting on a frictionless table. To what angle does the ball rebound?

MODEL This is a challenging problem. We can divide it into three parts. First the ball swings down as a pendulum. Second, the ball and block have a collision. Steel balls bounce off each other very well, so we will assume that the collision is perfectly elastic. Third, the ball, after it bounces off the block, swings back up as a pendulum.

VISUALIZE Figure 10.26 shows four distinct moments of time: as the ball is released, an instant before the collision, an instant after the collision but before the ball and block have had time to move, and as the ball reaches its highest point on the rebound. Call the ball A and the block B, so $m_A = 0.20$ kg and $m_B = 0.50$ kg.

SOLVE Part 1: The first part involves the ball only. Its initial height is

$$(y_0)_A = L - L\cos\theta_0 = L(1 - \cos\theta_0) = 0.293 \text{ m}$$

We can use conservation of mechanical energy to find the ball's velocity at the bottom, just before impact on the block:

$$\frac{1}{2}m_A(v_1)_A^2 + m_A g(y_1)_A = \frac{1}{2}m_A(v_0)_A^2 + m_A g(y_0)_A$$

Solving for the velocity at the bottom, where $(v_0)_A = 0$ and $(y_1)_A = 0$, gives

$$(v_1)_A = \sqrt{2g(y_0)_A} = 2.40 \text{ m/s}$$

Part 2: The ball and block undergo a perfectly elastic collision in which the block is initially at rest. These are the conditions for which Equations 10.43 were derived. The velocities *immediately* after the collision, prior to any further motion, are

$$(v_{2x})_A = \frac{m_A - m_B}{m_A + m_B}(v_{1x})_A = -1.03 \text{ m/s}$$

$$(v_{2x})_B = \frac{2m_A}{m_A + m_B}(v_{1x})_A = +1.37 \text{ m/s}$$

The ball rebounds toward the left with a speed of 1.03 m/s while the block moves to the right at 1.37 m/s. Kinetic energy has been conserved (you might want to check this), but it is now shared between the ball and the block.

Part 3: Now the ball is a pendulum with an initial speed of 1.03 m/s. Mechanical energy is again conserved, so we can find its maximum height at the point where $(v_3)_A = 0$:

$$\frac{1}{2}m_A(v_3)_A^2 + m_A g(y_3)_A = \frac{1}{2}m_A(v_2)_A^2 + m_A g(y_2)_A$$

Solving for the maximum height gives

$$(y_3)_A = \frac{(v_2)_A^2}{2g} = 0.0541 \text{ m}$$

The height $(y_3)_A$ is related to angle θ_3 by $(y_3)_A = L(1 - \cos\theta_3)$. This can be solved to find the angle of rebound:

$$\theta_3 = \cos^{-1}\left(1 - \frac{(y_3)_A}{L}\right) = 18.9°$$

The block speeds away at 1.37 m/s and the ball rebounds to an angle of 18.9°.

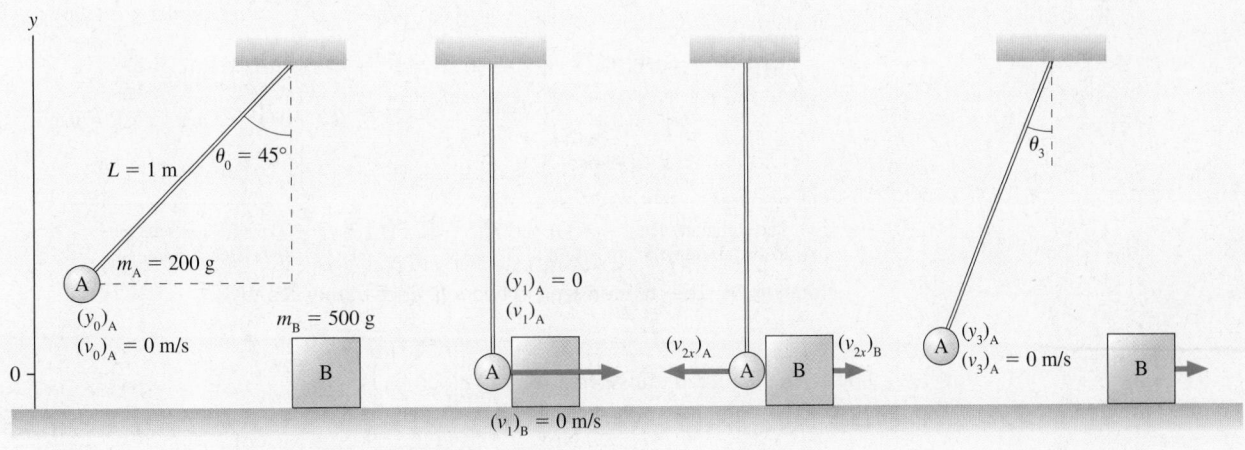

FIGURE 10.26 Four moments in the collision of a pendulum with a block.

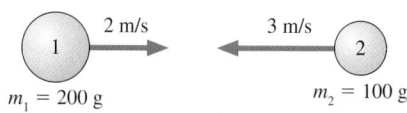

FIGURE 10.27 A perfectly elastic collision in which both balls have an initial velocity.

Using Reference Frames

Equations 10.43 assumed that ball 2 was at rest prior to the collision. Suppose, however, you need to analyze the perfectly elastic collision that is just about to take place in Figure 10.27. What are the direction and speed of each ball after the collision? You could solve the simultaneous momentum and energy equations, but the mathematics becomes quite messy when both balls have an initial velocity. Fortunately, there's an easier way.

You already know the answer—Equations 10.43—when ball 2 is initially at rest. And in Chapter 6 you learned the Galilean transformation of velocity. This transformation relates an object's velocity v as measured in reference frame S to its velocity v' in a different reference frame S′ that moves with velocity V relative to S. The Galilean transformation provides an elegant and straightforward way to analyze the collision of Figure 10.27.

> **TACTICS BOX 10.1 Analyzing elastic collisions**
>
> ❶ Use the Galilean transformation to transform the initial velocities of balls 1 and 2 from the "lab frame" S to a reference frame S′ in which ball 2 is at rest;
> ❷ Use Equations 10.43 to determine the outcome of the collision in frame S′; then
> ❸ Transform the final velocities back to the "lab frame" S.

Figure 10.28a shows the "before" situation in reference frame S, which we can think of as the lab frame. Notice, compared to Figure 10.27, that we've given $(v_{ix})_2$ as a *velocity* with an appropriate sign. The frame S′ in which ball 2 is at rest is a frame that is traveling alongside ball 2 with the same velocity: $V = -3$ m/s.

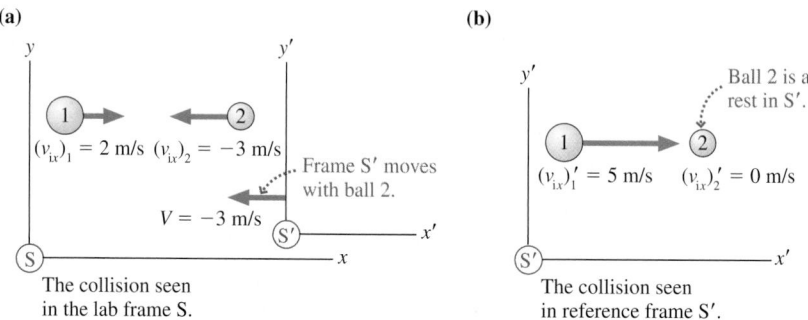

FIGURE 10.28 The collision seen in two reference frames, S and S′.

The Galilean transformation of velocities is

$$v' = v - V \tag{10.44}$$

where a prime represents a velocity measured in frame S′. We can apply this to find the initial velocities of the two balls in S′:

$$
\begin{aligned}
(v_{ix})'_1 &= (v_{ix})_1 - V = 2 \text{ m/s} - (-3 \text{ m/s}) = 5 \text{ m/s} \\
(v_{ix})'_2 &= (v_{ix})_2 - V = -3 \text{ m/s} - (-3 \text{ m/s}) = 0 \text{ m/s}
\end{aligned}
\tag{10.45}
$$

Figure 10.28b shows the "before" situation in reference frame S′, where ball 2 is at rest.

Now we can use Equations 10.43 to find the post-collision velocities in frame S':

$$(v_{fx})'_1 = \frac{m_1 - m_2}{m_1 + m_2}(v_{ix})'_1 = 1.67 \text{ m/s}$$

$$(v_{fx})'_2 = \frac{2m_1}{m_1 + m_2}(v_{ix})'_1 = 6.67 \text{ m/s}$$

(10.46)

Frame S' hasn't changed—it is still moving at $V = -3$ m/s—but the collision has caused both balls to have a velocity in S'.

Finally, we need to apply the reverse transformation $v = v' + V$, with the same V, to transform the post-collision velocities back to the lab frame:

$$(v_{fx})_1 = (v_{fx})'_1 + V = 1.67 \text{ m/s} + (-3 \text{ m/s}) = -1.33 \text{ m/s}$$

$$(v_{fx})_2 = (v_{fx})'_2 + V = 6.67 \text{ m/s} + (-3 \text{ m/s}) = 3.67 \text{ m/s}$$

(10.47)

Figure 10.29 shows the situation after the collision. It's not hard to check that these final velocities do, indeed, conserve both momentum and energy.

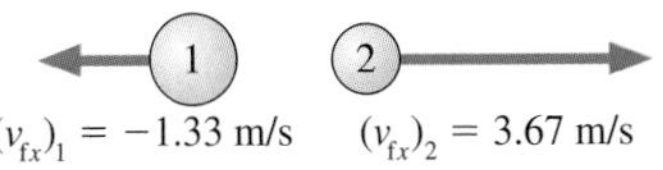

$(v_{fx})_1 = -1.33$ m/s $(v_{fx})_2 = 3.67$ m/s

FIGURE 10.29 The post-collision velocities in the lab frame.

10.7 Energy Diagrams

Potential energy is an energy of position. The gravitational potential energy depends on the height of an object and the elastic potential energy depends on a spring's displacement. Other potential energies you will meet in the future will depend in some way on position. Functions of position are easy to represent as graphs. A graph that shows a system's potential energy and total energy as a function of position is called an **energy diagram.** Energy diagrams allow you to visualize motion based on energy considerations. They can also be useful problem-solving tools, and they will play an important role when we get to quantum physics in Part VII.

Figure 10.30 is the energy diagram of a particle in free fall. The gravitational potential energy $U_g = mgy$ is graphed as a line through the origin with slope mg. The *potential-energy curve* is labeled PE. The line labeled TE is the *total energy line*, $E = K + U_g$. It is horizontal because the mechanical energy is conserved, meaning that the object's total mechanical energy E has the same value at every position.

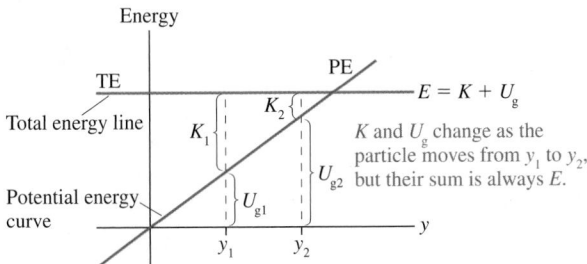

FIGURE 10.30 The energy diagram of a particle in free fall.

Suppose the particle is at position y_1. By definition, the distance from the axis to the potential-energy curve is the particle's potential energy U_{g1} at that position. Because $K_1 = E - U_{g1}$, the distance between the potential-energy curve and the total energy line is the particle's kinetic energy.

The four-frame "movie" of Figure 10.31 on the next page illustrates how an energy diagram is used to visualize motion. The first frame shows a particle projected upward from $y_a = 0$ with kinetic energy K_a. Initially the energy is entirely kinetic, with $U_{ga} = 0$. A pictorial representation and an energy bar chart help to illustrate what the energy diagram is showing.

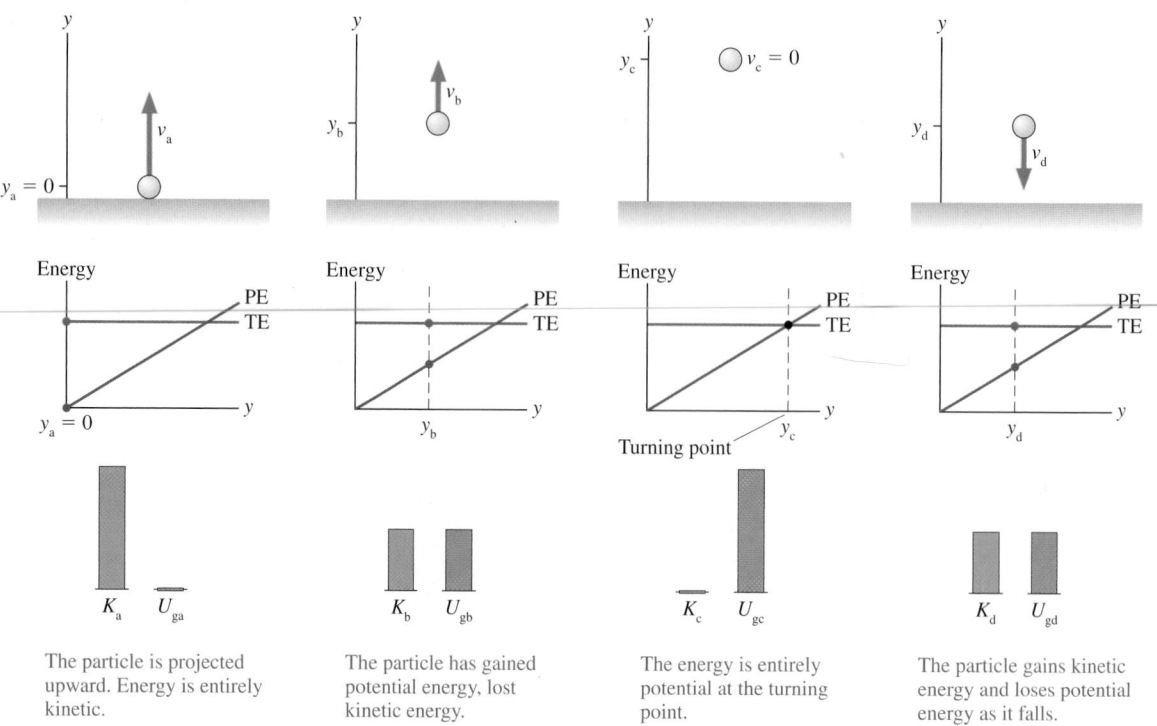

FIGURE 10.31   A four-frame "movie" of a particle in free fall.

In the second frame, the particle has gained height but lost speed. The potential-energy curve U_{gb} is higher, and the distance K_b between the potential-energy curve and the total energy line is less. The particle continues rising and slowing until, in the third frame, it reaches the y-value where the total energy line crosses the potential-energy curve. This point, where $K = 0$ and the energy is entirely potential, is a *turning point* where the particle reverses direction. Finally, we see the particle speeding up as it falls.

A particle with this amount of total energy would need negative kinetic energy to be to the right of the point, at y_c, where the total energy line crosses the potential-energy curve. Negative K is not physically possible, so **the particle cannot be at positions with $U > E$.** Now, it's certainly true that you could make the particle reach a larger value of y simply by throwing it harder. But that would increase E and move the total energy line higher.

NOTE ▶ The TE line is under your control. You can move the TE line as far up or down as you wish by changing the initial conditions, such as projecting the particle upward with a different speed or dropping it from a different height. Once you've determined the initial conditions, you can use the energy diagram to analyze the motion for that amount of total energy. ◀

Figure 10.32 shows the energy diagram of a mass on a horizontal spring. The potential-energy curve $U_s = \frac{1}{2}k(x - x_e)^2$ is a parabola centered at the equilibrium position x_e. The PE curve is determined by the spring constant; you can't change it. But you can set the TE to any height you wish simply by stretching the spring to the proper length. The figure shows one possible TE line.

Suppose you pull the mass out to position x_R and release it. Figure 10.33 is a four-frame movie of the subsequent motion. Initially, the energy is entirely potential. The restoring force of the spring pulls the mass toward x_e, increasing the kinetic energy as the potential energy decreases. The mass has maximum speed at position x_e, where $U_s = 0$, and then it slows down as the spring starts to compress.

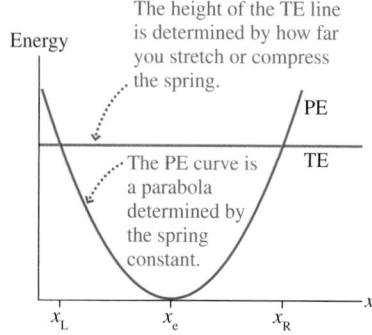

FIGURE 10.32   The energy diagram of a mass on a horizontal spring.

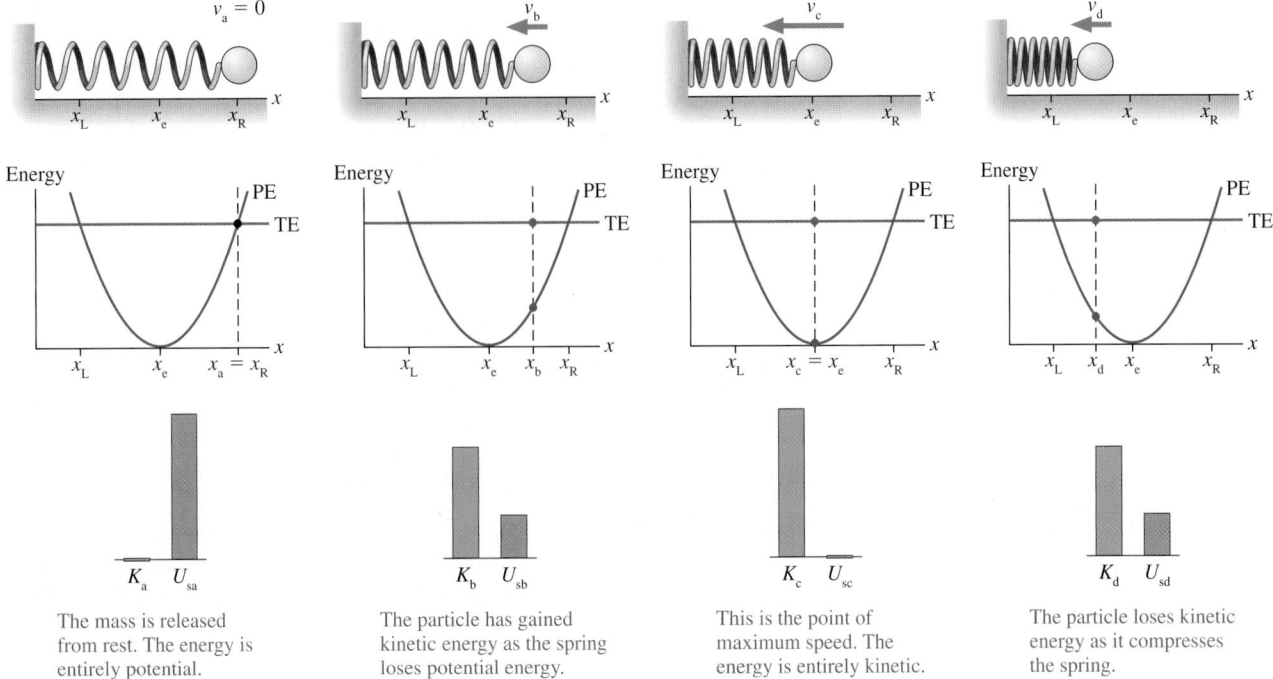

FIGURE 10.33 A four-frame movie of a mass oscillating on a spring.

If the movie were to continue, you should be able to visualize that position x_L is a turning point. The mass will instantaneously have $v_L = 0$ and $K_L = 0$, then reverse direction as the spring starts to expand. The mass will speed up until x_e, then slow down until reaching x_R, where it started. This is another turning point. It will reverse direction again and start the process over. In other words, the mass will *oscillate* back and forth between the left and right turning points at x_L and x_R where the TE line crosses the PE curve.

Figure 10.34 applies these ideas to a more general energy diagram. We don't know how this potential energy was created, but we can visualize the motion of a particle that has this potential energy. Suppose the particle is released from rest at position x_1. How will it then move?

The particle's kinetic energy at x_1 is zero, hence the TE line must cross the PE curve at this point. The particle cannot move to the left, because $U > E$, so it begins to move toward the right. The particle speeds up from x_1 to x_2 as U decreases and K increases, then slows down from x_2 to x_3 as it goes up the "potential energy hill." The particle doesn't stop at x_3 because it still has kinetic energy. It speeds up from x_3 to x_4, reaching its maximum speed at x_4, then slows down between x_4 and x_5. Position x_5 is a turning point, a point where the TE line crosses the PE curve. The particle is instantaneously at rest, then reverses direction. The particle will oscillate back and forth between x_1 and x_5, following the pattern of slowing down and speeding up that we've outlined.

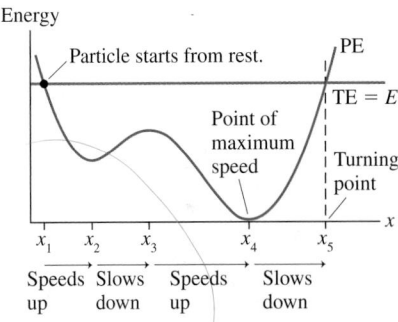

FIGURE 10.34 A more general energy diagram.

Equilibrium Positions

Positions x_2, x_3, and x_4, where the potential energy has a local minimum or maximum, are special positions. Consider a particle with the total energy E_2 shown in Figure 10.35. The particle can be at rest at x_2, with $K = 0$, but it cannot move away from x_2. In other words, a particle with energy E_2 is in *static equilibrium* at x_2. If you disturb the particle, giving it a small kinetic energy and a total energy just *slightly* larger than E_2, the particle will undergo a very small oscillation centered on x_2, like a marble in the bottom of a bowl. An equilibrium for which small

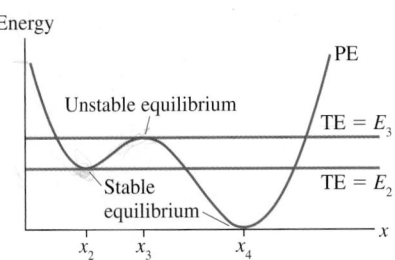

FIGURE 10.35 Points of stable and unstable equilibrium.

disturbances cause small oscillations is called a point of **stable equilibrium.** You should recognize that *any* minimum in the PE curve is a point of stable equilibrium. Position x_4 is also a point of stable equilibrium, in this case for a particle with $E = 0$.

Figure 10.35 also shows a particle with energy E_3 that is tangent to the curve at x_3. If a particle is placed *exactly* at x_3, it will stay there at rest ($K = 0$). But if you disturb the particle at x_3, giving it an energy only slightly more than E_3, it will speed up as it moves away from x_3. This is like trying to balance a marble on top of a hill. The slightest displacement will cause the marble to roll down the hill. A point of equilibrium for which a small disturbance causes the particle to move away is called a point of **unstable equilibrium.** Any maximum in the PE curve, such as x_3, is a point of unstable equilibrium.

We can summarize these lessons as follows:

TACTICS BOX 10.2 Interpreting an energy diagram

❶ The distance from the axis to the PE curve is the particle's potential energy. The distance from the PE curve to the TE line is its kinetic energy. These are transformed as the position changes, causing the particle to speed up or slow down, but the sum $K + U$ doesn't change.

❷ A point where the TE line crosses the PE curve is a turning point. The particle reverses direction.

❸ The particle cannot be at a point where the PE curve is above the TE line.

❹ The PE curve is determined by the properties of the system—mass, spring constant, and the like. You cannot change the PE curve. However, you can raise or lower the TE line simply by changing the initial conditions to give the particle more or less total energy.

❺ A minimum in the PE curve is a point of stable equilibrium. A maximum in the PE curve is a point of unstable equilibrium.

EXAMPLE 10.10 Balancing a mass on a spring

A spring of length L_0 and spring constant k is standing on one end. A block of mass m is placed on the spring, compressing it. What is the length of the compressed spring?

MODEL Assume an ideal spring obeying Hooke's law. The block + spring system has both gravitational potential energy U_g *and* elastic potential energy U_s. The block sitting on top of

the spring is at a point of stable equilibrium (small disturbances cause the block to oscillate slightly around the equilibrium position), so we can solve this problem by looking at the energy diagram.

VISUALIZE Figure 10.36a is a pictorial representation. We've used a coordinate system with the origin at ground level, so the equilibrium position of the uncompressed spring is $y_e = L_0$.

(a)

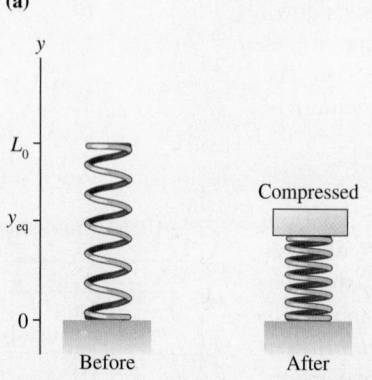

(b)

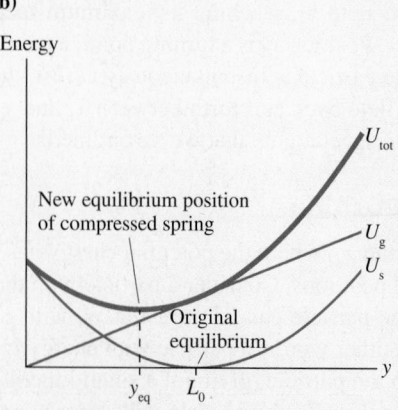

FIGURE 10.36 The block + spring system has both gravitational and elastic potential energy.

SOLVE Figure 10.36b shows the two potential energies separately and also shows the total potential energy

$$U_{tot} = U_g + U_s = mgy + \frac{1}{2}k(y - L_0)^2$$

The equilibrium position (the minimum of U_{tot}) has shifted from L_0 to a smaller value of y, closer to the ground. We can find the equilibrium by locating the position of the minimum in the PE curve. You know from calculus that the minimum of a function is at the point where the derivative (or slope) is zero. The derivative of U_{tot} is

$$\frac{dU_{tot}}{dy} = mg + k(y - L_0)$$

The derivative is zero at the point y_{eq}, so we can easily find

$$mg + k(y_{eq} - L_0) = 0$$

$$y_{eq} = L_0 - \frac{mg}{k}$$

The block compresses the spring by the length mg/k from its original length L_0, giving it a new equilibrium length $L_0 - mg/k$.

STOP TO THINK 10.6 A particle with the potential energy shown in the graph is moving to the right. It has 1 J of kinetic energy at $x = 1$ m. Where is the particle's turning point?

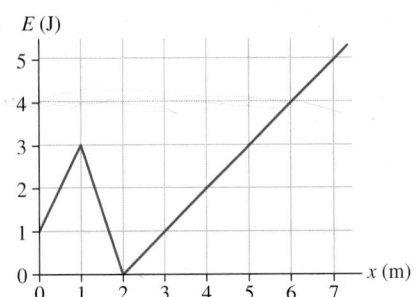

Molecular Bonds

Let's end this chapter by seeing how energy diagrams can allow us to understand something about molecular bonds. A *molecular bond* that holds two atoms together is an electric interaction between the charged electrons and nuclei. Figure 10.37 shows the potential-energy diagram for the diatomic molecule HCl (hydrogen chloride) as it has been experimentally determined. Distance x is the *atomic separation*, the distance between the hydrogen and the chlorine atoms. Note the very tiny distances: 1 nm = 10^{-9} m.

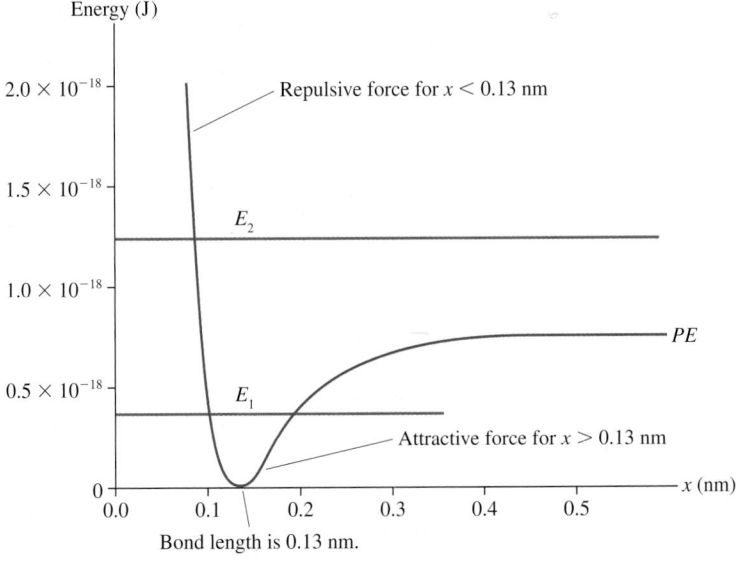

FIGURE 10.37 The energy diagram of the diatomic molecule HCl.

Although the potential energy is an electric energy, we can *interpret* the diagram using the steps in Tactics Box 10.2. The first thing you might notice is this potential-energy diagram has some similarities to a spring, with a deep potential-energy valley, but also some significant differences.

The molecule has a point of stable equilibrium at an atomic separation of $x_{eq} = 0.13$ nm. This is called the *bond length* of HCl, and you can find this value listed in chemistry books. If we try to push the atoms closer together (smaller x), the potential energy rises very rapidly. Physically, this is the repulsive electric force between the electrons orbiting each atom, preventing the atoms from getting too close.

There is also an attractive force between the atoms, called the *polarization force*. It is similar to the static electricity force by which a comb that has been brushed through your hair attracts small pieces of paper. If you try to pull the atoms apart (larger x), the attractive polarization force resists and is responsible for the increasing potential energy for $x > x_{eq}$. The equilibrium position is where the repulsive force between the electrons and the attractive polarization force are exactly balanced.

The repulsive force keeps getting stronger as you push the atoms together, and thus the potential-energy curve keeps getting steeper on the left. But the attractive polarization force gets *weaker* as the atoms get further apart. This is why the potential-energy curve becomes *less* steep as the atomic separation increases. Ultimately, at very large x, the potential energy no longer changes. This is not surprising, because two distant atoms do not interact with each other. This difference between the repulsive and attractive forces leads to an *asymmetric* curve.

It turns out that, for quantum physics reasons, a molecule cannot have $E = 0$ and thus cannot simply rest at the equilibrium position. By requiring the molecule to have some energy, such as E_1, we see that the atoms oscillate back and forth between two turning points. This is a *molecular vibration,* and atoms held together by molecular bonds are constantly vibrating. For a molecule having an energy $E_1 = 0.35 \times 10^{-18}$ J, as illustrated in Figure 10.37, the bond oscillates in length between roughly 0.10 nm and 0.18 nm.

Suppose we increase the molecule's energy to $E_2 = 1.25 \times 10^{-18}$ J. This could happen if the molecule absorbs some light. You can see from the energy diagram that atoms with this energy are not bound together at large values of x. There is no turning point on the right, so the atoms will keep moving apart. By raising the molecule's energy to E_2 we have *broken the molecular bond.* If the atoms happen to be moving together at the time the energy changes, they will "bounce" one last time (there is still a *left* turning point), then move away from each other and not return. The breaking of molecular bonds through the absorption of light is called *photodissociation.* It is an important process in the making of integrated circuits.

SUMMARY

The goal of Chapter 10 has been to introduce the ideas of kinetic and potential energy and to learn a new problem-solving strategy based on conservation of energy.

GENERAL PRINCIPLES

Law of Conservation of Mechanical Energy

If there are no friction or other energy-loss processes (to be explored more thoroughly in Chapter 11), then the mechanical energy $E_{mech} = K + U$ of a system is conserved. Thus

$$K_f + U_f = K_i + U_i$$

- K is the sum of the kinetic energies of all particles.
- U is the sum of all potential energies.

Solving Energy Conservation Problems

MODEL Choose a system without friction or other losses of mechanical energy.

VISUALIZE Draw a before-and-after pictorial representation.

SOLVE Use the law of conservation of energy

$$K_f + U_f = K_i + U_i$$

ASSESS Is the result reasonable?

IMPORTANT CONCEPTS

Kinetic energy is an energy of motion

$$K = \frac{1}{2}mv^2$$

Potential energy is an energy of position

- **Gravitational:** $U_g = mgy$

- **Elastic:** $U_s = \frac{1}{2}k\,(\Delta s)^2$

Basic Energy Model

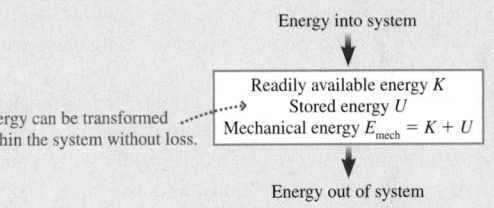

Energy into system

Energy can be transformed within the system without loss. ⋯⟶ Readily available energy K
Stored energy U
Mechanical energy $E_{mech} = K + U$

Energy out of system

Energy diagrams

These diagrams show the potential energy curve PE and the total mechanical energy line TE.

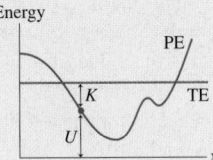

- The distance from the axis to the curve is PE.
- The distance from the curve to the TE line is KE.
- A point where the TE line crosses the PE curve is a **turning point.**
- Minima in the PE curve are points of **stable equilibrium.** Maxima are points of **unstable equilibrium.**

APPLICATIONS

Hooke's law

The restoring force of an ideal spring is

$$(F_{sp})_s = -k\Delta s$$

where k is the spring constant and $\Delta s = s - s_e$ is the displacement from equilibrium.

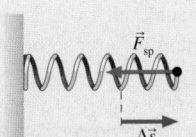

Perfectly elastic collisions

Both mechanical energy and momentum are conserved.

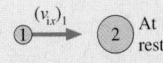

$$(v_{fx})_1 = \frac{m_1 - m_2}{m_1 + m_2}(v_{ix})_1 \qquad (v_{fx})_2 = \frac{2m_1}{m_1 + m_2}(v_{ix})_1$$

If ball 2 is moving, transform to a reference frame in which ball 2 is at rest.

TERMS AND NOTATION

energy
kinetic energy, K
gravitational potential energy, U_g
joule, J
mechanical energy
law of conservation of mechanical energy

restoring force
elastic
equilibrium length, L_0
displacement from equilibrium, Δs
spring constant, k
Hooke's law

elastic potential energy, U_s
perfectly elastic collision
energy diagram
stable equilibrium
unstable equilibrium

EXERCISES AND PROBLEMS

The 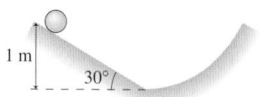 icon indicates that the problem can be done on an Energy Worksheet.

Exercises

Section 10.2 Kinetic Energy and Gravitational Potential Energy

1. Which has the larger kinetic energy, a 10 g bullet fired at 500 m/s or a 10 kg bowling ball rolled at 10 m/s?
2. The lowest point in Death Valley is 85 m below sea level. The summit of nearby Mt. Whitney has an elevation of 4420 m. What is the change in potential energy of an energetic 65 kg hiker who makes it from the floor of Death Valley to the top of Mt. Whitney?
3. At what speed does a 1000 kg compact car have the same kinetic energy as a 20,000 kg truck going 25 km/hr?
4. An oxygen atom is four times as massive as a helium atom. In an experiment, a helium atom and an oxygen atom have the same kinetic energy. What is the ratio v_{He}/v_O of their speeds?
5. a. What is the kinetic energy of a 1500 kg car traveling at a speed of 30 m/s ($\approx$65 mph)?
   b. From what height would the car have to be dropped to have this same amount of kinetic energy just before impact?
   c. Does your answer to part b depend on the car's mass?
6. A boy reaches out of a window and tosses a ball straight up with a speed of 10 m/s. The ball is 20 m above the ground as he releases it. Use energy to find
   a. The ball's maximum height above the ground.
   b. The ball's speed as it passes the window on its way down.
   c. The speed of impact on the ground.
7. a. With what minimum speed must you toss a 100 g ball straight up to hit the 10-m-high roof of the gymnasium if you release the ball 1.5 m above the ground? Solve this problem using energy.
   b. With what speed does the ball hit the ground?

Section 10.3 A Closer Look at Gravitational Potential Energy

8. What minimum speed does a 100 g puck need to make it to the top of a 3.0-m-long, 20° frictionless ramp?
9. A 55 kg skateboarder wants to just make it to the upper edge of a "quarter pipe," a track that is one-quarter of a circle with a radius of 3.0 m. What speed does he need at the bottom?
10. A 50 g ball is released from rest 1.0 m above the bottom of the track shown in Figure Ex10.10. It rolls down a straight 30° segment, then back up a parabolic segment whose shape is given by $y = \frac{1}{4}x^2$, where x and y are in m. How high will the ball go on the right before reversing direction and rolling back down?

FIGURE EX10.10

11. A pendulum is made by tying a 500 g ball to a 75-cm-long string. The pendulum is pulled 30° to one side, then released.
    a. What is the ball's speed at the lowest point of its trajectory?
    b. To what angle does the pendulum swing on the other side?
12. A 20 kg child is on a swing that hangs from 3.0-m-long chains. What is her maximum speed if she swings out to a 45° angle?
13. A 1500 kg car is approaching the hill shown in Figure Ex10.13 at 10.0 m/s when it suddenly runs out of gas.
    a. Can the car make it to the top of the hill by coasting?
    b. If your answer to (a) is yes, what is the car's speed after coasting down the other side?

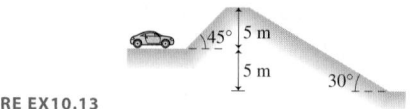

FIGURE EX10.13

Section 10.4 Restoring Forces and Hooke's Law

14. A 10-cm-long spring is attached to the ceiling. When a 2.0 kg mass is hung from it, the spring stretches to a length of 15 cm.
    a. What is the spring constant k?
    b. How long is the spring when a 3.0 kg mass is suspended from it?
15. A 5.0 kg mass hanging from a spring scale is slowly lowered onto a vertical spring, as shown in Figure Ex10.15.
    a. What does the spring scale read just before the mass touches the lower spring?
    b. The scale reads 20 N when the lower spring has been compressed by 2.0 cm. What is the value of the spring constant for the lower spring?
    c. At what compression length will the scale read zero?

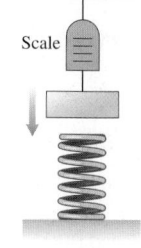

FIGURE EX10.15

16. A runner wearing spiked shoes pulls a 20 kg sled across frictionless ice using a horizontal spring with spring constant 150 N/m. The spring is stretched 20 cm from its equilibrium length. What is the acceleration of the sled?
17. You need to make a spring scale for measuring mass. You want each 1.0 cm length along the scale to correspond to a mass difference of 100 g. What should be the value of the spring constant?
18. A 60 kg student is standing atop a spring in an elevator that is accelerating upward at 3.0 m/s². The spring constant is 2500 N/m. By how much is the spring compressed?

Section 10.5 Elastic Potential Energy

19. How much energy can be stored in a spring with $k = 500$ N/m if the maximum possible stretch is 20 cm?
20. How far must you stretch a spring with $k = 1000$ N/m to store 200 J of energy?
21. A student places her 500 g physics book on a frictionless table. She pushes the book against a spring, compressing the spring by 4.0 cm, then releases the book. What is the book's speed as it slides away? The spring constant is 1250 N/m.

22. A block sliding along a horizontal frictionless surface with speed v collides with a spring and compresses it by 2.0 cm. What will be the compression if the same block collides with the spring at a speed of $2v$?

23. A 10 kg runaway grocery cart runs into a spring with spring constant 250 N/m and compresses it by 60 cm. What was the speed of the cart just before it hit the spring?

24. As a 15,000 kg jet plane lands on an aircraft carrier, its tail hook snags a cable to slow it down. The cable is attached to a spring with spring constant 60,000 N/m. If the spring stretches 30 m to stop the plane, what was the plane's landing speed?

Section 10.6 Elastic Collisions

25. A 50 g marble moving at 2.0 m/s strikes a 20 g marble at rest. What is the speed of each marble immediately after the collision?

26. A 50 g ball of clay traveling at speed v_0 hits and sticks to a 1.0 kg block sitting at rest on a frictionless surface.
    a. What is the speed of the block after the collision?
    b. Show that the mechanical energy is *not* conserved in this collision. What percentage of the ball's initial energy is "lost"?

27. Ball 1, with a mass of 100 g and traveling at 10 m/s, collides head on with ball 2, which has a mass of 300 g and is initially at rest. What are the final velocities of each ball if the collision is (a) perfectly elastic? (b) perfectly inelastic?

28. A proton is traveling to the right at 2.0×10^7 m/s. It has a head-on perfectly elastic collision with a carbon atom. The mass of the carbon atom is 12 times the mass of the proton. What are the speed and direction of each after the collision?

Section 10.7 Energy Diagrams

29. Figure Ex10.29 is the potential-energy diagram for a 20 g particle that is released from rest at $x = 1.0$ m.
    a. Will the particle move to the right or to the left? How can you tell?
    b. What is the particle's maximum speed? At what position does it have this speed?
    c. Where are the turning points of the motion?

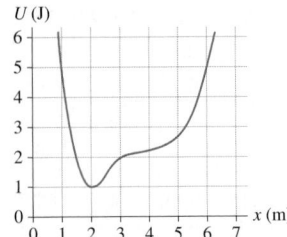

FIGURE EX10.29

30. Figure Ex10.30 is the potential-energy diagram for a 500 g particle that is released from rest at A. What are the particle's speeds at B, C, and D?

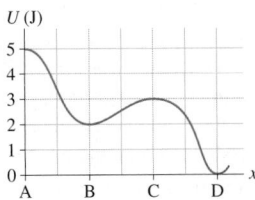

FIGURE EX10.30

31. What is the maximum speed of a 2.0 g particle that oscillates between $x = 2.0$ mm and $x = 8.0$ mm in Figure Ex10.31?

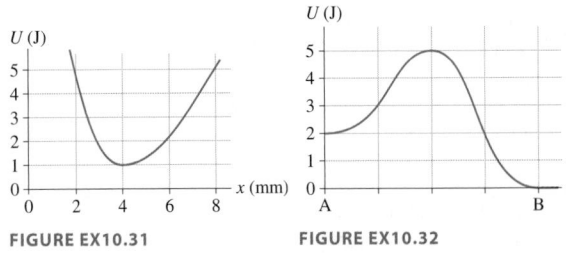

FIGURE EX10.31   **FIGURE EX10.32**

32. a. In Figure Ex10.32, what minimum speed does a 100 g particle need at point A to reach point B?
    b. What minimum speed does a 100 g particle need at point B to reach point A?

Problems

33. You're driving at 35 km/hr when the road suddenly descends 15 m into a valley. You take your foot off the accelerator and coast down the hill. Just as you reach the bottom you see the policeman hiding behind the speed limit sign that reads "70 km/hr." Are you going to get a speeding ticket?

34. A cannon tilted up at a 30° angle fires a cannon ball at 80 m/s from atop a 10-m-high fortress wall. What is the ball's impact speed on the ground below?

35. Your friend's Frisbee has become stuck 16 m above the ground in a tree. You want to dislodge the Frisbee by throwing a rock at it. The Frisbee is stuck pretty tight, so you figure the rock needs to be traveling at least 5.0 m/s when it hits the Frisbee.
    a. Does the speed with which you throw the rock depend on the angle at which you throw it? Explain.
    b. If you release the rock 2.0 m above the ground, with what minimum speed must you throw it?

36. A marble spins in a *vertical* plane around the inside of a smooth, 20-cm-diameter horizontal pipe. The marble's speed at the bottom of the circle is 3.0 m/s.
    a. What is the marble's speed at the top?
    b. Find an algebraic expression for the marble's speed when it is at angle θ, where the angle is measured from the bottom of the circle.
    c. Make a graph of v-versus-θ for one complete revolution.

37. A 50 g rock is placed in a slingshot and the rubber band is stretched. The force of the rubber band on the rock is shown by the graph in Figure P10.37.
    a. Is the rubber band stretched to the right or to the left? How can you tell?
    b. Does this rubber band obey Hooke's law? Explain.
    c. What is the rubber band's spring constant k?
    d. The rubber band is stretched 30 cm and then released. What is the speed of the rock?

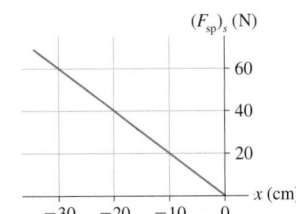

FIGURE P10.37

38. The spring in Figure P10.38a is compressed by length Δx. It launches the block across a frictionless surface with speed v_0. The two springs in Figure P10.38b are identical to the spring of Figure P10.38a. They are compressed by the same length Δx and used to launch the same block. What is the block's speed?

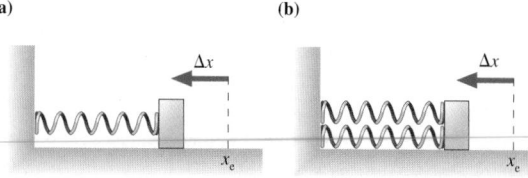

(a)   (b)

FIGURE P10.38

39. The spring in Figure P10.39a is compressed by length Δx. It launches the block across a frictionless surface with speed v_0. The two springs in Figure P10.39b are identical to the spring of Figure P10.39a. They are compressed the same *total* length Δx and used to launch the same block. What is the block's speed?

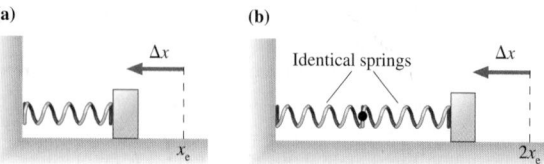

(a)   (b)

FIGURE P10.39

40. A 500 g rubber ball is dropped from a height of 10 m and undergoes a perfectly elastic collision with the earth.
    a. What is the earth's velocity after the collision? Assume the earth was at rest just before the collision.
    b. How many years would it take the earth to move 1.0 mm at this speed?

41. A 100 g marble rolls down a 40° incline. At the bottom, just after it exits onto a horizontal table, it collides with a 200 g steel ball at rest. How high above the table should the marble be released to give the steel ball a speed of 150 cm/s?

42. A package of mass m is released from rest at a warehouse loading dock and slides down a 3.0-m-high frictionless chute to a waiting truck. Unfortunately, the truck driver went on a break without having removed the previous package, of mass $2m$, from the bottom of the chute.
    a. Suppose the packages stick together. What is their common speed after the collision?
    b. Suppose the collision between the packages is elastic. To what height does the package of mass m rebound?

FIGURE P10.42

43. A 50 g ice cube can slide without friction up and down a 30° slope. The ice cube is pressed against a spring at the bottom of the slope, compressing the spring 10 cm. The spring constant is 25 N/m. When the ice cube is released, what distance will it travel up the slope before reversing direction?

44. A 1000 kg safe is 2.0 m above a heavy-duty spring when the rope holding the safe breaks. The safe hits the spring and compresses it 50 cm. What is the spring constant of the spring?

45. In a physics lab experiment, a spring clamped to the table is used to shoot a 20 g ball at a 30° angle. When the spring is compressed 20 cm, the ball travels horizontally 5.0 m and lands 1.5 m below the point at which it left the spring. What is the spring constant?

46. A vertical spring with $k = 490$ N/m is standing on the ground. You are holding a 5.0 kg block just above the spring, not quite touching it.
    a. How far does the spring compress if you let go of the block suddenly?
    b. How far does the spring compress if you slowly lower the block to the point where you can remove your hand without disturbing it?
    c. Why are your two answers different?

47. The desperate contestants on a TV survival show are very hungry. The only food they can see is some fruit hanging on a branch high in a tree. Fortunately, they have a spring they can use to launch a rock. The spring constant is 1000 N/m, and they can compress the spring a maximum of 30 cm. All the rocks on the island seem to have a mass of 400 g.
    a. With what speed does the rock leave the spring?
    b. If the fruit hangs 15 m above the ground, will they feast or go hungry?

48. A massless pan hangs from a spring that is suspended from the ceiling. When empty, the pan is 50 cm below the ceiling. If a 100 g clay ball is placed gently on the pan, the pan hangs 60 cm below the ceiling. Suppose the clay ball is dropped from the ceiling onto an empty pan. What is the pan's distance from the ceiling when the spring reaches its maximum length?

49. You have been hired to design a spring-launched roller coaster that will carry two passengers per car. The car goes up a 10-m-high hill, then descends 15 m to the track's lowest point. You've determined that the spring can be compressed a maximum of 2.0 m and that a loaded car will have a maximum mass of 400 kg. For safety reasons, the spring constant should be 10% larger than the minimum needed for the car to just make it over the top.
    a. What spring constant should you specify?
    b. What is the maximum speed of a 350 kg car if the spring is compressed the full amount?

50. It's been a great day of new, frictionless snow. Julie starts at the top of the 60° slope shown in Figure P10.50. At the bottom, a circular arc carries her through a 90° turn, and she then launches off a 3.0-m-high ramp. How far is her touchdown point from the base of the ramp?

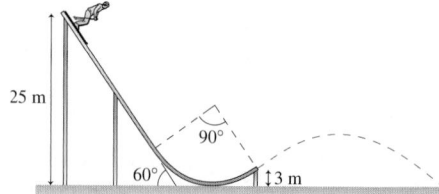

25 m   90°   60°   ‡3 m

FIGURE P10.50

51. A 100 g block on a frictionless table is firmly attached to one end of a spring with $k = 20$ N/m. The other end of the spring is anchored to the wall. A 20 g ball is thrown horizontally toward the block with a speed of 5.0 m/s.
    a. If the collision is perfectly elastic, what is the ball's speed immediately after the collision?
    b. What is the maximum compression of the spring?
    c. Repeat parts a and b for the case of a perfectly inelastic collision.

52. You have been asked to design a "ballistic spring system" to measure the speed of bullets. A bullet of mass m is fired into a block of mass M. The block, with the embedded bullet, then slides across a frictionless table and collides with a horizontal spring whose spring constant is k. The opposite end of the spring is anchored to a wall. The spring's maximum compression d is measured.
    a. Find an expression for the bullet's initial speed v_B in terms of m, M, k, and d.
    b. What was the speed of a 5.0 g bullet if the block's mass is 2.0 kg and if the spring, with $k = 50$ N/m, was compressed by 10 cm?
    c. What fraction of the bullet's energy is "lost"? Where did it go?

53. A roller coaster car on the frictionless track shown in Figure P10.53 starts from rest at height h. The track is straight until point A. Between points A and D, the track consists of circle-shaped segments of radius R.
    a. What is the *maximum* height h_{max} from which the car can start so as not to fly off the track when going over the hill at point C? Give your answer in terms of the radius R. **Hint:** This is a two-part problem. First find v_{max} at C.
    b. Evaluate h_{max} for a roller coaster that has $R = 10$ m.

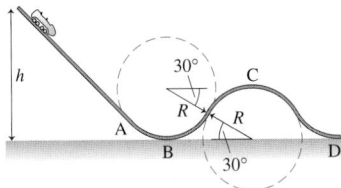

FIGURE P10.53

54. A pendulum is formed from a small ball of mass m on a string of length L. As Figure P10.54 shows, a peg is height $h = L/3$ above the pendulum's lowest point. From what minimum angle θ must the pendulum be released in order for the ball to go over the top of the peg without the string going slack?

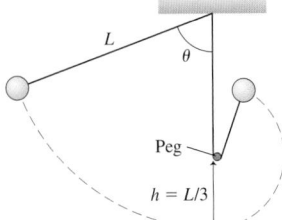

FIGURE P10.54

55. A block of mass m slides down a frictionless track, then around the inside of a circular loop-the-loop of radius R. From what minimum height h must the block start to make it around the loop without falling off? Give your answer as a multiple of R.

56. A new event has been proposed for the Winter Olympics. An athlete will sprint 100 m, starting from rest, then leap onto a 20 kg bobsled. The person and bobsled will then slide down a 50-m-long ice-covered ramp, sloped at 20°, and into a spring with a carefully calibrated spring constant of 2000 N/m. The athlete who compresses the spring the farthest wins the gold medal.

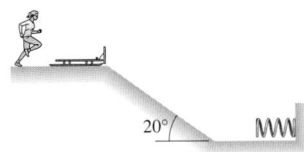

FIGURE P10.56

Lisa, whose mass is 40 kg, has been training for this event. She can reach a maximum speed of 12 m/s in the 100 m dash.
    a. How far will Lisa compress the spring?
    b. The Olympic committee has very exact specifications about the shape and angle of the ramp. Is this necessary? If the committee asks your opinion, what factors about the ramp will you tell them are important?

57. A 20 g ball is fired horizontally with initial speed v_0 toward a 100 g ball that is hanging motionless from a 1.0-m-long string. The balls undergo a head-on, perfectly elastic collision, after which the 100 g ball swings out to a maximum angle $\theta_{max} = 50°$. What was v_0?

58. A 100 g ball moving to the right at 4.0 m/s collides head-on with a 200 g ball that is moving to the left at 3.0 m/s.
    a. If the collision is perfectly elastic, what are the speed and direction of each ball after the collision?
    b. If the collision is perfectly inelastic, what are the speed and direction of the combined balls after the collision?

59. A 100 g ball moving to the right at 4.0 m/s catches up and collides with a 400 g ball that is moving to the right at 1.0 m/s. If the collision is perfectly elastic, what are the speed and direction of each ball after the collision?

60. Figure P10.60 shows the potential energy of a 500 g particle as it moves along the x-axis. Suppose the particle's mechanical energy is 12 J.
    a. Where are the particle's turning points?
    b. What is the particle's speed when it is at $x = 2.0$ m?
    c. What is the particle's maximum speed? At what position or positions does this occur?
    d. Give a written description of the motion of the particle as it moves from the left turning point to the right turning point.
    e. Suppose the particle's energy is lowered to 4.0 J. Describe the possible motions.

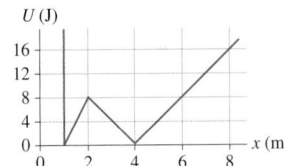

FIGURE P10.60

61. The ammonia molecule NH_3 has the tetrahedral structure shown in Figure P10.61a. The three hydrogen atoms form a triangle in the xy-plane at $z = 0$. The nitrogen atom is the apex of the pyramid. Figure P10.61b shows the potential energy of the nitrogen atom along a z-axis that is perpendicular to the H_3 triangle.
    a. At room temperature, the nitrogen atom has $\approx 0.4 \times 10^{-20}$ J of mechanical energy. Describe the position and motion of the nitrogen atom.
    b. The nitrogen atom can gain energy if the molecule absorbs light energy. Describe the motion of the nitrogen atom if its energy is 4×10^{-20} J.

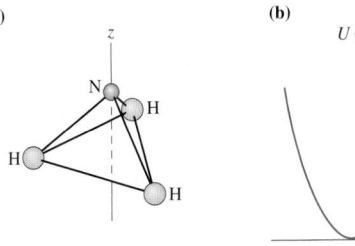

FIGURE P10.61

62. Protons and neutrons (together called *nucleons*) are held together in the nucleus of an atom by a force called the *strong force*. At very small separations, the strong force between two nucleons is larger than the repulsive electrical force between two protons—hence its name. But the strong force quickly weakens as the distance between the protons increases. A well-established model for the potential energy of two nucleons interacting via the strong force is

$$U = U_0[1 - e^{-x/x_0}]$$

where x is the distance between the centers of the two nucleons, x_0 is a constant having the value $x_0 = 2.0 \times 10^{-15}$ m, and $U_0 = 6.0 \times 10^{-11}$ J.

a. Calculate and draw an accurate potential-energy curve from $x = 0$ m to $x = 10 \times 10^{-15}$ m. Either calculate about 10 points by hand or use computer software.

b. Quantum effects are essential for a proper understanding of how nucleons behave. Nonetheless, let us innocently consider two neutrons *as if* they were small, hard, electrically neutral spheres of mass 1.67×10^{-27} kg and diameter 1.0×10^{-15} m. (We will consider neutrons rather than protons so as to avoid complications from the electric forces between protons.) You are going to hold two neutrons 5.0×10^{-15} m apart, measured between their centers, then release them. Draw the total energy line for this situation on your diagram of part a.

c. What speed do *each* of these neutrons have as they crash together? Keep in mind that *both* neutrons are moving.

63. Write a realistic problem for which the energy bar chart shown in Figure P10.63 correctly shows the energy at the beginning and end of the problem.

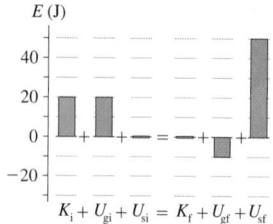

FIGURE P10.63

$K_i + U_{gi} + U_{si} = K_f + U_{gf} + U_{sf}$

In Problems 64 through 67 you are given the equation used to solve a problem. For each of these, you are to
a. Write a realistic problem for which this is the correct equation.
b. Draw the before-and-after pictorial representation.
c. Finish the solution of the problem.

64. $\frac{1}{2}(1500 \text{ kg})(5.0 \text{ m/s})^2 + (1500 \text{ kg})(9.80 \text{ m/s}^2)(10 \text{ m})$

$= \frac{1}{2}(1500 \text{ kg})(v_i)^2 + (1500 \text{ kg})(9.80 \text{ m/s}^2)(0 \text{ m})$

65. $\frac{1}{2}(0.20 \text{ kg})(2.0 \text{ m/s})^2 + \frac{1}{2}k(0 \text{ m})^2$

$= \frac{1}{2}(0.20 \text{ kg})(0 \text{ m/s})^2 + \frac{1}{2}k(-0.15 \text{ m})^2$

66. $(0.10 \text{ kg} + 0.20 \text{ kg})v_{1x} = (0.10 \text{ kg})(3.0 \text{ m/s})$

$\frac{1}{2}(0.30 \text{ kg})(0 \text{ m/s})^2 + \frac{1}{2}(3.0 \text{ N/m})(\Delta x_2)^2$

$= \frac{1}{2}(0.30 \text{ kg})(v_{1x})^2 + \frac{1}{2}(3.0 \text{ N/m})(0 \text{ m})^2$

67. $\frac{1}{2}(0.50 \text{ kg})(v_f)^2 + (0.50 \text{ kg})(9.80 \text{ m/s}^2)(0 \text{ m})$

$+ \frac{1}{2}(400 \text{ N/m})(0 \text{ m})^2$

$= \frac{1}{2}(0.50 \text{ kg})(0 \text{ m/s})^2$

$+ (0.50 \text{ kg})(9.80 \text{ m/s}^2)((-0.10 \text{ m})\sin 30°)$

$+ \frac{1}{2}(400 \text{ N/m})(-0.10 \text{ m})^2$

Challenge Problems

68. It's your birthday, and to celebrate you're going to make your first bungee jump. You stand on a bridge that is 100 m above a river and attach a 30-m-long bungee cord to your harness. A bungee cord, for practical purposes, is just a long spring, and this cord has a spring constant of 40 N/m. Assume that your mass is 80 kg. After a long hesitation, you dive off the bridge. How far are you above the water when the cord reaches its maximum elongation?

69. A 10 kg box slides 4.0 m down the frictionless ramp shown in Figure CP10.69, then collides with a spring whose spring constant is 250 N/m.
a. What is the maximum compression of the spring?
b. At what compression of the spring does the box have its maximum velocity?

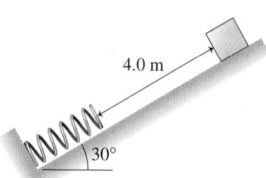

4.0 m

30°

FIGURE CP10.69

70. Old naval ships fired 10 kg cannon balls from a 200 kg cannon. It was very important to stop the recoil of the cannon, since otherwise the heavy cannon would go careening across the deck of the ship. In one design, a large spring with spring constant 20,000 N/m was placed behind the cannon. The other end of the spring braced against a post that was firmly anchored to the ship's frame. What was the speed of the cannon ball if the spring compressed 50 cm when the cannon was fired?

71. You have been asked to design a "ballistic spring system" to measure the speed of bullets. A spring whose spring constant is k is suspended from the ceiling. A block of mass M hangs from the spring. A bullet of mass m is fired vertically upward

into the bottom of the block. The spring's maximum compression d is measured.

a. Find an expression for the bullet's initial speed v_B in terms of m, M, k, and d.

b. What was the speed of a 10 g bullet if the block's mass is 2.0 kg and if the spring, with $k = 50$ N/m, was compressed by 45 cm?

72. A 2.0 kg cart has a spring with $k = 5000$ N/m attached to its front, parallel to the ground. This cart rolls at 4.0 m/s toward a stationary 1 kg cart.

a. What is the maximum compression of the spring during the collision?

b. What is the speed of each cart after the collision?

73. A 100 g steel ball and a 200 g steel ball each hang from 1-m-long strings. At rest, the balls hang side by side, barely touching. The 100 g ball is pulled to the left until its string is at a 45° angle. The 200 g ball is pulled to a 45° angle on the right. The balls are released so as to collide at the very bottom of their swings. To what angle does each ball rebound?

74. A sled starts from rest at the top of the frictionless, hemispherical, snow-covered hill shown in Figure CP10.74.

a. Find an expression for the sled's speed when it is at angle ϕ.

b. Use Newton's laws to find the maximum speed the sled can have at angle ϕ without leaving the surface.

c. At what angle ϕ_{max} does the sled "fly off" the hill?

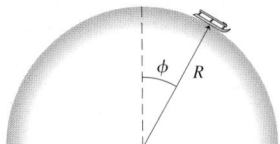

FIGURE CP10.74

Stop to Think 10.1: $(U_g)_3 > (U_g)_2 = (U_g)_4 > (U_g)_1$. Gravitational potential energy depends only on height, not speed.

Stop to Think 10.2: $v_A = v_B = v_C = v_D$. Her increase in kinetic energy depends only on the vertical height through which she falls, not the shape of the slide.

Stop to Think 10.3: b. Mechanical energy is conserved on a frictionless surface. Because $K_i = 0$ and $K_f = 0$, it must be true that $U_f = U_i$ and thus $y_f = y_i$. The final height matches the initial height.

Stop to Think 10.4: $k_1 > k_2 > k_3$. The spring constant is the slope of the force-versus-displacement graph.

Stop to Think 10.5: c. U_s depends on Δs^2, so doubling the compression increases U_s by a factor of 4. All the potential energy is converted to kinetic energy, so K increases by a factor of 4. But K depends on v^2, so v increases by only a factor of $(4)^{1/2} = 2$.

Stop to Think 10.6: $x = 6$ m. From the graph, the particle's potential energy at $x = 1$ m is $U = 3$ J. Its total energy is thus $E = K + U = 4$ J. A TE line at 4 J crosses the PE curve at $x = 6$ m.

11 Work

This bobsled team is increasing the sled's kinetic energy by pushing it forward. In the language of physics, they are doing *work* on the sled.

► Looking Ahead
The goal of Chapter 11 is to develop a more complete understanding of energy and its conservation. In this chapter you will learn to:

- Understand and apply the basic energy model.
- Calculate the work done on a system.
- Understand and use a more complete statement of conservation of energy.
- Use a general strategy for solving energy problems.
- Calculate the power supplied to or dissipated by a system.

◄ Looking Back
This chapter continues to develop energy ideas that were introduced in Chapter 10. Please review:

- Sections 10.2–10.3 Gravitational potential energy.
- Sections 10.4–10.5 Hooke's law and elastic potential energy.

Chapter 10 introduced the concept of energy. Although energy appears to be a useful idea, three major questions remain unanswered:

- How many kinds of energy are there?
- Under what conditions is energy conserved?
- How does a system gain or lose energy?

For example, this bobsled is gaining kinetic energy, but it's not doing so by losing potential energy. Instead, the runners are giving it kinetic energy by pushing it faster and faster. One of our goals in this chapter is to relate the energy gained by the sled to the strength of their push. Energy transferred by pushes and pulls is called *work*.

We will also explore how energy is *dissipated*. Because of friction, a bobsled sliding across a horizontal surface gradually slows and stops. What happens to its kinetic energy? By addressing these issues, we will put the concept of energy on a firmer foundation.

11.1 The Basic Energy Model

We will begin with an overview of where this chapter will take us, then come back to fill in the details. Consider a *system* of particles. The system can be characterized by two quantities that we call the *kinetic energy* and the *potential energy.* Kinetic energy is an energy due to the *motion* of the particles. The potential energy, which is often thought of as "stored energy," is due to *interactions* between the particles. For example, two particles connected by a stretched spring have a potential energy.

The sum of kinetic and potential energy is the *mechanical energy* $E_{mech} = K + U$. The term *mechanical* designates this form of energy as being due to motion and mechanical effects, such as stretching springs, rather than chemical effects or heat effects.

Mechanical energy is an energy of the object as a whole. That is, it is an energy *of the ball* or *of the rocket.* Suppose a ball is at rest, with zero mechanical energy. If you peered inside with a very powerful microscope, you would see the atoms inside the ball vibrating back and forth on their spring-like molecular bonds. This microscopic motion of the atoms and molecules within an object is a form of energy that is distinct from the object's mechanical energy. The total energy of the moving atoms is called the **thermal energy E_{th}.**

Thermal energy is associated with the system's *temperature.* A higher temperature means more microscopic motion and thus more thermal energy. Friction raises the temperature—think of rubbing your hands together briskly—so a system with friction "runs down" as its mechanical energy is transformed into thermal energy. We'll say more about thermal energy in Section 11.7.

We can define the **system energy E_{sys}** as the sum of the mechanical energy *of* the objects plus the thermal energy of the atoms *inside* the objects. That is,

$$E_{sys} = E_{mech} + E_{th} = K + U + E_{th} \qquad (11.1)$$

As Figure 11.1 shows, kinetic and potential energy can be changed back and forth into each other. You studied this process in Chapter 10. Kinetic and potential energy can also be changed into thermal energy, but, as we'll discuss later, thermal energy is not normally changed into kinetic or potential energy. These energy exchanges within the system are called **energy transformations.** Energy transformations within the system do not change the value of E_{sys}.

NOTE ▶ We will use an arrow $\rightarrow$ as a shorthand way to indicate an energy transformation. If the kinetic energy of a box is transformed into thermal energy as it slides across a table, we will indicate this by $K \rightarrow E_{th}$. ◀

A system of particles is always surrounded by a larger *environment.* Unless the system is completely isolated, it has the possibility of exchanging energy with the environment. An energy exchange between the system and the environment is called an **energy transfer.** There are two primary energy transfer processes. The first, and the only one we will be concerned with for now, is due to forces—pushes and pulls—exerted on the system by the environment. For example, you give a ball kinetic energy by pushing on it. This *mechanical* transfer of energy to or from the system is called **work.** The symbol for work is W.

The second means of transferring energy between the system and its environment is a *nonmechanical* process called *heat.* Heat is a crucial idea that we will add to the energy model when we study thermodynamics, but for now we want to concentrate on the mechanical transfer of energy via work.

Figure 11.2 shows a **basic energy model** in which energy can be *transferred* to or from the system and energy can be *transformed* within the system. Notice how similar Figure 11.2 is to John's model of the monetary system in Chapter 10. As a *basic* model, Figure 11.2 is certainly not complete, and we will add new features to the model as needed. Nonetheless, it is a good starting point.

As the arrows in Figure 11.2 show, energy can both enter and leave the system. We'll distinguish between the two directions of energy flow by allowing the work W to be either positive or negative.

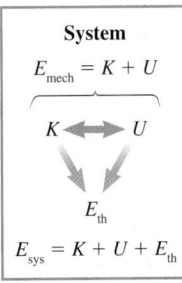

FIGURE 11.1 Energy can be transformed within the system.

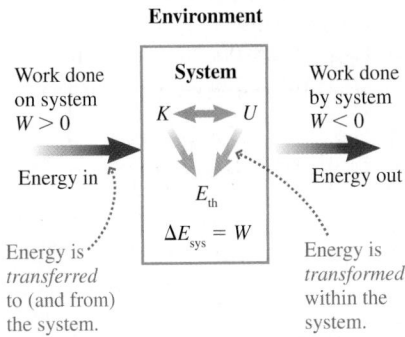

FIGURE 11.2 The basic energy model of a system interacting with its environment.

The sign of W is interpreted as follows:

$W > 0$ The environment does work on the system and the system's energy increases,

$W < 0$ The system does work on the environment and the system's energy decreases.

This is equivalent to considering expenditures (i.e., money out) to be negative income. In fact, this is how accountants really do handle expenditures.

Having established the quantities of the basic energy model, we can ask, what is the relationship among them? Our hypothesis, which is confirmed by experiment, is that

$$\Delta E_{sys} = \Delta K + \Delta U + \Delta E_{th} = W \qquad (11.2)$$

The two essential ideas of the basic energy model and Equation 11.2 are:

1. Energy can be *transferred* to a system by doing work on the system. This process changes the total energy of the system: $\Delta E_{sys} = W$.
2. If no work is done, energy can be *transformed* within the system between K, U, and E_{th} as long as the total energy of the system doesn't change: $\Delta E_{sys} = 0$.

This is the essence of the basic energy model. The rest of Chapter 11 will substantiate Equation 11.2 and look at its many implications.

STOP TO THINK 11.1   A child slides down a playground slide at constant speed. The energy transformation is

a. $U \rightarrow K$ b. $K \rightarrow U$
c. There is no transformation because energy is conserved.
d. $U \rightarrow E_{th}$ e. $K \rightarrow E_{th}$

11.2 Work and Kinetic Energy

Work is a common word in the English language, with many meanings. When you first think of work, you probably think of the first two definitions in this list. After all, we talk about "working out," or we say, "I just got home from work." But that is *not* what work means in physics.

The basic energy model uses *work* in the sense of definition 7: Energy transferred to or from a body or system by the application of force. The critical question we must answer is: *How much energy* does a force transfer?

We can answer this question by following the procedure we used in Chapter 10 to find the potential energy of gravity and of a spring. Figure 11.3 shows a force $\vec{F}$ acting on a particle of mass m as the particle moves along an s-axis from an initial position s_i, with kinetic energy K_i, to a final position s_f where the kinetic energy is K_f. As the figure suggests, the force may vary in magnitude and/or direction as the particle moves.

NOTE ▶ $\vec{F}$ may not be the only force acting on the particle. For the particle to move along a straight line, some other force must cancel any component of $\vec{F}$ perpendicular to the s-axis. However, for now we'll assume that $\vec{F}$ is the only force with a component parallel to the s-axis and hence is the only force capable of changing the particle's speed. ◀

The force component F_s parallel to the s-axis causes the particle to speed up or slow down, thus transferring energy to or from the particle. We say that force $\vec{F}$

One dictionary defines *work* as:

1. Physical or mental effort; labor.
2. The activity by which one makes a living.
3. A task or duty.
4. Something produced as a result of effort, such as a *work of art*.
5. Plural *works*: A factory or plant where industry is carried on, such as *steel works*.
6. Plural *works*: The essential or operating parts of a mechanism.
7. The transfer of energy to a body by application of a force.

does work on the particle. Our goal is to find a relationship between F_s and ΔK. The s-component of Newton's second law is

$$F_s = ma_s = m\frac{dv_s}{dt} \tag{11.3}$$

where the v_s is the s-component of $\vec{v}$. As we did in Chapter 10, we can use the chain rule to write

$$m\frac{dv_s}{dt} = m\frac{dv_s}{ds}\frac{ds}{dt} = mv_s\frac{dv_s}{ds} \tag{11.4}$$

where $ds/dt = v_s$. Substituting Equation 11.4 into Equation 11.3 gives

$$F_s = mv_s\frac{dv_s}{ds} \tag{11.5}$$

The crucial step here, as it was in Chapter 10, was changing from a derivative with respect to time to a derivative with respect to position. We're going to want to integrate, so first multiply through by ds to get

$$mv_s\,dv_s = F_s\,ds \tag{11.6}$$

Now we can integrate both sides from "before," where the position is s_i and the speed is v_i, to "after," giving

$$\int_{v_i}^{v_f} mv_s\,dv_s = \frac{1}{2}mv_s^2\Big|_{v_i}^{v_f} = \frac{1}{2}mv_f^2 - \frac{1}{2}mv_i^2 = \int_{s_i}^{s_f} F_s\,ds \tag{11.7}$$

The left side of Equation 11.7 is ΔK, the change in the particle's kinetic energy as it moves from s_i to s_f. The integral on the right apparently specifies the extent to which the applied force changes the particle's kinetic energy. We define the *work* done by force $\vec{F}$ as the particle moves from s_i to s_f as

$$W = \int_{s_i}^{s_f} F_s\,ds \tag{11.8}$$

The unit of work, that of force multiplied by distance, is the Nm. Using the definition of the newton gives

$$1\,\text{Nm} = 1\,(\text{kg m/s}^2)\,\text{m} = 1\,\text{kg m}^2/\text{s}^2 = 1\,\text{J}$$

Thus the unit of work is really the unit of energy. This is consistent with the idea that work is a transfer of energy. Rather than use Nm, we will measure work, just as we do energy, in joules.

Using Equation 11.8 as the definition of work, we can write Equation 11.7 as

$$\Delta K = W \tag{11.9}$$

Equation 11.9 is the quantitative statement that a force transfers kinetic energy to a particle, and thus *changes* the particle's kinetic energy, by pushing or pulling on it. Furthermore, **Equation 11.8 gives us a specific method to calculate *how much* energy is transferred by the push or pull.** This energy transfer, by mechanical means, is what we mean by the term *work*.

Notice that *no* work is done if $s_f = s_i$ because an integral that spans no interval is zero. **To change a particle's energy, a force must be applied as the particle undergoes a displacement.** If you were to hold a 200 lb weight over your head, you might break out in a sweat and your arms would tire. You might "feel" that you had done a lot of work, but you would have done *zero* work in the physics sense because the weight was not displaced while you were holding it and thus you transferred no energy to it.

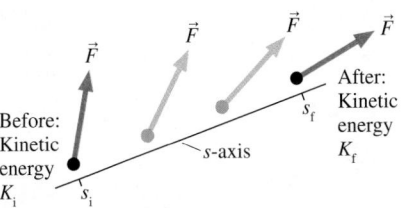

FIGURE 11.3 Force $\vec{F}$ does work as the particle moves from s_i to s_f.

This pitcher is increasing the ball's kinetic energy by doing work on it.

The Work-Kinetic Energy Theorem

Equation 11.8 is the work done by one force. Because $\vec{F}_{net} = \sum \vec{F}_i$, it's easy to see that the net work done on a particle by several forces is $W_{net} = \sum W_i$, where W_i is the work done by force $\vec{F}_i$. In that case, Equation 11.9 becomes

$$\Delta K = W_{net} \qquad (11.10)$$

This basic idea—that the net work done on a particle causes the particle's kinetic energy to change—is a general principle, one worth giving a name:

> **The work-kinetic energy theorem** When one or more forces act on a particle as it is displaced from an initial position to a final position, the net work done on the particle by these forces causes the particle's kinetic energy to *change* by $\Delta K = W_{net}$.

One of the questions that opened this chapter was "How does a system gain or lose energy?" The work-kinetic energy theorem begins to answer that question by saying that a system gains or loses kinetic energy when work (done by forces originating in the environment) transfers energy between the environment and the system.

An Analogy with the Impulse-Momentum Theorem

You might have noticed that there is a similarity between the work-kinetic energy theorem and the impulse-momentum theorem of Chapter 9:

Work-kinetic energy theorem: $\qquad \Delta K = W = \displaystyle\int_{s_i}^{s_f} F_s \, ds$

$$(11.11)$$

Impulse-momentum theorem: $\qquad \Delta p_s = J_s = \displaystyle\int_{t_i}^{t_f} F_s \, dt$

In both cases, a force acting on a particle changes the state of the system. If the force acts over a time interval from t_i to t_f, it creates an *impulse* that changes the particle's momentum. If the force acts over the spatial interval from s_i to s_f, it does *work* that changes the particle's kinetic energy. Figure 11.4 shows that the geometric interpretation of impulse as the area under the F-versus-t graph applies equally well to an interpretation of work as the area under the F-versus-s graph.

This does not mean that a force *either* creates an impulse *or* does work but does not do both. Quite the contrary. A force acting on a particle *both* creates an impulse *and* does work, changing both the momentum and the kinetic energy of the particle. Whether you use the work-kinetic energy theorem or the impulse-momentum theorem depends on the question you are trying to answer.

We can, in fact, express the kinetic energy in terms of the momentum as

$$K = \frac{1}{2}mv^2 = \frac{(mv)^2}{2m} = \frac{p^2}{2m} \qquad (11.12)$$

You cannot change a particle's kinetic energy without also changing its momentum.

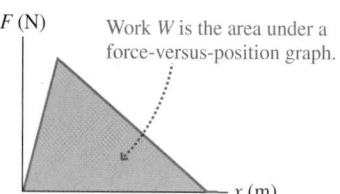

FIGURE 11.4 Impulse and work are both the area under a force graph, but it's very important to know what the horizontal axis is.

F (N) Impulse J_s is the area under a force-versus-time graph. t (s)

F (N) Work W is the area under a force-versus-position graph. x (m)

STOP TO THINK 11.2 A particle moving along the x-axis experiences the force shown in the graph. If the particle has 2.0 J of kinetic energy as it passes $x = 0$ m, what is its kinetic energy when it reaches $x = 4$ m?

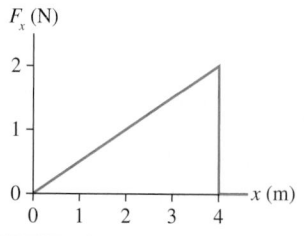

11.3 Calculating and Using Work

The work-kinetic energy theorem is a formal statement about the energy transferred to or from a particle by pushes and pulls. For this to be useful, we must be able to calculate the work. In this section we'll practice calculating work and using the work-kinetic energy theorem. We'll also introduce a new mathematical idea, the *dot product* of two vectors, that will allow us to write the work in a compact notation.

Constant Force

We'll begin by calculating the work done by a force $\vec{F}$ that acts with a *constant* strength and in a *constant* direction as a particle moves along a straight line through a displacement $\Delta\vec{r}$. As Figure 11.5a shows, we'll define the s-axis to point in the direction of motion. In that case, Δs is equal to Δr, the magnitude of the displacement vector $\Delta\vec{r}$.

Figure 11.5b shows the force acting on the particle as it moves along the line. The force vector $\vec{F}$ makes an angle θ with respect to the displacement $\Delta\vec{r}$, so the component of the force vector along the direction of motion is $F_s = F\cos\theta$. According to Equation 11.8, the work done on the particle by this force is

$$W = \int_{s_i}^{s_f} F_s\,ds = \int_{s_i}^{s_f} F\cos\theta\,ds$$

Both F and θ are constant, so they can be taken outside the integral. Thus

$$W = F\cos\theta \int_{s_i}^{s_f} ds = F\cos\theta(s_f - s_i) = F\cos\theta(\Delta s) = F(\Delta r)\cos\theta \quad (11.13)$$

We can use Equation 11.13 to calculate the work done by a constant force if we know the magnitude F of the force, the angle θ of the force from the line of motion, and the distance Δr through which the particle is displaced.

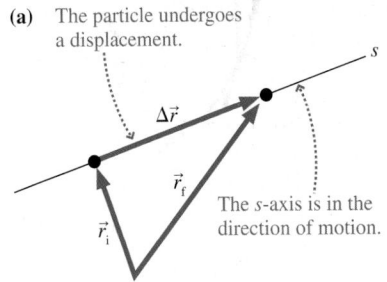

(a) The particle undergoes a displacement.

The s-axis is in the direction of motion.

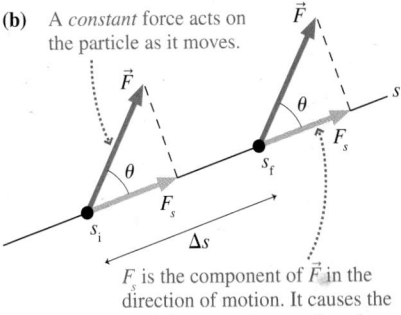

(b) A *constant* force acts on the particle as it moves.

F_s is the component of $\vec{F}$ in the direction of motion. It causes the particle to speed up or slow down.

FIGURE 11.5 Work being done by a *constant* force as a particle moves through displacement $\Delta\vec{r}$.

Activ
Physics   5.1

EXAMPLE 11.1 Pulling a suitcase
A rope inclined upward at a 45° angle pulls a suitcase through the airport. The tension in the rope is 20 N. How much work does the tension do if the suitcase is pulled 100 m?

MODEL Model the suitcase as a particle.

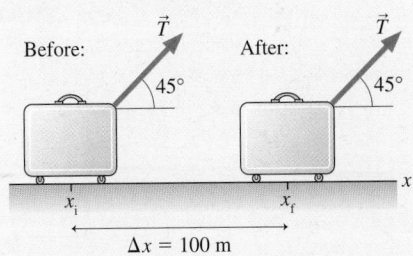

FIGURE 11.6 Pictorial representation of a suitcase pulled by a rope.

VISUALIZE Figure 11.6 shows a before-and-after pictorial representation.

SOLVE The motion is along the x-axis, so in this case $\Delta r = \Delta x$. We can use Equation 11.13 to find that the tension does work

$$W = T(\Delta x)\cos\theta = (20\text{ N})(100\text{ m})\cos 45° = 1410\text{ J}$$

ASSESS Because a person is pulling on the other end of the rope, we would say informally that the person does 1410 J of work on the suitcase.

According to the basic energy model, work can be either positive or negative to indicate energy transfer into or out of the system. The quantities F and Δr are always positive, so the sign of W is determined entirely by the angle θ between the force $\vec{F}$ and the displacement $\Delta\vec{r}$.

TACTICS BOX 11.1 **Calculating the work W of a constant force**

Force and displacement	θ	Work W	Sign	Energy transfer
	$0°$	$F(\Delta r)$	+	
				Energy is transferred to the system. The particle speeds up. K increases.
	$<90°$	$F(\Delta r)\cos\theta$	+	
	$90°$	0	0	No energy is transferred. Speed and K are constant.
	$>90°$	$F(\Delta r)\cos\theta$	−	
				Energy is transferred out of the system. The particle slows down. K decreases.
	$180°$	$-F(\Delta r)$	−	

NOTE ▶ You may have learned in an earlier physics course that work is "force times distance." This is *not* the definition of work, merely a special case. Work is "force times distance" only if the force is constant *and* parallel to the displacement (i.e., $\theta = 0°$). ◀

EXAMPLE 11.2 **Work and kinetic energy during a rocket launch**

A 150,000 kg rocket is launched straight up. The rocket motor generates a thrust of 4.0×10^6 N. What is the rocket's speed at a height of 500 m? Ignore air resistance and the slight mass loss due to burned fuel.

MODEL Model the rocket as a particle. Thrust and gravity are constant forces that do work on the rocket and change its energy.

VISUALIZE Figure 11.7 shows a pictorial representation and a free-body diagram.

SOLVE We can solve this problem with the work-kinetic energy theorem $\Delta K = W_{net}$. There are two forces, and both do work on the rocket. The thrust is in the direction of motion, with $\theta = 0°$, and thus

$$W_{thrust} = F_{thrust}(\Delta r) = (4.0 \times 10^6 \text{ N})(500 \text{ m}) = 2.00 \times 10^9 \text{ J}$$

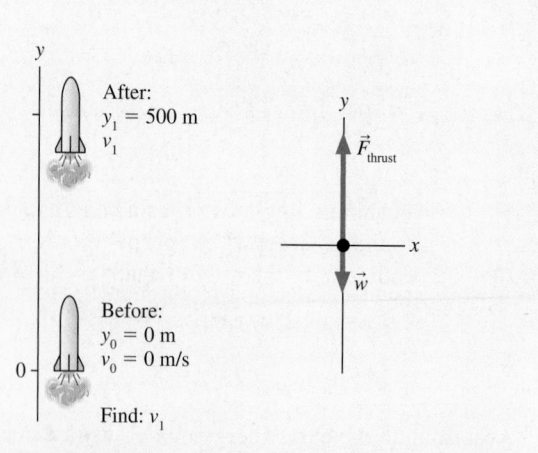

FIGURE 11.7 Pictorial representation and free-body diagram of a rocket launch.

The weight force points downward, opposite the displacement $\Delta\vec{r}$, so $\theta = 180°$. Thus the work done by gravity is

$$W_{grav} = -w(\Delta r) = -mg(\Delta r)$$
$$= -(1.5 \times 10^5 \text{ kg})(9.8 \text{ m/s}^2)(500 \text{ m}) = -0.74 \times 10^9 \text{ J}$$

The work done by the thrust is positive. By itself, the thrust would cause the rocket to speed up. The work done by gravity is negative. By itself, gravity would cause the rocket to slow down. The work-kinetic energy theorem, using $v_0 = 0$ m/s, is

$$\Delta K = \frac{1}{2}mv_1^2 - 0 = W_{net} = W_{thrust} + W_{grav} = 1.26 \times 10^9 \text{ J}$$

This is easily solved for the speed:

$$v_1 = \sqrt{\frac{2W_{net}}{m}} = 130 \text{ m/s}$$

ASSESS The net work is positive, meaning that energy is transferred *to* the rocket. In response, the rocket speeds up.

NOTE ▶ The work done by a force depends on the angle θ between the force $\vec{F}$ and the displacement $\Delta\vec{r}$, *not* on the direction the particle is moving. The work done on all four particles in Figure 11.8 is the same, despite the fact that they are moving in four different directions. ◀

Force Perpendicular to the Direction of Motion

Figure 11.9a shows a bird's-eye view of a car turning a corner. As you learned in Chapter 7, a friction force points toward the center of the circle. How much work does friction do on the car?

Zero! In Figure 11.9b we've "bent" the s-axis to follow the curve. You can see that the friction force is everywhere perpendicular to the small displacement ds. F_s, the component of the force parallel to the displacement, is everywhere zero. Thus static friction does *no* work on the car. This shouldn't be surprising. You know that the car's speed, and hence its kinetic energy, doesn't change as it rounds the curve. Thus, according to the work-kinetic energy theorem, $W = \Delta K = 0$.

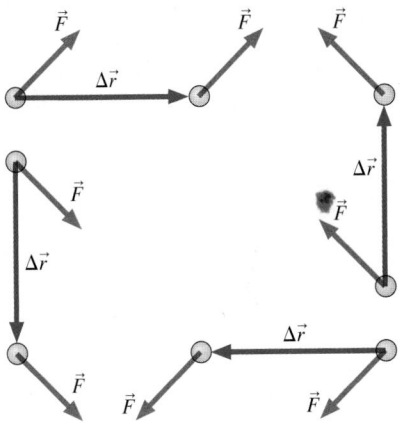

FIGURE 11.8 The same amount of work is done on each of these particles.

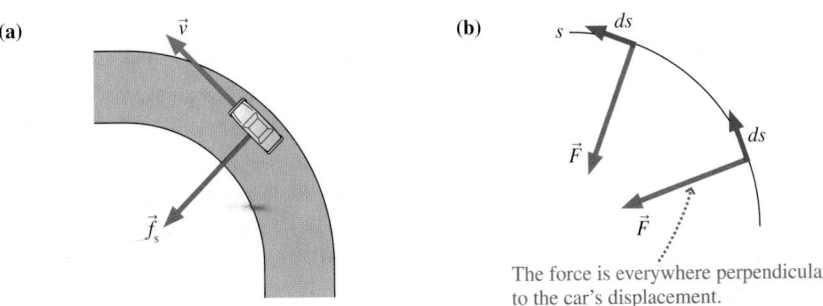

FIGURE 11.9 The force is everywhere perpendicular to the car's displacement. It does no work.

We used a car rounding a curve as a concrete example, but this is a general result: **A force that is everywhere perpendicular to the motion does no work.** A force perpendicular to the motion changes the *direction* of motion but not the particle's speed.

EXAMPLE 11.3 Pushing a puck
A 500 g ice hockey puck slides across frictionless ice with an initial speed of 2.0 m/s. A compressed-air gun can be used to exert a 1.0 N force on the puck. The air gun is aimed at the front edge of the puck with the compressed-air flow 30° below the horizontal. This force is applied continuously as the puck moves 50 cm. What is the puck's final speed?

MODEL Model the puck as a particle. Use the work-kinetic energy theorem to find its final speed.

VISUALIZE Figure 11.10 shows a pictorial representation. The angle between the compressed-air force $\vec{F}$ and the displacement $\Delta\vec{r}$ is $\theta = 150°$.

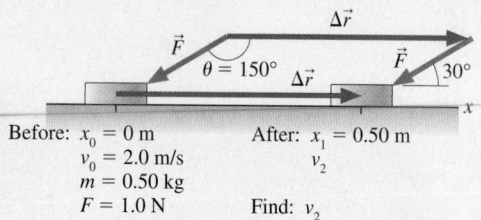

Before: $x_0 = 0$ m
$v_0 = 2.0$ m/s
$m = 0.50$ kg
$F = 1.0$ N

After: $x_1 = 0.50$ m
v_2

Find: v_2

FIGURE 11.10 Pictorial representation of the puck.

SOLVE Three forces act on the puck: its weight $\vec{w}$, the normal force $\vec{n}$, and the force $\vec{F}$ of the compressed air. The weight and the normal force are perpendicular to the direction of motion ($\theta = 90°$), hence they do no work on the puck. The only work W is done by the compressed-air force $\vec{F}$. The work is

$$W = F(\Delta r)\cos\theta = (1.0\text{ N})(0.50\text{ m})\cos 150° = -0.433\text{ J}$$

Now we can use the work-kinetic energy theorem to compute the final speed:

$$\Delta K = \frac{1}{2}mv_1^2 - \frac{1}{2}mv_0^2 = W$$

$$v_1 = \sqrt{v_0^2 + \frac{2W}{m}} = \sqrt{(2.0\text{ m/s})^2 + \frac{2(-0.433\text{ J})}{0.50\text{ kg}}}$$

$$= 1.51\text{ m/s}$$

ASSESS The work is negative, meaning that energy is transferred *from* the puck. In response, the puck slows down.

STOP TO THINK 11.3 A crane lowers a steel girder into place at a construction site. The girder moves with constant speed. Consider the work W_g done by gravity and the work W_T done by the tension in the cable. Which of the following is correct?

a. W_g is positive and W_T is positive.
b. W_g is positive and W_T is negative.
c. W_g is negative and W_T is positive.
d. W_g is negative and W_T is negative.
e. W_g and W_T are both zero.

The Dot Product of Two Vectors

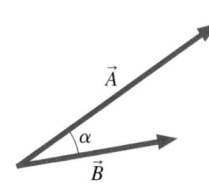

FIGURE 11.11 Vectors $\vec{A}$ and $\vec{B}$, with angle α between them.

There's something different about the quantity $F(\Delta r)\cos\theta$ in Equation 11.13. We've spent many chapters adding vectors, but this is the first time that we've *multiplied* two vectors. Multiplying vectors is not like multiplying scalars. In fact, there is more than one way to multiply vectors. We will introduce one way now, the *dot product*.

Figure 11.11 shows two vectors, $\vec{A}$ and $\vec{B}$, with angle α between them. We define the **dot product** of $\vec{A}$ and $\vec{B}$ as

$$\vec{A} \cdot \vec{B} = AB\cos\alpha \tag{11.14}$$

A dot product *must have* the dot symbol · between the vectors. The notation $\vec{A}\vec{B}$, without the dot, is *not* the same thing as $\vec{A} \cdot \vec{B}$. The dot product is also called the **scalar product** because the value is a scalar. Later, when we need it, we'll introduce a different way to multiply vectors called the *cross product* or the *vector product*.

The dot product of two vectors depends on the orientation of the vectors. Figure 11.12 shows five different situations, including the three "special cases" where $\alpha = 0°, 90°$, and $180°$.

NOTE ▶ The dot product of a vector with itself is well defined. If $\vec{B} = \vec{A}$ (i.e., $\vec{B}$ is a copy of $\vec{A}$), then $\alpha = 0°$. Thus $\vec{A} \cdot \vec{A} = A^2$. ◀

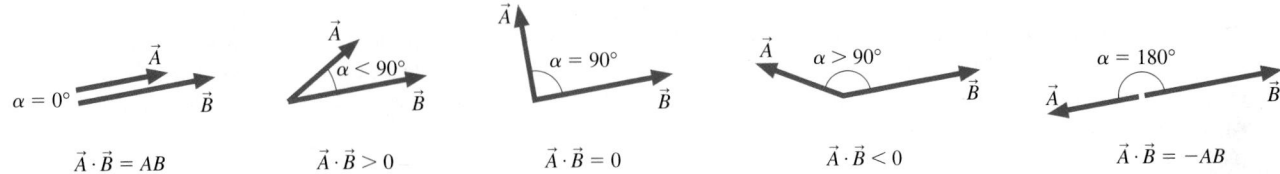

FIGURE 11.12 The dot product $\vec{A} \cdot \vec{B}$ as α ranges from $0°$ to $180°$.

EXAMPLE 11.4 Calculating a dot product
Compute the dot product of the two vectors shown in Figure 11.13.

SOLVE The angle *between* the vectors is $\alpha = 30°$, so

$$\vec{A} \cdot \vec{B} = AB\cos\alpha = (3)(4)\cos 30° = 10.4$$

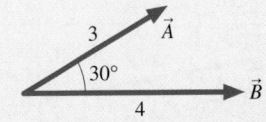

FIGURE 11.13 Vectors $\vec{A}$ and $\vec{B}$ of Example 11.4.

Like vector addition and subtraction, calculating the dot product of two vectors is often performed most easily using vector components. Figure 11.14 reminds you of the unit vectors $\hat{\imath}$ and $\hat{\jmath}$ that point in the positive *x*-direction and positive *y*-direction. The two unit vectors are perpendicular to each other, so their dot product is $\hat{\imath} \cdot \hat{\jmath} = 0$. Furthermore, because the magnitudes of $\hat{\imath}$ and $\hat{\jmath}$ are 1, $\hat{\imath} \cdot \hat{\imath} = 1$ and $\hat{\jmath} \cdot \hat{\jmath} = 1$.

In terms of components, we can write the dot product of vectors $\vec{A}$ and $\vec{B}$ as

$$\vec{A} \cdot \vec{B} = (A_x\hat{\imath} + A_y\hat{\jmath}) \cdot (B_x\hat{\imath} + B_y\hat{\jmath})$$

Multiplying this out, and using the results for the dot products of the unit vectors:

$$\vec{A} \cdot \vec{B} = A_xB_x\hat{\imath} \cdot \hat{\imath} + (A_xB_y + A_yB_x)\hat{\imath} \cdot \hat{\jmath} + A_yB_y\hat{\jmath} \cdot \hat{\jmath}$$
$$= A_xB_x + A_yB_y$$

(11.15)

That is, the dot product can be calculated as the sum of the products of the components.

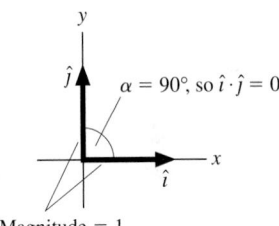

FIGURE 11.14 The unit vectors $\hat{\imath}$ and $\hat{\jmath}$.

EXAMPLE 11.5 Calculating a dot product using components
Compute the dot product of $\vec{A} = 3\hat{\imath} + 3\hat{\jmath}$ and $\vec{B} = 4\hat{\imath} - \hat{\jmath}$.

SOLVE Figure 11.15 shows vectors $\vec{A}$ and $\vec{B}$. We could calculate the dot product by first doing the geometry needed to find the angle between the vectors, then using Equation 11.14. But calculating the dot product from the vector components is much easier. It is

$$\vec{A} \cdot \vec{B} = A_xB_x + A_yB_y = (3)(4) + (3)(-1) = 9$$

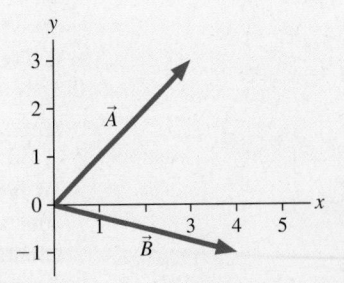

FIGURE 11.15 Vectors $\vec{A}$ and $\vec{B}$.

Looking at Equation 11.13, the work done by a constant force, you should recognize that it is the dot product of the force vector and the displacement vector:

$$W = \vec{F} \cdot \Delta\vec{r} \quad \text{(work done by a constant force)}$$

(11.16)

This definition of work is valid for a constant force.

EXAMPLE 11.6 **Calculating work using the dot product**

A 70 kg skier is gliding at 2.0 m/s when he starts down a very slippery 50-m-long, 10° slope. What is his speed at the bottom?

MODEL Model the skier as a particle and interpret "very slippery" to mean frictionless. Use the work-kinetic energy theorem to find his final speed.

VISUALIZE Figure 11.16 shows a pictorial representation.

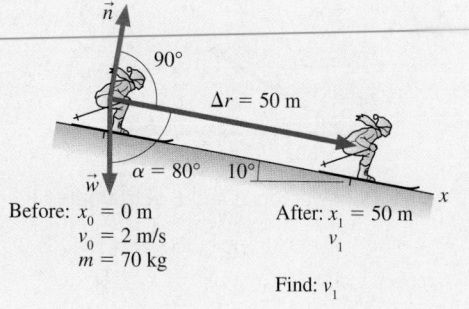

Before: $x_0 = 0$ m
$v_0 = 2$ m/s
$m = 70$ kg

After: $x_1 = 50$ m
v_1

Find: v_1

FIGURE 11.16 Pictorial representation of the skier.

SOLVE The only forces on the skier are $\vec{w}$ and $\vec{n}$. The normal force is perpendicular to the motion and thus does no work. The work done by gravity is easily calculated as a dot product:

$$W = \vec{w} \cdot \Delta\vec{r} = w(\Delta r)\cos\alpha$$
$$= (70 \text{ kg})(9.8 \text{ m/s}^2)(50 \text{ m})\cos 80° = 5960 \text{ J}$$

Notice that the angle *between* the vectors is 80°, not 10°. Then, from the work-kinetic energy theorem, we find

$$\Delta K = \frac{1}{2}mv_1^2 - \frac{1}{2}mv_0^2 = W$$

$$v_1 = \sqrt{v_0^2 + \frac{2W}{m}} = \sqrt{(2.0 \text{ m/s})^2 + \frac{2(5960 \text{ J})}{70 \text{ kg}}} = 13.2 \text{ m/s}$$

NOTE ▶ While in the midst of the mathematics of calculating work, do not lose sight of what the work-kinetic energy theorem is all about. It is a statement about *energy transfer*, saying that work is the energy transferred to or from the system due to forces exerted on the system. Work causes the system's kinetic energy to either increase or decrease. ◀

STOP TO THINK 11.4 Which force does the most work?

a. The 10 N force.
b. The 8 N force.
c. The 6 N force.
d. They all do the same amount of work.

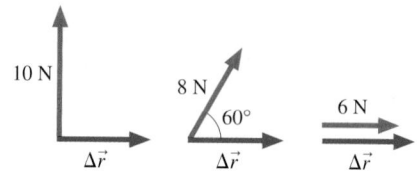

11.4 The Work Done by a Variable Force

We've learned how to calculate the work done on a particle by a force that has a constant magnitude and direction throughout the particle's displacement. But what about a force that changes in either magnitude or direction as the particle moves? Equation 11.8, the definition of work, is all we need:

$$W = \int_{s_i}^{s_f} F_s \, ds = \text{area under the force-versus-position graph} \quad (11.17)$$

The integral sums up the small amounts of work $F_s \, ds$ done in each step along the trajectory. The only new feature, since F_s varies with position, is that we cannot take F_s outside the integral. We must evaluate the integral either geometrically, by finding the area under the curve, or by actually doing the integration. We'll restrict our applications to motion along a straight line in order to avoid unnecessarily complex mathematics.

EXAMPLE 11.7 Using work to find the speed of a car

A 1500 kg car accelerates from rest. Figure 11.17 shows the net force on the car (propulsion force minus any drag forces) as a function of the car's position. What is the car's speed after traveling 200 m?

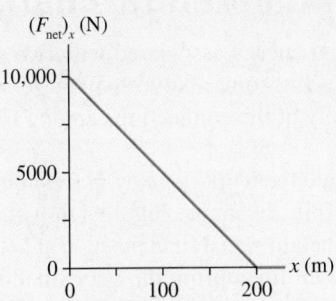

FIGURE 11.17 Force-versus-position graph for a car.

SOLVE The acceleration $a_x = (F_{net})_x/m$ is high as the car starts but decreases as the car picks up speed because of increasing drag. Figure 11.17 is a more realistic portrayal of the net force on a car than was our earlier model of a constant force. But a variable force means that we cannot use the familiar constant-acceleration kinematics. Instead, we can use the work-kinetic energy theorem. Because $v_0 = 0$ m/s, we have

$$\Delta K = \frac{1}{2}mv_1^2 - 0 = W_{net}$$

Starting from $x_0 = 0$ m, the work is

$$W_{net} = \int_{0\,m}^{x_1} (F_{net})_x \, dx$$

$$= \text{area under the } (F_{net})_x\text{-versus-}x \text{ graph from 0 m to } x_1$$

The area under the curve of Figure 11.17 is that of a triangle of width 200 m. Thus

$$W_{net} = \text{area} = \frac{1}{2}(10{,}000 \text{ N})(200 \text{ m}) = 1{,}000{,}000 \text{ J}$$

The work-kinetic energy theorem then gives

$$v_1 = \sqrt{\frac{2W_{net}}{m}} = \sqrt{\frac{2(1{,}000{,}000 \text{ J})}{1500 \text{ kg}}} = 36.5 \text{ m/s}$$

ASSESS Because 1 J = 1 kg m²/s², the quantity W/m has units m²/s². Thus the units of v_1 are m/s, as expected.

EXAMPLE 11.8 Using the work-kinetic energy theorem for a spring

The "pincube machine" was an ill-fated predecessor of the pinball machine. A 100 g cube is launched by pulling a spring back 20 cm and releasing it. What is the cube's launch speed, as it leaves the spring, if the spring constant is 20 N/m and the coefficient of kinetic friction is 0.10?

MODEL Model the cube as a particle and the spring as an ideal spring obeying Hooke's law. Use the work-kinetic energy theorem to find the launch speed.

VISUALIZE Figure 11.18 shows a before-and-after pictorial representation and a free-body diagram. We've placed the origin of the x-axis at the equilibrium position of the spring.

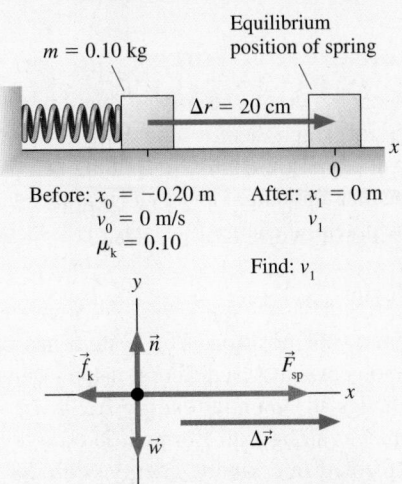

FIGURE 11.18 Pictorial representation of a cube being launched in the pincube machine.

SOLVE The normal force and weight are perpendicular to the motion and do no work. We can use the work-kinetic energy theorem, with $v_0 = 0$ m/s, to find the launch speed:

$$\Delta K = \frac{1}{2}mv_1^2 - 0 = W_{net} = W_{fric} + W_{sp}$$

Friction is a constant force $f_k = \mu_k mg$ in the direction *opposite* the motion, with $\theta = 180°$, so the work done by friction is

$$W_{fric} = \vec{f}_k \cdot \Delta\vec{r} = f_k(\Delta r)\cos 180° = -\mu_k mg\Delta x = -0.020 \text{ J}$$

The negative work of friction would, by itself, slow the block down.

The spring force is a variable force: $(F_{sp})_x = -k\Delta x = -kx$, where $\Delta x = x - x_e = x$ because we chose a coordinate system with $x_e = 0$ m. Despite the minus sign, $(F_{sp})_x$ is a positive quantity (force pointing to the right) because x is negative throughout the motion. The spring force points in the direction of motion, so W_{sp} is positive. We can use Equation 11.17 to evaluate W_{sp}:

$$W_{sp} = \int_{x_0}^{x_1} (F_{sp})_x \, dx = -k\int_{x_0}^{x_1} x \, dx = -\frac{1}{2}kx^2 \bigg|_{x_0}^{x_1}$$

$$= -\left(\frac{1}{2}kx_1^2 - \frac{1}{2}kx_0^2\right)$$

Evaluating W_{sp} for $x_0 = -0.20$ m and $x_1 = 0$ m gives

$$W_{sp} = \frac{1}{2}(20 \text{ N/m})(-0.20 \text{ m})^2 = 0.400 \text{ J}$$

The net work is $W_{net} = W_{fric} + W_{sp} = 0.380$ J, with which we can now find

$$v_1 = \sqrt{\frac{2W_{net}}{m}} = \sqrt{\frac{2(0.380 \text{ J})}{0.100 \text{ kg}}} = 2.76 \text{ m/s}$$

You might have noticed that the work done by the spring looks a lot like the spring's potential energy $U_{sp} = \frac{1}{2}k\Delta x^2$. The next section will find a connection between work and potential energy.

11.5 Force, Work, and Potential Energy

In Chapter 10 we found a potential energy associated with two specific forces, the gravitational force and the spring force. Now that we've related force to work, it's time to look more closely at the connections among force, work, and potential energy.

As a starting point, let's calculate the work done by gravity on an object sliding along a frictionless path of arbitrary shape. Figure 11.19 shows the object moving from an initial position at height y_i to a final position at height y_f. The displacement ds is essentially a straight line during the very small segment of the motion shown in the inset. The small amount of work dW_{grav} done by gravity as the object moves through ds is

$$dW_{grav} = \vec{w} \cdot \Delta\vec{r} = mg\cos\theta \, ds \tag{11.18}$$

where θ is the angle between $\vec{w}$ and the displacement vector $\Delta\vec{r}$.

It's easy to see that $\cos\theta \, ds$ is the vertical displacement, but we need to be careful with signs. The small displacement ds is positive because we earlier chose the s-axis to be positive in the direction of motion. But the y-axis points upward, so dy in Figure 11.19 is negative. In particular, $dy = -\cos\theta \, ds$. Thus the work done by gravity is

$$dW_{grav} = -mg \, dy \tag{11.19}$$

The main feature of Equation 11.19 is that dW_{grav} is independent of θ. It depends on the vertical displacement dy but *not* on the slope of the surface. The total work done by gravity is found by adding the work done in each segment of the motion (i.e., by integrating Equation 11.19). The result is

$$W_{grav} = -mg(y_f - y_i) = -mg\Delta y \tag{11.20}$$

Notice that **the work done by gravity is independent of the path followed by the object.** It depends on the initial and final heights, but not at all on how the object gets from height y_i to height y_f. For example, Figure 11.20 shows a particle moving along four different paths that have the same vertical displacement Δy. Despite the very different trajectories, the work done by gravity is the same in all four cases.

Conservative and Nonconservative Forces

The path independence of the work is perhaps surprising, but it turns out to be an essential ingredient of any force for which there is a potential energy. To see this, Figure 11.21 shows an object that can move from A to B along two possible paths while force $\vec{F}$ acts on it. Assume that a potential energy U is associated with force $\vec{F}$ in much the same way that the potential energy $U_g = mgy$ is associated with the gravitational force $\vec{F}_g = -mg\hat{\jmath}$.

There are three steps in the logic:

1. Potential energy is an energy of position. The system has one value of potential energy when the object is at A, a different value when the object is at B. Thus the overall change in potential energy $\Delta U = U_B - U_A$ is the same whether the object moves along path 1 or path 2.
2. Potential energy is transformed into kinetic energy, with $\Delta K = -\Delta U$. If ΔU is independent of the path, then ΔK is also independent of the path. The transformation of energy causes the object to have the same kinetic energy at B no matter which path it follows.

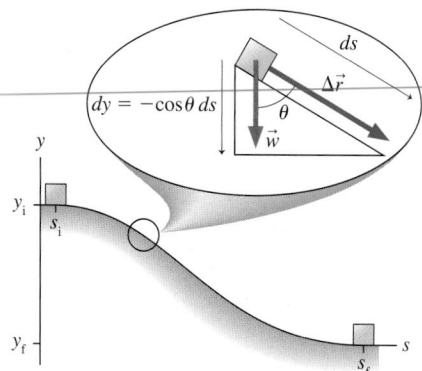

FIGURE 11.19 An object moves along an arbitrarily shaped path.

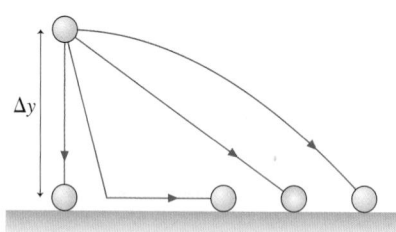

FIGURE 11.20 The work done by gravity is the same in all four cases.

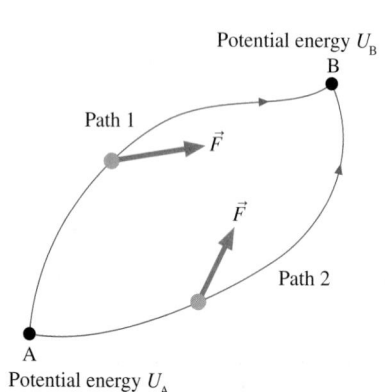

FIGURE 11.21 An object can move from A to B along either path 1 or path 2.

3. The change in a particle's kinetic energy is related to the work done on the particle by force $\vec{F}$. According to the work-kinetic energy theorem, $\Delta K = W$. Because ΔK is independent of the path, it *must* be the case that the work done by force $\vec{F}$ as the object moves from A to B is independent of the path followed.

A force for which the work done on a particle as it moves from an initial to a final position is independent of the path followed is called a **conservative force.** (The name, as you'll soon see, is related to the conditions under which mechanical energy is conserved.) The importance of conservative forces is that **a potential energy can be associated with any conservative force.** For example, our analysis of Figure 11.19 showed that gravity is a conservative force. Consequently, we can establish a gravitational potential energy.

To establish a general connection between work and potential energy, suppose a particle moves from initial position i to final position f under the influence of a conservative force $\vec{F}$. We'll denote the work done by the force as $W_c(\text{i} \rightarrow \text{f})$, where the notation i $\rightarrow$ f means "as the particle moves from position i to position f." Because $\Delta K = W$ and $\Delta K = -\Delta U$, the potential energy difference between these two points must be

$$\Delta U = U_f - U_i = -W_c(\text{i} \rightarrow \text{f}) \qquad (11.21)$$

Equation 11.21 is a general definition of the potential energy associated with a conservative force.

For example, in Equation 11.20 we found that the work due to gravity is independent of the path and is $W_{grav}(\text{i} \rightarrow \text{f}) = -mg(y_f - y_i)$. The gravitational potential energy is defined by

$$U_f - U_i = -W_{grav}(\text{i} \rightarrow \text{f}) = mgy_f - mgy_i$$

and thus $U_g = mgy$. This agrees with the gravitational potential energy of Chapter 10. That's not unexpected. The analysis of Chapter 10 that led to $U_g = mgy$ was really a calculation of work, although we didn't call it that at the time. What we've now done is to generalize that analysis to apply to *any* conservative force.

NOTE ▶ Strictly speaking, Equation 11.21 defines only the *change* in potential energy ΔU. We can add a constant to both U_f and U_i without changing ΔU. This was the basis for our discussion in Chapter 10 about the zero of potential energy. ◀

What about springs? A homework problem will let you show that Hooke's law is also a conservative force. In Example 11.8 we showed that the work done by a spring is

$$W_{sp}(\text{i} \rightarrow \text{f}) = \int_{x_i}^{x_f} F_{sp}\, dx = -\left(\frac{1}{2}kx_f^2 - \frac{1}{2}kx_i^2 \right)$$

from which it follows that $U_s = \frac{1}{2}kx^2$. Example 11.8 was a "special case" in that we defined the coordinate system to make $x_e = 0$. A more general analysis would give $U_s = \frac{1}{2}k(\Delta s)^2$, as you learned in Chapter 10.

Not all forces are conservative forces. For example, Figure 11.22 is a bird's-eye view of a box sliding across a floor from P to R. The friction force, with magnitude $\mu_k n = \mu_k mg$, points opposite the direction of motion and does *negative* work on the box. Specifically, the work done by the friction force as the box moves a distance Δs is

$$W_{fric} = f_k(\Delta r)\cos 180° = -\mu_k mg\Delta s \qquad (11.22)$$

W_{fric} depends on the distance traveled. If the box moves directly from P to R, $(\Delta s)_{PR} = \sqrt{2}L$. If it moves around the perimeter P $\rightarrow$ Q $\rightarrow$ R, the distance is

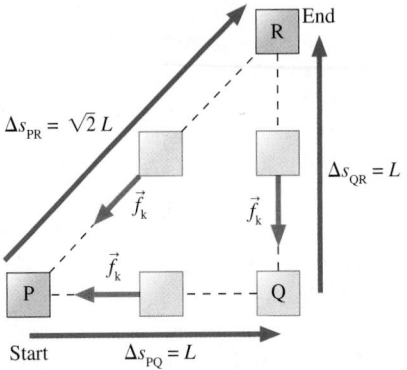

The work done by friction *does* depend on the path followed.

FIGURE 11.22 Top view of a box sliding from P to R.

$(\Delta s)_{PQR} = 2L$. More work (negative) is done on the longer path around the outside of the square. The work done by friction is *not* independent of the path followed.

A force for which the work is *not* independent of the path is called a **nonconservative force.** It is not possible to define a potential energy for a nonconservative force. Friction is a nonconservative force, so we cannot define a potential energy of friction.

This makes sense. If you toss a ball straight up, kinetic energy is transformed into gravitational potential energy. The ball has the potential to transform this energy back into kinetic energy, and it does so as the ball falls. But you cannot recover the kinetic energy lost to friction as a box slides to a halt. There's no "potential" that can be transformed back into kinetic energy.

Mechanical Energy

Consider a system of particles that interact via both conservative forces and nonconservative forces. The conservative forces do work W_c as the particles move from initial positions i to final positions f. The nonconservative forces do work W_{nc}. The total work done by *all* forces is $W_{net} = W_c + W_{nc}$. The change in the system's kinetic energy ΔK, as determined by the work-kinetic energy theorem, is

$$\Delta K = W_{net} = W_c(i \rightarrow f) + W_{nc}(i \rightarrow f) \tag{11.23}$$

The work done by the conservative forces can now be associated with a potential energy U. According to Equation 11.21, $W_c(i \rightarrow f) = -\Delta U$. With this definition, Equation 11.23 becomes

$$\Delta K + \Delta U = \Delta E_{mech} = W_{nc} \tag{11.24}$$

where, as in Chapter 10, the *mechanical energy* is $E_{mech} = K + U$.

Now we can see that **mechanical energy is conserved if there are no nonconservative forces.** That is

$$\Delta E_{mech} = 0 \text{ if } W_{nc} = 0 \tag{11.25}$$

This important conclusion is what we called the law of conservation of mechanical energy in Chapter 10. There we saw that friction prevents E_{mech} from being conserved, but we really didn't know why. Equation 11.24 tells us that any nonconservative force causes the mechanical energy to change. Friction and other "dissipative forces" lead to a loss of mechanical energy. Other outside forces, such as the pull of a rope, might increase the mechanical energy.

Equally important, Equation 11.24 tells us what to do if the mechanical energy isn't conserved. You can still use energy concepts to analyze the motion if you compute the work done by the nonconservative forces.

EXAMPLE 11.9 **Using work and potential energy together**

Use potential energy to find the launch speed of the "pincube machine" of Example 11.8. Recall that the 100 g cube is launched by pulling a spring back 20 cm and releasing it. The spring constant is 20 N/m and the coefficient of kinetic friction is 0.10.

MODEL This time let the system be the cube and the spring.

VISUALIZE Figure 11.18 shows the pictorial representation and free-body diagram..

SOLVE In solving this problem with the work-kinetic energy theorem, we had to explicitly calculate the work done by the

spring. Now let's use Equation 11.24. The spring force is a conservative force that we can associate with the elastic potential energy U_s. The friction force is nonconservative. Thus

$$\Delta K + \Delta U_s = W_{nc} = W_{fric}$$

The spring's potential energy with $x_e = 0$ m is $U_s = \frac{1}{2}kx^2$. In Equation 11.22 we found that the work done by friction on an object sliding across a horizontal surface is $W_{fric} = -\mu_k mg \Delta s$. Thus the energy equation becomes

$$\frac{1}{2}mv_1^2 - \frac{1}{2}mv_0^2 + \frac{1}{2}kx_1^2 - \frac{1}{2}kx_0^2 = -\mu_k mg(x_1 - x_0)$$

Using $v_0 = 0$ m/s, $x_1 = 0$ m, and $x_0 = -0.20$ m, we find

$$v_1 = \sqrt{\frac{kx_0^2}{m} + 2\mu_k g x_0} = 2.76 \text{ m/s}$$

ASSESS What appeared to be a difficult problem, with both friction and a spring, turned out to be straightforward when analyzed with energy and work.

Example 11.9 illustrates an important idea. When we associate a potential energy with a conservative force we

- Enlarge the system to include all objects that interact via conservative forces.
- "Precompute" the work. We can do this because we don't need to know what paths the objects are going to follow. This precomputed work becomes a potential energy and moves from the right side of $\Delta K = W$ to the left side of Equation 11.24.

In Example 11.8, the system consisted of just the cube. We treated the spring force as a force from the environment doing work on the system. In Example 11.9, where we revisited the same problem, we brought the spring into the system and represented the conservative spring-cube interaction with a potential energy.

NOTE ▶ When you use a potential energy, you've already taken the work of that force into account. Don't compute the work explicitly, or you'll be double counting it! ◀

To analyze a problem using work and energy, you can either

1. Use the work-kinetic energy theorem $\Delta K = W$ and explicitly compute the work done by *every* force. This was the method of Example 11.8. Or
2. Represent the work done by conservative forces as potential energies, then use $\Delta K + \Delta U = W_{nc}$. The only work that must be computed is the work of any nonconservative forces. This was the method of Example 11.9.

It's important to recognize that **these two methods yield the same result!** It's simply that method 2 "precomputes" some of the work and represents it as a potential energy. In practice, **method 2 is always easier and is the preferred method.**

11.6 Finding Force from Potential Energy

We know how to find the potential energy due to a conservative force. Now we need to learn how to go in reverse. That is, if we know a particle's potential energy, how do we find the force acting on it?

Figure 11.23a shows a particle moving through a *small* displacement Δs while being acted on by a conservative force $\vec{F}$. If Δs is sufficiently small, the force component F_s in the direction of motion is essentially constant during the displacement. The work done on the particle as it moves from s to $s + \Delta s$ is

$$W(s \rightarrow s + \Delta s) = F_s \Delta s \qquad (11.26)$$

This work is shown in Figure 11.23b as the area under the force curve in the narrow rectangle of width Δs.

Because $\vec{F}$ is a conservative force, the change in the particle's potential energy over this interval is defined in Equation 11.21 as

$$\Delta U = -W(s \rightarrow s + \Delta s) = -F_s \Delta s$$

which we can rewrite as

$$F_s = -\frac{\Delta U}{\Delta s} \qquad (11.27)$$

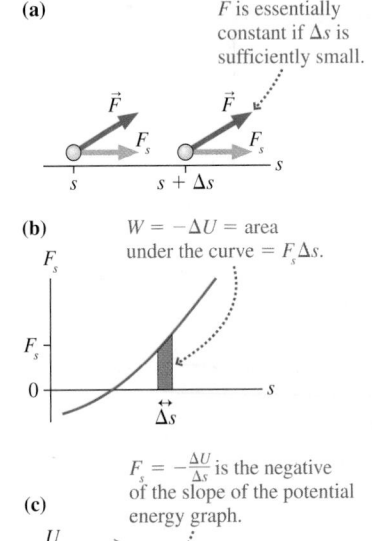

FIGURE 11.23 Relating force and potential energy.

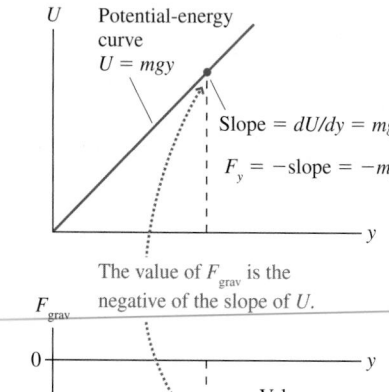

FIGURE 11.24 Gravitational potential energy and force diagrams.

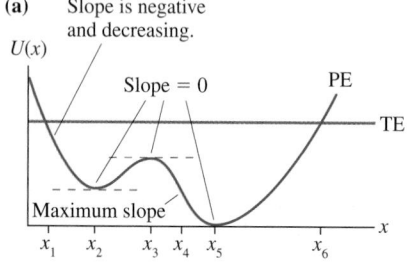

(b)

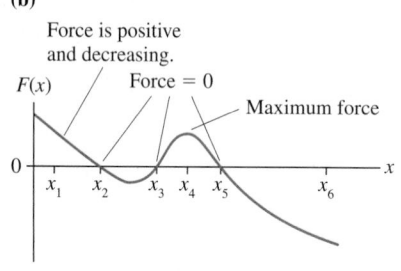

FIGURE 11.25 A potential-energy diagram and the corresponding force diagram.

In the limit $\Delta s \rightarrow 0$, we find that the force at position s is

$$F_s = \lim_{\Delta s \rightarrow 0}\left(-\frac{\Delta U}{\Delta s}\right) = -\frac{dU}{ds} \qquad (11.28)$$

We see that the force on the particle is the *negative* of the derivative of the potential energy with respect to position. Figure 11.23b shows that we can interpret this result graphically by saying

$$F_s = \text{the negative of the slope of the } U\text{-versus-}s \text{ graph at } s. \qquad (11.29)$$

In practice, of course, we will usually use either $F_x = -dU/dx$ or $F_y = -dU/dy$.

As an example, consider the gravitational potential energy $U_g = mgy$. Figure 11.24a shows the potential energy diagram U_g-versus-y. It is simply a straight-line graph passing through the origin. The force on the particle at position y, according to Equations 11.28 and 11.29, is simply

$$F_{\text{grav}} = -\frac{dU_g}{dy} = -(\text{slope of } U_g) = -mg$$

The negative sign, as always, indicates that the force points in the negative y-direction. Figure 11.24b shows the corresponding F-versus-y graph. At each point, the *value* of F is equal to the negative of the *slope* of the U-versus-y graph. This is similar to position and velocity graphs, where the value of v_x at any time t is equal to the slope of the x-versus-t graph.

We already knew that $F_{\text{grav}} = -mg$, of course, so the point of this particular example was to illustrate the meaning of Equation 11.29 rather than to find out anything new. Had we *not* known the force of gravity, we see is that it is possible to find it from the potential energy diagram.

Figure 11.25 is a more interesting example. You learned in Chapter 10 that a particle with the total energy TE will oscillate back and forth between turning points at x_1 and x_6 where the total energy line crosses the potential-energy curve. Now we can be more specific. The slope of the potential-energy graph is negative between x_1 and x_2. This means that the force on the particle, which is the negative of the slope of U, is *positive*. A particle between x_1 and x_2 experiences a force toward the right. The force decreases as the slope decreases until, at x_2, $F_x = 0$. This is consistent with our prior identification of x_2 as a point of stable *equilibrium*. The slope is positive (force negative and thus to the left) between x_2 and x_3, zero (zero force) at the unstable equilibrium point x_3, and so on. Point x_4, where the slope is most negative, is the point between x_1 and x_6 of maximum force.

Figure 11.25b is a plausible graph of F-versus-x. We don't know the exact shape, because we don't have an exact expression for U, but the force graph must look very much like this.

STOP TO THINK 11.5 A particle moves along the x-axis with the potential energy shown. The x-component of the force on the particle when it is at $x = 4$ m is

a. 4 N.
b. 2 N.
c. 1 N.
d. −4 N.
e. −2 N.
f. −1 N.

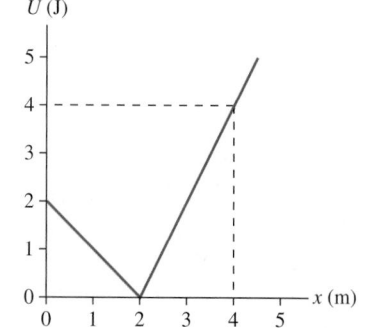

11.7 Thermal Energy

All of the objects we handle and use every day are extraordinarily large compared to the size of an atom. Each object is really a system consisting of vast numbers of particle-like atoms. We want to distinguish between the motion of the object as a whole and the motion of the atoms inside it. We will use the terms **macrophysics** to refer to the motion and dynamics of the object as a whole and **microphysics** to refer to the motion of atoms. You recognize the prefix *micro,* meaning "small." You may not be familiar with *macro,* which means "large" or "large-scale."

The connection between microscopic and macroscopic behavior is a topic we will explore in depth when we reach thermodynamics, in order to understand the bulk properties of materials, and again in electricity, to explain the electrical properties of conductors and insulators. For now, let's take a closer look at the microscopic energy in a system of atoms.

Kinetic and Potential Energy at the Microscopic Level

Figure 11.26 shows two different perspectives of an object. In the macrophysics perspective of Figure 11.26a you see an object of mass M moving as a whole with velocity v_{obj}. As a consequence of its motion, the object has macroscopic kinetic energy $K_{macro} = \frac{1}{2}Mv_{obj}^2$.

Figure 11.26b is a microphysics view of the same object, where now we see a *system of particles.* Each of these atoms is moving about, and in doing so they stretch and compress the spring-like molecular bonds between them. Consequently, there is a *microscopic potential energy* U_{micro} associated with molecular bonds. The energy stored in any one bond is very small, but there are incredibly many bonds.

Each moving atom has a kinetic energy $K_j = \frac{1}{2}mv_j^2$, where v_j is the speed of atom j. In this section we will use lowercase m to represent the mass of atoms within the system (which, for simplicity, we'll assume to all be the same) and uppercase M for the mass of the entire system. The kinetic energy of one atom is exceedingly small, but there are enormous numbers of atoms in a macroscopic object. The combined kinetic energy of all the atoms is what we call the *microscopic kinetic energy* K_{micro}. These microscopic energies associated with molecular bonds and moving atoms are quite distinct from the energies K_{macro} and U_{macro}.

Is the microscopic energy worth worrying about? To see, consider a 500 g ($\approx$1 lb) iron ball moving at the respectable speed of 20 m/s ($\approx$45 mph). Its macroscopic kinetic energy is

$$K_{macro} = \frac{1}{2}Mv_{obj}^2 = 100 \text{ J}$$

A periodic table of the elements shows that iron has atomic mass 56. Recall from chemistry that 56 g of iron is 1 gram-molecular weight and has Avogadro's number ($N_A = 6.02 \times 10^{23}$) atoms. Thus 500 g of iron is $\approx$9 gram-molecular weights and contains $N \approx 9N_A \approx 5.4 \times 10^{24}$ iron atoms. Thus the mass of each atom is

$$m = \frac{M}{N} \approx \frac{0.50 \text{ kg}}{5.4 \times 10^{24}} \approx 9 \times 10^{-26} \text{ kg}$$

How fast do atoms move? In Part IV you'll learn that the speed of sound in air at room temperature is $\approx$340 m/s. Sound travels by atoms bumping into each other, so the atoms in the air must have a speed of *at least* 340 m/s. The speed of sound in solids is even higher, usually >1000 m/s, but that's partially influenced by the spring-like molecular bonds. As a rough estimate, $v \approx$ 500 m/s is a reasonable guess. The kinetic energy of one iron atom at this speed is

$$K_{atom} = \frac{1}{2}mv^2 \approx 1.1 \times 10^{-20} \text{ J}$$

(a) The macroscopic motion of the sytem as a whole

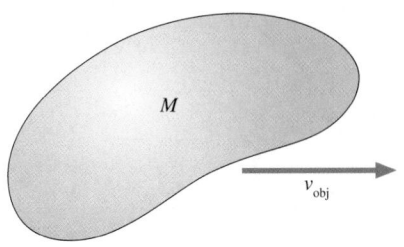

(b) The microscopic motion of the atoms inside

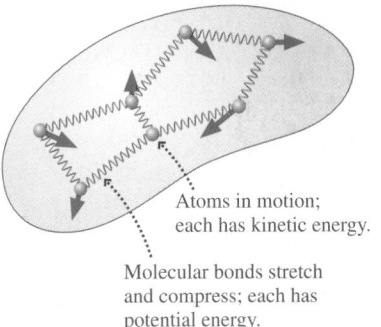

Atoms in motion; each has kinetic energy.

Molecular bonds stretch and compress; each has potential energy.

FIGURE 11.26 Two perspectives of motion and energy.

This is very tiny, but there are a great many atoms. If we assume, for our estimate, that all atoms move at this speed, the microscopic kinetic energy is

$$K_{\text{micro}} \approx N K_{\text{atom}} \approx 70{,}000 \text{ J}$$

The microscopic kinetic energy of the atoms moving inside the iron is much larger than the macroscopic kinetic energy of the object as a whole! We'll later see that, on average, U_{micro} for a solid is equal to K_{micro}, so the total microscopic energy is $\approx 140{,}000$ J.

The combined microscopic kinetic and potential energy of the atoms is called the *thermal energy* of the system:

$$E_{\text{th}} = K_{\text{micro}} + U_{\text{micro}} \tag{11.30}$$

This energy is usually hidden from view in our macrophysics perspective, but it is quite real. We will discover later, when we reach thermodynamics, that the thermal energy is proportional to the *temperature* of the system. Raising the temperature causes the atoms to move faster and the bonds to stretch more, giving the system more thermal energy.

NOTE ▶ The microscopic energy of the atoms in a system is *not* called "heat." The word *heat,* like the word *work,* has a narrow and precise meaning in physics that is much more restricted than its use in everyday language. We will introduce the concept of heat later, when we need it. For the time being we want to use the correct term *thermal energy* to describe the random, thermal motion of the particles in a system. If the temperature of a system goes up (i.e., it gets hotter), it is because the system's thermal energy has increased. ◀

Dissipative Forces

At the beginning of the chapter we asked "What happens to the energy?" when a system "runs down" because of friction. If you shove a book across the table, it gradually slows down and stops. Where did the energy go? The common answer "It went into heat" isn't quite right.

Figure 11.27 reminds you of the atomic-level model of friction that we introduced in Chapter 5. There we were interested in how the friction affects the macroscopic object. Now, think what friction does to the microscopic energy of the atoms. If two atoms temporarily stick together, the molecular bond gets stretched and U_{micro} increases. When the bond breaks and the atoms suddenly snap back into place, that potential energy gets transformed into microscopic kinetic energy K_{micro} of atoms bouncing around. Imagine having several balls connected by springs. If you pull one ball and then suddenly release it, you cause the whole system to jiggle and vibrate.

The forces of friction increase the microscopic kinetic and potential energy of the atoms. That is, they increase the object's thermal energy, and we perceive this as a higher temperature. That's why rubbing things together makes them warmer. Thus the correct answer to "What happens to the energy?" is "It is transformed into thermal energy."

Forces such as friction and drag that always oppose the motion are called **dissipative forces.** They cause the macroscopic kinetic energy of the system as a whole to be "dissipated" as thermal energy. Dissipative forces are always nonconservative forces.

Consider the book sliding to a halt. Call the work done by friction W_{diss} to indicate that it is a particular type of nonconservative work. The energy equation, Equation 11.24, is

$$\Delta K + \Delta U = \Delta K = W_{\text{diss}}$$

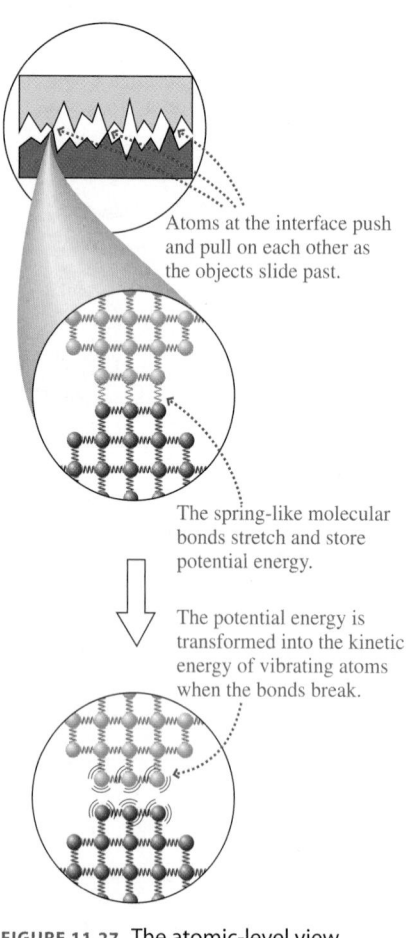

Atoms at the interface push and pull on each other as the objects slide past.

The spring-like molecular bonds stretch and store potential energy.

The potential energy is transformed into the kinetic energy of vibrating atoms when the bonds break.

FIGURE 11.27 The atomic-level view of friction.

where we've used $\Delta U = 0$ for horizontal motion. If the kinetic energy is transformed entirely into thermal energy, then

$$\Delta E_{\text{th}} = -\Delta K$$

ΔK is a negative number (loss of kinetic energy), so the explicit negative sign makes ΔE_{th} a positive number (gain of thermal energy).

Comparing these two equations, you can see that the relationship between the work done by dissipative forces and the change in thermal energy is

$$\Delta E_{\text{th}} = -W_{\text{diss}} \qquad (11.31)$$

W_{diss} is *always* a negative number because the force is always opposite the direction of motion. Consequently, ΔE_{th} is *always* a positive number. **Dissipative forces always increase the thermal energy, they never decrease it.**

The ballplayer's kinetic energy is being transformed into thermal energy.

EXAMPLE 11.10 **Calculating the increase in thermal energy**

A rope pulls a 10 kg wooden crate 3.0 m across a wood floor. What is the change in thermal energy?

SOLVE The change in thermal energy is $\Delta E_{\text{th}} = -W_{\text{diss}} = -W_{\text{fric}}$. From Equation 11.22,

$$\begin{aligned} W_{\text{fric}} &= f_k(\Delta r)\cos 180° = -\mu_k mg\Delta x \\ &= -(0.2)(10 \text{ kg})(9.8 \text{ m/s}^2)(3.0 \text{ m}) = -58.8 \text{ J} \end{aligned}$$

Thus

$$\Delta E_{\text{th}} = -(-58.8 \text{ J}) = 58.8 \text{ J}$$

11.8 Conservation of Energy

Let's return to the basic energy model and start pulling together the many ideas introduced in this chapter. Figure 11.28 shows a general system consisting of several macroscopic objects. These objects interact with each other, and they may be acted on by external forces from the environment. Both the interaction forces and the external forces do work on the objects. The change in the system's kinetic energy is given by the work-kinetic energy theorem, $\Delta K = W_{\text{net}}$.

We previously divided W_{net} into the work W_c done by conservative forces and the work W_{nc} done by nonconservative forces. The work done by the conservative forces can be represented by a potential energy U. Let's now make a further distinction by dividing the nonconservative forces into *dissipative forces* and *external forces*. That is,

$$W_{\text{nc}} = W_{\text{diss}} + W_{\text{ext}} \qquad (11.32)$$

To illustrate what we mean by an external force, suppose you pick up a box at rest on the floor and place it at rest on a table. The box gains gravitational potential energy, but $\Delta K = 0$. Or consider pulling the box across the table with a string. The box gains kinetic energy, but not by transforming potential energy. The force of your hand and the tension of the string are forces that "reach in" from the environment to change the system. Thus they are *external forces*.

We have to be careful choosing the system if we want this distinction to be valid. As you can imagine, we're going to associate W_{diss} with ΔE_{th}. We want the thermal energy E_{th} to be an energy *of the system*. Otherwise, it wouldn't make sense to talk about transforming kinetic energy into thermal energy. But for E_{th} to be an energy of the system, *both* objects involved in a dissipative interaction must

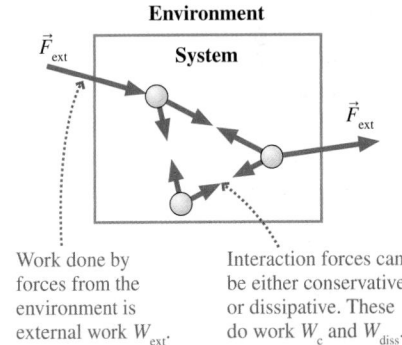

Environment

$\vec{F}_{\text{ext}}$ **System**

$\vec{F}_{\text{ext}}$

| Work done by forces from the environment is external work W_{ext}. | Interaction forces can be either conservative or dissipative. These do work W_c and W_{diss}. |

FIGURE 11.28 A system with both internal interaction forces and external forces.

be part of the system. The book sliding across the table raises the temperature of both the book *and the table*. Consequently, we must include both the book *and the table* in the system. The dissipative forces, like the conservative forces, are interaction forces *inside* the system.

With this distinction, the work-kinetic energy theorem is

$$\Delta K = W_c + W_{diss} + W_{ext} \tag{11.33}$$

As before, define the potential energy U such that $\Delta U = -W_c$. Remember that potential energy is really just the precomputed work of a conservative force. We've also seen that the work done by dissipative forces increases the system's thermal energy: $\Delta E_{th} = -W_{diss}$. With these substitutions, the work-kinetic energy theorem becomes

$$\Delta K = -\Delta U + -\Delta E_{th} + W_{ext}$$

We can write this more profitably as

$$\Delta K + \Delta U + \Delta E_{th} = \Delta E_{mech} + \Delta E_{th} = \Delta E_{sys} = W_{ext} \tag{11.34}$$

where $E_{sys} = E_{mech} + E_{th}$ is the total energy of the system. Equation 11.34 is the **energy equation** of the system.

Equation 11.34 is our most general statement about how the energy of a system changes, but we still need to give a clear interpretation as to what it says. In Chapter 9 we defined an *isolated system* as a system for which the *net* external force is zero. It follows that no external work is done on an isolated system: $W_{ext} = 0$. Thus one conclusion from Equation 11.34 is that **the total energy E_{sys} of an isolated system is conserved.** That is, $\Delta E_{sys} = 0$ for an isolated system. If, in addition, the system is also nondissipative (i.e., no friction forces), then $\Delta E_{th} = 0$. In that case, the mechanical energy E_{mech} is conserved.

These conclusions about energy can be summarized as the *law of conservation of energy:*

> **Law of conservation of energy** The total energy $E_{sys} = E_{mech} + E_{th}$ of an isolated system is a constant. The kinetic, potential, and thermal energy within the system can be transformed into each other, but their sum cannot change. Further, the mechanical energy $E_{mech} = K + U$ is conserved if the system is both isolated and nondissipative.

The law of conservation of energy is one of the most powerful statements in physics.

Figure 11.29a redraws the basic energy model of Figure 11.2. Now you can see that this picture is a pictorial representation of Equation 11.34. E_{sys}, the total energy of the system, changes only if external forces transfer energy in or out of the system by doing work on the system. The kinetic, potential, and thermal energy within the system can be transformed into each other by interaction forces within the system. As Figure 11.29b shows, $E_{sys} = K + U + E_{th}$ remains constant if the system is isolated. The *transfer* and *transformation* of energy are what the basic energy model is all about.

Energy Bar Charts

The energy bar charts of Chapter 10 can now be expanded to include the thermal energy and the work done by external forces. The energy equation, Equation 11.34, can be written

$$K_i + U_i + W_{ext} = K_f + U_f + \Delta E_{th} \tag{11.35}$$

(a) A system interacting with its environment

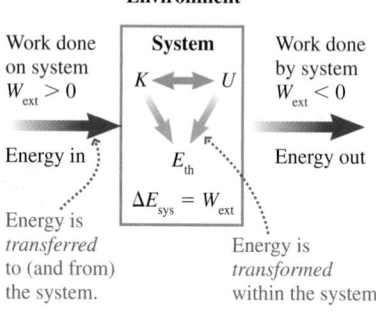

(b) An isolated system

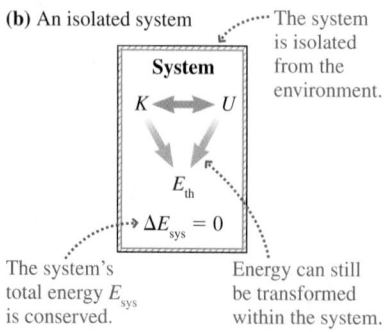

FIGURE 11.29 The basic energy model is a pictorial representation of the energy equation.

The left side is the "before" condition ($K_i + U_i$) plus any energy that is added to or removed from the system. The right side is the "after" situation. The "energy accounting" of Equation 11.35 can be represented by the bar chart of Figure 11.30.

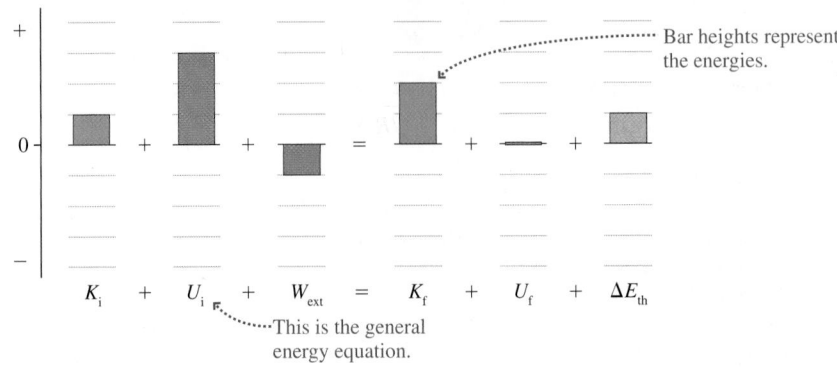

FIGURE 11.30 An energy bar chart shows how all the energy is accounted for.

NOTE ▶ We don't have any way to determine $(E_{th})_i$ or $(E_{th})_f$, but ΔE_{th} is always positive whenever the system contains dissipative forces. ◀

Let's look at a few examples.

EXAMPLE 11.11 Energy bar chart I

A box slides across a rough floor until it stops. Show the energy transfers and transformations on an energy bar chart.

SOLVE The box has an initial kinetic energy K_i. That energy is transformed into the thermal energy of the box and the floor. The potential energy doesn't change and no work is done by external forces, so the process is an energy transformation $K_i \rightarrow E_{th}$. This is shown in Figure 11.31. E_{sys} is conserved but E_{mech} is not.

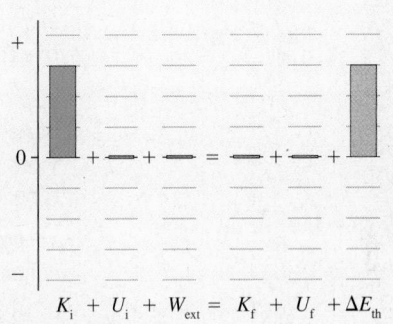

FIGURE 11.31 Energy bar chart for Example 11.11.

EXAMPLE 11.12 Energy bar chart II

A rope lifts a box at constant speed. Show the energy transfers and transformations on an energy bar chart.

SOLVE The tension in the rope is an external force that does work on the box, increasing the potential energy of the box. The kinetic energy is unchanged because the speed is constant. The process is an energy transfer $W_{ext} \rightarrow U_f$, as Figure 11.32 shows. This is not an isolated system, so E_{sys} is not conserved.

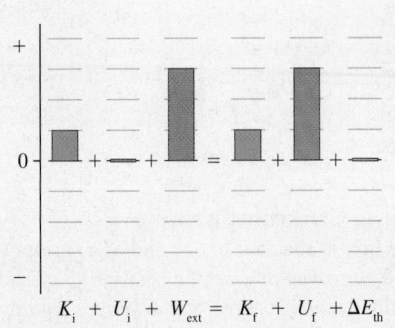

FIGURE 11.32 Energy bar chart for Example 11.12.

EXAMPLE 11.13 Energy bar chart III

The box that was lifted in Example 11.12 falls at a steady speed as the rope spins a generator and causes a light bulb to glow. Air resistance is negligible. Show the energy transfers and transformations on an energy bar chart.

SOLVE The initial potential energy decreases, but K does not change and $\Delta E_{th} = 0$. The tension in the rope is an external force that does work, but W_{ext} is negative in this case because $\vec{T}$ points up while the displacement $\Delta\vec{r}$ is down. Negative work means that energy is transferred from the system to the environment or, in more informal terms, that the *system does work on the environment*. The falling box does work on the generator to spin it and light the bulb. Energy is transferred out of the system and eventually ends up in the light bulb as electrical energy. The process is $U_i \rightarrow W_{ext}$. This is shown in Figure 11.33.

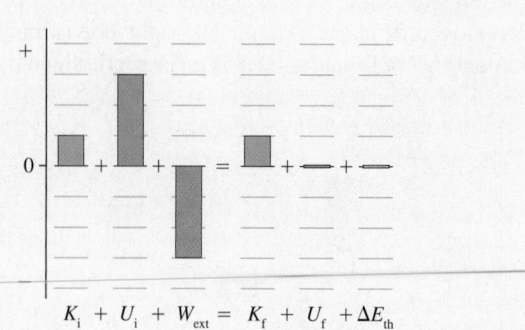

$$K_i + U_i + W_{ext} = K_f + U_f + \Delta E_{th}$$

FIGURE 11.33 Energy bar chart for Example 11.13.

Strategy for Energy Problems

  5.2–5.7, 6.5, 6.8, 6.9 Activ Physics

This is a good point at which to summarize the strategy we have been developing for using the concept of energy.

> **(MP) PROBLEM-SOLVING STRATEGY 11.1** **Solving energy problems**
>
> **MODEL** Identify which objects are part of the system and which are in the environment. If possible, choose a system without friction or other dissipative forces. Some problems may need to be subdivided into two or more parts.
>
> **VISUALIZE** Draw a before-and-after pictorial representation and an energy bar chart. A free-body diagram can be helpful if you're going to calculate work, although often the forces are simple enough to show on the pictorial representation.
>
> **SOLVE** If the system is both isolated and nondissipative, then the mechanical energy is conserved:
>
> $$K_f + U_f = K_i + U_i$$
>
> If there are external or dissipative forces, calculate W_{ext} and W_{diss}. Then use the more general energy equation
>
> $$K_f + U_f + \Delta E_{th} = K_i + U_i + W_{ext}$$
>
> Kinematics and/or other conservation laws may be needed for some problems.
>
> **ASSESS** Check that your result has the correct units, is reasonable, and answers the question.

EXAMPLE 11.14 Stretching a spring

The 5.0 kg box is attached to one end of a spring with spring constant 80 N/m. The other end of the spring is anchored to a wall. Initially the box is at rest at the spring's equilibrium position. A rope with a constant tension of 100 N then pulls the box away from the wall. What is the speed of the box after it has moved 50 cm? The coefficient of friction between the box and the floor is 0.30.

MODEL This is a complex situation, but one that we can analyze. First, identify the box, the spring, and the floor as the system. We need the floor inside the system because friction increases the temperature of the box *and* the floor. Friction is a dissipative force that does work W_{diss}. The tension in the rope is an external force. The work W_{ext} done by the rope's tension transfers energy into the system, causing K, U_s, and E_{th} all to increase.

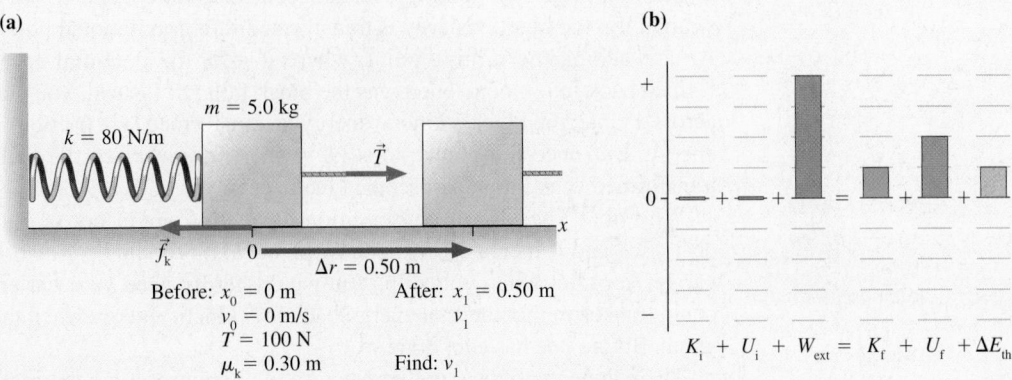

FIGURE 11.34 Pictorial representation and energy bar chart for Example 11.14.

VISUALIZE Figure 11.34a is a before-and-after pictorial representation. The energy transfers and transformations are shown in the energy bar chart of Figure 11.34b.

SOLVE We have an external force and a dissipative force for which we need to calculate the work. The work done by the rope's tension is

$$W_{ext} = \vec{T} \cdot \Delta\vec{r} = T(\Delta x)\cos 0° = (100 \text{ N})(0.50 \text{ m}) = 50 \text{ J}$$

The work done by friction is

$$W_{diss} = f_k(\Delta r)\cos 180° = -\mu_k mg\Delta x$$
$$= -(0.3)(5.0 \text{ kg})(9.8 \text{ m/s}^2)(0.50 \text{ m}) = -7.35 \text{ J}$$

and thus $\Delta E_{th} = +7.35$ J. The energy equation $K_f + U_f + \Delta E_{th} = K_i + U_i + W_{ext}$ is

$$\frac{1}{2}mv_1^2 + \frac{1}{2}kx_1^2 + \Delta E_{th} = \frac{1}{2}mv_0^2 + \frac{1}{2}kx_0^2 + W_{ext}$$

We know that $x_0 = 0$ m and $v_0 = 0$ m/s, so the energy equation simplifies to

$$\frac{1}{2}mv_1^2 = W_{ext} - \Delta E_{th} - \frac{1}{2}kx_1^2$$

Solving for the final speed v_1 gives

$$v_1 = \sqrt{\frac{2(W_{ext} - \Delta E_{th} - \frac{1}{2}kx_1^2)}{m}} = 3.61 \text{ m/s}$$

ASSESS We had to bring all the energy ideas together to solve this problem.

A Good Start, But . . .

We've made a good start at understanding energy, but there's still much to do. First, work is not the only way to transfer energy to a system. Suppose you put a pan of water on the stove and turn on the burner. The water temperature goes up, meaning that E_{th} is increasing, but no work is done. Instead, energy is transferred to the water by a *nonmechanical* means that we will call *heat*. Heat, which has the symbol Q, is the transfer of energy when there is a temperature difference between the system and the environment.

If we include heat as a second means of transferring energy, Equation 11.34, the energy equation, becomes

$$\Delta E_{sys} = W_{ext} + Q \tag{11.36}$$

This more general statement about energy is called the *first law of thermodynamics*. We will expand our basic energy model to include heat when we reach the study of thermodynamics in Part IV. Equation 11.36 will become essential for understanding thermal processes. The important point to notice for now is that we have developed a quite general model of energy that can later be expanded to include additional types of energy and energy transfers. Thermodynamics will be an extension of the ideas that we've established in Chapters 10 and 11, not an entirely new subject.

Second, there's something unusual about thermal energy. If you toss a block straight up, the kinetic energy is transformed into gravitational potential energy. After reaching the turning point, where $K = 0$, the potential energy is transformed back into kinetic energy as the block falls. If, instead, you push the block across a table, the block's kinetic energy is transformed (via friction) into thermal energy. But once the block stops ($K = 0$), you *never* see the thermal energy transformed back into macroscopic kinetic energy.

Why not? The law of conservation of energy would not be violated if E_{th} decreased and K increased. Think how practical this could be. The brakes of your car get very hot when you stop. You would hardly need your car engine if you could transform that thermal energy back into kinetic energy when the light turns green. But no one has ever done so.

There appears to be a one-way nature to the microscopic thermal energy that isn't true for an object's macroscopic kinetic or potential energy. It's easy to transform kinetic energy into thermal energy, difficult or impossible to transform it back. It seems as if some other law of physics is acting to prevent this. And indeed there is, a very important statement about energy transformation called the *second law of thermodynamics*.

So our basic energy model is a good start, but there's still much to do in Part IV as we expand these ideas into the full science of thermodynamics.

STOP TO THINK 11.6   A child at the playground slides down a pole at constant speed. This is a situation in which

a. $U \rightarrow K$. E_{mech} is not conserved but E_{sys} is.
b. $U \rightarrow E_{th}$. E_{mech} is conserved.
c. $U \rightarrow E_{th}$. E_{mech} is not conserved but E_{sys} is.
d. $K \rightarrow E_{th}$. E_{mech} is not conserved but E_{sys} is.
e. $U \rightarrow W_{ext}$. Neither E_{mech} nor E_{sys} are conserved.

11.9 Power

Work is a transfer of energy between the environment and a system. In many situations we would like to know *how fast* the energy is transferred. Does the force act quickly and transfer the energy very rapidly, or is it a slow and lazy transfer of energy? If you need to buy a motor to lift 2000 lb of bricks up 50 ft, it makes a *big* difference whether the motor has to do this in 30 s or 30 min!

The question "How fast?" implies that we are talking about a *rate*. For example, the velocity of an object—how fast it is going—is the *rate of change* of position. So when we raise the issue of how fast the energy is transferred, we are talking about the *rate of transfer* of energy. The rate at which energy is transferred or transformed is called the **power** P, and it is defined as

$$P \equiv \frac{dE_{sys}}{dt} \tag{11.37}$$

The English unit of power is the *horsepower*. The conversion factor to watts is

$$1 \text{ horsepower} = 1 \text{ hp} = 746 \text{ W}$$

Many common appliances, such as motors, are rated in hp.

The unit of power is the **watt,** which is defined as 1 watt = 1 W ≡ 1 J/s.

A force that is doing work (i.e., transferring energy) at a rate of 3 J/s has an "output power" of 3 W. The system gaining energy at the rate of 3 J/s is said to "consume" 3 W of power. Common prefixes used with power are mW (milliwatts), kW (kilowatts), and MW (megawatts).

EXAMPLE 11.15 Choosing a motor

What power motor is needed to lift a 2000 kg elevator at a steady 3.0 m/s?

SOLVE The tension in the cable does work on the elevator to lift it. Because the cable is pulled by the motor, we say that the motor does the work of lifting the elevator. The net force is zero, because the elevator moves at constant velocity, so the tension is simply $T = mg = 19,600$ N. The energy gained by the elevator is

$$\Delta E_{sys} = W_{ext} = T(\Delta y)$$

The power required to give the system this much energy in a time interval Δt is

$$P = \frac{\Delta E_{sys}}{\Delta t} = \frac{T(\Delta y)}{\Delta t}$$

But $\Delta y = v\Delta t$, so $P = Tv = (19,600 \text{ N})(3.0 \text{ m/s}) = 58,800 \text{ W} = 79 \text{ hp}$.

The idea of power as a *rate* of energy transfer applies no matter what the form of energy. Figure 11.35 shows several examples of the idea of power. For now, we primarily want to focus on *work* as the source of energy transfer. Within this more limited scope, power is simply the **rate of doing work:** $P = dW/dt$. If a particle moves through a small displacement $d\vec{r}$ while acted on by force $\vec{F}$, the force does a small amount of work dW given by

$$dW = \vec{F} \cdot d\vec{r}$$

Dividing both sides by dt, to give a rate of change, yields

$$\frac{dW}{dt} = \vec{F} \cdot \frac{d\vec{r}}{dt}$$

But $d\vec{r}/dt$ is the velocity $\vec{v}$, so we can write the power as

$$P = \vec{F} \cdot \vec{v} = Fv\cos\theta \qquad (11.38)$$

In other words, the power delivered to a particle by a force acting on it is the dot product of the force with the particle's velocity. These ideas will become clearer with some examples.

Highly trained athletes have a tremendous power output.

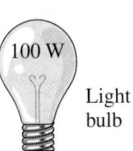

100 W

Light bulb

Electrical energy ⟶ light and heat at 100 J/s.

Athlete

$\frac{1}{2}$ hp

Chemical energy of glucose and fat ⟶ mechanical energy at ≈350 J/s ≈ $\frac{1}{2}$ hp.

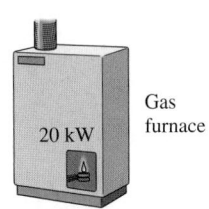

20 kW

Gas furnace

Chemical energy of gas ⟶ thermal energy at 20,000 J/s.

FIGURE 11.35 Examples of power.

EXAMPLE 11.16 Power output of a motor

A factory uses a motor and a cable to drag a 300 kg machine into the proper place on the factory floor. What power must the motor supply to drag the machine at a speed of 0.50 m/s? The coefficient of friction between the machine and the floor is 0.60.

SOLVE The force applied by the motor, through the cable, is the tension force $\vec{T}$. This force does work on the machine with power $P = Tv$. The machine is in dynamic equilibrium, because

the motion is at constant velocity, hence the tension in the rope balances the friction and is

$$T = f_k = \mu_k mg$$

The motor's power output is

$$P = Tv = \mu_k mgv = 882 \text{ W}$$

EXAMPLE 11.17 Power output of a car engine

A 1500 kg car has a front profile that is 1.6 m wide and 1.4 m high. The coefficient of rolling friction is 0.02. What power must the engine provide to drive at a steady 30 m/s ($\approx$65 mph) if 25% of the power is "lost" before reaching the drive wheels?

SOLVE The net force on a car moving at a steady speed is zero. The motion is opposed both by rolling friction and by air resistance. The forward force on the car $\vec{F}_{car}$ (recall that this is really $\vec{F}_{car\ on\ ground}$, a reaction to the drive wheels pushing backward on the ground with $\vec{F}_{car\ on\ ground}$) exactly balances the two opposing forces:

$$F_{car} = f_r + D$$

where $\vec{D}$ is the drag due to the air. Using the results of Chapter 5, where both rolling friction and drag were introduced, this becomes

$$F_{car} = \mu_r mg + \frac{1}{4}Av^2 = 294\ N + 504\ N = 798\ N$$

$A = (1.6\ m) \times (1.4\ m)$ is the front cross-section area of the car. The power required to push the car forward at this speed is

$$P_{car} = F_{car}v = (798\ N)(30\ m/s) = 23,900\ W = 32\ hp$$

This is the power *needed* at the drive wheels to push the car against the dissipative forces of friction and air resistance. The power output of the engine is larger because some energy is used to run the water pump, the power steering, and other accessories. In addition, energy is lost to friction in the drive train. If 25% of the power is lost (a typical value), leading to $P_{car} = 0.75P_{engine}$, the engine's power output is

$$P_{engine} = \frac{P_{car}}{0.75} = 31,900\ W = 43\ hp$$

ASSESS Automobile engines are typically rated at $\approx$200 hp. Most of that power is reserved for fast acceleration and climbing hills.

STOP TO THINK 11.7 Four students run up the stairs in the time shown. Rank in order, from largest to smallest, their power outputs P_a to P_d.

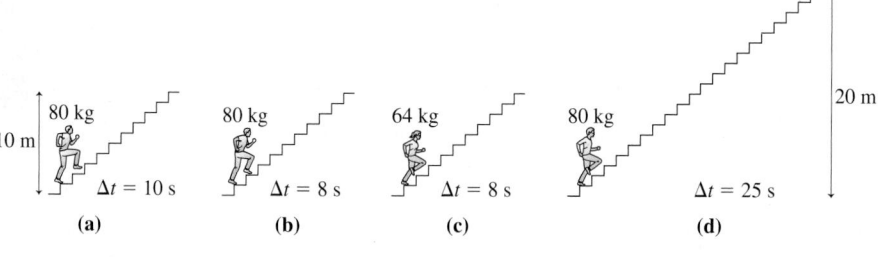

SUMMARY

The goal of Chapter 11 has been to develop a more complete understanding of energy and its conservation.

GENERAL PRINCIPLES

Basic Energy Model

- Energy is *transferred* to or from the system by work.
- Energy is *transformed* within the system.

Two versions of the energy equation are

$$\Delta E_{sys} = \Delta K + \Delta U + \Delta E_{th} = W_{ext}$$

$$K_f + U_f + \Delta E_{th} = K_i + U_i + W_{ext}$$

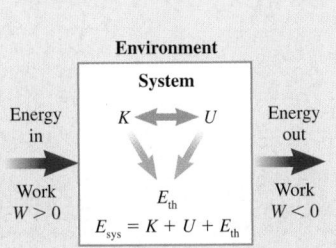

Solving Energy Problems

MODEL Identify objects in the system.

VISUALIZE Draw a before-and-after pictorial representation and an energy bar chart.

SOLVE Use the energy equation

$$K_f + U_f + \Delta E_{th} = K_i + U_i + W_{ext}$$

ASSESS Is the result reasonable?

Law of Conservation of Energy

- **Isolated system:** $W_{ext} = 0$. The total energy $E_{sys} = E_{mech} + E_{th}$ is conserved. $\Delta E_{sys} = 0$
- **Isolated, nondissipative system:** $W_{ext} = 0$ and $W_{diss} = 0$. The mechanical energy E_{mech} is conserved.

$$\Delta E_{mech} = 0 \text{ or } K_f + U_f = K_i + U_i$$

IMPORTANT CONCEPTS

The work-kinetic energy theorem is

$$\Delta K = W_{net} = W_c + W_{diss} + W_{ext}$$

Using $W_c = -\Delta U$ for conservative forces and $W_{diss} = -\Delta E_{th}$ for dissipative forces, this becomes the energy equation.

The work done by a force on a particle as it moves from s_i to s_f is

$$W = \int_{s_i}^{s_f} F_s \, ds = \text{area under the force curve}$$

$$= \vec{F} \cdot \Delta \vec{r} \text{ if } \vec{F} \text{ is a constant force}$$

Conservative forces are forces for which the work is independent of the path followed. The work done by a conservative force can be represented as a **potential energy**

$$\Delta U = U_f - U_i = -W_c(i \rightarrow f)$$

A conservative force is found from the potential energy by

$$F = -dU/ds = \text{negative of the slope of the PE curve}$$

Dissipative forces transform **macroscopic energy** into thermal energy, which is the **microscopic energy** of the atoms and molecules.

$$\Delta E_{th} = -W_{diss}$$

APPLICATIONS

Power is the rate at which energy is transferred or transformed:

$$P = \frac{dE_{sys}}{dt}$$

For a particle moving with velocity $\vec{v}$, the power delivered to the particle by force $\vec{F}$ is $P = \vec{F} \cdot \vec{v} = Fv\cos\theta$.

Dot product

$$\vec{A} \cdot \vec{B} = AB \cos \alpha = A_x B_x + A_y B_y$$

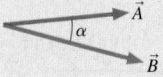

Energy bar charts display the energy equation

$$K_f + U_f + \Delta E_{th} = K_i + U_i + W_{ext} \text{ in graphical form.}$$

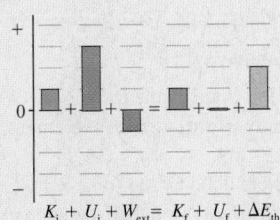

$$K_i + U_i + W_{ext} = K_f + U_f + \Delta E_{th}$$

TERMS AND NOTATION

thermal energy, E_{th}	work-kinetic energy theorem	microphysics
system energy, E_{sys}	dot product	dissipative force
energy transformation	scalar product	energy equation
energy transfer	conservative force	law of conservation of energy
work, W	nonconservative force	power, P
basic energy model	macrophysics	watt, W

EXERCISES AND PROBLEMS

The [✐] icon indicates that the problem can be done on an Energy Worksheet.

Exercises

Section 11.2 Work and Kinetic Energy

Section 11.3 Calculating and Using Work

1. Evaluate the dot product of the three pairs of vectors in Figure Ex11.1.

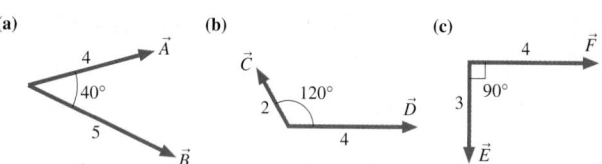

FIGURE EX11.1

2. Evaluate the dot product of the three pairs of vectors in Figure Ex11.2.

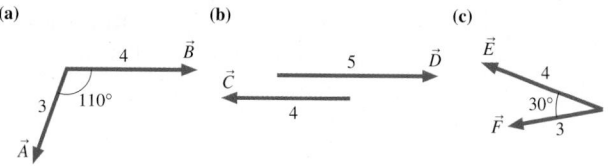

FIGURE EX11.2

3. Evaluate the dot product $\vec{A} \cdot \vec{B}$ if
   a. $\vec{A} = 3\hat{\imath} - 4\hat{\jmath}$ and $\vec{B} = -2\hat{\imath} + 6\hat{\jmath}$.
   b. $\vec{A} = 2\hat{\imath} + 3\hat{\jmath}$ and $\vec{B} = 6\hat{\imath} - 4\hat{\jmath}$.

4. Evaluate the dot product $\vec{A} \cdot \vec{B}$ if
   a. $\vec{A} = 5\hat{\imath} - 3\hat{\jmath}$ and $\vec{B} = 2\hat{\imath} - 2\hat{\jmath}$.
   b. $\vec{A} = 4\hat{\imath} - 2\hat{\jmath}$ and $\vec{B} = -3\hat{\imath} + 6\hat{\jmath}$.

5. How much work is done by the force $\vec{F} = (6.0\hat{\imath} - 3.0\hat{\jmath})$ N on a particle that moves through displacement (a) $\Delta\vec{r} = 2.0\hat{\imath}$ m and (b) $\Delta\vec{r} = 2.0\hat{\jmath}$ m?

6. How much work is done by the force $\vec{F} = (-5.0\hat{\imath} + 4.0\hat{\jmath})$ N on a particle that moves through displacement (a) $\Delta\vec{r} = 3.0\hat{\imath}$ m and (b) $\Delta\vec{r} = -3.0\hat{\jmath}$ m?

7. A 20 g particle is moving to the left at 30 m/s. How much work must be done on the particle to cause it to move to the right at 30 m/s?

8. A 2.0 kg book is lying on a 0.75-m-high table. You pick it up and place it on a bookshelf 2.25 m above the floor.
   a. How much work does gravity do on the book?
   b. How much work does your hand do on the book?

9. The two ropes seen in Figure Ex11.9 are used to lower a 255 kg piano 5.0 m from a second-story window to the ground. How much work is done by each of the three forces?

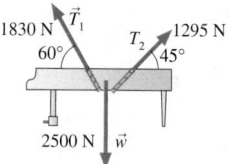

FIGURE EX11.9

10. The two ropes shown in the bird's-eye view of Figure Ex11.10 are used to drag a crate 3.0 m across the floor. How much work is done by each of the three forces?

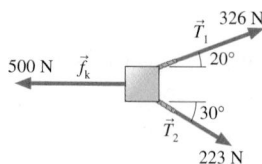

FIGURE EX11.10

11. Figure Ex11.11 is the velocity-versus-time graph for a 2.0 kg object moving along the x-axis. Determine the work done on the object during each of the five intervals AB, BC, CD, DE, and EF.

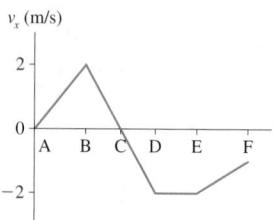

FIGURE EX11.11

Section 11.4 The Work Done by a Variable Force

12. Figure Ex11.12 is the force-versus-position graph for a particle moving along the *x*-axis. Determine the work done on the particle during each of the three intervals 0–1 m, 1–2 m, and 2–3 m.

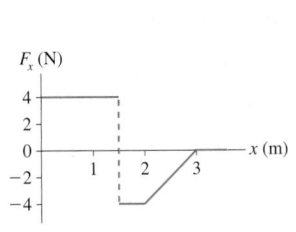

FIGURE EX11.12

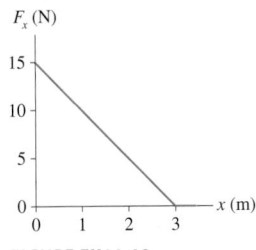

FIGURE EX11.13

13. A 500 g particle moving along the *x*-axis experiences the force shown in Figure Ex11.13. The particle's velocity is 2.0 m/s at *x* = 0 m. What is its velocity at *x* = 1 m, 2 m, and 3 m?

14. A 2.0 kg particle moving along the *x*-axis experiences the force shown in Figure Ex11.14. The particle's velocity is 4.0 m/s at *x* = 0 m. What is its velocity at *x* = 2 m and 4 m?

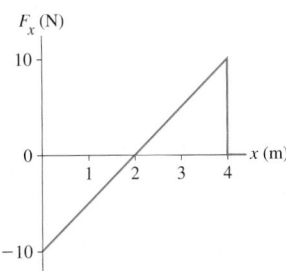

FIGURE EX11.14

15. A 500 g particle moving along the *x*-axis experiences the force shown in Figure Ex11.15. The particle goes from v_x = 2.0 m/s at *x* = 0 m to v_x = 6.0 m/s at *x* = 2 m. What is F_{max}?

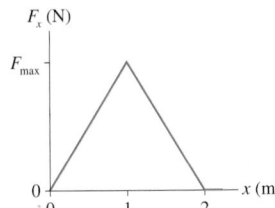

FIGURE EX11.15

Section 11.5 Force, Work, and Potential Energy

Section 11.6 Finding Force from Potential Energy

16. A particle has the potential energy shown in Figure Ex11.16. What is the *x*-component of the force on the particle at *x* = 5, 15, 25, and 35 cm?

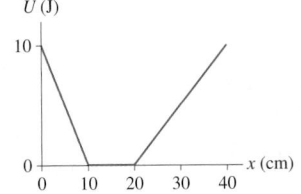

FIGURE EX11.16

17. A particle has the potential energy shown in Figure Ex11.17. What is the *x*-component of the force on the particle at *x* = 1 m and 3 m?

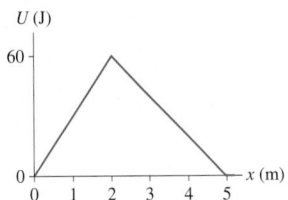

FIGURE EX11.17

18. A particle moving along the *y*-axis has the potential energy $U = 4y^3$ J, where *y* is in m.
    a. Graph the potential energy from *y* = 0 m to *y* = 2 m.
    b. What is the *y*-component of the force on the particle at *y* = 0 m, 1 m, and 2 m?

19. A particle moving along the *x*-axis has the potential energy $U = 10/x$ J, where *x* is in m.
    a. Graph the potential energy from *x* = 1 m to *x* = 10 m.
    b. What is the *x*-component of the force on the particle at *x* = 2 m, 5 m, and 8 m?

Section 11.7 Thermal Energy

20. The mass of a carbon atom is 2.00×10^{-26} kg.
    a. What is the kinetic energy of a carbon atom moving with a speed of 500 m/s?
    b. Two carbon atoms are joined by a spring-like carbon-carbon bond. The potential energy stored in the bond has the value you calculated in part a if the bond is stretched 0.050 nm. What is the value of the bond's spring constant?

21. In Part IV you'll learn to calculate that 1 mole (6.02×10^{23} atoms) of helium atoms in the gas phase has 3700 J of microscopic kinetic energy at room temperature. If we assume that all atoms move with the same speed, what is that speed? The mass of a helium atom is 6.68×10^{-27} kg.

22. A 1500 kg car traveling at 20 m/s skids to a halt.
    a. What energy transfers and transformations occur during the skid?
    b. What is the change in the thermal energy of the car and the road surface?

23. A 20 kg child slides down a 3.0-m-high playground slide. She starts from rest, and her speed at the bottom is 2.0 m/s.
    a. What energy transfers and transformations occur during the slide?
    b. What is the change in the thermal energy of the slide and the seat of her pants?

Section 11.8 Conservation of Energy

24. A system loses 1000 J of potential energy. In the process, it does 500 J of work on the environment and the thermal energy increases by 100 J. Show this process on an energy bar chart.

25. A system gains 500 J of kinetic energy while losing 200 J of potential energy. The thermal energy increases 100 J. Show this process on an energy bar chart.

26. How much work is done by the environment in the process shown in Figure Ex11.26? Is energy transferred from the environment to the system or from the system to the environment?

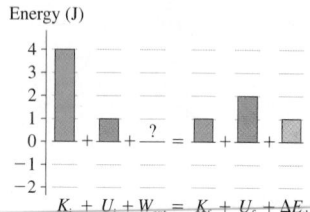

FIGURE EX11.26 $K_i + U_i + W_{ext} = K_f + U_f + \Delta E_{th}$

27. A cable with 20 N of tension pulls straight up on a 1.02 kg block that is initially at rest. What is the block's speed after being lifted 2.0 m?

Section 11.9 Power

28. a. How much work does an elevator motor do to lift a 1000 kg elevator a height of 100 m?
    b. How much power must the motor supply to do this in 50 s at constant speed?

29. a. How much work must you do to push a 10 kg block of steel across a steel table at a steady speed of 1.0 m/s for 3.0 s?
    b. What is your power output while doing so?

30. At midday, solar energy strikes the earth with an intensity of about 1 kW/m². What is the area of a solar collector that could collect 150 MJ of energy in 1 hr? This is roughly the energy content of 1 gallon of gasoline.

31. Which consumes more energy, a 1.2 kW hair dryer used for 10 min or a 10 W night light left on for 24 hr?

32. The electric company bills you in "kilowatt hours," abbreviated kwh.
    a. Is this energy, power, or force? Explain.
    b. Monthly electric use for a typical household is 500 kwh. What is this in basic SI units?

33. A 50 kg sprinter, starting from rest, runs 50 m in 7.0 s at constant acceleration.
    a. What is the magnitude of the horizontal force acting on the sprinter?
    b. What is the sprinter's power output at 2.0 s, 4.0 s, and 6.0 s?

Problems

34. A particle moves from A to D in Figure P11.34 while experiencing force $\vec{F} = (6\hat{i} + 8\hat{j})$ N. How much work does the force do if the particle follows path (a) ABD, (b) ACD, and (c) AD? Is this a conservative force? Explain.

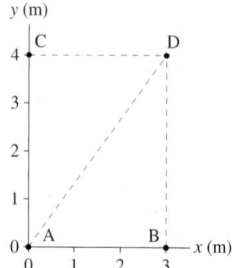

FIGURE P11.34

35. A 100 g particle experiences the one-dimensional, conservative force F_x shown in Figure P11.35.
    a. Draw a graph of the potential energy U from $x = 0$ m to $x = 5$ m. Let the zero of the potential energy be at $x = 0$ m. **Hint:** Think about the definition of potential energy *and* the geometric interpretation of the work done by a varying force.
    b. The particle is shot toward the right from $x = 1$ m with a speed of 25 m/s. What is the particle's mechanical energy?
    c. Draw the total energy line on your graph of part a.
    d. Where is the particle's turning point?

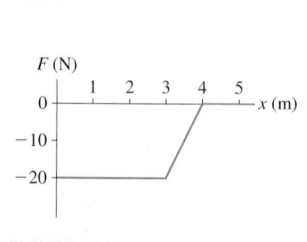

FIGURE P11.35

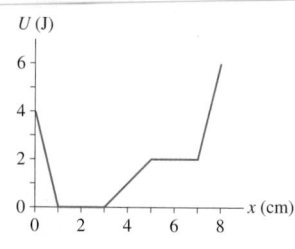

FIGURE P11.36

36. A 10 g particle has the potential energy shown in Figure P11.36.
    a. Draw a force-versus-position graph from $x = 0$ cm to $x = 8$ cm.
    b. How much work does the force do as the particle moves from $x = 2$ cm to $x = 6$ cm?
    c. What speed does the particle need at $x = 2$ cm to arrive at $x = 6$ cm with a speed of 10 m/s?

37. a. Figure P11.37a shows the force F_x exerted on a particle that moves along the x-axis. Draw a graph of the particle's potential energy as a function of position x. Let U be zero at $x = 0$ m.
    b. Figure P11.37b shows the potential energy U of a particle that moves along the x-axis. Draw a graph of the force F_x as a function of position x.

(a)
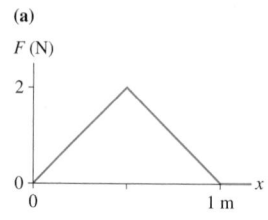

(b)
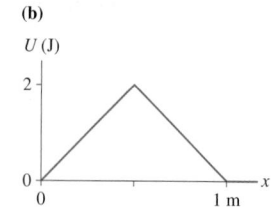

FIGURE P11.37

38. Figure P11.38 is the velocity-versus-time graph of a 500 g particle that starts at $x = 0$ m and moves along the x-axis. Draw graphs of the following by calculating and plotting numerical values at $t = 0, 1, 2, 3,$ and 4 s. Then sketch lines or curves of the appropriate shape between the points. Make sure you include appropriate scales on both axes of each graph.

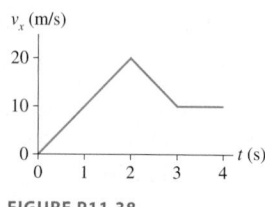

FIGURE P11.38

    a. Acceleration versus time.
    b. Position versus time.
    c. Kinetic energy versus time.
    d. Force versus time.
    e. Use your F_x-versus-t graph to determine the *impulse* delivered to the particle during the time interval 0–2 s and also the interval 2–4 s.

f. Use the impulse-momentum theorem to determine the particle's velocity at $t = 2$ s and at $t = 4$ s. Do your results agree with the velocity graph?

g. Now draw a graph of force versus *position*. This requires no calculations; just think carefully about what you learned in parts a to d.

h. Use your F_x-versus-x graph to determine the *work* done on the particle during the time interval 0–2 s and also the interval 2–4 s.

i. Use the work-kinetic energy theorem to determine the particle's velocity at $t = 2$ s and at $t = 4$ s. Do your results agree with the velocity graph?

39. A 1000 kg elevator accelerates upward at 1.0 m/s^2 for 10 m, starting from rest.
   a. How much work does gravity do on the elevator?
   b. How much work does the tension in the elevator cable do on the elevator?
   c. Use the work-kinetic energy theorem to find the kinetic energy of the elevator as it reaches 10 m.
   d. What is the speed of the elevator as it reaches 10 m?

40. Bob can throw a 500 g rock with a speed of 30 m/s. He moves his hand forward 1.0 m while doing so.
   a. How much work does Bob do on the rock?
   b. How much force, assumed to be constant, does Bob apply to the rock?
   c. What is Bob's maximum power output as he throws the rock?

41. Doug pushes a 5.0 kg crate up a 2.0-m-high 20° frictionless slope by pushing it with a constant *horizontal* force of 25 N. What is the speed of the crate as it reaches the top of the slope?
   a. Solve this problem using work and energy.
   b. Solve this problem using Newton's laws.

42. Sam, whose mass is 75 kg, straps on his skis and starts down a 50-m-high, 20° frictionless slope. A strong headwind exerts a *horizontal* force of 200 N on him as he skies. Find Sam's speed at the bottom (a) using work and energy, (b) using Newton's laws.

43. Susan's 10 kg baby brother Paul sits on a mat. Susan pulls the mat across the floor using a rope that is angled 30° above the floor. The tension is a constant 30 N and the coefficient of friction is 0.20. Use work and energy to find Paul's speed after being pulled 3.0 m.

44. A horizontal spring with spring constant 100 N/m is compressed 20 cm and used to launch a 2.5 kg box across a frictionless, horizontal surface. After the box travels some distance, the surface becomes rough. The coefficient of kinetic friction of the box on the surface is 0.15. Use work and energy to find how far the box slides across the rough surface before stopping.

45. A baggage handler throws a 15 kg suitcase horizontally along the floor of an airplane luggage compartment with an initial speed of 1.2 m/s. The suitcase slides 2.0 m before stopping. Use work and energy to find the suitcase's coefficient of kinetic friction on the floor.

46. Truck brakes can fail if they get too hot. In some mountainous areas, ramps of loose gravel are constructed to stop runaway trucks that have lost their brakes. The combination of a slight upward slope and a large coefficient of friction in the gravel brings the truck safely to a halt. Suppose a gravel ramp slopes upward at 6.0° and the coefficient of friction is 0.40. Use work and energy to find the length of a ramp that will stop a 15,000 kg truck that enters the ramp at 35 m/s ($\approx$75 mph).

47. A freight company uses a compressed spring to shoot 2.0 kg packages up a 1.0-m-high frictionless ramp into a truck, as Figure P11.47 shows. The spring constant is 500 N/m and the spring is compressed 30 cm.
   a. What is the speed of the package when it reaches the truck?
   b. A careless worker spills his soda on the ramp. This creates a 50-cm-long sticky spot with a coefficient of kinetic friction 0.30. Will the next package make it into the truck?

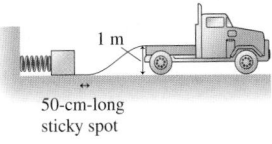

FIGURE P11.47

48. Use work and energy to find the speed of the 2.0 kg block just before it hits the floor if (a) the table is frictionless and if (b) the coefficient of kinetic friction of the 3.0 kg block is 0.15.

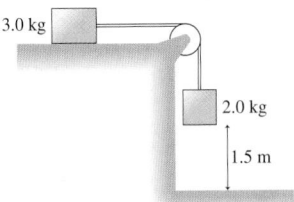

FIGURE P11.48

49. An 8.0 kg crate is pulled up a 30° incline by a person pulling on a rope that makes an 18° angle with the incline. The tension in the rope is 120 N and the crate's coefficient of kinetic friction on the incline is 0.25. If the crate moves 5.0 m, what is the work done by (a) the tension in the rope, (b) gravity, (c) friction, and (d) the normal force?

50. A 10.2 kg weather rocket generates a thrust of 200 N. The rocket, pointing upward, is clamped to the top of a vertical spring. The bottom of the spring, whose spring constant is 500 N/m, is anchored to the ground.
   a. Initially, before the engine is ignited, the rocket sits at rest on top of the spring. How much is the spring compressed?
   b. After the engine is ignited, what is the rocket's speed when the spring has stretched 40 cm? For comparison, what would be the rocket's speed after traveling this distance if it weren't attached to the spring?

51. A 50 kg ice skater is gliding along the ice, heading due north at 4.0 m/s. The ice has a small coefficient of static friction, to prevent the skater from slipping sideways, but $\mu_k = 0$. Suddenly, a wind from the northeast exerts a force of 4.0 N on the skater.
   a. Use work and energy to find the skater's speed after gliding 100 m in this wind.
   b. What is the minimum value of μ_s that allows her to continue moving straight north?

52. a. A 50 g ice cube can slide without friction up and down a 30° slope. The ice cube is pressed against a spring at the bottom of the slope, compressing the spring 10 cm. The spring constant is 25 N/m. When the ice cube is released, what total distance will it travel up the slope before reversing direction?
   b. The ice cube is replaced by a 50 g plastic cube whose coefficient of kinetic friction is 0.20. How far will the plastic cube travel up the slope?

53. A 5.0 kg box slides down a 5.0-m-high frictionless hill, starting from rest, across a 2.0-m-long horizontal surface, then hits a horizontal spring with spring constant 500 N/m. The other end of the spring is anchored against a wall. The ground under the spring is frictionless, but the 2.0-m-long horizontal surface is rough. The coefficient of kinetic friction of the box on this surface is 0.25.
    a. What is the speed of the box just before reaching the rough surface?
    b. What is the speed of the box just before hitting the spring?
    c. How far is the spring compressed?
    d. Including the first crossing, how many *complete* trips will the box make across the rough surface before coming to rest?

54. The spring shown in Figure P11.54 is compressed 50 cm and used to launch a 100 kg physics student. The track is frictionless until it starts up the incline. The student's coefficient of kinetic friction on the 30° incline is 0.15.
    a. What is the student's speed just after losing contact with the spring?
    b. How far up the incline does the student go?

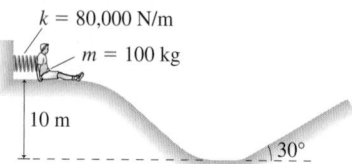

$k = 80,000$ N/m

$m = 100$ kg

10 m

30°

FIGURE P11.54

55. A block of mass m starts from rest at height h. It slides down a frictionless incline, across a rough horizontal surface of length L, then up a frictionless incline. The coefficient of kinetic friction on the rough surface is μ_k.
    a. What is the block's speed at the bottom of the first incline?
    b. How high does the block go on the second incline?
    Give your answers in terms of m, h, L, μ_k, and g.

56. Show that Hooke's law for an ideal spring is a conservative force. To do so, first calculate the work done by the spring as it expands from A to B. Then calculate the work done by the spring as it expands from A to point C, which is beyond B, then returns from C to B.

57. A clever engineer designs a "sprong" that obeys the force law $F_x = -q(x - x_e)^3$, where x_e is the equilibrium position of the end of the sprong and q is the sprong constant. For simplicity, we'll let $x_e = 0$ m. Then $F_x = -qx^3$.
    a. What are the units of q?
    b. Draw a graph of F_x versus x.
    c. Find an expression for the potential energy of a stretched or compressed sprong.
    d. A sprong-loaded toy gun shoots a 20 g plastic ball. What is the launch speed if the sprong constant is 40,000, with the units you found in part a, and the sprong is compressed 10 cm? Assume the barrel is frictionless.

58. A particle of mass m has initial conditions $x_0 = 0$ m and $v_0 > 0$ m/s. The particle experiences the variable force $F_x = F_0\sin(cx)$ as it moves to the right along the x-axis, where F_0 and c are constants.
    a. What are the units of F_0?
    b. What are the units of c?
    c. What is the value of F_x when the particle is at x_0?
    d. At what position x_{max} does the force first reach a maximum value? Your answer will be in terms of the constants F_0 and c and perhaps other numerical constants.
    e. Sketch a graph of F-versus-x from x_0 to x_{max}.
    f. What is the particle's velocity as it reaches x_{max}? Give your answer in terms of m, v_0, F_0, and c.

59. a. Estimate the height in meters of the two flights of stairs that go from the first to the third floor of a building.
    b. Estimate how long it takes you to *run* up these two flights of stairs.
    c. Estimate your power output in both watts and horsepower while running up the stairs.

60. A gardener pushes a 12 kg lawnmower whose handle is tilted up 37° above horizontal. The lawnmower's coefficient of rolling friction is 0.15. How much power does the gardener have to supply to push the lawnmower at a constant speed of 1.2 m/s?

61. A 5.0 kg cat leaps from the floor to the top of a 95-cm-high table. If the cat pushes against the floor for 0.20 s to accomplish this feat, what was her average power output during the pushoff period?

62. A 2.0 hp electric motor on a water well pumps water from 10 m below the surface. The density of water is 1.0 kg per liter. How many liters of water does the motor pump in 1 hr?

63. In a hydroelectric dam, water falls 25 m and then spins a turbine to generate electricity.
    a. What is ΔU of 1.0 kg of water?
    b. Suppose the dam is 80% efficient at converting the water's potential energy to electrical energy. How many kilograms of water must pass through the turbines each second to generate 50 MW of electricity? This is a typical value for a small hydroelectric dam.

64. The force required to tow a water skier at speed v is proportional to the speed. That is, $F_{tow} = Av$, where A is a proportionality constant. If a speed of 2.5 mph requires 2 hp, how much power is required to tow a water skier at 7.5 mph?

65. Estimate the maximum speed of a horse. Assume that a horse is 1.8 m tall and 0.5 m wide.

66. The engine in a 1500 kg car has a maximum power output of 200 hp, but 25% of the power is lost before reaching the drive wheels. The car has a front profile that is 1.6 m wide and 1.4 high. The coefficient of rolling friction is 0.02. What is the car's top speed? Is this answer reasonable?

67. A Porsche 944 Turbo has a rated power of 217 hp and a mass, with the driver, of 1480 kg. Two-thirds of the car's weight is over the drive wheels.
    a. What is the maximum acceleration of the Porsche on a concrete surface where $\mu_s = 1.0$?
    Hint: What force pushes the car forward?
    b. What is the speed of the Porsche at maximum power output?
    c. If the Porsche accelerates at a_{max}, how long does it take until it reaches the maximum power output?

In Problems 68 through 71 you are given the equation(s) used to solve a problem. For each of these, you are to
  a. Write a realistic problem for which this is the correct equation(s).
  b. Draw a pictorial representation.
  c. Finish the solution of the problem.

68. $\frac{1}{2}(2.0 \text{ kg})(4.0 \text{ m/s})^2 + 0$
    $+ (0.15)(2.0 \text{ kg})(9.8 \text{ m/s}^2)(2.0 \text{ m}) = 0 + 0 + T(2.0 \text{ m})$

69. $\frac{1}{2}(20 \text{ kg})v_1^2 + 0$
    $+ (0.15)(20 \text{ kg})(9.8 \text{ m/s}^2)\cos 40°((2.5 \text{ m})/\sin 40°)$
    $= 0 + (20 \text{ kg})(9.8 \text{ m/s}^2)(2.5 \text{ m}) + 0$

70. $F_{push} - (0.20)(30 \text{ kg})(9.8 \text{ m/s}^2) = 0$

    $75 \text{ W} = F_{push}v$

71. $T - (1500 \text{ kg})(9.8 \text{ m/s}^2) = (1500 \text{ kg})(1.0 \text{ m/s}^2)$

    $P = T(2.0 \text{ m/s})$

Challenge Problems

72. You've taken a summer job at a water park. In one stunt, a water skier is going to glide up the 2.0-m-high frictionless ramp shown in Figure CP11.72, then sail over a 5.0-m-wide tank filled with hungry sharks. You will be driving the boat that pulls her to the ramp. She'll drop the tow rope at the base of the ramp just as you veer away. What minimum speed must you have as you reach the ramp in order for her to live to do this again tomorrow?

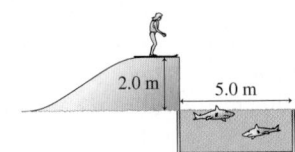

FIGURE CP11.72

73. The spring in Figure CP11.73 has a spring constant of 1000 N/m. It is compressed 15 cm, then launches a 200 g block. The horizontal surface is frictionless, but the block's coefficient of kinetic friction on the incline is 0.20. What distance *d* does the block sail through the air?

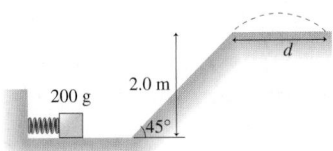

FIGURE CP11.73

74. As a hobby, you like to participate in reenactments of Civil War battles. Civil War cannons were "muzzle loaded," meaning that the gunpowder and the cannonball were inserted into the output end of the muzzle, then tamped into place with a long plunger. To recreate the authenticity of muzzle-loaded

cannons, but without the danger of real cannons, Civil War buffs have invented a spring-powered cannon that fires a 1.0 kg plastic ball. A spring, with spring constant 3000 N/m, is mounted at the back of the barrel. You place a ball in the barrel, then use a long plunger to press the ball against the spring and lock the spring into place, ready for firing. In order for the latch to catch, the ball has to be moving at a speed of at least 2.0 m/s when the spring has been compressed 30 cm. The coefficient of friction of the ball in the barrel is 0.30. The plunger doesn't touch the sides of the barrel.

 a. If you push the plunger with a constant force, what is the minimum force that you must use to compress and latch the spring? You can assume that no effort was required to push the ball down the barrel to where it first contacts the spring.

 b. What is the cannon's muzzle velocity if the ball travels a total distance of 1.5 m to the end of the barrel?

75. The equation *mgy* for gravitational potential energy is valid only for objects near the surface of a planet. Consider two very large objects of mass m_1 and m_2, such as stars or planets, whose centers are separated by the large distance *r*. These two large objects exert gravitational forces on each other. You'll learn in Chapter 12 that the gravitational potential energy is

$$U = -\frac{Gm_1m_2}{r}$$

where $G = 6.67 \times 10^{-11} \text{ Nm}^2/\text{kg}^2$ is called the *gravitational constant.*

 a. Sketch a graph of *U* versus *r*. The mathematical difficulty at $r = 0$ is not a physically significant problem because the masses will collide before they get that close together.

 b. What separation *r* has been chosen as the point of zero potential energy? Does this make sense? Explain.

 c. Two stars are at rest 1.0×10^{14} m apart. This is about 10 times the diameter of the solar system. The first star is the size of our sun, with a mass of 2.0×10^{30} kg and a radius of 7.0×10^8 m. The second star has mass 8.0×10^{30} kg and radius 11.0×10^8 m. Gravitational forces pull the two stars together. What is the speed of each star at the moment of impact?

Stop to Think 11.1: d. Constant speed means $\Delta K = 0$. Gravitational potential energy is lost and friction heats up the slide and the child's pants.

Stop to Think 11.2: 6.0 J. $K_f = K_i + W$. *W* is the area under the curve, which is 4 J.

Stop to Think 11.3: b. The weight force $\vec{w}$ is in the same direction as the displacement. It does positive work. The tension force $\vec{T}$ is opposite the displacement. It does negative work.

Stop to Think 11.4: c. $W = F(\Delta r)\cos\theta$. The 10 N force at 90° does no work at all. $\cos 60° = \frac{1}{2}$, so the 8 N force does less work than the 6 N force.

Stop to Think 11.5: e. Force is the negative of the slope of the potential energy diagram. At $x = 4$ m the potential energy has risen by 4 J over a distance of 2 m, so the slope is 2 J/m = 2 N.

Stop to Think 11.6: c. Constant speed means $\Delta K = 0$. Gravitational potential energy is lost, and friction heats up the pole and the child's hands.

Stop to Think 11.7: $P_b > P_a = P_c > P_d$. The work done is $mg\Delta y$, so the power is $mg\Delta y/\Delta t$. Runner b does the same work as a but in less time. The ratio $m/\Delta t$ is the same for a and c. Runner d does twice the work as a but takes more than twice as long.

PART

II

Conservation Laws

SUMMARY

In Part II we have discovered that we don't need to know all the details of an interaction to relate the properties of a system "before" the interaction to the system's properties "after" the interaction. Along the way, we found two important quantities, momentum and energy, that are often conserved. Momentum and energy are specific characteristics of a system of particles.

Momentum and energy have specific conditions under which they are conserved. The total momentum $\vec{P}$ and the total energy E_{sys} are conserved for an *isolated system,* one on which the net external force is zero. Further, the system's mechanical energy is conserved if the system is both isolated and nondissipative (i.e., no friction forces). These ideas are captured in the two most important conservation laws, the law of conservation of momentum and the law of conservation of energy.

Of course, not all systems are isolated. For both momentum and energy, it was useful to develop a *model* of a sys-

tem interacting with its environment. Interactions within the system do not change $\vec{P}$ or E_{sys}. The kinetic, potential, and thermal energy within the system can be transformed without changing E_{sys}. Interactions between the system and the environment *do* change the system's momentum and energy. In particular,

■ Impulse is the transfer of momentum to or from the system: $\Delta p_s = J_s$.
■ Work is the transfer of energy to or from the system: $\Delta E_{sys} = W_{ext}$.

The basic energy model is built around the twin ideas of the transfer and the transformation of energy.

The table below is a knowledge structure of conservation laws. You should compare this with the knowledge structure of Newtonian mechanics in the Part I Summary. Add the problem-solving strategies, and you now have a very powerful set of tools for understanding motion.

KNOWLEDGE STRUCTURE II **Conservation Laws**

ESSENTIAL CONCEPTS	Impulse, momentum, work, energy
BASIC GOALS	How is the system "after" an interaction related to the system "before"? What quantities are conserved, and under what conditions?

GENERAL PRINCIPLES	**Impulse-momentum theorem**	$\Delta p_s = J_s$
	Work-kinetic energy theorem	$\Delta K = W = W_c + W_{diss} + W_{ext}$
	Energy equation	$\Delta E_{sys} = \Delta K + \Delta U + \Delta E_{th} = W_{ext}$

CONSERVATION LAWS For an isolated system, with $\vec{F}_{net} = \vec{0}$ and $W_{net} = 0$
• The total momentum $\vec{P}$ is conserved.
• The total energy $E_{sys} = E_{mech} + E_{th}$ is conserved.

For an isolated and nondissipative system, with $W_{diss} = 0$
• The mechanical energy $E_{mech} = K + U$ is conserved.

BASIC PROBLEM-SOLVING STRATEGY Draw a before-and-after pictorial representation, then use the momentum or energy equations to relate "before" to "after." Where possible, choose a system for which momentum and/or energy are conserved. If necessary, calculate impulse and/or work.

Basic model of momentum and energy

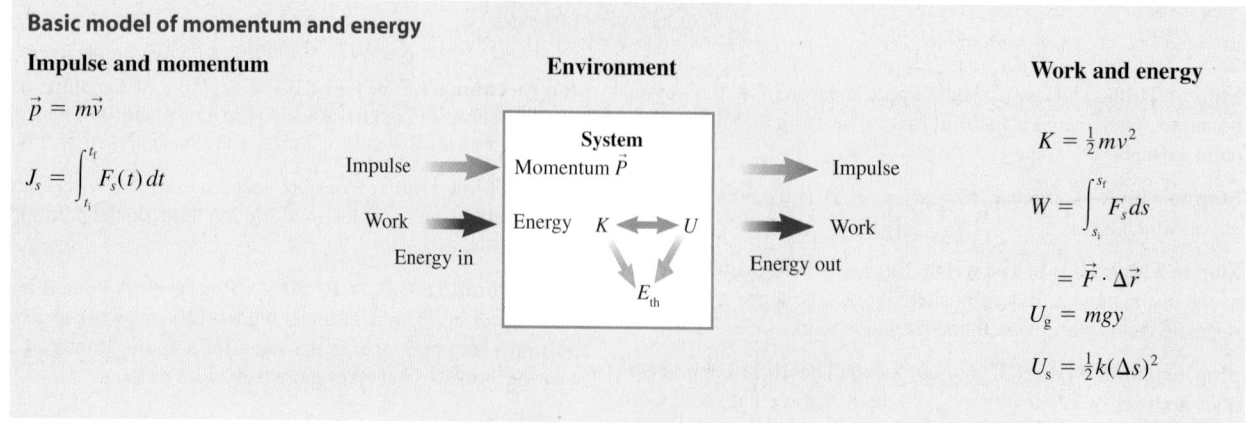

Impulse and momentum

$\vec{p} = m\vec{v}$

$J_s = \int_{t_i}^{t_f} F_s(t)\, dt$

Environment

Impulse →

Work →

Energy in

System
Momentum $\vec{P}$

Energy $K \longleftrightarrow U$

E_{th}

→ Impulse

→ Work

Energy out

Work and energy

$K = \frac{1}{2}mv^2$

$W = \int_{s_i}^{s_f} F_s\, ds$

$= \vec{F} \cdot \Delta \vec{r}$

$U_g = mgy$

$U_s = \frac{1}{2}k(\Delta s)^2$

Energy Conservation

You hear it all the time. Turn off lights. Buy a more fuel-efficient car. Conserve energy. But why conserve energy if energy is already conserved? Consider the earth as a whole. No work is done on the earth. And while heat energy flows from the sun to the earth, the earth radiates an equal amount of heat back into space. With no work and no net heat flow, the earth's total energy E_{earth} is conserved.

Pumping oil, driving your car, running a nuclear reactor, or turning on the lights are all interactions *within* the earth system. They transform energy from one type to another, but they don't affect the value of E_{earth}. Consider some examples.

■ Crude oil, stored in the earth, has chemical energy E_{chem}. Chemical energy, a form of microscopic potential energy, is released when chemical reactions rearrange the bonds. As you burn gasoline in your car engine, the chemical energy is transformed into the kinetic energy of the moving pistons. This kinetic energy, in turn, is transformed into the car's kinetic energy. The car's kinetic energy is ultimately dissipated as thermal energy in the brakes, air, tires, and road because of friction and drag. Overall, the energy process of driving looks like

$$E_{chem} \rightarrow K_{piston} \rightarrow K_{car} \rightarrow E_{th}$$

■ Water stored behind a dam has gravitational potential energy U_g. Potential energy is transformed into kinetic energy as the water falls, then into the spinning turbine's kinetic energy. The turbine converts mechanical energy into electric energy E_{elec}. The electric energy reaches a light bulb where it is transformed partly into thermal energy (light bulbs are hot!) and partly into light energy. The light is absorbed by surfaces, heating them slightly and thus transforming the light energy into thermal energy. The overall energy process is

$$U_g \rightarrow K_{water} \rightarrow K_{turbine} \rightarrow E_{elec} \rightarrow E_{light} \rightarrow E_{th}.$$

Do you notice a trend? Stored energy (fossil fuel, water behind a dam) is transformed through a series of steps, some of which are considered "useful," until the energy is ultimately dissipated as thermal energy. **The total energy has not changed, but its "usefulness" has.**

Thermal energy is rarely "useful" energy. A room full of moving air molecules has a huge thermal energy, but you can't run your lights or your air conditioner with it. You can't turn the thermal energy of your hot brakes back into the kinetic energy of the car. Energy may be conserved, but there's a one-way characteristic of the transformations.

The energy stored in fuels and the energy of the sun is "high-quality energy" because of its potential to be transformed into such useful forms of energy as moving your car or heating your house. But as Figure II.1 shows, high-quality energy becomes "degraded" into thermal energy, where it is no longer useful. Thus the phrase "conserve energy" isn't used literally. Instead, it means to conserve or preserve the earth's sources of high-quality energy.

Conserving high-quality energy is important because fossil fuels are a finite resource. Experts may disagree as to how long fossil fuels will last, but all agree that it won't be forever. Oil and natural gas will likely become scarce during your lifetime. In addition, burning fossil fuel generates carbon dioxide, a major contributor to global warming. Energy conservation helps fuels last longer and minimizes their side effects.

There are two paths to conserving energy. One is to use less high-quality energy. Turning off lights and bicycling rather than driving are actions that preserve high-quality energy. A second path is to use energy more efficiently. That is, get more of the useful activity (miles driven, rooms lit) for the same amount of high-quality energy.

Light bulbs offer a good example. A 100 W incandescent light bulb actually produces only about 10 W of light energy. Ninety watts of the high-quality electric energy is immediately degraded as thermal energy without doing anything useful. By contrast, a 25 W compact fluorescent bulb generates the same 10 W of light but only 15 W of thermal energy. The same amount of high-quality energy can light four times as many rooms if 100 W incandescent bulbs are replaced by 25 W compact fluorescent bulbs.

So why conserve energy if energy is already conserved? Because technological society needs a dependable and sustainable supply of high-quality energy. Both technology improvement and lifestyle choices will help us achieve a sustainable energy future.

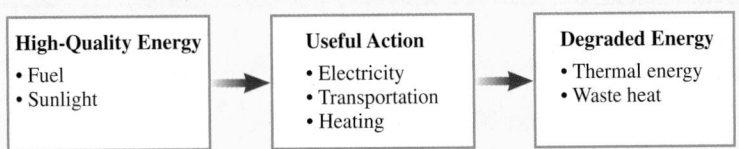

FIGURE II.1 "Using" energy transforms high-quality energy into thermal energy.

PART III

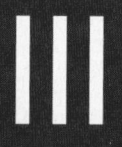

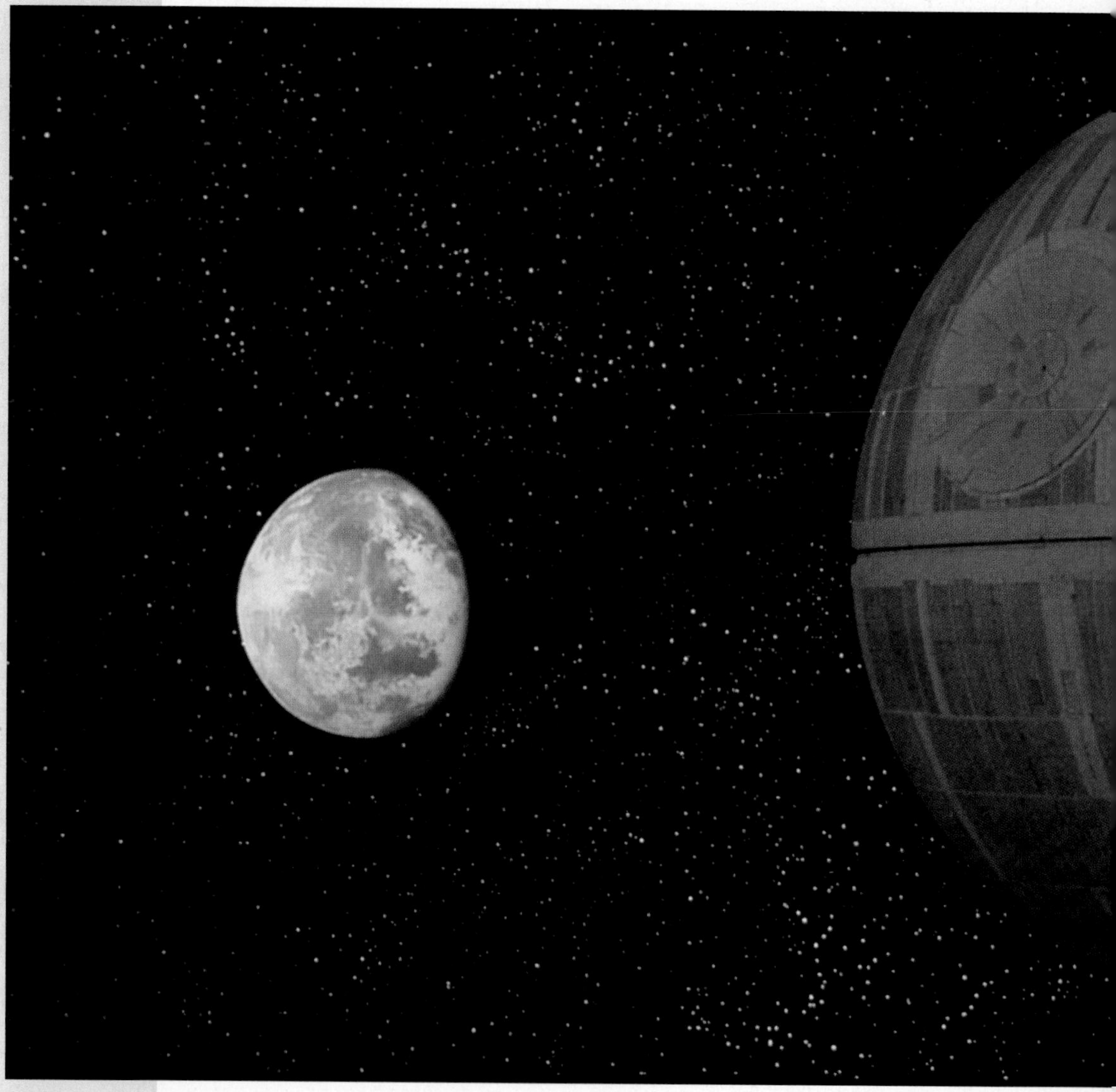

At what altitude above the surface of Alderaan does the Death Star orbit with a period of 10 hours? To find out, what properties of the planet do you need to estimate?

Applications of Newtonian Mechanics

From Nanostructures to the Stars

Engineering arose as a distinct discipline, separate from science, during the industrial revolution. Initially, the engineer's goal was to design new and better machines. Science and engineering in the 21st century may bear little resemblance to the practices of those early days, but the underlying principles remain the same. Whether your dream is to build nanomachines or to pilot spaceships to the stars, much of engineering and applied science is still built on a foundation of Newtonian mechanics.

Parts I and II of this textbook have established the *theory* of Newtonian mechanics. It is a theory of motion based on the concept of *forces* applied to *particles*. Our focus has been to develop the principal ideas of the theory, so you haven't seen too many examples of how this theory is applied to the "real world." In fact, most of the applications of Newtonian mechanics will be developed in other science and engineering courses. Nonetheless, we're now in a good position to examine some of the more practical aspects of Newtonian mechanics.

Our goal for Part III is to apply our newfound theory to four important topics:

- **Gravity.** By adding one more law, Newton's law of gravity, we'll be able to understand much about the physics of the space shuttle, communication satellites, the solar system, and interplanetary travel.
- **Rotation.** The primary limitation of the particle model is that we can't represent a spinning or rotating object as a particle. To understand rotation, a very important form of motion, we'll need to learn how to extend the particle model. We'll then be able to study the motion of objects such as rolling wheels and spinning space stations. Our study of rotation will also lead to another important conservation law, the law of conservation of angular momentum.
- **Oscillations.** Oscillations and vibrations are another important form of motion with many practical applications. Oscillations are seen in systems ranging from the pendulum in a grandfather's clock and the pistons in your car's engine to pieces of machinery that vibrate and the quartz crystal oscillator that provides the timing signals in sophisticated electronic circuits. The physics and mathematics of oscillations will also be the starting point for two other important topics later in this book: waves and AC circuits.
- **Fluids.** Liquids and gases are classified as fluids because they *flow*. Surprisingly, it takes no new physics to understand the basic mechanical properties of fluids. (The thermal properties of fluids, such as the ideal-gas law, will be studied in Part IV.) However, we will have to go well beyond the particle model to see how Newton's laws are applied to liquids and gases. By doing so we'll be able to understand what pressure is, how a steel ship can float, and how fluids flow through pipes.

atively few new concepts will be introduced in Part III. Instead, we will focus on *applying* our two primary sets of tools:

- Newton's laws of motion, and
- The conservation laws, especially conservation of energy.

These tools will help us analyze and understand a variety of interesting and practical applications.

Nanostructures

Nanostructures are motors, machines, and materials smaller than the width of a human hair. Some day, nanomachines introduced into a blood vessel may be able to perform microscopic surgery to repair a damaged heart. Nanostructures may become the basis for new materials that are stronger than steel but lighter than silk. Although nanostructures are mostly still in the research stage, they may come to dominate 21st-century technology.

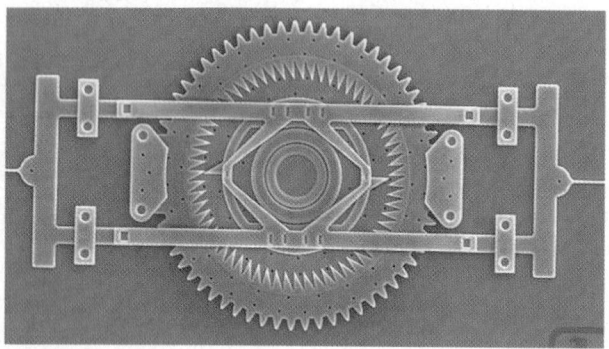

This microscopic machine is about the size of a human hair.

One interesting possibility is to develop self-assembling structures, not unlike mechanical DNA, wherein one type of nanomachine can build other nanomachines, including copies of itself, from raw materials. It's safe to say that nanostructures have megapotential.

But what do you need to know to design and use nanostructures? Mostly Newtonian mechanics. The forces and torques may be at the microscopic level, but you still have to analyze how the device will respond to those forces and torques. Friction, rather than being a minor nuisance, plays a crucial role that must be understood in detail. And many nanostructures will operate in fluids, so you must know the properties of both static and moving fluids.

Space

Humans have already visited the moon. It is entirely possible, perhaps even likely, that we'll begin to visit other planets during your lifetime. And it's undoubtedly true that

unmanned spacecraft will continue to explore the solar system and beyond. Astronauts may get the glory, but countless engineers and scientists make space flight possible. They design the rockets, the instruments, and the life-support systems. They also determine the trajectory that a spacecraft must follow to reach its destination. These are all applications of Newtonian mechanics.

How do you get a satellite into orbit? How do you move a spacecraft from earth orbit onto a trajectory toward Mars? Questions that have to be answered include

- What is the optimum trajectory through space?
- How long will it take?
- How much energy will be required?
- How will you stop the spacecraft when it gets to its destination?
- How will you get it back?

These are complex problems, but you should recognize that they are not fundamentally different from the types of problems you have learned to solve in Parts I and II.

Power over Our Environment

Early humans, for hundreds of thousands of years, had to endure whatever nature provided. Only within the last few thousand years have agriculture, fire, and simple machines provided some level of control over the environment. And it has been a mere couple of centuries since machines, and later electronics, began to do much of our work and provide us with "creature comforts."

Is it a coincidence that machines began to appear about a hundred years after Newton? Of course not. Galileo, Newton, and others in the 17th and 18th centuries ignited what we now call the *scientific revolution*. The machines and other devices we take for granted today are direct consequences of scientific knowledge and the scientific method.

But science and engineering are a two-edged sword. Science has given us the power to control our environment. At the same time, much of the progress of the last two hundred years has come at the expense of the environment. We humans have deforested much of the world, polluted our air and water, and driven many of our fellow travelers on Spaceship Earth to extinction. Now, at the beginning of the 21st century, the evidence is increasingly clear that humans are altering the earth's climate and causing other global changes.

Fortunately, science also gives us the ability to understand the consequences of our actions and to develop better techniques and procedures. It is more important than ever that scientists and engineers in the 21st century distinguish control that is beneficial from control that is harmful. We'll return to some of these ideas in the Summary to Part III.

12 Newton's Theory of Gravity

A solar eclipse occurs when the moon passes between the earth and the sun. The ancient Babylonians learned to predict eclipses, a remarkable feat of early science.

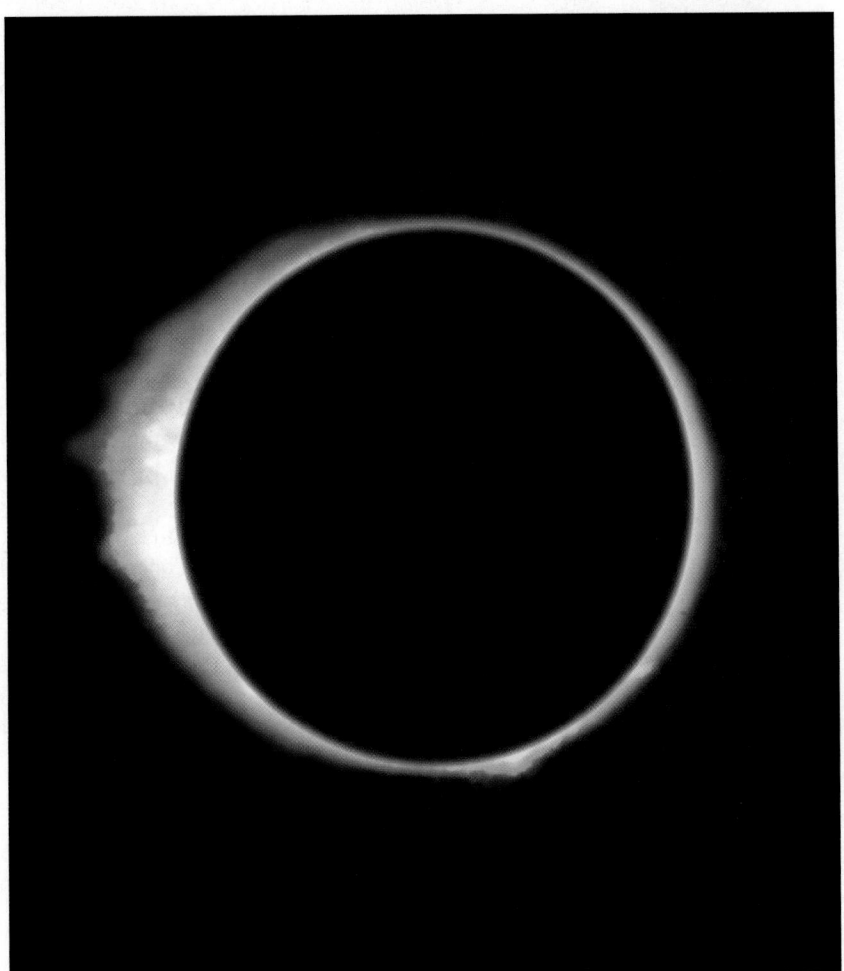

▶ **Looking Ahead**

The goal of Chapter 12 is to use Newton's theory of gravity to understand the motion of satellites and planets. In this chapter you will learn to:

- Place Newton's discovery of the law of gravity in historical context.
- Use Newton's theory of gravity to solve problems about orbital motion.
- Understand Kepler's laws of planetary orbits.
- Understand and use gravitational potential energy.

◀ **Looking Back**

Newton's theory of gravity depends on the properties of uniform circular motion. Please review:

- Sections 7.1–7.4 Uniform circular motion and circular orbits.
- Section 9.7 Angular momentum.
- Section 10.2 Gravitational potential energy.

Every ancient culture was fascinated with the motion of the heavens above. Without city lights or urban haze, the nighttime sky and the daytime sun were ever present, powerful experiences. The unknown people that built Stonehenge clearly used it as a solar observatory, and the ancient Babylonians learned to predict the occurrence of solar eclipses.

Our fascination with sky and the stars has not diminished in the 21st century. Today our interest may be in galaxies, black holes, and the Big Bang, but we're still exploring the heavens. One of the most important discoveries of science is that one pervasive force is dominant throughout the universe. This force is responsible for phenomena ranging from the orbiting space shuttle and solar eclipses to the dynamics of galaxies and the expansion of the universe. It is the force of gravity.

Newton's formulation of his theory of gravity was a pivotal event in the history of science. It was the first scientific theory to have broad explanatory and predictive power. Although Newton's theory is now over three centuries old, its importance has not diminished with time.

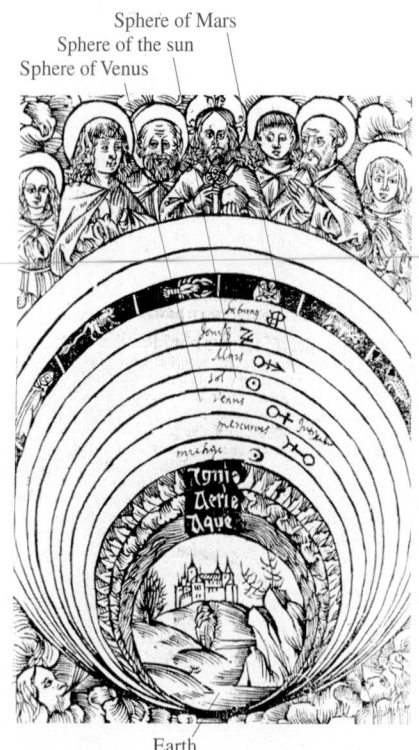

Sphere of Mars
Sphere of the sun
Sphere of Venus

Earth

FIGURE 12.1 The earth-centered cosmology of the ancient Greek and medieval periods.

12.1 A Little History

The study of the structure of the universe is called **cosmology.** The ancient Greeks developed a cosmological model, illustrated in Figure 12.1, that placed the earth at the center of the universe while the moon, the sun, the planets, and the stars were points of light that turned about the earth on large "celestial spheres." This viewpoint was further expanded by the second-century Egyptian astronomer Ptolemy (the P is silent). He developed an elaborate mathematical model of the solar system that quite accurately predicted the complex planetary motions. Ptolemy's earth-centered cosmology was accepted without question for more than a thousand years in medieval Europe.

Then, in the year 1543, the medieval world was turned on its head with the publication of Nicholas Copernicus's *De Revolutionibus.* Copernicus argued that it is not the earth at rest in the center of the universe—it is the sun! Furthermore, Copernicus asserted that all of the planets, including the earth, revolve about the sun (hence his title) in circular orbits.

Copernicus suspected that his ideas would not sit well with the authorities, and he tactfully waited until he was on his deathbed to have his book published. The Catholic Church did, indeed, denounce his views and for more than a hundred years persecuted astronomers who used Copernican ideas. After all, Copernicus's attempt to remove the earth from the center of the universe undermined the very foundations of medieval theology. Not until many decades later, when Galileo used a telescope to study the heavens, did the Copernican view become widely accepted.

The greatest astronomer of this period was Tycho Brahe, a Dane born in 1546, just three years after Copernicus's death. Tycho began observing the heavens as a teenager, and for 30 years, from 1570 to 1600, compiled the most accurate astronomical observations the world had known. The invention of the telescope was still to come, so Tycho's observations were all with the naked eye. However, Tycho did develop ingenious mechanical sighting devices that allowed him to determine the positions of stars and planets in the sky with unprecedented accuracy. Despite his careful observations, Tycho never accepted Copernicus's assertion that the earth was in motion.

Kepler

Tycho willed his records to a young mathematical assistant named Johannes Kepler. Kepler had become one of the first outspoken defenders of Copernicus, and one of his goals was to find evidence for circular planetary orbits in Tycho's records. To appreciate the difficulty of this task, keep in mind that Kepler was working before the development of graphs or of calculus—and certainly before calculators! His mathematical tools were algebra, geometry, and trigonometry, and he was faced with thousands upon thousands of individual observations of planetary positions measured as angles above the horizon.

Ten years of effort without success eventually led Kepler to think that Copernicus's insistence on circular orbits was unnecessary. Perhaps orbits could have other shapes. This was the mental barrier that had to be broken, and once beyond it Kepler was quickly able to deduce that the orbit of Mars is, indeed, not a circle but an *ellipse!* Furthermore, the speed of the planet is not constant but varies as it moves around the ellipse.

Kepler made three major discoveries that today we call **Kepler's laws** of planetary orbits:

1. The planets move in elliptical orbits, with the sun at one focus of the ellipse.
2. A line drawn between the sun and a planet sweeps out equal areas during equal intervals of time.

3. The square of a planet's orbital period is proportional to the cube of the semimajor-axis length.

Figure 12.2a shows that an ellipse has two *foci* (plural of *focus*), and the sun occupies one of these. The long axis of the ellipse is the *major axis,* and half the length of this axis is called the *semimajor-axis length.* As the planet moves, a line drawn from the sun to the planet "sweeps out" an area. Figure 12.2b shows two such areas. Kepler's discovery that the areas are equal for equal Δt implies that the planet moves faster when near the sun, slower when farther away.

(a)   **(b)**

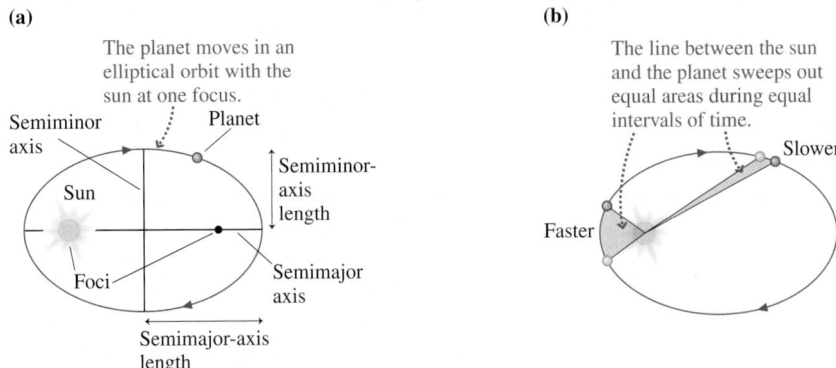

FIGURE 12.2 The elliptical orbit of a planet about the sun.

All the planets except Mercury and Pluto have elliptical orbits that are only very slightly distorted circles. As Figure 12.3 shows, a circle is an ellipse in which the two foci move to the center, effectively making one focus, and the semimajor-axis length becomes the radius. Because the mathematics of ellipses is difficult, this chapter will focus on circular orbits.

Kepler made an additional contribution that is less widely recognized but that was essential to prepare the way for Newton. For Ptolemy and, later, Copernicus, the role of the sun was merely to light and warm the earth and planets. The sun had no role in their movement. Kepler was the first to suggest that the sun was a center of force that somehow *caused* the planetary motions. Now, Kepler was working before Galileo and Newton, so he did not speak in terms of forces and centripetal accelerations. He thought that some type of rays or spirit emanated from the sun and pushed the planets around their orbits. The value of his contribution was not the specific mechanism he proposed but his introduction of the idea that the sun somehow exerts forces on the planets to determine their motion.

Kepler published the first two of his laws in 1609, the same year in which Galileo first turned a telescope to the heavens. Through his telescope Galileo could *see* moons orbiting Jupiter, just as Copernicus had suggested the planets orbit the sun. He could *see* that Venus has phases, like the moon, which implied its orbital motion about the sun. And he could *see* spots moving across the face of the sun, indicating that it was not a perfect and unchanging symbol of God. Galileo was tried and convicted of heresy (this was still the time of the Inquisition), forced to publicly recant his views, and left a broken man. But his evidence was too persuasive, and by the time of his death in 1642 the Copernican revolution was complete.

In hindsight, we can see that Kepler's analysis and Galileo's observations had set the stage for a major theoretical leap. All that was needed was a great intellect to recognize and pull together these ideas. Enter Isaac Newton, one of the most brilliant scientists ever to live.

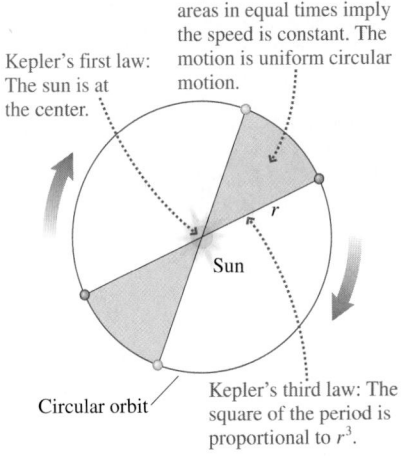

FIGURE 12.3 A circular orbit is a special case of an elliptical orbit.

12.2 Isaac Newton

Isaac Newton was born to a poor farming family in 1642, the year of Galileo's death. He entered Trinity College at Cambridge University at age 19 as a "sub-sizar," a poor student who had to work his way through school. Newton graduated in 1665, at age 23, just as an outbreak of the plague in England forced the universities to close for two years. He returned to his family farm for that period, during which he made important experimental discoveries in optics, laid the foundations for his theories of mechanics and gravitation, and made major progress toward his invention of calculus as a whole new branch of mathematics.

A popular image has Newton thinking of the idea of gravity after an apple fell on his head. This amusing story is at least close to the truth. Newton himself said that the "notion of gravitation" came to him as he "sat in a contemplative mood" and "was occasioned by the fall of an apple." It occurred to him that, perhaps, the apple was attracted to the *center* of the earth but was prevented from getting there by the earth's surface. And if the apple was so attracted, why not the moon?

Robert Hooke, discoverer of Hooke's law, had already suggested that the planets might be attracted to the sun with a strength proportional to the inverse square of the distance between the sun and the planet. This seems to have been a hunch rather than being based on any particular evidence, and Hooke failed to follow up on the idea. Newton's genius was not just his successful application of Hooke's suggestion, but his sudden realization that **the force of the sun on the planets was identical to the force of the earth on the apple.** In other words, gravitation is a *universal* force between all objects in the universe! This is not shocking today, but no one before Newton had ever thought that the mundane motion of objects on earth had any connection at all with the stately motion of the planets around the sun.

Newton reasoned along the following lines. Suppose the moon's circular motion around the earth is due to the pull of the earth's gravity. Then, as Figure 12.4a shows, the force on the moon has magnitude $F_m = m_m g_{at\ moon}$, where $g_{at\ moon}$ is the acceleration due to the earth's gravity at the distance of the moon. We cannot assume that $g_{at\ moon}$ is the familiar 9.80 m/s^2. If the strength of gravity diminishes with distance, as Hooke had suggested, then the acceleration due to gravity will also decrease.

> NOTE ▶ We need to be careful with notation. The symbol g_{moon} would be the acceleration caused by the *moon's* gravity—that is, the acceleration of a falling object on the moon. Here we're interested in the acceleration *of* the moon by the earth's gravity, which we'll call $g_{at\ moon}$. ◀

The moon's acceleration, which is found from Newton's second law, is $\vec{a}_m = \vec{F}_m/m_m = (g_{at\ moon}, \text{toward the earth})$. This centripetal acceleration, shown in Figure 12.4b, has only the radial component $a_r = g_{at\ moon}$. As you learned in Chapter 7, the centripetal acceleration of an object in uniform circular motion is

$$a_r = g_{at\ moon} = \frac{v_m^2}{r_m} \tag{12.1}$$

The moon's speed is related to the radius r_m and period T_m of its orbit by $v_m = $ circumference/period $= 2\pi r_m/T_m$. Combining these, Newton found

$$g_{at\ moon} = \frac{4\pi^2 r_m}{T_m^2} = \frac{4\pi^2(3.84 \times 10^8 \text{ m})}{(2.36 \times 10^6 \text{ s})^2} = 0.00272 \text{ m/s}^2$$

Astronomical measurements had established a reasonably good value for r_{moon} by the time of Newton, and the period $T_m = 27.3$ days was quite well known.

Isaac Newton, 1642–1727.

(a)

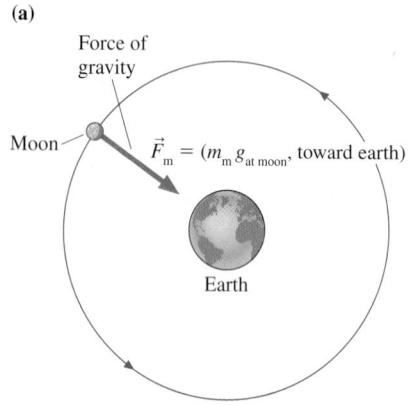

Force of gravity

Moon

$\vec{F}_m = (m_m g_{at\ moon}, \text{toward earth})$

Earth

(b)

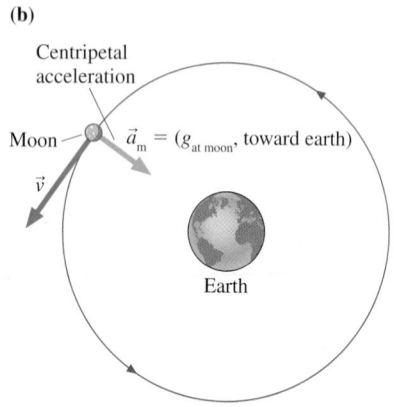

Centripetal acceleration

Moon

$\vec{a}_m = (g_{at\ moon}, \text{toward earth})$

$\vec{v}$

Earth

FIGURE 12.4 The centripetal acceleration of the moon is due to the pull of the earth's gravity.

The moon's centripetal acceleration is significantly less than the acceleration due to gravity on the earth's surface. In fact,

$$\frac{g_{\text{at moon}}}{g_{\text{on earth}}} = \frac{0.00272 \text{ m/s}^2}{9.80 \text{ m/s}^2} = \frac{1}{3600}$$

This is an interesting result, but it was Newton's next step that was critical. He compared the radius of the moon's orbit to the radius of the earth:

$$\frac{r_{\text{m}}}{R_{\text{e}}} = \frac{3.84 \times 10^8 \text{ m}}{6.37 \times 10^6 \text{ m}} = 60.2$$

NOTE ▶ We'll use a lowercase *r*, as in r_{m}, to indicate the radius of an orbit. We'll use an uppercase *R*, as in R_{e}, to indicate the radius of a star or planet. ◀

Newton recognized that 60.2^2 is almost exactly 3600. Thus, he reasoned:

- If *g* has the value 9.80 at the earth's surface, and
- If the force of gravity and *g* decrease in size depending inversely on the square of the distance from the center of the earth,
- Then *g* will have exactly the value it needs at the distance of the moon to cause the moon to orbit the earth with a period of 27.3 days.

His two ratios were not identical (because the earth isn't a perfect sphere and the moon's orbit isn't a perfect circle), but he found them to "answer pretty nearly" and knew that he had to be on the right track.

This flash of insight changed our most basic understanding of the universe. Copernicus displaced the earth from the center of the universe, and now Newton had shown that the laws of the heavens and the laws of earth are the same. Nonetheless, Newton did not publish his results for a long 22 years. The issue that troubled him was treating the sun, the earth, and the other planets as if they were single particles with all their mass at the center. If his idea about a universal force was correct, then *every atom* in the earth exerts a force on *every atom* in the moon. Newton had to show that all of these forces add up to give a result that is identical with treating the bodies as single particles. This is a problem in integral calculus, and Newton had first to develop the necessary mathematics. He did eventually succeed, and his theory of gravitation was published in 1687 along with his theory of mechanics (which we know as Newton's laws) in his great work *Philosophia Naturalis Principia Mathematica* (Mathematical Principles of Natural Philosophy). The rest is history.

I deduced that the forces which keep the planets in their orbs must be reciprocally as the squares of their distances from the centers about which they revolve; and thereby compared the force requisite to keep the Moon in her orb with the force of gravity at the surface of the Earth; and found them answer pretty nearly.

Isaac Newton

STOP TO THINK 12.1   A satellite orbits the earth with constant speed at a height above the surface equal to the earth's radius. The magnitude of the satellite's acceleration is

a. $4g_{\text{on earth}}$ b. $2g_{\text{on earth}}$ c. $g_{\text{on earth}}$
d. $\frac{1}{2}g_{\text{on earth}}$ e. $\frac{1}{4}g_{\text{on earth}}$ f. 0

12.3 Newton's Law of Gravity

Newton proposed that *every* object in the universe attracts *every other* object with a force that has the following properties:

1. The force is inversely proportional to the square of the distance between the objects.
2. The force is directly proportional to the product of the masses of the two objects.

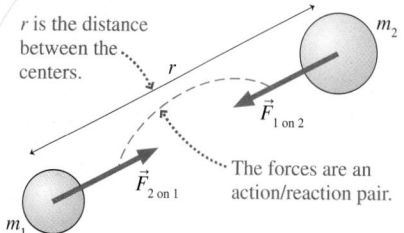

r is the distance between the centers.

$\vec{F}_{1 \text{ on } 2}$

$\vec{F}_{2 \text{ on } 1}$

The forces are an action/reaction pair.

m_1

m_2

r

FIGURE 12.5 The gravitational forces on masses m_1 and m_2.

To make these ideas more specific, Figure 12.5 shows masses m_1 and m_2 separated by distance r. Each mass exerts an attractive force on the other, a force that we call the **gravitational force.** These two forces form an action/reaction pair, so $\vec{F}_{1 \text{ on } 2}$ is equal and opposite to $\vec{F}_{2 \text{ on } 1}$. The magnitude of the forces is given by Newton's law of gravity.

Newton's law of gravity If two objects with masses m_1 and m_2 are a distance r apart, the objects exert attractive forces on each other of magnitude

$$F_{1 \text{ on } 2} = F_{2 \text{ on } 1} = \frac{Gm_1m_2}{r^2} \tag{12.2}$$

The forces are directed along the straight line joining the two objects.

The constant G, called the **gravitational constant,** is a proportionality constant necessary to relate the masses, measured in kilograms, to the force, measured in Newtons. In the SI system of units, G has the value

$$G = 6.67 \times 10^{-11} \text{ N m}^2/\text{kg}^2$$

Figure 12.6 is a graph of the gravitational force as a function of the distance between the two masses. As you can see, an inverse-square force decreases rapidly. Doubling the distance between the masses causes the force to decrease by a factor of 4.

Strictly speaking, Equation 12.2 is valid only for particles. As we noted, however, Newton was able to show that this equation also applies to spherical objects, such as planets, if r is the distance between their centers. Our intuition and common sense suggest this to us, as they did to Newton, but the proof is not essential, so we will omit it.

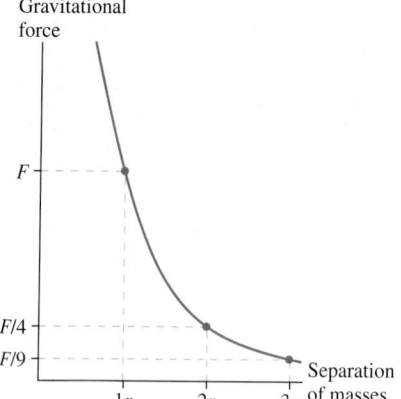

Gravitational force

F

$F/4$

$F/9$

$1r$ $2r$ $3r$ Separation of masses

FIGURE 12.6 The gravitational force is an inverse-square force.

Gravitational Force and Weight

Knowing G, we can calculate the size of the gravitational force. Consider two 1.0 kg masses that are 1.0 m apart. According to Newton's law of gravity, these two masses exert an attractive gravitational force on each other. The magnitude of the force is

$$F_{1 \text{ on } 2} = F_{2 \text{ on } 1} = \frac{Gm_1m_2}{r^2}$$

$$= \frac{(6.67 \times 10^{-11} \text{ N m}^2/\text{kg}^2)(1.0 \text{ kg})(1.0 \text{ kg})}{(1.0 \text{ m})^2} = 6.67 \times 10^{-11} \text{ N}$$

This is an exceptionally tiny force, especially when compared to the weight of one of the masses: $w = mg = 9.8$ N.

The fact that the gravitational force between two ordinary-size objects is so small is the reason that we are not aware of it. As you sit there reading, you are being attracted to this book, to the person sitting next to you, and to every object around you, but the forces are so tiny in comparison to your weight and to the normal forces and friction forces acting on you that they are completely undetectable. Only when one (or both) of the masses is exceptionally large—planet-size—does the force of gravity become important.

We find a more respectable result if we calculate the force *of the earth* on a 1.0 kg mass at the earth's surface:

$$F_{\text{earth on 1 kg}} = \frac{GM_e m_{1\,\text{kg}}}{R_e^2}$$

$$= \frac{(6.67 \times 10^{-11}\,\text{N m}^2/\text{kg}^2)(5.98 \times 10^{24}\,\text{kg})(1.0\,\text{kg})}{(6.37 \times 10^6\,\text{m})^2} = 9.8\,\text{N}$$

where the distance between the mass and the center of the earth is the earth's radius. The earth's mass M_e and radius R_e were taken from Table 12.2 in Section 12.6. This table, which is also printed inside the cover of the book, contains astronomical data that will be used for examples and homework.

The force $F_{\text{earth on 1 kg}} = 9.8\,\text{N}$ is exactly the weight of a 1.0 kg mass: $w = mg = 9.8\,\text{N}$. Is this a coincidence? Of course not. The weight $w = mg$ of an object *is* the "force of gravity" acting on it. This suggests that we should be able to use Newton's law of gravity to *predict* the value of g. We will return to this idea in Section 12.4.

Although weak, gravity is a *long-range* force. No matter how far apart two objects may be, there is a gravitational attraction between them given by Equation 12.2. Consequently, gravity is the most ubiquitous force in the universe. It not only keeps your feet on the ground, it also keeps the earth orbiting the sun, the solar system orbiting the center of the Milky Way galaxy, and the entire Milky Way galaxy performing an intricate orbital dance with other galaxies making up what is called the "local cluster" of galaxies.

The dynamics of stellar motions, spanning distances of hundreds of thousands of light years, are governed by Newton's law of gravity.

A galaxy of $\approx 10^{11}$ stars spanning a distance greater than 100,000 light years.

The Principle of Equivalence

Newton's law of gravity depends on a rather curious assumption. The concept of *mass* was introduced in Chapter 4 by considering the relationship between force and acceleration. The *inertial mass* of an object, which is the mass that appears in Newton's second law, is found by measuring the object's acceleration a in response to force F:

$$m_{\text{inert}} = \text{inertial mass} = \frac{F}{a} \qquad (12.3)$$

Gravity plays no role in this definition of mass.

The quantities m_1 and m_2 in Newton's law of gravity are being used in a very different way. Masses m_1 and m_2 govern the strength of the gravitational attraction between two objects. The mass used in Newton's law of gravity is called the **gravitational mass.** The gravitational mass of an object can be determined by measuring the attractive force exerted on it by another mass M a distance r away:

$$m_{\text{grav}} = \text{gravitational mass} = \frac{r^2 F_{M\,\text{on}\,m}}{GM} \qquad (12.4)$$

Acceleration does not enter into the definition of the gravitational mass.

These are two very different concepts of mass. Yet Newton, in his theory of gravity, asserts that the inertial mass in his second law is the very same mass that governs the strength of the gravitational attraction between two objects. The assertion that $m_{\text{grav}} = m_{\text{inert}}$ is called the **principle of equivalence.** It says that inertial mass is *equivalent to* gravitational mass.

As a hypothesis about nature, the principle of equivalence is subject to experimental verification or disproof. Many exceptionally clever experiments have looked for any difference between the gravitational mass and the inertial mass, and they have shown that any difference, if it exists at all, is less than 10 parts in a trillion! As far as we know today, the gravitational mass and the inertial mass are exactly the same thing.

But why should a quantity associated with the dynamics of motion, relating force to acceleration, have anything at all to do with the gravitational attraction? This is a question that intrigued Einstein and eventually led to his general theory of relativity, the theory about curved space-time and black holes. General relativity is beyond the scope of this textbook, but it explains the principle of equivalence as a property of the space itself.

Newton's Theory of Gravity

Newton's theory of gravity is more than just Equation 12.2. The *theory* of gravity consists of:

1. A specific force law for gravity, given by Equation 12.2, *and*
2. The principle of equivalence, *and*
3. An assertion that Newton's three laws of motion are universally applicable. These laws are as valid for heavenly bodies, the planets and stars, as for earthly objects.

Consequently, everything we have learned about forces, motion, and energy is relevant to the dynamics of satellites, planets, and galaxies.

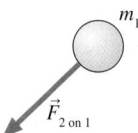
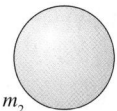

STOP TO THINK 12.2 The figure shows a binary star system. The mass of star 2 is twice the mass of star 1. Compared to $\vec{F}_{1\text{ on }2}$, the magnitude of the force $\vec{F}_{2\text{ on }1}$ is

a. four times as big.
b. twice as big.
c. the same size.
d. half as big.
e. one-quarter as big.

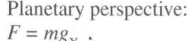
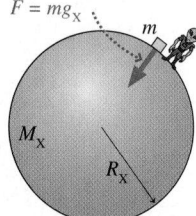

Planetary perspective:
$F = mg_X$

Planet X

Universal perspective:
$F = \dfrac{GM_X m}{R_X^2}$

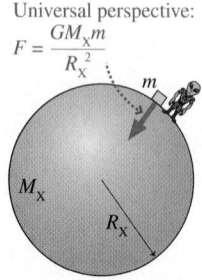

Planet X

FIGURE 12.7 Weighing an object of mass *m* on Planet X.

12.4 Little *g* and Big *G*

The familiar equation $w = mg$ works well when an object is on the surface of a planet, but mg will not help us find the force exerted on the same object if it is in orbit around the planet. Neither can we use mg to find the force of attraction between the earth and the moon. Newton's law of gravity provides a more fundamental starting point because it describes a *universal* force that exists between all objects.

To illustrate the connection between Newton's law of gravity and the familiar $w = mg$, Figure 12.7 shows an object of mass m on the surface of Planet X. Planet X inhabitant Mr. Xhzt places the object on a scale to weigh it and exclaims (translated), "Aha! This object has weight $w = mg_X$. The constant g_X is the acceleration due to gravity on my planet."

We, taking a more cosmic perspective, reply, "Yes, it has that weight *because* of a universal force of attraction between your planet and the object. The size of the force is determined by Newton's law of gravity."

We and Mr. Xhzt are both correct. Whether you choose to call it "weight" or "the force of gravitational attraction," we and Mr. Xhzt must arrive at the *same numerical value* for the magnitude of the force. Suppose an object of mass m is on the surface of a planet of mass M and radius R. The object's weight is

$$w = mg_{\text{surface}} \tag{12.5}$$

where $g_{surface}$ is the acceleration due to gravity at the planet's surface. The force of gravitational attraction, as given by Newton's law of gravity, is

$$F_{M \text{ on } m} = \frac{GMm}{R^2} \qquad (12.6)$$

Because these are two names and two expressions for the same force, we can equate the right-hand sides to find that

$$g_{surface} = \frac{GM}{R^2} \qquad (12.7)$$

We have used Newton's law of gravity to *predict* the value of *g* at the surface of a planet. The value depends on the mass and radius of the planet as well as on the value of *G*, which establishes the overall strength of the gravitational force.

The expression for $g_{surface}$ in Equation 12.7 is valid for any planet or star. Using the mass and radius of the earth, from Table 12.2, we can predict the earthly value of *g*:

$$g_{earth} = \frac{GM_e}{R_e^2} = \frac{(6.67 \times 10^{-11} \text{ N m}^2/\text{kg}^2)(5.98 \times 10^{24} \text{ kg})}{(6.37 \times 10^6 \text{ m})^2} = 9.83 \text{ m/s}^2$$

Galileo could measure g_{earth}, but he did not know *why* it was 9.8 m/s² rather than some other number. Now, with a more fundamental theory, we can see that the value for g_{earth} is a direct consequence of the size and mass of the earth. Similarly, you can show for homework that $g_{moon} = 1.62$ m/s² and $g_{jupiter} = 25.9$ m/s². A falling object on the moon would accelerate only about one-sixth as rapidly as on earth. On Jupiter, it would accelerate about 2.6 times more rapidly.

You probably noticed that the calculated value of g_{earth} is slightly larger than the "accepted" value of 9.80 m/s². Is there something wrong with the theory? No, the difference arises because we have not considered the earth's rotation. Figure 12.8 shows an object on the earth's equator as seen by an observer above the north pole. There are two forces on the object, the gravitational force $\vec{F}_{M \text{ on } m}$ and the normal force $\vec{n}$ of the ground. The *apparent weight* of the object, what a set of scales would read, is the magnitude *n* of the normal force. If the earth were not rotating, the object would be in static equilibrium and the two forces would exactly balance. Equation 12.7 gives the value of *g* on a nonrotating earth.

But the earth *is* rotating, and the object moves in a circle. An object in circular motion must have a *net* force acting on it, directed toward the center of the circle. Thus $n < F_{grav}$ for an object on a rotating planet. (This is the same reasoning we applied earlier to roller coasters doing loop-the-loops and buckets of water swinging overhead.) Because *n* is the apparent weight $w_{app} = mg_{app}$, the apparent value of *g* that we actually experience is slightly less than the value of *g* calculated from Newton's law of gravity. At midlatitudes, where most of us live, the reduction is just about 0.03 m/s², and hence the measured value for *g* on a rotating earth is 9.80 m/s².

Decrease of *g* with Distance

Equation 12.7 gives $g_{surface}$ at the surface of a planet. More generally, imagine an object of mass *m* at distance $r > R$ from the center of a planet. Further, suppose that gravity from the planet is the only force acting on the object. Then its acceleration, the acceleration due to gravity, is given by Newton's second law:

$$g = \frac{F_{M \text{ on } m}}{m} = \frac{GM}{r^2} \qquad (12.8)$$

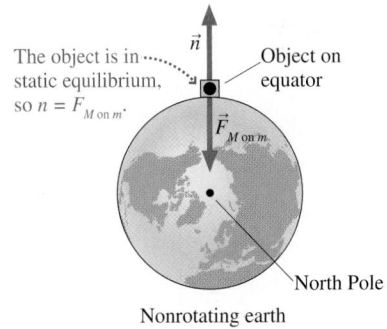

The object is in static equilibrium, so $n = F_{M \text{ on } m}$.

n — Object on equator

$\vec{F}_{M \text{ on } m}$

North Pole

Nonrotating earth

The object is in circular motion, so there is a net force toward the center. Thus $n < F_{M \text{ on } m}$.

$\vec{n}$

$\vec{F}_{M \text{ on } m}$

North Pole

Rotating earth

FIGURE 12.8 The earth's rotation affects the measured value of *g*.

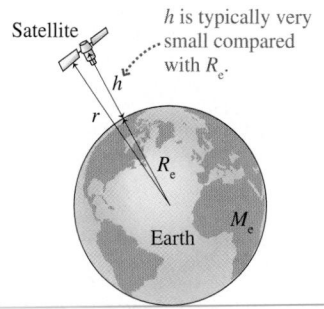

FIGURE 12.9 A satellite orbits the earth at height h.

This more general result agrees with Equation 12.7 if $r = R$, but it allows us to determine the "local" acceleration due to gravity at distances $r > R$. Equation 12.8 expresses Newton's discovery, with regard to the moon, that g decreases inversely with the square of the distance.

Figure 12.9 shows a satellite orbiting at height h above the earth's surface. Its distance from the center of the earth is $r = R_e + h$. Most people have a mental image that satellites orbit "far" from the earth, but in reality h is typically 200 miles $\approx 3 \times 10^5$ m while $R_e = 6.37 \times 10^6$ m. Thus the satellite is barely "skimming" the earth at a height only about 5% of the earth's radius!

The value of g at height h above the earth is

$$g = \frac{GM_e}{(R_e + h)^2} = \frac{GM_e}{R_e^2(1 + h/R_e)^2} = \frac{g_{\text{earth}}}{(1 + h/R_e)^2} \qquad (12.9)$$

where $g_{\text{earth}} = 9.83$ m/s^2 is the value calculated in Equation 12.7 for $h = 0$ on a nonrotating earth. Table 12.1 shows the value of g evaluated at several values of h.

TABLE 12.1 Variation of the earth's acceleration due to gravity with height above the ground

Height h	Example	g (m/s^2)
0 m	ground	9.83
4500 m	Mt. Whitney	9.82
10,000 m	jet airplane	9.80
300,000 m	space shuttle	8.90
3,590,000 m	communications satellite	0.22

NOTE ▶ The acceleration due to gravity on a satellite such as the space shuttle is only slightly less than the ground-level value. An object in orbit is not "weightless" because there is no gravity in space but because it is in free fall, as you learned in Sections 5.3 and 7.4. ◀

Weighing the Earth

We can predict g if we know the earth's mass. But how do we know the value of M_e? We cannot place the earth on a giant pan balance, so how is its mass known? Furthermore, how do we know the value of G? These are interesting and important questions.

Newton did not know the value of G. He could say that the gravitational force is proportional to the product $m_1 m_2$ and inversely proportional to r^2, but he had no means of knowing the value of the proportionality constant.

Determining G requires a *direct* measurement of the gravitational force between two known masses at a known separation. The small size of the gravitational force between ordinary-size objects makes this quite a feat. Yet the English scientist Henry Cavendish came up with an ingenious way of doing so with a device called a *torsion balance*. Two fairly small masses m, typically about 10 g, are placed on the ends of a lightweight rod. The rod is hung from a thin fiber, as shown in Figure 12.10a, and allowed to reach equilibrium.

If the rod is then rotated slightly and released, a *restoring force* will return it to equilibrium. This is analogous to displacing a spring from equilibrium, and in fact the restoring force and the angle of displacement obey a version of Hooke's law: $F_{\text{restore}} = k\Delta\theta$. The "torsion constant" k can be determined by timing the period of oscillations. Once k is known, a force that twists the rod slightly away from equilibrium can be measured by the product $k\Delta\theta$. It is possible to measure very small angular deflections, so this device can be used to determine very small forces.

Two larger masses M (typically lead spheres with $M \approx 10$ kg) are then brought close to the torsion balance, as shown in Figure 12.10b. The gravitational

(a)

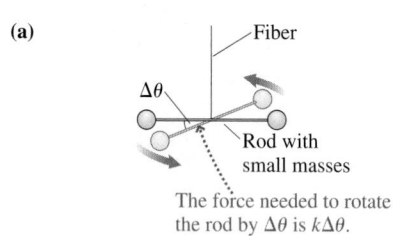

The force needed to rotate the rod by $\Delta\theta$ is $k\Delta\theta$.

(b)

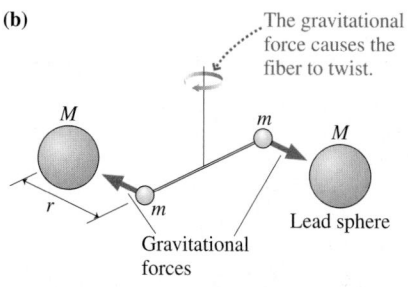

FIGURE 12.10 Cavendish's experiment to measure G.

attraction that they exert on the smaller hanging masses causes a small but measurable twisting of the balance, enough to measure $F_{M \text{ on } m}$. Because m, M, and r are all known, Cavendish was able to determine G from

$$G = \frac{F_{M \text{ on } m} r^2}{Mm} \tag{12.10}$$

His first results were not highly accurate, but improvements over the years in this and similar experiments have produced the value of G accepted today.

With an independently determined value of G, we can return to Equation 12.7 to find

$$M_e = \frac{g_{earth} R_e^2}{G} \tag{12.11}$$

We have weighed the earth! The value of g_{earth} at the earth's surface is known with great accuracy from kinematics experiments. The earth's radius R_e is determined by surveying techniques. Combining our knowledge from these very different measurements has given us a way to determine the mass of the earth.

The acceleration due to gravity g is constant on the surface of any given planet, but is different for each planet. The gravitational constant G is a constant of a different nature. It is what we call a *universal constant*. Its value establishes the strength of one of the fundamental forces of nature. As far as we know, the gravitational force between two masses would be the same anywhere in the universe. Universal constants tell us something about the most basic and fundamental properties of nature. You will soon meet other universal constants.

STOP TO THINK 12.3   A planet has four times the mass of the earth, but the acceleration due to gravity on the planet's surface is the same as on the earth's surface. The planet's radius is

a. $4R_e$.   b. $2R_e$.   c. R_e.   d. $\frac{1}{2}R_e$.   e. $\frac{1}{4}R_e$.

12.5 Gravitational Potential Energy

Gravitational problems are ideal for the conservation-law tools we developed in Chapters 9 through 11. Because gravity is the only force, and it is a conservative force, both the momentum and the mechanical energy of the system $m_1 + m_2$ are conserved. To employ conservation of energy, however, we need to determine an appropriate form for the gravitational potential energy for two particles interacting via Newton's law of gravity.

The definition of potential energy that we developed in Chapter 11 is

$$\Delta U = U_f - U_i = -W_c(i \rightarrow f) \tag{12.12}$$

where $W_c(i \rightarrow f)$ is the work done by a conservative force as a particle moves from position i to position f. Strictly speaking, this defines only ΔU, the *change* in potential energy. To find an explicit expression for U, we must choose a zero point of the potential energy.

For a flat earth, we used $F = -mg$ and the choice that $U = 0$ at the surface $(y = 0)$ to arrive at the now-familiar $U_g = mgy$. This result for U_g is valid only for $y \ll R_e$, when the earth's curvature and size are not apparent. We now need to find an expression for the gravitational potential energy of masses that interact over *large* distances.

Figure 12.11 shows two particles of mass m_1 and m_2. Let's calculate the work done on mass m_2 by the conservative force $\vec{F}_{1 \text{ on } 2}$ as m_2 moves from an initial position at distance r to a final position very far away. The force, which points to

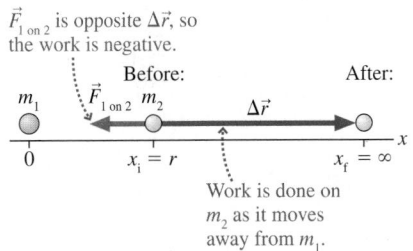

FIGURE 12.11 Calculating the work done by the gravitational force as mass m_2 moves from r to ∞.

the left, is opposite the displacement, hence this force does *negative* work. Consequently, due to the minus sign in Equation 12.12, ΔU is *positive*. A pair of masses *gains* potential energy as the masses move farther apart, just as a particle near the earth's surface gains potential energy as it moves to higher elevations.

Establish a coordinate system with m_1 at the origin and m_2 moving along the *x*-axis. The gravitational force is a variable force, so we need the full definition of work:

$$W(\text{i} \rightarrow \text{f}) = \int_{x_i}^{x_f} F_x \, dx \qquad (12.13)$$

The $\vec{F}_{1 \text{ on } 2}$ points toward the left, so its *x*-component is $(F_{1 \text{ on } 2})_x = -Gm_1m_2/x^2$. As mass m_2 moves from $x_i = r$ to $x_f = \infty$, the potential energy changes by

$$\Delta U = U_{\text{at} \, \infty} - U_{\text{at} \, r} = -\int_r^\infty (F_{1 \text{ on } 2})_x \, dx = -\int_r^\infty \left(\frac{-Gm_1m_2}{x^2} \right) dx$$

$$= +Gm_1m_2 \int_r^\infty \frac{dx}{x^2} = -\frac{Gm_1m_2}{x} \bigg|_r^\infty \qquad (12.14)$$

$$= \frac{Gm_1m_2}{r}$$

> **NOTE** ▶ We chose to integrate along the *x*-axis, but the fact that gravity is a conservative force means that ΔU will have this value if m_2 moves from r to ∞ along *any* path. ◀

To proceed further, we need to choose the point where $U = 0$. We would like our choice to be valid for any star or planet, regardless of its mass and radius. This will be the case if we set $U = 0$ at the point where the interaction between the masses vanishes. According to Newton's law of gravity, the strength of the interaction is zero only when $r = \infty$. Two masses infinitely far apart will have no tendency, or potential, to move together, so we will *choose* to place the zero point of potential energy at $r = \infty$. That is, $U_{\text{at} \, \infty} = 0$.

This choice gives us the gravitational potential energy of masses m_1 and m_2:

$$U_{\text{g}} = -\frac{Gm_1m_2}{r} \qquad (12.15)$$

This is the potential energy of masses m_1 and m_2 when their *centers* are separated by distance r. Figure 12.12 is a graph of U_{g} as a function of the distance r between the masses. Notice that it asymptotically approaches 0 as $r \rightarrow \infty$.

> **NOTE** ▶ Although Equation 12.15 looks rather similar to Newton's law of gravity, it depends only on $1/r$, *not* on $1/r^2$. ◀

It may seem disturbing that the potential energy is negative, but we encountered similar situations in Chapter 10. All a negative potential energy means is that the potential energy of the two masses at separation r is *less* than their potential energy at infinite separation. It is only the *change* in U that has physical significance, and the change will be the same no matter where we place the zero of potential energy.

Suppose two masses distance r_1 apart are released from rest. How will they move? From a force perspective, you would note that each mass experiences an attractive force and accelerates toward the other. The energy perspective of Figure 12.13 tells us the same thing. By moving toward smaller r (that is, $r_1 \rightarrow r_2$), the system *loses* potential energy and *gains* kinetic energy while conserving E_{mech}. The system is "falling downhill," although in a more general sense than we think about on a flat earth.

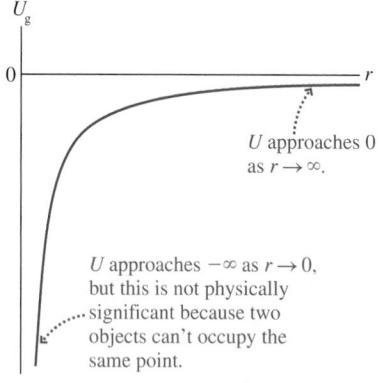

U approaches 0 as $r \rightarrow \infty$.

U approaches $-\infty$ as $r \rightarrow 0$, but this is not physically significant because two objects can't occupy the same point.

FIGURE 12.12 The gravitational potential-energy curve.

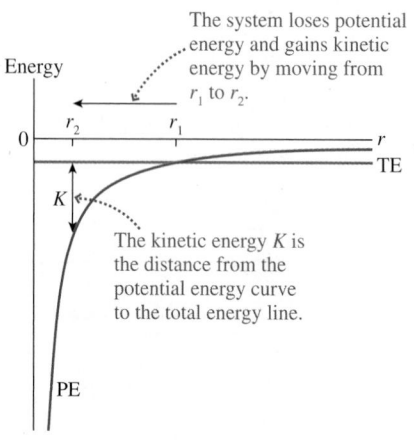

The system loses potential energy and gains kinetic energy by moving from r_1 to r_2.

The kinetic energy K is the distance from the potential energy curve to the total energy line.

FIGURE 12.13 Two masses gain kinetic energy as the separation between them changes from r_1 to r_2.

EXAMPLE 12.1 Crashing into the sun

Suppose the earth were suddenly to cease revolving around the sun. The gravitational force would then pull it directly into the sun. What would be the earth's speed as it crashed?

MODEL Model the earth and the sun as spherical masses. This is an isolated system, so its mechanical energy is conserved.

VISUALIZE Figure 12.14 is a before-and-after pictorial representation for this gruesome cosmic event. The "crash" occurs as the earth touches the sun, at which point the distance between their centers is $r_2 = R_s + R_e$. The initial separation r_1 is the radius of the earth's *orbit* about the sun, not the radius of the earth.

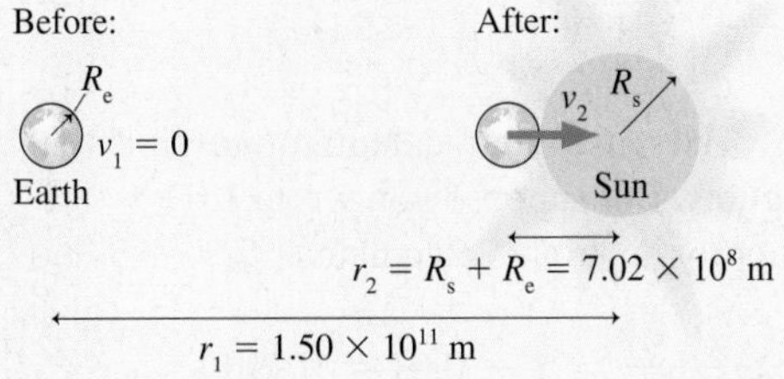

Before:

R_e
$v_1 = 0$
Earth

After:

v_2 R_s
Sun

$r_2 = R_s + R_e = 7.02 \times 10^8$ m

$r_1 = 1.50 \times 10^{11}$ m

FIGURE 12.14 Before-and-after pictorial representation of the earth crashing into the sun (not to scale).

SOLVE Strictly speaking, the kinetic energy is the sum $K = K_{earth} + K_{sun}$. However, the sun is so much more massive than the earth that the lightweight earth does almost all of the moving. It is a reasonable approximation to consider the sun as remaining at rest. In that case, the energy conservation equation $K_2 + U_2 = K_1 + U_1$ is

$$\frac{1}{2}M_e v_2^2 - \frac{GM_s M_e}{(R_s + R_e)} = 0 - \frac{GM_s M_e}{r_1}$$

This is easily solved for the earth's speed at impact. Using data from Table 12.2, we find

$$v_2 = \sqrt{2GM_s\left(\frac{1}{R_s + R_e} - \frac{1}{r_1}\right)} = 6.13 \times 10^5 \text{ m/s}$$

ASSESS The earth is really flying along at over 1 million miles per hour! It is worth noting that we do not have the mathematical tools to solve this problem using Newton's second law because the acceleration is not constant. But the solution is straightforward when we use energy conservation.

EXAMPLE 12.2 Escape speed

A 1000 kg rocket is fired straight away from the surface of the earth. What speed does the rocket need to "escape" from the gravitational pull of the earth and never return? Assume a non-rotating earth.

MODEL In a simple universe, consisting of only the earth and the rocket, an insufficient launch speed will cause the rocket eventually to fall back to earth. Once the rocket finally slows to a halt, gravity will ever so slowly pull it back. The only way the rocket can escape is to never stop ($v = 0$) and thus never have a turning point! That is, the rocket must continue moving away from the earth forever. The *minimum* launch speed for escape, which is called the **escape speed,** will cause the rocket to stop ($v = 0$) only as it reaches $r = \infty$. Now ∞, of course, is not a "place," so a statement like this means that we want the rocket's speed to approach $v = 0$ asymptotically as $r \rightarrow \infty$.

VISUALIZE Figure 12.15 is a before-and-after pictorial representation.

SOLVE The energy conservation equation $K_2 + U_2 = K_1 + U_1$ is

$$0 + 0 = \frac{1}{2}mv_1^2 - \frac{GM_e m}{R_e}$$

where we used the fact that both the kinetic and potential energy are zero at $r = \infty$. Thus the escape speed is

$$v_{escape} = v_1 = \sqrt{\frac{2GM_e}{R_e}} = 11{,}200 \text{ m/s} \approx 25{,}000 \text{ mph}$$

ASSESS The problem was mathematically easy; the difficulty was deciding how to interpret it. That is why—as you have now seen many times—the "physics" of a problem consists of thinking, interpreting, and modeling. We will see variations on this problem in the future, both with gravity and electricity, so you might want to review the *reasoning* involved. Notice that the answer does *not* depend on the rocket's mass, so this is the escape speed for any object.

Before:

$r_1 = R_e$
v_1

Earth

After:

$r_2 = \infty$
$v_2 = 0$

FIGURE 12.15 Pictorial representation of a rocket launched with sufficient speed to escape the earth's gravity.

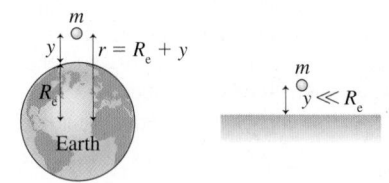

For a spherical earth:

$$U_g = -\frac{GM_em}{R_e + y}$$

We can treat the earth as flat if $y \ll R_e$:

$$U_g = mgy$$

FIGURE 12.16 We can treat the earth as flat if $y \ll R_e$.

The Flat-Earth Approximation

Equation 12.15 is the general form of the gravitational potential energy, but how is it related to our previous use of $U_g = mgy$ on a flat earth? Figure 12.16 shows an object of mass m located at height y above the surface of the earth. The object's distance from the earth's center is $r = R_e + y$ and its gravitational potential energy is

$$U_g = -\frac{GM_em}{r} = -\frac{GM_em}{R_e + y} = -\frac{GM_em}{R_e(1 + y/R_e)} \qquad (12.16)$$

where, in the last step, we factored R_e out of the denominator.

Suppose the object is very close to the earth's surface ($y \ll R_e$). In that case, the ratio $y/R_e \ll 1$. There is an approximation you will learn about in calculus, called the *binomial approximation*, that says

$$\frac{1}{1 + x} \approx 1 - x \quad \text{if } x \ll 1 \qquad (12.17)$$

As an illustration, you can easily use your calculator to find that $1/1.01 = 0.9901$, to four significant figures. But suppose you wrote $1.01 = 1 + 0.01$. You could then use the binomial approximation to calculate

$$\frac{1}{1.01} = \frac{1}{1 + 0.01} \approx 1 - 0.01 = 0.9900$$

You can see that the approximate answer is off by only 0.01%.

If we call $y/R_e = x$ in Equation 12.16 and use the binomial approximation, we find

$$U_g(\text{if } y \ll R_e) \approx -\frac{GM_em}{R_e}\left(1 - \frac{y}{R_e}\right) = -\frac{GM_em}{R_e} + m\left(\frac{GM_e}{R_e^2}\right)y \quad (12.18)$$

Now the first term is just the gravitational potential energy U_0 when the object is at ground level ($y = 0$). In the second term, you can recognize $GM_e/R_e^2 = g_{earth}$ from the definition of g in Equation 12.7. Thus we can write Equation 12.18 as

$$U_g(\text{if } y \ll R_e) = U_0 + mg_{earth}y \qquad (12.19)$$

Although we chose U_g to be zero when $r = \infty$, we are always free to change our minds. If we change the zero point of potential energy to be $U_0 = 0$ at the surface, which is the choice we made in Chapter 10, then Equation 12.19 becomes

$$U_g(\text{if } y \ll R_e) = mg_{earth}y \qquad (12.20)$$

We can sleep easier knowing that Equation 12.15 for the gravitational potential energy is consistent with our earlier "flat-earth" expression for the potential energy when $y \ll R_e$.

EXAMPLE 12.3 The speed of a projectile
A projectile is launched straight up from the earth.

a. With what speed should you launch it if it is to have a speed of 500 m/s at a height of 400 km? Ignore air resistance.
b. By what percentage would your answer be in error if you used a flat-earth approximation?

MODEL Mechanical energy is conserved.

VISUALIZE Figure 12.17 shows a pictorial representation.

SOLVE

a. Although the height is exaggerated in the figure, 400 km = 400,000 m is high enough that we cannot ignore

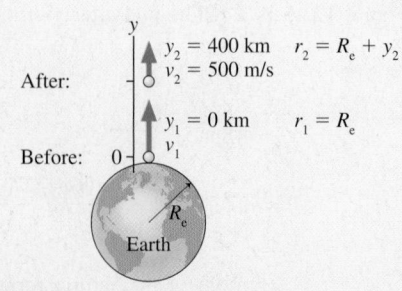

FIGURE 12.17 Pictorial representation of a projectile launched straight up.

the earth's spherical shape. The energy conservation equation $K_2 + U_2 = K_1 + U_1$ is

$$\frac{1}{2}mv_2^2 - \frac{GM_e m}{R_e + y_2} = \frac{1}{2}mv_1^2 - \frac{GM_e m}{R_e + y_1}$$

where we've written the distance between the projectile and the earth's center as $r = R_e + y$. The initial height is $y_1 = 0$. Notice that the projectile mass m cancels and is not needed. Solving for the launch speed,

$$v_1 = \sqrt{v_2^2 + 2GM_e\left(\frac{1}{R_e} - \frac{1}{R_e + y_2}\right)} = 2770 \text{ m/s}$$

This is about 6000 mph, much less than the escape speed.

b. The calculation is the same in the flat-earth approximation except that we use $U_g = mgy$ for the gravitational potential energy. Thus

$$\frac{1}{2}mv_2^2 + mgy_2 = \frac{1}{2}mv_1^2 + mgy_1$$

$$v_1 = \sqrt{v_2^2 + 2gy_2} = 2840 \text{ m/s}$$

The flat-earth value of 2840 m/s is 70 m/s too big. The error, as a percentage of the correct 2770 m/s, is

$$\text{error} = \frac{70}{2770} \times 100 = 2.5\%$$

ASSESS The true speed is less than the flat-earth approximation because the force of gravity decreases with height. Launching a rocket against a decreasing force takes less effort than it would with the flat-earth force of mg at all heights.

STOP TO THINK 12.4 Rank in order, from largest to smallest, the absolute values $|U_g|$ of the gravitational potential energies of these pairs of masses. The numbers give the relative masses and distances.

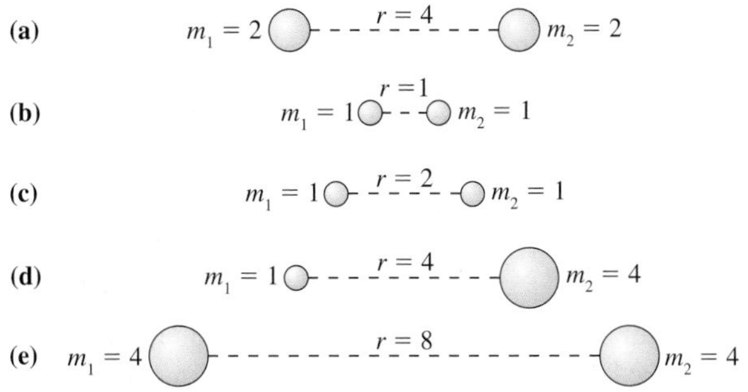

(a) $m_1 = 2$ ◯ - - - $r = 4$ - - - ◯ $m_2 = 2$

(b) $m_1 = 1$ ◦ - - $r = 1$ ◦ $m_2 = 1$

(c) $m_1 = 1$ ◦ - - $r = 2$ - ◦ $m_2 = 1$

(d) $m_1 = 1$ ◦ - - - $r = 4$ - - - ◯ $m_2 = 4$

(e) $m_1 = 4$ ◯ - - - - - - - $r = 8$ - - - - - - - ◯ $m_2 = 4$

12.6 Satellite Orbits and Energies

Solving Newton's second law to find the trajectory of a mass moving under the influence of gravity is mathematically beyond this textbook. It turns out that the solution is a set of elliptical orbits. This is Kepler's first law, which he discovered empirically by analyzing Tycho Brahe's observations. Kepler had no *reason* to think that orbits should be ellipses rather than some other shape. Newton was able to show that ellipses are a *consequence* of his theory of gravity.

The mathematics of ellipses is rather difficult, so we will restrict most of our analysis to the limiting case in which an ellipse becomes a circle. Most planetary orbits differ only very slightly from being circular. The earth's orbit, for example has a (semiminor axis/semimajor axis) ratio of 0.99986—very close to a true circle!

Figure 12.18 shows a massive body M, such as the earth or the sun, with a lighter body m orbiting it. The lighter body is called a **satellite**, even though it may be a planet orbiting the sun. Newton's second law for the satellite is

$$F_{M \text{ on } m} = \frac{GMm}{r^2} = ma_r = \frac{mv^2}{r} \qquad (12.21)$$

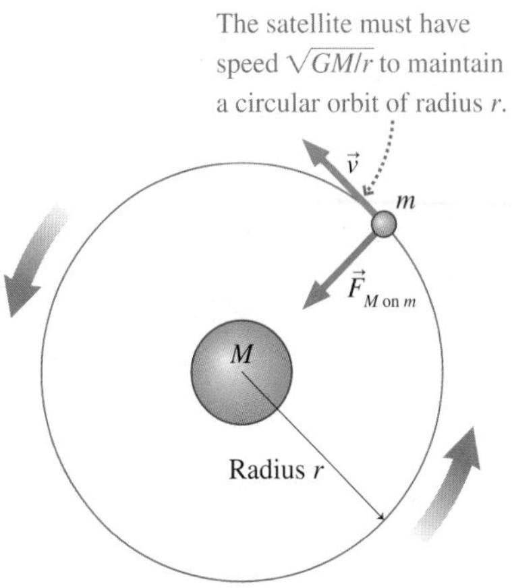

The satellite must have speed $\sqrt{GM/r}$ to maintain a circular orbit of radius r.

FIGURE 12.18 The orbital motion of a satellite due to the force of gravity.

The International Space Station appears to be floating, but it's actually traveling at nearly 8000 m/s as it orbits the earth.

Thus the speed of a satellite in a circular orbit is

$$v = \sqrt{\frac{GM}{r}} \qquad (12.22)$$

A satellite must have this specific speed in order to have a circular orbit of radius r about the larger mass M. If the velocity differs from this value, the orbit will become elliptical rather than circular. Notice that the orbital speed does *not* depend on the satellite's mass m. This is consistent with our previous discovery, for motion on a flat earth, that motion due to gravity is independent of the mass.

EXAMPLE 12.4 The speed of the space shuttle
The space shuttle in a 300-km-high orbit ($\approx$180 mi) wants to capture a smaller satellite for repairs. What are the speeds of the shuttle and the satellite in this orbit?

SOLVE Despite their different masses, the shuttle, the satellite, and the astronaut working in space to make the repairs all travel side-by-side with the same speed. They are simply in free-fall together. Using $r = R_e + h$ with $h = 300$ km $= 3.00 \times 10^5$ m, the speed is

$$v = \sqrt{\frac{(6.67 \times 10^{-11} \text{ N m}^2/\text{kg}^2)(5.98 \times 10^{24} \text{ kg})}{6.67 \times 10^6 \text{ m}}}$$

$$= 7730 \text{ m/s} \approx 16{,}000 \text{ mph}$$

ASSESS The answer depends on the mass of the earth but *not* on the mass of the satellite.

Kepler's Third Law

4.6 Activ Physics

An important parameter of circular motion is the *period*. Recall that the period T is the time to complete one full orbit. The relationship among speed, radius, and period is

$$v = \frac{1 \text{ circumference}}{1 \text{ period}} = \frac{2\pi r}{T} \qquad (12.23)$$

We can find a relationship between a satellite's period and the radius of its orbit by using Equation 12.22 for v:

$$v = \frac{2\pi r}{T} = \sqrt{\frac{GM}{r}} \qquad (12.24)$$

Squaring both sides and solving for T gives

$$T^2 = \left(\frac{4\pi^2}{GM}\right)r^3 \qquad (12.25)$$

In other words, the *square* of the period is proportional to the *cube* of the radius. This is Kepler's third law. You can see that Kepler's third law is a direct consequence of Newton's law of gravity.

Table 12.2 contains astronomical information about the sun, the earth, the moon, and other planets of the solar system. We can use these data to check the validity of Equation 12.25. Figure 12.19 is a graph of log T versus log r for all the planets in Table 12.2 except Mercury. Notice that the scales on each axis are increasing logarithmically—by *factors* of 10—rather than linearly. (Also, the vertical axis has converted T to the SI units of s.) As you can see, the graph is a very straight line with a statistical "best fit" equation

$$\log T = 1.500 \log r - 9.264$$

As a homework problem, you can show that the slope of 1.500 for this "log-log graph" confirms the prediction of Equation 12.25. The y-intercept value of this

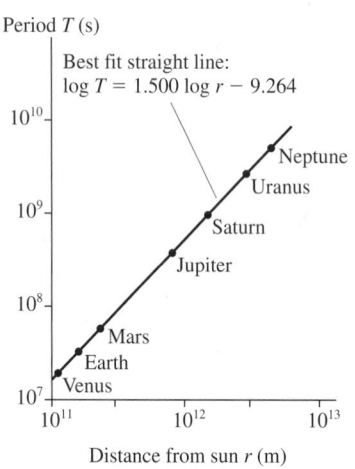

FIGURE 12.19 The graph of log T versus log r for the planetary data of Table 12.2.

TABLE 12.2 Useful astronomical data

Planetary body	Mean distance from sun (m)	Period (years)	Mass (kg)	Mean radius (m)
Sun	–	–	1.99×10^{30}	6.96×10^{8}
Moon	$3.84 \times 10^{8}*$	27.3 days	7.36×10^{22}	1.74×10^{6}
Mercury	5.79×10^{10}	0.241	3.18×10^{23}	2.43×10^{6}
Venus	1.08×10^{11}	0.615	4.88×10^{24}	6.06×10^{6}
Earth	1.50×10^{11}	1.00	5.98×10^{24}	6.37×10^{6}
Mars	2.28×10^{11}	1.88	6.42×10^{23}	3.37×10^{6}
Jupiter	7.78×10^{11}	11.9	1.90×10^{27}	6.99×10^{7}
Saturn	1.43×10^{12}	29.5	5.68×10^{26}	5.85×10^{7}
Uranus	2.87×10^{12}	84.0	8.68×10^{25}	2.33×10^{7}
Neptune	4.50×10^{12}	165	1.03×10^{26}	2.21×10^{7}

*Distance from earth.

line also contains useful information, and you can use it to determine the mass of the sun.

A particularly interesting application of Equation 12.25 is to communication satellites that are in **geosynchronous orbits** above the earth. These satellites have a period of 24 hours, making their orbital motion synchronous with the earth's rotation. As a result, a satellite in such an orbit appears to remain stationary over one point on the earth's equator. Equation 12.25 allows us to compute the radius of an orbit with this period:

$$r_{geo} = R_e + h_{geo} = \left[\left(\frac{GM}{4\pi^2} \right) T^2 \right]^{1/3}$$

$$= \left[\left(\frac{(6.67 \times 10^{-11} \text{ N m}^2/\text{kg}^2)(5.98 \times 10^{24} \text{ kg})}{4\pi^2} \right)(86{,}400 \text{ s})^2 \right]^{1/3}$$

$$= 4.225 \times 10^7 \text{ m}$$

The height of the orbit is

$$h_{geo} = r_{geo} - R_e = 3.59 \times 10^7 \text{ m} = 35{,}900 \text{ km} \approx 22{,}300 \text{ mi}$$

NOTE ▶ When using Equation 12.25, the period *must* be in SI units of s. ◀

Geosynchronous orbits are much higher than the low-earth orbits used by the space shuttle and remote-sensing satellites, where $h \approx 300$ km. Communications satellites in geosynchronous orbits were first proposed in 1948 by science fiction writer Arthur C. Clarke, 10 years before the first artificial satellite of any type!

EXAMPLE 12.5 Extrasolar planets

Astronomers using the most advanced telescopes have only recently seen evidence of planets orbiting nearby stars. These are called *extrasolar planets*. Suppose a planet is observed to have a 1200 day period as it orbits a star at the same distance that Jupiter is from the sun. What is the mass of the star in solar masses? (1 *solar mass* is defined to be the mass of the sun.)

SOLVE Here "day" means earth days, as used by astronomers to measure the period. Thus the planet's period in SI units is

$T = 1200$ days $= 1.037 \times 10^8$ s. The orbital radius is that of Jupiter, which we can find in Table 12.2 to be $r = 7.78 \times 10^{11}$ m. Solving Equation 12.25 for the mass of the star gives

$$M = \frac{4\pi^2 r^3}{GT^2} = 2.59 \times 10^{31} \text{ kg} \times \frac{1 \text{ solar mass}}{1.99 \times 10^{30} \text{ kg}}$$

$$= 13 \text{ solar masses}$$

ASSESS This is a large, but not extraordinary, star.

Two planets orbit a star. Planet 1 has orbital radius r_1 and planet 2 has $r_2 = 4r_1$. Planet 1 orbits with period T_1. Planet 2 orbits with period

a. $T_2 = 8T_1$.   b. $T_2 = 4T_1$.   c. $T_2 = 2T_1$.

d. $T_2 = \frac{1}{2}T_1$.   e. $T_2 = \frac{1}{4}T_1$.   f. $T_2 = \frac{1}{8}T_1$.

Kepler's Second Law

In Chapter 9 we defined a particle's *angular momentum* to be $L = mrv_t$, where v_t is the tangential component of the velocity vector. For a particle in circular motion, where v_t is simply the speed v, we showed that angular momentum is conserved if there is no tangential force on the particle. Although we won't prove it until Chapter 13, it turns out that angular momentum is conserved for a particle following a trajectory of *any* shape if the net tangential force is zero.

Figure 12.20a shows a satellite moving in an elliptical orbit around a star or planet at one focus. The tangential component of $\vec{v}$ is $v_t = v\sin\beta$, where β is the angle between $\vec{r}$ and $\vec{v}$. Consequently, the satellite's angular momentum is

$$L = mrv_t = mrv\sin\beta \qquad (12.26)$$

For a circular orbit, where β is always 90°, this reduces to simply $L = mrv$.

The only force on a satellite, the gravitational force, points directly toward the star or planet that the satellite is orbiting. The gravitational force has no tangential component, thus **the satellite's angular momentum is conserved as it orbits.**

The satellite moves forward a small distance $\Delta s = v\Delta t$ during the small interval of time Δt. This motion defines the triangle of area ΔA shown in Figure 12.20b. ΔA is the area "swept out" by the satellite during Δt. You can see that the height of the triangle is $h = \Delta s\sin\beta$, so the triangle's area is

$$\Delta A = \frac{1}{2} \times \text{base} \times \text{height} = \frac{1}{2} \times r \times \Delta s\sin\beta = \frac{1}{2}rv\sin\beta\Delta t \quad (12.27)$$

The *rate* at which the area is swept out by the satellite as it moves is

$$\frac{\Delta A}{\Delta t} = \frac{1}{2}rv\sin\beta = \frac{mrv\sin\beta}{2m} = \frac{L}{2m} \qquad (12.28)$$

The angular momentum L is conserved, so it has the same value at every point in the orbit. Consequently, the rate at which the area is swept out by the satellite is constant. This is Kepler's second law, which says that a line drawn between the sun and a planet sweeps out equal areas during equal intervals of time. We see that Kepler's second law is really a consequence of the conservation of angular momentum.

Kepler and Newton

Kepler's laws summarize observational data about the motions of the planets. They were an outstanding achievement, but they did not form a theory. Newton put forward a *theory,* a specific set of relationships between force and motion that allows *any* motion to be understood and calculated. Newton's theory of gravity has allowed us to *deduce* Kepler's laws and, thus, to understand them at a more fundamental level.

Furthermore, Kepler's laws are not perfectly accurate. The planets, in addition to being attracted to the sun, are also attracted toward each other and toward their orbiting moons. The consequences of these additional forces are small, but over time they provide measurable effects not contained in Kepler's laws. Newton's theory makes it possible to use the inverse-square law to calculate *all* of the forces, add them to get the net force acting on each planet, then proceed to solve Newton's

(a)

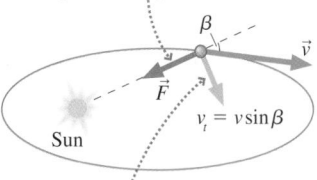

The gravitational force has no tangential component.

This is the component of $\vec{v}$ that would be tangent to a *circle* around the sun.

(b)

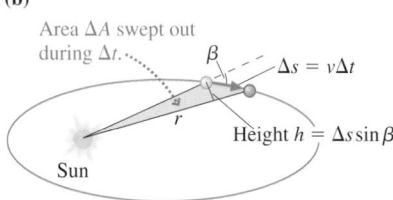

Area ΔA swept out during Δt.

FIGURE 12.20 Angular momentum is conserved for a planet in an elliptical orbit.

second law to determine the dynamics. The mathematics of the solution can be exceedingly difficult, and today is all done with computers, but even with hand calculations this procedure in the mid-19th century predicted the existence of an undiscovered planet that was having minor effects on the orbital motion of Uranus. The planet Neptune was discovered in 1846, just where the calculations predicted.

Orbital Energetics

Let us conclude this chapter by thinking about the energetics of orbital motion. We found, with Equation 12.24, that a satellite in a circular orbit must have $v^2 = GM/r$. A satellite's speed is determined entirely by the size of its orbit. The satellite's kinetic energy is thus

$$K = \frac{1}{2}mv^2 = \frac{GMm}{2r} \tag{12.29}$$

But $-GMm/r$ is the potential energy, U_g, so

$$K = -\frac{1}{2}U_g \tag{12.30}$$

This is an interesting result. In all our earlier examples, the kinetic and potential energy were two independent parameters. In contrast, a satellite can move in a circular orbit *only* if there is a very specific relationship between K and U. It is not that K and U *have* to have this relationship, but if they do not, the trajectory will be elliptical rather than circular.

Equation 12.30 gives us the mechanical energy of a satellite in a circular orbit:

$$E_{\text{mech}} = K + U_g = \frac{1}{2}U_g \tag{12.31}$$

The gravitational potential energy is negative, hence the *total* mechanical energy is also negative. Negative total energy is characteristic of a **bound system,** a system in which the satellite is bound to the central mass by the gravitational force and cannot get away. In an unbound system, the satellite can move infinitely far away to where $U = 0$. Because the kinetic energy K must be ≥ 0, the total energy of an unbound system must be ≥ 0. A negative value of E_{mech} tells us that the satellite is unable to escape the central mass.

Figure 12.21 shows the kinetic, potential, and total energy of a satellite in a circular orbit as a function of the orbit's radius. Notice how $E_{\text{mech}} = \frac{1}{2}U_g$. This figure can help us understand the energetics of transferring a satellite from one orbit to another. Suppose a satellite is in an orbit of radius r_1 and that we'd like it to be in a larger orbit of radius r_2. The kinetic energy at r_2 is less than at r_1 (the satellite moves more slowly in the larger orbit), but you can see that the total energy *increases* as r

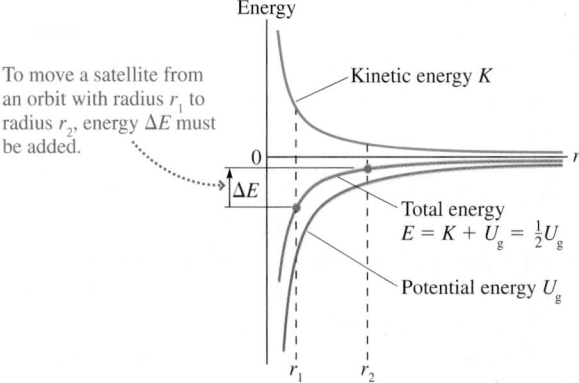

FIGURE 12.21 The kinetic, potential, and total energy of a satellite in a circular orbit.

increases. Consequently, transferring a satellite to a larger orbit requires a net energy increase $\Delta E > 0$. Where does this increase of energy come from?

Artificial satellites are raised to higher orbits by firing their rocket motors to create a forward thrust. This force does work on the satellite, and the energy equation of Chapter 11 tells us that this work increases the satellite's energy by $\Delta E_{\text{mech}} = W_{\text{ext}}$. Thus the energy to "lift" a satellite into a higher orbit comes from the chemical energy stored in the rocket fuel.

EXAMPLE 12.6 Raising a satellite

How much work must be done to boost a 1000 kg communications satellite from a low earth orbit with $h = 300$ km, where it is released by the space shuttle, to a geosynchronous orbit?

SOLVE The required work is $W_{\text{ext}} = \Delta E_{\text{mech}}$, and from Equation 12.31 we see that $\Delta E_{\text{mech}} = \frac{1}{2}\Delta U_{\text{g}}$. The initial orbit has radius $r_{\text{shuttle}} = R_{\text{e}} + h = 6.67 \times 10^6$ m. We earlier found the radius of a geosynchronous orbit to be 4.22×10^7 m. Thus

$$W_{\text{ext}} = \Delta E_{\text{mech}} = \frac{1}{2}\Delta U_{\text{g}} = \frac{1}{2}(-GM_{\text{e}}m)\left(\frac{1}{r_{\text{geo}}} - \frac{1}{r_{\text{shuttle}}}\right)$$

$$= 2.52 \times 10^{10} \text{ J}$$

ASSESS It takes a lot of energy to boost satellites to high orbits!

Firing the rocket tangentially to the circle here moves the satellite onto the elliptical orbit.

Kinetic energy is transformed into potential energy as the rocket moves "uphill."

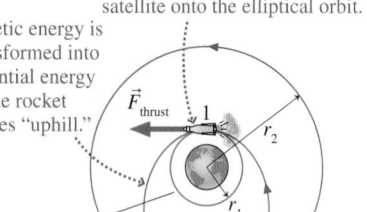

$\vec{F}_{\text{thrust}}$ 1 r_2

r_1

Initial orbit

Desired orbit

Elliptical transfer orbit

2 $\vec{F}_{\text{thrust}}$

A second firing here transfers it to the larger circular orbit.

FIGURE 12.22 Transferring a satellite to a larger circular orbit.

You might think that the way to get a satellite into a larger orbit would be to point the thrusters toward the earth and blast outward. That would work fine *if* the satellite were initially at rest and moved straight out along a linear trajectory. But an orbiting satellite is already moving and has significant inertia. A force directed straight outward would *change* the satellite's velocity vector in that direction but would not cause it to *move* along that line. (Remember all those earlier motion diagrams for motion along curved trajectories.) In addition, a force directed outward would be almost at right angles to the motion and would do essentially zero work. Navigating in space is not as easy as it appears in *Star Wars!*

To move the satellite in Figure 12.22 from the orbit with radius r_1 to the larger circular orbit of radius r_2, the thrusters are turned on at point 1 to apply a brief *forward* thrust force in the direction of motion, *tangent* to the circle. This force does a significant amount of work because the force is parallel to the displacement, so the satellite quickly gains kinetic energy ($\Delta K > 0$). But $\Delta U_{\text{g}} = 0$ because the satellite does not have time to change its distance from the earth during a thrust of short duration. With the kinetic energy increased, but not the potential energy, the satellite no longer meets the requirement $K = -\frac{1}{2}U_{\text{g}}$ for a circular orbit. Instead, it goes into an elliptical orbit.

In the elliptical orbit, the satellite moves "uphill" toward point 2 by transforming kinetic energy into potential energy. At point 2, the satellite has arrived at the desired distance from earth and has the "right" value of the potential energy. But it turns out that its kinetic energy is now *less* than that needed for a circular orbit. (The analysis is more complex than we want to pursue here. It will be left for a homework Challenge Problem.) If no action is taken, the satellite will continue on its elliptical orbit and "fall" back to point 1. But another *forward* thrust at point 2 increases its kinetic energy, without changing U_{g}, until the kinetic energy reaches the value $K = -\frac{1}{2}U_{\text{g}}$ required for a circular orbit. Presto! The second burn kicks the satellite into the desired circular orbit of radius r_2. The work $W_{\text{ext}} = \Delta E_{\text{mech}}$ is the *total* work done in both burns. It takes a more extended analysis to see how the work has to be divided between the two burns, but even without those details you now have enough knowledge about orbits and energy to understand the ideas that are involved.

SUMMARY

The goal of Chapter 12 has been to use Newton's theory of gravity to understand the motion of satellites and planets.

GENERAL PRINCIPLES

Newton's Theory of Gravity

1. Two objects with masses M and m a distance r apart exert attractive **gravitational forces** on each other of magnitude

$$F_{M \text{ on } m} = F_{m \text{ on } M} = \frac{GMm}{r^2}$$

where the **gravitational constant** is $G = 6.67 \times 10^{-11} \text{ N m}^2/\text{kg}^2$.

2. Gravitational mass and inertial mass are equivalent.

3. Newton's three laws of motion apply to satellites, planets, and stars.

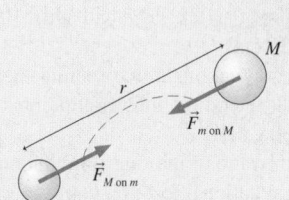

IMPORTANT CONCEPTS

Orbital motion of a planet (or satellite) is described by **Kepler's laws:**

1. Orbits are ellipses with the sun (or planet) at one focus.

2. A line between the sun and the planet sweeps out equal areas during equal intervals of time.

3. The square of the planet's period T is proportional to the cube of the orbit's semimajor axis.

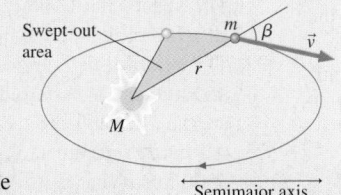

Circular orbits are a special case of an ellipse. For a circular orbit around a mass M,

$$v = \sqrt{\frac{GM}{r}} \quad \text{and} \quad T^2 = \left(\frac{4\pi^2}{GM}\right)r^3$$

Conservation of angular momentum

The angular momentum $L = mrv \sin \beta$ remains constant throughout the orbit. Kepler's second law is a consequence of this law.

Orbital energetics

A satellite's mechanical energy $E_{\text{mech}} = K + U_g$ is conserved, where the gravitational potential energy is

$$U_g = -\frac{GMm}{r}$$

For circular orbits, $K = -\frac{1}{2}U_g$ and $E_{\text{mech}} = \frac{1}{2}U_g$. Negative total energy is characteristic of a **bound system.**

APPLICATIONS

For a planet of mass M and radius R,

- The acceleration due to gravity on the surface is $g_{\text{surface}} = \dfrac{GM}{R^2}$

- The escape speed is $v_{\text{escape}} = \sqrt{\dfrac{2GM}{R}}$

- The radius of a geosynchronous orbit is $r_{\text{geo}} = \left(\dfrac{GM}{4\pi^2}T^2\right)^{1/3}$

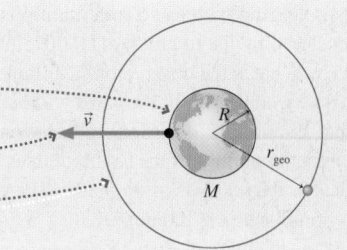

TERMS AND NOTATION

cosmology	gravitational constant, G	escape speed
Kepler's laws	gravitational mass	satellite
gravitational force	principle of equivalence	geosynchronous orbit
Newton's law of gravity	Newton's theory of gravity	bound system

EXERCISES AND PROBLEMS

Exercises

Section 12.3 Newton's Law of Gravity

1. a. What is the gravitational force of the sun on the earth?
   b. What is the gravitational force of the moon on the earth?
   c. The moon's force is what percent of the sun's force?

2. The centers of a 10 kg lead ball and a 100 g lead ball are separated by 10 cm.
   a. What gravitational force does each exert on the other?
   b. What is the ratio of this gravitational force to the weight of the 100 g ball?

3. What is the ratio of the sun's gravitational force on you to the earth's gravitational force on you?

4. What is the ratio of the sun's gravitational force on the moon to the earth's gravitational force on the moon?

5. A 1.0-m-diameter lead sphere has a mass of 5900 kg. A dust particle rests on the surface. What is the ratio of the gravitational force of the sphere on the dust particle to the weight of the dust particle?

6. Estimate the force of attraction between a 50 kg woman and a 70 kg man sitting 1.0 m apart.

7. The space shuttle orbits 300 km above the surface of the earth.
   a. What is the force of gravity on a 1.0 kg sphere inside the space shuttle?
   b. The sphere floats around inside the space shuttle, apparently "weightless." How is this possible?

Section 12.4 Little g and Big G

8. a. What is the acceleration due to gravity at the surface of the sun?
   b. What is the sun's acceleration due to gravity at the distance of the earth?

9. What is the acceleration due to gravity at the surface of (a) the moon and (b) Jupiter?

10. A starship is circling a distant planet of radius R. The astronauts find that the acceleration due to gravity at their altitude is half the value at the planet's surface. How far above the surface are they orbiting? Your answer will be a multiple of R.

11. A sensitive gravimeter at a mountain observatory finds that the acceleration due to gravity is 0.0075 m/s^2 less than that at sea level. What is the observatory's altitude?

12. Suppose we could shrink the earth without changing its mass. At what fraction of its current radius would the acceleration due to gravity at the surface be three times its present value?

13. Planet Z is 10,000 km in diameter. The acceleration due to gravity on Planet Z is 8.0 m/s^2.
    a. What is the mass of Planet Z?
    b. What is the acceleration due to gravity 10,000 km above Planet Z's north pole?

Section 12.5 Gravitational Potential Energy

14. a. What is the acceleration due to gravity on Mars?
    b. An astronaut on earth can throw a ball straight up to a height of 15 m. How high can he throw the ball on Mars?

15. A projectile is shot straight up from the earth's surface at a speed of 10,000 km/hr. How high does it go?

16. A rocket is launched straight up from the earth's surface at a speed of 15,000 m/s. What is its speed when it is very far away from the earth?

17. What is the escape speed from Jupiter?

18. A space station orbits the sun at the same distance as the earth but on the opposite side of the sun. A small probe is fired away from the station. What minimum speed does the probe need to escape the solar system?

19. You have been visiting a distant planet. Your measurements have determined that the planet's mass is twice that of earth but the acceleration due to gravity at the surface is only one-fourth as large.
    a. What is the planet's radius?
    b. To get back to earth, you need to escape the planet. What minimum speed does your rocket need?

Section 12.6 Satellite Orbits and Energies

20. A satellite orbits the sun with a period of 1.0 day. What is the radius of its orbit?

21. The space shuttle is in a 250-mile-high orbit. What are the shuttle's orbital period, in minutes, and its speed?

22. The *asteroid belt* circles the sun between the orbits of Mars and Jupiter. One asteroid has a period of 5.0 earth years. What are the asteroid's orbital radius and speed?

23. An earth satellite moves in a circular orbit at a speed of 5500 m/s. What is its orbital period?

24. What are the speed and altitude of a geosynchronous satellite orbiting Mars? Mars rotates on its axis once every 24.8 hours.

25. Use information about the earth and its orbit to determine the mass of the sun.

26. Three satellites orbit a planet of radius R, as shown in Figure Ex12.26. Satellites S_1 and S_3 have mass m. Satellite S_2 has mass $2m$. Satellite S_1 orbits in 250 minutes and the force on S_1 is 10,000 N.
    a. What are the periods of S_2 and S_3?
    b. What are the forces on S_2 and S_3?
    c. What is the kinetic-energy ratio K_1/K_3 for S_1 and S_3?

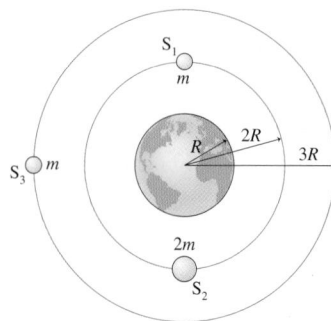

FIGURE EX12.26

27. A 4000 kg lunar lander is in orbit 50 km above the surface of the moon. It needs to move out to a 300-km-high orbit in order to link up with the mother ship that will take the astronauts home. How much work must the thrusters do?

28. The space shuttle is in a 250-km-high circular orbit. It needs to reach a 610-km-high circular orbit to catch the Hubble Space Telescope for repairs. The shuttle's mass is 75,000 kg. How much energy is required to boost it to the new orbit?

Problems

29. Two spherical objects have a combined mass of 150 kg. The gravitational attraction between them is 8.00×10^{-6} N when their centers are 20 cm apart. What is the mass of each?
30. Two 100 kg lead spheres are suspended from 100-m-long massless cables. The tops of the cables have been carefully anchored *exactly* 1 m apart. What is the distance between the centers of the spheres?
31. In a Cavendish balance experiment, 200 g masses are attached to the ends of a 40-cm-long rod. The rod is then suspended by a fiber whose torsion constant is 1.0×10^{-5} N/rad. That is, applying force F to *both* ends of the rod, in opposite directions, causes the rod to turn through angle $\Delta\theta$ such that $F = k\Delta\theta$. Once the rod is stationary, 20 kg lead spheres are brought close to each end of the rod. Through what angle does the rod turn if the distance between the center of each sphere and the corresponding 200 g mass is 15.0 cm? Give your answer in both radians and degrees. The angle is small, but it is measurable.
32. A new moon is almost exactly in line between the earth and the sun. A full moon is on the opposite side of the earth from the sun. What is the ratio of the *net* gravitational force on the moon when it is new to when it is full?
33. A 20 kg sphere is at the origin and a 10 kg sphere is at $(x, y) = (20$ cm, 0 cm$)$. At what point or points could you place a small mass such that the net gravitational force on it due to the spheres is zero?
34. Figure P12.34 shows three masses. What are the magnitude and the direction of the net gravitational force on (a) the 20.0 kg mass and (b) the 5.0 kg mass? Give the direction as an angle cw or ccw of the y-axis.

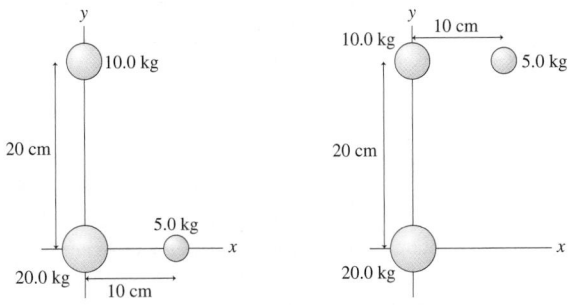

FIGURE P12.34   **FIGURE P12.35**

35. Figure P12.35 shows three masses. What are the magnitude and the direction of the net gravitational force on (a) the 10.0 kg mass and (b) the 20.0 kg mass? Give the direction as an angle cw or ccw of the y-axis.
36. What are the gravitational potential energies of (a) the 20.0 kg mass and (b) the 5.0 kg mass in Figure P12.34?
37. What are the gravitational potential energies of (a) the 10.0 kg mass and (b) the 20.0 kg mass in Figure P12.35?

38. Consider an object of mass m on the equator of a planet with mass M, radius R, and period of rotation T. If the planet did not rotate, the object would be in static equilibrium and its apparent weight would be $w_{app} = mg_{true}$, where $g_{true} = MG/R^2$. Because of the planet's rotation, the object is rather like a roller-coaster car going over the top of a hill. It has an effective weight $w_{app} = n = mg_{app}$, but $g_{app} \neq g_{true}$. Any experiment performed by a physics student to measure the acceleration due to gravity will determine the value of g_{app}, not g_{true}.
    a. Find an algebraic expression for g_{app} in terms of M, R, G, and T.
    b. Evaluate both g_{true} and g_{app} for the earth.
    c. Which of these g's have we been using in this text?
    d. How will the value of g measured at the North Pole compare to the value measured at the equator? Will it be larger, smaller, or the same? Explain.
39. a. At what height above the earth is the acceleration due to gravity 10% of its value at the surface?
    b. What is the speed of a satellite orbiting at that height?
40. A 1.0 kg object is released from rest 500 km ($\approx$300 miles) above the earth.
    a. What is its impact speed as it hits the ground? Ignore air resistance.
    b. What would the impact speed be if the earth were flat?
    c. By what percent is the flat-earth calculation in error?
    d. Suppose a space shuttle astronaut, while out for a stroll, "drops" a 1.0 kg hammer. Will it fall to earth? Explain why or why not.
41. A huge cannon is assembled on an airless planet. The planet has a radius of 5.0×10^6 m and a mass of 2.6×10^{24} kg. The cannon fires a projectile straight up at 5000 m/s.
    a. What height does the projectile reach above the surface?
    b. An observation satellite orbits the planet at a height of 1000 km. What is the projectile's speed as it passes the satellite?
42. Two meteoroids are heading for earth. Their speeds as they cross the moon's orbit are 2.0 km/s.
    a. The first meteoroid is heading straight for earth. What is its speed of impact?
    b. The second misses the earth by 5000 km. What is its speed at its closest point?
43. A binary star system has two stars, each with the same mass as our sun, separated by 1.0×10^{12} m. A comet is very far away and essentially at rest. Slowly but surely, gravity pulls the comet toward the stars. Suppose the comet travels along a straight line that passes through the midpoint between the two stars. What is the comet's speed at the midpoint?
44. Suppose that on earth you can jump straight up a distance of 50 cm. Can you escape from a 4.0-km-diameter asteroid with a mass of 1.0×10^{14} kg?
45. Figure P12.45 shows two identical planets of mass M and radius R spaced $6R$ apart. Ignore the rotation of the planets about each other. What is the escape speed of a rocket launched from (a) point A and (b) point B?

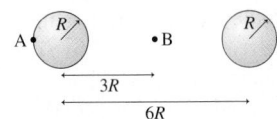

FIGURE P12.45

46. Two stars with the mass and radius of the sun are separated by distance $10R_s$, measured between their centers. A 10,000 kg space capsule moves along a line between the stars.
    a. Sketch a reasonably accurate graph of the space capsule's potential energy along a line passing through the two stars. Let one star be at $x = 0$ and the other at $x = 10R_s$.
    b. Suppose the space capsule is at rest exactly midway between the stars. Is this a point of equilibrium? If so, is it a stable or an unstable equilibrium? Explain.
    c. Suppose the space capsule is at rest 1.0 m closer to one star than the other. What will be the speed of the space capsule when it meets its ultimate fate?

47. a. At what distance from the center of the earth, along a line from the earth to the moon, is there no net force on a 10,000 kg spacecraft? This is the *crossover point* at which the attraction to the moon becomes larger than the attraction to the earth.
    b. The gravitational potential energy is the sum of the potential energies with respect to the earth and the moon. Calculate the spacecraft's potential energy at several points between the surface of the earth and the surface of the moon, including the crossover point. Then draw a reasonably accurate graph showing the potential energy from the earth to the moon.
    c. What is the mechanical energy of the spacecraft as it orbits 300 km above the moon? Don't forget to include a term due to the earth's gravitational potential energy.
    d. What is the minimum work the spacecraft engines must do to reach the crossover point and head home?
    e. If the spacecraft crosses the crossover point with the minimum possible energy, what will its speed be as it reaches the edge of the earth's 100-km-high atmosphere?

48. Two spherical asteroids have the same radius R. Asteroid 1 has mass M and asteroid 2 has mass $2M$. The two asteroids are released from rest with distance $10R$ between their centers. What is the speed of each asteroid just before they collide?
    Hint: You will need to use two conservation laws.

49. Two Jupiter-size planets are released from rest 1.0×10^{11} m apart. What are their speeds as they crash together?

50. a. How much energy must a 50,000 kg space shuttle lose to descend from a 500-km-high circular orbit to a 300-km-high orbit?
    b. Give a *qualitative* description, including a sketch, of how the shuttle would do this.

51. Suppose a cosmic accident increases the earth's rotation on its axis. At what rotational period, in hours, will objects at the equator begin to "fly off" the earth?

52. Mars has a small moon, Phobos, that orbits with a period of 7 h 39 min. The radius of Phobos' orbit is 9.4×10^6 m. What is the mass of Mars?

53. In 2000, NASA placed a satellite in orbit around an asteroid. Consider a spherical asteroid with a mass of 1.0×10^{16} kg and a radius of 8.8 km.
    a. What is the speed of a satellite orbiting 5.0 km above the surface?
    b. What is the escape speed from the asteroid?

54. You are the science officer on a visit to a distant solar system. Prior to landing on a planet you measure its diameter to be 1.8×10^7 m and its rotation period to be 22.3 hours. You have previously determined that the planet orbits 2.2×10^{11} m from its star with a period of 402 earth days. Once on the surface you find that the acceleration due to gravity is 12.2 m/s². What are the mass of (a) the planet and (b) the star?

55. NASA would like to place a satellite in orbit around the moon such that the satellite always remains in the same position over the lunar surface. What is the satellite's altitude?

56. A satellite orbiting the earth is directly over a point on the equator at 12:00 midnight every two days. It is not over that point at any time in between. What is the radius of the satellite's orbit?

57. Figure 12.19 showed a graph of log T versus log r for the planetary data given in Table 12.2. Such a graph is called a *log-log graph*. The scales in Figure 12.19 are logarithmic, not linear, meaning that each division along the axis corresponds to a *factor* of 10 increase in the value. Strictly speaking, the "correct" labels on the y-axis should be 7, 8, 9, and 10 because these are the logarithms of $10^7, \ldots, 10^{10}$.
    a. Consider two quantities u and v that are related by the expression $v^p = Cu^q$, where C is a constant. The exponents p and q are not necessarily integers. Define $x = \log u$ and $y = \log v$. Find an expression for y in terms of x.
    b. What *shape* will a graph of y versus x have? Explain.
    c. What *slope* will a graph of y versus x have? Explain.
    d. Figure 12.19 showed that the "best fit" line passing through all the planetary data points has the equation $\log T = 1.500 \log r - 9.264$. This is an *experimentally* determined relationship between $\log T$ and $\log r$, using measured data. Is this experimental result consistent with what you would expect from Newton's theory of gravity? Explain.
    e. Use the experimentally determined "best fit" line to find the mass of the sun.

58. Figure P12.58 shows two planets of mass m orbiting a star of mass M. The planets are in the same orbit, with radius r, but are always at opposite ends of a diameter. Find an expression for the orbital period T.

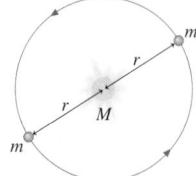

FIGURE P12.58

59. Humans currently use energy at the rate of about 10^{13} W. Suppose a future inventor finds a way to extract usable energy from the moon's orbital motion and beam that energy, without loss, to earth. If our energy use remains constant, by how much will the radius and period of the moon's orbit decrease after its energy has been tapped for 100 years? Give your answers as numerical values and as a percentage decrease.

60. Large stars can explode as they finish burning their nuclear fuel, causing a *supernova*. The explosion blows away the outer layers of the star. According to Newton's third law, the forces that push the outer layers away have *reaction forces* that are inwardly directed on the core of the star. These forces compress the core and can cause the core to undergo a *gravitational collapse*. The gravitational forces keep pulling all the matter together tighter and tighter, crushing atoms out

of existence. Under these extreme conditions, the protons and electrons can be squeezed together to form a neutron. If the collapse is halted when the neutrons all come into contact with each other, the result is an object called a *neutron star,* an entire star consisting of solid nuclear matter. Many neutron stars rotate about their axis with a period of ≈1 s and, as they do so, send out a pulse of electromagnetic waves once a second. These stars were discovered in the 1960s and are called *pulsars.*

   a. Consider a neutron star with a mass equal to the sun, a radius of 10 km, and a rotation period of 1.0 s. What is the speed of a point on the equator of the star?

   b. What is g at the surface of this neutron star?

   c. A 1.0 kg mass has a weight on earth of 9.8 N. What would be its weight on the star?

   d. How many revolutions per minute are made by a satellite orbiting 1.0 km above the surface?

   e. What is the radius of a geosynchronous orbit about the neutron star?

61. Astronomers discover a binary star system that has a period of 90 days. The binary star system consists of two equal-mass stars, each with a mass twice that of the sun, that rotate like a dumbbell about the *center of mass* at the midpoint between them. How far apart are the two stars?

62. Three stars, each with the mass of our sun, form an equilateral triangle with sides 1.0×10^{12} m long. (This triangle would just about fit within the orbit of Jupiter.) The triangle has to rotate, because otherwise the stars would crash together in the center. What is the period of rotation? Give your answer in years.

63. Pluto moves in a fairly elliptical orbit around the sun. Pluto's speed at its closest approach of 4.43×10^9 km is 6.12 km/s. What is Pluto's speed at the most distant point in its orbit, where it is 7.30×10^9 km from the sun?

64. Mercury moves in a fairly elliptical orbit around the sun. Mercury's speed is 38.8 km/s when it is at its most distant point, 6.99×10^9 m from the sun. How far is Mercury from the sun at its closest point, where its speed is 59.0 km/s?

65. Comets move around the sun in very elliptical orbits. At its closet approach, in 1986, Comet Halley was 8.79×10^7 km from the sun and moving with a speed of 54.6 km/s. What will the comet's speed be when it crosses Neptune's orbit in 2006?

66. A spaceship is in a circular orbit of radius r_0 about a planet of mass M. A brief but intense firing of its engine in the forward direction decreases the spaceship's speed by 50%. This causes the spaceship to move into an elliptical orbit.

   a. What is the spaceship's new speed, just after the rocket burn is complete, in terms of M, G, and r_0?

   b. In terms of r_0, what are the spaceship's maximum and minimum distance from the planet in its new orbit?

67. A planet is orbiting a star when, for no apparent reason, the star's gravity suddenly vanishes. As Figure P12.67 shows, the planet then obeys Newton's first law and heads outward along a straight line. Is Kepler's second law still obeyed? That is, are equal areas swept out in equal intervals of time as the planet moves away?

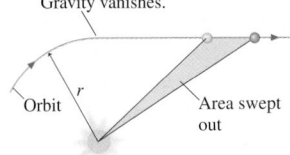

FIGURE P12.67

In Problems 68 through 71 you are given the equation(s) used to solve a problem. For each of these, you are to
   a. Write a realistic problem for which this is the correct equation(s).
   b. Draw a pictorial representation.
   c. Finish the solution of the problem.

68.
$$\frac{(6.67 \times 10^{-11} \text{ N m}^2/\text{kg}^2)(5.68 \times 10^{26} \text{ kg})}{r^2}$$
$$= \frac{(6.67 \times 10^{-11} \text{ N m}^2/\text{kg}^2)(5.98 \times 10^{24} \text{ kg})}{(6.37 \times 10^6 \text{ m})^2}$$

69.
$$\frac{(6.67 \times 10^{-11} \text{ N m}^2/\text{kg}^2)(5.98 \times 10^{24} \text{ kg})(1000 \text{ kg})}{r^2}$$
$$= \frac{(1000 \text{ kg})(1997 \text{ m/s})^2}{r}$$

70.
$$\frac{1}{2}(100 \text{ kg})v_2^2$$
$$- \frac{(6.67 \times 10^{-11} \text{ N m}^2/\text{kg}^2)(7.36 \times 10^{22} \text{ kg})(100 \text{ kg})}{(1.74 \times 10^6 \text{ m})}$$
$$= 0 - \frac{(6.67 \times 10^{-11} \text{ N m}^2/\text{kg}^2)(7.36 \times 10^{22} \text{ kg})(100 \text{ kg})}{(3.48 \times 10^6 \text{ m})}$$

71. $(2.0 \times 10^{30} \text{ kg})v_{f1} + (4.0 \times 10^{30} \text{ kg})v_{f2} = 0$
$$\frac{1}{2}(2.0 \times 10^{30} \text{ kg})v_{f1}^2 + \frac{1}{2}(4.0 \times 10^{30} \text{ kg})v_{f2}^2$$
$$- \frac{(6.67 \times 10^{-11} \text{ N m}^2/\text{kg}^2)(2.0 \times 10^{30} \text{ kg})(4.0 \times 10^{30} \text{ kg})}{(1.0 \times 10^9 \text{ m})}$$
$$= 0 + 0$$
$$- \frac{(6.67 \times 10^{-11} \text{ N m}^2/\text{kg}^2)(2.0 \times 10^{30} \text{ kg})(4.0 \times 10^{30} \text{ kg})}{(1.0 \times 10^{12} \text{ m})}$$

Challenge Problems

72. In 1996, the Solar and Heliospheric Observatory (SOHO) was "parked" in an orbit slightly inside the earth's orbit, as shown in Figure CP12.72. The satellite's period in this orbit is exactly one year, so it remains fixed relative to the earth. At this point, called a *Lagrange point,* the light from the sun is never blocked by the earth, yet the satellite remains "nearby" so that data are easily transmitted to earth. What is SOHO's distance from the earth?

Hint: Use the binomial approximation. SOHO's distance from the earth is much less than the earth's distance from the sun.

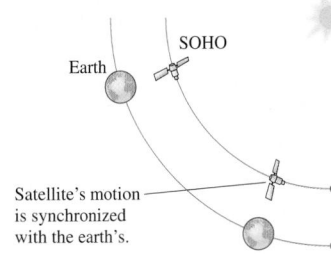

FIGURE CP12.72

73. The solar system is 25,000 light years from the center of our Milky Way galaxy. One *light year* is the distance light travels in one year at a speed of 3.0×10^8 m/s. Astronomers have determined that the solar system is orbiting the center of the galaxy at a speed of 230 km/s.
    a. Assuming the orbit is circular, what is the period of the solar system's orbit? Give your answer in years.
    b. Our solar system was formed roughly 5 billion years ago. How many orbits has it completed?
    c. The gravitational force on the solar system is the net force due to all the matter inside our orbit. Most of that matter is concentrated near the center of the galaxy. Assume that the matter has a spherical distribution, like a giant star. What is the approximate mass of the galactic center?
    d. Assume that the sun is a typical star with a typical mass. If galactic matter is made up of stars, approximately how many stars are in the center of the galaxy?

    Astronomers have spent many years trying to determine how many stars there are in the Milky Way. The number of stars seems to be only about 10% what you found in part (d). In other words, about 90% of the mass of the galaxy appears to be in some form other than stars. This is called the *dark matter* of the universe. No one knows what the dark matter is. This is one of the outstanding scientific questions of our day.

74. The space shuttle, in a 300-km-high orbit, needs to perform an experiment that has to take place well away from the space-craft. To do so, a 100 kg payload is "lowered" toward the earth on a 10-km-long massless rope. (We'll overlook the details of how they do this and simply assume they can.) The payload can be hauled back on board the shuttle after the experiment. Assume that any initial motions associated with lowering the payload have damped out and that the shuttle and payload are flying in steady-state conditions.
    a. What is the angle of the rope as measured from a line drawn from the center of the earth through the payload? Explain.
    b. What is the tension in the rope?

75. Your job with NASA is to monitor satellite orbits. One day, during a routine survey, you find that a 400 kg satellite in a 1000-km-high circular orbit is going to collide with a smaller 100 kg satellite traveling in the same orbit but in the opposite direction. Knowing the construction of the two satellites, you expect they will become enmeshed into a single piece of space debris. When you notify your boss of this impending collision, he asks you to quickly determine whether the space debris will continue to orbit or crash into the earth. What will the outcome be?

76. Let's look in more detail at how a satellite is moved from one circular orbit to another. Figure CP12.76 shows two circular orbits, of radii r_1 and r_2, and an elliptical orbit that connects them. Points 1 and 2 are at the ends of the semimajor axis of the ellipse.
    a. A satellite moving along the elliptical orbit has to satisfy two conservation laws. Use these two laws to prove that the velocities at points 1 and 2 are

    $$v_1' = \sqrt{\frac{2GM(r_2/r_1)}{r_1 + r_2}} \text{ and } v_2' = \sqrt{\frac{2GM(r_1/r_2)}{r_1 + r_2}}$$

    The prime indicates that these are the velocities on the ellip-tical orbit. Both reduce to Equation 12.22 if $r_1 = r_2 = r$.
    b. Consider a 1000 kg communication satellite that needs to be boosted from an orbit 300 km above the earth to a geo-synchronous orbit 35,900 km above the earth. Find the velocity v_1 on the lower circular orbit and the velocity v_1' at the low point on the elliptical orbit that spans the two cir-cular orbits.
    c. How much work must the rocket motor do to transfer the satellite from the circular orbit to the elliptical orbit?
    d. Now find the velocity v_2' at the top of the elliptical orbit and the velocity v_2 of the upper circular orbit.
    e. How much work must the rocket motor do to transfer the satellite from the elliptical orbit to the upper circular orbit?
    f. Compute the total work done and compare your answer to the result of Example 12.6.

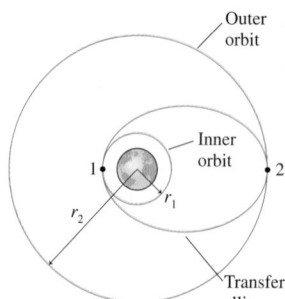

FIGURE CP12.76

STOP TO THINK ANSWERS

Stop to Think 12.1: e. The acceleration decreases inversely with the square of the distance. At height R_e, the distance from the center of the earth is $2R_e$.

Stop to Think 12.2: c. Newton's third law requires $F_{1 \text{ on } 2} = F_{2 \text{ on } 1}$.

Stop to Think 12.3: b. $g_{\text{surface}} = GM/R^2$. Because of the square, a radius twice as large balances a mass four times as large.

Stop to Think 12.4: In absolute value, $U_e > U_a = U_b = U_d > U_c$. $|U_g|$ is proportional to $m_1 m_2/r$.

Stop to Think 12.5: a. T^2 is proportional to r^3, or T is proportional to $r^{3/2}$. $4^{3/2} = 8$.

13 Rotation of a Rigid Body

Not all motion can be described as that of a particle. Rotation requires the idea of an extended object.

▶ **Looking Ahead**

The goal of Chapter 13 is to understand the physics of rotating objects. In this chapter you will learn to:

- Apply the rigid-body model to extended objects.
- Calculate torques and moments of inertia.
- Understand the rotation of a rigid body around a fixed axis.
- Understand rolling motion.
- Apply conservation of energy and angular momentum to rotational problems.
- Use vector mathematics to describe rotational motion.

◀ **Looking Back**

Rotational motion will revisit many of the major themes introduced in Parts I and II, especially the properties of circular motion. Please review:

- Section 4.5 Newton's second law.
- Sections 7.1 and 7.2 The mathematics of circular motion.
- Section 9.6 Angular momentum.
- Section 10.2 Kinetic and gravitational potential energy.

This diver is moving toward the water along a parabolic trajectory, much like a cannon ball. At the same time, she's rotating rapidly around her center of mass. This combination of two types of motion is what makes a great dive both interesting to watch and difficult to perform.

Our goal in this chapter is to understand rotational motion. We will focus our attention on what are called *rigid bodies*. Wheels, axles, and gyroscopes are examples of rigid bodies that rotate. Divers, gymnasts, and ice skaters also rotate, although the fact that they are *not* rigid bodies makes their motions more complex. Even so, we will be able to understand many aspects of their motion by modeling them as rigid bodies.

You will quickly discover that the physics of rotational motion is analogous to the physics of linear motion that you studied in Parts I and II. For example, the new concepts of torque and angular acceleration are the rotational analogs of force and acceleration, and we'll find a new version of Newton's second law that is the rotational equivalent of $\vec{F} = m\vec{a}$. Similarly, energy and conservation laws will continue to be important tools. Rotational motion is an important application of Newtonian mechanics to extended objects.

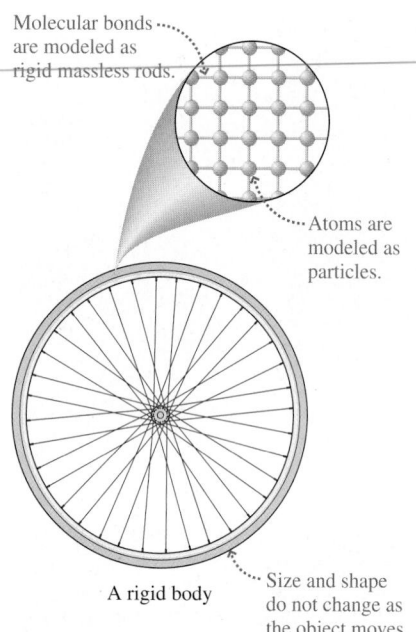

Molecular bonds are modeled as rigid massless rods.

Atoms are modeled as particles.

A rigid body

Size and shape do not change as the object moves.

FIGURE 13.1 The rigid-body model of an extended object.

13.1 Rotational Kinematics

Thus far, our study of physics has focused almost exclusively on the *particle model* in which an object is represented as a mass at a single point in space. The particle model is a perfectly good description of the physics in a vast number of situations, but there are other situations for which we need to consider the motion of an *extended object*—a system of particles for which the size and shape *do* make a difference and cannot be neglected.

A **rigid body** is an extended object whose size and shape do not change as it moves. For example, a bicycle wheel can be thought of as a rigid body. Figure 13.1 shows a rigid body as a collection of atoms held together by the rigid "massless rods" of molecular bonds.

Real molecular bonds are, of course, not perfectly rigid. That's why an object seemingly as rigid as a bicycle wheel can flex and bend. Thus Figure 13.1 is really a simplified *model* of an extended object, the **rigid-body model.** The rigid-body model is a very good approximation of many real objects of practical interest, such as wheels and axles. Even nonrigid objects can often be modeled as a rigid body during parts of their motion. For example, the diver in the opening photograph is well described as a rotating rigid body while she's in the tuck position.

Figure 13.2 illustrates the three basic types of motion of a rigid body: **translational motion, rotational motion,** and **combination motion.**

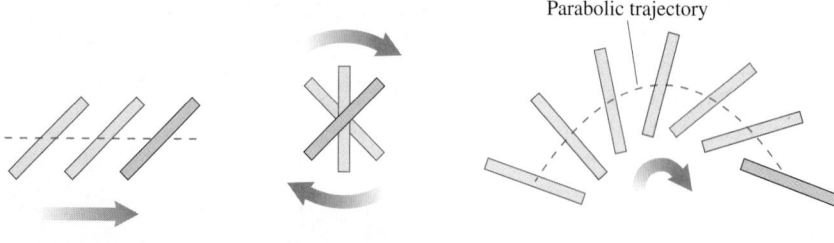

Parabolic trajectory

Translational motion:
The object as a whole moves along a trajectory but does not rotate.

Rotational motion:
The object rotates about a fixed point. Every point on the object moves in a circle.

Combination motion:
An object rotates as it moves along a trajectory.

FIGURE 13.2 Three basic types of motion of a rigid body.

Circular Motion

Rotation is an extension of circular motion, so we begin with a brief summary of some of the main results of Chapter 7. Figure 13.3 shows a particle of mass m rotating in a circle of radius r. Angle θ is the particle's angular position. The angular velocity

$$\omega = \frac{d\theta}{dt} \tag{13.1}$$

is the rate at which the particle moves around the circle. The SI units of ω are radians per second (rad/s). Revolutions per second (s^{-1}) and revolutions per minute (rpm) are frequently used units, but these need to be converted to rad/s for most calculations.

NOTE ▶ The sign convention is that ω is positive for counterclockwise (ccw) rotation, negative for clockwise (cw) rotation. ◀

A vector associated with the particle can be described in terms of its radial component, toward or away from the center, and its tangential component. Figure 13.3 reminds you of the radial and tangential components of $\vec{a}$ and $\vec{v}$ for circular motion.

You learned in Chapter 7 that the velocity, acceleration, and angular velocity of circular motion are related by

$$v_r = 0 \qquad a_r = \frac{v_t^2}{r} = \omega^2 r$$

$$v_t = \frac{ds}{dt} = r\omega \qquad a_t = \frac{dv_t}{dt} \tag{13.2}$$

Here s is the arc length around the circle ($s = r\theta$). The sign convention for ω implies that v_t and a_t are positive if they point in the ccw direction, negative if they point in the cw direction.

Angular Velocity and Angular Acceleration

Figure 13.4 shows a wheel rotating on an axle. Notice that two points on the wheel, marked with dots, turn through the *same angle* as the wheel rotates, even through their radii may be different. That is, $\Delta\theta_1 = \Delta\theta_2$ during some time interval Δt. As a consequence, the two points have equal angular velocities: $\omega_1 = \omega_2$. Thus we can refer to the angular velocity ω *of the wheel*.

NOTE ▶ Two points in a rotating object have different tangential velocities v_t if they have different distances from the point of rotation, but *all* points have the *same* angular velocity ω. Thus angular velocity is one of the most important parameters of a rotating object. ◀

We will want to consider situations in which an object's rotation speeds up or slows down; that is, situations in which the points on the object have a tangential acceleration a_t. You saw in Equation 13.2 that $a_t = dv_t/dt$ and that $v_t = r\omega$. Combining these two equations, we find

$$a_t = r\frac{d\omega}{dt} \tag{13.3}$$

In taking the derivative, we used the fact that r is a constant for circular motion.

We originally defined acceleration as $a = dv/dt$, the rate of change of the velocity. The derivative in Equation 13.3 is the rate of change of the *angular* velocity. By analogy, let's define the **angular acceleration** α (Greek alpha) to be

$$\alpha \equiv \frac{d\omega}{dt} \tag{13.4}$$

The units of angular acceleration are rad/s^2.

Angular acceleration is the *rate* at which the angular velocity ω changes, just as the linear acceleration is the rate at which the linear velocity v changes. If a wheel starts from rest with $\alpha = 2$ rad/s^2, its angular velocity will increase to 2 rad/s at $t = 1$ s, increase another 2 rad/s to 4 rad/s at $t = 2$ s, increase another 2 rad/s to 6 rad/s at $t = 3$ s, and so on. That is, the angular velocity increases by 2 rad/s per second.

NOTE ▶ Be careful with the sign of α. You learned in Chapter 2 that positive and negative values of the acceleration can't be interpreted as simply "speeding up" and "slowing down." ◀

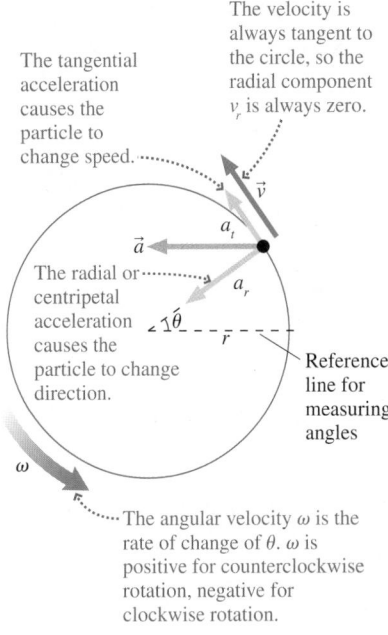

The tangential acceleration causes the particle to change speed.

The velocity is always tangent to the circle, so the radial component v_r is always zero.

The radial or centripetal acceleration causes the particle to change direction.

Reference line for measuring angles

The angular velocity ω is the rate of change of θ. ω is positive for counterclockwise rotation, negative for clockwise rotation.

FIGURE 13.3 A particle in circular motion.

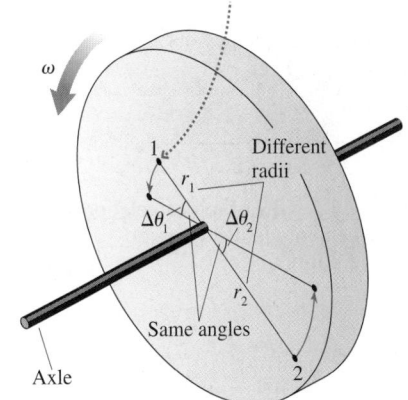

Every point on the wheel undergoes circular motion with the same angular velocity ω.

Different radii

Same angles

Axle

FIGURE 13.4 The two points on the wheel rotate with the same angular velocity.

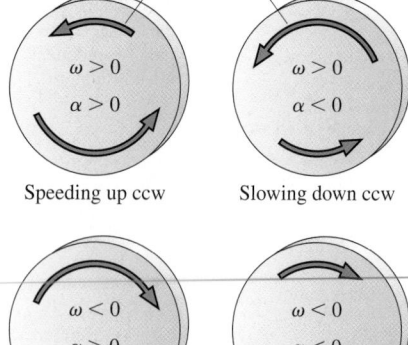

Initial angular velocity

Speeding up ccw

Slowing down ccw

Slowing down cw

Speeding up cw

FIGURE 13.5 The signs of angular velocity and acceleration.

Because $\vec{a}$ is a vector, positive a_x means that v_x is increasing to the right or decreasing to the left. Negative a_x means that v_x is increasing to the left or decreasing to the right. For rotational motion, α is positive if ω is increasing ccw or decreasing cw, negative if ω is increasing cw or decreasing ccw. These are illustrated in Figure 13.5.

Comparing Equations 13.3 and 13.4, we see that the tangential and angular accelerations are related by

$$a_t = r\alpha \qquad (13.5)$$

Two points on a rotating object have the *same* angular acceleration α, but in general they have *different* tangential accelerations because they are moving in circles of different radii. Notice the analogy between Equation 13.5 and the similar equation $v_t = r\omega$ for tangential and angular velocity.

Graphically, the angular acceleration α is the slope of the ω-versus-t graph. We can integrate Equation 13.4 to find

$$\omega_f = \omega_i + \text{area under the angular acceleration } \alpha \text{ curve between } t_i \text{ and } t_f \qquad (13.6)$$

These relationships involving slopes and areas, illustrated in Figure 13.6, are exactly the same for rotational motion as they were for linear motion.

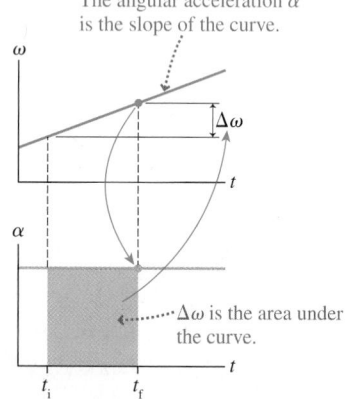

The angular acceleration α is the slope of the curve.

$\Delta\omega$ is the area under the curve.

FIGURE 13.6 The graphical relationships between angular velocity and acceleration.

EXAMPLE 13.1 A rotating wheel
Figure 13.7a is a graph of angular velocity versus time for a rotating wheel. Describe the motion and draw a graph of angular acceleration versus time.

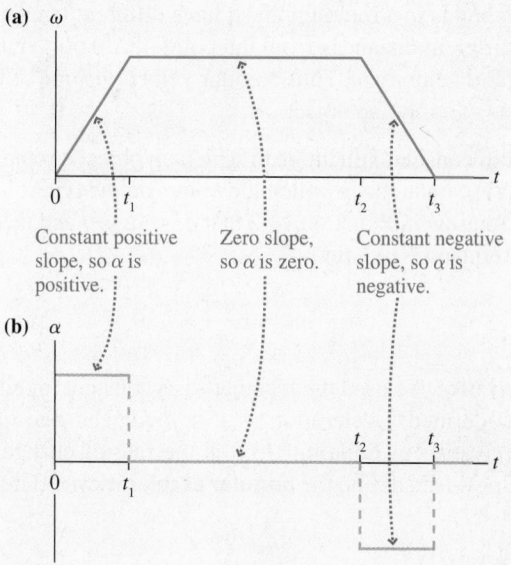

(a) ω

Constant positive slope, so α is positive.

Zero slope, so α is zero.

Constant negative slope, so α is negative.

(b) α

FIGURE 13.7 ω-versus-t and α-versus-t graphs for a rotating wheel.

MODEL The wheel is a rotating rigid body.

SOLVE This is a wheel that starts from rest, gradually speeds up *counterclockwise* until reaching top speed at t_1, maintains a constant angular velocity until t_2, then gradually slows down until stopping at t_3. The motion is always ccw because ω is always positive. The angular acceleration graph of Figure 13.7b is based on the fact that α is the slope of the ω-versus-t graph.

In Chapter 7 we developed kinematic equations for θ and ω. We can now replace the a_t/r that appeared in those equations with the angular acceleration α. Table 13.1 shows the resulting kinematic equations for constant angular acceleration. These equations apply to a particle in circular motion or to any rigid-body rotation. Notice that the rotational kinematic equations are exactly analogous to the linear kinematic equations.

Act|v
Phys|cs   7.7

TABLE 13.1 Rotational and linear kinematics for constant acceleration

Rotational kinematics	Linear kinematics
$\omega_f = \omega_i + \alpha\Delta t$	$v_f = v_i + a\Delta t$
$\theta_f = \theta_i + \omega_i\Delta t + \frac{1}{2}\alpha(\Delta t)^2$	$x_f = x_i + v_i\Delta t + \frac{1}{2}a(\Delta t)^2$
$\omega_f^2 = \omega_i^2 + 2\alpha\Delta\theta$	$v_f^2 = v_i^2 + 2a\Delta x$

EXAMPLE 13.2 A rotating crankshaft

A car's tachometer records the rotation frequency of the crankshaft. A car engine is idling at 500 rpm. When the light turns green, the crankshaft rotation speeds up at a constant rate to 2500 rpm over an interval of 3.0 s. How many revolutions does the crankshaft make during these 3 s?

MODEL The crankshaft is a rotating rigid body with constant angular acceleration.

SOLVE Imagine painting a dot on the crankshaft. Let the dot be at $\theta_i = 0$ rad at $t = 0$ s. Three seconds later the dot will have turned to angle

$$\theta_f = \omega_i\Delta t + \frac{1}{2}\alpha(\Delta t)^2$$

where $\Delta t = 3.0$ s. We can find the angular acceleration from the initial and final angular velocities, but first they must be converted to SI units:

$$\omega_i = 500\frac{\text{rev}}{\text{min}} \times \frac{1 \text{ min}}{60 \text{ s}} \times \frac{2\pi \text{ rad}}{1 \text{ rev}} = 52.4 \text{ rad/s}$$

$$\omega_f = 2500\frac{\text{rev}}{\text{min}} = 5\omega_i = 262.0 \text{ rad/s}$$

The angular acceleration α is

$$\alpha = \frac{\Delta\omega}{\Delta t} = \frac{(262.0 \text{ rad/s} - 52.4 \text{ rad/s})}{3.0 \text{ s}}$$

$$= \frac{209.6 \text{ rad/s}}{3.0 \text{ s}} = 69.9 \text{ rad/s}^2$$

During these 3.0 s, the dot turns through an angle

$$\Delta\theta = \omega_i\Delta t + \frac{1}{2}\alpha(\Delta t)^2$$

$$= (52.4 \text{ rad/s})(3.0 \text{ s}) + \frac{1}{2}(69.9 \text{ rad/s}^2)(3.0 \text{ s})^2 = 472 \text{ rad}$$

Because $472/2\pi = 75$, the crankshaft completes 75 revolutions as it spins up to 2500 rpm.

ASSESS This problem is solved just like the linear kinematics problems you learned to solve in Chapter 2.

STOP TO THINK 13.1   The fan blade is slowing down. What are the signs of ω and α?

a. ω is positive and α is positive.   b. ω is positive and α is negative.
c. ω is negative and α is positive.   d. ω is negative and α is negative.

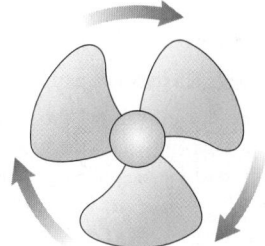

13.2 Rotation About the Center of Mass

(a)

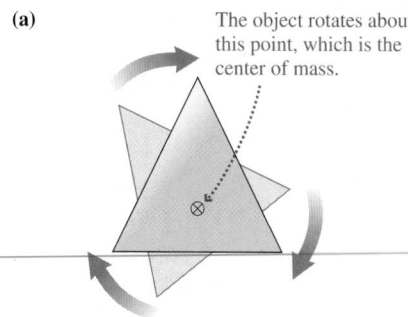

The object rotates about this point, which is the center of mass.

(b)

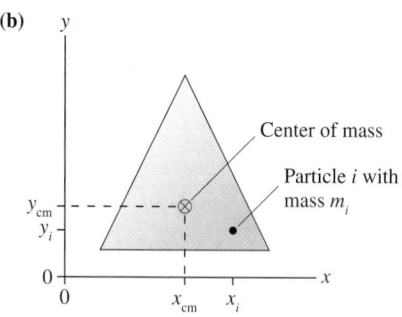

Center of mass

Particle i with mass m_i

FIGURE 13.8 Rotation about the center of mass.

Imagine yourself floating in a space capsule deep in space. Suppose you take an object like that shown in Figure 13.8a, push two corners in opposite directions to spin it, then let go. The object will rotate, but it will have no translational motion as it floats beside you. *About what point does it rotate?* That is the question we need to answer.

An unconstrained object (i.e., one not on an axle or a pivot) on which there is no net force rotates about a point called the **center of mass.** The center of mass remains motionless while every other point in the object undergoes circular motion around it. You need not go deep into space to demonstrate rotation about the center of mass. If you have an air table, a flat object rotating on the air table rotates about its center of mass.

To locate the center of mass, Figure 13.8b models the object as if it were constructed from particles numbered $i = 1, 2, 3, \ldots$ Particle i has mass m_i and is located at position (x_i, y_i). We'll prove later in this section that the center of mass is located at position

$$x_{\text{cm}} = \frac{1}{M}\sum_i m_i x_i = \frac{m_1 x_1 + m_2 x_2 + m_3 x_3 + \cdots}{m_1 + m_2 + m_3 + \cdots}$$

$$y_{\text{cm}} = \frac{1}{M}\sum_i m_i y_i = \frac{m_1 y_1 + m_2 y_2 + m_3 y_3 + \cdots}{m_1 + m_2 + m_3 + \cdots}$$

(13.7)

where $M = m_1 + m_2 + m_3 + \cdots$ is the object's total mass.

NOTE ▶ A three-dimensional object would need a similar equation for z_{cm}. For simplicity, we'll restrict ourselves to objects for which only the x- and y-coordinates are relevant. ◀

Let's see if Equations 13.7 make sense. Suppose you have an object consisting of N particles, all with the same mass m. That is, $m_1 = m_2 = \cdots = m_N = m$. We can factor the m out of the numerator and the denominator becomes simply Nm. The m cancels and x-coordinate of the center of mass is

$$x_{\text{cm}} = \frac{x_1 + x_2 + \cdots + x_N}{N} = x_{\text{average}}$$

In this case, x_{cm} is simply the *average* x-coordinate of all the particles. Likewise, y_{cm} will be the average of all the y-coordinates.

This *does* make sense! If the particle masses are all the same, the center of mass should be at the center of the object. And the "center of the object" is the average of the positions of all the particles. To allow for *unequal* masses, Equations 13.7 are called a *weighted average*. Particles of higher mass count more than particles of lower mass, but the basic idea remains the same. **The center of mass is the mass-weighted center of the object.**

EXAMPLE 13.3 The center of mass

A 500 g ball and a 2.0 kg ball are connected by a massless 50-cm-long rod.

a. Where is the center of mass?

b. What is the speed of each ball if they rotate about the center of mass at 40 rpm?

MODEL Model each ball as a particle.

VISUALIZE Figure 13.9 shows the two masses. We've chosen a coordinate system in which the masses are on the x-axis with the 2.0 kg mass at the origin.

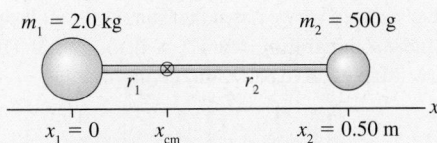

FIGURE 13.9 Finding the center of mass.

SOLVE

a. We can use Equation 13.7 to calculate that the center of mass is

$$x_{cm} = \frac{m_1 x_1 + m_2 x_2}{m_1 + m_2}$$

$$= \frac{(2.0 \text{ kg})(0.0 \text{ m}) + (0.50 \text{ kg})(0.50 \text{ m})}{2.0 \text{ kg} + 0.50 \text{ kg}} = 0.10 \text{ m}$$

$y_{cm} = 0$ because all the masses are on the x-axis. The center of mass is 20% of the way from the 2.0 g ball to the 0.50 kg ball.

b. Each ball rotates about the center of mass. The radii of the circles are $r_1 = 0.10$ m and $r_2 = 0.40$ m. The tangential velocities are $(v_i)_t = r_i \omega$, but this equation requires ω to be in rad/s. The conversion is

$$\omega = 40 \frac{\text{rev}}{\text{min}} \times \frac{1 \text{ min}}{60 \text{ s}} \times \frac{2\pi \text{ rad}}{1 \text{ rev}} = 4.19 \text{ rad/s}$$

Consequently,

$$(v_1)_t = r_1 \omega = (0.10 \text{ m})(4.19 \text{ rad/s}) = 0.419 \text{ m/s}$$

$$(v_2)_t = r_2 \omega = (0.40 \text{ m})(4.19 \text{ rad/s}) = 1.68 \text{ m/s}$$

ASSESS The center of mass is closer to the heavier ball than to the lighter ball. We expected this because x_{cm} is a mass-weighted average of the positions.

For any realistic object, carrying out the summations of Equations 13.7 over all the atoms in the object is not practical. Instead, as Figure 13.10 shows, we can divide an extended object into many small cells or boxes, each with the very small mass Δm. We will number the cells 1, 2, 3, ..., just as we did the particles. Cell i has coordinates (x_i, y_i) and mass $m_i = \Delta m$. The center-of-mass coordinates are then

$$x_{cm} = \frac{1}{M} \sum_i x_i \Delta m \quad \text{and} \quad y_{cm} = \frac{1}{M} \sum_i y_i \Delta m$$

Now, as you might expect, we'll let the cells become smaller and smaller, with the total number increasing. As each cell becomes infinitesimally small, we can replace Δm with dm and the sum by an integral. Then

$$x_{cm} = \frac{1}{M} \int x \, dm \quad \text{and} \quad y_{cm} = \frac{1}{M} \int y \, dm \qquad (13.8)$$

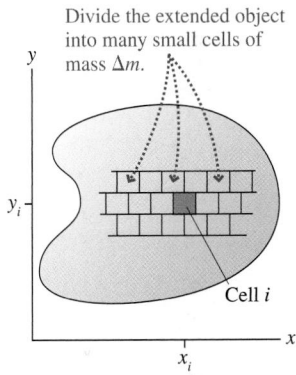

FIGURE 13.10 Calculating the center of mass of an extended object.

Equation 13.8 is a formal definition of the center of mass, but it is *not* ready to integrate in this form. First, integrals are carried out over *coordinates*, not over masses. Before we can integrate, we must replace dm by an equivalent expression involving a coordinate differential such as dx or dy. Second, no limits of integration have been specified. The procedure for using Equation 13.8 is best shown with an example.

EXAMPLE 13.4 The center of mass of a rod

Find the center of mass of a thin, uniform rod of length L and mass M. Use this result to find the tangential acceleration of one tip of a 1.60-m-long rod that rotates about its center of mass with an angular acceleration of 6.0 rad/s.

VISUALIZE Figure 13.11 shows the rod. We've chosen a coordinate system such that the rod lies along the x-axis from 0 to L. Because the rod is "thin," we'll assume that $y_{cm} = 0$.

SOLVE Our first task is to find x_{cm}, which lies somewhere on the x-axis. To do this, divide the rod into many small cells of

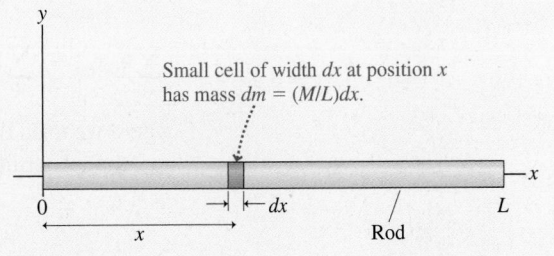

FIGURE 13.11 Finding the center of mass of a long, thin rod.

mass dm. One such cell, at position x, is shown. The cell's width is dx. Because the rod is *uniform*, the mass of this little cell is the *same fraction* of the total mass M that dx is of the total length L. That is,

$$\frac{dm}{M} = \frac{dx}{L}$$

Consequently, we can express dm in terms of the coordinate differential dx as

$$dm = \frac{M}{L} dx$$

NOTE ▶ The change of variables from dm to the differential of a coordinate is *the* key step in calculating the center of mass. ◀

With this expression for dm, Equation 13.8 for x_{cm} becomes

$$x_{cm} = \frac{1}{M}\left(\frac{M}{L}\int x\,dx\right) = \frac{1}{L}\int_0^L x\,dx$$

where in the last step we've noted that summing "all the mass in the rod" means integrating from $x = 0$ to $x = L$. This is a straightforward integral to carry out, giving

$$x_{cm} = \frac{1}{L}\left[\frac{x^2}{2}\right]_0^L = \frac{1}{L}\left[\frac{L^2}{2} - 0\right] = \frac{1}{2}L$$

The center of mass is at the center of the rod. For a 1.60-m-long rod, each tip of the rod rotates in a circle with $r = \frac{1}{2}L = 0.80$ m. The tangential acceleration, the rate at which the tip is speeding up, is

$$a_t = r\alpha = (0.80\text{ m})(6.0\text{ rad/s}^2) = 4.80\text{ m/s}^2$$

ASSESS You could have guessed that the center of mass is at the center of the rod, but now we've shown it rigorously.

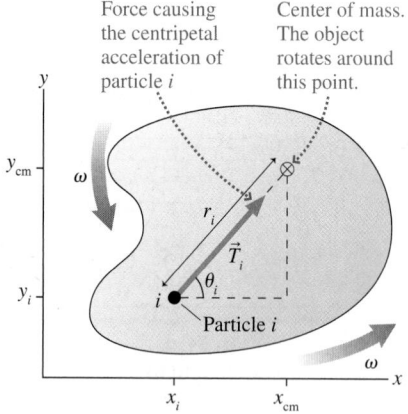

y

y_{cm}

ω

r_i

$\vec{T}_i$

θ_i

i Particle i

y_i

ω

x_i x_{cm} x

Force causing the centripetal acceleration of particle i

Center of mass. The object rotates around this point.

FIGURE 13.12 Finding the center of mass.

NOTE ▶ For any symmetrical object of uniform density, the center of mass is at the physical center of the object. ◀

To see where the center-of-mass equation comes from, Figure 13.12 shows an object rotating about its center of mass. Particle i is moving in a circle, so it *must* have a centripetal acceleration. Acceleration requires a force, and this force is due to tension in the molecular bonds that hold the object together. Force $\vec{T}_i$ on particle i has magnitude

$$T_i = m_i(a_i)_r = m_i r_i \omega^2 \tag{13.9}$$

where r_i is the distance of particle i from the center of mass and we used Equation 13.2 for a_r. All points in a rotating object have the *same* angular velocity, so ω doesn't need a subscript.

The internal tension forces are all paired as action/reaction forces, equal in magnitude but opposite in direction, so the sum of all the tension forces must be zero. That is, $\sum \vec{T}_i = \vec{0}$. The x-component of this sum is

$$\sum_i (T_i)_x = \sum_i T_i \cos\theta_i = \sum_i (m_i r_i \omega^2)\cos\theta_i = 0 \tag{13.10}$$

You can see from Figure 13.12 that $\cos\theta_i = (x_{cm} - x_i)/r_i$. Thus

$$\sum_i (T_i)_x = \sum_i (m_i r_i \omega^2)\frac{x_{cm} - x_i}{r_i} = \left(\sum_i m_i x_{cm} - \sum_i m_i x_i\right)\omega^2 = 0 \tag{13.11}$$

This equation will be true if the term in parentheses is zero. x_{cm} is a constant, so we can bring it outside the summation to write

$$\sum_i m_i x_{cm} - \sum_i m_i x_i = \left(\sum_i m_i\right)x_{cm} - \sum_i m_i x_i = M x_{cm} - \sum_i m_i x_i = 0 \tag{13.12}$$

where we used the fact that $\sum m_i$ is simply the object's total mass M. Solving for x_{cm}, the x-coordinate of the object's center of mass is

$$x_{cm} = \frac{1}{M}\sum_i m_i x_i = \frac{m_1 x_1 + m_2 x_2 + m_3 x_3 + \cdots}{m_1 + m_2 + m_3 + \cdots} \tag{13.13}$$

This was Equation 13.7. The y-equation is found similarly.

13.3 Torque

Newton's genius, summarized in his second law of motion, was to recognize force as the cause of acceleration. But what about *angular* acceleration? What do Newton's laws have to tell us about rotational motion? We need to find a rotational equivalent of force.

Consider the common experience of pushing open a door. Figure 13.13 is a top view of a door that is hinged on the left. Four pushing forces are shown, all of equal strength. Which of these will be most effective at opening the door?

Force $\vec{F}_1$ will open the door, but force $\vec{F}_2$, which pushes straight at the hinge, will not. Force $\vec{F}_3$ will open the door, but not as easily as $\vec{F}_1$. What about $\vec{F}_4$? It is perpendicular to the door, it has the same magnitude as $\vec{F}_1$, but you know from experience that pushing close to the hinge is not as effective as pushing at the outer edge of the door.

The ability of a force to cause a rotation or a twisting motion depends on three factors:

1. The magnitude F of the force.
2. The distance r from the point of application to the pivot.
3. The angle at which the force is applied.

To make these ideas specific, Figure 13.14 shows a force $\vec{F}$ applied at one point on a rigid body. For example, a string might be pulling on the object at that point, in which case the force would be a tension force. Figure 13.14 defines the distance r from the pivot to the point of application and the angle ϕ (Greek phi).

> **NOTE** ▶ Angle ϕ is measured *counterclockwise* from the dotted line that extends outward along the radial line. This is consistent with our sign convention for the angular position θ. ◀

Let's define a new quantity called the **torque** τ (Greek tau) as

$$\tau \equiv rF\sin\phi \tag{13.14}$$

Torque depends on the three properties we just listed: the magnitude of the force, its distance from the pivot, and its angle. Loosely speaking, τ measures the "effectiveness" of the force at causing an object to rotate about a pivot. **Torque is the rotational equivalent of force.**

The SI units of torque are newton-meters, abbreviated N m. Although we defined 1 N m = 1 J during our study of energy, torque is not an energy-related quantity and so we do *not* use joules as a measure of torque.

Torque, like force, has a sign. A torque that rotates the object in a ccw direction is positive while a negative torque gives a cw rotation. Figure 13.15 summarizes the signs. Notice that a force pushing straight toward the pivot or pulling straight out from the pivot exerts *no* torque.

Torque is to rotational motion as force is to linear motion.

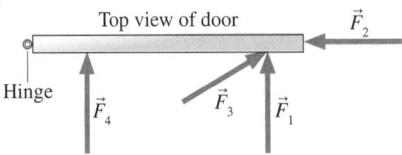

FIGURE 13.13 The four forces are the same strength, but they have different effects on the swinging door.

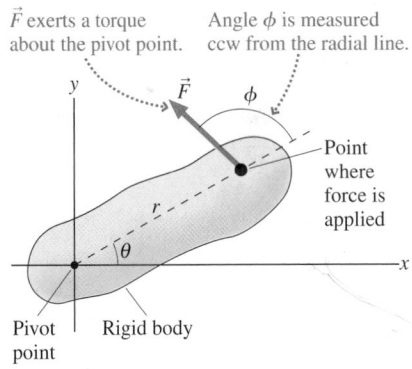

FIGURE 13.14 Force $\vec{F}$ exerts a torque about the pivot point.

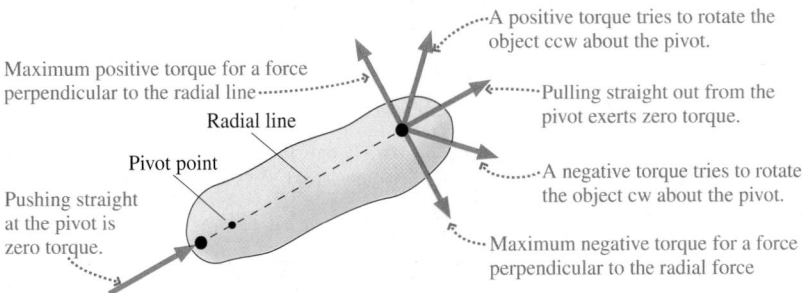

FIGURE 13.15 Signs and strengths of the torque.

7.1

NOTE ▶ Torque differs from force in a very important way. Torque is calculated or measured *about a pivot point*. To say that a torque is 20 Nm is meaningless. You need to say that the torque is 20 Nm about a particular point. Torque can be calculated about any pivot point, but its value depends on the point chosen. In practice, we measure or calculate torques about the same point from which we measure an object's angular position θ (and thus its angular velocity ω and angular acceleration α). This assumption is built into the equations of rotational dynamics. ◀

Returning to the door of Figure 13.13, you can see that $\vec{F}_1$ is most effective at opening the door because $\vec{F}_1$ exerts the largest torque *about the pivot point*. $\vec{F}_3$ has equal magnitude, but it is applied at an angle less than 90° and thus exerts less torque. $\vec{F}_2$, pushing straight at the hinge with $\phi = 0°$, exerts no torque at all. And $\vec{F}_4$, with a smaller value for r, exerts less torque than $\vec{F}_1$.

Interpreting Torque

Torque can be interpreted from two perspectives. First, Figure 13.16a shows that the quantity $F \sin\phi$ is the tangential force component F_t. Consequently, the torque can be written

$$\tau = rF_t \tag{13.15}$$

In other words, torque is the product of r with the force component F_t that is *perpendicular* to the radial line. This interpretation makes sense because the radial component of $\vec{F}$ points straight at the pivot point and cannot exert a torque.

Alternatively, Figure 13.16b shows that $d = r \sin\phi$ is the distance from the pivot to the **line of action,** the line along which force $\vec{F}$ acts. Thus the torque can also be written

$$|\tau| = dF \tag{13.16}$$

The distance d from the pivot to the line of action is called the **moment arm** (or the *lever arm*), so we can say that the torque is the product of the force with the moment arm. This second perspective on torque is widely used in applications.

NOTE ▶ Equation 13.16 gives only $|\tau|$, the magnitude of the torque; the sign has to be supplied by observing the direction in which the torque acts. ◀

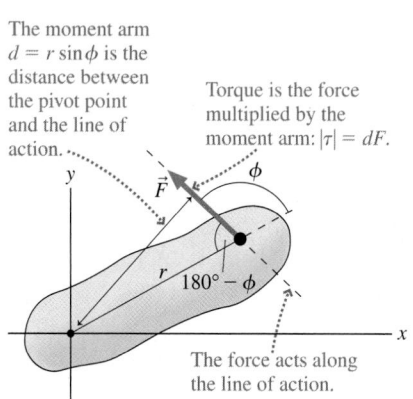

(a)
Torque is due to the tangential component of force: $\tau = rF_t$.

$F_t = F \sin\phi$

(b)
The moment arm $d = r \sin\phi$ is the distance between the pivot point and the line of action.

Torque is the force multiplied by the moment arm: $|\tau| = dF$.

$180° - \phi$

The force acts along the line of action.

FIGURE 13.16 Two useful interpretations of the torque.

EXAMPLE 13.5 Applying a torque

Luis uses a 20-cm-long wrench to turn a nut. The wrench handle is tilted 30° above the horizontal, and Luis pulls straight down on the end with a force of 100 N. How much torque does Luis exert on the nut?

VISUALIZE Figure 13.17 shows the situation. The angle is a negative $\phi = -120°$ because it is *clockwise* from the radial line.

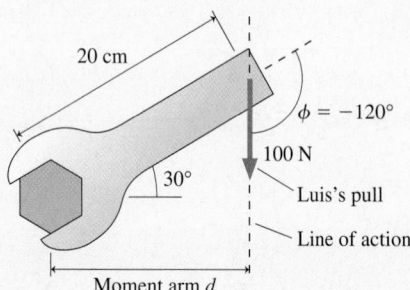

20 cm

$\phi = -120°$

100 N

Luis's pull

30°

Line of action

Moment arm d

FIGURE 13.17 A wrench being used to turn a nut.

SOLVE The tangential component of the force is

$$F_t = F \sin\phi = -86.6 \text{ N}$$

According to our sign convention, F_t is negative because it points in a cw direction. The torque, from Equation 13.15, is

$$\tau = rF_t = (0.20 \text{ m})(-86.6 \text{ N}) = -17.3 \text{ Nm}$$

The torque is negative because it tries to rotate the nut in a cw direction.

Alternatively, Figure 13.17 shows that the moment arm from the pivot to the line of action is

$$d = r \sin(60°) = 0.173 \text{ m}$$

Inserting the moment arm in Equation 13.16 gives

$$|\tau| = dF = (0.173 \text{ m})(100 \text{ N}) = 17.3 \text{ Nm}$$

The torque acts to give a cw rotation, so we insert a minus sign to end up with

$$\tau = -17.3 \text{ Nm}$$

ASSESS Notice that Luis could increase the torque by changing the angle so that his pull is perpendicular to the wrench ($\phi = 90°$).

STOP TO THINK 13.2 Rank in order, from largest to smallest, the five torques τ_a to τ_e. The rods all have the same length and are pivoted at the dot.

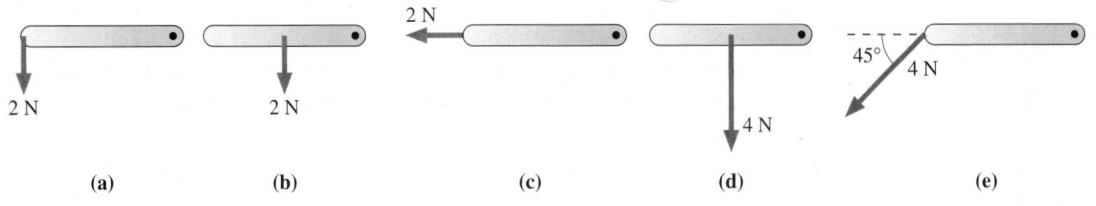

(a) (b) (c) (d) (e)

Net Torque

Figure 13.18 shows forces $\vec{F}_1$, $\vec{F}_2$, $\vec{F}_3$, ... applied to an extended object. The object is free to rotate about the axle, but the axle prevents the object from having any translational motion. It does so by exerting force $\vec{F}_{axle}$ on the object to balance the other forces and keep $\vec{F}_{net} = \vec{0}$.

Forces $\vec{F}_1$, $\vec{F}_2$, $\vec{F}_3$, ... exert torques τ_1, τ_2, τ_3, ... on the object, but $\vec{F}_{axle}$ does *not* exert a torque because it is applied at the pivot point and has zero moment arm. Thus the *net* torque about the axle is the sum of the torques due to the applied forces:

$$\tau_{net} = \tau_1 + \tau_2 + \tau_3 + \cdots = \sum_i \tau_i \qquad (13.17)$$

In practice, usually only a small number of forces exert torques. For example, the only torque we considered in Example 13.5 was due to Luis's pull.

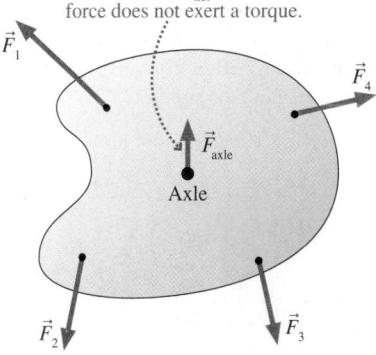

FIGURE 13.18 The forces exert a net torque about the pivot point.

Gravitational Torque

Gravity exerts a torque on many objects. If the object in Figure 13.19 is released, a torque due to the force of gravity will cause it to rotate around the axle. To calculate the torque about the axle, we start with the fact that gravity acts on *every* particle in the object, exerting a downward force of magnitude $F_i = m_i g$ on particle i. The *magnitude* of the gravitational torque on particle i is $|\tau_i| = d_i m_i g$, where d_i is the moment arm. But we need to be careful with signs.

A moment arm must be a positive number because it's a distance. If we establish a coordinate system with the origin at the axle, then you can see from Figure 13.19a that the moment arm d_i of particle i is $|x_i|$. A particle to the right of the axle (positive x_i) experiences a *negative* torque because gravity tries to rotate this particle in a clockwise direction. Similarly, a particle to the left of the axle (negative x_i) has a positive torque. The torque is opposite in sign to x_i, so we can get the sign of the torque right by writing

$$\tau_i = -x_i m_i g = -(m_i x_i)g \qquad (13.18)$$

The net torque due to gravity is found by summing Equation 13.18 over all particles:

$$\tau_{grav} = \sum_i \tau_i = \sum_i (-m_i x_i g) = -\left(\sum_i m_i x_i\right)g \qquad (13.19)$$

But according to the definition of center of mass, Equations 13.7, $\sum m_i x_i = M x_{cm}$. Thus the torque due to gravity is

$$\tau_{grav} = -M g x_{cm} \qquad (13.20)$$

where x_{cm} is the position of the center of mass *relative to the axis of rotation*.

(a)

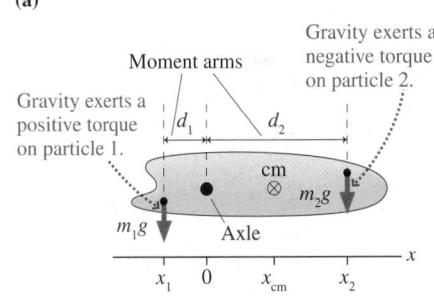

(b)

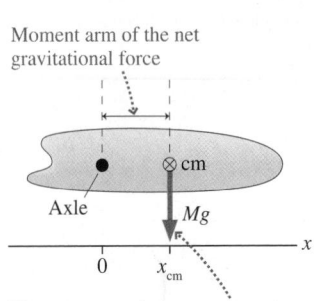

The net torque due to gravity acts as if all the mass is concentrated at the center of mass.

FIGURE 13.19 Gravity exerts a torque on this object.

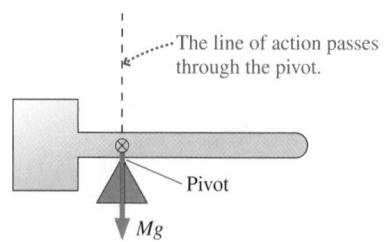

FIGURE 13.20 An object balances on a pivot that is directly under the center of mass.

Equation 13.20 has the simple interpretation shown in Figure 13.19b. Mg is the net gravitational force on the entire object, and x_{cm} is the moment arm between the rotation axis and the center of mass. The gravitational torque on an extended object of mass M is equivalent to the torque of a *single* force vector $\vec{F}_{grav} = -Mg\hat{j}$ acting at the object's center of mass.

In other words, **the gravitational torque is found by treating the object as if all its mass were concentrated at the center of mass.** This is the basis for the well-known technique of finding an object's center of mass by balancing it. An object will balance on a pivot, as shown in Figure 13.20, only if the center of mass is directly above the pivot point. If the pivot is *not* under the center of mass, the gravitational torque will cause the object to rotate.

EXAMPLE 13.6 The gravitational torque on a beam

The 4.00-m-long 500 kg steel beam shown in Figure 13.21 is supported 1.20 m from the right end. What is the gravitational torque about the support?

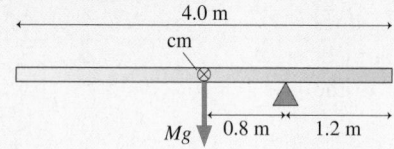

FIGURE 13.21 A steel beam supported at one point.

MODEL The center of mass of the beam is at the midpoint. $x_{cm} = -0.80$ m is measured from the pivot point.

SOLVE This is a straightforward application of Equation 13.20. The gravitational torque is

$$\tau_{grav} = -Mgx_{cm} = -(500 \text{ kg})(9.80 \text{ m/s}^2)(-0.80 \text{ m})$$
$$= 3920 \text{ N m}$$

ASSESS The torque is positive because gravity tries to rotate the beam ccw around the point of support. Notice that the beam in Figure 13.21 is *not* in equilibrium. It will fall over unless other forces, not shown, are supporting it.

Couples

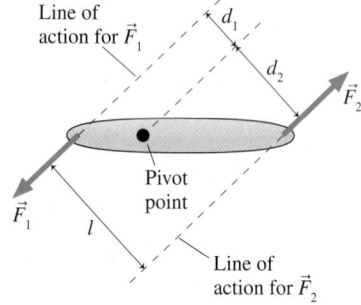

FIGURE 13.22 Two equal-but-opposite forces form a *couple*.

One way to rotate an extended object without translating it is to apply a pair of equal-but-opposite forces at two different points, as shown in Figure 13.22. This pair of forces is called a **couple.** The forces are in opposite directions, but the torques exerted by these forces act in the *same* direction. In this example, both torques are positive. Thus the two forces of a couple exert a *net torque*.

The forces are equal in magnitude, so we can write simply $F_1 = F_2 = F$. The moment arm of $\vec{F}_1$ is d_1, so the torque exerted by force $\vec{F}_1$ has magnitude $|\tau_1| = d_1 F$. Similarly, force $\vec{F}_2$ exerts a torque of magnitude $|\tau_2| = d_2 F$. The *net* torque is

$$|\tau_{net}| = d_1 F + d_2 F = (d_1 + d_2)F = lF \tag{13.21}$$

where l is the distance between the lines of action of the two forces. The sign of τ_{net} has to be supplied by observing the direction in which the torque acts.

There was nothing special about the point chosen as the pivot point in Figure 13.22. It could have been any point on the object. Hence a couple exerts the same net torque lF about *any* point on the object. This is not true of torques in general, just the torque exerted by a couple.

> **NOTE ▶** If an object is unconstrained (i.e., not on an axle), a couple causes the object to rotate about its center of mass. ◀

13.4 Rotational Dynamics

We've found that torque is the rotational equivalent of force. Now we need to learn what torque does. Figure 13.23 shows a model rocket engine attached to one end of a lightweight, rigid rod. When the engine is ignited, the thrust, which is *not* perpendicular to the rod, exerts force $\vec{F}_{\text{thrust}}$ on the rocket. The rod, which rotates around a pivot at the other end, exerts a tension force $\vec{T}$.

The tangential component of the thrust $F_t = F_{\text{thrust}} \sin \phi$ causes the rocket to speed up. In Chapter 7, we found that Newton's second law in the tangential direction is $F_t = ma_t$. This was perfectly adequate for single-particle motion, but our goal in this chapter is to understand the rotation of a rigid body. All points in a rotating object have the same angular acceleration so it will be useful to express Newton's second law in terms of angular acceleration rather than tangential acceleration.

The tangential and angular accelerations are related by $a_t = r\alpha$, allowing us to write Newton's second law as

$$F_t = ma_t = mr\alpha \tag{13.22}$$

Multiplying both sides by r gives us

$$rF_t = mr^2\alpha$$

But rF_t is the torque τ on the particle, hence Newton's second law for a *single particle* is

$$\tau = mr^2\alpha \tag{13.23}$$

At the beginning of Section 13.3 we asked the question, "What do Newton's laws have to tell us about rotational motion?" We now have the first part of an answer: **A torque causes an angular acceleration.** This is the rotational equivalent of our prior discovery, for motion along a line, that a force causes an acceleration. Now all that remains is to expand this idea from a single particle to an extended object.

Newton's Second Law

Figure 13.24 shows a rigid body that undergoes *pure rotational motion* about a fixed and unmoving axis. This might be an unconstrained rotation about the object's center of mass, such as we considered in Section 13.2. Or it might be an object, such as a pulley or a turbine, that is rotating on an axle. We'll assume that axles turn on frictionless bearings unless otherwise noted.

The forces $\vec{F}_1, \vec{F}_2, \vec{F}_3, \ldots$ in Figure 13.24 act on particles of masses $m_1, m_2, m_3 \ldots$ and exert torques $\tau_1, \tau_2, \tau_3, \ldots$ about the rotation axis. These torques cause the object to have an angular acceleration. The *net* torque on the object is the sum of the torques on all the individual particles. This is

$$\tau_{\text{net}} = \sum_i \tau_i = \sum_i (m_i r_i^2 \alpha) = \left(\sum_i m_i r_i^2 \right) \alpha \tag{13.24}$$

where we used Equation 13.23, $\tau_i = m_i r_i^2 \alpha$, to relate the individual torques to the angular acceleration. We're making explicit use of the fact that every particle in a rotating rigid body has the *same* angular acceleration α.

The quantity $\sum m_i r_i^2$, which is the proportionality constant between angular acceleration and net torque, is called the object's **moment of inertia** I:

$$I = m_1 r_1^2 + m_2 r_2^2 + m_3 r_3^2 + \cdots = \sum_i m_i r_i^2 \tag{13.25}$$

The units of moment of inertia are kg m^2. An object's moment of inertia, like torque, *depends on the axis of rotation*. Once the axis is specified, allowing the values of r_i to be determined, the moment of inertia *about that axis* can be calculated from Equation 13.25.

NOTE ▶ The *moment* in *moment of inertia* and *moment arm* has nothing to do with time. These terms stem from the Latin *momentum*, meaning "motion." ◄

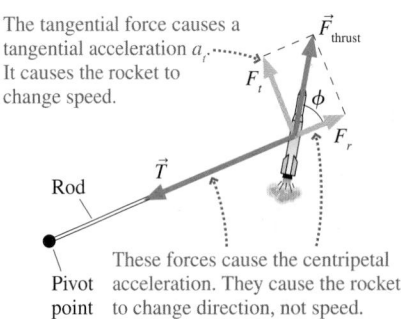

The tangential force causes a tangential acceleration a_t. It causes the rocket to change speed.

These forces cause the centripetal acceleration. They cause the rocket to change direction, not speed.

FIGURE 13.23 The thrust force exerts a torque on the rocket and causes an angular acceleration.

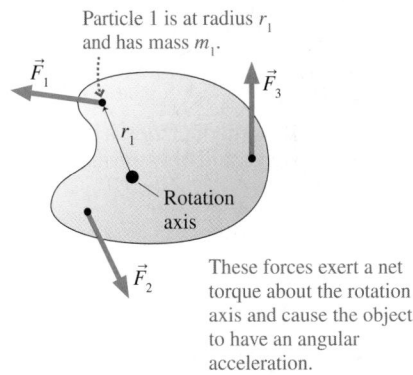

Particle 1 is at radius r_1 and has mass m_1.

Rotation axis

These forces exert a net torque about the rotation axis and cause the object to have an angular acceleration.

FIGURE 13.24 The forces on a rigid body exert a torque about the rotation axis.

Substituting the moment of inertia into Equation 13.24 puts the final piece of the puzzle into place. An object that experiences a net torque τ_{net} about the axis of rotation undergoes an angular acceleration

$$\alpha = \frac{\tau_{net}}{I} \quad \text{(Newton's second law for rotational motion)} \quad (13.26)$$

where I is the moment of inertia of the object *about the rotation axis*. This result, which is Newton's second law for rotation, is the fundamental equation of rigid-body dynamics.

In practice we often write $\tau_{net} = I\alpha$, but Equation 13.26 better conveys the idea that **torque is the cause of angular acceleration.** In the absence of a net torque ($\tau_{net} = 0$), the object either does not rotate ($\omega = 0$) or rotates with *constant* angular velocity ($\omega = $ constant).

Before rushing to calculate moments of inertia, let's get a better understanding of the meaning. First, notice that **moment of inertia is the rotational equivalent of mass.** It plays the same role in Equation 13.26 as mass m in the now-familiar $\vec{a} = \vec{F}_{net}/m$. Recall that the quantity we call *mass* was actually defined as the *inertial mass*. Objects with larger mass have a larger *inertia*, meaning that they're harder to accelerate. Similarly, an object with a larger moment of inertia is harder to rotate. Loosely speaking, it takes a larger torque to spin an object with a larger moment of inertia than an object with a smaller moment of inertia. The fact that *moment of inertia* retains the word *inertia* reminds us of this.

But why does the moment of inertia depend on the distances r_i from the rotation axis? Think about the two wheels shown in Figure 13.25. They have the same total mass M and the same radius R. As you probably know from experience, it's much easier to spin the wheel whose mass is concentrated at the center than to spin the one whose mass is concentrated around the rim. This is because having the mass near the center (smaller values of r_i) lowers the moment of inertia.

Thus an object's moment of inertia depends not only on the object's mass but on *how the mass is distributed* around the rotation axis. This is well known to bicycle racers. Every time a cyclist accelerates, she has to "spin up" the wheels and tires. The larger the moment of inertia, the more effort it takes and the slower her acceleration. For this reason, racers use the lightest possible tires, and they put those tires on wheels that have been designed to keep the mass as close as possible to the center without sacrificing the necessary strength and rigidity.

Table 13.2 summarizes the analogies between linear and rotational dynamics.

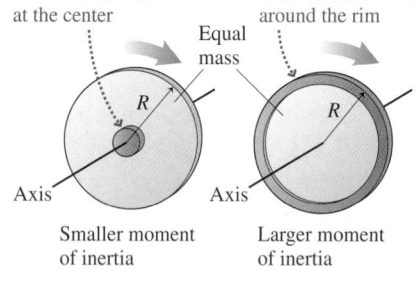

Mass concentrated at the center

Mass concentrated around the rim

Equal mass

R

R

Axis

Axis

Smaller moment of inertia

Larger moment of inertia

FIGURE 13.25 Moment of inertia depends both on the mass and how the mass is distributed.

TABLE 13.2 Rotational and linear dynamics

Rotational dynamics		Linear dynamics	
torque	τ_{net}	force	$\vec{F}_{net}$
moment of inertia	I	mass	m
angular acceleration	α	acceleration	$\vec{a}$
second law	$\alpha = \tau_{net}/I$	second law	$\vec{a} = \vec{F}_{net}/m$

EXAMPLE 13.7 Rotating rockets
Far out in space, a 100,000 kg rocket and a 200,000 kg rocket are docked at opposite ends of a motionless 90-m-long connecting tunnel. The tunnel is rigid and its mass is much less than that of either rocket. The rockets start their engines simultane-

ously, each generating 50,000 N of thrust in opposite directions. What is the structure's angular velocity after 30 s?

MODEL The entire structure can be modeled as two masses at the ends of a massless, rigid rod. The two thrust forces are a couple that exerts a torque on the structure. We'll assume the thrust forces are perpendicular to the connecting tunnel. This is an unconstrained rotation, so the structure will rotate about its center of mass.

VISUALIZE Figure 13.26 shows the rockets and defines distances r_1 and r_2 from the center of mass.

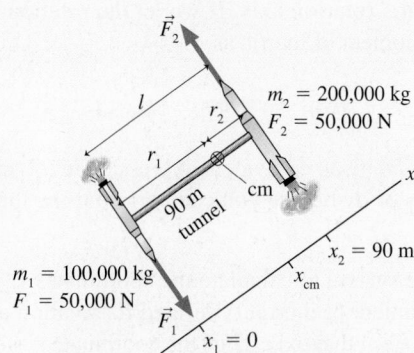

FIGURE 13.26 The thrusts are a couple that exerts a torque on the structure.

SOLVE Our strategy will be to use Newton's second law to find the angular acceleration, followed by rotational kinematics to find ω. We'll need to determine the moment of inertia, and that requires knowing the distances of the two rockets from the rotation axis. As we did in Example 13.3, choose a coordinate sys-

tem in which the masses are on the x-axis and in which m_1 is at the origin. Then

$$x_{cm} = \frac{m_1 x_1 + m_2 x_2}{m_1 + m_2}$$

$$= \frac{(100{,}000 \text{ kg})(0 \text{ m}) + (200{,}000 \text{ kg})(90 \text{ m})}{100{,}000 \text{ kg} + 200{,}000 \text{ kg}} = 60 \text{ m}$$

The structure's center of mass is $r_1 = 60$ m from the 100,000 kg rocket and $r_2 = 30$ m from the 200,000 kg rocket. The moment of inertia about the center of mass is

$$I = m_1 r_1^2 + m_2 r_2^2 = 540{,}000{,}000 \text{ kg m}^2$$

The two rocket thrusts form a couple that exerts torque

$$\tau_{net} = lF = (90 \text{ m})(50{,}000 \text{ N}) = 4{,}500{,}000 \text{ N m}$$

With I and τ_{net} now known, we can use Newton's second law to find the angular acceleration:

$$\alpha = \frac{\tau}{I} = \frac{4{,}500{,}000 \text{ N m}}{540{,}000{,}000 \text{ kg m}^2} = 0.00833 \text{ rad/s}^2$$

After 30 seconds, the structure's angular velocity is

$$\omega = \alpha \Delta t = 0.250 \text{ rad/s}$$

ASSESS Few of us have the experience to judge whether or not 0.25 rad/s is a reasonable answer to this problem. The significance of the example is to demonstrate the approach to a rotational dynamics problem.

STOP TO THINK 13.3   Rank in order, from largest to smallest, the angular accelerations α_a to α_e.

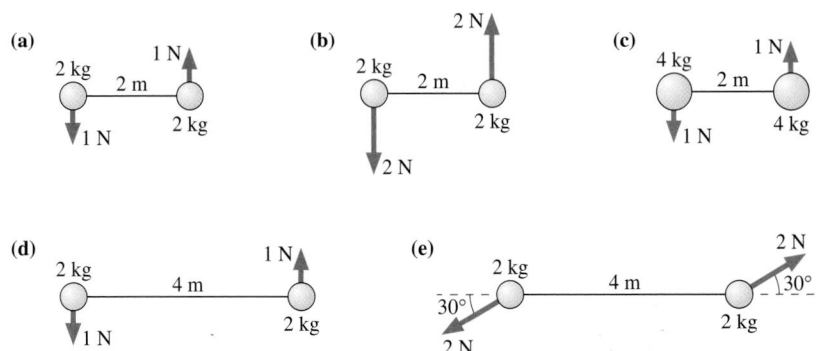

Moment of Inertia

Newton's second law for rotational motion is easy to write, but we can't make use of it without knowing an object's moment of inertia. Unlike mass, we can't measure moment of inertia by putting an object on a scale. And while we can guess that the center of mass of a symmetrical object is at the physical center of the object, we can *not* guess the moment of inertia of even a simple object. To find I, we really must carry through the calculation.

Activ
Physics   7.6

Equation 13.25 defines the moment of inertia as a sum over all the particles in the system. As we did for the center of mass, we can replace the individual particles with cells 1, 2, 3, . . . of mass Δm. Then the moment of inertia summation can be converted to an integration:

$$I = \sum_i r_i^2 \Delta m \xrightarrow[\Delta m \to 0]{} I = \int r^2\, dm \qquad (13.27)$$

where r is the distance from the rotation axis. If we let the rotation axis be the z-axis, then we can write the moment of inertia as

$$I = \int (x^2 + y^2)\, dm \qquad (13.28)$$

> **NOTE** ▶ You *must* replace dm by an equivalent expression involving a coordinate differential such as dx or dy before you can carry out the integration of Equation 13.28. ◀

You can use any coordinate system to calculate the coordinates x_{cm} and y_{cm} of the center of mass. But the moment of inertia is defined for rotation about a particular axis, and r is measured from that axis. Thus the coordinate system used for moment of inertia calculations *must* have its origin at the pivot point. Two examples will illustrate these ideas.

EXAMPLE 13.8 Moment of inertia of a rod about a pivot at one end

Find the moment of inertia of a thin, uniform rod of length L and mass M that rotates about a pivot at one end.

MODEL An object's moment of inertia depends on the axis of rotation. In this case, the rotation axis is at the end of the rod.

VISUALIZE Figure 13.27 defines an x-axis with the origin at the pivot point.

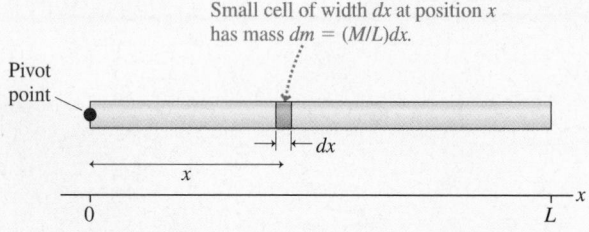

FIGURE 13.27 Finding the moment of inertia about one end of a long, thin rod.

SOLVE Because the rod is thin, we can assume that $y \approx 0$ for all points on the rod. Thus

$$I = \int x^2\, dm$$

The small amount of mass dm in the small length dx is $dm = (M/L)\, dx$, as we found in Example 13.4. The rod extends from $x = 0$ to $x = L$, so the moment of inertia for a rod about one end is

$$I_{end} = \frac{M}{L} \int_0^L x^2\, dx = \frac{M}{L}\left[\frac{x^3}{3}\right]_0^L = \frac{1}{3}ML^2$$

ASSESS The moment of inertia involves a product of the total mass M with the *square* of a length, in this case L. All moments of inertia have a similar form, although the fraction in front will vary.

EXAMPLE 13.9 Moment of inertia of a circular disk about an axis through the center

Find the moment of inertia of a circular disk of radius R and mass M that rotates on an axis passing through its center.

VISUALIZE Figure 13.28 shows the disk and defines distance r from the axis.

SOLVE This is a situation of great practical importance. To solve this problem, we need to use a two-dimensional integration scheme that you learned in calculus. Rather than dividing the disk into little boxes, let's divide it into narrow *rings* of

mass dm. Figure 13.28 shows one such ring, of radius r and width dr. Let dA represent the area of this ring. The mass dm in this ring is the same fraction of the total mass M as dA is of the total area A. That is,

$$\frac{dm}{M} = \frac{dA}{A}$$

Thus the mass in the small area dA is

$$dm = \frac{M}{A}\, dA$$

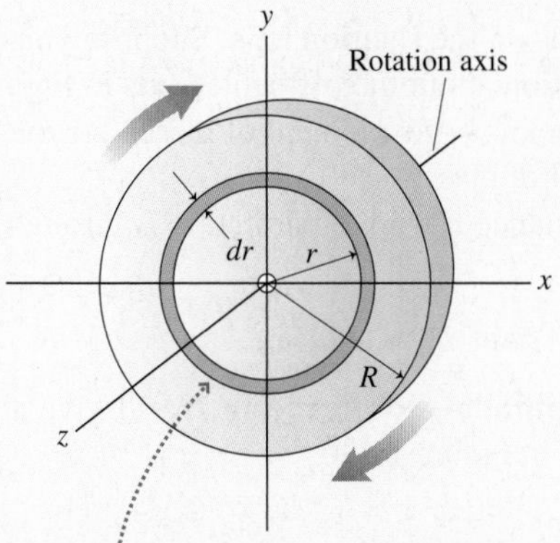

Narrow ring of width dr has mass $dm = (M/A)dA$.
Its area is $dA = $ width $\times$ circumference $= 2\pi r\, dr$.

FIGURE 13.28 Finding the moment of inertia of a disk about an axis through the center.

This is the reasoning we used to find the center of mass of the rod in Example 13.4, only now we're using it in two dimensions.

The total area of the disk is $A = \pi R^2$, but what is dA? If we imagine unrolling the little ring, it would form a long, thin rectangle of width $2\pi r$ and height dr. Thus the *area* of this little ring is $dA = 2\pi r\, dr$. With this information we can write

$$dm = \frac{M}{\pi R^2}(2\pi r\, dr) = \frac{2M}{R^2} r\, dr$$

Now we have an expression for dm in terms of a coordinate differential dr, so we can proceed to carry out the integration for I. Using Equation 13.27, we find

$$I_{\text{disk}} = \int r^2\, dm = \int r^2\left(\frac{2M}{R^2} r\, dr\right) = \frac{2M}{R^2}\int_0^R r^3\, dr$$

where in the last step we have used the fact that the disk extends from $r = 0$ to $r = R$. Performing the integration gives

$$I_{\text{disk}} = \frac{2M}{R^2}\left[\frac{r^4}{4}\right]_0^R = \frac{1}{2}MR^2$$

ASSESS Once again, the moment of inertia involves a product of the total mass M with the *square* of a length, in this case R.

Moments of inertia for most practical situations are tabulated and found in various science and engineering handbooks. You would need to compute I yourself only for an object of unusual shape. Table 13.3 is a short list of common moments of inertia.

TABLE 13.3 Moments of inertia of objects with uniform density

Object and axis	Picture	I	Object and axis	Picture	I
Thin rod, about center	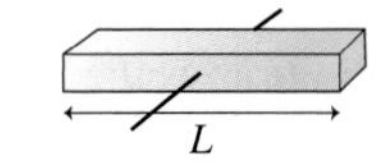	$\frac{1}{12}ML^2$	Cylinder or disk, about center	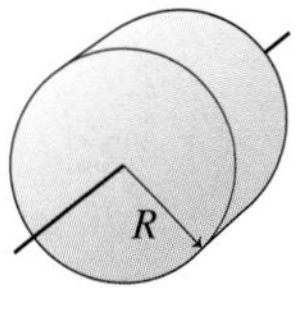	$\frac{1}{2}MR^2$
Thin rod, about end	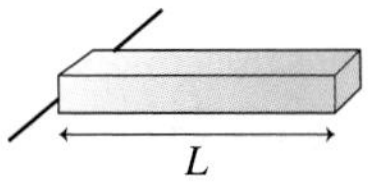	$\frac{1}{3}ML^2$	Cylindrical hoop, about center	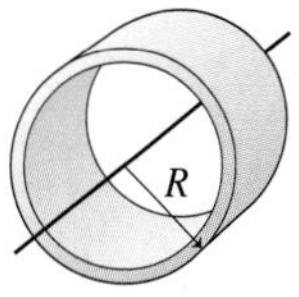	MR^2
Plane or slab, about center	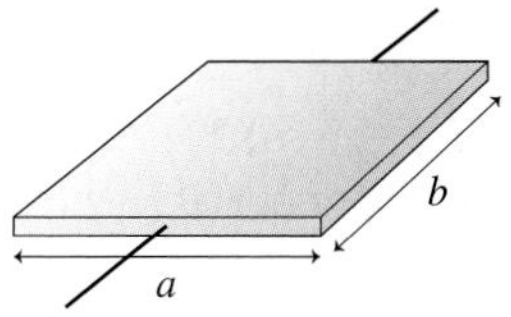	$\frac{1}{12}Ma^2$	Solid sphere, about diameter	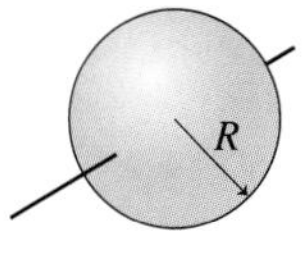	$\frac{2}{5}MR^2$
Plane or slab, about edge	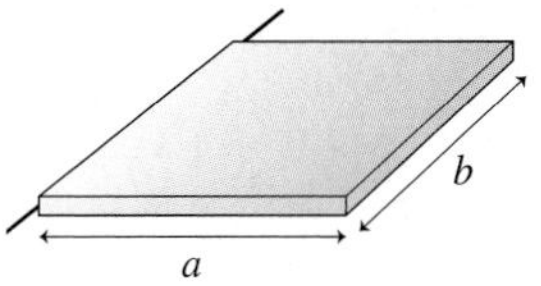	$\frac{1}{3}Ma^2$	Spherical shell, about diameter	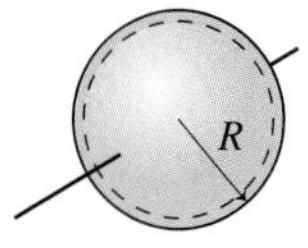	$\frac{2}{3}MR^2$

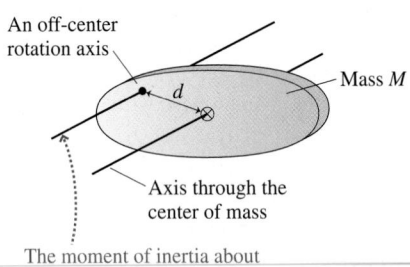

An off-center rotation axis

Mass M

Axis through the center of mass

The moment of inertia about this axis is $I = I_{cm} + Md^2$.

FIGURE 13.29 Rotation about an off-center axis.

The Parallel-Axis Theorem

The moment of inertia depends on the rotation axis. Suppose you need to know the moment of inertia for rotation about the off-center axis in Figure 13.29. You can find this quite easily if you know the moment of inertia for rotation around a *parallel axis* through the center of mass.

If the axis of interest is distance d from a parallel axis through the center of mass, the moment of inertia is

$$I = I_{cm} + Md^2 \qquad (13.29)$$

Equation 13.29 is called the **parallel-axis theorem.** We'll give a proof for the one-dimensional object shown in Figure 13.30.

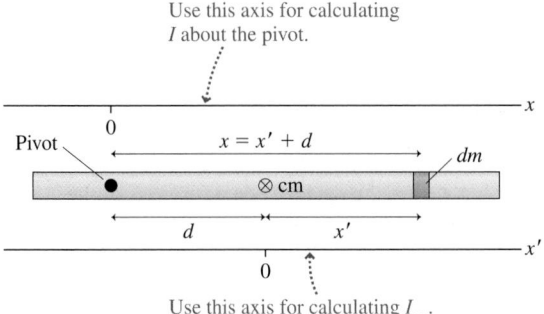

Use this axis for calculating I about the pivot.

Pivot

$x = x' + d$

dm

$\otimes$ cm

d

x'

Use this axis for calculating I_{cm}.

FIGURE 13.30 Proving the parallel-axis theorem.

The x-axis has its origin at the rotation axis and the x'-axis has its origin at the center of mass. You can see that the coordinates of dm along these two axes are related by $x = x' + d$. By definition, the moment of inertia about the rotation axis is

$$I = \int x^2 \, dm = \int (x' + d)^2 \, dm = \int (x')^2 \, dm + 2d \int x' \, dm + d^2 \int dm \quad (13.30)$$

The first of the three integrals on the right, by definition, is the moment of inertia I_{cm} about the center of mass. The third is simply Md^2 because adding up (integrating) all the dm gives the total mass M.

If you refer back to Equation 13.8, the definition of the center of mass, you'll see that the middle integral on the right is equal to Mx'_{cm}. But $x'_{cm} = 0$ because we specifically chose the x'-axis to have its origin at the center of mass. Thus the second integral is zero and we end up with Equation 13.29. The proof in two dimensions is similar.

EXAMPLE 13.10 The moment of inertia of a thin rod
Find the moment of inertia of a thin rod with mass M and length L about an axis one-third of the length from one end.

SOLVE From Table 13.3 we know the moment of inertia about the center of mass is $\frac{1}{12}ML^2$. The center of mass is at the center of the rod. An axis $\frac{1}{3}L$ from one end is $d = \frac{1}{6}L$ from the center of mass. Using the parallel-axis theorem,

$$I = I_{cm} + Md^2 = \frac{1}{12}ML^2 + M\left(\frac{1}{6}L\right)^2 = \frac{1}{9}ML^2$$

STOP TO THINK 13.4   Four Ts are made from two identical rods of equal mass and length. Rank in order, from largest to smallest, the moments of inertia I_a to I_d for rotation about the dotted line.

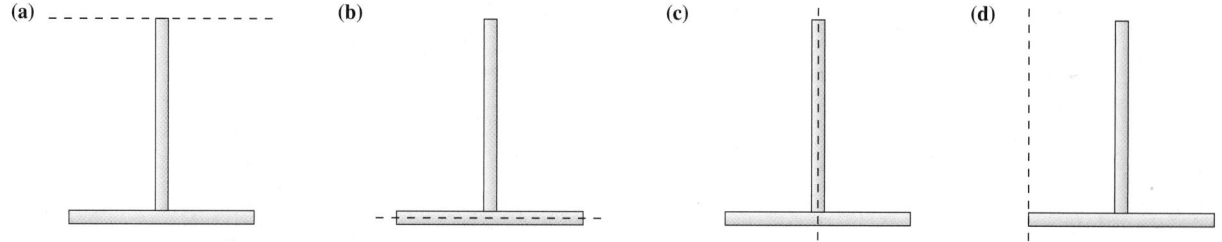

(a)   (b)   (c)   (d)

13.5 Rotation About a Fixed Axis

In this section we'll look at several examples of rotational dynamics for rigid bodies that rotate about a fixed axis. The restriction to a fixed axis avoids complications that arise for an object undergoing a combination of rotational and translational motion. The problem-solving strategy for rotational dynamics is very similar to that of linear dynamics.

 7.8–7.10

 PROBLEM-SOLVING STRATEGY 13.1 **Rotational dynamics problems**

MODEL Model the object as a simple shape.

VISUALIZE Draw a pictorial representation to clarify the situation, define coordinates and symbols, and list known information.

- Identify the axis about which the object rotates.
- Identify forces and determine their distance from the axis. For most problems it will be useful to draw a free-body diagram.
- Identify any torques caused by the forces and the signs of the torques.

SOLVE The mathematical representation is based on Newton's second law for rotational motion

$$\tau_{net} = I\alpha \quad \text{or} \quad \alpha = \frac{\tau_{net}}{I}$$

- Find the moment of inertia in Table 13.3 or, if needed, calculate it as an integral or by using the parallel-axis theorem.
- Use rotational kinematics to find angles and angular velocities.

ASSESS Check that your result has the correct units, is reasonable, and answers the question.

EXAMPLE 13.11 **Starting an airplane engine**
The engine in a small airplane is specified to have a torque of 60 N m. This engine drives a 2.0-m-long, 40 kg propeller. On start-up, how long does it take the propeller to reach 200 rpm?

MODEL The propeller can be modeled as a rod that rotates about its center. The engine exerts a torque on the propeller.

VISUALIZE Figure 13.31 on the next page shows the propeller and the rotation axis.

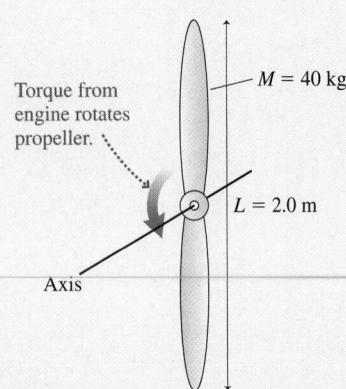

FIGURE 13.31 A rotating airplane propeller.

SOLVE The moment of inertia of a rod rotating about its center is found from Table 13.3:

$$I = \frac{1}{12}ML^2 = \frac{1}{12}(40 \text{ kg})(2.0 \text{ m})^2 = 13.33 \text{ kg m}^2$$

The 60 N m torque of the engine causes an angular acceleration

$$\alpha = \frac{\tau}{I} = \frac{60 \text{ N m}}{13.33 \text{ kg m}^2} = 4.50 \text{ rad/s}^2$$

The time needed to reach $\omega_f = 200$ rpm $= 3.33$ rev/s $= 20.9$ rad/s is

$$\Delta t = \frac{\Delta \omega}{\alpha} = \frac{\omega_f - \omega_i}{\alpha} = \frac{20.9 \text{ rad/s} - 0 \text{ rad/s}}{4.5 \text{ rad/s}^2} = 4.6 \text{ s}$$

ASSESS We've assumed a constant angular acceleration, which is reasonable for the first few seconds while the propeller is still turning slowly. Eventually, air resistance and friction will cause opposing torques and the angular acceleration will decrease. At full speed, the negative torque due to air resistance and friction cancels the torque of the engine. Then $\tau_{net} = 0$ and the propeller turns at *constant* angular velocity with no angular acceleration.

EXAMPLE 13.12 An off-center disk

Figure 13.32 shows a piece of a large machine. A 10.0-cm-diameter, 5.0 kg disk turns on an axle. A vertical cable attached to the edge of the disk exerts a 100 N force but, initially, a pin keeps the disk from rotating. What is the initial angular acceleration of the disk when the pin is removed?

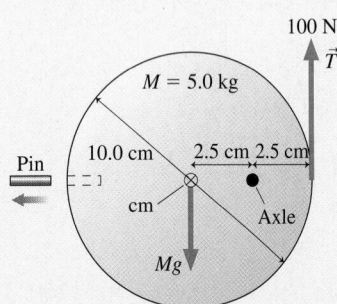

FIGURE 13.32 A disk rotates on an off-center axle after the pin is removed.

MODEL The disk has an off-center axle. Gravity exerts a torque about the axle.

VISUALIZE Both the cable and gravity rotate the disk ccw, so their torques are positive.

SOLVE After the pin is removed, the forces on the disk are a downward weight force, an upward force from the cable, and a force exerted by the axle. The axle force, which is exerted at the pivot, does not contribute to the torque and doesn't affect the rotation. The center of mass is to the *left* of the axle, at $x_{cm} = -\frac{1}{2}R$; thus the gravitational torque is

$$\tau_{grav} = -Mgx_{cm} = \frac{1}{2}MgR$$

This is a positive torque, as expected. The net torque, including the cable, is

$$\tau_{net} = \tau_{grav} + \tau_{cable} = \frac{1}{2}MgR + \frac{1}{2}RT = 3.73 \text{ N m}$$

Before finding the angular acceleration, we need to know the moment of inertia about the axle. This is where the parallel-axis theorem is useful. We know the moment of inertia about an axis through the center, from Table 13.3. The axle is offset by $d = \frac{1}{2}R$. Thus

$$I = I_{cm} + Md^2 = \frac{1}{2}MR^2 + M\left(\frac{1}{2}R\right)^2 = \frac{3}{4}MR^2$$

$$= 9.38 \times 10^{-3} \text{ kg m}^2$$

The torque causes an angular acceleration

$$\alpha = \frac{\tau_{net}}{I} = \frac{3.73 \text{ N m}}{9.38 \times 10^{-3} \text{ kg m}^2} = 397 \text{ rad/s}^2$$

The angular acceleration is positive, indicating that the disk begins rotating in a ccw direction.

ASSESS As the disk rotates, τ_{net} will change as the moment arms change. Consequently, the disk will *not* have constant angular acceleration. This is simply the *initial* value of α.

Constraints Due to Ropes and Pulleys

Many important applications of rotational dynamics involve objects, such as pulleys, that are connected via ropes or belts to other objects. Figure 13.33 shows a rope passing over a pulley and connected to an object in linear motion. If the rope turns on the pulley *without slipping*, then the rope's speed v_{rope} must exactly match the speed of the rim of the pulley, which is $v_{rim} = |\omega| R$. If the pulley has an angular acceleration, the rope's acceleration a_{rope} must match the *tangential* acceleration of the rim of the pulley, $a_t = |\alpha| R$.

The object attached to the other end of the rope has the same speed and acceleration as the rope. Consequently, an object connected to a pulley of radius R by a rope that does not slip must obey the constraints

$$v_{obj} = |\omega| R$$
$$a_{obj} = |\alpha| R \qquad \text{(motion constraints for a nonslipping rope)} \quad (13.31)$$

These constraints are very similar to the acceleration constraints introduced in Chapter 8 for two objects connected by a string or rope.

NOTE ▶ The constraints are given as magnitudes. Specific problems will need to introduce signs that depend on the direction of motion and on the choice of coordinate system. ◀

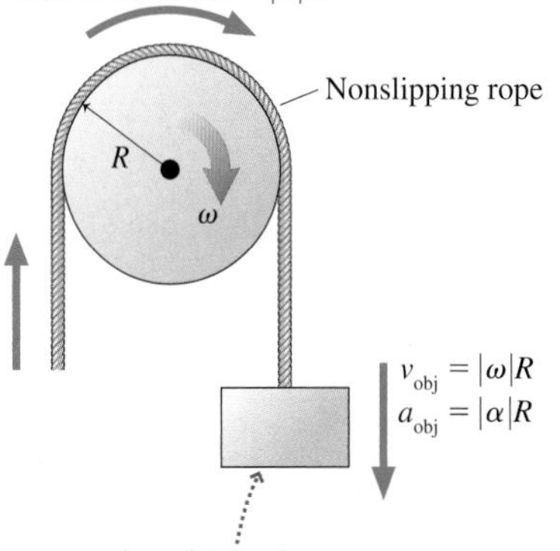

Rim speed $= |\omega| R$.
Rim acceleration $= |\alpha| R$.

Nonslipping rope

R

ω

$v_{obj} = |\omega| R$
$a_{obj} = |\alpha| R$

The motion of the object must match the motion of the rim.

FIGURE 13.33 The rope's motion must match the motion of the rim of the pulley.

EXAMPLE 13.13 **Lowering a block**

A 2.0 kg block is attached to a massless string that is wrapped around a 1.0 kg, 4.0-cm-diameter cylinder, as shown in Figure 13.34a. The cylinder rotates on an axle through the center. The block is released from rest 1.0 m above the floor. How long does it take to reach the floor?

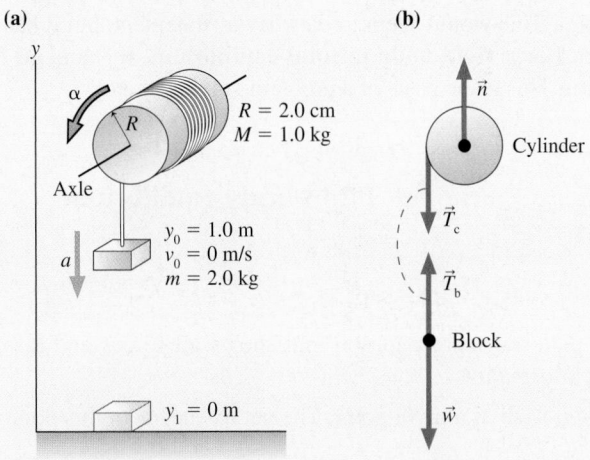

(a)

$R = 2.0$ cm
$M = 1.0$ kg

Axle

$y_0 = 1.0$ m
$v_0 = 0$ m/s
$m = 2.0$ kg

$y_1 = 0$ m

(b)

$\vec{n}$

Cylinder

$\vec{T}_c$

$\vec{T}_b$

Block

$\vec{w}$

FIGURE 13.34 The falling block turns the cylinder.

MODEL Assume the string does not slip.

VISUALIZE Figure 13.34b shows the free-body diagram for the cylinder and the block. The string tension exerts an upward force on the block and a downward force on the outer edge of the cylinder. The string is massless, so these two tension forces act as if they are an action/reaction pair: $T_b = T_c = T$.

SOLVE Newton's second law applied to the linear motion of the block is

$$ma_y = T - mg$$

where, as usual, the y-axis points upward. What about the cylinder? There's no net force on it, because its center of mass is stationary, so the axle must exert an upward normal force $\vec{n}$ to balance $\vec{T}_c$. However, $\vec{n}$ does not exert a torque because it passes through the rotation axis. The only torque comes from the string tension. The moment arm for the tension is $d = R$, and the torque is positive because the string turns the cylinder ccw. Thus $\tau_{string} = TR$ and Newton's second law for the rotational motion is

$$\alpha = \frac{\tau_{net}}{I} = \frac{TR}{\frac{1}{2}MR^2} = \frac{2T}{MR}$$

The moment of inertia of a cylinder rotating about a center axis was taken from Table 13.3.

The last piece of information we need is the constraint due to the fact that the string doesn't slip. Equation 13.31 relates only the absolute values, but in this problem α is positive (ccw acceleration) while a_y is negative (downward acceleration). Hence

$$a_y = -\alpha R$$

Using α from the cylinder's equation in the constraint, we find

$$a_y = -\alpha R = -\frac{2T}{MR}R = -\frac{2T}{M}$$

Thus the tension is $T = -\frac{1}{2}Ma_y$. If we use this value of the tension in the block's equation, we can solve for the acceleration:

$$ma_y = -\frac{1}{2}Ma_y - mg$$

$$a_y = -\frac{g}{(1 + M/2m)} = -7.84 \text{ m/s}^2$$

The time to fall through $\Delta y = -1.0$ m is found from kinematics:

$$\Delta y = \frac{1}{2}a_y(\Delta t)^2$$

$$\Delta t = \sqrt{\frac{2\Delta y}{a_y}} = \sqrt{\frac{2(-1.0\text{ m})}{-7.84\text{ m/s}^2}} = 0.505\text{ s}$$

ASSESS The expression for the acceleration gives $a_y = -g$ if $M = 0$. This makes sense because the block would be in free fall if there were no cylinder. When the cylinder has mass, the downward force of gravity on the block has to accelerate the block *and* spin the cylinder. Consequently, the acceleration is reduced and the block takes longer to fall.

Structures such as bridges are analyzed in engineering statics.

13.6 Rigid-Body Equilibrium

We now have two versions of Newton's second law: $\vec{F}_{net} = M\vec{a}$ for translational motion and $\tau_{net} = I\alpha$ for rotational motion. Our goal in this short section is to look at how Newton's laws apply to objects in equilibrium. There are three cases:

- If $\vec{F}_{net} = \vec{0}$, then the object is in *translational equilibrium* and the center of mass is either at rest or moving in a straight line with constant speed. However, the object may be rotating.
- If $\tau_{net} = 0$, then the object is in *rotational equilibrium* and it is either not rotating or is rotating with a constant angular velocity. However, the object may have a translational motion.
- The condition for a rigid body to be in *total equilibrium* is both $\vec{F}_{net} = \vec{0}$ *and* $\tau_{net} = 0$. That is, no net force *and* no net torque.

A motionless rigid body is in total static equilibrium. An important branch of engineering called *statics* analyzes buildings, dams, bridges, and other structures in total static equilibrium.

No matter which pivot point you choose, an object that is not rotating is not rotating about that point. This would seem to be a trivial statement, but it has an important implication: **For a rigid body in total equilibrium, there is no net torque about any point.** This is the basis of a problem-solving strategy.

 PROBLEM-SOLVING STRATEGY 13.2 Rigid-body equilibrium problems

---

MODEL Model the object as a simple shape.

VISUALIZE Draw a pictorial representation that shows all forces and distances. List known information.

- Pick any point you wish as a pivot point. The net torque about this point is zero.
- Determine the moment arms of all forces about this pivot point.
- Determine the sign of each torque about this pivot point.

SOLVE The mathematical representation is based on the fact that an object in total equilibrium has no net force and no net torque.

$$\vec{F}_{net} = \vec{0} \quad \text{and} \quad \tau_{net} = 0$$

- Write equations for $\sum F_x = 0$, $\sum F_y = 0$ and $\sum \tau = 0$.
- Solve the three simultaneous equations.

ASSESS Check that your result is reasonable and answers the question.

Although you can pick any point you wish as a pivot point, some choices make the problem easier than others. Often the best choice is a point at which several forces act because the torques exerted by those forces will be zero.

Activ
Physics ONLINE   7.2–7.5

EXAMPLE 13.14 **Will the ladder slip?**

A 3.0-m-long ladder leans against a frictionless wall at an angle of 60°. What is the minimum value of μ_s, the coefficient of static friction with the ground, that prevents the ladder from slipping?

MODEL The ladder is a rigid rod of length L. To not slip, it must be in both translational equilibrium ($\vec{F}_{net} = \vec{0}$) and rotational equilibrium ($\tau_{net} = 0$).

VISUALIZE Figure 13.35 shows the ladder and the forces acting on it.

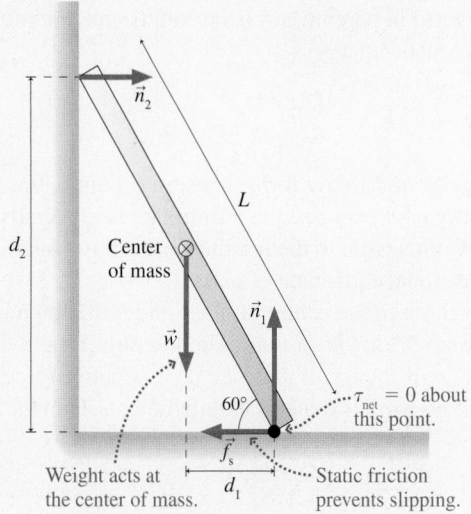

Weight acts at the center of mass.

Static friction prevents slipping.

FIGURE 13.35 A ladder in total equilibrium.

SOLVE The x- and y-components of $\vec{F}_{net} = \vec{0}$ are

$$\sum F_x = n_2 - f_s = 0$$

$$\sum F_y = n_1 - w = n_1 - Mg = 0$$

The net torque is zero about *any* point, so which should we choose? The bottom corner of the ladder is a good choice because two forces pass through this point and have no torque about it. The torque about the bottom corner is

$$\tau_{net} = d_1 w - d_2 n_2 = \frac{1}{2}(L\cos 60°)Mg - (L\sin 60°)n_2 = 0$$

The signs are based on the observation that $\vec{w}$ would cause the ladder to rotate ccw while $\vec{n}_2$ would cause it to rotate cw. All together, we have three equations in the three unknowns n_1, n_2, and f_s. If we solve the third for n_2,

$$n_2 = \frac{\frac{1}{2}(L\cos 60°)Mg}{L\sin 60°} = \frac{Mg}{2\tan 60°}$$

we can then substitute this into the first to find

$$f_s = \frac{Mg}{2\tan 60°}$$

Our model of friction is $f_s \leq f_{s\,max} = \mu_s n_1$. We can find n_1 from the second equation: $n_1 = Mg$. Using this, the model of static friction tells us that

$$f_s \leq \mu_s Mg$$

Comparing these two expressions for f_s, we see that μ_s must obey

$$\mu_s \geq \frac{1}{2\tan 60°} = 0.29$$

Thus the minimum value of the coefficient of static friction is 0.29.

ASSESS You know from experience that you can lean a ladder or other object against a wall if the ground is "rough," but it slips if the surface is too smooth. 0.29 is a "medium" value for the coefficient of static friction, which is reasonable.

STOP TO THINK 13.5   A student holds a meter stick straight out with one or more masses dangling from it. Rank in order, from most difficult to least difficult, how hard it will be for the student to keep the meter stick from rotating.

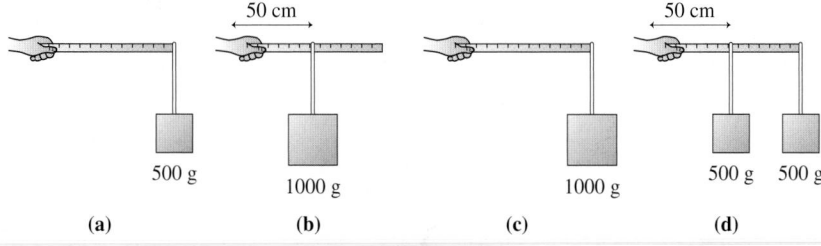

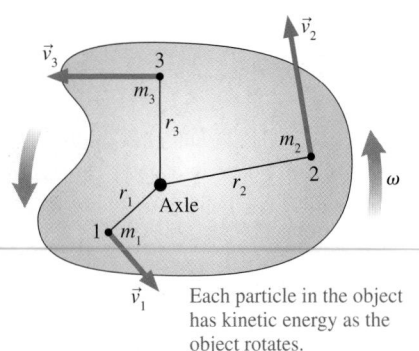

Each particle in the object has kinetic energy as the object rotates.

FIGURE 13.36 Rotational kinetic energy is due to the motion of the particles.

7.12, 7.13   Activ Physics

13.7 Rotational Energy

A rotating rigid body has kinetic energy because all atoms in the object are in motion. The kinetic energy due to rotation is called **rotational kinetic energy.** Rotational kinetic energy is *in addition* to any translational kinetic energy the object has if its center of mass is moving.

Figure 13.36 shows a few of the particles making up a solid object that rotates with angular velocity ω. Particle i, which rotates in a circle of radius r_i, moves with speed $v_i = r_i\omega$. The object's rotational kinetic energy is the sum of the kinetic energies of each of the particles:

$$K_{\text{rot}} = \frac{1}{2}m_1 v_1^2 + \frac{1}{2}m_2 v_2^2 + \cdots$$

$$= \frac{1}{2}m_1 r_1^2 \omega^2 + \frac{1}{2}m_2 r_2^2 \omega^2 + \cdots = \frac{1}{2}\left(\sum_i m_i r_i^2\right)\omega^2 \tag{13.32}$$

You will recognize that the term in parentheses is our old friend, the moment of inertia I. Thus the rotational kinetic energy is

$$K_{\text{rot}} = \frac{1}{2}I\omega^2 \tag{13.33}$$

Rotational kinetic energy is *not* a new form of energy. This is the familiar kinetic energy of motion, only now expressed in a form that is especially convenient for rotational motion. Comparison to the familiar $\frac{1}{2}mv^2$ shows again that the moment of inertia I is the rotational equivalent of mass.

If the rotation axis is not through the center of mass, then rotation may cause the center of mass to move up or down. In that case, the object's gravitational potential energy $U_g = Mgy_{\text{cm}}$ will change. If there are no dissipative forces (i.e., if the axle is frictionless) and if no work is done by external forces, then the object's mechanical energy

$$E_{\text{mech}} = K_{\text{rot}} + U_g = \frac{1}{2}I\omega^2 + Mgy_{\text{cm}} \tag{13.34}$$

is a conserved quantity.

EXAMPLE 13.15 The speed of a rotating rod
A 1.0-m-long, 200 g rod is hinged at one end and connected to a wall. It is held out horizontally, then released. What is the speed of the tip of the rod as it hits the wall?

MODEL The mechanical energy is conserved if we assume the hinge is frictionless. The rod's gravitational potential energy is transformed into rotational kinetic energy as it "falls."

VISUALIZE Figure 13.37 is a familiar before-and-after pictorial representation of the rod. We've placed the origin of the coordinate system at the pivot point.

SOLVE Mechanical energy is conserved, so we can equate the rod's final mechanical energy to its initial mechanical energy:

$$\frac{1}{2}I\omega_1^2 + Mgy_{\text{cm1}} = \frac{1}{2}I\omega_0^2 + Mgy_{\text{cm0}}$$

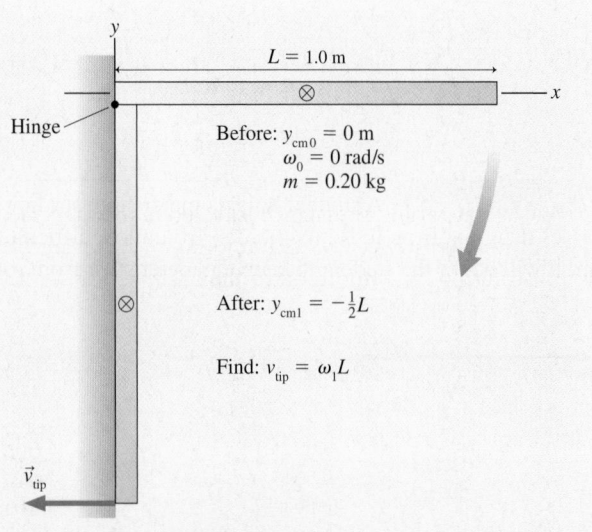

FIGURE 13.37 A before-and-after pictorial representation of the rod.

The initial conditions are $\omega_0 = 0$ and $y_{cm0} = 0$. The center of mass moves to $y_{cm1} = -\frac{1}{2}L$ as the rod hits the wall. Using $I = \frac{1}{3}ML^2$ for a rod rotating about one end, we have

$$\frac{1}{2}I\omega_1^2 + Mgy_{cm1} = \frac{1}{6}ML^2\omega_1^2 - \frac{1}{2}MgL = 0$$

We can solve this for the rod's angular velocity as it hits the wall:

$$\omega_1 = \sqrt{\frac{3g}{L}}$$

The tip of the rod is moving in a circle with radius $r = L$. Its final speed is

$$v_{tip} = \omega_1 L = \sqrt{3gL} = 5.42 \text{ m/s}$$

ASSESS Energy conservation is a powerful tool for rotational motion, just as it was for translational motion.

13.8 Rolling Motion

Rolling is a *combination motion* in which an object rotates about an axis that is moving along a straight-line trajectory. For example, Figure 13.38 is a time-exposure photo of a rolling wheel with one light bulb on the axis and a second light bulb at the edge. The axis light moves straight ahead, but the edge light follows a curve called a *cycloid*. Let's see if we can understand this interesting motion. We'll consider only objects that roll without slipping.

Figure 13.39 shows a round object—a wheel or a sphere—that rolls forward exactly one revolution. The point that had been on the bottom follows the cycloid, the curve you saw in Figure 13.38, to the top and back to the bottom. *Because the object doesn't slip*, the center of mass moves forward exactly one circumference: $\Delta x_{cm} = 2\pi R$.

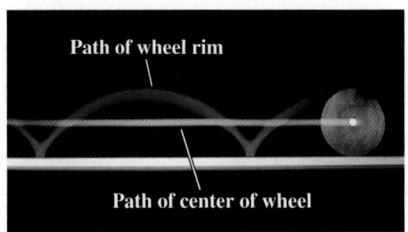

FIGURE 13.38 The trajectories of the center of a wheel and of a point on the rim are seen in a time-exposure photograph.

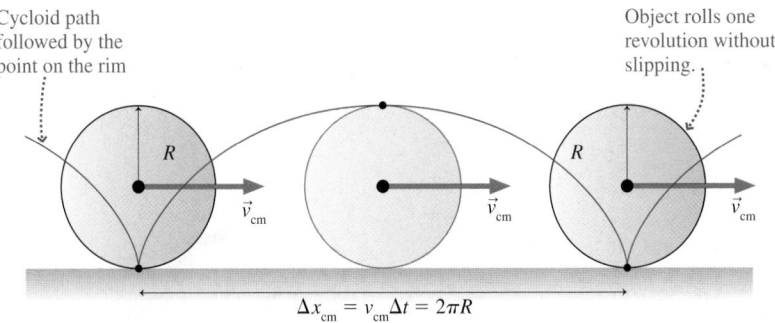

FIGURE 13.39 An object rolling through one revolution.

We can also write the distance traveled in terms of the velocity of the center of mass: $\Delta x_{cm} = v_{cm}\Delta t$. But Δt, the time it takes the object to make one complete revolution, is nothing other than the rotation period T. In other words, $\Delta x_{cm} = v_{cm}T$.

These two expressions for Δx_{cm} come from two perspectives on the motion, one looking at the rotation and the other at the translation of the center of mass. But it's the same distance no matter how you look at it, so these two expressions must be equal. Consequently,

$$\Delta x_{cm} = 2\pi R = v_{cm}T \qquad (13.35)$$

If we divide by T, we can write the center of mass velocity as

$$v_{cm} = \frac{2\pi}{T}R \qquad (13.36)$$

But $2\pi/T$ is the angular velocity ω, as you learned in Chapter 7, leading to

$$v_{cm} = R\omega \qquad (13.37)$$

Equation 13.37 is the **rolling constraint,** the basic link between translation and rotation for objects that roll without slipping.

(a)

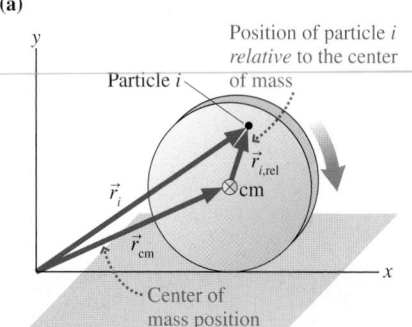

Position of particle i *relative* to the center of mass

> **NOTE** ▶ The rolling constraint is equivalent to Equation 13.31 for the speed of a rope that doesn't slip as it passes over a pulley. ◀

Let's look carefully at a particle in the rolling object. As Figure 13.40a shows, the position vector $\vec{r}_i$ for particle i is the vector sum $\vec{r}_i = \vec{r}_{cm} + \vec{r}_{i,\,rel}$. Taking the time derivative of this equation, we can write the velocity of particle i as

$$\vec{v}_i = \vec{v}_{cm} + \vec{v}_{i,\,rel} \qquad (13.38)$$

In other words, the velocity of particle i can be divided into two parts: the velocity $\vec{v}_{cm}$ of the object as a whole plus the velocity $\vec{v}_{i,\,rel}$ of particle i relative to the center of mass (i.e., the velocity that particle i would have if the object were only rotating and had no translational motion).

(b)

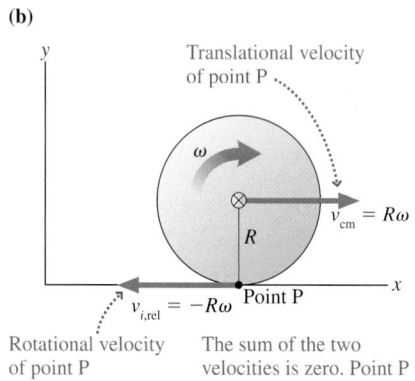

Translational velocity of point P

Rotational velocity of point P

The sum of the two velocities is zero. Point P is instantaneously at rest.

FIGURE 13.40 The motion of a particle in the rolling object.

Figure 13.40b applies this idea to point P at the very bottom of the rolling object, the point of contact between the object and the surface. This point is moving around the center of the object at angular velocity ω, so $v_{i,\,rel} = -R\omega$. The negative sign indicates that the motion is cw. At the same time, the center-of-mass velocity, Equation 13.37, is $v_{cm} = R\omega$. Adding these, we find that the velocity of point P, the lowest point, is $v_i = 0$. In other words, **the point on the bottom of a rolling object is instantaneously at rest.**

Although this seems surprising, it is really what we mean by "rolling without slipping." If the bottom point had a velocity, it would be moving horizontally relative to the surface. In other words, it would be slipping or sliding across the surface. To roll without slipping, the bottom point, the point touching the surface, must be at rest.

Figure 13.41 shows how the velocity vectors at the top, center, and bottom of a rotating wheel are found by adding the rotational velocity vectors to the center of mass velocity. You can see that $v_{bottom} = 0$ and that $v_{top} = 2R\omega = 2v_{cm}$.

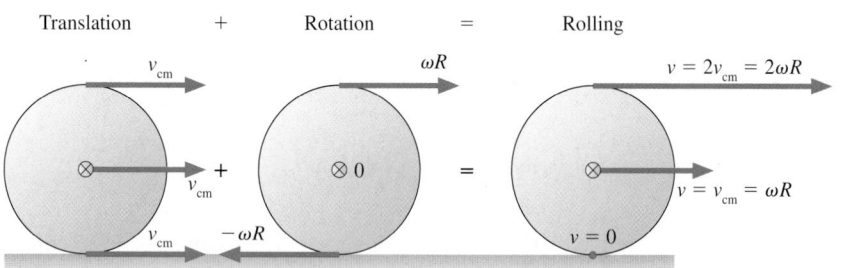

FIGURE 13.41 Rolling is a combination of translation and rotation.

Kinetic Energy of a Rolling Object

7.11 Activ Physics

In the last section we found that the rotational kinetic energy of a rigid body in pure rotational motion is $K_{rot} = \frac{1}{2}I\omega^2$. Now we would like to find the kinetic energy of an object that rolls, a combination of rotational and translation motion.

We begin with the observation that the bottom point in Figure 13.42 is instantaneously at rest. Consequently, we can think of an axis through P as an *instan-*

taneous axis of rotation. In other words, as Figure 13.42 shows, P is the pivot point for the entire object at this one instant of time.

The idea of an instantaneous axis of rotation seems a little far-fetched, but it is confirmed by looking at the instantaneous velocities of the center point and the top point. We found these in Figure 13.41 and they are shown again in Figure 13.42. They are exactly what you would expect as the tangential velocity $v_t = r\omega$ for rotation about P at distances R and $2R$.

From this perspective, the object's motion is pure rotation about point P. Thus the kinetic energy is that of pure rotation,

$$K = K_{\text{rotation about P}} = \frac{1}{2}I_P\omega^2 \qquad (13.39)$$

I_P is the moment of inertia for rotation about point P. We can use the parallel-axis theorem to write I_P in terms of the moment of inertia I_{cm} about the center of mass. Point P is displaced by distance $d = R$, thus

$$I_P = I_{cm} + MR^2$$

Using this expression in Equation 13.39 gives us the kinetic energy:

$$K = \frac{1}{2}I_{cm}\omega^2 + \frac{1}{2}M(R\omega)^2 \qquad (13.40)$$

We know from the rolling constraint that $R\omega$ is the center-of-mass velocity v_{cm}. Thus the kinetic energy of a rolling object is

$$K = \frac{1}{2}I_{cm}\omega^2 + \frac{1}{2}Mv_{cm}^2 = K_{rot} + K_{cm} \qquad (13.41)$$

In other words, **the rolling motion of a rigid body can be described as a translation of the center of mass (with kinetic energy K_{cm}) plus a rotation about the center of mass (with kinetic energy K_{rot}).**

The Great Downhill Race

Figure 13.43 shows a contest in which a sphere, a cylinder, and a circular hoop, all of mass M and radius R, are placed at height h on a slope of angle θ. All three are released from rest at the same instant of time and roll down the ramp without slipping. To make things more interesting, they are joined by a particle of mass M that slides down the ramp without friction. Which one will win the race to the bottom of the hill?

If the particle model were valid, all four would arrive at the bottom at the same time and with the same speed. Does the fact that three of them rotate change this conclusion? We can use conservation of energy to answer this question.

An object's initial gravitational potential energy is transformed into kinetic energy as it rolls (or slides, in the case of the particle). The kinetic energy, as we just discovered, is a combination of translational and rotational kinetic energy. If we choose the bottom of the ramp as the zero point of potential energy, the statement of energy conservation $K_f = U_i$ can be written

$$\frac{1}{2}I_{cm}\omega^2 + \frac{1}{2}Mv_{cm}^2 = Mgh \qquad (13.42)$$

The translational and rotational velocities are related by $\omega = v_{cm}/R$. In addition, notice from Table 13.3 that the moments of inertia of all the objects can be written in the form

$$I_{cm} = cMR^2 \qquad (13.43)$$

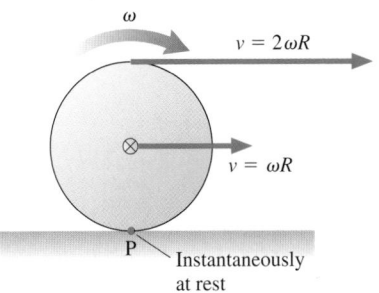

FIGURE 13.42 Rolling motion is an instantaneous rotation about point P.

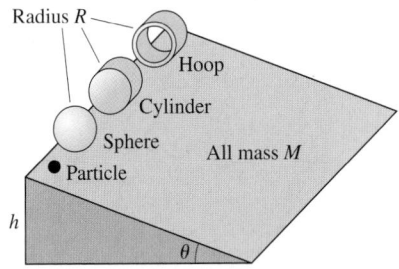

FIGURE 13.43 Which will win the downhill race?

where c is a constant that depends on the object's geometry. For example, $c = \frac{2}{5}$ for a sphere but $c = 1$ for a circular hoop. Even the particle can be represented by $c = 0$, which eliminates the rotational kinetic energy.

With this information, Equation 13.42 becomes

$$\frac{1}{2}(cMR^2)(v_{cm}/R)^2 + \frac{1}{2}Mv_{cm}^2 = \frac{1}{2}M(1 + c)v_{cm}^2 = Mgh$$

Thus the finishing speed of an object with $I = cMR^2$ is

$$v_{cm} = \sqrt{\frac{2gh}{(1 + c)}} \tag{13.44}$$

The final speed is independent of both M and R, but it does depend on the *shape* of the rolling object. The particle, with the smallest value of c, will finish with the highest speed while the circular hoop, with the largest c, will be the slowest. In other words, the rolling aspect of the motion *does* matter!

We can use Equation 13.44 to find the acceleration a_{cm} of the center of mass. The objects move through distance $\Delta x = h/\sin\theta$, so we can use constant-acceleration kinematics to find

$$v_{cm}^2 = 2a_{cm}\Delta x$$
$$a_{cm} = \frac{v_{cm}^2}{2\Delta x} = \frac{2gh/(1 + c)}{2h/\sin\theta} = \frac{g\sin\theta}{1 + c} \tag{13.45}$$

Recall, from Chapter 2, that $a_{particle} = g\sin\theta$ is the acceleration of a particle sliding down a frictionless incline. We can use this act to write Equation 13.45 in an interesting form:

$$a_{cm} = \frac{a_{particle}}{1 + c} \tag{13.46}$$

This analysis leads us to the conclusion that **the acceleration of a rolling object is less—in some cases significantly less—than the acceleration of a particle.** The reason is that the energy has to be shared between translational kinetic energy and rotational kinetic energy. A particle, by contrast, can put all its energy into translational kinetic energy.

Figure 13.44 shows the results of the race. The simple particle wins by a fairly wide margin. Of the solid objects, the sphere has the largest acceleration. Even so, its acceleration is only 71% the acceleration of a particle. The acceleration of the circular hoop, which comes in last, is a mere 50% that of a particle.

NOTE ▶ The objects that accelerate the fastest are those whose mass is most concentrated near the center. Placing the mass far from the center, as in the hoop, increases the moment of inertia. Thus it requires a larger effort to get a hoop rolling than to get a sphere of equal mass rolling. ◀

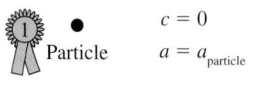

$c = 0$
$a = a_{particle}$
Particle

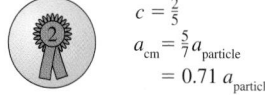

$c = \frac{2}{5}$
$a_{cm} = \frac{5}{7}a_{particle}$
$= 0.71\,a_{particle}$
Solid sphere

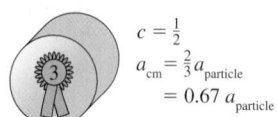

$c = \frac{1}{2}$
$a_{cm} = \frac{2}{3}a_{particle}$
$= 0.67\,a_{particle}$
Solid cylinder

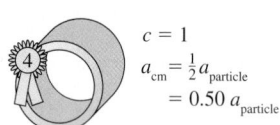

$c = 1$
$a_{cm} = \frac{1}{2}a_{particle}$
$= 0.50\,a_{particle}$
Circular hoop

FIGURE 13.44 And the winner is. . . .

13.9 The Vector Description of Rotational Motion

Thus far we have considered only rotation about a fixed axis, such as an axle, or in the case of rolling motion an axis that remains parallel to a fixed axis. In these situations, we can describe rotational motion in terms of a scalar angular velocity ω and a scalar torque τ, using a plus or minus sign to indicate the direction of rotation. This is very much analogous to the one-dimensional kinematics of Chapter 2. For more general rotational motion, angular velocity, torque, and other quantities must be treated as *vectors*. We won't go into much detail, because the subject rapidly gets very complicated, but we will sketch some important basic ideas.

The Angular Velocity Vector

Figure 13.45 shows a rotating rigid body. We can define an angular velocity vector $\vec{\omega}$ as follows:

- The magnitude of $\vec{\omega}$ is the object's angular velocity ω.
- $\vec{\omega}$ points along the axis of rotation in the direction given by the *right-hand rule* illustrated in Figure 13.45.

If the object rotates in the *xy*-plane, the vector $\vec{\omega}$ points along the *z*-axis. The scalar angular velocity $\omega = v_t/r$ that we've been using is now seen to be ω_z, the *z*-component of the vector $\vec{\omega}$. You should convince yourself that the sign convention for ω (positive for ccw rotation, negative for cw rotation) is equivalent to having the vector $\vec{\omega}$ pointing in the $+z$-direction or the $-z$-direction.

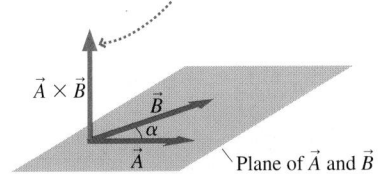

1. Using your right hand, curl your fingers in the direction of rotation with your thumb along the rotation axis.

2. Your thumb is then pointing in the direction of $\vec{\omega}$.

$\vec{\omega}$

Rotation axis

FIGURE 13.45 The angular velocity vector $\vec{\omega}$ is found using the right-hand rule.

The Cross Product of Two Vectors

We defined the torque exerted by force $\vec{F}$ to be $\tau = rF\sin\phi$. The quantity F is the magnitude of the force vector $\vec{F}$ and the distance r is really the magnitude of the position vector $\vec{r}$. Hence torque looks very much like a product of the two vectors $\vec{r}$ and $\vec{F}$. Previously, in conjunction with the definition of work, we introduced the dot product of two vectors: $\vec{A} \cdot \vec{B} = AB\cos\alpha$, where α is the angle between the vectors. $\tau = rF\sin\phi$ is a different way of multiplying vectors that depends on the *sine* of the angle between them.

Figure 13.46 shows two vectors, $\vec{A}$ and $\vec{B}$, with angle α between them. We define the **cross product** of $\vec{A}$ and $\vec{B}$ as the vector

$$\vec{A} \times \vec{B} \equiv (AB\sin\alpha, \text{ in the direction given by the right-hand rule}) \quad (13.47)$$

The symbol $\times$ between the vectors is *required* to indicate a cross product. The cross product is also called the **vector product** because the result is a vector.

The **right-hand rule,** which specifies the direction of $\vec{A} \times \vec{B}$, can be stated in three different but equivalent ways:

The cross product is perpendicular to the plane.

$\vec{A} \times \vec{B}$

$\vec{B}$

α

$\vec{A}$

Plane of $\vec{A}$ and $\vec{B}$

FIGURE 13.46 The cross product $\vec{A} \times \vec{B}$ is a vector perpendicular to the plane of vectors $\vec{A}$ and $\vec{B}$.

Using the right-hand rule

Spread your *right* thumb and index finger apart by angle α. Bend your middle finger so that it is *perpendicular* to your thumb and index finger. Orient your hand so that your thumb points in the direction of $\vec{A}$ and your index finger in the direction of $\vec{B}$. Your middle finger now points in the direction of $\vec{A} \times \vec{B}$.

Make a loose fist with your *right* hand with your thumb extended outward. Orient your hand so that your thumb is perpendicular to the plane of $\vec{A}$ and $\vec{B}$ and your fingers are curling *from* the line of vector $\vec{A}$ *toward* the line of vector $\vec{B}$. Your thumb now points in the direction of $\vec{A} \times \vec{B}$.

Imagine using a screwdriver to turn the slot in the head of a screw from the direction of $\vec{A}$ to the direction of $\vec{B}$. The screw will move either "in" or "out." The direction in which the screw moves is the direction of $\vec{A} \times \vec{B}$.

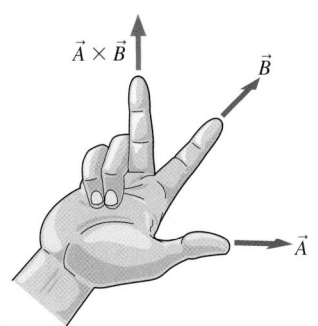

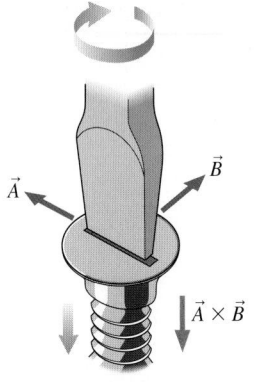

These are easier to demonstrate than to describe in words! Your instructor will show you how they work. Some individuals find one method of thinking about the direction of the cross product easier than the others, but they all work, and you'll soon find the method that works best for you.

Referring back to Figure 13.46, you should use the right-hand rule to convince yourself that the cross product $\vec{A} \times \vec{B}$ is a vector that points *upward*, perpendicular to the plane of $\vec{A}$ and $\vec{B}$. Figure 13.47 shows that the cross product, like the dot product, depends on the angle between the two vectors. Notice the two special cases: $\vec{A} \times \vec{B} = \vec{0}$ when $\alpha = 0°$ (parallel vectors) and $\vec{A} \times \vec{B}$ has its maximum magnitude AB when $\alpha = 90°$ (perpendicular vectors).

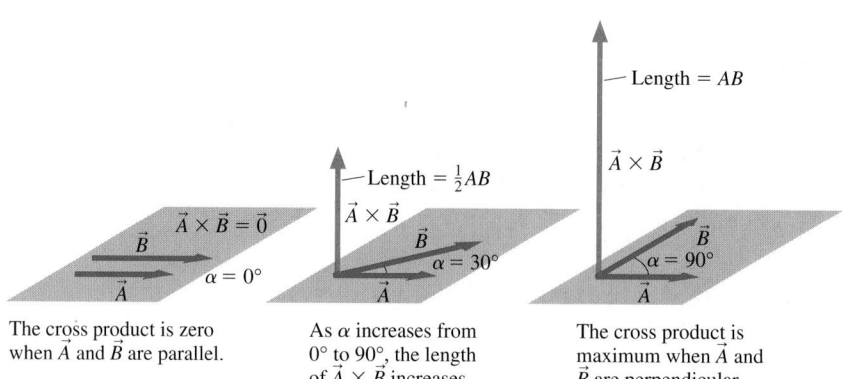

The cross product is zero when $\vec{A}$ and $\vec{B}$ are parallel.

As α increases from 0° to 90°, the length of $\vec{A} \times \vec{B}$ increases.

The cross product is maximum when $\vec{A}$ and $\vec{B}$ are perpendicular.

FIGURE 13.47 The cross-product vector increases from 0 to AB as α increases from 0° to 90°.

EXAMPLE 13.16 Calculating a cross product

Figure 13.48 shows vectors $\vec{C}$ and $\vec{D}$ in the plane of the page. What is the cross product $\vec{E} = \vec{C} \times \vec{D}$?

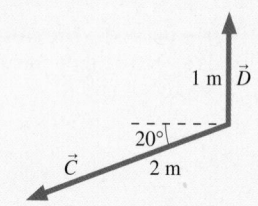

FIGURE 13.48 Vectors $\vec{C}$ and $\vec{D}$.

SOLVE The angle between the two vectors is $\alpha = 110°$. Consequently, the magnitude of the cross product is

$$E = CD\sin\alpha = (1 \text{ m})(2 \text{ m})\sin(110°) = 1.88 \text{ m}^2$$

The direction of $\vec{E}$ is given by the right-hand rule. To curl your right fingers from $\vec{C}$ to $\vec{D}$, you have to point your thumb *into* the page. Alternatively, if you turned a screwdriver from $\vec{C}$ to $\vec{D}$ you would be driving a screw *into* the page. Thus

$$\vec{E} = (1.88 \text{ m}^2, \text{ into page}).$$

ASSESS Notice that $\vec{E}$ has units of m².

(a)

(b)

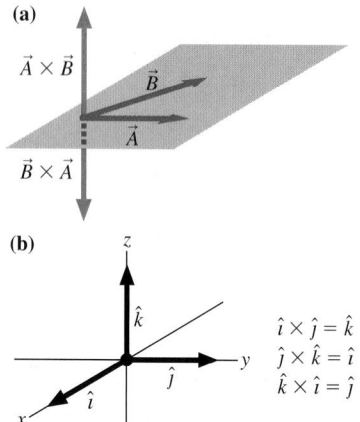

FIGURE 13.49 Properties of the cross product.

$\hat{i} \times \hat{j} = \hat{k}$
$\hat{j} \times \hat{k} = \hat{i}$
$\hat{k} \times \hat{i} = \hat{j}$

The cross product has three important properties:

1. The product $\vec{A} \times \vec{B}$ is *not* equal to the product $\vec{B} \times \vec{A}$. That is, the cross product does not obey the commutative rule $ab = ba$ that you know from arithmetic. In fact, you can see from the right-hand rule that the product $\vec{B} \times \vec{A}$ points in exactly the opposite direction from $\vec{A} \times \vec{B}$. Thus, as Figure 13.49a shows,

$$\vec{B} \times \vec{A} = -\vec{A} \times \vec{B}$$

2. In a *right-handed coordinate system*, which is the standard coordinate system of science and engineering, the z-axis is oriented relative to the xy-plane such that the unit vectors obey $\hat{i} \times \hat{j} = \hat{k}$. This is shown in Figure 13.49b. You can also see from this figure that and $\hat{j} \times \hat{k} = \hat{i}$ and $\hat{k} \times \hat{i} = \hat{j}$.

3. The derivative of a cross product is the same as the derivative of the product of two scalar quantities:

$$\frac{d}{dt}(\vec{A} \times \vec{B}) = \frac{d\vec{A}}{dt} \times \vec{B} + \vec{A} \times \frac{d\vec{B}}{dt} \qquad (13.48)$$

EXAMPLE 13.17 Calculating a cross product using unit vectors

What is the cross product of vectors $\vec{A} = 2\hat{\imath} + 3\hat{\jmath}$ and $\vec{B} = 2\hat{\imath} + 3\hat{\jmath} + 2\hat{k}$?

VISUALIZE Figure 13.50 shows vectors $\vec{A}$ and $\vec{B}$. Their cross product is perpendicular to the plane of $\vec{A}$ and $\vec{B}$.

SOLVE We could do the geometry to find the angle between $\vec{A}$ and $\vec{B}$, but it is easier to evaluate the cross product using unit vectors.

$$\vec{A} \times \vec{B} = (2\hat{\imath} + 3\hat{\jmath}) \times (2\hat{\imath} + 3\hat{\jmath} + 2\hat{k})$$

$$= (4\hat{\imath} \times \hat{\imath}) + (6\hat{\imath} \times \hat{\jmath}) + (4\hat{\imath} \times \hat{k})$$

$$\quad + (6\hat{\jmath} \times \hat{\imath}) + (9\hat{\jmath} \times \hat{\jmath}) + (6\hat{\jmath} \times \hat{k})$$

$$= \vec{0} + (6\hat{\imath} \times \hat{\jmath}) - (4\hat{k} \times \hat{\imath}) - (6\hat{\imath} \times \hat{\jmath})$$

$$\quad + \vec{0} + (6\hat{\jmath} \times \hat{k})$$

The fact that the cross product of two parallel vectors is zero allowed us to set $\hat{\imath} \times \hat{\imath} = \hat{\jmath} \times \hat{\jmath} = \vec{0}$. We also wrote $\hat{\imath} \times \hat{k} = -\hat{k} \times \hat{\imath}$ and $\hat{\jmath} \times \hat{\imath} = -\hat{\imath} \times \hat{\jmath}$. Now we can use the unit vector cross products shown in Figure 13.49b to complete the calculation and find

$$\vec{A} \times \vec{B} = 6\hat{\imath} - 4\hat{\jmath}$$

This vector is shown in Figure 13.50.

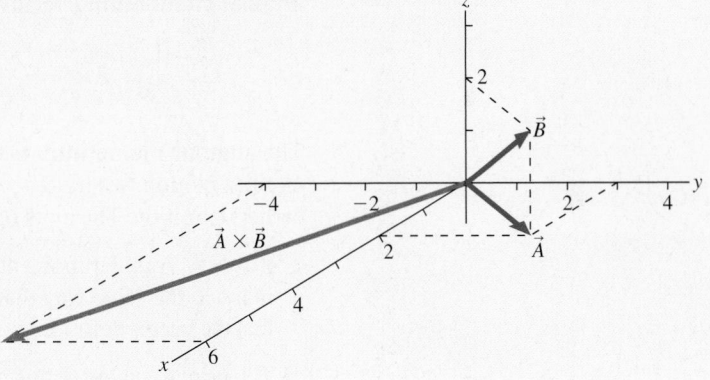

FIGURE 13.50 Vectors $\vec{A}$, $\vec{B}$, and their cross product.

Torque on a Particle

Now let's return to torque. Figure 13.51a shows force $\vec{F}$ acting on a particle located at position $\vec{r}$. This force exerts a torque about the origin. Let's define a *torque vector*

$$\vec{\tau} \equiv \vec{r} \times \vec{F} \qquad (13.49)$$

In Figure 13.51b we've placed the vector tails together in order to use the right-hand rule. You can see that the torque vector is perpendicular to the plane of $\vec{r}$ and $\vec{F}$. The angle between the vectors is ϕ, so the magnitude of the cross product is $\tau = rF|\sin\phi|$.

For motion in the xy-plane, the torque vector $\vec{\tau}$ points along the z-axis. The scalar torque $\tau = rF\sin\phi$ that we've been using is really τ_z, the z-component of the vector $\vec{\tau}$. This is the basis for our earlier sign convention for τ. In Figure 13.51, where the torque causes a ccw rotation, the torque vector $\vec{\tau}$ points in the positive z-direction, and thus τ_z is positive.

Angular Momentum of a Particle

We introduced angular momentum in Chapter 9, where for *circular* motion we defined $L = mrv_t$. In terms of the linear momentum $\vec{p} = m\vec{v}$, the angular momentum of a particle in circular motion is $L = rp$.

Circular motion is distinguished by the fact that $\vec{v}$ and $\vec{p}$ are perpendicular to $\vec{r}$. But suppose that $\vec{r}$ and $\vec{p}$ are *not* perpendicular. Figure 13.52 on the next page shows a particle moving along a trajectory. At this instant of time, the particle's

(a)

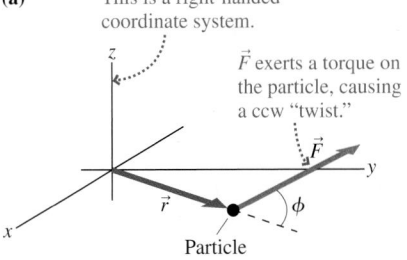

This is a right-handed coordinate system.

$\vec{F}$ exerts a torque on the particle, causing a ccw "twist."

(b)

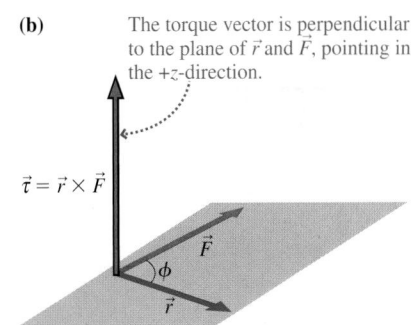

The torque vector is perpendicular to the plane of $\vec{r}$ and $\vec{F}$, pointing in the +z-direction.

$\vec{\tau} = \vec{r} \times \vec{F}$

FIGURE 13.51 The torque vector is perpendicular to the plane of $\vec{r}$ and $\vec{F}$.

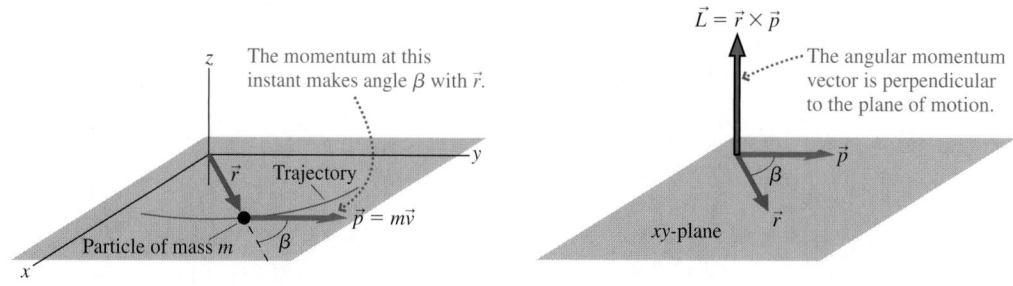

FIGURE 13.52 The angular momentum vector $\vec{L}$.

momentum vector $\vec{p}$ makes an angle β with the position vector $\vec{r}$. The particle's **angular momentum** $\vec{L}$ relative to the origin is a vector defined as

$$\vec{L} \equiv \vec{r} \times \vec{p} = (mrv\sin\beta, \text{ direction of right-hand rule}) \qquad (13.50)$$

The **angular momentum vector is perpendicular to the plane of motion.** For circular motion, where $\beta = 90°$, the magnitude $L = rp = rmv_t$ agrees with our earlier definition. The units of angular momentum are kg m²/s.

NOTE ▶ Angular momentum is the rotational equivalent of linear momentum in much the same way that torque is the rotational equivalent of force. Notice that the vector definitions are parallel: $\vec{\tau} \equiv \vec{r} \times \vec{F}$ and $\vec{L} \equiv \vec{r} \times \vec{p}$. ◀

The primary value of this more general definition of angular momentum lies in its connection to torque. To show this, take the time derivative of $\vec{L}$:

$$\frac{d\vec{L}}{dt} = \frac{d}{dt}(\vec{r} \times \vec{p}) = \left(\frac{d\vec{r}}{dt} \times \vec{p} + \vec{r} \times \frac{d\vec{p}}{dt}\right) \qquad (13.51)$$

$$= \vec{v} \times \vec{p} + \vec{r} \times \vec{F}_{net}$$

where we used Equation 13.48 for the derivative of a cross product. We also used the definition $\vec{v} = d\vec{r}/dt$ and, from Chapter 9, $\vec{F}_{net} = d\vec{p}/dt$.

Vectors $\vec{v}$ and $\vec{p}$ are parallel, and the cross product of two parallel vectors is $\vec{0}$. Thus the first term in Equation 13.51 vanishes. The second term $\vec{r} \times \vec{F}_{net}$ is the net torque, $\vec{\tau}_{net} = \vec{\tau}_1 + \vec{\tau}_2 + \cdots$, so we arrive at

$$\frac{d\vec{L}}{dt} = \vec{\tau}_{net} \qquad (13.52)$$

Equation 13.52, which says **a net torque causes the particle's angular momentum to change,** is the rotational equivalent of $d\vec{p}/dt = \vec{F}_{net}$.

13.10 Angular Momentum of a Rigid Body

Equation 13.52 is the angular momentum of a single particle. The angular momentum of a system consisting of particles with individual angular momenta $\vec{L}_1, \vec{L}_2, \vec{L}_3, \ldots$ is the vector sum

$$\vec{L} = \vec{L}_1 + \vec{L}_2 + \vec{L}_3 + \cdots = \sum_i \vec{L}_i \qquad (13.53)$$

We can combine Equations 13.52 and 13.53 to find the rate of change of the system's angular momentum:

$$\frac{d\vec{L}}{dt} = \sum_i \frac{d\vec{L}_i}{dt} = \sum_i \vec{\tau}_i = \vec{\tau}_{net} \qquad (13.54)$$

The net torque includes the torque due to internal forces, acting within the system, and the torque due to external forces. Because the internal forces are action/reaction pairs of forces, acting with the same strength in opposite directions, the net torque due to internal forces is zero. That is, the forces within a system of particles do not exert a net torque on the system. Thus the only forces that contribute to the net torque are external forces exerted on the system by the environment.

For a system of particles, **the rate of change of the system's angular momentum is the net torque on the system.** Equation 13.54 is analogous to the Chapter 9 result $d\vec{P}/dt = \vec{F}_{net}$, which says that the rate of change of a system's total linear momentum is the net force on the system. Table 13.4 summarizes the analogies between linear and angular momentum and energy.

TABLE 13.4 Angular and linear momentum and energy

Angular momentum	Linear momentum
$K_{rot} = \frac{1}{2}I\omega^2$	$K_{cm} = \frac{1}{2}Mv_{cm}^2$
$\vec{L} = I\vec{\omega}$ *	$\vec{P} = M\vec{v}_{cm}$
$d\vec{L}/dt = \vec{\tau}_{net}$	$d\vec{P}/dt = \vec{F}_{net}$
The angular momentum of a system is conserved if there is no net torque.	The linear momentum of a system is conserved if there is no net force.

*Rotation about an axis of symmetry.

Conservation of Angular Momentum

A net torque on a rigid body causes its angular momentum to change. Conversely, the angular momentum does *not* change—it is *conserved*—for a system with no net torque. This is the basis of the law of conservation of angular momentum.

Activ Physics ONLINE 7.14

> **Law of conservation of angular momentum** The angular momentum $\vec{L}$ of an isolated system ($\vec{\tau}_{net} = \vec{0}$) is conserved. The final angular momentum $\vec{L}_f$ is equal to the initial angular momentum $\vec{L}_i$.

If the angular momentum is conserved, both the magnitude *and* the direction of $\vec{L}$ are unchanged.

EXAMPLE 13.18 **An expanding rod**

Two equal masses are at the ends of a massless 50-cm-long rod. The rod spins at 2.0 rev/s about an axis through its midpoint. Suddenly, a compressed gas expands the rod out to a length of 160 cm. What is the rotation frequency after the expansion?

MODEL The forces that expand the rod push outward from the pivot and exert no torques. Thus the system's angular momentum is conserved.

VISUALIZE Figure 13.53 on the next page is a before-and-after pictorial representation. The angular momentum vectors $\vec{L}_1$ and $\vec{L}_2$ are perpendicular to the plane of motion.

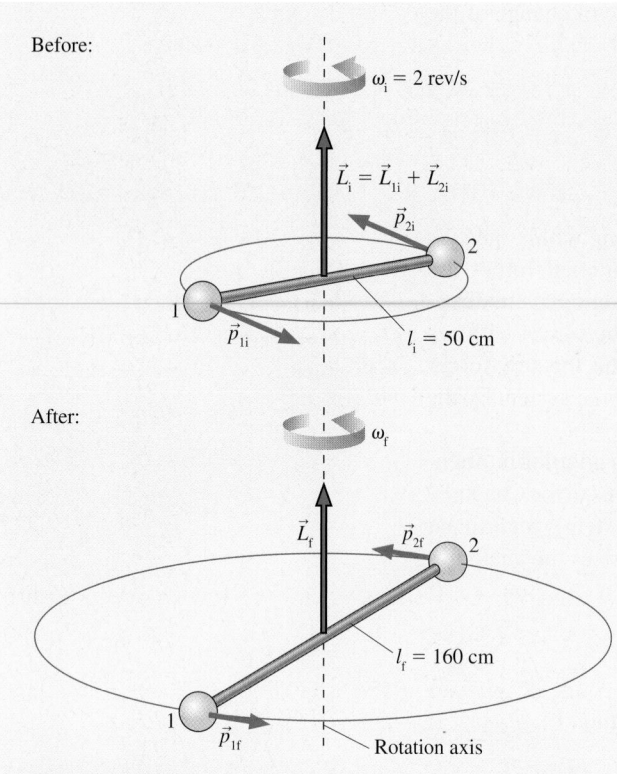

Before:

$\omega_i = 2$ rev/s

$\vec{L}_i = \vec{L}_{1i} + \vec{L}_{2i}$

$\vec{p}_{2i}$

2

1

$\vec{p}_{1i}$

$l_i = 50$ cm

After:

ω_f

$\vec{L}_f$

$\vec{p}_{2f}$

2

$l_f = 160$ cm

1

$\vec{p}_{1f}$

Rotation axis

FIGURE 13.53 The system before and after the rod expands.

SOLVE The particles are moving in circles, so the magnitude of each angular momentum is $L = rp = mrv_t = mr^2\omega = \frac{1}{4}ml^2\omega$, where we used $r = \frac{1}{2}l$. Because $\vec{L}_1$ and $\vec{L}_2$ point in the same direction, the magnitude of their sum is simply the sum of their magnitudes. Thus the initial angular momentum of the system is

$$L_i = \frac{1}{4}ml_i^2\omega_i + \frac{1}{4}ml_i^2\omega_i = \frac{1}{2}ml_i^2\omega_i$$

Similarly, the angular momentum after the expansion is $L_f = \frac{1}{2}ml_f^2\omega_f$. Angular momentum is conserved as the rod expands, thus

$$\frac{1}{2}ml_f^2\omega_f = \frac{1}{2}ml_i^2\omega_i$$

Solving for ω_f, we find

$$\omega_f = \left(\frac{l_i}{l_f}\right)^2\omega_i = \left(\frac{50 \text{ cm}}{160 \text{ cm}}\right)^2(2.0 \text{ rev/s}) = 0.195 \text{ rev/s}$$

ASSESS The values of the masses weren't needed. All that matters is the ratio of the lengths.

The expansion of the rod in Example 13.18 causes a dramatic slowing of the rotation. Similarly, the rotation would speed up if the weights were pulled in. This is how an ice skater controls her speed as she does a spin. Pulling in her arms decreases her moment of inertia and causes her angular velocity to increase. Similarly, extending her arms increases her moment of inertia and her angular velocity drops until she can skate out of the spin. It's all a matter of conserving angular momentum.

Angular Momentum and Angular Velocity

The analogy between linear and rotational motion has been so consistent that you might expect one more. The Chapter 9 result $\vec{P} = M\vec{v}_{cm}$ might give us reason to anticipate that angular momentum and angular velocity are related by $\vec{L} = I\vec{\omega}$. Unfortunately, the analogy breaks down at this point. For an arbitrarily shaped object, the angular momentum vector and the angular velocity vector don't necessarily point in the same direction. The general relationship between $\vec{L}$ and $\vec{\omega}$ is beyond the scope of this text.

The good news is that the analogy *does* continue to hold for the rotation of a *symmetrical* object about the symmetry axis. For example, the axis of a cylinder or disk is a symmetry axis, as is any diameter through a sphere. For the rotation of a symmetrical object about the symmetry axis, the angular momentum and angular velocity are related by

$$\vec{L} = I\vec{\omega} \quad \text{(rotation about an axis of symmetry)} \qquad (13.55)$$

This relationship is shown for a spinning disk in Figure 13.54. Equation 13.55 is particularly important for applying the law of conservation of momentum to symmetrical rigid bodies.

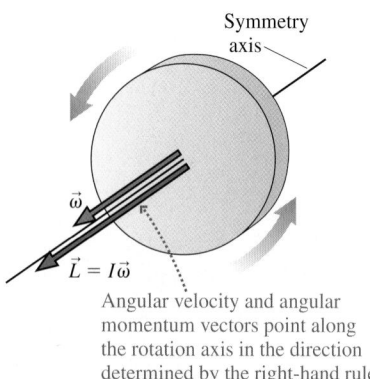

Symmetry axis

$\vec{\omega}$

$\vec{L} = I\vec{\omega}$

Angular velocity and angular momentum vectors point along the rotation axis in the direction determined by the right-hand rule.

FIGURE 13.54 The angular momentum vector of a rigid body rotating about an axis of symmetry.

EXAMPLE 13.19 Angular momentum of the earth

What is the angular momentum of the earth as it rotates on its axis?

MODEL The earth is a sphere that rotates about an axis of symmetry.

SOLVE This is a straightforward application of Equation 13.55. The earth's angular frequency is

$$\omega = \frac{1 \text{ rev}}{24 \text{ hr}} \times \frac{1 \text{ hr}}{3600 \text{ s}} \times \frac{2\pi \text{ rad}}{1 \text{ rev}} = 7.27 \times 10^{-5} \text{ rad/s}$$

The earth's moment of inertia is

$$I = \frac{2}{5} M_e R_e^2 = \frac{2}{5}(5.98 \times 10^{24} \text{ kg})(6.37 \times 10^6 \text{ m})^2$$

$$= 9.71 \times 10^{37} \text{ kg m}^2$$

Thus the magnitude of the earth's angular momentum is

$$L = I\omega = 7.06 \times 10^{33} \text{ kg m}^2/\text{s}$$

If you wrap your right fingers around a globe in the direction of the earth's rotation, from west to east, your thumb will point in the direction of the north pole. Consequently, the earth's angular momentum vector is

$$\vec{L} = (7.06 \times 10^{33} \text{ kg m}^2/\text{s, along the axis}$$
$$\text{through the North Pole})$$

ASSESS The earth's angular momentum is very large not because the earth is spinning rapidly, but because the earth has such a large moment of inertia.

EXAMPLE 13.20 Two interacting disks

A 20-cm-diameter, 2.0 kg solid disk is rotating at 200 rpm. A 20-cm-diameter, 1.0 kg circular loop is dropped straight down onto the rotating disk. Friction causes the loop to accelerate until it is "riding" on the disk. What is the final angular velocity of the combined system?

MODEL The friction between the two objects creates torques that speed up the loop and slow down the disk. But these torques are internal to the combined disk + loop system, so $\tau_{net} = 0$ and the *total* angular momentum of the disk + loop system is conserved.

VISUALIZE Figure 13.55 is a before-and-after pictorial representation. Initially only the disk is rotating, at angular velocity $\vec{\omega}_i$. The rotation is about an axis of symmetry, so the angular momentum $\vec{L} = I\vec{\omega}$ is parallel to $\vec{\omega}$. At the end of the problem, $\vec{\omega}_{disk} = \vec{\omega}_{loop} = \vec{\omega}_f$.

SOLVE Both angular momentum vectors point along the rotation axis. Conservation of angular momentum tells us that the magnitude of $\vec{L}$ is unchanged. Thus

$$L_f = I_{disk}\omega_f + I_{loop}\omega_f = L_i = I_{disk}\omega_i$$

Solving for ω_f gives

$$\omega_f = \frac{I_{disk}}{I_{disk} + I_{loop}}\omega_i$$

The moments of inertia for a disk and a loop can be found in Table 13.3, leading to

$$\omega_f = \frac{\frac{1}{2}M_{disk}R^2}{\frac{1}{2}M_{disk}R^2 + M_{loop}R^2}\omega_i = 100 \text{ rpm}$$

ASSESS What appeared to be a difficult problem turns out to be fairly easy once you recognize that the total angular momentum is conserved.

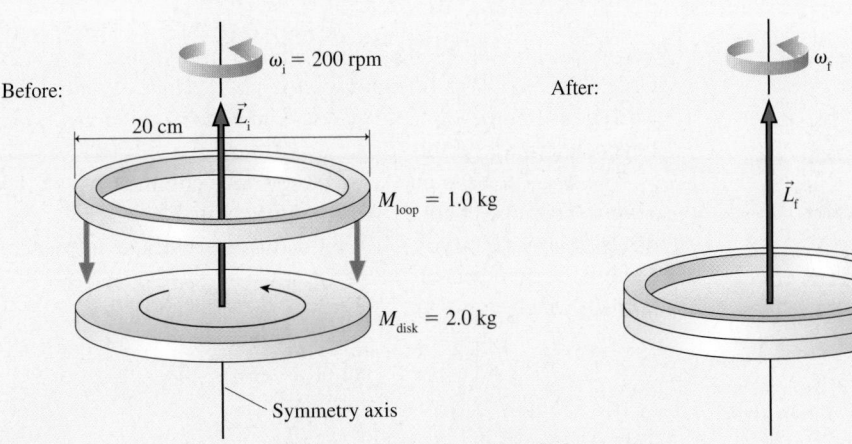

FIGURE 13.55 The circular hoop drops onto the rotating disk.

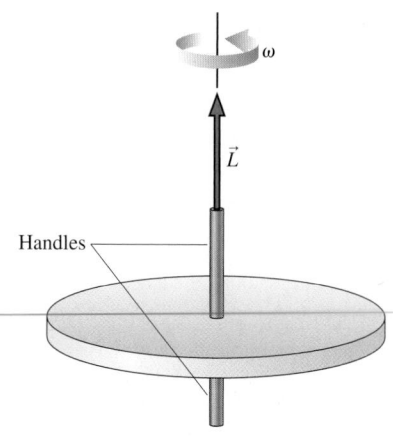

FIGURE 13.56 The vector nature of angular momentum makes it difficult to turn a rapidly spinning wheel.

When angular momentum is conserved, not only is its magnitude conserved, its direction—the direction of the rotation axis—must remain unchanged. This is often shown with the lecture demonstration illustrated in Figure 13.56. A bicycle wheel with two handles is given a spin, then handed to an unsuspecting student. The student is asked to turn the wheel 90°. Surprisingly, this is *very hard to do*.

The reason is that the wheel's angular momentum vector, which points straight up, is highly resistant to change. If the wheel is spinning fast, a *large* torque must be supplied to change $\vec{L}$. This directional stability of a rapidly spinning object is why gyroscopes are used as navigational devices on ships and planes. Once the axis of a spinning gyroscope is pointed north, it will maintain that direction as the ship or plane moves.

STOP TO THINK 13.6   Two buckets spin around in a horizontal circle on frictionless bearings. Suddenly, it starts to rain. As a result,

a. The buckets continue to rotate at constant angular velocity because the rain is falling vertically while the buckets move in a horizontal plane.
b. The buckets continue to rotate at constant angular velocity because the total mechanical energy of the bucket + rain system is conserved.
c. The buckets speed up because the potential energy of the rain is transformed into kinetic energy.
d. The buckets slow down because the angular momentum of the bucket + rain system is conserved.
e. Both a and b.
f. None of the above.

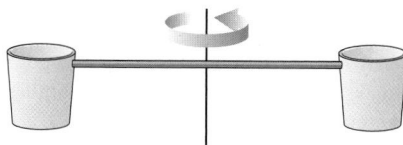

SUMMARY

The goal of Chapter 13 has been to understand the physics of rotating objects.

GENERAL PRINCIPLES

Rotational Dynamics

Every point on a **rigid body** rotating about a fixed axis has the same angular velocity ω and angular acceleration α.

Newton's second law for rotational motion is

$$\alpha = \frac{\tau_{net}}{I}$$

Use rotational kinematics to find angles and angular velocities.

Conservation Laws

Energy is conserved for an isolated system.

- Pure rotation $E = K_{rot} + U_g = \frac{1}{2}I\omega^2 + Mgy_{cm}$
- Rolling $E = K_{rot} + K_{cm} + U_g = \frac{1}{2}I\omega^2 + \frac{1}{2}Mv_{cm}^2 + Mgy_{cm}$

Angular momentum is conserved if $\vec{\tau}_{net} = \vec{0}$.

- Particle $\vec{L} = \vec{r} \times \vec{p}$
- Rigid body rotating about axis of symmetry $\vec{L} = I\vec{\omega}$

IMPORTANT CONCEPTS

Angular velocity

$$\omega = \frac{d\theta}{dt}$$

Angular acceleration is the rotational equivalent of acceleration

$$\alpha = \frac{d\omega}{dt}$$

Torque is the rotational equivalent of force

$$\tau = rF\sin\phi = rF_t = dF$$

Vector description of rotation

Torque $\vec{\tau} = \vec{r} \times \vec{F}$

Angular velocity $\vec{\omega}$ points along the rotation axis in the direction of the right-hand rule.

For a rigid body rotating about an axis of symmetry, the angular momentum is $\vec{L} = I\vec{\omega}$.

Newton's second law is

$$\frac{d\vec{L}}{dt} = \vec{\tau}_{net}$$

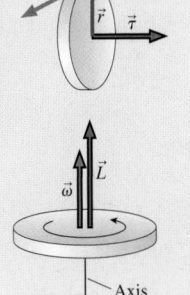

A system of particles on which there is no net force undergoes unconstrained rotation about the center of mass

$$x_{cm} = \frac{1}{M}\int x\, dm \qquad y_{cm} = \frac{1}{M}\int y\, dm$$

The gravitational torque on a body can be found by treating the body as a particle with all the mass M concentrated at the center of mass.

The moment of inertia

$$I = \int r^2\, dm$$

is the rotational equivalent of mass. The moment of inertia depends on how the mass is distributed around the axis. If I_{cm} is known, the I about a parallel axis distance d away is given by the **parallel-axis theorem** $I = I_{cm} + Md^2$.

APPLICATIONS

Rotational kinematics

$$\omega_f = \omega_i + \alpha\Delta t$$
$$\theta_f = \theta_i + \omega_i\Delta t + \frac{1}{2}\alpha(\Delta t)^2$$
$$\omega_f^2 = \omega_i^2 + 2\alpha\,\Delta\theta$$
$$v_t = r\omega \qquad a_t = r\alpha$$

Rigid-body equilibrium

An object is in total equilibrium only if both $\vec{F}_{net} = \vec{0}$ and $\vec{\tau}_{net} = \vec{0}$.

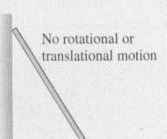

No rotational or translational motion

Rolling motion

For an object that rolls without slipping

$$v_{cm} = R\omega$$
$$K = K_{rot} + K_{cm}$$

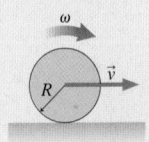

TERMS AND NOTATION

rigid body	torque, τ	rolling constraint
rigid-body model	line of action	cross product
translational motion	moment arm, d	vector product
rotational motion	couple	right-hand rule
combination motion	moment of inertia, I	angular momentum, $\vec{L}$
angular acceleration, α	parallel-axis theorem	law of conservation
center of mass	rotational kinetic energy, K_{rot}	of angular momentum

EXERCISES AND PROBLEMS

Exercises

Section 13.1 Rotational Kinematics

1. The graph shows the angular velocity of the crankshaft in a car. Draw a graph of the angular acceleration versus time. Include appropriate numerical scales on both axes.

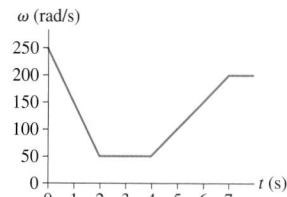

FIGURE EX13.1

2. The graph shows the angular acceleration of a turntable that starts from rest. Draw a graph of the angular velocity versus time. Include appropriate numerical scales on both axes.

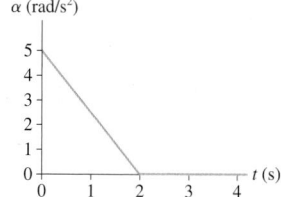

FIGURE EX13.2

3. The graph shows the angular velocity of one wheel on a car.
   a. Draw a graph of the angular acceleration versus time. Include appropriate numerical scales on both axes.
   b. Describe the car's motion.

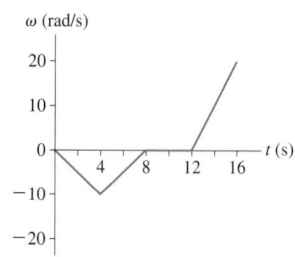

FIGURE EX13.3

4. a. Figure Ex13.4a shows angular velocity versus time. Draw the corresponding graph of angular acceleration versus time.
   b. Figure Ex13.4b shows angular acceleration versus time. Draw the corresponding graph of angular velocity versus time. Assume $\omega_0 = 0$.

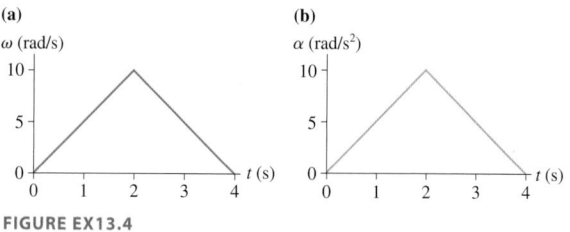

FIGURE EX13.4

5. A skater holds her arms outstretched as she spins at 180 rpm. What is the speed of her hands if they are 140 cm apart?

6. A magnetic computer disk 8.0 cm in diameter is initially at rest. A small dot is painted on the edge of the disk. The disk accelerates at 600 rad/s² for $\frac{1}{2}$ s, then coasts at a steady angular velocity for another $\frac{1}{2}$ s. What is the speed of the dot at $t = 1.0$ s? Through how many revolutions has it turned?

7. A high-speed drill rotating ccw at 2400 rpm comes to a halt in 2.5 s.
   a. What is the drill's angular acceleration?
   b. How many revolutions does it make as it stops?

8. An electric-generator turbine spins at 3600 rpm. Friction is so small that it takes the turbine 10 min to coast to a stop. How many revolutions does it make while stopping?

Section 13.2 Rotation About the Center of Mass

9. An equilateral triangle 5.0 cm on a side rotates about its center of mass at 120 rpm. What is the speed of one tip of the triangle?

10. The three masses shown in Figure Ex13.10 are connected by massless, rigid rods. What are the coordinates of the center of mass?

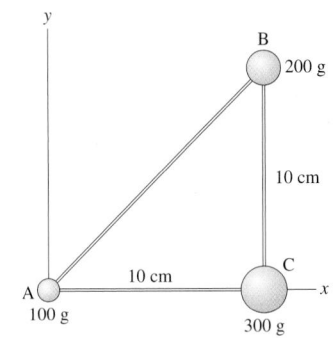

FIGURE EX13.10

Section 13.3 Torque

11. What is the net torque on the pulley about the axle?

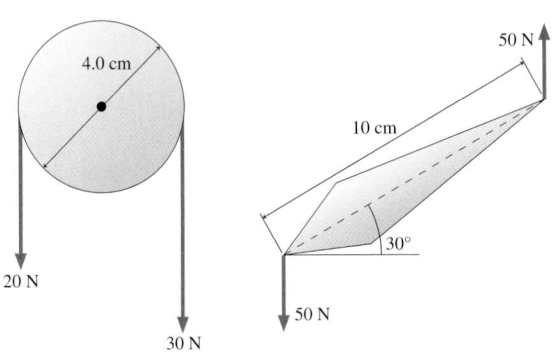

FIGURE EX13.11 **FIGURE EX13.12**

12. What is the net torque about the center of mass?
13. The tune-up specifications of a car call for the spark plugs to be tightened to a torque of 38 N m. You plan to tighten the plugs by pulling on the end of a 25-cm-long wrench. Because of the cramped space under the hood, you'll need to pull at an angle of 120° with respect to the wrench shaft. With what force must you pull?
14. The 20-cm-diameter disk in Figure Ex13.14 can rotate on an axle through its center. What is the net torque about the axle?

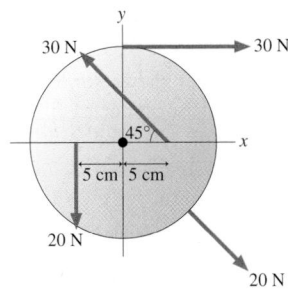

FIGURE EX13.14

15. A 4.0-m-long, 500 kg steel beam extends horizontally from the point where it has been bolted to the framework of a new building under construction. A 70 kg construction worker stands at the far end of the beam. What is the magnitude of the torque about the point where the beam is bolted into place?

16. An athlete at the gym holds a 3.0 kg steel ball in his hand. His arm is 70 cm long and has a mass of 4.0 kg. What is the magnitude of the torque about his shoulder if he holds his arm
    a. Straight out to his side, parallel to the floor?
    b. Straight, but 45° below horizontal?

Section 13.4 Rotational Dynamics

Section 13.5 Rotation About a Fixed Axis

17. The four masses shown in Figure Ex13.17 are connected by massless, rigid rods.
    a. Find the coordinates of the center of mass.
    b. Find the moment of inertia about an axis that passes through mass A and is perpendicular to the page.

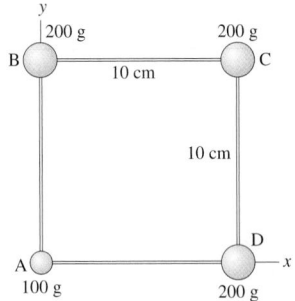

FIGURE EX13.17

18. The four masses shown in Figure Ex13.17 are connected by massless, rigid rods.
    a. Find the coordinates of the center of mass.
    b. Find the moment of inertia about a diagonal axis that passes through masses B and D.
19. The three masses shown in Figure Ex13.19 are connected by massless, rigid rods.
    a. Find the coordinates of the center of mass.
    b. Find the moment of inertia about an axis that passes through mass A and is perpendicular to the page.
    c. Find the moment of inertia about an axis that passes through masses B and C.

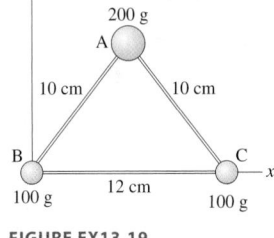

FIGURE EX13.19

20. An object's moment of inertia is 2.0 kg m². Its angular velocity is increasing at the rate of 4.0 rad/s per second. What is the torque on the object?
21. An object whose moment of inertia is 4.0 kg m² experiences the torque shown in Figure Ex13.21. What is the object's angular velocity at $t = 3.0$ s? Assume it starts from rest.

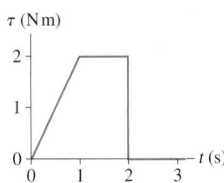

FIGURE EX13.21

22. A 1.0 kg ball and a 2.0 kg ball are connected by a 1.0-m-long rigid, massless rod. The rod is rotating cw about its center of mass at 20 rpm. What torque will bring the balls to a halt in 5.0 s?

23. A 200 g, 20-cm-diameter plastic disk is spun on an axle through its center by an electric motor. What torque must the motor supply to take the disk from 0 to 1800 rpm in 4.0 s?

24. The 200 g model rocket shown in Figure Ex13.24 generates 4.0 N of thrust. It spins in a horizontal circle at the end of a 100 g rigid rod. What is its angular acceleration?

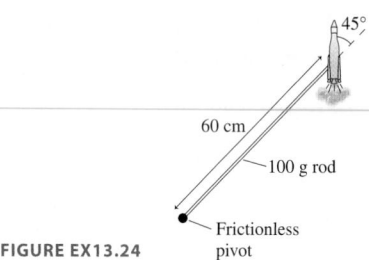

FIGURE EX13.24

Section 13.6 Rigid-Body Equilibrium

25. How much torque must the pin exert to keep the rod from rotating?

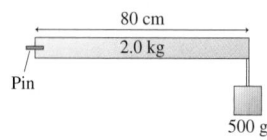

FIGURE EX13.25

26. Is the object in Figure Ex13.26 in equilibrium? Explain.

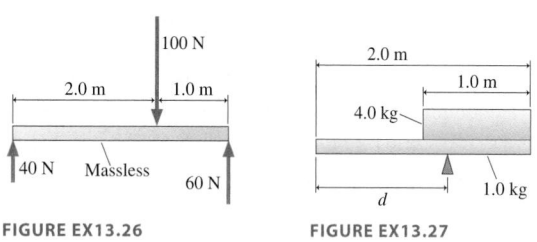

FIGURE EX13.26   FIGURE EX13.27

27. The two objects in Figure Ex13.27 are balanced on the pivot. What is distance d?

Section 13.7 Rotational Energy

28. What is the rotational kinetic energy of the earth? Assume the earth is a uniform sphere.

29. The three 200 g masses in Figure Ex13.29 are connected by massless, rigid rods to form a triangle. What is the triangle's rotational kinetic energy if it rotates at 5.0 rev/s about an axis through the center?

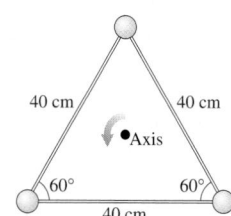

FIGURE EX13.29

30. A thin, 100 g disk with a diameter of 8.0 cm rotates about an axis through its center with 0.15 J of kinetic energy. What is the speed of a point on the rim?

31. A 300 g ball and a 600 g ball are connected by a 40-cm-long massless, rigid rod. The structure rotates about its center of mass at 100 rpm. What is its rotational kinetic energy?

Section 13.8 Rolling Motion

32. A car tire is 60 cm in diameter. The car is traveling at a speed of 20 m/s.
   a. What is the tire's rotation frequency, in rpm?
   b. What is the speed of a point at the top edge of the tire?
   c. What is the speed of a point at the bottom edge of the tire?

33. A 500 g, 8.0-cm-diameter can rolls across the floor at 1.0 m/s. What is the can's kinetic energy?

34. An 8.0-cm-diameter, 400 g sphere is released from rest at the top of a 2.1-m-long, 25° incline. It rolls, without slipping, to the bottom.
   a. What is the sphere's angular velocity at the bottom of the incline?
   b. What fraction of its kinetic energy is rotational?

Section 13.9 The Vector Description of Rotational Motion

35. Evaluate the cross products $\vec{A} \times \vec{B}$ and $\vec{C} \times \vec{D}$.

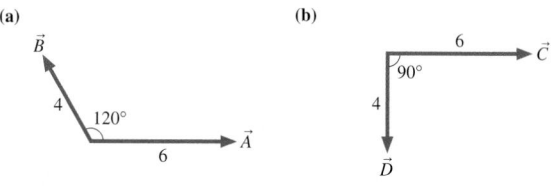

FIGURE EX13.35

36. Evaluate the cross products $\vec{A} \times \vec{B}$ and $\vec{C} \times \vec{D}$.

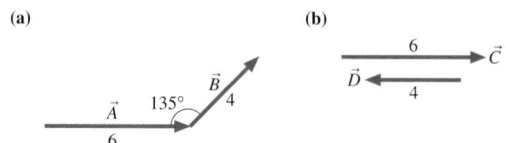

FIGURE EX13.36

37. a. What is $(\hat{\imath} \times \hat{\jmath}) \times \hat{\imath}$?
   b. What is $\hat{\imath} \times (\hat{\jmath} \times \hat{\imath})$?

38. a. What is $\hat{\imath} \times (\hat{\imath} \times \hat{\jmath})$?
   b. What is $(\hat{\imath} \times \hat{\jmath}) \times \hat{k}$?

39. Vector $\vec{A} = 3\hat{\imath} - \hat{\jmath}$ and vector $\vec{B} = 2\hat{\imath} + 3\hat{\jmath} - \hat{k}$.
   a. What is the cross product $\vec{A} \times \vec{B}$?
   b. Show vectors $\vec{A}$, $\vec{B}$, and $\vec{A} \times \vec{B}$ on a three-dimensional coordinate system.

40. Consider the vector $\vec{C} = 3\hat{\imath}$.
   a. What is a vector $\vec{D}$ such that $\vec{C} \times \vec{D} = \vec{0}$?
   b. What is a vector $\vec{E}$ such that $\vec{C} \times \vec{E} = 6\hat{k}$?
   c. What is a vector $\vec{F}$ such that $\vec{C} \times \vec{F} = -3\hat{\jmath}$?

41. Force $\vec{F} = -10\hat{\jmath}$ N is exerted on a particle at $\vec{r} = (5\hat{\imath} + 5\hat{\jmath})$ m. What is the torque on the particle about the origin?

42. Force $\vec{F} = (-10\hat{\imath} + 10\hat{\jmath})$N is exerted on a particle at $\vec{r} = 5\hat{\jmath}$ m. What is the torque on the particle about the origin?

Section 13.10 Angular Momentum of a Rigid Body

43. A 200 g block is attached to one end of a 40-cm-long massless rod. The rod rotates about a frictionless pivot at the other end. The block starts from rest and experiences a constant torque of 0.050 N m.
    a. How long does it take the block to complete 10 revolutions?
    b. What is the block's angular momentum at the time of completion?
    c. Show that $\Delta L / \Delta t = \tau$.

44. What are the magnitude and direction of the angular momentum of the 200 g particle in Figure Ex13.44?

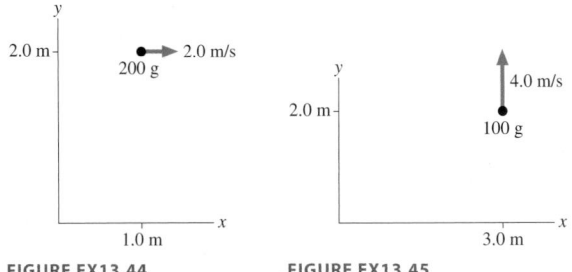

FIGURE EX13.44   FIGURE EX13.45

45. What are the magnitude and direction of the angular momentum of the 100 g particle in Figure Ex13.45?

46. What is the angular momentum of the 500 g rotating bar in Figure Ex13.46?

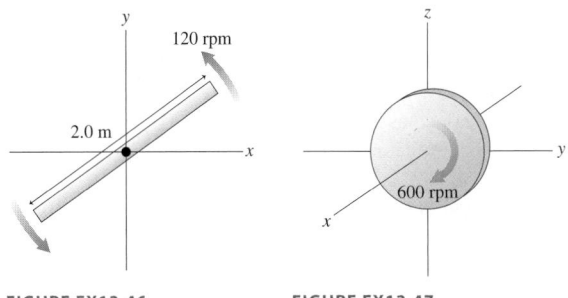

FIGURE EX13.46   FIGURE EX13.47

47. What is the angular momentum of the 2.0 kg, 4.0-cm-diameter rotating disk in Figure Ex13.47?

Problems

48. As the earth rotates, what is the speed of (a) a physics student in Miami, Florida, at latitude 26°, and (b) a physics student in Fairbanks, Alaska, at latitude 65°? Ignore the revolution of the earth around the sun.

49. A 60-cm-diameter wheel is rolling along at 20 m/s. What is the speed of a point at the front edge of the wheel?

50. An 800 g steel plate has the shape of the isosceles triangle shown in Figure P13.50. Locate the plate's center of mass.
    Hint: Divide the triangle into vertical strips of width dx, then relate the mass dm of a strip at position x to the values of x and dx.

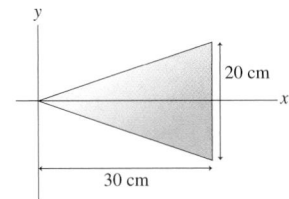

FIGURE P13.50

51. What is the moment of inertia of a 2.0 kg, 20-cm-diameter disk for rotation about an axis (a) through the center, and (b) through the edge of the disk.

52. Calculate by direct integration the moment of inertia for a thin rod of mass M and length L about an axis located distance d from one end. Confirm that your answer agrees with Table 13.3 when $d = 0$ and when $d = L/2$.

53. a. A disk of mass M and radius R has a hole of radius r centered on the axis. Calculate the moment of inertia of the disk.
    b. Confirm that your answer agrees with Table 13.3 when $r = 0$ and when $r = R$.
    c. A 4.0-cm-diameter disk with a 3.0-cm-diameter hole rolls down a 50-cm-long, 20° ramp. What is its speed at the bottom? What percent is this of the speed of a particle sliding down a frictionless ramp?

54. Determine the moment of inertia about the axis of the object shown in Figure P13.54.

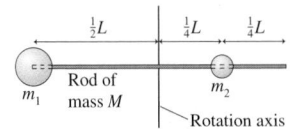

FIGURE P13.54

55. Calculate the moment of inertia about a center axis of a long rod with an $L \times L$ square cross section.

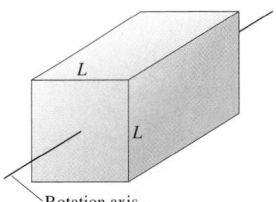

FIGURE P13.55   Rotation axis

56. A 3.0-m-long rigid beam with a mass of 100 kg is supported at each end. An 80 kg student stands 2.0 m from support 1. How much upward force does each support exert on the beam?

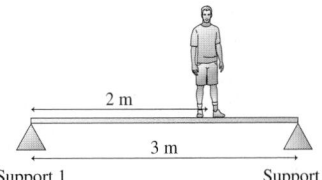

FIGURE P13.56   Support 1   Support 2

57. An 80 kg construction worker sits down 2.0 m from the end of a 1450 kg steel beam to eat his lunch. The cable supporting the beam is rated at 15,000 N. Should the worker be worried?

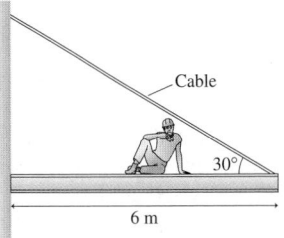

FIGURE P13.57

58. A forearm can be modeled as a 1.2 kg, 32-cm-long "beam" that pivots at the elbow and is supported by the biceps. How much force must the biceps exert to hold a 500 g ball with the forearm parallel to the floor?

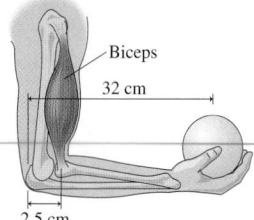

FIGURE P13.58

59. A 40 kg, 5.0-m-long beam is supported, but not attached to, the two posts in Figure P13.59. A 20 kg boy starts walking along the beam. How close can he get to the right end of the beam without it falling over?

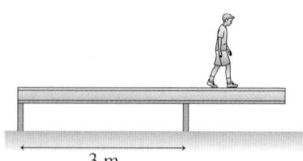

FIGURE P13.59

60. A 3.0-m-long ladder, as shown in Figure 13.35, leans against a frictionless wall. The coefficient of static friction between the ladder and the floor is 0.40. What is the minimum angle the ladder can make with the floor without slipping?

61. A 1.0 kg mass at $(x, y) = (20$ cm, 20 cm$)$ and a 3.0 kg mass at $(20$ cm, 100 cm$)$ are connected by a massless, rigid rod. They rotate about the center of mass.
    a. What are the coordinates of the center of mass?
    b. What is the moment of inertia about the center of mass?
    c. What constant torque will cause an angular velocity of 6.25 rad/s at the end of 3.0 s, starting from rest?
    d. At what angle, with respect to the axis of the rod, should 1.5 N forces be applied to each mass to give the torque you found in part c?

62. Starting from rest, a 12-cm-diameter compact disk takes 3.0 s to reach its operating angular velocity of 2000 rpm. Assume that the angular acceleration is constant. The disk's moment of inertia is 2.5×10^{-5} kg m^2.
    a. How much torque is applied to the disk?
    b. How many revolutions does it make before reaching full speed?

63. The two stars in a binary star system have masses 2.0×10^{30} kg and 6.0×10^{30} kg. They are separated by 2.0×10^{12} m. What are
    a. The system's rotation period, in years?
    b. The speed of each star?

64. A 60-cm-long, 500 g bar rotates in a horizontal plane on an axle that passes through the center of the bar. Compressed air is fed in through the axle, passes through a small hole down the length of the bar, and escapes as air jets from holes at the ends of the bar. The jets are perpendicular to the bar's axis. Starting from rest, the bar spins up to an angular velocity of 150 rpm at the end of 10 s.
    a. How much force does each jet of escaping air exert on the bar?

b. If the axle is moved to one end of the bar while the air jets are unchanged, what will be the bar's angular velocity at the end of 10 seconds?

65. Flywheels are large, massive wheels used to store energy. They can be spun up slowly, then the wheel's energy can be released quickly to accomplish a task that demands high power. An industrial flywheel has a 1.5 m diameter and a mass of 250 kg. Its maximum angular velocity is 1200 rpm.
    a. A motor spins up the flywheel with a constant torque of 50 Nm. How long does it take the flywheel to reach top speed?
    b. How much energy is stored in the flywheel?
    c. The flywheel is disconnected from the motor and connected to a machine to which it will deliver energy. Half the energy stored in the flywheel is delivered in 2 seconds. What is the average power delivered to the machine?
    d. How much torque does the flywheel exert on the machine?

66. A 3.0 kg block is attached to a string that is wrapped around a 2.0 kg, 4.0-cm-diameter *hollow* cylinder that is free to rotate. (Use Figure 13.34 but treat the cylinder as hollow.) The block is released 1.0 m above the ground.
    a. Use Newton's second law to find the speed of the block as it hits the ground.
    b. Use conservation of energy to find the speed of the block as it hits the ground.

67. The two blocks in Figure P13.67 are connected by a massless rope that passes over a pulley. The pulley is 12 cm in diameter and has a mass of 2.0 kg. As the pulley turns, friction at the axle exerts a torque of magnitude 0.50 Nm. If the blocks are released from rest, how long does it take the 4.0 kg block to reach the floor?

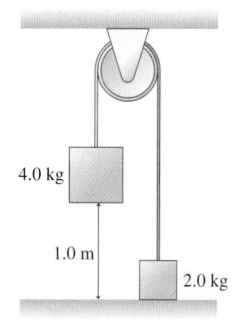

FIGURE P13.67

68. Blocks of mass m_1 and m_2 are connected by a massless string that passes over the frictionless pulley in Figure P13.68. Mass m_1 slides on a horizontal, frictionless surface. Mass m_2 is released while the blocks are at rest.
    a. Assume the pulley is massless. Find the acceleration of m_1 and the tension in the string. This is a Chapter 8 review problem.
    b. Suppose the pulley has mass m_p and radius R. Find the acceleration of m_1 and the tensions in the upper and lower portions of the string. Verify that your answers agree with part a if you set $m_p = 0$.

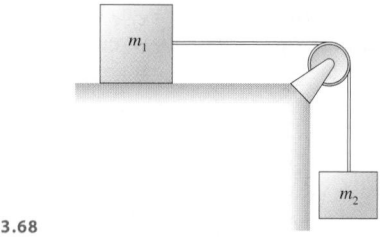

FIGURE P13.68

69. The 2.0 kg, 30-cm-diameter disk in Figure P13.69 is spinning at 300 rpm. How much friction force must the brake apply to the rim to bring the disk to a halt in 3.0 s?

FIGURE P13.69

70. Suppose the connecting tunnel in Example 13.7 has a mass of 50,000 kg.
    a. How far from the 100,000 kg rocket is the center of mass of the entire structure?
    b. What is the structure's angular velocity after 30 s?

71. A hollow sphere is rolling along a horizontal floor at 5.0 m/s when it comes to a 30° incline. How far up the incline does it roll before reversing direction?

72. A 5.0 kg, 60-cm-diameter disk rotates on an axle passing through one edge. The axle is parallel to the floor. The cylinder is held with the center of mass at the same height as the axle, then released.
    a. What is the cylinder's initial angular acceleration?
    b. What is the cylinder's angular velocity when it is directly below the axle?

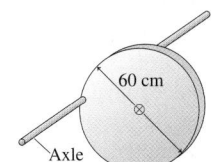

FIGURE P13.72

73. A hoop of mass M and radius R rotates about an axle at the edge of the hoop. The hoop starts at its highest position and is given a very small push to start it rotating. At its lowest position, what are (a) the angular velocity and (b) the speed of the lowest point on the hoop?

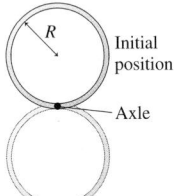

FIGURE P13.73

74. A long, thin rod of mass M and length L is standing straight up on a table. Its lower end rotates on a frictionless pivot. A very slight push causes the rod to fall over. As it hits the table, what are (a) the angular velocity and (b) the speed of the tip of the rod?

75. A sphere of mass M and radius R is rigidly attached to a thin rod that passes through the sphere at distance $\frac{1}{2}R$ from the center. A string wrapped around the rod exerts torque τ about the axis of the rod. Find an expression for the sphere's angular acceleration. The rod's moment of inertia is negligible.

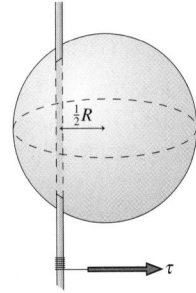

FIGURE P13.75

76. You've been given a pulley for your birthday. It's a fairly big pulley, 12 cm in diameter and with a mass of 2.0 kg. You get to wondering whether the pulley is uniform. That is, is the mass evenly distributed, or is it concentrated toward the center or near the rim? To find out, you hang the pulley on a hook, wrap a string around it several times, and suspend your 1.0 kg physics book 1.0 m above the floor. With your stopwatch, you find that it takes 0.71 s for your book to hit the floor. What can you conclude about the pulley?

77. A solid sphere of radius R is placed at a height of 30 cm on a 15° slope. It is released and rolls, without slipping, to the bottom.
    a. From what height should a circular hoop of radius R be released on the same slope in order to equal the sphere's speed at the bottom?
    b. Can a circular hoop of different diameter be released from a height of 30 cm and match the sphere's speed at the bottom? If so, what is the diameter? If not, why not?

78. A 2.0 kg, 20-cm-diameter turntable rotates at 100 rpm on frictionless bearings. Two 500 g blocks fall from above, hit the turntable simultaneously at opposite ends of a diagonal, and stick. What is the turntable's angular velocity, in rpm, just after this event?

79. A 200 g, 40-cm-diameter turntable rotates on frictionless bearings at 60 rpm. A 20 g block sits at the center of the turntable. A compressed spring shoots the block radially outward along a frictionless groove in the surface of the turntable. What is the turntable's rotation angular velocity when the block reaches the outer edge?

80. A merry-go-round is a common piece of playground equipment. A 3.0-m-diameter merry-go-round with a mass of 250 kg is spinning at 20 rpm. John runs tangent to the merry-go-round at 5.0 m/s, in the same direction that it is turning, and jumps onto the outer edge. John's mass is 30 kg. What is the merry-go-round's angular velocity, in rpm, after John jumps on?

81. A satellite follows the elliptical orbit shown. The only force on the satellite is the gravitational attraction of the planet. The satellite's speed at point a is 8000 m/s.
    a. Is there any torque on the satellite? Explain.
    b. What is the satellite's speed at point b?
    c. What is the satellite's speed at point c?

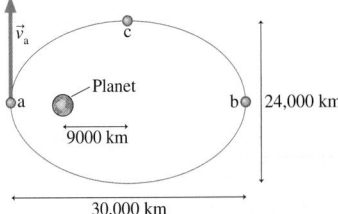

FIGURE P13.81

82. A 45 kg figure skater is spinning on the toes of her skates at 1.5 rpm. Her arms are outstretched as far as they will go. In this orientation, the skater can be modeled as a cylindrical torso (40 kg, 20 cm average diameter, 168 cm tall) plus two rod-like arms (2.5 kg each, 78 cm long) that each rotate about an axis through the inner end of the rod. The skater then raises her arms straight above her head, where she appears to be a 45 kg, 20-cm-diameter, 200-cm-tall cylinder. What is her new rotation frequency, in revolutions per second?

83. A billiard ball of mass m rolls without slipping across a table at speed v_0. It collides with a rod of length d and mass $M = 2m$. The rod is pivoted about a frictionless axle through its center, and it is initially hanging straight up and down at rest. After the collision, the billiard ball moves straight ahead with half its initial speed.

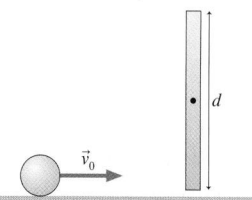

FIGURE P13.83

  a. What is the rod's angular velocity after the collision?
  b. Is the mechanical energy conserved?

Challenge Problems

84. The marble rolls down a track and around a loop-the-loop of radius R. The marble has mass m and radius r. What minimum height h must the track have for the marble to make it around the loop-the-loop without falling off?

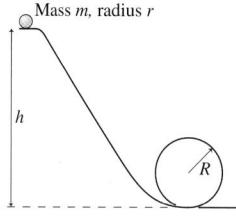

Mass m, radius r

FIGURE CP13.84

85. A 2.0 kg wood block hangs from the bottom of a 1.0 kg, 1.0-m-long rod. The block and rod form a pendulum that swings on a frictionless pivot at the top end of the rod. A 10 g bullet is fired into the block, where it sticks, causing the pendulum to swing out to a 30° angle. What was the speed of the bullet? You can treat the wood block as a particle.

86. A 10 g bullet traveling at 400 m/s strikes a 10 kg, 1.0-m-wide door at the edge opposite the hinge. The bullet embeds itself in the door, causing the door to swing open. What is the angular velocity of the door just after impact?

87. During most of its lifetime, a star maintains an equilibrium size in which the inward force of gravity on each atom is balanced by an outward pressure force due to the heat of the nuclear reactions in the core. But after all the hydrogen "fuel" is consumed by nuclear fusion, the pressure force drops and the star undergoes a *gravitational collapse* until it becomes a *neutron star*. In a neutron star, the electrons and protons of the atoms are squeezed together by gravity until they fuse into neutrons. Neutron stars spin very rapidly and emit intense pulses of radio and light waves, one pulse per rotation. These "pulsing stars" were discovered in the 1960s and are called *pulsars*.

  a. A star with the mass ($M = 2.0 \times 10^{30}$ kg) and size ($R = 3.5 \times 10^8$ m) of our sun rotates once every 30 days. After undergoing gravitational collapse, the star forms a pulsar that is observed by astronomers to emit radio pulses every 0.1 s. By treating the neutron star as a solid sphere, deduce its radius.
  b. What is the speed of a point on the equator of the neutron star?

  Your answer will be somewhat too large because a star cannot be accurately modeled as a solid sphere. Even so, you will be able to show that a star, whose mass is 10^6 larger than the earth's, can be compressed by gravitational forces to a size smaller than a typical state in the United States!

88. A physics professor stands at rest on a 5.0 kg, 50-cm-diameter frictionless turntable. His assistant has a 64-cm-diameter bicycle wheel to which 4.0 kg of lead weights have been added around the rim. Handles extend outward from the axis so that the wheel can be held as it spins. The assistant spins the wheel to 180 rpm and holds it in a horizontal plane (the rotation axis is vertical) such that the rotation is ccw as seen from the ceiling. He then hands the spinning wheel to the professor.

  a. When the professor takes the wheel by the handles and the assistant lets go, does anything happen to the professor? If so, *describe* the professor's motion and *calculate* any relevant numerical quantities. If not, explain why not.
  b. Then the professor turns the spinning wheel over 180° so that the handle that had been pointing toward the ceiling now points toward the floor. Does anything happen to the professor? If so, *describe* the professor's motion and *calculate* any relevant numerical quantities. If not, explain why not.

  Hint: You'll need to *model* both the professor and the wheel. The professor has a total mass of 75 kg. His legs and torso are 70 kg. They have an average diameter of 25 cm and a height of 180 cm. His arms are 2.5 kg each, and he hold the handles of the wheel 45 cm from the center of his body. Don't forget that the wheel both spins *and* moves with the professor.

STOP TO THINK ANSWERS

Stop to Think 13.1: c. ω is negative because the rotation is cw. Because ω is negative but becoming *less* negative, the change $\Delta\omega$ is *positive*. So α is positive.

Stop to Think 13.2: $\tau_e > \tau_a = \tau_d > \tau_b > \tau_c$. The tangential component in e is larger than 2 N.

Stop to Think 13.3: $\alpha_b > \alpha_a > \alpha_c = \alpha_d = \alpha_e$. Angular acceleration is proportional to torque and inversely proportional to the moment of inertia. The moment of inertia depends on the *square* of the radius. The tangential force component in e is the same as in d.

Stop to Think 13.4: $I_a > I_d > I_b > I_c$. The moment of inertia is smaller when mass is more concentrated near the rotation axis.

Stop to Think 13.5: c > d > a = b. To keep the meter stick in equilibrium, the student must supply a torque equal and opposite to the torque due to the hanging masses. Torque depends on the mass *and* on how far the mass is from the pivot point.

Stop to Think 13.6: d. There is no net torque on the bucket + rain system, so the angular momentum is conserved. The addition of mass on the outer edge of the circle increases I, so ω must decrease. Mechanical energy is not conserved because the raindrop collisions are inelastic.

14 Oscillations

A playground swing is just one example of oscillatory motion.

▶ Looking Ahead
The goal of Chapter 14 is to understand systems that oscillate with simple harmonic motion. In this chapter you will learn to:

- Understand the kinematics of simple harmonic motion.
- Use graphical and mathematical representations of oscillatory motion.
- Understand energy in oscillating systems.
- Understand the dynamics of oscillating systems.
- Recognize the importance of resonance and damping in oscillating systems.

◀ Looking Back
Simple harmonic motion is closely related to circular motion. Much of our analysis of oscillating systems will be based on the law of conservation of energy. Please review:

- Section 7.1 Uniform circular motion.
- Sections 10.4 and 10.5 Restoring forces and elastic potential energy.
- Section 10.7 Energy diagrams.

This girl is having a good time on the swing. At the same time, she is demonstrating an important type of motion—*oscillatory motion*. Examples of oscillatory motion abound. A marble rolling back and forth in the bottom of a bowl and a car bouncing up and down on its springs are oscillating. So are a piece of vibrating machinery, a ringing bell, and the current in an electric circuit used to drive an antenna. A vibrating guitar string pushes the air molecules back and forth to send out a sound wave, showing that oscillations are closely related to waves.

Oscillatory motion is a repetitive motion back and forth about an equilibrium position. Swinging motions and vibrations of all kinds are oscillatory motions. All oscillatory motion is *periodic*.

Our goal in this chapter is to study the physics of oscillations. Much of our analysis will be focused on the most basic form of oscillatory motion, *simple harmonic motion*. We will start with the kinematics of simple harmonic motion—a mathematical description of the motion. Then we will examine oscillatory motion from the twin perspectives of energy and Newton's laws. Finally, we will look at how oscillations are built up by driving forces and how they decay over time.

14.1 Simple Harmonic Motion

Objects or systems of objects that undergo oscillatory motion are called **oscillators.** Figure 14.1 shows position-versus-time graphs for several different oscillating systems. Although the shapes of the graphs are different, all these oscillators have two things in common:

1. The oscillation takes place about an equilibrium position, and
2. The motion is periodic.

The time to complete one full cycle, or one oscillation, is called the **period** of the motion. Period is given the symbol T.

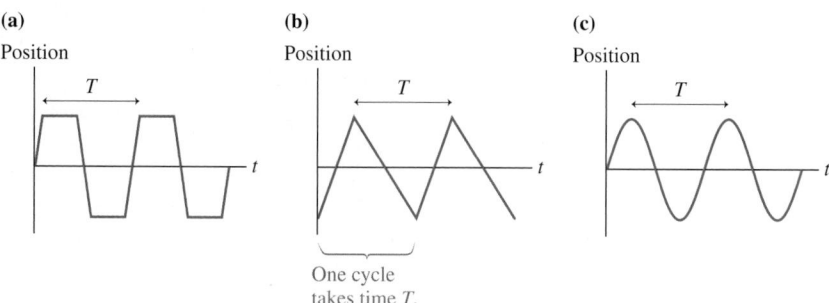

FIGURE 14.1 Several examples of position-versus-time graphs for oscillating systems.

A closely related piece of information is the number of cycles, or oscillations, completed per second. If the period is $\frac{1}{10}$ s, then the oscillator can complete 10 cycles in one second. Conversely, an oscillation period of 10 s allows only $\frac{1}{10}$ of a cycle to be completed per second. In general, T seconds per cycle implies that $1/T$ cycles will be completed each second. The number of cycles per second is called the **frequency** f of the oscillation. The relationship between frequency and period is

$$f = \frac{1}{T} \quad \text{or} \quad T = \frac{1}{f} \tag{14.1}$$

The units of frequency are **hertz,** abbreviated Hz, named in honor of the German physicist Heinrich Hertz who produced the first artificially generated radio waves in 1887. By definition,

$$1 \text{ Hz} \equiv 1 \text{ cycle per second} = 1 \text{ s}^{-1}$$

We will frequently deal with very rapid oscillations and make use of the units shown in Table 14.1.

NOTE ▶ Uppercase and lowercase letters *are* important. 1 MHz is 1 megahertz = 10^6 Hz, but 1 mHz would be 1 millihertz = 10^{-3} Hz! ◀

TABLE 14.1 Units of frequency

Frequency	Period
10^3 Hz = 1 kilohertz = 1 kHz	1 ms
10^6 Hz = 1 megahertz = 1 MHz	1 μs
10^9 Hz = 1 gigahertz = 1 GHz	1 ns

EXAMPLE 14.1 Frequency and period of a radio station
What is the oscillation period for the broadcast of a 100 MHz FM radio station?

SOLVE The frequency of current oscillations in the radio transmitter is 100 MHz = 1.0×10^8 Hz. The period is the inverse of the frequency, hence

$$T = \frac{1}{f} = \frac{1}{1.0 \times 10^8 \text{ Hz}} = 1.0 \times 10^{-8} \text{ s} = 10 \text{ ns}$$

A system can oscillate in many ways, but we will be especially interested in the smooth *sinusoidal* oscillation of Figure 14.1c. This sinusoidal oscillation, the most basic of all oscillatory motions, is called **simple harmonic motion,** often abbreviated SHM. Let's look at a graphical description before we dive into the mathematics of simple harmonic motion.

Figure 14.2a shows an air-track glider attached to a spring. If the glider is pulled out a few centimeters and released, it will oscillate back and forth on the nearly frictionless air track. Figure 14.2b shows actual results from an experiment in which a computer was used to measure the glider's position 20 times every second. This is a position-versus-time graph that has been rotated 90° from its usual orientation in order for the x-axis to match the motion of the glider.

The object's maximum displacement from equilibrium is called the **amplitude** A of the motion. The object's position oscillates between $x = -A$ and $x = +A$. When using a graph, notice that the amplitude is the distance from the *axis* to the maximum, *not* the distance from the minimum to the maximum.

Figure 14.3a shows the data with the graph axes in their "normal" positions. You can see that the amplitude in this experiment was $A = 0.17$ m, or 17 cm. You can also measure the period to be $T = 1.60$ s. Thus the oscillation frequency was $f = 1/T = 0.625$ Hz.

Figure 14.3b is a velocity-versus-time graph that the computer produced by using $\Delta x/\Delta t$ to find the slope of the position graph at each point. The velocity graph is also sinusoidal, oscillating between $-v_{max}$ (maximum speed to the left) and $+v_{max}$ (maximum speed to the right). As the figure shows,

- The instantaneous velocity is zero at the points where $x = \pm A$. These are the *turning points* in the motion.
- The maximum speed v_{max} is reached as the object passes through the equilibrium position at $x = 0$ m. The *velocity* is positive as the object moves to the right but *negative* as it moves to the left.

We can ask three important questions about this oscillating system:

1. How is the maximum speed v_{max} related to the amplitude A?
2. How are the period and frequency related to the object's mass m, the spring constant k, and the amplitude A?
3. Is the sinusoidal oscillation a consequence of Newton's laws?

A mass oscillating on a spring is the prototype of simple harmonic motion. Our analysis, in which we answer these questions, will be of a spring-mass system. Even so, most of what we learn will be applicable to other types of SHM.

Kinematics of Simple Harmonic Motion

Figure 14.4 on the next page redraws the position-versus-time graph of Figure 14.3 as a smooth curve. Although these are empirical data (we don't yet have any "theory" of oscillation) the position-versus-time graph is clearly a cosine function. We can write the object's position as

$$x(t) = A\cos\left(\frac{2\pi t}{T}\right) \tag{14.2}$$

where the notation $x(t)$ indicates that the position x is a *function* of time t. Because $\cos(2\pi) = \cos(0)$, it's easy to see that the position at time $t = T$ is the same as the position at $t = 0$. In other words, this is a cosine function with period T. Be sure to convince yourself that this function agrees with the five special points shown in Figure 14.4.

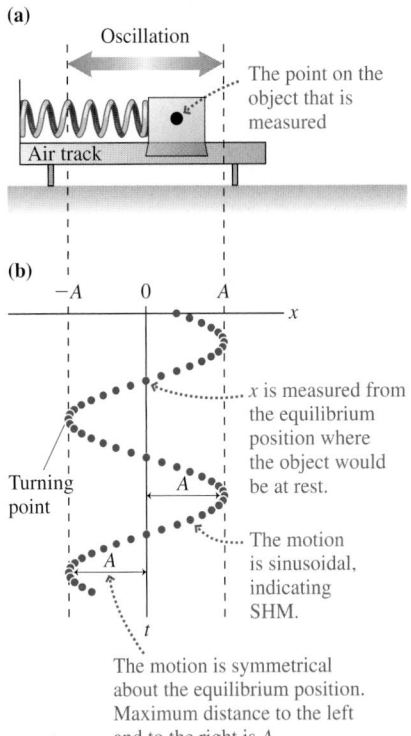

FIGURE 14.2 A prototype simple-harmonic-motion experiment.

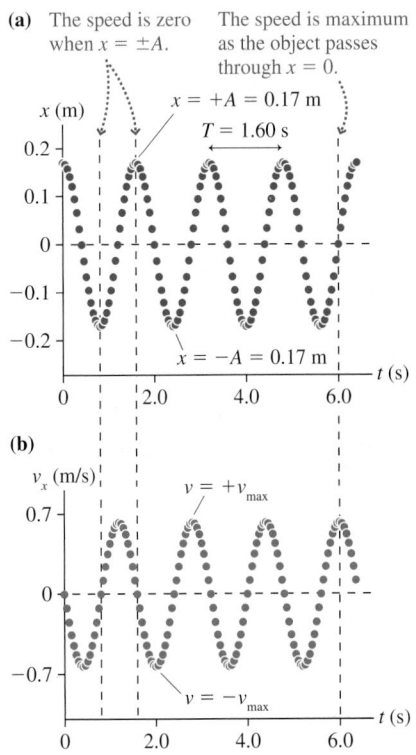

FIGURE 14.3 Position and velocity graphs of the experimental data.

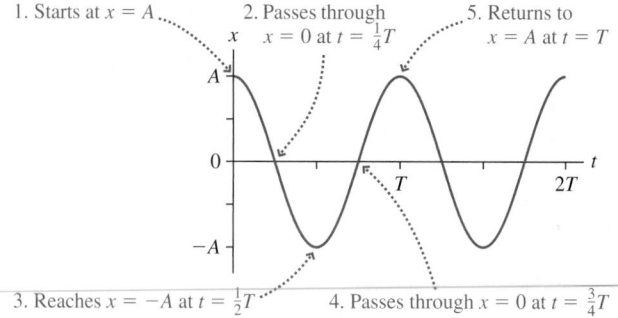

FIGURE 14.4 The position-versus-time graph for simple harmonic motion.

NOTE ▶ The argument of the cosine function is in *radians*. That will be true throughout this chapter. It's especially important to remember to set your calculator to radian mode before working oscillation problems. Leaving it in degree mode will lead to major errors. ◀

We can write Equation 14.2 in two alternative forms. Because the oscillation frequency is $f = 1/T$, we can write

$$x(t) = A\cos(2\pi ft) \tag{14.3}$$

Recall from Chapter 7 that a particle in circular motion has an *angular velocity* ω that is related to the period by $\omega = 2\pi/T$, where ω is in rad/s. Now that we've defined the frequency f, you can see that ω and f are related by

$$\omega \text{ (in rad/s)} = \frac{2\pi}{T} = 2\pi f \text{ (in Hz)} \tag{14.4}$$

In this context, ω is called the **angular frequency.** The position can be written in terms of ω as

$$x(t) = A\cos\omega t \tag{14.5}$$

Equations 14.2, 14.3, and 14.5 are equivalent ways to write the position of an object moving in simple harmonic motion.

Just as the position graph was clearly a cosine function, the velocity graph shown in Figure 14.5 is clearly an "upside-down" sine function with the same period T. The velocity v_x, which is a function of time, can be written

$$v_x(t) = -v_{\max}\sin\left(\frac{2\pi t}{T}\right) = -v_{\max}\sin(2\pi ft) = -v_{\max}\sin\omega t \tag{14.6}$$

NOTE ▶ $v_{\max}$ is the maximum *speed* and thus is inherently a *positive* number. The minus sign in Equation 14.6 is needed to turn the sine function upside down. ◀

We deduced Equation 14.6 from the experimental results, but we could equally well find it from the position function of Equation 14.2. After all, velocity is the time derivative of position. Table 14.2 reminds you of the derivatives of sine and cosine functions. Using the derivative of the cosine function, we find

$$v_x(t) = \frac{dx}{dt} = -\frac{2\pi A}{T}\sin\left(\frac{2\pi t}{T}\right) = -2\pi fA\sin(2\pi ft) = -\omega A\sin\omega t \tag{14.7}$$

We can draw an important conclusion by comparing Equation 14.7, the mathematical definition of velocity, to Equation 14.6, the empirical description of the

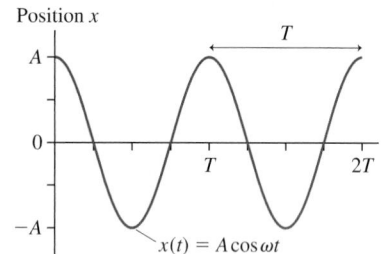

Position x

$x(t) = A\cos\omega t$

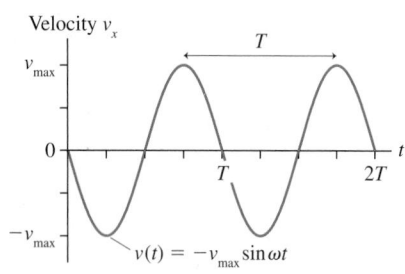

Velocity v_x

$v(t) = -v_{\max}\sin\omega t$

FIGURE 14.5 Position and velocity graphs for simple harmonic motion.

TABLE 14.2 Derivatives of sine and cosine functions

$$\frac{d}{dt}(a\sin(bt + c)) = +ab\cos(bt + c)$$

$$\frac{d}{dt}(a\cos(bt + c)) = -ab\sin(bt + c)$$

velocity. Namely, the maximum speed of an oscillation is related to the oscillation amplitude by

$$v_{max} = \frac{2\pi A}{T} = 2\pi f A = \omega A \qquad (14.8)$$

Equation 14.8 answers the first question we posed above, which was how the maximum speed v_{max} is related to the amplitude A. Not surprisingly, the object moves faster if you stretch the spring further and give the oscillation a larger amplitude.

EXAMPLE 14.2 A system in simple harmonic motion

An air-track glider is attached to a spring, pulled 20 cm to the right, and released at $t = 0$. It makes 15 oscillations in 10 s.

a. What is the period of oscillation?
b. What is the object's maximum speed?
c. What are the position and velocity at $t = 0.80$ s?

MODEL An object oscillating on a spring is in simple harmonic motion.

SOLVE

a. The oscillation frequency is

$$f = \frac{15 \text{ oscillations}}{10 \text{ s}} = 1.5 \text{ oscillations/s} = 1.5 \text{ Hz}$$

Thus the period is $T = 1/f = 0.667$ s.

b. The oscillation amplitude is $A = 0.20$ m. Thus the maximum speed is

$$v_{max} = \frac{2\pi A}{T} = \frac{2\pi(0.20 \text{ m})}{0.667 \text{ s}} = 1.88 \text{ m/s}$$

c. The object starts at $x = +A$ at $t = 0$. This is exactly the oscillation described by Equations 14.2 and 14.6. The position at $t = 0.80$ s is

$$x = A\cos\left(\frac{2\pi t}{T}\right) = (0.20 \text{ m})\cos\left(\frac{2\pi(0.80 \text{ s})}{0.667 \text{ s}}\right)$$

$$= (0.20 \text{ m})\cos(7.54 \text{ rad}) = 0.062 \text{ m} = 6.2 \text{ cm}$$

The velocity at this instant of time is

$$v_x = -v_{max}\sin\left(\frac{2\pi t}{T}\right) = -(1.88 \text{ m/s})\sin\left(\frac{2\pi(0.80 \text{ s})}{0.667 \text{ s}}\right)$$

$$= -(1.88 \text{ m/s})\sin(7.54 \text{ rad}) = -1.79 \text{ m/s} = -179 \text{ cm/s}$$

At $t = 0.80$ s, which is slightly more than one period, the object is 6.2 cm to the right of equilibrium and moving to the *left* at 179 cm/s. Notice the use of radians in the calculations.

EXAMPLE 14.3 Finding the time

A mass oscillating in simple harmonic motion starts at $x = A$ and has period T. At what time, as a fraction of T, does the object first pass through $x = \frac{1}{2}A$?

SOLVE Figure 14.4 noted that the object passes through the equilibrium position $x = 0$ at $t = \frac{1}{4}T$. This is one-quarter of the total distance in one-quarter of a period. You might expect it to take $\frac{1}{8}T$ to reach $\frac{1}{2}A$, but this is *not* the case because the SHM graph is not linear between $x = A$ and $x = 0$. We need to use $x(t) = A\cos(2\pi t/T)$. First, write the equation with $x = \frac{1}{2}A$:

$$x = \frac{A}{2} = A\cos\left(\frac{2\pi t}{T}\right)$$

Then solve for the time at which this position is reached:

$$t = \frac{T}{2\pi}\cos^{-1}\left(\frac{1}{2}\right) = \frac{T}{2\pi}\frac{\pi}{3} = \frac{1}{6}T$$

ASSESS The motion is slow at the beginning and then speeds up, so it takes longer to move from $x = A$ to $x = \frac{1}{2}A$ than it does to move from $x = \frac{1}{2}A$ to $x = 0$. Notice that the answer is independent of the amplitude A.

STOP TO THINK 14.1 An object moves with simple harmonic motion. If the amplitude and the period are both doubled, the object's maximum speed is

a. quadrupled.
d. halved.

b. doubled.
e. quartered.

c. unchanged.

14.2 Simple Harmonic Motion and Circular Motion

(a)

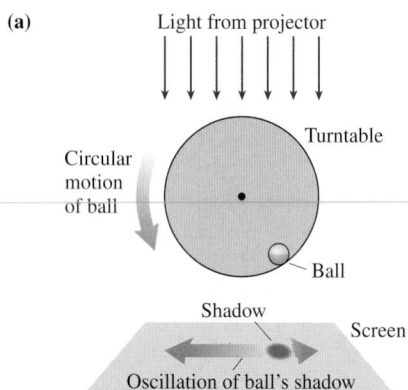

(b)

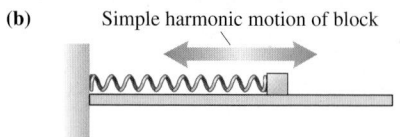

FIGURE 14.6 A projection of the circular motion of a rotating ball matches the simple harmonic motion of an object on a spring.

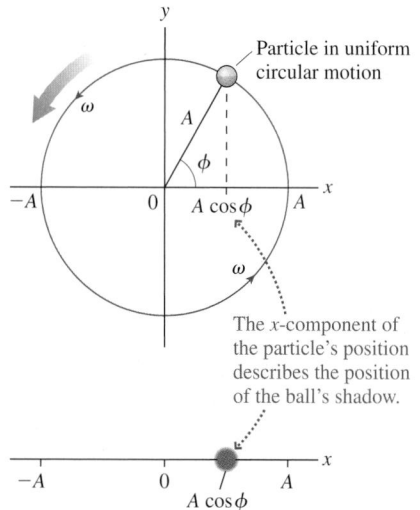

FIGURE 14.7 A particle in uniform circular motion with radius A and angular velocity ω.

The graphs of Figure 14.5 and the position function $x(t) = A\cos\omega t$ are for an oscillation in which the object just happened to be at $x_0 = A$ at $t = 0$. But you will recall that $t = 0$ is an arbitrary choice, the instant of time when you or someone else starts a stopwatch. What if you had started the stopwatch when the object was at $x_0 = -A$, or when the object was somewhere in the middle of an oscillation? In other words, what if the oscillator had different *initial conditions*. The position graph would still show an oscillation, but neither Figure 14.5 nor $x(t) = A\cos\omega t$ would describe the motion correctly.

To learn how to describe the oscillation for other initial conditions it will help to turn to a topic you studied in Chapter 7—circular motion. There's a very close connection between simple harmonic motion and circular motion.

Imagine you have a turntable with a small ball glued to the edge. Figure 14.6a shows how to make a "shadow movie" of the ball by projecting a light past the ball and onto a screen. The ball's shadow oscillates back and forth as the turntable rotates. This is certainly periodic motion, with the same period as the turntable, but is it simple harmonic motion?

To find out, you could place a real object on a real spring directly below the shadow, as shown in Fig, 14.6b. If you did so, and if you adjusted the turntable to have the same period as the spring, you would find that the shadow's motion exactly matches the simple harmonic motion of the object on the spring. **Uniform circular motion projected onto one dimension is simple harmonic motion.**

To understand this, consider the particle in Figure 14.7. It is in uniform circular motion, moving *counterclockwise* in a circle with radius A. As in Chapter 7, we can locate the particle by the angle ϕ measured ccw from the x-axis. Projecting the ball's shadow onto a screen in Figure 14.6 is equivalent to observing just the x-component of the particle's motion. Figure 14.7 shows that the x-component, when the particle is at angle ϕ, is

$$x = A\cos\phi \tag{14.9}$$

Recall that the particle's *angular velocity*, in rad/s, is

$$\omega = \frac{d\phi}{dt} \tag{14.10}$$

This is the rate at which the angle ϕ is increasing. If the particle starts from $\phi_0 = 0$ at $t = 0$, its angle at a later time t is simply

$$\phi = \omega t \tag{14.11}$$

As ϕ increases, the particle's x-component is

$$x(t) = A\cos\omega t \tag{14.12}$$

This is identical to Equation 14.5 for the position of a mass on a spring! Thus the x-component of a particle in uniform circular motion is simple harmonic motion.

NOTE ▶ When used to describe oscillatory motion, ω is called the *angular frequency* rather than the angular velocity. The angular frequency of an oscillator has the same numerical value, in rad/s, as the angular velocity of the corresponding particle in circular motion. ◀

The names and units can be a bit confusing until you get used to them. It may help to notice that *cycle* and *oscillation* are not true units. Unlike the "standard meter" or the "standard kilogram," to which you could compare a length or a mass, there is no "standard cycle" to which you can compare an oscillation. Cycles and oscillations are simply counted events. Thus the frequency f has units of hertz, where $1 \text{ Hz} = 1 \text{ s}^{-1}$. We may *say* "cycles per second" just to be clear, but the actual units are only "per second."

The radian is the SI unit of angle. However, the radian is a *defined* unit. Further, its definition as a ratio of two lengths ($\theta = s/r$) makes it a *pure number* without dimensions. As we noted in Chapter 7, the unit of angle, be it radians or degrees, is really just a *name* to remind us that we're dealing with an angle. The 2π in Equation 14.12 (and in many similar situations), which is stated without units, *means* 2π rad/cycle. When multiplied by the frequency f in cycles/s, it gives the frequency in rad/s. That is why, in this context, ω is called the angular *frequency*.

NOTE ▶ *Hertz* is specifically "cycles per second" or "oscillations per second." It is used for f but *not* for ω. We'll always be careful to use rad/s for ω, but you should be aware that many books give the units of ω as simply s^{-1}. ◀

The Phase Constant

Now we're ready to consider the issue of other initial conditions. The particle in Figure 14.7 started at $\phi_0 = 0$. This was equivalent to an oscillator starting at the far right edge, $x_0 = A$. Figure 14.8 shows a more general situation in which the initial angle ϕ_0 can have any value. The angle at a later time t is then

$$\phi = \omega t + \phi_0 \tag{14.13}$$

In this case, the particle's projection onto the x-axis at time t is

$$x(t) = A\cos(\omega t + \phi_0) \tag{14.14}$$

If Equation 14.14 describes the particle's projection, then it must also be the position of an oscillator in simple harmonic motion. The oscillator's velocity v_x is found by taking the derivative dx/dt. The resulting equations,

$$x(t) = A\cos(\omega t + \phi_0)$$
$$v_x(t) = -\omega A \sin(\omega t + \phi_0) = -v_{max}\sin(\omega t + \phi_0) \tag{14.15}$$

are the two primary kinematic equations of simple harmonic motion.

The quantity $\phi = \omega t + \phi_0$, which steadily increases with time, is called the **phase** of the oscillation. The phase is simply the *angle* of the circular-motion particle whose shadow matches the oscillator. The constant ϕ_0 is called the **phase constant**. It specifies the *initial conditions* of the oscillator.

To see what the phase constant means, set $t = 0$ in Equations 14.15:

$$x_0 = A\cos\phi_0$$
$$v_{0x} = -\omega A \sin\phi_0 \tag{14.16}$$

The position x_0 and velocity v_{0x} at $t = 0$ are the initial conditions. **Different values of the phase constant correspond to different starting points on the circle and thus to different initial conditions.**

The perfect cosine function of Figure 14.5 and the equation $x(t) = A\cos\omega t$ are for an oscillation with $\phi_0 = 0$ rad. You can see from Equations 14.16 that $\phi_0 = 0$ rad implies $x_0 = A$ and $v_0 = 0$. That is, the particle starts from rest at the point of maximum displacement.

Figure 14.9 on the next page illustrates these ideas by looking at three values of the phase constant: $\phi_0 = \pi/3$ rad (60°), $-\pi/3$ rad (−60°), and π rad (180°). For each value of ϕ_0 you see the oscillator at its starting position, the starting position shown on a circle, and both position and velocity graphs. All the graphs have the same amplitude and the same period, but they are *shifted* relative to the graphs of Figure 14.5 (which were for $\phi_0 = 0$ rad) so that the maximum displacement $x = A$ occurs at a time other than $t = 0$.

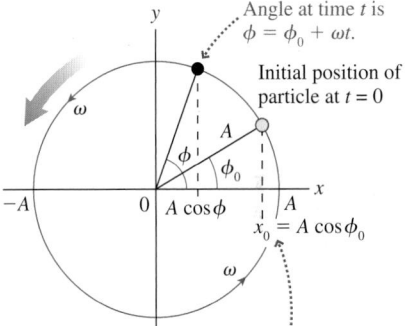

The initial x-component of the particle's position can be anywhere between $-A$ and A, depending on ϕ_0.

FIGURE 14.8 A particle in uniform circular motion with initial angle ϕ_0.

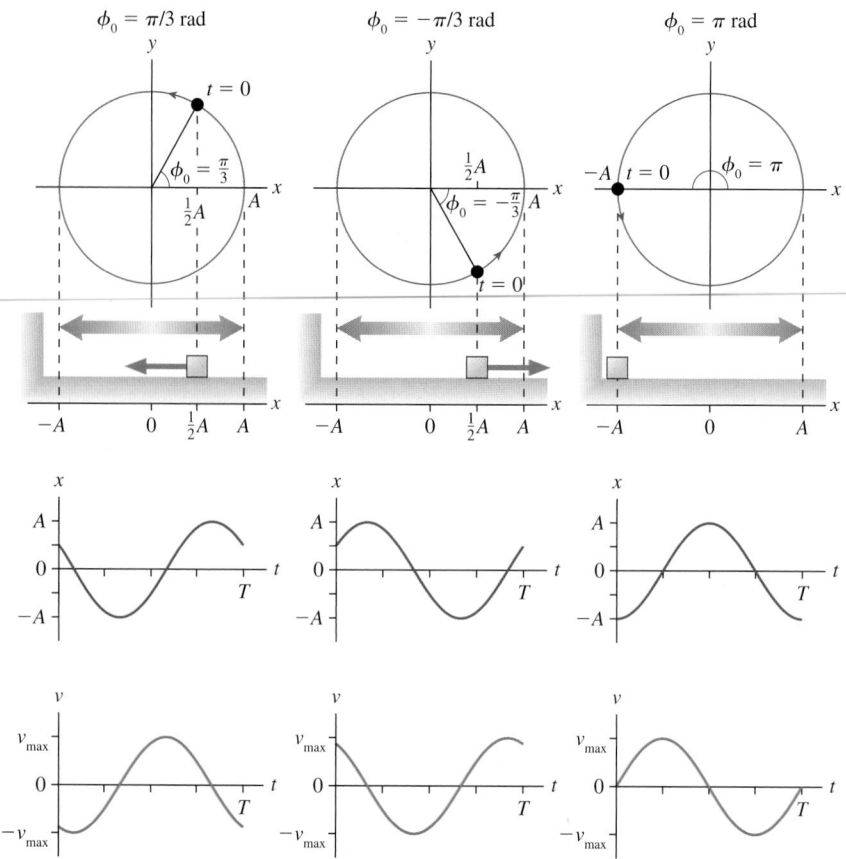

FIGURE 14.9 Oscillations described by the phase constants $\phi_0 = \pi/3$ rad, $-\pi/3$ rad, and π rad.

Notice that $\phi_0 = \pi/3$ rad and $\phi_0 = -\pi/3$ rad have the same starting position, $x_0 = \frac{1}{2}A$. This is a property of the cosine function in Equation 14.16. But these are *not* the same initial conditions. In one case the oscillator starts at $\frac{1}{2}A$ while moving to the right, in the other case it starts at $\frac{1}{2}A$ while moving to the left. You can distinguish between the two by visualizing the circular motion.

All values of the phase constant ϕ_0 between 0 and π rad correspond to a particle in the upper half of the circle and *moving to the left*. Thus v_{0x} is negative. All values of the phase constant ϕ_0 between π and 2π rad (or, as they are usually stated, between $-\pi$ and 0 rad) have the particle in the lower half of the circle and *moving to the right*. Thus v_{0x} is positive. If you're told that the oscillator is at $x = \frac{1}{2}A$ and moving to the right at $t = 0$, then the phase constant must be $\phi_0 = -\pi/3$ rad, not $+\pi/3$ rad.

EXAMPLE 14.4 Using the initial conditions

An object on a spring oscillates with a period of 0.80 s and an amplitude of 10 cm. At $t = 0$ s, it is 5.0 cm to the left of equilibrium and moving to the left. What are its position and direction of motion at $t = 2.0$ s?

MODEL An object oscillating on a spring is in simple harmonic motion.

SOLVE We can find the phase constant ϕ_0 from the initial condition $x_0 = -5.0$ cm $= A\cos\phi_0$. This condition gives:

$$\phi_0 = \cos^{-1}\left(\frac{x_0}{A}\right) = \cos^{-1}\left(-\frac{1}{2}\right) = \pm\frac{2}{3}\pi \text{ rad} = \pm 120°$$

Because the oscillator is moving to the *left* at $t = 0$, it is in the upper half of the circular-motion diagram and must have a phase constant between 0 and π rad. Thus ϕ_0 is $\frac{2}{3}\pi$ rad. The angular frequency is

$$\omega = \frac{2\pi}{T} = \frac{2\pi}{0.80 \text{ s}} = 7.854 \text{ rad/s}$$

Thus the object's position at time $t = 2.0$ s is

$$x(t) = A\cos(\omega t + \phi_0)$$

$$= (10 \text{ cm})\cos\left((7.854 \text{ rad/s})(2.0 \text{ s}) + \frac{2}{3}\pi\right)$$

$$= (10 \text{ cm})\cos(17.80 \text{ rad})$$

$$= 5.00 \text{ cm}$$

The object is now 5.0 cm to the right of equilibrium. But which way is it moving? There are two ways to find out. The direct way is to calculate the velocity at $t = 2.0$ s:

$$v_x = -\omega A \sin(\omega t + \phi_0) = +68.1 \text{ cm/s}$$

The velocity is positive, so the motion is to the right. Alternatively, we could note that the phase at $t = 2.0$ s is $\phi = 17.80$ rad. Dividing by π, you can see that

$$\phi = 17.80 \text{ rad} = 5.67\pi \text{ rad} = (4\pi + 1.67\pi) \text{ rad}$$

The 4π rad represents complete revolutions. The "extra" phase of 1.67π rad falls between π and 2π rad, so the particle in the circular-motion diagram is in the lower half of the circle and moving to the right.

NOTE ▶ The inverse-cosine function $\cos^{-1}$ is a *two-valued* function. Your calculator returns a single value, an angle between 0 rad and π rad. But the negative of this angle is also a solution. As Example 14.4 demonstrates, you must use additional information to choose between them. ◀

STOP TO THINK 14.2   The figure shows four oscillators at $t = 0$. Which one has the phase constant $\phi_0 = \pi/4$ rad?

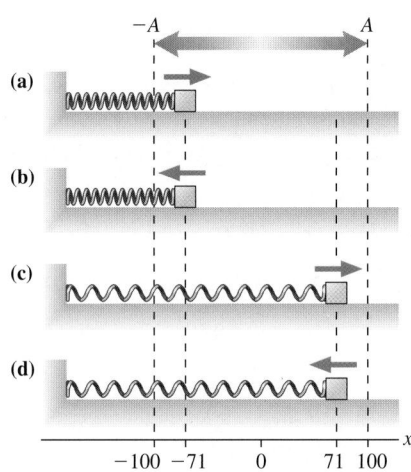

14.3 Energy in Simple Harmonic Motion

We've begun to develop the mathematical language of simple harmonic motion, but thus far we haven't included any physics. We've made no mention of the mass of the object or the spring constant of the spring. An energy analysis, using the tools of Chapters 10 and 11, is a good starting place. We already have all of the information we need because we found the elastic potential energy of a spring in Section 10.5.

 9.3

Figure 14.10a on the next page shows an object oscillating on a spring, our prototype of simple harmonic motion. Now we'll specify that the object has mass m, the spring has spring constant k, and the motion takes place on a frictionless surface. You learned in Chapter 10 that the elastic potential energy when the object is at position x is $U_s = \frac{1}{2}k(\Delta x)^2$, where $\Delta x = x - x_e$ is the displacement

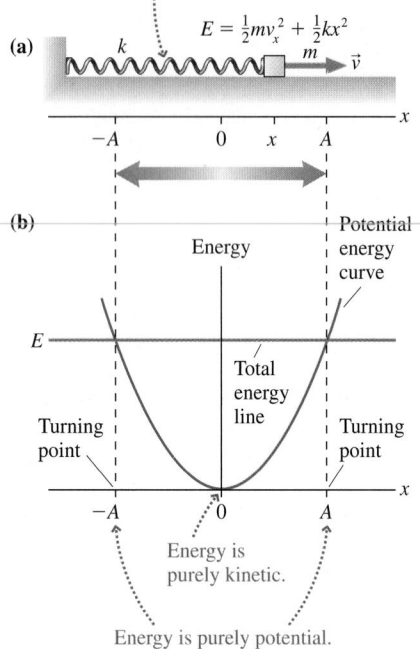

Energy is transformed between kinetic and potential, but the total mechanical energy E doesn't change.

(a)

$E = \frac{1}{2}mv_x^2 + \frac{1}{2}kx^2$

k m $\vec{v}$

$-A$ 0 x A x

(b)

Energy

Potential energy curve

E

Total energy line

Turning point

Turning point

$-A$ 0 A x

Energy is purely kinetic.

Energy is purely potential.

FIGURE 14.10 The energy is transformed between kinetic energy and potential energy as the object oscillates, but the mechanical energy $E = K + U$ doesn't change.

from the equilibrium position x_e. In this chapter we'll always use a coordinate system in which $x_e = 0$, making $\Delta x = x$. There's no chance for confusion with gravitational potential energy, so we can omit the subscript s and write the elastic potential energy as

$$U = \frac{1}{2}kx^2 \tag{14.17}$$

Thus the mechanical energy of an object oscillating on a spring is

$$E = K + U = \frac{1}{2}mv_x^2 + \frac{1}{2}kx^2 \tag{14.18}$$

Figure 14.10b is an energy diagram, showing the potential-energy curve $U = \frac{1}{2}kx^2$ as a parabola. Recall that a particle oscillates between the *turning points* where the total energy line E crosses the potential-energy curve. The left turning point is at $x = -A$ and the right turning point at $x = +A$. To go beyond these points would require a negative kinetic energy, which is physically impossible.

You can see that **the particle has purely potential energy at $x = \pm A$ and purely kinetic energy as it passes through the equilibrium point at $x = 0$.** At maximum displacement, with $x = \pm A$ and $v = 0$, the energy is

$$E(\text{at } x = \pm A) = U = \frac{1}{2}kA^2 \tag{14.19}$$

At $x = 0$, where $v_x = \pm v_{max}$, the energy is

$$E(\text{at } x = 0) = K = \frac{1}{2}mv_{max}^2 \tag{14.20}$$

These two cases are indicated on Figure 14.10.

The system's mechanical energy is conserved, because the surface is frictionless and there are no external forces, so the energy at maximum displacement and the energy at maximum speed, Equations 14.19 and 14.20, must be equal. That is

$$\frac{1}{2}m(v_{max})^2 = \frac{1}{2}kA^2 \tag{14.21}$$

From Equation 14.21 we can see that the maximum speed is related to the amplitude by

$$v_{max} = \sqrt{\frac{k}{m}}A \tag{14.22}$$

This is a relationship based on the physics of the situation.

Earlier, using kinematics, we found that

$$v_{max} = \frac{2\pi A}{T} = 2\pi fA = \omega A \tag{14.23}$$

Comparing Equations 14.22 and 14.23, we see that frequency and period of an oscillating spring are determined by the spring constant k and the object's mass m:

$$\omega = \sqrt{\frac{k}{m}} \qquad f = \frac{1}{2\pi}\sqrt{\frac{k}{m}} \qquad T = 2\pi\sqrt{\frac{m}{k}} \tag{14.24}$$

These three expressions are really only one equation. They say the same thing, but each expresses it in slightly different terms.

Equation 14.24 is the answer to the second question we posed at the beginning of the chapter, where we asked how the period and frequency are related to the object's mass m, the spring constant k, and the amplitude A. It is perhaps surprising, but **the period and frequency do not depend on the amplitude A.** A small oscillation and a large oscillation have the same period.

Because energy is conserved, we can combine Equations 14.18, 14.19, and 14.20 to write

$$E = \frac{1}{2}mv_x^2 + \frac{1}{2}kx^2 = \frac{1}{2}kA^2 = \frac{1}{2}m(v_{max})^2 \quad \text{(conservation of energy)} \quad (14.25)$$

Any pair of these may be useful, depending on the known information. For example, you can use the amplitude A to find the speed at any point x by combining the first and second expressions for E. The speed v at position x is

$$v = \sqrt{\frac{k}{m}(A^2 - x^2)} = \omega\sqrt{A^2 - x^2} \quad (14.26)$$

Similarly, you can use the first and second expressions to find the amplitude from the initial conditions x_0 and v_0:

$$A = \sqrt{x_0^2 + \frac{mv_0^2}{k}} = \sqrt{x_0^2 + \left(\frac{v_0}{\omega}\right)^2} \quad (14.27)$$

Figure 14.11 shows graphically how the kinetic and potential energy change with time. They both oscillate but remain *positive* because x and v_x are squared. Energy is continuously being transformed back and forth between the kinetic energy of the moving block and the stored potential energy of the spring, but their sum remains constant. Notice that K and U both oscillate *twice* each period; make sure you understand why.

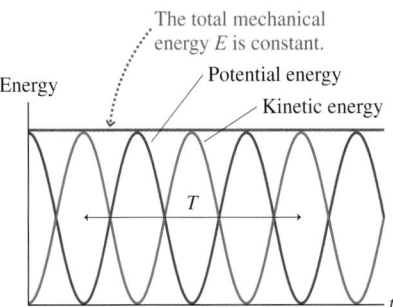

FIGURE 14.11 Kinetic energy, potential energy, and the total mechanical energy for simple harmonic motion.

EXAMPLE 14.5 Using conservation of energy
A 500 g block on a spring is pulled a distance of 20 cm and released. The subsequent oscillations are measured to have a period of 0.80 s. At what position or positions is the block's speed 1.0 m/s?

MODEL The motion is simple harmonic motion. Energy is conserved.

SOLVE The block starts from the point of maximum displacement, where $E = U = \frac{1}{2}kA^2$. At a later time, when the position is x and the velocity is v_x, energy conservation requires

$$\frac{1}{2}mv_x^2 + \frac{1}{2}kx^2 = \frac{1}{2}kA^2$$

Solving for x, we find

$$x = \sqrt{A^2 - \frac{mv_x^2}{k}} = \sqrt{A^2 - \left(\frac{v}{\omega}\right)^2}$$

where we used $k/m = \omega^2$ from Equation 14.24. The angular frequency is easily found from the period: $\omega = 2\pi/T = 7.85$ rad/s. Thus

$$x = \sqrt{(0.20 \text{ m})^2 - \left(\frac{1.0 \text{ m/s}}{7.85 \text{ rad/s}}\right)^2} = \pm0.154 \text{ m} = \pm15.4 \text{ cm}$$

There are two positions because the block has this speed on either side of equilibrium.

STOP TO THINK 14.3   Four springs have been compressed from their equilibrium position at $x = 0$ cm. When released, they will start to oscillate. Rank in order, from highest to lowest, the maximum speeds of the oscillators.

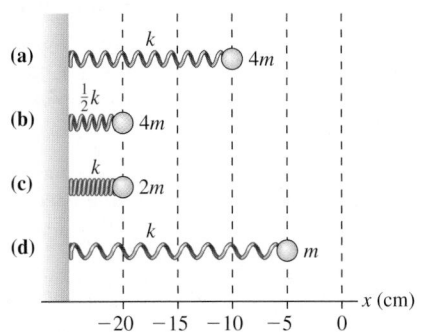

14.4 The Dynamics of Simple Harmonic Motion

9.1, 9.2 Actiｖ
Physｉcs
ONLINE

Our analysis thus far has been based on the experimental observation that the oscillation of a spring "looks" sinusoidal. It's time to show that Newton's second law *predicts* sinusoidal motion.

A motion diagram will help us visualize the object's acceleration. Figure 14.12 shows one cycle of the motion, separating motion to the left and motion to the right to make the diagram clear. As you can see, the object's velocity is large as it passes through the equilibrium point at $x = 0$, but $\vec{v}$ is *not changing* at that point. Acceleration measures the *change* of the velocity, hence $\vec{a} = \vec{0}$ at $x = 0$.

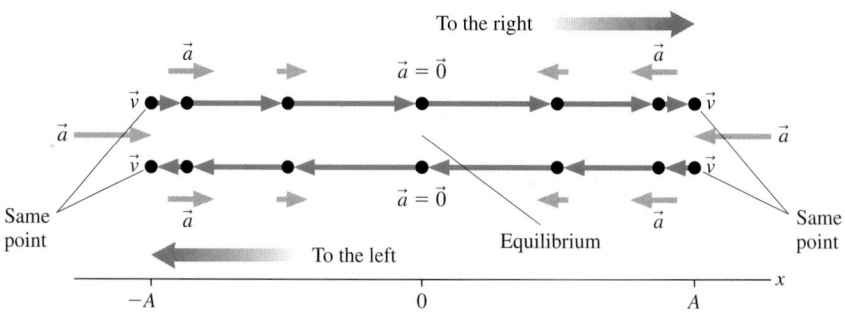

FIGURE 14.12 Motion diagram of simple harmonic motion. The left and right motions are separated for clarity but really occur along the same line.

On the other hand, the velocity is changing rapidly at the turning points. At the right turning point, $\vec{v}$ changes from a right-pointing vector to a left-pointing vector. Thus the acceleration $\vec{a}$ at the right turning point is large and *to the left*. In one-dimensional motion, the acceleration component a_x has a large *negative* value at the right turning point. Similarly, the acceleration $\vec{a}$ at the left turning point is large and *to the right*. Consequently, a_x has a large positive value at the left turning point.

NOTE ▶ This is exactly the same motion-diagram analysis we used in Chapters 1 and 2 to determine the acceleration at the turning point of a ball tossed straight up. ◀

Our motion diagram analysis suggests that the acceleration a_x is a maximum (most positive) when the displacement is most negative, a minimum (most negative) when the displacement is a maximum, and zero when $x = 0$. This is confirmed by taking the derivative of the velocity,

$$a_x = \frac{dv_x}{dt} = \frac{d}{dt}(-\omega A \sin \omega t) = -\omega^2 A \cos \omega t \qquad (14.28)$$

then graphing it.

Figure 14.13 shows the position graph that we started with in Figure 14.4 and the corresponding acceleration graph. Comparing the two, you can see that the acceleration graph looks like an upside-down position graph. In fact, because $x = A \cos \omega t$, Equation 14.28 for the acceleration can be written

$$a_x = -\omega^2 x \qquad (14.29)$$

That is, **the acceleration is proportional to the *negative* of the displacement.** The acceleration is, indeed, maximum when the displacement is most negative and minimum when the displacement is most positive.

Our interest in the acceleration is that the acceleration is related to the net force by Newton's second law. Consider again our prototype mass on a spring, shown in Figure 14.14. This is the simplest possible oscillation, with no distractions due

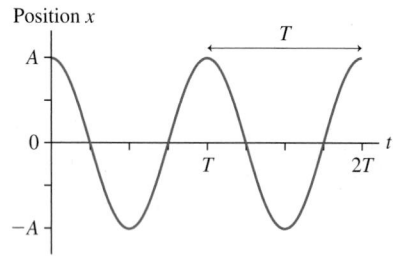

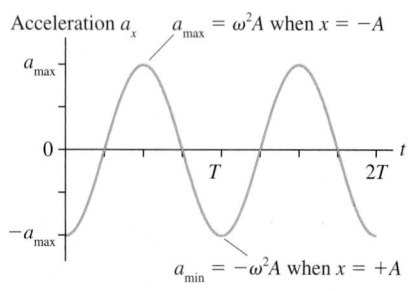

FIGURE 14.13 Position and acceleration graphs for an oscillating spring. We've chosen $\phi_0 = 0$.

to friction or gravitational forces. We will assume the spring itself to be massless, as we have done previously. The only force acting on the object in the x-direction is the spring force $\vec{F}_{sp}$.

As you learned in Chapter 10, the spring force is given by Hooke's law:

$$(F_{sp})_x = -k\Delta x \qquad (14.30)$$

The minus sign indicates that the spring force is a **restoring force,** a force that always points back toward the equilibrium position. If we place the origin of the coordinate system at the equilibrium position, as we've done throughout this chapter, then $\Delta x = x$ and Hooke's law is simply $(F_{sp})_x = -kx$.

The x-component of Newton's second law for the object attached to the spring is

$$(F_{net})_x = (F_{sp})_x = -kx = ma_x \qquad (14.31)$$

Equation 14.31 is easily rearranged to read

$$a_x = -\frac{k}{m}x \qquad (14.32)$$

You can see that Equation 14.32 is identical to Equation 14.29 if the system oscillates with angular frequency $\omega = \sqrt{k/m}$. We previously found this expression for ω from an energy analysis. Our experimental observation that the acceleration is proportional to the *negative* of the displacement is exactly what Hooke's law would lead us to expect. That's the good news.

The bad news is that a_x is not a constant. As the object's position changes, so does the acceleration. Nearly all of our kinematic tools have been based on constant acceleration. We can't use those tools to analyze oscillations, so we must go back to the very definition of acceleration:

$$a_x = \frac{dv_x}{dt} = \frac{d^2x}{dt^2}$$

Acceleration is the second derivative of position with respect to time. If we use this definition in Equation 14.32, it becomes

$$\frac{d^2x}{dt^2} = -\frac{k}{m}x \qquad \text{(equation of motion for a mass on a spring)} \qquad (14.33)$$

Equation 14.33, which is called the **equation of motion,** is a second-order differential equation. Unlike other equations we've dealt with, Equation 14.33 cannot be solved by direct integration. We'll need to take a different approach.

Solving the Equation of Motion

The solution to an algebraic equation such as $x^2 = 4$ is a number. The solution to a differential equation is a *function*. The x in Equation 14.33 is really $x(t)$, the position as a function of time. The solution to this equation is a function $x(t)$ whose second derivative is the function itself multiplied by $(-k/m)$.

One important property of differential equations that you will learn about in math is that the solutions are *unique*. That is, there is only *one* solution to Equation 14.33. If we were able to *guess* a solution, the uniqueness property would tell us that we had found the *only* solution. That might seem a rather strange way to solve equations, but in fact differential equations are frequently solved by using your knowledge of what the solution needs to look like to guess an appropriate functional form. Let us give it a try!

We know from experimental evidence that the oscillatory motion of a spring appears to be sinusoidal. Let us *guess* that the solution to Equation 14.33 should have the functional form

$$x(t) = A\cos(\omega t + \phi_0) \qquad (14.34)$$

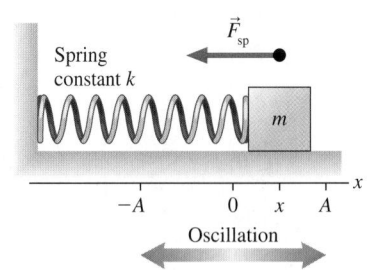

FIGURE 14.14 The prototype of simple harmonic motion: a mass oscillating on a horizontal spring without friction.

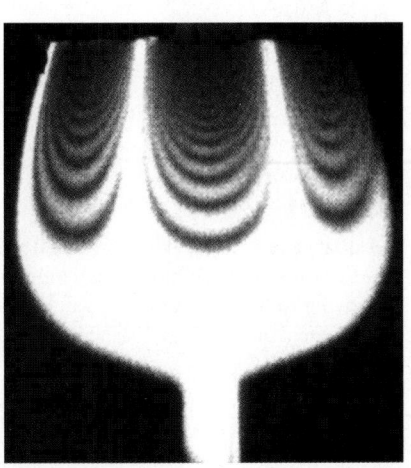

An optical technique called *interferometry* reveals the bell-like vibrations of a wine glass.

where A, ω, and ϕ_0 are unspecified constants that we can adjust to any values that might be necessary to satisfy the differential equation.

If you were to guess that a solution to the algebraic equation $x^2 = 4$ is $x = 2$, you would verify your guess by substituting it into the original equation to see if it works. We need to do the same thing here: substitute our guess for $x(t)$ into Equation 14.33 to see if, for an appropriate choice of the three constants, it works. To do so, we need the second derivative of $x(t)$. That is straightforward:

$$x(t) = A\cos(\omega t + \phi_0)$$

$$\frac{dx}{dt} = -\omega A\sin(\omega t + \phi_0) \qquad (14.35)$$

$$\frac{d^2x}{dt^2} = -\omega^2 A\cos(\omega t + \phi_0)$$

If we now substitute the first and third of Equations 14.35 into Equation 14.33 we find

$$-\omega^2 A\cos(\omega t + \phi_0) = -\frac{k}{m}A\cos(\omega t + \phi_0) \qquad (14.36)$$

Equation 14.36 will be true at all instants of time if and only if $\omega^2 = k/m$. There do not seem to be any restrictions on the two constants A and ϕ_0.

So we have found—by guessing!—that *the* solution to the equation of motion for a mass oscillating on a spring is

$$x(t) = A\cos(\omega t + \phi_0) \qquad (14.37)$$

where the angular frequency

$$\omega = 2\pi f = \sqrt{\frac{k}{m}} \qquad (14.38)$$

is determined by the mass and the spring constant.

> NOTE ▶ Once again we see that the oscillation frequency is independent of the amplitude A. ◀

Equations 14.37 and 14.38 seem somewhat anticlimactic because we've been using these results for the last several pages. But keep in mind that we had been *assuming* $x = A\cos\omega t$ simply because the experimental observations "looked" like a cosine function. We've now justified that assumption by showing that Equation 14.37 really is the solution to Newton's second law for a mass on a spring. **The *theory* of oscillation, based on Hooke's law for a spring and Newton's second law, is in good agreement with the experimental observations.** This conclusion gives an affirmative answer to the last of the three questions that we asked early in the chapter, which was whether the sinusoidal oscillation of SHM is a consequence of Newton's laws.

EXAMPLE 14.6 Analyzing an oscillator

At $t = 0$ s, a 500 g block oscillating on a spring is observed moving to the right at $x = 15$ cm. It reaches a maximum displacement of 25 cm at $t = 0.300$ s.

a. Draw a position-versus-time graph for one cycle of the motion.
b. At what times during the first cycle does the mass pass through $x = 20$ cm?

MODEL The motion is simple harmonic motion.

SOLVE

a. The position equation of the block is $x(t) = A\cos(\omega t + \phi_0)$. We know that the amplitude is $A = 0.25$ m and that $x_0 = 0.15$ m. From these two pieces of information we obtain the phase constant:

$$\phi_0 = \cos^{-1}\left(\frac{x_0}{A}\right) = \cos^{-1}(.60) = \pm 0.927 \text{ rad}$$

The object is initially moving to the right, which tells us that the phase constant must be between $-\pi$ and 0 rad. Thus $\phi_0 = -0.927$ rad. The block reaches its maximum displacement $x_{max} = A$ at time $t = 0.30$ s. At that instant of time

$$x_{max} = A = A\cos(\omega t + \phi_0)$$

This can be true only if $\cos(\omega t + \phi_0) = 1$, which requires $\omega t + \phi_0 = 0$. Thus

$$\omega = \frac{-\phi_0}{t} = \frac{-(-0.927\ \text{rad})}{(0.300\ \text{s})} = 3.09\ \text{rad/s}$$

Now that we know ω, it is straightforward to compute the period

$$T = \frac{2\pi}{\omega} = 2.03\ \text{s}$$

Figure 14.15 graphs $x(t) = (25\ \text{cm})\cos(3.09t - 0.927)$, where t is in s, from $t = 0$ s to $t = 2.03$ s.

b. From $x = A\cos(\omega t + \phi_0)$, the time at which the mass reaches position $x = 20$ cm is

$$t = \frac{1}{\omega}\left(\cos^{-1}\left(\frac{x}{A}\right) - \phi_0\right)$$

$$= \frac{1}{3.09\ \text{rad/s}}\left(\cos^{-1}\left(\frac{20\ \text{cm}}{25\ \text{cm}}\right) + 0.927\ \text{rad}\right) = 0.508\ \text{s}$$

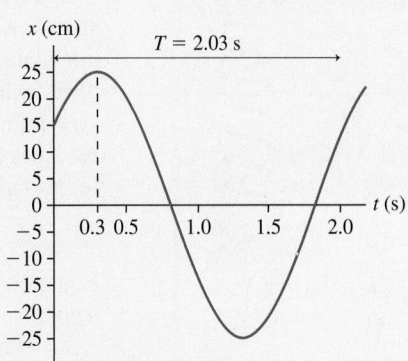

FIGURE 14.15 Position-versus-time graph for the oscillator of Example 14.6.

A calculator returns only one value of $\cos^{-1}$, in the range 0 to π rad, but we noted earlier that $\cos^{-1}$ actually has two values. Indeed, you can see in Figure 14.15 that there are two times at which the mass passes $x = 20$ cm. Because they are symmetrical on either side of $t = 0.300$ s, when $x = A$, the first point is $(0.508\ \text{s} - 0.300\ \text{s}) = 0.208$ s *before* the maximum. Thus the mass passes through $x = 20$ cm at $t = 0.092$ s and again at $t = 0.508$ s.

STOP TO THINK 14.4 This is the position graph of a mass on a spring. What can you say about the velocity and the force at the instant indicated by the dotted line?

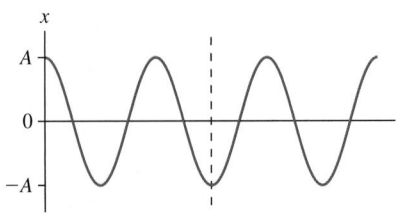

a. Velocity is positive; force is to the right.
b. Velocity is negative; force is to the right.
c. Velocity is zero; force is to the right.
d. Velocity is positive; force is to the left.
e. Velocity is negative; force is to the left.
f. Velocity is zero; force is to the left.
g. Velocity and force are both zero.

14.5 Vertical Oscillations

We have focused our analysis on a horizontally oscillating spring. But the typical demonstration you'll see in class is a mass bobbing up and down on a spring hung vertically from a support. Is it safe to assume that a vertical oscillation is the same as a horizontal oscillation? Or does the additional force of gravity change the motion? Let us look at this more carefully.

Activ Physics ONLINE 9.4, 9.5

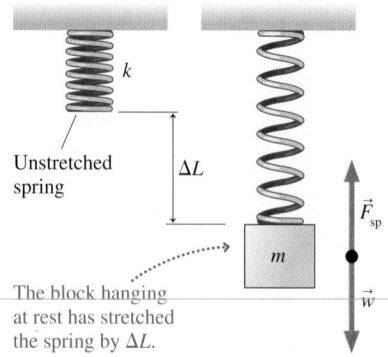

FIGURE 14.16 The weight of the block stretches the spring.

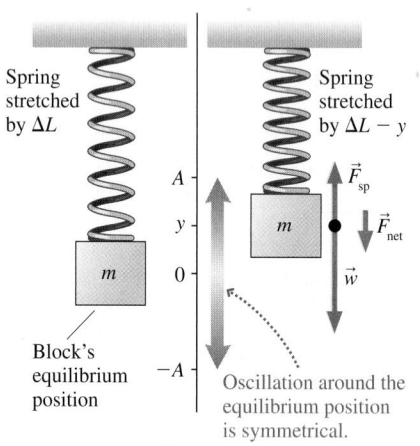

FIGURE 14.17 The block oscillates around the equilibrium position.

Figure 14.16 shows a block of mass m hanging from a spring of spring constant k. An important fact to notice is that the equilibrium position of the block is *not* where the spring is at its unstretched length. At the equilibrium position of the block, where it hangs motionless, the spring has stretched by ΔL.

Finding ΔL is a static-equilibrium problem in which the upward spring force balances the downward weight force of the block. The y-component of the spring force is given by Hooke's law,

$$(F_{sp})_y = -k\Delta y = +k\Delta L \qquad (14.39)$$

Equation 14.39 makes a distinction between ΔL, which is simply a *distance* and is a positive number, and the displacement Δy. The block is displaced downward, so $\Delta y = -\Delta L$. Newton's first law for the block in equilibrium is

$$(F_{net})_y = (F_{sp})_y + w_y = k\Delta L - mg = 0 \qquad (14.40)$$

from which we can find

$$\Delta L = \frac{mg}{k} \qquad (14.41)$$

This is the distance the spring stretches when the block is attached to it.

Let the block oscillate around this equilibrium position, as shown in Figure 14.17. We've now placed the origin of the y-axis at the block's equilibrium position in order to be consistent with our analyses of oscillations throughout this chapter. If the block moves upward, as the figure shows, the spring gets shorter compared to its equilibrium length, but the spring is still *stretched* compared to its unstretched length in Figure 14.16. When the block is at position y, the spring is stretched by an amount $\Delta L - y$ and hence exerts an *upward* spring force $F_{sp} = k(\Delta L - y)$. The net force on the block at this point is

$$(F_{net})_y = (F_{sp})_y + w_y = k(\Delta L - y) - mg = (k\Delta L - mg) - ky \quad (14.42)$$

But $k\Delta L - mg$ is zero, from Equation 14.41, so the net force on the block is simply

$$(F_{net})_y = -ky \qquad (14.43)$$

Equation 14.43 for vertical oscillations is *exactly* the same as Equation 14.31 for horizontal oscillations, where we found $(F_{net})_x = -kx$. That is, the restoring force for vertical oscillations is identical to the restoring force for horizontal oscillations. The role of gravity is to determine where the equilibrium position is, but it doesn't affect the motion around the equilibrium position.

Because the net force is the same, Newton's second law will have exactly the same oscillatory solution:

$$y(t) = A\cos(\omega t + \phi_0) \qquad (14.44)$$

with, again, $\omega = \sqrt{k/m}$. The vertical oscillations of a mass on a spring are the same simple harmonic motion as that of a block on a horizontal spring. This is an important finding because it was not obvious that the motion would still be simple harmonic motion when gravity was included. Because the motions are the same, **everything we have learned about horizontal oscillations is equally valid for vertical oscillations.**

EXAMPLE 14.7 **A vertical oscillation**

A 200 g block hangs from a spring with spring constant 10 N/m. The block is pulled down to a point where the spring is 30 cm longer than its unstretched length, then released. Where is the block and what is its velocity 3.0 s later?

MODEL Vertical oscillations are simple harmonic motion.

VISUALIZE Figure 14.18 shows the situation.

SOLVE Although the spring begins by being stretched 30 cm, this is *not* the amplitude of the oscillation. Oscillations occur around the equilibrium position, so we have to begin by finding

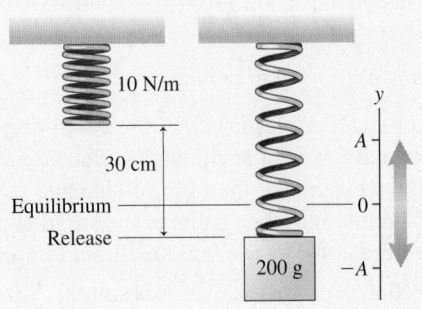

FIGURE 14.18 A block oscillates about its equilibrium position.

the equilibrium point where the block hangs motionless on the spring. The spring stretch at equilibrium is given by Equation 14.41,

$$\Delta L = \frac{mg}{k} = 0.196 \text{ m} = 19.6 \text{ cm}$$

Stretching the spring 30 cm pulls the block 10.4 cm below the equilibrium point, so $A = 10.4$ cm. That is, the block oscillates with an amplitude $A = 10.4$ cm about a point that is 19.6 cm beneath the spring's original end point. The block's position as a function of time, as measured from the equilibrium position, is

$$y(t) = (10.4 \text{ cm}) \cos(\omega t + \phi_0)$$

where $\omega = \sqrt{k/m} = 7.07 \sqrt{\text{s}}$. The initial condition

$$y_0 = -A = A \cos \phi_0$$

requires the phase constant to be $\phi_0 = \pi$ rad. At $t = 3.0$ s the block's position is

$$y = (10.4 \text{ cm}) \cos((7.07 \text{ rad/s})(3.0 \text{ s}) + \pi \text{ rad}) = 7.4 \text{ cm}$$

The block is 7.4 cm *above* the equilibrium position, or 12.2 cm *below* the original end of the spring. Its velocity at this instant is

$$v_x = -\omega A \sin(\omega t + \phi_0) = 52 \text{ cm/s}$$

14.6 The Pendulum

Now let's look at another very common oscillator: a pendulum. Figure 14.19a shows a mass m attached to a string of length L and free to swing back and forth. The pendulum's position can be described by the arc of length s, which is zero when the pendulum hangs straight down. Because angles are measured ccw, s and θ are positive when the pendulum is to the right of center, negative when it is to the left.

Two forces are acting on the mass: the string tension $\vec{T}$ and the weight $\vec{w}$. It will be convenient to do what we did in our study of circular motion: divide the forces into tangential components, parallel to the motion, and radial components parallel to the string. These are shown on the free-body diagram of Figure 14.19b.

Newton's second law for the tangential component, parallel to the motion, is

$$(F_{net})_t = \sum F_t = w_t = -mg \sin \theta = ma_t \qquad (14.45)$$

Using $a_t = d^2s/dt^2$ for acceleration "around" the circle, and noting that the mass cancels, we can write Equation 14.45 as

$$\frac{d^2s}{dt^2} = -g \sin \theta \qquad (14.46)$$

where angle θ is related to the arc length by $\theta = s/L$. This is the equation of motion for an oscillating pendulum. The sine function makes this equation more complicated than the equation of motion for an oscillating spring.

The Small-Angle Approximation

Suppose we restrict the pendulum's oscillations to *small angles* of less than about 10°. This restriction allows us to make use of an interesting and important piece of geometry.

Figure 14.20 on the next page shows an angle θ and a circular arc of length $s = r\theta$. A right triangle has been constructed by dropping a perpendicular from the top of the arc to the axis. The height of the triangle is $h = r \sin \theta$. Suppose that

(a)

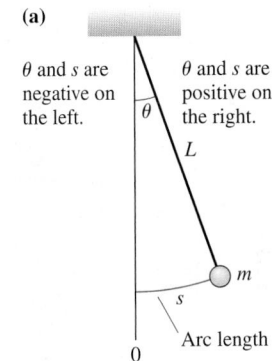

θ and s are negative on the left.

θ and s are positive on the right.

L

m

s

Arc length

0

(b)

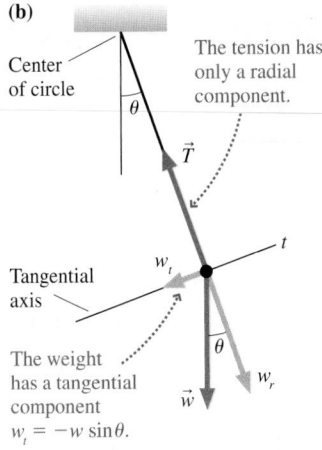

Center of circle

The tension has only a radial component.

$\vec{T}$

Tangential axis

t

w_t

The weight has a tangential component $w_t = -w \sin \theta$.

$\vec{w}$

w_r

FIGURE 14.19 The motion of a pendulum.

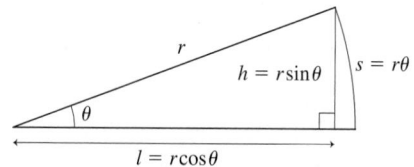

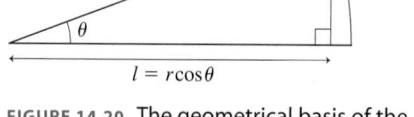

FIGURE 14.20 The geometrical basis of the small-angle approximation.

TABLE 14.3 Small-angle approximations. θ must be in radians.

$\sin\theta \approx \theta$
$\cos\theta \approx 1$
$\tan\theta \approx \sin\theta \approx \theta$

9.10–9.12   Activ Physics

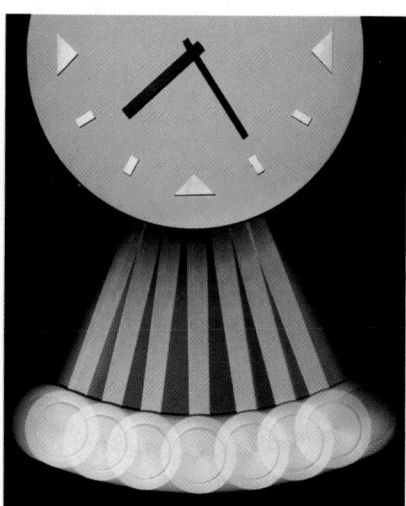

The pendulum clock has been used for hundreds of years.

the angle θ is "small." In that case there is very little difference between h and s. If $h \approx s$, then $r\sin\theta \approx r\theta$. It follows that

$$\sin\theta \approx \theta \qquad (\theta \text{ in radians})$$

The result that $\sin\theta \approx \theta$ for small angles is called the **small-angle approximation.** We can similarly note that $l \approx r$ for small angles. Because $l = r\cos\theta$, it follows that $\cos\theta \approx 1$. Finally, we can take the ratio of sine and cosine to find $\tan\theta \approx \sin\theta \approx \theta$. Table 14.3 summarizes the results of the small-angle approximation. We will have other occasions to use the small-angle approximation throughout the remainder of this text.

NOTE ▶ The small-angle approximation is valid *only* if the angle θ is in radians! ◀

How small does θ have to be to justify using the small-angle approximation? It's easy to use your calculator to find that the small-angle approximation is good to three significant figures, an error of $\leq 0.1\%$, up to angles of ≈ 0.10 rad ($\approx 5°$). In practice, we will use the approximation up to about $10°$, but for angles any larger it rapidly loses validity and produces unacceptable results.

If we restrict the pendulum to $\theta < 10°$, we can use the small-angle approximation to write $\sin\theta \approx \theta = s/L$. In that case, Equation 14.45 for the net force on the mass is

$$(F_{\text{net}})_t = -(mg/L)s$$

and the equation of motion becomes

$$\frac{d^2s}{dt^2} = -\frac{g}{L}s \qquad (14.47)$$

This is *exactly* the same equation as Equation 14.33 for a mass oscillating on a spring. The names are different, with x replaced by s and k/m by g/L, but that does not make it a different equation.

Because we know the solution to the spring problem, we can immediately write the solution to the pendulum problem just by changing variables and constants:

$$s(t) = A\cos(\omega t + \phi_0) \quad \text{or} \quad \theta(t) = \theta_{\text{max}}\cos(\omega t + \phi_0) \qquad (14.48)$$

The angular frequency

$$\omega = 2\pi f = \sqrt{\frac{g}{L}} \qquad (14.49)$$

is determined by the length of the string. The pendulum is interesting in that **the frequency, and hence the period, is independent of the mass.** It depends only on the length of the pendulum. The amplitude A and the phase constant ϕ_0 are determined by the initial conditions, just as they were for an oscillating spring.

EXAMPLE 14.8 A pendulum clock
What length pendulum has a period of exactly 1 s?

SOLVE The period is independent of the mass and depends only on the length. From Equation 14.49,

$$T = \frac{1}{f} = 2\pi\sqrt{\frac{L}{g}}$$

Solving for L, we find

$$L = g\left(\frac{T}{2\pi}\right)^2 = 0.248 \text{ m}$$

ASSESS This is a convenient length for a practical clock.

EXAMPLE 14.9 The maximum angle of a pendulum
A 300 g mass on a 30-cm-long string oscillates as a pendulum. It has a speed of 0.25 m/s as it passes through the lowest point. What maximum angle does the pendulum reach?

MODEL Assume that the angle remains small, in which case the motion is simple harmonic motion.

SOLVE The angular frequency of the pendulum is

$$\omega = \sqrt{\frac{g}{L}} = \sqrt{\frac{9.8 \text{ m/s}^2}{0.30 \text{ m}}} = 5.72 \text{ rad/s}$$

The speed at the lowest point is $v_{max} = \omega A$, so the amplitude is

$$A = s_{max} = \frac{v_0}{\omega} = \frac{0.25 \text{ m/s}}{5.72 \text{ rad/s}} = 0.0437 \text{ m}$$

The maximum angle, at the maximum arc length s_{max}, is

$$\theta_{max} = \frac{s_{max}}{L} = \frac{0.04347 \text{ m}}{0.30 \text{ m}} = 0.145 \text{ rad} = 8.3°$$

ASSESS Because the maximum angle is less than 10°, our analysis based on the small-angle approximation is valid.

The Conditions for Simple Harmonic Motion

You can begin to see how, in a sense, we have solved *all* simple-harmonic-motion problems once we have solved the problem of the horizontal spring. The restoring force of a spring, $F_{sp} = -kx$, is directly proportional to the displacement x from equilibrium. The pendulum's restoring force, in the small-angle approximation, is directly proportional to the displacement s. A restoring force that is directly proportional to the displacement from equilibrium is called a **linear restoring force.** For *any* linear restoring force, the equation of motion is identical to the spring equation (other than perhaps using different symbols). Consequently, **any system with a linear restoring force will undergo simple harmonic motion around the equilibrium position.**

This is why an oscillating spring is the prototype of SHM. Everything that we learn about an oscillating spring can be applied to the oscillations of any other linear restoring force, ranging from the vibration of airplane wings to the motion of electrons in electric circuits. Let's summarize this information with a Tactics Box.

Activ
Physics ONLINE 9.6–9.9

TACTICS BOX 14.1 Identifying and analyzing simple harmonic motion

❶ If the net force acting on a particle is a linear restoring force, the motion will be simple harmonic motion around the equilibrium position.

❷ The position as a function of time is $x(t) = A\cos(\omega t + \phi_0)$. The velocity as a function of time is $v_x(t) = -\omega A \sin(\omega t + \phi_0)$. The maximum speed is $v_{max} = \omega A$. The equations are given here in terms of x, but they can be written in terms of y, θ, or some other parameter if the situation calls for it.

❸ The amplitude A and the phase constant ϕ_0 are determined by the initial conditions through $x_0 = A\cos\phi_0$ and $v_{0x} = -\omega A \sin\phi_0$.

❹ The angular frequency ω (and hence the period $T = 2\pi/\omega$) depends on the physics of the particular situation. But ω does *not* depend on A or ϕ_0.

❺ Mechanical energy is conserved. Thus $\frac{1}{2}mv_x^2 + \frac{1}{2}kx^2 = \frac{1}{2}kA^2 = \frac{1}{2}m(v_{max})^2$. Energy conservation provides a relationship between position and velocity that is independent of time.

STOP TO THINK 14.5 One person swings on a swing and finds that the period is 3.0 s. A second person of equal mass joins him. With two people swinging, the period is

a. 6.0 s.
b. >3.0 s but not necessarily 6.0 s.
c. 3.0 s.
d. <3.0 s but not necessarily 1.5 s.
e. 1.5 s.
f. Can't tell without knowing the length.

The shock absorbers in cars and trucks are heavily damped springs. The vehicle's vertical motion, after hitting a rock or a pothole, is a damped oscillation.

14.7 Damped Oscillations

A pendulum left to itself gradually slows down and stops. The sound of a ringing bell gradually dies away. All real oscillators do run down—some very slowly but others quite quickly—as friction or other dissipative forces transform their mechanical energy into the thermal energy of the oscillator and its environment. An oscillation that runs down and stops is called a **damped oscillation.**

There are many possible reasons for the dissipation of energy: air resistance, friction, internal forces within the metal of the spring as it flexes, and so on. While it would be impractical to account for all of these, a reasonable model is to consider only air resistance because, in many cases, it will be the predominant dissipative force.

The drag force of air resistance is a complex force. There is no "law of air resistance" to tell us exactly how big air-resistance forces are. Chapter 5 introduced a *model* of air resistance in which the drag force was proportional to v^2. That's a good model when velocities are reasonably high, as they are for runners, baseballs, and cars, but a pendulum or an oscillating spring usually moves much more slowly. It is known from experiments that the drag force on *slowly* moving objects is *linearly* proportional to the velocity. Thus a reasonable model of the drag force for a slowly moving object is

$$\vec{D} = -b\vec{v} \qquad \text{(model of the drag force)} \qquad (14.50)$$

where the minus sign is the mathematical statement that the force is always opposite in direction to the velocity in order to slow the object.

The **damping constant** b depends in a complicated way on the shape of the object (long, narrow objects have less air resistance than wide, flat ones) *and* on the viscosity of the air or other medium in which the particle moves. The damping constant plays the same role in our model of air resistance that the coefficient of friction does in our model of friction.

The units of b need to be such that they will give units of force when multiplied by units of velocity. As you can confirm, these units are kg/s. A value $b = 0$ kg/s corresponds to the limiting case of no resistance, in which case the mechanical energy is conserved. A typical value of b for a spring or a pendulum in air is ≤ 0.10 kg/s. Poorly shaped objects or objects moving in a liquid (which is much more viscous than air) can have significantly larger values of b.

Figure 14.21 shows a mass oscillating on a spring in the presence of a drag force. With the drag included, Newton's second law is

$$(F_{\text{net}})_x = (F_{\text{sp}})_x + D_x = -kx - bv_x = ma_x \qquad (14.51)$$

Using $v_x = dx/dt$ and $a_x = d^2x/dt^2$, we can write Equation 14.51 as

$$\frac{d^2x}{dt^2} + \frac{b}{m}\frac{dx}{dt} + \frac{k}{m}x = 0 \qquad (14.52)$$

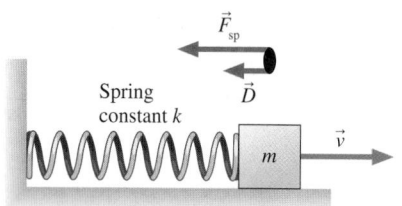

FIGURE 14.21 An oscillating mass in the presence of a drag force.

Equation 14.52 is the equation of motion of a damped oscillator. If you compare it to Equation 14.33, the equation of motion for a block on a frictionless surface, you'll see it differs by the inclusion of the term involving dx/dt.

Equation 14.52 is another second-order differential equation. We will simply assert (and, as a homework problem, you can confirm) that the solution is

$$x(t) = Ae^{-bt/2m}\cos(\omega t + \phi_0) \qquad \text{(damped oscillator)} \qquad (14.53)$$

where the angular frequency is given by

$$\omega = \sqrt{\frac{k}{m} - \frac{b^2}{4m^2}} = \sqrt{\omega_0^2 - \frac{b^2}{4m^2}} \qquad (14.54)$$

Here $\omega_0 = \sqrt{k/m}$ is the angular frequency of an undamped oscillator ($b = 0$). The constant e is the base of natural logarithms, so $e^{-bt/2m}$ is an *exponential function*.

Notice that Equation 14.53 reduces to our previous solution, $x(t) = A\cos(\omega t + \phi_0)$, when $b = 0$. This makes sense and gives us confidence in Equation 14.53. A *lightly damped* system, which oscillates many times before stopping, is one for which $b/2m \ll \omega_0$. In that case, $\omega \approx \omega_0$ is a good approximation. That is, light damping does not affect the oscillation frequency. (*Heavy damping*, which stops the motion within a few oscillations, causes the oscillation frequency to be significantly lowered.) We will focus on lightly damped systems for the rest of this section.

Figure 14.22 is a graph of the position $x(t)$ for a lightly damped oscillator, as given by Equation 14.53. Notice that the term $Ae^{-bt/2m}$, which is shown by the dashed line, acts as a slowly varying amplitude:

$$x_{max}(t) = Ae^{-bt/2m} \qquad (14.55)$$

where A is the *initial* amplitude, at $t = 0$. The oscillation keeps bumping up against this line, slowly dying out with time.

A slowly changing line that provides a border to a rapid oscillation is called the **envelope** of the oscillations. In this case, the oscillations have an *exponentially decaying envelope*. Figure 14.22 shows how real oscillations decay with time. Make sure you study Figure 14.22 long enough to see how both the oscillations and the decaying amplitude are related to Equation 14.53.

Changing the amount of damping, by changing the value of b, affects how quickly the oscillations decay. Figure 14.23 shows just the envelope $x_{max}(t)$ for several oscillators that are identical except for the value of the damping constant b. (You need to imagine a rapid oscillation within each envelope, as in Figure 14.22). Increasing b causes the oscillations to damp more quickly, while decreasing b makes them last longer.

Energy in Damped Systems

When considering the oscillator's mechanical energy, it is useful to define the **time constant** τ (Greek tau) to be

$$\tau = \frac{m}{b} \qquad (14.56)$$

Because b has units of kg/s, τ has units of seconds. With this definition, we can write the oscillation amplitude as $x_{max}(t) = Ae^{-t/2\tau}$.

Because of the drag force, the mechanical energy is no longer conserved. At any particular time we can compute the mechanical energy from

$$E(t) = \frac{1}{2}kx_{max}^2 = \frac{1}{2}k(Ae^{-t/2\tau})^2 = \left(\frac{1}{2}kA^2\right)e^{-t/\tau} = E_0 e^{-t/\tau} \qquad (14.57)$$

where $E_0 = \frac{1}{2}kA^2$ is the initial energy at $t = 0$ and where we used $(z^m)^2 = z^{2m}$. In other words, **the oscillator's mechanical energy decays exponentially with time constant τ.**

As Figure 14.24 shows, the time constant is the amount of time needed for the energy to decay to e^{-1}, or 37%, of its initial value. By $t = 2\tau$, the energy has been reduced to $E(\text{at } 2\tau) = E_0 e^{-2} = 0.13E_0$. We say that the time constant τ measures the "characteristic time" during which the energy of the oscillation is dissipated. Roughly two-thirds of the initial energy is gone after one time constant has elapsed and nearly 90% has dissipated after two time constants have gone by.

For practical purposes, we can speak of the time constant as the *lifetime* of the oscillation—about how long it lasts. An oscillator with $\tau = 3$ s will oscillate for

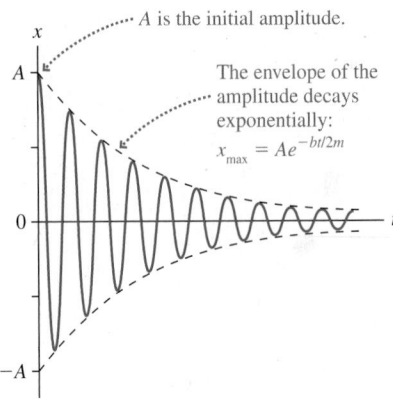

FIGURE 14.22 Position-versus-time graph for a damped oscillator.

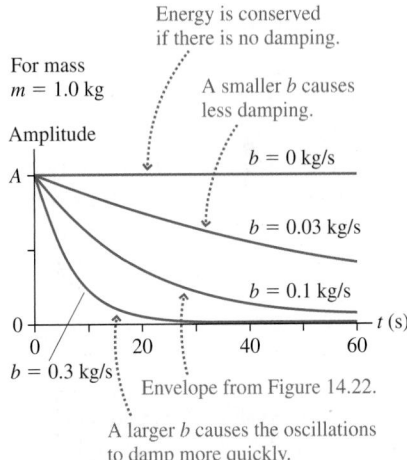

FIGURE 14.23 Several oscillation envelopes, corresponding to different values of the damping constant b.

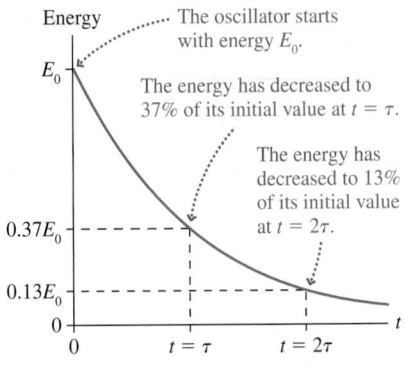

FIGURE 14.24 Exponential decay of the mechanical energy of an oscillator.

roughly 3 s, while one with $\tau = 30$ s will continue for roughly 30 s. Mathematically, there is never a time when the oscillation is "over." Equation 14.57 approaches zero asymptotically, but it never gets there in any finite time. Thus there is not any specific time t at which we can say "The oscillation is over." The best we can do is define a characteristic time when the motion is "almost over," and that is what the time constant τ does.

There are many, many examples of exponential decay in science and engineering. You will see several more in this text, and you will find others in more advanced courses. Wherever exponential decay occurs, the idea of a time constant is used to characterize how long the decay lasts. No matter what is decaying, it will have only 37% of its original value (roughly one-third) after a time interval of one time constant has elapsed.

EXAMPLE 14.10 A damped pendulum

A 500 g mass swings on a 60-cm-string as a pendulum. The amplitude is observed to decay to half its initial value after 35 s.

a. What is the time constant for this oscillator?
b. At what time will the *energy* have decayed to half its initial value?

MODEL The motion is a damped oscillation.

SOLVE

a. The initial amplitude at $t = 0$ is $x_{max} = A$. At $t = 35$ s the amplitude is $x_{max} = \frac{1}{2}A$. The amplitude of oscillation at time t is given by Equation 14.55:

$$x_{max}(t) = Ae^{-bt/2m} = Ae^{-t/2\tau}$$

In this case,

$$\frac{1}{2}A = Ae^{-(35\,s)/2\tau}$$

Notice that we do not need to know A itself because it cancels out. To solve for τ, take the natural logarithm of both sides of the equation:

$$\ln\left(\frac{1}{2}\right) = -\ln 2 = \ln e^{-(35\,s)/2\tau} = -\frac{(35\,s)}{2\tau}$$

This is easily rearranged to give

$$\tau = \frac{35\,s}{2\ln 2} = 25.2\,s$$

If desired, we could now determine the damping constant to be $b = m/\tau = 0.020$ kg/s.

b. The energy at time t is given by

$$E(t) = E_0 e^{-t/\tau}$$

The time at which an exponential decay is reduced to $\frac{1}{2}E_0$, half its initial value, has a special name. It is called the **half-life** and given the symbol $t_{1/2}$. The concept of the half-life is widely used in applications such as radioactive decay. To relate $t_{1/2}$ to τ, first write

$$E(\text{at } t = t_{1/2}) = \frac{1}{2}E_0 = E_0 e^{-t_{1/2}/\tau}$$

The E_0 cancels, giving

$$\frac{1}{2} = e^{-t_{1/2}/\tau}$$

Again, take the natural logarithm of both sides:

$$\ln\left(\frac{1}{2}\right) = -\ln 2 = \ln e^{-t_{1/2}/\tau} = -t_{1/2}/\tau$$

Finally, solve for $t_{1/2}$:

$$t_{1/2} = \tau \ln 2 = 0.693\tau$$

This result that $t_{1/2}$ is 69% of τ is valid for any exponential decay. In this particular problem, half the energy is gone at

$$t_{1/2} = (.693)(25.2\,s) = 17.5\,s$$

ASSESS The oscillator loses energy faster than it loses amplitude. This is what we should expect because the energy depends on the *square* of the amplitude.

STOP TO THINK 14.6 Rank in order, from largest to smallest, the time constants τ_a to τ_d of the decays shown in the figure.

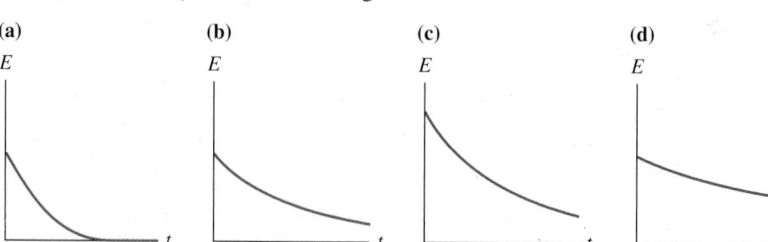

14.8 Driven Oscillations and Resonance

Thus far we have focused on the free oscillations of an isolated system. Some initial disturbance displaces the system from equilibrium, and it then oscillates freely until its energy is dissipated. These are very important situations, but they do not exhaust the possibilities. Another important situation is an oscillator that is subjected to a periodic external force. Its motion is called a **driven oscillation.**

A simple example of a driven oscillation is pushing a child on a swing, where your push is a periodic external force applied to the swing. A more complex example would be a car driving over a series of equally spaced bumps. Each bump causes a periodic upward force on the car's shock absorbers, which are big, heavily damped springs. The electromagnetic coil on the back of a loudspeaker cone provides a periodic magnetic force to drive the cone back and forth, causing it to send out sound waves. Air turbulence moving across the wings of an aircraft can exert periodic forces on the wings and other aerodynamic surfaces, causing them to vibrate if they are not properly designed.

As these examples suggest, driven oscillations have many important applications. However, driven oscillations are a mathematically complex subject. We will simply hint at some of the results, saving the details for more advanced classes.

Consider an oscillating system that, when left to itself, oscillates at a frequency f_0. We will call this the **natural frequency** of the oscillator. The natural frequency for a mass on a spring is $\sqrt{k/m}/2\pi$, but it might be given by some other expression for another type of oscillator. Regardless of the expression, f_0 is simply the frequency of the system if it is displaced from equilibrium and released.

Suppose that this system is subjected to a *periodic* external force of frequency f_{ext}. This frequency, which is called the **driving frequency,** is completely independent of the oscillator's natural frequency f_0. Somebody or something in the environment selects the frequency f_{ext} of the external force, causing the force to push on the system f_{ext} times every second.

Although it is possible to solve Newton's second law with an external driving force, we will be content to look at a graphical representation of the solution. The most important result is that the oscillation amplitude depends very sensitively on the frequency f_{ext} of the driving force. The response to the driving frequency is shown in Figure 14.25 for a system with $m = 1$ kg, a natural frequency $f_0 = 2$ Hz, and a damping constant $b = 0.20$ kg/s. This graph of amplitude versus driving frequency, called the **response curve,** occurs in many different applications.

When the driving frequency is substantially different from the oscillator's natural frequency, at the right and left edges of Figure 14.25, the system oscillates but its amplitude is very small. The system simply does not respond well to a driving frequency that differs much from f_0. As the driving frequency gets closer and closer to the natural frequency, the amplitude of the oscillation rises dramatically. After all, f_0 is the frequency at which the system "wants" to oscillate, so it is quite happy to respond to a driving frequency near f_0. Hence the amplitude reaches a maximum exactly when the driving frequency matches the system's natural frequency: $f_{ext} = f_0$.

You can understand this if you think about the energy. When f_{ext} matches f_0, the external force always pushes the oscillator at the same point in its cycle. For example, you always push a child on a swing just as the swing reaches its highest point on your side. Such push always *adds energy* to the system, pushing the amplitude higher.

By contrast, suppose you try to push a swing at some frequency other than its natural oscillation frequency. Sometimes you would push it as it goes forward, thus adding energy to the system, but in other cycles you would be trying to push it forward as it comes back. This would decelerate the swing and remove energy from the system. The net result would be a small amplitude. Only the frequency-matching condition builds up the amplitude.

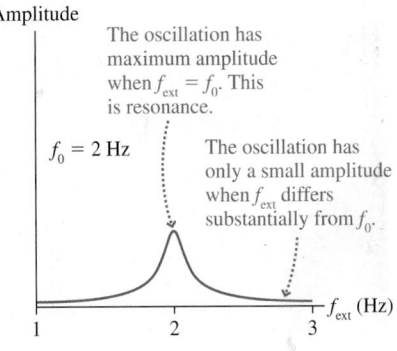

FIGURE 14.25 The response curve shows the amplitude of a driven oscillator at frequencies near its natural frequency $f_0 = 2$ Hz.

Amplitude

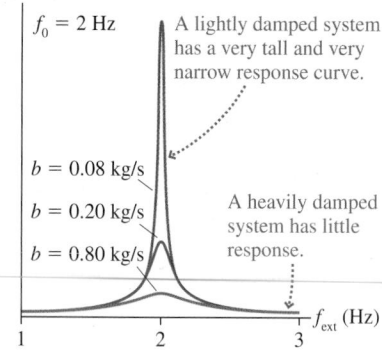

$f_0 = 2$ Hz

A lightly damped system has a very tall and very narrow response curve.

$b = 0.08$ kg/s

$b = 0.20$ kg/s

$b = 0.80$ kg/s

A heavily damped system has little response.

f_{ext} (Hz)

1   2   3

FIGURE 14.26 The resonance amplitude becomes higher and narrower as the damping constant decreases.

The amplitude can become exceedingly large when the frequencies match, especially if the damping constant is very small. Figure 14.26 shows the same oscillator with three different values of the damping constant. There's very little response if the damping constant is increased to 0.80 kg/s, but the amplitude for $f_{ext} = f_0$ becomes very large when the damping constant is reduced to 0.08 kg/s. This large-amplitude response to a driving force whose frequency matches the natural frequency of the system is a phenomenon called **resonance.** The condition for resonance is

$$f_{ext} = f_0 \quad \text{(resonance condition)} \quad (14.58)$$

Within the context of driven oscillations, the natural frequency f_0 is often called the **resonance frequency.**

There are many examples of resonance. We've seen that pushing a child on a swing is one. You may have heard that soldiers "break step" when crossing a footbridge, rather than marching in unison. The bridge has a natural frequency with which it bounces up and down. A group of soldiers marching across at exactly this frequency would cause a resonance, and potentially a disaster. A similar situation occurs when a singer is able to shatter a crystal goblet by singing the right note. The goblet has a natural frequency, which is the musical sound you hear if you lightly tap a wine glass. The singer causes a sound wave to impinge on the goblet, exerting a small driving force at the frequency of the note she is singing. If the singer's frequency matches the natural frequency of the goblet—resonance! The vibration amplitude may grow so large that the glass has a structural failure and shatters.

An important feature of Figure 14.26 is how the amplitude and width of the resonance depend on the damping constant. A heavily damped system responds fairly little, even at resonance, but it responds to a wide range of driving frequencies. Very lightly damped systems can reach exceptionally high amplitudes, but notice that the range of frequencies to which the system responds becomes narrower and narrower as b decreases.

This allows us to understand why singers can break crystal goblets but not inexpensive, everyday glasses. An inexpensive glass gives a "thud" when tapped, but a fine crystal goblet "rings" for several seconds. In physics terms, the goblet has a much longer time constant than the glass. That, in turn, implies that the goblet is very lightly damped while the ordinary glass is heavily damped (because the internal forces within the glass are not those of a high-quality crystal structure). Only the lightly damped goblet, like the top curve in Figure 14.26, can reach amplitudes large enough to shatter. The restriction, though, is that its natural frequency has to be matched very precisely, which is why such demonstrations are done by professional singers or musicians.

It is worth noting that there are many mechanical systems, such as the wings on airplanes, for which it is essential that resonances be avoided! It is important to understand the conditions of resonance in order to design structures without them.

A singer or musical instrument can shatter a crystal goblet by matching the goblet's natural oscillation frequency.

SUMMARY

The goal of Chapter 14 has been to understand systems that oscillate with simple harmonic motion.

GENERAL PRINCIPLES

Dynamics

SHM occurs when a linear restoring force acts to return a system to an equilibrium position.

Horizontal spring

$(F_{net})_x = -kx$

Vertical spring

The origin is at the equilibrium position $\Delta L = mg/k$.

$(F_{net})_y = -ky$

$\omega = \sqrt{\dfrac{k}{m}} \qquad T = 2\pi\sqrt{\dfrac{m}{k}}$

Pendulum

$(F_{net})_t = -\left(\dfrac{mg}{L}\right)s$

$\omega = \sqrt{\dfrac{g}{L}} \qquad T = 2\pi\sqrt{\dfrac{L}{g}}$

Energy

If there is **no friction** or dissipation, kinetic and potential energy are alternately transformed into each other, but the total mechanical energy $E = K + U$ is conserved.

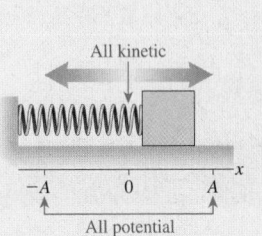

$$E = \frac{1}{2}mv_x^2 + \frac{1}{2}kx^2$$

$$= \frac{1}{2}m(v_{max})^2$$

$$= \frac{1}{2}kA^2$$

In a **damped system,** the energy decays exponentially

$$E = E_0 e^{-t/\tau}$$

where τ is the time constant.

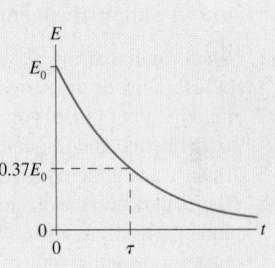

IMPORTANT CONCEPTS

Simple harmonic motion (SHM) is a sinusoidal oscillation with period T and amplitude A.

Frequency $f = \dfrac{1}{T}$

Angular frequency

$$\omega = 2\pi f = \frac{2\pi}{T}$$

Position $x(t) = A\cos(\omega t + \phi_0)$

$$= A\cos\left(\frac{2\pi t}{T} + \phi_0\right)$$

Velocity $v_x(t) = -v_{max}\sin(\omega t + \phi_0)$ with maximum speed $v_{max} = \omega A$

Acceleration $a_x = -\omega^2 x$

SHM is the projection onto the x-axis of **uniform circular motion.**

$\phi = \omega t + \phi_0$ is the phase

The position at time t is

$$x(t) = A\cos\phi$$
$$= A\cos(\omega t + \phi_0)$$

The phase constant ϕ_0 determines the initial conditions:

$$x_0 = A\cos\phi_0 \qquad v_{0x} = -\omega A\sin\phi_0$$

APPLICATIONS

Resonance

When a system is driven by a periodic external force, it responds with a large-amplitude oscillation if $f_{ext} \approx f_0$ where f_0 is the system's natural oscillation frequency, or **resonant frequency.**

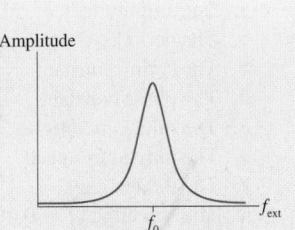

Damping

If there is a drag force $\vec{D} = -b\vec{v}$, where b is the damping constant, then (for lightly damped systems)

$$x(t) = Ae^{-bt/2m}\cos(\omega t + \phi_0)$$

The time constant for energy loss is $\tau = m/b$.

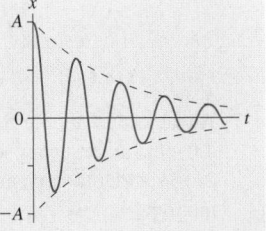

TERMS AND NOTATION

oscillatory motion	amplitude, A	linear restoring force	natural frequency, f_0
oscillator	angular frequency, ω	damped oscillation	driving frequency, f_{ext}
period, T	phase, ϕ	damping constant, b	response curve
frequency, f	phase constant, ϕ_0	envelope	resonance
hertz, Hz	restoring force	time constant, τ	resonance frequency, f_0
simple harmonic motion, SHM	equation of motion	half-life, $t_{1/2}$	
	small-angle approximation	driven oscillation	

EXERCISES AND PROBLEMS

Exercises

Section 14.1 Simple Harmonic Motion

1. When a guitar string plays the note "A," the string vibrates at 440 Hz. What is the period of the vibration?

2. In taking your pulse, you count 75 heartbeats in 1 min. What are the period and frequency of your heart's oscillations?

3. An air-track glider attached to a spring oscillates between the 10 cm mark and the 60 cm mark on the track. The glider completes 10 oscillations in 33 s. What are the (a) period, (b) frequency, (c) angular frequency, (d) amplitude, and (e) maximum speed of the glider?

4. An air-track glider is attached to a spring. The glider is pulled to the right and released from rest at $t = 0$ s. It then oscillates with a period of 2.0 s and a maximum speed of 40 cm/s.
   a. What is the amplitude of the oscillation?
   b. What is the glider's position at $t = 0.25$ s?

Section 14.2 Simple Harmonic Motion and Circular Motion

5. What are the (a) amplitude, (b) frequency, and (c) phase constant of the oscillation shown in Figure Ex14.5?

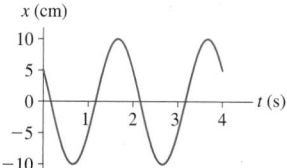

FIGURE EX14.5

6. What are the (a) amplitude, (b) frequency, and (c) phase constant of the oscillation shown in Figure Ex14.6?

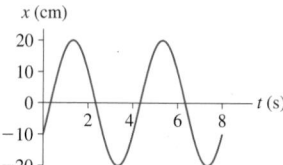

FIGURE EX14.6

7. An object in simple harmonic motion has an amplitude of 4.0 cm, a frequency of 2.0 Hz, and a phase constant of $2\pi/3$ rad. Draw a position graph showing two cycles of the motion.

8. An object in simple harmonic motion has an amplitude of 6.0 cm, a frequency of 0.50 Hz, and a phase constant of $-2\pi/3$ rad. Draw a position graph showing two cycles of the motion.

9. An object in simple harmonic motion has amplitude 4.0 cm and frequency 4.0 Hz, and at $t = 0$ s it passes through the equilibrium point moving to the right. Write the function $x(t)$ that describes the object's position.

10. An object in simple harmonic motion has amplitude 10.0 cm and frequency 0.50 Hz, and at $t = 0$ s it is at its largest distance to the left of the equilibrium position. Write the function $x(t)$ that describes the object's position.

11. An air-track glider attached to a spring oscillates with a period of 1.5 s. At $t = 0$ s the glider is 5.0 cm left of the equilibrium position and moving to the right at 36.3 cm/s.
    a. What is the phase constant?
    b. What is the phase at $t = 0$ s, 0.5 s, 1.0 s, and 1.5 s?

Section 14.3 Energy in Simple Harmonic Motion

Section 14.4 The Dynamics of Simple Harmonic Motion

12. A block attached to a spring with unknown spring constant oscillates with a period of 2.0 s. What is the period if
    a. The mass is doubled?
    b. The mass is halved?
    c. The amplitude is doubled?
    d. The spring constant is doubled?
    Parts a to d are independent questions, each referring to the initial situation.

13. A 200 g air-track glider is attached to a spring. The glider is pushed in 10 cm and released. A student with a stopwatch finds that 10 oscillations take 12.0 s. What is the spring constant?

14. The position of a 50 g oscillating mass is given by $x(t) = (2.0 \text{ cm}) \cos(10t - \pi/4)$, where t is in s. Determine:
    a. The amplitude.
    b. The period.
    c. The spring constant.
    d. The phase constant.
    e. The initial conditions.
    f. The maximum speed.
    g. The total energy.
    h. The velocity at $t = 0.40$ s.

15. A 200 g mass attached to a horizontal spring oscillates at a frequency of 2.0 Hz. At $t = 0$ s, the mass is at $x = 5.0$ cm and has $v_x = -30$ cm/s. Determine:
    a. The period.
    b. The angular frequency.
    c. The amplitude.
    d. The phase constant.
    e. The maximum speed.
    f. The maximum acceleration.
    g. The total energy.
    h. The position at $t = 0.40$ s.

16. A 507 g mass oscillates with an amplitude of 10 cm on a spring whose spring constant is 20 N/m. At $t = 0$ s the mass is 5.0 cm to the right of the equilibrium position and moving to the right. Determine:
    a. The period.
    b. The angular frequency.
    c. The phase constant.
    d. The initial velocity.
    e. The maximum speed.
    f. The total energy.
    g. The position at $t = 1.3$ s.
    h. The velocity at $t = 1.3$ s.

17. A 1.0 kg block is attached to a spring with spring constant 16 N/m. While the block is sitting at rest, a student hits it with a hammer and almost instantaneously gives it a speed of 40 cm/s. What are
    a. The amplitude of the subsequent oscillations?
    b. The block's speed at the point where $x = \frac{1}{2}A$?

Section 14.5 Vertical Oscillations

18. A spring is hanging from the ceiling. Attaching a 500 g physics book to the spring causes it to stretch 20 cm in order to come to equilibrium.
    a. What is the spring constant?
    b. From equilibrium, the book is pulled down 10 cm and released. What is the period of oscillation?
    c. What is the book's maximum speed? At what position or positions does it have this speed?

19. A spring with spring constant 15.0 N/m hangs from the ceiling. A ball is attached to the spring and allowed to come to rest. It is then pulled down 6.0 cm and released. If the ball makes 30 oscillations in 20 s, what are its (a) mass and (b) maximum speed?

20. A spring is hung from the ceiling. When a block is attached to its end, it stretches 2.0 cm before reaching its new equilibrium length. The block is then pulled down slightly and released. What is the frequency of oscillation?

Section 14.6 The Pendulum

21. Make a table with 4 columns and 8 rows. In row 1, label the columns θ (°), θ (rad), $\sin\theta$, and $\cos\theta$. In the left column, starting in row 2, write 0, 2, 4, 6, 8, 10, and 12.
    a. Convert each of these angles, in degrees, to radians. Put the results in column 2. Show four decimal places.
    b. Calculate the sines and cosines. Put the results, showing four decimal places, in columns 3 and 4.
    c. What is the first angle for which θ and $\sin\theta$ differ by more than 0.0010?

d. What is the first angle for which $\cos\theta$ is less than 0.9900?
    e. Over what range of angles does the small-angle approximation appear to be valid?

22. A mass on a string of unknown length oscillates as a pendulum with a period of 4.0 s. What is the period if
    a. The mass is doubled?
    b. The string length is doubled?
    c. The string length is halved?
    d. The amplitude is doubled?
    Parts a to d are independent questions, each referring to the initial situation.

23. A 200 g ball is tied to a string. It is pulled to an angle of 8.0° and released to swing as a pendulum. A student with a stopwatch finds that 10 oscillations take 12.0 s. How long is the string?

24. What is the period of a 1.0-m-long pendulum on (a) the earth and (b) Venus?

25. What is the length of a pendulum whose period on the moon matches the period of a 2.0-m-long pendulum on the earth?

26. The angle of a pendulum is $\theta(t) = (0.10 \text{ rad})\cos(5t + \pi)$, where t is in s. Determine:
    a. The amplitude.
    b. The frequency.
    c. The phase constant.
    d. The length of the string.
    e. The initial conditions.
    f. The angle at $t = 2.0$ s.

Section 14.7 Damped Oscillations

27. The amplitude of an oscillator decreases to 36.8% of its initial value in 10.0 s. What is the value of the time constant?

28. Calculate and draw an accurate position graph from $t = 0$ s to $t = 10$ s of a damped oscillator having a frequency of 1.0 Hz and a time constant of 4.0 s.

29. A 250 g air-track glider is attached to a spring with spring constant 4.0 N/m. The damping constant due to air resistance is 0.015 kg/s. The glider is pulled out 20 cm from equilibrium and released. How many oscillations will it make during the time in which the amplitude decays to e^{-1} of its initial value?

30. A spring with spring constant 15.0 N/m hangs from the ceiling. A 500 g ball is attached to the spring and allowed to come to rest. It is then pulled down 6.0 cm and released. What is the time constant if the ball's amplitude has decreased to 3.0 cm after 30 oscillations?

Problems

31. An object oscillating on a spring has the position graph shown in Figure P14.31. Draw a position graph if the following changes are made.
    a. The amplitude is halved and the frequency is halved.
    b. The mass is quadrupled.
    Parts a and b are independent questions, each starting from the graph shown.

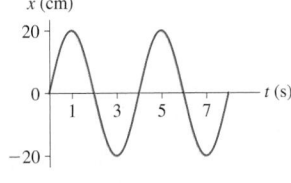

FIGURE P14.31

32. An object oscillating on a spring has the velocity graph shown in Figure P14.32. Draw a velocity graph if the following changes are made.
    a. The amplitude is doubled and the frequency is halved.
    b. The mass is quadrupled.
    Parts a and b are independent questions, each starting from the graph shown.

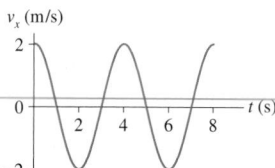

FIGURE P14.32

33. Figure P14.33 is the position-versus-time graph of a particle in simple harmonic motion.
    a. What is the phase constant?
    b. What is the velocity at $t = 0$ s?
    c. What is v_{max}?

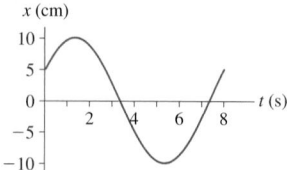

FIGURE P14.33

34. Figure P14.34 is the velocity-versus-time graph of a particle in simple harmonic motion.
    a. What is the amplitude of the oscillation?
    b. What is the phase constant?
    c. What is the position at $t = 0$ s?

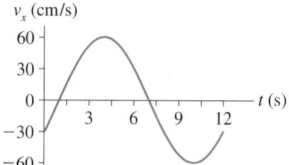

FIGURE P14.34

35. The two graphs in Figure P14.35 are for two different vertical mass/spring systems.
    a. What is the frequency of system A? What is the first time at which the mass has maximum speed while traveling in the upward direction?
    b. What is the period of system B? What is the first time at which the energy is all potential?
    c. If both systems have the same mass, what is the ratio k_A/k_B of their spring constants?

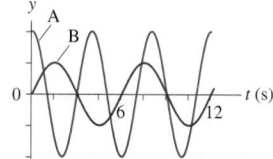

FIGURE P14.35

36. Astronauts in space cannot weigh themselves by standing on a bathroom scale. Instead, they determine their mass by oscillating on a large spring. Suppose an astronaut attaches one end of a large spring to her belt and the other end to a hook on the wall of the space capsule. A fellow astronaut then pulls her away from the wall and releases her. The spring's length as a function of time is shown in Figure P14.36.
    a. What is her mass if the spring constant is 240 N/m?
    b. What is her speed when the spring's length is 1.2 m?

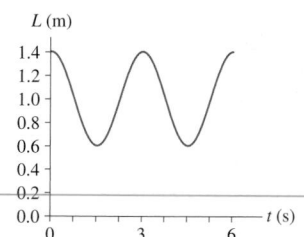

FIGURE P14.36

37. A 1.0 kg block oscillates on a spring with spring constant 20 N/m. At $t = 0$ s the block is 20 cm to the right of the equilibrium position and moving to the left at a speed of 100 cm/s. Determine the period of oscillation and draw a graph of position versus time.

38. An object in SHM oscillates with a period of 4.0 s and an amplitude of 10 cm. How long does the object take to move from $x = 0.0$ cm to $x = 6.0$ cm?

39. The motion of a particle is given by $x(t) = (25$ cm$)\cos(10t)$, where t is in s. At what time is the kinetic energy twice the potential energy?

40. a. When the displacement of a mass on a spring is $\frac{1}{2}A$, what fraction of the energy is kinetic energy and what fraction is potential energy?
    b. At what displacement, as a fraction of A, is the energy half kinetic and half potential?

41. For a particle in simple harmonic motion, show that $v_{max} = (\pi/2)v_{avg}$ where v_{avg} is the average speed during one cycle of the motion.

42. A block on a spring is pulled to the right and released at $t = 0$ s. It passes $x = 3.0$ cm at $t = 0.685$ s, and it passes $x = -3.0$ cm at $t = 0.886$ s.
    a. What is the angular frequency?
    b. What is the amplitude?
    Hint: $\cos(\pi - \theta) = -\cos\theta$.

43. A 100 g ball attached to a spring with spring constant 2.5 N/m oscillates horizontally on a frictionless table. Its velocity is 20 cm/s when $x = -5.0$ cm.
    a. What is the amplitude of oscillation?
    b. What is the ball's maximum acceleration?
    c. What is the ball's position when the acceleration is maximum?
    d. What is the speed of the ball when $x = 3.0$ cm?

44. The velocity of an object in simple harmonic motion is given by $v_x(t) = -(0.35$ m/s$)\sin(20t + \pi)$, where t is in s.
    a. What is the first time after $t = 0$ s at which the velocity is -0.25 m/s?
    b. What is the object's position at that time?

45. A 300 g oscillator has a speed of 95.4 cm/s when its displacement is 3.0 cm and 71.4 cm/s when its displacement is 6.0 cm. What is the oscillator's maximum speed?

46. An ultrasonic transducer, of the type used in medical ultrasound imaging, is a very thin disk ($m = 0.10$ g) driven back and forth in SHM at 1.0 MHz by an electromagnetic coil.
    a. The maximum restoring force that can be applied to the disk without breaking it is 40,000 N. What is the maximum oscillation amplitude that won't rupture the disk?
    b. What is the disk's maximum speed at this amplitude?

47. A 5.0 kg block hangs from a spring with spring constant 2000 N/m. The block is pulled down 5.0 cm from the equilibrium position and given an initial velocity of 1.0 m/s back toward equilibrium. What are the (a) frequency, (b) amplitude, and (c) total mechanical energy of the motion?

48. A 2.0 g spider is dangling at the end of a silk thread. You can make the spider bounce up and down on the thread by tapping lightly on his feet with a pencil. You soon discover that you can give the spider the largest amplitude on his little bungee cord if you tap exactly once every second. What is the spring constant of the silk thread?

49. A 200 g block hangs from a spring with spring constant 10 N/m. At $t = 0$ s the block is 20 cm below the equilibrium point and moving upward with a speed of 100 cm/s. What are the block's
    a. Oscillation frequency?
    b. Displacement from equilibrium when the speed is 50 cm/s?
    c. Position at $t = 1.0$ s?

50. A spring with spring constant k is suspended vertically from a support and a mass m is attached. The mass is held at the point where the spring is not stretched. Then the mass is released and begins to oscillate. The lowest point in the oscillation is 20 cm below the point where the mass was released. What is the oscillation frequency?

51. A compact car has a mass of 1200 kg. Assume that the car has one spring on each wheel, that the springs are identical, and that the mass is equally distributed over the four springs.
    a. What is the spring constant of each spring if the empty car bounces up and down 2.0 times each second?
    b. What will be the car's oscillation frequency while carrying four 70 kg passengers?

52. A block hangs in equilibrium from a vertical spring. When a second identical block is added, the original block sags by 5.0 cm. What is the oscillation frequency of the two-block system?

53. A 500 g block slides along a frictionless surface at a speed of 0.35 m/s. It runs into a horizontal massless spring with spring constant 50 N/m that extends outward from a wall. It compresses the spring, then is pushed back in the opposite direction by the spring, eventually losing contact with the spring.
    a. How long does the block remain in contact with the spring?
    b. How would your answer to part a change if the block's initial speed were doubled?

54. Figure P14.54 shows a 1.0 kg mass riding on top of a 5.0 kg mass as it oscillates on a frictionless surface. The spring constant is 50 N/m and the coefficient of static friction between the two blocks is 0.50. What is the maximum oscillation amplitude for which the upper block does not slip?

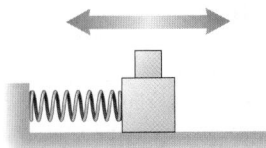

FIGURE P14.54

55. The two blocks in Figure P14.54 oscillate on a frictionless surface with a period of 1.5 s. The upper block just begins to slip when the amplitude is increased to 40 cm. What is the coefficient of static friction between the two blocks?

56. It is said that Galileo discovered a basic principle of the pendulum—that the period is independent of the amplitude—by using his pulse to time the period of swinging lamps in the cathedral as they swayed in the breeze. Suppose that one oscillation of a swinging lamp takes 5.5 s.
    a. How long is the lamp chain?
    b. What maximum speed does the lamp have if its maximum angle from vertical is 3.0°?

57. A 100 g mass on a 1.0-m-long string is pulled 8.0° to one side and released. How long does it take for the pendulum to reach 4.0° on the opposite side?

58. Astronauts on the first trip to Mars take along a pendulum that has a period on earth of 1.50 s. The period on Mars turns out to be 2.45 s. What is the Martian acceleration due to gravity?

59. The earth's acceleration due to gravity varies from 9.78 m/s² at the equator to 9.83 m/s² at the poles, because the earth is not a perfect sphere. A pendulum whose length is precisely 1.000 m can be used to measure g. Such a device is called a *gravimeter*.
    a. How long do 100 oscillations take at the equator?
    b. How long do 100 oscillations take at the north pole?
    c. Is the difference between your answers to parts a and b measurable? What kind of instrument could you use to measure the difference?
    d. Suppose you take your gravimeter to the top of a high mountain peak near the equator. There you find that 100 oscillations take 201.0 s. What is g on the mountain top?

60. In a science museum, a 110 kg brass pendulum bob swings at the end of a 15.0-m-long wire. The pendulum is started at exactly 8:00 A.M. every morning by pulling it 1.5 m to the side and releasing it. Because of its compact shape and smooth surface, the pendulum's damping constant is only 0.010 kg/s. At exactly 12:00 noon, how many oscillations will the pendulum have completed and what is its amplitude?

61. A 500 g air-track glider attached to a spring with spring constant 10 N/m is sitting at rest on a frictionless air track. A 250 g glider is pushed toward it from the far end of the track at a speed of 120 cm/s. It collides with and sticks to the 500 g glider. What are the amplitude and period of the subsequent oscillations?

62. A 200 g block attached to a horizontal spring is oscillating with an amplitude of 2.0 cm and a frequency of 2.0 Hz. Just as it passes through the equilibrium point, moving to the right, a sharp blow directed to the left exerts a 20 N force for 1.0 ms. What are the new (a) frequency and (b) amplitude?

63. A 1.00 kg block is attached to a horizontal spring with spring constant 2500 N/m. The block is at rest on a frictionless surface. A 10 g bullet is fired into the block, in the face opposite the spring, and sticks.
    a. What was the bullet's speed if the subsequent oscillations have an amplitude of 10.0 cm?
    b. Could you determine the bullet's speed by measuring the oscillation frequency? If so, how? If not, why not?

64. A pendulum consists of a massless, rigid rod with a mass at one end. The other end is pivoted on a frictionless pivot so that it can turn through a complete circle. The pendulum is inverted, so the mass is directly above the pivot point, then released. The speed of the mass as it passes through the lowest point is 5.0 m/s. If the pendulum undergoes small-amplitude oscillations at the bottom of the arc, what will the frequency be?

65. Figure P14.65 is a top view of an object of mass m connected between two stretched rubber bands of length L. The object rests on a frictionless surface. At equilibrium, the tension in each rubber band is T. Find an expression for the frequency of oscillations *perpendicular* to the rubber bands. Assume the amplitude is sufficiently small that the magnitude of the tension in the rubber bands is essentially unchanged as the mass oscillates.

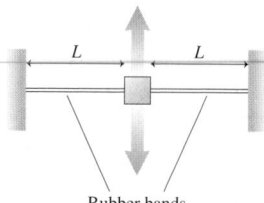

FIGURE P14.65   Rubber bands

66. A molecular bond can be modeled as a spring between two atoms that vibrate with simple harmonic motion. Figure P14.66 shows an SHM approximation for the potential energy of an HCl molecule. For $E < 4 \times 10^{-19}$ J it is a good approximation to the more accurate HCl potential energy curve that was shown in Figure 10.37. Because the chlorine atom is so much more massive than the hydrogen atom, it is reasonable to assume that the hydrogen atom ($m = 1.67 \times 10^{-27}$ kg) vibrates back and forth while the chlorine atom remains at rest. Use the graph to estimate the vibrational frequency of the HCl molecule.
Hint: How can you use the graph to determine the spring constant of the molecular bond?

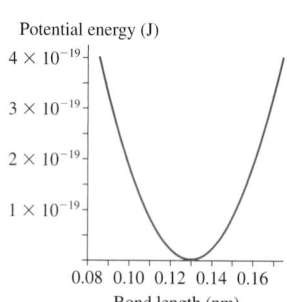

FIGURE P14.66   Bond length (nm)

67. A marble can roll around the inside of a vertical circular hoop of radius R. It undergoes small-amplitude oscillations if displaced slightly from the equilibrium position at the lowest point. Find an expression for the period of these small-amplitude oscillations.

68. A penny rides on top of a piston as it undergoes vertical simple harmonic motion with an amplitude of 4.0 cm. If the frequency is low, the penny rides up and down without difficulty. If the frequency is steadily increased, there comes a point at which the penny leaves the surface.
    a. At what point in the cycle does the penny first lose contact with the piston?
    b. What is the maximum frequency for which the penny just barely remains in place for the full cycle?

69. On your first trip to Planet X you happen to take along a 200 g mass, a 40-cm-long spring, a meter stick, and a stopwatch. You're curious about the acceleration due to gravity on Planet X, where ordinary tasks seem easier than on earth, but you can't find this information in your Visitor's Guide. One night you suspend the spring from the ceiling in your room and hang the mass from it. You find that the mass stretches the spring by 31.2 cm. You then pull the mass down 10.0 cm and release it. With the stopwatch you find that 10 oscillations take 14.5 s. Can you now satisfy your curiosity?

70. The 15 g head of a bobble-head doll oscillates in SHM at a frequency of 4.0 Hz.
    a. What is the spring constant of the spring on which the head is mounted?
    b. Suppose the head is pushed 2.0 cm against the spring, then released. What is the head's maximum speed as it oscillates?
    c. The amplitude of the head's oscillations decreases to 0.5 cm in 4.0 s. What is the head's damping constant?

71. An oscillator with a mass of 500 g and a period of 0.50 s has an amplitude that decreases by 2.0% during each complete oscillation.
    a. If the initial amplitude is 10 cm, what will be the amplitude after 25 oscillations?
    b. At what time will the energy be reduced to 60% of its initial value?

72. A 200 g oscillator in a vacuum chamber has a frequency of 2.0 Hz. When air is admitted, the oscillation decreases to 60% of its initial amplitude in 50 s. How many oscillations will have been completed when the amplitude is 30% of its initial value?

73. You've been hired by the circus to help design a new act. Samson, the great trapeze artist, will swing back and forth on a trapeze that hangs from 6.0-m-long ropes. Delilah, his beautiful young assistant, will stand precariously on a small platform that is 2.0 m below the pivot point at which the trapeze ropes are tied. The goal is for Delilah to dive off the platform, fall 4.0 m, and then be caught by Samson as he swings through the lowest point of his arc. Obviously, timing is everything. Samson will be swinging in an arc with an amplitude of 8.0°. At what angle should Samson be when Delilah leaps? There's no safety net, and Delilah, who is also the circus treasurer, won't write your check until after the first show. Keep in mind that Samson and Delilah aren't scientists, so they'll need your answer in units they can understand.

74. Invent a device, based on what you have learned in this chapter, with which a "weightless" astronaut in space can measure her mass. To recognize your invention, the patent office will need:
    a. A sketch.
    b. A description of the operating principles.
    c. Values for all major components.
    d. Sample calculations showing that these component values are "reasonable."

Challenge Problems

75. A block on a frictionless table is connected as shown in Figure CP14.75 to two springs having spring constants k_1 and k_2. Show that the block's oscillation frequency is given by
$$f = \sqrt{f_1^2 + f_2^2}$$
where f_1 and f_2 are the frequencies at which it would oscillate if attached to spring 1 or spring 2 alone.

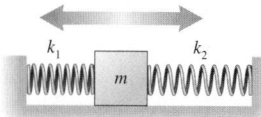

FIGURE CP14.75

76. A block on a frictionless table is connected as shown in Figure CP14.76 to two springs having spring constants k_1 and k_2. Find an expression for the block's oscillation frequency f in

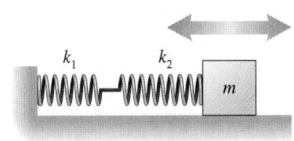

FIGURE CP14.76

terms of the frequencies f_1 and f_2 at which it would oscillate if attached to spring 1 or spring 2 alone.

77. A spring is standing upright on a table with its bottom end fastened to the table. A block is dropped from a height 3.0 cm above the top of the spring. The block sticks to the top end of the spring and then oscillates with an amplitude of 10 cm. What is the oscillation frequency?

78. Jose, whose mass is 75 kg, has just completed his first bungee jump and is now bouncing up and down at the end of the cord. His oscillations have an initial amplitude of 11.0 m and a period of 4.0 s.
    a. What is the spring constant of the bungee cord?
    b. What is Jose's maximum speed while oscillating?

c. From what height above the lowest point did Jose jump?
d. If the damping constant due to air resistance is 6.0 kg/s, how many oscillations will Jose make before his amplitude has decreased to 2.0 m?
Hint: Although not entirely realistic, treat the bungee cord as an ideal spring that can be compressed to a shorter length as well as stretched to a longer length.

79. A 1000 kg car carrying two 100 kg football players travels over a bumpy "washboard" road with the bumps spaced 3.0 m apart. The driver finds that the car bounces up and down with maximum amplitude when he drives at a speed of 5.0 m/s ($\approx$11 mph). The car then stops and picks up three more 100 kg passengers. By how much does the car body sag on its suspension when these three additional passengers get in?

80. Prove that the expression for $x(t)$ in Equation 14.53 is a solution to the equation of motion for a damped oscillator, Equation 14.52, if and only if the angular frequency ω is given by the expression in Equation 14.54.

STOP TO THINK ANSWERS

Stop to Think 14.1: c. $v_{max} = 2\pi A/T$. Doubling A and T leaves v_{max} unchanged.

Stop to Think 14.2: d. Think of circular motion. At 45°, the particle is in the first quadrant (positive x) and moving to the left (negative v_x).

Stop to Think 14.3: c > b > a = d. Energy conservation $\frac{1}{2}kA^2 = \frac{1}{2}m(v_{max})^2$ gives $v_{max} = \sqrt{k/m}A$. k or m have to be increased or decreased by a factor of 4 to have the same effect as increasing or decreasing A by a factor of 2.

Stop to Think 14.4: c. $v_x = 0$ because the slope of the position graph is zero. The negative value of x shows that the particle is left of the equilibrium position, so the restoring force is to the right.

Stop to Think 14.5: c. The period of a pendulum does not depend on its mass.

Stop to Think 14.6: $\tau_d > \tau_b = \tau_c > \tau_a$. The time constant is the time to decay to 37% of the initial height. The time constant is independent of the initial height.

15 Fluids and Elasticity

Kayaking through the rapids requires an intuitive understanding of fluids.

▶ **Looking Ahead**
The goal of Chapter 15 is to understand macroscopic systems that flow or deform. In this chapter you will learn to:

- Understand and use the concept of mass density.
- Understand pressure in liquids and gases.
- Use a wide variety of units for measuring pressure.
- Use Archimedes' principle to understand buoyancy.
- Use an ideal-fluid model to investigate how fluids flow.
- Calculate the elastic deformation of solids and liquids.

◀ **Looking Back**
The material in this chapter depends on the conditions of equilibrium. Please review:

- Section 4.6 Equilibrium and Newton's first law.
- Section 10.4 Hooke's law and restoring forces.

This kayak is floating on water, a fluid. The water itself is in motion. Surprisingly, we need no new laws of physics to understand how fluids flow or why some objects float while others sink. The physics of fluids, often called *fluid mechanics,* is an important application of Newton's laws and the law of conservation of energy—physics that you learned in Parts I and II.

Fluids are macroscopic systems, and our study of fluids will take us well beyond the particle model. Two new concepts, *density* and *pressure,* will be introduced to describe macroscopic systems. We'll begin with *fluid statics,* situations in which the fluid remains at rest. Suction cups and floating aircraft carriers are just two of the applications we'll explore. Then we'll turn to fluids in motion. Bernoulli's equation, the governing principle of *fluid dynamics,* will explain how water flows through fire hoses, how airplanes stay aloft, and many things in between. We'll then end this chapter with a brief look at a different but related property of macroscopic systems, the *elasticity* of solids.

15.1 Fluids

Quite simply, a **fluid** is a substance that flows. Because they flow, fluids take the shape of their container rather than retaining a shape of their own. You may think that gases and liquids are quite different, but both are fluids, and their similarities are often more important than their differences.

Gases and Liquids

A **gas,** shown in Figure 15.1a, is a system in which each molecule moves through space as a free, noninteracting particle until, on occasion, it collides with another molecule or with the wall of the container. The gas you are most familiar with is air, a mixture of mostly nitrogen and oxygen molecules. Gases are fairly simple macroscopic systems, and Part IV of this textbook will delve into the thermal properties of gases. For now, two properties of gases interest us:

1. Gases are *fluids*. They flow, and they exert pressure on the walls of their container.
2. Gases are *compressible*. That is, the volume of a gas is easily increased or decreased, a consequence of the "empty space" between the molecules in a gas.

Liquids are more complicated than either gases or solids. Liquids, like solids, are nearly *incompressible*. This property tells us that the molecules in a liquid, as in a solid, are about as close together as they can get without coming into contact with each other. At the same time, a liquid flows and deforms to fit the shape of its container. The fluid nature of a liquid tells us that the molecules are free to move around.

Together, these observations suggest the model of a **liquid** shown in Figure 15.1b. Here you see a system in which the molecules are loosely held together by weak molecular bonds. The bonds are strong enough that the molecules never get far apart but not strong enough to prevent the molecules from sliding around each other.

Volume and Density

One important parameter that characterizes a macroscopic system is its volume V, the amount of space the system occupies. The SI unit of volume is m^3. Nonetheless, both cm^3 and, to some extent, liters (L) are widely used metric units of volume. In most cases, you *must* convert these to m^3 before doing calculations.

While it is true that $1\ m = 100\ cm$, it is *not* true that $1\ m^3 = 100\ cm^3$. Figure 15.2 shows that the volume conversion factor is $1\ m^3 = 10^6\ cm^3$. You can think of this process as cubing the linear conversion factor:

$$1\ m^3 = 1\ m^3 \times \left(\frac{100\ cm}{1\ m}\right)^3 = 10^6\ cm^3$$

A liter is $1000\ cm^3$, so $1\ m^3 = 10^3$ L. A milliliter (1 mL) is the same as $1\ cm^3$.

A system is also characterized by its *density*. Suppose you have several blocks of copper, each of different size. Each block has a different mass m and a different volume V. Nonetheless, all the blocks are copper, so there should be some quantity that has the *same* value for all the blocks, telling us, "This is copper, not some

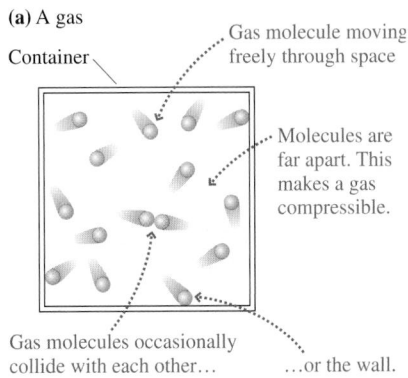

(a) A gas

Container

Gas molecule moving freely through space

Molecules are far apart. This makes a gas compressible.

Gas molecules occasionally collide with each other... ...or the wall.

(b) A liquid

A liquid has a well-defined surface.

Molecules are about as close together as they can get. This makes a liquid incompressible.

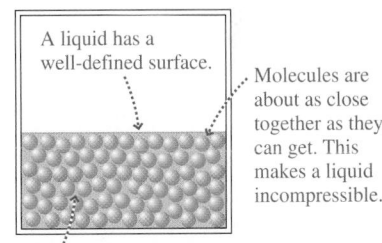

Molecules have weak bonds that keep them close together. But the molecules can slide around each other, allowing the liquid to flow and conform to the shape of its container.

FIGURE 15.1 Simple atomic-level models of gases and liquids.

Subdivide the 1 m × 1 m × 1 m cube into little cubes 1 cm on a side. You will get 100 subdivisions along each edge.

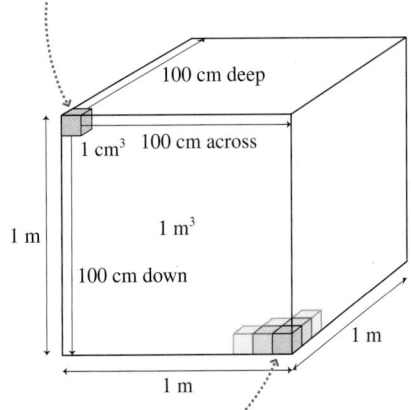

100 cm deep

$1\ cm^3$ 100 cm across

1 m

$1\ m^3$

100 cm down

1 m

1 m

There are $100 \times 100 \times 100 = 10^6$ little $1\ cm^3$ cubes in the big $1\ m^3$ cube.

FIGURE 15.2 There are $10^6\ cm^3$ in $1\ m^3$.

other material." The most important such parameter is the *ratio* of mass to volume, which we call the **mass density** ρ (lowercase Greek rho):

$$\rho = \frac{m}{V} \quad \text{(mass density)} \quad (15.1)$$

Conversely, an object of density ρ has mass

$$m = \rho V \quad (15.2)$$

The SI units of mass density are kg/m^3. Nonetheless, units of g/cm^3 are widely used. You need to convert these to SI units before doing most calculations. You must convert both the grams to kilograms and the cubic centimeters to cubic meters. The net result is the conversion factor

$$1 \text{ g/cm}^3 = 1000 \text{ kg/m}^3$$

TABLE 15.1 Densities of fluids at standard temperature (0°C) and pressure (1 atm)

Substance	ρ (kg/m³)
Air	1.28
Ethyl alcohol	790
Gasoline	680
Glycerin	1260
Helium gas	0.18
Mercury	13,600
Oil (typical)	900
Seawater	1030
Water	1000

The mass density is usually called simply "the density" if there is no danger of confusion. However, we will meet other types of density as we go along, and sometimes it is important to be explicit about which density you are using. Table 15.1 provides a short list of mass densities of various fluids. Notice the enormous difference between the densities of gases and liquids. Gases have lower densities because the molecules in gases are farther apart than in liquids.

What does it *mean* to say that the density of gasoline is 680 kg/m^3 or, equivalently, 0.68 g/cm^3? Density is a mass-to-volume ratio. It is often described as the "mass per unit volume," but for this to make sense you have to know what is meant by "unit volume." Regardless of which system of length units you use, a **unit volume** is one of those units cubed. For example, if you measure lengths in meters, a unit volume is 1 m^3. But 1 cm^3 is a unit volume if you measure lengths in cm, and 1 mi^3 is a unit volume if you measure lengths in miles.

Density is the mass of one unit of volume, whatever the units happen to be. To say that the density of gasoline is 680 kg/m^3 is to say that the mass of 1 m^3 of gasoline is 680 kg. The mass of 1 cm^3 of gasoline is 0.68 g, so the density of gasoline in those units is 0.68 g/cm^3.

The mass density is independent of the object's size. That is, mass and volume are parameters that characterize a *specific piece* of some substance—say copper—whereas the mass density characterizes the substance itself. All pieces of copper have the same mass density, which differs from the mass density of any other substance. Thus mass density allows us to talk about the properties of copper in general without having to refer to any specific piece of copper.

EXAMPLE 15.1 **Weighing the air**
What is the mass of air in a living room having dimensions 4.0 m × 6.0 m × 2.5 m?

MODEL Table 15.1 gives air density at a temperature of 0°C. The air density doesn't vary significantly over a small range of temperatures (we'll study this issue in the next chapter), so we'll use this value even though most people keep their living room warmer than 0°C.

SOLVE The room's volume is

$$V = (4.0 \text{ m}) \times (6.0 \text{ m}) \times (2.5 \text{ m}) = 60 \text{ m}^3$$

The mass of the air is

$$m = \rho V = (1.28 \text{ kg/m}^3)(60 \text{ m}^3) = 77 \text{ kg}$$

ASSESS This is perhaps more mass than you might have expected from a substance that hardly seems to be there. For comparison, a swimming pool this size would contain 60,000 kg of water.

STOP TO THINK 15.1 A piece of glass is broken into two pieces of different size. Rank order, from largest to smallest, the mass densities of pieces 1, 2, and 3.

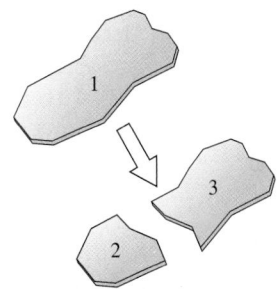

15.2 Pressure

Pressure is a word we all know and use. You probably have a commonsense idea of what pressure is. For example, you feel the effects of varying pressure against your eardrums when you swim underwater or take off in an airplane. Cans of whipped cream are "pressurized" to make the contents squirt out when you press the nozzle. It's hard to open a "vacuum sealed" jar of jelly the first time, but easy after the seal is broken.

You've undoubtedly seen water squirting out of a hole in the side of a container, as in Figure 15.3. Notice that the water emerges at greater speed from a hole at greater depth. And you've probably felt the air squirting out of a hole in a bicycle tire or inflatable air mattress. These observations suggest that

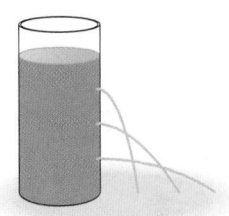

FIGURE 15.3 Water pressure pushes the water *sideways,* out of the holes.

- "Something" pushes the water or air *sideways,* out of the hole.
- In a liquid, the "something" is larger at greater depths. In a gas, the "something" appears to be the same everywhere.

Our goal is to turn these everyday observations into a precise definition of pressure.

Figure 15.4 shows a fluid—either a liquid or a gas—pressing against a small area A with force $\vec{F}$. This is the force that pushes the fluid out of a hole. In the absence of a hole, $\vec{F}$ pushes against the wall of the container. Let's define the **pressure** at this point in the fluid to be the ratio of the force to the area on which the force is exerted:

$$p = \frac{F}{A} \tag{15.3}$$

Notice that pressure is a scalar, not a vector. You can see, from Equation 15.3, that a fluid exerts a force of magnitude

$$F = pA \tag{15.4}$$

on a surface of area A. The force is *perpendicular* to the surface.

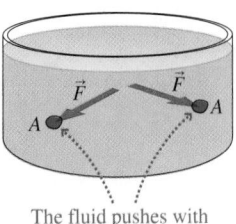

The fluid pushes with force $\vec{F}$ against area A.

FIGURE 15.4 The fluid presses against area A with force $\vec{F}$.

> **NOTE** ▶ Pressure itself is *not* a force, even though we sometimes talk informally about "the force exerted by the pressure." The correct statement is that the *fluid* exerts a force on a surface. ◀

From its definition, pressure has units of N/m². The SI unit of pressure is the **pascal,** defined as

$$1 \text{ pascal} = 1 \text{ Pa} \equiv 1 \frac{\text{N}}{\text{m}^2}$$

This unit is named for the 17th-century French scientist Blaise Pascal, who was one of the first to study fluids. Large pressures are often given in kilopascals, where 1 kPa = 1000 Pa.

Equation 15.3 is the basis for the simple pressure-measuring device shown in Figure 15.5a. Because the spring constant k and the area A are known, we can determine the pressure by measuring the compression of the spring. Once we've built such a device, we can place it in various liquids and gases to learn about pressure. Figure 15.5b shows what we can learn from a series of simple experiments.

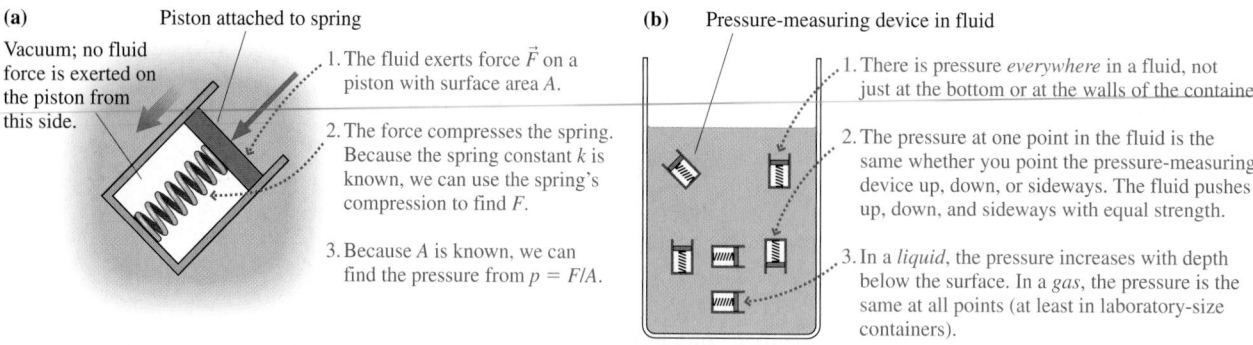

(a) Piston attached to spring

Vacuum; no fluid force is exerted on the piston from this side.

1. The fluid exerts force $\vec{F}$ on a piston with surface area A.

2. The force compresses the spring. Because the spring constant k is known, we can use the spring's compression to find F.

3. Because A is known, we can find the pressure from $p = F/A$.

(b) Pressure-measuring device in fluid

1. There is pressure *everywhere* in a fluid, not just at the bottom or at the walls of the container.

2. The pressure at one point in the fluid is the same whether you point the pressure-measuring device up, down, or sideways. The fluid pushes up, down, and sideways with equal strength.

3. In a *liquid*, the pressure increases with depth below the surface. In a *gas*, the pressure is the same at all points (at least in laboratory-size containers).

FIGURE 15.5 Learning about pressure.

The first statement in Figure 15.5b is especially important. Pressure exists at *all* points within a fluid, not just at the walls of the container. You may recall that tension exists at *all* points in a string, not only at its ends where it is tied to an object. We understood tension as the different parts of the string *pulling* against each other. Pressure is an analogous idea, except that the different parts of a fluid are *pushing* against each other.

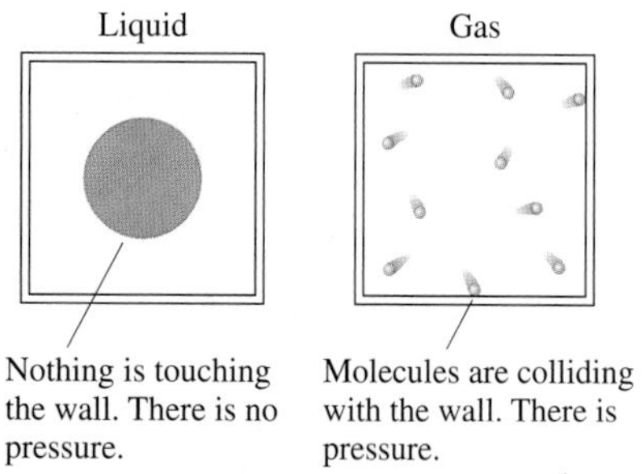

Causes of Pressure

Gases and liquids are both fluids, but they have some important differences. Liquids are nearly incompressible; gases are highly compressible. The molecules in a liquid attract each other via molecular bonds; the molecules in a gas do not interact other than through occasional collisions. These differences affect how we think about pressure in gases and liquids.

Imagine that you have two sealed jars, each containing a small amount of mercury and nothing else. All the air has been removed from the jars. Suppose you take the two jars into orbit on the space shuttle, where they are "weightless." One jar you keep cool, so that the mercury is a liquid. The other you heat until the mercury boils and becomes a gas. What can we say about the pressure in these two jars?

As Figure 15.6 shows, molecular bonds hold the liquid mercury together. It might quiver like Jello, but it remains a cohesive drop floating in the center of the jar. The liquid drop exerts no forces on the walls, so there's *no* pressure in the jar containing the liquid. (If we actually did this experiment, a very small fraction of the mercury would be in the vapor phase and create what is called *vapor pressure*. We can make the vapor pressure negligibly small by keeping the temperature low.)

The gas is different. Figure 15.1 introduced an atomic-level model of a gas in which a molecule moves freely until it collides with another molecule or with a wall of the container. Figure 15.7 shows a few of the gas molecules colliding with a wall. Recall, from our study of collisions in Chapter 9, that each molecule as it bounces exerts a tiny impulse on the wall. The impulse from any one collision is extremely small, but there are an extraordinarily large number of collisions every second. These collisions cause the gas to have a pressure.

The gas pressure can be calculated from the net force the molecules exert on the wall, divided by the area of the wall. We will do that calculation in Chapter 18. For now, we'll simply note that the pressure is proportional to the gas density in the container and to the absolute temperature.

Liquid

Gas

Nothing is touching the wall. There is no pressure.

Molecules are colliding with the wall. There is pressure.

FIGURE 15.6 Liquids and gases in a "weightless" environment.

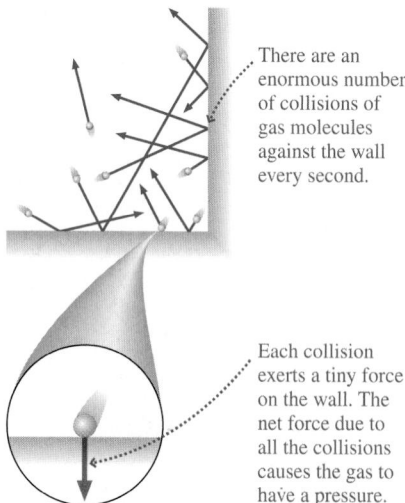

There are an enormous number of collisions of gas molecules against the wall every second.

Each collision exerts a tiny force on the wall. The net force due to all the collisions causes the gas to have a pressure.

FIGURE 15.7 The pressure in a gas is due to the net force of the molecules colliding with the walls.

Figure 15.8 shows the jars back on earth. Because of gravity, the liquid now fills the bottom of the jar and exerts a force on the bottom and the sides. Liquid mercury is incompressible, so the volume of liquid in Figure 15.8 is the same as in Figure 15.6. There is still no pressure on the top of the jar (other than the very small vapor pressure).

At first glance, the situation in the gas-filled jar seems unchanged from Figure 15.6. However, the earth's gravitational pull causes the gas density to be *slightly* more at the bottom of the jar than at the top. Because the pressure due to collisions is proportional to the density, the pressure is *slightly* larger at the bottom of the jar than at the top.

Thus there appear to be two contributions to the pressure in a container of fluid.

1. A *gravitational contribution* that arises from gravity pulling down on the fluid. Because a fluid can flow, forces are exerted on both the bottom and sides of the container. The gravitational contribution depends on the strength of the gravitational force.
2. A *thermal contribution* due to the collisions of freely moving gas molecules with the walls. The thermal contribution depends on the absolute temperature of the gas.

A detailed analysis finds that these two contributions are not entirely independent of each other, but the distinction is useful for a basic understanding of pressure. Let's see how these two contributions apply to different situations.

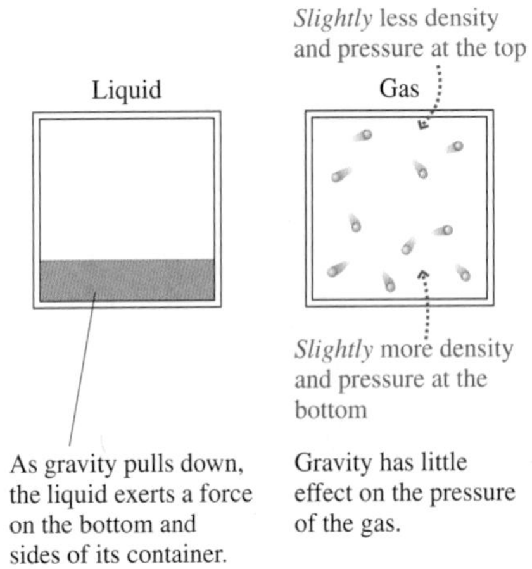

Pressure in Gases

The pressure in a laboratory-size container of gas is due almost entirely to the thermal contribution. A container would have to be ≈ 100 m tall for gravity to cause the pressure at the top to be even 1% less than the pressure at the bottom. Laboratory-size containers are much less than 100 m tall, so we can quite reasonably assume that p has the *same* value at all points in a laboratory-size container of gas. Homework problems will let you verify that the gravitational contribution to the pressure in a container of gas is negligible.

Decreasing the number of molecules in a container decreases the gas pressure simply because there are fewer collisions with the walls. If a container is completely empty, with no atoms or molecules, then the pressure is $p = 0$ Pa. This is a *perfect vacuum*. No perfect vacuum exists in nature, not even in the most remote depths of outer space, because it is impossible to completely remove every atom from a region of space. In practice, a **vacuum** is an enclosed space in which $p \ll 1$ atm. Using $p = 0$ Pa is then a very good approximation.

Atmospheric Pressure

The earth's atmosphere is *not* a laboratory-size container. The height of the atmosphere is such that the gravitational contribution to pressure *is* important. As Figure 15.9 shows, the density of air slowly decreases with increasing height until reaching zero in the vacuum of space. Consequently, the pressure of the air, what we call the *atmospheric pressure* p_{atmos}, decreases with height. The air pressure is less in Denver than in Miami.

The atmospheric pressure *at sea level* varies slightly with the weather, but the global average sea-level pressure is 101,300 Pa. Consequently, we define the **standard atmosphere** as

$$1 \text{ standard atmosphere} = 1 \text{ atm} \equiv 101,300 \text{ Pa} = 101.3 \text{ kPa}$$

The standard atmosphere, usually referred to simply as "atmospheres," is a commonly used unit of pressure. But it is not an SI unit, so you must convert atmospheres to pascals before doing most calculations with pressure.

Liquid   Gas

Slightly less density and pressure at the top

Slightly more density and pressure at the bottom

As gravity pulls down, the liquid exerts a force on the bottom and sides of its container.

Gravity has little effect on the pressure of the gas.

FIGURE 15.8 Gravity affects the pressure of the fluids.

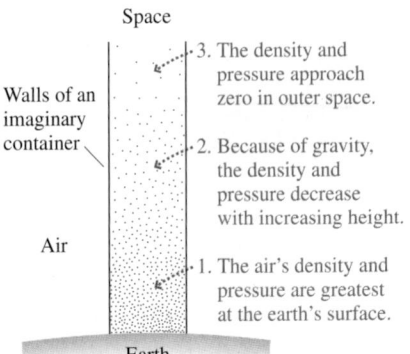

Space

3. The density and pressure approach zero in outer space.

Walls of an imaginary container

2. Because of gravity, the density and pressure decrease with increasing height.

Air

1. The air's density and pressure are greatest at the earth's surface.

Earth

FIGURE 15.9 The pressure and density decrease with increasing height in the atmosphere.

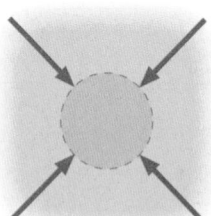

The forces of a fluid push in *all* directions.

FIGURE 15.10 Pressure forces in a fluid push with equal strength in all directions.

Removing the air from a container has very real consequences.

NOTE ▶ Unless you happen to live right at sea level, the atmospheric pressure around you is somewhat less than 1 atm. Pressure experiments use a barometer to determine the actual atmospheric pressure. For simplicity, this textbook will always assume that the pressure of the air is $p_{atmos} = 1$ atm unless stated otherwise. ◀

Given that the pressure of the air at sea level is 101.3 kPa, you might wonder why the weight of the air doesn't crush your forearm when you rest it on a table. Your forearm has a surface area of ≈ 200 cm$^2 = 0.02$ m^2, so the force of the air pressing against it is ≈ 2000 N (≈ 450 pounds). How can you even lift your arm?

The reason, as Figure 15.10 shows, is that a fluid exerts pressure forces in *all* directions. There *is* a downward force of ≈ 2000 N on your forearm, but the air underneath your arm exerts an *upward* force of the same magnitude. The *net* force is very close to zero. (To be accurate, there is net *upward* force called the buoyant force. We'll study buoyancy in Section 15.4. For most objects, the buoyant force of the air is too small to notice.)

But, you say, there isn't any air under my arm if I rest it on a table. Actually, there is. There would be a *vacuum* under your arm if there were no air. Imagine placing your arm on the top of a large vacuum cleaner suction tube. What happens? You feel a downward force as the vacuum cleaner "tries to suck your arm in." However, the downward force you feel is not a *pulling* force from the vacuum cleaner. It is the *pushing* force of the air above your arm *when the air beneath your arm is removed and cannot push back.* Air molecules do not have hooks! They have no ability to "pull" on your arm. The air can only push.

Vacuum cleaners, suction cups, and other similar devices are powerful examples of how strong atmospheric pressure forces can be *if* the air is removed from one side of an object so as to produce an unbalanced force. The fact that we are *surrounded* by the fluid allows us to move around in the air, just as we swim underwater, oblivious of these strong forces.

EXAMPLE 15.2 A suction cup
A 10.0-cm-diameter suction cup is pushed against a smooth ceiling. What is the maximum weight of an object that can be suspended from the suction cup without pulling it off the ceiling? The weight of the suction cup is negligible.

MODEL Pushing the suction cup against the ceiling pushes the air out. We'll assume that the volume enclosed between the suction cup and the ceiling is a perfect vacuum with $p = 0$ Pa. We'll also assume that the atmospheric pressure in the room is 1 atm.

VISUALIZE Figure 15.11 shows a free-body diagram of the suction cup stuck to the ceiling. The downward normal force of the ceiling is distributed around the rim of the suction cup, but in the particle model we can show this as a single force vector.

SOLVE The suction cup remains stuck to the ceiling, in static equilibrium, as long as $\vec{F}_{air} = \vec{n} + \vec{w}$. The magnitude of the upward force exerted by the air is

$$F_{air} = pA = p\pi r^2 = (101{,}300 \text{ Pa})\pi(0.050 \text{ m})^2 = 796 \text{ N}$$

There is no downward force from the air in this situation because there is no air inside the cup. Increasing the suspended

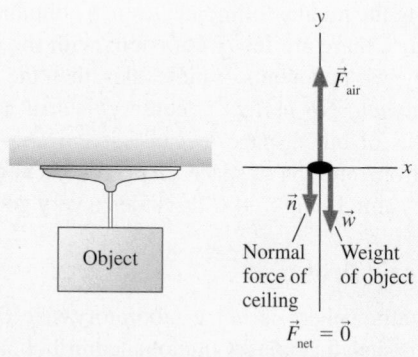

FIGURE 15.11 A suction cup is held to the ceiling by air pressure pushing upward on the bottom.

weight decreases the normal force n by an equal amount. The maximum weight has been reached when n is reduced to zero. This happens when

$$w_{max} = F_{air} = 796 \text{ N}$$

Hence this suction cup can support a weight of up to 796 N.

Pressure in Liquids

Whereas a gas fills a container, gravity causes a liquid to fill the bottom of a container. Thus it's not surprising that the pressure in a liquid is due almost entirely to the gravitational contribution. We'd like to determine the pressure at depth d below the surface of the liquid. We will assume that the liquid is at rest; flowing liquids will be considered later in this chapter.

The shaded cylinder of liquid in Figure 15.12 extends from the surface to depth d. This cylinder, like the rest of the liquid, is in static equilibrium with $\vec{F}_{net} = \vec{0}$. Three forces act on this cylinder: its weight mg, a downward force $p_0 A$ due to the pressure p_0 at the surface of the liquid, and an upward force pA due to the liquid beneath the cylinder pushing up on the bottom of the cylinder. This third force is a consequence of our earlier observation that different parts of a fluid push against each other. Pressure p, which is what we're trying to find, is the pressure at the bottom of the cylinder.

The upward force balances the two downward forces, so

$$pA = p_0 A + mg \tag{15.5}$$

The liquid is a cylinder of cross-section area A and height d. Its volume is $V = Ad$ and its mass is $m = \rho V = \rho Ad$. Substituting this expression for the mass of the liquid into Equation 15.5, we find that the area A cancels from all terms. The pressure at depth d in a liquid is

$$p = p_0 + \rho gd \qquad \text{(hydrostatic pressure at depth } d\text{)} \tag{15.6}$$

where ρ is the liquid's density. Because of our assumption that the fluid is at rest, the pressure given by Equation 15.6 is called the **hydrostatic pressure.** The fact that g appears in Equation 15.6 reminds us that this a gravitational contribution to the pressure.

As expected, $p = p_0$ at the surface, where $d = 0$. Pressure p_0 is often due to the air or other gas above the liquid. $p_0 = 1$ atm $= 101.3$ kPa for a liquid that is open to the air. However, p_0 can also be the pressure due to a piston or a closed surface pushing down on the top of the liquid.

NOTE ▶ Equation 15.6 assumes that the liquid is *incompressible;* that is, its density ρ doesn't increase with depth. This is an excellent assumption for liquids, but not a good one for a gas, which *is* compressible. Even so, Equation 15.6 can be used with gases over fairly small distances, a few tens of meters or less, because the density is nearly constant over these distances. Equation 15.6 should not be used for calculating the pressure at different heights in the atmosphere. (A homework problem will let you derive a different equation for the pressure of the atmosphere.) ◀

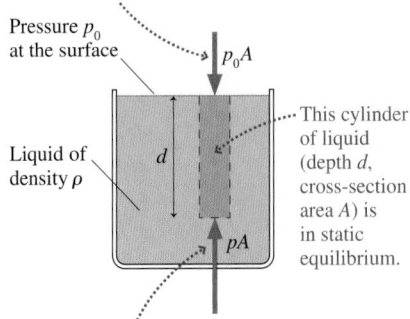

Whatever is above the liquid pushes down on the top of the cylinder.

Pressure p_0 at the surface

$p_0 A$

This cylinder of liquid (depth d, cross-section area A) is in static equilibrium.

Liquid of density ρ

d

pA

The liquid beneath the cylinder pushes up on the cylinder. The pressure at depth d is p.

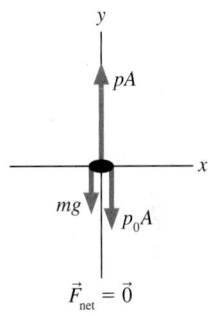

Free-body diagram of the column of liquid

FIGURE 15.12 Measuring the pressure at depth d in a liquid.

EXAMPLE 15.3 The pressure on a submarine
A submarine cruises at a depth of 300 m. What is the pressure at this depth? Give the answer both in pascals and atmospheres.

SOLVE The density of seawater, from Table 15.1, is $\rho = 1030$ kg/m³. The pressure at depth $d = 300$ m is found from Equation 15.6 to be

$$p = p_0 + \rho gd = 1.013 \times 10^5 \text{ Pa}$$
$$+ (1030 \text{ kg/m}^3)(9.80 \text{ m/s}^2)(300 \text{ m})$$
$$= 3.13 \times 10^6 \text{ Pa}$$

Converting the answer to atmospheres gives

$$p = 3.13 \times 10^6 \text{ Pa} \times \frac{1 \text{ atm}}{1.013 \times 10^5 \text{ Pa}} = 30.9 \text{ atm}$$

ASSESS The pressure deep in the ocean is very large. Windows on submersibles must be very thick to withstand the large forces.

(a)   (b)

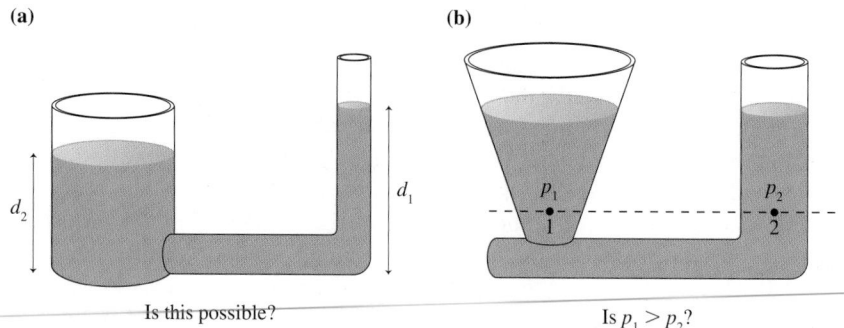

FIGURE 15.13 Some properties of a liquid in hydrostatic equilibrium are not what you might expect.

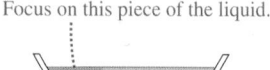

Focus on this piece of the liquid.

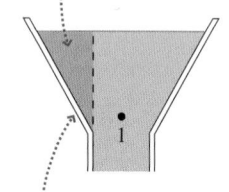

The wall of the tube supports the "extra" liquid.

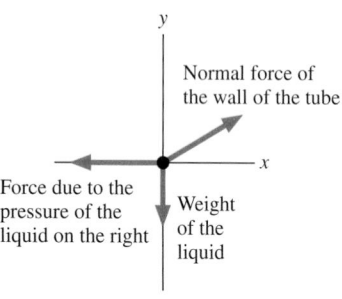

FIGURE 15.14 The weight of the liquid is supported by the wall of the tube.

The hydrostatic pressure in a liquid depends only on the depth and the pressure at the surface. This observation has some important implications. Figure 15.13a shows two connected tubes. It's certainly true that the larger volume of liquid in the wide tube weighs more than the liquid in the narrow tube. You might think that this extra weight would push the liquid in the narrow tube higher than in the wide tube. But it doesn't. If d_1 were larger than d_2, then, according to the hydrostatic pressure equation, the pressure at the bottom of the narrow tube would be higher than the pressure at the bottom of the wide tube. This *pressure difference* would cause the liquid to *flow* from right to left until the heights were equal.

Thus a first conclusion: **A connected liquid in hydrostatic equilibrium rises to the same height in all open regions of the container.**

Figure 15.13b shows two connected tubes of different shape. The conical tube holds more liquid above the dotted line, so you might think that $p_1 > p_2$. But it isn't. Both points are at the same depth, thus $p_1 = p_2$. You can arrive at the same conclusion by thinking about the pressure at the bottom of the tubes. If p_1 were larger than p_2, the pressure at the bottom of the left tube would be larger than the pressure at the bottom of the right tube. This would cause the liquid to flow until the pressures were equal.

If $p_1 = p_2$, you might be wondering what's holding up the "extra" liquid in the conical tube. Figure 15.14 shows that the weight of this extra liquid is supported by the wall of the tube. Only the liquid that's *directly above* point 1 needs to be supported by the pressure at point 1.

Thus a second conclusion: **The pressure is the same at all points on a horizontal line through a connected liquid in hydrostatic equilibrium.**

NOTE ▶ Both of these conclusions are restricted to liquids in hydrostatic equilibrium. The situation is entirely different for flowing fluids, as we'll see later in the chapter. ◀

EXAMPLE 15.4 Pressure in a closed tube
Water fills the tube shown in Figure 15.15. What is the pressure at the top of the closed tube?

MODEL This is a liquid in hydrostatic equilibrium. The closed tube is not an open region of the container, so the water cannot rise to an equal height. Nevertheless, the pressure is still the same at all points on a horizontal line. In particular, the pressure at the top of the closed tube equals the pressure in the open tube at the height of the dotted line. Assume $p_0 = 1$ atm.

FIGURE 15.15 What is the pressure at the top of the closed tube?

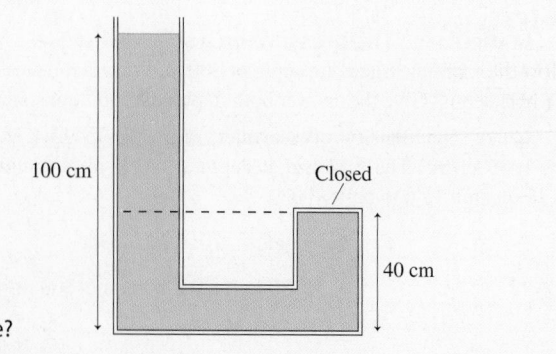

SOLVE A point 40 cm above the bottom of the open tube is at a depth of 60 cm. The pressure at this depth is

$$p = p_0 + \rho g d = 1.013 \times 10^5 \text{ Pa}$$

$$+ (1000 \text{ kg/m}^3)(9.80 \text{ m/s}^2)(0.60 \text{ m})$$

$$= 1.072 \times 10^5 \text{ Pa}$$

$$= 1.06 \text{ atm}$$

This is the pressure at the top of the closed tube.

ASSESS The water in the open tube *pushes* the water in the closed tube up against the top of the tube. Consequently, in accordance with Newton's third law, the top of the tube *presses down on the liquid* with a force of magnitude $F = pA$.

We can draw one more conclusion from the hydrostatic pressure equation $p = p_0 + \rho g d$. If we change the pressure p_0 at the surface to p_1, the pressure at depth d becomes $p' = p_1 + \rho g d$. The *change* in pressure $\Delta p = p_1 - p_0$ is the same at all points in the fluid, independent of the size or shape of the container. This idea, that **a change in the pressure at one point in an incompressible fluid appears undiminished at all points in the fluid,** was first recognized by Blaise Pascal, and is called **Pascal's principle.**

For example, if we compressed the air above the open tube in Example 15.4 to a pressure of 1.5 atm, an increase of 0.5 atm, the pressure at the top of the closed tube would increase to 1.56 atm. Pascal's principle is the basis for hydraulic systems, as we'll see in the next section.

STOP TO THINK 15.2 Water is slowly poured into the container until the water level has risen into tubes A, B, and C. The water doesn't overflow from any of the tubes. How do the water depths in the three columns compare to each other?

a. $d_A > d_B > d_C$.
b. $d_A < d_B < d_C$.
c. $d_A = d_B = d_C$.
d. $d_A = d_C > d_B$.
e. $d_A = d_C < d_B$.

Water

A   B   C

15.3 Measuring and Using Pressure

The pressure in a fluid is measured with a *pressure gauge,* which is often a device very similar to that in Figure 15.5. The fluid pushes against some sort of spring, usually a diaphragm, and the spring's displacement is registered by a pointer on a dial.

Many pressure gauges, such as tire gauges and the gauges on air tanks, measure not the actual or absolute pressure p but what is called **gauge pressure.** The gauge pressure, denoted p_g, is the pressure *in excess* of 1 atm. That is,

$$p_g = p - 1 \text{ atm} \tag{15.7}$$

You need to add 1 atm = 101.3 kPa to the reading of a pressure gauge to find the absolute pressure p that you need for doing most science or engineering calculations: $p = p_g + 1$ atm.

A tire-pressure gauge reads the gauge pressure p_g, not the absolute pressure p. The gauge reads zero when the tire is flat, but this doesn't mean there is a vacuum inside. Zero gauge pressure means the inside pressure is 1 atm.

EXAMPLE 15.5 **An underwater pressure gauge**

An underwater pressure gauge reads 60 kPa. What is its depth?

MODEL The gauge reads gauge pressure, not absolute pressure.

SOLVE The hydrostatic pressure at depth d, with $p_0 = 1$ atm, is $p = 1$ atm $+ \rho g d$. Thus the gauge pressure is

$$p_g = p - 1 \text{ atm} = (1 \text{ atm} + \rho g d) - 1 \text{ atm} = \rho g d$$

The term $\rho g d$ is the pressure *in excess* of atmospheric pressure and thus *is* the gauge pressure. Solving for d, we find

$$d = \frac{60{,}000 \text{ Pa}}{(1000 \text{ kg/m}^3)(9.80 \text{ m/s}^2)} = 6.1 \text{ m}$$

Solving Hydrostatic Problems

We now have enough information to formulate a set of rules for thinking about hydrostatic problems.

> TACTICS BOX 15.1 **Hydrostatics**
>
> ❶ **Draw a picture.** Show open surfaces, pistons, boundaries, and other features that affect the pressure. Include height and area measurements and fluid densities. Identify the points at which you need to find the pressure.
> ❷ **Determine the pressure at surfaces.**
> ■ **Surface open to the air:** $p_0 = p_{\text{atmos}}$, usually 1 atm.
> ■ **Surface covered by a gas:** $p_0 = p_{\text{gas}}$.
> ■ **Closed surface:** $p = F/A$ where F is the force the surface, such as a piston, exerts on the fluid.
> ❸ **Use horizontal lines.** Pressure in a connected fluid is the same at any point along a horizontal line.
> ❹ **Allow for gauge pressure.** Pressure gauges read $p_g = p - 1$ atm.
> ❺ **Use the hydrostatic pressure equation.** $p = p_0 + \rho g d$.

Manometers and Barometers

Gas pressure is sometimes measured with a device called a *manometer*. A manometer, shown in Figure 15.16, is a U-shaped tube connected to the gas at one end and open to the air at the other end. The tube is filled with a liquid—usually mercury—of density ρ. The liquid is in static equilibrium. A scale allows the user to measure the height h of the right side above the left side.

Steps 1–3 from Tactics Box 15.1 lead to the conclusion that the pressures p_1 and p_2 must be equal. Pressure p_1, at the surface on the left, is simply the gas pressure: $p_1 = p_{\text{gas}}$. Pressure p_2 is the hydrostatic pressure at depth $d = h$ in the liquid on the right: $p_2 = 1$ atm $+ \rho g h$. Equating these two pressures gives

$$p_{\text{gas}} = 1 \text{ atm} + \rho g h \tag{15.8}$$

Figure 15.16 assumed $p_{\text{gas}} > 1$ atm, so the right side of the liquid is higher than the left. Equation 15.8 is also valid for $p_{\text{gas}} < 1$ atm if the distance of the right side *below* the left side is considered to be a negative value of h.

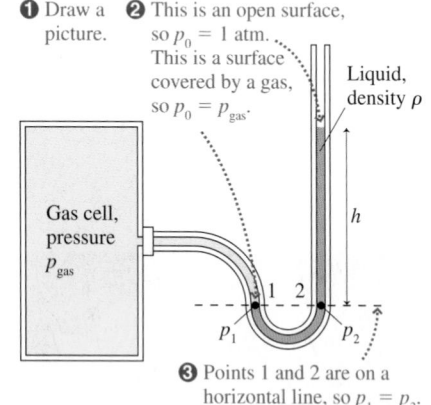

❶ Draw a picture. ❷ This is an open surface, so $p_0 = 1$ atm. This is a surface covered by a gas, so $p_0 = p_{\text{gas}}$.

Gas cell, pressure p_{gas}

Liquid, density ρ

h

1 2

p_1 p_2

❸ Points 1 and 2 are on a horizontal line, so $p_1 = p_2$.

FIGURE 15.16 A manometer is used to measure gas pressure.

EXAMPLE 15.6 **Using a manometer**

The pressure of a gas cell is measured with a mercury manometer. The mercury is 36.2 cm higher in the outside arm than in the arm connected to the gas cell.

a. What is the gas pressure?

b. What is the reading of a pressure gauge attached to the gas cell?

SOLVE

a. From Table 15.1, the density of mercury is $\rho = 13,600 \text{ kg/m}^3$. Equation 15.8 with $h = 0.362$ m gives

$$p_{\text{gas}} = 1 \text{ atm} + \rho g h = 149.5 \text{ kPa}$$

We had to change 1 atm to 101,300 Pa before adding. Converting the result to atmospheres, $p_{\text{gas}} = 1.476$ atm.

b. The pressure gauge reads gauge pressure: $p_{\text{g}} = p - 1 \text{ atm} = 0.476$ atm or 48.2 kPa.

ASSESS Manometers are useful over a pressure range from near vacuum up to ≈ 2 atm. For higher pressures, the mercury column would be too tall to be practical.

Another important pressure-measuring instrument is the *barometer,* which is used to measure the atmospheric pressure p_{atmos}. Figure 15.17a shows a glass tube, sealed at the bottom, that has been completely filled with a liquid. If we temporarily seal the top end, we can invert the tube and place it in a beaker of the same liquid. When the temporary seal is removed, some, but not all, of the liquid runs out, leaving a liquid column in the tube that is a height h above the surface of the liquid in the beaker. This device, shown in Figure 15.17b, is a barometer. What does it measure? And why doesn't *all* the liquid in the tube run out?

We can analyze the barometer much as we did the manometer. Points 1 and 2 in Figure 15.17b are on a horizontal line drawn even with the surface of the liquid. The liquid is in hydrostatic equilibrium, so the pressure at these two points must be equal. Liquid runs out of the tube only until a balance is reached between the pressure at the base of the tube and the pressure of the air.

You can think of a barometer as rather like a seesaw. If the pressure of the atmosphere increases, it presses down on the liquid in the beaker. This forces liquid up the tube until the pressures at points 1 and 2 are equal. If the atmospheric pressure falls, liquid has to flow out of the tube to keep the pressures equal at these two points.

The pressure at point 2 is the pressure due to the weight of the liquid in the tube plus the pressure of the gas above the liquid. But in this case there is no gas above the liquid! Because the tube had been completely full of liquid when it was inverted, the space left behind when the liquid ran out is a vacuum (ignoring a very slight *vapor pressure* of the liquid, negligible except in extremely precise measurements). Thus pressure p_2 is simply $p_2 = \rho g h$.

Equating p_1 and p_2 gives

$$p_{\text{atmos}} = \rho g h \qquad (15.9)$$

Thus we can measure the atmosphere's pressure by measuring the height of the liquid column in a barometer.

The average air pressure at sea level causes a column of mercury in a mercury barometer to stand 760 mm above the surface. Knowing that the density of mercury is 13,600 kg/m³ (at 0°C), we can use Equation 15.9 to find that the average atmospheric pressure is

$$p_{\text{atmos}} = \rho_{\text{Hg}} g h = (13,600 \text{ kg/m}^3)(9.80 \text{ m/s}^2)(0.760 \text{ m})$$

$$= 1.013 \times 10^5 \text{ Pa} = 101.3 \text{ kPa}$$

This is the value given earlier as "one standard atmosphere." Now you can see that 1 atm = 101.3 kPa *because,* on average, a mercury barometer gives a reading of 760 mm.

The barometric pressure varies slightly from day to day as the weather changes. Weather systems are called *high-pressure systems* or *low-pressure systems,* depending upon whether the local sea-level pressure is higher or lower than one standard atmosphere. Higher pressure is usually associated with fair weather while lower pressure portends rain.

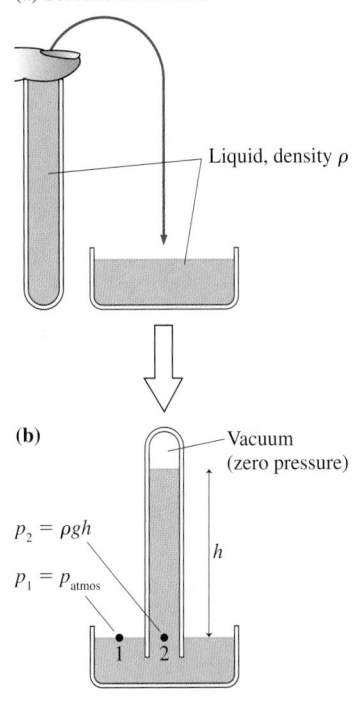

(a) Seal and invert tube.

Liquid, density ρ

(b)

Vacuum (zero pressure)

$p_2 = \rho g h$

$p_1 = p_{\text{atmos}}$

h

1 2

FIGURE 15.17 A barometer.

Pressure Units

In practice, pressure is measured in a number of different units. This plethora of units and abbreviations has arisen historically as scientists and engineers working on different subjects (liquids, high-pressure gases, low-pressure gases, weather, etc.) developed what seemed to them the most convenient units. These units continue in use through tradition, so it is necessary to become familiar with converting back and forth between them. Table 15.2 gives the basic conversions.

TABLE 15.2 Pressure units

Unit	Abbreviation	Conversion to 1 atm	Uses
pascal	Pa	101.3 kPa	SI unit: $1\text{ Pa} = 1\text{ N/m}^2$ use in most calculations
atmosphere	atm	1 atm	general
millimeters of mercury	mm of Hg	760 mm of Hg	gases and barometric pressure
inches of mercury	in	29.92 in	barometric pressure in U.S. weather forecasting
pounds per square inch	psi	14.7 psi	engineering and industry

Blood Pressure

The last time you had a medical checkup, the doctor may have told you something like, "Your blood pressure is 120 over 80." What does that mean?

About every 0.8 s, assuming a pulse rate of 75 beats per minute, your heart "beats." The heart muscles contract and push blood out into your aorta. This contraction, like squeezing a balloon, raises the pressure in your heart. The pressure increase, in accordance with Pascal's principle, is transmitted through all your arteries.

Figure 15.18 is a pressure graph showing how blood pressure changes during one cycle of the heartbeat. The medical condition of *high blood pressure* usually means that your systolic pressure is higher than necessary for blood circulation. The high pressure causes undue stress and strain on your entire circulatory system, often leading to serious medical problems. Low blood pressure can cause you to get dizzy if you stand up quickly because the pressure isn't adequate to pump the blood up to your brain.

Blood pressure is measured with a cuff that goes around your arm. The doctor or nurse pressurizes the cuff, places a stethoscope over the artery in your arm, then slowly releases the pressure while watching a pressure gauge. Initially, the cuff squeezes the artery shut and cuts off the blood flow. When the cuff pressure drops below the systolic pressure, the pressure pulse during each beat of your heart forces the artery open briefly and a squirt of blood goes through. You can feel this, and the doctor or nurse records the pressure when she hears the blood start to flow. This is your systolic pressure.

This pulsing of the blood through your artery lasts until the cuff pressure reaches the diastolic pressure. Then the artery remains open continuously and the blood flows smoothly. This transition is easily heard in the stethoscope, and the doctor or nurse records your diastolic pressure.

Blood pressure is measured in millimeters of mercury. And it is a gauge pressure, the pressure in excess of 1 atm. A fairly typical blood pressure of a healthy young adult is 120/80, meaning that the systolic pressure is $p_g = 120$ mm of Hg (absolute pressure $p = 880$ mm of Hg) and the diastolic pressure is 80 mm of Hg.

Now that you know, be sure to ask your doctor questions next time you have your blood pressure measured!

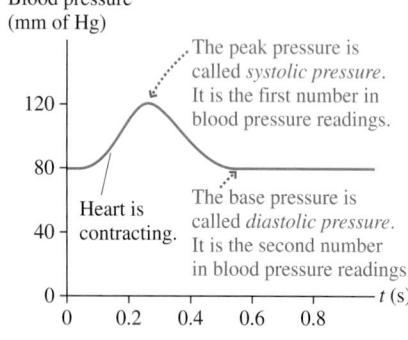

Blood pressure (mm of Hg)

The peak pressure is called *systolic pressure.* It is the first number in blood pressure readings.

Heart is contracting.

The base pressure is called *diastolic pressure.* It is the second number in blood pressure readings.

FIGURE 15.18 Blood pressure during one cycle of a heart beat.

The Hydraulic Lift

The use of pressurized liquids to do useful work is a technology known as **hydraulics.** Pascal's principle is the fundamental idea underlying hydraulic devices. If you increase the pressure at one point in a liquid by pushing a piston in, that pressure increase is transmitted to all points in the liquid. A second piston at some other point in the fluid can then push outward and do useful work.

The brake system in your car is a hydraulic system. Stepping on the brake pushes a piston into the *master brake cylinder* and increases the pressure in the *brake fluid.* The fluid itself hardly moves, but the pressure increase is transmitted to the four wheels where it pushes the brake pads against the spinning brake disk. You've used a pressurized liquid to achieve the useful goal of stopping your car.

One advantage of hydraulic systems over simple mechanical linkages is the possibility of *force multiplication.* To see how this works, we'll analyze a *hydraulic lift,* such as the one that lifts your car at the repair shop. Figure 15.19a shows force $\vec{F}_2$, perhaps due to a weight of mass m, pressing down on a liquid via a piston of area A_2. A much smaller force $\vec{F}_1$ presses down on a piston of area A_1. Can this system possibly be in equilibrium?

As you now know, the hydrostatic pressure is the same at all points along a horizontal line through a fluid. Consider the line passing through the liquid/piston interface on the left in Figure 15.19a. Pressures p_1 and p_2 must be equal, thus

$$p_0 + \frac{F_1}{A_1} = p_0 + \frac{F_2}{A_2} + \rho g h \qquad (15.10)$$

The atmosphere presses equally on both sides, so p_0 cancels. The system is in static equilibrium if

$$F_2 = \frac{A_2}{A_1}F_1 - \rho g h A_2 \qquad (15.11)$$

If the height h is very small, so that the term $\rho g h A_2$ is negligible, then F_2 (the weight of the heavy object) is larger than F_1 by the factor A_2/A_1. In other words, a small force applied to a small piston really can support a large car because both apply the *same pressure* to the fluid. The ratio A_2/A_1 is a force-multiplying factor.

> **NOTE** ▶ Force $\vec{F}_2$ is the force of the heavy object pushing *down* on the liquid. According to Newton's third law, the liquid pushes *up* on the object with a force of equal magnitude. Thus F_2 in Equation 15.11 is the "lifting force." ◀

Suppose we need to lift the car higher. If piston 1 is pushed down distance d_1, as in Figure 15.19b, it displaces volume $V_1 = A_1 d_1$ of liquid. Because the liquid is incompressible, V_1 must equal the volume $V_2 = A_2 d_2$ added beneath piston 2 as it rises distance d_2. That is,

$$d_2 = \frac{d_1}{A_2/A_1} \qquad (15.12)$$

The distance is *divided* by the same factor as that by which force is multiplied. A small force may be able to support a heavy weight, but you have to push the small piston a large distance to raise the heavy weight by a small amount.

This conclusion is really just a statement of energy conservation. Work is done *on* the liquid by a small force pushing the liquid through a large displacement. Work is done *by* the liquid when it lifts the heavy weight through a small distance. A full analysis must consider the fact that the gravitational potential energy of the liquid is also changing, so we can't simply equate the output work to the input work, but you can see that energy considerations require piston 1 to move further than piston 2.

Force $\vec{F}_1$ in Equation 15.11 is the force that balances the heavy object at height h. As a homework problem, you can show that the magnitude of force $\vec{F}_1$ must be increased by

$$\Delta F = \rho g (A_1 + A_2) d_2 \qquad (15.13)$$

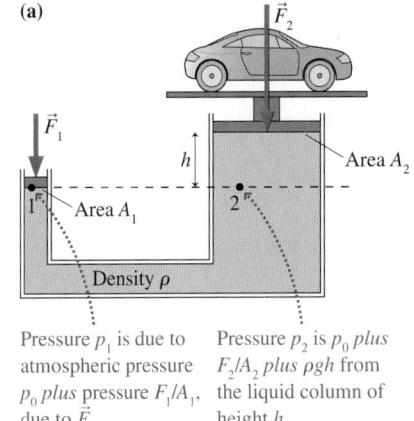

(a)

Pressure p_1 is due to atmospheric pressure p_0 plus pressure F_1/A_1, due to $\vec{F}_1$.

Pressure p_2 is p_0 *plus* F_2/A_2 *plus* $\rho g h$ from the liquid column of height h.

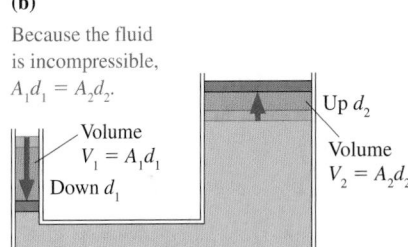

(b)

Because the fluid is incompressible, $A_1 d_1 = A_2 d_2$.

Volume $V_1 = A_1 d_1$

Down d_1

Up d_2

Volume $V_2 = A_2 d_2$

FIGURE 15.19 A hydraulic lift.

in order to lift the heavy object through distance d_2 to a new height $h + d_2$, where ρ is the density of the liquid. Surprisingly, ΔF is independent of the weight you're lifting.

EXAMPLE 15.7 Lifting a car

The hydraulic lift at a car repair shop is filled with oil. The car rests on a 25-cm-diameter piston. To lift the car, compressed air is used to push down on a 6.0-cm-diameter piston.

a. What air-pressure force will support a 1300 kg car level with the compressed-air piston?

b. By how much must the air-pressure force be increased to lift the car 2.0 m?

MODEL Assume that the oil is incompressible. Its density, from Table 15.1, is 900 kg/m^3.

SOLVE

a. The weight of the car pressing down on the piston is $F_2 = mg = 12{,}740$ N. The piston areas are $A_1 = \pi(0.030 \text{ m})^2 =$

0.00283 m^2 and $A_2 = \pi(0.125 \text{ m})^2 = 0.0491$ m^2. The force required to hold the car level with the compressed air piston, with $h = 0$ m, is

$$F_1 = \frac{F_2}{A_2/A_1} = \frac{12{,}740 \text{ N}}{(0.0491 \text{ m}^2)/(0.00283 \text{ m}^2)} = 734 \text{ N}$$

b. To raise the car $d_2 = 2.0$ m, the air-pressure force must be increased by

$$\Delta F = \rho g(A_1 + A_2)d_2 = 916 \text{ N}$$

ASSESS 734 N is roughly the weight of an average adult male. The multiplication factor $A_2/A_1 = (25 \text{ cm}/6 \text{ cm})^2 = 17$ makes it quite easy to hold up the car.

STOP TO THINK 15.3   Rank in order, from largest to smallest, the magnitudes of the forces $\vec{F}_1$, $\vec{F}_2$, and $\vec{F}_3$ required to balance the masses. The masses are in kilograms.

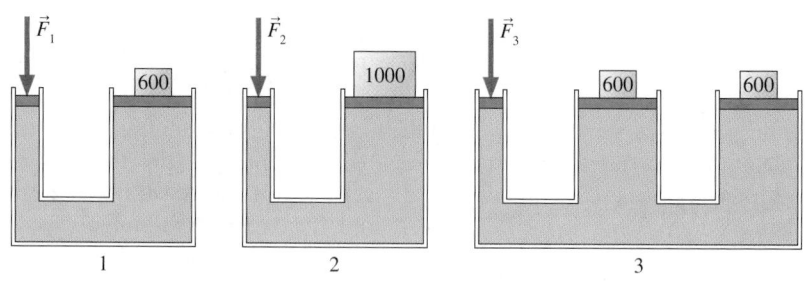

15.4 Buoyancy

A rock, as you know, sinks like a rock. Wood floats on the surface of a lake. A penny with a mass of a few grams sinks, but a massive steel aircraft carrier floats. How can we understand these diverse phenomena?

An air mattress floats effortlessly on the surface of a swimming pool. But if you've ever tried to push an air mattress underwater, you know it is nearly impossible. As you push down, the water pushes up. This net upward force of a fluid is called the **buoyant force.**

The basic reason for the buoyant force is easy to understand. Figure 15.20 shows a cylinder submerged in a liquid. The pressure in the liquid increases with depth, so the pressure at the bottom of the cylinder is larger than at the top. Both cylinder ends have equal area, so force $\vec{F}_{\text{up}}$ is larger than force $\vec{F}_{\text{down}}$. (Remember that pressure forces push in *all* directions.) Consequently, the pressure in the liquid exerts a *net upward force* on the cylinder of magnitude $F_{\text{net}} = F_{\text{up}} - F_{\text{down}}$. This is the buoyant force.

The submerged cylinder illustrates the idea in a simple way, but the result is not limited to cylinders or to liquids. Suppose we isolate a parcel of fluid of arbi-

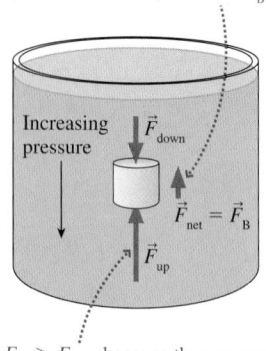

The net force of the fluid on the cylinder is the buoyant force $\vec{F}_{\text{B}}$.

$F_{\text{up}} > F_{\text{down}}$ because the pressure is greater at the bottom. Hence the fluid exerts a net upward force.

FIGURE 15.20 The buoyant force arises because the fluid pressure at the bottom of the cylinder is larger than at the top.

trary shape and volume by drawing an imaginary boundary around it, as shown in Figure 15.21a. This parcel is in static equilibrium. Consequently, the parcel's weight force pulling it down must be balanced by an upward force. The upward force, which is exerted on this parcel of fluid by the surrounding fluid, is the buoyant force $\vec{F}_B$. The buoyant force matches the weight of the fluid: $F_B = w$.

Now imagine that we could somehow remove this parcel of fluid and instantaneously replace it with an object having exactly the same shape and size, as shown in Figure 15.21b. Because the buoyant force is exerted by the *surrounding* fluid, and the surrounding fluid hasn't changed, the buoyant force on this new object is *exactly the same* as the buoyant force on the parcel of fluid that we removed.

When an object (or a portion of an object) is immersed in a fluid, it *displaces* fluid that would otherwise fill that region of space. This fluid is called the **displaced fluid.** The displaced fluid's volume is exactly the volume of the portion of the object that is immersed in the fluid. Figure 15.21 leads us to conclude that the magnitude of the upward buoyant force matches the weight of this displaced fluid.

This idea was first recognized by the ancient Greek mathematician and scientist Archimedes, perhaps the greatest scientist of antiquity, and today we know it as *Archimedes' principle.*

> **Archimedes' principle** A fluid exerts an upward buoyant force $\vec{F}_B$ on an object immersed in or floating on the fluid. The magnitude of the buoyant force equals the weight of the fluid displaced by the object.

Suppose the fluid has density ρ_f and the object displaces volume V_f of fluid. The mass of the displaced fluid is $m_f = \rho_f V_f$ and so its weight is $w = \rho_f V_f g$. Thus Archimedes' principle in equation form is

$$F_B = \rho_f V_f g \tag{15.14}$$

NOTE ▶ It is important to distinguish the density and volume of the displaced fluid from the density and volume of the object. To do so, we'll use subscript f for the fluid and o for the object. ◀

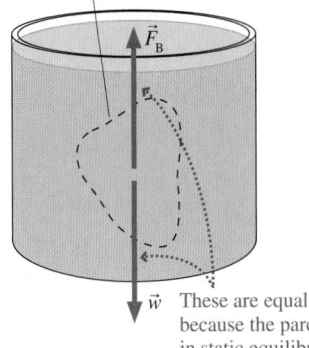

(a)

Imaginary boundary around a parcel of fluid

$\vec{w}$ These are equal because the parcel is in static equilibrium.

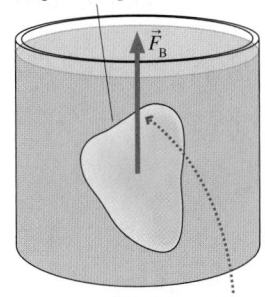

(b)

Real object with same size and shape as the parcel of fluid

The buoyant force on the object is the same as on the parcel of fluid because the *surrounding* fluid has not changed.

FIGURE 15.21 The buoyant force on an object is the same as the buoyant force on the fluid it displaces.

EXAMPLE 15.8 Holding a block of wood underwater
A 10 cm × 10 cm × 10 cm block of wood with a density 700 kg/m^3 is held underwater by a string tied to the bottom of the container. What is the tension in the string?

MODEL The buoyant force on the wood is given by Archimedes' principle.

VISUALIZE Figure 15.22 shows the forces acting on the wood.

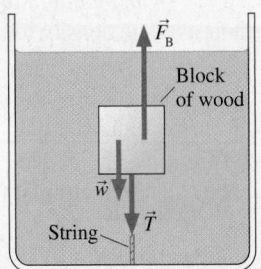

FIGURE 15.22 The forces acting on the submerged wood.

SOLVE The block is in static equilibrium, so

$$\sum F_y = F_B - T - w = 0$$

Thus the tension is $T = F_B - w$. The mass of the block is $m_o = \rho_o V_o$ and its weight is $w = \rho_o V_o g$. The buoyant force, given by Equation 15.14, is $F_B = \rho_f V_f g$. Thus

$$T = \rho_f V_f g - \rho_o V_o g = (\rho_f - \rho_o) V_o g$$

where we've used the fact that $V_f = V_o$ for a completely submerged object. The volume is $V_o = 1000 \text{ cm}^3 = 1.0 \times 10^{-3} \text{ m}^3$, and hence the tension in the string is

$$T = ((1000 \text{ kg/m}^3) - (700 \text{ kg/m}^3))$$
$$\times (1.0 \times 10^{-3} \text{ m}^3)(9.8 \text{ m/s}^2) = 2.94 \text{ N}$$

ASSESS The tension depends on the *difference* in densities. The tension would vanish if the wood density matched the water density.

Float or Sink?

If you *hold* an object underwater and then release it, it either floats to the surface, sinks, or remains "hanging" in the water. How can we predict which it will do? The net force on the object an instant after you release it is $\vec{F}_{net} = \vec{F}_B - \vec{w}_o$. Whether it heads for the surface or the bottom depends on whether the buoyancy force F_B is larger or smaller than the object's weight w_o.

The magnitude of the buoyant force is $\rho_f V_f g$. The weight of a uniform object, such as a block of steel, is simply $\rho_o V_o g$. But a compound object, such as a scuba diver, may have pieces of varying density. If we define the **average density** to be $\rho_{avg} = m_o / V_o$, the weight of a compound object can be written $w_o = \rho_{avg} V_o g$.

Comparing $\rho_f V_f g$ to $\rho_{avg} V_o g$, and noting that $V_f = V_o$ for an object that is submerged, we see that an object floats or sinks depending on whether the fluid density ρ_f is larger or smaller than the object's average density ρ_{avg}. If the densities are equal, the object is in static equilibrium and hangs motionless. This is called **neutral buoyancy.** These conditions are summarized in Tactics Box 15.2.

TACTICS BOX 15.2 Finding whether an object floats or sinks

❶ Object sinks

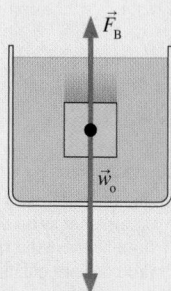

❷ Object floats

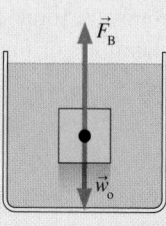

❸ Object has neutral buoyancy

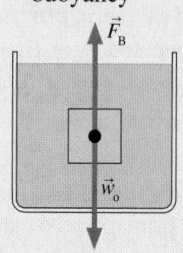

An object sinks if it weighs more than the fluid it displaces; that is, if its average density is greater than the density of the fluid:

$$\rho_{avg} > \rho_f$$

An object floats on the surface if it weighs less than the fluid it displaces; that is, if its average density is less than the density of the fluid.

$$\rho_{avg} < \rho_f$$

An object hangs motionless in the fluid if it weighs exactly the same as the fluid it displaces. It has neutral buoyancy if its average density equals the density of the fluid.

$$\rho_{avg} = \rho_f$$

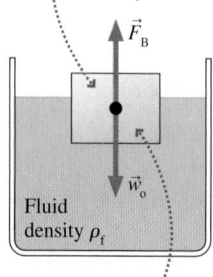

An object of density ρ_o and volume V_o is floating on a fluid of density ρ_f.

Fluid density ρ_f

The submerged volume of the object is equal to the volume V_f of displaced fluid.

FIGURE 15.23 A floating object is in static equilibrium.

As an example, steel is denser than water, so a chunk of steel sinks. Oil is less dense than water, so oil floats on water. Fish use *swim bladders* filled with air and scuba divers use weighted belts to adjust their average density to match the water. Both are examples of neutral buoyancy.

If you release a block of wood underwater, the net upward force causes the block to shoot to the surface. Then what? To understand floating, let's begin with a *uniform* object such as the block shown in Figure 15.23. This object contains nothing tricky, like indentations or voids. Because it's floating, it must be the case that $\rho_o < \rho_f$.

Now that the object is floating, it's in static equilibrium. The upward buoyant force, given by Archimedes' principle, exactly balances the downward weight of the object. That is,

$$F_B = \rho_f V_f g = w_o = \rho_o V_o g \qquad (15.15)$$

In this case, the volume of the displaced fluid is *not* the same as the volume of the object. In fact, we can see from Equation 15.15 that the volume of fluid displaced by a floating object of uniform density is

$$V_f = \frac{\rho_o}{\rho_f} V_o < V_o \qquad (15.16)$$

You've often heard it said that "90% of an iceberg is underwater." Equation 15.16 is the basis for that statement. Most icebergs break off glaciers and are fresh-water ice with a density of 917 kg/m^3. The density of seawater is 1030 kg/m^3. Thus

$$V_f = \frac{917 \text{ kg/m}^3}{1030 \text{ kg/m}^3} V_o = 0.89 V_o$$

V_f, the displaced water, is the volume of the iceberg that is underwater. You can see that, indeed, 89% of the volume of an iceberg is underwater.

NOTE ▶ Equation 15.16 applies only to *uniform* objects. It does not apply to boats, hollow spheres, or other objects of nonuniform composition. ◀

EXAMPLE 15.9 **Measuring the density of an unknown liquid**

You need to determine the density of an unknown liquid. You notice that a block floats in this liquid with 4.6 cm of the side of the block submerged. When the block is placed in water, it also floats but with 5.8 cm submerged. What is the density of the unknown liquid?

MODEL The block is an object of uniform composition.

VISUALIZE Figure 15.24 shows the block and defines the cross-section area A and submerged lengths h_u in the unknown liquid and h_w in water.

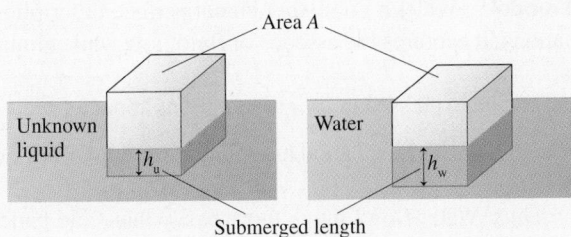

FIGURE 15.24 The block is floating.

SOLVE The block is floating, so Equation 15.16 applies. The block displaces volume $V_u = Ah_u$ of the unknown liquid. Thus

$$V_u = Ah_u = \frac{\rho_o}{\rho_u} V_o$$

Similarly, the block displaces volume $V_w = Ah_w$ of the water, leading to

$$V_w = Ah_w = \frac{\rho_o}{\rho_w} V_o$$

Because there are two fluids, we've used subscripts w for water and u the unknown in place of the fluid subscript f. The product $\rho_o V_o$ appears in both equations, hence

$$\rho_u Ah_u = \rho_w Ah_w$$

The unknown area A cancels, and the density of the unknown liquid is

$$\rho_u = \frac{h_w}{h_u} \rho_w = \frac{5.8 \text{ cm}}{4.6 \text{ cm}} 1000 \text{ kg/m}^3 = 1260 \text{ kg/m}^3$$

ASSESS Comparison with Table 15.1 shows that the unknown liquid is likely to be glycerin.

Boats

We'll conclude by designing a boat. Figure 15.25 is a physicist's idea of a boat. Four massless but rigid walls are attached to a solid steel plate of mass m_o and area A. The steel plate by itself would, of course, head straight for the bottom. But now, as the steel plate settles down into the water, the sides allow the boat to displace a volume of water much larger than that displaced by the steel alone. The boat will float if the weight of the displaced water equals the weight of the boat.

In terms of density, the boat will float if $\rho_{avg} < \rho_f$. If the sides of the boat are height h, the boat's volume is $V_o = Ah$ and its average density is $\rho_{avg} = m_o/V_o = m_o/Ah$. The boat will float if

$$\rho_{avg} = \frac{m_o}{Ah} < \rho_f \qquad (15.17)$$

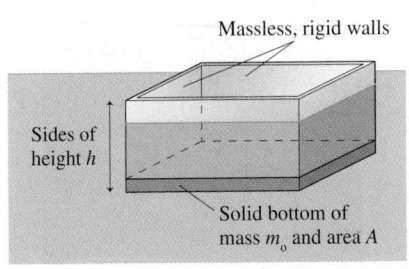

FIGURE 15.25 A physicist's boat.

Thus the minimum height of the sides, a height that would allow the boat to float (in perfectly still water!) with water right up to the rails, is

$$h_{min} = \frac{m_o}{\rho_f A} \tag{15.18}$$

As a quick example, a 5 m × 10 m steel "barge" with a 2-cm-thick floor has an area of 50 m^2 and a mass of 7900 kg. The minimum height of the massless walls, as given by Equation 15.18, is 16 cm.

Real ships and boats are more complicated, but the same idea holds true. Whether it's made of concrete, steel, or lead, **a boat will float if its geometry allows it to displace enough water to equal the weight of the boat.**

STOP TO THINK 15.4 An ice cube is floating in a glass of water that is filled entirely to the brim. When the ice cube melts, the water level will

   a. fall.
   b. stay the same, right at the brim.
   c. rise, causing the water to spill.

15.5 Fluid Dynamics

The wind blowing through your hair, a white-water river, and oil gushing from an oil well are examples of fluids in motion. We've focused thus far on fluid statics, but it's time to turn our attention to fluid dynamics.

Fluid flow is a complex subject. Many aspects, especially turbulence and the formation of eddies, are still not well understood and are areas of current science and engineering research. We will avoid these difficulties by using a simplified *model.* The **ideal-fluid model** provides a good, though not perfect, description of fluid flow in many situations. It captures the essence of fluid flow while eliminating unnecessary details.

The ideal-fluid model can be expressed in four assumptions about a fluid:

1. The fluid is *incompressible.* This is a good assumption for liquids but questionable for gases.
2. The fluid is *nonviscous.* Water flows much more easily than cold pancake syrup because the syrup is a very *viscous* fluid. **Viscosity** is a resistance to flow, and in a fluid is analogous to the kinetic friction of a solid object. Assuming that a fluid is nonviscous is equivalent to assuming that there's no friction. This is the weakest of the four assumptions for many liquids, but assuming a nonviscous liquid avoids major mathematical difficulties.
3. The flow is *steady.* That is, the fluid velocity at each point in the fluid is constant; it does not fluctuate or change with time. Flow under these conditions is called **laminar flow,** and it is distinguished from *turbulent flow.*
4. The flow is *irrotational.* The definition of *irrotational* is fairly technical, but there's a simple test. If a tiny paddle wheel anywhere in the fluid does not spin, then the flow is irrotational.

The rising smoke in the photograph of Figure 15.26 begins as laminar flow, recognizable by the smooth contours, but at some point undergoes a transition to turbulent flow. A laminar-to-turbulent transition is not uncommon in fluid flow. The ideal-fluid model can be applied to the laminar flow, but not to the turbulent flow.

Turbulent flow

Laminar flow

FIGURE 15.26 Rising smoke changes from laminar flow to turbulent flow.

The Equation of Continuity

Figure 15.27 is another interesting photograph. Here smoke is being used to help engineers visualize the air flow around a car in a wind tunnel. The smoothness of the flow tells us this is laminar flow. But notice also how the individual smoke trails retain their identity. They don't cross or get mixed together. Each smoke trail represents a *streamline* in the fluid.

Streamline

FIGURE 15.27 The laminar air flow around a car in a wind tunnel is made visible with smoke. Each smoke trail represents a streamline.

Imagine that we could inject a colored drop of water into a stream of water flowing as an ideal fluid. Because the flow is steady and frictionless, and the water is incompressible, this colored drop would maintain its identity as it flowed along. Its shape might change, becoming compressed or elongated, but it would not mix with the surrounding water.

The path or trajectory followed by this "particle of fluid" is called a **streamline.** Smoke particles mixed with the air allow you to see the streamlines in the wind-tunnel photograph of Figure 15.27. Notice also how the individual smoke trails retain their identity. Figure 15.28 illustrates three important properties of streamlines.

A bundle of neighboring streamlines, such as those shown in Figure 15.29a, form a **flow tube.** Because streamlines never cross, all the streamlines that cross plane 1 within area A_1 later cross plane 2 within area A_2. A flow tube is like an invisible pipe that keeps this portion of the flowing fluid distinct from other portions. Real pipes are also flow tubes.

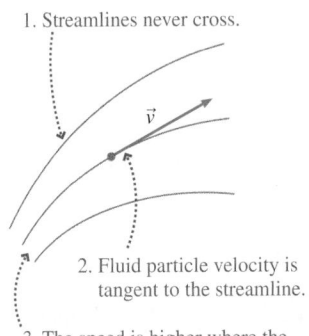

1. Streamlines never cross.

$\vec{v}$

2. Fluid particle velocity is tangent to the streamline.

3. The speed is higher where the streamlines are closer together.

FIGURE 15.28 Particles in an ideal fluid move along streamlines.

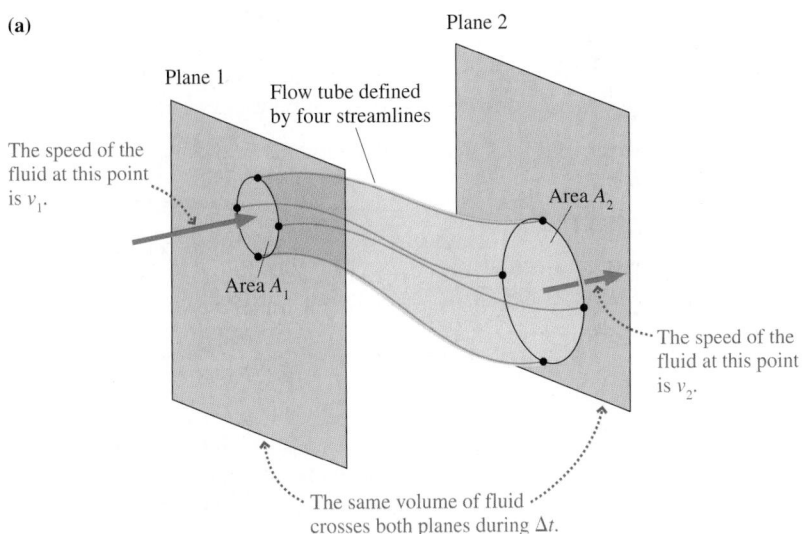

(a)

Plane 2

Plane 1

Flow tube defined by four streamlines

The speed of the fluid at this point is v_1.

Area A_2

Area A_1

The speed of the fluid at this point is v_2.

The same volume of fluid crosses both planes during Δt.

(b) The fluid moves this distance during Δt.

$\Delta x_1 = v_1 \Delta t$

v_1

A_1

$\Delta x_2 = v_2 \Delta t$

v_2

A_2

Volume $v_1 A_1 \Delta t$

Volume $v_2 A_2 \Delta t$

The fluid is incompressible, so these volumes must be equal.

FIGURE 15.29 A flow tube.

(a) Garden hose

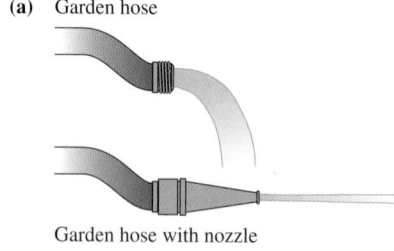

Garden hose with nozzle

(b)

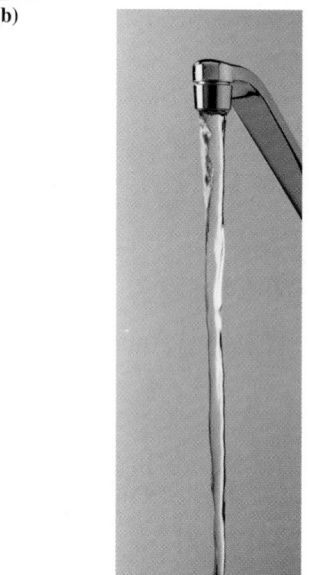

FIGURE 15.30 The speed of the fluid changes as the flow tube diameter changes. This is a consequence of the equation of continuity.

When you squeeze a toothpaste tube, the volume of toothpaste that emerges matches the amount by which you reduce the volume of the tube. An incompressible fluid in a flow tube acts the same way. Fluid is not created or destroyed within the flow tube, and it cannot be stored. If volume V enters the flow tube through area A_1 during some interval of time Δt, then an equal volume V must leave the flow tube through area A_2.

Figure 15.29b shows the flow crossing A_1 during a small interval of time Δt. If the fluid speed at this point is v_1, the fluid moves forward a small distance $\Delta x_1 = v_1 \Delta t$ and fills the volume $V_1 = A_1 \Delta x_1 = v_1 A_1 \Delta t$. The same analysis for the fluid crossing A_2 with fluid speed v_2 would find $V_2 = v_2 A_2 \Delta t$. These two volumes must be equal, leading to the conclusion that

$$v_1 A_1 = v_2 A_2 \qquad (15.19)$$

Equation 15.19 is called the **equation of continuity,** and it is one of two important equations for the flow of an ideal fluid. The equation of continuity says that **the volume of an incompressible fluid entering one part of a flow tube must be matched by an equal volume leaving downstream.**

An important consequence of the equation of continuity is that **flow is faster in narrower parts of a flow tube, slower in wider parts.** You're familiar with this conclusion from many everyday observations. The garden hose shown in Figure 15.30a squirts farther after you put a nozzle on it. This is because the narrower opening of the nozzle gives the water a higher exit speed. Water flowing from the faucet shown in Figure 15.30b picks up speed as it falls. As a result, the flow tube "necks down" to a smaller diameter.

The quantity

$$Q = vA \qquad (15.20)$$

is called the **volume flow rate.** The SI units of Q are m³/s, although in practice Q may be measured in cm³/s, liters per minute, or, in the United States, gallons per minute. Another way to express the meaning of the equation of continuity is to say that **the volume flow rate is constant at all points in a flow tube.**

EXAMPLE 15.10 **Water through a garden hose**
A garden hose has an inside diameter of 16 mm. The hose can fill a 10 L bucket in 20 s.

a. What is the speed of the water out of the end of the hose?
b. What diameter nozzle would increase the fluid speed by a factor of 4?

MODEL Treat the water as an ideal fluid. The hose itself is a flow tube, so the equation of continuity applies.

SOLVE

a. The volume flow rate is $Q = (10 \text{ L})/(20 \text{ s}) = 0.50$ L/s. To convert this to SI units you must recall that there are 1000 L in 1 m³, or 1 L $= 10^{-3}$ m³. Thus $Q = 5.0 \times 10^{-4}$ m³/s. We can find the speed of the water from Equation 15.20:

$$v = \frac{Q}{A} = \frac{Q}{\pi r^2} = \frac{5.0 \times 10^{-4} \text{ m}^3/\text{s}}{\pi(0.008 \text{ m})^2} = 2.5 \text{ m/s}$$

b. $Q = vA$ remains constant. To increase v by a factor of 4, A must be reduced by a factor of 4. The cross-section area depends on the square of the radius, so the area is reduced by a factor of 4 if the radius is reduced by a factor of 2. Thus the necessary nozzle diameter is 8 mm.

STOP TO THINK 15.5 The figure shows volume flow rates (in cm³/s) for all but one tube. What is the volume flow rate through the unmarked tube? Is the flow direction in or out?

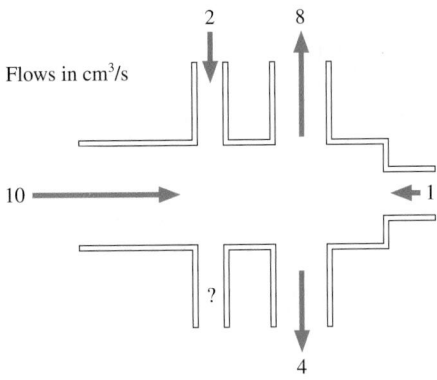

Flows in cm³/s

Bernoulli's Equation

The equation of continuity is one of two important relationships for ideal fluids. The other is a statement of energy conservation. The general statement of energy conservation that you learned in Chapter 11 is

$$\Delta K + \Delta U = W_{ext} \tag{15.21}$$

where W_{ext} is the work done by any external forces.

Let's see how this applies to the flow tube of Figure 15.31. Our system for analysis is the volume of fluid within the flow tube. Work is done on this volume of fluid by the pressure forces of the *surrounding* fluid. At point 1, the fluid to the left of the flow tube exerts force $\vec{F}_1$ on the system. This force points to the right. At the other end of the flow tube, at point 2, the fluid to the right of the flow tube exerts force $\vec{F}_2$ to the left. The pressure inside the flow tube is not relevant because those forces are internal to the system. Only external forces change the total energy.

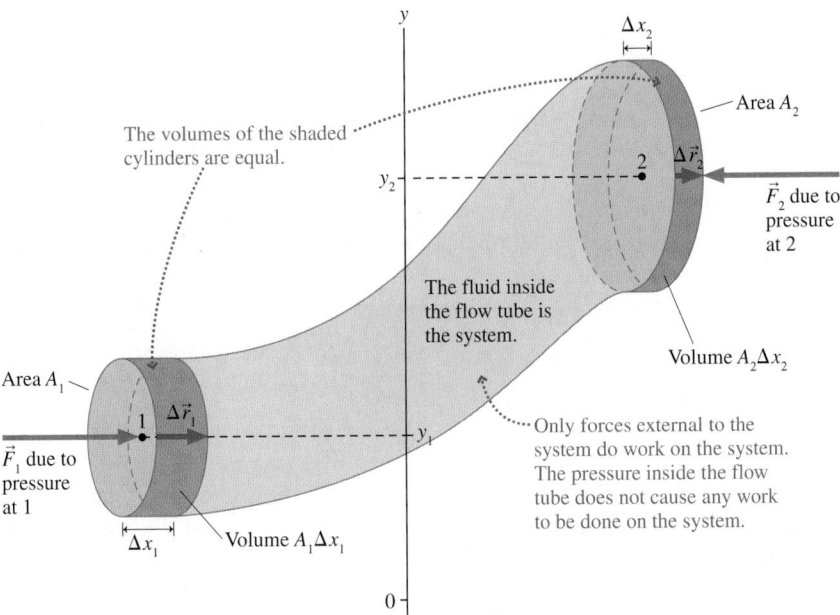

The volumes of the shaded cylinders are equal.

Area A_2

$\vec{F}_2$ due to pressure at 2

The fluid inside the flow tube is the system.

Volume $A_2\Delta x_2$

Only forces external to the system do work on the system. The pressure inside the flow tube does not cause any work to be done on the system.

Area A_1

$\vec{F}_1$ due to pressure at 1

Volume $A_1\Delta x_1$

FIGURE 15.31 Energy analysis of a flow tube.

At point 1, force $\vec{F}_1$ pushes the fluid through displacement $\Delta\vec{r}_1$. $\vec{F}_1$ and $\Delta\vec{r}_1$ are parallel, so the work done on the fluid at this point is

$$W_1 = \vec{F}_1 \cdot \Delta\vec{r}_1 = F_1\Delta r_1 = (p_1A_1)\Delta x_1 = p_1V \tag{15.22}$$

The A_1 and Δx_1 enter the equation from different terms, but they conveniently combine to give the fluid volume V.

The situation is much the same at point 2 except that $\vec{F}_2$ points opposite the displacement $\Delta\vec{r}_2$. This introduces a $\cos(180°) = -1$ into the dot product for the work, giving

$$W_2 = \vec{F}_2 \cdot \Delta\vec{r}_2 = -F_1\Delta r_1 = -(p_2A_2)\Delta x_2 = -p_2V \tag{15.23}$$

The pressure from the left at point 1 pushes the fluid ahead, a positive work. The pressure from the right at point 2 tries to slow the fluid down, a negative work. Together, the work by external forces is

$$W_{\text{ext}} = W_1 + W_2 = p_1V - p_2V \tag{15.24}$$

Now let's see how this work changes the kinetic and potential energy of the system. A small volume of fluid $V = A_1\Delta x_1$ passes point 1 and, at some later time, arrives at point 2, where the unchanged volume is $V = A_2\Delta x_2$. The change in gravitational potential energy for this volume of fluid is

$$\Delta U = mgy_2 - mgy_1 = \rho Vgy_2 - \rho Vgy_1 \tag{15.25}$$

where ρ is the fluid density. Similarly, the change in kinetic energy is

$$\Delta K = \frac{1}{2}mv_2^2 - \frac{1}{2}mv_1^2 = \frac{1}{2}\rho Vv_2^2 - \frac{1}{2}\rho Vv_1^2 \tag{15.26}$$

Combining Equations 15.24, 15.25, and 15.26 gives us the energy equation for the fluid in the flow tube:

$$\frac{1}{2}\rho Vv_2^2 - \frac{1}{2}\rho Vv_1^2 + \rho Vgy_2 - \rho Vgy_1 = p_1V - p_2V \tag{15.27}$$

The volume V cancels out of all the terms. If we regroup the terms, the energy equation becomes

$$p_1 + \frac{1}{2}\rho v_1^2 + \rho gy_1 = p_2 + \frac{1}{2}\rho v_2^2 + \rho gy_2 \tag{15.28}$$

Equation 15.28 is called **Bernoulli's equation.** It is named for the 18th-century Italian scientist Daniel Bernoulli, who made some of the earliest studies of fluid dynamics.

Bernoulli's equation is really nothing more than a statement about work and energy. It is sometimes useful to express Bernoulli's equation in the alternative form

$$p + \frac{1}{2}\rho v^2 + \rho gy = \text{constant} \tag{15.29}$$

This version of Bernoulli's equation tells us that the quantity $p + \frac{1}{2}\rho v^2 + \rho gy$ remains constant along a streamline.

One important implication of Bernoulli's equation is easily demonstrated. Before reading the next paragraph, try the simple experiment illustrated in Figure 15.32. Really, do try this!

What happened? You probably expected your breath to press the strip of paper down. Instead, the strip *rose*. In fact, the harder you blow, the more nearly the strip becomes parallel to the floor. This counterintuitive result is a consequence

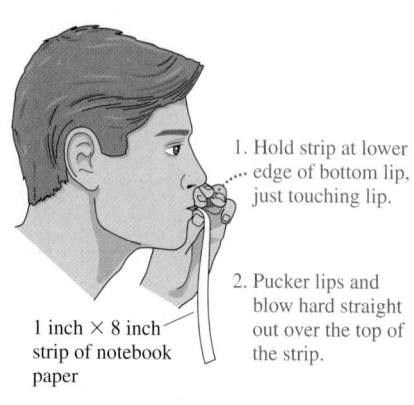

1. Hold strip at lower edge of bottom lip, just touching lip.

2. Pucker lips and blow hard straight out over the top of the strip.

1 inch × 8 inch strip of notebook paper

FIGURE 15.32 A simple demonstration of Bernoulli's equation.

of Bernoulli's equation. As the air speed above the strip of paper increases, the pressure has to *decrease* to keep the quantity $p + \frac{1}{2}\rho v^2 + \rho gy$ constant. The air-pressure force pressing down on the top surface is then *less* than the air-pressure force pushing up on the bottom surface, resulting in a net upward force on the paper.

NOTE ▶ Using Bernoulli's equation is very much like using the law of conservation of energy. Rather than identifying a "before" and "after," you want to identify two points on a streamline. As the following examples show, Bernoulli's equation is often used in conjunction with the equation of continuity. ◀

EXAMPLE 15.11 An irrigation system
Water flows through the pipes shown in Figure 15.33. The water's speed through the lower pipe is 5.0 m/s and a pressure gauge reads 75 kPa. What is the reading of the pressure gauge on the upper pipe?

MODEL Treat the water as an ideal fluid obeying Bernoulli's equation. Consider a streamline connecting point 1 in the lower pipe with point 2 in the upper pipe.

SOLVE Bernoulli's equation, Equation 15.28, relates the pressure, fluid speed, and heights at points 1 and 2. It is easily solved for the pressure p_2 at point 2:

$$p_2 = p_1 + \frac{1}{2}\rho v_1^2 - \frac{1}{2}\rho v_2^2 + \rho gy_1 - \rho gy_2$$

$$= p_1 + \frac{1}{2}\rho(v_1^2 - v_2^2) + \rho g(y_1 - y_2)$$

All quantities on the right are known except v_2, and that is where the equation of continuity will be useful. The cross-section areas and water speeds at points 1 and 2 are related by

$$v_1 A_1 = v_2 A_2$$

from which we find

$$v_2 = \frac{A_1}{A_2}v_1 = \frac{r_1^2}{r_2^2}v_1 = \frac{(0.030\text{ m})^2}{(0.020\text{ m})^2}(5.0\text{ m/s}) = 11.25\text{ m/s}$$

The pressure at point 1 is $p_1 = 0.75$ kPa $+ 1$ atm $= 176{,}300$ Pa. We can now use the above expression for p_2 to calculate $p_2 = 105{,}900$ Pa. This is the absolute pressure; the pressure gauge on the upper pipe will read

$$p_2 = 176{,}300\text{ Pa} - 1\text{ atm} = 4.6\text{ kPa}$$

ASSESS Reducing the pipe size decreases the pressure because it makes $v_2 > v_1$. Gaining elevation also reduces the pressure.

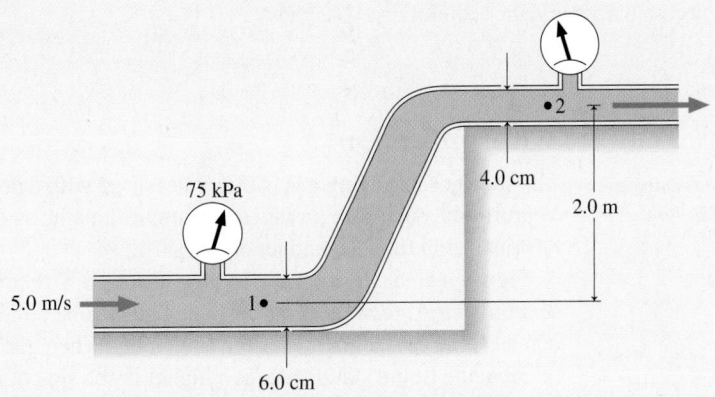

FIGURE 15.33 The water pipes of an irrigation system.

EXAMPLE 15.12 Hydroelectric power
Small hydroelectric plants in the mountains sometimes bring the water from a reservoir down to the power plant through enclosed tubes. In one such plant, the 100-cm-diameter intake tube in the base of the dam is 50 m below the reservoir surface. The water drops 200 m through the tube before flowing into the turbine through a 50-cm-diameter nozzle.

a. What is the water speed into the turbine?
b. By how much does the inlet pressure differ from the hydrostatic pressure at that depth?

MODEL Treat the water as an ideal fluid obeying Bernoulli's equation. Consider a streamline that begins at the surface of the reservoir and ends at the exit of the nozzle. The pressure at the surface is $p_1 = p_{atmos}$ and $v_1 \approx 0$ m/s. The water discharges into air, so $p_2 = p_{atm}$ at the exit.

VISUALIZE Figure 15.34 on the next page is a pictorial representation of the situation.

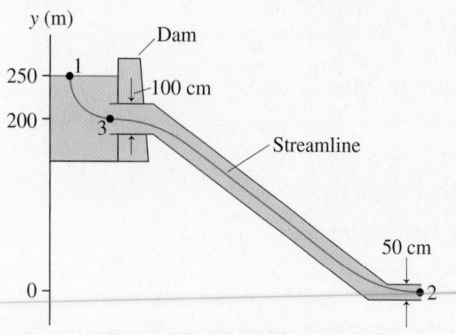

FIGURE 15.34 Pictorial representation of the water flow to a hydroelectric plant.

SOLVE

a. The power plant is in the mountains, where $p_{atmos} < 1$ atm, but p_{atmos} occurs on both sides of Bernoulli's equation and cancels. Bernoulli's equation, with $v_1 = 0$ m/s and $y_2 = 0$ m, is

$$p_{atmos} + \rho g y_1 = p_{atmos} + \frac{1}{2}\rho v_2^2$$

p_{atmos} cancels, as expected, as does the density ρ. Solving for v_2 gives

$$v_2 = \sqrt{2gy_1} = \sqrt{2(9.80 \text{ m/s}^2)(250 \text{ m})} = 70.0 \text{ m/s}$$

b. You might expect the pressure p_3 at the intake to be the hydrostatic pressure $p_{atmos} + \rho g d$ at depth d. But the water is *flowing* into the intake tube, so it's not in static equilibrium.

We can find the intake speed v_3 from the equation of continuity:

$$v_3 = \frac{A_2}{A_3}v_2 = \frac{r_2^2}{r_3^2}\sqrt{2gy_1}$$

The intake is along the streamline between points 1 and 2, so we can apply Bernoulli's equation to points 1 and 3:

$$p_{atmos} + \rho g y_1 = p_3 + \frac{1}{2}\rho v_3^2 + \rho g y_3$$

Solving this equation for p_3, and noting that $y_1 - y_3 = d$, we find

$$p_3 = p_{atmos} + \rho g(y_1 - y_3) - \frac{1}{2}\rho v_3^2$$

$$= p_{atmos} + \rho g d - \frac{1}{2}\rho \left(\frac{r_2}{r_3}\right)^4 (2gy_1)$$

$$= p_{static} - \rho g y_1 \left(\frac{r_2}{r_3}\right)^4$$

The intake pressure is *less* than hydrostatic pressure by the amount

$$\rho g y_1 \left(\frac{r_2}{r_3}\right)^4 = 153{,}000 \text{ Pa} = 1.5 \text{ atm}$$

ASSESS The water's exit speed from the nozzle is the same as if it fell 250 m from the surface of the reservoir. This isn't surprising because we've assumed a nonviscous (i.e., frictionless) liquid. "Real" water would have less speed but still flow very fast.

Two Applications

The speed of a flowing gas is often measured with a device called a **Venturi tube.** Venturi tubes measure gas speeds in environments as different as chemistry laboratories, wind tunnels, and jet engines.

Figure 15.35 shows gas flowing through a tube that changes from cross-section area A_1 to area A_2. A U-shaped glass tube containing liquid of density ρ_{liq} connects the two segments of the flow tube. When gas flows through the horizontal tube, the liquid stands height h higher in the side of the U tube connected to the narrow segment of the flow tube.

Figure 15.35 shows how a Venturi tube works. We can make this analysis quantitative and determine the gas-flow speed from the liquid height h. Two pieces of information we have to work with are Bernoulli's equation

$$p_1 + \frac{1}{2}\rho v_1^2 + \rho g y_1 = p_2 + \frac{1}{2}\rho v_2^2 + \rho g y_2 \qquad (15.30)$$

and the equation of continuity

$$v_2 A_2 = v_1 A_1 \qquad (15.31)$$

In addition, the hydrostatic equation for the liquid tells us that the pressure p_2 above the right tube differs from the pressure p_1 above the left tube by $\rho_{liq}gh$. That is,

$$p_2 = p_1 - \rho_{liq}gh \qquad (15.32)$$

1. As the gas flows into a smaller cross section, it speeds up (equation of continuity). As it speeds up, the pressure decreases (Bernoulli's equation).

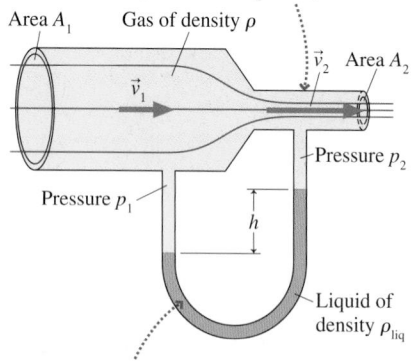

2. The U tube acts like a manometer. The liquid level is higher on the side where the pressure is lower.

FIGURE 15.35 A Venturi tube measures gas-flow speeds.

First use Equations 15.31 and 15.32 to eliminate v_2 and p_2 in Bernoulli's equation:

$$p_1 + \frac{1}{2}\rho v_1^2 = (p_1 - \rho_{liq}gh) + \frac{1}{2}\rho\left(\frac{A_1}{A_2}\right)^2 v_1^2 \qquad (15.33)$$

The potential energy terms have disappeared because $y_1 = y_2$ for a horizontal tube. Equation 15.33 can now be solved for v_1, then v_2 is obtained from Equation 15.31. We'll skip a few algebraic steps and go right to the result:

$$v_1 = A_2\sqrt{\frac{2\rho_{liq}gh}{\rho(A_1^2 - A_2^2)}}$$

$$v_2 = A_1\sqrt{\frac{2\rho_{liq}gh}{\rho(A_1^2 - A_2^2)}} \qquad (15.34)$$

In practice, the equations for the gas-flow speeds have to be corrected for the fact that the gas, which is compressible, is not an ideal liquid. But Equation 15.34 is reasonably accurate even without corrections as long as the flow speeds are much less than the speed of sound, about 340 m/s. For us, the Venturi tube is an example of the power of Bernoulli's equation.

As a final example, we can use Bernoulli's equation to understand, at least qualitatively, how airplane wings generate *lift*. Figure 15.36 shows the cross section of an airplane wing. This shape is called an *airfoil*.

Although you usually think of an airplane moving through the air, in the airplane's reference frame it is the air that flows across a stationary wing. As it does, the streamlines must separate. The flat bottom of the wing does not significantly alter the streamlines going under the wing. But the streamlines going over the wing get bunched together. This bunching reduces the cross-section area of a flow tube of streamlines. Consequently, in accordance with the equation of continuity, the air speed must increase as it flows across the top of the wing.

As you've seen several times, an increased air speed implies a decreased air pressure. This is the lesson of Bernoulli's equation. Because the air pressure above the wing is less than the air pressure below, the air exerts a net upward force on the wing, just as it did on the paper strip you blew across. The upward force of the air due to a pressure difference across the wing is called **lift.**

A complete analysis of the lift of a wing is quite complicated and involves many factors in addition to Bernoulli's equation. Nonetheless, you should now be able to understand one of the important physical principles that are involved.

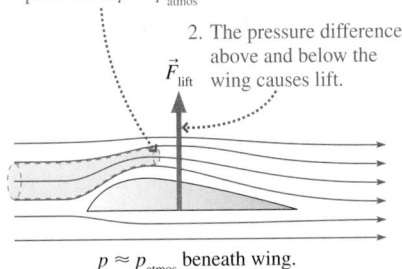

1. Flow tube decreases in size due to compression of streamlines. The higher speed lowers the pressure to $p < p_{atmos}$.

2. The pressure difference above and below the wing causes lift.

$\vec{F}_{lift}$

$p \approx p_{atmos}$ beneath wing.

FIGURE 15.36 Air flow over a wing generates lift by creating unequal pressures above and below.

STOP TO THINK 15.6   Rank in order, from highest to lowest, the liquid heights h_1 to h_4 in tubes 1 to 4. The air flow is from left to right.

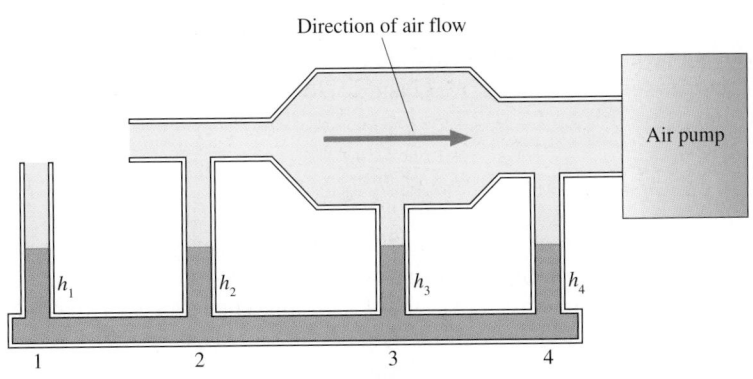

Direction of air flow

Air pump

h_1 h_2 h_3 h_4

1   2   3   4

15.6 Elasticity

The final subject to explore in this chapter is elasticity. Although elasticity applies primarily to solids rather than fluids, you will see that similar ideas come into play.

Tensile Stress and Young's Modulus

Suppose you clamp one end of a solid rod while using a strong machine to pull on the other with force $\vec{F}$. Figure 15.37a shows the experimental arrangement. We usually think of solids as being, well, solid. But any material, be it plastic, concrete, or steel, will stretch as the spring-like molecular bonds expand.

Figure 15.37b shows graphically the amount of force needed to stretch the rod by the amount ΔL. This graph contains several regions of interest. First is the *elastic region,* ending at the *elastic limit.* As long as ΔL is less than the elastic limit, the rod will return to its initial length L when the force is removed. Just such a reversible stretch is what we mean when we say a material is *elastic.* A stretch beyond the elastic limit will permanently deform the object; it will not return to its initial length when the force is removed. And, not surprisingly, there comes a point when the rod breaks.

For most materials, the graph begins with a *linear region,* which is where we will focus our attention. If ΔL is within the linear region, the force needed to stretch the rod is

$$F = k\Delta L \tag{15.35}$$

where k is the slope of the graph. You'll recognize Equation 15.35 as none other than Hooke's law.

The difficulty with Equation 15.35 is that the proportionality constant k depends both on the composition of the rod—whether it is, say, plastic or aluminum—and on the rod's length and cross-section area. It would be useful to characterize the elastic properties of plastic in general, or aluminum in general, without needing to know the dimensions of a specific rod.

We can meet this goal by thinking about Hooke's law at the atomic scale. The elasticity of a material is directly related to the spring constant of the molecular bonds between neighboring atoms. As Figure 15.38 shows, the force pulling each bond is proportional to the quantity F/A. This force causes each bond to stretch by an amount proportional to $\Delta L/L$. We don't know what the proportionality constants are, but we don't need to. Hooke's law applied to a molecular bond tells us that the force pulling on a bond is proportional to the amount that the bond stretches. Thus F/A must be proportional to $\Delta L/L$. We can write their proportionality as

$$\frac{F}{A} = Y\frac{\Delta L}{L} \tag{15.36}$$

The proportionality constant Y is called **Young's modulus.** It is directly related to the spring constant of the molecular bonds, so it depends on the material from which the object is made but *not* on the object's geometry.

A comparison of Equations 15.35 and 15.36 shows that Young's modulus can be written

$$Y = \frac{kL}{A} \tag{15.37}$$

This is not a definition of Young's modulus but simply an expression for making an experimental determination of the value of Young's modulus. This k is the spring constant of the bar seen in Figure 15.37. It is a quantity easily measured in the laboratory.

The quantity F/A, where A is the cross-section area, is called **tensile stress.** Notice that it is essentially the same definition as pressure. Even so, tensile stress

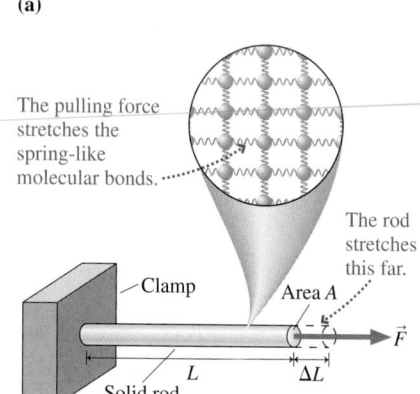

(a)

The pulling force stretches the spring-like molecular bonds.

The rod stretches this far.

Clamp

Area A

$\vec{F}$

L

ΔL

Solid rod

(b)

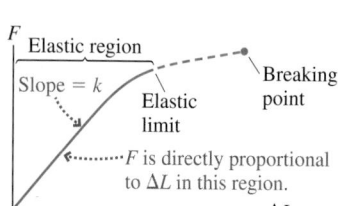

F

Elastic region

Slope = k

Elastic limit

Breaking point

F is directly proportional to ΔL in this region.

ΔL

Linear region

FIGURE 15.37 Stretching a solid rod.

The number of bonds is proportional to area A. If the rod is pulled with force F, the force pulling on each bond is proportional to F/A.

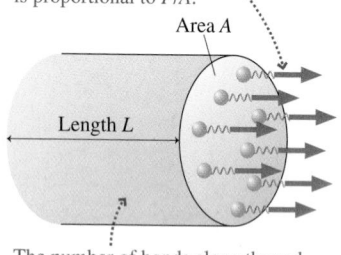

Area A

Length L

The number of bonds along the rod is proportional to length L. If the rod stretches by ΔL, the stretch of each bond is proportional to $\Delta L/L$.

FIGURE 15.38 A material's elasticity is directly related to the spring constant of the molecular bonds.

differs in that the stress is applied in a particular direction whereas pressure forces are exerted in all directions. Another difference is that stress is measured in N/m^2 rather than pascals. The quantity $\Delta L/L$, the fractional increase in the length, is called **strain.** Strain is dimensionless. The numerical values of strain are always very small because solids cannot be stretched very much before reaching the breaking point.

With these definitions, Equation 15.36 can be written

$$\text{stress} = Y \times \text{strain} \qquad (15.38)$$

Because strain is dimensionless, Young's modulus Y has the same dimensions as stress, namely N/m^2. Table 15.3 gives values of Young's modulus for several common materials. Large values of Y characterize materials that are stiff and rigid. "Softer" materials, at least relatively speaking, have smaller values of Y. You can see that steel has a larger Young's modulus than aluminum.

TABLE 15.3 Elastic properties of various materials

Substance	Young's modulus (N/m^2)	Bulk modulus (N/m^2)
Aluminum	7×10^{10}	7×10^{10}
Concrete	3×10^{10}	–
Copper	11×10^{10}	14×10^{10}
Mercury	–	3×10^{10}
Plastic (polystyrene)	0.3×10^{10}	–
Steel	20×10^{10}	16×10^{10}
Water	–	0.2×10^{10}
Wood (Douglas fir)	1×10^{10}	–

We introduced Young's modulus by considering how materials stretch. But Equation 15.38 and Young's modulus also apply to the compression of materials. Compression is particularly important in engineering applications, where beams, columns, and support foundations are compressed by the load they bear. Concrete is often compressed, as in columns that support highway overpasses, but rarely stretched.

NOTE ▶ Whether the rod is stretched or compressed, Equation 15.38 is valid only in the linear region of the graph in Figure 15.37b. The breaking point is usually well outside the linear region, so you can't use Young's modulus to compute the maximum possible stretch or compression. ◀

EXAMPLE 15.13   Stretching a wire
A 2.0-m-long, 1.0-mm-diameter wire is suspended from the ceiling. Hanging a 4.5 kg mass from the wire stretches the wire's length by 1.0 mm. What is Young's modulus for this wire? Can you identify the material?

MODEL The hanging mass creates tensile stress in the wire.

SOLVE The force pulling on the wire, which is simply the weight of the hanging mass, produces tensile stress

$$\frac{F}{A} = \frac{mg}{\pi r^2} = \frac{(4.5 \text{ kg})(9.80 \text{ m/s}^2)}{\pi (0.0005 \text{ m})^2} = 5.6 \times 10^7 \text{ N/m}^2$$

The resulting stretch of 1.0 mm is a strain of $\Delta L/L = (1.0 \text{ mm})/(2000 \text{ mm}) = 5.0 \times 10^{-4}$. Thus Young's modulus for the wire is

$$Y = \frac{F/A}{\Delta L/L} = 11 \times 10^{10} \text{ N/m}^2$$

Referring to Table 15.3, we see that the wire is made of copper.

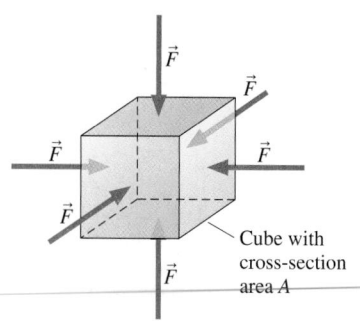

FIGURE 15.39 An object is compressed by pressure forces pushing equally on all sides.

Volume Stress and the Bulk Modulus

Young's modulus characterizes the response of an object to being pulled in one direction. Figure 15.39 shows an object being squeezed in all directions. For example, objects under water are squeezed from all sides by the water pressure. The force per unit area F/A applied to *all* surfaces of an object is called the **volume stress.** Because the force pushes equally on all sides, the volume stress (unlike the tensile stress) really is the same as pressure p.

No material is perfectly rigid. A volume stress applied to an object compresses its volume slightly. The **volume strain** is defined as $\Delta V/V$. The volume strain is a *negative* number because the volume stress *decreases* the volume.

Volume stress, or pressure, is linearly proportional to the volume strain, much as the tensile stress is linearly proportional to the strain in a rod. That is,

$$\frac{F}{A} = p = -B\frac{\Delta V}{V} \tag{15.39}$$

where B is called the **bulk modulus.** The negative sign in Equation 15.39 ensures that the pressure is a positive number. Table 15.3 gives values of the bulk modulus for several materials. Smaller values of B correspond to materials that are more easily compressed. Both solids and liquids can be compressed and thus have a bulk modulus, whereas Young's modulus applies only to solids.

EXAMPLE 15.14 **Compressing a sphere**
A 1.00-m-diameter solid steel sphere is lowered to a depth of 10,000 m in a deep ocean trench. By how much does its diameter shrink?

MODEL The water pressure applies a volume stress to the sphere.

SOLVE The water pressure at $d = 10,000$ m is

$$p = p_0 + \rho g d = 1.01 \times 10^8 \text{ Pa}$$

where we used the density of seawater. The bulk modulus of steel, taken from Table 15.3, is 16×10^{10} N/m². Thus the volume strain is

$$\frac{\Delta V}{V} = -\frac{p}{B} = -\frac{1.01 \times 10^8 \text{ Pa}}{16 \times 10^{10} \text{ Pa}} = -6.3 \times 10^{-4}$$

The volume of a sphere is $V = (4/3)\pi r^3$. For a very small change, we can use calculus to relate the volume change to the change in radius:

$$\Delta V = \frac{4\pi}{3}\Delta(r^3) = \frac{4\pi}{3} \cdot 3r^2\Delta r = 4\pi r^2\Delta r$$

Using this expression for ΔV gives the volume strain:

$$\frac{\Delta V}{V} = \frac{4\pi r^2\Delta r}{(4/3)\pi r^3} = \frac{3\Delta r}{r} = -6.3 \times 10^{-4}$$

Solving for Δr gives $\Delta r = -1.05 \times 10^{-4}$ m $= -0.105$ mm. The diameter changes by twice this, decreasing 0.21 mm.

ASSESS The immense pressure of the deep ocean causes only a tiny change in the sphere's diameter. You can see that treating solids and liquids as incompressible is an excellent approximation under nearly all circumstances.

SUMMARY

The goal of Chapter 15 has been to understand macroscopic systems that flow or deform.

GENERAL PRINCIPLES

Fluid Statics

Gases	Liquids
• Freely moving particles	• Loosely bound particles
• Compressible	• Incompressible
• Pressure primarily thermal	• Pressure primarily gravitational
• Pressure constant in a laboratory-size container	• Hydrostatic pressure at depth d is $p = p_0 + \rho g d$

Fluid Dynamics

Ideal-fluid model

• Incompressible

• Smooth, laminar flow

• Nonviscous

• Irrotational

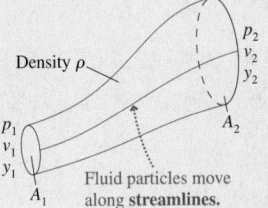

Density ρ

p_1, v_1, y_1 A_1 p_2, v_2, y_2 A_2

Fluid particles move along **streamlines.**

Equation of continuity

$$v_1 A_1 = v_2 A_2$$

Bernoulli's equation

$$p_1 + \tfrac{1}{2}\rho v_1^2 + \rho g y_1 = p_2 + \tfrac{1}{2}\rho v_2^2 + \rho g y_2$$

Bernoulli's equation is a statement of energy conservation.

IMPORTANT CONCEPTS

Density $\rho = m/V$, where m is mass and V is volume.

Pressure $p = F/A$, where F is the magnitude of the fluid force and A is the area on which the force acts.

• Exists at all points in a fluid

• Pushes equally in all directions

• Constant along a horizontal line

• Gauge pressure $p_g = p - 1$ atm

APPLICATIONS

Buoyancy is the upward force of a fluid on an object.

Archimedes' principle

The magnitude of the buoyant force equals the weight of the fluid displaced by the object.

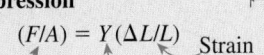

ρ_f $\vec{F}_B$ $\vec{w}_o$

Sink	$\rho_{avg} > \rho_f$	$F_B < w_o$
Rise to surface	$\rho_{avg} < \rho_f$	$F_B > w_o$
Neutrally buoyant	$\rho_{avg} = \rho_f$	$F_B = w_o$

Elasticity describes the deformation of solids and liquids under stress.

Linear stretch and compression

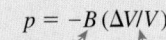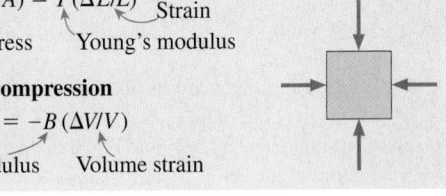

A $\vec{F}$ L ΔL

$$(F/A) = Y\,(\Delta L/L)$$

Tensile stress   Young's modulus   Strain

Volume compression

$$p = -B\,(\Delta V/V)$$

Bulk modulus   Volume strain

TERMS AND NOTATION

fluid	hydrostatic pressure	ideal-fluid model	lift
gas	Pascal's principle	viscosity	Young's modulus, Y
liquid	gauge pressure, p_g	laminar flow	tensile stress
mass density, ρ	hydraulics	streamline	strain
unit volume	buoyant force	flow tube	volume stress
pressure, p	displaced fluid	equation of continuity	volume strain
pascal, Pa	Archimedes' principle	volume flow rate, Q	bulk modulus, B
vacuum	average density, ρ_{avg}	Bernoulli's equation	
standard atmosphere, atm	neutral buoyancy	Venturi tube	

EXERCISES AND PROBLEMS

Exercises

Section 15.1 Fluids

1. A 100 mL beaker holds 120 g of liquid. What is the liquid's density in SI units?

2. Containers A and B have equal volumes. The mass of the liquid in B is 6300 times the mass of helium gas in container A. Identify the liquid in B.

3. A 6 m × 12 m swimming pool slopes linearly from a 1.0 m depth at one end to a 3.0 m depth at the other. What is the mass of water in the pool?

4. a. 50 g of gasoline are mixed with 50 g of water. What is the average density of the mixture?
   b. 50 cm³ of gasoline are mixed with 50 cm³ of water. What is the average density of the mixture?

Section 15.2 Pressure

5. The deepest point in the ocean is 11 km below sea level, deeper than Mt. Everest is tall. What is the pressure in atmospheres at this depth?

6. a. What volume of water has the same mass as 8.0 m³ of ethyl alcohol?
   b. If this volume of water is in a cubic tank, what is the pressure at the bottom?

7. A 1.0-m-diameter vat of liquid is 2.0 m deep. The pressure at the bottom of the vat is 1.3 atm. What is the mass of the liquid in the vat?

8. A 50-cm-thick layer of oil floats on a 120-cm-thick layer of water. What is the pressure at the bottom of the water layer?

9. A research submarine has a 20-cm-diameter window 8.0 cm thick. The manufacturer says the window can withstand forces up to 1.0×10^6 N. What is the submarine's maximum safe depth? The pressure inside the submarine is maintained at 1.0 atm.

10. A 20-cm-diameter circular cover is placed over a 10-cm-diameter hole that leads into an evacuated chamber. The pressure in the chamber is 20 kPa. How much force is required to pull the cover off?

Section 15.3 Measuring and Using Pressure

11. What is the gas pressure inside the box?

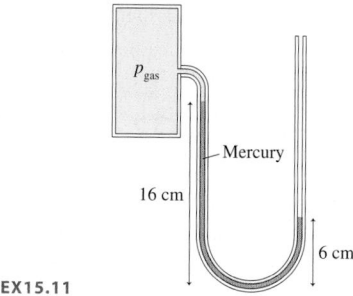

FIGURE EX15.11

12. What is the height of a water barometer at atmospheric pressure?

13. A 55 kg cheerleader uses an oil-filled hydraulic lift to hold four 110 kg football players at a height of 1.0 m. If her piston is 16 cm in diameter, what is the diameter of the football players' piston?

14. How far must a 2.0-cm-diameter piston be pushed down into one cylinder of a hydraulic lift to raise an 8.0-cm-diameter piston by 20 cm?

Section 15.4 Buoyancy

15. A 6.0-cm-diameter sphere with a mass of 89.3 g is neutrally buoyant in a liquid. Identify the liquid.

16. Two identical beakers are filled to the same height with water. Beaker B has a plastic sphere floating in it. Which beaker, with all its contents, weighs more? Or are they equal? Explain.

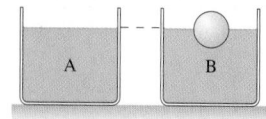

FIGURE EX15.16

17. Styrofoam has a density of 300 kg/m³. What is the maximum mass that can hang without sinking from a 50-cm-diameter Styrofoam sphere in water? Assume the volume of the mass is negligible compared to that of the sphere.

18. A 10 cm × 10 cm × 10 cm wood block with a density of 700 kg/m^3 floats in water. What is the distance from the top of the block to the water if the water is fresh? If it's seawater?

19. What is the tension in the string?

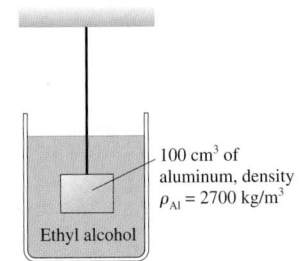

FIGURE EX15.19

100 cm^3 of aluminum, density ρ_{Al} = 2700 kg/m^3

Ethyl alcohol

Section 15.5 Fluid Dynamics

20. River Pascal with a volume flow rate of 5.0×10^5 L/s joins with River Archimedes, which carries 10.0×10^5 L/s, to form the Bernoulli River. The Bernoulli River is 150 m wide and 10 m deep. What is the speed of the water in the Bernoulli River?

21. Water flowing through a 2.0-cm-diameter pipe can fill a 300 L bathtub in 5.0 minutes. What is the speed of the water in the pipe?

22. A 1.0-cm-diameter pipe widens to 2.0 cm, then narrows to 0.5 cm. Liquid flows through the first segment at a speed of 4.0 m/s.
    a. What is the speed in the second and third segments?
    b. What is the volume flow rate through the pipe?

23. A long horizontal tube has a square cross section with sides of width L. A fluid moves through the tube with speed v_0. The tube then changes to a circular cross section with diameter L. What is the fluid's speed in the circular part of the tube?

24. What does the top pressure gauge read?

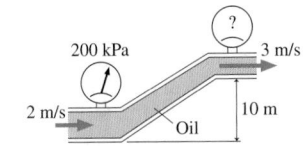

FIGURE EX15.24

200 kPa   ?   3 m/s   2 m/s   10 m   Oil

Section 15.6 Elasticity

25. What hanging mass will stretch a 2.0-m-long, 0.50-mm-diameter steel wire by 1.0 mm?

26. An 80-cm-long, 1.0-mm-diameter steel guitar string must be tightened to a tension of 2000 N by turning the tuning screws. By how much is the string stretched?

27. A 3.0-m-tall, 50-cm-diameter concrete column supports a 200,000 kg load. By how much is the column compressed?

28. At what ocean depth would the volume of an aluminum sphere be reduced by 0.10%?

Problems

29. A gymnasium is 16 m high. By what percent is the air pressure at the floor greater than the air pressure at the ceiling?

30. The two 60-cm-diameter cylinders in Figure P15.30, closed at one end, open at the other, are joined to form a single cylinder, then the air inside is removed.

    a. How much force does the atmosphere exert on the flat end of each cylinder?
    b. Suppose one cylinder is bolted to a sturdy ceiling. How many 100 kg football players would need to hang from the lower cylinder to pull the two cylinders apart?

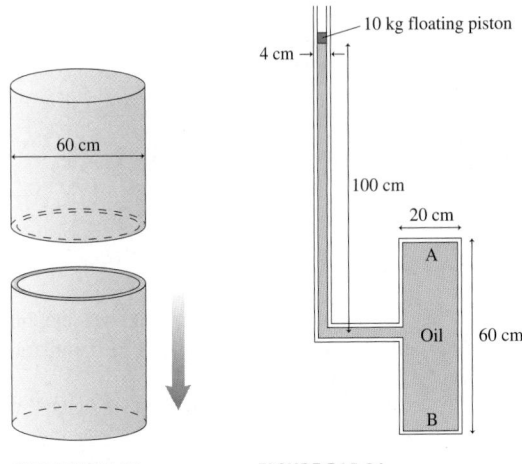

FIGURE P15.30   **FIGURE P15.31**

60 cm   10 kg floating piston   4 cm   100 cm   20 cm   A   Oil   60 cm   B

31. a. In Figure P15.31, how much force does the fluid exert on the end of the cylinder at A?
    b. How much force does the fluid exert on the end of the cylinder at B?

32. A friend asks you how much pressure is in your car tires. You know that the tire manufacturer recommends 30 psi, but it's been a while since you've checked. You can't find a tire gauge in the car, but you do find the owner's manual and a ruler. Fortunately, you've just finished taking physics, so you tell your friend, "I don't know, but I can figure it out." From the owner's manual you find that the car's mass is 1500 kg. It seems reasonable to assume that each tire supports one-fourth of the weight. With the ruler you find that the tires are 15 cm wide and the flattened segment of the tire in contact with the road is 13 cm long. What answer will you give your friend?

33. A 2.0 mL syringe has an inner diameter of 6.0 mm, a needle inner diameter of 0.25 mm, and a plunger pad diameter (where you place your finger) of 1.2 cm. A nurse uses the syringe to inject medicine into a patient whose blood pressure is 140/100.
    a. What is the minimum force the nurse needs to apply to the syringe?
    b. The nurse empties the syringe in 2.0 s. What is the flow speed of the medicine through the needle?

34. What is the longest vertical soda straw you could possibly drink from?

35. What is the total mass of the earth's atmosphere?

36. Suppose the density of the earth's atmosphere were a constant 1.3 kg/m^3, independent of height, until reaching the top. How thick would the atmosphere be?

37. Your science teacher has assigned you the task of building a water barometer. You've learned that the pressure of the atmosphere can vary by as much as 5% from 1 standard atmosphere as the weather changes.
    a. What minimum height must your barometer have?
    b. One stormy day the TV weather person says, "The barometric pressure this afternoon is a low 29.55 inches." What is the height of the water in your barometer?

38. The container shown in Figure P15.38 is filled with oil. It is open to the atmosphere on the left.
    a. What is the pressure at point A?
    b. What is the pressure difference between points A and B? Between points A and C?

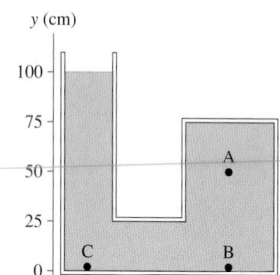

FIGURE P15.38

39. a. A 70 kg student balances a 1200 kg elephant on a hydraulic lift. What is the diameter of the piston the student is standing on?
    b. A second 70 kg student joins the first student. How high do they lift the elephant?

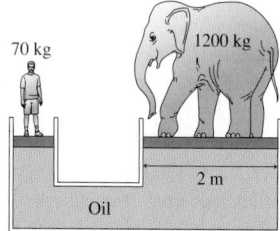

FIGURE P15.39

40. Figure 15.19 showed a hydraulic lift with force $\vec{F}_1$ balancing force $\vec{F}_2$. Assume that force $\vec{F}_2$ is the unchanging weight mg of an object of mass m. Derive Equation 15.13, which states that the force *increment* needed to lift the weight through distance d_2 is $\Delta F = \rho g (A_1 + A_2) d_2$, where ρ is the density of the liquid.

41. A U-shaped tube, open to the air on both ends, contains mercury. Water is poured into the left arm until the water column is 10.0 cm deep. How far upward from its initial position does the mercury in the right arm rise?

42. Glycerin is poured into an open U-shaped tube until the height in both sides is 20 cm. Ethyl alcohol is then poured into one arm until the height of the alcohol column is 20 cm. The two liquids do not mix. What is the difference in height between the top surface of the glycerin and the top surface of the alcohol?

43. Water stands at depth d behind a dam of width w.
    a. Find an expression for the net force of the water on the dam.
    b. Evaluate the net force on a 100-m-wide dam with a 60 m water depth.
    Hint: This problem requires an integration.

44. An aquarium tank is 100 cm long, 35 cm wide, and 40 cm deep. It is filled to the top.
    a. What is the force of the water on the bottom (100 cm × 35 cm) of the tank?
    b. What is the force of the water on the front window (100 cm × 40 cm) of the tank?
    Hint: This problem requires an integration.

45. It's possible to use the ideal gas law to show that the density of the earth's atmosphere decreases exponentially with height. That is, $\rho = \rho_0 \exp(-z/z_0)$, where z is the height above sea level, ρ_0 is the density at sea level (you can use the Table 15.1 value), and z_0 is called the *scale height* of the atmosphere. (See Challenge Problem 76.)
    a. Determine the value of z_0.
    b. What is the density of the air in Denver, at an elevation of 1600 m? What percent of sea-level density is this?
    Hint: This problem requires an integration. What is the weight of a column of air?

46. The average density of the body of a fish is 1080 kg/m³. To keep from sinking, the fish increases its volume by inflating an internal air bladder with air. By what percent must the fish increase its volume to be neutrally bouyant in fresh water? You can use the Table 15.1 value for the density of air.

47. A 6.0-cm-tall cylinder floats in water with its axis perpendicular to the surface. The length of the cylinder above water is 2.0 cm. What is the cylinder's mass density?

48. A sphere completely submerged in water is tethered to the bottom with a string. The tension in the string is one-third the weight of the sphere. What is the density of the sphere?

49. A 10 kg irregularly shaped rock with a density of 4200 kg/m³ rests on the bottom of a swimming pool. What is the normal force of the pool bottom on the rock?

50. You need to determine the density of a ceramic statue. If you suspend it from a spring scale, the scale reads 28.4 N. If you then lower the statue into a tub of water, so that it is completely submerged, the scale reads 17.0 N. What is the density?

51. A 5.0 kg rock whose density is 4800 kg/m³ is suspended by a string such that half of the rock's volume is under water. What is the tension in the string?

52. A 10 cm × 10 cm × 10 cm block of steel ($\rho_{steel} = 7900$ kg/m³) is suspended from a spring scale. The scale is in newtons.
    a. What is the scale reading if the block is in air?
    b. What is the scale reading after the block has been lowered into a beaker of oil and is completely submerged?

53. A 10-cm-diameter, 20-cm-tall steel cylinder ($\rho_{steel} = 7900$ kg/m³) floats in mercury. The axis of the cylinder is perpendicular to the surface. What length of steel is above the surface?

54. A cylinder with cross-section area A floats with its long axis vertical in a liquid of density ρ.
    a. Pressing down on the cylinder pushes it deeper into the liquid. Find an expression for the force needed to push the cylinder distance x deeper into the liquid and hold it there.
    b. A 4.0-cm-diameter cylinder floats in water. How much work must be done to push the cylinder 10 cm deeper into the water?
    Hint: An integration is required.

55. A less-dense liquid of density ρ_1 floats on top of a more-dense liquid of density ρ_2. A uniform cylinder of length l and density ρ, with $\rho_1 < \rho < \rho_2$, floats at the interface with its long axis vertical. What fraction of the length is in the more-dense liquid?

56. A 30-cm-tall, 4.0-cm-diameter plastic tube has a sealed bottom. 250 g of lead pellets are poured into the bottom of the tube, whose mass is 30 g, then the tube is lowered into a liquid. The tube floats with 5.0 cm extending above the surface. What is the density of the liquid?

57. A 355 mL soda can is 6.2 cm in diameter and has a mass of 20 g. Such a soda can half full of water is floating upright in water. What length of the can is above the water level?

58. The cylinder shown in the figure is completely submerged in a liquid of density ρ.

    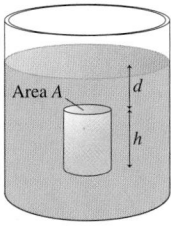
    FIGURE P15.58

    a. Find expressions for the forces F_{top} and F_{bot} that the liquid exerts on the top and bottom of the cylinder.
    b. Write an expression for the magnitude of the buoyant force on the cylinder.
    c. Can you write your expression of part b in terms of the weight of the liquid that has been displaced by the cylinder?
    Congratulations! You've just rediscovered Archimedes' principle.

59. The bottom of a steel "boat" is a 5 m × 10 m × 2 cm piece of steel ($\rho_{steel} = 7900$ kg/m^3). The sides are made of 0.50-cm-thick steel. What minimum height must the sides have for this boat to float in perfectly calm water?

60. Water flows at 5.0 L/s through a horizontal pipe that narrows smoothly from 10 cm diameter to 5.0 cm diameter. A pressure gauge in the narrow section reads 50 kPa. What is the reading of a pressure gauge in the wide section?

61. Water flows from the pipe shown in the figure with a speed of 4.0 m/s.
    a. What is the water pressure as it exits into the air?
    b. What is the height h of the standing column of water?

    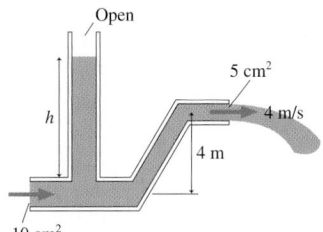
    FIGURE P15.61

62. Water flows out of a 16-mm-diameter sink faucet at 1.0 m/s. At what distance below the faucet has the water stream narrowed to 10 mm diameter?

63. A hurricane wind blows across a 6.0 m × 15.0 m flat roof at a speed of 130 km/hr.
    a. Is the air pressure above the roof higher or lower than the pressure inside the house? Explain.
    b. What is the pressure difference?
    c. How much force is exerted on the roof? If the roof cannot withstand this much force, will it "blow in" or "blow out"?

64. Air flows through this tube at a rate of 1200 cm^3/s. Assume that air is an ideal fluid. What is the height h of mercury in the right side of the U-tube?

    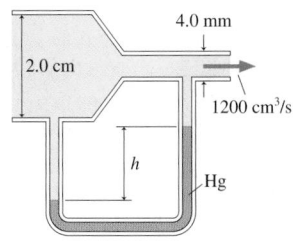
    FIGURE P15.64

65. Air flows through the tube shown in the figure. Assume that air is an ideal fluid.
    a. What are the air speeds v_1 and v_2 at points 1 and 2?
    b. What is the volume flow rate?

    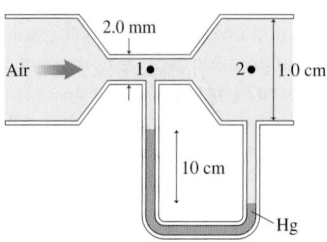
    FIGURE P15.65

66. A water tank of height h has a small hole at height y. The water is replenished to keep h from changing. The water squirting from the hole has range x. The range approaches zero as $y \to 0$ because the water squirts right onto the table. The range also approaches zero as $y \to h$ because the horizontal velocity becomes zero. Thus there must be some height y between 0 and h for which the range is a maximum.
    a. Find an algebraic expression for the flow speed v with which the water exits the hole at height y.
    b. Find an algebraic expression for the range of a particle shot horizontally from height y with speed v.
    c. Combine your expressions from parts a and b. Then find the maximum range x_{max} and the height y of the hole. "Real" water won't achieve quite this range because of viscosity, but it will be close.

    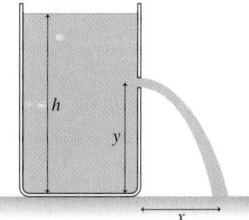
    FIGURE P15.66

67. A 4.0-mm-diameter hole is 1.0 m below the surface of a 2.0-m-diameter tank of water.
    a. What is the volume flow rate through the hole, in L/min?
    b. What is the rate, in mm/min, at which the water level in the tank will drop if the water is not replenished?

    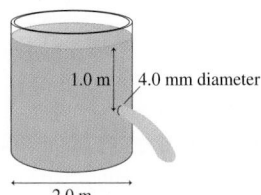
    FIGURE P15.67

68. A 70 kg mountain climber dangling in a crevasse stretches a 50-m-long, 1.0-cm-diameter rope by 8.0 cm. What is Young's modulus for the rope?

69. A large 10,000 L aquarium is supported by four wood posts (Douglas fir) at the corners. Each post has a square 4.0 cm × 4.0 cm cross section and is 80 cm tall. By how much is each post compressed by the weight of the aquarium?

70. a. What is the pressure at a depth of 5000 m in the ocean?
    b. What is the fractional volume change $\Delta V/V$ of seawater at this pressure?

c. What is the density of seawater at this pressure?

d. The hydrostatic pressure equation that you used in part a assumes the liquid is incompressible. That's an excellent assumption for many applications, but the slight compressibility of water means that the hydrostatic pressure equation doesn't give perfect results at great depths. Is the actual pressure at 5000 m more or less than you computed in part a? Explain.

71. A cylindrical steel pressure vessel with volume 1.30 m³ is to be tested. The vessel is entirely filled with water, then a piston at one end of the cylinder is pushed in until the pressure inside the vessel has increased by 2000 kPa. Suddenly, a safety plug on the top bursts. How many liters of water come out?

Challenge Problems

72. A 1.0-m-tall cylinder contains air at a pressure of 1 atm. A very thin, frictionless piston of negligible mass is placed at the top of the cylinder, to prevent any air from escaping, then mercury is slowly poured into the cylinder until no more can be added without the cylinder overflowing. What is the height h of the column of compressed air?

    Hint: Boyle's law, which you learned in chemistry, says $p_1V_1 = p_2V_2$ for a gas compressed at constant temperature, which we will assume to be the case.

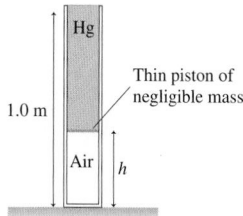

FIGURE CP15.72

73. A 1.0 g balloon is filled with helium gas until it becomes a 20-cm-diameter sphere. What maximum mass can be tied to the balloon (with a massless string) without the balloon sinking to the floor?

74. A cone of density ρ_0 and total height l floats in a liquid of density ρ_f. The height of the cone above the liquid is h. What is the ratio h/l of the exposed height to the total height?

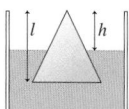

FIGURE CP15.74

75. A cylinder of density ρ_0, length l, and cross-section area A floats in a liquid of density ρ_f with its axis perpendicular to the surface. Length h of the cylinder is submerged when the cylinder floats at rest.

    a. Write expressions for the object's weight w and for the buoyant force F_B. Use these to show that $h = (\rho_0/\rho_f)l$.

    b. Suppose the cylinder is distance y above its equilibrium position. Find an expression for $(F_{net})_y$, the y-component of the net force. Use your expressions from part a to cancel some of the terms.

c. You should recognize your result of part b as a version of Hooke's law. What is the "spring constant" k?

d. If you push a floating object down and release it, it bobs up and down. So it is like a spring in the sense that it oscillates if displaced from equilibrium. Use your "spring constant" and what you know about simple harmonic motion to show that the cylinder's oscillation period is

$$T = 2\pi\sqrt{\frac{h}{g}}$$

e. What is the oscillation period for a 100-m-tall iceberg ($\rho_{ice} = 917$ kg/m³) in seawater?

76. The pressure of the atmosphere decreases with increasing elevation. Let's figure out how.

    a. Establish a z-axis that points up, with $z = 0$ at sea level. Suppose the pressure at height z is known to be p and the air density is ρ. Use the hydrostatic pressure equation to write an expression for the pressure at height $z + dz$, where dz is so small that the density has not changed. Your expression will be in terms of p, ρ, dz, and perhaps some constants. The pressure *decreases* as you gain elevation, so be careful with signs.

    b. Using your expression from part a, write an expression for dp, the amount by which the pressure *changes* in going from z to $z + dz$. Pressure is decreasing, so your expression should be negative.

    c. You need to integrate your expression from part b, but you can't because the density ρ is not a constant. *If* the temperature remains constant, which we will assume, then the ideal gas law implies that pressure is directly proportional to density. That is, $p/\rho = p_0/\rho_0$, where p_0 and ρ_0 are the sea-level values of pressure and density. Use this to rewrite your expression for dp in terms of p, dz, and various constants.

    d. Now you have an integrable expression, although you must first divide by p to get all the pressure terms on one side of the equation. Carry out the integration and use the fact that $p = p_0$ at $z = 0$ to determine the integration constant. Then solve for the pressure at height z. Your final result should be in the form $p = p_0\exp(-z/z_0)$.

    e. z_0 is called the *scale height* of the atmosphere. It is the height at which $p = e^{-1}p_0$, or about 37% of the sea-level pressure. Determine the numerical value of z_0.

    f. The lower layer of the atmosphere, called the troposphere, has a height of about 15,000 m. This is the region of the atmosphere where weather occurs. Above it is the stratosphere, where conditions are very different. Draw a graph of pressure versus height up to a height of 15,000 m.

    Comment: We assumed a constant-temperature atmosphere. In the real atmosphere, the temperature in the troposphere decreases with increasing height. This alters how the pressure changes, but not enormously. Your result is a reasonably good approximation.

Stop to Think 15.1: $\rho_1 = \rho_2 = \rho_3$. Density depends only on what the object is made of, not how big the pieces are.

Stop to Think 15.2: c. These are all open tubes, so the liquid rises to the same height in all three despite their different shapes.

Stop to Think 15.3: $F_2 > F_1 = F_3$. The masses in 3 do not add. The pressure underneath each of the two large pistons is mg/A_2, and the pressure under the small piston must be the same.

Stop to Think 15.4: b. The weight of the displaced water equals the weight of the ice cube. When the ice cube melts and turns into water, that amount of water will exactly fill the volume that the ice cube is now displacing.

Stop to Think 15.5: 1 cm³/s out. The fluid is incompressible, so the sum of what flows in must match the sum of what flows out. 13 cm³/s is known to be flowing in while 12 cm³/s flows out. An additional 1 cm³/s must flow out to achieve balance.

Stop to Think 15.6: $h_2 > h_4 > h_3 > h_1$. The liquid level is higher where the pressure is lower. The pressure is lower where the flow speed is higher. The flow speed is highest in the narrowest tube, zero in the open air.

Applications of Newtonian Mechanics

We have developed two parallel perspectives of motion, each with its own concepts and techniques. We focused on the first of these in Part I, where we dealt with the relationship between force and motion. Newton's second law is the principle most central to the force/motion perspective. Then, in Part II, we developed a before-and-after perspective based on the idea of conservation laws. Newton's laws were essential in the development of conservation laws, but they remain hidden in the background when the conservation laws are applied. Together, these two perspectives form the heart of Newtonian mechanics.

Our goal in Part III has been to see how Newtonian mechanics is applied to several diverse but important topics. We added only one new law of physics in Part III, Newton's law of gravity, and we introduced few completely new concepts. Instead, we've broadened our understanding of the force/motion perspective and the conservation-law perspective through our investigations of gravity, rotational motion, oscillations, and fluids. In reviewing Part III, pay close attention to the interplay between these two perspectives. Recognizing which is the best tool in a particular situation will help you improve your problem-solving ability.

Our knowledge of mechanics is now essentially complete. We will add a few additional ideas as we need them, but our journey into physics will be taking us in entirely new directions as we continue on. Hence this is an opportune moment to step back a bit to take a look at the "big picture." Newtonian mechanics may seem all very factual and straightforward to us today, but keep in mind that these ideas are all human inventions. There was a time when they did not exist and when our concepts of nature were quite different from what they are today.

KNOWLEDGE STRUCTURE III **Applications of Newtonian Mechanics**

Newton's Theory of Gravity

Any two masses exert attractive gravitational forces on each other.

Newton's law of gravity is

$$F_{m \text{ on } M} = F_{M \text{ on } m} = \frac{GMm}{r^2}$$

- Kepler's laws describe the elliptical orbits of satellites and planets.

- The gravitational potential energy is

$$U_g = -\frac{GMm}{r}$$

Rotation of a Rigid Body

A rigid body is a system of particles.
Rotational motion is analogous to linear motion.

Rotational motion	**Linear motion**
Angular acceleration α	Acceleration a
Torque τ	Force F
Moment of inertia I	Mass m
Angular momentum L	Momentum p

- **Newton's second law** $\tau_{\text{net}} = I\alpha$
- **Rotational kinetic energy** $K = \frac{1}{2}I\omega^2$

**NEWTON'S LAWS
+
CONSERVATION LAWS**

Oscillations

Systems with a linear restoring force exhibit simple harmonic oscillation.

- The **kinematic equations of SHM** are

$$x(t) = A\cos(\omega t + \phi_0)$$
$$v(t) = -v_{\max}\sin(\omega t + \phi_0)$$

where $v_{\max} = \omega A$ and the phase constant ϕ_0 describes the initial conditions.

- **Energy is transformed between kinetic and potential** as the system oscillates. In an undamped system, the total mechanical energy

$$E = \tfrac{1}{2}mv^2 + \tfrac{1}{2}kx^2 = \tfrac{1}{2}m(v_{\max})^2 = \tfrac{1}{2}kA^2$$

is conserved.

Fluids and Elasticity

Fluids are systems that flow. Gases and liquids are fluids. Fluids are better characterized by density and pressure than by mass and force.

- **Liquids** Pressure is primarily gravitational. The hydrostatic pressure is

$$p = p_0 + \rho g d$$

- **Gases** Pressure is primarily thermal. Pressure in a container is constant.

- **Archimedes' principle** The buoyant force is equal to the weight of the displaced liquid.

For fluid flow, **Bernoulli's equation**

$$p_1 + \tfrac{1}{2}\rho v_1^2 + \rho g y_1 = p_2 + \tfrac{1}{2}\rho v_2^2 + \rho g y_2$$

is really a statement of energy conservation.

The Newtonian Synthesis

Newton's achievements, praised by no less than Einstein as "perhaps the greatest advance in thought that a single individual was ever privileged to make," are often called the *Newtonian synthesis*. "Synthesis" means "the uniting or combining of separate elements to form a coherent whole." It is often said of Newton that he "united the heavens and the earth." In doing so, he changed forever the way we view ourselves and our relationship to the universe.

As we noted in Chapter 12, medieval cosmology considered the heavenly bodies to be perfect, unchanging objects quite unrelated to imperfect and changeable earthly matter. Their perfection and immortality symbolized the perfection of God above, while the material bodies of humans were imperfect and mortal. This cosmology was mirrored in medieval feudal society. The king—ordained by God and whose symbol was the sun—was surrounded by a small circle of nobles and a larger circle of serfs and peasants. Taken together, the ideas and institutions of science, religion, and society of this time form what we call the medieval *worldview*. Their worldview, in its many facets, was hierarchical and authoritarian, reflecting their understanding of "natural order" in the universe.

Copernicus weakened medieval cosmology by questioning the position of the earth in the universe. Galileo, with his telescope, found that the heavens are not perfect and unchanging. Now, at the end of the 17th century, the success of Newton's theories implied that the sun and the planets were merely ordinary matter, obeying the same natural laws as earthly matter. This uniting of earthly motions and heavenly motions—the *synthesis* in the Newtonian synthesis—dealt the final blow to the medieval worldview.

Newton's success changed the way we see and think about the universe. Rather than seeing whirling celestial spheres, people began to think of the universe in terms of the motion of material particles following rigid laws. This Newtonian conception of the cosmos is often called a "clockwork universe." The technology of clocks was progressing rapidly in the 18th century, and people everywhere admired the consistency and predictability of these little machines. The Newtonian universe is a very large machine, but one that is consistent, predictable, and law-abiding. In other words, a perfect clock.

Major thinkers of the 17th and 18th centuries soon concluded that God had created the world by placing all the particles in their original positions, then giving them a push to get them going. God, in this role, was called the "prime mover." But once the universe was started, it went along perfectly well just by obeying Newton's laws. No divine intervention or guidance was needed. This is certainly a very different view of our relationship to God and the universe than was contained in the medieval worldview.

Newton also influenced the way people think about themselves and their society. His theories clearly demonstrated that the universe is not random or capricious but, instead, follows natural laws. Others soon began to apply the concept of natural law to human nature, human behavior, and human institutions. The main protagonist in this school of thought was the English philosopher and political scientist John Locke, a contemporary of Newton. Locke developed a theory of human behavior from the ideas of natural laws and empirical evidence. We cannot go into Locke's theories here, but Newton's success helped to propel Locke's ideas into the mainstream of 18th-century political thought.

Locke's writings had a great influence on a young American colonial named Thomas Jefferson. The concept of natural laws, as they apply to individuals, is very much behind Jefferson's enunciation of "unalienable rights" in the Declaration of Independence. In fact, the first sentence of the Declaration refers explicitly to "the Laws of Nature and of Nature's God." The idea of *checks and balances*, built into the Constitution of the United States, is very much a mechanical and clock-like model of how political institutions function.

Just as medieval feudalism mirrored the medieval understanding of the universe, contemporary constitutional democracy mirrors, in many ways, the Newtonian cosmology. Hierarchy and authority have been replaced by equality and law because they now seem to us the "natural order" of things. Having grown up with this modern worldview, we find it difficult to imagine any other. Nonetheless, it is important to realize that vastly different worldviews have existed at other times and in other cultures.

Science has changed dramatically in the last hundred-odd years. Newton's clockwork universe has been superseded by relativity and quantum physics. Entirely new theories and sciences, such as evolution, ecology, and psychology, have appeared. These new ideas are slowly working their way into other areas of thought and human activity, and bit by bit they are changing the ways in which we see ourselves, our society, and our relationship to nature. A future worldview is in the making.

R2-D2's internal systems have been on standby while inside the ice cave on the planet Hoth. How much energy must R2-D2's fuel cells provide to raise the temperature of his internal systems from the standby temperature of −20°C to a working temperature of 30°C? To find out, what properties of R2-D2 do you need to estimate?

Thermodynamics

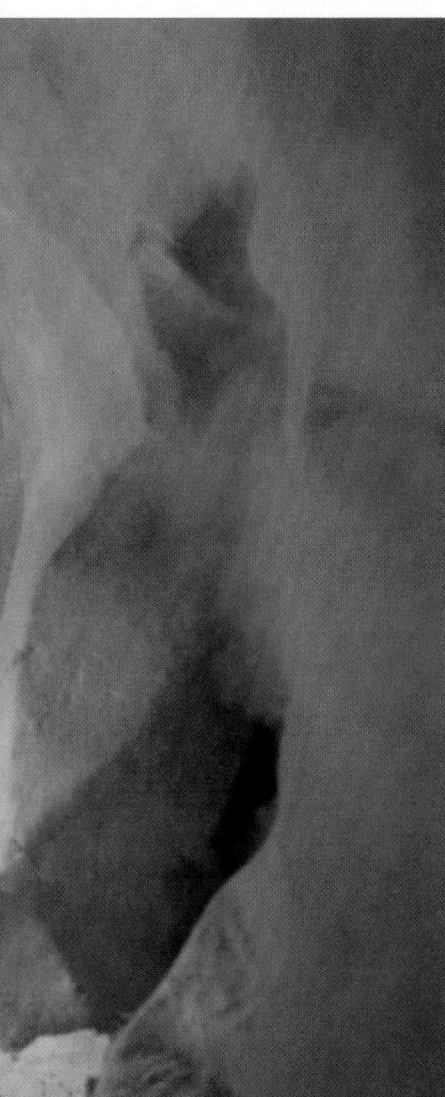

It's All About Energy

If one area of science is universally applicable, from astrophysics to zoology, it is thermodynamics. Thermodynamics is the science of energy in its broadest context, especially the conversion of energy from one form to another. The word energy itself comes from a Greek root meaning "to do work," and much of thermodynamics is concerned with converting stored energy of various forms, such as the energy in fuels, into the energy called useful work.

Thermodynamics arose hand-in-hand with the industrial revolution as the systematic study of converting heat energy into mechanical motion and work. Hence the name *thermo + dynamics*. Indeed, the analysis of engines and generators of various kinds remains the focus of engineering thermodynamics. But thermodynamics, as a science, now extends to all forms of energy conversions, including those involving living organisms. Here are a few examples:

- Gasoline, diesel, and jet engines convert the energy of a fuel into the mechanical energy of moving pistons, wheels, and gears.
- A generator converts the energy of a fuel into the mechanical energy of a spinning turbine and then into electrical energy.
- A fuel cell converts chemical energy into electrical energy.
- A photovoltaic cell converts the electromagnetic energy of light into electrical energy.
- A laser converts electrical energy into the electromagnetic energy of light.
- An organism converts the chemical energy of food into a variety of other forms of energy, including kinetic energy, sound energy, and thermal energy.

The major goal of Part IV will be to understand how energy transformations such as these take place. Although we can look at only a few applications, these chapters will prepare you to use the ideas of thermodynamics in your own field of study.

Thermodynamics deals with *macroscopic* systems, the prefix *macro* (the opposite of *micro*) meaning large. We will be concerned with systems that are solids, liquids, or gases, rather than the "particles" to which we have grown accustomed. The language of thermodynamics will be of temperature, pressure, volume, and moles rather than of position, velocity, and force. An important task will be to learn this new language.

The questions we will be asking are also different from those that have occupied us until now. Typical questions of thermodynamics are

- How do temperature and pressure change during a certain process?
- How does a physical system such as an engine do mechanical work? How is the work related to the temperature and pressure inside the engine?

- If an engine "burns fuel" to do work, how is the amount of work done related to the fuel energy that was used? Are there limits on how efficient an engine can be?

As you can see, there are many new ideas to learn.

The Micro/Macro Connection

Thermodynamics describes and characterizes a macroscopic system as a whole, regardless of whether the system is a solid crystal, a beaker of liquid, or a container of gas. But as you know, one of the major scientific discoveries of the last two hundred years is that solids, liquids, and gases all consist of atoms. These atoms, each a microscopic particle, are ceaselessly moving about and colliding with each other. Their motion and interactions are described by Newtonian mechanics and, in some cases, quantum physics.

A second important goal of Part IV will be to demonstrate that there is a connection between the *microphysics* of the atoms and molecules in the system and the *macrophysics* of the system as a whole. That is, the macroscopic quantities of thermodynamics can be understood in terms of the microscopic physics of moving atoms and molecules. This link between the realms of the small and the large is called the *micro/macro connection*.

The discovery of this link between the microscopic and the macroscopic is an exciting story and a major success of physics. You will find that the familiar concepts of thermodynamics, such as temperature and pressure, have their roots in atomic-level motion and collisions. You will also find it possible to learn a great deal about the properties of molecules, such as their speeds, on the basis of purely macroscopic measurements. The micro/macro connection will lead to the second law of thermodynamics, one of the most subtle but also one of the most profound and far-reaching statements in physics.

Practical Thermodynamics

Our ultimate destination in Part IV is an understanding of the thermodynamics of *heat engines*. A heat engine is a device, such as a gasoline-powered internal combustion engine, that transforms heat energy into useful work. These are the devices that power our modern society.

Understanding how to transform heat into work will be a significant achievement, but we first have many steps to take along the way. We need to understand the concepts of temperature and pressure. We need to learn about the properties of solids, liquids, and gases. Most important, we need to expand our view of energy to include *heat,* the energy that is transferred between two systems at different temperatures. And at a deeper level, we need to see how these concepts are connected to the underlying microphysics of randomly moving molecules. Only after all these steps have been taken will we be able to analyze a real heat engine.

This is an ambitious goal, but one we can achieve.

16 A Macroscopic Description of Matter

Liquid steel is poured from a giant crucible.

▶ **Looking Ahead**
The goal of Chapter 16 is to learn the characteristics of macroscopic systems. In this chapter you will learn to:

- Understand the basic properties of solids, liquids, and gases.
- Interpret a phase diagram.
- Work with different temperature scales.
- Use the ideal-gas law.
- Understand ideal-gas processes and represent them on a pV diagram.

◀ **Looking Back**
The material in this chapter depends on thermal energy and the properties of fluids. Please review:

- Section 11.7 Thermal energy.
- Sections 15.1–15.3 Fluids and pressure.

A room full of air, a beaker of water, and a crucible of red-hot molten steel are examples of **macroscopic systems,** systems that are large enough to see or touch. These are the systems of our everyday experience. Our goal in this chapter is twofold:

- To learn what kind of physical properties characterize macroscopic systems.
- To begin the process of connecting a system's macroscopic properties to the underlying motions of the atoms in the system.

The properties of a macroscopic system as a whole are called its **bulk properties.** One fairly obvious example is the system's mass. Other bulk properties to be discussed in this chapter include volume, density, temperature, and pressure. Macroscopic systems are also characterized as being either solid, liquid, or gas. These are called the *phases* of matter, and we'll be interested in when and how a system changes from one phase to another.

Ultimately we would like to understand the macroscopic properties of solids, liquids, and gases in terms of the microscopic motions of their atoms and molecules. Developing this **micro/macro connection** will take several chapters, but we'll start laying the foundations in this chapter. This effort to understand macroscopic properties in terms of the particle-like atoms will pay handsome dividends when we later come to electricity and then quantum physics.

16.1 Solids, Liquids, and Gases

The ice cube you take out of the freezer soon becomes a puddle of liquid water. Then, more slowly, it evaporates to become water vapor in the air. Water is unique. It is the only substance whose three **phases**—solid, liquid, and gas—are familiar from everyday experience.

Each of the elements and most compounds can exist as a solid, liquid, or gas. The change between liquid and solid (freezing or melting) or between liquid and gas (boiling or condensing) is called a **phase change.** Phase changes, which we'll investigate later in this chapter, occur at a well-defined temperature. We're familiar with only one, or perhaps two, of the phases of most substances because their melting point and/or the boiling point are far outside the range of normal human experience. Nonetheless, as the opening photo shows, substances as hard and solid as steel can melt and even boil if the temperature is sufficiently high.

The notion of three distinct phases is less useful for more complex systems. A piece of wood is solid, but liquid wood and gaseous wood don't exist. *Liquid crystals,* which are used to display the numbers on your digital watch, have characteristics of both solids *and* liquids. Complex systems have many interesting properties, but this text will focus on macroscopic systems for which the three phases are distinct.

NOTE ▶ This use of the word *phase* has no relationship at all to the *phase* or *phase constant* of simple harmonic motion and waves. ◀

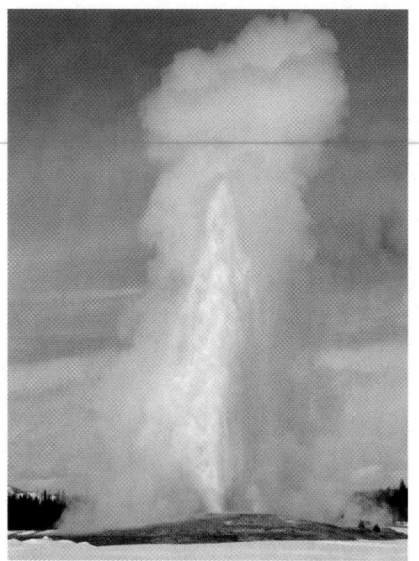

The three phases of water—solid, liquid, and gas—are familiar from everyday experience.

Solids, liquids, and gases

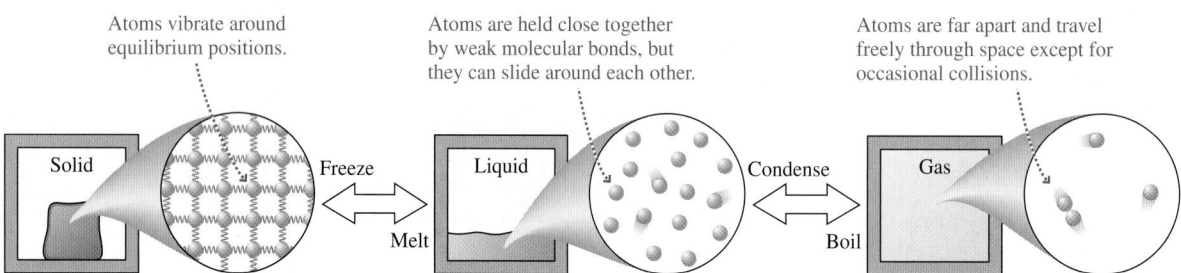

Atoms vibrate around equilibrium positions.

Atoms are held close together by weak molecular bonds, but they can slide around each other.

Atoms are far apart and travel freely through space except for occasional collisions.

Solid — Freeze / Melt — Liquid — Condense / Boil — Gas

A **solid** is a rigid macroscopic system with a definite shape and volume. It consists of particle-like atoms connected together by spring-like molecular bonds. Each atom jiggles and vibrates around an equilibrium position, but an atom is *not* free to move around inside the solid. Solids are nearly *incompressible,* telling us that the atoms in a solid are just about as close together as they can get.

The solid shown here is a **crystal,** meaning that the atoms are arranged in a periodic array. The elements and many compounds have a crystal structure when in their solid phase. In other solids, such as glass, the atoms are frozen into random positions. These are called **amorphous solids.**

A **liquid** is more complicated than either a solid or a gas. Like a solid, a liquid is nearly *incompressible.* This tells us that the molecules in a liquid are about as close together as they can get. Like a gas, a liquid flows and deforms to fit the shape of its container. The fluid nature of a liquid tells us that the molecules are free to move around.

Together, these observations suggest a model in which the molecules of the liquid are loosely held together by weak molecular bonds. The bonds are strong enough that the molecules never get far apart but not strong enough to prevent the molecules from sliding around each other.

A **gas** is a system in which each molecule moves through space as a free, noninteracting particle until, on occasion, it collides with another molecule or with the wall of the container. A gas is a *fluid.* A gas is also highly *compressible,* telling us that there is lots of space between the molecules.

Gases are fairly simple macroscopic systems, hence many of our examples in Part IV will be based on gases.

State Variables

The parameters used to characterize or describe a macroscopic system are known as **state variables** because, taken all together, they describe the *state* of the macroscopic system. You met some state variables in earlier chapters: Volume, pressure, mass, mass density, and thermal energy. We'll soon introduce several new state variables: moles, number density, and, most important, the temperature T.

The state variables are not all independent of each other. For example, you learned in Chapter 15 that a system's mass density ρ is defined in terms of the system's mass M and volume V as

$$\rho = \frac{M}{V} \quad \text{(mass density)} \tag{16.1}$$

In this chapter we'll use an uppercase M for the system mass and a lowercase m for the mass of an atom. Table 16.1 is a short list of mass densities of various substances.

If we change the value of any of the state variables, then we are changing the state of the system. For example, to *compress* a gas means to decrease its volume. Other state variables, such as pressure and temperature, may also change as the volume changes. We will use the symbol Δ to represent a *change* in the value of a state variable. That is, ΔT is a *change* of temperature and Δp is a *change* of pressure. **For any quantity X, ΔX is *always* $X_f - X_i$, the final value minus the initial value.**

A system is said to be in **thermal equilibrium** if its state variables are constant and not changing. As an example, a gas is in thermal equilibrium if it has been left undisturbed long enough for p, V, and T to reach steady values. One of the important goals of Part IV is to establish the conditions under which a macroscopic system reaches thermal equilibrium.

TABLE 16.1 Densities of materials

Substance	ρ (kg/m^3)
Air at STP*	1.2
Ethyl alcohol	790
Water (solid)	920
Water (liquid)	1000
Aluminum	2700
Copper	8920
Gold	19,300
Iron	7870
Lead	11,300
Mercury	13,600
Silicon	2330

*$T = 0°C$, $p = 1$ atm

EXAMPLE 16.1 The mass of a lead pipe
A project on which you are working uses a cylindrical lead pipe with outer and inner diameters of 4.0 cm and 3.5 cm, respectively, and a length of 50 cm. What is its mass?

SOLVE The mass density of lead is $\rho_{\text{lead}} = 11,300$ kg/m^3. The volume of a circular cylinder of length l is $V = \pi r^2 l$. In this case we need to find the volume of the outer cylinder, of radius

r_2, *minus* the volume of air in the inner cylinder, of radius r_1. The volume of the pipe is

$$V = \pi r_2^2 l - \pi r_1^2 l = \pi(r_2^2 - r_1^2)l = 1.47 \times 10^{-4} \text{ m}^3$$

Hence the pipe's mass is

$$M = \rho_{\text{lead}} V = 1.66 \text{ kg}$$

STOP TO THINK 16.1 The pressure in a system is measured to be 60 kPa. At a later time the pressure is 40 kPa. The value of Δp is

a. 60 kPa.   b. 40 kPa.   c. 20 kPa.   d. -20 kPa.

16.2 Atoms and Moles

The mass of a macroscopic system is directly related to the total number of atoms or molecules in the system, denoted N. Because N is determined simply by counting, it is a number with no units. A typical macroscopic system has $N \sim 10^{25}$ atoms, an incredibly large number.

The symbol $\sim$, if you are not familiar with it, stands for "has the order of magnitude." It means that the number is only known to within a factor of 10 or so. The

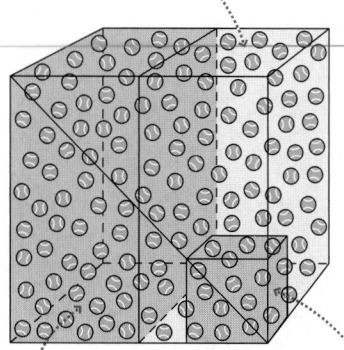

A 100 m³ room has 10,000 tennis balls bouncing around. The number density of tennis balls in the room is $N/V = 10{,}000/100$ m³ $= 100$ m⁻³.

∴ If we look at only half the room, we would find 5000 balls in 50 m³, again giving $N/V = 5000/50$ m³ $= 100$ m⁻³.

In one-tenth of the room, we would find 1000 balls in 10 m³, again giving $N/V = 1000/10$ m³ $= 100$ m⁻³.

FIGURE 16.1 The number density of a uniform system is independent of the volume.

statement $N \sim 10^{25}$, which is read "N is of order 10^{25}," implies that N is somewhere in the range 10^{24} to 10^{26}. It is far less precise than the "approximately equal" symbol $\approx$. As we begin to deal with large numbers it will often be necessary to distinguish "really large" numbers, such as 10^{25}, from "small" numbers such as a mere 10^{5}. Saying $N \sim 10^{25}$ gives us a rough idea of how large N is and allows us to know that it differs significantly from 10^{5} or even 10^{15}.

It is often useful to know the number of atoms or molecules per cubic meter in a system. We call this quantity the **number density.** It characterizes how densely the atoms are packed together within the system. In an N-atom system that fills volume V, the number density is

$$\frac{N}{V} \quad \text{(number density)} \tag{16.2}$$

The SI units of number density are m⁻³. The number density of atoms in a solid is $(N/V)_{\text{solid}} \sim 10^{29}$ m⁻³. The number density of a gas depends on the pressure, but is usually less than 10^{27} m⁻³.

The value of N/V in a *uniform* system is independent of the volume V. That is, the number density is the same whether you look at the whole system or just a portion of it. For example, Figure 16.1 shows a 100 m³ room with 10,000 tennis balls bouncing around. The number density of tennis balls in the room is $N/V = 10{,}000/100$ m³ $= 100$ m⁻³. If you were to look at only a one-tenth portion of the room, you would find 1000 balls in 10 m³, again giving $N/V = 1000/10$ m³ $= 100$ m⁻³. The number density is the same in both cases.

NOTE ▸ While we might say "There are 100 tennis balls per cubic meter," or "There are 10^{29} atoms per cubic meter," tennis balls and atoms are not units. The units of N/V are simply m⁻³. ◂

Atomic Mass and Atomic Mass Number

You will recall from chemistry that atoms of different elements have different masses. The mass of an atom is determined primarily by its most massive constituents, the protons and neutrons in its nucleus. The *sum* of the number of protons and neutrons is called the **atomic mass number** A:

$$A = \text{number of protons } + \text{ number of neutrons}$$

A, which by definition is an integer, is written as a leading superscript on the atomic symbol. For example, the common isotope of hydrogen, with one proton and no neutrons, is ¹H. The "heavy hydrogen" isotope called *deuterium,* which includes one neutron, is ²H. The primary isotope of carbon, with six protons (which makes it carbon) and six neutrons, is ¹²C. The radioactive isotope ¹⁴C, used for carbon dating of archeological finds, contains six protons and eight neutrons.

The **atomic mass** scale is established by defining the mass of ¹²C to be exactly 12 u, where u is the symbol for the **atomic mass unit.** That is, $m(^{12}\text{C}) = 12$ u. The atomic mass of any other atom is its mass relative to ¹²C. For example, careful experiments with hydrogen find that the mass *ratio* $m(^{1}\text{H})/m(^{12}\text{C})$ is $1.0078/12$. Thus the atomic mass of hydrogen is $m(^{1}\text{H}) = 1.0078$ u.

The numerical value of the atomic mass of ¹H is close to, but not exactly, its atomic mass number $A = 1$. The slight difference is due to the electron mass and to various relativistic effects. For our purposes, it will be sufficient to overlook the slight difference and use the integer atomic mass numbers as the values of the atomic mass. That is, we'll use $m(^{1}\text{H}) = 1$ u, $m(^{4}\text{He}) = 4$ u, and $m(^{16}\text{O}) = 16$ u. For molecules, the **molecular mass** is the sum of the atomic masses of the atoms forming the molecule. Thus the molecular mass of the diatomic molecule O_2, the constituent of oxygen gas, is $m(O_2) = 32$ u.

NOTE ▶ An element's atomic mass number is *not* the same as its atomic number. The *atomic number*, which is the element's position in the periodic table, is the number of protons. ◀

Table 16.2 shows the atomic mass numbers of some of the elements that we'll use for examples and homework problems. A complete periodic table, including atomic masses, is found in Appendix B.

Moles and Molar Mass

One way to specify the amount of substance in a macroscopic system is to give its mass. Another way, one connected to the number of atoms, is to measure the amount of substance in *moles*. By definition, one **mole** of matter, be it solid, liquid, or gas, is the amount of substance containing as many basic particles as there are atoms in 12 g of ^{12}C. Many decades of ingenious experiments have determined that there are 6.02×10^{23} atoms in 12 g of ^{12}C, so we can say that 1 mole of substance, abbreviated 1 mol, is 6.02×10^{23} basic particles.

The basic particle depends on the substance. Helium is a **monatomic gas,** meaning that the basic particle is the helium atom. Thus 6.02×10^{23} helium atoms are 1 mol of helium. But oxygen gas is a **diatomic gas** because the basic particle is the two-atom diatomic molecule O_2. 1 mol of oxygen gas contains 6.02×10^{23} *molecules* of O_2 and thus $2 \times 6.02 \times 10^{23}$ oxygen atoms. Table 16.3 lists the monatomic and diatomic gases that we will use for examples and homework problems.

The number of basic particles per mole of substance is called **Avogadro's number,** N_A. The value of Avogadro's number is

$$N_A = 6.02 \times 10^{23} \ \text{mol}^{-1}$$

Avogadro's number, like the gravitational constant G, is one of the basic constants of nature.

Despite its name, Avogadro's number is not simply "a number"; it has units. Because there are N_A particles per mole, the number of moles in a substance containing N basic particles is

$$n = \frac{N}{N_A} \tag{16.3}$$

where n is the symbol for moles.

Avogadro's number allows us to determine atomic masses in kilograms. Knowing that N_A ^{12}C atoms have a mass of 12 g, we know the mass of one ^{12}C atom must be

$$m(^{12}\text{C}) = \frac{12 \ \text{g}}{6.02 \times 10^{23}} = 1.993 \times 10^{-23} \ \text{g} = 1.993 \times 10^{-26} \ \text{kg}$$

We defined the atomic mass scale such that $m(^{12}\text{C}) = 12$ u. Thus the conversion factor between atomic mass units and kilograms is

$$1 \ \text{u} = \frac{m(^{12}\text{C})}{12} = 1.661 \times 10^{-27} \ \text{kg}$$

This conversion factor allows us to calculate the mass in kg of any atom. For example, a ^{20}Ne atom has atomic mass $m(^{20}\text{Ne}) = 20$ u. Multiplying by 1.661×10^{-27} kg/u gives $m(^{20}\text{Ne}) = 3.32 \times 10^{-26}$ kg. If the atomic mass is specified in kilograms, the number of atoms in a system of mass M can be found from

$$N = \frac{M}{m} \tag{16.4}$$

TABLE 16.2 Some atomic mass numbers

Element		A
^{1}H	Hydrogen	1
^{4}He	Helium	4
^{12}C	Carbon	12
^{14}N	Nitrogen	14
^{16}O	Oxygen	16
^{20}Ne	Neon	20
^{27}Al	Aluminum	27
^{40}Ar	Argon	40
^{207}Pb	Lead	207

One mole of helium, sulfur, copper, and mercury.

TABLE 16.3 Monatomic and diatomic gases

Monatomic		Diatomic	
He	Helium	H_2	Hydrogen
Ne	Neon	N_2	Nitrogen
Ar	Argon	O_2	Oxygen

The **molar mass** of a substance is the mass *in grams* of 1 mol of substance. The molar mass, which we'll designate M_{mol}, has units g/mol. By definition, the molar mass of ^{12}C is 12 g/mol. For other substances, whose atomic or molecular masses are given relative to ^{12}C, the numerical value of the molar mass equals the numerical value of the atomic or molecular mass. For example, the molar mass of He, with $m = 4$ u, is $M_{mol}(He) = 4$ g/mol and the molar mass of diatomic O_2 is $M_{mol}(O_2) = 32$ g/mol.

Equation 16.4 uses the atomic mass to find the number of atoms in a system. Similarly, you can use the molar mass to determine the number of moles. For a system of mass M consisting of atoms or molecules with molar mass M_{mol},

$$n = \frac{M \text{ (in grams)}}{M_{mol}} \tag{16.5}$$

NOTE ▶ Equation 16.5 is the one of the few instances where the proper units are *grams* rather than kilograms. ◄

EXAMPLE 16.2 Moles of oxygen

100 g of oxygen gas is how many moles of oxygen?

SOLVE We can do the calculation two ways. First, let's determine the number of molecules in 100 g of oxygen. The diatomic oxygen molecule O_2 has molecular mass $m = 32$ u. Converting this to kg, we get the mass of one molecule:

$$m = 32 \text{ u} \times \frac{1.661 \times 10^{-27} \text{ kg}}{1 \text{ u}} = 5.31 \times 10^{-26} \text{ kg}$$

Thus the number of molecules in 100 g = 0.10 kg is

$$N = \frac{M}{m} = \frac{0.100 \text{ kg}}{5.31 \times 10^{-26} \text{ kg}} = 1.88 \times 10^{24}$$

Knowing the number of molecules gives us the number of moles:

$$n = \frac{N}{N_A} = 3.13 \text{ mol}$$

Alternatively, we can use Equation 16.5 to find

$$n = \frac{M(\text{in grams})}{M_{mol}} = \frac{100 \text{ g}}{32 \text{ g/mol}} = 3.13 \text{ mol}$$

STOP TO THINK 16.2   Which system contains more atoms: 5 mol of helium $(A = 4)$ or 1 mol of neon $(A = 20)$?

a. Helium.   b. Neon.   c. They have the same number of atoms.

16.3 Temperature

We are all familiar with the idea of temperature. You hear the word used nearly every day. But just what is temperature a measure of? Mass is a measure of the amount of substance in a system. Velocity is a measure of how fast a system moves. What physical property of the system have you determined if you measure its temperature?

We will begin with the commonsense idea that temperature is a measure of how "hot" or "cold" a system is. These are properties that we can judge without needing an elaborate theory. As we develop these ideas, we'll find that **temperature** T is related to a system's *thermal energy*. We defined thermal energy in Chapter 10 as the kinetic and potential energy of the atoms and molecules in a system as they vibrate (a solid) or move around (a gas). A system has more thermal energy when it is "hot" than when it is "cold." We'll study temperature more carefully in Chapter 18 and replace these vague notions of hot and cold with a precise relationship between temperature and thermal energy.

To start, we need a means to measure the temperature of a system. This is what a *thermometer* does. A thermometer can be any small macroscopic system that undergoes a measurable change as it exchanges thermal energy with its surroundings. It is placed in contact with a larger system whose temperature it will measure. In a common glass-tube thermometer, for example, a small volume of mercury or alcohol expands or contracts when placed in contact with a "hot" or "cold" object. The object's temperature is determined by the length of the column of liquid.

Other thermometers include:

- Bimetallic strips (two strips of different metals sandwiched together) that curl and uncurl as the temperature changes. These are used in thermostats, such as the one in your house.
- Thermocouples that generate a small voltage depending on the temperature. Thermocouples are widely used for sensing temperatures in inhospitable environments, such as in your car's engine.
- Ideal gases, whose pressure varies with the temperature. We will look at an example of a gas thermometer in a minute.

A thermometer needs a *temperature scale* to be a useful measuring device. In 1742, the Swedish astronomer Anders Celsius sealed mercury into a small capillary tube and observed how it moved up and down the tube as the temperature changed. He selected two temperatures that anyone could reproduce, the freezing and boiling points of pure water, and labeled them 0 and 100. He then marked off the glass tube into one hundred equal intervals between these two reference points. By doing so, he invented the temperature scale that we today call the *Celsius scale.* The units of the Celsius temperature scale are "degrees Celsius," which we abbreviate °C. Note that the degree symbol ° is part of the unit, not part of the number.

NOTE ▶ Because of the 100 equal intervals, the Celsius scale is also called the *centigrade scale.* ◀

The *Fahrenheit scale,* still widely used in the United States, is related to the Celsius scale by

$$T_F = \frac{9}{5}T_C + 32°$$ (16.6)

Table 16.4 shows several temperatures measured on the Celsius and Fahrenheit scales and also on the Kelvin scale.

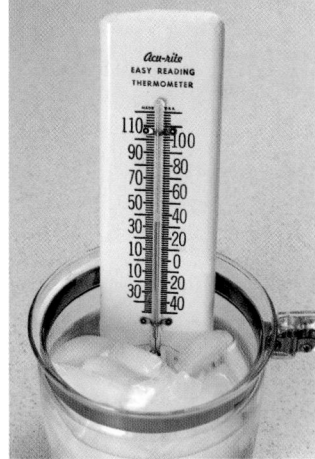

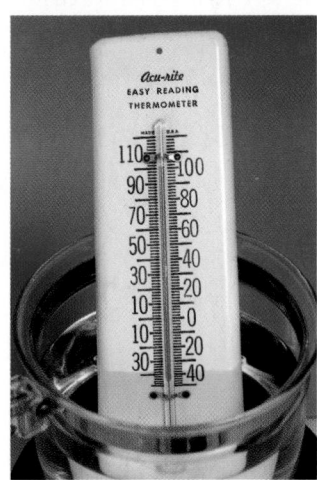

Thermal expansion of the liquid in the thermometer tube pushes it higher in the hot water than in the ice water.

TABLE 16.4 Temperatures measured with different scales

Temperature	T (°C)	T (K)	T (°F)
Melting point of iron	1538	1811	2800
Boiling point of water	100	373	212
Normal body temperature	37.0	310	98.6
Room temperature	20	293	68
Freezing point of water	0	273	32
Boiling point of nitrogen	−196	77	−321
Absolute zero	−273	0	−460

Absolute Zero and Absolute Temperature

Any physical property that changes with temperature can be used as a thermometer. In practice, the most useful thermometers have a physical property that changes *linearly* with temperature. One of the most important scientific thermometers is the **constant-volume gas thermometer** shown in Figure 16.2a. This thermometer depends on the fact that the *absolute* pressure (not the gauge pressure) of a gas in a sealed container increases linearly as the temperature increases.

A gas thermometer is first calibrated by recording the pressure at two reference temperatures, such as the boiling and freezing points of water. These two points are plotted on a pressure-versus-temperature graph and a straight line is drawn through them. The gas bulb is then brought into contact with the system whose temperature is to be measured. The pressure is measured, then the corresponding temperature is read off the graph.

Figure 16.2b shows the pressure-temperature relationship for three different gases. Notice two important things about this graph.

1. There is a *linear* relationship between temperature and pressure.
2. All gases extrapolate to *zero pressure* at the same temperature: $T_0 = -273°C$. No gas actually gets that cold without condensing, although helium comes very close, but it is surprising that you get the same zero-pressure temperature for any gas and any starting pressure.

The pressure in a gas is due to collisions of the molecules with each other and the walls of the container. A pressure of zero would mean that all motion, and thus all collisions, had ceased. If there were no atomic motion, the system's thermal energy would be zero. The temperature at which all motion would cease, and at which $E_{th} = 0$, is called **absolute zero.** Because temperature is related to thermal energy, absolute zero is the lowest temperature that has physical meaning. We see from the gas-thermometer data that $T_0 = -273°C$. We'll give a somewhat more precise definition of absolute zero in the next section.

It is useful to have a temperature scale with the zero point at absolute zero. Such a temperature scale is called an **absolute temperature scale.** Any system whose temperature is measured on an absolute scale will have $T > 0$. The absolute temperature scale having the same unit size as the Celsius scale is called the *Kelvin scale.* It is the SI scale of temperature. The units of the Kelvin scale are *kelvins,* abbreviated as K. The conversion between the Celsius scale and the Kelvin scale is

$$T_K = T_C + 273 \tag{16.7}$$

NOTE ▶ The units are simply "kelvins," *not* "degrees Kelvin." ◀

On the Kelvin scale, absolute zero is 0 K, the freezing point of water is 273 K, and the boiling point of water is 373 K. While most practical macroscopic devices utilize temperatures in the range ≈ 100 K to ≈ 1000 K, it is worth noting that scientists study the properties of matter from temperatures as low as $\approx 10^{-9}$ K (1 nK) on the one extreme to as high as $\approx 10^7$ K on the other!

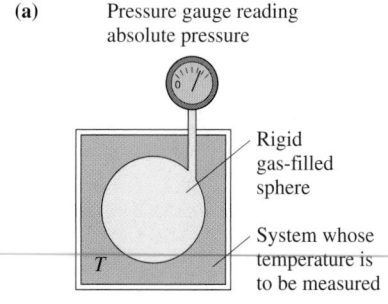

(a) Pressure gauge reading absolute pressure

Rigid gas-filled sphere

T

System whose temperature is to be measured

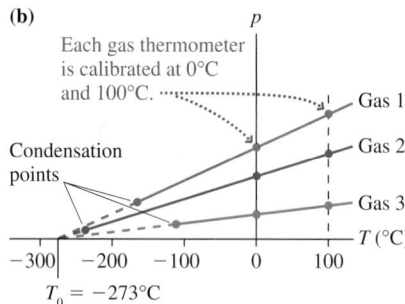

(b)

Each gas thermometer is calibrated at 0°C and 100°C.

Condensation points

Gas 1
Gas 2
Gas 3

T (°C)

-300 -200 -100 0 100

$T_0 = -273°C$

FIGURE 16.2 The pressure in a constant-volume gas thermometer extrapolates to zero at $T_0 = -273°C$. This is the basis for the concept of absolute zero.

STOP TO THINK 16.3 The temperature of a glass of water increases from 20°C to 30°C. What is ΔT?

a. 10 K
b. 283 K
c. 293 K
d. 303 K

16.4 Phase Changes

The temperature inside the freezer compartment of a refrigerator is typically about $-20°C$. Suppose you were to remove a few ice cubes from the freezer and place them in a sealed container with a thermometer. Then, as Figure 16.3a shows, put a steady flame under the container. We'll assume that the heating is done so slowly that the inside of the container always has a single, well-defined temperature.

Figure 16.3b shows the temperature as a function of time. After steadily rising from the initial $-20°C$, the temperature remains fixed at $0°C$ for an extended period of time. This is the interval of time during which the ice melts. As it's melting, the ice temperature is $0°C$ and the liquid water temperature is $0°C$. Even though the system is being heated, the liquid water temperature doesn't begin to rise until all the ice has melted. If you were to turn off the flame at any point, the system would remain a mixture of ice and liquid water at $0°C$.

NOTE ▶ In everyday language, the three phases of water are called *ice, water,* and *steam.* That is, the term *water* implies the liquid phase. Scientifically, these are the solid, liquid, and gas phases of the compound called *water.* To be clear, we'll use the term *water* in the scientific sense of a collection of H_2O molecules. We'll say either *liquid* or *liquid water* to denote the liquid phase. ◀

The thermal energy of a solid is the kinetic energy of the vibrating atoms plus the potential energy of the stretched and compressed molecular bonds. Melting occurs when the thermal energy gets so large that the molecular bonds begin to break, allowing the atoms to move around. The temperature at which a solid becomes a liquid or, if the thermal energy is reduced, a liquid becomes a solid is called the **melting point** or the **freezing point.** Melting and freezing are *phase changes.*

A system at the melting point is in **phase equilibrium,** meaning that any amount of solid can coexist with any amount of liquid. Raise the temperature ever so slightly and the entire system becomes liquid. Lower it slightly and it all becomes solid. But exactly at the melting point the system has no tendency to move one way or the other. That is why the temperature remains constant at the melting point until the phase change is complete.

You can see the same thing happening in Figure 16.3b at $100°C$, the boiling point. This is a phase equilibrium between the liquid phase and the gas phase, and any amount of liquid can coexist with any amount of gas at this temperature. Above this temperature, the thermal energy is too large for bonds to be established between molecules, so the system is a gas. If the thermal energy is reduced, the molecules begin to bond with each other and stick together. In other words, the gas condenses into a liquid. The temperature at which a gas becomes a liquid or, if the thermal energy is increased, a liquid becomes a gas is called the **condensation point** or the **boiling point.**

NOTE ▶ Liquid water becomes solid ice at $0°C$, but that doesn't mean the temperature of ice is always $0°C$. Ice reaches the temperature of its surroundings. If the air temperature in a freezer is $-20°C$, then the ice temperature is $-20°C$. Likewise, steam can be heated to temperatures above $100°C$. That doesn't happen when you boil water on the stove because the steam escapes, but steam can be heated far above $100°C$ in a sealed container. ◀

One difficulty with the experiment of Figure 16.3 is that the outcome depends on the pressure. If you carried out the experiment in Denver, rather than Miami, you would find that the melting point is slightly above $0°C$ and the boiling is several degrees below $100°C$. In other words, the phase-change temperatures are pressure dependent.

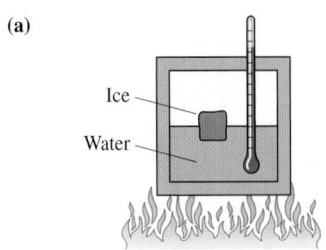

(a)

Ice

Water

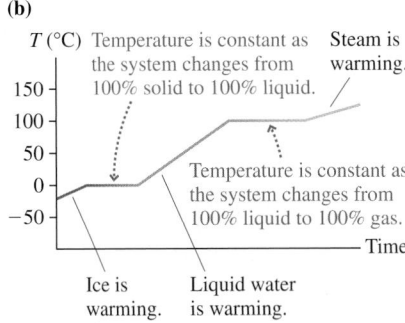

(b)

T (°C)   Temperature is constant as   Steam is
          the system changes from   warming.
          100% solid to 100% liquid.

150
100
50
0
-50
                                       Temperature is constant as
                                       the system changes from
                                       100% liquid to 100% gas.
                                                              Time

Ice is   Liquid water
warming.   is warming.

FIGURE 16.3 The temperature as a function of time as water is transformed from solid to liquid to gas.

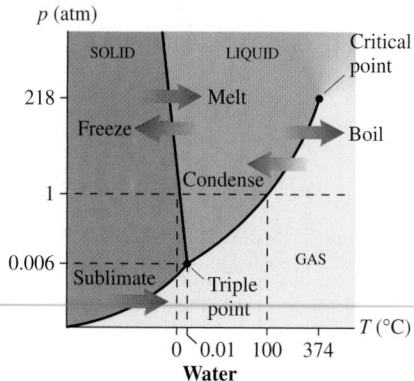

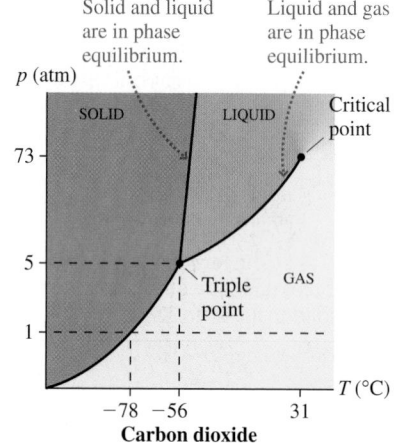

FIGURE 16.4 Phase diagrams (not to scale) for water and carbon dioxide.

A **phase diagram** is used to show how the phases and phase changes of a substance vary with both temperature and pressure. Figure 16.4 shows the phase diagrams for water and carbon dioxide. You can see that the diagram is divided into three regions corresponding to the solid, liquid, and gas phases. The boundary lines separating the regions indicate the phase transitions. The system is in phase equilibrium at a pressure-temperature point that falls on one of these lines.

Phase diagrams contain a great deal of information. Notice on the water phase diagram that the dotted line at $p = 1$ atm crosses the solid-liquid boundary at $0°C$ and the liquid-gas boundary at $100°C$. These well-known melting and boiling point temperatures of water apply only at standard atmospheric pressure. You can see that in Denver, where $p_{atmos} < 1$ atm, water melts at slightly above $0°C$ and boils at a temperature below $100°C$. A *pressure cooker* works by allowing the pressure inside to exceed 1 atm. This raises the boiling point, so foods that are in boiling water are at a temperature $>100°C$ and cook faster.

In general, crossing the solid-liquid boundary corresponds to melting or freezing while crossing the liquid-gas boundary corresponds to boiling or condensing. But you can see that there's another possibility—crossing the solid-gas boundary. The phase change in which a solid becomes a gas is called **sublimation.** It is not an everyday experience with water because water sublimates only at pressures far below atmospheric pressure. (It does happen in a vacuum chamber, which is how food products are *freeze dried.*) But you are familiar with the sublimation of dry ice. Dry ice is solid carbon dioxide. You can see on the carbon dioxide phase diagram that the dotted line at $p = 1$ atm crosses the solid-*gas* boundary, rather than the solid-liquid boundary, at $T = -78°C$. This is the *sublimation temperature* of dry ice.

Liquid carbon dioxide does exist, but only at pressures greater than 5 atm and temperatures greater than $-56°C$. A CO_2 fire extinguisher contains *liquid* carbon dioxide under high pressure. (You can hear the liquid slosh if you shake a CO_2 fire extinguisher.)

One important difference between the water and carbon dioxide phase diagrams is the slope of the solid-liquid boundary. For most substances, the solid phase is denser than the liquid phase and the liquid is denser than the gas. Pressurizing the substance compresses it and increases the density. If you start compressing CO_2 gas at room temperature, thus moving upward through the phase diagram along a vertical line, you'll first condense it to a liquid and eventually, if you keep compressing, change it into a solid.

Water is a very unusual substance in that the density of ice is *less* than the density of liquid water. That is why ice floats. If you compress ice, making it denser, you eventually cause a phase transition in which the ice turns to liquid water! Consequently, the solid-liquid boundary for water slopes to the left.

The liquid-gas boundary ends at a point called the **critical point.** Below the critical point, liquid and gas are clearly distinct and there is a phase change if you go from one to the other. But there is no clear distinction between liquid and gas at pressures or temperatures above the critical point. The system is a *fluid,* but it can be varied continuously between high density and low density without a phase change.

The final point of interest on the phase diagram is the **triple point** where the phase boundaries meet. Two phases are in phase equilibrium along the boundaries. The triple point is the *one* value of temperature and pressure for which all three phases can coexist in phase equilibrium. That is, any amounts of solid, liquid, and gas can happily coexist at the triple point. For water, the triple point occurs at $T_3 = 0.01°C$ and $p_3 = 0.006$ atm.

The significance of the triple point of water is its connection to the Kelvin temperature scale. The Celsius scale required two *reference points,* the boiling and melting points of water. We can now see that these are not very satisfactory reference points because their values vary as the pressure changes. The Kelvin scale requires only a single reference point. The low end, zero temperature at zero pressure, is fixed by the physical properties of gases. We need only one reference point to determine the slope of the pressure-temperature line of a gas thermometer. And it needs to be a reference that can always be reproduced *with no variation.*

No matter where you are or what the local conditions might be, there's only one temperature at which ice, liquid water, and water vapor will coexist in equilibrium. If you produce this equilibrium in the laboratory, then you *know* the system is at the triple-point temperature. The triple-point temperature of water is an ideal reference point, hence the Kelvin temperature scale is *defined* to be a linear temperature scale starting from 0 K at absolute zero and passing through 273.16 K at the triple point of water. Because $T_3 = 0.01°C$, absolute zero on the Celsius scale is $T_0 = -273.15°C$.

NOTE ▶ To be consistent with our use of significant figures, $T_0 = -273$ K is the appropriate value to use in calculations *unless* you know other temperatures with an accuracy of better than 1°C. ◄

STOP TO THINK 16.4 For which is there a sublimation temperature that is higher than a melting temperature?

a. Water   b. Carbon dioxide   c. Both   d. Neither

16.5 Ideal Gases

Gases are the simplest macroscopic systems. Our goal for the rest of this chapter will be to understand how the macroscopic properties of a gas change as the state of the gas changes.

While we today take it for granted that matter consists of atoms, the evidence for atoms is by no means obvious. The concept of atoms was formulated by two Greek philosophers, Leucippus and Democritus, who flourished about 440–420 BCE. They suggested that all matter consists of small, hard, indivisible, and indestructible particles they called *atoms*. Atoms, they thought, are always in motion, and the spaces between atoms are matter-free regions called the *void*.

Aristotle and most other "mainstream thinkers" were not atomists because they could not accept the idea of the void—or what we, today, would call a vacuum. Consequently, the idea of atoms essentially disappeared for nearly 2000 years. Only with the beginnings of modern chemistry, during the Renaissance, did atomic thinking reappear. Newton was an atomist, but he held the view that atoms in a gas are static, rather than in motion, and that they expand and contract to stay in contact with other atoms. Gases in motion, like the wind, would then be a collective motion of all the atoms moving together with the same velocity.

Daniel Bernoulli advanced a quite different model of gases in 1738. He suggested that a gas consists of small, hard atoms moving randomly at fairly high velocities and, on occasion, colliding with each other or the walls of the container. In other words, Bernoulli reintroduced Democritus' idea of atoms—but updated to include the scientific discoveries of Galileo and Newton! Surprisingly, Bernoulli's ideas were not accepted for nearly a century. The value of his postulates was not recognized until a complete understanding of energy conservation was achieved in the mid-19th century. Numerous scientists then developed Bernoulli's ideas into the kinetic theory of gases that we will study in Chapter 18.

What can macroscopic observations suggest to us about the properties of atoms? One observation, which we noted earlier in the chapter, is that solids and liquids are nearly incompressible. From this we can infer that atoms are fairly "hard" and cannot be pressed together once they come into contact with each other. Atoms also resist being pulled apart. Solids would not be solid if the atoms were not held

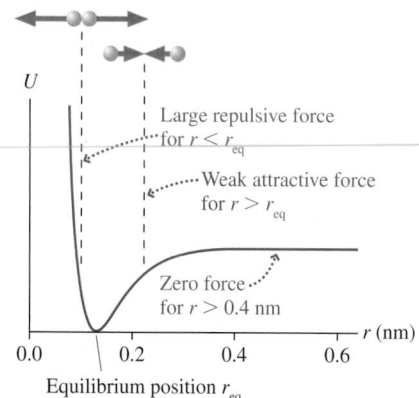

FIGURE 16.5 The potential-energy diagram for the interaction of two atoms.

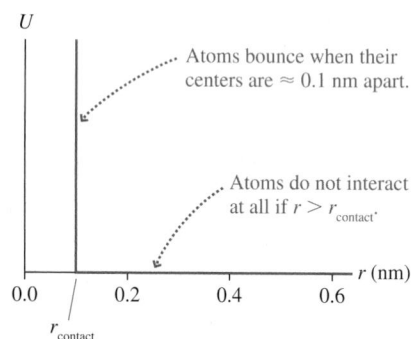

FIGURE 16.6 An idealized hard-sphere model of the interaction potential energy of two atoms.

together by attractive forces. These attractive forces are responsible for the *tensile strength* of solids—how hard you have to pull to break the solid—as well as for the cohesion of liquid droplets. Nonetheless, it is far easier to break a solid or disperse a liquid than it is to compress it, so these attractive forces must be weak in comparison to the repulsive forces that occur when we push the atoms too close together.

These observations imply that an atom is a small particle that is weakly attracted to other nearby atoms but strongly repelled by them if it gets too close. This is precisely the view of molecular bonds that we developed in Chapter 10. Figure 16.5 shows the potential-energy diagram of two atoms separated by distance r. Recall, from Chapter 11, that the force exerted by one atom on the other is the negative of the slope of this graph. The slope is large and negative for values of r less than the equilibrium value r_{eq}, so the force for $r < r_{eq}$ is large and repulsive. For r just slightly greater than r_{eq}, the modest positive slope indicates a weak attractive force. The slope has become nearly zero by $r \approx 0.4$ nm, hence the attractive force is restricted to atoms within about 0.4 nm of each other. Atoms separated by more than about 0.4 nm do not interact.

Solids and liquids are systems in which the atomic separation is very close to r_{eq}, thus the attractive and repulsive atomic forces are balanced. If you try to press the atoms closer together, the repulsive forces resist. If you try to pull them apart, the attractive forces resist.

A gas, by contrast, is much less dense and the average spacing of atoms is much larger than r_{eq}. Consequently, the atoms are usually *not interacting* with each other at all. Instead, they spend most of their time moving freely through space, only occasionally coming close enough to another atom to interact with it. When two atoms collide, it is the steep "wall" of the potential energy curve for $r < r_{eq}$ that is important. That wall represents the repulsive electrical force pushing the atoms apart as they collide. The small distance over which the atoms experience a weak attractive force is of essentially no importance because the atoms spend so little time at those distances.

The Ideal-Gas Model

With these ideas in mind, suppose we were to replace the actual potential-energy curve of Figure 16.5 with the approximate potential-energy curve of Figure 16.6. This is the potential-energy curve for the interaction of two "hard spheres" that have *no* interaction at all until they come into actual contact, at separation $r_{contact}$, and then bounce.

The *hard-sphere model* of the atom represents what we could call the *ideal atom*. It is Democritus' idea of a small, hard particle. A gas of such ideal atoms is called an **ideal gas.** It is a collection of small, hard, randomly moving atoms that occasionally collide and bounce off each other but otherwise do not interact. The ideal gas is a *model* of a real gas and, as with any other model, a simplified description. Nonetheless, experiments show that the ideal-gas model is quite good for gases if two conditions are met:

1. The density is low (i.e., the atoms occupy a volume much smaller than that of the container), and

2. The temperature is well above the condensation point.

If the density gets too high, or the temperature too low, then the attractive forces between the atoms begin to play an important role and our model, which ignores those attractive forces, fails. These are the forces that are responsible, under the right conditions, for the gas condensing into a liquid.

We've been using the term *atoms,* but many gases, as you know, consist of molecules rather than atoms. Only helium, neon, argon, and the other inert elements in the far-right column of the periodic table of the elements form monatomic gases. Hydrogen (H_2), nitrogen (N_2), and oxygen (O_2) are diatomic gases. As far as translational motion is concerned, the ideal-gas model does not distinguish between a monatomic gas and a diatomic gas; both are considered as simply small, hard spheres. Hence the terms *atoms* and *molecules* can be used interchangeably to mean the basic constituents of the gas.

Molecular Speeds

What kind of experimental evidence could distinguish between Newton's view of a gas of static atoms and Bernoulli's model of small atoms in random motion? Suppose we have a container of gas with a very small hole in the side, allowing some of the gas to flow out. If Newton is correct, the atoms should move out uniformly, all touching and moving with the *same speed*—like toothpaste out of a tube as you squeeze it! If Bernoulli is right, the individual atoms should shoot through the hole and emerge with many *different* speeds.

Figure 16.7 shows an experimental apparatus that measures the speeds of molecules in a gas. The molecules emerging from the source form what is called a *molecular beam*. At the right end, a detector counts the number of molecules making it through the apparatus each second. The entire experiment takes place inside a vacuum chamber so that the molecules can travel the distance from source to detector without undergoing any collisions.

The two rotating disks between the source and the detector form a *velocity selector*. Once every revolution, the slot in the first disk allows a small pulse of molecules to pass through. By the time these molecules reach the second disk, the slots have rotated. The molecules can pass through the slot in the second disk and be detected *only* if they have exactly the right speed to travel the distance L between the two disks in exactly the time interval Δt it takes the axle to complete one full revolution. These molecules, with $v = L/\Delta t$, can pass through the second slot as it reappears one revolution later. Molecules having any other speed are blocked by the second disk and are not detected. By changing the rotation frequency of the axle, and thus changing Δt, this apparatus can measure how many molecules have each of many possible speeds.

Figure 16.8 shows the results of an experiment performed with nitrogen gas (N_2) at $T = 20°C$. The results are presented in the form of a histogram. A **histogram** is a bar chart in which the height of each bar tells how many (or, in this case, what percent) of the molecules have a speed in the *range* of speeds shown below the bar. For example, 16% of the molecules have speeds in the range from 600 m/s to 700 m/s. All the bars added together sum to 100%, showing that this histogram describes *all* of the molecules leaving the hole.

Newton's view, in which all the molecules would have the same speed, is clearly incorrect. Instead, we see what is called a *distribution* of speeds, ranging from as low as ≈ 0 m/s to as high as ≈ 1100 m/s. This is exactly as expected for a Bernoulli gas of individual molecules moving randomly at many different speeds. Thus Figure 16.8 is convincing evidence that the basic assumptions behind the ideal-gas model are correct.

But we can learn more. You can see from Figure 16.8 that not all speeds are equally likely. There is a most probable speed of ≈ 550 m/s. This is really fast, ≈ 1200 mph! But also notice that the majority of molecular speeds do not differ much from the most probable speed. There are a few with very high or very low speeds, but well over 60% (sum of the center four bars) of the molecules have speeds within the range 300 m/s to 700 m/s.

The gas in the source consists of a very large number of molecules, each moving randomly and with a different speed. Yet the *collective* behavior of this sample of atoms is exceedingly stable and predictable. If we repeated this experiment over and over, we would get the results of Figure 16.8 every time. This idea—that the *average* properties of a very large number of molecules is a predictable quantity—will underlie our approach to the micro/macro connection in Chapter 18.

The only molecules that reach the detector are those whose speed allows them to travel distance L during the time it takes the disks to make one full revolution.

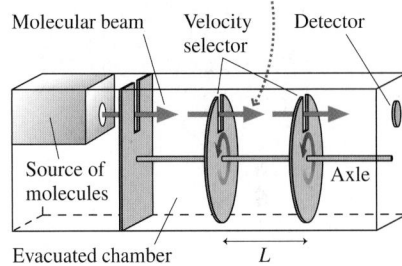

FIGURE 16.7 An apparatus to measure the speeds of molecules in a gas.

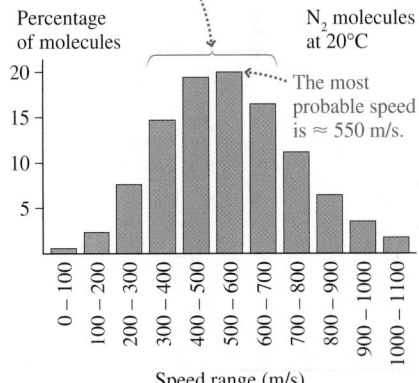

FIGURE 16.8 A histogram showing the distribution of speeds in a beam of N_2 molecules at $T = 20°C$.

The Ideal-Gas Law

Section 16.1 introduced the idea of *state variables,* those parameters that describe the state of a macroscopic system. The state variables for an ideal gas are the volume V of its container, the number of moles n of the gas present in the container, the temperature T of the gas and its container, and the pressure p that the gas exerts on the walls of the container. These four state parameters are not independent of each other. If you change the value of one—by, say, raising the temperature—then one or more of the others will change as well. Each change of the parameters is a *change of state* of the system.

Experiments during the 17th and 18th centuries showed that there is a very specific relationship between the four state variables. Suppose you measure p, V, n, and T of a gas, with the temperature measured in kelvins. Then change the state of the gas, by heating it or compressing it or doing something else to it, and measure p, V, n, and T again. Repeat this many times, changing the state of the gas each time, until you have a large table of p, V, n, and T values.

Then make a graph on which you plot pV, the product of the pressure and volume, on the vertical axis and nT, the product of the number of moles and temperature, on the horizontal axis. The very surprising result is that for *any* gas, whether it is hydrogen or helium or oxygen or methane, **you get exactly the same graph,** the linear graph shown in Figure 16.9. In other words, nothing about the graph indicates what gas was used because all gases give the same result.

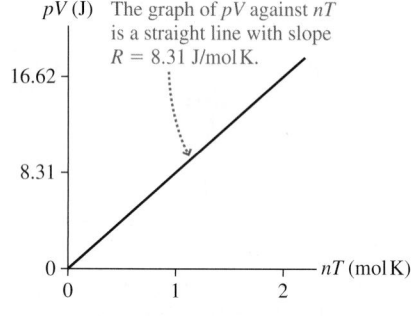

NOTE ▶ No real gas could extend to $nT = 0$ because it would condense. But an ideal gas never condenses because the only interactions among the molecules are hard-sphere collisions. ◀

As you can see, there is a very clear proportionality between the quantity pV and the quantity nT. If we designate the slope of the line in this graph as R, then we can write the relationship as

$$pV = R \times (nT)$$

It is customary to write this relationship in a slightly different form, namely

$$pV = nRT \qquad \text{(ideal-gas law)} \qquad (16.8)$$

FIGURE 16.9 A graph of pV versus nT for an ideal gas.

Equation 16.8 is the **ideal-gas law.**

The constant R, which is determined experimentally as the slope of the graph in Figure 16.9, is called the **universal gas constant.** Its value, in SI units, is

$$R = 8.31 \text{ J/mol K}$$

The units of R seem puzzling. The denominator mol K is clear because R multiplies nT. But what about the joules? The left side of the ideal-gas law, pV, has units

$$\text{Pa m}^3 = \frac{\text{N}}{\text{m}^2} \text{ m}^3 = \text{N m} = \text{joules}$$

The product pV has units of joules, as shown on the vertical axis in Figure 16.9.

NOTE ▶ You perhaps learned in chemistry to work gas problems using units of atmospheres and liters. To do so, you had a different numerical value of R that was expressed in those units. In physics, however, we always work gas problem in SI units. Pressures *must* be in Pa, volumes in m^3, and temperatures in K before you compute. Calculations using other units will give wildly incorrect answers. ◀

The surprising fact, and one worth commenting upon, is that *all* gases have the *same* graph and the *same* value of R. There is no obvious reason why a very simple atomic gas such as helium should have the same slope as a more complex gas such as methane (CH_4). Nonetheless, both turn out to have the same value for R. The ideal-gas law, within its limits of validity, describes *all* gases with a single value of the constant R.

EXAMPLE 16.3 Calculating a gas pressure
100 g of oxygen gas is distilled into an evacuated 600 cm³ container. What is the gas pressure at a temperature of 150°C?

MODEL The gas can be treated as an ideal gas. Oxygen is a diatomic gas of O_2 molecules.

SOLVE From the ideal-gas law, the pressure is $p = nRT/V$. In Example 16.2 we calculated the number of moles in 100 g of O_2 and found $n = 3.13$ mol. Gas problems typically involve several conversions to get quantities into the proper units, and this example is no exception. The SI units of V and T are m³ and K, respectively, thus

$$V = (600 \text{ cm}^3)\left(\frac{1 \text{ m}}{100 \text{ cm}}\right)^3 = 6.00 \times 10^{-4} \text{ m}^3$$

$$T = (150 + 273) \text{ K} = 423 \text{ K}$$

With this information, the pressure is

$$p = \frac{nRT}{V} = \frac{(3.13 \text{ mol})(8.31 \text{ J/mol K})(423 \text{ K})}{6.00 \times 10^{-4} \text{ m}^3}$$

$$= 1.83 \times 10^7 \text{ Pa} = 181 \text{ atm}$$

In this text we will consider only gases in sealed containers. The number of moles (and number of molecules) will not change during a problem. In that case,

$$\frac{pV}{T} = nR = \text{constant} \qquad (16.9)$$

If the gas is initially in state i, characterized by the state variables p_i, V_i, and T_i, and at some later time in a final state f, the state variables for these two states are related by

$$\frac{p_f V_f}{T_f} = \frac{p_i V_i}{T_i} \qquad \text{(ideal gas in a sealed container)} \qquad (16.10)$$

This before-and-after relationship between the two states, reminiscent of a conservation law, will be valuable for many problems.

EXAMPLE 16.4 Calculating a gas temperature
A cylinder of gas is at 0°C. A piston compresses the gas to half its original volume and three times its original pressure. What is the final gas temperature?

MODEL Treat the gas as an ideal gas in a sealed container.

SOLVE The before-and-after relationship of Equation 16.10 can be written

$$T_2 = T_1 \frac{p_2}{p_1} \frac{V_2}{V_1}$$

In this problem, the compression of the gas results in $V_2/V_1 = \frac{1}{2}$ and $p_2/p_1 = 3$. The initial temperature is $T_1 = 0°C = 273$ K. With this information,

$$T_2 = 273 \text{ K} \times 3 \times \frac{1}{2} = 409 \text{ K} = 136°C$$

ASSESS We did not need to know actual values of the pressure and volume, just the *ratios* by which they change.

We will often want to refer to the number of molecules N in a gas rather than the number of moles n. This is an easy change to make. Because $n = N/N_A$, the ideal-gas law in terms of N is

$$pV = nRT = \frac{N}{N_A}RT = N\frac{R}{N_A}T \qquad (16.11)$$

R/N_A, the ratio of two known constants, is known as **Boltzmann's constant** k_B:

$$k_B = \frac{R}{N_A} = 1.38 \times 10^{-23} \text{ J/K}$$

The subscript B distinguishes Boltzmann's constant from a spring constant or other uses of the symbol k.

Ludwig Boltzmann was an Austrian physicist who did some of the pioneering work in statistical physics during the mid-19th century. Boltzmann's constant k_B can be thought of as the "gas constant per molecule," whereas R is the "gas constant per mole." With this definition, the ideal-gas law in terms of N is

$$pV = Nk_BT \qquad \text{(ideal-gas law)} \tag{16.12}$$

Equations 16.8 and 16.12 are both the ideal-gas law, just expressed in terms of different state variables.

Recall that the number density (molecules per m³) was defined as N/V. A rearrangement of Equation 16.12 gives the number density as

$$\frac{N}{V} = \frac{p}{k_BT} \tag{16.13}$$

This is a useful consequence of the ideal-gas law, but keep in mind that the pressure *must* be in SI units of pascals and the temperature *must* be in SI units of kelvins.

EXAMPLE 16.5 The distance between molecules

"Standard temperature and pressure," abbreviated **STP**, are $T = 0°C$ and $p = 1$ atm. Estimate the average distance between gas molecules at STP.

SOLVE Imagine freezing all the molecules at some instant of time. After doing so, place an imaginary cube around each molecule to separate it from all its neighbors. This divides the total volume V of the gas into N small cubes of volume v_i such that the sum of all these small volumes v_i equals the full volume V. Although each of these volumes is somewhat different, we can define an *average* little volume:

$$v_{avg} = \frac{V}{N} = \frac{1}{N/V}$$

That is, the average volume per molecule (m³ per atom) is the inverse of the number of molecules per m³. Note that this is not the volume of the molecule itself, which is much smaller, but the average surrounding volume of space that each molecule can claim as its own, separating it from the other molecules. If we now use Equation 16.13, the number density is

$$\frac{N}{V} = \frac{p}{k_BT} = \frac{1.01 \times 10^5 \text{ Pa}}{(1.38 \times 10^{-23} \text{ J/K})(273 \text{ K})}$$

$$= 2.69 \times 10^{25} \text{ molecules/m}^3$$

where we have used the definition of STP in SI units. Thus the average volume per molecule is

$$v_{avg} = \frac{1}{N/V} = 3.72 \times 10^{-26} \text{ m}^3$$

The volume of a cube is $V = l^3$, where l is the length of each edge. Hence the average length of one of our little cubes is

$$l = (v_{avg})^{1/3} = 3.34 \times 10^{-9} \text{ m} = 3.34 \text{ nm}$$

Because each molecule sits at the center of a cube, the average distance between two molecules is the distance between opposite corners of the cube. As Figure 16.10 shows, this distance is

$$\text{average distance} = \sqrt{l^2 + l^2 + l^2} = \sqrt{3}l = 5.7 \text{ nm}$$

The average distance between molecules in a gas at STP is ≈ 5.7 nm.

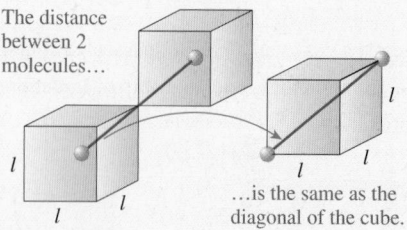

The distance between 2 molecules...

...is the same as the diagonal of the cube.

FIGURE 16.10 The distance between two molecules.

The results of this example are important. One of the basic assumptions of the ideal-gas model is that the atoms are "far apart" in comparison to the distance over which atoms exert attractive forces on each other. That distance, as was seen in Figure 16.6, is about 0.4 nm. A gas at STP has an average distance between atoms roughly 14 times the interaction distance. We can safely conclude that the ideal-gas model works well for gases under "typical" circumstances.

> **STOP TO THINK 16.5** You have two containers of equal volume. One is full of helium gas. The other holds an equal mass of nitrogen gas. Both gases have the same pressure. How does the temperature of the helium compare to the temperature of the nitrogen?
>
> a. $T_{helium} > T_{nitrogen}$ b. $T_{helium} = T_{nitrogen}$ c. $T_{helium} < T_{nitrogen}$

16.6 Ideal-Gas Processes

The ideal-gas law is the connection between the state variables pressure, temperature, and volume. If the state variables change, as they would from heating or compressing the gas, the state of the gas changes. An **ideal-gas process** is the means by which the gas changes from one state to another.

The pV Diagram

It will be very useful to represent ideal-gas processes on a graph called a pV **diagram.** This is nothing more than a graph of pressure versus volume. The important idea behind the pV diagram is that *each point* on the graph represents a single, unique state of the gas. That seems surprising at first, because a point on the graph only directly specifies the values of p and V. But knowing p and V, and assuming that n is known for a sealed container, we can find the temperature by using the ideal-gas law. Thus each point actually represents a triplet of values (p, V, T) specifying the state of the gas.

For example, Figure 16.11a is a pV diagram showing three states of a system consisting of 1 mol of gas. The values of p and V can be read from the axes for each point, then the temperature at that point determined from the ideal-gas law. An ideal-gas process can be represented as a "trajectory" in the pV diagram. The trajectory shows all the intermediate states through which the gas passes. Figures 16.11b and 16.11c show two different processes by which the gas of Figure 16.11a can be changed from state 1 to state 3.

There are infinitely many ways to change the gas from state 1 to state 3. Although the initial and final state are the same for each of them, the particular process by which the gas changes—that is, the particular trajectory—will turn out to have very real consequences. For example, you will soon learn that the work done by an expanding gas, a quantity of very practical importance in various devices, depends on the trajectory followed. The pV diagram is an important graphical representation of the process.

Quasi-Static Processes

Strictly speaking, the ideal-gas law applies only to gases in *thermal equilibrium.* We'll give a careful definition of thermal equilibrium later; for now it is sufficient to say that a system is in thermal equilibrium if its state variables are constant and not changing. Consider an ideal-gas process that changes a gas from state 1 to state 2. The initial and final states are states of thermal equilibrium, with steady values of p, V, and T. But the process, by definition, causes some of these state variables to change. The gas is *not* in thermal equilibrium while the process of changing from state 1 to state 2 is underway.

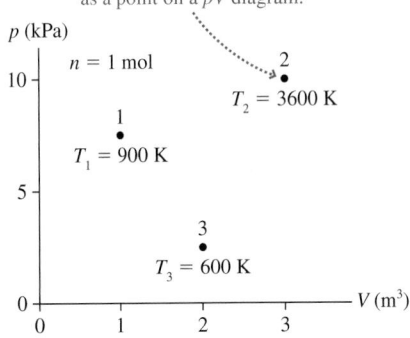

(a) Each state of an ideal gas is represented as a point on a pV diagram.

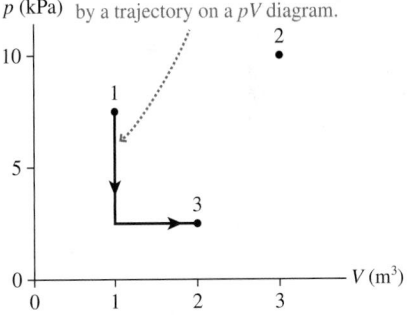

(b) A process that changes the gas from one state to another is represented by a trajectory on a pV diagram.

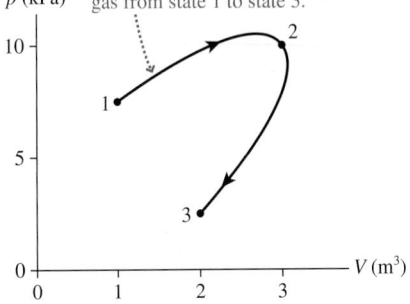

(c) This trajectory represents a different process that takes the gas from state 1 to state 3.

FIGURE 16.11 The state of the gas and ideal-gas processes can be shown on a pV diagram.

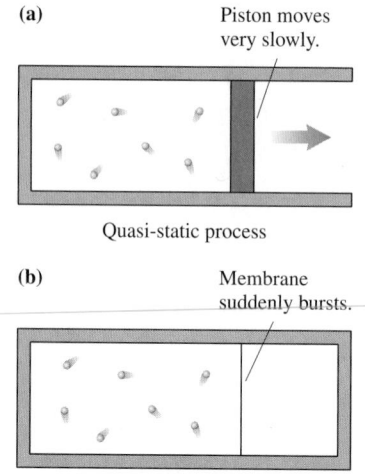

(a) Piston moves very slowly.

Quasi-static process

(b) Membrane suddenly bursts.

Irreversible process

FIGURE 16.12 The slow motion of the piston is a quasi-static process. The bursting of the membrane is not.

To use the ideal-gas law throughout, we will assume that the process occurs *so slowly* that the system is never far from equilibrium. In other words, the values of p, V, and T at any point in the process are essentially the same as the equilibrium values they would assume if we stopped the process at that point. A process that is essentially in thermal equilibrium at all times is called a **quasi-static process.** It is an idealization, like a frictionless surface, but one that is a very good approximation in many real situations.

An important characteristic of a quasi-static process is that the trajectory through the pV diagram can be *reversed.* If you quasi-statically expand a gas by slowly pulling a piston out, as shown in Figure 16.12a, you can reverse the process by slowing pushing the piston in. The gas retraces its pV trajectory until it has returned to its initial state. Contrast this with what happens when the membrane bursts in Figure 16.12b. That is a sudden process, not at all quasi-static. The *irreversible* process of Figure 16.12b cannot be represented on a pV diagram.

The critical question is: How slow must a process be to qualify as quasi-static? That turns out to be a difficult question to answer. This textbook will always assume that processes are quasi-static. It turns out to be a reasonable assumption for the types of examples and homework problems we will look at. Irreversible processes will be left to more advanced courses.

Constant-Volume Process

Many important gas processes take place in a container of constant, unchanging volume. A constant-volume process is called an **isochoric process,** where *iso* is a prefix meaning "constant" or "equal" while *choric* is from a Greek root meaning "volume." An isochoric process is one for which

$$V_f = V_i \tag{16.14}$$

For example, suppose that you have a gas in the closed, rigid container shown in Figure 16.13a. Warming the gas with a Bunsen burner will raise its pressure without changing its volume. This process is shown as the vertical line $1 \rightarrow 2$ on the pV diagram of Figure 16.13b. A constant-volume cooling, by placing the container on a block of ice, would lower the pressure and be represented as the vertical line from 2 to 1. **Any isochoric process appears on a pV diagram as a vertical line.**

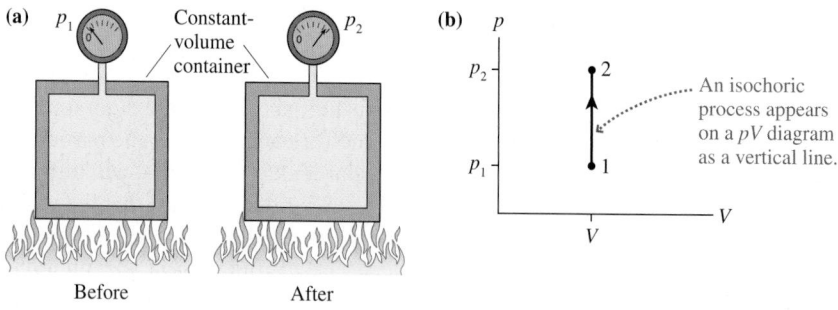

(a) p_1 Constant-volume container p_2

Before   After

(b) p

p_2 ⚫ 2

An isochoric process appears on a pV diagram as a vertical line.

p_1 ⚫ 1

V

FIGURE 16.13 A constant-volume (isochoric) process.

EXAMPLE 16.6 A constant-volume gas thermometer

A constant-volume gas thermometer is placed in contact with a reference cell containing water at the triple point. After reaching equilibrium, the gas pressure is recorded as 55.78 kPa. The thermometer is then placed in contact with a sample of unknown temperature. After the thermometer reaches a new

equilibrium, the gas pressure is 65.12 kPa. What is the temperature of this sample?

MODEL The thermometer's volume doesn't change, so this is an isochoric process.

SOLVE The temperature at the triple point of water is $T_1 = 0.01°C = 273.16$ K. The ideal-gas law for a closed system is $p_2V_2/T_2 = p_1V_1/T_1$. The volume doesn't change, so $V_2/V_1 = 1$. Thus

$$T_2 = T_1 \frac{V_2}{V_1} \frac{p_2}{p_1} = T_1 \frac{p_2}{p_1} = (273.16 \text{ K}) \frac{65.12 \text{ kPa}}{55.78 \text{ kPa}}$$

$$= 318.90 \text{ K} = 45.75°\text{C}$$

The temperature *must* be in kelvins to do this calculation, although it is common to convert the final answer to °C. The fact that the pressures were given to four significant figures justified using $T_K = T_C + 273.15$ rather than the usual $T_C + 273$.

ASSESS $T_2 > T_1$, which we expected from the increase in pressure.

Constant-Pressure Process

Other gas processes take place at a constant, unchanging pressure. A constant-pressure process is called an **isobaric process,** where *baric* is from the same root as "barometer" and means "pressure." An isobaric process is one for which

$$p_f = p_i \tag{16.15}$$

Figure 16.14a shows one method of changing the state of a gas while keeping the pressure constant. A cylinder of gas has a tight-fitting, massless piston that can slide up and down but seals the container so that no atoms enter or escape. The mass on top applies a constant downward force Mg on the piston. The air also presses down on the piston. In equilibrium, the upward force pA on the piston by the gas in the cylinder, where A is the piston's area, exactly balances the downward force Mg plus the air-pressure force $p_{atmos}A$. Consequently, the gas pressure inside the cylinder is

$$p = p_{atmos} + \frac{Mg}{A} \tag{16.16}$$

In other words, the gas pressure is determined by the requirement that the gas must support both the mass on top of the piston and the air pressing inward. This pressure is independent of the temperature of the gas or the height of the piston, so it stays constant as long as M is unchanged.

If the cylinder is warmed, the gas will expand and push the piston up. But the pressure, determined by mass M, will not change. This process is shown on the pV diagram of Figure 16.14b as the horizontal line $1 \rightarrow 2$. We call this an *isobaric expansion.* An *isobaric compression* occurs if the gas is cooled, lowering the piston. **Any isobaric process appears on a pV diagram as a horizontal line.**

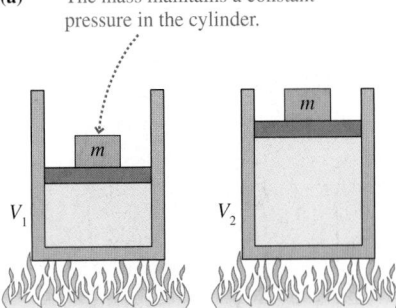

(a) The mass maintains a constant pressure in the cylinder.

Before   After

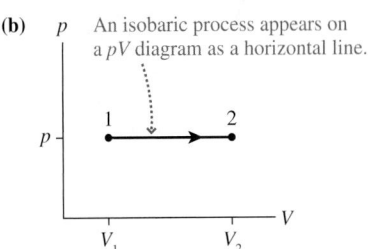

(b) An isobaric process appears on a pV diagram as a horizontal line.

FIGURE 16.14 A constant-pressure (isobaric) process.

EXAMPLE 16.7 A constant-pressure compression
A gas occupying 50 cm³ at 50°C is cooled at constant pressure until the temperature is 10° C. What is its final volume?

MODEL The pressure of the gas doesn't change, so this is an isobaric process.

SOLVE By definition, $p_1/p_2 = 1$ for an isobaric process. Using the ideal-gas law for constant n, we have

$$V_2 = V_1 \frac{p_1}{p_2} \frac{T_2}{T_1} = V_1 \frac{T_2}{T_1}$$

Temperatures *must* be in kelvins to use the ideal-gas law. Thus

$$V_2 = (50 \text{ cm}^3) \frac{(10 + 273) \text{ K}}{(50 + 273) \text{ K}} = 43.8 \text{ cm}^3$$

ASSESS As long as we use *ratios*, we do not need to convert volumes or pressures to SI units. That is because the conversion is a multiplicative factor that cancels. But the conversion of temperature is an *additive* factor that does *not* cancel. That is why you must always convert temperatures to kelvins in ideal-gas calculations.

Constant-Temperature Process

The last process we wish to look at for now is one that takes place at a constant temperature. A constant-temperature process is called an **isothermal process.** An isothermal process is one for which $T_f = T_i$. Because $pV = nRT$, a constant-

Activ
Physics ONLINE   8.4

(a)

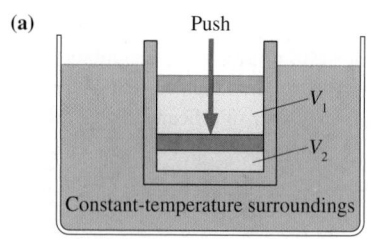

Push

V_1

V_2

Constant-temperature surroundings

(b)

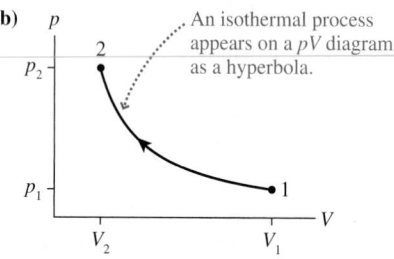

An isothermal process appears on a pV diagram as a hyperbola.

(c)

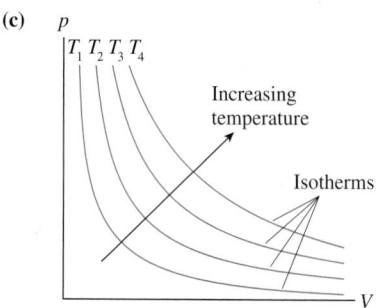

Increasing temperature

Isotherms

FIGURE 16.15 A constant-temperature (isothermal) process.

temperature process in a closed system (constant n) is one for which the product pV doesn't change. Thus

$$p_f V_f = p_i V_i \qquad (16.17)$$

in an isothermal process.

One possible isothermal process is illustrated in Figure 16.15a. A piston is being pushed down to compress a gas, but the gas cylinder is floating in a large container of liquid that is held at a constant temperature. If the piston is pushed *slowly*, then heat energy transfer through the walls of the cylinder will keep the gas at the same temperature as the surrounding liquid. This would be an *isothermal compression*. The reverse process, with the piston slowly pulled out, would be an *isothermal expansion*.

Representing an isothermal process on the pV diagram is a little more complicated than the two previous processes because both p and V change. As long as T remains fixed, we have the relationship

$$p = \frac{nRT}{V} = \frac{\text{constant}}{V} \qquad (16.18)$$

The inverse relationship between p and V causes the graph of an isothermal process to be a *hyperbola*.

The process shown as $1 \rightarrow 2$ in Figure 16.15b represents the *isothermal compression* shown in Figure 16.15a. An *isothermal expansion* would move in the opposite direction along the hyperbola.

The location of the hyperbola depends on the value of T. A lower-temperature process is represented by a hyperbola closer to the origin than a higher-temperature process. Figure 16.15c shows four hyperbolas representing the temperatures T_1 to T_4 where $T_4 > T_3 > T_2 > T_1$. These are called **isotherms**. A gas undergoing an isothermal process will move along the isotherm of the appropriate temperature.

EXAMPLE 16.8 A constant-temperature compression

A gas cylinder with a tight-fitting, moveable piston contains 200 cm^3 of air at 1.0 atm. It floats on the surface of a swimming pool of $15°C$ water. The cylinder is then slowly pulled underwater to a depth of 3.0 m. What is the volume of gas at this depth?

MODEL The gas's temperature doesn't change, so this is an isothermal compression.

SOLVE At the surface, the pressure inside the cylinder must exactly equal the outside air pressure of 1.0 atm. If the pressures were not equal, a net force would push the piston in or pull it out until the pressures balanced and equilibrium was achieved. As the cylinder is pulled underwater, the increasing water pressure pushes the piston in and compresses the gas. Equilibrium at depth d requires that the gas pressure inside the cylinder equal

the water pressure $p_{water} = p_0 + \rho g d$, where $p_0 = 1.0$ atm is the pressure at the surface. As long as the cylinder moves slowly, the gas will stay at the same temperature as the surrounding water. The value of T is not important; all we need to know is that the compression is isothermal. In that case, because $T_2/T_1 = 1$,

$$V_2 = V_1 \frac{T_2}{T_1} \frac{p_1}{p_2} = V_1 \frac{p_1}{p_2} = V_1 \frac{p_0}{p_0 + \rho g d}$$

The initial pressure p_0 must be in SI units: $p_0 = 1.0$ atm $= 1.013 \times 10^5$ Pa. Then a straightforward computation gives $V_2 = 155 \text{ cm}^3$.

ASSESS V_2 is less than V_1. This is expected because the gas is being compressed.

EXAMPLE 16.9 A multistep process

A gas at 2.0 atm pressure and a temperature of $200°C$ is first expanded isothermally until its volume has doubled. It then undergoes an isobaric compression until it returns to its original volume. First show this process on a pV diagram. Then find the final temperature and pressure.

MODEL Many practical ideal-gas processes consists of several basic steps performed in series. In this case, the final state of the isothermal expansion is the initial state for an isobaric compression.

VISUALIZE Figure 16.16 shows the process. The gas starts in state 1 at pressure $p_1 = 2.0$ atm and volume V_1. As the gas

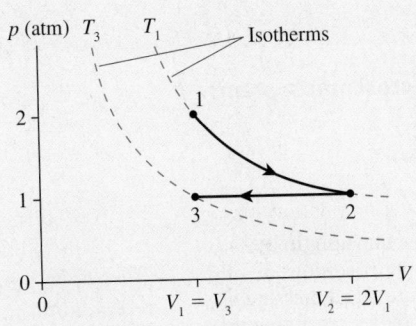

FIGURE 16.16 A pV diagram for the process of Example 16.9.

expands isothermally, it moves downward along an isotherm until it reaches volume $V_2 = 2V_1$. The pressure decreases during this process to a lower value, p_2. The gas is then compressed

at constant pressure p_2 until its final volume V_3 equals its original volume V_1. State 3 is on an isotherm closer to the origin, so we expect to find $T_3 < T_1$.

SOLVE $T_2/T_1 = 1$ during the isothermal expansion and $V_2 = 2V_1$, so the pressure at point 2 is

$$p_2 = p_1 \frac{T_2}{T_1} \frac{V_1}{V_2} = p_1 \frac{V_1}{2V_1} = \frac{1}{2}p_1 = 1.0 \text{ atm}$$

$p_3/p_2 = 1$ during the isobaric compression and $V_3 = V_1 = \frac{1}{2}V_2$, so

$$T_3 = T_2 \frac{p_3}{p_2} \frac{V_3}{V_2} = T_2 \frac{\frac{1}{2}V_2}{V_2} = \frac{1}{2}T_2 = 236.5 \text{ K} = -36.5°\text{C}$$

where we converted T_2 to 473 K before doing calculations and then converted T_3 back to °C. The final state, with $T_3 = -36.5°\text{C}$ and $p_3 = 1.0$ atm, is one in which both the pressure and the absolute temperature are half their original values.

What is the ratio T_f/T_i for this process?

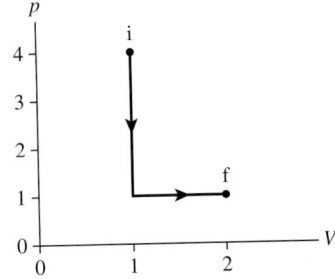

a. $\frac{1}{4}$

b. $\frac{1}{2}$

c. 1 (no change)

d. 2

e. 4

f. There's not enough information to tell.

SUMMARY

The goal of Chapter 16 has been to learn the characteristics of macroscopic systems.

GENERAL PRINCIPLES

Three Phases of Matter

Solid	Rigid, definite shape. Nearly incompressible.
Liquid	Molecules loosely held together by molecular bonds, but able to move around. Nearly incompressible.
Gas	Molecules move freely through space. Compressible.

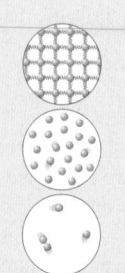

The different phases exist for different conditions of temperature T and pressure p. The boundaries separating the regions of a phase diagram are lines of phase equilibrium. Any amounts of the two phases can coexist in equilibrium. The **triple point** is the one value of temperature and pressure at which all three phases can coexist in equilibrium.

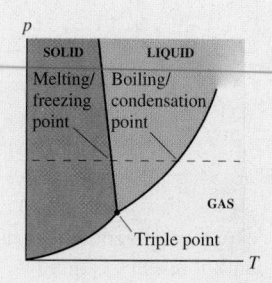

IMPORTANT CONCEPTS

Ideal-Gas Model

- Atoms and molecules are small, hard spheres that travel freely through space except for occasional collisions with each other or the walls.

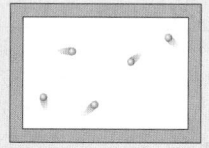

- The molecules have a distribution of speeds.

- The model is valid when the density is low and the temperature well above the condensation point.

Ideal-Gas Law

The **state variables** of an ideal gas are related by the ideal-gas law

$$pV = nRT \quad \text{or} \quad pV = Nk_\text{B}T$$

where $R = 8.31$ J/mol K is the universal gas constant and $k_\text{B} = 1.38 \times 10^{-23}$ J/K is Boltzmann's constant.

p, V, and T *must* be in SI units of Pa, m³, and K. For a gas in a sealed container, with constant n:

$$\frac{p_2 V_2}{T_2} = \frac{p_1 V_1}{T_1}$$

Counting atoms and moles

A macroscopic sample of matter consists of N atoms (or molecules), each of mass m (the **atomic** or **molecular mass**):

$$N = \frac{M}{m}$$

Alternatively, we can state that the sample consists of n **moles**

$$n = \frac{N}{N_\text{A}} \quad \text{or} \quad \frac{M(\text{in grams})}{M_\text{mol}}$$

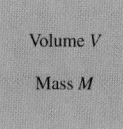
Volume V
Mass M

$N_\text{A} = 6.02 \times 10^{23}$ mol⁻¹ is **Avogadro's number.**

The numerical value of the molar mass M_mol, in g/mol, equals the numerical value of the atomic or molecular mass m in u. The atomic or molecular mass m, in atomic mass units u, is well approximated by the **atomic mass number** A.

$$1 \text{ u} = 1.661 \times 10^{-27} \text{ kg}$$

The **number density** of the sample is $\frac{N}{V}$.

APPLICATIONS

Temperature scales

$$T_\text{F} = \frac{9}{5}T_\text{C} + 32° \qquad T_\text{K} = T_\text{C} + 273$$

The Kelvin temperature scale is based on:

- Absolute zero at $T_0 = 0$ K
- The triple point of water at $T_3 = 273.16$ K

Three basic gas processes

1. **Isochoric,** or constant volume
2. **Isobaric,** or constant pressure
3. **Isothermal,** or constant temperature

pV diagram

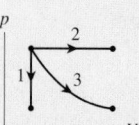

TERMS AND NOTATION

macroscopic system	atomic mass number, A	absolute temperature scale	universal gas constant, R
bulk properties	atomic mass	melting point	Boltzmann's constant, k_B
micro/macro connection	atomic mass unit, u	freezing point	STP
phase	molecular mass	phase equilibrium	ideal-gas process
phase change	mole, n	condensation point	pV diagram
solid	monatomic gas	boiling point	quasi-static process
crystal	diatomic gas	phase diagram	isochoric process
amorphous solid	Avogadro's number, N_A	sublimation	isobaric process
liquid	molar mass, M_{mol}	critical point	isothermal process
gas	temperature, T	triple point	isotherm
state variable	constant-volume gas	ideal gas	
thermal equilibrium	thermometer	histogram	
number density, N/V	absolute zero, T_0	ideal-gas law	

EXERCISES AND PROBLEMS

Exercises

Section 16.1 Solids, Liquids, and Gases

1. What volume of water has the same mass as 2.0 m³ of lead?
2. The nucleus of a uranium atom has a diameter of 1.5×10^{-14} m and a mass of 4.0×10^{-25} kg. What is the density of the nucleus?
3. What is the diameter of a copper sphere that has the same mass as a 10 cm × 10 cm × 10 cm cube of aluminum?
4. A hollow sphere, with an outer diameter of 10.0 cm and an inner diameter of 8.0 cm, has a mass of 690 g. Of what material is it made?

Section 16.2 Atoms and Moles

5. How many atoms are in a 2.0 cm × 2.0 cm × 2.0 cm cube of aluminum?
6. How many moles are in a 2.0 cm × 2.0 cm × 2.0 cm cube of copper?
7. What is the number density of (a) aluminum and (b) lead?
8. An element in its solid phase has mass density 1750 kg/m³ and number density 4.39×10^{28} atoms/m³. What is the element's atomic mass number?
9. 1.0 mol of gold is shaped into a cube. What is the length of each edge?
10. What volume of aluminum has the same number of atoms as 10 cm³ of mercury?

Section 16.3 Temperature

11. The lowest and highest natural temperatures ever recorded on earth are −127°F in Antarctica and 136°F in Libya. What are these temperatures in °C and in K?
12. At what temperature does the numerical value in °F match the numerical value in °C?

13. A demented scientist creates a new temperature scale, the "Z scale." He decides to call the boiling point of nitrogen 0°Z and the melting point of steel 1000°Z.
    a. What is the boiling point of water on the Z scale?
    b. Convert 500°Z to the Celsius scale and to the Kelvin scale.
14. The sample in an experiment is initially at 10°C. If the sample's temperature is doubled, what is the new temperature in °C?

Section 16.4 Phase Changes

15. What is the temperature in °F and the pressure in Pa at the triple point of (a) water and (b) carbon dioxide?
16. a. Is there a highest temperature at which ice can exist? If so, what is it? If not, why not?
    b. Is there a lowest temperature at which water vapor can exist? If so, what is it? If not, why not?
17. An aquanaut lives in an underwater apartment 100 m beneath the surface of the ocean. Compare the freezing and boiling points of water in the aquanaut's apartment to their values at the surface. Are they higher, lower, or the same? Explain.
18. a. A sample of water vapor in an enclosed cylinder has an initial pressure of 500 Pa at an initial temperature of −0.01°C. A piston squeezes the sample smaller and smaller, without limit. Describe what happens to the water as the squeezing progresses.
    b. Repeat part a if the initial temperature is 0.03°C warmer.

Section 16.5 Ideal Gases

19. A rigid container holds 2.0 mol of gas at a pressure of 1.0 atm and a temperature of 30°C.
    a. What is the container's volume?
    b. What is the pressure if the temperature is raised to 130°C?

20. A cylinder contains 4.0 g of nitrogen gas. A piston compresses the gas to half its initial volume. Afterward,
    a. Has the mass density of the gas changed? If so, by what factor? If not, why not?
    b. Has the number of moles of gas changed? If so, by what factor? If not, why not?

21. 3.0 mol of gas at a temperature of $-120°C$ fills a 2.0 L container. What is the gas pressure?

22. A gas at $100°C$ fills volume V_0. If the pressure is held constant, what is the volume if (a) the Celsius temperature is doubled and (b) the Kelvin temperature is doubled.

23. A 15-cm-diameter compressed-air tank is 50 cm tall. The pressure at $20°C$ is 150 atm.
    a. How many moles of air are in the tank?
    b. What volume would this air occupy at STP?

24. A 20-cm-diameter cylinder that is 40 cm long contains 50 g of oxygen gas at $20°C$.
    a. How many moles of oxygen are in the cylinder?
    b. How many oxygen molecules are in the cylinder?
    c. What is the number density of the oxygen?
    d. What is the reading of a pressure gauge attached to the tank?

25. A 10-cm-diameter cylinder of helium gas is 30 cm long and at $20°C$. The pressure gauge reads 120 psi.
    a. How many helium atoms are in the cylinder?
    b. What is the mass of the helium?
    c. What is the helium number density?
    d. What is the helium mass density?

26. A $1.0 \text{ m} \times 1.0 \text{ m} \times 1.0 \text{ m}$ cube of nitrogen gas is at $20°C$ and 1.0 atm. Estimate the number of molecules in the cube with a speed between 700 m/s and 1000 m/s.

Section 16.6 Ideal-Gas Processes

27. A rigid sphere has a valve that can be opened or closed. The sphere with the valve open is placed in boiling water in a room where the air pressure is 1.0 atm. After a long period of time has elapsed, the valve is closed. What will be the pressure inside the sphere if it is then placed in (a) a mixture of ice and water and (b) an insulated box filled with dry ice?

28. A gas with initial state variables p_1, V_1, and T_1 expands isothermally until $V_2 = 2V_1$. What are (a) T_2 and (b) p_2?

29. 0.10 mol of argon gas is admitted to an evacuated 50 cm³ container at $20°C$. The gas then undergoes an isochoric heating to a temperature of $300°C$.
    a. What is the final pressure of the gas?
    b. Show the process on a pV diagram. Include a proper scale on both axes.

30. 0.10 mol of argon gas is admitted to an evacuated 50 cm³ container at $20°C$. The gas then undergoes an isobaric heating to a temperature of $300°C$.
    a. What is the final volume of the gas?
    b. Show the process on a pV diagram. Include a proper scale on both axes.

31. 0.10 mol of argon gas is admitted to an evacuated 50 cm³ container at $20°C$. The gas then undergoes an isothermal expansion to a volume of 200 cm³.
    a. What is the final pressure of the gas?
    b. Show the process on a pV diagram. Include a proper scale on both axes.

32. 0.0040 mol of gas undergoes the process shown in Figure Ex16.32.
    a. What type of process is this?
    b. What are the initial and final temperatures?

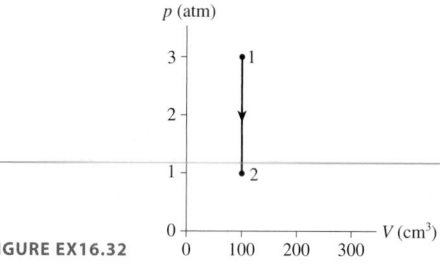

FIGURE EX16.32

33. 0.0040 mol of gas undergoes the process shown in Figure Ex16.33.
    a. What type of process is this?
    b. What are the initial and final temperatures?
    c. What is the final volume V_2?

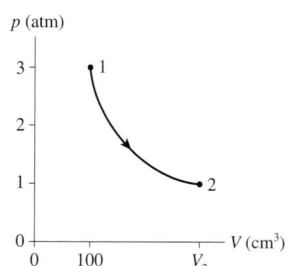

FIGURE EX16.33

34. A gas with an initial temperature of $900°C$ undergoes the process shown in Figure Ex16.34.
    a. What type of process is this?
    b. What is the final temperature?
    c. How many moles of gas are there?

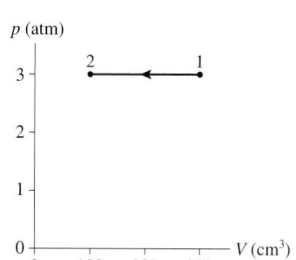

FIGURE EX16.34

Problems

35. The atomic mass number of copper is $A = 64$. Assume that atoms in solid copper form a cubic crystal lattice. To envision this, imagine that you place atoms at the centers of tiny sugar cubes, then stack the little sugar cubes to form a big cube. If you dissolve the sugar, the atoms left behind are in a cubic crystal lattice. What is the smallest distance between two copper atoms?

36. An element in its solid phase forms a cubic crystal lattice (see Problem 35) with mass density 7950 kg/m³. The smallest spacing between two adjacent atoms is 0.227 nm. What is the element's atomic mass number?

37. a. Use data from Figure 16.8 to estimate the average kinetic energy of a nitrogen molecule at room temperature.
    b. You will learn in Chapter 18 that the average molecular kinetic energy at temperature T is *independent* of the composition of the gas. Assuming this, estimate the average speed of a hydrogen molecule at room temperature.

38. Estimate the number density of gas molecules in the earth's atmosphere at sea level.

39. The solar corona is a very hot atmosphere surrounding the visible surface of the sun. X-ray emissions from the corona show that its temperature is about 2×10^6 K. The gas pressure in the corona is about 0.03 Pa. Estimate the number density of particles in the solar corona.

40. Current vacuum technology can achieve a pressure of 1.0×10^{-10} mm of Hg. At this pressure, and at a temperature of 20°C, how many molecules are in 1 cm³?

41. The semiconductor industry manufactures integrated circuits in large vacuum chambers where the pressure is 1.0×10^{-10} mm of Hg.
    a. What fraction is this of atmospheric pressure?
    b. At $T = 20°C$, how many molecules are in a cylindrical chamber 40 cm in diameter and 30 cm tall?

42. On average, each person in the industrialized world is responsible for the emission of 10,000 kg of carbon dioxide (CO_2) every year. This includes CO_2 that you generate directly, by burning fossil fuels to operate your car or your furnace, as well as CO_2 generated on your behalf by electric generating stations and manufacturing plants. CO_2 is a greenhouse gas that contributes to global warming. If you were to store your yearly CO_2 emissions in a cube at STP, how long would each edge of the cube be?

43. A gas at 25°C and atmospheric pressure fills a cylinder. The gas is transferred to a new cylinder with three times the volume, after which the pressure is half the original pressure. What is the new temperature of the gas?

44. What is the ratio of the volume occupied by 300°C steam at 1.0 atm pressure to the volume occupied by an equal mass of liquid water?

45. 10,000 cm³ of 200°C steam at a pressure of 20 atm is cooled until it condenses. What is the volume of the liquid water? Give your answer in cm³.

46. An electric generating plant boils water to produce high-pressure steam. The steam spins a turbine that is connected to the generator.
    a. How many liters of water must be boiled to fill a 5.0 m³ boiler with 50 atm of steam at 400°C?
    b. The steam has dropped to 2.0 atm pressure at 150°C as it exits the turbine. How much volume does it now occupy?

47. On a cool morning, when the temperature is 15°C, you measure the pressure in your car tires to be 30 psi. After driving 20 mi on the freeway, the temperature of your tires is 45°C. What pressure will your tire gauge now show?

48. The air temperature and pressure in a laboratory are 20°C and 1.0 atm. A 1.0 L container is open to the air. The container is then sealed and placed in a bath of boiling water. After reaching thermal equilibrium, the container is opened. How many moles of air escape?

49. The volume in a constant-*pressure* gas thermometer is directly proportional to the absolute temperature. A constant-pressure thermometer is calibrated by adjusting its volume to 1000 mL while it is in contact with a reference cell at the triple point of water. The volume increases to 1638 mL when the thermometer is placed in contact with a sample. What is the sample's temperature?

50. The mercury manometer shown in Figure P16.50 is attached to a gas cell. The mercury height h is 120 mm when the cell is placed in an ice-water mixture. The mercury height drops to 30 mm when the device is carried into an industrial freezer. What is the freezer temperature?
    Hint: The right tube of the manometer is much narrower than the left tube. What reasonable assumption can you make about the gas volume?

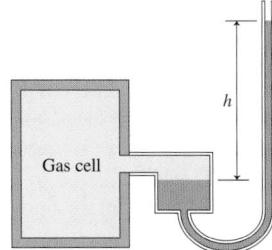

FIGURE P16.50

51. A 200 cm³ gas cell containing helium at 50°C is connected to a mercury manometer. The mercury in the outside arm of the manometer is 25.0 cm below the mercury in the arm attached to the gas cell.
    a. How many helium atoms are in the gas cell?
    b. What is the mass of the helium?

52. A diver 50 m deep in 10°C fresh water exhales a 1.0-cm-diameter bubble. What is the bubble's diameter just as it reaches the surface of the lake, where the water temperature is 20°C?
    Hint: Assume that the air bubble is always in thermal equilibrium with the surrounding water.

53. A compressed-air cylinder is known to fail if the pressure exceeds 110 atm. A cylinder that was filled to 25 atm at 20°C is stored in a warehouse. Unfortunately, the warehouse catches fire and the temperature reaches 950°C. Does the cylinder blow?

54. Reproduce Figure P16.54 on a piece of paper. A gas starts with pressure p_1 and volume V_1. Show on the figure the process in which the gas undergoes an isochoric process that doubles the pressure, then an isobaric process that doubles the volume, followed by an isothermal process that doubles the volume again. Label each of the three processes.

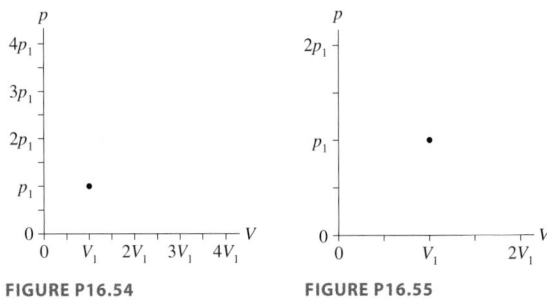

FIGURE P16.54   **FIGURE P16.55**

55. Reproduce Figure P16.55 on a piece of paper. A gas starts with pressure p_1 and volume V_1. Show on the figure the process in which the gas undergoes an isothermal process during which the volume is halved, then an isochoric process during which the pressure is halved, followed by an isobaric process during which the volume is doubled. Label each of the three processes.

56. 8.0 g of helium gas follows the process $1 \rightarrow 2 \rightarrow 3$ shown in Figure P16.56. Find the values of V_1, V_3, p_2, and T_3.

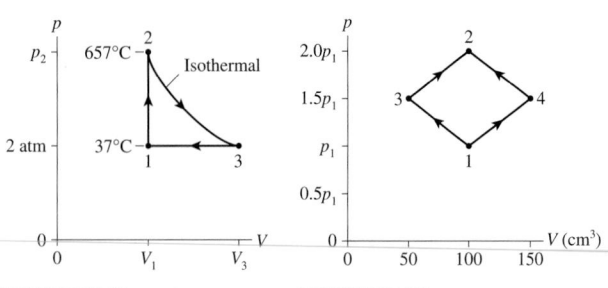

FIGURE P16.56   **FIGURE P16.57**

57. Figure P16.57 shows two different processes by which 1.0 g of nitrogen gas moves from state 1 to state 2. The temperature of state 1 is 25°C. What are (a) pressure p_1 and (b) temperatures (in °C) T_2, T_3, and T_4?

58. Figure P16.58 shows two different processes by which 80 mol of gas move from state 1 to state 2. The dashed line is an isotherm.
    a. What is the temperature of the isothermal process?
    b. What maximum temperature is reached along the straight-line process?

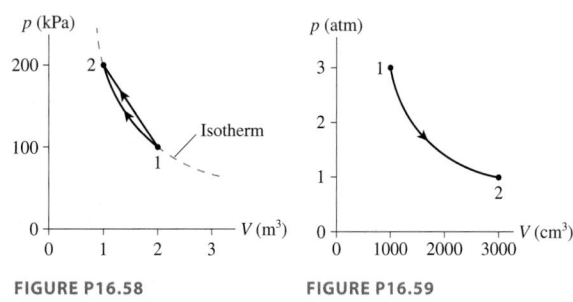

FIGURE P16.58   **FIGURE P16.59**

59. 0.10 mol of gas undergoes the process $1 \rightarrow 2$ shown in Figure P16.59.
    a. What are temperatures T_1 and T_2?
    b. What type of process is this?
    c. The gas undergoes an isochoric heating from point 2 until the pressure is restored to the value it had at point 1. What is the final temperature of the gas?

60. 0.0050 mol of gas undergoes the process $1 \rightarrow 2 \rightarrow 3$ shown in Figure P16.60. What are (a) temperature T_1, (b) pressure p_2, and (c) volume V_3?

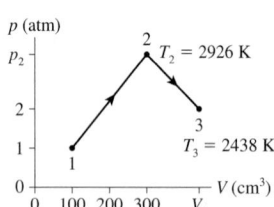

FIGURE P16.60

61. 10 g of dry ice (solid CO_2) is placed in a 10,000 cm³ container, then all the air is quickly pumped out and the container sealed. The container is warmed to 0°C, a temperature at which CO_2 is a gas.
    a. What is the gas pressure? Give your answer in atm.
    The gas then undergoes an isothermal compression until the pressure is 3.0 atm, immediately followed by an isobaric compression until the volume is 1000 cm³.

b. What is the final temperature of the gas?
c. Show the process on a pV diagram.

62. A container with 100 cm³ of gas is pressurized at constant volume until the gas pressure triples. It is then expanded at constant temperature until the pressure is one-half of the pressure the gas had before the experiment started.
    a. What is the final volume of the gas?
    b. Show this process on a pV diagram.

63. A container of gas at 2.0 atm pressure and 127°C is compressed at constant temperature until the volume is halved. It is then further compressed at constant pressure until the volume is halved again.
    a. What are the final pressure and temperature of the gas?
    b. Show this process on a pV diagram.

64. Five grams of nitrogen gas at an initial pressure of 3.0 atm and at 20°C undergo an isobaric expansion until the volume has tripled.
    a. What is the gas volume after the expansion?
    b. What is the gas temperature after the expansion?
       The gas pressure is then decreased at constant volume until the original temperature is reached.
    c. What is the gas pressure after the decrease?
       Finally, the gas is isothermally compressed until it returns to its initial volume.
    d. What is the final gas pressure?
    e. Show the full three-step process on a pV diagram. Use appropriate scales on both axes.

In Problems 65 through 68 you are given the equation(s) used to solve a problem. For each of these, you are to
    a. Write a realistic problem for which this is the correct equation(s).
    b. Draw a pV diagram.
    c. Finish the solution of the problem.

65. $p_2 = \dfrac{300 \text{ cm}^3}{100 \text{ cm}^3} \times 1 \times 2 \text{ atm}$

66. $(T_2 + 273) \text{ K} = \dfrac{200 \text{ kPa}}{500 \text{ kPa}} \times 1 \times (400 + 273) \text{ K}$

67. $V_2 = \dfrac{(400 + 273) \text{ K}}{(50 + 273) \text{ K}} \times 1 \times 200 \text{ cm}^3$

68. $(2.0 \times 101,300 \text{ Pa})(100 \times 10^{-6} \text{ m}^3) = n(8.31 \text{ J/mol K})T_1$

$n = \dfrac{0.12 \text{ g}}{20 \text{ g/mol}}$

$T_2 = \dfrac{200 \text{ cm}^3}{100 \text{ cm}^3} \times 1 \times T_1$

Challenge Problems

69. The 50 kg lead piston shown in Figure CP16.69 floats on 0.12 mol of compressed air.
    a. What is the piston height h if the temperature is 30°C?
    b. How far does the piston move if the temperature is increased by 100°C?

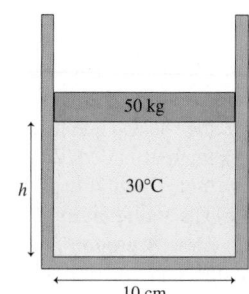

FIGURE CP16.69

70. The U-shaped tube in Figure CP16.70 has a total length of 1.0 m. It is open at one end, closed at the other, and is initially filled with air at 20°C and 1.0 atm pressure. Mercury is poured slowly into the open end without letting any air escape, thus compressing the air. This is continued until the open side of the tube is completely filled with mercury. What is the length L of the column of mercury?

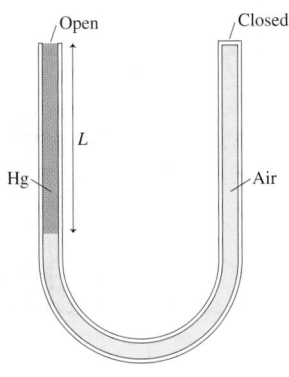

FIGURE CP16.70

71. A diving bell is a 3.0-m-tall cylinder closed at the upper end but open at the lower end. The temperature of the air in the bell is 20°C. The bell is lowered into the ocean until its lower end is 100 m deep. The temperature at that depth is 10°C.
    a. How high does the water rise in the bell after enough time has passed for the air to reach thermal equilibrium?
    b. A compressed-air hose from the surface is used to expel all the water from the bell. What minimum air pressure is needed to do this?

72. The 3.0-m-long pipe in Figure CP16.72 is closed at the top end. It is slowly pushed straight down into the water until the top end of the pipe is level with the water's surface. What is the length L of the trapped volume of air?

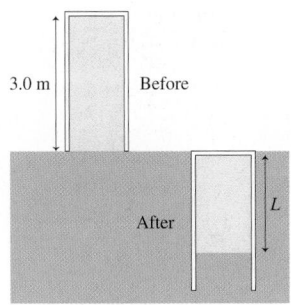

FIGURE CP16.72

73. The cylinder in Figure CP16.73 has a moveable piston attached to a spring. The cylinder's cross-section area is 10 cm², it contains 0.0040 mol of gas, and the spring constant is 1500 N/m. At 20°C the spring is neither compressed nor stretched. How far is the spring compressed if the gas temperature is raised to 100°C?

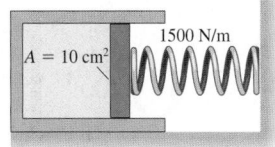

FIGURE CP16.73

74. Containers A and B in Figure CP16.74 hold the same gas. The volume of B is four times the volume of A. The two containers are connected by a thin tube (negligible volume) and a valve that is closed. The gas in A is at 300 K and pressure of 1.0×10^5 Pa. The gas in B is at 400 K and pressure of 5.0×10^5 Pa. Heaters will maintain the temperatures of A and B even after the valve is opened.
    a. After the valve is opened, gas will flow one way or the other until A and B have equal pressure. What is this final pressure?
    b. Is this a reversible or an irreversible process? Explain.

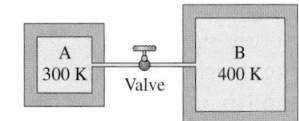

FIGURE CP16.74

STOP TO THINK ANSWERS

Stop to Think 16.1: d. The pressure *decreases* by 20 kPa.

Stop to Think 16.2: a. The number of atoms depends only on the number of moles, not the substance.

Stop to Think 16.3: a. The step size on the Kelvin scale is the same as the step size on the Celsius scale. A *change* of 10°C is a *change* of 10 K.

Stop to Think 16.4: a. On the water phase diagram, you can see that for a pressure just slightly below the triple-point pressure that the solid/gas transition occurs at a higher temperature than does the solid/liquid transition at high pressures. This is not true for carbon dioxide.

Stop to Think 16.5: c. $T = pV/nR$. Pressure and volume are the same, but n differs. The number of moles in mass M is $n = M/M_{mol}$. Helium, with the smaller molar mass, has a larger number of moles and thus a lower temperature.

Stop to Think 16.6: b. The temperature decreases by a factor of 4 during the isochoric process, where $p_f/p_i = \frac{1}{4}$. The temperature then increases by a factor of 2 during the isobaric expansion, where $V_f/V_i = 2$.

17 Work, Heat, and the First Law of Thermodynamics

The modern technological era was ushered in by a new understanding of the connection between work and heat.

▶ **Looking Ahead**

The goal of Chapter 17 is to expand our understanding of energy and to develop the first law of thermodynamics as a general statement of energy conservation. In this chapter you will learn to:

- Understand the energy transfers known as *work* and *heat*.
- Use the first law of thermodynamics.
- Calculate work and heat for ideal-gas processes.
- Use specific heats and heats of transformation in the practical application of calorimetry.
- Understand adiabatic processes.

◀ **Looking Back**

The material in this chapter continues the development of energy ideas from Chapter 11. Many of the examples depend on the properties of ideal gases. Please review:

- Section 11.4 Work.
- Sections 11.7–11.8 Conservation of energy.
- Sections 16.4–16.6 Phase changes and ideal gases.

The industrial revolution was powered by the steam engine. Heat from a wood or coal fire was used to boil water and produce high-pressure steam. The expanding steam pushed a piston that, through a series of gears and levers, turned paddle wheels, ran machinery, or even powered massive locomotives. Humans had used heat for thousands of years for activities ranging from cooking to metallurgy, but the steam engine marked the first time in human history that heat was used to do work.

Our goal in this chapter is to investigate the connection between work and heat in macroscopic systems. Work and heat are *energy transfers* between the system and its environment, so we will be continuing the development of energy concepts that we began in Chapters 10 and 11. In addition, we will want to understand how the state of a system *changes* in response to work and heat. These two ideas, the transfer of energy and the change in the system, are related to each other through the *first law of thermodynamics,* a powerful statement about energy conservation.

We have two primary tasks: first, to complete our ideas about energy and its transfer; second, to investigate the consequences of energy transfer to and from macroscopic systems. Many practical devices, ranging from the internal combustion engine in your car to nuclear reactors, depend upon these consequences.

17.1 It's All About Energy

A key idea of Chapter 11 was the work-kinetic energy theorem in the form

$$\Delta K = W_c + W_{diss} + W_{ext} \qquad (17.1)$$

Equation 17.1 tells us that the kinetic energy of a particle or a system of particles is changed when forces do work on the particles by pushing or pulling them through a distance. Here

1. W_c is the work done by conservative forces. This work can be represented as a change in the system's potential energy: $\Delta U = -W_c$.
2. W_{diss} is the work done by friction-like dissipative forces within the system. This work increases the system's thermal energy: $\Delta E_{th} = -W_{diss}$.
3. W_{ext} is the work done by external forces that originate in the environment. The tension in a rope would be an external force, as would the push of a piston rod.

With these definitions, Equation 17.1 becomes

$$\Delta K + \Delta U + \Delta E_{th} = W_{ext} \qquad (17.2)$$

The system's *mechanical energy* was defined as $E_{mech} = K + U$. Figure 17.1 reminds you that the mechanical energy is associated with the motion of the system as a whole while E_{th} is associated with the motion of the atoms and molecules within the system. In the language of Part IV, E_{mech} is the *macroscopic* energy of the system as a whole while E_{th} is the *microscopic* energy of the particle-like atoms and spring-like molecular bonds. This led to our final energy statement of Chapter 11:

$$\Delta E_{sys} = \Delta E_{mech} + \Delta E_{th} = W_{ext} \qquad (17.3)$$

where $E_{sys} = E_{mech} + E_{th}$ is the total energy of the system. We interpreted Equation 17.3 by saying that the total energy of an *isolated system,* one for which $W_{ext} = 0$, is constant. This was the essence of the law of conservation of energy as stated in Chapter 11.

The emphasis in Chapters 10 and 11 was on isolated systems, those with $W_{ext} = 0$. We were interested in learning how kinetic and potential energy were *transformed* into each other and, where there is friction, into thermal energy. Now we want to focus on how energy is *transferred* between the system and its environment. That is, on systems where W_{ext} is *not* zero. This section and the next will look more closely at thermal energy and work. Section 17.3 will then introduce *heat* as another form of energy transfer.

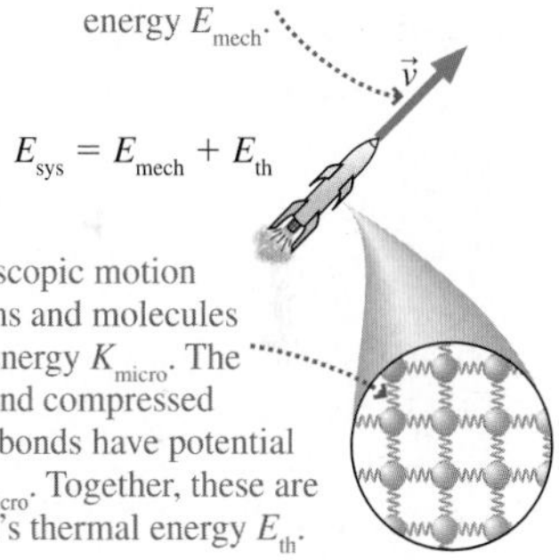

Thermal Energy

Thermal energy exists at the microscopic level of atoms and molecules. It is the sum of K_{micro}, the kinetic energy of all the moving atoms and molecules, and U_{micro}, the potential energy stored in the spring-like molecular bonds. That is,

$$E_{th} = K_{micro} + U_{micro} \qquad (17.4)$$

The thermal energy is hidden from our macroscopic view but nonetheless is quite real.

The macroscopic energy of the system as a whole is its mechanical energy E_{mech}.

$$E_{sys} = E_{mech} + E_{th}$$

$\vec{v}$

The microscopic motion of the atoms and molecules is kinetic energy K_{micro}. The stretched and compressed molecular bonds have potential energy U_{micro}. Together, these are the system's thermal energy E_{th}.

FIGURE 17.1 The total energy of a system consists of the macroscopic mechanical energy of the system as a whole plus the microscopic thermal energy of the atoms.

Strictly speaking, the thermal energy due to molecular motion is only one form of energy that can be stored within a system at the microscopic level. For example, a system might have *chemical energy* that can be released via chemical reactions between molecules in the system. Chemical energy is quite important in engineering thermodynamics, where it is needed to characterize combustion processes. *Nuclear energy* is stored in the atomic nuclei and can be released during radioactive decay. All the sources of microscopic energy taken together are called the system's **internal energy:**

$$E_{\text{int}} = E_{\text{th}} + E_{\text{chem}} + E_{\text{nuc}} + \cdots \qquad (17.5)$$

The total energy of the system is then $E_{\text{sys}} = E_{\text{mech}} + E_{\text{int}}$. This textbook will concentrate on simple thermodynamic systems in which the internal energy is entirely thermal: $E_{\text{int}} = E_{\text{th}}$. We'll leave other forms of internal energy to more advanced courses.

Chapter 16 noted that E_{th} is associated with the system's temperature. We now need to be a bit more specific.

- During a phase change, from solid to liquid or from liquid to gas, a system's thermal energy increases but its temperature does not. We'll look at phase changes later in this chapter.
- If there are no phase changes, increasing the system's temperature increases its thermal energy. **Conversely, a system's thermal energy does not change ($\Delta E_{\text{th}} = 0$) during an isothermal process ($\Delta T = 0$).** This simple idea will have major consequences.

EXAMPLE 17.1 **The thermal energy of a gas**
Estimate the thermal energy of 1 mol of nitrogen gas at 20°C.

MODEL For a gas, the thermal energy is the total kinetic energy of the moving molecules. That is, $E_{\text{th}} = K_{\text{micro}}$.

SOLVE Chapter 16 looked at the distribution of speeds of N_2 molecules at 20°C (Figure 16.8). While there was a range of speeds, the average speed was $v_{\text{avg}} \approx 550$ m/s. Consequently, an average molecule in the gas has kinetic energy $K_{\text{avg}} = \frac{1}{2}mv_{\text{avg}}^2$. Nitrogen atoms have atomic mass number $A = 14$, so the mass of each molecule is

$$m = 28 \text{ u} = 4.65 \times 10^{-26} \text{ kg}$$

Thus

$$K_{\text{avg}} = \frac{1}{2}mv_{\text{avg}}^2 \approx \frac{1}{2}(4.65 \times 10^{-26} \text{ kg})(550 \text{ m/s})^2$$

$$\approx 7.0 \times 10^{-21} \text{ J}$$

There are N_A molecules in 1 mol of gas, so the thermal energy is

$$E_{\text{th}} = K_{\text{micro}} = N_A K_{\text{avg}} \approx 4200 \text{ J}$$

ASSESS This is a substantial amount of energy, far more than most people would suspect.

The pistons in a car engine do work on the air-fuel mixture by compressing it.

Work

We introduced the idea of *work* in Chapter 11. Now, in preparation for the new concept of *heat,* we need to make our ideas about work more precise. **Work is the energy transferred between a system and the environment when a net force acts on the system over a distance.** The process itself is a **mechanical interaction,** meaning that the system and the environment interact via macroscopic pushes and pulls. Loosely speaking, we say that the environment (or a particular force from the environment) "does work" on the system. A system is in **mechanical equilibrium** with the environment or with another system if there is no net force on the system.

Figure 17.2 reminds you that work can be either positive or negative. **The sign of the work is *not* just an arbitrary convention, nor does it have anything to do with the choice of coordinate system.** The sign of the work tells us which way energy is being transferred.

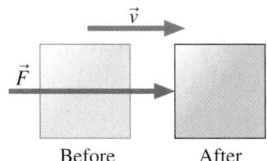
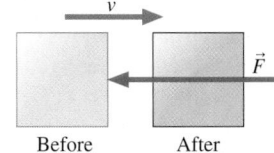

Work is *positive* when the force is in the direction of motion.
- Energy is transferred from the environment to the system.
- The system's energy increases.

Work is *negative* when the force is opposite to the motion.
- Energy is transferred from the system to the environment.
- The system's energy decreases.

FIGURE 17.2 The sign of work.

In contrast to the mechanical energy or the thermal energy, **work is not a state variable.** That is, work is not a number characterizing the system. Instead, work is the amount of energy that moves between the system and the environment during a mechanical interaction. We can measure the *change* in a state variable, such as a temperature change $\Delta T = T_f - T_i$, but it would make no sense to talk about a "change of work." Consequently, work appears in the energy equation simply as W_{ext}, never as ΔW_{ext}.

Energy Transfer

Doing work on a system can have very different consequences. Figure 17.3a shows an object being lifted at steady speed by a rope. The rope's tension is an external force doing work W_{ext} on the system. In this case, the energy transferred into the system goes entirely to increasing the system's macroscopic potential energy U_{grav}, part of the mechanical energy. The energy transfer process $W_{ext} \rightarrow E_{mech}$ is shown graphically in the energy bar chart of Figure 17.3a.

Contrast this with Figure 17.3b, where the same rope with the same tension now drags the object at steady speed across a rough surface. The tension does the same amount of work, but the mechanical energy does not change. Instead, friction increases the thermal energy of the object + surface system. The energy transfer process $W_{ext} \rightarrow E_{th}$ is shown in the energy bar chart of Figure 17.3b.

The point of this example is that the energy transferred to a system can go entirely to the system's mechanical energy, entirely to its thermal energy, or (imagine dragging the object up an incline) some combination of the two. The energy isn't lost, but where it ends up depends on the circumstances.

That Can't Be All

You can transfer energy into a system by the mechanical process of doing work on the system. But that can't be all there is to energy transfer. What happens when you place a pan of water on the stove and light the burner? The water temperature increases, so $\Delta E_{th} > 0$. But no work is done ($W_{ext} = 0$) and there is no change in the water's mechanical energy ($\Delta E_{mech} = 0$). This process clearly violates the energy equation $\Delta E_{mech} + \Delta E_{th} = W_{ext}$. What's wrong?

Nothing is wrong. The energy equation is correct as far as it goes, but it is incomplete. Work is energy transferred in a mechanical interaction, but that is not the only way a system can interact with its environment. Energy can also be transferred between the system and the environment if they have a *thermal interaction.* The energy transferred in a thermal interaction is called *heat.*

The symbol for heat is Q. When heat is included, the energy equation becomes

$$\Delta E_{sys} = \Delta E_{mech} + \Delta E_{th} = W + Q \qquad (17.6)$$

Heat and work, now on an equal footing, are both energy transferred between the system and the environment.

(a)

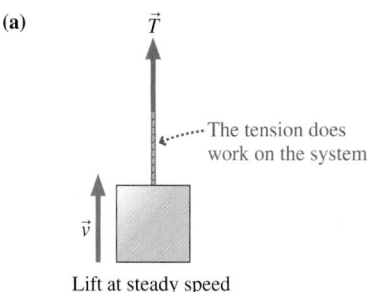

Lift at steady speed

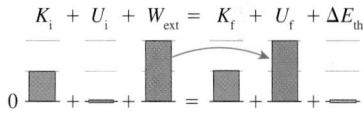

The energy transferred to the system goes entirely to the system's mechanical energy.

(b)

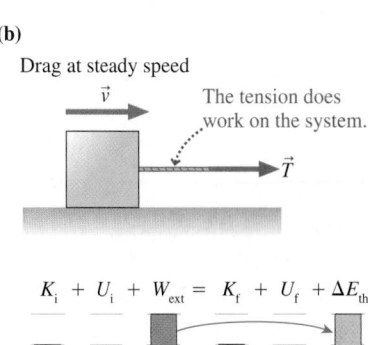

The energy transferred to the system goes entirely to the system's thermal energy.

FIGURE 17.3 The work done by the rope's tension can have very different consequences.

NOTE ▶ We've dropped the subscript "ext" from W. The work that we consider in thermodynamics is *always* the work done by the environment on the system. We won't need to distinguish this work from W_c or W_{diss}, so the subscript is superfluous. ◀

The next section looks at how work is calculated for ideal-gas processes. Heat will then be explored in Section 17.3.

STOP TO THINK 17.1   A gas cylinder and piston are covered with heavy insulation. The piston is pushed into the cylinder, compressing the gas. In this process, the gas temperature

a. increases.
b. decreases.
c. doesn't change.
d. there's not sufficient
   information to tell.

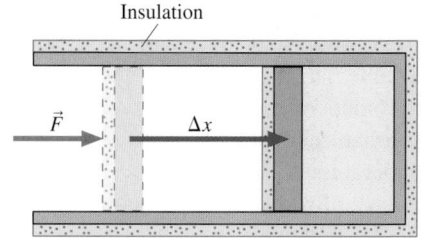

17.2 Work in Ideal-Gas Processes

You learned in Chapter 11 how to calculate work. The small amount of work dW done by force $\vec{F}$ as a system moves through the small displacement $d\vec{s}$ is $dW = \vec{F} \cdot d\vec{s}$. If we restrict ourselves to situations where $\vec{F}$ is either parallel or opposite to $d\vec{s}$, then the total work done on the system as it moves from s_i to s_f is

$$W = \int_{s_i}^{s_f} F_s\,ds \qquad (17.7)$$

Work is positive if the displacement is in the same direction as the force, negative if the force and the displacement are in opposite directions.

Let's apply these definitions to a gas as it expands or is compressed. Figure 17.4a shows a gas cylinder that is sealed at one end by a movable piston. Force $\vec{F}_{ext}$, perhaps a force supplied by a piston rod, is equal in magnitude and opposite in direction to $\vec{F}_{gas}$. The gas pressure would blow the piston out of the cylinder if the external force weren't there! Using the coordinate system of Figure 17.4a,

$$F_{ext} = -F_{gas} = -pA \qquad (17.8)$$

Suppose the piston moves the small distance dx shown in Figure 17.4b. As it does so, the external force (i.e., the environment) does work

$$dW = F_{ext}\,dx = -pA\,dx \qquad (17.9)$$

If dx is positive (the gas expands), then dW is negative. This is because the external force is opposite the displacement. dW is positive if the gas is slightly compressed (negative dx) because the force and the displacement are in the same direction. This is an important idea.

As the piston moves dx, the volume of the gas changes by $dV = A\,dx$. Consequently, Equation 17.9 can be written in terms of the cylinder's volume as

$$dW = -p\,dV \qquad (17.10)$$

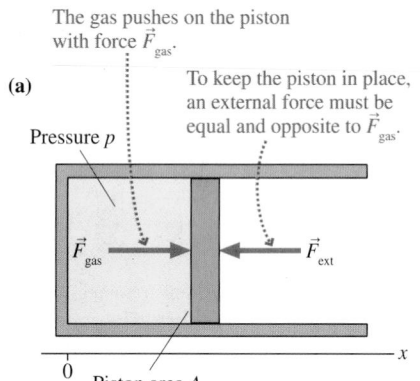

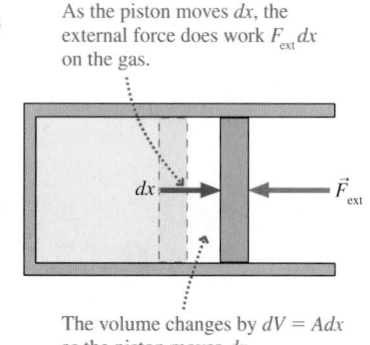

FIGURE 17.4 The external force does work on the gas as the piston moves.

If we let the piston move in a slow quasi-static process from initial volume V_i to final volume V_f, the total work done by the environment on the gas is found by integrating Equation 17.10:

$$W = -\int_{V_i}^{V_f} p \, dV \qquad \text{(work done on a gas)} \qquad (17.11)$$

Equation 17.11 is a key result of thermodynamics. Although we used a cylinder to derive Equation 17.11, it turns out to be true for a container of any shape.

NOTE ▶ The pressure of a gas usually changes as the gas expands or contracts. Consequently, p is *not* a constant that can be brought outside the integral. You need to know how the pressure changes with volume before you can carry out the integration. ◀

We can give the work done on a gas a nice geometric interpretation. You learned in Chapter 16 how to represent an ideal-gas process as a curve in the pV diagram. Figure 17.5 shows that the work done on a gas is the negative of the area under the pV curve as the volume changes from V_i to V_f. That is

$W = $ the negative of the area under the pV curve between V_i and V_f

Figure 17.5a shows a process in which a gas *expands* from V_i to a larger volume V_f. The area under the curve is positive, so the environment does a negative amount of work on an expanding gas. Figure 17.5b shows a process in which a gas is compressed to a smaller volume. This one is a little trickier because we have to integrate "backward" along the V-axis. You learned in calculus that integrating from a larger limit to a smaller limit gives a negative result, so the area in Figure 17.5b is a negative area. Consequently, as the minus sign in Equation 17.11 indicates, the environment does a positive amount of work on a gas to compress it.

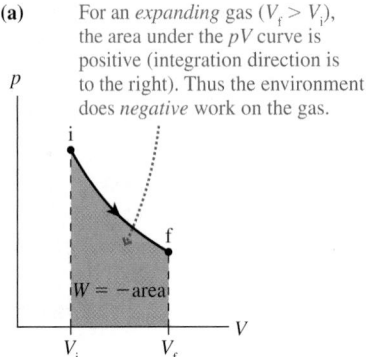

(a) For an *expanding* gas ($V_f > V_i$), the area under the pV curve is positive (integration direction is to the right). Thus the environment does *negative* work on the gas.

(b) For a *compressed* gas ($V_f < V_i$), the area is negative because the integration direction is to the left. Thus the environment does *positive* work on the gas.

FIGURE 17.5 The work done on a gas is the negative of the area under the curve.

EXAMPLE 17.2 The work done on an expanding gas
How much work is done on the gas in the ideal-gas process of Figure 17.6?

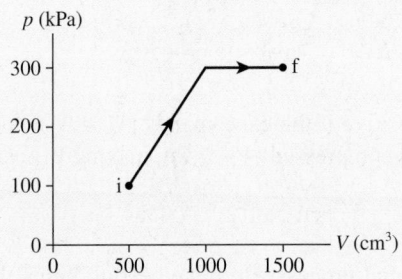

FIGURE 17.6 The ideal-gas process of Example 17.2.

MODEL The work done on a gas is the negative of the area under the pV curve. The gas is *expanding,* so we expect the work to be negative.

SOLVE The work W is the negative of the area under the curve from $V_i = 500$ cm³ to $V_f = 1500$ cm³. Volumes *must* be converted to SI units of m³. The area from 500 cm³ to 1000 cm³ can

be divided into a rectangle (between 0 kPa to 100 kPa) and a triangle (between 100 and 300 kPa). This area is

$$\text{Area}(500 \rightarrow 1000 \text{ cm}^3)$$
$$= ((1000 - 500) \times 10^{-6} \text{m}^3)(100{,}000 \text{ Pa} - 0 \text{ Pa})$$
$$+ \frac{1}{2}((1000 - 500) \times 10^{-6} \text{ m}^3)$$
$$\times (300{,}000 \text{ Pa} - 100{,}000 \text{ Pa})$$
$$= 100 \text{ J}$$

The area from 1000 cm³ to 1500 cm³ is a rectangle:

$$\text{Area}(1000 \rightarrow 1500 \text{ cm}^3)$$
$$= ((1500 - 1000) \times 10^{-6} \text{ m}^3)(300{,}000 \text{ Pa} - 0 \text{ Pa})$$
$$= 150 \text{ J}$$

The total area under the curve is 250 J, so the work done on the gas as it expands is

$$W = -(\text{area under the } pV \text{ curve}) = -250 \text{ J}$$

ASSESS We noted previously that the product Pa m³ is equivalent to joules. The work is negative, as expected, because the external force pushing on the piston is opposite the direction of the piston's displacement.

Equation 17.11 is the basis for a problem-solving strategy.

 PROBLEM-SOLVING STRATEGY 17.1 Work in ideal-gas processes

MODEL Assume the gas is ideal and the process is quasi-static.

VISUALIZE Show the process on a pV diagram. Note whether it happens to be one of the basic gas processes: isochoric, isobaric, or isothermal.

SOLVE Calculate the work as the area under the pV curve either geometrically or by carrying out the integration.

$$\text{Work done on the gas } W = -(\text{area under } pV \text{ curve}) = -\int p \, dV$$

ASSESS Check your signs.

- $W > 0$ when the gas is compressed. Energy is transferred from the environment to the gas.
- $W < 0$ when the gas expands. Energy is transferred from the gas to the environment.
- No work is done if the volume doesn't change. $W = 0$.

(a)

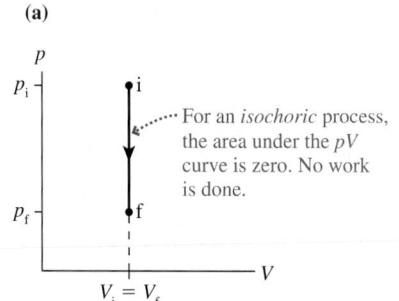

For an *isochoric* process, the area under the pV curve is zero. No work is done.

Isochoric Process

The isochoric process in Figure 17.7a is one in which the volume does not change. Consequently,

$$W = 0 \qquad \text{(isochoric process)} \tag{17.12}$$

An isochoric process is the *only* ideal-gas process in which no work is done.

(b)

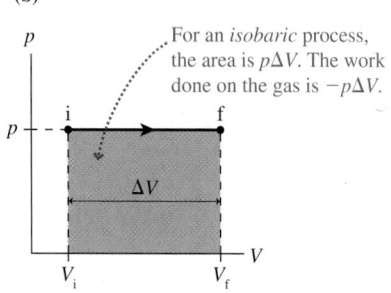

For an *isobaric* process, the area is $p\Delta V$. The work done on the gas is $-p\Delta V$.

Isobaric Process

Figure 17.7b shows an isobaric process in which the volume changes from V_i to V_f. The rectangular area under the curve is $p\Delta V$, so the work done during this process is

$$W = -p\Delta V \qquad \text{(isobaric process)} \tag{17.13}$$

where $\Delta V = V_f - V_i$. ΔV is positive if the gas expands ($V_f > V_i$), so W is negative. ΔV is *negative* if the gas is compressed ($V_f < V_i$), making W positive.

(c)

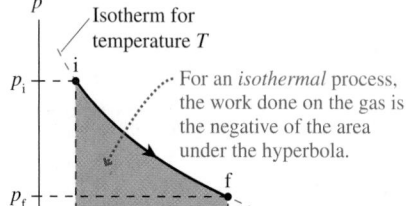

Isotherm for temperature T

For an *isothermal* process, the work done on the gas is the negative of the area under the hyperbola.

Isothermal Process

Figure 17.7c shows an isothermal process. Here we need to know the pressure as a function of volume before we can integrate Equation 17.11. From the ideal-gas law, $p = nRT/V$. Thus the work on the gas as the volume changes from V_i to V_f is

$$W = -\int_{V_i}^{V_f} p \, dV = -\int_{V_i}^{V_f} \frac{nRT}{V} dV = -nRT \int_{V_i}^{V_f} \frac{dV}{V} \tag{17.14}$$

FIGURE 17.7 Calculating the work done during ideal-gas processes.

where we could take the T outside the integral because temperature is constant during an isothermal process. This is a straightforward integration, giving

$$W = -nRT \int_{V_i}^{V_f} \frac{dV}{V} = -nRT \ln V \Big|_{V_i}^{V_f} \tag{17.15}$$

$$= -nRT(\ln V_f - \ln V_i) = -nRT \ln\left(\frac{V_f}{V_i}\right)$$

Because $nRT = p_i V_i = p_f V_f$ during an isothermal process, we can write the work three ways:

$$W = -nRT \ln\left(\frac{V_f}{V_i}\right) = -p_i V_i \ln\left(\frac{V_f}{V_i}\right) = -p_f V_f \ln\left(\frac{V_f}{V_i}\right) \tag{17.16}$$
$$\text{(isothermal process)}$$

Which version of Equation 17.16 is easiest to use will depend on the information you're given. The pressure, volume, and temperature *must* be in SI units.

EXAMPLE 17.3 The work of an isothermal compression
A cylinder contains 7.0 g of nitrogen gas. How much work must be done to compress the gas at a constant temperature of 80°C until the volume is halved?

MODEL This is an isothermal ideal-gas process.

SOLVE Nitrogen gas is N_2, with molar mass $M_{mol} = 28$ g/mol, so 7.0 g is 0.25 mol of gas. The temperature is $T = 353$ K. Although we don't know the actual volume, we do know that $V_f = \frac{1}{2} V_i$. The volume ratio is all we need to calculate the work:

$$W = -nRT \ln\left(\frac{V_f}{V_i}\right)$$

$$= -(0.25 \text{ mol})(8.31 \text{ J/mol K})(353 \text{ K}) \ln\left(\frac{1}{2}\right) = 508 \text{ J}$$

ASSESS The work is positive because a force from the environment pushes the piston inward to compress the gas.

Work Depends on the Path

Figure 17.8 shows two different processes that take a gas from an initial state i to a final state f. Although the initial and final states are the same, the work done during these two processes is *not* the same. The area under the curve of process A is larger than the area under the curve of process B, so $|W_A| > |W_B|$. (The work is negative, so W_A is more negative than W_B.) As this example shows, **the work done during an ideal-gas process depends on the path followed through the pV diagram.**

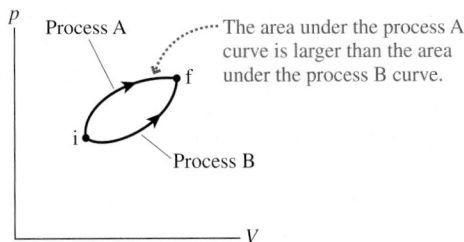

FIGURE 17.8 The work done during these two ideal-gas processes is not the same.

You may be thinking that work is supposed to be independent of the path, but there are two big reasons why that is not the case here. First, we found in Chapter 11 that work is independent of the path only for a *conservative* force.

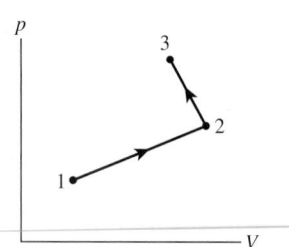

FIGURE 17.9 The work done during the process $1 \rightarrow 2 \rightarrow 3$ must be calculated in two steps.

In general, the forces that push and pull on pistons are not conservative forces. Second, the path that we considered in Chapter 11 was the trajectory of a particle from one point to another through space. For an ideal-gas process, the "path" is a sequence of thermodynamic states as they are represented on a pV diagram. It is a figurative path, because we can draw a picture of it on a pV diagram, but it is not a literal path.

The path dependence of work has an important implication for multistep processes such as the one shown in Figure 17.9. The total work done on the gas during the process $1 \rightarrow 2 \rightarrow 3$ must be calculated as $W_{1 \text{ to } 3} = W_{1 \text{ to } 2} + W_{2 \text{ to } 3}$. In this case, $W_{1 \text{ to } 2}$ is negative and $W_{2 \text{ to } 3}$ is positive. Trying to compute the work in a single step, using $\Delta V = V_3 - V_1$, would give you the work of a process that goes directly from 1 to 3. The initial and final states are the same, but the work is *not* the same because work depends on the path followed through the pV diagram.

STOP TO THINK 17.2 Two processes are shown that take an ideal gas from state 1 to state 3. Compare the work done by process A to the work done by process B.

a. $W_A = W_B = 0$
b. $W_A = W_B$ but neither is zero
c. $W_A > W_B$
d. $W_A < W_B$

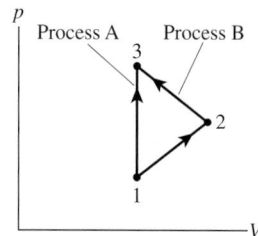

17.3 Heat

Heat is a more elusive concept than work. We use the word "heat" very loosely in the English language, often as synonymous with *hot*. We might say, on a very hot day, that "This heat is oppressive." If your apartment is cold you may say, "Turn up the heat." These expressions date to a time long ago when it was thought that heat was a *substance* with fluid-like properties.

If you place a hot object and a cold object together, they evolve toward a common final temperature. Common sense suggests that "something" flows from the hot object to the cold until equilibrium is achieved. This "heat fluid" was called *caloric*. The notion that objects somehow "contain" heat, and that heat can move around, lingers on in expressions like *heat flow, heat loss,* and *heat capacity.*

One of the first to disagree with this notion, in the late 1700s, was the American-born Benjamin Thompson. Thompson fled to Europe during the American Revolution, settling in Bavaria and later receiving the title Count Rumford. There, while watching the hot metal chips thrown off during the boring of cannons, he began to think about heat. If caloric is a substance, the cannon and borer should eventually run out of caloric and the heat generation ought to decrease with time. But it does not. Rumford noted that the heat generation appears to be "inexhaustible," which is not consistent with the idea of heat as a substance. He concluded that heat is not a substance—it is *motion!*

Rumford was beginning to think along the same lines as had Bernoulli. But Rumford's ideas were speculative and qualitative, hardly a scientific theory, and their implications were not immediately grasped by others. Like Bernoulli's, it would be some time before his insight was recognized and validated.

The turning point was the work of British physicist James Joule in the 1840s. Unlike Bernoulli and Count Rumford, Joule carried out careful experiments to

Heat is the energy transferred in a thermal interaction.

learn how it is that systems change their temperature. Using experiments like those shown in Figure 17.10, Joule found that you can raise the temperature of a beaker of water by two entirely different means:

1. Heating it with a flame, or
2. Doing work on it with a rapidly spinning paddle wheel.

The final state of the water is *exactly* the same in both cases. This implies that heat and work are essentially equivalent to each other. In other words, heat is not a substance. Instead, heat is the *energy* that is transferred from a hotter object to a colder object.

It was Joule who placed the idea of conservation of energy on a firm scientific footing and established it as a law of nature. Heat and work, which previously had been regarded as two completely different phenomena, were seen to be simply two different ways of transferring energy to or from a system. Joule's discoveries vindicated the earlier ideas of Bernoulli and Count Rumford, and they opened the door for rapid advancements in the subject of thermodynamics during the second half of the 19th century.

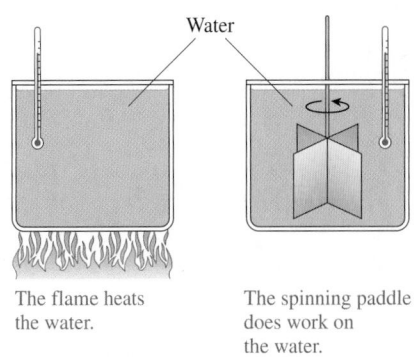

Water

The flame heats the water.

The spinning paddle does work on the water.

FIGURE 17.10 Joule's experiments to show the equivalence of heat and work.

Thermal Interactions

Heat is the energy transferred between a system and the environment as a consequence of a *temperature difference* between them. Unlike a mechanical interaction in which work is done, heat requires no macroscopic motion of the system. Instead (we'll look at the details in Chapter 18), energy is transferred when the *faster* molecules in the hotter object collide with the *slower* molecules in the cooler object. On average, these collisions cause the faster molecules to lose energy and the slower molecules to gain energy. The net result is that energy is transferred from the hotter object to the colder object. The process itself, whereby energy is transferred between the system and the environment via atomic-level collisions, is called a **thermal interaction.** *Heat* is the energy transferred during a thermal interaction.

When you place a pan of water on the stove, heat is the energy transferred *from* the hotter flame *to* the cooler water. If you place the water in a freezer, heat is the energy transferred from the warmer water to the colder air in the freezer. A system is in **thermal equilibrium** with the environment, or two systems are in thermal equilibrium with each other, if there is no temperature difference.

It is well worthwhile to compare this statement about heat and thermal interactions with the first paragraph about work in Section 17.1. The analogy would be complete if we were able to say that the environment (or an object in the environment) "does heat" on the system. Unfortunately, the English language doesn't work that way. Loosely speaking, we say that the environment "heats" the system.

Like work, **heat is not a state variable.** Heat is not a property of the system. Instead, heat is the amount of energy that moves between the system and the environment during a thermal interaction. It would not be meaningful to talk about a "change of heat." Thus heat appears in the energy equation simply as a value Q, never as ΔQ.

Figure 17.11 shows that Q is positive when energy is transferred *into* the system from the environment. This implies that $T_{env} > T_{sys}$. A negative Q represents heat transfer *from* the system back to the environment when $T_{env} < T_{sys}$. The system is in thermal equilibrium with the environment ($Q = 0$) when $T_{env} = T_{sys}$.

NOTE ▶ For both heat and work, a positive value indicates energy being transferred from the environment to the system. Table 17.1 on the next page summarizes the similarities and differences between work and heat. ◀

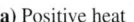

(a) Positive heat

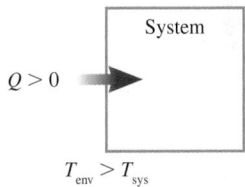

System

$Q > 0$

$T_{env} > T_{sys}$

(b) Negative heat

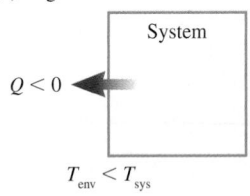

System

$Q < 0$

$T_{env} < T_{sys}$

(c) Thermal equilibrium

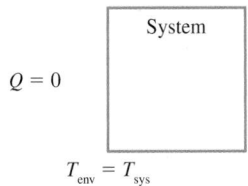

System

$Q = 0$

$T_{env} = T_{sys}$

FIGURE 17.11 The sign of heat.

TABLE 17.1 Understanding work and heat

	Work	Heat
Interaction:	Mechanical	Thermal
Requires:	Force and displacement	Temperature difference
Process:	Macroscopic pushes and pulls	Microscopic collisions
Positive value:	$W > 0$ when the force on the system is in the direction of the displacement. Gases are compressed. Energy is transferred in.	$Q > 0$ when the environment is at a higher temperature than the system. Energy is transferred in.
Negative value:	$W < 0$ when the force on the system is opposite the displacement. Gases expand. Energy is transferred out.	$Q < 0$ when the system is at a higher temperature than the environment. Energy is transferred out.
Equilibrium:	A system is in mechanical equilibrium when the environment exerts no net force on it.	A system is in thermal equilibrium when it is at the same temperature as the environment.

Units of Heat

Heat is energy transferred between the system and the environment. Consequently, the SI unit of heat is the joule. Historically, before the connection between heat and work had been recognized, a unit for measuring heat, the calorie, had been defined as

$$1 \text{ calorie} = 1 \text{ cal} = \text{the quantity of heat needed to change} \\ \text{the temperature of 1 g of water by } 1 \text{ °C.}$$

Once Joule established that heat is energy, it was apparent that the calorie is really a unit of energy. Joule used experiments to determine the conversion factor between heat units and more conventional energy units. In today's SI units, that conversion is

$$1 \text{ cal} = 4.186 \text{ J}$$

The calorie you know in relation to food is not the same as the heat calorie. The *food calorie,* abbreviated Cal with a capital C, is

$$1 \text{ food calorie} = 1 \text{ Cal} = 1000 \text{ cal} = 1 \text{ kcal} = 4186 \text{ J}$$

The food calorie measures the food's chemical energy, stored energy that is available for doing work or for keeping your body warm. That extra dessert you ate last night containing 300 Cal has a chemical energy

$$E_{\text{chem}} = 300 \text{ Cal} = 3 \times 10^5 \text{ cal} = 1.26 \times 10^6 \text{ J}$$

We will not use calories in this textbook, but there are some fields of science and engineering where calories are still widely used. All the calculations you learn to do with joules can equally well be done with calories.

The Trouble with Heat

The trouble with heat is twofold: conceptual and linguistic. At the conceptual level, it is important to distinguish between *heat, temperature,* and *thermal energy.* These three ideas are related, but the distinctions between them are crucial. Common language can easily mislead you. If an object slides to a halt because of friction, most people say that the object's kinetic energy is "converted into heat." In fact, heat is not involved in this process. Nowhere was there a transfer of energy due to a temperature difference. Instead, the object's mechanical energy is transformed into the *thermal energy* of the atoms and molecules. In brief,

■ Thermal energy is an energy *of the system* due to the motion of its atoms and molecules. It is a *form* of energy. Thermal energy is a state variable, and it makes sense to talk about how E_{th} changes during a process. The system's thermal energy continues to exist even if the system is isolated and not interacting thermally with its environment.

■ Heat is energy transferred *between the system* and the environment as they interact. Heat is *not* a particular form of energy, nor is it a state variable. It makes no sense to talk about how heat changes. $Q = 0$ if a system does not interact thermally with its environment. Heat may cause the system's thermal energy to change, but that doesn't mean that heat and thermal energy are the same thing.

■ Temperature is a state variable that quantifies the "hotness" or "coldness" of a system. We haven't given a precise definition of temperature, but it is related to the thermal energy *per molecule*. A temperature difference is a requirement for a thermal interaction in which heat energy is transferred between the system and the environment.

It is especially important not to associate an observed temperature increase with heat. Heating a system is one way to change its temperature, but, as Joule showed, not the only way. You can also change the system's temperature by doing work on the system. **Observing the system tells us nothing about the process by which energy enters or leaves the system.**

We have two problems on the linguistic front. One, already alluded to, are those terms such as "heat flow" and "heat capacity" that are vestiges of history. These phrases, which are used even in scientific and technical discourse, incorrectly suggest that heat is a substance that can flow from one object to another or be contained in an object. With experience, scientists and engineers learn to use these phrases without meaning what the phrase, interpreted literally, seems to suggest.

A second problem is that the phrase "to heat" uses the word "heat" as a verb, whereas our definition of "heat" uses the word as a noun. These two uses make no distinction between the energy transferred and the process of transferring the energy. With work, the phase "to *do* work" allows us to separate the process from the energy transferred (i.e., "the work") in the process.

Unfortunately, physics textbooks can't reinvent language. We will try to be very careful in our choice of words and phrases, and will highlight points were the language is potentially confusing or misleading. Being forewarned will help you avoid some of these pitfalls.

STOP TO THINK 17.3   Which one or more of the following processes involves heat?

a. The brakes in your car get hot when you stop.
b. A steel block is held over a candle.
c. You push a rigid cylinder of gas across a frictionless surface.
d. You push a piston into a cylinder of gas, increasing the temperature of the gas.
e. You place a cylinder of gas in hot water. The gas expands, causing a piston to rise and lift a weight. The temperature of the gas does not change.

17.4 The First Law of Thermodynamics

8.8–8.10   Activ
Physics
ONLINE

Heat was the missing piece that we needed to arrive at a completely general statement of the law of conservation of energy. Restating Equation 17.6,

$$\Delta E_{sys} = \Delta E_{mech} + \Delta E_{th} = W + Q$$

Work and heat, two ways of transferring energy between a system and the environment, cause the system's energy to change.

At this point in the text we are not interested in systems that have a macroscopic motion of the system as a whole. Moving macroscopic systems were important to us for many chapters, but now, as we investigate the thermal properties of a system, we would like the system as a whole to rest peacefully on the laboratory bench while we study it. So we will assume, throughout the remainder of Part IV, that $\Delta E_{mech} = 0$.

With this assumption clearly stated, the law of conservation of energy becomes

$$\Delta E_{th} = W + Q \qquad \text{(first law of thermodynamics)} \qquad (17.17)$$

The energy equation, in this form, is called the **first law of thermodynamics** or simply "the first law." The first law is nothing more than a statement about the conservation of energy.

Chapters 10 and 11 introduced the basic energy model. It was called *basic* because it included work but not heat. The first law of thermodynamics has included heat, but it excludes situations where the mechanical energy changes. Figure 17.12 is a pictorial representation of the **thermodynamic energy model** described by the first law. Work and heat are energies transferred between the system and the environment. Energy added to the system (W or Q positive) increases the system's thermal energy ($\Delta E_{th} > 0$). Likewise, the thermal energy decreases when energy is removed from the system.

Two comments are worthwhile:

1. The first law of thermodynamics doesn't tell us anything about the value of E_{th}, only how E_{th} changes. Doing 1 J of work changes the thermal energy by $\Delta E_{th} = 1$ J regardless of whether $E_{th} = 10$ J or 10,000 J.
2. The system's thermal energy isn't the only thing that changes. Work or heat that change the thermal energy also change the pressure, volume, temperature, and other state variables. The first law tells us only about ΔE_{th}. Other laws and relationships must be used to learn how the other state variables change.

The first law is one of the most important analytic tools of thermodynamics. It will be especially important when we look at practical thermodynamics in Chapter 19. We'll use the first law in the remainder of this chapter to study some of the thermal properties of matter.

Lifting Weights

A simple model can help us understand how the first law of thermodynamics is related to ideal-gas process and pV diagrams. Figure 17.13 shows a gas cylinder with several special properties:

■ You can fix the gas volume by inserting the locking pin into the piston. Without the pin, the piston can slide up or down. The piston is massless and insulated.
■ You can change the gas pressure by adding or removing masses on top of the piston. Work is done as the piston moves the masses up and down.

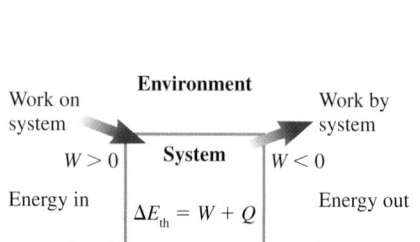

Environment

Work on system
$W > 0$

System
$\Delta E_{th} = W + Q$

Work by system
$W < 0$

Energy in

Energy out

$Q > 0$

$Q < 0$

Heat to system

Heat from system

FIGURE 17.12 The thermodynamic energy model.

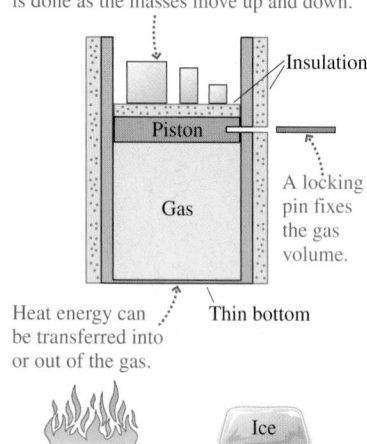

Masses determine the gas pressure. Work is done as the masses move up and down.

Insulation

Piston

A locking pin fixes the gas volume.

Gas

Heat energy can be transferred into or out of the gas.

Thin bottom

Flame

Ice

FIGURE 17.13 A gas that can be heated, have work done on it, or be maintained at a constant temperature.

■ You can warm or cool the gas by placing the cylinder above a flame or on a block of ice. The thin bottom of the cylinder is the only surface through which heat energy can be transferred.

■ All processes are assumed to take place quasi-statically.

When the piston is unlocked, the gas pressure is determined by the atmospheric pressure and by the total mass M on the piston:

$$p_{gas} = p_{atmos} + \frac{Mg}{A} \qquad (17.18)$$

The pressure doesn't change as the piston moves unless you change the mass. (This result is from our discussion of isobaric processes in Chapter 16.) Equation 17.18 is *not* valid when the piston is locked. The pressure with the piston locked could be either higher or lower than Equation 17.18.

EXAMPLE 17.4 **An isochoric cooling process**

Design and implement a process that will decrease the pressure in the gas cylinder of Figure 17.13 without changing the volume. Describe the steps, then show the process on a pV diagram and as a first-law bar chart.

MODEL To bring about this process:

■ Insert the locking pin so that the volume cannot change.

■ Place the cylinder on the block of ice. Heat energy will be transferred from the gas to the ice, causing the gas temperature and pressure to fall.

■ Remove the cylinder from the ice when the desired pressure is reached.

■ Remove masses from the piston until the total mass M balances the new gas pressure. This step must be done before removing the locking pin; otherwise the piston will move when the pin is removed.

■ Remove the locking pin.

VISUALIZE This is the isochoric process shown in Figure 17.14a. The final point is on a lower isotherm than the initial point, so $T_f < T_i$. No work is done ($W = 0$) in an isochoric process because the piston doesn't move. Heat energy was transferred out of the gas ($Q < 0$) and the thermal energy of the gas decreased ($\Delta E < 0$) as the temperature fell. From the first law, with $W = 0$, we can write explicitly that $E_{th\,f} = E_{th\,i} + Q$. Figure 17.14b shows this result on a first-law bar chart. We don't know the value of the initial thermal energy $E_{th\,i}$, so the height of the $E_{th\,i}$ bar is arbitrary. But we can show how the thermal energy *changes* as a consequence of the gas process.

In this example, the $E_{th\,f}$ bar shows that the thermal energy has been decreased by the amount of energy that left the system as heat.

(a)

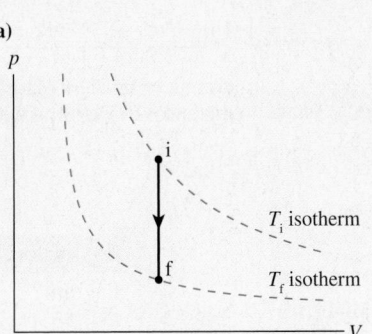

(b)

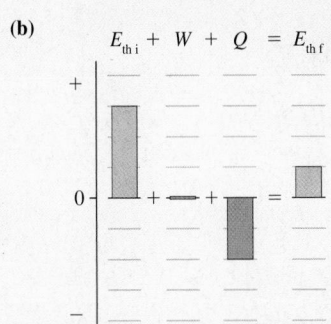

FIGURE 17.14 A pV diagram and a first-law bar chart for a process that decreases the pressure without changing the volume.

EXAMPLE 17.5 **An isothermal expansion**

Design and implement a process that will increase the volume in the gas cylinder of Figure 17.13 without changing the temperature. Describe the steps, then show the process on a pV diagram and as a first-law bar chart.

MODEL To bring about this process:

■ Place the cylinder over the flame. Heat energy will be transferred to the gas and the gas will begin to expand.

■ The product pV must remain constant during an isothermal process. Slowly remove masses from the piston to reduce the pressure as the volume increases. The temperature remains constant as heat energy from the flame balances the work done in the expansion.

■ Remove the cylinder from the flame when the gas reaches the desired volume.

VISUALIZE This is the isothermal process shown in Figure 17.15a. $\Delta E_{th} = 0$ in an isothermal process ($\Delta T = 0$), so the first law $\Delta E_{th} = W + Q$ can be satisfied only if $W = -Q$. Heat energy is transferred to the gas, but the temperature of the gas doesn't change. Instead, the energy causes the gas to expand and do the work of lifting the masses. The work done *on* the gas by the force holding the piston in place is negative when a gas expands. This information is shown on the first-law bar chart of Figure 17.15b.

ASSESS It is surprising, but true, that we can heat the system without changing its temperature. But to do so, we must have a process in which the energy coming into the system as heat is exactly balanced by energy leaving the system as work.

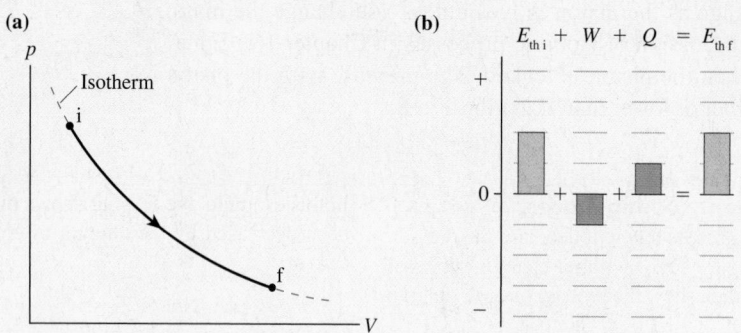

FIGURE 17.15 A pV diagram and a first-law bar chart for a process that increases the volume without changing the temperature.

STOP TO THINK 17.4 Which first-law bar chart describes the process shown in the pV diagram?

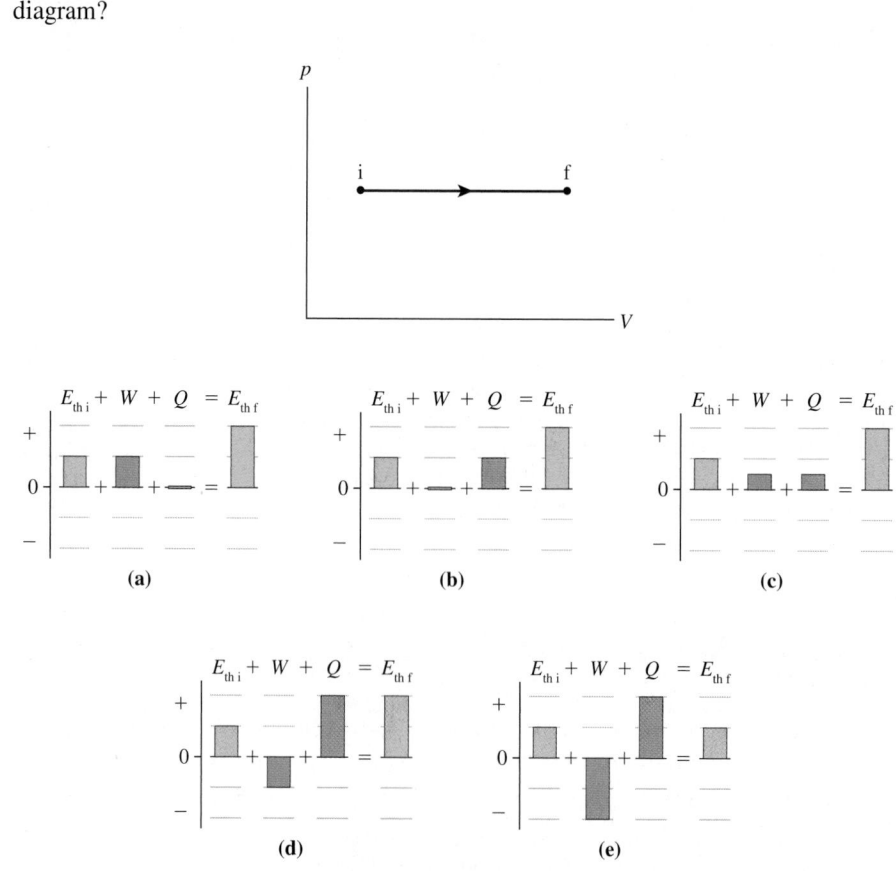

17.5 Thermal Properties of Matter

Joule established that heat and work are energy transferred between a system and its environment. Heat and work are equivalent in the sense that the change of the system is *exactly the same* whether you transfer heat energy to it or do an equal amount of work on it. Adding energy to the system, or removing it, changes the system's thermal energy.

What happens to a system when you change its thermal energy? In this section we'll consider two distinct possibilities:

- The temperature of the system changes.
- The system undergoes a phase change, such as melting or freezing.

Temperature Change and Specific Heat

Suppose you do an experiment in which you add energy to water, either by doing work on it or transferring heat to it. Either way, you will find that adding 4190 J of energy raises the temperature of 1 kg of water by 1 K. If you were fortunate enough to have 1 kg of gold, you need to add only 129 J of energy to raise its temperature by 1 K.

The amount of energy that raises the temperature of 1 kg of a substance by 1 K is called the **specific heat** of that substance. The symbol for specific heat is c. Water has specific heat $c_{water} = 4190$ J/kg K. The specific heat of gold is $c_{gold} = 129$ J/kg K. Specific heat depends only on the material from which an object is made. Table 17.2 provides some specific heats for common liquids and solids. Extensive tables can be found in handbooks of chemical and physical data.

> NOTE ▶ The term *specific heat* does not use the word *heat* in the way that we have defined it. Specific heat is an old idea, dating back to the days of the caloric theory when heat was thought to be a substance contained in the object. The term has continued in use even though our understanding of heat has changed. ◄

If energy c is required to raise the temperature of 1 kg of substance by 1 K, then energy Mc is needed to raise the temperature of mass M by 1 K and $(Mc)\Delta T$ is needed to raise the temperature of mass M by ΔT. In other words, the thermal energy of the system changes by

$$\Delta E_{th} = Mc\Delta T \quad \text{(temperature change)} \quad (17.19)$$

when its temperature changes by ΔT. ΔE_{th} can be either positive (thermal energy increases as the temperature goes up) or negative (thermal energy decreases as the temperature goes down). Recall that uppercase M is used for the mass of an entire system while lowercase m is reserved for the mass of an atom or molecule.

> NOTE ▶ In practice, ΔT is usually measured in °C. But the Kelvin and the Celsius temperature scales have the same step size, so ΔT in K has exactly the same numerical value as ΔT in °C. Thus
>
> - You do not need to convert temperatures from °C to K if you only need a temperature *change* ΔT.
> - You do need to convert anytime you need the actual temperature T. ◄

The first law of thermodynamics, $\Delta E_{th} = W + Q$, allows us to write Equation 17.19 as $Mc\Delta T = W + Q$. In other words, we can change the system's temperature either by heating it or by doing an equivalent amount of work on it. In working with solids and liquids, we almost always change the temperature by heating. If $W = 0$, which we will assume for the rest of this section, then the heat needed to bring about a temperature change ΔT is

$$Q = Mc\Delta T \quad \text{(temperature change)} \quad (17.20)$$

TABLE 17.2 Specific heats and molar specific heats of solids and liquids

Substance	c (J/kg K)	C (J/mol K)
Solids		
Aluminum	900	24.3
Copper	385	24.4
Iron	449	25.1
Gold	129	25.4
Lead	128	26.5
Ice	2090	37.6
Liquids		
Ethyl alcohol	2400	110.4
Mercury	140	28.1
Water	4190	75.4

Because $\Delta T = \Delta E_{th}/Mc$, it takes more energy to change the temperature of a substance with a large specific heat than to change the temperature of a substance with a small specific heat. You can think of specific heat as measuring the *thermal inertia* of a substance. Metals, with small specific heats, warm up and cool down quickly. A piece of aluminum foil can be safely held within seconds of removing it from a hot oven. Water, with a very large specific heat, is slow to warm up and slow to cool down. This is fortunate for us. The large thermal inertia of water is essential for the biological processes of life. We wouldn't be here studying physics if water had a small specific heat!

EXAMPLE 17.6 Heating gold

How much heat is required to raise the temperature of 500 g of gold from 10°C to 50°C?

MODEL Heating the gold changes its thermal energy and its temperature.

SOLVE The temperature change is $\Delta T = 40°C = 40$ K. Thus

$$Q = \Delta E_{th} = Mc\Delta T = (0.500 \text{ kg})(129 \text{ J/kg K})(40 \text{ K})$$

$$= 2580 \text{ J}$$

where we used $c_{gold} = 129$ J/kg K from Table 17.2. Thus 2580 J of heat is needed.

EXAMPLE 17.7 Quenching hot aluminum in ethyl alcohol

A 50 g aluminum disk at 300°C is placed in 200 cm^3 of ethyl alcohol at 10.0°C, then quickly removed. The aluminum temperature is found to have dropped to 120°C. What is the new temperature of the ethyl alcohol?

MODEL Heat is the energy transferred due to a temperature difference. If we assume that the container holding the alcohol is well insulated, then the disk and the alcohol interact with each other but nothing else. Conservation of energy tells us that the heat energy transferred out of the disk is the heat energy transferred into the alcohol.

SOLVE The temperature change of the disk is $\Delta T_{Al} = (120°C - 300°C) = -180°C = -180$ K. It is negative because the temperature decreases. The energy removed from the disk is

$$Q_{Al} = Mc\Delta T = (0.050 \text{ kg})(900 \text{ J/kg K})(-180 \text{ K})$$

$$= -8100 \text{ J}$$

Q_{Al} is negative because the energy is transferred out of the aluminum. The ethyl alcohol *gains* 8100 J of energy, thus $Q_{ethyl} = +8100$ J. We need to know the mass of the ethyl alcohol. Its density was given in Table 16.1 as $\rho = 790$ kg/m^3, hence its mass is

$$M = \rho V = (790 \text{ kg/m}^3)(200 \times 10^{-6} \text{ m}^3) = 0.158 \text{ kg}$$

The heat from the aluminum causes the alcohol's temperature to change by

$$\Delta T = \frac{Q_{ethyl}}{Mc} = \frac{8100 \text{ J}}{(0.158 \text{ kg})(2400 \text{ J/kg K})} = 21.4 \text{ K}$$

$$= 21.4°C$$

The ethyl alcohol ends up at temperature

$$T_f = T_i + \Delta T = 10.0°C + 21.4°C = 31.4°C$$

The **molar specific heat** is the amount of energy that raises the temperature of 1 mol of a substance by 1 K. We'll use an uppercase C for the molar specific heat. The heat needed to bring about a temperature change ΔT of n moles of substance is

$$Q = nC\Delta T \tag{17.21}$$

The molar specific heat C is not simply a conversion of c to other units because the number of moles per kilogram depends on the molar mass of a substance. The number of moles in mass M of a substance is

$$n = \frac{M(\text{in g})}{M_{mol}} = \frac{1000 \text{ g/kg}}{M_{mol}(\text{in g/mol})} M(\text{in kg})$$

Thus the molar specific heat is

$$C \text{ (in J/mol K)} = \frac{M_{\text{mol}} \text{(in g/mol)}}{1000 \text{ g/kg}} c \text{ (in J/kg K)} \qquad (17.22)$$

Molar specific heats are shown in Table 17.2. Look at the five elemental solids (excluding ice). All have C very near 25 J/mol K. If we were to expand the table, we would find that most elemental solids have $C \approx 25$ J/mol K. This can't be a coincidence, but what is it telling us? This is a puzzle we will address in Chapter 18, where we will explore thermal energy at the atomic level.

Phase Change and Heat of Transformation

Suppose you start with a system in its solid phase and heat it at a steady rate. Figure 17.16, which you saw in Chapter 16, shows how the system's temperature changes. At first, the temperature increases linearly. This is not hard to understand because Equation 17.20 can be written

$$\text{slope of the } T\text{-versus-}Q \text{ graph} = \frac{\Delta T}{Q} = \frac{1}{Mc} \qquad (17.23)$$

The slope of the graph depends inversely on the system's specific heat. A constant specific heat implies a constant slope and thus a linear graph. In fact, you can measure c from such a graph.

NOTE ▶ The different slopes indicate that the solid, liquid, and gas phases of a substance have different specific heats. ◀

But there are times, shown as horizontal line segments, during which heat is being transferred to the system but the temperature isn't changing. These are *phase changes*. The thermal energy continues to increase during a phase change, but the additional energy goes into breaking molecular bonds rather than speeding up the molecules. **A phase change is characterized by a change in thermal energy without a change in temperature.**

The amount of heat energy that causes 1 kg of a substance to undergo a phase change is called the **heat of transformation** of that substance. For example, laboratory experiments show that 333,000 J of heat are needed to melt 1 kg of ice at 0°C. The symbol for heat of transformation is L. The heat required for the entire system of mass M to undergo a phase change is

$$Q = ML \qquad \text{(phase change)} \qquad (17.24)$$

Heat of transformation is a generic term that refers to any phase change. Two specific heats of transformation are the **heat of fusion** L_f, the heat of transformation between a solid and a liquid, and the **heat of vaporization** L_v, the heat of transformation between a liquid and a gas. The heat needed for these phase changes is

$$Q = \begin{cases} \pm ML_f & \text{melt/freeze} \\ \pm ML_v & \text{boil/condense} \end{cases} \qquad (17.25)$$

where the $\pm$ indicates that heat must be *added* to the system during melting or boiling but *removed* from the system during freezing or condensing. **You must explicitly include the minus sign when it is needed.**

Table 17.3 on the next page gives the heats of transformation of a few substances. Notice that the heat of vaporization is always much larger than the heat of fusion. We can understand this. Melting breaks just enough molecular bonds to allow the system to lose rigidity and flow. Even so, the molecules in a liquid

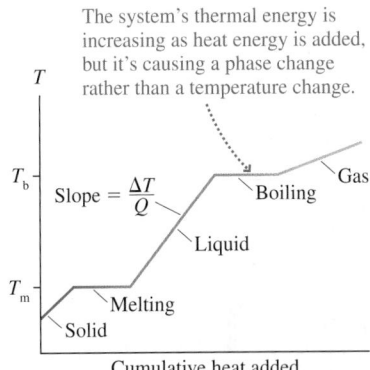

The system's thermal energy is increasing as heat energy is added, but it's causing a phase change rather than a temperature change.

FIGURE 17.16 The temperature of a system that is heated at a steady rate.

Lava—molten rock—undergoes a phase change when it contacts the much colder water. This is one way in which new islands are formed.

remain close together and loosely bonded. Vaporization breaks all bonds completely and sends the molecules flying apart. This process requires a larger increase in the thermal energy and thus a larger quantity of heat.

TABLE 17.3 Melting/boiling temperatures and heats of transformation

Substance	T_m (°C)	L_f (J/kg)	T_b (°C)	L_v (J/kg)
Nitrogen (N_2)	−210	0.26×10^5	−196	1.99×10^5
Ethyl alcohol	−114	1.09×10^5	78	8.79×10^5
Mercury	−39	0.11×10^5	357	2.96×10^5
Water	0	3.33×10^5	100	22.6×10^5
Lead	328	0.25×10^5	1750	8.58×10^5

EXAMPLE 17.8 Turning ice into steam
How much heat is required to change 200 mL of ice at −20°C (a typical freezer temperature) into steam?

MODEL Changing ice to steam requires four steps: Raise the temperature of the ice to 0°C, melt the ice to liquid water at 0°C, raise the water temperature to 100°C, then boil the water to produce steam at 100°C.

SOLVE The mass is $M = \rho V$. The density of ice (from Table 16.1) is 920 kg/m³, and $V = 200$ mL $= 200$ cm³ $= 2.00 \times 10^{-4}$ m³. Thus

$$M = \rho V = (920 \text{ kg/m}^3)(2.00 \times 10^{-4} \text{ m}^3) = 0.184 \text{ kg}$$

The heat needed for each step is

$$Q_1 = Mc_{ice}\Delta T_{ice} = (0.184 \text{ kg})(2090 \text{ J/kg K})(20 \text{ K})$$
$$= 7,700 \text{ J}$$

$$Q_2 = ML_f = (0.184 \text{ kg})(3.33 \times 10^5 \text{ J/kg}) = 61,300 \text{ J}$$

$$Q_3 = Mc_{water}\Delta T_{water} = (0.184 \text{ kg})(4190 \text{ J/kg K})(100 \text{ K})$$
$$= 77,100 \text{ J}$$

$$Q_4 = ML_v = (0.184 \text{ kg})(22.6 \times 10^5 \text{ J/kg}) = 415,800 \text{ J}$$

NOTE ▶ We used the specific heat of ice while warming the system in its solid phase. Then we used the specific heat of *water* while warming the system in its liquid phase. ◀

The total heat required is

$$Q = Q_1 + Q_2 + Q_3 + Q_4 = 562,000 \text{ J}$$

ASSESS Roughly 75% of the heat is used to change the water from a 100°C liquid to a 100°C gas. This is consistent with your experience that it takes much longer for a pan of water to boil away than it does to reach boiling.

STOP TO THINK 17.5 Objects A and B are brought into close thermal contact with each other, but they are well isolated from their surroundings. Initially $T_A = 0$°C and $T_B = 100$°C. The specific heat of A is less than the specific heat of B. The two objects will soon reach a common final temperature T_f. The final temperature is

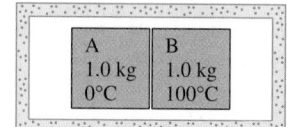

a. $T_f > 50$°C.
b. $T_f = 50$°C.
c. $T_f < 50$°C.

17.6 Calorimetry

At one time or another you've probably put an ice cube into a hot drink to cool it quickly. You were engaged, in a somewhat trial-and-error way, in a practical aspect of heat transfer known as **calorimetry.**

Figure 17.17 shows two systems thermally interacting with each other but isolated from everything else. Suppose they start at different temperatures T_1 and T_2. As you know from experience, heat energy will be transferred from the hotter to the colder system until they reach a common final temperature T_f. The systems will then be in thermal equilibrium and the temperature will not change further.

The insulation prevents any heat energy from being transferred to or from the environment, so energy conservation tells us that any energy leaving the hotter system must enter the colder system. That is, the systems *exchange* energy with no net loss or gain. The concept is straightforward, but to state the idea mathematically we need to be careful with signs.

Let Q_1 be the energy transferred to system 1 as heat. Q_1 is positive if energy *enters* system 1, negative if energy *leaves* system 1. Similarly, Q_2 is the energy transferred to system 2. The fact that the systems are merely exchanging energy can be written $|Q_1| = |Q_2|$. That is, the energy *lost* by the hotter system is the energy *gained* by the colder system. But Q_1 and Q_2 have opposite signs, so $Q_1 = -Q_2$. No energy is exchanged with the environment, hence it makes more sense to write this relationship as

$$Q_{net} = Q_1 + Q_2 = 0 \qquad (17.26)$$

This idea is not limited to the interaction of only two systems. If three or more systems are combined in isolation from the rest of their environment, each at a different initial temperature, they will all come to a common final temperature that can be found from the relationship

$$Q_{net} = Q_1 + Q_2 + Q_3 + \cdots = 0 \qquad (17.27)$$

NOTE ▶ The signs are very important in calorimetry problems. ΔT is always $T_f - T_i$, so ΔT and Q are negative for any system whose temperature decreases. The proper sign of Q for any phase change must be supplied *by you*, depending on the direction of the phase change. ◀

Heat energy is transferred from system 1 to system 2. Energy conservation requires

$$|Q_1| = |Q_2|$$

Opposite signs mean that

$$Q_{net} = Q_1 + Q_2 = 0$$

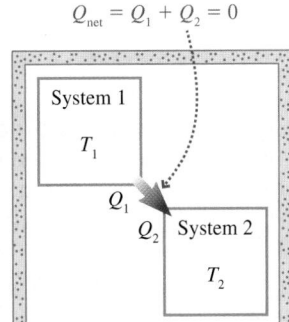

FIGURE 17.17 Two systems interact thermally.

 PROBLEM-SOLVING STRATEGY 17.2 Calorimetry problems

MODEL Identify the interacting systems. Assume that they are isolated from the larger environment.

VISUALIZE List known information and identify what you need to find. Convert all quantities to SI units.

SOLVE The mathematical representation, which is a statement of energy conservation, is

$$Q_{net} = Q_1 + Q_2 + \cdots = 0$$

- For systems that undergo a temperature change, $Q = Mc(T_f - T_i)$. Be sure to have the temperatures T_i and T_f in the correct order.
- For systems that undergo a phase change, $Q = \pm ML$. Supply the correct sign by observing whether energy enters or leaves the system during the transition.
- Some systems may undergo a temperature change *and* a phase change. Treat the changes separately. The heat energy is $Q = Q_{\Delta T} + Q_{phase}$.

ASSESS Is the final temperature in the middle? T_f that is higher or lower than all initial temperatures is an indication that something is wrong, usually a sign error.

NOTE ▶ You may have learned to solve calorimetry problems in other courses by writing $Q_{gained} = Q_{lost}$. That is, by balancing heat gained with heat lost. That approach works in simple problems, but it has two drawbacks. First, you often have to "fudge" the signs to make them work. Second, and more serious, you can't extend this approach to a problem with three or more interacting systems. Using $Q_{net} = 0$ is much preferred. ◀

EXAMPLE 17.9 Identifying a metal

200 g of an unknown metal is heated to a temperature of 200°C then dropped into 50 g of water at 20.0°C in an insulated container. The water temperature rises within a few seconds to 39.7°C, then changes no further. Identify the metal.

MODEL Two systems interact thermally. There are temperature changes but no phase changes.

VISUALIZE We know all the initial and final temperatures. For water, $c_w = 4190$ J/kg K is known from Table 17.2. Only the specific heat c_m of the metal is unknown.

SOLVE Energy conservation requires $Q_w + Q_m = 0$. Using $Q = Mc(T_f - T_i)$ for each, we have

$$Q_w + Q_m = M_w c_w (T_f - T_w) + M_m c_m (T_f - T_m) = 0$$

This is easily solved for the unknown specific heat:

$$\begin{aligned} c_m &= \frac{-M_w c_w (T_f - T_w)}{M_m (T_f - T_m)} \\ &= \frac{-(0.050 \text{ kg})(4190 \text{ J/kg K})(39.7°C - 20°C)}{(0.200 \text{ kg})(39.7°C - 200°C)} \\ &= 129 \text{ J/kg K} \end{aligned}$$

Referring to Table 17.2, we find we have either 200 g of gold or, if we made an ever-so-slight experimental error, 200 g of lead!

NOTE ▶ It's okay to compute ΔT in °C, even though the specific heat is in J/kg K, because a *change* of 1°C is equivalent to a *change* of 1 K. ◀

ASSESS The temperature of the unknown metal changed much more than the temperature of the water. That is, the unknown metal had much less thermal inertia, and thus, as we found, its specific heat should be much less than that of water.

EXAMPLE 17.10 Calorimetry with a phase change

Your 500 mL soda is at 20°C, room temperature, so you add 100 g of ice from the −20°C freezer. Does all the ice melt? If so, what is the final temperature? If not, what fraction of the ice melts? Assume that you have a well-insulated cup.

MODEL We have a thermal interaction between the soda, which is essentially water, and the ice. We need to distinguish between three possible outcomes. If all the ice melts, then $T_f > 0°C$. It's also possible that the soda will cool to 0°C before all the ice has melted, leaving the ice and liquid in equilibrium at 0°C. A third possibility is that the soda will freeze solid before the ice warms up to 0°C. That seems unlikely here, but there are situations, such as the pouring of molten metal out of furnaces, when all the liquid does solidify. We need to distinguish between these before knowing how to proceed.

VISUALIZE All the initial temperatures, masses, and specific heats are known. The final temperature of the combined soda + ice system is unknown.

SOLVE Let's first calculate the heat needed to melt all the ice and leave it as liquid water at 0°C. To do so, we must warm the ice to 0°C, then change it to water. The heat input for this two-stage process is

$$Q_{melt} = M_i c_i (20 \text{ K}) + M_i L_f = 37,500 \text{ J}$$

where L_f is the heat of fusion of water. It is used as a *positive* quantity because we must *add* heat to melt the ice. Next, let's calculate how much heat energy will leave the soda if it cools

all the way to 0°C. The volume is $V = 500 \text{ mL} = 5 \times 10^{-4} \text{ m}^3$ and thus the mass is $M_s = \rho V = 0.500 \text{ kg}$. The heat is

$$Q_{cool} = M_s c_w (-20 \text{ K}) = -41,900 \text{ J}$$

where $\Delta T = -20$ K because the temperature decreases. Because $|Q_{cool}| > Q_{melt}$, the soda has sufficient energy to melt all the ice. Hence the final state will be all liquid at $T_f > 0$. (Had we found $|Q_{cool}| < Q_{melt}$, then the final state would have been an ice-liquid mixture at 0°C.)

Energy conservation requires $Q_{ice} + Q_{soda} = 0$. The heat Q_{ice} consists of three terms: warming the ice to 0°C, melting the ice to water at 0°C, then warming the 0°C water to T_f. The mass will still be M_i in the last of these steps, because it is the "ice system," but we need to use the specific heat of *liquid water*. Thus

$$Q_{ice} + Q_{soda} = [M_i c_i (20 \text{ K}) + M_i L_f + M_i c_w (T_f - 0°C)]$$
$$+ M_s c_w (T_f - 20°C) = 0$$

We've already done part of the calculation, allowing us to write

$$37,500 \text{ J} + M_i c_w (T_f - 0°C) + M_s c_w (T_f - 20°C) = 0$$

Solving for T_f gives

$$T_f = \frac{20 M_s c_w - 37,500}{M_i c_w + M_s c_w} = 1.7°C$$

ASSESS As expected, the soda has been cooled to nearly the freezing point.

EXAMPLE 17.11 Three interacting systems

A 200 g piece of iron at 120°C and a 150 g piece of copper at −50°C are dropped into an insulated beaker containing 300 g of ethyl alcohol at 20°C. What is the final temperature?

MODEL Here you can't use a simple $Q_{gained} = Q_{lost}$ approach because you don't know whether the alcohol is going to warm up or cool down.

VISUALIZE All the initial temperatures, masses, and specific heats are known. We need to find the final temperature.

SOLVE Energy conservation requires

$$Q_i + Q_c + Q_e = M_i c_i (T_f − 120°C) + M_c c_c (T_f − (−50°C))$$

$$+ M_e c_e (T_f − 20°C) = 0$$

Solving for T_f gives

$$T_f = \frac{120 M_i c_i − 50 M_c c_c + 20 M_e c_e}{M_i c_i + M_c c_c + M_e c_e} = 25.7°C$$

ASSESS The temperature is between the initial iron and copper temperatures, as expected. It turns out that the alcohol warms up ($Q_e > 0$), but we had no way to know this without doing the calculation.

17.7 The Specific Heats of Gases

Specific heats are given in Table 17.2 for solids and liquids. Gases are harder to characterize because the heat required to cause a specified temperature change depends on the *process* by which the gas changes state.

Figure 17.18 shows two isotherms on the pV diagram for a gas. Processes A and B, which start on the T_i isotherm and end on the T_f isotherm, have the *same* temperature change $\Delta T = T_f − T_i$. But process A, which takes place at constant volume, requires a *different* amount of heat than does process B, which occurs at constant pressure. The reason is that work is done in process B but not in process A. This is a situation that we are now equipped to analyze.

It is useful to define two different versions of the specific heat of gases, one for constant volume (isochoric) processes and one for constant pressure (isobaric) processes. We will define these as molar specific heats because we usually do gas calculations using moles instead of mass. The quantity of heat needed to change the temperature of n moles of gas by ΔT is

$$Q = nC_V\Delta T \qquad \text{(temperature change at constant volume)}$$

$$Q = nC_P\Delta T \qquad \text{(temperature change at constant pressure)}$$

(17.28)

where C_V is the **molar specific heat at constant volume** and C_P is the **molar specific heat at constant pressure**. Table 17.4 gives the values of C_V and C_P for a few common monatomic and diatomic gases. The units are J/mol K.

NOTE ▶ Equation 17.28 applies to two specific ideal-gas processes. In a general gas process, for which neither p nor V are constant, we have no direct way to relate Q to ΔT. In that case, the heat must be found indirectly from the first law as $Q = \Delta E_{th} − W$. ◀

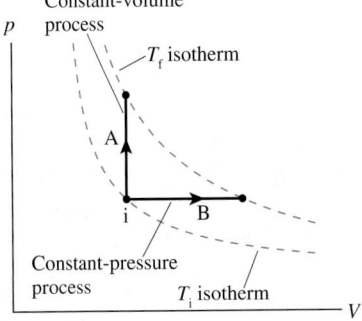

FIGURE 17.18 Processes A and B have the same ΔT and the same ΔE_{th}, but they require different amounts of heat.

TABLE 17.4 Molar specific heats of gases (J/mol K)

Gas	C_P	C_V	$C_P − C_V$
Monatomic Gases			
He	20.8	12.5	8.3
Ne	20.8	12.5	8.3
Ar	20.8	12.5	8.3
Diatomic Gases			
H_2	28.7	20.4	8.3
N_2	29.1	20.8	8.3
O_2	29.2	20.9	8.3

EXAMPLE 17.12 Heating and cooling a gas

Three moles of O_2 gas are at 20.0°C. 600 J of heat energy are transferred to the gas at constant pressure, then 600 J are removed at constant volume. What is the final temperature? Show the process on a pV diagram.

MODEL O_2 is a diatomic ideal gas. The gas is heated as an isobaric process, then cooled as an isochoric process.

SOLVE The heat transferred during the constant-pressure process causes a temperature rise:

$$\Delta T = T_2 − T_1 = \frac{Q}{nC_P} = \frac{600 \text{ J}}{(3.0 \text{ mol})(29.2 \text{ J/mol K})} = 6.8°C$$

where C_P for oxygen was taken from Table 17.4. Heating leaves the gas at temperature $T_2 = T_1 + \Delta T = 26.8°C$. The temperature then falls as heat is removed during the constant-volume process:

$$\Delta T = T_3 - T_2 = \frac{Q}{nC_V} = \frac{(-600 \text{ J})}{(3.0 \text{ mol})(20.9 \text{ J/mol K})} = -9.5°C$$

We used a *negative* value for Q because heat energy is transferred from the gas to the environment. The final temperature of the gas is $T_3 = T_2 + \Delta T = 17.3°C$. Figure 17.19 shows the process on a pV diagram. The gas expands (moves horizontally on the diagram) as heat is added, then cools at constant volume (moves vertically on the diagram) as heat is removed.

ASSESS The final temperature is lower than the initial temperature because $C_P > C_V$.

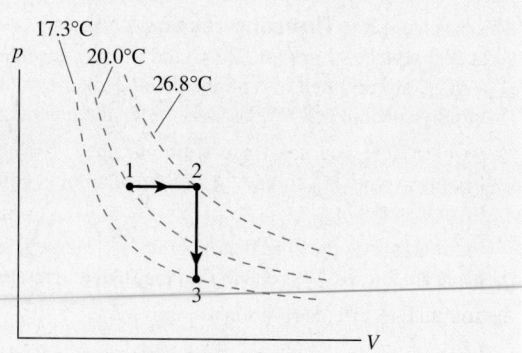

FIGURE 17.19 The pV diagram for Example 17.12.

EXAMPLE 17.13 **Calorimetry with a gas and a solid**

The interior volume of a 200 g hollow aluminum box is 800 cm³. The box contains nitrogen gas at STP. A 20 cm³ block of copper at a temperature of 300°C is placed inside the box, then the box is sealed. What is the final temperature?

MODEL This example has three interacting systems: the aluminum box, the nitrogen gas, and the copper block. They must all come to a common final temperature T_f.

VISUALIZE The box and gas have the same initial temperature: $T_{Al} = T_{N2} = 0°C$. The box doesn't change size, so this is a constant-volume process for the gas. The final temperature is unknown.

SOLVE Although one of the systems is now a gas, the calorimetry equation $Q_{net} = Q_{Al} + Q_{N2} + Q_{Cu} = 0$ is still appropriate. In this case,

$$Q_{net} = m_{Al}c_{Al}(T_f - T_{Al}) + n_{N2}C_V(T_f - T_{N2})$$
$$+ m_{Cu}c_{Cu}(T_f - T_{Cu}) = 0$$

Notice that we used masses and specific heats for the solids but moles and the molar specific heat for the gas. We used C_V because this is a constant-volume process. Solving for T_f gives

$$T_f = \frac{m_{Al}c_{Al}T_{Al} + n_{N2}C_V T_{N2} + m_{Cu}c_{Cu}T_{Cu}}{m_{Al}c_{Al} + n_{N2}C_V + m_{Cu}c_{Cu}}$$

The specific heat values are found in Tables 17.2 and 17.4. The mass of the copper is

$$m_{Cu} = \rho_{Cu}V_{Cu} = (8920 \text{ kg/cm}^3)(20 \times 10^{-6} \text{ m}^3) = 0.178 \text{ kg}$$

The number of moles of the gas is found from the ideal-gas law, using the initial conditions. Notice that inserting the copper block *displaces* 20 cm³ of gas, hence the gas volume is only $V = 780 \text{ cm}^3 = 7.80 \times 10^{-4} \text{ m}^3$. Thus

$$n_{N2} = \frac{pV}{RT} = 0.0348 \text{ mol}$$

Computing the final temperature from the above equation gives $T_f = 83°C$.

C_P and C_V

8.7

You may have noticed two curious features in Table 17.4. First, the molar specific heats of monatomic gases are *all alike*. And the molar specific heats of diatomic gases, while different from monatomic gases, are again *very nearly alike*. We saw a similar feature in Table 17.2 for the molar specific heats of solids. Second, the *difference* $C_P - C_V = 8.3 \text{ J/mol K}$ is the same in every case. And, most puzzling of all, the value of $C_P - C_V$ appears to be equal to the universal gas constant R! Why should this be?

The relationship between C_V and C_P hinges on one crucial idea: ΔE_{th}, **the change in the thermal energy of a gas, is the same for *any* two processes that have the same ΔT.** The thermal energy of a gas is associated with temperature, so any process that changes the gas temperature from T_i to T_f has the same ΔE_{th} as any other process that goes from T_i to T_f. Furthermore, the first law $\Delta E_{th} = Q + W$ tells us that a gas cannot distinguish between heat and work. The system's thermal energy changes in response to energy added to or removed from the system, but the response of the gas is the same whether you heat the system, do work on the system, or do some combination of both. Thus **any two processes that change the thermal energy of the gas by ΔE_{th} will cause the same temperature change ΔT.**

With that in mind, look back at Figure 17.18. Both gas processes have the same ΔT, so both have the same value of ΔE_{th}. Process A is an isochoric process in which no work is done (the piston doesn't move), so the first law for this process is

$$(\Delta E_{th})_A = W + Q = 0 + Q_{const\ vol} = nC_V\Delta T \qquad (17.29)$$

Process B is an isobaric process. You learned earlier that the work done on the gas during an isobaric process is $W = -p\Delta V$. Thus

$$(\Delta E_{th})_B = W + Q = -p\Delta V + Q_{const\ press} = -p\Delta V + nC_P\Delta T \quad (17.30)$$

$(\Delta E_{th})_B = (\Delta E_{th})_A$, because both have the same ΔT, so we can equate the right sides of Equations 17.29 and 17.30:

$$-p\Delta V + nC_P\Delta T = nC_V\Delta T \qquad (17.31)$$

For the final step, we can use the ideal-gas law $pV = nRT$ to relate ΔV and ΔT during process B. For any gas process,

$$\Delta(pV) = \Delta(nRT) \qquad (17.32)$$

For a constant-pressure process, where p is constant, Equation 17.32 becomes

$$p\Delta V = nR\Delta T \qquad (17.33)$$

Substituting this expression for $p\Delta V$ into Equation 17.31 gives

$$-nR\Delta T + nC_P\Delta T = nC_V\Delta T \qquad (17.34)$$

The $n\Delta T$ cancels, and we are left with

$$\boxed{C_P = C_V + R} \qquad (17.35)$$

This result, which applies to all ideal gases, is exactly what we see in the data of Table 17.4.

But that's not the only conclusion we can draw. Equation 17.29 found that $\Delta E_{th} = nC_V\Delta T$ for a constant-volume process. But we had just noted that ΔE_{th} is the same for *all* gas processes that have the same ΔT. Consequently, this expression for ΔE_{th} is equally true for any other process. That is

$$\boxed{\Delta E_{th} = nC_V\Delta T \qquad \text{(any ideal-gas process)}} \qquad (17.36)$$

Compare this result to Equation 17.28. We first made a distinction between constant-volume and constant-pressure processes, but now we're saying that Equation 17.36 is true for any process. Are we contradicting ourselves? No, the difference lies in what you need to calculate.

■ The change in thermal energy when the temperature changes by ΔT is the same for any process. That's Equation 17.36.
■ The *heat* required to bring about the temperature change depends on what the process is. That's Equation 17.28. An isobaric process requires more heat than an isochoric process that produces the same ΔT.

The reason for the difference is seen by writing the first law as $Q = \Delta E_{th} - W$. In an isochoric process, where $W = 0$, *all* the heat input is used to increase the gas temperature. But in an isobaric process, some of the energy that enters the system as heat then leaves the system as work ($W < 0$) done by the expanding gas. Thus more heat is needed to produce the same ΔT.

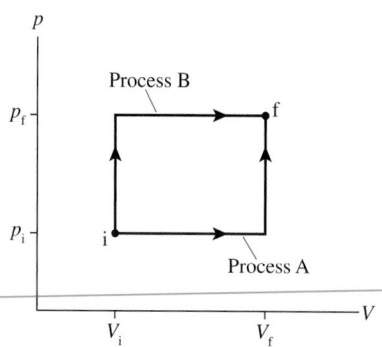

FIGURE 17.20 Is the heat input along these two paths the same or different?

Heat Depends on the Path

Consider the two ideal-gas processes shown in Figure 17.20. Even though the initial and final states are the same, the heat added during these two processes is *not* the same. We can use the first law $\Delta E_{th} = W + Q$ to see why.

The thermal energy is a state variable. That is, its value depends on the state of the gas, not the process by which the gas arrived at that state. Thus $\Delta E_{th} = E_{th\,f} - E_{th\,i}$ is the same for both processes. If ΔE_{th} is the same for processes A and B, then $W_A + Q_A = W_B + Q_B$.

You learned in Section 17.2 that the work done during an ideal-gas process depends on the path in the pV diagram. There's more area under the process B curve, so $|W_B| > |W_A|$. Both values of W are negative, because the gas expands, so W_B is more negative than W_A. Consequently, $W_A + Q_A$ can equal $W_B + Q_B$ only if $Q_B > Q_A$. **The heat added or removed during an ideal-gas process depends on the path followed through the pV diagram.**

Adiabatic Processes

We can conclude this section by introducing a fourth basic ideal-gas process. Of the three processes we've looked at thus far—isochoric, isobaric, and isothermal—an isochoric process does no work ($W = 0$) and the isothermal process has no change in thermal energy ($\Delta E_{th} = 0$). Work and thermal energy are two of the three quantities that appear in the first law of thermodynamics. This quite naturally leads us to wonder if there is a process for which no heat is transferred. There is.

An ideal-gas process in which no heat energy is transferred between the gas and the environment ($Q = 0$) is called an **adiabatic process.** The word comes from a Greek root meaning "heat does not pass through." Figure 17.21 compares an adiabatic process to isothermal and isochoric processes. In practice, there are two ways that an adiabatic process can come about. First, a gas cylinder can be completely surrounded by thermal insulation, such as thick pieces of Styrofoam. The environment can interact mechanically with the gas by pushing or pulling on the insulated piston, but there is no thermal interaction.

Second, the gas can be expanded or compressed very rapidly in what we call an *adiabatic expansion* or an *adiabatic compression.* In a rapid process there is essentially no time for heat to be transferred between the gas and the environment. We've already alluded to the idea that heat is transferred via atomic-level collisions. These collisions take time. If you stick one end of a copper rod into a flame, the other end will eventually get too hot to hold—but not instantly. Some amount of time—many seconds in this example—is required for heat to be transferred from one end to the other. A process that takes place faster than the heat can be transferred is adiabatic.

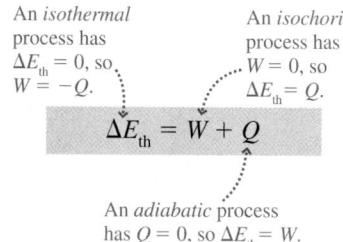

An *isothermal* process has $\Delta E_{th} = 0$, so $W = -Q$.

An *isochoric* process has $W = 0$, so $\Delta E_{th} = Q$.

$$\Delta E_{th} = W + Q$$

An *adiabatic* process has $Q = 0$, so $\Delta E_{th} = W$.

FIGURE 17.21 The relationship of three important processes to the first law of thermodynamics.

> NOTE ▶ You may recall reading in Chapter 16 that we are going to study only quasi-static processes, processes that proceed slowly enough to remain essentially in equilibrium at all times. Now we're proposing to study processes that take place very rapidly. Isn't this a contradiction? Yes, to some extent it is. What we need to establish are the appropriate time scales. How slow must a process go to be quasi-static? How fast must it go to be adiabatic? These types of calculations must be deferred to a more advanced course. It turns out—fortunately!—that many practical applications, such as the compression strokes in gasoline and diesel engines, are fast enough to be adiabatic yet slow enough to still be considered quasi-static. ◄

For an adiabatic process, with $Q = 0$, the first law of thermodynamics is $\Delta E_{th} = W$. Compressing a gas adiabatically ($W > 0$) increases the thermal energy. Thus **an adiabatic compression raises the temperature of a gas.** A gas that expands adiabatically ($W < 0$) gets colder as its thermal energy decreases.

Thus **an adiabatic expansion lowers the temperature of a gas.** You can use an adiabatic process to change the gas temperature without using heat!

The work done on a gas in an adiabatic process goes entirely to changing the thermal energy of the gas. But we just found that $\Delta E_{th} = nC_V\Delta T$ for *any* process. Thus

$$W = nC_V\Delta T \qquad \text{(adiabatic process)} \qquad (17.37)$$

Equation 17.37 joins with the equations we derived earlier for the work done in isochoric, isobaric, and isothermal processes.

Gas processes can be represented as trajectories in the pV diagram. For example, a gas moves along a hyperbola during an isothermal process. How does an adiabatic process appear in a pV diagram? The result is more important than the derivation, which is a bit tedious, so we'll begin with the answer and then, at the end of this section, show where it comes from.

First, define the **specific heat ratio** γ (lowercase Greek gamma) to be

$$\gamma = \frac{C_P}{C_V} = \begin{cases} 1.67 & \text{monatomic gas} \\ 1.40 & \text{diatomic gas} \end{cases} \qquad (17.38)$$

The specific heat ratio has many uses in thermodynamics. The values of γ for monatomic and diatomic gases are simple and independent of the gas. Values of γ for more complex gases, such as steam or gasoline vapor, are tabulated in books on engineering thermodynamics. Notice that γ is dimensionless.

An adiabatic process is one in which

$$pV^\gamma = \text{constant} \qquad \text{or} \qquad p_f V_f^\gamma = p_i V_i^\gamma \qquad (17.39)$$

This is similar to the isothermal $pV = $ constant, but somewhat more complex due to the exponent γ.

The curves found by graphing $p = $ constant$/V^\gamma$ are called **adiabats.** In Figure 17.22 you see that the two adiabats are steeper than the hyperbolic isotherms. An adiabatic process moves along an adiabat in the same way that an isothermal process moves along an isotherm. You can see that the temperature falls during an adiabatic expansion and rises during an adiabatic compression.

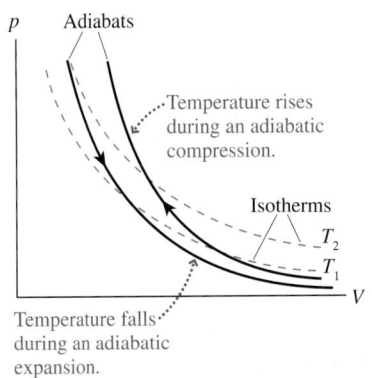

FIGURE 17.22 An adiabatic process moves along pV curves called *adiabats.*

EXAMPLE 17.14 An adiabatic compression

Air containing gasoline vapor is admitted into the cylinder of an internal combustion engine at 1.0 atm pressure and 30°C. The piston rapidly compresses the gas from 500 cm³ to 50 cm³, a *compression ratio* of 10.

a. What are the final temperature and pressure of the gas?
b. Show the compression process on a pV diagram.
c. How much work is done to compress the gas?

MODEL The compression is rapid, with insufficient time for heat to be transferred from the gas to the environment, so we will model it as an adiabatic compression. We'll treat the gas as if it were 100% air.

SOLVE

a. We know the initial pressure and volume, and we know the volume after the compression. For an adiabatic process, where pV^γ remains constant, the final pressure is

$$p_f = p_i \left(\frac{V_i}{V_f}\right)^\gamma = (1.0 \text{ atm})(10)^{1.40} = 25.1 \text{ atm}$$

Air is a mixture of N_2 and O_2, diatomic gases, so we used $\gamma = 1.40$. We can now find the temperature by using the ideal-gas law:

$$T_f = T_i \frac{p_f}{p_i} \frac{V_f}{V_i} = (303 \text{ K})(25.1)\left(\frac{1}{10}\right) = 761 \text{ K} = 488°\text{C}$$

Temperature *must* be in kelvins for doing gas calculations such as these.

b. Figure 17.23 shows the pV diagram. The 30°C and 488°C isotherms are included to show how the temperature changes during the process.

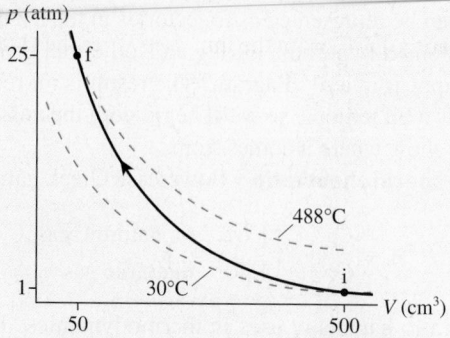

FIGURE 17.23 The adiabatic compression of the gas in an internal combustion engine.

c. The work done is $W = nC_V\Delta T$, with $\Delta T = 458$ K. The number of moles is found from the ideal-gas law and the initial conditions:

$$n = \frac{p_i V_i}{RT_i} = 0.0201 \text{ mol}$$

Thus the work done to compress the gas is

$$W = nC_V\Delta T = (0.0201 \text{ mol})(20.8 \text{ J/mol K})(458 \text{ K}) = 192 \text{ J}$$

ASSESS The temperature rises dramatically during the compression stroke of an engine. But the higher temperature has nothing to do with heat! **The temperature and thermal energy of the gas are increased not by heating the gas but by doing work on it.** This is an important idea to understand.

If we use the ideal-gas-law expression $p = nRT/V$ in the adiabatic equation $pV^\gamma = $ constant, we see that $TV^{\gamma-1}$ is also constant during an adiabatic process. Thus another useful equation for adiabatic processes is

$$T_f V_f^{\gamma-1} = T_i V_i^{\gamma-1} \tag{17.40}$$

Proof of Equation 17.39

Now let's see where Equation 17.39 comes from. Consider an adiabatic process in which an infinitesimal amount of work dW done on a gas causes an infinitesimal change in the thermal energy. For an adiabatic process, with $dQ = 0$, the first law of thermodynamics is

$$dE_{th} = dW \tag{17.41}$$

We can use Equation 17.36, which is valid for *any* gas process, to write $dE_{th} = nC_V dT$. Earlier in the chapter we found that the work done during a small volume change is $dW = -p\,dV$. With these substitutions, Equation 17.41 becomes

$$nC_V dT = -p\,dV \tag{17.42}$$

The ideal-gas law can now be used to write $p = nRT/V$. The n cancels, and the C_V can be moved to the other side of the equation to give

$$\frac{dT}{T} = -\frac{R}{C_V}\frac{dV}{V} \tag{17.43}$$

We're going to integrate Equation 17.43, but anticipating the need for $\gamma = C_P/C_V$ we can first use the fact that $C_P = C_V + R$ to write

$$\frac{R}{C_V} = \frac{C_P - C_V}{C_V} = \frac{C_P}{C_V} - 1 = \gamma - 1 \tag{17.44}$$

Now integrate Equation 17.43 from the initial state i to the final state f:

$$\int_{T_i}^{T_f}\frac{dT}{T} = -(\gamma - 1)\int_{V_i}^{V_f}\frac{dV}{V} \tag{17.45}$$

Carrying out the integration gives

$$\ln\!\left(\frac{T_f}{T_i}\right) = \ln\!\left(\frac{V_i}{V_f}\right)^{\gamma - 1} \tag{17.46}$$

where we used the logarithm properties $\log a - \log b = \log(a/b)$ and $c\log a = \log(a^c)$.

Taking the exponential of both sides now gives

$$\left(\frac{T_f}{T_i}\right) = \left(\frac{V_i}{V_f}\right)^{\gamma - 1} \tag{17.47}$$

$$T_f V_f^{\gamma - 1} = T_i V_i^{\gamma - 1}$$

This was Equation 17.40. Writing $T = pV/nR$ and canceling $1/nR$ from both sides of the equation gives Equation 17.39:

$$p_f V_f^{\gamma} = p_i V_i^{\gamma} \tag{17.48}$$

This was a lengthy derivation, but it is good practice at seeing how the ideal-gas law and the first law of thermodynamics can work together to yield results of great importance.

STOP TO THINK 17.6   For the two processes shown, which of the following is true:

a. $Q_A > Q_B$.
b. $Q_A = Q_B$.
c. $Q_A < Q_B$.

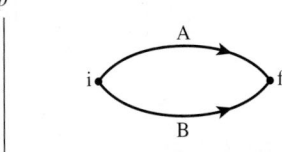

SUMMARY

The goal of Chapter 17 has been to expand our understanding of energy and to develop the first law of thermodynamics as a general statement of energy conservation.

GENERAL PRINCIPLES

First Law of Thermodynamics

$$\Delta E_{th} = W + Q$$

The first law is a general statement of energy conservation.

Work W and heat Q depend on the process by which the system is changed.

The change in the system depends only on the total energy exchanged $W + Q$, not on the process.

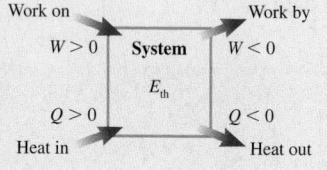

Energy

Thermal energy E_{th} Microscopic energy of moving molecules and stretched molecular bonds. ΔE_{th} depends on the initial/final states but is independent of the process.

Work W Energy transferred to the system by forces in a mechanical interaction.

Heat Q Energy transferred to the system via atomic-level collisions when there is a temperature difference. A thermal interaction.

IMPORTANT CONCEPTS

The work done on a gas is

$$W = -\int_{V_i}^{V_f} p\, dV$$

$$= -(\text{area under the } pV \text{ curve})$$

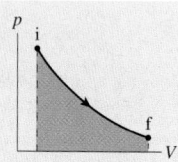

An adiabatic process is one for which $Q = 0$. Gases move along an **adiabat** for which $pV^\gamma = $ constant, where $\gamma = C_P/C_V$ is the **specific heat ratio.** An adiabatic process changes the temperature of the gas without heating it.

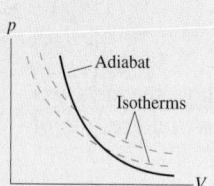

Calorimetry When two or more systems interact thermally, they come to a common final temperature determined by

$$Q_{net} = Q_1 + Q_2 + \ldots = 0$$

The heat of transformation L is the energy needed to cause 1 kg of substance to undergo a phase change

$$Q = \pm ML$$

The specific heat c of a substance is the energy needed to raise the temperature of 1 kg by 1 K.

$$Q = Mc\Delta T$$

The molar specific heat C is the energy needed to raise the temperature of 1 mol by 1 K.

$$Q = nC\Delta T$$

The molar specific heat of gases depends on the *process* by which the temperature is changed:

$C_V = $ molar specific heat at **constant volume.**

$C_P = $ molar specific heat at **constant pressure.**

$C_P = C_V + R$, where R is the universal gas constant.

SUMMARY OF BASIC GAS PROCESSES

Process	Definition	Stays constant	Work	Heat
Isochoric	$\Delta V = 0$	V and p/T	$W = 0$	$Q = nC_V\Delta T$
Isobaric	$\Delta p = 0$	p and V/T	$W = -p\Delta V$	$Q = nC_P\Delta T$
Isothermal	$\Delta T = 0$	T and pV	$W = -nRT \ln(V_f/V_i)$	$\Delta E_{th} = 0$
Adiabatic	$Q = 0$	pV^γ	$W = \Delta E_{th}$	$Q = 0$
All gas processes		Ideal-gas law	$pV = nRT$	
		First law	$\Delta E_{th} = W + Q = nC_V\Delta T$	

TERMS AND NOTATION

internal energy, E_{int}	first law of thermodynamics	calorimetry
work, W	thermodynamic energy model	molar specific heat at constant volume, C_V
mechanical interaction	specific heat, c	molar specific heat at constant pressure, C_P
mechanical equilibrium	molar specific heat, C	adiabatic process
heat, Q	heat of transformation, L	specific heat ratio, γ
thermal interaction	heat of fusion, L_f	adiabat
thermal equilibrium	heat of vaporization, L_v	

EXERCISES AND PROBLEMS

Exercises

Section 17.1 It's All About Energy

Section 17.2 Work in Ideal-Gas Processes

1. The average speed of the atoms in a 2.0 g sample of helium gas is 700 m/s. Estimate the thermal energy of the sample.
2. An 8.0 g sample of oxygen gas has 1700 J of thermal energy. Estimate the average speed of an oxygen molecule.
3. How much work is done on the gas in the process shown in Figure Ex17.3?

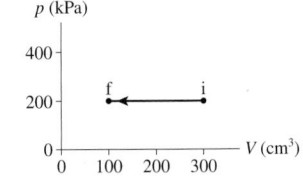

FIGURE EX17.3

4. How much work is done on the gas in the process shown in Figure Ex17.4?

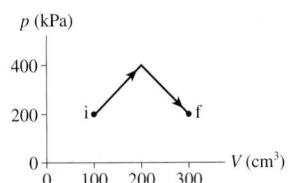

FIGURE EX17.4

5. 80 J of work are done on the gas in the process shown in Figure Ex17.5. What is V_1 in cm³?

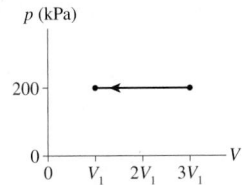

FIGURE EX17.5

6. A 2000 cm³ container holds 0.10 mol of helium gas at 300°C. How much work must be done to compress the gas to 1000 cm³ at (a) constant pressure and (b) constant temperature? (c) Show and label both processes on a single pV diagram.

Section 17.3 Heat

Section 17.4 The First Law of Thermodynamics

7. Draw a first-law bar chart (see Figure 17.14) for the gas process in Figure Ex17.7.

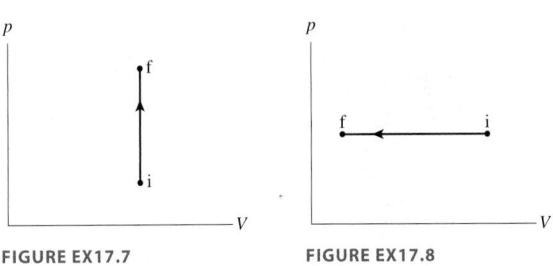

FIGURE EX17.7   FIGURE EX17.8

8. Draw a first-law bar chart (see Figure 17.14) for the gas process in Figure Ex17.8.
9. Draw a first-law bar chart (see Figure 17.14) for the gas process in Figure Ex17.9.

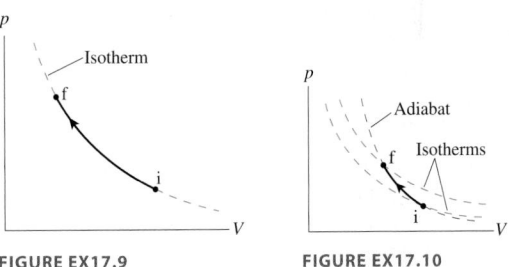

FIGURE EX17.9   FIGURE EX17.10

10. Draw a first-law bar chart (see Figure 17.14) for the gas process in Figure Ex17.10.
11. Design and implement a process that will use the gas cylinder of Figure 17.13 to carry out the process shown in Figure Ex17.11. Describe the steps, then show the process as a first-law bar chart.

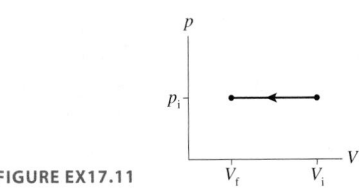

FIGURE EX17.11

12. Design and implement a process that will use the gas cylinder of Figure 17.13 to carry out the process shown in Figure Ex17.12. Describe the steps, then show the process as a first-law bar chart.

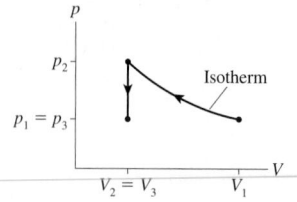

FIGURE EX17.12

13. 500 J of work are done on a system in a process that decreases the system's thermal energy by 200 J. How much heat energy is transferred to or from the system?

14. A gas is compressed from 600 cm³ to 200 cm³ at a constant pressure of 400 kPa. At the same time, 100 J of heat energy is transferred out of the gas. What is the change in thermal energy of the gas during this process?

Section 17.5 Thermal Properties of Matter

15. How much energy must be removed from a 6.0 cm × 6.0 cm × 6.0 cm block of ice to cool it from 0°C to −30°C?

16. A rapidly spinning paddle wheel raises the temperature of 200 mL of water from 21°C to 25°C. How much (a) work is done and (b) heat is transferred in this process?

17. How much heat is needed to change 20 g of mercury at 20°C into mercury vapor at the boiling point?

18. a. 100 J of heat energy are transferred to 20 g of mercury. By how much does the temperature increase?
    b. How much heat is needed to raise the temperature of 20 g of water by the same amount?

19. A beaker contains 200 mL of ethyl alcohol at 20°C. What is the minimum amount of energy that must be removed to produce solid ethyl alcohol?

20. What is the maximum mass of lead you could melt with 1000 J of heat, starting from 20°C?

Section 17.6 Calorimetry

21. 30 g of copper pellets are removed from a 300°C oven and immediately dropped into 100 mL of water at 20°C in an insulated cup. What will the new water temperature be?

22. A copper block is removed from a 300°C oven and dropped into 1.0 L of water at 20.0°C. The water quickly reaches 25.5°C and then remains at that temperature. What is the mass of the copper block?

23. A 50.0 g thermometer is used to measure the temperature of 200 mL of water. The specific heat of the thermometer, which is mostly glass, is 750 J/kg K, and it reads 20.0°C while lying on the table. After being completely immersed in the water, the thermometer's reading stabilizes at 71.2°C. What was the actual water temperature before it was measured?

24. A 750 g aluminum pan is removed from the stove and plunged into a sink filled with 10 L of water at 20.0°C. The water temperature quickly rises to 24.0°C. What was the initial temperature of the pan in °C and in °F?

25. A 500 g metal sphere is heated to 300°C, then dropped into a beaker containing 300 cm³ of mercury at 20.0°C. A short time later the mercury temperature stabilizes at 99.0°C. Identify the metal.

Section 17.7 The Specific Heats of Gases

26. A container holds 1.0 g of argon at a pressure of 8.0 atm.
    a. How much heat is required to increase the temperature by 100°C at constant volume?
    b. How much will the temperature increase if this amount of heat energy is transferred to the gas at constant pressure?

27. A container holds 1.0 g of oxygen at a pressure of 8.0 atm.
    a. How much heat is required to increase the temperature by 100°C at constant pressure?
    b. How much will the temperature increase if this amount of heat energy is transferred to the gas at constant volume?

28. The temperature of 2.0 g of helium is increased at constant volume by ΔT. What mass of oxygen can have its temperature increased by the same amount at constant volume using the same amount of heat?

29. A gas cylinder holds 0.10 mol of O_2 at 150°C and a pressure of 3.0 atm. The gas expands adiabatically until the volume is doubled. What are the final (a) pressure and (b) temperature?

30. A gas cylinder holds 0.10 mol of O_2 at 150°C and a pressure of 3.0 atm. The gas expands adiabatically until the pressure is halved. What are the final (a) volume and (b) temperature?

31. What compression ratios V_i/V_f will raise the temperature of (a) air and (b) argon from 30°C to 850°C in an adiabatic process?

Problems

32. A 1530 cm³ container holds 100°C steam at a pressure of 2.0 atm. How much heat is needed to change this water vapor into ice at −20°C?

33. A 5.0-m-diameter garden pond is 30 cm deep. Solar energy is incident on the pond at an average rate of 400 W/m². If the water absorbs all the solar energy and does not exchange energy with its surroundings, how many hours will it take to warm from 15°C to 25°C?

34. An 11 kg bowling ball at 0°C is dropped into a tub containing a mixture of ice and water. A short time later, when a new equilibrium has been established, there are 5.0 g less ice. From what height was the ball dropped? Assume no water or ice splashes out.

35. The burner on an electric stove has a power output of 2.0 kW. A 750 g stainless steel teakettle is filled with 20°C water and placed on the already hot burner. If it takes 3.0 min for the water to reach a boil, what volume of water, in cm³, was in the kettle? Stainless steel is mostly iron, so you can assume its specific heat is that of iron.

36. Two cars collide head on while each is traveling at 80 km/hr. Suppose all their kinetic energy is transformed into the thermal energy of the wrecks. What is the temperature increase of each car? You can assume that each car's specific heat is that of iron.

37. 10 g of aluminum at 200°C and 20 g of copper are dropped into 50 cm³ of ethyl alcohol at 15°C. The temperature quickly comes to 25°C. What was the initial temperature of the copper?

38. A 100 g ice cube at $-10°C$ is placed in an aluminum cup whose initial temperature is 70°C. The system comes to an equilibrium temperature of 20°C. What is the mass of the cup?

39. 512 g of an unknown metal at a temperature of 15°C is dropped into a 100 g aluminum container holding 325 g of water at 98°C. A short time later, the container of water and metal stabilizes at a new temperature of 78°C. Identify the metal.

40. An experiment finds that the specific heat at constant volume of a monatomic gas is 625 J/kg K. Identify the gas.

41. A 150 L (≈ 40 gal) electric hot-water tank has a 5.0 kW heater. How many minutes will it take to raise the water temperature from 65°F to 140°F?

42. An experiment measures the temperature of a 500 g substance while steadily supplying heat to it. Figure P17.42 shows the results of the experiment. What are the (a) specific heat of the solid phase, (b) specific heat of the liquid phase, (c) melting and boiling temperatures, and (d) heats of fusion and vaporization?

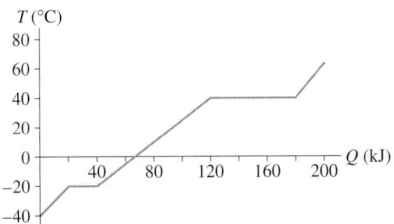

FIGURE P17.42

43. An experiment measures the temperature of a 200 g substance while steadily supplying heat to it. Figure P17.43 shows the results of the experiment. What are (a) the specific heat of the liquid phase and (b) the heat of vaporization?

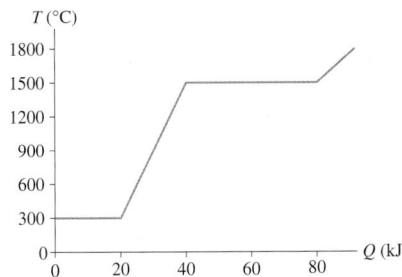

FIGURE P17.43

44. Liquid nitrogen is used in many low-temperature experiments. It is widely available, and cheaper than gasoline! How much heat must be removed from room temperature (20°C) nitrogen gas to produce 1.0 L of liquid nitrogen? The density of liquid nitrogen is 810 kg/m³.

45. Your 300 mL cup of coffee is too hot to drink when served at 90°C. What is the mass of an ice cube, taken from a $-20°C$ freezer, that will cool your coffee to a pleasant 60°C?

46. A typical nuclear reactor generates 1000 MW (1000 MJ/s) of electrical energy. In doing so, it produces 2000 MW of "waste heat" that must be removed from the reactor to keep it from melting down. Many reactors are sited next to large bodies of water so that they can use the water for cooling. Consider a reactor where the intake water is at 18°C. State regulations limit the temperature of the output water to 30°C so as not to harm aquatic organisms. How many liters of cooling water have to be pumped through the reactor each minute?

47. You find an empty cooler, the kind used to keep drinks cold, at a Fourth of July picnic. The cooler has aluminum walls surrounded by insulating material. This is a 20 L cooler that uses 2.0 kg of aluminum. Just for fun, you toss in a firecracker, slam the lid, and sit on it to keep the lid from blowing off. A minute later, when you open the lid, you see that a built-in thermometer has risen from 25°C to 28°C. How much energy was released by the firecracker when it exploded?

48. A beaker with a metal bottom is filled with 20 g of water at 20°C. It is brought into good thermal contact with a 4000 cm³ container holding 0.40 mol of a monatomic gas at 10 atm pressure. Both containers are well insulated from their surroundings. What is the gas pressure after a long time has elapsed? You can assume that the containers themselves are nearly massless and do not affect the outcome.

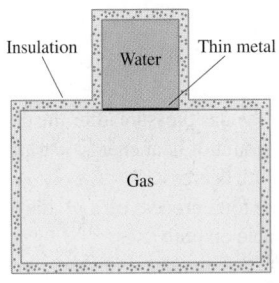

FIGURE P17.48

49. The average speed of molecules in 1.0 g of hydrogen gas is 700 m/s.
    a. What is the thermal energy of the gas?
    b. 500 J of heat are added to the gas, causing it to expand and do 300 J of work. Afterwards, what is the average molecular speed?

50. 2.0 mol of gas are at 30°C and a pressure of 1.5 atm. How much work must be done on the gas to compress it to one third of its initial volume at (a) constant temperature and (b) constant pressure? (c) Show both processes on a single pV diagram.

51. 500 J of work must be done to compress a gas to half its initial volume at constant temperature. How much work must be done to compress the gas by a factor of 10, starting from its initial volume?

52. A cylinder with a 16-cm-diameter piston contains gas at a pressure of 3.0 atm.
    a. How much force does the gas exert on the piston?
    b. How much force does the environment exert on the piston?
    c. The gas expands at constant pressure and pushes the piston out 10 cm. How much work is done by the environment?
    d. How much work is done by the gas?
    e. The thermal energy of the gas increases by 196 J in the expansion of part c. Was heat energy transferred to or from the gas in this process? How much?

53. A 10-cm-diameter cylinder contains argon gas at 10 atm pressure and a temperature of 50°C. A piston can slide in and out of the cylinder. The cylinder's initial length is 20 cm. 2500 J of heat are transferred to the gas, causing the gas to expand at constant pressure. What are (a) the final temperature and (b) the final length of the cylinder?

54. A cube 20 cm on each side contains 3.0 g of helium at 20°C. 1000 J of heat energy are transferred to this gas. What are (a) the final pressure if the process is at constant volume and (b) the final volume if the process is at constant pressure? (c) Show and label both processes on a single pV diagram.

55. 2.0 mol of an ideal monatomic gas are at a pressure of 1.0 atm and a temperature of 310 K. What is the gas's thermal energy? **Hint:** What is the thermal energy at $T = 0$ K?

56. n moles of an ideal gas at temperature T_1 and volume V_1 expand isothermally until the volume has doubled. In terms of n, T_1, and V_1, what are (a) the final temperature, (b) the work done on the gas, and (c) the heat energy transferred to the gas?

57. 5.0 g of nitrogen gas at 20°C and an initial pressure of 3.0 atm undergo an isobaric expansion until the volume has tripled.
    a. What is the gas volume and temperature after the expansion?
    b. How much heat energy is transferred to the gas to cause this expansion?
    The gas pressure is then decreased at constant volume until the original temperature is reached.
    c. What is the gas pressure after the decrease?
    d. What amount of heat energy is transferred from the gas as its pressure decreases?
    e. Show the total process on a pV diagram. Provide an appropriate scale on both axes.

58. Figure P17.58 shows two processes that take a gas from state i to state f. Show that $Q_A - Q_B = p_iV_i$.

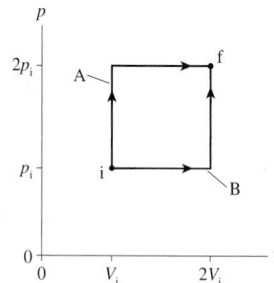

FIGURE P17.58

59. 0.10 mol of nitrogen gas follow the two processes shown in Figure P17.59. How much heat is required for each?

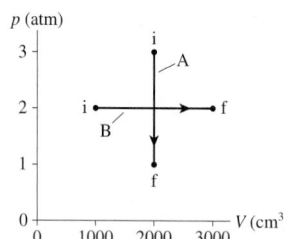

FIGURE P17.59

60. 0.10 mol of nitrogen gas follow the two processes shown in Figure P17.60. How much heat is required for each?

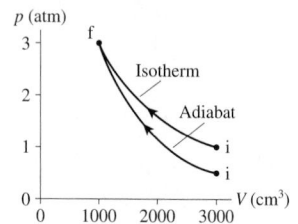

FIGURE P17.60

61. 0.10 mol of a monatomic gas follows the process shown in Figure P17.61.
    a. How much heat energy is transferred to or from the gas during process $1 \rightarrow 2$?
    b. How much heat energy is transferred to or from the gas during process $2 \rightarrow 3$?
    c. What is the total change in thermal energy of the gas?

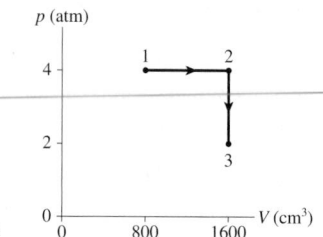

FIGURE P17.61

62. Two 800 cm³ containers hold identical amounts of a monatomic gas at 20°C. Container A is rigid. Container B has a 100 cm² piston with a mass of 10 kg that can slide up and down vertically without friction. Both containers are placed on identical heaters and heated for equal amounts of time.
    a. Will the final temperature of the gas in A be greater than, less than, or equal to the final temperature of the gas in B? Explain.
    b. Show both processes on a single pV diagram.
    c. What are the initial pressures in containers A and B?
    d. Suppose the heaters have 25 W of power and are turned on for 15 s. What is the final volume of container B?

63. Two cylinders each contain 0.10 mol of a diatomic gas at 300 K and a pressure of 3.0 atm. Cylinder A expands isothermally and cylinder B expands adiabatically until the pressure of each is 1.0 atm.
    a. What are the final temperature and volume of each?
    b. Show both processes on a single pV diagram. Use an appropriate scale on both axes.

64. A monatomic gas follows the process $1 \rightarrow 2 \rightarrow 3$ shown in Figure P17.64. How much heat is needed for (a) process $1 \rightarrow 2$ and (b) process $2 \rightarrow 3$?

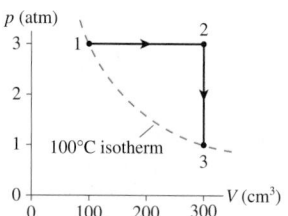

FIGURE P17.64

65. Figure P17.65 shows a thermodynamic process followed by 0.015 mol of hydrogen.
    a. How much work is done on the gas?
    b. By how much does the thermal energy of the gas change?
    c. How much heat energy is transferred to the gas?

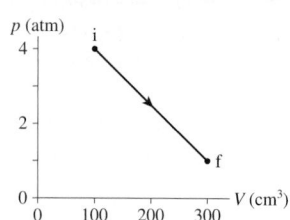

FIGURE P17.65

66. Figure P17.66 shows a thermodynamic process followed by 120 mg of helium.

    a. Determine the pressure (in atm), temperature (in °C), and volume (in cm^3) of the gas at points 1, 2, and 3. Put your results in a table for easy reading.

    b. How much work is done on the gas during each of the three segments?

    c. How much heat energy is transferred to or from the gas during each of the three segments?

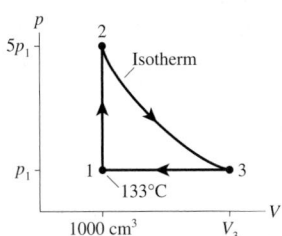

FIGURE P17.66

67. One cylinder in the diesel engine of a truck has an initial volume of 600 cm^3. Air is admitted to the cylinder at 30°C and a pressure of 1.0 atm. The piston rod then does 400 J of work to rapidly compress the air. What are its final temperature and volume?

68. The volume of a gas is halved during an adiabatic compression that increases the pressure by a factor of 2.5.

    a. What is the specific heat ratio γ?

    b. By what factor does the temperature increase?

69. a. What compression ratio V_{max}/V_{min} will raise the air temperature from 20°C to 1000°C in an adiabatic process?

    b. What pressure ratio p_{max}/p_{min} does this process have?

70. 2.0 g of helium at an initial temperature of 100°C and an initial pressure of 1.0 atm undergo an isobaric expansion until the volume has doubled. What are (a) the final temperature, (b) the work done on the gas, (c) the heat input to the gas, and (d) the change in thermal energy of the gas? (e) Show the process on a pV diagram, using proper scales on both axes.

71. 2.0 g of helium at an initial temperature of 100°C and an initial pressure of 1.0 atm undergo an isothermal expansion until the volume has doubled. What are (a) the final pressure, (b) the work done on the gas, (c) the heat input to the gas, and (d) the change in thermal energy of the gas? (e) Show the process on a pV diagram, using proper scales on both axes.

72. 14.0 g of nitrogen gas at STP are adiabatically compressed to a pressure of 20 atm. What are (a) the final temperature, (b) the work done on the gas, (c) the heat input to the gas, and (d) the compression ratio V_{max}/V_{min}? (e) Show the process on a pV diagram, using proper scales on both axes.

73. 14.0 g of nitrogen gas at STP are compressed in an isochoric process to a pressure of 20 atm. What are (a) the final temperature, (b) the work done on the gas, (c) the heat input to the gas, and (d) the pressure ratio p_{max}/p_{min}? (e) Show the process on a pV diagram, using proper scales on both axes.

74. When strong winds rapidly carry air down from mountains to a lower elevation, the air has no time to exchange heat with its surroundings. The air is compressed as the pressure rises, and its temperature can increase dramatically. These warm winds are called Chinook winds in the Rocky Mountains and Santa Ana winds in California. Suppose the air temperature high in the mountains behind Los Angeles is 0°C at an elevation where the air pressure is 60 kPa. What will the air temperature be, in °C and °F, when the Santa Ana winds have carried this air down to an elevation near sea level where the air pressure is 100 kPa?

In Problems 75 through 78 you are given the equation used to solve a problem. For each of these, you are to

    a. Write a realistic problem for which this is the correct equation.

    b. Finish the solution of the problem.

75. $50 \text{ J} = -n(8.31 \text{ J/mol K})(350 \text{ K})\ln\left(\dfrac{1}{3}\right)$

76. $(200 \times 10^{-6} \text{ m}^3)(13,600 \text{ kg/m}^3)$
    $\times (140 \text{ J/kg K})(90°C - 15°C)$
    $+ (0.50 \text{ kg})(449 \text{ J/kg K})(90°C - T_i) = 0$

77. $-200,000 \text{ J} = \dfrac{M \times 1000 \text{ g/kg}}{28 \text{ g/mol}}(29.1 \text{ J/mol K})$
    $\times (-196°C - 20°C) - M(199,000 \text{ J/kg})$

78. $(10 \text{ atm})V_2^{1.40} = (1.0 \text{ atm})V_1^{1.40}$

Challenge Problems

79. Figure CP17.79 shows a thermodynamic process followed by 120 mg of helium.

    a. Determine the pressure (in atm), temperature (in °C), and volume (in cm^3) of the gas at points 1, 2, and 3. Put your results in a table for easy reading.

    b. How much work is done on the gas during each of the three segments?

    c. How much heat is transferred to or from the gas during each of the three segments?

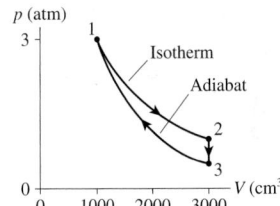

FIGURE CP17.79

80. A 6.0-cm-diameter cylinder of nitrogen gas has a 4.0-cm-thick movable copper piston. The cylinder is oriented vertically, as shown in Figure CP17.80, and the air above the piston is evacuated. When the gas temperature is 20°C, the piston floats 20 cm above the bottom of the cylinder.

    a. What is the gas pressure?

    b. How many gas molecules are in the cylinder?

    c. If the average molecular speed is 550 m/s, what is the thermal energy of the gas?

    Then 2.0 J of heat energy are transferred to the gas.

    d. What is the new equilibrium temperature of the gas?

    e. What is the final height of the piston?

    f. How much work is done on the gas as the piston rises?

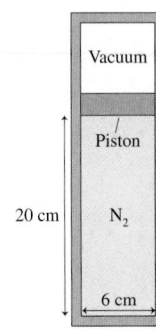

FIGURE CP17.80

81. You come into lab one day and find a well-insulated 2000 mL thermos bottle containing 500 mL of boiling liquid nitrogen. The remainder of the thermos has nitrogen gas at a pressure of 1.0 atm. The gas and liquid are in thermal equilibrium. While waiting for lab to start, you notice a piece of iron on the table with "197 g" written on it. Just for fun, you drop the iron into the thermos and seal the cap tightly so that no gas can escape. After a few seconds have passed, what is the pressure inside the thermos? The density of liquid nitrogen is 810 kg/m^3.

STOP TO THINK ANSWERS

Stop to Think 17.1: a. The piston does work W on the gas. There's no heat, because of the insulation, and $\Delta E_{mech} = 0$ because the gas as a whole doesn't move. Thus $\Delta E_{th} = W > 0$. The work increases the system's thermal energy and thus raises its temperature.

Stop to Think 17.2: d. $W_A = 0$ because A is an isochoric process. $W_B = W_{1\,to\,2} + W_{2\,to\,3}$. $|W_{2\,to\,3}| > |W_{1\,to\,2}|$ because there's more area under the curve, and $W_{2\,to\,3}$ is positive whereas $W_{1\,to\,2}$ is negative. Thus W_B is positive.

Stop to Think 17.3: b and e. The temperature rises in (d) from doing work on the gas ($\Delta E_{th} = W$), not from heat. (e) involves heat because there is a temperature difference. The temperature of the gas doesn't change because the heat is used to do the work of lifting a weight.

Stop to Think 17.4: d. The temperature increases so E_{th} must increase. W is negative in an expansion, so Q must be positive and larger than $|W|$.

Stop to Think 17.5: a. A has a smaller specific heat and thus less thermal inertia. The temperature of A will change more than the temperature of B.

Stop to Think 17.6: a. $W_A + Q_A = W_B + Q_B$. The area under process A is larger than the area under B, so W_A is *more negative* than W_B. Q_A has to be more positive than Q_B to maintain the equality.

18 The Micro/Macro Connection

Heating the air in a hot-air balloon increases the thermal energy of the air molecules. This causes the gas to expand, lowering its density and allowing the balloon to float in the cooler surrounding air.

▶ **Looking Ahead**

The goal of Chapter 18 is to understand the properties of a macroscopic system in terms of the microscopic behavior of its molecules. In this chapter you will learn to:

- Understand how molecular motions and collisions are responsible for macroscopic phenomena such as pressure and heat transfer.
- Establish a connection between temperature, thermal energy, and the average translational kinetic energy of the molecules in the system.
- Use the micro/macro connection to predict the molar specific heats of gases and solids.
- Use the second law of thermodynamics to understand how interacting systems come to thermal equilibrium.

◀ **Looking Back**

The material in this chapter depends on understanding heat, thermal energy, and the properties of ideal gases. Please review:

- Sections 16.5–16.6 Ideal gases.
- Sections 17.3–17.4 Heat and the first law of thermodynamics.
- Sections 17.5 and 17.7 Specific heats and molar specific heats.

A gas consists of a vast number of molecules ceaselessly colliding with each other and the walls of their container as they whiz about. A solid contains uncountable atoms vibrating around their equilibrium positions. Our goal in this chapter is to show how this turmoil at the microscopic level gives rise to predictable and steady values of macroscopic variables such as pressure, temperature, and specific heat.

This micro/macro connection, which goes by the more formal name **kinetic theory,** will help us elucidate several puzzles that we noted in the previous two chapters. For example, why do all elemental solids have the same molar specific heats, as do all monatomic gases and all diatomic gases? Kinetic theory will also give us a better understanding of *heat* and of how it is that two systems come to thermal equilibrium as they interact.

We'll also introduce a new law of nature, the second law of thermodynamics. The second law is rather subtle, but it has profound implications. We'll use the second law to understand why it is that heat energy "flows" from hot to cold rather than from cold to hot.

18.1 Molecular Collisions

Let's begin by thinking about gases at the atomic level. Chapter 16 presented experimental evidence that the molecules of a gas are moving with different speeds. Those data, seen again in Figure 18.1, show that the molecular speeds in N_2 at room temperature vary from ≈ 100 m/s to ≈ 1200 m/s with the most likely speed being ≈ 550 m/s. You can see that 16% of the molecules have speeds between 600 m/s and 700 m/s.

If you were to repeat the experiment a few seconds or a few hours later, you would again find that the most likely speed is ≈ 550 m/s and that 16% of the molecules have speeds between 600 m/s and 700 m/s. This is somewhat surprising when you think that each molecule in the gas undergoes millions of collisions every second (we'll derive a value later in the chapter). As Figure 18.2 shows, each collision causes one molecule to speed up and the other to slow down.

The "molecular deck of cards" is constantly being reshuffled, yet 16% of the molecules always have speeds between 600 m/s and 700 m/s. There's an important lesson here. We can't possibly keep track of what the individual molecules are doing, but *averages,* such as the average number of molecules in the speed range 600 to 700 m/s, can have precise, predictable values. **The micro/macro connection is built on the idea that macroscopic properties of a system, such as temperature or pressure, are related to the *average* behavior of the atoms and molecules.**

Mean Free Path

Imagine someone opening a bottle of strong perfume a few feet away from you. If molecular speeds are hundreds of meters per second, you might expect to smell the perfume almost instantly. But that isn't what happens. As you know, it takes many seconds for the molecules to *diffuse* across the room. Let's see why this is.

Figure 18.3 shows a "movie" of one molecule as it moves through a gas. Instead of zipping along in a straight line, as it would in a vacuum, the molecule follows a convoluted zig-zag path in which it frequently collides with other molecules. A molecule may have traveled hundreds of meters by the time it manages to get 1 or 2 m away from its starting point.

The random distribution of the molecules in the gas causes the straight-line segments between collisions to be of unequal lengths. A question we could ask is, "What is the *average* distance between collisions?" If a molecule has N_{coll} collisions as it travels distance L, the average distance between collisions, which is called the **mean free path** λ (lowercase Greek lambda), is

$$\lambda = \frac{L}{N_{coll}} \tag{18.1}$$

The concept of mean free path is used not only in gases but also to describe electrons moving through conductors and light passing through a medium that scatters the photons.

Our task is to determine the number of collisions. Figure 18.4a shows two molecules approaching each other. We will assume that the molecules are spherical and of radius *r.* We will also continue the ideal-gas assumption that the molecules undergo hard-sphere collisions, like billiard balls. In that case, the molecules will

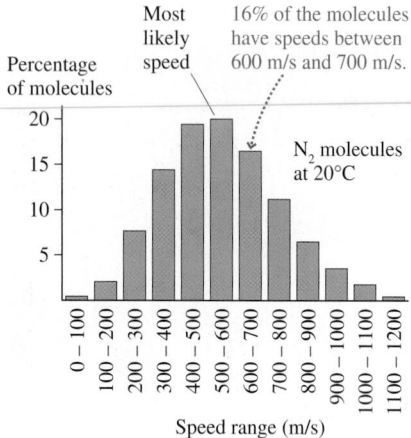

FIGURE 18.1 The distribution of molecular speeds in a sample of nitrogen gas.

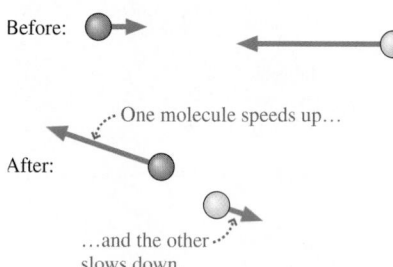

FIGURE 18.2 Frequent collisions change the speed and direction of molecules in a gas.

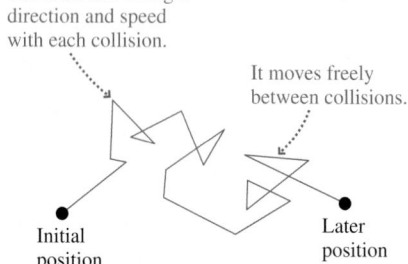

FIGURE 18.3 A single molecule follows a zig-zag path through a gas as it collides with other molecules.

(a) Two molecules will collide if
the distance between their
centers is less than $2r$.

(b) Target molecules

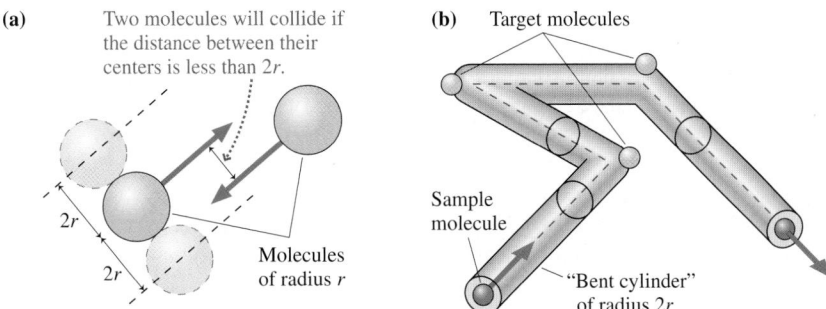

Sample
molecule

"Bent cylinder"
of radius $2r$

$2r$

$2r$

Molecules
of radius r

FIGURE 18.4 A sample molecule will collide with all target molecules whose centers are within a bent cylinder of radius $2r$ centered on its path.

collide if the distance between their *centers* is less than $2r$. They will miss if the distance is greater than $2r$.

Figure 18.4b shows a cylinder of radius $2r$ centered on the trajectory of a "sample" molecule. The sample molecule collides with any "target" molecule whose center is located within the cylinder, causing the cylinder to bend at that point. Hence the number of collisions N_{coll} is equal to the number of molecules in a cylindrical volume of length L.

The volume of a cylinder is $V_{cyl} = AL = \pi(2r)^2 L$. If the number density of the gas is N/V particles per m^3, then the number of collisions along a trajectory of length L is

$$N_{coll} = \frac{N}{V}V_{cyl} = \frac{N}{V}\pi(2r)^2 L = 4\pi\frac{N}{V}r^2 L \qquad (18.2)$$

Thus the mean free path between collisions is

$$\lambda = \frac{L}{N_{coll}} = \frac{1}{4\pi(N/V)r^2}$$

We made a tacit assumption in this derivation that the target molecules are at rest. While the general idea behind our analysis is correct, a more careful calculation in which all molecules move introduces an extra factor of $\sqrt{2}$, giving

$$\lambda = \frac{1}{4\sqrt{2}\pi(N/V)r^2} \qquad \text{(mean free path)} \qquad (18.3)$$

Measurements are necessary to determine precise values of atomic and molecular radii, but a reasonable rule of thumb is to assume that atoms in a monatomic gas have $r \approx 0.5 \times 10^{-10}$ m and diatomic molecules have $r \approx 1.0 \times 10^{-10}$ m.

EXAMPLE 18.1 **The mean free path at STP**
What is the mean free path of a nitrogen molecule at 1.0 atm pressure and room temperature (20°C)?

SOLVE Nitrogen is a diatomic molecule, so $r \approx 1.0 \times 10^{-10}$ m. We can use the ideal-gas law in the form $pV = Nk_BT$ to determine the number density:

$$\frac{N}{V} = \frac{p}{k_BT} = \frac{101,300 \text{ Pa}}{(1.38 \times 10^{-23} \text{ J/K})(293 \text{ K})} = 2.51 \times 10^{25} \text{ m}^{-3}$$

Thus the mean free path is

$$\lambda = \frac{1}{4\sqrt{2}\pi(N/V)r^2}$$

$$= \frac{1}{4\sqrt{2}\pi(2.51 \times 10^{25} \text{ m}^{-3})(1.0 \times 10^{-10} \text{ m})^2}$$

$$= 2.25 \times 10^{-7} \text{ m} = 225 \text{ nm}$$

ASSESS You learned in Example 16.5 that the average separation between gas molecules at STP is ≈5.7 nm. It seems that any given molecule can slip between its neighbors, which are spread out in three dimensions, and travel—on average— about 40 times the average spacing before it collides with another molecule.

STOP TO THINK 18.1 The table shows the properties of four gases, each having the same number of molecules. Rank in order, from largest to smallest, the mean free paths λ_A to λ_D of molecules in these gases.

Gas	A	B	C	D
Volume	V	$2V$	V	V
Atomic mass	m	m	$2m$	m
Atomic radius	r	r	r	$2r$

18.2 Pressure in a Gas

Why does a gas have pressure? In Chapter 15, where pressure was introduced, we suggested that the pressure in a gas is due to collisions of the molecules with walls of its container. The force due to one such collision may be unmeasurably tiny, but the steady rain of a vast number of molecules striking a wall each second exerts a measurable macroscopic force. The gas pressure is the force per unit area ($p = F/A$) resulting from these molecular collisions.

Our task in this section is to calculate the pressure by doing the appropriate averaging over molecular motions and collisions. This task can be divided into three main pieces:

1. Calculate the force a single molecule exerts on the wall during a collision.
2. Add the force due to all collisions.
3. Introduce an appropriate average speed.

Force Due to a Single Collision

Figure 18.5 shows a molecule with an x-component of velocity v_x approaching a wall. We will assume that the collision with the wall is perfectly elastic, an assumption we will justify later, in which case the molecule rebounds from the wall with its x-component of velocity changed from $+v_x$ to $-v_x$. This molecule experiences an impulse. We can use the impulse-momentum theorem from Chapter 9 to write

$$(J_x)_{\text{molecule}} = \Delta p = m(-v_x) - mv_x = -2mv_x \tag{18.4}$$

According to Newton's third law, the wall experiences the equal but opposite impulse

$$(J_x)_{\text{wall}} = +2mv_x \tag{18.5}$$

as a result of this single collision.

Impulse was defined in Chapter 9 as

$$J_x = \int_0^{\Delta t_{\text{coll}}} F_x(t)\, dt \tag{18.6}$$

where $F_x(t)$ is the time-varying x-component of the force the molecule exerts on the wall during the collision and Δt_{coll} is the duration of the collision. A typical collision force is shown in Figure 18.6a.

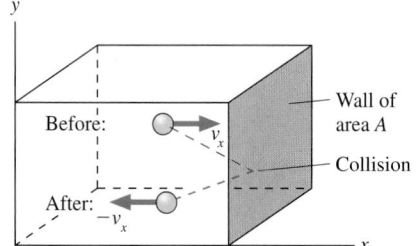

FIGURE 18.5 A molecule colliding with the wall exerts an impulse J_x on it.

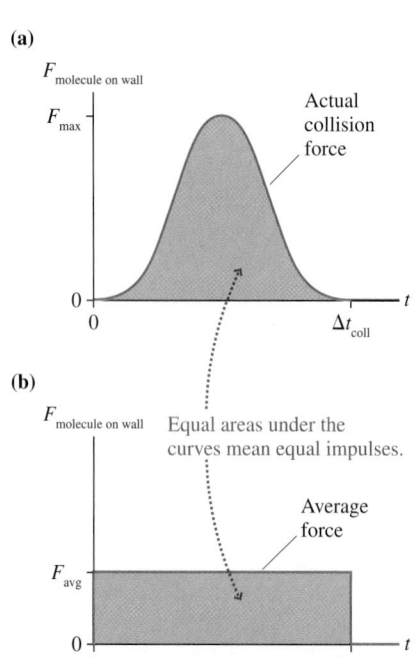

FIGURE 18.6 The force that a molecule exerts on the wall during a collision.

Suppose we replace the actual collision force with an *average* force F_{avg} defined, as Figure 18.6b shows, such that the area under the curve, or the impulse, is the same as for the true force. Then

$$(J_x)_{wall} = \int_0^{\Delta t_{coll}} F_{avg}\, dt = F_{avg}\Delta t_{coll} = 2mv_x$$

Thus the average force on the wall during *one* collision is

$$F_{avg} = \frac{2mv_x}{\Delta t_{coll}} \tag{18.7}$$

EXAMPLE 18.2 The average force in a collision

Estimate the average force a nitrogen molecule exerts on the wall of its container during a collision.

SOLVE We saw in the previous section that a typical molecular speed is ≈ 500 m/s. Although we don't know the details, the duration of a collision must be roughly the time it takes a molecule to move a distance equal to its radius. We noted that the radius of a typical diatomic molecule is 1.0×10^{-10} m, so a rough estimate for the duration of a collision is

$$\Delta t_{coll} \approx \frac{1.0 \times 10^{-10}\ \text{m}}{500\ \text{m/s}} \approx 2 \times 10^{-13}\ \text{s}$$

A nitrogen molecule has mass $m = 28$ u $= 4.65 \times 10^{-26}$ kg. Thus an estimate of the average force during a collision is

$$F_{avg} = \frac{2mv_x}{\Delta t_{coll}} \approx \frac{2(4.65 \times 10^{-26}\ \text{kg})(500\ \text{m/s})}{2 \times 10^{-13}\ \text{s}} \approx 2 \times 10^{-10}\ \text{N}$$

ASSESS The duration of a collision is extremely short, and the force exerted by one molecule is very tiny. But a very large number of molecules are all colliding essentially simultaneously.

Adding the Forces

The net force exerted on the wall by *all* the molecules is the average force per molecule, given by Equation 18.7, multiplied by the total number N_{coll} of such collisions that occur during the time interval Δt_{coll}. That is,

$$F_{net} = N_{coll}F_{avg} = 2mv_x\frac{N_{coll}}{\Delta t_{coll}} \tag{18.8}$$

The ratio $N_{coll}/\Delta t_{coll}$ is the *rate* at which molecules collide with the wall.

To find N_{coll}, the number of molecules that hit the wall during the time interval Δt_{coll}, assume for the moment that all molecules have the *same* x-component velocity v_x. These molecules all travel distance $\Delta x = v_x\Delta t_{coll}$ along the x-axis during interval Δt_{coll}. Figure 18.7 has shaded the gas that is within distance Δx of the wall. *Every one* of the molecules in this shaded region that is moving to the right will reach and collide with the wall during time Δt_{coll}. Molecules outside this region will not reach the wall during Δt_{coll} because they are not moving fast enough to travel distance Δx.

The shaded region has volume $A\Delta x$, where A is the surface area of the wall. Only half the molecules are moving to the right, hence the number of collisions during Δt_{coll} is

$$N_{coll} = \frac{1}{2}\frac{N}{V}A\Delta x = \frac{1}{2}\frac{N}{V}Av_x\Delta t_{coll} \tag{18.9}$$

Thus the *rate* of molecular collisions with a wall of area A is

$$\text{rate of collisions} = \frac{N_{coll}}{\Delta t_{coll}} = \frac{1}{2}\frac{N}{V}Av_x \tag{18.10}$$

Only molecules moving to the right in the shaded region will hit the wall during Δt_{coll}.

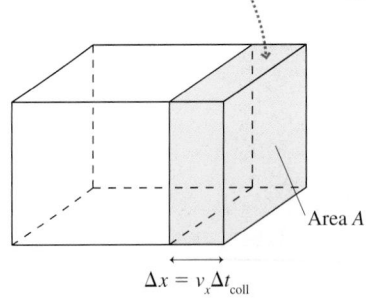

Area A

$\Delta x = v_x\Delta t_{coll}$

FIGURE 18.7 Determining the number of collisions.

The net force on the wall is found by substituting $N_{coll}/\Delta t_{coll}$ from Equation 18.10 into Equation 18.8:

$$F_{net} = (2mv_x)\left(\frac{1}{2}\frac{N}{V}Av_x\right) = \frac{N}{V}mv_x^2 A \qquad (18.11)$$

Notice that this expression for F_{net} does not depend on any details of the molecular collision.

Equation 18.11 is based on an assumption that all molecules have the same speed. This is a condition we need to relax if our result is to apply to real gases. We can correct for this by replacing the squared velocity v_x^2 in Equation 18.11 with its average value. That is,

$$F_{net} = \frac{N}{V}m(v_x^2)_{avg}A \qquad (18.12)$$

where $(v_x^2)_{avg}$ is the quantity v_x^2 averaged over all molecules in the container.

The Root-Mean-Square Speed

We need to be somewhat careful when averaging velocities. The velocity component v_x has a sign. At any instant of time, half the molecules in a container move to the right and have positive v_x while the other half move to the left and have negative v_x. Thus the *average velocity* is $(v_x)_{avg} = 0$. If this weren't true, the entire container of gas would move away!

The speed of a molecule is $v = (v_x^2 + v_y^2 + v_z^2)^{1/2}$. Thus the average of the speed squared is

$$(v^2)_{avg} = (v_x^2 + v_y^2 + v_z^2)_{avg} = (v_x^2)_{avg} + (v_y^2)_{avg} + (v_z^2)_{avg} \qquad (18.13)$$

The square root of $(v^2)_{avg}$ is called the **root-mean-square speed** v_{rms}:

$$v_{rms} = \sqrt{(v^2)_{avg}} \qquad \text{(root-mean-square speed)} \qquad (18.14)$$

This is usually called the *rms speed.* You can remember its definition by noting that its name is the *opposite* of the sequence of operations: first you square all the speeds, then you average the squares (find the mean), then you take the square root. Because the square root "undoes" the square, v_{rms} must, in some sense, give an average speed.

> **NOTE ▶** We could compute a true average speed v_{avg}, but that calculation turns out to be difficult. More important, the root-mean-square speed tends to arise naturally in many scientific and engineering calculations. It turns out that v_{rms} differs from v_{avg} by less than 10%, so for practical purposes we can interpret v_{rms} as being essentially the average speed of a molecule in a gas. ◀

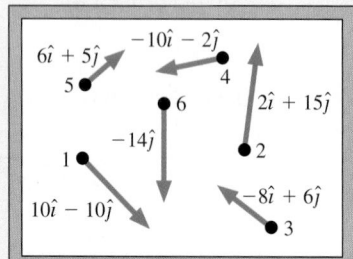

FIGURE 18.8 The molecular velocities of Example 18.3. Units are m/s.

EXAMPLE 18.3 Calculating the root-mean-square speed
Figure 18.8 shows the velocities of all the molecules in a six-molecule two-dimensional gas. Calculate and compare the average velocity $\vec{v}_{avg}$, the average speed v_{avg}, and the rms speed v_{rms}.

SOLVE Table 18.1 shows the velocity components v_x and v_y for each molecule, the squares v_x^2 and v_y^2, their sum $v^2 = v_x^2 + v_y^2$, and the speed $v = (v_x^2 + v_y^2)^{1/2}$. Averages of all the values in each column are shown at the bottom. You can see that the average velocity is $\vec{v}_{avg} = \vec{0}$ m/s and the average speed is $v_{avg} = 11.9$ m/s. The rms speed is

$$v_{rms} = \sqrt{(v^2)_{avg}} = \sqrt{148.3 \text{ m}^2/\text{s}^2} = 12.2 \text{ m/s}$$

ASSESS The rms speed is only 2.5% greater than the average speed.

TABLE 18.1 Calculation of rms speed and average speed for the molecules of Example 18.3

Molecule	v_x	v_y	v_x^2	v_y^2	v^2	v
1	10	−10	100	100	200	14.1
2	2	15	4	225	229	15.1
3	−8	6	64	36	100	10.0
4	−10	−2	100	4	104	10.2
5	6	5	36	25	61	7.8
6	0	−14	0	196	196	14.0
Average	0	0			148.3	11.9

There's nothing special about the x-axis. The coordinate system is something that *we* impose on the problem, so *on average* it must be the case that

$$(v_x^2)_{avg} = (v_y^2)_{avg} = (v_z^2)_{avg} \qquad (18.15)$$

Hence we can use Equation 18.13 and the definition of v_{rms} to write

$$v_{rms}^2 = (v_x^2)_{avg} + (v_y^2)_{avg} + (v_z^2)_{avg} = 3(v_x^2)_{avg}$$

Consequently, $(v_x^2)_{avg}$ is

$$(v_x^2)_{avg} = \frac{1}{3}v_{rms}^2 \qquad (18.16)$$

Using this result in Equation 18.12 gives us the net force on the wall of the container:

$$F_{net} = \frac{1}{3}\frac{N}{V}mv_{rms}^2 A \qquad (18.17)$$

Thus the pressure on the wall of the container due to all the molecular collisions is

$$p = \frac{F}{A} = \frac{1}{3}\frac{N}{V}mv_{rms}^2 \qquad (18.18)$$

We have met our goal. Equation 18.18 expresses the macroscopic pressure in terms of the microscopic physics. The pressure depends on the number density of molecules in the container and on how fast, on average, the molecules are moving.

EXAMPLE 18.4 The rms speed of helium atoms
A container holds helium at a pressure of 200 kPa and a temperature of 60°C. What is the rms speed of the helium atoms?

SOLVE The rms speed can be found from the pressure and the number density. Using the ideal-gas law gives us the number density:

$$\frac{N}{V} = \frac{p}{k_B T} = \frac{200{,}000 \text{ Pa}}{(1.38 \times 10^{-23} \text{ J/K})(333)} = 4.35 \times 10^{25} \text{ m}^{-3}$$

The mass of a helium atom is $m = 4$ u $= 6.64 \times 10^{-27}$ kg. Thus

$$v_{rms} = \sqrt{\frac{3p}{(N/V)m}} = 1440 \text{ m/s}$$

STOP TO THINK 18.2 The speed of every molecule in a gas is suddenly increased by a factor of 4. As a result, v_{rms} increases by a factor of

a. 2.

b. <4, but not necessarily 2.

c. 4.

d. >4 but not necessarily 16.

e. 16.

f. v_{rms} doesn't change.

18.3 Temperature

A molecule of mass m and velocity v has translational kinetic energy

$$\epsilon = \frac{1}{2}mv^2 \tag{18.19}$$

We'll use ϵ (lowercase Greek epsilon) to distinguish the energy of a molecule from the system energy E. Thus the average translational kinetic energy is

$$\epsilon_{avg} = \text{average translational kinetic energy per molecule}$$
$$= \frac{1}{2}m(v^2)_{avg} = \frac{1}{2}mv_{rms}^2 \tag{18.20}$$

We've included the word *translational* to distinguish ϵ from rotational kinetic energy, which we will consider later in this chapter.

We can write the gas pressure, Equation 18.18, in terms of the average translational kinetic energy as

$$p = \frac{2}{3}\frac{N}{V}\left(\frac{1}{2}mv_{rms}^2\right) = \frac{2}{3}\frac{N}{V}\epsilon_{avg} \tag{18.21}$$

The pressure is directly proportional to the average molecular translational kinetic energy. This makes sense. More-energetic molecules will hit the walls harder as they bounce and thus exert more force on the walls.

It's instructive to write Equation 18.21 as

$$pV = \frac{2}{3}N\epsilon_{avg} \tag{18.22}$$

We know, from the ideal-gas law, that

$$pV = Nk_BT \tag{18.23}$$

Comparing these two equations, we reach the significant conclusion that the average translational kinetic energy per molecule is

$$\epsilon_{avg} = \frac{3}{2}k_BT \qquad \text{(average translational kinetic energy)} \tag{18.24}$$

where the temperature T is in kelvins. For example, the average translational kinetic energy of a molecule at room temperature (20°C) is

$$\epsilon_{avg} = \frac{3}{2}(1.38 \times 10^{-23} \text{ J/K})(293 \text{ K}) = 6.1 \times 10^{-21} \text{ J}$$

NOTE ▶ The average translational kinetic energy per molecule depends *only* on the temperature, not on the molecule's mass. If two gases have the same temperature, their molecules have the same average translational kinetic energy. This will be an important idea when we look at the thermal interaction between two systems. ◀

Equation 18.24 is especially satisfying because it finally gives real meaning to the concept of temperature. Writing it as

$$T = \frac{2}{3k_B}\epsilon_{\text{avg}} \qquad (18.25)$$

we can see that, for a gas, **this thing we call *temperature* is a direct measure of the average translational kinetic energy.** A higher temperature corresponds to a larger value of ϵ_{avg} and thus to higher molecular speeds. This concept of temperature also gives meaning to *absolute zero* as the temperature at which $\epsilon_{\text{avg}} = 0$ and all molecular motion ceases. (Quantum effects at very low temperatures prevent the motions from actually stopping, but our classical theory predicts that they would.) Figure 18.9 summarizes what we've learned thus far about the micro/macro connection.

We can now justify our assumption that molecular collisions are perfectly elastic. Suppose they were not. That is, suppose that a small amount of kinetic energy was lost in each collision. If that were so, the average translational kinetic energy ϵ_{avg} of the gas would slowly decrease and we would see a steadily decreasing temperature. But that doesn't happen. The temperature of an isolated system remains perfectly constant, indicating that ϵ_{avg} is not changing with time. Consequently, the collisions must be perfectly elastic.

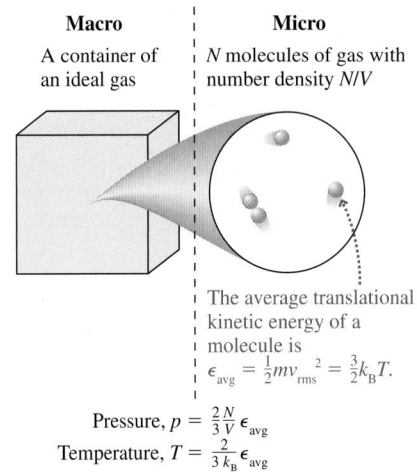

Macro	Micro
A container of an ideal gas	N molecules of gas with number density N/V

The average translational kinetic energy of a molecule is $\epsilon_{\text{avg}} = \frac{1}{2}mv_{\text{rms}}^2 = \frac{3}{2}k_B T$.

Pressure, $p = \frac{2}{3}\frac{N}{V}\epsilon_{\text{avg}}$

Temperature, $T = \frac{2}{3k_B}\epsilon_{\text{avg}}$

FIGURE 18.9 The micro/macro connection for pressure and temperature.

EXAMPLE 18.5 Total microscopic kinetic energy
What is the total translational kinetic energy of the molecules in 1.0 mol of gas at STP?

SOLVE The average translational kinetic energy per molecule is

$$\epsilon_{\text{avg}} = \frac{3}{2}k_B T = \frac{3}{2}(1.38 \times 10^{-23}\,\text{J/K})(273\,\text{K})$$

$$= 5.65 \times 10^{-21}\,\text{J}$$

1.0 mol of gas contains N_A molecules, hence the total kinetic energy is

$$K_{\text{micro}} = N_A \epsilon_{\text{avg}} = 3400\,\text{J}$$

ASSESS The energy of any one molecule is incredibly small. Nonetheless, a macroscopic system has substantial thermal energy because it consists of an incredibly large number of molecules.

By definition, $\epsilon_{\text{avg}} = \frac{1}{2}mv_{\text{rms}}^2$. Using the ideal-gas law, we found $\epsilon_{\text{avg}} = \frac{3}{2}k_B T$. By equating these two expressions we find that the rms speed of a molecule is

Activ
Physics   8.1–8.3

$$v_{\text{rms}} = \sqrt{\frac{3k_B T}{m}} \qquad (18.26)$$

The rms speed of a molecule depends on the square root of the temperature and inversely on the square root of the molecular mass.

EXAMPLE 18.6 Calculating an rms speed
What is the rms speed of a nitrogen molecule at room temperature?

SOLVE The molecular mass is $m = 28\,\text{u} = 4.68 \times 10^{-26}\,\text{kg}$ and $T = 20°\text{C} = 293\,\text{K}$. It is then a simple calculation to find

$$v_{\text{rms}} = \sqrt{\frac{3(1.38 \times 10^{-23}\,\text{J/K})(293\,\text{K})}{4.68 \times 10^{-26}\,\text{kg}}} = 509\,\text{m/s}$$

Some speeds will be greater than this and others smaller, but 509 m/s will be a typical or fairly average speed. This is in excellent agreement with the experimental results of Figure 18.1.

EXAMPLE 18.7 Laser cooling

It is possible to "cool" atoms by letting them interact with a laser beam under proper, carefully controlled conditions. Laser cooling is currently a subject of intense research activity, and it is now possible to cool a dilute gas of atoms to a temperature less than one *micro*kelvin! (The atoms are kept from solidifying by their extremely low density.) Various novel quantum effects appear at these incredibly low temperatures. What is the rms speed for a cesium atom at a temperature of 1.0 μK?

SOLVE Reference to the periodic table of the elements shows that the atomic mass of cesium is $m = 133$ u $= 2.22 \times 10^{-25}$ kg. At $T = 1\,\mu\text{K} = 1 \times 10^{-6}$ K the rms speed is

$$v_{\text{rms}} = \sqrt{\frac{3(1.38 \times 10^{-23}\ \text{J/K})(1.0 \times 10^{-6}\ \text{K})}{2.22 \times 10^{-25}\ \text{kg}}}$$

$$= 0.014\ \text{m/s} = 1.4\ \text{cm/s}$$

This is slow enough to enable us to "watch" the atoms moving about!

EXAMPLE 18.8 Mean time between collisions

Because v_{rms} is essentially the average molecular speed, the *mean time between collisions* is simply the time needed to travel distance λ, the mean free path, at speed v_{rms}. What is the mean time between collisions for a nitrogen molecule at 1.0 atm pressure and room temperature?

SOLVE We found $\lambda = 2.25 \times 10^{-7}$ m in Example 18.1 and $v_{\text{rms}} = 509$ m/s in Example 18.6. Thus the mean time between collisions is

$$\tau_{\text{coll}} = \frac{\lambda}{v_{\text{rms}}} = \frac{2.25 \times 10^{-7}\ \text{m}}{509\ \text{m/s}} = 4.42 \times 10^{-10}\ \text{s}$$

ASSESS The air molecules around us move very fast, they collide with their neighbors about two billion times every second, and they manage to move, on average, only about 225 nm between collisions.

STOP TO THINK 18.3 Which system (or systems) has the largest average translational kinetic energy per molecule?

  a. 1 mol of He at $p = 1$ atm, $T = 300$ K
  b. 2 mol of He at $p = 2$ atm, $T = 300$ K
  c. 1 mol of N_2 at $p = 0.5$ atm, $T = 600$ K
  d. 2 mol of N_2 at $p = 0.5$ atm, $T = 450$ K
  e. 1 mol of Ar at $p = 0.5$ atm, $T = 450$ K
  f. 2 mol of Ar at $p = 2$ atm, $T = 300$ K

18.4 Thermal Energy and Specific Heat

We defined the thermal energy of a system to be $E_{\text{th}} = K_{\text{micro}} + U_{\text{micro}}$, where K_{micro} is the microscopic kinetic energy of the moving molecules and U_{micro} is the potential energy of the stretched and compressed molecular bonds. We're now ready to take a microscopic look at thermal energy. In doing so, we'll be able to resolve the puzzle of the molar specific heats.

Monatomic Gases

Figure 18.10 shows a monatomic gas such as helium or neon. The atoms in an ideal gas have no molecular bonds with their neighbors, hence $U_{\text{micro}} = 0$. Furthermore, the kinetic energy of a monatomic gas particle is entirely translational kinetic energy ϵ. Thus the thermal energy of a monatomic gas of N atoms is

$$E_{\text{th}} = K_{\text{micro}} = \epsilon_1 + \epsilon_2 + \epsilon_3 + \cdots + \epsilon_N = N\epsilon_{\text{avg}} \tag{18.27}$$

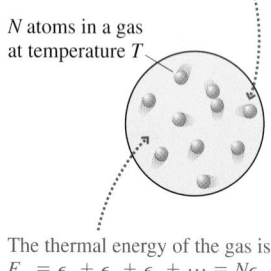

Atom i has translational kinetic energy ϵ_i but no potential energy or rotational kinetic energy.

N atoms in a gas at temperature T

The thermal energy of the gas is
$E_{\text{th}} = \epsilon_1 + \epsilon_2 + \epsilon_3 + \cdots = N\epsilon_{\text{avg}}$.

FIGURE 18.10 The atoms in a monatomic gas have only translational kinetic energy.

where ϵ_i is the translational kinetic energy of atom i. We found that $\epsilon_{avg} = \frac{3}{2}k_B T$, hence the thermal energy is

$$E_{th} = \frac{3}{2}Nk_B T = \frac{3}{2}nRT \qquad \text{(thermal energy of a monotomic gas)} \qquad (18.28)$$

where we used $N = nN_A$ and the definition of Boltzmann's constant, $k_B = R/N_A$.

We've noted for the last two chapters that thermal energy is associated with temperature. Now we have an explicit result for a monatomic gas: E_{th} is directly proportional to the temperature. Notice that E_{th} is independent of the atomic mass. Any two monatomic gases will have the same thermal energy if they have the same temperature and the same number of atoms (or moles).

If the temperature of a monatomic gas changes by ΔT, its thermal energy changes by

$$\Delta E_{th} = \frac{3}{2}nR\Delta T \qquad (18.29)$$

In Chapter 17 we found that the change in thermal energy for *any* ideal gas process is related to the molar specific heat at constant volume by

$$\Delta E_{th} = nC_V\Delta T \qquad (18.30)$$

Equation 18.29 is a microscopic result that we obtained by relating the temperature to the average translational kinetic energy of the atoms. Equation 18.30 is a macroscopic result that we arrived at from the first law of thermodynamics. We can make a micro/macro connection by combining these two equations. Doing so gives us a *prediction* for the molar specific heat:

$$C_V = \frac{3}{2}R = 12.5 \text{ J/mol K} \qquad \text{(monatomic gas)} \qquad (18.31)$$

This was exactly the value of C_V for all three monatomic gases in Table 17.4. Not only has the micro/macro connection shown that C_V is the same for all monatomic gases—the puzzle we had noted in Chapter 17—it has also made a specific prediction for the *value* of C_V. The perfect agreement of theory and experiment is strong evidence that gases really do consist of moving, colliding molecules.

The Equipartition Theorem

The particles of a monatomic gas are atoms. Their energy consists exclusively of their translational kinetic energy. A particle's translational kinetic energy can be written

$$\epsilon = \frac{1}{2}mv^2 = \frac{1}{2}mv_x^2 + \frac{1}{2}mv_y^2 + \frac{1}{2}mv_z^2 = \epsilon_x + \epsilon_y + \epsilon_z \qquad (18.32)$$

where we have written separately the energy associated with translational motion along the three axes. Because each axis in space is independent, we can think of ϵ_x, ϵ_y, and ϵ_z as independent *modes* of storing energy within the system.

Other systems have additional modes of energy storage. For example,

■ Two atoms joined by a spring-like molecular bond can vibrate back and forth. Both kinetic and potential energy are associated with this vibration.
■ A diatomic molecule, in additional to translational kinetic energy, has rotational kinetic energy if it rotates end-over-end like a dumbbell.

It will be useful to define the number of **degrees of freedom** as the number of distinct and independent modes of energy storage. A monatomic gas has three degrees of freedom, the three modes of translational kinetic energy. Systems that can vibrate or rotate have more degrees of freedom.

An important result of statistical physics says that the energy in a system is distributed so that all modes of energy storage have equal amounts of energy. This conclusion is known as the *equipartition theorem,* meaning that the energy is equally divided. The proof is beyond what we can do in this textbook, so we will state the theorem without proof:

> **Equipartition theorem** The thermal energy of a system of particles is equally divided among all the possible energy modes. For a system of N particles at temperature T, the energy stored in each mode (each degree of freedom) is $\frac{1}{2}Nk_BT$ or, in terms of moles, $\frac{1}{2}nRT$.

A monatomic gas has three degrees of freedom and thus, as we found above, $E_{th} = \frac{3}{2}Nk_BT$.

Solids

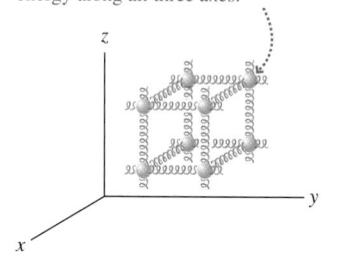

Each atom has microscopic translational kinetic energy *and* microscopic potential energy along all three axes.

FIGURE 18.11 A simple model of a solid.

Figure 18.11 reminds you of our "bedspring model" of a solid with particle-like atoms connected by a lattice of spring-like molecular bonds. How many degrees of freedom does a solid have? The kinetic energy of an atom as it vibrates around its equilibrium position is given by Equation 18.32. Three degrees of freedom are associated with the kinetic energy, just as in a monatomic gas. In addition, the molecular bonds can be compressed or stretched independently along the x-, y-, and z-axes. Three additional degrees of freedom are associated with these three modes of potential energy. Altogether, a solid has six degrees of freedom.

The energy stored in each of these six degrees of freedom is $\frac{1}{2}Nk_BT$. The thermal energy of a solid is the total energy stored in all six modes, or

$$E_{th} = 3Nk_BT = 3nRT \qquad \text{(thermal energy of a solid)} \qquad (18.33)$$

We can use this result to predict the molar specific heat of a solid. If the temperature changes by ΔT, then the thermal energy changes by

$$\Delta E_{th} = 3nR\Delta T \qquad (18.34)$$

In Chapter 17 we defined the molar specific heat of a solid such that

$$\Delta E_{th} = nC\Delta T \qquad (18.35)$$

By comparing Equations 18.34 and 18.35 we can predict that the molar specific heat of a solid is

$$C = 3R = 25.0 \text{ J/mol K} \qquad \text{(solid)} \qquad (18.36)$$

Not bad. The five elemental solids in Table 17.2 had molar specific heats clustered right around 25 J/mol K. They ranged from 24.3 J/mol K for aluminum to 26.5 J/mol K for lead. There are two reasons why the agreement between theory and experiment isn't quite as perfect as it was for monatomic gases. First, our simple bedspring model of a solid isn't quite as accurate as our model of a monatomic gas. Second, quantum effects are beginning to make their appearance. More on this shortly. Nonetheless, our ability to predict C to within a few percent from a simple model of a solid is further evidence for the atomic structure of matter.

Diatomic Molecules

Diatomic molecules are a bigger challenge. How many degrees of freedom does a diatomic molecule have? Figure 18.12 shows a diatomic molecule, such as molecular nitrogen N_2, oriented along the z-axis. Three degrees of freedom are associated with the molecule's translational kinetic energy. The molecule can have a dumbbell-like end-over-end rotation about either the x-axis or the y-axis. It can

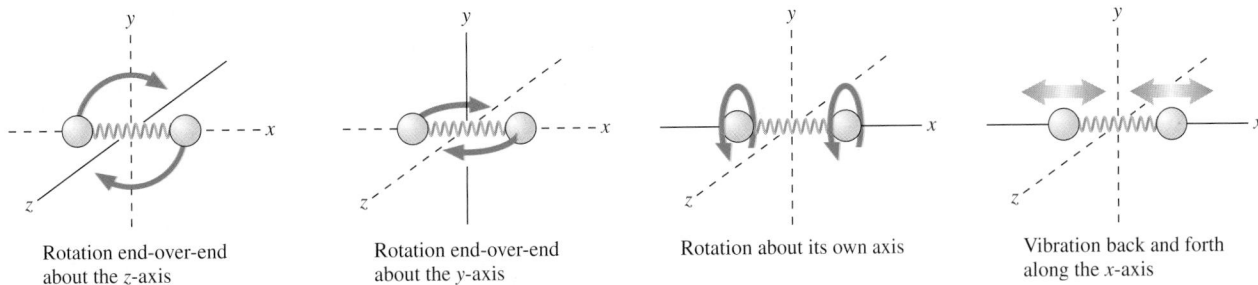

Rotation end-over-end about the z-axis

Rotation end-over-end about the y-axis

Rotation about its own axis

Vibration back and forth along the x-axis

FIGURE 18.12 A diatomic molecule can rotate or vibrate.

also rotate about its own axis. These are three rotational degrees of freedom. The two atoms can also vibrate back and forth, stretching and compressing the molecular bond. This vibrational motion has both kinetic and potential energy, thus two more degrees of freedom.

Altogether, then, a diatomic molecule has eight degrees of freedom, and we would expect the thermal energy of a gas of diatomic molecules to be $E_{th} = 4k_B T$. The analysis we followed for a monatomic gas would then lead to the prediction $C_V = 4R = 33.2$ J/mol K. As compelling as this reasoning seems to be, this is *not* the experimental value of C_V that was reported for diatomic gases in Table 17.4. Instead, we found $C_V = 20.8$ J/mol K.

Why should a theory that works so well for monatomic gases and solids fail so miserably for diatomic molecules? To see what's going on, notice that 20.8 J/mol K = $\frac{5}{2}R$. A monatomic gas, with three degrees of freedom, has $C_V = \frac{3}{2}R$. A solid, with six degrees of freedom, has $C = 3R$. A diatomic gas would have $C_V = \frac{5}{2}R$ if it had five degrees of freedom, not eight.

This discrepancy was a major conundrum as statistical physics developed in the late 19th century. Although it was not recognized as such at the time, we are here seeing our first evidence for the breakdown of classical Newtonian physics. Classically, a diatomic molecule has eight degrees of freedom. The equipartition theorem doesn't distinguish between them; all eight should have the same energy. But atoms and molecules are not classical particles. It took the development of quantum theory in the 1920s to accurately characterize the behavior of atoms and molecules. We don't yet have the tools needed to see why, but quantum effects prevent three of the modes—the two vibrational modes and the rotation of the molecule about its own axis—from being active at room temperature.

Figure 18.13 shows C_V as a function of temperature for hydrogen gas. C_V is right at $\frac{5}{2}R$ for temperatures from ≈ 200 K up to ≈ 800 K. But at very low temperatures C_V drops to the monatomic-gas value $\frac{3}{2}R$. The two rotational modes

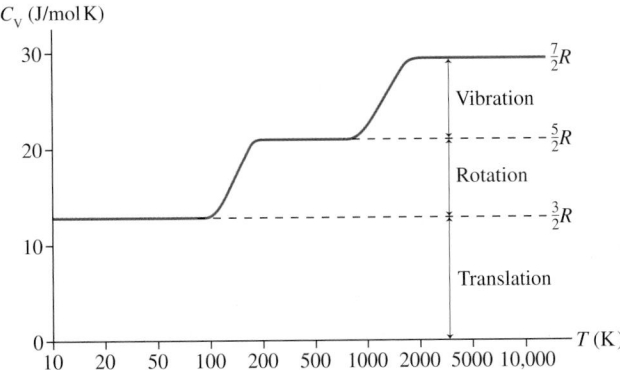

C_V (J/mol K)

Vibration — $\frac{7}{2}R$

Rotation — $\frac{5}{2}R$

Translation — $\frac{3}{2}R$

FIGURE 18.13 Hydrogen molar specific heat at constant volume as a function of temperature. The temperature scale is logarithmic.

become "frozen out" and the nonrotating molecule has only translational kinetic energy. Quantum physics can explain this, but not Newtonian physics. You can also see that the two vibrational modes *do* become active at very high temperatures, where C_V rises to $\frac{7}{2}R$. Thus the real answer to "What's wrong?" is that Newtonian physics is not the right physics for describing atoms and molecules. We are somewhat fortunate that Newtonian physics is adequate to understand monatomic gases and solids, at least at room temperature.

Accepting the quantum result that a diatomic gas has only five degrees of freedom at commonly used temperatures (the translational degrees of freedom and the two end-over-end rotations), we find

$$E_{th} = \frac{5}{2}Nk_BT = \frac{5}{2}nRT$$
$$C_V = \frac{5}{2}R = 20.8 \text{ J/mol K}$$
$$\text{(diatomic gases)} \qquad (18.37)$$

A diatomic gas has more thermal energy than a monatomic gas at the same temperature because the molecules have rotational as well as translational kinetic energy.

While the micro/macro connection firmly establishes the atomic structure of matter, it also heralds the need for a new theory of matter at the atomic level. That is a task we will take up in Part VII. For now, Table 18.2 summarizes what we have learned from kinetic theory about thermal energy and molar specific heats.

TABLE 18.2 Kinetic theory predictions for the thermal energy and the molar specific heat

System	Degrees of freedom	E_{th}	C_V
Monatomic gas	3	$\frac{3}{2}Nk_BT = \frac{3}{2}nRT$	$\frac{3}{2}R = 12.5 \text{ J/mol K}$
Diatomic gas	5	$\frac{5}{2}Nk_BT = \frac{5}{2}nRT$	$\frac{5}{2}R = 20.8 \text{ J/mol K}$
Elemental solid	6	$3Nk_BT = 3nRT$	$3R = 25.0 \text{ J/mol K}$

EXAMPLE 18.9 The rotational frequency of a molecule
The nitrogen molecule N_2 has a bond length of 0.12 nm. Estimate the rotational frequency of N_2 at 20°C.

MODEL Each mode of energy storage has the same energy.

SOLVE The molecule rotates end-over-end, as was shown in Figure 18.12. Because the molecule rotates about its midpoint, each of the nitrogen atoms undergoes uniform circular motion with radius $r = 0.06$ nm. We found in Chapter 7 that the speed of a particle in circular motion is $v = 2\pi r/T_{rot}$, where T_{rot} is the rotation period. Hence the rotational kinetic energy of the molecule is

$$\epsilon_{rot} = \frac{1}{2}mv^2 + \frac{1}{2}mv^2 = m\left(\frac{2\pi r}{T_{rot}}\right)^2 = \frac{4\pi^2 mr^2}{T_{rot}^2}$$

The energy associated with this mode is $\frac{1}{2}Nk_BT$, so the *average* rotational kinetic energy per molecule is

$$(\epsilon_{rot})_{avg} = \frac{1}{2}k_BT$$

Equating these two expressions for ϵ_{rot} gives us

$$\frac{4\pi^2 mr^2}{T_{rot}^2} = \frac{1}{2}k_BT$$

Thus the rotational period is

$$T_{rot} = \sqrt{\frac{8\pi^2 mr^2}{k_BT}} = 1.3 \times 10^{-12} \text{ s}$$

We evaluated T_{rot} at $T = 293$ K, using $m = 14 \text{ u} = 2.34 \times 10^{-26}$ kg for each *atom*. The molecule's rotational frequency is

$$f = \frac{1}{T_{rot}} = 7.7 \times 10^{11} \text{ rev/s}$$

ASSESS This is a *very* high frequency, but these values are typical of molecular rotations.

How many degrees of freedom does a bead on a fixed wire have?

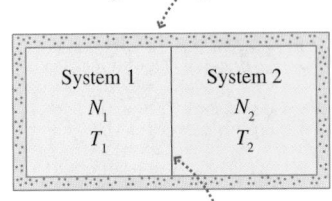

   a. 1   b. 2   c. 3   d. 4   e. 5   f. 6

18.5 Thermal Interactions and Heat

We can now look in more detail at what happens when two systems at different temperatures interact with each other. Figure 18.14 shows a rigid, insulated container that is divided into two sections by a very thin membrane. The left side, which we'll call system 1, has N_1 atoms at an initial temperature T_{1i}. System 2 on the right has N_2 atoms at an initial temperature T_{2i}. The membrane is so thin that atoms can collide at the boundary as if the membrane were not there, yet it is a barrier that prevents atoms from moving from one side to the other. The situation is analogous, on an atomic scale, to basketballs colliding through a shower curtain.

Suppose that system 1 is initially at a higher temperature: $T_{1i} > T_{2i}$. This is not an equilibrium situation. The temperatures will change with time until they eventually reach a common final temperature T_f. If you *watch* the gases as one warms and the other cools, you see nothing happening. This interaction is quite different from a mechanical interaction in which, for example, you might see a piston move from one side toward the other. The only way in which the gases can interact is via molecular collisions at the boundary. This is a *thermal interaction,* and our goal is to understand how thermal interactions bring the systems to thermal equilibrium.

System 1 and system 2 begin with thermal energies

$$E_{1i} = \frac{3}{2}N_1 k_B T_{1i} = \frac{3}{2}n_1 R T_{1i}$$

$$E_{2i} = \frac{3}{2}N_2 k_B T_{2i} = \frac{3}{2}n_2 R T_{2i}$$

(18.38)

We've written the energies for monatomic gases; you could do the same calculation if one or both of the gases is diatomic by replacing the $\frac{3}{2}$ with $\frac{5}{2}$. Notice that we've omitted "th" from the subscript to keep the notation manageable.

The total energy of the combined systems is $E_{tot} = E_{1i} + E_{2i}$. As systems 1 and 2 interact, their individual thermal energies E_1 and E_2 can change but their sum E_{tot} remains constant. The system will have reached thermal equilibrium when the individual thermal energies reach final values E_{1f} and E_{2f} that no longer change.

The Systems Exchange Energy

Figure 18.15 shows a fast atom and a slow atom approaching the barrier from opposite sides. They undergo a perfectly elastic collision at the barrier. Although no net energy is lost in a perfectly elastic collision, the faster atom loses energy while the slower one gains energy. In other words, there is an energy *transfer* from the faster atom's side to the slower atom's side.

The average translational kinetic energy per molecule is directly proportional to the temperature: $\epsilon_{avg} = \frac{3}{2}k_B T$. Because $T_{1i} > T_{2i}$, the atoms in system 1 are, on average, more energetic than the atoms in system 2. Thus *on average* the collisions transfer energy from system 1 to system 2. But not in every collision. Sometimes a fast atom in system 2 collides with a slow atom in system 1, transferring energy from 2 to 1. But the net energy transfer, from all collisions, is from the warmer system 1 to the cooler system 2. In other words, **heat is the energy transferred *via collisions*** between the more energetic (warmer) atoms on one side and the less energetic (cooler) atoms on the other.

Insulation prevents heat from entering or leaving the container.

FIGURE 18.14 Two gases can interact thermally through a very thin barrier.

A thin barrier prevents atoms from moving from system 1 to 2 but still allows them to collide.

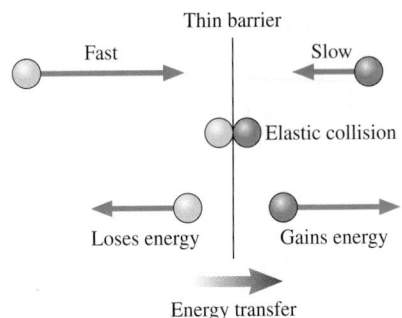

FIGURE 18.15 Collisions at the barrier transfer energy from faster molecules to slower molecules.

How do the systems "know" when they've reached thermal equilibrium? Energy transfer continues until the atoms on both sides of the barrier have the *same average translational kinetic energy*. Once the average translational kinetic energies are the same, there is no tendency for energy to flow in either direction. This is the state of thermal equilibrium, so the condition for thermal equilibrium is

$$(\epsilon_1)_{\text{avg}} = (\epsilon_2)_{\text{avg}} \qquad \text{(thermal equilibrium)} \qquad (18.39)$$

where, as before, ϵ is the translational kinetic energy of an atom.

Because the average energies are directly proportional to the final temperatures, $\epsilon_{\text{avg}} = \frac{3}{2}k_{\text{B}}T_{\text{f}}$, thermal equilibrium is characterized by the macroscopic condition

$$T_{1\text{f}} = T_{2\text{f}} = T_{\text{f}} \qquad \text{(thermal equilibrium)} \qquad (18.40)$$

In other words, **two thermally interacting systems reach a common final temperature *because* they exchange energy via collisions until the atoms on each side have, on average, equal translational kinetic energies.** This is a very important idea.

Equation 18.40 can be used to determine the equilibrium thermal energies. Because these are monatomic gases, $E_{\text{th}} = N\epsilon_{\text{avg}}$. Thus the equilibrium condition $(\epsilon_1)_{\text{avg}} = (\epsilon_2)_{\text{avg}} = (\epsilon_{\text{tot}})_{\text{avg}}$ implies

$$\frac{E_{1\text{f}}}{N_1} = \frac{E_{2\text{f}}}{N_2} = \frac{E_{\text{tot}}}{N_1 + N_2} \qquad (18.41)$$

from which we can conclude

$$E_{1\text{f}} = \frac{N_1}{N_1 + N_2} E_{\text{tot}}$$

$$E_{2\text{f}} = \frac{N_2}{N_1 + N_2} E_{\text{tot}} \qquad (18.42)$$

This result can also be written in terms of the number of moles. If we use $N = N_{\text{A}}n$ and note that the N_{A} cancels, Equation 18.42 becomes

$$E_{1\text{f}} = \frac{n_1}{n_1 + n_2} E_{\text{tot}}$$

$$E_{2\text{f}} = \frac{n_2}{n_1 + n_2} E_{\text{tot}} \qquad (18.43)$$

Notice that $E_{1\text{f}} + E_{2\text{f}} = E_{\text{tot}}$, verifying that energy has been conserved even while being redistributed between the systems.

No work is done on either system, because the barrier has no macroscopic displacement, so the first law of thermodynamics is

$$Q_1 = \Delta E_1 = E_{1\text{f}} - E_{1\text{i}}$$

$$Q_2 = \Delta E_2 = E_{2\text{f}} - E_{2\text{i}} \qquad (18.44)$$

As a homework problem you can show that $Q_1 = -Q_2$, as required by energy conservation. That is, the heat lost by one system is gained by the other. $|Q_1|$ is the quantity of heat that is transferred from the warmer gas to the cooler gas during the thermal interaction.

NOTE ▶ In general, the equilibrium thermal energies of the system are *not* equal. That is, $E_{1\text{f}} \neq E_{2\text{f}}$. They will be equal only if $N_1 = N_2$. Equilibrium is reached when the average translational kinetic energies in the two systems are equal; that is, when $(\epsilon_1)_{\text{avg}} = (\epsilon_2)_{\text{avg}}$, not when $E_{1\text{f}} = E_{2\text{f}}$. The distinction is important. Figure 18.16 summarizes these ideas. ◀

Collisions transfer energy from warmer system to cooler as more energetic atoms lose energy to less energetic atoms.

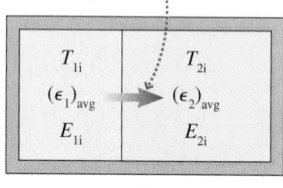

Thermal equilibrium occurs when the systems have the same average translational kinetic energy and thus the same temperature.

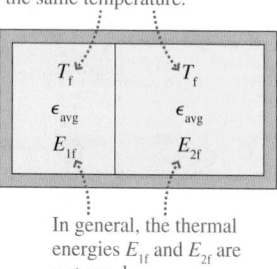

In general, the thermal energies $E_{1\text{f}}$ and $E_{2\text{f}}$ are *not* equal.

FIGURE 18.16 Equilibrium is reached when the atoms on each side have, on average, equal energies.

EXAMPLE 18.10 A thermal interaction

A sealed, insulated container has 2.0 g of helium at an initial temperature of 300 K on one side of a barrier and 10.0 g of argon at an initial temperature of 600 K on the other side.

a. How much heat energy is transferred, and in which direction?
b. What is the final temperature?

MODEL The systems start with different temperatures, so they are not in thermal equilibrium. Energy will be transferred via collisions from the argon to the helium until both systems have the same average molecular energy.

SOLVE

a. Let the helium be system 1. Helium has molar mass $M_{mol} = 4$ g/mol, so $n_1 = M/M_{mol} = 0.50$ mol. Similarly, argon has $M_{mol} = 40$ g/mol, so $n_2 = 0.25$ mol. The initial thermal energies of the two monatomic gases are

$$E_{1i} = \frac{3}{2}n_1 R T_{1i} = 225R = 1870 \text{ J}$$

$$E_{2i} = \frac{3}{2}n_2 R T_{2i} = 225R = 1870 \text{ J}$$

The systems start with *equal* thermal energies, but they are not in thermal equilibrium. The total energy is $E_{tot} = 3740$ J. In equilibrium, this energy is distributed between the two systems as

$$E_{1f} = \frac{n_1}{n_1 + n_2}E_{tot} = \frac{0.50}{0.75}3740 \text{ J} = 2493 \text{ J}$$

$$E_{2f} = \frac{n_2}{n_1 + n_2}E_{tot} = \frac{0.25}{0.75}3740 \text{ J} = 1247 \text{ J}$$

The heat entering or leaving each system is

$$Q_1 = Q_{He} = E_{1f} - E_{1i} = 623 \text{ J}$$

$$Q_2 = Q_{Ar} = E_{2f} - E_{2i} = -623 \text{ J}$$

The helium and the argon interact thermally via collisions at the boundary, causing 623 J of heat to be transferred from the warmer argon to the cooler helium.

b. These are constant-volume processes, thus $Q = nC_V\Delta T$. $C_V = \frac{3}{2}R$ for monatomic gases, thus the temperature changes are

$$\Delta T_{He} = \frac{Q_{He}}{\frac{3}{2}nR} = \frac{623 \text{ J}}{1.5(0.50 \text{ mol})(8.31 \text{ J/mol K})} = 100 \text{ K}$$

$$\Delta T_{Ar} = \frac{Q_{Ar}}{\frac{3}{2}nR} = \frac{-623 \text{ J}}{1.5(0.25 \text{ mol})(8.31 \text{ J/mol K})} = -200 \text{ K}$$

Both gases reach the common final temperature $T_f = 400$ K.

ASSESS $E_{1f} = E_{2f}$ because there are twice as many atoms in system 1.

The main idea of this section is that two systems reach a common final temperature not by magic or by a prearranged agreement but simply from the energy exchange of vast numbers of molecular collisions. Real interacting systems, of course, are separated by walls rather than our unrealistic thin membrane. As the systems interact, the energy is first transferred via collisions from system 1 into the wall and subsequently, as the cooler molecules collide with a warm wall, into system 2. That is, the energy transfer is $E_1 \rightarrow E_{wall} \rightarrow E_2$. This is still heat because the energy transfer is occurring via molecular collisions rather than mechanical motion.

STOP TO THINK 18.5   Systems A and B are interacting thermally. At this instant of time,

a. $T_A > T_B$.
b. $T_A = T_B$.
c. $T_A < T_B$.

A	B
$N = 1000$	$N = 2000$
$\epsilon_{avg} = 1.0 \times 10^{-20}$ J	$\epsilon_{avg} = 0.5 \times 10^{-20}$ J
$E_{th} = 1.0 \times 10^{-17}$ J	$E_{th} = 1.0 \times 10^{-17}$ J

18.6 Irreversible Processes and the Second Law of Thermodynamics

The previous section looked at the thermal interaction between a warm gas and a cold gas. Heat energy is transferred from the warm gas to the cold gas until they reach a common final temperature. But why isn't heat transferred from the cold gas to the warm gas, making the cold side colder and the warm side warmer? Such a process could still conserve energy, but it never happens. The transfer of heat energy from hot to cold is an example of an **irreversible process,** a process that can happen only in one direction.

Examples of irreversible processes abound. Stirring the cream in your coffee mixes the cream and coffee together. No amount of stirring ever unmixes them. If you shake a jar that has red marbles on the top and blue marbles on the bottom, the two colors are quickly mixed together. No amount of shaking ever separates them again. If you watched a movie of someone shaking a jar and saw the red and blue marbles separating, you would be certain that the movie was running backward. In fact, a reasonable definition of an irreversible process is one for which a backward-running movie shows a physically impossible process.

Figure 18.17a is a two-frame movie of a collision between two particles, perhaps two gas molecules. Suppose that sometime after the collision is over we could reach in and reverse the velocities of both particles. That is, replace vector $\vec{v}$ with vector $-\vec{v}$. Then, as in a movie playing backward, the collision would happen in reverse. This is the movie of Figure 18.17b.

(a) Forward movie   **(b)** The backward movie is equally plausible.

Before: After: Before: After:

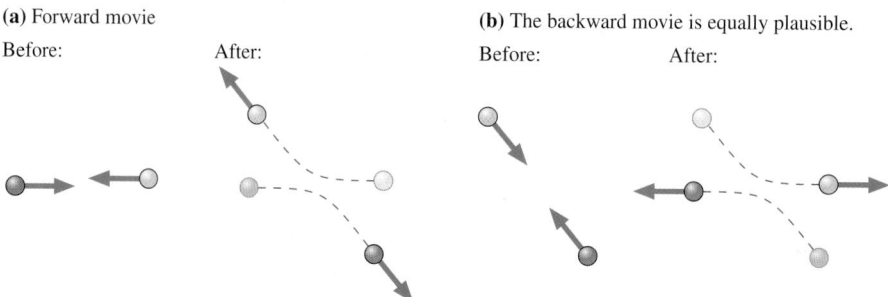

FIGURE 18.17 Molecular collisions are reversible.

You cannot tell, just by looking at the two movies, which is really going forward and which is being played backward. Maybe Figure 18.17b was the original collision and Figure 18.17a is the backward version. Nothing in either collision looks wrong, and no measurements you might make on either would reveal any violations of Newton's laws. Interactions at the molecular level are reversible processes.

Contrast this with the two-frame car crash movies in Figure 18.18. Past and future are clearly distinct in an irreversible process, and the backward movie of Figure 18.18b is obviously wrong. But what has been violated in the backward movie? To have the crumpled car spring away from the wall would not violate any laws of physics we have so far discovered. It would simply require transforming the thermal energy of the car and wall back into the macroscopic center-of-mass energy of the car as a whole.

The paradox stems from our assertion that macroscopic phenomena can be understood on the basis of microscopic molecular motions. If the microscopic motions are all reversible, how can the macroscopic phenomena end up being irreversible? If reversible collisions can cause heat to be transferred from hot to cold, why do they never cause heat to be transferred from cold to hot? There must be another law of physics preventing it. The law we seek must, in some sense, be able to distinguish the past from the future.

(a) Forward movie

Before: After:

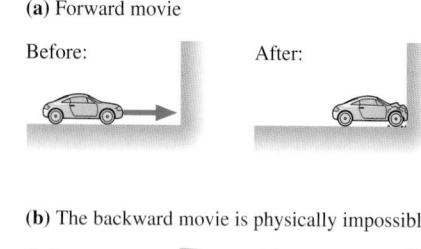

(b) The backward movie is physically impossible.

Before: After:

FIGURE 18.18 A car crash is irreversible.

Which Way to Equilibrium?

Stated another way, how do two systems initially at different temperatures "know" which way to go to reach equilibrium? Perhaps an analogy will help.

Figure 18.19 shows two boxes, numbered 1 and 2, containing identical balls. Box 1 starts with more balls than box 2, so $N_{1i} > N_{2i}$. Once every second, one ball is chosen at random and moved to the other box. This is a reversible process because a ball can move from box 2 to box 1 just as easily as from box 1 to box 2. What do you expect to see if you return several hours later?

Because balls are chosen at random, and because $N_{1i} > N_{2i}$, it's initially more likely that a ball will move from box 1 to box 2 than from box 2 to box 1. Sometimes a ball will move "backward" from box 2 to box 1, but overall there's a net movement of balls from box 1 to box 2. The system will evolve until $N_1 \approx N_2$. This is a stable situation—equilibrium!—with an equal number of balls moving in both directions.

But couldn't it go the other way, with N_1 getting even larger while N_2 decreases? In principle, any possible arrangement of the balls is possible in the same way that any number of heads are possible if you throw N coins in the air and let them fall. If you throw four coins, the odds are 1 in 2^4, or 1 in 16, of getting four heads. With four balls, the odds are 1 in 16 that, at a randomly chosen instant of time, you would find $N_1 = 4$. You wouldn't find that to be terribly surprising.

With 10 balls, the probability that $N_1 = 10$ is $0.5^{10} \approx 1/1000$. With 100 balls, the probability that $N_1 = 100$ has dropped to $\approx 10^{-30}$. With 10^{20} balls, the odds of finding all of them, or even most of them, in one box are so staggeringly small that it's safe to say it will "never" happen. Although each transfer is reversible, **the statistics of large numbers make it overwhelmingly more likely that the system will evolve toward a state in which $N_1 \approx N_2$ than toward a state in which $N_{1f} > N_{1i}$.**

The balls in our analogy represent energy. The total energy, like the total number of balls, is conserved, but molecular collisions can move energy between system 1 and system 2. Each collision is reversible, just as likely to transfer energy from 1 to 2 as from 2 to 1. But if $(\epsilon_{1i})_{avg} > (\epsilon_{2i})_{avg}$, and if we're dealing with two macroscopic systems where $N > 10^{20}$, then it's overwhelmingly likely that the net result of many, many collisions will be to transfer energy from system 1 to system 2 until $(\epsilon_{1f})_{avg} = (\epsilon_{2f})_{avg}$—in other words, for heat energy to be transferred from hot to cold.

The system reaches thermal equilibrium not by any plan or by outside intervention, but simply because **equilibrium is the *most probable* state in which to be.** It is *possible* that the system will move away from equilibrium, with heat moving from cold to hot, but remotely improbable in any realistic system. The consequence of a vast number of random events is that the system evolves in one direction, toward equilibrium, and not the other. **Reversible microscopic events lead to irreversible macroscopic behavior because some macroscopic states are vastly more probable than others.**

Order, Disorder, and Entropy

Figure 18.20 shows three different systems. At the top is a group of atoms arranged in a crystal-like lattice. This is a highly ordered and nonrandom system, with each atom's position precisely specified. Contrast this with the system on the bottom, where there is no order at all. The position of every atom was assigned entirely at random.

It is extremely improbable that the atoms in a container would *spontaneously* arrange themselves into the ordered pattern of the top picture. In a system of, say, 10^{20} atoms, the probability of this happening is similar to the probability that 10^{20} tossed coins will all be heads. We can safely say that it will never happen. By contrast, there are a vast number of arrangements like the one on the bottom that randomly fill the container.

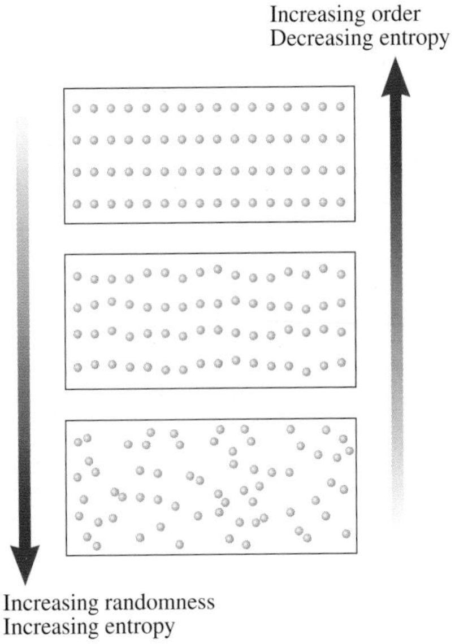

The middle picture of Figure 18.20 is an in-between situation. This situation might arise as a solid melts. The positions of the atoms are clearly not completely

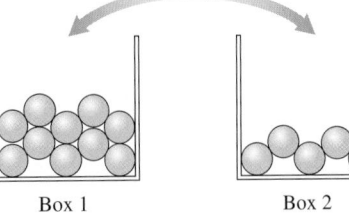

Balls are chosen at random and moved from one box to the other.

Box 1
N_1 balls

Box 2
N_2 balls

FIGURE 18.19 Two interacting systems. Balls are chosen at random and moved to the other box.

Increasing order
Decreasing entropy

Increasing randomness
Increasing entropy

FIGURE 18.20 Ordered and disordered systems.

random, so the system preserves some degree of order. This in-between situation is more likely to occur spontaneously than the highly ordered lattice on the top, but is less likely to occur than the completely random system on the bottom.

Scientists and engineers use a state variable called **entropy** to measure the probability that a macroscopic state will occur. The ordered lattice, which has a very small probability of occurrence, has a very low entropy. The entropy of the randomly filled container is high. The entropy of the middle picture is somewhere in between. It is often said that entropy measures the amount of *disorder* in a system. The entropy in Figure 18.20 increases as you move from the ordered system on the top to the disordered system on the bottom.

Similarly, two thermally interacting systems with different temperatures have a low entropy. These systems are ordered in the sense that the faster atoms are on one side of the barrier, the slower atoms on the other. The most random possible distribution of energy, and hence the least ordered system, corresponds to the situation where the two systems are in thermal equilibrium with equal temperatures. Entropy increases as two systems with initially different temperatures move toward equilibrium. Entropy would decrease if heat energy moved from cold to hot, making the hot system hotter and the cold system colder.

Entropy can be calculated, but we'll leave that to more advanced courses. For our purposes, the *concept* of entropy as a measure of the disorder in a system, or of the probability that a macroscopic state will occur, is more important than a numerical value.

The Second Law of Thermodynamics

The fact that macroscopic systems evolve irreversibly toward equilibrium is a statement about nature that is not contained in any of the laws of physics we have encountered. It is, in fact, a new law of physics, one known as the **second law of thermodynamics.**

The formal statement of the second law of thermodynamics is given in terms of entropy:

> **Second law, formal statement** The entropy of an isolated system never decreases. The entropy either increases, until the system reaches equilibrium, or, if the system began in equilibrium, stays the same.

The qualifier "isolated" is most important. We can order the system by reaching in from the outside, perhaps using little tweezers to place the atoms in a lattice. Similarly, we can transfer heat from cold to hot by using a refrigerator. The second law is about what a system can or cannot do *spontaneously,* on its own, without outside intervention.

The second law of thermodynamics tells us that an isolated system evolves such that

- Order turns into disorder and randomness.
- Information is lost rather than gained.
- The system "runs down."

An isolated system never spontaneously generates order out of randomness. It is not that the system "knows" about order or randomness, but rather that there are vastly more states corresponding to randomness than there are corresponding to order. As collisions occur at the microscopic level, the laws of probability dictate that the system will, on average, move inexorably toward the most probable and thus most random macroscopic state.

The second law of thermodynamics is often stated in several equivalent but more informal versions. One of these, and the one most relevant to our discussion, is

Second law, informal statement #1 When two systems at different temperatures interact, heat energy is transferred spontaneously from the hotter to the colder system, never from the colder to the hotter.

The second law of thermodynamics is an independent statement about nature, separate from the first law. The first law is a precise statement about energy conservation. The second law, by contrast, is a *probabilistic* statement, based on the statistics of very large numbers. While it is conceivable that heat could spontaneously move from cold to hot, it will never occur in any realistic macroscopic system.

The irreversible evolution from less-likely macroscopic states to more-likely macroscopic states is what gives us a macroscopic direction of time. Stirring blends your coffee and cream, it never unmixes them. Friction causes an object to stop while increasing its thermal energy; the random atomic motions of thermal energy never spontaneously organize themselves into a macroscopic motion of the entire object. A plant in a sealed jar dies and decomposes to carbon and various gases; the gases and carbon never spontaneously assemble themselves into a flower. These are all examples of irreversible processes. They each show a clear direction of time, a distinct difference between past and future.

Thus another statement of the second law is

Second law, informal statement #2 The time direction in which the entropy of an isolated macroscopic system increases is "the future."

Establishing the "arrow of time" is one of the most profound implications of the second law of thermodynamics.

The second law of thermodynamics has important implications for issues ranging from how we as a society use energy and resources to biological evolution and the future of the universe. We'll return to some of these issues in the Summary to Part IV. In the meantime, the second law will be used in Chapter 19 to understand some of the practical aspects of the thermodynamics of engines.

STOP TO THINK 18.6   Two identical boxes each contain 1,000,000 molecules. In box A, 750,000 molecules happen to be in the left half of the box while 250,000 are in the right half. In box B, 499,900 molecules happen to be in the left half of the box while 500,100 are in the right half. At this instant of time,

a. The entropy of box A is larger than the entropy of box B.
b. The entropy of box A is equal to the entropy of box B.
c. The entropy of box A is smaller than the entropy of box B.

SUMMARY

The goal of Chapter 18 has been to understand the properties of a macroscopic system in terms of the microscopic behavior of its molecules.

GENERAL PRINCIPLES

Kinetic theory, the **micro/macro connection,** relates the macroscopic properties of a system to the motion and collisions of its atoms and molecules.

The Equipartition Theorem

Tells us how collisions distribute the energy in the system. The energy stored in each mode of the system (each **degree of freedom**) is $\frac{1}{2}Nk_BT$ or, in terms of moles, $\frac{1}{2}nRT$.

The Second Law of Thermodynamics

Tells us how collisions move a system toward equilibrium. The entropy of an isolated system can only increase or, in equilibrium, stay the same.

• Order turns into disorder and randomness.

• Systems run down.

• Heat energy is transferred spontaneously from the hotter to the colder system, never from colder to hotter.

IMPORTANT CONCEPTS

Pressure is due to the force of the molecules colliding with the walls.

$$p = \frac{1}{3}\frac{N}{V}mv_{rms}^2 = \frac{2}{3}\frac{N}{V}\epsilon_{avg}$$

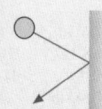

The average translational kinetic energy of a molecule is

$\epsilon_{avg} = \frac{3}{2}k_BT$. The temperature of the gas $T = \frac{2}{3k_B}\epsilon_{avg}$

measures the average translational kinetic energy.

Entropy measures the probability that a macroscopic state will occur or, equivalently, the amount of disorder in a system.

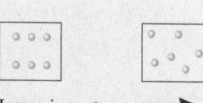

Increasing entropy

The thermal energy of a system is

E_{th} = translational kinetic energy + rotational kinetic energy + vibrational energy

• **Monatomic gas** $E_{th} = \frac{3}{2}Nk_BT = \frac{3}{2}nRT$

• **Diatomic gas** $E_{th} = \frac{5}{2}Nk_BT = \frac{5}{2}nRT$

• **Elemental solid** $E_{th} = 3Nk_BT = 3nRT$

Heat is energy transferred via collisions from more-energetic molecules on one side to less-energetic molecules on the other. Equilibrium is reached when $(\epsilon_1)_{avg} = (\epsilon_2)_{avg}$, which implies $T_{1f} = T_{2f}$.

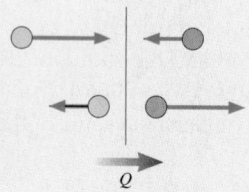

APPLICATIONS

The root-mean-square speed v_{rms} is the square root of the average of the squares of the molecular speeds:

$$v_{rms} = \sqrt{(v^2)_{avg}}$$

For molecules of mass m at temperature T,

$$v_{rms} = \sqrt{\frac{3k_BT}{m}}$$

Molar specific heats can be predicted from the thermal energy because $\Delta E_{th} = nC\Delta T$.

• **Monatomic gas** $C_V = \frac{3}{2}R$

• **Diatomic gas** $C_V = \frac{5}{2}R$

• **Elemental solid** $C = 3R$

TERMS AND NOTATION

kinetic theory	degrees of freedom	entropy
mean free path, λ	equipartition theorem	second law of thermodynamics
root-mean-square speed, v_{rms}	irreversible process	

EXERCISES AND PROBLEMS

Exercises

Section 18.1 Molecular Collisions

1. The number density of an ideal gas at STP is called the *Loschmidt number.* Calculate the Loschmidt number.
2. At what pressure will the mean free path in room-temperature (20°C) nitrogen be 1.0 m?
3. Integrated circuits are manufactured in vacuum chambers in which the air pressure is 1.0×10^{-10} mm of Hg. What are (a) the number density and (b) the mean free path of a molecule? Assume $T = 20°C$.
4. The mean free path of a molecule in a gas is 300 nm. What is the mean free path if the gas temperature is doubled at (a) constant volume and (b) constant pressure?
5. The pressure inside a tank of neon is 150 atm. The temperature is 25°C. On average, how many atomic diameters does a neon atom move between collisions?
6. A lottery machine uses blowing air to keep 2000 Ping-Pong balls bouncing around inside a 1.0 m × 1.0 m × 1.0 m box. The diameter of a Ping-Pong ball is 3.0 cm. What is the mean free path between collisions? Give your answer in cm.

Section 18.2 Pressure in a Gas

7. The molecules in a six-particle gas have velocities

    $\vec{v}_1 = (20\hat{\imath} + 30\hat{\jmath})$ m/s $\qquad \vec{v}_4 = (60\hat{\imath} - 20\hat{\jmath})$ m/s

    $\vec{v}_2 = (-40\hat{\imath} + 70\hat{\jmath})$ m/s $\qquad \vec{v}_5 = -50\hat{\jmath}$ m/s

    $\vec{v}_3 = (-80\hat{\imath} - 10\hat{\jmath})$ m/s $\qquad \vec{v}_6 = (40\hat{\imath} - 20\hat{\jmath})$ m/s

    Calculate (a) $\vec{v}_{avg}$, (b) v_{avg}, and (c) v_{rms}.
8. Eleven molecules have speeds 15, 16, 17, . . . , 25 m/s. Calculate (a) v_{avg}, and (b) v_{rms}.
9. The number density in a container of argon gas is 2.0×10^{25} m^{-3}. The atoms are moving with an rms speed of 455 m/s. What are (a) the pressure and (b) the temperature inside the container?
10. 5.0×10^{23} nitrogen molecules collide with a 10 cm^2 wall each second. Assume that the molecules all travel with a speed of 400 m/s and strike the wall head on. What is the pressure on the wall?
11. Oxygen molecules collide with a 10 cm^2 wall. Assume that the molecules all travel with a speed of 500 m/s and strike the wall head on. How many collisions are there per second if the oxygen pressure is 1.0 atm?

12. At 100°C the rms speed of nitrogen molecules is 576 m/s. Nitrogen at 100°C and a pressure of 2.0 atm is held in a container with a 10 cm × 10 cm square wall. Estimate the rate of molecular collisions (collisions/s) on this wall.

Section 18.3 Temperature

13. What are the rms speeds of helium and argon atoms at 1000°C?
14. 30 m/s is a typical highway speed for a car. At what temperature do the molecules of nitrogen gas have an rms speed of 30 m/s?
15. A gas consists of a mixture of neon and argon. The rms speed of the neon atoms is 400 m/s. What is the rms speed of the argon atoms?
16. At what temperature do hydrogen molecules have the same rms speed as nitrogen molecules at 100°C?
17. At what temperature is the rms speed of oxygen molecules (a) half and (b) twice its value at STP?
18. By what factor does the rms speed of a molecule change if the temperature is increased from 20°C to 100°C?
19. At what temperature would the rms speed of hydrogen molecules be the speed of light (3.0×10^8 m/s)? There is no upper limit to temperature, but Einstein's theory of relativity says that no material particle can attain the speed of light. Consequently, our results for ϵ_{avg} and v_{rms} would need to be modified for very high temperatures and speeds.
20. Suppose you double the temperature of a gas at constant volume. Do the following change? And if so, by what factor?
    a. The average translational kinetic energy of a molecule.
    b. The rms speed of a molecule.
    c. The mean free path.
21. What is the total translational kinetic energy at STP of 1.0 mol of (a) hydrogen, (b) helium, and (c) oxygen?
22. During a physics experiment, helium gas is cooled to a temperature of 10 K at a pressure of 0.10 atm. What are (a) the mean free path in the gas, (b) the rms speed of the atoms, and (c) the average energy per atom?
23. The temperature at the center of the sun is $\approx 2 \times 10^7$ K. What are (a) the average kinetic energy and (b) the rms speed of a proton in the center of the sun?
24. The atmosphere of the sun consists mostly of hydrogen *atoms* (not molecules) at a temperature of 6000 K. What are (a) the average translational kinetic energy per atom and (b) the rms speed of the atoms?

Section 18.4 Thermal Energy and Specific Heat

25. A 6.0 m × 8.0 m × 3.0 m room contains air at 20°C. What is the room's thermal energy?
26. What is the thermal energy of 100 cm^3 of lead at room temperature?
27. The thermal energy of 1.0 mol of a substance is increased by 1.0 J. What is the temperature change if the system is (a) a monatomic gas, (b) a diatomic gas, and (c) a solid?
28. 1.0 mol of a monatomic gas interacts thermally with 1.0 mol of an elemental solid. The gas temperature decreases by 50°C at constant volume. What is the temperature change of the solid?
29. A rigid container holds 0.20 g of hydrogen gas. How much heat is needed to change the temperature of the gas
    a. From 50 K to 100 K?
    b. From 250 K to 300 K?
    c. From 550 K to 600 K?
    d. From 2250 K to 2300 K?
30. A cylinder of nitrogen gas has a volume of 15,000 cm^3 and a pressure of 100 atm.
    a. What is the thermal energy of this gas at room temperature (20°C)?
    b. What is the mean free path in the gas?
    c. The valve is opened and the gas is allowed to expand slowly and isothermally until it reaches a pressure of 1.0 atm. What is the change in the thermal energy of the gas?

Section 18.5 Thermal Interactions and Heat

31. 2.0 mol of monatomic gas A initially has 5000 J of thermal energy. It interacts with 3.0 mol of monatomic gas B, which initially has 8000 J of thermal energy.
    a. Which gas has the higher initial temperature?
    b. What are the final thermal energies of each gas?
32. 4.0 mol of monatomic gas A initially has 9000 J of thermal energy. It interacts with 3.0 mol of monatomic gas B, which initially has 5000 J of thermal energy. How much heat energy is transferred between the systems, and in which direction, as they come to thermal equilibrium?

Problems

33. For a monatomic gas, what is the ratio of the volume per atom (V/N) to the volume *of* an atom when the mean free path equals the atomic diameter?
34. From what height must an oxygen molecule fall in a vacuum so that its kinetic energy at the bottom equals the average energy of an oxygen molecule at 300 K?
35. A gas at $p = 50$ kPa and $T = 300$ K has a mass density of 0.0802 kg/m^3.
    a. Identify the gas.
    b. What is the rms speed of the atoms in this gas?
    c. What is the mean free path of the atoms in the gas?
36. Interstellar space, far from any stars, is filled with a very low density of hydrogen atoms (H, not H_2). The number density is about 1 atom/cm^3 and the temperature is about 3 K.
    a. Estimate the pressure in interstellar space. Give your answer in Pa and in atm.

---

    b. What is the rms speed of the atoms?
    c. What is the edge length L of an $L \times L \times L$ cube of gas with 1.0 J of thermal energy?
37. Dust particles are ≈10 μm in diameter. They are pulverized rock, with $\rho \approx 2500$ kg/m^3. If you treat dust as an ideal gas, what is the rms speed of a dust particle at 20°C?
38. Uranium has two naturally occurring isotopes. ^{238}U has a natural abundance of 99.3% and ^{235}U has an abundance of 0.7%. It is the rarer ^{235}U that is needed for nuclear reactors. The isotopes are separated by forming uranium hexafluoride UF_6, which is a gas, then allowing it to diffuse through a series of porous membranes. $^{235}UF_6$ has a slightly larger rms speed than $^{238}UF_6$ and diffuses slightly faster. Many repetitions of this procedure gradually separate the two isotopes. What is the ratio of the rms speed of $^{235}UF_6$ to that of $^{238}UF_6$?
39. Equation 18.3 is the mean free path of a particle through a gas of identical particles of equal radius. An electron can be thought of as a point particle with zero radius.
    a. Find an expression for the mean free path of an electron through a gas.
    b. Electrons travel 3 km through the Stanford Linear Accelerator (SLAC). In order for scattering losses to be negligible, the pressure inside the accelerator tube must be reduced to the point where the mean free path is at least 50 km. What is the maximum possible pressure inside the accelerator tube, assuming $T = 20$°C? Give your answer both in Pa and atm.
40. Figure P18.40 is a histogram showing the speeds of the molecules in a very small gas. What are (a) the most probable speed, (b) the average speed, and (c) the rms speed?

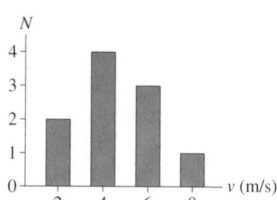

FIGURE P18.40

41. A 10-cm-diameter, 20-cm-long cylinder contains 2.0×10^{22} atoms of argon at a temperature of 50°C.
    a. What is the number density of the gas?
    b. What is the root-mean-square speed?
    c. What is $(v_x)_{rms}$, the rms value of the x-component of velocity?
    d. What is the rate at which atoms collide with one end of the cylinder?
    e. Determine the pressure in the cylinder using the results of kinetic theory.
    f. Determine the pressure in the cylinder using the ideal-gas law.
42. A 10 cm × 10 cm × 10 cm box contains 0.010 mol of nitrogen at 20°C. What is the rate of collisions (collisions/s) on one wall of the box?
43. A 100 cm^3 box contains helium at a pressure of 2.0 atm and a temperature of 100°C. It is placed in thermal contact with a 200 cm^3 box containing argon at a pressure of 4.0 atm and a temperature of 400°C.
    a. What is the initial thermal energy of each gas?
    b. What is the final thermal energy of each gas?

c. How much heat energy is transferred, and in which direction?

d. What is the final temperature?

e. What is the final pressure in each box?

44. 2.0 g of helium at an initial temperature of 300 K interacts thermally with 8.0 g of oxygen at an initial temperature of 600 K.

a. What is the initial thermal energy of each gas?

b. What is the final thermal energy of each gas?

c. How much heat energy is transferred, and in which direction?

d. What is the final temperature?

45. A gas of 1.0×10^{20} atoms or molecules has 1.0 J of thermal energy. Its molar specific heat at constant pressure is 20.8 J/mol K. What is the temperature of the gas?

46. How many degrees of freedom does a system have if $\gamma = 1.29$?

47. A system has f degrees of freedom. Show that $\gamma = (f + 2)/f$. Verify that this expression gives the values you learned in Chapter 17.

48. 1.0 mol of a monatomic gas and 1.0 mol of a diatomic gas are at 0°C. Both are heated at constant pressure until their volume doubles. What is the ratio $Q_{diatomic}/Q_{monatomic}$?

49. In the discussion following Equation 18.44 it was said that $Q_1 = -Q_2$. Prove that this is so.

50. A monatomic gas is adiabatically compressed to $\frac{1}{8}$ of its initial volume. Do each of the following quantities change? If so, does it increase or decrease, and by what factor? If not, why not?

a. The rms speed.

b. The mean free path.

c. The thermal energy of the gas.

d. The molar specific heat at constant volume.

51. Predict the molar specific heat at constant volume of (a) a two-dimensional monatomic gas and (b) a two-dimensional solid.

52. A diatomic molecule has only two rotational degrees of freedom because rotation about its own axis is "frozen out." This isn't a problem for polyatomic molecules with more than two atoms. They can rotate about any of three mutually perpendicular axes and thus have three rotational degrees of freedom.

a. Predict the molar specific heat at constant volume of a polyatomic molecule.

b. Water vapor has $\gamma = 1.31$. Is this consistent with your prediction?

53. Equal masses of hydrogen gas and oxygen gas are mixed together in a container and held at constant temperature. What is the hydrogen/oxygen ratio of (a) v_{rms}, (b) ϵ_{avg}, and (c) E_{th}?

54. A gas mixture consists of 2.0 mol of nitrogen and 1.0 mol of neon. What is the molar specific heat at constant volume of the mixture?

55. An experiment you're designing needs a gas with $\gamma = 1.50$. You recall from your physics class that no individual gas has this value, but it occurs to you that you could produce a gas with $\gamma = 1.50$ by mixing together a monatomic gas and a diatomic gas. What fraction of the molecules need to be monatomic?

56. n_1 moles of a monatomic gas and n_2 moles of a diatomic gas are mixed together in a container.

a. Derive an expression for the molar specific heat at constant volume of the mixture.

b. Show that your expression has the expected behavior if $n_1 \to 0$ or $n_2 \to 0$.

57. A 1.0 kg ball is at rest on the floor in a 2.0 m $\times$ 2.0 m $\times$ 2.0 m room of air at STP. Air is 80% nitrogen (N_2) and 20% oxygen (O_2) by volume.

a. What is the thermal energy of the air in the room?

b. What fraction of the thermal energy would have to be conveyed to the ball for it to be spontaneously launched to a height of 1.0 m?

c. By how much would the air temperature have to decrease to launch the ball?

d. Your answer to part c is so small as to be unnoticeable, yet this event never happens. Why not?

58. An inventor wants you to invest money with his company, offering you 10% of all future profits. He reminds you that the brakes on cars get extremely hot when they stop and that there is a large quantity of thermal energy in the brakes. He has invented a device, he tells you, that converts that thermal energy into the forward motion of the car. This device will take over from the engine after a stop and accelerate the car back up to its original speed, thereby saving a tremendous amount of gasoline. Now, you're a smart person, so he admits up front that this device is not 100% efficient, that there is some unavoidable heat loss to the air and to friction within the device, but the upcoming research for which he needs your investment will make those losses extremely small. You do also have to start the car with cold brakes after it has been parked awhile, so you'll still need a gasoline engine for that. Nonetheless, he tells you, his prototype car gets 500 miles to the gallon and he expects to be at well over 1000 miles to the gallon after the next phase of research. Should you invest? Base your answer on an analysis of the *physics* of the situation.

Challenge Problems

59. 1.0 mol of a diatomic gas with $C_V = \frac{5}{2}R$ has initial pressure p_i and volume V_i. The gas undergoes a process in which the pressure is directly proportional to the volume until the rms speed of the molecules has doubled.

a. Show this process on a pV diagram.

b. How much heat does this process require? Give your answer in terms of p_i and V_i.

60. At what temperature does the rms speed of (a) a nitrogen molecule and (b) a hydrogen molecule equal the escape speed from the earth's surface? (c) You'll find that these temperatures are very high, so you might think that the earth's gravity could easily contain both gases. But not all molecules move with v_{rms}. There is a distribution of speeds, and a small percentage of molecules have speeds several times v_{rms}. Bit by bit, a gas can slowly leak out of the atmosphere as its fastest molecules escape. A reasonable rule of thumb is that the earth's gravity can contain a gas only if the average translational kinetic energy per molecule is less than 1% of the kinetic energy needed to escape. Use this rule to show why the earth's atmosphere contains nitrogen but not hydrogen, even though hydrogen is the most abundant element in the universe.

61. Consider a container like that shown in Figure 18.14, with n_1 moles of a monatomic gas on one side and n_2 moles of a diatomic gas on the other. The monatomic gas has initial temperature T_{1i}. The diatomic gas has initial temperature T_{2i}.
    a. Show that the equilibrium thermal energies are

    $$E_{1f} = \frac{3n_1}{3n_1 + 5n_2}(E_{1i} + E_{2i})$$

    $$E_{2f} = \frac{5n_2}{3n_1 + 5n_2}(E_{1i} + E_{2i})$$

    b. Show that the equilibrium temperature is

    $$T_f = \frac{3n_1 T_{1i} + 5n_2 T_{2i}}{3n_1 + 5n_2}$$

    c. 2.0 g of helium at an initial temperature of 300 K interacts thermally with 8.0 g of oxygen at an initial temperature of 600 K. What is the final temperature? How much heat energy is transferred, and in which direction?

STOP TO THINK ANSWERS

Stop to Think 18.1: $\lambda_B > \lambda_A = \lambda_C > \lambda_D$. Increasing the volume makes the gas less dense, so λ increases. Increasing the radius makes the targets larger, so λ decreases. The mean free path doesn't depend on the atomic mass.

Stop to Think 18.2: c. Each v^2 increases by a factor of 16 but, after averaging, v_{rms} takes the square root.

Stop to Think 18.3: c. The average translational kinetic energy per molecule depends *only* on the temperature.

Stop to Think 18.4: b. The bead can slide along the wire (one degree of translational motion) and rotate around the wire (one degree of rotational motion).

Stop to Think 18.5: a. Temperature measures the average translational kinetic energy *per molecule,* not the thermal energy of the entire system.

Stop to Think 18.6: c. With 1,000,000 molecules, it's highly unlikely that 750,000 of them would spontaneously move into one side of the box. A state with a very small probability of occurrence has a very low entropy. Having an imbalance of only 100 out of 1,000,000 is well within what you might expect for random fluctuations. This is a highly probable situation and thus one of large entropy.

19 Heat Engines and Refrigerators

This coal-fired power plant is generating electricity by turning heat into work—but not very efficiently. Roughly two-thirds of the fuel's energy goes into the cooling water and is lost as waste heat.

▶ **Looking Ahead**

The goal of Chapter 19 is to investigate the physical principles that govern the operation of heat engines and refrigerators. In this chapter you will learn to:

- Understand and analyze heat engines and refrigerators.
- Understand the concept and significance of the Carnot engine.
- Characterize the performance of a heat engine in terms of its thermal efficiency and that of a refrigerator in terms of its coefficient of performance.
- Recognize that the second law of thermodynamics limits the efficiencies of heat engines and refrigerators.

◀ **Looking Back**

The material in this chapter depends on the first and second laws of thermodynamics. Most of the examples will be based on ideal gases. Please review:

- Sections 16.5–16.6 Ideal gases.
- Sections 17.2–17.4 Work, heat, and the first law of thermodynamics.
- Section 18.6 The second law of thermodynamics.

The earliest humans learned to use the heat from fires to warm themselves and cook their food. They were transforming heat energy into thermal energy. But is there a way to transform heat into *work?* Can we use the energy released by the fuel to grind corn, pump water, accelerate cars, launch rockets, or do any other task in which a force is exerted through a distance?

The first practical device for turning heat into work was the steam engine. A steam engine boils water to make high-pressure steam, then uses the steam to push a piston and do work. The 19th and 20th centuries saw the development of the steam turbine, the gasoline engine, the jet engine, and other devices that transform the heat from burning fuel into useful work. These are the devices that power modern society.

"Heat engine" is the generic term for *any* device that uses a cyclical process to transform heat energy into work. The power plant shown in the photo and the engine in your car are examples of heat engines. A closely related concept is that of a *refrigerator,* a device that uses work to move heat energy from a cold object to a hot object. Our goal in this chapter is to investigate the physical principles that *all* heat engines and *all* refrigerators must obey. We'll discover that the second law of thermodynamics places sharp constraints on the maximum possible efficiency of heat engines and refrigerators.

A car engine transforms the chemical energy stored in the fuel into work and ultimately into the car's kinetic energy.

19.1 Turning Heat into Work

Thermodynamics is the branch of physics that studies the transformation of energy. Many practical devices are designed to transform energy from one form, such as the heat from burning fuel, into another, such as work. Chapters 17 and 18 established two laws of thermodynamics that any such device must obey:

First law Energy is conserved; that is, $\Delta E_{th} = W + Q$.

Second law Most macroscopic processes are irreversible. In particular, heat energy is transferred spontaneously from a hotter to a colder system but never from a colder system to a hotter system.

Our goal in this chapter is to discover what these two laws, especially the second law, imply about devices that turn heat into work. The questions we want to address include:

- How does a practical device transform heat into work?
- What are the limitations and restrictions on these energy transformations?

Much of this chapter will be an exercise in logical deduction. The reasoning is subtle but important.

Work Done by the System

In mechanics, we used the term "work" to mean the work done *on* the system by an external force; this definition considers work and heat to be separate but equivalent ways of transferring energy to the system. However, it is useful in practical thermodynamics to turn things around and speak of the work done on the environment *by* the system.

Figure 19.1a shows a piston in a gas cylinder. The gas pressure pushes outward on the piston with force $\vec{F}_{gas}$. Some force in the environment, usually a force applied by a piston rod, pushes inward with force $\vec{F}_{ext}$. This external force is needed to keep the gas pressure from blowing the piston out of the cylinder. For any quasi-static process, where the system is essentially in equilibrium at all times, these two forces must balance: $\vec{F}_{gas} = -\vec{F}_{ext}$.

The work W done *on* the system is the work done by $\vec{F}_{ext}$ as the piston moves through a displacement Δx. You learned in Chapter 17 that W is the *negative* of the area under the pV curve of the process. But force $\vec{F}_{gas}$ also does work on the moving piston. Because $\vec{F}_{gas} = -\vec{F}_{ext}$, the work done by $\vec{F}_{gas}$, which we call the work W_s done *by* the system, has the same absolute value as the work W but the opposite sign. As Figure 19.1b shows, the work done *by* the system is

$$W_s = -W = \text{the area under the } pV \text{ curve} \qquad (19.1)$$

W_s is positive when energy is transferred *out* of the system by a mechanical interaction.

Work done by the environment and work done by the system are not mutually exclusive. Both $\vec{F}_{gas}$ and $\vec{F}_{ext}$ do work as the piston moves. Energy is transferred *into* the system as a gas is compressed, hence W is positive and W_s is negative. Energy is transferred *out* of the system as a gas expands, thus W is negative and W_s is positive.

NOTE ▶ When energy is transferred *into* a system, by compressing the gas, it is customary to say "the environment does work on the system." Similarly, when the gas pushes the piston out and transfers energy *out* of the system, we customarily say "the system does work on the environment." These expressions focus on the work that is positive. Neither is meant to imply that the "other" work isn't being done at the same time. ◀

(a)

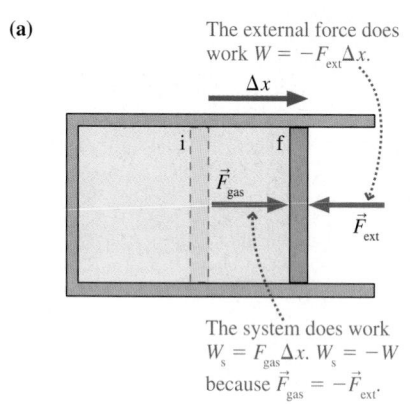

The external force does work $W = -F_{ext}\Delta x$.

The system does work $W_s = F_{gas}\Delta x$. $W_s = -W$ because $\vec{F}_{gas} = -\vec{F}_{ext}$.

(b)

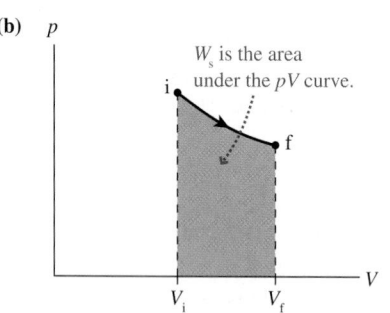

W_s is the area under the pV curve.

FIGURE 19.1 Forces $\vec{F}_{gas}$ and $\vec{F}_{ext}$ both do work as the piston moves.

The first law of thermodynamics $\Delta E_{th} = W + Q$ can be written in terms of W_s as

$$Q = W_s + \Delta E_{th} \qquad \text{(first law of thermodynamics)} \qquad (19.2)$$

It's easy to interpret this version of the first law. Because energy must be conserved, **any energy transferred into a system as heat is either used to do work or stored within the system as an increased thermal energy.**

Energy-Transfer Diagrams

Suppose you drop a hot rock into the ocean. Heat is transferred from the rock to the ocean until the rock and ocean are at the same temperature. Although the ocean warms up ever so slightly, ΔT_{ocean} is so small as to be completely insignificant. For all practical purposes, the ocean is infinite and unchangeable.

An **energy reservoir** is an object or a part of the environment so large that its temperature and thermal energy do not change when heat is transferred between the system and the reservoir. A reservoir at a higher temperature than the system is called a *hot reservoir*. A vigorously burning flame is a hot reservoir for small objects placed in the flame. A reservoir at a lower temperature than the system is called a *cold reservoir*. The ocean is a cold reservoir for the hot rock. We will use T_H and T_C to designate the temperatures of the hot and cold reservoirs.

Hot and cold reservoirs are idealizations, in the same category as frictionless surfaces and massless strings. No real object can maintain a perfectly constant temperature as heat is transferred in or out. Even so, an object can be accurately modeled as a reservoir if it is much larger than the system that thermally interacts with it.

Heat energy is transferred between a system and a reservoir if they have different temperatures. We will define

Q_H = amount of heat transferred to or from a hot reservoir.

Q_C = amount of heat transferred to or from a cold reservoir.

By definition, Q_H and Q_C are *positive* quantities. The direction of heat transfer, which determines the sign of Q in the first law, will always be clear as we deal with thermodynamic devices. For example, the heat transferred *from* the system to a cold reservoir is $Q = -Q_C$.

Figure 19.2a shows a heavy copper bar between a hot reservoir (at temperature T_H) and a cold reservoir (at temperature T_C). Heat Q_H is transferred from the hot reservoir into the copper and heat Q_C is transferred from the copper to the cold reservoir. Figure 19.2b is an **energy-transfer diagram** for this process. The hot reservoir is always drawn at the top, the cold reservoir at the bottom, and the system—the copper bar in this case—between them. The reservoirs and the system are connected by "pipes" that show the energy transfers. Figure 19.2b shows heat Q_H being transferred into the system and Q_C being transferred out.

These cooling towers are using evaporation to transfer heat energy to the atmosphere. The atmosphere is a cold reservoir.

(a)
Heat is transferred from hot to cold.

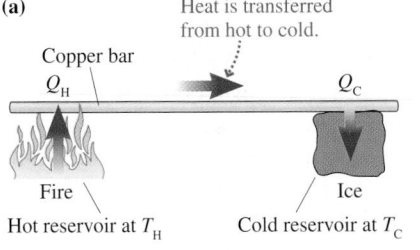

Copper bar
Q_H Q_C
Fire   Ice
Hot reservoir at T_H Cold reservoir at T_C

(b)
Heat energy is transferred from a hot reservoir to a cold reservoir. Energy conservation requires $Q_C = Q_H$.

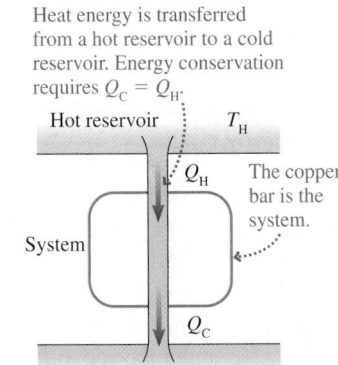

Hot reservoir T_H
Q_H
The copper bar is the system.
System
Q_C
Cold reservoir T_C

(c)
The second law forbids a process in which heat is spontaneously transferred from a colder object to a hotter object.

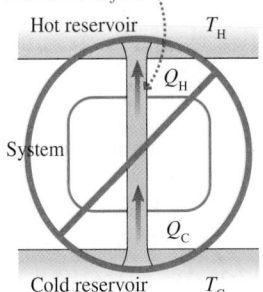

Hot reservoir T_H
Q_H
System
Q_C
Cold reservoir T_C

FIGURE 19.2 Energy-transfer diagrams.

The first law of thermodynamics $Q = W_s + \Delta E_{th}$ refers to the *system*. Q is the net heat to the system. In this case, because Q_C is the quantity of heat *leaving* the system, $Q = Q_H - Q_C$. The copper bar does no work, so $W_s = 0$. The bar warms up when first placed between the two reservoirs, but it soon comes to a steady state where its temperature no longer changes. Then $\Delta E_{th} = 0$. Thus the first law tells us that $Q = Q_H - Q_C = 0$, from which we conclude that

$$Q_C = Q_H \tag{19.3}$$

In other words, all of the heat transferred into the hot end of the rod is subsequently transferred out of the cold end. This isn't surprising. After all, we know that heat is transferred spontaneously from a hotter object to a colder object. Even so, there has to be some *means* by which the heat energy gets from the hotter object to the colder. The copper bar provides a route for the energy transfer, and $Q_C = Q_H$ is the statement that energy is conserved as it moves through the bar.

Contrast Figure 19.2b with Figure 19.2c. Figure 19.2c shows a system in which heat is being transferred from the cold reservoir to the hot reservoir. The first law of thermodynamics is not violated, because $Q_H = Q_C$, but the second law is. If there were such a system, it would allow the spontaneous (i.e., with no outside input or assistance) transfer of heat from a colder object to a hotter object. The process of Figure 19.2c is forbidden by the second law of thermodynamics.

Work into Heat and Heat into Work

Turning work into heat is easy. Take two rocks out of the ocean and rub them together vigorously until both are warmer. You're creating a mechanical interaction in which work increases the thermal energy of the rocks, or $W \rightarrow \Delta E_{th}$. Then toss both back into the ocean, where they return to their initial temperature as the excess thermal energy is transferred as heat from the slightly warmer rocks to the colder water ($\Delta E_{th} \rightarrow Q_C$). Figure 19.3 is the energy-transfer diagram for this process.

> **NOTE** ▶ Energy-transfer diagrams show the "work pipe" entering or leaving the system from the side. ◀

The conversion of work into heat is 100% efficient. That is, *all* of the energy supplied to the system as work W is transferred into the ocean as heat Q_C. This perfect transformation of work into heat can continue as long as there is motion. (It was this continual production of heat energy in the boring of cannons that Count Rumford recognized as being in conflict with the caloric theory.)

But the reverse—transforming heat into work—isn't as easy.

Figure 19.4 shows the special gas cylinder that we used in Figure 17.13. Heat energy is transferred from the flame into the gas, but the temperature remains constant because the heat energy from the flame is used to do the work of lifting the mass. $\Delta E_{th} = 0$ in an isothermal expansion, so the first law is

$$W_s = Q \tag{19.4}$$

The energy that is transferred into the gas as heat is transformed with 100% efficiency into work done by the gas as it lifts the mass. So why did we just say that transforming heat into work isn't as easy as transforming work into heat?

There's a difference. In Figure 19.3, where we transformed work into heat, the system *returned to its initial state*. We can repeat the process over and over, continuing to transform work into heat as long as there is motion. But Figure 19.4 is a one-time process. The gas does work once as it lifts the piston, but then the gas is no longer in its initial state. We cannot repeat the process. Extracting more and more work from the device of Figure 19.4 requires lifting the piston higher and higher until, ultimately, it reaches the end of the cylinder.

A one-time device for turning heat into work will eventually, when enough heat has been added, melt down or blow up! **To be practical, a device that transforms heat into work must return to its initial state at the end of the process**

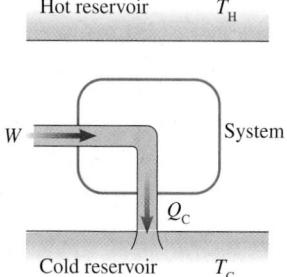

FIGURE 19.3 Work can be transformed into heat with 100% efficiency.

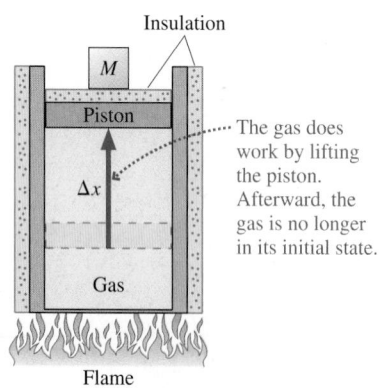

The gas does work by lifting the piston. Afterward, the gas is no longer in its initial state.

FIGURE 19.4 An isothermal process transforms heat into work, but only as a one-time event.

and be ready for continued use. You want your car engine to turn over and over as long as there is fuel.

Perhaps Figure 19.4 is just a bad idea for turning heat into work. Perhaps some other device can turn heat into work continuously. Interestingly, no one has ever invented a "perfect engine" that transforms heat into work with 100% efficiency *and returns to its initial state* so that it can continue to do work as long as there is fuel. Of course, that such a device has not been invented is not a proof that it can't be done. We'll provide a proof shortly, but for now we'll make the hypothesis that the process of Figure 19.5 is somehow forbidden.

Notice the asymmetry between Figures 19.3 and 19.5. The perfect transformation of work into heat is permitted, but the perfect transformation of heat into work is forbidden. This asymmetry parallels the asymmetry of the two processes in Figure 19.2. In fact, we'll soon see that the "perfect engine" of Figure 19.5 is forbidden for exactly the same reason: the second law of thermodynamics.

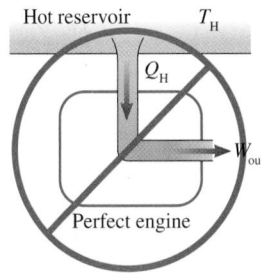

FIGURE 19.5 There are no perfect engines that turn heat into work with 100% efficiency.

19.2 Heat Engines and Refrigerators

We've alluded to the fact that your car engine operates in a *cycle* in which the temperature and pressure conditions inside the cylinders are repeated over and over. Another essential aspect of your car engine allows it to continue doing work. Your car engine has both a hot reservoir, the combustion of fuel, *and* a cold reservoir that consists of the radiator and the surrounding air. Heat is transferred *to* the engine when the spark plug ignites the fuel. Heat is transferred *out* of the engine to the cooling water and ultimately the air.

The steam generator at your local electric power plant works by boiling water to produce high-pressure steam that spins a turbine (which then spins a generator to produce electricity). The steam pressure and temperature drop as the steam goes through the turbine. The steam is then condensed to liquid water and pumped back to the boiler to start the process again. There are two crucial ideas here. First, the device works in a cycle, with the water returning to its initial conditions once a cycle. Second, heat is transferred to the water in the boiler, but heat is transferred *out* of the water in the condenser.

Car engines and steam generators are examples of what we call *heat engines*. A **heat engine** is any closed-cycle device that extracts heat Q_H from a hot reservoir, does useful work, and exhausts heat Q_C to a cold reservoir. A **closed-cycle device** is one that periodically *returns to its initial conditions,* repeating the same process over and over. That is, all state variables (pressure, temperature, thermal energy, and so on) return to their initial values once every cycle. Consequently, a heat engine can continue to do useful work for as long as it is attached to the reservoirs.

Figure 19.6 is the energy-transfer diagram of a heat engine. Unlike the forbidden "perfect engine" of Figure 19.5, a heat engine is connected both to a hot reservoir *and* to a cold reservoir. You can think of a heat engine as "siphoning off" some of the heat that moves from the hot reservoir to the cold reservoir and transforming that heat into work—some heat, but not all.

Because the temperature and thermal energy of a heat engine return to their initial values at the end of each cycle, an important characteristic of *any* heat engine is that

$$(\Delta E_{th})_{net} = 0 \quad \text{(any heat engine, over one full cycle)} \quad (19.5)$$

Consequently, the first law of thermodynamics *for a full cycle* of a heat engine is $(\Delta E_{th})_{net} = Q + W = Q - W_s = 0$.

Let's define W_{out} to be the useful work done *by* the heat engine *per cycle.* The net heat transfer per cycle is $Q_{net} = Q_H - Q_C$, hence the first law applied to a heat engine is

$$W_{out} = Q_{net} = Q_H - Q_C \quad \text{(work per cycle done by a heat engine)} \quad (19.6)$$

The steam turbine in a modern power plant is an enormous device. Expanding steam does work by spinning the turbine.

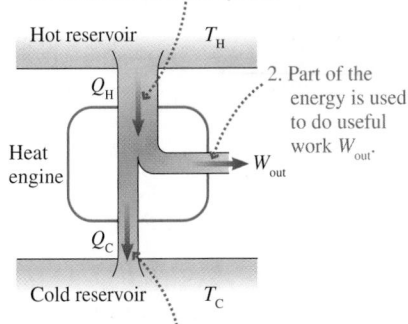

1. Heat energy Q_H is transferred from the hot reservoir to the system.

2. Part of the energy is used to do useful work W_{out}.

3. The remaining energy $Q_C = Q_H - W_{out}$ is exhausted to the cold reservoir as waste heat.

FIGURE 19.6 The energy-transfer diagram of a heat engine.

This is just energy conservation. The energy transferred into the engine (Q_H) and energy transferred out of the engine (Q_C and W_{out}) have to balance. The energy-transfer diagram of Figure 19.6 is a pictorial representation of Equation 19.6.

NOTE ▶ Equations 19.5 and 19.6 apply only to a *full cycle* of the heat engine. They are *not* valid for any of the individual processes that make up a cycle. ◀

For practical reasons, we would like an engine to do the maximum amount of work with the minimum amount of fuel. We can measure the performance of a heat engine in terms of its **thermal efficiency** η (lowercase Greek eta), defined as

$$\eta = \frac{W_{out}}{Q_H} = \frac{\text{what you get}}{\text{what you had to pay}} \qquad (19.7)$$

Using Equation 19.6 for W_{out}, we can also write the thermal efficiency as

$$\eta = 1 - \frac{Q_C}{Q_H} \qquad (19.8)$$

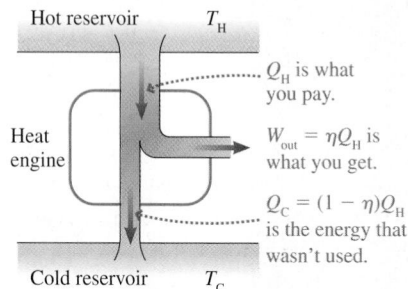

Q_H is what you pay.

$W_{out} = \eta Q_H$ is what you get.

$Q_C = (1 - \eta)Q_H$ is the energy that wasn't used.

FIGURE 19.7 η is the fraction of heat energy that is transformed into useful work.

Figure 19.7 illustrates the idea of thermal efficiency.

A *perfect* heat engine would have $\eta_{perfect} = 1$. That is, it would be 100% efficient at converting heat from the hot reservoir (the burning fuel) into work. You can see from Equation 19.8 that a perfect engine would have no exhaust ($Q_C = 0$) and would not need a cold reservoir. Figure 19.5 has already suggested that there are no perfect heat engines, that an engine with $\eta = 1$ is impossible. A heat engine *must* exhaust energy to a cold reservoir. This energy transfer to the cold reservoir is called **waste heat.** It is energy that was extracted from the hot reservoir but *not* transformed to useful work.

Practical heat engines, such as car engines and steam generators, have thermal efficiencies in the range $\eta \approx 0.1–0.4$. This is not large. Well over half the input heat energy is "wasted." Can a clever designer do better, or is this some kind of physical limitation?

STOP TO THINK 19.1   Rank in order, from largest to smallest, the work W_{out} performed by these four heat engines.

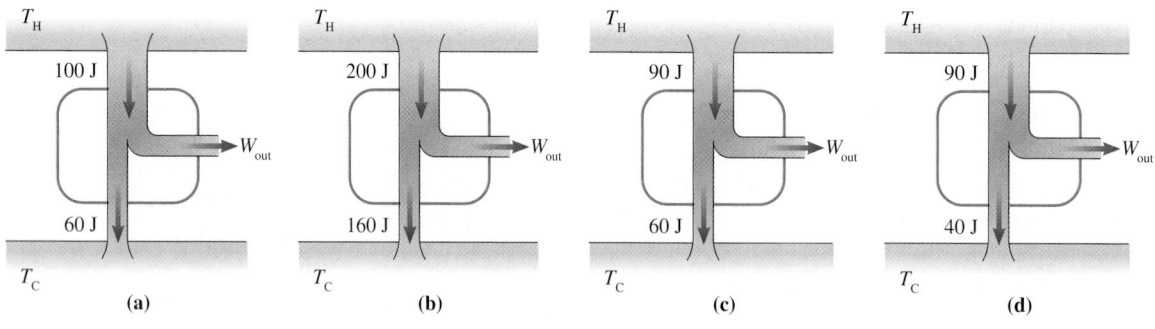

A Heat-Engine Example

This discussion has been somewhat abstract. To illustrate how these ideas actually work, Figure 19.8 shows a simple engine that converts heat into the work of lifting mass M. None of the state variables changes value as the mass is removed in step (c), but the piston must be locked in place to prevent the gas pressure from blowing it out once the mass is gone. The gas does work on the environment while lift-

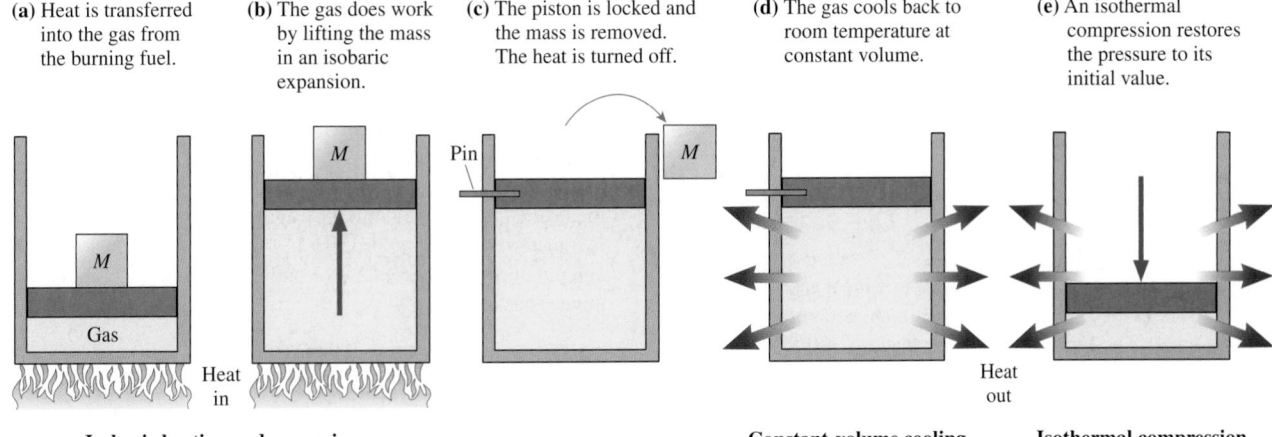

(a) Heat is transferred into the gas from the burning fuel.

(b) The gas does work by lifting the mass in an isobaric expansion.

(c) The piston is locked and the mass is removed. The heat is turned off.

(d) The gas cools back to room temperature at constant volume.

(e) An isothermal compression restores the pressure to its initial value.

Isobaric heating and expansion   Constant-volume cooling   Isothermal compression

FIGURE 19.8 A simple heat engine transforms heat into work.

ing the mass during step (b) (W_s is positive, W is negative). The environment does work on the gas during the compression of step (e) (W is positive, W_s is negative).

The net effect of this multistep process is to convert some of the fuel's energy into the useful work of lifting the mass. There has been no net change in the gas, which has returned to its initial pressure, volume, and temperature at the end of step (e). We can start the whole process over again and continue lifting masses (doing work) as long as we have fuel.

Figure 19.9 shows the heat-engine process on a pV diagram. It is a *closed cycle* because the gas returns to its initial conditions. No work is done during the isochoric process, so the *net* work done by the heat engine over the course of one full cycle is $W_{out} = (W_s)_{1\rightarrow2} + (W_s)_{3\rightarrow1}$.

Notice that the cyclical process of Figure 19.9 involves two *cooling processes* in which heat is transferred *from* the gas to the environment. Heat energy is transferred from hotter objects to colder objects, so the system *must* be connected to a cold reservoir with $T_C < T_{gas}$ during these two processes. A key to understanding heat engines is that they require both a heat source (burning fuel) *and* a heat sink (cooling water, the air, or something at a lower temperature than the system).

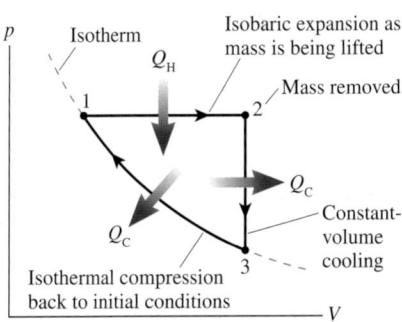

FIGURE 19.9 The closed-cycle pV diagram for the heat engine of Figure 19.8.

EXAMPLE 19.1 Analyzing a heat engine I

A heat engine follows the ideal-gas process shown in Figure 19.10. Analyze this engine to determine (a) the net work done per cycle, (b) the engine's thermal efficiency, and (c) the

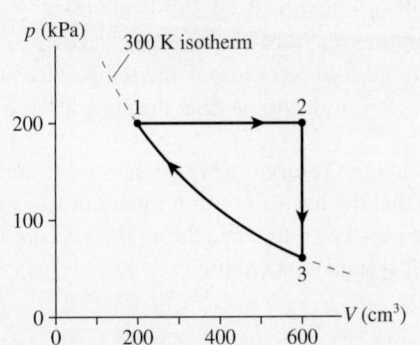

FIGURE 19.10 The heat engine of Example 19.1.

engine's power output if it runs at 600 rpm. Assume the gas is monatomic.

MODEL The gas follows a closed cycle, starting and ending in state 1. The cycle consists of three distinct processes, each of which was studied in Chapters 16 and 17. For each of the three we need to determine the work done and the heat transferred.

SOLVE To begin, we can use the initial conditions at state 1 and the ideal-gas law to determine the number of moles of gas:

$$n = \frac{p_1 V_1}{RT_1} = \frac{(200 \times 10^3 \text{ Pa})(2.0 \times 10^{-4} \text{ m}^3)}{(8.31 \text{ J/mol K})(300 \text{ K})} = 0.01604 \text{ mol}$$

Process $1 \rightarrow 2$: The work done *on* the gas in the isobaric expansion is

$$W_{12} = -p\Delta V = -(200 \times 10^3 \text{ Pa})(6.0 - 2.0) \times 10^{-4} \text{ m}^3$$
$$= -80.0 \text{ J}$$

Thus the work done *by* the gas is $(W_s)_{12} = +80.0$ J. We can use the ideal-gas law at constant pressure to find $T_2 = (V_2/V_1)T_1 = 3T_1 = 900$ K. The heat transfer during a constant-pressure process is

$$Q_{12} = nC_P\Delta T$$
$$= (0.01604 \text{ mol})(20.8 \text{ J/mol K})(900 \text{ K} - 300 \text{ K})$$
$$= 200.0 \text{ J}$$

where we used $C_P = \frac{5}{2}R$ for a monatomic ideal gas.

Process $2 \rightarrow 3$: No work is done in an isochoric process, so $(W_s)_{23} = 0$. The temperature drops back to 300 K, so the heat transfer is

$$Q_{23} = nC_V\Delta T$$
$$= (0.01604 \text{ mol})(12.5 \text{ J/mol K})(300 \text{ K} - 900 \text{ K})$$
$$= -120.0 \text{ J}$$

where we used $C_V = \frac{3}{2}R$.

Process $3 \rightarrow 4$: The gas returns to its initial state with volume V_1. The work done *on* the gas during an isothermal process is

$$W_{31} = -nRT \ln\left(\frac{V_1}{V_3}\right)$$
$$= -(0.01604 \text{ mol})(8.31 \text{ J/mol K})(300 \text{ K})\ln\left(\frac{1}{3}\right)$$
$$= 43.9 \text{ J}$$

Thus $(W_s)_{31} = -43.9$ J. W_s is negative because the environment does work on the gas to compress it. An isothermal process has $\Delta E_{th} = 0$ and hence, from the first law,

$$Q_{31} = -W_{31} = -43.9 \text{ J}$$

Q is negative because the gas must be cooled as it is compressed to keep the temperature constant.

a. The *net* work done by the engine during one cycle is

$$W_{out} = (W_s)_{12} + (W_s)_{23} + (W_s)_{31} = 36.1 \text{ J}$$

As a consistency check, notice that the net heat transfer is

$$Q_{net} = Q_{12} + Q_{23} + Q_{31} = 36.1 \text{ J}$$

Equation 19.6 told us that a heat engine *must* have $W_{out} = Q_{net}$, and we see that it does.

b. The efficiency depends not on the net heat transfer but on the heat Q_H transferred into the engine from the flame. Heat is transferred in during process $1 \rightarrow 2$, where Q is positive, and out during processes $2 \rightarrow 3$ and $3 \rightarrow 1$, where Q is negative. Thus

$$Q_H = Q_{12} = 200.0 \text{ J}$$
$$Q_C = |Q_{23}| + |Q_{31}| = 163.9 \text{ J}$$

Notice that $Q_H - Q_C = 36.1$ J $= W_{out}$. In this heat engine, 200.0 J of heat from the hot reservoir does 36.1 J of useful work. Thus the thermal efficiency is

$$\eta = \frac{W_{out}}{Q_H} = \frac{36.1 \text{ J}}{200.0 \text{ J}} = 0.18 \text{ or } 18\%$$

This heat engine is far from being a perfect engine!

c. An engine running at 600 rpm goes through 10 cycles per second. The power output is the work done *per second*:

$$P_{out} = (\text{work per cycle}) \times (\text{cycles per second})$$
$$= 361 \text{ J/s} = 361 \text{ W}$$

ASSESS Although we didn't need Q_{net}, verifying that $Q_{net} = W_{out}$ was a check of self-consistency. Heat-engine analysis requires many calculations and offers many opportunities to get signs wrong. However, there are a sufficient number of self-consistency checks so that you can almost always spot calculational errors *if you check for them*. Also notice that we carried an extra significant figure in some of the intermediate calculations, such as $Q_H = 200.0$ J. Heat-engine calculations often require subtracting one number from another number of similar size; hence keeping an extra significant figure during intermediate calculations helps to minimize round-off error.

Let's think about this example a bit more before going on. We've said that a heat engine operates between a hot reservoir and a cold reservoir. Figure 19.10 doesn't explicitly show the reservoirs. Nonetheless, we know that heat is transferred from a hotter object to a colder object. Heat Q_H is transferred into the system during process $1 \rightarrow 2$ as the gas warms from 300 K to 900 K. For this to be true, the hot-reservoir temperature T_H must be ≥ 900 K. Likewise, heat Q_C is transferred from the system to the cold reservoir as the temperature drops from 900 K to 300 K in process $2 \rightarrow 3$. For this to be true, the cold-reservoir temperature T_C must be ≤ 300 K.

So we really don't know what the reservoirs are or their exact temperatures, but we can say with certainty that the hot-reservoir temperature T_H must exceed the highest temperature reached by the system and the cold-reservoir temperature T_C must be less than the coldest system temperature.

Refrigerators

Your house or apartment has a refrigerator. Very likely it has an air conditioner. The purpose of these devices is to make air that is cooler than its environment even colder. The first does so by blowing hot air out into a warm room, the second by blowing it out to the hot outdoors. You've probably felt the hot air exhausted by an air conditioner compressor or coming out from beneath the refrigerator.

At first glance, a refrigerator or air conditioner may seem to violate the second law of thermodynamics. After all, doesn't the second law forbid heat from being transferred from a colder object to a hotter object? Not quite: The second law says that heat is not *spontaneously* transferred from a colder to a hotter object. A refrigerator or air conditioner requires electric power to operate. They do cause heat to be transferred from cold to hot, but the transfer is "assisted" rather than spontaneous.

We'll define a **refrigerator** as any closed-cycle device that uses external work W_{in} to remove heat Q_C from a cold reservoir and exhaust heat Q_H to a hot reservoir. Figure 19.11 is the energy-transfer diagram of a refrigerator. The cold reservoir is the air inside the refrigerator or the air inside your house on a summer day. To keep the air cold, in the face of inevitable "heat leaks," the refrigerator or air conditioner compressor continuously removes heat from the cold reservoir and exhausts heat into the room or outdoors. You can think of a refrigerator as "pumping heat uphill," much as a water pump lifts water uphill.

Because a refrigerator, like a heat engine, is a cyclical device, $\Delta E_{th} = 0$. Conservation of energy requires

$$Q_H = Q_C + W_{in} \qquad (19.9)$$

To move energy from a colder to a hotter reservoir, a refrigerator must exhaust *more* heat to the outside than it removes from the inside. This has significant implications as to whether or not you can cool a room by leaving the refrigerator door open.

The thermal efficiency of a heat engine was defined as "what you get (useful work W_{out})" versus "what you pay for (fuel to supply Q_H)." By analogy, we define the **coefficient of performance** K of a refrigerator to be

$$K = \frac{Q_C}{W_{in}} = \frac{\text{what you get}}{\text{what you had to pay}} \qquad (19.10)$$

What you get, in this case, is the removal of heat from the cold reservoir. But you have to pay the electric company for the work needed to run the refrigerator. A better refrigerator will require less work to remove a given amount of heat, thus having a larger coefficient of performance.

A perfect refrigerator would require no work at all ($W_{in} = 0$) and would have $K_{perfect} = \infty$. But if Figure 19.11 had no work input, it would look just like Figure 19.2b. That device was forbidden by the second law of thermodynamics because, with no work input, heat would be able to move *spontaneously* from cold to hot.

We noted in Chapter 18 that the second law of thermodynamics can be stated several different but equivalent ways. We can now give a third statement:

> **Second law, informal statement #3** There are no perfect refrigerators with $K = \infty$.

Any real refrigerator or air conditioner *must* use work to move energy from the cold reservoir to the hot reservoir, hence $K < \infty$.

No Perfect Heat Engines

We hypothesized above that there are no perfect heat engines—that is, no heat engines like the one shown in Figure 19.5 with $Q_C = 0$ and $\eta = 1$. Now we're ready to prove this hypothesis. Figure 19.12 on the next page shows a hot reservoir at temperature T_H and a cold reservoir at temperature T_C. An ordinary refrigerator, one that obeys all the laws of physics, is operating between these two reservoirs.

This air conditioner transfers heat energy *from* the cool indoors *to* the hot exterior.

The amount of heat exhausted to the hot reservoir is larger than the amount of heat extracted from the cold reservoir.

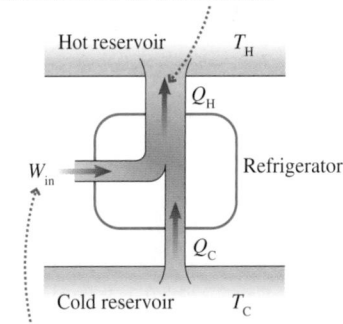

External work is used to remove heat from a cold reservoir and exhaust heat to a hot reservoir.

FIGURE 19.11 The energy-transfer diagram of a refrigerator.

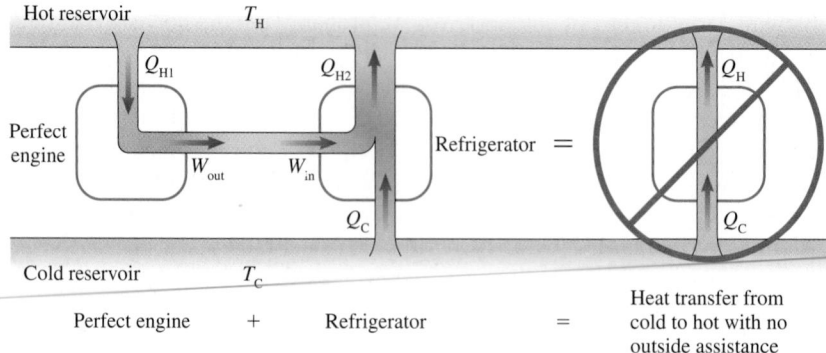

FIGURE 19.12 A perfect engine driving an ordinary refrigerator would be able to violate the second law of thermodynamics.

Suppose we had a perfect heat engine, one that takes in heat Q_H from the high-temperature reservoir and transforms that energy entirely into work W_{out}. If we had such a heat engine, we could use its output to provide the work input to the refrigerator. The two devices combined have no connection to the external world. That is, there's no net input or net output of work.

If we built a box around the heat engine and refrigerator, so that you couldn't see what was inside, the only thing you would observe is heat being transferred *with no outside assistance* from the cold reservoir to the hot reservoir. But a spontaneous or unassisted transfer of heat from a colder to a hotter object is exactly what the second law of thermodynamics forbids. Consequently, our assumption of a perfect heat engine must be wrong. Hence another statement of the second law of thermodynamics is:

> **Second law, informal statement #4** There are no perfect heat engines with $\eta = 1$.

Any real heat engine *must* exhaust waste heat Q_C to a cold reservoir.

Unanswered Questions

We noted that this chapter would be an exercise in logical deduction. By using only energy conservation and the fact that heat is not spontaneously transferred from cold to hot, we've been able to deduce that

- Heat engines and refrigerators exist.
- They must use a closed-cycle process, with $(\Delta E_{th})_{net} = 0$.
- There are no perfect heat engines. A heat engine *must* exhaust heat to a cold reservoir.
- There are no perfect refrigerators. A refrigerator *must* use external work.

This is a good start, but it leaves many unanswered questions. For example,

- With good design and good materials, can we make a heat engine whose thermal efficiency η approaches 1? Or is there an upper limit η_{max} that cannot be exceeded?
- If η has a maximum value, what is it?
- Likewise, is there an upper limit K_{max} for the coefficient of performance of a refrigerator? If so, what is it?

Section 19.6 will show that there is, indeed, an upper limit to η that no heat engine can exceed and an upper limit to K that no refrigerator can exceed. We'll be able to establish an actual value for η_{max} and will find that, for many practical engines, η_{max} is distressingly low.

STOP TO THINK 19.2   It's a really hot day and your air conditioner is broken. Your roommate says, "Let's open the refrigerator door and cool this place off." Will this work?

a. Yes.

b. No.

c. It might, but it will depend on how hot the room is.

19.3 Ideal-Gas Heat Engines

We will focus on heat engines that use a gas as the *working substance*. The gasoline engine in your car is an engine that alternately compresses and expands a gaseous fuel-air mixture. Diesel engines, jet engines, and gas turbines also fall into this category. The details of engines such as steam generators that rely upon phase changes will be deferred to more advanced courses.

A gas heat engine can be represented by a closed-cycle trajectory in the pV diagram, such as the one shown in Figure 19.13a. This observation leads to an interesting and important geometric interpretation of the work done by the system during one full cycle. You learned in Section 19.1 that the work done *by* the system is the area under the curve of a pV trajectory. As Figure 19.13b shows, the work done during a full cycle is the work W_{expand} done by the system as it expands from V_{min} to V_{max} plus the work $W_{compress}$ done by the system as it is compressed from V_{max} back to V_{min}. W_{expand} is positive and $W_{compress}$ is negative, thus

$$W_{out} = W_{expand} - |W_{compress}| = \text{area } inside \text{ the closed curve} \quad (19.11)$$

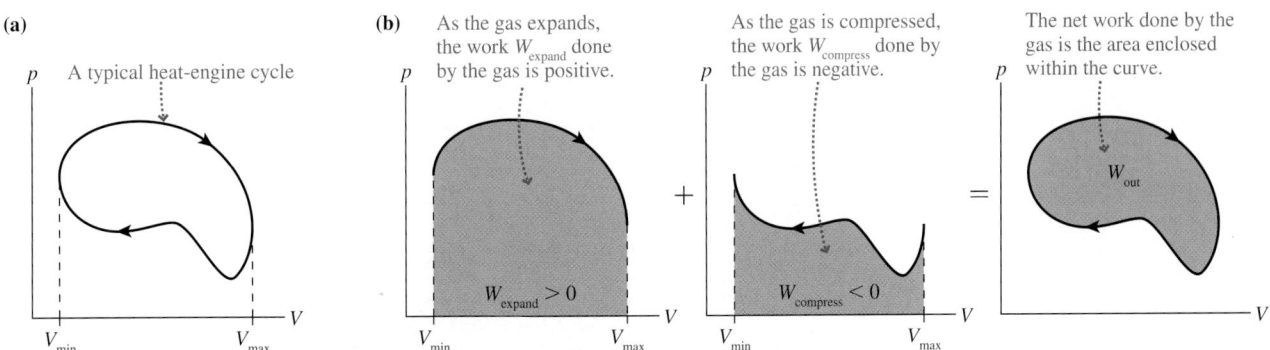

(a) A typical heat-engine cycle

(b) As the gas expands, the work W_{expand} done by the gas is positive. $W_{expand} > 0$

As the gas is compressed, the work $W_{compress}$ done by the gas is negative. $W_{compress} < 0$

The net work done by the gas is the area enclosed within the curve. W_{out}

FIGURE 19.13 The work W_{out} done by the system during one full cycle is the area enclosed within the curve.

You can see that **the net work done by a gas heat engine during one full cycle is the area enclosed by the pV curve for the cycle.** A thermodynamic cycle with a larger enclosed area does more work than one with a smaller enclosed area. Notice that the gas must go around the pV trajectory in a *clockwise* direction for W_{out} to be positive. We'll see later that a refrigerator uses a counterclockwise (ccw) cycle.

Ideal-Gas Summary

We've learned a lot about ideal gases in the last three chapters. It will be helpful to summarize the major relationships we need in order to analyze heat engines. All gas processes obey the ideal-gas law $pV = nRT$ and the first law of thermodynamics $\Delta E_{th} = Q - W_s$. Results for specific gas processes are shown in Table 19.1 on the next page. This table shows W_s, the work done *by* the system, so the signs are opposite those in Chapter 17.

TABLE 19.1 Summary of ideal-gas processes

Process	Gas law	Work W_s	Heat Q	Thermal energy
Isochoric	$p_i/T_i = p_f/T_f$	0	$nC_V\Delta T$	$\Delta E_{th} = Q$
Isobaric	$V_i/T_i = V_f/T_f$	$p\Delta V$	$nC_P\Delta T$	$\Delta E_{th} = Q - W_s$
Isothermal	$p_iV_i = p_fV_f$	$nRT\ln(V_f/V_i)$ $pV\ln(V_f/V_i)$	$Q = W_s$	$\Delta E_{th} = 0$
Adiabatic	$p_iV_i^\gamma = p_fV_f^\gamma$ $T_iV_i^{\gamma-1} = T_fV_f^{\gamma-1}$	$(p_fV_f - p_iV_i)/(1-\gamma)$ $-nC_V\Delta T$	0	$\Delta E_{th} = -W_s$
Any	$p_iV_i/T_i = p_fV_f/T_f$	area under curve		$\Delta E_{th} = nC_V\Delta T$

TABLE 19.2 Properties of monatomic and diatomic gases

	Monatomic	Diatomic
E_{th}	$\frac{3}{2}nRT$	$\frac{5}{2}nRT$
C_V	$\frac{3}{2}R$	$\frac{5}{2}R$
C_P	$\frac{5}{2}R$	$\frac{7}{2}R$
γ	$\frac{5}{3} = 1.67$	$\frac{7}{5} = 1.40$

There is one entry in this table that you haven't seen before. The expression

$$W_s = \frac{p_fV_f - p_iV_i}{1 - \gamma} \quad \text{(work in an adiabatic process)} \quad (19.12)$$

for the work done in an adiabatic process follows from writing $W_s = -\Delta E_{th} = -nC_V\Delta T$, which you learned in Chapter 17, then using $\Delta T = \Delta(pV)/R$ and the definition of γ. The proof will be left for a homework problem.

Finally, we learned in Chapter 18 that the thermal energy of an ideal gas depends only on its temperature. Table 19.2 summarizes the thermal energy, molar specific heats, and specific heat ratio $\gamma = C_P/C_V$ for monatomic and diatomic gases.

A Strategy for Heat-Engine Problems

8.12, 8.13 Activ Physics ONLINE

The engine of Example 19.1 was not a realistic heat engine, but it did illustrate the kinds of reasoning and computations involved in the analysis of a heat engine. A basic strategy for analyzing a heat engine is as follows.

(MP) **PROBLEM-SOLVING STRATEGY 19.1** **Heat-engine problems**

MODEL Identify each process in the cycle.

VISUALIZE Draw the pV diagram of the cycle.

SOLVE There are several steps in the mathematical analysis.

- Use the ideal-gas law to complete your knowledge of n, p, V, and T at one point in the cycle.
- Use the ideal-gas law and equations for specific gas processes to determine p, V, and T at the beginning and end of each process.
- Calculate Q, W_s, and ΔE_{th} for each process.
- Find W_{out} by adding W_s for each process in the cycle. If the geometry is simple you can confirm this value by finding the area enclosed within the pV curve.
- Add just the *positive* values of Q to find Q_H.
- Verify that $(\Delta E_{th})_{net} = 0$. This is a self-consistency check to verify that you haven't made any mistakes.
- Calculate the thermal efficiency η and any other quantities you need to complete the solution.

ASSESS Is $(\Delta E_{th})_{net} = 0$? Do all the signs of W_s and Q make sense? Does η have a reasonable value? Have you answered the question?

EXAMPLE 19.2 **Analyzing a heat engine II**

A heat engine with a diatomic gas as the working substance uses the closed cycle shown in Figure 19.14. How much work does this engine do per cycle and what is its thermal efficiency?

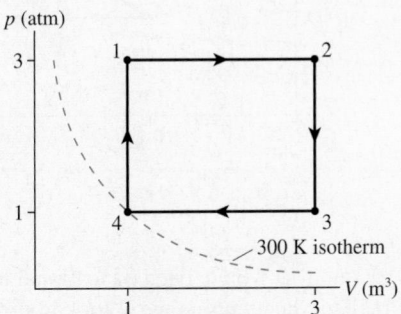

FIGURE 19.14 The pV diagram for the heat engine of Example 19.2.

MODEL Processes $1 \rightarrow 2$ and $3 \rightarrow 4$ are isobaric. Processes $2 \rightarrow 3$ and $4 \rightarrow 1$ are isochoric.

VISUALIZE The pV diagram has been drawn for us in the problem statement.

SOLVE We know the pressure, volume, and temperature at state 4. The number of moles of gas in the heat engine is

$$n = \frac{p_4 V_4}{RT_4} = \frac{(101,300 \text{ Pa})(1.0 \text{ m}^3)}{(8.31 \text{ J/mol K})(300 \text{ K})} = 40.6 \text{ mol}$$

$p/T =$ constant during an isochoric process and $V/T =$ constant during an isobaric process. These allow us to find that $T_1 = T_3 = 900$ K and $T_2 = 2700$ K. This completes our knowledge of the state variables at all four corners of the diagram.

Process $1 \rightarrow 2$ is an isobaric expansion, so

$$(W_s)_{12} = p\Delta V = (3.0 \times 101,300 \text{ Pa})(2.0 \text{ m}^3) = 6.08 \times 10^5 \text{ J}$$

where we converted the pressure to pascals. The heat transfer during an isobaric expansion is

$$Q_{12} = nC_P\Delta T = (40.6 \text{ mol})(29.1 \text{ J/mol K})(1800 \text{ K})$$
$$= 21.27 \times 10^5 \text{ J}$$

where we used $C_P = \frac{7}{2}R$ for a diatomic gas. Then, using the first law,

$$\Delta E_{12} = Q_{12} - (W_s)_{12} = 15.19 \times 10^5 \text{ J}$$

Process $2 \rightarrow 3$ is an isochoric process, so $(W_s)_{23} = 0$ and

$$\Delta E_{23} = Q_{23} = nC_V\Delta T = -15.19 \times 10^5 \text{ J}$$

Notice that ΔT is *negative*.

Process $3 \rightarrow 4$ is an isobaric compression. Now ΔV is negative, so

$$(W_s)_{34} = p\Delta V = -2.03 \times 10^5 \text{ J}$$

and

$$Q_{34} = nC_P\Delta T = -7.09 \times 10^5 \text{ J}$$

Then $\Delta E_{th} = Q_{34} - (W_s)_{34} = -5.06 \times 10^5$ J.

Process $4 \rightarrow 1$ is another constant-volume process, so again $(W_s)_{41} = 0$ and

$$\Delta E_{23} = Q_{23} = nC_V\Delta T = 5.06 \times 10^5 \text{ J}$$

The results of all four processes are shown in Table 19.3. The net results for W_{out}, Q_{net}, and $(\Delta E_{th})_{net}$ are found by summing the columns. As expected, $W_{out} = Q_{net}$ and $(\Delta E_{th})_{net} = 0$.

TABLE 19.3 Energy transfers in Example 19.2. All energies $\times 10^5$ J

Process	W_s	Q	ΔE_{th}
$1 \rightarrow 2$	6.08	21.27	15.19
$2 \rightarrow 3$	0	−15.19	−15.19
$3 \rightarrow 4$	−2.03	−7.09	−5.06
$4 \rightarrow 1$	0	5.06	5.06
Net	4.05	4.05	0

The work done during one cycle is $W_{out} = 4.05 \times 10^5$ J. Heat enters the system from the hot reservoir during processes $1 \rightarrow 2$ and $4 \rightarrow 1$, where Q is positive. Summing these gives $Q_H = 26.33 \times 10^5$ J. Thus the thermal efficiency of this engine is

$$\eta = \frac{W_{out}}{Q_H} = \frac{4.05 \times 10^5 \text{ J}}{26.33 \times 10^5 \text{ J}} = 0.15 = 15\%$$

ASSESS The verification that $W_{out} = Q_{net}$ and $(\Delta E_{th})_{net} = 0$ gives us great confidence that we didn't make any calculational errors. This engine may not seem very efficient, but η is quite typical of many real engines.

We noted in Example 19.1 that a heat engine's hot reservoir temperature T_H must exceed the highest temperature reached by the system and the cold reservoir temperature T_C must be less than the coldest system temperature. Although we don't know what the reservoirs are in Example 19.2, we can be sure that $T_H > 2700$ K and $T_C < 300$ K.

What is the thermal efficiency of this heat engine?

a. 0.10
b. 0.50
c. 0.25
d. 4
e. Can't tell without
   knowing Q_C.

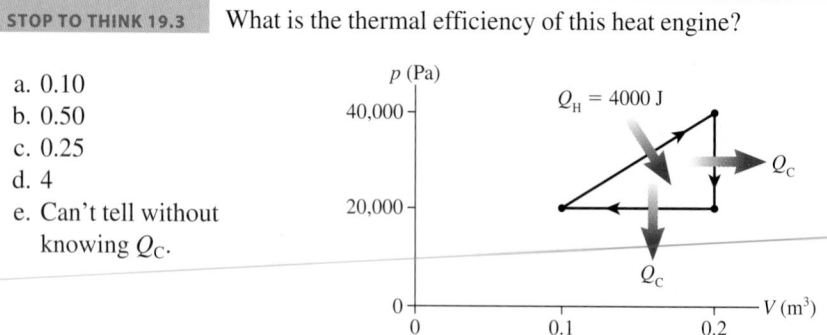

The Brayton Cycle

The heat engines of Examples 19.1 and 19.2 have been educational but not realistic. As an example of a more realistic heat engine we'll look at the thermodynamic cycle known as the *Brayton cycle*. It is a reasonable model of a *gas turbine engine*. Gas turbines are used for electric power generation and as the basis for jet engines in aircraft and rockets. The *Otto cycle*, which describes the gasoline internal combustion engine, and the *diesel cycle*, which, not surprisingly, describes the diesel engine, will be the subject of homework problems.

Figure 19.15a is a schematic look at a gas turbine engine and Figure 19.15b is the corresponding pV diagram. To begin the Brayton cycle, air at an initial pressure p_1 is rapidly compressed in a *compressor*. This is an *adiabatic process*, with $Q = 0$, because there is no time for heat to be exchanged with the surroundings. Recall that an adiabatic compression raises the temperature of a gas by doing work on it, not by heating it, so the air leaving the compressor is very hot.

The hot gas flows into a combustion chamber. Fuel is continuously admitted to the combustion chamber where it mixes with the hot gas and is ignited, transferring heat to the gas at constant pressure and raising the gas temperature yet further. The high-pressure gas then expands, spinning a turbine that does some form of useful work. This is an adiabatic expansion, with $Q = 0$, that drops the temperature and pressure of the gas. Maximum work is produced with the maximum pressure drop, so the pressure at the end of the expansion through the turbine is back to p_1. However, the gas is still quite hot. The gas completes the cycle by flowing through a device called a **heat exchanger** that transfers heat energy to a cooling fluid. Large power plants are often sited on rivers or oceans in order to use the water for the cooling fluid in the heat exchanger.

This thermodynamic cycle, called a Brayton cycle, has two adiabatic processes—the compression and the expansion through the turbine—plus a constant-pressure heating and a constant-pressure cooling. There's no heat transfer during the adiabatic processes. The hot reservoir temperature must be $T_H \geq T_3$ in order for heat to be transferred into the gas during process $2 \rightarrow 3$. Similarly, the heat exchanger will remove heat from the gas only if $T_C \leq T_1$.

The thermal efficiency of any heat engine is

$$\eta = \frac{W_{out}}{Q_H} = 1 - \frac{Q_C}{Q_H}$$

Heat is transferred into the gas only during process $2 \rightarrow 3$. This is an isobaric process, so $Q_H = nC_P\Delta T = nC_P(T_3 - T_2)$. Similarly, heat is transferred out only during the isobaric process $4 \rightarrow 1$.

We have to be careful with signs. Q_{41} is negative, because the temperature decreases, but Q_C was defined as the *amount* of heat exchanged with the cold reservoir, a positive quantity. Thus

$$Q_C = |Q_{41}| = |nC_P(T_1 - T_4)| = nC_P(T_4 - T_1) \tag{19.13}$$

A jet engine uses a modified Brayton cycle.

(a)

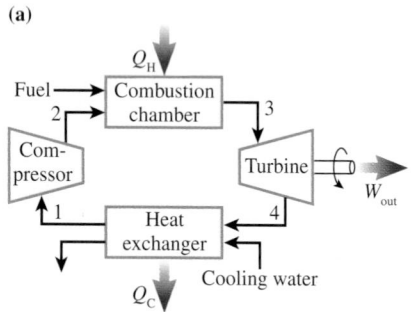

(b)

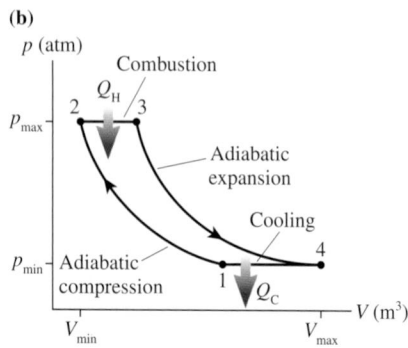

FIGURE 19.15 A gas turbine engine follows a Brayton cycle.

With these expressions for Q_H and Q_C, the thermal efficiency is

$$\eta_B = 1 - \frac{T_4 - T_1}{T_3 - T_2} \qquad (19.14)$$

The subscript B indicates that this result applies to the Brayton cycle; it is not a general result. The efficiency depends on the temperatures, not the pressures or volumes. However, this expression isn't useful unless we compute all four temperatures. We can cast Equation 19.14 into a more useful form.

You learned in Chapter 17 that $pV^\gamma = $ constant during an adiabatic process, where $\gamma = C_P/C_V$ is the specific heat ratio. If we use $V = nRT/p$ from the ideal gas law, $V^\gamma = (nR)^\gamma T^\gamma p^{-\gamma}$. $(nR)^\gamma$ is a constant, so we can write $pV^\gamma = $ constant as

$$p^{(1-\gamma)}T^\gamma = \text{constant} \qquad (19.15)$$

Equation 19.15 is a pressure-temperature relationship for an adiabatic process. Because $(T^\gamma)^{1/\gamma} = T$, we can simplify Equation 19.15 by raising both sides to the power $1/\gamma$. Doing so gives

$$p^{(1-\gamma)/\gamma}T = \text{constant} \qquad (19.16)$$

during an adiabatic process.

Process $1 \rightarrow 2$ is an adiabatic process, hence

$$p_1^{(1-\gamma)/\gamma}T_1 = p_2^{(1-\gamma)/\gamma}T_2 \qquad (19.17)$$

Isolating T_1 gives

$$T_1 = \frac{p_2^{(1-\gamma)/\gamma}}{p_1^{(1-\gamma)/\gamma}}T_2 = \left(\frac{p_2}{p_1}\right)^{(1-\gamma)/\gamma}T_2 = \left(\frac{p_{max}}{p_{min}}\right)^{(1-\gamma)/\gamma}T_2 \qquad (19.18)$$

If we define the **pressure ratio** r_p as $r_p = p_{max}/p_{min}$, then T_1 and T_2 are related by

$$T_1 = r_p^{(1-\gamma)/\gamma}T_2 \qquad (19.19)$$

The algebra of getting to Equation 19.19 was a bit tricky, but the final result is fairly simple.

Similarly, process $3 \rightarrow 4$ is an adiabatic process. The exact same reasoning leads to

$$T_4 = r_p^{(1-\gamma)/\gamma}T_3 \qquad (19.20)$$

If we substitute these expressions for T_1 and T_4 into Equation 19.14, the efficiency is

$$\eta_B = 1 - \frac{T_4 - T_1}{T_3 - T_2} = 1 - \frac{r_p^{(1-\gamma)/\gamma}T_3 - r_p^{(1-\gamma)/\gamma}T_2}{T_3 - T_2} = 1 - \frac{r_p^{(1-\gamma)/\gamma}(T_3 - T_2)}{T_3 - T_2}$$

$$= 1 - r_p^{(1-\gamma)/\gamma}$$

Remarkably, all the temperatures cancel and we're left with an expression that depends only on the pressure ratio. Noting that $(1 - \gamma)$ is negative, it's useful to make one final change and write

$$\eta_B = 1 - \frac{1}{r_p^{(\gamma-1)/\gamma}} \qquad (19.21)$$

Figure 19.16 is a graph of the efficiency of the Brayton cycle as a function of the pressure ratio, assuming $\gamma = 1.40$ for a diatomic gas such as air. You can see that η_B first grows quickly as the pressure ratio is increased, reaching $\approx 50\%$ at $r_p = 10$, then levels off. The pressure ratio is determined by the design of the compressor. A compressor with tighter-fitting parts and better seals will have a higher pressure ratio. However, any increase in efficiency beyond $\approx 50\%$ has to be weighed against the higher cost of a better compressor.

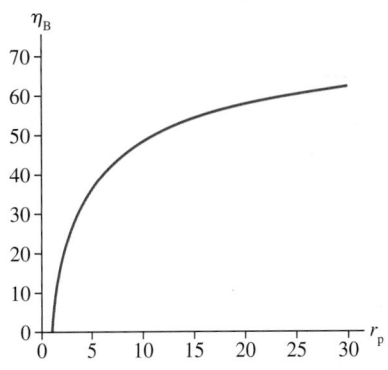

FIGURE 19.16 The efficiency of a Brayton cycle as a function of the pressure ratio r_p.

In Example 19.2 we found the thermal efficiency $\eta = W_{out}/Q_H$ by explicitly computing W_{out} and Q_H. Here, by contrast, we've determined the thermal efficiency of the Brayton cycle by using the relationship between the initial and final temperatures during an adiabatic process. The price we pay for this simplified analysis is that we didn't find an expression for the work done by a heat engine following the Brayton cycle. To calculate the work, which you can do as a homework problem, there's no avoiding the step-by-step analysis of the problem-solving strategy.

19.4 Ideal-Gas Refrigerators

Suppose that we were to operate a Brayton heat engine backward, going *ccw* in the *pV* diagram. Figure 19.17a shows a device for doing this. Figure 19.17b is its *pV* diagram and Figure 19.17c is the energy-transfer diagram. Starting from point 4, the gas is adiabatically compressed to increase its temperature and pressure. It then flows through a high-temperature heat exchanger where the gas *cools* at constant pressure from temperature T_3 to T_2. The gas then expands adiabatically, leaving it significantly colder at T_1 than it started at T_4. It completes the cycle by flowing through a low-temperature heat exchanger, where it *warms* back to its starting temperature.

Suppose that the low-temperature heat exchanger is a closed container of air surrounding a pipe through which the engine's cold gas is flowing. The heat-exchange process $1 \rightarrow 4$ *cools* the air in the container as it warms the gas flowing through the pipe. If you were to place eggs and milk inside this closed container, you would call it a refrigerator!

Going around a closed *pV* cycle in a ccw direction reverses the sign of W for each process in the cycle. Consequently, the area inside the curve of Figure 19.17b is W_{in}, the work done *on* the system. The work W_{in} done on the system is used to extract heat Q_C from the cold reservoir and exhaust a larger amount of heat $Q_H = Q_C + W_{in}$ to the hot reservoir. But where, in this situation, are the energy reservoirs?

Understanding a refrigerator is a little harder than understanding a heat engine. The key is to remember that **heat is always transferred from a hotter object to a colder object.** In particular,

- The gas in a refrigerator can extract heat Q_C *from* the cold reservoir only if the gas temperature is *lower* than the cold-reservoir temperature T_C. Heat energy is then transferred *from* the cold reservoir *into* the colder gas.
- The gas in a refrigerator can exhaust heat Q_H *to* the hot reservoir only if the gas temperature is *higher* than the hot-reservoir temperature T_H. Heat energy is then transferred *from* the warmer gas *into* the hot reservoir.

These two requirements place severe constraints on the thermodynamics of a refrigerator. Because there is no reservoir colder than T_C, the gas cannot reach a temperature lower than T_C by heat exchange. The gas in a refrigerator *must* use an adiabatic expansion ($Q = 0$) to lower the temperature below T_C. Likewise, a gas refrigerator requires an adiabatic compression to raise the gas temperature above T_H.

The reversed Brayton cycle of Figure 19.17b does, indeed, have two adiabatic processes. The adiabatic expansion lowers the temperature to T_1, then heat Q_C is transferred *from* the cold reservoir *to* the gas during process $1 \rightarrow 4$. Consequently, the cold reservoir temperature must be $T_C \geq T_4$. Contrast this with the same cycle run clockwise (cw) as a heat engine, where we saw that the cold reservoir must be $T_C \leq T_1$.

Similar reasoning applies on the hot side. In order for heat Q_H to be transferred *into* the hot reservoir during process $3 \rightarrow 2$, the hot reservoir temperature must be

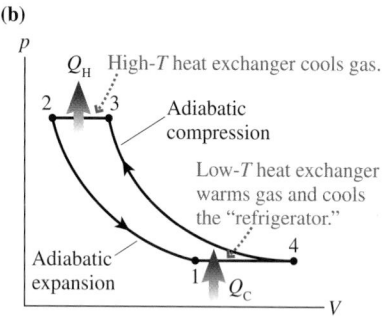

(a)

(b)

(c)

FIGURE 19.17 A refrigerator that extracts heat from the cold reservoir and exhausts heat to the hot reservoir.

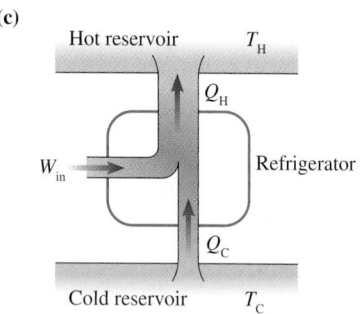

$T_H \leq T_2$. This requirement of the high-temperature reservoir differs distinctly from the Brayton heat engine, which required $T_H \geq T_3$. Figure 19.18 compares a Brayton-cycle heat engine to a Brayton-cycle refrigerator.

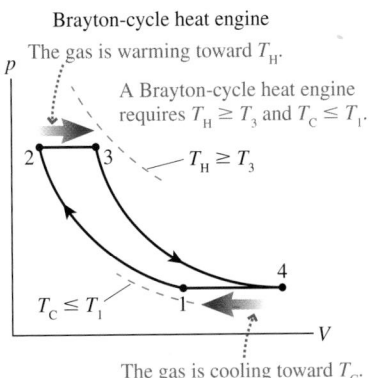
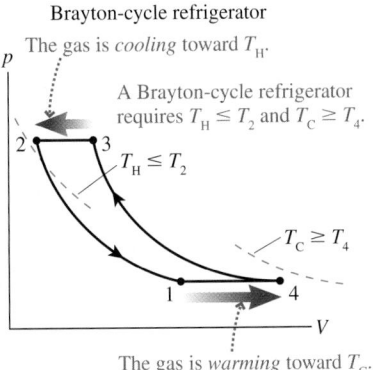

Brayton-cycle heat engine

The gas is warming toward T_H.

A Brayton-cycle heat engine requires $T_H \geq T_3$ and $T_C \leq T_1$.

$T_H \geq T_3$

$T_C \leq T_1$

The gas is cooling toward T_C.

Brayton-cycle refrigerator

The gas is *cooling* toward T_H.

A Brayton-cycle refrigerator requires $T_H \leq T_2$ and $T_C \geq T_4$.

$T_H \leq T_2$

$T_C \geq T_4$

The gas is *warming* toward T_C.

FIGURE 19.18 A comparison of a Brayton-cycle heat engine to a Brayton-cycle refrigerator.

These cooling coils are the refrigerator's high-temperature heat exchanger. Heat energy is being transferred from hot gas inside the coils to the cooler room air.

The important point—a point that will be important in the next section—is that a Brayton refrigerator is *not* simply a Brayton heat engine running backward. To make a Brayton refrigerator you must both reverse the cycle *and* change the hot and cold reservoirs.

NOTE ▶ Some heat engines cannot be converted to refrigerators under any circumstances. We'll leave it as a homework problem to show that the heat engine of Example 19.2, if run backward, is a total loser. Its energy-transfer diagram, shown in Figure 19.19, shows work being done to transfer energy "downhill" even faster than it would move spontaneously from hot to cold! ◀

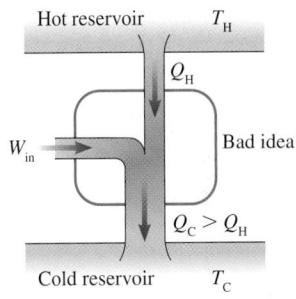

FIGURE 19.19 This is the energy-transfer diagram if the heat engine of Example 19.2 is run backward.

EXAMPLE 19.3 **Analyzing a refrigerator**

A refrigerator using helium gas operates on a reversed Brayton cycle with a pressure ratio of 5.0. Prior to compression, the gas occupies 100 cm³ at a pressure of 150 kPa and a temperature of $-23°C$. Its volume at the end of the expansion is 80 cm³. What are the refrigerator's coefficient of performance and its power input if it operates at 60 cycles per second?

MODEL The Brayton cycle has two adiabatic processes and two isobaric processes. The work per cycle needed to run the refrigerator is $W_{in} = Q_H - Q_C$, hence we can determine both the coefficient of performance and the power requirements from Q_H and Q_C. Heat energy is transferred only during the two isobaric processes.

VISUALIZE Figure 19.20 shows the pV cycle. We know from the pressure ratio of 5.0 that the maximum pressure is 750 kPa. Neither V_2 nor V_3 is known.

SOLVE To calculate heat we're going to need the temperatures at the four corners of the cycle. First, we can use the conditions of state 4 to find the number of moles of helium:

$$n = \frac{p_4 V_4}{RT_4} = 0.00722 \text{ mol}$$

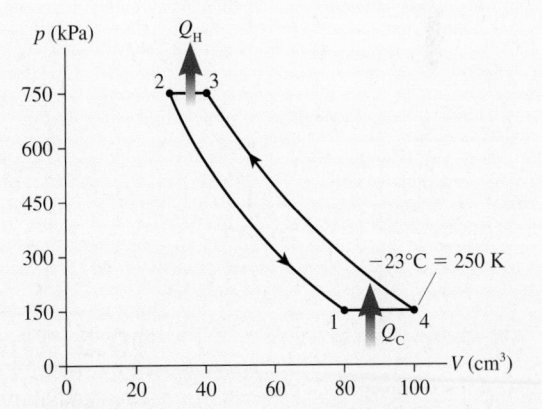

FIGURE 19.20 A Brayton-cycle refrigerator.

Process $1 \rightarrow 4$ is isobaric, hence temperature T_1 is

$$T_1 = \frac{V_1}{V_4} T_4 = (0.80)(250 \text{ K}) = 200 \text{ K} = -73°C$$

With Equation 19.16 we found that the quantity $p^{(1-\gamma)/\gamma} T$ remains constant during an isobaric process. Helium is a monatomic gas with $\gamma = \frac{5}{3}$, so $(1 - \gamma)/\gamma = -\frac{2}{5} = -0.40$. For the adiabatic compression $4 \rightarrow 3$,

$$p_3^{-0.40} T_3 = p_4^{-0.40} T_4$$

Solving for T_3 gives

$$T_3 = \left(\frac{p_4}{p_3}\right)^{-0.40} T_4 = \left(\frac{1}{5}\right)^{-0.40} (250 \text{ K}) = 476 \text{ K} = 203°C$$

The same analysis applied to the $2 \rightarrow 1$ adiabatic expansion gives

$$T_2 = \left(\frac{p_1}{p_2}\right)^{-0.40} T_1 = \left(\frac{1}{5}\right)^{-0.40} (200 \text{ K}) = 381 \text{ K} = 108°C$$

Now we can use $C_P = \frac{5}{2}R = 20.8$ J/molK for a monatomic gas to compute the heat transfers:

$$Q_H = |Q_{32}| = nC_P(T_3 - T_2)$$

$$= (0.00722 \text{ mol})(20.8 \text{ J/mol K})(95 \text{ K}) = 14.27 \text{ J}$$

$$Q_C = |Q_{14}| = nC_P(T_4 - T_1)$$

$$= (0.00722 \text{ mol})(20.8 \text{ J/mol K})(50 \text{ K}) = 7.51 \text{ J}$$

Thus the work *input* to the refrigerator is $W_{in} = Q_H - Q_C = 6.76$ J. During each cycle, 6.76 J of work are done *on* the gas to extract 7.51 J of heat from the cold reservoir. 14.27 J of heat are then exhausted into the hot reservoir.

The refrigerator's coefficient of performance is

$$K = \frac{Q_C}{W_{in}} = \frac{7.51 \text{ J}}{6.76 \text{ J}} = 1.1$$

The power input needed to run the refrigerator is

$$P_{in} = 6.76 \frac{\text{J}}{\text{cycle}} \times 60 \frac{\text{cycles}}{\text{s}} = 406 \frac{\text{J}}{\text{s}} = 406 \text{ W}$$

ASSESS These are fairly realistic values for a kitchen refrigerator. You pay your electric company for providing the work W_{in} that operates the refrigerator. The cold reservoir is the freezer compartment. The cold temperature T_C must be higher than T_4 ($T_C > -23°C$) in order for heat to be transferred *from* the cold reservoir *to* the gas. A typical freezer temperature is $-15°C$, so this condition is satisfied. The hot reservoir is the air in the room. The back and underside of a refrigerator have heat-exchanger coils where the hot gas, after compression, transfers heat to the air. The hot temperature T_H must be less than T_2 ($T_H < 108°C$) in order for heat to be transferred *from* the gas *to* the air. An air temperature $\approx 25°C$ under a refrigerator satisfies this condition.

STOP TO THINK 19.4 What, if anything, is wrong with this refrigerator?

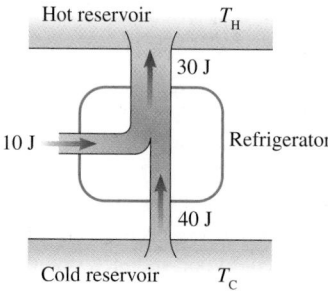

19.5 The Limits of Efficiency

Everyone knows that heat can produce motion. That it possesses vast motive power no one can doubt, in these days when the steam engine is everywhere so well known. . . . Notwithstanding the satisfactory condition to which they have been brought today, their theory is very little understood. The question has often been raised whether the motive power of heat is unbounded, or whether the possible improvements in steam engines have an assignable limit.

Sadi Carnot

Thermodynamics has its historical roots in the development of the steam engine and other machines of the early industrial revolution. Early steam engines, built on the basis of experience rather than scientific understanding, were not very efficient at converting fuel energy into work. The first major theoretical analysis of heat engines was published by the French engineer Sadi Carnot in 1824. Interestingly, his analysis *precedes* Joule's discovery of the full statement of conservation of energy. The question that Carnot raised was one we posed at the end of Section 19.3: Can we make a heat engine whose thermal efficiency η approaches 1, or is there an upper limit η_{max} that cannot be exceeded? To frame the question more clearly, imagine that we have a hot reservoir at temperature T_H and a cold reservoir at T_C. What is the most efficient heat engine (maximum η) that can operate between these two energy reservoirs? Similarly, what is the most efficient refrigerator (maximum K) that can operate between the two reservoirs?

We just saw that a refrigerator is, in some sense, a heat engine running backward. We might thus suspect that the most efficient heat engine may be related to the most efficient refrigerator. Suppose we have a heat engine that we can turn into a refrigerator by reversing the direction of operation, thus changing the direction of the energy transfers, and with *no other changes.* In particular, the heat engine and the refrigerator operate between the same two energy reservoirs at temperatures T_H and T_C.

> NOTE ▶ The heat engine we're looking for cannot be a Brayton-cycle heat engine. As we found, a Brayton-cycle refrigerator requires reversing the heat engine's direction of operation *and* changing the temperatures of the energy reservoirs. ◀

Figure 19.21a shows such a heat engine and its corresponding refrigerator. Notice that the refrigerator has *exactly the same* work and heat transfer as the heat engine, only in the opposite directions. A device that can be operated as either a heat engine or a refrigerator between the same two energy reservoirs and with the same energy transfers, with only their direction changed, is called a **perfectly reversible engine.** A perfectly reversible engine is an idealization, as was the concept of a perfectly elastic collision. Nonetheless, it will allow us to establish limits that no real engine can exceed.

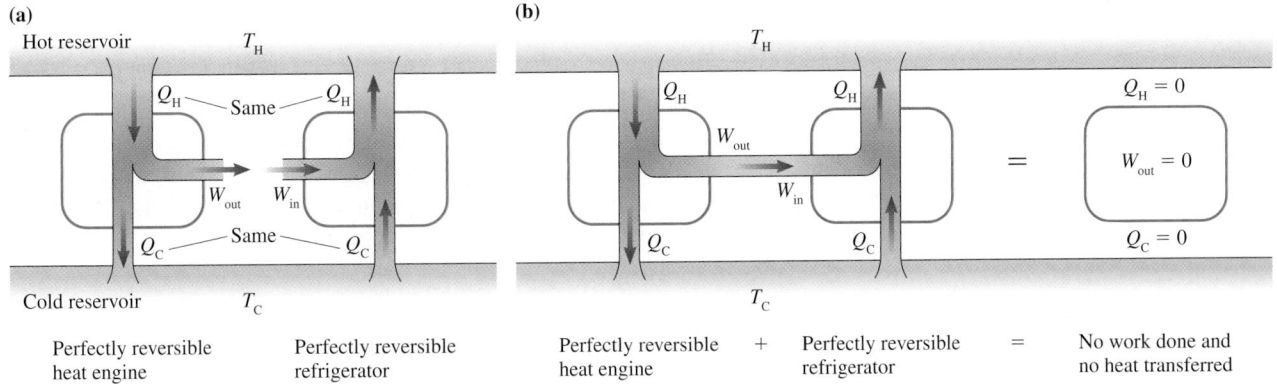

(a)
Hot reservoir T_H

Q_H — Same — Q_H

W_{out} W_{in}

Q_C — Same — Q_C

Cold reservoir T_C

Perfectly reversible heat engine   Perfectly reversible refrigerator

(b)
T_H

Q_H Q_H

W_{out}

W_{in}

Q_C Q_C

T_C

Perfectly reversible heat engine   +   Perfectly reversible refrigerator   =

$Q_H = 0$

$W_{out} = 0$

$Q_C = 0$

No work done and no heat transferred

FIGURE 19.21 If a perfectly reversible heat engine is used to operate a perfectly reversible refrigerator, the two devices exactly cancel each other. Consequently no net work is done and no net heat is transferred.

Suppose we have a perfectly reversible heat engine and a perfectly reversible refrigerator (the same device running backward) operating between a hot reservoir at temperature T_H and a cold reservoir at temperature T_C. Because the work W_{in} needed to operate the refrigerator is exactly the same as the useful work W_{out} done by the heat engine, we can use the heat engine, as shown in Figure 19.21b, to drive the refrigerator. The heat Q_C that the engine exhausts to the cold reservoir is exactly the same as the heat Q_C that the refrigerator extracts from the cold reservoir. Similarly, the heat Q_H that the engine extracts from the hot reservoir matches the heat Q_H that the refrigerator exhausts to the hot reservoir. Consequently, there is no net heat transfer in either direction. The refrigerator exactly replaces all the heat energy that had been transferred out of the hot reservoir by the heat engine.

You may want to compare the reasoning used here with the reasoning we used with Figure 19.12. There we tried to use the output of a "perfect" heat engine to run a refrigerator but did *not* succeed.

A Perfectly Reversible Engine Has Maximum Efficiency

Now we've arrived at the critical step in the reasoning. Suppose I claim to have a heat engine that can operate between temperatures T_H and T_C with *more* efficiency than a perfectly reversible engine. Figure 19.22 shows the output of this heat engine operating the same perfectly reversible refrigerator that we used in Figure 19.21b.

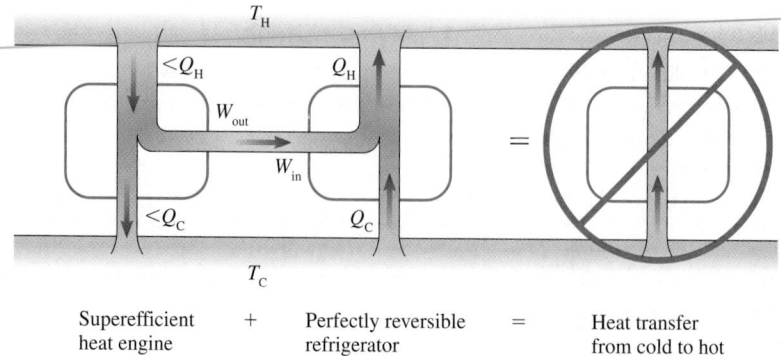

| Superefficient heat engine | + | Perfectly reversible refrigerator | = | Heat transfer from cold to hot |

FIGURE 19.22 A heat engine more efficient than a perfectly reversible engine could be used to violate the second law of thermodynamics.

Recall that the thermal efficiency and the work of a heat engine are

$$\eta = \frac{W_{out}}{Q_H} \quad \text{and} \quad W_{out} = Q_H - Q_C$$

If the new heat engine is more efficient than the perfectly reversible engine it replaces, it needs *less* heat Q_H from the hot reservoir to perform the *same* work W_{out}. If Q_H is less while W_{out} is the same, then Q_C must also be less. That is, the new heat engine exhausts less heat to the cold reservoir than does the perfectly reversible heat engine.

When this new heat engine drives the perfectly reversible refrigerator, the heat it exhausts to the cold reservoir is *less* than the heat extracted from the cold reservoir by the refrigerator. Similarly, this engine extracts *less* heat from the hot reservoir than the refrigerator exhausts. Thus the net result of using this heat engine to operate a perfectly reversible refrigerator is that heat is transferred from the cold reservoir to the hot reservoir *without outside assistance*.

But this can't happen. It would violate the second law of thermodynamics. Hence we have to conclude that no heat engine operating between reservoirs at temperatures T_H and T_C can be more efficient than a perfectly reversible engine. This very important conclusion is another version of the second law:

> **Second law, informal statement #5** No heat engine operating between reservoirs at temperatures T_H and T_C can be more efficient than a perfectly reversible engine operating between these temperatures.

The answer to our question "Is there a maximum η that cannot be exceeded?" is a clear "Yes!" The maximum possible efficiency η_{max} is that of a perfectly reversible engine. Because the perfectly reversible engine is an idealization, any real engine will have an efficiency less than η_{max}.

A similar argument shows that no refrigerator can be more efficient than a perfectly reversible refrigerator. If we had such a refrigerator, and if we ran it with the output of a perfectly reversible heat engine, we could transfer heat from cold to hot with no outside assistance. Thus:

Second law, informal statement #6 No refrigerator operating between reservoirs at temperatures T_H and T_C can have a coefficient of performance larger than that of a perfectly reversible refrigerator operating between these temperatures.

Conditions for a Perfectly Reversible Engine

This argument tells us that η_{max} and K_{max} exist, but it doesn't tell us what they are. Our final task will be to "design" and analyze a perfectly reversible engine. Under what conditions is an engine reversible?

An engine transfers energy by both mechanical and thermal interactions. Mechanical interactions are pushes and pulls. The environment does work on the system, transferring energy into the system by pushing in on a piston. The system transfers energy back to the environment by pushing out on the piston. This exchange of energy is perfectly reversible if two conditions are met:

1. The energy transferred out (W_{out}) equals the energy transferred in (W_{in}), and
2. The system returns to its initial state, with no change of temperature or pressure.

Both these conditions will be met only if the motion is *frictionless*. The slightest bit of friction will prevent the mechanical transfer of energy from being perfectly reversible.

The circumstances under which heat transfer can be *completely* reversed aren't quite so obvious. After all, Chapter 18 emphasized the *irreversible* nature of heat transfer. If objects A and B are in thermal contact, with $T_A > T_B$, then heat energy is transferred from A to B. But the second law of thermodynamics prohibits a heat transfer from B back to A. Likewise, heat can move from B to A, but not from A to B, if $T_A < T_B$. Heat transfer through a temperature *difference* is an irreversible process.

But suppose $T_A = T_B$. With no temperature difference, any heat that is transferred from A to B can, at a later time, be transferred from B back to A. This transfer wouldn't violate the second law, which prohibits only heat transfer from a colder object to a hotter object. Now you might object, and rightly so, that heat *can't* move from A to B if they are at the same temperature because heat, by definition, is the energy transferred between two objects at different temperatures.

This is true, so let's consider a limiting case in which $T_A = T_B + dT$. That is, the temperature difference is infinitesimal. Heat is transferred from A to B, but *very slowly!* If you later try to make the heat move from B back to A, the second law will prevent you from doing so with perfect precision. But because the temperature difference is infinitesimal, you'll be missing only an infinitesimal amount dQ of heat. You can transfer heat reversibly in the limit $dT \rightarrow 0$, but you must be prepared to spend an infinite amount of time doing so.

Thus the thermal transfer of energy is reversible if the heat is transferred infinitely slowly in an isothermal process. This is an idealization, but so are completely frictionless processes. Nonetheless, we can now say that a perfectly reversible engine must use only two types of processes:

1. Frictionless mechanical interactions with no heat transfer ($Q = 0$), and
2. Thermal interactions in which heat is transferred in an isothermal process ($\Delta E_{th} = 0$).

Any engine that uses only these two types of processes is called a **Carnot engine.** A Carnot engine is a perfectly reversible engine; thus it has the maximum possible thermal efficiency η_{max} and, if operated as a refrigerator, the maximum possible coefficient of performance K_{max}.

19.6 The Carnot Cycle

8.14   Activ
Physics

No real engine is perfectly reversible, so a Carnot engine is an idealization. Nonetheless, an analysis of the Carnot engine will allow us to establish a maximum possible thermal efficiency that no real heat engine can exceed.

The definition of a Carnot engine does not specify whether the engine's working substance is a gas or a liquid. It makes no difference. Our argument that a perfectly reversible engine is the most efficient possible heat engine depended only on the engine's reversibility. It did not depend on any details of how the engine is constructed or what it uses for a working substance. Consequently, **any Carnot engine operating between T_H and T_C must have exactly the same efficiency as any other Carnot engine operating between the same two energy reservoirs.** If we can determine the thermal efficiency of one Carnot engine, we'll know the efficiency of all Carnot engines. Because liquids and phase changes are complicated, we'll analyze a Carnot engine that uses an ideal gas.

The Carnot Cycle

The **Carnot cycle** is an ideal-gas cycle that consists of the two adiabatic processes ($Q = 0$) and two isothermal processes ($\Delta E_{th} = 0$) shown in Figure 19.23. These are the two types of processes allowed in a perfectly reversible gas engine. As a Carnot cycle operates,

1. The gas is isothermally compressed while in thermal contact with the cold reservoir at temperature T_C. Heat energy $Q_C = |Q_{12}|$ is removed from the gas as it is compressed in order to keep the temperature constant. The compression must take place extremely slowly because there can be only an infinitesimal temperature difference between the gas and the reservoir.
2. The gas is adiabatically compressed while thermally isolated from the environment. This compression increases the gas temperature until it matches temperature T_H of the hot reservoir. No heat is transferred during this process.
3. After reaching maximum compression, the gas expands isothermally at temperature T_H. Heat $Q_H = Q_{34}$ is transferred from the hot reservoir into the gas as it expands in order to keep the temperature constant.
4. Finally, the gas expands adiabatically, with $Q = 0$, until the temperature decreases back to T_C.

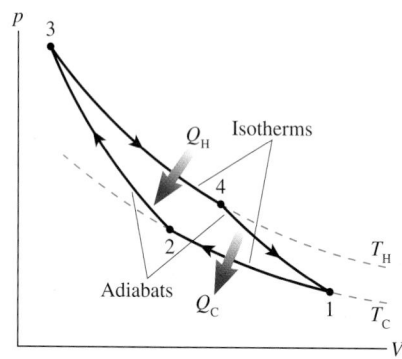

FIGURE 19.23 The Carnot cycle is perfectly reversible.

Work is done in all four processes of the Carnot cycle, but heat is transferred only during the two isothermal processes.

The thermal efficiency of any heat engine is

$$\eta = \frac{W_{out}}{Q_H} = 1 - \frac{Q_C}{Q_H}$$

We can determine η_{Carnot} by finding that heat transfer in the two isothermal processes.

Process $1 \rightarrow 2$: Table 19.1 gives us the heat transfer in an isothermal process at temperature T_C:

$$Q_{12} = (W_s)_{12} = nRT_C \ln\left(\frac{V_2}{V_1}\right) = -nRT_C \ln\left(\frac{V_1}{V_2}\right) \qquad (19.22)$$

$V_1 > V_2$, so the logarithm on the right is positive. Q_{12} is negative because heat is transferred out of the system, but Q_C is simply the *amount* of heat transferred to the cold reservoir:

$$Q_C = |Q_{12}| = nRT_C \ln\left(\frac{V_1}{V_2}\right) \qquad (19.23)$$

Process $3 \rightarrow 4$: Similarly, the heat transferred in the isothermal expansion at temperature T_H is

$$Q_H = Q_{34} = (W_s)_{34} = nRT_H \ln\frac{V_4}{V_3} \qquad (19.24)$$

Thus the thermal efficiency of the Carnot cycle is

$$\eta_{\text{Carnot}} = 1 - \frac{Q_C}{Q_H} = 1 - \frac{T_C \ln(V_1/V_2)}{T_H \ln(V_4/V_3)} \qquad (19.25)$$

We can simplify this expression. During the two adiabatic processes,

$$T_C V_2^{\gamma-1} = T_H V_3^{\gamma-1} \quad \text{and} \quad T_C V_1^{\gamma-1} = T_H V_4^{\gamma-1} \qquad (19.26)$$

An algebraic rearrangement gives

$$V_2 = V_3\left(\frac{T_H}{T_C}\right)^{1/(\gamma-1)} \quad \text{and} \quad V_1 = V_4\left(\frac{T_H}{T_C}\right)^{1/(\gamma-1)} \qquad (19.27)$$

from which it follows that

$$\frac{V_1}{V_2} = \frac{V_4}{V_3} \qquad (19.28)$$

Consequently, the two logarithms in Equation 19.25 cancel and we're left with the result that the thermal efficiency of a Carnot engine operating between a hot reservoir at temperature T_H and a cold reservoir at temperature T_C is

$$\eta_{\text{Carnot}} = 1 - \frac{T_C}{T_H} \qquad \text{(Carnot thermal efficiency)} \qquad (19.29)$$

This remarkably simple result, an efficiency that depends only on the ratio of the temperatures of the hot and cold reservoirs, is Carnot's legacy to thermodynamics.

NOTE ▶ Temperatures T_H and T_C are *absolute* temperatures. They *must* be in kelvins. ◀

EXAMPLE 19.4 A Carnot engine
A Carnot engine is cooled by water at $T_C = 10°C$. What temperature must be maintained in the hot reservoir of the engine to have a thermal efficiency of 70%?

MODEL The efficiency of a Carnot engine depends only on the temperatures of the hot and cold reservoirs.

SOLVE The thermal efficiency $\eta_{\text{Carnot}} = 1 - T_C/T_H$ can be rearranged to give

$$T_H = \frac{T_C}{1 - \eta_{\text{Carnot}}} = 943 \text{ K} = 670°C$$

ASSESS A "real" engine would need a higher temperature than this to provide 70% efficiency because no real engine will match the Carnot efficiency.

EXAMPLE 19.5 A real engine
The heat engine of Example 19.2 had a highest temperature of 2700 K, a lowest temperature of 300 K, and a thermal efficiency of 15%. What is the efficiency of a Carnot engine operating between these two temperatures?

SOLVE The Carnot efficiency is

$$\eta_{\text{Carnot}} = 1 - \frac{T_C}{T_H} = 1 - \frac{300 \text{ K}}{2700 \text{ K}} = 0.89 = 89\%$$

ASSESS The thermodynamic cycle used in the example doesn't come anywhere close to the Carnot efficiency.

The Maximum Efficiency

In Section 19.2 we tried to invent a perfect engine with $\eta = 1$ and $Q_C = 0$. We found that we could not do so without violating the second law, so no engine can have $\eta = 1$. However, that example didn't rule out an engine with $\eta = 0.9999$. Further analysis has now shown that no heat engine operating between energy reservoirs at temperatures T_H and T_C can be more efficient than a perfectly reversible engine operating between these temperatures.

We've now reached the endpoint of this line of reasoning by establishing an exact result for the thermal efficiency of a perfectly reversible engine, the Carnot engine. We can summarize our conclusions:

> **Second law, informal statement #7** No heat engine operating between energy reservoirs at temperatures T_H and T_C can exceed the Carnot efficiency $\eta_{\text{Carnot}} = 1 - T_C/T_H$.

As Example 19.5 showed, real engines usually fall well short of the Carnot limit.

We also found that no refrigerator can exceed the coefficient of performance of a perfectly reversible refrigerator. We'll leave the proof as a homework problem, but an analysis very similar to that above shows that the coefficient of performance of a Carnot refrigerator is

$$K_{\text{Carnot}} = \frac{T_C}{T_H - T_C} \qquad \text{(Carnot coefficient of performance)} \qquad (19.30)$$

Thus we can state:

> **Second law, informal statement #8** No refrigerator operating between energy reservoirs at temperatures T_H and T_C can exceed the Carnot coefficient of performance $K_{\text{Carnot}} = T_C/(T_H - T_C)$.

EXAMPLE 19.6 Brayton versus Carnot
The Brayton-cycle refrigerator of Example 19.3 had a coefficient of performance $K = 1.1$. How does this compare to the limit set by the second law of thermodynamics?

SOLVE Example 19.3 found that the reservoir temperatures had to be $T_C \geq 250$ K and $T_H \leq 381$ K. A Carnot refrigerator operating between 250 K and 381 K has

$$K_{\text{Carnot}} = \frac{T_C}{T_H - T_C} = \frac{250 \text{ K}}{381 \text{ K} - 250 \text{ K}} = 1.9$$

ASSESS This is the minimum value of K_{Carnot}. It will be even higher if $T_C > 250$ K or $T_H < 381$ K. The coefficient of performance of the reasonably realistic refrigerator of Example 19.3 is less than 60% of the limiting value.

Statements #7 and #8 of the second law are a major result of this chapter, one with profound implications. The efficiency limit of a heat engine is set by the temperatures of the hot and cold reservoirs. High efficiency requires $T_C/T_H \ll 1$ and thus $T_H \gg T_C$. However, practical realities often prevent T_H from being significantly larger than T_C, in which case the engine cannot possibly have a large efficiency. This limit on the efficiency of heat engines is a consequence of the second law of thermodynamics.

EXAMPLE 19.7 **Generating electricity**

An electric power plant boils water to produce high-pressure steam at 400°C. The high-pressure steam spins a turbine as it expands, then the turbine spins the generator. The steam is then condensed back to water in an ocean-cooled heat exchanger at 25°C. What is the *maximum* possible efficiency with which heat energy can be converted to electrical energy?

MODEL The maximum possible efficiency is that of a Carnot engine operating between these temperatures.

SOLVE The Carnot efficiency depends on absolute temperatures, so we must use $T_H = 400°C = 673$ K and $T_C = 25°C = 298$ K. Then

$$\eta_{max} = 1 - \frac{298}{673} = 0.56 = 56\%$$

ASSESS This is an upper limit. Real coal-, oil-, gas-, and nuclear-heated steam generators actually operate at $\approx 35\%$ thermal efficiency. (The heat *source* has nothing to do with the efficiency. All it does is boil water.) Thus, as the photo at the beginning to this chapter noted, electric power plants convert only about one-third of the fuel energy to electric energy while exhausting about two-thirds of the energy to the environment as waste heat. Not much can be done to alter the low-temperature limit. The high-temperature limit is determined by the maximum temperature and pressure the boiler and turbine can withstand. The efficiency of electricity generation is far less than most people imagine, but it is an unavoidable consequence of the second law of thermodynamics.

A limit on the efficiency of heat engines was not expected. We are used to thinking in terms of energy conservation, so it comes as no surprise that we cannot make an engine with $\eta > 1$. But the limits arising from the second law were not anticipated, nor are they obvious. Nonetheless, they are a very real fact of life and a very real constraint on any practical device. No one has ever invented a machine that exceeds the second-law limits, and we have seen that the maximum efficiency for realistic engines is surprisingly low.

STOP TO THINK 19.5 Could this heat engine be built?

a. Yes.
b. No.
c. It's impossible to tell without knowing what kind of cycle it uses.

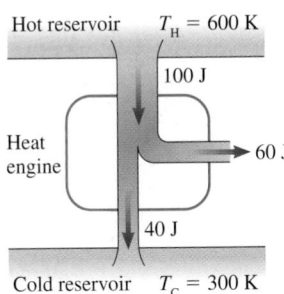

Hot reservoir $T_H = 600$ K

100 J

Heat engine   60 J

40 J

Cold reservoir $T_C = 300$ K

SUMMARY

The goal of Chapter 19 has been to investigate the physical principles that govern the operation of heat engines and refrigerators.

GENERAL PRINCIPLES

Heat Engines

Devices which transform heat into work. They require two energy reservoirs at different temperatures.

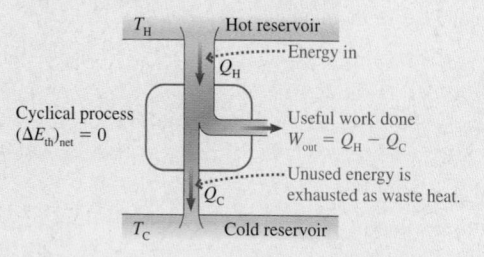

Thermal efficiency

$$\eta = \frac{W_{out}}{Q_H} = \frac{\text{what you get}}{\text{what you pay}}$$

Second law limit:

$$\eta \leq 1 - \frac{T_C}{T_H}$$

Refrigerators

Devices which use work to transfer heat from a colder object to a hotter object.

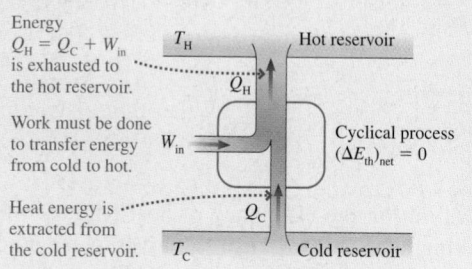

Coefficient of performance

$$K = \frac{Q_C}{W_{in}} = \frac{\text{what you get}}{\text{what you pay}}$$

Second law limit:

$$K \leq \frac{T_C}{T_H - T_C}$$

IMPORTANT CONCEPTS

A perfectly reversible engine (a **Carnot engine**) can be operated as either a heat engine or a refrigerator between the same two energy reservoirs by reversing the cycle and with no other changes.

- A **Carnot heat engine** has the maximum possible thermal efficiency of any heat engine operating between T_H and T_C.

$$\eta_{Carnot} = 1 - \frac{T_C}{T_H}$$

- A **Carnot refrigerator** has the maximum possible coefficient of performance of any refrigerator operating between T_H and T_C.

$$K_{Carnot} = \frac{T_C}{T_H - T_C}$$

The Carnot cycle for a gas engine consists of two isothermal processes and two adiabatic processes.

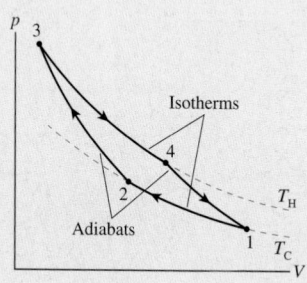

An energy reservoir is a part of the environment so large in comparison to the system that its temperature doesn't change as the system extracts heat energy from or exhausts heat energy to the reservoir. All heat engines and refrigerators operate between two energy reservoirs at different temperatures T_H and T_C.

The **work** W_s done *by* the system has the opposite sign to the work done *on* the system.

W_s = area under pV curve

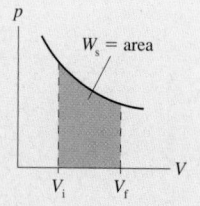

APPLICATIONS

To analyze a heat engine or refrigerator:

MODEL Identify each process in the cycle.

VISUALIZE Draw the pV diagram of the cycle.

SOLVE There are several steps:

- Determine p, V, and T at the beginning and end of each process.
- Calculate ΔE_{th}, W_s, and Q for each process.
- Determine W_{in} or W_{out}, Q_H, and Q_C.
- Calculate $\eta = W_{out}/Q_H$ or $K = Q_C/W_{in}$.

ASSESS Verify $(\Delta E_{th})_{net} = 0$. Check signs.

TERMS AND NOTATION

thermodynamics	closed-cycle device	refrigerator	pressure ratio, r_p
energy reservoir	thermal efficiency, η	coefficient of performance, K	perfectly reversible engine
energy-transfer diagram	waste heat	heat exchanger	Carnot engine
heat engine			Carnot cycle

EXERCISES AND PROBLEMS

Exercises

Section 19.1 Turning Heat into Work

Section 19.2 Heat Engines and Refrigerators

1. What are (a) W_{out} and Q_C and (b) the thermal efficiency for the heat engine shown in Figure Ex19.1?

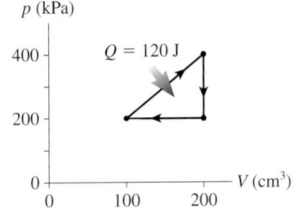

FIGURE EX19.1

2. What are (a) W_{out} and Q_H and (b) the thermal efficiency for the heat engine shown in Figure Ex19.2?

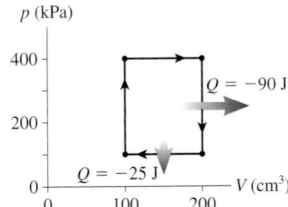

FIGURE EX19.2

3. A heat engine extracts 55 kJ of heat from the hot reservoir each cycle and exhausts 40 kJ of heat. What are (a) the thermal efficiency and (b) the work done per cycle?

4. A heat engine does 20 J of work per cycle while exhausting 30 J of waste heat. What is the engine's thermal efficiency?

5. A heat engine with a thermal efficiency of 40% does 100 J of work per cycle. How much heat is (a) extracted from the hot reservoir and (b) exhausted to the cold reservoir per cycle?

6. A refrigerator requires 20 J of work and exhausts 50 J of heat per cycle. What is the refrigerator's coefficient of performance?

7. 50 J of work are done per cycle on a refrigerator with a coefficient of performance of 4.0. How much heat is (a) extracted from the cold reservoir and (b) exhausted to the hot reservoir per cycle?

8. The power output of a car engine running at 2400 rpm is 500 kW. How much (a) work is done and (b) heat is exhausted per cycle if the engine's thermal efficiency is 20%? Give your answers in kJ.

9. A 32%-efficient electric power plant produces 900 MW of electric power and discharges waste heat into 20°C ocean water. Suppose the waste heat could be used to heat homes during the winter instead of being discharged into the ocean. A typical American house requires an average 20 kW for heating. How many homes could be heated with the waste heat of this one power plant?

10. 1.0 L of 20°C water is placed in a refrigerator. The refrigerator's motor must supply an extra 8.0 W power to chill the water to 5°C in 1.0 hr. What is the refrigerator's coefficient of performance?

Section 19.3 Ideal-Gas Heat Engines

Section 19.4 Ideal-Gas Refrigerators

11. The heat-engine cycle of Figure Ex19.11 consists of four processes. Make a chart with rows labeled A to D and columns labeled ΔE_{th}, W_s, and Q. Fill each box in the chart with $+$, $-$, or 0 to indicate whether the quantity increases, decreases, or stays the same during that process.

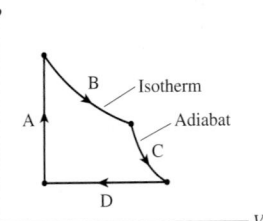

FIGURE EX19.11

12. The refrigerator cycle of Figure Ex19.12 consists of four processes. Make a chart with rows labeled A–D and columns labeled ΔE_{th}, W_s, and Q. Fill each box in the chart with $+$, $-$, or 0 to indicate whether the quantity increases, decreases, or stays the same during that process.

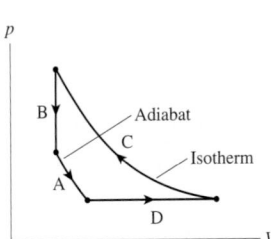

FIGURE EX19.12

13. How much work is done per cycle by a gas following the pV trajectory of Figure Ex19.13?

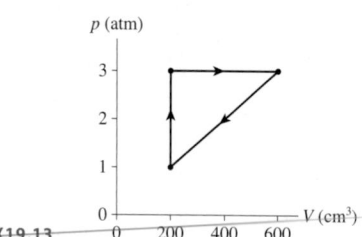

FIGURE EX19.13

14. A gas following the pV trajectory of Figure Ex19.14 does 60 J of work per cycle. What is p_{max}?

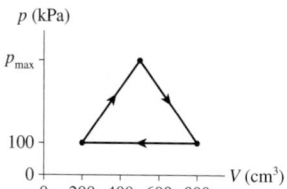

FIGURE EX19.14

15. What are (a) the thermal efficiency and (b) the heat exhausted to the cold reservoir for the heat engine shown in Figure Ex19.15?

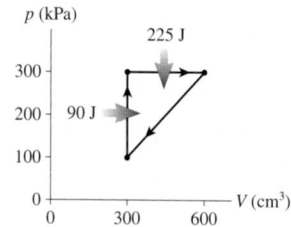

FIGURE EX19.15

16. What are (a) the thermal efficiency and (b) the heat extracted from the hot reservoir for the heat engine shown in Figure Ex19.16?

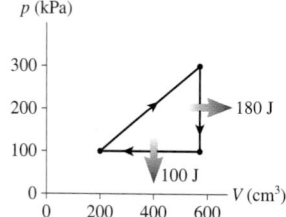

FIGURE EX19.16

17. At what pressure ratio would a heat engine operating with a Brayton cycle have an efficiency of 60%? Assume that the gas is diatomic.

18. A heat engine uses a diatomic gas in a Brayton cycle. What is the engine's thermal efficiency if the gas volume is halved during the compression?

Section 19.5 The Limits of Efficiency

Section 19.6 The Carnot Cycle

19. Which, if any, of the heat engines in Figure Ex19.19 violate (a) the first law of thermodynamics or (b) the second law of thermodynamics? Explain.

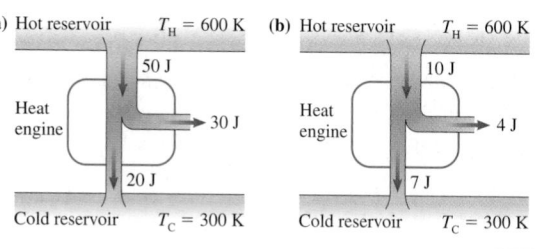

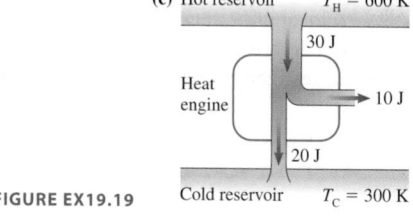

FIGURE EX19.19

20. Which, if any, of the refrigerators in Figure Ex19.20 violate (a) the first law of thermodynamics or (b) the second law of thermodynamics? Explain.

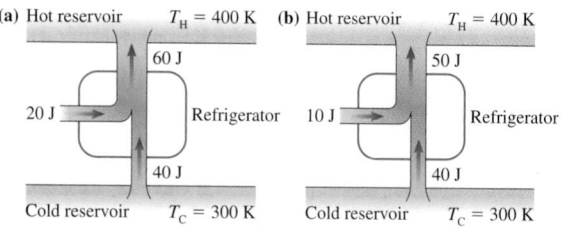

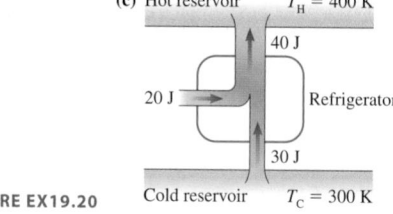

FIGURE EX19.20

21. At what cold-reservoir temperature (in °C) would a Carnot engine with a hot-reservoir temperature of 427°C have an efficiency of 60%?

22. a. A heat engine does 200 J of work per cycle while exhausting 600 J of heat to the cold reservoir. What is the engine's thermal efficiency?
    b. A Carnot engine with a hot-reservoir temperature of 400°C has the same thermal efficiency. What is the cold-reservoir temperature in °C?

23. A heat engine does 10 J of work and exhausts 15 J of waste heat during each cycle.
    a. What is the engine's thermal efficiency?
    b. If the cold-reservoir temperature is 20°C, what is the minimum possible temperature in °C of the hot reservoir?

24. A Carnot engine operating between energy reservoirs at temperatures 300 K and 500 K produces a power output of 1000 W. What are (a) the thermal efficiency of this engine, (b) the rate of heat input, in W, and (c) the rate of heat output, in W?

25. A Carnot engine whose cold-reservoir temperature is 7°C has a thermal efficiency of 40%. By how much should the temperature of the hot reservoir be increased to raise the efficiency to 60%?

26. A heat engine operating between energy reservoirs at 20°C and 600°C has 30% of the maximum possible efficiency. How much energy must this engine extract from the hot reservoir to do 1000 J of work?

27. A Carnot refrigerator operating between −20°C and +20°C extracts heat from the cold reservoir at the rate 200 J/s. What are (a) the coefficient of performance of this refrigerator, (b) the rate at which work is done on the refrigerator, and (c) the rate at which heat is exhausted to the hot side?

28. Liquid helium is kept in a refrigerator that removes heat energy from the helium at its boiling temperature of 4.2 K and exhausts heat energy into the 20°C laboratory. How much work must a Carnot refrigerator do to remove 1.0 J of heat energy from the liquid helium?

29. The coefficient of performance of a refrigerator is 5.0. The compressor uses 10 J of energy per cycle.
    a. How much heat energy is exhausted per cycle?
    b. If the hot-reservoir temperature is 27°C, what is the lowest possible temperature in °C of the cold reservoir?

30. A Carnot refrigerator with a cold-reservoir temperature of −13°C has a coefficient of performance of 5.0. To increase the coefficient of performance to 10.0, should the hot-reservoir temperature be increased or decreased, and by how much? Explain.

Problems

31. A 20%-efficient car engine accelerates the 1500 kg car from rest to 15 m/s. How much heat energy is transferred to the engine by burning gasoline?

32. The engine that powers a crane burns fuel at a flame temperature of 2000°C. It is cooled by 20°C air. The crane lifts a 2000 kg steel girder 30 m upward. How much heat energy is transferred to the engine by burning fuel if the engine is 40% as efficient as a Carnot engine?

33. 100 mL of water at 15°C is placed in the freezer compartment of a refrigerator with a coefficient of performance of 4.0. How much heat energy is exhausted into the room as the water is changed to ice at −15°C?

34. Prove that the work done in an adiabatic process i → f is $W_s = (p_f V_f - p_i V_i)/(1 - \gamma)$.

35. Statement #6 of the second law says that no refrigerator operating between reservoirs at temperatures T_H and T_C can have a coefficient of performance larger than that of a perfectly reversible refrigerator operating between these temperatures. Prove that this is true.

    Hint: The proof is a logical proof, similar to the proof of statement #5 of the second law, not a mathematical proof.

36. Prove that the coefficient of performance of a Carnot refrigerator is $K_{Carnot} = T_C/(T_H - T_C)$.

37. A Carnot heat engine with thermal efficiency η and a Carnot refrigerator operate between the same two energy reservoirs. Show that the refrigerator's coefficient of performance is $K = (1 - \eta)/\eta$.

38. You have a hot reservoir at temperature T_H, a cold reservoir at T_C, and a third energy reservoir with a temperature T_3 between T_C and T_H. A Carnot engine operating between T_H and T_C extracts heat Q_H from the hot reservoir and exhausts heat Q_C to the cold reservoir. Suppose you use two Carnot engines, one operating between T_H and T_3 and the other between T_3 and T_C. To make an appropriate comparison, the first must extract heat Q_H from the hot reservoir and the second must exhaust heat Q_C to the cold reservoir. Is the combined work done by the two engines more, less, or equal to the work of the single engine?

39. There has long been an interest in using the vast quantities of thermal energy in the oceans to run heat engines. A heat engine needs a temperature *difference*, a hot side and a cold side. Conveniently, the ocean surface waters are warmer than the deep ocean waters. Suppose you build a floating power plant in the tropics where the surface water temperature is ≈30°C. This would be the hot reservoir of the engine. For the cold reservoir, water would be pumped up from the ocean bottom where it is always ≈5°C. What is the maximum possible efficiency of such a power plant?

40. The ideal gas in a Carnot engine extracts 1000 J of heat energy during the isothermal expansion at 300°C. How much heat energy is exhausted during the isothermal compression at 50°C?

41. The hot-reservoir temperature of a Carnot engine with 25% efficiency is 80°C higher than the cold-reservoir temperature. What are the reservoir temperatures, in °C?

42. Which would make the larger increase in the thermal efficiency of a Carnot engine, raising the hot-reservoir temperature by ΔT or lowering the cold-reservoir temperature by ΔT? Explain.

43. The heat exhausted to the cold reservoir of a Carnot engine is two-thirds the heat extracted from the hot reservoir. What is the temperature ratio T_C/T_H?

44. Figure P19.44 shows a Carnot heat engine driving a Carnot refrigerator.
    a. Determine Q_1, Q_2, Q_3, and Q_4.
    b. Is Q_3 greater than, less than, or equal to Q_1?
    c. Do these two devices, when operated together in this way, violate the second law?

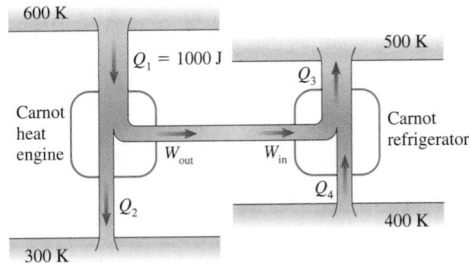

FIGURE P19.44

45. A heat engine running backward is called a refrigerator if its purpose is to extract heat from a cold reservoir. The same engine running backward is called a *heat pump* if its purpose is to exhaust warm air into the hot reservoir. Heat pumps are widely used for home heating. You can think of a heat pump as a refrigerator that is cooling the already cold outdoors and, with its exhaust heat Q_H, warming the indoors. Perhaps this seems a little silly, but consider the following. Electricity can be directly used to heat a home by passing an electric current through a heating coil. This is a direct, 100% conversion of work to heat. That is, 15 kW of electric power (generated by doing work at the rate 15 kJ/s at the power plant) produces

heat energy inside the home at a rate of 15 kJ/s. Suppose that the neighbor's home has a heat pump with a coefficient of performance of 5.0, a realistic value.

a. How much electric power (in kW) does the heat pump use to deliver 15 kJ/s of heat energy to the house?

b. An average price for electricity is about 40 MJ per dollar. A furnace or heat pump will run typically 200 hours per month during the winter. What does one month's heating cost in the home with a 15 kW electric heater and in the home of the neighbor who uses an equivalent heat pump?

46. Three engineering students submit their solutions to a design problem in which they were asked to design an engine that operates between temperatures 300 K and 500 K. The heat input/output and work done by their designs are shown in the following table:

Student	Q_H	Q_C	W_{out}
1	250 J	140 J	110 J
2	250 J	170 J	90 J
3	250 J	160 J	90 J

Critique their designs. Are they acceptable or not? Is one better than the others? Explain.

47. A typical coal-fired power plant burns 300 metric tons of coal *every hour* to generate 750 MW of electricity. 1 metric ton = 1000 kg. The density of coal is 1500 kg/m³ and its heat of combustion is 28 MJ/kg. Assume that *all* heat is transferred from the fuel to the boiler and that *all* the work done in spinning the turbine is transformed into electrical energy.

a. Suppose the coal is piled up in a 10 m × 10 m room. How tall must the pile be to operate the plant for one day?

b. What is the power plant's thermal efficiency?

48. A nuclear power plant generates 2000 MW of heat energy from nuclear reactions in the reactor's core. This energy is used to boil water and produce high-pressure steam at 300°C. The steam spins a turbine, which produces 700 MW of electric power, then the steam is condensed and the water is cooled to 30°C before starting the cycle again.

a. What is the maximum possible thermal efficiency of the power plant?

b. What is the plant's actual efficiency?

c. Cooling water from a river flows through the condenser (the low-temperature heat exchanger) at the rate of 1.2 × 10⁸ L/hr (≈30 million gallons per hour). If the river water enters the condenser at 18°C, what is its exit temperature?

49. The electric output of a power plant is 750 MW. Cooling water flows through the power plant at the rate 1.0 × 10⁸ L/hr. The cooling water enters the plant at 16°C and exits at 27°C. What is the power plant's thermal efficiency?

50. a. A large nuclear power plant has a power output of 1000 MW. In other words, it generates electric energy at the rate 1000 MJ/s. How much energy does this power plant supply in one day?

b. The oceans are vast. How much energy could be extracted from 1 km³ of water if its temperature is decreased by 1°C? For simplicity, assume fresh water.

c. A friend of yours who is an inventor comes to you with an idea. He has done the calculations that you just did in parts a and b, and he's concluded that a few cubic kilometers of

ocean water could meet most of the energy needs of the United States. This is an insignificant fraction of the U.S. coastal waters. In addition, the oceans are constantly being reheated by the sun, so energy obtained from the ocean is essentially solar energy. He has sketched out some design plans—highly secret, of course, because they're not patented—and now he needs some investors to provide money for a prototype. A working prototype will lead to a patent. As an initial investor, you'll receive a fraction of all future royalties. Time is of the essence because a rival inventor is working on the same idea. He needs $10,000 from you right away. You could make millions if it works out. Will you invest? If so, explain why. If not, why not? Either way, your explanation should be based on scientific principles. Sketches and diagrams are a reasonable part of an explanation.

51. Design a heat engine using a three-step closed cycle that meets the following specifications:

$$p_{min} = 1 \text{ atm} \qquad p_{max} = 3 \text{ atm}$$
$$V_{min} = 100 \text{ cm}^3 \qquad W_{out} = 40.5 \text{ J}$$

Give your answer as a pV diagram showing the cycle.

52. A heat engine using 1.0 mol of a monatomic gas follows the cycle shown in Figure P19.52. 3750 J of heat energy is transferred to the gas during process 1 → 2.

a. Determine W_s, Q, and ΔE for each of the four processes in this cycle. Display your results in a table.

b. What is the thermal efficiency of this heat engine?

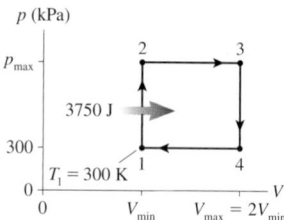

FIGURE P19.52

53. A heat engine using a diatomic gas follows the cycle shown in Figure P19.53. Its temperature at point 1 is 20°C.

a. Determine W_s, Q, and ΔE for each of the three processes in this cycle. Display your results in a table.

b. What is the thermal efficiency of this heat engine?

c. What is the power output of the engine if it runs at 500 rpm?

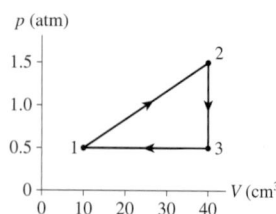

FIGURE P19.53

54. Figure P19.54 shows the cycle for a heat engine that uses a gas having $\gamma = 1.25$. The initial temperature is $T_1 = 300$ K, and this engine operates at 20 cycles per second.

a. What is the power output of the engine?

b. What is the engine's thermal efficiency?

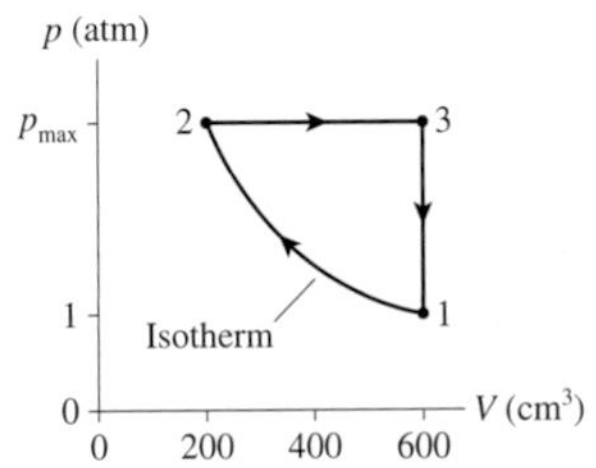

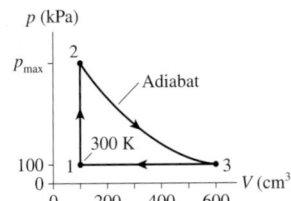

FIGURE P19.54

55. A heat engine using a monatomic gas follows the cycle shown in Figure P19.55.
    a. Find W_s, Q, and ΔE for each process in the cycle. Display your results in a table.
    b. What is the thermal efficiency of this heat engine?

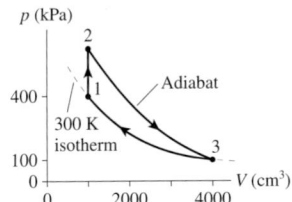

FIGURE P19.55

56. A heat engine uses a diatomic gas that follows the pV cycle in Figure P19.56.
    a. Determine the pressure, volume, and temperature at point 2.
    b. Determine ΔE_{th}, W_s, and Q for each of the three processes. Put your results in a table for easy reading.
    c. How much work does this engine do per cycle and what is its thermal efficiency?

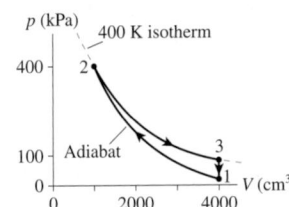

FIGURE P19.56

57. A heat engine uses a diatomic gas that follows the pV cycle in Figure P19.57.
    a. Determine the pressure, volume, and temperature at point 1.
    b. Determine ΔE_{th}, W_s, and Q for each of the three processes. Put your results in a table for easy reading.
    c. How much work does this engine do per cycle and what is its thermal efficiency?

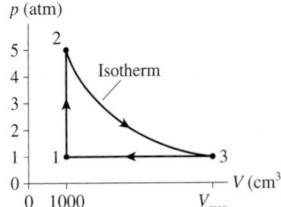

FIGURE P19.57

58. A Brayton-cycle heat engine follows the cycle shown in Figure P19.58. The heat input from the burning fuel is 2.0 MJ per cycle. Determine the engine's thermal efficiency by explicitly computing the work done per cycle. Compare your answer with the efficiency that you can determine from Equation 19.21.

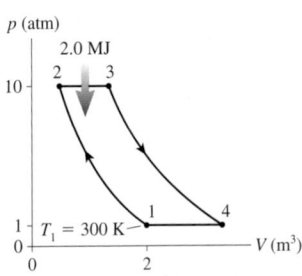

FIGURE P19.58

59. A heat engine using 120 mg of helium as the working substance follows the cycle shown in Figure P19.59.
    a. Determine the pressure, temperature, and volume of the gas at points 1, 2, and 3.
    b. What is the engine's thermal efficiency?
    c. What is the maximum possible efficiency of a heat engine that operates between T_{max} and T_{min}?

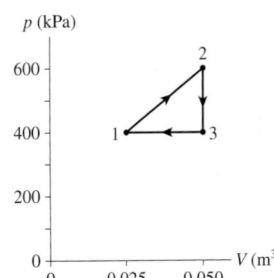

FIGURE P19.59

60. The heat engine shown in Figure P19.60 uses 2.0 mol of a monatomic gas as the working substance.
    a. Determine T_1, T_2, and T_3.
    b. Make a table that shows ΔE_{th}, W_s, and Q for each of the three processes.
    c. What is the engine's thermal efficiency?

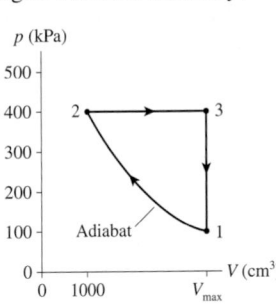

FIGURE P19.60

61. The heat engine shown in Figure P19.61 uses 0.020 mol of a diatomic gas as the working substance.
    a. Determine T_1, T_2, and T_3.
    b. Make a table that shows ΔE_{th}, W_s, and Q for each of the three processes.
    c. What is the engine's thermal efficiency?

FIGURE P19.61

62. A heat engine using 2.0 g of helium gas is initially at STP. The gas goes through the following closed cycle:
    - Isothermal compression until the volume is halved.
    - Isobaric expansion until the volume is restored to its initial value.
    - Isochoric cooling until the pressure is restored to its initial value.

    How much work does this engine do per cycle and what is its thermal efficiency?

63. A heat engine with 0.20 mol of a monatomic ideal gas initially fills a 2000 cm^3 cylinder at 600 K. The gas goes through the following closed cycle:
    - Isothermal expansion to 4000 cm^3.
    - Isochoric cooling to 300 K.
    - Isothermal compression to 2000 cm^3.
    - Isochoric heating to 600 K.

    How much work does this engine do per cycle and what is its thermal efficiency?

64. Figure P19.64 is the pV diagram of Example 19.2, but now the device is operated in reverse.
    a. During which processes is heat transferred into the gas?
    b. Is this Q_H, heat extracted from a hot reservoir, or Q_C, heat extracted from a cold reservoir? Explain.
    c. Determine the values of Q_H and Q_C.
    Hint: The calculations have been done in Example 19.2 and do not need to be repeated. Instead, you need to determine which processes now contribute to Q_H and which to Q_C.
    d. Is the area inside the curve W_{in} or W_{out}? What is its value?
    e. Show that Figure 19.19 is the energy-transfer diagram of this device.
    f. The device is now being operated in a ccw cycle. Is it a refrigerator? Explain.

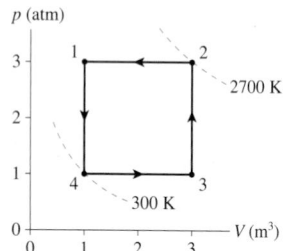

FIGURE P19.64

In Problems 65 through 68 you are given the equation(s) used to solve a problem. For each of these, you are to
  a. Write a realistic problem for which this is the correct equation(s).
  b. Finish the solution of the problem.

65. $0.80 = 1 - (0°C + 273)/(T_H + 273)$

66. $4.0 = Q_C/W_{in}$

    $Q_H = 100$ J

67. $0.20 = 1 - Q_C/Q_H$

    $W_{out} = Q_H - Q_C = 20$ J

68. 400 kJ $= \frac{1}{2}(p_{max} - 100$ kPa$)(3.0$ m$^3 - 1.0$ m$^3)$

Challenge Problems

69. Figure CP19.69 shows a heat engine going through one cycle. The gas is diatomic. The masses are such that when the pin is removed, in steps 3 and 6, the piston does not move.
    a. Draw the pV diagram for this heat engine.
    b. How much work is done per cycle?
    c. What is this engine's thermal efficiency?

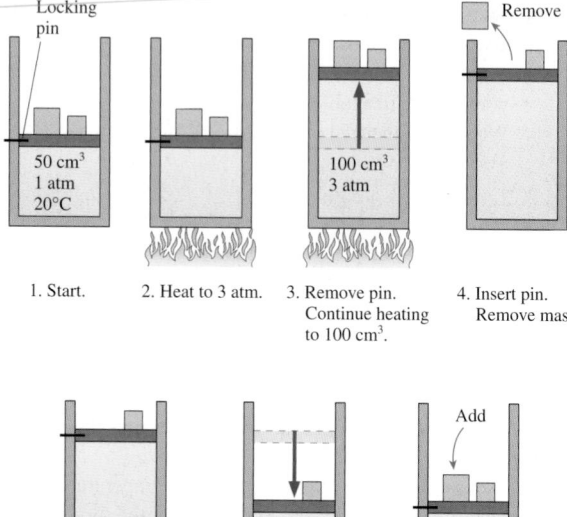

FIGURE CP19.69

70. Figure CP19.70 shows two compartments separated by a thin wall. The left side contains 0.060 mol of helium at an initial temperature of 600 K and the right side contains 0.030 mol of helium at an initial temperature of 300 K. The compartment on the right is attached to a vertical cylinder. A 2.0 kg piston can slide without friction up and down the cylinder. Neither the cylinder diameter nor the volumes of the compartments are known.
    a. What is the final temperature?
    b. How much heat is transferred from the left side to the right side?
    c. How high is the piston lifted due to this heat transfer?
    d. What fraction of the heat is converted into work?

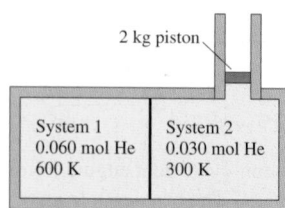

FIGURE CP19.70

71. The gasoline engine in your car can be modeled as the Otto cycle shown in Figure CP19.71. A fuel-air mixture is sprayed into the cylinder at point 1, where the piston is at its farthest distance from the spark plug. This mixture is compressed as the piston moves toward the spark plug during the *compression stroke*. The spark plug fires at point 2, releasing heat energy that had been stored in the gasoline. The fuel burns so quickly that the piston doesn't have time to move, so the heating is an isochoric process. The hot, high-pressure gas then pushes the piston outward during the *power stroke*. Finally, an exhaust value opens to allow the gas temperature and pressure to drop back to their initial values before starting the cycle over again.

a. Analyze the Otto cycle and show that the work done per cycle is

$$W_{out} = \frac{nR}{1 - \gamma}(T_2 - T_1 + T_4 - T_3)$$

b. Use the adiabatic connection between T_1 and T_2 and also between T_3 and T_4 to show that the thermal efficiency of the Otto cycle is

$$\eta = 1 - \frac{1}{r^{(\gamma - 1)}}$$

where $r = V_{max}/V_{min}$ is the engine's *compression ratio*.

c. Graph η versus r out to $r = 30$ for a diatomic gas.

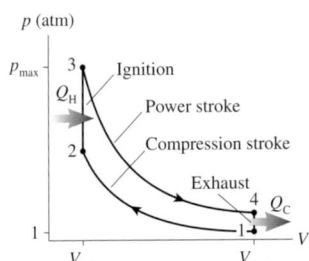

FIGURE CP19.71

72. Figure CP19.72 shows the diesel cycle. It is similar to the Otto cycle (see Problem CP19.71), but there are two important differences. First, the fuel is not admitted until the air is fully compressed at point 2. Because of the high temperature at the end of an adiabatic compression, the fuel begins to burn spontaneously. (There are no spark plugs in a diesel engine!) Second, combustion takes place more slowly, with fuel continuing to be injected. This makes the ignition stage a constant-pressure process. The cycle shown, for one cylinder of a diesel engine, has a *displacement* $V_{max} - V_{min}$ of 1000 cm^3 and a compression ratio $r = V_{max}/V_{min} = 21$. These are typical values for a diesel truck. The engine operates with intake air ($\gamma = 1.40$) at 25°C and 1.0 atm pressure. The quantity of fuel injected into the cylinder has a heat of combustion of 1000 J.

a. Find p, V, and T at each of the four corners of the cycle. Display your results in a table.

b. What is the net work done by the cylinder during one full cycle?

c. What is the thermal efficiency of this engine?

d. What is the power output in kW and horsepower (1 hp = 746 W) of an eight-cylinder diesel engine running at 2400 rpm?

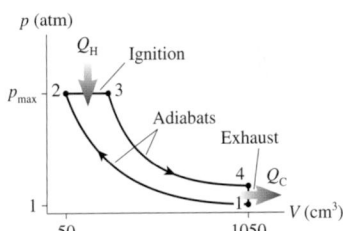

FIGURE CP19.72

Stop to Think 19.1: $W_d > W_a = W_b > W_c$. $W_{out} = Q_H - Q_C$.

Stop to Think 19.2: b. Energy conservation requires $Q_H = Q_C + W_{in}$. The refrigerator will exhaust more heat out the back than it removes from the front. A refrigerator with an open door will heat the room rather than cool it.

Stop to Think 19.3: c. W_{out} = area inside triangle = 1000 J. $\eta = W_{out}/Q_H = 1000/4000 = 0.25$.

Stop to Think 19.4: To conserve energy, the heat Q_H exhausted to the hot reservoir needs to be $Q_H = Q_C + W_{out} = 40 J + 10 J = 50 J$. The numbers shown here, with $Q_C = Q_H + W_{out}$, would be appropriate to a heat engine, not a refrigerator.

Stop to Think 19.5: b. The efficiency of this engine would be $\eta = W_{out}/Q_H = 0.6$. That exceeds the Carnot efficiency $\eta_{Carnot} = 1 - T_C/T_H = 0.5$, so it is not possible.

Thermodynamics

Part IV had two important goals: first, to learn how energy is transformed; second, to establish a micro/macro connection in which we can understand the macroscopic properties of solids, liquids, and gases in terms of the microscopic motions of atoms and molecules. We have been quite successful. You have learned that:

- Temperature is a measure of the thermal energy of the molecules in a system, and the average energy per molecule is simply $\frac{1}{2}k_B T$ per degree of freedom.
- The pressure of a gas is due to collisions of the molecules with the walls of the container.
- Heat is the energy transferred between two systems that have different temperatures. The *mechanism* of heat transfer is molecular collisions at the boundary between the two systems.
- Work, heat, and thermal energy can be transformed into each other in accord with the first law of thermodynamics, $\Delta E_{th} = W + Q$. This is a statement that energy is conserved.

- Practical devices for turning heat into work, called heat engines, are limited in their efficiency by the second law of thermodynamics.

The knowledge structure of thermodynamics below summarizes the basic laws, diagramming our energy model and presenting our model of a heat engine in pictorial form. Thermodynamics, more than most topics in physics, can seem very "equation oriented." It's undeniable that there are more equations than we used in earlier parts of this text and more things to remember. But focusing on the equations is seeing only the trees, not the forest. A better strategy is to focus on the ideas embedded in the knowledge structure. You can find the necessary equations if you know how the ideas are connected, but memorizing all the equations won't help if you don't know which are relevant to different situations.

KNOWLEDGE STRUCTURE IV **Thermodynamics**

ESSENTIAL CONCEPTS	Work, heat, and thermal energy.
BASIC GOALS	How is energy converted from one form to another?
	How are macroscopic properties related to microscopic behavior?

GENERAL PRINCIPLES	**First law of thermodynamics**	Energy is conserved, $\Delta E_{th} = W + Q$.
	Second law of thermodynamics	Heat is not spontaneously transferred from a colder object to a hotter object.

GAS LAWS AND PROCESSES   Ideal-gas law $pV = nRT = Nk_B T$

- Isochoric process $V = $ constant and $W = 0$
- Isothermal process $T = $ constant and $\Delta E_{th} = 0$
- Isobaric process $p = $ constant
- Adiabatic process $Q = 0$

Energy Transformation

Work on system	Environment	Work by system
W > 0	**System** Thermal energy E_{th} + Other state variables $p, V, T, n, M, \ldots$	W < 0
Energy in		Energy out
Q > 0	First law: $\Delta E_{th} = W + Q$	Q < 0
Heat to system		Heat out of system

Heat Engines

W_{out} = area inside pV curve
       = $Q_H - Q_C$

$$\eta = \frac{W_{out}}{Q_H}$$

$$\eta_{max} = \eta_{Carnot} = 1 - \frac{T_C}{T_H}$$

Work

Requires volume change

Gas: $W = -\int p\, dV$

      $= -(\text{area under } pV \text{ curve})$

Thermal Energy

$E_{th} = \frac{1}{2}Nk_B T$ per degree of freedom

Heat

Requires temperature difference

$Q = Mc\Delta T$ or $nC\Delta T$

$Q = \pm ML$ for phase changes

Order Out of Chaos

The second law predicts that systems will run down, that order will evolve toward disorder and randomness, and that complexity will give way to simplicity. But just look around you!

- Plants grow from simple seeds to complex entities.
- Single-cell fertilized eggs grow into complex adult organisms.
- Electric current passing through a "soup" of simple random molecules produces such complex chemicals as amino acids.
- Over the last billion or so years life has evolved from simple unicellular organisms to very complex forms.
- Knowledge and information seem to grow every year, not to fade away.

Everywhere we look, it seems, the second law is being violated. How can this be?

There is an important qualification in the second law of thermodynamics: It applies only to *isolated* systems, systems that do not exchange energy with their environment. The situation is entirely different if energy is transferred into or out of the system, and we cannot predict what will happen to the entropy of a nonisolated system. The popular-science literature is full of arguments and predictions that make incorrect use of the second law by trying to apply it to systems that are not isolated.

Systems that become *more* ordered as time passes, and in which the entropy decreases, are called *self-organizing systems*. All the examples listed above are self-organizing systems. One of the major characteristics of self-organizing systems is a substantial flow of energy *through* the system. For example, plants and animals take in energy from the sun or chemical energy from food, make use of that energy, and then give waste heat back to the environment via evaporation, decay, and other means. It is this energy flow that allows the systems to maintain, or even increase, a high degree of order and a very low entropy.

But—and this is the important point—the entropy of the *entire* system, including the earth and the sun, undergoes a significant *increase* so as to let selected subsystems decrease their entropy and become more ordered. The second law is not violated at all, but you must apply the second law to the combined systems that are interacting and not just to a single subsystem.

The snowflake in the photo is a beautiful example. As water freezes, the random motion of water molecules is transformed into a highly ordered crystal. The entropy of the water molecules certainly decreases, but water doesn't

A snowflake is a highly ordered arrangement of water molecules. The creation of a snowflake decreases the entropy of the water, but the second law of thermodynamics is not violated because the water molecules are not an isolated system.

freeze as an isolated system. For it to freeze, heat energy must be transferred from the water to the surrounding air. The entropy of the air increases by *more* than the entropy of the water decreases. Thus the *total* entropy of the water + air system increases when a snowflake is formed, just as the second law predicts.

Self-organization is closely related to nonlinear mechanics, chaos, and the geometry of fractals. It has important applications in fields ranging from ecology to computer science to aeronautical engineering. For example, the air flow across a wing gives rise to large-scale turbulence—eddies and whirlpools—in the wake behind an airplane. Their formation affects the aerodynamics of the plane and can also create hazards for following aircraft. Whirlpools are ordered, large-scale macroscopic structures with low entropy, but they are produced from disordered, random collisions of the air molecules.

Self-organizing systems are a very active field of research in both science and engineering. The 1977 Nobel Prize in chemistry was awarded to the Belgian scientist Ilya Prigogine for his studies of *nonequilibrium thermodynamics*, the basic science underlying self-organizing systems. Prigogine and others have shown how energy flow through a system can, when the conditions are right, "bring order out of chaos."

If violent storms drive waves across the ocean of planet Kamino at a speed of 75 m/s, what is the wavelength of an ocean wave whose period is 10 s?

Waves and Optics

Beyond the Particle Model

Parts I–IV of this text have been about the physics of particles. You saw that macroscopic systems, such as solids and gases, can be thought of as systems of particles. A *particle* is one of the two fundamental models of physics. The other, to which we now turn our attention, is a *wave*.

Waves are ubiquitous in nature. Familiar examples of waves include

- Ripples on a pond.
- Surf crashing on a beach.
- The swaying ground of an earthquake.
- A vibrating guitar string.
- The sweet sound of a flute.
- The colors of the rainbow.
- A laser beam shooting through space.

The physics of waves is the subject of Part V, the next stage of our journey. Despite the great diversity of types and sources of waves, there is a single, elegant physical theory that is capable of describing them all. Our exploration of wave phenomena will call upon water waves, sound waves, and light waves for examples, but we want to emphasize the unity and coherence of the ideas that are common to *all* types of waves.

We will start with waves that travel outward through some medium, like the ripples that spread out from a pebble hitting a pool of water or the sounds that emanate from a loudspeaker. These are called *traveling waves*. After arriving at an understanding of a single traveling wave, we will see what happens when several traveling waves are combined. This investigation will lead us to *standing waves*, which are essential for understanding music and lasers, and to the phenomenon called *interference*, one of the most important defining characteristics of waves.

Two chapters will be devoted to light and optics, perhaps the most important application of waves. Light is an elusive subject, and we will end up developing three different models of light:

- The *wave model*, which explains the interference and diffraction of light.
- The *ray model*, which provides a framework for understanding how lenses and mirrors form images.
- The *photon model*, which will bring us to the foothills of quantum physics.

Although light is an electromagnetic wave, these chapters depend on nothing more than the "waviness" of light waves for your understanding. You can study these chapters either before or after your study of electricity and magnetism in

609

Part VI. The electromagnetic aspects of light waves will be taken up in Chapter 34.

Finally, waves will give us new information about atoms and their properties. Much of the light and color in our environment is due to the emission and absorption of light by atoms and molecules. The atoms of each element in the periodic table emit a unique "fingerprint" of light, a fingerprint that can be read with optical instruments and that provides clues about the structure of the atoms. The search for a theory that could successfully decipher these clues ultimately led to the development of quantum physics at the beginning of the 20th century. Part V will conclude with an initial look at the connection between atoms and light. We will then return to this important topic in Part VII.

Waves and Particles

Our journey through Part V will lead us to a closer examination of the relationship between waves and particles. The classical physics of the 18th and 19th centuries made a fundamental distinction between waves and particles. There was a well-defined wave-particle dichotomy, an either-or situation, and each of the objects in nature could be characterized as one or the other. Planets, projectiles, and atoms were clearly particles, or systems of particles, while sound and light were waves. That is not to say that particles (air molecules) are not relevant to sound waves, but sound itself is a collective, wave-like behavior of the air. Sound does not follow a parabolic trajectory as a particle would.

Table OV.V.1 lists some of the basic, defining characteristics of the wave and particle models. The idea that a particle exists at a specific location is particularly important. It has a position coordinate x. You can put your finger on it. It is *here*. A wave, by contrast, is diffuse, spread out, not to be found at a single point in space. This distinction seems perfectly clear. If you consider a cork bobbing up and down on an ocean swell, there is no doubt which is a particle and which a wave.

TABLE OV.V.1 Basic characteristics of particles and waves

Particles	Waves
Discrete	Continuous
Localized (here)	Nonlocalized (everywhere)
Individual	Collective

Yet as we enter the atomic realm, we will be faced with evidence that its inhabitants are not so easily classified. Electrons, protons, even light itself will be found to have characteristics of both particles *and* waves. In the strange world of quantum physics, "either-or" is replaced with "both-and." Rather than the wave-particle dichotomy of classical physics, we will come to recognize a *wave-particle duality* in quantum physics. The breakdown of the distinction between waves and particles undermines the Newtonian worldview, but at the same time it provides us with a richer and deeper understanding of nature.

20 Traveling Waves

This surfer is "catching a wave." At the same time, he is seeing light waves and hearing sound waves.

▶ Looking Ahead

The goal of Chapter 20 is to learn the basic properties of traveling waves. In this chapter you will learn to:

- Use the wave model and understand how it differs from the particle model.
- Understand how a wave travels through a medium.
- Recognize the properties of sinusoidal waves.
- Understand the important characteristics of sound and light waves.
- Use the Doppler effect to find the speed of wave sources and observers.

◀ Looking Back

The material in this chapter depends on the concept of simple harmonic motion. Please review:

- Sections 14.1–14.2 The properties of simple harmonic motion.

You may not realize it, but you are surrounded by waves. The "waviness" of a water wave is readily apparent, from the ripples on a pond to ocean waves large enough to surf. It's less apparent that sound and light are also waves because their wave properties are discovered only by careful observations and experiments. We will even find, quite surprisingly, that matter exhibits wave-like behavior when we get to the microscopic scale of electrons and atoms.

Our overarching goal in Part V is to understand the properties and characteristics that are common to waves of all types. In other words, we want to find the "essence of waviness" that all waves possess. In this chapter we start with the idea of a *traveling wave*. When your friend speaks to you, a sound wave travels through the air to your ear. Light waves travel from the sun to the earth. A sudden fracture in the earth's crust sends out a shock wave that is felt far away as an earthquake. To understand phenomena such as these we need both new models and new mathematics.

The boat's wake is a wave moving across the surface of the lake.

20.1 The Wave Model

The *particle model* of Parts I–IV focused on those aspects of motion that are common to many systems. Balls, cars, and rockets obviously differ from one another, but the general features of their motions are well described by treating them as particles. In Part V we will explore the basic properties of waves with a **wave model** that emphasizes those aspects of wave behavior common to all waves. Although water waves, sound waves, and light waves are clearly different, the wave model will allow us to understand many of their important features.

The wave model is built around the idea of a **traveling wave,** which is an organized disturbance that travels with a well-defined wave speed. This definition seems straightforward, but several new terms must be understood before the concept of a traveling wave will make sense.

To begin, we can distinguish three types of waves:

1. **Mechanical waves** can travel only within a material *medium,* such as air or water. Two mechanical waves that you are familiar with are sound waves and water waves.
2. **Electromagnetic waves,** which include visible light as well as radio waves, microwaves, and x rays, are a self-sustaining oscillation of the *electromagnetic field.* Electromagnetic waves require no material medium and can travel through a vacuum. We'll look more closely at electromagnetic waves in Section 20.5.
3. **Matter waves** are the basis for quantum physics. One of the most significant discoveries of the 20th century was that material particles, such as electrons and atoms, have wave-like characteristics. Chapter 24 will introduce matter waves.

The **medium** of a mechanical wave is the substance through or along which the wave moves. For example, the medium of a water wave is the water, the medium of a sound wave is the air, and the medium of a wave on a stretched string is the string. A medium must be *elastic.* That is, a restoring force of some sort brings the medium back to equilibrium after it has been displaced or disturbed. The tension in a stretched string pulls the string back straight after you pluck it. Gravity restores the level surface of a lake after the wave generated by a boat has passed by.

As a wave passes through a medium, the atoms that make up the medium—we'll simply call them the particles of the medium—are displaced from equilibrium. This is a **disturbance** of the medium. The water ripples of Figure 20.1 are a disturbance of the water's surface. A pulse traveling down a string is a disturbance, as is the wake of a boat and the sonic boom created by a jet traveling faster than the speed of sound. The disturbance of a wave is an *organized* motion of the particles in the medium, in contrast with the *random* molecular motions of thermal energy.

A wave disturbance is created by a *source.* The source of a wave might be a rock thrown into water, your hand plucking a stretched string, or an oscillating loudspeaker cone pushing on the air. Once created, the disturbance travels outward through the medium at the **wave speed** *v.* This is the speed with which a ripple moves across the water or a pulse travels down a string.

NOTE ▶ The disturbance propagates through the medium, and a wave does transfer *energy,* but **the medium as a whole does not move!** The ripples on the pond (the disturbance) move outward from the splash of the rock, but there is no outward flow of water from the splash. Likewise, the particles of a string oscillate up and down but do not move in the direction of a pulse traveling along the string. **A wave transfers energy, but it does not transfer any material or substance outward from the source.** ◀

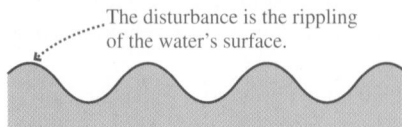

..........The disturbance is the rippling of the water's surface.

The water is the medium.

FIGURE 20.1 Ripples on a pond are a traveling wave.

We can identify two distinct types of wave motion: transverse and longitudinal.

Two types of wave motion

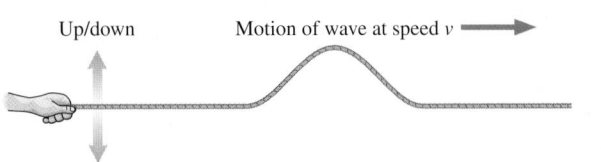

A transverse wave

For mechanical waves, a **transverse wave** is a wave in which the particles in the medium move *perpendicular* to the direction in which the wave travels. For example, a wave travels along a string in a horizontal direction while the particles that make up the string oscillate vertically. Electromagnetic waves are also transverse waves because the electromagnetic fields oscillate perpendicular to the direction in which the wave travels.

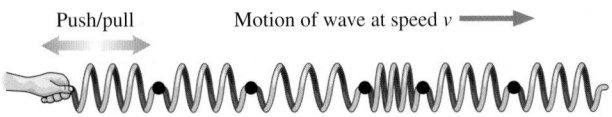

A longitudinal wave

In a **longitudinal wave,** the particles in the medium move *parallel* to the direction in which the wave travels. Here we see a chain of masses connected by springs. If you give the first mass in the chain a sharp push, a disturbance travels down the chain by compressing and expanding the springs. Sound waves in gases and liquids are the most well known examples of longitudinal waves. An oscillating loudspeaker cone compresses and expands the air much like the springs in this figure.

Some waves are more complex. For example, water waves have characteristics of both transverse and longitudinal waves. The surface of the water moves up and down vertically, but individual water molecules actually move both perpendicular *and* parallel to the direction of the wave. We will not analyze these more complex waves in this text.

Traveling Waves

How does a mechanical wave travel through a medium? In answering this question, we must be careful to distinguish the motion of the wave from the motion of the particles that make up the medium. The wave itself is not a particle, so we cannot apply Newton's laws to the wave. However, we can use Newton's laws to examine how the medium responds to a disturbance.

Figure 20.2 on the next page shows a transverse *wave pulse* traveling to the right along a stretched string. Imagine watching a little dot on the string as a wave pulse passes by. As the pulse approaches from the left, the string near the dot begins to curve. This is the **leading edge** of the pulse. Once the string curves, the tension forces pulling on a small segment of string no longer cancel each other. Instead, as you can see in the first part of the figure, the tension in the string exerts a net upward force on this segment of string, causing it to accelerate upward.

The string's curvature reverses near the top of the pulse, causing the net force and the acceleration to point downward. Thus this piece of string slows until reaching a turning point at the highest point of the pulse, then it speeds up in the downward direction. This is much like a mass on a spring as the mass approaches the highest point and then reverses direction. Finally, the force and acceleration turn upward on the **trailing edge** of the pulse. The piece of string decelerates and comes to rest when the pulse has completely passed by.

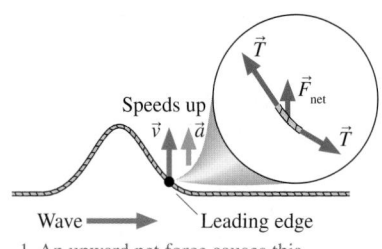

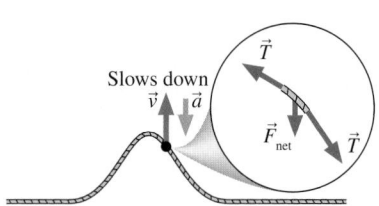

Speeds up $\vec{v}$ $\vec{a}$

Wave ⟶ Leading edge

1. An upward net force causes this piece of string to accelerate upward.

Slows down $\vec{v}$ $\vec{a}$

2. The change in curvature causes the net force to point down. $\vec{a}$ is opposite to $\vec{v}$, so the piece of string slows down.

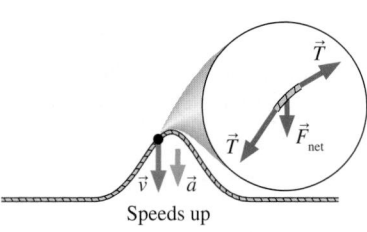

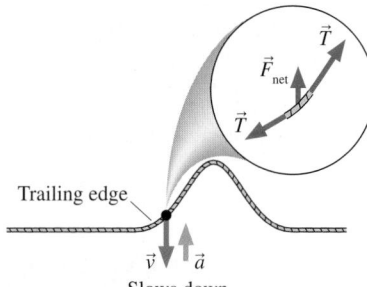

$\vec{v}$ $\vec{a}$

Speeds up

3. The downward net force causes the piece of string to speed up in the downward direction.

Trailing edge

$\vec{v}$ $\vec{a}$

Slows down

4. The net upward force causes the piece of string to slow down and stop.

FIGURE 20.2 The motion of a small segment of string as a wave pulse travels to the right.

The point of this analysis is that no new physical principles are required to understand how a wave moves. The motion of a pulse along a string is a direct consequence of the tension acting on the segments of the string. An external force may have been required to create the pulse, but once started the pulse continues to move because of the *internal dynamics* of the medium.

This discovery leads to the important and somewhat surprising conclusion that the wave speed is a property *of the medium*. **The wave speed depends on the restoring forces within the medium but not at all on the shape or size of the pulse, how the pulse was generated, or how far it has traveled.** In Section 20.3 we'll apply the ideas of Figure 20.2 to derive an explicit expression for the wave speed on a string.

20.2 One-Dimensional Waves

Some waves travel in one dimension (waves on a string), some in two dimensions (ripples on a pond), and yet others in three dimensions (sound from a loudspeaker or light from the sun). We will begin our study with an analysis of one-dimensional waves, which are the easiest to visualize. The ideas developed in this section will carry over to two- and three-dimensional waves.

Waves on a String

The prototype of a one-dimensional wave is a wave on a string. Consider a string of total length L and total mass m. The ratio of mass to length is called the **linear density** μ of the string:

$$\mu = \frac{m}{L} \qquad (20.1)$$

Linear density characterizes the *type* of string we are using. A fat string has a larger value of μ than a skinny string made of the same material. Similarly, a steel wire has a larger value of μ than a plastic string of the same diameter.

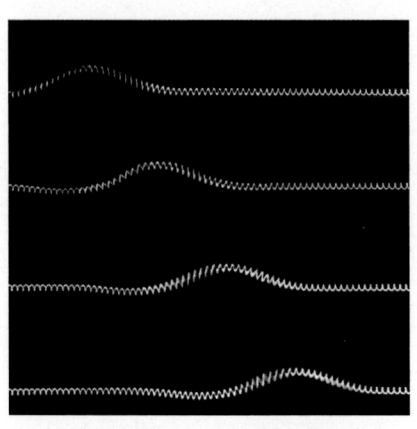

This sequence of photographs shows a wave pulse traveling along a string.

NOTE ▶ We'll assume that strings are *uniform,* meaning that the linear density is the same everywhere along the length of the string. ◀

As an example, the linear density of a 2.0-m-long string with a mass of 4.0 g is

$$\mu = \frac{0.0040 \text{ kg}}{2.0 \text{ m}} = 0.0020 \text{ kg/m}$$

The units tell us that the *numerical* value of μ is the mass in kg of a 1-m-long section of string; that is, the linear density is the mass *per meter* of string. But because μ is a ratio, we can apply it to a segment of string of any length. Thus the mass of any length L of string is $m = \mu L$.

The wave speed on a string depends on both the string's linear density μ and the tension T_s in the string. (The subscript s on the symbol T_s for the string's tension will distinguish it from the symbol T for the *period* of oscillation.) In Section 20.3 we will show that the wave speed on a stretched string is

$$v_{\text{string}} = \sqrt{\frac{T_s}{\mu}} \qquad \text{(wave speed on a stretched string)} \qquad (20.2)$$

This is the wave *speed,* not the wave velocity, so v_{string} always has a positive value.

Every point on a wave pulse travels with this speed. You can increase the wave speed either by *increasing* the string's tension (make it tighter) or by *decreasing* the string's linear density (make it skinnier). We'll examine the implications for stringed musical instruments in Chapter 21.

Actïv
Physïcs   ONLINE   10.2

EXAMPLE 20.1 **The speed of a wave pulse**

A 2.0-m-long string with a mass of 4.0 g is tied to a wall at one end, stretched horizontally to a pulley 1.5 m away, then tied to a physics book of mass M that hangs from the string. Experiments find that a wave pulse travels along the stretched string at 40 m/s. What is the mass of the book?

MODEL The wave pulse is a traveling wave on a stretched string. The hanging book is in static equilibrium.

VISUALIZE Figure 20.3 shows the situation and a free-body diagram of the book.

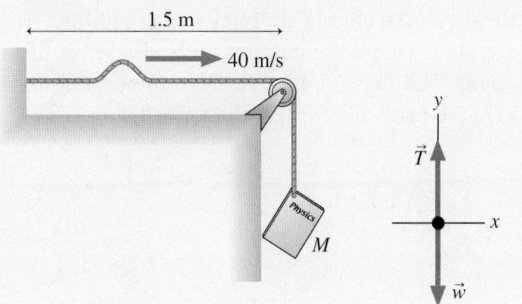

FIGURE 20.3 A wave pulse traveling on a string.

SOLVE The book is in static equilibrium, hence

$$(F_{\text{net}})_y = T_s - w = T_s - Mg = 0$$

Thus the tension in the string is $T_s = Mg$. The linear density of the 1.5-m-long segment of string between the wall and the pulley is exactly the same as μ for the entire string, which we computed in the calculation following Equation 20.1 to be 0.0020 kg/m. The length of the string between the wall and the pulley is not relevant. Squaring both sides of Equation 20.2 gives

$$v^2 = \frac{T_s}{\mu} = \frac{Mg}{\mu}$$

from which we find

$$M = \frac{\mu v^2}{g} = \frac{(0.0020 \text{ kg/m})(40 \text{ m/s})^2}{9.8 \text{ m/s}^2} = 0.327 \text{ kg} = 327 \text{ g}$$

ASSESS To be precise, 327 g is the combined mass of the book and the short length of the string that hangs from the pulley. This is one-quarter of the string, with a mass of 1 g, hence the book's mass is 326 g.

Snapshot Graphs and History Graphs

To understand waves we must deal with functions of *two* variables. Until now, we have been concerned with quantities that depend only on time, such as $x(t)$ or $v(t)$. Functions of the one variable t are all right for a particle, because a particle is only

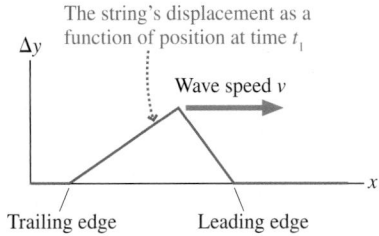

FIGURE 20.4 A snapshot graph of a wave pulse on a string.

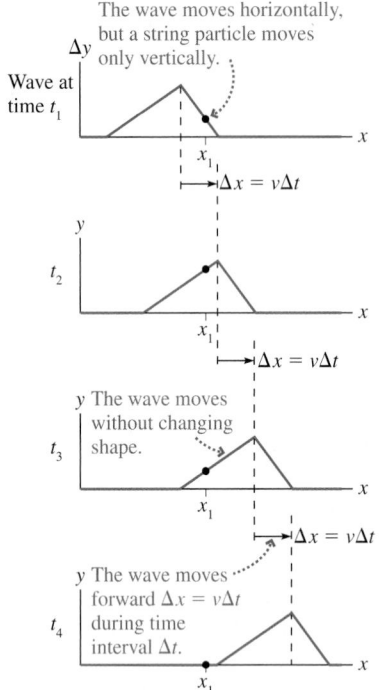

FIGURE 20.5 A sequence of snapshot graphs shows the wave in motion.

in one place at a time, but a wave is not localized. It is spread out through space at each instant of time. To describe a wave mathematically requires a function that specifies not only an instant of time (when) but also a point in space (where).

It will be helpful to think about waves graphically before we get into a mathematical analysis. Consider the wave pulse shown moving along a stretched string in Figure 20.4. (We will consider somewhat artificial triangular and square-shaped pulses in this section to make clear where the edges of the pulse are.) The graph shows the string's displacement Δy at a particular instant of time t_1 as a function of position x along the string. This is a "snapshot" of the wave, much like what you might make with a camera whose shutter is opened briefly at t_1. A graph that shows the wave's displacement as a function of position at a single instant of time is called a **snapshot graph.** For a wave on a string, a snapshot graph is literally a picture of the wave at this instant.

Figure 20.5 shows a sequence of snapshot graphs as the wave of Figure 20.4 continues to move. These are like successive frames from a movie. Notice that the wave pulse moves forward distance $\Delta x = v \Delta t$ during the time interval Δt. That is, the wave moves with constant speed.

A snapshot graph tells only half the story. It tells us *where* the wave is and how it varies with position, but only at one instant of time. It gives us no information about how the wave *changes* with time. As a different way of portraying the wave, suppose we follow the dot marked on the string in Figure 20.5 and produce a graph showing how the displacement of this dot changes with time. The result, shown below in Figure 20.6, is a displacement-versus-time graph at a single position in space. A graph that shows the wave's displacement as a function of time at a single position in space is called a **history graph.** It tells the history of that particular point in the medium.

You might think we have made a mistake; the graph of Figure 20.6 is reversed compared to Figure 20.5. It is not a mistake, but it requires careful thought to see why. As the wave moves toward the dot, the steep leading edge causes the dot to rise quickly. On the displacement-versus-time graph, *earlier* times (smaller values of t) are to the *left* and later times (larger t) to the right. Thus the leading edge of the wave is on the *left* side of the Figure 20.6 history graph. As you move to the right on Figure 20.6 you see the slowly falling trailing edge of the wave as it moves past the dot at later times.

The snapshot graph of Figure 20.4 and the history graph of Figure 20.6 portray complementary information. The snapshot graph tells us how things look throughout all of space, but at only one instant of time. The history graph tells us how things look at all times, but at only one position in space. We need them both to have the full story of the wave. An alternative representation of the wave is the series of graphs of Figure 20.7, where we can get a clearer sense of the wave moving forward. But graphs like these are essentially impossible to draw by hand, so it is necessary to move back and forth between snapshot graphs and history graphs.

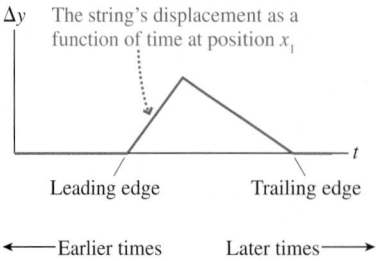

FIGURE 20.6 A history graph for the dot on the string in Figure 20.5.

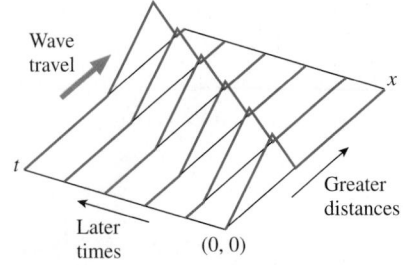

FIGURE 20.7 An alternative look at a traveling wave.

EXAMPLE 20.2 Finding a history graph from a snapshot graph

Figure 20.8 is a snapshot graph at $t = 0$ s of a wave moving to the right at a speed of 2 m/s. Draw a history graph for the position $x = 8$ m.

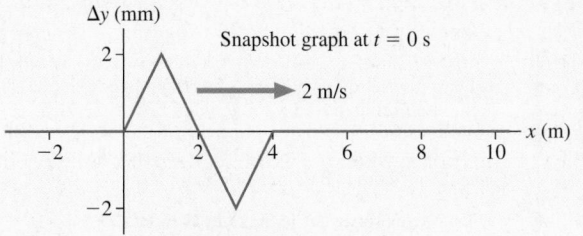

FIGURE 20.8 A snapshot graph at $t = 0$ s.

MODEL This is a wave traveling at constant speed. The pulse moves 2 m to the right every second.

VISUALIZE The snapshot graph of Figure 20.8 shows the wave at all points on the x-axis at $t = 0$ s. You can see that nothing is happening at $x = 8$ m at this instant of time because the wave has not yet reached $x = 8$ m. In fact, at $t = 0$ s the leading edge of the wave is still 4 m away from $x = 8$ m. Because the wave

is traveling at 2 m/s, it will take 2 s for the leading edge to reach $x = 8$ m. Thus the history graph for $x = 8$ m will be zero until $t = 2$ s. The first part of the wave causes a *downward* displacement of the medium, so immediately after $t = 2$ s the displacement at $x = 8$ m will be negative. The negative portion of the wave pulse is 2 m wide and takes 1 s to pass $x = 8$ m, so the midpoint of the pulse reaches $x = 8$ m at $t = 3$ s. The positive portion takes another 1 s to go past, so the trailing edge of the pulse arrives at $t = 4$ s. You could also note that the trailing edge was initially 8 m away from $x = 8$ m and needed 4 s to travel that distance at 2 m/s. The displacement at $x = 8$ m returns to zero at $t = 4$ s and remains zero for all later times. This information is all portrayed on the history graph of Figure 20.9.

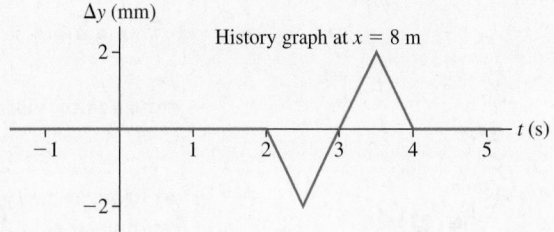

FIGURE 20.9 The corresponding history graph at $x = 8$ m.

STOP TO THINK 20.1   The graph at the right is the history graph at $x = 4$ m of a wave traveling to the right at a speed of 2 m/s. Which is the history graph of this wave at $x = 0$ m?

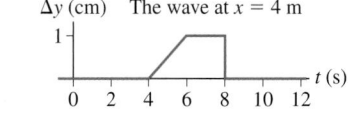

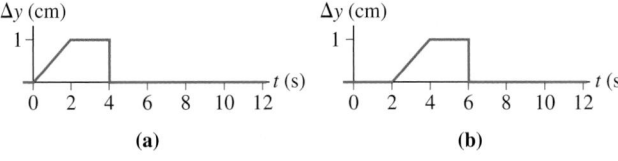

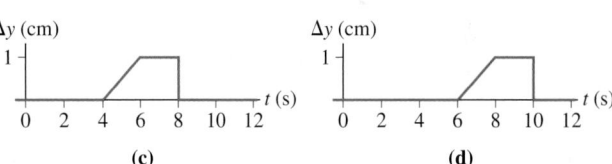

Longitudinal Waves

For a wave on a string, a transverse wave, the snapshot graph is literally a picture of the wave. Not so for a longitudinal wave, where the particles in the medium are displaced parallel to the direction in which the wave is traveling. Thus the displacement is Δx rather than Δy, and a snapshot graph is a graph of Δx versus x.

Figure 20.10a on the next page is a snapshot graph of a longitudinal wave, such as a sound wave. It's purposefully drawn to have the same shape as the string wave in Example 20.2. Without practice, what this graph tells us about the particles in the medium will not be obvious.

To help you find out, Figure 20.10b provides a tool for visualizing longitudinal waves. In the second row, we've used information from the graph to displace the particles in the medium to the right or to the left of their equilibrium positions. For example, the particle at $x = 1$ cm has been displaced 0.5 cm to the right because the snapshot graph shows $\Delta x = 0.5$ cm at $x = 1$ cm. We now have a picture of the longitudinal wave pulse at $t_1 = 0$ s. You can see that the medium is compressed to higher density at the center of the pulse and, to compensate, expanded

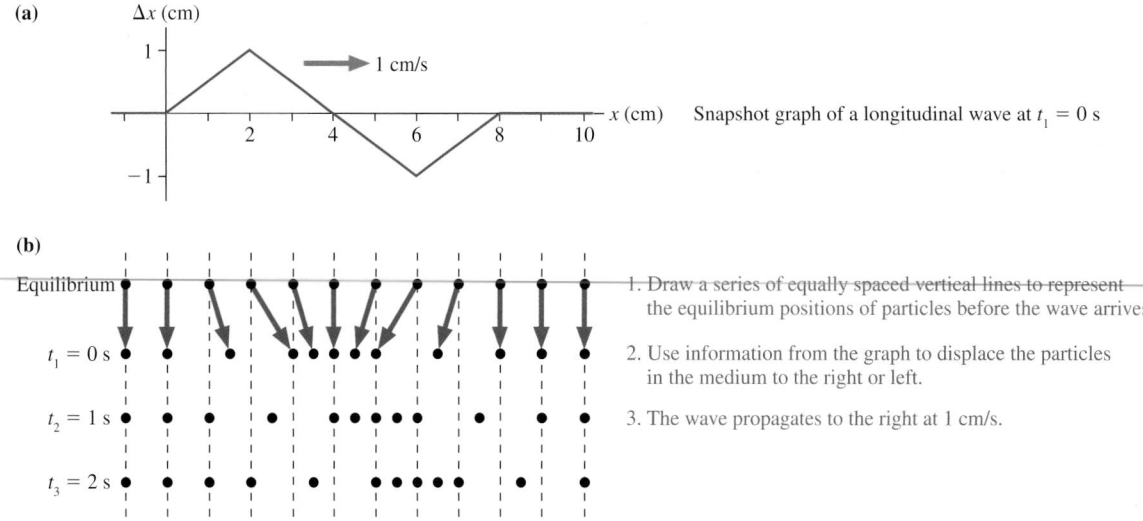

FIGURE 20.10 Visualizing a longitudinal wave.

to lower density at the leading and trailing edges. Two further lines show the medium at $t_2 = 1$ s and $t_3 = 2$ s so that you can see the wave propagating through the medium at 1 cm/s.

The Displacement

A traveling wave causes the particles of the medium to be displaced from their equilibrium positions. One of our goals is to develop a mathematical representation to describe all types of waves, so let's use the generic symbol D to stand for the *displacement* of a wave of any type. But what do we mean by a "particle" in the medium? And what about electromagnetic waves, for which there is no medium?

For a string, where the atoms stay fixed relative to each other, you can think of either the atoms themselves or very small segments of the string as being the particles of the medium. D is then the perpendicular displacement Δy of a point on the string. Sound waves and water waves require a bit more care because the atoms and molecules of a fluid have random thermal motions. Figure 20.10 is a nice way to visualize the motion of a sound wave, but no individual molecule in the air actually moves this way. For a fluid, we'll let a very small volume of the fluid be our particle. If the size of this little volume is large in comparison to the mean free path of the molecules, then the random thermal motions will average to zero. At the same time, the volume can be sufficiently small to be thought of as a "point" in the fluid.

Thus D for a sound wave is the longitudinal displacement Δx of a small volume of fluid. For any other mechanical wave, D is the appropriate displacement. Even electromagnetic waves can be described within the same mathematical representation if D is interpreted as a yet-undefined *electromagnetic field strength,* a "displacement" in a more abstract sense as an electromagnetic wave passes through a region of space.

NOTE ▶ For convenience, we'll use the terminology of mechanical waves traveling through a medium. Even so, the mathematical representation of waves that we develop can be used to describe electromagnetic waves, and we will do so in Chapter 22. ◀

Because the displacement of a particle in the medium depends both on *where* the particle is (position x) and on *when* you observe it (time t), D must be a function of the two variables x and t. That is,

$$D(x, t) = \text{the displacement at time } t \text{ of a particle at position } x \quad (20.3)$$

The values of *both* variables—where and when—must be specified before you can evaluate the displacement D.

Not every function of two variables describes a traveling wave. What we need is a function to describe a displacement that is translated along the x-axis at steady speed *without changing shape*. As a simple example, Figure 20.11a shows the parabola $f(x) = x^2$, which is centered at $x = 0$, and the *same parabola* drawn at $x = 2$ and at $x = 4$. The parabola at $x = 2$ can be written $f(x - 2) = (x - 2)^2$. Similarly, $f(x - 4) = (x - 4)^2$ is the parabola centered at $x = 4$.

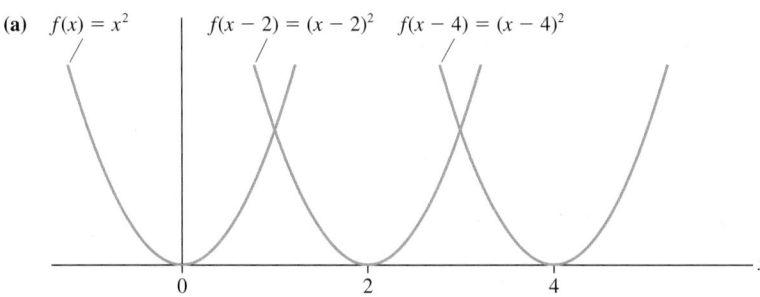

(a) $f(x) = x^2$ $f(x - 2) = (x - 2)^2$ $f(x - 4) = (x - 4)^2$

These are the same parabolas, but centered at different values of x.

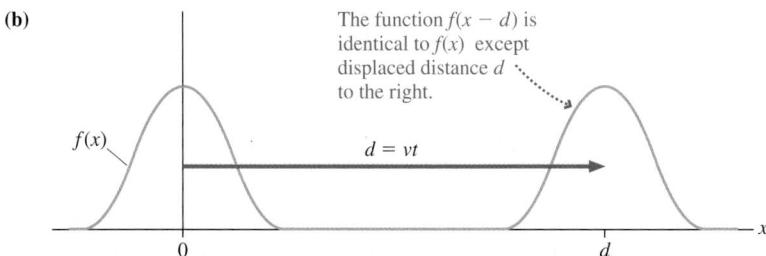

(b)

The function $f(x - d)$ is identical to $f(x)$ except displaced distance d to the right.

$f(x)$

$d = vt$

FIGURE 20.11 The graph of $f(x - d)$ looks exactly the same as the graph of $f(x)$ *except* that it is shifted distance d to the right.

In general, the graph of the function $f(x - d)$ looks exactly the same as the graph of $f(x)$ *except* that it is shifted distance d to the right. Whatever value $f(x)$ has at $x = 0$, the function $f(x - d)$ has the same value at $x = d$ where $x - d = 0$. Figure 20.11b applies this idea to a wave pulse. On the left is a snapshot graph of the wave at $t = 0$. At a later time t, the wave pulse looks *exactly the same* except that it has moved distance $d = vt$ to the right. If the wave at $t = 0$ is described by the function $f(x)$, then the shifted wave at time t is described by the function $f(x - d) = f(x - vt)$.

Thus we see that the displacement of a wave traveling in the positive x-direction with wave speed v must be a function of the form

$$D(x, t) = D(x - vt)$$

(wave traveling in the positive x-direction with speed v)　　(20.4)

That is, the two variables x and t always appear together as $x - vt$. Examples of such functions include $D(x, t) = (x - vt)^2$, $D(x, t) = e^{(x - vt)}$, and $D(x, t) = \sin(x - vt)$. Each and every wave has its own function, depending on the shape and speed of the wave, so there is not a single function that describes all waves. But all waves traveling in the positive x-direction, regardless of their shape, are described by *some* function of the form $D(x - vt)$.

EXAMPLE 20.3 A traveling wave pulse

Draw snapshot graphs at $t = 0$, 1, and 2 s to show that the displacement

$$D(x, t) = \frac{2 \text{ m}^3}{(x - (2 \text{ m/s})t)^2 + 1 \text{ m}^2}$$

where x is in m and t is in s, represents a traveling wave pulse.

VISUALIZE The snapshot graphs for $t = 0$, 1, and 2 s are shown in Figure 20.12. You can see that the function $D(x, t)$ is a pulse that travels in the positive x-direction without changing shape. That is, it is a traveling wave with speed $v = 2$ m/s.

ASSESS The displacement is a function of $(x - (2 \text{ m/s})t)$. According to Equation 20.4, this should represent a wave traveling in the positive x-direction with speed $v = 2$ m/s. The graphs show that it does.

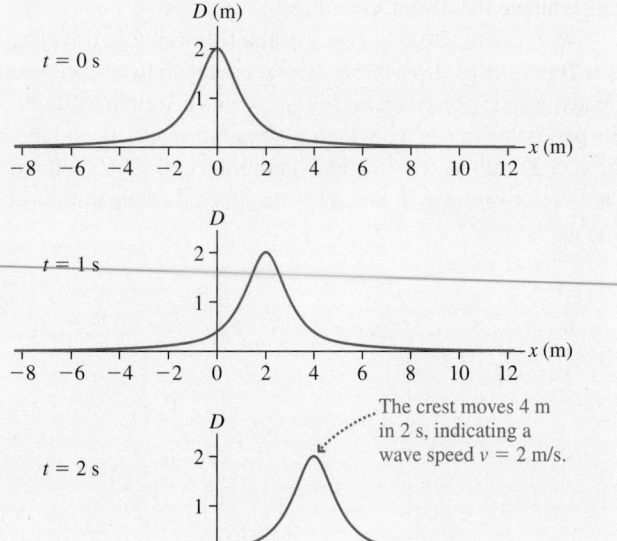

FIGURE 20.12 Three snapshot graphs of the displacement function of Example 20.3.

It should come as no surprise that the displacement of a wave traveling in the *negative x*-direction with wave speed v will have the form

$$D(x, t) = D(x + vt)$$

(wave traveling in the negative x-direction with speed v)   (20.5)

You can verify this with a homework problem.

> **NOTE** ▶ Not every wave moves along the x-axis. A change of variables will allow us to describe a wave traveling in the positive y-direction as $D(y - vt)$ and a wave traveling in the negative z-direction as $D(z + vt)$. We will usually let waves travel along the x-axis, just for convenience, but do not lose sight of the fact that we can equally well describe waves moving in any direction. ◀

STOP TO THINK 20.2 Which of the following actions would make a pulse travel faster down a stretched string? More than one answer may be correct. If so, give all that are correct.

a. Move your hand up and down more quickly as you generate the pulse.
b. Move your hand up and down a larger distance as you generate the pulse.
c. Use a heavier string of the same length, under the same tension.
d. Use a lighter string of the same length, under the same tension.
e. Stretch the string tighter to increase the tension.
f. Loosen the string to decrease the tension.
g. Put more force into the wave.

20.3 Sinusoidal Waves

10.1 Activ Physics ONLINE

A wave source that oscillates with simple harmonic motion (SHM) generates a **sinusoidal wave.** For example, a loudspeaker cone that oscillates in SHM radiates a sinusoidal sound wave. The sinusoidal electromagnetic waves broadcast by television and FM radio stations are generated by electrons oscillating back and forth in the antenna wire with SHM.

Figure 20.13 shows a sinusoidal wave moving through a medium. The source of the wave, which is undergoing vertical SHM, is located at $x = 0$. Notice how the wave crests move with steady speed toward larger values of x at later times t.

Figure 20.14a is a history graph for a sinusoidal wave, showing the displacement of the medium at one point in space. Each particle in the medium undergoes simple harmonic motion with frequency f, so this graph of SHM is identical to the graphs you learned to work with in Chapter 14. The *period* of the wave, shown on the graph, is the time interval for one cycle of the motion. The period is related to the wave *frequency f* by

$$T = \frac{1}{f} \tag{20.6}$$

exactly as in simple harmonic motion.

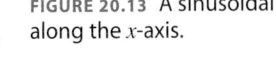

FIGURE 20.13 A sinusoidal wave moving along the x-axis.

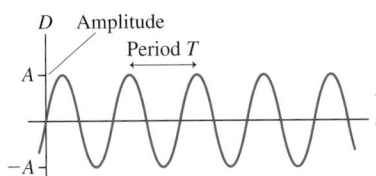 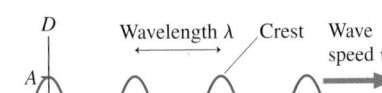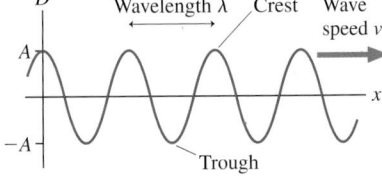

FIGURE 20.14 History and snapshot graphs for a sinusoidal wave.

Displacement-versus-time is only half the story. Figure 20.14b shows a snapshot graph for the same wave at one instant in time. Here we see the wave stretched out in space, moving to the right with speed v. The **amplitude** A of the wave is the maximum value of the displacement. The crests of the wave have displacement $D_{crest} = A$ and the troughs have displacement $D_{trough} = -A$.

An important characteristic of a sinusoidal wave is that it is periodic *in space* as well as in time. As you move from left to right along the "frozen" wave in the snapshot graph of Figure 20.14b, the disturbance repeats itself over and over. The distance spanned by one cycle of the motion is called the **wavelength** of the wave. Wavelength is symbolized by λ (lowercase Greek lambda) and, because it is a length, it is measured in units of meters. The wavelength is shown in Figure 20.14b as the distance between two crests, but it could equally well be the distance between two troughs.

NOTE ▶ Wavelength is the spatial analog of period. The period T is the *time* in which the disturbance at a single point in space repeats itself. The wavelength λ is the *distance* in which the disturbance at one instant of time repeats itself. ◀

The Fundamental Relationship for Sinusoidal Waves

There is an important relationship between the wavelength and the period of a wave. Figure 20.15 shows this relationship through five snapshot graphs of a sinusoidal wave at time increments of one-quarter of the period T. One full period has elapsed between the first graph and the last, which you can see by observing the motion at a fixed point on the x-axis. Each point in the medium has undergone exactly one complete oscillation.

The critical observation is that the wave crest marked by an arrow has moved one full wavelength between the first graph and the last. That is, **during a time interval of exactly one period T, each crest of a sinusoidal wave travels forward a distance of exactly one wavelength λ**. Because speed is distance divided by time, the wave speed must be

$$v = \frac{\text{distance}}{\text{time}} = \frac{\lambda}{T} \tag{20.7}$$

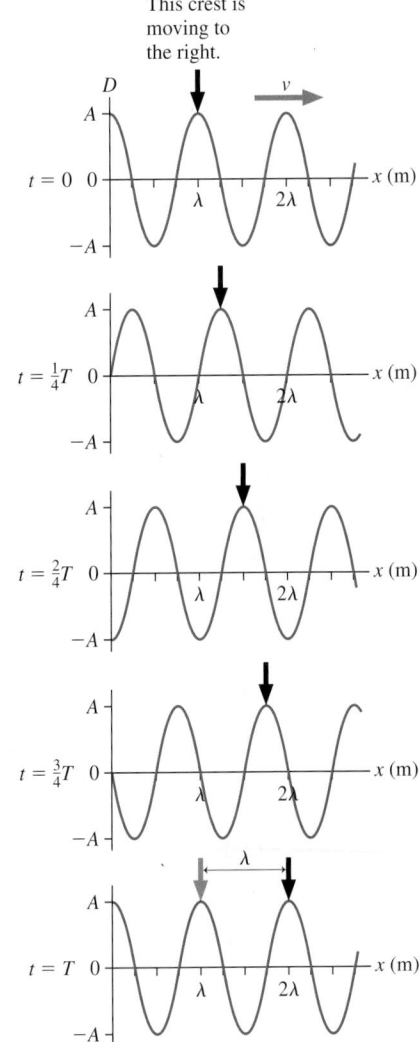

During a time interval of exactly one period, the crest has moved forward exactly one wavelength.

FIGURE 20.15 A series of snapshot graphs at time increments of one-quarter of the period T.

Because $f = 1/T$, it is customary to write Equation 20.7 in the form

$$v = \lambda f \tag{20.8}$$

Although Equation 20.8 has no special name, it is *the* fundamental relationship for periodic waves. When using it, keep in mind the *physical* meaning that **a wave moves forward a distance of one wavelength during a time interval of one period.**

> **NOTE** ▶ Wavelength and period are defined only for *periodic* waves, so Equations 20.7 and 20.8 apply only to periodic waves. A wave pulse has a wave speed, but it doesn't have a wavelength or a period. Hence Equations 20.7 and 20.8 cannot be applied to wave pulses. ◀

STOP TO THINK 20.3 What is the frequency of this traveling wave?

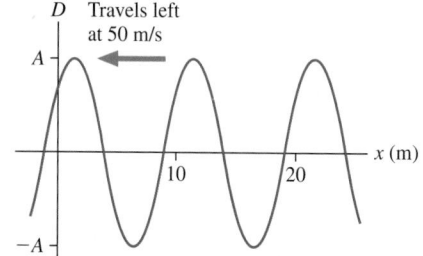

a. 0.1 Hz
b. 0.2 Hz
c. 2 Hz
d. 5 Hz
e. 10 Hz
f. 500 Hz

The Mathematics of Sinusoidal Waves

Section 20.2 introduced the function $D(x, t)$ that gives the displacement of a particle in the medium at position x and time t. We learned that the displacement must be a function of the form $D(x - vt)$ for a wave traveling in the positive x-direction with speed v. It's now straightforward to deduce the displacement function for a sinusoidal wave.

Figure 20.16 shows a snapshot graph at $t = 0$ of a sinusoidal wave. The sinusoidal function that describes the displacement of this wave is

$$D(x, t = 0) = A\sin\left(2\pi\frac{x}{\lambda} + \phi_0\right) \tag{20.9}$$

where the notation $D(x, t = 0)$ means that we've frozen the time at $t = 0$ to make the displacement a function only of x. The term ϕ_0 is a *phase constant* that characterizes the initial conditions. (We'll return to the phase constant momentarily.)

The function of Equation 20.9 is periodic with period λ. We can see this by writing

$$D(x + \lambda) = A\sin\left(2\pi\frac{(x + \lambda)}{\lambda} + \phi_0\right) = A\sin\left(2\pi\frac{x}{\lambda} + \phi_0 + 2\pi \text{ rad}\right)$$

$$= A\sin\left(2\pi\frac{x}{\lambda} + \phi_0\right) = D(x)$$

where we used the fact that $\sin(a + 2\pi \text{ rad}) = \sin a$. In other words, the disturbance created by the wave at $x + \lambda$ is exactly the same as the disturbance at x.

Now, just as we did in the previous section for Figure 20.11b, we can set the wave in motion by replacing x in Equation 20.9 with $x - vt$. This will cause the

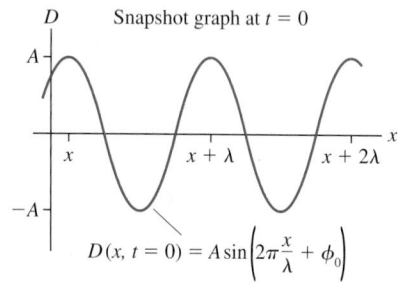

FIGURE 20.16 A sinusoidal wave is "frozen" at $t = 0$.

wave to be displaced distance $d = vt$ to the right at time t but with no change in shape. Thus the displacement equation of a sinusoidal wave is

$$D(x, t) = A\sin\left(2\pi\frac{x - vt}{\lambda} + \phi_0\right) = A\sin\left(2\pi\left(\frac{x}{\lambda} - \frac{t}{T}\right) + \phi_0\right) \quad (20.10)$$

In the last step we used $v = \lambda f = \lambda/T$ to write $v/\lambda = 1/T$. The function of Equation 20.10 is not only periodic in space with period λ, it is also periodic in time with period T. That is, $D(x, t + T) = D(x, t)$.

It will be useful to introduce two new quantities. First, recall from simple harmonic motion the *angular frequency*

$$\omega = 2\pi f = \frac{2\pi}{T} \quad (20.11)$$

The units of ω are rad/s, although many textbooks use simply s^{-1}.

You can see that ω is 2π times the reciprocal of the period in time. This suggests that we define an analogous quantity, called the **wave number** k, that is 2π times the reciprocal of the period in space:

$$k = \frac{2\pi}{\lambda} \quad (20.12)$$

The units of k are rad/m, although many textbooks use simply m^{-1}.

NOTE ▶ The wave number k is *not* a spring constant, even though it uses the same symbol. This is a most unfortunate use of symbols, but every major textbook and professional tradition uses the same symbol k for these two very different meanings, so we have little choice but to follow along. ◀

We can use the fundamental relationship $v = \lambda f$ to find an analogous relationship between ω and k:

$$v = \lambda f = \frac{2\pi}{k}\frac{\omega}{2\pi} = \frac{\omega}{k}$$

which is usually written

$$\omega = vk \quad (20.13)$$

Equation 20.13 contains no new information. It is a variation of Equation 20.8, but one that is convenient when working with k and ω.

If we use the definitions of Equations 20.11 and 20.12, Equation 20.10 for the displacement can be written

$$D(x, t) = A\sin(kx - \omega t + \phi_0)$$
(sinusoidal wave traveling in the positive x-direction) $\quad (20.14)$

A sinusoidal wave traveling in the negative x-direction, would be $A\sin(kx + \omega t + \phi_0)$. Equation 20.14 is graphed versus x and t in Figure 20.17.

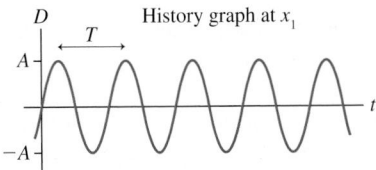

If x is fixed, $D(x_1, t) = A\sin(kx_1 - \omega t + \phi)$ gives a sinusoidal history graph at one point in space, x_1. It repeats every T s.

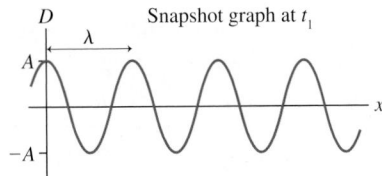

If t is fixed, $D(x, t_1) = A\sin(kx - \omega t_1 + \phi)$ gives a sinusoidal snapshot graph at one instant of time, t_1. It repeats every λ m.

FIGURE 20.17 Interpreting the equation of a sinusoidal traveling wave.

Just as it did for simple harmonic motion, the phase constant ϕ_0 characterizes the initial conditions. At $(x, t) = (0\text{ m}, 0\text{ s})$ Equation 20.14 becomes

$$D(0\text{ m}, 0\text{ s}) = A\sin\phi_0 \qquad (20.15)$$

Different values of ϕ_0 describe different initial conditions for the wave.

EXAMPLE 20.4 Analyzing a sinusoidal wave

A sinusoidal wave with an amplitude of 1.0 cm and a frequency of 100 Hz travels at 200 m/s in the positive x-direction. At $t = 0$ s, the point $x = 1.0$ m is on a crest of the wave.

a. Determine the values of A, v, λ, k, f, ω, T, and ϕ_0 for this wave.
b. Write the equation for the wave's displacement as it travels.
c. Draw a snapshot graph of the wave at $t = 0$ s.

VISUALIZE The snapshot graph will be sinusoidal, but we must do some numerical analysis before we know how to draw it.

SOLVE

a. There are several numerical values associated with a sinusoidal traveling wave, but they are not all independent. From the problem statement itself we learn that

$$A = 1.0\text{ cm} \qquad v = 200\text{ m/s} \qquad f = 100\text{ Hz}$$

We can then find:

$$\lambda = v/f = 2.00\text{ m}$$
$$k = 2\pi/\lambda = \pi\text{ rad/m} \quad \text{or} \quad 3.14\text{ rad/m}$$
$$\omega = 2\pi f = 628\text{ rad/s}$$
$$T = 1/f = 0.0100\text{ s} = 10.0\text{ ms}$$

The phase constant ϕ_0 is determined by the initial conditions. We know that a wave crest, with displacement $D = A$, is passing $x_0 = 1.0$ m at $t_0 = 0$ s. Equation 20.14 at x_0 and t_0 is

$$D(x_0, t_0) = A = A\sin(k(1.0\text{ m}) + \phi_0)$$

This equation is true only if $\sin(k(1.0\text{ m}) + \phi_0) = 1$, which requires

$$k(1.0\text{ m}) + \phi_0 = \frac{\pi}{2}\text{rad}$$

Solving for the phase constant gives

$$\phi_0 = \frac{\pi}{2}\text{ rad} - (\pi\text{ rad/m})(1.0\text{ m}) = -\frac{\pi}{2}\text{rad}$$

b. With the information gleaned from part a, the wave's displacement is

$$D(x, t) = 1.0\text{ cm} \times$$
$$\sin\left[(3.14\text{ rad/m})x - (628\text{ rad/s})t - \frac{\pi}{2}\text{ rad}\right]$$

Notice that we included units with A, k, ω, and ϕ_0.

c. We know that $x = 1.0$ m is a wave crest at $t = 0$ s and that the wavelength is $\lambda = 2.0$ m. Because the origin is $\lambda/2$ away from the crest at $x = 1.0$ m, we expect to find a wave trough at $x = 0$. This is confirmed by calculating $D(0\text{ m}, 0\text{ s}) = (1.0\text{ cm})\sin(-\pi/2\text{ rad}) = -1.0$ cm. Figure 20.18 is a snapshot graph that portrays this information.

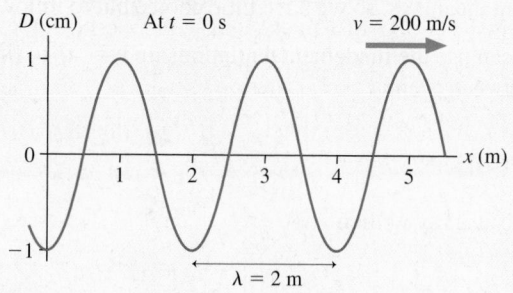

FIGURE 20.18 A snapshot graph at $t = 0$ s of the sinusoidal wave of Example 20.4.

Wave Motion on a String

The displacement equation, Equation 20.14, allows us to learn more about wave motion on a string. As a wave travels along the x-axis, the points on the string oscillate back and forth in the y-direction. The displacement D of a point on the string is simply that point's y-coordinate, so Equation 20.14 for a string wave is

$$y(x, t) = A\sin(kx - \omega t + \phi_0) \qquad (20.16)$$

The velocity of a particle on the string—**which is not the same as the velocity of the wave along the string**—is the time derivative of Equation 20.16:

$$v_y = \frac{dy}{dt} = -\omega A\cos(kx - \omega t + \phi_0) \qquad (20.17)$$

The maximum velocity of a small segment of the string is $v_{\max} = \omega A$. This is the same result we found for simple harmonic motion because the motion of the string particles *is* simple harmonic motion. Figure 20.19 shows velocity vectors *of the particles* at different points on a sinusoidal wave.

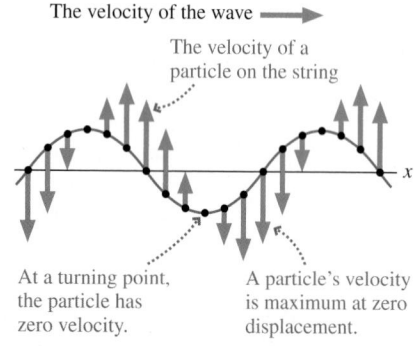

The velocity of the wave ⟶
The velocity of a particle on the string
At a turning point, the particle has zero velocity.
A particle's velocity is maximum at zero displacement.

FIGURE 20.19 A snapshot graph of a wave on a string with vectors showing the velocity *of the string* at various points.

NOTE ▶ Creating a wave of larger amplitude increases the speed of particles *in* the medium, but it does *not* change the speed of the wave *through* the medium. ◀

Pursuing this line of thought, we can derive an expression for the wave speed along the string. Figure 20.20 shows a small segment of the string with length $\Delta x \ll \lambda$ right at a crest of the wave. You can see that the string's tension exerts a downward force on this piece of the string, pulling it back to equilibrium. Newton's second law for this small segment of string is

$$(F_{net})_y = ma_y = (\mu \Delta x)a_y \tag{20.18}$$

where we used the string's linear density μ to write the mass as $m = \mu \Delta x$.

From simple harmonic motion, we know that this point of maximum displacement is also the point of maximum acceleration. The acceleration of a point on the string is the time derivative of Equation 20.17:

$$a_y = \frac{dv_y}{dt} = -\omega^2 A \sin(kx - \omega t + \phi_0) \tag{20.19}$$

Thus the acceleration at the crest of the wave is $a_y = -\omega^2 A$. But the angular frequency ω with which the particles of the string oscillate is related to the wave's speed v along the string by Equation 20.13, $\omega = vk$. Thus

$$a_y = -\omega^2 A = -v^2 k^2 A \tag{20.20}$$

A large wave speed causes the particles of the string to oscillate more quickly and thus to have a larger acceleration.

Our main task is to determine the net force acting on this segment of the string. You can see from Figure 20.20 that the y-component of the tension is $T_s \sin\theta$, where θ is the angle of the string at $x = \frac{1}{2}\Delta x$. θ is a *negative* angle because it is below the x-axis. This segment of string is pulled from both ends, so

$$(F_{net})_y = 2T_s \sin\theta \tag{20.21}$$

The angle θ is a very small angle, because $\Delta x \ll \lambda$, so we can use the small-angle approximation ($\sin u \approx \tan u$ if $u \ll 1$) to write

$$(F_{net})_y \approx 2T_s \tan\theta \tag{20.22}$$

where $\tan\theta$ is the slope of the string at $x = \frac{1}{2}\Delta x$.

At this specific instant, with the crest of the wave at $x = 0$, the equation of the string is

$$y = A\cos(kx)$$

The slope of the string at $x = \frac{1}{2}\Delta x$ is the derivative evaluated at that point:

$$\tan\theta = \frac{dy}{dx}\bigg|_{at \Delta x/2} = -kA\sin(kx)\big|_{at \Delta x/2} = -kA\sin\left(\frac{k\Delta x}{2}\right)$$

Now $\Delta x \ll \lambda$, so $k\Delta x/2 = \pi \Delta x/\lambda \ll 1$. Thus the small-angle approximation ($\sin u \approx u$ if $u \ll 1$) of the slope is

$$\tan\theta \approx -kA\left(\frac{k\Delta x}{2}\right) = -\frac{k^2 A \Delta x}{2} \tag{20.23}$$

If we substitute this expression for $\tan\theta$ into Equation 20.22, we find that the net force on this little piece of string is

$$(F_{net})_y = -k^2 A T_s \Delta x \tag{20.24}$$

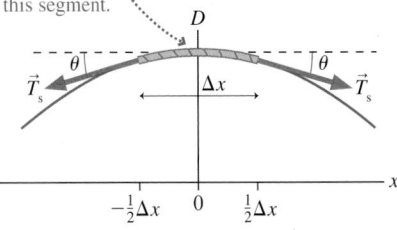

A small segment of the string at the crest of the wave. Because of the curvature of the string, the tension forces exert a net downward force on this segment.

FIGURE 20.20 A small segment of string at the crest of a wave.

Now we can use Equation 20.20 for a_y and Equation 20.24 for $(F_{net})_y$ in Newton's second law. With these substitutions, Equation 20.18 becomes

$$(F_{net})_y = -k^2 A T_s \Delta x = (\mu \Delta x) a_y = -v^2 k^2 A \mu \Delta x \qquad (20.25)$$

The term $-k^2 A \Delta x$ cancels, and we're left with

$$v = \sqrt{\frac{T_s}{\mu}} \qquad (20.26)$$

This was the result that we stated, without proof, in Equation 20.2. Although we've derived Equation 20.26 with the assumption of a sinusoidal wave, the wave speed does not depend on the shape of the wave. Thus any wave on a stretched string will have this wave speed.

EXAMPLE 20.5 Generating a sinusoidal wave

A very long string with $\mu = 2.0$ g/m is stretched along the x-axis with a tension of 5.0 N. At $x = 0$ m it is tied to a 100 Hz simple harmonic oscillator that vibrates perpendicular to the string with an amplitude of 2.0 mm. The oscillator is at its maximum positive displacement at $t = 0$ s.

a. Write the displacement equation for the traveling wave on the string.
b. At $t = 5.0$ ms, what is the string's displacement at a point 2.7 m from the oscillator?

MODEL The oscillator generates a sinusoidal traveling wave on a string. The displacement of the wave has to match the displacement of the oscillator at $x = 0$ m.

SOLVE

a. The equation for the displacement is

$$D(x, t) = A \sin(kx - \omega t + \phi_0)$$

with A, k, ω, and ϕ_0 to be determined. The wave amplitude is the same as the amplitude of the oscillator that generates the wave, so $A = 2.0$ mm. The oscillator has its maximum displacement $y_{osc} = A = 2.0$ mm at $t = 0$ s, thus

$$D(0 \text{ m}, 0 \text{ s}) = A \sin(\phi_0) = A$$

This requires the phase constant to be $\phi_0 = \pi/2$ rad. The wave's frequency is $f = 100$ Hz, the frequency of the source, therefore the angular frequency is $\omega = 2\pi f = 200\pi$ rad/s.

We still need $k = 2\pi/\lambda$, but we do not know the wavelength. However, we have enough information to determine the wave speed, and we can then use either $\lambda = v/f$ or $k = \omega/v$. The speed is

$$v = \sqrt{\frac{T_s}{\mu}} = \sqrt{\frac{5.0 \text{ N}}{0.0020 \text{ kg/m}}} = 50 \text{ m/s}$$

Using v, we find $\lambda = 0.50$ m and $k = 2\pi/\lambda = 4\pi$ rad/m. Thus the wave's displacement is

$$D(x, t) = (2.0 \text{ mm}) \times$$
$$\sin\left[2\pi((2.0 \text{ m}^{-1})x - (100 \text{ s}^{-1})t) + \frac{\pi}{2} \text{ rad}\right]$$

where x is in m and t in s. Notice that we have separated out the 2π. This step is not essential, but for some problems it makes subsequent steps easier.

b. The wave's displacement at $t = 5.0$ ms $= 0.0050$ s is

$$D(x, t = 5.0 \text{ ms}) = (2.0 \text{ mm}) \sin\left(4\pi x - \pi + \frac{\pi}{2} \text{ rad}\right)$$

$$= (2.0 \text{ mm}) \sin\left(4\pi x - \frac{\pi}{2} \text{ rad}\right)$$

At $x = 2.7$ m (calculator set to radians!), the displacement is

$$D(2.7 \text{ m}, 5.0 \text{ ms}) = 1.62 \text{ mm}$$

20.4 Waves in Two and Three Dimensions

Suppose you were to take a photograph of ripples spreading on a pond. If you mark the location of the *crests* on the photo, your picture would look like Figure 20.21a. The lines that locate the crests are called **wave fronts,** and they are spaced precisely one wavelength apart. The diagram shows only a single instant of time, but you can imagine a movie in which you would see the wave fronts moving outward from the source at speed v. A wave like this is called a **circular wave.** It is a two-dimensional wave that spreads across a surface.

Although the wave fronts are circles, you would hardly notice the curvature if you observed a small section of the wave front very, very far away from the source. The wave fronts would appear to be parallel lines, still spaced one wavelength apart and traveling at speed v. A good example is an ocean wave reaching a beach. Ocean waves are generated by storms and wind far out at sea, hundreds or thousands of miles away. By the time they reach the beach where you are working on your tan, the crests appear to be straight lines. An aerial view of the ocean would show a wave diagram like Figure 20.21b.

Many waves of interest, such as sound waves or light waves, move in three dimensions. For example, loudspeakers and light bulbs emit **spherical waves.** That is, the crests of the wave form a series of concentric spherical shells separated by the wavelength λ. In essence, the waves are three-dimensional ripples. It will still be useful to draw wave-front diagrams such as Figure 20.21, but now the circles are slices through the spherical shells locating the wave crests.

If you observe a spherical wave very, very far from its source, the small piece of the wave front that you can see is a little patch on the surface of a very large sphere. If the radius of the sphere is sufficiently large, you will not notice the curvature and this little patch of the wave front appears to be a plane. Figure 20.22 illustrates the idea of a **plane wave.**

To visualize a plane wave, imagine standing on the x-axis facing a sound wave as it comes toward you from a very distant loudspeaker. Sound is a longitudinal wave, so the particles of medium oscillate toward you and away from you. If you were to locate all of the particles that, at one instant of time, were at their maximum displacement toward you, they would all be located in a plane perpendicular to the travel direction. This is one of the wave fronts in Figure 20.22, and all the particles in this plane are doing exactly the same thing at that instant of time. This plane is moving toward you at speed v. There is another plane one wavelength behind it where the molecules are also at maximum displacement, yet another two wavelengths behind the first, and so on.

Because a plane wave's displacement depends on x but not on y or z, the displacement function $D(x, t)$ describes a plane wave just as readily as it does a one-dimensional wave. Once you specify a value for x, the displacement is the same at every point in the yz-plane that slices the x-axis at that value (i.e., one of the planes shown in Figure 20.22).

NOTE ▶ There are no perfect plane waves in nature, but it is a good model for many waves of practical interest. ◀

We can describe a circular wave or a spherical wave by changing the mathematical description from $D(x, t)$ to $D(r, t)$, where r is the distance measured outward from the source. Then the displacement of the medium will be the same at every point on a spherical surface. In particular, a sinusoidal spherical wave with wave number k and angular frequency ω is written

$$D(r, t) = A(r)\sin(kr - \omega t + \phi_0) \qquad (20.27)$$

Other than the change of x to r, the only difference is that the amplitude is now a function of r. A one-dimensional wave propagates with no change in the wave amplitude. But circular and spherical waves spread out to fill larger and larger volumes of space. To conserve energy, an issue we'll look at later in the chapter, the wave's amplitude has to decrease with increasing distance r. This is why sound and light decrease in intensity as you get farther from the source. We don't need to specify exactly how the amplitude decreases with distance, but you should be aware that it does.

Phase and Phase Difference

The quantity $(kx - \omega t + \phi_0)$ is called the **phase** of the wave, denoted ϕ. The phase of a wave will be an important concept in Chapters 21 and 22, where we will explore the consequences of adding various waves together. For now, we can

(a)

Wave fronts are the crests of the wave. They are spaced one wavelength apart.

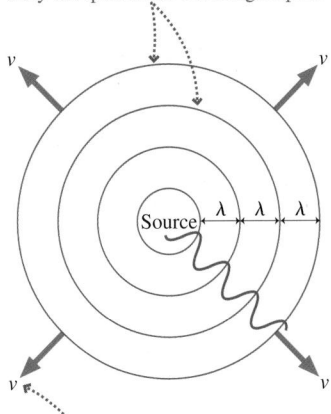

The circular wave fronts move outward from the source at speed v.

(b)

Very far away from the source, small sections of the wave fronts appear to be straight lines.

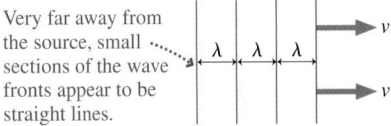

FIGURE 20.21 The wave fronts of a circular or spherical wave.

Very far from the source, small segments of spherical wave fronts appear to be planes. The wave is cresting at every point in these planes.

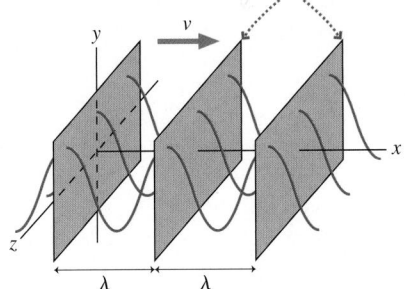

FIGURE 20.22 A plane wave.

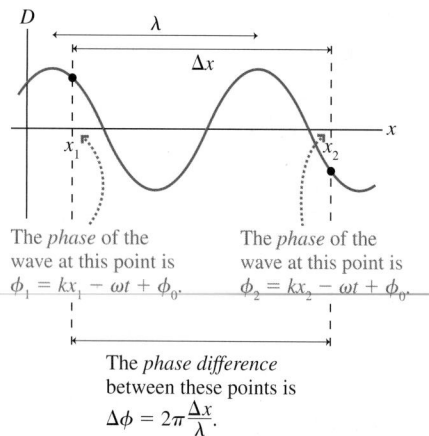

The *phase* of the wave at this point is $\phi_1 = kx_1 - \omega t + \phi_0$.

The *phase* of the wave at this point is $\phi_2 = kx_2 - \omega t + \phi_0$.

The *phase difference* between these points is $\Delta\phi = 2\pi\frac{\Delta x}{\lambda}$.

FIGURE 20.23 The phase difference between two points on a wave.

note that the wave fronts seen in Figures 20.21 and 20.22 are "surfaces of constant phase." To see this, notice that we can use the phase to write the displacement as simply $D(x, t) = A\sin\phi$. Because each point on a wave front has the same displacement, the phase must be the same at every point.

It will be useful to know the *phase difference* $\Delta\phi$ between two different points on a sinusoidal wave. Figure 20.23 shows two points on a sinusoidal wave at time t. The phase difference between these point is

$$\Delta\phi = \phi_2 - \phi_1 = (kx_2 - \omega t + \phi_0) - (kx_1 - \omega t + \phi_0)$$

$$= k(x_2 - x_1) = k\Delta x = 2\pi\frac{\Delta x}{\lambda} \tag{20.28}$$

That is, **the phase difference between two points on a wave depends on the ratio of their separation Δx to the wavelength λ**. For example, two points on a wave separated by $\Delta x = \frac{1}{2}\lambda$ have a phase difference $\Delta\phi = \pi$.

An important consequence of Equation 20.28 is that **the phase difference between two adjacent wave fronts is $\Delta\phi = 2\pi$**. This follows from the fact that two adjacent wave fronts are separated by $\Delta x = \lambda$. This is an important idea. Moving from one crest of the wave to the next corresponds to changing the *distance* by λ and to changing the *phase* by 2π.

EXAMPLE 20.6 The phase difference between two points on a sound wave

A 100 Hz sound wave travels with a wave speed of 343 m/s.

a. What is the phase difference between two points 60 cm apart along the direction the wave is traveling?

b. How far apart are two points whose phase differs by 90°?

MODEL Treat the wave as a plane wave traveling in the positive x-direction.

SOLVE

a. The phase difference between two points at positions x_1 and x_2 is

$$\Delta\phi = 2\pi\frac{\Delta x}{\lambda}$$

In this case, $\Delta x = 60$ cm $= 0.60$ m. The wavelength is

$$\lambda = \frac{v}{f} = \frac{343 \text{ m/s}}{100 \text{ Hz}} = 3.43 \text{ m}$$

and thus

$$\Delta\phi = 2\pi\frac{0.60 \text{ m}}{3.43 \text{ m}} = 0.350\pi \text{ rad} = 63.0°$$

b. A phase difference $\Delta\phi = 90°$ is $\pi/2$ radians. This will be the phase difference between two points when $\Delta x/\lambda = \frac{1}{4}$, or when $\Delta x = \lambda/4$. In this case, with $\lambda = 3.43$ m, $\Delta x = 85.8$ cm.

ASSESS The phase difference increases as Δx increases, so we expect the answer to part b to be larger than 60 cm.

STOP TO THINK 20.4   What is the phase difference between the crest of a wave and the adjacent trough?

a. -2π b. 0

c. $\pi/4$ d. $\pi/2$

e. π f. 3π

20.5 Sound and Light

Although there are many kinds of waves in nature, two are especially significant for us as humans. These are sound waves and light waves, the basis of hearing and seeing.

Sound Waves

We usually think of sound waves traveling in air, but sound can travel through any gas, through liquids, and even through solids. Figure 20.24 shows a loudspeaker cone vibrating back and forth in a fluid such as air or water. Each time the cone moves forward, it collides with the molecules and pushes them closer together. A half cycle later, as the cone moves backward, the fluid has room to expand and the density decreases a little. These regions of higher and lower density are called **compressions** and **rarefactions.**

This periodic sequence of compressions and rarefactions travels outward from the loudspeaker as a longitudinal sound wave. A similar type of sound wave is produced if you hit the end of a metal rod with a hammer, sending a compression pulse through the metal.

> NOTE ▶ Sound waves in gases and liquids are always longitudinal waves, but sound waves in solids can be either longitudinal or transverse. For a transverse wave to propagate, a plane of molecules oscillating perpendicular to the direction of motion has to be able to "drag" the neighboring planes of atoms along with it. Neighboring planes slip in a gas or liquid, so these media won't support a transverse wave. (Think how much easier it is to slide your hand sideways in water than to push against the water.) But the stronger molecular bonds in a solid do support transverse sound waves, sometimes called *shear waves*. Their speed is usually different from the speed of longitudinal sound waves. We'll assume that all sound waves are longitudinal waves unless otherwise noted. ◀

The speed of sound waves depends on the properties of the medium. A thermodynamic analysis of the compressions and expansions shows that the wave speed in a gas depends on the temperature and on the molecular mass of the gas. For air at room temperature (20°C),

$$v_{sound} = 343 \text{ m/s} \qquad \text{(sound speed in air at 20°C)}$$

The speed of sound is a little bit lower at lower temperatures and a little higher at higher temperatures. Liquids and solids are less compressible than air, and that makes the speed of sound in those media higher than in air. Table 20.1 gives the speed of sound in several substances.

A speed of 343 m/s is high, but not extraordinarily so. A distance as small as 100 m is enough to notice a slight delay between when you see something, such as a person hammering a nail, and when you hear it. The time required for sound to travel 1 km is $t = (1000 \text{ m})/(343 \text{ m/s}) \approx 3$ s. You may have learned to estimate the distance to a bolt of lightning by timing the number of seconds between when you see the flash and when you hear the thunder. Because sound takes 3 s to travel 1 km, the time divided by 3 gives the distance in kilometers. Or, in English units, the time divided by 5 gives the distance in miles.

Your ears are able to detect sinusoidal sound waves with frequencies between about 20 Hz and about 20,000 Hz, or 20 kHz. Low frequencies are perceived as a "low pitch" bass note while high frequencies are heard as a "high pitch" treble note. Your high-frequency range of hearing can deteriorate either with age or as a result of exposure to loud sounds that damage the ear.

Sound waves exist at frequencies well above 20 kHz, even though humans can't hear them. These are called *ultrasonic* frequencies. Oscillators vibrating at frequencies of many MHz generate the ultrasonic waves used in ultrasound medical imaging. A 3 MHz frequency traveling through water (which is basically what your body is) at a sound speed of 1480 m/s has a wavelength of about 0.5 mm. It is this very small wavelength that allows ultrasound to image very small objects. We'll see why when we study *diffraction* in Chapter 22.

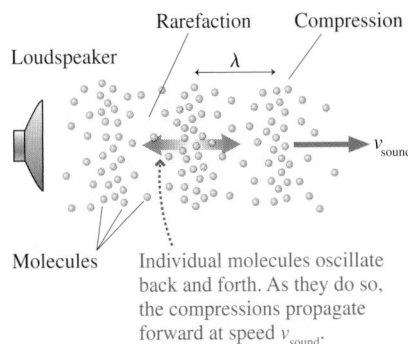

FIGURE 20.24 A sound wave in a fluid is a sequence of compressions and rarefactions that travels outward with speed v_{sound}. The variation in density and the amount of motion have been greatly exaggerated.

Activ
Physics
10.3

TABLE 20.1 The speed of sound

Medium	Speed (m/s)
Air (0°C)	331
Air (20°C)	343
Helium (0°C)	970
Ethyl alcohol	1170
Water	1480
Granite	6000
Aluminum	6420

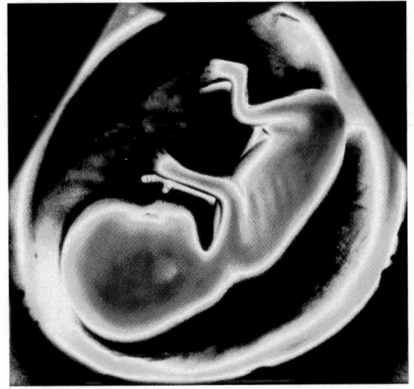

This ultrasound image is an example of how high-frequency sound waves can be used to "see" within the human body.

EXAMPLE 20.7 Sound wavelengths

What are the wavelengths of sound waves at the limits of human hearing and at the midrange frequency of 500 Hz? Notes sung by human voices are near 500 Hz, as are notes played by striking keys near the center of a piano keyboard.

MODEL Assume a room temperature 20°C.

SOLVE We can use the fundamental relationship $\lambda = v/f$ to find the wavelengths for sounds of various frequencies:

$$f = 20 \text{ Hz} \qquad \lambda = \frac{343 \text{ m/s}}{20 \text{ Hz}} = 17 \text{ m}$$

$$f = 500 \text{ Hz} \qquad \lambda = \frac{343 \text{ m/s}}{500 \text{ Hz}} = 0.69 \text{ m}$$

$$f = 20,000 \text{ Hz} \qquad \lambda = \frac{343 \text{ m/s}}{20,000 \text{ Hz}} = 0.017 \text{ m} = 1.7 \text{ cm}$$

ASSESS The wavelength of a 20 kHz note is a small 1.7 cm while, at the other extreme, a 20 Hz note has a huge wavelength of 17 m! This is because a wave moves forward one wavelength during a time interval of one period, and a wave traveling at 343 m/s can move 17 m during the $\frac{1}{20}$ s period of a 20 Hz note. The 69 cm wavelength of a 500 Hz note is more of a "human scale." You might note that most musical instruments are a meter or a little less in size. This is not a coincidence. You will see in the next chapter how the wavelength produced by a musical instrument is related to its size.

Electromagnetic Waves

A light wave is an *electromagnetic wave,* an oscillation of the electromagnetic field. Other electromagnetic waves, such as radio waves, microwaves, and ultraviolet light, have the same physical characteristics as light waves even though we cannot sense them with our eyes. It is easy to demonstrate that light will pass unaffected through a container from which all the air has been removed, and light reaches us from distant stars through the vacuum of interstellar space. Such observations raise interesting but difficult questions. If light can travel through a region in which there is no matter, then what is the *medium* of a light wave? What is it that is waving?

It took scientists over 50 years, most of the 19th century, to answer this question. We will examine the answers in more detail in Part VI after we introduce the ideas of electric and magnetic fields. For now we can say that light waves are a "self-sustaining oscillation of the electromagnetic field." Being self-sustaining means that electromagnetic waves require *no material medium* in order to travel, hence electromagnetic waves are not mechanical waves. Fortunately, we can learn about the wave properties of light without having to understand electromagnetic fields. In fact, the discovery that light propagates as a wave was made 60 years before it was realized that light is an electromagnetic wave. We, too, will be able to learn much about the wave nature of light without having to know just what it is that is waving.

It was predicted theoretically in the late 19th century, and has been subsequently confirmed experimentally with outstanding precision, that all electromagnetic waves travel through vacuum with the same speed, called the *speed of light.* The value of the speed of light is

$$v_{\text{light}} = c = 299,792,458 \text{ m/s} \qquad \text{(electromagnetic wave speed in vacuum)}$$

where the special symbol c is used to designate the speed of light. (This is the c in Einstein's famous formula $E = mc^2$.) Now *this* is really moving—about one million times faster than the speed of sound in air! At this speed, light could circle the earth 7.5 times in a mere one second—*if* there were some way to make it go in circles.

NOTE ▶ $c = 3.00 \times 10^8$ m/s is the appropriate value to use in most calculations. ◀

The wavelengths of light are extremely small. You will learn in Chapter 22 how these wavelengths are determined, but for now we will note that visible light is an electromagnetic wave with a wavelength (in air) in the range of roughly 400 nm (400×10^{-9} m) to 700 nm (700×10^{-9} m). Each wavelength is perceived as a different color, with the longer wavelengths seen as orange or red light and the shorter wavelengths seen as blue or violet light. A prism is able to spread the different wavelengths apart, from which we learn that "white light" is all the colors, or wavelengths, combined. The spread of colors seen with a prism, or seen in a rainbow, is called the *visible spectrum.*

If the wavelengths of light are unbelievably small, the oscillation frequencies are unbelievably large. The frequency for a 600 nm wavelength of light is

$$f = \frac{v}{\lambda} = \frac{3.00 \times 10^8 \text{ m/s}}{600 \times 10^{-9} \text{ m}} = 5.00 \times 10^{14} \text{ Hz}$$

The frequencies of light waves are roughly a factor of a trillion (10^{12}) higher than sound frequencies.

Electromagnetic waves exist at many frequencies other than the rather limited range that our eyes detect. One of the major technological advances of the twentieth century was learning to generate and detect electromagnetic waves at many frequencies, ranging from low-frequency radio waves to the extraordinarily high frequencies of x rays. Figure 20.25 shows that the visible spectrum is a small slice out of the much broader **electromagnetic spectrum.**

White light passing through a prism is spread out into a band of colors called the *visible spectrum.*

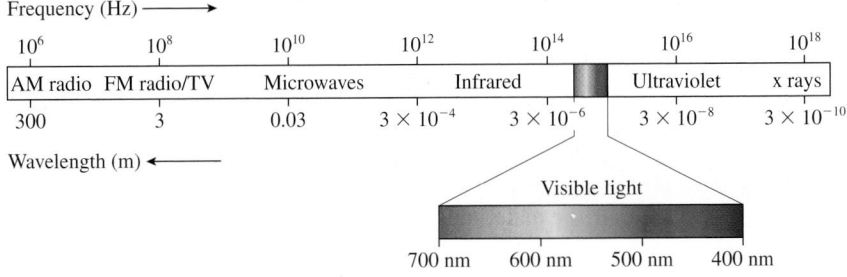

FIGURE 20.25 The electromagnetic spectrum from 10^6 Hz to 10^{18} Hz.

EXAMPLE 20.8 Traveling at the speed of light
A satellite exploring Jupiter transmits data to the earth as a radio wave with a frequency of 200 MHz. What is the wavelength of the electromagnetic wave, and how long does it take the signal to travel 800 million kilometers from Jupiter to the earth?

SOLVE Radio waves are sinusoidal electromagnetic waves that travel with speed c. Thus

$$\lambda = \frac{c}{f} = \frac{3.00 \times 10^8 \text{ m/s}}{2.00 \times 10^8 \text{ Hz}} = 1.5 \text{ m}$$

The time needed to travel 800×10^6 km $= 8.0 \times 10^{11}$ m is

$$\Delta t = \frac{\Delta x}{c} = \frac{8.0 \times 10^{11} \text{ m}}{3.00 \times 10^8 \text{ m/s}} = 2700 \text{ s} = 45 \text{ min}$$

The Index of Refraction

Light waves travel with speed c in a vacuum, but they slow down as they pass through transparent materials such as water or glass or even, to a very slight extent, air. The slowdown is a consequence of interactions between the electromagnetic field of the wave and the electrons in the material. The speed of light in a material is characterized by the material's **index of refraction** n, defined as

$$n = \frac{\text{speed of light in a vacuum}}{\text{speed of light in the material}} = \frac{c}{v} \qquad (20.29)$$

where v is the speed of light in the material. The index of refraction of a material is always greater than 1 because $v < c$. A vacuum has $n = 1$ exactly. Table 20.2 shows the index of refraction for several materials. You can see that liquids and solids have larger indices of refraction than gases.

TABLE 20.2 Typical indices of refraction

Material	Index of refraction
Vacuum	1 exactly
Air	1.0003
Water	1.33
Glass	1.50
Diamond	2.42

NOTE ▶ An accurate value for the index of refraction of air is relevant only in very precise measurements. We will assume $n_{\text{air}} = 1.00$ in this text. ◀

If the speed of a light wave changes as it enters into a transparent material, such as glass, what happens to the light's frequency and wavelength? Because $v = \lambda f$, either λ or f or both have to change when v changes.

As an analogy, think of a sound wave in the air as it impinges on the surface of a pool of water. As the air oscillates back and forth, it periodically pushes on the surface of the water. These pushes generate the compressions of the sound wave that continues on into the water. Because each push of the air causes one compression of the water, the frequency of the sound wave in the water must be *exactly the same* as the frequency of the sound wave in the air. In other words, **the frequency of a wave does not change as the wave moves from one medium to another.**

The same is true for electromagnetic waves, although the pushes are a bit more complex as the electric and magnetic fields of the wave interact with the atoms at the surface of the material. Nonetheless, the frequency does not change as the wave moves from one material to another.

Figure 20.26 shows a light wave passing through a transparent material with index of refraction n. As the wave travels through vacuum it has wavelength λ_{vac} and frequency f_{vac} such that $\lambda_{\text{vac}} f_{\text{vac}} = c$. In the material, $\lambda_{\text{mat}} f_{\text{mat}} = v = c/n$. The frequency does not change as the wave enters ($f_{\text{mat}} = f_{\text{vac}}$), so the wavelength must. The wavelength in the material is

$$\lambda_{\text{mat}} = \frac{v}{f_{\text{mat}}} = \frac{c}{n f_{\text{mat}}} = \frac{c}{n f_{\text{vac}}} = \frac{\lambda_{\text{vac}}}{n} \qquad (20.30)$$

The wavelength in the transparent material is less than the wavelength in vacuum. This makes sense. Suppose a marching band is marching at one step per second at a speed of 1 m/s. Suddenly they slow their speed to $\frac{1}{2}$ m/s but maintain their march at one step per second. The only way to go slower while marching at the same pace is to take *smaller steps*. When a light wave enters a material, the only way it can go slower while oscillating at the same frequency is to have a *smaller wavelength*.

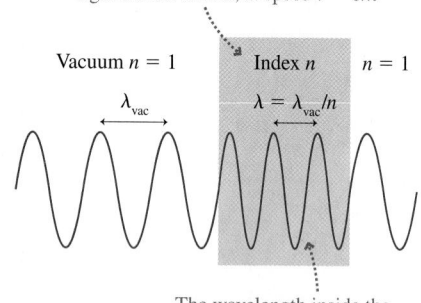

A transparent material in which light travels slower, at speed $v = c/n$

Vacuum $n = 1$ Index n $n = 1$

λ_{vac} $\lambda = \lambda_{\text{vac}}/n$

The wavelength inside the material decreases, but the frequency doesn't change.

FIGURE 20.26 Light passing through a transparent material with index of refraction n.

EXAMPLE 20.9 Light traveling through glass

Orange light with a wavelength of 600 nm is incident upon a 1.00-mm-thick glass microscope slide.

a. What is the light speed in the glass?
b. How many wavelengths of the light are inside the slide?

SOLVE

a. From Table 20.2 we see that the index of refraction of glass is $n_{\text{glass}} = 1.50$. Thus the speed of light in glass is

$$v_{\text{glass}} = \frac{c}{n_{\text{glass}}} = \frac{3.00 \times 10^8 \text{ m/s}}{1.50} = 2.00 \times 10^8 \text{ m/s}$$

b. Because $n_{\text{air}} = 1.00$, the wavelength of the light is the same in air and vacuum: $\lambda_{\text{vac}} = \lambda_{\text{air}} = 600$ nm. Thus the wavelength inside the glass is

$$\lambda_{\text{glass}} = \frac{\lambda_{\text{vac}}}{n_{\text{glass}}} = \frac{600 \text{ nm}}{1.50} = 400 \text{ nm} = 4.00 \times 10^{-7} \text{ m}$$

N wavelengths span a distance $d = N\lambda$, so the number of wavelengths in $d = 1.00$ mm is

$$N = \frac{d}{\lambda} = \frac{1.00 \times 10^{-3} \text{ m}}{4.00 \times 10^{-7} \text{ m}} = 2500$$

ASSESS The fact that 2500 wavelengths fit within 1 mm shows how small the wavelengths of light are.

STOP TO THINK 20.5 A light wave travels through three transparent materials of equal thickness. Rank in order, from the largest to smallest, the indices of refraction n_1, n_2, and n_3.

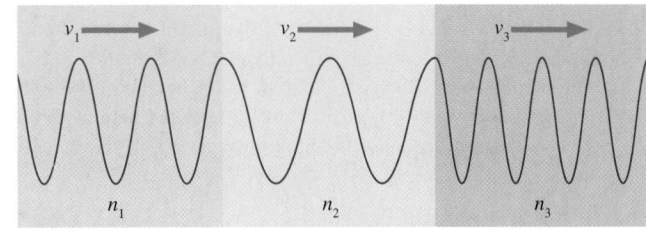

20.6 Power and Intensity

A traveling wave transfers energy from one point to another. The sound wave from a loudspeaker sets your eardrum into motion and, if at the proper frequency, can shatter a glass. Light waves from the sun warm the earth and, if focused with a lens, can start a fire. The *power* of a wave is the rate, in joules per second, at which the wave transfers energy. As you learned in Chapter 11, power is measured in watts. A loudspeaker might emit 2 W of power, meaning that energy in the form of sound waves is radiated at the rate of 2 joules per second. A light bulb might emit 2 W, or 2 J/s, of visible light. (In fact, this is about right for a so-called 100 watt bulb, with the other 98 W of power being emitted as heat, or infrared radiation, rather than as visible light.)

Imagine doing two experiments with a light bulb that emits 2 W of visible light. In the first, you hang the bulb in the center of a room and allow the light to illuminate the walls. In the second experiment, you use mirrors and lenses to "capture" the bulb's light and focus it onto a small spot on one wall. This is what a slide projector does. The energy emitted by the bulb is the same in both cases, but, as you know, the light is much brighter when focused onto a small area. We would say that the focused light is more *intense* than the diffuse light that goes in all directions. Similarly, a loudspeaker that beams its sound forward into a small area produces a louder sound in that area than a speaker of equal power that radiates the sound in all directions. Quantities such as brightness and loudness depend not only on the rate of energy transfer, or power, but also on the *area* that receives that power.

Figure 20.27 shows a wave impinging on a surface of area a. The surface is perpendicular to the direction in which the wave is traveling. This might be a real, physical surface, such as your eardrum or a photovoltaic cell, but it could equally well be a mathematical surface in space that the wave passes right through. If the wave has power P, we define the **intensity** I of the wave to be

$$I = \frac{P}{a} = \text{power-to-area ratio} \qquad (20.31)$$

The SI units of intensity are W/m². Because intensity is a power-to-area ratio, a wave focused into a small area will have a larger intensity than a wave of equal power that is spread out over a large area.

Energy from the sun is a practical and efficient way to heat water, as these solar panels are doing.

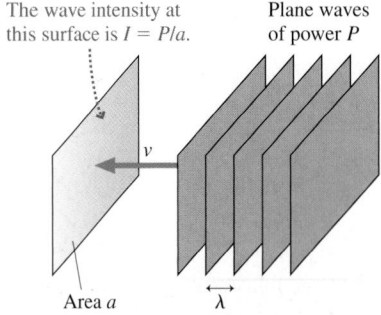

FIGURE 20.27 Plane waves of power P impinge on area a with intensity $I = P/a$.

EXAMPLE 20.10 The intensity of a laser beam

A helium-neon laser, the kind that provides the familiar red light of classroom demonstrations and supermarket checkout scanners, emits 1.0 mW of light power into a laser beam that is 1.0 mm in diameter. What is the intensity of the laser beam?

MODEL The laser beam is a light wave.

SOLVE The light waves of the laser beam pass through a mathematical surface that is a circle of diameter 1 mm. The intensity of the laser beam is

$$I = \frac{P}{a} = \frac{P}{\pi r^2} = \frac{0.0010 \text{ W}}{\pi (0.00050 \text{ m})^2} = 1270 \text{ W/m}^2$$

ASSESS This is roughly the intensity of sunlight at noon on a summer day. The difference between the sun and a small laser is not their intensities, which are about the same, but their powers. The laser has a small power of 1 mW. It can produce a very intense wave only because the area through which the wave passes is very small. The sun, by contrast, radiates a total power $P_{sun} \approx 4 \times 10^{26}$ W. This immense power is spread through *all* of space, producing an intensity of 1400 W/m² at a distance of 1.5×10^{11} m, the radius of the earth's orbit.

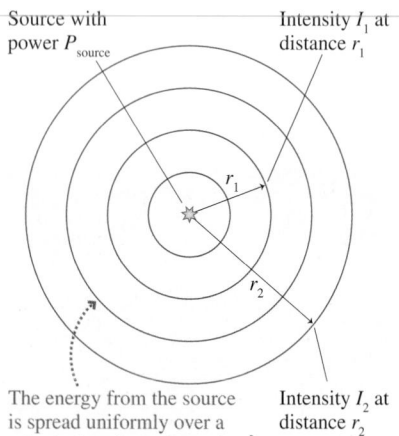

Source with power P_{source}

Intensity I_1 at distance r_1

r_1

r_2

The energy from the source is spread uniformly over a spherical surface of area $4\pi r^2$.

Intensity I_2 at distance r_2

FIGURE 20.28 A source emitting uniform spherical waves.

If a source of spherical waves radiates uniformly in all directions, then, as Figure 20.28 shows, the power at distance r is spread uniformly over the surface of a sphere of radius r. The surface area of a sphere is $a = 4\pi r^2$, so the intensity of a uniform spherical wave is

$$I = \frac{P_{source}}{4\pi r^2} \qquad \text{(intensity of a uniform spherical source)} \qquad (20.32)$$

The inverse-square dependence of r is really just a statement of energy conservation. The source emits energy at the rate P joules per second. The energy is spread over a larger and larger area as the wave moves outward. Consequently, the energy *per unit area* must decrease in proportion to the surface area of a sphere. If the intensity at distance r_1 is $I_1 = P_{source}/4\pi r_1^2$ and the intensity at r_2 is $I_2 = P_{source}/4\pi r_2^2$, then you can see that the intensity *ratio* is

$$\frac{I_1}{I_2} = \frac{r_2^2}{r_1^2} \qquad (20.33)$$

You can use Equation 20.33 to compare the intensities at two distances from a source without needing to know the power of the source.

> **NOTE** ▶ Wave intensities are strongly affected by reflections and absorption. Equations 20.32 and 20.33 apply to situations such as the light from a star or the sound from a firework exploding high in the air. Indoor sound does *not* obey a simple inverse-square law because of the many reflecting surfaces. ◀

For a sinusoidal wave, each particle in the medium oscillates back and forth in simple harmonic motion. You learned in Chapter 14 that a particle in SHM with amplitude A has energy $E = \frac{1}{2}kA^2$, where k is the spring constant of the medium, *not* the wave number. It is this oscillatory energy of the medium that is transferred, particle to particle, as the wave moves through the medium.

Because a wave's intensity is proportional to the rate at which energy is transferred through the medium, and because the oscillatory energy in the medium is proportional to the *square* of the amplitude, we can infer that for *any* wave

$$I = CA^2 \qquad (20.34)$$

where C is a proportionality constant that depends on the type of wave. That is, **the intensity of a wave is proportional to the square of its amplitude.** If you double the amplitude of a wave, you increase its intensity by a factor of four. This relationship between intensity and amplitude will be important when we study some of the properties of light waves.

20.7 The Doppler Effect

10.8, 10.9   Activ Physics ONLINE

Our final topic for this chapter is an interesting effect that occurs when you are in motion relative to a wave source. It is called the *Doppler effect*. You've likely noticed that the pitch of an ambulance's siren drops as it goes past you. A higher pitch suddenly becomes a lower pitch. Why?

Figure 20.29 shows a source of sound waves moving away from Pablo and toward Nancy at a steady speed v_s. The subscript s indicates that this is the speed of the source, not the speed of the waves. The source is emitting sound waves of frequency f_0 as it travels. Part a of the figure is a motion diagram showing the position of the source times $t = 0$, T, $2T$, and $3T$, where $T = 1/f_0$ is the period of the waves.

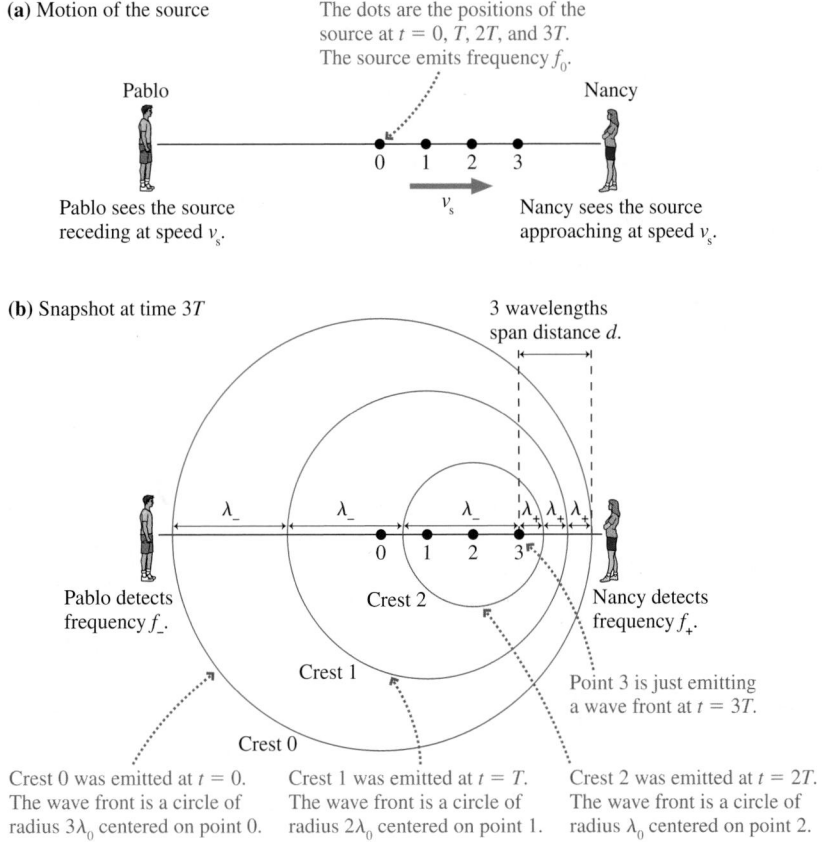

(a) Motion of the source

The dots are the positions of the source at $t = 0$, T, $2T$, and $3T$. The source emits frequency f_0.

Pablo

Nancy

0 1 2 3

v_s

Pablo sees the source receding at speed v_s.

Nancy sees the source approaching at speed v_s.

(b) Snapshot at time $3T$

3 wavelengths span distance d.

λ_- λ_- λ_- λ_+ λ_+ λ_+

0 1 2 3

Pablo detects frequency f_-.

Crest 2

Nancy detects frequency f_+.

Crest 1

Point 3 is just emitting a wave front at $t = 3T$.

Crest 0

Crest 0 was emitted at $t = 0$. The wave front is a circle of radius $3\lambda_0$ centered on point 0.

Crest 1 was emitted at $t = T$. The wave front is a circle of radius $2\lambda_0$ centered on point 1.

Crest 2 was emitted at $t = 2T$. The wave front is a circle of radius λ_0 centered on point 2.

FIGURE 20.29 A motion diagram showing the wave fronts emitted by a source as it moves to the right at speed v_s.

Nancy measures the frequency of the wave emitted by the *approaching source* to be f_+. At the same time, Pablo measures the frequency of the wave emitted by the *receding source* to be f_-. Our task is to relate f_+ and f_- to the source frequency f_0 and speed v_s.

After a wave crest leaves the source, its motion is governed by the properties of the medium. That is, the motion of the source cannot affect a wave that has already been emitted. Thus each circular wave front in Figure 20.29b is centered on the point from which it was emitted. The wave crest from point 3 was emitted just as this figure was made, but it hasn't yet had time to travel any distance.

You can see that the wave crests are bunched up in the direction the source is moving, stretched out behind it. The distance between one crest and the next is one wavelength, so the wavelength λ_+ that Nancy measures is *less* than the wavelength $\lambda_0 = v/f_0$ that would be emitted if the source were at rest. Similarly, λ_- behind the source is larger than λ_0.

These crests move through the medium at the wave speed v. Consequently, the frequency $f_+ = v/\lambda_+$ detected by the observer whom the source is approaching is *higher* than the frequency f_0 emitted by the source. Similarly, $f_- = v/\lambda_-$

detected behind the source is *lower* than frequency f_0. This change of frequency when a source moves relative to an observer is called the **Doppler effect.**

The wavelength detected by Nancy is $\lambda_+ = d/3$, where d is the difference between how far the wave has moved and how far the source has moved at time $t = 3T$. These distances are

$$\Delta x_{\text{wave}} = vt = 3vT$$
$$\Delta x_{\text{source}} = v_s t = 3v_s T \tag{20.35}$$

Thus the wavelength of the wave emitted by an approaching source is

$$\lambda_+ = \frac{d}{3} = \frac{\Delta x_{\text{wave}} - \Delta x_{\text{source}}}{3} = \frac{3vT - 3v_s T}{3} = (v - v_s)T \tag{20.36}$$

You can see that our arbitrary choice of three periods was not relevant because the 3 cancels. Thus the frequency detected in Nancy's direction is

$$f_+ = \frac{v}{\lambda_+} = \frac{v}{(v - v_s)T} = \frac{v}{(v - v_s)}f_0 \tag{20.37}$$

where $f_0 = 1/T$ is the frequency of the source and is the frequency you would detect if the source were at rest. We'll find it convenient to write the detected frequency as

$$f_+ = \frac{f_0}{1 - v_s/v} \qquad \text{(Doppler effect for an approaching source)}$$

$$\tag{20.38}$$

$$f_- = \frac{f_0}{1 + v_s/v} \qquad \text{(Doppler effect for a receding source)}$$

Proof of the second version, for the frequency f_- of a receding source, will be left for a homework problem. You can see that $f_+ > f_0$ in front of the source, because the denominator is less than 1, and $f_- < f_0$ behind the source.

EXAMPLE 20.11 How fast are the police traveling?
A police siren has a frequency of 550 Hz as the police car approaches you, 450 Hz after it has passed you and is receding. How fast are the police traveling? The temperature is 20°C.

MODEL The siren's frequency is altered by the Doppler effect. The frequency is f_+ as the car approaches and f_- as it moves away.

SOLVE To find v_s, rewrite Equations 20.38 as

$$f_0 = (1 + v_s/v)f_-$$
$$f_0 = (1 - v_s/v)f_+$$

Subtract the second equation from the first, giving

$$0 = f_- - f_+ + \frac{v_s}{v}(f_- + f_+)$$

This is easily solved to give

$$v_s = \frac{f_+ - f_-}{f_+ + f_-}v = \frac{100 \text{ Hz}}{1000 \text{ Hz}}343 \text{ m/s} = 34.3 \text{ m/s}$$

ASSESS If you now solve for the siren frequency when at rest, you will find $f_0 = 495$ Hz. Surprisingly, the at-rest frequency is *not* halfway between f_- and f_+.

NOTE ▶ The frequency of an approaching source is shifted upward, from f_0 to f_+, but the frequency *does not change* as the source gets closer. It's often said that the frequency *rises* as a source approaches, but you can see that is not the case. What does rise is the intensity, or loudness, of the sound. Interestingly, a sound of constant frequency but increasing loudness is often *perceived* to be increasing in pitch. You might perceive that the pitch of an approaching ambulance is rising, but measurements would show that the frequency remains constant as the intensity increases. ◀

A Stationary Source and a Moving Observer

Suppose the police car in Example 20.11 is at rest while you drive toward it at 34.3 m/s. You might think that this is equivalent to having the police car move toward you at 34.3 m/s, but it isn't. Mechanical waves move through a medium, and the Doppler effect depends not just on how the source and the observer move with respect to each other but also how they move with respect to the medium. We'll omit the proof, but it's not hard to show that the frequencies heard by an observer moving at speed v_o relative to a stationary source emitting frequency f_0 are

$$f_+ = (1 + v_o/v)f_0 \quad \text{(observer approaching a source)}$$

$$f_- = (1 - v_o/v)f_0 \quad \text{(observer receding from a source)}$$

(20.39)

A quick calculation shows that the frequency of the police siren as you approach it at 34.3 m/s is 545 Hz, not the 550 Hz you heard as it approached you at 34.3 m/s.

The Doppler Effect for Light Waves

The Doppler effect is observed for all types of waves, not just sound waves. If a source of light waves is receding from you, the wavelength λ_- that you detect is longer than the wavelength λ_0 emitted by the source. Because the wavelength is shifted toward the red end of the visible spectrum, the longer wavelengths of light, this effect is called the **red shift.** Similarly, the light you detect from a source moving toward you is **blue shifted** to shorter wavelengths.

Although the underlying reason for the Doppler shift for light is the same as for sound waves, there is one fundamental difference. We derived Equation 20.38 for the Doppler-shifted frequencies by measuring the wave speed v relative to the medium. For electromagnetic waves in empty space, there is no medium. Consequently, we would need to turn to Einstein's theory of relativity to determine the frequency of light waves from a moving source. The result, which we will state without proof, is

$$\lambda_{\text{red}} = \sqrt{\frac{1 + v_s/c}{1 - v_s/c}}\, \lambda_0$$

(Doppler effect for the light of a receding source)

(20.40)

$$\lambda_{\text{blue}} = \sqrt{\frac{1 - v_s/c}{1 + v_s/c}}\, \lambda_0$$

(Doppler effect for the light of an approaching source)

Here v_s is the speed of the source *relative to* the observer.

EXAMPLE 20.12 Measuring the velocity of a galaxy

Hydrogen atoms in the laboratory emit red light with wavelength 656 nm. In the light from a distant galaxy, this "spectral line" is observed at 691 nm. What is the speed of this galaxy relative to the earth?

MODEL The observed wavelength is longer than the wavelength emitted by atoms at rest with respect to the observer (i.e., red shifted), so we are looking at light emitted from a galaxy that is receding from us.

SOLVE Squaring the expression for λ_{red} in Equation 20.40 and solving for v_s gives

$$v_s = \frac{(\lambda_{\text{red}}/\lambda_0)^2 - 1}{(\lambda_{\text{red}}/\lambda_0)^2 + 1}\, c$$

$$= \frac{(691 \text{ nm}/656 \text{ nm})^2 - 1}{(691 \text{ nm}/656 \text{ nm})^2 + 1}\, c$$

$$= 0.052c = 1.56 \times 10^7 \text{ m/s}$$

ASSESS The galaxy is moving away from the earth at about 5% of the speed of light!

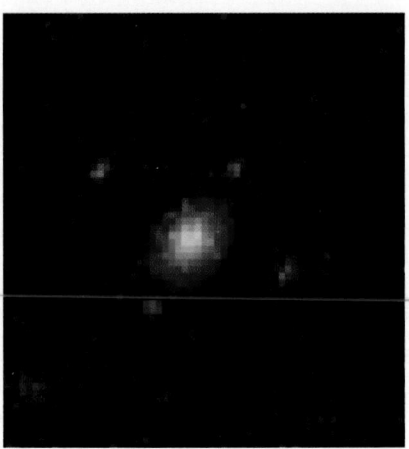

FIGURE 20.30 A Hubble Space Telescope picture of a quasar.

In the 1920s, an analysis of the red shifts of many galaxies led the astronomer Edwin Hubble to the conclusion that the galaxies of the universe are *all* moving apart from each other. Extrapolating backward in time must bring us to a point when all the matter of the universe—and even space itself, according to the theory of relativity—began rushing out of a primordial fireball. Many observations and measurements since have given support to the idea that the universe began in a *Big Bang* about 15 billion years ago.

As an example, Figure 20.30 is a Hubble Space Telescope picture of a *quasar,* short for *quasistellar object.* Quasars are extraordinarily powerful sources of light and radio waves. The light reaching us from quasars is highly red shifted, corresponding in some cases to objects that are moving away from us at greater than 90% of the speed of light. Astronomers have determined that some quasars are 10 to 12 *billion* light years away from the earth, hence the light we see was emitted when the universe was only about 25% of its present age. Today, the red shifts of distant quasars and supernovae (exploding stars) are being used to refine our understanding of the structure and evolution of the universe.

STOP TO THINK 20.6 Amy and Zack are both listening to the source of sound waves that is moving to the right. Compare the frequencies each hears.

a. $f_{Amy} > f_{Zack}$
b. $f_{Amy} = f_{Zack}$
c. $f_{Amy} < f_{Zack}$

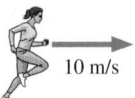
Amy
10 m/s

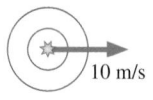
10 m/s
f_0

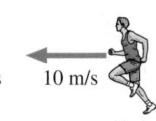
10 m/s
Zack

SUMMARY

The goal of Chapter 20 has been to learn the basic properties of traveling waves.

GENERAL PRINCIPLES

The Wave Model

This model is based on the idea of a traveling wave, which is an organized disturbance traveling at a well-defined **wave speed** v.

- In transverse waves the particles of the medium move perpendicular to the direction in which the wave travels.

- In longitudinal waves the particles of the medium move parallel to the direction in which the wave travels.

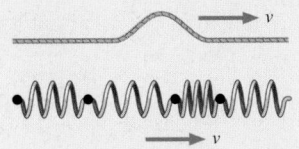

A wave transfers **energy,** but no material or substance is transferred outward from the source.

Three basic types of waves:

- Mechanical waves travel through a material medium such as water or air.

- Electromagnetic waves require no material medium and can travel through a vacuum.

- Matter waves describe the wavelike characteristics of atomic-level particles.

For mechanical waves, the speed of the wave is a property of the medium. Speed does not depend on the size or shape of the wave.

IMPORTANT CONCEPTS

The **displacement** D of a wave is a function of both position (where) and time (when).

- A snapshot graph shows the wave's displacement as a function of position at a single instant of time.

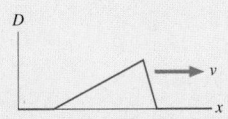

- A history graph shows the wave's displacement as a function of time at a single point in space.

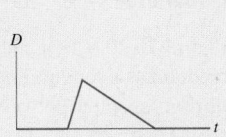

A wave traveling in the positive x-direction with speed v must be a function of the form $D(x - vt)$.

A wave traveling in the negative x-direction with speed v must be a function of the form $D(x + vt)$.

Sinusoidal waves are periodic in both time (period T) and space (wavelength λ).

$$D(x, t) = A\sin\left[2\pi(x/\lambda - t/T) + \phi_0\right]$$
$$= A\sin(kx - \omega t + \phi_0)$$

where A is the **amplitude,** $k = 2\pi/\lambda$ is the **wave number,** $\omega = 2\pi f = 2\pi/T$ is the **angular frequency,** and ϕ_0 is the **phase constant** that describes initial conditions.

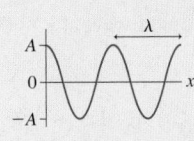

One-dimensional waves

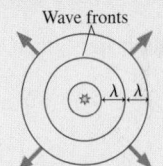

Two- and three-dimensional waves

The fundamental relationship for any sinusoidal wave is $v = \lambda f$.

APPLICATIONS

Wave speeds for some specific waves:

- **String** (transverse): $v = \sqrt{T_s/\mu}$
- **Sound** (longitudinal): $v = 343$ m/s in 20°C air
- **Light** (transverse): $v = c/n$, where $c = 3.00 \times 10^8$ m/s is the speed of light in a vacuum and n is the material's **index of refraction.**

The wave intensity is the power-to-area ratio
$$I = P/A$$
For a circular or spherical wave
$$I = P_{source}/4\pi r^2$$

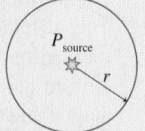

The Doppler effect occurs when a wave source and detector are moving with respect to each other: the frequency detected differs from the frequency f_0 emitted.

Approaching source

$$f_+ = \frac{f_0}{1 - v_s/v}$$

Observer approaching a source

$$f_+ = (1 + v_o/v)f_0$$

Receding source

$$f_- = \frac{f_0}{1 + v_s/v}$$

Observer receding from a source

$$f_- = (1 - v_o/v)f_0$$

The Doppler effect for light uses a result derived from the theory of relativity.

TERMS AND NOTATION

wave model	trailing edge	plane wave
traveling wave	linear density, μ	phase, ϕ
mechanical waves	snapshot graph	compression
electromagnetic waves	history graph	rarefaction
matter waves	sinusoidal wave	electromagnetic spectrum
medium	amplitude, A	index of refraction, n
disturbance	wavelength, λ	intensity, I
wave speed, v	wave number, k	Doppler effect
transverse wave	wave front	red shift
longitudinal wave	circular wave	blue shift
leading edge	spherical wave	

EXERCISES AND PROBLEMS

Exercises

Section 20.2 One-Dimensional Waves

1. Draw the history graph $D(x = 6 \text{ m}, t)$ of this wave at $x = 6$ m.

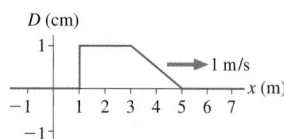

FIGURE EX20.1 Snapshot graph of a wave at $t = 0$ s

2. Draw the history graph $D(x = 0 \text{ m}, t)$ of this wave at $x = 0$ m.

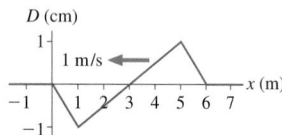

FIGURE EX20.2 Snapshot graph of a wave at $t = 2$ s

3. Draw the snapshot graph $D(x, t = 1 \text{ s})$ of this wave at $t = 1$ s.

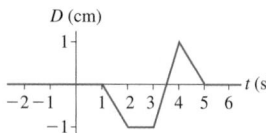

History graph of a wave at $x = 0$ m
FIGURE EX20.3 Wave moving to the right at 1 m/s

4. Draw the snapshot graph $D(x, t = 0 \text{ s})$ of this wave at $t = 0$ s.

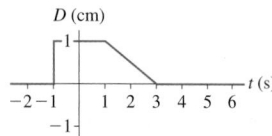

History graph of a wave at $x = 2$ m
FIGURE EX20.4 Wave moving to the left at 1 m/s

5. Figure Ex20.5 is the snapshot graph at $t = 0$ s of a *longitudinal* wave. Draw the corresponding picture of the particle positions, as was done in Figure 20.10. Let the equilibrium spacing between the particles be 1.0 cm.

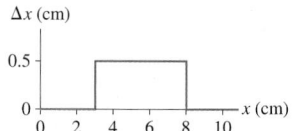

FIGURE EX20.5

6. Figure Ex20.6 is a picture at $t = 0$ s of the particles in a medium as a longitudinal wave is passing through. The equilibrium spacing between the particles is 1.0 cm. Draw the snapshot graph $D(x, t = 0 \text{ s})$ of this wave at $t = 0$ s.

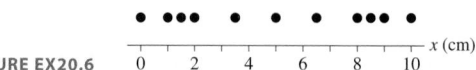

FIGURE EX20.6

7. The wave speed on a string under tension is 200 m/s. What is the speed if the tension is doubled?

8. The wave speed on a string is 150 m/s when the tension is 75 N. What tension will give a speed of 180 m/s?

9. A wave travels along a string at a speed of 280 m/s. What will be the speed if the string is replaced by one made of the same material and under the same tension but having twice the radius?

Section 20.3 Sinusoidal Waves

10. A wave has angular frequency 30 rad/s and wavelength 2.0 m. What are its (a) wave number and (b) wave speed?

11. A wave travels with speed 200 m/s. Its wave number is 1.5 rad/m. What are its (a) wavelength and (b) frequency?

12. The displacement of a wave traveling in the positive x-direction is $D(x, t) = (3.5 \text{ cm}) \sin(2.7x - 124t)$, where x is in m and t is in s. What are the (a) frequency, (b) wavelength, and (c) speed of this wave?

13. The displacement of a wave traveling in the negative y-direction is $D(y, t) = (5.2 \text{ cm}) \sin(5.5y + 72t)$, where y is in m and t is in s. What are the (a) frequency, (b) wavelength, and (c) speed of this wave?

14. Figure Ex20.14 is a snapshot graph at $t = \frac{1}{4}$ s of a traveling wave with displacement $D(x, t) = (2.0 \text{ cm}) \sin(2\pi x - 4\pi t)$, where x is in m and t is in s.
    a. Reproduce this graph on your page, then use a *dotted* line to show the snapshot graph at $t = 0$ s and a *dashed* line to show the snapshot graph at $t = \frac{1}{8}$ s.
    b. What is the speed of this wave?

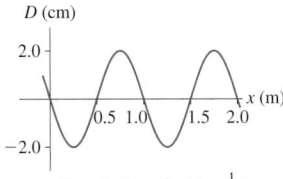

FIGURE EX20.14   Snapshot graph at $t = \frac{1}{4}$ s

15. What are the amplitude, wavelength, and frequency of this wave?

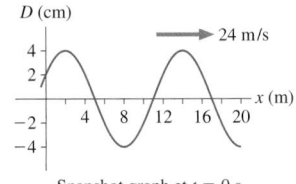

FIGURE EX20.15   Snapshot graph at $t = 0$ s

16. What are the amplitude, frequency, and wavelength of this wave?

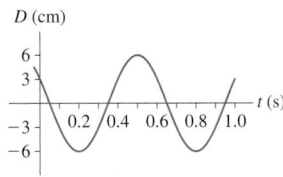

FIGURE EX20.16   History graph at $x = 0$ m
Wave traveling left at 2 m/s

Section 20.4 Waves in Two and Three Dimensions

17. A circular wave travels outward from the origin. At one instant of time, the phase at $r_1 = 20$ cm is 0 rad and the phase at $r_2 = 80$ cm is 3π rad. What is the wavelength of the wave?

18. A spherical wave with a wavelength of 2.0 m is emitted from the origin. At one instant of time, the phase at $r = 4.0$ m is π rad. At that instant, what is the phase at $r = 3.5$ m and at $r = 4.5$ m?

19. A loudspeaker at the origin emits sound waves on a day when the speed of sound is 340 m/s. A crest of the wave simultaneously passes listeners at the (x, y) coordinates (40 m, 0 m) and (0 m, 30 m). What are the lowest two possible frequencies of the sound?

20. A sound source is located somewhere along the x-axis. Experiments show that the same wave front simultaneously reaches listeners at $x = -7.0$ m and $x = +3.0$ m.
    a. What is the x-coordinate of the source?
    b. A third listener is positioned along the positive y-axis. What is her y-coordinate if the same wave front reaches her at the same instant it does the first two listeners?

Section 20.5 Sound and Light

21. The back wall of an auditorium is 26 m from the stage. If you are seated in the middle row, how much time elapses between a sound from the stage reaching your ear directly and the same sound reaching your ear after reflecting from the back wall?

22. A hammer taps on the end of a 4.0-m-long metal bar at room temperature. A microphone at the other end of the bar picks up two pulses of sound, one that travels through the metal and one that travels through the air. The pulses are separated in time by 11.0 ms. What is the speed of sound in this metal?

23. Oil explorers set off explosives to make loud sounds, then listen for the echoes from underground oil deposits. Geologists suspect that there is oil under 500-m-deep Lake Physics. It's known that Lake Physics is carved out of a granite basin. Explorers detect a weak echo 0.94 s after exploding dynamite at the lake surface. If it's really oil, how deep will they have to drill into the granite to reach it?

24. a. What is the wavelength of a 2.0 MHz ultrasound wave traveling through aluminum?
    b. What frequency of electromagnetic wave would have the same wavelength as the ultrasound wave of part a?

25. a. At 20°C, what is the frequency of a sound wave in air with a wavelength of 20 cm?
    b. What is the frequency of an electromagnetic wave with a wavelength of 20 cm?
    c. What would be the wavelength of a sound wave in water that has the same frequency as the electromagnetic wave of part b?

26. a. What is the frequency of blue light that has a wavelength of 450 nm?
    b. What is the frequency of red light that has a wavelength of 650 nm?
    c. What is the index of refraction of a material in which the red-light wavelength is 450 nm?

27. a. Telephone signals are often transmitted over long distances by microwaves. What is the frequency of microwave radiation with a wavelength of 3.0 cm?
    b. Microwave signals are beamed between two mountaintops 50 km apart. How long does it take a signal to travel from one mountaintop to the other?

28. a. An FM radio station broadcasts at a frequency of 101.3 MHz. What is the wavelength?
    b. What is the frequency of a sound source that produces the same wavelength in 20°C air?

29. a. How long does it take light to travel through a 3.0-mm-thick piece of window glass?
    b. Through what thickness of water could light travel in the same amount of time?

30. A light wave has a 670 nm wavelength in air. Its wavelength in a transparent solid is 420 nm.
    a. What is the speed of light in this solid?
    b. What is the light's frequency in the solid?

31. A helium-neon laser beam has a wavelength in air of 633 nm. It takes 1.38 ns for the light to travel through 30 cm of an unknown liquid. What is the wavelength of the laser beam in the liquid?

Section 20.6 Power and Intensity

32. A sound wave with intensity 2.0×10^{-3} W/m² is perceived to be modestly loud. Your eardrum is 6.0 mm in diameter. How much energy will be transferred to your eardrum while listening to this sound for 1.0 min?

33. The intensity of electromagnetic waves from the sun is 1.4 kW/m² just above the earth's atmosphere. Eighty percent of this reaches the surface at noon on a clear summer day. Suppose you think of your back as a 30 cm × 50 cm rectangle. How many joules of solar energy fall on your back as you work on your tan for 1.0 hr?

34. The sound intensity from a jack hammer breaking concrete is 2.0 W/m² at a distance of 2.0 m from the point of impact. This is sufficiently loud to cause permanent hearing damage if the operator doesn't wear ear protection. What is the sound intensity for a person watching from 50 m away?

35. A concert loudspeaker suspended high off the ground emits 35 W of sound power. A small microphone with a 1.0 cm² area is 50 m from the speaker.
    a. What is the sound intensity at the position of the microphone?
    b. How much sound energy impinges on the microphone each second?

Section 20.7 The Doppler Effect

36. An opera singer in a convertible sings a note at 600 Hz while cruising down the highway at 90 km/hr. What is the frequency heard by
    a. A person standing beside the road in front of the car?
    b. A person on the ground behind the car?

37. A mother hawk screeches as she dives at you. You recall from biology that female hawks screech at 800 Hz, but you hear the screech at 900 Hz. How fast is the hawk approaching?

38. A whistle you use to call your hunting dog has a frequency of 21 kHz, but your dog is ignoring it. You suspect the whistle may not be working, but you can't hear sounds above 20 kHz. To test it, you ask a friend to blow the whistle, then you hop on your bicycle. In which direction should you ride (toward or away from your friend) and at what minimum speed to know if the whistle is working?

39. A friend of yours is loudly singing a single note at 400 Hz while racing toward you at 25.0 m/s on a day when the speed of sound is 340 m/s.
    a. What frequency do you hear?
    b. What frequency does your friend hear if you suddenly start singing at 400 Hz?

Problems

40. The displacement of a traveling wave is
$$D(x, t) = \begin{cases} 1 \text{ cm} & \text{if } |x - 3t| \leq 1 \\ 0 \text{ cm} & \text{if } |x - 3t| > 1 \end{cases}$$
where x is in m and t in s.
    a. Draw displacement-versus-position graphs at 1 s intervals from $t = 0$ s to $t = 3$ s. Use an x-axis that goes from −2 to 12 m. Stack the four graphs vertically, similar to Figure 20.12.

    b. Determine the wave speed from the graphs. Explain how you did so.
    c. Determine the wave speed from the equation for $D(x, t)$. Does it agree with your answer to part b?

41. The displacement of a traveling wave is
$$D(x, t) = \begin{cases} 1 \text{ cm} & \text{if } |x + 2t| \leq \frac{1}{2} \\ 0 \text{ cm} & \text{if } |x + 2t| > \frac{1}{2} \end{cases}$$
where x is in m and t in s.
    a. Draw displacement-versus-position graphs at 1 s intervals from $t = -2$ s to $t = +2$ s. Use an x-axis that goes from −6 to 6 m. Stack the five graphs vertically, similar to Figure 20.12.
    b. Determine the wave speed from the graphs. Explain how you did so.
    c. Determine the wave speed from the equation for $D(x, t)$. Does it agree with your answer to part b?

42. This is a snapshot graph at $t = 0$ s of a 5.0 Hz wave traveling to the left.
    a. What is the wave speed?
    b. What is the phase constant of the wave?
    c. Write the displacement equation for this wave.

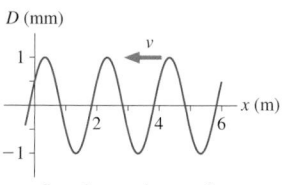

FIGURE P20.42   Snapshot graph at $t = 0$ s

43. This is a history graph at $x = 0$ m of a wave traveling in the positive x-direction at 4.0 m/s.
    a. What is the wavelength?
    b. What is the phase constant of the wave?
    c. Write the displacement equation for this wave.

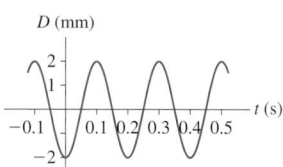

FIGURE P20.43   History graph at $x = 0$ m
Wave traveling right at 4.0 m/s

44. An ultrasound unit sends a 2.4 MHz sound wave into a 25-cm-long tube filled with an unknown liquid. A small microphone right next to the ultrasonic generator detects both the transmitted wave and the sound wave that has reflected off the far end of the tube. The two sound pulses are 4.4 divisions apart on an oscilloscope for which the horizontal time sweep is set to 100 μs/division. What is the speed of sound in the liquid?

45. A 2.0-m-long string is under 20 N of tension. A pulse travels the length of the string in 50 ms. What is the mass of the string?

46. Andy (mass 80 kg) uses a 3.0-m-long rope to pull Bob (mass 60 kg) across the floor ($\mu_k = 0.20$) at a constant speed of 1.0 m/s. Bob signals to Andy to stop by "plucking" the rope, sending a wave pulse forward along the rope. The pulse reaches Andy 150 ms later. What is the mass of the rope?

47. String 1 in Figure P20.47 has linear density 2.0 g/m and string 2 has linear density 4.0 g/m. A student sends pulses in both

directions by quickly pulling up on the knot, then releasing it. What should the string lengths L_1 and L_2 be if the pulses are to reach the ends of the strings simultaneously?

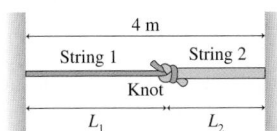

FIGURE P20.47

48. Ships measure the distance to the ocean bottom with *sonar*. A pulse of sound waves is aimed at the ocean bottom, then sensitive microphones listen for the echo. The graph shows the delay time as a function of the ship's position as it crosses 60 km of ocean. Draw a graph of the ocean bottom. Let the ocean surface define $y = 0$ and ocean bottom have negative values of y. This way your graph will be a picture of the ocean bottom. The speed of sound in ocean water varies slightly with temperature, but you can use 1500 m/s as an average value.

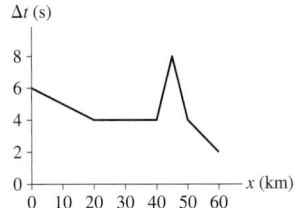

FIGURE P20.48

49. A long string is shaken at its left end at 5.0 Hz, sending a sinusoidal wave down the string at 2.0 m/s. After the wave has traveled 1.0 m, the linear density of the string suddenly decreases by a factor of four. Draw a snapshot graph of the first 3.0 m of the string at an instant when the left end is at its equilibrium position.

50. One cue your hearing system uses to localize a sound (i.e., to tell where a sound is coming from) is the slight difference in the arrival times of the sound at your ears. Your ears are spaced approximately 20 cm apart. Consider a sound source 5.0 m from the center of your head along a line 45° to your right. What is the *difference* in arrival times? Give your answer in microseconds.

    Hint: You are looking for the difference between two numbers that are nearly the same. What does this near equality imply about the necessary precision during intermediate stages of the calculation?

51. A 256 Hz sound wave in 20°C air propagates into the water of a swimming pool. What are the water-to-air ratios of the wave's frequency, wave speed, and wave length?

52. Earthquakes are essentially sound waves traveling through the earth. They are called seismic waves. Because the earth is solid, it can support both longitudinal and transverse seismic waves. These travel at different speeds. The speed of longitudinal waves, called P waves, is 8000 m/s. Transverse waves, called S waves, travel at a slower 4500 m/s. A seismograph records the two waves from a distant earthquake. If the S wave arrives 2.0 min after the P wave, how far away was the earthquake? You can assume that the waves travel in straight lines, although actual seismic waves follow more complex routes.

53. One way to monitor global warming is to measure the average temperature of the ocean. Researchers are doing this by mea-

suring the time it takes sound pulses to travel underwater over large distances. At a depth of 1000 m, where ocean temperatures hold steady near 4°C, the average sound speed is 1480 m/s. It's known from laboratory measurements that the sound speed increases 4.0 m/s for every 1.0°C increase in temperature. In one experiment, where sounds generated near California are detected in the South Pacific, the sound waves travel 8000 km. If the smallest time change that can be reliably detected is 1.0 s, what is the smallest change in average temperature that can be measured?

54. A sound wave is described by $D(y, t) = (0.020 \text{ mm}) \times \sin[(8.96 \text{ rad/m})y + (3140 \text{ rad/s})t + \pi/4 \text{ rad}]$, where y is in m and t is in s.
    a. In what direction is this wave traveling?
    b. Along which axis is the air oscillating?
    c. What are the wavelength, the wave speed, and the period of oscillation?
    d. Draw a displacement-versus-time graph $D(y = 1 \text{ m}, t)$ at $y = 1$ m from $t = 0$ s to $t = 0.004$ s.

55. A wave on a string is described by $D(x, t) = (3.0 \text{ cm}) \times \sin[2\pi(x/(2.4 \text{ m}) + t/(0.20 \text{ s}) + 1)]$, where x is in m and t is in s.
    a. In what direction is this wave traveling?
    b. What are the wave speed, the frequency, and the wave number?
    c. At $t = 0.50$ s, what is the displacement of the string at $x = 0.20$ m?

56. A wave on a string is described by $D(x, t) = (2.0 \text{ cm}) \times \sin[(12.57 \text{ rad/m})x - (638 \text{ rad/s})t]$, where x is in m and t is in s. The linear density of the string is 5.0 g/m. What are
    a. The string tension?
    b. The maximum displacement of a point on the string?
    c. The maximum speed of a point on the string?

57. Write the displacement equation for a sinusoidal wave that is traveling in the negative y-direction with wavelength 50 cm, speed 4.0 m/s, and amplitude 5.0 cm. Assume $\phi_0 = 0$.

58. Write the displacement equation for a sinusoidal wave that is traveling in the positive x-direction with frequency 200 Hz, speed 400 m/s, amplitude 0.010 mm, and phase constant $\pi/2$.

59. Show that a wave traveling in the *negative x*-direction with wave speed v must be a function of the form $D(x + vt)$.

60. Show that $D(x, t + T) = D(x, t)$ for a sinusoidal traveling wave. This shows that the wave is periodic with period T.

61. A spherical sound source at the origin emits a sound wave with frequency 13,100 Hz and wave speed 346 m/s. What is the phase difference in degrees and in rad between the two points with (x, y, z) coordinates (1.0 cm, 3.0 cm, 2.0 cm) and $(-1.0$ cm, 1.5 cm, 2.5 cm)?

62. A string with linear density 2.0 g/m is stretched along the positive x-axis with tension 20 N. One end of the string, at $x = 0$ m, is tied to a hook that oscillates up and down at a frequency of 100 Hz with a maximum displacement of 1.0 mm. At $t = 0$ s, the hook is at its lowest point.
    a. What are the wave speed on the string and the wavelength?
    b. What are the amplitude and phase constant of the wave?
    c. Write the equation for the displacement $D(x, t)$ of the traveling wave.
    d. What is the string's displacement at $x = 0.50$ m and $t = 15$ ms?

63. A sound wave of frequency 686 Hz is traveling in 20°C air as a plane wave in the positive y-direction. At time $t = 0$ s, one crest of the wave is located in the plane $y = 12.5$ cm.
    a. Draw a graph of $D(y, t = 0 \text{ s})$ as a function of y from $y = -150$ cm to $y = +150$ cm.
    b. What is the phase constant for this wave? There are *two* values that satisfy the mathematics, so justify your answer.
    c. Write the wave equation $D(y, t)$ for the wave, leaving the amplitude A unspecified.
    d. Draw an xy-coordinate system in which y ranges from -150 cm to $+150$ cm. Show the wave fronts of the wave at time $t = 0$ s. Number each of the wave fronts (1, 2, . . .) at the side of the graph.
    e. Repeat part d at time $t = 0.729$ ms. Keep the same number attached to each wave front so that you can see how the wave fronts have moved.
    f. At $t = 0.729$ ms, what is the phase of the wave at $y = +12.5$ cm and at $y = -12.5$ cm?

64. Figure P20.64 shows a snapshot graph of a wave traveling to the right along a string at 45 m/s. At this instant, what is the velocity of points 1, 2, and 3 on the string?

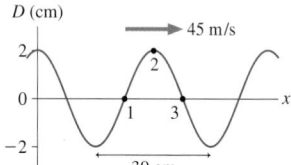

FIGURE P20.64

65. Figure P20.65 is a snapshot graph of the instantaneous *velocity* v_{particle} of the particles on a string. The wave is moving to the left at 50 cm/s. Draw a snapshot graph of the string's displacement at this instant of time.

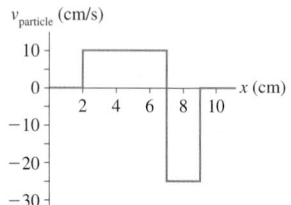

FIGURE P20.65

66. A string that is under 50.0 N of tension has linear density 5.0 g/m. A sinusoidal wave with amplitude 3.0 cm and wavelength 2.0 m travels along the string. What is the maximum velocity of a particle on the string?

67. A sinusoidal wave travels along a stretched string. A particle on the string has a maximum velocity of 2.0 m/s and a maximum acceleration of 200 m/s². What are the frequency and amplitude of the wave?

68. The sun emits electromagnetic waves with a power of 4×10^{26} W. Determine the intensity of electromagnetic waves from the sun just outside the atmospheres of Venus, the earth, and Mars.

69. a. A 100 W lightbulb produces 2.0 W of visible light. (The other 98 W are dissipated as heat and infrared radiation.) What is the light intensity on a wall 2.0 m away from the lightbulb?
    b. A krypton laser produces a cylindrical red laser beam 2.0 mm in diameter with 2.0 W of power. What is the light intensity on a wall 2.0 m away from the laser?

70. An AM radio station broadcasts with a power of 25 kW at a frequency of 920 kHz. Estimate the intensity of the radio wave at a point 10 km from the broadcast antenna.

71. The sound intensity 50 m from a wailing tornado siren is 0.10 W/m².
    a. What is the intensity at 1000 m?
    b. The weakest intensity likely to be heard over background noise is ≈ 1 μW/m². Estimate the maximum distance at which the siren can be heard.

72. Lasers can be used to drill or cut material. One such laser generates a series of high-intensity pulses rather than a continuous beam of light. Each pulse contains 500 mJ of energy and lasts 10 ns. The laser fires 10 such pulses per second.
    a. What is the *peak power* of the laser light? The peak power is the power output during one of the 10 ns pulses.
    b. What is the average power output of the laser? The average power is the total energy delivered per second.
    c. A lens focuses the laser beam to a 10-μm-diameter circle on the target. During a pulse, what is the light intensity on the target?
    d. The intensity of sunlight at midday is about 1100 W/m². What is the ratio of the laser intensity on the target to the intensity of the midday sun?

73. A bat locates insects by emitting ultrasonic "chirps" and then listening for echoes from the bugs. Suppose a bat chirp has a frequency of 25 kHz. How fast would the bat have to fly, and in what direction, for you to just barely be able to hear the chirp at 20 kHz?

74. A physics professor demonstrates the Doppler effect by tying a 600 Hz sound generator to a 1.0-m-long rope and whirling it around her head in a horizontal circle at 100 rpm. What are the highest and lowest frequencies heard by a student in the classroom?

75. Show that the Doppler frequency f_- of a receding source is $f_- = f_0/(1 + v_s/v)$.

76. A starship approaches its home planet at a speed of $0.1c$. When it is 54×10^6 km away, it uses its green laser beam ($\lambda = 540$ nm) to signal its approach.
    a. How long does the signal take to travel to the home planet?
    b. At what wavelength is the signal detected on the home planet?

77. You are cruising to Jupiter at the posted speed limit of $0.1c$ when suddenly a daredevil passes you, going in the same direction, at $0.3c$. At what wavelength does your rocket cruiser's light detector "see" his red tail lights? Is this wavelength ultraviolet, visible, or infrared? Use 650 nm for the wavelength of red light.

78. Wavelengths of light from a distant galaxy are found to be 0.5% longer than the corresponding wavelengths measured in a terrestrial laboratory. Is the galaxy approaching or receding from the earth? At what speed?

79. You have just been pulled over for running a red light, and the police officer has informed you that the fine will be $250. In desperation, you suddenly recall an idea that your physics professor recently discussed in class. In your calmest voice, you tell the officer that the laws of physics prevented you from knowing that the light was red. In fact, as you drove toward it, the light was Doppler shifted to where it appeared green to you. "OK," says the officer, "Then I'll ticket you for speeding. The fine is $1 for every 1 km/hr over the posted speed limit of 50 km/hr." How big is your fine? Use 650 nm as the wavelength of red light and 540 nm as the wavelength of green light.

Challenge Problems

80. Figure CP20.80 shows two masses hanging from a steel wire. The mass of the wire is 60.0 g. A wave pulse travels along the wire from point 1 to point 2 in 24.0 ms. What is mass m?

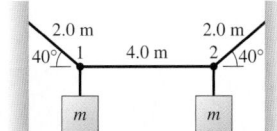

FIGURE CP20.80

81. A wire is made by welding together two metals having different densities. Figure CP20.81 shows a 2.0-m-long section of wire centered on the junction, but the wire extends much farther in both directions. The wire is placed under 2250 N tension, then a 1500 Hz wave with an amplitude of 3.0 mm is sent down the wire. How many wavelengths (complete cycles) of the wave are in this 2.0-m-long section of the wire?

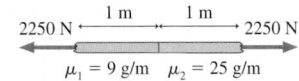

FIGURE CP20.81

82. A rope of mass m and length L hangs from a ceiling.
    a. Show that the wave speed on the rope a distance y above the lower end is $v = \sqrt{gy}$.
    b. Show that the time for a pulse to travel the length of the string is $\Delta t = 2\sqrt{L/g}$.

83. Some modern optical devices are made with glass whose index of refraction changes with distance from the front surface. Figure CP20.83 shows the index of refraction as a function of the distance into a slab of glass of thickness L. The index of refraction increases linearly from n_1 at the front surface to n_2 at the rear surface.
    a. Find an expression for the time light takes to travel through this piece of glass.
    b. Evaluate your expression for a 1.0-cm-thick piece of glass for which $n_1 = 1.50$ and $n_2 = 1.60$.

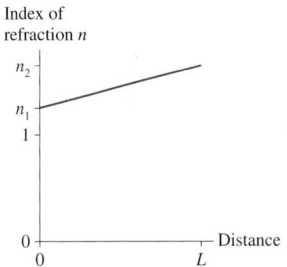

FIGURE CP20.83

STOP TO THINK ANSWERS

Stop to Think 20.1: b. The wave is traveling to the right at 2 m/s, so each point on the wave passed $x = 0$ m, the point of interest, 2 s before reaching $x = 4$ m. The graph has the same shape, but everything happens 2 s earlier.

Stop to Think 20.2: d and e. The wave speed depends on properties of the medium, not on how you generate the wave. For a string, $v = \sqrt{T_s/\mu}$. Increasing the tension or decreasing the linear density (lighter string) will increase the wave speed.

Stop to Think 20.3: d. The wavelength—the distance between two crests—is seen to be 10 m. The frequency is $f = v/\lambda = (50 \text{ m/s})/(10 \text{ m}) = 5$ Hz.

Stop to Think 20.4: e. A crest and an adjacent trough are separated by $\lambda/2$. This is a phase difference of π.

Stop to Think 20.5: $n_3 > n_1 > n_2$. $\lambda = \lambda_{vac}/n$, so a shorter wavelength corresponds to a larger index of refraction.

Stop to Think 20.6: c. Zack hears a higher frequency as he and the source approach. Amy is moving with the source, so $f_{Amy} = f_0$.

21 Superposition

This piece of glass has one color if you look through it, a different color if you look at light reflected from it. It is called *dichroic glass,* meaning two-color glass, and it is used both for jewelry and in the electro-optics industry. Ordinary colored glass, such as a green bottle, reflects and transmits the same color of light. Why is dichroic glass different? You'll learn in this chapter that the interesting optical properties of dichroic glass are due to the combination of *two* traveling waves.

The combination of two or more waves is called a *superposition* of waves. In this chapter we will explore how waves are superimposed and learn that superposition is important to applications ranging from musical instruments to lasers. This chapter also lays the groundwork for our study of light-wave optics in Chapter 22.

21.1 The Principle of Superposition

Figure 21.1a shows two baseball players, Alan and Bill, at batting practice. Unfortunately, someone has turned the pitching machines so that pitching machine A throws baseballs toward Bill while machine B throws toward Alan. If two baseballs are launched at the same time, and with the same speed, they collide at the crossing point and bounce away. Two particles cannot occupy the same point of space at the same time.

(a)

(b)

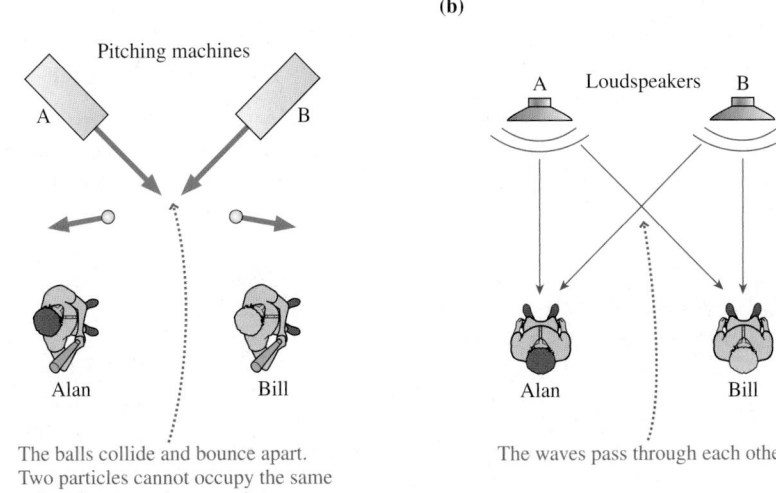

Pitching machines

The balls collide and bounce apart. Two particles cannot occupy the same point of space at the same time.

The waves pass through each other.

FIGURE 21.1 Unlike particles, two waves can pass directly through each other.

But waves, unlike particles, can pass directly through each other. In Figure 21.1b Alan and Bill are listening to the stereo system in the locker room after practice. Because both hear the music quite well, without distortion or missing sound, the sound wave that travels from loudspeaker A toward Bill must pass through the wave traveling from loudspeaker B toward Alan. What happens to the medium at a point where two waves are present simultaneously?

If wave 1 displaces a particle in the medium by D_1 and wave 2 *simultaneously* displaces it by D_2, the net displacement of the particle is simply $D_1 + D_2$. This is a very important idea because it tells us how to combine waves. It is known as the *principle of superposition*.

> **Principle of superposition** When two or more waves are *simultaneously* present at a single point in space, the displacement of the medium at that point is the sum of the displacements due to each individual wave.

When different objects are laid on top of each other, they are said to be *superimposed*. But through some quirk in the English language, the result of superimposing objects is called a *superposition*, without the syllable "im" in the middle. When one wave is "placed" on top of another wave, we have a superposition of waves, hence the name of the principle.

Mathematically, the net displacement of a particle in the medium is

$$D_{net} = D_1 + D_2 + \cdots = \sum_i D_i \qquad (21.1)$$

where D_i is the displacement that would be caused by wave i alone. We will make the simplifying assumption that the displacements of the individual waves are along the same line so that we can add displacements as scalars rather than vectors.

To use the principle of superposition you must know the displacement that each wave would cause if it traveled through the medium alone. Then you go through the medium *point by point* and add the displacements due to each wave *at that point* to find the net displacement at that point. The outcome will be different at each and every point in the medium because the displacements are different at each point.

To illustrate, Figure 21.2 shows five snapshot graphs taken 1 s apart of two waves traveling at the same speed (1 m/s) in opposite directions along a string.

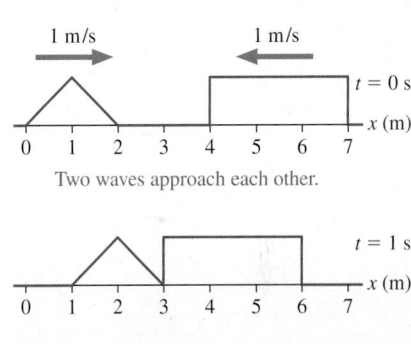

Two waves approach each other.

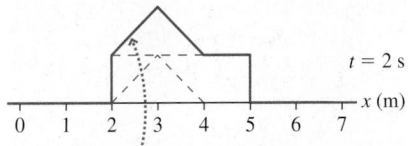

The net displacement is the point-by-point summation of the individual waves.

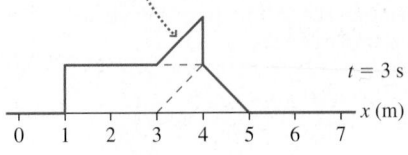

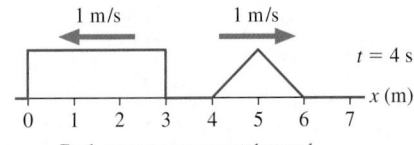

Both waves emerge unchanged.

FIGURE 21.2 The superposition of two waves on a string as they pass through each other.

The principle of superposition comes into play wherever the waves overlap. The displacement of each wave is shown as a dotted line. The solid line is the sum *at each point* of the two displacements at that point. This is the displacement that you would actually observe as the two waves pass through each other.

NOTE ▶ Superposition is the key idea for this chapter and the next, so make sure you understand how the solid line is the point-by-point addition of the displacements due to the individual waves. ◀

STOP TO THINK 21.1 Two pulses on a string approach each other at speeds of 1 m/s. What is the shape of the string at $t = 6$ s?

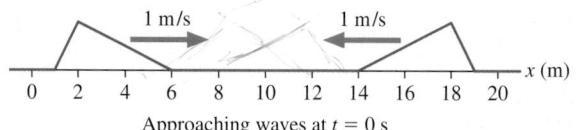

Approaching waves at $t = 0$ s

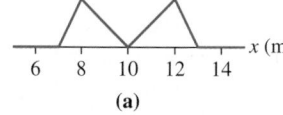

(a)

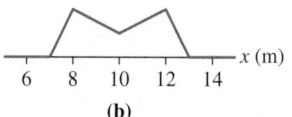

(b)

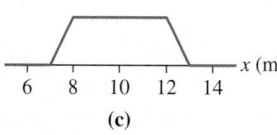

(c)

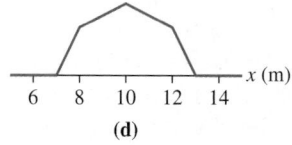

(d)

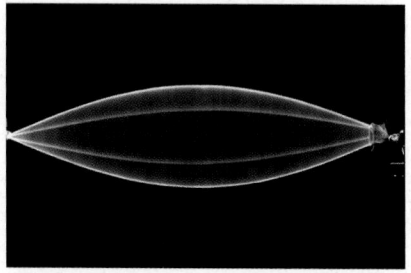

FIGURE 21.3 A vibrating string is an example of a standing wave.

21.2 Standing Waves

Figure 21.3 is a time-lapse photograph of a *standing wave* on a vibrating string. It's not obvious from the photograph, but a standing wave is actually the superposition of two waves. To understand this, let's begin by thinking about two sinusoidal waves traveling in opposite directions through a medium. For example, suppose you point two loudspeakers at each other or shake both ends of a very long string. Throughout this section, **we will assume that the two waves have the same frequency, the same wavelength, and the same amplitude.** In other words, they're identical waves except that one travels to the right and the other to the left. What happens as these two waves pass through each other?

Figure 21.4a shows nine snapshot graphs, at intervals of $\frac{1}{8}T$, of the two waves moving through each other. The dots identify two of the crests to help you see that the red wave is traveling to the right and the green wave to the left. At *each point,* the net displacement of the medium is found by adding the red displacement and the green displacement. The resulting blue wave is the superposition of the two traveling waves.

Figure 21.4a is rather complicated, so Figure 21.4b shows just the blue superposition of the two waves. This is the wave that you would actually observe in the medium. Interestingly, the blue dot shows that the wave in Figure 21.4b is moving neither right nor left. This is a wave, but it is not a traveling wave. The wave in Figure 21.4b is called a **standing wave** because the crests and troughs "stand in place" as it oscillates.

(a)

The red wave is traveling to the right.

The green wave is traveling to the left.

(b)

The superposition of the red and green waves is a standing wave.

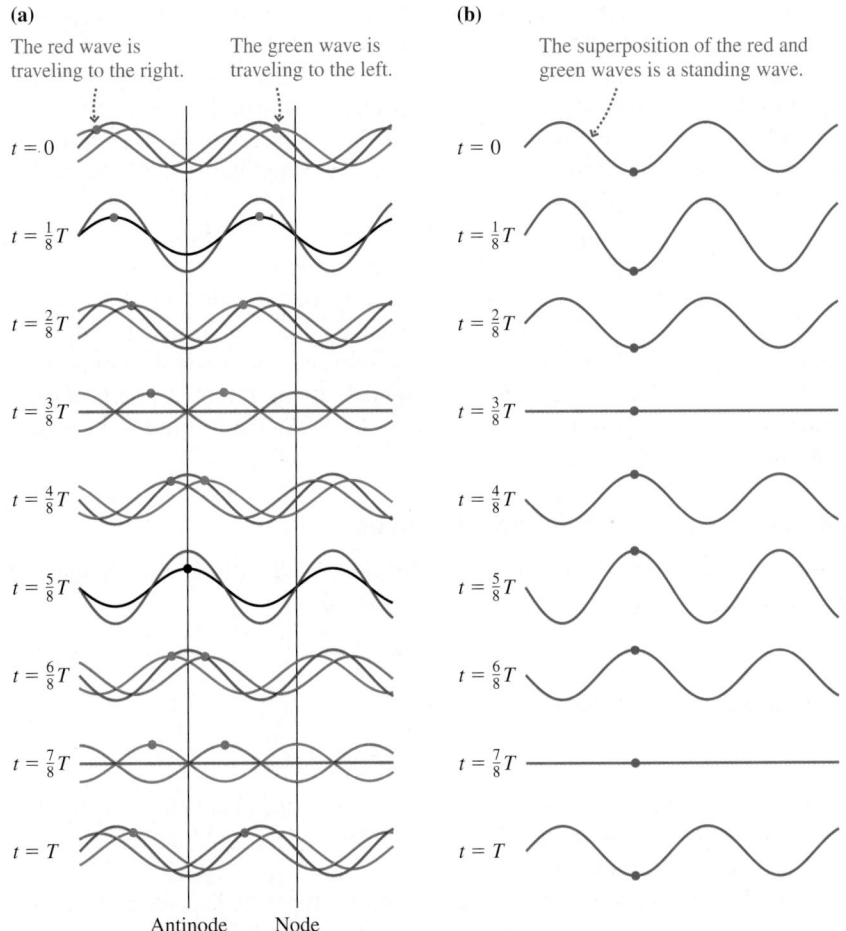

Antinode   Node

FIGURE 21.4 Two sinusoidal waves traveling in opposite directions.

Nodes and Antinodes

Figure 21.5 has collapsed the nine graphs of Figure 21.4b into a single graphical representation of a standing wave. Compare this to the Figure 21.3 photograph of a vibrating string and you can see that the vibrating string is a standing wave. A striking feature of a standing-wave pattern is the existence of points that *never move!* These points, which are spaced $\lambda/2$ apart, are called **nodes.** Halfway between the nodes are the points where the particles in the medium oscillate with maximum displacement. These points of maximum amplitude are called **antinodes,** and you can see that they are also spaced $\lambda/2$ apart.

It seems surprising and counterintuitive that some particles in the medium have no motion at all. To understand how this happens, look carefully at the two traveling waves in Figure 21.4a. You will see that the nodes occur at points where at *every instant* of time the displacements of the two traveling waves have equal magnitudes but *opposite signs.* Thus the superposition of the displacements at these points is always zero. The antinodes correspond to points where the two displacements have equal magnitudes and the *same sign* at all times. The superposition at these points gives a displacement twice that of each individual wave.

Two waves 1 and 2 are said to be *in phase* at a point where D_1 is *always* equal to D_2. The superposition at that point yields a wave whose amplitude is twice that of the individual waves. This is called a point of *constructive interference.* The antinodes of a standing wave are points of constructive interference between the two traveling waves.

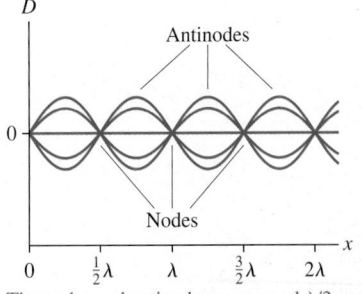

The nodes and antinodes are spaced $\lambda/2$ apart.

FIGURE 21.5 Standing waves are often represented as they would be seen in a time-lapse photograph.

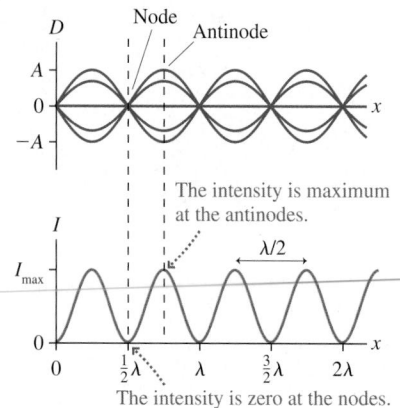

FIGURE 21.6 The intensity of a standing wave is maximum at the antinodes, zero at the nodes.

In contrast, two waves are said to be *out of phase* at points where D_1 is *always* equal to $-D_2$. Their superposition gives a wave with zero amplitude—no wave at all! This is a point of *destructive interference*. The nodes of a standing wave are points of destructive interference. We will defer the main discussion of constructive and destructive interference until later in this chapter, but you'll then recognize that you're seeing constructive and destructive interference at the antinodes and nodes of a standing wave.

In Chapter 20 you learned that the *intensity* of a wave is proportional to the square of the amplitude: $I = CA^2$. You can see in Figure 21.6 that the points of maximum intensity occur where the standing wave oscillates with the largest amplitude (i.e., the antinodes) and that the intensity is zero at the nodes. If this is a sound wave, the loudness periodically varies from zero (no sound) to a maximum and back to zero. The key idea is that **the intensity is maximum at points of constructive interference and zero (if the waves have equal amplitudes) at points of destructive interference.**

The Mathematics of Standing Waves

A sinusoidal wave traveling to the right along the x-axis with angular frequency $\omega = 2\pi f$, wave number $k = 2\pi/\lambda$, and amplitude a is

$$D_\text{R} = a\sin(kx - \omega t) \tag{21.2}$$

An equivalent wave traveling to the left is

$$D_\text{L} = a\sin(kx + \omega t) \tag{21.3}$$

We previously used the symbol A for the wave amplitude, but here we will use a lower case a to represent the amplitude of each individual wave and reserve A for the amplitude of the net wave.

According to the principle of superposition, the net displacement of the medium when both waves are present is the sum of D_R and D_L:

$$D(x, t) = D_\text{R} + D_\text{L} = a\sin(kx - \omega t) + a\sin(kx + \omega t) \tag{21.4}$$

We can simplify Equation 21.4 by using the trigonometric identity

$$\sin(\alpha \pm \beta) = \sin\alpha\cos\beta \pm \cos\alpha\sin\beta$$

Doing so gives

$$\begin{aligned} D(x, t) &= a(\sin kx\cos\omega t - \cos kx\sin\omega t) + a(\sin kx\cos\omega t + \cos kx\sin\omega t) \\ &= (2a\sin kx)\cos\omega t \end{aligned} \tag{21.5}$$

It is useful to write Equation 21.5 as

$$D(x, t) = A(x)\cos\omega t \tag{21.6}$$

where the **amplitude function** $A(x)$ is defined as

$$A(x) = 2a\sin kx \tag{21.7}$$

The amplitude reaches a maximum value $A_\text{max} = 2a$ at points where $\sin kx = 1$.

The displacement $D(x, t)$ given by Equation 21.6 is neither a function of $x - vt$ nor a function of $x + vt$, hence it is *not* a traveling wave. Instead, the $\cos\omega t$ term in Equation 21.6 describes a medium in which each point oscillates in simple harmonic motion with frequency $f = \omega/2\pi$. The function $A(x) = 2a\sin kx$ determines the amplitude of the oscillation for a particle at position x.

Figure 21.7 graphs Equation 21.6 at several different instants of time. Notice that the graphs are identical to those of Figure 21.5, showing us that Equation 21.6 is the mathematical description of a standing wave. You can see that the amplitude of oscillation, given by $A(x)$, varies from point to point in the medium.

The nodes of the standing wave are the points at which the amplitude is zero. They are located at positions x for which

$$A(x) = 2a \sin kx = 0 \qquad (21.8)$$

Equation 21.8 is true if

$$kx_m = \frac{2\pi x_m}{\lambda} = m\pi \qquad m = 0, 1, 2, 3, \ldots \qquad (21.9)$$

Thus the position x_m of the mth node is

$$x_m = m\frac{\lambda}{2} \qquad m = 0, 1, 2, 3, \ldots \qquad (21.10)$$

where m is an integer. You can see that the spacing between two adjacent nodes is $\lambda/2$, in agreement with Figure 21.6. The nodes are *not* spaced by λ, as you might have expected.

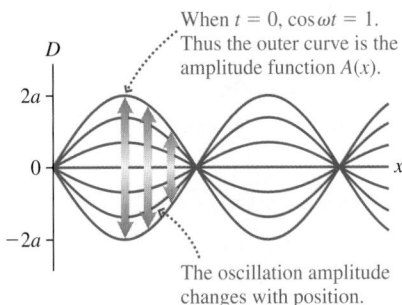

FIGURE 21.7 The net displacement due to two counter-propagating sinusoidal waves.

EXAMPLE 21.1 Node spacing on a string

A very long string has a linear density of 5.0 g/m and is stretched with a tension of 8.0 N. 100 Hz waves with amplitudes of 2.0 mm are generated at the ends of the string.

a. What is the node spacing along the resulting standing wave?
b. What is the maximum displacement of the string?

MODEL Two counter-propagating waves of equal frequency create a standing wave.

VISUALIZE The standing wave will look like Figure 21.5.

SOLVE

a. The speed of the waves on the string is

$$v = \sqrt{\frac{T_s}{\mu}} = \sqrt{\frac{8.0\ \text{N}}{0.0050\ \text{kg/m}}} = 40\ \text{m/s}$$

and thus the wavelength is

$$\lambda = \frac{v}{f} = \frac{40\ \text{m/s}}{100\ \text{Hz}} = 0.40\ \text{m} = 40\ \text{cm}$$

Consequently, the spacing between adjacent nodes is $\lambda/2 =$ 20 cm.

b. The maximum displacement, at the antinodes where $\sin kx = 1$, is

$$A_{\max} = 2a = 4.0\ \text{mm}$$

21.3 Transverse Standing Waves

Wiggling both ends of a very long string is not a practical way to generate standing waves. Instead, as in the photograph in Figure 21.3, standing waves are usually seen on a string that is fixed at both ends. To understand why this condition causes standing waves, we need to examine what happens when a traveling wave encounters a discontinuity or a boundary.

Figure 21.8a on the next page shows a *discontinuity* between a string with a larger linear density and one with a smaller linear density. The tension is the same in both strings, so the wave speed is slower on the left, faster on the right. Whenever a wave encounters a discontinuity, some of the wave's energy is *transmitted* forward and some is *reflected*. Here the larger fraction of the energy is transmitted, but you can see that neither the transmitted not the reflected pulse has the same amplitude as the initial wave pulse.

Light waves exhibit an analogous behavior when they encounter a piece of glass. Most of the light wave's energy is transmitted through the glass, which is

Activ
Physics ONLINE   10.4, 10.6

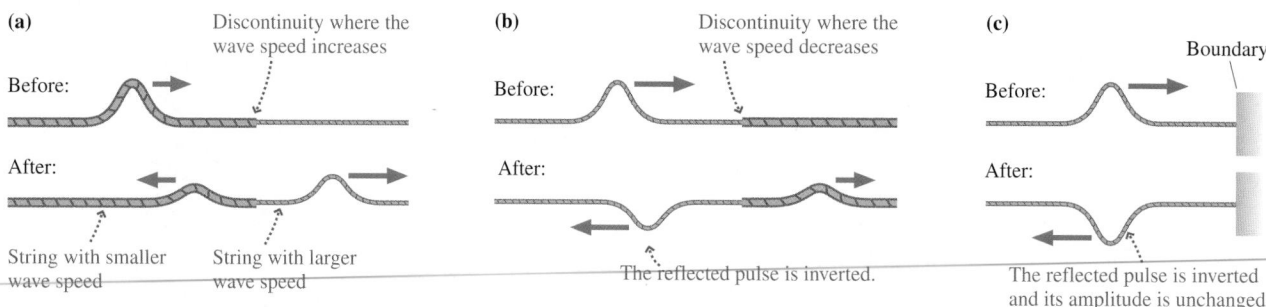

(a) Discontinuity where the wave speed increases

Before:

After:

String with smaller wave speed | String with larger wave speed

(b) Discontinuity where the wave speed decreases

Before:

After:

The reflected pulse is inverted.

(c) Boundary

Before:

After:

The reflected pulse is inverted and its amplitude is unchanged.

FIGURE 21.8 A wave reflects when it encounters a discontinuity or a boundary.

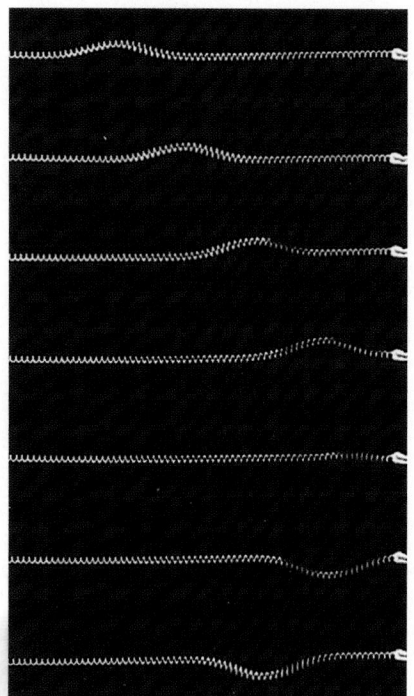

FIGURE 21.9 A strobe photo of a pulse traveling along a rope-like spring.

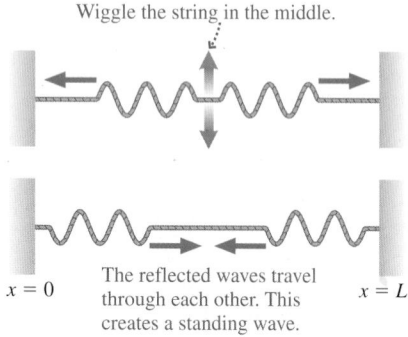

Wiggle the string in the middle.

$x = 0$ The reflected waves travel through each other. This creates a standing wave. $x = L$

FIGURE 21.10 Reflections at the two boundaries cause a standing wave on the string.

why glass is transparent, but a small amount of energy is reflected. That is how you see your reflection dimly in a storefront window.

In Figure 21.8b, an incident wave encounters a discontinuity at which the wave speed decreases. Once again, some of the wave's energy is transmitted and some is reflected. But notice something interesting. When a wave encounters a discontinuity at which the speed decreases, the reflected pulse is *inverted*. A displacement D on the incident wave becomes displacement $-D$ on the reflected wave. Because $\sin(\phi + \pi) = -\sin\phi$, we say that the reflected wave has a *phase change of π upon reflection*. This aspect of reflection will be important later in the chapter when we look at the interference of light waves.

The wave in Figure 21.8c reflects from a *boundary*. You can think of this as Figure 21.8b in the limit that the string on the right becomes infinitely massive. Thus the reflection in Figure 21.8c looks like that of Figure 21.8b with one exception: Because there is no transmitted wave, *all* the wave's energy is reflected. Hence **the amplitude of a wave reflected from a boundary is unchanged.** Figure 21.9 is a sequence of strobe photos in which you see a pulse on a rope-like spring reflecting from a boundary at the right of the photo. The reflected pulse is inverted but otherwise unchanged.

Standing Waves on a String

Figure 21.10 shows a string of length L that is tied at $x = 0$ and $x = L$. This string has *two* boundaries where reflections can occur. If you wiggle the string in the middle, sinusoidal waves travel outward in both directions and soon reach the boundaries. Because the speed of a reflected wave does not change, **the wavelength and frequency of a reflected sinusoidal wave are unchanged.** Consequently, reflections at the ends of the string cause two waves of *equal amplitude and wavelength* to travel in opposite directions along the string. As we've just seen, these are the conditions that cause a standing wave!

To connect the mathematical analysis of standing waves in Section 21.2 with the physical reality of a string tied down at the ends, we need to impose *boundary conditions*. A **boundary condition** is a mathematical statement of any constraint that *must* be obeyed at the boundary or edge of a medium. Because the string is tied down at the ends, the displacements at $x = 0$ and $x = L$ must be zero at all times. Thus the standing-wave boundary conditions are $D(x = 0, t) = 0$ and $D(x = L, t) = 0$. Stated another way, we require nodes at both ends of the string.

We found that the displacement of a standing wave is $D(x, t) = (2a \sin kx)\cos \omega t$. You can see that this equation already satisfies the boundary condition $D(x = 0, t) = 0$. That is, the origin has already been located at a node. The second boundary condition, at $x = L$, requires $D(x = L, t) = 0$. This condition will be met at all times if

$$2a \sin kL = 0 \quad \text{(boundary condition at } x = L) \quad (21.11)$$

Equation 21.11 will be true if $\sin kL = 0$, which in turn requires

$$kL = \frac{2\pi L}{\lambda} = m\pi \qquad m = 1, 2, 3, 4, \ldots \qquad (21.12)$$

kL must be a multiple of $m\pi$, but $m = 0$ is excluded because L can't be zero.

For a string of fixed length L, the only quantity in Equation 21.12 that can vary is λ. That is, the boundary condition can be satisfied only if the wavelength has one of the values

$$\lambda_m = \frac{2L}{m} \qquad m = 1, 2, 3, 4, \ldots \qquad (21.13)$$

In other words, **a standing wave can exist on the string *only* if its wavelength is one of the values given by Equation 21.13.** The mth possible wavelength $\lambda_m = 2L/m$ is just the right size so that its mth node is located at the end of the string (at $x = L$).

NOTE ▶ Other wavelengths, which would be perfectly acceptable wavelengths for a traveling wave, cannot exist as a *standing* wave of length L because they cannot meet the boundary conditions requiring a node at each end of the string. ◀

If standing waves are possible only for certain wavelengths, then only a few specific oscillation frequencies are allowed. Because $\lambda f = v$ for a sinusoidal wave, the oscillation frequency corresponding to wavelength λ_m is

$$f_m = \frac{v}{\lambda_m} = \frac{v}{2L/m} = m\frac{v}{2L} \qquad m = 1, 2, 3, 4, \ldots \qquad (21.14)$$

The lowest allowed frequency

$$f_1 = \frac{v}{2L} \qquad \text{(fundamental frequency)} \qquad (21.15)$$

which corresponds to wavelength $\lambda_1 = 2L$, is called the **fundamental frequency** of the string. The allowed frequencies can be written in terms of the fundamental frequency as

$$f_m = mf_1 \qquad m = 1, 2, 3, 4, \ldots \qquad (21.16)$$

The allowed standing-wave frequencies are all integer multiples of the fundamental frequency. The higher-frequency standing waves are called **harmonics,** with the $m = 2$ wave at frequency f_2 called the *second harmonic,* the $m = 3$ wave called the *third harmonic,* and so on.

Figure 21.11 graphs the first four possible standing waves on a string of fixed length L. These possible standing waves are called the **normal modes** of the string. Each mode, numbered by the integer m, has a unique wavelength and frequency. Keep in mind that these drawings simply show the *envelope,* or outer edge, of the oscillations. The string is continuously oscillating at all positions between these edges, as we showed in more detail in Figure 21.5.

There are several things to note about the normal modes of a string.

1. m is the number of *antinodes* on the standing wave, not the number of nodes. You can tell a string's mode of oscillation by counting the number of antinodes.
2. The *fundamental mode,* with $m = 1$, has $\lambda_1 = 2L$, not $\lambda_1 = L$. Only half of a wavelength is contained between the boundaries, a direct consequence of the fact that the spacing between nodes is $\lambda/2$.

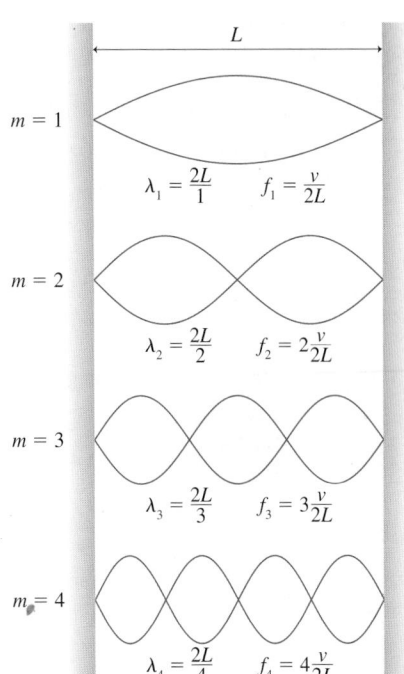

FIGURE 21.11 The first four normal modes for standing waves on a string of length L.

FIGURE 21.12 Time-exposure photograph of the $m = 4$ standing-wave mode on a stretched string.

3. The frequencies of the normal modes form an arithmetic series: f_1, $2f_1$, $3f_1$, $4f_1$, ... The fundamental frequency f_1 can be found as the *difference* between the frequencies of any two adjacent modes. That is, $f_1 = \Delta f = f_{m+1} - f_m$.

Figure 21.12 is a time-exposure photograph of the $m = 4$ standing wave on a string. The nodes and antinodes are quite distinct. The string vibrates four times faster for the $m = 4$ mode than for the fundamental $m = 1$ mode.

EXAMPLE 21.2 A standing wave on a string

A 2.50-m-long string vibrates as a 100 Hz standing wave with nodes 1.00 m and 1.50 m from one end of the string and at no points in between these two. Which harmonic is this, and what is the string's fundamental frequency?

MODEL The nodes of a standing wave are spaced $\lambda/2$ apart.

VISUALIZE The standing wave looks like Figure 21.5.

SOLVE If there are no nodes between the two at 1.0 m and 1.5 m, then the node spacing is $\lambda/2 = 0.50$ m. The number of 0.50-m-wide segments that fit into a 2.50 m length is five, so this is the $m = 5$ mode and 100 Hz is the fifth harmonic. The harmonic frequencies are $f_m = mf_1$, hence the fundamental frequency is

$$f_1 = \frac{f_5}{5} = \frac{100 \text{ Hz}}{5} = 20 \text{ Hz}$$

STOP TO THINK 21.2 A standing wave on a string vibrates as shown at the right. Suppose the tension is quadrupled while the frequency and the length of the string are held constant. Which standing-wave pattern is produced?

Original standing wave

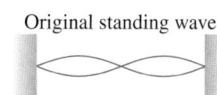

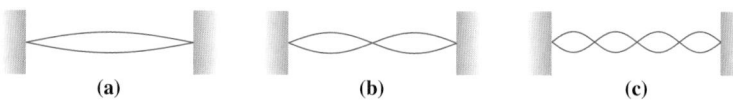

(a)   (b)   (c)

No standing wave for these conditions

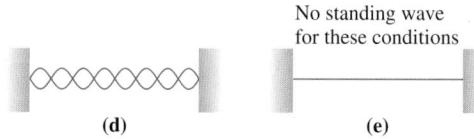

(d)   (e)

Standing Electromagnetic Waves

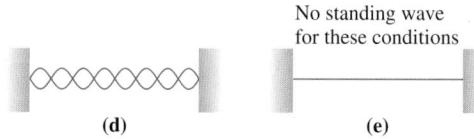

Laser cavity

Standing light wave

Laser beam

Full reflector   Partial reflector

FIGURE 21.13 A laser contains a standing light wave between two parallel mirrors.

A vibrating string is only one example of a transverse standing wave. Another transverse wave is an electromagnetic wave. Standing electromagnetic waves can be established between two parallel mirrors that reflect light back and forth. The mirrors are boundaries, analogous to the boundaries at the ends of a string. In fact, this is exactly how a laser operates. The two facing mirrors in Figure 21.13 form what is called a *laser cavity*.

Because the mirrors act exactly like the points to which a string is tied, the boundary condition is that the light wave must have a node at the surface of each mirror. (To allow some of the light to escape the laser cavity and form the *laser beam*, one of the mirrors is only partially reflective. This doesn't affect the boundary condition.)

Because the boundary conditions are the same, Equations 21.13 and 21.14 for λ_m and f_m apply to a laser just as they do to a vibrating string. The primary difference is the size of the wavelength. A typical laser cavity has a length $L \approx 30$ cm, and visible light has a wavelength $\lambda \approx 600$ nm. The standing light wave in a laser cavity has a mode number m that is approximately

$$m = \frac{2L}{\lambda} \approx \frac{2 \times 0.30 \text{ m}}{6.00 \times 10^{-7} \text{ m}} = 1,000,000$$

In other words, the standing light wave inside a laser cavity has approximately one million antinodes! This is a consequence of the very small wavelength of light.

EXAMPLE 21.3 The standing light wave inside a laser

Helium-neon lasers emit the red laser light commonly used in classroom demonstrations and supermarket checkout scanners. A helium-neon laser operates at a wavelength of precisely 632.9924 nm when the spacing between the mirrors is 310.372 mm.

a. In which mode does this laser operate?
b. What is the next longest wavelength that could form a standing wave in this laser cavity?

MODEL The light wave forms a standing wave between the two mirrors.

VISUALIZE The standing wave looks like Figure 21.13.

SOLVE

a. We can use $\lambda_m = 2L/m$ to find that m (the mode) is

$$m = \frac{2L}{\lambda_m} = \frac{2 \times 0.310372 \text{ m}}{6.329924 \times 10^{-7} \text{ m}} = 980,650$$

There are 980,650 antinodes in the standing light wave between the mirrors.

b. The next longest wavelength that can fit in this laser cavity will have one less node. It will be the $m = 980,649$ mode and its wavelength will be

$$\lambda = \frac{2L}{m} = \frac{2(0.310372 \text{ m})}{980,649} = 632.9930 \text{ nm}$$

ASSESS The mode number is very large, as expected. The wavelength increases by a mere 0.0006 nm when the mode number is decreased by 1.

Microwave radiation, which has a wavelength of a few centimeters, can also set up standing waves between reflective surfaces. This is not always good. If the microwaves in a microwave oven form a standing wave, there are nodes where the electromagnetic field intensity is always zero. These nodes cause cold spots where the food does not heat. Although designers of microwave ovens try to prevent standing waves, ovens usually do have cold spots spaced $\lambda/2$ apart at nodes in the microwave field. A turntable in a microwave oven keeps the food moving so that no part of your dinner remains at a node.

EXAMPLE 21.4 Cold spots in a microwave oven

Cold spots in a microwave oven are found to be 1.25 cm apart. What is the frequency of the microwaves?

MODEL A standing wave is created by microwaves reflecting from the walls.

SOLVE The cold spots are nodes in the microwave standing wave. Nodes are spaced $\lambda/2$ apart, so the wavelength of the microwave radiation must be $\lambda = 2.5$ cm $= 0.025$ m. The speed of microwaves is the speed of light, $v = c$, so the frequency is

$$f = \frac{c}{\lambda} = \frac{3.00 \times 10^8 \text{ m/s}}{0.025 \text{ m}} = 1.2 \times 10^{10} \text{ Hz} = 12 \text{ GHz}$$

21.4 Standing Sound Waves and Musical Acoustics

A long, narrow column of air, such as the air in a tube or a pipe, can support a *longitudinal* standing sound wave. As a sound wave travels through a tube, the air oscillates *parallel* to the tube. If the tube is closed at the end, with a cap, the air at the closed end cannot oscillate. Consequently, **the closed end of a column of air must be a node.**

Figure 21.14 shows the $m = 2$ standing wave inside a column of air closed at both ends. We call this a *closed-closed tube*. You can see that the displacement graph looks exactly like the $m = 2$ standing-wave mode on a string. Because there must be a node at both ends, the possible wavelengths and frequencies of a closed air column are the same as for a string of the same length.

The standing wave must have a node at a closed end.

Air molecules undergo longitudinal oscillations.

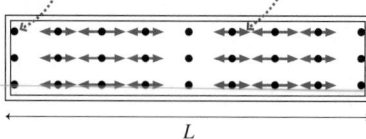

Physical representation of the $m = 2$ mode

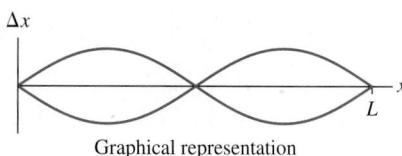

Graphical representation

FIGURE 21.14 The $m = 2$ longitudinal standing wave inside a closed column of air.

EXAMPLE 21.5 Singing in the shower

A shower stall is 2.45 m (8 ft) tall. For what frequencies less than 500 Hz are there standing sound waves in the shower stall?

MODEL The shower stall, at least to a first approximation, is a column of air 2.45 m long. It is closed at the ends by the ceiling and floor. Assume a room-temperature speed of sound.

VISUALIZE A standing sound wave will have nodes at the ceiling and the floor. The $m = 2$ mode will look like Figure 21.14 rotated 90°.

SOLVE The fundamental frequency for a standing sound wave in this air column is

$$f_1 = \frac{v}{2L} = \frac{343 \text{ m/s}}{2(2.45 \text{ m})} = 70 \text{ Hz}$$

The possible standing-wave frequencies are integer multiples of the fundamental frequency. These are 70 Hz, 140 Hz, 210 Hz, 280 Hz, 350 Hz, 420 Hz, and 490 Hz.

ASSESS The many possible standing waves in a shower cause the sound to *resonate*, which helps explain why some people like to sing in the shower. Our approximation of the shower stall as a one-dimensional tube is actually a bit too simplistic. A three-dimensional analysis would find additional modes, making the "sound spectrum" even richer.

Air columns closed at both ends are of limited interest unless, as in Example 21.5, you are inside the column. Columns of air that *emit* sound are open at one or both ends. Many musical instruments fit this description. For example, a flute is a tube of air that is open at both ends. The flutist blows across one end to create a standing wave inside the tube, and a note of that frequency is emitted from both ends of the flute. (The blown end of a flute is open on the side, rather than across the tube. That is necessary for practical reasons of how flutes are played, but from a physics perspective this is the "end" of the tube because it opens the tube to the atmospheric pressure of the surrounding air.) A trumpet, however, is open at the bell end but is *closed* by the player's lips at the other end.

You saw earlier that a wave is partially transmitted and partially reflected at a discontinuity. When a sound wave traveling through a tube of air reaches an open end, some of the wave's energy is transmitted out of the tube to become the sound that you hear and some portion of the wave is reflected back into the tube. These reflections, analogous to the reflection of a string wave from a boundary, allow standing sound waves to exist in a tube of air that is open at one or both ends.

Not surprisingly, the *boundary condition* at the open end of a column of air is not the same as the boundary condition at a closed end. We've seen that the closed

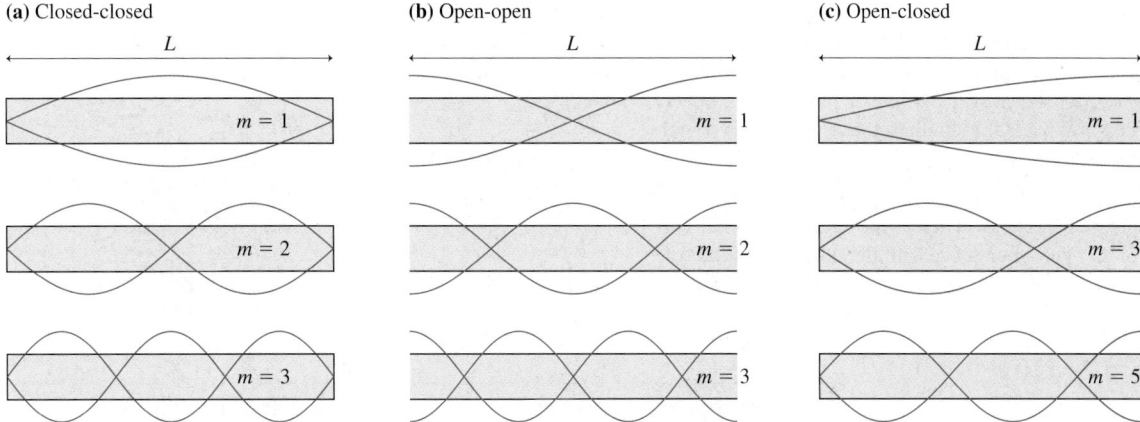

FIGURE 21.15 The first three standing sound wave modes in columns of air with different boundary conditions.

end must be a node because the air has no room to vibrate. By contrast, **an open end of an air column is required to be an antinode.** (A careful analysis shows that the antinode is actually just outside the open end, but for our purposes we'll assume the antinode is exactly at the open end.)

Figure 21.15 shows displacement graphs of the first three standing-wave modes of a tube closed at both ends (a *closed-closed tube*), a tube open at both ends (an *open-open tube*), and a tube open at one end but closed at the other (an *open-closed tube*), all with the same length L. The standing wave in the open-open tube looks like the closed-closed tube except that the positions of the nodes and antinodes are interchanged. In both cases there are m half-wavelength segments between the ends, thus the wavelengths and frequencies of an open-open tube and a closed-closed tube are the same as those of a string tied at both ends:

$$\begin{cases} \lambda_m = \dfrac{2L}{m} \\[2mm] f_m = m\dfrac{v}{2L} = mf_1 \end{cases} \quad \begin{aligned} & m = 1, 2, 3, 4, \dots \\ & \text{(open-open or closed-closed tube)} \end{aligned} \quad (21.17)$$

The open-closed tube is different. The fundamental mode has only one-quarter of a wavelength in a tube of length L, hence the $m = 1$ wavelength is $\lambda_1 = 4L$. This is twice the λ_1 wavelength of an open-open or a closed-closed tube. Consequently, **the fundamental frequency of an open-closed tube is half that of an open-open or a closed-closed tube of the same length.** It will be left as a homework problem for you to show that the possible wavelengths and frequencies of an open-closed tube of length L are

$$\begin{cases} \lambda_m = \dfrac{4L}{m} \\[2mm] f_m = m\dfrac{v}{4L} = mf_1 \end{cases} \quad \begin{aligned} & m = 1, 3, 5, 7, \dots \\ & \text{(open-closed tube)} \end{aligned} \quad (21.18)$$

Notice that m in Equation 21.18 takes on only *odd* values.

NOTE ▶ Because sound is a longitudinal wave, the graphs of Figure 21.15 are *not* "pictures" of the wave as they are for a string wave. The graphs show the displacement Δx, *parallel* to the axis, versus position x. The tube itself is shown merely to indicate the location of the open and closed ends, but the diameter of the tube is *not* related to the amplitude of the displacement. ◀

EXAMPLE 21.6 **The length of an organ pipe**

An organ pipe open at both ends sounds its second harmonic at a frequency of 523 Hz. (Musically, this is the note one octave above middle C.) What is the length of the pipe from the sounding hole to the end?

MODEL An organ pipe, similar to a flute, has a *sounding hole* where compressed air is blown across the edge of the pipe. This is one end of an open-open tube, with the other end at the true "end" of the pipe. Assume a room-temperature speed of sound.

SOLVE The second harmonic is the $m = 2$ mode, which for an open-open tube has frequency

$$f_2 = 2\frac{v}{2L}$$

Thus the length of the organ pipe is

$$L = \frac{v}{f_2} = \frac{343 \text{ m/s}}{523 \text{ Hz}} = 0.656 \text{ m} = 65.6 \text{ cm}$$

ASSESS This is a typical length for an organ pipe.

STOP TO THINK 21.3 An open-open tube of air supports standing waves at frequencies of 300 Hz and 400 Hz and at no frequencies between these two. The second harmonic of this tube has frequency

a. 100 Hz.　　b. 200 Hz.　　c. 400 Hz.　　d. 600 Hz.　　e. 800 Hz.

Musical Instruments

10.5　Activ Physics ONLINE

An important application of standing waves is to musical instruments. Think about stringed musical instruments, such as the guitar, the piano, and the violin. These instruments all have strings that are fixed at the ends and tightened to create tension. A disturbance is generated on the string by plucking, striking, or bowing it. Regardless of how it is generated, the disturbance creates a standing wave on the string.

The fundamental frequency of a vibrating string is

$$f_1 = \frac{v}{2L} = \frac{1}{2L}\sqrt{\frac{T_s}{\mu}}$$

where T_s is the tension in the string and μ is its linear density. The fundamental frequency is the musical note you hear when the string is sounded. Increasing the tension in the string raises the fundamental frequency, a fact known to anyone who has ever tuned a stringed instrument.

NOTE ▶ v is the wave speed *on the string,* not the speed of sound in air. ◀

For instruments like the guitar or the violin, the strings are all the same length and under approximately the same tension. Were that not the case, the neck of the instrument would tend to twist toward the side of higher tension. The strings have different frequencies because they differ in linear density. The lower-pitched strings are "fat" while the higher-pitched strings are "skinny." This difference changes the frequency by changing the wave speed. *Small* adjustments are then made in the tension to bring each string to the exact desired frequency. Once the instrument is tuned, you play it by using your finger tips to alter the effective length of the string. As you shorten the string's length, the frequency and pitch go up.

A piano covers a much wider range of frequencies than a guitar or violin. This range cannot be produced by changing only the linear densities of the strings. The high end would have strings too thin to use without breaking, and the low end would have solid rods rather than flexible wires! So a piano is tuned through a combination of changing the linear density *and* the length of the strings. The bass note strings are not only fatter, they are also longer.

With a wind instrument, blowing into the mouthpiece creates a standing sound wave inside a tube of air. The player changes the notes by using her fingers to cover holes or open valves, changing the length of the tube and thus its frequency.

The strings on a harp vibrate as standing waves. Their frequencies determine the notes that you hear.

As we noted about the flute, the fact that the holes are on the side makes very little difference. The first open hole becomes an antinode because the air is free to oscillate in and out of the opening.

According to Equations 21.17 and 21.18, a wind instrument's frequency depends on the speed of sound *inside* the instrument. But the speed of sound depends on the temperature of the air. When a wind player first blows into the instrument, the air inside starts to rise in temperature. This increases the sound speed, which in turn raises the instrument's frequency for each note until the air temperature reaches a steady state. Consequently, wind players must "warm up" before tuning their instrument. For strings, the speed appearing in Equation 21.17 is the wave speed on the string as determined by the tension, not the sound speed in air.

Many wind instruments have a "buzzer" at one end of the tube, such as a vibrating reed on a saxophone or clarinet or vibrating lips on a trumpet or trombone. Buzzers like these generate a continuous range of frequencies rather than single notes, which is why they sound like a "squawk" if you play on just the mouthpiece without the rest of the instrument. When connected to the body of the instrument, most of those frequencies cause no response of the air molecules. But the frequency from the buzzer that matches the fundamental frequency of the instrument causes the build-up of a large-amplitude response at just that frequency—a standing-wave resonance. This is the energy input that generates and sustains the musical note.

EXAMPLE 21.7 The notes on a clarinet

A clarinet is 66 cm long. The speed of sound in warm air is 350 m/s. What are the frequencies of the lowest note on a clarinet and of the next highest harmonic?

MODEL A clarinet is an open-closed tube because the player's lips and the reed seal the tube at the upper end.

SOLVE The lowest frequency is the fundamental frequency. For an open-closed tube, the fundamental frequency is

$$f_1 = \frac{v}{4L} = \frac{350 \text{ m/s}}{4(0.66 \text{ m})} = 133 \text{ Hz}$$

An open-closed tube has only *odd* harmonics. The next highest harmonic is $m = 3$, with frequency $f_3 = 3f_1 = 399$ Hz.

Except in unusual situations, a vibrating string plays only the musical note corresponding to the fundamental frequency f_1. Thus stringed instruments must use several strings in order to obtain a reasonable range of notes. By contrast, wind instruments can sound at the second or third harmonic of the tube of air (f_2 or f_3). These higher frequencies are sounded by *overblowing* (flutes, brass instruments) or with special *register keys* that open small holes in the side of the instrument to encourage the formation of an antinode at that point (clarinets, saxophones). The controlled use of these higher harmonics gives wind instruments a wide range of notes.

21.5 Interference in One Dimension

One of the most basic characteristics of waves is the ability of two waves to combine into a single wave whose displacement is given by the principle of superposition. This combination, or superposition, of waves is often called **interference.** A standing wave is the interference pattern produced when two waves of equal frequency travel in opposite directions. In this section we will look at the interference of two waves traveling in the *same* direction. Then, after the fundamental ideas are established, we will expand our discussion to interference in two and three dimensions.

Figure 21.16a shows two light waves impinging on a partially silvered mirror. Such a mirror partially transmits and partially reflects each wave, causing two *overlapped* light waves to travel along the *x*-axis to the right of the mirror. Or consider the two loudspeakers in Figure 21.16b. The sound wave from loudspeaker 2 passes just to the side of loudspeaker 1, hence two overlapped sound waves travel to the right along the *x*-axis. We want to find out what happens when two overlapped waves travel in the same direction along the same axis.

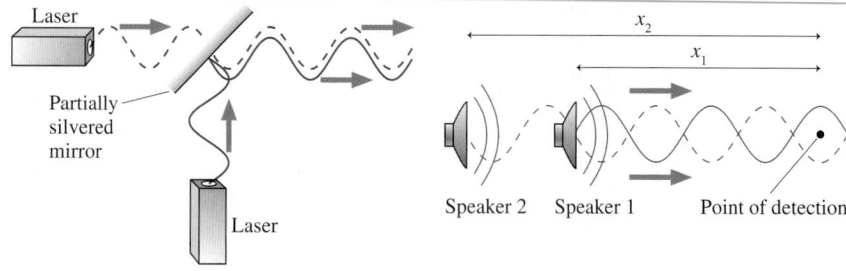

FIGURE 21.16 Two overlapped waves travel along the *x*-axis.

Figure 21.16b shows a point on the *x*-axis where the overlapped waves are detected, either by your ear or by a microphone. This point is distance x_1 from loudspeaker 1 and distance x_2 from loudspeaker 2. (We will use loudspeakers and sound waves for most of our examples, but our analysis will be valid for any wave.) What is the amplitude of the combined waves at this point?

Throughout this section, **we will assume that the waves are sinusoidal, have the same frequency and amplitude, and travel to the right along the *x*-axis.** Thus we can write the displacements of the two waves as

$$D_1(x_1, t) = a\sin(kx_1 - \omega t + \phi_{10}) = a\sin\phi_1$$
$$D_2(x_2, t) = a\sin(kx_2 - \omega t + \phi_{20}) = a\sin\phi_2 \tag{21.19}$$

where ϕ_1 and ϕ_2 are the *phases* of the waves. Both waves have the same amplitude a, the same wave number $k = 2\pi/\lambda$, and the same angular frequency $\omega = 2\pi f$.

The phase constants ϕ_{10} and ϕ_{20} are characteristics *of the sources*, not the medium. Figure 21.17 shows snapshot graphs at $t = 0$ of waves emitted by three sources with phase constants $\phi_0 = 0$ rad, $\phi_0 = \pi/2$ rad, and $\phi_0 = \pi$ rad. You can see that **the phase constant tells us what the source is doing at $t = 0$.** For example, a loudspeaker at its center position and moving backward at $t = 0$ has $\phi_0 = 0$ rad. Looking back at Figure 21.16b, you can see that loudspeaker 1 has phase constant $\phi_{10} = 0$ rad and loudspeaker 2 has $\phi_{20} = \pi$ rad.

NOTE ▶ We will often consider *identical sources*, by which we mean that $\phi_{20} = \phi_{10}$. ◀

Let's examine overlapped waves graphically before diving into the mathematics. Figure 21.18 shows two important situations. In part a, the crests of the two waves are aligned as they travel along the *x*-axis. In part b, the crests of one wave align with the troughs of the other wave. The graphs and the wave fronts are slightly displaced from each other so that you can see what each wave is doing, but the *physical situation* is one in which the waves are traveling *on top of* each other. Recall, from Chapter 20, that the wave fronts shown in the middle panel locate the crests of the waves.

The two waves of Figure 21.18a have the same displacement at every point: $D_1(x) = D_2(x)$. Consequently, they must have the same phase. That is, $\phi_1 = \phi_2$

(a) Snapshot graph at $t = 0$ for $\phi_0 = 0$ rad

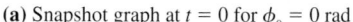
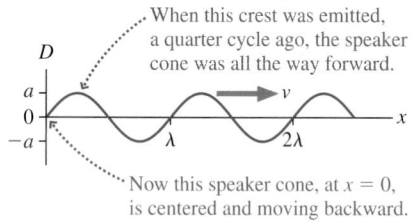

When this crest was emitted, a quarter cycle ago, the speaker cone was all the way forward.

Now this speaker cone, at $x = 0$, is centered and moving backward.

(b) Snapshot graph at $t = 0$ for $\phi_0 = \pi/2$ rad

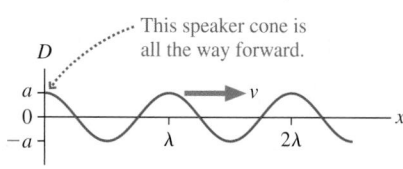

This speaker cone is all the way forward.

(c) Snapshot graph at $t = 0$ for $\phi_0 = \pi$ rad

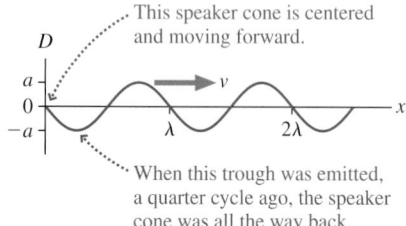

This speaker cone is centered and moving forward.

When this trough was emitted, a quarter cycle ago, the speaker cone was all the way back.

FIGURE 21.17 Waves from three sources having phase constants $\phi_0 = 0$ rad, $\phi_0 = \pi/2$ rad, and $\phi_0 = \pi$ rad.

or, more precisely, $\phi_1 = \phi_2 \pm 2\pi m$ where m is an integer. Two waves that are aligned crest-to-crest and trough-to-trough are said to be **in phase.** Waves that are in phase march along "in step" with each other.

When we combine two in-phase waves, using the principle of superposition, the net displacement at each point is twice the displacement of each individual wave. The superposition of two waves to create a wave with an amplitude *larger* than either individual wave is called **constructive interference.** When the waves are exactly in phase, giving $A = 2a$, we have *maximum constructive interference.*

In Figure 21.18b, where the crests of one wave align with the troughs of the other, the waves march along "out of step" with $D_1(x) = -D_2(x)$ at every point. Two waves that are aligned crest-to-trough are said to be *180° out of phase* or, more generally, just **out of phase.** A superposition of two waves that creates a wave with an amplitude smaller than either individual wave is called **destructive interference.** In this case, because $D_1 = -D_2$, the net displacement is *zero* at *every point* along the axis. The combination of two waves that cancel each other to give no wave is called *perfect destructive interference.*

> NOTE ▶ Perfect destructive interference occurs only if the two waves have equal amplitudes, as we're assuming. Two waves of unequal amplitudes can interfere destructively if they have equal wavelengths, but the cancellation won't be perfect. ◀

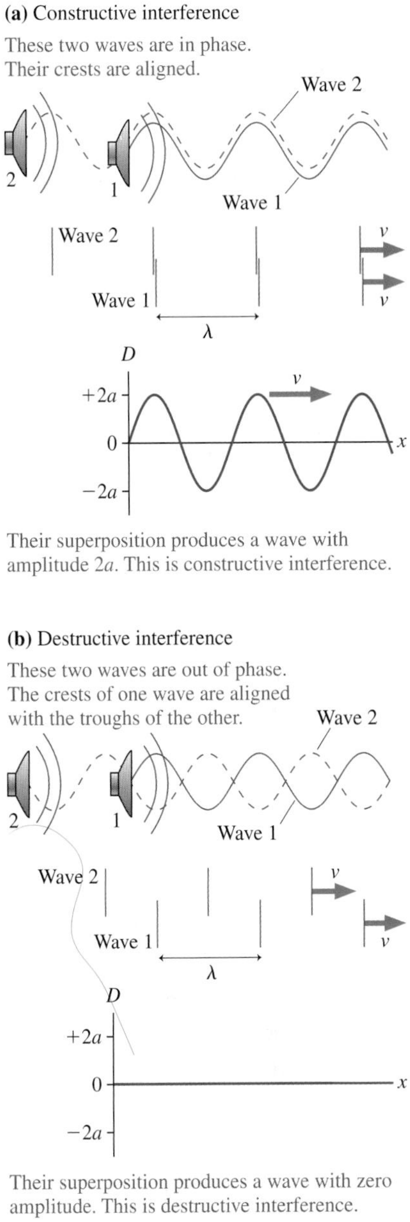

The Phase Difference

To understand interference, we need to focus on the *phases* of the two waves, which are

$$\phi_1 = kx_1 - \omega t + \phi_{10}$$
$$\phi_2 = kx_2 - \omega t + \phi_{20} \tag{21.20}$$

The difference between the two phases ϕ_1 and ϕ_2, called the **phase difference** $\Delta\phi$, is

$$\begin{aligned}\Delta\phi &= \phi_2 - \phi_1 = (kx_2 - \omega t + \phi_{20}) - (kx_1 - \omega t + \phi_{10}) \\ &= k(x_2 - x_1) + (\phi_{20} - \phi_{10}) \\ &= 2\pi\frac{\Delta x}{\lambda} + \Delta\phi_0\end{aligned} \tag{21.21}$$

You can see that there are two contributions to the phase difference. $\Delta x = x_2 - x_1$, the distance between the two sources, is called **path-length difference.** It is the extra distance traveled by wave 2 on the way to the point where the two waves are combined. $\Delta\phi_0 = \phi_{20} - \phi_{10}$ is the *inherent phase difference* between the sources. Although the individual phases are dependent on time, the time variable drops out of the phase difference.

The condition of being in phase, where crests are aligned with crests and troughs with troughs, is that $\Delta\phi = 0, 2\pi, 4\pi$, or any integer multiple of 2π. Thus the condition for maximum constructive interference is

Maximum constructive interference:

$$\Delta\phi = 2\pi\frac{\Delta x}{\lambda} + \Delta\phi_0 = 2m\pi \text{ rad} \qquad m = 0, 1, 2, 3, \ldots \tag{21.22}$$

For identical sources, which have $\Delta\phi_0 = 0$ rad, maximum constructive interference occurs when $\Delta x = m\lambda$. That is, **two identical sources produce maximum constructive interference when the path-length difference is an integer number of wavelengths.**

(a) Constructive interference

These two waves are in phase. Their crests are aligned.

Their superposition produces a wave with amplitude 2a. This is constructive interference.

(b) Destructive interference

These two waves are out of phase. The crests of one wave are aligned with the troughs of the other.

Their superposition produces a wave with zero amplitude. This is destructive interference.

FIGURE 21.18 Constructive and destructive interference of two waves traveling along the *x*-axis.

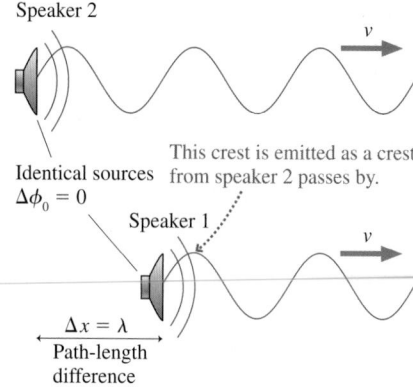

Speaker 2

v

This crest is emitted as a crest from speaker 2 passes by.

Identical sources
$\Delta\phi_0 = 0$

Speaker 1

v

$\Delta x = \lambda$
Path-length
difference

The two waves are in phase ($\Delta\phi = 2\pi$ rad) and interfere constructively.

FIGURE 21.19 Two identical sources one wavelength apart.

Figure 21.19 shows two identical sources (i.e., the two loudspeakers are doing the same thing at the same time), so $\Delta\phi_0 = 0$ rad. The path-length difference Δx is the extra distance traveled by the wave from loudspeaker 2 before it combines with loudspeaker 1. In this case, $\Delta x = \lambda$. Because a wave moves forward exactly one wavelength during a time interval of one period, loudspeaker 1 emits a crest exactly as a crest of wave 2 passes by. The two waves are "in step," with $\Delta\phi = 2\pi$ rad, so the two waves interfere constructively to produce a wave of amplitude $2a$.

Perfect destructive interference, where the crests of one wave are aligned with the troughs of the other, occurs when two waves are *out of phase,* meaning that $\Delta\phi = \pi$, 3π, 5π, or any odd multiple of π. Thus the condition for perfect destructive interference is

> **Perfect destructive interference:**
>
> $$\Delta\phi = 2\pi\frac{\Delta x}{\lambda} + \Delta\phi_0 = 2\left(m + \frac{1}{2}\right)\pi \text{ rad} \qquad m = 0, 1, 2, 3, \ldots$$
>
> (21.23)

For identical sources, which have $\Delta\phi_0 = 0$ rad, perfect destructive interference occurs when $\Delta x = (m + \frac{1}{2})\lambda$. **That is, two identical sources produce perfect destructive interference when the path-length difference is a half-integer number of wavelengths.**

Two waves can be out of phase because the sources are located at different positions, because the sources themselves are out of phase, or because of a combination of these two. Figure 21.20 illustrates these ideas by showing three different ways in which two waves interfere destructively. Each of these three arrangements creates waves with $\Delta\phi = \pi$ rad.

(a) The sources are out of phase.

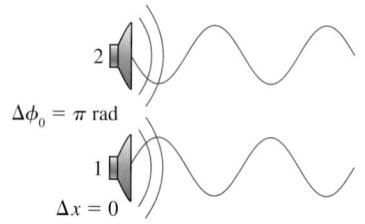

2

$\Delta\phi_0 = \pi$ rad

1

$\Delta x = 0$

(b) Identical sources are separated by half a wavelength.

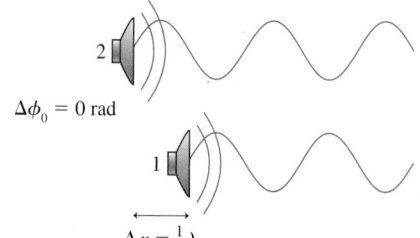

2

$\Delta\phi_0 = 0$ rad

1

$\Delta x = \frac{1}{2}\lambda$

(c) The sources are both separated and partially out of phase.

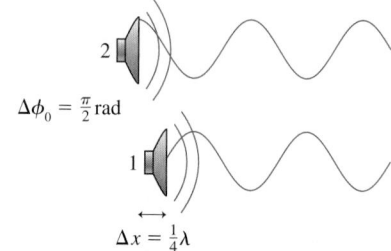

2

$\Delta\phi_0 = \frac{\pi}{2}$ rad

1

$\Delta x = \frac{1}{4}\lambda$

FIGURE 21.20 Destructive interference three ways.

NOTE ▶ Don't confuse the phase difference of the waves ($\Delta\phi$) with the phase difference of the sources ($\Delta\phi_0$). It is $\Delta\phi$, the phase difference of the waves, that governs interference. ◀

EXAMPLE 21.8 Interference between two sound waves
You are standing in front of two side-by-side loudspeakers playing sounds of the same frequency. Initially there is almost no sound at all. Then one of the speakers is moved slowly away from you. The sound intensity increases as the separation between the speakers increases, reaching a maximum when the speakers are 0.75 m apart. Then, as the speaker continues to move, the sound starts to decrease. What is the distance between the speakers when the sound intensity is again a minimum?

MODEL The changing sound intensity is due to the interference of two overlapped sound waves.

VISUALIZE Moving one speaker relative to the other, as in Figure 21.20, changes the phase difference between the waves.

SOLVE A minimum sound intensity implies that the two sound waves are interfering destructively. Initially the loudspeakers are side-by-side, so the situation is as shown in Figure 21.20a with $\Delta x = 0$ and $\Delta\phi_0 = \pi$ rad. That is, the speakers themselves are out of phase. Moving one of the speakers does not change $\Delta\phi_0$, but it does change the path-length difference Δx and thus increases the overall phase difference $\Delta\phi$. Constructive interference, causing maximum intensity, is reached when

$$\Delta\phi = 2\pi\frac{\Delta x}{\lambda} + \Delta\phi_0 = 2\pi\frac{\Delta x}{\lambda} + \pi = 2\pi \text{ rad}$$

where we used $m = 1$ because this is the first separation giving constructive interference. The speaker separation at which this occurs is $\Delta x = \lambda/2$. This is the situation shown in Figure 21.21.

Because $\Delta x = 0.75$ m is $\lambda/2$, the sound's wavelength is $\lambda = 1.50$ m. The next point of destructive interference, with $m = 1$, occurs when

$$\Delta\phi = 2\pi\frac{\Delta x}{\lambda} + \Delta\phi_0 = 2\pi\frac{\Delta x}{\lambda} + \pi = 3\pi \text{ rad}$$

Thus the distance between the speakers when the sound intensity is again a minimum is

$$\Delta x = \lambda = 1.50 \text{ m}$$

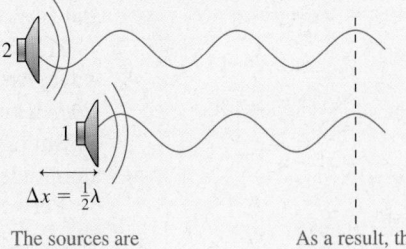

The sources are out of phase, $\Delta\phi_0 = \pi$ rad.

$\Delta x = \frac{1}{2}\lambda$

The sources are separated by half a wavelength.

As a result, the waves are in phase.

FIGURE 21.21 Two out-of-phase sources generate waves that are in phase if the sources are one half-wavelength apart.

ASSESS A separation of λ gives constructive interference for two *identical* speakers ($\Delta\phi_0 = 0$). Here the phase difference of π rad between the speakers (one is pushing forward as the other pulls back) gives destructive interference at this separation.

STOP TO THINK 21.4 Two loudspeakers emit waves with $\lambda = 2.0$ m. Speaker 2 is 1.0 m in front of speaker 1. What, if anything, must be done to cause constructive interference between the two waves?

a. Move speaker 1 forward (to the right) 1.0 m.
b. Move speaker 1 forward (to the right) 0.5 m.
c. Move speaker 1 backward (to the left) 0.5 m.
d. Move speaker 1 backward (to the left) 1.0 m.
e. Nothing. The situation shown already causes constructive interference.
f. Constructive interference is not possible for any placement of the speakers.

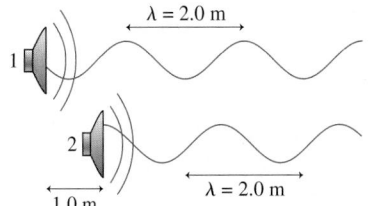

$\lambda = 2.0$ m

$\lambda = 2.0$ m

1.0 m

21.6 The Mathematics of Interference

Let's look more closely at the superposition of two waves . As two waves of equal amplitude and frequency travel together along the x-axis, the net displacement of the medium is

$$D = D_1 + D_2 = a\sin(kx_1 - \omega t + \phi_{10}) + a\sin(kx_2 - \omega t + \phi_{20})$$
$$= a\sin\phi_1 + a\sin\phi_2 \qquad (21.24)$$

where the phases ϕ_1 and ϕ_2 were defined in Equation 21.20.

A useful trigonometric identity is

$$\sin\alpha + \sin\beta = 2\cos\left[\frac{1}{2}(\alpha - \beta)\right]\sin\left[\frac{1}{2}(\alpha + \beta)\right] \qquad (21.25)$$

This identity is certainly not obvious, although it is easily proven by working backward from the right side. We can use this identity to write the net displacement of Equation 21.24 as

$$D = \left[2a\cos\frac{\Delta\phi}{2}\right]\sin(kx_{\text{avg}} - \omega t + (\phi_0)_{\text{avg}}) \qquad (21.26)$$

where $\Delta\phi = \phi_2 - \phi_1$ is the phase difference between the two waves, exactly as in Equation 21.21. $x_{avg} = (x_1 + x_2)/2$ is the average distance to the two sources and $(\phi_0)_{avg} = (\phi_{10} + \phi_{20})/2$ is the average phase constant of the sources.

The sine term shows that the superposition of the two waves is still a traveling wave. An observer would see a sinusoidal wave moving along the x-axis with the *same* wavelength and frequency as the original waves.

But how *big* is this wave compared to the two original waves? They each had amplitude a, but the amplitude of their superposition is

$$A = \left| 2a \cos\left(\frac{\Delta\phi}{2}\right) \right| \tag{21.27}$$

where we have used an absolute value sign because amplitudes must be positive. Depending upon the phase difference of the two waves, the amplitude of their superposition can be anywhere from zero (perfect destructive interference) to $2a$ (maximum constructive interference).

The amplitude has its maximum value $A = 2a$ if $\cos(\Delta\phi/2) = \pm 1$. This occurs when

$$\Delta\phi = 2m\pi \quad \text{(maximum amplitude } A = 2a) \tag{21.28}$$

where m is an integer. Similarly, the amplitude is zero if $\cos(\Delta\phi/2) = 0$, which occurs when

$$\Delta\phi = 2\left(m + \frac{1}{2}\right)\pi \quad \text{(minimum amplitude } A = 0) \tag{21.29}$$

Equations 21.28 and 21.29 are identical to the conditions of Equations 21.22 and 21.23 for constructive and destructive interference. We initially found these conditions by considering the alignment of the crests and troughs. Now we have confirmed them with an algebraic addition of the waves.

It is entirely possible, of course, that the two waves are neither exactly in phase nor exactly out of phase. Equation 21.27 allows us to calculate the amplitude of the superposition for any value of the phase difference. As an example, Figure 21.22 shows the calculated interference of two waves that differ in phase by 40°, by 90°, and by 160°.

For $\Delta\phi = 40°$, the interference is constructive but not maximum constructive.

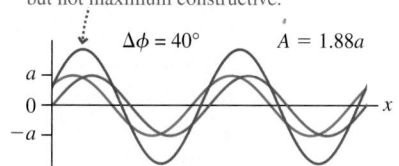

$\Delta\phi = 40°$ $A = 1.88a$

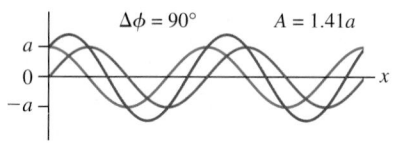

$\Delta\phi = 90°$ $A = 1.41a$

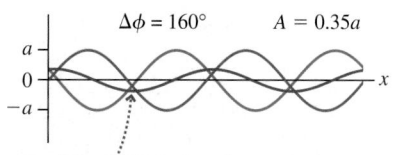

$\Delta\phi = 160°$ $A = 0.35a$

For $\Delta\phi = 160°$, the interference is destructive but not perfect destructive.

FIGURE 21.22 The interference of two waves for three different values of the phase difference.

EXAMPLE 21.9 More interference of sound waves

Two loudspeakers emit 500 Hz sound waves with an amplitude of 0.10 mm. Speaker 2 is 1.00 m behind speaker 1, and the phase difference between the speakers is 90°. What is the amplitude of the sound wave at a point 2.00 m in front of speaker 1?

MODEL The amplitude is determined by the interference of the two waves. Assume that the speed of sound has its room-temperature value of 343 m/s.

SOLVE The amplitude of the sound wave is

$$A = \left| 2a \cos(\Delta\phi/2) \right|$$

where $a = 0.10$ mm and the phase difference between the waves is

$$\Delta\phi = \phi_2 - \phi_1 = 2\pi\frac{\Delta x}{\lambda} + \Delta\phi_0$$

The sound's wavelength is

$$\lambda = \frac{v}{f} = \frac{343 \text{ m/s}}{500 \text{ Hz}} = 0.686 \text{ m}$$

Distances $x_1 = 2.0$ m and $x_2 = 3.0$ m are measured from the speakers, so the path-length difference is $\Delta x = 1.00$ m. We're given that the inherent phase difference between the speakers is $\Delta\phi_0 = \pi/2$ rad. Thus the phase difference at the observation point is

$$\Delta\phi = 2\pi\frac{\Delta x}{\lambda} + \Delta\phi_0 = 2\pi\frac{1.00 \text{ m}}{0.686 \text{ m}} + \frac{\pi}{2} \text{ rad} = 10.73 \text{ rad}$$

and the amplitude of the wave at this point is

$$A = \left| 2a \cos\left(\frac{\Delta\phi}{2}\right) \right| = (0.200 \text{ mm})\cos\left(\frac{10.73}{2}\right) = 0.121 \text{ mm}$$

ASSESS The interference is constructive, because $A > a$, but less than maximum constructive interference.

Application: Thin-Film Optical Coatings

Interference in one dimension has an important application in the optics industry, namely the use of **thin-film optical coatings.** These films, less than 1 μm (10^{-6} m) thick, are placed on glass surfaces, such as lenses, to control reflections from the glass. Dichroic glass, shown in the photograph that opens this chapter, is made with a thin-film coating on the glass. Thin films are also used for the antireflection coatings used on the lenses in cameras, microscopes, and other optical equipment.

Figure 21.23 shows a light wave of wavelength λ approaching a piece of glass that is coated with a transparent film whose index of refraction is n. The thickness d of this film has been greatly exaggerated in the figure. The air-film boundary is a discontinuity at which the wave speed suddenly decreases, and you saw earlier, in Figure 21.8, that a discontinuity causes a reflection. Most of the light is transmitted into the film, but a little bit is reflected.

Furthermore, you saw in Figure 21.8 that the wave reflected from a discontinuity at which the speed decreases is *inverted* with respect to the incident wave. For a sinusoidal wave, which we're now assuming, the inversion is represented mathematically as phase shift of π rad. The speed of a light wave decreases when it enters a material with a *larger* index of refraction. Thus **a light wave that reflects from a boundary at which the index of refraction increases has a phase shift of π rad.** There is no phase shift for the reflection from a boundary at which the index of refraction decreases. The reflection in Figure 21.23 is from a boundary between air ($n_{air} = 1.00$) and a transparent film with $n_{film} > n_{air}$, so the reflected wave is inverted due to the phase shift of π rad.

When the transmitted wave reaches the glass, most of it continues on into the glass but a portion is reflected back to the right. We'll assume that the index of refraction of the glass is larger than that of the film, $n_{glass} > n_{film}$, so this reflection also has a phase shift of π rad. This second reflection, after traveling back through the film, passes back into the air. There are now *two* equal-frequency waves traveling to the right, and these waves will interfere. If the two reflected waves are *in phase,* they will interfere constructively to cause a *strong reflection.* If the two reflected waves are *out of phase,* they will interfere destructively to cause a *weak reflection* or, if their amplitudes are equal, *no reflection* at all.

This suggests practical uses for thin-film optical coatings. The reflections from glass surfaces, even if weak, are often undesirable. For example, reflections degrade the performance of optical equipment. These reflections can be eliminated by coating the glass with a film whose thickness is chosen to cause *destructive* interference of the two reflected waves. This is called an *antireflection coating.*

Dichroic glass is created by applying a coating that causes strong *constructive* interference for one wavelength or one band of wavelengths. This produces the color you see in reflected light. To conserve energy, any light that is strongly reflected must be absent from the light that is transmitted through the glass. Consequently, the transmitted light has the complementary color—white light minus the reflected color.

The phase difference between the two reflected waves is

$$\Delta\phi = \phi_2 - \phi_1 = (kx_2 + \phi_{20} + \pi \text{ rad}) - (kx_1 + \phi_{10} + \pi \text{ rad})$$
$$= 2\pi\frac{\Delta x}{\lambda_f} - \Delta\phi_0 \tag{21.30}$$

where we explicitly included the reflection phase shift of each wave. In this case, because *both* waves had a phase shift of π rad, the reflection phase shifts cancel.

The wavelength λ_f is the wavelength *in the film* because that's where the path-length difference Δx occurs. You learned in Chapter 20 that the wavelength in a

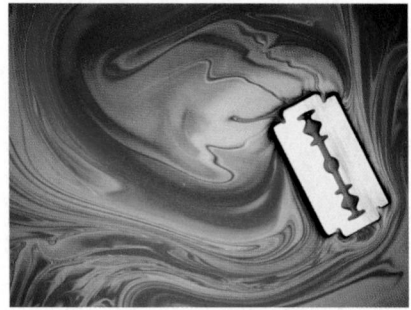

The colors of an oil slick are due to thin-film interference. Different wavelengths of light, and thus different colors, are strongly reflected as the thickness of the oil film slowly changes.

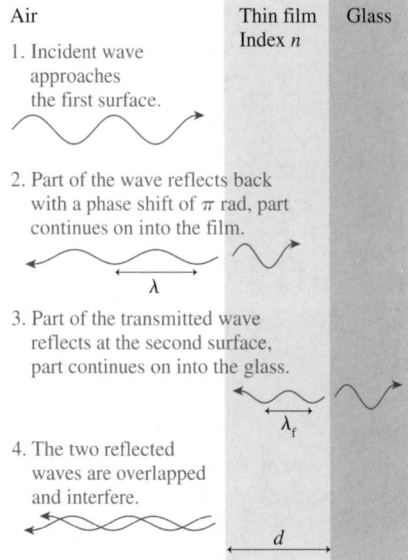

FIGURE 21.23 The two reflections, one from the coating and one from the glass, interfere.

transparent material with index of refraction n is $\lambda_f = \lambda/n$, where the unsubscripted λ is the wavelength in vacuum or air. That is, λ is the wavelength that we measure on "our" side of the air-film boundary.

The path-length difference between the two waves is $\Delta x = 2d$ because wave 2 travels through the film *twice* before rejoining wave 1. The two waves have a common origin—the initial division of the incident wave at the front surface of the film—so the inherent phase difference is $\Delta\phi_0 = 0$. Thus the phase difference of the two reflected waves is

$$\Delta\phi = 2\pi\frac{2d}{\lambda/n} = 2\pi\frac{2nd}{\lambda} \tag{21.31}$$

The interference is constructive, causing a strong reflection, when $\Delta\phi = 2m\pi$ rad. Constructive interference occurs for wavelengths

$$\lambda_C = \frac{2nd}{m} \qquad m = 1, 2, 3, \ldots \qquad \text{(constructive interference)} \tag{21.32}$$

You will notice that m starts with 1, rather than 0, in order to give meaningful results. Destructive interference, with minimum reflection, requires $\Delta\phi = 2\left(m - \frac{1}{2}\right)\pi$ rad. This occurs for wavelengths

$$\lambda_D = \frac{2nd}{m - \frac{1}{2}} \qquad m = 1, 2, 3, \ldots \qquad \text{(destructive interference)} \tag{21.33}$$

We've used $m - \frac{1}{2}$, rather than $m + \frac{1}{2}$, so that m can start with 1 to match the condition for constructive interference.

NOTE ▶ The exact condition for constructive or destructive interference is satisfied for only a few discrete wavelengths λ. Nonetheless, reflections are strongly enhanced (nearly constructive interference) for a range of wavelengths near λ_C. Likewise, there is a range of wavelengths near λ_D for which the reflection is nearly canceled. ◀

EXAMPLE 21.10 Designing an antireflection coating
Magnesium fluoride (MgF_2) is often used as an antireflection coating on lenses. The index of refraction of MgF_2 is 1.39. What is the thinnest film of MgF_2 that works as an antireflection coating at $\lambda = 510$ nm, near the center of the visible spectrum?

MODEL Reflection is minimized if the two reflected waves interfere destructively.

SOLVE The film thicknesses that cause destructive interference at wavelength λ are

$$d = \left(m - \frac{1}{2}\right)\frac{\lambda}{2n}$$

The thinnest film has $m = 1$. Its thickness is

$$d = \frac{\lambda}{4n} = \frac{510 \text{ nm}}{4(1.39)} = 92 \text{ nm}$$

The film thickness is significantly less than the wavelength of visible light!

ASSESS The reflected light is completely eliminated (perfect destructive interference) only if the two reflected waves have equal amplitudes. In practice, they don't. Nonetheless, the reflection is reduced from $\approx 4\%$ of the incident intensity for "bare glass" to well under 1%. Furthermore, the intensity of reflected light is much reduced across most of the visible spectrum (400–700 nm), even though the phase difference deviates more and more from π rad as the wavelength moves away from 510 nm. It is the increasing reflection at the ends of the visible spectrum ($\lambda \approx 400$ nm and $\lambda \approx 700$ nm), where $\Delta\phi$ deviates significantly from π rad, that gives a reddish-purple tinge to the lenses on cameras and binoculars.

Homework problems will let you "design" a piece of dichroic glass and explore situations where only one of the two reflections has a reflection phase shift of π rad.

21.7 Interference in Two and Three Dimensions

Ripples on a lake move in two dimensions. The glow from a light bulb spreads outward as a spherical wave. In Section 20.4, we noted that a circular or spherical wave can be written

$$D(r, t) = a\sin(kr - \omega t + \phi_0) \tag{21.34}$$

where r is the distance measured outward from the source. Equation 21.34 is our familiar wave equation with the one-dimensional coordinate x replaced by a more general radial coordinate r. Strictly speaking, the amplitude a of a circular or spherical wave diminishes as r increases. However, we will assume that a remains essentially constant over the region in which we study the wave. Figure 21.24 shows the wave-front diagram for a circular or spherical wave. Recall that the wave fronts represent the *crests* of the wave and are spaced by the wavelength λ.

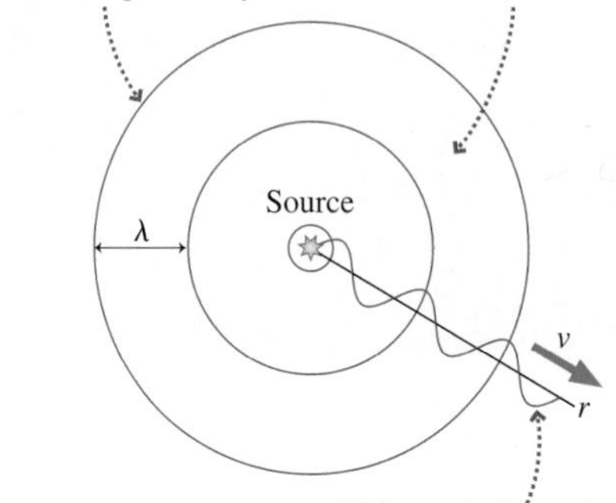

What happens when two circular or spherical waves overlap? For example, imagine two paddles oscillating up and down on the surface of a pond. We will assume that the two paddles oscillate with the same frequency and amplitude and that they are in phase. Figure 21.25 shows the wave fronts of the two waves. The ripples overlap as they travel, and, as was the case in one dimension, this causes interference.

Constructive interference with $A = 2a$ occurs where two crests align or two troughs align. Several locations of constructive interference are marked in Figure 21.25. Intersecting wave fronts are points where two crests are aligned. It's a bit harder to visualize, but two troughs are aligned when a midpoint between two wave fronts is overlapped with another midpoint between two wave fronts. Destructive interference with $A = 0$ occurs where the crest of one wave aligns with a trough of the other wave. Several points of destructive interference are also indicated in Figure 21.25.

A picture on a page is static, but the wave fronts are in motion. Try to imagine the wave fronts of Figure 21.25 expanding outward as new circular rings are born at the sources. The waves will move forward half a wavelength during half a period, causing the crests in Figure 21.25 to be replaced by troughs while the troughs become crests.

The important point to recognize is that **the motion of the waves does not affect the points of constructive and destructive interference.** Points in the figure where two crests overlap will become points where two troughs overlap, but this overlap is still constructive interference. Similarly, points in the figure where a crest and a trough overlap will become a point where a trough and a crest overlap—still destructive interference.

The mathematical description of interference in two or three dimensions is very similar to that of one-dimensional interference. The net displacement of a particle in the medium is

$$D = D_1 + D_2 = a\sin(kr_1 - \omega t + \phi_{10}) + a\sin(kr_2 - \omega t + \phi_{20}) \tag{21.35}$$

The only difference between Equation 21.35 and the earlier one-dimensional Equation 21.24 is that the linear coordinates x_1 and x_2 have been changed to radial coordinates r_1 and r_2. Thus our conclusions are unchanged. The superposition of the two waves yields a wave traveling outward with amplitude

$$A = \left| 2a\cos\left(\frac{\Delta\phi}{2}\right) \right| \tag{21.36}$$

where the phase difference, with x replaced by r, is now

$$\Delta\phi = 2\pi\frac{\Delta r}{\lambda} + \Delta\phi_0 \tag{21.37}$$

The wave fronts are crests, separated by λ.

Troughs are halfway between wave fronts.

Source

λ

v

r

This graph shows the displacement of the medium.

FIGURE 21.24 A circular or spherical wave.

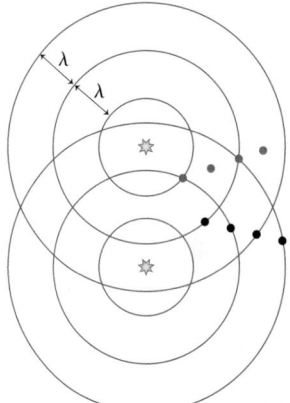

Two in-phase sources emit circular or spherical waves.

λ

λ

- Points of constructive interference. A crest is aligned with a crest, or a trough with a trough.

- Points of destructive interference. A crest is aligned with a trough of another wave.

FIGURE 21.25 The overlapping ripple patterns of two sources. A few points of constructive and destructive interference are noted.

Two overlapping water waves create an interference pattern.

The term $2\pi(\Delta r/\lambda)$ is the phase difference that arises when the waves travel different distances from the sources to the point at which they combine. Δr itself is the *path-length difference*. As before, $\Delta\phi_0$ is any inherent phase difference of the sources themselves.

Maximum constructive interference with $A = 2a$ occurs, just as in one dimension, at those points where $\cos(\Delta\phi/2) = \pm1$. Similarly, perfect destructive interference occurs at points where $\cos(\Delta\phi/2) = 0$. The conditions for constructive and destructive interference are

Maximum constructive interference:
$$\Delta\phi = 2\pi\frac{\Delta r}{\lambda} + \Delta\phi_0 = 2m\pi$$

Perfect destructive interference:
$$\Delta\phi = 2\pi\frac{\Delta r}{\lambda} + \Delta\phi_0 = 2\left(m + \frac{1}{2}\right)\pi$$

$$m = 0, 1, 2, \ldots \quad (21.38)$$

For two identical sources (i.e., sources that oscillate in phase with $\Delta\phi_0 = 0$), the conditions for constructive and destructive interference are very simple:

Constructive: $\Delta r = m\lambda$

Destructive: $\Delta r = \left(m + \dfrac{1}{2}\right)\lambda$

(identical sources)

$$(21.39)$$

The waves from two identical sources interfere constructively at points where the path-length difference is an integer number of wavelengths because, for these values of Δr, crests are aligned with crests and troughs with troughs. **The waves interfere destructively where the path-length difference is a half-integer number of wavelengths** because, for these values of Δr, crests are aligned with troughs. These two statements are the essence of interference.

NOTE ▶ Equation 21.39 applies only if the sources are in phase. If the sources are not in phase, you must use the more general Equation 21.38 to locate the points of constructive and destructive interference. ◀

Wave fronts are spaced exactly one wavelength apart, hence we can measure the distances r_1 and r_2 simply by counting the rings in the wave-front pattern. In Figure 21.26, which is based on Figure 21.25, Point A is distance $r_1 = 3\lambda$ from the first source and $r_2 = 2\lambda$ from the second. The path-length difference is $\Delta r_A = 1\lambda$, the condition for the maximum constructive interference of identical sources. Point B, on the other hand, has $\Delta r_B = \frac{1}{2}\lambda$, so it is a point of perfect destructive interference.

NOTE ▶ Interference is determined by Δr, the path-length *difference*, rather than by r_1 or r_2. ◀

• At A, $\Delta r_A = \lambda$, so this is a point of constructive interference.

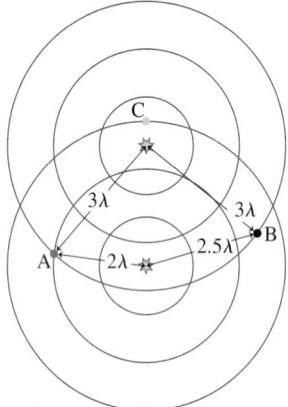

• At B, $\Delta r_B = \frac{1}{2}\lambda$, so this is a point of destructive interference.

FIGURE 21.26 The path-length difference Δr determines whether the interference at a particular point is constructive or destructive.

STOP TO THINK 21.5 The interference at point C in Figure 21.26 is

a. Maximum constructive.   b. Constructive, but less than maximum.
c. Perfect destructive.   d. Destructive, but not perfect.
e. There is no interference at point C.

We can now locate the points of maximum constructive interference, for which $\Delta r = m\lambda$, by drawing a line through *all* the points at which $\Delta r = 0$, another line through all the points at which $\Delta r = \lambda$, and so on. These lines, shown as red lines in Figure 21.27, are called **antinodal lines.** They are analogous to the antinodes of a standing wave, hence the name. An antinode is a *point* of maximum constructive interference; for circular waves, oscillation at maximum amplitude occurs along a continuous *line*. Similarly, destructive interference occurs along lines called **nodal lines.** The displacement is *always zero* along these lines, just as it is at a node in a standing-wave pattern.

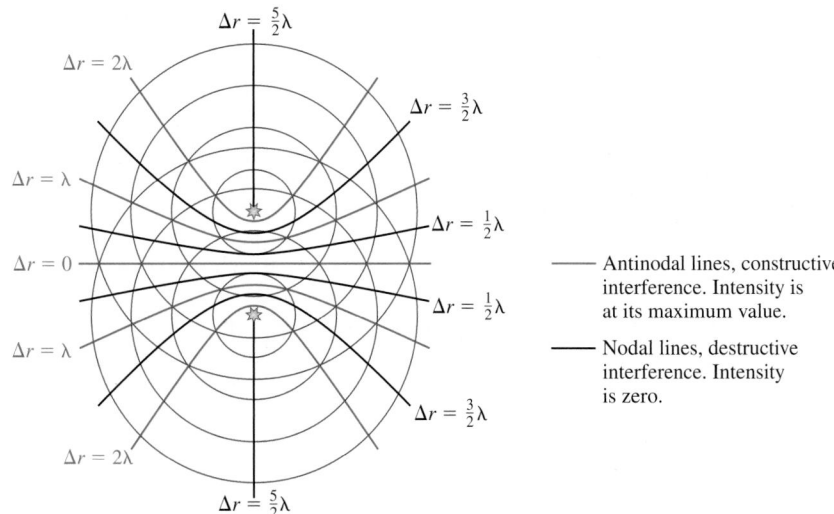

FIGURE 21.27 The points of constructive and destructive interference fall along antinodal and nodal lines.

A Problem-Solving Strategy for Interference Problems

The information in this section is the basis of a strategy for solving interference problems. This strategy applies equally well to interference in one dimension if you use Δx instead of Δr.

(MP) PROBLEM-SOLVING STRATEGY 21.1 **Interference of two waves**

MODEL Make simplifying assumptions, such as assuming waves are circular and of equal amplitude.

VISUALIZE Draw a picture showing the sources of the waves and the point where the waves interfere. Give relevant dimensions. Identify the distances r_1 and r_2 from the sources to the point. Note any phase difference $\Delta\phi_0$ between the two sources.

SOLVE The interference depends on the path-length difference $\Delta r = r_2 - r_1$ and the source phase difference $\Delta\phi_0$.

Constructive: $\quad \Delta\phi = 2\pi\dfrac{\Delta r}{\lambda} + \Delta\phi_0 = 2m\pi$

Destructive: $\quad \Delta\phi = 2\pi\dfrac{\Delta r}{\lambda} + \Delta\phi_0 = 2\left(m + \dfrac{1}{2}\right)\pi$

$$m = 0, 1, 2, \ldots$$

For identical sources ($\Delta\phi_0 = 0$), the interference is constructive if $\Delta r = m\lambda$, destructive if $\Delta r = (m + \frac{1}{2})\lambda$.

ASSESS Check that your result has the correct units, is reasonable, and answers the question.

EXAMPLE 21.11 Two-dimensional interference between two loudspeakers

Two loudspeakers are 2.0 m apart and in phase with each other. Both emit 700 Hz sound waves into a room where the speed of sound is 341 m/s. A listener stands 5.0 m in front of the loudspeakers and 2.0 m to one side of the center. Is the interference at this point constructive, destructive, or something in between? How will the situation differ if the loudspeakers are out of phase?

MODEL The two speakers are sources of in-phase, circular waves. The overlap of these waves causes interference.

VISUALIZE Figure 21.28 shows the loudspeakers and defines the distances r_1 and r_2 to the point of observation. The figure includes dimensions and notes that $\Delta\phi_0 = 0$ rad.

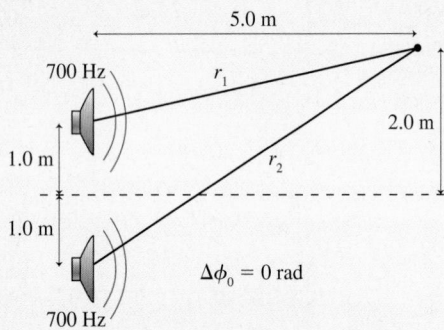

FIGURE 21.28 Pictorial representation of the interference between two loudspeakers.

SOLVE It's not r_1 or r_2 that matter, but the *difference* Δr between them. From the geometry of the figure we can calculate that

$$r_1 = \sqrt{(5.0 \text{ m})^2 + (1.0 \text{ m})^2} = 5.10 \text{ m}$$
$$r_2 = \sqrt{(5.0 \text{ m})^2 + (3.0 \text{ m})^2} = 5.83 \text{ m}$$

Thus the path-length difference is $\Delta r = r_2 - r_1 = 0.73$ m. The wavelength of the sound waves is

$$\lambda = \frac{v}{f} = \frac{341 \text{ m/s}}{700 \text{ Hz}} = 0.487 \text{ m}$$

In terms of wavelengths, the path-length difference is $\Delta r/\lambda = 1.50$, or

$$\Delta r = \frac{3}{2}\lambda$$

Because the sources are in phase ($\Delta\phi_0 = 0$), this is the condition for *destructive* interference. If the sources were out of phase ($\Delta\phi_0 = \pi$ rad), then the phase difference of the waves at the listener would be

$$\Delta\phi = 2\pi\frac{\Delta r}{\lambda} + \Delta\phi_0 = 2\pi\left(\frac{3}{2}\right) + \pi \text{ rad} = 4\pi \text{ rad}$$

This is an integer multiple of 2π rad, so in this case the interference would be *constructive*.

ASSESS Both the path-length difference *and* any inherent phase difference of the sources must be considered when evaluating interference.

Picturing Interference

A *contour map* is a useful way to visualize an interference pattern. Figure 21.29a shows the superposition of the waves from two identical sources ($\Delta\phi_0 = 0$) emitting waves with $\lambda = 1$ m. The sources, indicated with black dots, are located at $y = \pm 1$ m. Positive displacements are shown in red, with the deepest red representing the maximum displacement of the wave at this instant in time. These are the points where the crests of the individual waves interfere constructively to give $D = 2a$. Negative displacements are blue, with the darkest blue being the most negative displacement of the wave. These are also points of constructive interference, with two troughs overlapping to give $D = -2a$.

To understand this figure, try to visualize the waves expanding outward from center. The red-blue-red-blue-red-. . . pattern of crests and troughs moves outward along the antinodal lines as a *traveling wave* of amplitude $A = 2a$. Nothing ever happens along the nodal lines, where the amplitude is always zero. Between the nodal and antinodal lines are traveling waves whose amplitude varies between 0 and $2a$.

Suppose you were to observe the *intensity* of the wave as it crosses the vertical line at $x = 4$ m on the right edge of the figure. If, for example, these are sound waves, you could listen to (or measure, with a microphone) the sound intensity as you walk from $(x, y) = (4 \text{ m}, -4 \text{ m})$ at the bottom of the figure to $(x, y) = (4 \text{ m}, 4 \text{ m})$ at the top. The intensity is zero as you cross the nodal lines at $y \approx \pm 1$ m ($\Delta r = \frac{1}{2}\lambda$). The intensity is maximum at the antinodal lines at

(a) Two identical sources

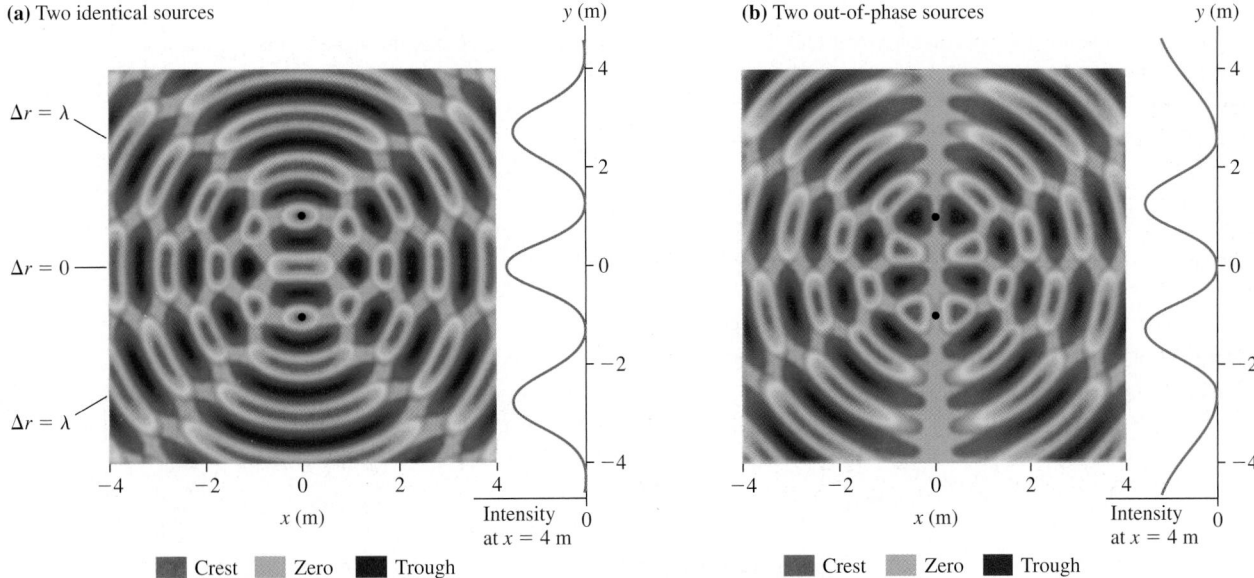

FIGURE 21.29 A contour map of the interference pattern of two sources. The graph on the right side of each figure shows the wave intensity along a vertical line at $x = 4$ m.

$y = 0$ ($\Delta r = 0$) and $y \approx \pm 2.5$ m ($\Delta r = \lambda$), where a wave of maximum amplitude streams out from the sources.

The intensity is shown in the rather unusual graph on the right side of Figure 21.29a. It is unusual in the sense that the intensity, the quantity of interest, is graphed to the left. The peaks are the points of constructive interference, where you would measure maximum amplitude. The zeros are points of destructive interference, where the intensity is zero.

Figure 21.29b is a contour map of the interference pattern produced by the same two sources but with the sources themselves now out of phase ($\Delta\phi_0 = \pi$ rad). We'll leave the investigation of this figure for you to study, but notice that the nodal and antinodal lines are reversed from those of Figure 21.29a.

EXAMPLE 21.12 The intensity of two interfering loudspeakers

Two loudspeakers are 6.0 m apart and in phase. They emit equal-amplitude sound waves with a wavelength of 1.0 m. Each speaker alone creates sound with intensity I_0. An observer at point A is 10 m in front of the plane containing the two loudspeakers and centered between them. A second observer at point B is 10 m directly in front of one of the speakers. In terms of I_0, what is the intensity I_A at point A and the intensity I_B at point B?

MODEL The two speakers are sources of in-phase, circular waves. The overlap of these waves causes interference.

VISUALIZE Figure 21.30 shows the two loudspeakers and the two points of observation. Distances r_1 and r_2 are defined for point B.

SOLVE Let the amplitude of the wave from each speaker be a. The intensity of a wave is proportional to the square of the amplitude, so the intensity of each speaker alone is $I_0 = ca^2$,

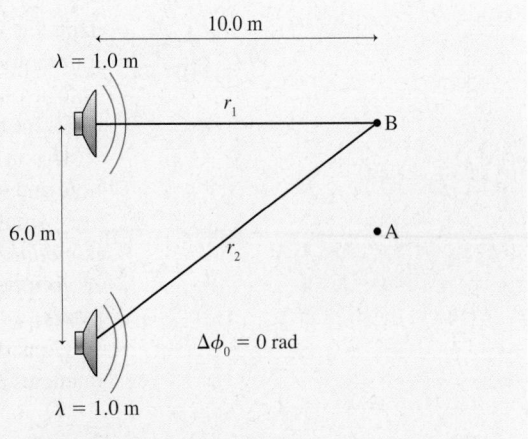

FIGURE 21.30 Pictorial representation of the interference between two loudspeakers.

where c is an unknown proportionality constant. Point A is a point of constructive interference because the speakers are in phase ($\Delta\phi_0 = 0$) and the path-length difference is $\Delta r = 0$. The amplitude at this point is given by Equation 21.36:

$$A_A = \left| 2a\cos\left(\frac{\Delta\phi}{2}\right) \right| = 2a\cos(0) = 2a$$

Consequently, the intensity at this point is

$$I_A = cA_A^2 = c(2a)^2 = 4ca^2 = 4I_0$$

The intensity at A is four times that of either speaker played alone. At point B, the path-length difference is

$$\Delta r = \sqrt{(10.0 \text{ m})^2 + (6.0 \text{ m})^2} - 10.0 \text{ m} = 1.662 \text{ m}$$

The phase difference of the waves at this point is

$$\Delta\phi = 2\pi\frac{\Delta r}{\lambda} = 2\pi\frac{1.662 \text{ m}}{1.0 \text{ m}} = 10.44 \text{ rad}$$

Consequently, the amplitude at B is

$$A_B = \left| 2a\cos\left(\frac{\Delta\phi}{2}\right) \right| = |2a\cos(5.22 \text{ rad})| = 0.972a$$

Thus the intensity at this point is

$$I_B = cA_B^2 = c(0.972a)^2 = 0.95ca^2 = 0.95I_0$$

ASSESS Although B is directly in front of one of the speakers, superposition of the two waves results in an intensity that is less than it would be if this speaker played alone.

STOP TO THINK 21.6 These two loudspeakers are in phase. They emit equal-amplitude sound waves with a wavelength of 1.0 m. At the point indicated, is the interference maximum constructive, perfect destructive, or something in between?

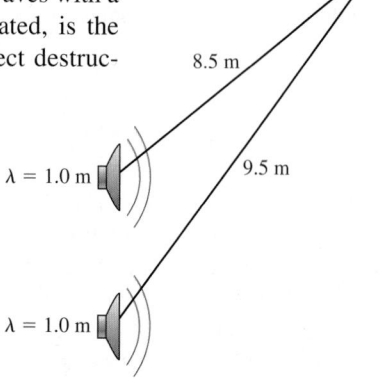

8.5 m

9.5 m

$\lambda = 1.0$ m

$\lambda = 1.0$ m

21.8 Beats

10.7   Activ Physics

Thus far we have looked at the superposition of sources having the same wavelength and frequency. We can also use the principle of superposition to investigate a phenomenon that is easily demonstrated with two sources of slightly different frequency.

If you listen to two sounds with very different frequencies, such as a high note and a low note, you hear two distinct tones. But if the frequency difference is very small, just one or two hertz, then you hear a single tone whose intensity is *modulated* once or twice every second. That is, the sound goes up and down in volume, loud, soft, loud, soft, . . . , making a distinctive sound pattern called **beats.**

Consider two sinusoidal waves traveling along the x-axis with angular frequencies $\omega_1 = 2\pi f_1$ and $\omega_2 = 2\pi f_2$. The two waves are

$$D_1 = a\sin(k_1 x - \omega_1 t + \phi_{10})$$
$$D_2 = a\sin(k_2 x - \omega_2 t + \phi_{20})$$

(21.40)

where the subscripts 1 and 2 indicate that the frequencies, wave numbers, and phase constants of the two waves may be different.

To simplify the analysis, let's make several assumptions:

1. The two waves have the same amplitude a,
2. A detector, such as your ear, is located at the origin ($x = 0$),
3. The two sources are in phase ($\phi_{10} = \phi_{20}$), and
4. The source phases happen to be $\phi_{10} = \phi_{20} = \pi$ rad.

None of these assumptions is essential to the outcome. All could be otherwise and we would still come to basically the same conclusion, but the mathematics would be far messier. Making these assumptions allows us to emphasize the physics with the least amount of mathematics.

With these assumptions, the two waves as they reach the detector at $x = 0$ are

$$D_1 = a\sin(-\omega_1 t + \pi) = a\sin\omega_1 t$$
$$D_2 = a\sin(-\omega_2 t + \pi) = a\sin\omega_2 t \tag{21.41}$$

where we've used the trigonometric identity $\sin(\pi - \theta) = \sin\theta$. The principle of superposition tells us that the *net* displacement of the medium at the detector is the sum of the displacements of the individual waves. Thus

$$D = D_1 + D_2 = a(\sin\omega_1 t + \sin\omega_2 t) \tag{21.42}$$

Earlier, in our analysis of the mathematics of interference, we used the trigonometric identity

$$\sin\alpha + \sin\beta = 2\cos\left[\frac{1}{2}(\alpha - \beta)\right]\sin\left[\frac{1}{2}(\alpha + \beta)\right]$$

We can use this identity again to write Equation 21.42 as

$$D = 2a\cos\left[\frac{1}{2}(\omega_1 - \omega_2)t\right]\sin\left[\frac{1}{2}(\omega_1 + \omega_2)t\right]$$
$$= \left[2a\cos(\omega_{\text{mod}}t)\right]\sin(\omega_{\text{avg}}t) \tag{21.43}$$

where $\omega_{\text{avg}} = \frac{1}{2}(\omega_1 + \omega_2)$ is the *average* angular frequency and $\omega_{\text{mod}} = \frac{1}{2}(\omega_1 - \omega_2)$ is called the *modulation frequency*.

We are interested in the situation when the two frequencies are very nearly equal: $\omega_1 \approx \omega_2$. In that case, ω_{avg} hardly differs from either ω_1 or ω_2 while ω_{mod} is very near to—but not exactly—zero. When ω_{mod} is very small, the term $\cos(\omega_{\text{mod}}t)$ oscillates *very* slowly. We have grouped it with the $2a$ term because, together, they provide a slowly changing "amplitude" for the rapid oscillation at frequency ω_{avg}.

Figure 21.31 is a history graph of the wave at the detector ($x = 0$). It shows the oscillation of the air against your ear drum at frequency $f_{\text{avg}} = \omega_{\text{avg}}/2\pi = \frac{1}{2}(f_1 + f_2)$. This oscillation determines the note you hear; it differs little from the two notes at frequencies f_1 and f_2. We are especially interested in the time-dependent amplitude, shown as a dotted line, that is given by the term $2a\cos(\omega_{\text{mod}}t)$. This periodically varying amplitude is called a **modulation** of the wave, which is where ω_{mod} gets its name.

As the amplitude rises and falls, the sound alternates as loud, soft, loud, soft, and so on. But that is exactly what you hear when you listen to beats! The alternating loud and soft sounds arise from the two waves being alternately in phase and out of phase, causing constructive and then destructive interference.

Imagine two people walking side-by-side at just slightly different paces. Initially both of their right feet hit the ground together, but after a while they get out of step. A little bit later they are back in step and the process alternates. The sound waves are doing the same. Initially the crests of each wave, of amplitude a, arrive together at your ear and the net displacement is doubled to $2a$. But after a while the two waves, being of slightly different frequency, get out of step and a crest of

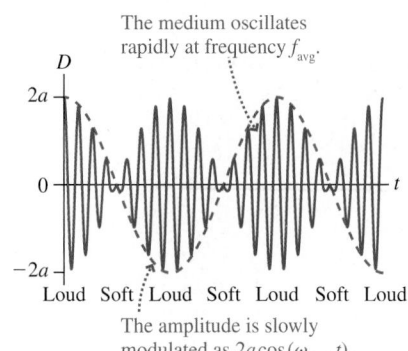

The medium oscillates rapidly at frequency f_{avg}.

Loud Soft Loud Soft Loud Soft Loud

The amplitude is slowly modulated as $2a\cos(\omega_{\text{mod}}t)$.

FIGURE 21.31 Beats are caused by the superposition of two waves of nearly identical frequency.

one arrives with a trough of the other. When this happens, the two waves cancel each other to give a net displacement of zero. This process alternates over and over, loud and soft.

Notice, from the figure, that the sound intensity rises and falls *twice* during one cycle of the modulation envelope. Each "loud-soft-loud" is one beat, so the **beat frequency** f_{beat}, which is the number of beats per second, is *twice* the modulation frequency $f_{mod} = \omega_{mod}/2\pi$. From the above definition of ω_{mod}, the beat frequency is

$$f_{beat} = 2f_{mod} = 2\frac{\omega_{mod}}{2\pi} = 2 \cdot \frac{1}{2}\left(\frac{\omega_1}{2\pi} - \frac{\omega_2}{2\pi}\right) = f_1 - f_2 \qquad (21.44)$$

where, to keep f_{beat} from being negative, we will always let f_1 be the larger of the two frequencies. The beat frequency is simply the *difference* between the two individual frequencies.

EXAMPLE 21.13 Listening to beats

One flutist plays a note of 510 Hz while a second plays a note of 512 Hz. What frequency do you hear? What is the beat frequency?

SOLVE You hear a note with frequency

$$f_{avg} = 511 \text{ Hz}$$

The beat frequency is

$$f_{beat} = f_1 - f_2 = 2 \text{ Hz}$$

You (and they) would hear two beats per second.

ASSESS If a 510 Hz note and a 512 Hz note were played separately, you would not be able to perceive the slight difference in frequency. But when the two notes are played together, the quite obvious beats tell you that the frequencies are slightly different. Musicians learn to make constant minor adjustments in their tuning as they play in order to eliminate beats between themselves and other players.

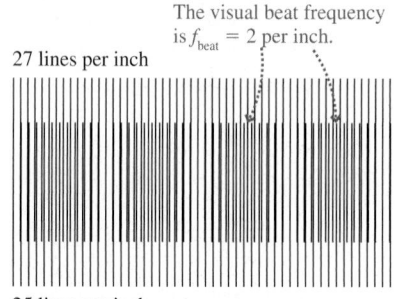

27 lines per inch

The visual beat frequency is f_{beat} = 2 per inch.

25 lines per inch

FIGURE 21.32 A graphical example of beats.

Beats aren't limited to sound waves. Figure 21.32 shows a graphical example of beats. Two "fences" of slightly different frequencies—25 lines per inch and 27 lines per inch—are superimposed on each other. The density of the lines varies as they alternate in step and out of step, giving a visual "loud/soft" alternation. The difference in the two frequencies is two lines per inch. You can confirm, with a ruler, that the figure has two "beats" per inch, in agreement with Equation 21.44.

Beats are important in many other situations. For example, you have probably seen movies where rotating wheels seem to turn slowly backward. Why is this? Suppose the movie camera is shooting at 30 frames per second but the wheel is rotating 32 times per second. The combination of the two produces a "beat" of 2 Hz, meaning that the wheel appears to rotate only twice per second. The same is true if the wheel is rotating 28 times per second, but in this case, where the wheel frequency slightly lags the camera frequency, it appears to rotate *backward* twice per second!

STOP TO THINK 21.7 You hear three beats per second when two sound tones are generated. The frequency of one tone is known to be 610 Hz. The frequency of the other is

a. 604 Hz.   b. 607 Hz.
c. 613 Hz.   d. 616 Hz.
e. Either a or d.   f. Either b or c.

SUMMARY

The goal of Chapter 21 has been to understand and use the idea of superposition.

GENERAL PRINCIPLES

Principle of Superposition

The displacement of a medium when more than one wave is present is the sum of the displacements due to each individual wave.

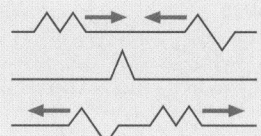

IMPORTANT CONCEPTS

Standing waves are due to the superposition of two traveling waves moving in opposite directions.

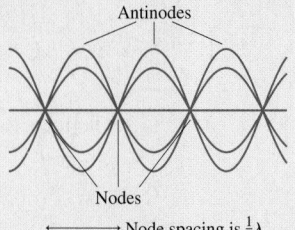

Antinodes

Nodes

Node spacing is $\frac{1}{2}\lambda$

The amplitude at position x is

$$A(x) = 2a \sin kx$$

where a is the amplitude of each wave.

The boundary conditions determine which standing wave frequencies and wavelengths are allowed.

Interference

In general, the superposition of two or more waves into a single wave is called interference.

Maximum constructive interference occurs where crests are aligned with crests and troughs with troughs. These waves are in phase. The maximum displacement is $A = 2a$.

Perfect destructive interference occurs where crests are aligned with troughs. These waves are out of phase. The amplitude is $A = 0$.

Interference depends on the phase difference $\Delta\phi$ between the two waves.

Constructive: $\Delta\phi = 2\pi\dfrac{\Delta r}{\lambda} + \Delta\phi_0 = 2m\pi$

Destructive: $\Delta\phi = 2\pi\dfrac{\Delta r}{\lambda} + \Delta\phi_0 = 2(m + \frac{1}{2})\pi$

Antinodal lines, constructive interference. $A = 2a$

Nodal lines, destructive interference. $A = 0$

Δr is the path-length difference of the two waves and $\Delta\phi_0$ is any phase difference between the sources. For identical sources (in phase, $\Delta\phi_0 = 0$):

Interference is constructive if the path-length difference $\Delta r = m\lambda$.

Interference is destructive if the path-length difference $\Delta r = (m + \frac{1}{2})\lambda$.

The amplitude at a point where the phase difference is $\Delta\phi$ is $A = \left| 2a \cos\left(\dfrac{\Delta\phi}{2}\right) \right|$

APPLICATIONS

Boundary conditions

Strings, electromagnetic waves, and sound waves in closed-closed tubes must have nodes at both ends.

$$\lambda_m = \frac{2L}{m} \qquad f_m = m\frac{v}{2L} = mf_1$$

where $m = 1, 2, 3, \ldots$

The frequencies and wavelengths are the same for a sound wave in an open-open tube, which has antinodes at both ends.

A sound wave in an open-closed tube must have a node at the closed end but an antinode at the open end. This leads to

$$\lambda_m = \frac{4L}{m} \qquad f_m = m\frac{v}{4L} = mf_1$$

where $m = 1, 3, 5, 7, \ldots$

Beats (loud-soft-loud-soft modulations of intensity) occur when two waves of slightly different frequency are superimposed.

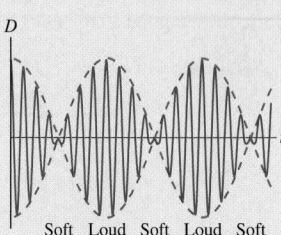

Soft Loud Soft Loud Soft

The beat frequency between waves of frequencies f_1 and f_2 is

$$f_{\text{beat}} = f_1 - f_2$$

TERMS AND NOTATION

principle of superposition	normal mode	thin-film optical coating
standing wave	interference	antinodal line
node	in phase	nodal line
antinode	constructive interference	beats
amplitude function, $A(x)$	out of phase	modulation
boundary condition	destructive interference	beat frequency, f_{beat}
fundamental frequency, f_1	phase difference, $\Delta\phi$	
harmonic	path-length difference, Δx or Δr	

EXERCISES AND PROBLEMS

Exercises

Section 21.1 The Principle of Superposition

1. Figure Ex21.1 is a snapshot graph at $t = 0$ s of two waves approaching each other at 1 m/s. Draw six snapshot graphs, stacked vertically, showing the string at 1 s intervals from $t = 1$ s to $t = 6$ s.

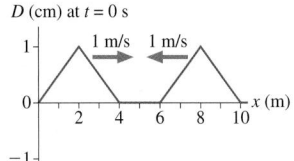

FIGURE EX21.1

2. Figure Ex21.2 is a snapshot graph at $t = 0$ s of two waves approaching each other at 1 m/s. Draw six snapshot graphs, stacked vertically, showing the string at 1 s intervals from $t = 1$ s to $t = 6$ s.

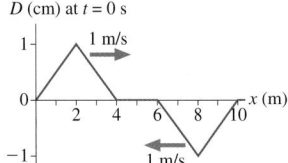

FIGURE EX21.2

3. Figure Ex21.3 is a snapshot graph at $t = 0$ s of two waves approaching each other at 1 m/s. Draw four snapshot graphs, stacked vertically, showing the string at $t = 2, 4, 6,$ and 8 s.

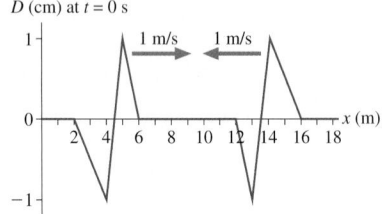

FIGURE EX21.3

4. Figure Ex21.4 is a snapshot graph at $t = 0$ s of two waves approaching each other at 1 m/s. Draw four snapshot graphs, stacked vertically, showing the string at $t = 2, 4, 6,$ and 8 s.

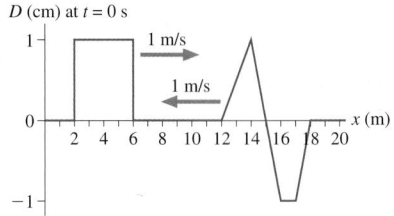

FIGURE EX21.4

5. Figure Ex21.5a is a snapshot graph at $t = 0$ s of two waves approaching each other at 1 m/s.
   a. At what time was the snapshot graph in Figure Ex21.5b taken?
   b. Draw a history graph of the string at $x = 5$ m from $t = 0$ s to $t = 6$ s.

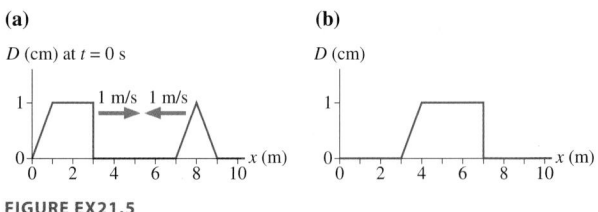

FIGURE EX21.5

Section 21.2 Standing Waves

Section 21.3 Transverse Standing Waves

6. Figure Ex21.6 is a snapshot graph at $t = 0$ s of two waves moving to the right at 1 m/s. The string is fixed at $x = 8$ m. Draw four snapshot graphs, stacked vertically, showing the string at $t = 2, 4, 6,$ and 8 s.

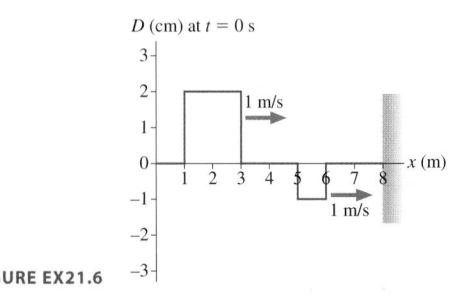

FIGURE EX21.6

7. A 2.0-m-long string is fixed at both ends and tightened until the wave speed is 40 m/s. What is the frequency of the standing wave shown in Figure Ex21.7?

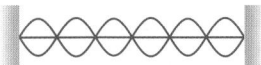
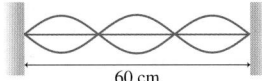
60 cm

FIGURE EX21.7   **FIGURE EX21.8**

8. Figure Ex21.8 shows a standing wave oscillating at 100 Hz on a string. What is the wave speed?
9. Figure Ex21.9 shows a standing wave that is oscillating at frequency f_0.
   a. How many antinodes will there be if the frequency is doubled to $2f_0$? Explain.
   b. If the tension in the string is increased by a factor of four, for what frequency, in terms of f_0, will the string continue to oscillate as a standing wave with three antinodes?

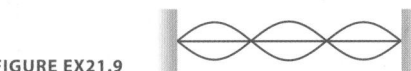

FIGURE EX21.9

10. a. What are the three longest wavelengths for standing waves on a 240-cm-long string that is fixed at both ends?
    b. If the frequency of the second-longest wavelength is 50 Hz, what is the frequency of the third-longest wavelength?
11. Standing waves on a 1.0-m-long string that is fixed at both ends are seen at successive frequencies of 24 Hz and 36 Hz.
    a. What are the fundamental frequency and the wave speed?
    b. Draw the standing-wave pattern when the string oscillates at 36 Hz.
12. A 121-cm-long, 4.0 g string oscillates in its $m = 3$ mode with a frequency of 180 Hz and a maximum amplitude of 5.0 mm. What are (a) the wavelength and (b) the tension in the string?
13. A guitar string with a linear density of 2.0 g/m is stretched between supports that are 60 cm apart. The string is observed to form a standing wave with three antinodes when driven at a frequency of 420 Hz. What are (a) the frequency of the fifth harmonic of this string and (b) the tension in the string?
14. A carbon dioxide laser is an infrared laser. A CO_2 laser with a cavity length of 53.00 cm oscillates in the $m = 100,000$ mode. What are the wavelength and frequency of the laser beam?

Section 21.4 Standing Sound Waves and Musical Acoustics

15. Figure Ex21.15 shows a standing sound wave in an 80-cm-long tube. The tube is filled with an unknown gas. What is the speed of sound in this gas?

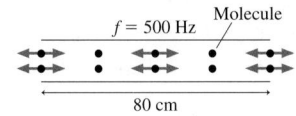
$f = 500$ Hz   Molecule

FIGURE EX21.15   80 cm

16. What are the three longest wavelengths for standing sound waves in a 121-cm-long tube that is (a) open at both ends and (b) open at one end, closed at the other?
17. The lowest pedal note on a large pipe organ has a fundamental frequency of 16.4 Hz. This extreme bass note, four octaves below middle C, is more felt than heard. What is the length of pipe between the sounding hole and the open end?

18. The fundamental frequency of an open-open tube is 1500 Hz when the tube is filled with 0°C helium. What is its frequency when filled with 0°C air?
19. A violin string is 30 cm long. It sounds the musical note A (440 Hz) when played without fingering. How far from the end of the string should you place your finger to play the note C (523 Hz)?
20. The lowest note on a grand piano has a frequency of 27.5 Hz. The entire string is 2.00 m long and has a mass of 400 g. The vibrating section of the string is 1.90 m long. What tension is needed to tune this string properly?

Section 21.5 Interference in One Dimension

Section 21.6 The Mathematics of Interference

21. Two loudspeakers in a 20°C room emit 686 Hz sound waves along the x-axis.
    a. If the speakers are in phase, what is the smallest distance between the speakers for which the interference of the sound waves is destructive?
    b. If the speakers are out of phase, what is the smallest distance between the speakers for which the interference of the sound waves is constructive?
22. Two loudspeakers emit sound waves along the x-axis. The sound has maximum intensity when the speakers are 20 cm apart. The sound intensity decreases as the distance between the speakers is increased, reaching zero at a separation of 60 cm.
    a. What is the wavelength of the sound?
    b. If the distance between the speakers continues to increase, at what separation will the sound intensity again be a maximum?
23. Two identical loudspeakers separated by distance d emit 170 Hz sound waves along the x-axis. As you walk along the axis, away from the speakers, you don't hear anything even though both speakers are on. What are three possible values for d? Assume a sound speed of 340 m/s.
24. What is the thinnest film of MgF_2 ($n = 1.39$) on glass that produces a strong reflection for orange light with a wavelength of 600 nm?
25. A very thin oil film ($n = 1.25$) floats on water ($n = 1.33$). What is the thinnest film that produces a strong reflection for green light with a wavelength of 500 nm?

Section 21.7 Interference in Two and Three Dimensions

26. Figure Ex21.26 shows the circular wave fronts emitted by two sources.
    a. Are these sources in phase or out of phase? Explain.
    b. Make a table with rows labeled P, Q, and R and columns labeled r_1, r_2, Δr, and C/D. Fill in the table for points P, Q, and R, giving the distances as multiples of λ and indicating, with a C or a D, whether the interference at that point is constructive or destructive. **FIGURE EX21.26**

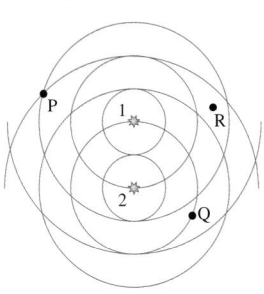

27. Figure Ex21.27 shows the circular wave fronts emitted by two sources.
    a. Are these sources in phase or out of phase? Explain.
    b. Make a table with rows labeled P, Q, and R and columns labeled r_1, r_2, Δr, and C/D. Fill in the table for points P, Q, and R, giving the distances as multiples of λ and indicating, with a C or a D, whether the interference at that point is constructive or destructive.

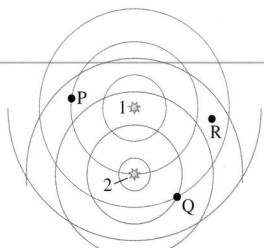

FIGURE EX21.27

28. Two identical loudspeakers 2.0 m apart are emitting 1800 Hz sound waves into a room where the speed of sound is 340 m/s. Is the point 4.0 m in front of one of the speakers, perpendicular to the plane of the speakers, a point of maximum constructive interference, perfect destructive interference, or something in between?

29. Two out-of-phase radio antennas at $x = \pm300$ m on the x-axis are emitting 3.0 MHz radio waves. Is the point $(x, y) = (300$ m, 800 m$)$ a point of maximum constructive interference, perfect destructive interference, or something in between?

Section 21.8 Beats

30. Two strings are adjusted to vibrate at exactly 200 Hz. Then the tension in one string is increased slightly. Afterward, three beats per second are heard when the strings vibrate at the same time. What is the new frequency of the string that was tightened?

31. A flute player hears four beats per second when she compares her note to a 523 Hz tuning fork (the note C). She can match the frequency of the tuning fork by pulling out the "tuning joint" to lengthen her flute slightly. What was her initial frequency?

Problems

32. Two wave pulses on a string travel in opposite directions at 100 m/s. Figure P21.32 shows a snapshot graph of the string at $t = 0$ s, when the two waves are overlapped, and a snapshot graph of the right-traveling wave at $t = 0.05$ s. Draw a snapshot graph of the left-traveling wave at $t = 0.05$ s.

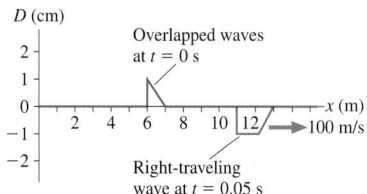

FIGURE P21.32

33. Two waves on a string travel in opposite directions at 100 m/s. Figure P21.33 shows a snapshot graph of the string at $t = 0$ s, when the two waves are overlapped, and a snapshot graph of the left-traveling wave at $t = 0.05$ s. Draw a snapshot graph of the right-traveling wave at $t = 0.05$ s.

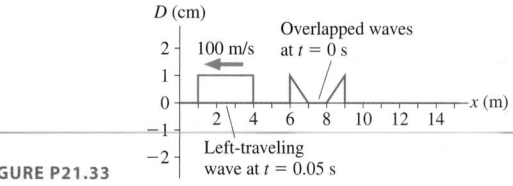

FIGURE P21.33

34. The superposition of two counter-propagating, equal-frequency sinusoidal waves is a standing wave. What about the superposition of two equal-frequency standing waves? Consider the superposition $D = a\sin(kx)\cos(\omega t) + a\cos(kx)\sin(\omega t)$. Each of these two standing waves has period $T = 2\pi/\omega$ and wavelength $\lambda = 2\pi/k$.
    a. Draw nine snapshot graphs, one every eighth of a period from $t = 0$ to $t = T$, showing the x-axis from $x = 0$ to $x = 2\lambda$. Sketch each of the standing waves as a dotted line, then show their superposition as a solid line. Stack the graphs vertically, similar to Figure 21.4a.
    b. Is the superposition a standing wave or a traveling wave? If it's a traveling wave, which way is it moving? Use your graphs to explain.
    c. Use a well-known trigonometric identity to write the displacement D in a form that is clearly a standing wave or a traveling wave. Show that your result agrees with your observations of part b.

35. A 2.0-m-long string vibrates at its second-harmonic frequency with a maximum amplitude of 2.00 cm. One end of the string is at $x = 0$ cm. Find the oscillation amplitude at $x = 10, 20, 30, 40,$ and 50 cm.

36. A string vibrates at its third-harmonic frequency. The amplitude at a point 30 cm from one end is half the maximum amplitude. How long is the string?

37. A string of length L vibrates at its fundamental frequency. The amplitude at a point $\frac{1}{4}L$ from one end is 2.0 cm. What is the amplitude of each of the traveling waves that form this standing wave?

38. Two sinusoidal waves with equal wavelengths travel along a string in opposite directions at 3.0 m/s. The time between two successive instants when the antinodes are at maximum height is 0.25 s. What is the wavelength?

39. An 80-cm-long guitar string with a linear density of 1.0 g/m is under 200 N tension. It is plucked and vibrates at its fundamental frequency. What is the wavelength of the sound wave that reaches your ear in a 20°C room?

40. A violinist places her finger so that the vibrating section of a 1.0 g/m string has a length of 30 cm, then she draws her bow across it. A listener nearby in a 20°C room hears a note with a wavelength of 40 cm. What is the tension in the string?

41. A particularly beautiful note reaching your ear from a rare Stradivarius violin has a wavelength of 39.1 cm. The room is slightly warm, so the speed of sound is 344 m/s. If the string's linear density is 0.60 g/m and the tension is 150 N, how long is the vibrating section of the violin string?

42. A heavy piece of hanging sculpture is suspended by a 90-cm-long, 5.0 g steel wire. When the wind blows hard, the wire hums at its fundamental frequency of 80 Hz. What is the mass of the sculpture?

43. Astronauts visiting Planet X have a 2.5-m-long string whose mass is 5.0 g. They tie the string to a support, stretch it horizontally over a pulley 2.0 m away, and hang a 1.0 kg mass on the free end. Then the astronauts begin to excite standing waves on the string. Their data show that standing waves exist at frequencies of 64 Hz and 80 Hz, but at no frequencies in between. What is the value of g, the acceleration due to gravity, on Planet X?

44. A 75 g bungee cord has an equilibrium length of 1.20 m. The cord is stretched to a length of 1.80 m, then vibrated at 20 Hz. This produces a standing wave with two antinodes. What is the spring constant of the bungee cord?

45. A steel wire is used to stretch a spring. An oscillating magnetic field drives the steel wire back and forth. A standing wave with three antinodes is created when the spring is stretched 8.0 cm. What stretch of the spring produces a standing wave with two antinodes?

FIGURE P21.45

46. The microwave generator in Figure P21.46 can produce microwaves at any frequency between 10 GHz and 20 GHz. The microwaves are aimed, through a small hole, into a "microwave cavity" that consists of a 10-cm-long cylinder with reflective ends.
   a. Which frequencies will create standing waves in the microwave cavity?
   b. For which of these frequencies is the cavity midpoint an antinode?

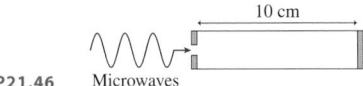

FIGURE P21.46   Microwaves

47. An open-open organ pipe is 78.0 cm long. An open-closed pipe has a fundamental frequency equal to the third harmonic of the open-open pipe. How long is the open-closed pipe?

48. A narrow column of air is found to have standing waves at frequencies of 390 Hz, 520 Hz, and 650 Hz and at no frequencies in between these. The behavior of the tube at frequencies less than 390 Hz or greater than 650 Hz is not known.
   a. Is this an open-open tube or an open-closed tube? Explain.
   b. How long is the tube?
   c. Draw a displacement graph of the 520 Hz standing wave in the tube.
   d. The air in the tube is replaced with carbon dioxide, which has a sound speed of 280 m/s. What are the new frequencies of these three modes?

49. In 1866, the German scientist Adolph Kundt developed a technique for accurately measuring the speed of sound in various gases. A long glass tube, known today as a Kundt's tube, has a vibrating piston at one end and is closed at the other. Very finely ground particles of cork are sprinkled in the bottom of the tube before the piston is inserted. As the vibrating piston is slowly moved forward, there are a few positions that cause the cork particles to collect in small, regularly spaced

piles along the bottom. Figure P21.49 shows an experiment in which the tube is filled with pure oxygen and the piston is driven at 400 Hz. What is the speed of sound in oxygen?

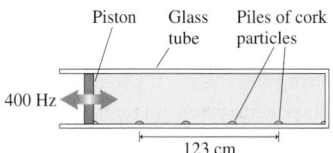

FIGURE P21.49

50. A 40-cm-long tube has a 40-cm-long insert that can be pulled in and out. A vibrating tuning fork is held next to the tube. As the insert is slowly pulled out, the sound from the tuning fork creates standing waves in the tube when the total length L is 42.5 cm, 56.7 cm, and 70.9 cm. What is the frequency of the tuning fork?

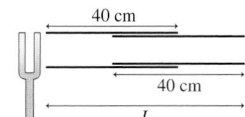

FIGURE P21.50

51. A 1.0-m-tall vertical tube is filled with 20°C water. A tuning fork vibrating at 580 Hz is held just over the top of the tube as the water is slowly drained from the bottom. At what water heights, measured from the bottom of the tube, will there be a standing wave in the tube?

52. A 50-cm-long wire with a mass of 1.0 g and a tension of 440 N passes across the open end of an open-closed tube of air. The wire, which is fixed at both ends, is bowed at the center so as to vibrate at its fundamental frequency and generate a sound wave. Then the tube length is adjusted until the fundamental frequency of the tube is heard. What is the length of the tube? Assume $v_{sound} = 340$ m/s.

53. A 25-cm-long wire with a linear density of 20 g/m passes across the open end of an 85-cm-long open-closed tube of air. If the wire, which is fixed at both ends, vibrates at its fundamental frequency, the sound wave it generates excites the second vibrational mode of the tube of air. What is the tension in the wire? Assume $v_{sound} = 340$ m/s.

54. A 50-cm-long wire with a mass of 1.0 g and a tension of 440 N passes across the open top of a vertical tube partially filled with water. The wire, which is fixed at both ends, is bowed at the center so as to vibrate at its fundamental frequency and generate a sound wave. The water level in the tube is slowly lowered until the sound wave from the wire sets up a standing wave in the tube. It is then lowered another 36.0 cm until the next standing wave is detected. Use this information to determine the speed of sound in air.

55. A longitudinal standing wave can be created in a long, thin aluminum rod by stroking the rod with very dry fingers. This is often done as a physics demonstration, creating a high-pitched, very annoying whine. From a wave perspective, the standing wave is equivalent to a sound standing wave in an open-open tube. In particular, both ends of the rod are antinodes. What is the fundamental frequency of a 2.0-m-long aluminum rod?

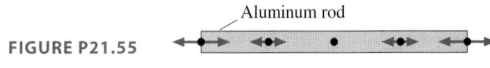

FIGURE P21.55

56. Figure P21.56 is a snapshot picture showing some of the air molecules in a tube of air at their equilibrium positions and when a standing sound wave is present inside the tube.
    a. Is the left end of the tube open or closed? The right end? Explain.
    b. If the tube is 1.5 m long, what is the wavelength of this wave?
    c. Draw a snapshot picture showing the air molecules one half cycle later.

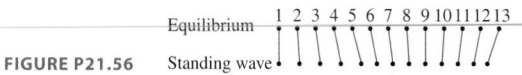

Equilibrium   1 2 3 4 5 6 7 8 9 10 11 12 13

FIGURE P21.56   Standing wave

57. Analyze the standing sound waves in an open-closed tube to show that the possible wavelengths and frequencies are given by Equation 21.18.
58. Two loudspeakers emit sound waves of the same frequency along the x-axis. The amplitude of each wave is a. The sound intensity is minimum when speaker 2 is 10 cm behind speaker 1. The intensity increases as speaker 2 is moved forward and first reaches maximum, with amplitude $2a$, when it is 30 cm in front of speaker 1. What is
    a. The wavelength of the sound?
    b. The phase difference between the two loudspeakers?
    c. The amplitude of the sound if the speakers are placed side by side?
59. Two in-phase loudspeakers emit identical 1000 Hz sound waves along the x-axis. What distance should one speaker be placed behind the other for the sound to have an amplitude 1.5 times that of each speaker alone?
60. Two loudspeakers emit sound waves along the x-axis. Speaker 2 is 2.0 m behind speaker 1. Both loudspeakers are connected to the same signal generator, which is oscillating at 340 Hz, but the wire to speaker 1 passes through a box that delays the signal by 1.47 ms. Is the interference along the x-axis maximum constructive interference, perfect destructive interference, or something in between? Assume $v_{sound} = 340$ m/s.
61. Two loudspeakers emit sound waves along the x-axis. A listener in front of both speakers hears a maximum sound intensity when speaker 2 is at the origin and speaker 1 is at $x = 0.50$ m. If speaker 1 is slowly moved forward, the sound intensity decreases and then increases, reaching another maximum when speaker 1 is at $x = 0.90$ m.
    a. What is the frequency of the sound? Assume $v_{sound} = 340$ m/s.
    b. What is the phase difference between the speakers?
62. A sheet of glass is coated with a 500-nm-thick layer of oil ($n = 1.42$).
    a. For what *visible* wavelengths of light do the reflected waves interfere constructively?
    b. For what *visible* wavelengths of light do the reflected waves interfere destructively?
    c. What is the color of reflected light? What is the color of transmitted light?
63. A jewelry maker has asked your glass studio to produce a sheet of dichroic glass that will appear red for transmitted light and blue for reflected light. You decide that "red light" should be centered at 640 nm and that "blue light" is 480 nm. If you use a MgF$_2$ coating ($n = 1.39$), how thick should the coating be?

64. Example 21.10 showed that a 92-nm-thick coating of MgF$_2$ ($n = 1.39$) on glass acts as an antireflection coating for light with a wavelength of 510 nm. Without the coating, the intensity of reflected light is $I_0 = Ca^2$, where a is the amplitude of the reflected light wave and C is an unknown proportionality constant.
    a. Let I_λ be the intensity of light reflected from the coated glass at wavelength λ. Find an expression for the ratio I_λ/I_0 as a function of the wavelength λ. This ratio is the reflection intensity from the coated glass relative to the reflection intensity from uncoated glass. A ratio less than 1 indicates that the coating is reducing the reflection intensity.

    Hint: The amplitude of the superposition of two waves depends on the phase difference between the waves. Although not entirely accurate, assume that both reflected waves have amplitude a.
    b. Evaluate I_λ/I_0 at $\lambda = 400, 450, 500, 550, 600, 650,$ and 700 nm. This spans the range of visible light.
    c. Draw a graph of I_λ/I_0 versus λ.
65. A manufacturing firm has hired your company, Acoustical Consulting, to help with a problem. Their employees are complaining about the annoying hum from a piece of machinery. Using a frequency meter, you quickly determine that the machine emits a rather loud sound at 1200 Hz. After investigating, you tell the owner that you cannot solve the problem entirely, but you can at least improve the situation by eliminating reflections of this sound from the walls. You propose to do this by installing mesh screens in front of the walls. A portion of the sound will reflect from the mesh; the rest will pass through the mesh and reflect from the wall. How far should the mesh be placed in front of the wall for this scheme to work?
66. A soap bubble is essentially a very thin film of water ($n = 1.33$) surrounded by air. The colors that you see in soap bubbles are produced by interference, much like the colors of dichroic glass.
    a. Derive an expression for the wavelengths λ_C for which constructive interference causes a strong reflection from a soap bubble of thickness d.

    Hint: Think about the reflection phase shifts at both boundaries.
    b. What visible wavelengths of light are strongly reflected from a 390-nm-thick soap bubble? What color would such a soap bubble appear to be?
67. Two radio antennas are separated by 2.0 m. Both broadcast identical 750 MHz waves. If you walk around the antennas in a circle of radius 10 m, how many maxima will you detect?
68. You are standing 2.5 m directly in front of one of the two loudspeakers shown in Figure P21.68. They are 3.0 m apart and both are playing a 686 Hz tone in phase. As you begin to walk directly away from the speaker, at what distances from the speaker do you hear a *minimum* sound intensity? The room temperature is 20°C.

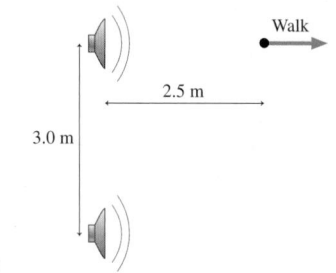

FIGURE P21.68

69. Two loudspeakers 5.0 m apart are playing the same frequency. If you stand 12.0 m in front of the plane of the speakers, centered between them, you hear a sound of maximum intensity. As you walk parallel to the plane of the speakers, staying 12.0 m in front of them, you first hear a minimum of sound intensity when you are directly in front of one of the speakers.
    a. What is the frequency of the sound? Assume a sound speed of 340 m/s.
    b. If you stay 12.0 m in front of one of the speakers, for what other frequencies between 100 Hz and 1000 Hz is there a minimum sound intensity at this point?

70. Two in-phase loudspeakers are located at (x, y) coordinates $(-3.0$ m, $+2.0$ m$)$ and $(-3.0$ m, -2.0 m$)$. They emit identical sound waves with a 2.0 m wavelength and amplitude a. Determine the amplitude of the sound at the five positions on the y-axis $(x = 0)$ with $y = 0.0$ m, 0.5 m, 1.0 m, 1.5 m, and 2.0 m.

71. Your firm has been hired to design a system that allows airplane pilots to make instrument landings in rain or fog. You've decided to place two radio transmitters 50 m apart on either side of the runway. These two transmitters will broadcast the same frequency, but out of phase with each other. This will cause a nodal line to extend straight off the end of the runway (see Figure 21.29b). As long as the airplane's receiver is silent, the pilot knows she's directly in line with the runway. If she drifts to one side or the other, the radio will pick up a signal and sound a warning beep. To have sufficient accuracy, the first intensity maxima need to be 60 m on either side of the nodal line at a distance of 3.0 km. What frequency should you specify for the transmitters?

72. Two radio antennas are 100 m apart along a north-south line. They broadcast identical radio waves at a frequency of 3.0 MHz. Your job is to monitor the signal strength with a handheld receiver. To get to your first measuring point, you walk 800 m east from the midpoint between the antennas, then 600 m north.
    a. What is the phase difference between the waves at this point?
    b. Is the interference at this point maximum constructive, perfect destructive, or somewhere in between? Explain.
    c. If you now begin to walk farther north, does the signal strength increase, decrease, or stay the same? Explain.

73. The three identical loudspeakers in Figure P21.73 play a 170 Hz tone in a room where the speed of sound is 340 m/s. You are standing 4 m in front of the middle speaker. At this point, the amplitude of the wave from each speaker is a.
    a. What is the amplitude at this point?
    b. How far must speaker 2 be moved to the left to produce a maximum amplitude at the point where you are standing?
    c. When the amplitude is maximum, by what factor is the sound intensity greater than the sound intensity from a single speaker?

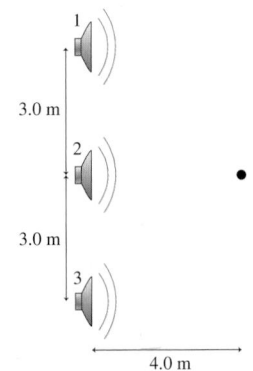

FIGURE P21.73

3.0 m

3.0 m

4.0 m

74. Piano tuners tune pianos by listening to the beats between the *harmonics* of two different strings. When properly tuned, the note A should have a frequency of 440 Hz and the note E should be at 659 Hz.
    a. What is the frequency difference between the third harmonic of the A and the second harmonic of the E?
    b. A tuner first tunes the A string very precisely by matching it to a 440 Hz tuning fork. She then strikes the A and E strings simultaneously and listens for beats between the harmonics. What beat frequency indicates that the E string is properly tuned?
    c. The tuner starts with the tension in the E string a little low, then tightens it. What is the frequency of the E string when she hears four beats per second?

75. A flutist assembles her flute in a room where the speed of sound is 342 m/s. When she plays the note A, it is in perfect tune with a 440 Hz tuning fork. After a few minutes, the air inside her flute has warmed to where the speed of sound is 346 m/s.
    a. How many beats per second will she hear if she now plays the note A as the tuning fork is sounded?
    b. How far does she need to extend the "tuning joint" of her flute to be in tune with the tuning fork?

76. Two lasers with very nearly the same wavelength can generate a beat frequency if both laser beams illuminate a photodetector with a very fast response. In an experiment, one laser's wavelength has been stabilized at 780.54510 nm. The second laser starts with a longer wavelength that is slowly decreased until the beat frequency between the two lasers is 98.5 MHz. What is the second laser's wavelength?

77. Two loudspeakers emit 400 Hz notes. One speaker sits on the ground. The other speaker is in the back of a pickup truck. You hear eight beats per second as the truck drives away from you. What is the truck's speed?

78. Two loudspeakers face each other from opposite walls of a room. Both are playing exactly the same frequency, thus setting up a standing wave with distance $\lambda/2$ between antinodes. Assume that λ is much less than the room width, so there are many antinodes.
    a. Yvette starts at one speaker and runs toward the other at speed v_Y. As the does so, she hears a loud-soft-loud modulation of the sound intensity. From your perspective, as you sit at rest in the room, Yvette is running through the nodes and antinodes of the standing wave. Find an expression for the number of sound maxima she hears per second.
    b. From Yvette's perspective, the two sound waves are Doppler shifted. They're not the same frequency, so they don't create a standing wave. Instead, she hears a loud-soft-loud modulation of the sound intensity because of beats. Find an expression for the beat frequency that Yvette hears.
    c. Are your answers to parts a and b the same or different? *Should* they be the same or different?

Challenge Problems

79. a. The frequency of a standing wave on a string is f when the string's tension is T. If the tension is changed by the *small* amount ΔT, without changing the length, show that the frequency changes by an amount Δf such that

$$\frac{\Delta f}{f} = \frac{1}{2}\frac{\Delta T}{T}$$

b. Two identical strings vibrate at 500 Hz when stretched with the same tension. What percentage increase in the tension of one of the strings will cause five beats per second when both strings vibrate simultaneously?

80. A 280 Hz sound wave is directed into one end of a trombone slide and a microphone is placed at the other end to record the intensity of sound waves that are transmitted through the tube. The straight sides of the slide are 80 cm in length and 10 cm apart with a semicircular bend at the end. For what slide extensions s will the microphone detect a maximum of sound intensity?

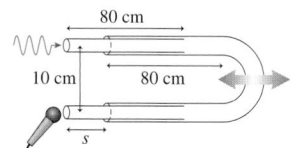

FIGURE CP21.80

81. As the captain of the scientific team sent to Planet Physics, one of your tasks is to measure g. You have a long, thin wire labeled 1.00 g/m and a 1.25 kg weight. You have your accurate space cadet chronometer but, unfortunately, you seem to have forgotten a meter stick. Undeterred, you first find the midpoint of the wire by folding it in half. You then attach one end of the wire to the wall of your laboratory, stretch it horizontally to pass over a pulley at the midpoint of the wire, then tie the 1.25 kg weight to the end hanging over the pulley. By vibrating the wire, and measuring time with your chronometer, you find that the wire's second harmonic frequency is 100 Hz. Next, with the 1.25 kg weight still tied to one end of the wire, you attach the other end to the ceiling to make a pendulum. You find that the pendulum requires 314 s to complete 100 oscillations. Pulling out your trusty calculator, you get to work. What value of g will you report back to headquarters?

82. A 22-cm-long, 1.0-mm-diameter copper wire is joined smoothly to a 60-cm-long, 1.0-mm-diameter aluminum wire. The resulting wire is stretched with 20 N of tension between fixed supports 82 cm apart. The densities of copper and aluminum are 8920 kg/m³ and 2700 kg/m³, respectively.
   a. What is the lowest-frequency standing wave for which there is a node at the junction between the two metals?
   b. At that frequency, how many antinodes are on the aluminum wire?

83. Ultrasound has many medical applications, one of which is to monitor fetal heartbeats by reflecting ultrasound off a fetus in the womb.
   a. Consider an object moving at speed v_0 toward an at-rest source that is emitting sound waves of frequency f_0. Show that the reflected wave (i.e., the echo) that returns to the source has a Doppler-shifted frequency

   $$f_{echo} = \left(\frac{v + v_0}{v - v_0}\right)f_0$$

   where v is the speed of sound in the medium.
   b. Suppose the object's speed is much less than the wave speed: $v_0 \ll v$. Then $f_{echo} \approx f_0$, and a microphone that is sensitive to these frequencies will detect a beat frequency if it listens to f_0 and f_{echo} simultaneously. Use the binomial

approximation and other appropriate approximations to show that the beat frequency is $f_{beat} \approx (2v_0/v)f_0$.
   c. The reflection of 2.40 MHz ultrasound waves from the surface of a fetus's beating heart is combined with the 2.40 MHz wave to produce a beat frequency that reaches a maximum of 65 Hz. What is the maximum speed of the surface of the heart? The speed of ultrasound waves within the body is 1540 m/s.
   d. Suppose the surface of the heart moves in simple harmonic motion at 90 beats/min. What is the amplitude in mm of the heartbeat?

84. A water wave is called a *deep-water wave* if the water's depth is more than one-quarter of the wavelength. Unlike the waves we've considered in this chapter, the speed of a deep water wave depends on its wavelength:

$$v = \sqrt{\frac{g\lambda}{2\pi}}$$

Longer wavelengths travel faster. Let's apply this to standing waves. Consider a diving pool that is 5.0 m deep and 10.0 m wide. Standing water waves can set up across the width of the pool. Because water sloshes up and down at the sides of the pool, the boundary conditions require antinodes at $x = 0$ and $x = L$. Thus a standing water wave resembles a standing sound wave in an open-open tube.
   a. What are the wavelengths of the first three standing-wave modes for water in the pool? Do they satisfy the condition for being deep-water waves? Draw a graph of each.
   b. What are the wave speeds for each of these waves?
   c. Derive a general expression for the frequencies f_m of the possible standing waves. Your expression should be in terms of m, g, and L.
   d. What are the oscillation *periods* of the first three standing-wave modes?

85. The broadcast antenna of an AM radio station is located at the edge of town. The station owners would like to beam all of the energy into town and none into the countryside, but a single antenna radiates energy equally in all directions. Figure CP21.85 shows two parallel antennas separated by distance L. Both antennas broadcast a signal at wavelength λ, but antenna 2 can delay its broadcast relative to antenna 1 by a time interval Δt in order to create a phase difference $\Delta\phi_0$ between the sources. Your task is to find values of L and Δt such that the waves interfere constructively on the town side and destructively on the country side.

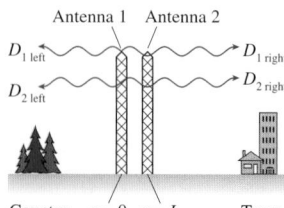

FIGURE CP21.85 Country $x = 0$ $x = L$ Town

Let antenna 1 be at $x = 0$. The wave that travels to the right is $a\sin[2\pi(x/\lambda - t/T)]$. The wave that travels to the left is $a\sin[2\pi(-x/\lambda - t/T)]$. (It must be this, rather than $a\sin[2\pi(x/\lambda + t/T)]$, so that the two waves match at $x = 0$.) Antenna 2 is at $x = L$. It broadcasts waves

$a \sin\left[2\pi((x - L)/\lambda - t/T) + \phi_{20}\right]$ to the right and $a \sin\left[2\pi(-(x - L)/\lambda - t/T) + \phi_{20}\right]$ to the left.

a. What is the smallest value of L for which you can create perfect constructive interference on the town side and perfect destructive interference on the country side? Your answer will be a multiple or fraction of the wavelength λ.

b. What phase constant ϕ_{20} of antenna 2 is needed?

c. What fraction of the oscillation period T must Δt be to produce the proper value of ϕ_{20}?

d. Evaluate both L and Δt for the realistic AM radio frequency of 1000 KHz.

Comment: This is a simple example of what is called a *phased array*, where phase differences between identical emitters are used to "steer" the radiation in a particular direction. Phased arrays are widely used in radar technology.

STOP TO THINK ANSWERS

Stop to Think 21.1: c. The figure shows the two waves at $t = 6$ s and their superposition. The superposition is the *point-by-point* addition of the displacements of the two individual waves.

Stop to Think 21.2: a. The allowed standing-wave frequencies are $f_m = m(v/2L)$, so the mode number of a standing wave of frequency f is $m = 2Lf/v$. Quadrupling T_s increases the wave speed v by a factor of two. The initial mode number was 2, so the new mode number is 1.

Stop to Think 21.3: b. 300 Hz and 400 Hz are allowed standing waves, but they are not f_1 and f_2 because 400 Hz $\neq 2 \times 300$ Hz. Because there's a 100 Hz difference between them, these must be $f_3 = 3 \times 100$ Hz and $f_4 = 4 \times 100$ Hz, with a fundamental frequency $f_1 = 100$ Hz. Thus the second harmonic is $f_2 = 2 \times 100$ Hz = 200 Hz.

Stop to Think 21.4: c. Shifting the top wave 0.5 m to the left aligns crest with crest and trough with trough.

Stop to Think 21.5: c. $r_1 = 0.5\lambda$ and $r_2 = 3.0\lambda$, so $\Delta r = 2.5\lambda$. This is the condition for perfect destructive interference.

Stop to Think 21.6: Maximum constructive. The path-length difference is $\Delta r = 1.0$ m $= \lambda$. For identical sources, interference is constructive when Δr is an integer multiple of λ.

Stop to Think 21.7: f. The beat frequency is the difference between the two frequencies.

22 Wave Optics

The rainbows of color from the surface of a CD are evidence that light is a wave.

You've probably noticed the rainbow splash of colors when a bright light reflects from the surface of a compact disk. You may be surprised to learn that the colors from a CD are closely related to the iridescence of bird feathers and to the technology underlying supermarket checkout scanners, holograms, and optical computers. All of these, in one way or another, depend on the interference of light waves. In some, such as the CD, the interference is incidental to the intended use of the product. Others, such as holograms, have been designed to make precise use of the wave-like properties of light.

The study of light is called **optics,** and this is the first of three chapters to explore optics and the nature of light. Light is an elusive topic. You will find, perhaps surprisingly, that there is no simple description of light. Light behaves quite differently in different situations, and we will ultimately need three different *models* of light to capture this behavior. We begin, in this chapter, with situations in which light acts as a wave. The groundwork for *wave optics* has been laid in Chapters 20 and 21, and we will now apply those ideas to light waves.

Although light is an electromagnetic wave, this chapter depends on nothing more than the "waviness" of light waves. You can study this chapter either before or after your study of electricity and magnetism in Part VI.

22.1 Light and Optics

What is light? The first Greek scientists and philosophers did not make a distinction between light and vision. Light, to them, was not something that existed apart from seeing. But gradually there arose a view that light actually "exists," that light is some sort of physical entity that is present regardless of whether or not someone is looking. But if light is a physical entity, what is it? What are its characteristics? Is it a wave, similar to sound? Or is light a collection of small particles that blows by like the wind?

Newton, in addition to doing pioneering work in mathematics and mechanics in the 1660s, was also one of the early investigators of the nature of light. Newton knew that a water wave, after passing through an opening, *spreads out* to fill the space behind the opening. You can see this in Figure 22.1a, where plane waves, approaching from the left, spread out in circular arcs after passing through a hole in a barrier. This inexorable spreading of waves is the phenomenon called **diffraction.** Diffraction is a sure sign that whatever is passing through the hole is a wave.

In contrast, Figure 22.1b shows that sunlight passing through a door makes a sharp-edged shadow as it falls upon the floor. We don't see sunlight light spreading out after it passes through a hole. Instead, this behavior is exactly what you would expect if light consists of particles traveling in straight lines. Some particles would pass through the window to make a bright area on the wall, others would be blocked and cause a well-defined shadow. This reasoning led Newton to the conclusion that light consists of very small, light, fast particles that he called *corpuscles.*

Newton was vigorously opposed by Robert Hooke (of Hooke's law) and the Dutch scientist Christian Huygens, who argued that light was some sort of wave. Although the debate was lively, and sometimes acrimonious, Newton eventually prevailed. The belief that light consists of corpuscles was not seriously questioned for more than a hundred years after Newton's death.

The situation changed dramatically in1801, when the English scientist Thomas Young announced that he had produced *interference* between two waves of light. Young's experiment, which we will analyze in the next section, was a painstakingly difficult experiment with the technology of his era. Nonetheless, Young's experiment quickly settled the debate in favor of a wave theory of light because interference is a distinctly wave-like phenomenon.

But if light is a wave, what is it that is waving? This was the question that Young posed to the 19th century. It was ultimately established, through theoretical and experimental efforts by numerous scientists, that light is an *electromagnetic wave,* an oscillation of the electromagnetic field that requires no material medium in which to travel. Further, as we have already seen, visible light is just one small slice out of a vastly broader *electromagnetic spectrum.*

That light is a wave, an electromagnetic wave, seemed well established by about 1880. But this satisfying conclusion was undermined within the next 25 years. A new discovery, called the photoelectric effect, seemed to be inconsistent with the theory of electromagnetic waves. In 1905, an unknown young physicist named Albert Einstein was able to explain the photoelectric effect by treating light as a novel type of wave having certain particle-like characteristics. These wave-like particles of light soon came to be known as *photons.*

Einstein's introduction of the concept of the photon can now be seen as the end of *classical physics* and the beginning of a new era called *quantum physics.* Equally important, Einstein's theory marked yet another shift in our age-old effort to understand light.

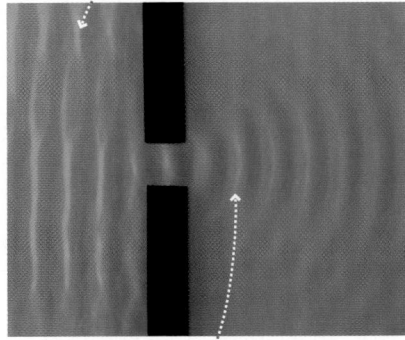

(a) Plane waves approach from the left.

Circular waves spread out on the right.

(b)

A beam of sunlight has a sharp edge.

FIGURE 22.1 Water waves spread out behind a small hole in a barrier, but light passing through a doorway makes a sharp-edged shadow.

Models of Light

Light is a real physical entity, but the nature of light is elusive. Light is the chameleon of the physical world. Under some circumstances, light acts like particles traveling in straight lines. But change the circumstances, and light shows the same kinds of wave-like behavior as sound waves or water waves. Change the circumstances yet again, and light exhibits behavior that is neither wave-like nor particle-like but has characteristics of both.

Rather than an all-encompassing "theory of light," it will be better to develop several **models of light.** Each model successfully explains the behavior of light within a certain domain—that is, within a certain range of physical situations. Our task will be twofold:

1. To develop clear and distinct models of light.
2. To learn the conditions and circumstances for which each model is valid.

The second task is especially important.

We'll begin with a brief summary of all three models, so that you will have a road map of where we're headed. Each of these models will be developed in the coming chapters.

The wave model: The wave model of light is the most widely applicable model, responsible for the widely known "fact" that light is a wave. It is certainly true that, under many circumstances, light exhibits the same behavior as sound or water waves. Lasers and electro-optical devices, critical technologies of the 21st century, are best understood in terms of the wave model of light. Some aspects of the wave model of light were introduced in Chapters 20 and 21, and the wave model is the primary focus of this chapter. The study of light as a wave is called **wave optics.**

The ray model: An equally well-known "fact" is that light travels in a straight line. These straight-line paths are called *light rays.* In Newton's view, light rays are the trajectories of particle-like corpuscles of light. The properties of prisms, mirrors, lenses, and optical instruments such as telescopes and microscopes are best understood in terms of light rays. Unfortunately, it's difficult to reconcile the statement "light travels in a straight line" with the statement "light is a wave." For the most part, waves and rays are mutually exclusive models of light. One of our most important tasks will be to learn when each model is appropriate. The ray model of light, the basis of **ray optics,** is the subject of the next chapter.

The photon model: Modern technology is increasingly reliant on quantum physics. In the quantum world, light behaves like neither a wave nor a particle. Instead, light consists of *photons* that have both wave-like and particle-like properties. Photons are the *quanta* of light. Much of the quantum theory of light is beyond the scope of this textbook, but we will take a peek at the important ideas in Chapter 24 and again in Part VII.

22.2 The Interference of Light

Suppose that Newton had seen the experiment depicted in Figure 22.2. Here light passes through a "window" that is only 0.1 mm wide, about twice the width of a human hair. The photograph shows how the light appears on a viewing screen 2 m behind the aperture. If light consists of corpuscles traveling in straight lines, as Newton thought, we should see a narrow strip of light, about 0.1 mm wide, with dark shadows on either side. Instead, we see a band of light extending over about 2.5 cm, a distance much wider than the aperture, with dimmer patches of light extending even farther on either side.

If you compare Figure 22.2 to the water wave of Figure 22.1, you see that *the light is spreading out* behind the 0.1-mm-wide hole. The light is exhibiting diffraction, the sure signature of waviness. The diffraction of light wasn't known in

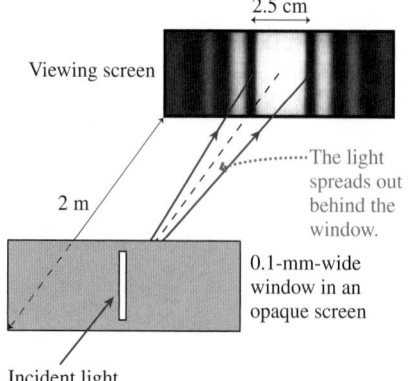

FIGURE 22.2 Light, just like a water wave, does spread out behind a hole *if* the hole is sufficiently small.

Newton's time because, as we shall see, diffraction is apparent only for very narrow apertures, typically less than 0.5 mm in width. Such apertures are hard to make, and, because they are so narrow, the light on the viewing screen is extremely dim unless you have a very bright source of light. Consequently, we can surmise that Newton would have reached a very different conclusion had he been able to perform this experiment.

We will look at diffraction in more detail later in the chapter. For now, we merely need the *observation* that light does, indeed, spread out behind a hole that is sufficiently small.

Young's Double-Slit Experiment

Rather than one small hole, suppose we use two. Figure 22.3a shows an experiment in which a laser beam is aimed at an opaque screen containing two long, narrow slits that are very close together. This pair of slits is called a **double slit,** and in a typical experiment they are ≈ 0.1 mm wide and spaced ≈ 0.5 mm apart. We will assume that the laser beam illuminates both slits equally, and any light passing through the slits impinges on a viewing screen. This is the essence of Young's experiment of 1801, although he used sunlight rather than a laser.

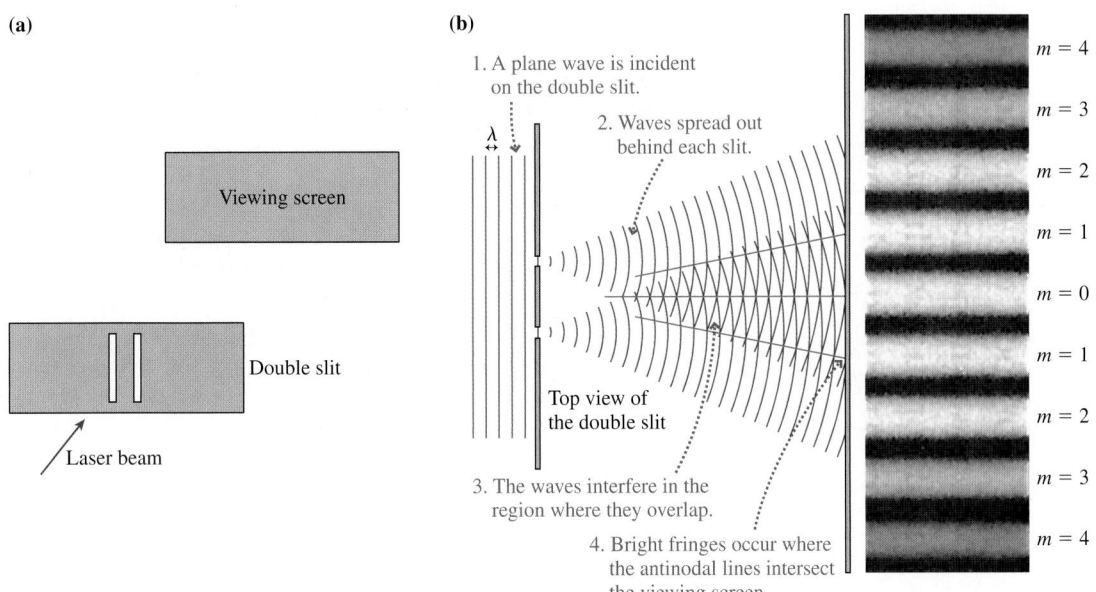

FIGURE 22.3 A double-slit interference experiment.

What should we expect to see on the screen? Figure 22.3b is a view from above the experiment, looking down on the top ends of the slits and the top edge of the viewing screen. Because the slits are very narrow, **light spreads out behind each slit** as it did in Figure 22.2, and these two spreading waves overlap in the region between the slits and the screen.

The primary conclusion of Chapter 21 was that two overlapped waves of equal length produce interference. In fact, Figure 22.3b is equivalent to the waves emitted by two loudspeakers, a situation we analyzed in Section 21.7. (It is very useful to compare Figure 22.3b with Figures 21.27 and 21.29a.) Nothing in that analysis depended on what type of wave it was, so the conclusions apply equally well to two overlapped light waves. If light really is a wave, we should see interference between the two light waves over the small region, typically 2 or 3 cm wide, where they overlap on the viewing screen.

The photograph in Figure 22.3b shows how the screen looks. As expected, the light is intense at points where an antinodal line intersects the screen. There is no light at all at points where a nodal line intersects the screen. These alternating bright and dark bands of light, due to constructive and destructive interference, are called **interference fringes.** The fringes are numbered $m = 0, 1, 2, 3, \ldots$, going outward from the center. The brightest fringe, at the midpoint of the viewing screen, with $m = 0$, is called the **central maximum.**

STOP TO THINK 22.1 Suppose the viewing screen in Figure 22.3 is moved closer to the double slit. What happens to the interference fringes?

    a. They get brighter but otherwise do not change.
    b. They get brighter and closer together.
    c. They get brighter and farther apart.
    d. They get out of focus.
    e. They fade out and disappear.

Analyzing Double-Slit Interference

16.1–16.3 Activ Physics

Figure 22.3 showed qualitatively that interference is produced behind a double slit by the overlap of the light waves spreading out behind each opening. Now let's analyze the experiment more carefully. Figure 22.4 shows the geometry of a double-slit experiment in which the spacing between the two slits is d and the distance to the viewing screen is L. We will assume that L is *very* much larger than d.

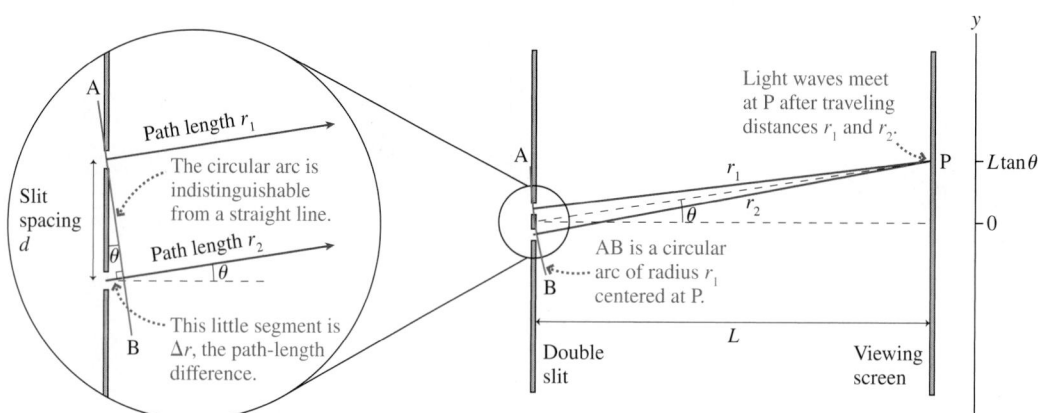

FIGURE 22.4 Geometry of the double-slit experiment.

Our goal, as it was with sound waves in Chapter 21, is to determine if the interference at a particular point is constructive, destructive, or in between. Let P be a point on the screen that is distance r_1 from one slit and r_2 from the other. We can specify point P either by the distance y from the midpoint on the viewing screen or by the angle θ from the midpoint between the slits. Angle θ and distance y are related by

$$y = L \tan\theta \qquad (22.1)$$

Figure 22.3b showed that both slits are illuminated by the *same* wave front from the laser. Consequently, the slits act as sources of identical waves ($\Delta\phi_0 = 0$). You learned in Chapter 21 that constructive interference between the waves from identical sources occurs at points for which the path-length difference $\Delta r = r_2 - r_1$ is an integer number of wavelengths:

$$\Delta r = m\lambda \qquad m = 0, 1, 2, 3, \ldots \qquad \text{(constructive interference)} \qquad (22.2)$$

Thus the interference at point P is constructive, producing a bright fringe, if $\Delta r = m\lambda$ at that point.

The midpoint on the viewing screen at $y = 0$ is equally distant from both slits ($\Delta r = 0$). Constructive interference at the midpoint produces the bright fringe identified as the central maximum in Figure 21.3b. The path-length difference increases as you move away from the center of the screen, and the $m = 1$ fringes occur at the positions where $\Delta r = 1\lambda$. That is, one wave has traveled exactly one wavelength farther than the other. In general, **the mth bright fringe occurs where one wave has traveled m wavelengths farther than the other and thus $\Delta r = m\lambda$.**

We need to find the specific positions on the screen where $\Delta r = m\lambda$. To do so, draw a circular arc of radius r_1 centered at point P. This arc, labeled AB in Figure 22.4, divides the path from the lower slit into a long segment of length r_1 and a much shorter segment of length $\Delta r = r_2 - r_1$.

Because L is very large in comparison to the slit spacing d, the two paths r_1 and r_2 are essentially parallel and the arc AB is indistinguishable from a straight line forming one side of a right triangle. This is shown in the magnified portion of Figure 22.4, where you can see that

$$\Delta r = d\sin\theta \qquad (22.3)$$

Bright fringes (constructive interference) occur at angles θ_m such that

$$\Delta r = d\sin\theta_m = m\lambda \qquad m = 0, 1, 2, 3, \ldots \qquad (22.4)$$

We have added the subscript m to denote that θ_m is the angle of the mth bright fringe, starting with $m = 0$ at the center.

In practice, the angle θ in a double-slit experiment is a very small angle ($<1°$) We can use the small-angle approximation $\sin\theta \approx \theta$, where θ must be in radians, to write Equation 22.4 as

$$\theta_m = m\frac{\lambda}{d} \qquad m = 0, 1, 2, 3, \ldots \qquad \text{(angles of bright fringes)} \qquad (22.5)$$

This gives the angular positions *in radians* of the bright fringes in the interference pattern.

It is usually more convenient to measure the *position* of the mth bright fringe, as measured from the center of the viewing screen. Using the small-angle approximation once again, this time in the form $\tan\theta \approx \theta$, we can substitute θ_m from Equation 22.5 for $\tan\theta_m$ in Equation 22.1 to find that the mth bright fringe occurs at position

$$y_m = \frac{m\lambda L}{d} \qquad m = 0, 1, 2, 3, \ldots \qquad \text{(positions of bright fringes)} \qquad (22.6)$$

The interference pattern is symmetrical, so there is an mth bright fringe at the same distance on both sides of the center. You can see this in Figure 22.3b. As we've already noted, **the $m = 1$ fringes occur at points on the screen where the light from one slit travels exactly one wavelength farther than the light from the other slit.**

NOTE ► Equations 22.5 and 22.6 do *not* apply to the interference of sound waves from two loudspeakers. The approximations we've used (small angles, $L \gg d$) are usually not valid for the much longer wavelengths of sound waves. ◄

Equation 22.6 predicts that **the interference pattern is a series of equally spaced bright lines** on the screen, exactly as shown in Figure 22.3b. How do we know the fringes are equally spaced? The **fringe spacing** between the m fringe and the $m + 1$ fringe is

$$\Delta y = y_{m+1} - y_m = \frac{(m + 1)\lambda L}{d} - \frac{m\lambda L}{d} = \frac{\lambda L}{d} \tag{22.7}$$

Because Δy is independent of m, *any* two bright fringes have the same spacing.

The dark fringes in the photograph are bands of destructive interference. You learned in Chapter 21 that destructive interference occurs at positions where the path-length difference of the waves is a half-integer number of wavelengths:

$$\Delta r = \left(m + \frac{1}{2}\right)\lambda \qquad m = 0, 1, 2, 3, \ldots \tag{22.8}$$
(destructive interference)

We can use Equation 22.4 for Δr and the small-angle approximation to find that the dark fringes are located at positions

$$y'_m = \left(m + \frac{1}{2}\right)\frac{\lambda L}{d} \qquad m = 0, 1, 2, 3, \ldots \tag{22.9}$$
(positions of dark fringes)

We have used y'_m, with a prime, to distinguish the location of the mth minimum from the mth maximum at y_m. You can see from Equation 22.9 that **the dark fringes are located exactly halfway between the bright fringes.**

EXAMPLE 22.1 Double-slit interference of a laser beam
Light from a helium-neon laser ($\lambda = 633$ nm) illuminates two slits spaced 0.40 mm apart. A viewing screen is 2.0 m behind the slits. What are the distances between the two $m = 2$ bright fringes and between the two $m = 2$ dark fringes?

MODEL Two closely spaced slits produce a double-slit interference pattern.

VISUALIZE The interference pattern looks like the photograph of Figure 22.3b. It is symmetrical, with $m = 2$ bright fringes at equal distances on both sides of the central maximum.

SOLVE The $m = 2$ bright fringe is located at position

$$y_m = \frac{m\lambda L}{d} = \frac{2(633 \times 10^{-9}\,\text{m})(2.0\,\text{m})}{4.0 \times 10^{-4}\,\text{m}}$$

$$= 6.3 \times 10^{-3}\,\text{m} = 6.3\,\text{mm}$$

Each of the $m = 2$ fringes is 6.3 mm from the central maximum, hence the distance between the two $m = 2$ bright fringes is 12.6 mm. The $m = 2$ dark fringe is located at

$$y'_m = \left(m + \frac{1}{2}\right)\frac{\lambda L}{d} = 7.9\,\text{mm}$$

Thus the distance between the two $m = 2$ dark fringes is 15.8 mm.

ASSESS As the fringes are counted outward from the center, the $m = 2$ bright fringe occurs *before* the $m = 2$ dark fringe.

EXAMPLE 22.2 Measuring the wavelength of light
A double-slit interference pattern is observed on a screen 1.0 m behind two slits spaced 0.30 mm apart. Ten bright fringes span a distance of 1.65 cm. What is the wavelength of the light?

MODEL It is not always obvious which fringe is the central maximum. Slight imperfections in the slits can make the interference fringe pattern less than ideal. However, you do not need to identify the $m = 0$ fringe because you can make use of the fact that the fringe spacing Δy is uniform. Ten bright fringes have *nine* spaces between them (not ten—be careful!).

VISUALIZE The interference pattern looks like the photograph of Figure 22.3b.

SOLVE The fringe spacing is

$$\Delta y = \frac{1.65\,\text{cm}}{9} = 1.833 \times 10^{-3}\,\text{m}$$

Using this fringe spacing in Equation 22.7, we find that the wavelength is

$$\lambda = \frac{d}{L}\Delta y = 5.50 \times 10^{-7} \text{ m} = 550 \text{ nm}$$

It is customary to express the wavelengths of light in nanometers. Be sure to do this as you solve problems.

ASSESS Young's double-slit experiment not only demonstrated that light is a wave, it provided a means for measuring the wavelength. You learned in Chapter 20 that the wavelengths of visible light span the range 400–700 nm. These lengths are smaller than we can easily comprehend. A wavelength of 550 nm, which is in the middle of the visible spectrum, is only about 1% of the diameter of a human hair.

STOP TO THINK 22.2 Light of wavelength λ_1 illuminates a double slit, and interference fringes are observed on a screen behind the slits. When the wavelength is changed to λ_2, the fringes get closer together. Is λ_2 larger or smaller than λ_1?

Intensity of the Double-Slit Interference Pattern

Equations 22.6 and 22.9 locate the positions of maximum and zero intensity. To complete our analysis we need to calculate the light *intensity* at every point on the screen. All the tools we need to do this calculation were developed in Chapters 20 and 21.

You learned in Chapter 20 that the wave intensity I is proportional to the square of the wave's amplitude. The light spreading out behind a *single* slit produces the wide band of light that you saw in Figure 22.2. The intensity in this band of light is $I_1 = Ca^2$, where a is the light-wave amplitude at the screen due to *one* wave and C is a proportionality constant.

If there were no interference, the light intensity due to two slits would be twice the intensity of one slit: $I_2 = 2I_1 = 2Ca^2$. In other words, two slits would cause the broad band of light on the screen to be twice as bright. But that's not what happens. Instead, the superposition of the two light waves creates bright and dark interference fringes.

We found in Chapter 21 (Equation 21.36) that the net amplitude of two superimposed waves is

$$A = \left| 2a \cos\left(\frac{\Delta\phi}{2}\right) \right| \tag{22.10}$$

where a is the amplitude of each individual wave. Because the sources are in phase, the phase difference $\Delta\phi$ at the point where the two waves are combined is $\Delta\phi = 2\pi(\Delta r/\lambda)$. Using Equation 22.3 for the path-length difference Δr, along with the small-angle approximation and Equation 22.2 for y gives us the phase difference at position y on the screen:

$$\Delta\phi = 2\pi\frac{\Delta r}{\lambda} = 2\pi\frac{d\sin\theta}{\lambda} \approx 2\pi\frac{d\tan\theta}{\lambda} = \frac{2\pi d}{\lambda L}y \tag{22.11}$$

Substituting Equation 22.11 into Equation 22.10, we find the wave amplitude at position y to be

$$A = \left| 2a \cos\left(\frac{\pi d}{\lambda L}y\right) \right| \tag{22.12}$$

Consequently, the light intensity at position y on the screen is

$$I = CA^2 = 4Ca^2\cos^2\left(\frac{\pi d}{\lambda L}y\right) \tag{22.13}$$

But Ca^2 is I_1, the light intensity of a single slit. Thus the intensity of the double-slit interference pattern at position y is

$$I_{\text{double}} = 4I_1 \cos^2\left(\frac{\pi d}{\lambda L}y\right) \tag{22.14}$$

Figure 22.5a is a graph of the double-slit intensity versus position y. Notice the unusual orientation of the graph, with the intensity increasing toward the *left* so that the y-axis can match the experimental layout. You can see that the intensity oscillates between dark fringes ($I_{\text{double}} = 0$) and bright fringes ($I_{\text{double}} = 4I_1$). The maxima occur at points where $y_m = m\lambda L/d$. This is what we found earlier for the positions of the bright fringes, so Equation 22.14 is consistent with our initial analysis.

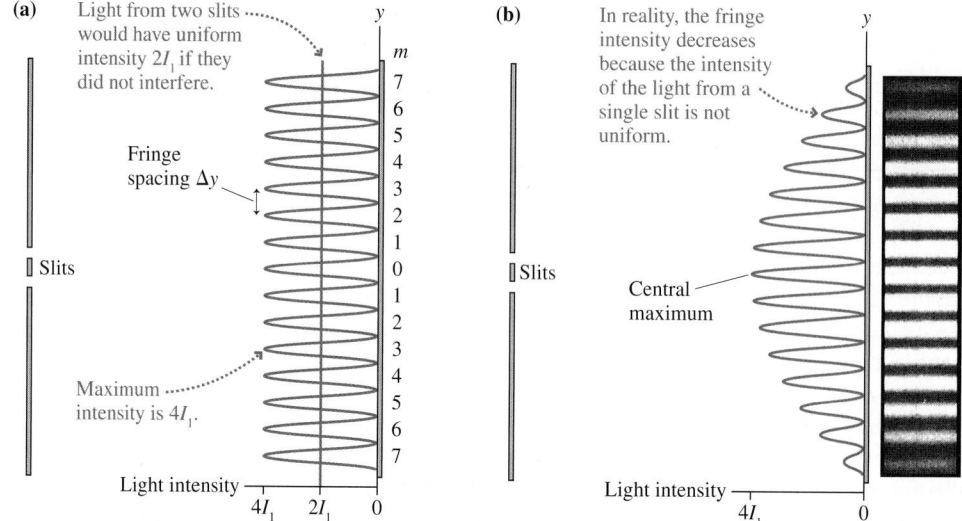

FIGURE 22.5 Intensity of the interference fringes in a double-slit experiment.

One curious feature is that the light intensity at the maxima is $I = 4I_1$, four times the intensity of the light from each slit alone. You might think that two slits would make the light twice as intense as one slit, but interference leads to a different result. Mathematically, two slits make the *amplitude* twice as big at points of constructive interference ($A = 2a$), so the intensity increases by a factor $2^2 = 4$. Physically, this is conservation of energy. The line labeled $2I_1$ in Figure 22.5a is the uniform intensity that two slits would produce *if* the waves did not interfere. Interference does not change the amount of light energy coming through the two slits, but it does redistribute the light energy on the viewing screen. You can see that the *average* intensity of the oscillating curve is $2I_1$, but the intensity of the bright fringes gets pushed up from $2I_1$ to $4I_1$ in order for the intensity of the dark fringes to drop from $2I_1$ to 0.

There is still one problem. Equation 22.14 predicts that all interference fringes are equally bright, but you saw in Figure 22.3b that the fringes decrease in brightness as you move away from the center. The erroneous prediction stems from our assumption that the amplitude a of the wave from each slit is constant across the screen. But this isn't really true. A more detailed calculation, in which the amplitude gradually decreases as you move away from the center, finds that Equation 22.14 is correct if I_1 slowly decreases as y increases.

Figure 22.5b summarizes this analysis by graphing the light intensity (Equation 22.14) with I_1 slowly decreasing as y increases. Comparing this graph to the photograph, you can see that the wave model of light has provided an excellent description of Young's double-slit interference experiment.

22.3 The Diffraction Grating

Suppose we were to replace the double slit with an opaque screen that has *N* closely spaced slits. When illuminated from one side, each of these slits becomes the source of a light wave that diffracts, or spreads out, behind the slit. Such a multi-slit device is called a **diffraction grating.** The light intensity pattern on a screen behind a diffraction grating is due to the interference of *N* overlapped waves.

Figure 22.6 shows a diffraction grating in which *N* slits are equally spaced a distance *d* apart. This is a top view of the grating, as we look down on the experiment, and the slits extend above and below the page. Only 10 slits are shown here, but a practical grating will have hundreds or even thousands of slits. Suppose a plane wave of wavelength λ approaches from the left. The crest of a plane wave arrives *simultaneously* at each of the slits, causing the wave emerging from each slit to be in phase with the wave emerging from every other slit. Each of these emerging waves spreads out, just like the light wave in Figure 22.2, and after a short distance they all overlap with each other and interfere.

We want to know how the interference pattern will appear on a screen behind the grating. The light wave at the screen is the superposition of *N* waves, from *N* slits, as they spread and overlap. As we did with the double slit, we'll assume that the distance *L* to the screen is very large in comparison with the slit spacing *d*, hence the path followed by the light from one slit to a point on the screen is *very nearly* parallel to the path followed by the light from neighboring slits. The paths cannot be perfectly parallel, of course, or they would never meet to interfere, but the slight deviation from perfect parallelism is too small to notice. You can see in Figure 22.6 that the wave from one slit travels distance $\Delta r = d \sin\theta$ more than the wave from the slit above it and $\Delta r = d \sin\theta$ less than the wave below it. This is the same reasoning we used in Figure 22.4 to analyze the double-slit experiment.

Figure 22.6 was a magnified view of the slits. Figure 22.7 steps back to where we can see the viewing screen. If the angle θ is such that $\Delta r = d \sin\theta = m\lambda$, where *m* is an integer, then the light wave arriving at the screen from one slit will be *exactly in phase* with the light waves arriving from the two slits next to it. But each of those waves is in phase with waves from the slits next to them, and so on until we reach the end of the grating. In other words, *N* **light waves, from *N* different slits, will** all **be in phase with each other when they arrive at a point on the screen at angle θ_m such that**

$$d \sin\theta_m = m\lambda \qquad m = 0, 1, 2, 3, \dots \qquad (22.15)$$

The screen will have bright constructive-interference fringes at the values of θ_m given by Equation 22.15. When this happens, we say that the light is "diffracted at angle θ_m." Because it's usually easier to measure distances rather than angles, the distance y_m from the center to the *m*th maximum is

$$y_m = L \tan\theta_m \qquad \text{(positions of bright fringes)} \qquad (22.16)$$

The integer *m* is called the **order** of the diffraction. For example, light diffracted at $\theta_2 = 60°$ would be the second-order diffraction. Practical gratings, with very small values for *d*, display only a few orders. Because *d* is usually very small, it is customary to characterize a grating by the number of *lines per millimeter.* Here "line" is synonymous with "slit," so the number of lines per millimeter is simply the inverse of the slit spacing *d* in millimeters.

NOTE ▶ The condition for constructive interference in a grating of *N* slits is identical to Equation 22.4 for just two slits. Equation 22.15 is simply the requirement that the path-length difference between adjacent slits, be they two

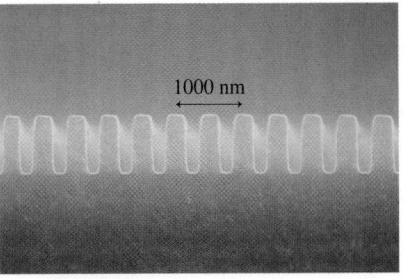

A microscopic side-on look at a diffraction grating.

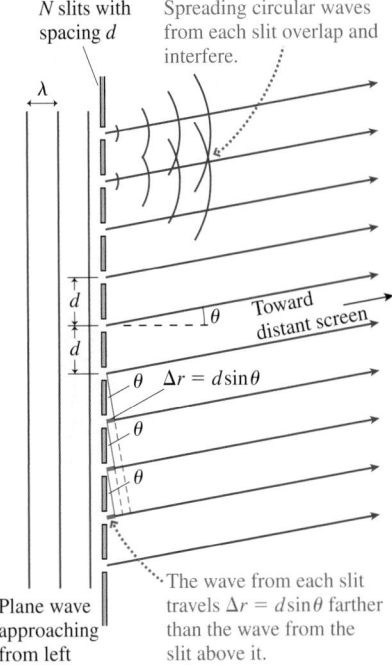

FIGURE 22.6 Top view of a diffraction grating with *N* = 10 slits.

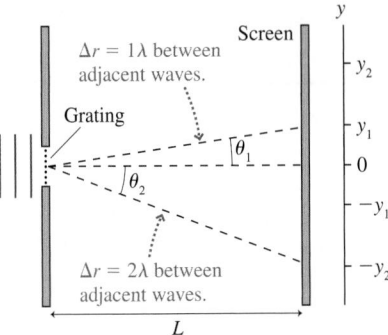

FIGURE 22.7 Angles of constructive interference.

or N, is $m\lambda$. But unlike the angles in double-slit interference, the angles of constructive interference from a diffraction grating are generally *not* small angles. The reason is that the slit spacing d in a diffraction grating is so small that λ/d is not a small number. Thus you *cannot* use the small-angle approximation to simplify Equations 22.15 and 22.16. ◄

The wave amplitude at the points of constructive interference is Na because N waves of amplitude a combine in phase. Because the intensity depends on the square of the amplitude, the intensities of the bright fringes of a diffraction grating are

$$I_{max} = N^2 I_1 \tag{22.17}$$

where, as before, I_1 is the intensity of the wave from a single slit. Equation 22.17 is consistent with our prior conclusion that the intensity of a bright fringe in a double-slit interference experiment is four times the intensity of the light from each slit alone. You can see that the fringe intensities increase rapidly as the number of slits increases.

Not only do the fringes get brighter as N increases, they also get narrower. This is again a matter of conservation of energy. If the light waves did not interfere, the intensity from N slits would be NI_1. Interference increases the intensity of the bright fringes by an extra factor of N, so to conserve energy the width of the bright fringes must be proportional to $1/N$. For a realistic diffraction grating, with $N > 100$, the interference pattern consists of a small number of *very* bright and *very* narrow fringes while most of the screen remains dark. Figure 22.8a shows the interference pattern behind a diffraction grating both graphically and with a simulation of the viewing screen. A comparison with Figure 22.5b shows that the bright fringes of a diffraction grating are much sharper and more distinct than the fringes of a double slit.

Because the bright fringes are so distinct, diffraction gratings are used for measuring the wavelengths of light. Suppose the incident light consists of two slightly different wavelengths. Each wavelength will be diffracted at a slightly different angle and, if N is sufficiently large, we'll see two distinct fringes on the screen. Figure 22.8b illustrates this idea. By contrast, the fringes in a double-slit experiment are so broad that it would not be possible to distinguish the fringes of one wavelength from those of the other.

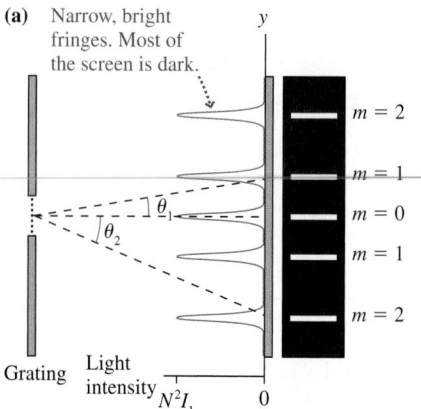

(a) Narrow, bright fringes. Most of the screen is dark.

Grating   Light intensity $N^2 I_1$

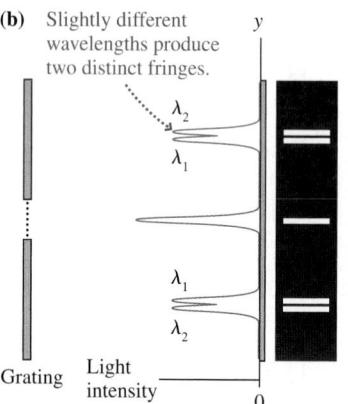

(b) Slightly different wavelengths produce two distinct fringes.

Grating   Light intensity

FIGURE 22.8 The interference pattern behind a diffraction grating.

EXAMPLE 22.3 Measuring wavelengths emitted by sodium atoms

Light from a sodium lamp passes through a diffraction grating having 1000 slits per millimeter. The interference pattern is viewed on a screen 1.000 m behind the grating. Two bright yellow fringes are visible 72.88 cm and 73.00 cm from the central maximum. What are the wavelengths of these two fringes?

VISUALIZE This is the situation shown in Figure 22.8b. The two fringes are very close together, so we expect the wavelengths to be only slightly different. No other yellow fringes are mentioned, so we will assume these two fringes are the first-order diffraction ($m = 1$).

SOLVE The distance y_m of a bright fringe from the central maximum is related to the diffraction angle by $y_m = L\tan\theta_m$. Thus the diffraction angles of these two fringes are

$$\theta_1 = \tan^{-1}\left(\frac{y_1}{L}\right) = \begin{cases} 36.08° & \text{fringe at 72.88 cm} \\ 36.13° & \text{fringe at 73.00 cm} \end{cases}$$

These angles must satisfy the interference condition $d\sin\theta_1 = \lambda$, so the wavelengths are

$$\lambda = d\sin\theta_1$$

What is d? If a 1 mm length of the grating has 1000 slits, then the spacing from one slit to the next must be 1/1000 mm, or $d = 1.00 \times 10^{-6}$ m. Thus the wavelengths creating the two bright fringes are

$$\lambda = d\sin\theta_1 = \begin{cases} 589.0 \text{ nm} & \text{fringe at 72.88 cm} \\ 589.6 \text{ nm} & \text{fringe at 73.00 cm} \end{cases}$$

ASSESS We had data accurate to four significant figures, and all four were necessary to distinguish the two wavelengths.

The science of measuring the wavelengths of atomic and molecular emissions is called **spectroscopy.** The two sodium wavelengths in this example are called the *sodium doublet,* a name given to two closely spaced wavelengths emitted by the atoms of one element. This doublet is an identifying characteristic of sodium. Because no other element emits these two wavelengths, the doublet can be used to identify the presence of sodium in a sample of unknown composition, even if sodium is only a very minor constituent. This procedure is called *spectral analysis.*

Reflection Gratings

We have analyzed what is called a *transmission grating,* with many parallel slits. In practice, most diffraction gratings are manufactured as *reflection gratings.* The simplest reflection grating, shown in Figure 22.9a, is a mirror with hundreds or thousands of narrow, parallel grooves cut into the surface. The grooves divide the surface into many parallel reflective stripes, each of which, when illuminated, becomes the source of a spreading wave. Thus an incident light wave is divided into *N* overlapped waves. The interference pattern is exactly the same as the interference pattern of light transmitted through *N* parallel slits.

Reflection gratings can also be created by carving or molding a series of narrow, parallel ridges into a reflective surface, as illustrated in Figure 22.9b. The *iridescence* of some bird feathers and insect shells arises from biological structures that have small, parallel ridges. These ridges form a reflection grating that diffracts the different wavelengths of sunlight at different angles. Multicolored hues of iridescence appear as the angle between the grating and your eye changes.

The rainbow of colors reflected from the surface of a CD is a similar display of interference. The surface of a CD is smooth plastic with a mirror-like reflective coating. Millions and millions of microscopic holes, each about 1 μm in diameter, are "burned" into the surface with a laser. The presence or absence of a hole at a particular location on the disk is interpreted as the 0 or 1 of digitally encoded information. But from an optical perspective, the array of holes in a shiny surface is a two-dimensional version of the reflection grating shown in Figure 22.9a. Less precise plastic reflection gratings can be manufactured at very low cost simply by stamping holes or grooves into a reflective surface, and these are widely sold as toys and novelty items. Rainbows of color are seen as each wavelength of white light is diffracted at a unique angle.

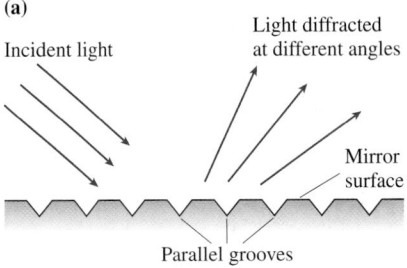

(a)

A reflection grating can be made by cutting parallel grooves in a mirror surface. These can be very precise, for scientific use, or mass produced in plastic.

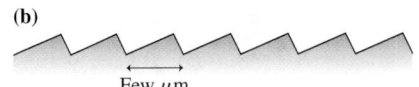

(b)

Few μm

Naturally occurring microscopic ridges are present in some bird feathers and insect shells. These cause iridescence when white light reflects off them.

FIGURE 22.9 Reflection gratings.

STOP TO THINK 22.3   White light passes through a diffraction grating and forms rainbow patterns on a screen behind the grating. For each rainbow,

a. the red side is on the right, the violet side on the left.
b. the red side is on the left, the violet side on the right.
c. the red side is closest to the center of the screen, the violet side is farthest from the center.
d. the red side is farthest from the center of the screen, the violet side is closest to the center.

22.4 Single-Slit Diffraction

We opened this chapter with a photograph of a water wave passing through a hole in a barrier, then spreading out on the other side. You then saw a photograph showing that light, after passing through a very narrow slit, also spreads out on the other side. This phenomenon is called *diffraction.* We're now ready to look at the details of diffraction.

Activ
Physics   16.6

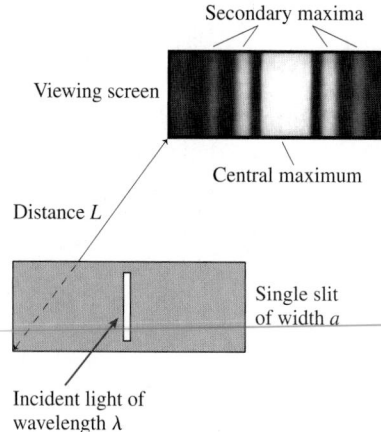

FIGURE 22.10 A single-slit diffraction experiment.

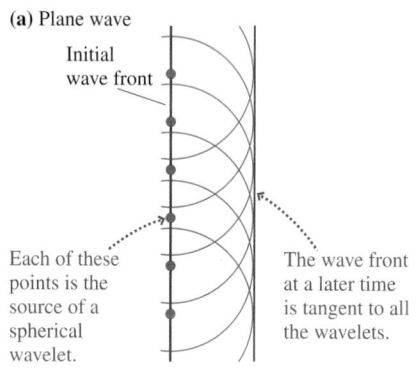

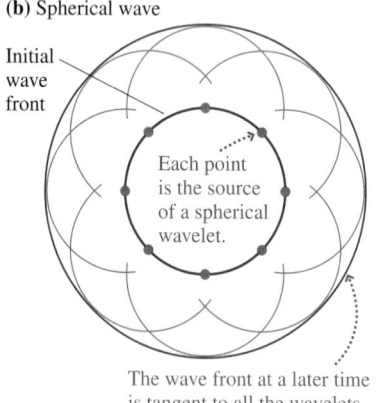

FIGURE 22.11 Huygens' principle applied to the propagation of plane waves and spherical waves.

Figure 22.10 shows the experimental arrangement for observing the diffraction of light through a narrow slit of width a. Diffraction through a tall, narrow slit is known as **single-slit diffraction.** A viewing screen is placed distance L behind the slit, and we will assume that $L \gg a$. The light pattern on the viewing screen consists of a *central maximum* flanked by a series of weaker **secondary maxima** and dark fringes. Notice that the central maximum is significantly broader than the secondary maxima. It is also significantly brighter than the secondary maxima, although that is hard to tell here because this photograph has been overexposed to make the secondary maxima show up better.

Huygens' Principle

Our analysis of the superposition of waves from distinct sources, such as two loudspeakers or the two slits in a double-slit experiment, has tacitly assumed that the sources are *point sources,* with no measurable extent. To understand diffraction, we need to think about the propagation of an *extended* wave front. This is a problem that was first considered by the Dutch scientist Christiaan Huygens, a contemporary of Newton who argued that light is a wave.

Huygens lived before a mathematical theory of waves had been developed, so he developed a geometrical model of wave propagation. His idea, which we now call **Huygens' principle,** has two steps:

1. Each point on a wave front is the source of a spherical *wavelet* that spreads out at the wave speed.
2. At a later time, the shape of the wave front is the line tangent to all the wavelets.

Figure 22.11 illustrates Huygens' principle for a plane wave and a spherical wave. As you can see, the line tangent to the wavelets of a plane wave is a plane that has propagated to the right. The line tangent to the wavelets of a spherical wave is a larger sphere. We've shown only a small number of wavelets, which leaves the new surface looking a bit bumpy, but you can easily imagine that the new surface would be perfectly smooth if we drew a wavelet for *every* point on the wave front.

Huygens' principle is a visual device, not a theory of waves. Nonetheless, the full mathematical theory of waves, as it developed in the 19th century, justifies Huygens' basic idea, although it is beyond the scope of this textbook to prove it.

Analyzing Single-Slit Diffraction

Figure 22.12a shows a wave front passing through a narrow slit of width a. According to Huygens' principle, each point on the wave front can be thought of as the source of a spherical wavelet. These wavelets overlap and interfere, producing the diffraction pattern seen on the viewing screen. The full mathematical analysis, using *every* point on the wave front, is a fairly difficult problem in calculus. We'll be satisfied with a geometrical analysis based on just a few wavelets.

Figure 22.12b shows several wavelets that travel straight ahead to the central point on the screen. (The screen is *very* far to the right in this magnified view of the slit.) The paths to the screen are very nearly parallel to each other, thus all the wavelets travel the same distance and arrive at the screen *in phase* with each other. The *constructive interference* between these wavelets produces the central maximum of the diffraction pattern at $\theta = 0$.

The situation is different at points away from the center. Wavelets 1 and 2 in Figure 22.12c start from points that are distance $a/2$ apart. Suppose that Δr_{12}, the extra distance traveled by wavelet 2, happens to be $\lambda/2$. In that case, wavelets 1 and 2 arrive out of phase and interfere destructively. But if Δr_{12} is $\lambda/2$, then the difference Δr_{34} between paths 3 and 4 and the difference Δr_{56} between paths 5 and 6 are also $\lambda/2$. Those pairs of wavelets also interfere destructively. The superposition of all the wavelets produces perfect destructive interference.

(a) Greatly magnified view of slit

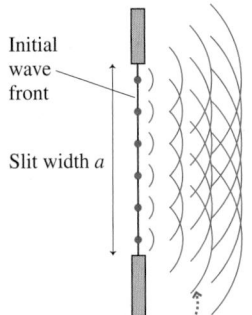

Initial wave front

Slit width a

The wavelets from each point on the initial wave front overlap and interfere, creating a diffraction pattern on the screen.

(b)

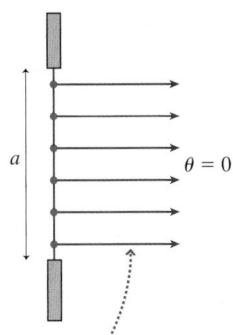

a

$\theta = 0$

The wavelets going straight forward all travel the same distance to the screen. Thus they arrive in phase and interfere constructively to produce the central maximum.

(c)

θ

1
3
5
2
4
6

Each point on the wave front is paired with another point distance $a/2$ away.

$\dfrac{a}{2}$

Δr_{12}

θ

These wavelets all meet on the screen at angle θ. Wavelet 2 travels distance $\Delta r_{12} = (a/2)\sin\theta$ farther than wavelet 1.

FIGURE 22.12 Each point on the wave front is a source of spherical wavelets. The superposition of these wavelets produces the diffraction pattern on the screen.

Figure 22.12c happens to show six wavelets, but our conclusion is valid for any number of wavelets. The key idea is that **every point on the wave front can be paired with another point that is distance $a/2$ away.** If the path-length difference is $\lambda/2$, the wavelets that originate at these two points will arrive at the screen out of phase and interfere destructively. When we sum the displacements of all N wavelets, they will—pair by pair—add to zero. The viewing screen at this position will be dark. This is the main idea of the analysis, one worth thinking about carefully.

You can see from Figure 22.12c that $\Delta r_{12} = (a/2)\sin\theta$. This path-length difference will be $\lambda/2$, the condition for destructive interference, if

$$\Delta r_{12} = \frac{a}{2}\sin\theta_1 = \frac{\lambda}{2} \tag{22.18}$$

or, equivalently, if $a\sin\theta_1 = \lambda$.

NOTE ▶ Equation 22.18 cannot be satisfied if the slit width a is less than the wavelength λ. If a wave passes through an opening smaller than the wavelength, the central maximum of the diffraction pattern expands to where it *completely* fills the space behind the opening. There are no minima or dark spots at any angle. This situation is uncommon for light waves, because λ is so small, but quite common in the diffraction of sound and water waves. ◀

We can extend this idea to find other angles of perfect destructive interference. Suppose each wavelet is paired with another wavelet from a point $a/4$ away. If Δr between these wavelets is $\lambda/2$, then all N wavelets will again cancel in pairs to give complete destructive interference. The angle θ_2 at which this occurs is found by replacing $a/2$ in Equation 22.18 with $a/4$, leading to the condition $a\sin\theta_2 = 2\lambda$. This process can be continued, and we find that the general condition for complete destructive interference is

$$a\sin\theta_p = p\lambda \qquad p = 1, 2, 3, \ldots \tag{22.19}$$

When $\theta_p \ll 1$ rad, which is almost always true for light waves, we can use the small-angle approximation to write

$$\theta_p = p\frac{\lambda}{a} \qquad p = 1, 2, 3, \ldots \qquad \text{(angles of dark fringes)} \tag{22.20}$$

(a)

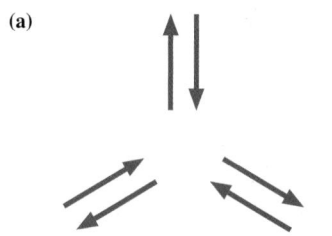

Each pair of vectors interferes destructively. The vector sum of all six vectors is zero.

(b)

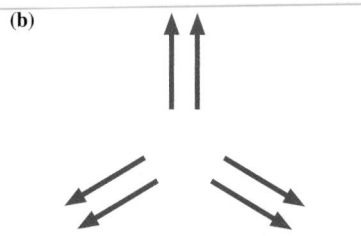

Each pair of vectors interferes constructively. Even so, the vector sum of all six vectors is zero.

FIGURE 22.13 Destructive interference by pairs leads to net destructive interference, but constructive interference by pairs does *not* necessarily lead to net constructive interference.

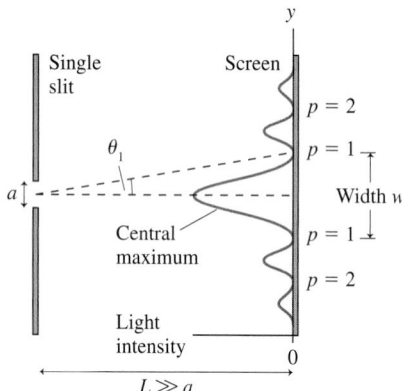

FIGURE 22.14 A graph of the intensity of a single-slit diffraction pattern.

Equation 22.20 gives the angles *in radians* to the dark minima in the diffraction pattern of Figure 22.10. Notice that $p = 0$ is explicitly *excluded*. $p = 0$ corresponds to the straight-ahead position at $\theta = 0$, but you saw in Figures 22.10 and 22.12b that $\theta = 0$ is the central *maximum,* not a minimum.

NOTE ▶ It is perhaps surprising that Equations 22.19 and 22.20 are *mathematically* the same as the condition for the mth *maximum* of the double-slit interference pattern. But the physical meaning here is quite different. Equation 22.20 locates the *minima* (dark fringes) of the single-slit diffraction pattern. ◀

You might think that we could use this method of pairing wavelets from different points on the wave front to find the maxima in the diffraction pattern. Why not take two points on the wave front that are distance $a/2$ apart, find the angle at which their wavelets are in phase and interfere constructively, then sum over all points on the wave front? There is a subtle but important distinction. Figure 22.13 shows six vector arrows. The arrows in Figure 22.13a are arranged in pairs such that the two members of each pair cancel. The sum of all six vectors is clearly the zero vector $\vec{0}$, representing destructive interference. This is the procedure we used in Figure 22.12c to arrive at Equation 22.18.

The arrows in Figure 22.13b are arranged in pairs such that the two members of each pair point in the same direction—constructive interference! Nonetheless, the sum of all six vectors is still $\vec{0}$. To have N waves interfere constructively requires more than simply having constructive interference between pairs. Each pair must also be in phase with every other pair, a condition not satisfied in Figure 22.13b. Stated another way, destructive interference by pairs leads to net destructive interference, but constructive interference by pairs does *not* necessarily lead to net constructive interference. It turns out that there is no simple formula to locate the maxima of a single-slit diffraction pattern.

It is possible, although beyond the scope of this textbook, to calculate the entire light intensity pattern. The results of such a calculation are shown graphically in Figure 22.14. You can see the bright central maximum at $\theta = 0$, the weaker secondary maxima, and the dark points of destructive interference at the angles given by Equation 22.20. Compare this graph to the photograph of Figure 22.10 and make sure you see the agreement between the two.

EXAMPLE 22.4 Diffraction of a laser through a slit
Light from a helium-neon laser ($\lambda = 633$ nm) passes through a narrow slit and is seen on a screen 2.0 m behind the slit. The first minimum in the diffraction pattern is 1.2 cm from the central maximum. How wide is the slit?

MODEL A narrow slit produces a single-slit diffraction pattern. A displacement of only 1.2 cm in a distance of 200 cm means that angle θ_1 is certainly a small angle.

VISUALIZE The intensity pattern will look like Figure 22.14.

SOLVE We can use the small-angle approximation to find that the angle to the first minimum is

$$\theta_1 = \frac{1.2 \text{ cm}}{200 \text{ cm}} = 0.00600 \text{ rad} = 0.344°$$

The first minimum is at angle $\theta_1 = \lambda/a$, from which we find that the slit width is

$$a = \frac{\lambda}{\theta_1} = \frac{633 \times 10^{-9} \text{ m}}{6.00 \times 10^{-3} \text{ rad}} = 1.06 \times 10^{-4} \text{ m} = 0.106 \text{ mm}$$

ASSESS This is typical of the slit widths used to observe single-slit diffraction. You can see that the small-angle approximation is well satisfied.

The Width of a Single-Slit Diffraction Pattern

We'll find it useful, as we did for the double slit, to measure positions on the screen rather than angles. The position of the pth dark fringe, at angle θ_p, is $y_p = L\tan\theta_p$, where L is the distance from the slit to the viewing screen. Using Equation 22.20 for θ_p and the small-angle approximation $\tan\theta_p \approx \theta_p$, we find that the dark fringes in the single-slit diffraction pattern are located at

$$y_p = \frac{p\lambda L}{a} \qquad p = 1, 2, 3, \ldots \qquad \text{(positions of dark fringes)} \qquad (22.21)$$

$p = 0$ is explicitly excluded because the midpoint on the viewing screen is the central maximum, not a dark fringe.

A diffraction pattern is dominated by the central maximum, which is much brighter than the secondary maxima. The width w of the central maximum, shown in Figure 22.14, is defined as the distance between the two $p = 1$ minima on either side of the central maximum. Because the pattern is symmetrical, the width is simply $w = 2y_1$. This is

$$w = \frac{2\lambda L}{a} \qquad (22.22)$$

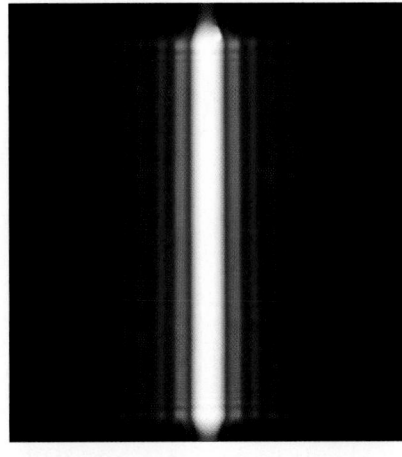

The central maximum of this single-slit diffraction pattern appears white because it is overexposed. The width of the central maximum is clear.

The width of the central maximum is *twice* the spacing $\lambda L/a$ between the dark fringes on either side. The farther back you move the screen (increasing L), the wider the pattern of light on it becomes. In other words, the light waves are *spreading out* behind the slit, and they fill a wider and wider region as they travel farther.

An important implication of Equation 22.22, one contrary to common sense, is that a narrower slit (smaller a) causes a *wider* diffraction pattern. **The smaller the opening you squeeze a wave through, the *more* it spreads out on the other side.**

EXAMPLE 22.5 **Determining the wavelength**
Light passes through a 0.12-mm-wide slit and forms a diffraction pattern on a screen 1.00 m behind the slit. The width of the central maximum is 0.85 cm. What is the wavelength of the light?

SOLVE From Equation 22.22, the wavelength is

$$\lambda = \frac{aw}{2L} = \frac{(1.2 \times 10^{-4}\,\text{m})(0.0085\,\text{m})}{2(1.00\,\text{m})}$$

$$= 5.10 \times 10^{-7}\,\text{m} = 510\,\text{nm}$$

STOP TO THINK 22.4   The figure shows two single-slit diffraction patterns. The distance between the slit and the viewing screen is the same in both cases. Which of the following could be true?

a. The slits are the same for both; $\lambda_1 > \lambda_2$.
b. The slits are the same for both; $\lambda_2 > \lambda_1$.
c. The wavelengths are the same for both; $a_1 > a_2$.
d. The wavelengths are the same for both; $a_2 > a_1$.
e. The slits and the wavelengths are the same for both; $p_1 > p_2$.
f. The slits and the wavelengths are the same for both; $p_2 > p_1$.

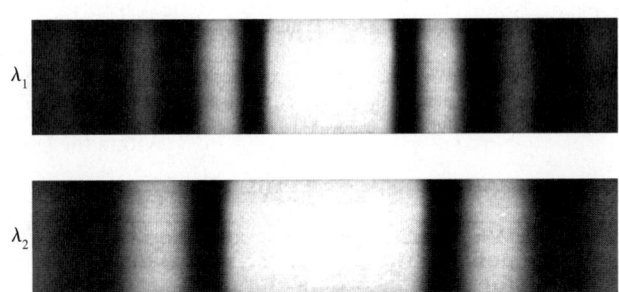

22.5 Circular-Aperture Diffraction

Diffraction occurs if a wave passes through an opening of any shape. Diffraction by a single slit establishes the basic ideas of diffraction, but a common situation of practical importance is diffraction of a wave by a **circular aperture.** Circular diffraction is mathematically more complex than diffraction from a slit, and we will present results without derivation.

Consider some examples. A loudspeaker cone generates sound by the rapid oscillation of a diaphragm, but the sound wave must pass through the circular aperture defined by the outer edge of the speaker cone before it travels into the room beyond. This is diffraction by a circular aperture. Telescopes and microscopes are the reverse. Light waves from outside need to enter the instrument. To do so, they must pass through a circular input lens. In fact, the performance limit of optical instruments is determined by the diffraction of the circular openings through which the waves must pass. This is an issue we'll look at more closely in the next chapter.

Figure 22.15 shows a circular aperture of diameter D. Light waves passing through this aperture spread out to generate a *circular* diffraction pattern. You should compare this to Figure 22.10 for a single slit to note the similarities and differences. The diffraction pattern still has a *central maximum,* now circular in shape, and it is surrounded by a series of secondary bright fringes. Most of the intensity is contained within the central maximum.

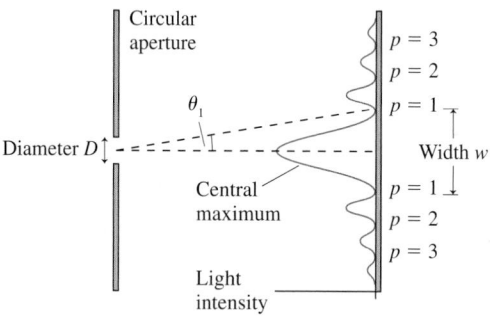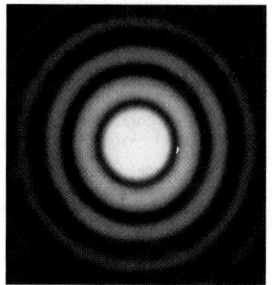

FIGURE 22.15 The diffraction of light by a circular opening.

Angle θ_1 locates the first minimum in the intensity, where there is complete destructive interference. A mathematical analysis of circular diffraction leads to the result

$$\theta_1 = \frac{1.22\lambda}{D} \tag{22.23}$$

where D is the *diameter* of the circular opening. This is very similar to the result for a single slit, but not quite the same. Equation 22.23 has assumed the small-angle approximation, which is almost always valid for the diffraction of light but usually is *not* valid for the diffraction of longer-wavelength sound waves.

Within the small-angle approximation, the width of the central maximum is

$$w = 2y_1 = 2L\tan\theta_1 \approx \frac{2.44\lambda L}{D} \tag{22.24}$$

This is similar to the width of the central maximum in a single-slit diffraction pattern, but not quite the same. The diameter of the diffraction pattern increases with distance L, showing that light spreads out behind a circular aperture, but it decreases if the size D of the aperture is increased.

EXAMPLE 22.6 Shining a laser through a circular hole
Light from a helium-neon laser ($\lambda = 633$ nm) passes through a 0.50-mm-diameter hole. How far away should a viewing screen be placed to observe a diffraction pattern whose central maximum is 3.0 mm in diameter?

SOLVE Equation 22.24 gives us the appropriate screen distance:

$$L = \frac{wD}{2.44\lambda} = \frac{(3.0 \times 10^{-3}\,\text{m})(5.0 \times 10^{-4}\,\text{m})}{2.44(633 \times 10^{-9}\,\text{m})} = 0.97\,\text{m}$$

The Wave and Ray Models of Light

We opened this chapter by noting that there are three models of light, each useful within a certain range of circumstances. We are now at a point where we can establish an important condition that separates the wave model of light from the ray model of light.

When light passes through an opening of size a, the angle of the first diffraction minimum is

$$\theta_1 = \sin^{-1}\left(\frac{\lambda}{a}\right) \tag{22.25}$$

Equation 22.25 is for a slit, but the result is very nearly the same if a is the diameter of a circular aperture. Regardless of the shape of the opening, the factor that determines how much a wave spreads out behind an opening is the ratio λ/a, the size of the wavelength compared to the size of the opening.

Figure 22.16 illustrates the difference between a wave whose wavelength is much smaller than the size of the opening and a second wave whose wavelength is comparable to the opening. A wave with $\lambda/a \approx 1$ quickly spreads to fill the region behind the opening. Light waves, because of their very short wavelength, almost always have $\lambda/a \ll 1$ and diffract to produce a slowly spreading "beam" of light.

Now we can better appreciate Newton's dilemma. With everyday-sized openings, sound and water waves have $\lambda/a \approx 1$ and diffract to fill the space behind the opening. Consequently, this is what we come to expect for the behavior of waves. Newton saw no evidence of this for light passing through openings. We see now that light really does spread out behind an opening, but the very small λ/a ratio usually makes the diffraction pattern too small to see. Diffraction begins to be discernible only when the size of the opening is a fraction of a millimeter or less. If we wanted the diffracted light wave to *fill* the space behind the opening ($\theta_1 \approx 90°$), as a sound wave does, we would need to reduce the size of the opening to $a \approx 0.001$ mm! Although holes this small can be made today, with the processes used to make integrated circuits, the amount of light able to pass through such a small opening is so small that its intensity is too weak to be seen by the eye.

Figure 22.17 shows light passing through a hole of diameter D. According to the ray model, light rays passing through the hole travel straight ahead to create a bright circular spot of diameter D on a viewing screen. This is the *geometric image* of the slit. In reality, diffraction causes the light to spread out behind the slit, but—and this is the important point—**we will not notice the spreading if it is less than the diameter D of the geometric image.** That is, we will not be aware of diffraction unless the bright spot on the screen increases in diameter.

This idea provides a reasonable criterion for when to use ray optics and when to use wave optics:

- If the spreading due to diffraction is less than the size of the opening, use the ray model and think of light as traveling in straight lines.
- If the spreading due to diffraction is larger than the size of the opening, use the wave model of light.

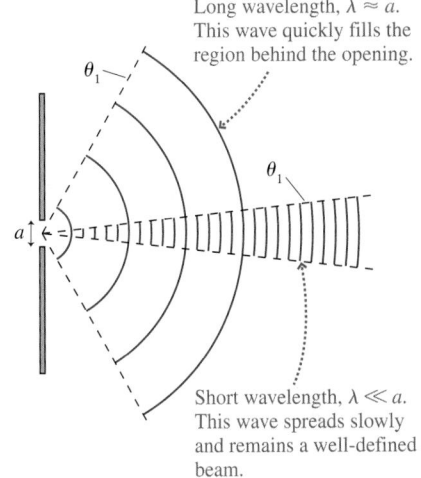

Long wavelength, $\lambda \approx a$. This wave quickly fills the region behind the opening.

Short wavelength, $\lambda \ll a$. This wave spreads slowly and remains a well-defined beam.

FIGURE 22.16 The diffraction of a long-wavelength wave and a short-wavelength wave through the same opening.

If light travels in straight lines, the image on the screen is the same size as the hole. Diffraction will not be noticed unless the light spreads over a diameter larger than D.

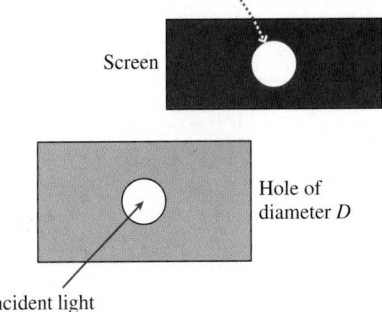

Screen

Hole of diameter D

Incident light

FIGURE 22.17 Diffraction will be noticed only if the bright spot on the screen is wider than D.

The cross-over point between these two regimes occurs when the spreading due to diffraction is equal to the size of the opening. The central-maximum width of a circular-aperture diffraction pattern is $w = 2.44\lambda L/D$. If we equate this diffraction width to the diameter of the aperture itself, we have

$$\frac{2.44\lambda L}{D_c} = D_c \tag{22.26}$$

where the subscript c on D_c indicates that this is the cross-over between the ray model and the wave model. Solving for D_c, we find

$$D_c = \sqrt{2.44\lambda L} \tag{22.27}$$

This is the diameter of a circular aperture whose diffraction pattern, at distance L, has width $w = D$. We know that visible light has $\lambda \approx 500$ nm, and a typical distance in laboratory work is $L \approx 1$ m. For these values,

$$D_c \approx 1 \text{ mm}$$

This brings us to an important and very practical conclusion, presented in Tactics Box 22.1.

TACTICS BOX 22.1 Choosing a model of light

❶ When light passes through openings < 1 mm in size, diffraction effects are usually important. Use the wave model of light.
❷ When light passes through openings > 1 mm in size, diffraction effects are usually not important. Use the ray model of light.

Openings ≈ 1 mm in size are a gray area. Whether one should use a ray model or a wave model will depend on the precise values of λ and L. We'll avoid such ambiguous cases in this book, sticking with examples and homework that fall clearly within the wave model or the ray model. Lenses and mirrors, in particular, are almost always 1 mm in size. We will study the optics of lenses and mirrors in the chapter on ray optics. This chapter on wave optics deals with objects and openings <1 mm in size.

22.6 Interferometers

Scientists and engineers have devised many ingenious methods for using interference to control the flow of light and to make very precise measurements with light waves. A device that makes practical use of interference is called an **interferometer.**

Interference requires two waves of *exactly* the same wavelength. One way of guaranteeing that two waves have exactly equal wavelengths is to divide a wave front into two pieces. This is what happens in double-slit interference. Another way to produce light waves with identical wavelengths is to divide one wave into two parts of smaller amplitude. Later, at a different point in space, the two parts are recombined. Interferometers are based on the division and recombination of a single wave.

To illustrate the idea, Figure 22.18 shows an *acoustical interferometer*. A sound wave is sent into the left end of the tube. The wave splits into two parts at the junction, and waves of smaller amplitude travel around each side. Distance L can be changed by sliding the upper tube in and out like a trombone. After traveling distances r_1 and r_2, the waves recombine and their superposition travels out to the microphone. The sound emerging from the right end has maximum intensity,

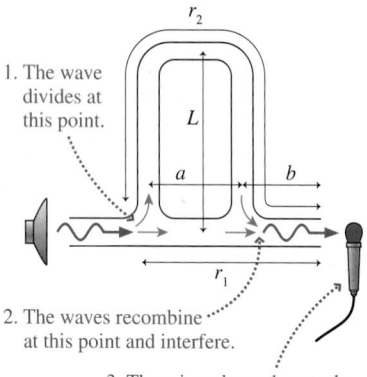

1. The wave divides at this point.

2. The waves recombine at this point and interfere.

3. The microphone detects the superposition of the two waves that traveled different distances.

FIGURE 22.18 An acoustical interferometer.

zero intensity, or somewhere in between depending on the phase difference between the two waves as they recombine.

The two waves traveling through the interferometer started from the *same* source, the loudspeaker, hence the phase difference $\Delta\phi_0$ between the wave sources is automatically zero. The phase difference $\Delta\phi$ between the recombined waves is due entirely to the different distances they travel. Consequently, the conditions for constructive and destructive interference are those we found in Chapter 21 for identical sources:

Constructive: $\Delta r = m\lambda$

Destructive: $\Delta r = \left(m + \dfrac{1}{2}\right)\lambda$ $m = 0, 1, 2, \ldots$ (22.28)

The distance each wave travels is easily found from Figure 22.18:

$$r_1 = a + b$$

$$r_2 = L + a + L + b = 2L + a + b$$

Thus the path-length difference between the waves is $\Delta r = r_2 - r_1 = 2L$, and the conditions for constructive and destructive interference are

Constructive: $L = m\dfrac{\lambda}{2}$

Destructive: $L = \left(m + \dfrac{1}{2}\right)\dfrac{\lambda}{2}$ $m = 0, 1, 2, \ldots$ (22.29)

The interference conditions involve $\lambda/2$ rather than just λ because the wave following the upper path travels distance L *twice,* once up and once down. The upper wave travels a full wavelength λ farther than the lower wave when $L = \lambda/2$.

The interferometer is used by recording the alternating maxima and minima in the sound as the top tube is pulled out and L changes. The interference changes from a maximum to a minimum and back to a maximum every time L increases by half a wavelength. Figure 22.19 is a graph of the sound intensity at the microphone as L is increased. You can see, from Equation 22.29, that the number Δm of maxima appearing as the length changes by ΔL is

$$\Delta m = \frac{\Delta L}{\lambda/2}$$ (22.30)

Equation 22.30 is the basis for measuring wavelengths very accurately.

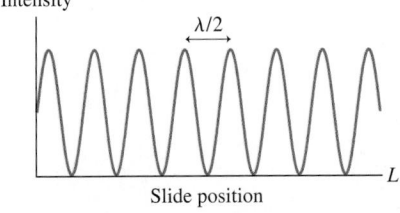

FIGURE 22.19 Interference maxima and minima alternate as the slide on an acoustical interferometer is withdrawn.

EXAMPLE 22.7 **Measuring the wavelength of sound**

A loudspeaker broadcasts a sound wave into an acoustical interferometer. The interferometer is adjusted so that the output sound intensity is a maximum, then the slide is slowly withdrawn. Exactly 10 new maxima appear as the slide moves 31.52 cm. What is the wavelength of the sound wave?

MODEL An interferometer produces a new maximum each time L increases by $\lambda/2$, causing the path-length difference Δr to increase by λ.

SOLVE Using Equation 22.30,

$$\lambda = \frac{2\Delta L}{\Delta m} = \frac{2(31.52 \text{ cm})}{10} = 6.304 \text{ cm}$$

ASSESS The wavelength can be determined to four significant figures because the distance was measured to four significant figures.

The Michelson Interferometer

Albert Michelson, the first American scientist to receive a Nobel prize, invented an optical interferometer analogous to the acoustical interferometer. The Michelson interferometer has been widely used for over a century to make precise measurements of wavelengths and distances. A contemporary version of a Michelson

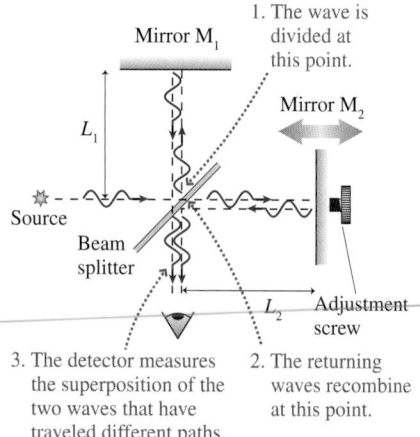

1. The wave is divided at this point.

Mirror M$_1$

Mirror M$_2$

L_1

Source

Beam splitter

L_2 Adjustment screw

3. The detector measures the superposition of the two waves that have traveled different paths.

2. The returning waves recombine at this point.

FIGURE 22.20 A Michelson interferometer.

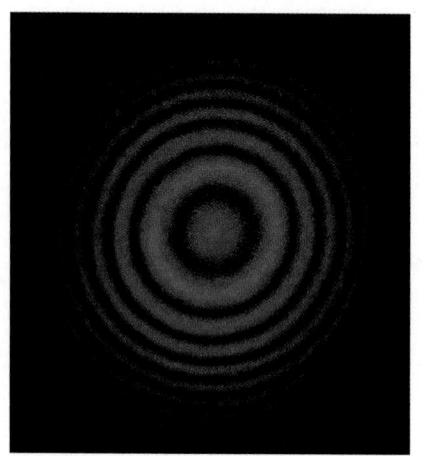

FIGURE 22.21 Photograph of the interference fringes produced by a Michelson interferometer.

interferometer several kilometers in length is being built underground, in old mine shafts, to search for *gravity waves* arriving at the earth from distant supernovae.

Figure 22.20 shows a Michelson interferometer. A light wave enters from the left and strikes a **beam splitter,** a partially silvered mirror that reflects half the light but transmits the other half. The beam splitter divides the wave in two, and the two waves then travel toward mirrors M$_1$ and M$_2$. Half of the wave reflected from M$_1$ is transmitted through the beam splitter, where it recombines with the reflected half of the wave returning from M$_2$. The superimposed waves travel on to a light detector, originally a human observer but now more likely an electronic photodetector.

Mirror M$_2$ can be moved forward or backward by turning a precision screw. This is equivalent to pulling out the slide on the acoustical interferometer. The waves travel distances $r_1 = 2L_1$ and $r_2 = 2L_2$, with the factors of two appearing because the waves travel to the mirrors and back again. Thus the path-length difference between the two waves is

$$\Delta r = 2L_2 - 2L_1 \qquad (22.31)$$

Δr is all we need to determine if the light seen by the detector exhibits constructive interference (bright light) or destructive interference (no light). The condition for constructive interference is $\Delta r = m\lambda$, hence constructive interference occurs when

$$\text{Constructive:} \quad L_2 - L_1 = m\frac{\lambda}{2} \qquad m = 0, 1, 2, \ldots \qquad (22.32)$$

This result is essentially identical to Equation 22.29 for an acoustical interferometer. Both divide a wave, send the two smaller waves along two paths that differ in length by Δr, then recombine the two waves at a detector.

You might expect the interferometer output to be either "bright" or "dark." Instead, a viewing screen shows the pattern of circular interference fringes seen in Figure 22.21. Our analysis was for light waves that strike the beam splitter at exactly 45° and impinge on the mirrors exactly perpendicular to the surface. In an actual experiment, some of the light waves enter the interferometer at slightly different angles and, as a result, the recombined waves have slightly altered path-length differences Δr. These waves cause the alternating bright and dark fringes as you move outward from the center of the pattern. Their analysis will be left to more advanced courses in optics. Equation 22.32 is valid at the *center* of the circular pattern, thus there is a bright central spot when Equation 22.32 is true.

If mirror M$_2$ is moved by turning the screw, the central spot in the fringe pattern alternates between bright and dark. The output recorded by a detector looks exactly like the alternating loud and soft sounds shown in Figure 22.19. Suppose the interferometer is adjusted to produce a bright central spot. The next bright spot will appear when M$_2$ has moved half a wavelength, increasing the path-length difference by one full wavelength. The number Δm of maxima appearing as M$_2$ moves through distance ΔL_2 is

$$\Delta m = \frac{\Delta L_2}{\lambda/2} \qquad (22.33)$$

Very precise wavelength measurements can be made by moving the mirror while counting the number of new bright spots that appear at the center of the pattern. The number Δm is counted and known exactly. The only limitation on how precisely λ can be measured this way is the precision with which distance ΔL_2 can be measured. Unlike λ, which is microscopic, ΔL_2 is typically a few millimeters, a macroscopic distance that can be measured very accurately using precision screws, micrometers, precision spacing blocks, and other techniques. Michelson's invention provided a way to transfer the precision of macroscopic distance measurements to an equal precision for the wavelength of light.

EXAMPLE 22.8 Measuring the wavelength of light

An experimenter uses a Michelson interferometer to measure one of the wavelengths of light emitted by neon atoms. She slowly moves mirror M_2 until 10,000 new bright central spots have appeared. (In a modern experiment, a photodetector and computer would be used to eliminate the possibility of experimenter error while counting.) She then measures that the mirror has moved a distance of 3.164 mm. What is the wavelength of the light?

MODEL An interferometer produces a new maximum each time L_2 increases by $\lambda/2$, causing the path-length difference Δr to increase by λ.

SOLVE The mirror moves $\Delta L_2 = 3.164$ mm $= 3.164 \times 10^{-3}$ m. We can use Equation 22.33 to find

$$\lambda = \frac{2\Delta L_2}{\Delta m} = 6.328 \times 10^{-7} \text{ m} = 632.8 \text{ nm}$$

ASSESS A measurement of ΔL_2 accurate to four significant figures allowed us to determine λ to four significant figures. This happens to be the neon wavelength that is emitted as the laser beam in a helium-neon laser.

The technology of wavelength measurements using a Michelson interferometer became so good that in 1960 an international scientific committee decided to use it as the *definition* of the meter. They defined the meter to be exactly 1,650,763.73 wavelengths of a particular orange color of light emitted by the krypton isotope ^{86}Kr. Scientists would aim this light at a Michelson interferometer, then move mirror M_2 while counting out a predetermined number of new bright spots. The total distance traveled by M_2 was then, *by definition,* exactly 1 meter (or, in actual practice, some fraction of 1 meter). This was a vast improvement over the previous definition of the meter as the distance between two scratches on a special bar kept in a safe in Paris, and it stood until 1983 when the meter was again refined in terms of the speed of light.

STOP TO THINK 22.5   A Michelson interferometer using light of wavelength λ has been adjusted to produce a bright spot at the center of the interference pattern. Mirror M_1 is then moved distance λ toward the beam splitter while M_2 is moved distance λ away from the beam splitter. How many bright-dark-bright fringe shifts are seen?

a. 0   b. 1
c. 2   d. 4
e. 8   f. It's not possible to say without knowing λ.

Measuring Indices of Refraction

The Michelson interferometer can be used to measure indices of refraction, especially of gases, as a function of wavelength. In Figure 22.22, a cell of accurately known thickness d has been inserted into one arm of the interferometer. To begin, all the air is pumped out of the cell. As light travels from the beam splitter to the mirror and back, the number of wavelengths inside the cell is

$$m_1 = \frac{2d}{\lambda_{\text{vac}}} \tag{22.34}$$

where the 2 appears because the light passes through the cell twice.

The cell is then filled with gas at 1 atm pressure. Light travels slightly slower in the gas, and its index of refraction is $n = c/v$. You learned in Chapter 20 that the wavelength of light in a material with index of refraction n is λ_{vac}/n. With the cell filled, the number of wavelengths spanning distance d is

$$m_2 = \frac{2d}{\lambda} = \frac{2d}{\lambda_{\text{vac}}/n} \tag{22.35}$$

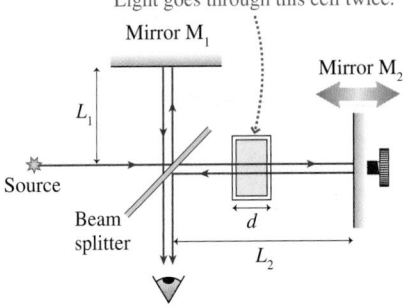

Gas-filled cell of thickness d. Light goes through this cell twice.

FIGURE 22.22 Measuring the index of refraction.

The physical distance has not changed, but the number of wavelengths along the lower path has. Filling the cell has increased the lower path by

$$\Delta m = m_2 - m_1 = (n - 1)\frac{2d}{\lambda_{\text{vac}}} \qquad (22.36)$$

wavelengths. Each increase of one wavelength causes one bright-dark-bright fringe shift at the output, so the index of refraction can be determined by counting fringe shifts as the cell is filled.

EXAMPLE 22.9 Measuring the index of refraction

A Michelson interferometer uses a helium-neon laser with wavelength $\lambda_{\text{vac}} = 633.0$ nm. As a 4.00-cm-thick cell is slowly filled with a gas, 43 bright-dark-bright fringe shifts are seen and counted. What is the index of refraction of the gas at this wavelength?

MODEL The gas increases the number of wavelengths in one arm of the interferometer. Each additional wavelength causes one bright-dark-bright fringe shift.

SOLVE We can rearrange Equation 22.36 to find that the index of refraction is

$$n = 1 + \frac{\lambda_{\text{vac}}\Delta m}{2d} = 1 + \frac{(6.330 \times 10^{-7}\text{ m})(43)}{2(0.0400\text{ m})} = 1.00034$$

ASSESS This may seem like a six-significant-figure result, but it's really only two. What we're measuring is not n but $n - 1$. We know the fringe count to two significant figures, and that has allowed us to compute $n - 1 = \lambda_{\text{vac}}\Delta m/2d = 3.4 \times 10^{-4}$.

Holography

No discussion of wave optics would be complete without mentioning holography, which has both scientific and artistic applications. The basic idea is a simple extension of interferometry.

Figure 22.23a shows how a **hologram** is made. A beam splitter divides a laser beam into two waves. One wave illuminates the object of interest. The light scattered by this object is a very complex wave, but it is the wave you would see if you looked at the object from the position of the film. The other wave, called the *reference beam,* is reflected directly toward the film. The scattered light and the reference beam meet at the film and interfere. The film records their interference pattern.

(a) Recording a hologram

The interference between the scattered light and the reference beam is recorded on the film.

Film

Plane waves

Reference beam

Laser →

Beam splitter

Object beam

The scattered light has a complex wave front.

Object

(b) A hologram

An enlarged photo of the developed film. This is the hologram.

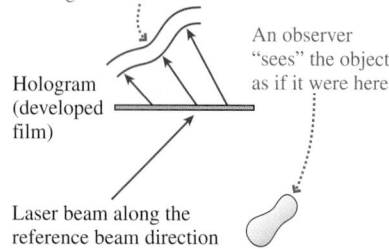

(c) Playing a hologram

The diffraction of the laser beam through the light and dark patches of the film reconstructs the original scattered wave.

Hologram (developed film)

An observer "sees" the object as if it were here.

Laser beam along the reference beam direction

FIGURE 22.23 Holography is an important application of wave optics.

The interference patterns we've looked at in this chapter have been simple geometrical patterns of stripes and circles because the light waves have been well-behaved plane waves and spherical waves. The light wave scattered by the object in Figure 22.23a is exceedingly complex. As a result, the interference pattern recorded on the film—the hologram—is a seemingly random pattern of whorls and blotches. Figure 22.23b is an enlarged photograph of a portion of a hologram. It's certainly not obvious that information is stored in this pattern, but it is.

The hologram is "played" by sending just the reference beam through it, as seen in Figure 22.23c. The reference beam diffracts through the transparent parts of the hologram, just as it would through the slits of a diffraction grating. Amazingly, the diffracted wave is *exactly the same* as the light wave that had been scattered by the object! In other words, the diffracted reference beam *reconstructs* the original scattered wave. As you look at this diffracted wave, from the far side of the hologram, you "see" the object exactly as if it were there. The view is three dimensional because, by moving your head with respect to the hologram, you can see different portions of the wave front.

SUMMARY

The goal of Chapter 22 has been to understand and apply the wave model of light.

GENERAL PRINCIPLES

Huygens' principle says that each point on a wave front is the source on a spherical wavelet. The wave front at a later time is tangent to all the wavelets.

Diffraction is the spreading of a wave after it passes through an opening.

Constructive and destructive interference are due to the overlap of two or more waves as they spread behind openings.

IMPORTANT CONCEPTS

The wave model of light considers light to be a wave propagating through space. Diffraction and interference are important.

The ray model of light considers light to travel in straight lines like little particles. Diffraction and interference are not important.

Diffraction is important when the width of the diffraction pattern of an aperture equals or exceeds the size of the aperture. For a circular aperture, the crossover between the ray and wave models occurs for an opening of diameter $D_c = \sqrt{2.44\lambda L}$.

In practice, $D_c \approx 1$ mm. Thus

- Use the wave model when light passes through openings <1 mm in size. Diffraction effects are usually important.
- Use the ray model when light passes through openings >1 mm in size. Diffraction is usually not important.

APPLICATIONS

Single slit of width a.
A bright **central maximum** of width

$$w = \frac{2\lambda L}{a}$$

is flanked by weaker **secondary maxima.**
Dark fringes are located at angles such that

$$a\sin\theta_p = p\lambda \qquad p = 1, 2, 3, \ldots$$

If $\lambda/p \ll 1$, then from the small-angle approximation

$$\theta_p = \frac{p\lambda}{a} \qquad y_p = \frac{p\lambda L}{a}$$

Circular aperture of diameter D.
A bright central maximum of diameter

$$w = \frac{2.44\lambda L}{D}$$

is surrounded by circular secondary maxima.
The first dark fringe is located at

$$\theta_1 = \frac{1.22\lambda}{D} \qquad y_1 = \frac{1.22\lambda L}{D}$$

For an aperture of any shape, a smaller opening causes a more rapid spreading of the wave behind the opening.

Interference due to wave-front division

Waves overlap as they spread out behind slits. Constructive interference occurs along antinodal lines. Bright fringes are seen where the antinodal lines intersect the viewing screen.

Double slit with separation d.
Equally spaced bright fringes are located at

$$\theta_m = \frac{m\lambda}{d} \qquad y_m = \frac{m\lambda L}{d} \qquad m = 0, 1, 2, \ldots$$

The **fringe spacing** is $\Delta y = \frac{\lambda L}{d}$

Diffraction grating with slit spacing d.
Very bright and narrow fringes are located at angles and positions

$$d\sin\theta_m = m\lambda \qquad y_m = L\tan\theta_m$$

Interference due to amplitude division

An interferometer divides a wave, lets the two waves travel different paths, then recombines them. Interference is constructive if one wave travels an integer number of wavelengths more or less than the other wave. The difference can be due to an actual path-length difference or to a different index of refraction.

Michelson interferometer

The number of bright-dark-bright fringe shifts as mirror M_2 moves distance ΔL_2 is

$$\Delta m = \frac{\Delta L_2}{\lambda/2}$$

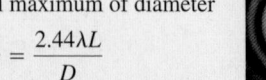

TERMS AND NOTATION

optics	double slit	order, m	circular aperture
diffraction	interference fringes	spectroscopy	interferometer
models of light	central maximum	single-slit diffraction	beam splitter
wave optics	fringe spacing, Δy	secondary maxima	hologram
ray optics	diffraction grating	Huygens' principle	

EXERCISES AND PROBLEMS

Exercises

Section 22.2 The Interference of Light

1. Figure Ex22.1 shows light waves passing through two closely spaced slits. The graph shows the light intensity on a screen behind the slits. Reproduce the graph axes on your page, including the zero and the tick marks that locate the fringes of the double-slit interference pattern.
   a. Draw a graph to show how the light intensity pattern will appear if the right slit is blocked, allowing light to go only through the left slit.
   b. Explain why you drew the graph this way.

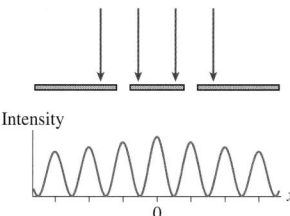

FIGURE EX22.1

2. Light of wavelength 500 nm illuminates a double slit, and the interference pattern is observed on a screen. At the position of the $m = 2$ bright fringe, how much farther is it to the more distant slit than to the nearer slit?
3. Two narrow slits 50 μm apart are illuminated with light of wavelength 500 nm. What is the angle of the $m = 2$ bright fringe in radians? In degrees?
4. Light from a sodium lamp ($\lambda = 589$ nm) illuminates two narrow slits. The fringe spacing on a screen 150 cm behind the slits is 4.0 mm. What is the spacing (in mm) between the two slits?
5. A double-slit interference pattern is created by two narrow slits spaced 0.20 mm apart. The distance between the first and the fifth minimum on a screen 60 cm behind the slits is 6.0 mm. What is the wavelength of the light used in this experiment?
6. A double-slit experiment is performed with light of wavelength 600 nm. The bright interference fringes are spaced 1.8 mm apart on the viewing screen. What will the fringe spacing be if the light is changed to a wavelength of 400 nm?
7. Light from a helium-neon laser ($\lambda = 633$ nm) is used to illuminate two narrow slits. The interference pattern is observed on a screen 3.0 m behind the slits. Twelve bright fringes are seen, spanning a distance of 52 mm. What is the spacing (in mm) between the slits?

Section 22.3 The Diffraction Grating

8. A 1.0-cm-wide diffraction grating has 1000 slits. It is illuminated by light of wavelength 550 nm. What are the angles of the first two diffraction orders?
9. Light of wavelength 600 nm illuminates a diffraction grating. The second-order maximum is at angle 39.5°. How many lines per millimeter does this grating have?
10. A helium-neon laser ($\lambda = 633$ nm) illuminates a diffraction grating. The distance between the two $m = 1$ bright fringes is 32 cm on a screen 2.0 m behind the grating. What is the spacing between slits of the grating?
11. A diffraction grating with 600 lines/mm is illuminated with light of wavelength 500 nm. A very wide viewing screen is 2.0 m behind the grating.
    a. What is the distance between the two $m = 1$ bright fringes?
    b. How many bright fringes can be seen on the screen?
12. A 500 line/mm diffraction grating is illuminated by light of wavelength 510 nm. How many diffraction orders are seen, and what is the angle of each?
13. The two most prominent wavelengths in the light emitted by a hydrogen discharge lamp are 656 nm (red) and 486 nm (blue). Light from a hydrogen lamp illuminates a diffraction grating with 500 lines/mm, and the light is observed on a screen 1.5 m behind the grating. What is the distance between the first-order red and blue fringes?

Section 22.4 Single-Slit Diffraction

14. A helium-neon laser ($\lambda = 633$ nm) illuminates a single slit and is observed on a screen 1.5 m behind the slit. The distance between the first and second minima in the diffraction pattern is 4.75 mm. What is the width (in mm) of the slit?
15. A 0.50-mm-wide slit is illuminated by light of wavelength 500 nm. What is the width of the central maximum on a screen 2.0 m behind the slit?
16. In a single-slit experiment, the slit width is 200 times the wavelength of the light. What is the width of the central maximum on a screen 2.0 m behind the slit?
17. The second minimum in the diffraction pattern of a 0.10-mm-wide slit occurs at 0.70°. What is the wavelength of the light?
18. You need to use your cell phone, which broadcasts an 800 MHz signal, but you're behind two massive, radio-wave-absorbing buildings that have only a 15 m space between them. What is the angular width, in degrees, of the electromagnetic wave after it emerges from between the buildings?

19. The opening to a cave is a tall, 30-cm-wide crack. A bat that is preparing to leave the cave emits a 30 kHz ultrasonic chirp. How wide is the "sound beam" 100 m outside the cave opening? Use $v_{sound} = 340$ m/s.

Section 22.5 Circular-Aperture Diffraction

20. A 0.50-mm-diameter hole is illuminated by light of wavelength 500 nm. What is the width of the central maximum on a screen 2.0 m behind the slit?
21. Light from a helium-neon laser ($\lambda = 633$ nm) passes through a circular aperture and is observed on a screen 4.0 m behind the aperture. The width of the central maximum is 2.5 cm. What is the diameter (in mm) of the hole?
22. You want to photograph a circular diffraction pattern whose central maximum has a diameter of 1.0 cm. You have a helium-neon laser ($\lambda = 633$ nm) and a 0.12-mm-diameter pinhole. How far behind the pinhole should you place the viewing screen?
23. Infrared light of wavelength 2.5 μm illuminates a 0.20-mm-diameter hole. What is the angle of the first dark fringe in radians? In degrees?

Section 22.6 Interferometers

24. Moving mirror M_2 of a Michelson interferometer a distance of 100 μm causes 500 bright-dark-bright fringe shifts. What is the wavelength of the light?
25. A Michelson interferometer uses red light with a wavelength of 656.45 nm from a hydrogen discharge lamp. How many bright-dark-bright fringe shifts are observed if mirror M_2 is moved exactly 1 cm?
26. A Michelson interferometer uses light whose wavelength is known to be 602.446 nm. Mirror M_2 is slowly moved while exactly 33,198 bright-dark-bright fringe shifts are observed. What distance has M_2 moved? Be sure to give your answer to an appropriate number of significant figures.
27. A Michelson interferometer operating at a 600 nm wavelength has a 2.00-cm-long glass cell in one arm. To begin, the air is pumped out of the cell and mirror M_2 is adjusted to produce a bright spot at the center of the interference pattern. Then a valve is opened and air is slowly admitted into the cell. The index of refraction of air at 1.0 atm pressure is 1.00028. How many bright-dark-bright fringe shifts are observed as the cell fills with air?

Problems

28. Figure P22.28 shows the light intensity on a screen 2.5 m behind an aperture. The aperture is illuminated with light of wavelength 600 nm.
    a. Is the aperture a single slit or a double slit? Explain.
    b. If the aperture is a single slit, what is its width? If it is a double slit, what is the spacing between the slits?

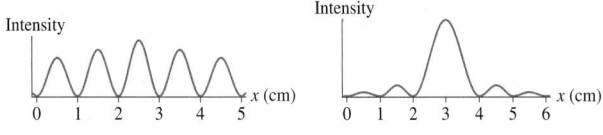

FIGURE P22.28   FIGURE P22.29

29. Figure P22.29 shows the light intensity on a screen 2.5 m behind an aperture. The aperture is illuminated with light of wavelength 600 nm.
    a. Is the aperture a single slit or a double slit? Explain.
    b. If the aperture is a single slit, what is its width? If it is a double slit, what is the spacing between the slits?
30. A double slit is illuminated simultaneously with orange light of wavelength 600 nm and light of an unknown wavelength. The $m = 4$ bright fringe of the unknown wavelength overlaps the $m = 3$ bright orange fringe. What is the unknown wavelength?
31. In a double-slit experiment, the slit separation is 200 times the wavelength of the light. What is the angular separation between two adjacent bright fringes?
32. What is the intensity of a double-slit interference pattern at a point on the screen halfway between the central maximum and the first minimum? Give your answer as a multiple of I_1, the intensity due to each individual wave.
33. A double-slit interference pattern has a fringe spacing of 4.0 mm. How far from the central maximum is the first position at which the intensity is equal to I_1?
34. Figure P22.34 shows the light intensity on a screen behind a double slit. The slit spacing is 0.20 mm and the wavelength of the light is 600 nm. What is the distance from the slits to the screen?

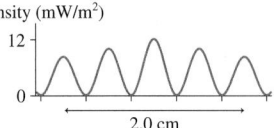

FIGURE P22.34

35. Figure P22.34 shows the light intensity on a screen behind a double slit. The slit spacing is 0.20 mm and the screen is 2.0 m behind the slits. What is the wavelength of the light?
36. Figure P22.34 shows the light intensity on a screen behind a double slit. Suppose one slit is covered. What will be the light intensity at the center of the screen due to the remaining slit?
37. A diffraction grating having 500 lines/mm diffracts visible light at 30°. What is the light's wavelength?
38. Light emitted by Element X passes through a diffraction grating having 1200 lines/mm. The diffraction pattern is observed on a screen 75.0 cm behind the grating. Bright fringes are *seen* on the screen at distances of 56.2 cm, 65.9 cm, and 93.5 cm from the central maximum. No other fringes are seen.
    a. What is the value of m for each of these diffracted wavelengths? Explain why only one value is possible.
    b. What are the wavelengths of light emitted by Element X?
39. Helium atoms emit light at several wavelengths. Light from a helium lamp illuminates a diffraction grating and is observed on a screen 50.0 cm behind the grating. The emission at wavelength 501.5 nm creates a first-order bright fringe 21.90 cm from the central maximum. What is the wavelength of the bright fringe that is 31.60 cm from the central maximum?
40. A diffraction grating produces a first-order maximum at an angle of 20.0°. What is the angle of the second-order maximum?
41. A diffraction grating is illuminated simultaneously with red light of wavelength 660 nm and light of an unknown wavelength. The fifth-order maximum of the unknown wavelength exactly overlaps the third-order maximum of the red light. What is the unknown wavelength?

42. White light (400–700 nm) is incident on a 600 line/mm diffraction grating. What is the width of the first-order rainbow on a screen 2.0 m behind the grating?

43. White light (400–700 nm) is incident on a 300 line/mm diffraction grating. What wavelengths in the third-order rainbow overlap wavelengths in the fourth-order rainbow?

44. For your science fair project you need to design a diffraction grating that will disperse the visible spectrum (400–700 nm) over 30.0° in first order.
    a. How many lines per millimeter does your grating need?
    b. What is the first-order diffraction angle of light from a sodium lamp ($\lambda = 589$ nm)?

45. Figure P22.45 shows the interference pattern on a screen 1.0 m behind an 800 line/mm diffraction grating. What is the wavelength of the light?

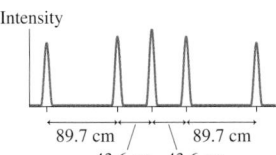

Intensity

| 89.7 cm | 89.7 cm |
43.6 cm 43.6 cm

FIGURE P22.45

46. Figure P22.45 shows the interference pattern on a screen 1.0 m behind a diffraction grating. The wavelength of the light is 600 nm. How many lines per millimeter does the grating have?

47. Light from a sodium lamp ($\lambda = 589$ nm) illuminates a narrow slit and is observed on a screen 75 cm behind the slit. The distance between the first and third dark fringes is 7.5 mm. What is the width (in mm) of the slit?

48. What is the width of a slit for which the first minimum is at 45° when the slit is illuminated by a helium-neon laser ($\lambda = 633$ nm)?

49. For what slit-width-to-wavelength ratio does the first minimum of a single-slit diffraction pattern appear at (a) 30°, (b) 60°, and (c) 90°?

50. Light from a helium-neon laser ($\lambda = 633$ nm) is incident on a single slit. What is the largest slit width for which there are no minima in the diffraction pattern?

51. Figure P22.51 shows the light intensity on a screen behind a single slit. The wavelength of the light is 500 nm and the screen is 1.0 m behind the slit. What is the width (in mm) of the slit?

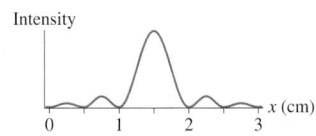

Intensity

x (cm)
0   1   2   3

FIGURE P22.51

52. Figure P22.51 shows the light intensity on a screen behind a single slit. The wavelength of the light is 600 nm and the slit width is 0.15 mm. What is the distance from the slit to the screen?

53. Figure P22.51 shows the light intensity on a screen behind a circular aperture. The wavelength of the light is 500 nm and the screen is 1.0 m behind the slit. What is the diameter (in mm) of the aperture?

54. Light from a helium-neon laser ($\lambda = 633$ nm) illuminates a circular aperture. It is noted that the diameter of the central maximum on a screen 50 cm behind the aperture matches the diameter of the geometric image. What is the aperture's diameter?

55. One day, after pulling down your window shade, you notice that sunlight is passing through a pinhole in the shade and making a small patch of light on the far wall. Having recently studied optics in your physics class, you're not too surprised to see that the patch of light seems to be a circular diffraction pattern. It appears that the central maximum is about 1 cm across, and you estimate that the distance from the window shade to the wall is about 3 m. Estimate (a) the average wavelength of sunlight and (b) the diameter of the pinhole.

56. A radar for tracking aircraft broadcasts a 12 GHz microwave beam from a 2.0-m-diameter circular radar antenna. From a wave perspective, the antenna is a circular aperture through which the microwaves diffract.
    a. What is the diameter of the radar beam at a distance of 30 km?
    b. If the antenna emits 100 kW of power, what is the average microwave intensity at 30 km?

57. A helium-neon laser ($\lambda = 633$ nm) is built with a glass tube of inside diameter 1.0 mm. One mirror is partially transmitting to allow the laser beam out. An electrical discharge in the tube causes it to glow like a neon light. From an optical perspective, the laser beam is a light wave that diffracts out through a 1.0-mm-diameter circular opening.
    a. Can a laser beam be *perfectly* parallel, with no spreading? Why or why not?
    b. The angle θ_1 to the first minimum is called the *divergence angle* of a laser beam. What is the divergence angle of this laser beam?
    c. What is the diameter (in mm) of the laser beam after it travels 3.0 m?

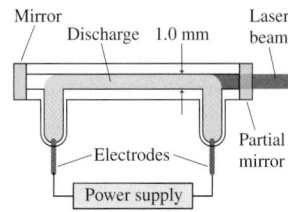

Mirror   Laser
    Discharge   1.0 mm   beam

                         Partial
    Electrodes   mirror

    Power supply

FIGURE P22.57

    d. What is the diameter of the laser beam after it travels 1.0 km?

58. Scientists use *laser range-finding* to measure the distance to the moon with great accuracy. A brief laser pulse is fired at the moon, then the time interval is measured until the "echo" is seen by a telescope. A laser beam spreads out as it travels because it diffracts through a circular exit as it leaves the laser. In order for the reflected light to be bright enough to detect, the laser spot on the moon must be no more than 1 km in diameter. Staying within this diameter is accomplished by using a special large-diameter laser. If $\lambda = 532$ nm, what is the minimum diameter of the circular opening from which the laser beam emerges? The earth-moon distance is 384,000 km.

59. Light of wavelength 600 nm passes though two slits separated by 0.20 mm and is observed on a screen 1.0 m behind the slits. The location of the central maximum is marked on the screen and labeled $y = 0$.
    a. At what distance, on either side of $y = 0$, are the $m = 1$ bright fringes?
    b. A very thin piece of glass is then placed in one slit. Because light travels slower in glass than in air, the wave passing through the glass is delayed by 5.0×10^{-16} s in comparison to the wave going through the other slit. What fraction of the period of the light wave is this delay?
    c. With the glass in place, what is the phase difference $\Delta\phi_0$ between the two waves as they leave the slits?
    d. The glass causes the interference fringe pattern on the screen to shift sideways. Which way does the central maximum move (toward or away from the slit with the glass) and by how far?

60. A 600 line/mm diffraction grating is in an empty aquarium tank. The index of refraction of the glass walls is $n_{glass} = 1.50$. A helium-neon laser ($\lambda = 633$ nm) is outside the aquarium. The laser beam passes through the glass wall and illuminates the diffraction grating.
    a. What is the first-order diffraction angle of the laser beam?
    b. What is the first-order diffraction angle of the laser beam after the aquarium is filled with water ($n_{water} = 1.33$)?

61. You've set up a Michelson interferometer with a helium-neon laser ($\lambda = 632.8$ nm). After adjusting mirror M_2 to produce a bright spot at the center of the pattern, you carefully move M_2 away from the beam splitter while counting 1200 new bright spots at the center. Then you put the laser away. Later another student wants to restore the interferometer to its starting condition, but he mistakenly sets up a hydrogen discharge lamp and uses the 656.5 nm emission from hydrogen atoms. He then counts 1200 new bright spots while slowly moving M_2 back toward the beam splitter. What is the net displacement of M_2 when he is done? Is M_2 now closer to or farther from the beam splitter?

62. A Michelson interferometer uses light from a sodium lamp. Sodium atoms emit light having wavelengths 589.0 nm and 589.6 nm. The interferometer is initially set up with both arms of equal length ($L_1 = L_2$), producing a bright spot at the center of the interference pattern. How far must mirror M_2 be moved so that one wavelength has produced one more new maxima than the other wavelength?

63. A light wave has wavelength 500 nm in vacuum.
    a. What is the wavelength of this light as it travels through water ($n_{water} = 1.33$)?
    b. Suppose that a 1.0-mm-thick layer of water is inserted into one arm of a Michelson interferometer. How many "extra" wavelengths does the light now travel in this arm?
    c. By how many fringes will this water layer shift the interference pattern?

64. A 0.10-mm-thick piece of glass is inserted into one arm of a Michelson interferometer that is using light of wavelength 500 nm. This causes the fringe pattern to shift by 200 fringes. What is the index of refraction of this piece of glass?

65. Transparent materials known as *electro-optic crystals* change their index of refraction when a voltage is applied to them. They are used to control the output of interferometers in optical computers and optical switching devices. To illustrate the basic idea, suppose one arm of a Michelson interferometer contains a 0.100-mm-thick electro-optic crystal with an initial index of refraction $n = 1.550$. The interferometer operates with light of wavelength 0.981 μm from a semiconductor laser. Mirror M_2 is first adjusted to make the output a bright fringe, which we can interpret as a 1 in binary arithmetic. What is the first index of refraction larger than 1.550 for which output goes dark—a binary 0? A voltage that can change the index of refraction by this amount can be used to switch binary signals from 1's to 0's or from 0's to 1's.

66. To illustrate one of the ideas of holography in a simple way, consider a diffraction grating with slit spacing d. The small-angle approximation is usually not valid for diffraction gratings, because d is only slightly larger than λ, but assume that the λ/d ratio of this grating is small enough to make the small-angle approximation valid.
    a. Use the small-angle approximation to find an expression for the fringe spacing on a screen at distance L behind the grating.
    b. Rather than a screen, suppose you place a piece of film at distance L behind the grating. The bright fringes will expose the film, but the dark spaces in between will leave the film unexposed. After being developed, the film will be a series of alternating light and dark stripes. What if you were to now "play" the film by using it as a diffraction grating? In other words, what happens if you shine the same laser through the film and look at the film's diffraction pattern on a screen at the same distance L? Demonstrate that the film's diffraction pattern is a reproduction of the original diffraction grating.

Challenge Problems

67. A double-slit experiment is set up using a helium-neon laser ($\lambda = 633$ nm). Then a very thin piece of glass ($n = 1.50$) is placed over one of the slits. Afterward, the central point on the screen is occupied by what had been the $m = 10$ dark fringe. How thick is the glass?

68. Light of wavelength 600 nm passes through a double slit and is viewed on a screen 2.0 m behind the slits. Each slit is 0.040 mm wide and they are separated by 0.200 mm. How many bright fringes are seen on the screen?
    Hint: What determines the width of the region on the screen where fringes are observed?

69. Light consisting of two nearly equal wavelengths $\lambda + \Delta\lambda$ and λ, where $\Delta\lambda \ll \lambda$, is incident on a diffraction grating. The slit separation of the grating is d.
    a. Show that the angular separation of these two wavelengths in the mth order is
    $$\Delta\theta = \frac{\Delta\lambda}{\sqrt{(d/m)^2 - \lambda^2}}$$
    b. Sodium atoms emit light at 589.0 nm and 589.6 nm. What are the first-order and second-order angular separations (in degrees) of these two wavelengths for a 600 line/mm grating?

70. Figure CP22.70 shows two nearly overlapped intensity peaks of the sort you might produce with a diffraction grating (see Figure 22.8b). As a practical matter, two peaks can just barely be resolved if their spacing Δy equals the width w of each peak, where w is measured at half of the peak's height. Two peaks closer together than w will merge into a single peak. We can use this idea to understand the *resolution* of a diffraction grating.

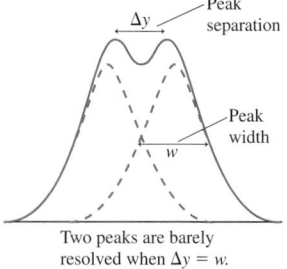

FIGURE CP22.70

    a. In the small-angle approximation, the position of the $m = 1$ peak of a diffraction grating falls at the same location as the $m = 1$ fringe of a double slit: $y_1 = \lambda L/d$. Suppose two wavelengths differing by $\Delta\lambda$ pass through a grating at the same time. Find an expression for Δy, the separation of their first-order peaks.
    b. We noted that the widths of the bright fringes are proportional to $1/N$, where N is the number of slits in the grating. Let's hypothesize that the fringe width is $w = y_1/N$. Show

that this is true for the double-slit pattern. We'll then assume it to be true as N increases.

c. Use your results from parts a and b together with the idea that $\Delta y_{min} = w$ to find an expression for $\Delta\lambda_{min}$, the minimum wavelength separation (in first order) for which the diffraction fringes can barely be resolved.

d. Ordinary hydrogen atoms emit red light with a wavelength of 656.45 nm. In deuterium, which is a "heavy" isotope of hydrogen, the wavelength is 656.27 nm. What is the minimum number of slits in a diffraction grating that can barely resolve these two wavelengths in the first-order diffraction pattern?

71. The diffraction grating analysis in this chapter assumed that the incident light is normal to the grating. Figure CP22.71 shows a plane wave approaching a diffraction grating at angle ϕ.

a. Show that the angles θ_m for constructive interference are given by the grating equation

$$d(\sin\theta_m + \sin\phi) = m\lambda$$

where $m = 0, \pm1, \pm2, \ldots$. Angles are considered positive if they are above the horizontal line, negative if below it.

b. The two first-order maxima, $m = +1$ and $m = -1$, are no longer symmetrical about the center. Find θ_1 and θ_{-1} for 500 nm light incident on a 600 line/mm grating at $\phi = 30°$.

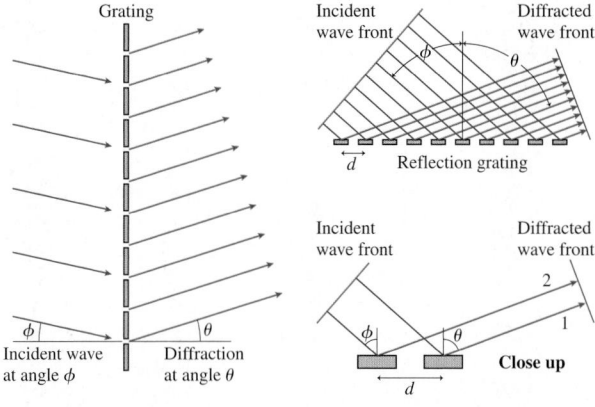

FIGURE CP22.71   **FIGURE CP22.72**

72. Figure CP22.72 shows light of wavelength λ incident at angle ϕ on a *reflection* grating of spacing d. We want to find the angles θ_m at which constructive interference occurs.

a. The figure shows paths 1 and 2 along which two waves travel and interfere. Find an expression for the path-length difference $\Delta r = r_2 - r_1$.

b. Using your result from part a, find an equation (analogous to Equation 22.15) for the angles θ_m at which diffraction occurs when the light is incident at angle ϕ. Notice that m can be a negative integer in your expression, indicating that path 2 is shorter than path 1.

c. Show that the zeroth-order diffraction is simply a "reflection." That is, $\theta_0 = \phi$.

d. Light of wavelength 500 nm is incident at $\phi = 40°$ on a reflection grating having 700 reflection lines/mm. Find all angles θ_m at which light is diffracted. Negative values of θ_m are interpreted as an angle left of the vertical.

e. Draw a picture showing a *single* 500 nm light ray incident at $\phi = 40°$ and showing all the diffracted waves at the correct angles.

73. The pinhole camera of Figure CP22.73 images distant objects by allowing only a narrow bundle of light rays to pass through the hole and strike the film. If light consisted of particles, you could make the image sharper and sharper (at the expense of getting dimmer and dimmer) by making the aperture smaller and smaller. In practice, diffraction of light by the circular aperture limits the maximum sharpness that can be obtained. Consider two distant points of light, such as two distant street-

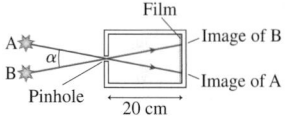

FIGURE CP22.73

lights. Each will produce a circular diffraction pattern on the film. The two images can just barely be resolved if the central maximum of one image falls on the first dark fringe of the other image. (This is called Rayleigh's criterion, and we will explore its implication for optical instruments in the next chapter.)

a. Optimum sharpness of one image occurs when the diameter of the central maximum equals the diameter of the pinhole. What is the optimum hole size for a pinhole camera in which the film is 20 cm behind the hole? Assume $\lambda = 550$ nm, an average value for visible light.

b. For this hole size, what is the angle α (in degrees) between two distant sources that can barely be resolved?

c. What is the distance between two street lights 1 km away that can barely be resolved?

STOP TO THINK ANSWERS

Stop to Think 22.1: b. The antinodal lines seen in Figure 22.3b are diverging.

Stop to Think 22.2: Smaller. Shorter-wavelength light doesn't spread as rapidly as longer-wavelength light. The fringe spacing Δy is directly proportional to the wavelength λ.

Stop to Think 22.3: d. Larger wavelengths have larger diffraction angles. Red light has a larger wavelength than violet light, so red light is diffracted farther from the center.

Stop to Think 22.4: b or c. The width of the central maximum, which is proportional to λ/a, has increased. This could occur either because the wavelength has increased or because the slit width has decreased.

Stop to Think 22.5: d. Moving M_1 in by λ decreases r_1 by 2λ. Moving M_2 out by λ increases r_2 by 2λ. These two actions together change the path length by $\Delta r = 4\lambda$.

23 Ray Optics

A focused laser beam can cut and drill the hardest metals.

▶ **Looking Ahead**

The goals of Chapter 23 are to understand and apply the ray model of light. In this chapter you will learn to:

- Use the ray model of light.
- Calculate angles of reflection and refraction.
- Understand color and dispersion.
- Use ray tracing to analyze lens systems.
- Use refraction theory to calculate the properties of lens systems.
- Understand how diffraction limits the resolution of optical systems.

◀ **Looking Back**

The material in this chapter depends on the wave model of light. Please review:

- Section 20.5 Light waves and the index of refraction.
- Sections 22.1 and 22.5 The wave and ray models of light.
- Sections 22.5 and 22.6 Diffraction, especially by a circular aperture.

Humans have always been fascinated by light. Simple mirrors are found in ancient archeological sites from Egypt to China. Our ancestors had learned by 1500 BCE to start fires by focusing sunlight with a simple lens. From there, it's only a small step to drilling holes with a focused laser beam.

Chapter 22 introduced the three models of light but then emphasized the wave optics of interference and diffraction. This chapter will analyze basic optical systems such as mirrors and lenses in terms of straight-line light trajectories. This is *ray optics,* and it is a subject of immense practical value. The ray model of light will take center stage, but we'll find that we can't entirely avoid the fact that light is a wave. It will turn out, perhaps surprisingly, that the waviness of light is what ultimately sets the performance limits of optical systems. You'll finish this chapter with a much more complete understanding of what light is and how it behaves.

23.1 The Ray Model of Light

A flashlight makes a beam of light through the night's darkness. Sunbeams stream into a darkened room through a small hole in the shade, and laser beams are even more well defined. Our everyday experience that light travels in straight lines is the basis of the *ray model* of light.

The ray model is an oversimplification of reality, but nonetheless is very useful within its range of validity. In particular, the ray model of light is valid as long as any apertures through which the light passes (lenses, mirrors, holes, and the like) are very large compared to the wavelength of light. With such apertures, diffraction and other wave aspects of light are negligible and can be ignored. The analysis of Section 22.5 found that the crossover between wave optics and ray optics occurs for apertures ≈1 mm in diameter. Lenses and mirrors are almost always larger than 1 mm, so the ray model of light is an excellent basis for the practical optics of image formation.

To begin, let us define a **light ray** as a line in the direction along which light energy is flowing. A light ray is an abstract idea, not a physical entity or a "thing." Any narrow beam of light, such as the laser beam in Figure 23.1, is actually a bundle of many parallel light rays. You can think of a single light ray as the limiting case of a laser beam whose diameter approaches zero. Laser beams are good approximations of light rays, certainly adequate for demonstrating ray behavior, but any real laser beam is a bundle of many parallel rays.

The following table outlines five basic ideas and assumptions of the ray model of light.

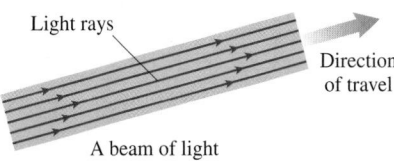

FIGURE 23.1 A laser beam or beam of sunlight is a bundle of parallel light rays.

The ray model of light

	Light rays travel in straight lines.
	Light travels through a transparent material in straight lines called light rays. The speed of light is $v = c/n$, where n is the index of refraction of the material.
	Light rays can cross.
	Light rays do not interact with each other. Two rays can cross without either being affected in any way.
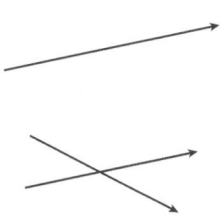	A light ray travels forever unless it interacts with matter.
	A light ray continues forever unless it has an interaction with matter that causes the ray to change direction or to be absorbed. Light interacts with matter in four different ways: ■ At an interface between two materials, light can be either *reflected* or *refracted*. ■ Within a material, light can be either *scattered* or *absorbed*. These interactions are discussed later in the chapter.
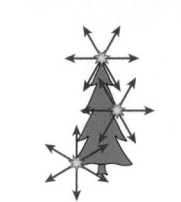	An object is a source of light rays.
	An **object** is a source of light rays. Rays originate from *every* point on the object, and each point sends rays in *all* directions. We make no distinction between self-luminous objects and reflective objects.
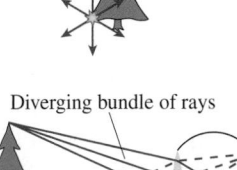	The eye sees by focusing a diverging bundle of rays.
	The eye "sees" an object when *diverging* bundles of rays from each point on the object enter the pupil and are focused to an image on the retina. (Imaging is discussed later in the chapter.) From the movements the eye's lens has to make to focus the image, your brain "computes" the distance d at which the rays originated, and you perceive the object as being at that point.

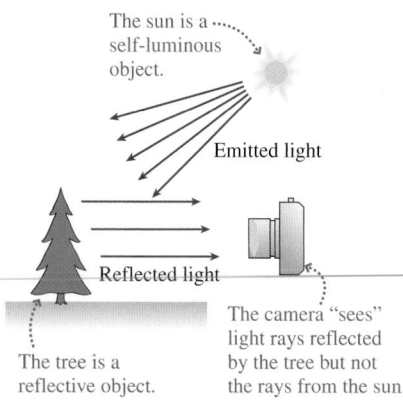

FIGURE 23.2 Self-luminous and reflective objects.

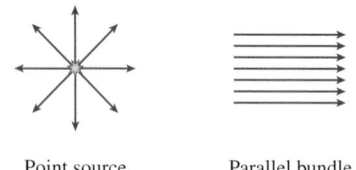

Point source   Parallel bundle

FIGURE 23.3 Point sources and parallel bundles represent idealized objects.

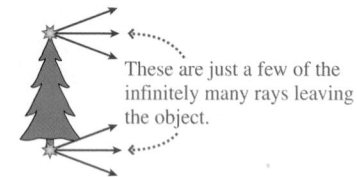

These are just a few of the infinitely many rays leaving the object.

FIGURE 23.4 A ray diagram simplifies the situation by showing only a few rays.

Objects

Figure 23.2 illustrates the idea that objects can be either *self-luminous,* such as the sun, flames, and light bulbs, or *reflective.* Most objects are reflective. A tree, unless it is on fire, is seen or photographed by virtue of reflected sunlight or reflected skylight. People, houses, and this page in the book reflect light from self-luminous sources. In this chapter we are concerned not with how the light originates but with how it behaves after leaving the object.

Light rays from an object are emitted in all directions, but you are not *aware* of light rays unless they enter the pupil of your eye. Consequently, most light rays go completely unnoticed. For example, light rays travel from the sun to the tree in Figure 23.2, but you're not aware of these unless the tree reflects some of them into your eye. Or consider a laser beam traveling through the room. You've probably noticed that it's almost impossible to see a laser beam from the side unless there's dust in the air. The dust scatters a few of the light rays toward your eye, but in the absence of dust you would be completely unaware of a very powerful light beam traveling past you. **Light rays exist everywhere, quite independently of whether you are seeing them.**

Figure 23.3 shows two idealized sets of light rays. The diverging rays from a **point source** are emitted in all directions. It will often be useful to think of each point on an object as a point source of light rays. A **parallel bundle** of rays could be a laser beam. Alternatively it could represent a *distant object,* an object such as a star so far away that the rays arriving at the observer are essentially parallel to each other.

Ray Diagrams

Rays originate from *every* point on an object and travel outward in *all* directions, but a diagram trying to show all these rays would be hopelessly messy and confusing. To simplify the picture, we usually use a **ray diagram** that shows only a few rays. For example, Figure 23.4 is a ray diagram showing only a few rays leaving the top and bottom points of the object and traveling to the right. These rays will be sufficient to show us how the object is imaged by lenses or mirrors.

NOTE ▶ Ray diagrams are the basis for a *pictorial representation* that we'll use throughout this chapter. Be careful not think that a ray diagram shows all of the rays. The rays shown on the diagram are just a subset of the infinitely many rays leaving the object. ◀

Apertures

A popular form of entertainment during ancient Roman times was a visit to a **camera obscura,** Latin for "dark room." As Figure 23.5a shows, a camera obscura was a darkened room with a single, small hole to the outside world. After their eyes became dark adapted, visitors could see a dim but full-color image of

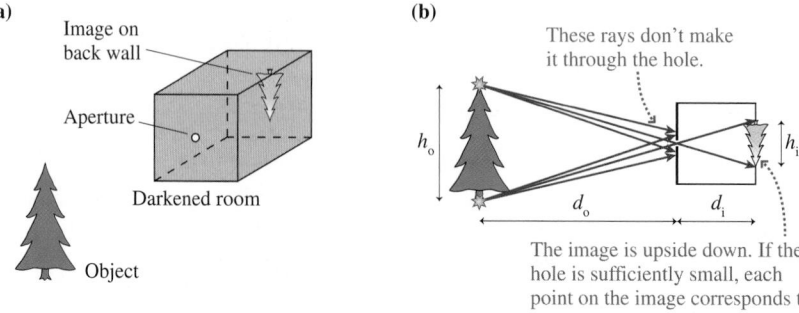

FIGURE 23.5 A camera obscura.

the outside world displayed on the back wall of the room. However, the image was upside down! The *pinhole camera* is a miniature and more modern version of the camera obscura.

A hole through which light passes is called an **aperture.** Figure 23.5b uses the ray model of light passing through a small aperture to explain how the camera obscura works. Each point on an object emits light rays in all directions, but only a very few of these rays pass through the aperture and reach the back wall. As the figure illustrates, the geometry of the rays causes the image to be upside down.

Actually, as you may have realized, each *point* on the object illuminates a small but finite *patch* on the wall. This is because the finite size of the aperture allows several rays from each point on the object to pass through at slightly different angles. As a result, the image is slightly blurred and out of focus. Maximum sharpness is achieved by making the hole smaller and smaller, which makes the image dimmer and dimmer. (Diffraction also becomes an issue if the hole gets too small.) A practical camera obscura or pinhole camera has to accept a small amount of blurring as the tradeoff for having an image bright enough to see.

An interesting aspect of the camera obscura is that the image size is not the same as the object size. We define the **magnification** m as the ratio of image height h_i to object height h_o:

$$m = \frac{h_i}{h_o} \tag{23.1}$$

While you may think of magnification as meaning "larger," our definition allows the image to be either larger than the object ($m > 1$) or smaller ($m < 1$). In the case of the camera obscura, you can see from the similar triangles in Figure 23.5b that

$$m = \frac{h_i}{h_o} = \frac{d_i}{d_o} \tag{23.2}$$

where d_o is the distance to the object and d_i is the depth of the camera obscura. Any realistic camera obscura has $d_i < d_o$ and magnification $m < 1$.

These ideas will all reappear when we consider forming images with lenses. There we will discover how a modern camera, with a lens, improves on the camera obscura.

We can apply the ray model to more complex apertures, such as the L-shaped aperture in Figure 23.6. The pattern of light on the screen is found by tracing all the straight-line paths—the ray trajectories—that start from the point source and pass through the aperture. We will see an enlarged L on the screen, with a sharp boundary between the image and the dark shadow.

FIGURE 23.6 A point source and an aperture.

STOP TO THINK 23.1 A long, thin light bulb illuminates a vertical aperture. Which pattern of light do you see on a viewing screen behind the aperture?

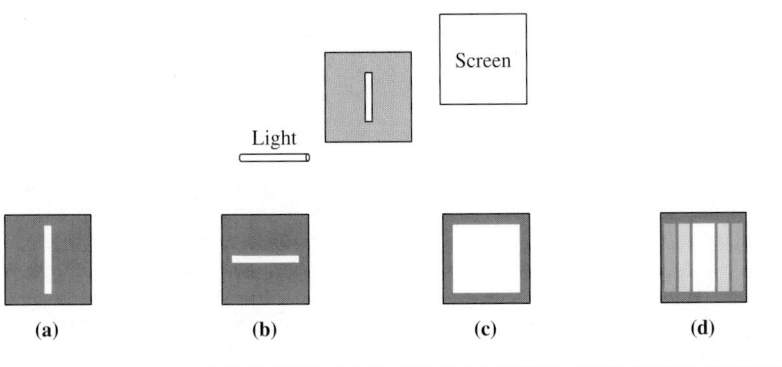

(a)   (b)   (c)   (d)

Reflection is an everyday experience.

23.2 Reflection

Reflection of light is a familiar, everyday experience. You see your reflection in the bathroom mirror first thing every morning, reflections in your car's rear-view mirror as you drive to school, and the sky reflected in puddles of standing water. Reflection from a flat, smooth surface, such as a mirror or a piece of polished metal, is called **specular reflection,** from *speculum,* the Latin word for "mirror."

Figure 23.7a shows a bundle of parallel light rays reflecting from a mirror-like surface. You can see that the incident and reflected rays are both in a plane that is normal, or perpendicular, to the reflective surface. A three-dimensional perspective accurately shows the relation between the light rays and the surface, but figures such as this are hard to draw by hand. Instead, it is customary to represent reflection with the simpler pictorial representation of Figure 23.7b. In this figure,

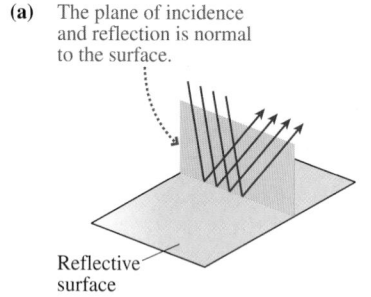

(a) The plane of incidence and reflection is normal to the surface.

Reflective surface

■ The plane of the page is the plane of incidence and reflection. The reflective surface extends into and out of the page.
■ A *single* light ray represents the entire bundle of parallel rays. This is oversimplified, but it keeps the figure and the analysis clear.

The angle θ_i between the ray and a line perpendicular to the surface—the *normal* to the surface—is called the **angle of incidence.** Similarly, the **angle of reflection** θ_r is the angle between the reflected ray and the normal to the surface. The **law of reflection,** easily demonstrated with simple experiments, states that

1. The incident ray and the reflected ray are in the same plane normal to the surface, and
2. The angle of reflection equals the angle of incidence: $\theta_r = \theta_i$.

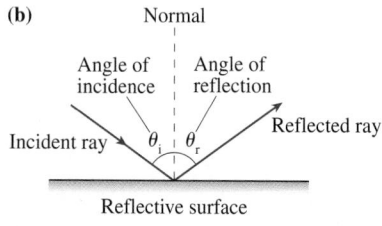

(b)   Normal

Angle of incidence   Angle of reflection

Incident ray θ_i θ_r Reflected ray

Reflective surface

FIGURE 23.7 Specular reflection of light.

NOTE ▶ Optics calculations *always* use the angle measured from the normal, not the angle between the ray and the surface. ◀

EXAMPLE 23.1 **Light reflecting from a mirror**

A dressing mirror on a closet door is 1.5 m tall. The bottom is 0.5 m above the floor. A bare light bulb hangs 1.0 m from the closet door, 2.5 m above the floor. How long is the streak of reflected light across the floor?

MODEL Treat the light bulb as a point source and use the ray model of light.

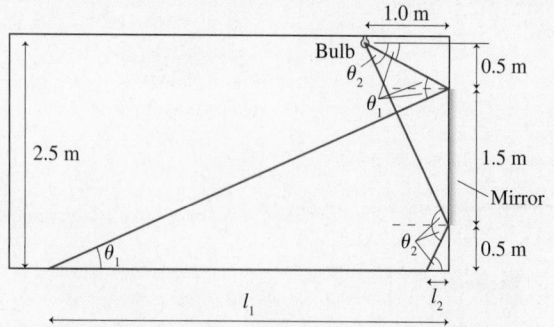

FIGURE 23.8 Pictorial representation of the light rays reflecting from a mirror.

VISUALIZE Figure 23.8 is a pictorial representation of the light rays. We need to consider only the two rays that strike the edges of the mirror. All other reflected rays will fall between these two.

SOLVE Figure 23.8 has used the law of reflection to set the angles of reflection equal to the angles of incidence. Other angles have been identified with simple geometry. The two angles of incidence are

$$\theta_1 = \tan^{-1}\left(\frac{0.5 \text{ m}}{1.0 \text{ m}}\right) = 26.6°$$

$$\theta_2 = \tan^{-1}\left(\frac{2.0 \text{ m}}{1.0 \text{ m}}\right) = 63.4°$$

The distances to the points where the rays strike the floor are then

$$l_1 = \frac{2.0 \text{ m}}{\tan \theta_1} = 4.00 \text{ m}$$

$$l_2 = \frac{0.5 \text{ m}}{\tan \theta_2} = 0.25 \text{ m}$$

Thus the length of the light streak is $l_1 - l_2 = 3.75$ m.

Diffuse Reflection

Most objects are seen by virtue of their reflected light. For a "rough" surface, the law of reflection $\theta_r = \theta_i$ is obeyed at each point but the irregularities of the surface cause the reflected rays to leave in many random directions. This situation, shown in Figure 23.9, is called **diffuse reflection.** It is how you see this page, the wall, your hand, your friend, and so on. Diffuse reflection is far more prevalent than the mirror-like specular reflection.

By a "rough" surface, we mean a surface that is rough or irregular in comparison to the wavelength of light. Because visible light wavelengths are $\approx 0.5 \ \mu$m, any surface with texture, scratches, or other irregularities larger than $1 \ \mu$m will cause diffuse reflection rather than specular reflection. A piece of paper may feel quite smooth to your hand, but a microscope would show that the surface consists of distinct fibers much larger than $1 \ \mu$m in size. By contrast, the irregularities on a mirror or a piece of polished metal are much smaller than $1 \ \mu$m. The law of reflection is equally valid for both specular and diffuse reflection, but the nature of the surface causes the outcomes to be quite different.

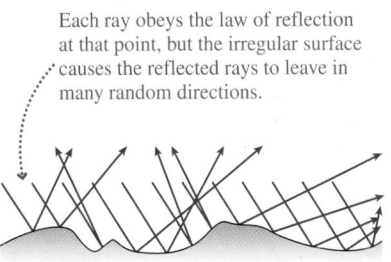

Each ray obeys the law of reflection at that point, but the irregular surface causes the reflected rays to leave in many random directions.

Magnified view of surface

FIGURE 23.9 Diffuse reflection from an irregular surface.

The Plane Mirror

One of the most commonplace observations is that you can see yourself in a mirror. How? Figure 23.10a shows rays from point source P reflecting from a mirror. Consider the particular ray shown in Figure 23.10b. The reflected ray travels along a line that passes through point P' on the "back side" of the mirror. Because $\theta_r = \theta_i$, simple geometry dictates that P' is the same distance behind the mirror as P is in front of the mirror. That is, $s' = s$.

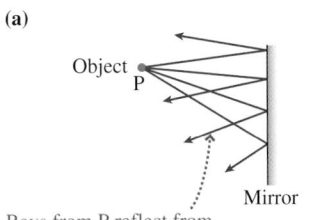

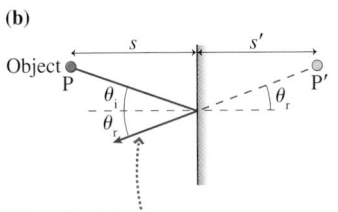

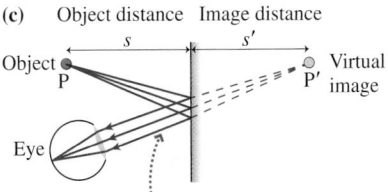

(a)

Rays from P reflect from the mirror. Each ray obeys the law of reflection.

(b)

This reflected ray appears to have come from point P'.

(c)

The reflected rays *all* diverge from P', which appears to be the source of the reflected rays. Your eye collects the bundle of diverging rays and "sees" the light coming from P'.

FIGURE 23.10 The light rays reflecting from a plane mirror.

The location of point P' in Figure 23.10b is independent of the value of θ_i. Consequently, *all* reflected rays travel along lines that pass through the *same* point P'. The original light rays diverged from point P, but the reflected rays now diverge from point P'. Consequently, as Figure 23.10c shows, **the reflected rays all *appear* to be coming from point P'.** For a plane mirror, the distance s' to point P' is equal to the object distance s:

$$s' = s \qquad \text{(plane mirror)} \qquad (23.3)$$

If rays diverge from an *object point* P and interact with a mirror so that the reflected rays diverge from point P' and *appear* to come from P', then we call P' a **virtual image** of point P. The image is "virtual" in the sense that no rays actually leave P', which is in darkness behind the mirror. But as far as your eye is concerned, the light rays act exactly *as if* the light really originated at P'. So while you may say "I see P in the mirror," what you are actually seeing is the virtual image of P. Distance s' is the *image distance*.

For an extended object, such as the one in Figure 23.11, each point on the object from which rays strike the mirror has a corresponding image point an equal

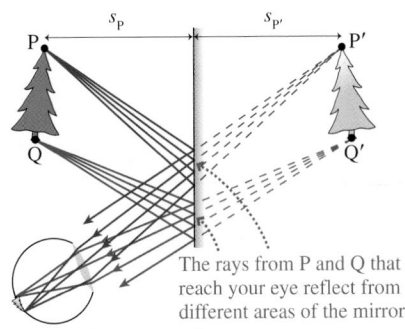

The rays from P and Q that reach your eye reflect from different areas of the mirror.

Your eye intercepts only a very small fraction of all the reflected rays.

FIGURE 23.11 Each point on the extended object has a corresponding image point an equal distance on the opposite side of the mirror.

distance on the opposite side of the mirror. The eye captures and focuses diverging bundles of rays from each point of the image in order to see the full image in the mirror. Two facts are worth noting:

1. Rays from each point on the object spread out in all directions and strike *every point* on the mirror. Only a very few of these rays enter your eye, but the other rays are very real and might be seen by other observers.
2. Rays from points P and Q enter your eye after reflecting from *different* areas of the mirror. This is why you can't always see the full image of an object in a very small mirror.

EXAMPLE 23.2 **How high is the mirror?**

If your height is h, what is the shortest mirror on the wall in which you can see your full image? Where must the top of the mirror be hung?

MODEL Use the ray model of light.

VISUALIZE Figure 23.12 is a pictorial representation of the light rays. We need to consider only the two rays that leave your head and feet and reflect into your eye.

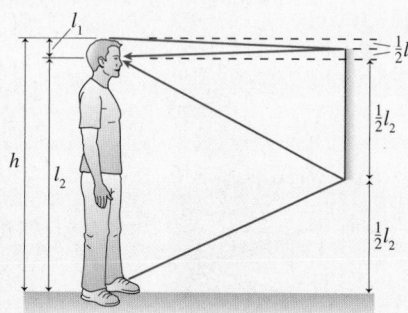

FIGURE 23.12 Pictorial representation of light rays from your head and feet reflecting into your eye.

SOLVE Let the distance from your eyes to the top of your head be l_1 and the distance to your feet be l_2. Your height is $h = l_1 + l_2$. A light ray from the top of your head that reflects from the mirror at $\theta_r = \theta_i$ and enters your eye must, by congruent triangles, strike the mirror a distance $\frac{1}{2}l_1$ above your eyes. Similarly, a ray from your foot to your eye strikes the mirror a distance $\frac{1}{2}l_2$ below your eyes. The distance between these two points on the mirror is $\frac{1}{2}l_1 + \frac{1}{2}l_2 = \frac{1}{2}h$. A ray from anywhere else on your body will reach your eye if it strikes the mirror between these two points. Pieces of the mirror outside these two points are irrelevant, not because rays don't strike them but because the reflected rays don't reach your eye. Thus the shortest mirror in which you can see your full reflection is $\frac{1}{2}h$. But this will only work if the top of the mirror is hung midway between your eyes and the top of your head.

ASSESS It is interesting that the answer does not depend on how far you are from the mirror.

Left and Right

It's common wisdom that a mirror "reverses left and right." But why, then, does it not also reverse up and down? What's special about left and right?

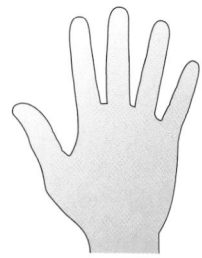

Right or left?

FIGURE 23.13 Can you tell which hand this is?

Hold your hands in front of you so that you can see the back of your right hand but the palm of your left hand. If the lighting were poor so that you could see only the outlines of your hands, as in Figure 23.13, could you tell which is a "right hand" and which is a "left hand"? No! Unlike "up" and "down," the terms "right" and "left" do not have an absolute, unambiguous meaning. Right and left are determined by the orientation of your thumb *relative to* your palm. Without knowing where the palm is, you can't assign a handedness to a hand.

In fact, a mirror does not "reverse right and left" any more than it reverses up and down. Instead, a mirror reverses *front and back*. Hold your right hand out, palm away from you and thumb pointing left. Imagine turning your hand inside-out in the sense that everything on the palm side is pulled through toward you and everything on the back side is pushed through away from you. Your fingers would still point up and your thumb would still point to the left, but you would see your palm rather than the back of your hand. Neither up/down nor left/right have reversed, but your "right hand" has become a "left hand."

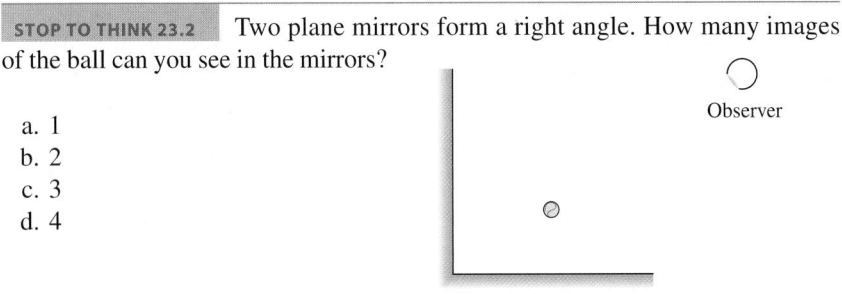

STOP TO THINK 23.2   Two plane mirrors form a right angle. How many images of the ball can you see in the mirrors?

a. 1
b. 2
c. 3
d. 4

Observer

23.3 Refraction

It has been known since antiquity that two things happen when a light ray is incident on a smooth boundary between two transparent materials, such as the boundary between air and glass:

Activ
Physics   15.1–15.3

1. Part of the light *reflects* from the boundary, obeying the law of reflection. This is how you see reflections from pools of water or storefront windows, even though water and glass are transparent.
2. Part of the light continues into the second medium. It is *transmitted* rather than reflected, but the transmitted ray changes direction as it crosses the boundary. The transmission of light from one medium to another, but with a change in direction, is called **refraction.**

The photograph of Figure 23.14 shows the refraction of a laser beam as it passes through a glass prism. Notice that the ray direction changes as the light enters and leaves the glass. You can also see two weak reflections leaving the top surface of the prism.

Reflection from the boundary between transparent media is usually weak. Typically 95% of the light is transmitted and only 5% is reflected. Our goal in this section is to understand refraction, so we will usually ignore the weak reflection and focus on the transmitted light.

NOTE ▶ A transparent material through which light travels is called the *medium* (plural *media*). This term has to be used with caution. The material does affect the light speed, but a transparent material differs from the medium of a sound or water wave in that particles of the medium do *not* oscillate as a light wave passes through. For a light wave it is the electromagnetic field that oscillates. ◀

Figure 23.15a shows the refraction of light rays in a parallel beam of light, such as a laser beam, and rays from a point source. It's good to remember that an infinite number of rays are incident on the boundary, but our analysis will be simplified if we focus on a single light ray. Figure 23.15b is a ray diagram showing

FIGURE 23.14 A laser beam refracts through a glass prism.

(a)

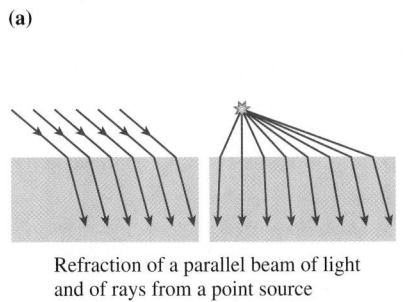

Refraction of a parallel beam of light and of rays from a point source

FIGURE 23.15 Refraction of light rays.

(b)

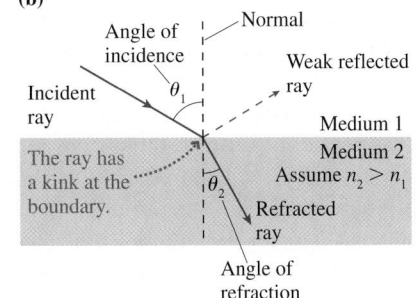

Angle of incidence

Normal

Incident ray

Weak reflected ray

θ_1

The ray has a kink at the boundary.

Medium 1

Medium 2
Assume $n_2 > n_1$

θ_2

Refracted ray

Angle of refraction

(c)

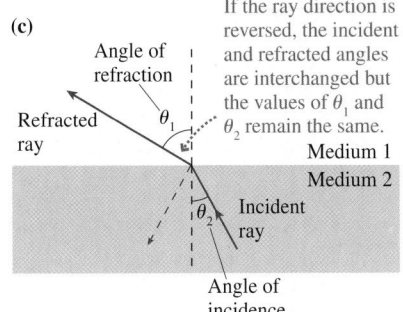

If the ray direction is reversed, the incident and refracted angles are interchanged but the values of θ_1 and θ_2 remain the same.

Angle of refraction

Refracted ray

θ_1

Medium 1

Medium 2

θ_2

Incident ray

Angle of incidence

the refraction of a single ray at a boundary between medium 1 and medium 2. Let the angle between the ray and the normal be θ_1 in medium 1 and θ_2 in medium 2. For the medium in which the ray is approaching the boundary, this is the *angle of incidence* as we've previously defined it. The angle on the transmitted side, *measured from the normal,* is called the **angle of refraction.** Notice that θ_1 is the angle of incidence in Figure 23.15b and the angle of refraction in Figure 23.15c, where the ray is traveling in the opposite direction, even though the value of θ_1 has not changed.

The relationship between angles θ_1 and θ_2 is generally believed to have been discovered in 1621 by the Dutch scientist Willebrod Snell, although some historians of science believe it may have been known before then. Regardless of its discoverer, we refer to the "law of refraction" as **Snell's law.** If a ray refracts between medium 1 and medium 2, having indices of refraction n_1 and n_2, the ray angles θ_1 and θ_2 in the two media are related by

$$n_1 \sin\theta_1 = n_2 \sin\theta_2 \qquad \text{(Snell's law of refraction)} \qquad (23.4)$$

Notice that Snell's law does not mention which is the incident angle and which the refracted angle.

The Index of Refraction

To Snell and his contemporaries, n was simply an "index of the refractive power" of a transparent substance. The relation between the index of refraction and the speed of light was not recognized until the development of a wave theory of light in the 19th century. Theory predicts, and experiment confirms, that light travels through a transparent medium, such as glass or water, at a speed *less* than its speed c in vacuum. We define the *index of refraction n* of a transparent medium as

$$n = \frac{c}{v_{\text{medium}}} \qquad (23.5)$$

where v_{medium} is the light speed in the medium. This implies, of course, that $v_{\text{medium}} = c/n$. The index of refraction of a medium is always $n > 1$ except for vacuum, which has $n = 1$ exactly.

Table 23.1 shows measured values of n for several materials. There are many types of glass, each with a slightly different index of refraction, so we will keep things simple by accepting $n = 1.50$ as a typical value. Notice that zircon, the material used to make "cubic zirconium" costume jewelry, has an index of refraction much higher than glass, although not nearly equal to diamond.

We can accept Snell's law as simply an empirical discovery about the behavior of light. Alternatively, and perhaps surprisingly, we can use the wave model of light to justify Snell's law. Figure 23.16 shows the wave front at time t of a plane wave approaching a boundary at speed $v_1 = c/n_1$. The light rays are perpendicular to the wave front, so the angle of incidence θ_1 is also the angle between the wave front and the boundary.

Huygens' principle, which you learned in Chapter 22, says that we can locate the wave front at a later time by thinking of each point on the wave front as the source of spherical wavelets. A spherical wavelet from point B travels distance $v_1 \Delta t$ during the time interval Δt. Because point A had just reached the boundary at time t, its wavelet travels through medium 2 at speed $v_2 = c/n_2$. This wavelet travels distance $v_2 \Delta t$ during the time interval Δt.

The wave front at time $t + \Delta t$ is tangent to the wavelets. You can see that a kink has developed because of the change in wave speed. The rays in medium 2 are still perpendicular to the wave front, but they now travel at the refracted angle θ_2. Notice that the two right triangles AB'B and AB'A' share a common

TABLE 23.1 Indices of refraction

Medium	n
Vacuum	1.00 exactly
Air (actual)	1.0003
Air (accepted)	1.00
Water	1.33
Ethyl alcohol	1.36
Oil	1.46
Glass (typical)	1.50
Polystyrene plastic	1.59
Zircon	1.96
Diamond	2.41
Silicon (infrared)	3.50

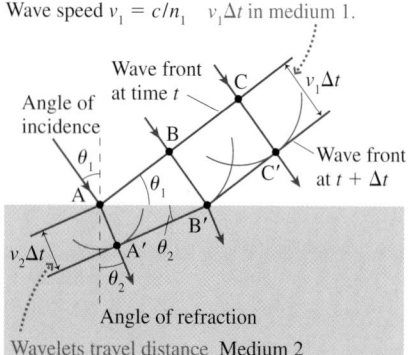

FIGURE 23.16 Snell's law is a consequence of Huygens' principle.

hypotenuse AB′. Calculating the length of the hypotenuse for both triangles, using trigonometry, and equating the results gives

$$\frac{v_1 \Delta t}{\sin\theta_1} = \frac{v_2 \Delta t}{\sin\theta_2} \qquad (23.6)$$

The Δt cancels and Equation 23.6 can be written

$$\frac{1}{v_1}\sin\theta_1 = \frac{1}{v_2}\sin\theta_2 \qquad (23.7)$$

If we now multiply both sides by c and use the definitions $n_1 = c/v_1$ and $n_2 = c/v_2$, we arrive at

$$n_1\sin\theta_1 = n_2\sin\theta_2 \qquad (23.8)$$

which is Snell's law.

Examples of Refraction

Look back at Figure 23.15. As the ray in Figure 23.15b moves from medium 1 to medium 2, where $n_2 > n_1$, it bends closer to the normal. In Figure 23.15c, where the ray moves from medium 2 to medium 1, it bends away from the normal. This is a general conclusion that follows from Snell's law:

- When a ray is transmitted into a material with a higher index of refraction, it bends toward the normal.
- When a ray is transmitted into a material with a lower index of refraction, it bends away from the normal.

This rule becomes a central idea in a procedure for analyzing refraction problems.

TACTICS BOX 23.1 Analyzing refraction

❶ **Draw a ray diagram.** Represent the light beam with one ray.
❷ **Draw a line normal to the boundary.** Do this at each point where the ray intersects a boundary.
❸ **Show the ray bending in the correct direction.** The angle is larger on the side with the smaller index of refraction. This is the qualitative application of Snell's law.
❹ **Label angles of incidence and refraction.** Measure all angles from the normal.
❺ **Use Snell's law.** Calculate the unknown angle or unknown index of refraction.

EXAMPLE 23.3 Deflecting a laser beam
A laser beam is aimed at a 1.0-cm-thick sheet of glass at an angle 30° above the glass.

a. What is the laser beam's direction of travel in the glass?
b. What is its direction in the air on the other side?
c. By what distance is the laser beam displaced?

MODEL Represent the laser beam with a single ray and use the ray model of light.

VISUALIZE Figure 23.17 is a pictorial representation in which the first four steps of Tactics Box 23.1 have been identified. Notice that the angle of incidence is $\theta_1 = 60°$, not the 30° value given in the problem.

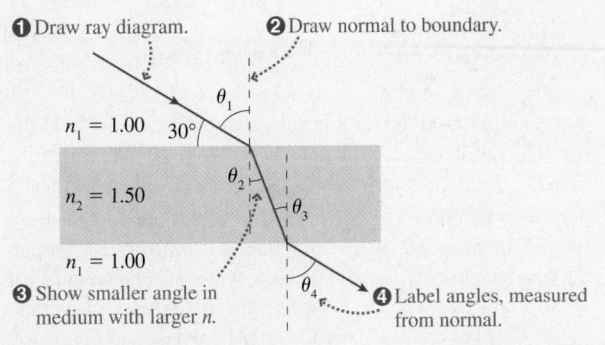

❶ Draw ray diagram.   ❷ Draw normal to boundary.
$n_1 = 1.00$ 30° θ_1
$n_2 = 1.50$ θ_2 θ_3
$n_1 = 1.00$
❸ Show smaller angle in medium with larger n. θ_4 ❹ Label angles, measured from normal.

FIGURE 23.17 The ray diagram of a laser beam passing through a sheet of glass.

SOLVE

a. Snell's law, the final step in the Tactics Box, is $n_1 \sin \theta_1 = n_2 \sin \theta_2$. Using $\theta_1 = 60°$, we find that the direction of travel in the glass is

$$\theta_2 = \sin^{-1} \left(\frac{n_1 \sin \theta_1}{n_2} \right) = \sin^{-1} \left(\frac{\sin 60°}{1.5} \right)$$

$$= \sin^{-1}(0.577) = 35.3°$$

b. Snell's law at the second boundary is $n_2 \sin \theta_3 = n_1 \sin \theta_4$. You can see from Figure 23.17 that the interior angles are equal: $\theta_3 = \theta_2 = 35.3°$. Thus the ray emerges back into the air traveling at angle

$$\theta_4 = \sin^{-1} \left(\frac{n_2 \sin \theta_3}{n_1} \right) = \sin^{-1}(1.5 \sin 35.3°)$$

$$= \sin^{-1}(0.867) = 60°$$

This is the same as θ_1, the original angle of incidence. The glass doesn't change the direction of the laser beam.

c. Although the exiting laser beam is parallel to the initial laser beam, it has been displaced sideways by distance d. Figure 23.18 shows the geometry for finding d. From trigonometry, $d = l \sin \phi$. Further $\phi = \theta_1 - \theta_2$ and $l = t/\cos \theta_2$, where t is the thickness of the glass. Combining these gives

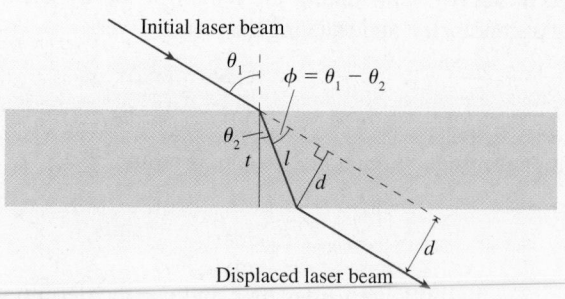

FIGURE 23.18 The laser beam is deflected sideways by distance d.

$$d = l \sin \phi = \frac{t}{\cos \theta_2} \sin(\theta_1 - \theta_2)$$

$$= \frac{(1.0 \text{ cm}) \sin 24.7°}{\cos 35.3°} = 0.51 \text{ cm}$$

The glass causes the laser beam to be displaced sideways by 0.51 cm.

ASSESS The laser beam exits the glass still traveling in the same direction as it entered. This is a general result for light traveling through a medium with parallel sides. Notice that the displacement d becomes zero in the limit $t \rightarrow 0$. This will be an important observation when we get to lenses.

EXAMPLE 23.4 **Measuring the index of refraction**

Figure 23.19 shows a laser beam deflected by a 30°-60°-90° prism. What is the prism's index of refraction?

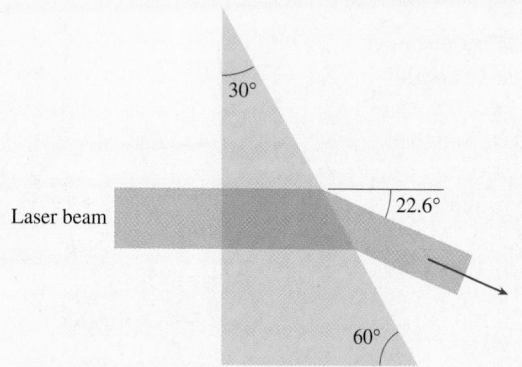

FIGURE 23.19 A prism deflects a laser beam.

MODEL Represent the laser beam with a single ray and use the ray model of light.

VISUALIZE Figure 23.20 uses the steps of Tactics Box 23.1 to draw a ray diagram. The ray is incident perpendicular to the front face of the prism ($\theta_{\text{incident}} = 0°$), thus it is transmitted through the first boundary without deflection. At the second boundary it is especially important to *draw the normal to the surface* at the point of incidence and to *measure angles from the normal*.

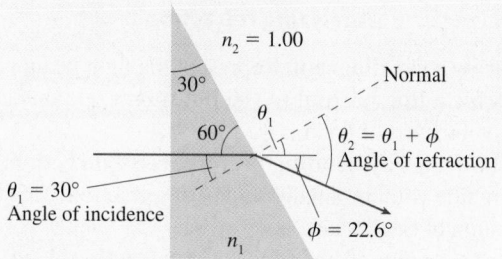

θ_1 and θ_2 are measured from the normal.

FIGURE 23.20 Pictorial representation of a laser beam passing through the prism.

SOLVE From the geometry of the triangle you can find that the laser's angle of incidence on the hypotenuse of the prism is $\theta_1 = 30°$, the same as the apex angle of the prism. The ray exits the prism at angle θ_2 such that the deflection is $\phi = \theta_2 - \theta_1 = 22.6°$. Thus $\theta_2 = 52.6°$. Knowing both angles and $n_2 = 1.00$ for air, we can use Snell's law to find n_1:

$$n_1 = \frac{n_2 \sin \theta_2}{\sin \theta_1} = \frac{1.00 \sin 52.6°}{\sin 30°} = 1.59$$

ASSESS Referring to the indices of refraction in Table 23.1, we see that the prism is made of plastic.

Total Internal Reflection

What would have happened in Example 23.4 if the prism angle had been 45° rather than 30°? The light rays would approach the rear surface of the prism at an angle of incidence $\theta_1 = 45°$. When we try to calculate the angle of refraction at which the ray emerges into the air, we find

$$\sin\theta_2 = \frac{n_1}{n_2}\sin\theta_1 = \frac{1.59}{1.00}\sin 45° = 1.12$$

$$\theta_2 = \sin^{-1}(1.12) = \text{???}$$

Angle θ_2 doesn't compute because the sine of an angle can't be larger than 1. The ray is unable to refract through the boundary. Instead, 100% of the light *reflects* from the boundary back into the prism. This process is called **total internal reflection,** often abbreviated TIR. That it really happens is illustrated in Figure 23.21. Here three laser beams enter a prism from the left. The bottom two refract out through the right side of the prism. The blue beam, which is incident on the prism's back face at a slightly larger angle of incidence, undergoes total internal reflection and then emerges through the right surface.

Figure 23.22 shows several rays leaving a point source in a medium with index of refraction n_1. The medium on the other side of the boundary has $n_2 < n_1$. As we've seen, crossing a boundary into a material with a lower index of refraction causes the ray to bend away from the normal. Two things happen as angle θ_1 increases. First, the refraction angle θ_2 approaches 90°. Second, the fraction of the light energy that is transmitted decreases while the fraction that is reflected increases.

A **critical angle** is reached when $\theta_2 = 90°$. Because $\sin 90° = 1$, Snell's law $n_1 \sin\theta_c = n_2 \sin 90°$ gives the critical angle of incidence as

$$\theta_c = \sin^{-1}\left(\frac{n_2}{n_1}\right) \tag{23.9}$$

The refracted light vanishes at the critical angle and the reflection becomes 100% for any angle $\theta_1 \geq \theta_c$. The critical angle is well defined because of our assumption that $n_2 < n_1$. **There is no critical angle and no total internal reflection if $n_2 > n_1$.**

As a quick example, the critical angle in a typical piece of glass at the glass-air boundary is

$$\theta_{c\ glass} = \sin^{-1}\left(\frac{1.00}{1.50}\right) = 42°$$

The fact that the critical angle is less than 45° has important applications. For example, Figure 23.23 shows a pair of binoculars. The lenses are much farther apart than your eyes, so the light rays need to be brought together before exiting the eyepieces. Rather than using mirrors, which get dirty, are easily scratched, and require alignment, binoculars use a pair of prisms on each side. Thus the light undergoes two total internal reflections and emerges from the eyepiece.

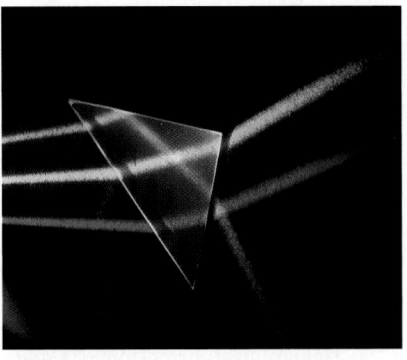

FIGURE 23.21 One of the three laser beams undergoes total internal reflection.

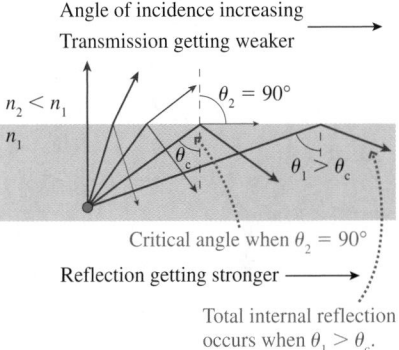

FIGURE 23.22 Refraction and reflection of rays as the angle of incidence increases.

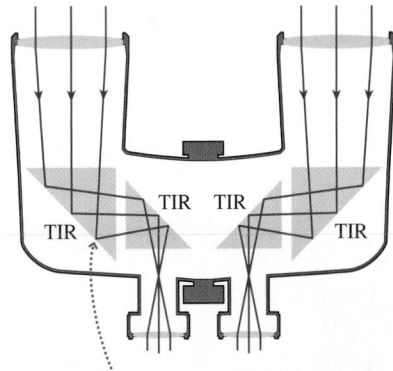

Angles of incidence exceed the critical angle.

FIGURE 23.23 Binoculars and other optical instruments make use of total internal reflection.

EXAMPLE 23.5 Total internal reflection

A light bulb is set in the bottom of a 3.0-m-deep swimming pool. What is the diameter of the circle of light seen on the water's surface from above?

MODEL Represent the light bulb as a point source and use the ray model of light.

VISUALIZE Figure 23.24 on the next page is a pictorial representation of the light rays. The light bulb emits rays at all angles,

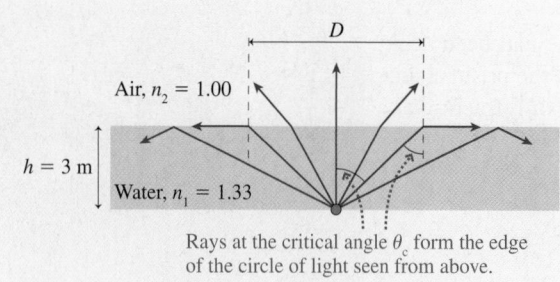

Rays at the critical angle θ_c form the edge of the circle of light seen from above.

FIGURE 23.24 Pictorial representation of the rays leaving a light bulb at the bottom of a swimming pool.

but only some of the rays refract into the air where they can be seen from above. Rays striking the surface at greater than the critical angle undergo TIR and remain within the water. The diameter of the circle of light is the distance between the two points at which rays strike the surface at the critical angle.

SOLVE From trigonometry, the circle diameter is $D = 2h\tan\theta_c$, where h is the depth of the water. The critical angle for a water-air boundary is $\theta_c = \sin^{-1}(1.00/1.33) = 48.7°$. Thus

$$D = 2(3.0\text{ m})\tan 48.7° = 6.83\text{ m}$$

Fiber Optics

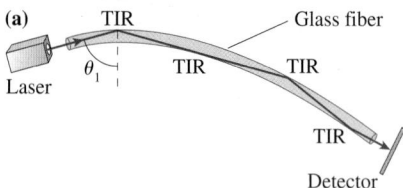

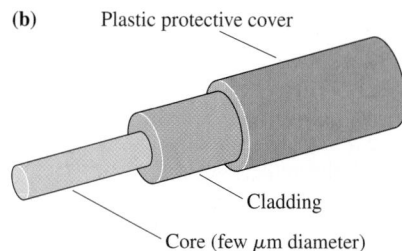

FIGURE 23.25 Light rays are confined within an optical fiber by total internal reflection.

The most important modern application of total internal reflection is the transmission of light through optical fibers. Figure 23.25a shows a laser beam shining into the end of a long, narrow-diameter glass tube. The light rays pass easily from the air into the glass, but they then impinge on the inside wall of the glass tube at an angle of incidence θ_1 approaching 90°. This is well above the critical angle, so the laser beam undergoes TIR and remains inside the glass. The laser beam continues to "bounce" its way down the tube as if the light were inside a pipe. Indeed, optical fibers are sometimes called "light pipes." The rays are *below* the critical angle ($\theta_1 \approx 0$) when they finally reach the end of the fiber, thus they refract out without difficulty and can be detected.

While a simple glass tube can transmit light, reliance on a glass-air boundary is not sufficiently reliable for commercial use. Any small scratch on the side of the tube alters the rays' angle of incidence and allows leakage of light. Figure 23.25b shows the construction of a practical optical fiber. A small-diameter glass *core* is surrounded by a layer of glass *cladding*. The glasses used for the core and the cladding have $n_{core} > n_{cladding}$, thus light undergoes TIR at the core-cladding boundary and remains confined within the core. This boundary is not exposed to the environment and hence retains its integrity even under adverse conditions.

Even glass of the highest purity is not perfectly transparent. Absorption in the glass, even if very small, causes a gradual decrease in light intensity. The glass used for the core of optical fibers has a minimum absorption at a wavelength of 1.3 μm, in the infrared, so this is the laser wavelength used for long-distance signal transmission. Light at this wavelength can travel hundreds of kilometers through a fiber without significant loss.

Fiber optics have replaced copper wires for carrying digital signals.

STOP TO THINK 23.3 A light ray travels from medium 1 to medium 3 as shown. For these media,

a. $n_3 > n_1$.
b. $n_3 = n_1$.
c. $n_3 < n_1$.
d. We can't compare n_1 to n_3 without knowing n_2.

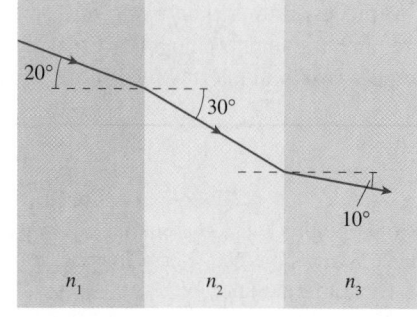

23.4 Image Formation by Refraction

You've likely made an interesting observation while looking at fish in an aquarium. First you see a fish that appears to be swimming close to the front window of the aquarium. But then, if you look through the side of the aquarium, you find that the fish is actually farther from the front window than you thought. Somehow, as you look through the front window, the fish appears to be closer than it really is. This is a puzzle crying out for an explanation.

To begin, recall how vision functions. A diverging bundle of rays leaves the object, enters the pupil of the eye, and is focused on the retina. By adjusting the eye's lens to achieve a good focus, your brain determines the distance d from which the rays originated. This is where you perceive the object to be. Figure 23.26a shows how you would see a fish out of water at distance d.

Now place the fish back in the aquarium at the same distance d. For simplicity, we'll ignore the glass wall of the aquarium and consider the water-air boundary. (The thin glass of a typical window has only a very small effect on the refraction of the rays and doesn't change the conclusions.) Light rays again leave the fish, but this time they refract at the water-air boundary. Because they're going from a higher to a lower index of refraction, the rays refract *away from* the normal. Figure 23.26b shows the consequences.

A bundle of diverging rays still enters your eye, but now these rays *seem* to be diverging from a closer point, at distance d'. As far as your eye and brain are concerned, it's exactly *as if* the rays really originate at distance d', and this is the location at which you "see" the fish. **The object appears closer than it really is because of the refraction of light at the boundary.**

We found that the rays reflected from a mirror diverge from a point that is not the object point. We called that point a *virtual image.* Similarly, if rays from an object point P refract at a boundary between two media such that the rays then diverge from a point P′ and *appear* to come from P′, we call P′ a virtual image of point P. The virtual image of the fish is what you see.

Let's examine this image formation a bit more carefully. Figure 23.27 shows a boundary between two transparent media having indices of refraction n_1 and n_2. Point P, a source of light rays, is the object. Point P′, from which the rays *appear* to diverge, is the virtual image of P. The figure assumes $n_1 > n_2$, but this assumption isn't necessary. Distance s is called the **object distance.** Our goal is to determine distance s', the **image distance.**

A line perpendicular to the boundary is called the **optical axis.** Consider a ray that leaves the object at angle θ_1 with respect to the optical axis. θ_1 is also the angle of incidence at the boundary, where the ray refracts into the second medium at angle θ_2. By tracing the refracted ray backward, you can see that θ_2 is also the angle between the refracted ray and the optical axis at point P′.

The distance l is common to both the incident and the refracted rays, and you can see that $l = s\tan\theta_1 = s'\tan\theta_2$. Thus

$$s' = \frac{\tan\theta_1}{\tan\theta_2}s \tag{23.10}$$

Snell's law relates the sines of angles θ_1 and θ_2. That is,

$$\frac{\sin\theta_1}{\sin\theta_2} = \frac{n_2}{n_1} \tag{23.11}$$

In practice, the angle between any of these rays and the optical axis is very small because the size of the pupil of your eye is very much less than the distance between the object and your eye. (The angles in the figure have been greatly exaggerated.) Rays that are nearly *parallel* to the *axis* are called **paraxial rays.**

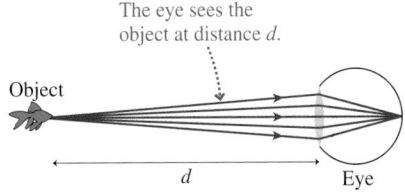

(a) A fish out of water

The eye sees the object at distance d.

Object

d Eye

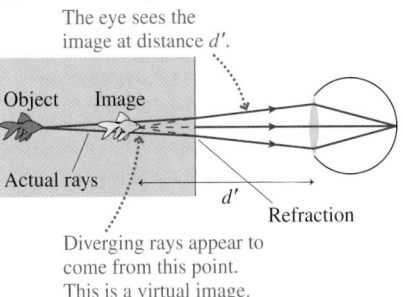

(b) A fish in the aquarium

The eye sees the image at distance d'.

Object   Image

Actual rays

d'

Refraction

Diverging rays appear to come from this point. This is a virtual image.

FIGURE 23.26 Refraction of the light rays causes a fish in the aquarium to be seen at distance d'.

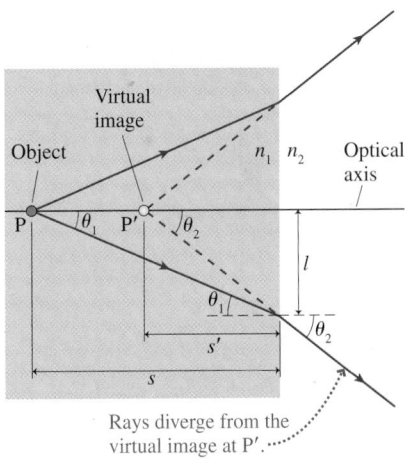

Virtual image

Object

n_1 n_2 Optical axis

P θ_1 P′ θ_2

l

θ_1 θ_2

s'

s

Rays diverge from the virtual image at P′.

FIGURE 23.27 Finding the virtual image P′ of an object at P.

The small-angle approximation $\sin\theta \approx \tan\theta \approx \theta$, where θ is in radians, can be applied to paraxial rays. Consequently,

$$\frac{\tan\theta_1}{\tan\theta_2} \approx \frac{\sin\theta_1}{\sin\theta_2} = \frac{n_2}{n_1} \tag{23.12}$$

Using this result in Equation 23.10, we find that the image distance is

$$s' = \frac{n_2}{n_1}s \tag{23.13}$$

NOTE ▶ The fact that the result for s' is independent of θ_1 implies that *all* paraxial rays appear to diverge from the same point P′. This property of the diverging rays is essential in order to have a well-defined image. ◀

This section has given us a first look at image formation via refraction. We will extend this idea to image formation with lenses in Section 23.6.

EXAMPLE 23.6 An air bubble in a window

A fish and a sailor look at each other through a 5.0-cm-thick glass porthole in a submarine. There happens to be a small air bubble right in the center of the glass. How far behind the surface of the glass does the air bubble appear to the fish? To the sailor?

MODEL Represent the air bubble as a point source and use the ray model of light.

VISUALIZE Paraxial light rays from the bubble refract into the air on one side and into the water on the other. The ray diagram looks like Figure 23.27.

SOLVE The index of refraction of the glass is $n_1 = 1.50$. The bubble is in the center of the window, so the object distance

from either side of the window is $s = 2.5$ cm. From the water side, the fish sees the bubble at an image distance

$$s' = \frac{n_2}{n_1}s = \frac{1.33}{1.50}(2.5 \text{ cm}) = 2.2 \text{ cm}$$

The sailor, in air, sees the bubble at an image distance

$$s' = \frac{n_2}{n_1}s = \frac{1.00}{1.50}(2.5 \text{ cm}) = 1.7 \text{ cm}$$

ASSESS The image distance is *less* for the sailor because of the *larger* difference between the two indices of refraction.

23.5 Color and Dispersion

One of the most obvious visual aspects of light is the phenomenon of color. Yet color, for all its vivid sensation, is not inherent in the light itself. Color is a *perception,* not a physical quantity. Color is associated with the wavelength of light, but the fact that we see light with a wavelength of 650 nm as "red" tells us how our visual system responds to electromagnetic waves of this wavelength. There is no "redness" associated with the light wave itself.

Most of the results of optics do not depend on color. Rays of red light pass through glass the same as rays of blue light, so we generally don't need to know the color of light—or, to be more precise, its wavelength—to use the laws of reflection and refraction. Nonetheless, color is an interesting subject, one worthy of a short digression.

Color

It has been known since antiquity that irregularly shaped glass and crystals cause sunlight to be broken into various colors. A common idea was that the glass or crystal somehow altered the properties of the light by *adding* color to the light. Newton suggested a different explanation. He first passed a sunbeam through a prism, producing the familiar rainbow of light. We say that the prism *disperses* the light. Newton's novel idea, shown in Figure 23.28a, was to use a second

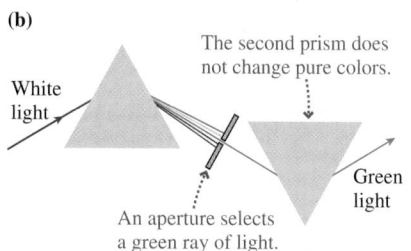

(a)

White light

A prism disperses white light into colors.

A second prism can combine the colors back into white light.

White light

(b)

White light

An aperture selects a green ray of light.

The second prism does not change pure colors.

Green light

FIGURE 23.28 Newton used prisms to study color.

prism, inverted with respect to the first, to "reassemble" the colors. He found that the light emerging from the second prism was a beam of pure, white light.

But the emerging light beam is white only if *all* the rays are allowed to move between the two prisms. Blocking some of the rays with small obstacles, as in Figure 23.28b, causes the emerging light beam to have color. This suggests that color is associated with the light itself, not with anything that the prism is "doing" to the light. Newton tested this idea by inserting a small aperture between the prisms to pass only the rays of a particular color, such as green. If the prism alters the properties of light, then the second prism should change the green light to other colors. Instead, the light emerging from the second prism is unchanged from the green light entering the prism.

These and similar experiments show that

1. What we perceive as white light is a mixture of all colors. White light can be dispersed into its various colors and, equally important, mixing all the colors produces white light.
2. The index of refraction of a transparent material differs slightly for different colors of light. Glass has a slightly larger index of refraction for violet light than for green light or red light. Consequently, different colors of light refract at slightly different angles and follow slightly different paths through a piece of glass. A prism does not alter the light or add anything to the light; it simply causes the different colors that are inherent in white light to follow slightly different trajectories.

Dispersion

It was Thomas Young, with his two-slit interference experiment, who showed that what we perceive as different colors are associated with light of different wavelengths. The longest wavelengths are perceived as red light and the shortest wavelengths are perceived as violet light. Table 23.2 is a brief summary of the *visible spectrum* of light. Visible-light wavelengths are used so frequently that it is well worth committing this short table to memory.

Newton's observation that the index of refraction varies slightly with color implies that the index of refraction varies slightly with wavelength. This is known as **dispersion.** Figure 23.29 shows the *dispersion curves* of two common glasses. Notice that *n* is *larger* when the wavelength is *shorter,* thus violet light refracts more than red light.

I procured me a triangular glass prism to try therewith the celebrated phenomena of colors.

Isaac Newton

TABLE 23.2 A brief summary of the visible spectrum of light

Color	Approximate wavelength
Deepest red	700 nm
Red	650 nm
Green	550 nm
Blue	450 nm
Deepest violet	400 nm

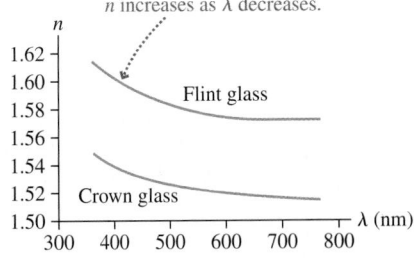

FIGURE 23.29 Dispersion curves show how the index of refraction varies with wavelength.

EXAMPLE 23.7 **Dispersing light with a prism**

Example 23.4 found that a ray incident on a 30° prism is deflected by 22.6° if the prism's index of refraction is 1.59. Suppose this is the index of refraction of deep violet light, and that deep red light has an index of refraction of 1.54.

a. What is the deflection angle for deep red light?
b. If a beam of white light is dispersed by this prism, how wide is the rainbow spectrum on a screen 2.0 m away?

VISUALIZE Figure 23.20 showed the geometry. A ray of any wavelength is incident on the rear surface of the prism at $\theta_1 = 30°$.

SOLVE

a. If $n_1 = 1.54$ for deep red light, the refraction angle is

$$\theta_2 = \sin^{-1}\left(\frac{n_1 \sin\theta_1}{n_2}\right) = \sin^{-1}\left(\frac{1.54 \sin 30°}{1.00}\right) = 50.4°$$

Example 23.4 showed that the deflection angle is $\phi = \theta_2 - \theta_1$, so deep red light is deflected by $\phi_{red} = 20.4°$. This angle is slightly smaller than the previously observed $\phi_{violet} = 22.6°$.

b. The entire spectrum is spread between $\phi_{red} = 20.4°$ and $\phi_{violet} = 22.6°$. The angular spread is

$$\delta = \phi_{violet} - \phi_{red} = 2.2° = 0.038 \text{ rad}$$

At distance r, the spectrum spans an arc length

$$s = r\delta = (2.0 \text{ m})(0.038 \text{ rad}) = 0.076 \text{ m} = 7.6 \text{ cm}$$

ASSESS The angle is so small that there's no appreciable difference between arc length and a straight line. The spectrum will be 7.6 cm wide at a distance of 2.0 m.

Rainbows

One of the most interesting sources of color in nature is the rainbow. The details get somewhat complicated, but Figure 23.30a shows that the basic cause of the rainbow is a combination of refraction, reflection, and dispersion.

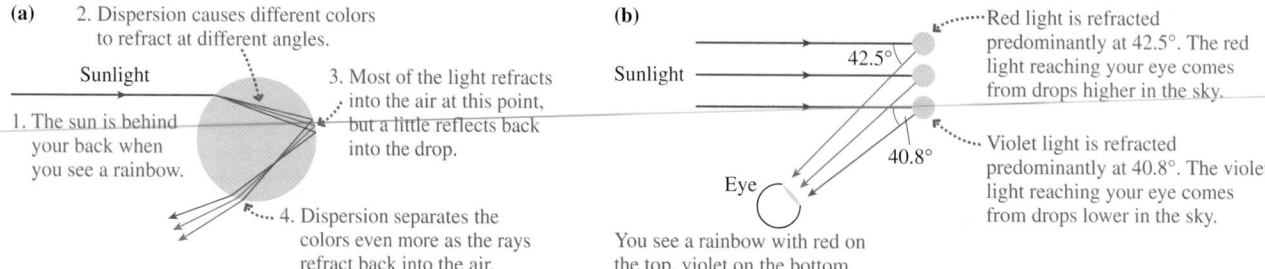

(a) 2. Dispersion causes different colors to refract at different angles.

Sunlight

1. The sun is behind your back when you see a rainbow.

3. Most of the light refracts into the air at this point, but a little reflects back into the drop.

4. Dispersion separates the colors even more as the rays refract back into the air.

(b) Sunlight

42.5°

Eye

40.8°

You see a rainbow with red on the top, violet on the bottom.

Red light is refracted predominantly at 42.5°. The red light reaching your eye comes from drops higher in the sky.

Violet light is refracted predominantly at 40.8°. The violet light reaching your eye comes from drops lower in the sky.

FIGURE 23.30 Light seen in a rainbow has undergone refraction + reflection + refraction in a raindrop.

Figure 23.30a might lead you to think that the top edge of a rainbow is violet. In fact, the top edge is red, and violet is on the bottom. The rays leaving the drop in Figure 23.30a are spreading apart, so they can't all reach your eye. As Figure 23.30b shows, a ray of red light reaching your eye comes from a drop *higher* in the sky than a ray of violet light. In other words, the colors you see in a rainbow refract toward your eye from different raindrops, not from the same drop. You have to look higher in the sky to see the red light than to see the violet light.

Colored Filters and Colored Objects

White light passing through a piece of green glass emerges as green light. A possible explanation would be that the green glass *adds* "greenness" to the white light, but Newton found otherwise. Green glass is green because it *removes* any light that is "not green." More precisely, a piece of colored glass *absorbs* all wavelengths except those of one color, and that color is transmitted through the glass without hindrance. We can think of a piece of colored glass or plastic as a *filter* that removes all wavelengths except a chosen few.

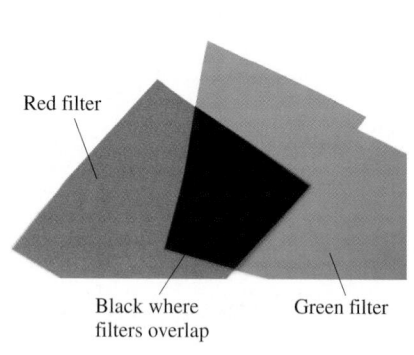

Red filter

Black where filters overlap

Green filter

No light at all passes through both a green and a red filter.

EXAMPLE 23.8 **Filtering light**

White light passes through a green filter and is observed on a screen. Describe how the screen will look if a second green filter is placed between the first filter and the screen. Describe how the screen will look if a red filter is placed between the green filter and the screen.

VISUALIZE The first filter removes all light except for wavelengths near 550 nm that we perceive as green light. A second green filter doesn't have anything to do. The nongreen wavelengths have already been removed, and the green light emerging from the first filter will pass through the second filter without difficulty. The screen will continue to be green and its intensity will not change. A red filter, by contrast, absorbs all wavelengths except those near 650 nm. The red filter will absorb the green light, and *no* light will reach the screen. The screen will be dark.

This behavior is true not just for glass filters, which transmit light, but for *pigments* that absorb light of some wavelengths but *reflect* light at other wavelengths. For example, red paint contains pigments that reflect light at wavelengths near 650 nm while absorbing all other wavelengths. Pigments in paints, inks, and natural objects are responsible for most of the color we observe in the world, from the red of lipstick to the blue of a bluebird's feathers.

As an example, Figure 23.31 shows the absorption curve of *chlorophyll*. Chlorophyll is essential for photosynthesis in green plants. The chemical reactions of photosynthesis are able to use red light and blue/violet light, thus chlorophyll has evolved to absorb red light and blue/violet light from sunlight and put it to use. But green and yellow light are not absorbed. Instead, to conserve energy, these wavelengths are mostly *reflected* to give the object a greenish-yellow color. When you look at the green leaves on a tree, you're seeing the light that was reflected because it *wasn't* needed for photosynthesis.

Light Scattering: Blue Skies and Red Sunsets

In the ray model of Section 23.1 we noted that light within a medium can be scattered or absorbed. As we've now seen, the absorption of light can be wavelength dependent and can create color in objects. What are the effects of scattering?

Light can scatter from small particles that are suspended in a medium. If the particles are large compared to the wavelengths of light—even though they may be microscopic and not readily visible to the naked eye—the light essentially reflects off the particles. The law of reflection doesn't depend on wavelength, so all colors are scattered equally. White light scattered from many small particles makes the medium appear cloudy and white. Two well-known examples are clouds, where micrometer-size water droplets scatter the light, and milk, which is a colloidal suspension of microscopic droplets of fats and proteins.

A more interesting aspect of scattering occurs at the atomic level. The atoms and molecules of a transparent medium are much smaller than the wavelengths of light, so they can't scatter light simply by reflection. Instead, the oscillating electric field of the light wave interacts with the electrons in each atom in such a way that the light is scattered. This atomic-level scattering is called **Rayleigh scattering.**

Unlike the scattering by small particles, Rayleigh scattering from atoms and molecules *does* depend on the wavelength. A detailed analysis shows that the intensity of scattered light depends inversely on the fourth power of the wavelength: $I_{\text{scattered}} \propto \lambda^{-4}$. This wavelength dependence explains why the sky is blue and sunsets are red.

As sunlight travels through the atmosphere, the λ^{-4} dependence of Rayleigh scattering causes the shorter wavelengths to be preferentially scattered. If we take 650 nm as a typical wavelength for red light and 450 nm for blue light, the intensity of scattered blue light relative to scattered red light is

$$\frac{I_{\text{blue}}}{I_{\text{red}}} = \left(\frac{650}{450}\right)^4 \approx 4$$

Four times more blue light is scattered toward us than red light and thus, as Figure 23.32 shows, the sky appears blue.

Because of the earth's curvature, sunlight has to travel much farther through the atmosphere when we see it at sunrise or sunset than it does during the midday hours. In fact, the path length through the atmosphere at sunset is so long that essentially all the short wavelengths have been lost due to Rayleigh scattering. Only the longer wavelengths remain—orange and red—and they make the colors of the sunset.

23.6 Thin Lenses: Ray Tracing

A camera obscura or a pinhole camera forms images on a screen, but the images are faint and not perfectly focused. The ability to create a bright, well-focused image is vastly improved by using a lens. A **lens** is a transparent material that uses refraction at *curved* surfaces to form an image from diverging light rays. We will

Chlorophyll absorbs most of the red and blue/violet light for use in photosynthesis.

The green and yellow light that is not absorbed is reflected and gives plants their green color.

FIGURE 23.31 The absorption curve of chlorophyll.

Sunsets are red because all the blue light has scattered as the sunlight passes through the atmosphere.

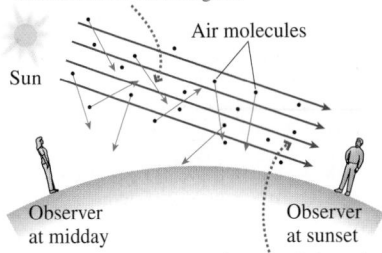

At midday the scattered light is mostly blue because molecules preferentially scatter shorter wavelengths.

At sunset, when the light has traveled much farther through the atmosphere, the light is mostly red because the shorter wavelengths have been lost to scattering.

FIGURE 23.32 Rayleigh scattering by molecules in the air gives the sky and sunsets their color.

defer a detailed analysis of the refraction of lenses until the next section. First, we want to establish a pictorial method of understanding image formation. This method is called **ray tracing.**

Figure 23.33 shows parallel light rays entering two different lenses. The left lens, called a **converging lens,** causes the rays to refract *toward* the optical axis. The common point through which initially parallel rays pass is called the **focal point** of the lens. The distance of the focal point from the lens is called the **focal length** *f* of the lens. The right lens, called a **diverging lens,** refracts parallel rays *away from* the optical axis. This lens also has a focal point, but it is not as obvious in the figure.

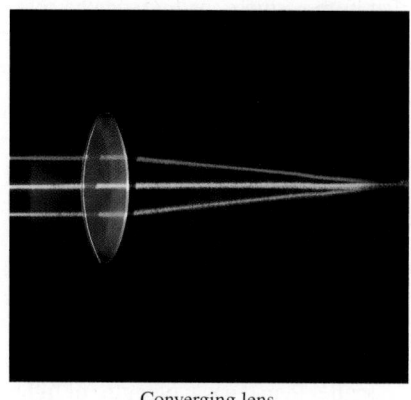 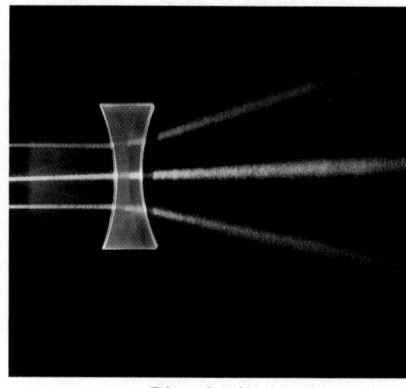

Converging lens   Diverging lens

FIGURE 23.33 Parallel light rays pass through a converging lens and a diverging lens.

NOTE ▶ A converging lens is thicker in the center than at the edges. A diverging lens is thicker at the edges than at the center. ◀

Figure 23.34 clarifies the situation. In the case of a diverging lens, a backward projection of the diverging rays shows that they all *appear* to have started from the same point. This is the focal point of a diverging lens, and its distance from the lens is the focal length of the lens. In the next section we'll relate the focal length to the curvature and index of refraction of the lens, but for now we'll use the practical definition that **the focal length is the distance from the lens at which rays parallel to the optical axis converge or from which they diverge.**

NOTE ▶ The focal length *f* is a property *of the lens,* independent of how the lens is used. The focal length characterizes a lens in much the same way that a mass *m* characterizes an object or a spring constant *k* characterizes a spring. ◀

Converging Lenses

These basic observations about lenses are enough to understand image formation by a thin lens. A **thin lens** is a lens whose thickness is very small in comparison to its focal length and in comparison to the object and image distances. We'll make the approximation that the thickness of a thin lens is zero and that the lens lies in a plane called the **lens plane.** Within this approximation, all refraction occurs as the rays cross the lens plane, and all distances are measured from the lens plane. Fortunately, the thin-lens approximation is quite good for most practical applications of lenses.

NOTE ▶ We'll *draw* lenses as if they have a thickness, because that is how we expect lenses to look, but our analysis will not depend on the shape or thickness of a lens. ◀

Focal length *f*

Parallel rays

Optical axis

Converging lens

This is the focal point. Rays actually converge at this point.

Focal length *f*

Parallel rays

Optical axis

This is the focal point. Rays appear to diverge from this point.

Diverging lens

FIGURE 23.34 The focal point and focal length of converging and diverging lenses.

Figure 23.35 shows three important situations of light rays passing through a thin, converging lens. Part a is familiar from Figure 23.34. If the direction of each of the rays in Figure 23.35a is reversed, Snell's law tells us that each ray will exactly retrace its path and emerge from the lens parallel to the optical axis. This leads to Figure 23.35b, which is the "mirror image" of part (a).

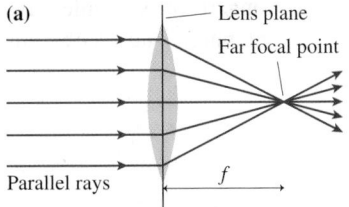

(a)

Any ray initially parallel to the optical axis will refract through the focal point on the far side of the lens.

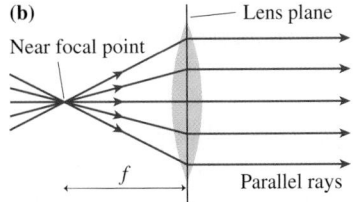

(b)

Any ray passing through the near focal point emerges from the lens parallel to the optical axis.

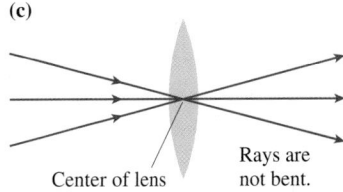

(c)

Any ray directed at the center of the lens passes through in a straight line.

FIGURE 23.35 Three important sets of rays passing through a thin, converging lens.

Notice that the lens actually has *two* focal points, located at distances f on either side of the lens. The focal point on the side from which the light rays are approaching is the *near focal point.* The focal point opposite the side from which the light rays are approaching is the *far focal point.*

Figure 23.35c shows several rays passing through the *center* of the lens. At the center, the two sides of a lens are very nearly parallel to each other. Earlier, in Example 23.3, we found that a ray passing through a piece of glass with parallel sides is *displaced* but *not bent* and that the displacement becomes zero as the thickness approaches zero. Consequently, a ray through the center of a thin lens, with zero thickness, is neither bent nor displaced but travels in a straight line.

These three situations form the basis for ray tracing.

Real Images

Figure 23.36 shows a lens and an object whose distance from the lens is larger than the focal length. Rays from point P on the object are refracted by the lens so as to converge at point P′ on the opposite side of the lens. If rays diverge from an object point P and interact with a lens such that the refracted rays *converge* at point P′, then we call P′ a **real image** of point P. Contrast this with our prior definition of a *virtual image* as a point from which rays *diverge.*

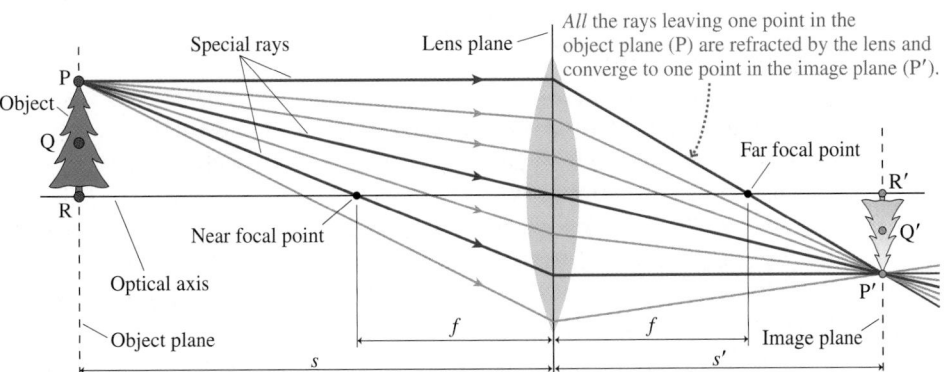

FIGURE 23.36 Rays from an object point P are refracted by the lens and converge to a real image at point P′.

All points on the object that are in the same plane, the **object plane,** converge to image points in the **image plane.** Points Q and R in the object plane of Figure 23.36 have image points Q′ and R′ in the same plane as point P′. Once we locate *one* point in the image plane, such as point P′, we know that the full image lies in the same plane.

There are two important observations to make about Figure 23.36. First, the image is upside down with respect to the object. This is called an **inverted image,** and it is a standard characteristic of real image formation with a converging lens. You have to put slides into a slide tray upside down so that the image seen on the screen is right side up. Second, rays from point P *fill* the entire lens surface, and all portions of the lens contribute to the image. A larger lens will "collect" more rays and thus make a brighter image. This is the big advantage of a lens over a camera obscura or a pinhole camera.

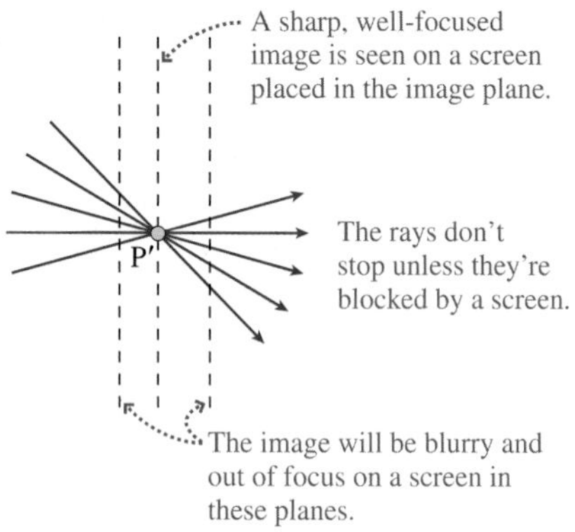

Figure 23.37 is a close-up view of the rays very near the image plane. The rays don't stop at P′ unless we place a screen in the image plane. When we do so, we see a sharp, well-focused image on the screen. To focus an image, you must either move the screen to coincide with the image plane or move the lens or object to make the image plane coincide with the screen. For example, the focus knob on a slide projector moves the lens closer to or farther from the slide until the image plane matches the screen position.

NOTE ▶ The ability to view and record *real* images, where the rays actually converge, sets real images apart from *virtual* images. But keep in mind that we need not *see* a real image in order to *have* an image. A real image exists at a point in space where the rays converge even if there's no viewing screen in the image plane. ◀

Figure 23.36 highlights three "special rays" that are based on the three situations of Figure 23.35. Notice that these three rays alone are sufficient to locate the image point P′. That is, we don't need to draw all the rays shown in Figure 23.36. The procedure known as *ray tracing* consists of locating the image by the use of just these three rays.

A sharp, well-focused image is seen on a screen placed in the image plane.

The rays don't stop unless they're blocked by a screen.

The image will be blurry and out of focus on a screen in these planes.

FIGURE 23.37 A close-up look at the rays near the image plane.

15.9   Actıv
Physıcs

TACTICS BOX 23.2 Ray tracing for a converging lens

❶ **Draw an optical axis.** Use graph paper or a ruler! Establish an appropriate scale.

❷ **Center the lens on the axis.** Mark and label the focal points at distance *f* on either side.

❸ **Represent the object with an upright arrow at distance *s*.** It's usually best to place the base of the arrow on the axis and to draw the arrow about half the radius of the lens.

❹ **Draw the three "special rays" from the tip of the arrow.** Use a straightedge.

   a. A ray parallel to the axis refracts through the far focal point.
   b. A ray that enters the lens along a line through the near focal point emerges parallel to the axis.
   c. A ray through the center of the lens does not bend.

❺ **Extend the rays until they converge.** This is the image point. Draw the rest of the image in the image plane. If the base of the object is on the axis, then the base of the image will also be on the axis.

❻ **Measure the image distance *s′*.** Also, if needed, measure the image height relative to the object height.

EXAMPLE 23.9 **Finding the image of a flower**

A 4.0-cm-diameter flower is 200 cm from the 50-cm-focal-length lens of a camera. How far should the film be placed behind the lens to record a well-focused image? What is the diameter of the image on the film?

MODEL The flower is in the object plane. Use ray tracing to locate the image.

VISUALIZE Figure 23.38 shows the ray-tracing diagram and the steps of Tactics Box 23.2. The image has been drawn in the plane where the three special rays converge. You can see *from the drawing* that the image distance is $s' \approx 67$ cm. This is where the film needs to be placed to record a focused image.

The heights of the object and image are labeled h and h'. The ray through the center of the lens is a straight line, thus the object and image both subtend the same angle θ. Using similar triangles,

$$\frac{h'}{s'} = \frac{h}{s}$$

Solving for h' gives

$$h' = h\frac{s'}{s} = (4.0 \text{ cm})\frac{67 \text{ cm}}{200 \text{ cm}} = 1.3 \text{ cm}$$

The flower's image has a diameter of 1.3 cm.

ASSESS We've been able to learn a great deal about the image from a simple geometric procedure.

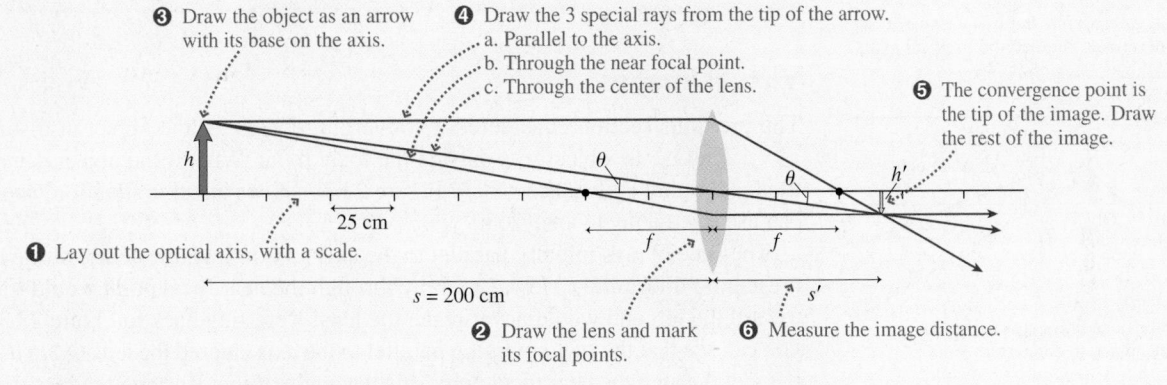

FIGURE 23.38 Ray-tracing diagram for Example 23.9.

Magnification

The image can be either larger or smaller than the object, depending on the location and focal length of the lens. But there's more to a description of the image than just its size. We also want to know its *orientation* relative to the object. That is, is the image upright or inverted?

Earlier in the chapter, we defined the magnification m as the ratio of image height to object height. It's now useful to expand that definition to include information about the orientation of the image. The revised definition of the **magnification,** which we'll call M, is

$$M = -\frac{h'}{h} = -\frac{s'}{s} \tag{23.14}$$

You just saw in Example 23.9 that the image-to-object height ratio is $h'/h = s'/s$. Consequently, we interpret the magnification M as follows:

1. A positive value of M indicates that the image is upright relative to the object. A negative value of M indicates that the image is inverted relative to the object.
2. The absolute value of M gives the size ratio of the image and object: $h'/h = |M|$.

The magnification in Example 23.9 would be $M = -0.33$, indicating that the image is inverted and 33% the size of the object.

NOTE ▶ "Magnification" can be less than 1, meaning that the image is smaller than the object (i.e., "demagnified"). ◀

STOP TO THINK 23.4 A lens produces a sharply focused, inverted image on a screen. What will you see on the screen if the lens is removed?

a. The image will be inverted and blurry.
b. The image will be upright and sharp.
c. The image will be upright and blurry.
d. The image will be much dimmer but otherwise unchanged.
e. There will be no image at all.

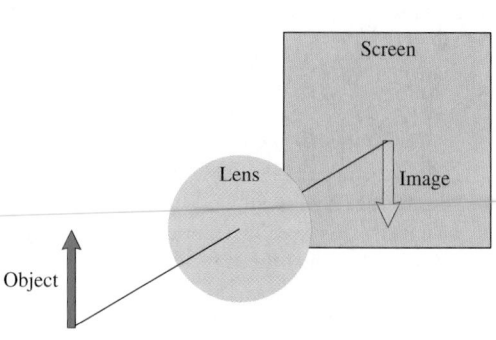

A ray *along a line* through the near focal point refracts parallel to the optical axis.

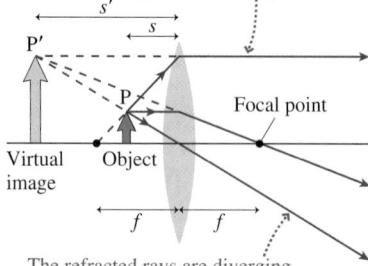

The refracted rays are diverging. They appear to come from point P′.

FIGURE 23.39 Rays from an object at distance $s < f$ are refracted by the lens and diverge to form a virtual image at point P′.

(a)

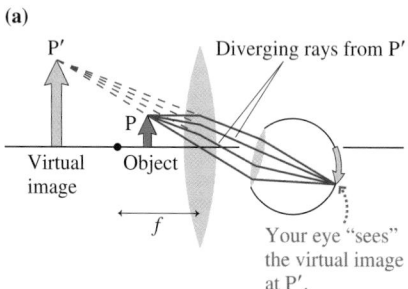

Your eye "sees" the virtual image at P′.

(b)

FIGURE 23.40 A converging lens is a magnifying glass when the object distance is less than f.

Virtual Images

The previous section considered a converging lens with the object at distance $s > f$. That is, the object was outside the focal point. What if the object is inside the focal point, at distance $s < f$? Figure 23.39 shows just this situation, and we can use ray tracing to analyze it.

The special rays initially parallel to the axis and through the center of the lens present no difficulties. However, a ray through the near focal point would travel toward the left and would never reach the lens! Referring back to Figure 23.35b, you can see that the rays emerging parallel to the axis entered the lens *along a line* passing through the near focal point. It's the angle of incidence on the lens that is important, not whether the light ray actually passes through the focal point. This was the basis for the wording of step 4b in Tactics Box 23.2 and is the third special ray shown in Figure 23.39.

You can see that the three refracted rays don't converge. Instead, all three rays appear to *diverge* from point P′. This is the situation we found for rays reflecting from a mirror and for the rays refracting out of an aquarium. Point P′ is a *virtual image* of the object point P. Furthermore, it is an **upright image**, having the same orientation as the object.

The refracted rays, which are all to the right of the lens, *appear* to come from P′, but none of the rays were ever at that point. No image would appear on a screen placed in the image plane at P′. So what good is a virtual image?

Your eye collects and focuses bundles of diverging rays, thus, as Figure 23.40a shows, you can "see" a virtual image by looking *through* the lens. This is exactly what you do with a magnifying glass, producing a scene like the one in Figure 23.40b. In fact, you view a virtual image anytime you look *through* the eyepiece of an optical instrument such as a microscope or binoculars.

The image distance s' for a virtual image is defined to be a *negative number* ($s' < 0$), indicating that the image is on the opposite side of the lens from a real image. With this choice of sign, the definition of magnification, $M = -s'/s$, is still valid. A virtual image with negative s' has $M > 0$, implying that the image is upright. This agrees with the ray tracing in Figure 23.39 and the photograph of Figure 23.40b.

NOTE ▶ A lens thicker in the middle than at the edges is classified as a converging lens. The light rays from an object *can* converge to form a real image after passing through such a lens, but only if the object distance is larger than the focal length of the lens: $s > f$. If $s < f$, the rays leaving a converging lens are diverging to produce a virtual image. ◀

EXAMPLE 23.10 Magnifying a flower

To see a flower better, a naturalist holds a 6.0-cm-focal-length magnifying glass 4.0 cm from the flower. What is the magnification?

MODEL The flower is in the object plane. Use ray tracing to locate the image.

VISUALIZE Figure 23.41 shows the ray-tracing diagram. The three special rays diverge from the lens, but we can use a straightedge to extend the rays backward to the point from which they diverge. This point, the image point, is seen to be 12 cm to the left of the lens. Because this is a virtual image, the image distance is $s' = -12$ cm. Thus the magnification is

$$M = -\frac{s'}{s} = -\frac{-12 \text{ cm}}{4.0 \text{ cm}} = 3.0$$

The image is three times as large as the object and, because M is positive, upright.

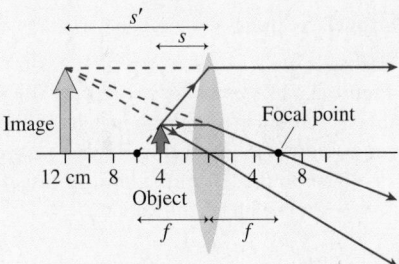

FIGURE 23.41 Ray-tracing diagram for Example 23.10.

Diverging Lenses

A lens thicker at the edges than in the middle is called a *diverging lens*. Figure 23.42 shows three important sets of rays passing through a diverging lens. These are based on Figures 23.33 and 23.34, where you saw that rays initially parallel to the axis diverge after passing through a diverging lens.

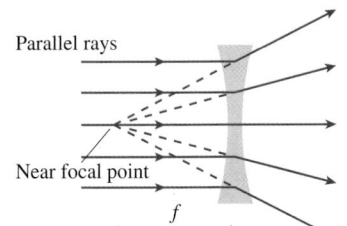

Any ray initially parallel to the optical axis diverges along a line through the near focal point.

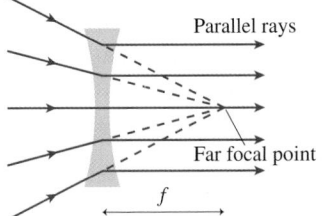

Any ray directed along a line toward the far focal point emerges from the lens parallel to the optical axis.

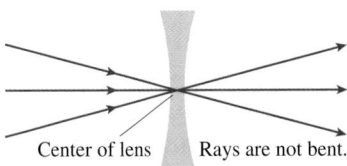

Any ray directed at the center of the lens passes through in a straight line.

FIGURE 23.42 Three important sets of rays passing through a thin, diverging lens.

Ray tracing follows the steps of Tactics Box 23.2 for a converging lens *except* that two of the three special rays in step 4 are different.

TACTICS BOX 23.3 Ray tracing for a diverging lens

❶–❸ **Follow steps 1 through 3 of Tactics Box 23.2.**
❹ **Draw the three "special rays" from the tip of the arrow.** Use a straightedge.

  a. A ray parallel to the axis diverges along a line through the near focal point.
  b. A ray along a line toward the far focal point emerges parallel to the axis.
  c. A ray through the center of the lens does not bend.

❺ **Trace the diverging rays backward.** The point from which they are diverging is the image point, which is always a virtual image.
❻ **Measure the image distance s'.** This will be a negative number.

EXAMPLE 23.11 Demagnifying a flower

A diverging lens with a focal length of 50 cm is placed 100 cm from a flower. Where is the image? What is its magnification?

MODEL The flower is in the object plane. Use ray tracing to locate the image.

VISUALIZE Figure 23.43 shows the ray-tracing diagram. The three special rays (labeled a, b, and c to match the Tactics Box) do not converge. However, they can be traced backward to an intersection ≈33 cm to the left of the lens. A virtual image is formed at $s' = -33$ cm with magnification

$$M = -\frac{s'}{s} = -\frac{-33 \text{ cm}}{100 \text{ cm}} = 0.33$$

The image, which can be seen by looking *through* the lens, is one-third the size of the object and upright.

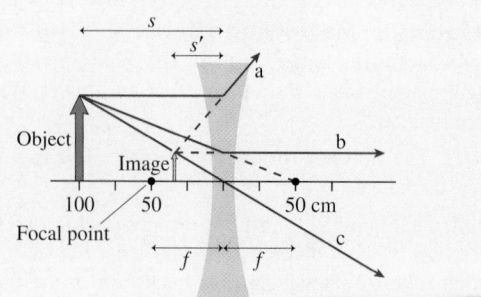

FIGURE 23.43 Ray-tracing diagram for Example 23.11.

ASSESS Ray tracing with a diverging lens is somewhat trickier than for a converging lens, so this example is worth careful study.

Diverging lenses *always* make virtual images and, for this reason, are rarely used alone. However, they have important applications when used in combination with other lenses. Cameras, eyepieces, and eyeglasses often incorporate diverging lenses.

Lens Combinations

Optical instruments, such as microscopes and cameras, are built with multiple lenses. There are many reasons for this having to do with image quality and the overall orientation and magnification of the image. A full analysis of optical instruments is beyond this text, but we can use the ideas of ray tracing to understand some of the basic ideas of lens combinations.

As an example, Figure 23.44 shows a telescope similar to the one with which Galileo discovered sunspots and the moons of Jupiter. It consists of a large converging lens, called the *objective,* and a smaller converging lens used as the *eyepiece.* The lenses are placed such that their focal points nearly coincide.

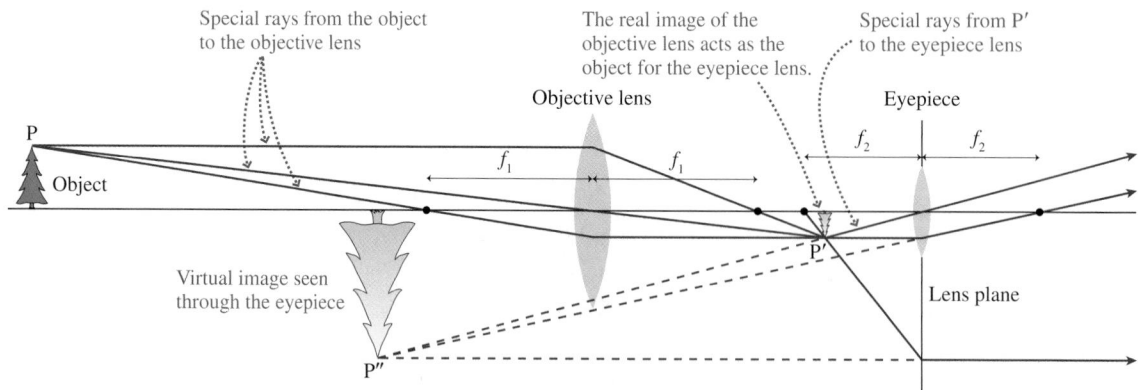

FIGURE 23.44 Ray-tracing diagram of a Galilean telescope.

The rays passing through the objective converge to a real image at P′, but they don't stop there. Instead, light rays *diverge* from P′ as they approach the second lens. As far as the eyepiece is concerned, the rays are coming from P′, and thus P′ is the object for the second lens. In other words, **the image of the first lens in a lens combination is the object for the second lens.**

The three special rays for the objective lens locate the image P', but only one of these (parallel to the axis) is a special ray for the eyepiece. However, these aren't the only rays. Other rays will leave P' at the correct angles to be the special rays for the eyepiece. That is, a new set of special rays is drawn from P' to the second lens and used to find the final image point P''.

NOTE ▶ One ray seems to "miss" the eyepiece lens, but this isn't really a problem. First, we don't know the actual diameter of the lens. The lens diameter in the figure was an arbitrary choice, and the actual lens might be larger or smaller than shown. Second, the purpose of the special rays is to locate the point where *all* rays converge (or from which they diverge). Whether each special ray actually makes it through the lens isn't relevant. We can let the special rays refract as they cross the *lens plane,* regardless of whether the lens itself extends that far. ◀

The eyepiece acts as a magnifier because its object, point P', is inside the focal point. Consequently, P'' is an enlarged, inverted, virtual image that is seen by looking through the eyepiece. The fact that a Galilean telescope produces an inverted image is not a problem in astronomy, but Galilean telescopes are not suitable for bird watching. Other telescope designs produce an upright image.

23.7 Thin Lenses: Refraction Theory

Ray tracing is a powerful visual approach for understanding image formation, but it doesn't provide precise information about the image location or image properties. We need to develop a quantitative relationship between the object distance s and the image distance s'. We will first analyze the refraction at a single spherical surface, using a method similar to our analysis of image formation by a plane refracting surface, in Section 23.4. Then we'll put two spherical surfaces together to form a lens.

To begin, Figure 23.45 shows a *spherical* boundary between two transparent media with indices of refraction n_1 and n_2. The sphere has radius of curvature R and is centered at point C. Consider a ray that leaves object point P at angle α and later, after refracting, reaches point P'. Figure 23.45 has exaggerated the angles to make the picture clear, but we will restrict our analysis to *paraxial rays* traveling nearly parallel to the axis. For paraxial rays, all the angles are small and we can use the small-angle approximation.

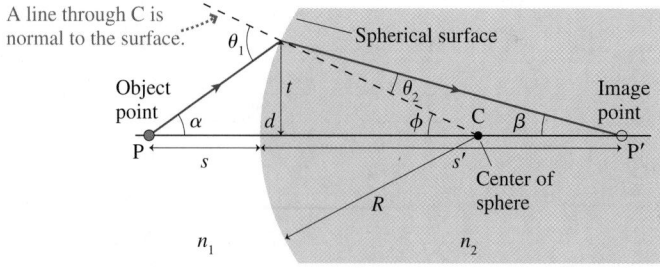

FIGURE 23.45 Image formation due to refraction at a spherical surface. The angles are exaggerated.

The ray from P is incident on the boundary at angle θ_1 and refracts into medium n_2 at angle θ_2, both measured from the normal to the surface at the point of incidence. Snell's law for the refraction is $n_1\sin\theta_1 = n_2\sin\theta_2$, which in the small-angle approximation is

$$n_1\theta_1 = n_2\theta_2 \qquad (23.15)$$

You can see from the geometry of Figure 23.45 that angles α, β, and ϕ are related by

$$\theta_1 = \alpha + \phi$$
$$\theta_2 = \phi - \beta \tag{23.16}$$

Using these expressions in Equation 23.15, we can write Snell's law as

$$n_1(\alpha + \phi) = n_2(\phi - \beta) \tag{23.17}$$

This is one important relationship between the angles.

The line of height t, from the axis to the point of incidence, is the vertical leg of three different right triangles having vertices at points P, C, and P′. Consequently,

$$\tan\alpha \approx \alpha = \frac{t}{s+d} \qquad \tan\beta \approx \beta = \frac{t}{s'-d} \qquad \tan\phi \approx \phi = \frac{t}{R-d} \tag{23.18}$$

But $d \to 0$ for paraxial rays, thus

$$\alpha = \frac{t}{s} \qquad \beta = \frac{t}{s'} \qquad \phi = \frac{t}{R} \tag{23.19}$$

This is the second important relationship that comes from the geometry of Figure 23.45.

If we use the angles of Equation 23.19 in Equation 23.17, we find

$$n_1\left(\frac{t}{s} + \frac{t}{R}\right) = n_2\left(\frac{t}{R} - \frac{t}{s'}\right) \tag{23.20}$$

The t cancels, and we can rearrange Equation 23.20 to read

$$\frac{n_1}{s} + \frac{n_2}{s'} = \frac{n_2 - n_1}{R} \tag{23.21}$$

Equation 23.21 is independent of angle α. Consequently, **all paraxial rays that leave point P later converge at point P′.** If an object is located at distance s from a spherical refracting surface, an image will be formed at distance s' given by Equation 23.21.

Equation 23.21 was derived for a surface that is convex toward the object point, and the image is real. However, the result is also valid for virtual images or for surfaces that are concave toward the object point as long as we adopt the *sign convention* shown in Table 23.3.

TABLE 23.3 Sign convention for refracting surfaces

	Positive	Negative
R	Convex toward the object	Concave toward the object
s'	Real image, opposite side from object	Virtual image, same side as object

Section 23.4 considered image formation due to refraction by a plane surface. There we found (in Equation 23.13) an image distance $s' = (n_2/n_1)s$. A plane can be thought of as a sphere in the limit $R \to \infty$, so we should be able to reach the same conclusion from Equation 23.21. As $R \to \infty$, the term $(n_2 - n_1)/R \to 0$ and Equation 23.21 becomes $s' = -(n_2/n_1)s$. This seems to differ from Equation 23.13, but it doesn't really. Equation 23.13 gives the actual distance to the image. Equation 23.21 is based on a sign convention in which virtual images have negative image distances, hence the minus sign.

EXAMPLE 23.12 Image formation inside a glass rod

One end of a 4.0-cm-diameter glass rod is shaped as a hemisphere. A small light bulb is 6.0 cm from the end of the rod. Where is the bulb's image located?

MODEL Model the light bulb as a point source of light and consider the paraxial rays that refract into the glass rod.

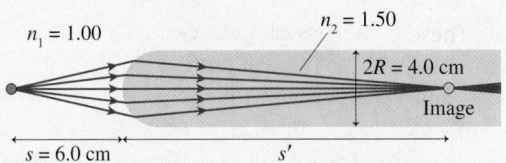

FIGURE 23.46 The curved surface refracts the light to form a real image.

VISUALIZE Figure 23.46 shows the situation. $n_1 = 1.00$ for air and $n_2 = 1.50$ for glass.

SOLVE The radius of the surface is half the rod diameter, so $R = 2.0$ cm. Equation 23.21 is

$$\frac{1.00}{6.0 \text{ cm}} + \frac{1.50}{s'} = \frac{1.50 - 1.00}{2.0 \text{ cm}} = \frac{0.50}{2.0 \text{ cm}}$$

Solving for the image distance s' gives

$$\frac{1.50}{s'} = \frac{0.50}{2.0 \text{ cm}} - \frac{1.00}{6.0 \text{ cm}} = 0.0833 \text{ cm}^{-1}$$

$$s' = \frac{1.50}{0.0833} = 18 \text{ cm}$$

ASSESS This is a real image located 18 cm inside the glass rod.

EXAMPLE 23.13 A goldfish in a bowl

A goldfish lives in a spherical fish bowl 50 cm in diameter. If the fish is 10 cm from the near edge of the bowl, where does the fish appear when viewed from the outside?

MODEL Model the fish as a point source and consider the paraxial rays that refract from the water into the air. The thin glass wall has little effect and will be ignored.

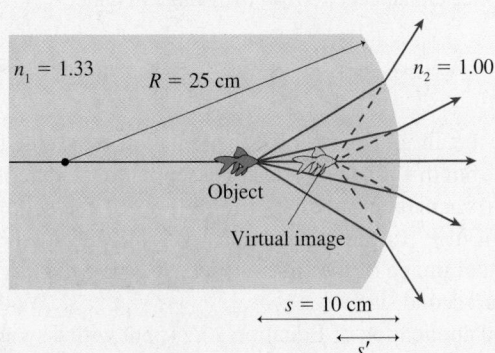

FIGURE 23.47 The curved surface of a fish bowl produces a virtual image of the fish.

VISUALIZE Figure 23.47 shows the rays refracting *away* from the normal as they move from the water into the air. We expect to find a virtual image at a distance less than 10 cm.

SOLVE The object is in the water, so $n_1 = 1.33$ and $n_2 = 1.00$. The inner surface is concave (you can remember "concave" because it's like looking into a cave), so $R = -25$ cm. The object distance is $s = 10$ cm. Thus Equation 23.21 is

$$\frac{1.33}{10 \text{ cm}} + \frac{1.00}{s'} = \frac{1.00 - 1.33}{-25 \text{ cm}} = \frac{0.33}{25 \text{ cm}}$$

Solving for the image distance s' gives

$$\frac{1.00}{s'} = \frac{0.33}{25} - \frac{1.33}{10} = -0.119 \text{ cm}^{-1}$$

$$s' = \frac{1.00}{-0.119} = -8.35 \text{ cm}$$

ASSESS The image is virtual, located to the left of the boundary. A person looking into the bowl will see a fish that appears to be 8.35 cm from the edge of the bowl.

STOP TO THINK 23.5 Which of these actions will move the image point P′ farther from the boundary? More than one may work.

a. Increase the radius of curvature R.
b. Increase the index of refraction n.
c. Increase the object distance s.
d. Decrease the radius of curvature R.
e. Decrease the index of refraction n.
f. Decrease the object distance s.

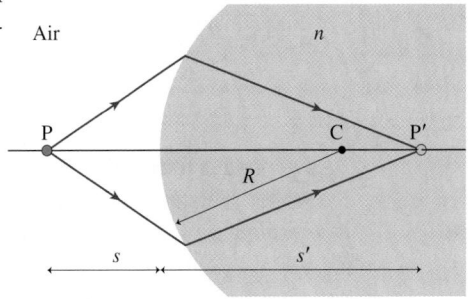

Lenses

15.10–15.12 Activ Physics ONLINE

A lens consists of *two* spherical surfaces having radii of curvature R_1 and R_2 and thickness t. The lens material has index of refraction n, and for simplicity we'll assume that the lens is surrounded by air. We'll analyze the converging lens shown in Figure 23.48, but our results will apply to any lens if we use the sign convention given above in Table 23.3.

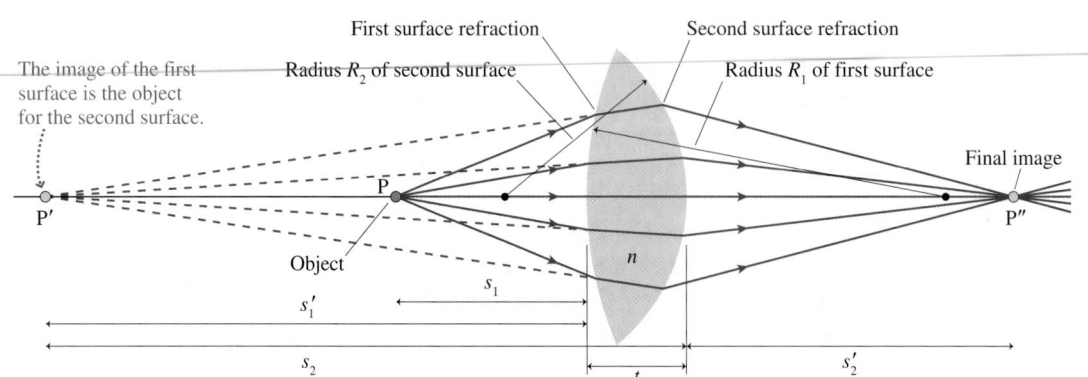

FIGURE 23.48 Image formation by a lens.

The object at point P is distance s_1 to the left of the lens. The first surface of the lens, of radius R_1, refracts the rays from P to create an image at point P′. We can use Equation 23.21 for a spherical surface to find the image distance s_1':

$$\frac{1}{s_1} + \frac{n}{s_1'} = \frac{n-1}{R_1} \tag{23.22}$$

where we used $n_1 = 1$ for the air and $n_2 = n$ for the lens. We'll assume that the image P′ is a virtual image, but this assumption isn't essential to the outcome.

The image P′ of the first surface becomes the object for the second surface. Object distance s_2 from P′ to the second surface looks like it should be $s_2 = s_1' + t$, but P′ is a virtual image of the first surface, so s_1' is a *negative* number. Thus the distance to the second surface is $s_2 = |s_1'| + t = t - s_1'$. We can find the image of P′ by a second application of Equation 23.21, but with a switch. The rays are incident on the surface from within the lens, so this time $n_1 = n$ and $n_2 = 1$. Consequently,

$$\frac{n}{t - s_1'} + \frac{1}{s_2'} = \frac{1-n}{R_2} \tag{23.23}$$

For a *thick lens*, where the thickness t is not negligible, we can solve Equations 23.22 and 23.23 in sequence to find the position of the image point P″. But our primary interest is the *thin lens*. In the limit $t \to 0$, Equation 23.23 becomes

$$-\frac{n}{s_1'} + \frac{1}{s_2'} = \frac{1-n}{R_2} = -\frac{n-1}{R_2} \tag{23.24}$$

Our goal is to find the distance s_2' to point P″, the image produced by the lens as a whole. This goal is easily reached if we simply add Equations 23.22 and 23.24, eliminating s_1' and giving

$$\frac{1}{s_1} + \frac{1}{s_2'} = \frac{n-1}{R_1} - \frac{n-1}{R_2} = (n-1)\left(\frac{1}{R_1} - \frac{1}{R_2}\right) \tag{23.25}$$

The numerical subscripts on s_1 and s_2' no longer serve a purpose. If we replace s_1 by s, the object distance of the lens, and s_2' by s', the image distance, Equation 23.25 becomes the *thin-lens equation*

$$\frac{1}{s} + \frac{1}{s'} = \frac{1}{f} \quad \text{(thin-lens equation)} \tag{23.26}$$

where the *focal length* of the lens is given by

$$\frac{1}{f} = (n-1)\left(\frac{1}{R_1} - \frac{1}{R_2}\right) \quad \text{(lens maker's equation)} \tag{23.27}$$

Equation 23.27 is known as the *lens maker's equation*. It allows you to determine the focal length from the shape of a lens and the material used to make it.

We can verify that this expression for f really is the focal length of the lens by recalling that rays initially parallel to the optical axis pass through the focal point on the far side. In fact, this was our *definition* of the focal length of a lens. Parallel rays must come from an object extremely far away, with object distance $s \to \infty$. In that case, $1/s = 0$ and Equation 23.26 tells us that the parallel rays will converge at distance $s' = f$ on the far side of the lens, exactly as expected.

Similarly, Equation 23.26 gives $1/s' = 0$, or $s' \to \infty$, for a point source of light at object distance $s = f$. In other words, an object at the near focal point produces an image infinitely far away, so the rays leave the lens traveling parallel to the axis. This is what we saw in Figure 23.35b. Thus the quantity f in Equation 23.27 really does represent the focal length of the lens.

We derived the thin-lens equation and the lens maker's equation from the specific lens geometry shown in Figure 23.48, but the results are valid for any lens as long as all quantities are given appropriate signs. The sign convention used with Equations 23.26 and 23.27 is given in Table 23.4.

TABLE 23.4 Sign convention for thin lenses

	Positive	Negative
R_1, R_2	Convex toward the object	Concave toward the object
f	Converging lens, thicker in center	Diverging lens, thinner in center
s'	Real image, opposite side from object	Virtual image, same side as object

EXAMPLE 23.14 Focal length of a meniscus lens
What is the focal length of the glass *meniscus lens* shown in Figure 23.49? Is this a converging or diverging lens?

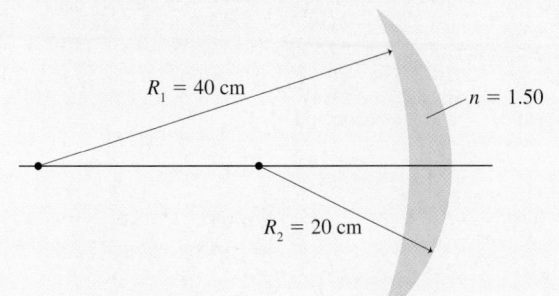

FIGURE 23.49 A meniscus lens.

SOLVE If the object is on the left, then the first surface has $R_1 = -40$ cm (concave toward the object) and the second surface has $R_2 = -20$ cm (also concave toward the object). The index of refraction of glass is $n = 1.50$, so the lens maker's equation is

$$\frac{1}{f} = (n-1)\left(\frac{1}{R_1} - \frac{1}{R_2}\right) = (1.50 - 1)\left(\frac{1}{-40 \text{ cm}} - \frac{1}{-20 \text{ cm}}\right)$$
$$= 0.0125 \text{ cm}^{-1}$$

Inverting this expression gives $f = 80$ cm. This is a converging lens, as seen both from the positive value of f and from the fact that the lens is thicker in the center.

EXAMPLE 23.15 Designing an eyeglass lens

A plastic eyeglass lens to correct for near-sightedness is a diverging lens with a planoconcave design. The patient needs a focal length of -1.5 m. What is the radius of the inner surface of the lens?

VISUALIZE Figure 23.50 clarifies the shape of the lens and defines radius R_2. The index of refraction for plastic was taken from Table 23.1.

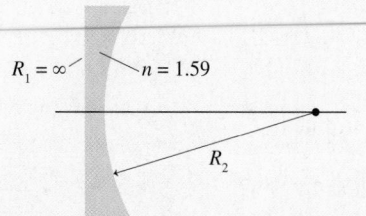

$R_1 = \infty$ $n = 1.59$

R_2

FIGURE 23.50 A planoconcave lens.

SOLVE The front surface of the lens is planar, which we can consider a portion of a sphere with $R_1 \to \infty$. R_2 is positive because the second surface is convex toward the object (assumed to be on the left). The lens maker's equation can be solved for R_2, giving

$$\frac{1}{R_2} = \frac{1}{R_1} - \frac{1}{(n-1)f} = 0 - \frac{1}{(1.59-1)(-1.5 \text{ m})}$$

$$= 1.13 \text{ m}^{-1}$$

$$R_2 = 0.885 \text{ m}$$

ASSESS Optometrists measure lenses not by their focal length but by their *power*, which is the inverse of the focal length: $P = 1/f$. The unit of power is the *diopter*, defined as 1 m^{-1}. A patient needing a focal length of -1.5 m would be given a prescription for a -0.67 diopter lens.

Thin-Lens Image Formation

In practice, the focal length of a lens is specified by the manufacturer. We can use the thin-lens equation to calculate exactly where an image of the lens is located. Alternatively, we can use the thin-lens equation to determine the focal length of a lens needed to create an image at a desired location. Although the thin-lens equation allows precise calculations, the lessons of ray tracing should not be forgotten. The most powerful tool of optical analysis is a combination of ray tracing, to gain an intuitive understanding of the ray trajectories, and the thin-lens equation.

EXAMPLE 23.16 A magnifying lens

A stamp collector uses a magnifying lens that sits 2.0 cm above the stamp. The magnification is 4. What is the focal length of the lens?

MODEL A magnifying lens is a converging lens with the object distance less than the focal length ($s < f$). Assume it is a thin lens.

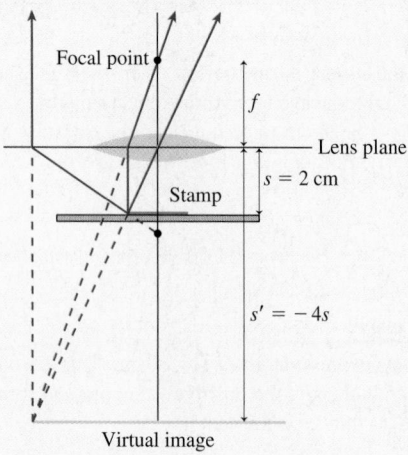

Focal point

f

Lens plane

$s = 2$ cm

Stamp

$s' = -4s$

Virtual image

FIGURE 23.51 Pictorial representation of a magnifying lens.

VISUALIZE Figure 23.51 shows the lens and a ray-tracing diagram. We don't know, nor do we need to know, the actual shape of the lens. Any combination of materials and surfaces that produces the desired focal length will function as a magnifying lens. The figure shows a generic converging lens.

SOLVE A virtual image is upright, so $M = +4$. The magnification is $M = -s'/s$, thus

$$s' = -4s = -4(2.0 \text{ cm}) = -8.0 \text{ cm}$$

We can use s and s' in the thin-lens equation to find the focal length:

$$\frac{1}{f} = \frac{1}{s} + \frac{1}{s'} = \frac{1}{2.0 \text{ cm}} + \frac{1}{-8.0 \text{ cm}} = 0.375 \text{ cm}^{-1}$$

$$f = 2.67 \text{ cm}$$

ASSESS $f > 2$ cm, as expected.

EXAMPLE 23.17 A combination camera lens

A camera "lens" is usually a combination of two or more single lenses. Consider a camera lens consisting of a diverging lens, with $f_1 = -120$ mm, and a converging lens, with $f_2 = 42$ mm, spaced 60 mm apart. A 10-cm-tall object is 500 mm from the first lens.

a. What are the location, size, and orientation of the image?
b. What is the focal length of a *single* lens that could produce an image in the same location if placed at the midpoint of the lens combination? This focal length is a reasonable definition of the *equivalent focal length* of the lens combination.

MODEL Each lens is a thin lens. The image of the first lens is the object for the second lens.

VISUALIZE Figure 23.52 shows the two lenses and a ray-tracing diagram. The ray tracing shows that the lens combination will produce a real, inverted image ≈ 55 mm behind the second lens. We can use the thin-lens equation to make this estimate more precise.

SOLVE

a. $s_1 = 500$ mm is the object distance of the first lens. Its image, which is a virtual image, is found from the thin-lens equation:

$$\frac{1}{s_1'} = \frac{1}{f_1} - \frac{1}{s_1} = \frac{1}{-120 \text{ mm}} - \frac{1}{500 \text{ mm}} = -0.0103 \text{ mm}^{-1}$$

$$s_1' = -97 \text{ mm}$$

This image distance is consistent with the ray-tracing diagram. The image of the first lens is now the object for the second lens. The object distance is $s_2 = 97$ mm + 60 mm = 157 mm. A second application of the thin-lens equation yields

$$\frac{1}{s_2'} = \frac{1}{f_2} - \frac{1}{s_2} = \frac{1}{42 \text{ mm}} - \frac{1}{157 \text{ mm}} = 0.0174 \text{ mm}^{-1}$$

$$s_2' = 57 \text{ mm}$$

The image of the lens combination is 57 mm behind the second lens. The magnifications of the two lenses are

$$M_1 = -\frac{s_1'}{s_1} = -\frac{-97 \text{ cm}}{500 \text{ cm}} = +0.194$$

$$M_2 = -\frac{s_2'}{s_2} = -\frac{57 \text{ cm}}{157 \text{ cm}} = -0.363$$

The second lens magnifies the image of the first lens, which magnifies the object, so the total magnification is

$$M = M_1 M_2 = -0.070$$

Thus the image is 57 mm behind the second lens, inverted (M is negative), and 0.70 cm tall.

b. If a single lens midway between these two lenses produced an image in the same plane, its object and image distances would be $s = 500$ mm + 30 mm = 530 mm and $s' = 57$ mm + 30 mm = 87 mm. A final application of the thin-lens equation gives the focal length:

$$\frac{1}{f_{eq}} = \frac{1}{s} + \frac{1}{s'} = \frac{1}{530 \text{ mm}} + \frac{1}{87 \text{ mm}} = 0.0134 \text{ mm}^{-1}$$

$$f_{eq} = 75 \text{ mm}$$

This is the equivalent focal length of the camera lens.

ASSESS These are reasonably accurate numbers for a simple two-lens system on a 35 mm camera. It would be sold as a "75 mm lens."

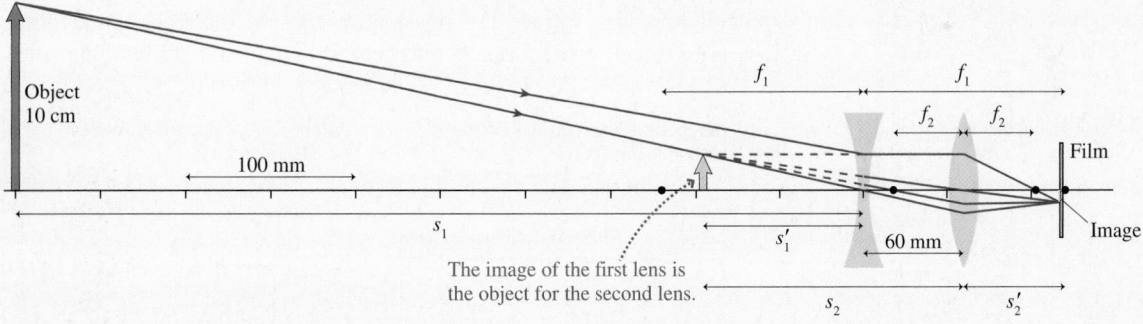

FIGURE 23.52 Pictorial representation of a combination camera lens.

STOP TO THINK 23.6 The image of a slide on a screen is blurry because the screen is in front of the image plane. To focus the image, should you move the lens toward the slide or away from the slide?

Lens Aberrations

The imaging in the last example could have been accomplished by a single lens. Why would anyone ever use a lens combination in place of a single lens? There are two primary reasons.

(a) Chromatic aberration

Different wavelengths focus at different points.

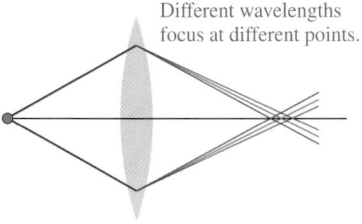

(b) Spherical aberration

Rays at different angles focus at different points.

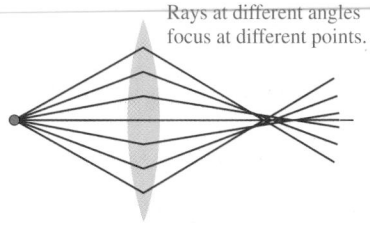

FIGURE 23.53 Chromatic aberration and spherical aberration prevent simple lenses from forming perfect images.

First, any lens has dispersion. That is, its index of refraction varies slightly with wavelength. Because the index of refraction for violet light is larger than for red light, a lens's focal length is shorter for violet light than for red light. Consequently, different colors of light come to a focus at slightly different distances from the lens. If red light is sharply focused on a viewing screen, then blue and violet wavelengths are not well focused. This imaging error, illustrated in Figure 23.53a, is called **chromatic aberration.**

Second, our analysis of image formation was based on paraxial rays traveling nearly parallel to the optical axis. This assumption allowed us to use the small-angle approximation. A more exact analysis, taking all the rays into account, finds that rays incident on the outer edges of a spherical surface are not focused at exactly the same point as rays incident near the center. This imaging error, shown in Figure 23.53b, is called **spherical aberration.** Spherical aberration, which causes the image to be slightly blurred, gets worse as the lens diameter increases.

Fortunately, the aberrations of a converging lens and a diverging lens are in opposite directions. When a converging lens and a diverging lens are used in combination, their aberrations tend to cancel. Thus a combination lens, such as the one in Example 23.17, can produce a much sharper focus than can a single lens with the equivalent focal length. Consequently, cameras, microscopes, and other optical equipment use combination lenses rather than single lenses.

23.8 The Resolution of Optical Instruments

16.8

According to the ray model of light, a perfect lens (one with no aberrations) should be able to form a perfect image. But the ray model of light, while a very good model for lenses, is not an absolutely correct description of light. If we look closely, the wave aspects of light haven't entirely disappeared. In fact, the performance of optical equipment is limited by the waviness of light.

Figure 23.54a shows a plane wave being focused by a lens of diameter D. Only those waves passing *through* the lens can be focused, so the lens acts like a circular aperture in an opaque barrier. In other words, the lens both *focuses and diffracts* light waves. Figure 23.54b separates these two effects by modeling a real lens as an "ideal" diffractionless lens behind a circular aperture of diameter D.

(a) A lens acts as a circular aperture.

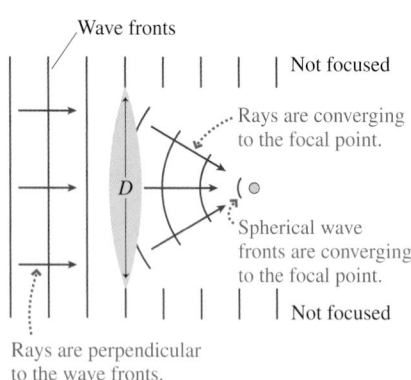

(b) The aperture and focusing effects can be separated.

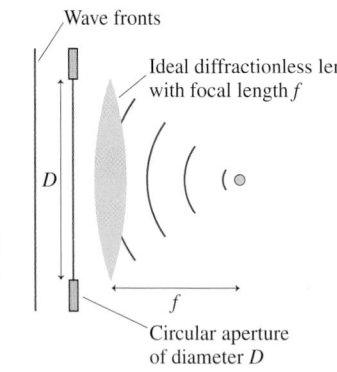

(c) The lens produces a diffraction pattern in the focal plane.

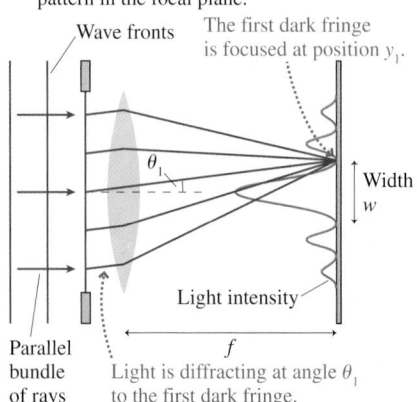

FIGURE 23.54 A lens both focuses and diffracts the light passing through.

You learned in Chapter 22 that a circular aperture produces a diffraction pattern with a bright central maximum surrounded by dimmer circular fringes. A converging lens brings parallel light rays to a focus at distance. Consequently, as

Figure 23.54c shows, a lens behind a circular aperture collects all the light rays diffracting at angle θ and brings these rays together in the focal plane of the lens. The net result is that the image of a parallel bundle of rays is not a perfect point but, instead, a circular diffraction pattern.

The angle to the first minimum of a circular diffraction pattern is $\theta_1 = 1.22\lambda/D$. The ray that passes through the center of a lens is not bent, so Figure 23.54c uses this ray to show that the position of the dark fringe is $y_1 = f\tan\theta_1 \approx f\theta_1$. Thus the width of the central maximum in the focal plane is

$$w = 2f\theta_1 = \frac{2.44\lambda f}{D} \qquad \text{(minimum spot size)} \qquad (23.28)$$

This is the **minimum spot size** to which a lens can focus light.

Lenses are often limited by aberrations, so not all lenses can focus light to a spot this small. A well-crafted lens, for which this is the minimum spot size, is called a *diffraction-limited lens.* No optical design can overcome the spreading of light due to diffraction, and it is because of this spreading that the image point has a minimum spot size.

For various reasons, it is difficult to produce a diffraction-limited lens having a focal length less than its diameter. That is, $f \geq D$ for any realistic lens. This implies that **the smallest diameter to which you can focus a spot of light, no matter how hard you try, is $w_{min} \approx 2.5\lambda$.** This is a fundamental limit on the performance of optical equipment. Diffraction has very real consequences!

One example of these consequences is found in the manufacturing of integrated circuits. Integrated circuits are made by creating a "mask" that shows all the components and their connections. A lens images this mask onto the surface of a semiconductor wafer that has been coated with a substance called *photoresist.* Bright areas in the mask expose the photoresist, and subsequent processing steps chemically etch away the exposed areas while leaving behind areas that had been in the shadows of the mask. This process is called photolithography.

The power of a microprocessor and the amount of memory in a memory chip depend on how small the circuit elements can be made. Diffraction dictates that a circuit element can be no smaller than the smallest spot to which light can be focused. That is, no feature on the chip can be smaller than roughly 2.5λ. If the mask is projected with ultraviolet light having $\lambda \approx 200$ nm $= 0.2$ μm, then the smallest elements on a chip are about 0.50 μm wide. This is, in fact, just about the current limit of technology.

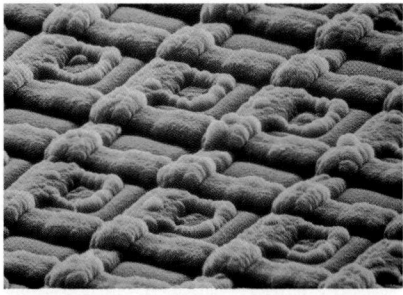

The size of the features in an integrated circuit is limited by the diffraction of light.

EXAMPLE 23.18 Looking at the stars
A 12-cm-diameter telescope lens has a focal length of 1.0 m. What is the diameter of the image of a star in the focal plane if the lens is diffraction limited *and* if the earth's atmosphere is not a limitation?

MODEL Stars are so far away that they appear as points in space. An ideal diffractionless lens would focus their light to an arbitrarily small point. Diffraction prevents this. Model the telescope lens as a 12-cm-diameter aperture in front of an ideal lens with a 1.0 m focal length.

SOLVE The minimum spot size in the focal plane of this lens is

$$w = \frac{2.44\lambda f}{D}$$

where D is the lens diameter. What is λ? Because stars emit white light, the *longest* wavelengths spread the most and determine the size of the image that is seen. If we use $\lambda = 700$ nm as the approximate upper limit of visible wavelengths, we find $w = 1.4 \times 10^{-5}$ m $= 14$ μm.

ASSESS This is certainly small, and it would appear as a point to your unaided eye. Nonetheless, the spot size would be easily noticed if it were recorded on film and enlarged. Turbulence and temperature effects in the atmosphere, the causes of the "twinkling" of stars, generally prevent ground-based telescopes from being this good, but space-based telescopes really are diffraction limited.

Resolution

Suppose you point a telescope at two nearby stars in a galaxy far, far away. If you use the best possible detector, will you be able to distinguish separate images for the two stars, or will they blur into a single blob of light? A similar question could be asked of a microscope. Can two microscopic objects, very close together, be distinguished if sufficient magnification is used? Or is there some size limit at which they will blur together and never be separated? These are important questions about the **resolution** of optical instruments.

Because of diffraction, the image of a distant star is not a point but a circular diffraction pattern. Our question, then, really amounts to asking how close together two diffraction patterns can be before you can no longer distinguish them. One of the major scientists of the 19th century, Lord Rayleigh, studied this problem and suggested a reasonable rule that today is called **Rayleigh's criterion.**

Figure 23.55 shows two objects being imaged by a lens of diameter D. The angular separation between the objects, as seen from the lens, is α. Rayleigh's criterion states that

- The two objects are resolvable if $\alpha > \theta_1$, where $\theta_1 = 1.22\lambda/D$ is the angle of the first dark fringe in the circular diffraction pattern.
- The two objects are not resolvable if $\alpha < \theta_1$ because their diffraction patterns are too overlapped.
- The two objects are marginally resolvable if $\alpha = \theta_1$. The central maximum of one image falls exactly on top of the first dark fringe of the other image. This is the situation shown in the figure

Figure 23.56 shows enlarged photographs of the images of two point sources. The images are circular diffraction patterns, not points. The two images are close but distinct where the objects are separated by $\alpha > \theta_1$. Two objects really were recorded in the photo on the right, but their separation is $\alpha < \theta_1$ and their images have blended together. In the middle photo, with $\alpha = \theta_1$, you can see that the two images are just barely resolved.

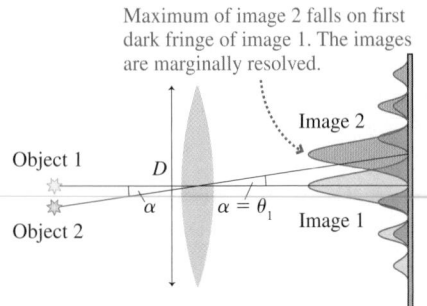

FIGURE 23.55 Two images that are marginally resolved.

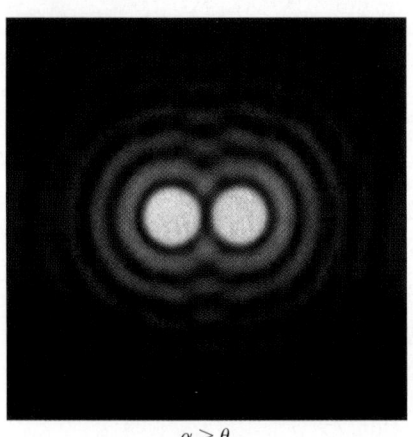

$\alpha > \theta_1$
Resolved

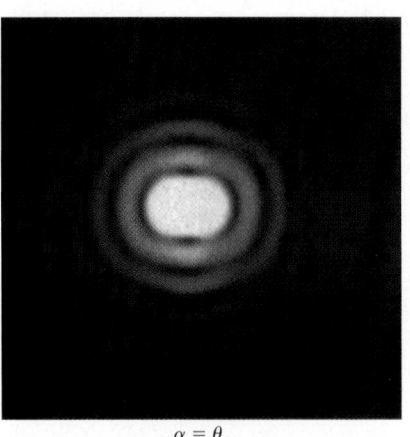

$\alpha = \theta_1$
Marginally resolved

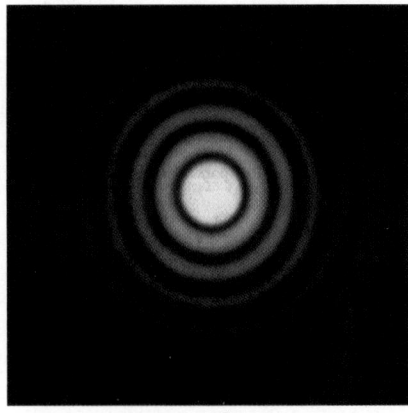

$\alpha < \theta_1$
Not resolved

FIGURE 23.56 Enlarged photographs of the images of two closely spaced objects.

For telescopes, the angle $\theta_1 = 1.22\lambda/D$ is called the *angular resolution* of the telescope. The angular resolution is a function of the lens diameter and the wavelength; the magnification is not a factor. Two overlapped, unresolved images will remain overlapped and unresolved no matter what the magnification. For visible light, where λ is pretty much fixed, the only parameter over which the astronomer has any control is the diameter of the lens or mirror of the telescope. The urge to build ever larger telescopes is motivated, in part, by a desire to improve the angular resolution. (Another motivation is to increase the light-gathering power so as to see objects farther away.)

A microscope is rather like a telescope in reverse. The object is located at distance $s \approx f$ in front of the lens and the image is formed much farther away behind the lens. (Think how close the objective of a microscope is placed to the object and how far away, by comparison, the eyepiece is.) A microscope with an objective lens of diameter D can marginally resolve two objects separated by angle $\alpha = \theta_1 = 1.22\lambda/D$. At distance f, the physical separation d between the two objects is $s = f\tan\alpha \approx f\alpha$, where we've used the small-angle approximation. Thus the smallest separation of two objects that can be resolved by a microscope is

$$d_{\min} = f\theta_1 = \frac{1.22\lambda f}{D} \approx \lambda \tag{23.29}$$

In the last step we have used the fact, previously noted, that for any real lens the minimum value of f/D is ≈ 1.

The ultimate performance of a microscope is limited by the diffraction of light through the objective lens. Objects smaller than about one wavelength of light, roughly 1 μm, *cannot* be resolved by an optical microscope. This is a conclusion of fundamental importance. Because atoms are approximately 0.1 nm in diameter, vastly smaller than the wavelength of visible or even ultraviolet light, there is no hope of ever seeing atoms with an optical microscope. This limitation is not simply a matter of needing a better design or more precise components. It is a fundamental limit set by the wave nature of the light with which we see.

STOP TO THINK 23.7   Four diffraction-limited lenses focus plane waves of light with the same wavelength λ. Rank in order, from largest to smallest, the spot sizes w_1 to w_4.

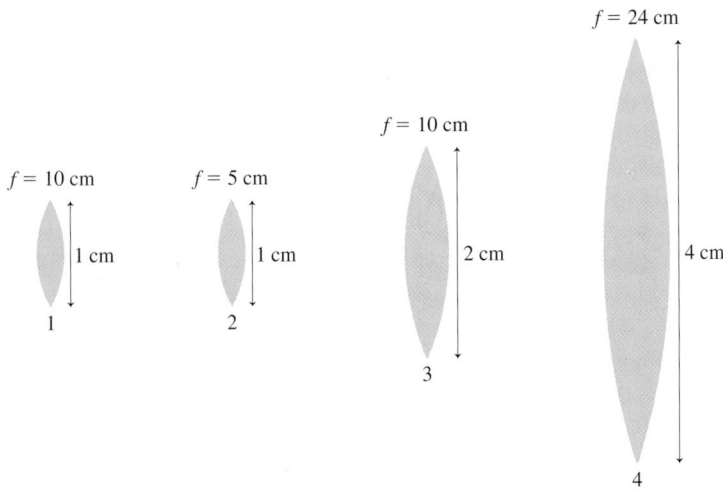

SUMMARY

The goal of Chapter 23 has been to understand and apply the ray model of light.

GENERAL PRINCIPLES

Reflection

Law of reflection: $\theta_r = \theta_i$

Reflection can be **specular** (mirror-like) or **diffuse** (from rough surfaces).

Plane mirrors: A virtual image is formed at P' with $s' = s$.

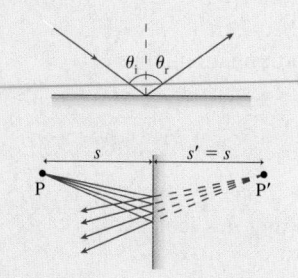

Refraction

Snell's law of refraction:
$$n_1 \sin \theta_1 = n_2 \sin \theta_2$$

Index of refraction is $n = c/v$. The ray is closer to the normal on the side with the larger index of refraction.

If $n_2 < n_1$, total internal reflection (TIR) occurs when the angle of incidence $\theta_1 > \theta_c = \sin^{-1}(n_2/n_1)$.

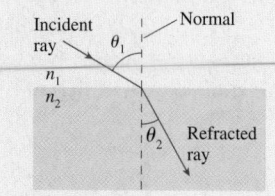

IMPORTANT CONCEPTS

The ray model of light

Light travels along straight lines, called **light rays**, at speed $v = c/n$.

A light ray continues forever unless an interaction with matter causes it to reflect, refract, scatter, or be absorbed.

Light rays come from **objects.** Each point on the object sends rays in all directions.

The eye sees an object (or an image) when diverging rays are collected by the pupil and focused on the retina.

▶ Ray optics is valid when lenses, mirrors, and apertures are larger than ≈ 1 mm in size.

Image formation

If rays diverge from P and interact with a lens or mirror so that the refracted/reflected rays *diverge* from P′ and appear to come from P′, then P′ is a virtual image of P.

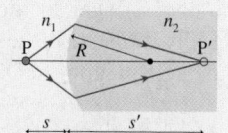

If rays diverge from P and interact with a lens so that the refracted rays *converge* at P′, then P′ is a real image of P.

Spherical surface: Object and image distances are related by

$$\frac{n_1}{s} + \frac{n_2}{s'} = \frac{n_2 - n_1}{R}$$

Plane surface: $R \to \infty$, so $|s'/s| = n_2/n_1$

APPLICATIONS

Ray tracing

3 special rays in 3 basic situations:

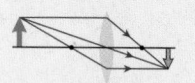  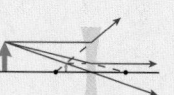

| Converging lens Real image | Converging lens Virtual image | Diverging lens Virtual image |

Magnification $M = -\dfrac{h'}{h} = -\dfrac{s'}{s}$

M is $+$ for an upright image, $-$ for inverted.

The height ratio is $h'/h = |M|$.

Thin lenses

The image and object distance are related by

$$\frac{1}{s} + \frac{1}{s'} = \frac{1}{f}$$

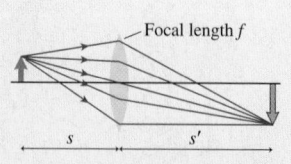

where the focal length is given by the lens maker's equation

$$\frac{1}{f} = (n - 1)\left(\frac{1}{R_1} - \frac{1}{R_2}\right)$$

R	$+$ for surface convex toward object	$-$ for concave
f	$+$ for a converging lens	$-$ for diverging
s'	$+$ for a real image	$-$ for virtual

Lens combinations: Image of first is object for second.

Resolution of optical instruments

Because of diffraction, the **minimum spot size** of a lens of diameter D is $w = 2.44 \lambda f/D$.

Rayleigh's criterion: Two objects separated by angle α are marginally resolvable if $\alpha = \theta_1 = 1.22 \lambda/D$.

TERMS AND NOTATION

light ray	law of reflection	optical axis	lens plane
object	diffuse reflection	paraxial rays	real image
point source	virtual image	dispersion	object plane
parallel bundle	refraction	Rayleigh scattering	image plane
ray diagram	angle of refraction	lens	inverted image
camera obscura	Snell's law	ray tracing	upright image
aperture	total internal reflection	converging lens	chromatic aberration
magnification, m or M	(TIR)	focal point	spherical aberration
specular reflection	critical angle, θ_c	focal length, f	minimum spot size
angle of incidence	object distance, s	diverging lens	resolution
angle of reflection	image distance, s'	thin lens	Rayleigh's criterion

EXERCISES AND PROBLEMS

Exercises

Section 23.1 The Ray Model of Light

1. a. How long (in ns) does it take light to travel 1.0 m in vacuum?
   b. What distance does light travel in water, glass, and zircon during the time that it travels 1.0 m in vacuum?

2. A 5.0-cm-thick layer of oil is sandwiched between a 1.0-cm-thick sheet of glass and a 2.0-cm-thick sheet of polystyrene plastic. How long (in ns) does it take light incident perpendicular to the glass to pass through this 8.0-cm-thick sandwich?

3. A point source of light illuminates an aperture 2.0 m away. A 12.0-cm-wide bright patch of light appears on a screen 1.0 m behind the aperture. How wide is the aperture?

4. Figure Ex23.4 is the top view of a room. Red and green light bulbs separated by 0.25 m shine through the door and illuminate the back wall. Over what range of x is the back wall illuminated by (a) the red and (b) the green light?

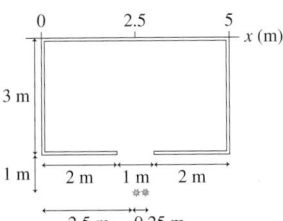

FIGURE EX23.4

5. A student has built a 20-cm-long pinhole camera for a science fair project. She wants to photograph the Washington Monument, which is 167 m (550 ft) tall, and to have the image on the film be 5.0 cm high. How far should she stand from the Washington Monument?

Section 23.2 Reflection

6. The mirror in Figure Ex23.6 deflects a horizontal laser beam by 60°. What is the angle ϕ?

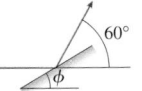

FIGURE EX23.6

7. A light ray leaves point A in Figure Ex23.7, reflects from the mirror, and reaches point B. How far below the top edge does the ray strike the mirror?

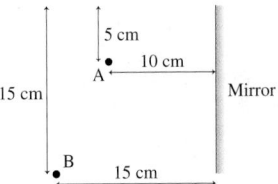

FIGURE EX23.7

8. You need to place a mirror on the left wall of Figure Ex23.8 so that the reflected light from the bulb exactly fills the right wall.
   a. What is the proper height of the mirror, and how far below the ceiling should the top edge be?
   b. A 10-cm-tall screen to the right of the bulb prevents the bulb from illuminating the right wall directly. What is the height of the screen's shadow on the right wall? Consider only the shadow due to light coming directly from the bulb, not light reflected by the mirror.

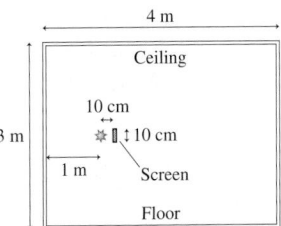

FIGURE EX23.8

9. At what angle ϕ should the laser beam in Figure Ex23.9 be aimed at the mirrored ceiling in order to hit the midpoint of the far wall?

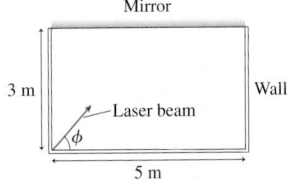

FIGURE EX23.9

10. Figure Ex23.10 is the top view of a room. As you walk along the wall opposite the mirror (i.e., along the *x*-axis), over what range of *x* can you see the entire green arrow in the mirror?

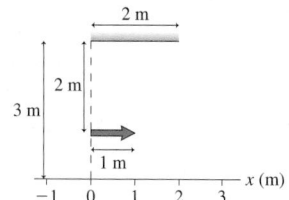

FIGURE EX23.10

11. It is 165 cm from your eyes to your toes. You're standing 200 cm in front of a tall mirror. How far is it from your eyes to the image of your toes?

Section 23.3 Refraction

12. A 1.0-cm-thick layer of water stands on a horizontal slab of glass. A light ray in the air is incident on the water 60° from the normal. What is the ray's direction of travel in the glass?

13. A diamond is underwater. A light ray enters one face of the diamond, then travels at an angle of 30° with respect to the normal. What was the ray's angle of incidence on the diamond?

14. An underwater diver sees the sun 50° above horizontal. How high is the sun above the horizon to a fisherman in a boat above the diver?

15. A laser beam in air is incident on a liquid at an angle of 37° with respect to the normal. The laser beam's angle in the liquid is 26°. What is the liquid's index of refraction?

16. The glass core of an optical fiber has an index of refraction 1.60. The index of refraction of the cladding is 1.48. What is the maximum angle a light ray can make with the wall of the core if it is to remain inside the fiber?

17. A thin glass rod is submerged in oil. What is the critical angle for light traveling inside the rod?

Section 23.4 Image Formation by Refraction

18. A fish in a flat-sided aquarium sees a can of fish food on the counter. To the fish's eye, the can looks to be 30 cm outside the aquarium. What is the actual distance between the can and the aquarium? (You can ignore the thin glass wall of the aquarium.)

19. A biologist keeps a specimen of his favorite beetle embedded in a cube of polystyrene plastic. The hapless bug appears to be 2.0 cm within the plastic. What is the beetle's actual distance beneath the surface?

20. A 150-cm-tall diver is standing completely submerged on the bottom of a swimming pool full of water. You are sitting on the end of the diving board, almost directly over her. How tall does the diver appear to be?

21. To a fish in an aquarium, the 4.00-mm-thick walls appear to be only 3.50 mm thick. What is the index of refraction of the walls?

Section 23.5 Color and Dispersion

22. A sheet of glass has $n_{red} = 1.52$ and $n_{violet} = 1.55$. A narrow beam of white light is incident on the glass at 30°. What is the angular spread of the light inside the glass?

23. A hydrogen discharge lamp emits light with two prominent wavelengths: 656 nm (red) and 486 nm (blue). The light enters a flint-glass prism perpendicular to one face and then refracts through the hypotenuse back into the air. The angle between these two faces is 35°.
    a. Use Figure 23.29 to estimate to ±0.002 the index of refraction of flint glass at these two wavelengths.
    b. What is the angle (in degrees) between the red and blue light as it leaves the prism?

24. A narrow beam of white light is incident on a sheet of quartz. The beam disperses in the quartz, with red light ($\lambda \approx 700$ nm) traveling at an angle of 26.3° with respect to the normal and violet light ($\lambda \approx 400$ nm) traveling at 25.7°. The index of refraction of quartz for red light is 1.45. What is the index of refraction of quartz for violet light?

25. Infrared telescopes, which use special infrared detectors, are able to peer farther into star-forming regions of the galaxy because infrared light is not scattered as strongly as is visible light by the tenuous clouds of hydrogen gas from which new stars are created. For what wavelength of light is the scattering only 1% that of light with a visible wavelength of 500 nm?

Section 23.6 Thin Lenses: Ray Tracing

26. An object is 20 cm in front of a converging lens with a focal length of 10 cm. Use ray tracing to determine the location of the image. Is the image upright or inverted?

27. An object is 30 cm in front of a converging lens with a focal length of 10 cm. Use ray tracing to determine the location of the image. Is the image upright or inverted?

28. An object is 6 cm in front of a converging lens with a focal length of 10 cm. Use ray tracing to determine the location of the image. Is the image upright or inverted?

29. An object is 15 cm in front of a diverging lens with a focal length of −10 cm. Use ray tracing to determine the location of the image. Is the image upright or inverted?

Section 23.7 Thin Lenses: Refraction Theory

30. Find the focal length of the glass lens in Figure Ex23.30.

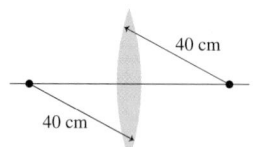

FIGURE EX23.30

31. Find the focal length of the planoconvex polystyrene plastic lens in Figure Ex23.31.

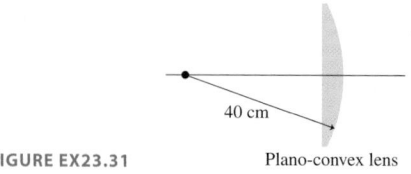

FIGURE EX23.31   Plano-convex lens

32. Find the focal length of the glass lens in Figure Ex23.32.

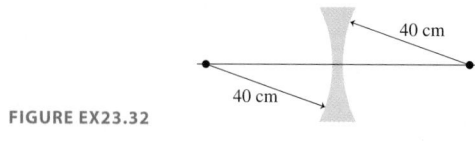

FIGURE EX23.32

33. Find the focal length of the meniscus polystyrene plastic lens in Figure Ex23.33.

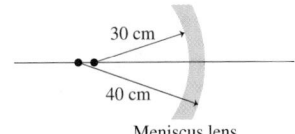

FIGURE EX23.33   Meniscus lens

34. A goldfish lives in a 50-cm-diameter spherical fish bowl. The fish sees a cat watching it. If the cat's face is 20 cm from the edge of the bowl, how far from the edge does the fish see it as being? (You can ignore the thin glass wall of the bowl.)

35. An air bubble inside an 8.0-cm-diameter plastic ball is 2.0 cm from the surface. As you look at the ball with the bubble turned toward you, how far beneath the surface does the bubble appear to be?

Section 23.8 The Resolution of Optical Instruments

36. A scientist needs to focus a helium-neon laser beam (λ = 633 nm) to a 10-μm-diameter spot 8.0 cm behind the lens. What focal-length lens should she use? What minimum diameter must the lens have?

37. Two light bulbs are 1.0 m apart. From what distance can these light bulbs be marginally resolved by a small telescope with a 4.0-cm-diameter objective lens? Assume that the lens is diffraction limited and λ = 600 nm.

Problems

38. An advanced computer sends information to its various parts via infrared light pulses traveling through silicon fibers. To acquire data from memory, the central processing unit sends a light-pulse request to the memory unit. The memory unit processes the request, then sends a data pulse back to the central processing unit. The memory unit takes 0.5 ns to process a request. If the information has to be obtained from memory in 2.0 ns, what is the maximum distance the memory unit can be from the central processing unit?

39. A red ball is placed at point A in Figure P23.39.
    a. How many images are seen by an observer at point O?
    b. Where is each image located?
    c. Draw a ray diagram showing the formation of each image.

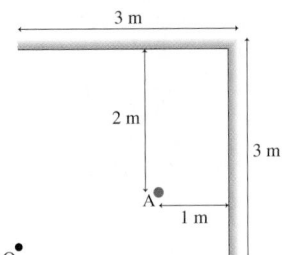

FIGURE P23.39

40. A laser beam is incident on the left mirror in Figure P23.40. Its initial direction is parallel to a line that bisects the mirrors. What is the angle ϕ of the reflected laser beam?

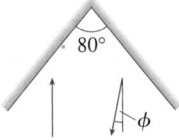

FIGURE P23.40

41. The place you get your hair cut has two nearly parallel mirrors 5.0 m apart. As you sit in the chair, your head is 2.0 m from the nearer mirror. Looking toward this mirror, you first see your face and then, farther away, the back of your head. (The mirrors need to be slightly nonparallel for you to be able to see the back of your head, but you can treat them as parallel in this problem.) How far away does the back of your head appear to be? Neglect the thickness of your head.

42. You're helping with an experiment in which a vertical cylinder will rotate about its axis by a very small angle. You need to devise a way to measure this angle. You decide to use what is called an *optical lever*. You begin by mounting a small mirror on top of the cylinder. A laser 5.0 m away shoots a laser beam at the mirror. Before the experiment starts, the mirror is adjusted to reflect the laser beam directly back to the laser. Later, you measure that the reflected laser beam, when it returns to the laser, has been deflected sideways by 2.0 mm. How many degrees has the cylinder rotated?

43. A 1.0-cm-thick layer of water stands on a horizontal slab of glass. Light from a source within the glass is incident on the glass-water boundary. What is the maximum angle of incidence for which the light ray can emerge into the air above the water?

44. A microscope is focused on a black dot. When a 1.00-cm-thick piece of plastic is placed over the dot, the microscope objective has to be raised 0.40 cm to bring the dot back into focus. What is the index of refraction of the plastic?

45. What is the angle of incidence in air of a light ray whose angle of refraction in glass is half the angle of incidence?

46. A meter stick lies on the bottom of a 100-cm-long tank with its zero mark against the left edge. You look into the tank at a 30° angle, with your line of sight just grazing the upper left edge of the tank. What mark do you see on the meter stick if the tank is (a) empty, (b) half full of water, and (c) completely full of water?

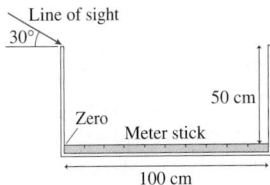

FIGURE P23.46

47. The 80-cm-tall, 65-cm-wide tank shown in Figure P23.47 is completely filled with water. The tank has marks every 10 cm along one wall, and the 0 cm mark is barely submerged. As you stand beside the opposite wall, your eye is level with the top of the water.
    a. Can you see the marks from the top of the tank (the 0 cm mark) going down, or from the bottom of the tank (the 80 cm mark) coming up? Explain.
    b. Which is the lowest or highest mark, depending on your answer to part a, that you can see?

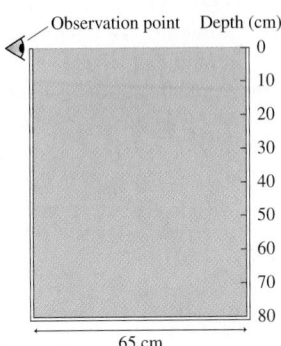

FIGURE P23.47

48. A 4.0-m-wide swimming pool is filled to the top. The bottom of the pool becomes completely shaded in the afternoon when the sun is 20° above the horizon. How deep is the pool?

49. It's nighttime, and you've dropped your goggles into a swimming pool that is 3.0 m deep. If you hold a laser pointer 1.0 m above the edge of the pool, you can illuminate the goggles if the laser beam enters the water 2.0 m from the edge. How far are the goggles from the edge of the pool?

50. Shown from above in Figure P23.50 is one corner of a rectangular box filled with water. A laser beam starts 10 cm from side A of the container and enters the water at position x. You can ignore the thin walls of the container.
    a. If x = 15 cm, does the laser beam refract back into the air through side B or reflect from side B back into the water? Determine the angle of refraction or reflection.
    b. Repeat part a for x = 25 cm.
    c. Find the minimum value of x for which the laser beam passes through side B and emerges into the air.

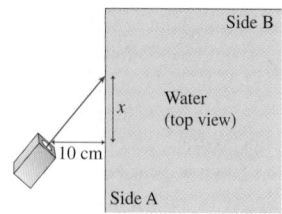

FIGURE P23.50

51. A fish is 20 m from the shore of a lake. A bonfire is burning on the edge of the lake nearest the fish.
    a. Does the fish need to be shallow (just below the surface) or very deep to see the light from the bonfire? Explain.
    b. What is the deepest or shallowest, depending on your answer to part a, that the fish can be and still see light from the fire?

52. One of the contests at the school carnival is to throw a spear at an underwater target lying flat on the bottom of a pool. The water is 1.0 m deep. You're standing on a small stool that places your eyes 3.0 m above the bottom of the pool. As you look at the target, your gaze is 30° below horizontal. At what angle below horizontal should you throw the spear in order to hit the target? Your raised arm brings the spear point to the level of your eyes as you throw it, and over this short distance you can assume that the spear travels in a straight line rather than a parabolic trajectory.

53. A narrow beam of white light is incident at 30° on a 10.0-cm-thick piece of glass. The rainbow of dispersed colors spans 1.00 mm on the bottom surface of the glass. The index of refraction for deep red light is 1.513. What is the index of refraction for deep violet light?

54. White light is incident onto a 30° prism at the 40° angle shown in Figure P23.54. Violet light emerges perpendicular to the rear face of the prism. The index of refraction of violet light in this glass is 2.0% larger than the index of refraction of red light. At what angle ϕ does red light emerge from the rear face?

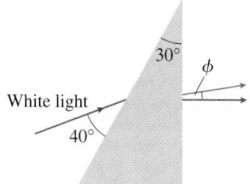

FIGURE P23.54

55. a. What is the smallest angle θ_1 for which a laser beam will undergo TIR on the hypotenuse of this glass prism?
    b. After reflecting from the hypotenuse at angle θ_c, the laser beam exits the prism through the bottom face. Does it exit to the right or to the left of the normal? At what angle?

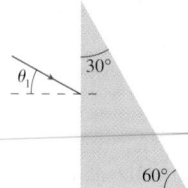

FIGURE P23.55

56. There's one angle of incidence β onto a prism for which the light inside an isosceles prism travels parallel to the base and emerges at angle β.
    a. Find an expression for β in terms of the prism's apex angle α and index of refraction n.
    b. A laboratory measurement finds that $\beta = 52.2°$ for a prism that is shaped as an equilateral triangle. What is the prism's index of refraction?

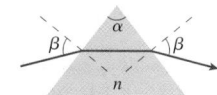

FIGURE 23.56

57. A 6.0-cm-diameter zircon sphere has an air bubble exactly in the center. As you look into the sphere, how far beneath the surface does the bubble appear to be?

58. Parallel light rays enter a transparent sphere along a line passing through the center of the sphere. The rays come to a focus on the far surface of the sphere. What is the sphere's index of refraction?

59. A 2.0-cm-tall object is 40 cm in front of a converging lens that has a 20 cm focal length.
    a. Use ray tracing to find the position and height of the image. To do this accurately use a ruler or paper with a grid. Determine the image distance and image height by making measurements on your diagram.
    b. Calculate the image position and height. Compare with your ray-tracing answers in part a.

60. A 1.0-cm-tall object is 10 cm in front of a converging lens that has a 30 cm focal length.
    a. Use ray tracing to find the position and height of the image. To do this accurately use a ruler or paper with a grid. Determine the image distance and image height by making measurements on your diagram.
    b. Calculate the image position and height. Compare with your ray-tracing answers in part a.

61. A 2.0-cm-tall object is 15 cm in front of a converging lens that has a 20 cm focal length.
    a. Use ray tracing to find the position and height of the image. To do this accurately use a ruler or paper with a grid. Determine the image distance and image height by making measurements on your diagram.
    b. Calculate the image position and height. Compare with your ray-tracing answers in part a.

62. A 1.0-cm-tall object is 75 cm in front of a converging lens that has a 30 cm focal length.
    a. Use ray tracing to find the position and height of the image. To do this accurately use a ruler or paper with a grid. Determine the image distance and image height by making measurements on your diagram.

b. Calculate the image position and height. Compare with your ray-tracing answers in part a.

63. A 2.0-cm-tall object is 15 cm in front of a diverging lens that has a −20 cm focal length.
    a. Use ray tracing to find the position and height of the image. To do this accurately use a ruler or paper with a grid. Determine the image distance and image height by making measurements on your diagram.
    b. Calculate the image position and height. Compare with your ray-tracing answers in part a.

64. A 1.0-cm-tall object is 60 cm in front of a diverging lens that has a −30 cm focal length.
    a. Use ray tracing to find the position and height of the image. To do this accurately use a ruler or paper with a grid. Determine the image distance and image height by making measurements on your diagram.
    b. Calculate the image position and height. Compare with your ray-tracing answers in part a.

65. A 2.0-cm-diameter spider is 2.0 m from a wall. Determine the focal length and position (measured from the wall) of a lens that will make a half-size image of the spider on the wall.

66. A 2.0-cm-tall candle flame is 2.0 m from a wall. You happen to have a lens with a focal length of 32 cm. How many places can you put the lens to form a well-focused image of the candle flame on the wall? For each location, what is the height and orientation of the image?

67. a. Estimate the diameter of your eyeball.
    b. Bring this page up to the closest distance at which the text is sharp—not the closest at which you can still read it, but the closest at which the letters remain sharp. If you wear glasses or contact lenses, leave them on. This distance is called the *near point* of your (possibly corrected) eye. Record it.
    c. Estimate the effective focal length of your eye. The effective focal length includes the focusing due to the lens, the curvature of the cornea, and any corrections you wear. Ignore the effects of the fluid in your eye.

68. A slide projector needs to create a 98-cm-high image of a 2.0-cm-tall slide. The screen is 300 cm from the slide.
    a. What focal length does the lens need? Assume that it is a thin lens.
    b. How far should you place the lens from the slide?

69. An object is 60 cm from a screen. What are the radii of a symmetric converging plastic lens (i.e., two equally curved surfaces) that will form an image on the screen twice the height of the object?

70. Two converging lenses with focal lengths of 40 cm and 20 cm are 10 cm apart. A 2.0-cm-tall object is 15 cm in front of the 40-cm-focal-length lens.
    a. Use ray tracing to find the position and height of the image. To do this accurately use a ruler or paper with a grid. Determine the image distance and image height by making measurements on your diagram.
    b. Calculate the image position and height. Compare with your ray-tracing answers in part a.

71. A converging lens with a focal length of 40 cm and a diverging lens with a focal length of −40 cm are 160 cm apart. A 2.0-cm-tall object is 60 cm in front of the converging lens.
    a. Use ray tracing to find the position and height of the image. To do this accurately use a ruler or paper with a grid. Determine the image distance and image height by making measurements on your diagram.

b. Calculate the image position and height. Compare with your ray-tracing answers in part a.

72. High-power lasers are used to cut and weld materials by focusing the laser beam to a very small spot. This is like using a magnifying lens to focus the sun's light to a small spot that can burn things. As an engineer, you have designed a laser cutting device in which the material to be cut is placed 5.0 cm behind the lens. You have selected a high-power laser with a wavelength of 1.06 μm. (This distance and wavelength are both very typical values.) Your calculations indicate that the laser must be focused to a 5.0-μm-diameter spot in order to have sufficient power to make the cut. What is the minimum diameter of the lens you must install?

73. Once dark adapted, the pupil of your eye is approximately 7 mm in diameter. The headlights of an oncoming car are 120 cm apart. If the lens of your eye is diffraction limited, at what distance are the two headlights marginally resolved? Assume a wavelength of 600 nm. (Your eye is not really good enough to resolve headlights at this distance, due both to aberrations in the lens and to the size of the receptors in your retina, but it comes reasonably close.)

74. Alpha Centauri, the nearest star to our solar system, is 4.3 light years away. Assume that Alpha Centauri has a planet with an advanced civilization. Professor Dhg, at the planet's Astronomical Institute, wants to build a telescope with which he can find out if there are planets orbiting the sun.
    a. What is the minimum diameter for an objective lens that will just barely resolve Jupiter and the sun? The radius of Jupiter's orbit is 780 million km. Assume $\lambda = 600$ nm.
    b. Building a telescope of the necessary size does not appear to be a major problem. What practical difficulties might prevent Professor Dhg's experiment from succeeding?

75. Optical disk storage uses a small infrared laser ($\lambda \approx 800$ nm) to read, via reflected light, "pits" that are burned into a plastic surface.
    a. What is the smallest spot size to which the laser beam can be focused?
    b. Assume the pits are located on a two-dimensional square grid with a spacing 25% larger than the laser spot size. (Spacing them any closer would risk reading errors.) Each pit records 1 bit of information, and it takes 8 bits to form 1 byte, the standard unit of data storage. An optical disk has a usable surface area with an inner diameter of 4 cm and an outer diameter of 11 cm. How many megabytes (MB) of data can be stored on an optical disk?

Challenge Problems

76. Figure CP23.76 on the next page shows a light ray that travels from point A to point B. The ray crosses the boundary at position x, making angles θ_1 and θ_2 in the two media. Suppose that you did *not* know Snell's law.
    a. Write an expression for the *time t* it takes the light ray to travel from A to B. Your expression should be in terms of the distances a, b, and w; the variable x; and the indices of refraction n_1 and n_2.
    b. The time depends on x. There's one value of x for which the light travels from A to B in the shortest possible time. We'll call it x_{min}. Write an expression (but don't try to solve it!) from which x_{min} could be found.

c. Now, by using the geometry of the figure, derive Snell's law from your answer to part b.

You've proven that Snell's law is equivalent to the statement that "light traveling between two points follows the path that requires the shortest time." This interesting way of thinking about refraction is called *Fermat's principle*.

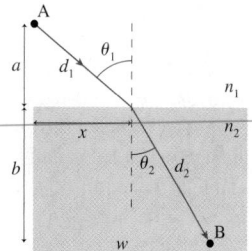

FIGURE CP23.76

77. A beam of white light enters a transparent material. Wavelengths for which the index of refraction is n are refracted at angle θ_2. Wavelengths for which the index of refraction is $n + \delta n$, where $\delta n \ll n$, are refracted at angle $\theta_2 + \delta\theta$.
   a. Show that the angular separation in radians is $\delta\theta = -\tan\theta_2(\delta n/n)$.
   b. A beam of white light is incident on a piece of glass at 30.0°. Deep violet light is refracted 0.28° more than deep red light. The index of refraction for deep red light is known to be 1.552. What is the index of refraction for deep violet light?

78. A fortune teller's "crystal ball" (actually just glass) is 10 cm in diameter. Her secret ring is placed 6 cm from the edge of the ball.
   a. An image of the ring appears on the opposite side of the crystal ball. How far is the image from the center of the ball?
   b. Draw a ray diagram showing the formation of the image.
   c. The crystal ball is removed and a thin lens is placed where the center of the ball had been. If the image is still in the same position, what is the focal length of the lens?

79. Consider a lens having index of refraction n_2 and surfaces with radii R_1 and R_2. The lens is immersed in a fluid that has index of refraction n_1.
   a. Derive a generalized lens maker's equation to replace Equation 23.27 when the lens is surrounded by a medium other than air. That is, when $n_1 \neq 1$.

   b. A symmetric converging glass lens (i.e., two equally curved surfaces) has two surfaces with radii of 40 cm. Find the focal length of this lens in air and the focal length of this lens in water.

80. The closest you can bring an object to your eye and still see it clearly is called the eye's *near point*. A near point of 25 cm is considered normal vision. If your near point is more than 25 cm, you are *far sighted* and may need to have your vision corrected. Call your actual near point d. This is the shortest distance at which your unaided eye can focus. The purpose of a corrective lens is to create a virtual image at distance d of an object held at 25 cm. If you hold an object 25 cm away that you would like to focus on but can't, the lens creates a virtual image that you can see. Suppose your actual near point is 100 cm.
   a. What is the focal length of an eyeglass lens that will restore normal vision?
   b. Draw a ray diagram of your eye viewing an object 25 cm away.

   You can neglect the small distance between your eye and the lens.

81. The farthest distance at which you can see an object clearly is called the eye's *far point*. A far point infinitely far away is considered normal. (Stars, which are essentially at infinity, should appear as small, well-focused points of light.) If your far point is less than infinity, you are *near sighted* and may need to have your vision corrected. Call your actual far point d. This is the farthest distance at which your unaided eye can focus. The purpose of a corrective lens is to create a virtual image at distance d of an object that is very far away. If you look toward a very distant object that you would like to focus on but can't, the lens creates a virtual image that you can see. Suppose your actual far point is 200 cm.
   a. What is the focal length of an eyeglass lens that will restore normal vision?
   b. Draw a ray diagram of your eye viewing an object very far away.

   You can neglect the small distance between your eye and the lens.

82. A 1.0-cm-tall object is 110 cm from a screen. A diverging lens with focal length −20 cm is 20 cm in front of the object. What are the focal length and location of a second lens that will produce a well-focused, 2.0-cm-tall image on the screen?

STOP TO THINK ANSWERS

Stop to Think 23.1: c. The light spreads vertically as it goes through the vertical aperture. The light spreads horizontally due to different points on the horizontal light bulb.

Stop to Think 23.2: c. There's one image behind the vertical mirror and a second behind the horizontal mirror. A third image in the corner arises from rays that reflect twice, once off each mirror.

Stop to Think 23.3: a. The ray travels closer to the normal in both media 1 and 3 than in medium 2, so n_1 and n_3 are both larger than n_2. The angle is smaller in medium 3 than in medium 1, so $n_3 > n_1$.

Stop to Think 23.4: e. The rays from the object are diverging. Without a lens, the rays cannot converge to form any kind of image on the screen.

Stop to Think 23.5: a, e, or f. Any of these will increase the angle of refraction θ_2.

Stop to Think 23.6: Away from. You need to decrease s' to bring the image plane on the screen. s' is decreased by increasing s.

Stop to Think 23.7: $w_1 > w_4 > w_2 = w_3$. The spot size is proportional to f/D.

24 Modern Optics and Matter Waves

This image from a scanning tunneling microscope shows individual silicon atoms at the surface of a silicon crystal.

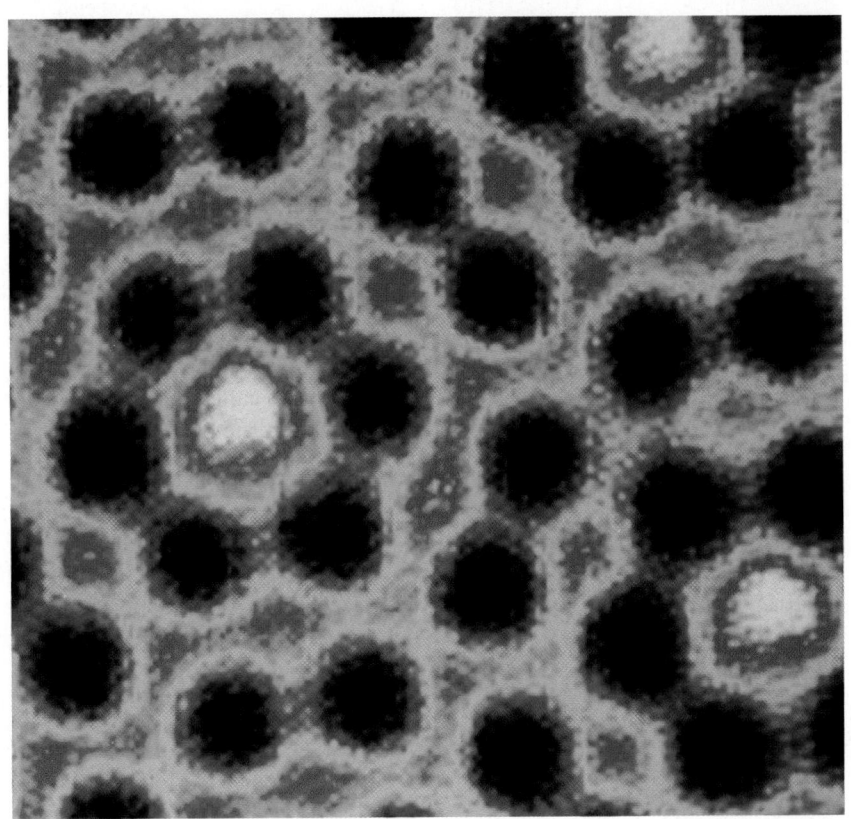

▶ Looking Ahead
The goal of Chapter 24 is to explore the limits of the wave and particle models. In this chapter you will learn to:

- Understand how light and x rays are used to study atoms and solids.
- Use the photon model of light.
- Recognize the experimental evidence for the wave nature of matter.
- Understand that energy quantization is a consequence of the wave-like properties of matter.

◀ Looking Back
The material in this chapter depends on the wave model of light. Please review:

- Section 22.1 Models of light.
- Sections 22.2–22.3 Double-slit interference and diffraction gratings.

The scanning tunneling microscope is one of the most important inventions of the last 30 years. For the first time, we can "see" the structure of materials at the atomic level. The scanning tunneling microscope works by exploiting the wave properties of electrons.

Wave properties? Aren't electrons particles?

To answer this question we must look more closely at the fundamental properties of matter and light. Our journey through physics has brought us to about 1890, a little over a century ago. The physics of particles and waves was well understood by then, and it seemed that Newtonian physics would soon succeed in explaining all the phenomena of nature in terms of the particle and wave models. But trouble was on the horizon. Discoveries made during the last decade of the 19th century and the opening years of the 20th century refused to yield to a Newtonian analysis.

At the heart of the crisis was a breakdown of the basic particle and wave models. As physicists probed more deeply into the nature of light, they began to make observations that couldn't be reconciled with the wave model. Sometimes, as you will see, light refuses to act like a wave and seems more like a collection of particles. Even more troubling experiments found that electrons sometimes behave like waves.

These discoveries eventually led to a radical new theory of light and matter called *quantum physics*. We will return to study quantum physics more thoroughly

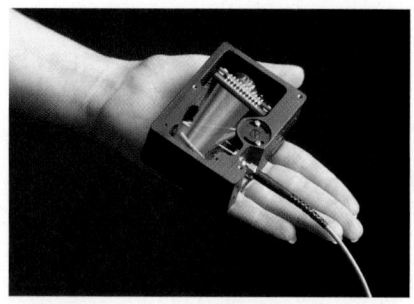

Some modern spectrometers are small enough to hold in your hand. (The rainbow has been painted onto this photograph.)

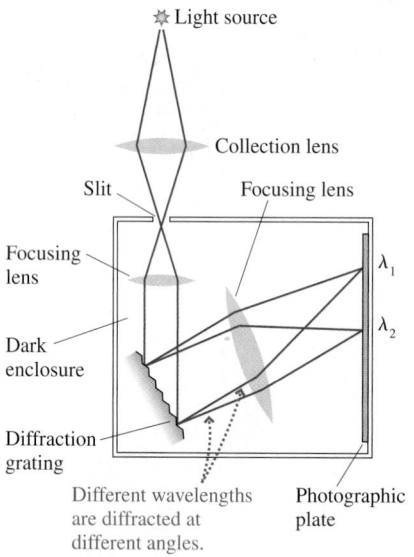

FIGURE 24.1 A diffraction spectrometer for the accurate measurement of wavelengths.

in Part VII. Our goal in this chapter, as we conclude our study of waves, is to use our knowledge of particles and waves to examine some of the experimental evidence that led to quantum physics. By doing so, we will discover the limits of the wave and particle models that we've developed.

24.1 Spectroscopy: Unlocking the Structure of Atoms

The basic discoveries of the interference and diffraction of light were made early in the 19th century. These phenomena were well understood by the end of the century, and the knowledge was used to design practical tools for measuring wavelengths with great accuracy. The primary instrument for measuring the wavelengths of light is a **spectrometer,** such as the one shown in Figure 24.1. The heart of a spectrometer is a diffraction grating that diffracts different wavelengths of light at different angles. A lens then focuses the interference fringes onto a *photographic plate* or (more likely today) an electronic array detector.

Each wavelength in the light is focused to a different position on the plate. The distinctive pattern of wavelengths emitted by a source of light and recorded on the detector is called the **spectrum** of the light. Spectroscopists discovered very early that there are two types of spectra, continuous spectra and discrete spectra:

- Hot, self-luminous objects, such as the sun or an incandescent light bulb, emit a *continuous spectrum* in which a rainbow is formed by light being emitted at every possible wavelength.
- In contrast, the light emitted by a gas discharge tube (such as those used to make neon signs) contains only certain discrete, individual wavelengths. Such a spectrum is called a **discrete spectrum.**

Figure 24.2 shows examples of spectra as they would appear on the photographic plate of a spectrometer. Each bright line, called a **spectral line,** represents *one* specific wavelength present in the light emitted by the source. A discrete spectrum is sometimes called a **line spectrum** because of its appearance on the plate. You can see that a neon light has its familiar reddish-orange color because nearly all of the wavelengths emitted by neon atoms fall within the wavelength range 600–700 nm that we perceive as orange and red.

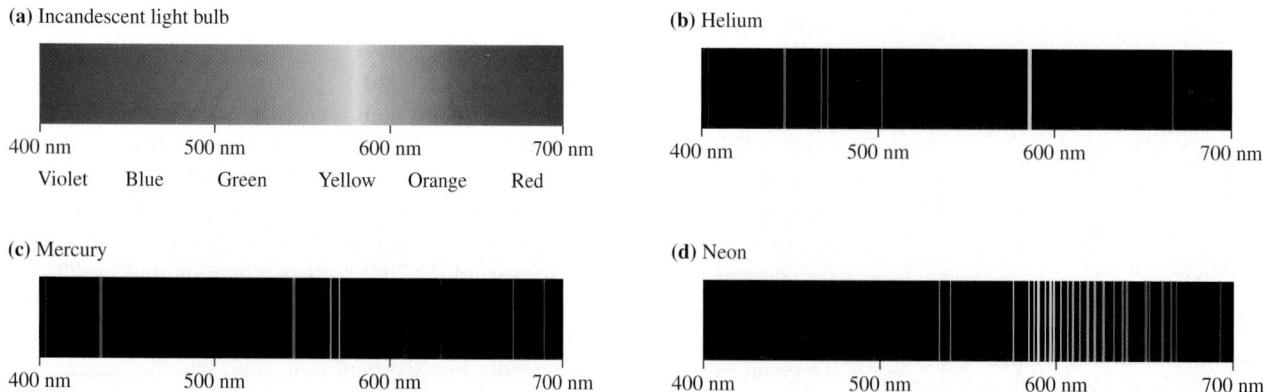

FIGURE 24.2 Examples of spectra in the visible wavelength range 400–700 nm.

Two important conclusions had been established by the end of the 19th century:

1. The light emitted by atoms in a gas discharge tube has a discrete spectrum.
2. Every element in the periodic table has its own unique spectrum.

The fact that each element emits a unique spectrum means that atomic spectra can be used as "fingerprints" to identify elements. Consequently, atomic spectroscopy is the basis of many contemporary technologies for analyzing the composition of unknown materials, monitoring air pollutants, and studying the atmospheres of the earth and other planets.

You know from chemistry that an element's *atomic number* specifies the number of protons and electrons within an atom. Hydrogen, with atomic number 1, has one electron and one proton while neon, at atomic number 10, has 10 electrons and 10 protons. It was soon recognized that an atom's internal structure determines the wavelengths the atom emits. If we only knew how to "decode" an element's spectrum, we would be able to determine the trajectories of the electrons within the atom.

Despite heroic attempts by some of the best scientists of the late 19th century, Newtonian mechanics and the (then) new theory of electromagnetism were completely unable to provide an explanation of atomic spectra or atomic structure. Not only did they fail to predict why one element's spectrum should differ from another, these classical theories predicted that atomic electrons should spiral into the nucleus, destroying the atoms and the universe in a small fraction of a second! This prediction is obviously incorrect.

Physics from the time of Newton through the mid-19th century had been spectacularly successful. But the physics of particles and waves was completely unable to explain the puzzle of discrete spectra. The first hint of a new direction in which to turn was made in 1885 by a Swiss school teacher named Johann Balmer.

Balmer and the Hydrogen Atom

Balmer was intrigued by the spectrum of hydrogen. Hydrogen is the simplest atom, with a single electron orbiting a proton, and it also has the simplest atomic spectrum. The *visible spectrum* of hydrogen, between 400 nm and 700 nm, consists of a mere four spectral lines. The wavelengths are given in Table 24.1. Physicists felt certain that such a simple spectrum must have a simple and straightforward explanation.

In 1885, Johann Balmer found by trial and error that the four wavelengths in the visible spectrum of hydrogen could be represented by the simple formula

$$\lambda = \frac{91.18 \text{ nm}}{\left(\dfrac{1}{2^2} - \dfrac{1}{n^2}\right)} \qquad n = 3, 4, 5, 6 \qquad (24.1)$$

Balmer's formula was able to reproduce the measured wavelengths with much better than 0.1% accuracy. Not only was his formula accurate, it was *simple,* in keeping with expectations that the hydrogen spectrum should have a simple explanation.

Balmer knew only the four *visible* wavelengths shown in Table 24.1, but an obvious question to ask was whether Equation 24.1 also predicts wavelengths for $n = 7$, 8, 9, and so on. The prediction for $n = 7$ is $\lambda = 397.1$ nm, an ultraviolet wavelength. Spectroscopists were just beginning to extend their craft into the ultraviolet and infrared regions of the spectrum, and it was soon confirmed that Balmer's formula does, indeed, work for *all* values of n.

Balmer's formula predicts a *series* of spectral lines of gradually decreasing wavelength, converging to the *series limit* wavelength of 364.7 nm as $n \to \infty$. Although there are an infinite number of spectral lines in this series, their intensities rapidly get weaker as n increases until, for large values of n, they blur together and cannot be resolved. This series of spectral lines is now called the **Balmer series.** Figure 24.3 shows a photograph of the Balmer series of hydrogen in which the series limit is quite obvious.

Activ Physics 18.2

TABLE 24.1 Wavelengths of visible lines in the hydrogen spectrum*

656.46 nm

486.27 nm

434.17 nm

410.29 nm

*Wavelengths in vacuum.

The spectral lines extend to the series limit at 364.7 nm.

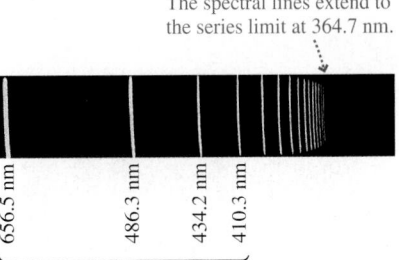

656.5 nm   486.3 nm   434.2 nm   410.3 nm

Four visible wavelengths known to Balmer

FIGURE 24.3 The Balmer series of hydrogen as seen on the photographic plate of a spectrometer.

With the success of Balmer's formula, it was natural to ask what happens if the 2^2 in Equation 24.1 is changed to 1^2 or 3^2 or m^2. It was easy to calculate that all spectral lines in the series with 1^2, if they existed, would have fairly extreme ultraviolet wavelengths while all those in the series with 3^2 would be in the infrared. Spectroscopists accepted the challenge and went to work developing the techniques for infrared and ultraviolet spectroscopy.

The $m = 1$ series was discovered by Theodore Lyman and is called the Lyman series, and the $m = 3$ series, found by Louis Paschen, is called the Paschen series. They provided confirmation, beyond any doubt, that Balmer's formula could be generalized to

$$\lambda = \frac{91.18 \text{ nm}}{\left(\dfrac{1}{m^2} - \dfrac{1}{n^2}\right)} \quad \begin{cases} m = 1 & \text{Lyman series} \\ m = 2 & \text{Balmer series} \\ m = 3 & \text{Paschen series} \\ \vdots \end{cases} \qquad (24.2)$$

$$n = m + 1, m + 2, \ldots$$

As spectroscopists acquired ever more data, it became increasingly clear that Equation 24.2 could predict *every* line in the hydrogen spectrum, from the extreme ultraviolet to the far infrared.

Surely Balmer's success was not a mere coincidence. There must be some underlying meaning to his formula. But what? Balmer did not present a *theory*. He simply said, "Here's a formula that accurately calculates the wavelengths in the hydrogen spectrum." In effect, Balmer's formula was a challenge. Any successful theory of atoms must be able to *derive* Equation 24.2 from the basic laws and principles of the theory. It was 30 years before a theory was proposed that could meet this challenge.

It is particularly striking that Equation 24.2 depends on two *integers*. Hydrogen atoms simply do not emit wavelengths for $m = 1.6$ or for $n = 3.4$. This must tell us *something* important about the structure of the hydrogen atom. Newtonian mechanics does not deal in such "discrete" quantities. Masses, forces, velocities, and energies can take on any value at all; they are not restricted to having only some values but not others.

However, we have seen one exception to this: standing waves. Standing waves exist only for certain frequencies and wavelengths that are described by an *integer* called the mode number. Could there, somehow, be a connection between standing waves and the structure of atoms? To answer this question, we must probe yet deeper into the nature of light and matter.

24.2 X-Ray Diffraction

In 1895, the German physicist Wilhelm Röntgen made a remarkable discovery. The late 19th century was a period in which the technology of vacuum tubes was being perfected, and Röntgen was studying how electrons (called *cathode rays* at the time) traveled through a vacuum. He sealed an electron-producing cathode and a metal target electrode into a vacuum tube, such as shown in Figure 24.4. A high voltage pulled electrons from the cathode and accelerated them to very high speed before they struck the target. Röntgen and others had done similar experiments previously, but one day he happened by chance to have left a sealed envelope containing film near the vacuum tube. He was later surprised to discover that the film had been exposed even though it had never been removed from the envelope. This serendipitous discovery was the beginning of the study of x rays.

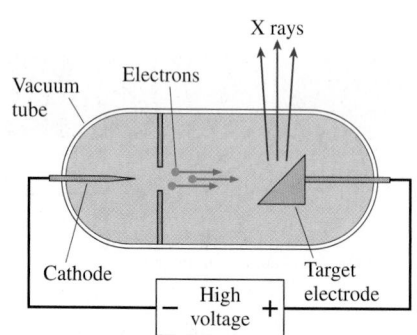

FIGURE 24.4 Röntgen's x-ray tube.

Röntgen quickly found that the vacuum tube was the source of whatever was exposing the film. But he had no idea what was coming from the tube, so he called them **x rays,** using the algebraic symbol *x* as meaning "unknown." X rays were unlike anything, particle or wave, ever discovered. Röntgen was not successful at reflecting the rays or at focusing them with a lens. He showed that they travel in straight lines, like particles, but they also pass right through most solid materials, something no known particle could do.

By the early 1900s, scientists suspected that x rays were an electromagnetic wave with a wavelength much less than that of visible light. At about the same time, scientists were first discovering that the size of an atom is ≈ 0.1 nm, and it was suggested that solids might consist of atoms arranged in crystalline lattices. In 1912, the German scientist Max von Laue noted that *if* x rays are waves with very short wavelengths, and *if* solids are atomic crystals with the atoms spaced about 0.1 nm apart, then x rays passing through a crystal ought to undergo diffraction from the "three-dimensional grating" of the crystal.

X-ray diffraction by crystals was soon confirmed experimentally. Measurements showed that x rays are indeed electromagnetic waves, not fundamentally different from visible light, with wavelengths in the range 0.01 nm to 10 nm.

To understand x-ray diffraction, we need to begin by looking at a crystal lattice. Figure 24.5a shows a simple cubic lattice of atoms. The crystal structure of most materials is more complex than this, but a cubic lattice will help you understand the ideas. We'll often draw just one *plane* of atoms, as in Figure 24.5b, so you'll have to visualize the three-dimensional structure of the crystal.

Suppose that a beam of x rays is incident at angle θ on the plane of atoms shown in Figure 24.6a. (Imagine the plane extending out of the page.) Most of the x rays are transmitted through the plane, because we know that x rays penetrate solids, but a small fraction of the wave may be reflected. The reflected wave obeys the law of reflection—the angle of reflection equals the angle of incidence—and the figure has been drawn accordingly.

A solid is not one plane of atoms but many parallel planes. As x rays pass through a solid, a small fraction of the wave reflects from each of the parallel planes of atoms shown in Figure 24.5b. The *net* reflection from the solid is the *superposition* of the waves reflected by each atomic plane. For most angles of incidence, the phases of the reflected waves are all different and their superposition is very near zero. In other words, as Röntgen observed, solids don't reflect x rays. However, there are a few specific angles of incidence for which the reflected waves all happened to be in phase. For these angles of incidence, the reflected waves interfere constructively to produce a strong reflection. This strong x-ray reflection at a few specific angles of incidence is called **x-ray diffraction.**

You can see from Figure 24.6b that the wave reflecting from any particular plane travels an extra distance $\Delta r = 2d\cos\theta$ before combining with the reflection from the plane immediately above it, where d is the spacing between the atomic planes. If $\Delta r = m\lambda$, these two waves will be in phase when they recombine. But the same geometry applies to all the planes of atoms. If the reflections from two neighboring planes are in phase, then *all* the reflections from *all* the planes are in phase and will interfere constructively to produce a strong reflection.

Consequently, x rays will reflect from the crystal when the angle of incidence θ_m satisfies

$$\Delta r = 2d\cos\theta_m = m\lambda \qquad m = 1, 2, 3, \ldots \qquad (24.3)$$

Equation 24.3 is called the **Bragg condition,** named for physicist W. L. Bragg who developed this technique for producing x-ray diffraction.

> **NOTE** ▶ Our reasoning is very similar to the reasoning we used in Chapter 21 to understand constructive and destructive interference in thin films. ◀

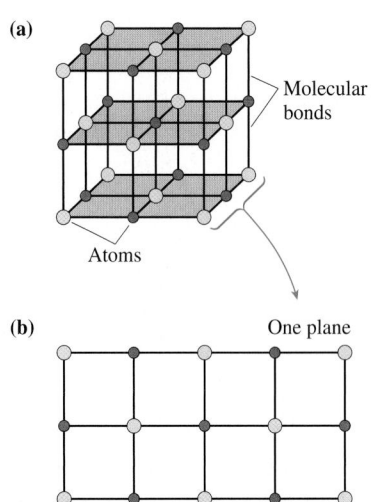

(a)

Molecular bonds

Atoms

(b)

One plane

The spacing between atomic planes is d.

FIGURE 24.5 Atoms arranged in a cubic lattice.

(a) X rays are transmitted and reflected at one plane of atoms.

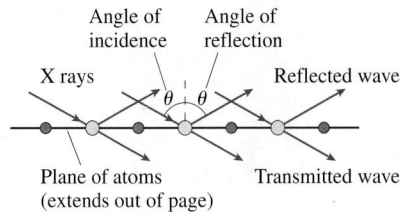

Angle of incidence   Angle of reflection

X rays θ | θ Reflected wave

Plane of atoms (extends out of page)   Transmitted wave

(b) The reflections from parallel planes interfere.

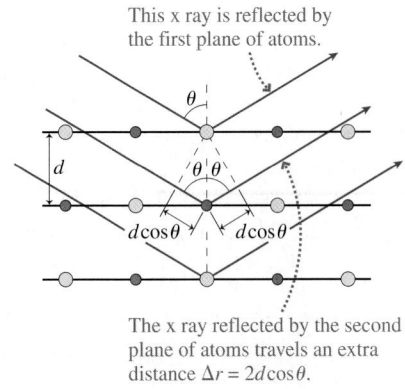

This x ray is reflected by the first plane of atoms.

θ

d

$\theta | \theta$

$d\cos\theta$ $d\cos\theta$

The x ray reflected by the second plane of atoms travels an extra distance $\Delta r = 2d\cos\theta$.

FIGURE 24.6 The x-ray reflections from parallel atomic planes interfere constructively to cause strong reflections for certain angles of incidence.

X-ray diffraction is measured by mounting a crystal so that it can be rotated through a range of angles, as shown in Figure 24.7a. A graph of the reflected x-ray intensity versus angle θ is called the *x-ray diffraction spectrum,* and it contains valuable information about the structure of the crystal.

(a)

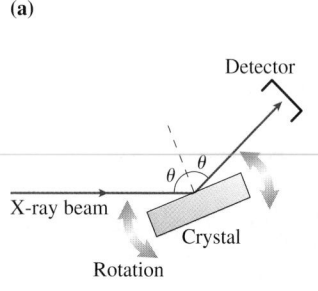

(b)

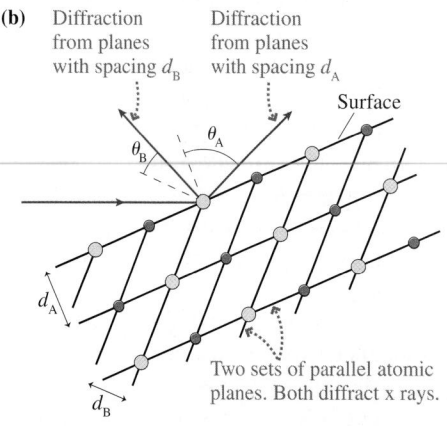

(c)

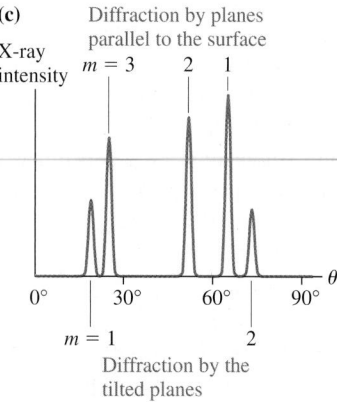

FIGURE 24.7 Producing and measuring an x-ray diffraction spectrum.

One complicating factor is that a crystal can be "sliced" into more than one set of parallel planes of atoms. Figure 24.7b shows a set of atomic planes with spacing d_A and another set of planes with spacing $d_B = d_A/\sqrt{2}$. The planes parallel to the surface cause diffraction if θ_A satisfies the Bragg condition for spacing d_A. Independently, the planes tilted at 45° cause diffraction if θ_B satisfies the Bragg condition for spacing d_B.

Figure 24.7c shows a simulated x-ray diffraction spectrum for a cubic lattice with atomic spacing $d_1 = 0.20$ nm and an x-ray wavelength $\lambda = 0.12$ nm. These are typical values. Real x-ray diffraction spectra are usually more complicated than this spectrum, but such spectra contain information with which scientists can deduce the crystalline structure of the solid.

Notice that the experimentally measured angle θ, which is measured from the surface of the crystal, is angle θ_A for the planes parallel to the surface. The experimental angle is *not* the same as angle θ_B, so it takes a little geometry to match the measured angles to the angles at which the tilted planes cause diffraction. The details will be left for a homework problem.

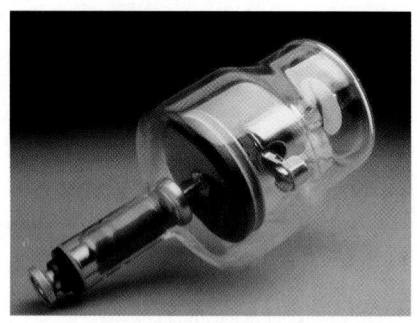

A modern x-ray tube that might be used for medical or dental x rays.

EXAMPLE 24.1 Analyzing x-ray diffraction

X rays with a wavelength of 0.105 nm are diffracted by a crystal. Diffraction maxima are observed at angles 31.6° and 55.4° and at no angles between these two. What is the spacing between the atomic planes causing this diffraction?

MODEL The angles must satisfy the Bragg condition. We don't know the values of m, but they are two consecutive values. Notice that θ_m *decreases* as m increases, so 31.6° corresponds to the larger value of m.

SOLVE d and λ are the same for both diffractions, so we can use the Bragg condition to find

$$\frac{m+1}{m} = \frac{\cos 31.6°}{\cos 55.4°} = 1.50 = \frac{3}{2}$$

Thus 55.4° is the second-order diffraction and 31.6° is the third-order diffraction. With this information we can use the Bragg condition again to find

$$d = \frac{2\lambda}{2\cos\theta_2} = \frac{0.105 \text{ nm}}{\cos 55.4°} = 0.185 \text{ nm}$$

ASSESS This is a reasonable value for the atomic spacing in a crystal.

Although the Bragg procedure is straightforward, most practical x-ray diffraction studies look at the diffraction of x rays that are *transmitted* through a crystal. Figure 24.8a shows a typical experimental arrangement. An x-ray tube generates several x-ray wavelengths, so Bragg diffraction is first used to select just one of these wavelengths by rotating a crystal to an angle meeting the Bragg condition. This part of the apparatus is called an *x-ray monochromator,* a device that selects one (mono) wavelength.

The known wavelength then passes through the sample and is diffracted by the three-dimensional grating of the crystal lattice. An x-ray film behind the sample records the locations of constructive interference. Because the grating is three-dimensional, the diffraction pattern consists of bright points rather than lines or fringes. Figure 24.8b shows a typical diffraction pattern. You can see that it is quite complicated. Nonetheless, crystallographers have developed many powerful analysis tools for deciphering such patterns. These techniques are computationally very intense, but modern supercomputers have made such analyses routine.

Today, x-ray diffraction is an essential tool for studying the atomic and molecular structure of solids. The most important properties of solids—their strength, chemical properties, ability to be cut or welded, optical properties, and so on—are consequences of their crystal structure. Modern engineering could not exist without the knowledge of materials gained through x-ray diffraction. Similarly, x-ray diffraction was used to deduce the double-helix structure of DNA, and it continues to elucidate the structure of biological molecules such as proteins. The techniques of x-ray diffraction are likely to become even more important in the future as physicists develop new superconducting materials, molecular biologists produce "designer drugs," and engineers design atomic-size nanostructures.

(a)

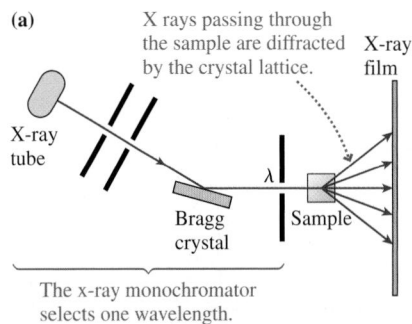

The x-ray monochromator selects one wavelength.

(b) X-ray diffraction pattern for niobium diboride

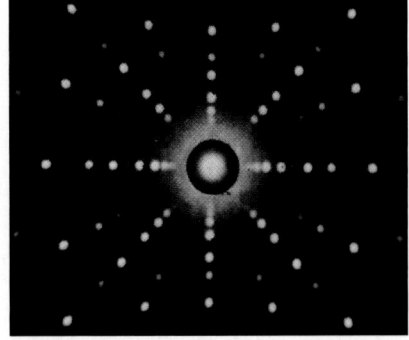

FIGURE 24.8 Using x-ray diffraction to study the atomic structure of a sample.

STOP TO THINK 24.1 The first-order diffraction of monochromatic x rays from crystal A occurs at an angle of 20°. The first-order diffraction of the same x rays from crystal B occurs at 30°. Which crystal has the larger atomic spacing?

24.3 Photons

Figure 24.9 shows three photographs made with a camera in which the film has been replaced by a special high-sensitivity detector. A correct exposure, at the right, shows a perfectly normal photograph of a woman. But with very faint illumination (left), the picture is *not* just a dim version of the properly exposed photo. Instead, it is a collection of dots. A few points on the detector have registered the presence of light, but most have not. As the illumination increases, the density of these dots increases until the dots form a full picture.

The photo at very low light levels shows individual points, as if particles are arriving at the detector.

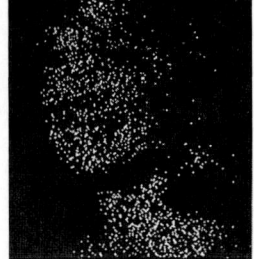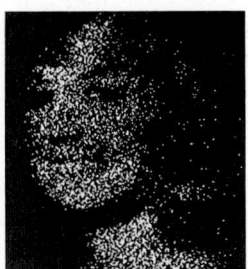

The particle-like behavior is not noticeable at higher light levels.

Increasing light intensity

FIGURE 24.9 Photographs made with an increasing level of light intensity.

This is not what we expected. If light is a wave, reducing its intensity should cause the picture to grow dimmer and dimmer until disappearing, but the entire picture would remain present. It should be like turning down the volume on your stereo until you can no longer hear the sound. Instead, the left photograph in Figure 24.9 looks as if someone randomly threw "pieces" of light at the detector, causing full exposure at a few *discrete* points but no exposure at others.

If we did not know that light is a wave, we would interpret the results of this experiment as evidence that light is a stream of some type of particle-like object. If these particles arrive frequently enough, they overwhelm the detector and it senses a steady "river" instead of the individual particles in the stream. Only at very low intensities do we become aware of the individual particles.

Double-Slit Interference Revisited

The particle-like behavior of light seen in Figure 24.9 was apparent only for very low-intensity light. Let's return to the experiment that showed most dramatically the wave nature of light—Young's double-slit interference experiment—and lower the light intensity by inserting filters between the light source and the slits. We cannot expect to see the interference fringes by eye for such a low intensity, so we will replace the viewing screen with the same detector used to make the photographs of Figure 24.9.

What might we predict for the outcome of this experiment? If light is a wave, there is no reason to think that the nature of the interference fringes will change. The detector should continue to show alternating light and dark bands that become dimmer and dimmer until they vanish.

Figure 24.10 shows the outcome of such an experiment at three low but increasing light levels. Contrary to our prediction, the detector does not show bands at all. Instead, it shows dots like those seen in Figure 24.9. The detector is registering particle-like objects. They arrive one-by-one, and they are localized at a specific point on the detector. This is particle-like behavior, not wave-like behavior. (Waves, you will recall, are not localized at a specific point in space.) But these dots of light are not entirely random. They are grouped into bands at *exactly* the positions where we expected to see bright constructive interference fringes.

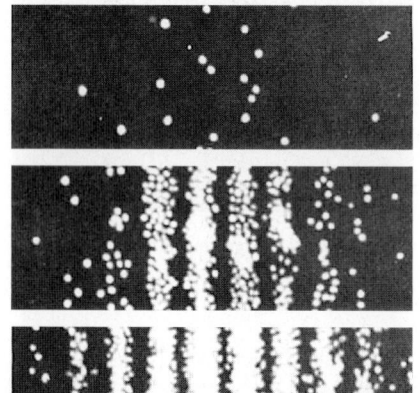

FIGURE 24.10 A simulation of a double-slit interference experiment with very low but increasing levels of light.

The Photon Model of Light

Figures 24.9 and 24.10 are our first evidence of the particle-like nature of light. These particle-like components of light are called **photons.** The concept of the photon was introduced by Albert Einstein to explain an experiment called the photoelectric effect, an experiment we will investigate in detail in Part VII.

The **photon model** of light consists of three basic postulates:

1. Light consists of discrete, massless units called photons. A photon travels in vacuum at the speed of light, 3.00×10^8 m/s.
2. Each photon has energy

$$E_{\text{photon}} = hf \tag{24.4}$$

where f is the frequency of the light and h is a *universal constant* called **Planck's constant.** The value of Planck's constant is

$$h = 6.63 \times 10^{-34} \, \text{J s}$$

In other words, the light comes in discrete "chunks" of energy hf.
3. The superposition of a sufficiently large number of photons has the characteristics of a classical light wave.

EXAMPLE 24.2 The energy of a photon

550 nm is the average wavelength of visible light.

a. What is the energy of a photon with a wavelength of 550 nm?
b. A typical incandescent light bulb emits about 1 J of visible light energy every second. Estimate the number of photons emitted per second.

SOLVE

a. The frequency of the photon is

$$f = \frac{c}{\lambda} = \frac{3.00 \times 10^8 \text{ m/s}}{550 \times 10^{-9} \text{ m}} = 5.45 \times 10^{14} \text{ Hz}$$

Equation 24.4 gives us the energy of this photon:

$$E_{\text{photon}} = hf = (6.63 \times 10^{-34} \text{ J s})(5.45 \times 10^{14} \text{ Hz})$$

$$= 3.61 \times 10^{-19} \text{ J}$$

This is an extremely small energy!

b. The photons emitted by a light bulb will span a range of energies, because the light spans a range of wavelengths, but the *average* photon energy will correspond to a wavelength near 550 nm. Thus we can estimate the number of photons in 1 J of light as

$$N \approx \frac{1 \text{ J}}{3.61 \times 10^{-19} \text{ J/photon}} \approx 3 \times 10^{18} \text{ photons}$$

A typical light bulb emits about 3×10^{18} photons every second.

ASSESS This is a staggeringly large number. It's not surprising that in our everyday life we would sense only the river and not the individual particles within the flow.

Most light sources with which you are familiar emit such vast numbers of photons that you are only aware of their wave-like superposition, just as you only notice the roar of a heavy rain on your roof and not the individual raindrops. But at extremely low intensities the light begins to appear as a stream of individual photons, like the random patter of raindrops when it is barely sprinkling. Each dot on the detector in Figures 24.9 and 24.10 signifies a point where one individual photon delivered its energy and caused a measurable signal.

Although photons are particle like, they are certainly not classical particles. Classical particles, such as Newton's corpuscles of light, would travel in straight lines through the two slits of a double-slit experiment and make just two bright areas on the detector. Instead, as Figure 24.10 shows, the *particle*-like photons seem to be landing at places where a *wave* undergoes constructive interference, thus forming the bands of dots.

Suppose that the detector in the double-slit interference experiment is 30 cm behind the slits and that the light intensity so low that only 10^6 photons arrive per second. This is experimentally quite feasible. On average, a new photon passes through the slits every 10^{-6} s. A photon moving at the speed of light travels distance $d = c\Delta t = 300$ m during 10^{-6} s. While one photon is traveling the 30 cm between the slits and the detector, the next photon is 300 m away. Or in the likely case that the light source is closer to the slits than 300 m, the next photon has not yet even been emitted by the light source! Under these conditions, only one photon at a time is passing through the double-slit apparatus.

If particle-like photons arrive at the detector in a banded pattern as a consequence of wave-like interference, as Figure 24.10 shows, but if only one photon at a time is passing through the experiment, what is it interfering with? The only possible answer is that the photon is somehow interfering *with itself.* Nothing else is present. But if each photon interferes with itself, rather than with other photons, then each photon, despite the fact that it is a particle-like object, must somehow go through *both* slits!

This all seems pretty crazy. But crazy or not, this is the way light behaves. Sometimes it exhibits particle-like behavior and sometimes it exhibits wave-like behavior. You may be expecting that we will now bring forth an "explanation" so that these observations will all "make sense." Sorry. This is simply how light really and truly behaves. The thing we call *light* is stranger and more complex than it first appeared, and there just is no way for these seemingly contradictory behaviors to make sense. We have to accept nature as it is rather than hoping that nature will conform to our expectations. Furthermore, this half-wave/half-particle behavior is not restricted to light.

STOP TO THINK 24.2 Does a photon of red light have more or less energy than a photon of blue light?

24.4 Matter Waves

An important experiment took place in 1927 at the Bell Telephone Laboratories in New York. Two physicists, Clinton Davisson and Lester Germer, were studying how electrons scatter from the surface of metals. They had been doing similar experiments for several years, but this time they happened to use a well-crystallized piece of nickel as their target. As they rotated the electron detector around the sample, as shown in Figure 24.11a, they discovered that the intensity of the scattered electron beam exhibited clear minima and maxima.

Notice that Davisson and Germer's experiment was very similar to the Bragg x-ray diffraction experiment that was shown in Figure 24.7a. And the scattered-electron intensity they observed was not unlike the x-ray intensity pattern shown in Figure 24.7c. Although we "know" that electrons are material particles, completely unlike light waves, suppose we were to analyze the Davisson-Germer experiment *as if* electrons were waves undergoing Bragg diffraction.

Davisson and Germer found that electrons incident normal to the crystal face at a speed of 4.35×10^6 m/s scattered at $\phi = 50°$. You can see in Figure 24.11b that this scattering can be interpreted as a mirror-like reflection from the atomic planes that slice diagonally through the crystal. The angle of incidence on this set of planes is $\theta = \phi/2 = 25°$. This is the angle in Equation 24.3, $2d\cos\theta_m = m\lambda$, the Bragg condition for diffraction.

You can also see that the spacing d between the atomic planes is related to the atomic spacing D by

$$d = D\sin\theta \tag{24.5}$$

Equation 24.5 allows us to write the Bragg condition in terms of the atomic spacing D, rather then the plane spacing d, as

$$2(D\sin\theta_m)\cos\theta_m = D(2\sin\theta_m\cos\theta_m) = D\sin(2\theta_m) = m\lambda \tag{24.6}$$

From x-ray diffraction, the atomic spacing of nickel was already known to be $D = 0.215$ nm. If we combine this value of D with the measured angle $\theta = 25°$, and if we assume $m = 1$, then we find that the "electron wavelength" is

$$\lambda = D\sin(2\theta) = 0.165 \text{ nm} \tag{24.7}$$

This seems like a pointless exercise. Yes, electrons reflect from a nickel surface with a scattering angle of 50°. But electrons are particles of matter, so there must be some explanation in terms of the collision of particles with the atoms at the surface of the crystal. Right? Nonetheless, Davisson and Germer searched for, and quickly found, 20 other reflections obeying the Bragg condition for *exactly* the same "wavelength" of 0.165 nm.

These results could not be a coincidence. Electrons, particles of matter with mass, were somehow, in some way, being *diffracted* by the grating of a crystal. Particles of matter, for the first time ever, were being observed to have wave-like properties!

(a)

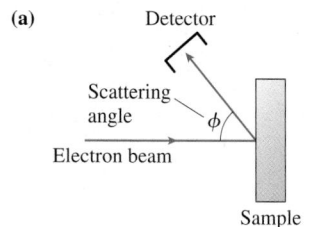

(b)

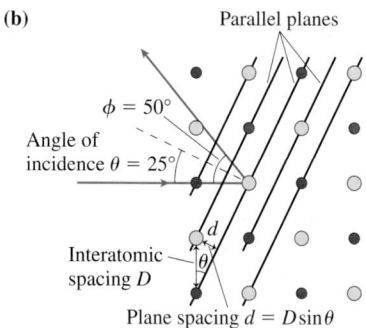

FIGURE 24.11 The Davisson-Germer experiment to study electrons scattered from metal surfaces.

The de Broglie Wavelength

Three years earlier, in 1924, a French graduate student named Louis-Victor de Broglie (Figure 24.12) was puzzling over the growing evidence that light seemed to have both wave-like and particle-like properties. Sometimes light acted like a classical wave, exhibiting interference and diffraction. Yet at other times, light seemed to come in small, localized pieces like a particle. Einstein had just won the Nobel prize in 1921 for his explanation of the photoelectric effect in terms of particle-like photons of light.

If light, something that we generally think of as a wave, can act like a particle, then it occurred to de Broglie that perhaps some object we generally think of as a particle would, in the right conditions, act like a wave. What are the most "particle-like" entities we can think of? Very likely electrons and protons, the basic building blocks of matter. Can an electron or a proton act like a wave? If they did so, how would we know? What behavior would they exhibit that is wave-like? And what is the "wavelength" of an electron—if it has one?

De Broglie postulated that a particle of mass m and momentum $p = mv$ has a wavelength

$$\lambda = \frac{h}{p} \tag{24.8}$$

where h is Planck's constant. This wavelength for material particles is now called the **de Broglie wavelength.** It depends *inversely* on the particle's momentum, so the largest wave effects will occur for particles having the smallest momentum.

What led de Broglie to this postulate? The constant that we now symbolize as h was first introduced into physics in the year 1900 by the German physicist Max Planck as part of an explanation of how incandescent objects emit continuous spectra. Planck's explanation was not widely accepted until 1905, when Einstein showed that the photoelectric effect could be understood if the energy E of a photon of light is related to its frequency f by $E_{\text{photon}} = hf$.

It was this relationship of energy to frequency that intrigued de Broglie. He reasoned that if matter has wave-like properties, it should also obey Einstein's $E = hf$. But he also knew that the kinetic energy of a particle of mass m is related to its momentum by

$$E = \frac{1}{2}mv^2 = \frac{1}{2}m\left(\frac{p}{m}\right)^2 = \frac{p^2}{2m} \tag{24.9}$$

What relationship between momentum and wavelength would allow these two statements about the particle's energy to be consistent with each other? The only possibility de Broglie could find was $\lambda = h/p$. The details of his reasoning, although not difficult, are not important to us. Our goal, instead, is to understand the experimental evidence for, and some of the implications of, de Broglie's bold and imaginative suggestion.

It is worth noting that there was absolutely *no* evidence for matter waves in 1924. Even so, de Broglie must have reasoned, perhaps the evidence was lacking only because no one had looked in the right places or used the right equipment and techniques. If Equation 24.8 is correct, what evidence would you expect to see? You know that the most obvious characteristic of waves is their ability to exhibit interference and diffraction, but you have also learned that diffraction effects are not easily observable unless the opening through which a wave passes is comparable in size to the wavelength. There is no obvious spreading when a wave passes through an opening of size $a \gg \lambda$. What wavelengths do material particles have, and is it likely that anyone would have seen their diffraction before 1924?

EXAMPLE 24.3 The de Broglie wavelength of a smoke particle

One of the smallest macroscopic particles we could imagine using for an experiment would be a very small smoke or soot particle. These are $\approx 1 \ \mu$m in diameter, too small to see with the naked eye and just barely at the limits of resolution of a microscope. A particle this size has mass $m \approx 10^{-18}$ kg. Estimate the de Broglie wavelength for a 1 μm diameter particle moving at the very slow speed of 1 mm/s.

SOLVE The particle's momentum is $p = mv \approx 10^{-21}$ kg m/s. The de Broglie wavelength of a particle with this momentum is

$$\lambda = \frac{h}{p} \approx 7 \times 10^{-13} \ \text{m}$$

ASSESS This wavelength is $\approx 1\%$ the size of an atom. We can't shoot a 1-μm-diameter particle though an atom-size hole, so we certainly don't expect to see any wave-like behavior. And if a 1 μm particle has a wavelength this small, the wavelength of a baseball must be smaller by many factors of 10. It is thus little wonder, if de Broglie's suggestion is correct, that we do not see macroscopic objects exhibiting wave-like behavior.

EXAMPLE 24.4 The de Broglie wavelength of an electron

Find the de Broglie wavelength of an electron with a speed of 4.35×10^6 m/s, the speed used in the Davisson-Germer experiment.

SOLVE The mass of an electron is 9.11×10^{-31} kg. Its de Broglie wavelength at this speed is

$$\lambda = \frac{h}{p} = \frac{h}{mv} = 0.167 \ \text{nm}$$

ASSESS This result is in near-perfect agreement with Davisson and Germer's experimentally determined wavelength of 0.165 nm! Electrons moving with speeds in this range have de Broglie wavelengths very similar to those of x rays. These wavelengths are exactly the right size to be diffracted by atomic crystals.

Davisson and Germer, who won the Nobel prize for their demonstration of the wave nature of electrons, had not set out to perform a breakthrough experiment. They were simply continuing research that had started years earlier, and they had never heard of de Broglie at the time they found unexpected and unexplainable results. However, being open-minded enough to seek out the advice and opinion of others, they learned that they might be able to demonstrate electron diffraction. A large element of chance and luck was involved; they just happened to be doing the right experiments at the right time. But their careful thought and study had also prepared them to recognize a unique opportunity when it came along. It was their willingness to give a fair test to a really crazy idea—that electrons might be waves!—that earned them a place in science history.

The Interference and Diffraction of Matter

17.5 Activ Physics

Further evidence in support of de Broglie's hypothesis was soon forthcoming. The English physicist G. P. Thompson performed a diffraction experiment with an electron beam transmitted *through* a crystal, an experiment exactly equivalent to Figure 24.8 for x-ray diffraction. Figures 24.13a and 24.13b show the diffraction patterns produced by x rays and electrons passing through an aluminum-foil target. (The foil is not a single crystal but, instead, thousands of tiny crystal grains at random orientations. As a consequence, the single-crystal diffraction spots of Figure 24.8b get rotated about the axis and form concentric diffraction circles.) The primary observation to make from Figure 24.13 is that **electrons diffract exactly like x rays.**

(a) X-ray diffraction pattern

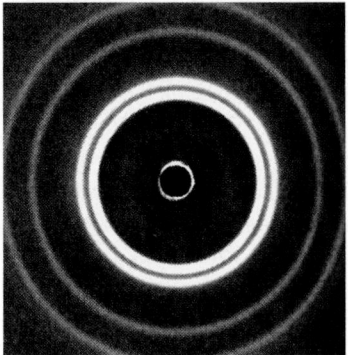

(b) Electron diffraction pattern

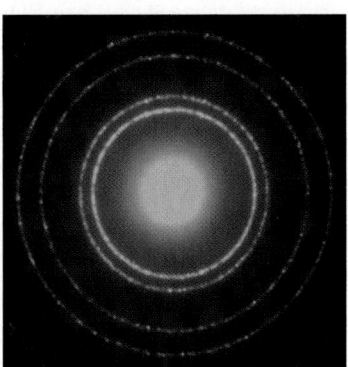
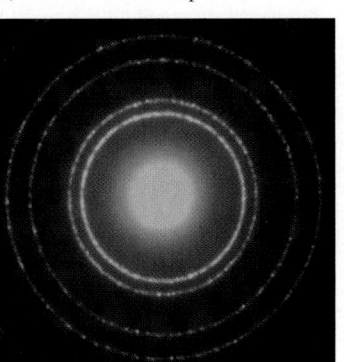

(c) Neutron diffraction pattern

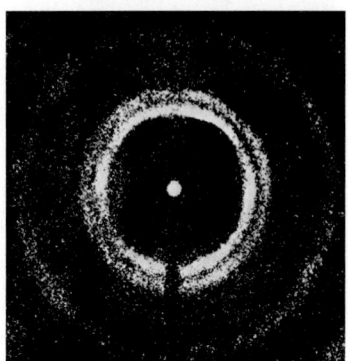

FIGURE 24.13 The diffraction patterns produced by x rays, electrons, and neutrons passing through an aluminum-foil target.

Later experiments demonstrated that de Broglie's hypothesis applies to other material particles as well. Neutrons have a much larger mass than electrons, which tends to decrease their de Broglie wavelength, but it is possible to generate very slow neutrons. The much smaller speed compensates for the heavier mass, so neutron wavelengths can be comparable to electron wavelengths. Figure 24.13c shows a neutron diffraction pattern. It is similar to the x ray and electron diffraction patterns, although of lower quality because neutrons are harder to detect. A neutron, too, is a matter wave. In fact, in recent years it has become possible to observe the interference and diffraction of entire atoms!

The classic test of "waviness" is Young's double-slit interference experiment. If an electron, or other material object, has wave-like properties, it should exhibit interference when passing through two slits. Does it? This experiment is not easy to do because the spacing between the two slits has to be very tiny. The technical challenges of such an experiment could not be met until around 1960, when it became possible to produce slits in a thin foil with a spacing of $\approx 2 \ \mu$m. Even then, various technical reasons required the electrons to have much higher velocities than Davisson and Germer used, reducing their de Broglie wavelength to ≈ 0.005 nm. This rather significant discrepancy between the wavelength and the slit spacing is equivalent to an optical double-slit experiment with a slit spacing of 20 cm. Nonetheless, the experiment was performed, and Figure 24.14a shows the highly enlarged electron pattern that was detected. Amazing as it seems, electrons, one of the basic building blocks of matter, produce interference fringes after passing through a double slit.

(a) Double-slit interference of electrons

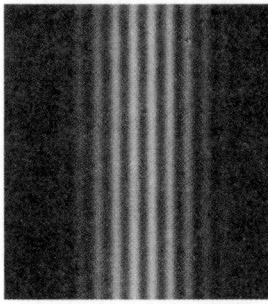

(b) Double-slit interference of neutrons

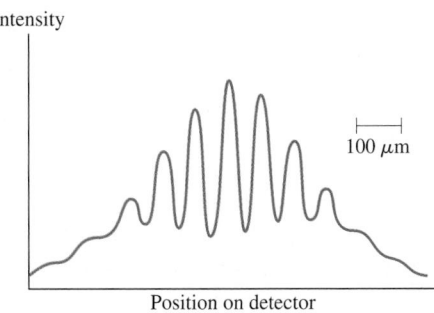

FIGURE 24.14 Double-slit interference patterns of electrons and neutrons.

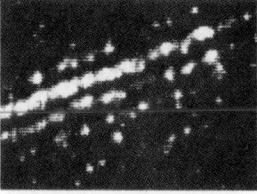

FIGURE 24.15 A double-slit interference pattern of electrons is built up electron by electron as they arrive at the detector.

Later, during the 1970s and 1980s, techniques were developed for observing the double-slit interference of neutrons. Figure 24.14b shows the pattern recorded when neutrons passed through two slits separated by 0.10 mm. The characteristic interference fringes are readily observed, despite the much larger mass of the neutron.

Figure 24.15 shows an electron double-slit experiment in which the intensity of the electron beam was reduced to only a few electrons per second. You can see that each electron is detected on the screen as a *particle,* a localized dot where the electron hits, but that the pattern of dots is the interference pattern of a *wave* with wavelength $\lambda = h/p$. Compare this picture to Figure 24.10 for photons. (Note that Figure 24.10 was a simulation, but Figure 24.15 is a photograph from a real experiment.) In both cases, electrons and photons, we see a combination of both wave-like and particle-like behaviors.

We noted earlier that each photon must in some sense interfere with itself. The same is true for electrons. If only a few electrons arrive per second, then only one electron at a time is in the region of the slits and the screen. Each electron somehow goes through both slits, has a wave-like interference with itself, but is finally detected at the screen as a particle-like dot.

NOTE ▶ We are *not* saying that photons and electrons are the same thing. We are saying that light and electrons are found to share both wave-like and particle-like properties, so under similar experimental conditions we can expect to see similar behavior. Nonetheless, electrons are matter. They are particles with mass and charge that obey $\lambda = h/p$. Photons have no mass, no charge, and obey $\lambda = c/f$. There are many situations in which the behaviors of electrons and photons are quite distinct. ◀

STOP TO THINK 24.3   A proton, an electron, and an oxygen atom each pass at the same speed through a 1-μm-wide slit. Which will produce a wider diffraction pattern on a detector behind the slit?

a. The proton.
b. The electron.
c. The oxygen atom.
d. All three will be the same.
e. None of them will produce a diffraction pattern.

24.5 Energy Is Quantized

De Broglie hypothesized that material particles have wave-like properties, and you've now seen experimental evidence that this must be true. This final section will explore one of the most important implications of the wave-like nature of matter.

You learned in Chapter 21 that waves confined between two boundaries form standing waves. Wave reflections cause the region between the two boundaries to be filled with waves traveling in both directions, and the superposition of two oppositely directed waves produces a standing wave.

To see how this applies to matter, let's consider what physicists call "a particle in a box." Figure 24.16 shows a particle of mass m confined inside a rigid box of length L. For simplicity, we'll consider only the one-dimensional motion parallel to the length of the box. Furthermore, we'll assume that collisions with the ends of the box are perfectly elastic, with no loss of kinetic energy.

(a) A classical particle of mass m bounces back and forth between the ends.

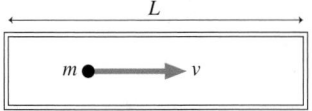

(b) Matter waves moving in opposite directions create standing waves.

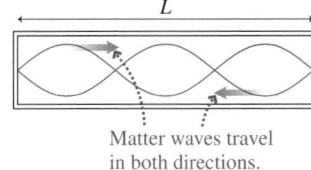

Matter waves travel in both directions.

FIGURE 24.16 A particle of mass m confined in a box of length L.

Figure 24.16a shows a classical particle, such as a ball or a dust particle, in the box. This particle will simply bounce back and forth at constant speed. But if particles have wave-like properties, perhaps we should consider a *wave* reflecting back and forth from the ends of the box. The reflections will create the standing wave shown in Figure 24.16b. This standing wave is analogous to the standing wave on a string that is tied at both ends.

In Chapter 21, we found that the wavelength of a standing wave is related to the length L of the confining region by

$$\lambda_n = \frac{2L}{n} \qquad n = 1, 2, 3, 4, \ldots \qquad (24.10)$$

The particle must also satisfy the de Broglie condition $\lambda = h/p$. Equating these two expressions for the wavelength gives

$$\frac{h}{p} = \frac{2L}{n} \qquad (24.11)$$

Solving Equation 24.11 for the particle's momentum p, we find

$$p_n = n\left(\frac{h}{2L}\right) \qquad n = 1, 2, 3, 4, \ldots \qquad (24.12)$$

Equation 24.12 is a most surprising result. It appears that the momentum of a wave-like particle can have only those *discrete* values given by Equation 24.12. Newtonian physics places no restrictions on the value of the momentum, hence Equation 24.12 represents a clear break with Newtonian physics.

The particle's energy, which is entirely kinetic energy, is related to its momentum by

$$E = \frac{1}{2}mv^2 = \frac{p^2}{2m} \qquad (24.13)$$

If we use Equation 24.12 for the momentum, we find that the particle's energy is restricted to the discrete values

$$E_n = \frac{1}{2m}\left(\frac{hn}{2L}\right)^2 = \frac{h^2}{8mL^2}n^2 \qquad n = 1, 2, 3, 4, \ldots \qquad (24.14)$$

This conclusion is one of the most profound discoveries of physics. Because of the wave nature of matter, which has ample experimental confirmation, **a confined particle can have only certain energies.** It is simply not possible for the particle to exist in the box with any energy other than one of the values given by Equation 24.14.

This result, that a confined particle can have only discrete values of energy, is called the **quantization** of energy. More informally, we say that energy is *quantized.* The number n is called the **quantum number,** and each value of n characterizes one **energy level** of the particle in the box.

Not only is the energy quantized, we see from Equation 24.14 that the energy of the particle in the box cannot be reduced below

$$E_1 = \frac{h^2}{8mL^2} \qquad (24.15)$$

E_1 is the *least* kinetic energy a particle can have. Because $E_1 > 0$, **the particle is *always* in motion;** it cannot be made to stay at rest! These properties of a wave-like particle in a box are in stark contrast to those of a classical Newtonian particle, for which the possible energies are continuous and the minimum kinetic energy is zero. In terms of E_1, the allowed energies are

$$E_n = n^2 E_1 \qquad (24.16)$$

This result is analogous to our earlier finding that standing waves can exist for only the discrete frequencies $f_n = n f_1$.

Notice that the allowed energies are inversely proportional to both m and L^2. The quantization of energy is not apparent with macroscopic objects, or else we would have known about it long ago, so both m and L have to be exceedingly small before energy quantization has any significance. This is an important observation because any new theory about matter and energy cannot be in conflict with our observations of macroscopic objects. Newtonian physics still works for baseballs.

EXAMPLE 24.5 The minimum energy of a smoke particle
What is the first allowed energy of the very small 1-μm-diameter particle of Example 24.3 if it is confined to a very small box 10 μm in length?

SOLVE This is about as small as we can easily imagine making macroscopic particles and boxes. Example 24.3 noted that such a particle has $m \approx 10^{-18}$ kg. The first allowed energy, $n = 1$, is

$$E_1 = \frac{h^2}{8mL^2} \approx 5 \times 10^{-40} \text{ J}$$

ASSESS This is an unimaginably small amount of energy. By comparison, the kinetic energy of a 1-μm-diameter particle moving at a barely perceptible speed of 1 mm/s is $K \approx 5 \times 10^{-25}$ J, a factor of 10^{15} larger. There is no way we could ever observe or measure discrete energies this small, so it is not surprising that we are unaware of energy quantization for macroscopic objects.

EXAMPLE 24.6 The minimum energy of an electron
What are the first three allowed energies of an electron confined to a 0.10-nm-long box?

SOLVE The mass of an electron is $m = 9.11 \times 10^{-31}$ kg. Thus the first allowed energy is

$$E_1 = \frac{h^2}{8mL^2} = 6.0 \times 10^{-18} \text{ J}$$

This is the lowest allowed energy. The next two allowed energies are

$$E_2 = 2^2 E_1 = 24.0 \times 10^{-18} \text{ J}$$

$$E_3 = 3^2 E_1 = 54.0 \times 10^{-18} \text{ J}$$

ASSESS An electron with energy E_1 has speed $v = 3.6 \times 10^6$ m/s, roughly 1% of the speed of light. A 0.10-nm-long box is about the size of an atom. The very large speed of an electron with the *minimum* electron energy in an atomic-size box suggests that the wave nature of electrons *is* important for the physics of atoms.

Energy quantization is simply not an issue for the physics of macroscopic objects. Newtonian physics works fine. But at the much smaller scale of atoms, the physics is very new and different. An atom is certainly more complicated than a simple one-dimensional box, but an electron is "confined" within an atom rather as a particle is in a box. Thus the electron orbits must, in some sense, be standing waves. This idea will be the key that unlocks the mystery of discrete atomic spectra. We will study the atom more carefully in Part VII, but you can see from this introduction that the quantization of energy *is* important at the microscopic scale of the atom.

These examples raise more questions than they answer. If matter is some kind of wave, what is waving? What is the medium of a matter wave? What kind of displacement does it undergo? De Broglie's hypothesis is not a *theory,* and it provides no answers to important questions such as these.

De Broglie's suggestion came nearly 40 years after Balmer's discovery, 40 years during which the atom was being explored and the failures of classical physics were becoming ever more apparent. His suggestion was the final spark, setting off a burst of activities and new ideas that led within a year to a complete and revolutionary new theory—quantum physics. We will revisit these issues later, but for now it is important to see just how far we have been able to come with our study of waves.

STOP TO THINK 24.4   A proton, an electron, and an oxygen atom are each confined in a 1-nm-long box. Rank in order, from largest to smallest, the minimum possible energies of these particles.

SUMMARY

The goal of Chapter 24 has been to explore the limits of the wave and particle models.

GENERAL PRINCIPLES

The two basic models of classical physics

The particle model

A particle is localized at one point in space.
A particle follows a well-defined trajectory.

The wave model

A wave is spread out through space.
A wave exhibits interference and diffraction.

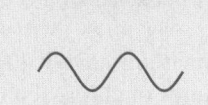

The breakdown of classical physics

A closer look at light and matter finds that these classical models are not sufficient. Light and matter are neither particles nor waves, but exhibit characteristics of both.

IMPORTANT CONCEPTS

Light

- Exhibits interference and diffraction
    Wave-like: $c = \lambda f$
- Detected at localized positions
    Particle-like: $E = hf$
- Particle-like "chunks" of light are called photons.

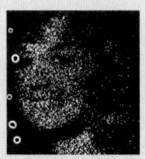

Matter

- Detected at localized positions
    Particle-like: $E = \frac{1}{2}mv^2$
- Exhibits interference and diffraction
    Wave-like: $\lambda = h/p$
- The wavelength is called the **de Broglie wavelength.**

Quantization

A "particle" confined to one-dimensional box of length L sets up a standing wave with the de Broglie wavelength. Because only certain wavelengths can oscillate, only certain discrete energies are allowed:

$$E_n = \frac{h^2}{8mL^2}n^2 \qquad n = 1, 2, 3, \ldots$$

Energy is quantized into discrete levels rather than being continuous as it is in classical physics. Quantization is not important for macroscopic objects, but energy quantization plays a very large role at the atomic level.

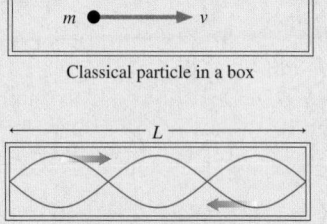

Classical particle in a box

Quantum particle in a box

APPLICATIONS

Hydrogen spectrum

The wavelengths in the spectrum of hydrogen atoms are

$$\lambda = \frac{91.18 \text{ nm}}{\left(\dfrac{1}{m^2} - \dfrac{1}{n^2}\right)} \qquad \begin{array}{l} m = 1, 2, 3, \ldots \\ n = m + 1, m + 2, \ldots \end{array}$$

The series of spectral lines with $m = 2$ is the Balmer series.

Diffraction by atomic crystals

X rays and matter particles with wavelength λ undergo strong reflections from atomic planes spaced by d when the angle of incidence satisfies the Bragg condition

$$2d\cos\theta = m\lambda \qquad m = 1, 2, 3, \ldots$$

TERMS AND NOTATION

spectrometer	line spectrum	Bragg condition	de Broglie wavelength
spectrum	Balmer series	photon	quantization
discrete spectrum	x ray	photon model	quantum number, n
spectral line	x-ray diffraction	Planck's constant, h	energy level, E_n

EXERCISES AND PROBLEMS

Data for Chapter 24: $m_{electron} = 9.11 \times 10^{-31}$ kg
$m_{proton} = m_{neutron} = 1.67 \times 10^{-27}$ kg

Exercises

Section 24.1 Spectroscopy:
Unlocking the Structure of Atoms

1. What are the wavelengths of spectral lines in the Balmer series with $n = 6, 8,$ and 10?
2. Show that the series limit of the Balmer series is 364.7 nm.
3. Which member of the Balmer series has wavelength 389.0 nm?

Section 24.2 X-Ray Diffraction

4. X rays with a wavelength of 0.12 nm undergo first-order diffraction from a crystal at a 68° angle of incidence. What is the angle of second-order diffraction?
5. X rays with a wavelength of 0.20 nm undergo first-order diffraction from a crystal at a 54° angle of incidence. At what angle does first-order diffraction occur for x rays with a wavelength of 0.15 nm?
6. X rays diffract from a crystal in which the spacing between atomic planes is 0.175 nm. The second-order diffraction occurs at 45.0°. What is the angle of the first-order diffraction?
7. X rays with a wavelength of 0.085 nm diffract from a crystal in which the spacing between atomic planes is 0.180 nm. How many diffraction orders are observed?

Section 24.3 Photons

8. What is the energy of a photon of visible light that has a wavelength of 500 nm?
9. What is the energy of an x-ray photon that has a wavelength of 1.0 nm?
10. What is the wavelength of a photon whose energy is twice that of a photon with a 600 nm wavelength?
11. What is the energy of 1 mol of photons that have a wavelength of 1.0 μm?

Section 24.4 Matter Waves

12. Estimate your de Broglie wavelength while walking at a speed of 1 m/s.

13. a. What is the speed of an electron with a de Broglie wavelength of 0.20 nm?
    b. What is the speed of a proton with a de Broglie wavelength of 0.20 nm?
14. What is the kinetic energy of an electron with a de Broglie wavelength of 1.0 nm?
15. a. What is the de Broglie wavelength of a 200 g baseball with a speed of 30 m/s?
    b. What is the speed of a 200 g baseball with a de Broglie wavelength of 0.20 nm?

Section 24.5 Energy Is Quantized

16. What is the smallest box in which you can confine an electron if you want to know for certain that the electron's speed is no more than 10 m/s?
17. What is the length of a box in which the minimum energy of an electron is 1.5×10^{-18} J?
18. The nucleus of an atom is 5.0 femtometers in diameter, where 1 femtometer = 1 fm = 10^{-15} m. A very simple model of the nucleus is a box in which protons are confined. Estimate the energy of a proton in the nucleus by finding the first three allowed energies of a proton in a 5.0-fm-long box.

Problems

19. a. Calculate the wavelengths of the first four members of the Lyman series in the spectrum of hydrogen.
    b. What is the series limit for the Lyman series?
    c. Light from a hydrogen discharge lamp passes through a diffraction grating and registers on a detector 1.5 m behind the grating. The $m = 1$ diffraction of the first member of the Lyman series is located 37.6 cm from the central maximum. What is the position of the second member of the Lyman series?
20. a. Calculate the wavelengths of the first four members of the Paschen series in the spectrum of hydrogen.
    b. What is the series limit for the Paschen series?
    c. Light from a hydrogen discharge passes through a diffraction grating and registers on a detector 1.5 m behind the grating. The $m = 1$ diffraction of the first member of the Paschen series is located 60.7 cm from the central maximum. What is the position of the second member of the Paschen series?

21. *Gamma rays* are photons with very high energy.
    a. What is the wavelength of a gamma-ray photon with energy 1.0×10^{-13} J?
    b. How many visible-light photons with a wavelength of 500 nm would you need to match the energy of this one gamma-ray photon?

22. A 1000 kHz AM radio station broadcasts with a power of 20 kW. How many photons does the transmitting antenna emit each second?

23. A helium-neon laser emits a light beam with a wavelength of 633 nm. The power of the laser beam is 1.0 mW.
    a. What is the energy of one photon of laser light?
    b. How many photons does the laser emit each second?

24. Example 24.2 found that a typical incandescent light bulb emits $\approx 3 \times 10^{18}$ visible-light photons per second. Your eye, when it is fully dark adapted, can barely see the light from an incandescent light bulb 10 km away. How many photons per second are incident at the image point on your retina? The diameter of a dark-adapted pupil is ≈ 6 mm.

25. X-ray photons with energies of 1.50×10^{-15} J are incident on a crystal. The spacing between the atomic planes in the crystal is 0.21 nm. At what angles of incidence will the x rays diffract from the crystal?

26. X rays with a wavelength of 0.070 nm diffract from a crystal. Two adjacent angles of x-ray diffraction are 45.6° and 21.0°. What is the distance between the atomic planes responsible for the diffraction?

27. a. Show that the Bragg condition for x-ray diffraction at normal incidence is equivalent to the condition for maximum reflectivity of a thin film.
    b. Researchers have recently learned how to fabricate thin-film coatings only a few atoms thick out of alternating layers of tungsten and boron carbide. These coatings are expected to greatly improve the x-ray telescopes used in astronomy. What are the two longest x-ray wavelengths that will reflect at normal incidence from a film with a thickness of 1.2 nm?

28. The basic idea of Bragg diffraction is not limited to x rays. One contemporary application is in optical fibers. It is sometimes useful to block one particular wavelength of light, by reflecting it, while transmitting all other wavelengths. Figure P24.28 shows that this can be done by building a short section of fiber, called a *fiber grating,* in which the index of refraction varies periodically. A small fraction of the light wave traveling through the fiber reflects from each little "bump" in the index of refraction. For most wavelengths, the reflected waves have random phases and their superposition is essentially zero. These wavelengths are transmitted through the fiber grating. If, however, the reflections are all in phase for some wavelength, that wavelength is strongly reflected and the transmitted light is strongly attenuated. Consider a fiber grating in a glass fiber ($n = 1.50$) with a spacing of 0.45 μm. What is the air wavelength of infrared light that is blocked by this fiber grating?

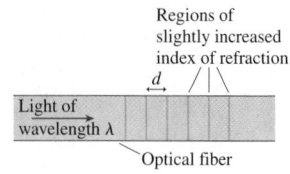

FIGURE P24.28

29. Electrons with a speed of 2.0×10^6 m/s pass through a double-slit apparatus. Interference fringes are detected with a fringe spacing of 1.5 mm.
    a. What will the fringe spacing be if the electrons are replaced by neutrons with the same speed?
    b. What speed must neutrons have to produce interference fringes with a fringe spacing of 1.5 mm?

30. Electrons pass through a 1.0-μm-wide slit with a speed of 1.5×10^6 m/s. How wide is the electron diffraction pattern on a detector 1.0 m behind the slit?

31. The double-slit neutron diffraction pattern shown in Figure 24.14b was measured 3.0 m behind two slits having a separation of 0.10 mm. From measurements that you can make *on the figure* (notice the scale on the figure), determine the speed of the neutrons.

32. In a Davisson-Germer experiment, electrons with a speed of 4.30×10^6 m/s are scattered at an angle of 60°. What is the atomic spacing D?

33. a. What are the energies of the first three energy levels of an electron confined in a one-dimensional box of length 0.70 nm?
    b. How much energy must the electron lose to move from the $n = 2$ energy level to the $n = 1$ energy level?
    c. Suppose that an electron can move from the $n = 2$ level to the $n = 1$ level by emitting a photon of light. If energy is conserved, what must the photon's wavelength be? Give your answer in nm.

34. a. What is the minimum energy of a 10 g Ping-Pong ball in a 10-cm-long box?
    b. At what speed does it have this energy?

35. What is the length of a box in which the difference between an electron's first and second allowed energies is 1.0×10^{-19} J?

36. Two adjacent allowed energies of an electron in a one-dimensional box are 3.6×10^{-19} J and 6.4×10^{-19} J. What is the length of the box?

37. a. Find an expression for the allowed velocities of a particle of mass m in a one-dimensional box of length L.
    b. What are the first three allowed velocities for an electron in a 0.20-nm-long box?

38. It can be shown that the allowed energies of a particle of mass m in a two-dimensional square box of side L are

$$E_{nm} = \frac{h^2}{8mL^2}(n^2 + m^2)$$

The energy depends on two quantum numbers, n and m, both of which must have an integer value 1, 2, 3, . . .
    a. What is the minimum energy for a particle in a two-dimensional square box of side L?
    b. What are the five lowest allowed energies? Give your values as multiples of E_{min}.

Challenge Problems

39. Let's further explore the x-ray diffraction spectrum of Figure 24.7c. This spectrum was for x rays of wavelength 0.12 nm incident on a cubic crystal with atomic spacing 0.20 nm.
    a. Calculate all the diffraction angles θ_A for diffraction from the atomic planes parallel to the surface. Do these angles agree with Figure 24.7c?

b. Calculate all the diffraction angles θ_B for diffraction from the tilted atomic planes.

c. At what measured angles θ should diffraction from the tilted atomic planes be observed? How does your answer compare to Figure 24.7c?

Hint: A second set of atomic planes is tilted in the opposite direction.

40. X rays with wavelength 0.10 nm are incident on a crystal with a hexagonal crystal structure. The x-ray diffraction spectrum is shown in Figure CP24.40. What is the atomic spacing D of this crystal?

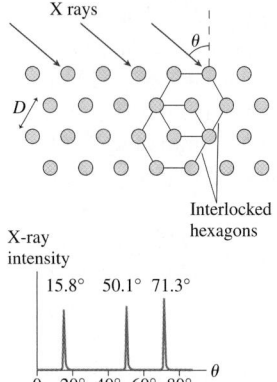

FIGURE CP24.40

41. A sound wave is easily transmitted through an open-open tube of air only if the sound's wavelength λ matches one of the possible standing waves in the tube. This is called a *standing-wave resonance*. The same idea applies to a modern semiconductor device called a *quantum-well device* in which electrons are incident on a thin layer of material that "confines" the electrons. Electrons are trapped within this layer, and it is not easy for electrons to enter or leave. However, electrons can easily flow through this layer if they are able to excite a de Broglie standing-wave resonance in the confinement layer.

a. The confinement layer in a quantum-well device is 5.0 nm thick. What are the four longest-wavelength de Broglie standing waves in this layer?

b. What are the four lowest electron speeds for which an electron current will flow through this layer?

We will study quantum well devices in more detail in Part VII. They are used to make light-emitting diodes and semiconductor lasers.

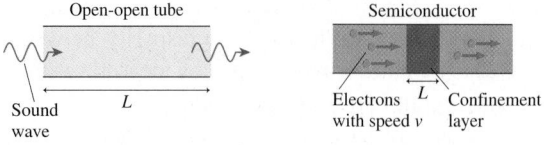

FIGURE CP24.41

42. In Chapter 22, where we studied diffraction gratings, you learned that light with wavelength λ is diffracted by a piece of matter (the grating) with a periodic structure (many slits with spacing d). Experiments done in the 1990s showed that the roles of light and matter can be reversed. That is, matter with a de Broglie wavelength λ can be diffracted by light with a periodic structure. A periodic structure of light is easily created by reflecting a laser beam back on itself to create a standing wave. The experimental challenge, which is quite difficult, is to create a "monochromatic" beam of atoms with a large de Broglie wavelength. Figure CP24.42 shows a beam of sodium atoms ($m = 3.84 \times 10^{-26}$ kg) that are all traveling with a uniform speed of 50 m/s. The atomic beam crosses a laser-beam standing wave with a wavelength of 600 nm. Assuming that the diffraction obeys the diffraction-grating equation (it does), how far will the first-order-diffracted atoms be deflected sideways on a detector 1.0 m behind the laser beam?

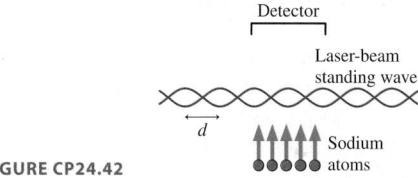

FIGURE CP24.42

STOP TO THINK ANSWERS

Stop to Think 24.1: A. The Bragg condition $2d\sin\theta_1 = \lambda$ tells us that larger values of d go with smaller values of θ_1.

Stop to Think 24.2: Less. $E = hf$, and red light, because of its longer wavelength, has a smaller frequency.

Stop to Think 24.3: b. The widest diffraction pattern occurs for the largest wavelength. The de Broglie wavelength is inversely proportional to the particle's mass.

Stop to Think 24.4: $E_{elec} > E_{proton} > E_{oxy}$. The minimum energy E_1 is inversely proportional to the particle's mass.

Waves and Optics

We end our study of waves a long distance from where we started. Who would have guessed, as we examined our first pulse on a string, that we would end up with quantum numbers? But despite the wide disparity between string waves and matter waves, a few key ideas have stayed with us throughout Part V: the principle of superposition, interference and diffraction, and standing waves. As part of your final study of waves, you should trace the influence of these ideas through the chapters of Part V.

One point we have tried to emphasize is the *unity* of wave physics. We did not need separate theories of string waves and sound waves and light waves. Instead, a few basic ideas enabled us to understand waves of all types. By focusing on similarities, we have been able to analyze sound and light as well as strings and electrons in a single part of this book.

Unfortunately, the physics of waves is not as easily summarized as the physics of particles. Newton's laws and the conservation laws are two very general sets of principles about particles, principles that allowed us to develop the powerful problem-solving strategies of Parts I and II. You probably noticed that we have not found any general problem-solving strategies for wave problems.

This is not to say that wave physics has no structure. Rather, the knowledge structure of waves and optics rests more heavily on *phenomena* than on general principles. Unlike the knowledge structure of Newtonian mechanics, which was a "pyramid of ideas," the knowledge structure of waves is a logical grouping of the major topics you studied. This is a different way of structuring knowledge, but it still provides you with a mental framework for analyzing and thinking about wave problems.

KNOWLEDGE STRUCTURE V **Waves and Optics**

ESSENTIAL CONCEPTS	Wave speed, wavelength, frequency, phase, wave front, and ray.
BASIC GOALS	What are the distinguishing features of waves?
	How does a wave travel through a medium?
	How does a medium respond to the presence of more than one wave?
	What is light and what are its properties?
GENERAL PRINCIPLES	Principle of superposition
	$v = \lambda f$ for periodic waves

Traveling Waves

- The wave speed v is a property of the medium.
- The motion of particles in the medium is distinct from the motion of the wave.
- Snapshot graphs and history graphs show the same wave from different perspectives.
- The Doppler effect of shifted frequencies is observed whenever the wave source or the detector is moving.

Standing Waves

- Standing waves are the superposition of waves moving in opposite directions.
- Nodes and antinodes are spaced by $\lambda/2$.
- Only certain discrete frequencies are allowed, depending on the boundary conditions.

Interference

- Interference is constructive if two waves are in phase: $\Delta\phi = 0, 2\pi, 4\pi, \ldots$
- Interference is destructive if two waves are out of phase: $\Delta\phi = \pi, 3\pi, 5\pi, \ldots$
- The phase difference depends on the path length difference Δr and on any phase difference of the sources.
- Beats occur when $f_1 \neq f_2$.

Light and Optics

- The wave model, used for interference and diffraction, is appropriate when apertures are comparable in size to the wavelength.
- The ray model, used for mirrors and lenses, is appropriate when apertures are much larger than the wavelength.
- Diffraction, a wave effect, limits the best possible resolution of a lens.
- The photon model (discrete chunks of energy) has both wave and particle aspects.

Matter Waves

- "Particles" are not waves, but they have wave-like aspects.
- The de Broglie wavelength is $\lambda = h/mv$.
- Standing matter waves for a confined particle lead to the quantization of energy.

Wave-Particle Duality

The various objects of classical physics are *either* particles *or* waves. There's no middle ground. Planets, projectiles, and atoms are particles or collections of particles, while sound and light are clearly waves. Particles follow trajectories given by Newton's laws; waves obey the principle of superposition, spread out, and exhibit interference. This wave-particle dichotomy seemed obvious until physicists encountered evidence that light sometimes acts like a particle and, even stranger, that matter sometimes acts like a wave. The wave-like aspects of matter continue even today to baffle physicist and nonphysicist alike, but the evidence that matter has wave-like properties is overwhelming and irrefutable.

You might at first think that light and matter are *both* a wave *and* a particle, but that idea does not work. The basic definitions of particleness and waviness are mutually exclusive. Something that is a wave—spread out, non-localized—cannot simultaneously be a discrete, localized particle. It is more profitable to conclude that light and matter are *neither* a wave *nor* a particle.

Although all the macroscopic phenomena with which we are familiar seem to be easily classified as either a wave or a particle, we have no reason to expect that phenomena outside the range of our experience have to be one or the other. And, indeed, it is at the microscopic scale of atoms and their constituents—a physical scale not directly accessible to our five senses—that the classical concepts of particles and waves turn out to be simply too limited to explain the subtleties of nature. Rather than forcing every phenomenon in nature into the mold of "wave" or "particle," we need to adopt a more open-minded approach. Let an electron be an electron and a photon a photon, and nature, rather than our assumptions, will guide us as to just what these entities are.

Although matter and light have both wave-like aspects and particle-like aspects, they show us only one face at a time. If we arrange an experiment to measure a wave-like property, such as interference, we find photons and electrons acting like waves, not particles. An experiment to look for particles will find photons and electrons acting like particles, not waves.

These two aspects of light and matter are *complementary* to each other, like a two-piece jigsaw puzzle. Neither the wave nor the particle model alone provides an adequate picture of light or matter, but taken together they provide us with a basis for understanding these elusive but most fundamental constituents of nature. This two-sided point of view is called *wave-particle duality*.

Wave-particle duality is an established and accepted principle of physics as we start the 21st century. But what does it all mean? What does wave-particle duality tell us about the nature of the universe in which we live? Scientists and nonscientists alike, for over two hundred years, felt that the clockwork universe of Newtonian physics was a fundamental description of reality. The Newtonian worldview provided a familiar and widely accepted understanding of the place and role of humans in the universe.

But wave-particle duality, along with Einstein's relativity, undermines the basic assumptions of the Newtonian worldview. The certainty and predictability of classical physics have given way to a new understanding of the universe in which chance and uncertainty play key roles. Will the 21st century see the development and acceptance of a "quantum worldview" that reflects a different understanding of nature and reality? It is interesting to speculate about what a quantum worldview would be and what implications it would have for society.

We will leave waves behind for a while as we head out in entirely new directions—but not forever. The discovery of matter waves has left us with unfinished business, and we will return in Part VII for a more in-depth study of the strange world of quantum physics. There we will find that electrons and other atomic particles are described by a completely new kind of wave called a *probability wave*.

Anakin's and Obi-Wan's speeder discharges the power coupler on the planet Coruscant by flying through it. If the discharge current through the speeder is 5000 A, what voltage is developed across the speeder? To find out, what properties of the speeder do you need to estimate?

Electricity and Magnetism

Charges, Currents, and Fields

Amber, or fossilized tree resin, has long been prized for the beauty of its lustrous yellow color. Amber is of scientific interest today because biologists have learned how to recover strands of DNA from million-year-old insects that were trapped in the resin. But amber has an ancient scientific connection as well. The Greek word for amber is *elektron*.

It has been known since at least the fifth century B.C. that a piece of amber that has been rubbed briskly can attract feathers or small pieces of straw—seemingly magical powers to a pre-scientific society. It was also known to Greeks of the same time period that certain stones from the region they called *Magnesia* (in present-day Turkey) could pick up pieces of iron. It is from these humble beginnings that we today have high-speed computers, lasers, fiber-optic communications, and magnetic-resonance imaging as well as such mundane modern-day miracles as the light bulb.

The story of electricity and magnetism is vast. The development of a successful electromagnetic theory, which occupied the leading physicists of Europe for most of the nineteenth century, led to sweeping revolutions in both science and technology. The complete formulation of the theory of the electromagnetic field has been called by no less than Einstein "the most important event in physics since Newton's time." Not surprisingly, all that we can do in this text is to develop some of the basic ideas and concepts, leaving many details and advanced applications to later courses. Even so, our study of electricity and magnetism will require learning many new and important ideas. Foremost among these will be the idea of a *field*.

Phenomena and Theories

The basic phenomena of electricity and magnetism are not as familiar to most people as those of mechanics. You have spent your entire life exerting forces on objects and watching them move, but your experience with electricity and magnetism is probably much more limited.

It is hard to motivate the need for a major new theory if you are not aware of what the theory is meant to explain. We will deal with this lack of experience by placing a large emphasis on the basic *phenomena* of electricity and magnetism. We will begin by looking in detail at the fundamental properties of *electric charge* and the process of *charging* an object. It is easy to make systematic observations of how charges behave, and we will be led to consider the forces between charges and how charges behave in different materials. The development of electrical technology and the dawn of the electronic age came about as scientists and

781

engineers learned to *control* the movement of charges. Electric current, whether it be for lighting a light bulb or changing the state of a computer memory element, is simply a controlled motion of charges through conducting materials. One of our goals will be to understand how charges move through electric circuits.

When we turn to magnetic behavior, we will start by observing how magnets stick to some metals but not others and how magnets affect compass needles. But our most important observation, which you may have seen, is that an electric current can affect a compass needle in exactly the same way as a magnet. This observation will suggest to us that there is a close connection between electricity and magnetism, and we will explore this relationship in detail. Our path will eventually lead to the discovery of electromagnetic waves.

Our goal in Part VI is to develop a theory that will explain the phenomena of electricity and magnetism. The linchpin of our theory will be the entirely new concept of a *field*. Electricity and magnetism are about the long-range interactions of charges, both static charges and moving charges, and the field concept will help us understand how these interactions take place. Much of our attention will be focused on the interplay between charges and fields: How fields are created by charges and how charges, in return, respond to the fields. Bit by bit, we will assemble a theory—based on the new concepts of electric and magnetic fields—that will allow us to understand, explain, and predict a wide range of electromagnetic behavior.

The Microscopic Model

There are two different aspects to the theory of electromagnetism. The field theory provides a macroscopic perspective on the phenomena, but we can also take a microscopic view. At the microscopic level, we want to know what charges are, how they are related to atoms and molecules, and how they move about through various kinds of materials. We will develop a microscopic model of electrons and ions moving in response to electric and magnetic forces. This microscopic perspective of electricity and magnetism is analogous to the kinetic theory of gases in our study of thermodynamics.

Likely *the* most important scientific discovery of the modern era is that matter consists of atoms. It was found near the end of the nineteenth century that the atoms themselves are not indestructible objects but, instead, have constituents that are *charged particles*. We know these today as electrons and protons. Much of the time these charged

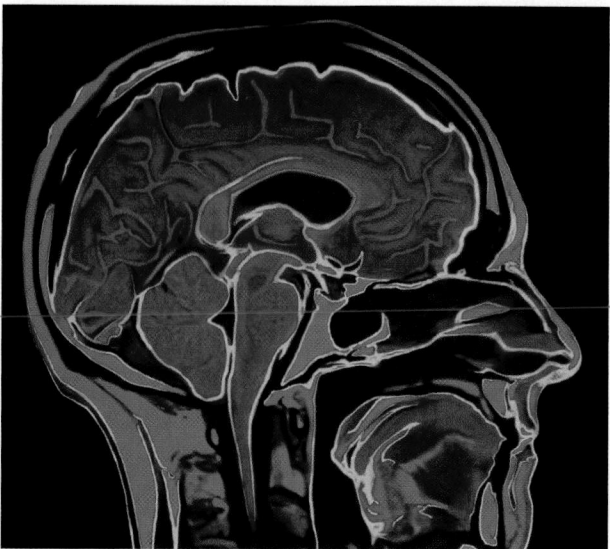

Magnetic-resonance imaging, or MRI, uses the magnetic properties of atoms as a non-invasive probe of the human body.

constituents all balance, and the atoms, as well as macroscopic objects built of these atoms, are electrically neutral. That has been the implicit assumption for all the physics we have done until now. But under some circumstances, the charges can become separated and move about. An important goal of our microscopic model will be to understand how charges become separated and how charged particles move through conductors as what we call a *current*.

But our interest at the microscopic level is more than simply how currents flow. Electricity and magnetism are of particular significance in our quest to understand the atomic structure of matter itself. The electric force is the "glue" that holds the atom together and that binds atoms into molecules and solids. The forces that we see in the macroscopic world as friction, tension, adhesion, and other contact forces are really electric forces acting at the atomic level. Magnetism plays a lesser, but not insignificant, role. While magnetism is not involved in the forces that bind atoms and molecules together, it does figure prominently in the macroscopic behavior of materials. In addition, the magnetic properties of atoms have become important probes into the interior of solids and liquids. Magnetic-resonance imaging in medicine is the most well known example of this technology, but many of the same techniques are used in science and engineering to characterize materials. For all these many reasons, acquiring a knowledge of electricity and magnetism is an essential part of science and engineering education.

25 Electric Charges and Forces

Lightning is a vivid manifestation of electric charges and forces.

▶ Looking Ahead

The goal of Chapter 25 is to develop a basic understanding of electric phenomena in terms of charges, forces, and fields. In this chapter you will learn to:

- Use a charge model to explain basic electric phenomena.
- Understand the electric properties of insulators and conductors.
- Use Coulomb's law to calculate the electric force between charges.
- Use a field model to explain the long-range interaction between charges.
- Calculate and display the electric field of a point charge.

◀ Looking Back

The mathematical analysis of electric forces and fields makes extensive use of vector addition. The electric force is in some ways analogous to gravity. Please review:

- Sections 3.2–3.4 Vector properties and vector addition.
- Sections 12.3–12.4 Newton's theory of gravity.

The electric force is one of the fundamental forces of nature. Sometimes, as in this lightning strike, electric forces can be wild and uncontrolled. On the other hand, controlled electricity is the cornerstone of our modern, technological society. Electric devices range from light bulbs and motors to computers and sophisticated medical equipment. Try imagining what it would be like to live without electricity!

But how do we control and manage this force? What are the properties of electricity and electric forces? How do we generate, transport, and use electricity? These are the questions we will explore throughout Part VI. Electricity is a big topic, and we cannot hope to answer all these questions at once.

A charged object, such as a comb that you've run through your hair, picks up small pieces of paper.

We will begin, in this chapter, by investigating some of the basic phenomena of electricity. It's hard to see what rubbing plastic rods with wool has to do with computers or generators, but only by starting at the very beginning, with simple observations, will we develop the understanding we need to use electricity in a controlled and precise manner.

25.1 Developing a Charge Model

You can receive a mildly unpleasant shock and produce a little spark if you touch a metal doorknob after walking across a carpet. Vigorously brushing your freshly washed hair makes all the hairs fly apart. A plastic comb that you've run through your hair will pick up bits of paper and other small objects, but a metal comb won't.

The common factor in these observations is that two objects are *rubbed* together. Why should rubbing an object cause forces and sparks? What kind of forces are these? Why do metallic objects behave differently from nonmetallic? These are the questions with which we begin our study of electricity.

Our first goal is to develop a model for understanding electric phenomena in terms of *charges* and *forces*. We will later use our contemporary knowledge of atoms to understand electricity on a microscopic level, but the basic concepts of electricity make *no* reference to atoms or electrons. The theory of electricity was well established long before the electron was discovered.

Experimenting with Charges

Let us enter a laboratory where we can make observations of electric phenomena. This is a modest laboratory, much like one you would have found in the year 1800. The major tools in the lab are:

- A variety of plastic, glass, and wood rods, each a few inches long. These can be held in your hand or suspended by threads from a support.
- A few metal rods with wood handles.
- Pieces of wool and silk.
- Small metal spheres, an inch or two in diameter, on wood stands.

We will manipulate and use these tools with the goal of developing a theory to explain the phenomena we see.

Discovering electricity I

Experiment 1	Experiment 2	Experiment 3	Experiment 4

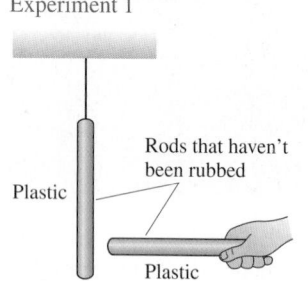

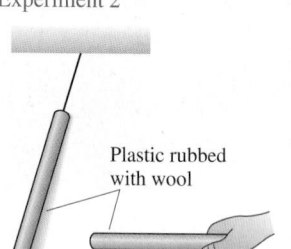

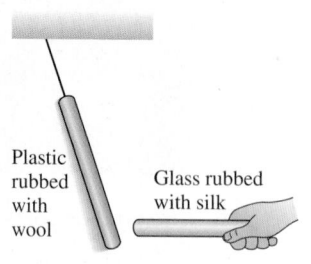

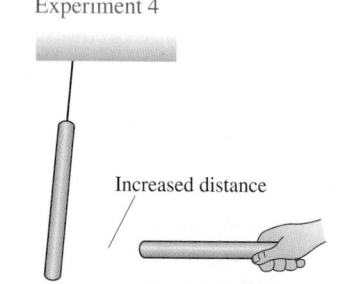

Take a plastic rod that has been undisturbed for a long period of time and hang it by a thread. Pick up another undisturbed plastic rod and bring it close to the hanging rod. Nothing happens to either rod.

Vigorously rub both the hanging plastic rod and the hand-held plastic rod with wool. Now the hanging rod tries to move away from the handheld rod when you bring the two close together. Rubbing two glass rods with silk produces the same result: The two rods repel each other.

Bring a glass rod that has been rubbed with silk close to a hanging plastic rod that has been rubbed with wool. These two rods *attract* each other.

Further observations show that:

- The strength of these forces is greater for rods that have been rubbed more vigorously.
- The strength of the forces decreases as the separation between the rods increases.

No forces were observed in Experiment 1. We will say that the original objects are **neutral.** Rubbing the rods (Experiments 2 and 3) somehow causes forces to be exerted between them. We will call the rubbing process **charging** and say that a rubbed rod is *charged.* For now, these are simply descriptive terms. The terms don't tell us anything about the process itself.

Experiment 2 shows that there is a *long-range repulsive force* between two identical objects that have been charged in the *same* way, such as two plastic rods both rubbed with wool. Furthermore, Experiment 4 finds that the force between two charged objects depends on the distance between them. This is the first long-range force we've encountered since gravity was introduced in Chapter 4. It is also the first time we've observed a repulsive force, so right away we see that new ideas will be needed to understand electricity.

Experiment 3 is a puzzle. Two rods *seem* to have been charged in the same way, by rubbing, but these two rods *attract* each other rather than repel. Why should the outcome of Experiment 3 differ from that of Experiment 2? Back to the lab.

Discovering electricity II

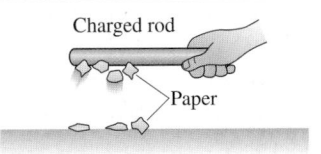

Experiment 5

Hold a charged (i.e., rubbed) plastic rod over small pieces of paper on the table. The pieces of paper leap up and stick to the rod. A charged glass rod does the same. However, a neutral rod has no effect on the pieces of paper.

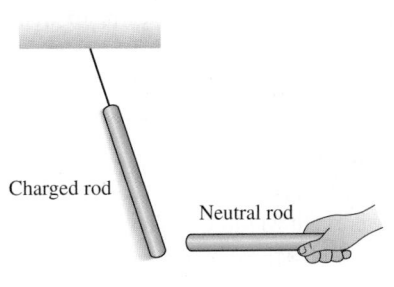

Experiment 6

Rub a plastic rod with wool and a glass rod with silk. Hang both by threads, some distance apart. Both rods are attracted to a *neutral* (i.e., unrubbed) plastic rod that is held close. Interestingly, both are also attracted to a *neutral* glass rod. In fact, the charged rods are attracted to *any* neutral object, such as a finger, a piece of paper, or a metal rod.

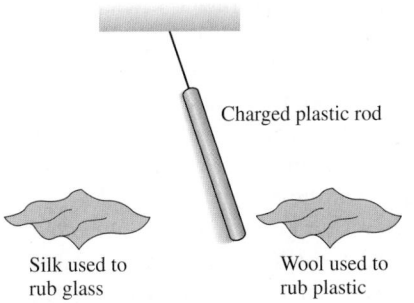

Experiment 7

Rub a hanging plastic rod with wool and then hold the *wool* close to the rod. The rod is weakly *attracted* to the wool. The plastic rod is *repelled* by a piece of silk that has been used to rub glass.

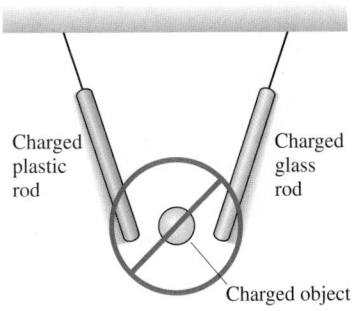

Experiment 8

Continued experiments show that:
- Other objects, after being rubbed, attract one of the hanging charged rods (plastic or glass) and repel the other. These objects always pick up small pieces of paper.
- There appear to be *no* objects that, after being rubbed, pick up pieces of paper and attract *both* the charged plastic and glass rods.

Our first set of experiments found that charged objects exert forces on each other. The forces are sometimes attractive, sometimes repulsive. Experiments 5 and 6 now show that there is an attractive force between a charged object and a *neutral* (uncharged) object. This discovery presents us with a bit of a problem: How can we tell if an object is charged or neutral? Because of the attractive force between a charged and a neutral object, simply observing an electric force does *not* imply that an object is charged.

However, an important characteristic of any *charged* object appears to be that **a charged object picks up small pieces of paper.** This behavior provides a straightforward test to answer the question, "Is this object charged?" An object that passes the test by picking up paper is charged; an object that fails the test is neutral.

These observations let us tentatively advance the first stages of a **charge model.** We can give our model these postulates:

1. Frictional forces, such as rubbing, add something called **charge** to an object or remove it from the object. The process itself is called *charging.* More vigorous rubbing produces a larger quantity of charge.
2. There are two and only two kinds of charge. For now we will call these "plastic charge" and "glass charge." Other objects can sometimes be charged by rubbing, but the charge they receive is either "plastic charge" or "glass charge."
3. Two **like charges** (plastic/plastic or glass/glass) exert repulsive forces on each other. Two **opposite charges** (plastic/glass), exert attractive forces on each other.
4. The force between two charges is a long-range force. The size of the force increases as the quantity of charge increases and decreases as the distance between the charges increases.
5. *Neutral* objects have an *equal mixture* of both "plastic charge" and "glass charge." The rubbing process somehow manages to separate the two.

Postulate 2 is based on Experiment 8. If an object is charged (i.e., picks up paper), it always attracts one charged rod and repels the other. That is, it acts either "like plastic" or "like glass." If there were a third kind of charge, different from the first two, an object with that charge should pick up paper and attract *both* the charged plastic and glass rods. No such objects have ever been found.

The basis for postulate 5 is the observation in Experiment 7 that a charged plastic rod is attracted to the wool used to rub it but repelled by silk that has rubbed glass. It appears that rubbing glass causes the silk to acquire "plastic charge." The easiest way to explain this is to hypothesize that the silk starts out with equal amounts of "glass charge" and "plastic charge" and that the rubbing somehow transfers "glass charge" from the silk to the rod. This leaves an excess of "glass charge" on the rod and an excess of "plastic charge" on the silk.

While the charge model is *consistent* with the observations, it is by no means proved. One could easily imagine other hypotheses that are just as consistent with the limited observations we have made so far. We still have some very large unexplained puzzles, such as why charged objects exert attractive forces on neutral objects.

STOP TO THINK 25.1 To determine if an object has "glass charge," you need to

a. See if the object attracts a charged plastic rod.
b. See if the object repels a charged glass rod.
c. Do both a and b.
d. Do either a or b.

Electric Properties of Materials

We still need to clarify how different types of materials respond to charges.

Discovering electricity III

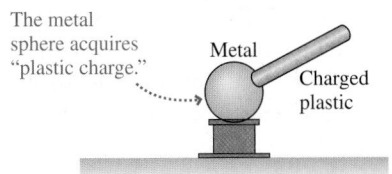

The metal sphere acquires "plastic charge."

Metal · Charged plastic

Experiment 9

Charge a plastic rod by rubbing it with wool. Touch a neutral metal sphere with the rubbed area of the rod. The metal sphere then picks up small pieces of paper and repels a charged, hanging plastic rod. The metal sphere appears to have acquired "plastic charge."

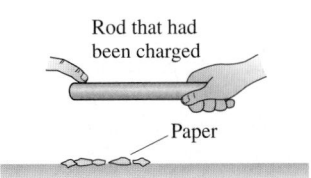

Rod that had been charged

Paper

Experiment 10

Charge a plastic rod, then run your finger along it. After you've done so, the rod no longer picks up small pieces of paper or repels a charged, hanging plastic rod. Similarly, the metal sphere of Experiment 9 no longer repels the plastic rod after you touch it with your finger.

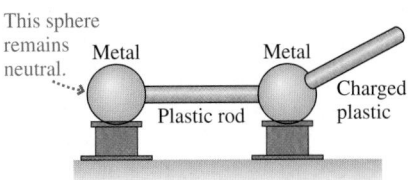

This sphere remains neutral.

Metal · Metal · Plastic rod · Charged plastic

Experiment 11

Place two metal spheres close together with a plastic rod connecting them. Charge a second plastic rod, by rubbing, and touch it to one of the metal spheres. Afterward, the metal sphere that was touched picks up small pieces of paper and repels a charged, hanging plastic rod. The other metal sphere does neither.

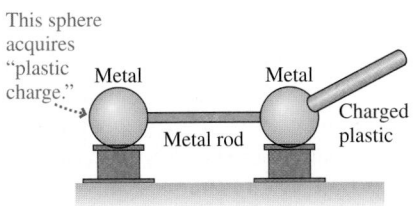

This sphere acquires "plastic charge."

Metal · Metal · Metal rod · Charged plastic

Experiment 12

Repeat Experiment 11 with a metal rod connecting the two metal spheres. Touch one metal sphere with a charged plastic rod. Afterward, *both* metal spheres pick up small pieces of paper and repel a charged, hanging plastic rod.

Our final set of experiments has shown that

- Charge can be *transferred* from one object to another, but only when the objects *touch*. Contact is required. Removing charge from an object, which you can do by touching it, is called **discharging.**
- There are two types or classes of materials with very different electric properties. We call these *conductors* and *insulators*.

Experiment 12, in which a metal rod is used, is in sharp contrast to Experiment 11. Charge somehow *moves through* or along a metal rod, from one sphere to the other, but remains *fixed in place* on a plastic or glass rod. Let us define **conductors** as those materials through or along which charge easily moves and **insulators** as those materials on or in which charges remain immobile. Glass and plastic are insulators; metal is a conductor.

This new information allows us to add two more postulates to our charge model:

6. There are two types of materials. Conductors are materials through or along which charge easily moves. Insulators are materials on or in which charges remain fixed in place.

7. Charge can be transferred from one object to another by contact.

NOTE ▶ Both insulators and conductors can be charged. They differ in the *mobility* of the charge. ◀

We have by no means exhausted the number of experiments and observations we might try. Early scientific investigators were faced with all of these results, plus many others. Moreover, many of these experiments are hard to reproduce with much accuracy. How should we make sense of it all? The charge model seems promising, but certainly not proven. We have not yet explained how charged objects exert attractive forces on *neutral* objects, nor have we explained what charge is, how it is transferred, or *why* it should move through some objects but not others. Nonetheless, we will take advantage of our historical hindsight and continue to pursue this model. Homework problems will let you practice using the model to explain other observations.

EXAMPLE 25.1 Transferring charge

In Experiment 12, touching one metal sphere with a charged plastic rod caused a second metal sphere to become charged with the same type of charge as the rod. Use the postulates of the charge model to explain this.

SOLVE We need the following ideas from the charge model:

1. Charge is transferred upon contact.
2. Metal is a conductor.
3. Like charges repel.

The plastic rod was charged by rubbing with wool. The charge doesn't move around on the rod, because it is an insulator, but some of the "plastic charge" is transferred to the metal upon contact. Once in the metal, which is a conductor, the charges are free to move around. Furthermore, because like charges repel, these plastic charges quickly move as far apart as they possibly can. Some move through the connecting metal rod to the second sphere. Consequently, the second sphere acquires "plastic charge."

25.2 Charge

As you probably know, the modern names for the two types of charge are *positive charge* and *negative charge.* You may be surprised to learn that the names were coined by Benjamin Franklin. Franklin was among the first to make quantitative studies of electric phenomena, as opposed to making qualitative observations, and he found that charge behaves like positive and negative numbers. If a plastic rod is charged twice, by rubbing, and twice transfers charge to a metal sphere, the electric forces exerted by the sphere are doubled. That is, $2 + 2 = 4$. But the sphere is found to be neutral after receiving equal amounts of "plastic charge" and "glass charge." This is like $2 + (-2) = 0$. These experiments establish an important property of charge.

So what is positive and what is negative? It's entirely up to us! Franklin established the convention that a glass rod that has been rubbed with silk is *positively* charged. That's it. Any other object that repels a charged glass rod is also positively charged. Any charged object that attracts a charged glass rod is negatively charged. It was only long afterward, with the discovery of electrons and protons, that electrons were found to be attracted to a charged glass rod while protons were repelled. Thus *by convention* electrons have a negative charge and protons a positive charge.

> **NOTE** ▶ In hindsight, it would have been better had Franklin made the opposite choice. Electrons are the carriers of electric currents in metals, and the convention of assigning a negative charge to electrons will later present us with some sign difficulties that could have been avoided with positive electrons. ◀

Atoms and Electricity

Now let's fast forward to the 21st century. The theory of electricity was developed without knowledge of atoms, but there is no reason for us to continue to overlook this important part of our contemporary perspective. For now, we will assert without proof some of the relevant characteristics of atoms and matter. You will have later opportunities to learn about the experimental evidence supporting these assertions.

Figure 25.1 shows that an atom consists of a very small and dense *nucleus* (diameter $\sim 10^{-14}$ m) surrounded by much less massive orbiting *electrons*. The electron orbital frequencies are so enormous ($\sim 10^{15}$ revolutions per second) that the electrons seem to form an **electron cloud** of diameter $\sim 10^{-10}$ m, a factor 10^4 larger than the nucleus. In fact, the wave-particle duality of quantum physics destroys any notion of a well-defined electron trajectory, and *all* we know of the electrons is the size and shape of the electron cloud.

Experiments at the end of the 19th century—experiments we will study in Part VII—revealed that electrons are *particles* with both mass and a negative charge. The nucleus is a composite structure consisting of *protons,* which are positively charged particles, and neutral *neutrons.* The atom is held together by the attractive electric force between the positive nucleus and the negative electrons.

One of the most important discoveries is that **charge, like mass, is an inherent property of electrons and protons.** It's no more possible to have an electron without charge than it is to have an electron without mass. As far as we know today, electrons and protons have charges of opposite sign but *exactly* equal magnitude. (Very careful experiments have never found any difference.) This atomic-level unit of charge, called the **fundamental unit of charge,** is represented by the symbol e. Table 25.1 shows the masses and charges of protons and electrons. We need to define a unit of charge, which we will do in Section 25.5, before we can specify how much charge e is.

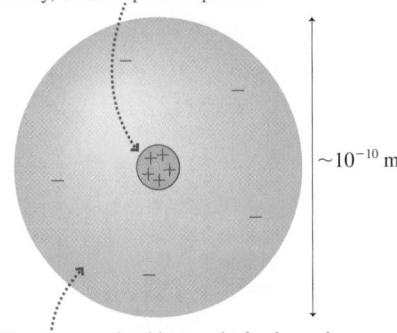

The nucleus, exaggerated for clarity, contains positive protons.

$\sim 10^{-10}$ m

The electron cloud is negatively charged.

FIGURE 25.1 An atom.

TABLE 25.1 Protons and electrons

Particle	Mass (kg)	Charge
Proton	1.67×10^{-27}	$+e$
Electron	9.11×10^{-31}	$-e$

The Micro/Macro Connection

Electrons and protons are the basic charges in nature. **There are no other sources of charge.** Consequently, the various observations we made in Section 25.1 need to be explained in terms of electrons and protons.

NOTE ▶ Electrons and protons are particles of matter. Their motion is governed by Newton's laws. Electrons can move from one object to another when the objects are in contact, but neither electrons nor protons can leap through the air from one object to another. An object does not become charged simply from being close to a charged object. ◀

Charge is represented by the symbol q (or sometimes Q). A macroscopic object, such as a plastic rod, has charge

$$q = N_p e - N_e e = (N_p - N_e)e \tag{25.1}$$

where N_p and N_e are the number of protons and electrons contained in the object. Most macroscopic objects have an *equal number* of protons and electrons and therefore have $q = 0$. An object with no *net* charge (i.e., $q = 0$) is said to be *electrically neutral.*

NOTE ▶ *Neutral* does *not* mean "no charges" but, instead, means that there is no *net* charge. A typical 1 cm^3 solid contains $\sim 10^{24}$ electrons and an equal number of protons. This is a tremendous number of charges, but most solids are electrically neutral or very close to it. A glass rod loses only $\sim 10^{10}$ electrons as it is charged by rubbing. This corresponds to only 1 electron out of every 10^{14}. ◀

A charged object has an unequal number of protons and electrons. An object is positively charged if $N_p > N_e$. It is negatively charged if $N_p < N_e$. Notice that an object's charge is always an integer multiple of e. That is, the amount of charge on

A positive ion with
net charge $q = +e$

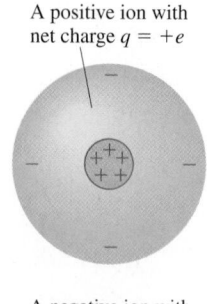

A negative ion with
net charge $q = -e$

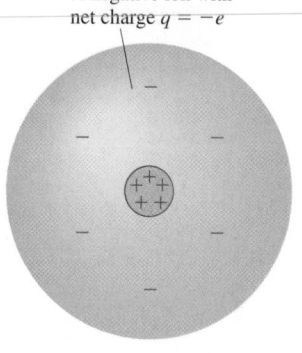

FIGURE 25.2 Positive and negative ions.

Electrically
neutral molecule

Atoms

Bond

Friction

These bonds were
broken by friction.

Positive
molecular
ion

Negative
molecular
ion

This half of the
molecule lost an
electron as the
bond broke.

This half of the
molecule gained an
extra electron as the
bond broke.

FIGURE 25.3 Charging by friction usually creates molecular ions as bonds are broken.

an object varies by small but discrete steps, not continuously. This is called **charge quantization.**

In practice, objects acquire a positive charge not by gaining protons, which you might expect, but by losing electrons. Protons are *extremely* tightly bound within the nucleus and cannot be added to or removed from atoms. Electrons, on the other hand, are bound rather loosely and can be removed without great difficulty. The process of removing an electron from the electron cloud of an atom is called **ionization.** An atom that is missing an electron is called a *positive ion.* Its *net* charge is $q = +e$.

It turns out that some atoms can accommodate an *extra* electron and thus become a *negative ion* with net charge $q = -e$. A saltwater solution is a good example. When table salt (the chemical sodium chloride, NaCl) dissolves, it separates into positive sodium ions Na^+ and negative chlorine ions Cl^-. Figure 25.2 shows positive and negative ions.

All the charging processes we observed in Section 25.1 involved rubbing and friction. The forces of friction cause molecular bonds at the surface to break as the two materials slide past each other. Molecules are electrically neutral, but Figure 25.3 shows that *molecular ions* can be created when one of the bonds in a large molecule is broken. The positive molecular ions remain on one material and the negative ions on the other, so one of the objects being rubbed ends up with a net positive charge and the other with a net negative charge. This is the way in which a plastic rod is charged by rubbing with wool or a comb is charged by passing through your hair.

Frictional charging via bond breaking works best with large organic molecules. This explains not only how plastic is charged by rubbing with wool but also such familiar experiences as the production of "static cling" in a clothes dryer. Metals usually can *not* be charged by rubbing them.

Charge Conservation and Charge Diagrams

One of the important discoveries about charge is the **law of conservation of charge:** Charge is neither created nor destroyed. Charge can be transferred from one object to another as electrons and ions move about, but the *total* amount of charge remains constant. For example, charging a plastic rod by rubbing it with wool transfers electrons from the wool to the plastic as the molecular bonds break. The wool is left with a positive charge equal in magnitude but opposite in sign to the negative charge of the rod: $q_{wool} = -q_{plastic}$. The *net* charge remains zero.

Diagrams are going to be an important tool for understanding and explaining charges and the forces on charged objects. As you begin to use diagrams, it will be important to make explicit use of charge conservation. The net number of plusses and minuses drawn on your diagrams should *not* change as you show them moving around.

TACTICS BOX 25.1 Drawing charge diagrams

❶ Draw a simplified two-dimensional cross section of the object.
❷ Draw *surface* charges *very close* to the object's boundary.
❸ Draw *interior* charges uniformly within the interior of the object.
❹ Show only the *net* charge. A neutral object should show *no* charges, not a lot of plusses and minuses.
❺ Conserve charge from one diagram to the next if you use a series of diagrams to explain a process.

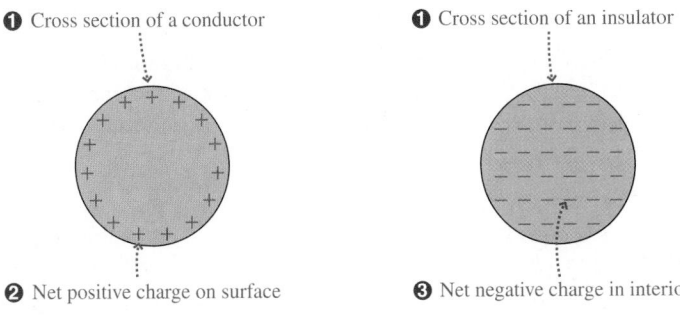

❶ Cross section of a conductor ❶ Cross section of an insulator

❷ Net positive charge on surface ❸ Net negative charge in interior

FIGURE 25.4 Charge diagrams.

Figure 25.4 shows two examples of charge diagrams: Step 5 will become clearer as you see it used in examples. Step 4 is especially important. For example, a positively charged object is missing electrons. Regardless of how the object became charged, the charge diagram should show plusses.

STOP TO THINK 25.2 Rank in order, from most positive to most negative, the charges q_a to q_e of these five systems.

Proton	Electron	17 protons 19 electrons	1,000,000 protons 1,000,000 electrons	Glass ball missing 3 electrons
•	•			○
(a)	**(b)**	**(c)**	**(d)**	**(e)**

25.3 Insulators and Conductors

You have seen that there are two classes of materials as defined by their electrical properties: insulators and conductors. It's time for a closer look at these materials.

Figure 25.5 looks inside an insulator and a metallic conductor. The electrons in the insulator are all tightly bound to the positive nuclei and not free to move around. Charging an insulator by friction leaves patches of molecular ions on the surface, but these patches are immobile.

In metals, the outer atomic electrons (called the *valence electrons* in chemistry) are only weakly bound to the nuclei. As the atoms come together to form a solid, these outer electrons become detached from their parent nuclei and are free to wander about through the entire solid. The solid *as a whole* remains electrically neutral, because we have not added or removed any electrons, but the electrons are now rather like a negatively charged gas or liquid—what physicists like to call a **sea of electrons**—permeating an array of positively charged **ion cores.**

The primary consequence of this structure is that electrons in a metal are highly mobile. They can quickly and easily move through the metal in response to electric forces. The motion of charges through a material is what we will later call a **current,** and the charges that physically move are called the **charge carriers.** The charge carriers in metals are electrons.

Metals aren't the only conductors. Ionic solutions, such as salt water, are also good conductors. But the charge carriers in an ionic solution are the ions, not electrons. We'll focus on metallic conductors because of their importance in applications of electricity.

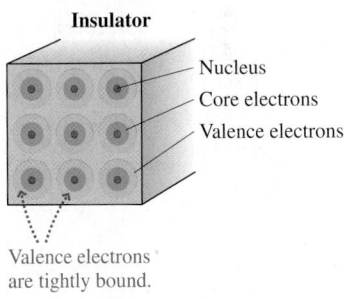

Insulator

Nucleus
Core electrons
Valence electrons

Valence electrons are tightly bound.

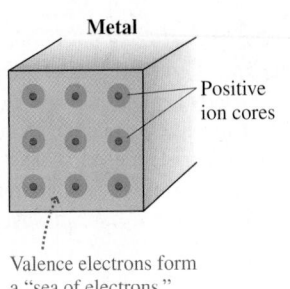

Metal

Positive ion cores

Valence electrons form a "sea of electrons."

FIGURE 25.5 A microscopic look at insulators and conductors.

Charging

Insulators are often charged by rubbing, as shown in Figure 25.6. The drawing shows that the charges on the rod are right at the surface and that charge is conserved. The charge on the rod is immobile. It can be transferred to another object upon contact, but it doesn't move around on the rod.

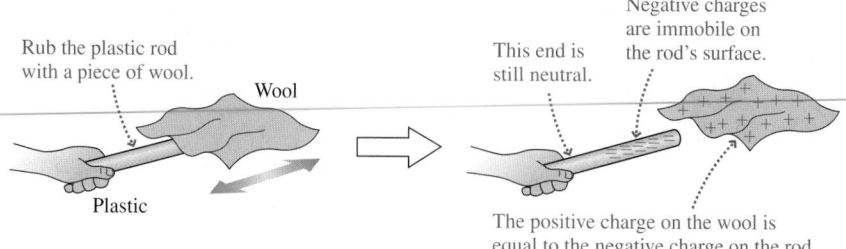

Rub the plastic rod with a piece of wool.

Wool

Plastic

This end is still neutral.

Negative charges are immobile on the rod's surface.

The positive charge on the wool is equal to the negative charge on the rod.

FIGURE 25.6 An insulating rod is charged by rubbing.

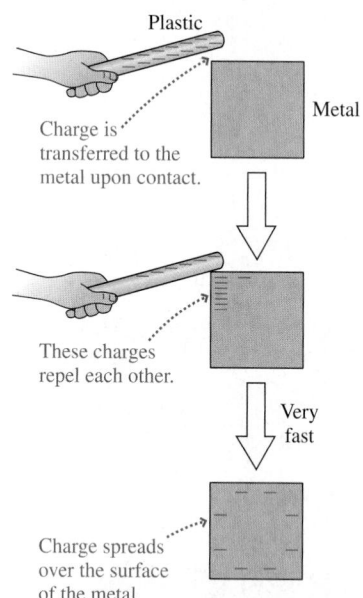

Plastic

Metal

Charge is transferred to the metal upon contact.

These charges repel each other.

Very fast

Charge spreads over the surface of the metal.

FIGURE 25.7 A conductor is charged by contact with a charged plastic rod.

Metals usually cannot be charged by rubbing, but Experiment 9 showed that a metal sphere can be charged by contact with a charged plastic rod. Figure 25.7 gives a pictorial explanation. An essential idea is that **the electrons in a conductor are free to move.** Once charge is transferred to the metal, repulsive forces between the negative charges cause the electrons to move apart from each other.

Note that the newly added electrons do not themselves need to move to the far corners of the metal. Because of the repulsive forces, the newcomers simply "shove" the entire electron sea a little to the side. The electron sea takes an extremely short time to adjust itself to the presence of the added charge, typically less than 10^{-9} s. For all practical purposes, a conductor responds *instantaneously* to the addition or removal of charge.

Other than this very brief interval during which the electron sea is adjusting, the charges in an *isolated* conductor are in static equilibrium. That is, the charges are at rest and there is no *net* force on any charge. This condition is called **electrostatic equilibrium.** If there *were* a net force on one of the charges, it would quickly move to an equilibrium point at which the force was zero.

Electrostatic equilibrium has an important consequence:

In an isolated conductor, any excess charge is located on the surface of the conductor.

To see this, suppose there *were* an excess electron in the interior of an isolated conductor. The extra electron would upset the electrical neutrality of the interior and exert forces on nearby electrons, causing them to move. But their motion would violate the assumption of static equilibrium, so we're forced to conclude that there cannot be any excess electrons in the interior. Because any excess electrons repel each other, they move as far apart as possible and spread out along the surface.

EXAMPLE 25.2 Charging an electroscope
Many electricity demonstrations are carried out with the help of an *electroscope* like the one shown in Figure 25.8. Touching the sphere at the top of an electroscope with a charged plastic rod causes the leaves to fly apart and remain hanging at an angle. Use charge diagrams to explain why.

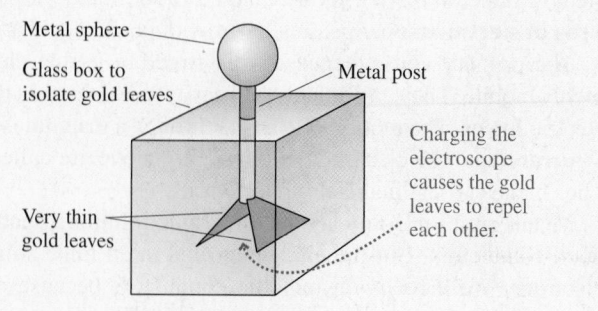

Metal sphere

Glass box to isolate gold leaves

Metal post

Charging the electroscope causes the gold leaves to repel each other.

Very thin gold leaves

FIGURE 25.8 A charged electroscope.

MODEL We'll use the charge model and the model of a conductor as a material through which electrons move.

VISUALIZE Figure 25.9 uses a series of charge diagrams to show the charging of an electroscope.

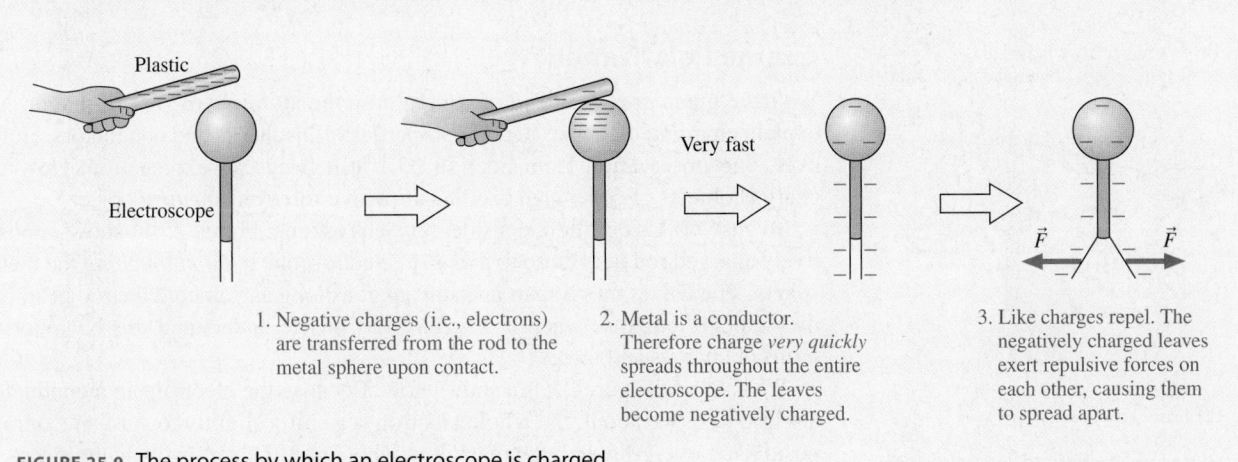

1. Negative charges (i.e., electrons) are transferred from the rod to the metal sphere upon contact.

2. Metal is a conductor. Therefore charge *very quickly* spreads throughout the entire electroscope. The leaves become negatively charged.

3. Like charges repel. The negatively charged leaves exert repulsive forces on each other, causing them to spread apart.

FIGURE 25.9 The process by which an electroscope is charged.

Discharging

Pure water is not a terribly good conductor, but nearly all water contains a variety of dissolved minerals that float around as ions. Dissolved table salt, as we noted previously, separates into Na^+ and Cl^- ions. These ions are the charge carriers, allowing salt water to be a fairly good conductor.

The human body consists largely of salt water. Consequently, and occasionally tragically, humans are reasonably good conductors. This fact allows us to understand how it is that *touching* a charged object discharges it, as we observed in Experiment 10. Figure 25.10 shows a person touching a positively charged metal, one that is missing electrons. Upon contact, some of the negative Cl^- ions on the skin surface transfer their extra electron to the metal, neutralizing both the metal and the chlorine atoms. This leaves the body with an excess of positive Na^+ ions and, thus, a net positive charge. As in any conductor, these excess positive charges quickly spread as far apart as possible over the surface of the conductor.

The net effect of touching a charged metal is that it and the conducting human together become a much larger conductor than the metal alone. Any excess charge that was initially confined to the metal can now spread over the larger metal + human conductor. This may not entirely discharge the metal, but in typical circumstances, where the human is much larger than the metal, the residual charge remaining on the metal is much reduced from the original charge. The metal, for most practical purposes, is discharged. In essence, two conductors in contact "share" the charge that was originally on just one of them.

Moist air is a conductor, although a rather poor one. Charged objects in air slowly lose their charge as the object shares its charge with the air. The earth itself is a giant conductor because of its water, moist soil, and a variety of ions—not, admittedly, as good a conductor as a piece of copper, but a conductor nonetheless. Any object that is physically connected to the earth through a conductor is said to be **grounded.** The effect of being grounded is that the object shares any excess charge it has with the entire earth! But the earth is so enormous that any conductor attached to the earth will be completely discharged.

The purpose of *grounding* objects, such as circuits and appliances, is to prevent the buildup of any charge on the objects. As you will see later, grounding has the effect of preventing a *voltage difference* between the object and the ground. The third prong on appliances and electronics that have a three-prong plug is the

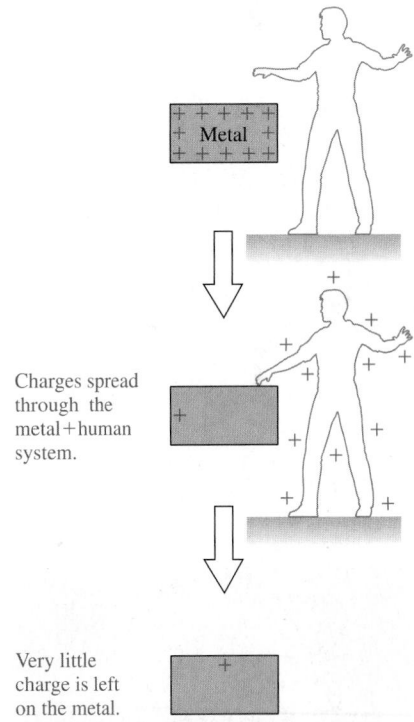

Charges spread through the metal + human system.

Very little charge is left on the metal.

FIGURE 25.10 Touching a charged metal discharges it.

ground connection. The building wiring physically connects that third wire deep into the ground somewhere just outside the building, often by attaching it to a metal water pipe that goes underground.

Charge Polarization

We have made great strides in learning how the atomic structure of matter can explain charging processes and the properties of insulators and conductors. However, one observation from Section 25.1 still needs an explanation. How do charged objects of either sign exert an attractive force on a *neutral* object?

To answer this question, consider an electroscope. Figure 25.11 shows a positively charged rod held close to a *neutral* electroscope *without touching* the metal sphere. The leaves move apart and stay apart as long as you hold the rod near, but they quickly collapse when it is removed. Can we understand this behavior in terms of charges and forces?

We can, and Figure 25.12a shows how. Because the electrons in a conductor are free to move about, the whole electron sea shifts slightly toward an external positive charge. Although the metal as a whole is still electrically neutral, we say that the object has been *polarized*. **Charge polarization** is a slight separation of the positive and negative charge in a neutral object.

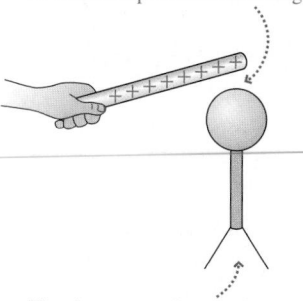

Bring a positively charged glass rod close to an electroscope without touching the sphere.

The electroscope is neutral, yet the leaves repel each other. Why?

FIGURE 25.11 A charged rod held close to an electroscope causes the leaves to repel each other.

(a)

The sea of electrons is attracted to the rod and shifts so that there is excess negative charge on the near surface.

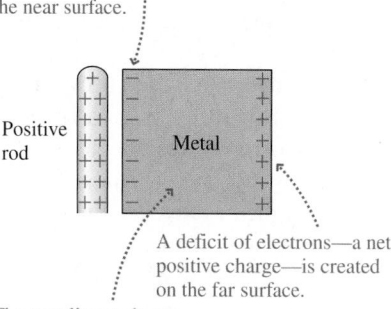

Positive rod

Metal

A deficit of electrons—a net positive charge—is created on the far surface.

The metal's net charge is still zero, but it has been *polarized* by the charged rod.

(b)

The electroscope is polarized by the charged rod. The sea of electrons shifts toward the rod.

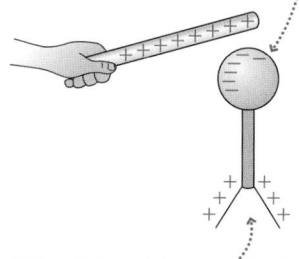

Although the net charge on the electroscope is still zero, the leaves have excess positive charge and repel each other.

FIGURE 25.12 A charged rod polarizes a metal.

Toner particles in a photocopy machine stick to charged *carrier beads* because of a polarization force. Later, the toner particles will be transferred to charged areas on a sheet of paper to form the photocopied image.

Charge polarization produces a net positive charge on the leaves of the electroscope shown in Figure 25.12b, so they repel each other. But because the electroscope has no *net* charge, the electron sea quickly readjusts to balance the positive ions once the rod is removed.

You might think that *all* the electrons in Figure 25.12a would rush to the side near the positive charge, but once the electron sea shifts slightly, the stationary positive ions begin to exert a force, a restoring force, pulling the electrons back to the right. The equilibrium position for the sea of electrons is just far enough to the left that the forces due to the external charge and the positive ions are in balance. In practice, the displacement of the electron sea is usually *less than 10^{-15} m*!

Charge polarization explains not only why the electroscope leaves deflect but also how a charged object exerts an attractive force on a neutral object. Figure 25.13 shows a positively charged rod near a neutral piece of metal. Because the electric force decreases with distance, the attractive force on the electrons at the top surface is *slightly greater* than the repulsive force on the ions. The net

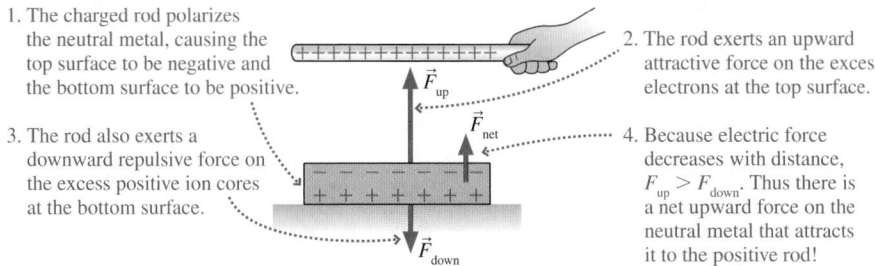

1. The charged rod polarizes the neutral metal, causing the top surface to be negative and the bottom surface to be positive.

2. The rod exerts an upward attractive force on the excess electrons at the top surface.

3. The rod also exerts a downward repulsive force on the excess positive ion cores at the bottom surface.

4. Because electric force decreases with distance, $F_{up} > F_{down}$. Thus there is a net upward force on the neutral metal that attracts it to the positive rod!

FIGURE 25.13 The polarization force on a neutral piece of metal is due to the slight charge separation.

force toward the charged rod is called a **polarization force.** The polarization force arises because the charges in the metal are separated, *not* because the rod and metal are oppositely charged.

A negatively charged rod would push the electron sea slightly away, polarizing the metal to have a positive upper surface charge and negative lower surface charge. Once again, these are the conditions for the charge to exert a *net attractive force* on the metal. Thus our charge model explains how a charged object of *either* sign attracts neutral pieces of metal.

The Electric Dipole

Now let's consider a slightly trickier situation. Why does a charged rod pick up paper, which is an insulator rather than a metal? First consider what happens if we bring a positive charge near an atom. As Figure 25.14a shows, the charge polarizes the atom. The electron cloud doesn't move far, because the force from the positive nucleus pulls it back, but the center of positive charge and the center of negative charge are now slightly separated.

(a)

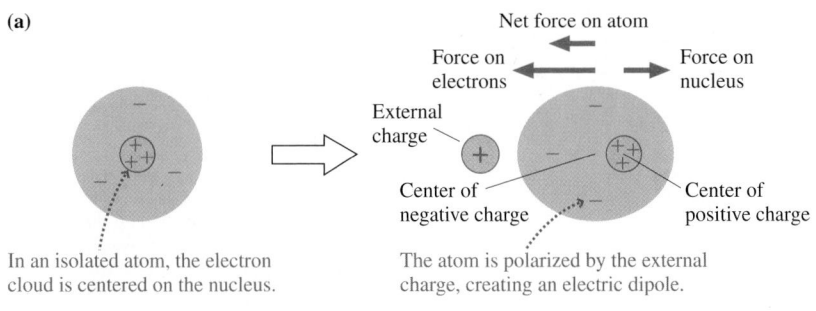

In an isolated atom, the electron cloud is centered on the nucleus.

Net force on atom

Force on electrons | Force on nucleus

External charge

Center of negative charge | Center of positive charge

The atom is polarized by the external charge, creating an electric dipole.

(b)

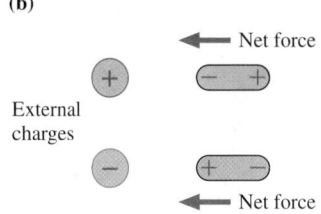

External charges

Net force

Net force

Electric dipoles can be created by either positive or negative charges. In both cases, there is an attractive net force toward the external charge.

FIGURE 25.14 A neutral atom is polarized by an external charge, forming an *electric dipole*.

Two opposite charges with a slight separation between them form what is called an **electric dipole.** Figure 25.14b shows that an external charge of either sign polarizes the atom to produce an electric dipole with the near end opposite in sign to the charge. (The actual distortion from a perfect sphere is minuscule, nothing like the distortion shown in the figure.) The attractive force on the dipole's near end *slightly* exceeds the repulsive force on its far end because the near end is closer to the charge. The net force, which is an *attractive* force between the charge and the atom, is another example of a polarization force.

An insulator has no sea of electrons to shift if an external charge is brought close. Instead, as Figure 25.15 shows, all the individual atoms inside the insulator become polarized. The polarization force acting *on each atom* produces a net

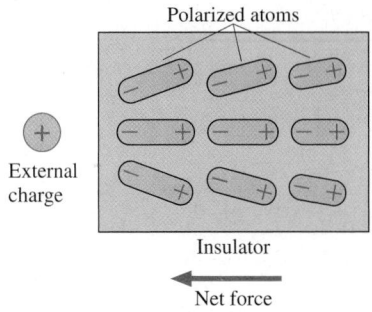

Polarized atoms

External charge

Insulator

Net force

FIGURE 25.15 The atoms in an insulator are polarized by an external charge.

polarization force toward the external charge. This solves the puzzle. A charged rod picks up pieces of paper by

- Polarizing the atoms in the paper,
- Then exerting an attractive polarization force on each atom.

This is a significant conclusion. Make sure you understand all the steps in the reasoning.

STOP TO THINK 25.3   An electroscope is positively charged by *touching* it with a positive glass rod. The electroscope leaves spread apart and the glass rod is removed. Then a negatively charged plastic rod is brought close to the top of the electroscope, but it doesn't touch. What happens to the leaves?

a. The leaves get closer together.   b. The leaves spread farther apart.
c. One leaf moves higher, the other lower.   d. The leaves don't move.

Charging by Induction

Charge polarization is responsible for an interesting and counterintuitive way of charging an electroscope. Figure 25.16 shows a positively charged glass rod held near an electroscope but not touching it, while a person touches the electroscope with a finger. Unlike what happens in Figure 25.11, the electroscope leaves do not move.

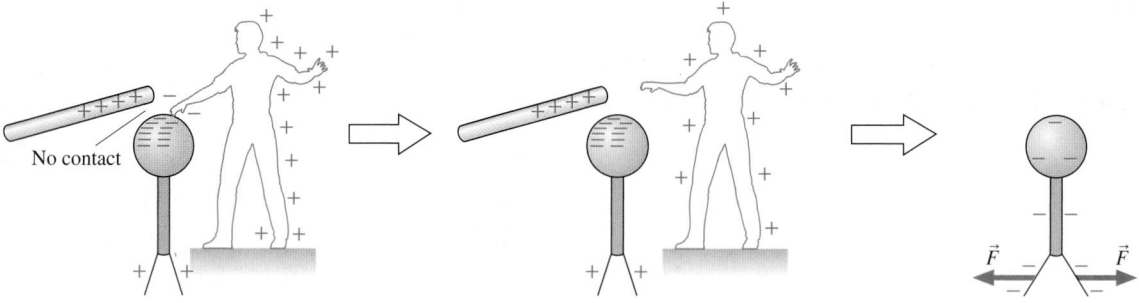

1. The charged rod polarizes the electroscope + person conductor. The leaves repel slightly, due to polarization within the electroscope, but overall the electroscope has an excess of electrons and the person has a deficit of electrons.

2. The negative charge on the electroscope is isolated when contact is broken.

3. When the rod is removed, the leaves first collapse, as the polarization vanishes, then repel as the excess negative charge spreads out. The electroscope has been *negatively* charged.

FIGURE 25.16 Charging by induction.

Charge polarization occurs, as it did in Figure 25.11, but this time in the much larger electroscope + person conductor. If the person removes his or her finger while the system is polarized, the electroscope is left with a *net* negative charge and the person has a net positive charge. When the rod is then removed, the electroscope leaves first collapse, as the polarization vanishes, then spread apart as the excess negative charge spreads out. The electroscope has been charged *opposite to the rod* in a process called **charging by induction.**

25.4 Coulomb's Law

The last few sections have established a *model* of charges and electric forces. This model is very good at explaining electric phenomena and providing a general understanding of electricity. Now we need to become quantitative. Experiment 4 in Section 25.1 found that the electric force increases for objects with more charge and decreases as charged objects are moved farther apart. The force law that describes this behavior is known as *Coulomb's law.*

Charles Coulomb was one of many scientists investigating electricity in the late 18th century. Coulomb had the idea of studying electric forces using the torsion balance scheme by which Cavendish had measured the value of the gravitational constant G (see Section 12.4). This was a difficult experiment. Cavendish's masses could be placed in position and did not change, but Coulomb was constantly having to recharge the ends of his balance. How could he do this reproducibly? How could he know if two objects were "equally charged"? How could he know for sure where the charge was located?

Despite these obstacles, Coulomb announced in 1785 that the electric force obeys an *inverse-square law* analogous to Newton's law of gravity. Historians of science debate whether Coulomb really discovered this law from his data or, perhaps, if he leapt to unwarranted conclusions because he so wanted his discovery to match that of the great Newton. Nonetheless, Coulomb's discovery or lucky guess, whichever it was, was subsequently confirmed, and the basic law of electric force bears his name.

Coulomb's law This law can be stated as follows:

1. If two charged particles having charges q_1 and q_2 are a distance r apart, the particles exert forces on each other of magnitude

$$F_{1 \text{ on } 2} = F_{2 \text{ on } 1} = \frac{K|q_1||q_2|}{r^2} \qquad (25.2)$$

where K is called the **electrostatic constant.** These forces are an action/reaction pair, equal in magnitude and opposite in direction.
2. The forces are directed along the line joining the two particles. The forces are *repulsive* for two like charges and *attractive* for two opposite charges.

We sometimes speak of the "force between charge q_1 and charge q_2," but keep in mind that we are really dealing with charged *objects* that also have a mass, a size, and other properties. Charge is not some disembodied entity that exists apart from matter. Coulomb's law describes the force between charged *particles,* which are also called **point charges.** A charged particle, which is an extension of the particle model we used in Part I, has a mass and a charge but has no size.

Coulomb's law looks much like Newton's law of gravity, but there is one important difference: The charge q can be either positive or negative. Consequently, the absolute value signs in Equation 25.2 are especially important. The first part of Coulomb's law gives only the *magnitude* of the force, which is always positive. The direction must be determined from the second part of the law. Figure 25.17 shows the forces between different combinations of positive and negative charges.

Units of Charge

Coulomb had no *unit* of charge, so he was unable to determine a value for K, whose numerical value depends upon the units of both charge and distance. The SI unit of charge, the **coulomb** (C), is derived from the SI unit of *current,* so we'll have to await the study of current in Chapter 28 before giving a precise definition. For now we'll note that the fundamental unit of charge e has been measured to have the value

$$e = 1.60 \times 10^{-19} \text{ C}$$

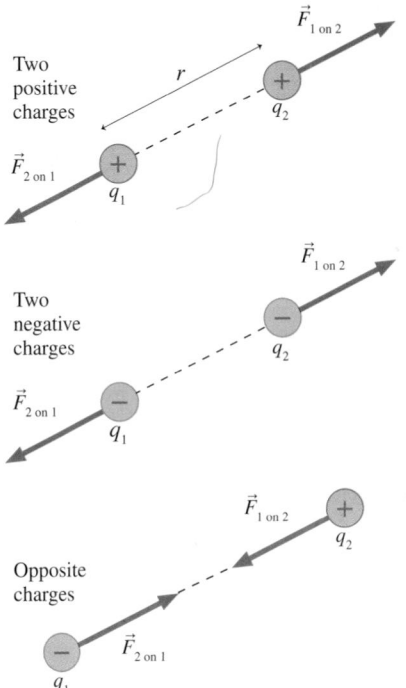

FIGURE 25.17 Attractive and repulsive forces between charges.

This is a very small amount of charge. Stated another way, 1 C is the net charge of roughly 6.25×10^{18} protons.

NOTE ▶ The amount of charge produced by rubbing plastic or glass rods is typically in the range 1 nC (10^{-9} C) to 100 nC (10^{-7} C). This corresponds to an excess or deficit of 10^{10} to 10^{12} electrons. ◀

Once the unit of charge is established, torsion balance experiments such as Coulomb's can be used to measure the electrostatic constant K. In SI units

$$K = 8.99 \times 10^9 \text{ N m}^2/\text{C}^2$$

It is customary to round this to $K = 9.0 \times 10^9$ N m^2/C^2 for all but extremely precise calculations, and we will do so.

Surprisingly, we will find that Coulomb's law is not explicitly used in much of the theory of electricity. While it *is* the basic force law, most of our future discussion and calculations will be of things called *fields* and *potentials*. It turns out that we can make many future equations easier to use if we rewrite Coulomb's law in a somewhat more complicated way. Let's define a new constant, called the **permittivity constant** ϵ_0 (pronounced "epsilon zero" or "epsilon naught"), as

$$\epsilon_0 = \frac{1}{4\pi K} = 8.85 \times 10^{-12} \text{ C}^2/\text{N m}^2$$

Rewriting Coulomb's law in terms of ϵ_0 gives us

$$F = \frac{1}{4\pi\epsilon_0} \frac{|q_1||q_2|}{r^2} \tag{25.3}$$

It will be easiest when using Coulomb's law directly to use the electrostatic constant K. However, in later chapters we will switch to the second version with ϵ_0.

Using Coulomb's Law

11.1–11.3   Activ
ONLINE
Physics

Coulomb's law is a force law, and forces are vectors. It has been many chapters since we made much use of vectors and vector addition, but these mathematical techniques will be essential in our study of electricity and magnetism. You may wish to review vector addition in Chapter 3.

There are three important observations regarding Coulomb's law:

1. **Coulomb's law applies only to point charges.** A point charge is an idealized material object with charge and mass but with no size or extension. For practical purposes, two charged objects can be modeled as point charges if they are much smaller than the separation between them.
2. Strictly speaking, **Coulomb's law applies only to electrostatics,** the electric forces between *static charges*. In practice, Coulomb's law is a good approximation to the electric force between two moving charged particles if their relative speed is much less than the speed of light.
3. **Electric forces, like other forces, can be superimposed.** If multiple charges 1, 2, 3, . . . are present, the *net* electric force on charge j due to all other charges is

$$\vec{F}_{\text{net}} = \vec{F}_{1 \text{ on } j} + \vec{F}_{2 \text{ on } j} + \vec{F}_{3 \text{ on } j} + \cdots \tag{25.4}$$

where each of the $\vec{F}_{i \text{ on } j}$ is given by Equation 25.2 or 25.3.

These conditions are the basis of a strategy for using Coulomb's law to solve electrostatic force problems.

PROBLEM-SOLVING STRATEGY 25.1 **Electrostatic forces and Coulomb's law**

MODEL Identify point charges or objects that can be modeled as point charges.

VISUALIZE Use a *pictorial representation* to establish a coordinate system, show the positions of the charges, show the force vectors on the charges, define distances and angles, and identify what the problem is trying to find. This is the process of translating words to symbols.

SOLVE The mathematical representation is based on Coulomb's law

$$F_{1 \text{ on } 2} = F_{2 \text{ on } 1} = \frac{K|q_1||q_2|}{r^2}$$

- Show the directions of the forces—repulsive for like charges, attractive for opposite charges—on the pictorial representation.
- When possible, do graphical vector addition on the pictorial representation. While not exact, it tells you the type of answer you should expect.
- Write each force vector in terms of its *x*- and *y*-components, then add the components to find the net force. Use the pictorial representation to determine which components are positive and which are negative.

ASSESS Check that your result has the correct units, is reasonable, and answers the question.

EXAMPLE 25.3 **The sum of two forces**

Two +10 nC charged particles are 2.0 cm apart on the *x*-axis. What is the net force on a +1.0 nC charge midway between them? What is the net force if the charged particle on the right is replaced by a −10 nC charge?

MODEL Model the charged particles as point charges.

VISUALIZE Figure 25.18 establishes a coordinate system and shows the forces $\vec{F}_{1 \text{ on } 3}$ and $\vec{F}_{2 \text{ on } 3}$.

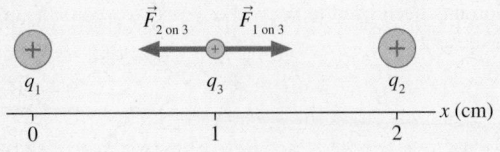

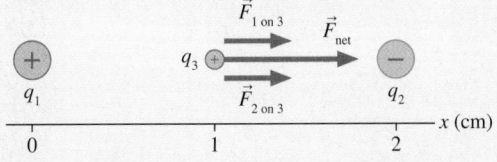

FIGURE 25.18 A pictorial representation of the charges and forces.

SOLVE Electric forces are vectors, and the net force on q_3 is the *vector* sum $\vec{F}_{\text{net}} = \vec{F}_{1 \text{ on } 3} + \vec{F}_{2 \text{ on } 3}$. Charges q_1 and q_2 each exert a repulsive force on q_3, but they are equal in magnitude and opposite in direction. Consequently, $\vec{F}_{\text{net}} = \vec{0}$. The situation changes if q_2 is negative. Now the two forces are equal in magnitude but in the *same* direction, so $\vec{F}_{\text{net}} = 2\vec{F}_{1 \text{ on } 3}$. The magnitude of the force is given by Coulomb's law:

$$F_{1 \text{ on } 3} = \frac{K|q_1||q_2|}{r_{13}^2}$$

$$= \frac{(9.0 \times 10^9 \text{ N m}^2/\text{C}^2)(10 \times 10^{-9} \text{ C})(1 \times 10^{-9} \text{ C})}{(0.010 \text{ m})^2}$$

$$= 9.0 \times 10^{-4} \text{ N}$$

Thus the net force on the 1.0 nC charge is $\vec{F}_{\text{net}} = 1.8 \times 10^{-3}\hat{\imath}$ N.

ASSESS This example illustrates the important idea that electric forces are *vectors*.

EXAMPLE 25.4 The point of zero force

Two positively charged particles q_1 and $q_2 = 3q_1$ are 10.0 cm apart. Where (other than at infinity) could a third charge q_3 be placed so as to experience no net force?

MODEL Model the charged particles as point charges.

VISUALIZE Figure 25.19 establishes a coordinate with q_1 at the origin. We first need to identify the region of space in which q_3 must be located. We have no information about the sign of q_3, so apparently the position for which we are looking will work for either sign. You can see from the figure that the forces at point A, above the axis, and at point B, outside the charges, cannot possibly add to zero. However, at some point C on the x-axis *between* the charges, the two forces will be oppositely directed.

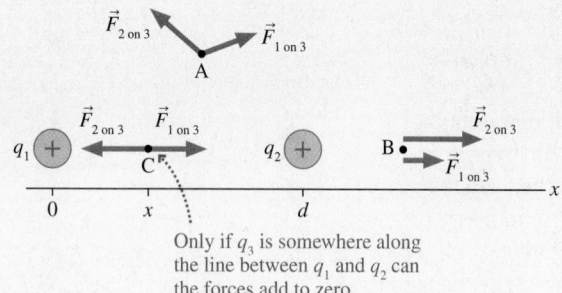

Only if q_3 is somewhere along the line between q_1 and q_2 can the forces add to zero.

FIGURE 25.19 A pictorial representation of the charges and forces.

SOLVE The mathematical problem is to find the position for which the forces $\vec{F}_{1\,\text{on}\,3}$ and $\vec{F}_{2\,\text{on}\,3}$ are equal in magnitude. If q_3 is distance x from q_1, it is distance $d - x$ from q_2. The *magnitudes* of the forces are

$$F_{1\,\text{on}\,3} = \frac{Kq_1|q_3|}{r_{13}^2} = \frac{Kq_1|q_3|}{x^2}$$

$$F_{2\,\text{on}\,3} = \frac{Kq_2|q_3|}{r_{23}^2} = \frac{K(3q_1)|q_3|}{(d-x)^2}$$

Charges q_1 and q_2 are positive and do not need absolute value signs. Equating the two forces gives

$$\frac{Kq_1|q_3|}{x^2} = \frac{3Kq_1|q_3|}{(d-x)^2}$$

The term $Kq_1|q_3|$ cancels. Multiplying by $x^2(d-x)^2$ gives

$$(d-x)^2 = 3x^2$$

which can be rearranged into the quadratic equation

$$2x^2 + 2dx - d^2 = 2x^2 + 20x - 100 = 0$$

where we used $d = 10$ cm and x is in cm. The solutions to this equation are

$$x = +3.66 \text{ cm} \quad \text{or} \quad -13.66 \text{ cm}$$

Both are points where the *magnitudes* of the two forces are equal, but $x = -13.66$ cm is a point where the magnitudes are equal but the directions are the same. The solution we want, which is between the charges, is $x = 3.66$ cm. Thus the point to place q_3 is 3.66 cm from q_1 along the line joining q_1 and q_2.

ASSESS q_1 is smaller than q_2, so we expect the point at which the forces balance to be closer to q_1 than to q_2. The solution seems reasonable. Note that the problem statement has no coordinates, so "$x = 3.66$ cm" is *not* an acceptable answer. You need to describe the position relative to q_1 and q_2.

EXAMPLE 25.5 Three charges

Three charged particles with $q_1 = -50$ nC, $q_2 = +50$ nC, and $q_3 = +30$ nC are placed on the corners of the 5.0 cm × 10.0 cm rectangle shown in Figure 25.20. What is the net force on charge q_3 due to the other two charges? Give your answer both in component form and as a magnitude and direction.

MODEL Model the charged particles as point charges.

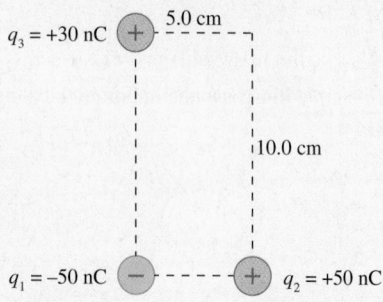

FIGURE 25.20 The three charges of Example 25.5.

VISUALIZE The pictorial representation of Figure 25.21 establishes a coordinate system. q_1 and q_3 are opposite charges, so force vector $\vec{F}_{1\,\text{on}\,3}$ is an attractive force toward q_1. q_2 and q_3 are like charges, so force vector $\vec{F}_{2\,\text{on}\,3}$ is a repulsive force away from q_2. q_1 and q_2 have equal magnitudes, but $\vec{F}_{2\,\text{on}\,3}$ has been drawn shorter than $\vec{F}_{1\,\text{on}\,3}$ because q_2 is farther from q_3. Vector addition has been used to draw the net force vector $\vec{F}_3$ and to define the angle ϕ.

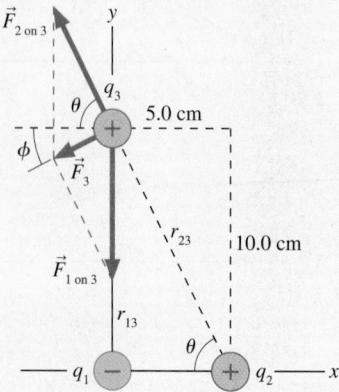

FIGURE 25.21 A pictorial representation of the charges and forces.

SOLVE The question asks for a *force,* so our answer will be the *vector* sum $\vec{F}_3 = \vec{F}_{1 \text{ on } 3} + \vec{F}_{2 \text{ on } 3}$. We need to write $\vec{F}_{1 \text{ on } 3}$ and $\vec{F}_{2 \text{ on } 3}$ in component form. The magnitude of force $\vec{F}_{1 \text{ on } 3}$ can be found using Coulomb's law:

$$F_{1 \text{ on } 3} = \frac{K|q_1||q_3|}{r_{13}^2}$$

$$= \frac{(9.0 \times 10^9 \text{ N m}^2/\text{C}^2)(50 \times 10^{-9} \text{ C})(30 \times 10^{-9} \text{ C})}{(0.100 \text{ m})^2}$$

$$= 1.35 \times 10^{-3} \text{ N}$$

where we used $r_{13} = 10$ cm. The pictorial representation shows that $\vec{F}_{1 \text{ on } 3}$ points down, in the negative y-direction, so

$$\vec{F}_{1 \text{ on } 3} = -1.35 \times 10^{-3} \hat{j} \text{ N}$$

To calculate $\vec{F}_{2 \text{ on } 3}$ we first need the distance r_{23} between the charges:

$$r_{23} = \sqrt{(5.0 \text{ cm})^2 + (10.0 \text{ cm})^2} = 11.18 \text{ cm}$$

The magnitude of $\vec{F}_{2 \text{ on } 3}$ is thus

$$F_{2 \text{ on } 3} = \frac{K|q_2||q_3|}{r_{23}^2}$$

$$= \frac{(9.0 \times 10^9 \text{ N m}^2/\text{C}^2)(50 \times 10^{-9} \text{ C})(30 \times 10^{-9} \text{ C})}{(0.1118 \text{ m})^2}$$

$$= 1.08 \times 10^{-3} \text{ N}$$

This is only a magnitude. The *vector* $\vec{F}_{2 \text{ on } 3}$ is

$$\vec{F}_{2 \text{ on } 3} = -F_{2 \text{ on } 3} \cos\theta \hat{i} + F_{2 \text{ on } 3} \sin\theta \hat{j}$$

where angle θ is defined in the figure and the signs (negative x-component, positive y-component) were determined from the pictorial representation. From the geometry of the rectangle,

$$\theta = \tan^{-1}\left(\frac{10.0 \text{ cm}}{5.0 \text{ cm}}\right) = \tan^{-1}(2.0) = 63.4°$$

Thus $\vec{F}_{2 \text{ on } 3} = (-4.83\hat{i} + 9.66\hat{j}) \times 10^{-4}$ N. Now we can add $\vec{F}_{1 \text{ on } 3}$ and $\vec{F}_{2 \text{ on } 3}$ to find

$$\vec{F}_3 = \vec{F}_{1 \text{ on } 3} + \vec{F}_{2 \text{ on } 3} = (-4.83\hat{i} - 3.84\hat{j}) \times 10^{-4} \text{ N}$$

This would be an acceptable answer for many problems, but sometimes we need the net force as a magnitude and direction. With angle ϕ as defined in the figure, these are

$$F_3 = \sqrt{F_{3x}^2 + F_{3y}^2} = 6.17 \times 10^{-4} \text{ N}$$

$$\phi = \tan^{-1}\left|\frac{F_{3y}}{F_{3x}}\right| = 38.5°$$

Thus $\vec{F}_3 = (6.17 \times 10^{-4} \text{ N}, 38.5°$ below the negative x-axis).

ASSESS The forces are not large, but they are typical of electrostatic forces. Even so, you'll soon see that these forces can produce very large accelerations because the masses of the charged objects are usually very small.

EXAMPLE 25.6 Lifting a glass bead

A small plastic sphere is charged to -10 nC. It is held 1.0 cm above a small glass bead at rest on a table. The bead has a mass of 15 mg and a charge of $+10$ nC. Will the glass bead "leap up" to the plastic sphere?

MODEL Model the plastic sphere and glass bead as point charges.

VISUALIZE Figure 25.22 establishes a y-axis, identifies the plastic sphere as q_1 and the glass bead as q_2, and shows a free-body diagram.

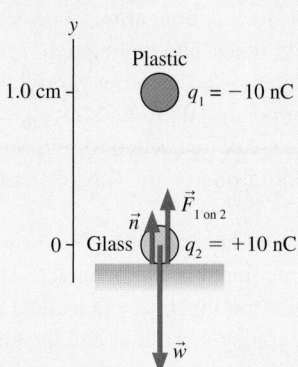

FIGURE 25.22 A pictorial representation of the charges and forces.

SOLVE If $F_{1 \text{ on } 2}$ is less than the bead's weight $w = m_2 g$, then the bead will remain at rest on the table with $\vec{F}_{1 \text{ on } 2} + \vec{w} + \vec{n} = \vec{0}$. But if $F_{1 \text{ on } 2}$ is greater than the bead's weight, the glass bead will accelerate upward from the table. Using the values provided,

$$F_{1 \text{ on } 2} = \frac{K|q_1||q_2|}{r^2} = 9.0 \times 10^{-3} \text{ N}$$

$$w = m_2 g = 1.5 \times 10^{-4} \text{ N}$$

$F_{1 \text{ on } 2}$ exceeds the weight by a factor of 60, so the glass bead will leap upward.

ASSESS The values used in this example are realistic for spheres ≈ 2 mm in diameter. In general, as in this example, electric forces are *significantly* larger than weight forces. Consequently, we can neglect weight forces when working electric-force problems unless the particles are fairly massive.

STOP TO THINK 25.4 Charges A and B exert repulsive forces on each other. $q_A = 4q_B$. Which statement is true?

a. $F_{A \text{ on } B} > F_{B \text{ on } A}$
b. $F_{A \text{ on } B} = F_{B \text{ on } A}$
c. $F_{A \text{ on } B} < F_{B \text{ on } A}$

A   B

25.5 The Concept of a Field

Electric and magnetic forces, like gravity, are *long-range forces.* No contact is required for one charged particle to exert a force on another charged particle. Somehow, in a process called *action at a distance,* the force is transmitted through empty space. The concept of action at a distance greatly troubled many of the leading thinkers of Newton's day, following the publication of his theory of gravity. Force, they believed, should have some *mechanism* by which it is exerted, and the idea of action at a distance, with no apparent mechanism, was more than most scientists could accept. Nonetheless, they could not dispute the success of Newton's theory.

The great prestige and success of Newton was able to keep scientists' doubts and reservations in check until the end of the 18th century, when investigations of electric and magnetic phenomena reopened the issue of action at a distance. For example, consider the charged particles A and B in Figure 25.23. If the particles have been at rest for a long period of time, then we can confidently use Coulomb's law to determine the force that A exerts on B. But suppose that A suddenly starts moving, as shown by the arrow. In response, the force vector on B must pivot to follow A. Does this happen *instantly?* Or is there some *delay* between when A moves and when the force $\vec{F}_{A \text{ on } B}$ responds?

Neither Coulomb's law nor Newton's law of gravity is dependent upon time, so the answer from the perspective of Newtonian physics has to be "instantly." Yet most scientists found this troubling. What if A is 100,000 light years from B? Will B respond *instantly* to an event 100,000 light years away? The idea of instantaneous transmission of forces was becoming unbelievable to most scientists by the beginning of the 19th century. But if there is a delay, how long is it? How does the information to "change force" get sent from A to B? These were the issues when a young Michael Faraday appeared on the scene.

Michael Faraday is one of the most interesting figures in the history of science. Born in 1791, the son of a poor blacksmith near London, Faraday was sent to work at an early age with almost no formal education. As a teenager, he found employment with a printer and bookbinder, and he began to read the books that came through the shop. By happenstance, a customer brought in a copy of the *Encyclopedia Britannica* to be rebound, and there Faraday discovered a lengthy article about electricity. It was all the spark he needed to set him on a course that, by his death in 1867, would make him one of the most esteemed scientists in Europe.

You will learn more about Faraday in later chapters. For now, suffice it to say that Faraday was never able to become fluent in mathematics. Apparently the late age at which he started his studies was too much of a detriment for mathematical learning. In place of mathematics, Faraday's brilliant and insightful mind developed many ingenious *pictorial* methods for thinking about and describing physical phenomena. By far the most important of these was the field.

Faraday was particularly impressed with the pattern that iron filings make when sprinkled around a magnet, as seen in Figure 25.24. The pattern's regularity

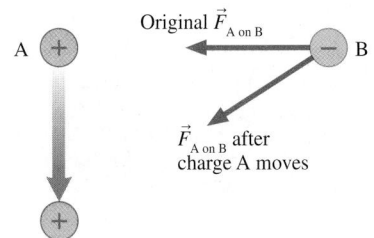

FIGURE 25.23 If charge A moves, how long does it take the force vector on B to respond?

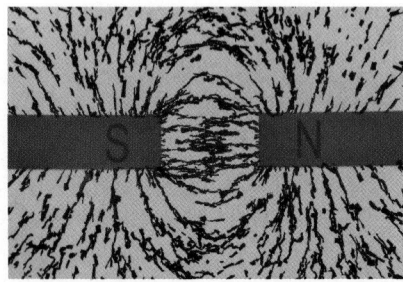
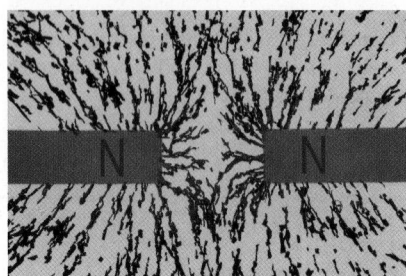

FIGURE 25.24 Iron filings sprinkled around the ends of a magnet suggest that the influence of the magnet extends into the space around it.

and the curved lines suggested to Faraday that the *space itself* around the magnet is filled with some kind of magnetic influence. Perhaps the magnet in some way alters the space around it. In this view, a piece of iron near the magnet responds not directly to the magnet but, instead, to the alteration of space caused by the magnet. This space alteration, whatever it is, is the *mechanism* by which the long-range force is exerted.

Figure 25.25 illustrates Faraday's idea. The Newtonian view was that A and B interact directly. In Faraday's view, A first alters or modifies the space around it, and Object B then comes along and interacts with this altered space. The alteration of space becomes the *agent* by which A and B interact. Furthermore, this alteration could easily be imagined to take a finite time to propagate outward from A, perhaps in a wave-like fashion. If A changes, B responds only when the new alteration of space reaches it. The interaction between B and this alteration of space is a *local* interaction, rather like a contact force.

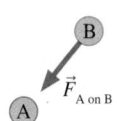

In the Newtonian view, A exerts a force directly on B.

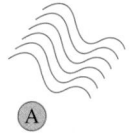

In Faraday's view, A alters the space around it. (The wavy lines are poetic license. We don't know what the alteration looks like.)

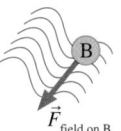

Particle B then responds to the altered space. The altered space is the agent that exerts the force on B.

FIGURE 25.25 Newton's and Faraday's ideas about long-range forces.

Faraday's idea came to be called a **field.** The term *field,* which comes from mathematics, describes a function $f(x, y, z)$ that assigns a value to every point in space. When used in physics, a field conveys the idea that the physical entity exists at every point in space. That is, indeed, what Faraday was suggesting about how long-range forces operate. The charge makes an alteration *everywhere* in space. Other charges then respond to the alteration at their position. The alteration of the space around a mass is called the **gravitational field.** Similarly, the space around a charge is altered to create the **electric field.**

NOTE ▶ The concept of a field is in sharp contrast to the concept of a particle. A particle exists at *one* point in space. The purpose of Newton's laws of motion is to determine how the particle moves from point to point along a trajectory. A field exists simultaneously at *all* points in space. A wave is an example of a field, although we didn't use the term during our study of waves. ◀

Faraday proposed a novel way to think about how one object exerts forces on another, but his idea was not taken seriously at first. It seemed too vague and non-mathematical to scientists steeped in the Newtonian tradition of particles and forces. But the significance of the concept of field grew as electromagnetic theory developed during the first half of the 19th century. What seemed at first a pictorial "gimmick" came to be seen as more and more essential for understanding electric and magnetic forces.

Faraday's field ideas were finally placed on a firm mathematical foundation in 1865 by James Clerk Maxwell, a young Scottish physicist possessing both great physical insight and mathematical ability. Maxwell was able to describe completely all the known behaviors of electric and magnetic fields in four equations, known today as Maxwell's equations. We will explore aspects of Maxwell's theory as we go along, then look at the full implications of Maxwell's equations in Chapter 34.

I have preferred to seek an explanation [of electric and magnetic phenomena] by supposing them to be produced by actions which go on in the surrounding medium as well as in the excited bodies, and endeavoring to explain the action between distant bodies without assuming the existence of forces capable of acting directly, . . . The theory I propose may therefore be called a theory of the Electromagnetic Field because it has to do with the space in the neighborhood of the electric and magnetic bodies.

James Clerk Maxwell, 1865

An Example of a Field: The Gravitational Field

Before we dive into investigating the electric field, we will take a slight detour to examine the gravitational field. The field concept is not as necessary for understanding the force of gravity as it will be for electric and magnetic forces.

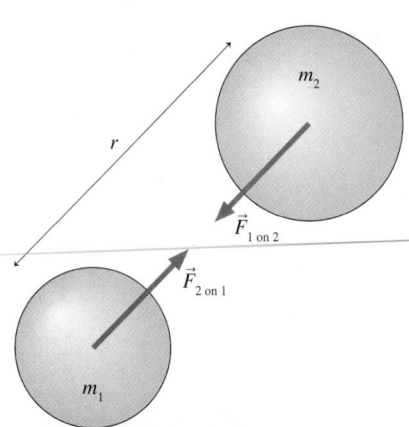

FIGURE 25.26 The gravitational forces between two masses.

Nonetheless, it will be useful to introduce the mathematical statement of the field concept in a familiar context before venturing into new and unexplored territory.

Figure 25.26 reminds you of Newton's law of gravity, which says that masses 1 and 2 exert equal but opposite attractive forces on each other of magnitude

$$F_{1 \text{ on } 2} = F_{2 \text{ on } 1} = \frac{Gm_1m_2}{r^2} \tag{25.5}$$

where G is the gravitational constant and r is the distance between their centers of mass. Let's focus our attention on the force that mass 1 exerts on mass 2. Because force is a vector, we can write the force on mass 2 as

$$\vec{F}_{1 \text{ on } 2} = \left(\frac{Gm_1m_2}{r^2}, \text{ toward mass 1}\right) = m_2\left(\frac{Gm_1}{r^2}, \text{ toward mass 1}\right) \tag{25.6}$$

On the right side we separated out the mass m_2 on which the force is exerted.

Faraday's hypothesis, applied to gravity, is that the *gravitational field* of mass 1 exerts the force on mass 2. Suppose we define the gravitational field, which we will give the symbol $\vec{g}$, so that

$$\vec{F}_{\text{on } m} = m\vec{g} \tag{25.7}$$

In other words, $\vec{F}_{\text{on } m}$ is the gravitational force exerted on a mass m placed at a point in space where, due to the presence of *other* masses, the gravitational field is $\vec{g}$. The field exists at this point in space whether or not mass m is present; mass m simply *responds* to the presence of the field. Equation 25.7 is the mathematical representation of Faraday's idea that **the field is the agent that exerts the force.**

NOTE ▶ Equation 25.7 defines the gravitational field to be a *vector,* a quantity with both a magnitude and a direction. ◀

Comparing Equations 25.6 and 25.7 shows you that the gravitational field of mass 1 is

$$\vec{g}_{\text{mass } 1} = \left(\frac{Gm_1}{r^2}, \text{ toward mass 1}\right) \tag{25.8}$$

Equation 25.8 tells us that there is a gravitational field at *every point* in space around mass 1, regardless of whether or not another mass is present. At any particular point,

1. The field strength g is proportional to the mass m_1. Larger masses create stronger gravitational fields.
2. The field strength is inversely proportional to the square of the distance from mass 1. The field is weaker at larger distances, but there is never a point at which the field is exactly zero. The gravitational field of a mass extends through *all* of space.
3. The field points directly toward mass 1.

NOTE ▶ The gravitational field is determined by mass 1, the object that *creates* the field. The value m_2 of the mass that *experiences* the field does not enter into Equation 25.8. ◀

Figure 25.27 is a pictorial representation of the gravitational field of mass 1. It shows the vector $\vec{g}$ at just a *few* points in space. It is impossible to draw a vector at every point, but you should keep in mind that such a vector *is* at every point, whether shown or not.

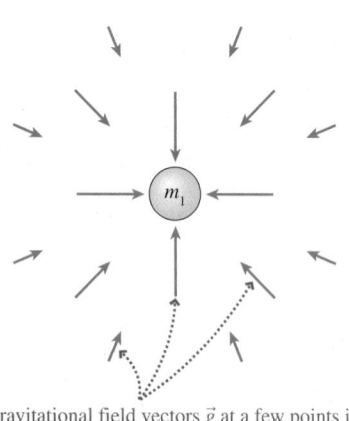

Gravitational field vectors $\vec{g}$ at a few points in space near mass m_1. All point toward A and have a magnitude inversely proportional to the square of the distance from the mass.

FIGURE 25.27 A pictorial representation of the gravitational field around mass 1.

EXAMPLE 25.7 The gravitational field of the sun
What is the sun's gravitational field at the position of the earth? Use the gravitational field to find the force of the sun on the earth.

MODEL The earth responds to the sun's gravitational field.

VISUALIZE Figure 25.27 shows what the sun's gravitational field looks like.

SOLVE The sun's gravitational field is given by Equation 25.8:

$$\vec{g} = \left(\frac{Gm_{sun}}{r_e^2}, \text{ toward sun}\right)$$

where r_e is the radius of the earth's orbit about the sun. Table 12.2 gives $m_{sun} = 1.99 \times 10^{30}$ kg and $r_e = 1.50 \times 10^{11}$ m. A straightforward calculation then gives

$$\vec{g} = (0.0059 \text{ N/kg, toward sun})$$

According to Equation 25.7, the force that this gravitational field exerts on the earth is

$$\vec{F}_{on \ earth} = m_e\vec{g} = (5.98 \times 10^{24} \text{ kg})(0.0059 \text{ N/kg, toward sun})$$
$$= (3.53 \times 10^{22} \text{ N, toward sun})$$

ASSESS It follows from the definition of $\vec{g}$ in Equation 25.7 that the units of $\vec{g}$ are N/kg.

The concept of a gravitational field is not limited to the gravitational field of a single mass. Figure 25.28 shows a mass placed in an arbitrary gravitational field $\vec{g}$ represented by the vector arrows. This is *not* the gravitational field of a single mass. In fact, we have no idea what masses are the source of this field because they are not shown. Nonetheless, knowing the *field* allows us to determine that the gravitational force exerted on this mass is $\vec{F}_{on \ m} = m\vec{g}$. The force is in the same direction as $\vec{g}$.

NOTE ▶ The field exists at the point where the mass is placed even though no field vector is drawn at that exact point. The field exists at *all* points in space. ◀

Our choice of the symbol $\vec{g}$ to represent the gravitational field was not arbitrary. Consider the situation near the surface of the earth. The gravitational force on mass m is then a simpler

$$\vec{F}_{on \ m} = -mg\hat{j} \qquad (25.9)$$

where the minus sign indicates a force vector pointing in the downward direction. This g is the familiar acceleration due to gravity, $g = 9.80$ m/s². According to Equation 25.7, the gravitational field just above the earth's surface is

$$\vec{g} = \frac{\vec{F}_{on \ m}}{m} = -g\hat{j} \qquad (25.10)$$

In other words, the gravitational field near the earth's surface is a downward vector with magnitude g. Our old friend g is nothing other than the magnitude of the gravitational field!

Figure 25.29 shows the gravitational field near the surface of the earth. All the vectors are of equal length $g = 9.8$ N/kg $= 9.8$ m/s², and they all point straight down. A mass m *responds* to the field by experiencing the force $\vec{F} = m\vec{g}$. It is important to keep in mind that mass m does not *cause* the field. The field is created by the earth and is there whether mass m is present or not.

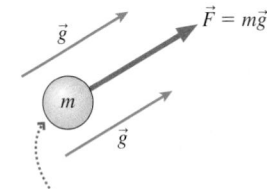

The gravitational field exerts a force on the mass even though no field vector is shown at this point.

FIGURE 25.28 A mass in a gravitational field experiences a force.

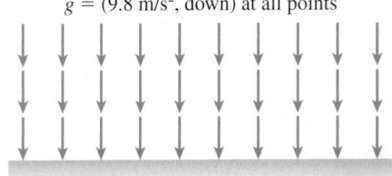

$\vec{g} = (9.8 \text{ m/s}^2, \text{ down})$ at all points

FIGURE 25.29 The gravitational field near the earth's surface.

25.6 The Field Model

Many of the electric phenomena you have seen in this chapter are more complex than gravitational phenomena. One reason is that whereas there is only one kind of mass, there are two kinds of charge, and thus both attractive and repulsive forces. Another is that there are two types of materials, conductors and insulators, with very different mobility for charge. These factors alone will make Faraday's concept of field useful for the study of electricity.

However, there's a more fundamental reason for introducing the electric field. For purely static situations, where electric and magnetic fields do not change with time, we *could* dispense with the concept of a field and base all our calculations on Coulomb's law. But later, when we get to the electro*dynamics* of fields that change with time, we'll find phenomena that can be understood *only* in terms of fields.

We begin our investigation of electric fields by postulating a **field model** that describes how two charges interact:

1. Some charges, which we will call the **source charges,** alter the space around them by creating an *electric field* $\vec{E}$.
2. A separate charge *in* the electric field experiences a force $\vec{F}$ exerted *by the field.*

We must accomplish two tasks to make this a useful model of electric interactions. First, we must learn how to calculate the electric field for a configuration of source charges. Second, we must determine the forces on and the motion of a charge in the electric field.

Suppose charge q experiences an electric force $\vec{F}_{\text{on } q}$ due to other charges. The strength and direction of this force vary from point to point in space, so $\vec{F}_{\text{on } q}$ is a continuous function of the charge's coordinates (x, y, z). This suggests that "something" is present at each point in space to cause the force that charge q experiences. Using Equation 25.7 for the gravitational field as an analogy, let us define the electric field $\vec{E}$ at the point (x, y, z) as

$$\vec{E}(x, y, z) = \frac{\vec{F}_{\text{on } q} \text{ at } (x, y, z)}{q} \tag{25.11}$$

We're *defining* the electric field as a force-to-charge ratio, hence the units of the electric field are newtons per coulomb, or N/C. The magnitude E of the electric field is called the **electric field strength.**

You can think of using charge q as a *probe* to determine if an electric field is present at a point in space. If charge q experiences an electric force at a point in space, as Figure 25.30a shows, we say that there is an electric field at that point causing the force. Further, we *define* the electric field at that point to be the vector given by Equation 25.11. Figure 25.30b shows the electric field at two points, but you can imagine "mapping out" the electric field by moving charge q all through space.

The basic idea of the field model is that **the field is the agent that exerts an electric force on charge** q. Notice three important things about the field:

1. Equation 25.11 assigns a *vector* to *every point* in space. That is, the electric field is a *vector field.* Electric field diagrams will show a sample of the vectors, but there is an electric field vector at every point whether one is shown or not.
2. If q is positive, the electric field vector points in the same direction as the force on the charge.
3. Because q appears in Equation 25.11, it may seem that the electric field depends on the size of the charge used to probe the field. It doesn't. We know from Coulomb's law that the force $\vec{F}_{\text{on } q}$ is proportional to q. Thus the electric field defined in Equation 25.11 is *independent* of the charge q that probes the field. The electric field depends only on the source charges that create the field.

In practice we will usually want to turn Equation 25.11 around and find the force exerted by a known field. That is, a charge q at a point in space where the electric field is $\vec{E}$ experiences an electric force

$$\vec{F}_{\text{on } q} = q\vec{E} \tag{25.12}$$

If q is positive, the force on charge q is in the direction of $\vec{E}$. The force on a negative charge is *opposite* the direction of $\vec{E}$.

(a)

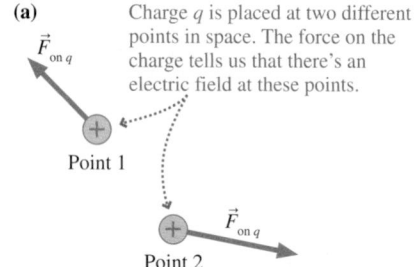

Charge q is placed at two different points in space. The force on the charge tells us that there's an electric field at these points.

Point 1

Point 2

(b)

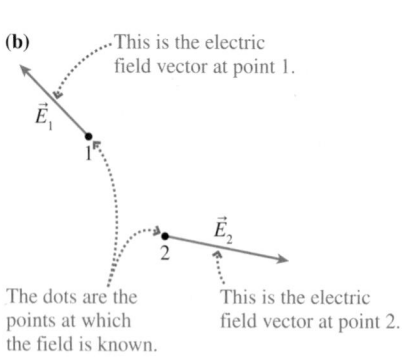

This is the electric field vector at point 1.

This is the electric field vector at point 2.

The dots are the points at which the field is known.

FIGURE 25.30 Charge q is a probe of the electric field.

STOP TO THINK 25.5   An electron is placed at the position marked by the dot. The force on the electron is

a. Zero.
b. To the right.
c. To the left.
d. There's not enough information to tell.

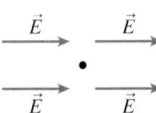

The Electric Field of a Point Charge

We will begin to put the definition of the electric field to full use in the next chapter. For now, to develop the ideas, we will determine the electric field of a single point charge q. Figure 25.31a shows charge q and a point in space at which we would like to know the electric field. We need a second charge, shown as q' in Figure 25.31b, to serve as a probe of the electric field.

For the moment, assume both charges are positive. The force on q', which is repulsive and directed straight away from q, is given by Coulomb's law:

$$\vec{F}_{\text{on } q'} = \left(\frac{1}{4\pi\epsilon_0} \frac{qq'}{r^2}, \text{ away from } q \right) \tag{25.13}$$

It's customary to use $1/4\pi\epsilon_0$ rather than K for field calculations. Equation 25.11 defined the electric field in terms of the force on a probe charge, thus the electric field at this point is

$$\vec{E} = \frac{\vec{F}_{\text{on } q'}}{q'} = \left(\frac{1}{4\pi\epsilon_0} \frac{q}{r^2}, \text{ away from } q \right) \tag{25.14}$$

The electric field is shown in Figure 25.31c.

NOTE ▶ The expression for the electric field is similar to Coulomb's law. To distinguish the two, remember that Coulomb's law has a product of two charges in the numerator. It describes the force between *two* charges. The electric field has a single charge in the numerator. It is the field of *a* charge. ◀

The field *strength* at distance r from a point charge depends inversely on the square of the distance: $E = q/4\pi\epsilon_0 r^2$. In Figure 25.32a, the field strength E_1 is larger than the field strength E_2 because $r_1 < r_2$. If we calculate the field at a sufficient number of points, we can draw a **field diagram** such as the one shown in Figure 25.32b. Notice that the field vectors all point straight away from charge q.

(a)

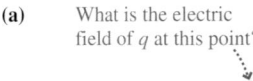

What is the electric field of q at this point?

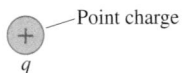

Point charge

(b)

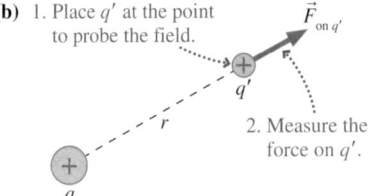

1. Place q' at the point to probe the field.

2. Measure the force on q'.

(c)

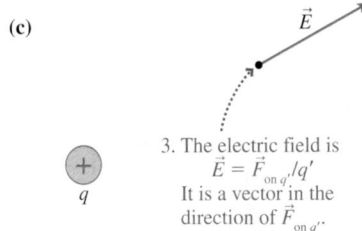

3. The electric field is $\vec{E} = \vec{F}_{\text{on } q'}/q'$. It is a vector in the direction of $\vec{F}_{\text{on } q'}$.

FIGURE 25.31 Charge q' is used to probe the electric field of point charge q.

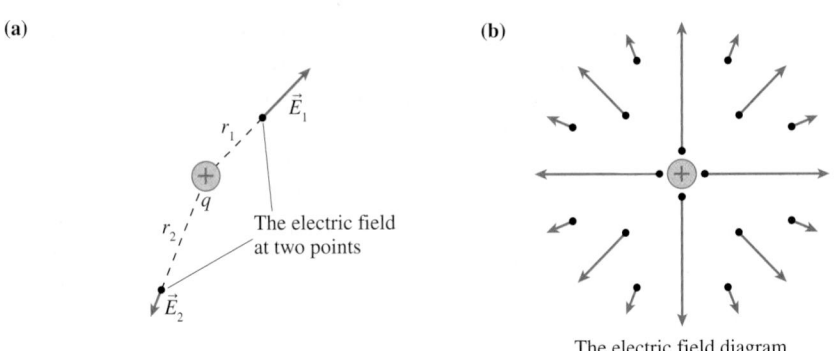

The electric field at two points

The electric field diagram of a positive point charge

FIGURE 25.32 The electric field near a positive charge.

11.4 Activ ONLINE Physics

Keep these three important points in mind when using field diagrams:

1. The diagram is just a representative sample of electric field vectors. The field exists at all the other points. A well-drawn diagram can tell you fairly well what the field would be like at a neighboring point.

2. The arrow indicates the direction and the strength of the electric field *at the point to which it is attached.* That is, at the point where the *tail* of the vector is placed. In this chapter, we indicate the point at which the electric field is measured with a dot. The length of any vector is significant only relative to the lengths of other vectors.

3. Although we have to draw a vector across the page, from one point to another, an electric field vector is *not* a spatial quantity. It does not "stretch" from one point to another. Each vector represents the electric field at *one point* in space.

Unit Vector Notation

Equation 25.14 is precise, but it's not terribly convenient. Furthermore, what happens if the source charge q is negative? We need a more concise notation to write the electric field, a notation that will allow q to be either positive or negative.

The basic need is to express "away from q" in mathematical notation. "Away from q" is a *direction* in space. To guide us, recall that we already have a notation for expressing certain directions; namely, the unit vectors $\hat{\imath}$, $\hat{\jmath}$, and $\hat{k}$. For example, unit vector $\hat{\imath}$ means "in the direction of the positive *x*-axis." With a minus sign, $-\hat{\imath}$ means "in the direction of the negative *x*-axis." Unit vectors, with a magnitude of 1 and no units, provide purely directional information.

With this in mind, let's define the unit vector $\hat{r}$ to be a vector of length 1 that points from the origin to a point of interest. Unit vector $\hat{r}$ provides no information at all about the distance to the point. It merely specifies the direction.

Figure 25.33a shows unit vectors $\hat{r}_1$, $\hat{r}_2$, and $\hat{r}_3$ pointing toward points 1, 2, and 3. Unlike $\hat{\imath}$ and $\hat{\jmath}$, unit vector $\hat{r}$ does not have a fixed direction. Instead, unit vector $\hat{r}$ specifies the direction "straight outward from this point." But that's just what we need to describe the electric field vector. Figure 25.33b shows the electric fields at points 1, 2, and 3 due to a positive charge at the origin. No matter which point you choose, the electric field at that point is "straight outward" from the charge. In other words, the electric field $\vec{E}$ points in the direction of the unit vector $\hat{r}$.

Using this notation, the electric field at distance r from a point charge q can be written

$$\vec{E} = \frac{1}{4\pi\epsilon_0}\frac{q}{r^2}\hat{r} \qquad \text{(electric field of a point charge)} \qquad (25.15)$$

where $\hat{r}$ is the unit vector from the charge to the point at which we want to know the field. Equation 25.15 is identical to Equation 25.14, but written in a notation in which the unit vector $\hat{r}$ expresses the idea "away from q."

Equation 25.15 works equally well if q is negative. A negative sign in front of a vector simply reverses its direction, so the unit vector $-\hat{r}$ points *toward* charge q. Figure 25.34 shows the electric field of a negative point charge. It looks like the electric field of a positive point charge except that the vectors point inward, toward the charge, instead of outward.

We'll end this chapter with two examples of the electric field of a point charge. Chapter 26 will expand these ideas to the electric fields of multiple charges and of extended objects.

(a)

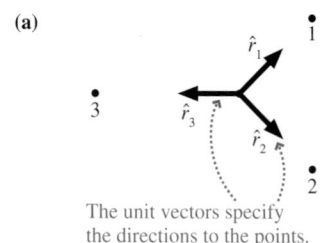

The unit vectors specify the directions to the points.

(b)   Electric field at point 1 is in the direction of $\hat{r}_1$.

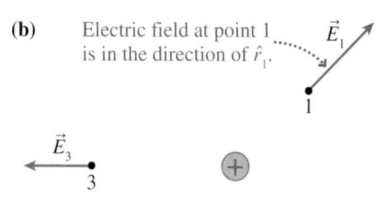

$\vec{E}_2$ is in the direction of $\hat{r}_2$.

FIGURE 25.33 Using the unit vector $\hat{r}$.

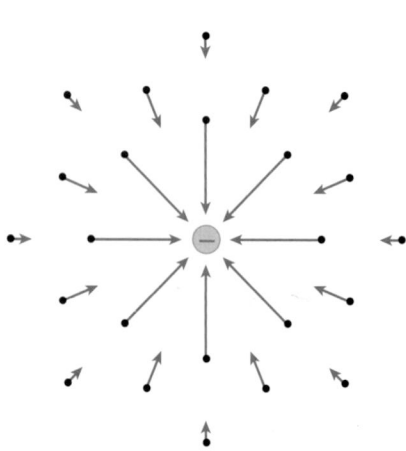

FIGURE 25.34 The electric field of a negative point charge.

EXAMPLE 25.8 Calculating the electric field

A -1.0 nC charged particle is located at the origin. Points 1 to 3 have (x, y) coordinates (1 cm, 0 cm), (0 cm, 1 cm), and (1 cm, 1 cm), respectively. Determine the electric field $\vec{E}$ at these three points, then show the vectors on an electric field diagram.

MODEL The electric field is that of a negative point charge.

VISUALIZE The electric field points straight *toward* the origin. It will be weaker at (1 cm, 1 cm), which is farther from the charge.

SOLVE The electric field is

$$\vec{E} = \frac{1}{4\pi\epsilon_0}\frac{q}{r^2}\hat{r}$$

where $q = -1.0$ nC $= 1.0 \times 10^{-9}$ C. The distance r is 1.0 cm $= 0.010$ m for points 1 and 2 and ($\sqrt{2} \times 1.0$ cm) $= 0.0141$ m for point 3. The *magnitude* of $\vec{E}$ at the three points is

$$E_1 = \frac{1}{4\pi\epsilon_0}\frac{|q|}{r_1^2}$$

$$= \frac{(9.0 \times 10^9 \, \text{N m}^2/\text{C}^2)(1.0 \times 10^{-9} \, \text{C})}{(0.010 \, \text{m})^2} = 90{,}000 \, \text{N/C}$$

$$E_2 = \frac{1}{4\pi\epsilon_0}\frac{|q|}{r_2^2}$$

$$= \frac{(9.0 \times 10^9 \, \text{N m}^2/\text{C}^2)(1.0 \times 10^{-9} \, \text{C})}{(0.010 \, \text{m})^2} = 90{,}000 \, \text{N/C}$$

$$E_3 = \frac{1}{4\pi\epsilon_0}\frac{|q|}{r_3^2}$$

$$= \frac{(9.0 \times 10^9 \, \text{N m}^2/\text{C}^2)(1.0 \times 10^{-9} \, \text{C})}{(0.0141 \, \text{m})^2} = 45{,}000 \, \text{N/C}$$

Because q is negative, the field at each of these positions points directly at charge q. The electric field vectors, in component form, are

$$\vec{E}_1 = -90{,}000\hat{i} \, \text{N/C}$$

$$\vec{E}_2 = -90{,}000\hat{j} \, \text{N/C}$$

$$\vec{E}_3 = -E_3\cos45°\hat{i} - E_3\sin45°\hat{j}$$

$$= (-31{,}800\hat{i} - 31{,}800\hat{j}) \, \text{N/C}$$

These vectors are shown on the electric field diagram of Figure 25.35.

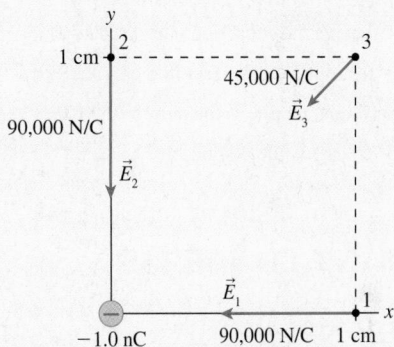

FIGURE 25.35 The electric field diagram of a -1.0 nC charged particle.

EXAMPLE 25.9 The electric field of a proton

The electron in a hydrogen atom orbits the proton at a radius of 0.053 nm.

a. What is the proton's electric field strength at the position of the electron?
b. What is the magnitude of the electric force on the electron?

SOLVE

a. The proton's charge is $q = e$. Its electric field strength at the distance of the electron is

$$E = \frac{1}{4\pi\epsilon_0}\frac{e}{r^2} = \frac{1}{4\pi\epsilon_0}\frac{1.60 \times 10^{-19} \, \text{C}}{(5.3 \times 10^{-11} \, \text{m})^2}$$

$$= 5.12 \times 10^{11} \, \text{N/C}$$

Notice how large this field is in comparison to the field of Example 25.8.

b. We could use Coulomb's law to find the force on the electron, but the whole point of knowing the electric field is that we can use it directly to find the force on a charge in the field. The magnitude of the force on the electron is

$$F_{\text{on elec}} = |q_e|E_{\text{of proton}}$$

$$= (1.60 \times 10^{-19} \, \text{C})(5.12 \times 10^{11} \, \text{N/C})$$

$$= 8.20 \times 10^{-8} \, \text{N}$$

STOP TO THINK 25.6 Rank in order, from largest to smallest, the electric field strengths E_1 to E_4 at points 1 to 4.

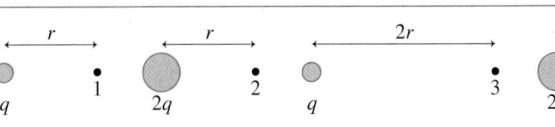

SUMMARY

The goal of Chapter 25 has been to develop a basic understanding of electric phenomena in terms of charges, forces, and fields.

GENERAL PRINCIPLES

Coulomb's Law

The forces between two charged particles q_1 and q_2 separated by distance r are

$$F_{1 \text{ on } 2} = F_{2 \text{ on } 1} = \frac{K|q_1||q_2|}{r^2}$$

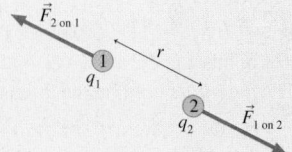

These forces are an action/reaction pair directed along the line joining the particles.

- The forces are repulsive for two like charges, attractive for two opposite charges.
- The net force on a charge is the sum of the forces from all other charges.
- The unit of charge is the coulomb (C).

IMPORTANT CONCEPTS

The Charge Model

There are **two kinds of charge,** called *positive* and *negative*.

- Fundamental charges are protons and electrons, with charge $\pm e$ where $e = 1.60 \times 10^{-19}$ C.
- Objects are charged by adding or removing electrons.
- The amount of charge is $q = (N_p - N_e)e$.
- An object with an equal number of protons and electrons is neutral, meaning no *net* charge.

Charged objects exert **electric forces** on each other.

- Like charges repel, opposite charges attract.
- The force increases as the charge increases.
- The force decreases as the distance increases.

There are **two types of material,** insulators and conductors.

- Charge remains fixed in or on an insulator.
- Charge moves easily through or along conductors.
- Charge is transferred by contact between objects.

Charged objects attract neutral objects.

- Charge polarizes metal by shifting the electron sea.
- Charge polarizes atoms, creating electric dipoles.
- The polarization force is always an attractive force.

The Field Model

Charges interact with each other via the electric field $\vec{E}$.

- Charge A alters the space around it by creating an electric field.

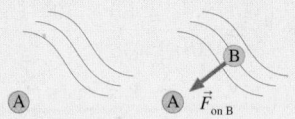

- The field is the agent that exerts a force. The force on charge q_B is $\vec{F}_{\text{on } B} = q_B \vec{E}$.

An electric field is identified and measured in terms of the force on a **probe charge** q.

$$\vec{E} = \vec{F}_{\text{on } q} / q$$

- The electric field exists at all points in space.
- An electric field vector shows the field only at one point, the point at the tail of the vector.

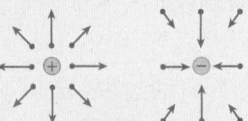

The electric field of a **point charge** is

$$\vec{E} = \frac{1}{4\pi\epsilon_0} \frac{q}{r^2} \hat{r}$$

TERMS AND NOTATION

neutral	ionization	Coulomb's law
charging	law of conservation of charge	electrostatic constant, K
charge model	sea of electrons	point charge
charge, q or Q	ion core	coulomb, C
like charges	current	permittivity constant, ϵ_0
opposite charges	charge carriers	field
discharging	electrostatic equilibrium	gravitational field, $\vec{g}$
conductor	grounded	electric field, $\vec{E}$
insulator	charge polarization	field model
electron cloud	polarization force	source charge
fundamental unit of charge, e	electric dipole	electric field strength, E
charge quantization	charging by induction	field diagram

EXERCISES AND PROBLEMS

Exercises

Section 25.1 Developing a Charge Model

Section 25.2 Charge

1. A glass rod is charged to +5 nC by rubbing.
   a. Have electrons been removed from the rod or protons added? Explain.
   b. How many electrons have been removed or protons added?

2. A plastic rod is charged to −20 nC by rubbing.
   a. Have electrons been added to the rod or protons removed? Explain.
   b. How many electrons have been added or protons removed?

3. A 2.0-mm-diameter copper ball is charged to +50 nC. What fraction of its electrons have been removed?

4. How many coulombs of positive charge are in 1.0 mol of O_2 gas?

Section 25.3 Insulators and Conductors

5. Figure 25.9 showed how an electroscope becomes negatively charged. The leaves will also repel each other if you touch the electroscope with a positively charged glass rod. Use a series of charge diagrams to explain what happens and why the leaves repel each other.

6. A plastic balloon that has been rubbed with wool will stick to a wall.
   a. Can you conclude that the wall is charged? If not, why not? If so, where does the charge come from?
   b. Draw a series of charge diagrams showing how the balloon is held to the wall.

7. Two neutral metal spheres on wood stands are touching. A negatively charged rod is held directly above the top of the left sphere, not quite touching it. While the rod is there, the right sphere is moved so that the spheres no longer touch. Then the rod is withdrawn. Afterward, what is the charge state of each sphere? Use charge diagrams to explain your answer.

8. You have two neutral metal spheres on wood stands. Devise a procedure for charging the spheres so that they will have opposite charges of *exactly* equal magnitude. Use charge diagrams to explain your procedure.

9. You have two neutral metal spheres on wood stands. Devise a procedure for charging the spheres so that they will have like charges of *exactly* equal magnitude. Use charge diagrams to explain your procedure.

10. An object passes the "Is it charged?" test by picking up small pieces of paper.
    a. Use a series of charge diagrams to explain how a charged object picks up a piece of paper.
    b. Write a paragraph explaining why this test works for both positively and negatively charged objects. Your explanation should be based on the properties of charges, forces, and insulators.

Section 25.4 Coulomb's Law

11. Two 1.0 kg masses are 1.0 m apart on a frictionless table. Each has +1.0 C of charge.
    a. What is the magnitude of the electric force on one of the masses?
    b. What is the initial acceleration of the mass if it is released and allowed to move?

12. Two small plastic spheres each have a mass of 2.0 g and a charge of −50.0 nC. They are placed 2.0 cm apart.
    a. What is the magnitude of the electric force between the spheres?
    b. By what factor is the electric force on a sphere larger than its weight?

13. A small glass bead has been charged to +20 nC. A metal ball bearing 1.0 cm above the bead feels a 0.018 N downward electric force. What is the charge on the ball bearing?

14. What is the net electric force on charge A in Figure Ex25.14?

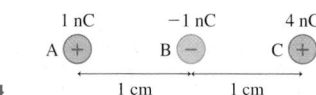

FIGURE EX25.14

15. Object A, which has been charged to $+10.0$ nC, is at the origin. Object B, which has been charged to -20.0 nC, is at $(x, y) = (0.0 \text{ cm}, 2.0 \text{ cm})$. Determine the electric force on each object. Write each force vector in component form.
16. A small glass bead has been charged to $+20$ nC. What are the magnitude and direction of the acceleration of (a) a proton and (b) an electron that is 1.0 cm from the center of the bead?

Section 25.5 The Concept of a Field

17. What is the strength of the gravitational field at a point where a 2.0 kg mass experiences a 60.0 N force?
18. Mass m experiences a gravitational force of magnitude 15 N. What will be the magnitude of the gravitational force on mass $3m$ at the same point in the gravitational field?
19. What are the magnitude and direction of the earth's gravitational field $\vec{g}$ at
    a. The surface of the earth?
    b. The radius of the moon's orbit?
20. A satellite orbits a planet. The gravitational field strength at the radius of the orbit is 12 N/kg. What will the gravitational field strength at the position of the satellite be if
    a. The radius of the orbit is doubled?
    b. The planet's density is doubled?
    c. The satellite's mass is doubled?

Section 25.6 The Field Model

21. What magnitude charge creates a 1.0 N/C electric field at a point 1.0 m away?
22. What are the strength and direction of the electric field 2.0 cm from a small glass bead that has been charged to $+6.0$ nC?
23. The electric field 2.0 cm from a small object points toward the object with a strength of 180,000 N/C. What is the object's charge?
24. What are the strength and direction of the electric field 1.0 mm from (a) a proton and (b) an electron?
25. What are the strength and direction of an electric field that will balance the weight of a 1.0 g plastic sphere that has been charged to -3.0 nC?
26. What are the strength and direction of an electric field that will balance the weight of (a) a proton and (b) an electron?
27. A $+10.0$ nC charge is located at the origin.
    a. What are the electric fields at the positions $(x, y) = (5 \text{ cm}, 0 \text{ cm})$, $(-5 \text{ cm}, 5 \text{ cm})$, and $(-5 \text{ cm}, -5 \text{ cm})$? Write each electric field vector in component form.
    b. Draw a field diagram showing the electric field vectors at these points.
28. A -10 nC charge is located at the origin.
    a. What are the electric fields at the positions $(x, y) = (0 \text{ cm}, 5 \text{ cm})$, $(-5 \text{ cm}, -5 \text{ cm})$, and $(-5 \text{ cm}, 5 \text{ cm})$? Write each electric field vector in component form.
    b. Draw a field diagram showing the electric field vectors at these points.

Problems

29. Pennies today are copper-covered zinc, but older pennies are 3.1 g of solid copper. What are the total positive charge and total negative charge in a solid copper penny that is electrically neutral?
30. A plastic rod that has been charged to -15 nC touches a metal sphere. Afterward, the rod's charge is -10 nC.
    a. What kind of charged particle was transferred between the rod and the sphere, and in which direction? That is, did it move from the rod to the sphere or from the sphere to the rod?
    b. How many charged particles were transferred?
31. A glass rod that has been charged to $+12$ nC touches a metal sphere. Afterward, the rod's charge is $+8$ nC.
    a. What kind of charged particle was transferred between the rod and the sphere, and in which direction? That is, did it move from the rod to the sphere or from the sphere to the rod?
    b. How many charged particles were transferred?
32. Two identical metal spheres A and B are connected by a metal rod. Both are initially neutral. 1.0×10^{12} electrons are added to sphere A, then the connecting rod is removed. Afterward, what are the charge of A and the charge of B?
33. Two identical metal spheres A and B are connected by a plastic rod. Both are initially neutral. 1.0×10^{12} electrons are added to sphere A, then the connecting rod is removed. Afterward, what are the charge of A and the charge of B?
34. Two protons are 2.0 fm apart.
    a. What is the magnitude of the electric force on one proton due to the other proton?
    b. What is the magnitude of the gravitational force on one proton due to the other proton?
    c. What is the ratio of the electric force to the gravitational force?
35. The nucleus of a ^{125}Xe atom (an isotope of the element xenon with mass 125 u) is 6.0 fm in diameter. It has 54 protons and charge $q = +54e$.
    a. What is the electric force on a proton 2.0 fm from the surface of the nucleus?
    Hint: Treat the spherical nucleus as a point charge.
    b. What is the proton's acceleration?
36. Two 1.0 g spheres are charged equally and placed 2.0 cm apart. When released, they begin to accelerate at 225 m/s^2. What is the magnitude of the charge on each sphere?
37. Objects A and B are both positively charged. Both have a mass of 100 g, but A has twice the charge of B. When A and B are placed 10 cm apart, B experiences an electric force of 0.45 N.
    a. How large is the force on A?
    b. What are the charges q_A and q_B?
    c. If the objects are released, what is the initial acceleration of A?
38. What is the force $\vec{F}$ on the 1 nC charge? Give your answer as a magnitude and a direction.

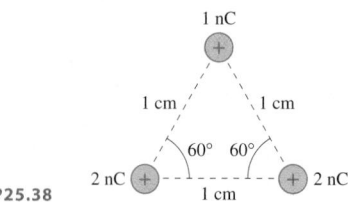

FIGURE P25.38

39. What is the force $\vec{F}$ on the 1 nC charge? Give your answer as a magnitude and a direction.

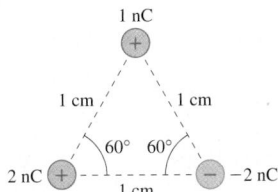

FIGURE P25.39

40. What is the force $\vec{F}$ on the -10 nC charge? Give your answer as a magnitude and a direction.

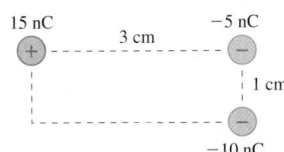

FIGURE P25.40

41. What is the force $\vec{F}$ on the -10 nC charge? Give your answer as a magnitude and a direction.

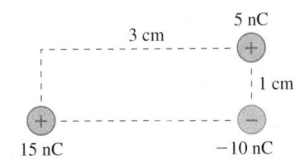

FIGURE P25.41

42. What is the force $\vec{F}$ on the 5 nC charge in Figure P25.42? Give your answer as a magnitude and a direction.

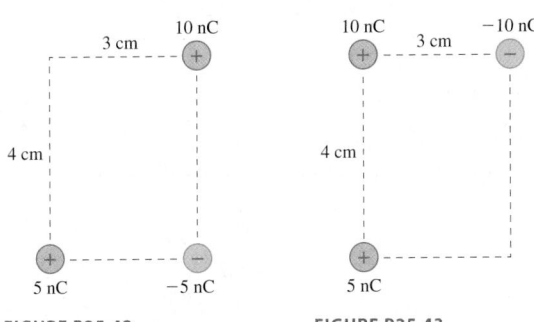

FIGURE P25.42   **FIGURE P25.43**

43. What is the force $\vec{F}$ on the 5 nC charge in Figure P25.43? Give your answer as a magnitude and a direction.
44. What is the force $\vec{F}$ on the 1 nC charge in the middle due to the four other charges? Give your answer in component form.

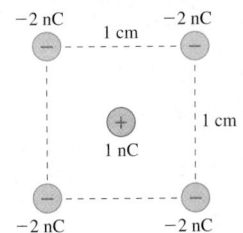

FIGURE P25.44

45. What is the force $\vec{F}$ on the 1 nC charge in the middle due to the four other charges? Give your answer in component form.

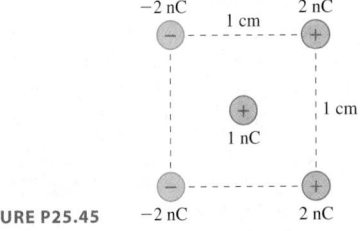

FIGURE P25.45

46. What is the force $\vec{F}$ on the 1 nC charge at the bottom? Give your answer in component form.

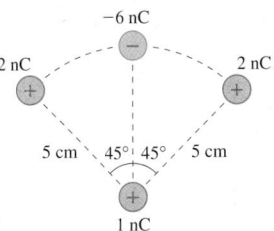

FIGURE P25.46

47. What is the force $\vec{F}$ on the 1 nC charge at the bottom? Give your answer in component form.

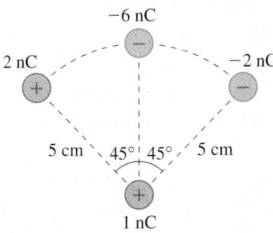

FIGURE P25.47

48. A $+2.0$ nC charge and a -4.0 nC charge are 1.0 cm apart.
    a. Where could you place a proton so that it would experience no net force?
    b. Would the net force be zero for an electron placed at the same position? Explain.
49. The net force on the 1 nC charge is zero. What is q?

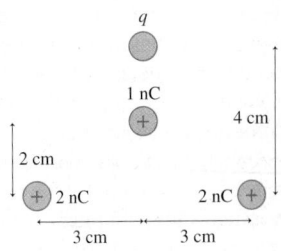

FIGURE P25.49

50. Charge q_2 is in static equilibrium. What is q_1?

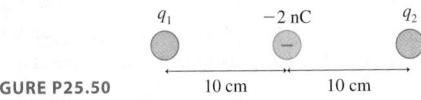

FIGURE P25.50

51. A positive point charge Q is located at $x = a$ and a negative point charge $-Q$ is at $x = -a$. A positive charge q can be placed anywhere on the y-axis. Find an expression for $(F_{net})_x$, the x-component of the net force on q.

52. A positive point charge Q is located at $x = a$ and a negative point charge $-Q$ is at $x = -a$. A positive charge q can be placed anywhere on the x-axis. Find an expression for $(F_{net})_x$, the x-component of the net force on q, when (a) $|x| < a$ and (b) $|x| > a$.

53. Figure P25.53 shows four charges at the corners of a square of side L. Assume q and Q are positive.
    a. Draw a diagram showing the three forces on charge q due to the other charges. Give your vectors the correct relative lengths.
    b. What is the magnitude of the net force on q?

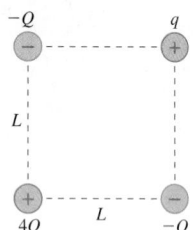

FIGURE P25.53

54. Two point charges q and $4q$ are distance L apart and free to move. A third charge is placed so that the entire three-charge system is in static equilibrium. What are the magnitude, sign, and position of the third charge?

55. Suppose that the magnitudes of the proton charge and the electron charge differ by a mere 1 part in 10^9.
    a. What would be the force between two 2.0-mm-diameter copper spheres 1.0 cm apart? Assume that each copper atom has an equal number of electrons and protons.
    b. Would this amount of force be detectable? What can you conclude from the fact that no such forces are observed?

56. In a simple model of the hydrogen atom, the electron moves in a circular orbit of radius 0.053 nm around a stationary proton. How many revolutions per second does the electron make?

57. As a science project, you've invented an "electron pump" that moves electrons from one object to another. To demonstrate your invention, you bolt a small metal plate to the ceiling, connect the pump between the metal plate and yourself, and start pumping electrons from the metal plate to you. How many electrons must be moved from the metal plate to you in order for you to hang suspended in the air 2.0 m below the ceiling? Your mass is 60 kg.
    Hint: Assume that both you and the plate can be modeled as point charges.

58. You have a lightweight spring whose unstretched length is 4.0 cm. You're curious to see if you can use this spring to measure charge. First, you attach one end of the spring to the ceiling and hang a 1.0 g mass from it. This stretches the spring to a length of 5.0 cm. You then attach two small plastic beads to the opposite ends of the spring, lay the spring on a frictionless table, and give each plastic bead the same charge. This stretches the spring to a length of 4.5 cm. What is the magnitude of the charge (in nC) on each bead?

59. Two 5.0 g spheres on 1.0-m-long threads repel each other after being charged to $+100$ nC. What is the angle θ? You can assume that θ is a small angle.

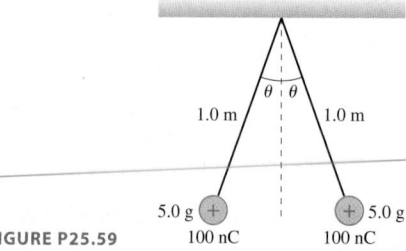

FIGURE P25.59

60. Two 3.0 g spheres on 1.0-m-long threads repel each other after being equally charged. What is the charge q?

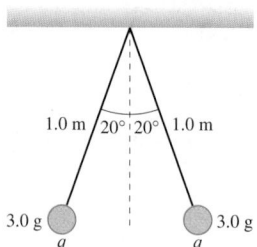

FIGURE P25.60

61. What are the electric fields at points 1, 2, and 3 in Figure P25.61? Give your answer in component form.

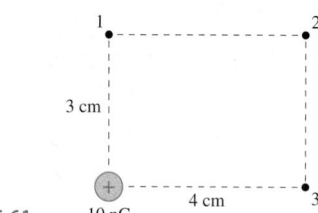

FIGURE P25.61

62. What are the electric fields at points 1 and 2 in Figure P25.62? Give your answer as a magnitude and direction.

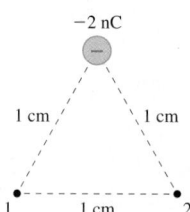

FIGURE P25.62

63. What are the electric fields at points 1, 2, and 3 in Figure P25.63? Give your answer in component form.

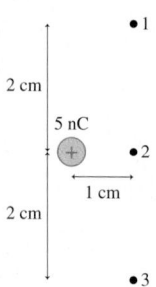

FIGURE P25.63

64. A -10.0 nC charge is located at position $(x, y) = (2.0$ cm, 1.0 cm$)$. At what (x, y) position(s) is the electric field
    a. $-225,000\hat{\imath}$ N/C?
    b. $(161,000\hat{\imath} - 80,500\hat{\jmath})$ N/C?
    c. $(28,800\hat{\imath} + 21,600\hat{\jmath})$ N/C?
65. A 10.0 nC charge is located at position $(x, y) = (1.0$ cm, 2.0 cm$)$. At what (x, y) position(s) is the electric field
    a. $-225,000\hat{\imath}$ N/C?
    b. $(161,000\hat{\imath} + 80,500\hat{\jmath})$ N/C?
    c. $(21,600\hat{\imath} - 28,800\hat{\jmath})$ N/C?
66. Three 1.0 nC charges are placed as shown in Figure P25.66. Each of these charges creates an electric field $\vec{E}$ at a point 3.0 cm in front of the middle charge.
    a. What are the three fields $\vec{E}_1$, $\vec{E}_2$, and $\vec{E}_3$ created by the three charges? Write your answer for each as a vector in component form.
    b. Do you think that electric fields obey a principle of superposition? That is, is there a "net field" at this point given by $\vec{E}_{net} = \vec{E}_1 + \vec{E}_2 + \vec{E}_3$? Use what you learned in this chapter and previously in our study of forces to argue why this is or is not true.
    c. If it is true, what is $\vec{E}_{net}$?

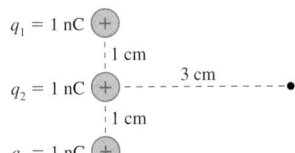

FIGURE P25.66

67. The electric field at a point in space is $\vec{E} = (200\hat{\imath} + 400\hat{\jmath})$ N/C.
    a. What is the electric force on a proton at this point? Give your answer in component form.
    b. What is the electric force on an electron at this point? Give your answer in component form.
    c. What is the magnitude of the proton's acceleration?
    d. What is the magnitude of the electron's acceleration?
68. An electric field $\vec{E} = 100,000\hat{\imath}$ N/C causes the 5.0 g ball in Figure P25.68 to hang at a $20°$ angle. What is the charge on the ball?

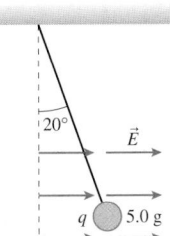

FIGURE P25.68

69. An electric field $\vec{E} = 200,000\hat{\imath}$ N/C causes the ball in Figure P25.69 to hang at an angle. What is θ?

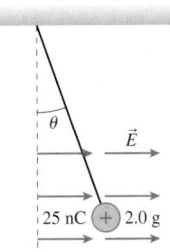

FIGURE P25.69

In Problems 70 through 73 you are given the equation(s) used to solve a problem. For each of these,
  a. Write a realistic problem for which this is the correct equation(s).
  b. Finish the solution of the problem.

70. $\dfrac{(9.0 \times 10^9 \text{ N m}^2/\text{C}^2) \times N \times (1.60 \times 10^{-19} \text{ C})}{(1.0 \times 10^{-6} \text{ m})^2}$

   $= 1.5 \times 10^6$ N/C

71. $\dfrac{(9.0 \times 10^9 \text{ N m}^2/\text{C}^2)q^2}{(0.0150 \text{ m})^2} = 0.020$ N

72. $\dfrac{(9.0 \times 10^9 \text{ N m}^2/\text{C}^2)(15 \times 10^{-9} \text{ C})}{r^2} = 54,000$ N/C

73. $\sum F_x = 2 \times \dfrac{(9.0 \times 10^9 \text{ N m}^2/\text{C}^2)(1.0 \times 10^{-9} \text{ C})q}{((0.020 \text{ m})/\sin 30°)^2} \times \cos 30°$

   $= 5.0 \times 10^{-5}$ N

   $\sum F_y = 0$ N

Challenge Problems

74. Three 3.0 g balls are tied to 80-cm-long threads and hung from a *single* fixed point. Each of the balls is given the same charge q. At equilibrium, the three balls form an equilateral triangle in a horizontal plane with 20.0 cm sides. What is q?
75. The identical small spheres shown in Figure CP25.75 are charged to $+100$ nC and -100 nC. They hang as shown in a $100,000$ N/C electric field. What is the mass of each sphere?

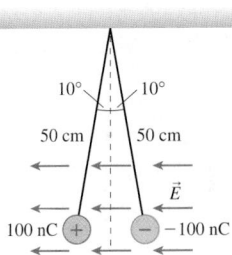

FIGURE CP25.75

76. You sometimes create a spark when you touch a doorknob after shuffling your feet on a carpet. Why? The air always has a few free electrons that have been kicked out of atoms by cosmic rays. If an electric field is present, a free electron is accelerated until it collides with an air molecule. It will transfer its kinetic energy to the molecule, then accelerate, then collide, then accelerate, collide, and so on. If the electron's kinetic energy just before a collision is 2.0×10^{-18} J or more, it has sufficient energy to kick an electron out of the molecule it hits. Where there was one free electron, now there are two! Each of these can then accelerate, hit a molecule, and kick out another electron. Then there will be four free electrons. In other words, as Figure CP25.76 on the next page shows, a sufficiently strong electric field causes a "chain reaction" of electron production. This is called a *breakdown* of the air. The current of moving electrons is what gives you the shock, and a spark is generated when the electrons recombine with the positive ions and give off excess energy as a burst of light.
    a. The average distance an electron travels between collisions is 2.0 μm. What acceleration must an electron have to gain 2.0×10^{-18} J of kinetic energy in this distance?

b. What force must act on an electron to give it the acceleration found in part a?

c. What strength electric field will exert this much force on an electron? This is the *breakdown field strength*.

d. Suppose a free electron in air is 1.0 cm away from a point charge. What minimum charge q_{min} must this point charge have to cause a breakdown of the air and create a spark?

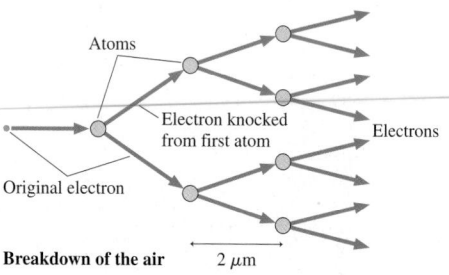

Atoms

Electron knocked from first atom

Original electron

Electrons

Breakdown of the air   2 μm

FIGURE CP25.76

77. In Section 25.3 we claimed that a charged object exerts a net attractive force on an electric dipole. Let's investigate this. The figure shows a *permanent* electric dipole consisting of charges $+q$ and $-q$ separated by the fixed distance s. Charge $+Q$ is distance r from the center of the dipole. We'll assume, as is usually the case in practice, that $s \ll r$.

a. Write an expression for the net force exerted on the dipole by charge $+Q$.

b. Is this force toward $+Q$ or away from $+Q$? Explain.

c. Use the *binomial approximation* $(1 + x)^{-n} \approx 1 - nx$ if $x \ll 1$ to show that your expression from part a can be written $F_{net} = 2KqQs/r^3$.

d. How can an electric force have an inverse-cube dependence? Doesn't Coulomb's law say that the electric force depends on the inverse square of the distance? Explain.

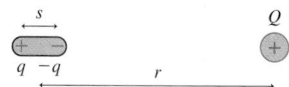

s

q $-q$

Q

r

FIGURE CP25.77

STOP TO THINK ANSWERS

Stop to Think 25.1: b. Charged objects are attracted to neutral objects, so an attractive force is inconclusive. Repulsion is the only sure test.

Stop to Think 25.2: $q_e(+3e) > q_a(+1e) > q_d(0) > q_b(-1e) > q_c(-2e)$.

Stop to Think 25.3: a. The negative plastic rod will polarize the electroscope by pushing electrons down toward the leaves. This will partially neutralize the positive charge the leaves had acquired from the glass rod.

Stop to Think 25.4: b. The two forces are an action/reaction pair, opposite in direction but *equal* in magnitude.

Stop to Think 25.5: c. There's an electric field at *all* points, whether an $\vec{E}$ vector is shown or not. The electric field at the dot is to the right. But an electron is a negative charge, so the force of the electric field on the electron is to the left.

Stop to Think 25.6: $E_2 > E_1 > E_4 > E_3$.

26 The Electric Field

► **Looking Ahead**

The goal of Chapter 26 is to learn how to calculate and use the electric field. In this chapter you will learn to:

- Calculate the electric field due to multiple point charges.
- Calculate the electric field due to a continuous distribution of charge.
- Use the electric field of dipoles, lines of charge, and planes of charge.
- Generate a uniform electric field with a parallel-plate capacitor.
- Calculate the motion of charges and dipoles in an electric field.

◄ **Looking Back**

This chapter builds on the ideas about electric forces and fields that were introduced in Chapter 25. Charged-particle motion in an electric field is similar to projectile motion. Please review:

- Section 6.3 Projectile motion.
- Section 25.4 Coulomb's law.
- Sections 25.5–25.7 The electric field of a point charge.

Liquid crystal displays work by using electric fields to align long polymer molecules.

You can't see them, but they are all around you—electric fields, that is. Electric fields line up polymer molecules to form the images in the liquid crystal display (LCD) of a wristwatch or a flat-panel computer screen. Electric fields are responsible for the electric currents that flow in your computer and your stereo, and they are essential to the functioning of your brain, your heart, and your DNA.

We introduced the idea of an electric field, in Chapter 25, in order to understand the long-range electric interaction between charges. The electric field of a point charge is pretty simple, but real-world charged objects have vast numbers of charges arranged in complex patterns. To make practical use of electric fields, we need to know how to calculate the electric field of a complicated distribution of charge. The major goal of this chapter is to develop a procedure for calculating the electric field of a specified configuration or arrangement of charge.

Chapter 25 made a distinction between those charged particles that are the *sources* of an electric field and other charged particles that *experience* and move in the electric field. This is a very important distinction. Most of this chapter will be concerned with the *sources* of the electric field. Only at the end, once we know how to calculate the electric field, will we look at what happens to charges that find themselves *in* an electric field.

26.1 Electric Field Models

The electric fields used in science and engineering are often caused by fairly complicated distributions of charge. Sometimes these fields require exact calculations, but much of the time we can understand the essential physics on the basis of simplified *models* of the electric field.

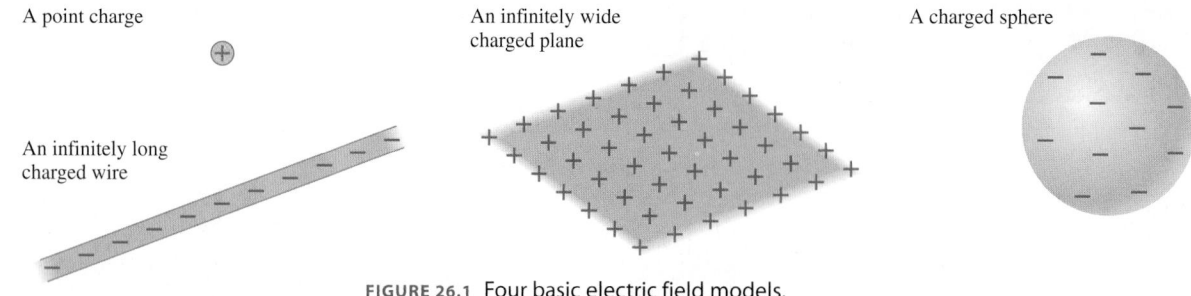

A point charge

An infinitely long charged wire

An infinitely wide charged plane

A charged sphere

FIGURE 26.1 Four basic electric field models.

Four widely used electric field models, illustrated in Figure 26.1, are:

- The electric field of a point charge.
- The electric field of an infinitely long charged wire.
- The electric field of an infinitely wide charged plane.
- The electric field of a charged sphere.

Small charged objects can often be modeled as point charges or charged spheres. Real wires aren't infinitely long, but in many practical situations this approximation is perfectly reasonable. As we derive and use these electric fields, we'll consider the conditions under which they are appropriate models.

Our starting point is the electric field of a point charge q. We found in Chapter 25 that

$$\vec{E} = \frac{1}{4\pi\epsilon_0} \frac{q}{r^2} \hat{r} \qquad \text{(electric field of a point charge)} \qquad (26.1)$$

where $\hat{r}$ is a unit vector pointing away from q. Figure 26.2 reminds you how the electric fields of positive and negative point charges look. Although we have to give each vector we draw a length, keep in mind that each arrow represents the electric field *at a point*. The electric field is not a spatial quantity that "stretches" from one end of the arrow to the other.

The electric field was defined as $\vec{E} = \vec{F}_{\text{on }q}/q$, where $\vec{F}_{\text{on }q}$ is the electric force on charge q. Forces add as vectors, so the net force on q due to a group of point charges is the vector sum

$$\vec{F}_{\text{on }q} = \vec{F}_{1 \text{ on }q} + \vec{F}_{2 \text{ on }q} + \cdots$$

Consequently, the net electric field due to a group of point charges is

$$\vec{E}_{\text{net}} = \frac{\vec{F}_{\text{on }q}}{q} = \frac{\vec{F}_{1 \text{ on }q}}{q} + \frac{\vec{F}_{2 \text{ on }q}}{q} + \cdots = \vec{E}_1 + \vec{E}_2 + \cdots = \sum_i \vec{E}_i \qquad (26.2)$$

where $\vec{E}_i$ is the field from point charge i.

Equation 26.2, which is the primary tool for calculating electric fields, tells us that **the net electric field is the *vector sum* of the electric fields due to each charge.** In other words, electric fields obey the *principle of superposition*. Figure 26.3 illustrates this important idea. Much of this chapter will be focused on the mathematical aspects of performing the sum.

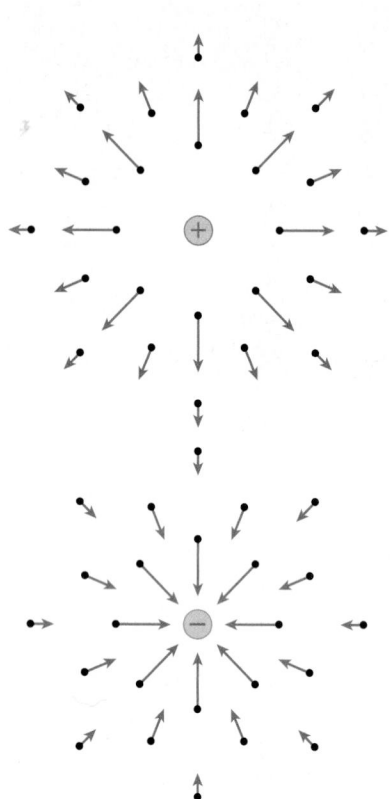

FIGURE 26.2 The electric field of a positive and a negative point charge.

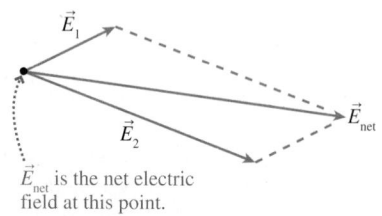

$\vec{E}_{\text{net}}$ is the net electric field at this point.

FIGURE 26.3 Electric fields obey the principle of superposition.

Limiting Cases and Typical Field Strengths

Suppose you have a small, charged object. The electric field very near the object depends on the object's shape and on how the charge is distributed. But from far away, the object appears to be a point charge in the distance. Thus the object's electric field should approach that of a point charge in the *limiting case* that the distance from the object approaches infinity.

We'll have occasion to look at limiting cases both very far from and very close to charged objects. Limiting cases allow us to

- Check a solution by seeing if it has the expected behavior as distances become very large or very small.
- Use simpler expressions for the electric field at points very close to or very far from a charged object.

We'll emphasize limiting cases throughout this chapter as we develop models of the electric field.

Knowing some typical electric field strengths will also be helpful. The values shown in Table 26.1 will help you judge the reasonableness of your solutions to problems.

TABLE 26.1 Typical electric field strengths

Field location	Field strength (N/C)
Inside a current-carrying wire	10^{-2}
Near the earth's surface	$10^2 – 10^4$
Near objects charged by rubbing	$10^3 – 10^6$
Electric breakdown in air, causing a spark	3×10^6
Inside an atom	10^{11}

26.2 The Electric Field of Multiple Point Charges

As we noted in Chapter 25, it is important to distinguish between those charges that are the sources of an electric field and those that experience and move in the electric field. Suppose the source of an electric field is a group of point charges $q_1, q_2, \ldots$. According to Equation 26.2, the electric field at each point in space is a superposition of the electric fields due to each individual charge.

The *vector* sum of Equation 26.2 can be written as the three simultaneous sums

$$(E_{net})_x = (E_1)_x + (E_2)_x + \cdots = \sum (E_i)_x$$
$$(E_{net})_y = (E_1)_y + (E_2)_y + \cdots = \sum (E_i)_y \qquad (26.3)$$
$$(E_{net})_z = (E_1)_z + (E_2)_z + \cdots = \sum (E_i)_z$$

The problem and the context will sometimes favor expressing $\vec{E}_{net}$ in component form: $\vec{E}_{net} = (E_{net})_x \hat{i} + (E_{net})_y \hat{j} + (E_{net})_z \hat{k}$. At other times you will want to write $\vec{E}_{net}$ as a magnitude and a direction.

 PROBLEM-SOLVING STRATEGY 26.1 **The electric field of multiple point charges**

MODEL Model charged objects as point charges.

VISUALIZE For the pictorial representation:

- Establish a coordinate system and show the locations of the charges.
- Identify the point P at which you want to calculate the electric field.
- Draw the electric field of each charge at P.
- Use symmetry to determine if any components of $\vec{E}_{net}$ are zero.

SOLVE The mathematical representation is $\vec{E}_{net} = \sum \vec{E}_i$.

- For each charge, determine its distance from P and the angle of $\vec{E}_i$ from the axes.
- Calculate the field strength of each charge's electric field.
- Write each vector $\vec{E}_i$ in component form.
- Sum the vector components to determine $\vec{E}_{net}$.
- If needed, determine the magnitude and direction of $\vec{E}_{net}$.

ASSESS Check that your result has the correct units, is reasonable, and agrees with any known limiting cases.

EXAMPLE 26.1 The electric field of three equal point charges

Three equal point charges q are located on the y-axis at $y = 0$ and at $y = \pm d$. What is the electric field at a point on the x-axis?

MODEL This problem is a step along the way to understanding the electric field of a charged wire. We'll assume that q is positive when drawing pictures, but the solution should allow for the possibility that q is negative. The question does not ask about any specific point, so we will be looking for a symbolic expression in terms of the unspecified position x.

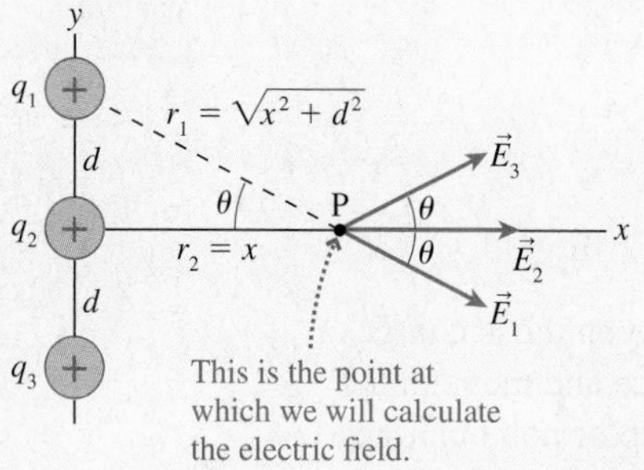

FIGURE 26.4
Calculating the electric field of three equal point charges.

VISUALIZE Figure 26.4 shows the charges, the coordinate system, and the three electric field vectors $\vec{E}_1$, $\vec{E}_2$, and $\vec{E}_3$. Each of these fields points *away from* its source charge because of the assumption that q is positive. We need to find the vector sum $\vec{E}_{net} = \vec{E}_1 + \vec{E}_2 + \vec{E}_3$.

Before rushing into a calculation, we can make our task *much* easier by first thinking qualitatively about the situation. For example, the fields $\vec{E}_1$, $\vec{E}_2$, and $\vec{E}_3$ all lie in the xy-plane, hence we can conclude without doing any calculations that $(E_{net})_z = 0$. Next, look at the y-components of the fields. Because the charges are located *symmetrically* on either side of the x-axis and are of equal value, the fields $\vec{E}_1$ and $\vec{E}_3$ have equal magnitudes and are tilted away from the x-axis by the same angle θ. Consequently, the y-components of $\vec{E}_1$ and $\vec{E}_3$ will *cancel* when added. $\vec{E}_2$ has no y-component, so we can conclude that $(E_{net})_y = 0$. The only component we need to calculate is $(E_{net})_x$.

SOLVE We're finally ready to calculate. The x-component of the electric field is

$$(E_{net})_x = (E_1)_x + (E_2)_x + (E_3)_x = 2(E_1)_x + (E_2)_x$$

where we used the fact that fields $\vec{E}_1$ and $\vec{E}_3$ have *equal* x-components. Vector $\vec{E}_2$ has *only* the x-component

$$(E_2)_x = E_2 = \frac{1}{4\pi\epsilon_0}\frac{q_2}{r_2^2} = \frac{1}{4\pi\epsilon_0}\frac{q}{x^2}$$

where $r_2 = x$ is the distance from q_2 to the point at which we are calculating the field. Vector $\vec{E}_1$ is at angle θ from the x-axis, so its x-component is

$$(E_1)_x = E_1\cos\theta = \frac{1}{4\pi\epsilon_0}\frac{q_1}{r_1^2}\cos\theta$$

where r_1 is the distance from q_1. This expression for $(E_1)_x$ is correct, but it is not yet sufficient. Both the distance r_1 and the angle θ vary with the position x and need to be expressed as functions of x. From the Pythagorean theorem, $r_1 = (x^2 + d^2)^{1/2}$. Then from trigonometry,

$$\cos\theta = \frac{x}{r_1} = \frac{x}{(x^2 + d^2)^{1/2}}$$

By combining these pieces, we see that $(E_1)_x$ is

$$(E_1)_x = \frac{1}{4\pi\epsilon_0}\frac{q}{x^2 + d^2}\frac{x}{(x^2 + d^2)^{1/2}} = \frac{1}{4\pi\epsilon_0}\frac{xq}{(x^2 + d^2)^{3/2}}$$

This expression is a bit complex, but notice that the dimensions of $x/(x^2 + d^2)^{3/2}$ are $1/m^2$, as they *must* be for the field of a point charge. Checking dimensions is a good way to verify that you haven't made algebra errors.

We can now combine $(E_1)_x$ and $(E_2)_x$ to write the x-component of $\vec{E}_{net}$ as

$$(E_{net})_x = 2(E_1)_x + (E_2)_x = \frac{q}{4\pi\epsilon_0}\left[\frac{1}{x^2} + \frac{2x}{(x^2 + d^2)^{3/2}}\right]$$

The other two components of $\vec{E}_{net}$ are zero, hence the electric field of the three charges at a point on the x-axis is

$$\vec{E}_{net} = \frac{q}{4\pi\epsilon_0}\left[\frac{1}{x^2} + \frac{2x}{(x^2 + d^2)^{3/2}}\right]\hat{\imath}$$

ASSESS This is the electric field only at points *on the x-axis*. Furthermore, this expression is valid only for $x > 0$. The electric field to the left of the charges points in the opposite direction, but our expression doesn't change sign for negative x. (This is a consequence of how we wrote $(E_2)_x$.) We would need to modify this expression to use it for negative values of x. The good news, though, is that our expression *is* valid for both positive and negative q. A negative value of q makes $(E_{net})_x$ negative, which would be an electric field pointing to the left, toward the negative charges.

Let's explore this example a bit more. If you think about it, there are two limiting cases for which we know what the result should be. First, let x become really, really small. As the point in Figure 26.4 approaches the origin, the fields $\vec{E}_1$ and $\vec{E}_3$ become opposite to each other and cancel. Thus as $x \to 0$, the field should be that of the single point charge q at the origin, a field we already know. Is it? Notice that

$$\lim_{x \to 0}\frac{2x}{(x^2 + d^2)^{3/2}} = 0 \tag{26.4}$$

Thus $E_{net} \to q/4\pi\epsilon_0 x^2$ as $x \to 0$, the expected field of a single point charge.

Now consider the opposite situation, where x becomes extremely large. From very far away, the three source charges will seem to merge into a single charge of size $3q$, just as three very distant light bulbs appear to be a single light. Thus the field for $x \gg d$ *should* be that of a point charge $3q$. Is it?

The field is zero in the limit $x \to \infty$. That doesn't tell us much, so we don't want to go *that* far away. We simply want x to be very large in comparison to the spacing d between the source charges. If $x \gg d$, then the denominator of the second term of $\vec{E}_{\text{net}}$ is well approximated by $(x^2 + d^2)^{3/2} \approx (x^2)^{3/2} = x^3$. Thus

$$\lim_{x \gg d}\left[\frac{1}{x^2} + \frac{2x}{(x^2 + d^2)^{3/2}}\right] = \frac{1}{x^2} + \frac{2x}{x^3} = \frac{3}{x^2} \tag{26.5}$$

Consequently, the net electric field far from the source charges is

$$\vec{E}_{\text{net}}(x \gg d) = \frac{1}{4\pi\epsilon_0}\frac{(3q)}{x^2}\hat{\imath} \tag{26.6}$$

As expected, this is the field of a point charge $3q$. These checks of limiting cases provide confidence in the result of the calculation.

It's often useful to represent the electric field strength as a graph. This is especially true when the algebraic expression is complicated. Figure 26.5 is a graph of E_{net} for the three charges of Example 26.1. Although we don't have any numerical values, we can specify x as a multiple of the charge separation d. Notice how the graph matches the field of a single point charge when $x \ll d$ and matches the field of a point charge $3q$ when $x \gg d$.

The Electric Field of a Dipole

Two opposite charges $\pm q$ separated by a small distance s form an *electric dipole*. You learned in Chapter 25 that an electric dipole is created when a neutral atom is polarized by an external charge. That was an example of an *induced electric dipole* because the polarization was caused, or induced, by the electric field of the external charge. There are also *permanent electric dipoles* in which two oppositely charged particles maintain a small permanent separation. Some molecules, such as the water molecule, are permanent dipoles due to the way in which the electrons are distributed between the atoms. Figure 26.6 shows two examples of electric dipoles.

Although a dipole has zero net charge, it *does* have a net electric field. Figure 26.7 shows an electric dipole aligned along the y-axis. It makes no difference whether this dipole is induced or permanent. You can see that $\vec{E}_{\text{dipole}}$ is in the negative y-direction at points along the x-axis because the x-components of $\vec{E}_+$ and $\vec{E}_-$ cancel. $\vec{E}_{\text{dipole}}$ is in the positive y-direction at points along the y-axis because E_+ is slightly larger than E_-.

Let's calculate the electric field of a dipole at a point on the axis of the dipole. This is the y-axis in Figure 26.7. The point is distance $r_+ = y - s/2$ from the positive charge and $r_- = y + s/2$ from the negative charge. The net electric field at this point has only a y-component, and the superposition of the fields of the two point charges gives

$$(E_{\text{dipole}})_y = (E_+)_y + (E_-)_y = \frac{1}{4\pi\epsilon_0}\frac{q}{(y - \frac{1}{2}s)^2} + \frac{1}{4\pi\epsilon_0}\frac{(-q)}{(y + \frac{1}{2}s)^2}$$

$$= \frac{q}{4\pi\epsilon_0}\left[\frac{1}{(y - \frac{1}{2}s)^2} - \frac{1}{(y + \frac{1}{2}s)^2}\right] \tag{26.7}$$

If we combine the two terms over a common denominator, Equation 26.7 becomes

$$(E_{\text{dipole}})_y = \frac{q}{4\pi\epsilon_0}\left[\frac{2ys}{(y - \frac{1}{2}s)^2(y + \frac{1}{2}s)^2}\right] \tag{26.8}$$

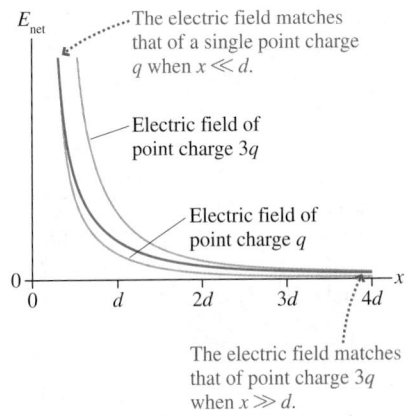

The electric field matches that of a single point charge q when $x \ll d$.

Electric field of point charge $3q$

Electric field of point charge q

The electric field matches that of point charge $3q$ when $x \gg d$.

FIGURE 26.5 The electric field strength along a line perpendicular to three equal point charges.

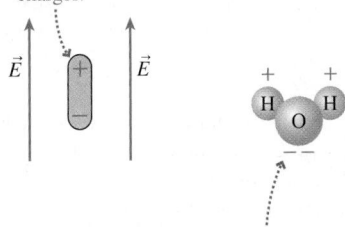

This dipole is *induced*, or stretched, by the electric field acting on the + and − charges.

A water molecule is a *permanent* dipole because the negative electrons spend more time with the oxygen atom.

FIGURE 26.6 Induced and permanent electric dipoles.

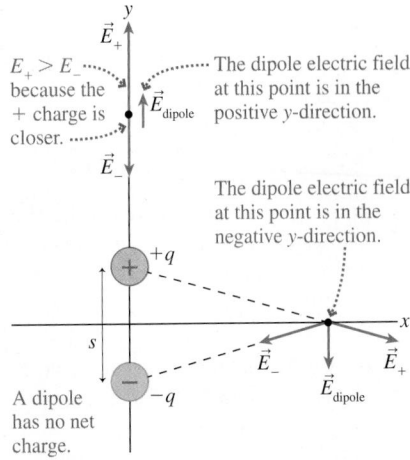

$E_+ > E_-$ because the + charge is closer.

The dipole electric field at this point is in the positive y-direction.

The dipole electric field at this point is in the negative y-direction.

A dipole has no net charge.

FIGURE 26.7 The dipole electric field at two points.

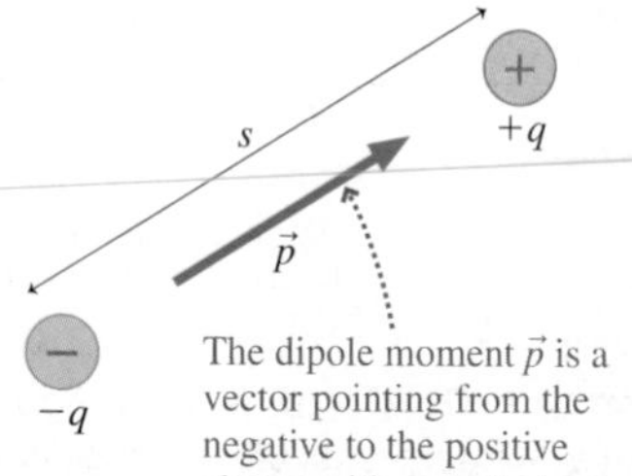

We omitted some of the algebraic steps, but be sure you can do this yourself. Some of the homework problems will require similar algebra.

In practice, we almost always observe the electric field of a dipole only for distances $y \gg s$; that is, only for distances much larger than the charge separation. In such cases, the denominator can be approximated $(y - \frac{1}{2}s)^2(y + \frac{1}{2}s)^2 \approx y^4$. With this approximation, Equation 26. 8 becomes

$$(E_{\text{dipole}})_y \approx \frac{1}{4\pi\epsilon_0}\frac{2qs}{y^3} \tag{26.9}$$

It is useful to define the **dipole moment** $\vec{p}$, shown in Figure 26.8, as the vector

$$\vec{p} = (qs, \text{ from the negative to the positive charge}) \tag{26.10}$$

The direction of $\vec{p}$ identifies the orientation of the dipole, and the dipole-moment magnitude $p = qs$ determines the electric field strength. The SI units of the dipole moment are C m.

We can use the dipole moment to write a succinct expression for the electric field at a point on the axis of a dipole:

$$\vec{E}_{\text{dipole}} \approx \frac{1}{4\pi\epsilon_0}\frac{2\vec{p}}{r^3} \quad \text{(on the axis of an electric dipole)} \tag{26.11}$$

where r is the distance measured from the *center* of the dipole. We've switched from y to r because we've now specified that Equation 26.11 is valid only along the axis of the dipole. Notice that the electric field along the axis points in the direction of the dipole moment $\vec{p}$.

A homework problem will let you calculate the electric field in the plane that bisects and is perpendicular to the dipole. This is the field shown on the x-axis in Figure 26.7, but it could equally well be the field on the z-axis as it comes out of the page. The field, for $r \gg s$, is

$$\vec{E}_{\text{dipole}} \approx -\frac{1}{4\pi\epsilon_0}\frac{\vec{p}}{r^3} \tag{26.12}$$

(in the plane perpendicular to an electric dipole)

This field is *opposite* to $\vec{p}$, and it is only half the strength of the on-axis field at the same distance.

NOTE ▶ Do these inverse-cube equations violate Coulomb's law? Not at all. Coulomb's law describes the force between two *point charges,* and from Coulomb's law we found that the electric field of a *point charge* varies with the inverse square of the distance. But a dipole is not a point charge. The field of a dipole decreases more rapidly than that of a point charge, which is to be expected because the dipole is, after all, electrically neutral. ◀

The dipole moment $\vec{p}$ is a vector pointing from the negative to the positive charge with magnitude qs.

FIGURE 26.8 The dipole moment.

EXAMPLE 26.2 The electric field of a water molecule
The water molecule H_2O has a permanent dipole moment of magnitude 6.2×10^{-30} C m. What is the electric field strength 1.0 nm from a water molecule at a point on the axis of the dipole?

MODEL The size of a molecule is ≈ 0.1 nm. Thus $r \gg s$, and we can use Equation 26.11 for the on-axis electric field of the molecule's dipole moment.

SOLVE The electric field strength at $r = 1.0$ nm on the dipole axis is

$$E \approx \frac{1}{4\pi\epsilon_0}\frac{2p}{r^3} = (9.0 \times 10^9 \,\text{N m}^2/\text{C}^2)\frac{2(6.2 \times 10^{-30}\,\text{C m})}{(1.0 \times 10^{-9}\,\text{m})^3}$$

$$= 1.1 \times 10^8 \,\text{N/C}$$

ASSESS By referring to Table 26.1 you can see that the field strength is "strong" compared to our everyday experience with charged objects but "weak" compared to the electric field inside the atoms themselves. This seems reasonable.

Picturing the Electric Field

We can't see the electric field. Consequently, we need pictorial tools to help us visualize it in a region of space. One method, which is introduced in Chapter 25, is to picture the electric field by drawing electric field vectors at various points in space. Another way to picture the field is to draw **electric field lines.** These are imaginary lines drawn through a region of space so that

- The tangent to a field line at any point is in the direction of the electric field $\vec{E}$ at that point, and
- The field lines are closer together where the electric field strength is larger.

Figure 26.9 shows the relationship between electric field lines and electric field vectors in one region of space.

Electric field lines have two other important properties:

- Every field line starts on a positive charge and ends on a negative charge.
- Field lines cannot cross.

The first of these properties follows from the fact that the electric field is created by charges. The second property is required to make sure that $\vec{E}$ has a unique direction at every point in space.

Figure 26.10a represents the electric field of a dipole as a field-vector diagram. Figure 26.10b shows the same field using electric field lines. Notice how the on-axis field points in the direction of $\vec{p}$, both above and below the dipole, while the field in the bisecting plane points opposite to $\vec{p}$. At most points, however, $\vec{E}$ has components both parallel to $\vec{p}$ and perpendicular to $\vec{p}$.

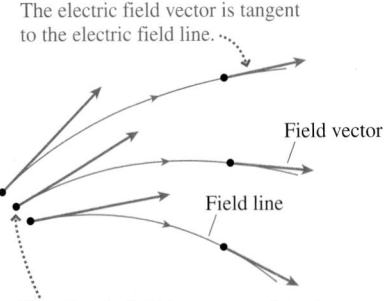

The electric field vector is tangent to the electric field line.

Field vector

Field line

The electric field is stronger where the electric field vectors are longer and where the electric field lines are closer together.

FIGURE 26.9 Electric field lines and electric field vectors.

11.5, 11.6

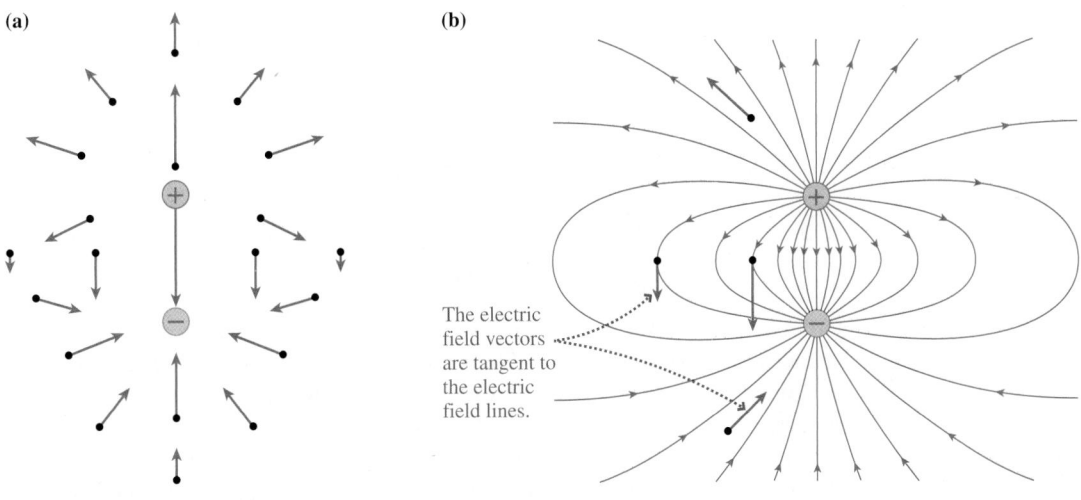

The electric field vectors are tangent to the electric field lines.

FIGURE 26.10 The electric field of a dipole.

A somewhat similar electric field is produced by two same-sign charges, as Figure 26.11 on the next page shows, using two positive charges. Figure 26.11a represents the field as vectors, Figure 26.11b represents it as electric field lines.

Neither field-vector diagrams nor field-line diagrams are perfect pictorial representations of an electric field. The field vectors are somewhat harder to draw, and they show the field at only a few points, but they do clearly indicate the direction and strength of the electric field at those points. Field-line diagrams perhaps look more elegant, and they're sometimes easier to sketch, but there's no formula for knowing where to draw the lines and it's harder to infer the actual direction and strength of the electric field.

(a)   **(b)**

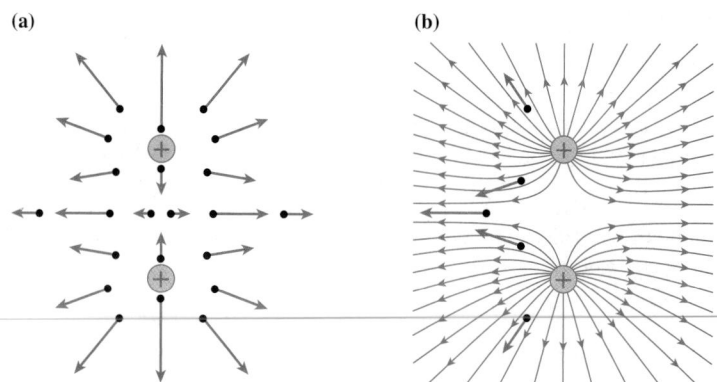

FIGURE 26.11 The electric field of two equal positive charges.

There simply is no pictorial way to show *exactly* what the field is. Only the mathematical representation is exact. We'll use both field-vector diagrams and field-line diagrams, depending upon the circumstances, but you'll see that the preference of this text is usually to use a field-vector diagram.

STOP TO THINK 26.1   At the position of the dot,
the electric field points

   a. Left.
   b. Right.
   c. Up.
   d. Down.
   e. The electric field is zero.

(a)   Charge Q on a rod of
length L. The linear
charge density is
$\lambda = Q/L$.

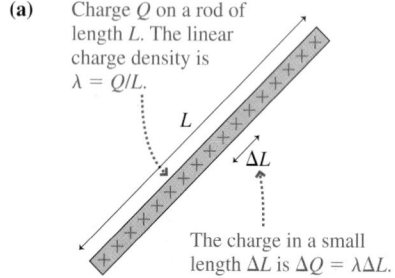

The charge in a small
length ΔL is $\Delta Q = \lambda \Delta L$.

(b)   Charge Q on a surface of area A. The
surface charge density is $\eta = Q/A$.

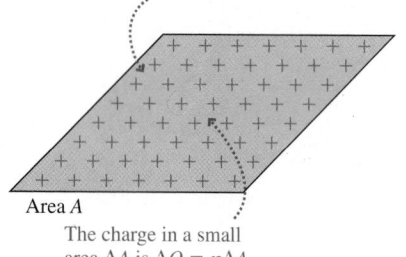

Area A

The charge in a small
area ΔA is $\Delta Q = \eta \Delta A$.

FIGURE 26.12 One-dimensional and
two-dimensional continuous charge
distributions.

26.3 The Electric Field of a Continuous Charge Distribution

Ordinary objects—tables, chairs, beakers of water, flowing streams—seem to our senses to be continuous distributions of matter. There is no obvious evidence for an atomic structure, even though we have good reasons to believe that we would find atoms if we subdivided the matter sufficiently far. Thus it is easier, for many practical purposes, to consider matter to be continuous and to talk about the *density* of matter. Density—the number of kilograms of matter per cubic meter—allows us to describe the *distribution* of matter *as if* the matter were continuous rather than atomic.

Much the same situation occurs with charge. If a charged object contains a large number of excess electrons—for example, 10^{12} extra electrons on a metal rod—it is not practical to keep track of every electron. It makes more sense to consider the charge to be *continuous* and to describe how it is *distributed* over the object.

Figure 26.12a shows an object of length L, such as a plastic rod or a metal wire, with charge Q spread uniformly along it. (We will use an uppercase Q for the total charge of an object, reserving lowercase q for individual point charges.) The **linear charge density** λ is defined to be

$$\lambda = \frac{Q}{L} \qquad (26.13)$$

Linear charge density, which has units of C/m, is the amount of charge *per meter* of length. The linear charge density of a 20-cm-long wire with 40 nC of charge is 2.0 nC/cm or 2.0×10^{-7} C/m.

NOTE ▶ The linear charge density λ is analogous to the linear mass density μ that you used in Chapter 20 to find the speed of a wave on a string. ◀

We'll also be interested in charged surfaces. Figure 26.12b shows a two-dimensional distribution of charge across a surface of area A. We define the **surface charge density** η (lowercase Greek eta) to be

$$\eta = \frac{Q}{A} \tag{26.14}$$

Surface charge density, with units of C/m², measures the amount of charge *per square meter*. A 1.0 mm $\times$ 1.0 mm square on a surface with $\eta = 2.0 \times 10^{-4}$ C/m² contains 2.0×10^{-10} C or 0.20 nC of charge.

Figure 26.12 and the definitions of Equations 26.13 and 26.14 assume that the object is **uniformly charged,** meaning that the charges are evenly spread over the object. We will assume objects to be uniformly charged unless noted otherwise.

NOTE ▶ Some textbooks represent the surface charge density with the symbol σ. Because σ is also used to represent *conductivity,* an idea we'll introduce in Chapter 28, we've selected a different symbol for surface charge density. ◀

STOP TO THINK 26.2 A piece of plastic is uniformly charged with surface charge density η_1. The plastic is then broken into a large piece with surface charge density η_2 and a small piece with surface charge density η_3. Rank in order, from largest to smallest, the surface charge densities η_1 to η_3.

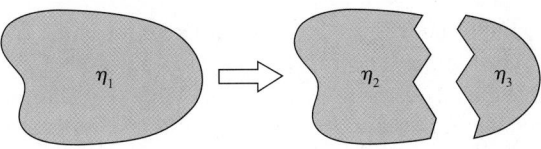

A Problem-Solving Strategy

Our goal is to find the electric field of a continuous distribution of charge, such as a charged rod or a charged disk. We have two basic tools to work with:

- The electric field of a point charge, and
- The principle of superposition.

We can apply these tools to a continuous distribution of charge if we follow a three-part strategy:

1. Divide the total charge Q into many small point-like charges ΔQ.
2. Use our knowledge of the electric field of a point charge to find the electric field of each ΔQ.
3. Calculate the net field $\vec{E}_{\text{net}}$ by summing the fields of all the ΔQ.

In practice, as you may have guessed, we'll let the sum become an integral.

The difficulty with electric field calculations is not the summation or integration itself, which is the last step, but in setting up the calculation and knowing *what* to integrate. We will go step-by-step through several examples to illustrate the procedures. However, we first need to flesh out the steps of the problem-solving strategy. The aim of this problem-solving strategy is to break a difficult problem down into small steps that are individually manageable.

PROBLEM-SOLVING STRATEGY 26.2 **The electric field of a continuous distribution of charge**

MODEL Model the distribution as a simple shape, such as a line of charge or a disk of charge. Assume the charge is uniformly distributed.

VISUALIZE For the pictorial representation:

❶ Draw a picture and establish a coordinate system.
❷ Identify the point P at which you want to calculate the electric field.
❸ Divide the total charge Q into small pieces of charge ΔQ, using shapes for which you *already know* how to determine $\vec{E}$. This is often, but not always, a division into point charges.
❹ Draw the electric field vector at P for one or two small pieces of charge. This will help you identify distances and angles that need to be calculated.
❺ Look for symmetries of the charge distribution that simplify the field. You may conclude that some components of $\vec{E}$ are zero.

SOLVE The mathematical representation is $\vec{E}_{net} = \sum \vec{E}_i$.

■ Use superposition to form an algebraic expression for *each* of the three components of $\vec{E}$ (unless you are sure one or more is zero) at point P.
■ Let the (x, y, z) coordinates of the point remain as variables.
■ Replace the small charge ΔQ with an equivalent expression involving a charge density and a coordinate, such as dx, that describes the shape of charge ΔQ. **This is the critical step in making the transition from a sum to an integral** because you need a coordinate to serve as the integration variable.
■ All angles and distances must be expressed in terms of the coordinates.
■ Let the sum become an integral. The integration will be over the coordinate variable that is related to ΔQ. The integration limits for this variable will depend on your choice of coordinate system. Carry out the integration and simplify the result as much as possible.

ASSESS Check that your result is consistent with any limits for which you know what the field should be.

EXAMPLE 26.3 **The electric field of a line of charge**

Figure 26.13 shows a thin, uniformly charged rod of length L with total charge Q that can be either positive or negative. Find the electric field strength at distance r in the plane that bisects the rod.

MODEL The rod is thin, so we'll assume the charge lies along a line and forms what we call a *line of charge*. This is an important charge distribution that models the electric field of a charged rod or a charged metal wire. The rod's linear charge density is $\lambda = Q/L$.

VISUALIZE Figure 26.14 illustrates the five steps of the problem-solving strategy. We've chosen a coordinate system in which the rod lies along the y-axis and point P, in the bisecting plane, is on the x-axis. We've then divided the rod into N small segments of charge ΔQ, each of which can be modeled as a point charge. For

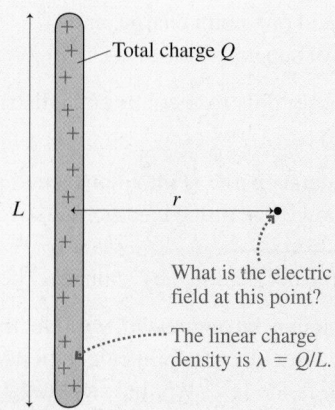

FIGURE 26.13 A thin, uniformly charged rod.

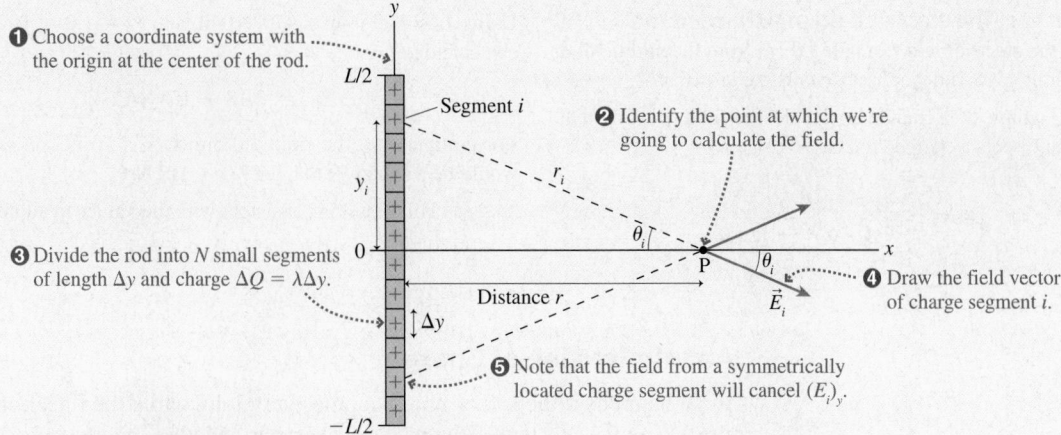

❶ Choose a coordinate system with the origin at the center of the rod.

Segment i

❷ Identify the point at which we're going to calculate the field.

❸ Divide the rod into N small segments of length Δy and charge $\Delta Q = \lambda \Delta y$.

❹ Draw the field vector of charge segment i.

Distance r

❺ Note that the field from a symmetrically located charge segment will cancel $(E_i)_y$.

FIGURE 26.14 Calculating the electric field of a line of charge.

every ΔQ in the bottom half of the wire with a field that points to the right and up, there's a matching ΔQ in the top half whose field points to the right and down. The y-components of these two fields cancel, hence the net electric field on the x-axis points straight away from the rod. The only component we need to calculate is E_x. (This is the same reasoning on the basis of symmetry that we used in Example 26.1.)

SOLVE Each of the little segments of charge can be modeled as a point charge. We know the electric field of a point charge, so we can write the x-component of $\vec{E}_i$, the electric field of segment i, as

$$(E_i)_x = E_i \cos \theta_i = \frac{1}{4\pi\epsilon_0} \frac{\Delta Q}{r_i^2} \cos \theta_i$$

where r_i is the distance from charge i to point P. You can see from the figure that $r_i = (y_i^2 + r^2)^{1/2}$ and $\cos \theta_i = r/r_i = r/(y_i^2 + r^2)^{1/2}$. With these, $(E_i)_x$ is

$$(E_i)_x = \frac{1}{4\pi\epsilon_0} \frac{\Delta Q}{y_i^2 + r^2} \frac{r}{\sqrt{y_i^2 + r^2}}$$

$$= \frac{1}{4\pi\epsilon_0} \frac{r \Delta Q}{(y_i^2 + r^2)^{3/2}}$$

Compare this result to the very similar calculation we did in Example 26.1. If we now sum this expression over all the charge segments, the net x-component of the electric field is

$$E_x = \sum_{i=1}^{N} (E_i)_x = \frac{1}{4\pi\epsilon_0} \sum_{i=1}^{N} \frac{r \Delta Q}{(y_i^2 + r^2)^{3/2}}$$

This is the same superposition we did for the $N = 3$ case in Example 26.1. The only difference is that we have now written the result as an explicit summation so that N can have any value. We want to let $N \to \infty$ and to replace the sum with an integral, but we can't integrate over Q; it's not a geometric quantity. This is where the linear charge density enters. The quantity of charge in each segment is related to its length Δy by $\Delta Q = \lambda \Delta y = (Q/L)\Delta y$. In terms of the linear charge density, the electric field is

$$E_x = \frac{Q/L}{4\pi\epsilon_0} \sum_{i=1}^{N} \frac{r \Delta y}{(y_i^2 + r^2)^{3/2}}$$

Now we're ready to let the sum become an integral. If we let $N \to \infty$, then each segment becomes an infinitesimal length $\Delta y \to dy$ while the discrete position variable y_i becomes the continuous integration variable y. The sum from $i = 1$ at the bottom end of the line of charge to $i = N$ at the top end will be replaced with an integral from $y = -L/2$ to $y = +L/2$. Thus in the limit $N \to \infty$,

$$E_x = \frac{Q/L}{4\pi\epsilon_0} \int_{-L/2}^{L/2} \frac{r \, dy}{(y^2 + r^2)^{3/2}}$$

This is a standard integral that you have learned to do in calculus and that can be found in any table of integrals. Note that r is a *constant* as far as this integral is concerned. Integrating gives

$$E_x = \frac{Q/L}{4\pi\epsilon_0} \frac{y}{r\sqrt{y^2 + r^2}} \bigg|_{-L/2}^{L/2}$$

$$= \frac{Q/L}{4\pi\epsilon_0} \left[\frac{L/2}{r\sqrt{(L/2)^2 + r^2}} - \frac{-L/2}{r\sqrt{(-L/2)^2 + r^2}} \right]$$

$$= \frac{1}{4\pi\epsilon_0} \frac{Q}{r\sqrt{r^2 + (L/2)^2}}$$

Because E_x is the *only* component of the field, the electric field strength E_{rod} at distance r from the center of a charged rod is

$$E_{\text{rod}} = \frac{1}{4\pi\epsilon_0} \frac{|Q|}{r\sqrt{r^2 + (L/2)^2}}$$

The field strength must be positive, so we added absolute value signs to Q to allow for the possibility that the charge could be negative. The only restriction is to remember that this is the electric field at a point in the plane that bisects the rod.

ASSESS Suppose we are at a point *very* far from the rod. If $r \gg L$, the length of the rod is not relevant and the rod appears to be a point charge Q in the distance. Thus in the *limiting case* $r \gg L$, we expect the rod's electric field to be that of a point charge. If $r \gg L$, the square root becomes $(r^2 + (L/2)^2)^{1/2} \approx (r^2)^{1/2} = r$ and the electric field strength at distance r becomes $E_{\text{rod}} \approx Q/4\pi\epsilon_0 r^2$. The fact that our expression of E_{rod} has the correct limiting behavior gives us confidence that we haven't made any mistakes in its derivation.

EXAMPLE 26.4   The electric field of a charged rod

What is the electric field strength 1.0 cm from the middle of an 8.0-cm-long glass rod that has been charged to 10 nC?

SOLVE Example 26.3 found that the electric field strength in the plane that bisects a charged rod is

$$E_{\text{rod}} = \frac{1}{4\pi\epsilon_0} \frac{|Q|}{r\sqrt{r^2 + (L/2)^2}}$$

Using $L = 0.080$ m, $r = 0.010$ m, and $Q = 1.0 \times 10^{-8}$ C, we can calculate

$$E_{\text{rod}} = 2.18 \times 10^5 \text{ N/C}.$$

For comparison, the field 1.0 cm from a 10 nC *point* charge would be a somewhat larger 9.0×10^5 N/C.

ASSESS This result is consistent with the values in Table 26.1.

An Infinite Line of Charge

What happens if the rod or wire becomes very long while the linear charge density λ remains constant? That is, more charge is added so that the ratio $\lambda = |Q|/L$ stays constant as L increases. In the limit that L approaches infinity, the electric field strength becomes

$$E_{\text{line}} = \lim_{L\to\infty} \frac{1}{4\pi\epsilon_0} \frac{|Q|}{r\sqrt{r^2 + (L/2)^2}} = \frac{1}{4\pi\epsilon_0} \frac{|Q|}{rL/2} = \frac{1}{4\pi\epsilon_0} \frac{2|\lambda|}{r} \qquad (26.15)$$

This is the field strength of an infinitely long **line of charge** having linear charge density λ. The linear charge density measures how close together or far apart the charges are on the rod, and this distance did not change as the rod's length L was increased.

NOTE ▶ Unlike a point charge, for which the field decreases as $1/r^2$, the field of an infinitely long charged wire decreases more slowly—as only $1/r$. ◀

Equation 26.15 is of considerable practical significance. Although no real wire is infinitely long, the fact that the field of a point charge decreases inversely with the square of the distance means that the electric field at a point near the wire is determined primarily by the nearest charges on the wire. Over most of the length of a wire, the ends of the wire are too far away to make any significant contribution. Consequently, the field of a realistic finite-length wire is well approximated by Equation 26.15, the field of an infinitely long line of charge, except at points near the end of wire.

Figure 26.15 shows the electric field vectors of an infinite line of positive charge. Notice that the field points straight away from the line at all points and the field strength decreases as the distance increases. The vectors would point inward for a negative line of charge.

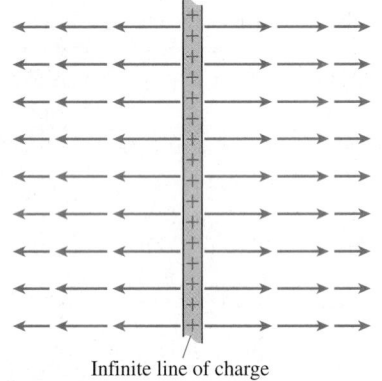

Infinite line of charge

FIGURE 26.15 The electric field of an infinite line of charge.

STOP TO THINK 26.3   Which of the following actions will increase the electric field strength at the position of the dot?

a. Make the rod longer without changing the charge.
b. Make the rod shorter without changing the charge.
c. Make the rod wider without changing the charge.
d. Make the rod narrower without changing the charge.
e. Add charge to the rod.
f. Remove charge from the rod.
g. Move the dot farther from the rod.
h. Move the dot closer to the rod.

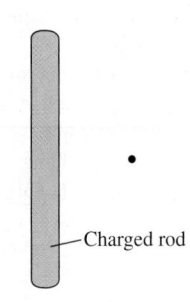

Charged rod

26.4 The Electric Fields of Rings, Planes, and Spheres

In this section we'll derive the electric fields for three important and related charge distributions: a ring of charge, a disk of charge, and a plane of charge. The ring of charge is the most fundamental and will be the basis for determining the other two. We'll also look at the electric field of a sphere of charge.

EXAMPLE 26.5 The electric field of a ring of charge
Figure 26.16 shows a thin, uniformly charged ring of radius R with total charge Q. Find the electric field at a point on the axis of the ring (perpendicular to the page).

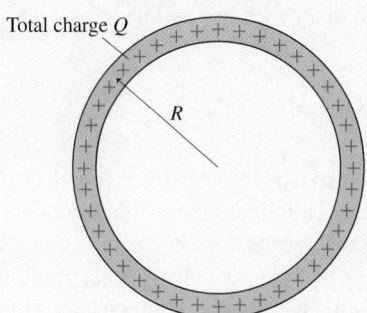

Total charge Q

R

FIGURE 26.16 A thin, uniformly charged ring.

MODEL Because the ring is thin, we'll assume the charge lies along circle of radius R. You can think of this as a line of charge of length $2\pi R$ wrapped into a circle. The linear charge density along the ring is $\lambda = Q/2\pi R$.

VISUALIZE Figure 26.17 illustrates the five steps of the problem-solving strategy. We've chosen a coordinate system in which

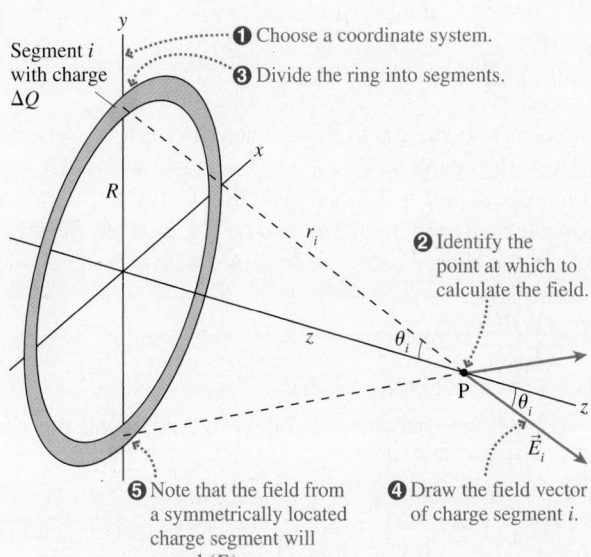

❶ Choose a coordinate system.
Segment i with charge ΔQ
❸ Divide the ring into segments.
y
x
R
r_i
❷ Identify the point at which to calculate the field.
z
θ_i
P
θ_i
z
$\vec{E}_i$
❺ Note that the field from a symmetrically located charge segment will cancel $(E_i)_y$.
❹ Draw the field vector of charge segment i.

FIGURE 26.17 Calculating the on-axis electric field of a ring of charge.

the ring lies in the xy-plane and point P is on the z-axis. We've then divided the ring into N small segments of charge ΔQ, each of which can be modeled as a point charge. For every pair of segments that are diametrically opposed, such as segments i and j in the figure, the components of their fields perpendicular to the axis will cancel. The symmetry of the ring tells us that the field at a point on the axis will point along the z-axis; outward for a positive ring, inward for a negative ring. Thus we need to calculate only the z-component E_z.

SOLVE The z-component of the electric field due to segment i is

$$(E_i)_z = E_i \cos\theta_i = \frac{1}{4\pi\epsilon_0} \frac{\Delta Q}{r_i^2} \cos\theta_i$$

You can see from the figure that *every* segment of the ring, independent of i, has

$$r_i = \sqrt{z^2 + R^2}$$
$$\cos\theta_i = \frac{z}{r_i} = \frac{z}{\sqrt{z^2 + R^2}}$$

Consequently, the field of segment i is

$$(E_i)_z = \frac{1}{4\pi\epsilon_0} \frac{\Delta Q}{z^2 + R^2} \frac{z}{\sqrt{z^2 + R^2}} = \frac{1}{4\pi\epsilon_0} \frac{z}{(z^2 + R^2)^{3/2}} \Delta Q$$

The net electric field is found by summing $(E_i)_z$ due to all N segments:

$$E_z = \sum_{i=1}^{N} (E_i)_z = \frac{1}{4\pi\epsilon_0} \frac{z}{(z^2 + R^2)^{3/2}} \sum_{i=1}^{N} \Delta Q$$

We were able to bring all terms involving z to the front because z is a constant as far as the summation is concerned. Surprisingly, we don't need to convert the sum to an integral to complete this calculation. The sum of all the ΔQ around the ring is simply the ring's total charge, $\sum \Delta Q = Q$, hence the electric field on the axis of a charged ring is

$$(E_{\text{ring}})_z = \frac{1}{4\pi\epsilon_0} \frac{zQ}{(z^2 + R^2)^{3/2}}$$

This expression is valid for both positive and negative z (i.e., on either side of the ring) and for both positive and negative charge.

ASSESS It will be left as a homework problem to show that this result gives the expected limit when $z \gg R$.

(a)

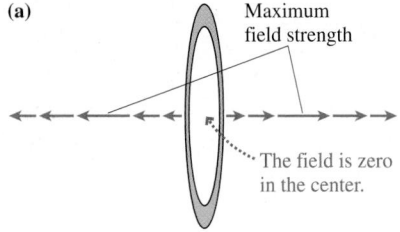

Maximum field strength

The field is zero in the center.

(b)

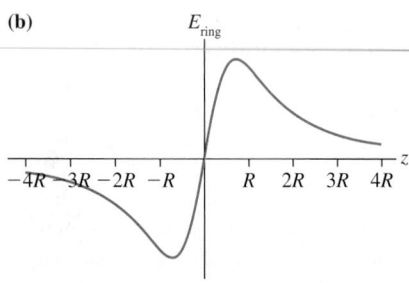

E_{ring}

$-4R$ $-3R$ $-2R$ $-R$ R $2R$ $3R$ $4R$ z

FIGURE 26.18 The on-axis electric field of a ring of charge.

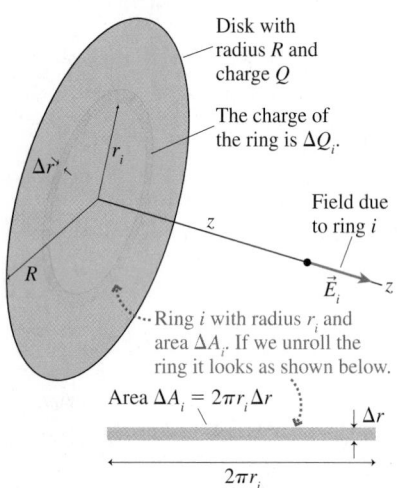

Disk with radius R and charge Q

The charge of the ring is ΔQ_i.

Δr

r_i

z

Field due to ring i

$\vec{E}_i$ z

R

Ring i with radius r_i and area ΔA_i. If we unroll the ring it looks as shown below.

Area $\Delta A_i = 2\pi r_i \Delta r$

Δr

$2\pi r_i$

FIGURE 26.19 Calculating the on-axis field of a charged disk.

Figure 26.18 shows two representations of the on-axis electric field of a positively charged ring. Figure 26.18a shows that the electric field vectors point away from the ring, increasing in length until reaching a maximum when $|z| \approx R$, then decreasing. The graph of $(E_{ring})_z$ in Figure 26.18b confirms that the field strength has a maximum on either side of the ring. Notice that the electric field at the center of the ring is zero, even though this point is surrounded by charge. You might want to spend a minute thinking about why this has to be the case.

A Disk of Charge

Our goal is to find the electric field of an infinitely wide charged plane because it is one of our basic electric field models. Most of the work is now done. We'll first use the result for a ring of charge to find the on-axis electric field of a *disk* of charge. Then we'll let the disk expand until it becomes a plane of charge.

Figure 26.19 shows a disk having radius R that is uniformly charged with charge Q. This is a mathematical disk, with no thickness, and its surface charge density is

$$\eta = \frac{Q}{A} = \frac{Q}{\pi R^2} \tag{26.16}$$

We would like to calculate the on-axis electric field of this disk. Our problem-solving strategy tells us to divide a continuous charge into segments for which we already know how to find $\vec{E}$. Because we now know the on-axis electric field of a ring of charge, let's divide the disk into N very narrow rings of radius r and width Δr. One such ring, with radius r_i and charge ΔQ_i, is shown.

We need to be careful with notation. The R in Example 26.5 was the radius of the ring. Now we have many rings, and the radius of ring i is r_i. Similarly, Q was the charge on the ring. Now the charge on ring i is ΔQ_i, a small fraction of the total charge on the disk. With these changes, the electric field of ring i, with radius r_i, is

$$(E_i)_z = \frac{1}{4\pi\epsilon_0} \frac{z\Delta Q_i}{(z^2 + r_i^2)^{3/2}} \tag{26.17}$$

The on-axis electric field of the charged disk is the sum of the electric fields of all of the rings:

$$(E_{disk})_z = \sum_{i=1}^{N} (E_i)_z = \frac{z}{4\pi\epsilon_0} \sum_{i=1}^{N} \frac{\Delta Q_i}{(z^2 + r_i^2)^{3/2}} \tag{26.18}$$

The critical step, as always, is to relate ΔQ to a coordinate. Because we now have a surface, rather than a line, the charge in ring i is $\Delta Q = \eta \Delta A_i$, where ΔA_i is the area of ring i. We can find ΔA_i, as you've learned to do in calculus, by "unrolling" the ring to form a narrow rectangle of length $2\pi r_i$ and height Δr. Thus the area of ring i is $\Delta A_i = 2\pi r_i \Delta r$ and the charge is $\Delta Q_i = 2\pi \eta r_i \Delta r$. With this substitution, Equation 26.18 becomes

$$(E_{disk})_z = \frac{\eta z}{2\epsilon_0} \sum_{i=1}^{N} \frac{r_i \Delta r}{(z^2 + r_i^2)^{3/2}} \tag{26.19}$$

As $N \rightarrow \infty$, $\Delta r \rightarrow dr$ and the sum becomes an integral. Adding all the rings means integrating from $r = 0$ to $r = R$, thus

$$(E_{disk})_z = \frac{\eta z}{2\epsilon_0} \int_0^R \frac{r\,dr}{(z^2 + r^2)^{3/2}} \tag{26.20}$$

All that remains is to carry out the integration. This is straightforward if we make the variable change $u = z^2 + r^2$. Then $du = 2r\,dr$ or, equivalently, $r\,dr = \frac{1}{2}du$. At the lower integration limit $r = 0$, our new variable is $u = z^2$. At the upper limit $r = R$, the new variable is $u = z^2 + R^2$.

NOTE ▶ When changing variables in a definite integral, you *must* also change the limits of integration. ◀

With this variable change the integral becomes

$$(E_{\text{disk}})_z = \frac{\eta z}{2\epsilon_0} \frac{1}{2} \int_{z^2}^{z^2+R^2} \frac{du}{u^{3/2}} = \frac{\eta z}{4\epsilon_0} \frac{-2}{u^{1/2}} \Big|_{z^2}^{z^2+R^2} = \frac{\eta z}{2\epsilon_0} \left[\frac{1}{z} - \frac{1}{\sqrt{z^2 + R^2}} \right] \quad (26.21)$$

If we multiply through by z, the on-axis electric field of a charged disk with $\eta = Q/\pi R^2$ is

$$(E_{\text{disk}})_z = \frac{\eta}{2\epsilon_0} \left[1 - \frac{z}{\sqrt{z^2 + R^2}} \right] \quad (26.22)$$

It's a bit difficult see what Equation 26.22 is telling us, so let's compare it to what we already know. First, you can see that the quantity in square brackets is dimensionless. The surface charge density $\eta = Q/A$ has the same units as q/r^2, so $\eta/2\epsilon_0$ has the same units as $q/4\pi\epsilon_0 r^2$. This tells us that $\eta/2\epsilon_0$ really is an electric field.

Next, let's move very far away from the disk. At distance $z \gg R$, the disk appears to be a point charge Q in the distance and the field of the disk should approach that of a point charge. If we let $z \to \infty$ in Equation 26.22, so that $z^2 + R^2 \approx z^2$, we find $(E_{\text{disk}})_z \to 0$. This is true, but not quite what we wanted. We need to let z be very large in comparison to R, but not so large as to make E_{disk} vanish. That requires a little more care in taking the limit.

We can cast Equation 26.22 into a somewhat more useful form by factoring the z^2 out of the square root to give

$$(E_{\text{disk}})_z = \frac{\eta}{2\epsilon_0} \left[1 - \frac{1}{\sqrt{1 + R^2/z^2}} \right] \quad (26.23)$$

Now $R^2/z^2 \ll 1$ if $z \gg R$, so the second term in the square brackets is of the form $(1 + x)^{-1/2}$ where $x \ll 1$. We can then use the *binomial approximation*

$$(1 + x)^n \approx 1 + nx \quad \text{if} \quad x \ll 1 \quad \text{(binomial approximation)}$$

to simplify the expression in square brackets:

$$1 - \frac{1}{\sqrt{1 + R^2/z^2}} = 1 - (1 + R^2/z^2)^{-1/2} \approx 1 - \left(1 + \left(-\frac{1}{2} \right) \frac{R^2}{z^2} \right) = \frac{R^2}{2z^2} \quad (26.24)$$

This is a good approximation when $z \gg R$. Substituting this approximation into Equation 26.23, we find that the electric field of the disk for $z \gg R$ is

$$(E_{\text{disk}})_z \approx \frac{\eta}{2\epsilon_0} \frac{R^2}{2z^2} = \frac{Q/\pi R^2}{4\epsilon_0} \frac{R^2}{z^2} = \frac{1}{4\pi\epsilon_0} \frac{Q}{z^2} \quad \text{if} \quad z \gg R \quad (26.25)$$

This is, indeed, the field of a point charge Q, giving us confidence in Equation 26.22 for the on-axis electric field of a disk of charge.

NOTE ▶ The binomial approximation is an important tool for looking at the limiting cases of electric fields. ◀

EXAMPLE 26.6 The electric field of a charged disk

A 10-cm-diameter plastic disk is charged uniformly with an extra 10^{11} electrons. What is the electric field 1.0 mm above the surface at a point near the center?

MODEL Model the plastic disk as a uniformly charged disk. We are seeking the on-axis electric field. Because the charge is negative, the field will point *toward* the disk.

SOLVE The total charge on the plastic square is $Q = N(-e) = -1.60 \times 10^{-8}$ C. The surface charge density is

$$\eta = \frac{Q}{A} = \frac{Q}{\pi R^2} = \frac{-1.60 \times 10^{-8} \text{ C}}{\pi (0.050 \text{ m})^2} = -2.04 \times 10^{-6} \text{ C/m}^2$$

The electric field at $z = 0.0010$ m, given by Equation 26.22, is

$$E_z = \frac{\eta}{2\epsilon_0}\left[1 - \frac{1}{\sqrt{1 + R^2/z^2}}\right] = \frac{-2.04 \times 10^{-6} \text{ C/m}^2}{2 \times 8.85 \times 10^{-12} \text{ C}^2/\text{N m}^2}$$

$$\times \left[1 - \frac{1}{\sqrt{1 + (0.050 \text{ m}/0.001 \text{ m})^2}}\right]$$

$$= -1.13 \times 10^5 \text{ N/C}$$

The minus sign indicates that the field points *toward,* rather than away from, the disk. As a vector,

$$\vec{E} = (1.13 \times 10^5 \text{ N/C, toward the disk})$$

ASSESS The total charge, -16 nC, is typical of the amount of charge produced on a small plastic object by rubbing or friction. Thus 10^5 N/C is a typical electric field strength near an object that has been charged by rubbing.

A Plane of Charge

Many electronic devices used charged, flat surfaces—disks, squares, rectangles, and so on—to steer electrons along the proper paths. These charged surfaces are called **electrodes.** Although any real electrode is finite in extent, we can often model an electrode as an infinite **plane of charge.** As long as the distance z to the plane is small in comparison to the distance to the edges, we can reasonably treat the edges *as if* they are infinitely far away.

The electric field of a plane of charge is found from the on-axis field of a charged disk by letting the radius $R \to \infty$. That is, a disk with infinite radius is an infinite plane. From Equation 26.22, we see that the electric field of a plane of charge with surface charge density η is:

$$E_{\text{plane}} = \frac{\eta}{2\epsilon_0} = \text{constant} \qquad (26.26)$$

This is a simple result, but what does it tell us? First, the field strength is directly proportional to the charge density η: More charge, bigger field. Second, and more interesting, the field strength is the same at *all* points in space, independent of the distance z. The field strength 1000 m from the plane is the same as the field strength 1 mm from the plane.

How can this be? It seems that the field should get weaker as you move away from the plane of charge. But remember that we are dealing with an *infinite* plane of charge. What does it mean to be "close to" or "far from" an infinite object? For a disk of finite radius R, whether a point at distance z is "close to" or "far from" the disk is a comparison of z to R. If $z \ll R$, the point is close to the disk. If $z \gg R$, the point is far from the disk. But as $R \to \infty$, we have no *scale* for distinguishing near and far. In essence, *every* point in space is "close to" a disk of infinite radius.

No real plane is infinite in extent, but we can interpret Equation 26.26 as saying that the field of a surface of charge, regardless of its shape, is a constant $\eta/2\epsilon_0$ for those points whose distance z to the surface is much smaller than their distance to the edge. Eventually, when $z \gg R$, the charged surface will begin to look like a point charge Q and the field will have to decrease as $1/z^2$.

Electrodes a few millimeters in size guided electrons through old-fashioned vacuum tubes. Modern field-effect transistors use an electrode, called a *gate,* that is only about 1 μm wide.

We do need to note that the derivation leading to Equation 26.26 considered only $z > 0$. For a positively charged plane, with $\eta > 0$, the electric field points *away from* the plane on both sides of the plane. This requires $E_z < 0$ ($\vec{E}$ pointing in the negative z-direction) on the side with $z < 0$. Thus a complete description of the electric field, valid for both sides of the plane and for either sign of η, is

$$(E_{\text{plane}})_z = \begin{cases} +\dfrac{\eta}{2\epsilon_0} & z > 0 \\[2ex] -\dfrac{\eta}{2\epsilon_0} & z < 0 \end{cases} \qquad (26.27)$$

Figure 26.20 shows two views of the electric field of a positively charged plane. All the arrows would be reversed for a negatively charged plane. Notice that the electric field on either side of the plane looks like the gravitational field near the surface of the earth. It would have been very difficult to anticipate this result from Coulomb's law or from the electric field of a single point charge, but step by step we have been able to use the concept of the electric field to look at increasingly complex distributions of charge.

The Electric Field of a Sphere of Charge

The one last charge distribution for which we need to know the electric field is a **sphere of charge.** This problem is analogous to wanting to know the gravitational field of a spherical planet or star. The procedure for calculating the field of a sphere of charge is the same as we used for lines and planes, but the integrations are significantly more difficult. We will skip the details of the calculations and, for now, simply assert the result without proof. In Chapter 27 we'll use an alternative procedure to find the field of a sphere of charge.

A sphere of charge Q and radius R, be it a uniformly charged sphere or just a spherical shell, has an electric field *outside* the sphere ($r \geq R$) that is exactly the same as that of a point charge Q located at the center of the sphere:

$$\vec{E}_{\text{sphere}} = \frac{Q}{4\pi\epsilon_0 r^2}\hat{r} \qquad \text{for } r \geq R \qquad (26.28)$$

This assertion is analogous to our earlier assertion that the gravitational force between stars and planets can be computed as if all the mass is at the center.

Figure 26.21 shows the electric field of a sphere of positive charge. The field of a negative sphere would point inward. Note that the field inside the sphere ($r < R$) is *not* given by Equation 26.28.

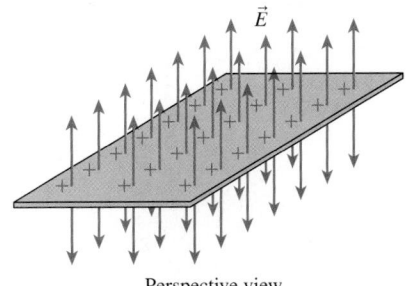

FIGURE 26.20 Perspective and edge views of the electric field of a positive plane of charge.

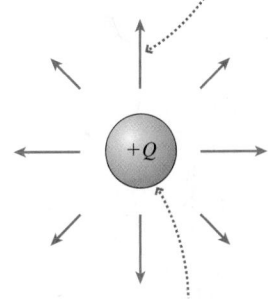

The electric field outside the sphere is the same as the field of a point charge Q at the center.

A positive sphere or spherical shell. We don't know what the electric field inside the sphere is.

FIGURE 26.21 The electric field of a sphere of positive charge.

STOP TO THINK 26.4   Rank in order, from largest to smallest, the electric field strengths E_a to E_e at these five points near a plane of charge.

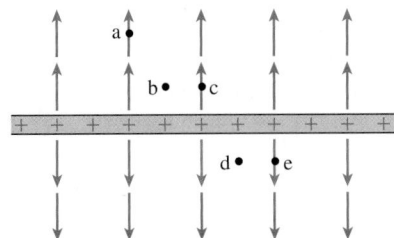

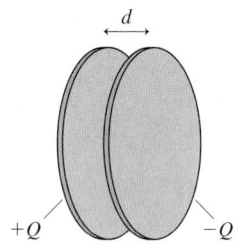

FIGURE 26.22 A parallel-plate capacitor.

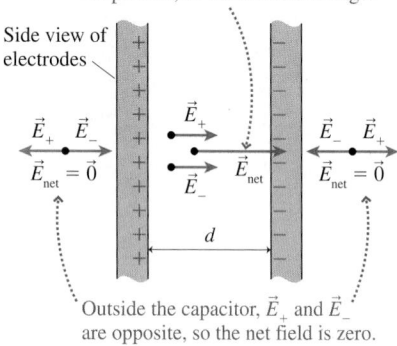

Inside the capacitor, $\vec{E}_+$ and $\vec{E}_-$ are parallel, so the net field is large.

Side view of electrodes

Outside the capacitor, $\vec{E}_+$ and $\vec{E}_-$ are opposite, so the net field is zero.

FIGURE 26.23 The electric fields inside and outside a parallel-plate capacitor.

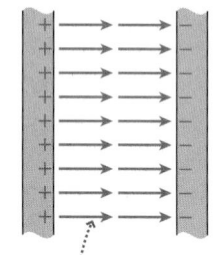

(a) Ideal capacitor

The field is constant, pointing from the positive to the negative electrode.

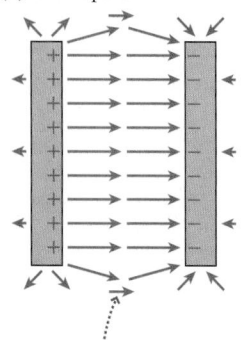

(b) Real capacitor

A weak fringe field extends outside the electrodes.

FIGURE 26.24 The electric field of a capacitor.

26.5 The Parallel-Plate Capacitor

Figure 26.22 shows two electrodes, one with charge $+Q$ and the other with $-Q$, placed face-to-face a distance d apart. This arrangement of two electrodes, charged equally but oppositely, is called a **parallel-plate capacitor.** Capacitors play important roles in many electric circuits. Our goal is to find the electric field both inside the capacitor (i.e., between the plates) and outside the capacitor.

NOTE ▶ The *net* charge of a capacitor is zero. Capacitors are charged by transferring electrons from one plate to the other. The plate that gains N electrons has charge $-Q = N(-e)$. The plate that loses electrons has charge $+Q$. ◄

Let's begin with a qualitative investigation. Figure 26.23 is an enlarged view of the capacitor plates, seen from the side. Because opposite charges attract, all of the charge is on the *inner* surfaces of the two plates. Thus the inner surfaces can be modeled as *charged planes* with equal but opposite surface charge densities. The electric field $\vec{E}_+$ of the positive plate points away from the charged surface. The field $\vec{E}_-$ of the negative plate points toward the surface. The figure shows the fields both between the plates and to the left and right of the capacitor.

NOTE ▶ You might think the right capacitor plate would somehow "block" the electric field created by the positive plate and prevent the presence of an $\vec{E}_+$ field to the right of the capacitor. To see that it doesn't have this effect, consider an analogous situation with gravity. The strength of gravity above a table is the same as its strength below it. Just as the table doesn't block the earth's gravitational field, intervening matter or charges do not alter or block an object's electric field. ◄

Inside the capacitor, $\vec{E}_+$ and $\vec{E}_-$ are parallel and of equal strength. Their superposition creates a net electric field inside the capacitor that points from the positive plate to the negative plate. Outside the capacitor, $\vec{E}_+$ and $\vec{E}_-$ point in opposite directions and, because the field of a plane of charge is independent of the distance from the plane, have equal magnitudes. Consequently, the fields $\vec{E}_+$ and $\vec{E}_-$ add to zero outside the capacitor plates.

We can calculate the fields between the capacitor plates from the field of an infinite charged plane. Between the electrodes, $\vec{E}_+$ is of magnitude $\eta/2\epsilon_0$ and points from the positive toward the negative side. The field $\vec{E}_-$ is *also* of magnitude $\eta/2\epsilon_0$ and *also* points from positive to negative. Thus the electric field inside the capacitor is

$$\vec{E}_{\text{capacitor}} = \vec{E}_+ + \vec{E}_- = \left(\frac{\eta}{\epsilon_0}, \text{ from positive to negative}\right)$$

$$= \left(\frac{Q}{\epsilon_0 A}, \text{ from positive to negative}\right)$$

(26.29)

Outside the capacitor plates, where $\vec{E}_+$ and $\vec{E}_-$ have equal magnitudes but *opposite* directions, $\vec{E} = \vec{0}$.

Figure 26.24a shows the electric field of an ideal parallel-plate capacitor constructed from two infinite charged planes. Now, it's true that no real capacitor is infinite in extent, but the ideal parallel-plate capacitor is a very good approximation for all but the most precise calculations as long as the electrode separation d is much smaller than the electrodes' size—that is, their edge length or radius. Figure 26.24b shows that the interior field of a real capacitor is virtually identical to that of an ideal capacitor but that the exterior field isn't quite zero. This weak field outside the capacitor is called the **fringe field.** We will keep things simple by always assuming the plates are very close together and using Equation 26.29 for the field inside a parallel-plate electrode.

NOTE ▶ The shape of the electrodes—circular or square or any other shape—is not relevant as long as the electrodes are very close together. ◀

Uniform Electric Fields

Figure 26.25 shows an electric field that is the *same*—in strength and direction—at every point in a region of space. This is called a **uniform electric field.** A uniform electric field is analogous to the uniform gravitational field near the surface of the earth. Uniform fields are of great practical significance because, as you will see in the next section, computing the trajectory of a charged particle moving in a uniform electric field is a straightforward process.

The easiest way to produce a uniform electric field is with a parallel-plate capacitor. Indeed, our interest in capacitors is due in large measure to the fact that the electric field is uniform. Many electric field problems refer to a uniform electric field. Such problems carry an implicit assumption that the action is taking place *inside* a parallel-plate capacitor.

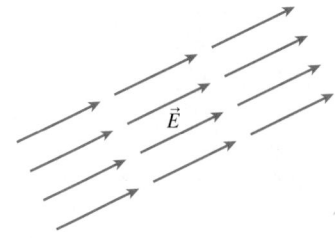

FIGURE 26.25 A uniform electric field.

EXAMPLE 26.7 The electric field inside a capacitor
Two 1.0 cm × 2.0 cm rectangular electrodes are 1.0 mm apart. What charge must be placed on each electrode to create a uniform electric field of strength 2.0×10^6 N/C? How many electrons must be moved from one electrode to the other to accomplish this?

MODEL The electrodes can be modeled as a parallel-plate capacitor because the spacing between them is much smaller than their lateral dimensions.

SOLVE The electric field strength inside the capacitor is $E = Q/\epsilon_0 A$. Thus the charge needed to produce a field of strength E is

$$Q = \epsilon_0 AE$$
$$= (8.85 \times 10^{-12} \text{ C}^2/\text{Nm}^2)(2.0 \times 10^{-4} \text{ m}^2)(2.0 \times 10^6 \text{ N/C})$$
$$= 3.5 \times 10^{-9} \text{ C} = 3.5 \text{ nC}$$

The positive plate must be charged to +3.5 nC and the negative plate to −3.5 nC. In practice, the plates are charged by using a *battery* to move electrons from one plate to the other. The number of electrons in 3.5 nC is

$$N = \frac{Q}{e} = \frac{3.5 \times 10^{-9} \text{ C}}{1.60 \times 10^{-19} \text{ C/electron}} = 2.2 \times 10^{10} \text{ electrons}$$

Thus 2.2×10^{10} electrons are moved from one electrode to the other. Note that the capacitor *as a whole* has no net charge.

ASSESS The plate spacing does not enter the result. As long as the spacing is much smaller than the plate dimensions, as is true in this example, the field is independent of the spacing.

STOP TO THINK 26.5 Rank in order, from largest to smallest, the forces F_a to F_e a proton would experience if placed at points a to e in this parallel-plate capacitor.

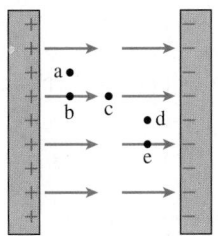

26.6 Motion of a Charged Particle in an Electric Field

Our motivation for introducing the concept of the electric field was to understand the long-range electric interaction of charges. We said that some charges, the *source charges,* create an electric field. Other charges then respond to that electric field. The first five sections of this chapter have focused on the electric field of the source charges. Now we turn our attention to the second half of the interaction.

Activ
Physics
ONLINE
11.9, 11.10

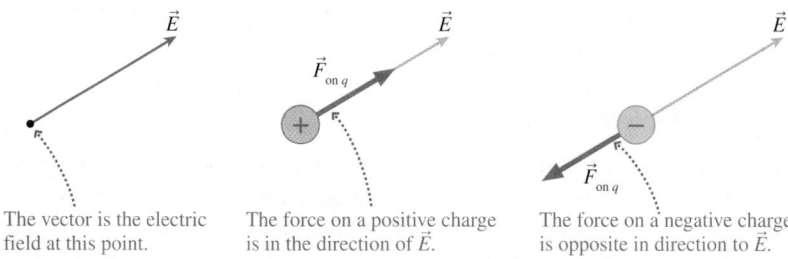

The vector is the electric field at this point.

The force on a positive charge is in the direction of $\vec{E}$.

The force on a negative charge is opposite in direction to $\vec{E}$.

FIGURE 26.26 The electric field exerts a force on a charged particle.

Figure 26.26 shows a particle of charge q and mass m at a point where an electric field $\vec{E}$ has been produced by *other* charges, the source charges. The electric field exerts a force

$$\vec{F}_{\text{on }q} = q\vec{E}$$

on the charged particle. This relationship between field and force was the *definition* of the electric field. Notice that the force on a negatively charged particle is *opposite* in direction to the electric field vector. Signs are important!

The force on the charged particle causes an acceleration

$$\vec{a} = \frac{\vec{F}_{\text{on }q}}{m} = \frac{q}{m}\vec{E} \tag{26.30}$$

This acceleration is the *response* of the charged particle to the source charges that created the electric field. The ratio q/m is especially important for the dynamics of charged-particle motion. It is called the **charge-to-mass ratio.** Two *equal* charges, say a proton and a Na^+ ion, will experience *equal* forces $\vec{F} = q\vec{E}$ if placed at the same point in an electric field, but their accelerations will be *different* because they have different masses and thus different charge-to-mass ratios. Two particles having different charges and masses *but* with the same charge-to-mass ratio will undergo the same acceleration and follow the same trajectory.

Motion in a Uniform Field

The motion of a charged particle in a *uniform* electric field is especially important for its basic simplicity and because of its many valuable applications. A uniform field is *constant* at all points—constant in both magnitude and direction—within the region of space where the charged particle is moving. It follows, from Equation 26.30, that **a charged particle in a uniform electric field will move with constant acceleration.** The magnitude of the acceleration is

$$a = \frac{qE}{m} = \text{constant} \tag{26.31}$$

where E is the electric field strength, and the direction of $\vec{a}$ is parallel or antiparallel to $\vec{E}$, depending on the sign of q.

Identifying the motion of a charged particle in a uniform field as being one of constant acceleration brings into play all the kinematic machinery that we developed in Chapter 2 for constant-acceleration motion. The basic trajectory of a charged particle in a uniform field is a *parabola,* analogous to the projectile motion of a mass in the near-earth uniform gravitational field. In the special case of a charged particle moving parallel to the electric field vectors, the motion is one-dimensional, analogous to the one-dimensional vertical motion of a mass tossed straight up or falling straight down.

NOTE ▶ The gravitational acceleration $\vec{a}_{\text{grav}}$ always points straight down. The electric field acceleration $\vec{a}_{\text{elec}}$ can point in *any* direction. You must determine the electric field $\vec{E}$ in order to learn the direction of $\vec{a}$. ◄

EXAMPLE 26.8 **An electron moving across a capacitor**

Two 6.0-cm-diameter electrodes are spaced 5.0 mm apart. They are charged by transferring 1.0×10^{11} electrons from one electrode to the other. An electron is released from rest at the surface of the negative electrode. How long does it take the electron to cross to the positive electrode? What is its speed as it collides with the positive electrode? Assume the space between the electrodes is a vacuum.

MODEL The electrodes form a parallel-plate capacitor. The electric field inside a parallel-plate capacitor is a uniform field, so the electron will have constant acceleration.

VISUALIZE Figure 26.27 shows the capacitor and the electron. The force on the negative electron is *opposite* the electric field, so the electron is repelled by the negative electrode as it accelerates across the gap of width d.

The capacitor was charged by transferring 10^{11} electrons from the right electrode to the left electrode.

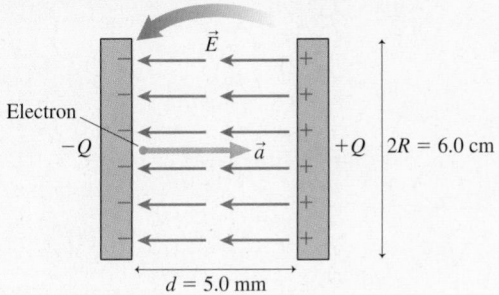

Electron

$-Q$ $\vec{a}$ $+Q$ $2R = 6.0$ cm

$d = 5.0$ mm

FIGURE 26.27 An electron accelerates across a capacitor.

SOLVE The electrodes are not point charges, so we cannot use Coulomb's law to find the force on the electron. Instead, we must analyze the electron's motion in terms of the electric field inside the capacitor. The field is the agent that exerts the force on the electron, causing it to accelerate. The electric field strength inside a parallel-plate capacitor with charge $Q = Ne$ is

$$E = \frac{\eta}{\epsilon_0} = \frac{Q}{\epsilon_0 A} = \frac{Ne}{\epsilon_0 \pi R^2} = 639{,}000 \text{ N/C}$$

The electron's acceleration in this field is

$$a = \frac{eE}{m} = 1.12 \times 10^{17} \text{ m/s}^2$$

where we used the electron mass $m = 9.11 \times 10^{-31}$ kg. This is an enormous acceleration compared to accelerations we're familiar with for macroscopic objects. We can use one-dimensional kinematics, with $x_i = 0$ and $v_i = 0$, to find the time required for the electron to cross the capacitor:

$$x_f = d = \frac{1}{2}a(\Delta t)^2$$

$$\Delta t = \sqrt{\frac{2d}{a}} = 2.98 \times 10^{-10} \text{ s} = 0.298 \text{ ns}$$

The electron's speed as it reaches the positive electrode is

$$v = a\Delta t = 3.34 \times 10^7 \text{ m/s}.$$

ASSESS We used e rather than $-e$ to find the acceleration because we already knew the direction; we only needed the magnitude. The electron's speed, after traveling a mere 5 mm, is approximately 10% the speed of light.

Parallel electrodes such as those in Example 26.8 are often used to accelerate charged particles. If the positive plate has a small hole in the center, a *beam* of electrons will pass through the hole, after accelerating across the capacitor gap, and emerge with a speed of 3.34×10^7 m/s. This is the basic idea of the *electron gun* used in televisions, oscilloscopes, computer display terminals, and other *cathode-ray tube* (CRT) devices. (A negatively charged electrode is called a *cathode*, so the physicists who first learned to produce electron beams in the late 19th century called them *cathode rays*.) The following example shows that parallel electrodes can also be used to deflect charged particles sideways.

EXAMPLE 26.9 **Deflecting an electron beam**

An electron gun creates a beam of electrons moving horizontally with a speed of 3.34×10^7 m/s. The electrons enter a 2.0-cm-long gap between two parallel electrodes where the electric field is $\vec{E} = (5.0 \times 10^4 \text{ N/C, down})$. In which direction, and by what angle, is the electron beam deflected by these electrodes?

MODEL The electric field between the electrodes is uniform. Assume that the electric field outside the electrodes is zero.

VISUALIZE Figure 26.28 shows an electron moving through the electric field. The electric field points down, so the force on the (negative) electrons is upward. The electrons will follow a parabolic trajectory, analogous to that of a ball thrown horizontally, except that the electrons "fall up" rather than down.

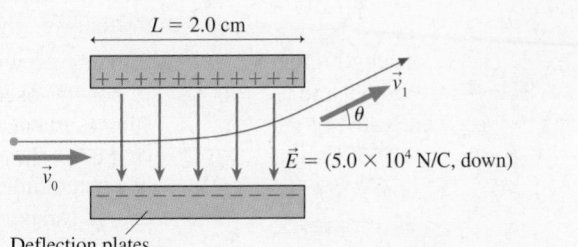

$L = 2.0$ cm

$+ + + + + + + + + + +$

$\vec{v}_1$

θ

$\vec{v}_0$

$\vec{E} = (5.0 \times 10^4 \text{ N/C, down})$

Deflection plates

FIGURE 26.28 The deflection of an electron beam in a uniform electric field.

SOLVE This is a two-dimensional motion problem. The electron enters the capacitor with velocity *vector* $\vec{v}_0 = v_{0x}\hat{\imath} = 3.34 \times 10^7 \hat{\imath}$ m/s and leaves with velocity $\vec{v}_1 = v_{1x}\hat{\imath} + v_{1y}\hat{\jmath}$. The electron's angle of travel upon leaving the electric field is

$$\theta = \tan^{-1}\left(\frac{v_{1y}}{v_{1x}}\right)$$

This is the *deflection angle*. To find θ we must compute the final velocity vector $\vec{v}_1$.

There is no horizontal force on the electron, so $v_{1x} = v_{0x} = 3.34 \times 10^7$ m/s. The electron's upward acceleration has magnitude

$$a = \frac{eE}{m} = \frac{(1.60 \times 10^{-19}\,\text{C})(5.0 \times 10^4\,\text{N/C})}{9.11 \times 10^{-31}\,\text{kg}}$$

$$= 8.78 \times 10^{15}\,\text{m/s}^2$$

We can use the fact that the horizontal velocity is constant to determine the time interval Δt needed to travel length 2.0 cm:

$$\Delta t = \frac{L}{v_{0x}} = \frac{0.020\,\text{m}}{3.34 \times 10^7\,\text{m/s}} = 5.99 \times 10^{-10}\,\text{s}$$

Vertical acceleration will occur during this time interval, resulting in a final vertical velocity

$$v_{1y} = v_{0y} + a\,\Delta t = 0 + (8.78 \times 10^{15}\,\text{m/s}^2)(5.99 \times 10^{-10}\,\text{s})$$

$$= 5.26 \times 10^6\,\text{m/s}$$

The electron's velocity as it leaves the capacitor is thus

$$\vec{v}_1 = (3.34 \times 10^7 \hat{\imath} + 5.26 \times 10^6 \hat{\jmath})\,\text{m/s}$$

and the deflection angle θ is

$$\theta = \tan^{-1}\left(\frac{v_{1y}}{v_{1x}}\right) = 8.95°$$

ASSESS The accelerations of charged particles in electric fields are enormous in comparison to the gravitational acceleration g. Thus it is rarely necessary to include the weight force when calculating the trajectories of charged particles. The only exception might be for a macroscopic charged object, such as a charged plastic bead, in a weak electric field.

Example 26.9 demonstrates how an electron beam is steered to a point on the screen of a cathode-ray tube. First, a high-speed electron beam is created by an electron gun like that of Example 26.8. The beam then passes first through a set of *vertical deflection plates,* as in Example 26.9, then through a second set of *horizontal deflection plates.* After leaving the deflection plates, it travels in a straight line (through vacuum, to eliminate collisions with air molecules) to the screen of the CRT, where it strikes a phosphor coating on the inside surface and makes a dot of light. Properly choosing the electric fields within the deflection plates, which is done by controlling their voltage, steers the electron beam to any point on the screen.

Motion in a Nonuniform Field

The motion of a charged particle in a nonuniform electric field can be quite complicated. Sophisticated mathematical techniques and computers are used to determine the trajectories. However, one type of motion in a nonuniform field is easy to analyze: the circular orbit of a charged particle around a charged sphere or wire.

Figure 26.29 shows a negatively charged particle orbiting a positively charged sphere, much as the moon orbits the earth. You will recall from Chapter 7 that Newton's second law for circular motion is $(F_{\text{net}})_r = mv^2/r$. Here the radial force has magnitude $|q|E$, where E is the electric field strength at distance r. Thus the charge can move in a circular orbit if

$$|q|E = \frac{mv^2}{r} \tag{26.32}$$

Specific examples of circular orbits will be left for homework problems.

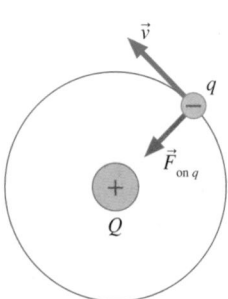

FIGURE 26.29 The circular motion of a charged particle around a charged sphere.

STOP TO THINK 26.6   Which electric field is responsible for the trajectory of the proton?

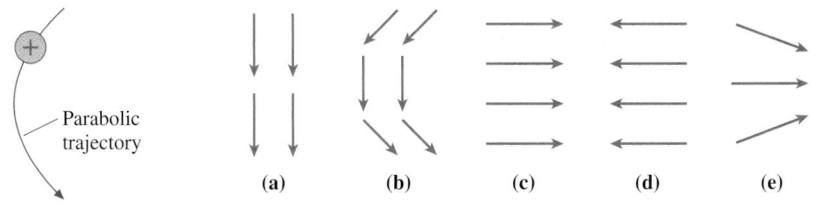

26.7 Motion of a Dipole in an Electric Field
===

Let us conclude this chapter by returning to one of the more striking puzzles we faced when making the observations at the beginning of Chapter 25. There you found that charged objects of *either* sign exert forces on neutral objects, such as when a comb used to brush your hair picks up pieces of paper. Our qualitative understanding of the *polarization force* was that it required two steps:

- The charge polarizes the neutral object, creating an induced electric dipole.
- The charge then exerts an attractive force on the near end of the dipole that is slightly stronger than the repulsive force on the far end.

We are now in a position to make that understanding more quantitative. We will analyze the force on a *permanent* dipole. A homework problem will let you think about *induced* dipoles.

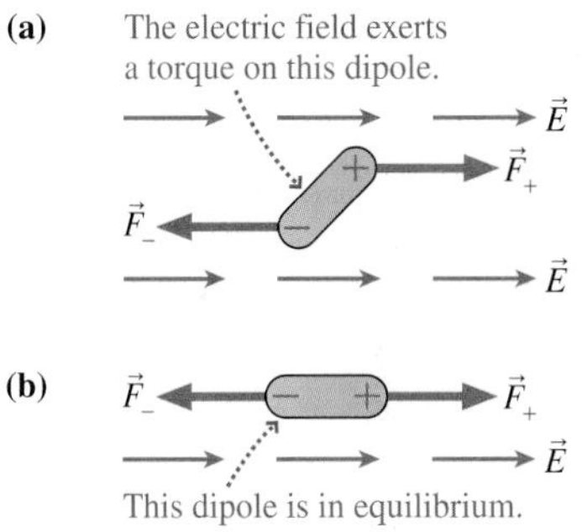

Dipoles in a Uniform Field

Figure 26.30a shows an electric dipole in a *uniform* external electric field $\vec{E}$ that has been created by source charges we do not see. That is, $\vec{E}$ is *not* the field of the dipole but, instead, is a field to which the dipole is responding. In this case, because the field is uniform, the dipole is presumably inside an unseen parallel-plate capacitor.

The net force on the dipole is the sum of the forces on the two charges forming the dipole. Because the charges $\pm q$ are equal in magnitude but opposite in sign, the two forces $\vec{F}_+ = +q\vec{E}$ and $\vec{F}_- = -q\vec{E}$, are also equal but opposite. Thus the net force on the dipole is

$$\vec{F}_{net} = \vec{F}_+ + \vec{F}_- = \vec{0} \tag{26.33}$$

There is no net force on a dipole in a uniform electric field.

There may be no net force, but the electric field *does* affect the dipole. Because the two forces in Figure 26.30a are in opposite directions but not aligned with each other, the electric field causes the dipole to *rotate*. The "twist" that the field exerts on the dipole is called a *torque*.

The torque causes the dipole to rotate until it is aligned with the electric field, as shown in Figure 26.30b. In this position, the dipole experiences not only no net force but also no torque. Thus Figure 26.30b represents the *equilibrium position* for a dipole in a uniform electric field. Notice that the positive end of the dipole is in the direction in which $\vec{E}$ points.

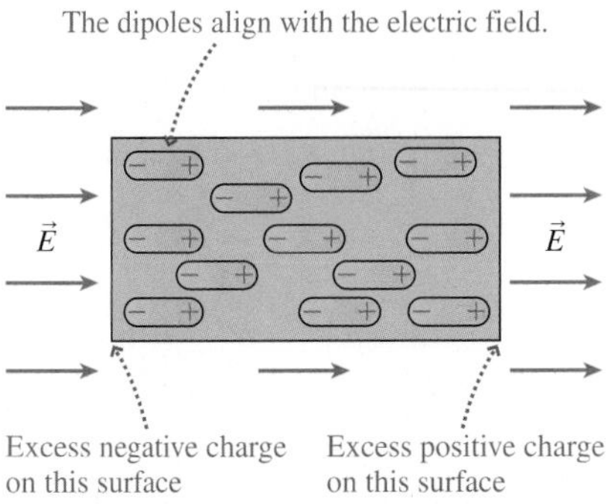

Figure 26.31 shows a sample of permanent dipoles, such as water molecules, in an external electric field. All the dipoles rotate until they are aligned with the electric field. This is the mechanism by which the sample becomes *polarized*. Once the dipoles are aligned, there is an excess of positive charge at one end of

(a) The electric field exerts a torque on this dipole.

(b) This dipole is in equilibrium.

FIGURE 26.30 A dipole in a uniform electric field.

The dipoles align with the electric field.

Excess negative charge on this surface   Excess positive charge on this surface

FIGURE 26.31 A sample of permanent dipoles is *polarized* in an electric field.

The torque due to a couple is $\tau = lF = pE\sin\theta$.

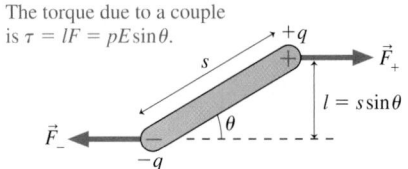

In terms of vectors, $\vec{\tau} = \vec{p} \times \vec{E}$.

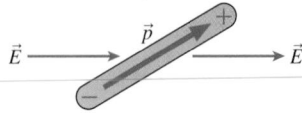

FIGURE 26.32 Calculating the torque on a dipole.

the sample and an excess of negative charge at the other end. The excess charges at the ends of the sample are the basis of the polarization forces we discussed in Section 25.3.

It's not hard to calculate the torque on a dipole. The two forces on the dipole in Figure 26.32 form what we called a *couple* in Chapter 13. There you learned that the torque τ on a couple is the product of the force F with the distance l between the lines along which the forces act. You can see that $l = s\sin\theta$, where θ is the angle the dipole makes with the electric field $\vec{E}$. Thus the torque on the dipole is

$$\tau = lF = (s\sin\theta)(qE) = pE\sin\theta \tag{26.34}$$

where $p = qs$ was our definition of the dipole moment. The torque is zero when the dipole is aligned with the field, making $\theta = 0$.

If you studied torque in Chapter 13, you will recall that the torque can be written in a compact mathematical form as the cross product between two vectors. The terms p and E in Equation 26.34 are the magnitudes of vectors, and θ is the angle between them. Thus in vector notation, the torque exerted on a dipole moment $\vec{p}$ by an electric field $\vec{E}$ is

$$\vec{\tau} = \vec{p} \times \vec{E} \tag{26.35}$$

The torque is greatest when $\vec{p}$ is perpendicular to $\vec{E}$, zero when $\vec{p}$ is aligned with or opposite to $\vec{E}$.

EXAMPLE 26.10 The angular acceleration of a dipole dumbbell

Two 1.0 g balls are connected by a 2.0-cm-long insulating rod of negligible mass. One ball has a charge of $+10$ nC, the other a charge of -10 nC. The rod is held in a 1.0×10^4 N/C uniform electric field at an angle of 30° with respect to the field, then released. What is its initial angular acceleration?

MODEL The two oppositely charged balls form an electric dipole. The electric field exerts a torque on the dipole, causing an angular acceleration.

VISUALIZE Figure 26.33 shows the dipole in the electric field. The angle between the dipole moment and the field is $\theta = 30°$.

1.0 g
+10 nC

$s = 2.0$ cm

$\vec{p}$

30°

$E = 1.0 \times 10^4$ N/C

1.0 g
-10 nC

FIGURE 26.33 The dipole of Example 26.10.

SOLVE The dipole moment is $p = qs = (1.0 \times 10^{-8}\,\text{C}) \times (0.020\,\text{m}) = 2.0 \times 10^{-10}$ Cm. The torque exerted on the dipole moment by the electric field is

$$\tau = pE\sin\theta = (2.0 \times 10^{-10}\,\text{Cm})(1.0 \times 10^4\,\text{N/C})\sin 30°$$

$$= 1.0 \times 10^{-6}\,\text{Nm}$$

You learned in Chapter 13 that a torque causes an angular acceleration $\alpha = \tau/I$, where I is the moment of inertia. The dipole rotates about its center of mass, which is at the center of the rod, so the moment of inertia is

$$I = m_1 r_1^2 + m_2 r_2^2 = 2m\left(\frac{1}{2}s\right)^2 = \frac{1}{2}ms^2 = 2.0 \times 10^{-7}\,\text{kg m}^2$$

Thus the rod's angular acceleration is

$$\alpha = \frac{\tau}{I} = \frac{1.0 \times 10^{-6}\,\text{Nm}}{2.0 \times 10^{-7}\,\text{kg m}^2} = 5.0\,\text{rad/s}^2$$

ASSESS This value of α is the initial angular acceleration, when the rod is first released. The torque and the angular acceleration will decrease as the rod rotates toward alignment with $\vec{E}$.

Dipoles in a Nonuniform Field

Suppose that a dipole is placed in a *nonuniform* electric field, one in which the field strength changes with position. For example, Figure 26.34 shows a dipole in the nonuniform field of a point charge. The first response of the dipole is to rotate until it is aligned with the field, with the dipole's positive end pointing in the same direction as the field. Now, however, there is a slight *difference* between the

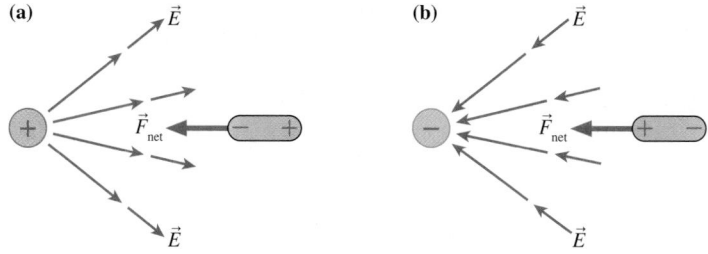

FIGURE 26.34 An aligned dipole is drawn toward a point charge.

forces acting on the two ends of the dipole. This difference occurs because the electric field, which depends on the distance from the point charge, is stronger at the end of the dipole nearest the charge. This causes a *net force* to be exerted on the dipole.

Which way does the force point? Figure 26.34a shows a positive point charge. Once the dipole is aligned, the leftward attractive force on its negative end will be slightly stronger than the rightward repulsive force on its positive end. This causes a net force to the *left,* toward the point charge. The dipole in Figure 26.34b aligns in the opposite orientation in the field of a negative point charge, but the net force is still to the left.

As you can see, **the net force on a dipole is toward the direction of the strongest field.** Because any finite-size charged object, such as a charged rod or a charged disk, has a field strength that increases as you get closer to the object, we can conclude that **a dipole will experience a net force toward any charged object.**

EXAMPLE 26.11 **The force on a water molecule**

The water molecule H_2O has a permanent dipole moment of magnitude 6.2×10^{-30} C m. A water molecule is located 10 nm from a Na^+ ion in a saltwater solution. What force does the ion exert on the water molecule?

VISUALIZE Figure 26.35 shows the ion and the dipole. The forces are an action/reaction pair.

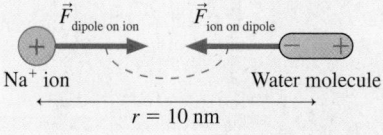

FIGURE 26.35 The interaction between an ion and a permanent dipole.

SOLVE A Na^+ ion has charge $q = +e$. The electric field of the ion aligns the water's dipole moment and exerts a net force on it. We could calculate the net force on the dipole as the small difference between the attractive force on its negative end and the repulsive force on its positive end. Alternatively, we know from Newton's third law that the force $\vec{F}_{\text{dipole on ion}}$ has the

same magnitude as the force $\vec{F}_{\text{ion on dipole}}$ that we are seeking. We calculated the on-axis field of a dipole in Section 26.2. An ion of charge $q = e$ will experience a force of magnitude $F = qE_{\text{dipole}} = eE_{\text{dipole}}$ when placed in that field. The dipole's electric field, which we found in Equation 26.11, is

$$E_{\text{dipole}} = \frac{1}{4\pi\epsilon_0}\frac{2p}{r^3}$$

The force on the ion at distance $r = 1.0 \times 10^{-8}$ m is

$$F_{\text{dipole on ion}} = eE_{\text{dipole}} = \frac{1}{4\pi\epsilon_0}\frac{2ep}{r^3} = 1.79 \times 10^{-14}\ \text{N}$$

Thus the force on the water molecule is $F_{\text{ion on dipole}} = 1.79 \times 10^{-14}$ N.

ASSESS While 1.79×10^{-14} N may seem like a very small force, it is $\approx 10^{11}$ times larger than the size of the earth's gravitational force on these atomic particles. Forces such as these cause water molecules to cluster around any ions that are in solution. This clustering, a phenomenon studied in a physical chemistry course, plays an important role in the microscopic physics of solutions.

SUMMARY

The goal of Chapter 26 has been to learn how to calculate and use the electric field.

GENERAL PRINCIPLES

Sources of $\vec{E}$

Electric fields are created by charges.

Two major tools for calculating $\vec{E}$ are

- The field of a point charge

$$\vec{E} = \frac{1}{4\pi\epsilon_0} \frac{q}{r^2} \hat{r}$$

- The principle of superposition

Multiple point charges

Use superposition: $\vec{E} = \vec{E}_1 + \vec{E}_2 + \vec{E}_3 + \dots$

Continuous distribution of charge

- Divide the charge into point-like ΔQ
- Find the field of each ΔQ
- Find $\vec{E}$ by summing the fields of all ΔQ

The summation usually becomes an integral. A critical step is replacing ΔQ with an expression involving a **charge density** (λ or η) and an integration coordinate.

Consequences of $\vec{E}$

The electric field exerts a force on a charged particle.

$$\vec{F} = q\vec{E}$$

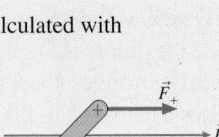

The force causes acceleration

$$\vec{a} = (q/m)\vec{E}$$

Trajectories of charged particles are calculated with kinematics.

The electric field exerts a torque on a dipole.

$$\tau = pE\sin\theta$$

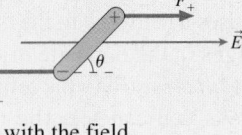

The torque tends to align the dipoles with the field.

In a nonuniform electric field, a dipole has a net force in the direction of increasing field strength.

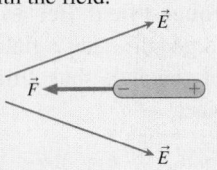

APPLICATIONS

The following fields are important models of the electric field:

Electric dipole

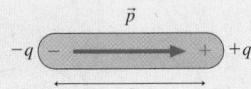

The electric dipole moment is

$\vec{p} = (qs,$ from negative to positive$)$

Field on axis $\vec{E} = \frac{1}{4\pi\epsilon_0} \frac{2\vec{p}}{r^3}$

Field in bisecting plane $\vec{E} = -\frac{1}{4\pi\epsilon_0} \frac{\vec{p}}{r^3}$

Infinite line of charge with linear charge density λ

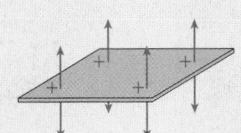

$$\vec{E} = \left(\frac{1}{4\pi\epsilon_0} \frac{2\lambda}{r}, \text{perpendicular to line} \right)$$

Infinite plane of charge with surface charge density η

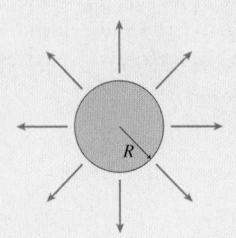

$$\vec{E} = \left(\frac{\eta}{2\epsilon_0}, \text{perpendicular to plane} \right)$$

Sphere of charge

Same as a point charge Q for $r > R$

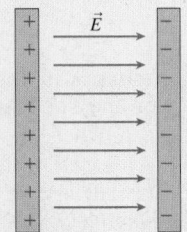

Parallel-plate capacitor

The electric field inside an ideal capacitor is a **uniform electric field**

$$\vec{E} = \left(\frac{\eta}{\epsilon_0}, \text{from positive to negative} \right)$$

A real capacitor has a weak **fringe field** around it.

TERMS AND NOTATION

dipole moment, $\vec{p}$	uniformly charged	plane of charge	fringe field
electric field line	line of charge	sphere of charge	uniform electric field
linear charge density, λ	electrode	parallel-plate capacitor	charge-to-mass ratio, q/m
surface charge density, η			

EXERCISES AND PROBLEMS

Exercises

Section 26.2 The Electric Field of Multiple Point Charges

1. What are the strength and direction of the electric field at the position indicated by the dot in Figure Ex26.1? Specify the direction as an angle above or below horizontal.

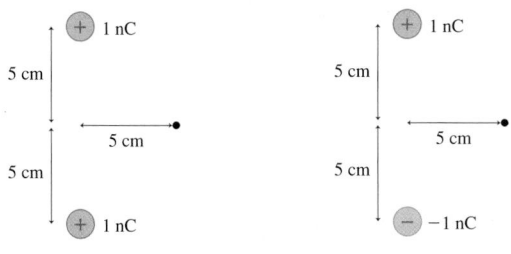

FIGURE EX26.1   **FIGURE EX26.2**

2. What are the strength and direction of the electric field at the position indicated by the dot in Figure Ex26.2? Specify the direction as an angle above or below horizontal.

3. What are the strength and direction of the electric field at the position indicated by the dot in Figure Ex26.3? Specify the direction as an angle above or below horizontal.

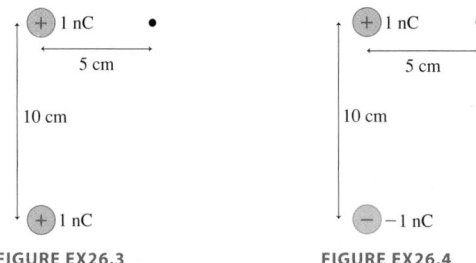

FIGURE EX26.3   **FIGURE EX26.4**

4. What are the strength and direction of the electric field at the position indicated by the dot in Figure Ex26.4? Specify the direction as an angle above or below horizontal.

5. An electric dipole is formed from ± 1.0 nC charges spaced 2.0 mm apart. The dipole is at the origin, oriented along the y-axis. What is the electric field strength at the points (a) $(x, y) = (10 \text{ cm}, 0 \text{ cm})$ and (b) $(x, y) = (0 \text{ cm}, 10 \text{ cm})$?

6. An electric dipole is formed from two charges, $\pm q$, spaced 1.0 cm apart. The dipole is at the origin, oriented along the y-axis. The electric field strength at the point $(x, y) = (0 \text{ cm}, 10 \text{ cm})$ is 360 N/C.
   a. What is the charge q? Give your answer in nC.
   b. What is the electric field strength at the point $(x, y) = (10 \text{ cm}, 0 \text{ cm})$?

Section 26.3 The Electric Field of a Continuous Charge Distribution

7. A 10-cm-long thin glass rod uniformly charged to $+10$ nC and a 10-cm-long thin plastic rod uniformly charged to -10 nC are placed side by side, 4.0 cm apart. What are the electric field strengths E_1 to E_3 at distances 1.0 cm, 2.0 cm, and 3.0 cm from the glass rod along the line connecting the midpoints of the two rods?

8. Two 10-cm-long thin glass rods uniformly charged to $+10$ nC are placed side by side, 4.0 cm apart. What are the electric field strengths E_1 to E_3 at distances 1.0 cm, 2.0 cm, and 3.0 cm from the rod on the left along the line connecting the midpoints of the two rods?

9. A 10-cm-long thin glass rod is uniformly charged to $+50$ nC. A small plastic bead, charged to -5.0 nC, is 4.0 cm from the center of the rod. What is the force (magnitude and direction) on the bead?

10. The electric field strength 5.0 cm from a very long charged wire is 2000 N/C. What is the electric field strength 10.0 cm from the wire?

11. The electric field 5.0 cm from a very long charged wire is (2000 N/C, toward the wire). What is the charge (in nC) on a 1.0-cm-long segment of the wire?

Section 26.4 The Electric Fields of Rings, Planes, and Spheres

12. Two 10-cm-diameter charged rings face each other, 20 cm apart. The left ring is charged to -20 nC and the right ring is charged to $+20$ nC.
   a. What is the electric field $\vec{E}$, both magnitude and direction, at the midpoint between the two rings?
   b. What is the force $\vec{F}$ on a 1.0 nC charge placed at the midpoint?

13. Two 10-cm-diameter charged rings face each other, 20 cm apart. Both rings are charged to $+20$ nC. What is the electric field strength at (a) the midpoint between the two rings and (b) the center of the left ring?

14. Two 10-cm-diameter charged disks face each other, 20 cm apart. The left disk is charged to -50 nC and the right disk is charged to $+50$ nC.
   a. What is the electric field $\vec{E}$, both magnitude and direction, at the midpoint between the two disks?
   b. What is the force $\vec{F}$ on a -1.0 nC charge placed at the midpoint?

15. Two 10-cm-diameter charged disks face each other, 20 cm apart. Both disks are charged to -50 nC. What is the electric field strength at (a) the midpoint between the two disks and (b) the center of the left disk?

16. A 20 cm $\times$ 20 cm metal electrode is uniformly charged to $+80$ nC. What is the electric field strength 2.0 mm above the center of the electrode?

17. The electric field strength 5.0 cm from a very wide charged electrode is 1000 N/C. What is the charge (in nC) on a 1.0-cm-diameter circular segment of the electrode?

18. The electric field strength 2.0 cm from a 10-cm-diameter metal ball is 50,000 N/C. What is the charge (in nC) on the ball?

19. Two 2.0-cm-diameter insulating spheres have a 6.0 cm space between them. One sphere is charged to $+10$ nC, the other to -15 nC. What is the electric field strength at the midpoint between the two spheres?

Section 26.5 The Parallel-Plate Capacitor

20. A parallel-plate capacitor is formed from two 4.0 cm $\times$ 4.0 cm electrodes spaced 2.0 mm apart. The electric field strength inside the capacitor is 1.0×10^6 N/C. What is the charge (in nC) on each electrode?

21. Two closely spaced circular disks form a parallel-plate capacitor. Transferring 1.5×10^9 electrons from one disk to the other causes the electric field strength to be 1.0×10^5 N/C. What are the diameters of the disks?

22. Air "breaks down" when the electric field strength reaches 3×10^6 N/C, causing a spark. A parallel-plate capacitor is made from two 4.0-cm-diameter disks. How many electrons must be transferred from one disk to the other to create a spark between the disks?

Section 26.6 Motion of a Charged Particle in an Electric Field

23. A 0.10 g plastic bead is charged by the addition of 1.0×10^{10} excess electrons. What electric field $\vec{E}$ (strength and direction) will cause the bead to hang suspended in the air?

24. Two 2.0-cm-diameter disks face each other, 1.0 mm apart. They are charged to ± 10 nC.
    a. What is the electric field strength between the disks?
    b. A proton is shot from the negative disk toward the positive disk. What launch speed must the proton have to just barely reach the positive disk?

25. The electron gun in a television tube uses a uniform electric field to accelerate electrons from rest to 5.0×10^7 m/s in a distance of 1.2 cm. What is the electric field strength?

26. An electron is released from rest 2.0 cm from an infinite charged plane. It accelerates toward the plane and collides with a speed of 1.0×10^7 m/s. What are (a) the surface charge density of the plane and (b) the time required for the electron to travel the 2.0 cm?

27. The surface charge density on an infinite charged plane is -2.0×10^{-6} C/m². A proton is shot straight away from the plane at 2.0×10^6 m/s. How far does the proton travel before reaching its turning point?

Section 26.7 Motion of a Dipole in an Electric Field

28. A point charge Q is distance r from the center of a dipole consisting of charges $\pm q$ separated by distance s. The charge is located in the plane that bisects the dipole. At this instant, what are (a) the force (magnitude and direction) and (b) the magnitude of the torque on the dipole? You can assume $r \gg s$.

29. An ammonia molecule (NH_3) has a permanent electric dipole moment 5.0×10^{-30} Cm. A proton is 2.0 nm from the molecule in the plane that bisects the dipole. What is the electric force of the molecule on the proton?

30. The permanent electric dipole moment of the water molecule (H_2O) is 6.2×10^{-30} Cm. What is the maximum possible torque on a water molecule in a 5.0×10^8 N/C electric field?

Problems

31. What are the strength and direction of the electric field at the position indicated by the dot? Give your answer in both component form and as a magnitude and direction.

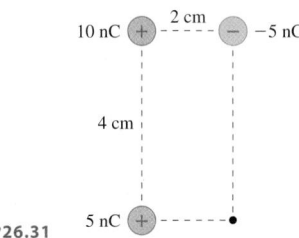

FIGURE P26.31

32. What are the strength and direction of the electric field at the position indicated by the dot? Give your answer in both component form and as a magnitude and direction.

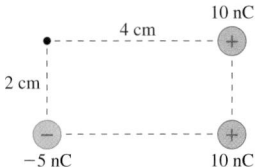

FIGURE P26.32

33. What are the strength and direction of the electric field at the position indicated by the dot in Figure P26.33? Give your answer in both component form and as a magnitude and direction.

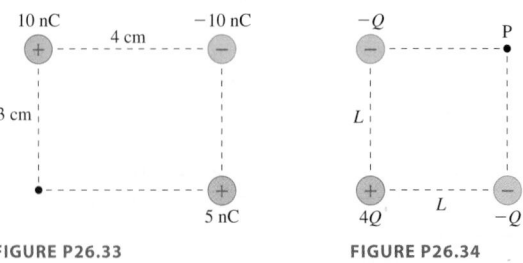

FIGURE P26.33   **FIGURE P26.34**

34. Figure P26.34 shows three charges at the corners of a square.
    a. Write the electric field at point P in component form.
    b. A particle with positive charge q and mass m is placed at point P and released. What is the initial magnitude of its acceleration?

35. Charges $-q$ and $+2q$ in Figure P26.35 are located at $x = \pm a$.
    a. Determine the electric field at points 1 to 4. Write each field in component form.
    b. Reproduce Figure P26.35, then draw the four electric field vectors on the figure.

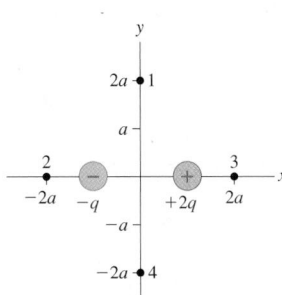

FIGURE P26.35

36. Two positive charges q are distance s apart on the y-axis.
    a. Find an expression for the electric field strength at distance x on the axis that bisects the two charges.
    b. For $q = 1.0$ nC and $s = 6.0$ mm, evaluate E at $x = 0, 2, 4, 6,$ and 10 mm.
    c. Draw a graph of E versus x for $0 \le x \le \infty$.
37. Derive Equation 26.12 for the field $\vec{E}_{dipole}$ in the plane that bisects an electric dipole.
38. Three charges are on the y-axis. Charges $-q$ are at $y = \pm d$ and charge $+2q$ is at $y = 0$.
    a. Determine the electric field $\vec{E}$ along the x-axis.
    b. Verify that your answer to part a has the expected behavior as x becomes very small and very large.
    c. Sketch a graph of E_x versus x for $0 \le x \le \infty$.
39. Three 10-cm-long rods form an equilateral triangle. Two of the rods are charged to $+10$ nC, the third to -10 nC. What is the electric field strength at the center of the triangle?
40. Figure P26.40 is a cross section of two infinite lines of charge that extend out of the page. Both have linear charge density λ.
    a. Find an expression for the electric field strength E at height y above the midpoint between the lines.
    b. Draw a graph of E versus y.

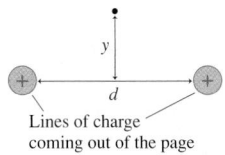

FIGURE P26.40   **FIGURE P26.41**

41. Figure P26.41 is a cross section of two infinite lines of charge that extend out of the page. The linear charge densities are $\pm\lambda$.
    a. Find an expression for the electric field strength E at height y above the midpoint between the lines.
    b. Draw a graph of E versus y.
42. Two infinite lines of charge, each with linear charge density λ, lie along the x- and y-axes, crossing at the origin. What is the electric field strength at position (x, y)?
43. A proton orbits a long charged wire, making 1.0×10^6 revolutions per second. The radius of the orbit is 1.0 cm. What is the wire's linear charge density?

44. Figure P26.44 shows a thin rod of length L with total charge Q.
    a. Find an expression for the electric field strength on the axis of the rod at distance r from the center.
    b. Verify that your expression has the expected behavior if $r \gg L$.
    c. Evaluate E at $r = 3.0$ cm if $L = 5.0$ cm and $Q = 3.0$ nC.

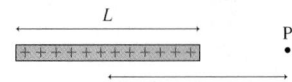

FIGURE P26.44

45. Figure P26.45 shows a thin rod of length L with total charge Q. Find an expression for the electric field $\vec{E}$ at distance x from the end of the rod. Give your answer in component form.

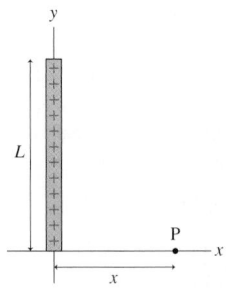

FIGURE P26.45

46. Show that the on-axis electric field of a ring of charge has the expected behavior when $z \ll R$ and when $z \gg R$.
47. a. Show that the maximum electric field strength on the axis of a ring of charge occurs at $z = R/\sqrt{2}$.
    b. What is the electric field strength at this point?
48. Charge Q is uniformly distributed along a thin, flexible rod of length L. The rod is then bent into the semicircle shown in Figure P26.48.
    a. Find an expression for the electric field $\vec{E}$ at the center of the semicircle.
    Hint: A small piece of arc length Δs spans a small angle $\Delta\theta = \Delta s/R$, where R is the radius.
    b. Evaluate the field strength if $L = 10$ cm and $Q = 30$ nC.

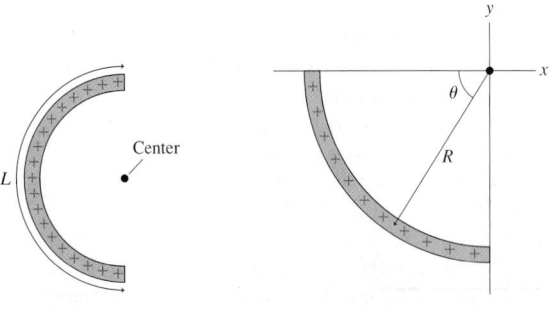

FIGURE P26.48   **FIGURE P26.49**

49. A plastic rod with linear charge density λ is bent into the quarter circle shown in Figure P26.49. We want to find the electric field at the origin.
    a. Write expressions for the x- and y-components of the electric field at the origin due to a small piece of charge at angle θ.
    b. Write, but do not evaluate, definite integrals for the x- and y-components of the net electric field at the origin.
    c. Evaluate the integrals and write $\vec{E}_{net}$ in component form.

50. You've hung two very large sheets of plastic facing each other with distance d between them, as shown in Figure P26.50. By rubbing them with wool and silk, you've managed to give one sheet a uniform surface charge density $\eta_1 = -\eta_0$ and the other a uniform surface charge density $\eta_2 = +3\eta_0$. What is the electric field vector at points 1, 2, and 3?

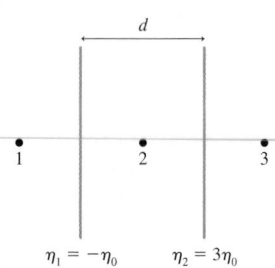

FIGURE P26.50 $\eta_1 = -\eta_0$ $\eta_2 = 3\eta_0$

51. Two parallel plates 1.0 cm apart are equally and oppositely charged. An electron is released from rest at the surface of the negative plate and simultaneously a proton is released from rest at the surface of the positive plate. How far from the negative plate is the point at which the electron and proton pass each other?

52. A proton traveling at a speed of 1.0×10^6 m/s enters the gap between the plates of a 2.0-cm-wide parallel-plate capacitor. The surface charge densities on the plates are $\pm 1.0 \times 10^{-6}$ C/m². How far has the proton been deflected sideways when it reaches the far edge of the capacitor? Assume the electric field is uniform inside the capacitor and zero outside.

53. The two parallel plates in Figure P26.53 are 2.0 cm apart and the electric field strength between them is 1.0×10^4 N/C. An electron is launched at a 45° angle from the positive plate. What is the maximum initial speed v_0 the electron can have without hitting the negative plate?

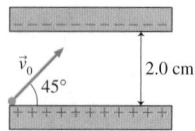

FIGURE P26.53

54. An electron is launched at a 45° angle and a speed of 5.0×10^6 m/s from the positive plate of the parallel-plate capacitor shown in Figure P26.54. The electron lands 4.0 cm away.
    a. What is the electric field strength inside the capacitor?
    b. What is the minimum spacing between the plates?

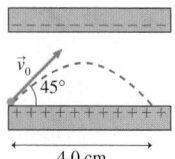

FIGURE P26.54   4.0 cm

55. A problem of practical interest is to make a beam of electrons turn a 90° corner. This can be done with the parallel-plate capacitor shown in Figure P26.55. An electron with kinetic energy 3.0×10^{-17} J moves up through a small hole in the bottom plate of the capacitor.
    a. Should the bottom plate be charged positive or negative relative to the top plate if you want the electron to turn to the right? Explain.

b. What strength electric field is needed if the electron is to emerge from an exit hole 1.0 cm away from the entrance hole, traveling at right angles to its original direction?
   Hint: The difficulty of this problem depends on how you choose your coordinate system.
c. What minimum separation d_{min} must the capacitor plates have?

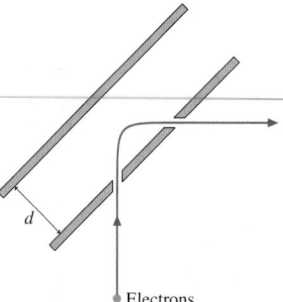

FIGURE P26.55   • Electrons

56. You have a summer intern position at a laboratory that uses a high-speed proton beam. The protons exit the machine at a speed of 2.0×10^6 m/s, and you've been asked to design a device to stop the protons safely. You know that protons will embed themselves in a metal target, but protons traveling faster than 2.0×10^5 m/s emit dangerous x rays when they hit. You decide to slow the protons to an acceptable speed, then let them hit a target. You take two metal plates, space them 2.0 cm apart, then drill a small hole through the center of one plate to let the proton beam enter. The opposite plate is the target in which the protons will embed themselves.
    a. What are the minimum surface charge densities you need to place on each plate? Which plate, positive or negative, faces the incoming proton beam?
    b. What happens if you charge the plates to $\pm 1.0 \times 10^{-5}$ C/m²? Does your device still work?

57. One type of ink-jet printer, called an electrostatic ink-jet printer, forms the letters by using deflecting electrodes to steer charged ink drops up and down vertically as the ink jet sweeps horizontally across the page. The ink jet forms 30-μm-diameter drops of ink, charges them by spraying 800,000 electrons on the surface, and shoots them toward the page at a speed of 20 m/s. Along the way, the drops pass through two parallel electrodes that are 6.0 mm long, 4.0 mm wide, and spaced 1.0 mm apart. The distance from the center of the plates to the paper is 2.0 cm. To form the letters, which have a maximum height of 6.0 mm, the drops need to be deflected up or down a maximum of 3.0 mm. Ink, which consists of dye particles suspended in alcohol, has a density of 800 kg/m³.
    a. Estimate the maximum electric field strength needed in the space between the electrodes.
    b. What amount of charge is needed on each electrode to produce this electric field?

58. A 2.0-mm-diameter glass sphere has a charge of +1.0 nC. What speed does an electron need to orbit the sphere 1.0 mm above the surface?

59. A proton orbits a 1.0-cm-diameter metal ball 1.0 mm above the surface. The orbital period is 1.0 μs. What is the charge on the ball?

60. In a classical model of the hydrogen atom, the electron orbits the proton in a circular orbit of radius 0.050 nm. What is the orbital frequency? The proton is so much more massive than the electron that you can assume the proton is at rest.

61. In a classical model of the hydrogen atom, the electron orbits a stationary proton in a circular orbit. What is the radius of the orbit for which the orbital frequency is $1.0 \times 10^{12} \text{ s}^{-1}$?

62. A *positron* is an elementary particle identical to an electron except that its charge is $+e$. An electron and a positron can rotate about their center of mass as if they were a dumbbell connected by a massless rod. What is the orbital frequency for an electron and a positron 1.0 nm apart?

63. An electric field can *induce* an electric dipole in a neutral atom or molecule by pushing the positive and negative charge in opposite directions. The dipole moment of an induced dipole is directly proportional to the electric field. That is, $\vec{p} = \alpha \vec{E}$, where α is called the *polarizability* of the molecule. A bigger field stretches the molecule farther and causes a larger dipole moment.
   a. What are the units of α?
   b. An ion with charge q is distance r from a molecule with polarizability α. Find an expression for the force $\vec{F}_{\text{ion on dipole}}$.

64. Show that a line of charge with linear charge density λ exerts an attractive force on an electric dipole with magnitude $F = 2\lambda p/4\pi\epsilon_0 r^2$. Assume that r is much larger than the charge separation in the dipole.

In Problems 65 through 68 you are given the equation(s) used to solve a problem. For each of these
   a. Write a realistic problem for which this is the correct equation(s).
   b. Finish the solution of the problem.

65. $(9.0 \times 10^9 \text{ Nm}^2/\text{C}^2)\dfrac{(2.0 \times 10^{-9} \text{ C}) s}{(0.025 \text{ m})^3} = 1150 \text{ N/C}$

66. $(9.0 \times 10^9 \text{ Nm}^2/\text{C}^2)\dfrac{2(2.0 \times 10^{-7} \text{ C/m})}{r} = 25{,}000 \text{ N/C}$

67. $\dfrac{\eta}{2\epsilon_0}\left[1 - \dfrac{z}{\sqrt{z^2 + R^2}}\right] = \dfrac{1}{2}\dfrac{\eta}{2\epsilon_0}$

68. $2.0 \times 10^{12} \text{ m/s}^2 = \dfrac{(1.60 \times 10^{-19} \text{ C}) E}{(1.67 \times 10^{-27} \text{ kg})}$

$E = \dfrac{Q}{(8.85 \times 10^{-12} \text{ C}^2/\text{Nm}^2)(0.020 \text{ m})^2}$

Challenge Problems

69. Your physics assignment is to figure out a way to use electricity to launch a small 6.0-cm-long plastic drink stirrer. You decide that you'll charge the little plastic rod by rubbing it with fur, then hold it near a long, charged wire. When you let go, the electric force of the wire on the plastic rod will shoot it away. Suppose you can charge the plastic stirrer to 10 nC and that the linear charge density of the long wire is $1.0 \times$ 10^{-7} C/m. What is the electric force on the plastic stirrer if the end closest to the wire is 2.0 cm away?

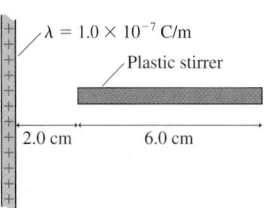

$\lambda = 1.0 \times 10^{-7} \text{ C/m}$

Plastic stirrer

2.0 cm   6.0 cm

FIGURE CP26.69

70. A rod of length L lies along the y-axis with its center at the origin. The rod has a nonuniform linear charge density $\lambda = a|y|$, where a is a constant with the units C/m^2.
   a. Draw a graph of λ versus y over the length of the rod.
   b. Determine the constant a in terms of L and the rod's total charge Q.
   Hint: This requires an integration. Think about how to handle the absolute value sign.
   c. Find the electric field strength of the rod at distance x on the x-axis.

71. a. An infinitely long *sheet* of charge of width L lies in the xy-plane between $x = -L/2$ and $x = L/2$. The surface charge density is η. Derive an expression for the electric field $\vec{E}$ at height z above the centerline of the sheet.
   b. Verify that your expression has the expected behavior if $z \ll L$ and if $z \gg L$.
   c. Draw a graph of field strength E versus z.

72. a. An infinitely long *sheet* of charge of width L lies in the xy-plane between $x = -L/2$ and $x = L/2$. The surface charge density is η. Derive an expression for the electric field $\vec{E}$ along the x-axis for points outside the sheet ($x > L/2$).
   b. Verify that your expression has the expected behavior if $x \gg L$.
   Hint: $\ln(1 + u) \approx u$ if $u \ll 1$.
   c. Draw a graph of field strength E versus x for $x > L/2$.

73. You have a summer intern position with a company that designs and builds nanomachines. An engineer with the company is designing a microscopic oscillator to help keep time, and you've been assigned to help him analyze the design. He wants to place a negative charge at the center of a very small, positively charged metal loop. His claim is that the negative charge will undergo simple harmonic motion at a frequency determined by the amount of charge on the loop.
   a. Consider a negative charge near the center of a positively charged ring. Show that there is a restoring force on the charge if it moves along the z-axis but stays close to the center. That is, show there's a force that tries to keep the charge at $z = 0$.
   b. Show that for *small* oscillations, with amplitude $\ll R$, a particle of mass m with charge $-q$ undergoes simple harmonic motion with frequency

$$f = \frac{1}{2\pi}\sqrt{\frac{qQ}{4\pi\epsilon_0 mR^3}}$$

R and Q are the radius and charge of the ring.
   c. Evaluate the oscillation frequency for an electron at the center of a 2.0-μm-diameter ring charged to 1.0×10^{-13} C.

74. We want to analyze how a charged object picks up a neutral piece of metal. Figure CP26.74 shows a small circular disk of aluminum foil lying flat on a table. The foil disk has radius R and thickness t. A glass ball with positive charge Q is at height h above the foil. Assume that $R \ll h$ and $t \ll R$. These assumptions imply that the electric field of the ball is very nearly constant over the volume of the foil disk.
    a. What are the magnitude and direction of the ball's electric field at the position of the foil? Your answer will be an expression involving Q and h.
    b. The ball's electric field polarizes the foil. The foil surfaces, with charges $+q$ and $-q$, then act as the plates of a parallel-plate capacitor with separation t. But the foil is a conductor in electrostatic equilibrium, so the electric field E_{in} *inside* the foil must be zero. $E_{in} = 0$ seems to be inconsistent with the surfaces of the foil acting as a parallel-plate capacitor. Use words *and* pictures to explain how $E_{in} = 0$ even though the surfaces of the foil are charged.
    c. Now write the condition that $E_{in} = 0$ as a mathematical statement and use it to find an expression for the charge q on the upper surface of the foil.
    d. Suppose $Q = 50$ nC, $R = 1$ mm, and $t = 0.01$ mm. These are all typical values. The density of aluminum is $\rho = 2700$ kg/m^3. How close must the ball be to lift the foil?

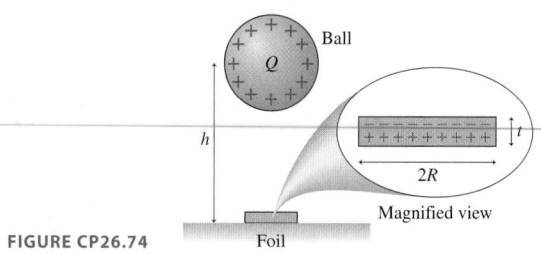

FIGURE CP26.74

STOP TO THINK ANSWERS

Stop to Think 26.1: c. From symmetry, the fields of the positive charges cancel. The net field is that of the negative charge, which is toward the charge.

Stop to Think 26.2: $\eta_3 = \eta_2 = \eta_1$. All pieces of a uniformly charged surface have the same surface charge density.

Stop to Think 26.3: b, e, and **h.** b and e both increase the linear charge density λ.

Stop to Think 26.4: $E_a = E_b = E_c = E_d = E_e$. The field strength of a charged plane is the same at all distances from the plane. An electric field diagram shows the electric field vectors at only a few points; the field exists at all points.

Stop to Think 26.5: $F_a = F_b = F_c = F_d = F_e$. The field strength inside a capacitor is the same at all points, hence the force on a charge is the same at all points. The electric field exists at all points whether or not a vector is shown at that point.

Stop to Think 26.6: c. Parabolic trajectories require *constant* acceleration and thus a *uniform* electric field. The proton has an initial velocity component to the left, but it's being pushed back to the right.

27 Gauss's Law

The nearly spherical shape of the girl's head determines the shape of the electric field that causes her hair to stream outwards.

The electric field of this charged sphere points straight out because that's the only field direction compatible with the *symmetry* of the sphere. Spheres, cylinders, and planes—common shapes for electrodes—all have a high degree of symmetry. As you'll see in this chapter, their symmetry determines the geometry of their electric fields.

You learned in Chapter 26 how to calculate electric fields by starting from Coulomb's law for the electric field of a point charge. This is a foolproof method in principle, but in practice it often requires excessive mathematical gymnastics to carry through the necessary integrals. In this chapter you'll learn how some important electric fields, those with a high degree of symmetry, can be deduced simply from the shape of the charge distribution. The principle underlying this approach to calculating electric fields is called *Gauss's law*.

Gauss's law and Coulomb's law are equivalent in the sense that each can be derived from the other. But Gauss's law gives us a very different perspective on electric fields, much as conservation laws give us a perspective on mechanics different from that of Newton's laws. In practice, Gauss's law will allow us to find some static electric fields that would be difficult to find using Coulomb's law. Ultimately, we'll find that Gauss's law is more general in that it applies not only to electrostatics but also to the electrodynamics of moving charges and fields that change with time.

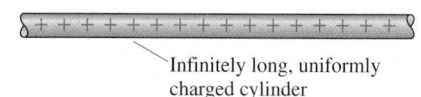

Infinitely long, uniformly charged cylinder

FIGURE 27.1 A charge distribution with cylindrical symmetry.

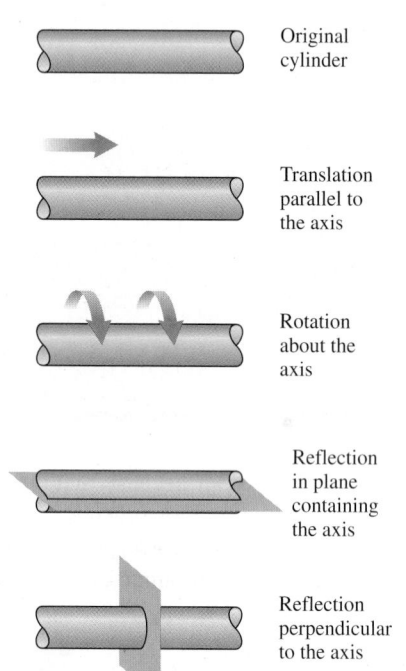

Original cylinder

Translation parallel to the axis

Rotation about the axis

Reflection in plane containing the axis

Reflection perpendicular to the axis

FIGURE 27.2 Transformations that don't change an infinite cylinder of charge.

27.1 Symmetry

Suppose we knew only two things about electric fields:

1. An electric field points away from positive charges, toward negative charges, and
2. An electric field exerts a force on a charged particle.

From this information alone, what can we deduce about the electric field of the infinitely long charged cylinder shown in Figure 27.1?

We don't know if the cylinder's diameter is large or small. We don't know if the charge density is the same at the outer edge as along the axis. All we know is that the charge is positive and the charge distribution has *cylindrical symmetry*.

Symmetry is an especially important idea in science and mathematics. We say that a charge distribution is **symmetric** if there are a group of *geometrical transformations* that don't cause any *physical* change. To make this idea concrete, suppose you close your eyes while a friend transforms a charge distribution in one of the following three ways. He or she can

■ *Translate* (that is, displace) the charge parallel to an axis,
■ *Rotate* the charge about an axis, or
■ *Reflect* the charge in a mirror.

When you open your eyes, will you be able to tell if the charge distribution has been changed? You might tell by observing a visual difference in the distribution. Or the results of an experiment with charged particles could reveal that the distribution has changed. If nothing you can see or do reveals any change, then we say that the charge distribution is symmetric under that particular transformation.

Figure 27.2 shows that the charge distribution of Figure 27.1 is symmetrical with respect to

■ Translations parallel to the cylinder axis. Shifting an infinitely long cylinder by 1 mm or 1000 m makes no noticeable or measurable change.
■ Rotation by any angle about the cylinder axis. Turning a cylinder about its axis by 1° or 100° makes no detectable change.
■ Reflections in any plane containing or perpendicular to the cylinder axis. Exchanging top and bottom, front and back, or left and right makes no detectable change.

But this isn't to say that *all* transformations are undetectable. You would notice immediately if your friend rotated the cylinder by 45° about an axis perpendicular to the cylinder axis.

A charge distribution that is symmetrical under these three groups of geometrical transformations is said to be *cylindrically symmetric*. Other charge distributions have other types of symmetries. Some charge distributions have no symmetry at all. Our interest in symmetry can be summed up in a single statement:

The symmetry of the electric field must match the symmetry of the charge distribution.

The proof of this statement is seen by assuming its converse. Suppose the electric field *did not* match the symmetry of the charge distribution. In that case, you could test whether or not the charge distribution had undergone a transformation by observing the motion of charged particles in the field. But the definition of symmetry requires that you not be able to tell. Thus the electric field must have the same symmetry as the charge distribution.

Now we're ready to see what we can learn about the electric field in Figure 27.1. Could the field look like Figure 27.3a? (Imagine this picture rotated about the axis.

Field vectors are also coming out of the page and going into the page.) That is, is this a *possible* field? This field looks the same if it's translated parallel to the cylinder axis, if up and down are exchanged by reflecting the field in a plane coming out of the page, or if you rotate the cylinder about its axis.

(a) Is this a possible electric field of an infinitely long charged cylinder? Suppose the charge and the field are reflected in a plane perpendicular to the axis.

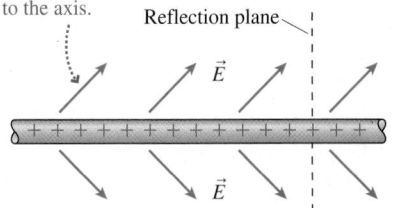 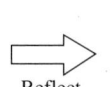

(b) The charge distribution is not changed by the reflection, but the field is. This field doesn't match the symmetry of the cylinder, so the cylinder's field can't look like this.

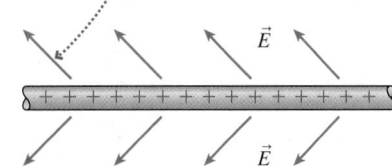

FIGURE 27.3 Could the field of a cylindrical charge distribution look like this?

But the proposed field fails one test. Suppose we reflect the field in a plane perpendicular to the axis, a reflection that exchanges left and right. This reflection, which would *not* make any change in the charge distribution itself, produces the field shown in Figure 27.3b. This change in the field is detectable because a positively charged particle would now have a component of motion to the left instead of to the right.

The field of Figure 27.3a, which makes a distinction between left and right, is not cylindrically symmetric and thus is *not* a possible field. In general, **the electric field of a cylindrically symmetric charge distribution cannot have a component parallel to the cylinder axis.**

Well then, what about the electric field shown in Figure 27.4a? Here we're looking down the axis of the cylinder. The electric field vectors are restricted to planes perpendicular to the cylinder and thus do not have any component parallel to the cylinder axis. This field is symmetric for rotations about the axis, but it's *not* symmetric for a reflection in a plane containing the axis.

The field of Figure 27.4b, after such a reflection, is easily distinguishable from the field of Figure 27.4a. Thus **the electric field of a cylindrically symmetric charge distribution cannot have a component tangent to the circular cross section.**

Figure 27.5 shows the only remaining possible field shape. The electric field is radial, pointing straight out from the cylinder like the bristles on a bottle brush. This is the one electric field shape that matches the symmetry of the charge distribution.

(a)

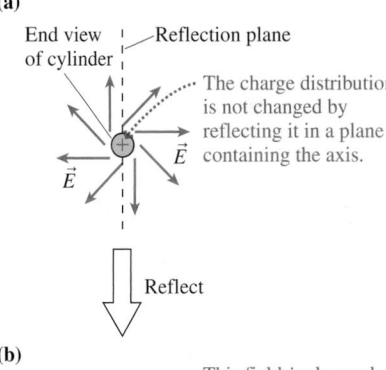

End view of cylinder — Reflection plane

The charge distribution is not changed by reflecting it in a plane containing the axis.

Reflect

(b)

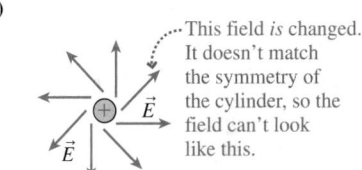

This field *is* changed. It doesn't match the symmetry of the cylinder, so the field can't look like this.

FIGURE 27.4 Or might the field of a cylindrical charge distribution look like this?

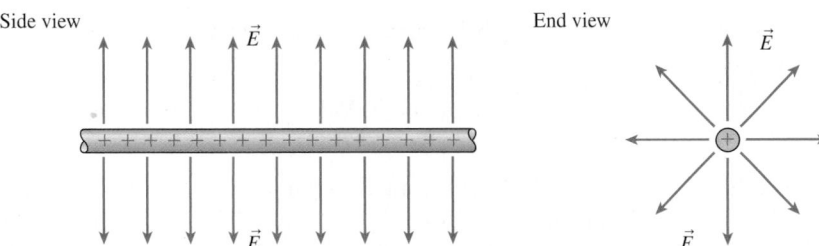

FIGURE 27.5 This is the only shape for the electric field that matches the symmetry of the charge distribution.

What Good Is Symmetry?

Given how little we assumed about Figure 27.1—that the charge distribution is cylindrically symmetric and that electric fields point away from positive charges—we've been able to deduce a great deal about the electric field. In particular, we've deduced the *shape* of the electric field.

Now, shape is not everything. We've learned nothing about the strength of the field or how strength changes with distance. Is E constant? Does it decrease like $1/r$ or $1/r^2$? We don't yet have a complete description of the field, but knowing what shape the field *has* to have will make finding the field strength a much easier task.

That's the good of symmetry. Symmetry arguments allow us to *rule out* many conceivable field shapes as simply being incompatible with the symmetry of the charge distribution. Knowing what doesn't happen, or can't happen, is often as useful as knowing what does happen. By the process of elimination, we're led to the one and only shape the field can possibly have. Reasoning on the basis of symmetry is a sometimes subtle but always powerful means of reasoning.

Three Fundamental Symmetries

Three fundamental symmetries appear frequently in electrostatics. The first row of Figure 27.6 shows the simplest form of each symmetry. The second row shows a more complex, but more realistic, situation with the same symmetry. We may not know the field strength, but the field *shape* in these more complex situations must match the symmetry of the charge distribution.

NOTE ▶ Figures must be finite in extent, but the planes and cylinders in Figure 27.6 are assumed to be infinite. ◀

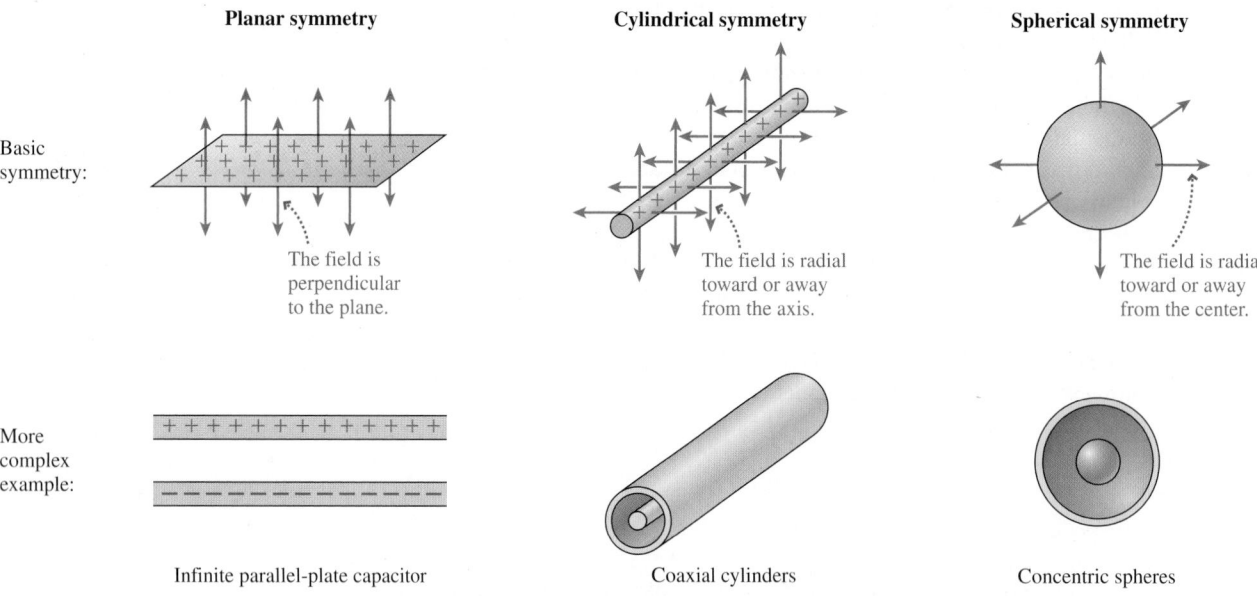

FIGURE 27.6 Three fundamental symmetries.

Objects do exist that are extremely close to being perfect spheres, but no real cylinder or plane can be infinite in extent. Even so, the fields of infinite planes and cylinders are good models for the fields of finite planes and cylinders at points not too close to an edge or an end. Planar and cylindrical electrodes are common in a vast number of practical devices, so the fields that we'll study in this chapter, even if idealized, are not without important applications.

STOP TO THINK 27.1   A uniformly charged rod has a *finite* length *L*. The rod is symmetric under rotations about the axis and under reflection in any plane containing the axis. It is *not* symmetric under translations or under reflections in a plane perpendicular to the axis unless that plane bisects the rod. Which field shape or shapes match the symmetry of the rod?

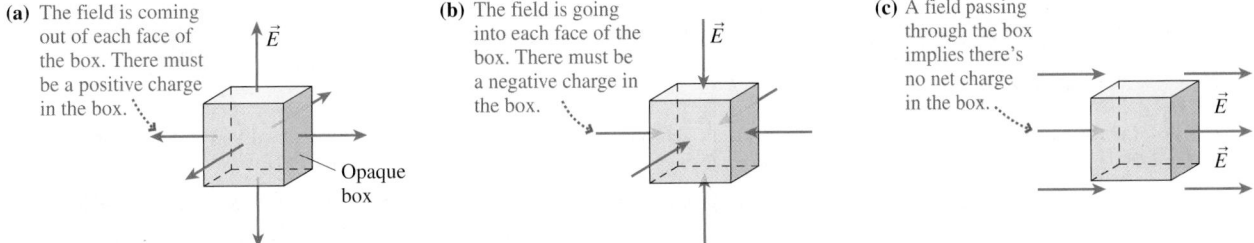

27.2 The Concept of Flux

Figure 27.7a shows an opaque box surrounding a region of space. We can't see what's in the box, but there's an electric field vector coming out of each face of the box. Can you figure out what's in the box?

(a) The field is coming out of each face of the box. There must be a positive charge in the box. $\vec{E}$ — Opaque box

(b) The field is going into each face of the box. There must be a negative charge in the box. $\vec{E}$

(c) A field passing through the box implies there's no net charge in the box. $\vec{E}$ $\vec{E}$

FIGURE 27.7 Although we can't see into the boxes, the electric fields passing through the faces tell us something about what's in them.

Of course you can. Because electric fields point away from positive charges, and the electric field is coming out of every face of the box, it seems clear that the box contains a positive charge or charges. Similarly, the box in Figure 27.7b certainly contains a negative charge.

What can we tell about the box in Figure 27.7c? The electric field points into the box on the left. An equal electric field points out on the right. This might be the electric field between a large positive electrode somewhere out of sight on the left and a large negative electrode off to the right. An electric field passes through the box, but we see no evidence there's any charge (or at least any net charge) inside the box.

These examples suggest that the electric field as it passes into, out of, or through the box is in some way connected to the charge within the box. However, these simple pictures don't tell us how much charge there is or where within the box the charge is located. Perhaps a better box would be more informative.

(a)

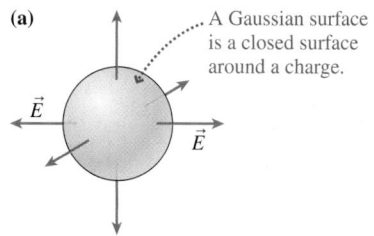

A Gaussian surface is a closed surface around a charge.

(b)

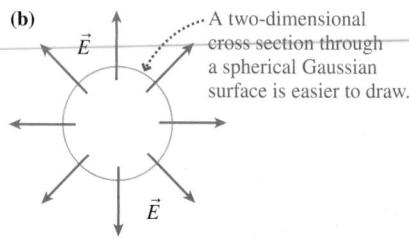

A two-dimensional cross section through a spherical Gaussian surface is easier to draw.

FIGURE 27.8 Gaussian surface surrounding a charge. A two-dimensional cross section is usually easier to draw.

Suppose we surround a region of space with a *closed surface,* a surface that divides space into distinct inside and outside regions. Within the context of electrostatics, a closed surface through which an electric field passes is called a **Gaussian surface,** named after the 19th-century mathematician Karl Gauss who developed the mathematical foundations of geometry. This is an imaginary, mathematical surface, not a physical surface, although it might coincide with a physical surface. For example, Figure 27.8a shows a spherical Gaussian surface surrounding a charge.

A closed surface must, of necessity, be a surface in three dimensions. But three-dimensional pictures are hard to draw, so we'll often look at two-dimensional cross sections through a Gaussian surface, such as the one shown in Figure 27.8b. Now, a better choice of box makes it more clear what's inside. We can tell from the *spherical symmetry* of the electric field vectors poking through the surface that the positive charge inside must be spherically symmetric and centered at the *center* of the sphere. Notice two features that will soon be important: The electric field is everywhere *perpendicular* to the spherical surface and has the *same magnitude* at each point on the surface.

Figure 27.9 shows another example. An electric field emerges from four sides of the cube in Figure 27.9a but not from the top or bottom. We might be able to guess what's within the box, but we can't be sure. Figure 27.9b uses a different Gaussian surface, a *closed* cylinder (i.e., the cylindrical wall *and* the flat ends), and Figure 27.9c simplifies the drawing by showing two-dimensional end and side views. Now, with a better choice of surface, we can tell that the cylindrical Gaussian surface surrounds some kind of cylindrical charge distribution, such as a charged wire. Again, the electric field is everywhere *perpendicular* to the cylindrical surface and has the *same magnitude* at each point on the surface.

(a)

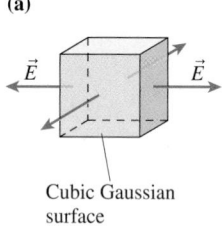

Cubic Gaussian surface

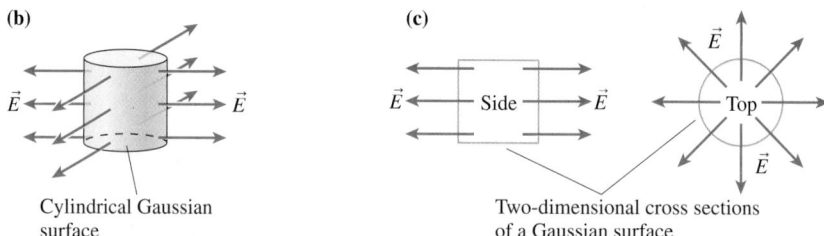

FIGURE 27.9 Gaussian surface is most useful when it matches the shape of the field.

(a)

A Gaussian surface that doesn't match the symmetry of the electric field isn't very useful.

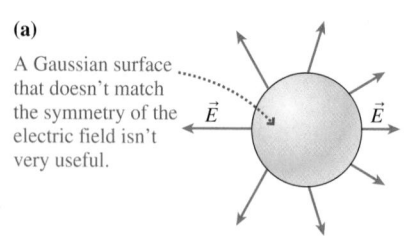

(b)

A nonclosed surface doesn't provide enough information about the charges.

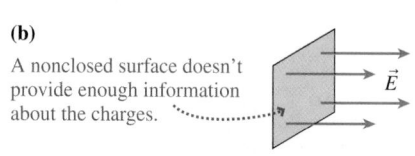

FIGURE 27.10 Not every surface is useful for learning about charge.

For contrast, consider the spherical surface in Figure 27.10a. This is also a Gaussian surface, and the protruding electric field tells us there's a positive charge inside. It might be a point charge located on the left side, but we can't really say. A Gaussian surface that doesn't match the symmetry of the charge distribution isn't terribly useful.

The nonclosed surface of Figure 27.10b doesn't provide much help either. What appears to be a uniform electric field to the right could be due to a large positive plate on the left, a large negative plate on the right, or both. A nonclosed surface doesn't provide enough information.

These examples lead us to two conclusions:

1. The electric field, in some sense, "flows" *out of* a closed surface surrounding a region of space containing a net positive charge and *into* a closed surface surrounding a net negative charge. The electric field may flow *through* a closed surface surrounding a region of space in which there is no net charge, but the *net flow* is zero.

2. The electric field pattern through the surface is particularly simple if the closed surface matches the symmetry of the charge distribution inside.

Now, it's true that the electric field doesn't flow like a fluid, but the metaphor is a useful one. The Latin word for flow is *flux,* and the amount of electric field passing through a surface is called the **electric flux.** Our first conclusion, stated in terms of electric flux, is

- There is an outward flux through a closed surface around a net positive charge.
- There is an inward flux through a closed surface around a net negative charge.
- There is no net flux through a closed surface around a region of space in which there is no net charge.

This chapter has been entirely qualitative thus far as we've established pictorially what we mean by symmetry, the idea of flux, and the fact that the electric flux through a closed surface has something to do with the charge inside. Understanding these qualitative ideas is essential, but to go further we need to make these ideas quantitative and precise. In the next section, you'll learn how to calculate the electric flux through a surface. Then, in the section following that, we'll establish a precise relationship between the net flux through a Gaussian surface and the enclosed charge. That relationship, Gauss's law, will allow us to determine the electric fields of some interesting and useful charge distributions.

STOP TO THINK 27.2 This box contains

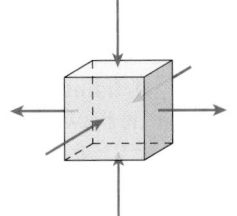

a. A positive charge.
b. A negative charge.
c. No charge.
d. A net positive charge.
e. A net negative charge.
f. No net charge.

27.3 Calculating Electric Flux

Let's start with a brief overview of where this section will take us. We'll begin with a definition of flux that is easy to understand, then we'll turn that simple definition into a formidable-looking integral. We need the integral because the simple definition applies only to uniform electric fields and flat surfaces. Those are a good starting point, but we'll soon need to calculate the flux of nonuniform fields through curved surfaces.

Mathematically, the flux of a nonuniform field through a curved surface is described by a special kind of integral called a *surface integral.* It's quite possible that you have not yet encountered surface integrals in your calculus course, and the "novelty factor" contributes to making this integral look worse than it really is. We will emphasize over and over the idea that an integral is just a fancy way of doing a sum, in this case the sum of the small amounts of flux through many small pieces of a surface.

The good news is that *every* surface integral we need to evaluate in this chapter, or that you will need to evaluate for the homework problems, is either zero or is so easy that you will be able to do it in your head. This seems like an astounding claim, but you will soon see it is true. The key will be to make effective use of the *symmetry* of the electric field.

Now that you've been warned, you needn't panic at the sight of the mathematical notation that will be introduced. We'll go step by step, and you'll see that, at least as far as electrostatics is concerned, calculating the electric flux is not difficult.

The Basic Definition of Flux

Imagine holding a rectangular wire loop of area A in front of a fan. As Figure 27.11 shows, the volume of air flowing through the loop each second depends on the angle between the loop and the direction of flow. The flow is maximum through a loop that is perpendicular to the air flow; no air goes through the same loop if it lies parallel to the flow.

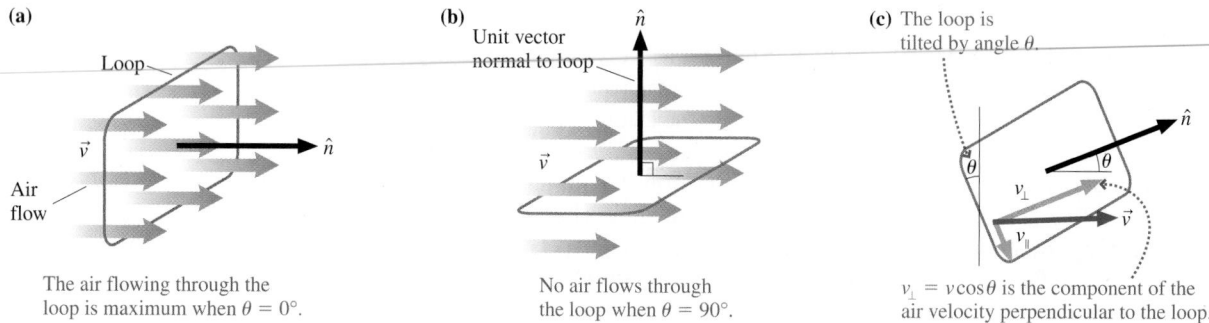

(a)

Loop

$\vec{v}$

Air flow

$\hat{n}$

The air flowing through the loop is maximum when $\theta = 0°$.

(b)

$\hat{n}$

Unit vector normal to loop

$\vec{v}$

No air flows through the loop when $\theta = 90°$.

(c) The loop is tilted by angle θ.

$\hat{n}$

θ

$v_\perp$

$v_\parallel$

$\vec{v}$

θ

$v_\perp = v\cos\theta$ is the component of the air velocity perpendicular to the loop.

FIGURE 27.11 The amount of air flowing through a loop depends on the angle between $\vec{v}$ and $\hat{n}$.

The flow direction is identified by the velocity vector $\vec{v}$. We can identify the loop's orientation by defining a unit vector $\hat{n}$ normal to the plane of the loop. Angle θ is then the angle between $\vec{v}$ and $\hat{n}$. The loop perpendicular to the flow in Figure 27.11a has $\theta = 0°$; the loop parallel to the flow in Figure 27.11b has $\theta = 90°$. You can think of θ as the angle by which a loop has been tilted away from perpendicular.

NOTE ▶ A surface has two sides, so $\hat{n}$ could point either way. We'll choose the side that makes $\theta \leq 90°$. ◀

You can see from Figure 27.11c that the velocity vector $\vec{v}$ can be decomposed into components $v_\perp = v\cos\theta$ perpendicular to the loop and $v_\parallel = v\sin\theta$ parallel to the loop. Only the perpendicular component $v_\perp$ carries air *through* the loop. Consequently, the volume of air flowing through the loop each second is

$$\text{volume of air per second (m}^3\text{/s)} = v_\perp A = vA\cos\theta \qquad (27.1)$$

$\theta = 0°$ is the orientation for maximum flow through the loop, as expected, and no air flows through the loop if it is tilted $90°$.

An electric field doesn't flow in a literal sense, but we can apply the same idea to an electric field passing through a surface. Figure 27.12 shows a surface of area A in a uniform electric field $\vec{E}$. Unit vector $\hat{n}$ is normal to the surface and θ is the angle between $\hat{n}$ and $\vec{E}$. Only the component $E_\perp = E\cos\theta$ passes *through* the surface.

With this in mind, and using Equation 27.1 as an analog, let's define the *electric flux* Φ_e as

$$\Phi_e = E_\perp A = EA\cos\theta \qquad (27.2)$$

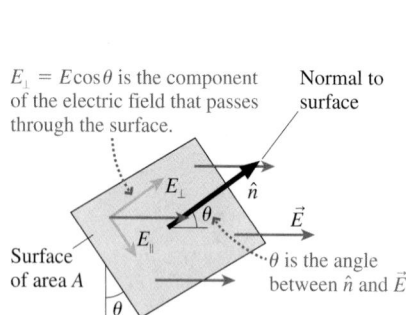

$E_\perp = E\cos\theta$ is the component of the electric field that passes through the surface.

Normal to surface

$E_\perp$

$\hat{n}$

θ

$\vec{E}$

$E_\parallel$

Surface of area A

θ is the angle between $\hat{n}$ and $\vec{E}$.

θ

FIGURE 27.12 An electric field passing through a surface.

The electric flux measures the amount of electric field passing through a surface of area A if the normal to the surface is tilted at angle θ from the field.

Equation 27.2 looks very much like a vector dot product: $\vec{E} \cdot \vec{A} = EA\cos\theta$. For this idea to work, let's define an **area vector** $\vec{A} = A\hat{n}$ to be a vector in the direction of $\hat{n}$—that is, *perpendicular* to the surface—with a magnitude A equal to the area of the surface. Vector $\vec{A}$ has units of m^2. Figure 27.13a shows two area vectors.

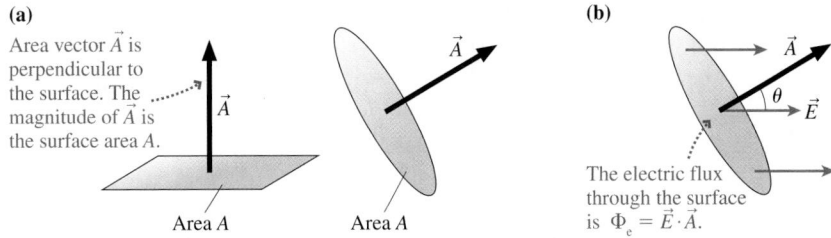

(a)

Area vector $\vec{A}$ is perpendicular to the surface. The magnitude of $\vec{A}$ is the surface area A.

$\vec{A}$

Area A

Area A

(b)

$\vec{A}$

θ

$\vec{E}$

The electric flux through the surface is $\Phi_e = \vec{E} \cdot \vec{A}$.

FIGURE 27.13 The electric flux can be defined in terms of the area vector $\vec{A}$.

Figure 27.13b shows an electric field passing through a surface of area A. The angle between vectors $\vec{A}$ and $\vec{E}$ is the same angle used in Equation 27.2 to define the electric flux, so Equation 27.2 really is a dot product. We can define the electric flux more concisely as

$$\Phi_e = \vec{E} \cdot \vec{A} \qquad \text{(electric flux of a constant electric field)} \qquad (27.3)$$

Writing the flux as a dot product helps make clear how angle θ is defined: θ is the angle between the electric field and a line *perpendicular* to the plane of the surface.

NOTE ▶ Figure 27.13b shows a circular area, but the shape of the surface is not relevant. However, Equation 27.3 is restricted to a *constant* electric field passing through a *planar* surface. ◀

EXAMPLE 27.1 The electric flux inside a parallel-plate capacitor

Two 100 cm^2 parallel electrodes are spaced 2.0 cm apart. One is charged to $+5$ nC, the other to -5 nC. A $1.0 \text{ cm} \times 1.0 \text{ cm}$ surface between the electrodes is tilted to where its normal makes a $45°$ angle with the electric field. What is the electric flux through this surface?

MODEL Assume the surface is located near the center of the capacitor where the electric field is uniform. The electric flux doesn't depend on the shape of the surface.

VISUALIZE The surface is square, rather than circular, but otherwise the situation looks like Figure 27.13b.

SOLVE In Chapter 26, we found the electric field inside a parallel-plate capacitor to be

$$E = \frac{Q}{\epsilon_0 A_{\text{plates}}} = \frac{5.0 \times 10^{-9} \text{ C}}{(8.85 \times 10^{-12} \text{ C}^2/\text{N m}^2)(1.0 \times 10^{-2} \text{ m}^2)}$$

$$= 5.65 \times 10^4 \text{ N/C}$$

A $1.0 \text{ cm} \times 1.0 \text{ cm}$ surface has $A = 1.0 \times 10^{-4} \text{ m}^2$. The electric flux through this surface is

$$\Phi_e = \vec{E} \cdot \vec{A} = EA \cos\theta$$

$$= (5.65 \times 10^4 \text{ N/C})(1.0 \times 10^{-4} \text{ m}^2) \cos 45°$$

$$= 4.00 \text{ N m}^2/\text{C}$$

ASSESS The units of electric flux are the product of electric field and area units: $\text{N m}^2/\text{C}$.

The Electric Flux of a Nonuniform Electric Field

Our initial definition of the electric flux assumed that the electric field $\vec{E}$ was constant over the surface. How should we calculate the electric flux if $\vec{E}$ varies from point to point on the surface? We can answer this question by returning to the analogy of air flowing through a loop. Suppose the air flow varies from point to point. We can still find the total volume of air passing through the loop each second by dividing the loop into many small areas, finding the flow through each small area, then adding them. Similarly, **the electric flux through a surface can be calculated as the sum of the fluxes through smaller pieces of the surface.** Because flux is a scalar, adding fluxes is easier than adding electric fields.

Figure 27.14 on the next page shows a surface in a nonuniform electric field. Imagine dividing the surface into many small pieces of area δA. Each little area

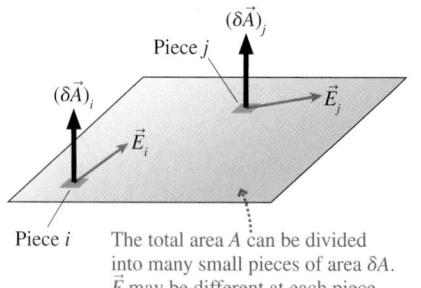

Piece i The total area A can be divided into many small pieces of area δA. $\vec{E}$ may be different at each piece.

FIGURE 27.14 A surface in a nonuniform electric field.

has an area vector $\delta\vec{A}$ perpendicular to the surface. Two of the little pieces, labeled i and j, are shown in the figure. The electric fluxes through these two pieces differ because the electric fields are different.

Consider the small piece i where the electric field is $\vec{E}_i$. The small electric flux $\delta\Phi_i$ through area $(\delta\vec{A})_i$ is

$$\delta\Phi_i = \vec{E}_i \cdot (\delta\vec{A})_i \tag{27.4}$$

The flux through every other little piece of the surface is found the same way. The total electric flux through the entire surface is then the sum of the fluxes through each of the small areas:

$$\Phi_e = \sum_i \delta\Phi_i = \sum_i \vec{E}_i \cdot (\delta\vec{A})_i \tag{27.5}$$

Now let's go to the limit $\delta\vec{A} \to d\vec{A}$. That is, the little areas become infinitesimally small, and there are infinitely many of them. Then the sum becomes an integral, and the electric flux through the surface is

$$\Phi_e = \int_{\text{surface}} \vec{E} \cdot d\vec{A} \tag{27.6}$$

The integral in Equation 27.6 is called a **surface integral.**

Equation 27.6 may look rather frightening if you haven't seen surface integrals before. Despite its appearance, a surface integral is no more complicated than integrals you know from calculus. After all, what does $\int f(x)\,dx$ really mean? This expression is a shorthand way to say "Divide the x-axis into many little segments of length δx, evaluate the function $f(x)$ in each of them, then add up $f(x)\,\delta x$ for all the segments along the line." The integral in Equation 27.6 differs only in that we're dividing a surface into little pieces instead of a line into little segments. In particular, we're summing the fluxes through a vast number of very tiny pieces

You may be thinking, "OK, I understand the idea, but I don't know what to *do*. In calculus, I learned formulas for evaluating integrals such as $\int x^2 dx$. How do I evaluate a surface integral?" This is a good question. We'll deal with evaluation shortly, and it will turn out that the surface integrals in electrostatics are quite easy to evaluate. But don't confuse *evaluating* the integral with understanding what the integral *means*. The surface integral in Equation 27.6 is simply a shorthand notation for the summation of the electric fluxes through a vast number of very tiny pieces of a surface.

The electric field might be different at every point on the surface, but suppose it isn't. That is, suppose the surface is in a uniform electric field $\vec{E}$. A field that is the same at every single point on a surface is a constant as far as the integration of Equation 27.6 is concerned, so we can take it outside the integral. In that case,

$$\Phi_e = \int_{\text{surface}} \vec{E} \cdot d\vec{A} = \int_{\text{surface}} E\cos\theta\, dA = E\cos\theta \int_{\text{surface}} dA \tag{27.7}$$

The integral that remains in Equation 27.7 tells us to add up all the little areas into which the full surface was subdivided. But the sum of all the little areas is simply the area of the surface:

$$\int_{\text{surface}} dA = A \tag{27.8}$$

This idea—that the surface integral of dA is the area of the surface—is one we'll use to evaluate most of the surface integrals of electrostatics. If we substitute Equation 27.8 into Equation 27.7, we find that the electric flux in a uniform electric field is $\Phi_e = EA\cos\theta$. We already knew this, from Equation 27.2, but it was important to see that the surface integral of Equation 27.6 gives the correct result for the case of a uniform electric field.

The Flux Through a Curved Surface

Most of the Gaussian surfaces we considered in the last section were curved surfaces. Figure 27.15 shows an electric field passing through a curved surface. How do we find the electric flux through this surface? Just as we did for a flat surface!

Divide the surface into many small pieces of area δA. For each, define the area vector $\vec{\delta A}$ perpendicular to the surface *at that point*. Compared to Figure 27.14, the only difference that the curvature of the surface makes is that the $\vec{\delta A}$ are no longer parallel to each other. Find the small electric flux $\delta\Phi_i = \vec{E}_i \cdot (\vec{\delta A})_i$ through each little area, then add them all up. The result, once again, is

$$\Phi_e = \int_{surface} \vec{E} \cdot d\vec{A} \qquad (27.9)$$

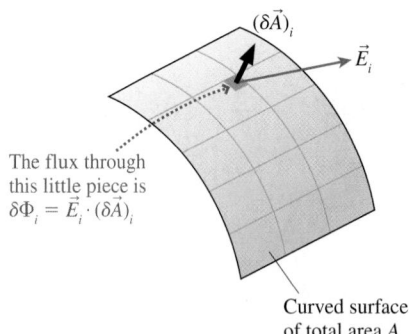

FIGURE 27.15 A curved surface in an electric field.

We *assumed,* in deriving this expression the first time, that the surface was flat and that all the $\vec{\delta A}$ were parallel to each other. But that assumption wasn't necessary. The *meaning* of Equation 27.9—a summation of the fluxes through a vast number of very tiny pieces—is unchanged if the pieces happen to lie on a curved surface.

We seem to be getting more and more complex, using surface integrals first for nonuniform fields and now for curved surfaces. But consider the two situations shown in Figure 27.16. The electric field $\vec{E}$ in Figure 27.16a is everywhere tangent, or parallel, to the curved surface. We don't need to know the magnitude of $\vec{E}$ to recognize that $\vec{E} \cdot d\vec{A}$ is *zero at every point* on the surface because $\vec{E}$ is perpendicular to $d\vec{A}$ at every point. Thus $\Phi_e = 0$. A tangent electric field never pokes through the surface, so it has no flux through the surface.

The electric field in Figure 27.16b is everywhere perpendicular to the surface *and* has the same magnitude E at every point. $\vec{E}$ differs in direction at different points on a curved surface, but at any particular point $\vec{E}$ is parallel to $d\vec{A}$ and $\vec{E} \cdot d\vec{A}$ is simply EdA. In this case,

$$\Phi_e = \int_{surface} \vec{E} \cdot d\vec{A} = \int_{surface} E\,dA = E \int_{surface} dA = EA \qquad (27.10)$$

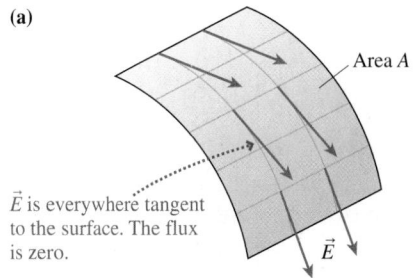

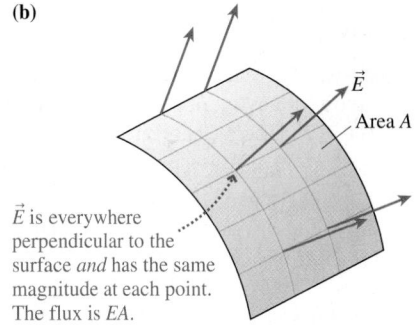

FIGURE 27.16 Electric fields that are everywhere tangent to or everywhere perpendicular to a curved surface.

In evaluating the integral, the fact that E has the same magnitude at every point on the surface allowed us to bring the constant value out of the integral. We then made use of the fact that the integral of dA over the surface is the surface area A.

We can summarize these two situations with a Tactics Box.

TACTICS BOX 27.1 Evaluating surface integrals

❶ If the electric field is everywhere tangent to a surface, the electric flux through the surface is $\Phi_e = 0$.

❷ If the electric field is everywhere perpendicular to a surface *and* has the same magnitude E at every point, the electric flux through the surface is $\Phi_e = EA$.

These two results will be of immeasurable value for using Gauss's law because *every* flux we'll need to calculate will be one of these situations. This is the basis for our earlier claim that the evaluation of surface integrals is not going to be difficult.

The Electric Flux through a Closed Surface

Our final step, to calculate the electric flux through a closed surface such as a box, a cylinder, or a sphere, requires nothing new. We've already learned how to calculate the electric flux through flat and curved surfaces, and a closed surface is nothing more than a surface that happens to be closed.

However, the mathematical notation for the surface integral over a closed surface differs slightly from what we've been using. It is customary to use a little circle on the integral sign to indicate that the surface integral is to be performed over a closed surface. With this notation, the electric flux through a closed surface is

$$\Phi_e = \oint \vec{E} \cdot d\vec{A} \qquad (27.11)$$

Only the notation has changed. The electric flux is still simply the summation of the fluxes through a vast number of very tiny pieces, pieces that now cover a closed surface.

NOTE ▶ A closed surface has a distinct inside and outside. The area vector $d\vec{A}$ is defined to always point *toward the outside*. This removes an ambiguity that was present for a single surface, where $d\vec{A}$ could point to either side. ◀

EXAMPLE 27.2 Calculating the electric flux through a closed cylinder

A cylindrical charge distribution has created the electric field $\vec{E} = E_0(r^2/r_0^2)\hat{r}$, where E_0 and r_0 are constants and where unit vector $\hat{r}$ lies in the xy-plane. Calculate the electric flux through a closed cylinder of length L and radius R that is centered along the z-axis.

MODEL The electric field extends radially outward from the z-axis with cylindrical symmetry. The z-component is $E_z = 0$. The cylinder is a Gaussian surface.

VISUALIZE Figure 27.17a is a view of the electric field looking along the z-axis. The field strength increases with increasing radial distance, and it's symmetric about the z-axis. Figure 27.17b is the closed Gaussian surface for which we need to calculate the electric flux. We can place the cylinder anywhere along the z-axis because the electric field extends forever in that direction.

SOLVE To calculate the flux, divide the closed cylinder into three surfaces: the top, the bottom, and the cylindrical wall. The electric field is tangent to the surface at every point on the top and bottom surfaces. Hence, according to Step 1 in Tactics Box 27.1, the flux through those two surfaces is zero. For the cylindrical wall, the electric field is perpendicular to the surface at every point *and* has the constant magnitude $E = E_0(R^2/r_0^2)$ at every point on the surface. Thus, from Step 2 in Tactics Box 27.1,

$$\Phi_{wall} = EA_{wall}$$

If we add up the three pieces, the net flux through the closed surface is

$$\Phi_e = \oint \vec{E} \cdot d\vec{A} = \Phi_{top} + \Phi_{bottom} + \Phi_{wall} = 0 + 0 + EA_{wall}$$

$$= EA_{wall}$$

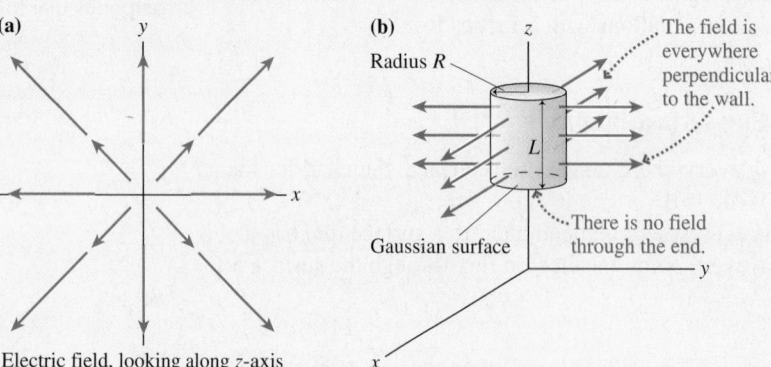

(a) Electric field, looking along z-axis

(b) Radius R — The field is everywhere perpendicular to the wall. — There is no field through the end. — Gaussian surface

FIGURE 27.17 The electric field and the closed surface through which we will calculate the electric flux.

We've evaluated the surface integral, using the two steps in Tactics Box 27.1, and there was nothing to it! To finish, all we need to recall is that the surface area of a cylindrical wall is circumference × height, or $A_{wall} = 2\pi RL$. Thus

$$\Phi_e = \left(E_0 \frac{R^2}{r_0^2}\right)(2\pi RL) = \frac{2\pi LR^3}{r_0^2}E_0$$

ASSESS LR^3/r_0^2 has units of m^2, an area, so this expression for Φ_e has units of Nm^2/C. These are the correct units for electric flux, giving us confidence in our answer. Notice the important role played by symmetry. The electric field was perpendicular to the wall and of constant value at every point on the wall *because* the Gaussian surface had the same symmetry as the charge distribution. We would not have been able to evaluate the surface integral in such an easy way for a surface of any other shape. Symmetry is the key.

Example 27.2 illustrated a two-step approach to performing a flux integral over a closed surface. In summary:

TACTICS BOX 27.2 Finding the flux through a closed surface

❶ Divide the closed surface into pieces that are everywhere tangent to the electric field and everywhere perpendicular to the electric field.
❷ Use Tactics Box 27.1 to evaluate the surface integrals over these surfaces, then add the results.

STOP TO THINK 27.3 The total electric flux through this box is

a. $0\,Nm^2/C$.   b. $1\,Nm^2/C$.   c. $2\,Nm^2/C$.
d. $4\,Nm^2/C$.   e. $6\,Nm^2/C$.   f. $8\,Nm^2/C$.

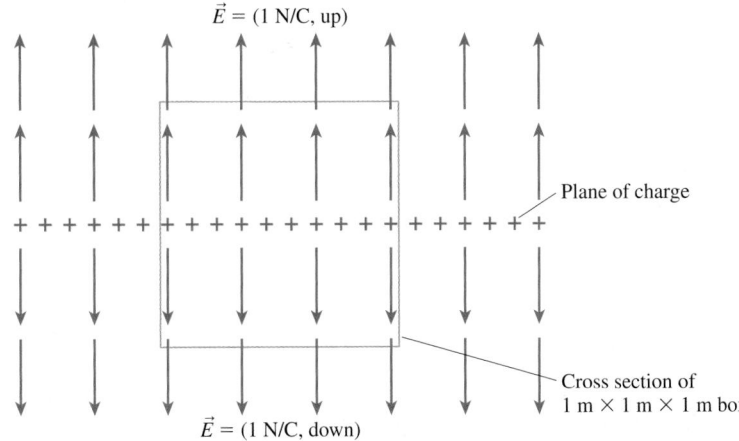

$\vec{E} = (1\,N/C,\ up)$

Plane of charge

$\vec{E} = (1\,N/C,\ down)$

Cross section of $1\,m \times 1\,m \times 1\,m$ box

27.4 Gauss's Law

The last section was long, but knowing how to calculate the electric flux through a closed surface is essential for the main topic of this chapter: Gauss's law. Gauss's law is equivalent to Coulomb's law for static charges, although Gauss's law will look very different.

Activ Physics ONLINE 11.8

The purpose of learning Gauss's law is twofold:

- Gauss's law allows the electric fields of some continuous distributions of charge to be found much more easily than does Coulomb's law.
- Gauss's law is valid for *moving* charges, but Coulomb's law is not (although it's a very good approximation for velocities that are much less than the speed of light). Thus Gauss's law is ultimately a more fundamental statement about electric fields than is Coulomb's law.

Let's start with Coulomb's law for the electric field of a point charge. Figure 27.18 shows a spherical Gaussian surface of radius r centered on a positive charge q. Keep in mind that this is an imaginary, mathematical surface, not a physical surface. There is a net flux through this surface because the electric field points outward at every point on the surface. To evaluate the flux, given formally by the surface integral of Equation 27.11, notice that the electric field is perpendicular to the surface at every point on the surface *and*, from Coulomb's law, it has the same magnitude $E = q/4\pi\epsilon_0 r^2$ at every point on the surface. This simple situation arises because **the Gaussian surface has the same symmetry as the electric field.**

Thus we know, without having to do any hard work, that the flux integral is

$$\Phi_e = \oint \vec{E} \cdot d\vec{A} = EA_{\text{sphere}} \qquad (27.12)$$

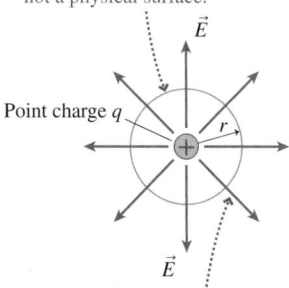

Cross section of a Gaussian sphere of radius r. This is a mathematical surface, not a physical surface.

$\vec{E}$

Point charge q

r

$\vec{E}$

The electric field is everywhere perpendicular to the surface *and* has the same magnitude at every point.

FIGURE 27.18 A spherical Gaussian surface surrounding a point charge.

The surface area of a sphere of radius r is $A_{\text{sphere}} = 4\pi r^2$. If we use A_{sphere} and the Coulomb-law expression for E in Equation 27.12, we find that the electric flux through the spherical surface is

$$\Phi_e = \frac{q}{4\pi\epsilon_0 r^2} 4\pi r^2 = \frac{q}{\epsilon_0} \qquad (27.13)$$

You should examine the logic of this calculation closely. We really did evaluate the surface integral of Equation 27.11, although it may appear, at first, as if we didn't do much. The integral was easily evaluated, we reiterate for emphasis, because the closed surface on which we performed the integration matched the *symmetry* of the charge distribution. In such cases, the surface integral for the flux is simply a field strength multiplied by a surface area.

NOTE ▶ We found Equation 27.13 for a positive charge, but it applies equally to negative charges. According to Equation 27.13, Φ_e is negative if q is negative. And that's what we would expect from the basic definition of flux, $\vec{E} \cdot \vec{A}$. The electric field of a negative charge points inward while the area vector of a closed surface points outward, making the dot product negative. ◀

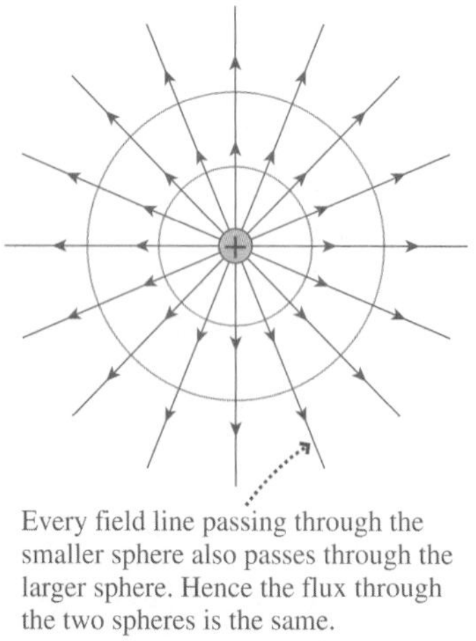

The Electric Flux Is Independent of the Surface Shape and Radius

Notice something interesting about Equation 27.13. The electric flux depends on the amount of charge but *not* on the radius of the sphere. Although this may seem a bit surprising, it's really a direct consequence of what we *mean* by flux. Think of the fluid analogy with which we introduced the term "flux." If fluid flows outward from a central point, all the fluid crossing a small-radius spherical surface will, at some later time, cross a large-radius spherical surface. No fluid is lost along the way, and no new fluid is created. Similarly, the point charge in Figure 27.19 is the only source of electric field. Every electric field line passing through a small-radius spherical surface also passes through a large-radius spherical surface. Hence the electric flux is independent of r.

NOTE ▶ This argument hinges on the fact that Coulomb's law is an inverse-square force law. The electric field strength, which is proportional to $1/r^2$, decreases with distance. But the surface area, which increases in proportion

Every field line passing through the smaller sphere also passes through the larger sphere. Hence the flux through the two spheres is the same.

FIGURE 27.19 The electric flux is the same through *every* sphere centered on a point charge.

to r^2, exactly compensates for this decrease. Consequently, the electric flux of a point charge through a spherical surface is independent of the radius of the sphere. ◄

This conclusion about the flux has an extremely important generalization. Figure 27.20a shows a point charge and a closed Gaussian surface of arbitrary shape and dimensions. All we know is that the charge is *inside* the surface. What is the electric flux through this surface?

One way to answer the question is to approximate the surface as a patchwork of spherical and radial pieces. The spherical pieces are centered on the charge and the radial pieces lie along lines extending outward from the charge. The figure, of necessity, has shown fairly large pieces that don't match the actual surface all that well. However, we can make this approximation as good as we want by letting the pieces become sufficiently small.

The electric field is everywhere tangent to the radial pieces. Hence the electric flux through the radial pieces is zero. The spherical pieces, although at varying distances from the charge, form a *complete sphere*. That is, any line drawn radially outward from the charge will pass through exactly one spherical piece, and no radial lines can avoid passing through a spherical piece. You could even imagine, as Figure 27.20b shows, sliding the spherical pieces in and out *without changing the angle they subtend* until they come together to form a complete sphere.

Consequently, the electric flux through these spherical pieces that, when assembled, form a complete sphere must be exactly the same as the flux q/ϵ_0 through a spherical Gaussian surface. In other words, **the flux through *any* closed surface surrounding a point charge q is**

$$\Phi_e = \oint \vec{E} \cdot d\vec{A} = \frac{q}{\epsilon_0} \tag{27.14}$$

This surprisingly simple result is a consequence of the fact that Coulomb's law is an inverse-square force law. Even so, the reasoning that got us to Equation 27.14 is rather subtle and well worth reviewing.

Charge Outside the Surface

The closed surface shown in Figure 27.21a has a point charge q outside the surface but no charges inside. Now what can we say about the flux? By approximating this surface with spherical and radial pieces *centered on the charge,* as we did in Figure 27.20, we can reassemble the surface into the equivalent surface of Figure 27.21b. This closed surface consists of sections of two spherical shells, and it is equivalent in the sense that the electric flux through this surface is the same as the electric flux through the original surface of Figure 27.21a.

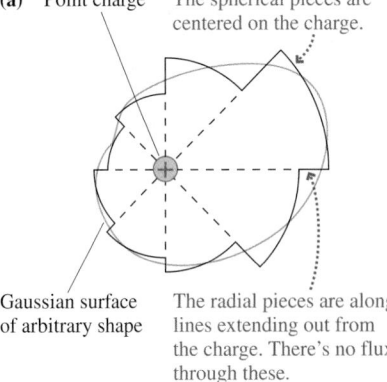

(a) Point charge · The spherical pieces are centered on the charge.

Gaussian surface of arbitrary shape · The radial pieces are along lines extending out from the charge. There's no flux through these.

The approximation with spherical and radial pieces can be as good as desired by letting the pieces become sufficiently small.

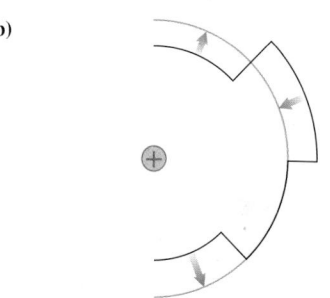

(b)

The spherical pieces can slide in or out to form a complete sphere. Hence the flux through the pieces is the same as the flux through a sphere.

FIGURE 27.20 An arbitrary Gaussian surface can be approximated with spherical and radial pieces.

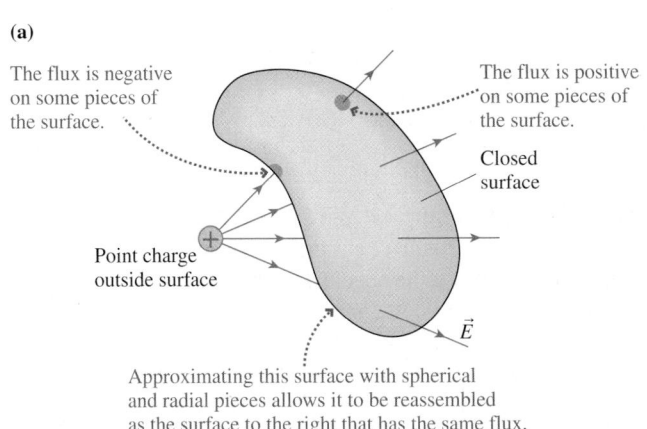

(a)

The flux is negative on some pieces of the surface.

The flux is positive on some pieces of the surface.

Closed surface

Point charge outside surface

$\vec{E}$

Approximating this surface with spherical and radial pieces allows it to be reassembled as the surface to the right that has the same flux.

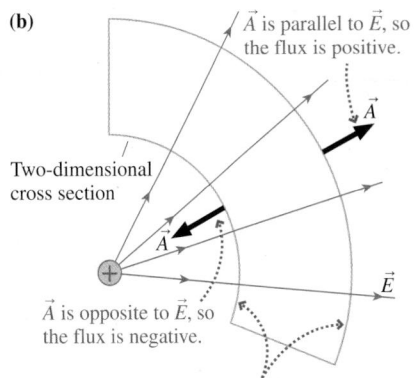

(b)

$\vec{A}$ is parallel to $\vec{E}$, so the flux is positive.

$\vec{A}$

Two-dimensional cross section

$\vec{A}$

$\vec{A}$ is opposite to $\vec{E}$, so the flux is negative.

$\vec{E}$

The fluxes through these surfaces are equal but opposite. The net flux is zero.

FIGURE 27.21 A point charge outside a Gaussian surface.

If the electric field were a fluid flowing outward from the charge, all the fluid *entering* the closed region through the first spherical surface would later *exit* the region through the second spherical surface. There is no *net* flow into or out of the closed region. Similarly, every electric field line entering this closed volume through one spherical surface exits through the other spherical surface.

Mathematically, the electric fluxes through the two spherical surfaces have the same magnitude because Φ_e is independent of r. But they have *opposite signs* because the outward-pointing area vector $\vec{A}$ is parallel to $\vec{E}$ on one surface but opposite to $\vec{E}$ on the other. The sum of the fluxes through the two surfaces is zero, and we are led to the conclusion that **the net electric flux is zero through a closed surface that does not contain any net charge.** Charges outside the surface do not produce a net flux through the surface.

This isn't to say that the flux through a small piece of the surface is zero. In fact, as Figure 27.21a shows, nearly every piece of the surface has an electric field either entering or leaving and thus has a nonzero flux. But some of these are positive and some are negative. When summed over the *entire* surface, the positive and negative contributions exactly cancel to give no *net* flux.

Multiple Charges

Finally, consider an arbitrary Gaussian surface and a group of charges q_1, q_2, q_3, . . . such as those shown in Figure 27.22. Some of these charges are inside the surface, others outside. The charges can be either positive or negative. What is the net electric flux through the closed surface?

By definition, the net flux is

$$\Phi_e = \oint \vec{E} \cdot d\vec{A}$$

From the principle of superposition, the electric field is $\vec{E} = \vec{E}_1 + \vec{E}_2 + \vec{E}_3 + \cdots$, where $\vec{E}_1, \vec{E}_2, \vec{E}_3, \ldots$ are the fields of the individual charges. Thus the flux can be written

$$\Phi_e = \oint \vec{E}_1 \cdot d\vec{A} + \oint \vec{E}_2 \cdot d\vec{A} + \oint \vec{E}_3 \cdot d\vec{A} + \cdots$$
$$= \Phi_1 + \Phi_2 + \Phi_3 + \cdots \tag{27.15}$$

where $\Phi_1, \Phi_2, \Phi_3, \ldots$ are the fluxes through the Gaussian surface due to the individual charges. That is, the net flux is the sum of the fluxes due to individual charges. But we know what those are: q/ϵ_0 for the charges inside the surface and zero for the charges outside. Thus

$$\Phi_e = \left(\frac{q_1}{\epsilon_0} + \frac{q_2}{\epsilon_0} + \cdots + \frac{q_i}{\epsilon_0} \text{ for all charges inside the surface} \right.$$
$$\left. + (0 + 0 + \cdots + 0 \text{ for all charges outside the surface}) \right. \tag{27.16}$$

Define

$$Q_{in} = q_1 + q_2 + \cdots + q_i \text{ for all charges inside the surface} \tag{27.17}$$

as the total charge enclosed *within* the surface. With this definition, we can write our result for the net electric flux in a very neat and compact fashion. For any *closed* surface enclosing total charge Q_{in}, the net electric flux through the surface is

$$\Phi_e = \oint \vec{E} \cdot d\vec{A} = \frac{Q_{in}}{\epsilon_0} \tag{27.18}$$

This result for the electric flux is known as **Gauss's law.**

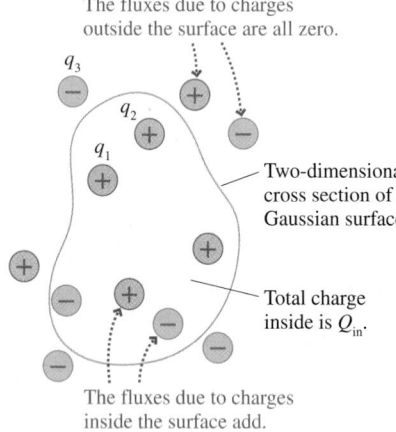

The fluxes due to charges outside the surface are all zero.

q_3

q_2

q_1

Two-dimensional cross section of a Gaussian surface

Total charge inside is Q_{in}.

The fluxes due to charges inside the surface add.

FIGURE 27.22 Charges both inside and outside a Gaussian surface.

What Does Gauss's Law Tell Us?

In one sense, Gauss's law doesn't say anything new or anything that we didn't already know from Coulomb's law. After all, we derived Gauss's law from Coulomb's law. But in another sense, Gauss's law is more important than Coulomb's law. Gauss's law states a very general property of electric fields—namely, that charges create electric fields in just such a way that the net flux of the field is the same through *any* surface surrounding the charges, no matter what its size and shape may be. This fact may have been implied by Coulomb's law, but it was by no means obvious. And Gauss's law will turn out to be particularly useful later when we combine it with other electric and magnetic field equations.

Gauss's law is the mathematical statement of our observations in Section 27.2. There we noticed a net "flow" of electric field out of a closed surface containing charges. Gauss's law quantifies this idea by making a specific connection between the "flow," now measured as electric flux, and the amount of charge.

But is it useful? Although to some extent Gauss's law is a formal statement about electric fields, not a tool for solving practical problems, there are exceptions: Gauss's law will allow us to find the electric fields of some very important and very practical charge distributions much more easily than if we had to rely on Coulomb's law. We'll consider some examples in the next section.

STOP TO THINK 27.4   These are two-dimensional cross sections through three-dimensional closed spheres and a cube. Rank order, from largest to smallest, the electric fluxes Φ_a to Φ_e through surfaces a to e.

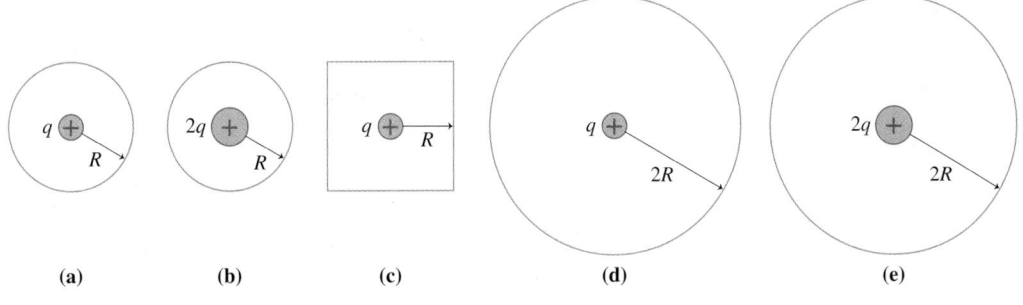

(a)   (b)   (c)   (d)   (e)

27.5 Using Gauss's Law

In this section, we'll use Gauss's law to determine the electric fields of several important charge distributions. Some of these you already know, from Chapter 26; others will be new. Several important observations can be made about using Gauss's law:

1. Gauss's law applies only to a *closed* surface, called a Gaussian surface.
2. A Gaussian surface is not a physical surface. It need not coincide with the boundary of any physical object (although it could if we wished). It is an imaginary, mathematical surface in the space surrounding one or more charges.
3. We can't find the electric field from Gauss's law alone. We need to apply Gauss's law in situations where, from symmetry and superposition, we already can guess the *shape* of the field.

These observations and our previous discussion of symmetry and flux lead to the following strategy for solving electric field problems with Gauss's law.

(MP) PROBLEM-SOLVING STRATEGY 27.1 **Gauss's law**

MODEL Model the charge distribution as a distribution with symmetry.

VISUALIZE Draw a picture of the charge distribution.

- Determine the symmetry of its electric field.
- Choose and draw a Gaussian surface with the *same symmetry*.
- You need not enclose all the charge within the Gaussian surface.
- Be sure every part of the Gaussian surface is either tangent to or perpendicular to the electric field.

SOLVE The mathematical representation is based on Gauss's law

$$\Phi_e = \oint \vec{E} \cdot d\vec{A} = \frac{Q_{in}}{\epsilon_0}$$

- Use Tactics Boxes 27.1 and 27.2 to evaluate the surface integral.

ASSESS Check that your result has the correct units, is reasonable, and answers the question.

EXAMPLE 27.3 Outside a sphere of charge

In Chapter 26 we asserted, without proof, that the electric field outside a sphere of total charge Q is the same as the field of a point charge Q at the center. Use Gauss's law to prove this result.

MODEL The charge distribution within the sphere need not be uniform (i.e., the charge density might increase or decrease with r), but it must have spherical symmetry in order for us to use Gauss's law. We will assume that it does.

VISUALIZE Figure 27.23 shows a sphere of charge Q and radius R. We want to find $\vec{E}$ outside this sphere, for distances $r > R$. The spherical symmetry of the charge distribution tells us that the electric field must point *radially outward* from the sphere. Although Gauss's law is true for any surface surrounding the charged sphere, it is useful only if we choose a Gaussian surface to match the spherical symmetry of the charge distribution and the field. Thus a spherical surface of radius $r > R$ *concentric with* the charged sphere will be our Gaussian surface. Because this surface surrounds the entire sphere of charge, the enclosed charge is simply $Q_{in} = Q$.

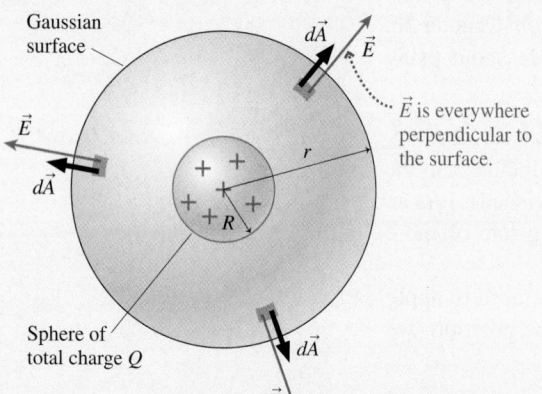

FIGURE 27.23 A spherical Gaussian surface surrounding a sphere of charge.

SOLVE Gauss's law is

$$\Phi_e = \oint \vec{E} \cdot d\vec{A} = \frac{Q_{in}}{\epsilon_0} = \frac{Q}{\epsilon_0}$$

To calculate the flux, notice that the electric field is everywhere perpendicular to the spherical surface. And although we don't know the electric field magnitude E, spherical symmetry dictates that E must have the same value at all points equally distant from the center of the sphere. Thus we have the simple result that the net flux through the Gaussian surface is

$$\Phi_e = EA_{sphere} = 4\pi r^2 E$$

where we used the fact that the surface area of a sphere is $A_{sphere} = 4\pi r^2$. With this result for the flux, Gauss's law is

$$4\pi r^2 E = \frac{Q}{\epsilon_0}$$

Thus the electric field at distance r outside a sphere of charge is

$$E_{outside} = \frac{1}{4\pi\epsilon_0} \frac{Q}{r^2}$$

Or in vector form, making use of the fact that $\vec{E}$ is radially outward,

$$\vec{E}_{outside} = \frac{1}{4\pi\epsilon_0} \frac{Q}{r^2} \hat{r}$$

where $\hat{r}$ is a radial unit vector.

ASSESS The field is exactly that of a point charge Q, which is what we wanted to show.

The derivation of the electric field of a sphere of charge depended crucially on a proper choice of the Gaussian surface. We would not have been able to evaluate the flux integral so simply for any other choice of surface. It's worth noting that the result of Example 27.3 can also be proven by the superposition of point-charge fields, but it requires a difficult three-dimensional integral and about a page of algebra. We obtained the answer using Gauss's law in just a few lines. Where Gauss's law works, it works *extremely* well! However, it works only in situations, such as this, with a very high degree of symmetry.

EXAMPLE 27.4 Inside a sphere of charge

What is the electric field *inside* a uniformly charged sphere?

MODEL We haven't considered a situation like this before. To begin, we don't know if the field strength is increasing or decreasing as we move outward from the center of the sphere. But the field inside must have spherical symmetry. That is, the field must point radially inward or outward, and the field strength can depend only on r. This is sufficient information to solve the problem because it allows us to choose a Gaussian surface.

VISUALIZE Figure 27.24 shows a spherical Gaussian surface with radius $r \leq R$ inside, and *concentric with*, the sphere of charge. This surface matches the symmetry of the charge distribution, hence $\vec{E}$ is perpendicular to this surface and the field strength E has the same value at all points on the surface.

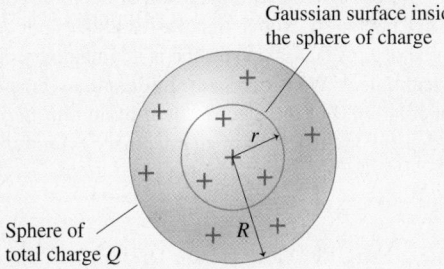

Gaussian surface inside the sphere of charge

Sphere of total charge Q

FIGURE 27.24 A spherical Gaussian surface inside a uniform sphere of charge.

SOLVE The flux integral is identical to that of Example 27.3:

$$\Phi_e = EA_{\text{sphere}} = 4\pi r^2 E$$

Consequently, Gauss's law is

$$\Phi_e = 4\pi r^2 E = \frac{Q_{\text{in}}}{\epsilon_0}$$

The difference between this example and Example 27.3 is that Q_{in} is no longer the total charge of the sphere. Instead, Q_{in} is the amount of charge *inside* the Gaussian sphere of radius r. Because the charge distribution is *uniform*, the ratio Q_{in}/Q must be the same as the ratio V_r/V_R of the volumes of the spheres of radii r and R. That is,

$$\frac{Q_{\text{in}}}{Q} = \frac{V_r}{V_R} = \frac{\frac{4}{3}\pi r^3}{\frac{4}{3}\pi R^3} = \frac{r^3}{R^3}$$

from which we find Q_{in} to be

$$Q_{\text{in}} = \frac{r^3}{R^3} Q$$

The amount of enclosed charge increases with the cube of the distance r from the center and, as expected, equals $Q_{\text{in}} = Q$ if $r = R$. With this expression for Q_{in}, Gauss's law is

$$4\pi r^2 E = \frac{(r^3/R^3)Q}{\epsilon_0}$$

Thus the electric field at radius r inside a uniformly charged sphere is

$$E_{\text{inside}} = \frac{1}{4\pi\epsilon_0} \frac{Q}{R^3} r$$

The electric field strength inside the sphere increases *linearly* with the distance r from the center.

ASSESS The field inside and the field outside a sphere of charge match at the boundary of the sphere, $r = R$, where both give $E = Q/4\pi\epsilon_0 R^2$. In other words, the field strength is *continuous* as we cross the boundary of the sphere. These results are shown graphically in Figure 27.25.

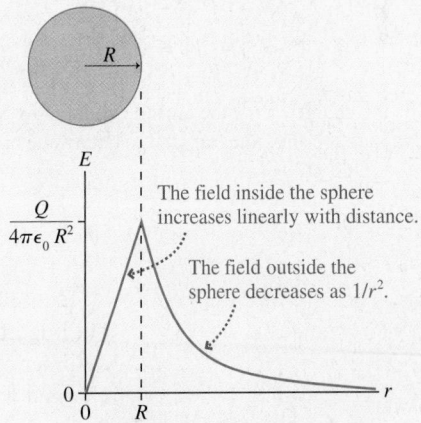

The field inside the sphere increases linearly with distance.

The field outside the sphere decreases as $1/r^2$.

FIGURE 27.25 The electric field strength of a uniform sphere of charge of radius R.

EXAMPLE 27.5 The electric field of a long, charged wire

In Chapter 26, we used superposition to find the electric field of an infinitely long line of charge with linear charge density (C/m) λ. It was not an easy derivation. Find the electric field using Gauss's law.

MODEL A long, charged wire can be modeled as an infinitely long line of charge.

VISUALIZE Figure 27.26 shows an infinitely long line of charge. We can use the symmetry of the situation to see that the only possible shape of the electric field is to point straight into or out from the wire, rather like the bristles on a bottle brush. The shape of the field suggests that we choose our Gaussian surface to be a cylinder of radius r and length L, centered on the wire. Because Gauss's law refers to *closed* surfaces, we must include the ends of the cylinder as part of the surface.

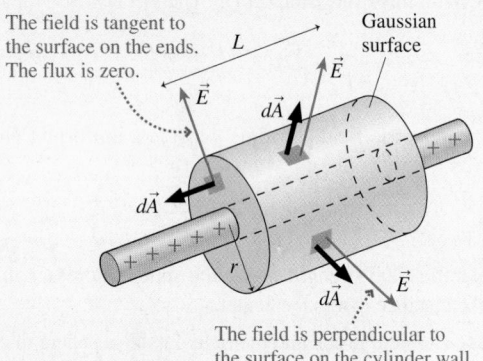

The field is tangent to the surface on the ends. The flux is zero.

Gaussian surface

The field is perpendicular to the surface on the cylinder wall.

FIGURE 27.26 A Gaussian surface around a charged wire.

SOLVE Gauss's law is

$$\Phi_e = \oint \vec{E} \cdot d\vec{A} = \frac{Q_{in}}{\epsilon_0}$$

where Q_{in} is the charge *inside* the closed cylinder. We have two tasks: to evaluate the flux integral, and to determine how much charge is inside the closed surface. The wire has linear charge density λ, so the amount of charge inside a cylinder of length L is simply

$$Q_{in} = \lambda L$$

Finding the net flux is just as straightforward. We can divide the flux through the entire closed surface into the flux through each end plus the flux through the cylindrical wall. The electric field $\vec{E}$, pointing straight out from the wire, is tangent to the end surfaces at every point. Thus the flux through these two surfaces is zero. On the wall, $\vec{E}$ is perpendicular to the surface and has the same strength E at every point. Thus

$$\Phi_e = \Phi_{top} + \Phi_{bottom} + \Phi_{wall} = 0 + 0 + EA_{cyl} = 2\pi rLE$$

where we used $A_{cyl} = 2\pi rL$ as the surface area of a cylindrical wall of radius r and length L. Once again, the proper choice of the Gaussian surface reduces the flux integral merely to finding a surface area. With these expressions for Q_{in} and Φ_e, Gauss's law is

$$\Phi_e = 2\pi rLE = \frac{Q_{in}}{\epsilon_0} = \frac{\lambda L}{\epsilon_0}$$

Thus the electric field at distance r from a long, charged wire is

$$E_{wire} = \frac{\lambda}{2\pi\epsilon_0 r}$$

ASSESS This agrees exactly with the result of the more complex derivation in Chapter 26. Notice that the result does not depend on our choice of L. A Gaussian surface is an imaginary device, not a physical object. We needed a finite-length cylinder to do the flux calculation, but the electric field of an *infinitely* long wire can't depend on the length of an imaginary cylinder.

Example 27.5, for the electric field of a long, charged wire contains a subtle but important idea, one that often occurs when using Gauss's law. The Gaussian cylinder of length L encloses only some of the wire's charge. The pieces of the charged wire outside the cylinder are not enclosed by the Gaussian surface and consequently do not contribute anything to the net flux. Even so, *they are essential* to the use of Gauss's law because it takes the *entire* charged wire to produce an electric field with cylindrical symmetry. In other words, the wire outside the cylinder may not contribute to the flux, but it affects the *shape* of the electric field. Our ability to write $\Phi_e = EA_{cyl}$ depended on knowing that E is the same at every point on the wall of the cylinder. That would not be true for a charged wire of finite length, so we cannot use Gauss's law to find the electric field of a finite-length charged wire.

EXAMPLE 27.6 The electric field of a plane of charge

Use Gauss's law to find the electric field of an infinite plane of charge with surface charge density (C/m^2) η.

MODEL A uniformly charged flat electrode can be modeled as an infinite plane of charge.

VISUALIZE Figure 27.27 shows a uniformly charged plane with surface charge density η. We will assume that the plane extends infinitely far in all directions, although we obviously have to show "edges" in our drawing. The planar symmetry allows the electric field to point only straight toward or away from the plane. With this in mind, choose as a Gaussian surface a cylinder with length L and faces of area A centered on the plane of charge. Although we've drawn them as circular, the shape of the faces is not relevant.

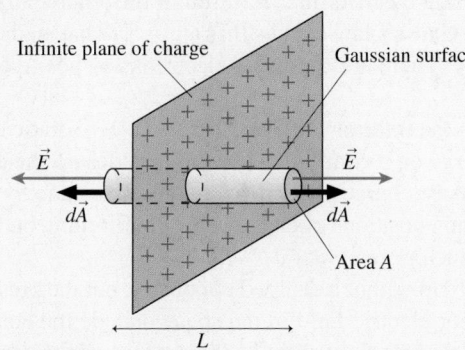

Infinite plane of charge   Gaussian surface $\vec{E}$ $d\vec{A}$ $\vec{E}$ $d\vec{A}$ Area A L

FIGURE 27.27 The Gaussian surface extends to both sides of a plane of charge.

SOLVE The electric field is perpendicular to both faces of the cylinder, so the total flux through both faces is $\Phi_{\text{faces}} = 2EA$. (The fluxes add rather than cancel because the area vector $\vec{A}$ points *outward* on each face.) There's *no* flux through the wall of the cylinder because the field vectors are tangent to the wall. Thus the net flux is simply

$$\Phi_e = 2EA$$

The charge inside the cylinder is the charge contained in area A of the plane. This is

$$Q_{\text{in}} = \eta A$$

With these expressions for Q_{in} and Φ_e, Gauss's law is

$$\Phi_e = 2EA = \frac{Q_{\text{in}}}{\epsilon_0} = \frac{\eta A}{\epsilon_0}$$

Thus the electric field of an infinite charged plane is

$$E_{\text{plane}} = \frac{\eta}{2\epsilon_0}$$

This agrees with the result of Chapter 26.

ASSESS Again, the details of how we chose the Gaussian surface don't affect the result. Notice that this is another example of a Gaussian surface enclosing only some of the charge. Most of the plane's charge is outside the Gaussian surface and does not contribute to the flux, but it does affect the shape of the field. We wouldn't have planar symmetry, with the electric field exactly perpendicular to the plane, without all the rest of the charge on the plane.

The plane of charge is an especially good example of how powerful Gauss's law can be. Finding the electric field of a plane of charge via superposition was a difficult and tedious derivation. With Gauss's law, once you see how to apply it, the problem is simple enough to solve in your head!

You might wonder, then, why we bothered with superposition at all. The reason is that Gauss's law, powerful though it may be, is effective only in a very limited number of situations where the field is highly symmetrical. Superposition always works, even if the derivation is messy, because superposition goes directly back to the fields of individual point charges. It's good to use Gauss's law when you can, but superposition is often the only way to attack real-world charge distributions.

STOP TO THINK 27.5 Which Gaussian surface would allow you to use Gauss's law to determine the electric field outside a uniformly charged cube?

a. A sphere whose center coincides with the center of the charged cube.
b. A cube whose center coincides with the center of the charged cube and which has parallel faces.
c. Either a or b.
d. Neither a nor b.

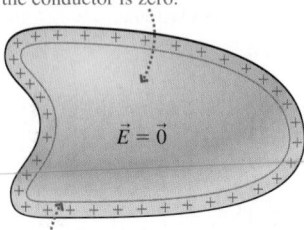

The electric field inside the conductor is zero.

$\vec{E} = \vec{0}$

The flux through the Gaussian surface is zero. There's no net charge inside the conductor. Hence all the excess charge is on the surface.

FIGURE 27.28 A Gaussian surface just inside a conductor that's in electrostatic equilibrium.

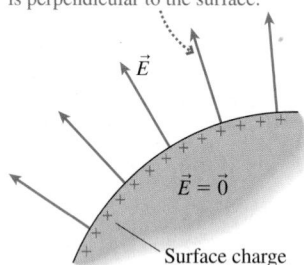

The electric field at the surface is perpendicular to the surface.

$\vec{E}$

$\vec{E} = \vec{0}$

Surface charge

FIGURE 27.29 The electric field at the surface of a charged conductor.

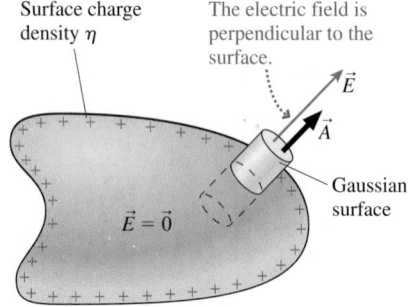

Surface charge density η

The electric field is perpendicular to the surface.

$\vec{E}$

$\vec{A}$

$\vec{E} = \vec{0}$

Gaussian surface

FIGURE 27.30 A Gaussian surface extending through the surface of the conductor has a flux only through the outer face.

27.6 Conductors in Electrostatic Equilibrium

Consider a charged conductor, such as a charged metal electrode, in electrostatic equilibrium. That is, there is no current through the conductor and the charges are all stationary. In Chapter 25, you learned that **the electric field is zero at all points within a conductor in electrostatic equilibrium.** That is, $\vec{E}_{in} = \vec{0}$. If this weren't true, the electric field would cause the charge carriers to move and thus violate the assumption that all the charges are at rest. Let's use Gauss's law to see what else we can learn.

At the Surface of a Conductor

Figure 27.28 shows a Gaussian surface just barely inside the physical surface of a conductor that's in electrostatic equilibrium. The electric field is zero at all points within the conductor, hence the electric flux Φ_e through this Gaussian surface must be zero. But if $\Phi_e = 0$, Gauss's law tells us that $Q_{in} = 0$. That is, there's no net charge within this surface. There are charges—electrons and positive ions—but no *net* charge.

If there's no net charge in the interior of a conductor in electrostatic equilibrium, then **all the excess charge on a charged conductor resides on the exterior surface of the conductor.** Any charges added to a conductor quickly spread across the surface until reaching positions of electrostatic equilibrium, but there is no net charge *within* the conductor.

There may be no electric field within a charged conductor, but the presence of net charge requires an exterior electric field in the space outside the conductor. Figure 27.29 shows that **the electric field right at the surface of the conductor has to be perpendicular to the surface.** To see that this is so, suppose $\vec{E}_{surface}$ had a component tangent to the surface. This component of $\vec{E}_{surface}$ would exert a force on the surface charges and cause a surface current, thus violating the assumption that all charges are at rest. The only exterior electric field consistent with electrostatic equilibrium is one that is perpendicular to the surface.

We can use Gauss's law to relate the field strength at the surface to the charge density on the surface. Figure 27.30 shows a small Gaussian cylinder with faces very slightly above and below the surface of a charged conductor. The charge inside this Gaussian cylinder is ηA, where η is the surface charge density at this point on the conductor. There's a flux $\Phi = AE_{surface}$ through the outside face of this cylinder but, unlike Example 27.6 for the plane of charge, *no* flux through the inside face because $\vec{E}_{in} = \vec{0}$ within the conductor. Furthermore, there's no flux through the wall of the cylinder because $\vec{E}_{surface}$ is perpendicular to the surface. Thus the net flux is $\Phi_e = AE_{surface}$. Gauss's law is

$$\Phi_e = AE_{surface} = \frac{Q_{in}}{\epsilon_0} = \frac{\eta A}{\epsilon_0} \tag{27.19}$$

from which we can conclude that the electric field at the surface of a charged conductor is

$$\vec{E}_{surface} = \left(\frac{\eta}{\epsilon_0}, \text{ perpendicular to surface} \right) \tag{27.20}$$

In general, the surface charge density η is *not* constant on the surface of a conductor but varies in a complicated way that depends on the shape of the conductor. If we can determine η, either by calculating it or measuring it, then Equation 27.20 tells us the electric field at that point on the surface. Alternatively, we can use

Equation 27.20 to deduce the charge density on the conductor's surface if we know the electric field just outside the conductor.

Charges and Fields within a Conductor

Figure 27.31 shows a charged conductor with a hole inside. Can there be charge on this interior surface? To find out, place a Gaussian surface around the hole, infinitesimally close but entirely within the conductor. The electric flux Φ_e through this Gaussian surface is zero because the electric field is zero everywhere inside the conductor. Thus we must conclude that $Q_{in} = 0$. There's no net charge inside this Gaussian surface and thus no charge on the surface of the hole. Any excess charge resides on the *exterior* surface of the conductor, not on any interior surfaces.

Furthermore, because there's no electric field inside the conductor and no charge inside the hole, the electric field inside the hole must also be zero. This conclusion has an important practical application. For example, suppose we need to exclude the electric field from the region in Figure 27.32a enclosed within dotted lines. We can do so by surrounding this region with the neutral conducting box of Figure 27.32b.

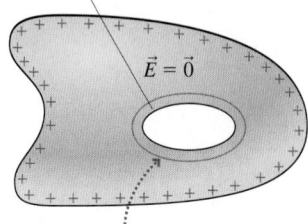

A hollow completely enclosed by the conductor

The flux through the Gaussian surface is zero. There's no net charge inside, hence no charge on this interior surface.

FIGURE 27.31 A Gaussian surface surrounding a hole inside a conductor in electrostatic equilibrium.

(a) Parallel-plate capacitor

$\vec{E}$

We want to exclude the electric field from this region.

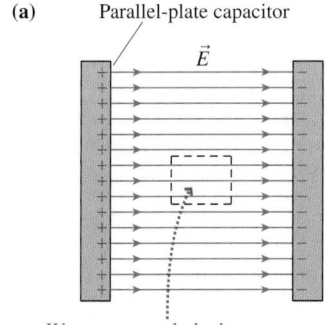

(b) The conducting box has been polarized and has induced surface charges.

$\vec{E} = \vec{0}$

The electric field is perpendicular to all conducting surfaces.

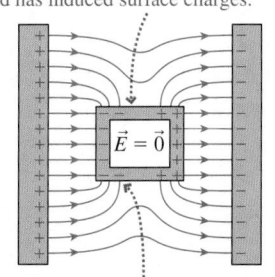

FIGURE 27.32 The electric field can be excluded from a region of space by surrounding it with a conducting box.

This region of space is now a hole inside a conductor, thus the interior electric field is zero. The use of a conducting box to exclude electric fields from a region of space is called **screening.** Solid metal walls are ideal, but in practice wire screen or wire mesh provides sufficient screening for all but the most sensitive applications. The price we pay is that the exterior field is now very complicated.

Finally, Figure 27.33 shows a charge q inside a hole within a neutral conductor. The electric field *within* the conductor is still zero, hence the electric flux through the Gaussian surface is zero. But $\Phi_e = 0$ requires $Q_{in} = 0$. Consequently, the charge inside the hole attracts an equal charge of opposite sign, and charge $-q$ now lines the inner surface of the hole.

The conductor as a whole is neutral, so moving $-q$ to the surface of the hole must leave $+q$ behind somewhere else. Where is it? It can't be in the interior of the conductor, as we've seen, and that leaves only the exterior surface. In essence, an internal charge polarizes the conductor just as an external charge would. Net charge $-q$ moves to the inner surface and net charge $+q$ is left behind on the exterior surface.

In summary, conductors in electrostatic equilibrium have the properties described in Tactics Box 27.3 on the next page.

The flux through the Gaussian surface is zero, hence there's no *net* charge inside this surface. There must be charge $-q$ on the inside surface to balance point charge q.

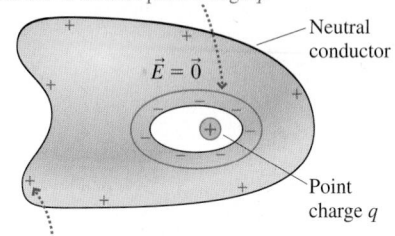

Neutral conductor

$\vec{E} = \vec{0}$

Point charge q

The outer surface must have charge $+q$ in order that the conductor remain neutral.

FIGURE 27.33 A charge in the hole causes a net charge on the interior and exterior surfaces.

TACTICS BOX 27.3 **Finding the electric field of a conductor in electrostatic equilibrium**

❶ The electric field is zero at all points within the volume of the conductor.

❷ Any excess charge resides entirely on the *exterior* surface.

❸ The external electric field at the surface of a charged conductor is perpendicular to the surface and of magnitude η/ϵ_0, where η is the surface charge density at that point.

❹ The electric field is zero inside any hole within a conductor unless there is a charge in the hole.

EXAMPLE 27.7 **The electric field at the surface of a charged metal sphere**

A 2.0-cm-diameter brass sphere has been given a charge of 2.0 nC. What is the electric field strength at the surface?

MODEL Brass is a conductor. The excess charge resides on the surface.

VISUALIZE The charge distribution has spherical symmetry. The electric field points radially outward from the surface.

SOLVE We can solve this problem two ways. One uses the fact that a sphere is the one shape for which any excess charge will spread out to a *uniform* surface charge density. Thus

$$\eta = \frac{q}{A_{\text{sphere}}} = \frac{q}{4\pi R^2} = \frac{2.0 \times 10^{-9}\,\text{C}}{4\pi(0.010\,\text{m})^2} = 1.59 \times 10^{-6}\,\text{C/m}^2$$

From Equation 27.20, we know the electric field at the surface has strength

$$E_{\text{surface}} = \frac{\eta}{\epsilon_0} = \frac{1.59 \times 10^{-6}\,\text{C/m}^2}{8.85 \times 10^{-12}\,\text{C}^2/\text{N\,m}^2} = 180{,}000\,\text{N/C}$$

Alternatively, we could have used the result, obtained earlier in the chapter, that the electric field strength outside a sphere of charge Q is $E_{\text{outside}} = Q_{\text{in}}/(4\pi\epsilon_0 r^2)$. But $Q_{\text{in}} = q$ and, at the surface, $r = R$. Thus

$$E_{\text{surface}} = \frac{1}{4\pi\epsilon_0}\frac{q}{R^2} = (9.0 \times 10^9\,\text{N\,m}^2/\text{C}^2)\frac{2.0 \times 10^{-9}\,\text{C}}{(0.010\,\text{m})^2}$$

$$= 180{,}000\,\text{N/C}$$

As we can see, both methods lead to the conclusion that $E_{\text{surface}} = 180{,}000\,\text{N/C}$.

SUMMARY

The goal of Chapter 27 has been to understand and apply Gauss's law.

GENERAL PRINCIPLES

Gauss's Law

For any *closed* surface enclosing net charge Q_{in}, the net electric flux through the surface is

$$\Phi_e = \oint \vec{E} \cdot d\vec{A} = \frac{Q_{in}}{\epsilon_0}$$

The electric flux Φ_e is the same for *any* closed surface enclosing charge Q_{in}.

Symmetry

The symmetry of the electric field must match the symmetry of the charge distribution.

In practice, Φ_e is computable only if the symmetry of the Gaussian surface matches the symmetry of the charge distribution.

IMPORTANT CONCEPTS

Charge creates the electric field that is responsible for the electric flux.

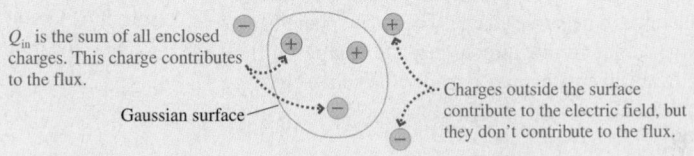

Q_{in} is the sum of all enclosed charges. This charge contributes to the flux.

Gaussian surface

Charges outside the surface contribute to the electric field, but they don't contribute to the flux.

Flux is the amount of electric field passing through a surface of area A.

$$\Phi_e = \vec{E} \cdot \vec{A}$$

where $\vec{A}$ is the **area vector**.

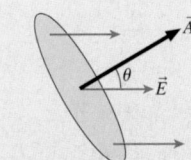

For closed surfaces:
A net flux in or out indicates that the surface encloses a net charge. Field lines through but with no *net* flux mean that the surface encloses no *net* charge.

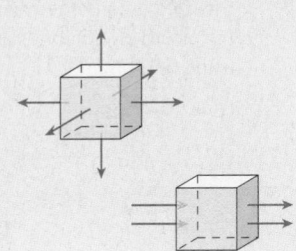

Surface integrals calculate the flux by summing the fluxes through many small pieces of the surface.

$$\Phi_e = \sum \vec{E} \cdot \delta\vec{A}$$
$$\rightarrow \int \vec{E} \cdot \delta\vec{A}$$

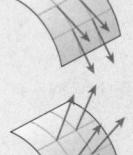

Two important situations:
If the electric field is everywhere tangent to the surface, then

$$\Phi_e = 0$$

If the electric field is everywhere perpendicular to the surface *and* has the same strength E at all points, then

$$\Phi_e = EA$$

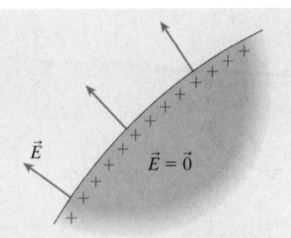

APPLICATIONS

Conductors in electrostatic equilibrium

- The electric field is zero at all points within the conductor.
- Any excess charge resides entirely on the exterior surface.
- The external electric field is perpendicular to the surface and of magnitude η/ϵ_0, where η is the surface charge density.
- The electric field is zero inside any hole within a conductor unless there is a charge in the hole.

TERMS AND NOTATION

symmetric	electric flux, Φ_e	surface integral	screening
Gaussian surface	area vector, $\vec{A}$	Gauss's law	

EXERCISES AND PROBLEMS

Exercises

Section 27.1 Symmetry

1. Figure Ex27.1 shows two cross sections of two infinitely long coaxial cylinders. The inner cylinder has a positive charge, the outer cylinder has an equal negative charge. Draw this figure on your paper, then draw electric field vectors showing the shape of the electric field.

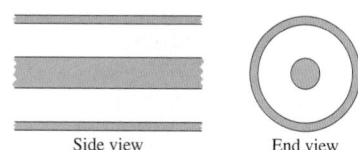

FIGURE EX27.1   Side view   End view

2. Figure Ex27.2 shows a cross section of two concentric spheres. The inner sphere has a negative charge. The outer sphere has a positive charge larger in magnitude than the charge on the inner sphere. Draw this figure on your paper, then draw electric field vectors showing the shape of the electric field.

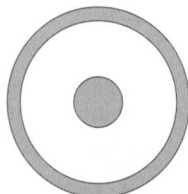

FIGURE EX27.2

3. Figure Ex27.3 shows a cross section of two infinite parallel planes of charge. Draw this figure on your paper, then draw electric field vectors showing the shape of the electric field.

FIGURE EX27.3

Section 27.2 The Concept of Flux

4. The electric field is constant over each face of the cube shown in Figure Ex27.4. Does the box contain positive charge, negative charge, or no charge? Explain.

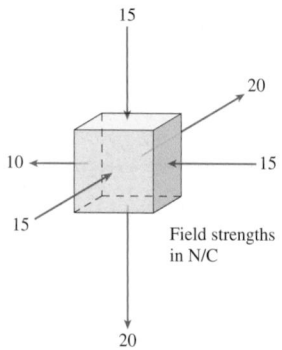

FIGURE EX27.4

5. The electric field is constant over each face of the cube shown in Figure Ex27.5. Does the box contain positive charge, negative charge, or no charge? Explain.

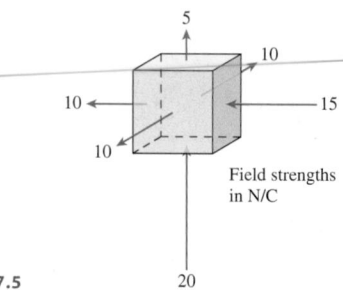

FIGURE EX27.5

6. The cube in Figure Ex27.6 contains negative charge. The electric field is constant over each face of the cube. Does the missing electric field vector on the front face point in or out? What is the minimum possible field strength?

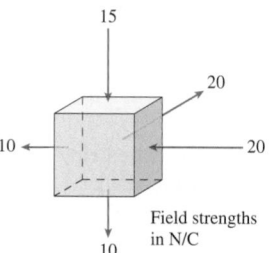

FIGURE EX27.6

7. The cube in Figure Ex27.7 contains no net charge. The electric field is constant over each face of the cube. Does the missing electric field vector on the front face point in or out? What is the field strength?

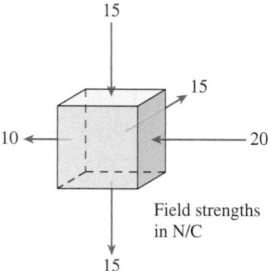

FIGURE EX27.7

Section 27.3 Calculating Electric Flux

8. What is the electric flux through the surface shown in Figure Ex27.8?

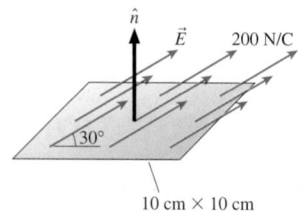

FIGURE EX27.8   10 cm × 10 cm

9. What is the electric flux through the surface shown in Figure Ex27.9?

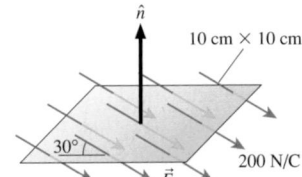

FIGURE EX27.9

10. The electric flux through the surface shown in Figure Ex27.10 is 15.0 Nm^2/C. What is the electric field strength?

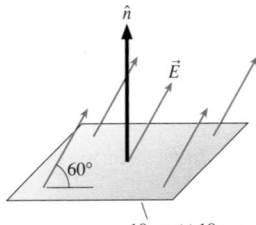

FIGURE EX27.10

11. A 2.0 cm × 3.0 cm rectangle lies in the xy-plane. What is the electric flux through the rectangle if
    a. $\vec{E} = (50\hat{\imath} + 100\hat{k})$ N/C?
    b. $\vec{E} = (50\hat{\imath} + 100\hat{\jmath})$ N/C?
12. A 2.0 cm × 3.0 cm rectangle lies in the xz-plane. What is the electric flux through the rectangle if
    a. $\vec{E} = (50\hat{\imath} + 100\hat{k})$ N/C?
    b. $\vec{E} = (50\hat{\imath} + 100\hat{\jmath})$ N/C?
13. A 4.0-cm-diameter circle lies in the xy-plane in a region where the electric field is $\vec{E} = (1000\hat{\imath} + 1000\hat{\jmath} + 1000\hat{k})$ N/C. What is the electric flux through the circle?
14. A 1.0 cm × 1.0 cm × 1.0 cm box is between the plates of a parallel-plate capacitor with two faces of the box perpendicular to $\vec{E}$. The electric field strength is 1000 N/C. What is the net electric flux through the box?
15. What is the net electric flux through the two cylinders shown in Figure Ex27.15?

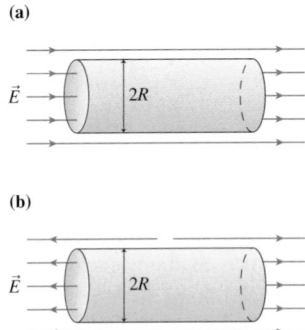

FIGURE EX27.15

Section 27.4 Gauss's Law

Section 27.5 Using Gauss's Law

16. The net electric flux through a closed surface is −1000 Nm^2/C. How much charge is enclosed within the surface?
17. 55.3 million excess electrons are inside a closed surface. What is the net electric flux through the surface?

18. A 10 nC point charge is at the center of a 2.0 m × 2.0 m × 2.0 m cube. What is the electric flux through the top surface of the cube?
19. The electric flux through each face of a 2.0 m × 2.0 m × 2.0 m cube is 100 Nm^2/C. How much charge is inside the cube?
20. Figure Ex27.20 shows three charges. Draw these charges on your paper four times. Then draw two-dimensional cross sections of three-dimensional closed surfaces through which the electric flux is (a) $2q/\epsilon_0$, (b) $3q/\epsilon_0$, (c) 0, and (d) $-q/\epsilon_0$.

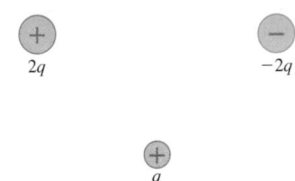

FIGURE EX27.20

21. Figure Ex27.21 shows three charges. Draw these charges on your paper four times. Then draw two-dimensional cross sections of three-dimensional closed surfaces through which the electric flux is (a) $-q/\epsilon_0$, (b) q/ϵ_0, (c) $3q/\epsilon_0$, and (d) $4q/\epsilon_0$.

FIGURE EX27.21

22. What is the net electric flux through the torus (i.e., doughnut shape) of Figure Ex27.22?

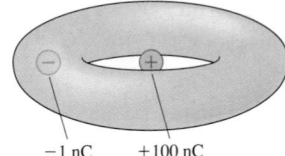

FIGURE EX27.22   −1 nC   +100 nC

23. What is the net electric flux through the cylinder of Figure Ex27.23?

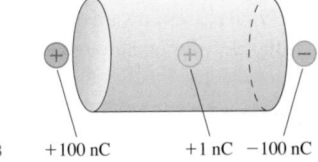

FIGURE EX27.23   +100 nC   +1 nC −100 nC

Section 27.6 Conductors in Electrostatic Equilibrium

24. The electric field strength just above one face of a copper penny is 2000 N/C. What is the surface charge density on this face of the penny?
25. A spark occurs at the tip of a metal needle if the electric field strength exceeds 3×10^6 N/C, the field strength at which air breaks down. What is the minimum surface charge density for producing a spark?

26. The conducting box in Figure Ex27.26 has been given an excess negative charge. The surface density of excess electrons at the center of the top surface is 5.0×10^{10} electrons/m². What are the electric field strengths E_1 to E_3 at points 1 to 3?

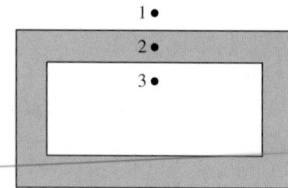

FIGURE EX27.26

Problems

27. Figure P27.27 shows the four sides of a 3.0 cm × 3.0 cm × 3.0 cm cube.
    a. What are the electric fluxes Φ_1 to Φ_4 through sides 1 to 4?
    b. What is the net flux through these four sides?

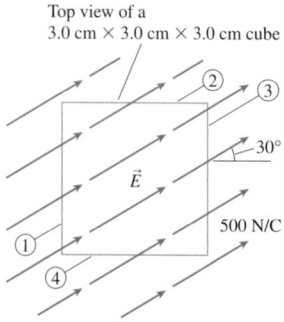

Top view of a
3.0 cm × 3.0 cm × 3.0 cm cube

FIGURE P27.27

28. Find the electric fluxes Φ_1 to Φ_5 through surfaces 1 to 5 in Figure P27.28.

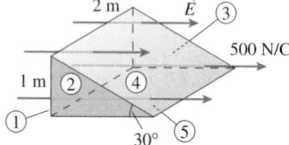

FIGURE P27.28

29. A tetrahedron has an equilateral triangle base with 20-cm-long edges and three equilateral triangle sides. The base is parallel to the ground, and a vertical uniform electric field of strength 200 N/C passes upward through the tetrahedron.
    a. What is the electric flux through the base?
    b. What is the electric flux through each of the three sides?
30. Figure P27.30 shows three Gaussian surfaces and the electric flux through each. What are the three charges q_1, q_2, and q_3?

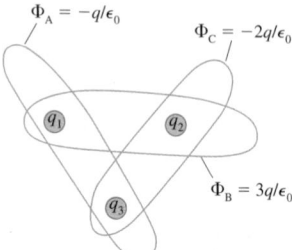

FIGURE P27.30

31. Charges $q_1 = -4Q$ and $q_2 = +2Q$ are located at $x = -a$ and $x = +a$, respectively. What is the net electric flux through a sphere of radius $2a$ centered at the origin?
32. A spherically symmetric charge distribution produces the electric field $\vec{E} = (5000r^2)\hat{r}$ N/C, where r is in m.
    a. What is the electric field strength at $r = 20$ cm?
    b. What is the electric flux through a 40-cm-diameter spherical surface that is concentric with the charge distribution?
    c. How much charge is inside this 40-cm-diameter spherical surface?
33. A spherically symmetric charge distribution produces the electric field $\vec{E} = (200/r)\hat{r}$ N/C, where r is in m.
    a. What is the electric field strength at $r = 10$ cm?
    b. What is the electric flux through a 20-cm-diameter spherical surface that is concentric with the charge distribution?
    c. How much charge is inside this 20-cm-diameter spherical surface?
34. A thin, horizontal 10 cm × 10 cm copper plate is charged with 1.0×10^{10} electrons. If the electrons are uniformly distributed on the surface, what are the strength and direction of the electric field
    a. 0.1 mm above the center of the top surface of the plate?
    b. at the plate's center of mass?
    c. 0.1 mm below the center of the bottom surface of the plate?
35. An initially neutral conductor contains a hollow cavity in which there is a +100 nC point charge. A charged rod transfers −50 nC to the conductor. Afterward, what is the charge (a) on the inner wall of the cavity wall, and (b) on the exterior surface of the conductor?
36. Figure P27.36 shows a hollow cavity within a neutral conductor. A point charge Q is inside the cavity. What is the net electric flux through the closed surface that surrounds the conductor?

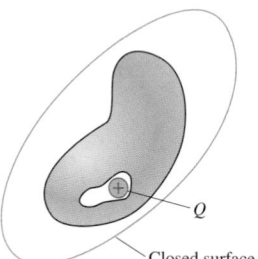

FIGURE P27.36   Closed surface

37. A 20-cm-radius ball is uniformly charged to 80 nC.
    a. What is the ball's charge density (C/m³)?
    b. How much charge is enclosed by spheres of radii 5, 10, and 20 cm?
    c. What is the electric field strength at points 5, 10, and 20 cm from the center?
38. A hollow metal sphere has inner radius a and outer radius b. The hollow sphere has charge $+2Q$. A point charge $+Q$ sits at the center of the hollow sphere.
    a. Determine the electric fields in the three regions $r \leq a$, $a < r < b$, and $r \geq b$.
    b. How much charge is on the inside surface of the hollow sphere? On the exterior surface?
39. Figure P27.39 shows a solid metal sphere at the center of a hollow metal sphere. What is the total charge on (a) the

exterior of the inner sphere, (b) the inside surface of the hollow sphere, and (c) the exterior surface of the hollow sphere?

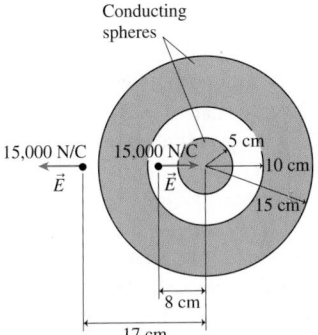

FIGURE P27.39

40. The earth has a vertical electric field at the surface, pointing down, that averages 100 N/C. This field is maintained by various atmospheric processes, including lightning. What is the excess charge on the surface of the earth?

41. Figure 27.32b showed a conducting box inside a parallel-plate capacitor. The electric field inside the box is $\vec{E} = \vec{0}$. Suppose the surface charge on the exterior of the box could be frozen. Draw a picture of the electric field inside the box after the box, with its frozen charge, is removed from the capacitor. **Hint:** Superposition.

42. A hollow metal sphere has 6 cm and 10 cm inner and outer radii, respectively. The surface charge density on the inside surface is -100 nC/m². The surface charge density on the exterior surface is $+100$ nC/m². What are the strength and direction of the electric field at points 4, 8, and 12 cm from the center?

43. A positive point charge q sits at the center of a hollow spherical shell. The shell, with radius R and negligible thickness, has net charge $-2q$. Find an expression for the electric field strength (a) inside the sphere, $r < R$, and (b) outside the sphere, $r > R$. In what direction does the electric field point in each case?

44. Find the electric field inside and outside a hollow plastic ball of radius R that has charge Q uniformly distributed on its outer surface.

45. A uniformly charged ball of radius a and charge $-Q$ is at the center of a hollow metal shell with inner radius b and outer radius c. The hollow sphere has net charge $+2Q$.
    a. Determine the electric field strength in the four regions $r \le a$, $a < r < b$, $b \le r \le c$, and $r > c$.
    b. Draw a graph of E versus r from $r = 0$ to $r = 2c$.

46. The three parallel planes of charge shown in Figure P27.46 have surface charge densities $-\frac{1}{2}\eta$, η, and $-\frac{1}{2}\eta$. Find the electric fields $\vec{E}_1$ to $\vec{E}_4$ in regions 1 to 4.

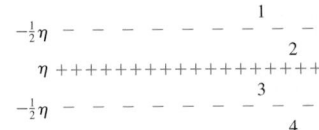

FIGURE P27.46

47. An infinite slab of charge of thickness $2z_0$ lies in the xy-plane between $z = -z_0$ and $z = +z_0$. The charge density ρ_0 (C/m³) is a constant.
    a. Use Gauss's law to find an expression for the electric field strength inside the slab ($-z_0 \le z \le z_0$).
    b. Find an expression for the electric field strength above the slab ($z \ge z_0$).
    c. Sketch a graph of E from $z = 0$ to $z = 3z_0$.

48. Figure P27.48 shows an infinitely wide conductor parallel to and distance d from an infinitely wide plane of charge with surface charge density η. What are the electric fields $\vec{E}_1$ to $\vec{E}_4$ in regions 1 to 4?

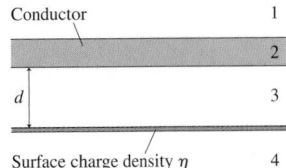

FIGURE P27.48   Surface charge density η

49. Figure P27.49 shows two very large slabs of metal that are parallel and distance d apart. Each slab has a total surface area (top + bottom) A. The thickness of each slab is so small in comparison to its lateral dimensions that the surface area around the sides is negligible. Metal 1 has total charge $Q_1 = Q$ and metal 2 has total charge $Q_2 = 2Q$. Assume Q is positive. In terms of Q and A, determine
    a. The electric field strengths E_1 to E_5 in regions 1 to 5.
    b. The surface charge densities η_a to η_d on the four surfaces a to d.

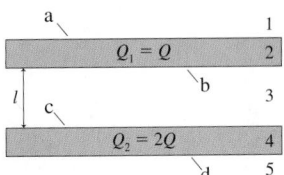

FIGURE P27.49

50. A long, thin straight wire with linear charge density λ runs down the center of a thin, hollow metal cylinder of radius R. The cylinder has a net linear charge density 2λ. Assume λ is positive. Find expressions for the electric field strength (a) inside the cylinder, $r < R$, and (b) outside the cylinder, $r > R$. In what direction does the electric field point in each of the cases?

51. A very long, uniformly charged cylinder has radius R and linear charge density λ. Find the cylinder's electric field (a) outside the cylinder, $r \ge R$, and (b) inside the cylinder, $r \le R$. (c) Show that your answers to parts a and b match at the boundary, $r = R$.

52. A spherical shell has inner radius R_{in} and outer radius R_{out}. The shell contains total charge Q, uniformly distributed. The interior of the shell is empty of charge and matter.
    a. Find the electric field outside the shell, $r \ge R_{out}$.
    b. Find the electric field in the interior of the shell, $r \le R_{in}$.
    c. Find the electric field within the shell, $R_{in} \le r \le R_{out}$.
    d. Show that your solutions match at both the inner and outer boundaries.
    e. Draw a graph of E versus r.

53. An early model of the atom, proposed by Rutherford after his discovery of the atomic nucleus, had a positive point charge $+Ze$ (the nucleus) at the center of a sphere of radius R with uniformly distributed negative charge $-Ze$. Z is the atomic number, the number of protons in the nucleus and the number of electrons in the negative sphere.
    a. Show that the electric field inside this atom is

$$E_{in} = \frac{Ze}{4\pi\epsilon_0}\left(\frac{1}{r^2} - \frac{r}{R^3}\right)$$

    b. What is E at the surface of the atom? Is this the expected value? Explain.
    c. A uranium atom has $Z = 92$ and $R = 0.10$ nm. What is the electric field strength at $r = \frac{1}{2}R$?

Challenge Problems

54. All examples of Gauss's law have used highly symmetrical surfaces where the flux integral is either zero or EA. Yet we've claimed that the net $\Phi_e = Q_{in}/\epsilon_0$ is independent of the surface. This is worth checking. Figure CP27.54 shows a cube of edge length L centered on a long thin wire with linear charge density λ. The flux through one face of the cube is *not* simply EA because, in this case, the electric field varies in both strength and direction. But you can calculate the flux by actually doing the flux integral.

    a. Consider the face parallel to the yz-plane. Define area $d\vec{A}$ as a strip of width dy and height L with the vector pointing in the x-direction. One such strip is located at position y. Use the known electric field of a wire to calculate the electric flux $d\Phi$ through this little area. Your expression should be written in terms of y, which is a variable, and various constants. It should not explicitly contain any angles.

    b. Now integrate $d\Phi$ to find the total flux through this face.

    c. Finally, show that the net flux through the cube is $\Phi_e = Q_{in}/\epsilon_0$.

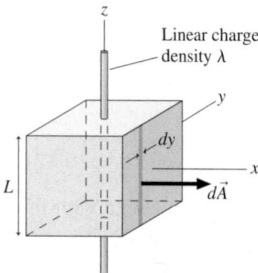

FIGURE CP27.54

55. An infinite cylinder of radius R has a linear charge density λ. The volume charge density (C/m³) within the cylinder $(r \leq R)$ is $\rho(r) = r\rho_0/R$, where ρ_0 is a constant to be determined.

    a. Draw a graph of ρ versus x for an x-axis that crosses the cylinder perpendicular to the cylinder axis. Let x range from $-2R$ to $2R$.

    b. The charge within a small volume dV is $dq = \rho dV$. The integral of ρdV over a cylinder of length L is the total charge $Q = \lambda L$ within the cylinder. Use this fact to show that $\rho_0 = 3\lambda/2\pi R^2$.

       Hint: Let dV be a cylindrical shell of length L, radius r, and thickness dr. What is the volume of such a shell?

    c. Use Gauss's law to find an expression for the electric field E inside the cylinder, $r \leq R$.

    d. Does your expression have the expected value at the surface, $r = R$? Explain.

56. A sphere of radius R has total charge Q. The volume charge density (C/m³) within the sphere is $\rho(r) = C/r^2$, where C is a constant to be determined.

    a. The charge within a small volume dV is $dq = \rho dV$. The integral of ρdV over the entire volume of the sphere is the total charge Q. Use this fact to determine the constant C in terms of Q and R.

       Hint: Let dV be a spherical shell of radius r and thickness dr. What is the volume of such a shell?

    b. Use Gauss's law to find an expression for the electric field E inside the sphere, $r \leq R$.

    c. Does your expression have the expected value at the surface, $r = R$? Explain.

57. A sphere of radius R has total charge Q. The volume charge density (C/m³) within the sphere is

$$\rho = \rho_0\left(1 - \frac{r}{R}\right)$$

This charge density decreases linearly from ρ_0 at the center to zero at the edge of the sphere.

    a. Show that $\rho_0 = 3Q/\pi R^3$.

       Hint: You'll need to do a volume integral.

    b. Show that the electric field inside the sphere points radially outward with magnitude

$$E = \frac{Qr}{4\pi\epsilon_0 R^3}\left(4 - 3\frac{r}{R}\right)$$

    c. Show that your result of part b has the expected value at $r = R$.

58. A spherical ball of charge has radius R and total charge Q. The electric field strength inside the ball $(r \leq R)$ is $E(r) = E_{max}(r^4/R^4)$.

    a. What is E_{max} in terms of Q and R?

    b. Find an expression for the volume charge density (C/m³) $\rho(r)$ inside the ball as a function of r.

    c. Verify that your charge density gives the total charge Q when integrated over the volume of the ball.

Stop to Think 27.1: a and d. Symmetry requires the electric field to be unchanged if front and back are reversed, if left and right are reversed, or if the field is rotated about the wire's axis. Fields a and d both have the proper symmetry. Other factors would now need to be considered to determine the correct field.

Stop to Think 27.2: e. The net flux is into the box.

Stop to Think 27.3: c. There's no flux through the four sides. The flux is positive 1 N m²/C through both the top and bottom because $\vec{E}$ and $\vec{A}$ both point outward.

Stop to Think 27.4: $\Phi_b = \Phi_e > \Phi_a = \Phi_c = \Phi_d$. The flux through a closed surface depends only on the amount of enclosed charge, not the size or shape of the surface.

Stop to Think 27.5: d. A cube doesn't have enough symmetry to use Gauss's law. The electric field of a charged cube is *not* constant over the face of a cubic Gaussian surface, so we can't evaluate the surface integral for the flux.

28 Current and Conductivity

The transition from candles to lightbulbs was both a technical and a social revolution.

Thomas Edison's invention of the lightbulb revolutionized the way people work and play. Can you imagine doing your physics homework by candle light?

Lights, stereos, microwave ovens, and computers are an important part of our contemporary lives. Devices such as these are connected by wires to a battery or an electrical outlet. What is happening *inside* the wire that makes the light come on or the stereo play? And *why* is it happening? We say that "electricity flows through the wire," but what does that statement mean? And equally important, *how do we know?* Simply looking at a wire between a battery and a lightbulb does not reveal whether anything is moving or flowing. As far as visual appearance is concerned, the wire is absolutely the same whether it is "conducting electricity" or not.

The objective of this chapter is to learn about current. We want to understand what moves through a current-carrying wire, and why. We'll also need to establish a connection between an electric current and the charging and electrostatic processes we have studied in the last three chapters.

28.1 The Electron Current

Let's start our exploration of current with a very simple system. Figure 28.1a shows a charged parallel-plate capacitor. We can verify that the plates are charged—one positively and the other negatively—by holding them, one at a time, near charged glass and plastic rods that are hanging from threads. If we now connect the two capacitor plates to each other with a metal wire, as shown in Figure 28.1b, the plates quickly become neutral. We say that the capacitor has been *discharged.*

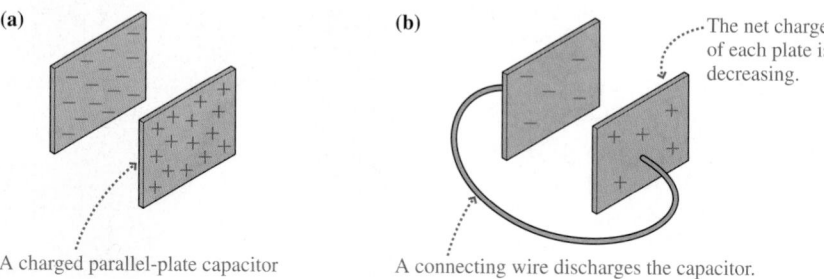

(a)   **(b)**   The net charge of each plate is decreasing.

A charged parallel-plate capacitor   A connecting wire discharges the capacitor.

FIGURE 28.1 A capacitor is discharged by a metal wire.

The wire is a conductor, a material through which charge easily moves. Apparently the excess charge on one capacitor plate is able to move through the wire to the other plate, neutralizing both plates. In Chapter 25, we defined a **current** as the motion of charge through a conductor. It would seem that the capacitor is discharged by a current in the connecting wire. Our goal in this chapter is to understand how it happens.

First, let's see what else we can observe. Figure 28.2 shows that the connecting wire gets warmer. If the wire is very thin in places, such as the thin filament in a lightbulb, the wire gets hot enough to glow. The current-carrying wire also deflects a compass needle. We will explore the connection between currents and magnetism in Chapter 32. For now, we will use "makes the wire warmer" and "deflects a compass needle" as *indicators* that a current is present in a wire.

Charge Carriers

Opposite charges attract, but the oppositely charged plates of a capacitor don't spontaneously discharge because the charges can't leap from one plate to the other. A connecting wire discharges the capacitor by providing a pathway for charge to move from one side of the capacitor to the other. However, merely observing that a wire discharges a capacitor doesn't answer an important question: Does positive charge move toward the negative plate, or does negative charge move toward the positive plate? *Either* motion would explain the observations we have made.

The charges that move in a current-carrying conductor are called the *charge carriers.* In Chapter 25 we simply *asserted* that the charge carriers in a metal are electrons, but we offered no evidence. One of the first clues was found by J. J. Thompson, the discoverer of the electron. In the 1890s, Thompson found that metals heated until they glow emit electrons. (This *thermal emission* from hot tungsten filaments is now the source of electrons in electron guns.) Thompson's observation suggested that the electrons are moving around inside the metal and can escape if they have sufficient thermal energy.

The *Tolman-Stewart experiment* of 1916 was the first direct evidence that electrons are the charge carriers in metals. Tolman and Stewart caused a metal rod to accelerate very rapidly. As Figure 28.3 shows, the inertia of the charge carriers within the metal (and Newton's first law) causes them to be "thrown" to the rear

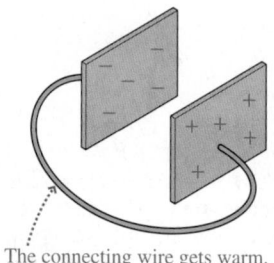

The connecting wire gets warm.

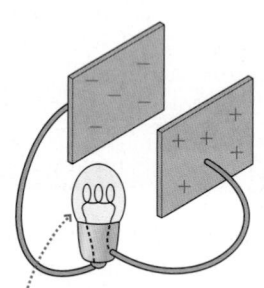

A light bulb glows. The light bulb filament is part of the connecting wire.

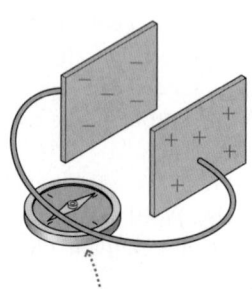

A compass needle is deflected.

FIGURE 28.2 Properties of a current.

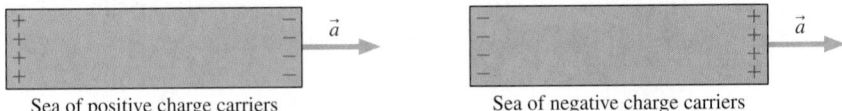

When a metal bar accelerates to the right, inertia causes the charge carriers to be displaced to the rear surface. The front surface becomes oppositely charged.

Sea of positive charge carriers   Sea of negative charge carriers

FIGURE 28.3 The Tolman-Stewart experiment to determine the sign of the charge carriers in a metal.

surface of the metal rod as it accelerates away. If the charge carriers are positive, their displacement relative to the metal should cause the rear surface to become positively charged and leave the front surface negatively charged, much as if the metal were polarized by an electric field. Negative charge carriers should give the rear surface a negative charge while leaving the front surface positive.

Tolman and Stewart found that the rear surface of a metal rod becomes negatively charged as it accelerates. The only negatively charged particles are electrons, thus *experimental evidence* tells us that **the charge carriers in metals are electrons.** We noted in Chapter 25 that the electrons act rather like a negatively charged gas or liquid in between the atoms of the lattice. This *model,* called the *sea of electrons,* is illustrated in Figure 28.4. It's not a perfect model, because it overlooks some quantum effects, but it is the basis for a reasonably good description of current in a metal. Notice that the conduction electrons are not attached to any particular atom in the metal.

NOTE ▶ Electrons are the charge carriers in *metals.* Other conductors, such as ionic solutions or semiconductors, have different charge carriers. We will focus on metals, because of their importance to circuits, but don't think that electrons are *always* the charge carrier. ◀

The Electron Current

The conduction electrons in a metal, like molecules in a gas, are moving around quite rapidly, but there is no *net* motion. We can change that by pushing on the sea of electrons, causing the entire sea of electrons to move in one direction like a gas or liquid flowing through a pipe. This net motion, which takes place at what we'll call the **drift speed** v_d, is superimposed on top of the random thermal motions of the individual electrons. The drift speed is quite small, with 10^{-4} m/s being a fairly typical value. We'll develop this model more fully in Section 28.2.

Figure 28.5 shows a cross section through a current-carrying wire in which the entire sea of electrons is moving from left to right at the drift speed. Suppose a microscopic observer could count the electrons that pass through this cross section. Let's define the **electron current** i to be the number of electrons *per second* that pass through a cross section of a wire or other conductor. The units of electron current are s^{-1}. Stated another way, the number N_e of electrons that pass through the cross section during the time interval Δt is

$$N_e = i\,\Delta t \tag{28.1}$$

It seems likely that the electron current is related to the drift speed. After all, increasing the electron speed should increase the number of electrons that are able to pass through a wire each second. To help us establish a relationship, Figure 28.6 on the next page shows the sea of electrons moving through a wire at the drift speed v_d. This is the *net* speed with which the electrons move, not the speed at which any one electron is bouncing around. The electrons that pass through a particular cross section of the wire during the interval Δt are shaded. How many of them are there?

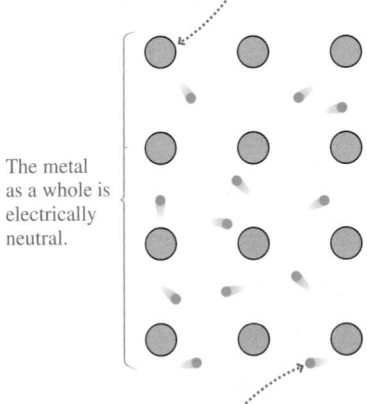

Ions (the metal atoms minus one valence electron) occupy fixed positions in the lattice.

The metal as a whole is electrically neutral.

The conduction electrons (one per atom) are free to move around. They are bound to the solid as a whole, not to any particular atom.

FIGURE 28.4 The sea of electrons is a model of how conduction electrons behave in a metal.

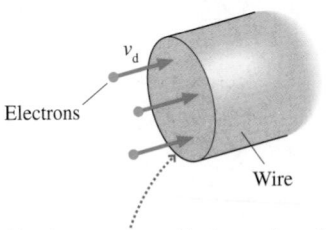

v_d

Electrons

Wire

The electron current i is the number of electrons passing through this cross section of the wire per second.

FIGURE 28.5 The electron current.

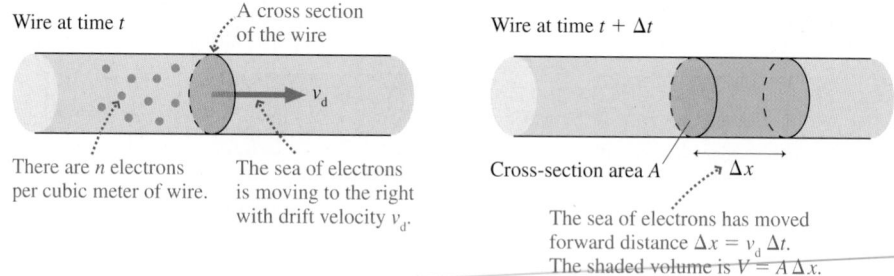

Wire at time t

A cross section of the wire

Wire at time $t + \Delta t$

There are n electrons per cubic meter of wire.

The sea of electrons is moving to the right with drift velocity v_d.

Cross-section area A

Δx

The sea of electrons has moved forward distance $\Delta x = v_d \Delta t$. The shaded volume is $V = A \Delta x$.

FIGURE 28.6 The sea of electrons moves to the right with velocity v_d.

The electrons travel distance $\Delta x = v_d \Delta t$ to the right during the interval Δt. This section of wire is a cylinder with volume $V = A \Delta x$. If the *number density* of conduction electrons is n electrons per cubic meter, then the total number of electrons in the cylinder is

$$N_e = nV = nA \Delta x = nAv_d \Delta t \qquad (28.2)$$

Comparing Equation 28.2 to Equation 28.1, you can see that the electron current in the wire is

$$i = nAv_d \qquad (28.3)$$

In most metals, each atom contributes one valence electron to the sea of electrons. Thus the number of conduction electrons per cubic meter is the same as the number of atoms per cubic meter, a quantity that can be determined from the metal's mass density. Table 28.1 gives values of the conduction-electron density n for several common metals. There is not a great deal of variation.

TABLE 28.1 Conduction-electron density in metals

Metal	Electron density (m^{-3})
Aluminum	6.0×10^{28}
Copper	8.5×10^{28}
Iron	8.5×10^{28}
Gold	5.9×10^{28}
Silver	5.8×10^{28}

EXAMPLE 28.1 The size of the electron current
What is the electron current in a 2.0-mm-diameter copper wire if the electron drift speed is 1.0×10^{-4} m/s?

SOLVE The wire's cross-section area is $A = \pi r^2 = 3.14 \times 10^{-6}$ m^2. Table 28.1 gives the electron density for copper as 8.5×10^{28} m^{-3}. Thus

$i = nAv_d$

$= (8.5 \times 10^{28} \text{ m}^{-3})(3.14 \times 10^{-6} \text{ m}^2)(1.0 \times 10^{-4} \text{ m/s})$

$= 2.7 \times 10^{19} \text{ s}^{-1}$

ASSESS This is an incredible number of electrons to pass through a section of the wire every second. The number is high not because the sea of electrons moves fast—in fact, it moves at literally a snail's pace—but because the density of electrons is so enormous. This is a fairly typical electron current.

The electron density n and cross-section area A are properties of the wire. We can't change those. Once we choose a particular piece of wire, the only parameter we can vary to control the size of a current is the drift speed v_d. As you'll soon see, the drift speed is determined by the electric field inside the wire.

STOP TO THINK 28.1 These four wires are made of the same metal. Rank in order, from largest to smallest, the electron currents i_a to i_d.

v

r

(a)

$2v$

r

(b)

$2r$

v

(c)

$\frac{1}{2}r$

$2v$

(d)

Conservation of Current

Returning to our capacitor, we can now say with confidence that the wire discharges the capacitor by allowing an electron current to flow from the negative plate, where there is an excess of electrons, to the positive plate. The flow of electrons continues until both plates are electrically neutral.

Figure 28.7 shows a lightbulb in the wire connecting two capacitor plates. The bulb glows while the electron current is discharging the capacitor. How do you think the electron current at point A compares to the electron current at point B? Are the electron currents at these points the same? Or is one larger than the other? Think about this before going on.

You might have predicted that the electron current at B is less than the electron current at A because the bulb, in order to be glowing, must use up some of the current. It's easy to test this prediction. We can compare the electron currents at A and B by comparing how far two compass needles are deflected. Or we could insert two small lightbulbs at A and B, small enough not to significantly influence the main lightbulb in Figure 28.7, and compare their brightness. All methods give the same result: The electron current at point B is *exactly equal* to the electron current at point A. The current leaving a lightbulb is exactly the same as the current entering the lightbulb.

This is an important observation, one that demands an explanation. After all, "something" makes the bulb glow, so why don't we observe a decrease in the electron current? The electron current *i* is the number of electrons moving through the wire per second. There are only two ways to decrease *i*: either decrease the number of electrons, or decrease their drift speed through the wire. Electrons are charged particles. The lightbulb can't destroy electrons without violating both the law of conservation of mass and the law of conservation of charge. Thus the *number* of electrons is not changed by the lightbulb.

Can the electrons slow down as they pass through the bulb? This is a little trickier, so first consider the analogous situation shown in Figure 28.8 where we see water flowing through a hose of constant diameter. Suppose the water flows into one end at a speed of 2.0 m/s. Is it possible for the water to flow out the other end at a speed of only 1.5 m/s?

We can't destroy water molecules any more than we can destroy electrons, nor can we increase the density of water by pushing the molecules closer together. Water can go in faster than it flows out only if water molecules are somehow stored inside the hose. If the hose is made of rubber, perhaps it's expanding in diameter as more and more water is stored inside. But this isn't sustainable because eventually the hose will burst. A sustainable flow exists only if the water flows out at the same speed it flows in.

The same is true for electrons in a constant-diameter wire. The only way the drift speed at B can be less than the drift speed at A is if electrons are being stored inside the bulb, making the bulb increasingly negative. But this isn't sustainable, and there are two reasons we know it doesn't happen. First, we can use charged plastic and glass rods to test whether a glowing bulb accumulates a negative charge. It doesn't. Second, if the bulb were to become increasingly negative, there would come a point when the repulsive force would stop the flow of new electrons into the bulb and the light would go out. This doesn't happen. **The rate of electrons going through the wire at B is the same as at A.**

The lightbulb doesn't "use up" current, but it *does* use energy. Imagine pushing a block along a rough surface at a constant speed. You have to do *work* on the block to keep it moving at a steady speed, and the energy you supply as work is dissipated by friction, making the block and the surface warmer. We've not yet identified what it is, but something is doing work by "pushing" the electrons through the wire *at constant speed*. The energy is dissipated by atomic-level friction as the electrons move through the atoms, making the wire hotter until, in the case of the light-bulb filament, it glows.

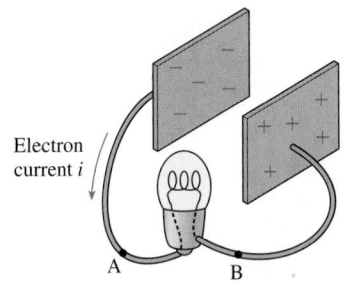

FIGURE 28.7 How does the electron current at A compare to the electron current at B?

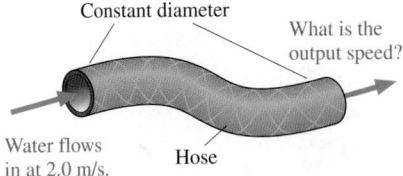

FIGURE 28.8 Water flowing through a hose.

There are many issues here that we'll need to look at before we can say that we understand how currents work, and we'll take them one at a time in this and subsequent chapters. For now, we draw a first important conclusion:

> **Law of conservation of current** The electron current is the same at all points in a current-carrying wire.

A Puzzle

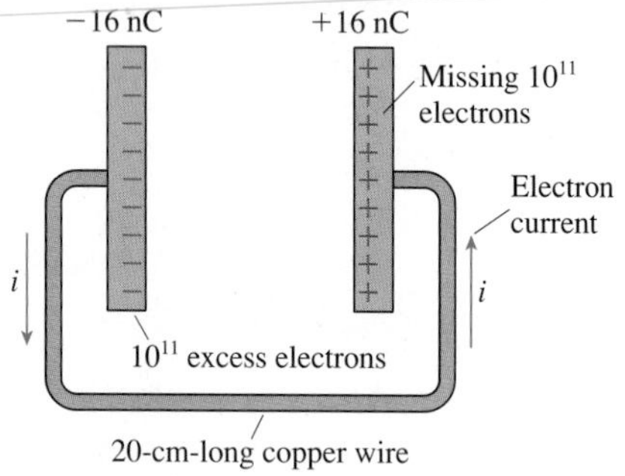

Let's end this section with a puzzle. Figure 28.9 shows a capacitor charged to ± 16 nC as it is being discharged by a 2.0-mm-diameter, 20-cm-long copper wire. *How long does it take* to discharge the capacitor?

We've noted that a fairly typical drift speed of the electron current through a wire is 10^{-4} m/s. At this rate, it would take 2000 s, or about a half hour, for an electron to travel 20 cm. We should have time to go for a cup of coffee while we wait for the discharge to occur!

But this isn't what happens. As far as our senses are concerned, the discharge of a capacitor by a copper wire is instantaneous. So what's wrong with our simple calculation?

The important point we overlooked is that the wire is *already full* of electrons. Return to the water-in-a-hose analogy. If the hose is already full of water, adding a drop to one end immediately (or very nearly so) pushes a drop out the other end. Likewise with the wire. As soon as the excess electrons move from the negative capacitor plate into the wire, they immediately (or very nearly so) push an equal number of electrons out the other end of the wire and onto the positive plate, thus neutralizing it. We don't have to wait for electrons to move all the way through the wire from one plate to the other. Instead, we just need to slightly rearrange the charges on the plates *and* in the wire.

Let's do a rough estimate of how much rearrangement is needed and how long the discharge takes. The negative plate in Figure 28.10, with $Q = -16$ nC, has 10^{11} excess electrons. Using the conduction-electron density of copper in Table 28.1, we can calculate that there are 5×10^{22} conduction electrons in the wire, a vastly larger number. The length of copper wire needed to hold 10^{11} electrons is a mere 4×10^{-13} m, only about 1% the diameter of an atom.

The instant the wire joins the capacitor plates, the repulsive forces between the excess 10^{11} electrons on the negative plate cause them to push their way into the wire. As they do so, 10^{11} electrons are squeezed out of the final 4×10^{-13} m of the wire and onto the positive plate. If the electrons all move together, and if they move at the typical drift speed of 10^{-4} m/s, both less than perfect assumptions but OK for the purpose of making an estimate, it takes 4×10^{-9} s, or 4 ns, to move 4×10^{-13} m and to discharge the capacitor. And, indeed, this is the right order of magnitude for how long the electrons take to rearrange themselves so that the capacitor plates are neutral.

FIGURE 28.9 How long does it take to discharge a capacitor?

−16 nC +16 nC

Missing 10^{11} electrons

Electron current

i i

10^{11} excess electrons

20-cm-long copper wire

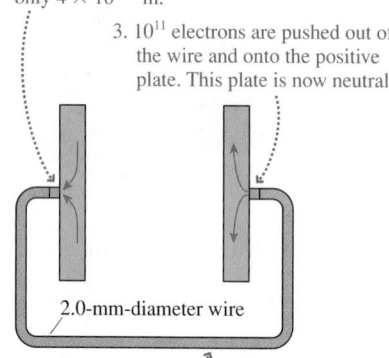

1. The 10^{11} excess electrons on the negative plate move into the wire. The length of wire needed to accommodate these electrons is only 4×10^{-13} m.

3. 10^{11} electrons are pushed out of the wire and onto the positive plate. This plate is now neutral.

2.0-mm-diameter wire

2. The sea of 5×10^{22} electrons in the wire is pushed to the side. It moves only 4×10^{-13} m, taking almost no time.

FIGURE 28.10 The sea of electrons needs only a minuscule rearrangement to discharge the capacitor.

STOP TO THINK 28.2 Why does the light in a room come on instantly when you flip a switch several meters away?

28.2 Creating a Current

Suppose you want to slide a book across the table to your friend. You give it a quick push to start it moving, but it begins slowing down because of friction as soon as you take your hand off. The book's kinetic energy is transformed into thermal energy, leaving the book and the table slightly warmer. The only way to keep the book moving at a *constant* speed is to continue pushing it.

As Figure 28.11 shows, the sea of electrons is similar to the book. If you push the sea of electrons, you create a current of electrons moving through the conductor. But the electrons aren't moving in a vacuum. Collisions between the electrons and the atoms of the metal transform the electrons' kinetic energy into the thermal energy of the metal, making the metal warmer. (Recall that "makes the wire warmer" is one of our indicators of a current.) Consequently, the sea of electrons will quickly slow down and stop *unless you continue pushing*. How do you push on electrons? With an electric field!

One of the important conclusions of Chapter 25 was that $\vec{E} = \vec{0}$ inside a conductor in electrostatic equilibrium. But a conductor with electrons moving through it is *not* in electrostatic equilibrium. **An electron current is a nonequilibrium motion of charges sustained by an internal electric field.**

Thus the quick answer to "What creates a current?" is "An electric field." But why is there an electric field in a current-carrying wire? How does it get established? What is the relationship between the strength of the electric field and the size of the electron current? These are the questions we need to answer.

Establishing the Electric Field in a Wire

Figure 28.12a shows two metal wires attached to the plates of a charged capacitor. The ends of the wires are close together, but not touching. The wires are conductors, so some of the charges on the capacitor plates spread out along the wires as a surface charge. (Remember that all excess charge on a conductor is located on the surface.)

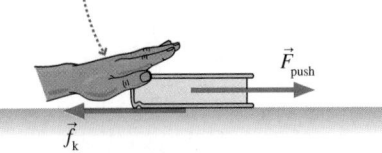

Because of friction, a steady push is needed to move the book at steady speed.

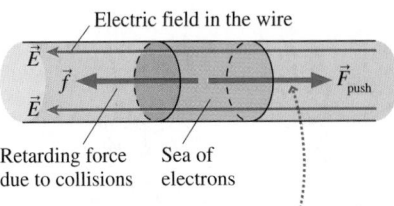

Electric field in the wire

Retarding force due to collisions   Sea of electrons

Because of collisions with atoms, a steady push is needed to move the sea of electrons at steady speed. Electrons are negative, so $\vec{F}_{push}$ is opposite to $\vec{E}$.

FIGURE 28.11 An electron current is sustained by pushing on the sea of electrons with an electric field.

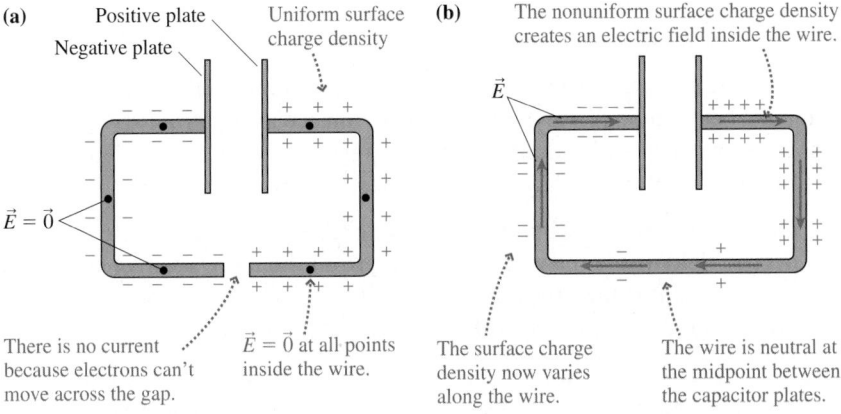

(a) Positive plate   Uniform surface charge density
Negative plate

$\vec{E} = \vec{0}$

There is no current because electrons can't move across the gap.

$\vec{E} = \vec{0}$ at all points inside the wire.

(b) The nonuniform surface charge density creates an electric field inside the wire.

$\vec{E}$

The surface charge density now varies along the wire.

The wire is neutral at the midpoint between the capacitor plates.

FIGURE 28.12 The surface charge on the wires before and after they are connected.

This is an electrostatic situation, with no charges in motion. Consequently, the electric field must be zero at *every* point inside the wire. Symmetry requires that there be equal amounts of charge to either side of a point in order to make $\vec{E} = \vec{0}$, hence the surface charge density must be uniform along each wire except near the ends (where the details need not concern us). We implied this uniform density in Figure 28.12a by drawing equally spaced + and − symbols along the wire. Remember that a positively charged surface is a surface that is *missing* electrons.

Now connect the ends of the wires together. What happens? The excess electrons on the surface of the negative wire suddenly have an opportunity to move onto the positive wire that is missing electrons. And, because opposite charges attract, they do. Within a *very* brief interval of time ($\approx 10^{-9}$ s), the sea of electrons shifts slightly and the surface charge is rearranged into a *nonuniform* distribution like that shown in Figure 28.12b. The surface charge near the positive and negative plates remains strongly positive and negative because of the large amount of charge on the capacitor plates, but the midpoint of the wire, halfway between the positive and negative plates, becomes electrically neutral. The new surface charge density on the wire varies from positive at the positive capacitor plate through zero at the midpoint to negative at the negative plate.

The new distribution of surface charge has an *extremely* important consequence. Figure 28.13 shows a section from a wire on which the surface charge density becomes more positive toward the left and more negative toward the right. Calculating the exact electric field is complicated, but we can understand the basic idea if we *model* this section of wire with four circular rings of charge.

The four rings A through D model the nonuniform charge distribution on the wire.

$\vec{E}_A$ points away from A and $\vec{E}_B$ points away from B, but A has more charge so the net field points to the right.

The nonuniform charge distribution creates a net field to the right at all points inside the wire.

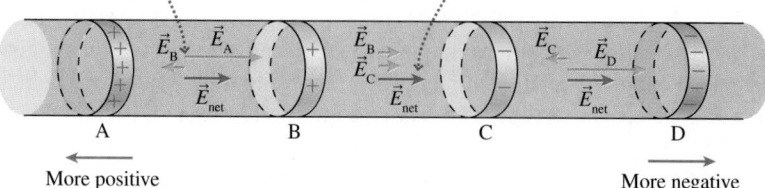

FIGURE 28.13 A varying surface charge distribution creates an internal electric field inside the wire.

In Chapter 26, we found that the on-axis field of a ring of charge

1. Points away from a positive ring, toward a negative ring;
2. Is proportional to the amount of charge on the ring; and
3. Decreases with distance away from the ring.

Because the field strength decreases with distance from the ring, the field at the midpoint between rings A and B is well approximated as $\vec{E}_{net} \approx \vec{E}_A + \vec{E}_B$. Ring A has more charge than ring B, so $\vec{E}_{net}$ points away from A.

The analysis of Figure 28.13 leads to a very important conclusion:

> The *nonuniform* distribution of surface charges along a wire creates a net electric field *inside* the wire that points from the more positive end of the wire toward the more negative end of the wire. This is the internal electric field $\vec{E}$ that pushes the electron current through the wire.

The following example shows that the electric field inside a current-carrying wire can be established with an extremely small amount of surface charge.

EXAMPLE 28.2 The surface charge on a current-carrying wire

Table 26.1 in Chapter 26 gave the typical electric field strength in a current-carrying wire as 0.01 N/C. (We'll verify this value later in this chapter.) Two 2.0-mm-diameter rings are 2.0 mm apart. They are charged to $\pm Q$. What value of Q causes the electric field at the midpoint to be 0.010 N/C?

MODEL Use the on-axis electric field of a ring of charge from Chapter 26.

VISUALIZE Figure 28.14 shows the two rings. Both contribute equally to the field strength, so the electric field strength of the positive ring is $E_+ = 0.0050$ N/C. The distance $z = 1.0$ mm is half the ring spacing.

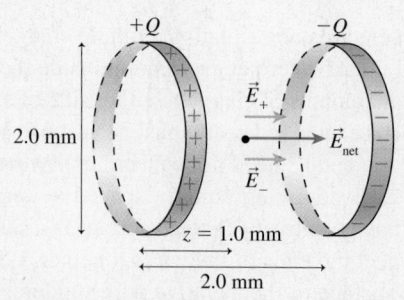

FIGURE 28.14 The electric field of two charged rings.

SOLVE Chapter 26 gives the on-axis electric field of a ring of charge Q as

$$E_+ = \frac{1}{4\pi\epsilon_0}\frac{zQ}{(z^2+R^2)^{3/2}}$$

Thus the charge needed to produce the desired field is

$$Q = \frac{4\pi\epsilon_0(z^2+R^2)^{3/2}}{z}E_+$$

$$= \frac{((0.0010 \text{ m})^2 + (0.0010 \text{ m})^2)^{3/2}}{(9.0\times10^9\text{ Nm}^2/\text{C}^2)(0.0010 \text{ m})}(0.0050 \text{ N/C})$$

$$= 1.6\times10^{-18}\text{ C}$$

ASSESS The electric field of a ring of charge is largest when $z \approx R$, so these two rings are a simple but reasonable model for estimating the electric field inside a 2.0-mm-diameter wire. We find that the surface charge needed to establish the electric field is *very small*. A mere 10 electrons have to be moved from one ring to the other to charge them to $\pm1.6\times10^{-18}$ C. This is sufficient to drive a sizable electron current through the wire.

Figure 28.13 showed how a varying surface charge density creates an electric field inside a straight wire. You might wonder how an electron current turns a corner. As Figure 28.15 shows, all it takes is just a little extra negative surface charge on the outside edge of a corner. When the electron current first begins, the electrons try to move straight ahead and thus pile up on the outside edge of the corner. This little bit of excess negative charge then exerts a repulsive force on the oncoming sea of electrons, just enough to turn them around the corner. Detailed calculations show that two or three extra electrons on the surface are usually all that are needed to make the current follow the wire around a bend.

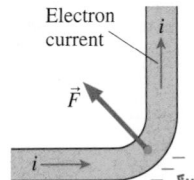

A few extra negative charges on the outside corner exert a repulsive force on the electrons, forcing the current to turn the corner.

FIGURE 28.15 The electron current turning a corner.

STOP TO THINK 28.3 The two charged rings are a model of the surface charge distribution along a wire. Rank in order, from largest to smallest, the electron currents E_a to E_e at the midpoint between the rings.

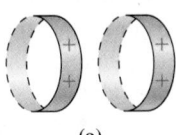

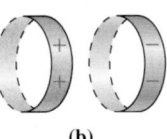
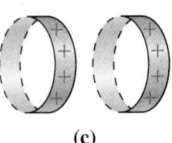
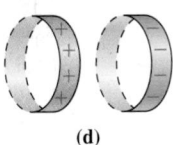
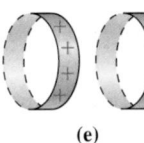

(a)   (b)   (c)   (d)   (e)

A Model of Conduction

An electric field inside a conductor pushes on the sea of electrons to create the electron current. The field has to *keep* pushing because the electrons continuously lose energy in collisions with the positive ions that form the structure of the solid. These collisions provide a drag force, much like friction.

The conduction electrons are analogous to the molecules in a gas. We characterized gases by their macroscopic parameters of temperature and pressure, but we needed an atomic-level perspective of colliding molecules in order to understand what temperature and pressure really are. The result was the kinetic theory of gases. We need a similar micro/macro link to help us understand how metals conduct electricity.

We will treat the conduction electrons—those electrons that make up the sea of electrons—as free particles moving through the crystal lattice of the metal. In the absence of an electric field, the electrons, like the molecules in a gas, move randomly in all directions with a distribution of speeds. If we assume that the average thermal energy of the electrons is given by the same $\frac{3}{2}kT$ that applies to an ideal gas, we can find that the average electron speed at room temperature is

(a) No electric field   Ions in the lattice
                             of the metal

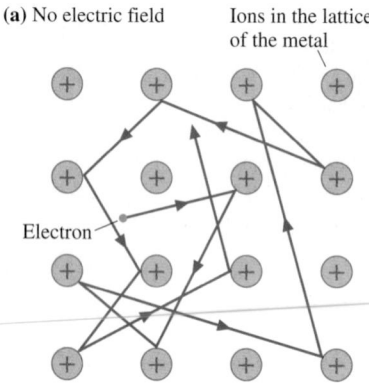

Electron

The electron has frequent collisions with
ions, but it undergoes no net displacement.

(b) With an electric field   Parabolic trajectories
                                  in the electric field

$\vec{E}$

$\vec{E}$

$\vec{E}$

Net displacement

A net displacement in the direction
opposite to $\vec{E}$ is superimposed on the
random thermal motion.

FIGURE 28.16 A microscopic view of a
conduction electron moving through
a metal.

$\approx 10^5$ m/s. This estimate turns out, for quantum physics reasons, to be not quite right, but it correctly indicates that the conduction electrons are moving very fast.

However, an individual electron does not travel far before colliding with an ion and being scattered to a new direction. Figure 28.16a shows that an electron moves in straight lines between collisions. Its *average* velocity is zero, as it bounces back and forth between collisions, and it undergoes no *net* displacement. This is similar to molecules in a container of gas.

Suppose we now turn on an electric field. Figure 28.16b shows that the steady electric force causes the electrons to move along *parabolic trajectories* between collisions. Because of the curvature of the trajectories, the negatively charged electrons begin to drift slowly in the direction opposite the electric field. The motion is similar to a ball moving in a pinball machine that has a slight downward tilt. An individual electron ricochets back and forth between the ions at a high rate of speed, but now there is a slow *net* motion in the "downhill" direction. Even so, this net displacement is a *very* small effect superimposed on top of the much larger thermal motion. Figure 28.16b has greatly exaggerated the rate at which the drift would occur.

Suppose an electron just had a collision with an ion and has rebounded with velocity $\vec{v}_i$. The acceleration of the electron between collisions is

$$a_x = \frac{F}{m} = \frac{eE}{m} \tag{28.4}$$

where E is the electric field strength inside the wire and m is the mass of the electron. (We'll assume that $\vec{E}$ points in the negative x-direction.) The field causes the x-component of the electron's velocity to increase linearly with time:

$$v_x = v_{ix} + a_x \Delta t = v_{ix} + \frac{eE}{m} \Delta t \tag{28.5}$$

The electron speeds up, with increasing kinetic energy, until its next collision with an ion. The collision transfers much of the electron's kinetic energy to the ion and thus to the thermal energy of the metal. **This energy transfer is the "friction" that raises the temperature of the wire.** The electron then rebounds, in a random direction, with a new initial velocity $\vec{v}_i$ and starts the process all over.

Figure 28.17a shows how the electron velocity abruptly changes due to a collision. Notice that the acceleration (the slope of the line) is the same before and after the collision. Figure 28.17b then follows an electron through a series of collisions. You can see that each collision "resets" the velocity. The primary observation we can make from Figure 28.17b is that this repeated process of speeding up and colliding gives the electron a nonzero *average* velocity. This average velocity, due to the electric field, is the *drift speed* v_d of the electron.

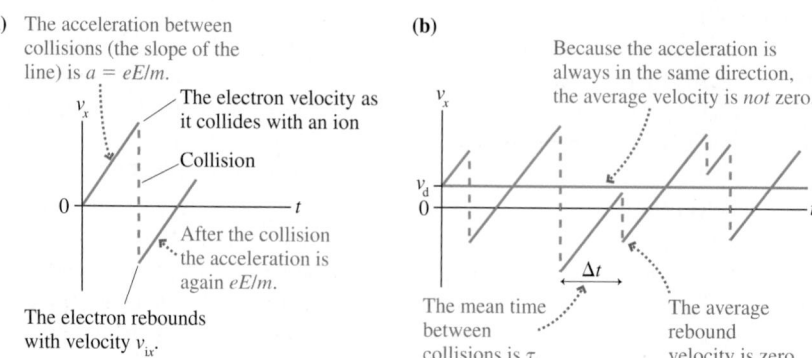

FIGURE 28.17 The electron velocity as a function of time.

If we observe all the electrons in the metal at one instant of time, their average velocity is

$$v_d = \overline{v_x} = \overline{v_{ix}} + \frac{eE}{m}\overline{\Delta t} \qquad (28.6)$$

where a bar over a quantity indicates an average value. The average value of v_{ix}, the velocity with which an electron rebounds after a collision, is zero. We know this because, in the absence of an electric field, the sea of electrons moves neither right nor left.

At any one instant of time, some electrons will have just recently collided and their acceleration time Δt will be shorter than average. Other electrons will be "overdue" for a collision and have Δt longer than average. When averaged over all electrons, the average value of Δt is the **mean time between collisions,** which we designate τ. The mean time between collisions is analogous to the mean free path between collisions in the kinetic theory of gases.

Thus the average speed at which the electrons are pushed along by the electric field is

$$v_d = \frac{e\tau}{m}E \qquad (28.7)$$

We can complete our model of conduction by using Equation 28.7 for v_d in the electron-current equation $i = nAv_d$. Upon doing so, we find that an electric field strength E in a wire of cross-section area A causes an electron current

$$i = \frac{ne\tau A}{m}E \qquad (28.8)$$

The electron density n and the mean time between collisions τ are properties of the metal.

Equation 28.8 is the main result of this model of conduction. We've found that **the electron current is directly proportional to the electric field strength.** A stronger electric field pushes the electrons faster and thus increases the electron current.

EXAMPLE 28.3 The electron current in a copper wire
The mean time between collisions for electrons in room-temperature copper is 2.5×10^{-14} s. What is the electron current in a 2.0-mm-diameter copper wire where the internal electric field strength is 0.010 N/C?

MODEL Use the model of conduction to relate the drift speed to the field strength.

SOLVE The electron current is $i = nAv_d$. The electron drift speed in a 0.010 N/C electric field can be found from Equation 28.7:

$$v_d = \frac{e\tau}{m}E = \frac{(1.60 \times 10^{-19}\text{ C})(2.5 \times 10^{-14}\text{ s})(0.010\text{ N/C})}{9.11 \times 10^{-31}\text{ kg}}$$

$$= 4.4 \times 10^{-5}\text{ m/s}$$

Copper has an electron density $n = 8.5 \times 10^{28}$ m^{-3}, and a 2.0-mm-diameter wire has a cross-section area $A = \pi r^2 = 3.14 \times 10^{-6}$ m^2. Thus the electron current is

$$i = nAv_d = 1.2 \times 10^{19}\text{ electrons/s}$$

ASSESS A *lot* of electrons are going past each second!

28.3 Batteries

We've focused our attention on the discharge of a capacitor because we can easily understand where all the charges are and how they move. By contrast, we can't easily see what's happening to the charges inside a battery. Nonetheless, batteries are the primary source of current in circuits and other practical applications, so it's important to understand how our ideas about the electron current apply to a battery.

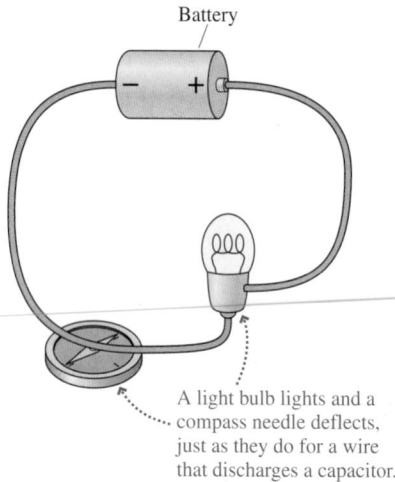

A light bulb lights and a compass needle deflects, just as they do for a wire that discharges a capacitor.

FIGURE 28.18 There is a current in a wire connecting the terminals of a battery.

The "charge escalator" inside a battery continuously "lifts" electrons from the positive to the negative terminal. This renewal of charge sustains the electron current.

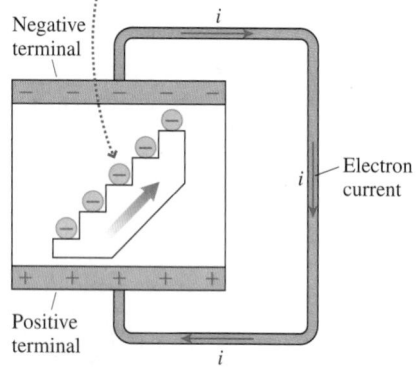

FIGURE 28.19 A battery can be thought of as a charge escalator.

Surface charges have created an electric field inside the wire.

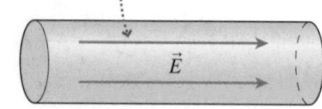

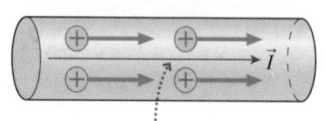

The current $\vec{I}$ is the rate at which the electric field seems to push *positive* charge through the wire. $\vec{I}$ is in the direction of $\vec{E}$.

FIGURE 28.20 The current I.

Figure 28.18 shows a wire connecting the two terminals of a battery, much like the wire that connected the capacitor plates in Figure 28.1. Just like that wire, the wire connecting the battery terminals gets warm, deflects a compass needle, and makes a lightbulb inserted into it glow brightly. These indicators tell us that charges flow through the wire from one terminal to the other. The current *in the wire* is the same whether it is supplied by a capacitor or a battery. Everything you've learned in this chapter about the electron current applies equally well to the current supplied by a battery—with one important difference.

That difference is the duration of the current. The current that discharges a capacitor is transient. It ceases as soon as the excess charge on the capacitor plates is removed; the lightbulb then goes out and the compass needle returns to its initial position. By contrast, the wire connecting the battery terminals *continues* to deflect the compass needle and *continues* to light the lightbulb. There is a *sustained* current in the wire—a sustained motion of charges—when it is connected to the battery. The capacitor quickly runs out of excess charge, but the battery can keep the motion going.

To understand why, we need to peer inside a battery. Figure 28.19 shows that the inner workings of a battery act like a *charge escalator* between two capacitor plates. Electrons are removed from the positive terminal and "lifted" to the negative terminal. It is the charge escalator that sustains the electron current in the wire by providing a continually renewed supply of electrons at the negative terminal.

The charge escalator does, however, have to be powered by some external source of energy. It is, in a very real sense, lifting the electrons "uphill" against the electric field. As you know, the energy to run the charge escalator comes from chemical reactions within the battery. A dead battery is one in which the supply of chemicals has been exhausted.

Our interest is in what happens to the charges, so we can conveniently ignore the details of the chemical reactions. All we care about is that the chemical reactions move electrons from the positive to the negative terminal of the battery. However, we can't overlook the fact that a description of the battery is going to require the ideas of work and energy. We will return to batteries and the charge escalator in Chapter 30. For now we need only to see that the role of a battery is to *maintain a charge separation* and thus to sustain a current.

28.4 Current and Current Density

We have developed the idea of a current as the motion of electrons through metals. But the properties of currents were known and used for a century before the discovery that electrons are the charge carriers in metals. We need to connect our ideas about the electron current to the conventional definition of current.

Because the coulomb is the SI unit of charge, and because currents are charges in motion, it seemed quite natural in the 19th century to define current as the *rate*, in coulombs per second, at which charge moves through a wire. Figure 28.20 shows a wire in which the electric field is $\vec{E}$. We define the current $\vec{I}$ in the wire to be

$$\vec{I} \equiv \left(\frac{dQ}{dt}, \text{ in the direction of } \vec{E} \right) \tag{28.9}$$

Strictly speaking, current is a vector. It has both a magnitude and a direction, and we will draw current arrows on diagrams to indicate the direction. For calculations, however, we are usually interested only in the magnitude $I = dQ/dt$.

The SI unit for current is the coulomb per second, which is called the **ampere** A:

1 ampere = 1 A ≡ 1 coulomb per second = 1 C/s

The current unit is named after the French scientist André Marie Ampère, who made major contributions to the study of electricity and magnetism in the early 19th century. The *amp* is an informal abbreviation of ampere. Household currents are typically ≈1 A. For example, the current through a 100 watt lightbulb is 0.85 A. The current in an electric hair dryer is ≈10 A. Currents in consumer electronics, such as stereos and computers are much less. They are typically measured in milliamps ($1\ mA = 10^{-3}\ A$) or microamps ($1\ \mu A = 10^{-6}\ A$).

For a *steady current,* which will be our primary focus, the amount of charge delivered by current I during the time interval Δt is

$$Q = I\Delta t \tag{28.10}$$

Equation 28.10 is closely related to Equation 28.1, which said that the number of electrons delivered during a time interval Δt is $N_e = i\Delta t$. Each electron has charge of magnitude e, hence the total charge of N_e electrons is $Q = eN_e$. Consequently, the conventional current I and the electron current i are related by

$$I = \frac{Q}{\Delta t} = \frac{eN_e}{\Delta t} = ei \tag{28.11}$$

Because electrons are the charge carriers, the rate at which charge moves is e times the rate at which the electrons move.

EXAMPLE 28.4 The current in a copper wire
The electron current in the copper wire of Example 28.3 was 1.2×10^{19} electrons/s. What is the current I? How much charge flows through a cross section of the wire each hour?

SOLVE The current in the wire is

$$I = ei = (1.60 \times 10^{-19}\ C)(1.2 \times 10^{19}\ s^{-1}) = 1.9\ A$$

The amount of charge passing through the wire in 1 hr = 3600 s is

$$Q = I\Delta t = (1.9\ A)(3600\ s) = 6840\ C$$

In one sense, the current I and the electron current i merely differ by a scale factor. The electron current i is the rate at which electrons move through a wire. The current I is the rate at which the charge of the electrons moves through the wire. The electron current i is more *fundamental* because it looks directly at the charge carriers. The current I is more *practical* because we can measure charge more easily than we can count electrons.

Despite the close connection between i and I, there's one important distinction. Because currents were known and studied before it was known what the charge carriers are, the direction of current was *defined* to be the direction in which positive charges *seem* to move. Thus the direction of the current $\vec{I}$ is the same as that of the internal electric field $\vec{E}$. But because the charge carriers turned out to be negative, at least for a metal, **the direction of the current $\vec{I}$ in a metal is opposite the direction of motion of the electrons.**

The situation shown in Figure 28.21 may seem disturbing, but it makes no real difference. A capacitor is discharged regardless of whether positive charges move toward the negative plate or negative charges move toward the positive plate. The primary application of the current is to the analysis of circuits, and in a circuit—a macroscopic device—we simply can't tell what is moving through the wires. All of our calculations will be correct and all of our circuits will work perfectly well if we choose to think of current as the flow of positive charge. The distinction is important only at the microscopic level.

Because the current I is in the direction of the electric field $\vec{E}$, **the current direction in a wire is from the positive terminal of a battery to the negative terminal.**

The current $\vec{I}$ is defined to point in the direction of $\vec{E}$. It is the direction in which positive charge carriers would move.

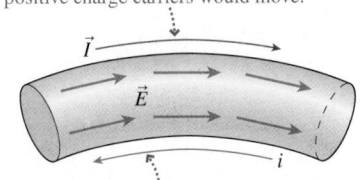

The electron current i is the motion of actual charge carriers. It is opposite to $\vec{E}$ and $\vec{I}$.

FIGURE 28.21 The current $\vec{I}$ is opposite the direction of motion of the electrons in a metal.

The Current Density in a Wire

We found that the electron current in a wire of cross-section area A is $i = nAv_d$. Thus the current I is

$$I = ei = nev_dA \qquad (28.12)$$

The quantity nev_d depends on the charge carriers and on the internal electric field that determines the drift speed, whereas A is simply a physical dimension of the wire. It will be useful to separate these quantities by defining the **current density** J in a wire as the current per square meter of cross section:

$$J = \text{current density} = \frac{I}{A} = nev_d \qquad (28.13)$$

The current density has units of A/m^2.

You learned earlier that the mass density ρ characterizes all pieces of a particular material, such as lead. A *specific* piece of the material, with known dimensions, is then characterized by its mass $m = \rho V$. Similarly, the current density J describes how charge flows through *any* piece of a particular kind of metal in response to an electric field. A *specific* piece of metal, shaped into a wire with cross-section area A, then has the specific current

$$I = JA \qquad (28.14)$$

EXAMPLE 28.5 Finding the electron drift speed
A 1.0 A current passes through a 1.0-mm-diameter aluminum wire. What is the drift speed of the electrons in the wire?

SOLVE We can find the drift speed from the current density. The current density is

$$J = \frac{I}{A} = \frac{I}{\pi r^2} = \frac{1.0 \text{ A}}{\pi (0.00050 \text{ m})^2} = 1.3 \times 10^6 \text{ A/m}^2$$

The electron drift speed is thus

$$v_d = \frac{J}{ne} = 1.3 \times 10^{-4} \text{ m/s} = 0.13 \text{ mm/s}$$

where the conduction-electron density for aluminum was taken from Table 28.1.

ASSESS We earlier used 1.0×10^{-4} m/s as a typical electron drift speed. This example shows where that value comes from.

Conservation of Current Revisited

Figure 28.22a shows a current-carrying wire. We've already seen that the electron current is the same at all points in a current-carrying wire. The conventional current I is measured differently, in terms of charge rather than the number of electrons, but its properties are the same. The law of conservation of current tells us that the current I is the same at all points in a current-carrying wire.

(a)

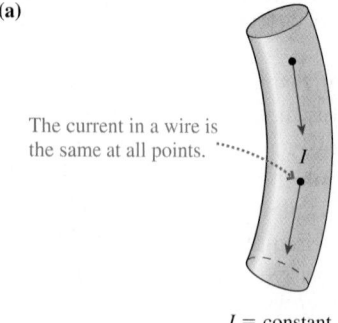

The current in a wire is the same at all points.

I

$I = \text{constant}$

(b)

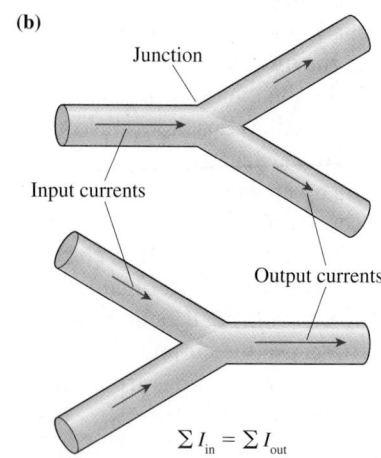

Junction

Input currents

Output currents

$\Sigma I_{in} = \Sigma I_{out}$

FIGURE 28.22 The sum of the currents into a junction must equal the sum of the currents leaving the junction.

Figure 28.22b shows two wires merging into one and one wire splitting into two. A point where a wire branches is called a **junction.** The presence of a junction doesn't change our basic reasoning. We cannot create or destroy electrons in the wire, and neither can we store them in the junction. The rate at which electrons flow into one *or many* wires must be exactly balanced by the rate at which they flow out of others. For a *junction,* the law of conservation of charge requires that

$$\sum I_{\text{in}} = \sum I_{\text{out}} \qquad (28.15)$$

where, as usual, the $\sum$ symbol means summation.

This basic conservation statement, that the sum of the currents into a junction equals the sum of the currents leaving, is called **Kirchhoff's junction law.** The junction law, which will play an important role in circuit analysis, is a very important consequence of the conservation of charge.

STOP TO THINK 28.4   What are the magnitude and the direction of the current in the fifth wire?

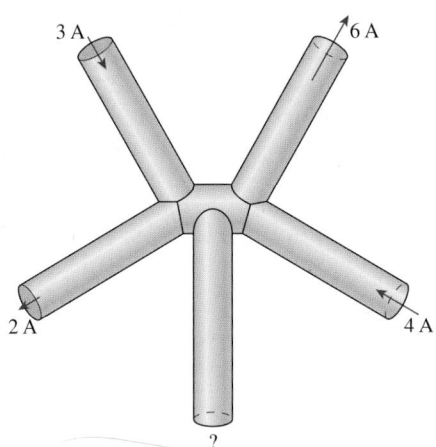

28.5 Conductivity and Resistivity

The current density $J = nev_{\text{d}}$ is directly proportional to the electron drift speed v_{d}. We used the microscopic model of conduction to find that the drift speed is $v_{\text{d}} = e\tau E/m$, where τ is the mean time between collisions. Thus we can write the current density as

$$J = nev_{\text{d}} = ne\left(\frac{e\tau E}{m}\right) = \frac{ne^2\tau}{m}E \qquad (28.16)$$

The quantity $ne^2\tau/m$ in Equation 28.16 depends *only* on the conducting material. It contains the electron density n and the mean time between collisions τ. Materials with larger values of these numbers conduct current better than materials with smaller values. That is, a given electric field strength will generate a larger current density in a material that has larger values of n and τ.

It makes sense, then, to define the **conductivity** σ of a material as

$$\sigma = \text{conductivity} = \frac{ne^2\tau}{m} \qquad (28.17)$$

where m is the mass of the electron. Conductivity, like density, characterizes a material as a whole. All pieces of copper (at the same temperature) have the same value of σ, but the conductivity of copper is different from that of aluminum. Notice that the mean time between collisions τ can be inferred from measured values of the conductivity.

With this definition of conductivity, Equation 28.16 becomes

$$J = \sigma E \tag{28.18}$$

This is a result of fundamental importance. Equation 28.18 tells us three things:

1. Current is caused by an electric field exerting forces on the charge carriers.
2. The current density, and hence the current $I = JA$, depends linearly on the strength of the electric field.
3. The current density also depends on the *conductivity* of the material. Different conducting materials have different conductivities because they have different values of the electron density n and, especially, different values of the mean time between electron collisions with the lattice of atoms.

The value of the conductivity is affected by the crystalline structure, by any impurities in the metal, and by the temperature. As the temperature increases, so do the thermal vibrations of the lattice atoms. This makes them "bigger targets" and causes collisions to be more frequent, thus lowing τ and decreasing the conductivity. Metals conduct better at low temperatures than they do at high temperatures.

For many practical applications of current it will be convenient to use the inverse of the conductivity, $1/\sigma$. This is called the **resistivity** and has the symbol ρ. Thus

$$\rho = \text{resistivity} = \frac{1}{\sigma} = \frac{m}{ne^2\tau} \tag{28.19}$$

The resistivity of a material tells us how reluctantly the electrons move in response to an electric field. Table 28.2 gives values of the conductivity and resistivity for several metals and for carbon. You can see that they do vary quite a bit, with copper and silver being the best two conductors.

The units of conductivity, from Equation 28.18, are those of J/E, namely $A\,C/N\,m^2$. These are clearly awkward. In Chapter 31 we will introduce a new unit called the *ohm*, symbolized by Ω (uppercase Greek omega). It will then turn out that resistivity has units of Ω m and conductivity has $\Omega^{-1}m^{-1}$.

NOTE ▶ *Resistivity* is not the same as *resistance,* a related quantity that you will meet in Chapter 31. ◀

TABLE 28.2 Resistivity and conductivity of conducting materials

Material	Resistivity $(\Omega\,m)$	Conductivity $(\Omega^{-1}m^{-1})$
Aluminum	2.8×10^{-8}	3.5×10^{7}
Copper	1.7×10^{-8}	6.0×10^{7}
Gold	2.4×10^{-8}	4.1×10^{7}
Iron	9.7×10^{-8}	1.0×10^{7}
Silver	1.6×10^{-8}	6.2×10^{7}
Tungsten	5.6×10^{-8}	1.8×10^{7}
Nichrome*	1.5×10^{-6}	6.7×10^{5}
Carbon	3.5×10^{-5}	2.9×10^{4}

*Nickel-chromium alloy used for heating wires

EXAMPLE 28.6 The electric field in a wire
A 2.0-mm-diameter aluminum wire carries a current of 800 mA. What is the electric field strength inside the wire?

SOLVE The electric field strength is

$$E = \frac{J}{\sigma} = \frac{I}{\sigma A} = \frac{I}{\sigma \pi r^2} = \frac{0.80\ \text{A}}{(3.5 \times 10^7\ \Omega^{-1}\text{m}^{-1})\pi(0.0010\ \text{m})^2}$$
$$= 0.0072\ \text{N/C}$$

where the conductivity of aluminum was taken from Table 28.2.

ASSESS This is a *very* small field in comparison with those we calculated in Chapters 25 and 26 for point charges and charged objects. This calculation justifies the claim in Table 26.1 that a typical electric field strength inside a current-carrying wire is 0.01 N/C.

EXAMPLE 28.7 **Mean time between collisions**

What is the mean time between collisions for electrons in copper?

SOLVE The mean time between collisions is related to a material's conductivity by

$$\tau = \frac{m\sigma}{ne^2}$$

The electron density of copper is found in Table 28.1 and the measured conductivity is found in Table 28.2. With this information,

$$\tau = \frac{(9.11 \times 10^{-31}\ \text{kg})(6.0 \times 10^7\ \Omega^{-1}\text{m}^{-1})}{(8.5 \times 10^{28}\ \text{m}^{-3})(1.60 \times 10^{-19}\ \text{C})^2} = 2.5 \times 10^{-14}\ \text{s}$$

This is the value that was given in Example 28.1.

The electric field strength found in Example 28.7 is roughly the size of the electric field 1 mm from a *single* electron. The lesson to be learned from this example is that it takes *very few* surface charges on a wire to create the internal electric field necessary for the wire to carry considerable current. Just a few excess electrons every centimeter are sufficient. The reason, once again, is the enormous value of the charge-carrier density n. Even through the electric field is very tiny and the drift speed is agonizingly slow, a wire can carry a substantial current due to the vast number of charge carriers able to move.

Superconductivity

In 1911, the Dutch physicist Kamerlingh Onnes was studying the conductivity of metals at very low temperatures. Scientists had just recently discovered how to liquefy helium, and this opened a whole new field of *low-temperature physics.* As we noted above, metals become better conductors (i.e., they have higher conductivity and lower resistivity) at lower temperatures. But the effect is gradual. Onnes, however, found that mercury suddenly and dramatically loses *all* resistance to current when cooled below a temperature of 4.2 K. This complete loss of resistance at low temperatures is called **superconductivity.**

Later experiments established that the resistivity of a superconducting metal is not just small, it is truly zero. The electrons are moving in a frictionless environment, and charge will continue to move through a superconductor *without an electric field.* Superconductivity was not understood until the 1950s, when it was explained as being a specific quantum effect.

Superconducting wires can carry enormous currents because the wires are not heated by electrons colliding with the atoms. Very strong magnetic fields can be created with superconducting electromagnets, but applications remained limited for many decades because all known superconductors required temperatures less than 20 K. This situation changed dramatically in 1986 with the discovery of *high-temperature superconductors.* These ceramic-like materials are superconductors at temperatures as "high" as 125 K. Although $-150°C$ may not seem like a high temperature to you, the technology for producing such temperatures is simple and inexpensive. Thus many new superconductor applications are likely to appear in coming years.

Superconductors have unusual magnetic properties. Here a small permanent magnet levitates above a disk of the high-temperature superconductor $YBa_2Cu_3O_7$ that has been cooled to liquid-nitrogen temperature.

STOP TO THINK 28.5   Rank in order, from largest to smallest, the current densities J_a to J_d in these four wires.

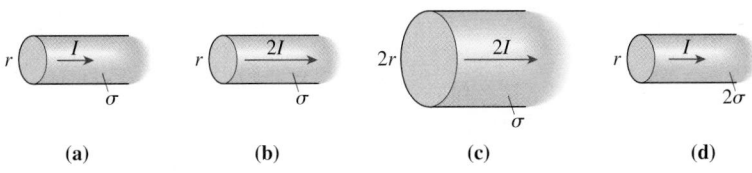

SUMMARY

The goal of Chapter 28 has been to learn how and why charge moves through a conductor as a current.

GENERAL PRINCIPLES

Current is a nonequilibrium motion of charges sustained by an electric field. Nonuniform surface charge density creates an electric field in a wire. The electric field pushes the electron current i in a direction opposite to $\vec{E}$. The conventional current I is in the direction in which positive charge *seems* to move.

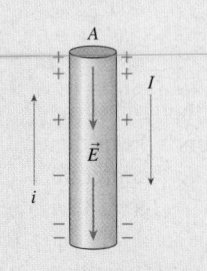

Electron current

   i = rate of electron flow

   $N_e = i\Delta t$

Conventional current

   I = rate of charge flow = ei

   $Q = I\Delta t$

Current density

   $J = I/A$

Conservation of Current

The current is the same at any two points in a wire.

At a junction,

   $$\sum I_{in} = \sum I_{out}$$

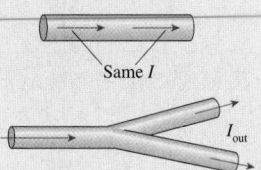

Batteries

The role of a battery is to maintain a charge separation and thus sustain a current. Chemical reactions power the "charge escalator" that moves electrons from the positive terminal to the negative terminal.

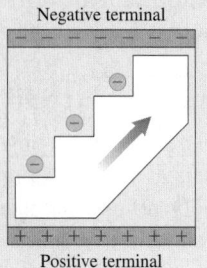

IMPORTANT CONCEPTS

Sea of electrons

Conduction electrons move freely around the positive ions that form the atomic lattice.

Conduction

An electric field causes a slow drift at speed v_d to be superimposed on the rapid but random thermal motions of the electrons.

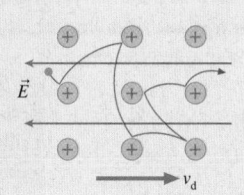
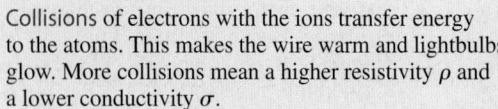

Collisions of electrons with the ions transfer energy to the atoms. This makes the wire warm and lightbulbs glow. More collisions mean a higher resistivity ρ and a lower conductivity σ.

The drift speed is $v_d = \dfrac{e\tau}{m}E$

where τ is the mean time between collisions. v_d is related to the electron current by

   $$i = nAv_d$$

where n is the electron density.

An electric field E in a conductor causes a current density $J = nev_d = \sigma E$ where the conductivity is

   $$\sigma = \frac{ne^2\tau}{m}$$

The resistivity is $\rho = 1/\sigma$.

TERMS AND NOTATION

current, I	mean time between collisions, τ	Kirchhoff's junction law
drift speed, v_d	ampere, A	conductivity, σ
electron current, i	current density, J	resistivity, ρ
law of conservation of current	junction	superconductivity

EXERCISES AND PROBLEMS

Exercises

Section 28.1 The Electron Current

1. 1.0×10^{20} electrons flow through a cross section of a 2.0-mm-diameter iron wire in 5.0 s. What is the electron drift speed?

2. Estimate how long it takes an electron to go from the wall switch to the overhead lightbulb in your bedroom.

3. The electron drift speed in a 1.0-mm-diameter gold wire is 5.0×10^{-5} m/s. How long does it take 1 mole of electrons to flow through a cross section of the wire?

4. 1.0×10^{16} electrons flow through a cross section of silver wire in 320 μs with a drift speed of 8.0×10^{-4} m/s. What is the diameter of the wire?

5. 1.44×10^{14} electrons flow through a cross section of a 2.0 mm $\times$ 2.0 mm square wire in 3.0 μs. The electron drift speed is 2.0×10^{-4} m/s. What metal is the wire made of?

Section 28.2 Creating a Current

6. What is the surface charge density of a 1.0-mm-diameter wire with 1000 excess electrons per centimeter of length?

7. a. How many conduction electrons are there in a 1.0-mm-diameter gold wire that is 10 cm long?
   b. How far must the sea of electrons in the wire move to deliver -32 nC of charge to an electrode?

8. The electron drift speed is 2.0×10^{-4} m/s in a metal with a mean free time between collisions of 5.0×10^{-14} s. What is the electric field strength?

9. The mean free time between collisions in iron is 4.2×10^{-15} s. What electric field strength causes a 5.0×10^{19} s^{-1} electron current in a 1.8-mm-diameter iron wire?

Section 28.3 Batteries

10. Draw Figure Ex28.10 on your paper.
    a. Use plusses and minuses to show how the surface charge is distributed along the wire.
    b. Show the electric field vector $\vec{E}$ at each of the seven points in the wire marked with a dot. The length of each vector should be proportional to E at that point.

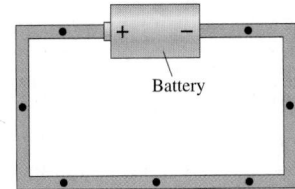

FIGURE EX28.10

11. A battery supplies a steady 1.5 A current to a circuit. How many electrons per second does the charge escalator transport from the positive terminal to the negative terminal?

Section 28.4 Current and Current Density

12. The current in an electric hair dryer is 10.0 A. How much charge and how many electrons flow through the hair dryer in 5.0 min?

13. 2.0×10^{13} electrons flow through a transistor in 1.0 ms. What is the current through the transistor?

14. In an ionic solution, 5.0×10^{15} positive ions with charge $+2e$ pass to the right each second while 6.0×10^{15} negative ions with charge $-e$ pass to the left. What is the current in the solution?

15. The current in a 2.0 mm $\times$ 2.0 mm square aluminum wire is 2.5 A. What are (a) the current density and (b) the electron drift speed?

16. The wires leading to and from a 0.12-mm-diameter lightbulb filament are 1.5 mm in diameter. The wire to the filament carries a current with a current density of 450,000 A/m^2. What are (a) the current and (b) the current density in the filament?

17. The current in a 100 watt lightbulb is 0.85 A. The filament inside the bulb is 0.25 mm in diameter.
    a. What is the current density in the filament?
    b. What is the electron current in the filament?

18. The electron drift speed in a gold wire is 3.0×10^{-4} m/s.
    a. What is the current density in the wire?
    b. What is the current if the wire diameter is 0.50 mm?

19. In an integrated circuit, the current density in a 2.5-μm-thick $\times$ 75-μm-wide gold film is 750,000 A/m^2. What is the current in the film?

20. A hollow copper wire with an inner diameter of 1.0 mm and an outer diameter of 2.0 mm carries a current of 10 A. What is the current density in the wire?

Section 28.5 Conductivity and Resistivity

21. What is the mean free time between collisions for electrons in an aluminum wire and in an iron wire?

22. What is the mean free time between collisions for electrons in silver and in gold?

23. The electric field in a 2.0 mm $\times$ 2.0 mm square aluminum wire is 0.012 N/C. What is the current in the wire?

24. What electric field strength is needed to create a 5.0 A current in a 2.0-mm-diameter iron wire?

25. A 3.0-mm-diameter wire carries a 12 A current when the electric field is 0.085 N/C. What is the wire's resistivity?

26. A 0.0075 N/C electric field creates a 3.9 mA current in a 1.0-mm-diameter wire. What material is the wire made of?

27. A 0.50-mm-diameter silver wire carries a 20 mA current. What are (a) the electric field and (b) the electron drift speed in the wire?

28. A metal cube 1.0 cm on each side is sandwiched between two electrodes. The electrodes create a 0.0050 N/C electric field in the metal. A current of 9.0 A passes through the cube, from the positive electrode to the negative electrode. Identify the metal.

Problems

29. The density of aluminum is 2700 kg/m^3. Verify the conduction-electron density for aluminum given in Table 28.1. The mass of an aluminum atom is 27 u.

30. For what electric field strength would the current in a 2.0-mm-diameter nichrome wire be the same as the current in a 1.0-mm-diameter aluminum wire in which the electric field strength is 0.0080 N/C?

31. The current in a wire is doubled. What happens to
    a. The current density?
    b. The conduction-electron density?
    c. The mean time between collisions?
    d. The electron drift speed?
32. The electric field strength inside a wire is doubled. What happens to
    a. The current?
    b. The conduction-electron density?
    c. The mean time between collisions?
    d. The electron drift speed?
33. The electron beam inside a television picture tube is 0.4 mm in diameter and carries a current of 50 μA. This electron beam impinges on the inside of the picture tube screen.
    a. How many electrons strike the screen each second?
    b. What is the current density in the electron beam?
    c. The electrons move with a velocity of 4.0×10^7 m/s. What electric field strength is needed to accelerate electrons from rest to this velocity in a distance of 5.0 mm?
    d. Each electron transfers its kinetic energy to the picture tube screen upon impact. What is the *power* delivered to the screen by the electron beam?
34. Figure P28.34 shows a 4.0-cm-wide plastic film being wrapped onto a 2.0-cm-diameter roller that turns at 90 rpm. The plastic has a uniform surface charge density -2.0 nC/cm².
    a. What is the current of the moving film?
    b. How long does it take the roller to accumulate a charge of -10 μC?

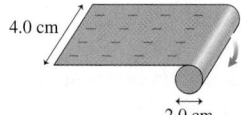

4.0 cm

FIGURE P28.34   2.0 cm

35. A sculptor has asked you to help electroplate gold onto a brass statue. You know that the charge carriers in the ionic solution are gold ions, and you've calculated that you must deposit 0.50 g of gold to reach the necessary thickness. How much current do you need, in mA, to plate the statue in 3.0 hours?
36. The biochemistry that takes place inside cells depends on various elements, such as sodium, potassium, and calcium, that are dissolved in water as ions. These ions enter cells through narrow pores in the cell membrane known as *ion channels.* Each ion channel, which is formed from a specialized protein molecule, is selective for one type of ion. Measurements with microelectrodes have shown that a 0.30-nm-diameter potassium ion (K^+) channel carries a current of 1.8 pA.
    a. How many potassium ions pass through if the ion channel opens for 1.0 ms?
    b. What is the current density in the ion channel?
37. The starter motor of a car engine draws a current of 150 A from the battery. The copper wire to the motor is 5.0 mm in diameter and 1.2 m long. The starter motor runs for 0.80 s until the car engine starts.
    a. How much charge passes through the starter motor?
    b. How far does an electron travel along the wire while the starter motor is on?
38. A car battery is rated at 90 A hr, meaning that it can supply a 90 A current for 1 hr before being completely discharged. If you leave your headlights on until the battery is completely dead, how much charge leaves the battery?

39. What fraction of the current in a wire of radius R flows in the part of the wire with radius $r \leq \frac{1}{2}R$?
40. You need to design a 1.0 A fuse that "blows" if the current exceeds 1.0 A. The fuse material in your stockroom melts at a current density of 500 A/cm². What diameter wire of this material will do the job?
41. A 62 g hollow copper cylinder is 10 cm long and has an inner diameter of 1.0 cm. The current density along the length of the cylinder is 150,000 A/m². What is the current in the cylinder?
42. A hollow metal cylinder has inner radius a, outer radius b, length L, and conductivity σ. The current I is *radially* outward from the inner surface to the outer surface.
    a. Find an expression for the electric field strength inside the metal as a function of the radius r from the cylinder's axis.
    b. Evaluate the electric field strength at the inner and outer surfaces of an iron cylinder if $a = 1.0$ cm, $b = 2.5$ cm, $L = 10$ cm, and $I = 25$ A.
43. A hollow metal sphere has inner radius a, outer radius b, and conductivity σ. The current I is *radially* outward from the inner surface to the outer surface.
    a. Find an expression for the electric field strength inside the metal as a function of the radius r from the center.
    b. Evaluate the electric field strength at the inner and outer surfaces of a copper sphere if $a = 1.0$ cm, $b = 2.5$ cm, and $I = 25$ A.
44. The total amount of charge in coulombs that has entered a wire at time t is given by the expression $Q = 4t - t^2$, where t is in seconds and $t \geq 0$.
    a. Graph Q versus t for the interval $0 \leq t \leq 4$ s.
    b. Find an expression for the current in the wire at time t.
    c. Graph I versus t for the interval $0 \leq t \leq 4$ s.
    d. Explain *why* I has the value at $t = 2.0$ s that you observe.
45. The total amount of charge that has entered a wire at time t is given by the expression $Q = (20 \text{ C})(1 - e^{-t/(2.0 \text{ s})})$, where t is in seconds and $t \geq 0$.
    a. Graph Q versus t for the interval $0 \leq t \leq 10$ s.
    b. Find an expression for the current in the wire at time t.
    c. What is the maximum value of the current?
    d. Graph I versus t for the interval $0 \leq t \leq 10$ s.
46. The current in a wire at time t is given by the expression $I = (2.0 \text{ A})e^{-t/(2.0 \text{ μs})}$, where t is in microseconds and $t \geq 0$.
    a. Graph I versus t for the interval $0 \leq t \leq 10$ μs.
    b. Find an expression for total amount of charge (in coulombs) that has entered the wire at time t. The initial conditions are $Q = 0$ C at $t = 0$ μs.
    c. Graph Q versus t for the interval $0 \leq t \leq 10$ μs.
47. The electric field in a current-carrying wire can be modeled as the electric field at the midpoint between two charged rings. Model a 3.0-mm-diameter aluminum wire as two 3.0-mm-diameter rings 2.0 mm apart. What is the current in the wire after 20 electrons are transferred from one ring to the other?
48. The two wires in Figure P28.48 are made of the same material. What are the current and the electron drift speed in the 2.0-mm-diameter segment of the wire?

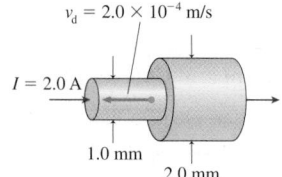

$v_d = 2.0 \times 10^{-4}$ m/s

$I = 2.0$ A

1.0 mm   2.0 mm

FIGURE P28.48

49. What is the electron drift speed at the 3.0-mm-diameter end (the left end) of the wire in Figure P28.49?

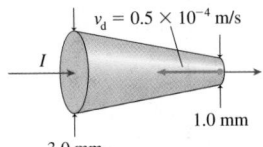

FIGURE P28.49
3.0 mm

50. What diameter should the nichrome wire in Figure P28.50 be in order for the electric field strength to be the same in both wires?

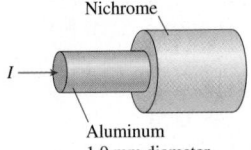

FIGURE P28.50
Aluminum
1.0 mm diameter

51. The two segments of the wire in Figure P28.51 have equal diameters but different conductivities σ_1 and σ_2. Current I passes through this wire. If the conductivities have the ratio $\sigma_2/\sigma_1 = 2$, what is the ratio E_2/E_1 of the electric field strengths in the two segments of the wire?

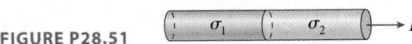

FIGURE P28.51

52. An aluminum wire consists of the three segments shown in Figure P28.52. The current in the top segment is 10 A. For each of these three segments, find the
   a. Current I.
   b. Current density J.
   c. Electric field E.
   d. Drift velocity v_d.
   e. Mean time between collisions τ.
   f. Electron current i.
   Place your results in a table for easy viewing.

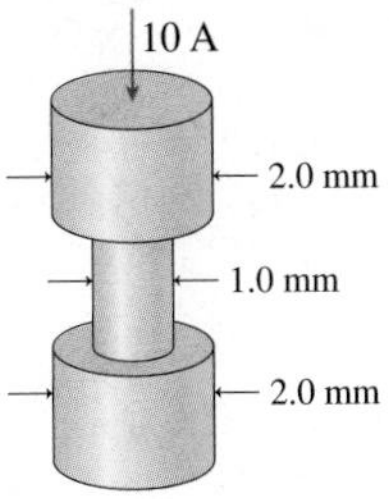

10 A

2.0 mm

1.0 mm

2.0 mm

FIGURE P28.52

Challenge Problems

53. The current supplied by a battery slowly decreases as the battery runs down. Suppose that the current as a function of time is $I = (0.75 \text{ A})e^{-t/(6 \text{ hr})}$. What is the total number of electrons transported from the positive electrode to the negative electrode by the charge escalator from the time the battery is first used until it is completely dead?

54. In a classical model of the hydrogen atom, the electron moves around the proton in a circular orbit of radius 0.053 nm.
   a. What is the electron's orbital frequency?
   b. What is the effective current of the electron?

55. Assume the conduction electrons in a metal can be treated as classical particles in an ideal gas.
   a. What is the rms velocity of electrons in room-temperature copper?
   b. How far, on average, does an electron move between collisions?

56. A 5.0-mm-diameter proton beam carries a total current of 1.5 mA. The current density in the proton beam, which increases with distance from the center, is given by $J = J_{edge}(r/R)$, where R is the radius of the beam and J_{edge} is the current density at the edge.
   a. How many protons per second are delivered by this proton beam?
   b. Determine the value of J_{edge}.

57. Figure CP28.57 shows a wire that is made of two equal-diameter segments with conductivities σ_1 and σ_2. When current I passes through the wire, a thin layer of charge appears at the boundary between the segments.
   a. Find an expression for the surface charge density η on the boundary. Give your result in terms of I, σ_1, σ_2, and the wire's cross-section area A.
   b. A 1.0-mm-diameter wire made of copper and iron segments carries a 5 A current. How much charge accumulates at the boundary between the segments?

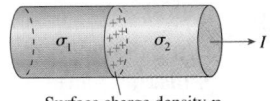

FIGURE CP28.57
Surface charge density η

STOP TO THINK ANSWERS

Stop to Think 28.1: $i_c > i_b > i_a > i_d$. The electron current is proportional to r^2v_d. Changing r by a factor of 2 has more influence than changing v_d by a factor of 2.

Stop to Think 28.2: The electrons don't have to move from the switch to the bulb, which could take hours. Because the wire between the switch and the bulb is already full of electrons, a flow of electrons from the switch into the wire immediately causes electrons to flow from the other end of the wire into the lightbulb.

Stop to Think 28.3: $E_d > E_b > E_e > E_a = E_c$. The electric field strength depends on the *difference* in the charge on the two wires.

The electric fields of the rings in a and c are opposed to each other, so the net field is zero. The rings in d have the largest charge *difference*.

Stop to Think 28.4: 1 A into the junction. The total current entering the junction must equal the total current leaving the junction.

Stop to Think 28.5: $J_b > J_a = J_d > J_c$. The current density $J = I/\pi r^2$ is independent of the conductivity σ, so a and d are the same. Changing r by a factor of 2 has more influence than changing I by a factor of 2.

29 The Electric Potential

Millions of lightbulbs transform electric energy into light and thermal energy.

▶ Looking Ahead

The goal of Chapter 29 is to calculate and use the electric potential and electric potential energy. In this chapter you will learn to:

- Use electric potential energy and conservation of energy to analyze the motion of charged particles.
- Use the electric potential to find the electric potential energy.
- Calculate the electric potential of useful and important charge distributions.
- Represent the electric potential graphically.

◀ Looking Back

This chapter depends heavily on the concepts of work, energy, and conservation of energy. Please review:

- Sections 10.2–10.5 Kinetic, gravitational, and elastic energy.
- Section 10.7 Energy diagrams.
- Sections 11.2–11.5 Work and potential energy.
- Section 26.3 Calculating the electric field of a continuous distribution of charge.

The sparkling lights of a big city are an awesome spectacle. These lights use a tremendous amount of energy. Where does all that energy come from?

Energy has been a theme throughout most of this textbook. Energy allows things to happen. A system without a source of energy is not terribly interesting; it just sits there. You want your lights to light, your computer to compute, and your stereo to keep your neighbors awake. In other words, you want devices that use electricity to *do* something, and that takes energy. It is time to see how the concept of energy helps us to understand and analyze electric phenomena.

In Chapter 25, we introduced the idea of the *electric field* to understand how one set of charges, the source charges, exerts electric forces on other charges. Now, to understand electric energy, we will introduce a new concept called the *electric potential*. In this chapter, you will study the basic properties of the electric potential and learn how it is connected to electric energy. In Chapter 30, we'll explore the relationship between the electric potential and the electric field. These two chapters will lead us directly into electric circuits, which are a practical application of the ideas of the electric potential and electric field.

29.1 Electric Potential Energy

We've alluded to energy many times in the last four chapters. We noted that a parallel-plate capacitor is charged by transferring electrons from one plate to the other, but the transfer doesn't happen by magic. It takes energy. A current flows in a wire only as long as energy is expended to push the sea of electrons forward against the "friction" of collisions with the atoms in the metal.

Our study of electric energy has two practical, and related, goals:

- To understand the motion of charged particles.
- To understand the fundamental ideas of electric circuits.

To meet these goals, we need to find out how electric energy is related to electric charges, forces, and fields.

Mechanical Energy

We will begin our investigation of electric energy by exploiting the close analogy between gravitational forces and electric forces. The gravitational force between two masses depends inversely on the square of the distance between them, as does the electric force between two point charges. Similarly, the uniform gravitational field near the earth's surface looks very much like the uniform electric field inside a parallel-plate capacitor.

You will recall that a system's mechanical energy $E_{\text{mech}} = K + U$ is conserved for particles that interact with each other via *conservative forces*, where K and U are the kinetic and potential energy. That is,

$$\Delta E_{\text{mech}} = \Delta K + \Delta U = 0 \qquad (29.1)$$

We need to be careful with notation because we are now using E to represent the electric field strength. To avoid confusion, we will represent mechanical energy either as the explicit sum $K + U$ or as E_{mech}, with an explicit subscript.

> **NOTE** ▶ Recall that for any quantity X, the *change* in X is $\Delta X = X_{\text{final}} - X_{\text{initial}}$. ◀

The kinetic energy $K = \sum K_i$, where $K_i = \frac{1}{2} m_i v_i^2$, is the sum of the kinetic energies of all the particles in the system. The potential energy U is the *interaction energy* of the system. In particular, we defined the *change* in potential energy in terms of the work W done by the forces of interaction as the system moves from an initial position or configuration i to a final position or configuration f:

$$\Delta U = U_{\text{f}} - U_{\text{i}} = -W_{\text{interaction forces}} \qquad \text{(position i} \rightarrow \text{position f)} \qquad (29.2)$$

This formal definition of ΔU is rather abstract and will make more sense when we see specific applications.

> **NOTE** ▶ The potential energy is an energy *of the system,* not of a particular particle in the system. That is, the familiar $U_{\text{grav}} = mgy$ is the energy of the earth + particle system due to their mutual gravitational interaction. Even so, we often speak of the *particle's* gravitational potential energy because the earth remains essentially at rest while the much less massive particle moves. ◀

A *constant* force does work

$$W = \vec{F} \cdot \Delta \vec{r} = F \Delta r \cos\theta \qquad (29.3)$$

on a particle that undergoes a linear displacement $\Delta \vec{r}$, where θ is the angle between the force $\vec{F}$ and $\Delta \vec{r}$. Figure 29.1 reminds you of the three special cases $\theta = 0°, 90°,$ and $180°$. It also shows that, in general, the work is done by the force component F_r in the direction of motion.

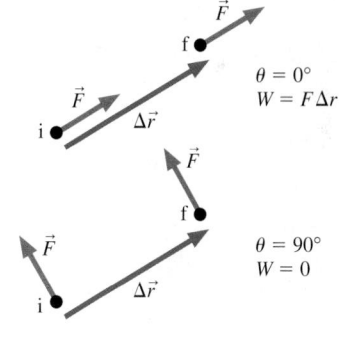

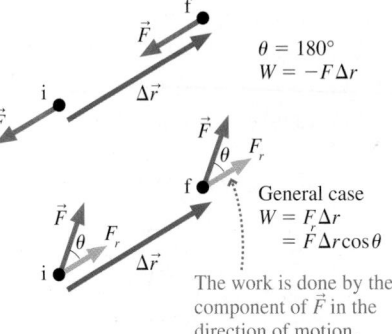

The work is done by the component of $\vec{F}$ in the direction of motion.

FIGURE 29.1 The work done by a constant force.

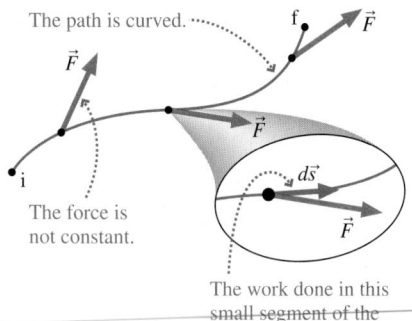

The path is curved. ····

$\vec{F}$

f

$\vec{F}$

i

The force is not constant.

$\vec{F}$

$d\vec{s}$

$\vec{F}$

The work done in this small segment of the motion is $F_s ds = \vec{F} \cdot d\vec{s}$.

FIGURE 29.2 The work done along a curved path or by a variable force.

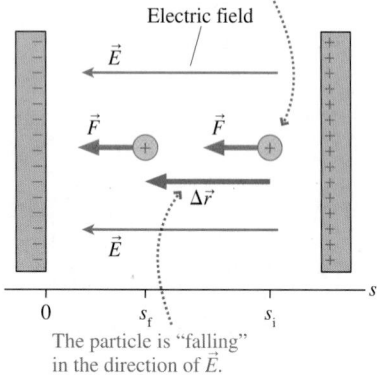

The gravitational field does work on the particle. We can express the work as a change in gravitational potential energy.

y

y_i

$\Delta\vec{r}$ $\vec{w}$

Gravitational field

y_f $\vec{g}$ $\vec{g}$ $\vec{w}$ $\vec{g}$

0

The net force on the particle is down. It gains kinetic energy (i.e., speeds up) as it loses potential energy.

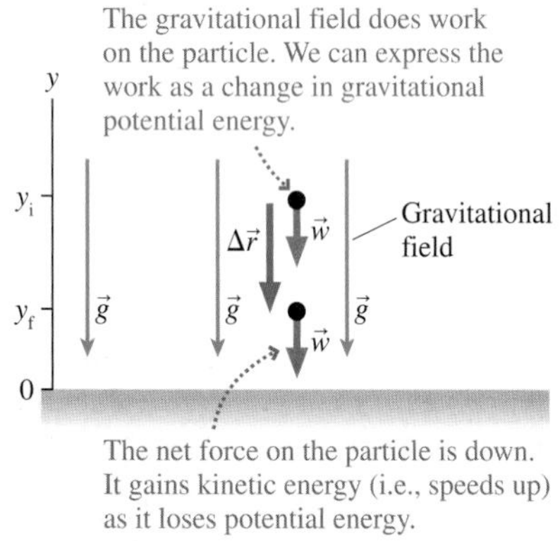

FIGURE 29.3 Potential energy is transformed into kinetic energy as a particle moves in a gravitational field.

The electric field does work on the particle. We can express the work as a change in electric potential energy.

Electric field

$\vec{E}$

$\vec{F}$ $\vec{F}$

$\Delta\vec{r}$

$\vec{E}$

0 s_f s_i s

The particle is "falling" in the direction of $\vec{E}$.

FIGURE 29.4 The electric field does work on the charged particle.

NOTE ▶ Work is *not* the oft-remembered "force times distance." Work is force times distance only in the one very special case in which the force is both constant *and* parallel to the displacement. ◀

If the force is *not* constant or the displacement is *not* along a linear path, we can calculate the work by dividing the path into many small segments. Figure 29.2 shows how this is done. The work done as the particle moves distance ds is $F_s ds$, where F_s is the force component parallel to ds (i.e., the component in the direction of motion). The total work done on the particle is

$$W = \sum_j (F_s)_j \Delta s_j \rightarrow \int_{s_i}^{s_f} F_s \, ds = \int_i^f \vec{F} \cdot d\vec{s} \qquad (29.4)$$

The second integral recognizes that $F_s ds = F\cos\theta\, ds$ is equivalent to the dot product $\vec{F} \cdot d\vec{s}$, allowing us to write the work in vector notation. As with Gauss's law, this integral looks more formidable than it really is. We'll look at examples shortly.

Finally, recall that a *conservative force* is one for which the work done as a particle moves from position i to position f is *independent of the path followed*. In other words, the integral in Equation 29.4 gives the same value for *any* path between points i and f. We'll assert for now, and prove later, that **the electric force is a conservative force.**

A Uniform Field

The gravitational field $\vec{g}$ near the earth's surface is a *uniform* field, $\vec{g} = (g, \text{down})$. Figure 29.3 shows a particle of mass m falling in the gravitational field. You learned in Chapter 10 that the particles *loses* potential energy and *gains* kinetic energy as it moves in the direction of the gravitational field. This is a fancy way of saying that the particle speeds up as it falls.

The gravitational force is in the same direction as the particle's displacement, so the gravitational field does a *positive* amount of work on the particle. The gravitational force is constant, so the work done by gravity is

$$W_{\text{grav}} = w\,\Delta r\cos 0° = mg\,|y_f - y_i| = mgy_i - mgy_f \qquad (29.5)$$

We have to be careful with signs because the displacement Δr must be a positive number.

Now we can see how the definition of ΔU in Equation 29.2 makes sense. The *change* in gravitational potential energy is

$$\Delta U_{\text{grav}} = U_f - U_i = -W_{\text{grav}}(i \rightarrow f) = mgy_f - mgy_i \qquad (29.6)$$

Comparing the initial and final terms on the two sides of the equation, we see that the gravitational potential energy near the earth is

$$U_{\text{grav}} = U_0 + mgy \qquad (29.7)$$

where U_0 is the value of U_{grav} at $y = 0$. We usually choose $U_0 = 0$, in which case $U_{\text{grav}} = mgy$, but such a choice is not necessary. The zero point of potential energy is an arbitrary choice because we have defined ΔU rather than U.

The uniform electric field between the plates of the parallel-plate capacitor of Figure 29.4 is analogous to the uniform gravitational field near the earth's surface. The one difference is that $\vec{g}$ always points down whereas the electric field inside a capacitor can point in any direction. To deal with this, let's define a coordinate axis s that points *from* the negative plate, which we define to be $s = 0$, *toward* the positive plate. The electric field $\vec{E}$ then points in the negative s-direction, just as the gravitational field $\vec{g}$ points in the negative y-direction. This s-axis, which is valid no matter how the capacitor is oriented, is analogous to the y-axis used for gravitational potential energy.

A positive charge q inside the capacitor speeds up and gains kinetic energy as it "falls" toward the negative plate. Is the charge losing potential energy as it gains kinetic energy? Indeed it is, and the calculation of the potential energy is just like the calculation of gravitational potential energy. The electric field exerts a *constant* force $F = qE$ on the charge in the direction of motion, thus the work done on the charge by the electric field is

$$W_{\text{elec}} = F\Delta r\cos 0° = qE|s_{\text{f}} - s_{\text{i}}| = qEs_{\text{i}} - qEs_{\text{f}} \tag{29.8}$$

where we again have to be careful with the signs because $s_{\text{f}} < s_{\text{i}}$.

The work done by the electric field causes the charge to experience a change in *electric* potential energy given by

$$\Delta U_{\text{elec}} = U_{\text{f}} - U_{\text{i}} = -W_{\text{elec}}(\text{i} \to \text{f}) = qEs_{\text{f}} - qEs_{\text{i}} \tag{29.9}$$

Comparing the initial and final terms on the two sides of the equation, we see that the **electric potential energy** of charge q in a uniform electric field is

$$U_{\text{elec}} = U_0 + qEs$$
(potential energy of charge q in a uniform electric field) $\tag{29.10}$

where s is the distance from the negative plate and U_0 is the potential energy at the negative plate $(s = 0)$. It will often be convenient to choose $U_0 = 0$, but the choice has no physical consequences because it doesn't affect ΔU_{elec}, the *change* in potential energy.

Equation 29.10 was derived with the assumption that q is positive, but it is valid for either sign of q. A negative value for q in Equation 29.10 causes the potential energy U_{elec} to become *more negative* as s increases. As Figure 29.5 shows, a negative charge speeds up and gains kinetic energy as it moves *away from* the negative plate of the capacitor.

NOTE ▶ Equation 29.10 is called "the potential energy of charge q," but we noted above that this is really the potential energy of the charge + capacitor system. However, to the extent that the charges on the capacitor plate stay fixed and only charge q moves, we're justified in thinking of this as the potential energy of just the charge. ◀

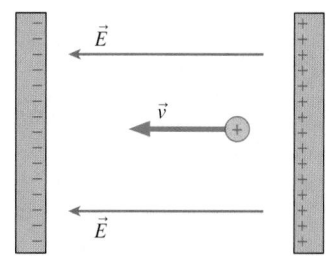

The potential energy of a positive charge decreases in the direction of $\vec{E}$. The charge gains kinetic energy as it moves toward the negative plate.

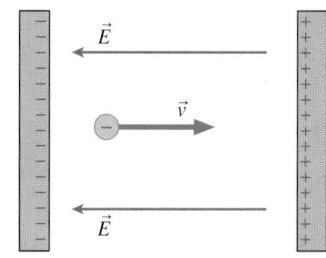

The potential energy of a negative charge decreases in the direction opposite to $\vec{E}$. The charge gains kinetic energy as it moves away from the negative plate.

FIGURE 29.5 A charged particle of either sign gains kinetic energy as it moves in the direction of decreasing potential energy.

EXAMPLE 29.1 Conservation of energy inside a capacitor

A 2.0 cm × 2.0 cm parallel-plate capacitor with a 2.0 mm spacing is charged to ± 1.0 nC. First a proton, then an electron are released from rest at the midpoint of the capacitor.

a. What is each particle's change in electric potential energy from its release until it collides with one of the plates?
b. What is each particle's speed as it reaches the plate?

MODEL The mechanical energy of each particle is conserved. A parallel-plate capacitor has a uniform electric field.

VISUALIZE Figure 29.6 is a before-and-after pictorial representation similar to those you learned to draw in Part II. The proton moves toward the negative plate, the electron toward the positive plate.

SOLVE

a. The s-axis was defined to point from the negative toward the positive plate of the capacitor. Both charged particles have $s_{\text{i}} = \frac{1}{2}d$, where $d = 2.0$ mm is the plate separation. The

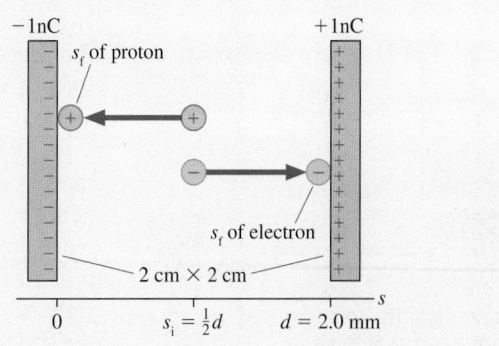

FIGURE 29.6 A proton and an electron in a capacitor.

positive proton loses potential energy and gains kinetic energy as it moves toward the negative plate. For the proton, with $q = +e$ and $s_{\text{f}} = 0$, the change in potential energy is

$$\Delta U_{\text{p}} = U_{\text{f}} - U_{\text{i}} = (U_0 + 0) - \left(U_0 + eE\frac{d}{2}\right) = -\frac{1}{2}eEd$$

where we used the electric potential energy for a charge in a uniform electric field. ΔU_p is negative, as expected. Notice that U_0 cancels when ΔU is calculated, which is why the value of U_0 has no physical consequences.

The electron moves toward the positive plate, which is the direction of decreasing potential energy for a negative charge. The electron has $q = -e$ and ends at $s_f = d$. Thus

$$\Delta U_e = U_f - U_i = (U_0 + (-e)Ed) - \left(U_0 + (-e)E\frac{d}{2}\right)$$

$$= -\frac{1}{2}eEd$$

Both particles have the *same* change in potential energy. The capacitor's electric field is

$$E = \frac{\eta}{\epsilon_0} = \frac{Q}{\epsilon_0 A} = 2.82 \times 10^5 \text{ N/C}$$

Using $d = 0.0020$ m, we find

$$\Delta U_p = \Delta U_e = -4.52 \times 10^{-17} \text{ J}$$

b. The law of conservation of energy is $\Delta K + \Delta U = 0$. Both particles are released from rest, hence $\Delta K = K_f - 0 = \frac{1}{2}mv_f^2$. Thus $\frac{1}{2}mv_f^2 = -\Delta U$, or

$$v_f = \sqrt{\frac{-2\Delta U}{m}} = \begin{cases} 2.33 \times 10^5 \text{ m/s for the proton} \\ 9.96 \times 10^6 \text{ m/s for the electron} \end{cases}$$

where we used the masses of the proton and the electron.

ASSESS Even though both particles have the same ΔU, the electron reaches a much larger final speed due to its much smaller mass.

STOP TO THINK 29.1   The positive charge is the end view of a positively charged glass rod. A negatively charged particle moves in a circular arc around the glass rod. Is the work done on the charged particle by the rod's electric field positive, negative, or zero?

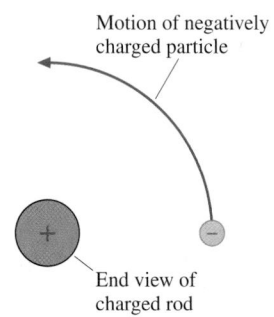

Motion of negatively charged particle

End view of charged rod

29.2 The Potential Energy of Point Charges

Now that we've introduced the idea of electric potential energy, let's look at *the fundamental interaction of electricity*—the force between two point charges. This force, given by Coulomb's law, varies with the distance between the two charges, so we'll begin by recalling how we dealt with variable forces in mechanics.

Figure 29.7 shows an object being pushed to the right by an expanding spring that has spring constant k. The force exerted by the spring decreases as it expands, hence we need to use the integral expression of Equation 29.4 to calculate the work done by the spring. The force, given by Hooke's law, is entirely in the direction of motion, so $F_s ds = F dx = -kx dx$. (The force is positive and to the right because, in this coordinate system, x is negative.) Thus the work done by the spring is

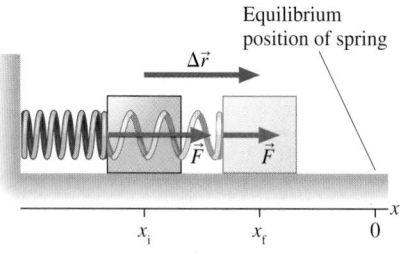

Equilibrium position of spring

$\Delta \vec{r}$

x_i x_f 0

FIGURE 29.7 The force exerted by a spring is not a constant force.

$$W_{sp} = \int_{x_i}^{x_f} F dx = -k \int_{x_i}^{x_f} x dx = -\frac{1}{2}kx_f^2 + \frac{1}{2}kx_i^2 \qquad (29.11)$$

The potential energy of the mass + spring system is related to the work done by

$$\Delta U_{sp} = U_f - U_i = -W_{sp}(i \rightarrow f) = \frac{1}{2}kx_f^2 - \frac{1}{2}kx_i^2 \qquad (29.12)$$

By comparing the left and right sides of this equation, we see that the potential energy of the mass + spring system is

$$U_{sp} = \frac{1}{2}kx^2 \qquad (29.13)$$

You should recall this result from Chapter 10. Here, where we can see the spring stretching and compressing, we really do need to think of this as the energy of the mass + spring system, not the energy of either the spring or the mass alone.

Figure 29.8 is the *energy diagram* for a mass + spring system. Recall that an energy diagram is a graphical representation of how the kinetic and potential energy are transformed as a particle moves. The parabola is the potential energy as a function of the object's position x, and the total energy line E_{mech} is the system's total mechanical energy. The particle oscillates between the *turning points* x_L and x_R where the total energy line crosses the potential-energy curve.

Now let's apply the same procedure to the interaction between two point charges. Figure 29.9a shows two charges q_1 and q_2, which we will assume to be like charges, exerting repulsive forces on each other. The potential energy of their interaction can be found by calculating the work done by q_1 on q_2 as q_2 moves from position x_i to position x_f. We'll assume that q_1 has been glued down and is unable to move.

We'll calculate the work just as we did for the spring, only we'll use Coulomb's law rather than Hooke's law for the force:

$$W_{elec} = \int_{x_i}^{x_f} F_{1 \text{ on } 2}\, dx = \int_{x_i}^{x_f} \frac{Kq_1q_2}{x^2}\, dx = Kq_1q_2 \frac{-1}{x}\Big|_{x_i}^{x_f} = -\frac{Kq_1q_2}{x_f} + \frac{Kq_1q_2}{x_i} \quad (29.14)$$

The potential energy of the two charges is related to the work done by

$$\Delta U_{elec} = U_f - U_i = -W_{elec}(i \rightarrow f) = \frac{Kq_1q_2}{x_f} - \frac{Kq_1q_2}{x_i} \qquad (29.15)$$

By comparing the left and right sides of the equation, as we did before, we see that the potential energy of the two-point-charge system is

$$U_{elec} = \frac{Kq_1q_2}{x} \qquad (29.16)$$

We chose to integrate along the x-axis for convenience, but what is really important is the *distance* between the charges. Thus a more general expression for the electric potential energy is

$$U_{elec} = \frac{Kq_1q_2}{r} = \frac{1}{4\pi\epsilon_0}\frac{q_1q_2}{r} \quad \text{(two point charges)} \qquad (29.17)$$

This is explicitly the energy *of the system,* not the energy of just q_1 or q_2.

NOTE ▶ The electric potential energy of two point charges looks *almost* the same as the force between the charges. The difference is the r in the denominator of the potential energy compared to the r^2 in Coulomb's law. Make sure you remember which is which! ◀

Two important points need to be noted:

■ We derived Equation 29.17 for two like charges, but it is equally valid for two opposite charges. The potential energy of two like charges is *positive* and of two opposite charges is *negative.*

■ Because the electric field outside a *sphere of charge* is the same as that of a point charge at the center, Equation 29.17 is also the electric potential energy of two charged spheres. Distance r is the distance between their centers.

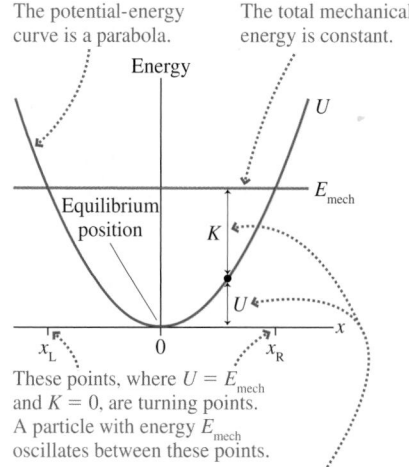

The potential-energy curve is a parabola. The total mechanical energy is constant.

These points, where $U = E_{mech}$ and $K = 0$, are turning points. A particle with energy E_{mech} oscillates between these points.

The mechanical energy is divided into kinetic and potential energy.

FIGURE 29.8 The energy diagram for a mass + spring system.

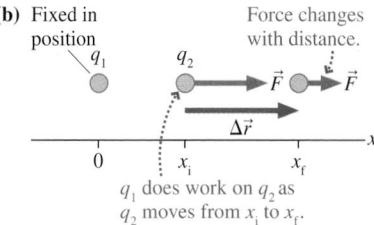

(a) $\vec{F}_{2 \text{ on } 1}$ q_1 q_2 $\vec{F}_{1 \text{ on } 2}$

(b) Fixed in position q_1 Force changes with distance. q_2 $\vec{F}$ $\vec{F}$ $\Delta\vec{r}$

q_1 does work on q_2 as q_2 moves from x_i to x_f.

FIGURE 29.9 The interaction between two point charges.

(a) Like charges

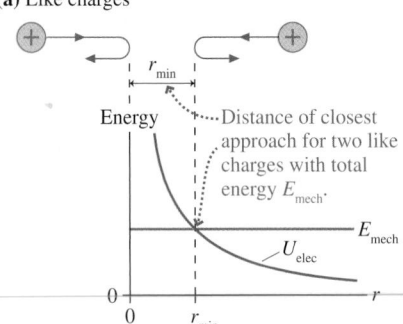

(b) Opposite charges

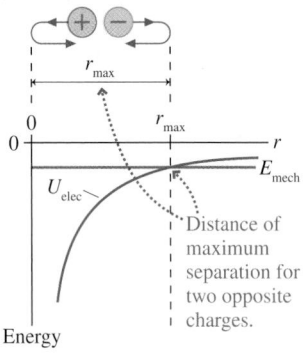

FIGURE 29.10 The potential-energy diagrams for two like charges and two opposite charges.

Figure 29.10 shows the potential-energy curves for two like charges and two opposite charges as a function of the distance r between them. Both curves are hyperbolas. Distance must be a positive number, so these graphs show only $r > 0$.

A charge released from rest must move in the direction of *decreasing* potential energy. Figure 29.10 shows that two like charges will move apart, because U decreases as r increases, and that two opposite charges will move toward one another. This is the expected behavior, based on Coulomb's law, but energy conservation provides an alternative perspective on the process.

More interesting, perhaps, is to consider two like charges that are shot toward each other with total mechanical energy E_{mech}. You can see in Figure 29.10a that the total energy line crosses the potential-energy curve at r_{min}. Two like charges shot toward each other will gradually slow down, because of the repulsive force between them, until the distance between them is r_{min}. At this point the kinetic energy is zero and both charged particles are instantaneously at rest. Both then reverse direction and move apart, speeding up as they go. r_{min} is the *distance of closest approach*. It is a quantity determined by energy conservation, not by analyzing the forces.

Similarly, you can see in Figure 29.10b that two oppositely charged particles shot apart from each other will slow down, losing kinetic energy, until reaching *maximum separation* r_{max}. Both reverse directions at the same instant, then they "fall" back together.

The Electric Force Is a Conservative Force

Potential energy can be defined only if the force is *conservative,* meaning that the work done on the particle as it moves from position i to position f is independent of the path followed between i and f. We *asserted* earlier that the electric force is a conservative force; now it's time to show it.

The work calculation in Equation 29.14 was based on Figure 29.9, where charge q_2 moved *straight out* from position i to position f. Figure 29.11 shows an alternative path between i and f. We can calculate the work done by the electric field as q_2 moves along this curved path by breaking the path into many small segments that are either radially outward from q_1 or circular arcs around q_1. The version shown in the figure is rather crude, with only a few segments, but you can imagine that we can approximate the true path arbitrarily closely by letting the number of segments approach infinity.

An alternative path for q_2 to move from i to f

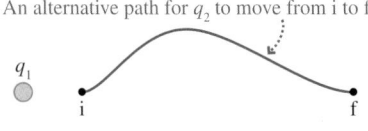

Approximate the path using circular arcs and radial lines centered on q_1.

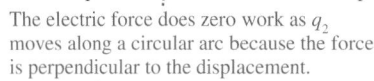

The electric force does zero work as q_2 moves along a circular arc because the force is perpendicular to the displacement.

All the work is done along the radial line segments, which are equivalent to a straight line from i to f.

FIGURE 29.11 Calculating the work done as q_2 moves along a curved path from i to f.

The electric force is a *central force,* directed straight at q_1. The electric force does *zero work* as q_2 moves along any of the circular arcs because the displacement is perpendicular to $\vec{F}$. All the work is done during the motion along the radial segments. The fact that the segments are displaced from each other doesn't affect the amount of work done along each segment. The total work, found by adding the work done along each radial segment, is equal to the total work that we calculated in Equation 29.14. The work is independent of the path, thus the electric force is a conservative force.

The Zero of Potential Energy

You can see both from Equation 29.17 for U_{elec} and from the graphs of Figure 29.10 that the potential energy of two charged particles approaches zero as $r \to \infty$. As you know, we can choose any point we wish as the zero of potential energy. The fact that only *changes* of potential energy ΔU appear in energy equations means that the zero of potential energy has no physical consequences.

Even if the zero point has no physical consequences, it's useful to think that $U = 0$ means "no interaction." A ball on the floor still has a weight, but its "ability to fall" is zero. Hence it makes sense to put the zero of gravitational potential energy at the floor—not essential, but useful.

For the two point charges, the zero point at infinity appeared from the integral in Equation 29.14 without our having made a conscious choice, but it's a good choice because two charged particles cease interacting only if they are infinitely far apart. A zero of potential energy at infinity allows us to think of U_{elec} as "the amount of interaction."

A zero point at infinity presents the one slight difficulty of interpreting negative energies. All a negative energy means is that the system has *less* energy than two charged particles that are infinitely far apart ($U_{\text{elec}} = 0$) *and* at rest ($K = 0$). Figure 29.12 shows that a system with the negative total energy $E_1 < 0$ is a *bound* system. The two charged particles cannot escape from each other. The electron and proton of a hydrogen atom are an example of a bound system.

Two opposite charges with the positive total energy $E_2 > 0$ *can* escape. They'll slow down as they move apart, but eventually the potential energy vanishes and they continue to coast apart with kinetic energy K_∞. The threshold condition is a system with $E = 0$. Two charged particles with $E = 0$ can escape, but it will take infinitely long because the kinetic energy approaches zero as the particles get far apart. The initial velocity that allows a particle to reach $r_f = \infty$ with $v_f = 0$ is called the **escape velocity.**

NOTE ▶ Real particles can't be infinitely far apart, but because U_{elec} decreases with distance, there comes a point when $U_{\text{elec}} = 0$ is an excellent approximation. Two charged particles for which $U_{\text{elec}} \approx 0$ are sometimes described as "far apart" or "far away." ◀

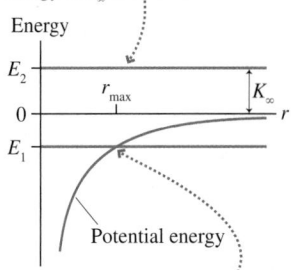

Two particles with total energy $E_2 > 0$ can move apart forever. Their kinetic energy is K_∞ as $r \to \infty$.

Two particles with total energy $E_1 < 0$ are a bound system. They can't get farther apart than r_{max}.

FIGURE 29.12 A system with $E_{\text{mech}} < 0$ is a *bound system.*

EXAMPLE 29.2 Approaching a charged sphere
A proton is fired from far away at a 1.0-mm-diameter glass sphere that has been charged to +100 nC. What initial speed must the proton have to just reach the surface of the glass?

MODEL Energy is conserved. The glass sphere can be treated as a charged particle, so the potential energy is that of two point charges. The proton starts "far away," which we interpret as sufficiently far to make $U_i \approx 0$.

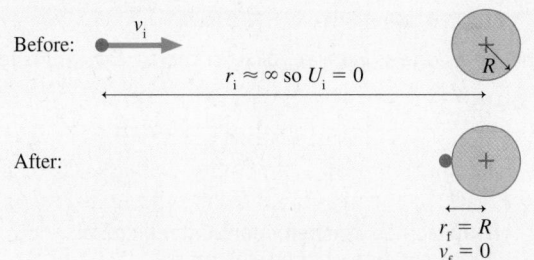

FIGURE 29.13 The before-and-after pictorial representation of a proton approaching a glass sphere.

VISUALIZE Figure 29.13 shows the before-and-after pictorial representation. To "just reach" the glass sphere means that the proton comes to rest, $v_f = 0$, as it reaches $r_f = 0.50$ mm, the *radius* of the sphere.

SOLVE Conservation of energy $K_f + U_f = K_i + U_i$ is

$$0 + \frac{Kq_p q_{\text{sphere}}}{r_f} = \frac{1}{2}mv_i^2 + 0$$

The proton charge is $q_p = e$. With this, we can solve for the proton's initial speed:

$$v_i = \sqrt{\frac{2Keq_{\text{sphere}}}{mr_f}} = 1.86 \times 10^7 \text{ m/s}$$

EXAMPLE 29.3 Escape velocity

An interaction between two elementary particles causes an electron and a positron (a positive electron) to be shot out back-to-back with equal speeds. What minimum speed must each particle have when they are 100 fm apart in order to escape each other?

MODEL Energy is conserved. The particles end "far apart," which we interpret as sufficiently far to make $U_f \approx 0$.

VISUALIZE Figure 29.14 shows the before-and-after pictorial representation. The minimum speed to escape is the speed that allows the particles to reach $r_f = \infty$ with $v_f = 0$.

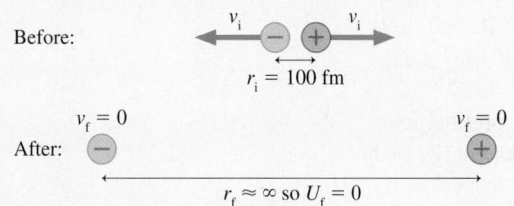

Before:

$r_i = 100$ fm

$v_f = 0$ $v_f = 0$

After:

$r_f \approx \infty$ so $U_f = 0$

FIGURE 29.14 The before-and-after pictorial representation of an electron and a positron flying apart.

SOLVE Here it is essential to interpret U_{elec} as the potential energy of the electron+positron system. Similarly, K is the *total* kinetic energy of the system. The electron and the positron, with equal masses and equal speeds, have equal kinetic energies. Conservation of energy $K_f + U_f = K_i + U_i$ is

$$0 + 0 + 0 = \frac{1}{2}mv_i^2 + \frac{1}{2}mv_i^2 + \frac{Kq_eq_p}{r_i} = mv_i^2 - \frac{Ke^2}{r_i}$$

Using $r_i = 100$ fm $= 1.0 \times 10^{-13}$ m, we can calculate the minimum initial speed to be

$$v_i = \sqrt{\frac{Ke^2}{mr_i}} = 5.0 \times 10^7 \text{ m/s}$$

ASSESS v_i is a little more than 10% the speed of light, just about the limit of what a "classical" calculation can predict. We would need to use the theory of relativity if v_i were any larger.

Multiple Point Charges

If more than two charges are present, the potential energy is the sum of the potential energies due to all pairs of charges:

$$U_{elec} = \sum_{i<j} \frac{Kq_iq_j}{r_{ij}} \tag{29.18}$$

where r_{ij} is the distance between q_i and q_j. The summation contains the $i < j$ restriction to ensure that each pair of charges is counted only once.

NOTE ▶ If two or more charges don't move, their potential energy doesn't change and can be thought of as an additive constant with no physical consequences. It's necessary to calculate only the potential energy for those pairs of charges for which the distance r_{ij} changes. ◀

EXAMPLE 29.4 Launching an electron

Three electrons are spaced 1.0 mm apart along a vertical line. The outer two electrons are fixed in position.

a. Is the center electron at a point of stable or unstable equilibrium?

b. If the center electron is displaced horizontally by an infinitesimal distance, what will be its speed when it is very far away?

MODEL Energy is conserved. The outer two electrons don't move, so we don't need to include the potential energy of their interaction.

VISUALIZE Figure 29.15 shows the before-and-after pictorial representation.

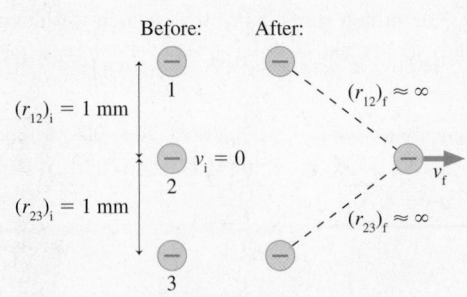

FIGURE 29.15 The before-and-after pictorial representation of three electrons.

SOLVE

a. The center electron is in equilibrium *exactly* in the center because the two electric forces on it balance. But if it moves a little to the right or left, no matter how little, then the horizontal components of the forces from both outer electrons will push the center electron farther away. This is an unstable equilibrium, like being on the top of a hill.

b. An infinitesimal displacement will cause the electron to move away. If the displacement is only infinitesimal, the initial conditions are $(r_{12})_i = (r_{23})_i = 1.0$ mm and $v_i = 0$. "Far away" is interpreted as $r_f \rightarrow \infty$, where $U_f \approx 0$. There are now *two* terms in the potential energy, so conservation of energy $K_f + U_f = K_i + U_i$ gives

$$\frac{1}{2}mv_f^2 + 0 + 0 = 0 + \left[\frac{Kq_1q_2}{(r_{12})_i} + \frac{Kq_2q_3}{(r_{23})_i} \right]$$

$$= \left[\frac{Ke^2}{(r_{12})_i} + \frac{Ke^2}{(r_{23})_i} \right]$$

This is easily solved to give

$$v_f = \sqrt{\frac{2}{m}\left[\frac{Ke^2}{(r_{12})_i} + \frac{Ke^2}{(r_{23})_i} \right]} = 1006 \text{ m/s}$$

STOP TO THINK 29.2 Rank in order, from largest to smallest, the potential energies U_a to U_d of these four pairs of charges. Each + symbol represents the same amount of charge.

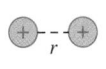 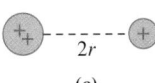 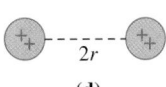

    (a)   (b)   (c)   (d)

29.3 The Potential Energy of a Dipole

The electric dipole has been our model for understanding how charged objects interact with neutral objects. In Chapter 26 we found that an electric field exerts a *torque* on a dipole. We can complete the picture by calculating the potential energy of an electric dipole in a uniform electric field.

Figure 29.16a shows a dipole in an electric field $\vec{E}$. Recall that the dipole moment $\vec{p}$ is a vector that points from $-q$ to q with magnitude $p = qd$. The forces $\vec{F}_+$ and $\vec{F}_-$ exert a torque on the dipole, but now we're interested in calculating the *work* done by these forces as the dipole rotates from angle ϕ_i to angle ϕ_f. The work on the positive and negative charges is the same, so we'll analyze the positive charge and then multiply by 2.

Figure 29.16b looks more closely at the force on the positive charge. As the dipole rotates, the charge's small displacement is $ds = r\,d\phi = (\frac{1}{2}d)d\phi$. The small amount of work done by force $\vec{F}_+$ is

$$dW_+ = \vec{F}_+ \cdot d\vec{s} = F_+(ds)\cos\theta = \left(\frac{1}{2}q\,dE\right)\cos\theta\,d\phi = \left(\frac{1}{2}pE\right)\cos\theta\,d\phi \quad (29.19)$$

You can see from the figure that $\theta = \phi + 90°$, hence $\cos\theta = \cos(\phi + 90°) = -\sin\phi$. Thus the work done by the electric field on the dipole as it rotates through the small angle $d\phi$ is

$$dW_{elec} = 2\,dW_+ = -pE\sin\phi\,d\phi \quad (29.20)$$

where the factor of 2 includes the work done on the negative charge.

The total work done by the electric field as the dipole turns from ϕ_i to ϕ_f is

$$W_{elec} = -pE\int_{\phi_i}^{\phi_f} \sin\phi\,d\phi = pE\cos\phi_f - pE\cos\phi_i \quad (29.21)$$

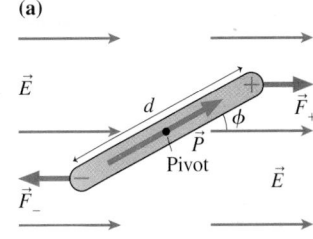

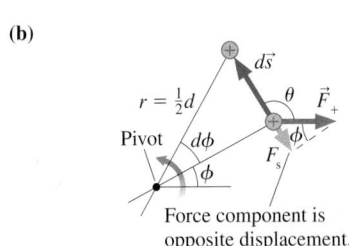

FIGURE 29.16 The electric field does work as a dipole rotates.

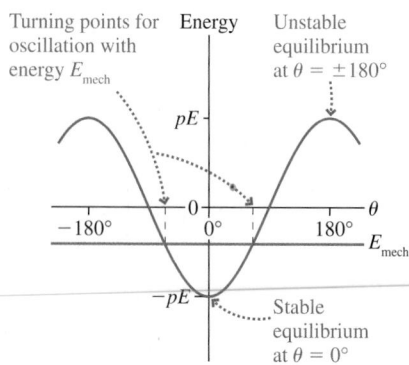

Turning points for
oscillation with
energy E_{mech}

Unstable
equilibrium
at $\theta = \pm 180°$

Stable
equilibrium
at $\theta = 0°$

FIGURE 29.17 The energy of a dipole in an electric field.

The potential energy associated with the work done on the dipole is

$$\Delta U_{dipole} = U_f - U_i = -W_{elec}(i \rightarrow f) = -pE\cos\phi_f + pE\cos\phi_i \quad (29.22)$$

By comparing the left and right sides of Equation 29.22, we see that the potential energy of an electric dipole $\vec{p}$ in a uniform electric field $\vec{E}$ is

$$U_{dipole} = -pE\cos\phi = -\vec{p} \cdot \vec{E} \quad (29.23)$$

Figure 29.17 shows the energy diagram of a dipole. The potential energy is minimum at $\theta = 0°$ where the dipole is aligned with the electric field. This is a point of stable equilibrium. A dipole exactly opposite $\vec{E}$, at $\theta = \pm 180°$, is at a point of unstable equilibrium. The slightest disturbance will cause it to flip around. A frictionless dipole with mechanical energy E_{mech} will oscillate back and forth between turning points on either side of $\theta = 0°$.

EXAMPLE 29.5 Rotating a molecule

The water molecule is a permanent electric dipole with dipole moment 6.2×10^{-30} Cm. A water molecule is aligned in an electric field with field strength 1.0×10^7 N/C. How much energy is needed to rotate the molecule 90°?

MODEL The molecule is at the point of minimum energy. It won't spontaneously rotate 90°. However, an external force that supplies energy, such as a collision with another molecule, can cause the water molecule to rotate.

SOLVE The molecule starts at $\theta_i = 0°$ and ends at $\theta_f = 90°$. The increase in potential energy is

$$\Delta U_{dipole} = U_f - U_i = -pE\cos 90° - (-pE\cos 0°)$$

$$= pE = 6.2 \times 10^{-23} \text{ J}$$

This is the energy needed to rotate the molecule 90°.

ASSESS ΔU_{dipole} is significantly less than $k_B T$ at room temperature. Thus collisions with other molecules can easily supply the energy to rotate the water molecules and keep them from staying aligned with the electric field.

29.4 The Electric Potential

11.11 Activ Physics

We introduced the concept of the *electric field* in Chapter 25 because action at a distance raised concerns and difficulties. The field provides an intermediary through which two charges exert forces on each other. Charge q_1 somehow alters the space around it by creating an electric field $\vec{E}_1$. Charge q_2 then responds to the field, experiencing force $\vec{F} = q_2\vec{E}_1$.

When we try to understand electric potential energy we face the same kinds of difficulties that action at a distance presented to an understanding of electric force. For a mass on a spring, we can *see* how the energy is stored in the stretched or compressed spring. But when we say two charged particles have a potential energy, an energy that can be converted to a tangible kinetic energy of motion, *where is the energy?* It's indisputable that two positive charges fly apart when you release them, gaining kinetic energy, but there's no obvious place that the energy had been stored.

In defining the electric field, we chose to separate the charges that are the *source* of the field from the charge *in* the field. The force on charge q is related to the electric field of the source charges by

force on q = [charge q] × [alteration of space by the source charges]

Let's try a similar procedure for the potential energy. The electric potential energy is due to the interaction of charge q with other charges, so let's divide up the potential energy of the system such that

potential energy of q + sources

= [charge q] × [*potential* for interaction of the source charges]

Figure 29.18 shows this idea schematically.

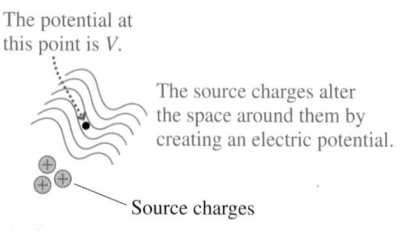

The potential at this point is V.

The source charges alter the space around them by creating an electric potential.

Source charges

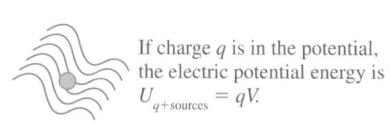

If charge q is in the potential, the electric potential energy is $U_{q+sources} = qV$.

FIGURE 29.18 Source charges alter the space around them by creating an electric potential.

Equations 29.10 and 29.17 show that the charge q can indeed be separated out from the expressions for the potential energy of a charge in a parallel-plate capacitor and for the potential energy of a point charge interacting with another point charge. Thus we will define the **electric potential** V (or, for brevity, just *the potential*) as

$$V \equiv \frac{U_{q+\text{sources}}}{q} \qquad (29.24)$$

Charge q is used as a probe to determine the electric potential, but the value of V is *independent of q*. **The electric potential is a property of the source charges.**

In practice, we're usually more interested in knowing the potential energy if a charge q happens to be at a point in space where the electric potential of the source charges is V. Turning Equation 29.24 around, we see that the electric potential energy is

$$U_{q+\text{sources}} = qV \qquad (29.25)$$

Once the potential has been determined, it's very easy to find the potential energy.

The unit of electric potential is the joule per coulomb, which is called the **volt** V:

$$1 \text{ volt} = 1 \text{ V} \equiv 1 \text{ J/C}$$

This unit is named for Alessandro Volta, who invented the electric battery in the year 1800. Microvolts (μV), millivolts (mV), and kilovolts (kV) are commonly used units because the electric potentials used in practical applications differ significantly in magnitude.

NOTE ▶ Once again, commonly used symbols and notation are in conflict. The symbol V is widely used to represent *volume,* and now we're introducing the same symbol to represent *potential.* To make matters more confusing, V is the abbreviation for *volts.* In printed text, V for potential is italicized and V for volts is not, but you can't make such a distinction in handwritten work. This is not a pleasant state of affairs, but these are the commonly accepted symbols. It's incumbent upon you to be especially alert to the *context* in which a symbol is used. ◀

What Good Is the Electric Potential?

The electric potential is an abstract idea, and it will take some practice to see just what it means and how it is useful. We'll use multiple representations—words, pictures, graphs, and analogies—to explain and describe the electric potential. We start with two essential ideas:

- The electric potential depends only on the source charges and their geometry. The potential is the "ability" of the source charges to have an interaction *if* a charge q shows up. This ability, or potential, is present throughout space **regardless of whether or not charge q is there to experience it.**
- If we know the electric potential V throughout a region of space, we'll immediately know the potential energy $U = qV$ of any charge that enters that region of space.

NOTE ▶ It is unfortunate that the terms *potential* and *potential energy* are so much alike. It is easy to confuse the two. Despite the similar names, they are very different concepts and are not interchangeable. Table 29.1 will help you to distinguish between the two. ◀

The source charges exert influence through the electric potential they establish throughout space. Once we know the potential—and the rest of this chapter deals

TABLE 29.1 Distinguishing electric potential and potential energy

The *electric potential* is a property of the source charges and, as you'll soon see, is related to the electric field. The electric potential is present whether or not a charged particle is there to experience it. Potential is measured in J/C, or V.

The *electric potential energy* is the interaction energy of a charged particle with the source charges. Potential energy is measured in J.

with how to calculate the potential—we can ignore the source charges and work just with the potential. The source charges remain hidden offstage.

The potential energy of a charged particle is determined by the electric potential: $U = qV$. Consequently, charged particles speed up or slow down as they move through a region of changing potential. Figure 29.19 illustrates this idea. It's useful to say that a particle moves through a **potential difference,** which is the difference $\Delta V = V_f - V_i$ between the potential at a starting point i and an ending point f. The potential difference between two points is often called the **voltage.** In illustrations, the potential difference between two points is represented by a double-headed green arrow.

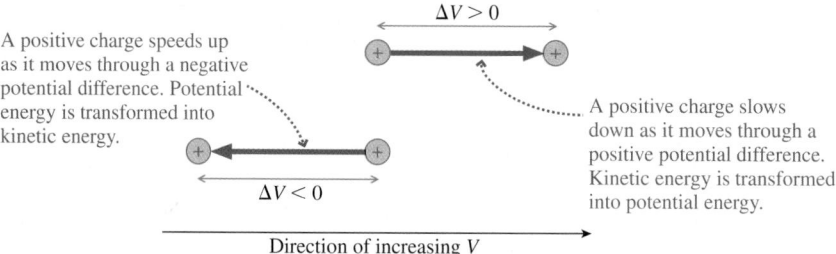

A positive charge speeds up as it moves through a negative potential difference. Potential energy is transformed into kinetic energy.

$\Delta V > 0$

A positive charge slows down as it moves through a positive potential difference. Kinetic energy is transformed into potential energy.

$\Delta V < 0$

Direction of increasing V

FIGURE 29.19 A charged particle speeds up or slows down as it moves through a potential difference.

If a particle moves through a potential difference ΔV, its electric potential energy is $\Delta U = q\Delta V$. We can write the conservation of energy equation in terms of the electric potential as $\Delta K + \Delta U = \Delta K + q\Delta V = 0$ or, as is often more practical,

$$K_f + qV_f = K_i + qV_i \qquad (29.26)$$

Conservation of energy is the basis of a powerful problem-solving strategy.

 PROBLEM-SOLVING STRATEGY 29.1 Conservation of energy in charge interactions

MODEL Check whether there are any dissipative forces that would keep the mechanical energy from being conserved.

VISUALIZE Draw a before-and-after pictorial representation. Define symbols that will be used in the problem, list known values, and identify what you're trying to find.

SOLVE The mathematical representation is based on the law of conservation of mechanical energy:

$$K_f + qV_f = K_i + qV_i$$

- Is the electric potential given in the problem statement? If not, you'll need to use a known potential, such as that of a point charge, or calculate the potential using the procedure given later, in Problem-Solving Strategy 29.2.
- K_i and K_f are the sums of the kinetic energies of all moving particles.
- Some problems may need additional conservation laws, such as conservation of charge or conservation of momentum.

ASSESS Check that your result has the correct units, is reasonable, and answers the question.

EXAMPLE 29.6 Moving through a potential difference

A proton with a speed of 2.0×10^5 m/s enters a region of space in which source charges have created an electric potential. What is the proton's speed after it moves through a potential difference of 100 V? What will be the final speed if the proton is replaced by an electron?

MODEL Energy is conserved. The electric potential determines the potential energy.

VISUALIZE Figure 29.20 is a before-and-after pictorial representation of a charged particle moving through a potential difference. A positive charge *slows down* as it moves into a region of higher potential ($K \to U$). A negative charge *speeds up* ($U \to K$).

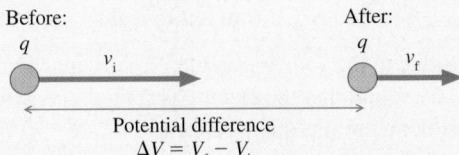

FIGURE 29.20 A charged particle moving through a potential difference.

SOLVE The potential energy of charge q is $U = qV$. Conservation of energy, now expressed in terms of the electric potential V, is $K_f + qV_f = K_i + qV_i$, or

$$K_f = K_i - q\Delta V$$

where $\Delta V = V_f - V_i$ is the potential difference. In terms of the speeds, energy conservation is

$$\frac{1}{2}mv_f^2 = \frac{1}{2}mv_i^2 - q\Delta V$$

We can solve this for the final speed:

$$v_f = \sqrt{v_i^2 - \frac{2q}{m}\Delta V}$$

For a proton, with $q = e$, the final speed is

$$(v_f)_{proton} = \sqrt{(2.0 \times 10^5 \text{ m/s})^2 - \frac{2(1.60 \times 10^{-19} \text{ C})(100 \text{ V})}{(1.67 \times 10^{-27} \text{ kg})}}$$

$$= 1.44 \times 10^5 \text{ m/s}$$

An electron, though, with $q = -e$ and a different mass, speeds up to $(v_f)_{electron} = 5.93 \times 10^6$ m/s.

ASSESS The electric potential *already existed* in space due to other charges that are not explicitly seen in the problem. The electron and proton have nothing to do with creating the potential. Instead, they *respond* to the potential by having potential energy $U = qV$.

STOP TO THINK 29.3   A proton is released from rest at point B, where the potential is 0 V. Afterward, the proton

a. Remains at rest at B.
b. Moves toward A with a steady speed.
c. Moves toward A with an increasing speed.
d. Moves toward C with a steady speed.
e. Moves toward C with an increasing speed.

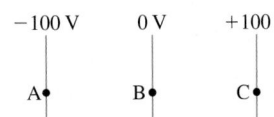

29.5 The Electric Potential Inside a Parallel-Plate Capacitor

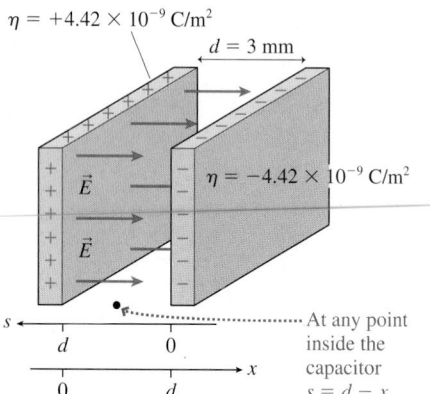

$\eta = +4.42 \times 10^{-9}$ C/m^2

$d = 3$ mm

$\eta = -4.42 \times 10^{-9}$ C/m^2

$\vec{E}$

$\vec{E}$

At any point inside the capacitor $s = d - x$.

FIGURE 29.21 A parallel-plate capacitor.

We began this chapter with the potential energy of a charge inside a parallel-plate capacitor. Now let's investigate the electric potential. Figure 29.21 shows two parallel electrodes, separated by distance d, with surface charge density $\pm\eta$. As a specific example, we'll let $d = 3.0$ mm and $\eta = 4.42 \times 10^{-9}$ C/m^2. The electric field inside the capacitor, as you learned in Chapter 26, is

$$\vec{E} = \left(\frac{\eta}{\epsilon_0}, \text{from positive toward negative}\right) \quad (29.27)$$

$$= (500 \text{ N/C, from left to right})$$

This electric field is due to the *source charges* on the capacitor plates.

In Section 29.1, we found that the electric potential energy of a charge q in the uniform electric field of a parallel-plate capacitor is

$$U_{\text{elec}} = U_{q+\text{sources}} = qEs \quad (29.28)$$

We've set the constant term U_0 to zero. U_{elec} is the energy of q interacting with the source charges on the capacitor plates.

Our new view of the interaction is to separate the role of charge q from the role of the source charges by defining the electric potential $V = U_{q+\text{sources}}/q$. Thus the electric potential inside a parallel-plate capacitor is

$$V = Es \quad \text{(electric potential inside a parallel-plate capacitor)} \quad (29.29)$$

where s is the distance from the *negative* electrode. The electric potential, like the electric field, exists at *all points* inside the capacitor. The electric potential is created by the source charges on the capacitor plates and exists whether or not charge q is inside the capacitor.

A first point to notice is that the electric potential increases linearly from the negative plate, where $V_- = 0$, to the positive plate, where $V_+ = Ed$. Let's define the *potential difference* ΔV_C between the two capacitor plates to be

$$\Delta V_C = V_+ - V_- = Ed \quad (29.30)$$

In our specific example, $\Delta V_C = (500 \text{ N/C})(0.003 \text{ m}) = 1.5$ V. The units work out because 1.5 (N m)/C = 1.5 J/C = 1.5 V.

NOTE ▶ People who work with circuits would call ΔV_C "the voltage across the capacitor" or simply "the voltage." ◀

Equation 29.30 has an interesting implication. Thus far, we've determined the electric field inside a capacitor by specifying the surface charge density η on the plates. Alternatively, we could specify the capacitor voltage ΔV_C (i.e., the potential difference between the capacitor plates) and then determine the electric field strength as

$$E = \frac{\Delta V_C}{d} \quad (29.31)$$

In fact, this is how E is determined in practical applications because it's easy to measure ΔV_C with a voltmeter but difficult, in practice, to know the value of η.

Equation 29.31 implies that the units of electric field are volts per meter, or V/m. We have been using electric field units of newtons per coulomb. In fact, as you can show as a homework problem, these units are equivalent to each other. That is,

$$1 \text{ N/C} = 1 \text{ V/m}$$

NOTE ▶ Volts per meter are the electric field units used by scientists and engineers in practice. We will now adopt them as our standard electric field units. ◀

Returning to the electric potential, we can substitute Equation 29.31 for E into Equation 29.29 for V. It will also be useful to measure the distance along the usual left-to-right x-axis, and you can see in Figure 29.21 that $s = d - x$. Thus the electric potential inside the capacitor is

$$V = Es = \frac{\Delta V_C}{d}(d - x) = \left(1 - \frac{x}{d}\right)\Delta V_C \qquad (29.32)$$

The potential *decreases* linearly from $V_+ = \Delta V_C = 1.5$ V at the positive plate ($x = 0$) to $V_- = 0$ V at the negative plate ($x = d = 3.0$ mm).

Let's explore the electric potential inside the capacitor by looking at several different, but related, ways that the potential can be represented graphically.

Graphical representations of the electric potential inside a capacitor

A graph of potential versus x. You can see the potential decreasing from 1.5 V at the positive plate to 0 V at the negative plate.

A three-dimensional view showing **equipotential surfaces.** These are mathematical surfaces, not physical surfaces, that have the same value of V at every point. The equipotential surfaces of a capacitor are planes parallel to the capacitor plates. The capacitor plates are also equipotential surfaces.

A two-dimensional **contour map.** The capacitor plates and the equipotential surfaces are seen edge on, so you need to imagine them extending above and below the plane of the page.

A three-dimensional **elevation graph.** The potential is graphed vertically versus the x-coordinate on one axis and a generalized "yz-coordinate" on the other axis. The face-on view of the elevation graph gives you the potential graph.

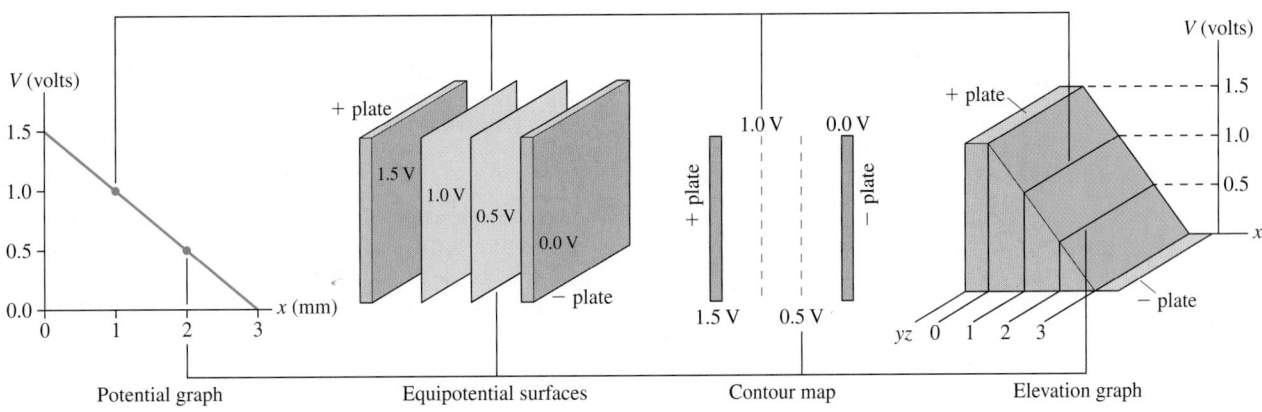

Potential graph   Equipotential surfaces   Contour map   Elevation graph

These four graphical representations show the same information from different perspectives, and the connecting lines help you see how they are related. If you think of the elevation graph as a "mountain," then the contour lines on the contour map are like the lines of a topographic map.

The potential graph and the contour map are the two representations most widely used in practice because they are easy to draw. Their limitation is that they are trying to convey three-dimensional information in a two-dimensional presentation. When you see graphs or contour maps, you need to imagine the three-dimensional equipotential surfaces or the three-dimensional elevation graph.

There's nothing special about showing equipotential surfaces or contour lines every 0.5 V. We chose these intervals because they were convenient. As an alternative, Figure 29.22 shows how the contour map looks if the contour lines are spaced every 0.3 V. Contour lines and equipotential surfaces are *imaginary* lines and surfaces drawn to help us visualize how the potential changes in space. Drawing the

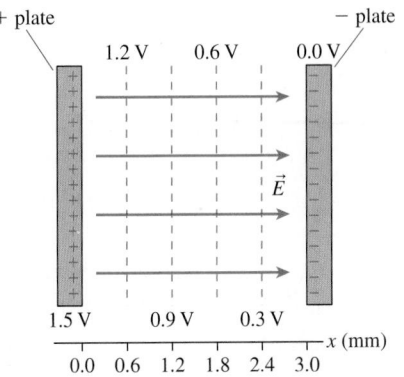

FIGURE 29.22 The contour lines of the electric potential and the electric field vectors inside a parallel-plate capacitor.

map more than one way reinforces the idea that there is an electric potential at *every* point inside the capacitor, not just at the points where we happened to draw a contour line or an equipotential surface.

Figure 29.22 also shows the electric field vectors. Notice that

- The electric field vectors are perpendicular to the equipotential surfaces.
- The potential decreases along the direction in which the electric field points. In other words, the electric field points "downhill" on a graph or map of the electric potential.

Chapter 30 will present a more in-depth exploration of the connection between the electric field and the electric potential. There you will find that these observations are always true. They are not unique to the parallel-plate capacitor.

Finally, you might wonder how we can arrange a capacitor to have a surface charge density of precisely 4.42×10^{-9} C/m². It turns out to be simple. As Figure 29.23 shows, we merely use wires to attach the capacitor plates to a 1.5 V battery. This is another topic that we'll look at more carefully in Chapter 30, but it's worth noting now that **a battery is a source of potential.** That's why batteries are labeled in volts, and it's a major reason that we need to thoroughly understand the concept of potential.

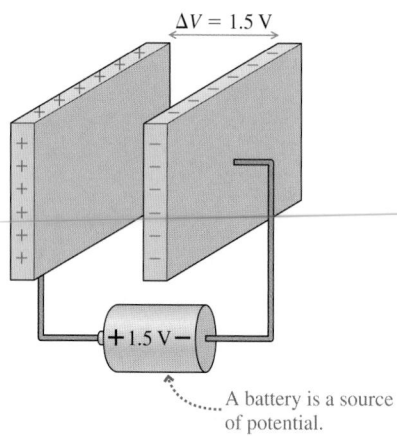

$\Delta V = 1.5$ V

$+1.5$ V $-$

A battery is a source of potential.

FIGURE 29.23 Using a battery to charge a capacitor to a precise value of ΔV_C.

EXAMPLE 29.7 A proton in a capacitor

A parallel-plate capacitor is constructed of two 2.0-cm-diameter disks spaced 2.0 mm apart. It is charged to a potential difference of 500 V.

a. What is the electric field strength inside?
b. How much charge is on each plate?
c. A proton is shot through a small hole in the negative plate with a speed of 2.0×10^5 m/s. Does it reach the other side? If not, where is the turning point?

MODEL Energy is conserved. The proton's potential energy inside the capacitor can be found from the capacitor's electric potential.

VISUALIZE Figure 29.24 is a before-and-after pictorial representation of the proton in the capacitor. Notice the *terminal symbols* where the potential is applied to the capacitor plates.

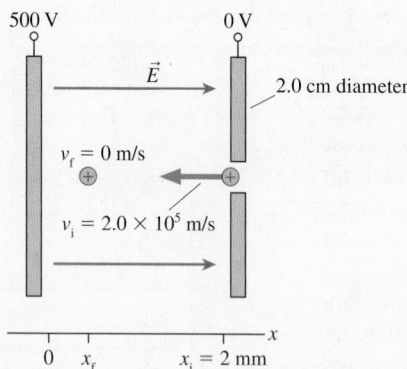

500 V   0 V

$\vec{E}$

2.0 cm diameter

$v_f = 0$ m/s

$v_i = 2.0 \times 10^5$ m/s

$0 \quad x_f \qquad x_i = 2$ mm

FIGURE 29.24 A before-and-after pictorial representation of a proton moving in a capacitor.

SOLVE

a. The electric field strength inside the capacitor is

$$E = \frac{\Delta V_C}{d} = \frac{500 \text{ V}}{0.0020 \text{ m}} = 2.5 \times 10^5 \text{ V/m}$$

b. Because $E = \eta/\epsilon_0$ for a parallel-plate capacitor, with $\eta = Q/A = Q/\pi R^2$, we find

$$Q = \pi R^2 \epsilon_0 E = 6.95 \times 10^{-10} \text{ C} = 0.695 \text{ nC}$$

c. The proton has charge $q = e$, and its potential energy at a point where the capacitor's potential is V is $U = eV$. It will gain potential energy $\Delta U = e\Delta V_C$ if it moves all the way across the capacitor. The increase in potential energy comes at the expense of kinetic energy, so the proton has sufficient kinetic energy to make it all the way across only if

$$K_i \geq e\Delta V_C$$

We can calculate that $K_i = 3.34 \times 10^{-17}$ J and that $e\Delta V_C = 8.00 \times 10^{-17}$ J. The proton does *not* have sufficient kinetic energy to be able to gain 8.00×10^{-17} J of potential energy, so it will not make it across. Instead, the proton will reach a turning point and reverse direction.

The proton starts at the negative plate, where $x_i = d = 2.0$ mm. Let the turning point be at x_f. The potential inside the capacitor is given by $V = \Delta V_C(1 - x/d)$ with $d = 0.0020$ m and $\Delta V_C = 500$ V. Conservation of energy requires $K_f + eV_f = K_i + eV_i$. This is

$$0 + e\Delta V_C\left(1 - \frac{x_f}{d}\right) = \frac{1}{2}mv_i^2 + 0$$

where we used $V_i = 0$ V at the negative plate ($x_i = d$). The solution for the turning point is

$$x_f = d - \frac{mdv_i^2}{2e\Delta V_C} = 1.16 \text{ mm}$$

The proton travels 0.84 mm, less than halfway across, before being turned back.

ASSESS We were able to use the electric potential inside the capacitor to determine the proton's potential energy. Notice that we used V/m as the electric field units.

We've assumed that $V = 0$ V at the negative plate, but that is not the only possible choice. Figure 29.25 shows three views of a parallel-plate capacitor that has a potential *difference* $\Delta V_C = 100$ V. These are contour maps, showing the edges of the equipotential surfaces.

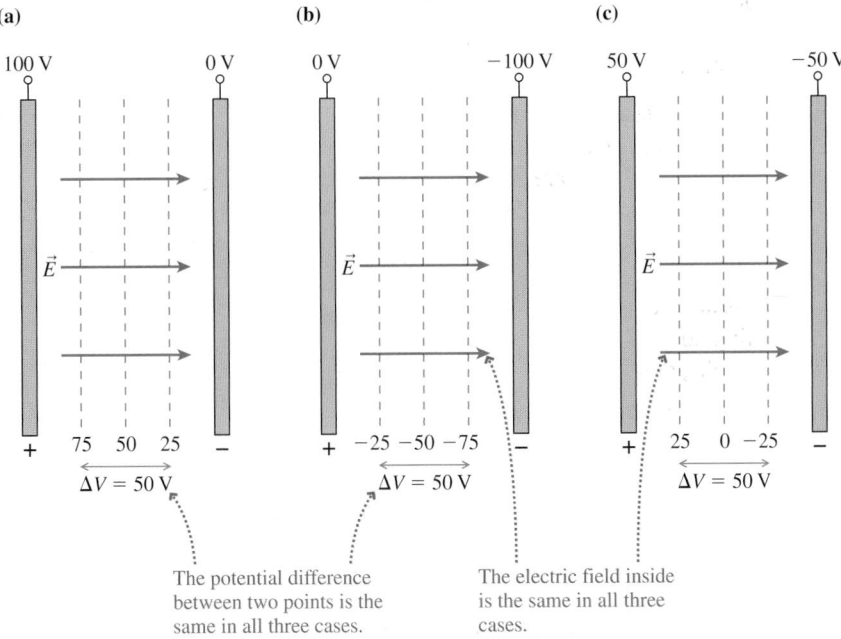

(a)

(b)

(c)

The potential difference between two points is the same in all three cases.

The electric field inside is the same in all three cases.

FIGURE 29.25 These three choices for $V = 0$ represent the same physical situation.

In Figure 29.25a we have chosen to let $V_{neg} = 0$ V. In this case, the potential of the positive plate $V_{pos} = \Delta V_C = 100$ V. At a point midway between the plates, at $x = d/2$, $V_{mid} = 50$ V.

Suppose, as Figure 29.25b shows, we choose the positive plate to have a potential of zero: $V_{pos} = 0$ V. After all, we're free to put the zero of potential and potential energy wherever we wish. Now the negative plate has potential $V_{neg} = -100$ V and the potential at the midpoint is $V_{mid} = -50$ V. Figure 29.25c shows yet a third choice, the symmetrical choice with $V = 0$ V in the center.

The important point to be made is that the three contour maps in Figure 29.25 represent the *same physical situation*. The potential difference between any two points is the same in all three maps. No matter which choice of zero we use, a charge q moving from $x = 0.75d$ to $x = 0.25d$ experiences a change of potential energy $\Delta U = 50q$ joules. We may *prefer* one of these figures over the other, but there is no measurable, physical difference between them.

STOP TO THINK 29.4   Rank in order, from largest to smallest, the potentials V_a to V_e at the points a to e.

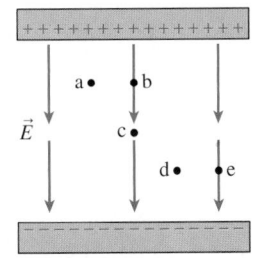

29.6 The Electric Potential of a Point Charge

To determine the potential of q at this point . . .

q

q'

r

. . . place charge q' at the point as a probe and measure the potential energy $U_{q'+q}$.

q

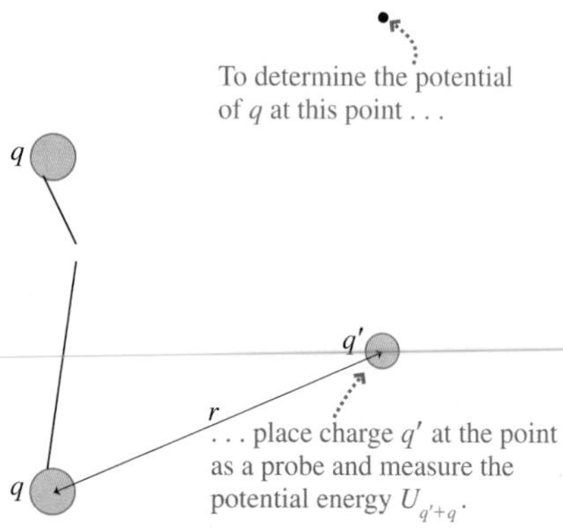

FIGURE 29.26 Measuring the electric potential of charge q.

Another important electric potential is that of a point charge. Let q in Figure 29.26 be the source charge, and let a second charge q' probe the electric potential of q. The potential energy of the two point charges is

$$U_{q'+q} = \frac{1}{4\pi\epsilon_0} \frac{qq'}{r} \tag{29.33}$$

Thus, by definition, the electric potential of charge q is

$$V = \frac{U_{q'+q}}{q'} = \frac{1}{4\pi\epsilon_0} \frac{q}{r} \quad \text{(electric potential of a point charge)} \tag{29.34}$$

The potential of Equation 29.34 extends through all of space, showing the influence of charge q, but it weakens with distance as $1/r$. This expression for V assumes that we have chosen $V = 0$ V to be at $r = \infty$. This is the most logical choice for a point charge because the influence of charge q ends at infinity.

The expression for the electric potential of charge q is similar to that for the electric field of charge q. The difference most quickly seen is that V depends on $1/r$ whereas $\vec{E}$ depends on $1/r^2$. But it is also important to notice that **the potential is a scalar** whereas the field is a *vector*. Thus the mathematics of using the potential are much easier than the vector mathematics using the electric field requires.

EXAMPLE 29.8 Calculating the potential of a point charge

What is the electric potential 1.0 cm from a $+1.0$ nC charge? What is the potential difference between a point 1.0 cm away and a second point 3.0 cm away?

SOLVE The potential at $r = 1.0$ cm is

$$V_{1\,cm} = \frac{1}{4\pi\epsilon_0} \frac{q}{r} = (9.0 \times 10^9 \,\text{N m}^2/\text{C}^2) \frac{1.0 \times 10^{-9}\,\text{C}}{0.010\,\text{m}}$$

$$= 900\,\text{V}$$

We can similarly calculate $V_{3\,cm} = 300$ V. Thus the potential difference between these two points is $\Delta V = V_{1\,cm} - V_3$ cm = 600 V.

ASSESS 1 nC is typical of the electrostatic charge produced by rubbing, and you can see that such a charge creates a fairly large potential nearby. Why are we not shocked and injured when working with the "high voltages" of such charges? The sensation of being shocked is a result of current, not potential. Some high-potential sources simply do not have the ability to generate much current. We will look at this issue in Chapter 31.

Visualizing the Potential of a Point Charge

Figure 29.27 shows four graphical representations of the electric potential of a point charge. These match the four representations of the electric potential inside a capacitor, and a comparison of the two is worthwhile. This figure assumes that q is positive; you may want to think about how the representations would change if q were negative.

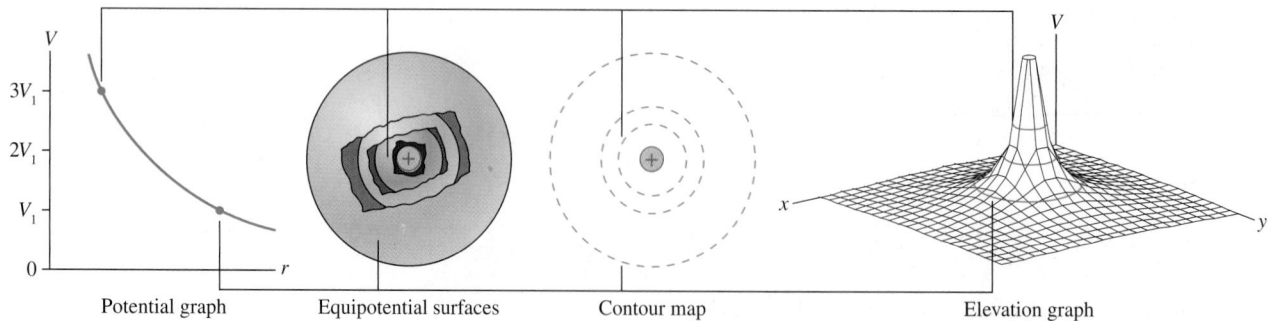

Potential graph   Equipotential surfaces   Contour map   Elevation graph

FIGURE 29.27 Four graphical representations of the electric potential of a point charge.

STOP TO THINK 29.5   Rank in order, from largest to smallest, the potential differences ΔV_{12}, ΔV_{13}, and ΔV_{23} between points 1 and 2, points 1 and 3, and points 2 and 3.

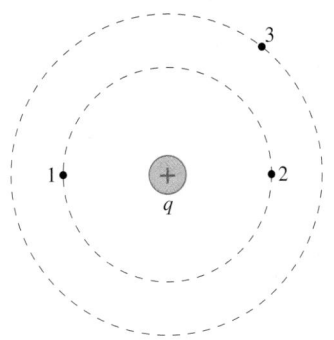

The Electric Potential of a Charged Sphere

In practice, you are more likely to work with a charged sphere, of radius R and total charge Q, than with a point charge. Outside a uniformly charged sphere, the electric potential is identical to that of a point charge Q at the center. That is,

$$V = \frac{1}{4\pi\epsilon_0}\frac{Q}{r} \qquad \text{(sphere of charge, } r \geq R) \qquad (29.35)$$

We can cast this result in a more useful form. It is customary to speak of charging an electrode, such as a sphere, "to" a certain potential, as in "Bob charged the sphere to a potential of 3000 volts." This potential, which we will call V_0, is the potential right on the surface of the sphere. We can see from Equation 29.35 that

$$V_0 = V(\text{at } r = R) = \frac{Q}{4\pi\epsilon_0 R} \qquad (29.36)$$

Consequently, a sphere of radius R that is charged to potential V_0 has total charge

$$Q = 4\pi\epsilon_0 R V_0 \qquad (29.37)$$

If we substitute this expression for Q into Equation 29.35, we can write the potential outside a sphere that is charged to potential V_0 as

$$V = \frac{R}{r}V_0 \qquad \text{(sphere charged to potential } V_0) \qquad (29.38)$$

Equation 29.38 tells us that the potential of a sphere is V_0 on the surface and decreases inversely with the distance. The potential at $r = 3R$ is $\frac{1}{3}V_0$.

EXAMPLE 29.9 A proton and a charged sphere

A proton is released from rest at the surface of a 1.0-cm-diameter sphere that has been charged to +1000 V.

a. What is the charge of the sphere?
b. What is the proton's speed when it is 1.0 cm from the sphere?

MODEL Energy is conserved. The potential outside the charged sphere is the same as the potential of a point charge at the center.

VISUALIZE Figure 29.28 is a before-and-after pictorial representation.

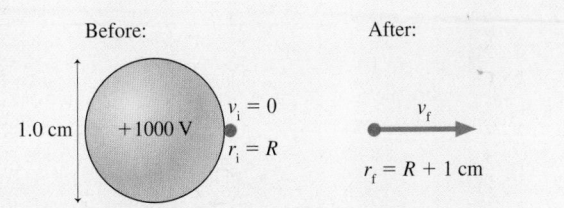

FIGURE 29.28 A before-and-after pictorial representation of a sphere and a proton.

SOLVE

a. We can use the sphere's potential to find that the charge of the sphere is

$$Q = 4\pi\epsilon_0 R V_0 = 0.56 \times 10^{-9} \text{ C} = 0.56 \text{ nC}$$

b. A sphere charged to $V_0 = +1000$ V is positively charged. The proton will be repelled by this charge and move away from the sphere. The conservation of energy equation $K_f + eV_f = K_i + eV_i$, with Equation 29.38 for the potential of a sphere, is

$$\frac{1}{2}mv_f^2 + \frac{eR}{r_f}V_0 = \frac{1}{2}mv_i^2 + \frac{eR}{r_i}V_0$$

The proton starts from the surface of the sphere, $r_i = R$, with $v_i = 0$. When the proton is 1.0 cm from the *surface* of the sphere, it has $r_f = 1.0$ cm $+ R = 1.5$ cm. Using these, we can solve for v_f:

$$v_f = \sqrt{\frac{2eV_0}{m}\left(1 - \frac{R}{r_f}\right)} = 3.57 \times 10^5 \text{ m/s}$$

ASSESS This example illustrates how the ideas of electric potential and potential energy work together, yet they are *not* the same thing.

29.7 The Electric Potential of Many Charges

Suppose there are many source charges $q_1, q_2, \ldots$ The electric potential V at a point in space is the sum of the potentials due to each charge:

$$V = \sum_i \frac{1}{4\pi\epsilon_0} \frac{q_i}{r_i} \tag{29.39}$$

where r_i is the distance from charge q_i to the point in space where the potential is being calculated. In other words, **the electric potential, like the electric field, obeys the principle of superposition.**

As an example, the contour map and elevation graph in Figure 29.29 show that the potential of an electric dipole is the sum of the potentials of the positive and negative charges. Potentials such as these have many practical applications. For example, electrical activity within the body can be monitored by measuring equipotential lines on the skin. Figure 29.29c shows that the equipotentials near the heart are a slightly distorted but recognizable electric dipole.

(a) Contour map

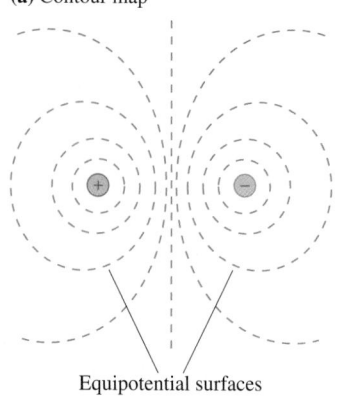

Equipotential surfaces

(b) Elevation graph

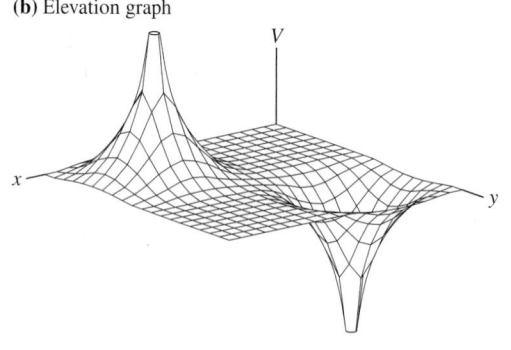

(c)

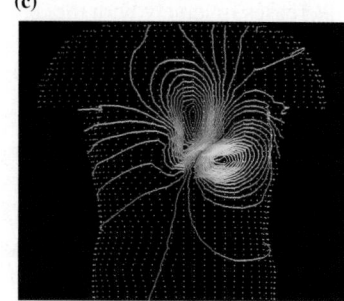

Equipotentials on the chest of a human are a slightly distorted electric dipole.

FIGURE 29.29 The electric potential of an electric dipole.

EXAMPLE 29.10 The potential of two charges

What is the electric potential at the point indicated in Figure 29.30?

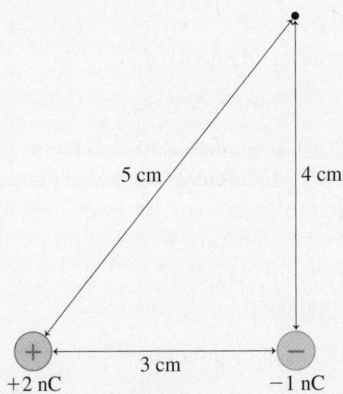

FIGURE 29.30 Finding the potential of two charges.

MODEL The potential is the sum of the potentials due to each charge.

SOLVE The potential at the indicated point is

$$V = \frac{1}{4\pi\epsilon_0} \frac{q_1}{r_1} + \frac{1}{4\pi\epsilon_0} \frac{q_2}{r_2}$$

$$= (9.0 \times 10^9 \, \text{N m}^2/\text{C}^2)\left(\frac{2.0 \times 10^{-9} \, \text{C}}{0.050 \, \text{m}} + \frac{-1.0 \times 10^{-9} \, \text{C}}{0.040 \, \text{m}}\right)$$

$$= 135 \, \text{V}$$

ASSESS The potential is a *scalar*, so we found the net potential by adding two scalars. We don't need any angles or components to calculate the potential.

A Continuous Distribution of Charge

Equation 29.39 is the basis for determining the potential of a continuous distribution of charge, such as a charged rod or a charged disk. The procedure is much like the one you learned in Chapter 26 for calculating the electric field of a continuous distribution of charge, but *easier* because the potential is a scalar. We will continue to assume that the object is *uniformly charged,* meaning that the charges are evenly spaced over the object.

The strategy uses three basic steps:

1. Divide the total charge Q into many small point-like charges ΔQ.
2. Use our knowledge of the potential of a point charge to find the electric potential of each ΔQ.
3. Calculate the net potential by summing the potentials of all the ΔQ.

In practice, as with electric field calculations, we usually will let the sum become an integral.

Before looking at examples, let's flesh out the steps of the problem-solving strategy on the next page. The goal of the strategy is to break a problem down into small steps that are individually manageable.

PROBLEM-SOLVING STRATEGY 29.2 The electric potential of a continuous distribution of charge

MODEL Model the charges as a simple shape, such as a line or a disk. Assume the charge is uniformly distributed.

VISUALIZE For the pictorial representation:

❶ Draw a picture and establish a coordinate system.

❷ Identify the point P at which you want to calculate the electric potential.

❸ Divide the total charge Q into small pieces of charge ΔQ, using shapes for which you *already know* how to determine V. This division is often, but not always, into point charges.

❹ Identify distances that need to be calculated.

SOLVE The mathematical representation is $V = \sum V_i$.

■ Use superposition to form an algebraic expression for the potential at P.

■ Let the (x, y, z) coordinates of the point remain as variables.

■ Replace the small charge ΔQ with an equivalent expression involving a *charge density* and a *coordinate,* such as dx, that describes the shape of charge ΔQ. **This is the critical step in making the transition from a sum to an integral** because you need a coordinate to serve as the integration variable.

■ All distances must be expressed in terms of the coordinates.

■ Let the sum become an integral. The integration will be over the coordinate variable that is related to ΔQ. The integration limits for this variable will depend on the coordinate system you have chosen. Carry out the integration and simplify the result.

ASSESS Check that your result is consistent with any limits for which you know what the potential should be.

EXAMPLE 29.11 **The potential of a ring of charge**

Figure 29.31 shows a thin, uniformly charged ring of radius R with total charge Q. Find the potential at a point on the axis of the ring.

MODEL Because the ring is thin, we'll assume the charge lies along a circle of radius R.

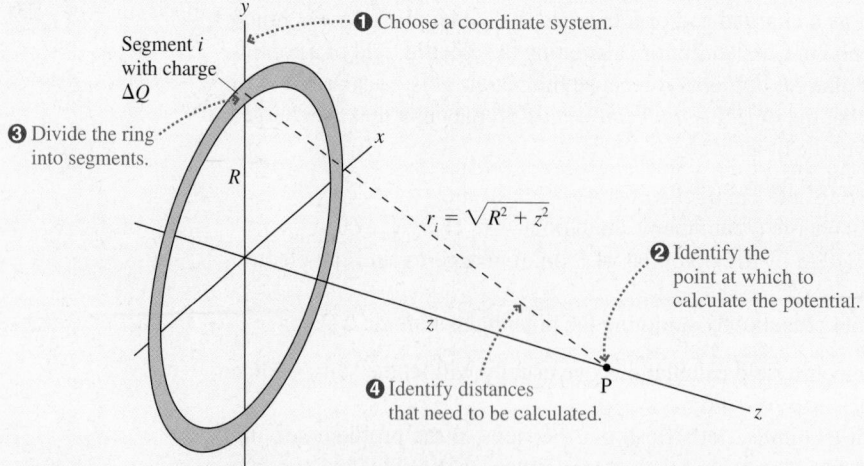

FIGURE 29.31 Finding the potential of a ring of charge.

VISUALIZE Figure 26.31 illustrates the four steps of the problem-solving strategy. We've chosen a coordinate system in which the ring lies in the xy-plane and point P is on the z-axis. We've then divided the ring into N small segments of charge ΔQ, each of which can be modeled as a point charge. The distance r_i between segment i and point P is

$$r_i = \sqrt{R^2 + z^2}$$

Note that r_i is a constant distance, the same for every charge segment. The potential V at P is the sum of the potentials due to each segment of charge:

$$V = \sum_{i=1}^{N} V_i = \sum_{i=1}^{N} \frac{1}{4\pi\epsilon_0} \frac{\Delta Q}{r_i} = \frac{1}{4\pi\epsilon_0} \frac{1}{\sqrt{R^2 + z^2}} \sum_{i=1}^{N} \Delta Q$$

We were able to bring all terms involving z to the front because z is a constant as far as the summation is concerned. Surprisingly, we don't need to convert the sum to an integral to complete this calculation. The sum of all the ΔQ charge segments around the ring is simply the ring's total charge, $\sum(\Delta Q) = Q$, hence the electric potential on the axis of a charged ring is

$$V_{\text{ring on axis}} = \frac{1}{4\pi\epsilon_0} \frac{Q}{\sqrt{R^2 + z^2}}$$

ASSESS From far away, the ring appears as a point charge Q in the distance. Thus we expect the potential of the ring to be that of a point charge when $z \gg R$. You can see that $V_{\text{ring}} \approx Q/4\pi\epsilon_0 z$ when $z \gg R$, which is, indeed, the potential of a point charge Q.

EXAMPLE 29.12 The potential of a disk of charge
Figure 29.32 shows a thin, uniformly charged disk of radius R with total charge Q. Find the potential at a point on the axis of the disk.

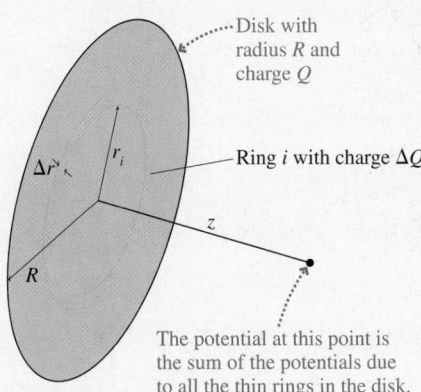

Disk with radius R and charge Q

Ring i with charge ΔQ_i

The potential at this point is the sum of the potentials due to all the thin rings in the disk.

FIGURE 29.32 Finding the potential of a disk of charge.

MODEL The disk has uniform surface charge density $\eta = Q/A = Q/\pi R^2$. We can take advantage of now knowing the on-axis potential of a ring of charge.

VISUALIZE Orient the ring in the xy-plane, as shown in Figure 29.32, with point P at distance z. Then divide the disk into *rings* of equal width Δr. Ring i has radius r_i and charge ΔQ_i. We can use the result of Example 29.11 to write the potential at distance z of ring i as

$$V_i = \frac{1}{4\pi\epsilon_0} \frac{\Delta Q_i}{\sqrt{r_i^2 + z^2}}$$

The potential at P due to all the rings is the sum

$$V = \sum_i V_i = \frac{1}{4\pi\epsilon_0} \sum_{i=1}^{N} \frac{\Delta Q_i}{\sqrt{r_i^2 + z^2}}$$

The critical step is to relate ΔQ_i to a coordinate. Because we now have a surface, rather than a line, the charge in ring i is $\Delta Q_i = \eta \Delta A_i$, where ΔA_i is the area of ring i. We can find ΔA_i, as you've learned to do in calculus, by "unrolling" the ring to form a narrow rectangle of length $2\pi r_i$ and height Δr. Thus the area of ring i is $\Delta A_i = 2\pi r_i \Delta r$ and the charge is

$$\Delta Q_i = \eta \Delta A_i = \frac{Q}{\pi R^2} 2\pi r_i \Delta r = \frac{2Q}{R^2} r_i \Delta r$$

With this substitution, the potential at P is

$$V = \frac{1}{4\pi\epsilon_0} \sum_{i=1}^{N} \frac{2Q}{R^2} \frac{r_i \Delta r_i}{\sqrt{r_i^2 + z^2}} \rightarrow \frac{Q}{2\pi\epsilon_0 R^2} \int_0^R \frac{r\,dr}{\sqrt{r^2 + z^2}}$$

where, in the last step, we let $N \rightarrow \infty$ and the sum become an integral. This integral can be found in a table of integrals, but it's not hard to evaluate with a change of variables. Let $u = r^2 + z^2$, in which case $r\,dr = \frac{1}{2}du$. Changing variables requires that we also change the integration limits. You can see that $u = z^2$ when $r = 0$, and $u = R^2 + z^2$ when $r = R$. With these changes, the on-axis potential of a charged disk is

$$V_{\text{disk on axis}} = \frac{Q}{2\pi\epsilon_0 R^2} \int_{z^2}^{R^2+z^2} \frac{\frac{1}{2}du}{u^{1/2}} = \frac{Q}{2\pi\epsilon_0 R^2} u^{1/2} \bigg|_{z^2}^{R^2+z^2}$$

$$= \frac{Q}{2\pi\epsilon_0 R^2} \left(\sqrt{R^2 + z^2} - z \right)$$

ASSESS Although we had to go through a number of steps, this procedure is easier than evaluating the electric field because we do not have to worry about components. Similar procedures can be followed to find the potential for any continuous distribution of charge.

We can find the potential V_0 of the disk itself by setting $z = 0$, giving $V_0 = Q/2\pi\epsilon_0 R$. In other words, placing charge Q on a disk of radius R charges it to potential V_0. The on-axis potential of the disk can be written in terms of V_0 as

$$V_{\text{disk on axis}} = V_0\left[\frac{\sqrt{R^2 + z^2} - z}{R}\right]$$

$$= V_0\left[\sqrt{1 + (z/R)^2} - (z/R)\right] \tag{29.40}$$

Figure 29.33 shows a graph of $V_{\text{disk on axis}}$ as a function of distance z along the axis. The potential of a point charge Q is shown for comparison. You can see that the charged disk begins to look like a point charge for $z \gg R$ but differs significantly from a point charge for $z \lesssim R$.

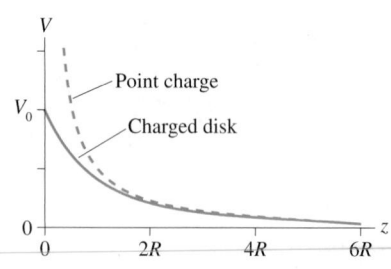

FIGURE 29.33 The potential of a charged disk and a point charge with the same Q.

EXAMPLE 29.13 **The potential of a dime**

A dime, which is 17.5 mm in diameter, is given a charge of +5.0 nC.

a. What is the potential of the dime?
b. What is the potential energy of an electron 1.0 cm above the dime?

MODEL The dime is a thin charged disk.

SOLVE

a. The potential of the dime itself is the potential of a disk at $z = 0$:

$$V_0 = \frac{Q}{2\pi\epsilon_0 R} = 10{,}300 \text{ V}$$

b. To calculate the potential energy $U = qV$ of charge q, we first need to determine the potential of the disk at $z = 1.0$ cm. Using Equation 29.40 for the potential on the axis of the dime, we find

$$V = V_0\left[\sqrt{1 + (z/R)^2} - (z/R)\right] = 3870 \text{ V}$$

The electron's charge is $q = -e = -1.60 \times 10^{-19}$ C, so its potential energy at $z = 1.0$ cm is $U = -6.2 \times 10^{-16}$ J.

SUMMARY

The goal of Chapter 29 has been to calculate and use the electric potential and electric potential energy.

GENERAL PRINCIPLES

Sources of V

The electric potential, like the electric field, is created by charges.

Two major tools for calculating V are

- The potential of a point charge $V = \dfrac{1}{4\pi\epsilon_0}\dfrac{q}{r}$

- The principle of superposition

Multiple point charges

Use superposition: $V = V_1 + V_2 + V_3 + \ldots$

Continuous distribution of charge

- Divide the charge into point-like ΔQ.
- Find the potential of each ΔQ.
- Find V by summing the potentials of all ΔQ.

The summation usually becomes an integral. A critical step is replacing ΔQ with an expression involving a charge density and an integration coordinate. Calculating V is usually easier than calculating $\vec{E}$ because the potential is a scalar.

Consequences of V

A charged particle has potential energy

$$U = qV$$

at a point where source charges have created an electric potential V.

The electric force is a conservative force, so the mechanical energy is conserved for a charged particle in an electric potential:

$$K_f + U_f = K_i + U_i$$

The potential energy of **two point charges** separated by distance r is

$$U_{q_1+q_2} = \frac{Kq_1q_2}{r} = \frac{1}{4\pi\epsilon_0}\frac{q_1q_2}{r}$$

The **zero point** of potential and potential energy is chosen to be convenient. For point charges, we let $U = 0$ when $r \to \infty$.

The potential energy in an electric field of an **electric dipole** with dipole moment $\vec{p}$ is

$$U_{\text{dipole}} = -pE\cos\theta = -\vec{p}\cdot\vec{E}$$

APPLICATIONS

Graphical representations of the potential:

Potential graph   **Equipotential surfaces**

Contour map   **Elevation graph**

Sphere of charge Q

Same as a point charge if $r \geq R$.

Parallel-plate capacitor

$V = Es$, where s is measured from the negative plate. The electric field inside is

$$E = \Delta V_C/d$$

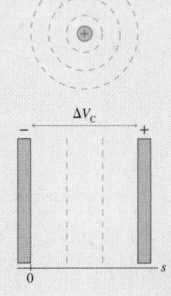

Units

Electric potential: $1\ \text{V} = 1\ \text{J/C}$

Electric field: $1\ \text{V/m} = 1\ \text{N/C}$

TERMS AND NOTATION

electric potential energy, U	volt, V	equipotential surface
escape velocity	potential difference, ΔV	contour map
electric potential, V	voltage, ΔV	elevation graph

EXERCISES AND PROBLEMS

Exercises

Section 29.1 Electric Potential Energy

1. The electric field strength is 50,000 N/C inside a parallel-plate capacitor with a 2.0 mm spacing. A proton is released from rest at the positive plate. What is the proton's speed when it reaches the negative plate?
2. The electric field strength is 20,000 N/C inside a parallel-plate capacitor with a 1.0 mm spacing. An electron is released from rest at the negative plate. What is the electron's speed when it reaches the positive plate?
3. A proton is released from rest at the positive plate of a parallel-plate capacitor. It crosses the capacitor and reaches the negative plate with a speed of 50,000 m/s. What will be the proton's final speed if the experiment is repeated with double the amount of charge on each capacitor plate?
4. A proton is released from rest at the positive plate of a parallel-plate capacitor. It crosses the capacitor and reaches the negative plate with a speed of 50,000 m/s. The experiment is repeated with a He$^+$ ion (charge e, mass 4 u). What is the ion's speed at the negative plate?

Section 29.2 The Potential Energy of Point Charges

5. What is the electric potential energy of the group of charges in Figure Ex29.5?

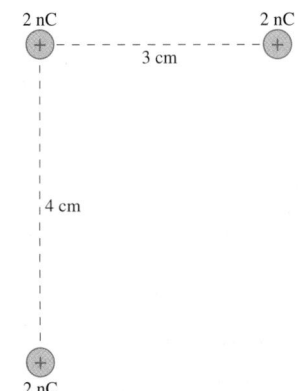

FIGURE EX29.5

6. What is the electric potential energy of the group of charges in Figure Ex29.6?

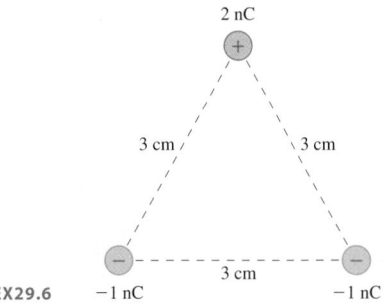

FIGURE EX29.6

7. What is the electric potential energy of the electron in Figure Ex29.7?

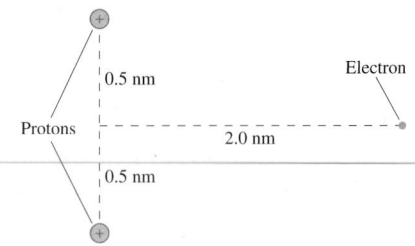

FIGURE EX29.7

8. Three electrons form an equilateral triangle 1.0 nm on each side. A proton is at the center of the triangle. What is the potential energy of this group of charges?

Section 29.3 The Potential Energy of a Dipole

9. A water molecule perpendicular to an electric field has 1.0×10^{-21} J more potential energy than a water molecule aligned with the field. The dipole moment of a water molecule is 6.2×10^{-30} C m. What is the strength of the electric field?
10. The graph shows the potential energy of an electric dipole. Consider a dipole that oscillates between $\pm 60°$.
    a. What is the dipole's mechanical energy?
    b. What is the dipole's kinetic energy when it is aligned with the electric field?

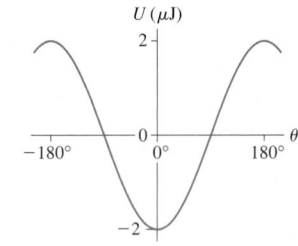

FIGURE EX29.10

Section 29.4 The Electric Potential

11. What is the speed of an electron that has been accelerated from rest through a potential difference of 1000 V?
12. What is the speed of a proton that has been accelerated from rest through a potential difference of -1000 V?
13. What potential difference is needed to accelerate a He$^+$ ion (charge $+e$, mass 4 u) from rest to a speed of 1.0×10^6 m/s?
14. What potential difference is needed to accelerate an electron from rest to a speed of 1.0×10^6 m/s?
15. An electron with an initial speed of 500,000 m/s is brought to rest by an electric field.
    a. Did the electron move into a region of higher potential or lower potential?
    b. What was the potential difference that stopped the electron?
16. A proton with an initial speed of 800,000 m/s is brought to rest by an electric field.
    a. Did the proton move into a region of higher potential or lower potential?
    b. What was the potential difference that stopped the proton?

Section 29.5 The Electric Potential Inside a Parallel-Plate Capacitor

17. Show that 1 V/m = 1 N/C.
18. a. What is the potential of an ordinary AA or AAA battery? (If you're not sure, find one and look at the label.)
    b. An AA battery is connected to a parallel-plate capacitor having 4.0-cm-diameter plates spaced 2 mm apart. How much charge does the battery supply to each plate?
19. A 2.0 cm × 2.0 cm parallel-plate capacitor has a 2.0 mm spacing. The electric field strength inside the capacitor is 1.0×10^5 V/m.
    a. What is the potential difference across the capacitor?
    b. How much charge is on each plate?
20. Two 2.0 cm × 2.0 cm plates that form a parallel-plate capacitor are charged to ±0.708 nC. What are the electric field strength inside and the potential difference across the capacitor if the spacing between the plates is (a) 1.0 mm and (b) 2.0 mm?
21. a. In Figure Ex29.21, which capacitor plate, left or right, is the positive plate?
    b. What is the electric field strength inside the capacitor?
    c. What is the potential energy of a proton at the midpoint of the capacitor?

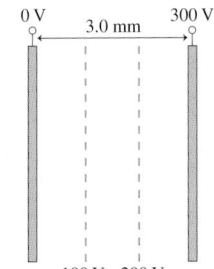

FIGURE EX29.21

Section 29.6 The Electric Potential of a Point Charge

22. A +25 nC charge is at the origin.
    a. What are the radii of the 1000 V, 2000 V, 3000 V, and 4000 V equipotential surfaces?
    b. Draw a contour map in the *xy*-plane showing the charge and these four surfaces.
23. a. What is the electric potential at points A, B, and C in Figure Ex29.23?
    b. What is the potential energy of an electron at each of these points?
    c. What are the potential differences ΔV_{AB} and ΔV_{BC}?

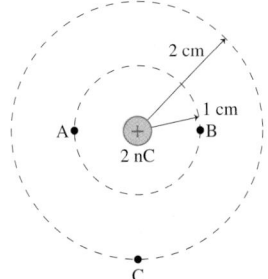

FIGURE EX29.23

24. A 1.0-mm-diameter ball bearing has 2.0×10^9 excess electrons. What is the ball bearing's potential?
25. A 2.0-mm-diameter glass bead is positively charged. The potential difference between a point 2.0 mm from the bead and a point 4.0 mm from the bead is 500 V. What is the charge on the bead?
26. In a semiclassical model of the hydrogen atom, the electron orbits the proton at a distance of 0.053 nm.
    a. What is the electric potential of the proton at the position of the electron?
    b. What is the electron's potential energy?

Section 29.7 The Electric Potential of Many Charges

27. What is the electric potential at the point indicated with the dot in Figure Ex29.27?

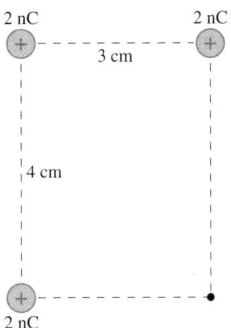

FIGURE EX29.27

28. What is the electric potential at the point indicated with the dot in Figure Ex29.28?

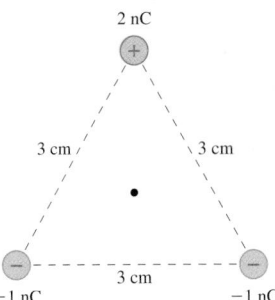

FIGURE EX29.28

29. a. What is the electric potential at point A in Figure Ex29.29?
    b. What is the potential energy of a proton at point A?

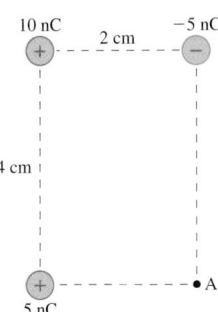

FIGURE EX29.29

30. a. What is the electric potential at point B in Figure Ex29.30?
    b. What is the potential energy of an electron at point B?

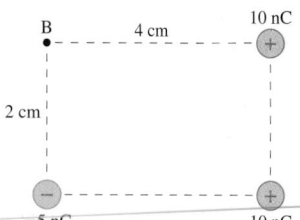

FIGURE EX29.30

31. A +3.0 nC charge is at $x = 0$ cm and a -1.0 nC charge is at $x = 4$ cm. At what point or points on the x-axis is the electric potential zero?

32. A -3.0 nC charge is on the x-axis at $x = -9$ cm and a $+4.0$ nC charge is on the x-axis at $x = 16$ cm. At what point or points on the y-axis is the electric potential zero?

33. Two point charges are located on the x-axis at $x = a$ and $x = b$. Figure Ex29.33 is a graph of E_x, the x-component of the electric field.
    a. What can you conclude about the signs and the relative magnitudes of the two charges?
    b. Draw a graph of the electric potential as a function of x.

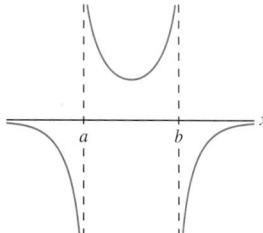

FIGURE EX29.33

34. Two point charges are located on the x-axis at $x = a$ and $x = b$. Figure Ex29.34 is a graph of V, the electric potential.
    a. What can you conclude about the signs and the relative magnitudes of the two charges?
    b. Draw a graph of E_x, the x-component of the electric field, as a function of x.

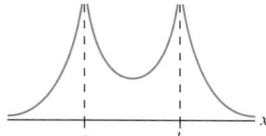

FIGURE EX29.34

35. The two halves of the rod are uniformly charged to $\pm Q$. What is the electric potential at the point indicated by the dot?

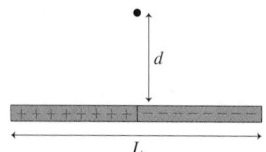

FIGURE EX29.35

Problems

36. Two point charges 2.0 cm apart have an electric potential energy $-180\ \mu$J. The total charge is 30 nC. What are the two charges?

37. Two positive point charges are 5.0 cm apart. If the electric potential energy is 72 μJ, what is the magnitude of the force between the two charges?

38. Bead A has a mass of 15 g and a charge of -5.0 nC. Bead B has a mass of 25 g and a charge of -10.0 nC. The beads are held 12 cm apart (measured between their centers) and released. What maximum speed and maximum acceleration are achieved by each bead?
    Hint: There are *two* conserved quantities. Make use of both.

39. A -2.0 nC charge and a $+2.0$ nC charge are located on the x-axis at $x = -1.0$ cm and $x = +1.0$ cm, respectively.
    a. At what position or positions on the x-axis is the electric field zero?
    b. At what position or positions on the x-axis is the electric potential zero?
    c. Draw graphs of the electric field strength and the electric potential along the x-axis.

40. A -10.0 nC point charge and a $+20.0$ nC point charge are 15 cm apart on the x-axis.
    a. What is the electric potential at the point on the x-axis where the electric field is zero?
    b. What are the magnitude and direction of the electric field at the point on the x-axis where the electric potential is zero?

41. Two small metal spheres with masses 2.0 g and 4.0 g are tied together by a 5.0-cm-long massless string and are at rest on a frictionless surface. Each is charged to $+2.0\ \mu$C.
    a. What is the energy of this system?
    b. What is the tension in the string?
    c. The string is cut. What is the speed of each sphere when they are far apart?
    Hint: There are *two* conserved quantities. Make use of both.

42. A proton and an alpha particle ($q = +2e$, $m = 4$ u) are fired directly toward each other from far away, each with an initial speed of $0.01c$. What is their distance of closest approach, as measured between their centers?
    Hint: There are *two* conserved quantities. Make use of both.

43. Figure P29.43 shows four charged particles at the corners of a rectangle. What is the total kinetic energy a long time after these charges have been released?

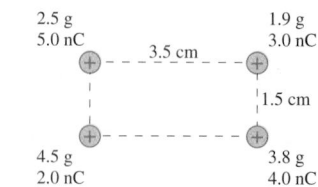

FIGURE P29.43

44. A proton's speed as it passes point A is 50,000 m/s. It follows the trajectory shown in Figure P29.44. What is the proton's speed at point B?

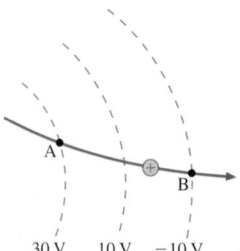

FIGURE P29.44   30 V   10 V -10 V

45. Two 2.0-cm-diameter disks spaced 2.0 mm apart form a parallel-plate capacitor. The electric field between the disks is 5.0×10^5 V/m.
    a. What is the voltage across the capacitor?
    b. How much charge is on each disk?
    c. An electron is launched from the negative plate. It strikes the positive plate at a speed of 2.0×10^7 m/s. What was the electron's speed as it left the negative plate?

46. What is the ratio $\Delta V_p / \Delta V_e$ of the potential differences that will accelerate a proton and an electron from rest to (a) the same final speed and (b) the same final kinetic energy?

47. An arrangement of source charges produces the electric potential $V = 5000x^2$ along the x-axis, where V is in volts and x is in meters.
    a. Graph the potential between $x = -10$ cm and $x = +10$ cm.
    b. Describe the motion of a positively charged particle in this potential.
    c. What is the mechanical energy of a 1.0 g, 10 nC charged particle if its turning points are at ± 8.0 cm?
    d. What is the particle's maximum speed?

48. In Figure P29.48, a proton is fired with a speed of 200,000 m/s from the midpoint of the capacitor toward the positive plate.
    a. Show that this is insufficient speed to reach the positive plate.
    b. What is the proton's speed as it collides with the negative plate?

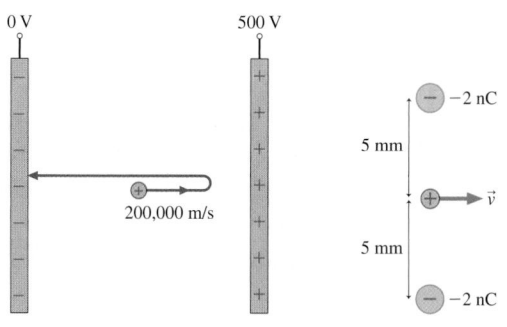

FIGURE P29.48   **FIGURE P29.49**

49. What is the escape speed of the proton in Figure P29.49?

50. What is the escape speed of an electron launched from the surface of a 1.0-cm-diameter plastic sphere that has been charged to 10 nC?

51. Your lab assignment is to use positive charge Q to launch a proton, starting from rest, so that it acquires the maximum possible speed. You can launch the proton from the surface of a sphere of positive charge Q and radius R, or from the center of a ring of charge Q and radius R, or from the center of a disk of charge Q and radius R. Which will you choose?

52. An electric dipole consists of 1.0 g spheres charged to ± 2.0 nC at the ends of a 10-cm-long massless rod. The dipole rotates on a frictionless pivot at its center. The dipole is held perpendicular to a uniform electric field with field strength 1000 V/m, then released. What is the dipole's angular velocity at the instant it is aligned with the electric field?

53. A proton is fired from far away toward the nucleus of an iron atom. Iron is element number 26, and the diameter of the nucleus is 9.0 fm. What initial speed does the proton need to just reach the surface of the nucleus? Assume the nucleus remains at rest.

54. A proton is fired from far away toward the nucleus of a mercury atom. Mercury is element number 80, and the diameter of the nucleus is 14.0 fm. If the proton is fired at a speed of 4.0×10^7 m/s, what is its closest approach to the surface of the nucleus? Assume the nucleus remains at rest.

55. In the form of radioactive decay known as *alpha decay*, an unstable nucleus emits a helium-atom nucleus, which is called an *alpha particle*. An alpha particle contains two protons and two neutrons, thus having mass $m = 4$ u and charge $q = 2e$. Suppose a uranium nucleus with 92 protons decays into thorium, with 90 protons, and an alpha particle. The alpha particle is initially at rest at the surface of the thorium nucleus, which is 15 fm in diameter. What is the speed of the alpha particle when it is detected in the laboratory? Assume the thorium nucleus remains at rest.

56. One form of nuclear radiation, *beta decay*, occurs when a neutron changes into a proton, an electron, and a chargeless particle called a *neutrino*: $n \rightarrow p^+ + e^- + \nu$ where ν is the symbol for a neutrino. When this change happens to a neutron within the nucleus of an atom, the proton remains behind in the nucleus while the electron and neutrino are ejected from the nucleus. The ejected electron is called a *beta particle*. One nucleus that exhibits beta decay is the isotope of hydrogen ^{3}H, called *tritium*, whose nucleus consists of one proton (making it hydrogen) and two neutrons (giving tritium an atomic mass $m = 3$ u). Tritium is radioactive, and it decays to helium: ^{3}H $\rightarrow$ ^{3}He $+ e^- + \nu$.
    a. Is charge conserved in the beta decay process? Explain.
    b. Why is the final product a helium atom? Explain.
    c. The nuclei of both ^{3}H and ^{3}He have radii of 1.5×10^{-15} m. With what minimum speed must the electron be ejected if it is to escape from the nucleus and not fall back?

57. Two 10.0-cm-diameter electrodes 0.5 cm apart form a parallel-plate capacitor. The electrodes are attached by metal wires to the terminals of a 15 V battery. After a long time, the capacitor is disconnected from the battery but is not discharged. What are the charge on each electrode, the electric field strength inside the capacitor, and the potential difference between the electrodes
    a. Right after the battery is disconnected?
    b. After insulating handles are used to pull the electrodes away from each other until they are 1.0 cm apart?
    c. After the original electrodes (not the modified electrodes of part b) are expanded until they are 20.0 cm in diameter?

58. Two 10.0-cm-diameter electrodes 0.5 cm apart form a parallel-plate capacitor. The electrodes are attached by metal wires to the terminals of a 15 V battery. What are the charge on each electrode, the electric field strength inside the capacitor, and the potential difference between the electrodes
    a. While the capacitor is attached to the battery?
    b. After insulating handles are used to pull the electrodes away from each other until they are 1.0 cm apart? The electrodes remain connected to the battery during this process.
    c. After the original electrodes (not the modified electrodes of part b) are expanded until they are 20.0 cm in diameter while remaining connected to the battery?

59. a. Find an algebraic expression for the electric field strength E_0 at the surface of a charged sphere in terms of the sphere's potential V_0 and radius R.
    b. What is the electric field strength at the surface of a 1.0-cm-diameter marble charged to 500 V?

60. Two spherical drops of mercury each have a charge of 0.1 nC and a potential of 300 V at the surface. The two drops merge to form a single drop. What is the potential at the surface of the new drop?

61. A Van de Graaff generator is a device for generating a large electric potential by building up charge on a hollow metal sphere. A typical classroom-demonstration model has a diameter of 30.0 cm.
    a. The generator is charged up by placing charge on the *inside* surface of the metal sphere. What happens to the charge after it is placed there? Explain.
    b. How much charge is needed on the sphere for its potential to be 500,000 V?
    c. What is the electric field strength just *inside* and just *outside* the surface of the sphere when it is charged to 500,000 V?

62. A thin spherical shell of radius R has total charge Q. What is the electric potential at the center of the shell?

63. Figure P29.63 shows two uniformly charged spheres. What is the potential difference between points a and b? Which point is at the higher potential?
    Hint: The potential at any point is the superposition of the potentials due to *all* charges.

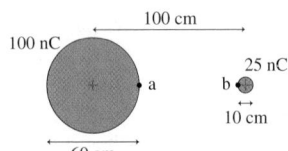

FIGURE P29.63

64. An electric dipole with dipole moment p is oriented along the y-axis.
    a. Find an expression for the electric potential on the y-axis at a point where y is much larger than the charge spacing s. Write your expression in terms of the dipole moment p.
    b. The dipole moment of a water molecule is 6.2×10^{-30} C m. What is the electric potential 1.0 nm from a water molecule along the axis of the dipole?

65. Two positive point charges q are located on the x-axis at $x = \pm\frac{1}{2}s$.
    a. Find an expression for the electric potential at a point on the x-axis where $|x| \ll s$.
    b. Describe the motion of a proton that moves along the x-axis near the origin.

66. Two positive point charges q are located on the y-axis at $y = \pm\frac{1}{2}s$.
    a. Find an expression for the potential along the x-axis.
    b. Draw a graph of V versus x for $-\infty < x < \infty$. For comparison, use a dotted line to show the potential of a point charge $2q$ located at the origin.

67. The arrangement of charges shown in Figure P29.67 is called a *linear electric quadrupole*. The positive charges are located at $y = \pm s$. Notice that the net charge is zero. Find an expression for the electric potential on the y-axis at distances $y \gg s$.

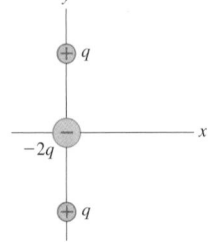

FIGURE P29.67

68. Show that the on-axis potential of a charged disk reduces to the potential of a point charge Q when $z \gg R$.

69. Figure P29.69 shows a thin rod of length L and charge Q. Find an expression for the electric potential a distance x away from the center of the rod on the axis of the rod.

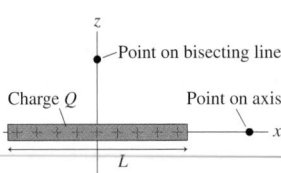

FIGURE P29.69

70. Figure P29.69 showed a thin rod of length L and charge Q. Find an expression for the electric potential a distance z away from the center of rod on the line that bisects the rod.

71. Figure P29.71 shows a thin rod with charge Q that has been bent into a semicircle of radius R. Find an expression for the electric potential at the center.

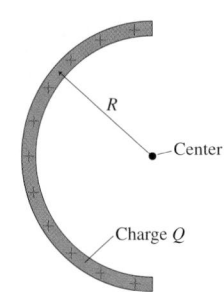

FIGURE P29.71

72. A disk with a hole has inner radius R_{in} and outer radius R_{out}. The disk is uniformly charged with total charge Q. Find an expression for the on-axis electric potential at distance z from the center of the disk. Verify that your expression has the correct behavior when $R_{in} \rightarrow 0$.

In Problems 73 through 76 you are given the equation(s) used to solve a problem. For each of these,
    a. Write a realistic problem for which this is the correct equation(s).
    b. Finish the solution of the problem.

73. $\dfrac{(9.0 \times 10^9 \text{ N m}^2/\text{C}^2)q_1 q_2}{0.030 \text{ m}} = 90 \times 10^{-6}$ J

    $q_1 + q_2 = 40$ nC

74. $\frac{1}{2}(1.67 \times 10^{-27} \text{ kg})(2.5 \times 10^6 \text{ m/s})^2 + 0 =$

    $\frac{1}{2}(1.67 \times 10^{-27} \text{ kg})v_i^2 +$

    $\dfrac{(9.0 \times 10^9 \text{ N m}^2/\text{C}^2)(2.0 \times 10^{-9} \text{ C})(1.60 \times 10^{-19} \text{ C})}{0.0010 \text{ m}}$

75. $\dfrac{100 \text{ V}}{0.0010 \text{ m}} = \dfrac{Q/(0.020 \text{ m})(0.020 \text{ m})}{8.85 \times 10^{-12} \text{ C}^2/\text{N m}^2}$

76. $\dfrac{(9.0 \times 10^9 \text{ N m}^2/\text{C}^2)(3.0 \times 10^{-9} \text{ C})}{0.030 \text{ m}} +$

    $\dfrac{(9.0 \times 10^9 \text{ N m}^2/\text{C}^2)(3.0 \times 10^{-9} \text{ C})}{(0.030 \text{ m}) + d} = 1200 \text{ V}$

Challenge Problems

77. A 1.0 nC charge is at the origin. A -3.0 nC charge is on the x-axis at $x = 4.0$ cm. Find *all* the points in the xy-plane at which the potential is zero. Give your answer as a contour map showing the $V = 0$ V equipotential line.

78. The four 1.0 g spheres shown in Figure CP29.78 are released simultaneously and allowed to move away from each other. What is the speed of each sphere when they are very far apart?

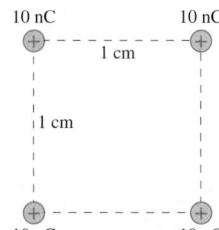

FIGURE CP29.78

79. The 2.0-mm-diameter spheres in Figure CP29.79 are released from rest. What are their speeds v_A and v_B when they are very far apart?

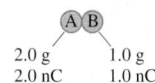

FIGURE CP29.79

80. The 2.0-mm-diameter spheres in Figure CP29.80 are released from rest. What are their speeds v_C and v_D at the instant they collide?

FIGURE CP29.80

81. An electric dipole has dipole moment p. If $r \gg s$, where s is the separation between the charges, show that the electric potential of the dipole can be written

$$V = \frac{1}{4\pi\epsilon_0}\frac{p\cos\theta}{r^2}$$

where r is the distance from the center of the dipole and θ is the angle from the dipole axis.

82. Electrodes of area A are spaced distance d apart to form a parallel-plate capacitor. The electrodes are charged to $\pm q$.
    a. What is the infinitesimal increase in electric potential energy dU if an infinitesimal amount of charge dq is moved from the negative electrode to the positive electrode?
    b. An uncharged capacitor can be charged to $\pm Q$ by transferring charge dq over and over and over. Use your answer to part a to show that the potential energy of a capacitor charged to $\pm Q$ is $U_{cap} = \frac{1}{2}Q\Delta V_C$.

83. A sphere of radius R has charge q.
    a. What is the infinitesimal increase in electric potential energy dU if an infinitesimal amount of charge dq is brought from infinity to the surface of the sphere?
    b. An uncharged sphere can acquire total charge Q by the transfer of charge dq over and over and over. Use your answer to part a to find an expression for the potential energy of a sphere of radius R with total charge Q.
    c. Your answer to part b is the amount of energy needed to assemble a charged sphere. It is often called the *self-energy* of the sphere. What is the self-energy of a proton, assuming it to be a charged sphere with a diameter of 1.0×10^{-15} m?

84. A hollow cylindrical shell of length L and radius R has charge Q uniformly distributed along its length. What is the electric potential at the center of the cylinder?

Stop to Think 29.1: Zero. The motion is always perpendicular to the electric force.

Stop to Think 29.2: $U_b = U_d > U_a = U_c$. The potential energy depends inversely on r. The effects of doubling the charge and doubling the distance cancel each other.

Stop to Think 29.3: c. The proton gains speed by losing potential energy. It loses potential energy by moving in the direction of decreasing electric potential.

Stop to Think 29.4: $V_a = V_b > V_c > V_d = V_e$. The potential decreases steadily from the positive to the negative plate. It depends only on the distance from the positive plate.

Stop to Think 29.5: $\Delta V_{13} = \Delta V_{23} > \Delta V_{12}$. The potential depends only on the *distance* from the charge, not the direction. $\Delta V_{12} = 0$ because these points are at the same distance.

30 Potential and Field

One way to establish a potential difference is with a battery.

▶ Looking Ahead

The goal of Chapter 30 is to understand how the electric potential is connected to the electric field. In this chapter you will learn to:

- Calculate the electric potential from the electric field.
- Calculate the electric field from the electric potential.
- Understand the geometry of the potential and the field.
- Understand and use sources of electric potential.
- Understand and use capacitors.

◀ Looking Back

This chapter continues our exploration of topics that we began in Chapters 28 and 29. Please review:

- Sections 28.3–28.5 Batteries, current, and resistivity.
- Sections 29.4–29.6 The electric potential and its graphical representation.

Everyone is familiar with batteries. You probably have several near you as you read this, in your watch, your calculator, your computer, and your CD player. Batteries are essential for electronics, but just what does a battery actually *do?*

Batteries are just one of several topics that we'll explore as we continue our investigation of the electric potential. The larger issue that we must first address is the connection between the electric potential and the electric field. The potential and the field are not two independent ideas, merely two different perspectives on how the source charges alter the space around them. Exploring the connection between the potential and the field will strengthen your understanding of both.

Our discussion of the electric potential will lead naturally into important applications, including batteries, capacitors, and the current in wires. These applications will set the stage for Chapter 31, on electric circuits.

30.1 Connecting Potential and Field

Chapter 29 introduced the concept of the *electric potential*. To continue our investigation of this important idea, Figure 30.1 shows schematically the four key ideas of force, field, potential energy, and potential. We explored the connection between the force on a particle and the particle's potential energy in Chapters 10 and 11. In Chapter 25, we dealt with the long-range nature of the electric force by generalizing the idea of a force to that of the electric field, and we've based the idea of electric potential on that of potential energy.

The "missing link" is the connection between the electric potential and the electric field, and that is the focus of this chapter. **The electric potential and electric field are not two distinct entities but, instead, two different perspectives or two different mathematical representations of how source charges alter the space around them.**

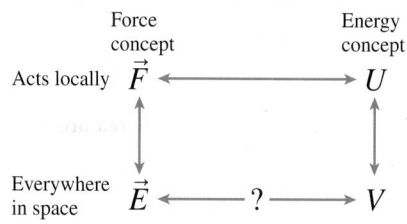

FIGURE 30.1 The four key ideas.

Finding the Potential from the Electric Field

Suppose we know the electric field in some region of space. If the potential and the field are really the same thing seen from two perspectives, we should be able to find the electric potential from the electric field. Chapter 29 introduced all the pieces for doing so, we just need to assemble them into a general statement.

First, we used the potential energy of charge q and the source charges to define the electric potential as

$$V \equiv \frac{U_{q+\text{sources}}}{q} \tag{30.1}$$

Second, we defined the potential energy in terms of the work done by force $\vec{F}$ on charge q as it moves from position i to position f:

$$\Delta U = -W(\text{i} \rightarrow \text{f}) = -\int_{s_i}^{s_f} F_s \, ds \tag{30.2}$$

Third, we found that the force exerted on charge q by the electric field is $\vec{F} = q\vec{E}$. Putting these three pieces together, you can see that the charge q cancels out and the potential difference between two points in space is

$$\Delta V = V(s_f) - V(s_i) = -\int_{s_i}^{s_f} E_s \, ds \tag{30.3}$$

where s is the position along a line from point i to point f.

As a simple example, suppose the electric field is uniform. A constant E_s can be taken outside the integral, and we find that the potential difference between two points is

$$\Delta V = -E_s \Delta s \quad \text{(uniform electric field)} \tag{30.4}$$

The potential difference depends only on the distance between the two points and on the component of $\vec{E}$ along the line between them.

NOTE ▶ Equation 30.3 determines only the potential *difference* ΔV. If you want to assign a specific value of V to a point in space, you must first make a choice of the zero point of the potential. ◀

Equation 30.3 is the basis for a procedure, shown in Tactics Box 30.1 on the next page, that can be used to find the potential difference between two points whenever you already know the electric field.

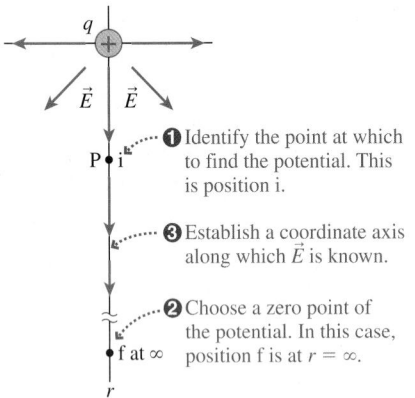

❶ Identify the point at which to find the potential. This is position i.

❸ Establish a coordinate axis along which $\vec{E}$ is known.

❷ Choose a zero point of the potential. In this case, position f is at $r = \infty$.

FIGURE 30.2 Finding the potential of a point charge.

To see how this works, let's use the electric field of a point charge to find its electric potential. Figure 30.2 identifies a point P at which we want to know the potential and calls this position i. In this case, because we want to assign a specific value of V to point P, we've chosen position f to be at $r = \infty$ and identified that as the zero point of the potential. The integration of Equation 30.3 is straight outward along the radial line from i to f:

$$\Delta V = V(\infty) - V(r) = -\int_r^\infty E_r dr \tag{30.5}$$

The electric field is radially outward, parallel to the line of integration, hence

$$E_r = \frac{1}{4\pi\epsilon_0}\frac{q}{r^2}$$

Thus the potential at distance r from a point charge q is

$$V(r) = V(\infty) + \frac{q}{4\pi\epsilon_0}\int_r^\infty \frac{dr}{r^2} = V(\infty) + \frac{q}{4\pi\epsilon_0}\frac{-1}{r}\bigg|_r^\infty = 0 + \frac{1}{4\pi\epsilon_0}\frac{q}{r} \tag{30.6}$$

We've rediscovered the potential of a point charge that you learned in Chapter 29:

$$V_{\text{point charge}} = \frac{1}{4\pi\epsilon_0}\frac{q}{r}$$

EXAMPLE 30.1 **The potential of a disk of charge**
In Chapter 26, the electric field strength at distance z on the axis of a disk of charge Q and radius R was found to be

$$E = \frac{Q}{2\pi R^2\epsilon_0}\left[1 - \frac{z}{(z^2 + R^2)^{1/2}}\right]$$

Find the electric potential on the axis of the disk.

VISUALIZE Figure 30.3 shows the disk, the z-axis, and the point P where we want to know the potential. We've chosen $z = \infty$ as the zero point of the potential.

SOLVE We'll integrate along the z-axis to point f at infinity. The electric field is parallel to the line of integration, so $E_z = E$. Equation 30.3 is

$$V(z) = V(\infty) + \int_z^\infty E_z(z)\,dz$$

$$= 0 + \frac{Q}{2\pi R^2\epsilon_0}\int_z^\infty\left[1 - \frac{z}{(z^2 + R^2)^{1/2}}\right]dz$$

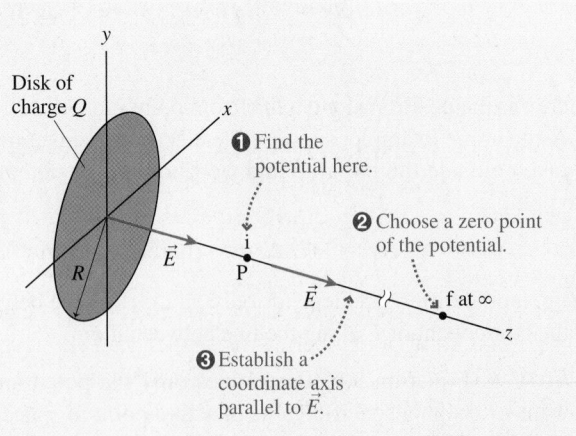

Disk of charge Q

❶ Find the potential here.

❷ Choose a zero point of the potential.

f at ∞

❸ Establish a coordinate axis parallel to $\vec{E}$.

FIGURE 30.3 Finding the on-axis potential of a charged disk.

The integral requires some care with the upper limit. The integral doesn't diverge, as it might appear, because the integrand becomes zero as $z \to \infty$. The second integration is done by changing variables to $u = z^2 + R^2$. Then, with $z\,dz = \frac{1}{2}du$, we find

$$V(z) = \frac{Q}{2\pi R^2 \epsilon_0}\left[\int dz - \int \frac{\frac{1}{2}du}{u^{1/2}}\right]_z^\infty = \frac{Q}{2\pi R^2 \epsilon_0}[z - u^{1/2}]_z^\infty$$

$$= \frac{Q}{2\pi R^2 \epsilon_0}[z - \sqrt{z^2 + R^2}]_z^\infty$$

The two halves of the integral cancel when $z \to \infty$, leaving

$$V_{\text{disk}} = \frac{Q}{2\pi R^2 \epsilon_0}(\sqrt{z^2 + R^2} - z)$$

ASSESS This agrees with Example 29.12, where we found the on-axis potential of a disk of charge by working directly with the continuous distribution of charge. Here we found the potential by explicitly recognizing the connection between the potential and the field.

30.2 Finding the Electric Field from the Potential

Now let's look at the reverse operation: finding the electric field when we know the potential. Figure 30.4 shows two points i and f separated by a very small distance Δs, so small that the electric field is essentially constant over this very short distance. The work done by the electric field as a charge q moves through this small distance is $W = F_s \Delta s = qE_s \Delta s$. Consequently, the potential difference between these two points is

$$\Delta V = \frac{\Delta U_{q+\text{sources}}}{q} = \frac{-W}{q} = -E_s \Delta s \qquad (30.7)$$

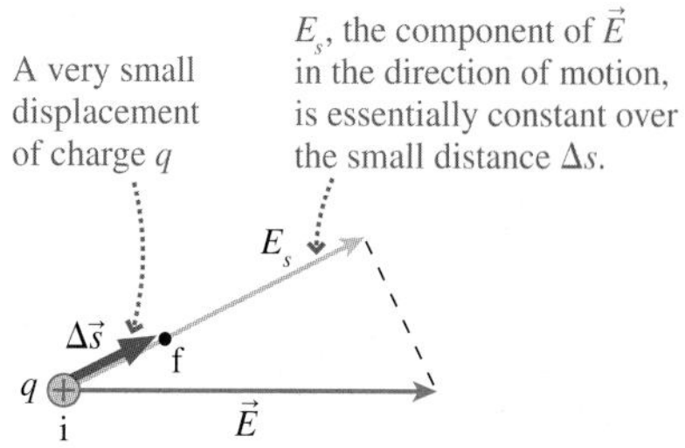

In terms of the potential, the component of the electric field in the s-direction is $E_s = -\Delta V/\Delta s$. In the limit $\Delta s \to 0$,

$$E_s = -\frac{dV}{ds} \qquad (30.8)$$

Now we have reversed Equation 30.3 and have a way to find the electric field from the potential. We'll begin with examples where the field is parallel to a coordinate axis, then we'll look at what Equation 30.8 tells us about the geometry of the field and the potential.

FIGURE 30.4 The electric field does work on charge q.

Actıv
Physıcs   11.12, 11.13

Field Parallel to a Coordinate Axis

The derivative of Equation 30.8 gives E_s, the component of the electric field parallel to the displacement $\Delta \vec{s}$. It doesn't tell us about the electric field component perpendicular to $\Delta \vec{s}$. Thus Equation 30.8 is most useful if we can use symmetry to select a coordinate axis that is parallel to $\vec{E}$ and along which the perpendicular component of $\vec{E}$ is known to be zero.

For example, suppose we knew the potential of a point charge to be $V = q/4\pi\epsilon_0 r$ but didn't remember the electric field. Symmetry requires that the field point straight outward from the charge, with only a radial component E_r. If we choose the s-axis to be in the radial direction, parallel to $\vec{E}$, we can use Equation 30. 8 to find

$$E = E_r = -\frac{dV}{dr} = -\frac{d}{dr}\left(\frac{q}{4\pi\epsilon_0 r}\right) = \frac{1}{4\pi\epsilon_0}\frac{q}{r^2} \qquad (30.9)$$

This is, indeed, the well-known electric field of a point charge.

Equation 30.8 is especially useful for a continuous distribution of charge because calculating V, which is a scalar, is usually much easier than calculating the vector $\vec{E}$ directly from the charge. Once V is known, $\vec{E}$ is found simply by taking a derivative.

EXAMPLE 30.2 The electric field of a ring of charge

In Chapter 29, we found the on-axis potential of a ring of radius R and charge Q to be

$$V_{\text{ring}} = \frac{1}{4\pi\epsilon_0}\frac{Q}{\sqrt{z^2 + R^2}}$$

Find the on-axis electric field of a ring of charge.

SOLVE Symmetry requires the electric field along the axis to point straight outward from the ring with only a z-component E_z. The electric field at distance z is

$$E = E_z = -\frac{dV}{dz} = -\frac{d}{dz}\left(\frac{1}{4\pi\epsilon_0}\frac{Q}{\sqrt{z^2 + R^2}}\right)$$

$$= \frac{1}{4\pi\epsilon_0}\frac{zQ}{(z^2 + R^2)^{3/2}}$$

ASSESS This result is in perfect agreement with the electric field we found in Chapter 26, but this calculation was easier because, unlike in Chapter 26, we didn't have to deal with angles.

A geometric interpretation of Equation 30.8 is that the electric field is the negative of the *slope* of the V-versus-s graph. This interpretation should be familiar. You learned in Chapter 11 that the force on a particle is the negative of the slope of the potential-energy graph: $F = -dU/ds$. In fact, Equation 30.8 is simply $F = -dU/ds$ with both sides divided by q to yield E and V. This geometric interpretation is an important step in developing an understanding of potential.

EXAMPLE 30.3 Finding E from the slope of V

Figure 30.5a is a graph of the electric potential in a region of space where $\vec{E}$ is parallel to the x-axis. Draw a graph of E_x versus x.

MODEL The electric field is the *negative* of the slope of the potential graph.

VISUALIZE Figure 30.5a shows the graph of the potential.

SOLVE There are three regions of different slope:

$0 < x < 2$ cm $\begin{cases} \Delta V/\Delta x = (20 \text{ V})/(0.020 \text{ m}) = 1000 \text{ V/m} \\ E_x = 1{,}000 \text{ V/m} \end{cases}$

$2 < x < 4$ cm $\begin{cases} \Delta V/\Delta x = 0 \text{ V/m} \\ E_x = 0 \text{ V/m} \end{cases}$

$4 < x < 8$ cm $\begin{cases} \Delta V/\Delta x = (-20 \text{ V})/(0.040 \text{ m}) = -500 \text{ V/m} \\ E_x = 500 \text{ V/m} \end{cases}$

The results are shown in Figure 30.5b.

ASSESS The electric field $\vec{E}$ points to the left (E_x is negative) for $0 < x < 2$ cm and to the right (E_x is positive) for $4 < x < 8$ cm. Notice that **the electric field is zero in a region of space where the potential is not changing.**

(a)

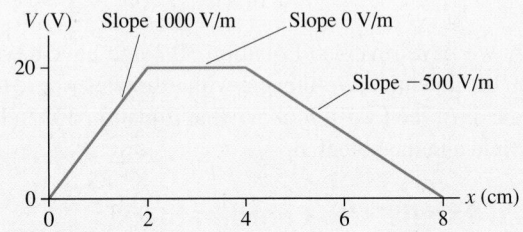

(b)

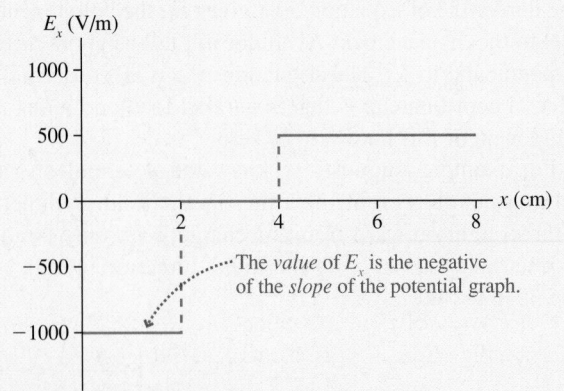

FIGURE 30.5 Graphs of V and E_x versus position x.

STOP TO THINK 30.1   Which potential-energy graph describes this electric field?

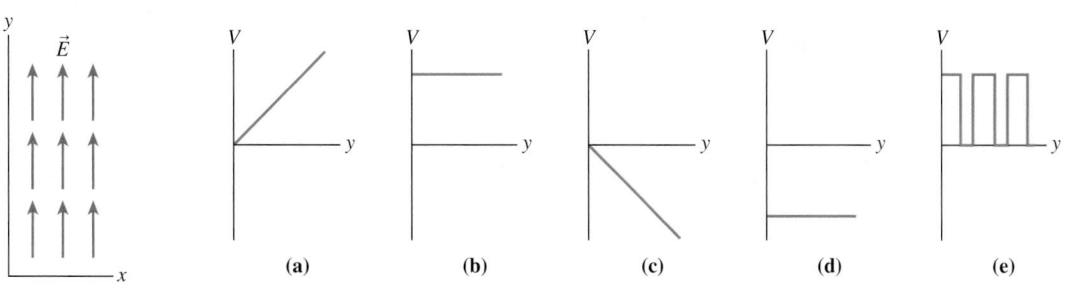

The Geometry of Potential and Field

Equations 30.3 for V in terms of E_s and 30.8 for E_s in terms of V have profound implications for the geometry of the potential and the field. Figure 30.6 shows two equipotential surfaces, with V_+ positive relative to V_-. To learn about the electric field $\vec{E}$ at point P, allow a charge to move through the two displacements $\Delta\vec{s}_1$ and $\Delta\vec{s}_2$. Displacement $\Delta\vec{s}_1$ is *tangent* to the equipotential surface, hence a charge moving in this direction experiences *no* potential difference. According to Equation 30.8, the electric field component along a direction of *constant* potential is $E_s = -dV/ds = 0$. In other words, the electric field component tangent to the equipotential is $E_{\parallel} = 0$.

Displacement $\Delta\vec{s}_2$ is *perpendicular* to the equipotential surface. There is a potential difference along $\Delta\vec{s}_2$, hence the electric field component is

$$E_{\perp} = -\frac{dV}{ds} \approx -\frac{\Delta V}{\Delta s} = -\frac{V_+ - V_-}{\Delta s}$$

You can see that the electric field is inversely proportional to Δs_2, the spacing between the equipotential surfaces. Furthermore, because $(V_+ - V_-) > 0$, the minus sign tells us that the electric field is *opposite* in direction to $\Delta\vec{s}_2$. In other words, $\vec{E}$ points in the direction of *decreasing* potential.

These important ideas about the geometry of the potential and the field are summarized in Figure 30.7.

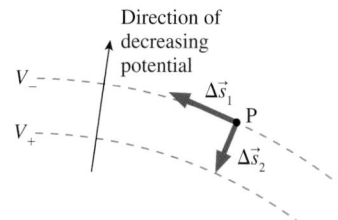

FIGURE 30.6 The electric field at P is related to the shape of the equipotential surfaces.

1. $\vec{E}$ is everywhere perpendicular to the equipotential surfaces.

2. $\vec{E}$ points "downhill," in the direction of decreasing V.

3. The field strength is inversely proportional to the spacing Δs between the equipotential surfaces.

Equipotential surfaces

Direction of decreasing potential

FIGURE 30.7 The geometry of the potential and the field.

Mathematically, we can calculate the individual components of $\vec{E}$ at any point by extending Equation 30.8 to three dimensions:

$$E_x = -\frac{\partial V}{\partial x} \qquad E_y = -\frac{\partial V}{\partial y} \qquad E_z = -\frac{\partial V}{\partial z} \qquad (30.10)$$

where $\partial V/\partial x$ is the partial derivative of V with respect to x while y and z are held constant. More advanced treatments of the electric field make extensive use of this mathematical relationship, but for the most part we'll limit our investigations to those we can analyze graphically.

EXAMPLE 30.4 Finding the electric field from the equipotential surfaces

In Figure 30.8 a 1 cm × 1 cm grid is superimposed on a contour map of the potential. Estimate the strength and direction of the electric field at points 1, 2, and 3. Show your results graphically by drawing the electric field vectors on the contour map.

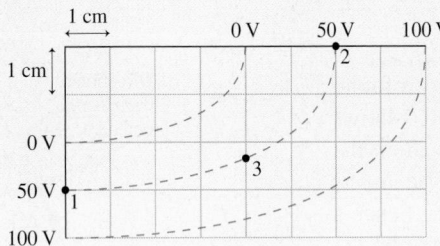

FIGURE 30.8 Equipotential lines.

MODEL The electric field is perpendicular to the equipotential lines, points "downhill," and depends on the slope of the potential hill.

VISUALIZE The potential is highest on the bottom and the right. An elevation graph of the potential would look like the lower-right quarter of a bowl or a football stadium.

SOLVE Some distant but unseen source charges have created an electric field and potential. We do not need to see the source charges to relate the field to the potential. Because $E \approx -\Delta V/\Delta s$, the electric field is stronger where the equipotential lines are closer together and weaker where they are far-

ther apart. If Figure 30.8 were a topographic map, you would interpret the closely spaced contour lines at the bottom of the figure as a steep slope.

Figure 30.9 shows how measurements of Δs from the grid are combined with values of ΔV to determine $\vec{E}$. Point 3 requires an estimate of the spacing between the 0 V and the 100 V surfaces. Notice that we're using the 0 V and 100 V equipotential surfaces to determine $\vec{E}$ at a point on the 50 V equipotential.

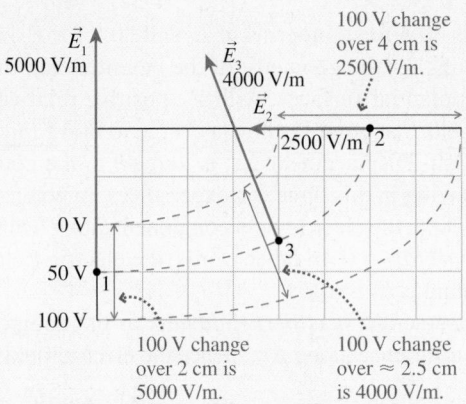

FIGURE 30.9 The electric field at points 1 to 3.

ASSESS The *directions* of $\vec{E}$ are found by drawing downhill vectors perpendicular to the equipotentials. The distances between the equipotential surfaces are needed to determine the field strengths.

Kirchhoff's Loop Law

Chapter 28 introduced *Kirchhoff's junction law,* which said that the sum of the currents flowing into a junction must equal the sum of the currents flowing out. The junction law is really a statement about conservation of charge.

A second law, called *Kirchhoff's loop law,* is a statement about conservation of energy. Figure 30.10 shows two points, 1 and 2, in a region of electric field and potential. You learned in Chapter 29 that the work done in moving a charge between points 1 and 2 is *independent of the path.* Consequently, the potential difference between points 1 and 2 along any two paths that join them is $\Delta V = 20$ V. This must be true in order for the idea of an equipotential surface to make sense.

Now consider the path 1–a–b–c–2–d–1 that ends where it started. What is the potential difference "around" this closed path? The potential increases by 20 V in moving from 1 to 2, but then decreases by 20 V in moving from 2 back to 1. Thus $\Delta V = 0$ V around the closed path.

The numbers are specific to this example, but the idea applies to any loop (i.e., a closed path) through an electric field. The situation is analogous to hiking on the side of a mountain. You may walk uphill during parts of your hike and downhill during other parts, but if you return to your starting point your *net* change of

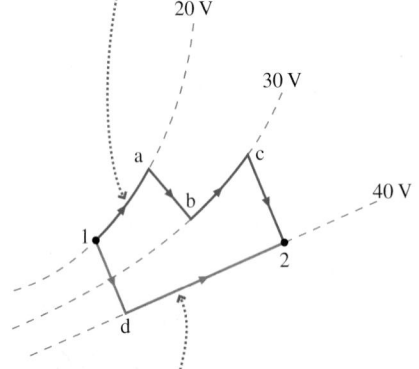

The potential difference along path 1-a-b-c-2 is $\Delta V = 0$ V + 10 V + 0 V + 10 V = 20 V.

The potential difference along path 1-d-2 is $\Delta V = 20$ V + 0 V = 20 V.

FIGURE 30.10 The potential difference between points 1 and 2 is the same along either path.

elevation is zero. So for any path that starts and ends at the same point, we can conclude that

$$\Delta V_{\text{loop}} = \sum_i (\Delta V)_i = 0 \qquad (30.11)$$

Stated in words, **the sum of all the potential differences encountered while moving around a loop or closed path is zero.** This is **Kirchhoff's loop law.**

Kirchhoff's loop law is a statement of energy conservation, because a charge that moves around a loop and returns to its starting point has $\Delta U = q\Delta V = 0$. Kirchhoff's loop law and Kirchhoff's junction law will turn out to be the two fundamental principles of circuit analysis.

STOP TO THINK 30.2   Which set of equipotential surfaces matches this electric field?

$\vec{E}$

(a)   0 V .... 50 V

(b)   0 V .... 50 V

(c)   0 V .... 50 V

(d)   50 V .... 0 V

(e)   50 V .... 0 V

(f)   50 V .... 0 V

30.3 A Conductor in Electrostatic Equilibrium

The basic relationships between potential and field allow us to draw some interesting and important conclusions about conductors. Consider a conductor, such as a metal, that is in electrostatic equilibrium. The conductor may be charged, but all the charges are at rest.

You learned in Chapter 25 that any excess charges on a conductor in electrostatic equilibrium are always located on the *surface* of the conductor. Using similar reasoning, we can conclude that **the electric field is zero at any interior point of a conductor in electrostatic equilibrium.** Why? If the field were other than zero, then there would be a force $\vec{F} = q\vec{E}$ on the charge carriers and they would move, creating a current. But there are no currents in a conductor in electrostatic equilibrium, so it must be that $\vec{E} = \vec{0}$ at all interior points.

NOTE ▶ $\vec{E} = \vec{0}$ inside a conductor in electrostatic equilibrium is in contrast with a current-carrying conductor, where there most definitely *is* an interior electric field to create the current density $J = \sigma E$. ◀

The two points inside the conductor in Figure 30.11 are connected by a line that remains entirely inside the conductor. We can find the potential difference $\Delta V = V_2 - V_1$ between these points by using Equation 30.3 to integrate E_s along the line from 1 to 2. But $E_s = 0$ at all points along the line, because $\vec{E} = \vec{0}$, thus the value of the integral is zero and $\Delta V = 0$. In other words, **any two points inside a conductor in electrostatic equilibrium are at the same potential.**

When a conductor is in electrostatic equilibrium, the *entire conductor* is at the same potential. If we charge a metal sphere, then the entire sphere is at a single potential. Similarly, a charged metal rod or wire is at a single potential *if* it is in electrostatic equilibrium. (This conclusion does *not* apply to a current-carrying wire because it is not in electrostatic equilibrium.)

A corona discharge, with crackling noises and glimmers of light, occurs at pointed metal tips where the electric field can be very strong.

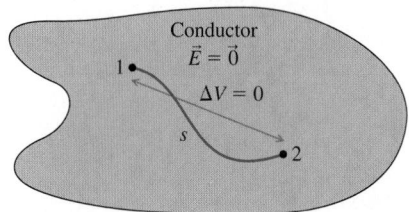

FIGURE 30.11 All points inside a conductor in electrostatic equilibrium are at the same potential.

If $\vec{E} = \vec{0}$ inside a charged conductor but $\vec{E} \neq \vec{0}$ outside, what happens right at the surface? If the entire conductor is at the same potential, then the surface is an equipotential surface. You have seen that the electric field is always perpendicular to an equipotential surface, hence **the exterior electric field $\vec{E}$ of a charged conductor is perpendicular to the surface.**

Figure 30.12 summarizes what we know about conductors in electrostatic equilibrium. Item 6, that the charge density and thus the electric field strength are largest at "sharp points," we'll assert without proof. These are important and practical conclusions because conductors are the primary components of electrical devices.

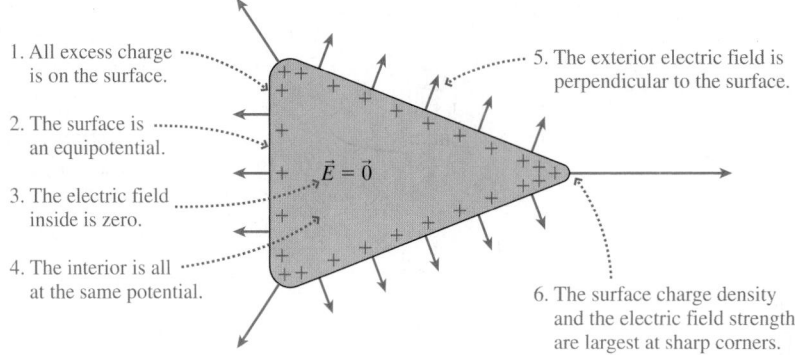

1. All excess charge is on the surface.

2. The surface is an equipotential.

3. The electric field inside is zero.

4. The interior is all at the same potential.

5. The exterior electric field is perpendicular to the surface.

6. The surface charge density and the electric field strength are largest at sharp corners.

$\vec{E} = \vec{0}$

FIGURE 30.12 Electric properties of a conductor in electrostatic equilibrium.

We can use similar reasoning to estimate the electric field and potential in between two charged conductors. As an example, Figure 30.13 shows a positively charged metal sphere near a flat metal plate. The surfaces of the sphere and the flat plate are equipotentials, hence the electric field must be perpendicular to both. Close to a surface, the electric field is still *nearly* perpendicular to the surface. Consequently, **an equipotential surface close to an electrode must roughly match the shape of the electrode.**

In between, the equipotential surfaces *gradually* change as they "morph" from one electrode shape to the other. It's not hard to sketch a contour map showing a plausible set of equipotential surfaces. You can then draw electric field lines (field lines are usually easier to draw than field vectors in these situations) that are perpendicular to the equipotentials, point "downhill," and are closer together where the contour line spacing is smaller.

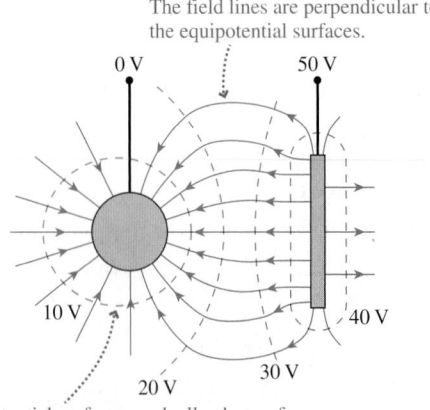

The field lines are perpendicular to the equipotential surfaces.

0 V   50 V

10 V   40 V

20 V   30 V

The equipotential surfaces gradually change from the shape of one electrode to the shape of the other.

FIGURE 30.13 Estimating the electric field and potential between two charged conductors.

STOP TO THINK 30.3   Three charged metal spheres of different radii are connected by a thin metal wire. The potential and electric field at the surface of each sphere are V and E. Which of the following is true?

a. $V_1 = V_2 = V_3$ and $E_1 = E_2 = E_3$ b. $V_1 = V_2 = V_3$ and $E_1 > E_2 > E_3$

c. $V_1 > V_2 > V_3$ and $E_1 = E_2 = E_3$ d. $V_1 > V_2 > V_3$ and $E_1 > E_2 > E_3$

e. $V_3 > V_2 > V_1$ and $E_3 = E_2 = E_1$ f. $V_3 > V_2 > V_1$ and $E_3 > E_2 > E_1$

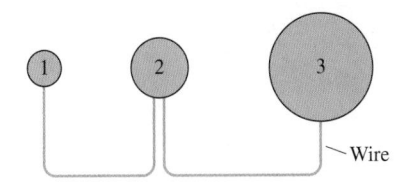

30.4 Sources of Electric Potential

A *separation of charge* creates an electric potential difference. Shuffling your feet on the carpet transfers electrons from the carpet to you, creating a potential difference between you and a doorknob that causes a spark and a shock as you touch it. Charging a capacitor by moving electrons from one plate to the other creates a potential difference across the capacitor.

In fact, as Figure 30.14 shows, *any* separation of charge causes a potential difference. The charge separation between the two electrodes creates an electric field $\vec{E}$ pointing from the positive toward the negative electrode. As a consequence, there is a potential difference between the electrodes that is given by

$$\Delta V = V_{\text{pos}} - V_{\text{neg}} = -\int_{\text{neg}}^{\text{pos}} E_s \, ds$$

where the integral runs from any point on the negative electrode to any point on the positive. The key idea is that **we can create a potential difference by creating a charge separation.**

One source of electric potential is a charged capacitor. However, you learned in Chapter 28 that the charge separation cannot be maintained if we "use" the capacitor. In Figure 30.15, each charge in the discharge current I loses potential energy as it "falls" through the wire, reaching the negative plate with $U_{\text{bottom}} = 0$. A charge at the negative plate has no means to return to the other side.

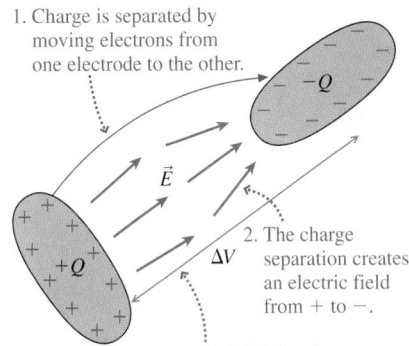

1. Charge is separated by moving electrons from one electrode to the other.

2. The charge separation creates an electric field from + to −.

3. Because of the electric field, there's a potential difference between the electrodes.

FIGURE 30.14 A charge separation creates a potential difference.

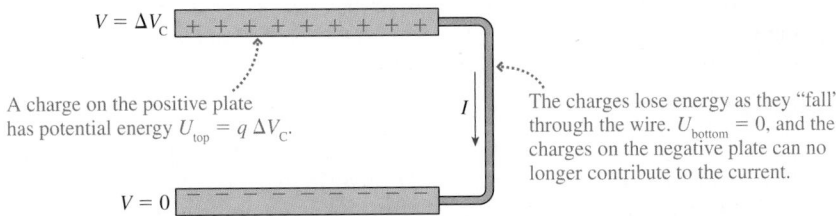

A charge on the positive plate has potential energy $U_{\text{top}} = q \, \Delta V_C$.

The charges lose energy as they "fall" through the wire. $U_{\text{bottom}} = 0$, and the charges on the negative plate can no longer contribute to the current.

FIGURE 30.15 A charged capacitor is a source of potential difference, but it cannot sustain a current.

To be practical, a source of potential needs a way to "lift" charges from the negative electrode back to the positive electrode where they can be reused. Once there, they could again "fall" through the wire and sustain the current. A device that continuously separates charge, moving it from the negative to the positive electrode, can *sustain* the potential difference and thus sustain a current in the wire.

The **Van de Graaff generator** shown in Figure 30.16a on the next page is a mechanical charge separator—essentially a fancy foot shuffler. A moving plastic or leather belt is charged, then the charge is mechanically transported via the conveyor belt to the spherical electrode at the top of the insulating column. The charging of the belt could be done by friction, but in practice a *corona discharge* due to the strong electric field at the tip of a needle is more efficient and reliable.

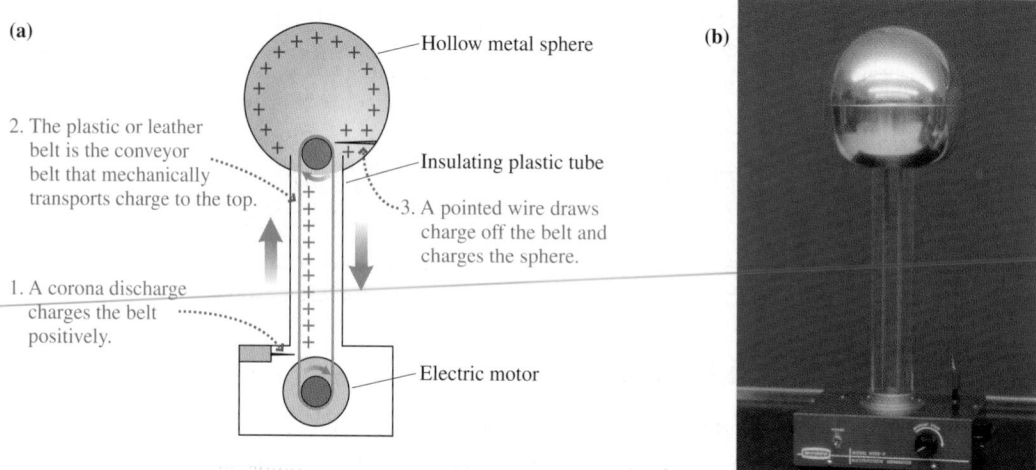

(a)

Hollow metal sphere

2. The plastic or leather belt is the conveyor belt that mechanically transports charge to the top.

Insulating plastic tube

3. A pointed wire draws charge off the belt and charges the sphere.

1. A corona discharge charges the belt positively.

Electric motor

(b)

FIGURE 30.16 A Van de Graaff generator.

A Van de Graaff generator has two noteworthy features:

■ Charge is mechanically transported from the negative side to the positive side. This charge separation creates *and sustains* a potential difference between the spherical electrode and its surroundings.

■ The electric field of the spherical electrode exerts a downward force on the positive charges moving up the belt. Consequently, *work must be done* to "lift" the positive charges. The work is done by the electric motor that runs the belt.

A classroom demonstration Van de Graaff generator like the one shown in Figure 30.16b creates a potential difference of several hundred thousand volts between the upper sphere and its surroundings. The maximum potential is reached when the electric field near the sphere becomes large enough to cause a breakdown of the air. This produces a spark and temporarily discharges the sphere. A large Van de Graaff generator surrounded by vacuum can reach a potential of 20 MV or more. These generators are used to accelerate protons for nuclear physics experiments.

For us, the Van de Graaff generator is important because it shows that a potential difference can be created and sustained by a device that separates charge.

Batteries and EMF

The most common source of electric potential is a **battery.** A battery consists of chemicals, called *electrolytes,* sandwiched between two electrodes made of different metals. Chemical reactions in the electrolytes separate charge by moving positive ions to one electrode and negative ions to the other. In other words, chemical reactions, rather than a mechanical conveyor belt, transport charge from one electrode to the other. The procedure is different, but the outcome is the same: a potential difference.

Figure 30.17 reminds you of the **charge escalator** model of a battery that we introduced in Section 28.3. The escalator "lifts" positive charges from the negative terminal to the positive terminal, where they can be reused. Lifting positive charges to a positive terminal requires that work be done, and the chemical reactions within the battery provide the energy to do this work. When the chemicals are used up, the reactions cease, and the battery is dead.

By separating the charge, the charge escalator establishes a potential difference ΔV_{bat} between the terminals. The value of ΔV_{bat} is determined by the specific chemical reactions employed by the battery. To see how, suppose the chemical

The charge "falls downhill" through the wire, but it can be sustained because of the charge escalator.

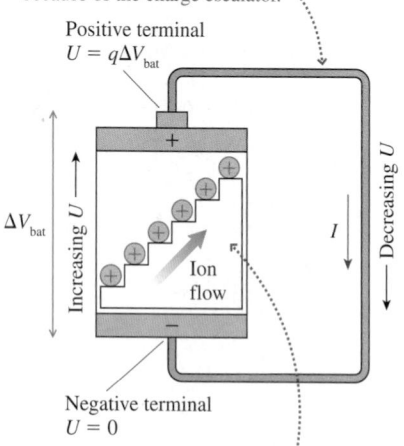

Positive terminal
$U = q\Delta V_{bat}$

ΔV_{bat}

Increasing U

Decreasing U

I

Ion flow

Negative terminal
$U = 0$

The charge escalator "lifts" charge from the negative side to the positive side. Charge q gains energy $\Delta U = q\Delta V_{bat}$.

FIGURE 30.17 The charge escalator model of a battery.

reactions do work W_{chem} to move charge q from the negative to the positive terminal. In an **ideal battery,** in which there are no internal energy losses, the charge gains electric potential energy $\Delta U = W_{chem}$. This is analogous to a book gaining gravitational potential energy as you do work to lift it from the floor to a shelf.

The quantity W_{chem}/q, which is the work done *per charge* by the charge escalator, is called the **emf** of the battery, pronounced as the sequence of three letters "e-m-f." The symbol for emf is $\mathcal{E}$, a script E, and the units are those of the electric potential: joules per coulomb, or volts. The *rating* of a battery, such as 1.5 V or 9 V, is the battery's emf. Originally the term emf was an abbreviation of "electromotive force." That is an outdated term (work per charge is not a force!), so today we just call it emf and it is not an abbreviation of anything.

NOTE ▶ The term *emf,* often capitalized as EMF, is widely used in popular science articles in newspapers and magazines to mean "electromagnetic field." You may have seen the abbreviation EMF if you've read about the debate over whether electric transmission lines, which generate electromagnetic fields, are a health hazard. This is *not* how we will use the term *emf.* ◀

By definition, the electric potential is related to the electric potential energy of charge q by $\Delta V = \Delta U/q$. But $\Delta U = W_{chem}$ for the charges in a battery, hence the potential difference between the terminals of an ideal battery is

$$\Delta V_{bat} = \frac{W_{chem}}{q} = \mathcal{E} \qquad \text{(ideal battery)} \qquad (30.12)$$

In other words, a battery constructed to have an emf of 1.5 V (i.e., the chemical reactions do 1.5 J of work to separate 1 C of charge) creates a 1.5 V potential difference. In practice, the measured potential difference ΔV_{bat} between the terminals of a real battery, called the **terminal voltage,** is usually slightly less than $\mathcal{E}$. You will learn the reason for this in the next chapter.

Electric generators, photocells, and other sources of potential difference use different means to separate charges, but otherwise they function exactly the same as a battery. The common feature of all such devices is that they use a *nonelectrical* means to separate charge and, thus, to create a potential difference. The emf $\mathcal{E}$ of any device is the work done per charge to separate the charge.

30.5 Connecting Potential and Current

An important consequence of the charge escalator model is that **a battery is a source of potential difference.** It is true that charges flow through a wire that connects the battery terminals, but this current is a *consequence* of the battery's potential difference. You can think of the battery's emf as being the *cause.* Current, heat, light, sound, and so on are all *effects* that happen when the battery is used in certain ways. Distinguishing between cause and effect will be vitally important for understanding how a battery functions in a circuit.

Our goal in this section is to find the connection between potential and current. Let's start by connecting a wire between the two terminals of a battery, as shown in Figure 30.18. You learned in Section 30.2 that the potential difference between any two points is independent of the path between them. Consequently, the potential difference between the two ends of the wire, along a path through the wire, is equal to the potential difference between the two terminals of the battery:

$$\Delta V_{wire} = \Delta V_{bat} \qquad (30.13)$$

Thus the battery is the *source* of a potential difference between the ends of the wire.

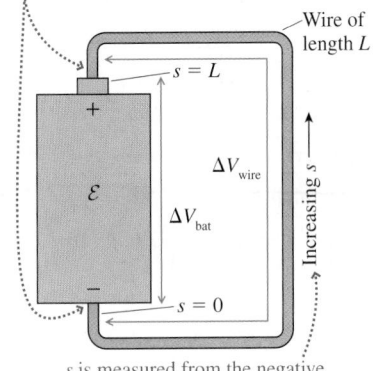

The potential difference between these two points is the same whether you go through the battery or through the wire.

Wire of length L

$s = L$

ΔV_{wire}

ΔV_{bat}

$\mathcal{E}$

Increasing s

$s = 0$

s is measured from the negative terminal, where $s = 0$, to the positive terminal, where $s = L$.

FIGURE 30.18 The potential difference along the wire is the same as the potential difference between the battery terminals.

(a) The potential difference between the ends of the wire establishes an electric field inside the wire.

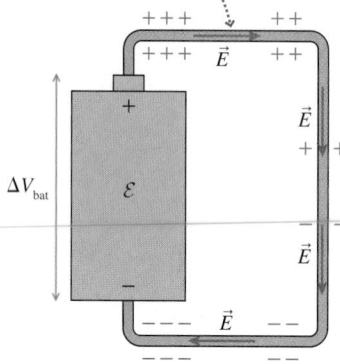

(b) The electric field drives a current through the wire.

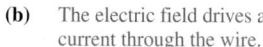

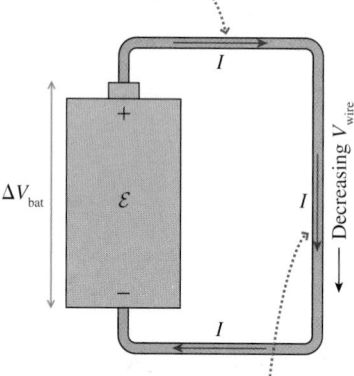

The current is in the direction of decreasing potential.

FIGURE 30.19 The electric field and the current inside the wire.

TABLE 30.1 Resistivity of materials

Material	Resistivity (Ω m)
Aluminum	2.8×10^{-8}
Copper	1.7×10^{-8}
Gold	2.4×10^{-8}
Iron	9.7×10^{-8}
Silver	1.6×10^{-8}
Tungsten	5.6×10^{-8}
Nichrome*	1.5×10^{-6}
Carbon	3.5×10^{-5}

*Nickel-chromium alloy used for heating wires

If there is a potential difference between the ends of the wire, there must be an electric field *inside* the wire. And if there's an internal electric field, a current will flow! The wire's electric field is related to the potential difference ΔV_{wire} by Equation 30.3:

$$\Delta V_{\text{wire}} = -\int_0^L E_s\, ds \qquad (30.14)$$

where L is the length of the wire and s is measured along the wire from $s = 0$ at the battery's negative end to $s = L$ at the positive end.

The field vector $\vec{E}_{\text{wire}}$ inside the wire points "downhill," from the positive toward the negative end of the wire. Thus the s-component of $\vec{E}_{\text{wire}}$, along the line of integration, is $E_s = -E_{\text{wire}}$. You learned in Chapter 28 that the electric field inside a constant-diameter wire is constant (a consequence of conservation of current), so E_{wire} can be taken outside the integral. Equation 30.14 is then

$$\Delta V_{\text{wire}} = -\int_0^L (-E_{\text{wire})}\, ds = E_{\text{wire}}\int_0^L ds = E_{\text{wire}}L \qquad (30.15)$$

Hence, the electric field strength inside the wire is

$$E_{\text{wire}} = \frac{\Delta V_{\text{wire}}}{L} \qquad (30.16)$$

Equation 30.16 is an important result. The electric field strength inside a constant-diameter wire—the field that drives the current forward—is simply the potential difference between the ends of the wire divided by the length of the wire. And ΔV_{wire} can be established by connecting the wire to a battery.

Figure 30.19a shows the electric field $\vec{E}_{\text{wire}}$ inside the wire. Notice how the surface charges on the wire—charges supplied by the battery—cause $\vec{E}$ to follow the wire, as you learned in Chapter 28. The electric field drives the current I shown in Figure 30.19b. Notice that the field points in the direction of *decreasing* potential, implying that **the current is in the direction of decreasing potential.**

Now we can use E_{wire} to find the current I in the wire. You learned in Chapter 28 that the current density in a wire of conductivity σ is $J = \sigma E_{\text{wire}}$. The current density is $J = I/A$, where A is the cross-section area of the wire. Thus

$$I = AJ = A\sigma E_{\text{wire}} = \frac{A}{\rho}E_{\text{wire}} \qquad (30.17)$$

where $\rho = 1/\sigma$ is the resistivity. Values of σ and ρ were tabulated for several metals in Table 28.2; Table 30.1 reproduces only the resistivity values.

The electric field strength in the wire is $E_{\text{wire}} = \Delta V_{\text{wire}}/L$, hence the current is

$$I = \frac{A}{\rho L}\Delta V_{\text{wire}} \qquad (30.18)$$

You can see that the **current in the wire is proportional to the potential difference between the ends of the wire.** Equation 30.18 is the connection we were seeking between potential and current, but we can cast it into a more useful form if we define the **resistance** of the wire as

$$R = \frac{\rho L}{A} \qquad (30.19)$$

The resistance is a property of a *specific* wire or conductor because it depends on the conductor's length and diameter as well as on the resistivity of the material from which it is made.

The SI unit of resistance is the **ohm,** defined as

$$1 \text{ ohm} = 1 \, \Omega \equiv 1 \text{ V/A}$$

where Ω is an uppercase Greek omega. The ohm is the basic unit of resistance, although kilohms ($1 \text{ k}\Omega = 10^3 \, \Omega$) and megohms ($1 \text{ M}\Omega = 10^6 \, \Omega$) are widely used. You can see from Equation 30.19 why the resistivity ρ has units of Ω m.

The resistance of a wire increases as the length increases. This seems reasonable, because it should be harder to push electrons through a long wire than a short one. Decreasing the cross-section area also increases the resistance. This again seems reasonable, because the same electric field can push more electrons through a fat wire than a skinny one.

NOTE ▶ It is important to distinguish between resistivity and resistance. *Resistivity* describes just the *material,* not any particular piece of it. *Resistance* characterizes a specific piece of the conductor having a specific geometry. The relationship between resistivity and resistance is analogous to that between mass density and mass. ◀

The definition of resistance allows us to write the current in a wire as

$$I = \frac{\Delta V_{\text{wire}}}{R} \qquad \text{(current in a wire)} \qquad (30.20)$$

In other words, establishing a potential difference ΔV_{wire} between the ends of a wire of resistance R creates an electric field that, in turn, causes a current $I = \Delta V_{\text{wire}}/R$ in the wire. The smaller the resistance, the larger the current.

EXAMPLE 30.5 **The current in a wire**
A 20-cm-long, 1.0-mm-diameter nichrome wire is connected to the terminals of a 1.5 V battery. What are the electric field and the current in the wire?

MODEL Assume the battery is an ideal battery, with $\Delta V_{\text{bat}} = \mathcal{E} = 1.5$ V.

SOLVE Connecting the wire to the battery makes $\Delta V_{\text{wire}} = \Delta V_{\text{bat}} = 1.5$ V. The electric field inside the wire is

$$E_{\text{wire}} = \frac{\Delta V_{\text{wire}}}{L} = \frac{1.5 \text{ V}}{0.20 \text{ m}} = 7.5 \text{ V/m}$$

This is not a very large field. The wire's resistance is

$$R = \frac{\rho L}{A} = \frac{\rho L}{\pi r^2}$$

From Table 30.1 we find that $\rho = 1.5 \times 10^{-6} \, \Omega$ m for nichrome. After converting L and r to meters, we find $R = 0.382 \, \Omega$. Thus the current in the wire is

$$I = \frac{\Delta V}{R} = \frac{1.5 \text{ V}}{0.382 \, \Omega} = 3.93 \text{ A}$$

The next chapter will look more carefully at currents in circuits. Our purpose here was to establish the connection between a source of potential difference and the current in a wire. The cause-and-effect sequence is the main idea.

1. A battery is a source of potential difference ΔV_{bat}. An ideal battery has $\Delta V_{\text{bat}} = \mathcal{E}$.
2. The battery creates a potential difference $\Delta V_{\text{wire}} = \Delta V_{\text{bat}}$ between the ends of a wire.
3. The potential difference ΔV_{wire} causes an electric field $E = \Delta V_{\text{wire}}/L$ in the wire.
4. The electric field establishes a current $I = JA = \sigma AE$ in the wire.
5. The magnitude of the current is determined *jointly* by the battery and the wire's resistance R to be $I = \Delta V_{\text{wire}}/R$.

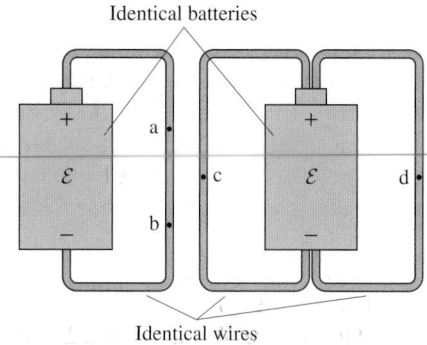

Identical batteries

Identical wires

30.6 Capacitance and Capacitors

We introduced the parallel-plate capacitor in Chapter 26 and have made frequent use of it since. We've assumed that the capacitor is charged, but we haven't really addressed the issue of *how* it gets charged. Figure 30.20 shows the two plates of a capacitor connected with conducting wires to the two terminals of a battery. What happens? And how is the potential difference ΔV_C across the capacitor related to the battery's potential difference ΔV_{bat}?

Figure 30.20a shows the situation shortly after the capacitor is connected and before it is fully charged. According to Kirchhoff's loop law, Equation 30.11, the sum of the four potential differences around the closed path must be zero. However, we have to be careful with signs. The potential increases (becomes more positive) from the negative to the positive terminal of the battery (ΔV_{bat}) as the charges move "uphill" on the charge escalator. The potential then decreases (becomes more negative) from the positive plate of the battery to the capacitor (ΔV_1), from the positive to the negative plate of the capacitor (ΔV_C), and from

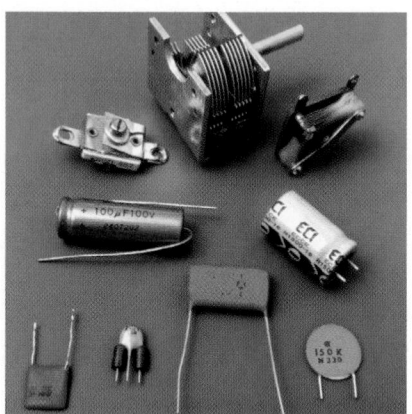

Capacitors are important elements in electric circuits. They come in a variety of sizes and shapes.

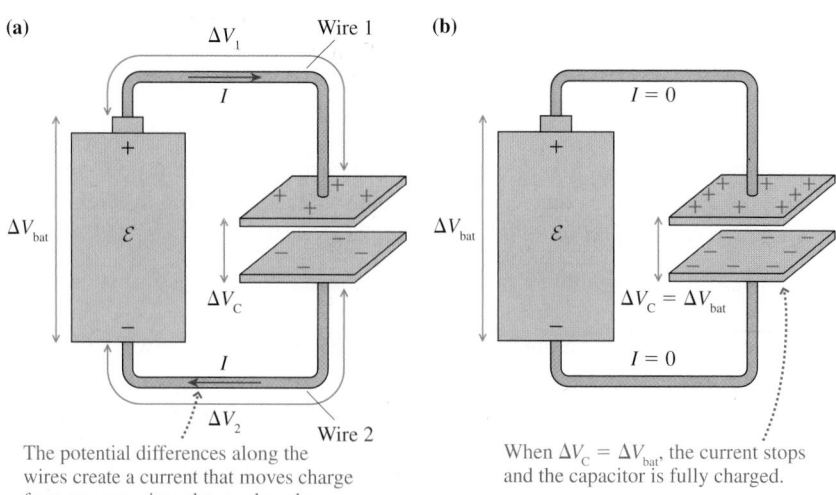

The potential differences along the wires create a current that moves charge from one capacitor plate to the other.

When $\Delta V_C = \Delta V_{bat}$, the current stops and the capacitor is fully charged.

FIGURE 30.20 A parallel-plate capacitor is charged by a battery.

the negative plate of the capacitor back to the battery (ΔV_2). Thus the requirement that the potential return to its starting point after a complete loop is

$$\Delta V_{\text{bat}} - \Delta V_1 - \Delta V_C - \Delta V_2 = 0 \tag{30.21}$$

Initially, when the capacitor is uncharged, $\Delta V_C = 0$. The potential differences along the wires are $\Delta V_1 = \Delta V_2 = \frac{1}{2}\Delta V_{\text{bat}}$ if we assume the wires are equal. These potential differences establish currents in the wires, and charges begin to flow through them. The current in wire 2 carries positive charge from the lower capacitor plate to the negative terminal of the battery. The charge escalator in the battery lifts this charge to the positive terminal and sends it as a current through wire 1 to the upper capacitor plate. The net effect is that the battery removes charge from one plate of the capacitor and transfers it to the other plate. In other words, the battery "charges the capacitor."

As the charge on the capacitor plates increases, the potential difference ΔV_C increases and the potential differences along the wire decrease. As ΔV_1 and ΔV_2 decrease, so does the current in the wire. When ΔV_1 and ΔV_2 reach zero, the current stops and the capacitor is fully charged. This is the situation in Figure 30.20b, where $\Delta V_C = \Delta V_{\text{bat}}$. The capacitor stops charging when $\Delta V_C = \Delta V_{\text{bat}}$ because there is no longer a potential *difference* along the wires to cause any further current.

You learned in Chapter 29 that a parallel-plate capacitor's potential difference is related to the electric field inside by $\Delta V_C = Ed$, where d is the separation between the plates. And you know from Chapter 26 that a capacitor's electric field is

$$E = \frac{Q}{\epsilon_0 A} \tag{30.22}$$

where A is the surface area of the plates. Combining these gives

$$Q = \frac{\epsilon_0 A}{d}\Delta V_C \tag{30.23}$$

In other words, **the charge on the capacitor plates is directly proportional to the potential difference between the plates.**

The ratio of the charge Q to the potential difference ΔV_C is called the **capacitance** C:

$$C = \frac{Q}{\Delta V_C} = \frac{\epsilon_0 A}{d} \qquad \text{(parallel-plate capacitor)} \tag{30.24}$$

Capacitance is a purely *geometric* property of two electrodes because it depends only on their surface area and spacing. The SI unit of capacitance is the **farad,** named in honor of Michael Faraday. One farad is defined as

$$1 \text{ farad} = 1 \text{ F} \equiv 1 \text{ C/V}$$

One farad turns out to be an enormous amount of capacitance. Practical capacitors are usually measured in units of microfarads (μF) or picofarads (1 pF $= 10^{-12}$ F).

With this definition of capacitance, Equation 30.24 can be written

$$Q = C\,\Delta V_C \qquad \text{(charge on a capacitor)} \tag{30.25}$$

Equation 30.25 for the charge on a capacitor is analogous to Equation 30.20 for the current in a wire, $I = \Delta V_{\text{wire}}/R$. There we found that the current in a wire is determined jointly by the potential difference supplied by a battery *and* a property of the wire called resistance. Now we see that the charge on a capacitor is determined jointly by the potential difference supplied by a battery *and* a property of the electrodes called capacitance.

EXAMPLE 30.6 **Charging a capacitor**

The spacing between the plates of a 1.0 μF capacitor is 0.050 mm.

a. What is the surface area of the plates?
b. How much charge is on the plates if this capacitor is attached to a 1.5 V battery?

MODEL Assume the battery is ideal and that the capacitor is a parallel-plate capacitor.

SOLVE

a. From the definition of capacitance,

$$A = \frac{dC}{\epsilon_0} = 5.65 \text{ m}^2$$

b. The charge is $Q = C\Delta V_C = 1.5 \times 10^{-6} \text{ C} = 1.5 \mu\text{C}$.

ASSESS The surface area needed to construct a 1.0 μF capacitor (a fairly typical value) is enormous. In practice, two very large sheets of thin metal foil are separated by a thin insulator ($d \approx 0.05$ mm). The sheets are then rolled up to form a cylinder. The insulator and the rolling do have some effect on the capacitance, which you can learn about in more advanced courses.

Forming a Capacitor

The parallel-plate capacitor is important because it is straightforward to analyze and it produces a uniform electric field. But capacitors and capacitance are not limited to flat, parallel electrodes. *Any* two electrodes, regardless of their shape, form a capacitor.

Figure 30.21 shows two arbitrary electrodes charged to $\pm Q$. The net charge, as was the case with a parallel-plate capacitor, is zero. By definition, the capacitance of the two electrodes is

$$C = \frac{Q}{\Delta V_C} \tag{30.26}$$

where ΔV_C is the potential difference between the positive and negative electrode. It might appear that the capacitance depends on the amount of charge, but the potential difference is proportional to Q. Consequently, **the capacitance depends only on the geometry of the electrodes.**

To make use of Equation 30.26, we must be able to determine the potential difference between the electrodes when they are charged to $\pm Q$. The following example shows how this is done.

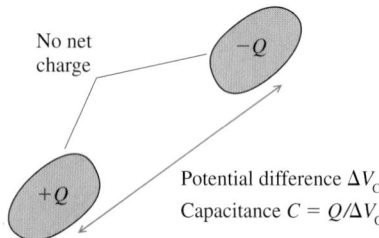

No net charge

Potential difference ΔV_C
Capacitance $C = Q/\Delta V_C$

FIGURE 30.21 Any two electrodes form a capacitor.

EXAMPLE 30.7 **A spherical capacitor**

A metal sphere of radius R_1 is inside and concentric with a hollow metal sphere of radius R_2. What is the capacitance of this spherical capacitor?

MODEL Assume the inner sphere is negative and the outer is positive.

VISUALIZE Figure 30.22 shows the two spheres. The electric field between them points from the positive outer sphere to the inner negative sphere.

SOLVE You might think we could find the potential difference between the spheres by using the Chapter 29 result for the potential of a charged sphere. However, that was the potential of an *isolated* charged sphere. To find the potential difference between *two* spheres we need to use Equation 30.3:

$$\Delta V = V(s_f) - V(s_i) = -\int_{s_i}^{s_f} E_s \, ds$$

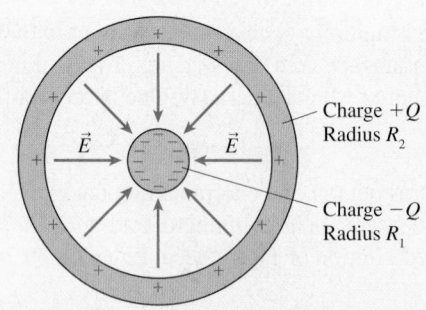

Charge $+Q$
Radius R_2

Charge $-Q$
Radius R_1

FIGURE 30.22 A spherical capacitor.

The electric field between the spheres is the superposition of the fields of the inner and outer sphere. The field of the inner sphere is that of point charge $-Q$, while, from Gauss's law, the interior field of the outer sphere is zero. If we integrate along a radial line from $r_i = R_1$ on the inner sphere to $r_f = R_2$ on the outer sphere, the potential difference between them is

$$\Delta V_C = -\int_{R_1}^{R_2}\left(\frac{-Q}{4\pi\epsilon_0 r^2}\right)dr = \frac{Q}{4\pi\epsilon_0}\int_{R_1}^{R_2}\frac{dr}{r^2} = \frac{Q}{4\pi\epsilon_0}\left(\frac{1}{R_1} - \frac{1}{R_2}\right)$$

Then, from the definition of capacitance,

$$C = \frac{Q}{\Delta V_C} = 4\pi\epsilon_0\left(\frac{1}{R_1} - \frac{1}{R_2}\right)^{-1} = 4\pi\epsilon_0\frac{R_1 R_2}{R_2 - R_1}$$

ASSESS As expected, the capacitance depends on the geometry but not on the charge Q. Note that we did not need to assume a negative inner sphere, but a positive inner sphere would have required us to integrate inward, from R_2 to R_1, to get a positive ΔV_C.

Combinations of Capacitors

In practice, two or more capacitors are sometimes joined together. Figure 30.23 illustrates two basic combinations: **parallel capacitors** and **series capacitors.** Notice that a capacitor, no matter what its actual geometrical shape, is represented in *circuit diagrams* by two parallel lines.

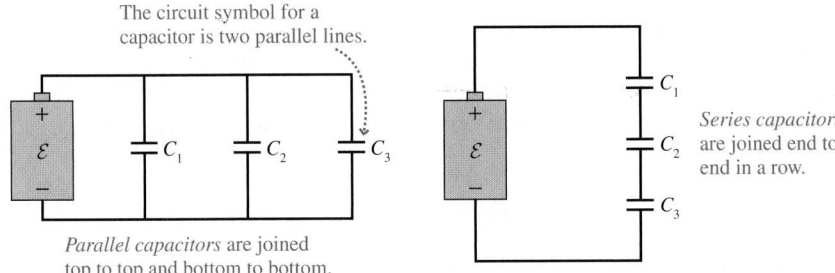

FIGURE 30.23 Parallel and series capacitors.

NOTE ▶ The terms "parallel capacitors" and "parallel-plate capacitor" do not describe the same thing. The former term describes how two or more capacitors are connected to each other, the latter describes how a particular capacitor is constructed. ◀

As we'll show, parallel or series capacitors (or, as is sometimes said, capacitors "in parallel" or "in series") can be represented by a single **equivalent capacitance.** We'll demonstrate this first with the two parallel capacitors C_1 and C_2 of Figure 30.24a. Because the two top electrodes are connected by a conducting wire, they form a single conductor in electrostatic equilibrium. Thus the two top electrodes are at the same potential. Similarly, the two connected bottom electrodes are at the same potential. Consequently, two (or more) capacitors in parallel each have the *same* potential difference ΔV_C between the two electrodes.

The charges on the two capacitors are $Q_1 = C_1\Delta V_C$ and $Q_2 = C_2\Delta V_C$. Altogether, the battery's charge escalator moved total charge $Q = Q_1 + Q_2$ from the negative electrodes to the positive electrodes. Suppose, as in Figure 30.24b, we replaced the two capacitors with a single capacitor having charge $Q = Q_1 + Q_2$ and potential difference ΔV_C. This capacitor is equivalent to the original two in the sense that the battery can't tell the difference. In either case, the battery has to establish the same potential difference and move the same amount of charge.

By definition, the capacitance of this equivalent capacitor is

$$C_{eq} = \frac{Q}{\Delta V_C} = \frac{Q_1 + Q_2}{\Delta V_C} = \frac{Q_1}{\Delta V_C} + \frac{Q_2}{\Delta V_C} = C_1 + C_2 \qquad (30.27)$$

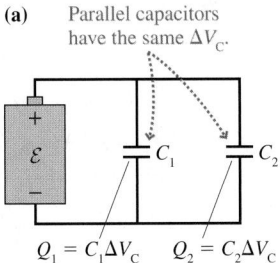

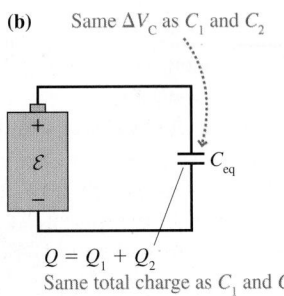

FIGURE 30.24 Replacing two parallel capacitors with an equivalent capacitor.

This analysis hinges on the fact that **parallel capacitors each have the same potential difference ΔV_C.** We could easily extend this analysis to more than two capacitors. If capacitors $C_1, C_2, C_3, \ldots$ are in parallel, their equivalent capacitance is

$$C_{eq} = C_1 + C_2 + C_3 + \cdots \quad \text{(parallel capacitors)} \quad (30.28)$$

Neither the battery nor any other part of a circuit can tell if the parallel capacitors are replaced by a single capacitor having capacitance C_{eq}.

Now consider the two series capacitors in Figure 30.25a. The center section, consisting of the bottom plate of C_1, the top plate of C_2, and the connecting wire, is electrically isolated. The battery cannot remove charge from or add charge to this section. If it starts out with no net charge, it must end up with no net charge. As a consequence, the two capacitors in series have equal charges $\pm Q$. The battery transfers Q from the bottom of C_2 to the top of C_1. This transfer polarizes the center section, as shown, but it still has $Q_{net} = 0$.

The potential differences across the two capacitors are $\Delta V_1 = Q/C_1$ and $\Delta V_2 = Q/C_2$. The total potential difference across both capacitors is $\Delta V_C = \Delta V_1 + \Delta V_2$. Suppose, as in Figure 30.25b, we replaced the two capacitors with a single capacitor having charge Q and potential difference $\Delta V_C = \Delta V_1 + \Delta V_2$. This capacitor is equivalent to the original two because the battery has to establish the same potential difference and move the same amount of charge in either case.

By definition, the *inverse* of the capacitance of this equivalent capacitor is

$$\frac{1}{C_{eq}} = \frac{\Delta V_C}{Q} = \frac{\Delta V_1 + \Delta V_2}{Q} = \frac{\Delta V_1}{Q} + \frac{\Delta V_2}{Q} = \frac{1}{C_1} + \frac{1}{C_2} \quad (30.29)$$

This analysis hinges on the fact that **series capacitors each have the same charge Q.** We could easily extend this analysis to more than two capacitors. If capacitors $C_1, C_2, C_3, \ldots$ are in series, their equivalent capacitance is

$$C_{eq} = \left(\frac{1}{C_1} + \frac{1}{C_2} + \frac{1}{C_3} + \cdots\right)^{-1} \quad \text{(series capacitors)} \quad (30.30)$$

NOTE ▶ Be careful to avoid the common error of adding the inverses but forgetting to invert the sum. ◀

Let's summarize the key facts before looking at a numerical example:

- Parallel capacitors all have the same potential difference ΔV_C. Series capacitors all have the same amount of charge $\pm Q$.
- The equivalent capacitance of a parallel combination of capacitors is *larger* than any single capacitor in the group. The equivalent capacitance of a series combination of capacitors is *smaller* than any single capacitor in the group.

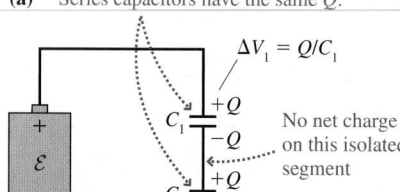

(a) Series capacitors have the same Q.

$\Delta V_1 = Q/C_1$

No net charge on this isolated segment

$\Delta V_2 = Q/C_2$

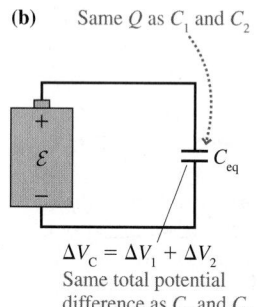

(b) Same Q as C_1 and C_2

$\Delta V_C = \Delta V_1 + \Delta V_2$
Same total potential difference as C_1 and C_2

FIGURE 30.25 Replacing two series capacitors with an equivalent capacitor.

EXAMPLE 30.8 A capacitor circuit
Find the charge on and the potential difference across each of the three capacitors in Figure 30.26.

MODEL Assume the battery is ideal, with $\Delta V_{bat} = \mathcal{E} = 12$ V. Use the results for parallel and series capacitors.

SOLVE The three capacitors are neither in parallel nor in series, but we can break them into smaller groups that are. A useful method of *circuit analysis* is first to combine elements until reaching a single equivalent element, then to reverse the process

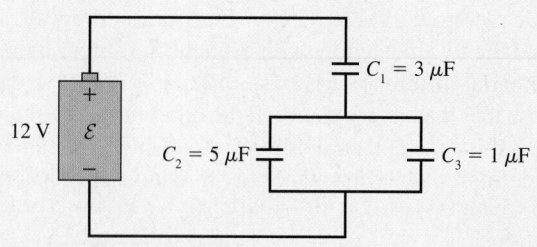

FIGURE 30.26 A capacitor circuit.

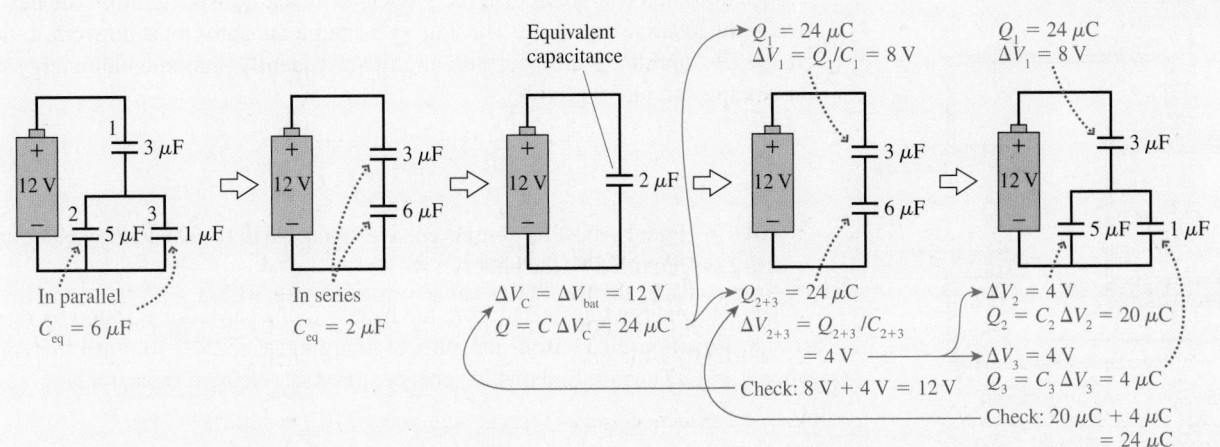

FIGURE 30.27 Analyzing the capacitor circuit.

and calculate values for each element. Figure 30.27 shows the analysis of this circuit. Notice that we redraw the circuit after every step. The equivalent capacitance of the 3 μF and 6 μF capacitors in series is found from

$$C_{eq} = \left(\frac{1}{3 \ \mu F} + \frac{1}{6 \ \mu F}\right)^{-1} = \left(\frac{2}{6} + \frac{1}{6}\right)^{-1} \mu F$$

$$= \left(\frac{1}{2}\right)^{-1} \mu F = 2 \ \mu F$$

Once we get to the single equivalent capacitance, we find that $\Delta V_C = \Delta V_{bat} = 12$ V and $Q = C\Delta V_C = 24 \ \mu$C. Now we can reverse direction. Capacitors in series all have the same charge, so the charge on C_1 and on C_{2+3} is $\pm 24 \ \mu$C. This is enough to

determine that $\Delta V_1 = 8$ V and $\Delta V_{2+3} = 4$ V. Capacitors in parallel all have the same potential difference, so $\Delta V_2 = \Delta V_3 = 4$ V. This is enough to find that $Q_2 = 20 \ \mu$C and $Q_3 = 4 \ \mu$C. The charge on and the potential difference across each of the three capacitors is shown in the final step of Figure 30.27.

ASSESS Notice that we had two important checks of internal consistency. $\Delta V_1 + \Delta V_{2+3} = 8$ V + 4 V add up to the 12 V we had found for the 2 μF equivalent capacitor. Then $Q_2 + Q_3 = 20 \ \mu$C + 4 μC add up to the 24 μC we had found for the 6 μF equivalent capacitor. We'll do much more circuit analysis of this type in the next chapter, but it's worth noting now that circuit analysis becomes nearly foolproof *if* you make use of these checks of internal consistency.

STOP TO THINK 30.5   Rank in order, from largest to smallest, the equivalent capacitance $(C_{eq})_a$ to $(C_{eq})_d$ of circuits a to d.

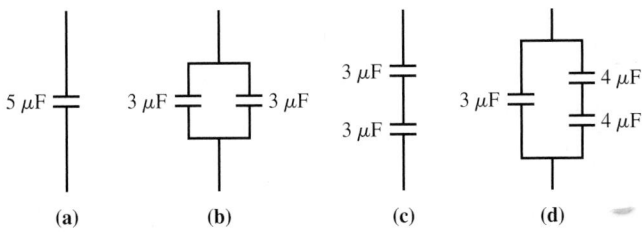

30.7 The Energy Stored in a Capacitor

Capacitors are important elements in electric circuits because of their ability to store energy. Figure 30.28 on the next page shows a capacitor being charged. The instantaneous value of the charge on the two plates is $\pm q$, and this charge separation has established a potential difference $\Delta V = q/C$ between the two electrodes.

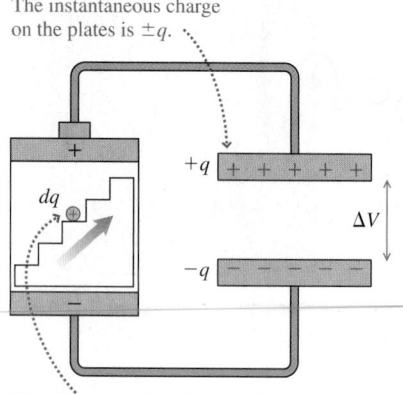

The instantaneous charge on the plates is $\pm q$.

dq

$+q$

$-q$

ΔV

The charge escalator does work $dq\,\Delta V$ to move charge dq from the negative plate to the positive plate.

FIGURE 30.28 The charge escalator does work on charge dq as the capacitor is being charged.

An additional charge dq is in the process of being transferred from the negative to the positive electrode. The battery's charge escalator must do work to lift charge dq "uphill" to a higher potential. Consequently, the potential energy of dq + capacitor increases by

$$dU = dq\,\Delta V = \frac{q\,dq}{C} \tag{30.31}$$

NOTE ▶ Energy must be conserved. This increase in the capacitor's potential energy is provided by the battery. ◀

The total energy transferred from the battery to the capacitor is found by integrating Equation 30.31 from the start of charging, when $q = 0$, until the end, when $q = Q$. Thus we find that the energy stored in a charged capacitor is

$$U_C = \frac{1}{C}\int_0^Q q\,dq = \frac{Q^2}{2C} \tag{30.32}$$

In practice, it is often easier to write the stored energy in terms of the capacitor's potential difference $\Delta V_C = Q/C$. This is

$$U_C = \frac{Q^2}{2C} = \frac{1}{2}C(\Delta V_C)^2 \tag{30.33}$$

The potential energy stored in a capacitor depends on the *square* of the potential difference across it. This result is reminiscent of the potential energy $U = \frac{1}{2}k(\Delta x)^2$ stored in a spring, and a charged capacitor really is analogous to a stretched spring. A stretched spring holds the energy until we release it, then that potential energy is transformed into kinetic energy. Likewise, a charged capacitor holds energy until we discharge it. Then the potential energy is transformed first into the kinetic energy of moving electrons (the current) and ultimately, as the electrons collide with atoms in the metal, into the thermal energy of the wire.

EXAMPLE 30.9 Storing energy in a capacitor

How much energy is stored in a 2.0 μF capacitor that has been charged to 5000 V? What is the average power dissipation if this capacitor is discharged in 10 μs?

SOLVE The energy stored in the charged capacitor is

$$U_C = \frac{1}{2}C(\Delta V_C)^2 = \frac{1}{2}(2.0 \times 10^{-6}\,\text{F})(5000\,\text{V})^2 = 25\,\text{J}$$

If this energy is released in 10 μs, the average power dissipation is

$$P = \frac{\Delta E}{\Delta t} = \frac{25\,\text{J}}{1.0 \times 10^{-5}\,\text{s}} = 2.5 \times 10^6\,\text{W} = 2.5\,\text{MW}$$

ASSESS The stored energy is equivalent to raising a 1 kg mass 2.5 m. This is a rather large amount of energy, which you can see by imagining the damage a 1 kg mass could do after falling 2.5 m. When this energy is released very quickly, which is possible in an electric circuit, it provides an *enormous* amount of power.

The usefulness of a capacitor stems from the fact that it can be charged very slowly, over many seconds, and then can release the energy very quickly. A mechanical analogy would be using a crank to slowly stretch the spring of a catapult, then quickly releasing the energy to launch a massive rock.

The capacitor described in Example 30.9 is typical of the capacitors used in high-power pulsed lasers. The capacitor is charged relatively slowly, in about 0.1 s, then quickly discharged into the laser tube to generate a high-power laser pulse.

Exactly the same thing occurs, only on a smaller scale, in the flash unit of a camera. The camera batteries charge a capacitor, then the energy stored in the capacitor is quickly discharged into a *flashlamp*. The charging process in a camera takes several seconds, which is why you can't fire a camera flash twice in quick succession.

An important medical application of capacitors is the *defibrillator*. A heart attack or a serious injury can cause the heart to enter a state known as *fibrillation* in which the heart muscles twitch randomly and cannot pump blood. A strong electric shock through the chest can sometimes restore the proper heart rhythm. A defibrillator has a large capacitor that can store up to 360 J of energy. This energy is released in about 2 ms through two "paddles" pressed against the patient's chest. It takes several seconds to charge the capacitor, which is why, on television medical shows, you hear an emergency room doctor or nurse shout, "Charging!"

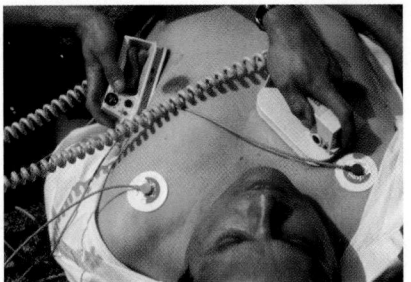

A defibrillator, which can restart the heart, discharges a capacitor through the patient's chest.

The Energy in the Electric Field

We can "see" the potential energy of a stretched spring in the tension of the coils. If a charged capacitor is analogous to a stretched spring, where is the stored energy? It's in the electric field!

Figure 30.29 shows a parallel-plate capacitor in which the plates have area A and are separated by distance d. The potential difference across the capacitor is related to the electric field inside the capacitor by $\Delta V_C = Ed$. The capacitance, which we found in Equation 30.24, is $C = \epsilon_0 A/d$. Substituting these into Equation 30.33, we find that the energy stored in the capacitor is

$$U_C = \frac{1}{2}C(\Delta V_C)^2 = \frac{1}{2}\frac{\epsilon_0 A}{d}(Ed)^2 = \frac{\epsilon_0}{2}(Ad)E^2 \tag{30.34}$$

The quantity Ad is the volume *inside* the capacitor, the region in which the capacitor's electric field exists. (Recall that an ideal capacitor has $\vec{E} = \vec{0}$ everywhere except between the plates.) Although we talk about "the energy stored in the capacitor," Equation 30.34 suggests that, strictly speaking, **the energy is stored in the capacitor's electric field.**

Because Ad is the volume in which the energy is stored, we can define an **energy density** u_E of the electric field:

$$u_E = \frac{\text{energy stored}}{\text{volume in which it is stored}} = \frac{U_C}{Ad} = \frac{\epsilon_0}{2}E^2 \tag{30.35}$$

The energy density has units J/m^3. We've derived Equation 30.35 for a parallel-plate capacitor, but it turns out to be the correct expression for any electric field.

From this perspective, charging a capacitor stores energy in the capacitor's electric field as the field grows in strength. Later, when the capacitor is discharged, the energy is released as the field collapses.

We first introduced the electric field as a way to visualize how a long-range force operates. But if the field can store energy, the field must be real, not merely a pictorial device. We'll explore this idea further in Chapter 34, where we'll find that the energy transported by a light wave—the very real energy of warm sunshine—is the energy of electric and magnetic fields.

Capacitor plate with area A

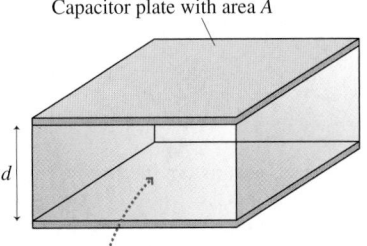

The capacitor's energy is stored in the electric field in volume Ad between the plates.

FIGURE 30.29 A capacitor's energy is stored in the electric field.

EXAMPLE 30.10 The energy density of the electric field
The plates of a parallel-plate capacitor are separated by 1.0 mm. What is the energy density in the capacitor's electric field if the capacitor is charged to 500 V?

SOLVE The electric field inside the capacitor is

$$E = \frac{\Delta V_C}{d} = \frac{500 \text{ V}}{0.0010 \text{ m}} = 5.0 \times 10^5 \text{ V/m}$$

Consequently, the energy density in the electric field is

$$u_E = \frac{\epsilon_0}{2}E^2 = \frac{1}{2}(8.85 \times 10^{-12} \text{ C}^2/\text{N m}^2)(5.0 \times 10^5 \text{ V/m})^2$$

$$= 1.1 \text{ J/m}^3$$

SUMMARY

The goal of Chapter 30 has been to understand how the electric potential is connected to the electric field.

GENERAL PRINCIPLES

Connecting V and $\vec{E}$

The electric potential and the electric field are two different perspectives of how source charges alter the space around them. V and $\vec{E}$ are related by

$$\Delta V = V(s_f) - V(s_i) = -\int_{s_i}^{s_f} E_s \, ds$$

where s is measured from point i to point f and E_s is the component of $\vec{E}$ parallel to the line of integration.

Graphically

ΔV = the negative of the area under the E_s graph

and

$$E_s = -\frac{dV}{ds}$$

= the negative of the slope of the potential graph.

The Geometry of Potential and Field

The electric field

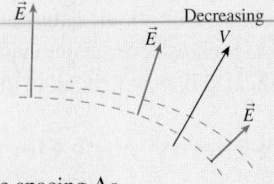

- Is perpendicular to the equipotential surfaces.
- Points "downhill" in the direction of decreasing V.
- Is inversely proportional to the spacing Δs between the equipotential surfaces.

Conservation of Energy

The sum of all potential differences around a closed path is zero. $\sum (\Delta V)_i = 0$.

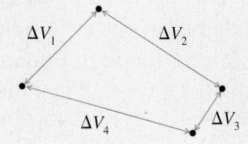

IMPORTANT CONCEPTS

A **battery** is a source of potential. The charge escalator in a battery uses chemical reactions to move charges from the negative terminal to the positive terminal.

$$\Delta V_{bat} = \mathcal{E}$$

where the emf $\mathcal{E}$ is the work per charge done by the charge escalator.

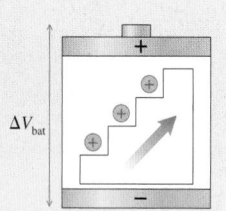

For a conductor in electrostatic equilibrium

- The interior electric field is zero.
- The exterior electric field is perpendicular to the surface.
- The surface is an equipotential.
- The interior is at the same potential as the surface.

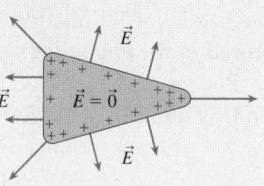

APPLICATIONS

Resistors

A potential difference ΔV_{wire} between the ends of a wire creates an electric field inside the wire

$$E_{wire} = \frac{\Delta V_{wire}}{L}$$

The electric field causes a current

$$I = \frac{\Delta V_{wire}}{R}$$

where $R = \dfrac{\rho L}{A}$ is the wire's **resistance.**

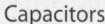

Capacitors

The **capacitance** of two conductors charged to $\pm Q$ is

$$C = \frac{Q}{\Delta V_C}$$

The energy stored in a capacitor is

$$U_C = \frac{1}{2} C (\Delta V_C)^2$$

Series capacitors

$$C_{eq} = \left(\frac{1}{C_1} + \frac{1}{C_2} + \frac{1}{C_3} + \cdots \right)^{-1}$$

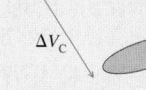

Parallel capacitors

$$C_{eq} = C_1 + C_2 + C_3 + \cdots$$

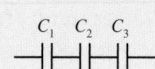

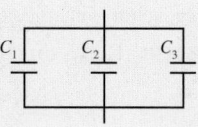

TERMS AND NOTATION

Kirchhoff's loop law	emf, $\mathcal{E}$	farad, F
Van de Graaff generator	terminal voltage, ΔV_{bat}	parallel capacitors
battery	resistance, R	series capacitors
charge escalator	ohm, Ω	equivalent capacitance, C_{eq}
ideal battery	capacitance, C	energy density, u_E

EXERCISES AND PROBLEMS

Exercises

Section 30.1 Connecting Potential and Field

1. What is the potential difference between $x_i = 10$ cm and $x_f = 30$ cm in the uniform electric field $E_x = 1000$ V/m?
2. What is the potential difference between $y_i = -5$ cm and $y_f = 5$ cm in the uniform electric field $\vec{E} = (20,000\hat{i} - 50,000\hat{j})$ V/m?

Section 30.2 Finding the Electric Field from the Potential

3. The electric potential in a region of uniform electric field is -1000 V at $x = -1.0$ m and $+1000$ V at $x = +1.0$ m. What is E_x?
4. The electric potential along the x-axis is $V = 100x^2$ V, where x is in meters. What is E_x at $x = 0$ m? At $x = 1$ m?
5. Figure Ex30.5 is a graph of V versus x. Draw the corresponding graph of E_x versus x.

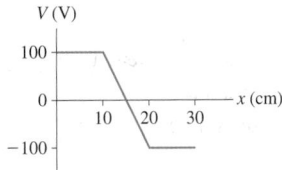

FIGURE EX30.5

6. Figure Ex30.6 is a graph of V versus x. Draw the corresponding graph of E_x versus x.

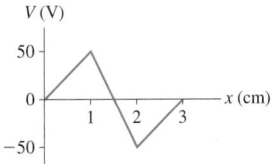

FIGURE EX30.6

7. What are the magnitude and direction of the electric field at the dot in Figure Ex30.7?

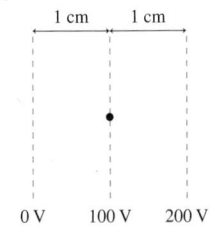

FIGURE EX30.7

8. What are the magnitude and direction of the electric field at the dot in Figure Ex30.8?

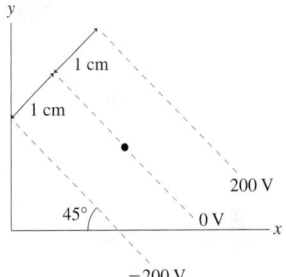

FIGURE EX30.8

9. What is the potential difference ΔV_{34} in Figure Ex30.9?

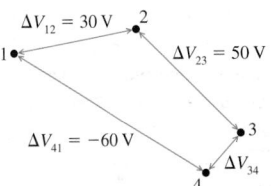

FIGURE EX30.9

Section 30.3 A Conductor in Electrostatic Equilibrium

10. The excess charge on a 20-cm-diameter metal sphere is 10 nC. Draw a graph of (a) E versus r and (b) V versus r over the range $0 < r < 40$ cm.

Section 30.4 Sources of Electric Potential

11. How much work does the charge escalator do to move 1.0 μC of charge from the negative terminal to the positive terminal of a 1.5 V battery?
12. How much work does the electric motor of a Van de Graaff generator do to lift a positive ion ($q = e$) if the potential of the spherical electrode is 1.0 MV?
13. What is the emf of a battery that does 0.60 J of work to transfer 0.050 C of charge from the negative to the positive terminal?

Section 30.5 Connecting Potential and Current

14. Wires 1 and 2 are made of the same metal. Wire 2 has twice the length and twice the diameter of wire 1. What are the ratios (a) ρ_2/ρ_1 of the resistivities and (b) R_2/R_1 of the resistances of the two wires?

15. What is the resistance of
    a. A 1.0-m-long copper wire that is 0.50 mm in diameter?
    b. A 10-cm-long piece of carbon with a 1.0 mm × 1.0 mm square cross section?

16. A 10-m-long wire with a diameter of 0.80 mm has a resistance of 1.1 Ω. Of what material is the wire made?

17. The electric field inside a 30-cm-long copper wire is 10 V/m. What is the potential difference between the ends of the wire?

18. a. How long must a 0.60-mm-diameter aluminum wire be to have a 0.50 A current when connected to the terminals of a 1.5 V flashlight battery?
    b. What is the current if the wire is half this length?

19. The terminals of a 0.70 V watch battery are connected by a 100-m-long gold wire with a diameter of 0.10 mm. What is the current in the wire?

Section 30.6 Capacitance and Capacitors

20. Two 2.0 cm × 2.0 cm square aluminum electrodes are spaced 0.50 mm apart. The electrodes are connected to a 100 V battery.
    a. What is the capacitance?
    b. What is the charge on each electrode?

21. You need to construct a 100 pF capacitor for a science project. You plan to cut two $L \times L$ metal squares and place spacers between them. The thinnest spacers you have are 0.20 mm thick. What is the proper value of L?

22. A switch that connects a battery to a 10 μF capacitor is closed. Several seconds later you find that the capacitor plates are charged to ± 30 μC. What is the emf of the battery?

23. What is the emf of a battery that will charge a 2.0 μF capacitor to ± 48 μC?

24. Two electrodes connected to a 9.0 V battery are charged to ± 45 nC. What is the capacitance of the electrodes?

25. A 6 μF capacitor, a 10 μF capacitor, and a 16 μF capacitor are connected in parallel. What is their equivalent capacitance?

26. A 6 μF capacitor, a 10 μF capacitor, and a 16 μF capacitor are connected in series. What is their equivalent capacitance?

27. What is the capacitance of the two metal spheres shown in Figure Ex30.27?

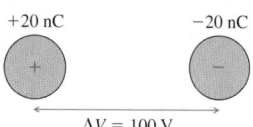

FIGURE EX30.27 $\Delta V = 100$ V

28. Initially, the switch in Figure Ex30.28 is open and the capacitor is uncharged. How much charge flows through the switch after the switch is closed?

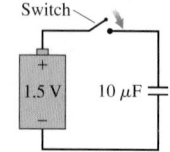

FIGURE EX30.28

Section 30.7 The Energy Stored in a Capacitor

29. To what potential should you charge a 1.0 μF capacitor to store 1.0 J of energy?

30. Figure Ex30.30 shows Q versus t for a 2.0 μF capacitor. Draw a graph showing U_C versus t.

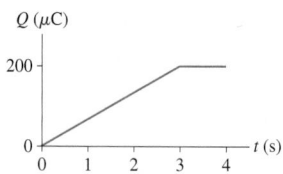

FIGURE EX30.30

31. Capacitor 2 has half the capacitance and twice the potential difference as capacitor 1. What is the ratio U_{C1}/U_{C2}?

32. 50 pJ of energy is stored in a 2.0 cm × 2.0 cm × 2.0 cm region of uniform electric field. What is the electric field strength?

33. A 2.0-cm-diameter parallel-plate capacitor with a spacing of 0.50 mm is charged to 200 V. What are (a) the total energy stored in the electric field and (b) the energy density?

Problems

34. a. Which point, A or B, has a larger electric potential?
    b. What is the potential difference between A and B?

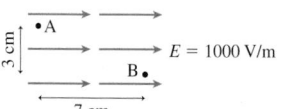

FIGURE P30.34   7 cm

35. The electric field in a region of space is $E_x = -1000x$ V/m, where x is in meters.
    a. Graph E_x versus x over the region -1 m $\leq x \leq 1$ m.
    b. What is the potential difference between $x_i = -20$ cm and $x_f = 30$ cm?

36. The electric field in a region of space is $E_x = 5000x$ V/m, where x is in meters.
    a. Graph E_x versus x over the region -1 m $\leq x \leq 1$ m.
    b. Find an expression for the potential V at position x. As a reference, let $V = 0$ V at the origin.
    c. Graph V versus x over the region -1 m $\leq x \leq 1$ m.

37. An infinitely long cylinder of radius R has linear charge density λ. The potential on the surface of the cylinder is V_0, and the electric field outside the cylinder is $E_r = \lambda/2\pi\epsilon_0 r$. Find the potential relative to the surface at a point that is distance r from the axis, assuming $r > R$.

38. Figure P30.38 shows E_x, the x-component of the electric field, as a function of position along the x-axis. Find and graph V versus x over the region 0 cm $\leq x \leq 3$ cm. As a reference, let $V = 0$ V at $x = 3$ cm.

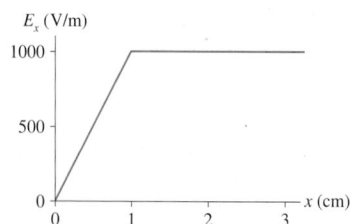

FIGURE P30.38

39. The three metal electrodes in Figure P30.39 are charged as shown. Draw a graph of (a) E_x versus x and (b) V versus x over the region $0 \leq x \leq 3$ cm.

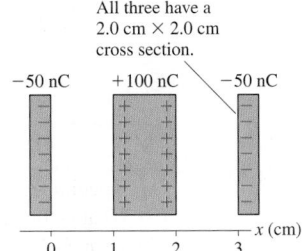

All three have a
2.0 cm $\times$ 2.0 cm
cross section.

-50 nC $+100$ nC -50 nC

FIGURE P30.39

40. Figure P30.40 shows a graph of V versus x in a region of space. The potential is independent of y and z.
    a. Draw a graph of E_x versus x.
    b. Draw a contour map of the potential in the xy-plane in the square-shaped region -3 m $\leq x \leq 3$ m and -3 m $\leq y \leq 3$ m. Show and label the -10 V, -5 V, 0 V, $+5$ V, and $+10$ V equipotential surfaces.
    c. Draw electric field vectors on your contour map of part b.

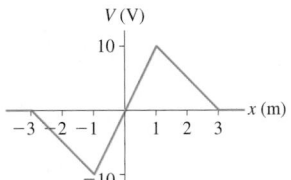

FIGURE P30.40

41. Use the on-axis potential of a charged disk from Chapter 29 to find the on-axis electric field of a charged disk.
42. a. Use the methods of Chapter 29 to find the potential at distance x on the axis of the charged rod shown in Figure P30.42.
    b. Use the result of part a to find the electric field at distance x on the axis of a rod.

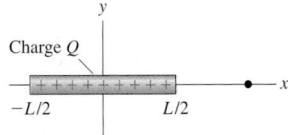

Charge Q

$-L/2$ $L/2$

FIGURE P30.42

43. Determine the magnitude and direction of the electric field at points 1 and 2 in Figure P30.43.

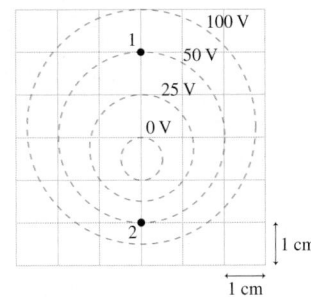

FIGURE P30.43

44. Figure P30.44 shows a set of equipotential lines and five labeled points.
    a. From measurements made on this figure with a ruler, using the scale on the figure, estimate the electric field strength E at the five points indicated.

    b. Trace the figure on your paper, then show the electric field vectors $\vec{E}$ at the five points.

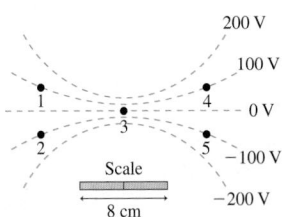

200 V
100 V
0 V
-100 V
-200 V
Scale
8 cm

FIGURE P30.44

45. The electric potential in a region of space is $V = (150x^2 - 200y^2)$ V, where x and y are in meters. What are the strength and the direction of the electric field at $(2.0$ m, 2.0 m$)$?
46. Figure P30.46 shows the electric potential at points on a 5.0 cm $\times$ 5.0 cm grid.
    a. Reproduce this figure on your paper, then draw the 50 V, 75 V, and 100 V equipotential surfaces.
    b. Determine the electric field (strength and direction) at the points A, B, C, and D.
    c. Draw the electric field vectors at points A, B, C, and D on your diagram.

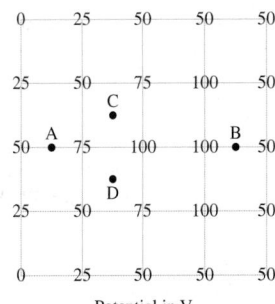

FIGURE P30.46   Potential in V

47. Metal sphere 1 has a positive charge of 6.0 nC. Metal sphere 2, which is twice the diameter of sphere 1, is initially uncharged. The spheres are then connected together by a long, thin metal wire. What are the final charges on each sphere?
48. The metal spheres in Figure P30.48 are charged to ± 300 V. Draw this figure on your paper, then draw a plausible contour map of the potential, showing and labeling the -300 V, -200 V, -100 V, . . . , 300 V equipotential surfaces.

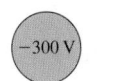

-300 V $+300$ V

FIGURE P30.48

49. The potential at the center of a 4.0-cm-diameter copper sphere is 500 V, relative to $V = 0$ V at infinity. How much excess charge is on the sphere?
50. A 15-cm-long nichrome wire is connected across the terminals of a 1.5 V battery.
    a. What is the electric field inside the wire?
    b. What is the current density inside the wire?
    c. If the current in the wire is 2.0 A, what is the wire's diameter?
51. A 20-cm-long hollow nichrome tube of inner diameter 2.8 mm, outer diameter 3.0 mm is connected to a 3.0 V battery. What is the current in the tube?

52. A 1.5 V battery provides 0.50 A of current.
    a. At what rate (C/s) is charge lifted by the charge escalator?
    b. How much work does the charge escalator do to lift 1.0 C of charge?
    c. What is the power output of the charge escalator?

53. A 1.5 V flashlight battery is connected to a wire with a resistance of 3.0 Ω. Figure P30.53 shows the battery's potential difference as a function of time. What is the total charge lifted by the charge escalator?

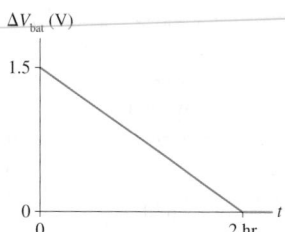

FIGURE P30.53

54. Two 10-cm-diameter metal plates are 1.0 cm apart. They are charged to ±12.5 nC. They are suddenly connected together by a 0.224-mm-diameter copper wire stretched taut from the center of one plate to the center of the other.
    a. What is the maximum current in the wire?
    b. What is the largest electric field in the wire?
    c. Does the current increase with time, decrease with time, or remain steady? Explain.
    d. What is the total amount of energy dissipated in the wire?

55. Two 2.0 cm × 2.0 cm metal electrodes are spaced 1.0 mm apart and connected by wires to the terminals of a 9.0 V battery.
    a. What are the charge on each electrode and the potential difference between them?
    The wires are disconnected, and insulated handles are used to pull the plates apart to a new spacing of 2.0 mm.
    b. What are the charge on each electrode and the potential difference between them?

56. Two 2.0 cm × 2.0 cm metal electrodes are spaced 1.0 mm apart and connected by wires to the terminals of a 9.0 V battery.
    a. What are the charge on each electrode and the potential difference between them?
    While the plates are still connected to the battery, insulated handles are used to pull them apart to a new spacing of 2.0 mm.
    b. What are the charge on each electrode and the potential difference between them?

57. A spherical capacitor with a 1.0 mm gap between the spheres has a capacitance of 100 pF. What are the diameters of the two spheres?

58. You need a capacitance of 50 μF, but you don't happen to have a 50 μF capacitor. You do have a 30 μF capacitor. What additional capacitor do you need to produce a total capacitance of 50 μF? Should you join the two capacitors in parallel or in series?

59. You need a capacitance of 50 μF, but you don't happen to have a 50 μF capacitor. You do have a 75 μF capacitor. What additional capacitor do you need to produce a total capacitance of 50 μF? Should you join the two capacitors in parallel or in series?

60. Find expressions for the equivalent capacitance of (a) N identical capacitors C in parallel and (b) N identical capacitors C in series.

61. What is the equivalent capacitance of the three capacitors in Figure P30.61?

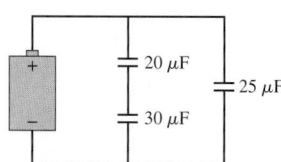

FIGURE P30.61

62. What is the equivalent capacitance of the three capacitors in Figure P30.62?

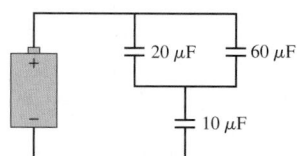

FIGURE P30.62

63. What are the charge on and the potential difference across each capacitor in Figure P30.63?

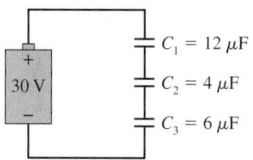

FIGURE P30.63

64. What are the charge on and the potential difference across each capacitor in Figure P30.64?

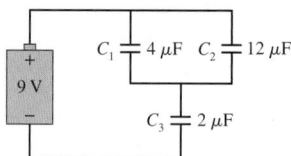

FIGURE P30.64

65. What are the charge on and the potential difference across each capacitor in Figure P30.65?

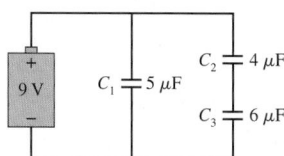

FIGURE P30.65

66. You have three 12 μF capacitors. Draw diagrams showing how you could arrange all three so that their equivalent capacitance is (a) 4.0 μF, (b) 8.0 μF, (c) 18 μF, and (d) 36 μF.

67. What is the capacitance of the three concentric metal spherical shells in Figure P30.67?
    Hint: Can you think of this as a combination of capacitors?

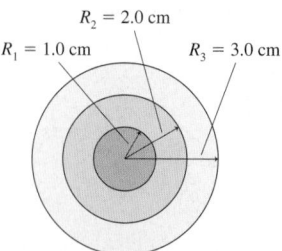

FIGURE P30.67

68. Six identical capacitors with capacitance C are connected as shown in Figure P30.68.
    a. What is the equivalent capacitance of these six capacitors?
    b. What is the potential difference between points a and b?

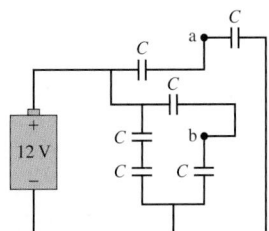

FIGURE P30.68

69. What is the capacitance of the two electrodes in Figure P30.69? **Hint:** Can you think of this as a combination of capacitors?

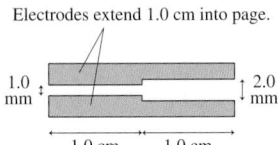

FIGURE P30.69

70. Initially, the switch in Figure P30.70 is in position A and capacitors C_2 and C_3 are uncharged. Then the switch is flipped to position B. Afterward, what are the charge on and the potential difference across each capacitor?

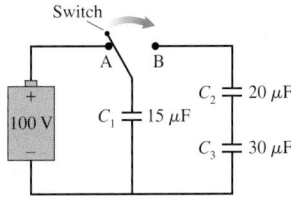

FIGURE P30.70

71. A battery with an emf of 60 V is connected to the two capacitors shown in Figure P30.71. Afterward, the charge on capacitor 2 is 450 μC. What is the capacitance of capacitor 2?

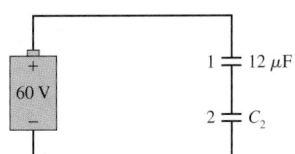

FIGURE P30.71

72. Capacitors $C_1 = 10\ \mu$F and $C_2 = 20\ \mu$F are each charged to 10 V, then disconnected from the battery without changing the charge on the capacitor plates. The two capacitors are then connected in parallel, with the positive plate of C_1 connected to the negative plate of C_2 and vice versa. Afterward, what are the charge on and the potential difference across each capacitor?

73. An isolated 5.0 μF parallel-plate capacitor has 4.0 mC of charge. An external force changes the distance between the electrodes until the capacitance is 2.0 μF. How much work is done by the external force?

74. A parallel-plate capacitor is constructed from two 10 cm × 10 cm electrodes spaced 1.0 mm apart. The capacitor plates are charged to ±10 nC, then disconnected from the battery.
    a. How much energy is stored in the capacitor?
    b. Insulating handles are used to pull the capacitor plates apart until the spacing is 2.0 mm. Now how much energy is stored in the capacitor?

c. Energy must be conserved. How do you account for the difference between a and b?

75. What is the energy density in the electric field at the surface of a 1.0-cm-diameter sphere charged to a potential of 1000 V?

76. The flash unit in a camera uses a 3.0 V battery to charge a capacitor. The capacitor is then discharged through a flashlamp. The discharge takes 10 μs, and the average power dissipated in the flashlamp is 10 W. What is the capacitance of the capacitor?

77. You need to melt a 0.50 kg block of ice at −10°C in a hurry. The stove isn't working, but you do have a 50 V battery. It occurs to you that you could build a capacitor from a couple of pieces of sheet metal that are nearby, charge the capacitor with the battery, then discharge it through the block of ice. If you use square sheets spaced 2.0 mm apart, what must the dimensions of the sheets be to accomplish your goal? Is this feasible?

In Problems 78 through 81 you are given the equation(s) used to solve a problem. For each of these, you are to
    a. Write a realistic problem for which this is the correct equation(s).
    b. Finish the solution of the problem.

78. $2z$ V/m $= -\dfrac{dV}{dz}$
    $V(z = 0) = 10$ V

79. $\dfrac{3.0\ \text{V}}{R} = 5.0$ A
    $R = \dfrac{(9.7 \times 10^{-8}\ \Omega\,\text{m})L}{\pi(0.00050\ \text{m})^2}$

80. 400 nC $= (100$ V$)C$
    $C = \dfrac{(8.85 \times 10^{-12}\ \text{C}^2/\text{Nm}^2)(0.10\ \text{m} \times 0.10\ \text{m})}{d}$

81. $\left(\dfrac{1}{3\ \mu\text{F}} + \dfrac{1}{6\ \mu\text{F}}\right)^{-1} + C = 4\ \mu$F

Challenge Problems

82. The electric potential in a region of space is $V = 100(x^2 - y^2)$ V, where x and y are in meters.
    a. Draw a contour map of the potential, showing and labeling the −400 V, −100 V, 0 V, +100 V, and +400 V equipotential surfaces.
    b. Find an expression for the electric field $\vec{E}$ at position (x, y).
    c. Draw the electric field lines on your diagram of part a.

83. Charge is uniformly distributed with charge density ρ inside a very long cylinder of radius R. Find the potential difference between the surface and the axis of the cylinder.

84. An electric dipole at the origin consists of two charges $\pm q$ spaced distance s apart along the y-axis.
    a. Find an expression for the potential $V(x, y)$ at an arbitrary point in the xy-plane. Your answer will be in terms of q, s, x, and y.
    b. Use the binomial approximation to simplify your result of part a when $s \ll x$ and $s \ll y$.
    c. Assuming $s \ll x$ and y, find expressions for E_x and E_y, the components of $\vec{E}$ for a dipole.
    d. What is the on-axis field $\vec{E}$? Does your result agree with Equation 26.11?
    e. What is the field $\vec{E}$ on the bisecting axis? Does your result agree with Equation 26.12?

85. Consider a uniformly charged sphere of radius R and total charge Q. The electric field E_{out} *outside* the sphere ($r \geq R$) is simply that of a point charge Q. In Chapter 27, we used Gauss's law to find that the electric field E_{in} *inside* the sphere ($r \leq R$) is radially outward with field strength

$$E_{in} = \frac{1}{4\pi\epsilon_0}\frac{Q}{R^3}r$$

a. Graph E versus R for $0 \leq r \leq 3R$.
b. The electric potential V_{out} *outside* the sphere is that of a point charge Q. Find an expression for the electric potential V_{in} at position r *inside* the sphere. As a reference, let $V_{in} = V_{out}$ at the surface of the sphere.
c. What is the ratio $V_{center}/V_{surface}$?
d. Graph V versus R for $0 \leq r \leq 3R$.

86. High-frequency signals are often transmitted along a *coaxial cable,* such as the one shown in Figure CP30.86. For example, the cable TV hookup coming into your home is a coaxial cable. The signal is carried on a wire of radius R_1 while the outer conductor of radius R_2 is grounded (i.e., at $V = 0$ V). An insulating material fills the space between them, and an insulating plastic coating goes around the outside.

a. Find an expression for the capacitance per meter of a coaxial cable. Assume that the insulating material between the cylinders is air.
b. Evaluate the capacitance per meter of a cable having $R_1 = 0.50$ mm and $R_2 = 3.0$ mm.

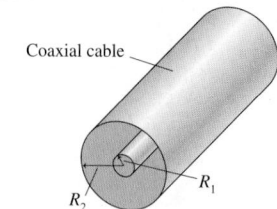

Coaxial cable

FIGURE CP30.86

R_2

R_1

STOP TO THINK ANSWERS

Stop to Think 30.1: c. E_y is the negative of the slope of the V-versus-y graph. E_y is positive, because $\vec{E}$ points up, so the graph has a negative slope. E_y has constant magnitude, so the slope has a constant value.

Stop to Think 30.2: c. $\vec{E}$ points "downhill," so V must decrease from right to left. E is larger on the left than on the right, so the contour lines must be closer together on the left.

Stop to Think 30.3: b. Because of the connecting wire, the three spheres form a single conductor in electrostatic equilibrium. Thus all points are at the same potential. The electric field of a sphere is related to the sphere's potential by $E = V/R$, so a smaller-radius sphere has a larger E.

Stop to Think 30.4: $I_a = I_b = I_c = I_d$. Conservation of current requires $I_a = I_b$. The current in each wire is $I = \Delta V/R$. All the wires have the same resistance, because they are identical, and they all have the same potential difference, because the battery is a *source of potential.*

Stop to Think 30.5: $(C_{eq})_b > (C_{eq})_a = (C_{eq})_d > (C_{eq})_c$. $(C_{eq})_b = 3\,\mu\text{F} + 3\,\mu\text{F} = 6\,\mu\text{F}$. The equivalent capacitance of series capacitors is less than any capacitor in the group, so $(C_{eq})_c < 3\,\mu\text{F}$. Only d requires any real calculation. The two $4\,\mu\text{F}$ capacitors are in series and are equivalent to a single $2\,\mu\text{F}$ capacitor. The $2\,\mu\text{F}$ equivalent capacitor is in parallel with $3\,\mu\text{F}$, so $(C_{eq})_d = 5\,\mu\text{F}$.

31 Fundamentals of Circuits

This microprocessor, the heart of a computer, is an extraordinarily complex electric circuit. Even so, its operations can be understood on the basis of a few fundamental physical principles.

▶ Looking Ahead

The goal of Chapter 31 is to understand the fundamental physical principles that govern electric circuits. In this chapter you will learn to:

- Understand the conducting and insulating materials used in circuits.
- Draw and use basic circuit diagrams.
- Analyze circuits containing resistors in series and in parallel.
- Calculate power dissipation in circuit elements.
- Understand the growth and decay of current in *RC* circuits.

◀ Looking Back

This chapter is based on our earlier development of the ideas of current and potential. Please review:

- Sections 28.4–28.5 Current and conductivity.
- Sections 30.4–30.5 Batteries, emf, and current.
- Section 30.6 Capacitors.

A computer is an incredible device. Surprising as it may seem, the power of a computer is achieved simply by the controlled flow of charges through tiny wires and circuit elements. The most powerful supercomputer is a direct descendant of the charged rods with which we began Part VI.

This chapter will bring together many of the ideas you have learned about the electric field and potential and apply them to the analysis of electric circuits. This single chapter will not pretend to be a full course on circuit analysis. Instead, as the title implies, our more modest goal is to describe the *fundamental physical principles* by which circuits operate. An understanding of these basic principles will prepare you to undertake a course in circuit analysis and design at a later time.

Our primary interest is with circuits in which the battery's potential difference is unchanging and all the currents in the circuit are constant. Such circuits are called **direct current,** or DC, circuits. We'll consider alternating-current (AC) circuits, in which the potential difference oscillates sinusoidally, in Chapter 35.

31.1 Resistors and Ohm's Law

In Part VI we have developed a theoretical understanding of charges and their interactions. Now we are ready to see how this information can be put to practical use. Figure 31.1 summarizes several key ideas that are of particular relevance to electric circuits.

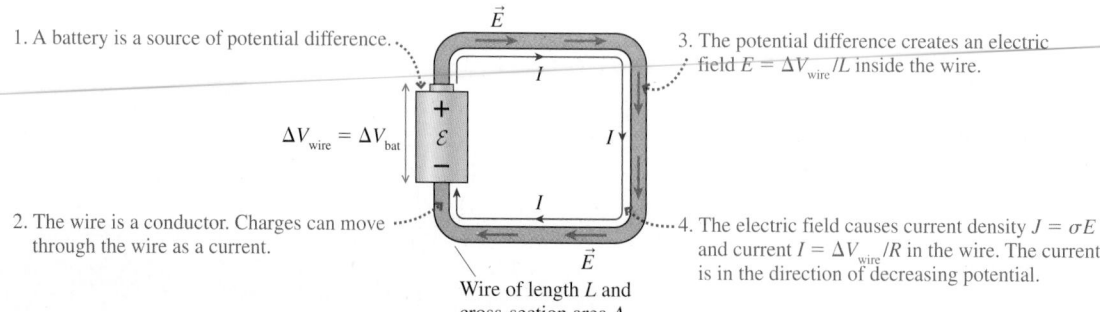

1. A battery is a source of potential difference.

$\Delta V_{wire} = \Delta V_{bat}$

3. The potential difference creates an electric field $E = \Delta V_{wire}/L$ inside the wire.

2. The wire is a conductor. Charges can move through the wire as a current.

4. The electric field causes current density $J = \sigma E$ and current $I = \Delta V_{wire}/R$ in the wire. The current is in the direction of decreasing potential.

Wire of length L and cross-section area A

FIGURE 31.1 A summary of the properties of charges, fields, and potentials that are relevant to electric circuits.

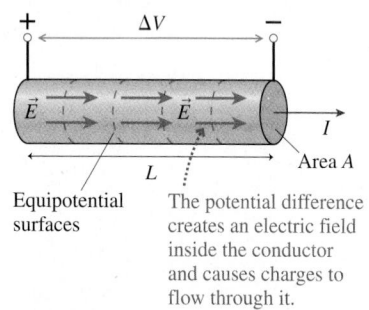

Equipotential surfaces

The potential difference creates an electric field inside the conductor and causes charges to flow through it.

FIGURE 31.2 The current I is related to the potential difference ΔV.

You learned in Chapter 30 that there is a simple relationship between the current I in a conducting element and ΔV, the potential difference between its ends. Figure 31.2 shows a conductor of length L and cross-section area A. The current density in the conductor is

$$J = \sigma E = \frac{E}{\rho} \tag{31.1}$$

where $E = \Delta V/L$ is the field strength created by the potential difference and $\rho = 1/\sigma$ is the material's resistivity. The actual current in the conductor, $I = JA$, is

$$I = JA = \frac{E}{\rho}A = \frac{\Delta V/L}{\rho}A = \frac{\Delta V}{(\rho L/A)} \tag{31.2}$$

The quantity

$$R = \frac{\rho L}{A} \tag{31.3}$$

is the *resistance* of the conductor. The unit of resistance is the *ohm,* where 1 ohm = 1 Ω = 1 V/A.

In terms of resistance, Equation 31.2 is

$$I = \frac{\Delta V}{R} \tag{31.4}$$

This relationship between I and ΔV is known as **Ohm's law.**

Circuit textbooks often write Ohm's law as $V = IR$ rather than $I = \Delta V/R$. This can be misleading until you have sufficient experience with circuit analysis. First, Ohm's law relates the current to the potential *difference* between the ends of the conductor. Engineers and circuit designers *mean* "potential difference" when they use the symbol V, but this use of the symbol is easily overlooked by beginners who think that it means "the potential."

Second, $V = IR$ or even $\Delta V = IR$ suggests that a current I causes a potential difference ΔV. As you have seen, the proper cause-and-effect sequence is the other way around. Current is a *consequence* of a potential difference, hence $I = \Delta V/R$ is a better description of cause and effect.

What is the current in a 1.0-mm-diameter, 10-cm-long copper wire that is attached to the terminals of a 1.5 V battery?

SOLVE The resistivity of copper, from Table 30.1, is $\rho = 1.7 \times 10^{-8} \; \Omega \, \text{m}$. Thus the resistance of the copper wire is

$$R = \frac{\rho L}{A} = \frac{\rho L}{\pi r^2} = 2.2 \times 10^{-3} \; \Omega$$

The current in the wire is then found from Ohm's law:

$$I = \frac{\Delta V}{R} = 680 \; \text{A}$$

ASSESS This is an enormous current because the resistance of a copper wire is extremely small. In fact, as you will learn later in this chapter, no real 1.5 V battery can deliver a current anywhere near this large.

Because the resistance of metals is so small, a circuit made exclusively of copper wires would have enormous currents and would quickly deplete the battery. It is useful to limit the current by using circuit elements that, although they are conductors, have a resistance significantly larger than the wires. Such devices are called **resistors.** Resistors are made either by using poorly conducting materials, such as carbon or nichrome, or by depositing very thin metal films on an insulating substrate. The cross-section area A of a thin film is so small that a film's resistance can be much larger than that of a solid wire. The resistors used in circuits typically range from 10 Ω to 1 MΩ. The resistance of the connecting metal wires, by comparison, is essentially zero.

The resistors used in circuits range from a few ohms to millions of ohms of resistance.

Ohmic and Nonohmic Materials

Despite its name, Ohm's law is *not* a law of nature. It is limited to those materials whose resistance R remains constant—or very nearly so—during use. The materials to which Ohm's law applies are called **ohmic.** Figure 31.3a shows that the current through an ohmic material is directly proportional to the potential difference. Doubling the potential difference results in a doubling of the current. Resistors and metal wires are ohmic devices.

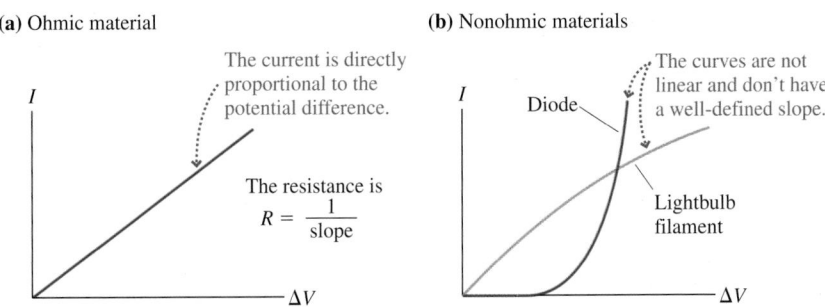

(a) Ohmic material

The current is directly proportional to the potential difference.

The resistance is $R = \dfrac{1}{\text{slope}}$

(b) Nonohmic materials

The curves are not linear and don't have a well-defined slope.

Diode

Lightbulb filament

FIGURE 31.3 Current-versus-potential-difference graphs for ohmic and nonohmic materials.

Many materials and devices are **nonohmic,** meaning that the current through the device is *not* directly proportional to the potential difference. Three important examples of nonohmic devices are

1. Batteries, where $\Delta V = \mathcal{E}$ is determined by chemical reactions, independent of I.
2. Semiconductors, where the I-versus-ΔV curve is very nonlinear.
3. Capacitors, where, as you'll learn later in this chapter, the relationship between I and ΔV differs from that of a resistor.

Figure 31.3b shows the I-versus-ΔV graphs of a semiconductor device called a *diode* and of a lightbulb filament. You can see that the inverse of the slope does not have a unique value that we can call "resistance."

NOTE ▶ Ohm's law is an important part of circuit analysis because resistors are essential components of almost any circuit. However, it is important that you apply Ohm's law *only* to the resistors and not to anything else. ◀

We can identify three basic classes of ohmic circuit materials:

1. *Wires* are metals with very small resistivities ρ and thus very small resistances ($R \ll 1\ \Omega$). An **ideal wire** has $R = 0\ \Omega$, hence the potential difference between the ends of an ideal wire is $\Delta V = 0$ V *even if there is a current in it*. We will usually adopt the *ideal-wire model* of assuming that any connecting wires in a circuit are ideal.

2. *Resistors* are poor conductors that have resistances in the range 10^1 to $10^6\ \Omega$. They are used to control the current in a circuit. Most resistors in a circuit have a specified value of R, such as 500 Ω. The filament in a lightbulb (a tungsten wire with a high resistance due to an extremely small cross-section area A) functions as a resistor as long as it is glowing, but the filament is slightly nonohmic because the value of its resistance when hot is larger than its room-temperature value.

3. *Insulators* are materials such as glass, plastic, or air. An **ideal insulator** has $R = \infty\ \Omega$, hence there is no current in an insulator even if there is a potential difference across it ($I = \Delta V/R = 0$ A). This is why insulators can be used to hold apart two conductors that are at different potentials. All practical insulators have $R \gg 10^9\ \Omega$ and can be treated, for our purposes, as ideal.

The Ideal-Wire Model

Figure 31.4a illustrates the ideal-wire model. Current must be conserved, hence the current I in the resistor is the same as the current in each wire. However, the resistor's resistance is *much* larger than the resistance of the wires: $R_{\text{resist}} \gg R_{\text{wire}}$. Consequently, the potential difference across the resistor $\Delta V_{\text{resist}} = IR_{\text{resist}}$ is *much* larger than the potential difference $\Delta V_{\text{wire}} = IR_{\text{wire}}$ between the ends of each wire.

Figure 31.4b shows the potential along the wire-resistor-wire combination. You can see a large *voltage drop,* or potential difference, across the resistor. The voltage drops across the two wires are extremely small. A very reasonable approximation is to assume that *all* the voltage drop occurs along the resistor, none along the wires. This is the approximation made by the ideal-wire model, which assumes that $R_{\text{wire}} = 0\ \Omega$ and $\Delta V_{\text{wire}} = 0$ V. With this approximation, shown in Figure 31.4c, the segments of the graph corresponding to the wires are horizontal.

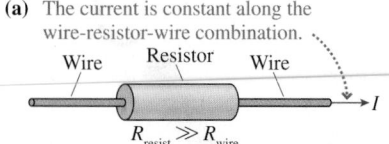

(a) The current is constant along the wire-resistor-wire combination.

$R_{\text{resist}} \gg R_{\text{wire}}$

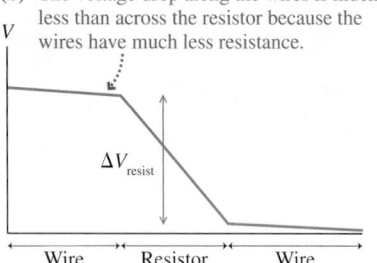

(b) The voltage drop along the wires is much less than across the resistor because the wires have much less resistance.

ΔV_{resist}

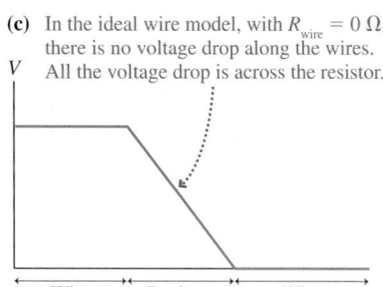

(c) In the ideal wire model, with $R_{\text{wire}} = 0\ \Omega$, there is no voltage drop along the wires. All the voltage drop is across the resistor.

FIGURE 31.4 The potential along a wire-resistor-wire combination.

STOP TO THINK 31.1 Conductors a to d are all made of the same material. Rank in order, from largest to smallest, the resistances R_a to R_d.

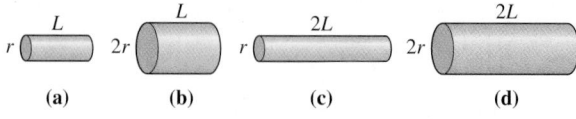

31.2 Circuit Elements and Diagrams

Figure 31.5 shows an electric circuit in which a resistor and a capacitor are connected by wires to a battery. To understand the functioning of this circuit, we do not need to know whether the wires are bent or straight, or whether the battery is to the right or to the left of the resistor. The literal picture of Figure 31.5 provides many irrelevant details. It is customary when describing or analyzing circuits to use a more abstract picture called a **circuit diagram.** A circuit diagram is a *logical* pic-

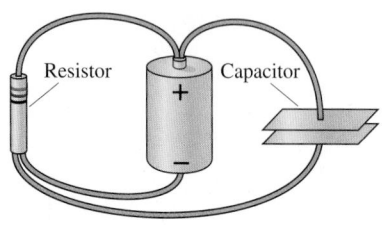

FIGURE 31.5 An electric circuit.

ture of what is connected to what. The actual circuit, once it is built, may *look* quite different from the circuit diagram, but it will have the same logic and connections.

A circuit diagram also replaces pictures of the circuit elements with symbols. Figure 31.6 shows the basic symbols that we will need. Notice that the *longer* line at one end of the battery symbol represents the positive terminal of the battery.

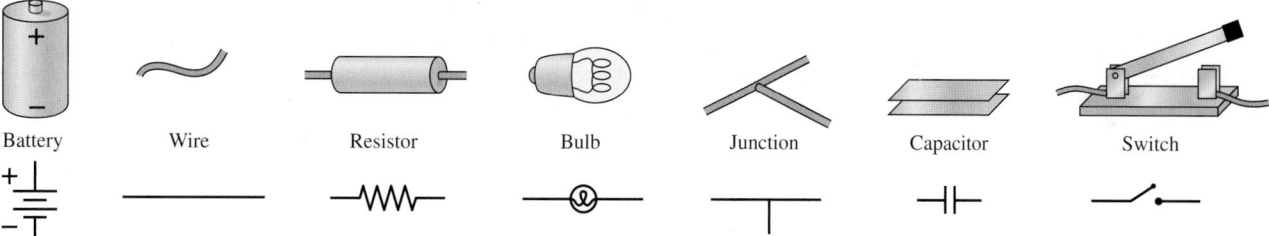

FIGURE 31.6 A library of basic symbols used for electric circuit drawings.

Figure 31.7 is a circuit diagram of the circuit shown in Figure 31.5. Notice how circuit elements are labeled. The battery's emf $\mathcal{E}$ is shown beside the battery, and + and − symbols, even though somewhat redundant, are shown beside the terminals. The resistance R of the resistor and capacitance C of the capacitor are written beside them. We would use numerical values for $\mathcal{E}$ and R if we knew them. The wires, which in practice may bend and curve, are shown as straight-line connections between the circuit elements. You should get into the habit of drawing your own circuit diagrams in a similar fashion.

Lightbulbs are important circuit elements, and Figure 31.8 gives more information about the anatomy of a light bulb. The important point to understand is that a lightbulb, like a wire or a resistor, has two "ends," and that current passes *through* the bulb. It is often useful to think of a lightbulb as a resistor that happens to give off light when a current is present. Even though, as Figure 31.3b showed, a lightbulb filament is not a perfectly ohmic material, the resistance of a *glowing* lightbulb remains reasonably constant if you don't change ΔV by much. A lightbulb's resistance is typically in the range from 10 Ω to 500 Ω.

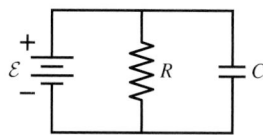

FIGURE 31.7 A circuit diagram for the circuit of Figure 31.5.

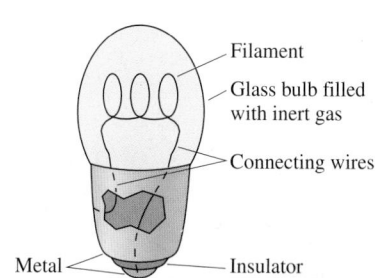

FIGURE 31.8 The anatomy of a lightbulb.

STOP TO THINK 31.2 Which of these diagrams represent the same circuit?

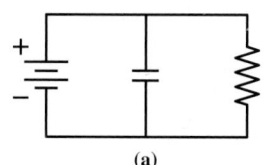 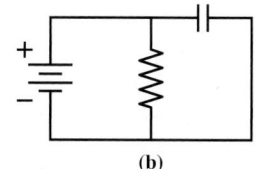 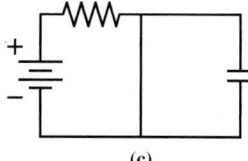 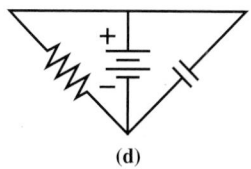

(a)	(b)	(c)	(d)

31.3 Kirchhoff's Laws and the Basic Circuit

We are now ready to begin analyzing circuits. To analyze a circuit means to find:

1. The potential difference across each circuit component.
2. The current in each circuit component.

Circuit analysis is based on *Kirchhoff's laws,* which we introduced in Chapters 28 and 30.

(a)

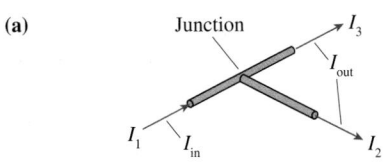

Junction law: $I_1 = I_2 + I_3$

(b)

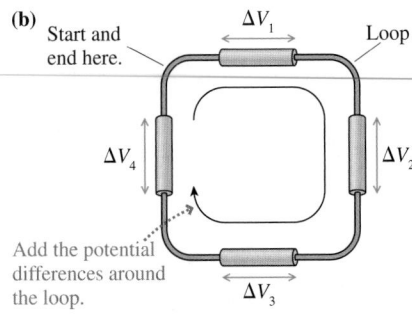

Loop law: $\Delta V_1 + \Delta V_2 + \Delta V_3 + \Delta V_4 = 0$

FIGURE 31.9 Kirchhoff's laws apply to junctions and loops.

You learned in Chapter 28 that charge and current are conserved. Consequently, the total current into the junction of Figure 31.9a must equal the total current leaving the junction. That is,

$$\sum I_{\text{in}} = \sum I_{\text{out}} \qquad (31.5)$$

This statement is **Kirchhoff's junction law.**

An important property of the electric potential that you learned in Chapter 30 is that the sum of the potential differences around any loop or closed path is zero. This is a statement of energy conservation, because a charge that moves around a closed path and returns to its starting point has $\Delta U = 0$. We apply this idea to the circuit of Figure 31.9b by adding all of the potential differences *around* the loop formed by the circuit. Doing so gives

$$\Delta V_{\text{loop}} = \sum_i (\Delta V)_i = 0 \qquad (31.6)$$

where $(\Delta V)_i$ is the potential difference of the ith component in the loop. This statement is **Kirchhoff's loop law.**

Kirchhoff's loop law can be true only if at least one of the $(\Delta V)_i$ is negative. To apply the loop law, we need to explicitly identify which potential differences are positive and which are negative.

TACTICS BOX 31.1   Using Kirchhoff's loop law

❶ **Draw a circuit diagram.** Label all known and unknown quantities.

❷ **Assign a direction to the current.** Draw and label a current arrow I to show your choice.
  - If you know the actual current direction, choose that direction.
  - If you don't know the actual current direction, make an arbitrary choice. All that will happen if you choose wrong is that your value for I will end up negative.

❸ **"Travel" around the loop.** Start at any point in the circuit, then go all the way around the loop in the direction you assigned to the current in step 2. As you go through each circuit element, ΔV is interpreted to mean

$$\Delta V = V_{\text{downstream}} - V_{\text{upstream}}$$

  - For an ideal battery in the negative-to-positive direction:
$$\Delta V_{\text{bat}} = +\mathcal{E}$$

  - For an ideal battery in the positive-to-negative direction:
$$\Delta V_{\text{bat}} = -\mathcal{E}$$

  - For a resistor:
$$\Delta V_{\text{R}} = -IR$$

❹ **Apply the loop law:**

$$\sum (\Delta V)_i = 0$$

NOTE ▶ ΔV_{R} for a resistor seems to be opposite Ohm's law, but Ohm's law was only concerned with the *magnitude* of the potential difference. Kirchhoff's law requires us to recognize that the electric potential inside a resistor *decreases* in the direction of the current. Thus $\Delta V = V_{\text{downstream}} - V_{\text{upstream}} < 0$. ◀

The Basic Circuit

The most basic electric circuit is that of a single resistor connected to the two terminals of a battery. Figure 31.10a shows a literal picture of the circuit elements and the connecting wires; Figure 31.10b is the circuit diagram. Notice that this is a **complete circuit,** forming a continuous path between the battery terminals.

(a)    **(b)**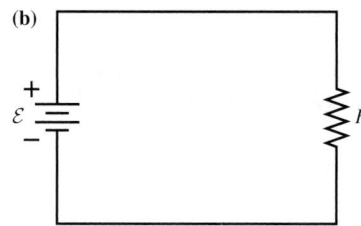

FIGURE 31.10 The basic circuit of a resistor connected to a battery.

The resistor might be a known resistor, such as "a 10 Ω resistor," or it might be some other resistive device, such as a lightbulb. Regardless of what the resistor is, it is called the **load.** The battery is called the **source.**

Figure 31.11 shows the use of Kirchhoff's loop law to analyze this circuit. Two things are worth noting:

1. This circuit has no junctions, so the current I is the same in all four sides of the circuit. Kirchhoff's junction law is not needed.
2. We're assuming the ideal-wire model, in which there are *no* potential differences along the connecting wires.

Kirchhoff's loop law, with two circuit elements, is

$$\Delta V_{loop} = \sum_i (\Delta V)_i = \Delta V_{bat} + \Delta V_R = 0 \qquad (31.7)$$

Let's look at each of the two terms in Equation 31.7:

1. The potential *increases* as we travel through the battery on our clockwise journey around the loop. We enter the negative terminal and, farther downstream, exit the positive terminal after having gained potential $\mathcal{E}$. Thus

$$\Delta V_{bat} = +\mathcal{E}$$

2. The *magnitude* of the potential difference across the resistor is $\Delta V = IR$, but Ohm's law does not tell us whether this should be positive or negative—and the difference is crucial. The potential of a conductor *decreases* in the direction of the current, which we've indicated with the + and − signs in Figure 31.11. Thus

$$\Delta V_R = V_{downstream} - V_{upstream} = -IR$$

NOTE ▶ Determining which potential differences are positive and which negative is perhaps *the* most important step in circuit analysis. ◀

With this information about ΔV_{bat} and ΔV_R, the loop equation becomes

$$\mathcal{E} - IR = 0 \qquad (31.8)$$

We can solve the loop equation to find that the current in the circuit is

$$I = \frac{\mathcal{E}}{R} \qquad (31.9)$$

We can then use the current to find that the resistor's potential difference is

$$\Delta V_R = -IR = -\mathcal{E} \qquad (31.10)$$

❶ Draw circuit diagram.

❷ The orientation of the battery indicates a clockwise current, so assign a clockwise direction to I.

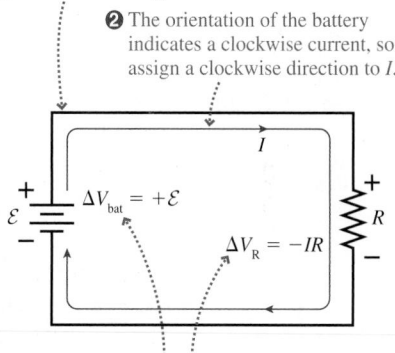

❸ Determine ΔV for each circuit element.

FIGURE 31.11 Analysis of the basic circuit using Kirchhoff's loop law.

This result should come as no surprise. The potential energy that the charges gain in the battery is subsequently lost as they "fall" through the resistor.

Equations 31.9 and 31.10 are the primary results of our circuit analysis. Notice that the current depends on the size of the resistance. The emf of a battery is a fixed quantity; the current that the battery delivers depends jointly on the emf and on the load.

EXAMPLE 31.2 A single-resistor circuit

A 15 Ω resistor is connected to the terminals of a 1.5 V battery.

a. What is the current in the circuit?
b. Draw a graph showing the potential as a function of distance traveled through the circuit, starting from $V = 0$ V at the negative terminal of the battery.

MODEL Assume ideal connecting wires and an ideal battery for which $\Delta V_{bat} = \mathcal{E}$.

VISUALIZE Figure 31.12 shows the circuit. We'll choose a clockwise (cw) direction for I.

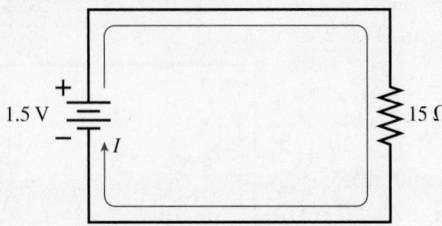

FIGURE 31.12 The circuit of Example 31.2.

SOLVE

a. This is the basic circuit of a single resistor connected to a single battery. The current is given by Equation 31.9:

$$I = \frac{\mathcal{E}}{R} = \frac{1.5 \text{ V}}{15 \text{ Ω}} = 0.10 \text{ A}$$

b. The battery's potential difference is $\Delta V_{bat} = \mathcal{E} = 1.5$ V. The resistor's potential difference is $\Delta V_R = -\mathcal{E} = -1.5$ V. Based on this, Figure 31.13 shows the potential experienced by charges flowing around the circuit. The distance s is measured from the battery's negative terminal, and we have chosen to let $V = 0$ V at that point. The potential ends at the value from which it started.

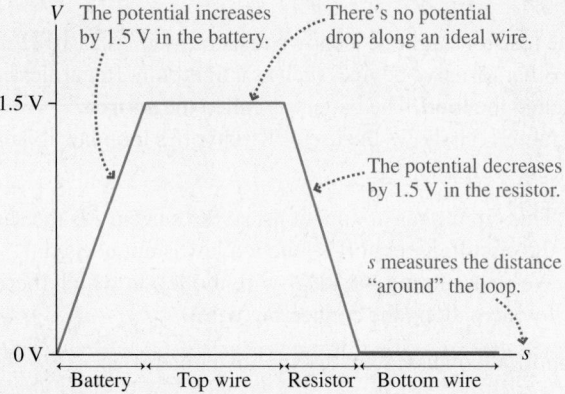

FIGURE 31.13 A graphical presentation of how the potential changes around the loop of the circuit in Figure 31.12.

ASSESS The value of I is positive. This tells us that the *actual* current direction is cw.

EXAMPLE 31.3 A more complex circuit

Analyze the circuit shown in Figure 31.14a.

a. Find the current in and the potential difference across each resistor.
b. Draw a graph showing how the potential changes around the circuit, starting from $V = 0$ V at the negative terminal of the 6 V battery.

MODEL Assume ideal connecting wires and ideal batteries, for which $\Delta V_{bat} = \mathcal{E}$.

VISUALIZE In Figure 31.14b the circuit has been redrawn; $\mathcal{E}_1$, $\mathcal{E}_2$, R_1, and R_2 defined; and the cw direction chosen for the current. This direction is an arbitrary choice because, with two batteries, we may not be sure of the actual current direction.

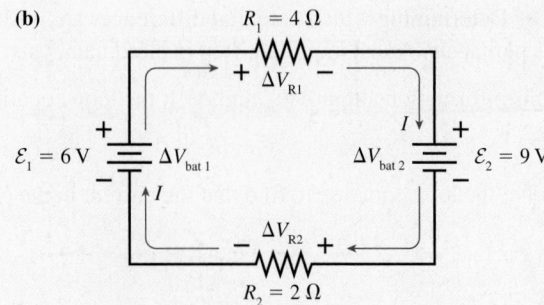

FIGURE 31.14 Circuit for Example 31.3.

SOLVE

a. How do we deal with *two* batteries? Can charge flow "backward" through a battery, from positive to negative? Consider the charge-escalator analogy. Left to itself, a charge escalator lifts charge from lower to higher potential. But it *is* possible to run down an up escalator, as many of you have probably done. If two escalators are placed "head to head," whichever is stronger will, indeed, force the charge to run down the up escalator of the other battery. The current in a battery *can* be from positive to negative if driven in that direction by a larger emf from a second battery. Indeed, this is how rechargeable batteries are recharged.

Because there are no junctions, the current is the same through *each* component in the circuit. With some thought, we might deduce whether the current is cw or ccw, but we do not need to know in advance of our analysis. We will choose a clockwise direction for the current and solve for the value of I. If our solution is positive, then the current really is cw. If the solution should turn out to be negative, we will know that the current is ccw. Kirchhoff's loop law, going clockwise from the negative terminal of battery 1, is

$$\Delta V_{\text{closed loop}} = \sum_i (\Delta V)_i = \Delta V_{\text{bat 1}} + \Delta V_{\text{R1}}$$
$$+ \Delta V_{\text{bat 2}} + \Delta V_{\text{R2}} = 0$$

All the signs are $+$ because this is a formal statement of *adding* potential differences around the loop. Next we can evaluate each ΔV. As we go cw, the charges *gain* potential in battery 1 but *lose* potential in battery 2. Thus $\Delta V_{\text{bat 1}} = +\mathcal{E}_1$ and $\Delta V_{\text{bat 2}} = -\mathcal{E}_2$. There is a *loss* of potential in traveling through each resistor, because we're traversing them in the direction we assigned to the current, so $\Delta V_{\text{R1}} = -IR_1$ and $\Delta V_{\text{R2}} = -IR_2$. Thus Kirchhoff's loop law becomes

$$\sum (\Delta V)_i = \mathcal{E}_1 - IR_1 - \mathcal{E}_2 - IR_2$$
$$= \mathcal{E}_1 - \mathcal{E}_2 - I(R_1 + R_2) = 0$$

We can solve this equation to find the current in the loop:

$$I = \frac{\mathcal{E}_1 - \mathcal{E}_2}{R_1 + R_2} = \frac{6\text{ V} - 9\text{ V}}{4\ \Omega + 2\ \Omega} = -0.50\text{ A}$$

The value of I is negative, hence the actual current in this circuit is 0.50 A *counterclockwise*. You perhaps anticipated this from the orientation of the larger 9 V battery.

b. The potential difference across the 4 Ω resistor is

$$\Delta V_{\text{R1}} = -IR_1 = -(-0.50\text{ A})(4\ \Omega) = +2.0\text{ V}$$

Because the current is actually ccw, the resistor's potential *increases* in the cw direction of our travel around the loop. Similarly, the potential difference across the 2 Ω resistor is $\Delta V_{\text{R2}} = 1.0$ V. Figure 31.15 is a graph of potential versus position, following a cw path around the loop starting from $V = 0$ V at the negative terminal of the 6 V battery.

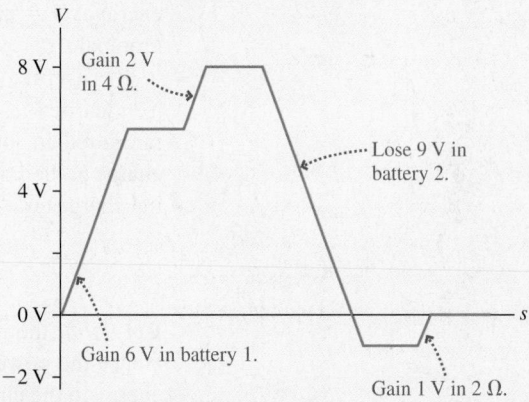

FIGURE 31.15 A graphical presentation of how the potential changes around the loop.

ASSESS Notice how the potential *drops* 9 V upon passing through battery 2 in the cw direction. It then gains 2 V upon passing through R_2 to end at the starting potential.

STOP TO THINK 31.3 What is ΔV across the unspecified circuit element? Does the potential increase or decrease when traveling through this element in the direction assigned to I?

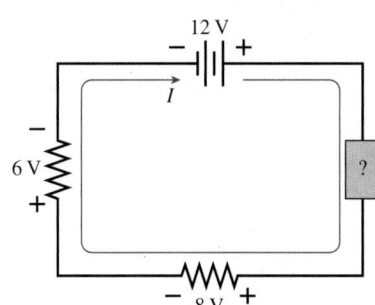

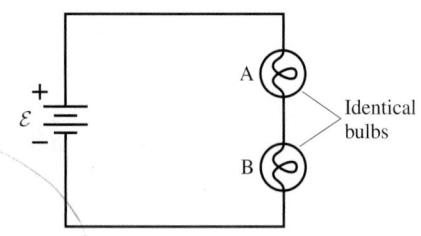

FIGURE 31.16 Which lightbulb is brighter?

31.4 Energy and Power

The circuit of Figure 31.16 has two identical lightbulbs, A and B. Which is brighter? Or are they equally bright? Think about this before going on.

You might have been tempted to say that A is brighter. After all, the current gets to A first, so A might "use up" some of the current and leave less for B. But this would violate the laws of conservation of charge and conservation of current. There are no junctions between A and B, so the current through the two bulbs must be the same. Hence the bulbs are equally bright.

It's not current that the bulbs use up, it's *energy*. A battery not only supplies a potential difference, it also supplies the energy to a circuit. The charge escalator is an energy-transfer process, transferring chemical energy E_{chem} stored in the battery to the potential energy U of the charges. That energy is then dissipated as the charges move through the wires and resistors, increasing their thermal energy until, in the case of the lightbulb filaments, they glow.

A charge gains potential energy $\Delta U = q\Delta V_{bat}$ as it moves up the charge escalator in the battery. For an ideal battery, with $\Delta V_{bat} = \mathcal{E}$, the battery supplies energy $\Delta U = q\mathcal{E}$ to charge q as it lifts the charge from the negative to the positive terminal.

It is useful to know the *rate* at which the battery supplies energy to the charges. You learned in Chapter 11 that the rate at which energy is transferred is *power*, measured in joules per second or *watts*. If energy $\Delta U = q\mathcal{E}$ is transferred to charge q, then the *rate* at which energy is transferred from the battery to the moving charges is

$$P_{bat} = \text{rate of energy transfer} = \frac{dU}{dt} = \frac{dq}{dt}\mathcal{E} \qquad (31.11)$$

But dq/dt, the rate at which charge moves through the battery, is the current I. Hence the power supplied by a battery, or the rate at which the battery transfers energy to the charges passing through it, is

$$P_{bat} = I\mathcal{E} \qquad \text{(power delivered by an emf)} \qquad (31.12)$$

$I\mathcal{E}$ has units of J/s, or W.

EXAMPLE 31.4 Delivering power
A 90 Ω load is connected to a 120 V battery. How much power is delivered by the battery?

SOLVE This is our basic battery-and-resistor circuit, which we analyzed earlier. In this case

$$I = \frac{\mathcal{E}}{R} = \frac{120\ \text{V}}{90\ \Omega} = 1.33\ \text{A}$$

Thus the power delivered by the battery is

$$P_{bat} = I\mathcal{E} = (1.33\ \text{A})(120\ \text{V}) = 160\ \text{W}$$

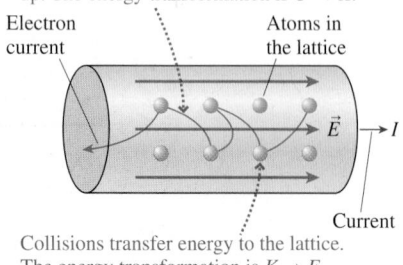

The electric field causes electrons to speed up. The energy transformation is $U \rightarrow K$.

Electron current

Atoms in the lattice

$\vec{E}$ → I

Current

Collisions transfer energy to the lattice. The energy transformation is $K \rightarrow E_{th}$.

FIGURE 31.17 A current-carrying resistor dissipates power because the electric force does work on the charges.

P_{bat} is the energy transferred per second from the battery's store of chemicals to the moving charges that make up the current. But what happens to this energy? Where does it end up? Figure 31.17, a section of a current-carrying resistor, reminds you of our microscopic model of conduction. The electrons accelerate in the electric field, then collide with atoms in the lattice. The acceleration phase is a transformation of potential to kinetic energy. The collisions then transfer the electron's kinetic energy to the *thermal* energy of the lattice. The potential energy was acquired in the battery, from the conversion of chemical energy, so the entire energy transfer process looks like

$$E_{chem} \rightarrow U \rightarrow K \rightarrow E_{th}$$

The net result is that **the battery's chemical energy is transferred to the thermal energy of the resistors,** raising their temperature.

Let's look more closely at the energy transfer in the resistor. Suppose the average distance between collisions is d. The electric force $\vec{F} = q\vec{E}$ exerted on charge q does work as it pushes the charge through distance d. The field is constant inside the resistor, so the work is simply

$$W = F\Delta s = qEd \qquad (31.13)$$

According to the work-kinetic energy theorem, this work increases the kinetic energy of charge q by $\Delta K = W = qEd$. This kinetic energy is transferred to the lattice when charge q collides with a lattice atom, causing the energy of the lattice to increase by

$$\Delta E_{\text{per collision}} = \Delta K = qEd$$

Collisions occur over and over as the charge makes its way through a resistor of length L. The total energy transferred while traveling distance L is

$$\Delta E_{\text{th}} = qEL \qquad (31.14)$$

But EL is the potential difference ΔV_R between the two ends of the resistor. Thus *each* charge q, as it travels the length of the resistor, transfers energy to the atomic lattice in the amount

$$\Delta E_{\text{th}} = q\Delta V_R \qquad (31.15)$$

The *rate* at which energy is transferred from the current to the resistor is thus

$$P_R = \frac{dE_{\text{th}}}{dt} = \frac{dq}{dt}\Delta V_R = I\Delta V_R \qquad (31.16)$$

We say that this power—so many joules per second—is *dissipated* by the resistor as charge flows through it. The resistor, in turn, transfers this energy to the air and to the circuit board on which it is mounted, causing the circuit and all its surroundings to heat up.

From our analysis of the basic circuit, in which a single resistor is connected to a battery, we learned that $\Delta V_R = \mathcal{E}$. That is, the potential difference across the resistor is exactly the emf supplied by the battery. But then Equations 31.12 and 31.16, for P_{bat} and P_R, are numerically equal, and we find that

$$P_R = P_{\text{bat}} \qquad (31.17)$$

The answer to the question "What happens to the energy supplied by the battery?" is "The battery's chemical energy is transformed into the thermal energy of the resistor." The *rate* at which the battery supplies energy is exactly equal to the *rate* at which the resistor dissipates energy. This is, of course, exactly what we would have expected from energy conservation.

EXAMPLE 31.5 The power of light
How much current is "drawn" by a 100 W lightbulb connected to a 120 V outlet?

MODEL Most household appliances, such as a 100 W lightbulb or a 1500 W hair dryer, have a power rating. The rating does *not* mean that these appliances *always* dissipate that much power. These appliances are intended for use at a standard household voltage of 120 V, and their rating is the power they will dissipate *if* operated with a potential difference of 120 V. Their power consumption will differ from the rating if they are operated at any other potential difference.

SOLVE Because the lightbulb is operating as intended, it will dissipate 100 W of power. Thus

$$I = \frac{P}{\Delta V_R} = \frac{100\ \text{W}}{120\ \text{V}} = 0.833\ \text{A}$$

ASSESS A current of 0.833 A in this lightbulb transfers 100 J/s to the thermal energy of the filament which, in turn, dissipates 100 J/s as heat and light to its surroundings.

A resistor obeys Ohm's law, $\Delta V_R = IR$. This gives us two alternative ways of writing the power dissipated by a resistor. We can either substitute IR for ΔV_R or substitute $\Delta V_R/R$ for I. Thus

$$P_R = I\Delta V_R = I^2 R = \frac{(\Delta V_R)^2}{R} \qquad \text{(power dissipated by a resistor)} \qquad (31.18)$$

If the same current I passes through several resistors in series, then $P_R = I^2 R$ tells us that most of the power will be dissipated by the largest resistance. This is why a lightbulb filament glows but the connecting wires do not. Essentially *all* of the power supplied by the battery is dissipated by the high-resistance lightbulb filament and essentially no power is dissipated by the low-resistance wires. The filament gets very hot, but the wires do not.

EXAMPLE 31.6 The power of sound

Most loudspeakers are designed to have a resistance of 8 Ω. If an 8 Ω loudspeaker is connected to a stereo amplifier with a rating of 100 W, what is the maximum possible current to the loudspeaker?

MODEL The rating of an amplifier is the *maximum* power it can deliver. Most of the time it delivers far less, but the maximum might be reached for brief, intense sounds like cymbal clashes.

SOLVE The loudspeaker is a resistive load. The maximum current to the loudspeaker occurs when the amplifier delivers maximum power $P_{max} = (I_{max})^2 R$. Thus

$$I_{max} = \sqrt{\frac{P_{max}}{R}} = \sqrt{\frac{100 \text{ W}}{8 \text{ Ω}}} = 3.5 \text{ A}$$

EXAMPLE 31.7 A dim bulb

How much power is dissipated by a 60 W (120 V) lightbulb when operated, using a dimmer switch, at 100 V?

MODEL The 60 W rating is for operation at 120 V.

SOLVE The lightbulb dissipates 60 W at $\Delta V_R = 120$ V. Thus the filament's resistance is

$$R = \frac{(\Delta V_R)^2}{P_R} = \frac{(120 \text{ V})^2}{60 \text{ W}} = 240 \text{ Ω}$$

The power dissipation when operated at $\Delta V_R = 100$ V is

$$P_R = \frac{(\Delta V_R)^2}{R} = \frac{(100 \text{ V})^2}{240 \text{ Ω}} = 42 \text{ W}$$

ASSESS Actually, this result is not quite correct. As noted previously, a filament is not a true ohmic material because the resistance changes somewhat with temperature. The filament's resistance at 100 V, where it glows less brightly, is not quite the same as its resistance at 120 V. The voltage in this example is still near 120 V, so the temperature of the filament will decrease only slightly, and our answer should be fairly accurate. However, this calculation would not give a good result for the power dissipation at 20 V, where the filament's temperature would be much less than at 120 V.

The electric meter on the side of your house or apartment records the kilowatt hours of electric energy that you use.

It is the stored chemical energy of the battery that is used up by a lightbulb. The energy is used by being converted first to the energy of the charges, then to the thermal energy of the filament, thus heating it, and lastly to the heat and light energy that we feel and see coming from the bulb. Conservation of energy is of paramount importance for understanding electric circuits.

Kilowatt Hours

The energy dissipated (i.e., transformed into thermal energy) by a resistor during time Δt is $E_{th} = P_R \Delta t$. The product of watts and seconds is joules, the SI unit of energy. However, your local electric company prefers to use a different unit, called *kilowatt hours,* to measure the energy you use each month.

A load that consumes P_R kW of electricity in Δt hours has used $P_R \Delta t$ **kilowatt hours** of energy, abbreviated kWh. For example, a 4000 W electric water heater uses 40 kWh of energy in 10 hours. A 1500 W hair dryer uses 0.25 kWh of energy in 10 minutes. Despite the rather unusual name, a kilowatt hour is a unit of energy. A homework problem will let you find the conversion factor from kilowatt hours to joules.

Your monthly electric bill specifies the number of kilowatt hours you used last month. This is the amount of energy that the electric company delivered to you, via an electric current, and that you transformed into light and thermal energy inside your home. The cost of electricity varies throughout the country, but the average cost of electricity in the United States is approximately 10¢ per kWh ($0.10/kWh). Thus it costs about $4.00 to run your water heater for 10 hours, about 2.5¢ to dry your hair.

STOP TO THINK 31.4   Rank in order, from largest to smallest, the powers P_a to P_d dissipated in resistors a to d.

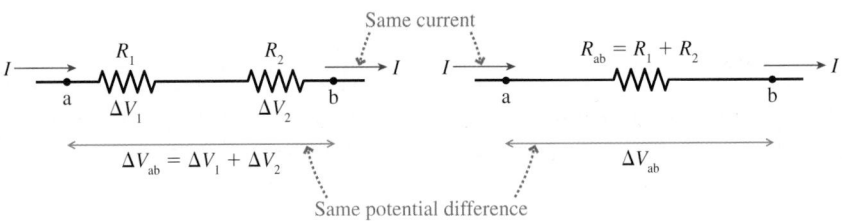

$$+ \; \Delta V \; -$$
$$R$$
(a)

$$+ \; 2\Delta V \; -$$
$$R$$
(b)

$$+ \; \Delta V \; -$$
$$2R$$
(c)

$$+ \; \Delta V \; -$$
$$\tfrac{1}{2}R$$
(d)

31.5 Series Resistors

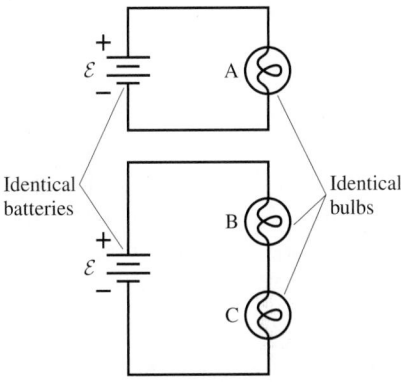

FIGURE 31.18 How does the brightness of bulb B compare to that of A?

Many circuits contain two or more resistors connected to each other in various ways. Thus much of circuit analysis consists of analyzing different *combinations* of resistors. As an example, consider the three lightbulbs in Figure 31.18. The batteries are identical and the bulbs are identical. You learned in the previous section that B and C are equally bright, because of conservation of current, but how does the brightness of B compare to that of A? Think about this before going on.

Figure 31.19a shows two resistors placed end to end between points a and b. Resistors that are aligned end to end, *with no junctions between them,* are called **series resistors** or, sometimes, resistors "in series." Because there are no junctions, and because current is conserved, the current I must be the same through each of these resistors. That is, the current out of the last resistor in a series is equal to the current into the first resistor.

(a) Two resistors in series   **(b)** An equivalent resistor

Same current

$$I \xrightarrow{} \quad R_1 \qquad R_2 \qquad \xrightarrow{} I \qquad I \xrightarrow{} \quad R_{ab} = R_1 + R_2 \qquad \xrightarrow{} I$$
$$a \quad \Delta V_1 \qquad \Delta V_2 \quad b \qquad\qquad a \qquad\qquad b$$

$$\Delta V_{ab} = \Delta V_1 + \Delta V_2 \qquad\qquad \Delta V_{ab}$$

Same potential difference

FIGURE 31.19 Replacing two series resistors with an equivalent resistor.

The potential differences across the two resistors are $\Delta V_1 = IR_1$ and $\Delta V_2 = IR_2$. The total potential difference ΔV_{ab} between points a and b is the sum of the individual potential differences:

$$\Delta V_{ab} = \Delta V_1 + \Delta V_2 = IR_1 + IR_2 = I(R_1 + R_2) \qquad (31.19)$$

Suppose, as in Figure 31.19b, we replaced the two resistors with a single resistor having current I and potential difference $\Delta V_{ab} = \Delta V_1 + \Delta V_2$. We can then use Ohm's law to find that the resistance R_{ab} between points a and b is

$$R_{ab} = \frac{\Delta V_{ab}}{I} = \frac{I(R_1 + R_2)}{I} = R_1 + R_2 \qquad (31.20)$$

Because the battery has to establish the same potential difference and provide the same current in both cases, the two resistors R_1 and R_2 act exactly the same as a *single* resistor of value $R_1 + R_2$. We can say that the single resistor R_{ab} is *equivalent* to the two resistors in series.

There was nothing special about having only two resistors. If we have N resistors in series, their **equivalent resistance** is

$$R_{eq} = R_1 + R_2 + \cdots + R_N \qquad \text{(series resistors)} \qquad (31.21)$$

The behavior of the circuit will be unchanged if the N series resistors are replaced by the single resistor R_{eq}. The key idea in this analysis is the fact that **resistors in series all have the same current.**

EXAMPLE 31.8 **A series resistor circuit**

a. What is the current in the circuit of Figure 31.20a?
b. Draw a graph of potential versus position in the circuit, going cw from $V = 0$ V at the battery's negative terminal.

MODEL The three resistors are end to end, with no junctions between them, and thus are in series. Assume ideal connecting wires and an ideal battery.

SOLVE

a. Nothing about the circuit's behavior will change if we replace the three series resistors by their equivalent resistance

$$R_{eq} = 15\,\Omega + 4\,\Omega + 8\,\Omega = 27\,\Omega$$

This is shown as an equivalent circuit in Figure 31.20b. Now we have a circuit with a single battery and a single resistor, for which we know the current to be

$$I = \frac{\mathcal{E}}{R_{eq}} = \frac{9\text{ V}}{27\,\Omega} = 0.333\text{ A}$$

b. $I = 0.333$ A is the current in each of the three resistors in the original circuit. Thus the potential differences across the resistors are $\Delta V_{R1} = -IR_1 = -5.00$ V, $\Delta V_{R2} = -IR_2 = -1.33$ V, and $\Delta V_{R3} = -IR_3 = -2.67$ V for the 15 Ω, the 4 Ω, and the 8 Ω resistors, respectively. Figure 31.20c shows that the potential increases by 9 V due to the battery's emf, then decreases by 9 V in three steps.

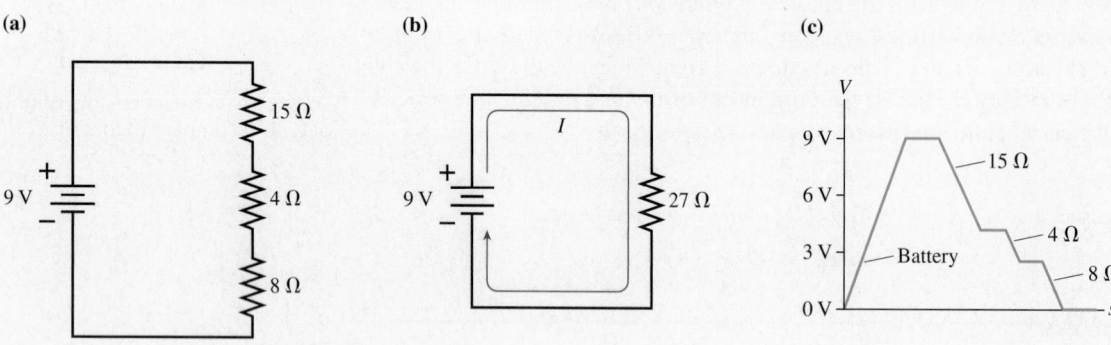

(a)   **(b)**   **(c)**

FIGURE 31.20 Analyzing a circuit with series resistors.

Now we can answer the lightbulb question posed at the beginning of this section. Suppose the resistance of each lightbulb is R. The battery drives current $I_A = \mathcal{E}/R$ through bulb A. Bulbs B and C are in series, with an equivalent resistance $R_{eq} = 2R$, but the battery has the same emf $\mathcal{E}$. Thus the current through bulbs B and C is $I_{B+C} = \mathcal{E}/R_{eq} = \mathcal{E}/2R = \frac{1}{2}I_A$. Bulb B has only half the current of bulb A, so B is dimmer.

Many people predict that A and B should be equally bright. It's the same battery, so shouldn't it provide the same current to both circuits? No! A battery is a source of emf, *not* a source of current. In other words, the battery's emf is the same no matter how the battery is used. When you buy a 1.5 V battery you're buying a device that provides a specified amount of potential difference, not a specified amount of current. The battery does provide the current to the circuit, but the *amount* of current depends on the resistance of the load. Your 1.5 V battery causes 1 A to pass through a 1.5 Ω load but only 0.1 A to pass through a 15 Ω load. As an analogy, think about a water faucet. The pressure in the water main underneath the street is a fixed and unvarying quantity set by the water company. It's like the emf of a battery. But the amount of water coming out of a faucet depends on how far you open it. A faucet opened slightly has a "high resistance," so only a little water flows. A wide-open faucet has a "low resistance," and the water flow is large.

We're spending a lot of time on this property of a battery because it's a critical idea for understanding circuits. In summary, **a battery provides a fixed and unvarying emf (potential difference). It does *not* provide a fixed and unvarying current. The amount of current depends jointly on the battery's emf *and* the resistance of the circuit attached to the battery.**

Ammeters

A device that measures the current in a circuit element is called an **ammeter.** Because charge flows *through* circuit elements, an ammeter must be placed *in series* with the circuit element whose current is to be measured.

Figure 31.21a shows a simple one-resistor circuit with an unknown emf $\mathcal{E}$. We can measure the current in the circuit by inserting the ammeter as shown in Figure 31.21b. Notice that we have to *break the connection* between the battery and the resistor in order to insert the ammeter. Now the current in the resistor has to first pass through the ammeter.

(a)   **(b)**   Ammeter

FIGURE 31.21 An ammeter measures the current in a circuit element.

Because the ammeter is now in series with the resistor, the total resistance seen by the battery is $R_{eq} = 6\ \Omega + R_{ammeter}$. In order that the ammeter measure the current without changing the current, the ammeter's resistance must, in this case, be $\ll 6\ \Omega$. Indeed, an ideal ammeter has $R_{ammeter} = 0\ \Omega$ and thus has no effect on the current. Real ammeters come very close to this ideal.

The ammeter in Figure 31.21b reads 0.50 A, meaning that the current in the 6 Ω resistor is $I = 0.50$ A. Thus the resistor's potential difference is $\Delta V_R = -IR = -3.0$ V. If the ammeter is ideal, as we will assume, then, from Kirchhoff's loop law, the battery's emf is $\mathcal{E} = -\Delta V_R = 3.0$ V.

STOP TO THINK 31.5   What are the current and the potential at points a to e?

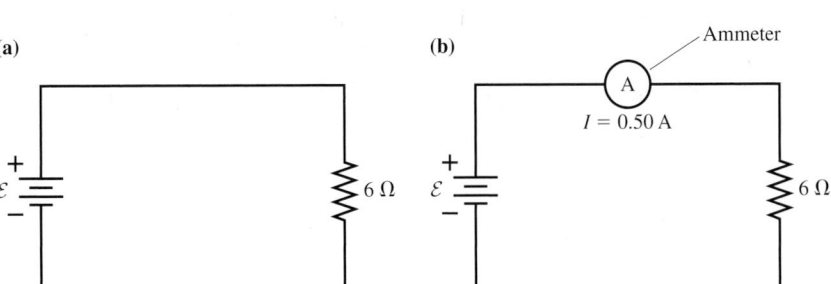

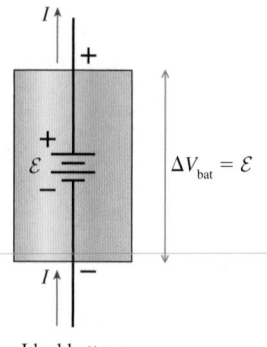

Ideal battery

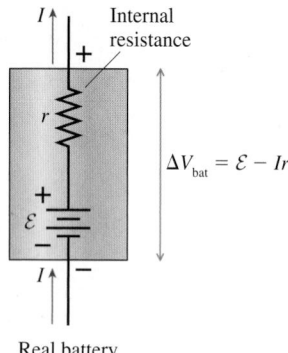

Real battery

FIGURE 31.22 An ideal battery and a real battery.

31.6 Real Batteries

Let's look at how real batteries differ from the ideal battery we have been assuming. Real batteries, like ideal batteries, separate charge and create a potential difference. However, real batteries also provide a slight resistance to the charges on the charge escalator. They have what is called an **internal resistance,** which is symbolized by r. Figure 31.22 shows both an ideal and a real battery.

From our vantage point outside a battery, we cannot see $\mathcal{E}$ and r separately. To the user, the battery provides a potential difference ΔV_{bat} called the **terminal voltage.** $\Delta V_{bat} = \mathcal{E}$ for an ideal battery, but the presence of the internal resistance affects ΔV_{bat}. Suppose the current in the battery is I. As charges travel from the negative to the positive terminal, they gain potential $\mathcal{E}$ but *lose* potential $\Delta V_{int} = -Ir$ due to the internal resistance. Thus the terminal voltage of the battery is

$$\Delta V_{bat} = \mathcal{E} - Ir \le \mathcal{E} \tag{31.22}$$

Only when $I = 0$, meaning that the battery is not being used, is $\Delta V_{bat} = \mathcal{E}$.

Figure 31.23 shows a single resistor R connected to the terminals of a battery having emf $\mathcal{E}$ and internal resistance r. Resistances R and r are in series, so we can replace them, for the purpose of circuit analysis, with a single equivalent resistor $R_{eq} = R + r$. Hence the current in the circuit is

$$I = \frac{\mathcal{E}}{R_{eq}} = \frac{\mathcal{E}}{R + r} \tag{31.23}$$

If $r \ll R$, so that the internal resistance of the battery is negligible, then $I = \mathcal{E}/R$, exactly the result we found before. But the current decreases significantly if $r \approx R$.

Although physically separated, the internal resistance r is electrically in series with R.

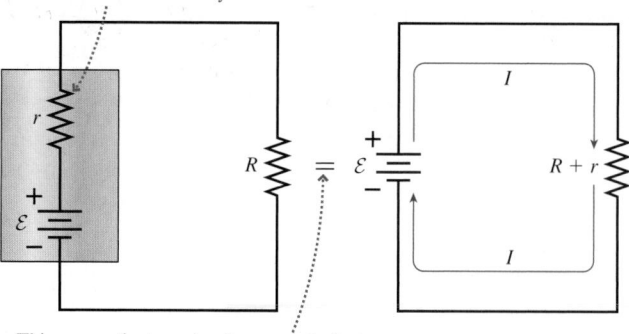

This means the two circuits are equivalent.

FIGURE 31.23 A single resistor connected to a real battery is in series with the battery's internal resistance, giving $R_{eq} = R + r$.

We can use Ohm's law to find that the potential difference across the load resistor R is

$$\Delta V_R = IR = \frac{R}{R + r}\mathcal{E} \tag{31.24}$$

Similarly, the potential difference across the terminals of the battery is

$$\Delta V_{bat} = \mathcal{E} - Ir = \mathcal{E} - \frac{r}{R + r}\mathcal{E} = \frac{R}{R + r}\mathcal{E} \tag{31.25}$$

The potential difference across the resistor is equal to the potential difference between the *terminals* of the battery, where the resistor is attached, *not* equal to the battery's emf.

EXAMPLE 31.9 Lighting up a flashlight

A 6 Ω flashlight bulb is powered by a 3 V battery having an internal resistance of 1 Ω. What are the power dissipation of the bulb and the terminal voltage of the battery?

MODEL Assume ideal connecting wires but not an ideal battery.

VISUALIZE The circuit diagram looks like Figure 31.23. R is the resistance of the bulb's filament.

SOLVE Equation 31.23 gives us the current:

$$I = \frac{\mathcal{E}}{R + r} = \frac{3\text{ V}}{6\ \Omega + 1\ \Omega} = 0.43\text{ A}$$

This is 15% less than the 0.5 A an ideal battery would supply. The potential difference across the resistor is $\Delta V_R = IR = 2.57$ V, thus the power dissipation is

$$P_R = I\Delta V = 1.1\text{ W}$$

The battery's terminal voltage is

$$\Delta V_{\text{bat}} = \frac{R}{R + r}\mathcal{E} = \frac{6\ \Omega}{6\ \Omega + 1\ \Omega}\,3\text{ V} = 2.6\text{ V}$$

ASSESS 1 Ω is a typical internal resistance for a flashlight battery. The internal resistance causes the battery's terminal voltage to be 0.4 V less than its emf in this circuit.

A Short Circuit

In Figure 31.24 we've replaced the resistor with an ideal wire having $R_{\text{wire}} = 0\ \Omega$. When a connection of very low or zero resistance is made between two points in a circuit that are normally separated by a higher resistance, then we have what is called a **short circuit.** The wire in Figure 31.24 is *shorting out* the battery.

If the battery were ideal, shorting it with an ideal wire ($R = 0\ \Omega$) would cause the current to be $I = \mathcal{E}/0 = \infty$. The current, of course, cannot really become infinite. Instead, the battery's internal resistance r becomes the only resistance in the circuit. If we use $R = 0\ \Omega$ in Equation 31.23, we find that the *short-circuit current* is

$$I_{\text{short}} = \frac{\mathcal{E}}{r} \tag{31.26}$$

A 3 V battery with 1 Ω internal resistance generates a short circuit current of 3 A. This is the *maximum possible current* that this battery can produce. Adding any external resistance R will decrease the current to a value less than 3 A.

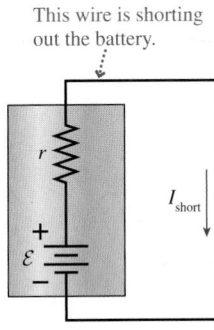

This wire is shorting out the battery.

FIGURE 31.24 The short-circuit current of a battery.

EXAMPLE 31.10 A short-circuited battery

What is the short-circuit current of a 12 V car battery with an internal resistance of 0.020 Ω? What happens to the power supplied by the battery?

SOLVE The short-circuit current is

$$I_{\text{short}} = \frac{\mathcal{E}}{r} = \frac{12\text{ V}}{0.02\ \Omega} = 600\text{ A}$$

Power is generated by chemical reactions in the battery and dissipated by the load resistance. But with a short-circuited battery, the load resistance is *inside* the battery! The "shorted" battery has to dissipate power $P = I^2 r = 7200$ W *internally*.

ASSESS This value is realistic. Car batteries are designed to drive the starter motor, which has a very small resistance and can draw a current of a few hundred amps. That is why the battery cables are so thick. A shorted car battery can produce an *enormous* amount of current. The normal response of a shorted car battery is to explode; it simply cannot dissipate this much power. Shorting a flashlight battery can make it rather hot, but your life is not in danger. Although the voltage of a car battery is relatively small, a car battery can be dangerous and should be treated with great respect.

Most of the time a battery is used under conditions in which $r \ll R$ and the internal resistance is negligible. The ideal battery model is fully justified in that case. Thus we will assume that batteries are ideal *unless stated otherwise.* But keep in mind that batteries (like all other sources of emf) do have an internal resistance, and that this internal resistance limits the maximum possible current of the battery.

31.7 Parallel Resistors

Figure 31.25 is another lightbulb puzzle. Initially the switch is open. The current is the same through bulbs A and B, because of conservation of current, and they are equally bright. Bulb C is not glowing. What happens to the brightness of A and B when the switch is closed? And how does the brightness of C then compare to that of A and B? Think about this before going on.

Figure 31.26a shows two resistors aligned side by side with their ends connected at junctions c and d. Resistors that are connected *at both ends* are called **parallel resistors** or, sometimes, resistors "in parallel." The left ends of both resistors are at the same potential V_c. Likewise, the right ends are at the same potential V_d. Thus the potential *differences* ΔV_1 and ΔV_2 are the *same* and are simply ΔV_{cd}.

Kirchhoff's junction law applies at the junctions. The input current I splits into currents I_1 and I_2 at the left junction. On the right, the two currents are recombined into current I. According to the junction law,

$$I = I_1 + I_2 \tag{31.27}$$

We can apply Ohm's law to each resistor, along with $\Delta V_1 = \Delta V_2 = \Delta V_{cd}$, to find that the current is

$$I = \frac{\Delta V_1}{R_1} + \frac{\Delta V_2}{R_2} = \frac{\Delta V_{cd}}{R_1} + \frac{\Delta V_{cd}}{R_2} = \Delta V_{cd}\left[\frac{1}{R_1} + \frac{1}{R_2}\right] \tag{31.28}$$

Suppose, as in Figure 31.26b, we replaced the two resistors with a single resistor having current I and potential difference ΔV_{cd}. This resistor is equivalent to the original two because the battery has to establish the same potential difference and provide the same current in either case. A second application of Ohm's law shows that the resistance between points c and d is

$$R_{cd} = \frac{\Delta V_{cd}}{I} = \left[\frac{1}{R_1} + \frac{1}{R_2}\right]^{-1} \tag{31.29}$$

The two resistors R_1 and R_2 act exactly the same as the single resistor R_{cd}. Resistor R_{cd} is *equivalent* to the two resistors in parallel.

There is nothing special about having chosen two resistors to be in parallel. If we have N resistors in parallel, the *equivalent resistance* is

$$R_{eq} = \left(\frac{1}{R_1} + \frac{1}{R_2} + \cdots + \frac{1}{R_N}\right)^{-1} \quad \text{(parallel resistors)} \tag{31.30}$$

The behavior of the circuit will be unchanged if the N parallel resistors are replaced by the single resistor R_{eq}. The key idea of this analysis is the fact that **resistors in parallel all have the same potential difference.**

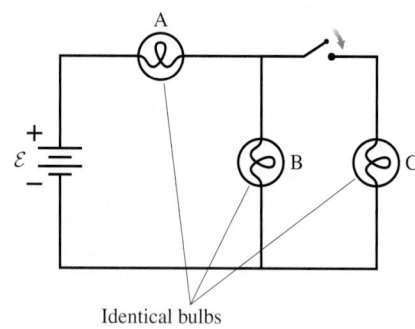

FIGURE 31.25 What happens to the brightness of the bulbs when the switch is closed?

(a) Two resistors in parallel

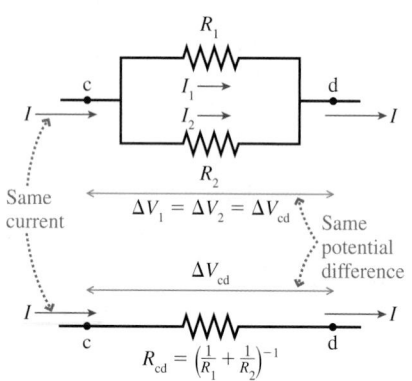

(b) An equivalent resistor

FIGURE 31.26 Replacing two parallel resistors with an equivalent resistor.

12.2

Two useful results of our analysis are the equivalent resistances of two identical resistors $R_1 = R_2 = R$ in series and in parallel:

Two identical resistors in series $R_{eq} = 2R$

Two identical resistors in parallel $R_{eq} = \dfrac{R}{2}$ (31.31)

EXAMPLE 31.11 **A parallel resistor circuit**

The three resistors of Figure 31.27 are connected to a 9 V battery. Find the potential difference across and current through each resistor.

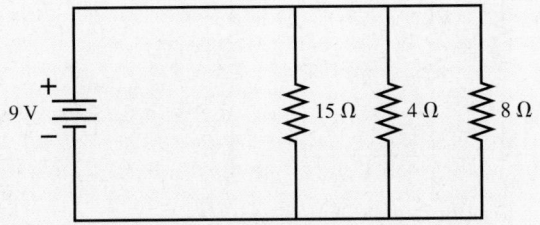

FIGURE 31.27 Parallel resistor circuit of Example 31.11.

MODEL The resistors are in parallel. Assume an ideal battery and ideal connecting wires.

SOLVE The three parallel resistors can be replaced by a single equivalent resistor

$$R_{eq} = \left(\frac{1}{15\ \Omega} + \frac{1}{4\ \Omega} + \frac{1}{8\ \Omega} \right)^{-1} = (0.4417\ \Omega^{-1})^{-1} = 2.26\ \Omega$$

The equivalent circuit is shown in Figure 31.28a, from which we find the current to be

$$I = \frac{\mathcal{E}}{R_{eq}} = \frac{9\ \text{V}}{2.26\ \Omega} = 3.98\ \text{A}$$

The potential difference across R_{eq} is $\Delta V_{eq} = \mathcal{E} = 9.0$ V. Now we have to be careful. Current I divides at the junction into the smaller currents I_1, I_2 and I_3 shown in Figure 31.28b. However, the division is *not* into three equal currents. According to Ohm's law, resistor i has current $I_i = \Delta V_i / R_i$. Because the resistors are in parallel, their potential differences are equal:

$$\Delta V_1 = \Delta V_2 = \Delta V_3 = \Delta V_{eq} = 9.0\ \text{V}$$

Thus the currents are

$$I_1 = \frac{9\ \text{V}}{15\ \Omega} = 0.60\ \text{A} \qquad I_2 = \frac{9\ \text{V}}{4\ \Omega} = 2.25\ \text{A}$$

$$I_3 = \frac{9\ \text{V}}{8\ \Omega} = 1.13\ \text{A}$$

ASSESS The *sum* of the three currents is 3.98 A, as required by Kirchhoff's junction law.

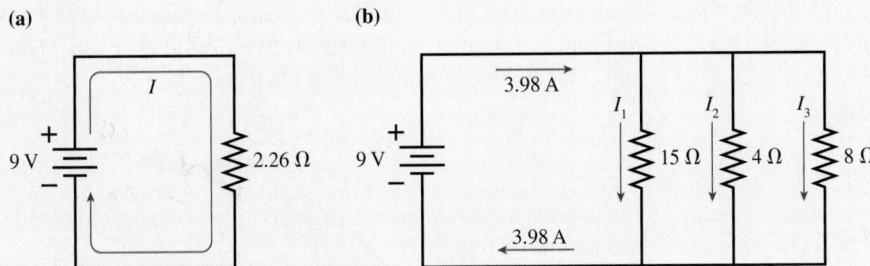

FIGURE 31.28 The parallel resistors can be replaced by a single equivalent resistor.

The result of Example 31.11 seems surprising. The equivalent of a parallel combination of 15 Ω, 4 Ω, and 8 Ω was found to be 2.26 Ω. How can the equivalent of a group of resistors be *less* than any single resistance in the group? Should not more resistors imply more resistance? The answer is yes for resistors in series but not for resistors in parallel. Even though a resistor is an obstacle to the flow of charge, parallel resistors provide more pathways for charge to get through. Consequently, the equivalent of several resistors in parallel is always *less* than any single resistor in the group.

Complex combinations of resistors can often be reduced to a single equivalent resistance through a step-by-step application of the series and parallel rules. The final example in this section illustrates this idea.

EXAMPLE 31.12 A combination of resistors

What is the equivalent resistance of the group of resistors shown in Figure 31.29a?

MODEL This circuit contains both series and parallel resistors.

SOLVE Reduction to a single equivalent resistance is best done in a series of steps, with the circuit being redrawn after each step. The procedure is shown in Figure 31.29b. Note that the

10 Ω and 25 Ω resistors are *not* in parallel. They are connected at their top ends but not at their bottom ends. Resistors must be connected at *both* ends to be in parallel. Similarly, the 10 Ω and 45 Ω resistors are *not* in series because of the junction between them. If the original group of four resistors occurred within a larger circuit, they could be replaced with a single 15.4 Ω resistor without having any effect on the rest of the circuit.

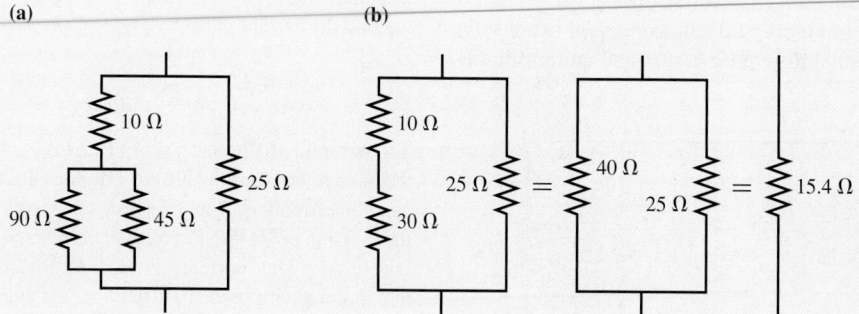

FIGURE 31.29 A combination of resistors is reduced to a single equivalent resistor.

Returning to the lightbulb question at the beginning of this section, suppose the resistance of each bulb in Figure 31.25 is R. Initially, before the switch is closed, bulbs A and B are in series with equivalent resistance $2R$. The current from the battery is

$$I_{before} = \frac{\mathcal{E}}{2R} = \frac{1}{2}\frac{\mathcal{E}}{R}$$

This is the current in both bulbs.

Closing the switch places bulbs B and C in parallel with each other. The equivalent resistance of two identical resistors in parallel is $R_{eq} = \frac{1}{2}R$. This equivalent resistance of B and C is in series with bulb A, hence the total resistance of the circuit is $\frac{3}{2}R$ and the current leaving the battery is

$$I_{after} = \frac{\mathcal{E}}{(3R/2)} = \frac{2}{3}\frac{\mathcal{E}}{R} > I_{before}$$

Closing the switch *decreases* the circuit resistance and thus *increases* the current leaving the battery.

All the charge flows through A, so A *increases* in brightness when the switch is closed. The current I_{after} then splits at the junction. Bulbs B and C have equal resistance, so the current splits equally. The current in B is $\frac{1}{3}(\mathcal{E}/R)$, which is *less* than I_{before}. Thus B *decreases* in brightness when the switch is closed. Bulb C has the same brightness as bulb B.

Voltmeters

A device that measures the potential difference across a circuit element is called a **voltmeter.** Because potential difference is measured *across* a circuit element, from one side to the other, a voltmeter is placed in *parallel* with the circuit element whose potential difference is to be measured.

Figure 31.30a shows a simple circuit in which a 17 Ω resistor is connected across a 9 V battery with an unknown internal resistance. By connecting a voltmeter across the resistor, as shown in Figure 31.30b, we can measure the potential difference across the resistor. Unlike an ammeter, using a voltmeter does *not* require that we break the connections.

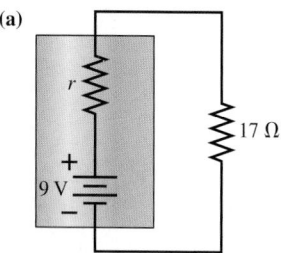

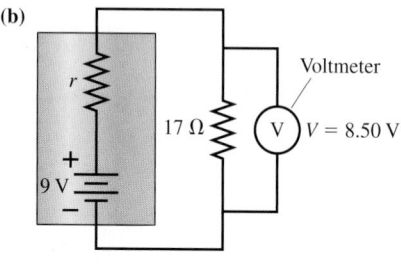

FIGURE 31.30 A voltmeter measures the potential difference across a circuit element.

Because the voltmeter is now in parallel with the resistor, the total resistance seen by the battery is $R_{eq} = (1/17\ \Omega + 1/R_{voltmeter})^{-1}$. In order that the voltmeter measure the voltage without changing the voltage, the voltmeter's resistance must, in this case, be $\gg 17\ \Omega$. Indeed, an *ideal voltmeter* has $R_{voltmeter} = \infty\ \Omega$, and thus has no effect on the voltage. Real voltmeters come very close to this ideal, and we will always assume them to be so.

The voltmeter in Figure 31.30b reads 8.50 V. This is less than $\mathcal{E}$ because of the battery's internal resistance. Equation 31.24 found an expression for the resistor's potential difference ΔV_R. That equation is easily solved for the internal resistance r:

$$r = \frac{\mathcal{E} - \Delta V_R}{\Delta V_R} R = \frac{0.5\ \text{V}}{8.5\ \text{V}}\ 17\ \Omega = 1.0\ \Omega$$

Here a voltmeter reading was the one piece of experimental data we needed in order to determine the battery's internal resistance.

STOP TO THINK 31.6 Rank in order, from brightest to dimmest, the identical bulbs A to D.

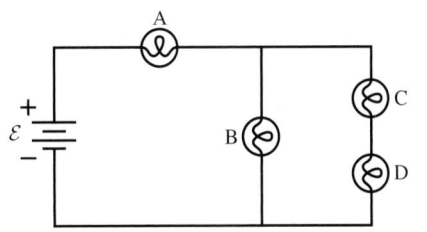

31.8 Resistor Circuits

We can use the information in this chapter to analyze a variety of more complex but more realistic circuits. We will thus have a chance to bring together the many ideas of this chapter and to see how they are used in practice.

Activ Physics   12.3–12.5

 PROBLEM-SOLVING STRATEGY 31.1   Resistor circuits

MODEL Assume that wires are ideal and, where appropriate, that batteries are ideal.

VISUALIZE Draw a circuit diagram. Label all known and unknown quantities.

SOLVE Base your mathematical analysis on Kirchhoff's laws and on the rules for series and parallel resistors.

- Step by step, reduce the circuit to the smallest possible number of equivalent resistors.
- Determine the current through and potential difference across the equivalent resistors.
- Rebuild the circuit, using the facts that the current is the same through all resistors in series and the potential difference is the same for all parallel resistors.

ASSESS Use two important checks as you rebuild the circuit.

- Verify that the sum of the potential differences across series resistors matches ΔV for the equivalent resistor.
- Verify that the sum of the currents through parallel resistors matches I for the equivalent resistor.

EXAMPLE 31.13 Analyzing a complex circuit
Find the current through and the potential difference across each of the four resistors in the circuit shown in Figure 31.31.

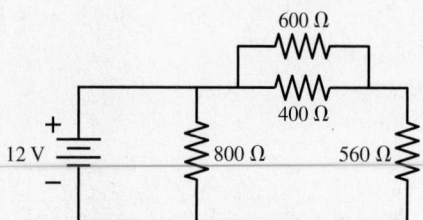

FIGURE 31.31 A complex resistor circuit for analysis.

MODEL Assume an ideal battery, with no internal resistance, and ideal connecting wires.

VISUALIZE Figure 31.31 shows the circuit diagram. We'll keep redrawing the diagram as we analyze the circuit.

SOLVE First, break the circuit down, step-by-step, into one with a single resistor. Figure 31.32a shows this done in four steps, using series and parallel resistors. The final circuit battery-and-resistor circuit is our basic circuit, which we know how to analyze. The current is

$$I = \frac{\mathcal{E}}{R} = \frac{12 \text{ V}}{400 \text{ }\Omega} = 0.030 \text{ A} = 30 \text{ mA}$$

The potential difference across the 400 Ω resistor is $\Delta V_{400} = \Delta V_{\text{bat}} = \mathcal{E} = 12$ V.

Second, rebuild the circuit, step-by-step, finding the currents and potential differences at each step. Figure 31.32b repeats the steps of Figure 31.32a exactly, but in reverse order. The 400 Ω resistor came from two 800 Ω resistors in parallel. Because $\Delta V_{400} = 12$ V, it must be true that each $\Delta V_{800} = 12$ V. The current through each 800 Ω is then $I = \Delta V/R = 15$ mA. The checkpoint is to note that 15 mA + 15 mA = 30 mA.

The right 800 Ω resistor was formed by 240 Ω and 560 Ω in series. Because $I_{800} = 15$ mA, it must be true that $I_{240} = I_{560} = 15$ mA. The potential difference across each is $\Delta V = IR$, so $\Delta V_{240} = 3.6$ V and $\Delta V_{560} = 8.4$ V. Here the checkpoint is to note that 3.6 V + 8.4 V = 12 V = ΔV_{800}, so the potential differences add as they should.

Finally, the 240 Ω resistor came from 600 Ω and 400 Ω in parallel, so they each have the same 3.6 V potential difference as their 240 Ω equivalent. The currents are $I_{600} = 6$ mA and $I_{400} = 9$ mA. Note that 6 mA + 9 mA = 15 mA, which is our third checkpoint. We now know all currents and potential differences.

ASSESS We *checked our work* at each step of the rebuilding process by verifying that currents summed properly at junctions and that potential differences summed properly along a series of resistances. This "check as you go" procedure is extremely important. It provides you, the problem solver, with a built-in error finder that will immediately inform you if a mistake has been made.

(a) Break down the circuit ⟶

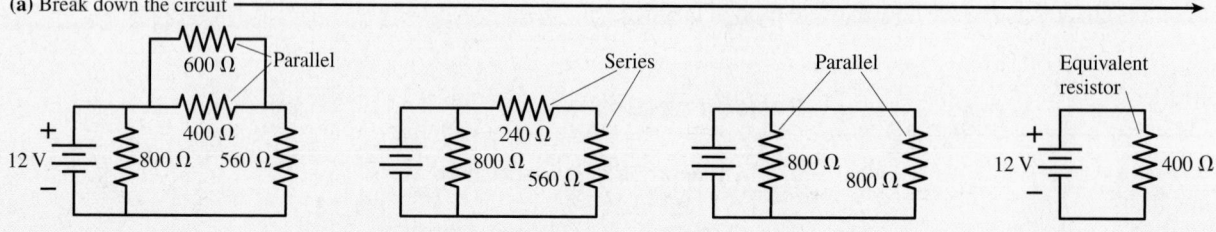

(b) Rebuild the circuit ⟵

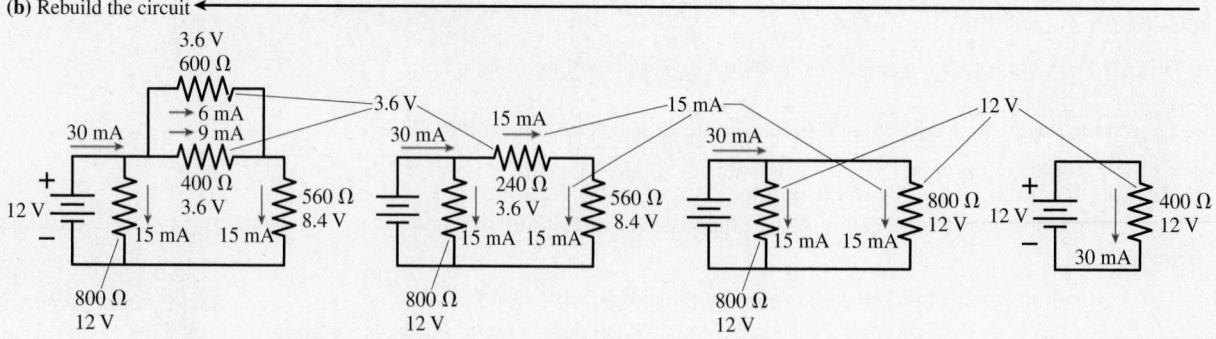

FIGURE 31.32 The step-by-step circuit analysis.

31.9 Getting Grounded

People who work with electronics are often heard to talk about things being "grounded." It always sounds quite serious, perhaps somewhat mysterious. What is it? Why do it?

The circuit analysis procedures we have discussed so far deal only with potential *differences*. Although we are free to choose the zero point of potential anywhere that is convenient, our analysis of circuits has not revealed any need to establish a zero point. Potential differences are all we have needed.

Difficulties can begin to arise, however, if you want to connect two *different* circuits together. Perhaps you would like to connect your CD player to your amplifier or your computer monitor to the computer itself. Incompatibilities can arise unless all the circuits to be connected have a *common* reference point for the potential.

You learned previously that the earth itself is a conductor. Suppose we have two circuits. If we connect *one* point of each circuit to the earth by an ideal wire, and we also agree to call the potential of the earth $V_{earth} = 0$ V, then both circuits have a common reference point. But notice something very important: *one* wire connects the circuit to the earth, but there is not a second wire returning to the circuit. That is, the wire connecting the circuit to the earth is not part of a complete circuit, so there is *no current* in this wire! Because the wire is an equipotential, it gives one point in the circuit the same potential as the earth, but it does *not* in any way change how the circuit functions. A circuit connected to the earth in this way is said to be **grounded,** and the wire is called the *ground wire*.

Figure 31.33a shows a fairly simple circuit with a 10 V battery and two resistors in series. The symbol beneath the circuit is the *ground symbol*. In this circuit, the symbol indicates that a wire has been connected between the negative battery terminal and the earth. This ground wire does not make a complete circuit, so there is no current in it. Consequently, the presence of the ground wire does not affect the circuit's behavior. The total resistance is $8 \, \Omega + 12 \, \Omega = 20 \, \Omega$, so the current in the loop is $I = (10 \text{ V})/(20 \, \Omega) = 0.5$ A. The potential differences across the two resistors are found, using Ohm's law, to be $\Delta V_8 = 4$ V and $\Delta V_{12} = 6$ V. These are the same values of the current and the potential differences that we would find if the ground wire were *not* present. So what has grounding the circuit accomplished?

Figure 31.33b shows the potential at several points in the circuit. By definition, $V_{earth} = 0$ V. The negative battery terminal and the bottom of the 12 Ω resistor are connected by ideal wires to the earth, so *the* potential at these two points must also be zero. The positive terminal of the battery is 10 V more positive than the negative terminal, so $V_{neg} = 0$ V implies $V_{pos} = +10$ V. Similarly, the fact that the potential *decreases* by 6 V as charge flows through the 12 Ω resistor now implies that *the* potential at the junction of the resistors must be $+6$ V. The potential difference across the 8 Ω resistor is 4 V, so the top has to be at $+10$ V. This agrees with the potential at the positive battery terminal, as it must because these two points are connected by an ideal wire.

Grounding the circuit has not changed the current or any of the potential differences. All that grounding the circuit does is allow us to have *specific values* for the potential at each point in the circuit. Now we can say "The voltage at the resistor junction is 6 V," whereas before all we could say was that "there is a 6 V potential difference across the 12 Ω resistor."

There is one important lesson from this: Nothing happens in a circuit "because" it is grounded. You cannot use "because it is grounded" to *explain* anything about a circuit's behavior. **Being grounded does not affect the circuit's behavior under normal conditions.**

The circular prong of a three-prong plug is a connection to ground.

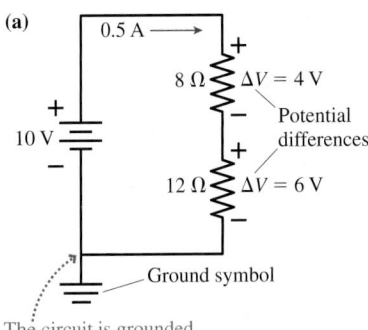

The circuit is grounded at this point.

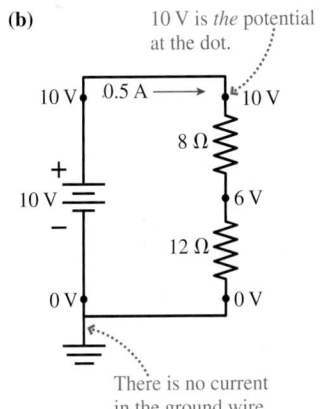

There is no current in the ground wire.

FIGURE 31.33 A circuit that is grounded at one point.

We added "under normal conditions" because there is one exception. Most circuits are enclosed in a case of some sort that is held away from the circuit with insulators. Sometimes, however, a circuit breaks or malfunctions in some way such that the case comes into electrical contact with the circuit. If the circuit uses high voltage, or even ordinary 120 V household voltage, anyone touching the case could be injured or killed by electrocution. To prevent this, many appliances or electrical instruments have the case itself grounded. Grounding ensures that the potential of the case will always remain at 0 V and be safe. If a malfunction occurs that connects the case to the circuit, a large current will pass through the ground wire to the earth and cause a fuse to blow. This is the *only* time the ground wire would ever have a current, and it is *not* a normal operation of the circuit.

Thus grounding a circuit serves two functions. First, it provides a common reference potential so that different circuits or instruments can be correctly interconnected. Second, it is an important safety feature to prevent injury or death from a defective circuit. For this reason you should *never* tamper with or try to defeat the ground connection (the third prong) on an electrical instrument's plug. If it has a ground connection, then it *needs* a ground connection and you should not try to plug it into a two-prong ungrounded outlet. Grounding the instrument does not affect its operation *under normal conditions,* but the abnormal and the unexpected are always with us. Play it safe.

EXAMPLE 31.14 A grounded circuit
Suppose the circuit of Figure 31.33 were grounded at the junction between the two resistors instead of at the bottom. Find the potential at each corner of the circuit.

VISUALIZE Figure 31.34 shows the new circuit. (It is customary to draw the ground symbol so that its "point" is always down.)

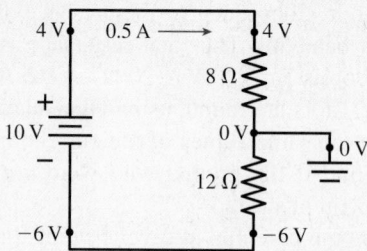

FIGURE 31.34 Circuit of Figure 31.33 grounded at the point between the resistors.

SOLVE Changing the ground point does not affect the circuit's behavior. The current is still 0.5 A, and the potential differences across the two resistors are still 4 V and 6 V. All that has happened is that we have moved the $V = 0$ V reference point. Because the earth has $V_{earth} = 0$ V, the junction itself now has a potential of 0 V. The potential decreases by 4 V as charge flows through the 8 Ω resistor. Because it *ends* at 0 V, the potential at the top of the 8 Ω must be +4 V. Similarly, the potential decreases by 6 V through the 12 Ω resistor. Because it *starts* at 0 V, the bottom of the resistor must be at −6 V. The negative battery terminal is at the same potential as the bottom of the 12 Ω resistor, because they are connected by a wire, so $V_{neg} = -6$ V. Finally, the potential increases by 10 V as the charge flows through the battery, so $V_{pos} = +4$ V, in agreement, as it should be, with the potential at the top of the 8 Ω.

You may wonder about the negative voltages. A negative voltage means only that the potential at that point is less than the potential at some other point that we chose to call $V = 0$ V. Only potential *differences* are physically meaningful, and only potential differences enter into Ohm's law: $I = \Delta V/R$. The potential difference across the 12 Ω resistor in this example is 6 V, decreasing from top to bottom, regardless of which point we choose to call $V = 0$ V.

31.10 *RC* Circuits

Thus far we've considered only circuits in which the current is steady and continuous. There are many circuits in which the time dependence of the current is a crucial feature. Charging and discharging a capacitor is an important example.

Figure 31.35a shows a charged capacitor, a switch, and a resistor. The capacitor has charge Q_0 and potential difference $\Delta V_C = Q_0/C$. There is no current, so the potential difference across the resistor is zero. Then, at $t = 0$, the switch closes and the capacitor begins to discharge through the resistor. A circuit such as this, with resistors and capacitors, is called an **RC circuit.**

How long does the capacitor take to discharge? How does the current through the resistor vary as a function of time? To answer these questions, Figure 31.35b shows the circuit after the switch has closed. Now the potential difference across the resistor is $\Delta V_R = -IR$, where I is the current discharging the capacitor.

Kirchhoff's loop law is valid for any circuit, not just circuits with batteries. The loop law applied to the circuit of Figure 31.33b, going around the loop cw, is

$$\Delta V_C + \Delta V_R = \frac{Q}{C} - IR = 0 \qquad (31.32)$$

Q and I in this equation are the *instantaneous* values of the capacitor charge and the resistor current.

The current I is the rate at which charge flows through the resistor: $I = dq/dt$. But the charge flowing through the resistor is charge that was *removed* from the capacitor. That is, an infinitesimal charge dq flows through the resistor when the capacitor charge *decreases* by dQ. Thus $dq = -dQ$, and the resistor current is related to the instantaneous capacitor charge by

$$I = -\frac{dQ}{dt} \qquad (31.33)$$

Now I is positive when Q is decreasing, as we would expect. The reasoning that has led to Equation 31.33 is rather subtle but very important. You'll see the same reasoning later in other contexts.

If we substitute Equation 31.33 into Equation 31.32 and then divide by R, the loop law for the *RC* circuit becomes

$$\frac{dQ}{dt} + \frac{Q}{RC} = 0 \qquad (31.34)$$

Equation 31.34 is a first-order differential equation for the capacitor charge Q, but one that we can solve by direct integration. First, rearrange Equation 31.34 to get all the charge terms on one side of the equation:

$$\frac{dQ}{Q} = -\frac{1}{RC}dt$$

The product RC is a constant for any particular circuit.

We know the capacitor charge was Q_0 at $t = 0$ when the switch was closed. We want to integrate from these starting conditions to charge Q at the unspecified time t. That is,

$$\int_{Q_0}^{Q} \frac{dQ}{Q} = -\frac{1}{RC} \int_0^t dt \qquad (31.35)$$

Activ Physics ONLINE   12.6–12.8

(a) Before the switch closes

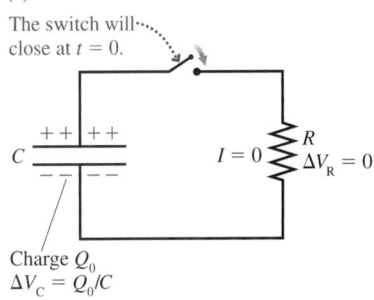

The switch will close at $t = 0$.

C $I = 0$ R $\Delta V_R = 0$

Charge Q_0
$\Delta V_C = Q_0/C$

(b) After the switch closes

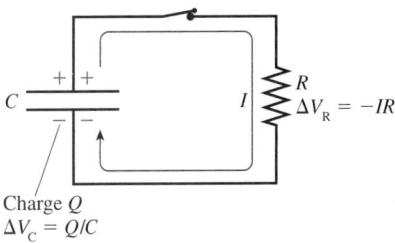

C I R $\Delta V_R = -IR$

Charge Q
$\Delta V_C = Q/C$

FIGURE 31.35 An *RC* circuit.

Both are well-known integrals, giving

$$\ln Q \Big|_{Q_0}^{Q} = \ln Q - \ln Q_0 = \ln\left(\frac{Q}{Q_0}\right) = -\frac{t}{RC}$$

We can solve for the capacitor charge Q by taking the exponential of both sides, then multiplying by Q_0. Doing so gives

$$Q = Q_0 e^{-t/RC} \tag{31.36}$$

Notice that $Q = Q_0$ at $t = 0$, as expected.

The argument of an exponential function must be dimensionless, so the quantity RC must have dimensions of time. It is useful to define the **time constant** τ of the RC circuit to be

$$\tau = RC \tag{31.37}$$

We can then write Equation 31.36 as

$$Q = Q_0 e^{-t/\tau} \tag{31.38}$$

The meaning of Equation 31.38 is easier to understand if we portray it graphically. Figure 31.36a shows the capacitor charge as a function of time. The charge decays exponentially, starting from Q_0 at $t = 0$ and asymptotically approaching zero as $t \to \infty$. The time constant τ is the time at which the charge has decreased to e^{-1} (about 37%) of its initial value. At time $t = 2\tau$, the charge has decreased to e^{-2} (about 13%) of its initial value.

NOTE ▶ The *shape* of the graph of Q is always the same, regardless of the specific value of the time constant τ. ◀

We find the resistor current by using Equation 31.33:

$$I = -\frac{dQ}{dt} = \frac{Q_0}{\tau} e^{-t/\tau} = \frac{Q_0}{RC} e^{-t/\tau} = \frac{\Delta V_0}{R} e^{-t/\tau} = I_0 e^{-t/\tau} \tag{31.39}$$

where I_0 is the initial current, immediately after the switch closes. Figure 31.36b is a graph of the resistor current versus t. You can see that the current undergoes the same exponential decay, with the same time constant, as the capacitor charge.

NOTE ▶ There's no specific time at which the capacitor has been discharged, because Q approaches zero asymptotically, but the charge and current have dropped to less than 1% of their initial values at $t = 5\tau$. Thus 5τ is a reasonable answer to the question, "How long does it take to discharge a capacitor?" ◀

(a)

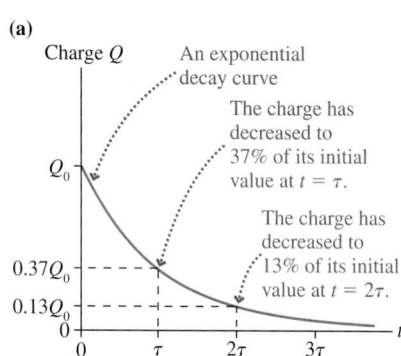

(b)

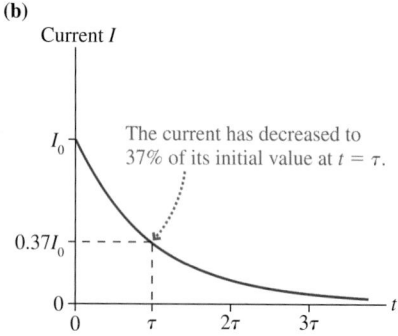

FIGURE 31.36 The decay curves of the capacitor charge and the resistor current.

EXAMPLE 31.15 Exponential decay in an RC circuit

The switch in Figure 31.37 has been in position a for a long time. It is changed to position b at $t = 0$ s. What are the charge on the capacitor and the current through the resistor at $t = 5.0 \, \mu\text{s}$?

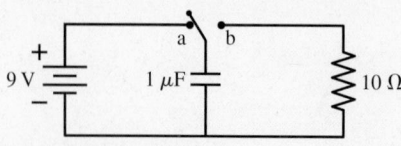

FIGURE 31.37 An RC circuit.

MODEL The battery charges the capacitor to 9.0 V. Then, when the switch is changed to position b, the capacitor discharges through the 10 Ω resistor. Assume ideal wires.

SOLVE The time constant of the RC circuit is

$$\tau = RC = (10 \, \Omega)(1.0 \times 10^{-6} \, \text{F}) = 10 \times 10^{-6} \, \text{s} = 10 \, \mu\text{s}$$

The capacitor is initially charged to 9.0 V, giving $Q_0 = C\Delta V_C = 9.0 \, \mu\text{C}$. The capacitor charge at $t = 5.0 \, \mu\text{s}$ is

$$Q = Q_0 e^{-t/RC} = (9.0 \, \mu\text{C})e^{-(5.0 \, \mu\text{s})/(10 \, \mu\text{s})}$$

$$= (9.0 \, \mu\text{C})e^{-0.5} = 5.5 \, \mu\text{C}$$

The initial current, immediately after the switch is closed, is $I_0 = Q_0/\tau = 0.90$ A. The resistor current at $t = 5.0 \, \mu\text{s}$ is

$$I = I_0 e^{-t/RC} = (0.90 \, \text{A})e^{-0.5} = 0.55 \, \text{A}$$

ASSESS This capacitor will be almost entirely discharged $5\tau = 50 \, \mu\text{s}$ after the switch is closed.

Charging a Capacitor

Figure 31.38a shows a circuit that charges a capacitor. After the switch is closed, the battery's charge escalator moves charge from the bottom electrode of the capacitor to the top electrode. The resistor, by limiting the current, slows the process but doesn't stop it. The capacitor charges until $\Delta V_C = \mathcal{E}$, then the charging current ceases. The full charge of the capacitor is $Q_{max} = C(\Delta V_C)_{max} = C\mathcal{E}$.

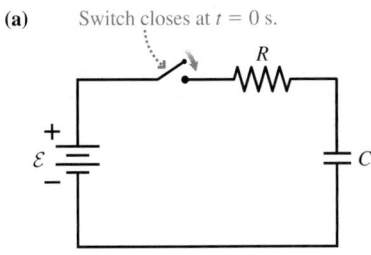

(a)   Switch closes at $t = 0$ s.

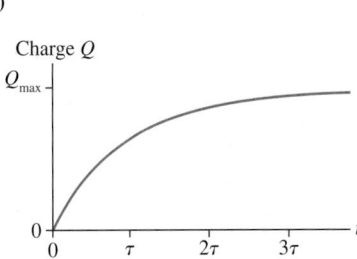

(b)

FIGURE 31.38 A circuit for charging a capacitor.

As a homework problem, you can show that the capacitor charge at time t is

$$Q = Q_{max}(1 - e^{-t/\tau}) \qquad (31.40)$$

where again $\tau = RC$. This "upside-down decay" to Q_{max} is shown graphically in Figure 31.38b. *RC* circuits that alternately charge and discharge a capacitor are at the heart of time-keeping circuits in computers and other digital electronics.

STOP TO THINK 31.7   The time constant for the discharge of this capacitor is

a. 5 s.
b. 4 s.
c. 2 s.
d. 1 s.
e. The capacitor doesn't discharge because the resistors cancel each other.

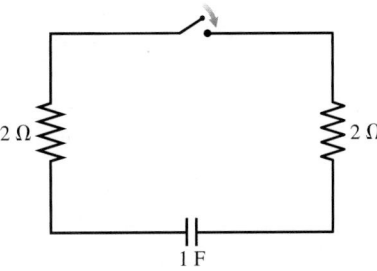

SUMMARY

The goal of Chapter 31 has been to understand the fundamental physical principles that govern electric circuits.

GENERAL STRATEGY

MODEL Assume that wires and, where appropriate, batteries are ideal.

VISUALIZE Draw a circuit diagram. Label known and unknown quantities.

SOLVE The solution is based on Kirchhoff's laws.

- Reduce the circuit to the smallest possible number of equivalent resistors.
- Find the current and the potential difference.
- "Rebuild" the circuit to find I and ΔV for each resistor.

ASSESS Verify that

- The sum of potential differences across series resistors matches ΔV for the equivalent resistor.
- The sum of the currents through parallel resistors matches I for the equivalent resistor.

Kirchhoff's loop law

For a closed loop:

- Assign a direction to the current I.
- $\sum_i (\Delta V)_i = 0$

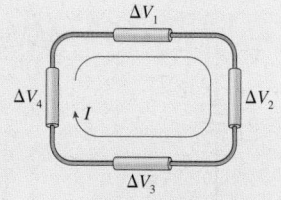

Kirchhoff's junction law

For a junction:

- $\sum I_{\text{in}} = \sum I_{\text{out}}$

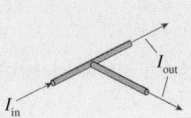

IMPORTANT CONCEPTS

Ohm's law

A potential difference ΔV between the ends of a conductor with resistance R creates a current

$$I = \frac{\Delta V}{R}$$

Signs of ΔV

$$\Delta V_{\text{bat}} = +\mathcal{E} \qquad \Delta V_{\text{bat}} = -\mathcal{E} \qquad \Delta V_R = -IR$$

The energy used by a circuit is supplied by the emf $\mathcal{E}$ of the battery through the energy transformations

$$E_{\text{chem}} \rightarrow U \rightarrow K \rightarrow E_{\text{th}}$$

The battery *supplies* energy at the rate

$$P_{\text{bat}} = I\mathcal{E}$$

The resistors *dissipate* energy at the rate

$$P_R = I\Delta V_R = I^2 R = \frac{(\Delta V_R)^2}{R}$$

APPLICATIONS

Series resistors

$$R_{\text{eq}} = R_1 + R_2 + R_3 + \dots$$

Parallel resistors

$$R_{\text{eq}} = \left(\frac{1}{R_1} + \frac{1}{R_2} + \frac{1}{R_3} + \dots \right)^{-1}$$

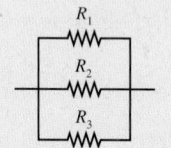

RC circuits

The discharge of a capacitor through a resistor satisfies:

$$Q = Q_0 e^{-t/\tau}$$

$$I = -\frac{dQ}{dt} = \frac{Q_0}{\tau} e^{-t/\tau} = I_0 e^{-t/\tau}$$

where $\tau = RC$ is the **time constant.**

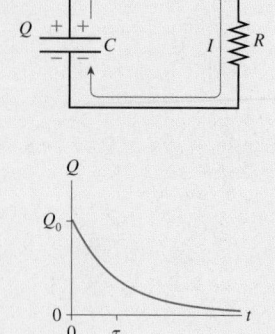

TERMS AND NOTATION

direct current	circuit diagram	kilowatt hour, kWh	short circuit
Ohm's law	Kirchhoff's junction law	series resistors	parallel resistors
resistor	Kirchhoff's loop law	equivalent resistance, R_{eq}	voltmeter
ohmic	complete circuit	ammeter	grounded
nonohmic	load	internal resistance, r	RC circuit
ideal wire	source	terminal voltage, ΔV_{bat}	time constant, τ
ideal insulator			

EXERCISES AND PROBLEMS

Exercises

Section 31.1 Resistors and Ohm's Law

1. Pencil "lead" is actually carbon. What is the resistance of the 0.70-mm-diameter, 6.0-cm-long lead from a mechanical pencil?
2. The resistance of a very fine aluminum wire with a 10 μm $\times$ 10 μm square cross section is 1000 Ω.
   a. How long is the wire?
   b. A 1000 Ω resistor is made by wrapping this wire in a spiral around a 3.0-mm-diameter glass core. How many turns of wire are needed?
3. A 3.0 V potential difference is applied between the ends of a 0.80-mm-diameter, 50-cm-long nichrome wire. What is the current in the wire?
4. A 1.0-mm-diameter, 20-cm-long copper wire carries a 3.0 A current. What is the potential difference between the ends of the wire?
5. A circuit calls for a 0.50-mm-diameter copper wire to be stretched between two points. You don't have any copper wire, but you do have aluminum wire in a wide variety of diameters. What diameter aluminum wire will provide the same resistance?
6. Figure Ex31.6 is a current-versus-potential-difference graph for a material. What is the material's resistance?

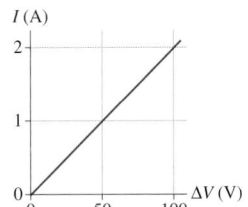

FIGURE EX31.6

Section 31.2 Circuit Elements and Diagrams

7. Draw a circuit diagram for the circuit of Figure Ex31.7.

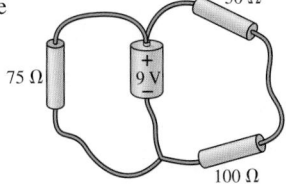

FIGURE EX31.7

8. Draw a circuit diagram for the circuit of Figure Ex31.8.

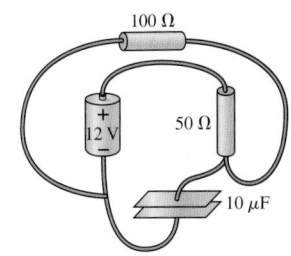

FIGURE EX31.8

Section 31.3 Kirchhoff's Laws and the Basic Circuit

9. In Figure Ex31.9, what is the current in the wire above the junction? Does charge flow toward or away from the junction?

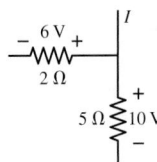

FIGURE EX31.9

10. a. What are the magnitude and direction of the current in the 30 Ω resistor in Figure Ex31.10?
    b. Draw a graph of the potential as a function of the distance traveled through the circuit, traveling cw from $V = 0$ V at the lower left corner.

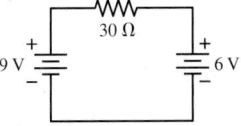

FIGURE EX31.10

11. a. What are the magnitude and direction of the current in the 18 Ω resistor in Figure Ex31.11?
    b. Draw a graph of the potential as a function of the distance traveled through the circuit, traveling cw from $V = 0$ V at the lower left corner.

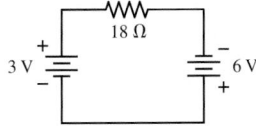

FIGURE EX31.11

12. a. What is the potential difference across each resistor in Figure Ex31.12?
    b. Draw a graph of the potential as a function of the distance traveled through the circuit, traveling cw from $V = 0$ V at the lower left corner.

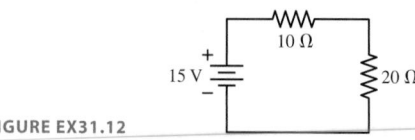

FIGURE EX31.12

Section 31.4 Energy and Power

13. What is the resistance of a 1500 W (120 V) hair dryer? What is the current in the hair dryer when it is used?
14. How much power is dissipated by each resistor in Figure Ex31.14?

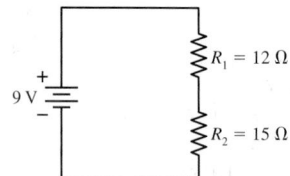

FIGURE EX31.14

15. A standard 100 W (120 V) lightbulb contains a 7.0-cm-long tungsten filament. The high-temperature resistivity of tungsten is 9.0×10^{-7} Ω m. What is the diameter of the filament?
16. How many joules are in 1 kWh?
17. A typical American family uses 1000 kWh of electricity a month.
    a. What is the average current in the 120 V power line to the house?
    b. On average, what is the resistance of a household?
18. A 60 W (120 V) night light is turned on for an average 12 hr a day year round. What is the annual cost of electricity at a billing rate of $0.10/kWh?

Section 31.5 Series Resistors

19. An 80-cm-long wire is made by welding a 1.0-mm-diameter, 20-cm-long copper wire to a 1.0-mm-diameter, 60-cm-long iron wire. What is the resistance of the composite wire?
20. Two of the three resistors in Figure Ex31.20 are unknown but equal. Is the total resistance between points a and b less than, greater than, or equal to 50 Ω? Explain.

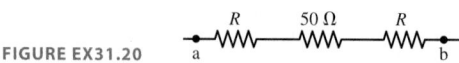

FIGURE EX31.20

21. What is the value of resistor R in Figure Ex31.21?

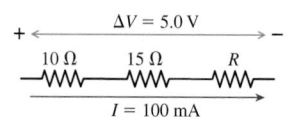

FIGURE EX31.21

22. Two 75 W (120 V) lightbulbs are wired in series, then the combination is connected to a 120 V supply. How much power is dissipated by each bulb?
23. The corroded contacts in a lightbulb socket have 5.0 Ω resistance. How much actual power is dissipated by a 100 W (120 V) lightbulb screwed into this socket?

Section 31.6 Real Batteries

24. The voltage across the terminals of a 9.0 V battery is 8.5 V when the battery is connected to a 20 Ω load. What is the battery's internal resistance?
25. Compared to an ideal battery, by what percentage does the battery's internal resistance reduce the potential difference across the 20 Ω resistor in Figure Ex31.25?

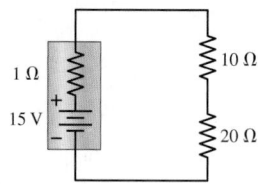

FIGURE EX31.25

26. What is the internal resistance of the battery in Figure Ex31.26? How much power is dissipated inside the battery?

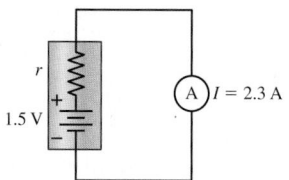

FIGURE EX31.26

Section 31.7 Parallel Resistors

27. A metal wire of resistance R is cut into two pieces of equal length. The two pieces are connected together side by side. What is the resistance of the two connected wires?
28. Two of the three resistors in Figure Ex31.28 are unknown but equal. Is the total resistance between points a and b less than, greater than, or equal to 200 Ω? Explain.

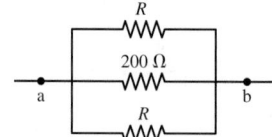

FIGURE EX31.28

29. What is the value of resistor R in Figure Ex31.29?

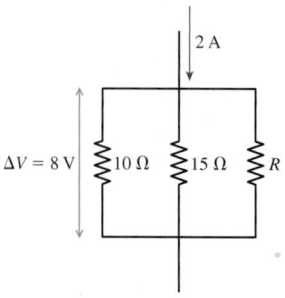

FIGURE EX31.29

30. What is the equivalent resistance between points a and b in Figure Ex31.30?

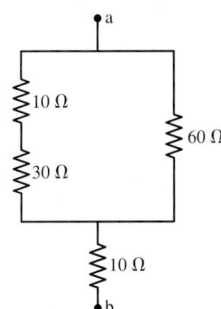

FIGURE EX31.30

31. What is the equivalent resistance between points a and b in Figure Ex31.31?

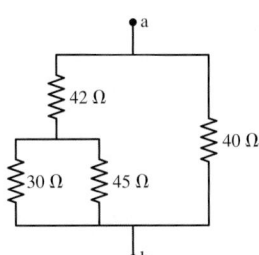

FIGURE EX31.31

32. What is the equivalent resistance between points a and b in Figure Ex31.32?

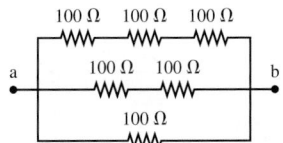

FIGURE EX31.32

33. What is the equivalent resistance between points a and b in Figure Ex31.33?

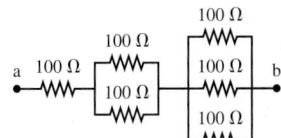

FIGURE EX31.33

Section 31.9 Getting Grounded

34. Determine the value of the potential at points a to d in Figure Ex31.34.

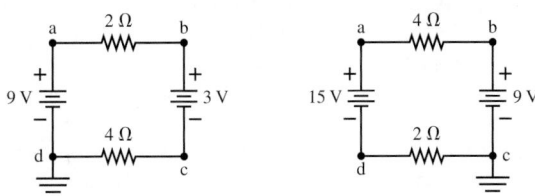

FIGURE EX31.34   **FIGURE EX31.35**

35. Determine the value of the potential at points a to d in Figure Ex31.35.

Section 31.10 *RC* Circuits

36. Show that the product *RC* has units of s.
37. What is the time constant for the discharge of the capacitors in Figure Ex31.37?

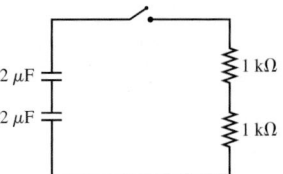

FIGURE EX31.37

38. What is the time constant for the discharge of the capacitors in Figure Ex31.38?

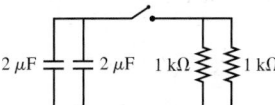

FIGURE EX31.38

39. A 10 μF capacitor initially charged to 20 μC is discharged through a 1.0 kΩ resistor. How long does it take to reduce the capacitor's charge to 10 μC?
40. The switch in Figure Ex31.40 has been in position a for a long time. It is changed to position b at $t = 0$ s. What are the charge Q on the capacitor and the current I through the resistor (a) immediately after the switch is closed? (b) at $t = 50$ μs? (c) at $t = 200$ μs?

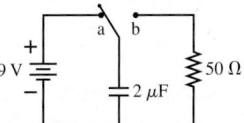

FIGURE EX31.40

41. What value resistor will discharge a 1.0 μF capacitor to 10% of its initial charge in 2.0 ms?
42. A capacitor is discharged through a 100 Ω resistor. The discharge current decreases to 25% of its initial value in 2.5 ms. What is the value of the capacitor?

Problems

43. Figure P31.43 shows five identical bulbs connected to an ideal battery. All the bulbs are glowing. Rank in order, from brightest to dimmest, the brightness of bulbs A to E. Explain.

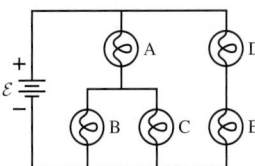

FIGURE P31.43

44. Figure P31.44 shows six identical bulbs connected to an ideal battery. All the bulbs are glowing. Rank in order, from brightest to dimmest, the brightness of bulbs A to F. Explain.

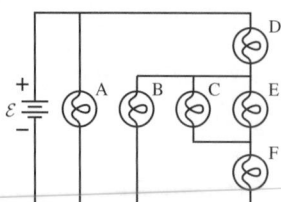

FIGURE P31.44

45. The battery in Figure P31.45 is ideal. Initially, bulbs A and B are both glowing. Bulb B is then removed from its socket. Does removing bulb B cause the potential difference ΔV_{12} between points 1 and 2 to increase, decrease, or become zero? Explain.

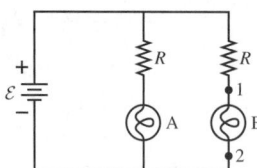

FIGURE P31.45

46. You've made the finals of the Science Olympics! As one of your tasks, you're given 1.0 g of aluminum and asked to make a wire, using all the aluminum, with a resistance of 1.0 Ω. What length and diameter will you choose for your wire?

47. Not too long ago houses were protected from excessive currents by fuses rather than circuit breakers. Sometimes a fuse blew out and a replacement wasn't at hand. Because a copper penny happens to have almost the same diameter as a fuse, some people replaced the fuse with a penny. Unfortunately, a penny never blows out, no matter how large the current, and the use of pennies in fuse boxes caused many house fires. Make the appropriate measurements on a penny, then calculate the resistance between the two faces of a copper penny.

48. You have three 12 Ω resistors. Draw diagrams showing how you could arrange all three so that their equivalent resistance is (a) 4.0 Ω, (b) 8.0 Ω, (c) 18 Ω, and (d) 36 Ω.

49. What is the equivalent resistance between points a and b in Figure P31.49?

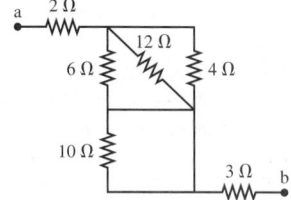

FIGURE P31.49

50. There is a current of 0.25 A in the circuit of Figure P31.50.
    a. What is the direction of the current? Explain.
    b. What is the value of the resistance R?
    c. What is the power dissipated by R?
    d. Make a graph of potential versus position, starting from $V = 0$ V in the lower left corner and proceeding cw.

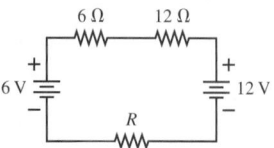

FIGURE P31.50

51. A variable resistor R is connected across the terminals of a battery. Figure P31.51 shows the current in the circuit as R is varied. What are the emf and internal resistance of the battery?

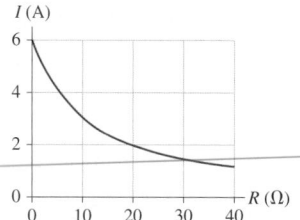

FIGURE P31.51

52. The 10 Ω resistor in Figure P31.52 is dissipating 40 W of power. How much power are the other two resistors dissipating?

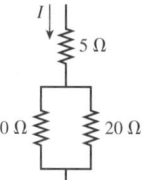

FIGURE P31.52

53. What are the emf and internal resistance of the battery in Figure P31.53?

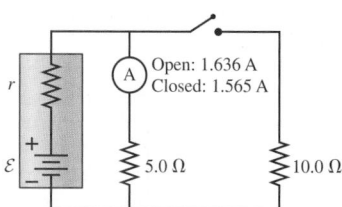

FIGURE P31.53

54. What are the resistance R and the emf of the battery in Figure P31.54?

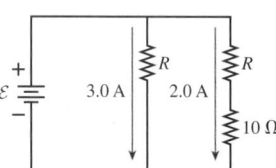

FIGURE P31.54

55. A 2.5 V battery and a 1.5 V battery, each with an internal resistance of 1 Ω, are connected in parallel. That is, their positive terminals are connected by a wire and their negative terminals are connected by a wire. What is the terminal voltage of each battery in this configuration?

56. a. Load resistor R is attached to a battery of emf $\mathcal{E}$ and internal resistance r. For what value of the resistance R, in terms of $\mathcal{E}$ and r, will the power dissipated by the load resistor be a maximum?
    b. What is the maximum power that the load can dissipate if the battery has $\mathcal{E} = 9.0$ V and $r = 1.0$ Ω?
    c. *Why* should the power dissipated by the load have a maximum value? Explain.
    Hint: What happens to the power dissipation when R is either very small or very large?

57. The ammeter in Figure P31.57 reads 3.0 A. Find I_1, I_2, and $\mathcal{E}$.

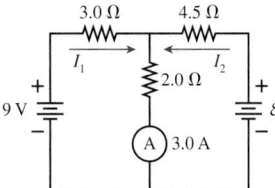

FIGURE P31.57

58. a. Suppose the circuit in Figure P31.58 is grounded at point d. Find the potential at each of the four points a, b, c, and d.
    b. Make a graph of potential versus position, starting from point d and proceeding cw.
    c. Repeat parts a and b for the same circuit grounded at point a instead of d.

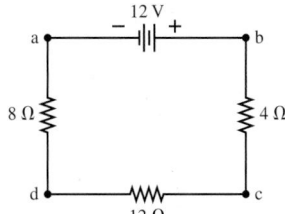

FIGURE P31.58

59. What is the current in the 2 Ω resistor in Figure P31.59?

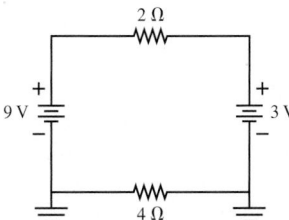

FIGURE P31.59

60. Energy experts tell us to replace regular incandescent light-bulbs with compact fluorescent bulbs, but it seems hard to justify spending $15 on a lightbulb. A 60 W incandescent bulb costs 50¢ and has a lifetime of 1000 hours. A 15 W compact fluorescent bulb produces the same amount of light as a 60 W incandescent bulb and is intended as a replacement. It costs $15 and has a lifetime of 10,000 hours. Compare the *life-cycle costs* of 60 W incandescent bulbs to 15 W compact fluorescent bulbs. The life-cycle cost of an object is the cost of purchasing it plus the cost of fueling and maintaining it over its useful life. Which is the cheaper source of light and which the more expensive? Assume that electricity costs $0.10/kWh.
    Hint: Be sure to compare the two over equal time spans.

61. A refrigerator has a 1000 W compressor, but the compressor runs only 20% of the time.
    a. If electricity costs $0.10/kWh, what is the monthly (30 day) cost of running the refrigerator?
    b. A more energy efficient refrigerator with an 800 W compressor costs $100 more. If you buy the more expensive refrigerator, how many months will it take to recover your additional cost?

62. For an ideal battery ($r = 0\ \Omega$), closing the switch in Figure P31.62 does not affect the brightness of bulb A. In practice, bulb A dims *just a little* when the switch closes. To see why, assume that the 1.5 V battery has an internal resistance

$r = 0.50\ \Omega$ and that the resistance of a glowing bulb is $R = 6\ \Omega$.
    a. What is the current through bulb A when the switch is open?
    b. What is the current through bulb A after the switch has closed?
    c. By what percent does the current through A change when the switch is closed?
    d. Would the current through A change if $r = 0\ \Omega$?

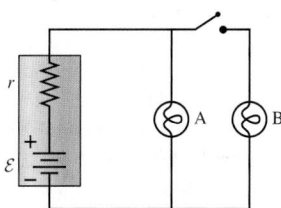

FIGURE P31.62

63. What are the battery current I_{bat} and the potential difference ΔV_{ab} between points a and b when the switch in Figure P31.63 is (a) open and (b) closed?

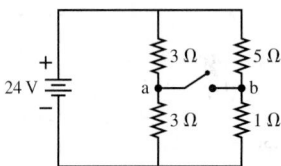

FIGURE P31.63

64. The circuit in Figure P31.64 is called a *voltage divider*. What value of R will make $V_{out} = V_{in}/10$?

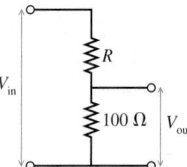

FIGURE P31.64

65. A circuit you're building needs an ammeter that goes from 0 mA to a full-scale reading of 50 mA. Unfortunately, the only ammeter in the storeroom goes from 0 μA to a full-scale reading of only 500 μA. Fortunately, you've just finished a physics class, and you realize that you can make this ammeter work by putting a resistor in parallel with it, as shown in Figure P31.65. You've measured that the resistance of the ammeter is 50 Ω, not the 0 Ω of an ideal ammeter.
    a. What value of R must you use so that the meter will go to full scale when the current I is 50 mA?
    b. What is the effective resistance of your ammeter?

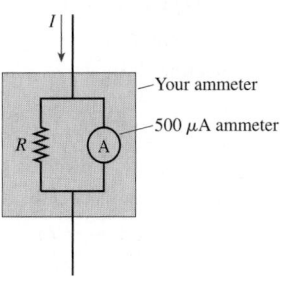

FIGURE P31.65

66. A circuit you're building needs a voltmeter that goes from 0 V to a full-scale reading of 5.0 V. Unfortunately, the only meter in the storeroom is an *ammeter* that goes from 0 μA to a full-scale reading of 500 μA. Fortunately, you've just finished a physics class, and you realize that you can convert this

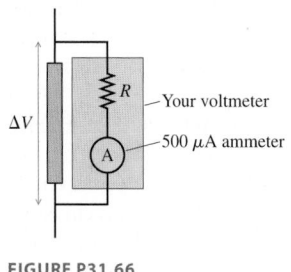

FIGURE P31.66

meter to a voltmeter by putting a resistor in series with it, as shown in Figure P31.66. You've measured that the resistance of the ammeter is 50 Ω, not the 0 Ω of an ideal ammeter. What value of R must you use so that the meter will go to full scale when the potential difference across the resistor is 5.0 V?

67. For the circuit shown in Figure P31.67, find the current through and the potential difference across each resistor. Place your results in a table for ease of reading.

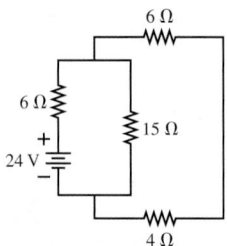

FIGURE P31.67

68. For the circuit shown in Figure P31.68, find the current through and the potential difference across each resistor. Place your results in a table for ease of reading.

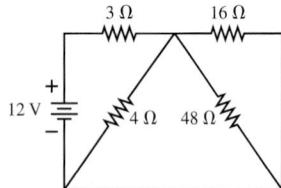

FIGURE P31.68

69. For the circuit shown in Figure P31.69, find the current through and the potential difference across each resistor. Place your results in a table for ease of reading.

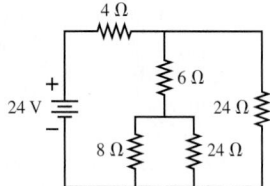

FIGURE P31.69

70. For the circuit shown in Figure P31.70, find the current through and the potential difference across each resistor. Place your results in a table for ease of reading.

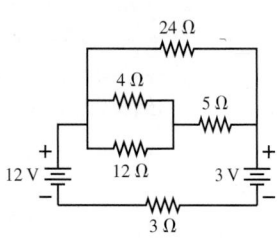

FIGURE P31.70

71. For the circuit in Figure P31.71, what are (a) the current through the 2 Ω resistor, (b) the power dissipated by the 20 Ω resistor, and (c) the potential at point a?

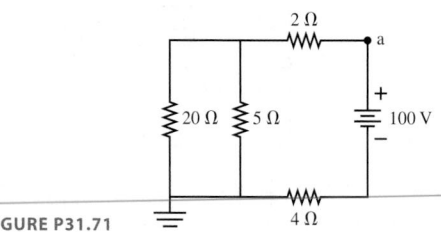

FIGURE P31.71

72. The capacitor in an *RC* circuit is discharged with a time constant of 10 ms. At what time after the discharge begins are (a) the charge on the capacitor reduced to half its initial value and (b) the energy stored in the capacitor reduced to half its initial value?

73. A 50 μF capacitor that had been charged to 30 V is discharged through a resistor. Figure P31.73 shows the capacitor voltage as a function of time. What is the value of the resistance?

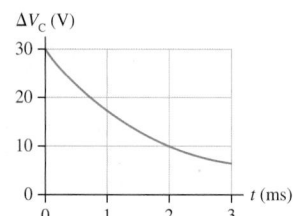

FIGURE P31.73

74. A 0.25 μF capacitor is charged to 50 V. It is then connected in series with a 25 Ω resistor and a 100 Ω resistor and allowed to discharge completely. How much energy is dissipated by the 25 Ω resistor?

75. The capacitors in Figure P31.75 are charged and the switch closes at $t = 0$ s. At what time has the current in the 8 Ω resistor decayed to half the value it had immediately after the switch was closed?

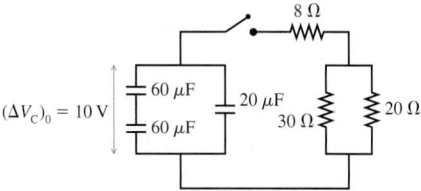

FIGURE P31.75

76. The capacitor in Figure P31.76 begins to charge after the switch closes at $t = 0$ s.
    a. What is ΔV_C a very long time after the switch has closed? Explain.
    b. What is Q_{max} in terms of $\mathcal{E}$, R, and C?
    c. In this circuit, does $I = +dQ/dt$ or $-dQ/dt$? Explain.
    d. Find an expression for the current I at time t. Graph I from $t = 0$ to $t = 5\tau$.

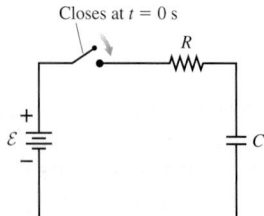

FIGURE P31.76

77. The switch in Figure P31.77 has been closed for a very long time.
    a. What is the charge on the capacitor?
    b. The switch is opened at $t = 0$ s. At what time has the charge on the capacitor decreased to 10% of its initial value?

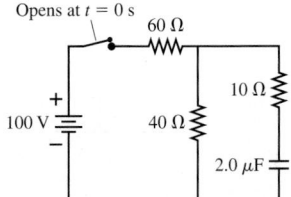

FIGURE P31.77

Challenge Problems

78. The switch in Figure CP31.78 has been in position a for a very long time. It is suddenly flipped to position b for 1.25 ms, then back to a. How much energy is dissipated by the 50 Ω resistor?

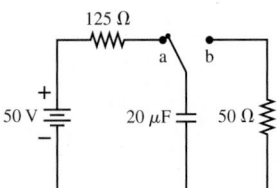

FIGURE CP31.78

79. The capacitor in Figure 31.38a begins to charge after the switch closes at $t = 0$ s. Analyze this circuit and show that $Q = Q_{max}(1 - e^{-t/\tau})$, where $Q_{max} = C\mathcal{E}$.

80. The switch in Figure 31.38a closes at $t = 0$ s and, after a very long time, the capacitor is fully charged. Find expressions for (a) the total energy supplied by the battery as the capacitor is being charged, (b) total energy dissipated by the resistor as the capacitor is being charged, and (c) the energy stored in the capacitor when it is fully charged. Your expressions will be in terms of $\mathcal{E}$, R, and C. (d) Do your results for parts a to c show that energy is conserved? Explain.

81. An *oscillator circuit* is important to many applications. A simple oscillator circuit can be built by adding a neon gas tube to an RC circuit, as shown in Figure CP31.81. Gas is normally a good insulator, and the resistance of the gas tube is essentially infinite when the light is off. This allows the capacitor to charge. When the capacitor voltage reaches a value V_{on}, the electric field inside the tube becomes strong enough to ionize the neon gas. Visually, the tube lights with an orange glow. Electrically, the ionization of the gas provides a very-low-resistance path through the tube. The capacitor very rapidly (we can think of it as instantaneously) discharges through the tube and the capacitor voltage drops. When the capacitor voltage has dropped to a value V_{off}, the electric field inside the tube becomes too weak to sustain the ionization and the neon light turns off. The capacitor then starts to charge again. The capacitor voltage oscillates between V_{off}, when it starts charging, and V_{on}, when the light comes on to discharge it.
    a. Show that the oscillation period is

$$T = RC\ln\left(\frac{\mathcal{E} - V_{off}}{\mathcal{E} - V_{on}}\right)$$

    b. A neon gas tube has $V_{on} = 80$ V and $V_{off} = 20$ V. What resistor value should you choose to go with a 10 μF capacitor and a 90 V battery to make a 10 Hz oscillator?

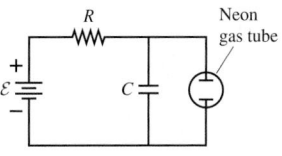

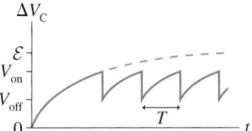

FIGURE CP31.81

Stop to Think 31.1: $R_c > R_a > R_d > R_b$. The resistance is proportional to L/r^2. Increasing r decreases R; increasing L increases R. But the radius has a larger effect because R depends on the square of r.

Stop to Think 31.2: a, b, and **d.** These three are the same circuit because the logic of the connections is the same. In c, the functioning of the circuit is changed by the extra wire connecting the two sides of the capacitor.

Stop to Think 31.3: ΔV increases by 2 V in the direction of I. Kirchhoff's loop law, starting on the left side of the battery, is then $+12$ V $+ 2$ V $- 8$ V $- 6$ V $= 0$ V.

Stop to Think 31.4: $P_b > P_d > P_a > P_c$. The power dissipated by a resistor is $P_R = (\Delta V_R)^2/R$. Increasing R decreases P_R; increasing ΔV_R increases P_R. But the potential has a larger effect because P_R depends on the square of ΔV_R.

Stop to Think 31.5: $I = 2$ A for all. $V_a = 20$ V, $V_b = 16$ V, $V_c = 10$ V, $V_d = 8$ V, $V_e = 0$ V. Current is conserved. The potential is 0 V on the right and increases by IR for each resistor going to the left.

Stop to Think 31.6: A > B > C = D. All the current from the battery goes through A, so it is brightest. The current divides at the junction, but not equally. Because B is in parallel with C + D but has half the resistance, twice as much current travels through B as through C + D. So B is dimmer than A but brighter than C and D. C and D are equal because of conservation of current.

Stop to Think 31.7: b. The two 2 Ω resistors are in series and equivalent to a 4 Ω resistor. Thus $\tau = RC = 4$ s.

32 The Magnetic Field

The beautiful aurora borealis, the northern lights, is due to the earth's magnetic field.

▶ Looking Ahead

The goal of Chapter 32 is to learn how to calculate and use the magnetic field. In this chapter you will learn to:

- Recognize basic magnetic phenomena.
- Calculate the magnetic field of moving charged particles and currents.
- Use the right-hand rule to find the directions of magnetic forces and fields.
- Understand the motion of a charged particle in a magnetic field.
- Calculate magnetic forces and torques on wires and current loops.
- Understand the magnetic properties of materials.

◀ Looking Back

This chapter uses what you have learned about circular motion, rotation, and dipoles to understand motion in a magnetic field. Please review:

- Sections 7.1–7.2 Uniform circular motion.
- Sections 13.3 and 13.9 Torque and the cross product of two vectors.
- Sections 25.5–25.6 Basic properties of fields.
- Sections 26.2 and 26.7 The properties of an electric dipole.

The shimmering aurora is one of the most beautiful of earth's natural displays. These lights, which mystified travelers and the inhabitants of northern realms for centuries, are a consequence of the earth's magnetism.

Magnetism, like electricity, has been known since antiquity. The ancient Greeks knew that certain minerals called *lodestones* could attract iron objects. Chinese navigators were using lodestone compasses by the year 1000, but compasses were not known in the West until nearly 1200. Later, in about 1600, William Gilbert recognized that compasses work because the earth itself is a magnet. The same forces that align compass needles are also responsible for the aurora.

Our task for this chapter is to investigate magnets and magnetism. Magnets are all around you. In addition to holding shopping lists and cartoons on refrigerators, magnets allow you to run electric motors, produce a picture on your television screen, store information on computer disks, cook food in a microwave oven, and listen to music using loudspeakers. Magnets are used in magnetic-resonance imaging to produce images of the interior of the human body, in high-energy physics experiments to identify subatomic particles, and in magnetic levitation trains.

Just what is magnetism? How are magnetic fields created? What are their properties? How are they used? These are the questions we will address.

32.1 Magnetism

We began our investigation of electricity in Chapter 25 by looking at the results of simple experiments with charged rods. Let's follow a similar approach with magnetism.

Discovering magnetism

Experiment 1

If a bar magnet is taped to a piece of cork and allowed to float in a dish of water, it always turns to align itself in an approximate north-south direction. The end of a magnet that points north is called the *north-seeking pole,* or simply, the **north pole.** The other end is the **south pole.**

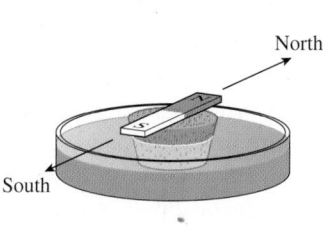

Experiment 2

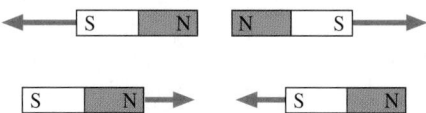

If the north pole of one magnet is brought near the north pole of another magnet, they exert repulsive forces on each other. Two south poles also repel each other, but the north pole of one magnet exerts an attractive force on the south pole of another magnet.

Experiment 3

The north pole of a bar magnet attracts one end of a compass needle and repels the other. Apparently the compass needle itself is a little bar magnet with a north pole and a south pole.

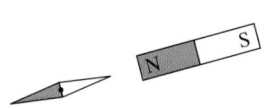

Experiment 4

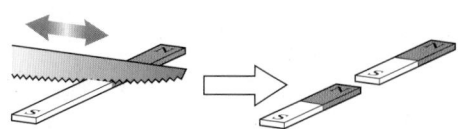

Cutting a bar magnet in half produces two weaker but still complete magnets, each with a north pole and a south pole. No matter how small the magnets are cut, even down to microscopic sizes, each piece remains a complete magnet with two poles.

Experiment 5

Magnets can pick up some objects, such as paper clips, but not all. If an object is attracted to one end of a magnet, it is also attracted to the other end. Most materials, including copper, aluminum, glass, and plastic, experience no force from a magnet.

Experiment 6

A magnet does not affect an electroscope. A charged rod exerts a weak *attractive* force on *both* ends of a magnet. However, the force is the same as the force on a metal bar that isn't a magnet, so it is simply a polarization force like the ones we studied in Chapter 25. Other than polarization forces, charges have *no effects* on magnets.

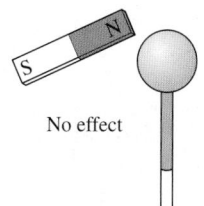

No effect

What do these experiments tell us?

1. Experiment 6 reveals that **magnetism is not the same as electricity.** Magnetic poles and electric charges share some similar behavior, but they are not the same. The magnetic force is a force of nature that we have not previously encountered.
2. Magnetism is a long-range force. Paper clips leap up to a magnet. You can feel the pull as you bring a refrigerator magnet close to the refrigerator.
3. Magnets have two poles, called north and south poles. The names are merely descriptive; they tell us nothing about how magnetism works. Two like poles exert repulsive forces on each other; two opposite poles exert attractive forces on each other. The behavior is *analogous* to electric charges, but, as noted, magnetic poles and electric charges are *not* the same.
4. The poles of a bar magnet can be identified by using it as a compass. Other magnets, such as flat refrigerator magnets or horseshoe magnets, aren't so easily made into a compass, but their poles can be identified by testing them against a bar magnet. A pole that attracts a known north pole and repels a known south pole must be a south magnetic pole.

5. Materials that are attracted to a magnet, or that a magnet sticks to, are called **magnetic materials.** The most common magnetic material is iron. Others include nickel and cobalt. Magnetic materials are attracted to *both* poles of a magnet. This attraction is analogous to how neutral objects are attracted to both positively and negatively charged rods by the polarization force. The difference is that *all* neutral objects are attracted to a charged rod whereas only a few materials are attracted to a magnet.

Our goal is to develop a theory of magnetism that will enable us to explain these observations.

Monopoles and Dipoles

It is a strange observation that cutting a magnet in half yields two weaker but still complete magnets, each with a north pole and a south pole. Every magnet that has ever been observed has both a north pole and south pole, thus forming a permanent **magnetic dipole.** A magnetic dipole is analogous to an electric dipole, but the two charges in an electric dipole can be separated and used individually. This appears *not* to be true for a magnetic dipole.

An isolated magnetic pole, such as a north pole in the absence of a south pole, would be called a **magnetic monopole.** No one has ever observed a magnetic monopole. On the other hand, no one has ever given a convincing reason why isolated magnetic poles should not exist, and some theories of subatomic particles say they should. Whether or not magnetic monopoles exist in nature remains an unanswered question at the most fundamental level of physics.

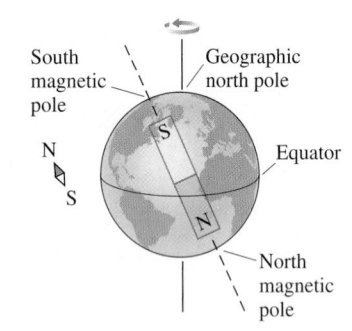

FIGURE 32.1 The earth is a large magnet.

Compasses and Geomagnetism

The north pole of a compass needle is attracted toward the geographic north pole of the earth and repelled by the earth's geographic south pole. Apparently the earth itself is a large magnet, as shown in Figure 32.1. The reasons for the earth's magnetism are complex, but geophysicists generally agree that the earth's magnetic poles arise from currents in its molten iron core. Two interesting facts about the earth's magnetic field are one, that the magnetic poles are offset slightly from the geographic poles of the earth's rotation axis, and two, that the geographic north pole is actually a *south* magnetic pole! You should be able to use what you have learned thus far to convince yourself that this is the case.

STOP TO THINK 32.1   Does the compass needle rotate clockwise (cw), counterclockwise (ccw), or not at all?

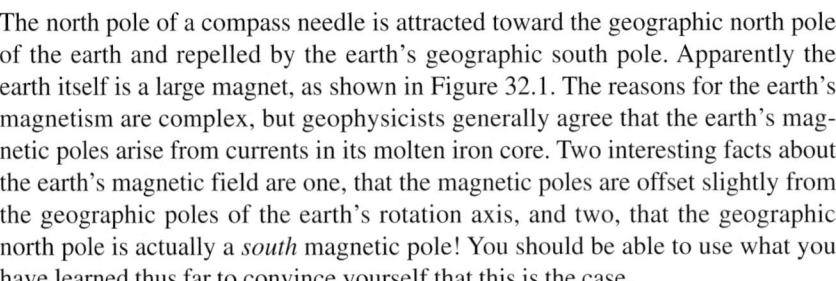

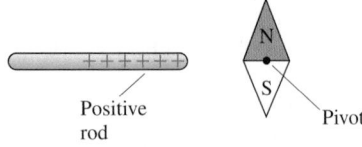

32.2 The Discovery of the Magnetic Field

As electricity began to be seriously studied in the 18th century, some scientists speculated that there might be a connection between electricity and magnetism. Interestingly, the link between electricity and magnetism was discovered *in the midst of a classroom lecture demonstration* in 1819 by the Danish scientist Hans Christian Oersted. Oersted was using a battery to produce a large current in a wire. By chance, a compass was sitting next to the wire, and Oersted noticed that the current caused the compass needle to turn. In other words, the compass responded as if a magnet had been brought near.

Oersted had long been interested in a possible connection between electricity and magnetism, so the significance of this serendipitous observation was immediately apparent to him. Oersted's discovery that **magnetism is caused by an electric current** will be our starting point for developing a theory of magnetism.

The Effect of a Current on a Compass

Let us use compasses to probe the magnetism created when a current passes through a long, straight wire. In Figure 32.2a, before the current is turned on, the compasses are aligned along a north-south line. You can see in Figure 32.2b that a strong current in the wire causes the compass needles to pivot until they are *tangent* to a circle around the wire. Figure 32.2c illustrates a **right-hand rule** that relates the orientation of the compass needles to the direction of the current.

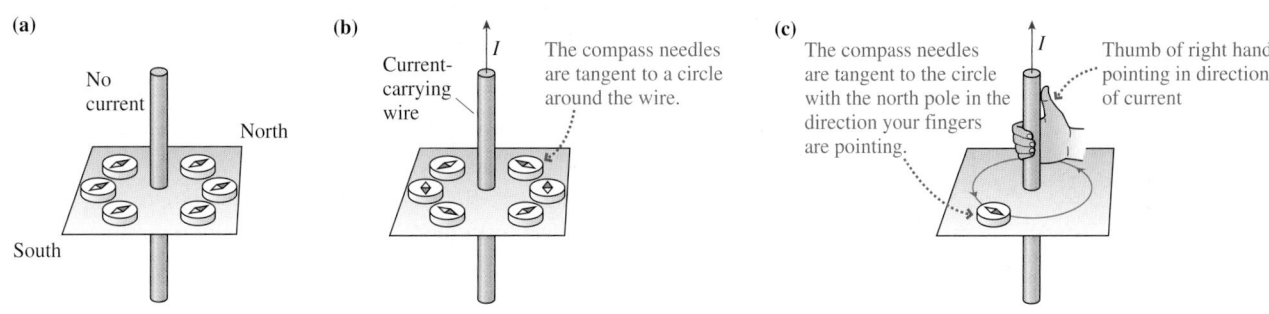

FIGURE 32.2 Response of compass needles to a current in a straight wire.

Magnetism is more demanding than electricity in requiring a three-dimensional perspective, of the sort shown in Figure 32.2. But since two-dimensional figures are easier to draw, we will make as much use of them as we can. Consequentially, we will often need to indicate field vectors or currents that are perpendicular to the page. Figure 32.3a shows the notation we will use. Figure 32.3b demonstrates this notation by showing the compasses around a current that is directed into the page. To use the right-hand rule with this drawing, point your right thumb into the page. Your fingers will curl cw, and that is the direction in which the north poles of the compass needles point.

The Magnetic Field

We introduced the idea of a *field* as a way to understand the long-range electric force. A charge alters the space around it by creating an electric field. A second charge then experiences a force due to the presence of the electric field. The electric field is the *means* by which charges interact with each other. Although this idea appeared rather far-fetched, it turned out to be very useful. We need a similar idea to understand the long-range force exerted by a current on a compass needle.

Let us define the **magnetic field** $\vec{B}$ as having the following properties:

1. A magnetic field is created at *all* points in space surrounding a current-carrying wire.
2. The magnetic field at each point is a vector. It has both a magnitude, which we call the *magnetic field strength B*, and a direction.
3. The magnetic field exerts forces on magnetic poles. The force on a north magnetic pole is parallel to $\vec{B}$; the force on a south magnetic pole is opposite to $\vec{B}$.

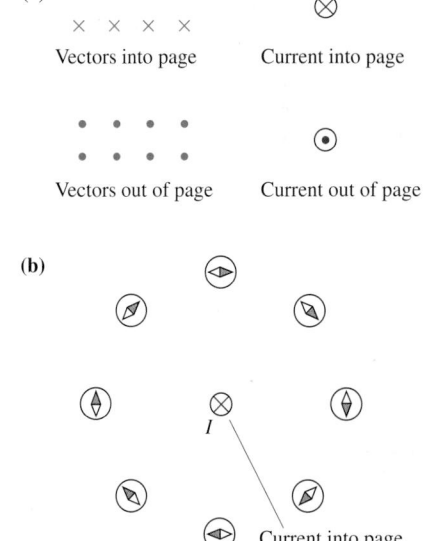

FIGURE 32.3 The notation for vectors and currents that are perpendicular to the page.

The magnetic force on the north pole is parallel to the magnetic field.

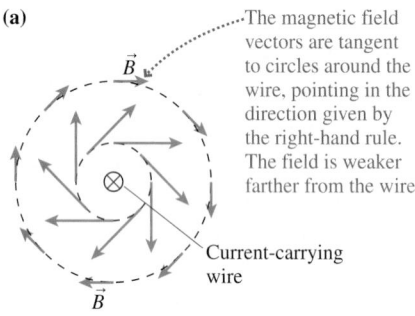

FIGURE 32.4 The magnetic field exerts forces on the poles of a compass, causing the needle to align with the field.

Figure 32.4 shows a compass needle in a magnetic field. The field vectors are shown at a few points, but keep in mind that the field is present at *all* points in space. A magnetic force is exerted on each of the two poles of the compass, parallel to $\vec{B}$ for the north pole and opposite to $\vec{B}$ for the south pole. This pair of opposite forces exerts a torque on the needle, rotating the needle until it is parallel to the magnetic field at that point.

Notice that the north pole of the compass needle, when it reaches the equilibrium position, is in the direction of the magnetic field. Thus a compass needle can be used as a probe of the magnetic field, just as a charge was a probe of the electric field. **Magnetic forces cause a compass needle to become aligned parallel to a magnetic field, with the north pole of the compass showing the direction of the magnetic field at that point.**

Look back at the compass alignments around the current-carrying wire in Figure 32.3b. Because compass needles align with the magnetic field, the magnetic field at each point must be tangent to a circle around the wire. Figure 32.5a shows the magnetic field by drawing field vectors. Notice that the field is weaker (shorter vectors) at greater distances from the wire.

Another way to picture the field is with the use of **magnetic field lines.** These are imaginary lines drawn through a region of space so that

- A tangent to a field line is in the direction of the magnetic field, and
- The field lines are closer together where the magnetic field strength is larger.

Figure 32.5b shows the magnetic field lines around a current-carrying wire. Notice that magnetic field lines form loops, with no beginning or ending point. This is in contrast to electric field lines, which stop and start on charges.

(a)

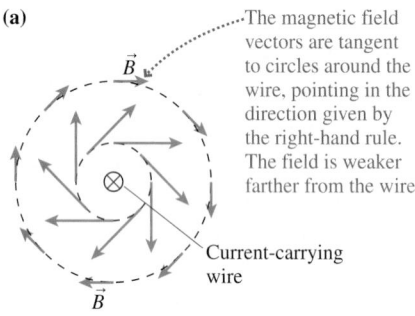

The magnetic field vectors are tangent to circles around the wire, pointing in the direction given by the right-hand rule. The field is weaker farther from the wire.

Current-carrying wire

(b)

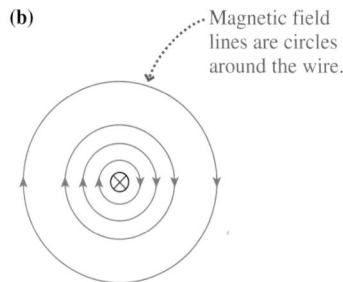

Magnetic field lines are circles around the wire.

(c)

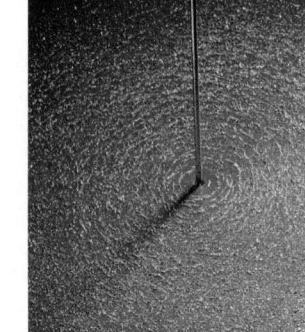

The magnetic field is revealed by the pattern of iron filings around the current-carrying wire.

FIGURE 32.5 The magnetic field around a current-carrying wire.

In the photograph of Figure 32.5c, iron filings that have been sprinkled around a current-carrying wire allow us to visualize the circular magnetic field pattern. It was patterns such as these that first suggested the field concept to Faraday.

NOTE ▶ The magnetic field of a current-carrying wire is very different from the electric field of a charged wire. The electric field of a charged wire points radially outward (positive wire) or inward (negative wire). ◀

Two Kinds of Magnetism?

You might be concerned that we have introduced two kinds of magnetism. We opened this chapter discussing permanent magnets and their forces. Then, without warning, we switched to the magnetic forces caused by a current. It is not at all obvious that these forces are the same kind of magnetism as that exhibited by stationary chucks of metal called "magnets." Perhaps there are two different types of magnetic forces, one having to do with currents and the other being responsible for permanent magnets. One of the major goals for our study of magnetism is to see that these two quite different ways of producing magnetic effects are really just two different aspects of a *single* magnetic force.

STOP TO THINK 32.2   The magnetic field at position P points

a. Up.
b. Down.
c. Into the page.
d. Out of the page.

• P

$\longrightarrow I$

32.3 The Source of the Magnetic Field: Moving Charges

Figure 32.5 is a qualitative picture of the wire's magnetic field. Our first task is to turn that picture into a quantitative description. Because current in a wire generates a magnetic field, and a current is a collection of moving charges, it's natural to wonder if *any* moving charge would do the same. Oersted's discovery encouraged the general assumption among scientists that this was the case, although confirmation was not to come until 1875, 55 years later, when a rapidly spinning charged disk was shown to produce the same magnetic effects as the current in a circular loop of wire.

Thus our starting point is the idea that **moving charges are the source of the magnetic field.** Figure 32.6 shows a charged particle q moving with velocity $\vec{v}$. The magnetic field of this moving charge is found to be

$$\vec{B} = \left(\frac{\mu_0}{4\pi} \frac{qv\sin\theta}{r^2}, \text{ direction given by the right-hand rule} \right) \quad (32.1)$$

where r is the distance from the charge and θ is the angle between $\vec{v}$ and $\vec{r}$.

Equation 32.1 is called the **Biot-Savart law** for a point charge (rhymes with *Leo* and *bazaar*), named for two French scientists whose investigations were motivated by Oersted's observations. It is analogous to Coulomb's law for the electric field of a point charge. Notice that the Biot-Savart law, like Coulomb's law, is an inverse-square law. However, the Biot-Savart law is somewhat more complex than Coulomb's law because the magnetic field depends on the angle θ between the charge's velocity and the line to the point where the field is evaluated.

NOTE ▶ The magnetic field of a moving charge is *in addition* to the charge's electric field. The charge has an electric field whether it is moving or not. ◀

The SI unit of magnetic field strength is the **tesla,** abbreviated as T. The tesla is defined as

$$1 \text{ tesla} = 1 \text{ T} \equiv 1 \text{ N/A m}$$

You will see later in the chapter that this definition is based on the magnetic force on a current-carrying wire. One tesla is quite a large field. Table 32.1 shows some typical magnetic field strengths. Most magnetic fields are a small fraction of a tesla.

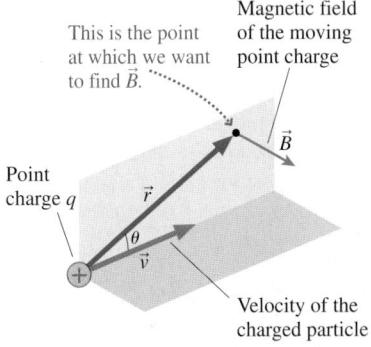

FIGURE 32.6 The magnetic field of a moving point charge.

TABLE 32.1 Typical magnetic field strengths

Field location	Field strength (T)
Surface of the earth	5×10^{-5}
Refrigerator magnet	5×10^{-3}
Laboratory magnet	0.1 to 1
Superconducting magnet	10

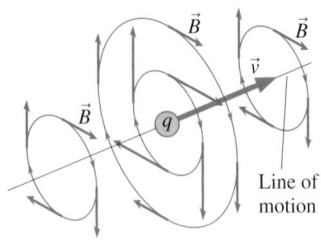

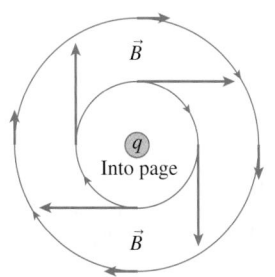

FIGURE 32.7 Two views of the magnetic field of a positive charge moving with velocity $\vec{v}$.

The constant μ_0 in Equation 32.1 is called the **permeability constant.** Its value is

$$\mu_0 = 4\pi \times 10^{-7}\,\text{Tm/A} = 1.257 \times 10^{-6}\,\text{Tm/A}$$

This constant plays a role in magnetism similar to that of the permittivity constant ϵ_0 in electricity.

The right-hand rule for finding the direction of $\vec{B}$ is similar to that used for a current-carrying wire: Point your right thumb in the direction of $\vec{v}$. The magnetic field vector $\vec{B}$ is perpendicular to the plane of $\vec{r}$ and $\vec{v}$, pointing in the direction in which your fingers curl. In other words, the $\vec{B}$ vectors are tangent to circles drawn about the charge's line of motion. Figure 32.7 shows a more complete view than Figure 32.6 of the magnetic field of a positive moving charge. Notice that $\vec{B}$ is zero along the line of motion, where $\theta = 0°$ or $180°$, due to the $\sin\theta$ term in Equation 32.1.

NOTE ▶ The vector arrows in Figure 32.7 would have the same lengths but be reversed in direction for a negative charge. ◀

The requirement that a charge be moving to generate a magnetic field is explicit in Equation 32.1. If the speed v of the particle is zero, the magnetic field (but not the electric field!) is zero. This helps to emphasize a fundamental distinction between electric and magnetic fields: **Charges create electric fields, but only *moving* charges create magnetic fields.**

EXAMPLE 32.1 **The magnetic field of a proton**

A proton moves along the x-axis with velocity $v_x = 1.0 \times 10^7$ m/s. As it passes the origin, what is the magnetic field at the (x, y, z) positions (1 mm, 0 mm, 0 mm), (0 mm, 1 mm, 0 mm), and (1 mm, 1 mm, 0 mm)?

MODEL The magnetic field is that of a moving charged particle.

VISUALIZE Figure 32.8 shows the geometry. The first point is on the x-axis, directly in front of the proton, with $\theta_1 = 0°$. The second point is on the y-axis, with $\theta_2 = 90°$, and the third is in the xy-plane.

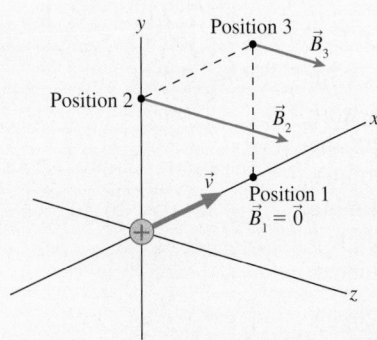

FIGURE 32.8 The magnetic field of Example 32.1.

SOLVE Position 1, which is along the line of motion, has $\theta = 0°$. Thus $\vec{B}_1 = 0$. Position 2 (at 0 mm, 1 mm, 0 mm) is at distance $r_2 = 1$ mm $= 0.001$ m. Equation 32.1, the Biot-Savart law, gives us the magnetic field strength at this point as

$$B = \frac{\mu_0}{4\pi}\frac{qv\sin\theta_2}{r_2^2}$$

$$= \frac{4\pi \times 10^{-7}\,\text{Tm/A}}{4\pi}\frac{(1.60 \times 10^{-19}\,\text{C})(1.0 \times 10^7\,\text{m/s})\sin 90°}{(0.0010\,\text{m})^2}$$

$$= 1.60 \times 10^{-13}\,\text{T}$$

According to the right-hand rule, the field points in the positive z-direction. Thus

$$\vec{B}_2 = 1.60 \times 10^{-13}\,\hat{k}\,\text{T}$$

where $\hat{k}$ is the unit vector in the positive z-direction. The field at position 3, at (1 mm, 1 mm, 0 mm), also points in the z-direction, but it is weaker than at position 2 both because r is larger *and* because θ is smaller. From geometry we know $r_3 = \sqrt{2}$ mm $= 0.00141$ m and $\theta_3 = 45°$. Another calculation using Equation 32.1 gives

$$\vec{B}_3 = 0.57 \times 10^{-13}\,\hat{k}\,\text{T}$$

ASSESS The magnetic field of a single moving charge is *very* small.

Superposition

The Biot-Savart law is the starting point for generating all magnetic fields, just as our earlier expression for the electric field of a point charge was the starting point for generating all electric fields. You learned in Chapter 26 that the total electric field caused by several charges $q_1, q_2, \ldots, q_n$ is the superposition of the electric fields of each separate charge.

Magnetic fields have been found experimentally to also obey the principle of superposition. If there are *n* moving point charges, the net magnetic field is given by the vector sum

$$\vec{B}_{\text{total}} = \vec{B}_1 + \vec{B}_2 + \cdots + \vec{B}_n \qquad (32.2)$$

where each individual $\vec{B}$ is calculated with Equation 32.1. The principle of superposition will be the basis for calculating the magnetic fields of several important current distributions.

The Vector Cross Product

In Chapter 25, we found that the electric field of a point charge could be written concisely and accurately as

$$\vec{E} = \frac{1}{4\pi\epsilon_0} \frac{q}{r^2} \hat{r}$$

where $\hat{r}$ is a *unit vector* that points from the charge to the point at which we wish to calculate the field. Unit vector $\hat{r}$ expresses the idea "away from q."

The unit vector $\hat{r}$ also allows us to write the Biot-Savart law more concisely and more accurately, but we'll need to use the form of vector multiplication called the *cross product*. To remind you, Figure 32.9 shows two vectors, $\vec{C}$ and $\vec{D}$, with angle α between them. The **cross product** of $\vec{C}$ and $\vec{D}$ is defined to be the vector

$$\vec{C} \times \vec{D} = (CD\sin\alpha, \text{ direction given by the right-hand rule}) \qquad (32.3)$$

The symbol $\times$ between the vectors is *required* to indicate a cross product.

> NOTE ▶ The cross product of two vectors and the right-hand rule used to determine the direction of the cross product were introduced in Section 13.9 to describe torque and angular momentum. If you omitted that section, you will want to turn to it now to read about the cross product. A review would be worthwhile even if you did learn about the cross product earlier. Study the examples in that section carefully if cross products are new to you. ◀

The Biot-Savart law, Equation 32.1, can be written in terms of the cross product as

$$\vec{B} = \frac{\mu_0}{4\pi} \frac{q\vec{v} \times \hat{r}}{r^2} \qquad \text{(magnetic field of a moving point charge)} \qquad (32.4)$$

where unit vector $\hat{r}$, shown in Figure 32.10, points from charge q to the point at which we want to evaluate the field. This expression for $\vec{B}$ has magnitude $(\mu_0/4\pi)(qv\sin\theta/r^2)$ (because the magnitude of unit vector $\hat{r}$ is 1) and points in the correct direction (given by the right-hand rule), so it agrees completely with Equation 32.1.

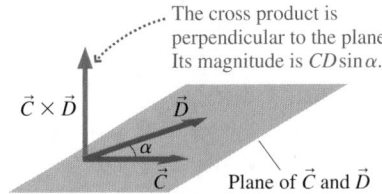

FIGURE 32.9 The cross product $\vec{C} \times \vec{D}$ is a vector perpendicular to the plane of vectors $\vec{C}$ and $\vec{D}$.

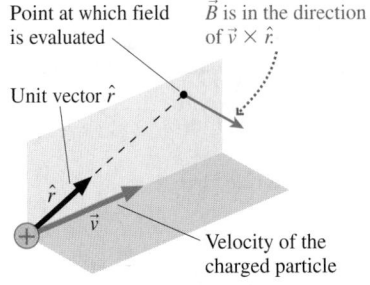

FIGURE 32.10 Unit vector $\hat{r}$ defines the direction from the moving charge to the point at which we want to evaluate the magnetic field.

EXAMPLE 32.2 The magnetic field direction of an electron

The electron in Figure 32.11 is moving to the right. What is the direction of the electron's magnetic field at the position indicated with a dot?

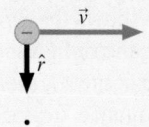

FIGURE 32.11 What is the direction of the electron's magnetic field at the dot?

VISUALIZE Because the charge is negative, the magnetic field points in the direction of $-(\vec{v} \times \hat{r})$, or opposite the direction of $\vec{v} \times \hat{r}$. Unit vector $\hat{r}$ points from the charge toward the dot. We can use the right-hand rule to find that $\vec{v} \times \hat{r}$ points *into* the page. Thus the electron's magnetic field at the dot points *out of* the page.

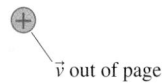

STOP TO THINK 32.3 The positive charge is moving straight out of the page. What is the direction of the magnetic field at the position of the dot?

$\vec{v}$ out of page

a. Up   b. Down   c. Left   d. Right

32.4 The Magnetic Field of a Current

13.1 Activ Physics ONLINE

Moving charges are the source of the magnetic field, but in practice we're more interested in the magnetic field of a current—a collection of moving charges—than in the very small magnetic fields of individual charges. The Biot-Savart law and the principle of superposition will be our primary tools for calculating magnetic fields. First, however, it will be useful to rewrite the Biot-Savart law in terms of current.

Figure 32.12a shows a current-carrying wire. The wire as a whole is electrically neutral, but current I represents the motion of positive charge carriers through the wire. Suppose the small amount of moving charge ΔQ spans the small length Δs. The charge has velocity $\vec{v} = \Delta\vec{s}/\Delta t$, where the vector $\Delta\vec{s}$, which is parallel to $\vec{v}$, is the charge's displacement vector. If ΔQ is small enough to treat as a point charge, the magnetic field it creates at a point in space is proportional to $(\Delta Q)\vec{v}$. We can write $(\Delta Q)\vec{v}$ in terms of the wire's current I as

$$(\Delta Q)\vec{v} = \Delta Q\frac{\Delta\vec{s}}{\Delta t} = \frac{\Delta Q}{\Delta t}\Delta\vec{s} = I\Delta\vec{s} \qquad (32.5)$$

where we used the definition of current, $I = \Delta Q/\Delta t$.

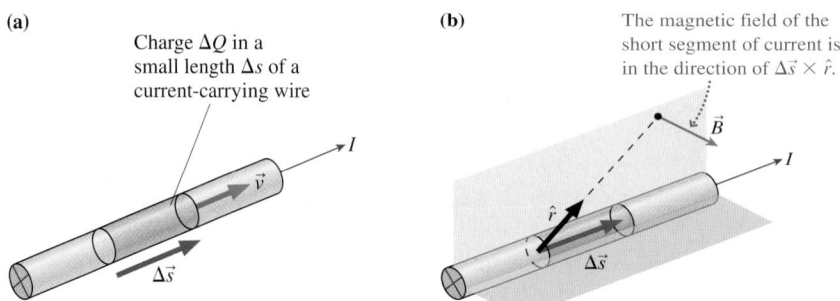

(a) Charge ΔQ in a small length Δs of a current-carrying wire

I
$\vec{v}$
$\Delta\vec{s}$

(b) The magnetic field of the short segment of current is in the direction of $\Delta\vec{s} \times \hat{r}$.

$\vec{B}$
I
$\hat{r}$
$\Delta\vec{s}$

FIGURE 32.12 Relating the magnetic field of wire to the current I.

If we replace $q\vec{v}$ in the Biot-Savart law with $I\Delta\vec{s}$, we find that the magnetic field of a very short segment of wire carrying current I is

$$\vec{B} = \frac{\mu_0}{4\pi}\frac{I\Delta\vec{s} \times \hat{r}}{r^2} \qquad (32.6)$$

(magnetic field of a very short segment of current)

Equation 32.6 is still the Biot-Savart law, only now written in terms of current rather than the motion of an individual charge. Figure 32.12b shows the direction of the current segment's magnetic field as determined by using the right-hand rule.

Equation 32.6 is the basis of a strategy for calculating the magnetic field of a current-carrying wire. You will recognize that it is the same basic strategy you learned for calculating the electric field of a continuous distribution of charge. The goal is to break a problem down into small steps that are individually manageable.

(MP) PROBLEM-SOLVING STRATEGY 32.1 **The magnetic field of a current**

MODEL Model the wire as a simple shape, such as a straight line or a loop.

VISUALIZE For the pictorial representation:

❶ Draw a picture and establish a coordinate system.

❷ Identify the point P at which you want to calculate the magnetic field.

❸ Divide the current-carrying wire into segments for which you *already know* how to determine $\vec{B}$. This is usually, though not always, a division into very short segments of length Δs.

❹ Draw the magnetic field vector for one or two segments. This will help you identify distances and angles that need to be calculated.

❺ Look for symmetries that simplify the field. You may conclude that some components of $\vec{B}$ are zero.

SOLVE The mathematical representation is $\vec{B}_{\text{net}} = \sum \vec{B}_k$.

■ Use superposition to form an algebraic expression for *each* of the three components of $\vec{B}$ (unless you are sure one or more is zero) at point P.

■ Let the (x, y, z) coordinates of the point remain as variables.

■ Express all angles and distances in terms of the coordinates.

■ Let $\Delta s \to ds$ and the sum become an integral. Think carefully about the integration limits for this variable; they will depend on the boundaries of the wire and on the coordinate system you have chosen to use. Carry out the integration and simplify the results as much as possible.

ASSESS Check that your result is consistent with any limits for which you know what the field should be.

EXAMPLE 32.3 The magnetic field of a long, straight wire
A long, straight wire carries current I in the positive x-direction. Find the magnetic field at a point that is distance d from the wire.

MODEL Because the wire is "long," let's model it as being infinitely long.

VISUALIZE Figure 32.13 illustrates the steps in the problem-solving strategy. We've chosen a coordinate system with point P on the y-axis. We've then divided the rod into small segments, each containing a small amount ΔQ of *moving charge*. Unit

vector $\hat{r}$ and angle θ_k are shown for segment k. You should use the right-hand rule to convince yourself that $\vec{B}_k$ points *out of the page,* in the positive z-direction. This is the direction no matter where segment k happens to be along the x-axis. Consequently, B_x (the component of $\vec{B}$ parallel to the wire) and B_y (the component of $\vec{B}$ straight away from the wire) are zero. The only component of $\vec{B}$ we need to evaluate is B_z, the component tangent to a circle around the wire.

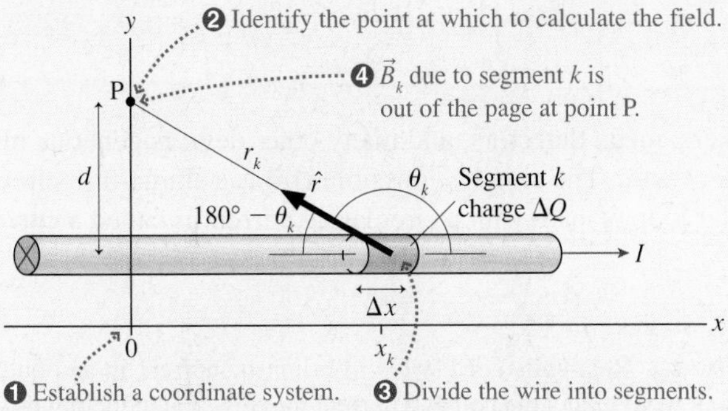

SOLVE We can use the Biot-Savart law to find the field $(B_k)_z$ of segment k. The cross product $\Delta \vec{s}_k \times \hat{r}$ has magnitude $(\Delta x)(1)\sin\theta_k$, hence

$$(B_k)_z = \frac{\mu_0}{4\pi} \frac{I \Delta x \sin\theta_k}{r_k^2} = \frac{\mu_0}{4\pi} \frac{I \sin\theta_k}{r_k^2} \Delta x = \frac{\mu_0}{4\pi} \frac{I \sin\theta_k}{x_k^2 + d^2} \Delta x$$

where we wrote the distance r_k in terms of x_k and d. We also need to express θ_k in terms of x_k and d. Because $\sin(180° - \theta) = \sin\theta$, this is

$$\sin\theta_k = \sin(180° - \theta_k) = \frac{d}{r_k} = \frac{d}{\sqrt{x_k^2 + d^2}}$$

With this expression for $\sin\theta_k$, the magnetic field of segment k is

$$(B_k)_z = \frac{\mu_0}{4\pi} \frac{Id}{(x_k^2 + d^2)^{3/2}} \Delta x$$

❷ Identify the point at which to calculate the field.

❹ $\vec{B}_k$ due to segment k is out of the page at point P.

P

y

r_k

d $\hat{r}$ θ_k Segment k

$180° - \theta_k$ charge ΔQ

$\longrightarrow I$

Δx

❶ Establish a coordinate system.   ❸ Divide the wire into segments.

0 x_k x

FIGURE 32.13 Calculating the magnetic field of a long, straight wire carrying current I.

Now we're ready to sum the magnetic field of all the segments. The superposition is a vector sum, but in this case only the z-components are nonzero. Thus

$$B_{wire} = \sum_k (B_k)_z$$

$$= \frac{\mu_0 Id}{4\pi} \sum_k \frac{\Delta x}{(x_k^2 + d^2)^{3/2}} \to \frac{\mu_0 Id}{4\pi} \int_{-\infty}^{\infty} \frac{dx}{(x^2 + d^2)^{3/2}}$$

Only at the very last step did we convert the sum to an integral. Then our model of the wire as being infinitely long sets the integration limits at $\pm\infty$. This is a standard integral that can be found in integral tables. Evaluation gives

$$B_{wire} = \frac{\mu_0 Id}{4\pi} \frac{x}{d^2 (x^2 + d^2)^{1/2}} \Big|_{-\infty}^{\infty} = \frac{\mu_0}{2\pi} \frac{I}{d}$$

This is the magnitude of the field. The field direction is determined by using the right-hand rule. We can combine these two pieces of information to write

$$\vec{B}_{wire} = \left(\frac{\mu_0}{2\pi} \frac{I}{d}, \text{ tangent to a circle around the wire in the right-hand direction} \right)$$

ASSESS Figure 32.14 shows the magnetic field of a current-carrying wire. Compare this to Figure 32.2 and convince yourself that the direction shown is that of the right-hand rule.

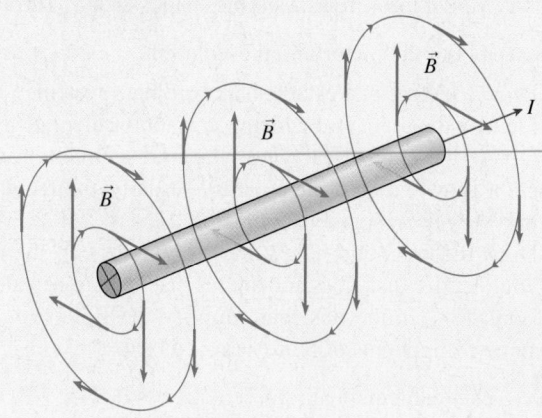

FIGURE 32.14 The magnetic field of a long, straight wire carrying current I.

NOTE ▶ The difficulty magnetic field calculations present is not doing the integration itself, which is the last step, but setting up the calculation and knowing *what* to integrate. The purpose of the problem-solving strategy is to guide you through the process of setting up the integral. ◀

EXAMPLE 32.4 The magnetic field strength near a heater wire

A 1.0-m-long, 1.0-mm-diameter nichrome heater wire is connected to a 12 V battery. What is the magnetic field strength 1.0 cm away from the wire?

MODEL 1 cm is much less than the 1 m length of the wire, so model the wire as infinitely long.

SOLVE The current through the wire is $I = \Delta V_{bat}/R$, where the wire's resistance R is

$$R = \frac{\rho L}{A} = \frac{\rho L}{\pi r^2} = 1.91 \,\Omega$$

The nichrome resistivity $\rho = 1.50 \times 10^{-6}\,\Omega\,m$ was taken from Table 28.2. Thus the current is $I = (12\,V)/(1.91\,\Omega) = 6.28\,A$. The magnetic field strength at distance $d = 1.0\,cm = 0.010\,m$ from the wire is

$$B_{wire} = \frac{\mu_0}{2\pi} \frac{I}{d} = (2.0 \times 10^{-7}\,T\,m/A)\frac{6.28\,A}{0.010\,m}$$

$$= 1.26 \times 10^{-4}\,T$$

ASSESS The magnetic field of the wire is slightly more than twice the strength of the earth's magnetic field.

13.2 Act∤v Phys∤cs

Motors, loudspeakers, metal detectors, and many other devices generate magnetic fields with *coils* of wire. The simplest possible coil is a single-turn circular loop of wire. A circular loop of wire with a circulating current is called a **current loop.**

EXAMPLE 32.5 The magnetic field of a current loop

Figure 32.15a shows a current loop, a circular loop of wire with radius R that carries current I. Find the magnetic field of the current loop at distance z on the axis of the loop.

MODEL Real coils need wires to bring the current in and out, but we'll model the coil as a current moving around the full circle shown in Figure 32.15b.

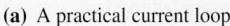

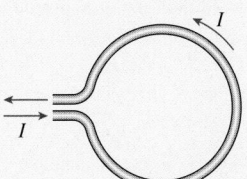 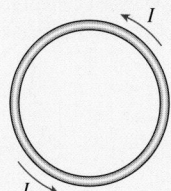

FIGURE 32.15 A current loop.

VISUALIZE Figure 32.16 shows a loop for which we've assumed that the current is circulating ccw. We've chosen a coordinate system in which the loop lies at $z = 0$ in the xy-plane. Let segment k be the segment at the top of the loop. Vector $\Delta\vec{s}_k$ is parallel to the x-axis and unit vector $\hat{r}$ is in the yz-plane, thus angle θ_k, the angle between $\Delta\vec{s}_k$ and $\hat{r}$, is 90°.

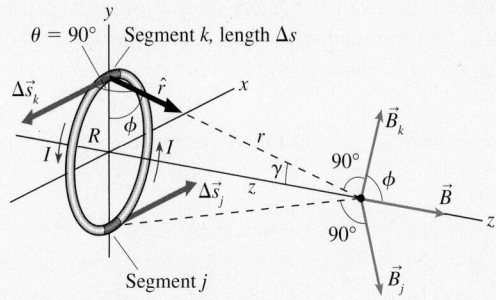

FIGURE 32.16 Calculating the magnetic field of a current loop.

The direction of $\vec{B}_k$, the magnetic field due to the current in segment k, is given by the cross product $\Delta\vec{s}_k \times \hat{r}$. $\vec{B}_k$ must be perpendicular to $\Delta\vec{s}_k$ *and* perpendicular to $\hat{r}$. You should convince yourself that $\vec{B}_k$ in Figure 32.16 points in the correct

direction. Notice that the xy-component of $\vec{B}_k$ is canceled by the xy-component of magnetic field $\vec{B}_j$ due to the current segment at the bottom of the loop, 180° away. In fact, *every* current segment on the loop can be paired with a segment 180° away, on the opposite side of the loop, such that the xy-components of $\vec{B}$ cancel and the components of $\vec{B}$ parallel to the z-axis add. The symmetry of the loop requires the on-axis magnetic field to point along the z-axis. Knowing that we need to sum only the z-components will simplify our calculation.

SOLVE We can use the Biot-Savart law to find the z-component $(B_k)_z = B_k \cos\phi$ of the magnetic field of segment k. The cross product $\Delta\vec{s}_k \times \hat{r}$ has magnitude $(\Delta s)(1)\sin 90° = \Delta s$, thus

$$(B_k)_z = \frac{\mu_0}{4\pi}\frac{I\Delta s}{r^2}\cos\phi = \frac{\mu_0 I \cos\phi}{4\pi(z^2 + R^2)}\Delta s$$

where we wrote distance r in terms of z and R. You can see, because $\phi + \gamma = 90°$, that angle ϕ is also the angle between $\hat{r}$ and the radius of the loop. Hence $\cos\phi = R/r$, and $(B_k)_z$ is

$$(B_k)_z = \frac{\mu_0 IR}{4\pi(z^2 + R^2)^{3/2}}\Delta s$$

The final step is to sum the magnetic fields due to all the segments:

$$B_{\text{loop}} = \sum_k (B_k)_z = \frac{\mu_0 IR}{4\pi(z^2 + R^2)^{3/2}}\sum_k \Delta s$$

In this case, unlike the straight wire, none of the terms multiplying Δs depends on the position of segment k, so all these terms can be factored out of the summation. We're left with a summation that adds up the lengths of all the small segments. But this is just the total length of the wire, which is the circumference $2\pi R$. Thus the on-axis magnetic field of a current loop is

$$B_{\text{loop}} = \frac{\mu_0 IR}{4\pi(z^2 + R^2)^{3/2}}2\pi R = \frac{\mu_0}{2}\frac{IR^2}{(z^2 + R^2)^{3/2}}$$

In practice, a coil often has N *turns* of wire. If the turns are all very close together, so that the magnetic field of each is essentially the same, then the magnetic field of a coil is N times the magnetic field of a current loop. The magnetic field at the center ($z = 0$) of an N-turn coil is

$$B_{\text{coil center}} = \frac{\mu_0}{2}\frac{NI}{R} \qquad (32.7)$$

EXAMPLE 32.6 Matching the earth's magnetic field
What current is needed in a 5-turn, 10-cm-diameter coil to cancel the earth's magnetic field at the center of the coil?

MODEL Scientists sometimes need to carry out measurements in zero magnetic field. One way to create a field-free region of space is to generate a magnetic field equal to the earth's field but pointing in the opposite direction. The vector sum of the two fields is zero.

VISUALIZE Figure 32.17 shows a five-turn coil of wire. The magnetic field is five times that of a single current loop.

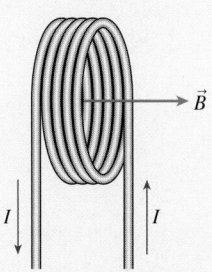

FIGURE 32.17 A coil of wire.

SOLVE The earth's magnetic field, from Table 32.1, is 5×10^{-5} T. We can use Equation 32.7 to find that the current needed to generate a 5×10^{-5} T field is

$$I = \frac{2RB}{\mu_0 N} = \frac{2(0.050 \text{ m})(5.0 \times 10^{-5} \text{ T})}{5(4\pi \times 10^{-7} \text{ Tm/A})} = 0.80 \text{ A}$$

ASSESS A 0.80 A current is easily produced. Although there are better ways to cancel the earth's field than using a simple current loop, this illustrates the idea.

32.5 Magnetic Dipoles

We were able to calculate the on-axis magnetic field of a current loop, but determining the field at other points requires either numerical integrations or an experimental mapping of the field. Figure 32.18 shows the full magnetic field of a current loop. This is a field with *rotational symmetry,* so to picture the full three-dimensional field, imagine Figure 32.18a rotated about the axis of the loop. Figure 32.18b shows the magnetic field in the plane of the loop as seen from the right. There is a clear sense that the magnetic field leaves the loop on one side, "flows" around the outside, then returns to the loop.

(a) Cross section through the current loop

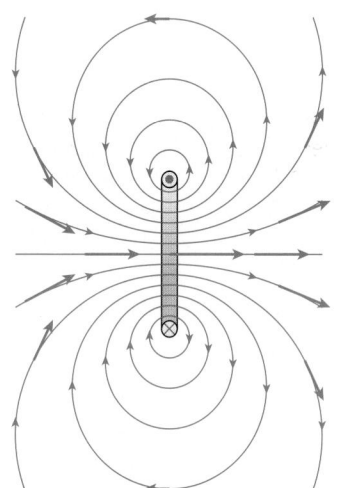

(b) The current loop seen from the right

The field emerges from the center of the loop.

The field returns around the outside of the loop.

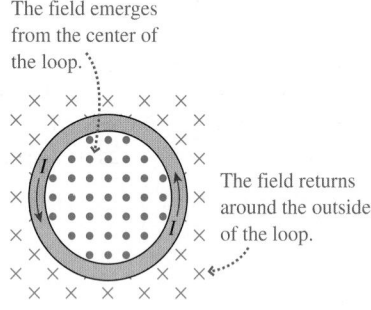

FIGURE 32.18 The magnetic field of a current loop.

There are two versions of the right-hand rule that you can use to determine which way a loop's field points. Try these in Figure 32.18. Being able to quickly ascertain the field direction of a current loop is an important skill.

TACTICS BOX 32.1 **Finding the magnetic field direction of a current loop**

Use either of the following methods to find the magnetic field direction:

❶ Point your right thumb in the direction of the current at any point on the loop and let your fingers curl through the center of the loop. Your fingers are then pointing in the direction in which $\vec{B}$ leaves the loop.

❷ Curl the fingers of your right hand around the loop in the direction of the current. Your thumb is then pointing in the direction in which $\vec{B}$ leaves the loop.

A Current Loop Is a Magnetic Dipole

A current loop has two distinct sides. Bar magnets and flat refrigerator magnets also have two distinct sides or ends, so you might wonder if current loops are related to these permanent magnets. Consider the following experiments with a current loop. Notice that we're using a simplified picture that shows the magnetic field only in the plane of the loop.

Investigating current loops

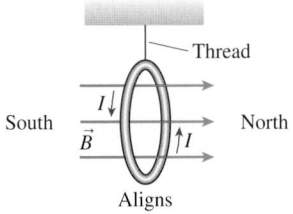
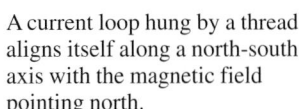

A current loop hung by a thread aligns itself along a north-south axis with the magnetic field pointing north.

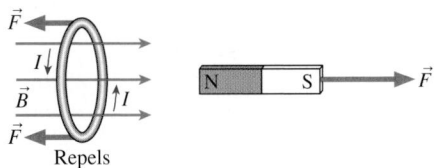

The north pole of a permanent magnet repels the side of a current loop from which the magnetic field is emerging.

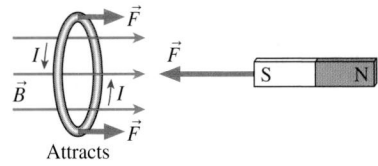

The south pole of a permanent magnet attracts the side of a current loop from which the magnetic field is emerging.

These investigations show that **a current loop is a magnet,** just like a permanent magnet. A magnet created by a current in a coil of wire is called an **electromagnet.** An electromagnet picks up small pieces of iron, influences a compass needle, and acts in every way like a permanent magnet.

In fact, Figure 32.19 shows that **a flat permanent magnet and a current loop generate the same magnetic field.** It is the field of a magnetic dipole, irrespective of how the dipole was produced. For both, you can identify the north pole as the face or end *from which* the magnetic field emerges. The magnetic field of both point *into* the south pole.

NOTE ▶ The magnetic field *inside* a permanent magnet differs from the magnetic field at the center of a current loop. Only the exterior field of a magnet matches the field of a current loop. ◀

One of the goals of this chapter is to show that magnetic forces exerted by currents and magnetic forces exerted by permanent magnets are just two different aspects of a single magnetism. We've now found a strong connection between permanent magnets and current loops, and this connection will turn out to be a big piece of the puzzle.

The Magnetic Dipole Moment

The expression for the electric field of an electric dipole was considerably simplified when we considered the field at distances significantly larger than the size of the charge separation s. The on-axis field of an electric dipole when $z \gg s$ is

$$\vec{E}_{\text{dipole}} = \frac{1}{4\pi\epsilon_0} \frac{2\vec{p}}{z^3}$$

where the electric dipole moment $\vec{p}$ is the vector $\vec{p} = (qs$, from the negative to the positive charge).

The on-axis magnetic field of a current loop, which we calculated in Example 32.5, is

$$B_{\text{loop}} = \frac{\mu_0}{2} \frac{IR^2}{(z^2 + R^2)^{3/2}}$$

(a) Current loop

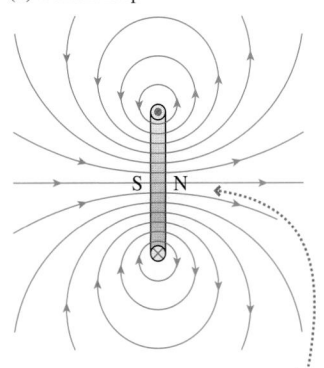

Whether it's a current loop or a permanent magnet, the magnetic field emerges from the north pole.

(b) Permanent magnet

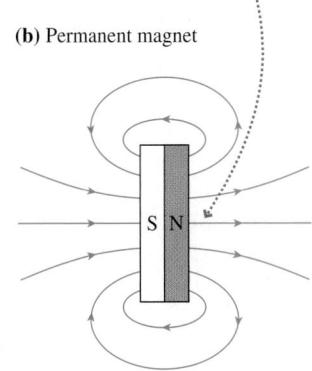

FIGURE 32.19 A current loop has magnetic poles and generates the same magnetic field as a flat permanent magnet.

If z is much larger than the diameter of the current loop, $z \gg R$, we can make the approximation $(z^2 + R^2)^{3/2} \rightarrow z^3$. Then the loop's field is

$$B_{\text{loop}} \approx \frac{\mu_0}{2} \frac{IR^2}{z^3} = \frac{\mu_0}{4\pi} \frac{2(\pi R^2)I}{z^3} = \frac{\mu_0}{4\pi} \frac{2AI}{z^3} \qquad (32.8)$$

where $A = \pi R^2$ is the area of the loop.

A more advanced treatment of current loops shows that, if z is much larger than the size of the loop, Equation 32.8 is the on-axis magnetic field of a current loop of *any* shape, not just a circular loop. The shape of the loop affects the nearby field, but the distant field depends only on the current I and the area A enclosed within the loop. With this in mind, let's define the **magnetic dipole moment** $\vec{\mu}$ of a current loop enclosing area A to be

$$\vec{\mu} = (AI, \text{ from the south pole to the north pole})$$

The SI units of the magnetic dipole moment are A m^2.

> **NOTE** ▶ Don't confuse the magnetic dipole moment $\vec{\mu}$ with the constant μ_0 in the Biot-Savart law. ◀

The magnetic dipole moment, like the electric dipole moment, is a vector. It has the same direction as the on-axis magnetic field. Thus the right-hand rule used for determining the direction of $\vec{B}$ also shows the direction of $\vec{\mu}$. Figure 32.20 shows the magnetic dipole moment of a circular current loop.

Because the on-axis magnetic field of a current loop points in the same direction as $\vec{\mu}$, we can combine Equation 32.8 and the definition of $\vec{\mu}$ to write the on-axis field of a magnetic dipole as

$$\vec{B}_{\text{dipole}} = \frac{\mu_0}{4\pi} \frac{2\vec{\mu}}{z^3} \qquad \text{(on the axis of a magnetic dipole)} \qquad (32.9)$$

If you compare $\vec{B}_{\text{dipole}}$ to $\vec{E}_{\text{dipole}}$, you can see that the magnetic field of a magnetic dipole has the same basic shape as the electric field of an electric dipole.

Because a permanent bar magnet is also a magnetic dipole, a permanent magnet also has a magnetic dipole moment. Its on-axis magnetic field is given by Equation 32.9 when z is much larger than the size of the magnet. Equation 32.9 and laboratory measurements of the on-axis magnetic field can be used to determine a permanent magnet's dipole moment.

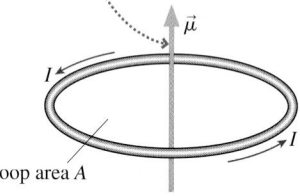

The magnetic dipole moment is perpendicular to the loop, in the direction of the right-hand rule. The magnitude of $\vec{\mu}$ is AI.

Loop area A

FIGURE 32.20 The magnetic dipole moment of a circular current loop.

EXAMPLE 32.7 The field of a magnetic dipole

a. The on-axis magnetic field strength 10 cm from a magnetic dipole is 1.0×10^{-5} T. What is the size of the magnetic dipole moment?

b. If the magnetic dipole is created by a 4.0-mm-diameter current loop, what is the current?

MODEL Assume that the distance 10 cm is much larger than the size of the dipole.

SOLVE

a. If $z \gg R$, we can use Equation 32.9 to find the magnetic dipole moment:

$$\mu = \frac{4\pi}{\mu_0} \frac{z^3 B}{2}$$

$$= \frac{4\pi}{4\pi \times 10^{-7} \text{ Tm/A}} \frac{(0.10 \text{ m})^3 (1.0 \times 10^{-5} \text{ T})}{2}$$

$$= 0.050 \text{ A m}^2$$

b. The magnetic dipole moment of a current loop is $\mu = AI$, so the necessary current is

$$I = \frac{\mu}{\pi R^2}$$

$$= \frac{(0.050 \text{ A m}^2)}{\pi (0.0020 \text{ m})^2} = 4000 \text{ A}$$

ASSESS Only a superconducting ring could carry a 4000 A current, so producing this magnetic field with a current loop is not very feasible. But 0.050 A m^2 is a quite modest dipole moment for a bar magnet, so this field could be produced with a permanent magnet.

STOP TO THINK 32.4   What is the current direction in this loop? And which side of the loop is the north pole?

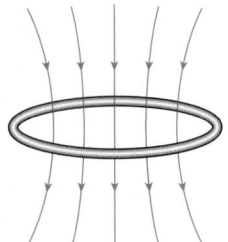

a. Current cw; north pole on top
b. Current cw; north pole on bottom
c. Current ccw; north pole on top
d. Current ccw; north pole on bottom

32.6 Ampère's Law and Solenoids

In principle, the Biot-Savart law can be used to calculate the magnetic field of any current distribution. In practice, the integrals are very difficult to evaluate for anything other than very simple situations. We faced a similar situation for calculating electric fields, but we discovered an alternative method—Gauss's law—for calculating the electric field of charge distributions with a high degree of symmetry. Gauss's law doesn't work in every situation, but it is simple and elegant where it does.

Likewise, there's an alternative method, called *Ampère's law,* for calculating the magnetic fields of current distributions with a high degree of symmetry. Ampère's law, like Gauss's law, doesn't work in all situations, but it is simple and elegant where it does. Whereas Gauss's law is written in terms of a surface integral, Ampère's law is based on the mathematical procedure called a *line integral.*

Line Integrals

We've flirted with the idea of a line integral ever since introducing the concept of work in Chapter 11, but now we need to take a more serious look at what a line integral represents and how it is used. Figure 32.21a shows a curved line that goes from an initial point i to a final point f.

(a)

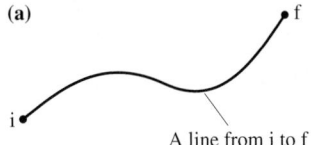

A line from i to f

(b)

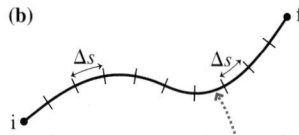

The line can be divided into many small segments. The sum of all the Δs's is the length L of the line.

FIGURE 32.21 Integrating along a line from i to f.

Suppose, as shown in Figure 32.21b, we divide the line up into many small segments of length Δs. The first segment is Δs_1, the second is Δs_2, and so on. The sum of all the Δs's is just the length L of the line between i and f. We can write this mathematically as

$$L = \sum_k \Delta s_k \rightarrow \int_i^f ds \tag{32.10}$$

where, in the last step, we let $\Delta s \rightarrow ds$ and the sum become an integral.

This integral is called a **line integral.** All we've done is to subdivide a line into infinitely many infinitesimal pieces, then add them up. This is exactly what you do in calculus when you evaluate an integral such as $\int x \, dx$. In fact, an integration along the *x*-axis *is* a line integral, one that happens to be along a straight line. Figure 32.21 differs only in that the line is curved. The underlying idea in both cases is that an integral is just a fancy way of doing a sum.

(a)

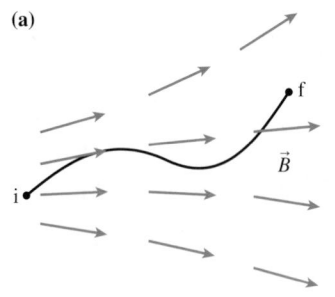

The line passes through a magnetic field.

(b)

Magnetic field at segment k

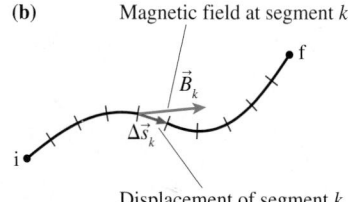

Displacement of segment k

FIGURE 32.22 Integrating $\vec{B}$ along a line from i to f.

The line integral of Equation 32.10 is not a terribly exciting one. Figure 32.22a makes things more interesting by allowing the line to pass through a magnetic field. Figure 32.22b again divides the line into small segments, but this time $\Delta \vec{s}_k$ is the displacement vector of segment k. The magnetic field at this point in space is $\vec{B}_k$.

Suppose we were to evaluate the dot product $\vec{B}_k \cdot \Delta \vec{s}_k$ at each segment, then add the values of $\vec{B}_k \cdot \Delta \vec{s}_k$ due to every segment. Doing so, and again letting the sum become an integral, we have

$$\sum_k \vec{B}_k \cdot \Delta \vec{s}_k \rightarrow \int_i^f \vec{B} \cdot d\vec{s} = \text{the line integral of } \vec{B} \text{ from i to f}$$

Once again, the integral is just a shorthand way to say "Divide the line into lots of little pieces, evaluate $\vec{B}_k \cdot \Delta \vec{s}_k$ for each piece, then add them up."

Although this process of evaluating the integral could be difficult, the only line integrals we'll need to deal with fall into two simple cases. If the magnetic field is *everywhere perpendicular* to the line, then $\vec{B} \cdot d\vec{s} = 0$ at every point along the line and the integral is zero. If the magnetic field is *everywhere tangent* to the line *and* has the same magnitude B at every point, then $\vec{B} \cdot d\vec{s} = B \, ds$ at every point and

$$\int_i^f \vec{B} \cdot d\vec{s} = \int_i^f B \, ds = B \int_i^f ds = BL \tag{32.11}$$

We used Equation 32.10 in the last step to integrate ds along the line.

Tactics Box 32.2 summarizes these two situations.

TACTICS BOX 32.2 Evaluating line integrals

❶ If $\vec{B}$ is everywhere perpendicular to a line, the line integral of $\vec{B}$ is

$$\int_i^f \vec{B} \cdot d\vec{s} = 0$$

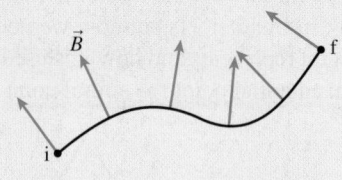

❷ If $\vec{B}$ is everywhere tangent to a line of length L *and* has the same magnitude B at every point, the line integral of $\vec{B}$ is

$$\int_i^f \vec{B} \cdot d\vec{s} = BL$$

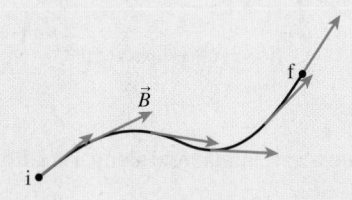

Ampère's Law

The French scientist André-Marie Ampère, for whom the SI unit of current is named, studied the properties of magnetism in the 1820s. Ampère noted that the magnetic field of a current-carrying wire is everywhere tangent to a circle around the wire *and* has the same magnitude $\mu_0 I/2\pi d$ at all points on the circle. According to Tactics Box 32.2, these conditions allow us to easily evaluate the line integral of $\vec{B}$ along a circular path around the wire.

Figure 32.23 shows a wire carrying current I into the page and the magnetic field at distance d. Suppose we were to integrate the magnetic field *all the way around* the circle. That is, the initial point i of the integration path and the final point f will be the same point. This would be a line integral around a *closed curve,* which is denoted

$$\oint \vec{B} \cdot d\vec{s}$$

The little circle on the integral sign indicates that the integration is performed around a closed curve. The notation has changed, but the meaning has not.

Because $\vec{B}$ is tangent to the circle *and* of constant magnitude at every point on the circle, we can use Option 2 from Tactics Box 32.2 to write

$$\oint \vec{B} \cdot d\vec{s} = BL = B(2\pi d) \tag{32.12}$$

where, in this case, the path length L is the circumference $2\pi d$ of the circle. We know that the magnetic field strength is $B = \mu_0 I / 2\pi d$, thus

$$\oint \vec{B} \cdot d\vec{s} = \mu_0 I \tag{32.13}$$

The interesting result is that the line integral of $\vec{B}$ around the current-carrying wire is independent of the radius of the circle. Any circle, from one touching the wire to one far away, would give the same result. The integral depends only on the amount of current passing *through* the circle that we integrated around.

This is reminiscent of Gauss's law. In our investigation of Gauss's law, we started with the observation that electric flux Φ_e through a sphere surrounding a point charge depends only on the amount of charge inside, not on the radius of the sphere. After examining several cases, we concluded that the shape of the surface wasn't relevant. The electric flux through *any* closed surface enclosing total charge Q_{in} turned out to be $\Phi_e = Q_{in}/\epsilon_0$.

Although we'll skip the details, the same type of reasoning that we used to prove Gauss's law shows that the result of Equation 32.13

- Is independent of the shape of the curve around the current.
- Is independent of where the current passes through the curve.
- Depends only on the total amount of current through the area enclosed by the integration path.

Thus whenever total current $I_{through}$ passes through an area bounded by a *closed curve,* the line integral of the magnetic field around the curve is

$$\oint \vec{B} \cdot d\vec{s} = \mu_0 I_{through} \tag{32.14}$$

This result for the magnetic field is known as **Ampère's law.**

To make practical use of Ampère's law, we need to determine which currents are positive and which are negative. The right-hand rule is once again the proper tool. If you curl your right fingers around the closed path in the direction in which you are going to integrate, then any current passing though the bounded area in the direction of your thumb is a positive current. Any current in the opposite direction is a negative current. In Figure 32.24, for example, currents I_1 and I_2 are positive, I_3 is negative. Thus $I_{through} = I_1 + I_2 - I_3$.

NOTE ▶ The integration path of Ampère's law is a mathematical curve through space. It does not have to match a physical surface or boundary, although it could if we want it to. ◀

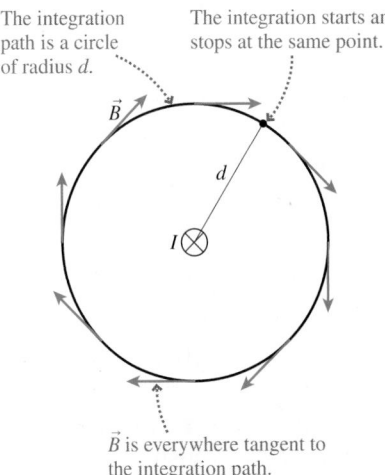

The integration path is a circle of radius d.

The integration starts and stops at the same point.

$\vec{B}$ is everywhere tangent to the integration path.

FIGURE 32.23 Integrating the magnetic field around a wire.

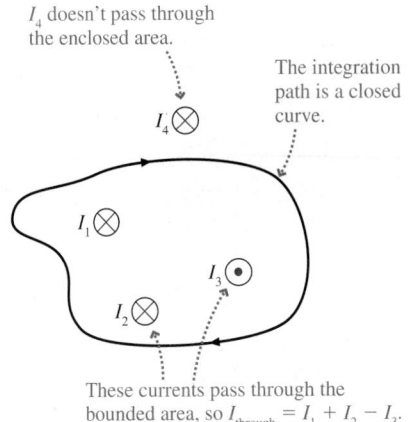

I_4 doesn't pass through the enclosed area.

The integration path is a closed curve.

These currents pass through the bounded area, so $I_{through} = I_1 + I_2 - I_3$.

FIGURE 32.24 Using Ampère's law.

In one sense, Ampère's law doesn't tell us anything new. After all, we derived Ampère's law from the Biot-Savart law for the magnetic field of a current. But in another sense, Ampère's law is more important than the Biot-Savart law because it states a very general property about magnetic fields. Ampère's law will turn out to be especially useful in Chapter 34 when we combine it with other electric and magnetic equations to form Maxwell's equations of the electromagnetic field. In the meantime, Ampère's law will allow us to find the magnetic fields of some important current distributions that have a high degree of symmetry.

EXAMPLE 32.8 The magnetic field inside a current-carrying wire

A wire of radius R carries current I. Find the magnetic field inside the wire at distance $r < R$ from the axis.

MODEL Assume the current density is uniform over the cross section of the wire.

VISUALIZE Figure 32.25 shows a cross section through the wire. The wire has perfect cylindrical symmetry, with all the charges moving parallel to the wire, so the magnetic field *must* be tangent to circles that are concentric with the wire. We don't know how the strength of the magnetic field depends on the distance from the center of the wire—that's what we're going to find—but the symmetry of the situation dictates the *shape* of the magnetic field.

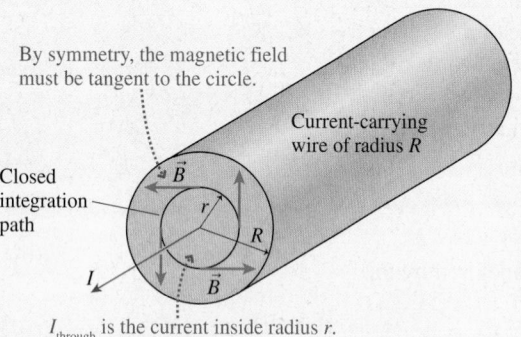

By symmetry, the magnetic field must be tangent to the circle.

Current-carrying wire of radius R

Closed integration path

$I_{through}$ is the current inside radius r.

FIGURE 32.25 Using Ampère's law inside a current-carrying wire.

SOLVE To find the field strength at radius r, draw a circle of radius r. The amount of current passing through this circle is

$$I_{through} = JA_{circle} = \pi r^2 J$$

where J is the current density. Our assumption of a uniform current density allows us to use the full current I passing though a wire of radius R to find that

$$J = \frac{I}{A} = \frac{I}{\pi R^2}$$

Thus the current through the circle of radius r is related to the total current I by

$$I_{through} = \frac{r^2}{R^2} I$$

Let's integrate $\vec{B}$ around the circumference of this circle. According to Ampère's law,

$$\oint \vec{B} \cdot d\vec{s} = \mu_0 I_{through} = \frac{\mu_0 r^2}{R^2} I$$

We know from the symmetry of the wire that $\vec{B}$ is everywhere tangent to the circle *and* has the same magnitude at all points on the circle. Consequently, the line integral of $\vec{B}$ around the circle can be evaluated using Option 2 of Tactics Box 32.2:

$$\oint \vec{B} \cdot d\vec{s} = BL = 2\pi r B$$

where $L = 2\pi r$ is the path length. If we substitute this expression into Ampère's law, we find that

$$2\pi r B = \frac{\mu_0 r^2}{R^2} I$$

Solving for B, we find that the magnetic field strength at radius r *inside* a current-carrying wire is

$$B = \frac{\mu_0 I}{2\pi R^2} r$$

ASSESS The magnetic field strength increases linearly with distance from the center of the wire until, at the surface of the wire, $B = \mu_0 I / 2\pi R$ matches our earlier solution for the magnetic field outside a current-carrying wire. This agreement at $r = R$ gives us confidence in our result. The magnetic field strength both inside and outside the wire is shown graphically in Figure 32.26.

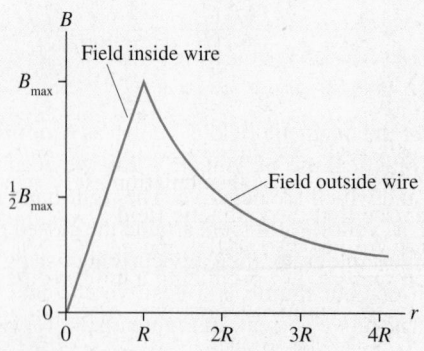

FIGURE 32.26 Graphical representation of the magnetic field inside and outside a current-carrying wire.

The Magnetic Field of a Solenoid

In our study of electricity, we made extensive use of the idea of a uniform electric field: a field that is the same at every point in space. We found that two closely spaced, parallel charged plates generate a uniform electric field between them, and this uniform field was one reason why we focused so much attention on learning about the parallel-plate capacitor.

Similarly, there are many applications of magnetism for which we would like to generate a **uniform magnetic field,** a field that has the same magnitude and the same direction at every point within some region of space. None of the sources we have looked at thus far produces a uniform magnetic field.

In practice, a uniform magnetic field is generated with a **solenoid.** A solenoid, shown in Figure 32.27, is a helical coil of wire with the same current I passing through each loop in the coil. Solenoids may have hundreds or thousands of coils, often called *turns,* sometimes wrapped in several layers.

We can understand a solenoid by thinking of it as a stack of current loops. Figure 32.28a shows the magnetic field of a single current loop at three points on the axis and three points equally distant from the axis. The field directly above the loop is opposite in direction to the field inside the loop. Figure 32.28b then shows three parallel loops. We can use information from Figure 32.28a to draw the magnetic fields of each loop at the center of loop 2 and at a point above loop 2.

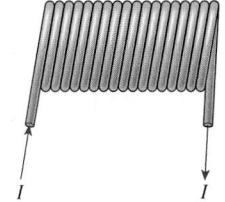

FIGURE 32.27 A solenoid.

(a) A single loop   **(b)** A stack of three loops

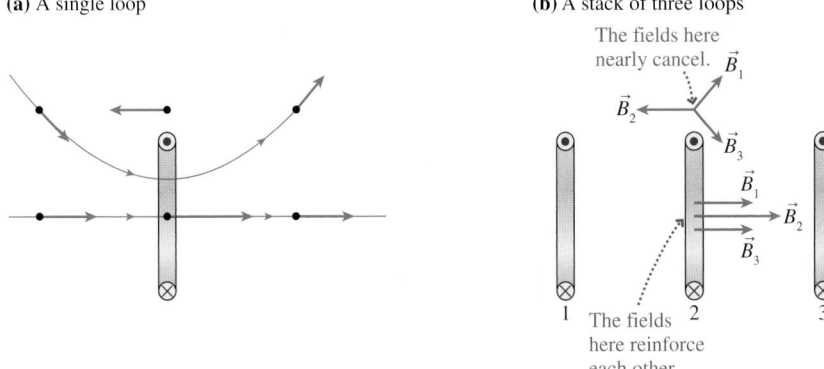

FIGURE 32.28 Using superposition to find the magnetic field of a stack of current loops.

The superposition of the fields at the center of loop 2 produces a *stronger* field than that of loop 2 alone. But the fields at the point above loop 2 tend to cancel, producing a net magnetic field that is either zero or very much weaker than the field at the center of the loop. We've used only three current loops to illustrate the idea, but these tendencies are reinforced by including more loops. With many current loops along the same axis, **the field in the center is strong and roughly parallel to the axis, whereas the field outside the loops is very weak.**

Figure 32.29 is a numerical calculation of the magnetic field of a 15-turn solenoid. You can see that the magnetic field inside the coils is nearly uniform (i.e., the field lines are nearly parallel and equally spaced) but the field outside is much weaker. Our goal of producing a uniform magnetic field can be achieved by increasing the number of coils until we have an *ideal solenoid* that is infinitely long and in which the coils are as close together as possible. **The magnetic field inside an ideal solenoid is uniform and parallel to the axis; the magnetic field outside is zero.** No real solenoid is infinitely long (just as no real capacitor is infinitely wide), but a very uniform magnetic field can be produced near the center of a tightly wound solenoid whose length is much larger than its diameter.

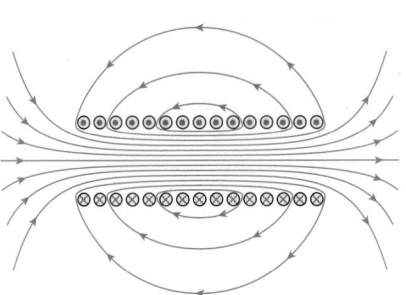

FIGURE 32.29 The magnetic field of a 15-turn solenoid.

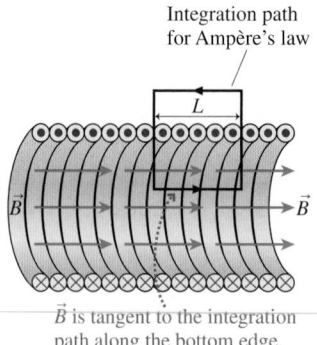

Integration path for Ampère's law

$\vec{B}$ is tangent to the integration path along the bottom edge.

FIGURE 32.30 A closed path inside and outside an ideal solenoid.

We can use Ampère's law to calculate the field of an ideal solenoid if we choose the integration path carefully. Figure 32.30 shows a cross section through an infinitely long solenoid. The integration path that we'll use is a rectangle of width L, enclosing N turns of the solenoid coil. Because this is a mathematical curve, not a physical boundary, there's no difficulty with letting it protrude through the wall of the solenoid wherever we wish. The solenoid's magnetic field direction, given by the right-hand rule, is left to right, so we'll integrate around this path in the ccw direction.

Each of the N wires enclosed by the integration path carries current I, so the total current passing through the rectangle is $I_{\text{through}} = NI$. Ampère's law is thus

$$\oint \vec{B} \cdot d\vec{s} = \mu_0 I_{\text{through}} = \mu_0 NI \tag{32.15}$$

The line integral around this path is the sum of the line integrals along each side. Along the bottom, where $\vec{B}$ is parallel to $d\vec{s}$ and of constant value B, the integral is simply BL. The integral along the top is zero because the magnetic field outside an ideal solenoid is zero.

The left and right sides sample the magnetic field both inside and outside the solenoid. The magnetic field outside is zero, but the magnetic field inside is not. However, the interior magnetic field is everywhere *perpendicular* to the line of integration. Consequently, as we recognized in Option 1 of Tactics Box 32.2, the line integral is zero.

Only the integral along the bottom path is nonzero, leading to

$$\oint \vec{B} \cdot d\vec{s} = BL = \mu_0 NI$$

Thus the strength of the uniform magnetic field inside a solenoid is

$$B_{\text{solenoid}} = \frac{\mu_0 NI}{L} = \mu_0 nI \tag{32.16}$$

where $n = N/L$ is the number of turns per unit length.

Objects inserted into the center of a solenoid are in a uniform magnetic field. Measurements that need a uniform magnetic field are often conducted inside a solenoid, which can be built quite large. The cylinder that surrounds a patient undergoing magnetic resonance imaging (MRI) contains a large solenoid made of superconducting wire, allowing it to carry the very large currents needed to generate a strong uniform magnetic field.

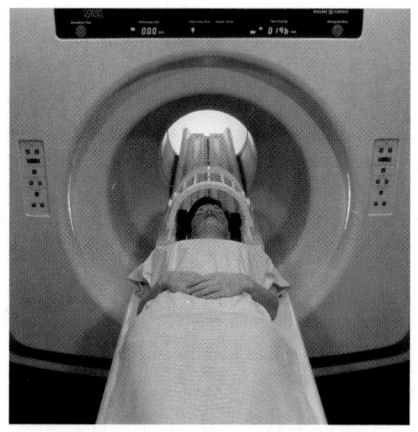

This patient is undergoing magnetic resonance imaging. The large cylinder surrounding the patient contains a solenoid to generate a uniform magnetic field.

EXAMPLE 32.9 Generating a uniform magnetic field

We wish to generate a 0.10 T uniform magnetic field near the center of a 10-cm-long solenoid. How many turns are needed if the wire can carry a maximum current of 10 A?

MODEL Assume that the solenoid is an ideal solenoid.

SOLVE Generating a magnetic field with a solenoid is a trade-off between current and turns of wire. A larger current requires fewer turns. However, wires have resistance, and too large a current can overheat the solenoid. Maximum safe currents are established on the basis of the wire's cross-section area. For a wire that

can carry 10 A, we can use Equation 32.16 to find the required number of turns:

$$N = \frac{LB}{\mu_0 I} = \frac{(0.10 \text{ m})(0.10 \text{ T})}{(4\pi \times 10^{-7} \text{ Tm/A})(10 \text{ A})} = 800 \text{ turns}$$

ASSESS A wire that can carry 10 A without overheating is about 1 mm in diameter, so only 100 turns can be placed in a 10 cm length. Thus it takes eight layers to reach the required number of turns.

For a real solenoid, of finite length, the magnetic field is approximately uniform *inside* the solenoid and weak, but not zero, outside. As Figure 32.31 shows, the magnetic field outside the solenoid looks like that of a bar magnet. Thus a

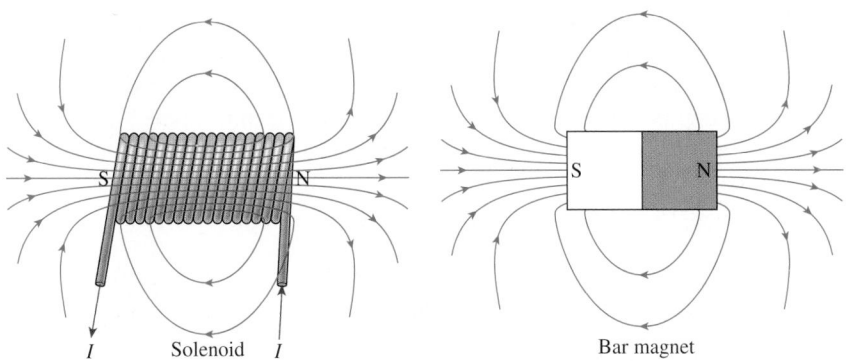

FIGURE 32.31 The magnetic fields of a finite-length solenoid and of a bar magnet.

solenoid is an electromagnet, with one end of the solenoid being a north magnetic pole and the other a south magnetic pole. You can use the right-hand rule to identify the north-pole end. A solenoid with many turns and a large current can be a very powerful magnet.

32.7 The Magnetic Force on a Moving Charge

It's time to switch our attention from how magnetic fields are generated to how magnetic fields exert forces and torques. Oersted discovered that a current passing through a wire causes a magnetic torque to be exerted on a nearby compass needle. Upon hearing of Oersted's discovery, Ampère reasoned that the current was acting like a magnet and, if this were true, that two current-carrying wires should exert magnetic forces on each other.

To find out, Ampère set up two parallel wires that could carry large currents in either the same direction or in opposite (or "antiparallel") directions. Figure 32.32 shows the outcome of his experiment. Notice that, for currents, "likes" attract and "opposites" repel. This is the opposite of what would have happened had the wires been charged and thus exerting electric forces on each other. Ampère's experiment showed that **a magnetic field exerts a force on a current,** but before we can analyze Ampère's experiment we must first investigate the magnetic force on a moving charge.

Magnetic Force

A current consists of moving charges. Ampère's experiment implied that a magnetic field exerts a force on a *moving* charge. This is true, although the exact form of the force law was not discovered until later in the 19th century. The magnetic force turns out to depend not only on the charge and the charge's velocity, but also on how the velocity vector is oriented relative to the magnetic field. Figure 32.33 shows the outcome of three experiments to observe the magnetic force on a charged particle.

"Like" currents attract.

"Opposite" currents repel.

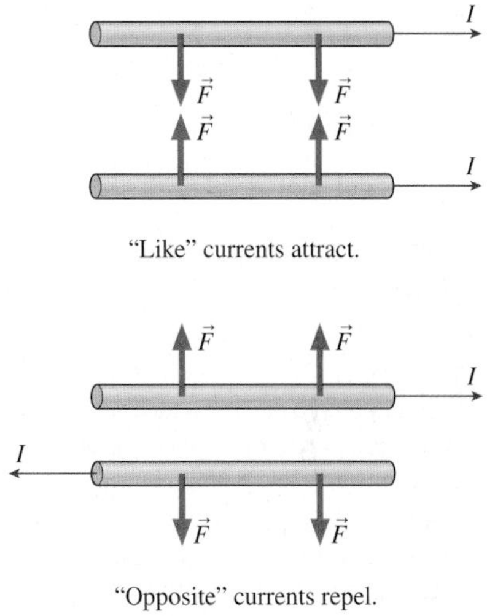

FIGURE 32.32 Ampère's experiment to observe the forces between parallel current-carrying wires.

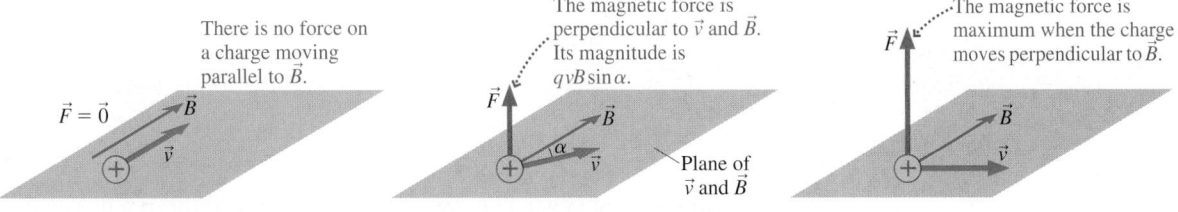

There is no force on a charge moving parallel to $\vec{B}$.

$\vec{F} = \vec{0}$

The magnetic force is perpendicular to $\vec{v}$ and $\vec{B}$. Its magnitude is $qvB\sin\alpha$.

Plane of $\vec{v}$ and $\vec{B}$

The magnetic force is maximum when the charge moves perpendicular to $\vec{B}$.

FIGURE 32.33 The relationship between $\vec{v}$, $\vec{B}$, and $\vec{F}$.

If you compare the experiment on the right in Figure 32.33 to Figure 32.9, you'll see that the relationship between $\vec{v}$, $\vec{B}$, and $\vec{F}$ is exactly the same as the geometric relationship between $\vec{C}$, $\vec{D}$, and $\vec{C} \times \vec{D}$. The magnetic force on a charge q as it moves through a magnetic field $\vec{B}$ with velocity $\vec{v}$ depends on the cross product between $\vec{v}$ and $\vec{B}$. The magnetic force on a moving charged particle can be written

$$\vec{F}_{\text{on } q} = q\vec{v} \times \vec{B} = (qvB\sin\alpha, \text{ direction of right-hand rule}) \quad (32.17)$$

where α is the angle between $\vec{v}$ and $\vec{B}$.

The right-hand rule is that of the cross product. Spread your right thumb and index finger apart by angle α, then bend your middle finger so that it is perpendicular to your thumb and index finger. Orient your hand so that your thumb points in the direction of $\vec{v}$ and your index finger in the direction of $\vec{B}$. Your middle finger is then pointing in the direction of the force $\vec{F}$. **The magnetic force on a moving charged particle is perpendicular to both $\vec{v}$ and $\vec{B}$.**

The magnetic force has several important properties:

1. Only a *moving* charge experiences a magnetic force. There is no magnetic force on a charge at rest ($v = 0$) in a magnetic field.
2. There is no force on a charge moving parallel ($\alpha = 0°$) or antiparallel ($\alpha = 180°$) to a magnetic field.
3. When there is a force, the force is perpendicular to *both* $\vec{v}$ and $\vec{B}$.
4. The force on a negative charge is in the direction *opposite* to $\vec{v} \times \vec{B}$.
5. For a charge moving perpendicular to $\vec{B}$ ($\alpha = 90°$), the magnitude of the magnetic force is $F = |q|vB$.

Figure 32.34 shows the relationship between $\vec{v}$, $\vec{B}$, and $\vec{F}$ for several moving charges. (The *source* of the magnetic field isn't shown, only the field itself.) You can see the inherent three-dimensionality of magnetism, with the force perpendicular to both $\vec{v}$ and $\vec{B}$. Thus the magnetic force is very different from the electric force, which is parallel to the electric field.

13.4 Activ Physics ONLINE

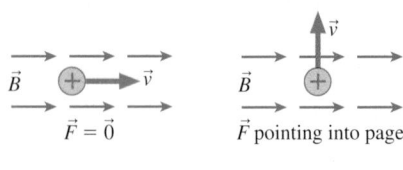

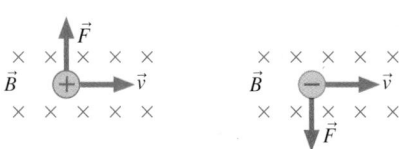

FIGURE 32.34 Magnetic forces on moving charges.

EXAMPLE 32.10 The magnetic force on an electron

A long wire carries a 10 A current from left to right. An electron 1.0 cm above the wire is traveling to the right at a speed of 1.0×10^7 m/s. What are the magnitude and the direction of the magnetic force on the electron?

MODEL The magnetic field is that of a long, straight wire.

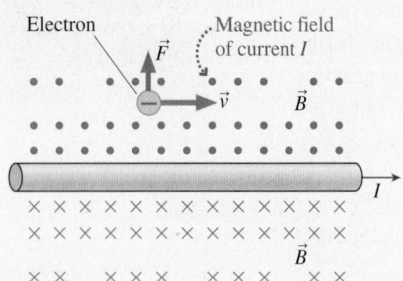

FIGURE 32.35 An electron moving parallel to a current-carrying wire.

VISUALIZE Figure 32.35 shows the current and an electron moving to the right. The right-hand rule tells us that the wire's magnetic field above the wire is out of the page, so the electron is moving perpendicular to the field.

SOLVE The electron charge is negative, thus the direction of the force is opposite the direction of $\vec{v} \times \vec{B}$. The right-hand rule shows that $\vec{v} \times \vec{B}$ points down, toward the wire, so $\vec{F}$ points up, away from the wire. The magnitude of the force is $|q|vB = evB$. The field is that of a long, straight wire,

$$B = \frac{\mu_0 I}{2\pi d} = 2.0 \times 10^{-4} \text{ T}$$

Thus the magnitude of the force on the electron is

$$F = evB = (1.60 \times 10^{-19} \text{ C})(1.0 \times 10^7 \text{ m/s})(2.0 \times 10^{-4} \text{ T})$$

$$= 3.2 \times 10^{-16} \text{ N}$$

The force on the electron is $\vec{F} = (3.2 \times 10^{-16} \text{ N, up})$.

ASSESS This force will cause the electron to curve away from the wire.

We can draw an interesting and important conclusion at this point. You have seen that the magnetic field is *created by* moving charges. Now you also see that magnetic forces are *exerted on* moving charges. Thus it appears that **magnetism** is an interaction between moving charges. Any two charges, whether moving or stationary, interact with each other through the electric field. In addition, two *moving* charges also interact with each other through the magnetic field. This fundamental observation is easy to lose sight of when we talk about currents, magnets, torques, and all the other phenomena of magnetism. But the most basic feature underlying of all these phenomena is an interaction between moving charges.

Cyclotron Motion

Many important applications of magnetism involve the motion of charged particles in a magnetic field. Your television picture tube functions by using magnetic fields to steer electrons as they move through a vacuum from the electron gun to the screen. Microwave generators, which are used in applications ranging from ovens to radar, use a device called a *magnetron* in which electrons oscillate rapidly in a magnetic field.

You've just seen that there is no force on a charge that has velocity $\vec{v}$ parallel or antiparallel to a magnetic field. Consequently, **a magnetic field has no effect on a charge moving parallel or antiparallel to the field.** To understand the motion of charged particles in magnetic fields, we need to consider only motion *perpendicular* to the field.

Figure 32.36 shows a positive charge q moving with a velocity $\vec{v}$ in a plane that is perpendicular to a *uniform* magnetic field $\vec{B}$. According to the right-hand rule, the magnetic force on this particle is *perpendicular* to the velocity $\vec{v}$. A force that is always perpendicular to $\vec{v}$ changes the *direction* of motion, by deflecting the particle sideways, but it cannot change the particle's speed. Thus **a particle moving perpendicular to a uniform magnetic field undergoes uniform circular motion at constant speed.** This motion is called the **cyclotron motion** of a charged particle in a magnetic field.

NOTE ▶ A negative charge will orbit in the opposite direction from that shown in Figure 32.36 for a positive charge. ◀

You've seen analogies to cyclotron motion in Parts I and II of this text. For a mass moving in a circle at the end of a string, the tension force is always perpendicular to $\vec{v}$. For a satellite moving in a circular orbit, the gravitational force is always perpendicular to $\vec{v}$. Now, for a charged particle moving in a magnetic field, it is the magnetic force of strength $F = qvB$ that points toward the center of the circle and causes the particle to have a centripetal acceleration.

Newton's second law for circular motion, which you learned in Chapter 7, is

$$F = qvB = ma_r = \frac{mv^2}{r} \tag{32.18}$$

Thus the radius of the cyclotron orbit is

$$r_{\text{cyc}} = \frac{mv}{qB} \tag{32.19}$$

The inverse dependence on B indicates that the size of the orbit can be decreased by increasing the magnetic field strength.

We can also determine the frequency of the cyclotron motion. Recall from your earlier study of circular motion that the frequency of revolution f is related to the speed and radius by $f = v/2\pi r$. A rearrangement of Equation 32.19 gives the **cyclotron frequency:**

$$f_{\text{cyc}} = \frac{qB}{2\pi m} \tag{32.20}$$

An electron beam undergoing circular motion in a magnetic field.

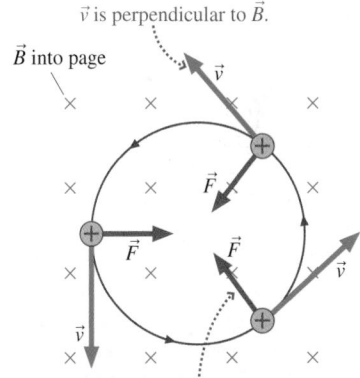

$\vec{v}$ is perpendicular to $\vec{B}$.

$\vec{B}$ into page

The magnetic force is always perpendicular to $\vec{v}$, causing the particle to move in a circle.

FIGURE 32.36 Cyclotron motion of a charged particle moving in a magnetic field.

Activ
Physics   13.7, 13.8

where the ratio q/m is the particle's *charge-to-mass ratio*. Notice that the cyclotron frequency depends on the charge-to-mass ratio and the magnetic field strength but *not* on the charge's velocity.

EXAMPLE 32.11 The radius of cyclotron motion

In Figure 32.37, an electron is accelerated from rest through a potential difference of 500 V, then injected into a uniform magnetic field. Once in the magnetic field, it completes half a revolution in 2.0 ns. What is the radius of its orbit?

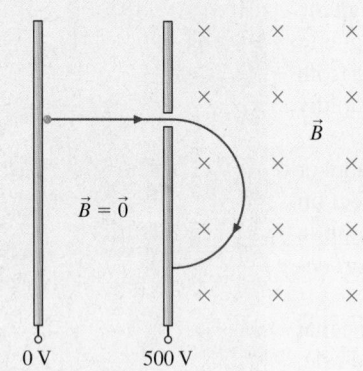

FIGURE 32.37 An electron is accelerated, then injected into a magnetic field.

MODEL Energy is conserved as the electron is accelerated by the potential difference. The electron then undergoes cyclotron motion in the magnetic field, although it completes only half a revolution before hitting the back of the acceleration electrode.

SOLVE The electron accelerates from rest ($v_i = 0$ m/s) at $V_i = 0$ V to speed v_f at $V_f = 500$ V. We can use conservation of energy $K_f + qV_f = K_i + qV_i$ to find the speed v_f with which it enters the magnetic field:

$$\frac{1}{2}mv_f^2 + (-e)V_f = 0 + 0$$

$$v_f = \sqrt{\frac{2eV_f}{m}} = \sqrt{\frac{2(1.60 \times 10^{-19}\,\text{C})(500\,\text{V})}{9.11 \times 10^{-31}\,\text{kg}}}$$

$$= 1.33 \times 10^7\,\text{m/s}$$

The cyclotron radius in the magnetic field is $r_{cyc} = mv/eB$, but we first need to determine the field strength. Were it not for the electrode, the electron would undergo circular motion with period $T = 4.0$ ns. Hence the cyclotron frequency is $f = 1/T = 2.5 \times 10^8$ Hz. We can use the cyclotron frequency to determine that the magnetic field strength is

$$B = \frac{2\pi m f_{cyc}}{e} = \frac{2\pi(9.11 \times 10^{-31}\,\text{kg})(2.50 \times 10^8\,\text{Hz})}{1.60 \times 10^{-19}\,\text{C}}$$

$$= 8.94 \times 10^{-3}\,\text{T}$$

Thus the radius of the electron's orbit is

$$r = \frac{mv}{qB} = 8.47 \times 10^{-3}\,\text{m} = 8.47\,\text{mm}$$

Figure 32.38a shows a more general situation in which the charged particle's velocity $\vec{v}$ is neither parallel to nor perpendicular to $\vec{B}$. The component of $\vec{v}$ parallel to $\vec{B}$ is not affected by the field, so the charged particle spirals around the magnetic field vectors in a helical trajectory. The radius of the helix is determined by $\vec{v}_\perp$, the component of $\vec{v}$ perpendicular to $\vec{B}$.

(a) Charged particles spiral around the magnetic field lines.

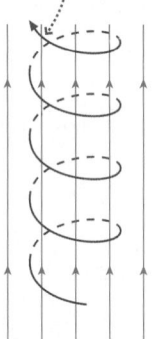

(b) The earth's magnetic field leads particles into the atmosphere near the poles, causing the aurora.

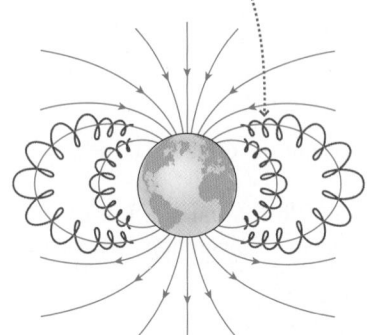

(c) The aurora seen from space

FIGURE 32.38 In general, charged particles spiral along helical trajectories around the magnetic field lines. This motion is responsible for the earth's aurora.

The motion of charged particles in a magnetic field is responsible for the earth's aurora, which you saw in the photograph at the beginning of the chapter. High-energy particles and radiation streaming out from the sun, called the *solar wind,* create ions and electrons as they strike molecules high in the atmosphere. Some of these charged particles become trapped in the earth's magnetic field, creating what is known as the *Van Allen radiation belt.*

As Figure 32.38b shows, the electrons spiral along the magnetic field lines until the field leads them into the atmosphere. The shape of the earth's magnetic field is such that most electrons enter the atmosphere in a circular region around the north magnetic pole and another around the south magnetic pole. There they collide with oxygen and nitrogen atoms, exciting the atoms and causing them to emit auroral light. Figure 32.38c shows a false-color image from space of the ultraviolet light emitted by the aurora.

STOP TO THINK 32.5   An electron moves perpendicular to a magnetic field. What is the direction of $\vec{B}$?

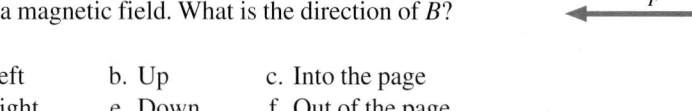

a. Left   b. Up   c. Into the page
d. Right   e. Down   f. Out of the page

The Cyclotron

Physicists studying the structure of the atomic nucleus and of elementary particles usually use a device called a *particle accelerator.* Charged particles, typically protons or electrons, are accelerated to very high speeds, close to the speed of light, and then collide with a target. The very large impact energies are sufficient to disrupt the nuclear forces, ejecting elementary particles that can be tracked and studied. The first practical particle accelerator, invented in the 1930s, was the **cyclotron.** Although cyclotrons have been superseded by newer accelerators for studying elementary particles, they remain important for many applications of nuclear physics.

A cyclotron, shown in Figure 32.39, consists of an evacuated chamber within a large, uniform magnetic field. Inside the chamber are two hollow conductors shaped like the letter D and hence called "dees." The dees are made of copper, which doesn't affect the magnetic field; are open along the straight sides; and are separated by a small gap. A charged particle, typically a proton, is injected into the magnetic field from a source near the center of the cyclotron, and it begins to move in and out of the dees in a circular cyclotron orbit.

The cyclotron operates by taking advantage of the fact that the cyclotron frequency f_{cyc} of a charged particle is independent of the particle's speed. An *oscillating* potential difference ΔV is connected across the dees and adjusted until its frequency is exactly the cyclotron frequency. There is almost no electric field inside the dees (you learned in Chapter 27 that the electric field inside a hollow conductor is zero), but a strong electric field points from the positive to the negative dee in the gap between them.

Suppose the proton emerges into the gap from the positive dee. The electric field in the gap *accelerates* the proton across the gap into the negative dee, and it gains kinetic energy $e\Delta V$. A half cycle later, when it next emerges into the gap, the potential of the dees (whose potential difference is oscillating at f_{cyc}) will have changed sign. The proton will *again* be emerging from the positive dee and will *again* accelerate across the gap and gain kinetic energy $e\Delta V$.

Because the dees change potential in time with the proton's orbit, and the proton's cyclotron frequency doesn't change as it speeds up, this pattern will continue orbit after orbit. The proton's kinetic energy increases by $2e\Delta V$ every orbit,

Cyclotrons are used in applied nuclear physics to accelerate ions to very high speeds.

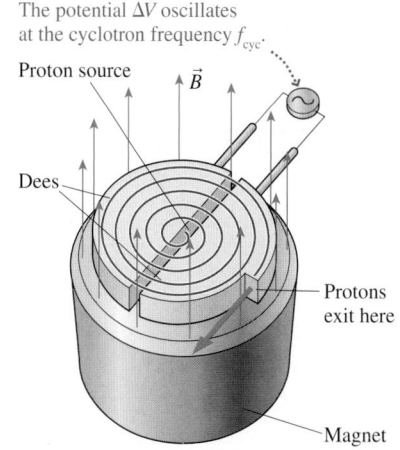

The potential ΔV oscillates at the cyclotron frequency f_{cyc}.

Proton source

$\vec{B}$

Dees

Protons exit here

Magnet

FIGURE 32.39 A cyclotron.

FIGURE 32.40 An electron and a positron moving in a bubble chamber. The magnetic field is perpendicular to the page.

so after N orbits its kinetic energy will be $K = 2Ne\,\Delta V$ (assuming that its initial kinetic energy was near zero). The radius of its orbit increases as it speeds up, hence the proton follows the *spiral* path shown in Figure 32.38 until it finally reaches the outer edge of the dee. It is then directed out of the cyclotron and aimed at a target. Although ΔV is modest, usually a few hundred volts, the fact that the proton can undergo many thousands of orbits before reaching the outer edge allows it to acquire a very large kinetic energy.

Magnetic fields are also important in the analysis of the elementary particles produced in these high-energy collisions. Notice that Equation 32.19 for $r_{\rm cyc}$ can be written $mv = p = r_{\rm cyc}qB$. In other words, the momentum of a charged particle can be determined by measuring the radius of its orbit in a known magnetic field. This is done inside a device called a *bubble chamber,* where the particles leave a map of their trajectories in the form of a string of tiny bubbles in liquid hydrogen. This bubble pattern is photographed, and the radius of the particle's orbit is measured from the photograph. Figure 32.40 is a photograph of a collision in which an electron and a positron (an *antielectron,* having the mass of an electron but charge $+e$) were created. Notice that they spiral in opposite directions, because of their opposite charges, with slowly decreasing radii as they lose energy in collisions with the hydrogen atoms.

The Hall Effect

A charged particle moving through a vacuum is deflected sideways, perpendicular to $\vec{v}$, by a magnetic field. In 1879, a graduate student named Edwin Hall showed that the same is true for the charges moving through a conductor as part of a current. This phenomenon—now called the **Hall effect**—is used to gain information about the charge carriers in a conductor. It is also the basis of a widely used technique for measuring magnetic field strengths.

Figure 32.41a shows a magnetic field perpendicular to a flat, current-carrying conductor. You learned in Chapter 28 that the charge carriers move through a conductor at the drift speed $v_{\rm d}$. Their motion is perpendicular to $\vec{B}$, so each charge carrier experiences a magnetic force $F_{\rm m} = ev_{\rm d}B$ perpendicular to both $\vec{B}$ and the current I. However, for the first time we have a situation in which it *does* matter whether the charge carriers are positive or negative.

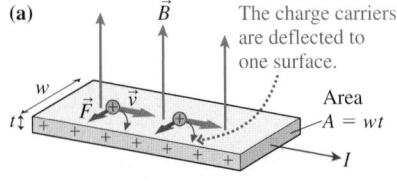

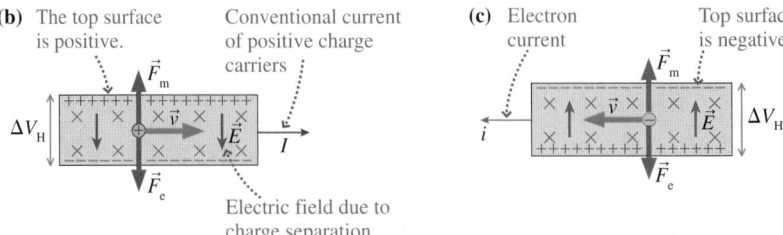

FIGURE 32.41 The charge carriers in a current are deflected to one surface of a conductor, creating the Hall voltage $\Delta V_{\rm H}$.

Figure 32.41b, which is looking along the magnetic field direction, shows that positive charge carriers moving in the direction of I are pushed toward the top surface of the conductor. This will create an excess positive charge on the top surface and leave an excess negative charge on the bottom. Figure 32.41c, where the electrons in an electron current i move opposite the direction of I, shows that electrons would be pushed toward the top surface. (Be sure to use the right-hand rule and the sign of the electron charge to confirm the deflections shown in these figures.) Thus the sign of the excess charge on the top surface is the same as the sign of the charge carriers. Experimentally, the top surface is negative when the conductor is a metal, and this is one more piece of evidence that the charge carriers in metals are electrons.

Electrons are deflected toward the top surface once the current starts flowing, but the process can't continue indefinitely. The excess charges on the surfaces, like the charges on the plates of a capacitor, create a potential difference ΔV between the two surfaces and an electric field $E = \Delta V/w$ *inside* the conductor. Charge builds up on the surface until the downward electric force $\vec{F}_e$ on the charge carriers exactly balances the upward magnetic force $\vec{F}_m$. Once the forces are balanced, a steady state is reached in which the charge carriers move in the direction of the current and no additional charge is deflected to the surface.

The steady state condition, in which $F_m = F_e$, is

$$F_m = ev_dB = F_e = eE = e\frac{\Delta V}{w} \tag{32.21}$$

Thus the steady-state potential difference between the two surfaces of the conductor, which is called the **Hall voltage** ΔV_H, is

$$\Delta V_H = wv_dB \tag{32.22}$$

You learned in Chapter 28 that the drift speed is related to the current density J by $J = nev_d$, where n is the charge-carrier density (charge carriers per m^3). Thus

$$v_d = \frac{J}{ne} = \frac{I/A}{ne} = \frac{I}{wtne} \tag{32.23}$$

where $A = wt$ is the cross-section area of the conductor. If we use this expression for v_d in Equation 32.23, we find that the Hall voltage is

$$\Delta V_H = \frac{IB}{tne} \tag{32.24}$$

The Hall voltage is very small for metals in laboratory-sized magnetic fields, typically in the microvolt range. Even so, measurements of the Hall voltage in a known magnetic field are used to determine the charge-carrier density n. Interestingly, the Hall voltage is larger for *poor* conductors that have smaller charge-carrier densities. A laboratory probe for measuring magnetic field strengths, called a *Hall probe*, measures ΔV_H for a poor conductor whose charge-carrier density is known. The magnetic field is then determined from Equation 32.24.

EXAMPLE 32.12 Measuring the magnetic field
A Hall probe consists of a strip of the metal bismuth that is 0.15 mm thick and 5.0 mm wide. Bismuth is a poor conductor with a charge-carrier density 1.35×10^{25} m^{-3}. The Hall voltage on the probe is 2.5 mV when the current through it is 1.50 A. What is the strength of the magnetic field, and what is the electric field strength inside the bismuth?

MODEL Assume the magnetic field is uniform over the Hall probe.

VISUALIZE The bismuth strip looks like Figure 32.41a. The thickness is $t = 1.5 \times 10^{-4}$ m and the width is $w = 5.0 \times 10^{-3}$ m.

SOLVE Equation 32.24 gives the Hall voltage. We can rearrange the equation to find that the magnetic field is

$$B = \frac{tne}{I}\Delta V_H$$

$$= \frac{(1.5 \times 10^{-4}\,\text{m})(1.35 \times 10^{25}\,\text{m}^{-3})(1.60 \times 10^{-19}\,\text{C})}{1.5\,\text{A}}\,0.0025\,\text{V}$$

$$= 0.54\,\text{T}$$

The electric field created inside the bismuth by the excess charge on the surface is

$$E = \frac{\Delta V_H}{w} = \frac{0.0025\,\text{V}}{5.0 \times 10^{-3}\,\text{m}} = 0.50\,\text{V/m}$$

ASSESS 0.54 T is a fairly typical strength for a laboratory magnet.

32.8 Magnetic Forces on Current-Carrying Wires

13.5 

We were motivated to look at the magnetic force on moving charges by the experiment in which Ampère observed magnetic forces between current-carrying wires. We're now ready to analyze Ampère's experiment. As a first step, let us find the force that a uniform magnetic field exerts on a long, straight wire that carries current I through the field. If a current-carrying wire is *parallel* to a magnetic field, the force on it is zero. This follows from the fact that there is no force on a charged particle moving parallel to $\vec{B}$.

It's more interesting to consider the wire in Figure 32.42 that is *perpendicular* to the magnetic field. Note that the field is an external magnetic field, created by a permanent magnet or by other currents; it is *not* the field of the current I. The direction of the force on the current is found by considering the force on each charge in the current. By the right-hand rule, each charge has a force of magnitude qvB directed to the left. Consequently, the entire length of wire within the magnetic field experiences a force to the left, perpendicular to both the current direction and the field direction.

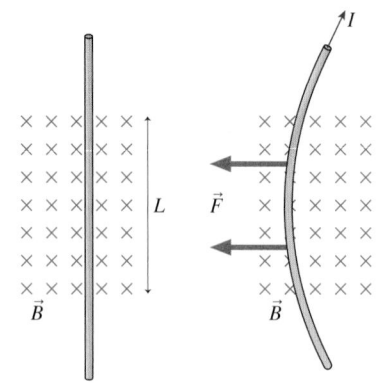

A wire is perpendicular to an externally created magnetic field.

A current through a wire that is fixed at the ends causes the wire to be bent sideways.

FIGURE 32.42 Magnetic force on a current-carrying wire.

NOTE ▶ The familiar right-hand rule can be applied directly to a current-carrying wire. Point your right thumb in the direction of the current (parallel to $\vec{v}$) and your index finger in the direction of $\vec{B}$. Your middle finger is then pointing in the direction of the force $\vec{F}$ on the wire. ◀

To find the magnitude of the force, we must relate qv of the charges to the current I in the wire. Consider a section of wire of length L in which charge carriers with total charge q move with speed v. The current I, by definition, is the charge q divided by the time Δt it takes the charge to flow out of this section of the wire: $I = q/\Delta t$. The time required is $\Delta t = L/v$, giving

$$I = \frac{q}{\Delta t} = \frac{q}{L/v} = \frac{qv}{L}$$

Thus $qv = IL$. If we substitute IL for qv in the force equation $F = qvB$, we find that the magnetic force on length L of a current-carrying wire is

$$F_{\text{wire}} = ILB \quad \text{(force on a current perpendicular to the field)} \quad (32.25)$$

Equation 32.25 is a simple result, but remember the two assumptions behind it: The wire must be perpendicular to the field, and the field must be constant over the length L of the wire. As an aside, you can see from Equation 32.25 that the magnetic field B must have units of N/A m. This is why we defined 1 T = 1 N/A m in Section 32.3.

EXAMPLE 32.13 Magnetic levitation

The 0.10 T uniform magnetic field of Figure 32.43 is horizontal, parallel to the floor. A straight segment of 1.0-mm-diameter copper wire, also parallel to the floor, is perpendicular to the magnetic field. What current through the wire, and in which direction, will allow the wire to "float" in the magnetic field?

MODEL The wire will float in the magnetic field if the magnetic force on the wire points upward and has magnitude mg, allowing it to balance the downward gravitational force.

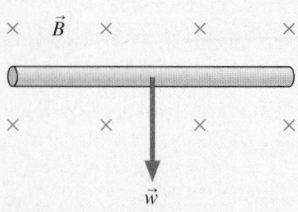

FIGURE 32.43 Magnetic levitation.

SOLVE We can use the right-hand rule to determine which current direction experiences an upward force. With $\vec{B}$ pointing away from us, the direction of the current needs to be from left to right. The forces will balance when

$$F = ILB = mg = \rho(\pi r^2 L)g$$

where $\rho = 8920 \text{ kg/m}^3$ is the density of copper. The length of the wire cancels, leading to

$$I = \frac{\rho \pi r^2 g}{B} = \frac{(8920 \text{ kg/m}^3)\pi(0.00050 \text{ m})^2(9.80 \text{ m/s}^2)}{0.10 \text{ T}}$$

$$= 0.687 \text{ A}$$

A 0.687 A current from left to right will levitate the wire in the magnetic field.

ASSESS A 0.687 A current is quite reasonable, but this idea is useful only if we can get the current into and out of this segment of wire. In practice, we could do so with wires that come in from below the page. These input and output wires would be parallel to $\vec{B}$ and not experience a magnetic force. Although this example is very simple, it is the basis for applications such as magnetic levitation trains.

Force Between Two Parallel Wires

Now consider Ampère's experimental arrangement of two parallel wires of length L, distance d apart. Figure 32.44a shows the currents I_1 and I_2 in the same directions; Figure 32.44b shows the currents in opposite directions. We will assume that the wires are sufficiently long to allow us to use the earlier result for the magnetic field of a long straight wire: $B = \mu_0 I/2\pi d$.

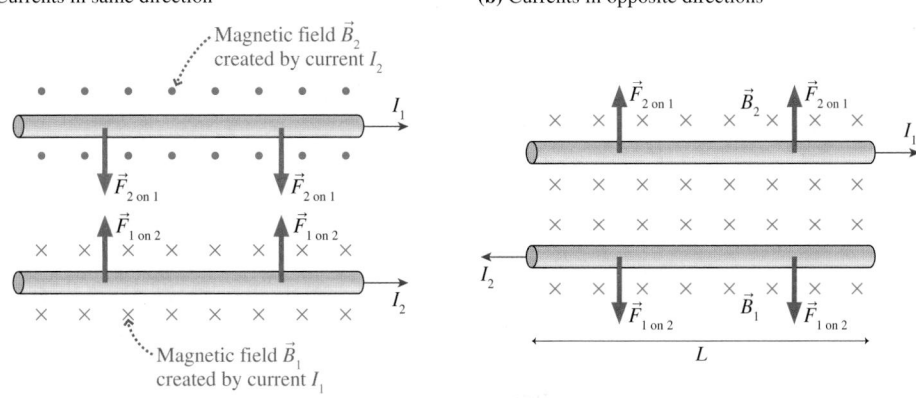

(a) Currents in same direction

(b) Currents in opposite directions

FIGURE 32.44 Magnetic forces between parallel current-carrying wires.

As Figure 32.44a shows, the current I_2 in the lower wire creates a magnetic field $\vec{B}_2$ at the position of the upper wire. $\vec{B}_2$ points out of the page, perpendicular to current I_1. **It is field $\vec{B}_2$, due to the lower wire, that exerts a magnetic force on the upper wire.** Using the right-hand rule, you can see that the force on the upper wire is downward, thus attracting it toward the lower wire. The field of the lower current is not a uniform field, but it is the *same* at all points along the upper wire because the two wires are parallel. Consequently, we can use the field of a long, straight wire to determine the magnetic force exerted by the lower wire on the upper wire when they are separated by distance d:

$$F_{\text{parallel wires}} = I_1 L B_2 = I_1 L \frac{\mu_0 I_2}{2\pi d} = \frac{\mu_0 L I_1 I_2}{2\pi d} \tag{32.26}$$

(force between two parallel wires)

As an exercise, you should convince yourself that the current in the upper wire exerts an upward-directed magnetic force on the lower wire with exactly the same magnitude. You should also convince yourself, using the right-hand rule, that the forces are repulsive and tend to push the wires apart if the two currents are in opposite directions.

Thus two parallel wires exert equal but opposite forces on each other, as required by Newton's third law. **Parallel wires carrying currents in the same direction attract each other; parallel wires carrying currents in opposite directions repel each other.**

EXAMPLE 32.14 A current balance

Two stiff, 50-cm-long parallel wires are connected at the ends by metal springs. Each spring has an unstretched length of 5.0 cm and a spring constant of 0.020 N/m. The wires push each other apart when a current travels around the loop. How much current is required to stretch the springs to lengths of 6.0 cm?

MODEL Two parallel wires carrying currents in opposite directions exert repulsive magnetic forces on each other.

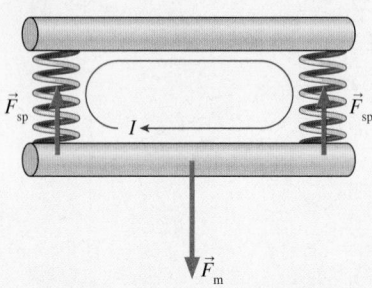

FIGURE 32.45 The current-carrying wires of Example 32.14.

VISUALIZE Figure 32.45 shows the "circuit." The springs are conductors, allowing a current to travel around the loop. In equilibrium, the repulsive magnetic forces between the wires are balanced by the restoring forces $F_{\text{sp}} = k\,\Delta y$ of the springs.

SOLVE Figure 32.45 shows the forces on the lower wire. The net force is zero, hence $F_{\text{m}} = 2F_{\text{sp}}$. The repulsive force between the wires is given by Equation 32.26 with $I_1 = I_2 = I$:

$$F_{\text{m}} = \frac{\mu_0 L I^2}{2\pi d} = 2F_{\text{sp}} = 2k\,\Delta y$$

where k is the spring constant and $\Delta y = 1.0$ cm is the amount by which each spring stretches. Solving for the current, we find

$$I = \sqrt{\frac{4\pi k d\,\Delta y}{\mu_0 L}} = 15.5 \text{ A}$$

ASSESS Devices similar to the one described, in which a magnetic force balances a mechanical force, are called *current balances*. They can be used to make very accurate current measurements.

32.9 Forces and Torques on Current Loops

13.6 **Activ Physics** ONLINE

You have seen that a current loop is a magnetic dipole, much like a permanent magnet. We will now look at some important features of how current loops behave in magnetic fields. This discussion will be largely qualitative, but it will highlight some of the important properties of magnets and magnetic fields. We will then use these ideas in the next section to make the connection between electromagnets and permanent magnets.

Figure 32.46a shows two current loops. Using what we've learned about the forces between parallel and antiparallel currents, you can see that **parallel cur-**

(a) Parallel currents attract, opposite currents repel.

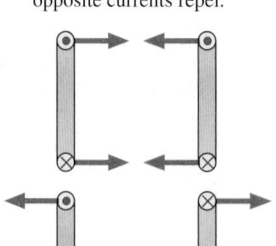

(b) Opposite poles attract, like poles repel.

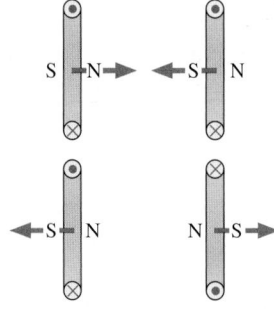

FIGURE 32.46 Magnetic forces between two parallel current loops.

rent loops exert attractive magnetic forces on each other if the currents circulate in the same direction; they repel each other when the currents circulate in opposite directions.

It is convenient to think of these forces in terms of magnetic poles. Figure 32.46b shows the north and south magnetic poles of the current loops. If the currents circulate in the same direction, a north and a south pole face each other and exert attractive forces on each other. If the currents circulate in opposite directions, the two like poles repel each other.

Here, at last, we have a real connection to the behavior of magnets that opened our discussion of magnetism—namely, that like poles repel and opposite poles attract. Now we have an *explanation* for this behavior, at least for electromagnets. **Magnetic poles attract or repel because the moving charges in one current exert attractive or repulsive magnetic forces on the moving charges in the other current.** Our tour through interacting moving charges is finally starting to show some practical results!

Now let's consider the forces on a current loop in a *uniform* magnetic field. Figure 32.47 shows a square current loop in a uniform magnetic field. The current in each of the four sides experiences a magnetic force due to the field $\vec{B}$. The forces $\vec{F}_{\text{front}}$ and $\vec{F}_{\text{back}}$ are opposite to each other and cancel. Forces $\vec{F}_{\text{top}}$ and $\vec{F}_{\text{bottom}}$ also add to give no net force, but because $\vec{F}_{\text{top}}$ and $\vec{F}_{\text{bottom}}$ don't act along the same line they will *rotate* the loop by exerting a torque on it.

The forces on the top and bottom segments form what we called a *couple* in Chapter 13. The torque due to a couple is the magnitude of the force multiplied by the distance d between the two lines of action. You can see that $d = L\sin\theta$, hence the torque on the loop—a torque exerted by the magnetic field—is

$$\tau = Fd = (ILB)(L\sin\theta) = (IL^2)B\sin\theta = \mu B\sin\theta \qquad (32.27)$$

where $\mu = IL^2 = IA$ is the size of the loop's magnetic dipole moment.

Although we derived Equation 32.27 for a square loop, the result is valid for a current loop of any shape. Notice that Equation 32.27 looks like another example of a cross product. We earlier defined the magnetic dipole moment vector $\vec{\mu}$ to be a vector perpendicular to the current loop in a direction given by the right-hand rule. Figure 32.47 shows that θ is the angle between $\vec{B}$ and $\vec{\mu}$, hence the torque on a magnetic dipole is

$$\vec{\tau} = \vec{\mu} \times \vec{B} \qquad (32.28)$$

The torque is zero when the magnetic dipole moment $\vec{\mu}$ is aligned parallel or antiparallel to the magnetic field, and is maximum when $\vec{\mu}$ is perpendicular to the field. It is this magnetic torque that causes a compass needle—a magnetic moment—to rotate until it is aligned with the magnetic field.

An Electric Motor

The torque on a current loop in a magnetic field is the basis for how an electric motor works. As Figure 32.48 on the next page shows, the *armature* of a motor is a coil of wire wound on an axle that is free to rotate. When a current passes through the coil, the magnetic field exerts a torque on the armature and causes it to rotate. If the current were steady, the armature would oscillate back and forth around the equilibrium position until (assuming there's some friction or damping) it stops with the plane of the coil perpendicular to the field. To keep the motor turning, a device called a *commutator* reverses the current direction in the coils every 180°. (Notice that the commutator is split, so the positive terminal of the battery sends current into whichever wire touches the bottom half of the commutator.) The current reversal prevents the armature from ever reaching an equilibrium position, so the magnetic torque keeps the motor spinning as long as there is a current.

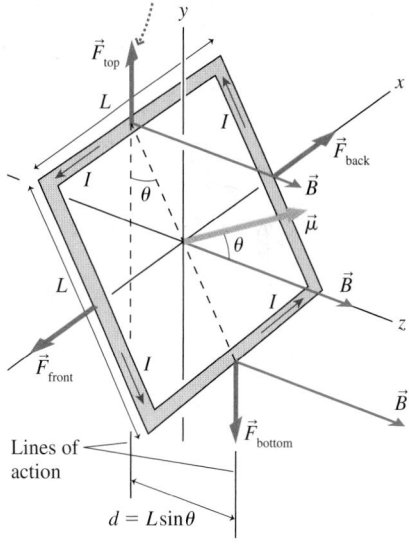

$\vec{F}_{\text{top}}$ and $\vec{F}_{\text{bottom}}$ exert a torque that rotates the loop about the *x*-axis.

FIGURE 32.47 A uniform magnetic field exerts a torque on a current loop.

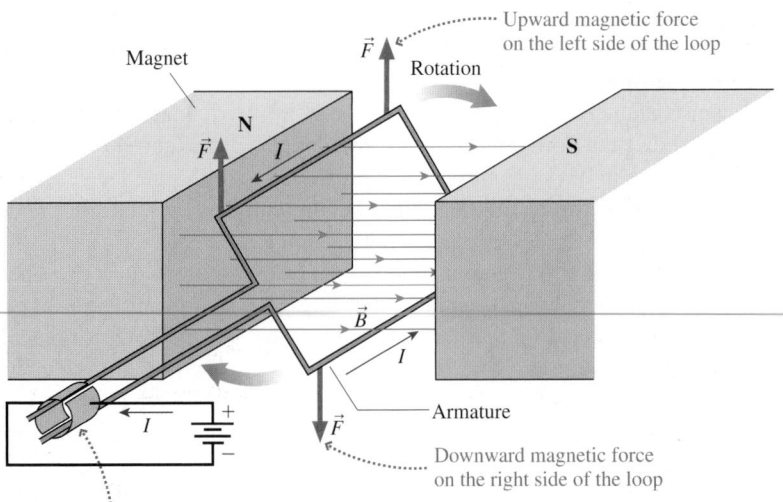

FIGURE 32.48 A simple electric motor.

STOP TO THINK 32.6 What is the current direction in the loop?

a. Out of the page at the top of the loop, into the page at the bottom.
b. Out of the page at the bottom of the loop, into the page at the top.

32.10 Magnetic Properties of Matter

Our theory has focused mostly on the magnetic properties of currents, yet our everyday experience is mostly with permanent magnets. We have seen that current loops and solenoids have magnetic poles and exhibit behaviors like those of permanent magnets, but we still lack a specific connection between electromagnets and permanent magnets. The goal of this section is to complete our understanding by developing an atomic-level view of the magnetic properties of matter.

Atomic Magnets

A plausible explanation for the magnetic properties of materials is the orbital motion of the atomic electrons. Figure 32.49 shows a simple, classical model of an atom in which a negative electron orbits a positive nucleus. In this picture of the atom, the electron's motion is that of a current loop! It is a microscopic current loop, to be sure, but a current loop nonetheless. Consequently, an orbiting electron acts as a tiny magnetic dipole, with a north pole and a south pole. You can think of the magnetic dipole as an atomic-size magnet. Experiments with *individual* hydrogen atoms verify that they are, indeed, tiny magnets.

However, the atoms of most elements contain many electrons. Unlike the solar system, where all of the planets orbit in the same direction, electron orbits are arranged to oppose each other: one electron moves counterclockwise for every electron that moves clockwise. Thus the magnetic moments of individual orbits tend to cancel each other and the *net* magnetic moment is either zero or very small.

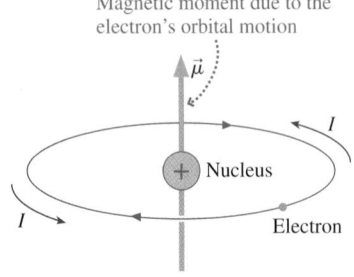

FIGURE 32.49 A classical orbiting electron is a tiny magnetic dipole.

The cancellation continues as atoms are joined into molecules and the molecules into solids. When all is said and done, the net magnetic moment of any bulk matter due to the orbiting electrons is so small as to be negligible. There are various subtle magnetic effects that can be observed under laboratory conditions, but orbiting electrons cannot explain the very strong magnetic effects of a piece of iron.

The Electron Spin

The key to understanding atomic magnetism was the 1922 discovery that electrons have an *inherent magnetic moment.* Perhaps this shouldn't be surprising. An electron has a *mass,* which allows it to interact with gravitational fields, and a *charge,* which allows it to interact with electric fields. There's no reason an electron shouldn't also interact with magnetic fields, and to do so it comes with a magnetic moment.

An electron's inherent magnetic moment is often called the electron *spin* because, in a classical picture, a spinning ball of charge would have a magnetic moment. This classical picture is not a realistic portrayal of how the electron really behaves, but its inherent magnetic moment makes it seem *as if* the electron were spinning. While it may not be spinning in a literal sense, each electron really is a microscopic bar magnet.

We must appeal to the results of quantum physics to find out what happens in an atom with many electrons. The spin magnetic moments, like the orbital magnetic moments, tend to oppose each other as the electrons are placed into their shells, causing the net magnetic moment of a *filled* shell to be zero. However, atoms containing an odd number of electrons must have at least one valence electron with an unpaired spin. These atoms have net magnetic moment due to the electron's spin.

But atoms with magnetic moments don't necessarily form a solid with magnetic properties. For most elements, the magnetic moments of the atoms are randomly arranged when the atoms join together to form a solid. As Figure 32.50 shows, this random arrangement produces a solid whose net magnetic moment is very close to zero. This agrees with our common experience that most materials are not magnetic; you cannot pick them up with a magnet or make a magnet from them. On the other hand, there are those materials such as iron that do exhibit strong magnetic properties, so we need to discover why these magnetic materials are different.

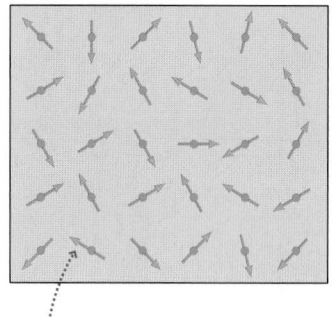

The atomic magnetic moments due to unpaired spins point in random directions. The sample has no net magnetic moment.

FIGURE 32.50 The random magnetic moments of the atoms in a typical solid produce no net magnetic moment.

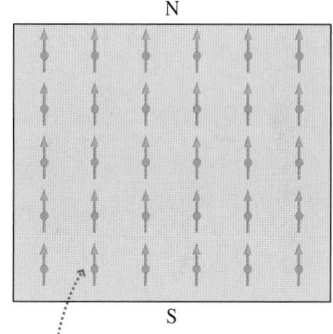

The atomic magnetic moments are aligned. The sample has a north and south magnetic pole.

FIGURE 32.51 The aligned atomic magnetic moments in a ferromagnetic material create a macroscopic magnetic dipole.

Ferromagnetism

It happens that in iron, and a few other substances, the spins interact with each other in such a way that atomic magnetic moments tend to all line up in the *same* direction. Materials that behave in this fashion are called **ferromagnetic,** with the prefix *ferro* meaning "iron-like." Figure 32.51 shows how the spin magnetic moments are aligned for the atoms making up a ferromagnetic solid.

In ferromagnetic materials, the individual magnetic moments add together to create a *macroscopic* magnetic dipole. The material has a north and a south magnetic pole, generates a magnetic field, and aligns parallel to an external magnetic field. In other words, it is a magnet!

Although iron is a magnetic material, a typical piece of iron is not a strong permanent magnet. You need not worry that a steel nail, which is mostly iron and is easily lifted with a magnet, will leap from your hands and pin itself against the hammer because of its own magnetism. It turns out, as shown in Figure 32.52, that a piece of iron is divided into small regions called **magnetic domains.** A typical domain size is roughly 0.1 mm—small, but not unreasonably so. The magnetic moments of all of the iron atoms within each domain are perfectly aligned, so each individual domain, like Figure 32.51, is a strong magnet.

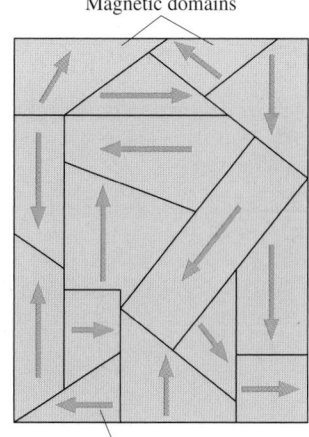

Magnetic domains

Magnetic moment of the domain

FIGURE 32.52 Magnetic domains in a ferromagnetic material. The net magnetic dipole is nearly zero.

However, the various magnetic domains that form a larger solid, such as you might hold in your hand, are randomly arranged. Their magnetic dipoles largely cancel, much like the cancellation that occurs on the atomic scale for nonferromagnetic substances, so the solid as a whole has only a small magnetic moment. That is why the nail is not a strong permanent magnet.

Induced Magnetic Dipoles

If a ferromagnetic substance is subjected to an *external* magnetic field, the external field exerts a torque on the magnetic dipole of each domain. The torque causes many of the domains to rotate and become aligned with the external field, just as a compass needle aligns with a magnetic field, although internal forces between the domains generally prevent the alignment from being perfect. In addition, atomic-level forces between the spins can cause the *domain boundaries* to move. Domains that are aligned along the external field become larger at the expense of domains that are opposed to the field. These changes in the size and orientation of the domains cause the material to develop a *net magnetic dipole* that is aligned with the external field. This magnetic dipole has been *induced* by the external field, so it is called an **induced magnetic dipole.**

NOTE ▶ The induced magnetic dipole is analogous to the polarization forces and induced electric dipoles that you studied in Chapter 26. ◀

Figure 32.53 shows a ferromagnetic material near the end of a solenoid. The magnetic moments of the domains align with the solenoid's field, creating an induced magnetic dipole whose south pole faces the solenoid's north pole. Consequently, the magnetic force between the poles pulls the ferromagnetic object to the electromagnet.

The fact that a magnet attracts and picks up ferromagnetic objects was one of the basic observations about magnetism with which we started the chapter. Now we have an *explanation* of how it works, based on three ideas:

1. Electrons are microscopic magnets due to their spin.
2. A ferromagnetic material in which the spins are aligned is organized into magnetic domains.
3. The individual domains align with an external magnetic field to produce an induced magnetic dipole moment for the entire object.

The object's magnetic dipole may not return to zero when the external field is removed because some domains remain "frozen" in the alignment they had in the external field. Thus a ferromagnetic object that has been in an external field may be left with a net magnetic dipole moment after the field is removed. In other words, the object has become a **permanent magnet.** A permanent magnet is simply a ferromagnetic material in which a majority of the magnetic domains are aligned with each other to produce a net magnetic dipole moment.

Whether or not a ferromagnetic material can be made into a permanent magnet depends on the internal crystalline structure of the material. *Steel* is an alloy of iron with other elements. An alloy of mostly iron with the right percentages of chromium and nickel produces *stainless steel,* which has virtually no magnetic properties at all because its particular crystalline structure is not conducive to the formation of domains. A very different steel alloy called Alnico V is made with 51% iron, 24% cobalt, 14% nickel, 8% aluminum, and 3% copper. It has extremely prominent magnetic properties and is used to make high-quality permanent magnets. You can see from the complex formula that developing good magnetic materials requires a lot of engineering skill as well as a lot of patience!

So we've come full circle. One of our initial observations about magnetism was that a permanent magnet can exert forces on some materials but not others.

Ferromagnetic material

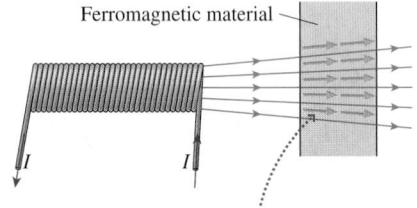

The magnetic domains align with the solenoid's magnetic field to produce an induced magnetic dipole.

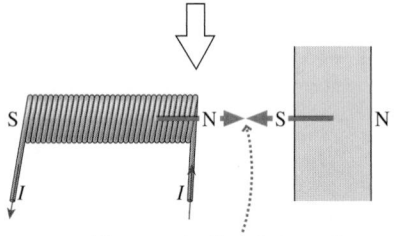

The attractive force between the opposite poles pulls the ferromagnetic material toward the solenoid.

FIGURE 32.53 The magnetic field of the solenoid creates an induced magnetic dipole in the iron.

The *theory* of magnetism that we then proceeded to develop was a theory about the interactions between moving charges. What moving charges had to do with permanent magnets was not obvious. But finally, by considering magnetic effects at the atomic level, we find that properties of permanent magnets and magnetic materials can be traced to the interactions of vast numbers of electron spins.

STOP TO THINK 32.7 Which magnet or magnets produced this induced magnetic dipole?

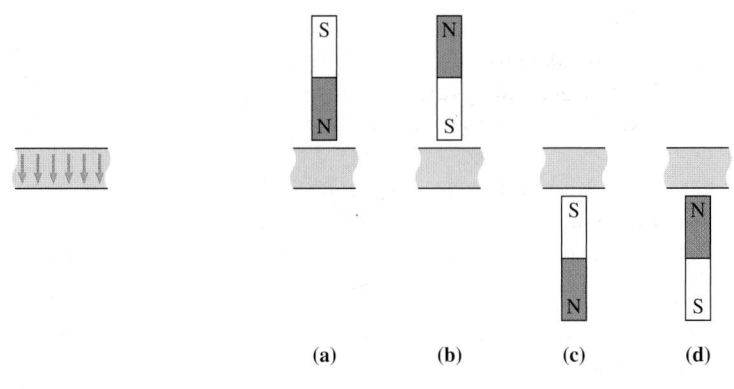

SUMMARY

The goal of Chapter 32 has been to learn how to calculate and use the magnetic field.

GENERAL PRINCIPLES

At its most fundamental level, magnetism is an interaction between moving charges. The magnetic field of one moving charge exerts a force on another moving charge.

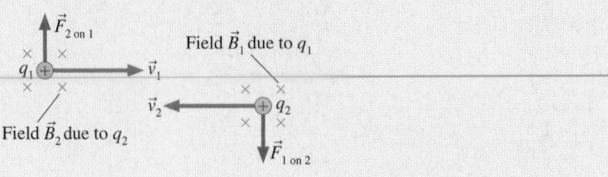

Magnetic Fields

The Biot-Savart law

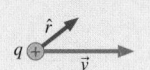

- A **point charge**, $\vec{B} = \dfrac{\mu_0}{4\pi} \dfrac{q\vec{v} \times \hat{r}}{r^2}$

- A **short current element**, $\vec{B} = \dfrac{\mu_0}{4\pi} \dfrac{I\Delta\vec{s} \times \hat{r}}{r^2}$

To find the magnetic field of a current

- Divide the wire into many short segments.
- Find the field of each segment Δs.
- Find $\vec{B}$ by summing the fields of all Δs, usually as an integral.

An alternative method for fields with a high degree of symmetry is Ampère's law

$$\oint \vec{B} \cdot d\vec{s} = \mu_0 I_{\text{through}}$$

where I_{through} is the current through the area bounded by the integration path.

Magnetic Forces

The magnetic force on a moving charge is

$$\vec{F} = q\vec{v} \times \vec{B}$$

The force is perpendicular to $\vec{v}$ and $\vec{B}$.

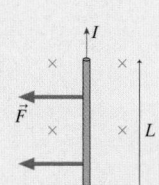

The magnetic force on a current-carrying wire perpendicular to a magnetic field is $F = ILB$.

$\vec{F} = \vec{0}$ for a charge or current moving parallel to $\vec{B}$.

The magnetic torque on a magnetic dipole is

$$\vec{\tau} = \vec{\mu} \times \vec{B}$$

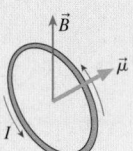

APPLICATIONS

Wire

Loop

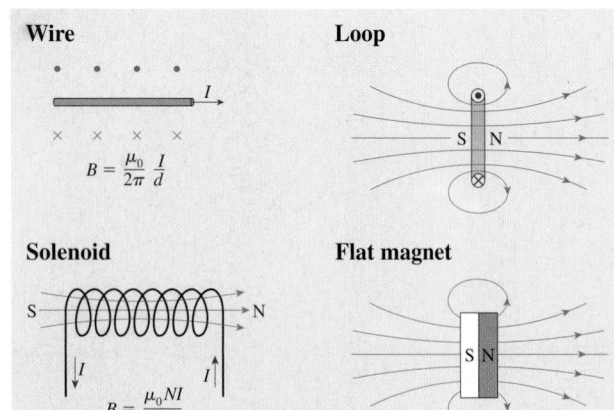

$B = \dfrac{\mu_0}{2\pi} \dfrac{I}{d}$

Solenoid

Flat magnet

$B = \dfrac{\mu_0 NI}{L}$

Right-hand rule

Point your right thumb in the direction of I. Your fingers curl in the direction of $\vec{B}$. For a dipole, $\vec{B}$ emerges from the side that is the north pole.

Charged-particle motion

No force if $\vec{v}$ is parallel to $\vec{B}$.

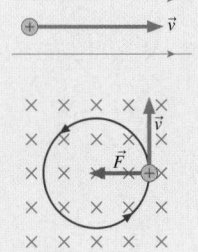

Circular motion at the cyclotron frequency $f_{\text{cyc}} = qB/2\pi m$ if $\vec{v}$ is perpendicular to $\vec{B}$.

Parallel wires and current loops

Parallel currents attract.
Opposite currents repel.

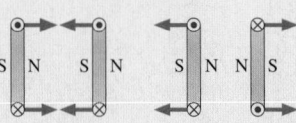

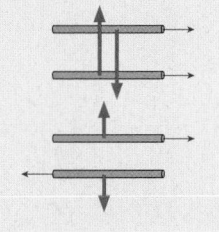

TERMS AND NOTATION

north pole	magnetic field lines	magnetic dipole moment, $\vec{\mu}$	cyclotron
south pole	Biot-Savart law	line integral	Hall effect
magnetic material	tesla, T	Ampère's law	Hall voltage, ΔV_H
magnetic dipole	permeability constant, μ_0	uniform magnetic field	ferromagnetic
magnetic monopole	cross product	solenoid	magnetic domain
right-hand rule	current loop	cyclotron motion	induced magnetic dipole
magnetic field, $\vec{B}$	electromagnet	cyclotron frequency, f_{cyc}	permanent magnet

EXERCISES AND PROBLEMS

Exercises

Section 32.2 The Magnetic Field

1. What is the current direction in the wire of Figure Ex32.1? Explain.

FIGURE EX32.1   **FIGURE EX32.2**

2. What is the current direction in the wire of Figure Ex32.2? Explain.

Section 32.3 The Source of the Magnetic Field: Moving Charges

3. Points 1 and 2 in Figure Ex32.3 are the same distance from the wires as the point where $B = 2.0$ mT. What are the strength and direction of $\vec{B}$ at points 1 and 2?

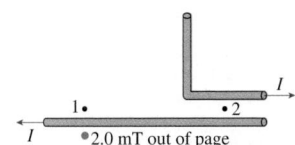

FIGURE EX32.3

4. What is the magnetic field strength at points 2 to 4 in Figure Ex32.4? Assume that the wires overlap closely and that points 1 to 4 are equally distant from the wires.

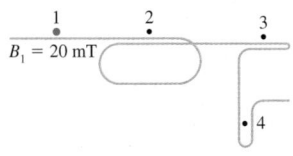

FIGURE EX32.4

5. A proton moves along the y-axis with $v_y = -1.0 \times 10^7$ m/s. As it passes the origin, what are the strength and direction of the magnetic field at the (x, y) positions (a) (1 cm, 0 cm), (b) (0 cm, 1 cm), and (c) (0 cm, −2 cm)?

6. An electron moves along the y-axis with $v_y = 1.0 \times 10^7$ m/s. As it passes the origin, what are the strength and direction of the magnetic field at the (x, y) positions (a) (0 cm, 1 cm), (b) (1 cm, 0 cm), and (c) (1 cm, 1 cm)?

7. What are the magnetic field strength and direction at the dot in Figure Ex32.7?

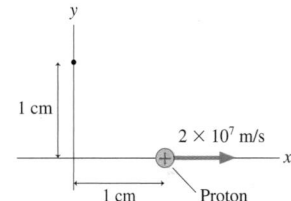

FIGURE EX32.7

8. What are the magnetic field strength and direction at the dot in Figure Ex32.8?

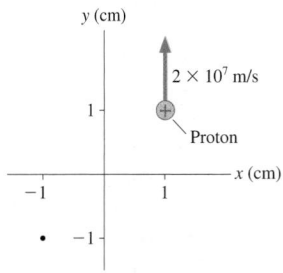

FIGURE EX32.8

9. A proton is passing the origin. The magnetic field at the (x, y, z) position (1 mm, 0 mm, 0 mm) is $1.0 \times 10^{-13} \, \hat{j}$ T. The field at (0 mm, 1 mm, 0 mm) is $-1.0 \times 10^{-13} \, \hat{i}$ T. What are the speed and direction of the proton?

Section 32.4 The Magnetic Field of a Current

10. What currents are needed to generate the magnetic field strengths of Table 32.1 at a point 1.0 cm from a long, straight wire?

11. At what distances from a very thin, straight wire carrying a 10 A current would the magnetic field strengths of Table 32.1 be generated?

12. For a current loop, what is the ratio of the magnetic field strength at $z = R$ to the magnetic field strength at the center of the loop?

13. The magnetic field at the center of a 1.0-cm-diameter loop is 2.5 mT.
    a. What is the current in the loop?
    b. A long straight wire carries the same current you found in part a. At what distance from the wire is the magnetic field 2.5 mT?

14. What are the magnetic field strength and direction at points 1 to 3 in Figure Ex32.14?

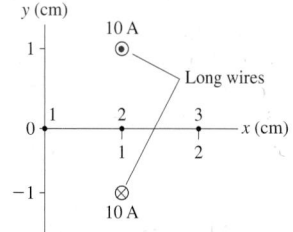

FIGURE EX32.14

15. What is the magnetic field $\vec{B}$ at points 1 to 3 in Figure Ex32.15?

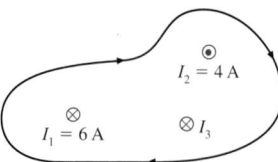

FIGURE EX32.15

Section 32.5 Magnetic Dipoles

16. A 100 A current circulates around a 2.0-mm-diameter super-conducting ring.
    a. What is the ring's magnetic dipole moment?
    b. What is the on-axis magnetic field strength 5.0 cm from the ring?

17. The on-axis magnetic field strength 10 cm from a small bar magnet is 5.0 μT.
    a. What is the bar magnet's magnetic dipole moment?
    b. What will be the magnetic field strength if the bar magnet is rotated end-over-end by 180°?

18. The earth's magnetic dipole moment is 8.0×10^{22} A m^2.
    a. What is the magnetic field strength on the surface of the earth at the earth's north magnetic pole? How does this compare to the value in Table 32.1? You can assume that the current loop is deep inside the earth.
    b. Astronauts discover an earth-size planet without a magnetic field. To create a magnetic field, so that compasses will work, they propose running a current through a wire around the equator. What size current would be needed?

19. A current loop with a 25 A current produces an on-axis magnetic field strength of 3.5 nT at a point 50 cm from the loop. What is the area of the loop? You can assume that the loop's radius is much less than 50 cm.

Section 32.6 Ampère's Law and Solenoids

20. What is the line integral of $\vec{B}$ between points i and f in Figure Ex32.20?

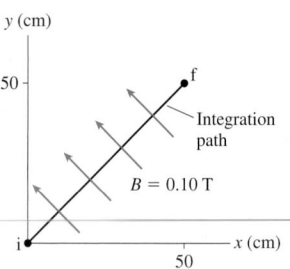

FIGURE EX32.20

21. What is the line integral of $\vec{B}$ between points i and f in Figure Ex32.21?

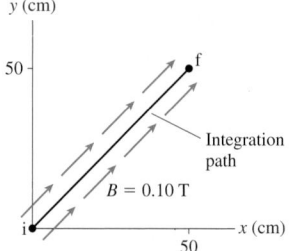

FIGURE EX32.21

22. What is the line integral of $\vec{B}$ between points i and f in Figure Ex32.22?

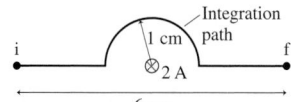

FIGURE EX32.22

23. The value of the line integral of $\vec{B}$ around the closed path in Figure Ex32.23 is 3.77×10^{-6} T m. What is I_3?

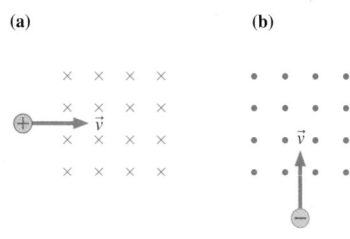

FIGURE EX32.23

24. A 2.0-cm-diameter, 15-cm-long solenoid is tightly wound from 1.0-mm-diameter wire. What current is needed to generate a 3.0 mT field inside the solenoid?

25. Magnetic resonance imaging needs a magnetic field strength of 1.5 T. The solenoid is 1.8 m long and 75 cm in diameter. It is tightly wound with a single layer of 2.0-mm-diameter superconducting wire. What size current is needed?

Section 32.7 The Magnetic Force on a Moving Charge

26. What is the *initial* direction of deflection for the charged particles entering the magnetic fields shown in Figure Ex32.26?

(a)

(b)

FIGURE EX32.26

27. What is the *initial* direction of deflection for the charged particles entering the magnetic fields shown in Figure Ex32.27?

(a)   (b)

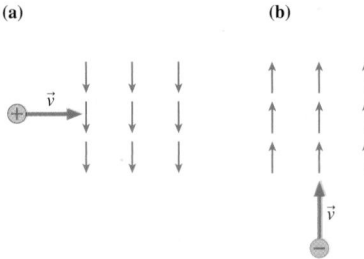

FIGURE EX32.27

28. Determine the magnetic field direction that causes the charged particles shown in Figure Ex32.28 to experience the indicated magnetic force.

(a)   (b)

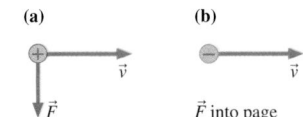

FIGURE EX32.28 $\vec{F}$ $\vec{F}$ into page

29. Determine the magnetic field direction that causes the charged particles shown in Figure Ex32.29 to experience the indicated magnetic force.

(a)   (b)

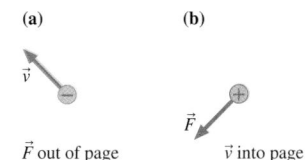

FIGURE EX32.29 $\vec{F}$ out of page $\vec{v}$ into page

30. A proton moves in the magnetic field $\vec{B} = 0.50\,\hat{\imath}$ T with a speed of 1.0×10^7 m/s in the directions shown in Figure Ex32.30. For each, what is magnetic force $\vec{F}$ on the proton? Give your answers in component form.

(a)   (b)

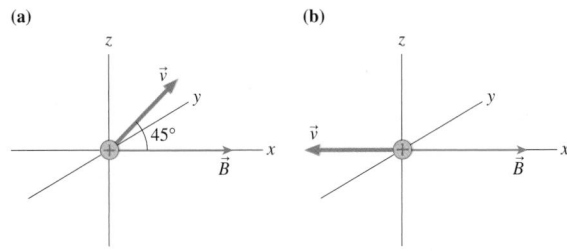

FIGURE EX32.30

31. An electron moves in the magnetic field $\vec{B} = 0.50\,\hat{\imath}$ T with a speed of 1.0×10^7 m/s in the directions shown in Figure Ex32.31. For each, what is magnetic force $\vec{F}$ on the electron? Give your answers in component form.

(a)   (b)

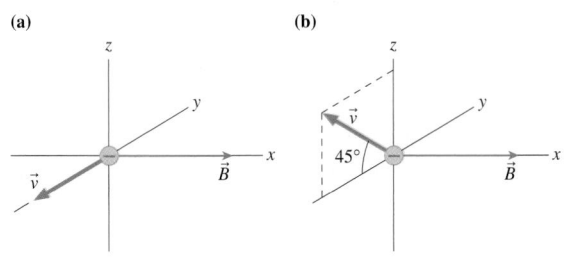

FIGURE EX32.31

32. What is the cyclotron frequency in a 3.00 T magnetic field of the ions (a) N_2^+, (b) O_2^+, and (c) CO^+? Give your answers in MHz. The masses of the atoms are shown in the table. The accuracy of your answers should reflect the accuracy of the data. (For this problem, assume that all the data you need are good to six significant figures. Although N_2^+ and CO^+ both have a *nominal* molecular mass of 28, they are easily distinguished by virtue of their different cyclotron resonance frequencies.)

Atomic masses

^{12}C	12.0000 u
^{14}N	14.0031 u
^{16}O	15.9949 u

33. Radio astronomers detect electromagnetic radiation at 45 MHz from an interstellar gas cloud. They suspect this radiation is emitted by electrons spiraling in a magnetic field. What is the magnetic field strength inside the gas cloud?

34. The aurora is caused when electrons and protons, moving in the earth's magnetic field of $\approx 5 \times 10^{-5}$ T, collide with molecules of the atmosphere and cause them to glow. What is the radius of the cyclotron orbit for
   a. An electron with speed 1.0×10^6 m/s?
   b. A proton with speed 5.0×10^4 m/s?

35. The Hall voltage across a 1.0-mm-thick conductor in a 1.0 T magnetic field is 3.2 μV when the current is 15 A. What is the charge-carrier density in this conductor?

Section 32.8 Magnetic Forces on Current-Carrying Wires

36. What magnetic field strength and direction will levitate the 2.0 g wire in Figure Ex32.36?

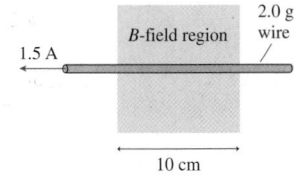

FIGURE EX32.36   10 cm

37. The right edge of the circuit in Figure Ex32.37 extends into a 50 mT uniform magnetic field. What are the magnitude and direction of the net force on the circuit?

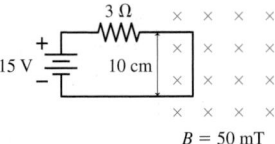

FIGURE EX32.37 $B = 50$ mT

38. What is the net force (magnitude and direction) on each wire in Figure Ex32.38?

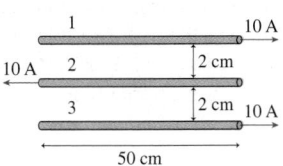

FIGURE EX32.38   50 cm

39. The two 10-cm-long parallel wires in Figure Ex32.39 are separated by 5.0 mm. For what value of the resistor R will the force between the two wires be 5.4×10^{-5} N?

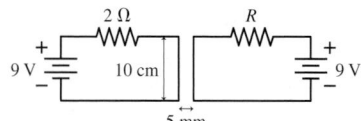

FIGURE EX32.39

Section 32.9 Forces and Torques on Current Loops

40. Figure Ex32.40 shows two square current loops. The loops are far apart and do not interact with each other.
    a. Use force diagrams to show that both loops are in equilibrium, having a net force of zero and no torque.
    b. One of the loop positions is stable. That is, the forces will return it to equilibrium if it is rotated slightly. The other position is unstable, like an upside-down pendulum. Which is which? Explain.

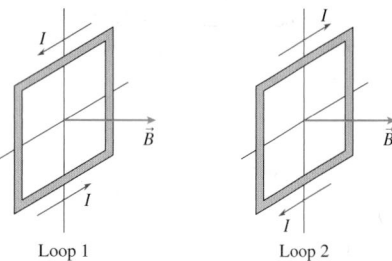

FIGURE EX32.40

41. A square current loop 5.0 cm on each side carries a 500 mA current. The loop is in a 1.2 T uniform magnetic field. The axis of the loop, perpendicular to the plane of the loop, is 30° away from the field direction. What is the magnitude of the torque on the current loop?

42. A small bar magnet experiences a 0.020 N m torque when the axis of the magnet is at 45° to a 0.10 T magnetic field. What is the magnitude of its magnetic dipole moment?

43. a. What is the magnitude of the torque on the current loop in Figure Ex32.43?
    b. What is the loop's equilibrium position?

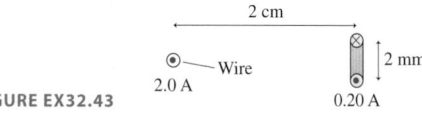

FIGURE EX32.43

Problems

44. You have a horizontal cathode ray tube (CRT) for which the controls have been adjusted such that the electron beam *should* make a single spot of light exactly in the center of the screen. You observe, however, that the spot is deflected to the right. It is possible that the CRT is broken. But as a clever scientist, you realize that your laboratory might be in either an electric or a magnetic field. Assuming that you do not have a compass, any magnets, or any charged rods, how can you use the CRT itself to determine whether the CRT is broken, is in an electric field, or is in a magnetic field? You cannot remove the CRT from the room.

45. Although the evidence is weak, there has been concern in recent years over possible health effects from the magnetic fields generated by transmission lines, the wiring in your house, and electrical appliances.
    a. The current carried by the wiring in the walls of your house rarely exceeds 10 A. What is the magnetic field strength 2 m from a long, straight wire carrying a current of 10 A?
    b. What percentage of the earth's magnetic field is your answer to (a)?
    Because the percentage is small, and because we live in the earth's field with no harmful effects, it is assumed that any possible health effects are due not to the field strength alone but to the 60 Hz oscillatory nature of fields from power lines.
    c. High-voltage transmission lines, on tall towers, typically carry 200 A at voltages of up to 500,000 V. Although this is much larger than household currents, the lines are roughly 20 m overhead. Estimate the magnetic field strength on the ground underneath such lines.
    d. A common electrical appliance is an electric blanket. Some consumer groups urge pregnant women not to use electric blankets, just in case there is a health risk. The current through the heater wires is approximately 1 A. Estimate, stating any assumptions you make, the magnetic field strength a fetus might experience. How does this compare to your answer to part a?

46. A wire carries current I into the junction shown in Figure P32.46. What is the magnetic field at the dot?

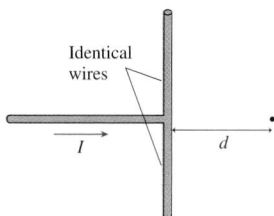

FIGURE P32.46

47. The two insulated wires in Figure P32.47 cross at a 30° angle but do not make electrical contact. Each wire carries a 5.0 A current. Points 1 and 2 are each 4.0 cm from the intersection and equally distant from both wires. What are the magnitude and direction of the magnetic fields at points 1 and 2?

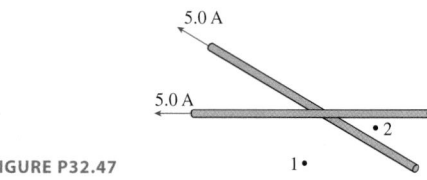

FIGURE P32.47

48. A long wire carrying a 5.0 A current perpendicular to the xy-plane intersects the x-axis at $x = -2.0$ cm. A second, parallel wire carrying a 3.0 A current intersects the x-axis at $x = +2.0$ cm. At what point or points on the x-axis is the magnetic field zero if (a) the two currents are in the same direction and (b) the two currents are in opposite directions?

49. The capacitor in Figure P32.49 is charged to 50 V. The switch closes at $t = 0$ s. Draw a graph showing the magnetic field strength as a function of time at the position of the dot. On your graph indicate the maximum field strength, and provide an appropriate numerical scale on the horizontal axis.

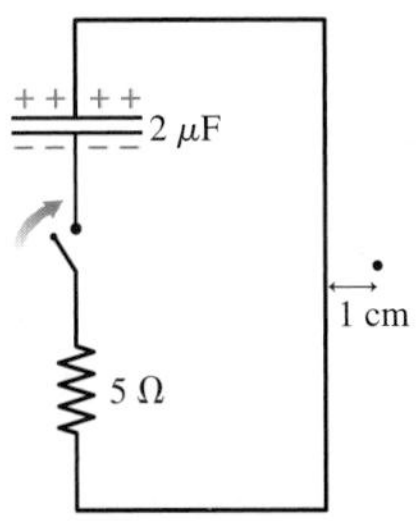

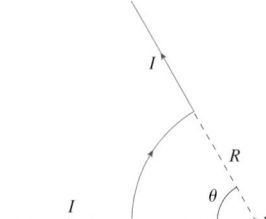

FIGURE P32.49

50. a. Find an expression for the magnetic field at the center (point P) of the circular arc in Figure P32.50.
    b. Does your result agree with the magnetic field of a current loop when $\theta = 2\pi$?

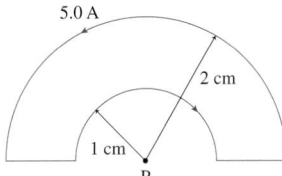

FIGURE P32.50

51. The element niobium, which is a metal, is a superconductor (i.e., no electrical resistance) at temperatures below 9 K. However, the superconductivity is destroyed if the magnetic field at the surface of the metal reaches or exceeds 0.10 T. What is the maximum current in a straight, 3.0-mm-diameter superconducting niobium wire?

52. What are the strength and direction of the magnetic field at point P in Figure P32.52?

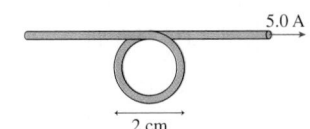

FIGURE P32.52

53. The earth's magnetic field, with a magnetic dipole moment of $8.0 \times 10^{22} \text{ A m}^2$, is generated by currents within the molten iron of the earth's outer core. (The inner core is solid iron.) As a simple model, consider the outer core to be a current loop made of a 1000-km-diameter "wire" of molten iron. The loop diameter, measured between the centers of the "wires," is 3000 km.
    a. What is the current in the current loop?
    b. What is the current density J in the current loop?
    c. To decide whether this is a large or a small current density, compare it to the current density of a 1.0 A current in a 1.0-mm-diameter wire.

54. What is the magnetic field at the center of the loop in Figure P32.54?

FIGURE P32.54

55. Your employer asks you to build a 20-cm-long solenoid with an interior field of 5.0 mT. The specifications call for a single layer of wire, wound with the coils as close together as possible. You have two spools of wire available. Wire with a #18

gauge has a diameter of 1.02 mm and has a maximum current rating of 6 A. Wire with a #26 gauge is 0.41 mm in diameter and can carry up to 1 A. Which wire should you use, and what current will you need?

56. The magnetic field strength at the north pole of a 2.0-cm-diameter, 8-cm-long Alnico magnet is 0.10 T. To produce the same field with a solenoid of the same size, carrying a current of 2.0 A, how many turns of wire would you need? Does this seem feasible? (See Problem 55 for information about wire sizes and maximum current.)

57. Two identical coils are parallel to each other on the same axis. They are separated by a distance equal to their radius. They each have N turns and carry equal currents I in the same direction.
    a. Find an expression for the magnetic field strength at the midpoint between the loops.
    b. Calculate the field strength if the loops are 10 cm in diameter, have 10 turns, and carry a 1.0 A current.

58. You have a 1.0-m-long copper wire. You want to make an N-turn current loop that generates a 1.0 mT magnetic field at the center when the current is 1.0 A. You must use the entire wire. What will be the diameter of your coil?

59. Use the Biot-Savart law to find the magnetic field strength at the center of the semicircle in Figure P32.59.

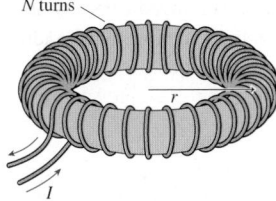

FIGURE P32.59

60. The *toroid* of Figure P32.60 is a coil of wire wrapped around a doughnut-shaped ring (a *torus*) made of nonconducting material. Toroidal magnetic fields are used to confine fusion plasmas.
    a. From symmetry, what must be the *shape* of the magnetic field in this toroid? Explain.
    b. Use Ampère's law to find an expression for the magnetic field strength at a distance r from the axis of a toroid with N closely spaced turns carrying current I.
    c. Is a toroidal magnetic field a uniform field? Explain.

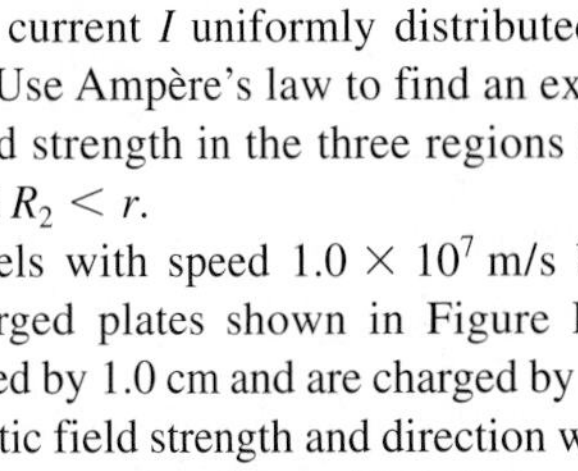

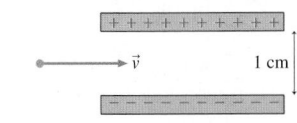

FIGURE P32.60

61. A long, hollow wire has inner radius R_1 and outer radius R_2. The wire carries current I uniformly distributed across the area of the wire. Use Ampère's law to find an expression for the magnetic field strength in the three regions $0 < r < R_1$, $R_1 < r < R_2$, and $R_2 < r$.

62. An electron travels with speed 1.0×10^7 m/s between the two parallel charged plates shown in Figure P32.62. The plates are separated by 1.0 cm and are charged by a 200 V battery. What magnetic field strength and direction will allow the electron to pass between the plates without being deflected?

FIGURE P32.62

63. An electron in a cathode-ray tube is accelerated through a potential difference of 10 kV, then passes through the 2.0-cm-wide region of uniform magnetic field in Figure P32.63. What field strength will deflect the electron by 10°?

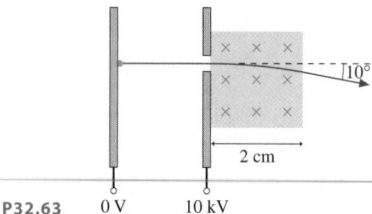

FIGURE P32.63   0 V   10 kV

64. The microwaves in a microwave oven are produced in a special tube called a *magnetron*. The electrons orbit the magnetic field at 2.4 GHz, and as they do so they emit 2.4 GHz electromagnetic waves.
    a. What is the magnetic field strength?
    b. If the maximum diameter of the electron orbit before the electron hits the wall of the tube is 2.5 cm, what is the maximum electron kinetic energy?

65. An antiproton (same properties as a proton except that $q = -e$) is moving in the combined electric and magnetic fields of Figure P32.65.
    a. What are the magnitude and direction of the antiproton's acceleration at this instant?
    b. What would be the magnitude and direction of the acceleration if $\vec{v}$ were reversed?

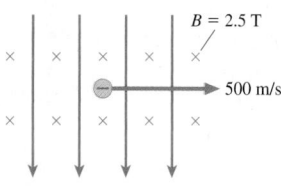

FIGURE P32.65 $E = 1000$ V/m

66. The uniform 30 mT magnetic field in Figure P32.66 points in the positive z-direction. An electron enters the region of magnetic field with a speed of 5.0×10^6 m/s and at an angle of 30° above the xy-plane. Find the radius r and the pitch p of the electron's spiral trajectory.

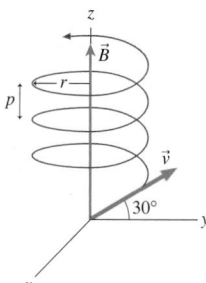

FIGURE P32.66

67. a. A 65-cm-diameter cyclotron uses a 500 V oscillating potential difference between the dees. What is the maximum kinetic energy of a proton if the magnetic field strength is 0.75 T?
    b. How many revolutions does the proton make before leaving the cyclotron?

68. For your senior project, you would like to build a cyclotron that will accelerate protons to 10% of the speed of light. The largest vacuum chamber you can find is 50 cm in diameter. What magnetic field strength will you need?

69. A Hall-effect probe to measure magnetic field strengths needs to be calibrated in a known magnetic field. Although it is not easy to do, magnetic fields can be precisely measured by measuring the cyclotron frequency of protons. A testing laboratory adjusts a magnetic field until the proton's cyclotron frequency is 10.0 MHz. At this field strength, the Hall voltage on the probe is 0.543 mV when the current through the probe is 0.150 mA. Later, when an unknown magnetic field is measured, the Hall voltage at the same current is 1.735 mV. What is the strength of this magnetic field?

70. Figure P32.70 shows a *mass spectrometer*, an analytical instrument used to identify the various molecules in a sample by measuring their charge-to-mass ratio e/m. The sample is ionized, the positive ions are accelerated (starting from rest) through a potential difference ΔV, and they then enter a region of uniform magnetic field. The field bends the ions into circular trajectories, but after just half a circle they either strike the wall or pass through a small opening to a detector. As the accelerating voltage is slowly increased, different ions reach the detector and are measured. Typical design values are a magnetic field strength $B = 0.200$ T and a spacing between the entrance and exit holes $d = 8.00$ cm. What accelerating potential difference ΔV is required to detect (a) N_2^+, (b) O_2^+, and (c) CO^+? See Exercise 32 for atomic data, and note there the comment about accuracy and significant figures.

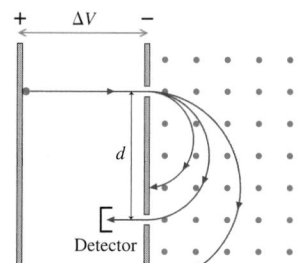

FIGURE P32.70

71. The two springs in Figure P32.71 each have a spring constant of 10 N/m. They are stretched by 1.0 cm when a current passes through the wire. How big is the current?

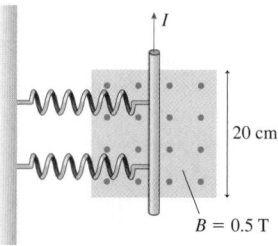

FIGURE P32.71

72. Figure P32.72 is a cross section through three long wires with linear mass density 50 g/m. They each carry equal currents in the directions shown. The lower two wires are 4.0 cm apart and are attached to a table. What current I will allow the upper wire to "float" so as to form an equilateral triangle with the lower wires?

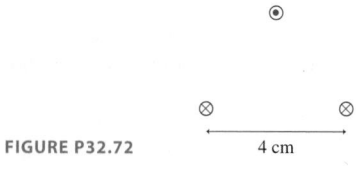

FIGURE P32.72   4 cm

73. A long, straight wire with linear mass density of 50 g/m is suspended by threads, as shown in Figure P32.73. A 10 A current in the wire experiences a horizontal magnetic force that deflects it to an equilibrium angle of 10°. What are the strength and direction of the magnetic field $\vec{B}$?

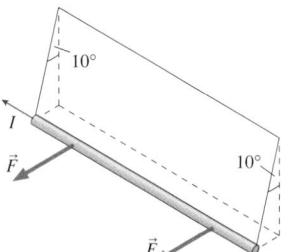

FIGURE P32.73

74. A bar magnet experiences a torque of magnitude 0.075 Nm when it is perpendicular to a 0.50 T external magnetic field. What is the strength of the bar magnet's on-axis magnetic field at a point 20 cm from the center of the magnet?

75. In the semiclassical Bohr model of the hydrogen atom, the electron moves in a circular orbit of radius 5.3×10^{-11} m with speed 2.2×10^6 m/s. According to this model, what is the magnetic field at the center of a hydrogen atom?
    Hint: Determine the *average* current of the orbiting electron.

76. A *nonuniform* magnetic field exerts a net force on a current loop of radius R. Figure P32.76 shows a magnetic field that is diverging from the end of a bar magnet. The magnetic field at the position of the current loop makes an angle θ with respect to the vertical.
    a. Find an expression for the net magnetic force on the current.
    b. Calculate the force if $R = 2.0$ cm, $I = 0.50$ A, $B = 200$ mT, and $\theta = 20°$.

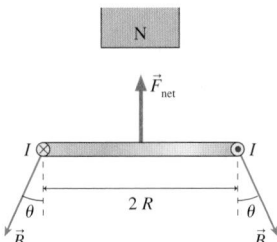

FIGURE P32.76

77. A computer diskette is a plastic disk coated with a ferromagnetic paint. A single magnetic domain can have its magnetic moment oriented to point either up or down, and these two orientations can be interpreted as a binary 0 (up) or 1 (down). Each 0 or 1 is called a *bit* of information. A diskette stores roughly 500,000 *bytes* of data on one side, and each byte contains eight bits. Estimate the width of a magnetic domain, and compare your answer to the typical domain size given in the text. List any assumptions you use in your estimate.

78. The ends of two permanent bar magnets can either attract or repel each other. Give a step-by-step description, using both words and picture, of how these magnetic forces result from the interaction between the electron spins. Consider both the attractive and the repulsive situations.

79. A permanent magnet can pick up a piece of nonmagnetized iron. Give a step-by-step description, using both words and picture, of how the magnetic force on the iron results from the interaction between the electron spins.

Challenge Problems

80. The 10-turn loop of wire shown in Figure CP32.80 lies in a horizontal plane, parallel to a uniform horizontal magnetic field, and carries a 2.0 A current. The loop is free to rotate about a nonmagnetic axle through the center. A 50 g mass hangs from one edge of the loop. What magnetic field strength will prevent the loop from rotating about the axle?

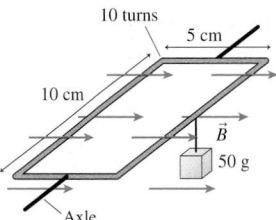

FIGURE CP32.80

81. a. Derive an expression for the magnetic field strength at distance d from the center of a straight wire of finite length L that carries current I.
    b. Determine the field strength at the center of a current-carrying *square* loop having sides of length $2R$.
    c. Compare your answer to part b to the field at the center of a *circular* loop of diameter $2R$. Do so by computing the ratio $B_{\text{square}}/B_{\text{circle}}$.

82. A long, straight conducting wire of radius R has a nonuniform current density $J = J_0 r/R$, where J_0 is a constant. The wire carries total current I.
    a. Find an expression for J_0 in terms of I and R.
    b. Find an expression for the magnetic field strength inside the wire at radius r.
    c. At the boundary, $r = R$, does your solution match the known field outside a long, straight current-carrying wire?

83. The coaxial cable shown in Figure CP32.83 consists of a solid inner conductor of radius R_1 surrounded by a hollow, very thin outer conductor of radius R_2. The two carry equal currents I, but in *opposite* directions. The current density is uniformly distributed over each conductor.
    a. Find expressions for three magnetic fields: within the inner conductor, in the space between the conductors, and outside the outer conductor.
    b. Draw a graph of B versus r from $r = 0$ to $r = 2R_2$ if $R_1 = \frac{1}{3}R_2$.

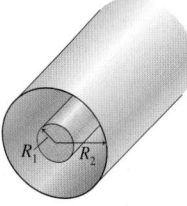

FIGURE CP32.83

84. An infinitely wide flat sheet of charge flows out of the page in Figure CP32.84. The current per unit width along the sheet (amps per meter) is given by the linear current density J_s.
    a. What is the *shape* of the magnetic field? To answer this question, you may find it helpful to approximate the current sheet as many parallel, closely spaced current-carrying wires. Give your answer as a picture showing magnetic field vectors.

b. Find the magnetic field strength at distance d above or below the current sheet.

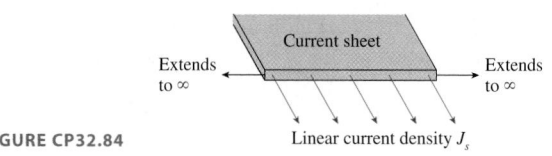

FIGURE CP32.84

STOP TO THINK ANSWERS

Stop to Think 32.1: Not at all. The charge exerts weak, attractive polarization forces on both ends of the compass needle, but in this configuration the forces will balance and have no net effect.

Stop to Think 32.2: d. Point your right thumb in the direction of the current and curl your fingers around the wire.

Stop to Think 32.3: b. Point your right thumb out of the page, in the direction of $\vec{v}$. Your fingers are pointing down as they curl around the left side.

Stop to Think 32.4: b. The right-hand rule gives a downward $\vec{B}$ for a clockwise current. The north pole is on the side from which the field emerges.

Stop to Think 32.5: c. For a field pointing into the page, $\vec{v} \times \vec{B}$ is to the right. But the electron is negative, so the force is in the direction of $-(\vec{v} \times \vec{B})$.

Stop to Think 32.6: b. Repulsion indicates that the south pole of the loop is on the right, facing the bar magnet; the north pole is on the left. Then the right-hand rule gives the current direction.

Stop to Think 32.7: a or c. Any downward magnetic field will align the magnetic domains as shown.

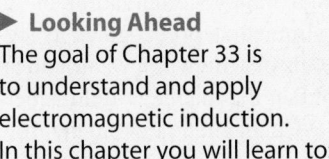

33 Electromagnetic Induction

Electromagnetic induction is the scientific principle that underlies many modern technologies, from the generation of electricity to communications and data storage.

▶ Looking Ahead

The goal of Chapter 33 is to understand and apply electromagnetic induction. In this chapter you will learn to:

- Calculate induced current.
- Calculate magnetic flux.
- Use Lenz's law and Faraday's law to determine the direction and size of induced currents.
- Understand how induced electric and magnetic fields lead to electromagnetic waves.
- Analyze circuits with inductors.

◀ Looking Back

This chapter will join together ideas about magnetic fields and electric potential. Please review:

- Section 11.3 The vector dot product.
- Section 30.4 Sources of electric potential.
- Sections 32.4–32.8 Magnetic fields and magnetic forces.

What do electric generators, metal detectors, video recorders, computer hard disks, and cell phones have in common? Surprisingly, these diverse technologies all stem from a single scientific principle, electromagnetic induction. **Electromagnetic induction** is the process of generating an electric current by varying the magnetic field that passes through a circuit.

The many applications of electromagnetic induction make it an important topic for study. But more fundamentally, electromagnetic induction establishes an important link between electricity and magnetism. We've been studying electric and magnetic fields as if they were separate, independent fields. Electromagnetic induction forms a link between $\vec{E}$ and $\vec{B}$, a link with important implications for understanding light as an electromagnetic wave.

Electromagnetic induction is a subtle topic, so we will build up to it gradually. We'll first examine different aspects of induction and become familiar with its basic characteristics. Section 33.5 will then introduce Faraday's law, a new law of physics not derivable from any previous laws you have studied. The remainder of the chapter will explore its implications and applications.

33.1 Induced Currents

Oersted's 1820 discovery that a current creates a magnetic field generated enormous excitement. Dozens of scientists immediately began to explore the implications of this discovery. One question they hoped to answer was whether the converse of Oersted's discovery was true. That is, can a magnet be used to create

FIGURE 33.1 Michael Faraday.

a current? There was not yet a good understanding of the origins or properties of electricity and magnetism, so scientists hoping to generate a current from magnetism had little to guide them. Many experiments were reported in which wires and coils were placed in or around magnets of various sizes and shapes, but no one was able to generate a current.

On the other side of the Atlantic, the American scientist Joseph Henry read of these new discoveries with great interest. American professors at the time were expected to devote all of their time to teaching, so Henry had little opportunity for research. It was during a one-month vacation in 1831 that Henry became the first to discover how to produce a current from magnetism, a process that we now call *electromagnetic induction.* But Henry had no time for follow-up studies, and he was not able to publish his discovery until the following year.

At about the same time, in England, Michael Faraday (Figure 33.1) made the same discovery and immediately published his findings. You met Faraday in Chapter 25 as the inventor of the concept of a *field.* The idea came to him as he observed that a compass needle stays tangent to a circle around a current-carrying wire. Faraday ascribed the needle's behavior to "circular lines of force," an idea that soon came to be known as the *magnetic field.* This pictorial representation played a crucial role in Faraday's discovery of the law of electromagnetic induction.

Credit in science usually goes to the first to publish, so today we study *Faraday's law* rather than *Henry's law.* The situation, however, is not entirely unjust. Henry had discovered an *effect,* but he was not able to do the research needed to understand the implications of his discovery. Even if Faraday did not have priority of discovery, it was Faraday who studied the new phenomenon of electromagnetic induction, established its properties, and realized that he had discovered a new law of nature.

Faraday's Discovery

Faraday's 1831 discovery, like Oersted's, was a happy combination of an unplanned event and a mind that was prepared to immediately recognize its significance. Faraday was experimenting with two coils of wire wrapped around an iron ring, as shown in Figure 33.2. He had hoped that the magnetic field generated by a current in the coil on the left would induce a magnetic field in the iron, and that the magnetic field in the iron might then somehow create a current in the circuit on the right.

Like all his previous attempts, this technique failed to generate a current. But Faraday happened to notice that the needle of the current meter jumped ever so slightly at the instant when he closed the switch in the circuit on the left. After the switch was closed, the needle immediately returned to zero. The needle again jumped when he later opened the switch, but this time in the opposite direction. Faraday recognized that the motion of the needle indicated a very slight current in the circuit on the right. But the effect happened only during the very brief interval when the current on the left was starting or stopping, not while it was steady.

Faraday applied his mental picture of lines of force to this discovery. The current on the left first magnetizes the iron ring, then the magnetic field of the iron ring passes through the coil on the right. Faraday's observation that the current-meter needle jumped only when the switch was opened and closed suggested to him that a current was generated only if the magnetic field was *changing* as it passed through the coil. This would explain why all the previous attempts to generate a current were unsuccessful: they had used only steady, unchanging magnetic fields.

Faraday set out to test this hypothesis. If the critical issue was *changing* the magnetic field through the loop, then the iron ring should not be necessary. That is, any method that changes the magnetic field should work. Faraday began a series of experiments to find out if this was true.

Closing the switch in the left circuit... ...causes a momentary current in the right circuit.

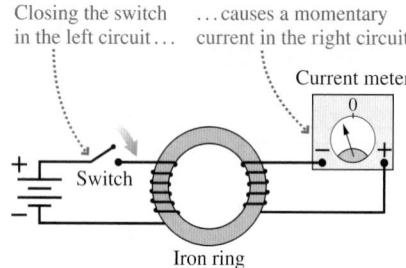

No current flows while the switch stays closed.

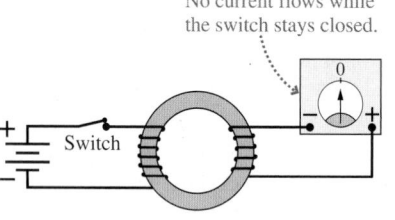

Opening the switch in the left circuit... ...causes a momentary current in the opposite direction.

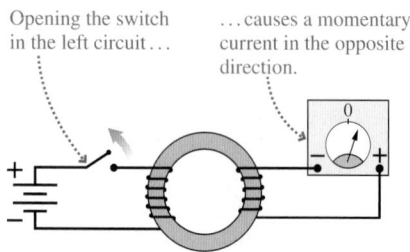

FIGURE 33.2 Faraday's discovery of electromagnetic induction.

Faraday investigates electromagnetic induction

Faraday placed one coil directly above the other, without the iron ring. There was no current in the lower circuit while the switch was in the closed position, but a momentary current appeared whenever the switch was opened or closed.

He pushed a bar magnet into a coil of wire. This action caused a momentary deflection of the current-meter needle, although *holding* the magnet inside the coil had no effect. A quick withdrawal of the magnet deflected the needle in the other direction.

Must the magnet move? Faraday created a momentary current by rapidly pulling a coil of wire out of a magnetic field, although there was no current if the coil was stationary in the magnetic field. Pushing the coil *into* the magnet caused the needle to deflect in the opposite direction.

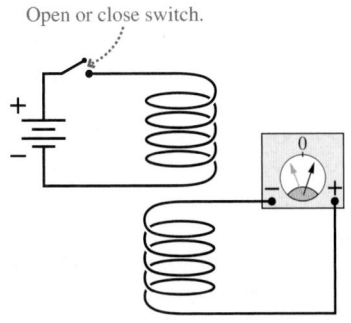

Opening or closing the switch creates a momentary current.

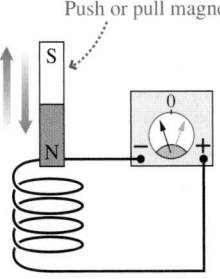

Pushing the magnet into the coil or pulling it out creates a momentary current.

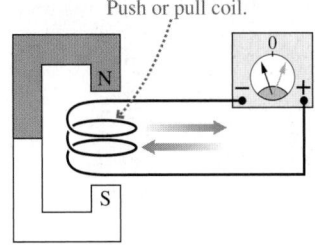

Pushing the coil into the magnet or pulling it out creates a momentary current.

To summarize

Faraday found that there is a current in a coil of wire if and only if the magnetic field passing through the coil is *changing*. This is an informal statement of what we'll soon call *Faraday's law*.

It makes no difference what causes the magnetic field to change: current stopping or starting in a nearby circuit, moving a magnet through the coil, or moving the coil in and out of a magnet. The effect is the same in all cases. There is no current if the field through the coil is not changing, so it's not the magnetic field itself that is responsible for the current but, instead, it is the *changing of the magnetic field.*

The current in a circuit due to a changing magnetic field is called an **induced current.** Opening the switch or moving the magnet *induces* a current in a nearby circuit. An induced current is not caused by a battery. It is a completely new way to generate a current, and we will have to discover how it is similar to and how it is different from currents we have studied previously.

The first induced currents were small, barely noticeable effects. Neither Faraday nor Henry could have answered the question, "What good is it?" Yet electromagnetic induction has became the basis of commercial electricity generation, of radio and television broadcasting, of computer memories and data storage, of cell phones, and much more.

33.2 Motional emf

An induced current can be created two different ways:

1. By changing the size or orientation of a circuit in a stationary magnetic field, or
2. By changing the magnetic field through a stationary circuit.

Although the effects are the same, the causes turn out to be different. We'll start our investigation of electromagnetic induction by looking at situations in which the magnetic field is fixed while the circuit moves or changes.

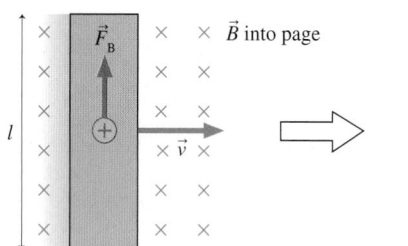

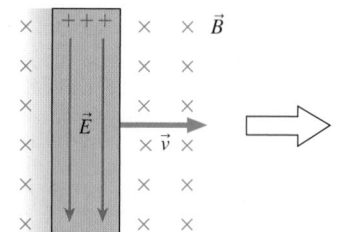

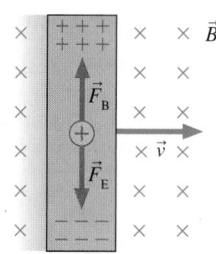

Charge carriers in the wire experience an upward force of magnitude $F_B = qvB$. Being free to move, positive charges flow upward (or, if you prefer, negative charges downward).

The charge separation creates an electric field in the conductor. $\vec{E}$ increases as more charge flows.

The charge flow continues until the downward electric force $\vec{F}_E$ is large enough to balance the upward magnetic force $\vec{F}_B$. Then the net force on a charge is zero and the current ceases.

FIGURE 33.3 The magnetic force on the charge carriers in a moving conductor creates an electric field inside the conductor.

To begin, consider a conductor of length l that moves with velocity $\vec{v}$ through a uniform magnetic field $\vec{B}$, as shown in Figure 33.3. The charge carriers inside the wire also move with velocity $\vec{v}$, so they each experience a magnetic force $\vec{F}_B = q\vec{v} \times \vec{B}$. For simplicity, we will assume that $\vec{v}$ is perpendicular to $\vec{B}$, in which case the strength of the force is $F_B = qvB$. This force causes the charge carriers to move, separating the positive and negative charges and thus creating an electric field inside the conductor.

The charge carriers continue to move until the electric force $F_E = qE$ exactly balances the magnetic force. This balance happens when the electric field strength is

$$E = vB \tag{33.1}$$

In other words, the magnetic force on the charge carriers in a moving conductor creates an electric field $E = vB$ inside the conductor.

The electric field, in turn, creates an electric potential difference between the two ends of the moving conductor. Figure 33.4a defines a coordinate system in which $\vec{E} = -vB\hat{j}$. Using the connection between the electric field and the electric potential that we found in Chapter 30,

$$\Delta V = V_{\text{top}} - V_{\text{bottom}} = -\int_0^l E_y \, dy = -\int_0^l (-vB) \, dy = vlB \tag{33.2}$$

Thus the motion of the wire through a magnetic field *induces* a potential difference vlB between the ends of the conductor. The potential difference depends on the strength of the magnetic field and on the wire's speed through the field.

There's an important analogy between this potential difference and the potential difference of a battery. Figure 33.4b reminds you that a battery uses a non-electric force—the charge escalator—to separate positive and negative charges. The emf $\mathcal{E}$ of the battery was defined as the work performed per charge (W/q) to separate the charges. An isolated battery, with no current, has a potential difference $\Delta V_{\text{bat}} = \mathcal{E}$. We could refer to a battery, where the charges are separated by chemical reactions, as a source of *chemical emf*.

The moving conductor develops a potential difference because of the work done by magnetic forces to separate the charges. You can think of the moving conductor as a "battery" that stays charged only as long as it keeps moving but "runs down" if it stops. The emf of the conductor is due to its motion, rather than to chemical reactions inside, so we can define the **motional emf** of a conductor moving with velocity $\vec{v}$ to be

$$\mathcal{E} = vlB \tag{33.3}$$

(a) Magnetic forces separate the charges and cause a potential difference between the ends. This is a motional emf.

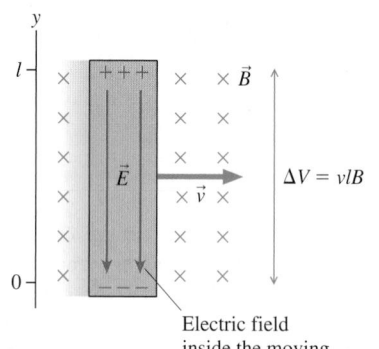

Electric field inside the moving conductor

(b) Chemical reactions separate the charges and cause a potential difference between the ends. This is a chemical emf.

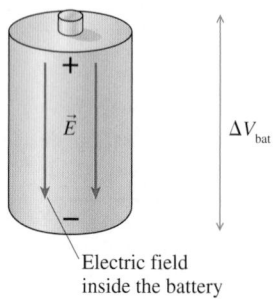

Electric field inside the battery

FIGURE 33.4 Two different ways to generate an emf.

STOP TO THINK 33.1   A square conductor moves through a uniform magnetic field. Which of the figures shows the correct charge distribution on the conductor?

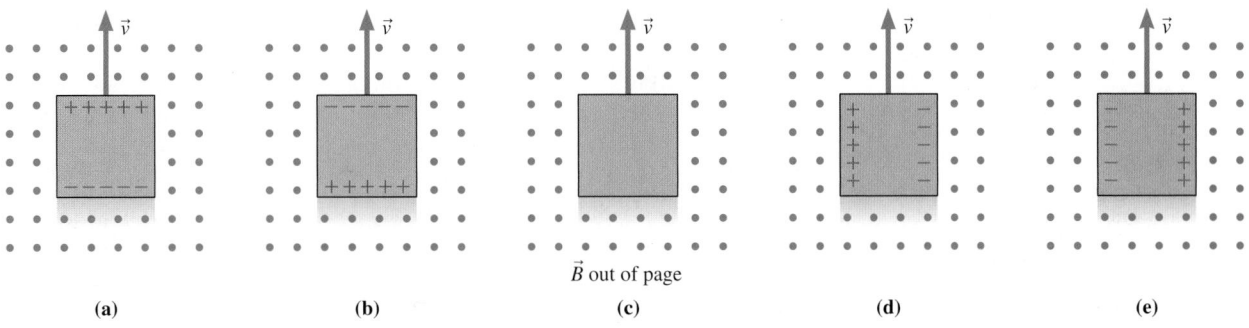

$\vec{B}$ out of page

(a)   (b)   (c)   (d)   (e)

EXAMPLE 33.1 **A battery substitute**
A 6.0-cm-long flashlight battery has an emf of 1.5 V. With what speed must a 6.0-cm-long wire move through a 0.10 T magnetic field to create a motional emf of 1.5 V?

SOLVE 1.5V is the motional emf. We can use equation 33.3 to find

$$v = \frac{\mathcal{E}}{lB} = \frac{1.5 \text{ V}}{(0.060 \text{ m})(0.10 \text{ T})} = 250 \text{ m/s}$$

ASSESS 250 m/s ≈ 500 mph. This might not be a very practical substitute for a battery, but it would work as long as the wire continued to move through the field with this speed.

EXAMPLE 33.2 **Potential difference along a rotating bar**
A metal bar of length *l* rotates with angular velocity ω about a pivot at one end of the bar. A uniform magnetic field $\vec{B}$ is perpendicular to the plane of rotation. What is the potential difference between the ends of the bar?

VISUALIZE Figure 33.5 is a pictorial representation of the bar. The magnetic forces on the charge carriers will cause the outer end to be positive with respect to the pivot.

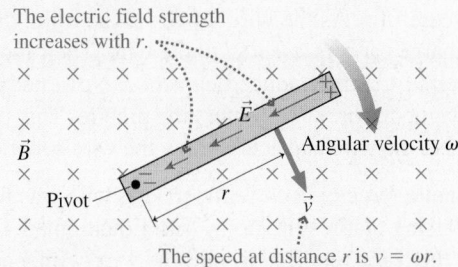

The electric field strength increases with *r*.

$\vec{B}$
Pivot
Angular velocity ω
r
$\vec{v}$

The speed at distance *r* is $v = \omega r$.

FIGURE 33.5 Pictorial representation of a metal bar rotating in a magnetic field.

SOLVE Even though the bar is rotating, rather than moving in a straight line, the velocity of each charge carrier is perpendicular to $\vec{B}$. Consequently, the electric field created inside the bar is exactly that given in Equation 33.1, $E = vB$. But v, the speed of the charge carrier, now depends on its distance from the pivot. Recall that in rotational motion the tangential speed at radius r from the center of rotation is $v = \omega r$. Thus the electric field at distance r from the pivot is $E = \omega r B$. The electric field increases in strength as you move outward along the bar.

The electric field $\vec{E}$ points toward the pivot, so its radial component is $E_r = -\omega r B$. If we integrate outward from the center, the potential difference between the ends of the bar is

$$\Delta V = V_{\text{tip}} - V_{\text{pivot}} = -\int_0^l E_r \, dr$$

$$= -\int_0^l (-\omega r B) \, dr = \omega B \int_0^l r \, dr = \frac{1}{2}\omega l^2 B$$

ASSESS $\frac{1}{2}\omega l$ is the speed at the midpoint of the bar. Thus ΔV is $v_{\text{mid}} l B$, which seems reasonable.

1. The charge carriers in the wire
are pushed upward by the
magnetic force.

Positive end
of wire

Moving wire

$\vec{B}$

$\vec{v}$

l

Conducting rail. Fixed
to table and doesn't move.

Negative end
of wire

2. The charge carriers flow
around the conducting loop
as an induced current.

FIGURE 33.6 A current is induced in the circuit as the wire moves through a magnetic field.

Induced Current in a Circuit

The moving conductor of Figure 33.3 had an emf, but it couldn't sustain a current because the charges had nowhere to go. It's like a battery that is disconnected from a circuit. We can change this by including the moving conductor in a circuit.

Figure 33.6 shows a conducting wire sliding with speed v along a U-shaped conducting rail. We'll assume that the rail is attached to a table and cannot move. The wire and the rail together form a closed conducting loop—a circuit.

Suppose a magnetic field $\vec{B}$ is perpendicular to the plane of the circuit. Charges in the moving wire will be pushed to the ends of the wire by the magnetic force, just as they were in Figure 33.3, but now the charges can continue to flow around the circuit. That is, the moving wire acts like a battery in a circuit.

The current in the circuit is an *induced current*. In this example, the induced current is counterclockwise (ccw). If the total resistance of the circuit is R, the induced current is given by Ohm's law as

$$I = \frac{\mathcal{E}}{R} = \frac{vlB}{R} \tag{33.4}$$

In this situation, the induced current is due to magnetic forces on moving charges.

STOP TO THINK 33.2 Is there an induced current in this circuit? If so, what is its direction?

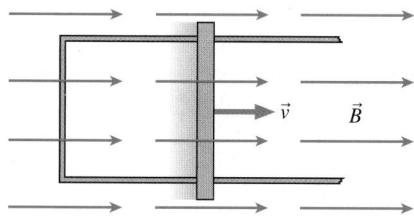

We've assumed that the wire is moving along the rail at constant speed. It turns out that we must apply a continuous pulling force $\vec{F}_{pull}$ to make this happen. Figure 33.7 shows why. The moving wire, which now carries induced current I, is in a magnetic field. You learned in Chapter 32 that a magnetic field exerts a force on a current-carrying wire. According to the right-hand rule, the magnetic force $\vec{F}_{mag}$ on the moving wire points to the left. This "magnetic drag" will cause the wire to slow down and stop *unless* we exert an equal but opposite pulling force $\vec{F}_{pull}$ to keep the wire moving.

NOTE ▶ Think about this carefully. As the wire moves to the right, the magnetic force $\vec{F}_B$ pushes the charge carriers *parallel* to the wire. Their motion, as they continue around the circuit, is the induced current I. Now, because we have a current, a second magnetic force $\vec{F}_{mag}$ enters the picture. This force on the current is *perpendicular* to the wire and acts to slow the wire's motion. ◀

The magnitude of the magnetic force on a current-carrying wire was found in Chapter 32 to be $F_{mag} = IlB$. Using that result, along with Equation 33.4 for the induced current, we find that the force required to pull the wire with a constant speed v is

$$F_{pull} = F_{mag} = IlB = \left(\frac{vlB}{R}\right)lB = \frac{vl^2B^2}{R} \tag{33.5}$$

A pulling force to the right must balance the magnetic force to keep the wire moving at constant speed. This force does work on the wire.

The induced current flows through the moving wire.

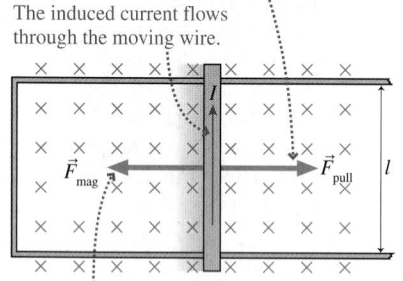

$\vec{F}_{mag}$

$\vec{F}_{pull}$

l

The magnetic force on the current-carrying wire is opposite the motion.

FIGURE 33.7 A pulling force is needed to move the wire to the right.

Energy Considerations

The environment must do work on the wire to pull it. What happens to the energy transferred to the wire by this work? Is energy conserved as the wire moves along the rail? It will be easier to answer this question if we think about power rather than work. Power is the *rate* at which work is done on the wire. You learned in Chapter 11 that the power exerted by a force pushing or pulling an object with velocity v is $P = Fv$. The power provided to the circuit by pulling on the wire is

$$P_{input} = F_{pull}v = \frac{v^2l^2B^2}{R} \tag{33.6}$$

This is the rate at which energy is added to the circuit by the pulling force.

But the circuit also dissipates energy by transforming electric energy into the thermal energy of the wires and components, heating them up. You learned in Chapter 31 that the power dissipated by current I as it passes through resistance R is $P = I^2R$. Equation 33.4 for the induced current I gives us the power dissipated by the circuit of Figure 33.6:

$$P_{dissipated} = I^2R = \frac{v^2l^2B^2}{R} \tag{33.7}$$

You can see that Equations 33.6 and 33.7 are identical. **The rate at which work is done on the circuit exactly balances the rate at which energy is dissipated.** Thus *energy is conserved.*

If you have to *pull* on the wire to get it to move to the right, you might think that it would spring back to the left on its own. Figure 33.8 shows the same circuit with the wire moving to the left. In this case, you must *push* the wire to the left to keep it moving. The magnetic force is always opposite to the wire's direction of motion.

In both Figure 33.7, where the wire is pulled, and Figure 33.8, where it is pushed, a mechanical force is used to create a current. In other words, we have a conversion of *mechanical* energy to *electric* energy. A device that converts mechanical energy to electric energy is called a **generator.** The slide-wire circuits of Figure 33.7 and 33.8 are simple examples of a generator. We will look at more practical examples of generators later in the chapter.

We can summarize our analysis as follows:

1. Pulling or pushing the wire through the magnetic field at speed v creates a motional emf $\mathcal{E}$ in the wire and induces a current $I = \mathcal{E}/R$ in the circuit.
2. To keep the wire moving at constant speed, a pulling or pushing force must balance the magnetic force on the wire. This force does work on the circuit.
3. The work done by the pulling or pushing force exactly balances the energy dissipated by the current as it passes through the resistance of the circuit.

1. The magnetic force on the charge carriers is down, so the induced current flows clockwise.

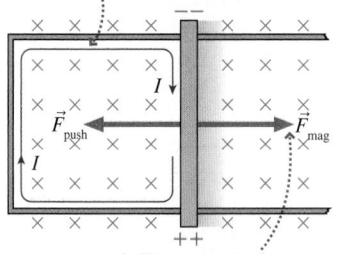

2. The magnetic force on the current-carrying wire is to the right.

FIGURE 33.8 A pushing force is needed to move the wire to the left.

EXAMPLE 33.3 Lighting a bulb

Figure 33.9 shows a circuit consisting of a flashlight bulb, rated 3.0 V/1.5 W, and ideal wires with no resistance. The right wire of the circuit, which is 10 cm long, is pulled at constant speed v through a perpendicular magnetic field of strength 0.10 T.

a. What speed must the wire have to light the bulb to full brightness?

b. What force is needed to keep the wire moving?

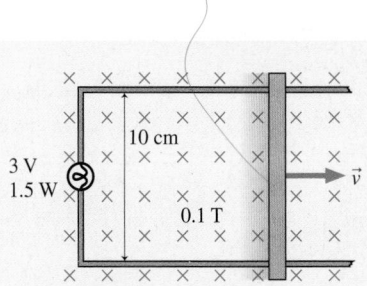

FIGURE 33.9 Circuit of Example 33.3.

MODEL Treat the moving wire as a source of motional emf.

VISUALIZE The direction of the magnetic force on the charge carriers, $\vec{F}_B = q\vec{v} \times \vec{B}$, will cause a counterclockwise (ccw) induced current.

SOLVE

a. The bulb's rating of 3.0 V/1.5 W means that at full brightness it will dissipate 1.5 W at a potential difference of 3.0 V. Because the power is related to the voltage and current by $P = I\Delta V$, the current causing full brightness is

$$I = \frac{P}{\Delta V} = \frac{1.5 \text{ W}}{3.0 \text{ V}} = 0.50 \text{ A}$$

The bulb's resistance, which is the total resistance of the circuit, is

$$R = \frac{\Delta V}{I} = \frac{3.0 \text{ V}}{0.50 \text{ A}} = 6.0 \text{ } \Omega$$

Equation 33.5 gives the speed needed to induce this current:

$$v = \frac{IR}{lB} = \frac{(0.50 \text{ A})(6.0 \text{ } \Omega)}{(0.10 \text{ m})(0.10 \text{ T})} = 300 \text{ m/s}$$

You can confirm from Equation 33.6 that the input power at this speed is 1.5 W.

b. From Equation 33.5, the pulling force must be

$$F_{\text{pull}} = \frac{vl^2B^2}{R} = 5.0 \times 10^{-3} \text{ N}$$

You can also obtain this result from $F_{\text{pull}} = P/v$.

ASSESS Example 33.1 showed that high speeds are needed to produce significant potential difference. Thus 300 m/s is not surprising. The pulling force is not very large, but even a small force can deliver large amounts of power $P = Fv$ when v is large.

Eddy Currents

Figure 33.10 shows a *rigid* square loop of wire between the poles of a magnet. The upper pole is a north pole, so the magnetic field points downward and is confined to the region between the poles. The magnetic field in Figure 33.10a passes through the loop, but the wires are not in the field. None of the charge carriers in the wire experience a magnetic force, so there is no induced current and it takes no force to pull the loop to the right.

But when the left edge of the loop enters the field, as shown in Figure 33.10b, the magnetic force on the charge carriers induces a current in the loop. The magnetic field then exerts a retarding magnetic force on this current, so **a pulling force must be exerted to pull the loop out of the magnetic field.** Note that the wire, typically copper, is *not* a magnetic material. A piece of the wire held near the magnet would feel no force. Nor would a force be required to pull the wire out if there were a gap in the loop, breaking the circuit and preventing a current. It is the *induced current* in the complete loop that causes the wire to experience a retarding force.

Figure 33.11 is an alternative way of viewing the situation. Pulling the loop out of the field is like pulling a magnet off the refrigerator door. Regardless of which way you look at it, a force is required to pull the loop out of the magnetic field.

(a)

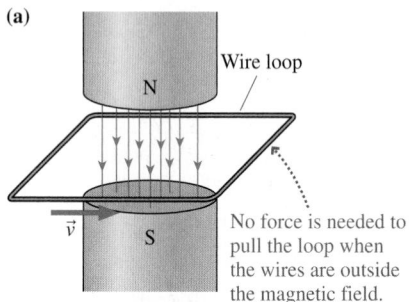

No force is needed to pull the loop when the wires are outside the magnetic field.

(b)

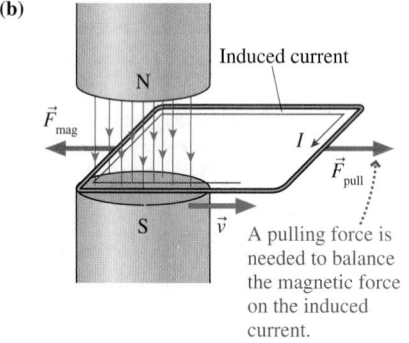

A pulling force is needed to balance the magnetic force on the induced current.

FIGURE 33.10 Pulling a loop of wire out of a magnetic field.

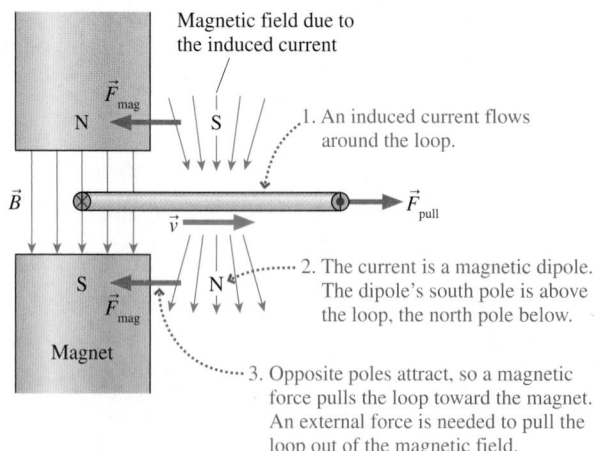

Magnetic field due to the induced current

1. An induced current flows around the loop.

2. The current is a magnetic dipole. The dipole's south pole is above the loop, the north pole below.

3. Opposite poles attract, so a magnetic force pulls the loop toward the magnet. An external force is needed to pull the loop out of the magnetic field.

FIGURE 33.11 Another way to think about pulling a loop out of a magnetic field.

(a) Eddy currents are induced when a metal sheet is pulled through a magnetic field.

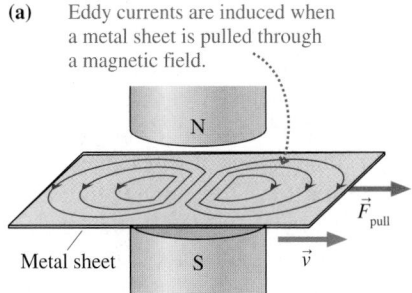

Metal sheet

(b) The magnetic force on the eddy currents is opposite in direction to $\vec{v}$.

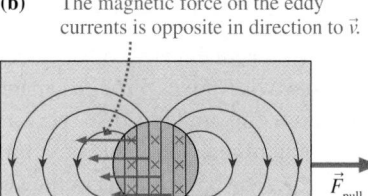

Metal sheet

Region between the permanent magnet's poles

FIGURE 33.12 Eddy currents.

These ideas have interesting implications. Consider pulling a *sheet* of metal through a magnetic field, as shown in Figure 33.12a. The metal, we will assume, is not a magnetic material, so it experiences no magnetic force if it is at rest. The charge carriers in the metal experience a magnetic force as the sheet is dragged between the pole tips of the magnet. A current is induced, just as in the loop of wire, but here the currents do not have wires to define their path. As a consequence, two "whirlpools" of current begin to circulate in the metal. These spread-out current whirlpools in a solid metal are called **eddy currents.**

Figure 33.12b shows the situation if we look down from the north pole of the magnet toward the south pole. There is a magnetic force on the eddy current as it passes between the pole tips. This force is to the left, acting as a retarding force. Thus **an external force is required to pull a metal through a magnetic field.** If the pulling force ceases, the retarding magnetic force quickly causes the metal to decelerate until it stops.

Eddy currents are often undesirable. The power dissipation of eddy currents can cause unwanted heating, and the magnetic forces on eddy currents means that extra energy must be expended to move metals in magnetic fields. But eddy currents also have important useful applications. A good example is magnetic braking, which is used in some trains and transit-system vehicles.

The moving train car has an electromagnet that straddles the rail, as shown in Figure 33.13. During normal travel, there is no current through the electromagnet and no magnetic field. To stop the car, a current is switched into the electromagnet. The current creates a strong magnetic field that passes *through* the rail, and the motion of the rail relative to the magnet induces eddy currents in the rail. The magnetic force between the electromagnet and the eddy currents acts as a braking force on the magnet and, thus, on the car. Magnetic braking systems are very efficient, and they have the added advantage that they heat the rail rather than the brakes.

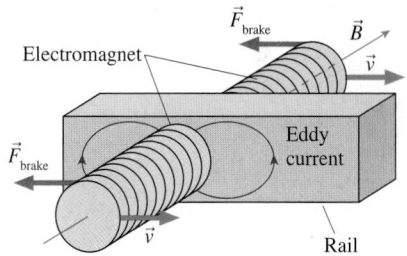

FIGURE 33.13 Magnetic braking systems are an application of eddy currents.

STOP TO THINK 33.3 A square loop of copper wire is pulled through a region of magnetic field. Rank in order, from strongest to weakest, the pulling forces $\vec{F}_1$, $\vec{F}_2$, $\vec{F}_3$, and $\vec{F}_4$ that must be applied to keep the loop moving at constant speed.

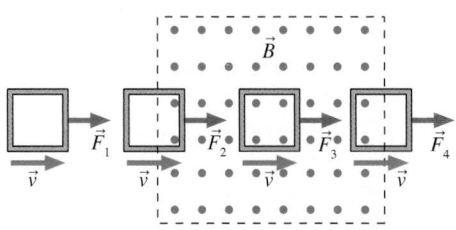

33.3 Magnetic Flux

We've begun our exploration of electromagnetic induction by analyzing a circuit in which one wire moves through a magnetic field. You might be wondering what this has to do with Faraday's discovery. Faraday found that a current is induced when the amount of magnetic field passing through a coil or a loop of wire changes. But that's exactly what happens as the slide wire moves down the rail in Figure 33.6! As the circuit expands, more magnetic field passes through. It's time to define more clearly what we mean by "the amount of field passing through a loop."

Imagine holding a rectangular loop of wire in front of a fan, as shown in Figure 33.14. The amount of air that flows through the loop depends on the effective area of the loop as seen along the direction of flow. You can see from the figure that the effective area (i.e., as seen facing the fan) is

$$A_{\text{eff}} = ab\cos\theta = A\cos\theta \qquad (33.8)$$

where A is the area of the loop and θ is the tilt angle of the loop. A loop perpendicular to the flow, with $\theta = 0°$, has $A_{\text{eff}} = A$, the full area of the loop. This is the orientation for maximum flow through the loop. No air at all flows through the loop if it is tilted 90°, and you can see that $A_{\text{eff}} = 0$ in this case.

FIGURE 33.14 The amount of air flowing through a loop depends on the effective area of the loop.

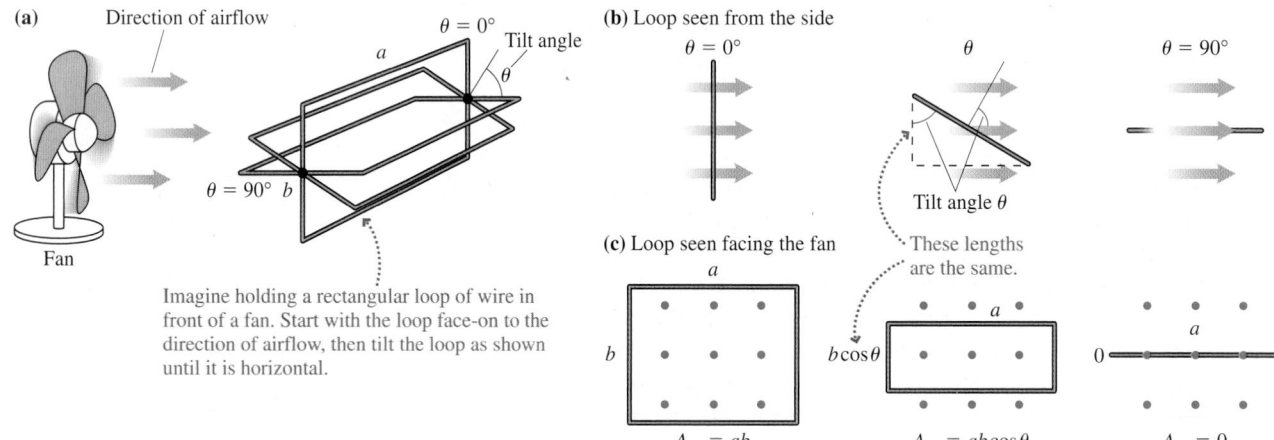

We can apply this idea to a magnetic field passing through a loop. Figure 33.15 shows a loop of area $A = ab$ in a uniform magnetic field. Think of these field vectors, seen here from behind, as if they were arrows shot into the page. The density of arrows (arrows per m²) is proportional to the strength B of the magnetic field; a stronger field would be represented by arrows spaced closer together. The number of arrows passing through a loop of wire depends on two factors:

1. The density of arrows, which is proportional to B, and
2. The effective area $A_{\text{eff}} = A\cos\theta$ of the loop.

The angle θ is the angle between the magnetic field and the axis of the loop. The maximum number of arrows passes through the loop when it is perpendicular to the magnetic field ($\theta = 0°$). No arrows pass through the loop if it is tilted 90°.

With this in mind, let's define the **magnetic flux** Φ_{m} as

$$\Phi_{\text{m}} = A_{\text{eff}}B = AB\cos\theta \qquad (33.9)$$

The magnetic flux measures the amount of magnetic field passing through a loop of area A if the loop is tilted at angle θ from the field. The SI unit of magnetic flux is the **weber.** From Equation 33.9 you can see that

$$1 \text{ weber} = 1 \text{ Wb} = 1 \text{ T m}^2$$

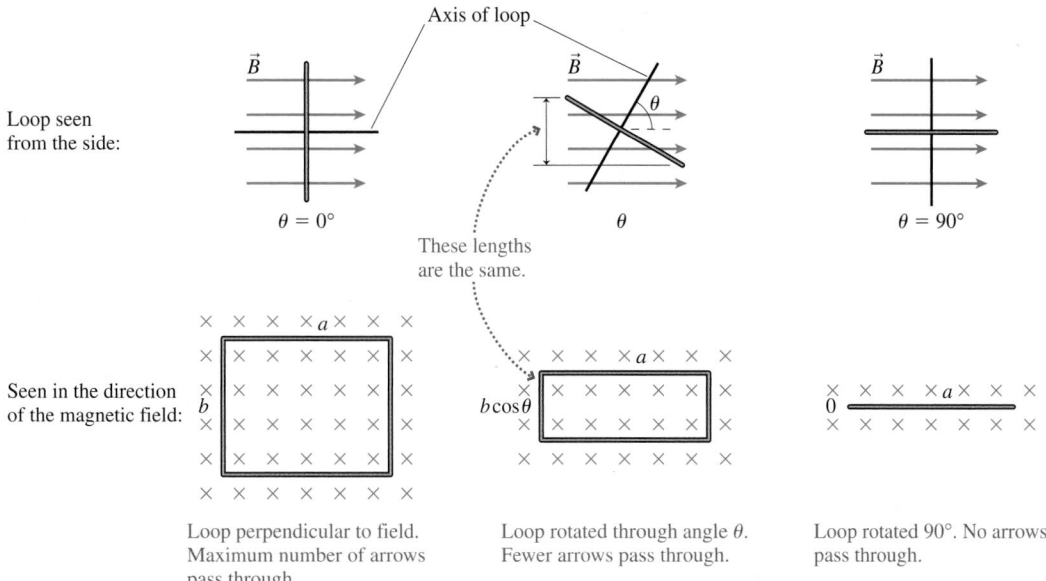

FIGURE 33.15 Magnetic field through a loop that is tilted at various angles.

Equation 33.9 is reminiscent of the vector dot product: $\vec{A} \cdot \vec{B} = AB \cos \theta$. With that in mind, let's define an **area vector** $\vec{A}$ to be a vector that is *perpendicular* to a loop and whose magnitude is equal to the area A of the loop. Vector $\vec{A}$ has units of m^2. Figure 33.16a shows the area vector $\vec{A}$ for a circular loop of area A.

Figure 33.16b shows a magnetic field passing through a loop. The angle between vectors $\vec{A}$ and $\vec{B}$ is the same angle used in Equations 33.8 and 33.9 to define the effective area and the magnetic flux. So Equation 33.9 really is a dot product, and we can define the magnetic flux more concisely as

$$\Phi_m = \vec{A} \cdot \vec{B} \tag{33.10}$$

Writing the flux as a dot product helps make clear how angle θ is defined: θ is the angle between the magnetic field and a line *perpendicular* to the plane of the loop.

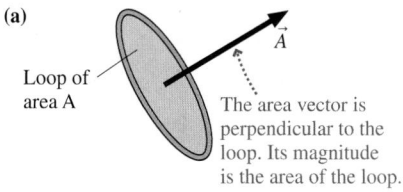

(a)
Loop of area A

The area vector is perpendicular to the loop. Its magnitude is the area of the loop.

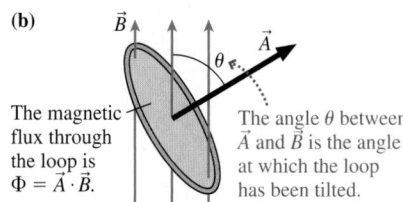

(b)

The magnetic flux through the loop is $\Phi = \vec{A} \cdot \vec{B}$.

The angle θ between $\vec{A}$ and $\vec{B}$ is the angle at which the loop has been tilted.

FIGURE 33.16 Magnetic flux can be defined in terms of an area vector $\vec{A}$.

EXAMPLE 33.4 A circular loop rotating in a magnetic field

Figure 33.17 is an edge view of a 10-cm-diameter circular loop rotating in a uniform 0.050 T magnetic field. What is the magnetic flux through the loop when θ is 0°, 30°, 60°, and 90°?

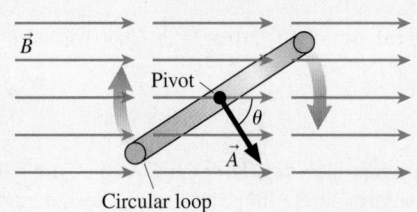

FIGURE 33.17 A circular loop in a magnetic field.

SOLVE Angle θ is the angle between the loop's area vector $\vec{A}$, which is perpendicular to the plane of the loop, and the magnetic field $\vec{B}$. Vector $\vec{A}$ has magnitude $A = \pi r^2 = 7.85 \times 10^{-3} \ m^2$. Thus the magnetic flux is

$$\Phi_m = \vec{A} \cdot \vec{B} = AB \cos \theta = \begin{cases} 3.93 \times 10^{-4} \ \text{Wb} & \theta = 0° \\ 3.40 \times 10^{-4} \ \text{Wb} & \theta = 30° \\ 1.96 \times 10^{-4} \ \text{Wb} & \theta = 60° \\ 0 \ \text{Wb} & \theta = 90° \end{cases}$$

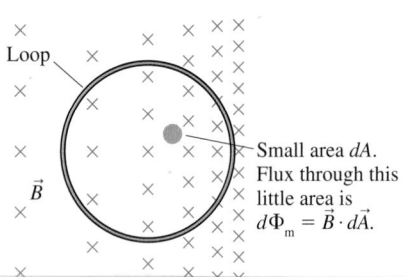

FIGURE 33.18 A loop in a nonuniform magnetic field.

Magnetic Flux in a Nonuniform Field

Equation 33.10 for the magnetic flux assumes that the field is uniform over the area of the loop. We can calculate the flux in a nonuniform field, one where the field strength changes from one edge of the loop to the other, but we'll need to use calculus.

Figure 33.18 shows a loop in a nonuniform magnetic field. Imagine dividing the loop into many small pieces of area dA. The infinitesimal flux $d\Phi_m$ through one such area, where the magnetic field is $\vec{B}$, is

$$d\Phi_m = \vec{B} \cdot d\vec{A} \qquad (33.11)$$

The total magnetic flux through the loop is the sum of the fluxes through each of the small areas. We find that sum by integrating. Thus the total magnetic flux through the loop is

$$\Phi_m = \int_{\text{area of loop}} \vec{B} \cdot d\vec{A} \qquad (33.12)$$

Equation 33.12 is a more general definition of magnetic flux. It may look rather formidable, so we'll illustrate its use with an example.

EXAMPLE 33.5 Magnetic flux from the current in a long straight wire

The 1.0 cm × 4.0 cm rectangular loop of Figure 33.19a is 1.0 cm away from a long straight wire. The wire carries a current of 1.0 A. What is the magnetic flux through the loop?

MODEL We'll treat the wire as if it were infinitely long. The magnetic field strength of a wire decreases with distance from the wire, so the field is *not* uniform over the area of the loop.

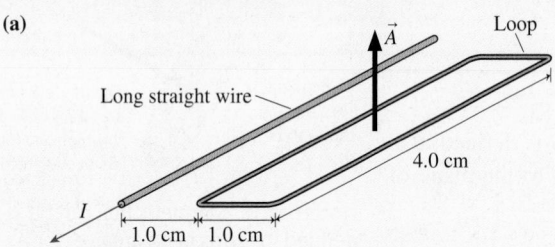

(a)

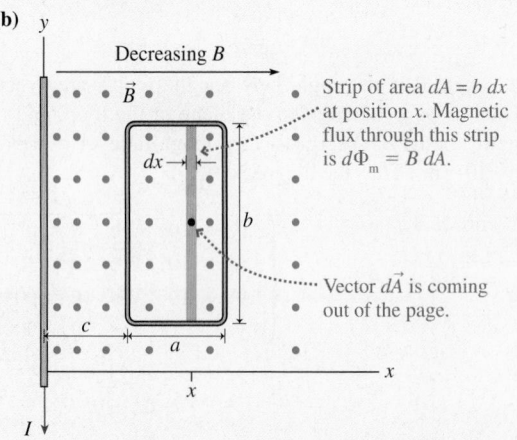

(b)

FIGURE 33.19 Magnetic flux through a loop due to the magnetic field of a long straight wire.

VISUALIZE Using the right-hand rule, we see that the field, as it circles the wire, is perpendicular to the plane of the loop. Figure 33.19b redraws the loop with the field coming out of the page and establishes a coordinate system.

SOLVE Let the loop have dimensions a and b, as shown, with the near edge at distance c from the wire. The magnetic field varies with distance x from the wire, but the field is constant along a line parallel to the wire. This suggests dividing the loop into many narrow rectangular strips of length b and width dx, each forming a small area $dA = b\,dx$. The magnetic field has the same strength at all points within this small area. One such strip is shown in the figure at position x.

The area vector $d\vec{A}$ is perpendicular to the strip (coming out of the page), which makes it parallel to $\vec{B}$ ($\theta = 0°$). Thus the infinitesimal flux through this little area is

$$d\Phi_m = \vec{B} \cdot d\vec{A} = B\,dA = Bb\,dx = \frac{\mu_0 Ib}{2\pi x}dx$$

where, from Chapter 32, we've used $B = \mu_0 I/2\pi x$ as the magnetic field at distance x from a long straight wire. Integrating "over the area of the loop" means to integrate from the near edge of the loop at $x = c$ to the far edge at $x = c + a$. Thus

$$\Phi_m = \frac{\mu_0 Ib}{2\pi}\int_c^{c+a}\frac{dx}{x} = \frac{\mu_0 Ib}{2\pi}\ln x\Big|_c^{c+a} = \frac{\mu_0 Ib}{2\pi}\ln\left(\frac{c+a}{c}\right)$$

Evaluating for $a = c = 0.010$ m, $b = 0.040$ m, and $I = 1.0$ A gives

$$\Phi_m = 5.55 \times 10^{-9}\text{ Wb}$$

ASSESS The flux measures how much of the wire's magnetic field passes through the loop, but we had to integrate, rather than simply using Equation 33.10, because the field is stronger at the near edge of the loop than at the far edge.

33.4 Lenz's Law

We started out by looking at a situation in which a moving wire caused a loop to expand in a magnetic field. This is one way to change the magnetic flux through the loop. But Faraday found that a current can be induced by any change in the magnetic flux, no matter how it's accomplished.

For example, a momentary current is induced in the loop of Figure 33.20 as the bar magnet is pushed toward the loop, increasing the flux through the loop. Pulling the magnet back out of the loop causes the current meter to deflect in the opposite direction. The conducting wires aren't moving, so this is not a motional emf. Nonetheless, the induced current is very real.

The German physicist Heinrich Lenz began to study electromagnetic induction after learning of Faraday's discovery. Three years later, in 1834, Lenz announced a rule for determining the direction of the induced current. We now call his rule **Lenz's law,** and it can be stated as follows:

> **Lenz's law** There is an induced current in a closed, conducting loop if and only if the magnetic flux through the loop is changing. The direction of the induced current is such that the induced magnetic field opposes the *change* in the flux.

Lenz's law is rather subtle, and it takes some practice to see how to apply it.

NOTE ▶ One difficulty with Lenz's law is the term *flux*. In everyday language, the word *flux* already implies that something is changing. Think of the phrase, "The situation is in flux." Not so in physics, where *flux* means "passes through." A steady magnetic field through a loop creates a steady, *un*changing magnetic flux. ◀

Lenz's law tells us to look for situations where the flux is *changing*. This can happen in three ways.

1. The magnetic field through the loop changes (increases or decreases),
2. The loop changes in area or angle, or
3. The loop moves into or out of a magnetic field.

Lenz's law depends on an idea that we hinted at in our discussion of eddy currents. If a current is induced in a loop, that current generates its own magnetic field $\vec{B}_{\text{induced}}$. This is the *induced magnetic field* of Lenz's law. You learned in Chapter 32 how to use the right-hand rule to determine the direction of this induced magnetic field.

In Figure 33.20, pushing the bar magnet into the loop causes the magnetic flux to *increase* in the downward direction. To oppose the *change* in flux, which is what Lenz's law requires, the loop itself needs to generate the *upward*-pointing magnetic field of Figure 33.21. The induced magnetic field at the center of the loop will point upward if the current is ccw. Thus pushing the north end of a bar magnet toward the loop induces a ccw current around the loop. The induced current ceases as soon as the magnet stops moving.

Now suppose the bar magnet is pulled back away from the loop, as shown in Figure 33.22a on the next page. There is a downward magnetic flux through the loop, but the flux *decreases* as the magnet moves away. According to Lenz's law, the induced magnetic field of the loop will *oppose this decrease*. To do so, the induced field needs to point in the *downward* direction, as shown in Figure 33.22b. Thus as the magnet is withdrawn, the induced current is clockwise (cw), opposite to the induced current of Figure 33.21.

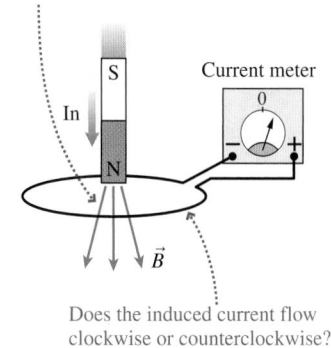

A bar magnet pushed into a loop increases the flux through the loop and induces a current to flow.

Does the induced current flow clockwise or counterclockwise?

FIGURE 33.20 Pushing a bar magnet toward the loop induces a current in the loop.

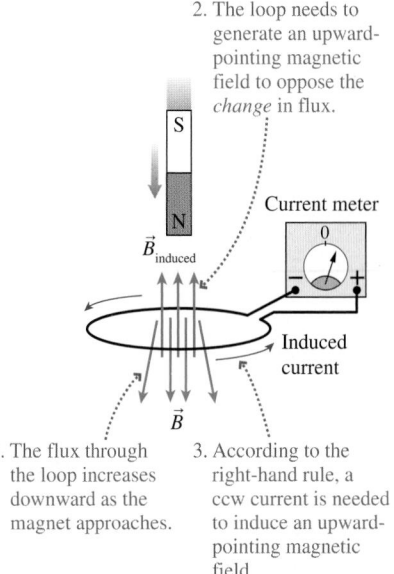

2. The loop needs to generate an upward-pointing magnetic field to oppose the *change* in flux.

1. The flux through the loop increases downward as the magnet approaches.

3. According to the right-hand rule, a ccw current is needed to induce an upward-pointing magnetic field.

FIGURE 33.21 The induced current is ccw.

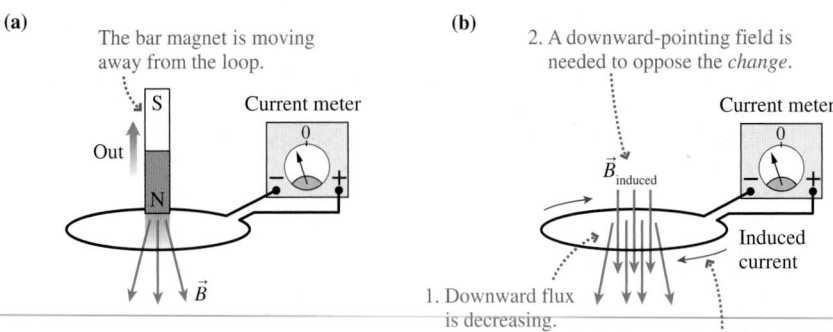

FIGURE 33.22 Pulling the magnet away induces a cw current.

NOTE ▶ Notice that the magnetic field of the bar magnet is pointing downward in both Figures 33.21 and 33.22. It is not the *flux* due to the magnet that the induced current opposes, but the *change* in the flux. This is a subtle but critical distinction. If the induced current opposed the flux itself, the current in both Figures 33.21 and 33.22 would be ccw to generate an upward magnetic field. But that's not what happens. When the field of the magnet points down and is increasing, the induced current opposes the increase by generating an upward field. When the field of the magnet points down but is decreasing, the induced current opposes the decrease by generating a downward field. ◀

Figure 33.23 shows six basic situations. The magnetic field can point either up or down through the loop. For each, the flux can either increase, hold steady, or decrease in strength. These observations form the basis for a set of rules about using Lenz's law.

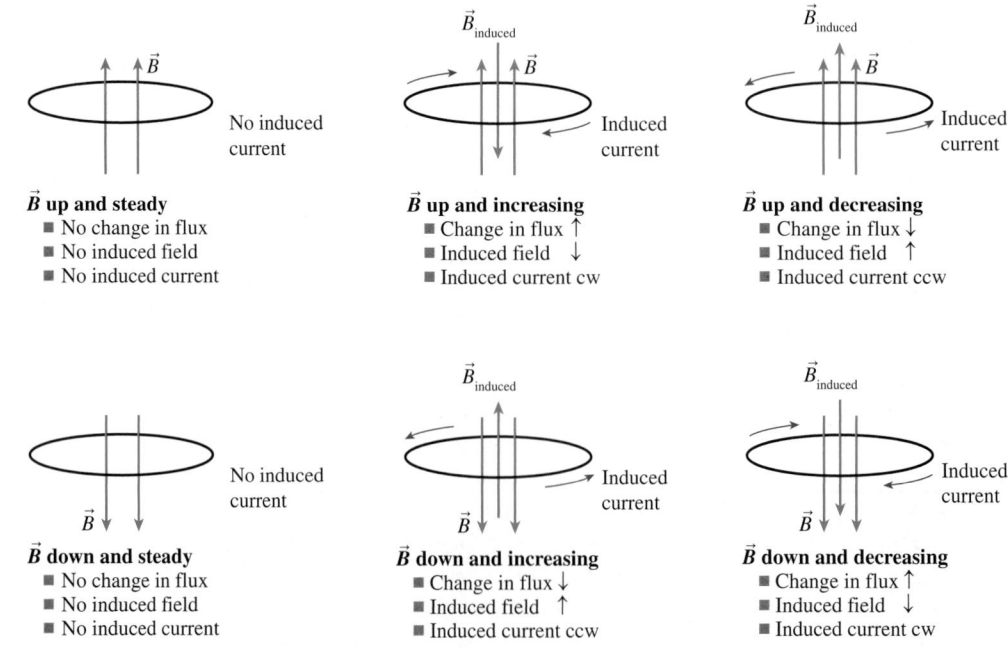

FIGURE 33.23 The induced current for six different situations.

TACTICS BOX 33.1 **Using Lenz's law**

❶ **Determine the direction of the applied magnetic field.** The field must pass through the loop.

❷ **Determine how the flux is changing.** Is it increasing, decreasing, or staying the same?

❸ **Determine the direction of an induced magnetic field that will oppose the *change* in the flux.**

- Increasing flux: the induced magnetic field points opposite the applied magnetic field.

- Decreasing flux: the induced magnetic field points in the same direction as the applied magnetic field.

- Steady flux: there is no induced magnetic field.

❹ **Determine the direction of the induced current.** Use the right-hand rule to determine the current direction in the loop that generates the induced magnetic field you found in step 3.

Let's look at some examples.

EXAMPLE 33.6 Lenz's law 1

The switch in the circuit of Figure 33.24a has been closed for a long time. What happens in the lower loop when the switch is opened?

MODEL We'll use the right-hand rule to find the magnetic fields of current loops.

SOLVE Figure 33.24b shows the four steps of using Lenz's law. Opening the switch induces a ccw current in the lower loop. This is a momentary current, lasting only until the magnetic field of the upper loop drops to zero.

ASSESS The conclusion is consistent with Figure 33.23.

(a)

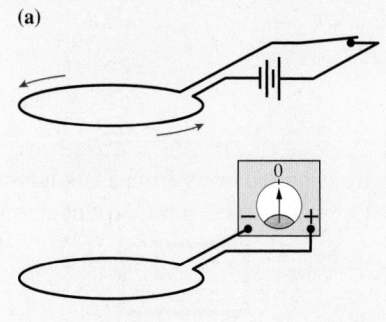

(b) ❶ The magnetic field of the upper loop points up, but it is decreasing as the current in the circuit rapidly decreases. $\vec{B}$ Switch opens.

❷ The flux through the loop is up and decreasing.

❸ The induced field needs to point upward to oppose the *change* in flux. $\vec{B}_{induced}$

Induced current

❹ A ccw current induces an upward magnetic field.

FIGURE 33.24 Circuits of Example 33.6.

EXAMPLE 33.7 Lenz's law 2

Figure 33.25a on the next page shows two solenoids facing each other. When the switch for coil 1 is closed, does the induced current in coil 2 pass from right to left or from left to right through the current meter?

MODEL We'll use the right-hand rule to find the magnetic fields of solenoids.

VISUALIZE It is very important to look at the *direction* in which a solenoid is wound around the cylinder. Notice that the two solenoids in Figure 33.25a are wound in opposite directions.

SOLVE Figure 33.25b shows the four steps of using Lenz's law. Closing the switch induces a current that passes from right to left through the current meter. The induced current is only momentary. It lasts only until the field from coil 1 reaches full strength and is no longer changing.

ASSESS The conclusion is consistent with Figure 33.23.

(a)

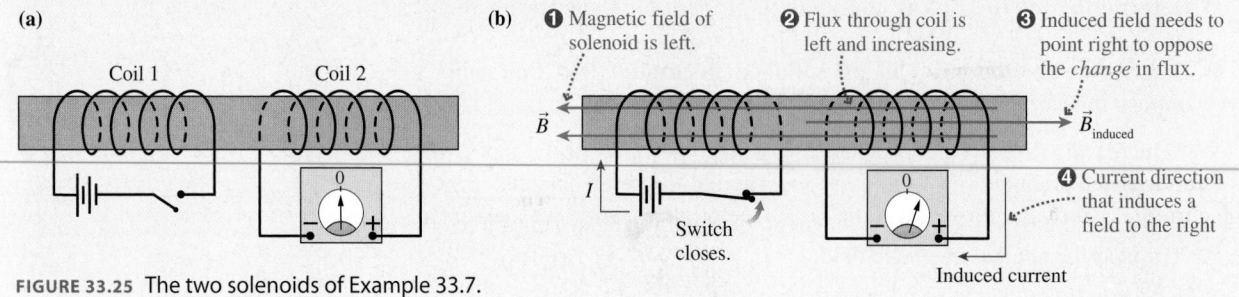

(b) ❶ Magnetic field of solenoid is left. ❷ Flux through coil is left and increasing. ❸ Induced field needs to point right to oppose the *change* in flux.

❹ Current direction that induces a field to the right

FIGURE 33.25 The two solenoids of Example 33.7.

EXAMPLE 33.8 A rotating loop

The loop of wire in Figure 33.26 was initially in the *xy*-plane, parallel to the magnetic field. It is suddenly rotated 90° about the *y*-axis until it is in the *yz*-plane, perpendicular to the magnetic field. In what direction is the induced current as the loop rotates?

SOLVE Unlike the magnetic fields in the previous examples, this magnetic field is constant and unchanging. Nonetheless, the *flux* through the loop changes as it rotates. To use Lenz's law,

1. The applied magnetic field points to the right.
2. Initially the flux is $\Phi = 0$, but after rotating the flux is $\Phi = AB$ toward the right, where A is the loop area. This is an increasing flux through the loop to the right.
3. To oppose this increase in the flux, the induced magnetic field of the loop must have an *x*-component toward the left.
4. This will be the case if an induced current is in a cw direction, as seen from the perspective of Figure 33.26.

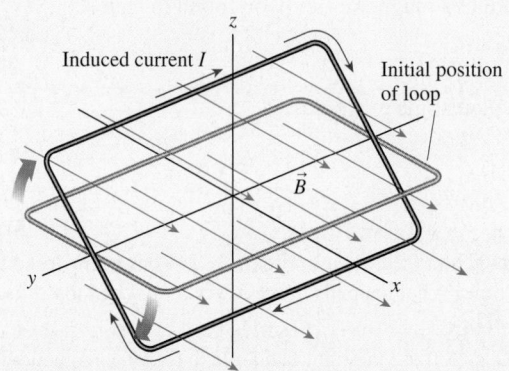

FIGURE 33.26 A current is induced in a loop as the loop rotates in a constant magnetic field.

STOP TO THINK 33.4 A current-carrying wire is pulled away from a conducting loop in the direction shown. As the wire is moving, is there a cw current around the loop, a ccw current, or no current?

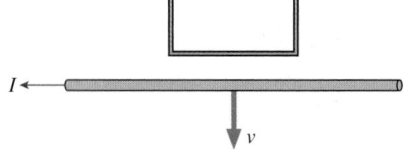

33.5 Faraday's Law

13.9, 13.10 ActivPhysics

Faraday discovered that a current is induced when the magnetic flux through a conducting loop changes. Lenz's law allows us to find the direction of the induced current. To put electromagnetic induction to practical use, we also need to know the *size* of the induced current.

Charges don't start moving spontaneously. A current requires an emf to provide the energy. We started our analysis of induced currents with circuits in which there is a *motional emf*. The motional emf can be understood in terms of magnetic forces on moving charges. But we've also seen that a current can be induced by changing the magnetic field through a stationary circuit, a circuit in which there is no motion. There *must* be an emf in this circuit, even though the mechanism for this emf is not yet clear.

The emf associated with a changing magnetic flux, regardless of what causes the change, is called an **induced emf** $\mathcal{E}$. Then, if there is a complete circuit having resistance R, a current

$$I_{\text{induced}} = \frac{\mathcal{E}}{R} \tag{33.13}$$

is established in the wire as a *consequence* of the induced emf. The direction of the current is given by Lenz's law. The last piece of information we need is the size of the induced emf $\mathcal{E}$.

The research of Faraday and others eventually led to the discovery of the basic law of electromagnetic induction, which we now call **Faraday's law.** Faraday's law is a new law of physics, not derivable from any previous laws you have studied. It states:

> **Faraday's law** An emf $\mathcal{E}$ is induced in a conducting loop if the magnetic flux through the loop changes. The magnitude of the emf is
>
> $$\mathcal{E} = \left| \frac{d\Phi_{\text{m}}}{dt} \right| \tag{33.14}$$
>
> and the direction of the emf is such as to drive an induced current in the direction given by Lenz's law.

In other words, the induced emf is the *rate of change* of the magnetic flux through the loop.

As a corollary to Faraday's law, a coil of wire consisting of N turns in a changing magnetic field acts like N batteries in series. The induced emf of each of the coils adds, so the induced emf of the entire coil is

$$\mathcal{E}_{\text{coil}} = N \left| \frac{d\Phi_{\text{per coil}}}{dt} \right| \quad \text{(Faraday's law for an } N\text{-turn coil)} \tag{33.15}$$

As a first example of using Faraday's law, return to the situation of Figure 33.6, where a wire moves through a magnetic field by sliding on a U-shaped conducting rail. Figure 33.27 shows the circuit again. The magnetic field $\vec{B}$ is perpendicular to the plane of the conducting loop, so $\theta = 0°$ and the magnetic flux is $\Phi = AB$, where A is the area of the loop. If the slide wire is distance x from the end, the area is $A = xl$ and the flux at that instant of time is

$$\Phi_{\text{m}} = AB = xlB \tag{33.16}$$

The flux through the loop increases as the wire moves. According to Faraday's law, the induced emf is

$$\mathcal{E} = \left| \frac{d\Phi_{\text{m}}}{dt} \right| = \frac{d}{dt}(xlB) = \frac{dx}{dt} lB = vlB \tag{33.17}$$

where the wire's velocity is $v = dx/dt$. We can now use Equation 33.13 to find that the induced current is

$$I = \frac{\mathcal{E}}{R} = \frac{vlB}{R} \tag{33.18}$$

The flux is increasing into the loop, so the induced magnetic field will oppose this increase by pointing out of the loop. This requires a ccw induced current in the loop. Faraday's law leads us to the conclusion that the loop will have a ccw induced

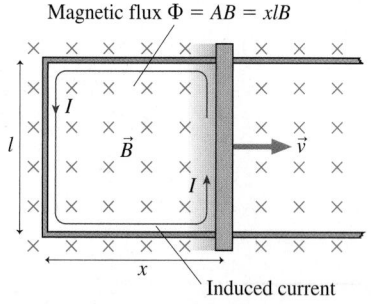

Magnetic flux $\Phi = AB = xlB$

FIGURE 33.27 The magnetic flux through the loop increases as the slide wire moves.

current $I = vlB/R$. This is exactly the conclusion we reached in Section 33.2, where we analyzed the situation from the perspective of magnetic forces on moving charge carriers. Faraday's law confirms what we already knew but, at least in this case, doesn't seem to offer anything new.

Using Faraday's Law

Most electromagnetic induction problems can be solved with a four-step strategy.

PROBLEM-SOLVING STRATEGY 33.1 Electromagnetic induction

MODEL Make simplifying assumptions about wires and magnetic fields.

VISUALIZE Draw a picture or a circuit diagram. Use Lenz's law to determine the direction of the induced current.

SOLVE The mathematical representation is based on Faraday's law

$$\mathcal{E} = \left| \frac{d\Phi_m}{dt} \right|$$

For an N-turn coil, multiply by N. The size of the induced current is $I = \mathcal{E}/R$.

ASSESS Check that your result has the correct units, is reasonable, and answers the question.

EXAMPLE 33.9 Electromagnetic induction in a circular loop

The magnetic field of Figure 33.28 decreases from 1.0 T to 0.4 T in 1.2 s. A 6.0-cm-diameter conducting loop with a resistance of 0.010 Ω is perpendicular to $\vec{B}$. What are the size and direction of the current induced in the loop?

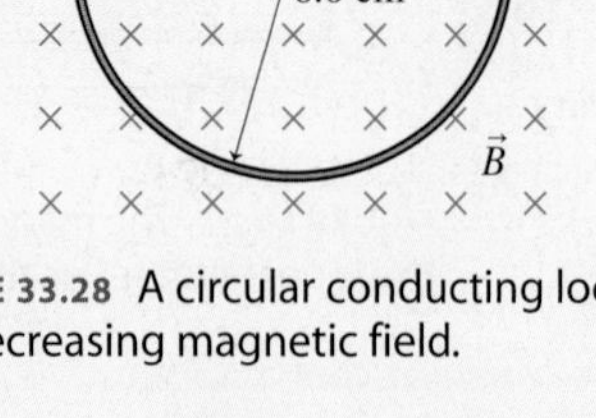

B decreases from 1.0 T to 0.4 T in 1.2 s.

$R = 0.010\ \Omega$

6.0 cm

$\vec{B}$

FIGURE 33.28 A circular conducting loop in a decreasing magnetic field.

MODEL Assume that B decreases linearly with time.

VISUALIZE The magnetic flux is into the page and decreasing. To oppose the *change* in the flux, the induced field needs to point into the page. This will be true if the induced current in the loop is cw.

SOLVE The magnetic field is perpendicular to the plane of the loop, hence $\theta = 0°$ and the magnetic flux is $\Phi_m = AB = \pi r^2 B$. The radius doesn't change with time, but B does. According to Faraday's law, the induced emf is

$$\mathcal{E} = \left| \frac{d\Phi_m}{dt} \right| = \left| \frac{d(\pi r^2 B)}{dt} \right| = \pi r^2 \left| \frac{dB}{dt} \right|$$

The *rate* at which the magnetic field changes is

$$\frac{dB}{dt} = \frac{\Delta B}{\Delta t} = \frac{-0.60\ \text{T}}{1.2\ \text{s}} = -0.50\ \text{T/s}$$

dB/dt is negative because the field is decreasing, but all we need for Faraday's law is the absolute value. Thus

$$\mathcal{E} = \pi r^2 \left| \frac{dB}{dt} \right| = \pi (0.030\ \text{m})^2 (0.50\ \text{T/s}) = 0.00141\ \text{V}$$

The current induced by this emf is

$$I = \frac{\mathcal{E}}{R} = \frac{0.00141\ \text{V}}{0.010\ \Omega} = 0.141\ \text{A}$$

The decreasing magnetic field causes a 0.141 A cw current that lasts for 1.2 s.

ASSESS We don't have much to go on for assessing the result. The emf is quite small, but, because the resistance of metal wires is also very small, the current is respectable. We know that electromagnetic induction produces currents large enough for practical applications, so this result seems plausible.

EXAMPLE 33.10 Electromagnetic induction in a solenoid

A 2.0-cm-diameter loop of wire with a resistance of 0.010 Ω is placed in the center of a solenoid. The solenoid, shown in Figure 33.29a, is 4.0 cm in diameter, 20 cm long, and wrapped with 1000 turns of wire. Figure 33.29b shows the current through the solenoid as a function of time as the solenoid is "powered up." A positive current is defined to be cw when seen from the left. Find the current in the loop as a function of time and show the result as a graph.

(a)

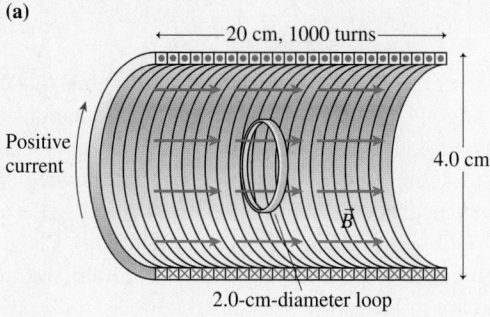

(b)

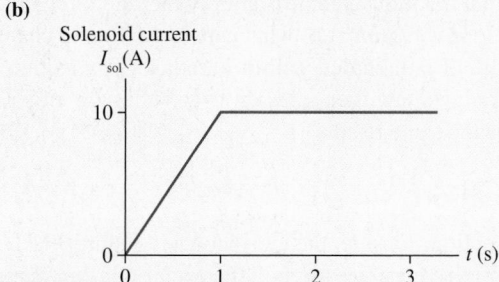

FIGURE 33.29 A loop inside a solenoid.

MODEL The solenoid's length is much greater than its diameter, so the field near the center should be nearly uniform.

VISUALIZE The magnetic field of the solenoid creates a magnetic flux through the loop of wire. The solenoid current is always positive, meaning that it is cw as seen from the left. Consequently, from the right-hand rule, the magnetic field inside the solenoid always points to the right. During the first second, while the solenoid current is increasing, the flux through the loop is to the right and increasing. To oppose the *change* in the flux, the loop's induced magnetic field must point to the left. Thus, again using the right-hand rule, the induced current must flow ccw as seen from the left. This is a *negative* current. There's no *change* in the flux for $t > 1$ s, so the induced current is zero.

SOLVE Now we're ready to use Faraday's law to find the magnitude of the current. Because the field is uniform inside the solenoid and perpendicular to the loop ($\theta = 0°$), the flux is $\Phi_m = AB$, where $A = \pi r^2 = 3.14 \times 10^{-4} \text{ m}^2$ is the area of the loop (*not* the area of the solenoid). The field of a long solenoid of length l was found in Chapter 32 to be

$$B = \frac{\mu_0 N I_{\text{sol}}}{l}$$

The flux through the loop when the solenoid current is I_{sol} is thus

$$\Phi_m = \frac{\mu_0 A N I_{\text{sol}}}{l}$$

The changing flux creates an induced emf $\mathcal{E}$ that is given by Faraday's law:

$$\mathcal{E} = \left| \frac{d\Phi_m}{dt} \right| = \frac{\mu_0 A N}{l} \left| \frac{dI_{\text{sol}}}{dt} \right| = 1.97 \times 10^{-6} \left| \frac{dI_{\text{sol}}}{dt} \right|$$

From the slope of the graph, we find

$$\left| \frac{dI_{\text{sol}}}{dt} \right| = \begin{cases} 10 \text{ A/s} & 0.0 \text{ s} < t < 1.0 \text{ s} \\ 0 & 1.0 \text{ s} < t < 3.0 \text{ s} \end{cases}$$

Thus the induced emf is

$$\mathcal{E} = \begin{cases} 1.97 \times 10^{-5} \text{ V} & 0.0 \text{ s} < t < 1.0 \text{ s} \\ 0 \text{ V} & 1.0 \text{ s} < t < 3.0 \text{ s} \end{cases}$$

Finally, the current induced in the loop is

$$I_{\text{loop}} = \frac{\mathcal{E}}{R} = \begin{cases} -1.97 \text{ mA} & 0.0 \text{ s} < t < 1.0 \text{ s} \\ 0 \text{ mA} & 1.0 \text{ s} < t < 3.0 \text{ s} \end{cases}$$

where the negative sign comes from Lenz's law. This result is shown in Figure 33.30.

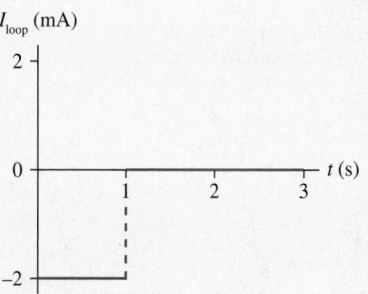

FIGURE 33.30 The induced current in the loop.

What Does Faraday's Law Tell Us?

The induced current in the slide-wire circuit of Figure 33.27 can be understood as a motional emf due to magnetic forces on moving charges. We had not anticipated this kind of current in Chapter 32, but it takes no new laws of physics to understand it.

The induced currents in Examples 33.9 and 33.10 are different. We cannot explain or predict these induced currents on the basis of previous laws or principles. This is new physics.

Faraday recognized that all induced currents are associated with a changing magnetic flux. There are two fundamentally different ways to change the magnetic flux through a conducting loop:

1. The loop can move or expand or rotate, creating a motional emf.
2. The magnetic field can change.

We can see both of these if we write Faraday's law as

$$\mathcal{E} = \left| \frac{d\Phi_m}{dt} \right| = \left| \vec{B} \cdot \frac{d\vec{A}}{dt} + \vec{A} \cdot \frac{d\vec{B}}{dt} \right| \tag{33.19}$$

The first term on the right side represents a motional emf. The magnetic flux changes because the loop itself is changing. This term includes not only situations like the slide-wire circuit, where the area A changes, but also loops that rotate in a magnetic field. The physical area of a rotating loop does not change, but the area *vector* $\vec{A}$ does. The loop's motion causes magnetic forces on the charge carriers in the loop.

The second term on the right side is the new physics in Faraday's law. It says that an emf can also be created simply by changing a magnetic field, even if nothing is moving. This was the case in Examples 33.9 and 33.10.

Faraday's law tells us that the induced emf is simply the rate of change of the magnetic flux through the loop, *regardless* of what causes the flux to change. The "old physics" of motional emf is included within Faraday's law as one way of changing the flux, but Faraday's law then goes on to say that any other way of changing the flux will have the same result.

An Unanswered Question

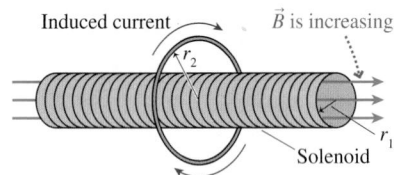

Induced current $\vec{B}$ is increasing.

r_2

Solenoid

r_1

FIGURE 33.31 A changing current in the solenoid induces a current in the loop.

As a final example in this section, consider the loop shown in Figure 33.31. A long, tightly wound solenoid of radius r_1 passes through the center of a conducting loop having a larger radius r_2. What happens to the loop if the solenoid current changes?

You learned in Chapter 32 that the magnetic field is strong inside the solenoid but, if the solenoid is sufficiently long, essentially zero outside the solenoid. Even so, changing the current through the solenoid *does* cause an induced current in the loop. The solenoid's magnetic field establishes a flux through the loop, and changing the solenoid current causes the magnetic flux to change. This is the essence of Faraday's law.

But the loop is completely outside the solenoid. How can the charge carriers in the conducting loop possibly know that the magnetic field inside the solenoid is changing? How do they know which way to move?

In the case of a motional emf, the *mechanism* that causes an induced current is the magnetic force on the moving charges. But here, where there's no motion, what is the mechanism that creates a current when the magnetic flux changes? This is an important question, one that we will answer in the next section.

STOP TO THINK 33.5 A conducting loop is halfway into a magnetic field. Suppose the magnetic field begins to increase rapidly in strength.
What happens to the loop?

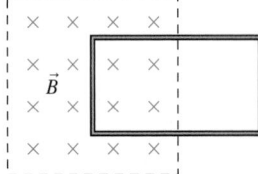

$\vec{B}$

a. The loop is pushed upward, toward the top of the page.
b. The loop is pushed downward, toward the bottom of the page.
c. The loop is pulled to the left, into the magnetic field.
d. The loop is pushed to the right, out of the magnetic field.
e. The tension in the wires increases but the loop does not move.

33.6 Induced Fields and Electromagnetic Waves

Faraday's law is a tool for calculating the strength of an induced current, but one important piece of the puzzle is still missing. What *causes* the current? That is, what *force* pushes the charges around the loop against the resistive forces of the metal?

The only agents that exert forces on charges are electric fields and magnetic fields. Magnetic forces are responsible for motional emfs, but magnetic forces cannot explain the current induced in a *stationary* loop by a changing magnetic field.

Figure 33.32a shows a conducting loop in an increasing magnetic field. According to Lenz's law, there is an induced current in the ccw direction. But something has to act on the charge carriers to make them move, so we infer that there must be an *electric* field tangent to the loop at all points. This electric field is *caused* by the changing magnetic field and is called an **induced electric field.** The induced electric field is the *mechanism* we were seeking that creates a current when there's a changing magnetic field inside a stationary loop.

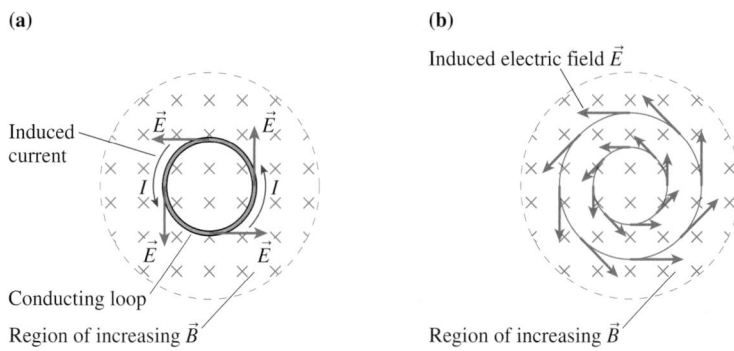

(a) **(b)**

Induced electric field $\vec{E}$

Induced current

I $\quad I$

Conducting loop

Region of increasing $\vec{B}$ Region of increasing $\vec{B}$

FIGURE 33.32 An induced electric field creates a current in the loop.

The conducting loop isn't necessary. The space in which the magnetic field is changing is filled with the pinwheel pattern of induced electric fields shown in Figure 33.32b. Charges will move if a conducting path is present, but the induced electric field is there as a direct consequence of the changing magnetic field.

But this is a rather peculiar electric field. All the electric fields we have examined until now have been created by charges. Electric field vectors pointed away from positive charges and toward negative charges. An electric field created by charges is called a **Coulomb electric field.** The induced electric field of Figure 33.32b is caused not by charges but by a changing magnetic field. It is called a **non-Coulomb electric field.**

So it appears that there are two different ways to create an electric field:

1. A Coulomb electric field is created by positive and negative charges.
2. A non-Coulomb electric field is created by a changing magnetic field.

Both exert a force $\vec{F} = q\vec{E}$ on a charge, and both create a current in a conductor. However, the origins of the fields are very different. Figure 33.33 is a quick summary of the two ways to create an electric field.

We first introduced the idea of a field as a way of thinking about how two charges exert long-range forces on each other through the emptiness of space. The field may have seemed like a useful pictorial representation of charge interactions, but we had little evidence that fields are *real,* that they actually exist. Now we do. The electric field has shown up in a completely different context, independent of charges, as the explanation of the very real existence of induced currents.

The electric field is not just a pictorial representation; it is real.

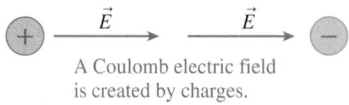

$\vec{E}$ $\qquad$ $\vec{E}$

A Coulomb electric field is created by charges.

$\vec{B}$ increasing or decreasing

$\vec{E}$ $\qquad$ $\vec{E}$

A non-Coulomb electric field is created by a changing magnetic field.

FIGURE 33.33 Two ways to create an electric field.

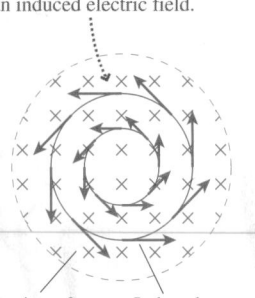

A changing magnetic field creates an induced electric field.

Region of increasing $\vec{B}$ Induced electric field E

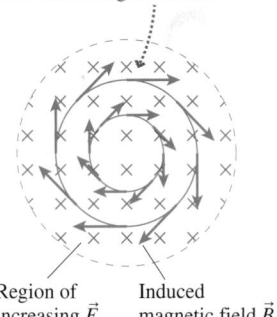

A changing electric field creates an induced magnetic field.

Region of increasing $\vec{E}$ Induced magnetic field $\vec{B}$

FIGURE 33.34 Maxwell hypothesized the existence of induced magnetic fields.

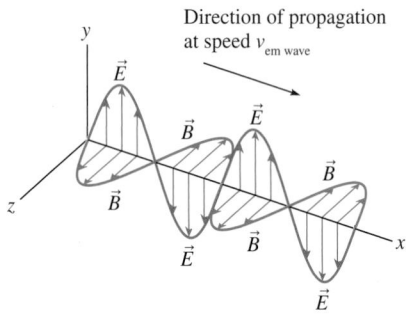

FIGURE 33.35 A self-sustaining electromagnetic wave.

The velocity of transverse undulations in our hypothetical medium, calculated from the electromagnetic experiments of Kohlrausch and Weber [who had measured ϵ_0 and μ_0], agrees so exactly with the velocity of light calculated from the optical experiments of Fizeau that we can scarcely avoid the inference that light consists of the transverse undulations of the same medium which is the cause of electric and magnetic phenomena.

James Clerk Maxwell

Maxwell's Theory

Faraday's field concept was capable of explaining the phenomena of electricity and magnetism as they were known in the 1830s and 1840s. But Faraday, despite his intuitive genius, lacked the mathematical skills to develop a true *theory* of electric and magnetic fields. It was not easy to predict new phenomena or develop applications without a theory.

In 1855, less than two years after receiving his undergraduate degree, the English physicist James Clerk Maxwell presented a paper titled "On Faraday's Lines of Force." In this paper, he began to sketch out how Faraday's pictorial ideas about fields could be given a rigorous mathematical basis. Maxwell then spent the next 10 years developing the mathematical theory of electromagnetism.

Maxwell was troubled by a certain lack of symmetry. Faraday had found that a changing magnetic field creates an induced electric field, a non-Coulomb electric field not tied to charges. But what, Maxwell began to wonder, about a changing *electric* field?

To complete the symmetry, Maxwell proposed that a changing electric field creates an **induced magnetic field,** a new kind of magnetic field not tied to the existence of currents. Figure 33.34 shows a region of space where the *electric* field is increasing. This region of space, according to Maxwell, is filled with a pinwheel pattern of induced magnetic fields. The induced magnetic field looks like the induced electric field, with $\vec{E}$ and $\vec{B}$ interchanged, except that—for technical reasons—the induced $\vec{B}$ points the opposite way from the induced $\vec{E}$. Although there was no experimental evidence that induced magnetic fields existed, Maxwell went ahead and included them in his electromagnetic field theory. This was an inspired hunch, soon to be vindicated.

Maxwell soon realized that it might be possible to establish self-sustaining electric and magnetic fields that would be entirely independent of any charges or currents. That is, a changing electric field $\vec{E}$ creates a magnetic field $\vec{B}$, which then changes in just the right way to recreate the electric field, which then changes in just the right way to again recreate the magnetic field, and so on. The fields are continually recreated through electromagnetic induction without any reliance on charges or currents.

The mathematics of Maxwell's theory is difficult, but eventually Maxwell was able to predict that electric and magnetic fields would be able to sustain themselves, free from charges and currents, if they took the form of an **electromagnetic wave.** The wave would have to have a very specific geometry, shown in Figure 33.35, in which $\vec{E}$ and $\vec{B}$ are perpendicular to each other as well as perpendicular to the direction of travel. That is, an electromagnetic wave would be a *transverse* wave.

Furthermore, Maxwell's theory predicted that the wave would travel with speed

$$v_{\text{em wave}} = \frac{1}{\sqrt{\epsilon_0 \mu_0}} \qquad (33.20)$$

where ϵ_0 is the permittivity constant from Coulomb's law and μ_0 is the permeability constant from the law of Biot and Savart. Maxwell computed that an electromagnetic wave, if it existed, would travel with speed $v_{\text{em wave}} = 3.00 \times 10^8$ m/s.

We don't know Maxwell's immediate reaction, but it must have been both shock and excitement. His predicted speed for electromagnetic waves, a prediction that came directly from his theory, was none other than the speed of light! This agreement could be just a coincidence, but Maxwell didn't think so. Making a bold leap of imagination, Maxwell concluded that **light is an electromagnetic wave.**

It took 25 more years for Maxwell's predictions to be tested. In 1886, the German physicist Heinrich Hertz discovered how to generate and transmit radio waves. Two years later, in 1888, he was able to show that radio waves travel at the speed of light. Maxwell, unfortunately, did not live to see his triumph. He had died in 1879, at the age of 48.

Chapter 34 will develop some of the mathematical details of Maxwell's theory and show how the ideas contained in Faraday's law lead to electromagnetic waves.

33.7 Induced Currents: Three Applications

There are many applications of Faraday's law and induced currents in modern technology. In this section we will look at three: generators, transformers, and metal detectors.

A generator inside a hydroelectric dam uses electromagnetic induction to convert the mechanical energy of a spinning turbine into electric energy.

Generators

We noted in Section 33.2 that a slide wire pulled through a magnetic field on a U-shaped track is a simple generator because it transforms mechanical energy into electric energy. Figure 33.36 shows a more practical generator. Here a coil of wire rotates in a magnetic field. Both the field and the area of the loop are constant, but the magnetic flux through the loop changes continuously as the loop rotates. The induced current is removed from the rotating loop by *brushes* that press up against rotating *slip rings*.

The flux through the coil is

$$\Phi_{\text{m}} = \vec{A} \cdot \vec{B} = AB\cos\theta = AB\cos\omega t \tag{33.21}$$

where ω is the angular frequency ($\omega = 2\pi f$) with which the coil rotates. The induced emf is given by Faraday's law,

$$\mathcal{E}_{\text{coil}} = N\frac{d\Phi_{\text{m}}}{dt} = ABN\frac{d}{dt}(\cos\omega t) = -\omega ABN\sin\omega t \tag{33.22}$$

where N is the number of turns on the coil. We've dropped the absolute value signs to demonstrate that the sign of $\mathcal{E}_{\text{coil}}$ alternates between positive and negative.

Because the emf alternates in sign, the current through resistor R alternates back and forth in direction. Hence the generator of Figure 33.36 is an alternating-current generator, producing what we call an *AC voltage*.

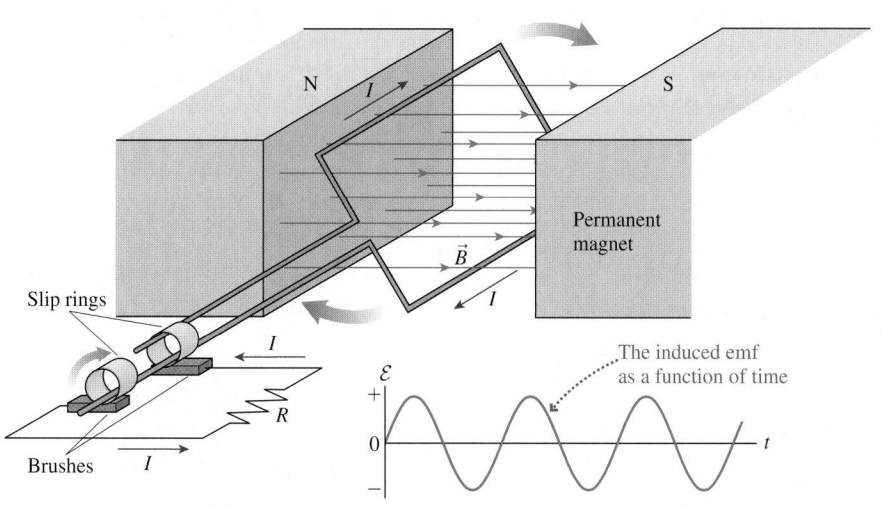

FIGURE 33.36 An alternating-current generator.

EXAMPLE 33.11 An AC generator

A coil with area 2.0 m^2 rotates in a 0.010 T magnetic field at a frequency of 60 Hz. How many turns are needed to generate a peak voltage of 160 V?

SOLVE The coil's maximum voltage is found from Equation 33.22:

$$\mathcal{E}_{max} = \omega ABN = 2\pi fABN$$

The number of turns needed to generate $\mathcal{E}_{max} = 160$ V is

$$N = \frac{\mathcal{E}_{max}}{2\pi fAB} = \frac{160 \text{ V}}{2\pi(60 \text{ Hz})(2.0 \text{ m}^2)(0.010 \text{ T})}$$

$$= 21 \text{ turns}$$

ASSESS A 0.010 T field is modest, so you can see that generating large voltages is not difficult with large (2 m^2) coils. Commercial generators use water flowing through a dam or turbines spun by expanding steam to rotate the generator coils. Work is required to rotate the coil, just as work was required to pull the slide wire in Section 33.2, because the magnetic field exerts retarding forces on the currents in the coil. Thus a generator is a device that turns motion (mechanical energy) into a current (electric energy). A generator is the opposite of a motor, which turns a current into motion.

Transformers

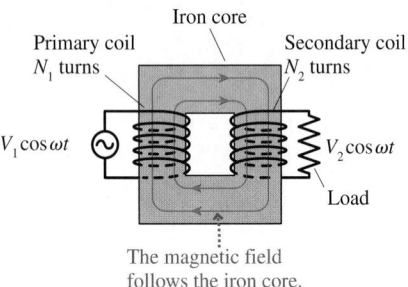

Primary coil N_1 turns — Iron core — Secondary coil N_2 turns

$V_1 \cos\omega t$ $V_2 \cos\omega t$

Load

The magnetic field follows the iron core.

FIGURE 33.37 A transformer.

Transformers are essential for transporting electric energy from the power plant to cities and homes.

Figure 33.37 shows two coils wrapped on an iron core. The left coil is called the **primary coil.** It has N_1 turns and is driven by an oscillating voltage $V_1 \cos \omega t$. The magnetic field of the primary follows the iron core and passes through the right coil, which has N_2 turns and is called the **secondary coil.** The alternating current through the primary coil causes an oscillating magnetic flux through the secondary coil and, hence, an induced emf. The induced emf of the secondary coil is delivered to resistance R as the oscillating voltage $V_2 \cos \omega t$.

The current through the primary coil is inversely proportional to the number of turns: $I_{prim} \propto 1/N_1$. (This relation is a consequence of the coil's inductance, an idea discussed later in the chapter. But, roughly speaking, the coil's resistance is proportional to the number of turns, and the current, according to Ohm's law, is inversely proportional to the resistance.) According to Faraday's law, the emf induced in the secondary coil is directly proportional to the number of turns: $\mathcal{E}_{sec} \propto N_2$. Combining these two proportionalities, the secondary voltage of an ideal transformer is related to the primary voltage by

$$V_2 = \frac{N_2}{N_1}V_1 \tag{33.23}$$

Depending on the ratio N_2/N_1, the voltage V_2 across the load can be *transformed* to a higher or a lower voltage than V_1. Consequently, this device is called a **transformer.** Transformers are widely used in the commercial generation and transmission of electricity. A *step-up transformer,* with $N_2 \gg N_1$, boosts the voltage of a generator up to several hundred thousand volts. Delivering power with smaller currents at higher voltages reduces losses due to the resistance of the wires. High-voltage transmission lines carry electric power to urban areas, where *step-down transformers* ($N_2 \ll N_1$) lower the voltage to 120 V.

Metal Detectors

Metal detectors, such as those used in airports for security, seem fairly mysterious. How can they detect the presence of *any* metal—not just magnetic materials such as iron—but not detect plastic or other materials? Metal detectors work because of induced currents.

A metal detector, shown in Figure 33.38, consist of two coils: a *transmitter coil* and a *receiver coil.* A high-frequency alternating current in the transmitter coil generates an alternating magnetic field along the axis. This magnetic field creates a changing flux through the receiver coil and causes an alternating induced current. The transmitter and receiver are similar to a transformer.

Suppose a piece of metal is placed between the transmitter and the receiver. The alternating magnetic field through the metal induces eddy currents in a plane parallel to the transmitter and receiver coils. The receiver coil then responds to the *superposition* of the transmitter's magnetic field and the magnetic field of the eddy currents. Because the eddy currents attempt to prevent the flux from changing, in accordance with Lenz's law, the net field at the receiver *decreases* when a piece of metal is inserted between the coils. Electronic circuits detect the current decrease in the receiver coil and set off an alarm. Eddy currents can't flow in an insulator, so this device detects only metals.

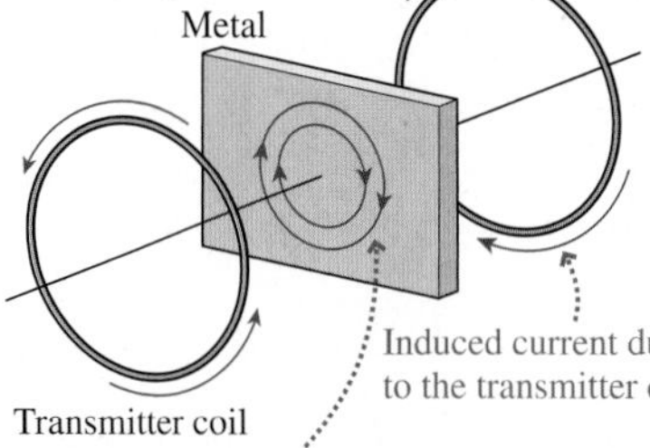

Eddy currents in the metal reduce the induced current in the receiver coil.

FIGURE 33.38 A metal detector.

33.8 Inductors

Capacitors were first introduced as devices that produce a uniform electric field. The capacitance (i.e., the *capacity* to store charge) was defined as the charge-to-voltage ratio $C = Q/\Delta V$. We later found that a capacitor stores potential energy $U_C = \frac{1}{2}C(\Delta V)^2$ and that this energy is released when the capacitor is discharged.

A coil of wire in the form of a solenoid is a device that produces a uniform magnetic field. Do solenoids in circuits have practical uses, as capacitors do? As a starting point to answering this question, notice that the charge on a capacitor is analogous to the magnetic flux through a solenoid. That is, a larger diameter capacitor plate holds more charge just as a larger diameter solenoid contains more flux. Using the definition of capacitance as an analog, let's define the **inductance** L of a magnetic-field device as its flux-to-current ratio

$$L = \frac{\Phi_m}{I} \tag{33.24}$$

Strictly speaking, this is called *self-inductance* because the flux we're considering is the magnetic flux the device creates in itself when there is a current.

The units of inductance are Wb/A. Recalling that $1\ \text{Wb} = 1\ \text{T m}^2$, this is equivalent to $\text{T m}^2/\text{A}$. It's convenient to define an SI unit of inductance called the **henry,** in honor of Joseph Henry, as

$$1\ \text{henry} = 1\ \text{H} \equiv 1\ \text{T m}^2/\text{A}$$

Practical inductances are usually in the range of millihenries (mH) or microhenries (μH).

Any circuit element can have an inductance by virtue of the fact that currents produce magnetic fields. In practice, however, inductance is usually negligible unless the magnetic field is concentrated, as it is in a solenoid. Consequently, a solenoid or coil of wire is our prototype of inductance. A coil of wire used in a circuit for the purpose of providing inductance is called an **inductor.** An *ideal inductor* is one for which the wire forming the coil has no electric resistance. The circuit symbol for an inductor is ⎯⎞⎛⎞⎛⎯ .

It's not hard to find the inductance of a solenoid. In Chapter 32 we found that the magnetic field inside a solenoid having N turns and length l is

$$B = \frac{\mu_0 NI}{l} \tag{33.25}$$

The magnetic flux through *each* coil is $\Phi_{\text{per coil}} = AB$, where A is the cross-section area of the solenoid. The total flux through all N coils is

$$\Phi_m = N\Phi_{\text{per coil}} = \frac{\mu_0 N^2 A}{l}I \tag{33.26}$$

Thus the inductance of the solenoid, using the definition of Equation 33.24, is

$$L_{\text{solenoid}} = \frac{\Phi_{\text{m}}}{I} = \frac{\mu_0 N^2 A}{l} \tag{33.27}$$

The inductance of a solenoid depends only on its geometry, not at all on the current. You may recall that the capacitance of two parallel plates depends only on their geometry, not at all on their potential difference.

EXAMPLE 33.12 **The length of an inductor**
An inductor is made by tightly wrapping 0.30-mm-diameter wire around a 4.0-mm-diameter cylinder. What length cylinder has an inductance of 10 μH?

SOLVE The cross-section area of the solenoid is $A = \pi r^2$. If the wire diameter is d, the number of turns of wire on a cylinder of length l is $N = l/d$. Thus the inductance is

$$L = \frac{\mu_0 N^2 A}{l} = \frac{\mu_0 (l/d)^2 \pi r^2}{l} = \frac{\mu_0 \pi r^2 l}{d^2}$$

The length needed to give inductance $L = 10^{-5}$ H is

$$l = \frac{d^2 L}{\mu_0 \pi r^2} = \frac{(0.00030 \text{ m})^2 (1.0 \times 10^{-5} \text{ H})}{(4\pi \times 10^{-7} \text{ Tm/A}) \pi (0.0020 \text{ m})^2}$$

$$= 0.057 \text{ m} = 5.7 \text{ cm}$$

The Potential Difference Across an Inductor

An inductor is not very interesting when the current through it is steady. If the inductor is ideal, with $R = 0 \ \Omega$, the potential difference due to a steady current is zero. Inductors become important circuit elements when currents are changing. Figure 33.39a shows a steady current into the left side of an inductor. The solenoid's magnetic field passes through the coils of the solenoid, establishing a flux.

In Figure 33.39b, the current into the solenoid is increasing. This creates an increasing flux to the left. According to Lenz's law, an induced current in the coils will oppose this increase by creating an induced magnetic field pointing to the right. This requires the induced current to be *opposite* the current into the solenoid. This induced current will carry positive charge carriers to the left until a potential difference ΔV is established across the solenoid.

You saw a similar situation in Section 33.2. The induced current in a conductor moving through a magnetic field carried positive charge carriers to the top of the wire and established a potential difference across the conductor. The induced current in the moving wire was due to magnetic forces on the moving charges. Now, in Figure 33.39b, the induced current is due to the non-Coulomb electric field induced by the changing magnetic field. Nonetheless, the outcome is the same: a potential difference across the conductor.

We can use Faraday's law to find the potential difference. The emf induced in a coil is

$$\mathcal{E}_{\text{coil}} = N \left| \frac{d\Phi_{\text{per coil}}}{dt} \right| = \left| \frac{d\Phi_{\text{m}}}{dt} \right| \tag{33.28}$$

where $\Phi = N\Phi_{\text{per coil}}$ is the total flux through all the coils. The inductance was defined such that $\Phi_{\text{m}} = LI$, so Equation 33.28 becomes

$$\mathcal{E}_{\text{coil}} = L \left| \frac{dI}{dt} \right| \tag{33.29}$$

The induced emf is directly proportional to the *rate of change* of current through the coil. We'll consider the appropriate sign in a moment, but Equation 33.29 gives us the size of the potential difference that is developed across a coil as the current through the coil changes. Note that $\mathcal{E}_{\text{coil}} = 0$ for a steady, unchanging current.

(a)

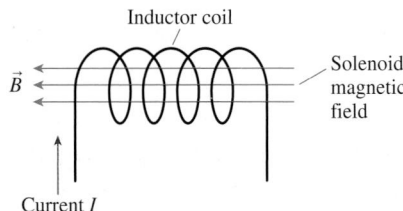

Inductor coil

$\vec{B}$

Solenoid magnetic field

Current I

(b)

The induced current is opposite the solenoid current.

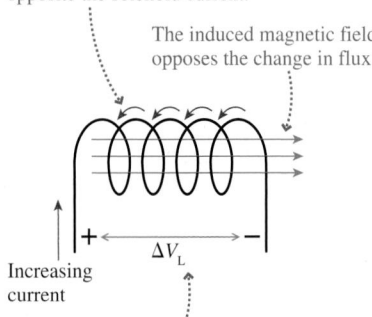

The induced magnetic field opposes the change in flux.

ΔV_{L}

Increasing current

The induced current carries positive charge carriers to the left and establishes a potential difference across the inductor.

FIGURE 33.39 Increasing the current through an inductor.

Figure 33.40 shows the same inductor, but now the current (still *in* to the left side) is decreasing. To oppose the decrease in flux, the induced current is in the *same* direction as the input current. The induced current carries charge to the right and establishes a potential difference opposite that in Figure 33.39b.

NOTE ▶ Notice that the induced current does not oppose the current through the inductor, which is from left to right in both Figures 33.39 and 33.40. Instead, in accordance with Lenz's law, the induced current opposes the *change* in the current in the solenoid. The practical result is that it is hard to change the current through an inductor. Any effort to increase or decrease the current is met with opposition in the form of an opposing induced current. You can think of the current in an inductor as having inertia, trying to continue what it was doing without change. ◀

Before we can use inductors in a circuit we need to establish a rule about signs that is consistent with our earlier circuit analysis. Figure 33.41 first shows current *I* passing through a resistor. You learned in Chapter 31 that the potential difference across a resistor is $\Delta V_R = -IR$, where the minus sign indicates that the potential *decreases* in the direction of the current.

We'll use the same convention for an inductor. The potential difference across an inductor, measured along the direction of the current, is

$$\Delta V_L = -L\frac{dI}{dt} \tag{33.30}$$

If the current is increasing ($dI/dt > 0$), the input side of the inductor is more positive than the output side and the potential decreases in the direction of the current ($\Delta V_L < 0$). This was the situation in Figure 33.39b. If the current is decreasing ($dI/dt < 0$), the input side is more negative and the potential increases in the direction of the current ($\Delta V_L > 0$). This was the situation in Figure 33.40.

The potential difference across an inductor can be very large if the current changes very abruptly (large *dI/dt*). Figure 33.42 shows an inductor connected across a battery. There is a large current through the inductor, limited only by the internal resistance of the battery. Suppose the switch is suddenly opened. A very large induced voltage is created across the inductor as the current rapidly drops to zero. This potential difference (plus ΔV_{bat}) appears across the gap of the switch as it is opened. A large potential difference across a small gap often creates a spark.

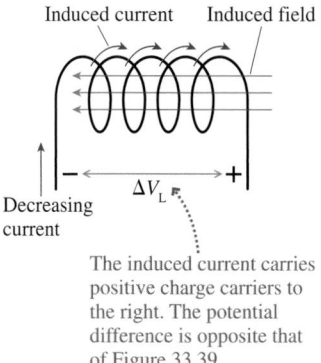

The induced current carries positive charge carriers to the right. The potential difference is opposite that of Figure 33.39.

FIGURE 33.40 Decreasing the current through an inductor.

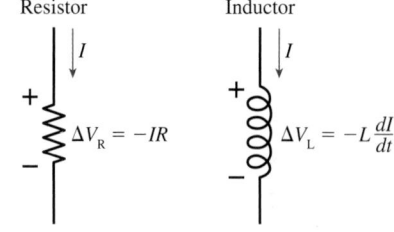

The potential always decreases.

The potential decreases if the current is increasing.

The potential increases if the current is decreasing.

FIGURE 33.41 The potential difference across a resistor and an inductor.

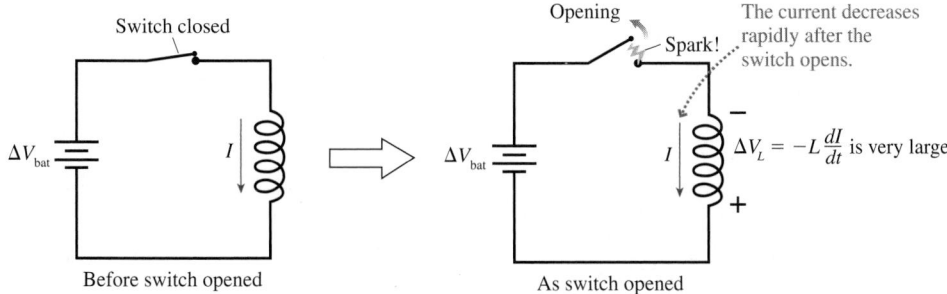

FIGURE 33.42 Creating sparks.

Indeed, this is exactly how the spark plugs in your car work. The car's generator sends a large current through the *coil,* which is a big inductor. A switch in the *distributor* is suddenly opened, breaking the current. The induced voltage, typically a few thousand volts, appears across the terminals of the spark plug, creating the spark that ignites the gasoline. A similar phenomenon happens if you unplug appliances such as toaster ovens or hair dryers while they are running. The heating coils in these devices have quite a bit of inductance. Suddenly pulling the plug is like opening a switch. The large induced voltage often causes a spark between the plug and the electric outlet.

EXAMPLE 33.13 Large voltage across an inductor

A 1.0 A current passes through a 10 mH inductor coil. What potential difference is induced across the coil if the current drops to zero in 5 μs?

MODEL Assume this is an ideal inductor, with $R = 0 \, \Omega$, and that the current decrease is linear with time.

SOLVE The rate of current decrease is

$$\frac{dI}{dt} \approx \frac{\Delta I}{\Delta t} = \frac{-1.0 \text{ A}}{5.0 \times 10^{-6} \text{ s}} = -2.0 \times 10^5 \text{ A/s}$$

The induced voltage is

$$\Delta V_{\text{L}} = -L\frac{dI}{dt} \approx -(0.010 \text{ H})(-2.0 \times 10^5 \text{ A/s})$$

$$= 2000 \text{ V}$$

ASSESS Inductors may be physically small, but they can pack a punch if you try to change the current through them too quickly.

STOP TO THINK 33.6   The potential at a is higher than the potential at b. Which of the following statements about the inductor current I could be true?

a. I is from a to b and steady.
b. I is from a to b and increasing.
c. I is from a to b and decreasing.
d. I is from b to a and steady.
e. I is from b to a and increasing.
f. I is from b to a and decreasing.

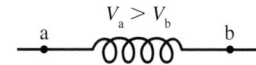

Energy in Inductors and Magnetic Fields

An inductor, like a capacitor, stores energy that can later be released. It is energy released from the coil in your car that becomes the spark of the spark plug. You learned in Chapter 31 that electric power is $P_{\text{elec}} = I\Delta V$. As current passes through an inductor, for which $\Delta V_{\text{L}} = -L(dI/dt)$, the electric power is

$$P_{\text{elec}} = I\Delta V_{\text{L}} = -LI\frac{dI}{dt} \tag{33.31}$$

P_{elec} is negative because the current is *losing* electric energy. That energy is being transferred to the inductor, which is *storing* energy U_{L} at the rate

$$\frac{dU_{\text{L}}}{dt} = +LI\frac{dI}{dt} \tag{33.32}$$

where we've noted that power is the rate of change of energy.

We can find the total energy stored in an inductor by integrating Equation 33.32 from $I = 0$, where $U_{\text{L}} = 0$, to a final current I. Doing so gives

$$U_{\text{L}} = L\int_0^I I\, dI = \frac{1}{2}LI^2 \tag{33.33}$$

The potential energy stored in an inductor depends on the square of the current through it. Notice the analogy with the energy $U_{\text{C}} = \frac{1}{2}C(\Delta V)^2$ stored in a capacitor.

In working with circuits we say that the energy is "stored in the inductor." Strictly speaking, the energy is stored in the inductor's magnetic field, analogous to how a capacitor stores energy in the electric field. We can use the inductance of a solenoid, Equation 33.27, to relate the inductor's energy to the magnetic field strength:

$$U_L = \frac{1}{2}LI^2 = \frac{\mu_0 N^2 A}{2l}I^2 = \frac{1}{2\mu_0}Al\left(\frac{\mu_0 NI}{l}\right)^2 \qquad (33.34)$$

We made the last rearrangement in Equation 33.34 because $\mu_0 NI/l$ is the magnetic field inside the solenoid. Thus

$$U_L = \frac{1}{2\mu_0}AlB^2 \qquad (33.35)$$

But Al is the volume inside the solenoid. Dividing by Al, the magnetic field *energy density* inside the solenoid (energy per m³) is

$$u_B = \frac{1}{2\mu_0}B^2 \qquad (33.36)$$

We've derived this expression for energy density based on the properties of a solenoid, but it turns out to be the correct expression for the energy density anywhere there's a magnetic field. Compare this to the energy density of an electric field $u_E = \frac{1}{2}\epsilon_0 E^2$ that we found in Chapter 30.

Energy in electric and magnetic fields

Electric fields	Magnetic fields
A capacitor stores energy $U_C = \frac{1}{2}C(\Delta V)^2$	An inductor stores energy $U_L = \frac{1}{2}LI^2$
Energy density in the field is $u_E = \frac{\epsilon_0}{2}E^2$	Energy density in the field is $u_B = \frac{1}{2\mu_0}B^2$

EXAMPLE 33.14 Energy stored in an inductor
The 10 μH inductor of Example 33.12 was 5.7 cm long and 4.0 mm in diameter. Suppose it carries a 100 mA current. What are the energy stored in the inductor, the magnetic energy density, and the magnetic field strength?

SOLVE The stored energy is

$$U_L = \frac{1}{2}LI^2 = \frac{1}{2}(10^{-5}\,\text{H})(0.10\,\text{A})^2 = 5.0 \times 10^{-8}\,\text{J}$$

The solenoid volume is $(\pi r^2)l = 7.16 \times 10^{-7}\,\text{m}^3$. Using this gives the energy density of the magnetic field:

$$u_B = \frac{5.0 \times 10^{-8}\,\text{J}}{7.16 \times 10^{-7}\,\text{m}^3} = 0.070\,\text{J/m}^3$$

From Equation 33.36, the magnetic field with this energy density is

$$B = \sqrt{2\mu_0 u_B} = 4.2 \times 10^{-4}\,\text{T}$$

33.9 *LC* Circuits

Telecommunication—radios, televisions, cell phones—is based on electromagnetic signals that *oscillate* at a well-defined frequency. These oscillations are generated and detected by a simple circuit consisting of an inductor and a capacitor in parallel. This is called an **LC circuit.** In this section we will learn why an *LC* circuit oscillates and determine the oscillation frequency.

Figure 33.43 shows a capacitor with initial charge Q_0, an inductor, and a switch. The switch has been open for a long time, so there is no current in the circuit. Then, at $t = 0$, the switch is closed. How does the circuit respond? Let's think it through qualitatively before getting into the mathematics.

As Figure 33.44 on the next page shows, the inductor provides a conducting path for discharging the capacitor. However, the discharge current has to pass through the inductor, and, as we've seen, an inductor resists changes in current. Consequently, the current doesn't stop when the capacitor charge reaches zero.

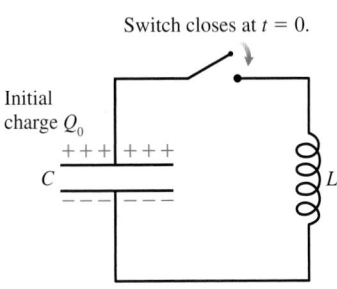

FIGURE 33.43 An *LC* circuit.

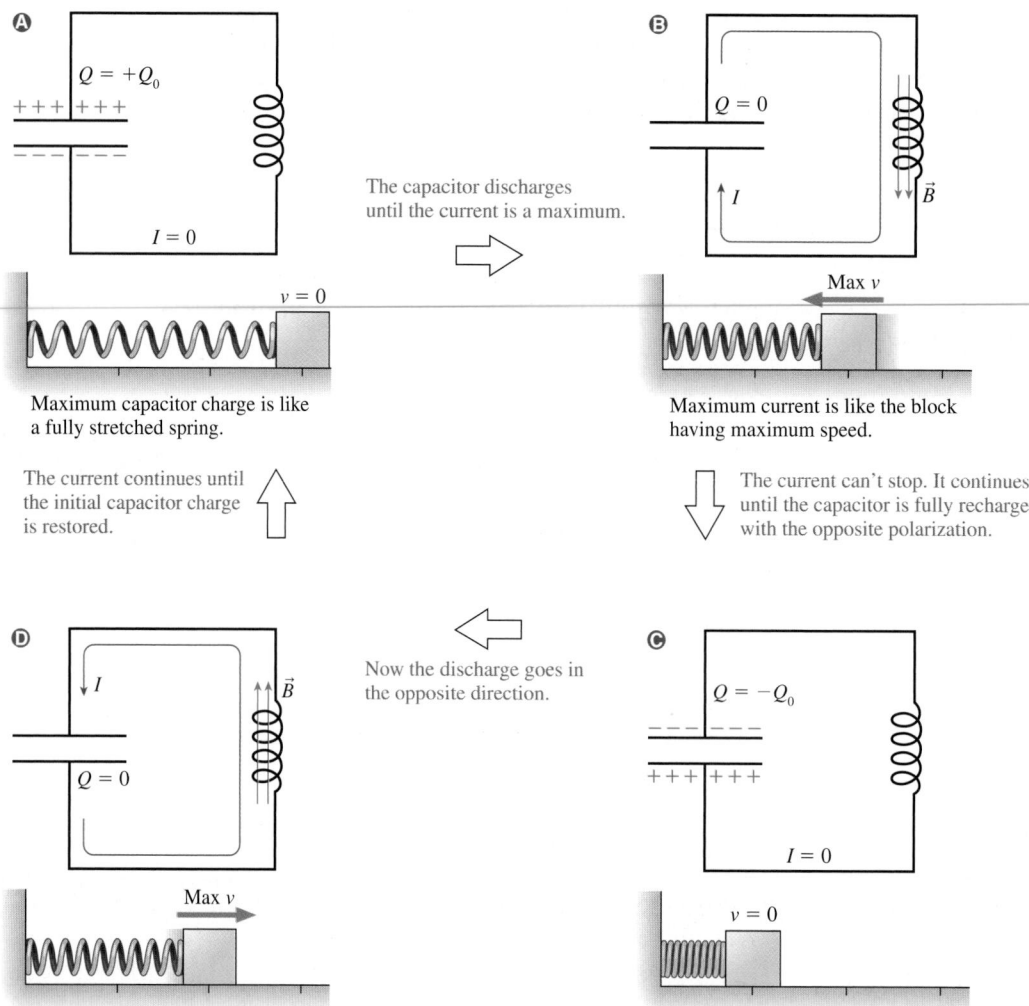

FIGURE 33.44 The capacitor charge oscillates much like a block attached to a spring.

A block attached to a stretched spring is a useful mechanical analogy. Closing the switch to discharge the capacitor is like releasing the block. But the block doesn't stop when it reaches the origin. Its momentum keeps it going until the spring is fully compressed. Likewise, the current continues until it has recharged the capacitor with the opposite polarization. This process repeats over and over, charging the capacitor first one way, then the other. That is, the charge and current *oscillate*.

The goal of our circuit analysis will be to find expressions showing how the capacitor charge Q and the inductor current I change with time. As always, our starting point for circuit analysis is Kirchhoff's voltage law, which says that all the potential differences around a closed loop must sum to zero. Choosing a cw direction for I, Kirchhoff's law is

$$\Delta V_C + \Delta V_L = 0 \tag{33.37}$$

You learned in Chapter 30 that the potential difference across a capacitor is $\Delta V_C = Q/C$, where Q is the charge on the top plate of the capacitor, and we found the potential difference across an inductor in Equation 33.30 above. Using these, Kirchhoff's law becomes

$$\frac{Q}{C} - L\frac{dI}{dt} = 0 \tag{33.38}$$

Equation 33.38 has two unknowns, Q and I. We need to eliminate one of the unknowns, and we can do so by finding another relation between Q and I. Current is the rate at which charge moves: $I = dq/dt$. But the charge flowing through the inductor is charge that was *removed* from the capacitor. That is, an infinitesimal charge dq flows through the inductor when the capacitor charge changes by $dQ = -dq$. Thus the current through the inductor is related to the charge on the capacitor by

$$I = -\frac{dQ}{dt} \qquad (33.39)$$

Now I is positive when Q is decreasing, as we would expect. This is a subtle but important step in the reasoning, one worth thinking about because it appears in other contexts.

Equations 33.38 and 33.39 are two equations in two unknowns. To solve them, we'll first take the time derivative of Equation 33.39:

$$\frac{dI}{dt} = \frac{d}{dt}\left(-\frac{dQ}{dt}\right) = -\frac{d^2Q}{dt^2} \qquad (33.40)$$

We can substitute this result into Equation 33.38:

$$\frac{Q}{C} + L\frac{d^2Q}{dt^2} = 0 \qquad (33.41)$$

Now we have an equation for the capacitor charge Q.

Equation 33.41 is a second-order differential equation for Q. Fortunately, it is an equation we've seen before and already know how to solve. To see this, rewrite Equation 33.41 as

$$\frac{d^2Q}{dt^2} = -\frac{1}{LC}Q \qquad (33.42)$$

Recall, from Chapter 14, that the equation of motion for an undamped mass on a spring is

$$\frac{d^2x}{dt^2} = -\frac{k}{m}x \qquad (33.43)$$

Equation 33.42 is *exactly the same equation*, with x replaced by Q and k/m replaced by $1/LC$. This should be no surprise because we've already seen that a mass on a spring is a mechanical analog of the *LC* circuit.

We know the solution to Equation 33.43. It is simple harmonic motion $x(t) = x_0 \cos\omega t$ with angular frequency $\omega = \sqrt{k/m}$. Thus the solution to Equation 33.42 must be

$$Q(t) = Q_0 \cos\omega t \qquad (33.44)$$

where Q_0 is the initial charge, at $t = 0$, and the angular frequency is

$$\omega = \sqrt{\frac{1}{LC}} \qquad (33.45)$$

The charge on the upper plate of the capacitor oscillates back and forth between $+Q_0$ (as shown in Figure 33.43) and $-Q_0$ (the opposite polarization) with period $T = 2\pi/\omega$.

As the capacitor charge oscillates, so does the current through the inductor. Using Equation 33.39 gives the current through the inductor:

$$I = -\frac{dQ}{dt} = \omega Q_0 \sin\omega t = I_{max} \sin\omega t \qquad (33.46)$$

where $I_{max} = \omega Q_0$ is the maximum current.

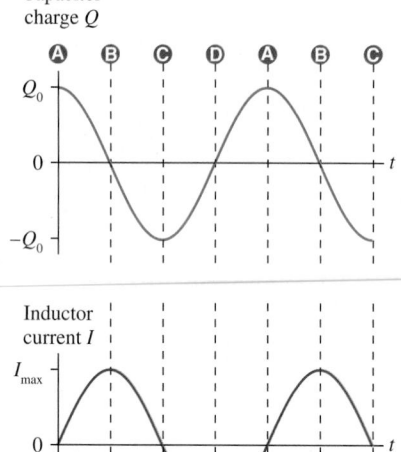

Capacitor charge Q

Inductor current I

FIGURE 33.45 The oscillations of an LC circuit.

An LC circuit is an *electric oscillator*, oscillating at frequency $f = \omega/2\pi$. Figure 33.45 shows graphs of the capacitor charge Q and the inductor current I as functions of time. The letters over the graph match the labels in Figure 33.44, and you should compare the two. Notice that Q and I are 90° out of phase. The current is zero when the capacitor is fully charged, as expected, and the charge is zero when the current is maximum.

EXAMPLE 33.15 An AM radio oscillator

You have a 1.0 mH inductor. What capacitor should you choose to make an oscillator with a frequency of 920 KHz? (This frequency is near the center of the AM radio band.)

SOLVE The angular frequency is $\omega = 2\pi f = 5.78 \times 10^6$ rad/s. Using Equation 33.45 for ω gives the required capacitor:

$$C = \frac{1}{\omega^2 L} = \frac{1}{(5.78 \times 10^6 \text{ rad/s})^2 (0.0010 \text{ H})}$$

$$= 3.0 \times 10^{-11} \text{ F} = 30 \text{ pF}$$

An LC circuit, like a mass on a spring, wants to respond only at its natural oscillation frequency $\omega = 1/\sqrt{LC}$. In Chapter 14 we defined a strong response at the natural frequency as a *resonance,* and resonance is the basis for all telecommunications. The input circuit in radios, televisions, and cell phones is an LC circuit driven by the signal picked up by the antenna. This signal is the superposition of hundreds of sinusoidal waves at different frequencies, one from each transmitter in the area, but the circuit responds only to the *one* signal that matches the circuit's natural frequency. That particular signal generates a large-amplitude current that can be further amplified and decoded to become the output that you hear.

Turning the dial on your radio or television changes a *variable capacitor,* thus changing the resonance frequency so that you pick up a different station. Cell phones are a bit more complicated. You don't change the capacitance yourself, but a "smart" circuit inside can change its capacitance in response to command signals it receives from the transmitter. The result is the same. Your cell phone responds to the one signal being broadcast to you and ignores the hundreds of other signals that are being broadcast simultaneously at different frequencies.

33.10 LR Circuits

14.1 Activ Physics ONLINE

A circuit consisting of an inductor, a resistor, and (perhaps) a battery is called an **LR circuit.** Figure 33.46a is an example of an LR circuit. We'll assume that the switch has been in position a for such a long time that the current is steady and unchanging. There's no potential difference across the inductor, because $dI/dt = 0$, so it simply acts like a piece of wire. The current flowing around the circuit is determined entirely by the battery and the resistor: $I_0 = \Delta V_{bat}/R$.

What happens if, at $t = 0$, the switch is suddenly moved to position b? With the battery no longer in the circuit, you might expect the current to stop immediately. But the inductor won't let that happen. The current will continue for some period of time as the inductor's magnetic field drops to zero. In essence, the energy stored in the inductor allows it to act like a battery for a short period of time. Our goal is to determine how the current decays after the switch is moved.

Figure 33.46b shows the circuit after the switch is changed. Our starting point, once again, is Kirchhoff's voltage law. The potential differences around a closed loop must sum to zero. For this circuit, Kirchhoff's law is

$$\Delta V_R + \Delta V_L = 0 \tag{33.47}$$

The potential differences in the direction of the current are $\Delta V_R = -IR$ for the resistor and $\Delta V_L = -L(dI/dt)$ for the inductor. Substituting these into Equation 33.47 gives

$$-RI - L\frac{dI}{dt} = 0 \tag{33.48}$$

We're going to need to integrate to find the current I as a function of time. Before doing so, we need to get all the current terms on one side of the equation and all the time terms on the other. A simple rearrangement of Equation 33.48 gives

$$\frac{dI}{I} = -\frac{R}{L}dt = -\frac{dt}{(L/R)} \tag{33.49}$$

We know that the current at $t = 0$, when the switch was moved, was I_0. We want to integrate from these starting conditions to current I at the unspecified time t. That is,

$$\int_{I_0}^{I} \frac{dI}{I} = -\frac{1}{(L/R)}\int_{0}^{t} dt \tag{33.50}$$

Both are common integrals, giving

$$\ln I \Big|_{I_0}^{I} = \ln I - \ln I_0 = \ln\left(\frac{I}{I_0}\right) = -\frac{t}{(L/R)} \tag{33.51}$$

We can solve for the current I by taking the exponential of both sides, then multiplying by I_0. Doing so gives I, the current as a function of time:

$$I = I_0 e^{-t/(L/R)} \tag{33.52}$$

Notice that $I = I_0$ at $t = 0$, as expected.

The argument of the exponential function must be dimensionless, so L/R must have dimensions of time. If we define the **time constant** τ of the LR circuit to be

$$\boxed{\tau = \frac{L}{R}} \tag{33.53}$$

then we can write Equation 33.52 as

$$I = I_0 e^{-t/\tau} \tag{33.54}$$

The time constant is the time at which the current has decreased to e^{-1} (about 37%) of its initial value. We can see this by computing the current at the time $t = \tau$.

$$I(\text{at } t = \tau) = I_0 e^{-\tau/\tau} = e^{-1} I_0 = 0.37 I_0 \tag{33.55}$$

Thus the time constant for an LR circuit functions in exactly the same way as the time constant for the RC circuit we analyzed in Chapter 31. At time $t = 2\tau$, the current has decreased to $e^{-2}I_0$, or about 13% of its initial value.

The current is graphed in Figure 33.47. You can see that the current decays exponentially. The *shape* of the graph is always the same, regardless of the specific value of the time constant τ.

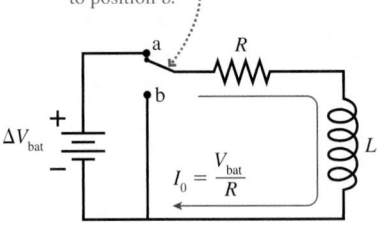

(a) The switch has been in this position for a long time. At $t = 0$ it is moved to position b.

(b)

This is the circuit with the switch in position b. The inductor prevents the current from stopping instantly.

FIGURE 33.46 An *LR* circuit.

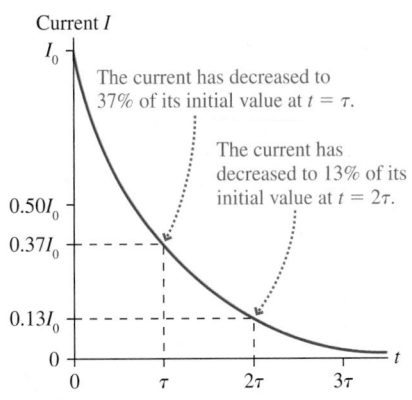

FIGURE 33.47 The current decay in an *LR* circuit.

EXAMPLE 33.16 Exponential decay in an *LR* circuit

The switch in Figure 33.48 has been in position a for a long time. It is changed to position b at $t = 0$ s.

a. What is the current in the circuit at $t = 5.0 \, \mu s$?

b. At what time has the current decayed to 1% of its initial value?

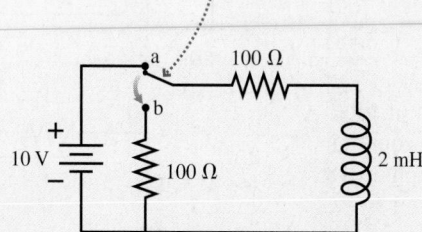

The switch moves from a to b at $t = 0$.

FIGURE 33.48 The *LR* circuit of Example 33.16.

MODEL This is an *LR* circuit. We'll assume ideal wires and an ideal inductor.

VISUALIZE The two resistors will be in series after the switch is thrown.

SOLVE Before the switch is thrown, while $\Delta V_L = 0$, the current is $I_0 = (10 \text{ V})/(100 \, \Omega) = 0.10 \text{ A} = 100 \text{ mA}$. This will be the initial current after the switch is thrown because the current through an inductor can't change instantaneously. The circuit resistance after the switch is thrown is $R = 200 \, \Omega$, so the time constant is

$$\tau = \frac{L}{R} = \frac{2.0 \times 10^{-3} \text{ H}}{200 \, \Omega} = 1.0 \times 10^{-5} \text{ s} = 10 \, \mu s$$

a. The current at $t = 5.0 \, \mu s$ is

$$I = I_0 e^{-t/\tau} = (100 \text{ mA}) e^{-(5 \, \mu s)/(10 \, \mu s)} = 61 \text{ mA}$$

b. To find the time at which a particular current is reached we need to go back to Equation 33.52 and solve for t:

$$t = -\frac{L}{R} \ln\left(\frac{I}{I_0}\right) = -\tau \ln\left(\frac{I}{I_0}\right)$$

The time at which the current has decayed to 1 mA (1% of I_0) is

$$t = -(10 \, \mu s) \ln\left(\frac{1 \text{ mA}}{100 \text{ mA}}\right) = 46 \, \mu s$$

ASSESS For all practical purposes, the current has decayed away in $\approx 50 \, \mu s$. The inductance in this circuit is not large, so a short decay time is not surprising.

STOP TO THINK 33.7 Rank in order, from largest to smallest, the time constants $\tau_1, \tau_2,$ and τ_3 of these three circuits.

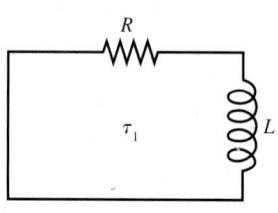

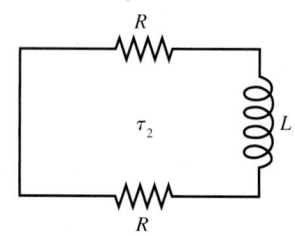

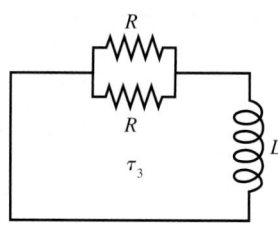

SUMMARY

The goal of Chapter 33 has been to understand and apply electromagnetic induction.

GENERAL PRINCIPLES

Faraday's Law

MODEL Make simplifying assumptions.

VISUALIZE Use Lenz's law to determine the direction of the **induced current.**

SOLVE The **induced emf** is

$$\mathcal{E} = \left| \frac{d\Phi}{dt} \right|$$

Multiply by N for an N-turn coil.
The size of the induced current is $I = \mathcal{E}/R$.

ASSESS Is the result reasonable?

Lenz's Law

There is an induced current in a closed conducting loop if and only if the magnetic flux through the loop is changing. The direction of the induced current is such that the induced magnetic field opposes the *change* in the flux.

Magnetic flux

Magnetic flux measures the amount of magnetic field passing through a surface.

$$\Phi = \vec{A} \cdot \vec{B} = AB \cos\theta$$

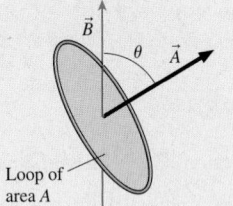

Loop of area A

IMPORTANT CONCEPTS

Three ways to change the flux

1. A loop moves into or out of a magnetic field.

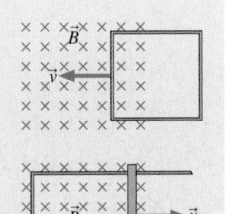

2. The loop changes area or rotates.

3. The magnetic field through the loop increases or decreases.

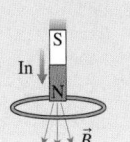

Two ways to create an induced current

1. A **motional emf** due to magnetic forces on moving charge carriers.

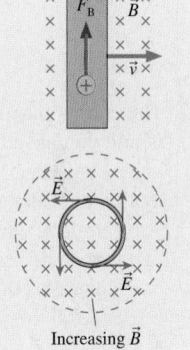

2. An induced electric field due to a changing magnetic field.

Increasing $\vec{B}$

APPLICATIONS

Inductors

Solenoid inductance $L_{\text{solenoid}} = \dfrac{\mu_0 N^2 A}{l}$

Potential difference $\Delta V_{\text{L}} = -L\dfrac{dI}{dt}$

Energy stored $U_{\text{L}} = \frac{1}{2}LI^2$

Magnetic energy density $u_{\text{B}} = \dfrac{1}{2\mu_0}B^2$

LC circuit

Oscillates at $\omega = \sqrt{\dfrac{1}{LC}}$

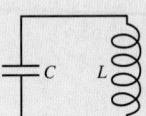

LR circuit

Exponential change with $\tau = \dfrac{L}{R}$

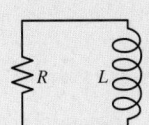

TERMS AND NOTATION

electromagnetic induction	induced emf, $\mathcal{E}$	transformer
induced current	Faraday's law	inductance, L
motional emf	induced electric field	henry, H
generator	Coulomb electric field	inductor
eddy current	non-Coulomb electric field	LC circuit
magnetic flux, Φ_m	induced magnetic field	LR circuit
weber, Wb	electromagnetic wave	time constant, τ
area vector, $\vec{A}$	primary coil	
Lenz's law	secondary coil	

EXERCISES AND PROBLEMS

Exercises

Section 33.2 Motional emf

1. A potential difference of 0.050 V is developed across a 10-cm-long wire as it moves through a magnetic field at 5.0 m/s. The magnetic field is perpendicular to the axis of the wire. What are the direction and strength of the magnetic field?

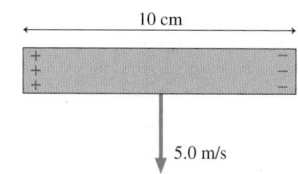

FIGURE EX33.1

2. The earth's magnetic field strength is 5×10^{-5} T. How fast would you have to drive your car to create a 1.0 V motional emf along your 1.0-m-long radio antenna? Assume that the motion of the antenna is perpendicular to $\vec{B}$.

3. A 10-cm-long wire is pulled along a U-shaped conducting rail in a perpendicular magnetic field. The total resistance of the wire and rail is 0.20 Ω. Pulling the wire with a force of 1.0 N causes 4.0 W of power to be dissipated in the circuit.
   a. What is the speed of the wire when pulled with 1.0 N?
   b. What is the strength of the magnetic field?

Section 33.3 Magnetic Flux

4. What is the magnetic flux through the loop shown in Figure Ex33.4?

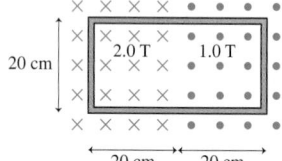

FIGURE EX33.4

5. A 2.0-cm-diameter solenoid passes through the center of a 6.0-cm-diameter loop. The magnetic field inside the solenoid is 0.20 T. What is the magnetic flux through the loop when it is perpendicular to the solenoid and when it is tilted at a 60° angle?

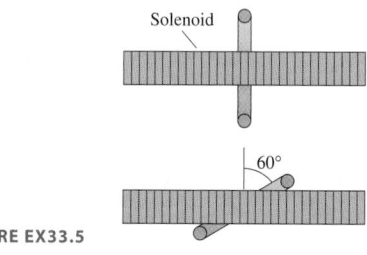

FIGURE EX33.5

Section 33.4 Lenz's Law

6. There is a ccw induced current in the conducting loop shown in Figure Ex33.6. Is the magnetic field inside the loop increasing in strength, decreasing in strength, or steady?

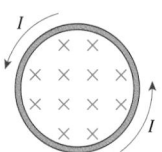

FIGURE EX33.6

7. A solenoid is wound as shown in Figure Ex33.7.
   a. Is there an induced current as magnet 1 is moved away from the solenoid? If so, what is the current direction through resistor R?
   b. Is there an induced current as magnet 2 is moved away from the solenoid? If so, what is the current direction through resistor R?

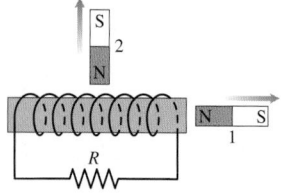

FIGURE EX33.7

8. The metal equilateral triangle in Figure Ex33.8, 20 cm on each side, is halfway into a 0.10 T magnetic field.
   a. What is the magnetic flux through the triangle?
   b. If the magnetic field strength decreases, what is the direction of the induced current in the triangle?

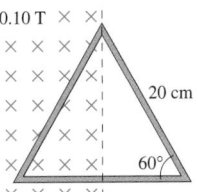

FIGURE EX33.8

9. The current in the solenoid of Figure Ex33.9 is decreasing. The solenoid is surrounded by a conducting loop. Is there a current in the loop? If so, is the loop current cw or ccw?

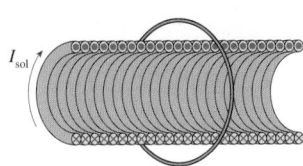

FIGURE EX33.9

Section 33.5 Faraday's Law

10. Figure Ex33.10 shows a 10-cm-diameter loop in three different magnetic fields. The loop's resistance is 0.10 Ω. For each case, determine the induced emf, the induced current, and the direction of the current.

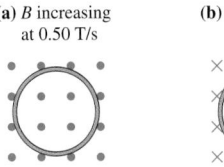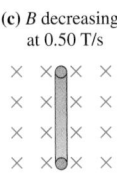

FIGURE EX33.10

11. A 1000-turn coil of wire that is 2.0 cm in diameter is in a magnetic field that drops from 0.10 T to 0 T in 10 ms. The axis of the coil is parallel to the field. What is the emf of the coil?

12. The loop in Figure Ex33.12 is being pushed into the 0.20 T magnetic field at 50 m/s. The resistance of the loop is 0.10 Ω. What are the direction and the magnitude of the current in the loop?

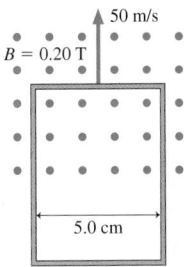

FIGURE EX33.12

13. The resistance of the loop in Figure Ex33.13 is 0.10 Ω. Is the magnetic field strength increasing or decreasing? At what rate (T/s)?

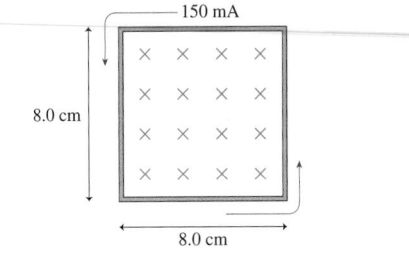

FIGURE EX33.13

Section 33.8 Inductors

14. You need to make a 100 μH inductor on a cylinder that is 5.0 cm long and 1.0 cm in diameter. You plan to wrap four layers of wire around the cylinder. What diameter wire should you use if the coils are tightly wound with no space between them? The wire diameter will be small enough that you don't need to consider the slight change in the coil's diameter for the outer layers.

15. What is the potential difference across a 10 mH inductor if the current through the inductor drops from 150 mA to 50 mA in 10 μs? What is the direction of this potential difference? That is, does the potential increase or decrease along the direction of the current?

16. The maximum allowable potential difference across a 200 mH inductor is 400 V. You need to raise the current through the inductor from 1.0 A to 3.0 A. What is the minimum time you should allow for changing the current?

17. How much energy is stored in a 3.0-cm-diameter, 12-cm-long solenoid that has 200 turns of wire and carries a current of 0.80 A?

Section 33.9 *LC* Circuits

18. A 2.0 mH inductor is connected in parallel with a variable capacitor. The capacitor can be varied from 100 pF to 200 pF. What is the range of oscillation frequencies for this circuit?

19. An FM radio station broadcasts at a frequency of 100 MHz. What inductance should be paired with a 10 pF capacitor to build a receiver circuit for this station?

20. An electric oscillator is made with a 0.10 μF capacitor and a 1.0 mH inductor. The capacitor is initially charged to 5.0 V. What is the maximum current through the inductor as the circuit oscillates?

Section 33.10 *LR* Circuits

21. What value of resistor R gives the circuit in Figure Ex33.21 a time constant of 10 μs?

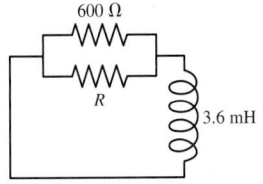

FIGURE EX33.21

22. At $t = 0$ s, the current in the circuit in Figure Ex33.22 is I_0. At what time is the current $\frac{1}{2}I_0$?

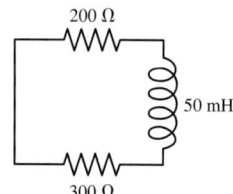

FIGURE EX33.22

Problems

23. A 10 cm × 10 cm square is bent at a 90° angle. A uniform 0.050 T magnetic field points downward at a 45° angle. What is the magnetic flux through the loop?

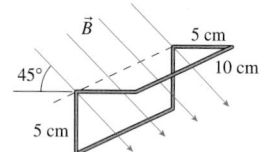

FIGURE P33.23

24. What is the magnetic flux through the loop shown in Figure P33.24?

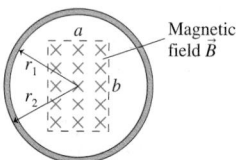

FIGURE P33.24

25. A 20 cm × 20 cm square loop has a resistance of 0.10 Ω. A magnetic field perpendicular to the loop is $B = 4t - 2t^2$, where B is in tesla and t is in seconds.
    a. Determine B, $\mathcal{E}$, and I at half-second intervals from 0 s to 2 s.
    b. Use your results of part a to draw graphs of B and I versus time.

26. A 5.0-cm-diameter coil has 20 turns and a resistance of 0.50 Ω. A magnetic field perpendicular to the coil is $B = 0.020t + 0.010t^2$, where B is in tesla and t is in seconds.
    a. Draw a graph of B as a function of time from $t = 0$ s to $t = 10$ s.
    b. Find an expression for the induced current I as a function of time.
    c. Evaluate I at $t = 5$ s and $t = 10$ s.

27. A 50-turn, 4.0-cm-diameter coil has a resistance of 1.0 Ω. A magnetic field perpendicular to the coil is $B = t - \frac{1}{4}t^2$, where B is in tesla and t is in seconds.
    a. Draw a graph of B as a function of time from $t = 0$ s to $t = 4$ s.
    b. Find an expression for the induced current $I(t)$ as a function of time.
    c. Evaluate I at $t = 1$, 2, and 3 s.

28. A 100-turn, 2.0-cm-diameter coil is at rest in a horizontal plane. A uniform magnetic field 60° away from vertical increases from 0.50 T to 1.50 T in 0.60 s. What is the induced emf in the coil?

29. A 25-turn, 10.0-cm-diameter coil is oriented in a vertical plane with its axis aligned east-west. A magnetic field pointing to the northeast decreases from 0.80 T to 0.20 T in 2.0 s. What is the emf induced in the coil?

30. A 100-turn, 8.0-cm-diameter coil is made of 0.50-mm-diameter copper wire. A magnetic field is perpendicular to the coil. At what rate must B increase to induce a 2.0 A current in the coil?

31. A 2.0 cm × 2.0 cm square loop of wire with resistance 0.010 Ω is parallel to a long straight wire. The near edge of the loop is 1.0 cm from the wire. The current in the wire is increasing at the rate of 100 A/s. What is the current in the loop?

32. A 4.0-cm-diameter loop with resistance 0.10 Ω surrounds a 2.0-cm-diameter solenoid. The solenoid is 10 cm long, has 100 turns, and carries the current shown in the graph. A positive current is cw when seen from the left. Determine the current in the loop as a function of time. Give your answer as a current-versus-time graph from $t = 0$ s to $t = 3$ s.

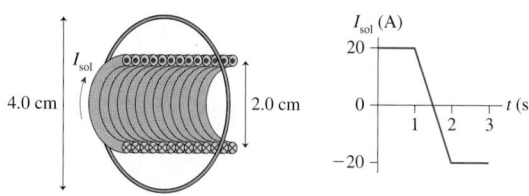

FIGURE P33. 32

33. A 20 cm × 20 cm square loop of wire lies in the xy-plane with its bottom edge on the x-axis. The resistance of the loop is 0.50 Ω. A magnetic field parallel to the z-axis is given by $B = 0.80y^2t$, where B is in tesla, y in meters, and t in seconds. What is the size of the induced current in the loop at $t = 0.50$ s?

34. Figure P33.34 shows a five-turn, 1.0-cm-diameter coil with $R = 0.10$ Ω inside a 2.0-cm-diameter solenoid. The solenoid is 8.0 cm long, has 120 turns, and carries the current shown in the graph. A positive current is cw when seen from the left. Determine the current in the coil as a function of time. Give your answer as a current-versus-time graph from $t = 0$ s to $t = 0.02$ s.

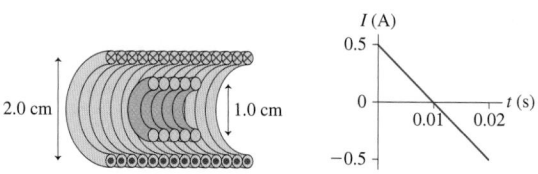

FIGURE P33.34

35. Two 20-turn coils are tightly wrapped on the same 2.0-cm-diameter cylinder with 1.0-mm-diameter wire. The current through coil 1 is shown in the graph. A positive current is into the page at the top of a loop. Determine the current in coil 2 as a function of time. Give your answer as a current-versus-time graph from $t = 0$ s to $t = 0.4$ s. Assume that the magnetic field of coil 1 passes entirely through coil 2.

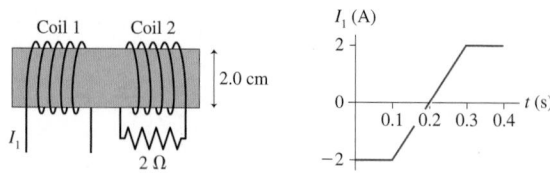

FIGURE P33.35

36. A loop antenna, such as is used on a television to pick up UHF broadcasts, is 25 cm in diameter. The plane of the loop is perpendicular to the oscillating magnetic field of a 150 MHz electromagnetic wave. The magnetic field through the loop is $B = (20 \text{ nT})\sin\omega t$.
    a. What is the maximum emf induced in the antenna?
    b. What is the maximum emf if the loop is turned 90° to be perpendicular to the oscillating electric field?

37. A 50-turn, 4.0-cm-diameter coil with $R = 0.50\ \Omega$ surrounds a 2.0-cm-diameter solenoid. The solenoid is 20 cm long and has 200 turns. The 60 Hz current through the solenoid is $I_{\text{sol}} = (0.50 \text{ A})\sin(2\pi ft)$. Find an expression for I_{coil}, the induced current in the coil as a function of time.

38. A 40-turn, 4.0-cm-diameter coil with $R = 0.40\ \Omega$ surrounds a 3.0-cm-diameter solenoid. The solenoid is 20 cm long and has 200 turns. The 60 Hz current through the solenoid is $I = I_0 \sin(2\pi ft)$. What is I_0 if the maximum current in the coil is 0.20 A?

39. Electricity is distributed from electrical substations to neighborhoods at 15,000 V. This is a 60 Hz oscillating (AC) voltage. Neighborhood transformers, such as those seen on utility poles, step this voltage down to the 120 V that is delivered to your house.
    a. How many turns does the primary coil on the transformer have if the secondary coil has 100 turns?
    b. No energy is lost in an ideal transformer, so the output power P_{out} from the secondary coil equals the input power P_{in} to the primary coil. Suppose a neighborhood transformer delivers 250 A at 120 V. What is the current in the 15,000 V line from the substation?

40. The square loop shown in Figure P33.40 moves into a 0.80 T magnetic field at a constant speed of 10 m/s. The loop is 10.0 cm on each side and has a resistance of 0.10 Ω. It enters the field at $t = 0$ s.
    a. Find the induced current in the loop as a function of time. Give your answer as a graph of I versus t from $t = 0$ s to $t = 0.020$ s.
    b. What is the maximum current? What is the position of the loop when the current is maximum?

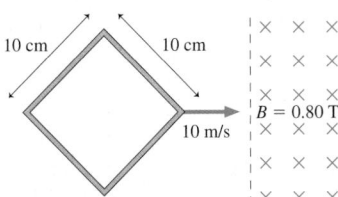

FIGURE P33.40

41. A small, 2.0-mm-diameter circular loop with $R = 0.020\ \Omega$ is at the center of a large 100-mm-diameter circular loop. Both loops lie in the same plane. The current in the outer loop changes from +1.0 A to −1.0 A in 0.10 s. What is the induced current in the inner loop?

42. A 4.0-cm-long slide wire moves outward with a speed of 100 m/s in a 1.0 T magnetic field. (See Figure 33.27.) At the instant the circuit forms a 4.0 cm × 4.0 cm square, with $R = 0.010\ \Omega$ on each side, what are
    a. The induced emf?
    b. The induced current?
    c. The potential difference between the two ends of the moving wire?

43. A 20-cm-long, zero-resistance slide wire moves outward, on zero-resistance rails, at a steady speed of 10 m/s in a 0.10 T magnetic field. (See Figure 33.27.) On the opposite side, a 1.0 Ω carbon resistor completes the circuit by connecting the two rails. The mass of the resistor is 50 mg.
    a. What is the induced current in the circuit?
    b. How much force is needed to pull the wire at this speed?
    c. If the wire is pulled for 10 s, what is the temperature increase of the carbon? The specific heat of carbon is 710 J/kgC°.

44. The 10-cm-wide, zero-resistance slide wire shown in Figure P33.44 is pushed toward the 2.0 Ω resistor at a steady speed of 0.50 m/s. The magnetic field strength is 0.50 T.
    a. How big is the pushing force?
    b. How much power does the pushing force supply to the wire?
    c. What are the direction and magnitude of the induced current?
    d. How much power is dissipated in the resistor?

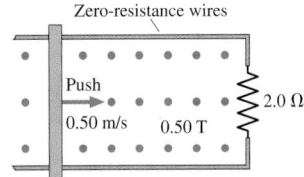

FIGURE P33.44

45. Your camping buddy has an idea for a light to go inside your tent. He happens to have a powerful (and heavy!) horseshoe magnet that he bought at a surplus store. This magnet creates a 0.20 T field between two pole tips 10 cm apart. His idea is to build a hand-cranked generator with a rotating 5.0-cm-radius semicircle between the pole tips. He thinks you can make enough current to fully light a 1.0 Ω lightbulb rated at 4.0 W. That's not super bright, but it should be plenty of light for routine activities in the tent.

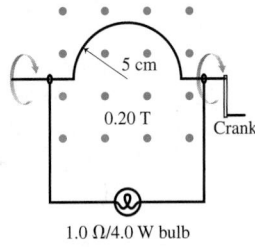

FIGURE P33.45

    a. Find an expression for the induced current as a function of time if you turn the crank at frequency f. Assume that the semicircle is at its highest point at $t = 0$.
    b. With what frequency will you have to turn the crank for the maximum current to fully light the bulb? Is this feasible?

46. You've decided to make a magnetic projectile launcher for your science project. An aluminum bar of length l slides along metal rails through a magnetic field B. The switch closes at $t = 0$ s, while the bar is at rest, and a battery of emf $\mathcal{E}_{\text{bat}}$ starts a current flowing around the loop. The battery has internal resistance r. The resistance of the rails and the bar are effectively zero.
    a. Show that the bar reaches a terminal velocity v_{term}, and find an expression for v_{term}.
    b. Evaluate v_{term} for $\mathcal{E}_{\text{bat}} = 1.0$ V, $r = 0.10\ \Omega$, $l = 6.0$ cm, and $B = 0.50$ T.

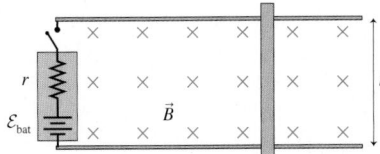

FIGURE P33.46

47. A slide wire of length l, mass m, and resistance R slides down a U-shaped metal track that is tilted upward at angle θ. The track has zero resistance and no friction. A vertical magnetic field B fills the loop formed by the track and the slide wire.
    a. Find an expression for the induced current I when the slide wire moves at speed v.
    b. Show that the slide wire reaches a terminal velocity v_{term}, and find an expression for v_{term}.

48. The plane of a 20 cm × 20 cm metal loop with a mass of 10 g and a resistance of 0.010 Ω is oriented vertically. A 1.0 T horizontal magnetic field, perpendicular to the loop, fills the top half of the loop. There is no magnetic field through the bottom half of the loop. The loop is released from rest and allowed to fall.
    a. Show that the loop reaches a terminal velocity v_{term}, and find a value for v_{term}.
    b. How long will it take the loop to leave the field? Assume that the time needed to reach v_{term} is negligible. How does this compare to the time it would take the loop to fall the same distance in the absence of a field?

49. Figure P33.49 shows a U-shaped conducting rail that is oriented vertically in a horizontal magnetic field. The rail has no electric resistance and does not move. A slide wire with mass m and resistance R can slide up and down without friction while maintaining electrical contact with the rail. The slide wire is released from rest.
    a. Show that the slide wire reaches a terminal velocity v_{term}, and find an expression for v_{term}.
    b. Determine the value of v_{term} if $l = 20$ cm, $m = 10$ g, $R = 0.10$ Ω, and $B = 0.50$ T.

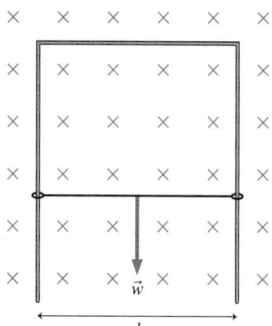

FIGURE P33.49

50. A 10-turn coil of wire having a diameter of 1.0 cm and a resistance of 0.20 Ω is in a 1.0 mT magnetic field, with the coil oriented for maximum flux. The coil is connected to an uncharged 1.0 μF capacitor rather than to a current meter. The coil is quickly pulled out of the magnetic field. Afterward, what is the voltage across the capacitor?
    Hint: Use $I = dq/dt$ to relate the *net* change of flux to the amount of charge that flows to the capacitor.

51. The magnetic field at one place on the earth's surface is 55 μT in strength and tilted 60° down from horizontal. A 200-turn coil having a diameter of 4.0 cm and a resistance of 2.0 Ω is connected to a 1.0 μF capacitor rather than to a current meter. The coil is held in a horizontal plane and the capacitor is discharged. Then the coil is quickly rotated 180° so that the side that had been facing up is now facing down. Afterward, what is the voltage across the capacitor? See the Hint in Problem 50.

52. A 100 mH inductor whose windings have a resistance of 4.0 Ω is connected across a 12 V battery having an internal resistance of 2.0 Ω. How much energy is stored in the inductor?

53. A solenoid inductor carries a current of 200 mA. It has a magnetic flux of 20 μWb per turn and stores 1.0 mJ of energy. How many turns does the inductor have?

54. A solenoid inductor has an emf of 0.20 V when the current through it changes at the rate 10.0 A/s. A steady current of 0.10 A produces a flux of 5.0 μWb per turn. How many turns does the inductor have?

55. a. What is the magnetic energy density at the surface of a 1.0-mm-diameter wire carrying a current of 1.0 A?
    b. A 2.0-cm-diameter tightly wound solenoid is made with 1.0-mm-diameter wire. What is the magnetic energy density inside the solenoid if the current is 1.0 A?

56. a. What is the magnetic energy density at the center of a 4.0-cm-diameter loop carrying a current of 1.0 A?
    b. What current in a straight wire gives the magnetic energy density you found in part a at a point 2.0 cm from the wire?

57. The earth's magnetic field at the earth's surface is approximately 50 μT. The earth's atmosphere is approximately 20 km thick.
    a. What is the total energy of the earth's magnetic field in the atmosphere? You can assume that the field strength within the atmosphere is constant.
    b. The world's total energy use is approximately 4×10^{18} J per year. If the magnetic energy in the atmosphere could be "harvested," what percentage of the world's annual energy use could it supply?

58. MRI (magnetic resonance imaging) is a medical technique that produces detailed "pictures" of the interior of the body. The patient is placed into a solenoid that is 40 cm in diameter and 1.0 m long. A 100 A current creates a 5.0 T magnetic field inside the solenoid. To carry such a large current, the solenoid wires are cooled with liquid helium until they become superconducting (no electric resistance).
    a. How much magnetic energy is stored in the solenoid? Assume that the magnetic field is uniform within the solenoid and quickly drops to zero outside the solenoid.
    b. How many turns of wire does the solenoid have?

59. One possible concern with MRI (see Problem 58) is turning the magnetic field on or off too quickly. Bodily fluids are conductors, and a changing magnetic field could cause electric currents to flow through the patient. Suppose a typical patient has a maximum cross-section area of 0.060 m². What is the smallest time interval in which a 5.0 T magnetic field can be turned on or off if the induced emf around the patient's body must be kept to less than 0.10 V?

60. Experiments to study vision often need to track the movements of a subject's eye. One way of doing so is to have the subject sit in a magnetic field while wearing special contact lenses that have a coil of very fine wire circling the edge. A current is induced in the coil each time the subject rotates his eye. Consider an experiment in which a 20-turn, 6.0-mm-diameter coil of wire circles the subject's cornea while a 1.0 T magnetic field is directed as shown. The subject begins by looking straight ahead. What emf is induced in the coil if the subject shifts his gaze by 5° in 0.20 s?

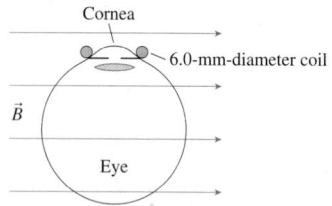

FIGURE P33.60

61. Figure P33.61 shows the current through a 10 mH inductor. Draw a graph showing the potential difference ΔV_L across the inductor for these 6 ms.

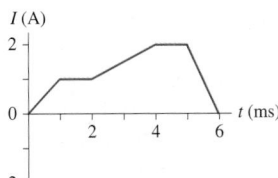

FIGURE P33.61

62. Figure P33.62 shows the current through a 10 mH inductor. Draw a graph showing the potential difference ΔV_L across the inductor for these 6 ms.

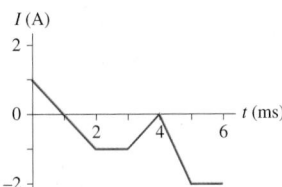

FIGURE P33.62

63. Figure P33.63 shows the potential difference across a 50 mH inductor. The current through the inductor at $t = 0$ s is 0.20 A. Draw a graph showing the current through the inductor from $t = 0$ s to $t = 40$ ms.

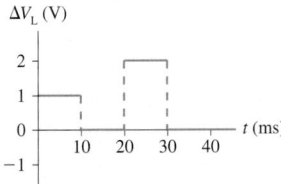

FIGURE P33.63

64. Figure P33.64 shows the potential difference across a 100 mH inductor. The current through the inductor at $t = 0$ s is 0.10 A. Draw a graph showing the current through the inductor from $t = 0$ s to $t = 40$ ms.

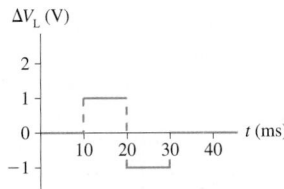

FIGURE P33.64

65. The current through inductance L is given by $I = I_0 e^{-t/\tau}$.
    a. Find an expression for the potential difference ΔV_L across the inductor.
    b. Evaluate ΔV_L at $t = 0$ s, 1, 2, and 3 ms if $L = 20$ mH, $I_0 = 50$ mA, and $\tau = 1.0$ ms.
    c. Draw a graph of ΔV_L versus time from $t = 0$ s to $t = 3$ ms.
66. The current through inductance L is given by $I = I_0 \sin \omega t$.
    a. Find an expression for the potential difference ΔV_L across the inductor.
    b. The maximum voltage across the inductor is 0.20 V when $L = 50 \ \mu$H and $f = 500$ kHz. What is I_0?

67. An LC circuit is built with a 20 mH inductor and an 8.0 μF capacitor. The current has its maximum value of 0.50 A at $t = 0$ s.
    a. How long is it until the capacitor is fully charged?
    b. What is the voltage across the capacitor at that time?
68. An LC circuit has a 10 mH inductor. The current has its maximum value of 0.60 A at $t = 0$ s. A short time later the capacitor reaches its maximum potential difference of 60 V. What is the value of the capacitance?
69. The maximum charge on the capacitor in an oscillating LC circuit is Q_0. What is the capacitor charge, in terms of Q_0, when the energy in the capacitor's electric field equals the energy in the inductor's magnetic field?
70. In recent years it has been possible to buy a 1.0 F capacitor. This is an enormously large amount of capacitance. Suppose you want to build a 1.0 Hz oscillator with a 1.0 F capacitor. You have a spool of 0.25-mm-diameter wire and a long 4.0-cm-diameter plastic cylinder. Design an inductor that will accomplish your goal.
71. The switch in Figure P33.71 has been in position 1 for a long time. It is changed to position 2 at $t = 0$ s.
    a. What is the maximum current through the inductor?
    b. What is the first time at which the current is maximum?

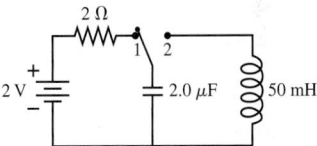

FIGURE P33.71

72. The 300 μF capacitor in Figure P33.72 is initially charged to 100 V, the 1200 μF capacitor is uncharged, and the switches are both open.
    a. What is the maximum voltage to which you can charge the 1200 μF capacitor by the proper closing and opening of the two switches?
    b. How would you do it? Describe the sequence in which you would close and open switches and the times at which you would do so. The first switch is closed at $t = 0$ s.

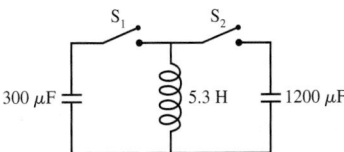

FIGURE P33.72

73. The switch in Figure P33.73 has been open for a long time. It is closed at $t = 0$ s.
    a. What is the current through the battery immediately after the switch is closed?
    b. What is the current through the battery after the switch has been closed a long time?

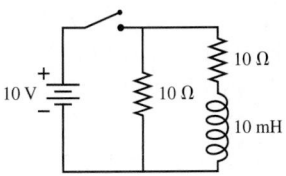

FIGURE P33.73

74. The switch in Figure P33.74 has been open for a long time. It is closed at $t = 0$ s. What is the current through the 20 Ω resistor
    a. immediately after the switch is closed?
    b. after the switch has been closed a long time?
    c. immediately after the switch is reopened?

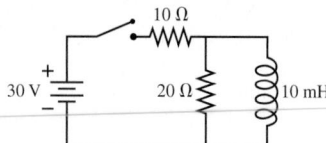

FIGURE P33.74

75. The switch in Figure P33.75 has been open for a long time. It is closed at $t = 0$ s.
    a. After the switch has been closed for a long time, what is the current in the circuit? Call this current I_0.
    b. Find an expression for the current I as a function of time. Write your expression in terms of I_0, R, and L.
    c. Sketch a current-versus-time graph from $t = 0$ s until the current is no longer changing.

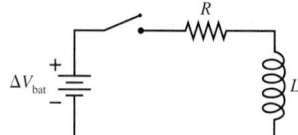

FIGURE P33.75

Challenge Problems

76. The L-shaped conductor in Figure CP33.76 moves at 10 m/s across a stationary L-shaped conductor in a 0.10 T magnetic field. The two vertices overlap, so that the enclosed area is zero, at $t = 0$ s. The conductor has a resistance of 0.010 ohms *per meter.*
    a. What is the direction of the induced current?
    b. Find expressions for the induced emf and the induced current as functions of time.
    c. Evaluate $\mathcal{E}$ and I at $t = 0.10$ s.

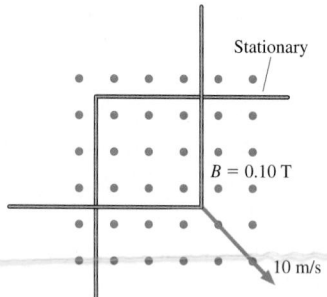

FIGURE CP33.76

77. The metal wire in Figure CP33.77 moves with speed v parallel to a straight wire that is carrying current I. The distance between the two wires is d. Find an expression for the potential difference between the two ends of the moving wire.

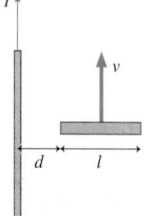

FIGURE CP33.77

78. A closed, square loop is formed with 40 cm of wire having $R = 0.10\ \Omega$, as shown in Figure CP33.78. A 0.50 T magnetic field is perpendicular to the loop. At $t = 0$ s, two diagonally opposite corners of the loop begin to move apart at 0.293 m/s.
    a. How long does it take the loop to collapse to a straight line?
    b. Find an expression for the induced current I as a function of time while the loop is collapsing. Assume that the sides remain straight lines during the collapse.
    c. Evaluate I at four or five times during the collapse, then draw a graph of I versus t.

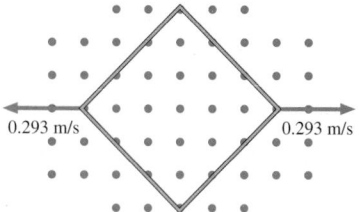

FIGURE CP33.78

79. Let's look at the details of eddy-current braking. A square loop, length l on each side, is shot with velocity v_0 into a uniform magnetic field B. The field is perpendicular to the plane of the loop. The loop has mass m and resistance R, and it enters the field at $t = 0$. Assume that the loop is moving to the right along the x-axis and that the field begins at $x = 0$.
    a. Find an expression for the loop's velocity as a function of time as it enters the magnetic field. You can ignore gravity, and you can assume that the back edge of the loop has not entered the field.
    b. Calculate and draw a graph of v over the interval $0\ \text{s} \le t \le 0.04$ s for the case that $v_0 = 10$ m/s, $l = 10$ cm, $m = 1.0$ g, $R = 0.0010\ \Omega$, and $B = 0.10$ T. The back edge of the loop does not reach the field during this time interval.

80. An 8.0 cm × 8.0 cm square loop is halfway into a magnetic field that is perpendicular to the plane of the loop. The loop's mass is 10 g and its resistance is 0.010 Ω. A switch is closed at $t = 0$ s, causing the magnetic field to increase from 0 to 1.0 T in 0.010 s.
    a. What is the induced current in the square loop?
    b. What is the force on the loop when the magnetic field is 0.50 T? Is the force directed into the magnetic field or away from the magnetic field?
    c. What is the loop's acceleration at $t = 0.005$ s, when the field strength is 0.50 T? If this acceleration stayed constant, how far would the loop move in 0.010 s?
    d. Because 0.50 T is the average field strength, your answer to c is an estimate of how far the loop moves during the 0.010 s in which the field increases to 1.0 T. If your answer is ≪8 cm, then it is reasonable to neglect the movement of the loop during the 0.010 s that the field ramps up. Is neglecting the movement reasonable?
    e. With what speed is the loop "kicked" away from the magnetic field?
    Hint: What is the impulse on the loop?

81. High-frequency signals are often transmitted along a *coaxial cable,* such as the one shown in Figure CP33.81. For example, the cable TV hookup coming into your home is a coaxial cable. The signal is carried on a wire of radius r_1 while the outer conductor of radius r_2 is grounded. A soft, flexible insulating material fills the space between them, and an insulating plastic coating goes around the outside.

    a. Find an expression for the inductance per meter of a coaxial cable. To do so, consider the flux through a rectangle of length l that spans the gap between the inner and outer conductor.

    b. Evaluate the inductance per meter of a cable having $r_1 = 0.50$ mm and $r_2 = 3.0$ mm.

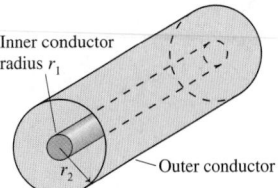

Inner conductor radius r_1

r_2

Outer conductor

FIGURE CP33.81

STOP TO THINK ANSWERS

Stop to Think 33.1: e. According to the right-hand rule, the magnetic force on a positive charge carrier is to the right.

Stop to Think 33.2: No. The charge carriers in the wire move parallel to $\vec{B}$. There's no magnetic force on a charge moving parallel to a magnetic field.

Stop to Think 33.3: $F_2 = F_4 > F_1 = F_3$. $\vec{F}_1$ is zero because there's no field. $\vec{F}_3$ is also zero because there's no current around the loop. The charge carriers in both the right and left edges are pushed to the bottom of the loop, creating a motional emf but no current. The currents at 2 and 4 are in opposite directions, but the forces on the segments in the field are both to the left and of equal magnitude.

Stop to Think 33.4: Clockwise. The wire's magnetic field as it passes through the loop is into the page. The flux through the loop

decreases into the page as the wire moves away. To oppose this decrease, the induced magnetic field needs to point into the page.

Stop to Think 33.5: d. The flux is increasing into the loop. To oppose this increase, the induced magnetic field needs to point out of the page. This requires a cw induced current. Using the right-hand rule, the magnetic force on the current in the left edge of the loop is to the right, away from the field. The magnetic forces on the top and bottom segments of the loop are in opposite directions and cancel each other.

Stop to Think 33.6: b or **f.** The potential decreases in the direction of increasing current and increases in the direction of decreasing current.

Stop to Think 33.7: $\tau_3 > \tau_1 > \tau_2$. $\tau = L/R$, so smaller total resistance gives a larger time constant. The parallel resistors have total resistance $R/2$. The series resistors have total resistance $2R$.

34 Electromagnetic Fields and Waves

A laser beam is a subtle interplay of oscillating electric and magnetic fields.

▶ Looking Ahead

The goal of Chapter 34 is to study the properties of electromagnetic fields and waves. In this chapter you will learn that:

- The electric field $\vec{E}$ and the magnetic field $\vec{B}$ are *real,* not just convenient fictions.
- The electric field and magnetic field are interdependent. Furthermore, the fields can exist independently of charges and currents.
- The fields obey four general laws, called Maxwell's equations.
- Maxwell's equations predict the existence of electromagnetic waves that travel at speed c, the speed of light.

◀ Looking Back

This chapter will synthesize many ideas about fields and motion. Please review:

- Section 6.4 Relative motion.
- Section 20.3 Sinusoidal traveling waves.
- Sections 27.3–27.4 The electric flux and Gauss's law.
- Sections 32.3 and 32.6 Magnetic fields and Ampère's law.
- Sections 33.5–33.6 Faraday's law and induced electric fields.

We've now spent nine chapters on electricity and magnetism. You might wonder what more there could be to learn. Surprisingly, there's much more. Our study of the basic properties of charges and currents has been limited mostly to *static* electric and magnetic fields, fields that don't change with time. To understand a laser beam, we need to know how electric and magnetic fields change with time. Other important examples of time-dependent electromagnetic phenomena include high-speed circuits, transmission lines, radar, and optical communications.

Faraday's law has been our one example thus far of time-dependent fields. In Chapter 33, you learned that a *changing* magnetic field creates an electric field. Our goal in this chapter is to explore further the dynamic relationships of electromagnetic fields. We'll also investigate how fields appear to someone moving through them. In the end, we'll find that it makes more sense to think of a single *electromagnetic field* rather than independent electric and magnetic fields.

Our study of electromagnetic fields will culminate in Maxwell's equations for the electromagnetic field. These equations play a role in electricity and magnetism analogous to Newton's laws in mechanics. Maxwell's realization that light is an electromagnetic wave was perhaps the most important discovery of the 19th century.

34.1 Electromagnetic Fields and Forces

We've looked at many examples of electric and magnetic phenomena, but now we want to return to the most basic ideas about electromagnetic fields and forces. The central idea of electricity and magnetism is that a charge alters the space around it by creating in that space an electric field and also, if the charge is moving, a magnetic field. These fields exert electric and magnetic forces on other charges. In other words, fields are the means by which charges interact with each other.

(a)

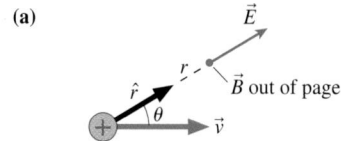

The electric and magnetic fields of a single point charge, we know from Chapters 25 and 32, are

$$\vec{E} = \frac{1}{4\pi\epsilon_0}\frac{q}{r^2}\hat{r} = \left(\frac{1}{4\pi\epsilon_0}\frac{q}{r^2}, \text{away from } q\right)$$

$$\vec{B} = \frac{\mu_0}{4\pi}\frac{q\vec{v}\times\hat{r}}{r^2} = \left(\frac{\mu_0}{4\pi}\frac{qv\sin\theta}{r^2}, \text{direction given by right-hand rule}\right)$$

(34.1)

(b)

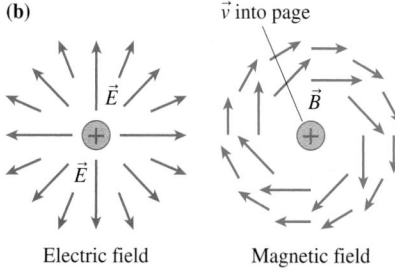

Electric field   Magnetic field

FIGURE 34.1 The electric and magnetic fields of a positive point charge.

where $\hat{r}$ is a unit vector pointing from the charge to the point at which the field is calculated. Figure 34.1a is the basic geometry, and Figure 34.1b reminds you of the fields of a positive charge.

Electric and magnetic fields can be portrayed using either field vectors or field lines. Both are useful visual aides, but neither is a perfect representation of a field. We'll find it most convenient in this chapter to use the field-line representation. Tactics Box 34.1 reminds you how to draw and use field lines.

TACTICS BOX 34.1 Drawing and using field lines

❶ The lines are continuous curves drawn tangent to the field vectors. Conversely, the field vector at any point is tangent to the field line at that point.

❷ Field lines never cross. This conclusion follows from the fact that the field vector has a unique direction at every point in space.

❸ The density of the lines indicates the field strength. Closely spaced lines indicate a large field strength, widely spaced lines a small field strength.

❹ Coulomb electric field lines start or stop only on charges. The lines start from positive charges and end on negative charges.

❺ Magnetic field lines form continuous, noncrossing loops. They do so because there are no magnetic monopoles on which lines could start or stop.

NOTE ▶ Rule 4 refers to *Coulomb electric fields,* which are those electric fields created by source charges. We'll modify rule 4 in Section 34.3 when we discuss the *induced electric fields* associated with Faraday's law. ◀

As examples, Figure 34.2 shows the electric field lines of a point charge and the magnetic field lines of a magnetic dipole. Notice how the magnetic field lines form continuous, noncrossing loops, although some of the field lines exceed the size of the picture. In both examples, the field lines are closer together in regions where the field is stronger.

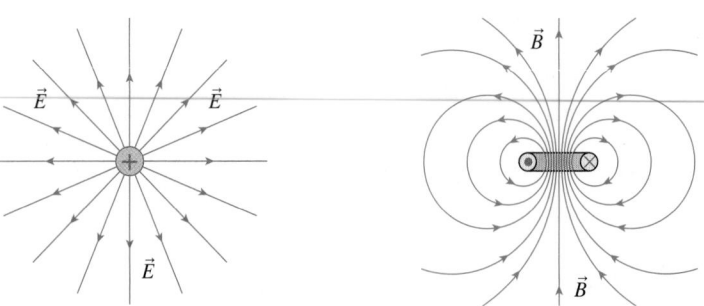

Electric field of a point charge   Magnetic field of a dipole

FIGURE 34.2 Electric and magnetic field-line diagrams.

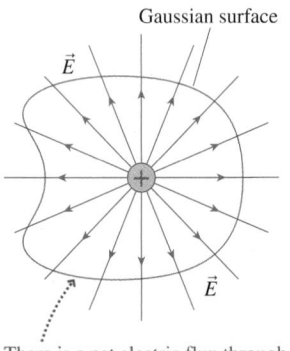

Gaussian surface

There is a net electric flux through this surface that encloses a charge.

FIGURE 34.3 A Gaussian surface enclosing a charge.

A *uniform* electric or magnetic field, which would be represented by parallel, equally spaced field lines, is an idealization that can only approximately be realized in practice. After all, magnetic field lines must eventually curve around and reconnect, although we might be able to make the curvature extremely small in a limited region of space. And electric field lines must start and stop on charges, although the lines might be very close to parallel if the charges are carefully arranged. The uniform field is an important model, one that we will continue to use in this chapter, but it is a good approximation only over a limited region of space.

Gauss's Law Revisited

Gauss's law, which you studied in Chapter 27, states a very general property of the electric field. It says that charges create electric fields in just such a way that the electric flux of the field is the same through *any* closed surface surrounding the charges. Figure 34.3 illustrates this idea by showing the field lines passing through a Gaussian surface that encloses a charge.

The mathematical statement of Gauss's law for the electric field says that for any *closed* surface enclosing total charge Q_{in}, the net electric flux through the surface is

$$\Phi_e = \oint \vec{E} \cdot d\vec{A} = \frac{Q_{in}}{\epsilon_0} \tag{34.2}$$

The circle on the integral sign indicates that the integration is over a closed surface.

We introduced the idea of magnetic flux Φ_m in Chapter 33, but for Faraday's law we considered the flux only through flat surfaces. What is the net magnetic flux over a closed surface? That is, what is the magnetic version of Gauss's law, analogous to Equation 34.2?

Figure 34.4 shows a Gaussian surface around a magnetic dipole. Magnetic field lines form continuous curves, without starting or stopping, so every field line leaving the surface at some point reenters it at another. Consequently, the net magnetic flux over a *closed* surface is zero.

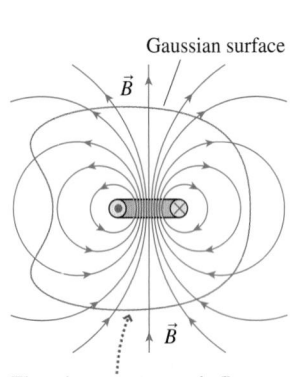

Gaussian surface

There is no net magnetic flux through this closed surface.

FIGURE 34.4 There is no net flux through a Gaussian surface around a magnetic dipole.

We've shown only one surface and one magnetic field, but this conclusion turns out to be a general property of magnetic fields. Because every north pole is always accompanied by a south pole, we can't enclose a "net pole" within a surface. Thus Gauss's law for magnetic fields is

$$\Phi_m = \oint \vec{B} \cdot d\vec{A} = 0 \tag{34.3}$$

Equation 34.2 is the mathematical statement that Coulomb electric field lines start and stop on charges. Equation 34.3 is the mathematical statement that magnetic field lines form closed loops; they don't start or stop (i.e., there are no magnetic monopoles). These two versions of Gauss's law are important statements about what types of fields can and cannot exist. They will become two of Maxwell's equations.

The Lorentz Force

A charge experiences a force in a region of electric or magnetic fields. The force is an interaction with the source charges and currents that created the fields, but one of the wonderful and important aspects of fields is that we don't need to know anything about the sources. Once we've determined their fields, we can forget about the sources and focus on what happens to a charge in the field.

You've learned that the force of an electric field on a charge is

$$\vec{F}_E = q\vec{E}$$

and that the force of a magnetic field on a moving charge is

$$\vec{F}_B = q\vec{v} \times \vec{B}$$

The magnitude of the magnetic force varies from 0 when $\vec{v}$ and $\vec{B}$ are parallel to qvB when the charge moves perpendicular to $\vec{B}$.

If a charge moves through a region of space in which there are both an electric *and* a magnetic field, the net force on the charge is

$$\vec{F} = q(\vec{E} + \vec{v} \times \vec{B}) \tag{34.4}$$

Equation 34.4 is called the **Lorentz force law.** It is the most general statement of the electric and magnetic forces on a charge.

EXAMPLE 34.1 **The motion of a proton**

A proton is launched with velocity $\vec{v}_0 = v_0\hat{j}$ into a region of space in which an electric field $\vec{E} = E_0\hat{i}$ and a magnetic field $\vec{B} = B_0\hat{i}$ are parallel. How many cyclotron orbits will the proton make while traveling distance L along the x-axis? Find an algebraic expression, then evaluate your answer if $E_0 = 10,000$ V/m, $B_0 = 0.10$ T, $v_0 = 1.0 \times 10^5$ m/s, and $L = 10$ cm.

MODEL Assume that the electric and magnetic fields are uniform fields.

VISUALIZE Figure 34.5 shows the proton moving in the parallel fields. The component of $\vec{v}$ perpendicular to $\vec{B}$ causes the proton to undergo cyclotron motion around the magnetic field. Simultaneously, the electric field causes the proton to accelerate along the x-axis. This component of $\vec{v}$ is parallel to $\vec{B}$ and not affected by the magnetic field. The combined motion is a *helix*

whose loops become increasingly stretched out as the proton gains speed.

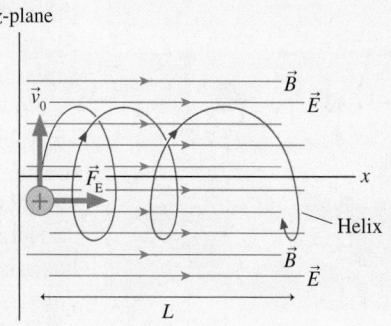

FIGURE 34.5 A proton moving in parallel electric and magnetic fields.

SOLVE The force on the proton is $\vec{F} = q(\vec{E} + \vec{v} \times \vec{B})$. The electric force points along the x-axis and causes the proton to accelerate, starting from $v_{0x} = 0$ m/s, with acceleration

$$a_x = \frac{F_x}{m} = \frac{eE_0}{m}$$

where we used $q = e$ for the charge of a proton. The kinematic equation is

$$\Delta x = L = \frac{1}{2}a_x(\Delta t)^2 = \frac{eE_0}{2m}(\Delta t)^2$$

Thus the time required to travel distance L is

$$\Delta t = \sqrt{\frac{2mL}{eE_0}}$$

The magnetic force is perpendicular to both $\vec{v}$ and $\vec{B}$, causing cyclotron motion in the yz-plane with cyclotron frequency

$$f_{cyc} = \frac{eB_0}{2\pi m}$$

The cyclotron frequency is independent of v_0. The number of cyclotron orbits during the time Δt that it takes the proton to move distance L is

$$N_{orbits} = f_{cyc}\Delta t = \frac{eB_0}{2\pi m}\sqrt{\frac{2mL}{eE_0}} = \frac{B_0}{2\pi}\sqrt{\frac{2eL}{mE_0}}$$

With the values given, $N_{orbits} = 15.6$.

STOP TO THINK 34.1   What is the direction of the net force on the moving charge?

a. Left
b. Right
c. Into the page
d. Out of the page
e. Up and left at 45°
f. Down and left at 45°

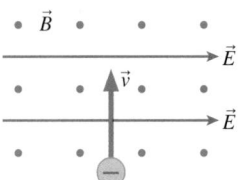

34.2 *E* or *B*? It Depends on Your Perspective

It seems clear, after the last nine chapters, that charges create an electric field $\vec{E}$ and that moving charges, or currents, create a magnetic field $\vec{B}$. Charges other than the source charges always respond to $\vec{E}$, but only moving charges respond to $\vec{B}$. Consider the following, however.

Figure 34.6a shows Sharon running past Bill with velocity $\vec{v}$ while carrying charge q. Bill sees a moving charge, and he knows that this charge creates a magnetic field given by Equation 34.1. But from Sharon's perspective, the charge is at rest. Stationary charges don't create magnetic fields, so Sharon claims that the magnetic field is zero. Is there or is there not a magnetic field?

Or what about the situation in Figure 34.6b? This time Sharon runs through an external magnetic field $\vec{B}$ that Bill has created. Bill sees a charge moving through a magnetic field, so he knows there's a force $\vec{F} = q\vec{v} \times \vec{B}$ on the charge. Using the right-hand rule, Bill determines that the force points straight up. But for Sharon the charge is still at rest. Stationary charges don't experience magnetic forces, so Sharon claims that $\vec{F} = \vec{0}$.

Now, we may be a bit uncertain about magnetic fields, because they are an abstract concept, but surely there can be no disagreement over forces. After all, the charge is either going to accelerate upward or it isn't, and Bill and Sharon should be able to agree on the outcome.

Here we have a genuine paradox, not merely faulty reasoning. This paradox has arisen because we have fields and forces that depend on velocity. The difficulty is that we haven't looked at the issue of velocity *with respect to what* or velocity *as measured by whom*. A closer look at how electromagnetic fields are viewed by two experimenters moving relative to each other will lead us to conclude that $\vec{E}$ and $\vec{B}$ are not, as we've been assuming, separate and independent entities. They are closely intertwined.

(a)

Charge q moves with velocity $\vec{v}$ relative to Bill.

(b)

Charge q moves through a magnetic field $\vec{B}$ established by Bill.

FIGURE 34.6 Sharon carries a charge past Bill.

Galilean Relativity

We introduced reference frames and relative motion in Chapter 6, and a review of Section 6.4 is highly recommended. Figure 34.7 shows two reference frames that we'll call frame S and frame S'. Frame S' moves with velocity $\vec{V}$ with respect to frame S. That is, an experimenter at rest in S sees the origin of S' go past with velocity $\vec{V}$. Of course, an experimenter at rest in S' would say that frame S has velocity $-\vec{V}$. We'll use an uppercase V for the velocity of reference frames, reserving lowercase v for the velocity of objects moving in the reference frames. There's no implication that either frame is "at rest." All we know is that the two frames move relative to each other.

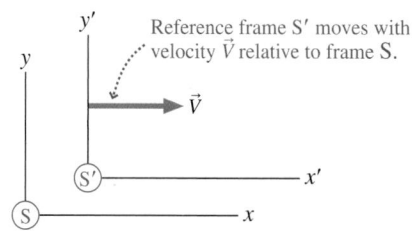

FIGURE 34.7 Reference frames S and S'.

NOTE ▶ We will consider only reference frames that move with respect to each other at *constant* velocity—unchanging speed in a straight line. You learned in Chapter 6 that these are called *inertial reference frames,* and they are the reference frames in which Newton's laws are valid. ◀

Figure 34.8 shows a physical object, such as a charged particle. Experimenters in frame S measure the motion of the particle and find that its velocity *relative to frame S* is $\vec{v}$. At the same time, experimenters in S' find that the particle's velocity *relative to frame S'* is $\vec{v}'$. In Chapter 6, we found that $\vec{v}$ and $\vec{v}'$ are related by

$$\vec{v}' = \vec{v} - \vec{V} \quad \text{or} \quad \vec{v} = \vec{v}' + \vec{V} \tag{34.5}$$

Equation 34.5, the *Galilean transformation of velocity,* allows us to transform a velocity measured in one reference frame into the velocity that would be measured by an experimenter in a different reference frame.

Suppose the particle in Figure 34.8 is accelerating. How does its acceleration $\vec{a}$, as measured by experimenters in frame S, compare to the acceleration $\vec{a}'$ measured in frame S'? We can answer this question by taking the time derivative of Equation 34.5:

$$\frac{d\vec{v}'}{dt} = \frac{d\vec{v}}{dt} - \frac{d\vec{V}}{dt}$$

The derivatives of $\vec{v}$ and $\vec{v}'$ are the particle's accelerations $\vec{a}$ and $\vec{a}'$ in frames S and S'. But $\vec{V}$ is a *constant* velocity, so $d\vec{V}/dt = 0$. Thus the Galilean transformation of acceleration is simply

$$\vec{a}' = \vec{a} \tag{34.6}$$

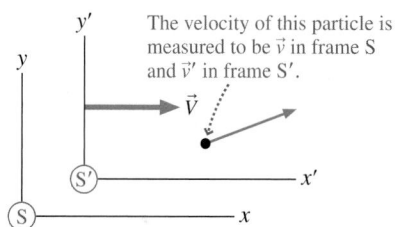

FIGURE 34.8 The particle's velocity is measured in both frame S and frame S'.

Sharon and Bill may measure different positions and velocities for a particle, but they *agree* on its acceleration. This agreement is important because acceleration is directly related to force. An experimenter in frame S would find that a force $\vec{F} = m\vec{a}$ is acting on the particle. Similarly, the force measured in frame S' is $\vec{F}' = m\vec{a}'$. But $\vec{a}' = \vec{a}$, hence

$$\vec{F}' = \vec{F} \tag{34.7}$$

Experimenters in all inertial reference frames agree about the force acting on a particle. This conclusion is the key to understanding how different experimenters see electric and magnetic fields.

The Transformation of Electric and Magnetic Fields

Now we're ready to return to the paradox that opened this section. Imagine that Bill has measured the electric field $\vec{E}$ and the magnetic field $\vec{B}$ in frame S. Our investigations thus far give us no reason to think that Sharon's measurements of the fields will differ from Bill's. After all, it seems like the fields are just "there," waiting to be measured. Thus our expectation is that Sharon, in frame S', will measure $\vec{E}' = \vec{E}$ and $\vec{B}' = \vec{B}$.

In S, the force on q is
due to a magnetic field.

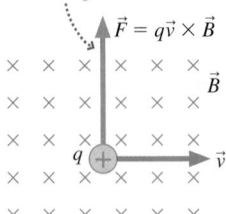

FIGURE 34.9 A charge in frame S moves through a magnetic field and experiences a magnetic force.

In S', the force on q is
due to an electric field.

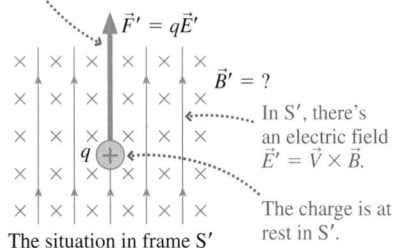

FIGURE 34.10 In frame S' the charge experiences an electric force.

To find out if this is true, Bill establishes a region of space in which there is a uniform magnetic field $\vec{B}$ but where $\vec{E} = \vec{0}$. Then, as shown in Figure 34.9, he shoots a positive charge q through the magnetic field. At an instant when q is moving horizontally with velocity $\vec{v}$, the Lorentz force $\vec{F} = q(\vec{E} + \vec{v} \times \vec{B}) = q\vec{v} \times \vec{B}$ is straight up.

Suppose that Sharon, in frame S', moves alongside the charge with velocity $\vec{V} = \vec{v}$. In other words, the charge is at rest in S'. We've just seen that experimenters in S and S' agree about forces, so if Bill finds an upward force in S then Sharon *must* observe an upward force in S'. But there is *no* magnetic force on a stationary charge, so how can this be?

Because Sharon in S' sees a stationary charge with an upward force that depends on the size of q, her only possible conclusion is that there is an upward-pointing *electric field*. After all, the electric field was initially defined in terms of the force experienced by a stationary charge. If the electric field in frame S' is $\vec{E}'$, then the force on the charge is $\vec{F}' = q\vec{E}'$. But we know that $\vec{F}' = \vec{F}$, and Bill has already measured $\vec{F} = q\vec{v} \times \vec{B} = q\vec{V} \times \vec{B}$. Thus we're led to the conclusion that

$$\vec{E}' = \vec{V} \times \vec{B} \tag{34.8}$$

As Sharon runs past Bill, she finds that at least part of Bill's magnetic field has become an electric field! **Whether a field is seen as "electric" or "magnetic" depends on the motion of the reference frame relative to the sources of the field.**

Figure 34.10 shows the situation from Sharon's perspective. The force on charge q is the same as that measured by Bill in Figure 34.9, but Sharon attributes this force to an electric field rather than a magnetic field. (Sharon needs a moving charge to measure magnetic forces, so we can't determine from this experiment whether or not Sharon experiences a magnetic field $\vec{B}'$. We'll return to this issue.)

More generally, suppose that an experimenter in S creates both an electric field $\vec{E}$ and a magnetic field $\vec{B}$. A charge moving with velocity $\vec{v}$ experiences the Lorentz force $\vec{F} = q(\vec{E} + \vec{v} \times \vec{B})$ shown in Figure 34.11a. The charge is at rest in frame S' that moves with velocity $\vec{V} = \vec{v}$, so the force in S' can be due only to an electric field, $\vec{F}' = q\vec{E}'$. Equating $\vec{F}$ and $\vec{F}'$, because experimenters in all inertial reference frames agree about forces, we find that

$$\vec{E}' = \vec{E} + \vec{V} \times \vec{B} \tag{34.9}$$

(a) The electric and magnetic
fields in frame S

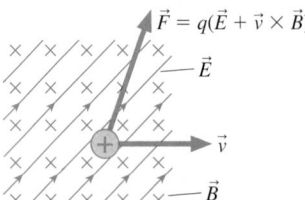

(b) The electric field in frame S',
where the charged particle is at rest

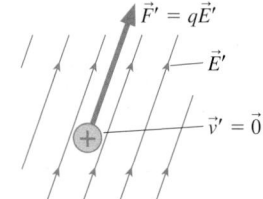

FIGURE 34.11 A charge in frame S experiences electric and magnetic forces. The charge experiences the same force in frame S', but it is due only to an electric field.

Equation 34.9 transforms the electric and magnetic fields in S into the electric field measured in frame S'. Figure 34.11b shows the outcome.

EXAMPLE 34.2 **Transforming the electric field**

Earlier, in Example 34.1, we considered a proton moving in the fields $\vec{E} = 10{,}000\hat{\imath}$ V/m and $\vec{B} = 0.10\hat{\imath}$ T. These are the fields in the laboratory. What is the electric field in a reference frame moving through the laboratory with velocity $\vec{V} = 1.0 \times 10^5 \hat{\jmath}$ m/s?

VISUALIZE Figure 34.12 shows the geometry. $\vec{E}$ and $\vec{B}$ are parallel to each other, along the *x*-axis, while velocity $\vec{V}$ of frame S′ is in the *y*-direction. Thus $\vec{V} \times \vec{B}$ points in the negative *z*-direction.

SOLVE $\vec{V}$ and $\vec{B}$ are perpendicular, so the magnitude of $\vec{V} \times \vec{B}$ is $VB = (1.0 \times 10^5$ m/s$)(0.10$ T$) = 10{,}000$ V/m. Thus the electric field in frame S′ is

$$\vec{E}' = \vec{E} + \vec{V} \times \vec{B} = (10{,}000\hat{\imath} - 10{,}000\hat{k}) \text{ V/m}$$

$$= (14{,}100 \text{ V/m}, 45° \text{ below the } x\text{-axis})$$

ASSESS A stationary positive charge in frame S′ experiences an electric force directed 45° below the *x*-axis. The force in frame S is the same, but, because the charge is moving in S, the force is attributed to a combination of electric and magnetic forces.

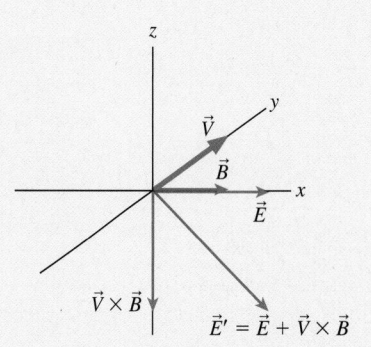

FIGURE 34.12 Finding the direction of field $\vec{E}'$.

Equation 34.9 transforms the fields $\vec{E}$ and $\vec{B}$ of frame S into the electric field $\vec{E}'$ of frame S′. In order to find a transformation equation for $\vec{B}'$, Figure 34.13a shows charge *q* at rest in frame S. Bill measures the fields of a stationary point charge, which we know are

$$\vec{E} = \frac{1}{4\pi\epsilon_0}\frac{q}{r^2}\hat{r} \qquad \vec{B} = \vec{0}$$

What are the fields at this point in space as measured by Sharon in frame S′? We can use Equation 34.9 to find $\vec{E}'$. Because $\vec{B} = \vec{0}$, the electric field in frame S′ is

$$\vec{E}' = \vec{E} = \frac{1}{4\pi\epsilon_0}\frac{q}{r^2}\hat{r} \tag{34.10}$$

In other words, Coulomb's law is still valid in a frame in which the point charge is moving. We needed to confirm that this is so, rather than just assuming it, because Coulomb's law was introduced in a frame in which the charges were at rest.

But Sharon also measures a magnetic field $\vec{B}'$ because, as seen in Figure 34.13b, charge *q* is moving away from her with velocity $\vec{v}' = -\vec{V}$. The magnetic field of a moving point charge is given by the Biot-Savart law, thus

$$\vec{B}' = \frac{\mu_0}{4\pi}\frac{q}{r^2}\vec{v}' \times \hat{r} = -\frac{\mu_0}{4\pi}\frac{q}{r^2}\vec{V} \times \hat{r} \tag{34.11}$$

where we used the fact that the charge's velocity in frame S′ is $\vec{v}' = -\vec{V}$.

It will be useful to rewrite Equation 34.11 as

$$\vec{B}' = -\frac{\mu_0}{4\pi}\frac{q}{r^2}\vec{V} \times \hat{r} = -\epsilon_0\mu_0\vec{V} \times \left(\frac{1}{4\pi\epsilon_0}\frac{q}{r^2}\hat{r}\right)$$

The expression in parentheses is simply $\vec{E}$, the electric field in frame S, so we find

$$\vec{B}' = -\epsilon_0\mu_0\vec{V} \times \vec{E} \tag{34.12}$$

Equation 34.12 expresses the remarkable idea that **the Biot-Savart law for the magnetic field of a moving point charge is nothing other than the Coulomb electric field of a stationary point charge transformed into a moving reference frame.**

(a) In frame S, the static charge creates an electric field but no magnetic field.

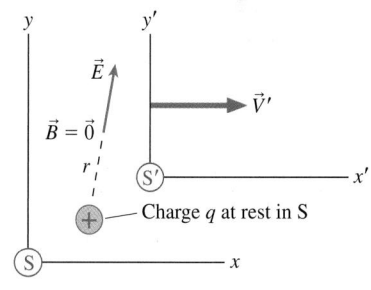

(b) In frame S′, the moving charge creates both an electric and a magnetic field.

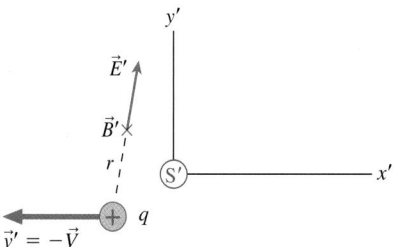

FIGURE 34.13 A charge at rest in frame S is moving in frame S′ and creates a magnetic field $\vec{B}'$.

We will assert without proof that if the experimenters in frame S create a magnetic field $\vec{B}$ in addition to the electric field $\vec{E}$, then the field $\vec{B}'$ measured in frame S′ is

$$\vec{B}' = \vec{B} - \epsilon_0\mu_0 \vec{V} \times \vec{E} \tag{34.13}$$

This is a more general transformation that matches Equation 34.9 for the electric field $\vec{E}'$.

Notice something interesting. The constant μ_0 has units of Tm/A; those of ϵ_0 are C²/Nm². By definition, 1 T = 1 N/Am and 1 A = 1 C/s. Consequently, the units of $\epsilon_0\mu_0$ turn out to be s²/m². In other words, the quantity $1/\sqrt{\epsilon_0\mu_0}$, with units of m/s, is a velocity. But what velocity? The constants are well known from measurements of static electric and magnetic fields, so it is straightforward to compute

$$\frac{1}{\sqrt{\epsilon_0\mu_0}} = \frac{1}{\sqrt{(1.26 \times 10^{-6}\,\text{Tm/A})(8.85 \times 10^{-12}\,\text{C}^2/\text{Nm}^2)}} = 3.00 \times 10^8 \text{ m/s}$$

Can this be a coincidence? Of all the possible values you might get from evaluating $1/\sqrt{\epsilon_0\mu_0}$, what are the chances it would come out to equal c, the speed of light? Maxwell was the first to discover this unexpected connection between the speed of light and the constants that govern the sizes of electric and magnetic forces, and he knew at once that this couldn't be a random coincidence. In Section 34.6 we'll show that electric and magnetic fields can exist as a *traveling wave,* and that the wave speed is predicted by the theory to be none other than

$$v_{\text{em}} = c = \frac{1}{\sqrt{\epsilon_0\mu_0}} \tag{34.14}$$

For now, we'll go ahead and write $\epsilon_0\mu_0 = 1/c^2$. With this, our **Galilean field transformation equations** are

$$\vec{E}' = \vec{E} + \vec{V} \times \vec{B} \qquad \qquad \vec{E} = \vec{E}' - \vec{V} \times \vec{B}'$$
$$\text{or}$$
$$\vec{B}' = \vec{B} - \frac{1}{c^2}\vec{V} \times \vec{E} \qquad \vec{B} = \vec{B}' + \frac{1}{c^2}\vec{V} \times \vec{E}' \tag{34.15}$$

where $\vec{V}$ is the velocity of frame S′ relative to frame S and where, to reiterate, the fields are measured *at the same point in space* by experimenters *at rest* in each reference frame.

NOTE ▶ We'll see shortly that these equations are valid only if $V \ll c$. ◀

We can no longer believe that electric and magnetic fields have a separate, independent existence. Changing from one reference frame to another mixes and rearranges the fields. Different experimenters watching an event will agree on the outcome, such as the deflection of a charged particle, but they will ascribe it to different combinations of fields. Our conclusion is that **there is just a single electromagnetic field that presents different faces, in terms of $\vec{E}$ and $\vec{B}$, to different viewers.** The whole concept of fields is beginning to look more complex, but also more interesting, than we first would have guessed!

EXAMPLE 34.3 **Two views of a magnetic field**
The 1.0 T magnetic field of a laboratory magnet points upward. A rocket flies past the laboratory, parallel to the ground, at 1000 m/s. What are the fields between the magnet's pole tips as measured by a scientist on board the rocket?

MODEL Assume that the laboratory and rocket reference frames are inertial reference frames.

VISUALIZE Figure 34.14 shows the magnet and establishes the reference frames.

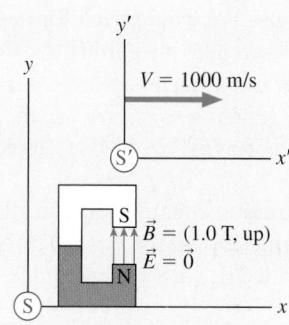

FIGURE 34.14 A rocket flies past a laboratory magnet.

SOLVE The fields in the laboratory frame are $\vec{B} = 1.0\hat{j}$ T and $\vec{E} = \vec{0}$. Frame S′, the frame of the rocket, moves with velocity $\vec{V} = 1000\hat{i}$ m/s. We can use Equations 34.15 to transform the

fields measured in the laboratory to the rocket frame S′. We find

$$\vec{E}' = \vec{E} + \vec{V} \times \vec{B} = \vec{V} \times \vec{B}$$

$$\vec{B}' = \vec{B} - \frac{1}{c^2}\vec{V} \times \vec{E} = \vec{B} = 1.0\hat{j}\ \text{T}$$

From the right-hand rule, $\vec{V} \times \vec{B}$ is out of the page, or in the $\hat{k}$ direction. $\vec{V}$ and $\vec{B}$ are perpendicular, so

$$\vec{E}' = VB\hat{k} = 1000\hat{k}\ \text{V/m} = (1000\ \text{V/m, out of page})$$

Thus the rocket scientist measures

$$\vec{B}' = 1.0\hat{j}\ \text{T} \quad\text{and}\quad \vec{E}' = 1000\hat{k}\ \text{V/m}$$

ASSESS The transformation equations apply only to fields measured at the *same* point in space. Thus these results apply to measurements made between the magnet's pole tips, where $\vec{B}$ is known, but not to other points in the laboratory.

Almost Relativity

Figure 34.15 shows two positive charges moving side by side through frame S with velocity $\vec{v}$. Charge q_1 creates an electric field and a magnetic field at the position of charge q_2. These are

$$\vec{E}_1 = \frac{1}{4\pi\epsilon_0}\frac{q_1}{r^2}\hat{j} \quad\text{and}\quad \vec{B}_1 = \frac{\mu_0}{4\pi}\frac{q_1 v}{r^2}\hat{k}$$

where r is the distance between the charges and we've used $\hat{r} = \hat{j}$ and $\vec{v} \times \hat{r} = v\hat{k}$.

How are the fields seen in frame S′, which moves with $\vec{V} = \vec{v}$ and in which the charges are at rest? From the field transformation equations,

$$\vec{B}_1' = \vec{B}_1 - \frac{1}{c^2}\vec{V} \times \vec{E}_1 = \frac{\mu_0}{4\pi}\frac{q_1 v}{r^2}\hat{k} - \frac{1}{c^2}\left(v\hat{i} \times \frac{1}{4\pi\epsilon_0}\frac{q_1}{r^2}\hat{j}\right)$$

$$= \frac{\mu_0}{4\pi}\frac{q_1 v}{r^2}\left(1 - \frac{1}{\epsilon_0\mu_0 c^2}\right)\hat{k} \tag{34.16}$$

where we used $\hat{i} \times \hat{j} = \hat{k}$. But $\epsilon_0\mu_0 = 1/c^2$, so the term in parentheses is zero and $\vec{B}' = \vec{0}$. This result was expected because q_1 is at rest in S′ and shouldn't create a magnetic field.

The transformation of the electric field is

$$\vec{E}_1' = \vec{E}_1 + \vec{V} \times \vec{B}_1 = \frac{1}{4\pi\epsilon_0}\frac{q_1}{r^2}\hat{j} + v\hat{i} \times \frac{\mu_0}{4\pi}\frac{q_1 v}{r^2}\hat{k}$$

$$= \frac{1}{4\pi\epsilon_0}\frac{q_1}{r^2}(1 - \epsilon_0\mu_0 v^2)\hat{j} = \frac{1}{4\pi\epsilon_0}\frac{q_1}{r^2}\left(1 - \frac{v^2}{c^2}\right)\hat{j} \tag{34.17}$$

where we used $\hat{i} \times \hat{k} = -\hat{j}$ and $\epsilon_0\mu_0 = 1/c^2$.

But now we have a problem. In frame S′, where the two charges are at rest and separated by distance r, the electric field due to charge q_1 should be simply

$$\vec{E}_1' = \frac{1}{4\pi\epsilon_0}\frac{q_1}{r^2}\hat{j}$$

The field transformation equations have given a "wrong" result for the electric field $\vec{E}'$.

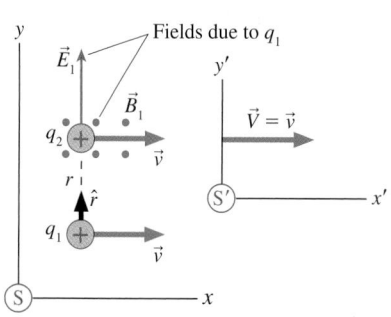

Fields seen in frame S

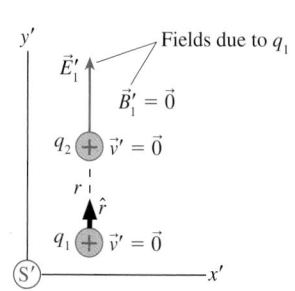

Fields seen in frame S′

FIGURE 34.15 Two charges moving parallel to each other.

It turns out that the field transformations of Equation 34.15, which are based on Galilean relativity, aren't quite right. We would need Einstein's relativity—a topic that we'll take up in Chapter 36—to give the correct transformations. However, the *Galilean* transformations in Equation 34.15 are equivalent to the relativistically correct transformations when $v \ll c$, in which case $v^2/c^2 \ll 1$. You can see that the two expressions for $\vec{E}_1'$ do, in fact, agree if v^2/c^2 can be neglected.

Thus our use of the field transformation equations has an additional rule: Set v^2/c^2 to zero. This is an acceptable rule for speeds $v < 10^7$ m/s. Even with this limitation, our investigation has provided us with a deeper understanding of electric and magnetic fields.

STOP TO THINK 34.2   Which diagram shows the fields in frame S′?

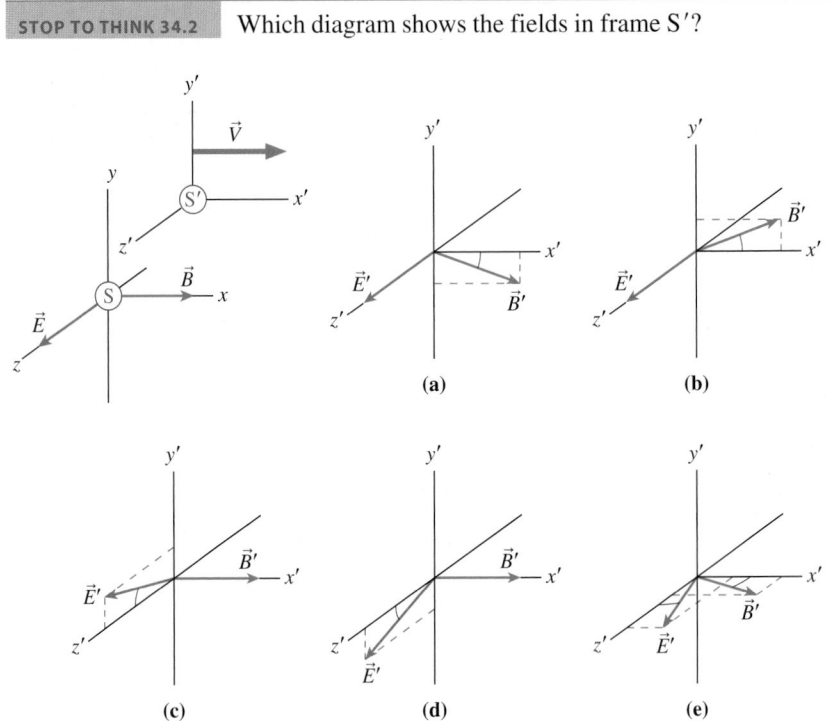

34.3 Faraday's Law Revisited

The transformation of electric and magnetic fields can give us new insight into Faraday's law. Figure 34.16a shows a reference frame S, which we can call the laboratory frame, in which a conducting loop is moving with velocity $\vec{v}$ into a magnetic field. You learned in Chapter 33 that the magnetic field exerts an upward force $\vec{F}_B = q\vec{v} \times \vec{B} = (qvB, \text{upward})$ on the charges in the leading edge of the wire, creating an emf $\mathcal{E} = vLB$ and an induced current in the loop. We called this a *motional emf.*

How do things appear to an experimenter who is in frame S′ that moves with the loop at velocity $\vec{V} = \vec{v}$ and for whom the loop is at rest? An important lesson of the previous section was that experimenters in different inertial reference frames agree about the outcome of any experiment, hence an experimenter in S′ agrees that there is an induced current in the loop. But the charges are at rest in frame S′, so there cannot be any magnetic force on them. How is the emf established in frame S′?

We couldn't have answered this question in Chapter 33, but now we've learned that the experimenter in frame S' doesn't see the same fields as the laboratory experimenter in S. In fact, we can use the field transformations to determine that the fields in S' are

$$\vec{E}' = \vec{E} + \vec{v} \times \vec{B} = \vec{v} \times \vec{B}$$

$$\vec{B}' = \vec{B} - \frac{1}{c^2}\vec{v} \times \vec{E} = \vec{B} \qquad (34.18)$$

where we used the fact that $\vec{E} = \vec{0}$ in the laboratory frame.

An experimenter in the loop's frame sees not only a magnetic field but also the electric field $\vec{E}'$ shown in Figure 34.16b. The magnetic field exerts no force on the charges, because they're at rest in this frame, but the electric field does. The force on charge q is $\vec{F} = q\vec{E}' = q\vec{v} \times \vec{B} = (qvB,$ upward). This is the same force as was measured in the laboratory frame, so it will cause the same emf and the same current. The outcome is identical, as we knew it had to be, but the experimenter in S' attributes the emf to an electric field whereas the experimenter in S attributes it to a magnetic field.

Field $\vec{E}'$ is, in fact, the *induced electric field*. Faraday's law, fundamentally, is a statement that **a changing magnetic field creates an electric field.** But only in frame S', the frame of the loop, is the magnetic field changing. Thus the induced electric field is seen in the loop's frame but not in the laboratory frame. The induced electric field is a *non-Coulomb* field because it is not created by static charges. It is a field that has been created in a new way.

Calculating the emf

The emf was defined in Chapter 30 as the work required per unit charge to separate the charge and thereby establish a potential difference. That is,

$$\mathcal{E} = \frac{W}{q} \qquad (34.19)$$

In batteries, a familiar source of emf, this work is done by chemical forces. But the emf that appears in Faraday's law arises when work is done by the forces of an induced electric field.

Figure 34.17 shows a charged particle moving through an electric field from a to b. We can calculate the work that the electric field does on the charge by dividing the path into many small displacement vectors $d\vec{s}$. The small amount of work done by the electric field as the charge moves through $d\vec{s}$ is $dW = \vec{F} \cdot d\vec{s} = q\vec{E} \cdot d\vec{s}$. The total work done on the particle as it moves from a to b is the sum of all the dW, or

$$W = q\int_a^b \vec{E} \cdot d\vec{s} \qquad (34.20)$$

This is a *line integral,* just like the line integral in Ampère's law except that it is an integral of $\vec{E} \cdot d\vec{s}$ rather than $\vec{B} \cdot d\vec{s}$. In Chapter 32 you learned that there are two situations in which evaluating the line integral is very simple:

- If $\vec{E}$ is everywhere perpendicular to the integration path, then $W = 0$.
- If $\vec{E}$ is everywhere tangent to an integration path of length L *and* has the same strength at all points, then $W = qEL$.

These two rules are sufficient for all the calculations we'll need to do *if* we choose the integration path carefully.

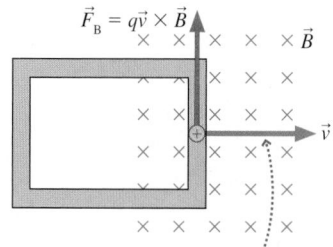

(a) Laboratory frame S

The loop is moving to the right.

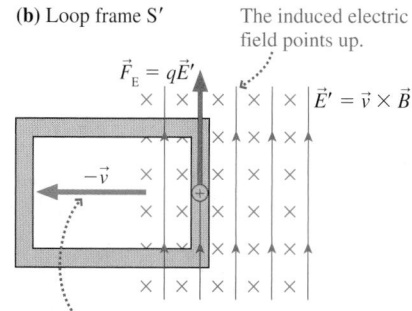

(b) Loop frame S'

The induced electric field points up.

The magnetic field is moving to the left.

FIGURE 34.16 A motional emf as seen in two different reference frames.

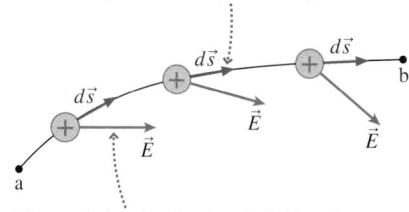

Divide the path into many small displacements.

The work done by the electric field as the charge moves through $d\vec{s}$ is $dW = q\vec{E} \cdot d\vec{s}$.

FIGURE 34.17 The electric field does work on a charge as it moves from a to b.

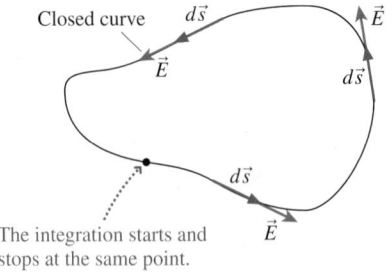

FIGURE 34.18 This electric field can do work as a charged particle moves around the closed curve.

The emf of Faraday's law is an emf around a *closed curve,* such as the one in Figure 34.18, through which the magnetic flux Φ_m is changing. We can calculate the work done by the electric field as a charge moves around a closed curve if we let the end points a and b in Figure 34.17 merge to a single point. The work done is

$$W_{\text{closed curve}} = q \oint \vec{E} \cdot d\vec{s} \tag{34.21}$$

where the integration symbol with the circle is the same as the one we used in Ampère's law to indicate an integral around a closed curve. If we use this expression for W in Equation 34.19, we find that the emf around a closed loop is

$$\mathcal{E} = \frac{W_{\text{closed curve}}}{q} = \oint \vec{E} \cdot d\vec{s} \tag{34.22}$$

Equation 34.22 relates the induced electric field to the induced emf of Faraday's law.

NOTE ▶ For a Coulomb electric field, created by charges, $W_{\text{closed curve}} = 0$ because a charge that moves around a closed path experiences no net change in potential energy. That's Kirchhoff's loop law. By contrast, an induced electric field is *not* a conservative field. We cannot associate an electric potential with an induced electric field, and a charge *does* gain energy by traversing a closed path. ◀

Establishing the Sign

Suppose the magnetic flux Φ_m through a loop is changing. According to Faraday's law, the emf around the loop is

$$\mathcal{E} = \left| \frac{d\Phi_m}{dt} \right|$$

in a direction given by Lenz's law. We've related the emf $\mathcal{E}$ to $\vec{E}$ and the flux Φ_m to $\vec{B}$, but we still need to incorporate Lenz's law into the mathematics. The difficulty is that the sign of Φ_m is ambiguous. If you have a surface, such as a circle enclosed by a current loop, how do you know if the flux is going "into" or "out of" of the surface?

We can deal with this ambiguity by establishing a *sign convention.* This convention works for both the magnetic flux Φ_m and the electric flux Φ_e.

TACTICS BOX 34.2 **Determining the sign of the flux**

❶ For a surface S bounded by a closed curve C, choose either the clockwise (cw) or counterclockwise (ccw) direction around C.

❷ Curl the fingers of your *right* hand around the curve in the chosen direction with your thumb perpendicular to the surface. Your thumb defines the positive direction. The loop's area vector $\vec{A}$ points in the direction of your thumb. The flux Φ through the surface is positive if the field is in the same direction as your thumb, negative if the field is in the opposite direction.

❸ A positive emf creates an induced current in the direction of your fingers; a negative emf creates a current in the opposite direction.

Figure 34.19 applies these rules to a loop moving into a magnetic field. You can see that $\mathcal{E}$ and $d\Phi_m/dt$ have opposite signs. In fact, if you consider all the possible combinations of increasing and decreasing fields, you find that $\mathcal{E}$ **and**

$d\Phi_m/dt$ **always have opposite signs.** This is Lenz's law at work, saying that the induced current (proportional to $\mathcal{E}$) *opposes* the change in the flux ($d\Phi_m/dt$). We can capture this idea mathematically by writing Faraday's law as

$$\mathcal{E} = -\frac{d\Phi_m}{dt} \tag{34.23}$$

where the minus sign is the mathematical statement of Lenz's law.

Now we can complete our task by using the Equation 34.22 expression for $\mathcal{E}$ in Equation 34.23:

$$\oint \vec{E} \cdot d\vec{s} = -\frac{d\Phi_m}{dt} = -\frac{d}{dt}\left[\int \vec{B} \cdot d\vec{A}\right] \tag{34.24}$$

where the line integral of $\vec{E}$ is around the closed curve that bounds the surface through which the magnetic flux is calculated. Equation 34.24 is Faraday's law written in terms of the fields $\vec{E}$ and $\vec{B}$.

But $\vec{E}$ and $\vec{B}$ according to which experimenter? **These are the fields in the reference frame of the loop,** where the loop is at rest. There is an induced electric field in the loop's reference frame if the magnetic flux through the loop is changing. The induced electric field is responsible for the emf that drives the induced current around the loop.

The Induced Electric Field

The solenoid in Figure 34.20a, whose upward magnetic field is increasing as the current increases, provides a good example of the connection between $\vec{E}$ and $\vec{B}$. You learned in Chapter 33 that the changing flux induces a current in a conducting loop inside the solenoid, and we could use Lenz's law to determine that the direction of the induced current would be clockwise. But Faraday's law, in the form of Equation 34.24, tells us that **an induced electric field is present whether there's a conducting loop or not.** The electric field is induced simply due to the fact that $\vec{B}$ is changing.

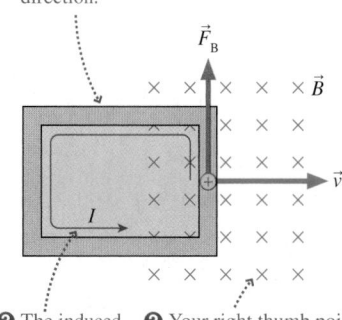

❶ The loop is closed curve C. The shaded area inside the loop is surface S. Choose a ccw direction.

❸ The induced current is in the direction of your fingers, so the emf $\mathcal{E}$ is positive.

❷ Your right thumb points *out* of the page. This magnetic field is into the page, so the flux Φ_m is negative. The flux is becoming more negative as the loop moves into the field, so $d\Phi_m/dt$ is also negative.

FIGURE 34.19 The emf is positive and the flux is negative.

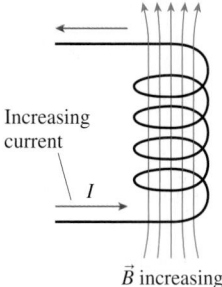

(a) The current through the solenoid is increasing.

Increasing current

I

$\vec{B}$ increasing

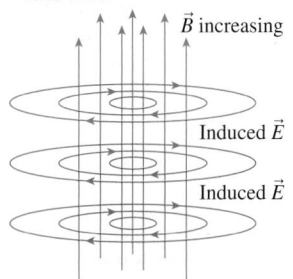

(b) The induced electric field circulates around the magnetic field lines.

$\vec{B}$ increasing

Induced $\vec{E}$

Induced $\vec{E}$

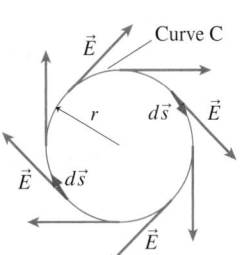

(c) Top view into the solenoid. $\vec{B}$ is coming out of the page.

$\vec{E}$ Curve C

r $d\vec{s}$ $\vec{E}$

$\vec{E}$ $d\vec{s}$

$\vec{E}$

FIGURE 34.20 The induced electric field circulates around the changing magnetic field inside a solenoid.

The shape of the induced electric field has to be such that it *could* drive a current around a conducting loop, if one were present, and it has to be consistent with the cylindrical symmetry of the solenoid. The only possible choice, shown in Figure 34.20b, is an electric field that circulates around the magnetic field lines.

NOTE ▶ Circular electric field lines violate rule 4 of Tactics Box 34.1 for drawing field lines, which said that electric field lines have to start and stop on charges. But we noted there that rule 4 applies only to Coulomb fields that are created by source charges. An induced electric field is a non-Coulomb field created not by source charges but by a changing magnetic field. Without source charges, induced electric field lines *must* form closed loops. ◄

To use Faraday's law, choose a *clockwise* direction around a circle of radius r as the closed curve for evaluating the emf and the flux. Figure 34.20c shows that the electric field vectors are everywhere tangent to the curve. In this case, the line integral of $\vec{E}$ is

$$\oint \vec{E} \cdot d\vec{s} = EL = 2\pi rE \tag{34.25}$$

where $L = 2\pi r$ is the length of the closed curve.

What about the flux? The closed curve surrounds area $A = \pi r^2$. The magnetic field inside a solenoid is uniform, so the flux has magnitude $|\Phi_m| = AB$. To determine the sign, curl your right fingers around the circle in the cw direction. Your right thumb points *into* the page, but $\vec{B}$ is coming out of the page. Thus the flux is a *negative*

$$\Phi_m = -AB = -\pi r^2 B$$

and the rate of change of the flux is

$$\frac{d\Phi_m}{dt} = -\pi r^2 \frac{dB}{dt} \tag{34.26}$$

If we substitute Equations 34.25 and 34.26 into Faraday's law, Equation 34.24, we find

$$\oint \vec{E} \cdot d\vec{s} = 2\pi rE = -\frac{d\Phi_m}{dt} = \pi r^2 \frac{dB}{dt}$$

Thus the induced electric field inside the solenoid ($r < R$) is

$$E = \frac{r}{2} \frac{dB}{dt} \tag{34.27}$$

This result shows very directly that the induced electric field is created by a *changing* magnetic field. A constant $\vec{B}$, with $dB/dt = 0$, would give $E = 0$.

We defined the positive direction to be clockwise. If the solenoid current is increasing, then dB/dt is positive as the magnetic field grows stronger, and E is positive. Thus the induced electric field lines circulate cw for an increasing solenoid current, which is the case shown in Figure 34.20. Conversely, a decreasing solenoid current would make dB/dt negative, hence E would be negative and the field lines would circulate ccw.

We have now extended Faraday's law to make an explicit connection between electric and magnetic fields. Faraday's law, in the form of Equation 34.24, joins Gauss's law and Gauss's law for magnetism as one of the fundamental equations of electromagnetic fields. We have one more equation to go.

EXAMPLE 34.4 An induced electric field

A 4.0-cm-diameter solenoid is wound with 2000 turns per meter. The current through the solenoid oscillates at 60 Hz with an amplitude of 2.0 A. What is the maximum strength of the induced electric field inside the solenoid?

MODEL Assume that the magnetic field inside the solenoid is uniform.

VISUALIZE The electric field lines are concentric circles around the magnetic field lines, as was shown in Figure 34.20. They will reverse direction twice every period as the current oscillates.

SOLVE You learned in Chapter 32 that the magnetic field strength inside a solenoid with n turns per meter is $B = \mu_0 nI$. In this case, the current through the solenoid is $I = I_0 \sin \omega t$, where $I_0 = 2.0$ A is the peak current and $\omega = 2\pi(60$ Hz$) = 377$ rad/s. Thus the induced electric field strength at radius r is

$$E = \frac{r}{2}\frac{dB}{dt} = \frac{r}{2}\frac{d}{dt}(\mu_0 n I_0 \sin \omega t) = \frac{1}{2}\mu_0 n r \omega I_0 \cos \omega t$$

The field strength is maximum at maximum radius ($r = R$) *and* at the instant when $\cos \omega t = 1$ (I changing at the maximum rate). That is,

$$E_{max} = \frac{1}{2}\mu_0 n R \omega I_0 = 0.019 \text{ V/m}$$

ASSESS This electric field strength, although not large, is similar to the field strength that the emf of a battery creates in a wire. Hence this induced electric field has the ability to drive a substantial induced current through a conducting loop *if* a loop is present. But the induced electric field exists inside the solenoid whether or not there is a conducting loop.

34.4 The Displacement Current

We introduced Ampère's law in Chapter 32 as an alternative method for calculating the magnetic field of a current. Whenever total current $I_{through}$ passes through an area bounded by a closed curve, the line integral of the magnetic field around the curve is

$$\oint \vec{B} \cdot d\vec{s} = \mu_0 I_{through} \qquad (34.28)$$

Figure 34.21 illustrates the geometry of Ampère's law. The sign of each current can be determined by using Tactics Box 34.2. In this case, $I_{through} = I_1 - I_2$.

Ampère's law, which is equivalent to the Biot-Savart law for the magnetic field of a moving charge, is the formal statement that **currents create magnetic fields.** Although Ampère's law can be used to calculate magnetic fields in situations with a high degree of symmetry, it is more important as a statement about what types of magnetic field can and cannot exist.

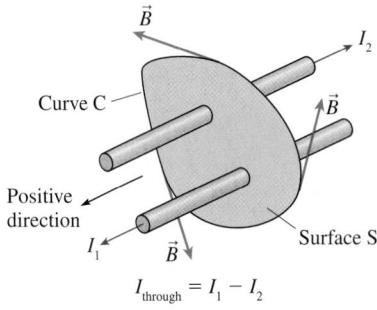

FIGURE 34.21 Ampère's law relates the line integral of $\vec{B}$ around curve C to the current passing through surface S.

Something Is Missing

Nothing restricts the bounded surface of Ampère's law to being flat. It's not hard to see that any current passing through surface S_1 in Figure 34.22 must also pass through the curved surface S_2. To interpret Ampère's law properly, we have to say that the current $I_{through}$ is the net current passing through *any* surface S that is bounded by curve C.

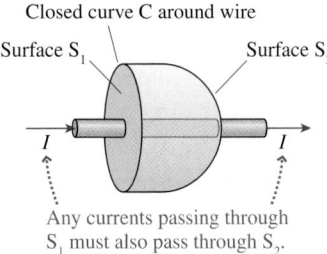

Any currents passing through S_1 must also pass through S_2.

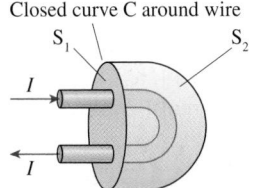

Even in this case, the *net* current through S_1, namely zero, matches the net current through S_2.

FIGURE 34.22 The *net* current passing through the flat surface S_1 also passes through the curved surface S_2.

But this leads to an interesting puzzle. Figure 34.23a on the next page shows a capacitor being charged. Current I, from the left, brings positive charge to the left capacitor plate. The same current carries charges away from the right capacitor plate, leaving the right plate negatively charged. This is a perfectly ordinary current in a conducting wire, and you can use the right-hand rule to verify that its magnetic field is as shown.

(a)

Cross section through a closed curve C around the wire

Current I passes through surface S_1.

No current passes through surface S_2.

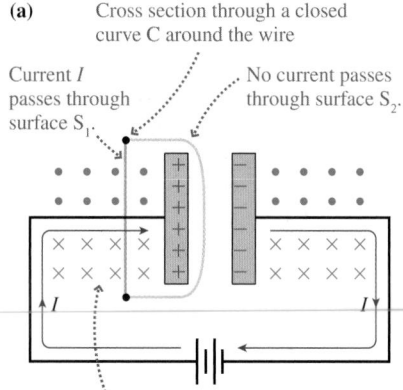

This is the magnetic field of the current I that is charging the capacitor.

(b)

Curve C

S_2

S_1

$I = \dfrac{dQ}{dt}$

$I = \dfrac{dQ}{dt}$

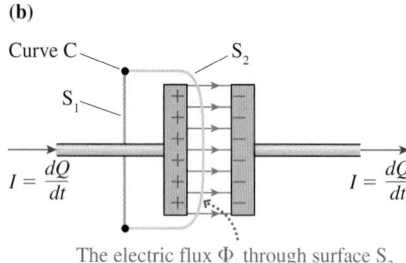

The electric flux Φ_e through surface S_2 increases as the capacitor charges.

FIGURE 34.23 There is no current through surface S_2 as the capacitor charges, but there is a changing electric flux.

Curve C is a closed curve encircling the wire on the left. The current passes through surface S_1, a flat surface across C, and we could use Ampère's law to find that the magnetic field is that of a straight wire. But what happens if we try to use surface S_2 to determine I_{through}? Ampère's law says that we can consider *any* surface bounded by curve C, and surface S_2 certainly qualifies. But *no* current passes through S_2. Charges are brought to the left plate of the capacitor and charges are removed from the right plate, but *no* charge moves across the gap between the plates. Surface S_1 has $I_{\text{through}} = I$, but surface S_2 has $I_{\text{through}} = 0$. Another dilemma!

It would appear that Ampère's law is either wrong or incomplete. Maxwell was the first to recognize the seriousness of this problem. He noted that there may be no current passing through S_2, but, as Figure 34.23b shows, there is an electric flux Φ_e through S_2 due to the electric field inside the capacitor. Furthermore, this flux is *changing* with time as the capacitor charges and the electric field strength grows. Faraday had discovered the significance of a changing magnetic flux, but no one had considered a changing electric flux.

The current I passes through S_1, so Ampère's law applied to S_1 gives

$$\oint \vec{B} \cdot d\vec{s} = \mu_0 I_{\text{through}} = \mu_0 I$$

We believe this result because it gives the correct magnetic field for a current-carrying wire. Now the line integral depends only on the magnetic field at points on curve C, so its value won't change if we choose a different surface S to evaluate the current. The problem is with the right side of Ampère's law, which would incorrectly give zero if applied to surface S_2. We need to modify the right side of Ampère's law to recognize that an electric flux rather than a current passes through S_2.

The electric flux between two capacitor plates of surface area A is

$$\Phi_e = EA$$

The capacitor's electric field is $E = Q/\epsilon_0 A$, hence the flux is actually independent of the plate size:

$$\Phi_e = \frac{Q}{\epsilon_0 A} A = \frac{Q}{\epsilon_0} \tag{34.29}$$

The *rate* at which the electric flux is changing is

$$\frac{d\Phi_e}{dt} = \frac{1}{\epsilon_0} \frac{dQ}{dt} = \frac{I}{\epsilon_0} \tag{34.30}$$

where we used $I = dQ/dt$. The flux is changing with time at a rate directly proportional to the charging current I.

Equation 34.30 suggests that the quantity $\epsilon_0(d\Phi_e/dt)$ is in some sense "equivalent" to current I. Maxwell called the quantity

$$I_{\text{disp}} = \epsilon_0 \frac{d\Phi_e}{dt} \tag{34.31}$$

the **displacement current.** He had started with a fluid-like model of electric and magnetic fields, and the displacement current was analogous to the displacement of a fluid. The fluid model has since been abandoned, but the name lives on despite the fact that nothing is actually being displaced.

Maxwell hypothesized that the displacement current was the "missing" piece of Ampère's law, so he modified Ampère's law to read

$$\oint \vec{B} \cdot d\vec{s} = \mu_0(I_{\text{through}} + I_{\text{disp}}) = \mu_0\left(I_{\text{through}} + \epsilon_0 \frac{d\Phi_e}{dt}\right) \tag{34.32}$$

Equation 34.32 is now known as the Ampère-Maxwell law. When applied to Figure 34.23b, the Ampère-Maxwell law gives

$$S_1: \quad \oint \vec{B} \cdot d\vec{s} = \mu_0 \left(I_{\text{through}} + \epsilon_0 \frac{d\Phi_e}{dt} \right) = \mu_0 (I + 0) = \mu_0 I$$

$$S_2: \quad \oint \vec{B} \cdot d\vec{s} = \mu_0 \left(I_{\text{through}} + \epsilon_0 \frac{d\Phi_e}{dt} \right) = \mu_0 (0 + I) = \mu_0 I$$

where, for surface S_2, we used Equation 34.30 for $d\Phi_e/dt$. Surfaces S_1 and S_2 now both give the same result for the line integral of $\vec{B} \cdot d\vec{s}$ around the closed curve C.

NOTE ▶ The displacement current I_{disp} between the capacitor plates is numerically equal to the current I in the wires leading to and from the capacitor, so in some sense it allows "current" to be conserved all the way through the capacitor. Nonetheless, the displacement current is *not* a flow of charge. The displacement current is equivalent to a real current in the sense that it creates the same magnetic field, but it does so with a changing electric flux rather than a flow of charge. ◀

The Induced Magnetic Field

Ordinary Coulomb electric fields are created by charges, but a second way to create an electric field is by having a changing magnetic field. That's Faraday's law. Ordinary magnetic fields are created by currents, but now we see that a second way to create a magnetic field is by having a changing electric field. Just as the electric field created by a changing $\vec{B}$ is called an induced electric field, the magnetic field created by a changing $\vec{E}$ is called an *induced magnetic field.*

Figure 34.24 shows the close analogy between induced electric fields, governed by Faraday's law, and induced magnetic fields, governed by the second term in the Ampère-Maxwell law. An increasing solenoid current causes an increasing magnetic field. The changing magnetic field, in turn, induces a circular electric field. The negative sign in Faraday's law dictates that the induced electric field direction is ccw when seen looking along the magnetic field direction.

An increasing capacitor charge causes an increasing electric field. The changing electric field, in turn, induces a circular magnetic field. But the sign of the Ampère-Maxwell law is positive, the opposite of the sign of Faraday's law, so the induced magnetic field direction is cw when you're looking along the electric field direction.

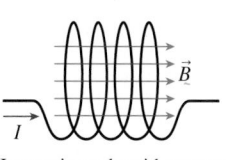
Increasing solenoid current

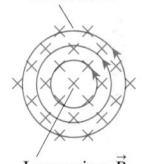
Induced $\vec{E}$

Increasing $\vec{B}$

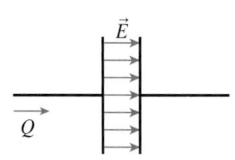
Increasing capacitor charge

Induced $\vec{B}$

Increasing $\vec{E}$

FIGURE 34.24 The close analogy between an induced electric field and an induced magnetic field.

EXAMPLE 34.5 The fields inside a charging capacitor

A 2.0-cm-diameter parallel-plate capacitor with a 1.0 mm spacing is being charged at the rate 0.50 C/s. What is the magnetic field strength inside the capacitor at a point 0.50 cm from the axis?

MODEL The electric field inside a parallel-plate capacitor is uniform. As the capacitor is charged, the changing electric field induces a magnetic field.

VISUALIZE Figure 34.25 shows the fields. The induced magnetic field lines are circles concentric with the capacitor.

SOLVE The electric field of a parallel-plate capacitor is $E = Q/\epsilon_0 A = Q/\epsilon_0 \pi R^2$. The electric flux through the circle of radius r (not the full flux of the capacitor) is

$$\Phi_e = \pi r^2 E = \pi r^2 \frac{Q}{\epsilon_0 \pi R^2} = \frac{r^2}{R^2} \frac{Q}{\epsilon_0}$$

Thus the Ampère-Maxwell law is

$$\oint \vec{B} \cdot d\vec{s} = \epsilon_0 \mu_0 \frac{d\Phi_e}{dt} = \epsilon_0 \mu_0 \frac{d}{dt} \left(\frac{r^2}{R^2} \frac{Q}{\epsilon_0} \right) = \mu_0 \frac{r^2}{R^2} \frac{dQ}{dt}$$

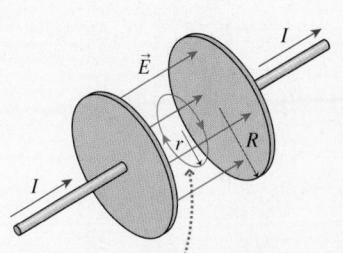

The magnetic field line is a circle concentric with the capacitor. The electric flux through this circle is $\pi r^2 E$.

FIGURE 34.25 The magnetic field strength is found by integrating around a closed curve of radius r.

The magnetic field is everywhere tangent to the circle of radius r, so the integral of $\vec{B} \cdot d\vec{s}$ around the circle is simply $BL = 2\pi rB$. With this value for the line integral, the Ampère-Maxwell law becomes

$$2\pi rB = \mu_0 \frac{r^2}{R^2} \frac{dQ}{dt}$$

and thus

$$B = \frac{\mu_0}{2\pi} \frac{r}{R^2} \frac{dQ}{dt} = (2.0 \times 10^{-7} \text{ T m/A}) \frac{0.0050 \text{ m}}{(0.010 \text{ m})^2} (0.50 \text{ C/s})$$

$$= 5.0 \times 10^{-6} \text{ T}$$

If a changing magnetic field can induce an electric field and a changing electric field can induce a magnetic field, what happens when both fields change simultaneously? That is the question that Maxwell was finally able to answer after he modified Ampère's law to include the displacement current, and it is the subject to which we turn next.

STOP TO THINK 34.3 The electric field in four identical capacitors is shown as a function of time. Rank in order, from largest to smallest, the magnetic field strength at the outer edge of the capacitor at time T.

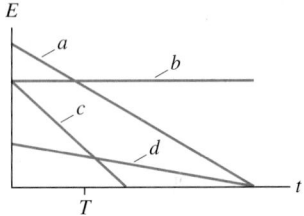

34.5 Maxwell's Equations

James Clerk Maxwell was a young, mathematically brilliant Scottish physicist. In 1855, barely 24 years old and having graduated from Cambridge University just two years earlier, he presented a paper to the Cambridge Philosophical Society entitled "On Faraday's Lines of Force." It had been 30 years and more since the major discoveries of Oersted, Ampère, Faraday, and others, but electromagnetism remained a loose collection of facts and "rules of thumb" without a consistent theory to link these ideas together.

Maxwell's goal, first enunciated in his paper of 1855, was to synthesize this body of knowledge and to place it in a proper mathematical framework. His desire was nothing less than to form a complete *theory* of electromagnetic fields. It took 10 years, until papers published in 1865 and 1868 laid out the theory in a form that looks familiar to us today. The critical step along the way was his recognition of the need to include a displacement current term in Ampère's law.

Maxwell's theory of electromagnetism is embodied in four equations that we today call **Maxwell's equations.** These are

$$\oint \vec{E} \cdot d\vec{A} = \frac{Q_{\text{in}}}{\epsilon_0} \qquad \text{Gauss's law}$$

$$\oint \vec{B} \cdot d\vec{A} = 0 \qquad \text{Gauss's law for magnetism}$$

$$\oint \vec{E} \cdot d\vec{s} = -\frac{d\Phi_{\text{m}}}{dt} \qquad \text{Faraday's law}$$

$$\oint \vec{B} \cdot d\vec{s} = \mu_0 I_{\text{through}} + \epsilon_0 \mu_0 \frac{d\Phi_{\text{e}}}{dt} \qquad \text{Ampère-Maxwell law}$$

You've seen all of these equations earlier in this chapter. It was Maxwell who first wrote them in a consistent mathematical form similar to this. (Not exactly the same, because our present-day vector notation wasn't developed until the 1890s, but Maxwell's versions were mathematically equivalent.) Neither Gauss nor Faraday nor Ampère would recognize these equations, but Maxwell had succeeded in capturing their physical ideas in a concise mathematical form.

Maxwell's claim is that these four equations are a *complete* description of electric and magnetic fields. They tell us how fields are created by charges and currents, and also how fields can be induced by the changing of other fields. We need one more equation for total completeness, an equation that tells us how matter responds to electromagnetic fields. But that's the Lorentz force law, another equation that we already have:

$$\vec{F} = q(\vec{E} + \vec{v} \times \vec{B}) \qquad \text{(Lorentz force law)}$$

Maxwell's equations for the fields, together with the Lorentz force law to tell us how matter responds to the fields, form the complete theory of electromagnetism.

Maxwell's equations bring us to the pinnacle of classical physics. Except at the quantum level of photons, these equations describe everything that is known about electromagnetic phenomena. In fact, they predict many new phenomena not known to Maxwell or his contemporaries, and they're the basis for all of modern circuit theory, electrical engineering, and other electromagnetic technology. When combined with Newton's three laws of motion, his law of gravity, and the first and second laws of thermodynamics, we have all of classical physics—a total of just 11 equations.

While some physicists might quibble over whether all 11 are truly fundamental, the important point is not the exact number but how few equations we need to describe the overwhelming majority of our experience of the physical world. It seems as if we could have written them all on page one of this book and been finished, but it doesn't work that way. Each of these equations is the synthesis of a tremendous number of physical phenomena and conceptual developments. To know physics isn't just to know the equations, but to know what the equations *mean* and how they're used. That's why it's taken us so many chapters and so much effort to get to this point. Each equation is simply a shorthand way to summarize a book's worth of information!

Let's summarize the physical meaning embodied in the five electromagnetic equations:

- **Gauss's law:** Charged particles create an electric field.
- **Faraday's law:** An electric field can also be created by a changing magnetic field.
- **Gauss's law for magnetism:** There are no magnetic monopoles.
- **Ampère-Maxwell law, first half:** Currents create a magnetic field.
- **Ampère-Maxwell law, second half:** A magnetic field can also be created by a changing electric field.
- **Lorentz force law, first half:** An electric force is exerted on a charged particle in an electric field.
- **Lorentz force law, second half:** A magnetic force is exerted on a moving charge in a magnetic field.

These are the *fundamental ideas* of electromagnetism. Other important ideas, such as Ohm's law, Kirchhoff's laws, and Lenz's law, despite their practical importance, are not fundamental ideas. They can be derived from Maxwell's equations, sometimes with the addition of empirically based concepts such as that of resistance.

Classical physics

Newton's first law
Newton's second law
Newton's third law
Newton's law of gravity
Gauss's law
Gauss's law for magnetism
Faraday's law
Ampère-Maxwell law
Lorentz force law
First law of thermodynamics
Second law of thermodynamics

Maxwell's equations can be used to understand motors, generators, antennas and receivers, the transmission of electrical signals through circuits, power lines, microwaves, the electromagnetic properties of materials, and much more. It's true that Maxwell's equations are mathematically more complex than Newton's laws and that their solution, for many problems of practical interest, requires advanced mathematics. Fortunately, we have the mathematical tools to get just far enough into Maxwell's equations to discover their most startling and revolutionary implication—the prediction of electromagnetic waves.

34.6 Electromagnetic Waves

It had been known since the early 19th century, from experiments on interference and diffraction, that light is a wave. We studied the wave properties of light in Part V, but at that time we were not able to determine just what is "waving."

Faraday speculated that light was somehow connected with electricity and magnetism, but Maxwell, using his equations of the electromagnetic field, was the first to understand that light is an oscillation of the electromagnetic field. Maxwell was able to predict that

- Electromagnetic waves can exist at any frequency, not just at the frequencies of visible light. This prediction was the harbinger of radio waves.
- All electromagnetic waves travel in a vacuum with the same speed, a speed that we now call the *speed of light.*

A general wave equation can be derived from Maxwell's equations, but the necessary mathematical techniques are beyond the level of this textbook. We'll adopt a simpler approach in which we *assume* an electromagnetic wave of a certain form and then show that it's consistent with Maxwell's equations. After all, the wave can't exist *unless* it's consistent with Maxwell's equations.

To begin, we're going to assume that electric and magnetic fields can exist independent of charges and currents in a *source-free* region of space. This is a very important assumption because it makes the statement that **fields are real entities.** They're not just cute pictures that tell us about charges and currents, but are real things that can exist all by themselves. Our assertion is that the fields can exist in a self-sustaining mode in which a changing magnetic field creates an electric field (Faraday's law) that in turn changes in just the right way to re-create the original magnetic field (the Ampère-Maxwell law).

The source-free Maxwell's equations, with no charges or currents, are

$$\oint \vec{E} \cdot d\vec{A} = 0 \qquad \oint \vec{E} \cdot d\vec{s} = -\frac{d\Phi_\mathrm{m}}{dt}$$

$$\oint \vec{B} \cdot d\vec{A} = 0 \qquad \oint \vec{B} \cdot d\vec{s} = \epsilon_0 \mu_0 \frac{d\Phi_\mathrm{e}}{dt} \tag{34.33}$$

Any electromagnetic wave traveling in empty space must be consistent with these equations.

Let's postulate that an electromagnetic plane wave traveling with speed v_em has the characteristics shown in Figure 34.26. It's a useful picture, and one that you'll see in any textbook, but a picture that can be very misleading if you don't think about it carefully. $\vec{E}$ and $\vec{B}$ are *not* spatial vectors. That is, they don't stretch spatially in the y- or z-direction for a certain distance. Instead, these vectors are showing the values of the electric and magnetic fields at *points* along a single line, the x-axis. An $\vec{E}$ vector pointing in the y-direction says that *at that point* on the x-axis, where the vector's tail is, the electric field points in the y-direction and has a certain strength. Nothing is "reaching" to a point in space above the x-axis. In fact, this picture contains no information about any points in space other than those right on the x-axis.

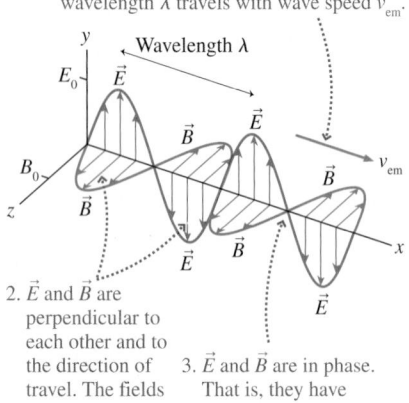

1. A sinusoidal wave with frequency *f* and wavelength λ travels with wave speed v_em.

2. $\vec{E}$ and $\vec{B}$ are perpendicular to each other and to the direction of travel. The fields have amplitudes E_0 and B_0.

3. $\vec{E}$ and $\vec{B}$ are in phase. That is, they have matching crests, troughs, and zeros.

FIGURE 34.26 A sinusoidal electromagnetic wave.

However, we are assuming that this is a *plane wave,* which, you'll recall from Chapter 20, is a wave for which the fields are the same at *all points* in any *yz*-plane, perpendicular to the *x*-axis. Figure 34.27a shows a small section of the *xy*-plane where, at this instant of time, $\vec{E}$ is pointing up and $\vec{B}$ is pointing toward you. The field strengths vary with *x*, the direction of travel, but not with *y*. As the wave moves forward, the fields that are now in the x_1-plane will soon arrive in the x_2-plane, and those now in the x_2-plane will move to x_3.

Figure 34.27b shows a section of the *yz*-plane that slices the *x*-axis at x_2. These fields are moving out of the page, coming toward you. The fields are the same at *every point* in this plane, which is what we mean by a plane wave. If you watched a movie of the event, you would see the $\vec{E}$ and $\vec{B}$ fields at each point in this plane *oscillating* in time, but always synchronized with all the other points in the plane. Thus you have to use your imagination to see that the $\vec{E}$ and $\vec{B}$ fields in Figure 34.26 are also the $\vec{E}$ and $\vec{B}$ fields *everywhere* in any *yz*-plane.

Gauss's Laws

Now that we understand the shape of the electromagnetic field, we can check its consistency with Maxwell's equations. This field is a sinusoidal wave, so the components of the fields are

$$E_x = 0 \quad E_y = E_0 \sin(2\pi(x/\lambda - ft)) \quad E_z = 0$$
$$B_x = 0 \quad B_y = 0 \qquad\qquad\qquad B_z = B_0 \sin(2\pi(x/\lambda - ft))$$

(34.34)

where E_0 and B_0 are the amplitudes of the oscillating electric and magnetic fields.

Figure 34.28 shows an imaginary box—a Gaussian surface—centered on the *x*-axis. Both electric and magnetic field vectors exist at each point in space, but the figure shows them separately for clarity. $\vec{E}$ oscillates along the *y*-axis, so all electric field lines enter and leave the box through the top and bottom surfaces; no electric field lines pass through the sides of the box.

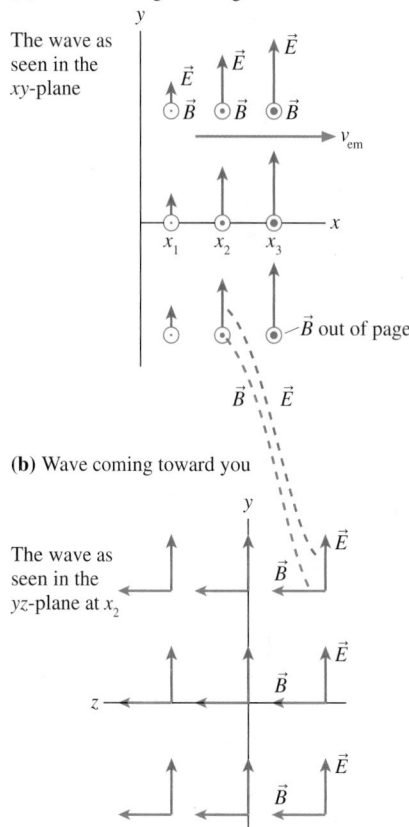

(a) Wave traveling to the right

The wave as seen in the *xy*-plane

(b) Wave coming toward you

The wave as seen in the *yz*-plane at x_2

FIGURE 34.27 Interpreting the electromagnetic wave of Figure 34.26.

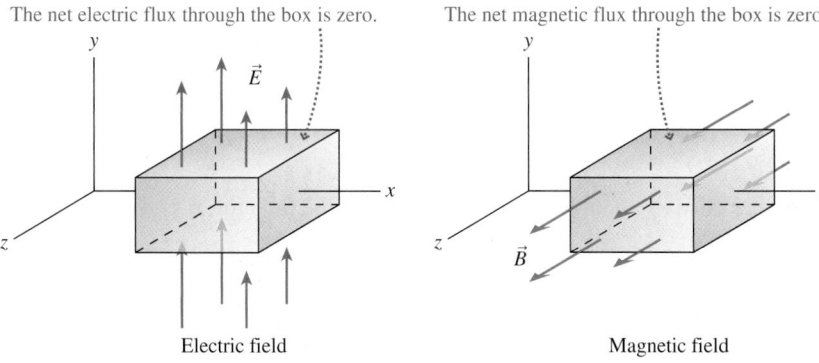

The net electric flux through the box is zero.

Electric field

The net magnetic flux through the box is zero.

Magnetic field

FIGURE 34.28 A closed surface can be used to check Gauss's law for the electric and magnetic fields.

Because this is a plane wave, the magnitude of each electric field vector entering the bottom of the box is exactly matched by the electric field vector leaving the top. The electric flux through the top of the box is equal in magnitude but opposite in sign to the flux through the bottom, and the flux through the sides is zero. Thus the *net* electric flux is $\Phi_e = 0$. There is no charge inside the box, because there are no sources in this region of space, so we also have $Q_{in} = 0$. Hence the electric field of a plane wave is consistent with the first of the source-free Maxwell's equations, Gauss's law.

The exact same argument applies to the magnetic field. The net magnetic flux is $\Phi_m = 0$, thus the magnetic field is consistent with the second of Maxwell's equations.

Faraday's Law

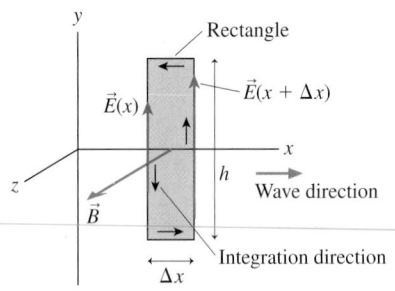

FIGURE 34.29 Faraday's law can be applied to a narrow rectangle in the xy-plane.

Faraday's law is concerned with the changing magnetic flux through a closed curve. We'll apply Faraday's law to a narrow rectangle in the xy-plane, shown in Figure 34.29, with height h and width Δx. We'll assume Δx to be so small that $\vec{B}$ is essentially constant over the width of the rectangle.

The magnetic field $\vec{B}$ points in the z-direction, perpendicular to the rectangle. The magnetic flux through the rectangle is $\Phi_m = B_z A_{rectangle} = B_z h \Delta x$, hence the flux *changes* at the rate

$$\frac{d\Phi_m}{dt} = \frac{d}{dt}(B_z h \Delta x) = \frac{\partial B_z}{\partial t} h \Delta x \tag{34.35}$$

The ordinary derivative dB_z/dt, which is the full rate of change of B from all possible causes, becomes a partial derivative $\partial B_z/\partial t$ in this situation because the change in magnetic flux is due entirely to the change of B with time and not at all to the spatial variation of B.

According to our sign convention, we have to go around the rectangle in a ccw direction to make the flux positive. Thus we must also use a ccw direction to evaluate the line integral

$$\oint \vec{E} \cdot d\vec{s} = \int_{right} \vec{E} \cdot d\vec{s} + \int_{top} \vec{E} \cdot d\vec{s} + \int_{left} \vec{E} \cdot d\vec{s} + \int_{bottom} \vec{E} \cdot d\vec{s} \tag{34.36}$$

The electric field $\vec{E}$ points in the y-direction, hence $\vec{E} \cdot d\vec{s} = 0$ at all points on the top and bottom edges, and these two integrals are zero.

Along the left edge of the loop, at position x, $\vec{E}$ has the same value at every point. Figure 34.27a shows that the direction of $\vec{E}$ is *opposite* to $d\vec{s}$, thus $\vec{E} \cdot d\vec{s} = -E_y(x)ds$. On the right edge of the loop, at position $x + \Delta x$, $\vec{E}$ is *parallel* to $d\vec{s}$ and $\vec{E} \cdot d\vec{s} = E_y(x + \Delta x)ds$. Thus the line integral of $\vec{E} \cdot d\vec{s}$ around the rectangle is

$$\oint \vec{E} \cdot d\vec{s} = -E_y(x)h + E_y(x + \Delta x)h = [E_y(x + \Delta x) - E_y(x)]h \tag{34.37}$$

NOTE ▶ $E_y(x)$ indicates that E_y is a function of the position x. It is *not* E_y multiplied by x. ◀

You learned in calculus that the derivative of the function $f(x)$ is

$$\frac{df}{dx} = \lim_{\Delta x \to 0}\left[\frac{f(x + \Delta x) - f(x)}{\Delta x}\right]$$

We've assumed that Δx is very small. If we now let the width of the rectangle go to zero, $\Delta x \to 0$, Equation 34.37 becomes

$$\oint \vec{E} \cdot d\vec{s} = \frac{\partial E_y}{\partial x} h \Delta x \tag{34.38}$$

We've used a partial derivative because E_y is a function of both position x and time t.

Now, using Equations 34.35 and 34.38, we can write Faraday's law as

$$\oint \vec{E} \cdot d\vec{s} = \frac{\partial E_y}{\partial x} h \Delta x = -\frac{d\Phi_m}{dt} = -\frac{\partial B_z}{\partial t} h \Delta x$$

The area $h \Delta x$ of the rectangle cancels, and we're left with

$$\frac{\partial E_y}{\partial x} = -\frac{\partial B_z}{\partial t} \tag{34.39}$$

Equation 34.39, which compares the rate at which E_y varies with position to the rate at which B_z varies with time, is a *required condition* that an electromagnetic wave must satisfy to be consistent with Maxwell's equations. We can use Equation 34.34 for E_y and B_z to evaluate the partial derivatives:

$$\frac{\partial E_y}{\partial x} = \frac{2\pi E_0}{\lambda}\cos(2\pi(x/\lambda - ft))$$

$$\frac{\partial B_z}{\partial t} = -2\pi f B_0 \cos(2\pi(x/\lambda - ft))$$

Thus the required condition of Equation 34.39 is

$$\frac{\partial E_x}{\partial x} = \frac{2\pi E_0}{\lambda}\cos(2\pi(x/\lambda - ft)) = -\frac{\partial B_z}{\partial t} = 2\pi f B_0 \cos(2\pi(x/\lambda - ft))$$

Canceling the many common factors, and multiplying by λ, we're left with

$$E_0 = (\lambda f)B_0 = v_{em}B_0 \qquad (34.40)$$

where we used the fact that $\lambda f = v$ for any sinusoidal wave.

Equation 34.40, which came from applying Faraday's law, tells us that the field amplitudes E_0 and B_0 of an electromagnetic wave are not arbitrary. **Once the amplitude B_0 of the magnetic field wave is specified, the electric field amplitude E_0 must be $E_0 = v_{em}B_0$.** Otherwise the fields won't satisfy Maxwell's equations.

The Ampère-Maxwell Law

We have one equation to go, but this one will now be easier. The Ampère-Maxwell law is concerned with the changing electric flux through a closed curve. Figure 34.30 shows a very narrow rectangle of width Δx and length l in the xz-plane. The electric field is perpendicular to this rectangle, hence the electric flux through it is $\Phi_e = E_y A_{rectangle} = E_y l\Delta x$. This flux is changing at the rate

$$\frac{d\Phi_e}{dt} = \frac{d}{dt}(E_y l\Delta x) = \frac{\partial E_y}{\partial t}l\Delta x \qquad (34.41)$$

The line integral of $\vec{B} \cdot d\vec{s}$ around this closed rectangle is calculated just like the line integral of $\vec{E} \cdot d\vec{s}$ in Figure 34.29. $\vec{B}$ is perpendicular to $d\vec{s}$ on the narrow ends, so $\vec{B} \cdot d\vec{s} = 0$. The field at *all* points on the left edge, at position x, is $\vec{B}(x)$, and this field is parallel to $d\vec{s}$ to make $\vec{B} \cdot d\vec{s} = B_z(x)ds$. Similarly, $\vec{B} \cdot d\vec{s} = -B_z(x + \Delta x)ds$ at all points on the right edge, where $\vec{B}$ is opposite to $d\vec{s}$. Thus, if we let $\Delta x \to 0$,

$$\oint \vec{B} \cdot d\vec{s} = B_z(x)l - B_z(x + \Delta x)l = -[B_z(x + \Delta x) - B_z(x)]l$$
$$= -\frac{\partial B_z}{\partial x}l\Delta x \qquad (34.42)$$

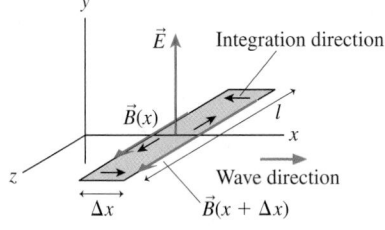

FIGURE 34.30 The Ampère-Maxwell law can be applied to a narrow rectangle in the xz-plane.

Equations 34.41 and 34.42 can now be used in the Ampère-Maxwell law:

$$\oint \vec{B} \cdot d\vec{s} = -\frac{\partial B_z}{\partial x}l\Delta x = \epsilon_0\mu_0\frac{d\Phi_e}{dt} = \epsilon_0\mu_0\frac{\partial E_y}{\partial t}l\Delta x$$

The area of the rectangle cancels, and we're left with

$$\frac{\partial B_z}{\partial x} = -\epsilon_0\mu_0\frac{\partial E_y}{\partial t} \qquad (34.43)$$

Equation 34.43 is a second required condition that the fields must satisfy. If we again evaluate the partial derivatives, using Equation 34.43 for E_y and B_z, we find that

$$\frac{\partial E_y}{\partial t} = -2\pi f E_0 \cos\left(2\pi(x/\lambda - ft)\right)$$

$$\frac{\partial B_z}{\partial x} = \frac{2\pi B_0}{\lambda} \cos\left(2\pi(x/\lambda - ft)\right)$$

With these, Equation 34.43 becomes

$$\frac{\partial B_z}{\partial x} = \frac{2\pi B_0}{\lambda} \cos\left(2\pi(x/\lambda - ft)\right) = -\epsilon_0\mu_0\frac{\partial E_y}{\partial t} = 2\pi\epsilon_0\mu_0 f E_0 \cos\left(2\pi(x/\lambda - ft)\right)$$

A final round of cancellations, and another use of $\lambda f = v_{em}$, leaves us with

$$E_0 = \frac{B_0}{\epsilon_0\mu_0\lambda f} = \frac{B_0}{\epsilon_0\mu_0 v_{em}} \tag{34.44}$$

The last of Maxwell's equations gives us another constraint between E_0 and B_0.

The Speed of Light

But how can Equation 34.40, which required $E_0 = v_{em}B_0$, and Equation 34.44 both be true at the same time? The one and only way is if

$$\frac{1}{\epsilon_0\mu_0 v_{em}} = v_{em}$$

from which we find

$$v_{em} = \frac{1}{\sqrt{\epsilon_0\mu_0}} = 3.00 \times 10^8 \text{ m/s} = c \tag{34.45}$$

This is a remarkable conclusion. The constants ϵ_0 and μ_0 are from electrostatics and magnetostatics, where they determine the size of $\vec{E}$ and $\vec{B}$ due to point charges. Coulomb's law and the Biot-Savart law, where ϵ_0 and μ_0 first appeared, have nothing to do with waves. Yet Maxwell's theory of electromagnetism ends up predicting that electric and magnetic fields can form a self-sustaining electromagnetic wave *if* that wave travels at the specific speed $v_{em} = 1/\sqrt{\epsilon_0\mu_0}$. No other speed will satisfy Maxwell's equations.

We've made no assumption about the frequency of the wave, so apparently all electromagnetic waves, regardless of their frequency, travel (in a vacuum) at the same speed $v_{em} = 1/\sqrt{\epsilon_0\mu_0}$. We now call this speed c, the "speed of light," but it applies equally well from low-frequency radio waves to ultrahigh-frequency x rays.

STOP TO THINK 34.4 An electromagnetic wave is propagating in the positive x-direction. At this instant of time, what is the direction of $\vec{E}$ at the center of the rectangle?

a. In the positive x-direction
b. In the negative x-direction
c. In the positive y-direction
d. In the negative y-direction
e. In the positive z-direction
f. In the negative z-direction

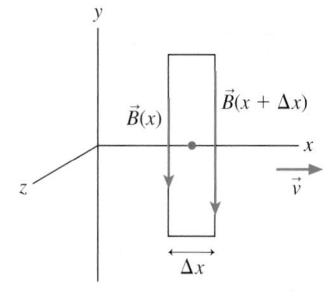

34.7 Properties of Electromagnetic Waves

We've demonstrated that one very specific sinusoidal wave is consistent with Maxwell's equations. It's possible to show that *any* electromagnetic wave, whether it's sinusoidal or not, must satisfy four basic conditions:

1. The fields $\vec{E}$ and $\vec{B}$ are perpendicular to the direction of propagation $\vec{v}_{em}$. Thus an electromagnetic wave is a transverse wave.
2. $\vec{E}$ and $\vec{B}$ are perpendicular to each other in a manner such that $\vec{E} \times \vec{B}$ is in the direction of $\vec{v}_{em}$.
3. The wave travels in a vacuum at speed $v_{em} = 1/\sqrt{\epsilon_0 \mu_0} = c$.
4. $E = cB$ at any point on the wave.

In this section, we'll look at some other properties of electromagnetic waves.

Energy and Intensity

Waves transfer energy. Ocean waves erode beaches, sound waves set your eardrum to vibrating, and light from the sun warms the earth. The energy flow of an electromagnetic wave is described by the **Poynting vector** $\vec{S}$, defined as

$$\vec{S} \equiv \frac{1}{\mu_0} \vec{E} \times \vec{B} \tag{34.46}$$

The Poynting vector, shown in Figure 34.31, has two important properties:

1. At any point, the Poynting vector points in the direction in which an electromagnetic wave is traveling. You can see this by looking back at Figure 34.26.
2. The magnitude S of the Poynting vector measures the rate of energy transfer per unit area of the wave. As a homework problem, you can show that the units of S are W/m^2, or power (joules per second) per unit area.

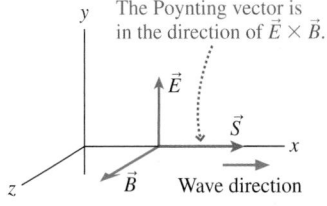

FIGURE 34.31 The Poynting vector.

Because $\vec{E}$ and $\vec{B}$ of an electromagnetic wave are perpendicular to each other, and $E = cB$, the magnitude of the Poynting vector is

$$S = \frac{EB}{\mu_0} = \frac{E^2}{c\mu_0}$$

The Poynting vector is a function of time, oscillating from zero to $S_{max} = E_0^2/c\mu_0$ and back to zero twice during each period of the wave's oscillation. That is, the energy flow in an electromagnetic wave is not smooth. It "pulses" as the electric and magnetic fields oscillate in intensity. We're unaware of this pulsing because the electromagnetic waves that we can sense—light waves—have such high frequencies.

Of more interest is the *average* energy transfer, averaged over one cycle of oscillation, which is the wave's **intensity** I. In our earlier study of waves, we defined the intensity of a wave to be $I = P/A$, where P is the power (energy transferred per second) of a wave that impinges on area A. Because $E = E_0\sin(2\pi(x/\lambda - ft))$, and the average over one period of $\sin^2(2\pi(x/\lambda - ft))$ is $\frac{1}{2}$, the intensity of an electromagnetic wave is

$$I = \frac{P}{A} = S_{avg} = \frac{1}{2c\mu_0} E_0^2 = \frac{c\epsilon_0}{2} E_0^2 \tag{34.47}$$

Equation 34.47 relates the intensity of an electromagnetic wave, a quantity that is easily measured, to the amplitude of the wave's electric field.

EXAMPLE 34.6 The electric field of a laser beam

A helium-neon laser, the laser commonly used for classroom demonstrations, emits a 1.0-mm-diameter laser beam with a power of 1.0 mW. What is the amplitude of the oscillating electric field in the laser beam?

MODEL The laser beam is an electromagnetic plane wave. Assume that the energy is uniformly distributed over the diameter of the laser beam.

SOLVE 1.0 mW, or 1.0×10^{-3} J/s, is the energy transported per second by the light wave. This energy is carried within a 1.0-mm-diameter beam, so the light intensity is

$$I = \frac{P}{A} = \frac{P}{\pi r^2} = \frac{1.0 \times 10^{-3} \text{ W}}{\pi (0.00050 \text{ m})^2} = 1270 \text{ W/m}^2$$

We can use Equation 34.47 to relate this intensity to the electric field amplitude:

$$E_0 = \sqrt{\frac{2I}{c\epsilon_0}} = \sqrt{\frac{2(1270 \text{ W/m}^2)}{(3.00 \times 10^8 \text{ m/s})(8.85 \times 10^{-12} \text{ C}^2/\text{Nm}^2)}}$$

$$= 978 \text{ V/m}$$

ASSESS This is a sizable electric field, comparable to the electric field near a charged glass or plastic rod.

The intensity of a plane wave, with constant electric field amplitude E_0, would not change with distance. But a plane wave is an idealization; there are no true plane waves in nature. You learned in Chapter 20 that, in order to conserve energy, the intensity of a wave far from its source decreases with the inverse square of the distance. If a source with power P_{source} emits electromagnetic waves *uniformly* in all directions, the electromagnetic wave intensity at distance r from the source is

$$I = \frac{P_{\text{source}}}{4\pi r^2} \tag{34.48}$$

Equation 34.48 simply expresses the recognition that the energy of the wave is spread over a sphere of surface area $4\pi r^2$.

Some sources, such as antennas, emit waves in some directions but not in others. Although Equation 34.48 does not apply to such nonuniform sources, it remains true that the intensity along any line from the source decreases with the inverse square of the distance.

STOP TO THINK 34.5 An electromagnetic wave is traveling in the positive y-direction. The electric field at one instant of time is shown at one position. The magnetic field at this position points

a. In the positive x-direction.
b. In the negative x-direction.
c. In the positive y-direction.
d. In the negative y-direction.
e. Toward the origin.
f. Away from the origin.

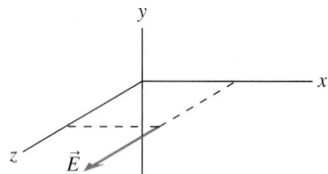

Radiation Pressure

Electromagnetic waves transfer not only energy but also momentum. An object gains momentum when it absorbs electromagnetic waves, much as a ball at rest gains momentum when struck by a ball in motion.

Suppose we shine a beam of light on an object that completely absorbs the light energy. If the object absorbs energy during a time interval Δt, its momentum changes by

$$\Delta p = \frac{\text{energy absorbed}}{c}$$

This is a consequence of Maxwell's theory that we'll state without proof.

The momentum change implies that the light is exerting a force on the object. Newton's second law, in terms of momentum, is $F = \Delta p / \Delta t$. The radiation force due to the beam of light is

$$F = \frac{\Delta p}{\Delta t} = \frac{(\text{energy absorbed})/\Delta t}{c} = \frac{P}{c}$$

where P is the power (joules per second) of the light.

It's more interesting to consider the force exerted on an object per unit area, which is called the **radiation pressure** p_{rad}. The radiation pressure on an object that absorbs all the light is

$$p_{rad} = \frac{F}{A} = \frac{(P/A)}{c} = \frac{I}{c} \qquad (34.49)$$

where I is the intensity of the light wave. The subscript on p_{rad} is important in this context to distinguish the radiation pressure from the momentum p.

EXAMPLE 34.7 Solar sailing

A low-cost way of sending spacecraft to other planets would be to use the radiation pressure on a solar sail. The intensity of the sun's electromagnetic radiation at distances near the earth's orbit is about 1300 W/m². What size sail would be needed to accelerate a 10,000 kg spacecraft toward Mars at 0.010 m/s²?

MODEL Assume that the solar sail is perfectly absorbing.

SOLVE The force that will create a 0.010 m/s² acceleration is $F = ma = 100$ N. We can use Equation 34.49 to find the sail area that, by absorbing light, will receive a 100 N force from the sun:

$$A = \frac{cF}{I} = \frac{(3.00 \times 10^8 \text{ m/s})(100 \text{ N})}{1300 \text{ W/m}^2} = 2.3 \times 10^7 \text{ m}^2$$

ASSESS If the sail is a square, it would need to be 4.8 km × 4.8 km, or roughly 3 mi × 3 mi. This is large, but not entirely out of the question with thin films that can be unrolled in space. But how will the crew return from Mars?

Antennas

We've seen that an electromagnetic wave is self-sustaining, independent of charges or currents. However, charges and currents are needed at the *source* of an electromagnetic wave. We'll take a brief look at how an electromagnetic wave is generated by an antenna.

Figure 34.32 is the electric field of an electric dipole. If the dipole is vertical, the electric field $\vec{E}$ at points along a horizontal line is also vertical. Reversing the dipole, by switching the charges, reverses $\vec{E}$. If the charges were to oscillate back and forth, switching position at frequency f, then $\vec{E}$ would oscillate in a vertical plane. The changing $\vec{E}$ would then create an induced magnetic field $\vec{B}$, which could then create an $\vec{E}$, which could then create a $\vec{B}$, . . . and an electromagnetic wave at frequency f would radiate out into space.

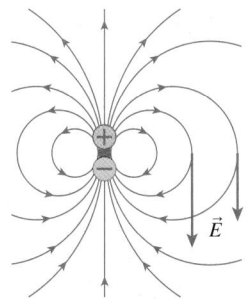

Positive charge on top

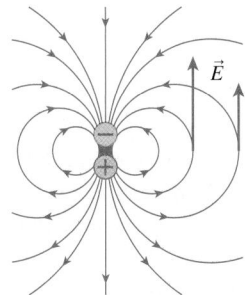

Negative charge on top

FIGURE 34.32 An electric dipole creates an electric field that reverses direction if the dipole charges are switched.

An oscillating voltage causes the dipole to oscillate.

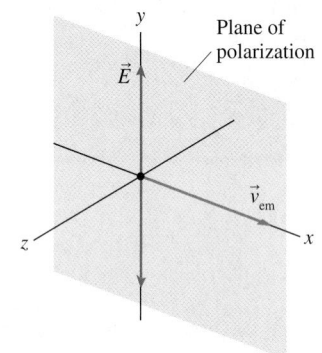

Antenna wire

The oscillating dipole causes an electromagnetic wave to move away from the antenna at speed $v_{em} = c$.

FIGURE 34.33 An antenna generates a self-sustaining electromagnetic wave.

This is exactly what an **antenna** does. Figure 34.33 shows two metal wires attached to the terminals of an oscillating voltage source. The figure shows an instant when the top wire is negative and the bottom is positive, but these will reverse in half a cycle. The wire is basically an oscillating dipole, and it creates an oscillating electric field. The oscillating $\vec{E}$ induces an oscillating $\vec{B}$, and they take off as an electromagnetic wave at speed $v_{em} = c$. The wave does need oscillating charges as a *wave source,* but once created it is self-sustaining and independent of the source. The antenna might be destroyed, but the wave could travel billions of light years across the universe, bearing the legacy of James Clerk Maxwell.

STOP TO THINK 34.6 The amplitude of the oscillating electric field at your cell phone is 4.0 μV/m when you are 10 km east of the broadcast antenna. What is the electric field amplitude when you are 20 km east of the antenna?

a. 1.0 μV/m
b. 2.0 μV/m
c. 4.0 μV/m
d. There's not enough information to tell.

34.8 Polarization

The plane of the electric field vector $\vec{E}$ and the Poynting vector $\vec{S}$ (the direction of propagation) is called the **plane of polarization** of an electromagnetic wave. Figure 34.34 shows just the electric field of two waves moving along the *x*-axis. The magnetic field, not shown, is perpendicular to $\vec{E}$. The electric field in Figure 34.34a oscillates vertically, so we would say that this wave is *vertically polarized.* Similarly the wave in Figure 34.34b is *horizontally polarized.* Other polarizations are possible, such as a wave polarized 30° away from horizontal.

(a) Vertical polarization

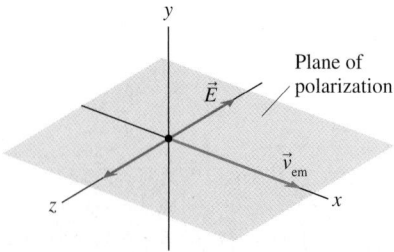

(b) Horizontal polarization

FIGURE 34.34 The plane of polarization is the plane in which the electric field vector oscillates.

NOTE ▶ This use of the term "polarization" is completely independent of the idea of *charge polarization* that you learned about in Chapter 25. ◄

Some wave sources, such as lasers and radio antennas, emit *polarized* electromagnetic waves with a well-defined plane of polarization. By contrast, most natural sources of electromagnetic radiation are unpolarized. Each atom in the sun's hot atmosphere emits light independently of all the other atoms, as does each tiny piece of metal in the incandescent filament of a light bulb. An electromagnetic wave that you see or measure is a superposition of waves from each of these tiny emitters. Although the wave from each individual emitter is polarized, it is polarized in a random direction with respect to the waves from all its neighbors. The net result is what we call an *unpolarized* wave, a wave whose electric field oscillates randomly with all possible orientations.

A few natural sources are *partially polarized,* meaning that one direction of polarization is more prominent than others. The light of the sky at right angles to the sun is partially polarized, because of how the sun's light scatters from air molecules to create skylight. Bees and other insects make use of this partial polarization to navigate. Light reflected from a flat, horizontal surface, such as a road or the surface of a lake, has a predominantly horizontal polarization. This is the rationale for using polarizing sunglasses.

The most common way of artificially generating polarized visible light is to send unpolarized light through a *polarizing filter.* The first widely used polarizing filter was invented by Edwin Land in 1928, while he was still an undergraduate

16.9

student. He developed an improved version, called Polaroid, in 1938. Polaroid, as shown in Figure 34.35, is a plastic sheet containing very long organic molecules known as polymers. The sheets are formed in such a way that the polymers are all aligned to form a grid, rather like the metal bars in a barbecue grill. The sheet is then chemically treated to make the polymer molecules somewhat conducting.

As a light wave travels through Polaroid, the component of the electric field oscillating parallel to the polymer grid drives the conduction electrons up and down the molecules. The electrons absorb energy from the light wave, so the parallel component of $\vec{E}$ is absorbed in the filter. But the conduction electrons can't oscillate perpendicular to the molecules, so the component of $\vec{E}$ perpendicular to the polymer grid passes through without absorption. Thus the light wave emerging from a polarizing filter is polarized perpendicular to the polymer grid.

Malus's Law

Suppose a *polarized* light wave of intensity I_0 approaches a polarizing filter. What is the intensity of the light that passes through the filter? Figure 34.36 shows that an oscillating electric field can be decomposed into components parallel and perpendicular to the polarizer's axis (i.e., the polarization direction transmitted by the polarizer). If we call the polarizer axis the y-axis, then the incident electric field is

$$\vec{E}_{\text{incident}} = E_\perp \hat{i} + E_\parallel \hat{j} = E_0 \sin\theta \hat{i} + E_0 \cos\theta \hat{j} \qquad (34.50)$$

where θ is the angle between the incident plane of polarization and the polarizer axis.

If the polarizer is ideal, meaning that light polarized parallel to the axis is 100% transmitted and light perpendicular to the axis is 100% blocked, then the electric field of the light transmitted by the filter is

$$\vec{E}_{\text{transmitted}} = E_\parallel \hat{j} = E_0 \cos\theta \hat{j} \qquad (34.51)$$

Because the intensity depends on the square of the electric field amplitude, you can see that the transmitted intensity is related to the incident intensity by

$$I_{\text{transmitted}} = I_0 \cos^2\theta \qquad \text{(incident light polarized)} \qquad (34.52)$$

This result, which was discovered experimentally in 1809, is called **Malus's law.**

Figure 34.37a shows that Malus's law can be demonstrated with two polarizing filters. The first, called the *polarizer,* is used to produce polarized light of intensity I_0. The second, called the *analyzer,* is rotated by angle θ relative to the polarizer. As the photographs of Figure 34.37b show, the transmission of the analyzer is (ideally) 100% when $\theta = 0°$ and steadily decreases to zero when $\theta = 90°$. Two polarizing filters with perpendicular axes, called *crossed polarizers,* block all the light.

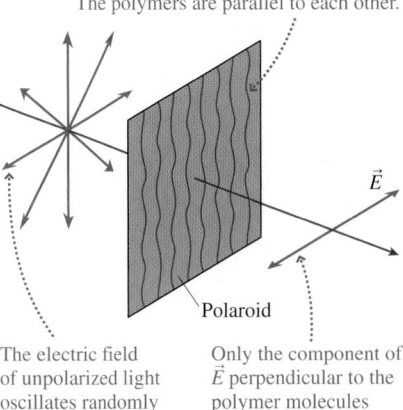

The polymers are parallel to each other.

The electric field of unpolarized light oscillates randomly in all directions.

Only the component of $\vec{E}$ perpendicular to the polymer molecules is transmitted.

FIGURE 34.35 A polarizing filter.

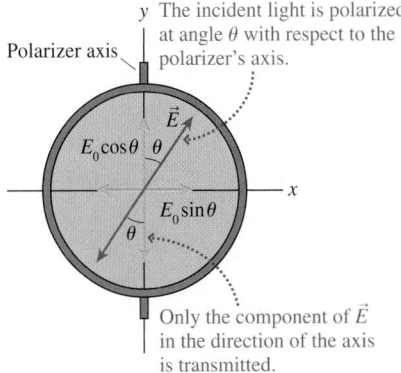

y The incident light is polarized at angle θ with respect to the polarizer's axis.

Polarizer axis

$E_0\cos\theta$ θ

$E_0\sin\theta$

θ

x

Only the component of $\vec{E}$ in the direction of the axis is transmitted.

FIGURE 34.36 An incident electric field can be decomposed into components parallel and perpendicular to a polarizer's axis.

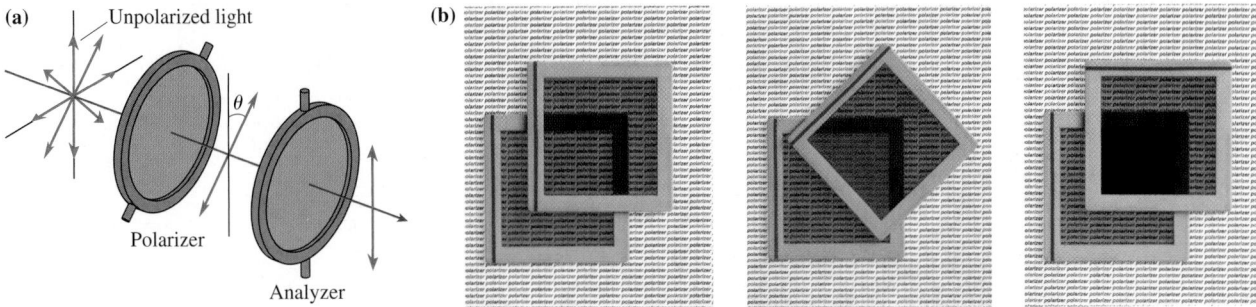

(a) Unpolarized light

θ

Polarizer

Analyzer

(b)

FIGURE 34.37 The intensity of the transmitted light depends on the angle between the polarizing filters.

Suppose the light incident on a polarizing filter is *unpolarized,* as is the light incident from the left on the polarizer in Figure 34.37a. The electric field of unpolarized light varies randomly through all possible values of θ. Because the *average* value of $\cos^2\theta$ is $\frac{1}{2}$, the intensity transmitted by a polarizing filter is

$$I_{transmitted} = \frac{1}{2}I_0 \text{ (incident light unpolarized)} \qquad (34.53)$$

In other words, a polarizing filter passes 50% of unpolarized light and blocks 50%.

In polarizing sunglasses, the polymer grid is aligned horizontally (when the glasses are in the normal orientation) so that the glasses transmit vertically polarized light. Most natural light is unpolarized, so the glasses reduce the light intensity by 50%. But *glare*—the reflection of the sun and the skylight from roads and other horizontal surfaces—has a strong horizontal polarization. This light is almost completely blocked by the Polaroid, so the sunglasses "cut glare" without affecting the main scene you wish to see.

You can test whether your sunglasses are polarized by holding them in front of you and rotating them as you look at the glare reflecting from a horizontal surface. Polarizing sunglasses will substantially reduce the glare when the glasses are "normal" but not when the glasses are 90° from normal. (You can also test them against a pair of sunglasses known to be polarizing by seeing if all light is blocked when the lenses of the two pairs are crossed.)

If you do have polarizing sunglasses, look at the sky about 90° from the sun in the early morning or late afternoon. You can detect the sky's polarization by rotating the glasses. Bees can automatically sense the polarization of skylight, but humans can't.

STOP TO THINK 34.7   Unpolarized light of equal intensity is incident on four pairs of polarizing filters. Rank in order, from largest to smallest, the intensities I_a to I_d transmitted through the second polarizer of each pair.

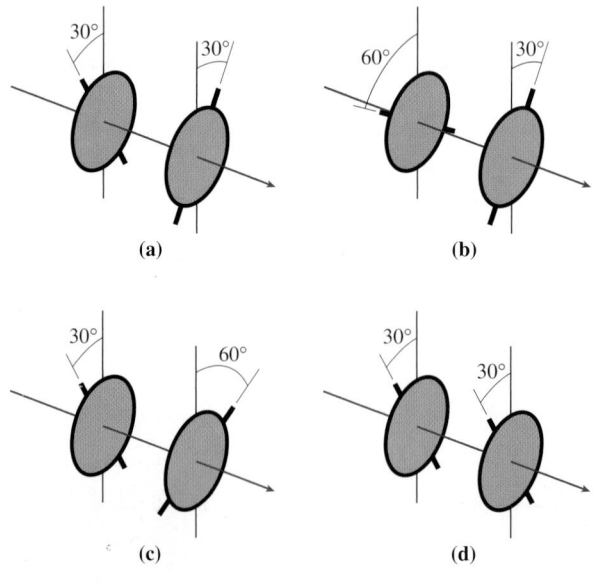

(a)   (b)

(c)   (d)

SUMMARY

The goal of Chapter 34 has been to study the properties of electromagnetic fields and waves.

GENERAL PRINCIPLES

Maxwell's Equations

These equations govern electromagnetic fields:

$$\oint \vec{E} \cdot d\vec{A} = \frac{Q_{in}}{\epsilon_0} \qquad \text{Gauss's law}$$

$$\oint \vec{B} \cdot d\vec{A} = 0 \qquad \text{Gauss's law for magnetism}$$

$$\oint \vec{E} \cdot d\vec{s} = -\frac{d\Phi_m}{dt} \qquad \text{Faraday's law}$$

$$\oint \vec{B} \cdot d\vec{s} = \mu_0 I_{through} + \epsilon_0 \mu_0 \frac{d\Phi_e}{dt} \qquad \text{Ampère-Maxwell law}$$

Maxwell's equations tell us that:

An electric field can be created by

- Charged particles
- A changing magnetic field

A magnetic field can be created by

- A current
- A changing electric field

Lorentz Force

This force law governs the interaction of charged particles with electromagnetic fields:

$$\vec{F} = q(\vec{E} + \vec{v} \times \vec{B})$$

- An electric field exerts a force on any charged particle.
- A magnetic field exerts a force on a moving charged particle.

Field Transformations

Fields measured in frame S to be $\vec{E}$ and $\vec{B}$ are found in frame S' to be

$$\vec{E}' = \vec{E} + \vec{V} \times \vec{B}$$

$$\vec{B}' = \vec{B} - \frac{1}{c^2}\vec{V} \times \vec{E}$$

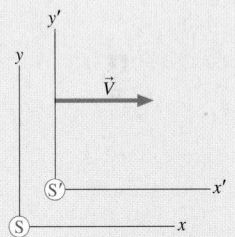

IMPORTANT CONCEPTS

Induced fields

An induced electric field is created by a changing magnetic field.

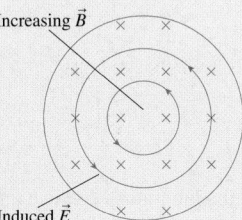

An induced magnetic field is created by a changing electric field.

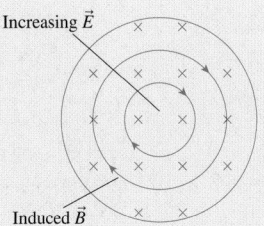

These fields can exist independently of charges and currents.

An electromagnetic wave is a self-sustaining electromagnetic field.

- An em wave is a transverse wave with $\vec{E}$, $\vec{B}$, and $\vec{v}$ mutually perpendicular.
- An em wave propagates with speed $v_{em} = c = 1/\sqrt{\epsilon_0 \mu_0}$.
- The electric and magnetic field strengths are related by $E = cB$.
- The **Poynting vector** $\vec{S} = (\vec{E} \times \vec{B})/\mu_0$ is the energy transfer in the direction of travel.
- The wave **intensity** is $I = P/A = (1/2c\mu_0)E_0^2 = (c\epsilon_0/2)E_0^2$.

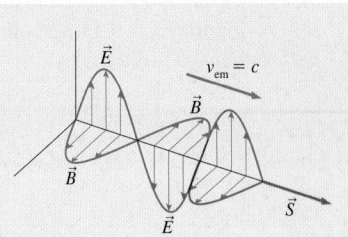

APPLICATIONS

Polarization

The electric field and the Poynting vector define the **plane of polarization.** The intensity of polarized light transmitted through a polarizing filter is given by Malus's law

$$I = I_0 \cos^2\theta$$

where θ is the angle between the electric field and the polarizer axis.

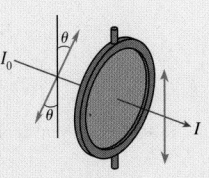

TERMS AND NOTATION

Lorentz force law	Maxwell's equations	antenna
Galilean field transformation	Poynting vector, $\vec{S}$	plane of polarization
equations	intensity, I	Malus's law
displacement current	radiation pressure, p_{rad}	

EXERCISES AND PROBLEMS

Exercises

Section 34.1 Electromagnetic Fields and Forces

1. The magnetic field is uniform over each face of the box shown in Figure Ex34.1. What are the magnetic field strength and direction on the front surface?

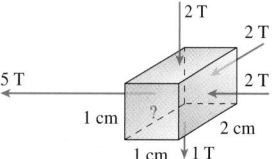

FIGURE EX34.1

2. What is the force (magnitude and direction) on the proton in Figure Ex34.2?

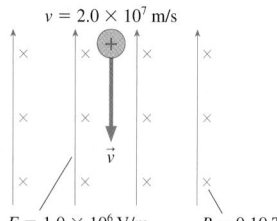

FIGURE EX34.2 $E = 1.0 \times 10^6$ V/m $B = 0.10$ T

3. An electron travels with $\vec{v} = 5.0 \times 10^6 \hat{\imath}$ m/s through a point in space where $\vec{E} = (2.0 \times 10^5 \hat{\imath} - 2.0 \times 10^5 \hat{\jmath})$ V/m and $\vec{B} = -0.10\hat{k}$ T. What is the force on the electron?

4. What electric field strength and direction will allow the electron in Figure Ex34.4 to pass through this region of space without being deflected?

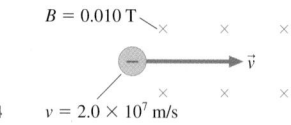

FIGURE EX34.4 $v = 2.0 \times 10^7$ m/s

5. What are the electric field strength and direction at the position of the proton in Figure Ex34.5?

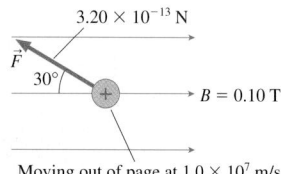

FIGURE EX34.5   Moving out of page at 1.0×10^7 m/s

6. An electron travels with $\vec{v} = 5.0 \times 10^6 \hat{\imath}$ m/s through a point in space where $\vec{B} = 0.10\hat{\jmath}$ T. The force on the electron at this point is $\vec{F} = (9.6 \times 10^{-14}\hat{\imath} - 9.6 \times 10^{-14}\hat{k})$ N. What is the electric field?

Section 34.2 *E* or *B*? It Depends on Your Perspective

7. A rocket cruises past a laboratory at 1.0×10^6 m/s in the positive *x*-direction just as a proton is launched with velocity (in the laboratory frame) $\vec{v} = (1.41 \times 10^6 \hat{\imath} + 1.41 \times 10^6 \hat{\jmath})$ m/s. What are the proton's speed and its angle from the *y*-axis (or *y′*-axis) in (a) the laboratory frame and (b) the rocket frame?

8. Figure Ex34.8 shows the electric and magnetic field in Frame S. A rocket travels parallel to one of the axes of the S coordinate system. Along which axis must the rocket travel, and in which direction (or directions), in order for the rocket scientists to measure (a) $B' > B$, (b) $B' = B$, and (c) $B' < B$?

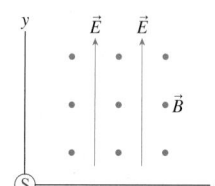

FIGURE EX34.8

9. Scientists in the laboratory create a uniform electric field $\vec{E} = -1.0 \times 10^6 \hat{\jmath}$ V/m in a region of space where $\vec{B} = \vec{0}$. What are the fields in the reference frame of a rocket traveling in the positive *x*-direction at 1.0×10^6 m/s?

10. A rocket zooms past the earth at $v = 2.0 \times 10^6$ m/s. Scientists on the rocket have created the electric and magnetic fields shown in Figure Ex34.10. What are the fields measured by an earthbound scientist?

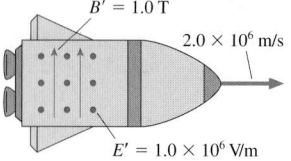

FIGURE EX34.10 $E' = 1.0 \times 10^6$ V/m

11. Laboratory scientists have created the electric and magnetic fields shown in Figure Ex34.11. These fields are also seen by scientists that zoom past in a rocket traveling in the *x*-direction at 1.0×10^6 m/s. According to the rocket scientists, what angle does the electric field make with the axis of the rocket?

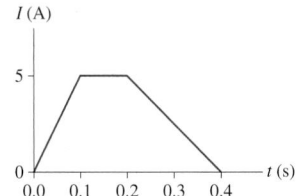

FIGURE EX34.11 $B = 0.50$ T $E = 1.0 \times 10^6$ V/m

Section 34.3 Faraday's Law Revisited

12. Figure Ex34.12 shows the current as a function of time through a 20-cm-long, 4.0-cm-diameter solenoid with 400 turns. Draw a graph of the induced electric field strength as a function of time at a point 1.0 cm from the axis of the solenoid.

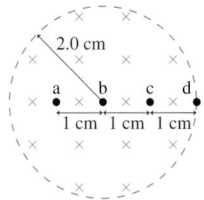

FIGURE EX34.12

13. The magnetic field inside a 5.0-diameter solenoid is 2.0 T and decreasing at 4.0 T/s. What is the electric field strength inside the solenoid at point (a) on the axis and (b) 2.0 cm from the axis?

14. The magnetic field in Figure Ex34.14 is decreasing at the rate 0.10 T/s. What is the acceleration (magnitude and direction) of a proton at rest at points a to d?

FIGURE EX34.14

Section 34.4 The Displacement Current

15. Show that the quantity $\epsilon_0(d\Phi_e/dt)$ has units of current.

16. Show that the displacement current inside a parallel-plate capacitor can be written $C(dV_C/dt)$.

17. At what rate must the potential difference increase across a 1.0 μF capacitor to create a 1.0 A displacement current in the capacitor?

18. A 10-cm-diameter parallel-plate capacitor has a 1.0 mm spacing. The electric field between the plates is increasing at the rate 1.0×10^6 V/m s. What is the magnetic field strength (a) on the axis, (b) 3.0 cm from the axis, and (c) 7.0 cm from the axis?

19. A square parallel-plate capacitor 5.0 cm on a side has a 0.50 mm gap. What is the displacement current in the capacitor if the potential difference across the capacitor is increasing at 500,000 V/s?

Section 34.6 Electromagnetic Waves

20. What is the magnetic field amplitude of an electromagnetic wave whose electric field amplitude is 10 V/m?

21. What is the electric field amplitude of an electromagnetic wave whose magnetic field amplitude is 2.0 mT?

22. The magnetic field of an electromagnetic wave in a vacuum is $B_z = (3.0\ \mu\text{T})\sin((1.00 \times 10^7)x - \omega t)$, where x is in m and t is in s. What are the wave's (a) wavelength, (b) frequency, and (c) electric field amplitude?

23. The electric field of an electromagnetic wave in a vacuum is $E_y = (20\ \text{V/m})\cos((6.28 \times 10^8)x - \omega t)$, where x is in m and t is in s. What are the wave's (a) wavelength, (b) frequency, and (c) magnetic field amplitude?

Section 34.7 Properties of Electromagnetic Waves

24. A radio wave is traveling in the negative y-direction. What is the direction of $\vec{E}$ at a point where $\vec{B}$ is in the positive x-direction?

25. Show that:
    a. The quantity cB has the same units as E.
    b. The Poynting vector has units W/m^2.

26. a. What is the magnetic field amplitude of an electromagnetic wave whose electric field amplitude is 100 V/m?
    b. What is the intensity of the wave?

27. A radio receiver can detect signals with electric field amplitudes as small as 300 μV/m. What is the intensity of the smallest detectable signal?

28. A 200 MW laser pulse is focused with a lens to a diameter of 2.0 μm.
    a. What is the laser beam's electric field amplitude at the focal point?
    b. How does this electric field compare to the electric field that keeps the electron bound to the proton of a hydrogen atom? The radius of the electron's orbit is 0.053 nm.

29. A radio antenna broadcasts a 1.0 MHz radio wave with 25 kW of power. Assume that the radiation is emitted uniformly in all directions.
    a. What is the wave's intensity 30 km from the antenna?
    b. What is the electric field amplitude at this distance?

30. At what distance from a 10 W point source of electromagnetic waves is the electric field amplitude (a) 100 V/m and (b) 0.010 V/m?

31. A 1000 W carbon-dioxide laser emits light with a wavelength of 10 μm into a 3.0-mm-diameter laser beam. What force does the laser beam exert on a completely absorbing target?

Section 34.8 Polarization

32. Figure Ex34.32 shows a vertically polarized radio wave of frequency 1.0×10^6 Hz traveling into the page. The maximum electric field strength is 1000 V/m. What are
    a. The maximum magnetic field strength?
    b. The magnetic field strength and direction at a point where $\vec{E} = (500\ \text{V/m, down})$?
    c. The smallest distance between a point on the wave having the magnetic field of part b and a point where the magnetic field is at maximum strength?

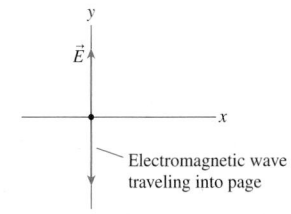

FIGURE EX34.32

33. Only 25% of the intensity of a polarized light wave passes through a polarizing filter. What is the angle between the electric field and the axis of the filter?

34. A 200 mW horizontally polarized laser beam passes through a polarizing filter whose axis is 25° from vertical. What is the power of the laser beam as it emerges from the filter?

35. Unpolarized light with intensity 350 W/m² passes first through a polarizing filter with its axis vertical, then through a polarizing filter with its axis 30° from vertical. What light intensity emerges from the second filter?

Problems

36. Figure P34.36 is an overhead view of children on a playground. Mary (M), who is at the origin of reference frame S, throws a ball horizontally to Tom (T) at 5.0 m/s. At the instant she releases the ball, Carlos (C) runs past her at 3.0 m/s.
    a. Write the ball's velocity vector $\vec{v}$ in component form, $\vec{v} = v_x \hat{i} + v_y \hat{j}$.
    b. What are Carlos's xy-coordinates in frame S at the instant Mary tosses the ball and at the instant Tom catches it?
    c. Carlos is at the origin of reference frame S′. Draw a picture of Carlos's coordinate axes, then show the position of the ball, as seen by Carlos, at the instant Mary tosses it and at the instant Tom catches it.
    d. Use Carlos's measurements of position and time to find the ball's velocity $\vec{v}'$ in Carlos's reference frame.
    e. Use the Galilean velocity transformation to find $\vec{v}'$ and show that it agrees with your answer to part d.

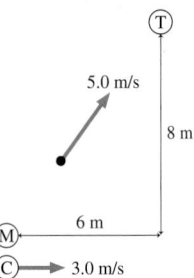

FIGURE P34.36

37. A proton is fired with a speed of 1.0×10^6 m/s through the parallel-plate capacitor shown in Figure P34.37. The capacitor's electric field is $\vec{E} = (1.0 \times 10^5$ V/m, down).
    a. What magnetic field $\vec{B}$, both strength and direction, must be applied to allow the proton to pass through the capacitor with no change in speed or direction?
    b. Find the electric and magnetic fields in the proton's reference frame.
    c. How does an experimenter in the proton's frame explain that the proton experiences no force as the charged plates fly by?

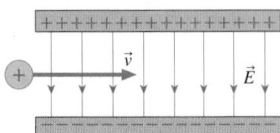

FIGURE P34.37

38. In Figure P34.38, a circular loop of radius r travels with speed v along a charged wire having linear charge density λ. The wire is at rest in the laboratory frame, and it passes through the center of the loop.
    a. What are $\vec{E}$ and $\vec{B}$ at a point on the loop as measured by a scientist in the laboratory? Include both strength and direction.
    b. What are the fields $\vec{E}'$ and $\vec{B}'$ at a point on the loop as measured by a scientist in the frame of the loop?
    c. Show that an experimenter in the loop's frame sees a current $I = \lambda v$ passing through the center of the loop.
    d. What electric and magnetic fields would an experimenter in the loop's frame calculate at distance r from the current of part c?
    e. Show that your field of parts b and d are the same.
    f. If the loop is made of a conducting material, will it have an induced current? Explain.

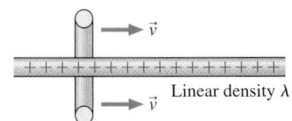

FIGURE P34.38

39. A very long, 1.0-mm-diameter wire carries a 2.5 A current from left to right. Thin plastic insulation on the wire is positively charged with linear charge density 2.5 nC/cm. A mosquito 1.0 cm from the center of the wire would like to move in such a way as to experience an electric field but no magnetic field. How fast and which direction should she fly?

40. The magnetic field inside a 4.0-cm-diameter superconducting solenoid varies sinusoidally between 8.0 T and 12.0 T at a frequency of 10 Hz.
    a. What is the maximum electric field strength at a point 1.5 cm from the solenoid axis?
    b. What is the value of B at the instant E reaches its maximum value?

41. Equation 34.27 is an expression for the induced electric field inside a solenoid ($r < R$). Find an expression for the induced electric field outside a solenoid ($r > R$) in which the magnetic field is changing at the rate dB/dt.

42. A simple series circuit consists of a 150 Ω resistor, a 25 V battery, a switch, and a 2.5 pF parallel-plate capacitor (initially uncharged) with plates 5.0 mm apart. The switch is closed at $t = 0$ s.
    a. After the switch is closed, find the maximum electric flux and the maximum displacement current through the capacitor.
    b. Find the electric flux and the displacement current at $t = 0.50$ ns.

43. A wire with conductivity σ carries current I. The current is increasing at the rate dI/dt.
    a. Show that there is a displacement current in the wire equal to $(\epsilon_0/\sigma)(dI/dt)$.
    b. Evaluate the displacement current for a copper wire in which the current is increasing at 1.0×10^6 A/s.

44. A 10 A current is charging a 1.0-cm-diameter parallel-plate capacitor.
    a. What is the magnetic field strength at a point 2.0 mm from the center of the wire leading to the capacitor?
    b. What is the magnetic field strength at a point 2.0 mm from the center of the capacitor?

45. Figure P34.45 shows the electric field inside a cylinder of radius $R = 3.0$ mm. The field strength is increasing with time as $E = 1.0 \times 10^8 t^2$ V/m, where t is in s. The electric field

outside the cylinder is always zero, and the field inside the cylinder was zero for $t < 0$.

a. Find an expression for the electric flux Φ_e through the entire cylinder as a function of time.

b. Draw a picture showing the magnetic field lines inside and outside the cylinder. Be sure to include arrowheads showing the field's direction.

c. Find an expression for the magnetic field strength as a function of time at a distance $r < R$ from the center. Evaluate the magnetic field strength at $r = 2.0$ mm, $t = 2.0$ s.

d. Find an expression for magnetic field strength as a function of time at a distance $r > R$ from the center. Evaluate the magnetic field strength at $r = 4.0$ mm, $t = 2.0$ s.

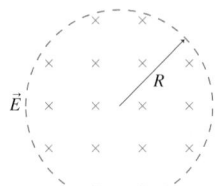

FIGURE P34.45

46. Assume that a 100 W light bulb radiates all its energy as a single wavelength of visible light. Estimate the electric and magnetic field strength at the surface of the bulb.

47. The intensity of sunlight reaching the earth is 1360 W/m².
    a. What is the power output of the sun?
    b. What is the intensity of sunlight on Mars?

48. When the Voyager 2 spacecraft passed Neptune in 1989, it was 4.5×10^9 km from the earth. Its radio transmitter, with which it sent back data and images, broadcast with a mere 21 W of power. Assuming that the transmitter broadcast equally in all directions,
    a. What signal intensity was received on the earth?
    b. What electric field amplitude was detected?
    The received signal was somewhat stronger than your result because the spacecraft used a directional antenna, but not by much.

49. In reading the instruction manual that came with your garage-door opener, you see that the transmitter unit in your car produces a 250 mW signal and that the receiver unit is supposed to respond to a radio wave of the correct frequency if the electric field amplitude exceeds 0.10 V/m. You wonder if this is really true. To find out, you put fresh batteries in the transmitter and start walking away from your garage while opening and closing the door. Your garage door finally fails to respond when you're 42 m away. Are the manufacturer's claims true?

50. The intensity of sunlight reaching the earth is 1360 W/m². Assuming all the sunlight is absorbed, what is the radiation-pressure force on the earth? Give your answer in newtons and as a percentage of the sun's gravitational force on the earth.

51. A laser beam shines straight up onto a flat, black foil with a mass of 25 μg. What laser power is needed to levitate the foil?

52. For a science project, you would like to horizontally suspend an 8.5 by 11 inch sheet of black paper in a vertical beam of light whose dimensions exactly match the paper. If the mass of the sheet is 1.0 g, what light intensity will you need?

53. You've recently read about a chemical laser that generates a 20-cm-diameter, 25 MW laser beam. One day, after physics class, you start to wonder if you could use the radiation pres-

sure from this laser beam to launch small payloads into orbit. To see if this might be feasible, you do a quick calculation of the acceleration of a 20-cm-diameter, 100 kg, perfectly absorbing block. What speed would such a block have if pushed *horizontally* 100 m along a frictionless track by such a laser?

54. An 80 kg astronaut has gone outside his space capsule to do some repair work. Unfortunately, he forgot to lock his safety tether in place, and he has drifted 5.0 m away from the capsule. Fortunately, he has a 1000 W portable laser with fresh batteries that will operate it for 1.0 hr. His only chance is to accelerate himself toward the space capsule by firing the laser in the opposite direction. He has a 10-hr supply of oxygen. Can he make it?

55. Unpolarized light of intensity I_0 is incident on three polarizing filters. The axis of the first is vertical, that of the second is 45° from vertical, and that of the third is horizontal. What light intensity emerges from the third filter?

Challenge Problems

56. A 4.0-cm-diameter parallel-plate capacitor with a 1.0 mm spacing is charged to 1000 V. A switch closes at $t = 0$ s, and the capacitor is discharged through a wire with 0.20 Ω resistance.
    a. Find an expression for the magnetic field strength inside the capacitor at $r = 1.0$ cm as a function of time.
    b. Draw of graph of B versus t.

57. a. Show that u_E and u_B, the energy densities of the electric and magnetic fields, are equal to each other in an electromagnetic wave. In other words, show that the wave's energy is divided equally between the electric field and the magnetic field.
    b. What is the total energy density in an electromagnetic wave of intensity 1000 W/m²?

58. Large quantities of dust should have been left behind after the creation of the solar system. Larger dust particles, comparable in size to soot and sand grains, are common. They create shooting stars when they collide with the earth's atmosphere. But very small dust particles are conspicuously absent. Astronomers believe that the very small dust particles have been blown out of the solar system by the sun. By comparing the forces on dust particles, determine the diameter of the smallest dust particles that can remain in the solar system over long periods of time. Assume that the dust particles are spherical, black, and have a density of 2000 kg/m³. The sun emits electromagnetic radiation with power 3.9×10^{26} W.

59. Consider current I passing through a resistor of radius r, length L, and resistance R.
    a. Determine the electric and magnetic fields at the surface of the resistor. Assume that the electric field is uniform throughout, including at the surface.
    b. Determine the strength and direction of the Poynting vector at the surface of the resistor.
    c. Show that the flux of the Poynting vector (i.e., the integral of $\vec{S} \cdot d\vec{A}$) over the surface of the resistor is I^2R. Then give an interpretation of this result.

60. Unpolarized light of intensity I_0 is incident on a stack of 7 polarizing filters, each with its axis rotated 15° cw with respect to the previous filter. What light intensity emerges from the last filter?

Stop to Think 34.1: a. The charge is negative, so the electric force is to the left. $\vec{v} \times \vec{B}$ is to the right, so the magnetic force on a negative charge is also to the left.

Stop to Think 34.2: b. $\vec{V}$ is parallel to $\vec{B}$, hence $\vec{V} \times \vec{B}$ is zero. Thus $\vec{E}' = \vec{E}$ and points in the positive z-direction. $\vec{V} \times \vec{E}$ points down, in the negative y direction, so $-\vec{V} \times \vec{E}/c^2$ points in the positive y-direction and causes $\vec{B}'$ to be angled upward.

Stop to Think 34.3: $B_c > B_a > B_d > B_b$. The induced magnetic field strength depends on the *rate dE/dt* at which the electric field is changing. Steeper slopes on the graph correspond to larger magnetic fields.

Stop to Think 34.4: e. $\vec{E}$ is perpendicular to $\vec{B}$ and to $\vec{v}$, so it can only be along the z-axis. According to the Ampère-Maxwell law, $d\Phi_e/dt$ has the same sign as the line integral of $\vec{B} \cdot d\vec{s}$ around the closed curve. The integral is positive for a cw integration. Thus, from the right-hand rule, $\vec{E}$ is either into the page (negative z-direction) and increasing, or out of the page (positive z-direction) and decreasing.

We can see from the figure that B is decreasing as the wave moves left to right, so E must also be decreasing. Thus $\vec{E}$ points along the positive z-axis.

Stop to Think 34.5: a. The Poynting vector $\vec{S} = (\vec{E} \times \vec{B})/\mu_0$ points in the direction of travel, which is the positive y-direction. $\vec{B}$ must point in the positive x-direction in order for $\vec{E} \times \vec{B}$ to point upward.

Stop to Think 34.6: b. The intensity along a line from the antenna decreases inversely with the square of the distance, so the intensity at 20 km is $\frac{1}{4}$ that at 10 km. But the intensity depends on the square of the electric field amplitude, or, conversely, E_0 is proportional to $I^{1/2}$. Thus E_0 at 20 km is $\frac{1}{2}$ that at 10 km.

Stop to Think 34.7: $I_d > I_a > I_b = I_c$. The intensity depends upon $\cos^2\theta$, where θ is the angle *between* the axes of the two filters. The filters in d have $\theta = 0°$. The two filters in both b and c are crossed ($\theta = 90°$) and transmit no light at all.

35 AC Circuits

Transmission lines carry alternating current at voltages as high as 500,000 V.

▶ **Looking Ahead**

The goal of Chapter 35 is to understand and apply basic techniques of AC circuit analysis. In this chapter you will learn to:

- Use phasors to analyze an AC circuit with resistors, capacitors, and inductors.
- Understand *RC* filter circuits.
- Understand resonance in an *RLC* circuit.
- Calculate the power loss in an AC circuit.

◀ **Looking Back**

The material in this chapter depends on the fundamentals of circuits and on the properties of resistors, capacitors, and inductors. The mathematical representation of AC circuits is based on simple harmonic motion. Please review:

- Sections 14.1–14.2 and 14.8 Simple harmonic motion and resonance.
- Section 30.6 Capacitors.
- Sections 31.1–31.4 Fundamentals of circuit analysis.
- Sections 33.8–33.10 Inductors.

Thomas Edison built the first large-scale electric generating station in 1882, in New York City. His entrepreneurial motive was to sell light bulbs, which he had invented a few years earlier. Edison's company, which later became General Electric, is still one of the largest manufacturers of electrical equipment.

Edison soon had competition from George Westinghouse, another name that probably looks familiar. Whereas Edison's system used *direct current* (DC), Westinghouse favored *alternating current* (AC). The technological debate between these two men and their companies lasted nearly 20 years, but eventually alternating current proved to be superior for the long-distance transmission of electric energy.

Today, more than a century later, a "grid" of AC electrical distribution systems spans the United States and other countries. Any device that plugs into an electric outlet uses an AC circuit. In this chapter, you will learn some of the basic techniques for analyzing AC circuits. However, these ideas are not limited to

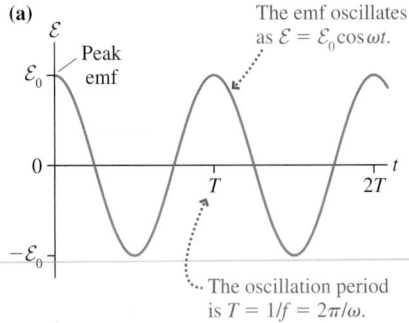

(a)

Peak emf

The emf oscillates as $\mathcal{E} = \mathcal{E}_0 \cos \omega t$.

The oscillation period is $T = 1/f = 2\pi/\omega$.

(b)

The length of the phasor is $\mathcal{E}_0$.

The phasor rotates ccw at angular frequency ω.

The *phase angle* is ωt.

The tip of the phasor goes once around the circle in time T.

The instantaneous emf value $\mathcal{E}_0 \cos \omega t$ is the projection of the phasor onto the horizontal axis.

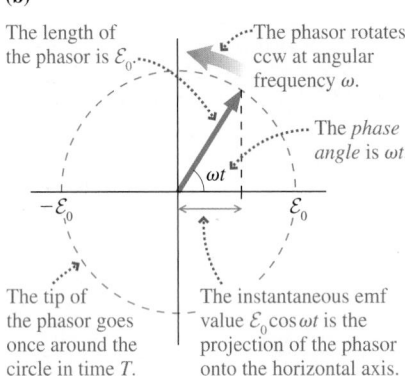

FIGURE 35.1 An oscillating quantity such as emf can be represented either as a graph or as a phasor diagram.

Graphical representation of the emf

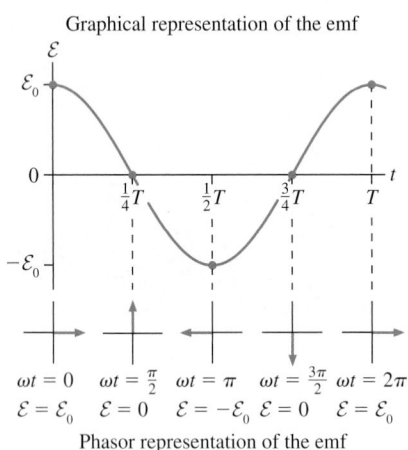

| $\omega t = 0$ | $\omega t = \frac{\pi}{2}$ | $\omega t = \pi$ | $\omega t = \frac{3\pi}{2}$ | $\omega t = 2\pi$ |
| $\mathcal{E} = \mathcal{E}_0$ | $\mathcal{E} = 0$ | $\mathcal{E} = -\mathcal{E}_0$ | $\mathcal{E} = 0$ | $\mathcal{E} = \mathcal{E}_0$ |

Phasor representation of the emf

FIGURE 35.2 The correspondence between a rotating phasor and points on a graph.

power-line circuits. Audio, radio, television, and telecommunication electronics are based on circuits that use oscillating voltages and currents. Any practical understanding of modern electronics is grounded, so to speak, in AC circuit analysis.

35.1 AC Sources and Phasors

One of the examples of Faraday's law cited in Chapter 33 was an electric generator. A turbine, which might be powered by expanding steam or falling water, causes a coil of wire to rotate in a magnetic field. As the coil spins, the emf and the induced current oscillate sinusoidally. The emf is alternately positive and then negative, causing the charges to flow in one direction and then, a half cycle later, in the other. The oscillation frequency in North and South America is $f = 60$ Hz, whereas most of the rest of the world uses a 50 Hz oscillation.

The generator's emf—the voltage—is determined by the magnetic field strength and the number of turns in the generator coil. The emf is a fixed, unvarying quantity, so it might seem logical to call a generator an *alternating voltage source*. Nonetheless, circuits powered by a sinusoidal emf are called **AC circuits,** where AC stands for *alternating current*. By contrast, the steady-current circuits you studied in Chapter 31 are called **DC circuits,** for *direct current*.

AC circuits are not limited to the use of 50 Hz or 60 Hz power-line voltages. Audio, radio, television, and telecommunication equipment all make extensive use of AC circuits, with frequencies ranging from approximately 10^2 Hz in audio circuits to approximately 10^9 Hz in cell phones. These devices use *electrical oscillators* rather than generators to produce a sinusoidal emf, but the basic principles of circuit analysis are the same.

You can think of an AC generator or oscillator as a battery whose output voltage undergoes sinusoidal oscillations. The instantaneous emf of an AC generator or oscillator, shown graphically in Figure 35.1a, can be written

$$\mathcal{E} = \mathcal{E}_0 \cos \omega t \qquad (35.1)$$

where $\mathcal{E}_0$ is the peak or maximum emf and $\omega = 2\pi f$ is the angular frequency in radians per second. Recall that the units of emf are volts. As you can imagine, the mathematics of AC circuit analysis are going to be very similar to the mathematics of simple harmonic motion.

An alternative way to represent the emf and other oscillatory quantities is with the *phasor diagram* of Figure 35.1b. A **phasor** is a vector that rotates *counterclockwise* (ccw) around the origin at angular frequency ω. The length or magnitude of the phasor is the maximum value of the quantity. For example, the length of an emf phasor is $\mathcal{E}_0$. The angle ωt is the *phase angle*, an idea you learned about in Chapter 14, where we made a connection between circular motion and simple harmonic motion.

The quantity's instantaneous value, the value you would measure at time t, is the projection of the phasor onto the horizontal axis. This is also analogous to the connection between circular motion and simple harmonic motion. Figure 35.2 helps you visualize the phasor rotation by showing how the phasor corresponds to the more familiar graph at several specific points in the cycle.

STOP TO THINK 35.1 The magnitude of the instantaneous value of the emf represented by this phasor is

a. Increasing.
b. Decreasing.
c. Constant.
d. It's not possible to tell without knowing t.

Resistor Circuits

In Chapter 31 you learned to analyze a circuit in terms of the current I, voltage V, and potential difference ΔV. Now, because the current and voltage are oscillating, we will use lowercase i to represent the *instantaneous* current through a circuit element and v for the circuit element's *instantaneous* voltage.

Figure 35.3 shows the instantaneous current i_R through a resistor R. The potential difference across the resistor, which we call the *resistor voltage* v_R, is given by Ohm's law:

$$v_R = i_R R \qquad (35.2)$$

The potential *decreases* in the direction of the current.

Figure 35.4 shows a resistor R connected across an AC emf $\mathcal{E}$. Notice that the circuit symbol for an AC generator is —⊙—. We can analyze this circuit in exactly the same way we analyzed a DC resistor circuit. Kirchhoff's loop law says that the sum of all the potential differences around a closed path is zero:

$$\sum \Delta V = \Delta V_{source} + \Delta V_R = \mathcal{E} - v_R = 0 \qquad (35.3)$$

The minus sign appears, just as it did in the equation for a DC circuit, because the potential *decreases* when we travel through a resistor in the direction of the current. We find from the loop law that $v_R = \mathcal{E} = \mathcal{E}_0 \cos\omega t$. This isn't surprising because the resistor is connected directly across the terminals of the emf.

The resistor voltage is a sinusoidal voltage at angular frequency ω. It will be useful to write

$$v_R = V_R \cos\omega t \qquad (35.4)$$

where V_R is the peak or maximum voltage. You can see that $V_R = \mathcal{E}_0$ in the single-resistor circuit of Figure 35.4. Thus the current through the resistor is

$$i_R = \frac{v_R}{R} = \frac{V_R \cos\omega t}{R} = I_R \cos\omega t \qquad (35.5)$$

where $I_R = V_R/R$ is the peak current.

NOTE ▶ Ohm's law applies to both the instantaneous *and* peak currents and voltages. ◀

The resistor's instantaneous current and voltage are in phase, both oscillating as $\cos\omega t$. Figure 35.5 shows the voltage and the current simultaneously on a graph and as a phasor diagram. The fact that the current phasor is shorter than the voltage phasor has no significance. Current and voltage are measured in different units, so you can't compare the length of one to the length of the other. Showing the two different quantities on a single graph—a tactic that can be misleading if you're not careful—is simply to show that they oscillate in phase and that their phasors rotate together at the same angle and frequency.

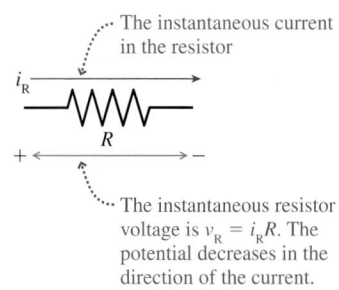

FIGURE 35.3 Instantaneous current i_R through a resistor.

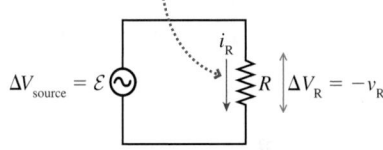

FIGURE 35.4 An AC resistor circuit.

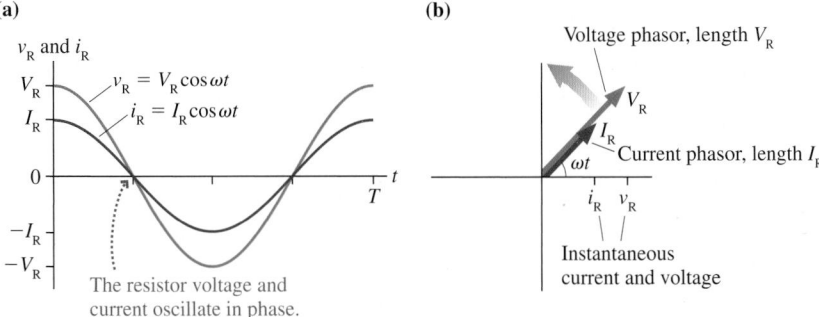

FIGURE 35.5 Graph and phasor diagram of the resistor current and voltage. The current and voltage are in phase.

EXAMPLE 35.1 Finding resistor voltages

In the circuit of Figure 35.6, what are (a) the peak voltage across each resistor and (b) the instantaneous resistor voltages at $t = 20$ ms?

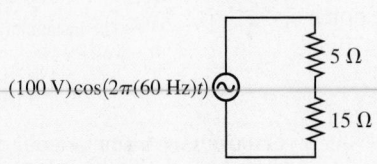

FIGURE 35.6 An AC resistor circuit.

VISUALIZE Figure 35.6 shows the circuit diagram. The two resistors are in series.

SOLVE

a. The equivalent resistance of the two series resistors is $R_{eq} = 5\ \Omega + 15\ \Omega = 20\ \Omega$. The instantaneous current through the equivalent resistance is

$$i_R = I_R \cos \omega t = \frac{v_R}{R_{eq}} = \frac{\mathcal{E}_0 \cos \omega t}{R_{eq}}$$

$$= \frac{(100\ \text{V})\cos(2\pi(60\ \text{Hz})t)}{20\ \Omega}$$

$$= (5.0\ \text{A})\cos(2\pi(60\ \text{Hz})t)$$

The peak current is $I_R = 5.0$ A, and this is also the peak current through the two resistors that form the 20 Ω equivalent resistance. Hence the peak voltage across each resistor is

$$V_R = I_R R = \begin{cases} 25\ \text{V} & 5\ \Omega\ \text{resistor} \\ 75\ \text{V} & 15\ \Omega\ \text{resistor} \end{cases}$$

b. The instantaneous current at $t = 0.020$ s is

$$i_R = (5.0\ \text{A})\cos(2\pi(60\ \text{Hz})(0.020\ \text{s})) = 1.545\ \text{A}$$

The resistor voltages at this time are

$$v_R = i_R R = \begin{cases} 7.7\ \text{V} & 5\ \Omega\ \text{resistor} \\ 23.2\ \text{V} & 15\ \Omega\ \text{resistor} \end{cases}$$

ASSESS The sum of the instantaneous voltages, 30.9 V, is what you would find by calculating $\mathcal{E}$ at $t = 20$ ms. This self-consistency gives us confidence in the answer.

STOP TO THINK 35.2 The resistor whose voltage and current phasors are shown here has resistance R

a. $>1\ \Omega$.
b. $<1\ \Omega$.
c. It's not possible to tell.

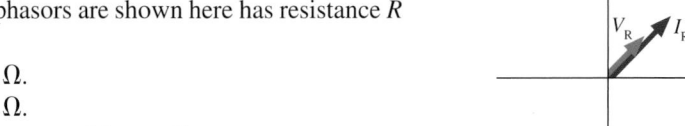

35.2 Capacitor Circuits

Figure 35.7a shows current i_C charging a capacitor with capacitance C. The instantaneous capacitor voltage is $v_C = q/C$, where $\pm q$ is the charge on the two capacitor plates at this instant of time. It is useful to compare Figure 35.7a to Figure 35.3 for a resistor.

Figure 35.7b, where capacitance C is connected across an AC source of emf $\mathcal{E}$, is the most basic capacitor circuit. The capacitor is in parallel with the source, so the capacitor voltage equals the emf: $v_C = \mathcal{E} = \mathcal{E}_0 \cos \omega t$. It will be useful to write

$$v_C = V_C \cos \omega t \tag{35.6}$$

where V_C is the peak or maximum voltage across the capacitor. You can see that $V_C = \mathcal{E}_0$ in this single-capacitor circuit.

To find the current to and from the capacitor, first write the charge

$$q = C v_C = C V_C \cos \omega t \tag{35.7}$$

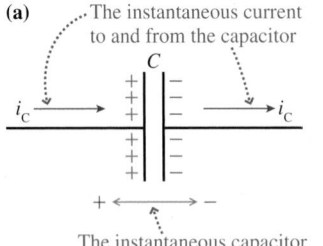

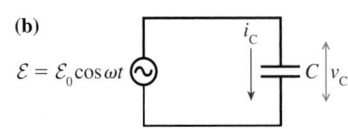

FIGURE 35.7 An AC capacitor circuit.

The current is the *rate* at which charge flows through the wires, $i_C = dq/dt$, thus

$$i_C = \frac{dq}{dt} = \frac{d}{dt}(CV_C\cos\omega t) = -\omega CV_C\sin\omega t \tag{35.8}$$

We can most easily see the relationship between the capacitor voltage and current if we use the trigonometric identity $-\sin(x) = \cos(x + \pi/2)$ to write

$$i_C = \omega CV_C\cos\left(\omega t + \frac{\pi}{2}\right) \tag{35.9}$$

In contrast to a resistor, a capacitor's current and voltage are *not* in phase. In Figure 35.8a, which shows a graph of the instantaneous voltage v_C and current i_C, you can see that the current peaks one-quarter of a period *before* the voltage peaks. The phase angle of the current phasor on the phasor diagram of Figure 35.8b is $\pi/2$ rad—a quarter of a circle—larger than the phase angle of the voltage phasor.

We can summarize this finding by saying

The AC current through a capacitor *leads* the capacitor voltage by $\pi/2$ rad, or 90°.

The current reaches its peak value I_C at the instant the capacitor is fully discharged and $v_C = 0$. The current is zero at the instant the capacitor is fully charged. You saw a similar behavior in the oscillation of an *LC* circuit in Chapter 33.

A simple harmonic oscillator provides a mechanical analogy of the 90° phase difference between current and voltage. You learned in Chapter 14 (refer to Section 14.1 and Figure 14.5) that the position and velocity of a simple harmonic oscillator are

$$x = A\cos\omega t$$

$$v = \frac{dx}{dt} = -\omega A\sin\omega t = -v_{max}\sin\omega t = v_{max}\cos\left(\omega t + \frac{\pi}{2}\right)$$

You can see in Figure 35.9 that the velocity leads the position by 90° in the same way that the capacitor current (which is proportional to the charge velocity) leads the voltage.

Capacitive Reactance

We can use Equation 35.9 to see that the peak current to and from a capacitor is $I_C = \omega CV_C$. This relationship between the peak voltage and peak current looks much like Ohm's law for a resistor if we define the **capacitive reactance** X_C to be

$$X_C \equiv \frac{1}{\omega C} \tag{35.10}$$

With this definition,

$$I_C = \frac{V_C}{X_C} \quad \text{or} \quad V_C = I_C X_C \tag{35.11}$$

The units of reactance, like those of resistance, are ohms.

NOTE ▶ Reactance relates the *peak* voltage V_C and current I_C. But reactance differs from resistance in that it does *not* relate the instantaneous capacitor voltage and current because they are out of phase. That is, $v_C \neq i_C X_C$. ◀

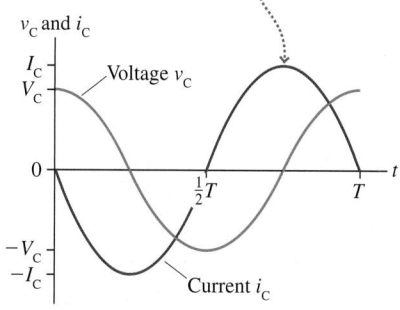

(a) i_C peaks $\frac{1}{4}T$ before v_C peaks. We say that the current *leads* the voltage by 90°.

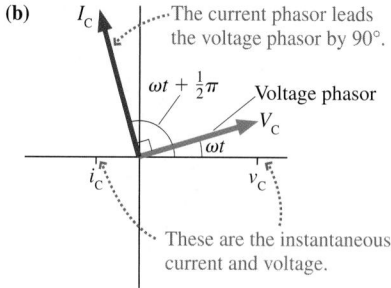

(b) The current phasor leads the voltage phasor by 90°. These are the instantaneous current and voltage.

FIGURE 35.8 Graph and phasor diagrams of the capacitor current and voltage.

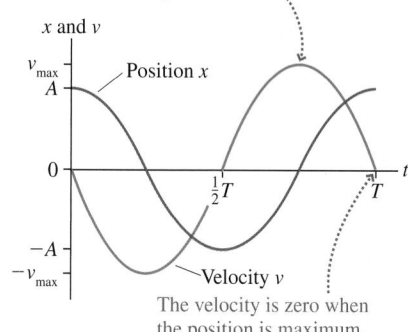

The velocity is maximum as the oscillator passes through equilibrium. The velocity is zero when the position is maximum.

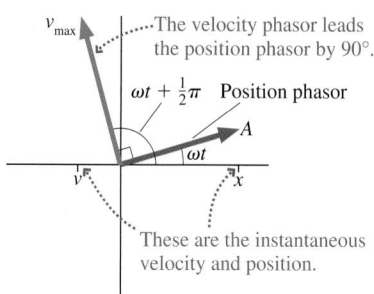

The velocity phasor leads the position phasor by 90°. These are the instantaneous velocity and position.

FIGURE 35.9 In a mechanical analogy, the velocity of a simple harmonic oscillator leads the position by 90°.

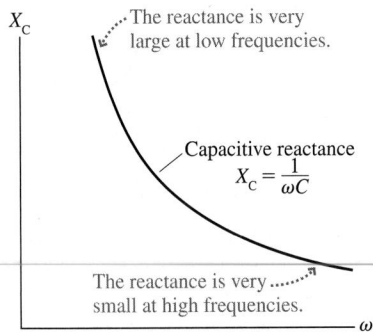

FIGURE 35.10 The capacitive reactance as a function of frequency.

A resistor's resistance R is independent of the emf frequency. In contrast, as Figure 35.10 shows, a capacitor's reactance X_C depends inversely on the frequency. The reactance becomes very large at low frequencies (i.e., the capacitor is a large impediment to current). This makes sense, because $\omega = 0$ would be a nonoscillating, DC circuit, and we know that a steady DC current cannot pass through a capacitor. The reactance decreases as the frequency increases until, at very high frequencies, $X_C \approx 0$ and the capacitor begins to act like an ideal wire. This result has important consequences for how capacitors are used in many circuits.

EXAMPLE 35.2 Capacitive reactance

What is the capacitive reactance of a 0.10 μF capacitor at a 100 Hz audio frequency and at a 100 MHz FM-radio frequency?

SOLVE At 100 Hz,

$$X_C(\text{at } 100 \text{ Hz}) = \frac{1}{\omega C} = \frac{1}{2\pi(100 \text{ Hz})(1.0 \times 10^{-7} \text{ F})}$$

$$= 15{,}900 \ \Omega$$

Increasing the frequency by a factor of 10^6 decreases X_C by a factor of 10^6, giving

$$X_C(\text{at } 100 \text{ MHz}) = 0.0159 \ \Omega$$

ASSESS A capacitor with a substantial reactance at audio frequencies has virtually no reactance at FM-radio frequencies.

EXAMPLE 35.3 Capacitor current

A 10 μF capacitor is connected to a 1000 Hz oscillator with a peak emf of 5.0 V. What is the peak current to the capacitor?

VISUALIZE Figure 35.7b showed the circuit diagram. It is a simple one-capacitor circuit.

SOLVE The capacitive reactance at $\omega = 2\pi f = 6280$ rad/s is

$$X_C = \frac{1}{\omega C} = \frac{1}{(6280 \text{ rad/s})(10 \times 10^{-6} \text{ F})} = 15.9 \ \Omega$$

The peak voltage across the capacitor is $V_C = \mathcal{E}_0 = 5.0$ V, hence the peak current is

$$I_C = \frac{V_C}{X_C} = \frac{5.0 \text{ V}}{15.9 \ \Omega} = 0.314 \text{ A}$$

ASSESS Using reactance is just like using Ohm's law, but don't forget it applies only to the *peak* current and voltage, not the instantaneous values.

STOP TO THINK 35.3 What is the capacitive reactance of "no capacitor," just a continuous wire?

a. 0
b. ∞
c. Undefined

35.3 *RC* Filter Circuits

You learned in Chapter 31 that a resistance R causes a capacitor to be charged or discharged with time constant $\tau = RC$. We called this an *RC* circuit. Now that we've looked at resistors and capacitors individually, let's explore what happens if an *RC* circuit is driven continuously by an alternating current source.

Figure 35.11 shows a circuit in which a resistor R and capacitor C are in series with an emf $\mathcal{E}$ that oscillates at angular frequency ω. Before launching into a formal analysis, let's try to understand qualitatively how this circuit will respond as the frequency is varied. If the frequency is very low, the capacitive reactance will be very large, and thus the peak current I_C will be very small. The peak current through the resistor is the same as the peak current to and from the capacitor (conservation of current requires $I_R = I_C$), hence we expect the resistor's peak voltage $V_R = I_R R$ to be very small at very low frequencies.

On the other hand, suppose the frequency is very high. Then the capacitive reactance approaches zero and the peak current, determined by the resistance alone, will be $I_R = \mathcal{E}_0/R$. The resistor's peak voltage $V_R = IR$ will approach the peak source voltage $\mathcal{E}_0$ at very high frequencies.

This reasoning leads us to expect that V_R will *increase* steadily from 0 to $\mathcal{E}_0$ as ω is increased from 0 to very high frequencies. Kirchhoff's loop law has to be obeyed, so the capacitor voltage V_C will *decrease* from $\mathcal{E}_0$ to 0 during the same change of frequency. A quantitative analysis will show us how this behavior can be used as a *filter*.

The goal of a quantitative analysis is to determine the peak current I and the two peak voltages V_R and V_C as functions of the emf amplitude $\mathcal{E}_0$ and frequency ω. Although this goal can be reached with a purely algebraic analysis, using a phasor diagram is easier and more informative. Our analytic procedure is based on the fact that the current i is the same for two circuit elements in series.

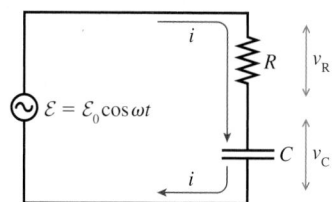

FIGURE 35.11 An *RC* circuit driven by an AC source.

Analyzing an *RC* circuit

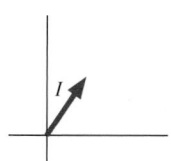

Begin by drawing a current phasor of length I. This is the starting point because the series circuit elements have the same current i. The angle at which the phasor is drawn is not relevant.

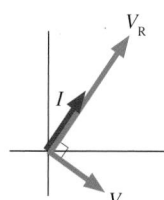

The current and voltage of a resistor are in phase, so draw a resistor voltage phasor of length V_R parallel to the current phasor I. The capacitor current leads the capacitor voltage by 90°, so draw a capacitor voltage phasor of length V_C that is 90° behind [i.e., clockwise (cw) from] the current phasor.

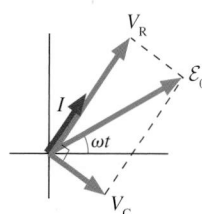

The series resistor and capacitor are in parallel with the emf, so their *instantaneous* voltages satisfy $v_R + v_C = \mathcal{E}$. This is a *vector* addition of phasors, so draw the emf phasor as the vector sum of the two voltage phasors. The emf is $\mathcal{E} = \mathcal{E}_0 \cos \omega t$, hence the emf phasor is at angle ωt.

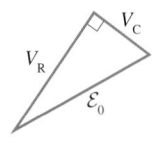

The length of the emf phasor, $\mathcal{E}_0$, is the hypotenuse of a right triangle formed by the resistor and capacitor phasors. Thus $\mathcal{E}_0^2 = V_R^2 + V_C^2$.

The relationship $\mathcal{E}_0^2 = V_R^2 + V_C^2$ is based on the peak values, not the instantaneous values, because the peak values are the lengths of the sides of the right triangle. The peak voltages are related to the peak current I via $V_R = IR$ and $V_C = IX_C$, thus

$$\mathcal{E}_0^2 = V_R^2 + V_C^2 = (IR)^2 + (IX_C)^2 = (R^2 + X_C^2)I^2$$
$$= (R^2 + 1/\omega^2 C^2)I^2 \tag{35.12}$$

Consequently, the peak current in the *RC* circuit is

$$I = \frac{\mathcal{E}_0}{\sqrt{R^2 + X_C^2}} = \frac{\mathcal{E}_0}{\sqrt{R^2 + 1/\omega^2 C^2}} \tag{35.13}$$

Knowing I gives us the two peak voltages:

$$V_R = IR = \frac{\mathcal{E}_0 R}{\sqrt{R^2 + X_C^2}} = \frac{\mathcal{E}_0 R}{\sqrt{R^2 + 1/\omega^2 C^2}}$$

$$V_C = IX_C = \frac{\mathcal{E}_0 X_C}{\sqrt{R^2 + X_C^2}} = \frac{\mathcal{E}_0/\omega C}{\sqrt{R^2 + 1/\omega^2 C^2}}$$

(35.14)

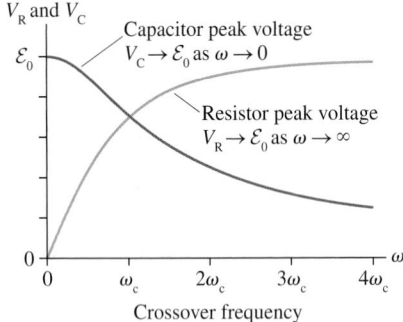

FIGURE 35.12 Graph of the resistor and capacitor peak voltages as functions of the emf frequency ω.

Frequency Dependence

Our goal was to see how the peak current and voltages varied as functions of the frequency ω. Equations 35.13 and 35.14 are rather complex and best interpreted by looking at graphs. Figure 35.12 is a graph of V_R and V_C versus ω.

You can see that our qualitative predictions have been borne out. That is, V_R increases from 0 to $\mathcal{E}_0$ as ω is increased while V_C decreases from $\mathcal{E}_0$ to 0. The explanation for this behavior is that the capacitive reactance X_C decreases as ω increases. For low frequencies, where $X_C \gg R$, the circuit is primarily capacitive. For high frequencies, where $X_C \ll R$, the circuit is primarily resistive.

The frequency at which $V_R = V_C$ is called the **crossover frequency** ω_c. The *crossover* frequency is easily found by setting the two expressions in Equation 35.14 equal to each other. The denominators are the same and cancel, as does $\mathcal{E}_0$, leading to

$$\omega_c = \frac{1}{RC}$$

(35.15)

In practice, $f_c = \omega_c/2\pi$ is also called the crossover frequency.

We'll leave it as a homework problem to show that $V_R = V_C = \mathcal{E}_0/\sqrt{2}$ when $\omega = \omega_c$. This may seem surprising. After all, shouldn't V_R and V_C add up to $\mathcal{E}_0$?

No! V_R and V_C are the *peak values* of oscillating voltages, not the instantaneous values. The instantaneous values do, indeed, satisfy $v_R + v_C = \mathcal{E}$ at all instants of time. But the resistor and capacitor voltages are out of phase with each other, as the phasor diagram shows, so the two circuit elements don't reach their peak values at the same time. The peak values are related by $\mathcal{E}_0^2 = V_R^2 + V_C^2$, and you can see that $V_R = V_C = \mathcal{E}_0/\sqrt{2}$ satisfies this equation.

> **NOTE** ▶ It's very important in AC circuit analysis to make a clear distinction between instantaneous values and peak values of voltages and currents. Relations that are true for one set of values may not be true for the other. ◀

Filters

Figure 35.13a is the circuit we've just analyzed, the only difference is that the capacitor voltage v_C is now identified as the *output voltage* v_{out}. This is a voltage you might measure or, perhaps, send to an amplifier for use elsewhere in an electronic instrument. You can see from the capacitor voltage graph in Figure 35.12 that the peak output voltage is $V_{out} \approx \mathcal{E}_0$ if $\omega \ll \omega_c$, but $V_{out} \approx 0$ if $\omega \gg \omega_c$. In other words,

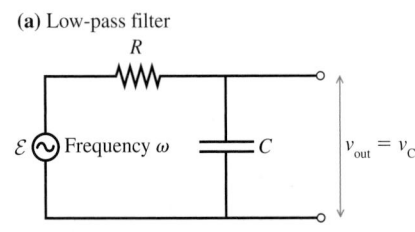

(a) Low-pass filter

Transmits frequencies $\omega < \omega_c$ and blocks frequencies $\omega > \omega_c$.

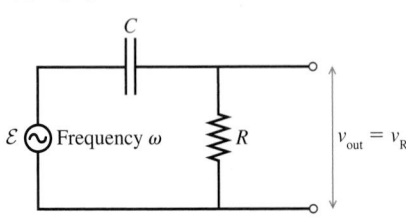

(b) High-pass filter

Transmits frequencies $\omega > \omega_c$ and blocks frequencies $\omega < \omega_c$.

FIGURE 35.13 Low-pass and high-pass filter circuits.

- If the frequency of an input signal is well below the crossover frequency, the input signal is transmitted with little loss to the output.
- If the frequency of an input signal is well above the crossover frequency, the input signal is strongly attenuated and the output is very nearly zero.

This circuit is called a **low-pass filter.**

The circuit of Figure 35.13b, which instead uses the resistor voltage v_R for the output v_{out}, is a **high-pass filter.** The output is $V_{out} \approx 0$ if $\omega \ll \omega_c$, but $V_{out} \approx \mathcal{E}_0$ if $\omega \gg \omega_c$. That is, an input signal whose frequency is well above the crossover frequency is transmitted without loss to the output.

Filter circuits are widely used in electronics. For example, a high-pass filter designed to have $f_c = 100$ Hz would pass the audio frequencies associated with speech ($f > 200$ Hz) while blocking 60 Hz "noise" that can be picked up from power lines. Similarly, the high-frequency hiss from old vinyl records can be attenuated with a low-pass filter that allows the lower-frequency audio signal to pass through.

A simple *RC* filter suffers from the fact that the crossover region where $V_R \approx V_C$ is fairly broad. More sophisticated filters have a sharper transition from off ($V_{out} \approx 0$) to on ($V_{out} \approx \mathcal{E}_0$), but they're based on the same principles as the *RC* filter analyzed here.

EXAMPLE 35.4 Designing a filter

For a science project, you've built a radio to listen to AM radio broadcasts at frequencies near 1 MHz. The basic circuit is an antenna, which produces a very small oscillating voltage when it absorbs the energy of an electromagnetic wave, and an amplifier. Unfortunately, your neighbor's short-wave radio broadcast at 10 MHz interferes with your reception. Having just finished physics, you decide to solve this problem by placing a filter between the antenna and the amplifier. You happen to have a 500 pF capacitor. What frequency should you select as the filter's crossover frequency? What value of resistance will you need to build this filter?

MODEL You want to block signals at 10 MHz while passing the lower-frequency AM signal at 1 MHz. Thus you need a low-pass filter.

VISUALIZE The circuit will look like the low-pass filter in Figure 35.13a. The oscillating voltage generated by the antenna will be the emf, and v_{out} will be sent to the amplifier.

SOLVE You might think that a crossover frequency near 5 MHz, about halfway between 1 MHz and 10 MHz, would work best. But 5 MHz is a factor of 5 higher than 1 MHz while only a factor of 2 less than 10 MHz. A crossover frequency that is the same factor above 1 MHz as it is below 10 MHz will give the best results. In practice, choosing $f_c = 3$ MHz would be sufficient. You can then use Equation 35.15 to select the proper resistor value:

$$R = \frac{1}{\omega_c C} = \frac{1}{2\pi(3 \times 10^6 \text{ Hz})(500 \times 10^{-12} \text{ F})}$$
$$= 106 \ \Omega \approx 100 \ \Omega$$

ASSESS Rounding to 100 Ω is appropriate because the crossover frequency was determined only to one significant figure. Such "sloppy design" is quite adequate when the two frequencies you need to distinguish are well separated.

STOP TO THINK 35.4 Rank in order, from largest to smallest, the crossover frequencies $(\omega_c)_a$ to $(\omega_c)_d$ of these four circuits.

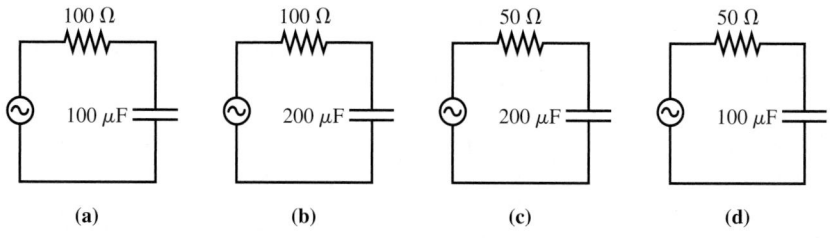

(a)   (b)   (c)   (d)

35.4 Inductor Circuits

Figure 35.14a on the next page shows the instantaneous current i_L through an inductor. If the current is changing, the instantaneous inductor voltage is

$$v_L = L\frac{di_L}{dt} \tag{35.16}$$

You learned in Chapter 33 that the potential decreases in the direction of the current if current is increasing ($di_L/dt > 0$) and increases if current is decreasing ($di_L/dt < 0$).

(a) The instantaneous current through the inductor

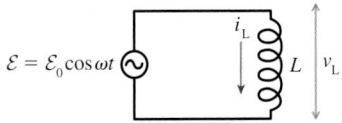

The instantaneous inductor voltage is $v_L = L(di_L/dt)$.

(b)

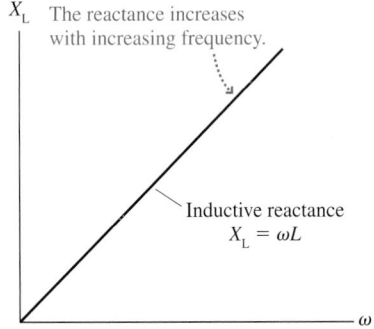

FIGURE 35.14 Using an inductor in an AC circuit.

Figure 35.14b, where inductance L is connected across an AC source of emf $\mathcal{E}$, is the simplest inductor circuit. The inductor is in parallel with the source, so the inductor voltage equals the emf: $v_L = \mathcal{E} = \mathcal{E}_0 \cos \omega t$. We can write

$$v_L = V_L \cos \omega t \tag{35.17}$$

where V_L is the peak or maximum voltage across the inductor. You can see that $V_L = \mathcal{E}_0$ in this single-inductor circuit.

We can find the inductor current i_L by integrating Equation 35.17. First, use Equation 35.17 to write Equation 35.16 as

$$di_L = \frac{v_L}{L} dt = \frac{V_L}{L} \cos \omega t \, dt \tag{35.18}$$

Integrating gives

$$i_L = \frac{V_L}{L} \int \cos \omega t \, dt = \frac{V_L}{\omega L} \sin \omega t = \frac{V_L}{\omega L} \cos\left(\omega t - \frac{\pi}{2}\right)$$

$$= I_L \cos\left(\omega t - \frac{\pi}{2}\right) \tag{35.19}$$

where $I_L = V_L/\omega L$ is the peak or maximum inductor current.

NOTE ▶ Mathematically, Equation 35.19 could have an integration constant i_0. An integration constant would represent a constant DC current through the inductor, but there is no DC source of potential in an AC circuit. Hence, on physical grounds, we set $i_0 = 0$ for an AC circuit. ◀

Define the **inductive reactance,** analogous to the capacitive reactance, to be

$$X_L \equiv \omega L \tag{35.20}$$

Then the peak current $I_L = V_L/\omega L$ and the peak voltage are related by

$$I_L = \frac{V_L}{X_L} \quad \text{or} \quad V_L = I_L X_L \tag{35.21}$$

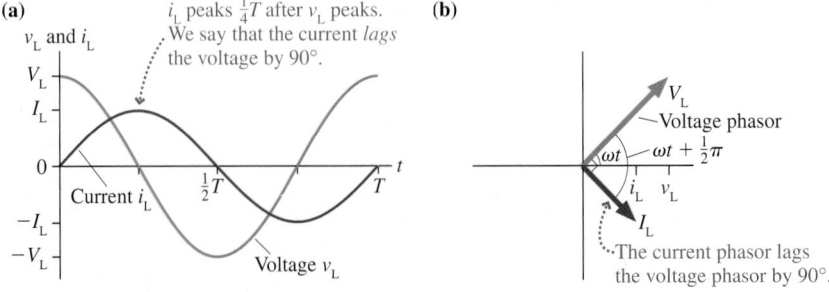

X_L The reactance increases with increasing frequency.

Inductive reactance $X_L = \omega L$

FIGURE 35.15 The inductive reactance as a function of frequency.

Figure 35.15 shows that the inductive reactance increases as the frequency increases. This makes sense. Faraday's law tells us that the induced voltage across a coil increases as the time rate of change of $\vec{B}$ increases, and $\vec{B}$ is directly proportional to the inductor current. For a given peak current I_L, $\vec{B}$ changes more rapidly at higher frequencies than at lower frequencies, and thus V_L is larger at higher frequencies than at lower frequencies.

Figure 35.16a is a graph of the inductor voltage and current. You can see that the current peaks one-quarter of a period *after* the voltage peaks. The angle of the current phasor on the phasor diagram of Figure 35.16b is $\pi/2$ rad less than the angle of the voltage phasor. We can summarize this finding by saying

The AC current through an inductor *lags* the inductor voltage by $\pi/2$ rad, or 90°.

(a) i_L peaks $\frac{1}{4}T$ after v_L peaks. We say that the current *lags* the voltage by 90°. **(b)**

The current phasor lags the voltage phasor by 90°.

FIGURE 35.16 Graphs and phasor diagrams of the inductor current and voltage.

EXAMPLE 35.5 Current and voltage of an inductor
A 25 μH inductor is used in a circuit that oscillates at 100 kHz . The current through the inductor reaches a peak value of 20 mA at $t = 5.0\ \mu$s. What is the peak inductor voltage, and when, closest to $t = 5.0\ \mu$s, does it occur?

MODEL The inductor current lags the voltage by 90°, or, equivalently, the voltage reaches its peak value one-quarter period *before* the current.

VISUALIZE The circuit looks like Figure 35.14b.

SOLVE The inductive reactance at $f = 100$ kHz is

$$X_C = \omega L = 2\pi(1.0 \times 10^5\ \text{Hz})(25 \times 10^{-6}\ \text{H}) = 16\ \Omega$$

Thus the peak voltage is $V_L = I_L X_L = (20\ \text{mA})(16\ \Omega) = 320$ mV. The voltage peak occurs one-quarter period before the current peaks, and we know that the current peaks at $t = 5.0\ \mu$s. The period of a 100 kHz oscillation is 10.0 μs, so the voltage peaks at

$$t = 5.0\ \mu\text{s} - \frac{10.0\ \mu\text{s}}{4} = 2.5\ \mu\text{s}$$

35.5 The Series *RLC* Circuit

The circuit of Figure 35.17, where a resistor, inductor, and capacitor are in series, is called a **series *RLC* circuit.** The series *RLC* circuit has many important applications because, as you will see, it exhibits resonance behavior.

The analysis, which is very similar to our analysis of the *RC* circuit in Section 35.3, will be based on a phasor diagram. Notice that the three circuit elements are in series with each other and, together, are in parallel with the emf. We can draw two conclusions that form the basis of our analysis:

1. The instantaneous current of all three elements is the same: $i = i_R = i_L = i_C$.
2. The sum of the instantaneous voltages matches the emf: $\mathcal{E} = v_R + v_L + v_C$.

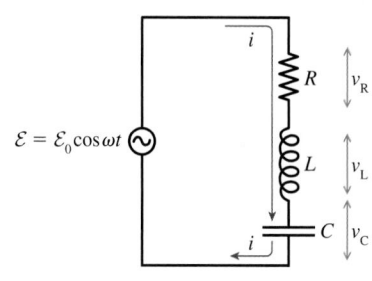

FIGURE 35.17 A series *RLC* circuit.

Analyzing an *RLC* circuit

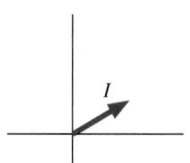

Begin by drawing a current phasor of length I. This is the starting point because the series circuit elements have the same current i.

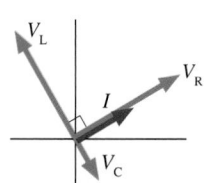

The current and voltage of a resistor are in phase, so draw a resistor voltage phasor parallel to the current phasor I. The capacitor current leads the capacitor voltage by 90°, so draw a capacitor voltage phasor that is 90° behind the current phasor. Finally, the inductor current *lags* the voltage by 90°, so draw an inductor voltage phasor 90° ahead of the current phasor.

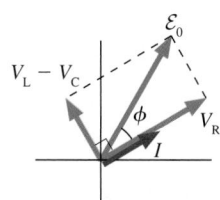

The instantaneous voltages satisfy $\mathcal{E} = v_R + v_L + v_C$. In terms of phasors, this is a *vector* addition. We can do the addition in two steps. Because the capacitor and inductor phasors are in opposite directions, their vector sum has length $V_L - V_C$. Adding the resistor phasor, at right angles, then gives the emf phasor $\mathcal{E}$ at angle ωt.

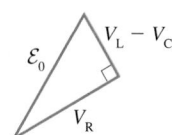

The length $\mathcal{E}_0$ of the emf phasor is the hypotenuse of a right triangle. Thus $\mathcal{E}_0^2 = V_R^2 + (V_L - V_C)^2$.

If $V_L > V_C$, which we've assumed, then the instantaneous current i lags the emf by a phase angle ϕ. We can write the current, in terms of ϕ, as

$$i = I\cos(\omega t - \phi) \tag{35.22}$$

Of course, there's no guarantee that V_L will be larger than V_C. If the opposite is true, $V_L < V_C$, the emf phasor is on the other side of the current phasor. Our analysis is still valid if we consider ϕ to be negative when i is ccw from $\mathcal{E}$. Thus ϕ can be anywhere between $-90°$ and $+90°$.

Now we can continue much as we did with the RC circuit. Based on the right triangle, $\mathcal{E}_0^2$ is

$$\mathcal{E}_0^2 = V_R^2 + (V_L - V_C)^2 = [R^2 + (X_L - X_C)^2]I^2 \tag{35.23}$$

where we wrote each of the peak voltages in terms of the peak current I and a resistance or a reactance. Consequently, the peak current in the RLC circuit is

$$I = \frac{\mathcal{E}_0}{\sqrt{R^2 + (X_L - X_C)^2}} = \frac{\mathcal{E}_0}{\sqrt{R^2 + (\omega L - 1/\omega C)^2}} \tag{35.24}$$

The three peak voltages, if you need them, are then found from $V_R = IR$, $V_L = IX_L$ and $V_C = IX_C$.

Impedance

The denominator of Equation 35.24 is called the **impedance** Z of the circuit:

$$Z = \sqrt{R^2 + (X_L - X_C)^2} \tag{35.25}$$

Impedance, like resistance and reactance, is measured in ohms. The circuit's peak current can be written in terms of the source emf and the circuit impedance as

$$I = \frac{\mathcal{E}_0}{Z} \tag{35.26}$$

Equation 35.26 is a compact way to write I, but it doesn't add anything new to Equation 35.24.

Phase Angle

It is often useful to know the phase angle ϕ between the emf and the current. You can see from Figure 35.18 that

$$\tan\phi = \frac{V_L - V_C}{V_R} = \frac{(X_L - X_C)I}{RI}$$

The current I cancels, and we're left with

$$\phi = \tan^{-1}\left(\frac{X_L - X_C}{R}\right) \tag{35.27}$$

The current lags the emf by
$$\phi = \tan^{-1}\left(\frac{V_L - V_C}{V_R}\right) = \tan^{-1}\left(\frac{X_L - X_C}{R}\right)$$

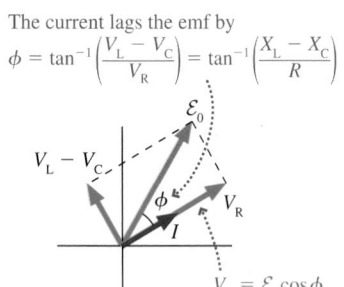

$$V_R = \mathcal{E}_0\cos\phi$$

FIGURE 35.18 The current is not in phase with the emf.

We can check that Equation 35.27 agrees with our analyses of single-element circuits. A resistor-only circuit has $X_L = X_C = 0$ and thus $\phi = \tan^{-1}(0) = 0$ rad. In other words, as we discovered previously, the emf and current are in phase. An AC inductor circuit has $R = X_C = 0$ and thus $\phi = \tan^{-1}(\infty) = \pi/2$ rad, agreeing with our earlier finding that the inductor current lags the voltage by 90°. A homework problem will let you check that Equation 35.27 gives the correct result for an AC capacitor circuit.

Other relationships can be found from the phasor diagram and written in terms of the phase angle. For example, it is frequently useful to write the peak resistor voltage as

$$V_R = \mathcal{E}_0 \cos\phi \qquad (35.28)$$

Notice that the resistor voltage oscillates in phase with the emf only if $\phi = 0$ rad.

Resonance

Suppose we vary the emf frequency ω while keeping everything else constant. There is very little current at very low frequencies because the capacitive reactance $X_C = 1/\omega C$ is very large. Similarly, there is very little current at very high frequencies because the inductive reactance $X_L = \omega L$ becomes very large.

If I approaches zero at very low and very high frequencies, there should be some intermediate frequency where I is a maximum. Indeed, you can see from Equation 35.24 that the denominator will be a minimum, making I a maximum, when $X_L = X_C$, or

$$\omega L = \frac{1}{\omega C} \qquad (35.29)$$

The frequency ω_0 that satisfies Equation 35.39 is called the **resonance frequency:**

$$\omega_0 = \frac{1}{\sqrt{LC}} \qquad (35.30)$$

This is the frequency for *maximum current* in the series *RLC* circuit. The maximum current

$$I_{max} = \frac{\mathcal{E}_0}{R} \qquad (35.31)$$

is that of a purely resistive circuit because the impedance is $Z = R$ at resonance.

You'll recognize ω_0 as the oscillation frequency of the *LC* circuit that we analyzed in Chapter 33. The current in an ideal *LC* circuit oscillates forever as energy is transferred back and forth between the capacitor and the inductor. This is analogous to an ideal, frictionless simple harmonic oscillator in which the energy is transformed back and forth between kinetic and potential.

Adding a resistor to the circuit is like adding damping to a mechanical oscillator. The emf is then a sinusoidal driving force, and the series *RLC* circuit is directly analogous to the driven, damped oscillator that you studied in Chapter 14. A mechanical oscillator exhibits *resonance* by having a large-amplitude response when the driving frequency matches the system's natural frequency. Equation 35.30 is the natural frequency of the series *RLC* circuit, the frequency at which the current would like to oscillate. Consequently, the circuit has a large current response when the oscillating emf matches this frequency.

Figure 35.19 shows the peak current I of a series *RLC* circuit as the emf frequency ω is varied. Notice how the current increases until reaching a maximum at frequency ω_0, then decreases. This is the hallmark of a resonance.

As R decreases, causing the damping to decrease, the maximum current becomes larger and the curve in Figure 35.19 becomes narrower. You saw exactly the same behavior for a driven mechanical oscillator. The emf frequency must be very close to ω_0 in order for a lightly damped system to respond, but the response at resonance is very large.

For a different perspective, Figure 35.20 on the next page graphs the instantaneous emf $\mathcal{E} = \mathcal{E}_0 \cos\omega t$ and current $i = I\cos(\omega t - \phi)$ for frequencies below, at, and above ω_0. The current and the emf are in phase at resonance ($\phi = 0$ rad)

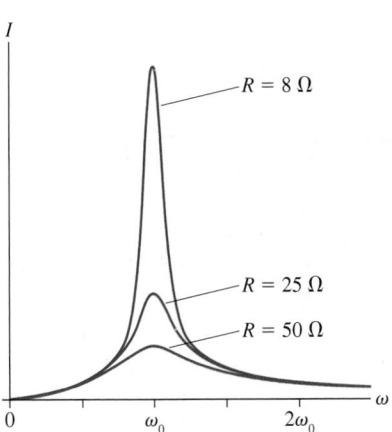

FIGURE 35.19 A graph of the current I versus emf frequency for a series *RLC* circuit.

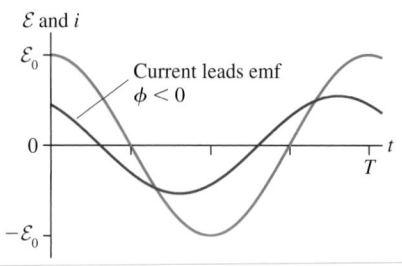

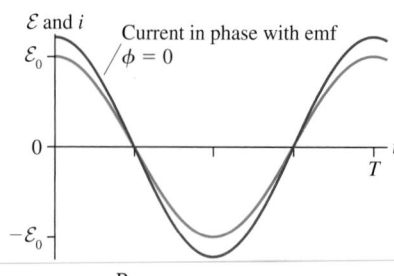

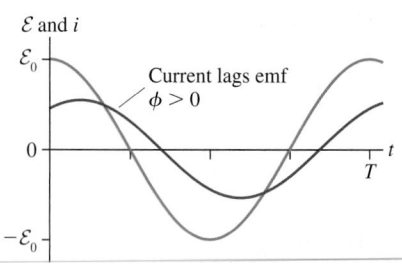

Below resonance: $\omega < \omega_0$

Resonance: $\omega = \omega_0$
Maximum current

Above resonance: $\omega > \omega_0$

FIGURE 35.20 Graphs of the emf $\mathcal{E}$ and the current I at frequencies below, at, and above the resonance frequency ω_0.

because the capacitor and inductor essentially cancel each other to give a purely resistive circuit. Away from resonance, the current decreases *and* begins to get out of phase with the emf. You can see, from Equation 35.27, that the phase angle ϕ is negative when $X_L < X_C$ (i.e., the frequency is below resonance) and positive when $X_L > X_C$ (the frequency is above resonance).

Resonance circuits are widely used in radio, television, and communication equipment because of their ability to respond to one particular frequency (or very narrow range of frequencies) while suppressing others. The selectivity of a resonance circuit improves as the resistance decreases, but the inherent resistance of the wires and the inductor coil keep R from being 0 Ω.

EXAMPLE 35.6 Designing a radio receiver

An AM radio antenna picks up a 1000 kHz signal with a peak voltage of 5.0 mV. The tuning circuit consists of a 60 μH inductor in series with a variable capacitor. The inductor coil has a resistance of 0.25 Ω, and the resistance of the rest of the circuit is negligible.

a. To what value should the capacitor be tuned to listen to this radio station?
b. What is the peak current through the circuit at resonance?
c. A stronger station at 1050 kHz produces a 10 mV antenna signal. What is the current at this frequency when the radio is tuned to 1000 MHz?

MODEL The inductor's 0.25 Ω resistance can be modeled as a resistance in series with the inductance, hence we have a series *RLC* circuit. The antenna signal at $\omega = 2\pi \times 1000$ kHz is the emf.

VISUALIZE The circuit looks like Figure 35.17.

SOLVE

a. The capacitor needs to be tuned to where it and the inductor are resonant at $\omega_0 = 2\pi \times 1000$ kHz. The appropriate value is

$$ C = \frac{1}{L\omega_0^2} = \frac{1}{(60 \times 10^{-6}\text{ H})(6.28 \times 10^6\text{ rad/s})^2} $$
$$ = 4.23 \times 10^{-10}\text{ F} = 423\text{ pF} $$

b. $X_L = X_C$ at resonance, so the peak current is

$$ I = \frac{\mathcal{E}_0}{R} = \frac{5.0 \times 10^{-3}\text{ V}}{0.25\text{ Ω}} = 0.020\text{ A} = 20\text{ mA} $$

c. The 1050 kHz signal is "off resonance," so we need to compute $X_L = \omega L = 396$ Ω and $X_C = 1/\omega C = 358$ Ω at $\omega = 2\pi \times 1050$ kHz. The peak voltage of this signal is $\mathcal{E}_0 = 10$ mV. With these values, Equation 35.24 for the peak current is

$$ I = \frac{\mathcal{E}_0}{\sqrt{R^2 + (X_L - X_C)^2}} = 0.26\text{ mA} $$

ASSESS These are realistic values for the input stage of an AM radio. You can see that the signal from the 1050 kHz station is strongly suppressed when the radio is tuned to 1000 kHz.

STOP TO THINK 35.5 A series *RLC* circuit has $V_C = 5.0$ V, $V_R = 7.0$ V, and $V_L = 9.0$ V. Is the frequency above, below, or equal to the resonance frequency?

35.6 Power in AC Circuits

A primary role of the emf is to supply energy. Some circuit devices, such as motors and lightbulbs, use the energy to perform useful tasks. Other circuit devices dissipate the energy as an increased thermal energy in the components and the surrounding air. Chapter 31 examined the topic of power in DC circuits. Now we can perform a similar analysis for AC circuits.

The emf supplies energy to a circuit at the rate

$$p_{\text{source}} = i\mathcal{E} \tag{35.32}$$

where i and $\mathcal{E}$ are the instantaneous current from and potential difference across the emf. We've used a lowercase p to indicate that this is the instantaneous power. We need to look at the power losses in individual circuit elements.

Resistors

You learned in Chapter 31 that the current through a resistor causes the resistor to dissipate energy at the rate

$$p_{\text{R}} = i_{\text{R}}v_{\text{R}} = i_{\text{R}}^2 R \tag{35.33}$$

We can use $i_{\text{R}} = I_{\text{R}}\cos\omega t$ to write the resistor's instantaneous power loss as

$$p_{\text{R}} = i_{\text{R}}^2 R = I_{\text{R}}^2 R\cos^2\omega t \tag{35.34}$$

Figure 35.21 shows the instantaneous power graphically. You can see that, because the cosine is squared, the power oscillates twice during every cycle of the emf. The energy dissipation peaks both when $i_{\text{R}} = I_{\text{R}}$ and when $i_{\text{R}} = -I_{\text{R}}$. This shows that energy dissipation doesn't depend on the current's direction through the resistor, a result that is hardly surprising.

In practice, we're usually more interested in the *average power* than in the instantaneous power. The **average power** P is the total energy dissipated per second. We can find P_{R} for a resistor by using the trigonometric identity $\cos^2(x) = \frac{1}{2}(1 + \cos 2x)$ to write

$$P_{\text{R}} = I_{\text{R}}^2 R\cos^2\omega t = I_{\text{R}}^2 R\left[\frac{1}{2}(1 + \cos 2\omega t)\right] = \frac{1}{2}I_{\text{R}}^2 R + \frac{1}{2}I_{\text{R}}^2 R\cos 2\omega t$$

The $\cos 2\omega t$ term oscillates positive and negative twice during each cycle of the emf. Its average, over one cycle of the emf, is therefore zero. Thus the average power loss in a resistor is

$$P_{\text{R}} = \frac{1}{2}I_{\text{R}}^2 R \qquad \text{(average power loss in a resistor)} \tag{35.35}$$

It is useful to write Equation 35.35 as

$$P_{\text{R}} = \left(\frac{I_{\text{R}}}{\sqrt{2}}\right)^2 R = (I_{\text{rms}})^2 R \tag{35.36}$$

where the quantity

$$I_{\text{rms}} = \frac{I_{\text{R}}}{\sqrt{2}} \tag{35.37}$$

is called the **root-mean-square current,** or rms current, I_{rms}. Technically, an rms quantity is the square root of the average, or mean, of the quantity squared. For quantities that oscillate sinusoidally, the rms value turns out to be the peak value divided by $\sqrt{2}$.

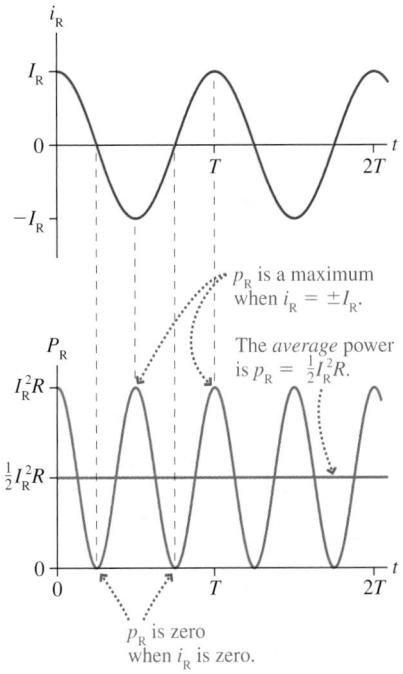

FIGURE 35.21 The instantaneous power loss in a resistor.

The rms current allows us to compare Equation 35.36 directly to the energy dissipated by a resistor in a DC circuit: $P = I^2R$. You can see that the average power loss of a resistor in an AC circuit with $I_{rms} = 1$ A is the same as in a DC circuit with $I = 1$ A. **As far as power is concerned, an rms current is equivalent to an equal DC current.**

Similarly, we can define the root-mean-square voltage and emf:

$$V_{rms} = \frac{V_R}{\sqrt{2}} \qquad \mathcal{E}_{rms} = \frac{\mathcal{E}_0}{\sqrt{2}} \tag{35.38}$$

The resistor's average power loss in terms of the rms quantities is

$$P_R = (I_{rms})^2 R = \frac{(V_{rms})^2}{R} = I_{rms} V_{rms} \tag{35.39}$$

and the average power supplied by the emf is

$$P_{source} = I_{rms} \mathcal{E}_{rms} \tag{35.40}$$

The single-resistor circuit that we analyzed in Section 35.1 had $V_R = \mathcal{E}$ or, equivalently, $V_{rms} = \mathcal{E}_{rms}$. You can see from Equations 35.39 and 35.40 that the power loss in the resistor exactly matches the power supplied by the emf. This must be the case in order to conserve energy.

NOTE ▶ Voltmeters, ammeters, and other AC electronic measuring instruments are almost always calibrated to give the rms value. An AC voltmeter would show that the "line voltage" of an electrical outlet in the United States is 120 V. This is $\mathcal{E}_{rms}$. The peak voltage $\mathcal{E}_0$ is larger by a factor of $\sqrt{2}$, or $\mathcal{E}_0 = 170$ V. The power-line voltage is sometimes specified as "120 V/60 Hz," showing the rms voltage and the frequency. ◀

EXAMPLE 35.7 Lighting a bulb

A 100 W incandescent lightbulb is plugged into a 120 V/60 Hz outlet. What is the resistance of the bulb's filament? What is the peak current through the bulb?

MODEL The filament in a lightbulb acts as a resistor.

VISUALIZE Figure 35.22 is a diagram of a simple one-resistor circuit.

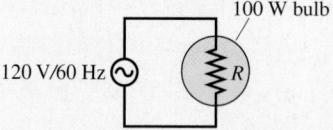

FIGURE 35.22 An AC circuit with a lightbulb as a resistor.

SOLVE A bulb labeled 100 W is designed to dissipate an average 100 W at $V_{rms} = 120$ V. We can use Equation 35.39 to find

$$R = \frac{(V_{rms})^2}{P_{avg}} = \frac{(120\text{ V})^2}{100\text{ W}} = 144\ \Omega$$

The rms current is then found from

$$I_{rms} = \frac{P_{avg}}{V_{rms}} = \frac{100\text{ W}}{120\text{ V}} = 0.833\text{ A}$$

The peak current is $I_R = \sqrt{2} I_{rms} = 1.18$ A.

ASSESS Calculations with rms values are just like the calculations you did for DC circuits.

Capacitors and Inductors

In Section 35.2, we found that the instantaneous current to a capacitor is $i_C = -\omega C V_C \sin \omega t$. Thus the instantaneous energy dissipation in a capacitor is

$$p_C = v_C i_C = (V_C \cos \omega t)(-\omega C V_C \sin \omega t) = -\frac{1}{2} \omega C V_C^2 \sin 2\omega t \quad (35.41)$$

where we used $\sin(2x) = 2\sin(x)\cos(x)$.

Figure 35.23 shows Equation 35.41 graphically. Energy is transferred into the capacitor (positive power) as it is charged, but, instead of being dissipated, as it would be by a resistor, the energy is stored as potential energy in the capacitor's electric field. Then, as the capacitor discharges, this energy is given back to the circuit. Power is the rate at which energy is *removed* from the circuit, hence p is negative as the capacitor transfers energy back into the circuit.

Returning to our mechanical analogy, a capacitor is like an ideal, frictionless simple harmonic oscillator. Kinetic and potential energy are constantly being exchanged, but there is no dissipation because none of the energy is transformed into thermal energy. The important conclusion is that **a capacitor's average power loss is zero: $P_C = 0$.**

The same is true of an inductor. An inductor alternately stores energy in the magnetic field, as the current is increasing, then transfers energy back to the circuit as the current decreases. The instantaneous power oscillates between positive and negative, but **an inductor's average power loss is zero: $P_L = 0$.**

NOTE ▶ We're assuming ideal capacitors and inductors. Real capacitors and inductors inevitably have a small amount of resistance and dissipate a small amount of energy. However, their energy dissipation is negligible compared to that of the resistors in most practical circuits. ◀

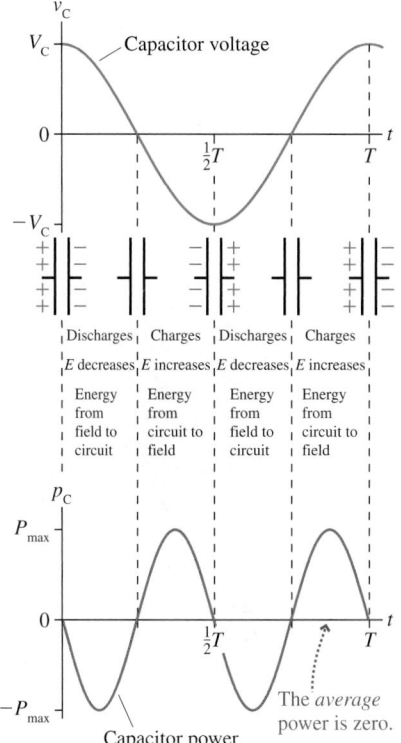

FIGURE 35.23 Energy flows into and out of a capacitor as it is charged and discharged.

The Power Factor

In an *RLC* circuit, energy is supplied by the emf and dissipated by the resistor. But an *RLC* circuit is unlike a purely resistive circuit in that the current is not in phase with the potential difference of the emf.

We found in Equation 35.22 that the instantaneous current in an *RLC* circuit is $i = I\cos(\omega t - \phi)$, where ϕ is the angle by which the current lags the emf. Thus the instantaneous power supplied by the emf is

$$p_{\text{source}} = i\mathcal{E} = (I\cos(\omega t - \phi))(\mathcal{E}_0 \cos \omega t) = I\mathcal{E}_0 \cos \omega t \cos(\omega t - \phi) \quad (35.42)$$

We can use the expression $\cos(x - y) = \cos(x)\cos(y) + \sin(x)\sin(y)$ to write the power as

$$p_{\text{source}} = (I\mathcal{E}_0 \cos \phi)\cos^2 \omega t + (I\mathcal{E}_0 \sin \phi)\sin \omega t \cos \omega t \quad (35.43)$$

In our analysis of the power loss in a resistor and a capacitor, we found that the average of $\cos^2 \omega t$ is $\frac{1}{2}$ and the average of $\sin \omega t \cos \omega t$ is zero. Thus we can immediately write that the *average* power supplied by the emf is

$$P_{\text{source}} = \frac{1}{2} I\mathcal{E}_0 \cos \phi = I_{\text{rms}} \mathcal{E}_{\text{rms}} \cos \phi \quad (35.44)$$

The rms values, you will recall, are $I/\sqrt{2}$ and $\mathcal{E}_0/\sqrt{2}$.

The term $\cos \phi$, which is called the **power factor,** arises because the current and the emf in a series *RLC* circuit are not in phase. Because the current and the emf aren't pushing and pulling together, the source delivers less energy to the circuit.

We'll leave it as a homework problem for you to show that the peak current in an *RLC* circuit can be written

$$I = I_{max} \cos\phi \tag{35.45}$$

where $I_{max} = \mathcal{E}_0/R$ was given in Equation 35.31. In other words, the current term in Equation 35.44 is a function of the power factor. Consequently, the average power is

$$P_{source} = P_{max} \cos^2\phi \tag{35.46}$$

where $P_{max} = \frac{1}{2}I_{max}\mathcal{E}_0$ is the *maximum* power the source can deliver to the circuit.

The source delivers maximum power only when $\cos\phi = 1$. This is the case when $X_L - X_C = 0$, requiring either a purely resistive circuit or an *RLC* circuit operating at the resonance frequency ω_0. The average power loss is zero for a purely capacitive or purely inductive load with, respectively, $\phi = -90°$ or $\phi = +90°$, which is in agreement with our analysis above.

Motors of various types, especially large industrial motors, use a significant fraction of the electric energy generated in industrialized nations. Motors operate most efficiently, doing the maximum work per second, when the power factor is as close to 1 as possible. But motors are inductive devices, due to their electromagnet coils, and if too many motors are attached to the electric grid, the power factor is pulled away from 1. To compensate, the electric company places large capacitors throughout the transmission system. The capacitors dissipate no energy, but they allow the electric system to deliver energy more efficiently by keeping the power factor close to 1.

Finally, we found in Equation 35.28 that the resistor's peak voltage in an *RLC* circuit is related to the emf peak voltage by $V_R = \mathcal{E}_0 \cos\phi$ or, dividing both sides by $\sqrt{2}$, $V_{rms} = \mathcal{E}_{rms} \cos\phi$. We can use this result to write the energy loss in the resistor as

$$P_R = I_{rms}V_{rms} = I_{rms}\mathcal{E}_{rms}\cos\phi \tag{35.47}$$

But this expression is P_{source}, as we found in Equation 35.44. Thus we see that the energy supplied to an *RLC* circuit by the emf is ultimately dissipated by the resistor.

Industrial motors use a significant fraction of the electric energy generated in the United States.

EXAMPLE 35.8 The power used by a motor

A motor attached to a 120 V/60 Hz power line uses 600 W of power at a power factor of 0.80.

a. What is the rms current to the motor?
b. What is the motor's resistance?

MODEL The motor has inductance. Otherwise, the power factor would be 1. Treat it as a series *RLC* circuit without the capacitor ($X_C = 0$).

SOLVE

a. The average power is $P_{source} = I_{rms}\mathcal{E}_{rms}\cos\phi$. Thus

$$I_{rms} = \frac{P_{avg}}{\mathcal{E}_{rms}\cos\phi} = \frac{600 \text{ W}}{(120 \text{ V})(0.80)} = 6.25 \text{ A}$$

b. We can use $V_{rms} = \mathcal{E}_{rms}\cos\phi$ and Ohm's law to find that the resistance is

$$R = \frac{V_{rms}}{I_{rms}} = \frac{\mathcal{E}_{rms}\cos\phi}{I_{rms}} = \frac{(120 \text{ V})(0.80)}{6.25 \text{ A}} = 15.4 \text{ } \Omega$$

STOP TO THINK 35.6   The emf and the current in a series *RLC* circuit oscillate as shown. Which of the following (perhaps more than one) would increase the rate at which energy is supplied to the circuit?

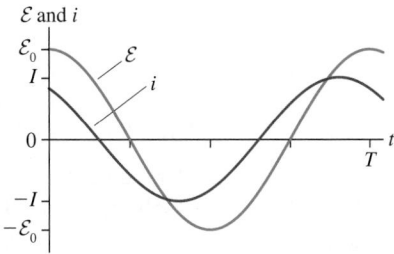

a. Increase $\mathcal{E}_0$ b. Increase L c. Increase C

d. Decrease $\mathcal{E}_0$ e. Decrease L f. Decrease C

SUMMARY

The goal of Chapter 35 has been to understand and apply basic techniques of AC circuit analysis.

IMPORTANT CONCEPTS

AC circuits are driven by an emf

$$\mathcal{E} = \mathcal{E}_0 \cos \omega t$$

that oscillates with angular frequency $\omega = 2\pi f$.

Phasors can be used to represent the oscillating emf, current, and voltage.

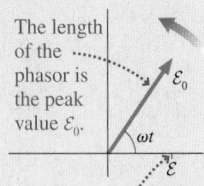

The length of the phasor is the peak value $\mathcal{E}_0$.

The horizontal projection is the instantaneous value $\mathcal{E}$.

Basic circuit elements

Element	i and v	Resistance/reactance	I and V	Power
Resistor	In phase	R is fixed	$V = IR$	$V_{rms}I_{rms}$
Capacitor	i leads v by 90°	$X_C = 1/\omega C$	$V = IX_L$	0
Inductor	i lags v by 90°	$X_L = \omega L$	$V = IX_C$	0

For many purposes, especially calculating power, the **root-mean-square** (rms) quantities

$$V_{rms} = V/\sqrt{2} \qquad I_{rms} = I/\sqrt{2} \qquad \mathcal{E}_{rms} = \mathcal{E}/\sqrt{2}$$

are equivalent to the corresponding DC quantities.

KEY SKILLS

Phasor diagrams

- Start with a phasor (v or i) common to two or more circuit elements.

- The sum of instantaneous quantities is vector addition.

- Use the Pythagorean theorem to relate peak quantities.

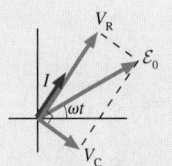

For an RC circuit, shown here,

$$v_R + v_C = \mathcal{E}$$

$$V_R^2 + V_C^2 = \mathcal{E}_0^2$$

Kirchhoff's laws

Loop law The sum of the potential differences around a loop is zero.

Junction law The sum of currents entering a junction equals the sum leaving the junction.

Instantaneous and peak quantities

Instantaneous quantities v and i generally obey different relationships than peak quantities V and I.

APPLICATIONS

RC filter circuits

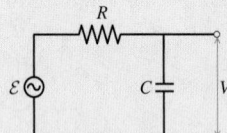

$$V_C = \mathcal{E}_0 X_C / \sqrt{R^2 + X_C^2}$$

$$V_C \rightarrow \mathcal{E}_0 \text{ as } \omega \rightarrow 0$$

A **low-pass filter** transmits low frequencies and blocks high frequencies.

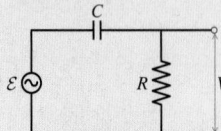

$$V_R = \mathcal{E}_0 R / \sqrt{R^2 + X_C^2}$$

$$V_R \rightarrow \mathcal{E}_0 \text{ as } \omega \rightarrow \infty$$

A **high-pass filter** transmits high frequencies and blocks low frequencies.

Series RLC circuits

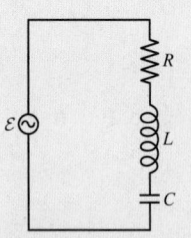

$I = \mathcal{E}_0/Z$ where Z is the **impedance**

$$Z = \sqrt{R^2 + (X_L - X_C)^2}$$

$$V_R = IR \qquad V_L = IX_L \qquad V_C = IX_C$$

When $\omega = \omega_0 = 1/\sqrt{LC}$ (the **resonance frequency**), the current in the circuit is a maximum $I_{max} = \mathcal{E}_0/R$.

In general, the current i lags behind $\mathcal{E}$ by the **phase angle** $\phi = \tan^{-1}((X_L - X_C)/R)$.

The power supplied by the emf is $P_{source} = I_{rms}\mathcal{E}_{rms} \cos\phi$, where $\cos\phi$ is called the **power factor.**

The power lost in a resistor is $P_R = I_{rms}V_{rms} = (I_{rms})^2 R$.

TERMS AND NOTATION

AC circuit	low-pass filter	resonance frequency, ω_0
DC circuit	high-pass filter	average power, P
phasor	inductive reactance, X_L	root-mean-square current, I_{rms}
capacitive reactance, X_C	series RLC circuit	power factor, $\cos\phi$
crossover frequency, ω_c	impedance, Z	

EXERCISES AND PROBLEMS

Exercises

Section 35.1 AC Sources and Phasors

1. The emf phasor in Figure Ex35.1 is shown at $t = 15$ ms.
   a. What is the angular frequency ω? Assume this is the first rotation.
   b. What is the instantaneous value of the emf?

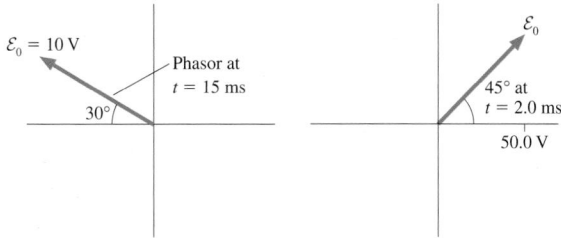

FIGURE EX35.1   FIGURE EX35.2

2. The emf phasor in Figure Ex35.2 is shown at $t = 2.0$ ms.
   a. What is the angular frequency ω? Assume this is the first rotation.
   b. What is the peak value of the emf?
3. A 111 Hz source of emf has a peak voltage of 50 V. Draw the emf phasor at $t = 3.0$ ms.
4. Draw the phasor for the emf $\mathcal{E} = (170 \text{ V})\cos((2\pi \times 60 \text{ Hz})t)$ at $t = 60$ ms.
5. A 200 Ω resistor is connected to an AC source with $\mathcal{E}_0 = 10$ V. What is the peak current through the resistor if the emf frequency is (a) 100 Hz? (b) 100 kHz?
6. Figure Ex35.6 shows voltage and current graphs for a resistor.
   a. What is the value of the resistance R?
   b. What is the emf frequency f?
   c. Draw the resistor's voltage and current phasors at $t = 15$ ms.

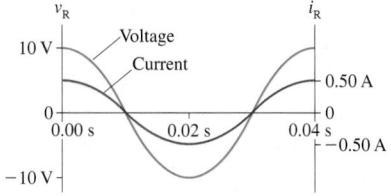

FIGURE EX35.6

Section 35.2 Capacitor Circuits

7. A 0.30 μF capacitor is connected across an AC generator that produces a peak voltage of 10.0 V. What is the peak current through the capacitor if the emf frequency is (a) 100 Hz? (b) 100 kHz?
8. The peak current through a capacitor is 10.0 mA. What is the current if
   a. The emf frequency is doubled?
   b. The emf peak voltage is doubled (at the original frequency)?
   c. The frequency is halved and, at the same time, the emf is doubled?
9. A 20 nF capacitor is connected across an AC generator that produces a peak voltage of 5.0 V.
   a. At what frequency f is the peak current 50 mA?
   b. What is the instantaneous value of the emf at the instant when $i_C = I_C$?
10. A capacitor is connected to a 15 kHz oscillator. The peak current is 65 mA when the rms voltage is 6.0 V. What is the value of the capacitance C?
11. The peak current through a capacitor is 8.0 mA when connected to an AC source with a peak voltage of 1.0 V. What is the capacitive reactance of the capacitor?

Section 35.3 RC Filter Circuits

12. A high-pass RC filter is connected to an AC source with a peak voltage of 10.0 V. The peak capacitor voltage is 6.0 V. What is the peak resistor voltage?
13. What are V_R and V_C if the emf frequency in Figure Ex35.13 is 10 kHz?

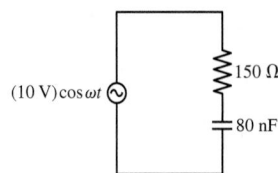

FIGURE EX35.13

14. A high-pass RC filter with a crossover frequency of 1000 Hz uses a 100 Ω resistor. What is the value of the capacitor?
15. A low-pass RC filter with a crossover frequency of 1000 Hz uses a 100 Ω resistor. What is the value of the capacitor?
16. A low-pass RC filter has a crossover frequency of 6280 rad/s. What is the crossover frequency if the value of the capacitor is doubled?

Section 35.4 Inductor Circuits

17. A 20 mH inductor is connected across an AC generator that produces a peak voltage of 10.0 V. What is the peak current through the inductor if the emf frequency is (a) 100 Hz? (b) 100 kHz?

18. The peak current through an inductor is 10.0 mA. What is the current if
    a. The emf frequency is doubled?
    b. The emf peak voltage is doubled (at the original frequency)?
    c. The frequency is halved and, at the same time, the emf is doubled?

19. A 500 μH inductor is connected across an AC generator that produces a peak voltage of 5.0 V.
    a. At what frequency f is the peak current 50 mA?
    b. What is the instantaneous value of the emf at the instant when $i_L = I_L$?

20. An inductor is connected to a 15 kHz oscillator. The peak current is 65 mA when the rms voltage is 6.0 V. What is the value of the inductance L?

21. The peak current through an inductor is 12.5 mA when connected to an AC source with a peak voltage of 1.0 V. What is the inductive reactance of the inductor?

Section 35.5 The Series *RLC* Circuit

22. At what frequency f do a 1.0 μF capacitor and a 1.0 μH inductor have the same reactance? What is the value of the reactance at this frequency?

23. What capacitor in series with a 100 Ω resistor and a 20 mH inductor will give a resonance frequency of 1000 Hz?

24. What inductor in series with a 100 Ω resistor and a 2.5 μF capacitor will give a resonance frequency of 1000 Hz?

25. A series *RLC* circuit has a 200 kHz resonance frequency. What is the resonance frequency if
    a. The resistor value is doubled?
    b. The capacitor value is doubled?

26. A series *RLC* circuit has a 200 kHz resonance frequency. What is the resonance frequency if
    a. The resistor value is doubled?
    b. The capacitor value is doubled and, at the same time, the inductor value is halved?

27. What is the phase angle of the current, in degrees, when the emf frequency in Figure Ex35.27 is (a) 14 kHz? (b) 18 kHz?

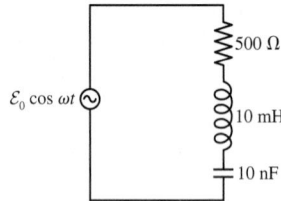

FIGURE EX35.27

Section 35.6 Power in AC Circuits

28. A resistor dissipates 2.0 W when the rms voltage of the emf is 10.0 V. At what rms voltage will the resistor dissipate 10.0 W?

29. The heating element of a hair drier dissipates 1500 W when connected to a 120 V/60 Hz power line. What is its resistance?

30. A 100 Ω resistor is connected to a 120 V/60 Hz power line. What is its average power loss?

31. A series *RLC* circuit attached to a 120 V/60 Hz power line draws 2.4 A of current with a power factor of 0.87. What is the value of the resistor?

32. A series *RLC* circuit with a 100 Ω resistor dissipates 80 W when attached to a 120 V/60 Hz power line. What is the power factor?

33. The motor of an electric drill draws a 3.5 A current at the power-line voltage of 120 V rms. What is the motor's power if the current lags the voltage by 20°?

Problems

34. a. For an *RC* circuit, find an expression for the angular frequency ω_{cap} at which $V_C = \frac{1}{2}\mathcal{E}_0$.
    b. What is V_R at this frequency?
    c. What is ω_{cap} if the crossover frequency is 6280 rad/s?

35. a. For an *RC* circuit, find an expression for the angular frequency ω_{res} at which $V_R = \frac{1}{2}\mathcal{E}_0$.
    b. What is V_C at this frequency?
    c. What is ω_{res} if the crossover frequency is 6280 rad/s?

36. a. Evaluate V_R in Figure P35.36 at emf frequencies 100, 300, 1000, 3000, and 10,000 Hz.
    b. Graph V_R versus frequency. Draw a smooth curve through your five points.

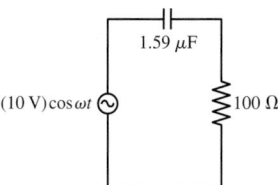

FIGURE P35.36

37. a. Evaluate V_C in Figure P35.37 at emf frequencies 1, 3, 10, 30, and 100 kHz.
    b. Graph V_C versus frequency. Draw a smooth curve through your five points.

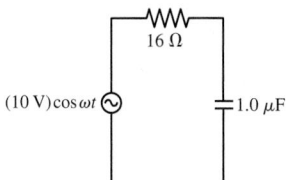

FIGURE P35.37

38. When two capacitors are connected in parallel across a 10.0 V rms, 1.0 kHz oscillator, the oscillator supplies a total rms current of 545 mA. When the same two capacitors are connected to the oscillator in series, the oscillator supplies an rms current of 126 mA. What are the values of the two capacitors?

39. For an *RC* filter circuit, show that $V_R = V_C = \mathcal{E}_0/\sqrt{2}$ at $\omega = \omega_c$.

40. Show that Equation 35.27 for the phase angle ϕ of a series *RLC* circuit gives the correct result for a capacitor-only circuit.

41. A low-pass filter consists of a 100 μF capacitor in series with a 159 Ω resistor. The circuit is driven by an AC source with a peak voltage of 5.0 V.
    a. What is the crossover frequency f_c?
    b. What is V_C when $f = \frac{1}{2}f_c$, f_c, and $2f_c$?

42. A high-pass filter consists of a 1.59 μF capacitor in series with a 100 Ω resistor. The circuit is driven by an AC source with a peak voltage of 5.0 V.
    a. What is the crossover frequency f_c?
    b. What is V_R when $f = \frac{1}{2}f_c$, f_c, and $2f_c$?

43. a. What is the peak current supplied by the emf in Figure P35.43?
    b. What is the peak voltage across the 3 μF capacitor?

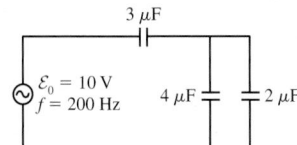

FIGURE P35.43

44. a. What is the average power supplied by the emf in Figure P35.44?
    b. What is the energy dissipated by each resistor?

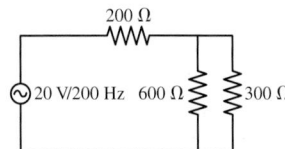

FIGURE P35.44

45. You have a resistor and a capacitor of unknown values. First, you charge the capacitor and discharge it through the resistor. By monitoring the capacitor voltage on an oscilloscope, you see that the voltage decays to half its initial value in 2.5 ms. You then use the resistor and capacitor to make a low-pass filter. What is the crossover frequency f_c?

46. Figure P35.46 shows a parallel RC circuit.
    a. Use a phasor-diagram analysis to find expressions for the peak currents I_R and I_C.
    Hint: What do the resistor and capacitor have in common? Use that as the initial phasor.
    b. Complete the phasor analysis by finding an expression for the peak emf current I.

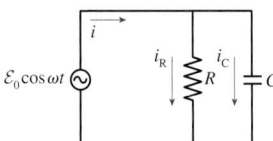

FIGURE P35.46

47. Use a phasor diagram to analyze the RL circuit of Figure P35.47. In particular,
    a. Find expressions for I, V_R, and V_L.
    b. What is V_R in the limits $\omega \rightarrow 0$ and $\omega \rightarrow \infty$?
    c. If the output is taken from the resistor, is this a low-pass or a high-pass filter? Explain.
    d. Find an expression for the crossover frequency ω_c.

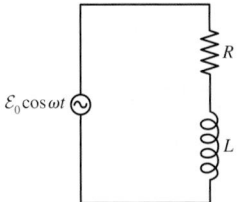

FIGURE P35.47

48. A series RLC circuit consists of a 100 Ω resistor, a 0.10 H inductor, and a 100 μF capacitor. It is attached to a 120 V/60 Hz power line. What are (a) the peak current I, (b) the phase angle ϕ, and (c) the average power loss?

49. A series RLC circuit consists of a 100 Ω resistor, a 0.15 H inductor, and a 30 μF capacitor. It is attached to a 120 V/60 Hz power line. What are (a) the peak current I, (b) the phase angle ϕ, and (c) the average power loss?

50. For the circuit of Figure P35.50,
    a. What is the resonance frequency, in both rad/s and Hz?
    b. Find V_R and V_L.
    c. How can V_L be larger than $\mathcal{E}_0$? Explain.

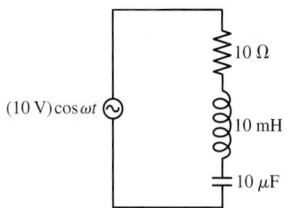

FIGURE P35.50

51. For the circuit of Figure P35.51,
    a. What is the resonance frequency, in both rad/s and Hz?
    b. Find V_R and V_C at resonance.
    c. How can V_C be larger than $\mathcal{E}_0$? Explain.

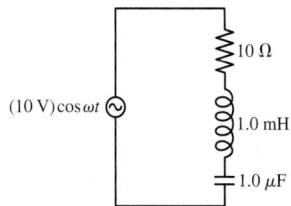

FIGURE P35.51

52. In Figure P35.52, what is the current supplied by the emf when (a) the frequency is very small and (b) the frequency is very large?

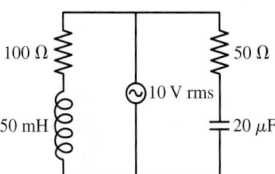

FIGURE P35.52

53. A series RLC circuit consists of a 50 Ω resistor, a 3.3 mH inductor, and a 480 nF capacitor. It is connected to an oscillator with a peak voltage of 5.0 V. Determine the impedance, the peak current, and the phase angle at frequencies (a) 3000 Hz, (b) 4000 Hz, and (c) 5000 Hz.

54. A series RLC circuit consists of a 50 Ω resistor, a 3.3 mH inductor, and a 480 nF capacitor. It is connected to a 5.0 kHz oscillator with a peak voltage of 5.0 V. What is the instantaneous current i when
    a. $\mathcal{E} = \mathcal{E}_0$?
    b. $\mathcal{E} = 0$ V and is decreasing?

55. A series RLC circuit consists of a 50 Ω resistor, a 3.3 mH inductor, and a 480 nF capacitor. It is connected to a 3.0 kHz oscillator with a peak voltage of 5.0 V. What is the instantaneous emf $\mathcal{E}$ when
    a. $i = I$?
    b. $i = 0$ A and is decreasing?
    c. $i = -I$?

56. A series *RLC* circuit consists of a 100 Ω resistor, a 10 mH inductor, and a 1.0 nF capacitor. It is connected to an oscillator with an rms voltage of 10 V. What is the power supplied to the circuit if (a) $\omega = \frac{1}{2}\omega_0$? (b) $\omega = \omega_0$? (c) $\omega = 2\omega_0$?

57. Show that the impedance of a series *RLC* circuit can be written

$$Z = \sqrt{R^2 + \omega^2 L^2 (1 - \omega_0^2/\omega^2)^2}$$

58. For a series *RLC* circuit, show that
   a. The peak current can be written $I = I_{max}\cos\phi$.
   b. The average power dissipation can be written $P_{avg} = P_{max}\cos^2\phi$.

59. The tuning circuit in an FM radio receiver is a series *RLC* circuit with a 0.200 μH inductor.
   a. The receiver is tuned to a station at 104.3 MHz. What is the value of the capacitor in the tuning circuit?
   b. FM radio stations are assigned frequencies every 0.2 MHz, but two nearby stations cannot use adjacent frequencies. What is the maximum resistance the tuning circuit can have if the peak current at a frequency of 103.9 MHz, the closest frequency that can be used by a nearby station, is to be no more than 0.10% of the peak current at 104.3 MHz? The radio is still tuned to 104.3 MHz, and you can assume the two stations have equal strength.

60. A television channel is assigned the frequency range from 54 MHz to 60 MHz. A series *RLC* tuning circuit in a TV receiver resonates in the middle of this frequency range. The circuit uses a 16 pF capacitor.
   a. What is the value of the inductor?
   b. In order to function properly, the current throughout the frequency range must be at least 50% of the current at the resonance frequency. What is the minimum possible value of the circuit's resistance?

61. Lightbulbs labeled 40 W, 60 W, and 100 W are connected to a 120 V/60 Hz power line as shown in Figure P35.61. What is the rate at which energy is dissipated in each bulb?

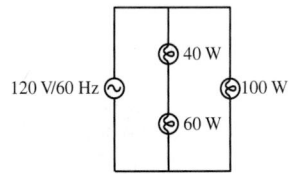

FIGURE P35.61

62. Commercial electricity is generated and transmitted as *three-phase electricity*. Instead of a single emf, three separate wires carry currents for the emfs $\mathcal{E}_1 = \mathcal{E}_0\cos\omega t$, $\mathcal{E}_2 = \mathcal{E}_0\cos(\omega t + 120°)$, and $\mathcal{E}_3 = \mathcal{E}_0\cos(\omega t - 120°)$ over three parallel wires, each of which supplies one-third of the power. This is why the long-distance transmission lines you see in the countryside have three wires. Suppose the transmission lines into a city supply a total of 450 MW of electric power, a realistic value.
   a. What would be the current in each wire if the transmission voltage were $\mathcal{E}_0 = 120$ V rms?
   b. In fact, transformers are used to step the transmission-line voltage up to 500 kV rms. What is the current in each wire?
   c. Big transformers are expensive. Why does the electric company use step-up transformers?

63. A motor attached to a 120 V/60 Hz power line draws an 8.0 A current. Its average energy dissipation is 800 W.
   a. What is the power factor?
   b. What is the rms resistor voltage?
   c. What is the motor's resistance?
   d. How much series capacitance needs to be added to increase the power factor to 1.0?

Challenge Problems

64. Commercial electricity is generated and transmitted as *three-phase electricity*. Instead of a single emf $\mathcal{E} = \mathcal{E}_0\cos\omega t$, three separate wires carry currents for the emfs $\mathcal{E}_1 = \mathcal{E}_0\cos\omega t$, $\mathcal{E}_2 = \mathcal{E}_0\cos(\omega t + 120°)$, and $\mathcal{E}_3 = \mathcal{E}_0\cos(\omega t - 120°)$. This is why the long-distance transmission lines you see in the countryside have three parallel wires, as do many distribution lines within a city.
   a. Draw a phasor diagram showing phasors for all three phases of a three-phase emf.
   b. Show that the sum of the three phases is zero, producing what is referred to as *neutral*. In *single-phase* electricity, provided by the familiar 120 V/60 Hz electric outlets in your home, one side of the outlet is neutral, as established at a nearby electrical substation. The other, called the *hot side*, is one of the three phases. (The round opening is connected to ground.)
   c. Show that the potential difference between any two of the phases has the rms value $\sqrt{3}\,\mathcal{E}_{rms}$, where $\mathcal{E}_{rms}$ is the familiar single-phase rms voltage. Evaluate this potential difference for $\mathcal{E}_{rms} = 120$ V. Some high-power home appliances, especially electric clothes dryers and hot-water heaters, are designed to operate between two of the phases rather than between one phase and neutral. Heavy-duty industrial motors are designed to operate from all three phases, but full three-phase power is rare in residential or office use.

65. The small transformers that power many consumer products produce a 12.0 V rms, 60 Hz emf. Design a circuit using resistors and capacitors that uses the transformer voltage as an input and produces a 6.0 V rms output that leads the input voltage by 45°.

66. You're the operator of a 15,000 V rms, 60 Hz electrical substation. When you get to work one day, you see that the station is delivering 6.0 MW of power with a power factor of 0.90.
   a. What is the rms current leaving the station?
   b. How much series capacitance should you add to bring the power factor up to 1.0?
   c. How much power will the station then be delivering?

67. a. Show that the peak inductor voltage in a series *RLC* circuit is maximum at frequency

$$\omega_L = \left(\frac{1}{\omega_0^2} - \frac{1}{2}R^2 C^2\right)^{-1/2}$$

   b. A series *RLC* circuit with $\mathcal{E}_0 = 10.0$ V consists of a 1.0 Ω resistor, a 1.0 μH inductor, and a 1.0 μF capacitor. What is V_L at $\omega = \omega_0$ and at $\omega = \omega_L$?

68. a. Show that the average power loss in a series RLC circuit is

$$P_{\text{avg}} = \frac{\omega^2 \mathcal{E}_{\text{rms}}^2 R}{\omega^2 R^2 + L^2(\omega^2 - \omega_0^2)^2}$$

b. Prove that the energy dissipation is a maximum at $\omega = \omega_0$.

69. Consider the parallel RLC circuit shown in Figure CP35.69.
   a. Show that the current drawn from the emf is

$$I = \mathcal{E}_0 \sqrt{\frac{1}{R^2} + \left(\frac{1}{\omega L} - \omega C\right)^2}$$

   Hint: Start with a phasor that is common to all three circuit elements.

   b. What is I in the limits $\omega \to 0$ and $\omega \to \infty$?
   c. Find the frequency for which I is a minimum.
   d. Sketch a graph of I versus ω.

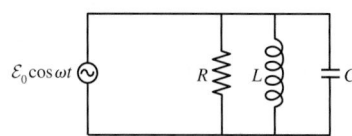

FIGURE CP35.69

70. The telecommunication circuit shown in Figure CP35.70 has a parallel inductor and capacitor in series with a resistor.
   a. Use a phasor diagram to show that the peak current through the resistor is

$$I = \frac{\mathcal{E}_0}{\sqrt{R^2 + \left(\dfrac{1}{X_L} - \dfrac{1}{X_C}\right)^{-2}}}$$

   Hint: Start with the inductor phasor v_L. Which voltages are shared? What does the junction law tell you about the currents? How do currents and voltages add on a phasor diagram?

   b. What is I in the limits $\omega \to 0$ and $\omega \to \infty$?
   c. What is the resonance frequency ω_0? What is I at this frequency?

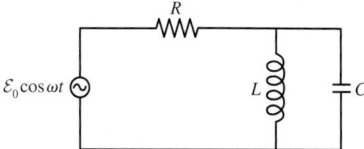

FIGURE CP35.70

STOP TO THINK ANSWERS

Stop to Think 35.1: a. The instantaneous emf value is the projection down onto the horizontal axis. The emf is negative but increasing in magnitude as the phasor, which rotates ccw, approaches the horizontal axis.

Stop to Think 35.2: c. Voltage and current are measured using different scales and units. You can't compare the length of a voltage phasor to the length of a current phasor.

Stop to Think 35.3: a. There is "no capacitor" when the separation between the two capacitor plates becomes zero and the plates touch. Capacitance C is inversely proportional to the plate spacing d, hence $C \to \infty$ as $d \to 0$. The capacitive reactance is inversely proportional to C, so $X_C \to 0$ as $C \to \infty$.

Stop to Think 35.4: $(\omega_c)_d > (\omega_c)_c = (\omega_c)_a > (\omega_c)_b$. The crossover frequency is $1/RC$.

Stop to Think 35.5: Above. $V_L > V_C$ tells us that $X_L > X_C$. This is the condition above resonance, where X_L is increasing with ω while X_C is decreasing.

Stop to Think 35.6: a, b, and **f.** You can always increase power by turning up the voltage. The current leads the emf, telling us that the circuit is primarily capacitive. The current can be brought into phase with the emf, thus maximizing the power, by decreasing C or increasing L.

Electricity and Magnetism

Mass and charge are the two most fundamental properties of matter. The first five parts of this text were investigations of the properties and interactions of masses. Part VI has been a study of the physics of charge—what charge is and how charges interact.

Electric and magnetic fields were introduced to enable us to understand the long-range forces of electricity and magnetism. The field concept is subtle, but it is an essential part of our modern understanding of the physical universe. One charge—the source charge—alters the space around it by creating an electric field and, if the charge is moving, a magnetic field. Other charges experience forces exerted *by* the fields. Thus the electric and magnetic fields are the agents by which charges interact.

Faraday's discovery of electromagnetic induction led scientists to recognize that the fields are *real* and can exist independently of charges. The most vivid confirmation of this reality was Maxwell's discovery of electromagnetic waves—the quintessential electromagnetic phenomenon.

Part VI has introduced many new phenomena, concepts and laws. The knowledge structure table draws together the major ideas about charges and fields, and it briefly summarizes some of the most important applications of electricity and magnetism.

KNOWLEDGE STRUCTURE VI **Electricity and Magnetism**

ESSENTIAL CONCEPTS	Charge, dipole, field, potential, emf
BASIC GOALS	How do charged particles interact?
	What are the properties and characteristics of electromagnetic fields?

GENERAL PRINCIPLES		
Coulomb's law	$\vec{E}_{\text{point charge}} = \dfrac{1}{4\pi\epsilon_0}\dfrac{q}{r^2}\hat{r} = \left(\dfrac{1}{4\pi\epsilon_0}\dfrac{q}{r^2}, \text{away from } q\right)$	
Biot-Savart law	$\vec{B}_{\text{point charge}} = \dfrac{\mu_0}{4\pi}\dfrac{q\vec{v}\times\hat{r}}{r^2} = \left(\dfrac{\mu_0}{4\pi}\dfrac{qv\sin\theta}{r^2}, \text{direction of right-hand rule}\right)$	
Faraday's law	$\mathcal{E} = \left|d\Phi_{\text{m}}/dt\right|$ $\qquad I_{\text{induced}} = \mathcal{E}/R$ in the direction of Lenz's law	
Lenz's law	An induced current flows around a conducting loop in the direction such that the induced magnetic field opposes the *change* in the magnetic flux.	
Lorentz force law	$\vec{F}_{\text{on } q} = q(\vec{E} + \vec{v}\times\vec{B})$	
Superposition	The electric or magnetic field due to multiple charges is the vector sum of the field of each charge. This principle was used to derive the fields of many special charge distributions, such as wires, planes, and loops.	

FIELD AND POTENTIAL The electric field of charges can also be described in terms of an electric potential V.

$$V_{\text{point charge}} = \frac{q}{4\pi\epsilon_0 r}$$

- The electric field is perpendicular to equipotential surfaces and in the direction of decreasing potential.
- The potential energy of charge q is $U = qV$. The total energy $K + U$ of a group of charges is conserved.

ELECTROMAGNETIC WAVES All the properties of electromagnetic fields are summarized mathematically in four equations called *Maxwell's equations*. From Maxwell's equations we learn that electromagnetic fields can exist independently of charges as an *electromagnetic wave*.

- An em wave travels at speed $c = 1/\sqrt{\epsilon_0\mu_0}$.
- $\vec{E}$ and $\vec{B}$ are perpendicular to each other and to the direction of travel, with $E = cB$.

Electric and magnetic properties of materials

- Charges move through conductors but not through insulators.
- Conductors and insulators are *polarized* in an electric field.
- A magnetic moment in a magnetic field experiences a torque.

Model of current and conductivity

- The charge carriers in metals are electrons.
- Emf → electric field → current density $J = \sigma E \rightarrow I = JA$.

Applications to circuits

- Circuits obey Kirchhoff's loop law (conservation of energy) and junction law (conservation of current).
- Resistors control the current: $I = \Delta V/R$ (Ohm's law).
- Capacitors store charge $Q = C\Delta V$ and energy $V_C = \frac{1}{2}C(\Delta V_C)^2$.

The Telecommunications Revolution

In 1800, the year that Volta invented the battery and that Thomas Jefferson was elected president, the fastest a message could travel was the speed of a man or woman on horseback. News took three days to travel from New York to Boston, and well over a month to reach the frontier outpost of Cincinnati.

But Oersted's 1820 discovery that a current creates a magnetic field soon introduced revolutionary changes to communications. The American scientist Joseph Henry, who shares with Faraday the credit for the discovery of electromagnetic induction, saw a simple electromagnet in 1825. Inspired, he set about improving the device. By 1830, Henry was able to send current through more than a mile of wire to activate an electromagnet and strike a bell.

In 1835, Henry met an entrepreneur interested in the commercial development of electric technology—Samuel F. B. Morse. Morse was one of the most prominent American artists of the early 19th century, but he also had an abiding interest in technology. In the 1830s he invented the famous code that bears his name—Morse code—and began to experiment with electromagnets.

With advice and encouragement from Henry, Morse developed the first practical telegraph. The first telegraph line, between Washington, D.C., and Baltimore, began operating in 1844; the first message sent was, "What hath God wrought?" For the first time, long-distance communications could take place essentially instantaneously.

Telegraph communication advanced as quickly as wire could be strung, and a worldwide network had been established by 1875. But the telegraph didn't hold its monopoly for long, as other inventors began to think about using electromagnetic devices to transmit speech. The first to succeed was Alexander Graham Bell, who invented the telephone in 1876.

The telegraph and telephone provided electromagnetic communication over wires, but the discovery of electromagnetic waves opened up another possibility—wireless communication at the speed of light. Radio technology developed rapidly in the late 19th century, and in 1901 the Italian inventor Guglielmo Marconi sent and received the first transatlantic radio message. World War I prompted further development of radio, because of the need to communicate with military units as they moved about, and by 1925 more than 1000 radio stations were operating in the United States.

Radio and, later, television spanned the globe by 1960, but radio stations reached a few hundred miles at best, and television transmission was pretty much limited to each city. National broadcasts within the United States required the signal to be transmitted via microwave relays to local stations for rebroadcast. This system made possible network television shows, but not live-from-the-scene broadcasts. News journalists had to film events, then return the film to the studio for broadcast. Television images from overseas could only be seen the next day, after film was flown back to the United States.

The first communications satellite was launched by NASA in 1960, followed two years later by a more practical satellite, Telstar, that used solar power to amplify signals received from earth and beam them back down. The first live transatlantic television transmission was made on July 11, 1962, and was broadcast throughout the United States.

Plans were made for a system of roughly 100 satellites, so that one would always be overhead, but another idea soon proved more practical. In 1945, 12 years before space flight began, the science-fiction writer Arthur C. Clarke proposed placing satellites in orbits 22,300 miles above the earth. A satellite at this altitude orbits with a 24-hour period, so from the ground it appears to hang stationary in space. We now call this a *geosynchronous orbit*. One such satellite would allow microwave communication between two points one-third of a world apart, so just three geosynchronous satellites would span the entire earth.

Much more energy is required to reach geosynchronous orbit than to reach low-earth orbit, but rocket technology was advancing faster than NASA could build Telstar satellites. The first commercial communications satellite was placed in geosynchronous orbit in 1965, and, for the first time, television images could be broadcast live to anywhere in the world. Today all of the world's intercontinental television and much of the intercontinental telephone traffic travels via microwaves to and from a cluster of these artificial stars floating high above the earth.

As we enter the 21st century, information and images span the world as quickly as or more quickly than they once moved through a small village. You can pick up the phone and talk to friends or relatives anywhere around the globe, and each evening's news brings live images from remote places. Telecommunication unites our world, and the technologies of telecommunications are direct descendants of Coulomb, Ampère, Oersted, Henry, and—most of all—Michael Faraday.

If the fusion reactor in the core of the Death Star generates 10^{15} W of power, how much mass does the Death Star lose each year as it converts mass into energy?

Relativity and Quantum Physics

Contemporary Physics

Our journey into physics is nearing its end. We began roughly 350 years ago with Newton's discovery of the laws of motion. The conclusion of Part VI brought us to the end of the 19th century, just over 100 years ago. Along the way you've learned about the motion of particles, the conservation of energy, the physics of waves, and the electromagnetic interactions that hold atoms together and generate light waves. We can begin the last phase of our journey with confidence.

Newton's mechanics and Maxwell's electromagnetism were the twin pillars of science at the end of the 19th century and the basis for much of engineering and applied science in the 20th century. Despite the successes of these theories, a series of discoveries starting around 1900 and continuing into the first few decades of the 20th century profoundly altered our understanding of the universe at the most fundamental level. These discoveries forced scientists to reconsider the very nature of space and time and to develop new models of light and matter.

The discoveries and new ideas of the early 20th century led to two new theories: relativity and quantum physics. These two theories form the basis for physics as it is practiced today and are already having a significant impact on 21st-century engineering. We will end our journey into physics with a look at these contemporary topics and some of their applications.

Relativity

The idea of measuring distance with a meter stick and time with a clock or stopwatch has been with us since Chapter 1. The basic notions of space and time seem so self-evident that no one had seriously questioned them. No one, that is, until an unknown young scientist named Albert Einstein began to ponder these issues in the years right around 1900.

It wasn't space and time that first troubled Einstein. Instead, he was bothered by what he saw as paradoxes and difficulties in Maxwell's theory of electromagnetism. Einstein was able to show that electromagnetism is a self-consistent theory only if the speed of electromagnetic waves—the speed of light—is the same in all inertial reference frames, no matter how the reference frames might be moving with respect to each other or to the source of the wave. But this strange behavior of a traveling wave can be true only if *space and time are different* for two experimenters moving relative to each other.

We'll need to explore what it means for space and time to be different for different experimenters. In doing so we'll discover some of the well-known puzzles of relativity, such as length contraction, the twin paradox, and a cosmic speed limit for particles. Our exploration of these fascinating ideas will end with what is perhaps the most famous equation in physics: Einstein's $E = mc^2$.

Quantum Physics

We ended Part V with experimental evidence that light sometimes acts like a particle and that electrons can exhibit wave-like behavior. These were observations only; we did not offer any explanation at the time. We now want to return to that thread of thought and see how it leads to the new ideas of quantum physics.

We'll begin by looking at *evidence* about the atomic world. What do we know about electrons and atoms, and how do we know it? Atoms were first thought to be tiny, indivisible pieces of matter that moved in accordance with Newton's laws of motion. But it became increasingly clear toward the end of the 19th century that atoms *can* be divided into smaller pieces. Furthermore, the classical physics of Newton and Maxwell was unable to explain the behavior of atoms or the light they emit.

We will focus our attention on a key experiment called the *photoelectric effect*. The experiment is straightforward and the data are unambiguous, but difficulties will arise when we try to explain the outcome. It was, once again, Albert Einstein who offered a fresh and original interpretation of the photoelectric effect in terms of the *quantization* of energy—in particular, the idea that the energy of a light wave is bundled into small, discrete packages that we now call *photons*.

We will then look at Niels Bohr's efforts to apply the ideas of energy quantization to atoms. Bohr's model of the atom was the first to explain the discrete spectra emitted by atoms, but he was unable to extend his model beyond the simplest hydrogen atom. Bohr was on the right track, but his ideas were still too classical. The missing ingredient in Bohr's model was de Broglie's hypothesis that matter has wave-like properties—a hypothesis we looked at in Chapter 24.

The complete theory of quantum physics was developed by Erwin Schrödinger soon after he learned of de Broglie's hypothesis. Schrödinger's new theory describes atomic particles in terms of an entirely new concept called a *wave function*. One of our most important tasks in Part VII will be to answer the questions:

- What is a wave function?
- How is the wave function interpreted and connected to experimental measurements?
- What new law of nature governs the behavior of the wave function?

Once we've answered these questions, we'll begin to see how Schrödinger's *quantum mechanics*—the quantum-physics analysis of motion—successfully explains the behavior of electrons and atoms.

We will concentrate on quantum mechanics in one dimension. This restriction will allow us to focus on physical phenomena without becoming sidetracked by the mathematics of quantum mechanics in three dimensions. Although one-dimensional models aren't perfect, they will be adequate for understanding the essential features of scanning tunneling microscopes, various kinds of semiconductor devices, radioactive decay, and other applications.

Quantum physics will give you an entirely new perspective on the nature of matter and light. The quantum world with its wave functions can seem strange and mysterious, yet quantum mechanics gives the most definitive and accurate predictions of any physical theory ever devised.

Atoms and Nuclei

You learned in Chapter 18 how macroscopic measurements of density, pressure, and temperature allow us to learn about the sizes and speeds of atoms. You've also learned how atoms can be polarized (Chapter 25), how electrons and ions can transfer charge as an electric current (Chapter 28), and how atoms act as tiny magnets (Chapter 32). And in Chapter 24 you saw the rather strange evidence that matter on the atomic scale sometimes acts like a wave.

But when you get right down to it, what *is* an atom? What makes an atom of carbon different from an atom of gold? And what's inside an atom? As important as these questions are, there's an even deeper question: How do we *know* about atoms? Atomic phenomena lie beyond the immediate realm of our senses, so learning to interpret macroscopic data in terms of atomic- and nuclear-level events is an essential aspect of learning about atoms and nuclei.

An understanding of atoms and nuclei depends on both classical electromagnetism and quantum mechanics. Although we can give only an introduction to atomic and nuclear physics, you will learn where the electron shell model of chemistry comes from, how atoms emit and absorb light, what's inside the nucleus, and why some nuclei undergo radioactive decay.

A scanning tunneling microscope image of individual iron atoms on a copper surface. The atoms form the Japanese character for "atom."

36 Relativity

These are the fundamental tools with which we learn about space and time.

▶ Looking Ahead
The goal of Chapter 36 is to understand how Einstein's theory of relativity changes our concepts of space and time. In this chapter you will learn to:

- Use the principle of relativity.
- Understand how time dilation and length contraction change our concepts of space and time.
- Use the Lorentz transformations of positions and velocities.
- Calculate relativistic momentum and energy.
- Understand how mass and energy are equivalent.

◀ Looking Back
The material in this chapter depends on an understanding of relative motion in Newtonian mechanics. Please review:

- Section 6.4 Inertial reference frames and the Galilean transformations.

Space and time seem like straightforward ideas. You can measure lengths with a ruler or meter stick. You can time events with a stop watch. Nothing could be simpler.

So it seemed to everyone until 1905, when an unknown young scientist had the nerve to suggest that this simple view of space and time was in conflict with other principles of physics. In the century since, Einstein's theory of relativity has radically altered our understanding of some of the most fundamental ideas in physics.

Relativity, despite its esoteric reputation, has very real implications for modern technology. Global positioning system (GPS) satellites depend on relativity, as do the navigation systems used by airliners. Nuclear reactors make tangible use of Einstein's famous equation $E = mc^2$ to generate 20% of the electricity used in the United States. The annihilation of matter in positron-emission tomography (PET scanners) has given neuroscientists an unprecedented ability to monitor activity within the brain.

The theory of relativity is fascinating, perplexing, and challenging. It is also vital to our contemporary understanding of the universe in which we live.

36.1 Relativity: What's It All About?

What do you think of when you hear the phrase "theory of relativity"? A white-haired Einstein? $E = mc^2$? Black holes? Time travel? Perhaps you've heard that the theory of relativity is so complicated and abstract that only a handful of people in the whole world really understand it.

There is, without doubt, a certain mystique associated with relativity, an aura of the strange and exotic. The good news is that understanding the ideas of relativity is well within your grasp. Einstein's *special theory of relativity,* the portion of relativity we'll study, is not mathematically difficult at all. The challenge is conceptual because relativity questions deeply held assumptions about the nature of space and time. In fact, that's what relativity is all about—space and time.

In one sense, relativity is not a new idea at all. Certain ideas about relativity are part of Newtonian mechanics. You had an introduction to these ideas in Chapter 6, where you learned about reference frames and the Galilean transformations. Einstein, however, thought that relativity should apply to *all* the laws of physics, not just mechanics. The difficulty, as you'll see, is that some aspects of relativity appear to be incompatible with the laws of electromagnetism, particularly the laws governing the propagation of light waves.

Lesser scientists might have concluded that relativity simply doesn't apply to electromagnetism. Einstein's genius was to see that the incompatibility arises from *assumptions* about space and time, assumptions no one had ever questioned because they seem so obviously true. Rather than abandon the ideas of relativity, Einstein changed our understanding of space and time.

Fortunately, you need not be a genius to follow a path that someone else has blazed. However, we will have to exercise the utmost care with regard to logic and precision. We will need to state very precisely just how it is that we know things about the physical world, then ruthlessly follow the logical consequences. The challenge is to stay on this path, not to let our prior assumptions—assumptions that are deeply ingrained in all of us—lead us astray.

What's Special About Special Relativity?

Einstein's first paper on relativity, in 1905, dealt exclusively with inertial reference frames, reference frames that move relative to each other with constant velocity. Ten years later, Einstein published a more encompassing theory of relativity that considers accelerated motion and its connection to gravity. The second theory, because it's more general in scope, is called *general relativity.* General relativity is the theory that describes black holes, curved spacetime, and the evolution of the universe. It is a fascinating theory but, alas, very mathematical and outside the scope of this textbook. If you're interested, many popular science books provide a nontechnical introduction to general relativity.

Motion at constant velocity is a "special case" of motion; namely, motion for which the acceleration is zero. Hence Einstein's first theory of relativity has come to be known as **special relativity.** It is special in the sense of being a restricted, special case of his more general theory, not special in the everyday sense of meaning distinctive or exceptional. Special relativity, with its conclusions about time dilation and length contraction, is what we will study.

36.2 Galilean Relativity

A firm grasp of Galilean relativity is necessary if we are to appreciate and understand what is new in Einstein's theory. Thus we begin with the ideas of relativity that are embodied in Newtonian mechanics.

Albert Einstein (1879–1955) was one of the most influential thinkers in history.

Reference Frames

Suppose you're passing me as we both drive in the same direction along a freeway. My car's speedometer reads 55 mph while your speedometer shows 60 mph. Is 60 mph your "true" speed? That is certainly your speed relative to someone standing beside the road, but your speed relative to me is only 5 mph. Your speed is 120 mph relative to a driver approaching from the other direction at 60 mph.

An object does not have a "true" speed or velocity. The very definition of velocity, $v = \Delta x/\Delta t$, assumes the existence of a coordinate system in which, during some time interval Δt, the displacement Δx is measured. The best we can manage is to specify an object's velocity relative to, or with respect to, the coordinate system in which it is measured.

Let's define a **reference frame** to be a coordinate system in which experimenters equipped with meter sticks, stopwatches, and any other needed equipment make position and time measurements on moving objects. Three ideas are implicit in our definition of a reference frame:

- A reference frame extends infinitely far in all directions.
- The experimenters are at rest in the reference frame.
- The number of experimenters and the quality of their equipment are sufficient to measure positions and velocities to any level of accuracy needed.

The first two bullets are especially important. It is often convenient to say "the laboratory reference frame" or "the reference frame of the rocket." These are shorthand expressions for "a reference frame, infinite in all directions, in which the laboratory (or the rocket) and a set of experimenters happen to be at rest."

NOTE ▶ A reference frame is not the same thing as a "point of view." That is, each person or each experimenter does not have his or her own private reference frame. **All experimenters at rest relative to each other share the same reference frame.** ◀

Figure 36.1 shows two reference frames called S and S′. The coordinate axes in S are x, y, z and those in S′ are x', y', z'. Reference frame S′ moves with velocity v relative to S or, equivalently, S moves with velocity $-v$ relative to S′. There's no implication that either reference frame is "at rest." Notice that the zero of time, when experimenters start their stopwatches, is the instant that the origins of S and S′ coincide.

We will restrict our attention to *inertial reference frames,* implying that the relative velocity v is constant. You should recall from Chapter 6 that an **inertial reference frame** is a reference frame in which Newton's first law, the law of inertia, is valid. In particular, an inertial reference frame is one in which an isolated particle, one on which there are no forces, either remains at rest or moves in a straight line at constant speed.

Any reference frame that moves at constant velocity with respect to an inertial reference frame is itself an inertial reference frame. Conversely, a reference frame that accelerates with respect to an inertial reference frame is *not* an inertial reference frame. Our restriction to reference frames moving with respect to each other at constant velocity—with no acceleration—is the "special" part of special relativity.

NOTE ▶ An inertial reference frame is an idealization. A true inertial reference would need to be floating in deep space, far from any gravitational influence. In practice, an earthbound laboratory is a good approximation of an inertial reference frame because the accelerations associated with the earth's rotation and motion around the sun are too small to influence most experiments. ◀

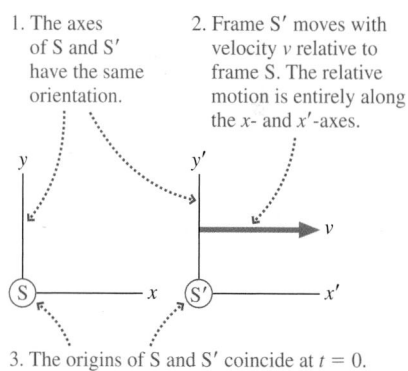

1. The axes of S and S′ have the same orientation.

2. Frame S′ moves with velocity v relative to frame S. The relative motion is entirely along the x- and x'-axes.

3. The origins of S and S′ coincide at $t = 0$. This is our definition of $t = 0$.

FIGURE 36.1 The standard reference frames S and S′.

Which of these is an inertial reference frame (or a very good approximation)?

a. Your bedroom
b. A car rolling down a steep hill
c. A train coasting along a level track
d. A rocket being launched
e. A roller coaster going over the top of a hill
f. A sky diver falling at terminal speed

The Galilean Transformations

Suppose a firecracker explodes at time t. The experimenters in reference frame S determine that the explosion happened at position x. Similarly, the experimenters in S' find that the firecracker exploded at x' in their reference frame. What is the relationship between x and x'?

Figure 36.2 shows the explosion and the two reference frames. You can see from the figure that $x = x' + vt$, thus

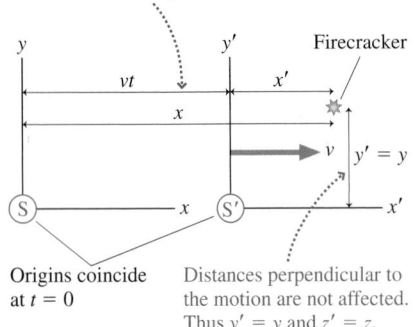

At time t, the origin of S' has moved distance vt to the right. Thus $x = x' + vt$.

Origins coincide at $t = 0$

Distances perpendicular to the motion are not affected. Thus $y' = y$ and $z' = z$.

FIGURE 36.2 The position of an exploding firecracker is measured in reference frames S and S'.

$$
\begin{array}{lll}
x = x' + vt & & x' = x - vt \\
y = y' & \text{or} & y' = y \\
z = z' & & z' = z
\end{array}
\tag{36.1}
$$

These equations, which you met in Chapter 6, are the *Galilean transformations of position*. If you know a position measured by the experimenters in one inertial reference frame, you can calculate the position that would be measured by experimenters in any other inertial reference frame.

Suppose the experimenters in both reference frames now track the motion of the object in Figure 36.3 by measuring its position at many instants of time. The experimenters in S find that the object's velocity is $\vec{u}$. During the *same time interval* Δt, the experimenters in S' measure the velocity to be $\vec{u}'$.

NOTE ▶ In this chapter, we will use v to represent the velocity of one reference frame relative to another. We will use $\vec{u}$ and $\vec{u}'$ to represent the velocities of objects with respect to reference frames S and S'. This notation differs from the notation of Chapter 6, where we used V to represent the relative velocity. ◀

We can find the relationship between $\vec{u}$ and $\vec{u}'$ by taking the time derivatives of Equation 36.1 and using the definition $u_x = dx/dt$:

$$u_x = \frac{dx}{dt} = \frac{dx'}{dt} + v = u_x' + v$$

$$u_y = \frac{dy}{dt} = \frac{dy'}{dt} = u_y'$$

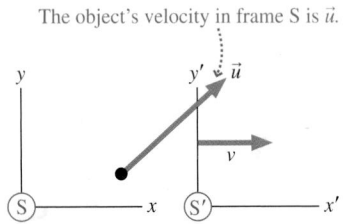

The object's velocity in frame S is $\vec{u}$.

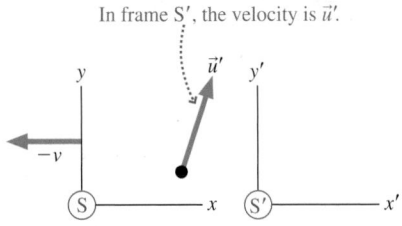

In frame S', the velocity is $\vec{u}'$.

FIGURE 36.3 The velocity of a moving object is measured in reference frames S and S'.

The equation for u_z is similar. The net result is

$$
\begin{array}{lll}
u_x = u_x' + v & & u_x' = u_x - v \\
u_y = u_y' & \text{or} & u_y' = u_y \\
u_z = u_z' & & u_z' = u_z
\end{array}
\tag{36.2}
$$

Equations 36.2 are the *Galilean transformations of velocity*. If you know the velocity of a particle as measured by the experimenters in one inertial reference frame, you can use Equations 36.2 to find the velocity that would be measured by experimenters in any other inertial reference frame.

EXAMPLE 36.1 The speed of sound

An airplane is flying at speed 200 m/s with respect to the ground. Sound wave 1 is approaching the plane from the front, sound wave 2 is catching up from behind. Both waves travel at 340 m/s relative to the ground. What is the speed of each wave relative to the plane?

MODEL Assume that the earth (frame S) and the airplane (frame S′) are inertial reference frames. Frame S′, in which the airplane is at rest, moves with velocity $v = 200$ m/s relative to frame S.

VISUALIZE Figure 36.4 shows the airplane and the sound waves.

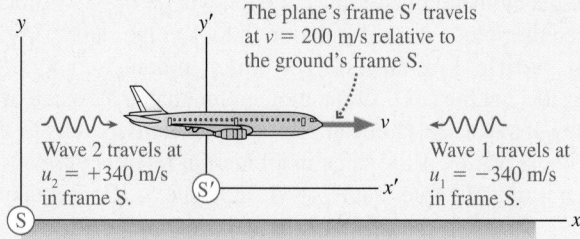

FIGURE 36.4 Experimenters in the plane measure different speeds for the sound waves than do experimenters on the ground.

SOLVE The speed of a mechanical wave, such as a sound wave or a wave on a string, is its speed *relative to its medium.* Thus the *speed of sound* is the speed of a sound wave through a reference frame in which the air is at rest. This is reference frame S, where wave 1 travels with velocity $u_1 = -340$ m/s and wave 2 travels with velocity $u_2 = +340$ m/s. Notice that the Galilean transformations use *velocities,* with appropriate signs, not just speeds.

The airplane travels to the right with reference frame S′ at velocity v. We can use the Galilean transformations of velocity to find the velocities of the two sound waves in frame S′:

$$u_1' = u_1 - v = -340 \text{ m/s} - 200 \text{ m/s} = -540 \text{ m/s}$$
$$u_2' = u_2 - v = 340 \text{ m/s} - 200 \text{ m/s} = 140 \text{ m/s}$$

ASSESS This isn't surprising. If you're driving 50 mph, a car coming the other way at 55 mph is approaching you at 105 mph. A car coming up behind you at 55 mph seems to be gaining on you at the rate of only 5 mph. Wave speeds behave the same. Notice that a mechanical wave would appear to be stationary to a person moving at the wave speed. To a surfer, the crest of the ocean wave remains at rest under his or her feet.

STOP TO THINK 36.2 Ocean waves are approaching the beach at 10 m/s. A boat heading out to sea travels at 6 m/s. How fast are the waves moving in the boat's reference frame?

a. 16 m/s   b. 10 m/s   c. 6 m/s   d. 4 m/s

The Galilean Principle of Relativity

Experimenters in reference frames S and S′ measure different values for position and velocity. What about the force on and the acceleration of the particle in Figure 36.5? The strength of a force can be measured with a spring scale. The experimenters in reference frames S and S′ both see the *same reading* on the scale (we'll assume the scale has a bright digital display easily seen by all experimenters), leading them to conclude that the force is the same in both frames. That is, $F' = F$.

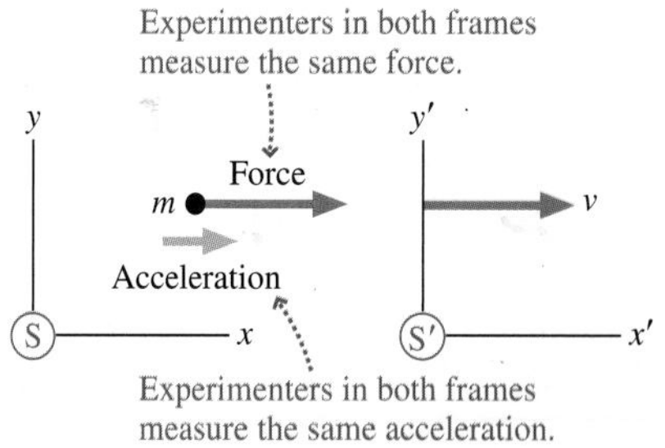

We can compare the accelerations measured in the two reference frames by taking the time derivative of the velocity transformation equation $u' = u - v$. (We'll assume, for simplicity, that the velocities and accelerations are all in the x-direction.) The relative velocity v between the two reference frames is *constant,* thus

$$a' = \frac{du'}{dt} = \frac{du}{dt} = a \tag{36.3}$$

Experimenters in reference frames S and S′ measure different values for an object's position and velocity, but they *agree* on its acceleration.

If $F = ma$ in reference frame S, then $F' = ma'$ in reference frame S′. Stated another way, if Newton's second law is valid in one inertial reference frame, then it is valid in all inertial reference frames. Because other laws of mechanics, such

Experimenters in both frames measure the same force.

FIGURE 36.5 Experimenters in both reference frames test Newton's second law by measuring the force on a particle and its acceleration.

as the conservation laws, follow from Newton's laws of motion, we can state this conclusion as the *Galilean principle of relativity:*

> **Galilean principle of relativity** The laws of mechanics are the same in all inertial reference frames.

(a) Collision seen in frame S

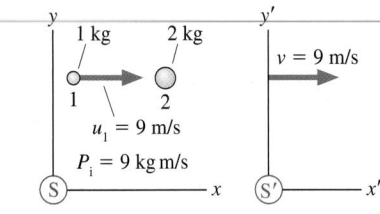

(b) Collision seen in frame S′

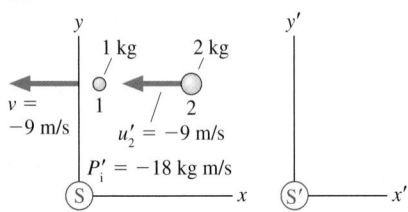

FIGURE 36.6 Total momentum measured in two reference frames.

The Galilean principle of relativity is easy to state, but to understand it we must understand what is and is not "the same." To take a specific example, consider the law of conservation of momentum. Figure 36.6a shows two particles about to collide. Their total momentum in frame S, where particle 2 is at rest, is $P_i = 9$ kg m/s. This is an isolated system, hence the law of conservation of momentum tells us that the momentum after the collision will be $P_f = 9$ kg m/s.

Figure 36.6b has used the velocity transformation to look at the same particles in frame S′ in which particle 1 is at rest. The initial momentum in S′ is $P_i' = -18$ kg m/s. Thus it is not the *value* of the momentum that is the same in all inertial reference frames. Instead, the Galilean principle of relativity tells us that the *law* of momentum conservation is the same in all inertial reference frames. If $P_f = P_i$ in frame S, then it must be true that $P_f' = P_i'$ in frame S′. Consequently, we can conclude that P_f' will be -18 kg m/s after the collision in S′.

Using Galilean Relativity

The principle of relativity is concerned with the laws of mechanics, not with the values that are needed to satisfy the laws. If momentum is conserved in one inertial reference frame, it is conserved in all inertial reference frames. Even so, a problem may be easier to solve in one reference frame than in others.

Elastic collisions provide a good example of using reference frames. You learned in Chapter 10 how to calculate the outcome of a perfectly elastic collision between two particles in the reference frame in which particle 2 is initially at rest. We can use that information together with the Galilean transformations to solve elastic-collision problems in any inertial reference frame.

> **TACTICS BOX 36.1 Analyzing elastic collisions**
>
> ❶ Use the Galilean transformations to transform the initial velocities of particles 1 and 2 from frame S to a reference frame S′ in which particle 2 is at rest.
> ❷ The outcome of the collision in S′ is given by
>
> $$u_{1f}' = \frac{m_1 - m_2}{m_1 + m_2}u_{1i}'$$
>
> $$u_{2f}' = \frac{2m_1}{m_1 + m_2}u_{1i}'$$
>
> ❸ Transform the two final velocities from frame S′ back to frame S.

EXAMPLE 36.2 An elastic collision

A 300 g ball moving to the right at 2 m/s has a perfectly elastic collision with a 100 g ball moving to the left at 4 m/s. What are the direction and speed of each ball after the collision?

MODEL The velocities are measured in the laboratory frame, which we call frame S.

VISUALIZE Figure 36.7a shows both the balls and reference frame S′ in which ball 2 is at rest.

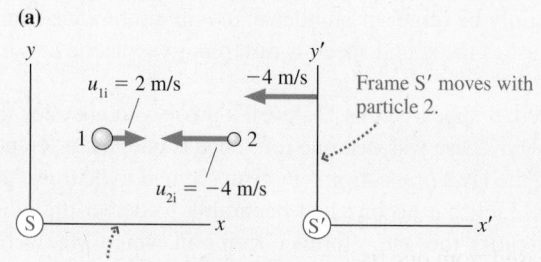

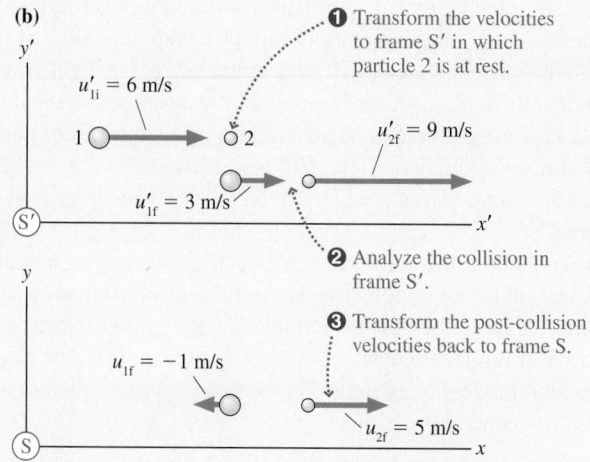

FIGURE 36.7 Using reference frames to solve an elastic-collision problem.

SOLVE The three steps of Tactics Box 36.1 are illustrated in Figure 36.7b. We're given u_{1i} and u_{2i}. The Galilean transformations of these velocities to frame S′, using $v = -4$ m/s, are

$$u'_{1i} = u_{1i} - v = (2 \text{ m/s}) - (-4 \text{ m/s}) = 6 \text{ m/s}$$
$$u'_{2i} = u_{2i} - v = (-4 \text{ m/s}) - (-4 \text{ m/s}) = 0 \text{ m/s}$$

The 100 g ball is at rest in frame S′, which is what we wanted. The velocities after the collision are

$$u'_{1f} = \frac{m_1 - m_2}{m_1 + m_2} u'_{1i} = 3 \text{ m/s}$$
$$u'_{2f} = \frac{2m_1}{m_1 + m_2} u'_{1i} = 9 \text{ m/s}$$

We've finished the collision analysis, but we're not done because these are the post-collision velocities in frame S′. Another application of the Galilean transformations tells us that the post-collision velocities in frame S are

$$u_{1f} = u'_{1f} + v = (3 \text{ m/s}) + (-4 \text{ m/s}) = -1 \text{ m/s}$$
$$u_{2f} = u'_{2f} + v = (9 \text{ m/s}) + (-4 \text{ m/s}) = 5 \text{ m/s}$$

Thus the 300 g ball rebounds to the left at a speed of 1 m/s and the 100 g ball is knocked to the right at a speed of 5 m/s.

ASSESS You can easily verify that momentum is conserved: $P_f = P_i = 0.20$ kg m/s. The calculations in this example were easy. The important point of this example, and one worth careful thought, is the *logic* of what we did and why we did it.

36.3 Einstein's Principle of Relativity

The 19th century was an era of optics and electromagnetism. Thomas Young demonstrated in 1801 that light is a wave, and by midcentury scientists had devised techniques for measuring the speed of light. Faraday discovered electromagnetic induction in 1831, setting in motion a train of events leading to Maxwell's conclusion, in 1864, that light is an electromagnetic wave.

If light is a wave, what is the medium in which it travels? This was perhaps *the* most important scientific question of the second half of the 19th century. The medium in which light waves were assumed to travel was called the **ether.** Experiments to measure the speed of light were assumed to be measuring its speed through the ether. But just what *is* the ether? What are its properties? Can we collect a jar full of ether to study? Despite the significance of these questions, experimental efforts to detect the ether or measure its properties kept coming up empty handed.

Maxwell's theory of electromagnetism didn't help the situation. The crowning success of Maxwell's theory was his prediction that light waves travel with speed

$$c = \frac{1}{\sqrt{\epsilon_0 \mu_0}} = 3.00 \times 10^8 \text{ m/s}$$

This is a very specific prediction with no wiggle room. The difficulty with such a specific prediction was the implication that Maxwell's laws of electromagnetism

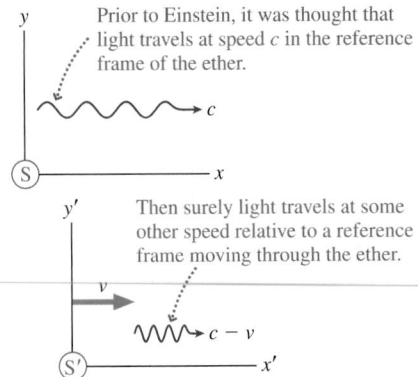

FIGURE 36.8 It seems as if the speed of light should differ from c in a reference frame moving through the ether.

are valid *only* in the reference frame of the ether. After all, as Figure 36.8 shows, the light speed should certainly be larger or smaller than c in a reference frame moving through the ether, just as the sound speed is different to someone moving through the air.

As the 19th century closed, it appeared that Maxwell's theory did not obey the classical principle of relativity. There was just one reference frame, the reference frame of the ether, in which the laws of electromagnetism seemed to be true. And to make matters worse, the fact that no one had been able to detect the ether meant that no one could identify the one reference frame in which Maxwell's equations "worked."

It was in this muddled state of affairs that a young Albert Einstein made his mark on the world. Even as a teenager, Einstein had wondered how a light wave would look to someone "surfing" the wave, traveling alongside the wave at the wave speed. You can do that with a water wave or a sound wave, but light waves seemed to present a logical difficulty. An electromagnetic wave sustains itself by virtue of the fact that a changing magnetic field induces an electric field and a changing electric field induces a magnetic field. But to someone moving with the wave, *the fields would not change.* How could there be an electromagnetic wave under these circumstances?

Several years of thinking about the connection between electromagnetism and reference frames led Einstein to the conclusion that *all* the laws of physics, not just the laws of mechanics, should obey the principle of relativity. In other words, the principle of relativity is a fundamental statement about the nature of the physical universe. Thus we can remove the restriction in the Galilean principle of relativity and state a much more general principle:

> **Principle of relativity** All the laws of physics are the same in all inertial reference frames.

All of the results of Einstein's theory of relativity flow from this one simple statement.

The Constancy of the Speed of Light

If Maxwell's equations of electromagnetism are laws of physics, and there's every reason to think they are, then, according to the principle of relativity, Maxwell's equations must be true in *every* inertial reference frame. On the surface this seems to be an innocuous statement, equivalent to saying that the law of conservation of momentum is true in every inertial reference frame. But follow the logic:

1. Maxwell's equations are true in all inertial reference frames.
2. Maxwell's equations predict that electromagnetic waves, including light, travel at speed $c = 3.00 \times 10^8$ m/s.
3. Therefore, **light travels at speed c in all inertial reference frames.**

Figure 36.9 shows the implications of this conclusion. *All* experimenters, regardless of how they move with respect to each other, find that *all* light waves, regardless of the source, travel in their reference frame with the *same* speed c. If Cathy's velocity toward Bill and away from Amy is $v = 0.9c$, Cathy finds, by making measurements in her reference frame, that the light from Bill approaches her at speed c, not at $c + v = 1.9c$. And the light from Amy, which left Amy at speed c, catches up from behind at speed c *relative to Cathy,* not the $c - v = 0.1c$ you would have expected.

Although this prediction goes against all shreds of common sense, the experimental evidence for it is strong. Laboratory experiments are difficult because

This light wave leaves Amy at speed c relative to Amy. It approaches Cathy at speed c relative to Cathy.

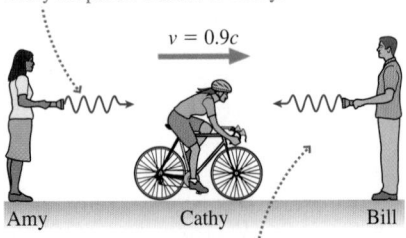

This light wave leaves Bill at speed c relative to Bill. It approaches Cathy at speed c relative to Cathy.

FIGURE 36.9 Light travels at speed c in all inertial reference frames, regardless of how the reference frames are moving with respect to the light source.

even the highest laboratory speed is insignificant in comparison to c. In the 1930s, however, the physicists R. J. Kennedy and E. M. Thorndike realized that they could use the earth itself as a laboratory. The earth's speed as it circles the sun is about 30,000 m/s. The *relative* velocity of the earth in January differs by 60,000 m/s from its velocity in July, when the earth is moving in the opposite direction. Kennedy and Thorndike were able to use a very sensitive and stable interferometer to show that the numerical values of the speed of light in January and July differ by less than 2 m/s.

More recent experiments have used unstable elementary particles, called π mesons, that decay into high-energy photons of light. The π mesons, created in a particle accelerator, move through the laboratory at 99.975% the speed of light, or $v = 0.99975c$ as they emit photons at the speed c in the π meson's reference frame. As Figure 36.10 shows, you would expect the photons to travel through the laboratory with speed $c + v = 1.99975c$. Instead, the measured speed of the photons in the laboratory was, within experimental error, 3.00×10^8 m/s.

In summary, *every* experiment designed to compare the speed of light in different reference frames has found that light travels at 3.00×10^8 m/s in every inertial reference frame, regardless of how the reference frames are moving with respect to each other.

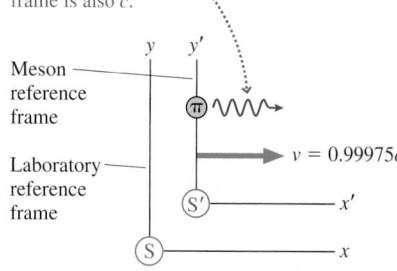

A photon is emitted at speed c relative to the π meson. Measurements find that the photon's speed in the laboratory reference frame is also c.

FIGURE 36.10 Experiments find that the photons travel through the laboratory with speed c, not the speed $1.99975c$ that you might expect.

How Can This Be?

You're in good company if you find this impossible to believe. Suppose I shot a ball forward at 50 m/s while driving past you at 30 m/s. You would certainly see the ball traveling at 80 m/s relative to you and the ground. What we're saying with regard to light is equivalent to saying that the ball travels at 50 m/s relative to my car and *at the same time* travels at 50 m/s relative to the ground, even though the car is moving across the ground at 30 m/s. It seems logically impossible.

You might think that this is merely a matter of semantics. If we can just get our definitions and use of words straight, then the mystery and confusion will disappear. Or perhaps the difficulty is a confusion between what we "see" versus what "really happens." In other words, a better analysis, one that focuses on what really happens, would find that light "really" travels at different speeds in different reference frames.

Alas, what "really happens" is that light travels at 3.00×10^8 m/s in every inertial reference frame, regardless of how the reference frames are moving with respect to each other. It's not a trick. There remains only one way to escape the logical contradictions.

The definition of velocity is $u = \Delta x / \Delta t$, the ratio of a distance traveled to the time interval in which the travel occurs. Suppose you and I both make measurements on an object as it moves, but you happen to be moving relative to me. Perhaps I'm standing on the corner, you're driving past in your car, and we're both trying to measure the velocity of a bicycle. Further, suppose we have agreed in advance to measure the bicycle as it moves from the tree to the lamppost in Figure 36.11 on the next page. Your $\Delta x'$ differs from my Δx because of your motion relative to me, causing you to calculate a bicycle velocity u' in your reference frame that differs from its velocity u in my reference frame. This is just the Galilean transformations showing up again.

Now let's repeat the measurements, but this time let's measure the velocity of a light wave as it travels from the tree to the lamppost. Once again, your $\Delta x'$ differs from my Δx, although the difference will be pretty small unless your car is moving at well above the legal speed limit. The obvious conclusion is that your light speed u' differs from my light speed u. But it doesn't. The experiments show that, for a light wave, we'll get the *same* values: $u' = u$.

The only way this can be true is if your Δt is not the same as my Δt. If the time it takes the light to move from the tree to the lamppost in your reference frame, a

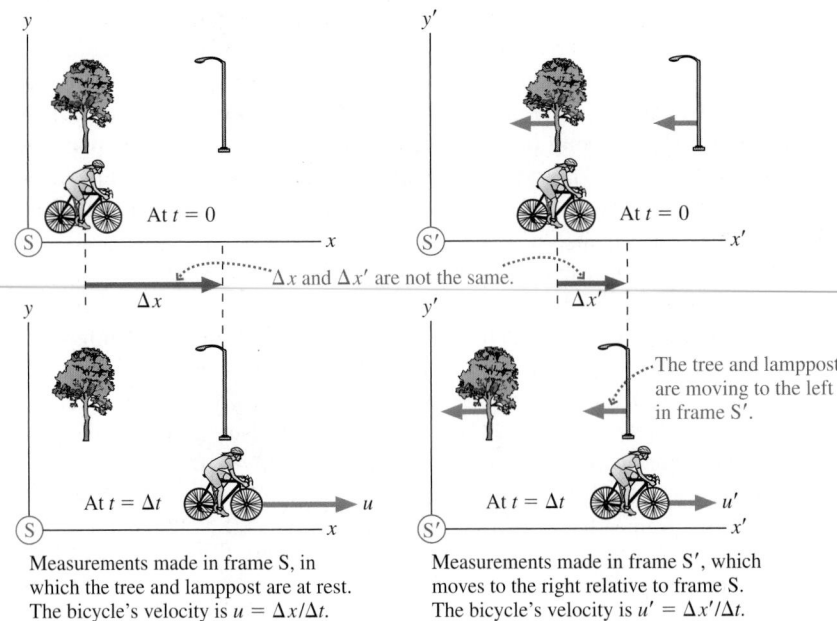

Measurements made in frame S, in which the tree and lamppost are at rest. The bicycle's velocity is $u = \Delta x / \Delta t$.

Measurements made in frame S', which moves to the right relative to frame S. The bicycle's velocity is $u' = \Delta x' / \Delta t$.

FIGURE 36.11 Measuring the velocity of an object by appealing to the basic definition $u = \Delta x / \Delta t$.

time we'll now call $\Delta t'$, differs from the time Δt it takes the light to move from the tree to the lamppost in my reference frame, then we might find that $\Delta x' / \Delta t' = \Delta x / \Delta t$. That is, $u' = u$ even though you are moving with respect to me.

We've assumed, since the beginning of this textbook, that time is simply time. It flows along like a river, and all experimenters in all reference frames simply use it. For example, suppose the tree and the lamppost both have big clocks that we both can see. Shouldn't we be able to agree on the time interval Δt the light needs to move from the tree to the lamppost?

Perhaps not. It's demonstrably true that $\Delta x' \neq \Delta x$. It's experimentally verified that $u' = u$ for light waves. Something must be wrong with *assumptions* that we've made about the nature of time. The principle of relativity has painted us into a corner, and our only way out is to reexamine our understanding of time.

36.4 Events and Measurements

To question some of our most basic assumptions about space and time requires extreme care. We need to be certain that no assumptions slip into our analysis unnoticed. Our goal is to describe the motion of a particle in a clear and precise way, making the barest minimum of assumptions.

Events

The fundamental entity of relativity is called an **event.** An event is a physical activity that takes place at a definite point in space and at a definite instant of time. A firecracker exploding is an event. A collision between two particles is an event. A light wave hitting a detector is an event.

Events can be observed and measured by experimenters in different reference frames. An exploding firecracker is as clear to you as you drive by in your car as it is to me standing on the street corner. We can quantify where and when an event occurs with four numbers: the coordinates (x, y, z) and the instant of time t. These four numbers, illustrated in Figure 36.12, are called the **spacetime coordinates** of the event.

An event has spacetime coordinates (x, y, z, t) in frame S and different spacetime coordinates (x', y', z', t') in frame S'.

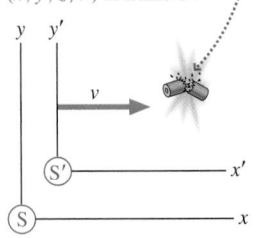

FIGURE 36.12 The location and time of an event are described by its spacetime coordinates.

The spatial coordinates of an event measured in reference frames S and S′ may differ. It now appears that the instant of time recorded in S and S′ may also differ. Thus the spacetime coordinates of an event measured by experimenters in frame S are (x, y, z, t) and the spacetime coordinates of the *same event* measured by experimenters in frame S′ are (x', y', z', t').

The motion of a particle can be described as a sequence of two or more events. We introduced this idea in the previous section when we agreed to measure the velocity of a bicycle and then of a light wave by comparing the object passing the tree (first event) to the object passing the lamppost (second event).

Measurements

Events are what "really happen," but how do we learn about an event? That is, how do the experimenters in a reference frame determine the spacetime coordinates of an event? This is a problem of *measurement.*

We defined a reference frame to be a coordinate system in which experimenters can make position and time measurements. That's a good start, but now we need to be more precise as to *how* the measurements are made. Imagine that a reference frame is filled with a cubic lattice of meter sticks, as shown in Figure 36.13. At every intersection is a clock, and all the clocks in a reference frame are *synchronized*. We'll return in a moment to consider how to synchronize the clocks, but assume for the moment it can be done.

Now, with our meter sticks and clocks in place, we can use a two-part measurement scheme:

■ The (x, y, z) coordinates of an event are determined by the intersection of meter sticks closest to the event.
■ The event's time t is the time displayed on the clock nearest the event.

You can imagine, if you wish, that each event is accompanied by a flash of light to illuminate the face of the nearest clock and make its reading known.

Several important issues need to be noted:

1. The clocks and meter sticks in each reference frame are imaginary, so they have no difficulty passing through each other.
2. Measurements of position and time made in one reference frame must use only the clocks and meter sticks in that reference frame.
3. There's nothing special about the sticks being 1 m long and the clocks 1 m apart. The lattice spacing can be altered to achieve whatever level of measurement accuracy is desired.
4. We'll assume that the experimenters in each reference frame have assistants sitting beside every clock to record the position and time of nearby events.
5. Perhaps most important, t is the time at which the event *actually happens,* not the time at which an experimenter sees the event or at which information about the event reaches an experimenter.
6. All experimenters in one reference frame agree on the spacetime coordinates of an event. In other words, **an event has a unique set of spacetime coordinates in each reference frame.**

The spacetime coordinates of this event are measured by the nearest meter stick intersection and the nearest clock.

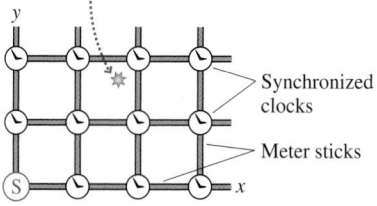

Synchronized clocks

Meter sticks

Reference frame S

Reference frame S′ has its own meter sticks and its own clocks.

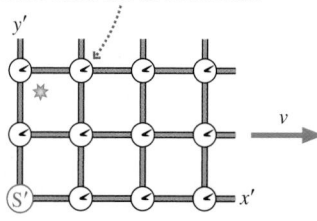

Reference frame S′

FIGURE 36.13 The spacetime coordinates of an event are measured by a lattice of meter sticks and clocks.

STOP TO THINK 36.3 A carpenter is working on a house two blocks away. You notice a slight delay between seeing the carpenter's hammer hit the nail and hearing the blow. At what time does the event "hammer hits nail" occur?

a. At the instant you hear the blow.
b. At the instant you see the hammer hit.
c. Very slightly before you see the hammer hit.
d. Very slightly after you see the hammer hit.

Clock Synchronization

It's important that all the clocks in a reference frame be **synchronized,** meaning that all clocks in the reference frame have the same reading at any one instant of time. We would not be able to use a sequence of events to track the motion of a particle if the clocks differed in their readings. Thus we need a method of synchronization. One idea that comes to mind is to designate the clock at the origin as the *master clock.* We could then carry this clock around to every clock in the lattice, adjust that clock to match the master clock, and finally return the master clock to the origin.

This would be a perfectly good method of clock synchronization in Newtonian mechanics, where time flows along smoothly, the same for everyone. But we've been driven to reexamine the nature of time by the possibility that time is different in reference frames moving relative to each other. Because the master clock would *move,* we cannot assume that the master clock keeps time in the same way as the stationary clocks.

We need a synchronization method that does not require moving the clocks. Fortunately, such a method is easy to devise. Each clock is resting at the intersection of meter sticks, so by looking at the meter sticks, the assistant knows, or can calculate, exactly how far each clock is from the origin. Once the distance is known, the assistant can calculate exactly how long a light wave will take to travel from the origin to each clock. For example, light will take 1.00 μs to travel to a clock 300 m from the origin.

> NOTE ▶ It's handy for many relativity problems to know that the speed of light is $c = 300$ m/μs. ◄

To synchronize the clocks, the assistants begin by setting each clock to display the light travel time from the origin, but they don't start the clocks. Next, as Figure 36.14 shows, a light flashes at the origin and, simultaneously, the clock at the origin starts running from $t = 0$ s. The light wave spreads out in all directions at speed c. A photodetector on each clock recognizes the arrival of the light wave and, without delay, starts the clock. The clock had been preset with the light travel time, so each clock as it starts reads exactly the same as the clock at the origin. Thus all the clocks will be synchronized after the light wave has passed by.

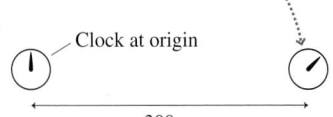

1. This clock is preset to 1.00 μs, the time it takes light to travel 300 m.

Clock at origin

300 m

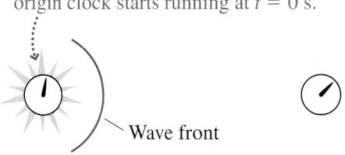

2. A light flashes at the origin and the origin clock starts running at $t = 0$ s.

Wave front

3. The clock starts when the light wave reaches it. It is now synchronized with the origin clock.

FIGURE 36.14 Synchronizing the clocks.

Events and Observations

We noted above that t is the time the event *actually happens.* This is an important point, one that bears further discussion. Light waves take time to travel. Messages, whether they're transmitted by light pulses, telephone, or courier on horseback, take time to be delivered. An experimenter *observes* an event, such as an exploding firecracker, only *at a later time* when light waves reach his or her eyes. But our interest is in the event itself, not the experimenter's observation of the event. The time at which the experimenter sees the event or receives information about the event is not when the event actually occurred.

Suppose at $t = 0$ s a firecracker explodes at $x = 300$ m. The flash of light from the firecracker will reach an experimenter at the origin at $t_1 = 1.0$ μs. The sound of the explosion will reach a sightless experimenter at $t_2 = 0.88$ s. Neither of these is the time t_{event} of the explosion, although the experimenter can work backward from these times, using known wave speeds, to determine t_{event}. In this example, the spacetime coordinates of the event—the explosion—are (300 m, 0 m, 0 m, 0 s).

EXAMPLE 36.3 Finding the time of an event

Experimenter A in reference frame S stands at the origin looking in the positive x-direction. Experimenter B stands at $x = 900$ m looking in the negative x-direction. A firecracker explodes somewhere between them. Experimenter B sees the light flash at $t = 3.00$ μs. Experimenter A sees the light flash at $t = 4.00$ μs. What are the spacetime coordinates of the explosion?

MODEL Experimenters A and B are in the same reference frame and have synchronized clocks.

VISUALIZE Figure 36.15 shows the two experimenters and the explosion at unknown position x.

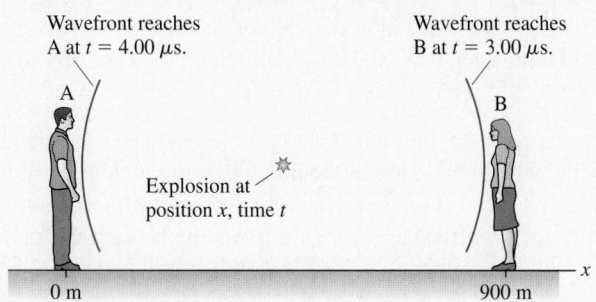

Wavefront reaches A at $t = 4.00$ μs.

Wavefront reaches B at $t = 3.00$ μs.

A

B

Explosion at position x, time t

0 m   900 m x

FIGURE 36.15 The light wave reaches the experimenters at different times. Neither of these is the time at which the event actually happened.

SOLVE The two experimenters observe light flashes at two different instants, but there's only one event. Light travels 300 m/μs, so the additional 1.00 μs needed for the light to reach experimenter A implies that distance $(x - 0$ m$)$ is 300 m longer than distance $(900$ m $- x)$. That is,

$$(x - 0 \text{ m}) = (900 \text{ m} - x) + 300 \text{ m}$$

This is easily solved to give $x = 600$ m as the position coordinate of the explosion. The light takes 1.00 μs to travel 300 m to experimenter B, 2.00 μs to travel 600 m to experimenter A. The light is received at 3.00 μs and 4.00 μs, respectively, hence it was emitted by the explosion at $t = 2.00$ μs. The spacetime coordinates of the explosion are (600 m, 0 m, 0 m, 2.00 μs).

ASSESS Although the experimenters *see* the explosion at different times, they agree that the explosion actually *happened* at $t = 2.00$ μs.

Simultaneity

Two events 1 and 2 that take place at different positions x_1 and x_2 but at the *same time* $t_1 = t_2$, as measured in some reference frame, are said to be **simultaneous** in that reference frame. Simultaneity is determined by when the events actually happen, not when they are seen or observed. In general, simultaneous events are *not* seen at the same time because of the difference in light travel times from the events to an experimenter.

EXAMPLE 36.4 Are the explosions simultaneous?

An experimenter in reference frame S stands at the origin looking in the positive x-direction. At $t = 3.0$ μs she sees firecracker 1 explode at $x = 600$ m. A short time later, at $t = 5.0$ μs, she sees firecracker 2 explode at $x = 1200$ m. Are the two explosions simultaneous? If not, which firecracker exploded first?

MODEL Light from both explosions travels toward the experimenter at 300 m/μs.

SOLVE The experimenter *sees* two different explosions, but perceptions of the events are not the events themselves. When did the explosions *actually* occur? Using the fact that light travels 300 m/μs, it's easy to see that firecracker 1 exploded at $t_1 = 1.0$ μs and firecracker 2 also exploded at $t_2 = 1.0$ μs. The events *are* simultaneous.

STOP TO THINK 36.4   A tree and a pole are 3000 m apart. Each is suddenly hit by a bolt of lightning. Mark, who is standing at rest midway between the two, sees the two lightning bolts at the same instant of time. Nancy is at rest under the tree. Define event 1 to be "lightning strikes tree" and event 2 to be "lightning strikes pole." For Nancy, does event 1 occur before, after, or at the same time as event 2?

36.5 The Relativity of Simultaneity

We've now established a means for measuring the time of an event in a reference frame, so let's begin to investigate the nature of time. The following "thought experiment" is very similar to one suggested by Einstein.

Figure 36.16 shows a long railroad car traveling to the right with a velocity v that may be an appreciable fraction of the speed of light. A firecracker is tied to each end of the car, right above the ground. Each firecracker is powerful enough that, when it explodes, it will make a burn mark on the ground at the position of the explosion.

Ryan is standing on the ground, watching the railroad car go by. Peggy is standing in the exact center of the car with a special box at her feet. This box has two light detectors, one facing each way, and a signal light on top. The box works as follows:

1. If a flash of light is received at the right detector before a flash is received at the left detector, then the light on top of the box will turn green.
2. If a flash of light is received at the left detector before a flash is received at the right detector, or if two flashes arrive simultaneously, the light on top will turn red.

The firecrackers explode as the railroad car passes Ryan, and he sees the two light flashes from the explosions simultaneously. He then measures the distances to the two burn marks and finds that he was standing exactly halfway between the marks. Because light travels equal distances in equal times, Ryan concludes that the two explosions were simultaneous in his reference frame, the reference frame of the ground. Further, because he was midway between the two ends of the car, he was directly opposite Peggy when the explosions occurred.

Figure 36.17a shows the sequence of events in Ryan's reference frame. Light travels at speed c in all inertial reference frames, so, although the firecrackers were moving, the light waves are spheres centered on the burn marks. Ryan determines that the light wave coming from the right reaches Peggy and the box before the light wave coming from the left. Thus, according to Ryan, the signal light on top of the box turns green.

How do things look in Peggy's reference frame, a reference frame moving to the right at velocity v relative to the ground? As Figure 36.17b shows, Peggy sees Ryan moving to the left with speed v. Light travels at speed c in all inertial reference frames, so the light waves are spheres centered on the ends of the car. If the explosions are simultaneous, as Ryan has determined, the two light waves reach her and the box simultaneously. Thus, according to Peggy, the signal light on top of the box turns red!

Now the light on top must be either green or red. *It can't be both!* Later, after the railroad car has stopped, Ryan and Peggy can place the box in front of them. Either it has a red light or a green light. Ryan can't see one color while Peggy sees the other. Hence we have a paradox. It's impossible for Peggy and Ryan both to be right. But who is wrong, and why?

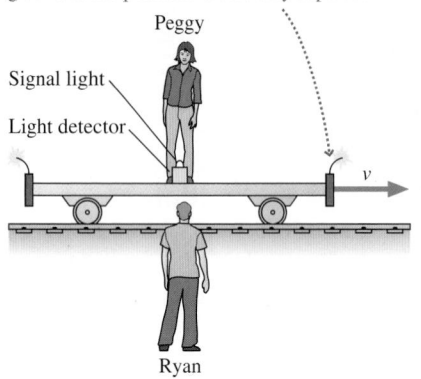

The firecrackers will make burn marks on the ground at the positions where they explode.

Peggy

Signal light

Light detector

v

Ryan

FIGURE 36.16 A railroad car traveling to the right with velocity v.

(a) The events in Ryan's frame

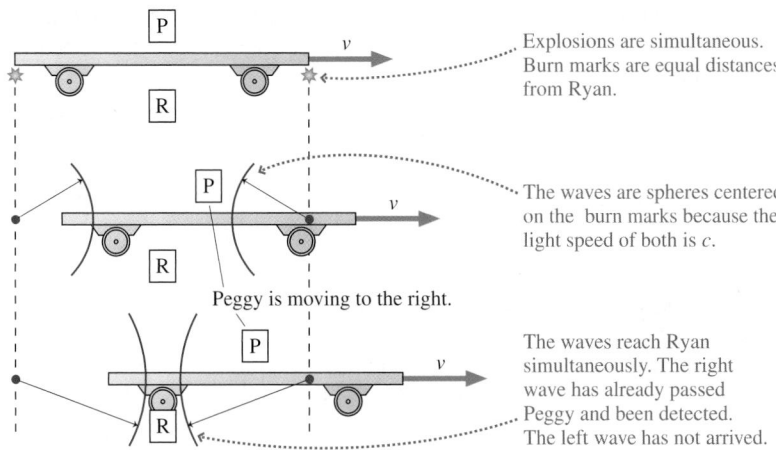

Explosions are simultaneous. Burn marks are equal distances from Ryan.

The waves are spheres centered on the burn marks because the light speed of both is c.

Peggy is moving to the right.

The waves reach Ryan simultaneously. The right wave has already passed Peggy and been detected. The left wave has not arrived.

(b) The events in Peggy's frame

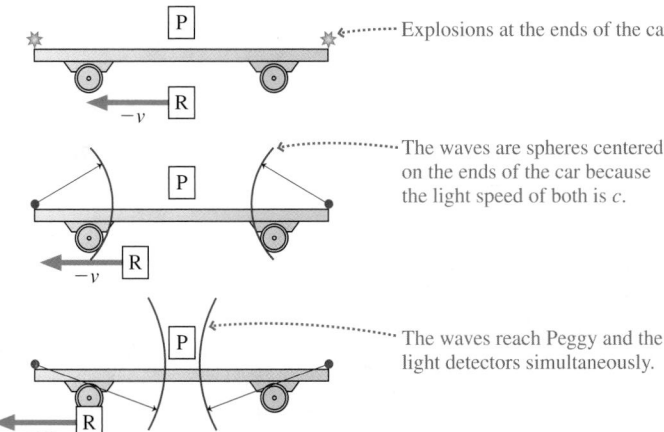

Explosions at the ends of the car

The waves are spheres centered on the ends of the car because the light speed of both is c.

The waves reach Peggy and the light detectors simultaneously.

FIGURE 36.17 Exploding firecrackers seen in two different reference frames.

What do we know with absolute certainty?

1. Ryan detected the flashes simultaneously.
2. Ryan was halfway between the firecrackers when they exploded.
3. The light from the two explosions traveled toward Ryan at equal speeds.

The conclusion that the explosions were simultaneous in Ryan's reference frame is unassailable. The light is green.

Peggy, however, made an assumption. It's a perfectly ordinary assumption, one that seems sufficiently obvious that you probably didn't notice, but an assumption nonetheless. Peggy assumed that the explosions were simultaneous.

Didn't Ryan find them to be simultaneous? Indeed, he did. Suppose we call Ryan's reference frame S, the explosion on the right event R, and the explosion on the left event L. Ryan found that $t_R = t_L$. But Peggy has to use a different set of clocks, the clocks in her reference frame S′, to measure the times t_R' and t_L' at which the explosions occurred. The fact that $t_R = t_L$ in frame S does *not* allow us to conclude that $t_R' = t_L'$ in frame S′.

In fact, the right firecracker must explode *before* the left firecracker in frame S′. Figure 36.17b, with its assumption about simultaneity, was incorrect. Figure 36.18 shows the situation in Peggy's reference frame with the right firecracker exploding first. Now the wave from the right reaches Peggy and the box first, as Ryan had concluded, and the light on top turns green.

The right firecracker explodes first.

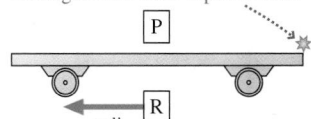

The left firecracker explodes later.   The right wave reaches Peggy first.

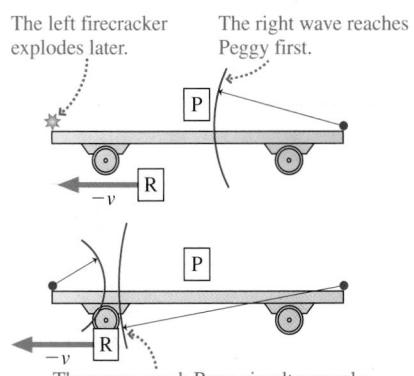

The waves reach Ryan simultaneously. The left wave has not reached Peggy.

FIGURE 36.18 The real sequence of events in Peggy's reference frame.

One of the most disconcerting conclusions of relativity is that **two events occurring simultaneously in reference frame S are *not* simultaneous in any reference frame S′ that is moving relative to S.** This is called the **relativity of simultaneity.**

The two firecrackers *really* explode at the same instant of time in Ryan's reference frame. And the right firecracker *really* explodes first in Peggy's reference frame. It's not a matter of when they see the flashes. Our conclusion refers to the times at which the explosions actually occur.

The paradox of Peggy and Ryan contains the essence of relativity, and it's worth careful thought. First, review the logic until you're certain that there *is* a paradox, a logical impossibility. Then convince yourself that the only way to resolve the paradox is to abandon the assumption that the explosions are simultaneous in Peggy's reference frame. If you understand the paradox and its resolution, you've made a big step toward understanding what relativity is all about.

STOP TO THINK 36.5 A tree and a pole are 3000 m apart. Each is suddenly hit by a bolt of lightning. Mark, who is standing at rest midway between the two, sees the two lightning bolts at the same instant of time. Nancy is flying her rocket at $v = 0.5c$ in the direction from the tree toward the pole. The lightning hits the tree just as she passes by it. Define event 1 to be "lightning strikes tree" and event 2 to be "lightning strikes pole." For Nancy, does event 1 occur before, after, or at the same time as event 2?

36.6 Time Dilation

The principle of relativity has driven us to the logical conclusion that time is not the same for two reference frames moving relative to each other. Our analysis thus far has been mostly qualitative. It's time to start developing some quantitative tools that will allow us to compare measurements in one reference frame to measurements in another reference frame.

Figure 36.19a shows a special clock called a **light clock.** The light clock is a box of height h with a light source at the bottom and a mirror at the top. The light source emits a very short pulse of light that travels to the mirror and reflects back to a light detector beside the source. The clock advances one "tick" each time the detector receives a light pulse, and it immediately, with no delay, causes the light source to emit the next light pulse.

Our goal is to compare two measurements of the interval between two ticks of the clock: one taken by an experimenter standing next to the clock and the other by an experimenter moving with respect to the clock. To be specific, Figure 36.19b shows the clock at rest in reference frame S′. We call this the **rest frame** of the clock. Reference frame S′ moves to the right with velocity v relative to reference frame S.

Relativity requires us to measure *events,* so let's define event 1 to be the emission of a light pulse and event 2 to be the detection of that light pulse. Experimenters in both reference frames are able to measure where and when these events occur *in their frame.* In frame S, the time interval $\Delta t = t_2 - t_1$ is one tick of the clock. Similarly, one tick in frame S′ is $\Delta t' = t_2' - t_1'$.

To be sure we have a clear understanding of the relativity result, let's first do a classical analysis. In frame S′, the clock's rest frame, the light travels straight up and down, a total distance $2h$, at speed c. The time interval is $\Delta t' = 2h/c$.

Figure 36.20a shows the operation of the light clock as seen in frame S. The clock is moving to the right at speed v in S, thus the mirror moves distance $\frac{1}{2}v(\Delta t)$

(a) A light clock

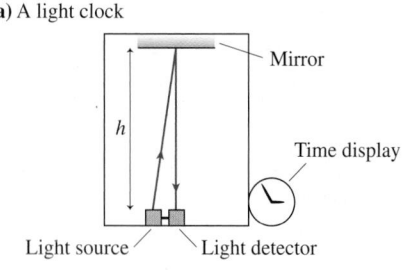

(b) The clock is at rest in frame S′.

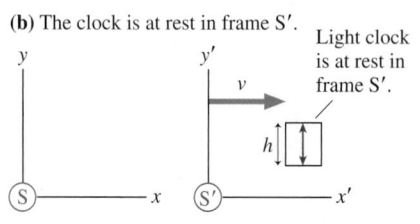

FIGURE 36.19 The ticking of a light clock can be measured by experimenters in two different reference frames.

during the time $\frac{1}{2}(\Delta t)$ in which the light pulse moves from the source to the mirror. The distance traveled by the light during this interval is $\frac{1}{2}u_{light}(\Delta t)$, where u_{light} is the speed of light in frame S. You can see from the vector addition in Figure 36.20b that the speed of light in frame S′ is $u_{light} = (c^2 + v^2)^{1/2}$. (Remember, this is a classical analysis in which the speed of light *does* depend on the motion of the reference frame relative to the light source.)

The Pythagorean theorem applied to the right triangle in Figure 36.20a is

$$h^2 + \left(\frac{1}{2}v\Delta t\right)^2 = \left(\frac{1}{2}u_{light}\Delta t\right)^2 = \left(\frac{1}{2}\sqrt{c^2 + v^2}\,\Delta t\right)^2$$
$$= \left(\frac{1}{2}c\,\Delta t\right)^2 + \left(\frac{1}{2}v\,\Delta t\right)^2 \qquad (36.4)$$

The term $(\frac{1}{2}v\Delta t)^2$ is common to both sides and cancels. Solving for Δt gives $\Delta t = 2h/c$, identical to $\Delta t'$. In other words, a classical analysis finds that the clock ticks at exactly the same rate in both frame S and frame S′. This shouldn't be surprising. There's only one kind of time in classical physics, measured the same by all experimenters independent of their motion.

The principle of relativity changes only one thing, but that change has profound consequences. According to the principle of relativity, light travels at the same speed in *all* inertial reference frames. In frame S′, the rest frame of the clock, the light simply goes straight up and back. The time of one tick,

$$\Delta t' = \frac{2h}{c} \qquad (36.5)$$

is unchanged from the classical analysis.

Figure 36.21 shows the light clock as seen in frame S. The difference from Figure 36.20a is that the light now travels along the hypotenuse at speed c. We can again use the Pythagorean theorem to write

$$h^2 + \left(\frac{1}{2}v\Delta t\right)^2 = \left(\frac{1}{2}c\,\Delta t\right)^2 \qquad (36.6)$$

Solving for Δt gives

$$\Delta t = \frac{2h/c}{\sqrt{1 - v^2/c^2}} = \frac{\Delta t'}{\sqrt{1 - v^2/c^2}} \qquad (36.7)$$

The time interval between two ticks in frame S is *not* the same as in frame S′.

It's useful to define $\beta = v/c$, the velocity as a fraction of the speed of light. For example, a reference frame moving with $v = 2.4 \times 10^8$ m/s has $\beta = 0.80$. In terms of β, Equation 36.7 is

$$\Delta t = \frac{\Delta t'}{\sqrt{1 - \beta^2}} \qquad (36.8)$$

NOTE ▶ The expression $(1 - v^2/c^2)^{1/2} = (1 - \beta^2)^{1/2}$ occurs frequently in relativity. The value of the expression is 1 when $v = 0$, and it steadily decreases to 0 as $v \to c$ (or $\beta \to 1$). The square root is an imaginary number if $v > c$, which would make Δt imaginary in Equation 36.8. Time intervals certainly have to be real numbers, suggesting that $v > c$ is not physically possible. One of the predictions of the theory of relativity, as you've undoubtedly heard, is that nothing can travel faster than the speed of light. Now you can begin to see why. We'll examine this topic more closely in Section 36.9. In the meantime, we'll require v to be less than c. ◀

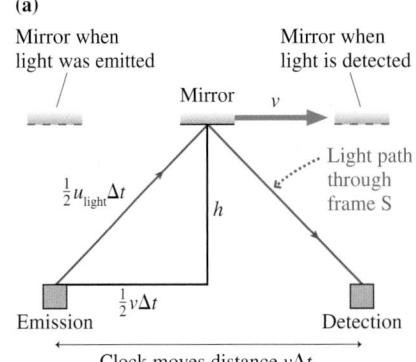

(a)

Mirror when light was emitted

Mirror when light is detected

Mirror v

$\frac{1}{2}u_{light}\Delta t$ h

Light path through frame S

Emission $\frac{1}{2}v\Delta t$ Detection

Clock moves distance $v\Delta t$.

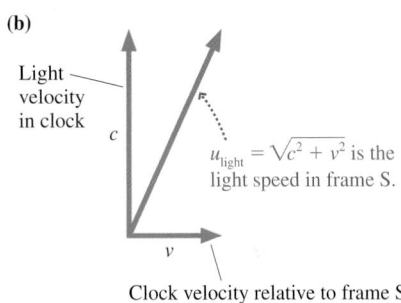

(b)

Light velocity in clock c

$u_{light} = \sqrt{c^2 + v^2}$ is the light speed in frame S.

v

Clock velocity relative to frame S

FIGURE 36.20 A classical analysis of the light clock.

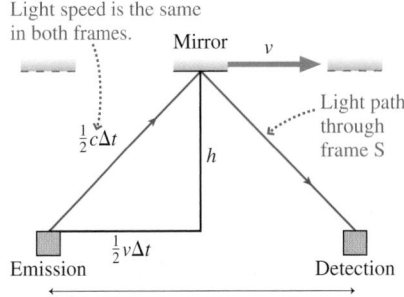

Light speed is the same in both frames.   Mirror v

$\frac{1}{2}c\Delta t$ h

Light path through frame S

Emission $\frac{1}{2}v\Delta t$ Detection

Clock moves distance $v\Delta t$.

FIGURE 36.21 A light clock analysis in which the speed of light is the same in all reference frames.

Proper Time

Frame S' has one important distinction. It is the *one and only* inertial reference frame in which the clock is at rest. Consequently, it is the one and only inertial reference frame in which the times of both events—the emission of the light and the detection of the light—are measured by the *same* clock. You can see that the light pulse in Figure 36.19, the rest frame of the clock, starts and ends at the same position and can be measured by one clock. In Figure 36.21, the emission and detection take place at different positions in frame S and must be measured by different clocks.

The time interval between two events that occur at the *same position* is called the **proper time** $\Delta\tau$. Only one inertial reference frame measures the proper time, and it does so with a single clock that is present at both events. An inertial reference frame moving with velocity $v = \beta c$ relative to the proper time frame must use two clocks to measure the time interval because the two events occur at different positions. The time interval in this frame is

$$\Delta t = \frac{\Delta\tau}{\sqrt{1 - \beta^2}} \geq \Delta\tau \qquad \text{(time dilation)} \qquad (36.9)$$

The "stretching out" of the time interval implied by Equation 36.9 is called **time dilation.** Time dilation is sometimes described by saying that "moving clocks run slow." This is not an accurate statement because it implies that some reference frames are "really" moving while others are "really" at rest. The whole point of relativity is that all inertial reference frames are equally valid, that all we know about reference frames is how they move relative to each other. A better description of time dilation is the statement that **the time interval between two ticks is the shortest in the reference frame in which the clock is at rest.** The time interval between two ticks is longer (i.e., the clock "runs slower") when it is measured in any reference frame in which the clock is moving.

NOTE ▶ Equation 36.9 was derived using a light clock because the operation of a light clock is clear and easy to analyze. But the conclusion is really about time itself. *Any* clock, regardless of how it operates, behaves the same. ◀

EXAMPLE 36.5 **From the sun to Saturn**

Saturn is 1.43×10^{12} m from the sun. A rocket travels along a line from the sun to Saturn at a constant speed of $0.9c$ relative to the solar system. How long does the journey take as measured by an experimenter on earth? As measured by an astronaut on the rocket?

MODEL Let the solar system be in reference frame S and the rocket be in reference frame S' that travels with velocity $v = 0.9c$ relative to S. Relativity problems must be stated in terms of *events.* Let event 1 be "the rocket and the sun coincide" (the experimenter on earth says that the rocket passes the sun; the astronaut on the rocket says that the sun passes the rocket) and event 2 be "the rocket and Saturn coincide."

VISUALIZE Figure 36.22 shows the two events as seen from the two reference frames. Notice that the two events occur at the *same position* in S', the position of the rocket, and consequently can be measured by *one* clock.

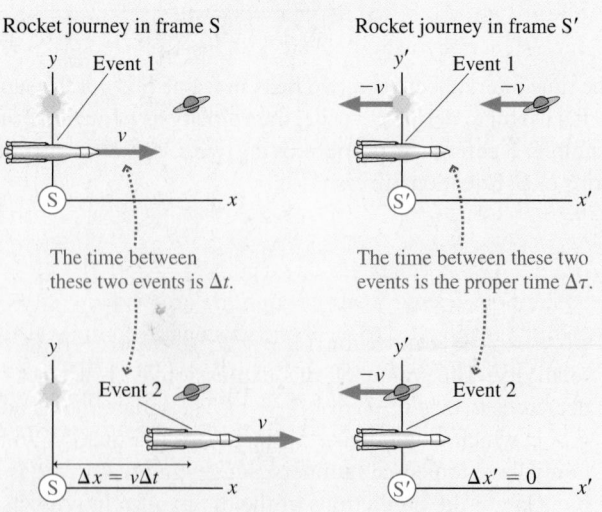

FIGURE 36.22 Pictorial representation of the trip as seen in frames S and S'.

SOLVE The time interval measured in the solar system reference frame, which includes the earth, is simply

$$\Delta t = \frac{\Delta x}{v} = \frac{1.43 \times 10^{12} \text{ m}}{0.9 \times (3.00 \times 10^8 \text{ m/s})} = 5300 \text{ s}$$

Relativity hasn't abandoned the basic definition $v = \Delta x/\Delta t$, although we do have to be sure that Δx and Δt are measured in just one reference frame and refer to the same two events.

How are things in the rocket's reference frame? The two events occur at the *same position* in S' and can be measured by *one* clock, the clock at the origin. Thus the time measured by the astronauts is the *proper time* $\Delta \tau$ between the two events. We can use Equation 36.9 with $\beta = 0.9$ to find

$$\Delta \tau = \sqrt{1 - \beta^2} \, \Delta t = \sqrt{1 - 0.9^2}(5300 \text{ s}) = 2310 \text{ s}$$

ASSESS The time interval measured between these two events by the astronauts is less than half the time interval measured by experimenters on earth. The difference has nothing to do with when earthbound astronomers *see* the rocket pass the sun and Saturn. Δt is the time interval from when the rocket actually passes the sun, as measured by a clock at the sun, until it actually passes Saturn, as measured by a synchronized clock at Saturn. The interval between *seeing* the events from earth, which would have to allow for light travel times, would be something other than 5300 s. Δt and $\Delta \tau$ are different because *time is different* in two reference frames moving relative to each other.

STOP TO THINK 36.6 Molly flies her rocket past Nick at constant velocity v. Molly and Nick both measure the time it takes the rocket, from nose to tail, to pass Nick. Which of the following is true?

a. Both Molly and Nick measure the same amount of time.
b. Molly measures a shorter time interval than Nick.
c. Nick measures a shorter time interval than Molly.

Experimental Evidence

Is there any evidence for the crazy idea that clocks moving relative to each other tell time differently? Indeed, there's plenty. An experiment in 1971 sent an atomic clock around the world on a jet plane while an identical clock remained in the laboratory. This was a difficult experiment because the traveling clock's speed was so small compared to c, but measuring the small differences between the time intervals was just barely within the capabilities of atomic clocks. It was also a more complex experiment than we've analyzed because the clock accelerated as it moved around a circle. Nonetheless, the traveling clock, upon its return, was 200 ns behind the clock that stayed at home, which was exactly as predicted by relativity.

Very detailed studies have been done on unstable particles called *muons* that are created at the top of the atmosphere, at a height of about 60 km, when high-energy cosmic rays collide with air molecules. It is well known, from laboratory studies, that stationary muons decay with a *half-life* of 1.5 μs. That is, half the muons decay within 1.5 μs, half of those remaining decay in the next 1.5 μs, and so on. The decays can be used as a clock.

The muons travel down through the atmosphere at very nearly the speed of light. The time needed to reach the ground, assuming $v \approx c$, is $\Delta t \approx$ (60,000 m)/ $(3 \times 10^8$ m/s$) = 200$ μs. This is 133 half lives, so the fraction of muons reaching the ground should be $\approx (1/2)^{133} = 10^{-40}$. That is, only 1 out of every 10^{40} muons should reach the ground. In fact, experiments find that about 1 in 10 muons reach the ground, an experimental result that differs by a factor of 10^{39} from our prediction!

The discrepancy is due to time dilation. In Figure 36.23, the two events "muon is created" and "muon hits ground" take place at two different places in the earth's reference frame. However, these two events occur at the *same position* in the muon's reference frame. (The muon is like the rocket in Example 36.5.) Thus

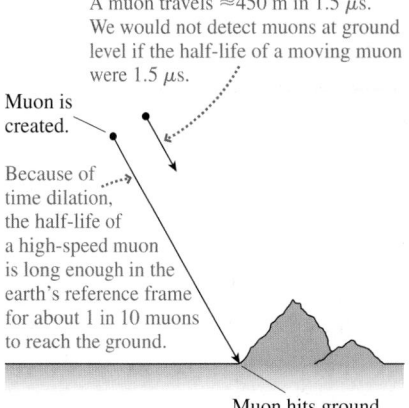

A muon travels ≈450 m in 1.5 μs. We would not detect muons at ground level if the half-life of a moving muon were 1.5 μs.

Muon is created.

Because of time dilation, the half-life of a high-speed muon is long enough in the earth's reference frame for about 1 in 10 muons to reach the ground.

Muon hits ground.

FIGURE 36.23 We wouldn't detect muons at the ground if not for time dilation.

the muon's internal clock measures the proper time. The time-dilated interval $\Delta t = 200$ μs in the earth's reference frame corresponds to a proper time $\Delta\tau \approx 5$ μs in the muon's reference frame. That is, in the muon's reference frame it takes only 5 μs from creation at the top of the atmosphere until the ground runs into it. This is 3.3 half-lives, so the fraction of muons reaching the ground is $(1/2)^{3.3} = 0.1$, or 1 out of 10. We wouldn't detect muons at the ground at all if not for time dilation.

The details are beyond the scope of this textbook, but dozens of high-energy particle accelerators around the world that study quarks and other elementary particles have been designed and built on the basis of Einstein's theory of relativity. The fact that they work exactly as planned is strong testimony to the reality of time dilation.

The Twin Paradox

The most well-known relativity paradox is the twin paradox. George and Helen are twins. On their 25th birthday, Helen departs on a starship voyage to a distant star. Let's imagine, to be specific, that her starship accelerates almost instantly to a speed of $0.95c$ and that she travels to a star that is 9.5 light years (9.5 ly) from earth. Upon arriving, she discovers that the planets circling the star are inhabited by fierce aliens, so she immediately turns around and heads home at $0.95c$.

A **light year,** abbreviated ly, is the distance that light travels in one year. A light year is vastly larger than the diameter of the solar system. The distance between two neighboring stars is typically a few light years. For our purpose, we can write the speed of light as $c = 1$ ly/year. That is, light travels 1 light year per year.

This value for c allows us to determine how long, according to George and his fellow earthlings, it takes Helen to travel out and back. Her total distance is 19 ly and, due to her rapid acceleration and rapid turn around, she travels essentially the entire distance at speed $v = 0.95c = 0.95$ ly/year. Thus the time she's away, as measured by George, is

$$\Delta t_{\text{G}} = \frac{19 \text{ ly}}{0.95 \text{ ly/year}} = 20 \text{ years} \tag{36.10}$$

George will be 45 years old when his sister Helen returns with tales of adventure.

While she's away, George takes a physics class and studies Einstein's theory of relativity. He realizes that time dilation will make Helen's clocks run more slowly than his clocks, which are at rest relative to him. Her heart—a clock—will beat fewer times and the minute hand on her watch will go around fewer times. In other words, she's aging more slowly than he is. Although she is his twin, she will be younger than he is when she returns.

Calculating Helen's age is not hard. We simply have to identify Helen's clock, because it's always with Helen as she travels, as the clock that measures proper time $\Delta\tau$. From Equation 36.9,

$$\Delta t_{\text{H}} = \Delta\tau = \sqrt{1 - \beta^2}\,\Delta t_{\text{G}} = \sqrt{1 - 0.95^2}\,(20 \text{ years}) = 6.25 \text{ years} \tag{36.11}$$

George will have just celebrated his 45th birthday as he welcomes home his 31-year-and-3-month-old twin sister.

This may be unsettling, because it violates our commonsense notion of time, but it's not a paradox. There's no logical inconsistency in this outcome. So why is it called "the twin paradox"? Read on.

Helen, knowing that she had quite of bit of time to kill on her journey, brought along several physics books to read. As she learns about relativity, she begins to think about George and her friends back on earth. Relative to her, they are all moving away at $0.95c$. Later they'll come rushing toward her at $0.95c$. Time dilation will cause their clocks to run more slowly than her clocks, which are at rest relative to her. In other words, as Figure 36.24 shows, Helen concludes that people on earth are aging more slowly than she is. Alas, she will be much older than they when she returns.

Finally, the big day arrives. Helen lands back on earth and steps out of the starship. George is expecting Helen to be younger than he is. Helen is expecting George to be younger than she is.

Here's the paradox! It's logically impossible for each to be younger than the other at the time when they are reunited. Where, then, is the flaw in our reasoning? It seems to be a symmetrical situation—Helen moves relative to George and George moves relative to Helen—but symmetrical reasoning has led to a conundrum.

But are the situations really symmetrical? George goes about his business day after day without noticing anything unusual. Helen, on the other hand, experiences three distinct periods during which the starship engines fire, she's crushed into her seat, and free dust particles that had been floating inside the starship are no longer, in the starship's reference frame, at rest or traveling in a straight line at constant speed. In other words, George spends the entire time in an inertial reference frame, *but Helen does not.* The situation is *not* symmetrical.

The principle of relativity applies *only* to inertial reference frames. Our discussion of time dilation was for inertial reference frames. Thus George's analysis and calculations are correct. Helen's analysis and calculations are *not* correct because she was trying to apply an inertial reference frame result to a noninertial reference frame.

Helen is younger than George when she returns. This is strange, but not a paradox. It is a consequence of the fact that time flows differently in two reference frames moving relative to each other.

> Helen is moving relative to me at 0.95c. Her clocks are running more slowly than mine, and when she returns she'll be younger than I am.

9.5 ly   0.95c

> George is moving relative to me at 0.95c. His clocks are running more slowly than mine, and when I return he'll be younger than I am.

FIGURE 36.24 The twin paradox.

36.7 Length Contraction

We've seen that relativity requires us to rethink our idea of time. Now let's turn our attention to the concepts of space and distance. Consider the rocket that traveled from the sun to Saturn in Example 36.5. Figure 36.25a shows the rocket moving with velocity v through the solar system reference frame S. Define $L = \Delta x = x_{\text{Saturn}} - x_{\text{sun}}$ as the distance between the sun and Saturn in frame S or, more generally, the *length* of the spatial interval between two points. The rocket's speed is $v = L/\Delta t$, where Δt is the time measured in frame S for the journey from the sun to Saturn.

Act**i**v
Physi**c**s   17.2

(a) Reference frame S: The solar system is stationary.

The rocket moves distance L in time Δt. This is the distance between the sun and Saturn in S.

L

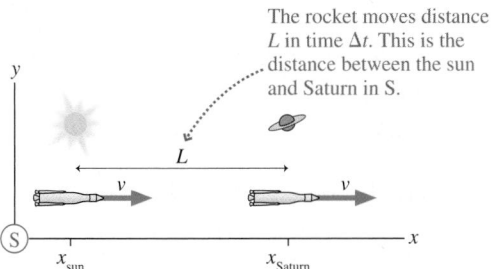

(b) Reference frame S′: The rocket is stationary.

Saturn moves distance L' in time $\Delta t' = \Delta \tau$. This is the distance between the sun and Saturn in S′.

L'

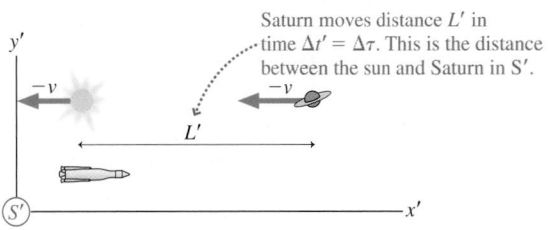

FIGURE 36.25 L and L' are the distances between the sun and Saturn in frames S and S′.

Figure 36.25b shows the situation in reference frame S′, where the rocket is at rest. The sun and Saturn move to the left at speed $v = L'/\Delta t'$, where $\Delta t'$ is the time measured in frame S′ for Saturn to travel distance L'.

Speed v is the relative speed between S and S′ and is the same for experimenters in both reference frames. That is,

$$v = \frac{L}{\Delta t} = \frac{L'}{\Delta t'} \tag{36.12}$$

The time interval $\Delta t'$ measured in frame S′ is the proper time $\Delta \tau$ because both events occur at the same position in frame S′ and can be measured by one clock. We can use the time-dilation result, Equation 36.9, to relate $\Delta \tau$ measured by the astronauts to Δt measured by the earthbound scientists. Then Equation 36.12 becomes

$$\frac{L}{\Delta t} = \frac{L'}{\Delta \tau} = \frac{L'}{\sqrt{1 - \beta^2} \, \Delta t} \tag{36.13}$$

The Δt cancels, and the distance L' in frame S′ is

$$L' = \sqrt{1 - \beta^2} L \tag{36.14}$$

Surprisingly, we find that **the distance between two objects in reference frame S′ is** *not the same* **as the distance between the same two objects in reference frame S.**

Frame S, in which the distance is L, has one important distinction. It is the *one and only* inertial reference frame in which the objects are at rest. Experimenters in frame S can take all the time they need to measure L because the two objects aren't going anywhere. The distance L between two objects or two points in space measured in the reference frame in which the objects are at rest is called the **proper length** ℓ. Only one inertial reference frame can measure the proper length.

We can use the proper length ℓ to write Equation 36.14 as

$$L' = \sqrt{1 - \beta^2}\, \ell \le \ell \tag{36.15}$$

This "shrinking" of the distance between two objects, as measured by an experiment moving with respect to the objects, is called **length contraction.** Although we derived length contraction for the distance between two distinct objects, it applies equally well to the length of any physical object that stretches between two points along the x- and x'-axes. The length of an object is greatest in the reference frame in which the object is at rest. The object's length is less (i.e., the length is contracted) when it is measured in any reference frame moving relative to the object.

The Stanford Linear Accelerator (SLAC) is a 2-mi-long electron accelerator. The accelerator's length is less than 1 m in the reference frame of the electrons.

EXAMPLE 36.6 The distance from the sun to Saturn
In Example 36.5 a rocket traveled along a line from the sun to Saturn at a constant speed of $0.9c$ relative to the solar system. The Saturn-to-sun distance was given as 1.43×10^{12} m. What is the distance between the sun and Saturn in the rocket's reference frame?

MODEL Saturn and the sun are, at least approximately, at rest in the solar system reference frame S. Thus the given distance is the proper length ℓ.

SOLVE We can use Equation 36.15 to find the distance in the rocket's frame S′:

$$L' = \sqrt{1 - \beta^2}\, \ell = \sqrt{1 - 0.9^2}\,(1.43 \times 10^{12}\text{ m})$$
$$= 0.62 \times 10^{12}\text{ m}$$

ASSESS The sun-to-Saturn distance measured by the astronauts is less than half the distance measured by experimenters on earth. L' and ℓ are different because *space is different* in two reference frames moving relative to each other.

The conclusion that space is different in reference frames moving relative to each other is a direct consequence of the fact that time is different. Experimenters in both reference frames agree on the relative velocity v, leading to Equation 36.12: $v = L/\Delta t = L'/\Delta t'$. We had already learned that $\Delta t' < \Delta t$ because of time dilation. Thus L' *has* to be less than L. That is the only way experimenters in the two reference frames can reconcile their measurements.

To be specific, the earthly experimenters in Examples 36.5 and 36.6 find that the rocket takes 5300 s to travel the 1.43×10^{12} m between the sun and Saturn. The rocket's speed is $v = L/\Delta t = 2.7 \times 10^6$ m/s $= 0.9c$. The astronauts in the rocket find that it takes only 2310 s for Saturn to reach them after the sun has passed by. But there's no conflict, because they also find that the distance is only 0.62×10^{12} m. Thus Saturn's speed toward them is $v = L'/\Delta t' = (0.62 \times 10^{12}$ m$)/(2310$ s$) = 2.7 \times 10^6$ m/s $= 0.9c$.

Another Paradox?

Carmen and Dan are in their physics lab room. They each select a meter stick, lay the two side by side, and agree that the meter sticks are exactly the same length. Then, for an extra-credit project, they go outside and run past each other, in opposite directions, at a relative speed $v = 0.9c$. Figure 36.26 shows their experiment and a portion of their conversation.

Now, Dan's meter stick can't be both longer and shorter than Carmen's meter stick. Is this another paradox? No! Relativity allows us to compare the *same* events as they're measured in two different reference frames. This did lead to a real paradox when Peggy rolled past Ryan on the train. There the signal light on the box turns green (a single event) or it doesn't, and Peggy and Ryan have to agree about it. But the events by which Dan measures the length (in Dan's frame) of Carmen's meter are *not the same events* as those that Carmen uses to measure the length (in Carmen's frame) of Dan's meter stick.

There's no conflict between their measurements. In Dan's reference frame, Carmen's meter stick has been length contracted and is less than 1 m in length. In Carmen's reference frame, Dan's meter stick has been length contracted and is less than 1 m in length. If this weren't the case, if both agreed that one of the meter sticks was shorter than the other, then we could tell which reference frame was "really" moving and which was "really" at rest. But the principle of relativity doesn't allow us to make that distinction. Each is moving relative to the other, so each should make the same measurement for the length of the other's meter stick.

The Spacetime Interval

Forget relativity for a minute and think about ordinary geometry. Figure 36.27 shows two ordinary coordinate systems. They are identical except for the fact that one has been rotated relative to the other. A student using the xy-system would measure coordinates (x_1, y_1) for point 1 and (x_2, y_2) for point 2. A second student, using the $x'y'$-system, would measure (x_1', y_1') and (x_2', y_2').

The students soon find that none of their measurements agree. That is, $x_1 \neq x_1'$ and so on. Even the intervals are different: $\Delta x \neq \Delta x'$ and $\Delta y \neq \Delta y'$. Each is a perfectly valid coordinate system, giving no reason to prefer one over the other, but each yields different measurements.

Is there *anything* on which the two students can agree? Yes, there is. The distance d between points 1 and 2 is independent of the coordinates. We can state this mathematically as

$$d^2 = (\Delta x)^2 + (\Delta y)^2 = (\Delta x')^2 + (\Delta y')^2 \tag{36.16}$$

The quantity $(\Delta x)^2 + (\Delta y)^2$ is called an **invariant** in geometry because it has the same value in any Cartesian coordinate system.

Returning to relativity, is there an invariant in the spacetime coordinates, some quantity that has the *same value* in all inertial reference frames? There is, and to find it let's return to the light clock that we analyzed in Figure 36.21. Figure 36.28 on the next page shows the light clock as seen in reference frames S' and S''. The speed of light is the same in both frames, even though both are moving with respect to each other and with respect to the clock.

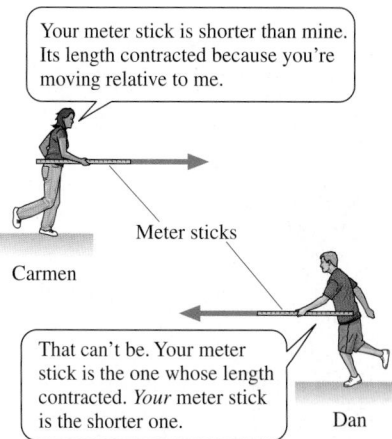

FIGURE 36.26 Carmen and Dan each measure the length of the other's meter stick as they move relative to each other.

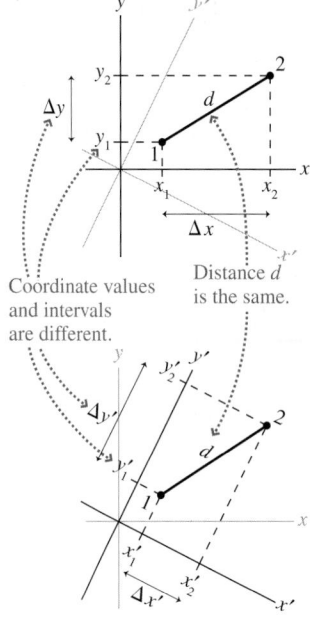

Measurements in the $x'y'$-system

FIGURE 36.27 Distance d is the same in both coordinate systems.

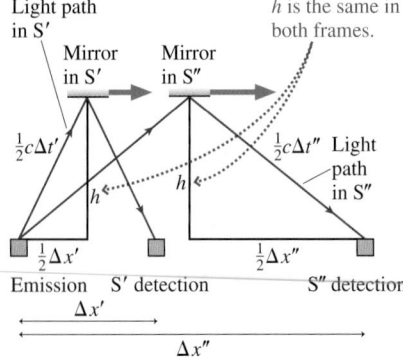

FIGURE 36.28 The light clock seen by experimenters in reference frames S′ and S″.

Light path in S′
Mirror in S′
Mirror in S″
h is the same in both frames.
$\frac{1}{2}c\Delta t'$
$\frac{1}{2}c\Delta t''$ Light path in S″
h
h
$\frac{1}{2}\Delta x'$
$\frac{1}{2}\Delta x''$
Emission   S′ detection   S″ detection
$\Delta x'$
$\Delta x''$

Notice that the clock's height h is common to both reference frames. Thus

$$h^2 = \left(\frac{1}{2}c\Delta t'\right)^2 - \left(\frac{1}{2}\Delta x'\right)^2 = \left(\frac{1}{2}c\Delta t''\right)^2 - \left(\frac{1}{2}\Delta x''\right)^2 \quad (36.17)$$

The factor $\frac{1}{2}$ cancels, allowing us to write

$$c^2(\Delta t')^2 - (\Delta x')^2 = c^2(\Delta t'')^2 - (\Delta x'')^2 \quad (36.18)$$

Let us define the **spacetime interval** s between two events to be

$$s^2 = c^2(\Delta t)^2 - (\Delta x)^2 \quad (36.19)$$

What we've shown in Equation 36.18 is that **the spacetime interval s has the same value in all inertial reference frames.** That is, the spacetime interval between two events is an invariant. It is a value that all experimenters, in all reference frames, can agree upon.

EXAMPLE 36.7 Using the spacetime interval
A firecracker explodes at the origin of an inertial reference frame. Then, 2.0 μs later, a second firecracker explodes 300 m away. Astronauts in a passing rocket measure the distance between the explosions to be 200 m. According to the astronauts, how much time elapses between the two explosions?

MODEL The spacetime coordinates of two events are measured in two different inertial reference frames. Call the reference frame of the ground S and the reference frame of the rocket S′. The spacetime interval between these two events is the same in both reference frames.

SOLVE The spacetime interval (or, rather, its square) in frame S is

$$s^2 = c^2(\Delta t)^2 - (\Delta x)^2 = (600 \text{ m})^2 - (300 \text{ m})^2$$

$$= 270{,}000 \text{ m}^2$$

where we used $c = 300 \text{ m}/\mu s$ to determine that $c\Delta t = 600$ m. The spacetime interval has the same value in frame S′. Thus

$$s^2 = 270{,}000 \text{ m}^2 = c^2(\Delta t')^2 - (\Delta x')^2$$

$$= c^2(\Delta t')^2 - (200 \text{ m})^2$$

This is easily solved to give $\Delta t' = 1.85 \mu$s.

ASSESS The two events are closer together in both space and time in the rocket's reference frame than in the reference frame of the ground.

Einstein's legacy, according to popular culture, was the discovery that "everything is relative." But it's not so. Time intervals and space intervals may be relative, as were the intervals Δx and Δy in the purely geometric analogy with which we opened this section, but some things are *not* relative. In particular, the spacetime interval s between two events is not relative. It is a well-defined number, agreed to by experimenters in each and every inertial reference frame.

STOP TO THINK 36.7   Beth and Charles are at rest relative to each other. Anjay runs past at velocity v while holding a long pole parallel to his motion. Anjay, Beth, and Charles each measure the length of the pole at the instant Anjay passes Beth. Rank in order, from largest to smallest, the three lengths L_A, L_B, and L_C.

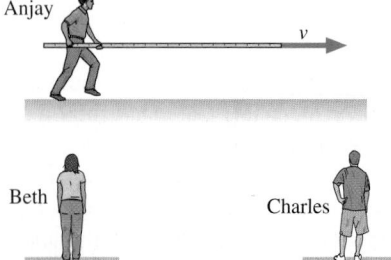

36.8 The Lorentz Transformations

The Galilean transformation $x' = x - vt$ of classical relativity lets us calculate the position x' of an event in frame S' if we know its position x in frame S. Classical relativity, of course, assumes that $t' = t$. Is there a similar transformation in relativity that would allow us to calculate an event's spacetime coordinates (x', t') in frame S' if we know their values (x, t) in frame S? Such a transformation would need to satisfy three conditions. It must

1. Agree with the Galilean transformations in the low-speed limit $v \ll c$.
2. Transform not only spatial coordinates but also time coordinates.
3. Ensure that the speed of light is the same in all reference frames.

We'll continue to use reference frames in the standard orientation of Figure 36.29. The motion is parallel to the x- and x'-axes, and we *define* $t = 0$ and $t' = 0$ as the instant when the origins of S and S' coincide.

The requirement that a new transformation agree with the Galilean transformation when $v \ll c$ suggests that we look for a transformation of the form

$$x' = \gamma(x - vt) \quad \text{and} \quad x = \gamma(x' + vt') \qquad (36.20)$$

where γ is a dimensionless function of velocity that satisfies $\gamma \to 1$ as $v \to 0$.

To determine γ, consider the following two events:

> Event 1: A flash of light is emitted from the origin of both reference frames $(x = x' = 0)$ at the instant they coincide $(t = t' = 0)$.

> Event 2: The light strikes a light detector. The spacetime coordinates of this event are (x, t) in frame S and (x', t') in frame S'.

Light travels at speed c in both reference frames, so the positions of event 2 are $x = ct$ in S and $x' = ct'$ in S'. Substituting these expressions for x and x' into Equation 36.20 gives

$$
\begin{aligned}
ct' &= \gamma(ct - vt) = \gamma(c - v)t \\
ct &= \gamma(ct' + vt') = \gamma(c + v)t'
\end{aligned}
\qquad (36.21)
$$

Solve the first for t', by dividing by c, then substitute this result for t' into the second:

$$ct = \gamma(c + v)\frac{\gamma(c - v)t}{c} = \gamma^2(c^2 - v^2)\frac{t}{c}$$

The t cancels, leading to

$$\gamma^2 = \frac{c^2}{c^2 - v^2} = \frac{1}{1 - v^2/c^2}$$

Thus the γ that "works" in the proposed transformation of Equation 36.20 is

$$\gamma = \frac{1}{\sqrt{1 - v^2/c^2}} = \frac{1}{\sqrt{1 - \beta^2}} \qquad (36.22)$$

You can see that $\gamma \to 1$ as $v \to 0$, as expected.

The transformation between t and t' is found by requiring that $x = x$ if you use Equation 36.20 to transform a position from S to S' and then back to S. The details will be left for a homework problem. Another homework problem will let you demonstrate that the y and z measurements made perpendicular to the relative motion are not affected by the motion. We tacitly assumed this condition in our analysis of the light clock.

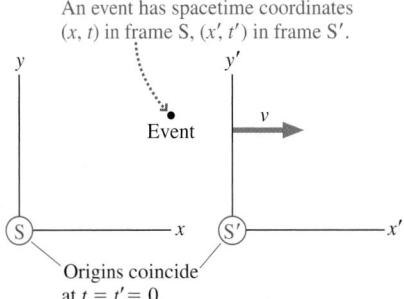

An event has spacetime coordinates (x, t) in frame S, (x', t') in frame S'.

Origins coincide at $t = t' = 0$.

FIGURE 36.29 The spacetime coordinates of an event are measured in inertial reference frames S and S'.

The full set of equations are called the **Lorentz transformations.** They are

$$
\begin{aligned}
x' &= \gamma(x - vt) & x &= \gamma(x' + vt') \\
y' &= y & y &= y' \\
z' &= z & z &= z' \\
t' &= \gamma(t - vx/c^2) & t &= \gamma(t' + vx'/c^2)
\end{aligned}
\tag{36.23}
$$

The Lorentz transformations transform the spacetime coordinates of *one* event. You should compare these to the Galilean transformation equations in Equation 36.1.

NOTE ▶ These transformations are named after the Dutch physicist H. A. Lorentz, who derived them prior to Einstein. Lorentz was close to discovering special relativity, but he didn't recognize that our concepts of space and time have to be changed before these equations can be properly interpreted. ◀

Using Relativity

Relativity is phrased in terms of *events,* hence relativity problems are solved by interpreting the problem statement in terms of specific events.

PROBLEM-SOLVING STRATEGY 36.1 Relativity

MODEL Frame the problem in terms of events, things that happen at a specific place and time.

VISUALIZE A pictorial representation defines the reference frames.

- Sketch the reference frames, showing their motion relative to each other.
- Show events. Identify objects that are moving with respect to the reference frames.
- Identify any proper time intervals and proper lengths. These are measured in an object's rest frame.

SOLVE The mathematical representation is based on the Lorentz transformations, but not every problem requires the full transformation equations.

- Problems about time intervals can often be solved using time dilation: $\Delta t = \gamma \Delta \tau$.
- Problems about distances can often be solved using length contraction: $L = \ell/\gamma$.

ASSESS Are the results consistent with Galilean relativity when $v \ll c$?

EXAMPLE 36.8 Ryan and Peggy revisited

Peggy is standing in the center of a long, flat railroad car that has firecrackers tied to both ends. The car moves past Ryan, who is standing on the ground, with velocity $v = 0.8c$. Flashes from the exploding firecrackers reach him simultaneously 1.0 μs after the instant that Peggy passes him, and he later finds burn marks on the track 300 m to either side of where he had been standing.

a. According to Ryan, what is the distance between the two explosions and at what times do the explosions occur relative to the time that Peggy passes him?

b. According to Peggy, what is the distance between the two explosions and at what times do the explosions occur relative to the time that Ryan passes her?

MODEL Let the explosion on Ryan's right, the direction in which Peggy is moving, be event R. The explosion on his left is event L.

VISUALIZE Peggy and Ryan are in inertial reference frames. Figure 36.30 shows Peggy's frame S′ moving with $v = 0.8c$ relative to Ryan's frame S. We've defined the reference frames such that Peggy and Ryan are at the origins. The instant they pass, by definition, is $t = t' = 0$ s. The two events are shown in Ryan's reference frame.

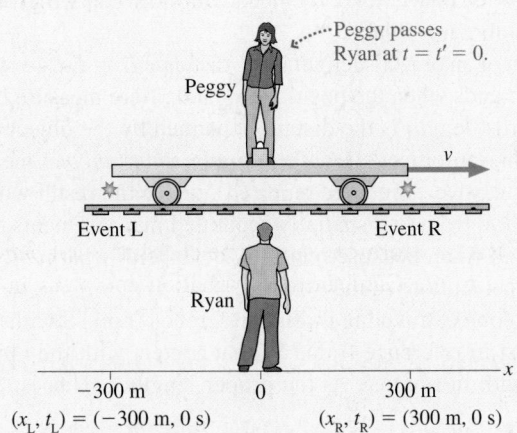

Peggy passes Ryan at $t = t' = 0$.

Peggy

v

Event L   Event R

Ryan

-300 m   0   300 m

$(x_L, t_L) = (-300$ m, 0 s$)$ $(x_R, t_R) = (300$ m, 0 s$)$

FIGURE 36.30 A pictorial representation of the reference frames and events.

SOLVE

a. The two burn marks tell Ryan that the distance between the explosions was $L = 600$ m. Light travels at $c = 300$ m/μs, and the burn marks are 300 m on either side of him, so Ryan can determine that each explosion took place 1.0 μs before he saw the flash. But this was the instant of time that Peggy passed him, so Ryan concludes that the explosions were simultaneous with each other and with Peggy's passing him. The spacetime coordinates of the two events in frame S are $(x_R, t_R) = (300$ m, 0 μs$)$ and $(x_L, t_L) = (-300$ m, 0 μs$)$.

b. We already know, from our qualitative analysis in Section 36.5, that the explosions are *not* simultaneous in Peggy's reference frame. Event R happens before event L in S′, but we don't know how they compare to the time at which Ryan passes Peggy. We can now use the Lorentz transformations to relate the spacetime coordinates of these events as measured by Ryan to the spacetime coordinates as measured by Peggy. Using $v = 0.8c$, we find that γ is

$$\gamma = \frac{1}{\sqrt{1 - v^2/c^2}} = \frac{1}{\sqrt{1 - 0.8^2}} = 1.667$$

For event L, the Lorentz transformations are

$$x'_L = 1.667((-300 \text{ m}) - (0.8c)(0 \text{ } \mu\text{s})) = -500 \text{ m}$$

$$t'_L = 1.667((0 \text{ } \mu\text{s}) - (0.8c)(-300 \text{ m})/c^2) = 1.33 \text{ } \mu\text{s}$$

And for event R,

$$x'_R = 1.667((300 \text{ m}) - (0.8c)(0 \text{ } \mu\text{s})) = 500 \text{ m}$$

$$t'_R = 1.667((0 \text{ } \mu\text{s}) - (0.8c)(300 \text{ m})/c^2) = -1.33 \text{ } \mu\text{s}$$

According to Peggy, the two explosions occur 1000 m apart. Furthermore, the first explosion, on the right, occurs 1.33 μs before Ryan passes her at $t' = 0$ s. The second, on the left, occurs 1.33 μs after Ryan goes by.

ASSESS Events that are simultaneous in frame S are *not* simultaneous in frame S′. The results of the Lorentz transformations agree with our earlier qualitative analysis.

A follow-up discussion of Example 36.8 is worthwhile. Because Ryan moves at speed $v = 0.8c = 240$ m/μs relative to Peggy, he moves 320 m during the 1.33 μs between the first explosion and the instant he passes Peggy, then another 320 m before the second explosion. Gathering this information together, Figure 36.31 shows the sequence of events in Peggy's reference frame.

The firecrackers define the ends of the railroad car, so the 1000 m distance between the explosions in Peggy's frame is the car's length L' in frame S′. The car is at rest in frame S′, hence length L' is the proper length: $\ell = 1000$ m. Ryan is measuring the length of a moving object, so he should see the car length contracted to

$$L = \sqrt{1 - \beta^2}\ell = \frac{\ell}{\gamma} = \frac{1000 \text{ m}}{1.667} = 600 \text{ m}$$

And, indeed, that is exactly the distance Ryan measured between the burn marks.

Finally, we can calculate the spacetime interval s between the two events. According to Ryan,

$$s^2 = c^2(\Delta t^2) - (\Delta x)^2 = c^2(0 \text{ } \mu\text{s})^2 - (600 \text{ m})^2 = -(600 \text{ m})^2$$

Peggy computes the spacetime interval to be

$$s^2 = c^2(\Delta t')^2 - (\Delta x')^2 = c^2(2.67 \text{ } \mu\text{s})^2 - (1000 \text{ m})^2 = -(600 \text{ m})^2$$

Their calculations of the spacetime interval agree, showing that s really is an invariant, but notice that s itself is an imaginary number.

Length

We've introduced the idea of length contraction, but we didn't precisely define just what we mean by the *length* of a moving object. The length of an object at rest is clear because we can take all the time we need to measure it with

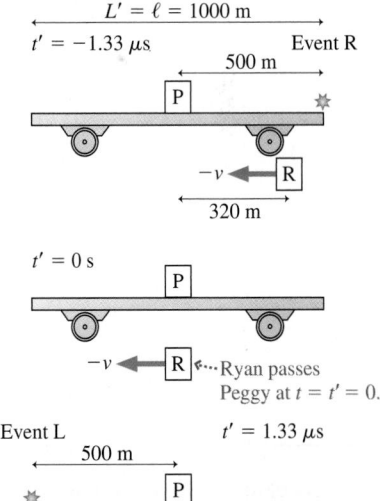

FIGURE 36.31 The sequence of events as seen in Peggy's reference frame.

meter sticks, surveying tools, or whatever we need. But how can we give clear meaning to the length of a moving object?

A reasonable definition of an object's length is the distance $L = \Delta x = x_R - x_L$ between the right and left ends when the positions x_R and x_L are measured *at the same time t*. In other words, length is the distance spanned by the object at *one instant* of time. Measuring an object's length requires *simultaneous* measurements of two positions (i.e., two events are required), hence the result won't be known until the information from two spatially separated measurements can be brought together. That's all right, because relativity deals with what *really happens* rather than with our perceptions of the events.

Figure 36.32 shows an object traveling through reference frame S with velocity v. The object is at rest in reference frame S' that travels with the object at velocity v, hence the length in frame S' is the proper length ℓ. That is, $\Delta x' = x_R' - x_L' = \ell$ in frame S'.

At time t, an experimenter (and his or her assistants) in frame S makes simultaneous measurements of the positions x_R and x_L of the ends of the object. The difference $\Delta x = x_R - x_L = L$ is the length in frame S. The Lorentz transformations of x_R and x_L are

$$x_R' = \gamma(x_R - vt)$$
$$x_L' = \gamma(x_L - vt)$$
(36.24)

where, it is important to note, t is the *same* for both because the measurements are simultaneous.

Subtracting the second equation from the first, we find

$$x_R' - x_L' = \ell = \gamma(x_R - x_L) = \gamma L = \frac{L}{\sqrt{1 - \beta^2}}$$

Solving for L, we find, in agreement with Equation 36.15, that

$$L = \sqrt{1 - \beta^2}\,\ell$$
(36.25)

This analysis has accomplished two things. First, by giving a precise definition of length, we've put our length-contraction result on a firmer footing. Second, we've had good practice at relativistic reasoning using the Lorentz transformation.

NOTE ▶ Length contraction does not tell us how an object would *look*. The visual appearance of an object is determined by light waves that arrive simultaneously at the eye. These waves left points on the object at different times (i.e., *not* simultaneously) because they had to travel different distances to the eye. The analysis needed to determine an object's visual appearance is considerably more complex. Length and length contraction are concerned only with the *actual* length of the object at one instant of time. ◀

The Binomial Approximation

You've met the binomial approximation earlier in this text and in your calculus class. The binomial approximation is useful when we need to calculate a relativistic expression for a nonrelativistic velocity $v \ll c$. Because $v^2/c^2 \ll 1$ in these cases, we can write

$$\sqrt{1 - \beta^2} = (1 - v^2/c^2)^{1/2} \approx 1 - \frac{1}{2}\frac{v^2}{c^2}$$

$$\gamma = \frac{1}{\sqrt{1 - \beta^2}} = (1 - v^2/c^2)^{-1/2} \approx 1 + \frac{1}{2}\frac{v^2}{c^2}$$
(36.26)

The following example illustrates the use of the binomial approximation.

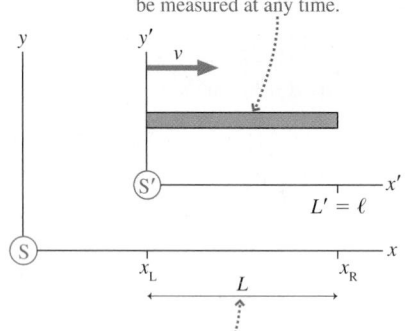

The object is at rest in frame S'. It's length is $L' = \ell$, which can be measured at any time.

$L' = \ell$

x_L L x_R

Since the object is moving in frame S, simultaneous measurements of its ends must be made in order to find its length L in frame S.

FIGURE 36.32 The length of an object is the distance between *simultaneous* measurements of the positions of the end points.

The binomial approximation

If $x \ll 1$, then $(1 + x)^n \approx 1 + nx$

EXAMPLE 36.9 The shrinking school bus

An 8.0-m-long school bus drives past at 30 m/s. By how much is its length contracted?

MODEL The school bus is at rest in an inertial reference frame S' moving at velocity $v = 30$ m/s relative to the ground frame S. The given length, 8.0 m, is the proper length ℓ in frame S'.

SOLVE In frame S, the school bus is length contracted to

$$L = \sqrt{1 - \beta^2}\,\ell$$

The bus's velocity v is much less than c, so we can use the binomial approximation to write

$$L \approx \left(1 - \frac{1}{2}\frac{v^2}{c^2}\right)\ell = \ell - \frac{1}{2}\frac{v^2}{c^2}\ell$$

The *amount* of the length contraction is

$$\ell - L = \frac{1}{2}\frac{v^2}{c^2}\ell = \left(\frac{30 \text{ m/s}}{3.0 \times 10^8 \text{ m/s}}\right)^2 (4.0 \text{ m})$$

$$= 4.0 \times 10^{-14} \text{ m} = 40 \text{ fm}$$

where 1 fm = 1 femtometer = 10^{-15} m.

ASSESS The amount the bus "shrinks" is only slightly larger than the diameter of the nucleus of an atom. It's no wonder that we're not aware of length contraction in our everyday lives. If you had tried to calculate this number exactly, your calculator would have shown $\ell - L = 0$. The difficulty is that the difference between ℓ and L shows up only in the 14th decimal place. A scientific calculator determines numbers to 10 or 12 decimal places, but that isn't sufficient to show the difference. The binomial approximation provides an invaluable tool for finding the very tiny difference between two numbers that are nearly identical.

The Lorentz Velocity Transformations

Figure 36.33 shows an object that is moving in both reference frame S and reference frame S'. Experimenters in frame S determine that the object's velocity is u while experimenters in S' find it to be u'. For simplicity, we'll assume that the object moves parallel to the x- and x'-axes.

The Galilean velocity transformation $u' = u - v$ was found by taking the time derivative of the position transformation. We can do the same with the Lorentz transformation if we take the derivative with respect to the time in each frame. Velocity u' in frame S' is

$$u' = \frac{dx'}{dt'} = \frac{d(\gamma(x - vt))}{d(\gamma(t - vx/c^2))} \tag{36.27}$$

where we've used the Lorentz transformations for position x' and time t'.

Carrying out the differentiation gives

$$u' = \frac{\gamma(dx - v\,dt)}{\gamma(dt - v\,dx/c^2)} = \frac{dx/dt - v}{1 - v(dx/dt)/c^2} \tag{36.28}$$

But dx/dt is u, the object's velocity in frame S, leading to

$$u' = \frac{u - v}{1 - uv/c^2} \tag{36.29}$$

You can see that Equation 36.29 reduces to the Galilean transformation $u' = u - v$ when $v \ll c$, as expected.

The reverse transformation, from S' to S, is found by reversing the sign of v. Altogether,

$$u' = \frac{u - v}{1 - uv/c^2} \quad \text{and} \quad u = \frac{u' + v}{1 + u'v/c^2} \tag{36.30}$$

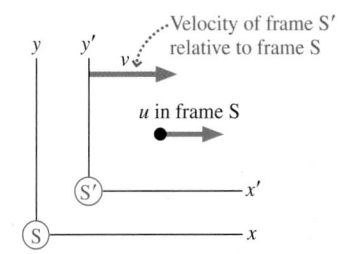

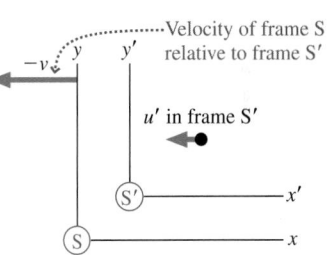

FIGURE 36.33 The velocity of a moving object is measured to be u in frame S and u' in frame S'.

Equations 36.30 are the Lorentz velocity transformation equations.

NOTE ▶ It is important to distinguish carefully between v, which is the relative velocity of the reference frames in which measurements are carried out, and u and u', which are the velocities of an *object* as measured in two different reference frames. ◀

EXAMPLE 36.10 A really fast bullet

A rocket flies past the earth at $0.9c$. As it goes by, the rocket fires a bullet in the forward direction at $0.95c$ with respect to the rocket. What is the bullet's speed with respect to the earth?

MODEL The rocket and the earth are inertial reference frames. Let the earth be frame S and the rocket be frame S'. The velocity of frame S' relative to frame S is $v = 0.9c$. The bullet's velocity in frame S' is $u' = 0.95c$.

SOLVE We can use the Lorentz velocity transformation to find

$$u = \frac{u' + v}{1 + u'v/c^2} = \frac{0.95c + 0.90c}{1 + (0.95c)(0.90c)/c^2} = 0.997c$$

The bullet's speed with respect to the earth is 99.7% of the speed of light.

NOTE ▶ Many relativistic calculations are much easier when velocities are specified as a fraction of c. ◀

ASSESS In Newtonian mechanics, the Galilean transformation of velocity would give $u = 1.85c$. Now, despite the very high speed of the rocket and of the bullet with respect to the rocket, the bullet's speed with respect to the earth remains less than c. This is yet more evidence that objects cannot exceed the speed of light.

Suppose the rocket in Example 36.10 fired a laser beam in the forward direction as it traveled past the earth at velocity v. The laser beam would travel away from the rocket at speed $u' = c$ in the rocket's reference frame S'. What is the laser beam's speed in the earth's frame S? According to the Lorentz velocity transformation, it must be

$$u = \frac{u' + v}{1 + u'v/c^2} = \frac{c + v}{1 + cv/c^2} = \frac{c + v}{1 + v/c} = \frac{c + v}{(c + v)/c} = c \qquad (36.31)$$

Light travels at speed c in both frame S and frame S'. This important consequence of the principle of relativity is "built into" the Lorentz transformations.

36.9 Relativistic Momentum

In Newtonian mechanics, the total momentum of a system is a conserved quantity. Further, as we've seen, the law of conservation of momentum, $P_f = P_i$, is true in all inertial reference frames *if* the particle velocities in different reference frames are related by the Galilean velocity transformations.

The difficulty, of course, is that the Galilean transformations are not consistent with the principle of relativity. It is a reasonable approximation when all velocities are much less than c, but the Galilean transformations fail dramatically as velocities approach c. It's not hard to show that $P_f' \neq P_i'$ if the particle velocities in frame S' are related to the particle velocities in frame S by the Lorentz transformations.

There are two possibilities:

1. The so-called law of conservation of momentum is not really a law of physics. It is approximately true at low velocities but fails as velocities approach the speed of light.
2. The law of conservation of momentum really is a law of physics, but the expression $p = mu$ is not the correct way to calculate momentum when the particle velocity u becomes a significant fraction of c.

Momentum conservation is such a central and important feature of mechanics that it seems unlikely to fail in relativity. How else might the momentum of a particle be defined?

The classical momentum, for one-dimensional motion, is $p = mu = m(\Delta x/\Delta t)$. Δt is the time needed to move distance Δx. That seemed clear enough within a Newtonian framework, but now we've learned that experimenters in different reference frames disagree about the amount of time needed. So whose Δt should we use?

One possibility is to use the time measured *by the particle*. This is the proper time $\Delta \tau$ because the particle is at rest in its own reference frame and needs only one clock. With this in mind, let's redefine the momentum of a particle of mass m moving with velocity $u = \Delta x/\Delta t$ to be

$$p = m\frac{\Delta x}{\Delta \tau} \qquad (36.32)$$

We can relate this new expression for p to the familiar Newtonian expression by using the time-dilation result $\Delta \tau = (1 - u^2/c^2)^{1/2}\Delta t$ to relate the proper time interval measured by the particle to the more practical time interval Δt measured by experimenters in frame S. With this substitution, Equation 36.32 becomes

$$p = m\frac{\Delta x}{\Delta \tau} = m\frac{\Delta x}{\sqrt{1 - u^2/c^2}\,\Delta t} = \frac{mu}{\sqrt{1 - u^2/c^2}} \qquad (36.33)$$

You can see that Equation 36.33 reduces to the classical expression $p = mu$ when the particle's speed $u \ll c$. That is an important requirement, but whether this is the "correct" expression for p depends on whether the total momentum P is conserved when the velocities of a system of particles are transformed with the Lorentz velocity transformation equations. The proof is rather long and tedious, so we will assert, without actual proof, that the momentum defined in Equation 36.33 does, indeed, transform correctly. **The law of conservation of momentum is still valid in all inertial reference frames** *if* **the momentum of each particle is calculated with Equation 36.33.**

The factor that multiplies mu in Equation 36.33 looks much like the factor γ in the Lorentz transformation equations for x and t, but there's one very important difference. The v in the Lorentz transformation equations is the velocity of a *reference frame*. The u in Equation 36.33 is the velocity of a particle moving *in* a reference frame.

With this distinction in mind, let's define the quantity

$$\gamma_p = \frac{1}{\sqrt{1 - u^2/c^2}} \qquad (36.34)$$

where the subscript p indicates that this is γ for a particle, not for a reference frame. In frame S′, where the particle moves with velocity u', the corresponding expression would be called γ_p'. With this definition of γ_p, the momentum of a particle is

$$p = \gamma_p mu \qquad (36.35)$$

EXAMPLE 36.11 Momentum of a subatomic particle
Electrons in a particle accelerator reach a speed of $0.999c$ relative to the laboratory. One collision of an electron with a target produces a muon that moves forward with a speed of $0.95c$ relative to the laboratory. The electron mass is $m_e = 9.11 \times 10^{-31}$ m/s and the muon mass is 1.90×10^{-28} m/s. What is the muon's momentum in the laboratory frame and in the frame of the electron beam?

MODEL Let the laboratory be reference frame S. The reference frame S′ of the electron beam (i.e., a reference frame in which the electrons are at rest) moves in the direction of the electrons at $v = 0.999c$. The muon velocity in frame S is $u = 0.95c$.

SOLVE γ_p for the muon in the laboratory reference frame is

$$\gamma_p = \frac{1}{\sqrt{1 - u^2/c^2}} = \frac{1}{\sqrt{1 - 0.95^2}} = 3.20$$

Thus the muon's momentum in the laboratory is

$$p = \gamma_p mu = (3.20)(1.90 \times 10^{-28}\text{ kg})(0.95 \times 3.00 \times 10^8\text{ m/s})$$
$$= 1.73 \times 10^{-19}\text{ kg m/s}$$

The momentum is a factor of 3.2 larger than the Newtonian momentum mu. To find the momentum in the electron-beam reference frame, we must first use the velocity transformation equation to find the muon's velocity in frame S′:

$$u' = \frac{u - v}{1 - uv/c^2} = \frac{0.95c - 0.999c}{1 - (0.95c)(0.999c)/c^2} = -0.962c$$

In the laboratory frame, the faster electrons are overtaking the slower muon. Hence the muon's velocity in the electron-beam frame is negative. γ_p' for the muon in frame S′ is

$$\gamma_p' = \frac{1}{\sqrt{1 - u'^2/c^2}} = \frac{1}{\sqrt{1 - 0.962^2}} = 3.66$$

The muon's momentum in the electron-beam reference frame is

$$p' = \gamma'_p m u'$$

$$= (3.66)(1.90 \times 10^{-28}\text{ kg})(-0.962 \times 3.00 \times 10^8\text{ m/s})$$

$$= -2.01 \times 10^{-19}\text{ kg m/s}$$

ASSESS From the laboratory perspective, the muon moves only slightly slower than the electron beam. But it turns out that the muon moves faster with respect to the electrons, although in the opposite direction, than it does with respect to the laboratory.

The Cosmic Speed Limit

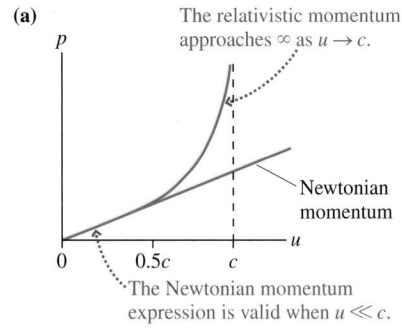

(a)

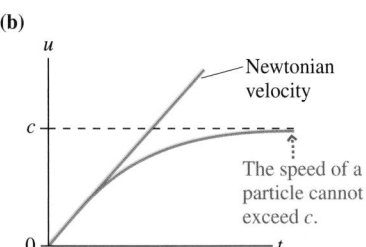

(b)

FIGURE 36.34 The speed of a particle cannot reach the speed of light.

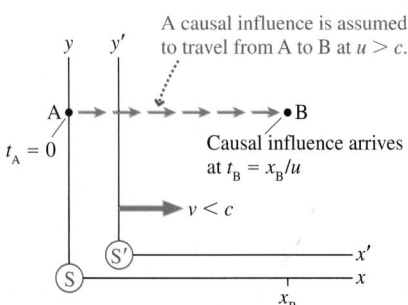

FIGURE 36.35 Assume that a causal influence can travel from A to B at a speed $u > c$.

Figure 36.34a is a graph of momentum versus velocity. For a Newtonian particle, with $p = mu$, the momentum is directly proportional to the velocity. The relativistic expression for momentum agrees with the Newtonian value if $u \ll c$, but p approaches ∞ as $u \to c$.

The implications of this graph become clear when we relate momentum to force. Consider a particle subjected to a constant force, such as a rocket that never runs out of fuel. If F is constant, we can see from $F = dp/dt$ that the momentum is $p = Ft$. If Newtonian physics were correct, a particle would go faster and faster as its velocity $u = p/m = (F/m)t$ increased without limit. But the relativistic result, shown in Figure 36.34b, is that the particle's velocity asymptotically approaches the speed of light ($u \to c$) as p approaches ∞. Relativity gives a very different outcome than Newtonian mechanics.

The speed c is a "cosmic speed limit" for material particles. A force cannot accelerate a particle to a speed higher than c because the particle's momentum becomes infinitely large as the speed approaches c. The amount of effort required for each additional increment of velocity becomes larger and larger until no amount of effort can raise the velocity any higher.

Actually, at a more fundamental level, c is a speed limit for *any* kind of **causal influence.** If I throw a rock and break a window, my throw is the *cause* of the breaking window and the rock is the *causal influence*. If I shoot a laser beam at a light detector that is wired to a firecracker, the light wave is the *causal influence* that leads to the explosion. A causal influence can be any kind of particle, wave, or information that travels from A to B and allows A to be the cause of B.

For two unrelated events—a firecracker explodes in Tokyo and a balloon bursts in Paris—the relativity of simultaneity tells us that they may be simultaneous in one reference frame but not in others. Or in one reference frame the firecracker may explode before the balloon bursts but in some other reference frame the balloon may burst first. These possibilities violate our commonsense view of time, but they're not in conflict with the principle of relativity.

For two causally related events—A *causes* B—it would be nonsense for an experimenter in any reference frame to find that B occurs before A. No experimenter in any reference frame, no matter how it is moving, will find that you are born before your mother is born. If A causes B, then it must be the case that $t_A < t_B$ in *all* reference frames.

Suppose there exists some kind of causal influence that *can* travel at speed $u > c$. Figure 36.35 shows a reference frame S in which event A is at the origin ($x_A = 0$). The faster-than-light causal influence—perhaps some yet-to-be-discovered "z ray"—leaves A at $t_A = 0$ and travels to the point at which it will cause event B. It arrives at x_B at time $t_B = x_B/u$.

How do events A and B appear in a reference frame S' that travels at an ordinary speed $v < c$ relative to frame S? We can use the Lorentz transformations to find out. Because $x_A = 0$ and $t_A = 0$, it's easy to see that $x'_A = 0$ and $t'_A = 0$. That is, the origins of S and S' overlap at the instant the causal influence leaves event A. More interesting is the time at which this influence reaches B in frame S′. The Lorentz time transformation for event B is

$$t_B' = \gamma\left(t_B - \frac{vx_B}{c^2}\right) = \gamma t_B\left(1 - \frac{v(x_B/t_B)}{c^2}\right) = \gamma t_B\left(1 - \frac{vu}{c^2}\right) \quad (36.36)$$

where we first factored out t_B, then made use of the fact that $u = x_B/t_B$ in frame S.

We're assuming $u > c$, so let $u = \alpha c$ where $\alpha > 1$ is a constant. Then $vu/c^2 = \alpha v/c$. Now follow the logic:

1. If $v > c/\alpha$, which is possible because $\alpha > 1$, then $vu/c^2 > 1$.
2. If $vu/c^2 > 1$, then the term $(1 - vu/c^2)$ is negative and $t_B' < 0$.
3. If $t_B' < 0$, then event B happens *before* event A in reference frame S′.

In other words, if a causal influence can travel faster than c, then there exist reference frames in which the effect happens before the cause. We know this can't happen, so our assumption $u > c$ must be wrong. **No causal influence of any kind—particle, wave, or yet-to-be-discovered z rays—can travel faster than c.**

The existence of a cosmic speed limit is one of the most interesting consequences of the theory of relativity. "Warp drive," in which a spaceship suddenly leaps to faster-than-light velocities, is simply incompatible with the theory of relativity. Rapid travel to the stars will remain in the realm of science fiction unless future scientific discoveries find flaws in Einstein's theory and open the doors to yet-undreamed-of theories. While we can't say with certainty that a scientific theory will never be overturned, there is currently not even a hint of evidence that disagrees with the special theory of relativity.

36.10 Relativistic Energy

Energy is our final topic in this chapter on relativity. Space, time, velocity, and momentum are changed by relativity, so it seems inevitable that we'll need a new view of energy.

In Newtonian mechanics, a particle's kinetic energy $K = \frac{1}{2}mu^2$ can be written in terms of its momentum $p = mu$ as $K = p^2/2m$. This suggests that a relativistic expression for energy will likely involve both the square of p and the particle's mass. We also hope that energy will be conserved in relativity, so a reasonable starting point is with the one quantity we've found that is the same in all inertial reference frames: the spacetime interval s.

Let a particle of mass m move through distance Δx during a time interval Δt, as measured in reference frame S. The spacetime interval is

$$s^2 = c^2(\Delta t)^2 - (\Delta x)^2 = \text{invariant}$$

We can turn this into an expression involving momentum if we multiply by $(m/\Delta\tau)^2$, where $\Delta\tau$ is the proper time (i.e., the time measured by the particle). Doing so gives

$$(mc)^2\left[\frac{\Delta t}{\Delta\tau}\right]^2 - \left[\frac{m\Delta x}{\Delta\tau}\right]^2 = (mc)^2\left[\frac{\Delta t}{\Delta\tau}\right]^2 - p^2 = \text{invariant} \quad (36.37)$$

where we used $p = m(\Delta x/\Delta\tau)$ from Equation 36.32.

Now Δt, the time interval in frame S, is related to the proper time by the time-dilation result $\Delta t = \gamma_p\Delta\tau$. With this change, Equation 36.37 becomes

$$(\gamma_p mc)^2 - p^2 = \text{invariant}$$

Finally, for reasons that will be clear in a minute, multiply by c^2, to get

$$(\gamma_p mc^2)^2 - (pc)^2 = \text{invariant} \quad (36.38)$$

To say that the right side is an *invariant* means it has the same value in all inertial reference frames. We can easily determine the constant by evaluating it in the reference frame in which the particle is at rest. In that frame, where $p = 0$ and $\gamma_p = 1$, we find that

$$(\gamma_p mc^2)^2 - (pc)^2 = (mc^2)^2 \tag{36.39}$$

Let's reflect on what this means before taking the next step. The spacetime interval s has the same value in all inertial reference frames. In other words, $c^2(\Delta t)^2 - (\Delta x)^2 = c^2(\Delta t')^2 - (\Delta x')^2$. Equation 36.39 was derived from the definition of the spacetime interval, hence the quantity mc^2 is also an invariant having the same value in all inertial reference frames. In other words, if experimenters in frames S and S' both make measurements on this particle of mass m, they will find that

$$(\gamma_p mc^2)^2 - (pc)^2 = (\gamma_p' mc^2)^2 - (p'c)^2 \tag{36.40}$$

Experimenters in different reference frames measure different values for the momentum, but experimenters in all reference frames agree that momentum is a conserved quantity. Equations 36.39 and 36.40 suggest that the quantity $\gamma_p mc^2$ is also an important property of the particle, a property that changes along with p in just the right way to satisfy Equation 36.39. But what is this property?

The first clue comes from checking the units. γ_p is dimensionless and c is a velocity, so $\gamma_p mc^2$ has the same units as the classical expression $\frac{1}{2}mv^2$; namely, units of energy. For a second clue, let's examine how $\gamma_p mc^2$ behaves in the low-velocity limit $u \ll c$. We can use the binomial approximation expression for γ_p to find

$$\gamma_p mc^2 = \frac{mc^2}{\sqrt{1 - u^2/c^2}} \approx \left(1 + \frac{1}{2}\frac{u^2}{c^2}\right)mc^2 = mc^2 + \frac{1}{2}mu^2 \tag{36.41}$$

The second term, $\frac{1}{2}mu^2$, is the low-velocity expression for the kinetic energy K. This is an energy associated with motion. But the first term suggests that the concept of energy is more complex than we originally thought. It appears that **there is an inherent energy associated with mass itself.**

With that as a possibility, subject to experimental verification, let's define the **total energy** E of a particle to be

$$E = \gamma_p mc^2 = E_0 + K = \text{rest energy} + \text{kinetic energy} \tag{36.42}$$

This total energy consists of a **rest energy**

$$E_0 = mc^2 \tag{36.43}$$

and a relativistic expression for the *kinetic energy*

$$K = (\gamma_p - 1)mc^2 = (\gamma_p - 1)E_0 \tag{36.44}$$

This expression for the kinetic energy is very nearly $\frac{1}{2}mu^2$ when $u \ll c$ but, as Figure 36.36 shows, differs significantly from the classical value for very high velocities.

Equation 36.43 is, of course, Einstein's famous $E = mc^2$, perhaps the most famous equation in all of physics. Before discussing its significance, we need to tie up some loose ends. First, notice that the right-hand side of Equation 36.39 is the square of the rest energy E_0. Thus we can write a final version of that equation:

$$E^2 - (pc)^2 = E_0^2 \tag{36.45}$$

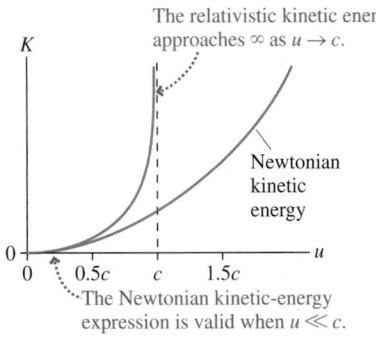

The relativistic kinetic energy approaches ∞ as $u \to c$.

Newtonian kinetic energy

The Newtonian kinetic-energy expression is valid when $u \ll c$.

FIGURE 36.36 The relativistic kinetic energy.

The quantity E_0 is an *invariant* with the same value mc^2 in *all* inertial reference frames.

Second, notice that we can write

$$pc = (\gamma_p mu)c = \frac{u}{c}(\gamma_p mc^2)$$

But $\gamma_p mc^2$ is the total energy E and $u/c = \beta_p$, where the subscript p, as on γ_p, implies that we're referring to the motion of a particle within a reference frame, not the motion of two reference frames relative to each other. Thus

$$pc = \beta_p E \qquad (36.46)$$

Figure 36.37 shows the "velocity-energy-momentum triangle," a convenient way to remember the relationships between the three quantities.

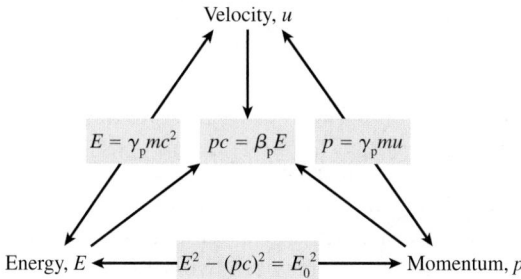

FIGURE 36.37 The velocity-energy-momentum triangle.

EXAMPLE 36.12 Kinetic energy and total energy
Calculate the rest energy and the kinetic energy of (a) a 100 g ball moving with a speed of 100 m/s and (b) an electron with a speed of 0.999c.

MODEL The ball, with $u \ll c$, is a classical particle. We don't need to use the relativistic expression for its kinetic energy. The electron is highly relativistic.

SOLVE

a. For the ball, with $m = 0.10$ kg,

$$E_0 = mc^2 = 9.0 \times 10^{15} \text{ J}$$

$$K = \frac{1}{2}mu^2 = 500 \text{ J}$$

b. For the electron, we start by calculating

$$\gamma_p = 1/(1 - u^2/c^2)^{1/2} = 22.4$$

Then, using $m_e = 9.11 \times 10^{-31}$ kg,

$$E_0 = mc^2 = 8.2 \times 10^{-14} \text{ J}$$

$$K = (\gamma_p - 1)E_0 = 170 \times 10^{-14} \text{ J}$$

ASSESS The ball's kinetic energy is a typical kinetic energy. Its rest energy, by contrast, is a staggeringly large number. For a relativistic electron, on the other hand, the kinetic energy is more important than the rest energy.

STOP TO THINK 36.8 An electron moves through the lab at 99% the speed of light. The lab reference frame is S and the electron's reference frame is S′. In which reference frame is the electron's rest mass larger?

a. Frame S, the lab frame
b. Frame S′, the electron's frame
c. It is the same in both frames.

FIGURE 36.38 An inelastic collision between two balls of clay does not seem to conserve the total energy E.

The tracks of elementary particles in a bubble chamber show the creation of an electron-positron pair. The negative electron and positive positron spiral in opposite directions in the magnetic field.

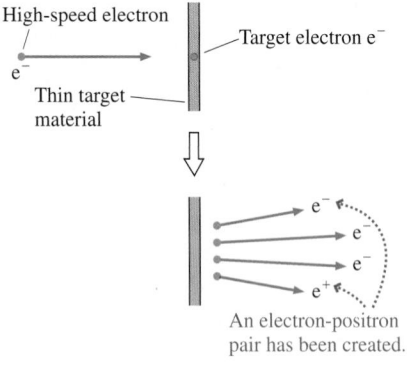

FIGURE 36.39 An inelastic collision between electrons can create an electron-positron pair.

Mass-Energy Equivalence

Now we're ready to explore the significance of Einstein's famous equation $E = mc^2$. Figure 36.38 shows two balls of clay approaching each other. They have equal masses, equal kinetic energies, and slam together in a perfectly inelastic collision to form one large ball of clay at rest. In Newtonian mechanics, we would say that the initial energy $2K$ is dissipated by being transformed into an equal amount of thermal energy, raising the temperature of the coalesced ball of clay. But Equation 36.42, $E = E_0 + K$, doesn't say anything about thermal energy. The total energy before the collision is $E_i = 2mc^2 + 2K$, with the factor of 2 appearing because there are two masses. It seems like the total energy after the collision, when the clay is at rest, should be $2mc^2$, but this value doesn't conserve total energy.

There's ample experimental evidence that energy is conserved, so there must be a flaw in our reasoning. The statement of energy conservation is

$$E_f = Mc^2 = E_i = 2mc^2 + 2K \qquad (36.47)$$

where M is the mass of clay after the collision. But, remarkably, this requires

$$M = 2m + \frac{2K}{c^2} \qquad (36.48)$$

In other words, **mass is not conserved.** The mass of clay after the collision is larger than the mass of clay before the collision. Total energy can be conserved only if kinetic energy is transformed into an "equivalent" amount of mass.

The mass increase in a collision between two balls of clay is incredibly small, far beyond any scientist's ability to detect. So how do we know if such a crazy idea is true?

Figure 36.39 shows an experiment that has been done countless times in the last 50 years at particle accelerators around the world. An electron that has been accelerated to $u \approx c$ is aimed at a target material. When a high-energy electron collides with an atom in the target, it can easily knock one of the electrons out of the atom. Thus we would expect to see two electrons leaving the target: the incident electron and the ejected electron. Instead, *four* particles emerge from the target: three electrons and a positron. A *positron*, or positive electron, is the antimatter version of an electron, identical to an electron in all respects other than having charge $q = +e$.

In chemical-reaction notation, the collision is

$$e^-(\text{fast}) + e^-(\text{at rest}) \rightarrow e^- + e^- + e^- + e^+$$

An electron and a positron have been *created*, apparently out of nothing. Mass $2m_e$ before the collision has become mass $4m_e$ after the collision. (Notice that charge has been conserved in this collision.)

Although the mass has increased, it wasn't created "out of nothing." This is an inelastic collision, just like the collision of the balls of clay, because the kinetic energy after the collision is less than before. In fact, if you measured the energies before and after the collision, you would find that the decrease in kinetic energy is exactly equal to the energy equivalent of the two particles that have been created: $\Delta K = 2m_e c^2$. The new particles have been created *out of energy!*

Particles can be created from energy and particles can return to energy. Figure 36.40 shows an electron colliding with a positron, its antimatter partner. When a particle and its antiparticle meet, they *annihilate* each other. The mass disappears, and the energy equivalent of the mass is transformed into two high-energy photons of light. Momentum conservation requires two photons, rather than one, and specifies that the two photons have equal energies and be emitted back to back.

If the electron and positron are fairly slow, so that $K \ll mc^2$, then $E_i \approx E_0 = mc^2$. In that case, energy conservation requires

$$E_f = 2E_{photon} = E_i \approx 2m_e c^2 \qquad (36.49)$$

You learned in Chapter 24 that the energy of a photon of light is $E_{photon} = hc/\lambda$, where h is Planck's constant. (Photons and their properties will be discussed again in Chapter 38.) Hence the wavelength of the emitted photons is

$$\lambda = \frac{hc}{m_e c^2} \approx 0.0024 \text{ nm} \qquad (36.50)$$

This is an extremely short wavelength, even shorter than the wavelengths of x rays. Photons in this wavelength range are called *gamma rays*. And, indeed, the emission of 0.0024 nm gamma rays is observed in many laboratory experiments in which positrons are able to collide with electrons and thus annihilate. In recent years, with the advent of gamma-ray telescopes on satellites, astronomers have found 0.0024 nm photons coming from many places in the universe, especially galactic centers—evidence that positrons are abundant throughout the universe.

Positron-electron annihilation is also the basis of the medical procedure known as a positron-emission tomography, or PET scans. A patient ingests a very small amount of a radioactive substance that decays by the emission of positrons. This substance is taken up by certain tissues in the body, especially those tissues with a high metabolic rate. As the substance decays, the positrons immediately collide with electrons, annihilate, and create two gamma-ray photons that are emitted back to back. The gamma rays, which easily leave the body, are detected, and their trajectories are traced backward into the body. The overlap of many such trajectories shows quite clearly the tissue where the positron emission is occurring. The results are usually shown as false-color photographs, with redder areas indicating regions of higher positron emission.

Conservation of Energy

The creation and annihilation of particles with mass, processes strictly forbidden in Newtonian mechanics, are vivid proof that neither mass nor the Newtonian definition of energy are conserved. Even so, the *total* energy—the kinetic energy *and* the energy equivalent of mass—remains a conserved quantity.

> **Law of conservation of total energy** The energy $E = \Sigma E_i$ of an isolated system is conserved, where $E_i = (\gamma_p)_i m_i c^2$ is the total energy of particle i.

Mass and energy are not the same thing, but, as the last few examples have shown, they are *equivalent* in the sense that mass can be transformed into energy and energy can be transformed into mass as long as the total mass is conserved.

Probably the most well-known application of the conservation of total energy is nuclear fission. The uranium isotope ^{236}U, containing 236 protons and neutrons, does not exist in nature. It can be created when a ^{235}U nucleus absorbs a neutron, increasing its atomic mass from 235 to 236. The ^{236}U nucleus quickly fragments into two smaller nuclei and several extra neutrons, a process known as **nuclear fission.** The nucleus can fragment in several ways, but one is

$$n + {}^{235}U \rightarrow {}^{236}U \rightarrow {}^{144}Ba + {}^{89}Kr + 3n$$

Ba and Kr are the atomic symbols for barium and krypton.

This reaction seems like an ordinary chemical reaction—until you check the masses. The masses of atomic isotopes are known with great precision from many decades of measurement in instruments called mass spectrometers. If you add up

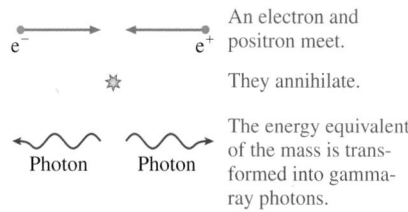

FIGURE 36.40 The annihilation of an electron-positron pair.

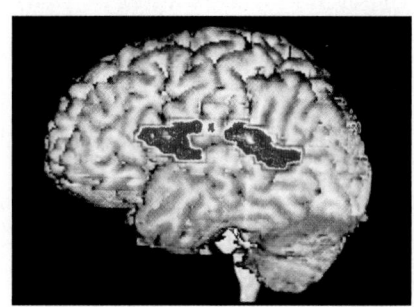

Positron-electron annihilation (a PET scan) provides a noninvasive look into the brain.

The mass of the reactants is 0.185 u more than the mass of the products.

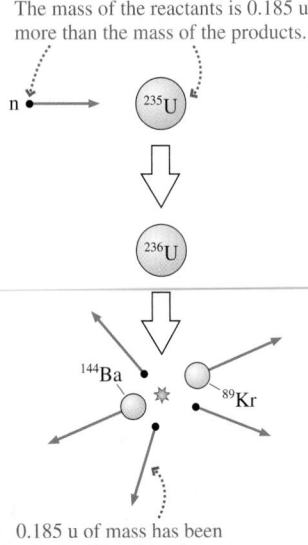

0.185 u of mass has been converted into kinetic energy.

FIGURE 36.41 In nuclear fission, the energy equivalent of lost mass is converted into kinetic energy.

the masses on both sides, you find that the mass of the products is 0.185 u smaller than the mass of the initial neutron and ^{235}U, where, you will recall, 1 u = 1.66 × 10^{-27} kg is the unified atomic mass unit. Converting to kilograms gives us the mass loss of 3.07 × 10^{-28} kg.

Mass has been lost, but the energy equivalent of the mass has not. As Figure 36.41 shows, the mass has been converted to kinetic energy, causing the two product nuclei and three neutrons to be ejected at very high speeds. The kinetic energy is easily calculated: $\Delta K = m_{lost}c^2 = 2.8 \times 10^{-11}$ J.

This is a very tiny amount of energy, but it is the energy released from *one* fission. The number of nuclei in a macroscopic sample of uranium is on the order of N_A, Avogadro's number. Hence the energy available if *all* the nuclei fission is enormous. This energy, of course, is the basis for both nuclear power reactors and nuclear weapons.

We started this chapter with an expectation that relativity would challenge our basic notions of space and time. We end by finding that relativity changes our understanding of mass and energy. Most remarkable of all is that each and every one of these new ideas flows from one simple statement: The laws of physics are the same in all inertial reference frames.

SUMMARY

The goal of Chapter 36 has been to understand how Einstein's theory of relativity changes our concepts of space and time.

GENERAL PRINCIPLES

Principle of Relativity All the laws of physics are the same in all inertial reference frames.

- The speed of light c is the same in all inertial reference frames.
- No particle or causal influence can travel at a speed greater than c.

IMPORTANT CONCEPTS

Space

Spatial measurements depend on the motion of the experimenter relative to the events. An object's length is the difference between *simultaneous* measurement of the positions of both ends.

Proper length ℓ is the length of an object measured in a reference frame in which the object is at rest. The L in a frame in which the object moves with velocity v is

$$L = \sqrt{1 - \beta^2}\,\ell \leq \ell$$

This is called **length contraction.**

Time

Time measurements depend on the motion of the experimenter relative to the events. Events that are simultaneous in reference frame S are not simultaneous in frame S′ moving relative to S.

Proper time $\Delta\tau$ is the time interval between two events measured in a reference frame in which the events occur at the same position. The time interval Δt in a frame moving with relative velocity v is

$$\Delta t = \Delta\tau/\sqrt{1 - \beta^2} \geq \Delta\tau$$

This is called **time dilation.**

Momentum

The law of conservation of momentum is valid in all inertial reference frames if the momentum of a particle with velocity u is $p = \gamma_p m u$, where

$$\gamma_p = 1/\sqrt{1 - u^2/c^2}$$

The momentum approaches ∞ as $u \to c$.

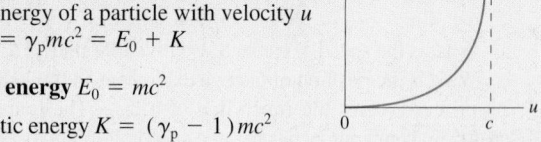

Invariants are quantities that have the same value in all inertial reference frames.

Spacetime interval: $s^2 = (c\Delta t)^2 - (\Delta x)^2$

Particle rest energy: $E_0^2 = (mc^2)^2 = E^2 - (pc)^2$

Energy

The law of conservation of energy is valid in all inertial reference frames if the energy of a particle with velocity u is $E = \gamma_p mc^2 = E_0 + K$

Rest energy $E_0 = mc^2$

Kinetic energy $K = (\gamma_p - 1)mc^2$

Mass-energy equivalence

Mass m can be transformed into energy $E = mc^2$.

Energy can be transformed into mass $m = \Delta E/c^2$.

APPLICATIONS

An event happens at a specific place in space and time. Spacetime coordinates are (x, t) in frame S and (x', t') in frame S′.

A reference frame is a coordinate system with meter sticks and clocks for measuring events. Experimenters at rest relative to each other share the same reference frame.

The Lorentz transformations transform spacetime coordinates and velocities between reference frames S and S′.

$$x' = \gamma(x - vt) \qquad x = \gamma(x' + vt')$$
$$y' = y \qquad\qquad y = y'$$
$$z' = z \qquad\qquad z = z'$$
$$t' = \gamma(t - vx/c^2) \qquad t = \gamma(t' + vx'/c^2)$$
$$u' = \frac{u - v}{1 - uv/c^2} \qquad u = \frac{u' + v}{1 + u'v/c^2}$$

where u and u' are the x- and x'-components of velocity.

$$\beta = \frac{v}{c} \quad \text{and} \quad \gamma = 1/\sqrt{1 - v^2/c^2} = 1/\sqrt{1 - \beta^2}$$

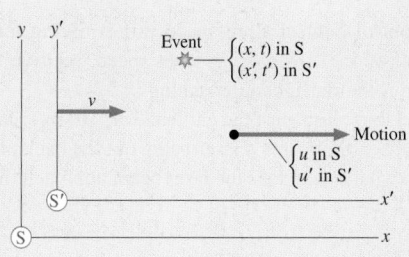

TERMS AND NOTATION

special relativity	simultaneous	invariant
reference frame	relativity of simultaneity	spacetime interval, s
inertial reference frame	light clock	Lorentz transformations
Galilean principle of relativity	rest frame	causal influence
ether	proper time, $\Delta\tau$	total energy, E
principle of relativity	time dilation	rest energy, E_0
event	light year, ly	law of conservation of total energy
spacetime coordinates, (x, y, z, t)	proper length, ℓ	nuclear fission
synchronized	length contraction	

EXERCISES AND PROBLEMS

Exercises

Section 36.2 Galilean Relativity

1. At $t = 1$ s, a firecracker explodes at $x = 10$ m in reference frame S. Four seconds later, a second firecracker explodes at $x = 20$ m. Reference frame S′ moves in the x-direction at a speed of 5 m/s. What are the positions and times of these two events in frame S′?

2. A firecracker explodes in reference frame S at $t = 1$ s. A second firecracker explodes at the same position at $t = 3$ s. In reference frame S′, which moves in the x-direction at speed v, the first explosion is detected at $x' = 4$ m and the second at $x' = -4$ m.
   a. What is the speed of frame S′ relative to frame S?
   b. What is the position of the two explosions in frame S?

3. A sprinter crosses the finish line of a race. The roar of the crowd in front approaches her at a speed of 360 m/s. The roar from the crowd behind her approaches at 330 m/s. What are the speed of sound and the speed of the sprinter?

4. A baseball pitcher can throw a ball with a speed of 40 m/s. He is in the back of a pickup truck that is driving away from you. He throws the ball in your direction, and it floats toward you at a lazy 10 m/s. What is the speed of the truck?

5. A boy on a skateboard coasts along at 5 m/s. He has a ball that he can throw at a speed of 10 m/s. What is the ball's speed relative to the ground if he throws the ball
   a. forward?
   b. backward?
   c. to the side?

Section 36.3 Einstein's Principle of Relativity

6. An out-of-control alien spacecraft is diving into a star at a speed of 1×10^8 m/s. At what speed, relative to the spacecraft, is the starlight approaching?

7. A starship blasts past the earth at 2×10^8 m/s. Just after passing the earth, it fires a laser beam out the back of the starship. With what speed does the laser beam approach the earth?

8. A positron moving in the positive x-direction at 2×10^8 m/s collides with an electron at rest. The positron and electron annihilate, producing two gamma-ray photons. Photon 1 travels in the positive x-direction and photon 2 travels in the negative x-direction. What is the speed of each photon?

Section 36.4 Events and Measurements

Section 36.5 The Relativity of Simultaneity

9. Your job is to synchronize the clocks in a reference frame. You are going to do so by flashing a light at the origin at $t = 0$ s. To what time should the clock at $(x, y, z) = (30$ m, 40 m, 0 m$)$ be preset?

10. Bjorn is standing at $x = 600$ m. Firecracker 1 explodes at the origin and firecracker 2 explodes at $x = 900$ m. The flashes from both explosions reach Bjorn's eye at $t = 3$ μs. At what time did each firecracker explode?

11. Bianca is standing at $x = 600$ m. Firecracker 1, at the origin, and firecracker 2, at $x = 900$ m, explode simultaneously. The flash from firecracker 1 reaches Bianca's eye at $t = 3$ μs. At what time does she see the flash from firecracker 2?

12. You are standing at $x = 9$ km. Lightning bolt 1 strikes at $x = 0$ km and lightning bolt 2 strikes at $x = 12$ km. Both flashes reach your eye at the same time. Your assistant is standing at $x = 3$ km. Does your assistant see the flashes at the same time? If not, which does she see first and what is the time difference between the two?

13. You are standing at $x = 9$ km and your assistant is standing at $x = 3$ km. Lightning bolt 1 strikes at $x = 0$ km and lightning bolt 2 strikes at $x = 12$ km. You see the flash from bolt 2 at $t = 10$ μs and the flash from bolt 1 at $t = 50$ μs. According to your assistant, were the lightning strikes simultaneous? If not, which occurred first and what was the time difference between the two?

14. Jose is looking to the east. Lightning bolt 1 strikes a tree 300 m from him. Lightning bolt 2 strikes a barn 900 m from him in the same direction. Jose sees the tree strike 1 μs before he sees the barn strike. According to Jose, were the lightning strikes simultaneous? If not, which occurred first and what was the time difference between the two?

15. You are flying your personal rocketcraft at 0.9c from Star A toward Star B. The distance between the stars, in the stars' reference frame, is 1 ly. Both stars happen to explode simultaneously in your reference frame at the instant you are exactly halfway between them. Do you see the flashes simultaneously? If not, which do you see first and what is the time difference between the two?

Section 36.6 Time Dilation

16. A cosmic ray travels 60 km through the earth's atmosphere in 400 μs, as measured by experimenters on the ground. How long does the journey take according to the cosmic ray?

17. At what speed, as a fraction of c, does a moving clock tick at half the rate of an identical clock at rest?

18. An astronaut travels to a star system 4.5 ly away at a speed of 0.9c. Assume that the time needed to accelerate and decelerate is negligible.
    a. How long does the journey take according to Mission Control on earth?
    b. How long does the journey take according to the astronaut?
    c. How much time elapses between the launch and the arrival of the first radio message from the astronaut saying that she has arrived?

19. A starship voyages to a distant planet 10 ly away. The explorers stay 1 yr, return at the same speed, and arrive back on earth 26 yr after they left. Assume that the time needed to accelerate and decelerate is negligible.
    a. What is the speed of the starship?
    b. How much time has elapsed on the astronauts' chronometers?

20. You fly 5000 km across the United States on an airliner at 250 m/s. You return two days later.
    a. Have you aged more or less than your friends at home?
    b. By how much?
    Hint: Use the binomial approximation.

21. Two clocks are synchronized. One is placed in a race car that drives around a 2.0-km-diameter track at 100 m/s for 24 hours. Afterward, by how much do the two clocks differ?
    Hint: Use the binomial approximation.

Section 36.7 Length Contraction

22. At what speed, as a fraction of c, will a moving rod have a length 60% that of an identical rod at rest?

23. Jill claims that her new rocket is 100 m long. As she flies past your house, you measure the rocket's length and find that it is only 80 m. Should Jill be cited for exceeding the 0.5c speed limit?

24. A muon travels 60 km through the atmosphere at a speed of 0.9997c. According to the muon, how thick is the atmosphere?

25. The Stanford Linear Accelerator (SLAC) accelerates electrons to $c = 0.99999997c$ in a 3.2-km-long tube. If they travel the length of the tube at full speed (they don't, because they are accelerating), how long is the tube in the electrons' reference frame?

26. Our Milky Way galaxy is 100,000 ly in diameter. A spaceship crossing the galaxy measures the galaxy's diameter to be a mere 1 ly.
    a. What is the speed of the spaceship relative to the galaxy?
    b. How long is the crossing time as measured in the galaxy's reference frame?

27. An optical interferometer can detect a displacement of 0.1λ, or $\approx$50 nm. At what speed would a meter stick "shrink" by 50 nm?
    Hint: Use the binomial approximation.

Section 36.8 The Lorentz Transformations

28. An event has spacetime coordinates $(x, t) = (1200 \text{ m}, 2.0 \text{ } \mu\text{s})$ in reference frame S. What are the event's spacetime coordinates (a) in reference frame S' that moves in the positive x-direction at 0.8c and (b) in reference frame S'' that moves in the negative x-direction at 0.8c?

29. A rocket travels in the x-direction at speed 0.6c with respect to the earth. An experimenter on the rocket observes a collision between two comets and determines that the spacetime coordinates of the collision are $(x', t') = (3 \times 10^{10} \text{ m}, 200 \text{ s})$. What are the spacetime coordinates of the collision in earth's reference frame?

30. In the earth's reference frame, a tree is at the origin and a pole is at $x = 30$ km. Lightning strikes both the tree and the pole at $t = 10$ μs. The lightning strikes are observed by a rocket traveling in the x-direction at 0.5c.
    a. What are the spacetime coordinates for these two events in the rocket's reference frame?
    b. Are the events simultaneous in the rocket's frame? If not, which occurs first?

31. A rocket cruising past earth at 0.8c shoots a bullet out the back door, opposite the rocket's motion, at 0.9c relative to the rocket. What is the bullet's speed relative to the earth?

32. A laboratory experiment shoots an electron to the left at 0.9c. What is the electron's speed relative to a proton moving to the right at 0.9c?

33. A distant quasar is found to be moving away from the earth at 0.8c. A galaxy closer to the earth and along the same line of sight is moving away from us at 0.2c. What is the recessional speed of the quasar as measured by astronomers in the other galaxy?

Section 36.9 Relativistic Momentum

34. A proton is accelerated to 0.999c.
    a. What is the proton's momentum?
    b. By what factor does the proton's momentum exceed its Newtonian momentum?

35. A 1 g particle has momentum 400,000 kg m/s. What is the particle's speed?

36. At what speed is a particle's momentum twice its Newtonian value?

37. What is the speed of a particle whose momentum is mc?

Section 36.10 Relativistic Energy

38. What are the kinetic energy, the rest energy, and the total energy of a 1 g particle with a speed of 0.8c?

39. A quarter-pound hamburger with all the fixings has a mass of 200 g. The food energy of the hamburger (480 food calories) is 2 MJ.
    a. What is the energy equivalent of the mass of the hamburger?
    b. By what factor does the energy equivalent exceed the food energy?

40. How fast must an electron move so that its total energy is 10% more than its rest mass energy?

41. At what speed is a particle's kinetic energy twice its rest energy?

42. At what speed is a particle's total energy twice its rest energy?

Problems

43. A 50 g ball moving to the right at 4.0 m/s overtakes and collides with a 100 g ball moving to the right at 2.0 m/s. The collision is perfectly elastic. Use reference frames and the Chapter 10 result for perfectly elastic collisions to find the speed and direction of each ball after the collision.

44. A 300 g ball moving to the right at 2 m/s has a perfectly elastic collision with a 100 g ball moving to the left at 8 m/s. Use reference frames and the Chapter 10 result for perfectly elastic collisions to find the speed and direction of each ball after the collision.

45. A billiard ball has a perfectly elastic collision with a second billiard ball of equal mass. Afterward, the first ball moves to the left at 2.0 m/s and the second to the right at 4.0 m/s. Use reference frames and the Chapter 10 result for perfectly elastic collisions to find the speed and direction of each ball before the collision.

46. A 9.0 kg artillery shell is moving to the right at 100 m/s when suddenly it explodes into two fragments, one twice as heavy as the other. Measurements reveal that 900 J of energy are released in the explosion and that the heavier fragment was in front of the lighter fragment. Find the velocity of each fragment relative to the ground by analyzing the explosion in the reference frame of (a) the ground and (b) the shell. (c) Is the problem easier to solve in one reference frame?

47. The diameter of the solar system is 10 light hours. A spaceship crosses the solar system in 15 hours, as measured on earth. How long, in hours, does the passage take according to passengers on the spaceship?
    Hint: c = 1 light hour per hour.

48. A 30-m-long rocket train car is traveling from Los Angeles to New York at 0.5c when a light at the center of the car flashes. When the light reaches the front of the car, it immediately rings a bell. Light reaching the back of the car immediately sounds a siren.
    a. Are the bell and siren simultaneous events for a passenger seated in the car? If not, which occurs first and by how much time?
    b. Are the bell and siren simultaneous events for a bicyclist waiting to cross the tracks? If not, which occurs first and by how much time?

49. The star Alpha goes supernova. Ten years later and 100 ly away, as measured by astronomers in the galaxy, star Beta explodes.
    a. Is it possible that the explosion of Alpha is in any way responsible for the explosion of Beta? Explain.
    b. An alien spacecraft passing through the galaxy finds that the distance between the two explosions is 120 ly. According to the aliens, what is the time between the explosions?

50. Two events in reference frame S occur 10 μs apart at the same point in space. The distance between the two events is 2400 m in reference frame S'.
    a. What is the time interval between the events in reference frame S'?
    b. What is the velocity of S' relative to S?

51. a. How fast must a rocket travel on a journey to and from a distant star so that the astronauts age 10 years while the Mission Control workers on earth age 120 years?
    b. As measured by Mission Control, how far away is the distant star?

52. In Section 36.6 we explained that muons can reach the ground because of time dilation. But how do things appear in the muon's reference frame, where the muon's half-life is only 1.5 μs? How can a muon travel the 60 km to reach the earth's surface before decaying? Resolve this apparent paradox. Be as quantitative as you can in your answer.

53. A cube has a density of 2000 kg/m³ while at rest in the laboratory. What is the cube's density as measured by an experimenter in the laboratory as the cube moves through the laboratory at 90% of the speed of light in a direction perpendicular to one of its faces?

54. In an attempt to reduce the extraordinarily long travel times for voyaging to distant stars, some people have suggested traveling at close to the speed of light. Suppose you wish to visit the red giant star Betelgeuse, which is 430 ly away, and that you want your 20,000 kg rocket to move so fast that you age only 20 years during the round trip.
    a. How fast must the rocket travel relative to earth?
    b. How much energy is needed to accelerate the rocket to this speed?
    c. Compare this amount of energy to the total energy used by the United States in the year 2000, which was roughly 1.0×10^{20} J.

55. A rocket traveling at 0.5c sets out for the nearest star, Alpha Centauri, which is 4.25 ly away from earth. It will return to earth immediately after reaching Alpha Centauri. What distance will the rocket travel and how long will the journey last according to (a) stay-at-home earthlings and (b) the rocket crew? (c) Which answers are the correct ones, those in part a or those in part b?

56. The star Delta goes supernova. One year later and 2 ly away, as measured by astronomers in the galaxy, star Epsilon explodes. Let the explosion of Delta be at $x_D = 0$ and $t_D = 0$. The explosions are observed by three spaceships cruising through the galaxy in the direction from Delta to Epsilon at velocities $v_1 = 0.3c$, $v_2 = 0.5c$, and $v_3 = 0.7c$.
    a. What are the times of the two explosions as measured by scientists on each of the three spaceships?
    b. Does one spaceship find that the explosions are simultaneous? If so, which one?
    c. Does one spaceship find that Epsilon explodes before Delta? If so, which one?
    d. Do your answers to parts b and c violate the idea of causality? Explain.

57. Two rockets approach each other. Each is traveling at 0.75c in the earth's reference frame. What is the speed of one rocket relative to the other?

58. A military jet traveling at 1500 m/s has engine trouble and the pilot must bail out. Her ejection seat shoots her forward at 300 m/s relative to the jet. How fast is she traveling relative to the ground according to the relativistic velocity transformation? Would the pilot or anyone else be able to tell the difference between the relativistic and the nonrelativistic result?

59. What is the speed of an electron after being accelerated from rest through a 20×10^6 V potential difference?

60. What is the speed of a proton after being accelerated from rest through a 50×10^6 V potential difference?

61. The half-life of a muon at rest is $1.5 \ \mu s$. Muons that have been accelerated to a very high speed and are then held in a circular storage ring have a half-life of $7.5 \ \mu s$.
    a. What is the speed of the muons in the storage ring?
    b. What is the total energy of a muon in the storage ring? The mass of a muon is 207 times the mass of an electron.

62. A solar flare blowing out from the sun at $0.9c$ is overtaking a rocket as it flies away from the sun at $0.8c$. According to the crew on board, with what speed is the flare gaining on the rocket?

63. This chapter has assumed that lengths perpendicular to the direction of motion are not affected by the motion. That is, motion in the x-direction does not cause length contraction along the y- or z-axes. To find out if this is really true, consider two spray-paint nozzles attached to rods perpendicular to the x-axis. It has been confirmed that, when both rods are at rest, both nozzles are exactly 1 m above the base of the rod. One rod is placed in the S reference frame with its base on the x-axis; the other is placed in the S′ reference frame with its base on the x'-axis. The rods then swoop past each other and, as Figure P36.63 shows, each paints a stripe across the other rod.

    We will use proof by contradiction. Assume that objects perpendicular to the motion *are* contracted. An experimenter in frame S finds that the S′ nozzle, as it goes past, is less than 1 m above the x-axis. The principle of relativity says that an experiment carried out in two different inertial reference frames will have the same outcome in both.
    a. Pursue this line of reasoning and show that you end up with a logical contradiction, two mutually incompatible situations.
    b. What can you conclude from this contradiction?

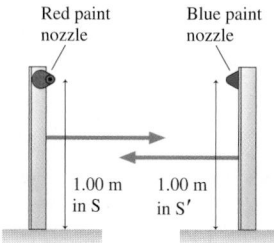

FIGURE P36.63

64. Derive the Lorentz transformations for t and t'.
    Hint: See the comment following Equation 36.22.

65. a. Derive a velocity transformation equation for u_y and u_y'. Assume that the reference frames are in the standard orientation with motion parallel to the x- and x'-axes.
    b. A rocket passes the earth at $0.8c$. As it goes by, it launches a projectile at $0.6c$ perpendicular to the direction of motion. What is the projectile's speed in the earth's reference frame?

66. What is the momentum of a particle with speed $0.95c$ and total energy 2.0×10^{-10} J?

67. What is the momentum of a particle whose total energy is four times its rest energy? Give your answer as a multiple of mc.

68. a. What are the momentum and total energy of a proton with speed $0.99c$?
    b. What is the proton's momentum in a different reference frame in which $E' = 5.0 \times 10^{-10}$ J?

69. At what speed is the kinetic energy of a particle twice its Newtonian value?

70. What is the speed of an electron whose total energy equals the rest mass of a proton?

71. A typical nuclear power plant generates electricity at the rate of 1000 MW. The efficiency of transforming thermal energy into electrical energy is $\frac{1}{3}$ and the plant runs at full capacity for 80% of the year. (Nuclear power plants are down about 20% of the time for maintenance and refueling.)
    a. How much thermal energy does the plant generate in one year?
    b. What mass of uranium is transformed into energy in one year?

72. The sun radiates energy at the rate 3.8×10^{26} W. The source of this energy is fusion, a nuclear reaction in which mass is transformed into energy. The mass of the sun is 2.0×10^{30} kg.
    a. How much mass does the sun lose each year?
    b. What percentage is this of the sun's total mass?
    c. Estimate the lifetime of the sun.

73. The radioactive element radium (Ra) decays by a process known as *alpha decay,* in which the nucleus emits a helium nucleus. (These high-speed helium nuclei were named alpha particles when radioactivity was first discovered, long before the identity of the particles was established.) The reaction is ^{226}Ra $\rightarrow$ ^{222}Rn $+$ ^{4}He, where Rn is the element radon. The accurately measured atomic masses of the three atoms are 226.015, 222.017, and 4.003. How much energy is released in each decay? (The energy released in radioactive decay is what makes nuclear waste "hot.")

74. The nuclear reaction that powers the sun is the fusion of four protons into a helium nucleus. The process involves several steps, but the net reaction is simply $4p \rightarrow {}^4$He $+$ energy. The mass of a helium nucleus is known to be 6.64×10^{-27} kg.
    a. How much energy is released in each fusion?
    b. What fraction of the initial rest mass energy is this energy?

75. An electron moving to the right at $0.9c$ collides with a positron moving to the left at $0.9c$. The two particles annihilate and produce two gamma-ray photons. What is the wavelength of the photons?

76. Section 36.10 looked at the inelastic collision e^- (fast) $+$ e^- (at rest) $\rightarrow e^- + e^- + e^- + e^+$.
    a. What is the threshold kinetic energy of the fast electron? That is, what minimum kinetic energy must the electron have to allow this process to occur?
    b. What is the speed of an electron with the threshold kinetic energy?

Challenge Problems

77. Two rockets, A and B, approach the earth from opposite directions at speed $0.8c$. The length of each rocket measured in its rest frame is 100 m. What is the length of rocket A as measured by the crew of rocket B?

78. Two rockets are each 1000 m long in their rest frame. Rocket Orion, traveling at $0.8c$ relative to the earth, is overtaking rocket Sirius, which is poking along at a mere $0.6c$. According to the crew on Sirius, how long does Orion take to completely pass? That is, how long is it from the instant the nose of Orion is at the tail of Sirius until the tail of Orion is at the nose of Sirius?

79. Some particle accelerators allow protons (p^+) and antiprotons (p^-) to circulate at equal speeds in opposite directions in a device called a *storage ring*. The particle beams cross each other at various points to cause $p^+ + p^-$ collisions. In one collision, the outcome is $p^+ + p^- \rightarrow e^+ + e^- + \gamma + \gamma$, where γ represents a high-energy gamma-ray photon. The electron and positron are ejected from the collision at $0.9999995c$ and the gamma-ray photon wavelengths are found to be 1.0×10^{-6} nm. What were the proton and antiproton speeds prior to the collision?

80. The rockets of the Goths and the Huns are each 1000 m long in their rest frame. The rockets pass each other, virtually touching, at a relative speed of $0.8c$. The Huns have a laser cannon at the rear of their rocket that shoots a deadly laser beam at right angles to the motion. The captain of the Hun rocket wants to send a threatening message to the Goths by "firing a shot across their bow." He tells his first mate, "The Goths' rocket is length contracted to 600 m. Fire the laser cannon at the instant the nose of our rocket passes the tail of their rocket. The laser beam will cross 400 m in front of them." But things are different in the Goths' reference frame. The Goth captain muses, "The Huns' rocket is length contracted to 600 m, 400 m shorter than our rocket. If they fire the laser cannon as their nose passes the tail of our rocket, the lethal laser blast will go right through our side."

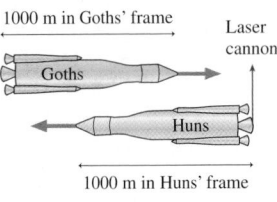

1000 m in Goths' frame

Laser cannon

Goths

Huns

1000 m in Huns' frame

FIGURE CP36.80

The first mate on the Hun rocket fires as ordered. Does the laser beam blast the Goths or not? Resolve this paradox. Show that, when properly analyzed, the Goths and the Huns agree on the outcome. Your analysis should contain both quantitative calculations and written explanation.

81. A very fast pole vaulter lives in the country. One day, while practicing, he notices a 10-m-long barn with the doors open at both ends. He decides to run through the barn at $0.866c$ while carrying his 16-m-long pole. The farmer, who sees him coming, says. "Aha! This guy's pole is length contracted to 8 m. There will be a short interval of time when the pole is entirely inside the barn. If I'm quick, I can simultaneously close both barn doors while the pole vaulter and his pole are inside." The pole vaulter, who sees the farmer beside the barn, thinks to himself, "That farmer is crazy. The barn is length contracted and is only 5 m long. My 16-m-long pole cannot fit into a 5-m-long barn. If the farmer closes the doors just as the tip of my pole reaches the back door, the front door will break off the last 11 m of my pole."

Can the farmer close the doors without breaking the pole? Show that, when properly analyzed, the farmer and the pole vaulter agree on the outcome. Your analysis should contain both quantitative calculations and written explanation. It's obvious that the pole vaulter cannot stop quickly, so you can assume that the doors are paper thin and that the pole breaks through without slowing down.

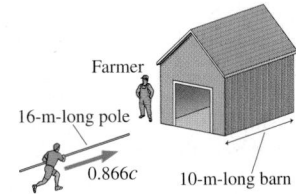

Farmer

16-m-long pole

0.866c

10-m-long barn

FIGURE CP36.81

STOP TO THINK ANSWERS

Stop to Think 36.1: a, c, and **f.** These move at constant velocity, or very nearly so. The others are accelerating.

Stop to Think 36.2: a. $u' = u - v = -10$ m/s $- 6$ m/s $= -16$ m/s. The *speed* is 16 m/s.

Stop to Think 36.3: c. Even the light has a slight travel time. The event is the hammer hitting the nail, not your seeing the hammer hit the nail.

Stop to Think 36.4: At the same time. Mark is halfway between the tree and the pole, so the fact that he *sees* the lightning bolts at the same time means they *happened* at the same time. It's true that Nancy *sees* event 1 before event 2, but the events actually occurred before she sees them. Mark and Nancy share a reference frame, because they are at rest relative to each other, and all experimenters in a reference frame, after correcting for any signal delays, *agree* on the spacetime coordinates of an event.

Stop to Think 36.5: After. This is the same as the case of Peggy and Ryan. In Mark's reference frame, as in Ryan's, the events are simultaneous. Nancy *sees* event 1 first, but the time when an event is seen is not when the event actually happens. Because all experi-

menters in a reference frame agree on the spacetime coordinates of an event, Nancy's position in her reference frame cannot affect the order of the events. If Nancy had been passing Mark at the instant the lightning strikes occur in Mark's frame, then Nancy would be equivalent to Peggy. Event 2, like the firecracker at the front of Peggy's railroad car, occurs first in Nancy's reference frame.

Stop to Think 36.6: c. Nick measures proper time because Nick's clock is present at both the "nose passes Nick" event and the "tail passes Nick" event. Proper time is the smallest measured time interval between two events.

Stop to Think 36.7: $L_A > L_B = L_C$. Anjay measures the pole's proper length because it is at rest in his reference frame. Proper length is the longest measured length. Beth and Charles may *see* the pole differently, but they share the same reference frame and their *measurements* of the length agree.

Stop to Think 36.8: c. The rest energy E_0 is an invariant, the same in all inertial reference frames. Thus $m = E_0/c^2$ is independent of speed.

37 The End of Classical Physics

Studies of the light emitted by gas discharge tubes helped bring classical physics to an end.

Except for relativity, and a brief preview of quantum physics in Chapter 24, everything we have studied until now was known by 1900. Newtonian mechanics, thermodynamics, and Maxwell's theory of electromagnetism form what we call *classical physics*. It is an impressive body of knowledge, with immense explanatory power. Many scientists of the late 1800s felt that they could use these theories to explain just about anything, and some even felt there was nothing significant left to discover.

But within the span of just a few years, right around 1900, investigations into the structure of matter led to many astonishing discoveries that were at odds with classical physics. Discoveries that defied explanation came from investigations as simple as measuring the spectrum of light emitted by gas discharge tubes. It was soon recognized that the laws of classical physics break down when applied to atomic systems. Physicists in the early years of the 20th century had to reexamine their most basic assumptions about the nature of matter and light.

Our goal in this chapter is twofold. The first is to learn how scientists in the 19th and early 20th centuries discovered the properties of atoms. Michael Faraday noted long ago that it is "easy to talk of atoms" but quite another thing to have real knowledge of atoms. We cannot see atoms, so what is the *evidence* that leads us to our current understanding of the atomic theory of matter?

Our second goal is to recognize that many of the newly found atomic properties could not be reconciled with classical physics. Before we launch into quantum physics, it is important to recognize where classical physics failed and why a new theory of light and matter was needed.

37.1 Physics in the 1800s

Scientists in 1800 had three major realms of inquiry. They wanted to understand the nature of matter, of electricity, and of light.

Matter

The idea that matter consists of small, indivisible particles is traceable to Leucippus and his student Democritus, who flourished in ancient Greece around 440–420 BCE. They called these particles *atoms,* Greek for "not divisible." Atomism was not widely accepted, due in no little part to the complete lack of evidence for atoms, but atomic ideas did manage to maintain a minority status throughout the Middle Ages. Then, at about the time of Newton and the beginnings of a mechanistic conception of the world, interest in atoms revived.

Robert Boyle began his study of gases in the 1660s, from which we today know Boyle's law as the fact that pV remains constant for an isothermal process. Newton noted that Boyle's law could be explained if a gas consisted of particles. In 1738, Daniel Bernoulli advanced the idea that gases are composed of small, atom-like particles in random motion. However, the *evidence* for atoms was still far too weak for Bernoulli's ideas to be more than a curiosity.

Things began to change in the early years of the 19th century. The English chemist John Dalton argued that much of what was known about chemical reactions, in particular the law of definite proportions, could be understood if all matter of a particular chemical element consisted of identical, indestructible atoms. The unique feature of Dalton's work, which made it more science than speculation, was his attempt to determine the relative masses of the atoms of different elements. These ideas were further extended by the Italian chemist Amedeo Avogadro, who postulated that atoms could stick together to form more complex entities he called *molecules* and that equal volumes of gases at equal temperatures contain equal numbers of molecules.

The evidence for atoms grew stronger as thermodynamics and the kinetic theory of gases developed in the mid-19th century. Two lines of inquiry led to rough estimates of atomic sizes. First, slight deviations from the ideal-gas law at high pressures could be understood if the atoms were beginning to come into close proximity to one another. Second, the viscosity of a gas, which is readily measured, could be related to the mean free paths of the molecules and the mean free paths, in turn, to the sizes of the atoms. The existence of atoms with diameters of approximately 10^{-10} m was widely accepted by 1890.

Electricity

The generation of static electricity by rubbing amber with fur had been known since antiquity, but the discovery of electric currents in the 17th century raised interesting new questions. For example, is the "electric substance" a continuous fluid or does it consist of granular particles of electricity? There was no direct evidence, but the flow of current suggested a fluid of some sort. This supposition was analogous to the prevailing belief that heat was a fluid called *caloric.*

Eighteenth-century studies of electricity relied on electrostatic generators. These mechanical devices used friction to produce large but mostly uncontrolled potential differences. Volta's invention of the battery in 1800 was a major improvement for investigators because a battery provides a controlled and reproducible potential difference. Further, it was perfect for creating currents in wires. Volta's battery stimulated a wave of research on electricity during the opening decades of the 19th century.

It took only two months from Volta's invention of the battery until the discovery that an electric current through water decomposes the water into hydrogen and oxygen, a process that came to be called **electrolysis.** The basic experiment, as it is done today in chemistry classes, is shown in Figure 37.1. The positive and negative terminals of a battery are connected to pieces of metal called *electrodes*. The negative electrode is called the *cathode* and the positive one is the *anode*. Bubbles of gas appear at the electrodes—hydrogen at one and oxygen at the other—and can be collected in tubes.

The outcome of this experiment does not surprise us today, but at the time of its first performance water had long been regarded as one of the basic elements. The decomposition of water forced scientists to reconsider the basic building blocks of matter. Furthermore, these newly discovered effects suggested a previously unsuspected connection between electricity and matter.

FIGURE 37.1 A current through water decomposes it into hydrogen and oxygen.

Light

The question "What is light?" had long been debated. Newton, as we have noted previously, favored a *corpuscular* theory whereby small particles of light travel in straight lines. His conviction was based largely on the sharp shadows cast by sunlight, in contrast to the diffraction of water waves as they pass barriers. Newton's view was dominant throughout the 18th century.

The situation changed quickly as the 19th century opened. In 1801, the English linguist, physician, and scientist Thomas Young demonstrated the interference of light with his celebrated double-slit experiment, shown in Figure 37.2. The wave model of light was given a more rigorous mathematical foundation in 1818 by the French physicist Augustin Fresnel. Fresnel's theory predicted a number of diffraction effects that had not been previously observed. These predicted effects were initially criticized as being contrary to common sense, but their subsequent experimental verification validated the wave model of light.

But if light is a wave, what is waving? What is the medium? How can light travel through a vacuum? The corpuscular theory had not faced these difficulties, but they were brought to the forefront by the overwhelming evidence that light must be some kind of wave.

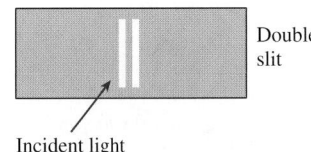

FIGURE 37.2 Young's double-slit experiment showed that light is a wave.

37.2 Faraday

The three lines of inquiry—matter, electricity, and light—came together during the 1820s in the person of Michael Faraday, one of the most remarkable scientific geniuses in history. Faraday conducted three investigations of particular interest to us.

Electrical Conduction in Liquids

Others had already begun to study electrolysis, the conduction of electric current through liquids, but it was Faraday's systematic and careful measurements that revealed the laws governing electrolysis. Faraday showed that electrolysis is most easily understood on the basis of an atomic theory of matter, and he found that there is a *charge* associated with each atom in the solution. Today these charges are called positive and negative *ions*.

In fuel cells, which will power cars in the near future, oxygen and hydrogen are combined to produce water and an electric current. This is the reverse of the electrolysis process shown in Figure 37.1.

Faraday's discoveries implied that

1. Atoms exist.
2. Electric charges are somehow (Faraday did not know how) associated with atoms.
3. There are two different kinds of charge, positive and negative.
4. Electricity is "granular" rather than a continuous fluid. That is, it comes in discrete amounts with a basic unit of charge.

Electrical Conduction in Gases

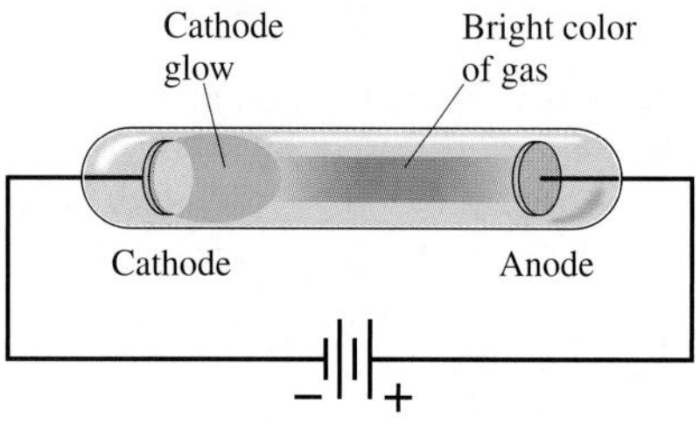

Faraday also investigated whether electric currents could pass through air. He sealed metal electrodes into a glass tube, lowered the pressure with a primitive vacuum pump, then attached an electrostatic generator. When he started the generator, the tube began to glow with a bright purple color! Faraday's device, called a **gas discharge tube,** is shown in Figure 37.3. The photo that opened this chapter is a modern gas discharge tube.

Faraday's investigations showed that

1. Current flows through a low-pressure gas, creating an electric discharge.
2. The color of the discharge depends on the type of gas in the tube.
3. Regardless of the type of gas, there is a separate, constant glow around the negative electrode (i.e., the cathode). This is called the **cathode glow.**

FIGURE 37.3 Faraday's gas discharge tube.

Today we know the purple color to be characteristic of nitrogen, the primary component of air. You are more likely familiar with the reddish-orange color of the neon discharge tubes used for signs, but neon was not discovered until long after Faraday's time. His investigations, though, showed an unexpected connection between the color of the light and type of atoms in the tube.

Electromagnetic Fields

Perhaps Faraday's most important contributions to physics were in the realms of magnetism and light. You will recall that it was Faraday who introduced the concept of electric and magnetic *fields.* Although these fields were first devised simply as a way of envisioning electric and magnetic processes, Faraday's later studies of electromagnetic induction showed that they have a real existence and real properties. These investigations paved the way for the discovery, about 30 years later, that light is an electromagnetic wave.

Faraday's discoveries were a major step toward providing real *evidence* for the existence of atoms. Altogether, Faraday established that atoms are associated with electricity, he demonstrated that different colors of light are associated with different kinds of atoms, and he prepared the way for showing that light is associated with electricity and magnetism.

Matter, electricity, and light—previously three separate ideas—had been intertwined. Even so, Faraday recognized that he had barely scratched the surface, that far more research was needed before atoms were understood.

Although we know nothing of what an atom is, yet we cannot resist forming some idea of a small particle which represents it to the mind; and though we are in equal, if not greater, ignorance of electricity, so as to be unable to say whether it is a particular matter or matters, or mere motion of ordinary matter, or some third kind of power or agent, yet there is an immensity of facts which justify us in believing that the atoms of matter are in some way endowed or associated with electric powers to which they owe their most striking qualities, and amongst them their mutual chemical affinity. . . .

Michael Faraday

37.3 Cathode Rays

Faraday's invention of the gas discharge tube had two major repercussions. One set of investigations, to which we will return in Section 37.8, led to the development of spectroscopy and eventually to quantum physics. Another set of investigations led to the discovery of the electron.

An important technological breakthrough came in the 1850s with the development of much-improved vacuum pumps. The German scientist Julius Plücker began a study of Faraday's gas discharge tube using lower gas pressures, and made two important observations:

1. As he reduced the pressure, the colored glow of the gas diminished and the cathode glow became more extended.
2. If the cathode glow extended to the wall of the glass tube, the glass itself emitted a greenish glow at that point.

A few years later, one of Plücker's students found that a solid object sealed inside the tube casts a *shadow* on the glass wall, as shown in Figure 37.4. This discovery suggested that the cathode emits *rays* of some form that travel in straight lines but are easily blocked by solid objects. These rays, which are invisible but cause the glass to glow where they strike it, were quickly dubbed **cathode rays.** This name lives on today in the *cathode-ray tube* that forms the picture tube in televisions and most computer display terminals. But naming the rays did nothing to explain them. What were they?

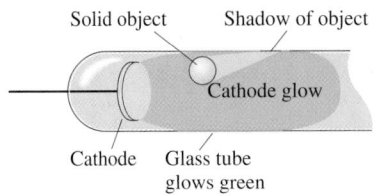

FIGURE 37.4 A solid object in the cathode glow casts a shadow.

Crookes Tubes

The most systematic studies on the new cathode rays were carried out during the 1870s by the English scientist Sir William Crookes. Crookes devised a set of glass tubes, such as the one shown in Figure 37.5, that could be used to make careful studies of cathode rays. His primary innovations were to elongate the tube, use yet lower pressure, and introduce a collimating hole for the rays to pass through. The net result was to generate a well-defined beam of cathode rays that created a small glowing spot where they struck the end of the tube. Today we call his design a **Crookes tube.**

The work of Crookes and others demonstrated that

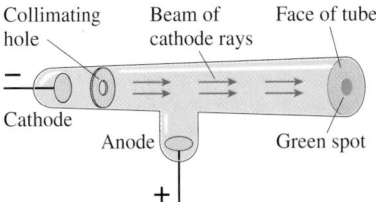

FIGURE 37.5 A Crookes tube.

1. There is an electric current in a tube in which cathode rays are emitted.
2. The rays are deflected by a magnetic field *as if* they are negative charges.
3. Cathodes made of any metal produce cathode rays. Furthermore, the ray properties are independent of the cathode material.
4. The rays can exert forces on objects and can transfer energy to objects. For example, a thin metal foil in the cathode-ray beam glows red hot.

Crookes' experiments led to more questions than they answered. Were the cathode rays some sort of particles? Or a wave? Were the rays themselves the carriers of the electric current, or were they something else that happened to be emitted whenever there was a current? Item 3 is worthy of note because it suggests that the cathode rays are a *fundamental* entity, not a part of the element from which they are emitted.

Although you can read the final answers in a book today, it is important to realize how difficult these questions were at the time and how experimental evidence was used to answer them. Crookes suggested that molecules in the gas collided with the cathode, somehow acquired a negative charge (i.e., became negative ions), and then "rebounded" with great speed as they were repelled by the negative cathode. These "charged molecules" would travel in a straight line, carry energy and momentum, be deflected by a magnetic field, and cause the tube to glow, or *fluoresce,* where they struck the glass. Crookes' theory predicted, of course, that the negative ions should also be deflected by an electric field. Crookes attempted to demonstrate this deflection by sealing electrodes into the tube and creating an electric field, but his efforts were inconclusive. Other than this troublesome difficulty, Crookes' model seemed to explain the observations.

However, Crookes' theory was immediately attacked. Critics noted that the cathode rays could travel the length of a 90-cm-long tube with no discernible deviation from a straight line. But the mean free path for molecules, due to collisions with other molecules, is only about 6 mm at the pressure in Crookes' tubes.

There was no chance at all that molecules could travel in a straight line for 150 times their mean free path! It was later discovered that the cathode rays could even penetrate very thin (≈ 2 μm thick) metal foils, something that no atomic-size particle could do. Crookes' theory, seemingly adequate when it was proposed, was wildly inconsistent with subsequent observations.

But if cathode rays were not particles, what were they? An alternative theory was that the cathode rays were electromagnetic waves. After all, light travels in straight lines, casts shadows, carries energy and momentum, and can, under the right circumstances, cause materials to fluoresce. It was known that hot metals emit light—incandescence—so it seemed plausible that the cathode could be emitting waves. A long path through the gas would present no problem, and it was known by 1890 that radio waves could penetrate thin foils. The major obstacle for the wave theory was the deflection of cathode rays by a magnetic field. But the theory of electromagnetic waves was quite new at the time, and many characteristics of these waves were still unknown. Visible light was not deflected by a magnetic field, but it was easy to think that some other form of electromagnetic waves might be so influenced.

The controversy over particles versus waves was intense. British scientists generally favored particles, but their continental counterparts preferred waves. Such controversies are an integral part of science, for they stimulate the best minds to come forward with new ideas and new experiments.

37.4 J. J. Thomson and the Discovery of the Electron

Shortly after Wilhelm Röntgen's 1895 discovery of x rays, the young English physicist J. J. Thomson began using them to study electrical conduction in gases. He found that x rays could discharge an electroscope and concluded that they must be ionizing the air molecules, thereby making the air conductive. That is, the x rays were splitting the molecules into charged fragments—ions!

This simple observation was of profound significance. Until then, the only form of ionization known was the creation of positive and negative ions in solutions where, for example, a molecule such as NaCl splits into two smaller charged pieces. Although the underlying process was not yet understood, the fact that two atoms could acquire charge as a molecule splits apart did not jeopardize the idea that the atoms themselves were indivisible. But after observing that even monatomic gases, such as helium, could be ionized by x rays, Thomson realized that **the atom itself must have charged constituents that could be split apart!** This was the first direct evidence that the atom is a complex structure, not a fundamental, indivisible unit of matter.

Thomson was also conducting experiments to investigate the nature of cathode rays. One of his first goals was to establish, once and for all, that cathode rays are charged particles. Other scientists, using a Crookes tube like the one shown in Figure 37.6a, had measured an electric current in a cathode-ray beam. Although its presence seemed to demonstrate that the rays are charged particles, proponents of the wave model argued that the current might be a separate, independent event that just happened to be following the same straight line as the rays.

Thomson realized that he could use magnetic deflection of the cathode rays to settle the issue. He built a modified tube, shown in Figure 37.6b, in which the collecting electrode was off to the side. Under normal operation, the cathode rays struck the center of the tube face and created a greenish spot on the glass. No current was measured by the electrode under these circumstances. Thomson then placed the tube in a magnetic field that deflected the cathode rays to the side. He could determine their trajectory by the location of the green spot as it moved

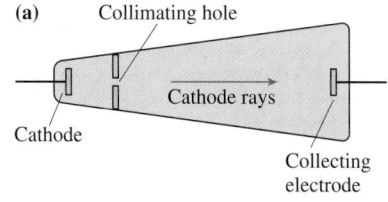

(a) Collimating hole

Cathode rays

Cathode

Collecting electrode

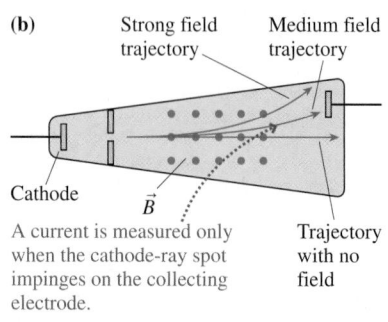

(b) Strong field trajectory   Medium field trajectory

Cathode

$\vec{B}$

A current is measured only when the cathode-ray spot impinges on the collecting electrode.

Trajectory with no field

FIGURE 37.6 Experiments to measure the electric current in a cathode-ray tube.

across the face of the tube. Just at the point when the field was strong enough to deflect the cathode rays onto the electrode, a current was detected! At an even stronger field, when the cathode rays were deflected completely to the other side of the electrode, the current ceased.

This was the first conclusive demonstration that cathode rays really are negatively charged particles. But why were they not deflected by an electric field? Thomson's first efforts to deflect the cathode rays met with the same inconclusive results that others had found, but his experience with the x-ray ionization of gases soon led him to recognize the difficulty. He realized that the rapidly moving cathode-ray particles must be colliding with the few remaining gas molecules in the tube with sufficient energy to *ionize* them by splitting them into charged pieces. The electric field created by these charges neutralized the field of the electrodes, hence there was no deflection.

Fortunately, vacuum technology was getting ever better. By using the most sophisticated techniques of his day, Thomson was able to lower the pressure to a point where ionization of the gas was not a problem. Then, just as he had expected, the cathode rays *were* deflected by an electric field!

Thomson's experiment was a decisive victory for the charged-particle model, but it still did not indicate anything about the nature of the particles. What were they?

J. J. Thomson.

Thomson's Crossed-Field Experiment

Thomson could measure the deflection of cathode-ray particles for various strengths of the magnetic field, but magnetic deflection depends both on the particle's charge-to-mass ratio q/m *and* on its speed. Measuring the charge-to-mass ratio, and thus learning something about the particles themselves, requires some means of measuring their velocity. To do so, Thomson devised the experiment for which he is most remembered.

Thomson built a tube containing the parallel-plate electrodes visible in the photo in Figure 37.7a. He then placed the tube between the poles of a magnet. Figure 37.7b shows that the electric and magnetic fields were perpendicular to each other, thus creating what came to be known as a **crossed-field experiment.**

The magnetic field, which is perpendicular to the particle's velocity $\vec{v}$, exerts a magnetic force on the charged particle of magnitude

$$F_B = qvB \qquad (37.1)$$

The magnetic field alone would cause a negatively charged particle to move along a *downward* circular arc. The particle doesn't move in a complete circle because the velocity is large and because the magnetic field is limited in extent. As you learned in Chapter 32, the radius of the arc is

$$r = \frac{mv}{qB} \qquad (37.2)$$

The net result is to *deflect* the beam of particles downward. This deflection is easily measured by observing the green spot where the particles strike the glass at the end of the tube. It is then a straightforward geometry problem to determine the radius of curvature r from the measured deflection.

Thomson's new idea was to create an electric field between the parallel-plate electrodes that would exert an *upward* force on the negative charges, pushing them back toward the center of the tube. The magnitude of the electric force on each particle is

$$F_E = qE \qquad (37.3)$$

Thomson adjusted the electric field strength until the cathode-ray beam, in the presence of both electric and magnetic fields, was exactly in the center of the tube.

(a)

(b)

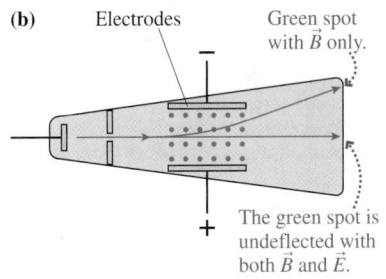

(c)

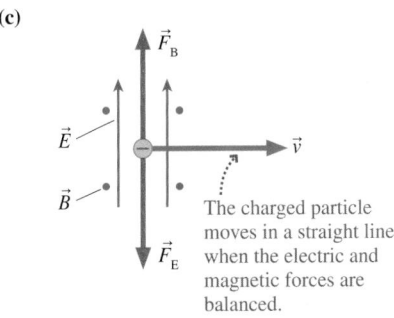

FIGURE 37.7 Thomson's crossed-field experiment to measure the velocity of cathode rays. The photograph shows his original tube and the coils he used to produce the magnetic field.

Zero deflection occurs when the magnetic and electric forces exactly balance each other, as Figure 37.7c shows. The force vectors point in opposite directions, and their magnitudes are equal when

$$F_B = qvB = F_E = qE$$

Notice that the charge q cancels. Once E and B are set, a charged particle can pass undeflected through the crossed fields only if its speed is

$$v = \frac{E}{B} \tag{37.4}$$

By balancing the magnetic force against the electric force, Thomson could determine the velocity of the charged-particle beam. Once he knew v, he could then use Equation 37.2 to find the charge-to-mass ratio:

$$\frac{q}{m} = \frac{v}{rB} \tag{37.5}$$

Thomson found that the charge-to-mass ratio of cathode rays is $q/m \approx 1 \times 10^{11}$ C/kg. This seems not terribly accurate in comparison to a modern value of 1.76×10^{11} C/kg, but keep in mind both the experimental limitations of his day and the fact that, prior to his work, no one had *any* idea of the charge-to-mass ratio.

EXAMPLE 37.1 A crossed-field experiment

An electron is fired between two parallel-plate electrodes that are 5.0 mm apart and 3.0 cm long. A potential difference ΔV between the electrodes establishes an electric field between them. A 3.0-cm-wide, 1.0 mT magnetic field overlaps the electrodes and is perpendicular to the electric field. When $\Delta V = 0$ V, the electron is deflected by 2.0 mm as it passes between the plates. What value of ΔV will allow the electron to pass through the plates without deflection?

MODEL Assume the fields between the electrodes are uniform and that they are zero outside the electrodes.

VISUALIZE Figure 37.8 shows an electron passing through the magnetic field between the plates when $\Delta V = 0$ V. The curvature has been exaggerated to make the geometry clear.

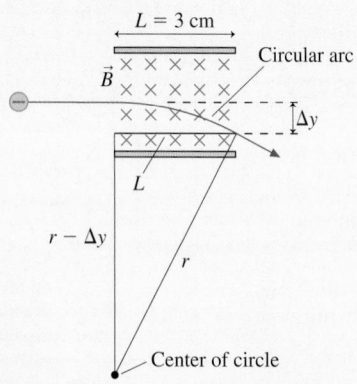

FIGURE 37.8 The electron's trajectory in Example 37.1.

SOLVE We can find the needed electric field, and thus ΔV, if we know the electron's speed. We can find the electron's speed from the radius of curvature of its circular arc in a magnetic field. Figure 37.8 shows a right triangle with hypotenuse r and width L. We can use the Pythagorean theorem to write

$$(r - \Delta y)^2 + L^2 = r^2$$

where Δy is the electron's deflection in the magnetic field. This is easily solved to find the radius of the arc:

$$r = \frac{(\Delta y)^2 + L^2}{2\Delta y} = \frac{(0.0020 \text{ m})^2 + (0.030 \text{ m})^2}{2(0.0020 \text{ m})} = 0.226 \text{ m}$$

The speed of an electron traveling along an arc with this radius is found from Equation 37.2:

$$v = \frac{erB}{m} = 4.0 \times 10^7 \text{ m/s}$$

Thus the electric field that will allow the electron to pass through without deflection is

$$E = vB = 40,000 \text{ V/m}$$

The electric field of a parallel-plate capacitor of spacing d is related to the potential difference by $E = \Delta V/d$, so the necessary potential difference is

$$\Delta V = Ed = (40,000 \text{ V/m})(0.0050 \text{ m}) = 200 \text{ V}$$

ASSESS A fairly small potential difference is sufficient to counteract the magnetic deflection.

The Electron

Notable as this accomplishment was, Thomson did not stop there. Next he measured q/m for different cathode materials. They were all the same: Whatever the charged particles were, they were identical for all the different elements. Thomson then compared his result to the charge-to-mass ratio of the hydrogen ion, known from electrolysis to have a value of $\approx 1 \times 10^8$ C/kg. This value was roughly 1000 times smaller than for the cathode-ray particles, which could imply that a cathode-ray particle has a much larger charge than a hydrogen ion, or a much smaller mass, or some combination of these.

Electrolysis experiments suggested the existence of a basic unit of charge, so it was tempting to assume that the cathode-ray charge was the same as the charge of a hydrogen ion. However, cathode rays were so different from the hydrogen ion that such an assumption could not be justified without some other evidence. To provide that evidence, Thomson called attention to previous experiments showing that cathode rays can penetrate thin metal foils but atoms cannot. This can be true, Thomson argued, only if cathode-ray particles are vastly smaller and thus much less massive than atoms.

In a paper published in 1897, J. J. Thomson assembled all of the evidence to announce the discovery that cathode rays are negatively charged particles, that they are much less massive ($\approx 0.1\%$) than atoms, and that they are identical when generated by different elements. In other words, Thomson had discovered a **subatomic particle,** one of the constituents of which atoms themselves are constructed. In recognition of the role this particle plays in electricity, it was later named the **electron.**

Thomson's discovery did not immediately convince everyone that the cathode-ray particles were a ubiquitous component of all atoms. But experiments by Thomson and others over the next few years showed that negative particles emitted from hot metal wires (a process discovered by Thomas Edison in his development of the lightbulb) had the same q/m; that one type of radioactive decay (today called *beta radiation*) consisted of particles with the same q/m; and that certain changes in the spectra of atoms when placed in a magnetic field could be understood if the atoms had a charged constituent with the same q/m. By 1900 it was clear to all that electrons were a fundamental building block of atoms. J. J. Thomson was awarded the Nobel prize in 1906.

STOP TO THINK 37.1   J. J. Thomson's conclusion that cathode-ray particles are *fundamental* constituents of atoms was based primarily on which observation?

a. They have a negative charge.
b. They are the same from all cathode materials.
c. Their mass is much less than that of hydrogen.
d. They penetrate very thin metal foils.

37.5 Millikan and the Fundamental Unit of Charge

Thomson measured the electron's charge-to-mass ratio and *surmised* that the mass must be much smaller than that of an atom, but clearly it was desirable to measure the charge q or the mass m directly. Thomson had discovered that air can be ionized by x rays, and he subsequently found that the water vapor in moist air condensed to form small droplets around the ions. By measuring the mass and charge of water droplets, Thomson and his students were able to determine that the unit of charge involved in the ionization of gases was roughly 1×10^{-19} C.

These were not very accurate experiments, but the value obtained was close to the charge of the hydrogen ion as determined (also rather crudely) from electrolysis. Thomson interpreted this information—that the same unit of charge is responsible for conduction in both liquids and gases—as implying the existence of a *fundamental unit of charge.* This fundamental unit of charge is designated e. A hydrogen ion has a charge $q_{\text{H ion}} = +e$ and an electron has $q_{\text{elec}} = -e$.

In 1906, the American scientist Robert Millikan began making his own measurements of e. He based his technique on his discovery that he could "catch" a charged oil droplet and then accurately measure its motion using the combined influence of gravity and an electric field.

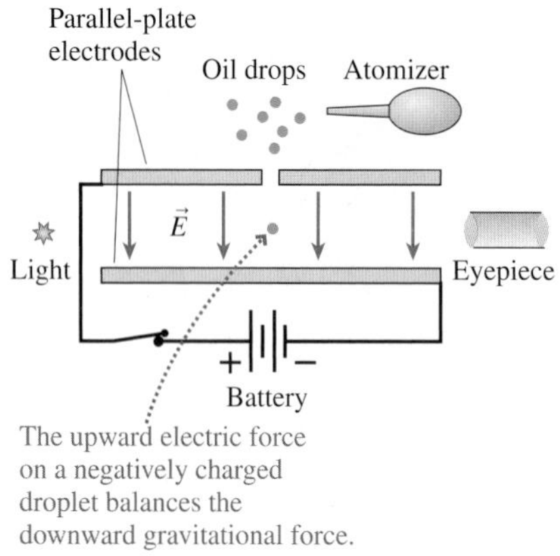

The **Millikan oil-drop experiment,** as we call it today, is illustrated in Figure 37.9. A squeeze-bulb atomizer sprayed out a very fine mist of oil droplets, some of which were charged from friction in the sprayer. These slowly settled toward a horizontal pair of parallel-plate electrodes, where a few droplets passed through a small hole in the top plate. Millikan observed the drops by shining a bright light between the plates and using an eyepiece to see the droplets' reflections. He then established an electric field by applying a voltage to the plates.

A drop will remain suspended between the plates, moving neither up nor down, if the electric field exerts an upward force on a charged drop that exactly balances the downward gravitational force. The forces balance when

$$m_{\text{drop}}g = q_{\text{drop}}E \tag{37.6}$$

and thus the charge on the drop is measured to be

$$q_{\text{drop}} = \frac{m_{\text{drop}}g}{E} \tag{37.7}$$

Parallel-plate electrodes
Oil drops   Atomizer
$\vec{E}$
Light   Eyepiece
Battery

The upward electric force on a negatively charged droplet balances the downward gravitational force.

FIGURE 37.9 Millikan's oil-drop apparatus to measure the fundamental unit of charge.

Notice that m and q are the mass and charge of the oil droplet, not that of an electron. But because the droplet is charged by acquiring (or losing) electrons, the charge of the droplet should be related to the fundamental unit of charge.

The field strength E could be determined accurately from the voltage applied to the plates, so the limiting factor in measuring q_{drop} was Millikan's ability to determine the mass of these small drops. Ideally, the mass could be found by measuring a drop's diameter and using the known density of the oil. However, the drops were too small (≈ 1 μm) to measure accurately by viewing through the eyepiece.

Instead, Millikan devised an ingenious method to find the size of the droplets. Objects this small are *not* in free fall. The air resistance forces are so large that the drops fall with a very small but constant speed. The motion of a sphere through a viscous medium is a problem that had been solved in the 19th century, and it was known that the sphere's terminal speed depends on its radius and on the viscosity of air. So rather than holding the droplets motionless, Millikan used the electric field to cause them to move slowly up and down through a known distance. He could determine the droplets' velocities by timing them with a stopwatch. Then, using the known viscosity of air, he could calculate their radii, compute their masses, and, finally, arrive at a value for their charge. Although it was a somewhat roundabout procedure, Millikan was able to measure the charge on a droplet with an accuracy of $\pm 0.1\%$ (one part in a thousand).

Millikan measured many hundreds of droplets, some for hours at a time, under a wide variety of conditions. He found that some of his droplets were positively charged and some negatively charged, but **all had charges that were integer multiples of a certain minimum charge value.** Millikan concluded that "the electric charges found on ions all have either exactly the same value or else some small exact multiple of that value." That value, the fundamental unit of charge that we now call e, is measured to be

$$e = 1.602 \times 10^{-19} \, \text{C}$$

We can then combine the measured e with the measured charge-to-mass ratio e/m to find that the mass of the electron is

$$m_{\text{elec}} = 9.11 \times 10^{-31} \, \text{kg}$$

Taken together, the experiments of Thomson, Millikan, and others provided overwhelming evidence that electric charge comes in discrete units and that *all* charges found in nature are multiples of a fundamental unit of charge we call e.

EXAMPLE 37.2 Suspending an oil drop

Oil has a density of 860 kg/m^3. A 1.0-μm-diameter oil droplet acquires 10 extra electrons as it is sprayed. What potential difference between two parallel plates 1.0 cm apart will cause the droplet to be suspended in air?

MODEL Assume a uniform electric field $E = \Delta V/d$ between the plates.

SOLVE The magnitude of the charge on the drop is $q_{\text{drop}} = 10e$. The mass of the charge is related to its density ρ and volume V by

$$m_{\text{drop}} = \rho V = \frac{4}{3}\pi R^3 \rho = 4.50 \times 10^{-16} \, \text{kg}$$

where the droplet's radius is $R = 5.0 \times 10^{-7}$ m. The electric field that will suspend this droplet against the force of gravity is

$$E = \frac{m_{\text{drop}}g}{q_{\text{drop}}} = 2760 \, \text{V/m}$$

Establishing this electric field between two plates spaced by $d = 0.010$ m requires a potential difference

$$\Delta V = Ed = 27.6 \, \text{V}$$

37.6 Rutherford and the Discovery of the Nucleus

By 1900, it was clear that atoms are not indivisible but, instead, are constructed of charged particles. Atomic sizes were known to be $\approx 10^{-10}$ m, but the electrons common to all atoms are much smaller and much less massive than the smallest atom. How do they "fit" into the larger atom? What is the positive charge of the atom? Where are the charges located inside the atoms?

J. J. Thomson proposed the first model of an atom. Because the electrons are very small and light compared to the whole atom, it seemed reasonable to think that the positively charged part would take up most of the space. Thomson suggested that the atom consists of a spherical "cloud" of positive charge, roughly 10^{-10} m in diameter, in which the smaller negative electrons are embedded. The positive charge exactly balances the negative, so the atom as a whole has no net charge. This model of the atom has often been called the "plum-pudding model" or the "raisin-cake model" for reasons that should be clear from the picture of Figure 37.10.

Thomson proposed that small, negative electrons are embedded in a sphere of positive charge.

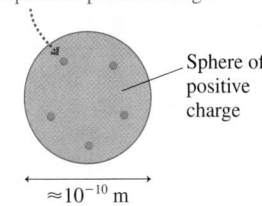

Sphere of positive charge

$\approx 10^{-10}$ m

FIGURE 37.10 Thomson's raisin-cake model of the atom.

Thomson was never able to make any predictions that would enable his model to be tested, and the Thomson atom did not stand the tests of time. His model is of interest today primarily to remind us that our current models of the atom are by no means obvious. Science has many side-steps and dead ends as it progresses.

One of Thomson's students was a New Zealander named Ernest Rutherford. While Rutherford and Thomson were studying the ionizing effects of x rays, in 1896, the French physicist Antoine Henri Becquerel announced the discovery that some new form of "rays" were emitted by crystals of uranium. These rays, like x rays, could expose film, pass through objects, and ionize the air. Yet they were emitted continuously from the uranium without having to "do" anything to it. This was the discovery of **radioactivity,** a topic we'll study in Chapter 42.

With x rays only a year old and cathode rays not yet completely understood, it was not known whether all these various kinds of rays were truly different or merely variations of a single type. Rutherford immediately began a study of these new rays. He quickly discovered that at least two *different* rays are emitted by a uranium crystal. The first, which he called **alpha rays,** were easily absorbed by a piece of paper. The second, **beta rays,** could penetrate through at least 0.1 inch of metal and through much greater thicknesses of soft materials.

As we have already noted, Thomson soon found that beta rays have the same charge-to-mass ratio as cathode rays. The beta rays turned out to be high-speed electrons emitted by the uranium crystal. Rutherford, using similar techniques, showed that alpha rays are *positively* charged particles. By 1906 he had measured their charge-to-mass ratio to be

$$\frac{q}{m} = \frac{1}{2}\frac{e}{m_{\text{H}}}$$

where m_{H} is the mass of a hydrogen atom. This value could indicate either a singly ionized hydrogen molecule H_2^+ ($q = e, m = 2m_{\text{H}}$) *or* a doubly ionized helium atom He^{++} ($q = 2e, m = 4m_{\text{H}}$).

In an ingenious experiment, Rutherford sealed a sample of radium—an emitter of alpha radiation—into a glass tube. Alpha rays could not penetrate the glass, so the particles were contained within the tube. Several days later, Rutherford used electrodes in the tube to create a discharge and observed the spectrum of the emitted light. He found the characteristic wavelengths of helium, but not those of hydrogen. Alpha rays (or alpha particles, as we now call them) consist of doubly ionized helium atoms (bare helium nuclei) emitted at high speed ($\approx 3 \times 10^7$ m/s) from the sample.

It had been quite a shock to discover that atoms are not indivisible, that they have an inner structure. Now, with the discovery of radioactivity, it appeared that some atoms were not even stable but could spit out various kinds of charged particles! Physics had come a long way from the simple atomic idea of Democritus.

The First Nuclear Physics Experiment

Rutherford soon realized that he could use these high-speed particles to probe inside other atoms. In 1909, Rutherford and his students Hans Geiger and Ernest Marsden set up the experiment shown in Figure 37.11 to shoot alpha particles at very thin metal foils. Some of the alpha particles penetrated the foil, but the beam of alpha particles that did so became somewhat spread out. This was not surprising. The alpha particle is charged, and it experiences forces from the positive and negative charges of the atoms as it passes through the foil. According to Thomson's raisin-cake model of the atom, the forces exerted on the alpha particle by the positive atomic charges were expected to roughly cancel the forces from the negative electrons, causing the alpha particles to be deflected only slightly. Indeed, this was the experimenters' initial observation.

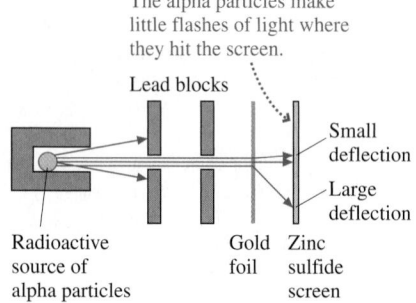

The alpha particles make little flashes of light where they hit the screen.

Lead blocks

Small deflection

Large deflection

Radioactive source of alpha particles

Gold foil

Zinc sulfide screen

FIGURE 37.11 Rutherford's experiment to shoot high-speed alpha particles through a thin gold foil.

At Rutherford's suggestion, Geiger and Marsden set up the apparatus to see if any alpha particles were deflected at *large* angles. It took only a few days to find the answer. Not only were alpha particles deflected at large angles, a very few were reflected almost straight backward toward the source!

How can we understand this result? Figure 37.12a shows that only a small deflection is expected for an alpha particle passing through a Thomson atom. But if an atom has a small, positive core, such as the one in Figure 37.12b, a few of the alpha particles can come very close to the core. Because the electric force varies with the inverse square of the distance, the very large force of this very close approach can cause a large-angle scattering or a backward deflection of the alpha particle. This is what Geiger and Marsden were observing.

I remember two or three days later Geiger coming to me in great excitement and saying, "We have been able to get some of the alpha particles coming backward." It was quite the most incredible event that has ever happened to me in my life. It was almost as if you fired a 15-inch shell at a piece of tissue paper and it came back and hit you. . . . It was then that I had the idea of an atom with a minute massive center, carrying a charge.

Ernest Rutherford

(a)

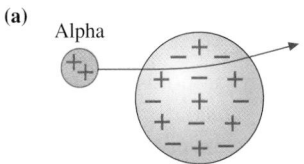

The alpha particle is only slightly deflected by a Thomson atom because forces from the spread-out positive and negative charges nearly cancel.

(b)

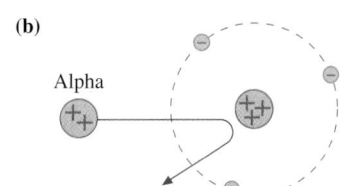

If the atom has a concentrated positive nucleus, some alpha particles will be able to come very close to the nucleus and thus feel a very strong repulsive force.

FIGURE 37.12 Alpha particles interact differently with a concentrated positive nucleus than they would with the spread-out charge in Thomson's model.

Thus the discovery of large-angle scattering of alpha particles led Rutherford to envision an atom in which negative electrons orbit an unbelievably small, massive, positive **nucleus,** rather like a miniature solar system. This is the **nuclear model of the atom.** Notice that nearly all of the atom is merely empty space—the void!

 19.1

EXAMPLE 37.3 **A nuclear physics experiment**

An alpha particle is shot with a speed of 2.0×10^7 m/s directly toward the nucleus of a gold atom. What is the distance of closest approach to the nucleus?

MODEL Energy is conserved in electric interactions. Assume that the gold nucleus, which is much more massive than the alpha particle, does not move. Also recall that the exterior electric field and potential of a sphere of charge can be found by treating the total charge as a point charge at the center.

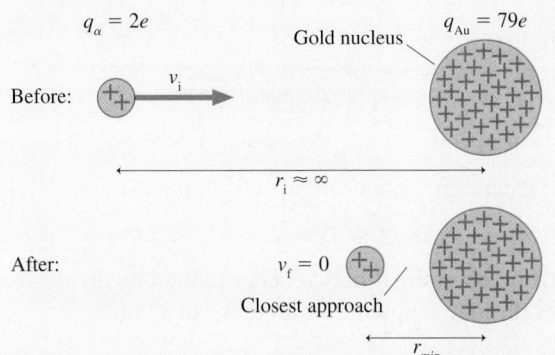

FIGURE 37.13 A before-and-after pictorial representation of an alpha particle colliding with a nucleus.

VISUALIZE Figure 37.13 is a pictorial representation. The motion is in and out along a straight line.

SOLVE We are not interested in how long the collision takes or any of the details of the trajectory, so using conservation of energy rather than Newton's laws is appropriate. Initially, when the alpha particle is very far away, the system has only kinetic energy. At the moment of closest approach, just before the alpha particle is reflected, the charges are at rest and the system has only potential energy. The conservation of energy statement $K_f + U_f = K_i + U_i$ is

$$0 + \frac{1}{4\pi\epsilon_0} \frac{q_\alpha q_{Au}}{r_{min}} = \frac{1}{2}mv_i^2 + 0$$

where q_α is the alpha-particle charge and we've treated the gold nucleus as a point charge q_{Au}. The mass m is that of the alpha particle. The solution for r_{min} is

$$r_{min} = \frac{1}{4\pi\epsilon_0} \frac{2q_\alpha q_{Au}}{mv_i^2}$$

The alpha particle is a helium nucleus, so $m = 4$ u $= 6.64 \times 10^{-27}$ kg and $q_\alpha = 2e = 3.20 \times 10^{-19}$ C. Gold has atomic number 79, so $q_{Au} = 79e = 1.26 \times 10^{-17}$ C. We can then calculate

$$r_{min} = 2.7 \times 10^{-14} \text{ m}$$

This is only about 1/10,000 the size of the atom itself!

ASSESS We ignored the atom's electrons in this example. In fact, they make almost no contribution to the alpha particle's trajectory. The alpha particle is exceedingly massive compared to the electrons, and the electrons are spread out over a distance very large compared to the size of the nucleus. Hence the alpha particle easily pushes them aside without any noticeable change in its velocity.

Rutherford went on to make careful experiments of how the alpha particles scattered at different angles. From these experiments he deduced that the diameter of the atomic nucleus is $\approx 1 \times 10^{-14}$ m $= 10$ fm (1 fm = 1 femtometer = 10^{-15} m), increasing a little for elements of higher atomic number and atomic mass.

It may seem surprising to you that the Rutherford model of the atom, with its solar system analogy, was not Thomson's original choice. However, scientists at the time could not imagine matter having the extraordinarily high density implied by a small nucleus. Neither could they understand what holds the nucleus together, why the positive charges do not push each other apart. Thomson's model, in which the positive charge was spread out and balanced by the negative electrons, actually made more sense. It would be several decades before the forces holding the nucleus together began to be understood, but Rutherford's evidence for a very small nucleus was indisputable.

> **STOP TO THINK 37.2**   If the alpha particle has a positive charge, which way will it be deflected in the magnetic field?
>
> a. Up   b. Down
> c. Into the page   d. Out of the page

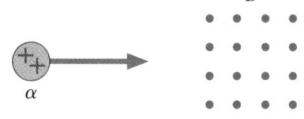

The Electron Volt

The joule is a unit of appropriate size in mechanics and thermodynamics, where we dealt with macroscopic objects, but it is poorly matched to the needs of atomic physics. It will be very useful to have an energy unit appropriate to atomic and nuclear events.

Figure 37.14 shows an electron accelerating from rest across a parallel-plate capacitor with a 1.0 V potential difference. What is the electron's kinetic energy when it reaches the positive plate? We know from energy conservation that $K_f + qV_f = K_i + qV_i$, where $U = qV$ is the electric potential energy. $K_i = 0$ because the electron starts from rest, and the electron's charge is $q = -e$. Thus

$$K_f = -q(V_f - V_i) = -q\Delta V = e\Delta V = (1.60 \times 10^{-19} \text{ C})(1.0 \text{ V})$$

$$= 1.60 \times 10^{-19} \text{ J}$$

Let us define a new unit of energy, called the **electron volt,** as

$$1 \text{ electron volt} = 1 \text{ eV} \equiv 1.60 \times 10^{-19} \text{ J}$$

With this definition, the kinetic energy gained by the electron in our example is

$$K_f = 1 \text{ eV}$$

In other words, **1 electron volt is the kinetic energy gained by an electron (or proton) if it accelerates through a potential difference of 1 volt.**

> **NOTE** ▶ The abbreviation eV uses a lower case e but an upper case V. Units of keV (10^3 eV), MeV (10^6 eV), and GeV (10^9 eV) are common. ◀

The electron volt can be a troublesome unit. One difficulty is its unusual name, which looks less like a unit than, say, "meter" or "second." A more significant

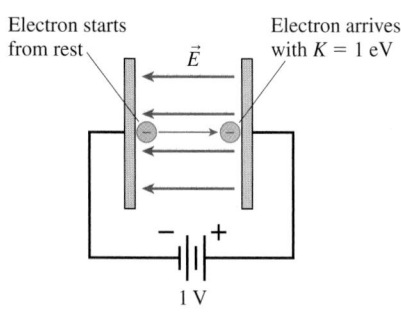

FIGURE 37.14 An electron accelerating across a 1 V potential difference gains 1 eV of kinetic energy.

difficulty is that the name suggests a relationship to volts. But *volts* are units of electric potential, whereas this new unit—with an admittedly confusing name—is a unit of energy! It is crucial to distinguish between the *potential V*, measured in volts, and an *energy* that can be measured either in joules or in electron volts. You can now use electron volts anywhere that you would previously have used joules. Doing so is no different from converting back and forth between pressure units of pascals and atmospheres.

NOTE ▶ To reiterate, the electron volt is a unit of *energy*, convertible to joules, and not a unit of potential. Potential is always measured in volts. However, the joule remains the SI unit of energy. It will be useful to express energies in eV, but you *must* convert this energy to joules before doing most calculations. ◀

EXAMPLE 37.4 The speed of an alpha particle
Alpha particles are usually characterized by their kinetic energy in MeV. What is the speed of an 8.30 MeV alpha particle?

SOLVE Alpha particles are helium nuclei, having $m = 4\,u = 6.64 \times 10^{-27}$ kg. The kinetic energy of this alpha particle is 8.30×10^6 eV. First, convert the energy to joules:

$$K = 8.30 \times 10^6 \text{ eV} \times \frac{1.60 \times 10^{-19} \text{ J}}{1.0 \text{ eV}} = 1.33 \times 10^{-12} \text{ J}$$

Now we can find the speed:

$$K = \frac{1}{2}mv^2 = 1.33 \times 10^{-12} \text{ J}$$

$$v = \sqrt{\frac{2K}{m}} = 2.0 \times 10^7 \text{ m/s}$$

This was the speed of the alpha particle in Example 37.3.

EXAMPLE 37.5 Energy of an electron
In a simple model of the hydrogen atom, the electron orbits the proton at 2.19×10^6 m/s in a circle with radius 5.29×10^{-11} m. What is the atom's energy in eV?

MODEL The electron has a kinetic energy of motion, and the electron + proton system (i.e., the atom) has an electric potential energy.

SOLVE The potential energy is that of two point charges, with $q_{\text{proton}} = +e$ and $q_{\text{elec}} = -e$. Thus

$$E = K + U = \frac{1}{2}m_{\text{elec}}v^2 + \frac{1}{4\pi\epsilon_0}\frac{(e)(-e)}{r} = -2.17 \times 10^{-18} \text{ J}$$

Conversion to eV gives

$$E = -2.17 \times 10^{-18} \text{ J} \times \frac{1 \text{ eV}}{1.60 \times 10^{-19} \text{ J}} = -13.6 \text{ eV}$$

ASSESS The negative energy reflects the fact that the electron is *bound* to the proton. You would need to *add* energy to remove the electron.

Using the Nuclear Model

The nuclear model of the atom makes it easy to understand and picture such processes as ionization. Because electrons orbit a positive nucleus, an x-ray photon or a rapidly moving particle, such as another electron, can knock one of the orbiting electrons away, creating a positive ion. Removing one electron makes a singly charged ion, with $q = +e$. Removing two electrons creates a doubly charged ion, with $q = +2e$. This is shown for lithium (atomic number 3) in Figure 37.15.

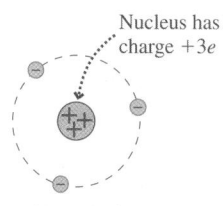
Neutral Li

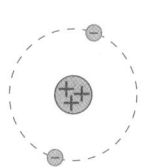
Singly charged Li$^+$

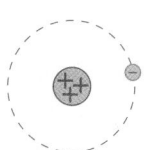
Doubly charged Li^{++}

FIGURE 37.15 Different ionization stages of the lithium atom ($Z = 3$).

The nuclear model also allows us to understand why, during chemical reactions and when an object is charged by rubbing, electrons are easily transferred but protons are not. The protons are tightly bound in the nucleus, shielded by all the electrons, but outer electrons are easily stripped away. Rutherford's nuclear model has explanatory power that was lacking in Thomson's model.

EXAMPLE 37.6 The ionization energy of hydrogen

What is the minimum energy required to ionize a hydrogen atom? The electron orbits the proton at 2.19×10^6 m/s in a circle with radius 5.29×10^{-11} m.

SOLVE In Example 37.5 we found that the atom's energy is $E_i = -13.6$ eV. Ionizing the atom means removing the electron and taking it very far away. As $r \rightarrow \infty$, the potential energy becomes zero. Further, using the least possible energy to ionize the atom will leave the electron, when it is very far

away, very nearly at rest. Thus the atom's energy after ionization is $E_f = K_f + U_f = 0 + 0 = 0$ eV. This is *larger* than E_i by 13.6 eV, so the minimum energy that is required to ionize a hydrogen atom is 13.6 eV. This is called the atom's *ionization energy*. If the electron receives ≥ 13.6 eV (2.17×10^{-18} J) of energy from a photon, or in a collision with another electron, or by any other means, it will be knocked out of the atom and leave a H^+ ion behind.

STOP TO THINK 37.3 Carbon is the sixth element in the periodic table. How many electrons are in a C^{++} ion?

37.7 Into the Nucleus

Chapter 42 will discuss nuclear physics in more detail, but it will be helpful to give a brief overview of the nucleus. The relative masses of many of the elements were known from chemistry experiments by the mid-19th century. By arranging the elements in order of ascending mass, and noting recurring regularities in their chemical properties, the Russian chemist Dmitri Mendeleev first proposed the periodic table of the elements in 1872. But what did it mean to say that hydrogen was atomic number 1, helium number 2, lithium number 3, and so on?

It soon became known that hydrogen atoms can only be singly ionized, producing H^+. A doubly ionized H^{++} is never observed. Helium, by contrast, can be both singly and doubly ionized, creating He^+ and He^{++}, but He^{+++} is not observed. Once Thomson discovered the electron and Millikan established the fundamental unit of charge, it seemed fairly clear that a hydrogen atom contains only one electron and one unit of positive charge, helium has two electrons and two units of positive charge, and so on. Thus the **atomic number** of an element, which is always an integer, describes the number of electrons (of a neutral atom) and the number of units of positive charge in the nucleus. The atomic number is represented by Z, so hydrogen is $Z = 1$, helium $Z = 2$, and lithium $Z = 3$. Elements are listed in the periodic table by their atomic number.

Rutherford's discovery of the nucleus quickly led to the recognition that the positive charge is associated with a positive subatomic particle called the **proton.** The proton's charge is $+e$, equal in magnitude but opposite in sign to the electron's charge. Further, because nearly all the atomic mass is associated with the nucleus, the proton is much more massive than the electron. According to Rutherford's nuclear model, atoms with atomic number Z consist of Z negative electrons, with net charge $-Ze$, orbiting a massive nucleus that contains protons and has net charge $+Ze$. The Rutherford atom went a large way toward explaining the periodic table.

But there was a problem. Helium, with atomic number 2, has twice as many electrons as hydrogen. Lithium, $Z = 3$, has three electrons. But it was known from chemistry measurements that helium is *four times* as massive as hydrogen and lithium is *seven times* as massive. If a nucleus contains Z protons to balance the Z orbiting electrons, and if nearly all the atomic mass is contained in the nucleus, then helium should be simply twice as massive as hydrogen and lithium three times as massive. Something else must be present in the nucleus to make the atoms more massive than our simple nuclear model predicts.

The Neutron

About 1910, J. J. Thomson and his student Francis Aston developed a device called a **mass spectrometer** for measuring the charge-to-mass ratios of atomic ions. (A mass spectrometer was the subject of homework problem 70 in Chapter 32.) As Aston and others began collecting data, they soon found that many elements consist of atoms of *differing* mass! Neon, for example, had been assigned an atomic mass of 20. But Aston found, as the data of Figure 37.16 show, that while 91% of neon atoms have mass $m = 20$ u, 9% have $m = 22$ u and a very small percentage have $m = 21$ u. Chlorine was found to be a mixture of 75% chlorine atoms with $m = 35$ u and 25% chlorine atoms with $m = 37$ u, both having atomic number $Z = 17$.

These difficulties were not resolved until the discovery, in 1932, of a third subatomic particle. This particle has essentially the same mass as a proton but *no* electric charge. It is called the **neutron.** Neutrons reside in the nucleus, with the protons, where they contribute to the mass of the atom but not to its charge. As you'll see in Chapter 42, neutrons help provide the "glue" that holds the nucleus together.

The neutron was the missing link needed to explain why atoms of the same element can have different masses. We now know that every atom with atomic number Z has a nucleus containing Z protons with charge $+Ze$. In addition, as shown in Figure 37.17, the nucleus contains N neutrons. There are a *range* of neutron numbers that happily form a nucleus with Z protons, creating a series of nuclei having the same Z-value (i.e., they are all the same chemical element) but different masses. Such a series of nuclei are called **isotopes.**

Chemical behavior is determined by the orbiting electrons. All isotopes of one element have the same number Z of orbiting electrons (if the atoms are electrically neutral) and have the same chemical properties. But different isotopes of the same element can have quite different nuclear properties. In addition, macroscopic behavior that depends on mass, such as the diffusion of a gas, can slightly favor one isotope over another.

An atom's **mass number** A is defined to be $A = Z + N$. It is the total number of protons and neutrons in a nucleus. The mass number, which is dimensionless, is *not* the same thing as the atomic mass m. By definition, A is an integer. But because the proton and neutron masses are both ≈ 1 u, the mass number A is *approximately* the mass in atomic mass units.

The notation used to label isotopes is ^{A}Z, where the mass number A is given as a *leading* superscript. The proton number Z is not specified by an actual number but, equivalently, by the chemical symbol for that element. The most common isotope of neon has $Z = 10$ protons and $N = 10$ neutrons. Thus it has mass number $A = 20$ and it is labeled ^{20}Ne. The neon isotope ^{22}Ne has $Z = 10$ protons (that's what makes it neon) and $N = 12$ neutrons. Helium has the two isotopes shown in Figure 37.18. The rare ^{3}He is only 0.0001% abundant, but it can be isolated and has important uses in scientific research.

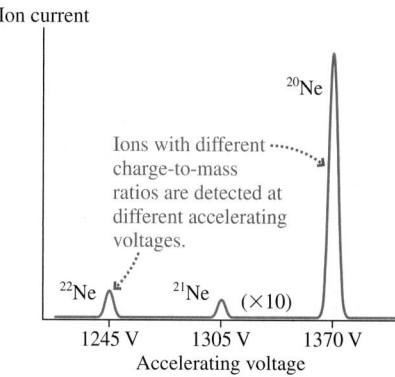

FIGURE 37.16 The mass spectrum of neon.

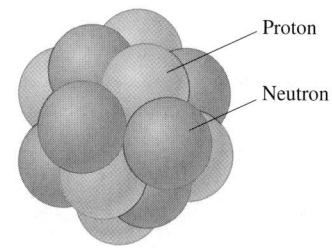

FIGURE 37.17 The nucleus of an atom contains protons and neutrons.

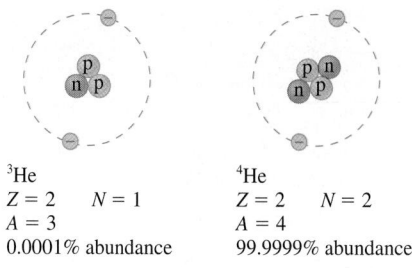

^{3}He
$Z = 2 \quad N = 1$
$A = 3$
0.0001% abundance

^{4}He
$Z = 2 \quad N = 2$
$A = 4$
99.9999% abundance

FIGURE 37.18 The two isotopes of helium. ^{3}He is only 0.0001% abundant.

STOP TO THINK 37.4 Carbon is the sixth element in the periodic table. How many protons and how many neutrons are there in a nucleus of the isotope ^{14}C?

37.8 The Emission and Absorption of Light

The investigations of cathode rays that led to Thomson's discovery of the electron all followed from Faraday's invention of the gas discharge tube. At the same time, a separate group of scientists was using the gas discharge tube for different purposes. Their discoveries about the emission and absorption of light would also, in the early years of the 20th century, come to bear on the issue of atomic structure.

A gas discharge tube exhibits both a *cathode glow* and a *positive column,* as it is called, that glows with bright color and is different for every gas. The positive column was a hindrance to the study of cathode rays, and those investigators learned that they could eliminate the bright glow by reducing the gas pressure. But other scientists were intrigued by the brightly colored light of the positive column. Why does every gas emit a different color? Can these colors tell us anything about the nature of the atoms and molecules? Fortunately, Faraday's discovery came just at the time that the interference and diffraction of light were first being understood. The production of diffraction gratings was well underway by mid century, and they were the ideal tool to study the light emitted by a discharge tube.

Figure 37.19a shows a typical experimental arrangement for recording the spectrum of light emitted by a gas. The light is focused on the entrance slit of a *spectrometer,* then diffracted by a grating. Different wavelengths in the light are diffracted at different angles, as you learned in Chapter 22, then focused on a film or a photographic plate. A modern spectrometer, widely used today in physics, chemistry, and astronomy, is little changed except that the film has been replaced by an electronic photodetector.

Examples of *emission spectra* are shown in Figure 37.19b. Each line represents one of the wavelengths of light coming from the discharge. These wavelengths can be measured with extremely high accuracy in a well-calibrated instrument. Scientists quickly learned that

1. Gases emit a **discrete spectrum,** consisting of discrete, specific wavelengths of light. This is in contrast to the *continuous* rainbow-like spectrum of the sun or an incandescent light source. Each wavelength in a spectrum is commonly called a **spectral line** because of its appearance in photographs such as Figure 37.19b.
2. Every element in the periodic table emits a unique spectrum.

Substances not only emit light, they can also absorb light. If you look at a lightbulb through a piece of red glass, the bulb looks red because the glass *absorbs* the yellow, green, and blue wavelengths of the white light. Only the red wavelengths make it through to be seen. Similarly, grass and leaves appear green because they absorb both red and blue wavelengths (red and blue wavelengths are the ones that drive photosynthesis), reflecting only the green and yellow wavelengths in the center of the visible spectrum.

NOTE ▶ A peculiarity of the English language is that the process of *absorbing* light is called *absorption,* not "absorbtion." ◀

Do gases absorb light? Indeed they do, although less strongly than solids or liquids because they are less dense. An absorption experiment is shown in Figure 37.20a. Here a white-light source emits a continuous spectrum that, in the absence of a gas, exposes the film completely and uniformly. When a sample of gas is placed in the light's path, any wavelengths absorbed by the gas are missing and the film is dark at that wavelength.

It was discovered that gases not only emit discrete wavelengths, they also absorb discrete wavelengths. But there is an important difference between the emission spectrum and the absorption spectrum of a gas: **Every wavelength that is absorbed by the gas is also emitted, but *not* every emitted wavelength is absorbed.** The wavelengths in the absorption spectrum appear as a subset of the

(a) Measuring an emission spectrum

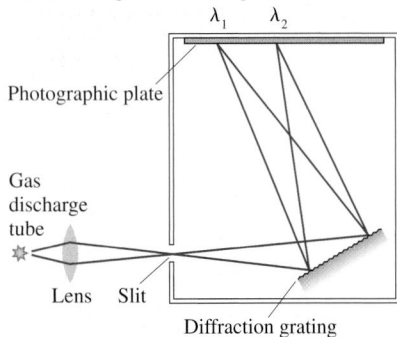

(b) The spectral lines extend to the series limit at 364.7 nm.

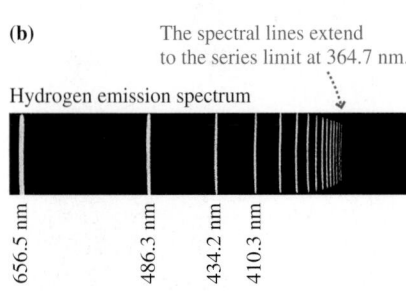

Neon emission spectrum

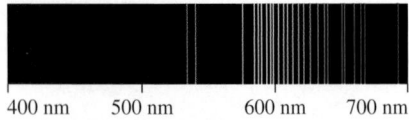

FIGURE 37.19 A grating spectrometer is used to study the emission of light.

wavelengths in the emission spectrum. As an example, Figure 37.20b shows both the emission and the absorption spectra of sodium atoms. All of the absorption wavelengths are prominent in the emission spectrum, but there are many emission lines for which no absorption occurs.

What causes atoms to emit or absorb light? Why a discrete spectrum? Why are some wavelengths emitted but not absorbed? Why is each element different? Nineteenth-century physicists struggled with these questions but could not answer them. Ultimately, their inability to understand the emission and absorption of light forced scientists to the unwelcome realization that classical physics was simply incapable of providing an understanding of atoms.

The only encouraging sign came from an unlikely source. While the spectra of other atoms have dozens or even hundreds of wavelengths, the emission spectrum of hydrogen (Figure 37.19b) is very simple and regular. If any spectrum could be understood, it should be that of the first element in the periodic table. The breakthrough came in 1885, not by an established and recognized scientist but by a Swiss school teacher, Johann Balmer. Balmer showed that the wavelengths in the hydrogen spectrum could be represented by the simple formula

$$\lambda = \frac{91.18 \text{ nm}}{\left(\frac{1}{2^2} - \frac{1}{n^2}\right)}, \qquad n = 3, 4, 5, \ldots \tag{37.8}$$

Balmer's story was told more completely in Section 24.1, to which you should refer.

Later experimental evidence, as ultraviolet and infrared spectroscopy developed, showed that Balmer's result could be generalized to

$$\lambda = \frac{91.18 \text{ nm}}{\left(\frac{1}{m^2} - \frac{1}{n^2}\right)}, \qquad m = 1, 2, 3, \ldots \qquad n = m + 1, m + 2, \ldots \tag{37.9}$$

We now refer to Equation 37.9 as the **Balmer formula,** although Balmer himself only suggested the original version of Equation 37.8 in which $m = 2$. Other than at the very highest levels of resolution, where new details appear that need not concern us in this text, the Balmer formula accurately describes *every* wavelength in the emission spectrum of hydrogen.

The Balmer formula is what we call *empirical knowledge.* It is an accurate mathematical representation found empirically—that is, through experimental evidence—but it does not rest on any physical principles or physical laws. Balmer's formula was useful, but no one was able to *derive* Balmer's formula from Newtonian mechanics or the theory of electromagnetism. Yet the formula was so simple that it must, everyone agreed, have a simple explanation. It would take 30 years to find it.

STOP TO THINK 37.5   These spectra are due to the same element. Which one is an emission spectrum and which is an absorption spectrum?

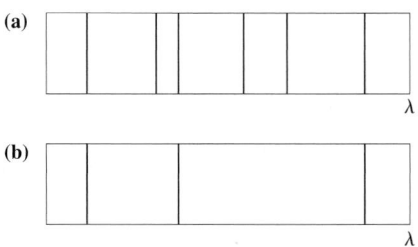

(a) Measuring an absorption spectrum

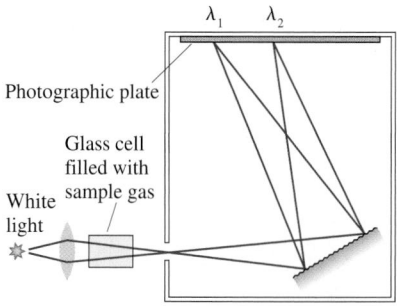

(b) Absorption and emission spectra of sodium

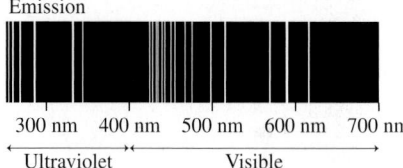

300 nm   400 nm   500 nm   600 nm   700 nm

Ultraviolet   Visible

FIGURE 37.20 Measuring an absorption spectrum. The images in part b are replicas of photographic plates.

37.9 Classical Physics at the Limit

At the start of the 19th century, only a few scientists believed that matter consists of atoms. By century's end, there was substantial evidence not only for atoms but for the existence of charged subatomic particles. The explorations into atomic structure culminated with Rutherford's nuclear model.

Rutherford's nuclear model of the atom matched the experimental evidence about the *structure* of atoms, but it had two serious shortcomings. Electrons orbiting the nucleus in a Rutherford atom are oscillating charged particles. According to Maxwell's theory of electricity and magnetism, these orbiting electrons should act as small antennas and radiate electromagnetic waves. That sounds encouraging, because we know that atoms can emit light, but it was easy to show that a Rutherford atom would radiate a *continuous* rainbow-like spectrum. Thus one failure of Rutherford's model was an inability to predict the discrete nature of emission and absorption spectra.

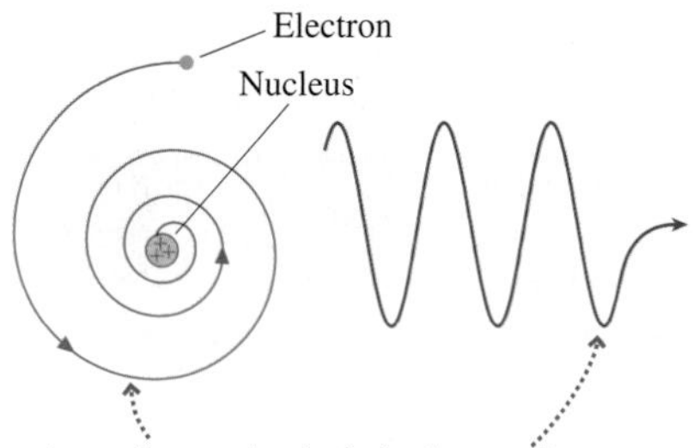

In addition, the atoms would continuously lose energy as they radiated electromagnetic waves. As Figure 37.21 shows, this would cause the electrons to spiral into the nucleus! Calculations showed that a Rutherford atom can last no more than about a microsecond. In other words, classical Newtonian mechanics and electromagnetism predict that an atom in which electrons orbit a nucleus would be highly unstable and would immediately self-destruct. This clearly does not happen.

The experimental efforts of the late 19th century had been impressive, and there could be no doubt about the existence of electrons, about the small positive nucleus, and about the unique discrete spectrum emitted by each atom. But the theoretical framework for understanding such observations had lagged behind. As the new century dawned, physicists could not explain the structure of atoms, could not explain the stability of matter, could not explain discrete spectra or why an element's absorption spectrum differs from its emission spectrum, and could not explain the origin of x rays or radioactivity.

Yet few physicists were willing to abandon the successful and long-cherished theories of classical physics. Despite the attention we have focused on the search for atomic structure, the large majority of scientists were working in other fields—electricity, acoustics, thermodynamics—for which classical physics remained completely satisfactory. Most considered these "problems" with atoms to be minor discrepancies that would soon be resolved. But classical physics had, indeed, reached its limit, and a whole new generation of brilliant young physicists, with new ideas, was about to take the stage. Among the first was an unassuming young man in Berne, Switzerland. His scholastic record had been mediocre, and the best job he could find upon graduation was as a clerk in the patent office, examining patent applications. He needed the job, having recently married a fellow student due, at least in part, to a child conceived out of wedlock. His name was Albert Einstein.

According to classical physics, an electron would spiral into the nucleus while radiating energy as an electromagnetic wave.

FIGURE 37.21 The fate of a Rutherford atom.

SUMMARY

The goal of Chapter 37 has been to understand how scientists discovered the properties of atoms and how these discoveries led to the need for a new theory of light and matter.

IMPORTANT CONCEPTS/EXPERIMENTS

Nineteenth-century scientists were focused on understanding matter, electricity, and light. Faraday's invention of the gas discharge tube launched two important avenues of inquiry.

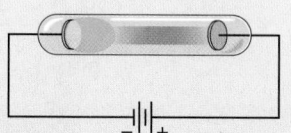

Cathode Rays and Atomic Structure

Thomson found that cathode rays are negative, subatomic particles. These were soon named electrons. Electrons are

- Constituents of atoms.
- The fundamental unit of negative charge.

Rutherford discovered the atomic nucleus. His nuclear model of the atom proposes

- A very small, dense positive nucleus.
- Orbiting negative electrons.

Later, different isotopes were recognized to contain different numbers of **neutrons** in a nucleus with the same number of **protons.**

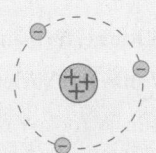

Atomic Spectra and the Nature of Light

The spectra emitted by the gas in a discharge tube consist of discrete wavelengths.

- Every element has a unique spectrum.
- Every spectral line in an element's absorption spectrum is present in its emission spectrum, but not all emission lines are seen in absorption.

Absorption

Emission

Balmer found that the wavelengths of the hydrogen emission spectrum are

$$\lambda = \frac{91.18 \text{ nm}}{\left(\dfrac{1}{m^2} - \dfrac{1}{n^2}\right)}, \qquad m = 1, 2, 3, \ldots \qquad n = m + 1, m + 2, \ldots$$

The end of classical physics . . .

Atomic spectra had to be related to atomic structure, but no one could understand how. According to all that was known,

- Rutherford's nuclear atom should not be stable.
- Atoms should radiate continuous rather than discrete spectra.

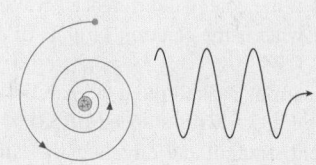

APPLICATIONS

Millikan's oil-drop experiment measured the fundamental unit of charge:

$$e = 1.60 \times 10^{-19} \text{ C}$$

One electron volt (1 eV) is the energy an electron or proton (charge $\pm e$) gains by accelerating through a potential difference of 1 V.

$$1 \text{ eV} = 1.60 \times 10^{-19} \text{ J}$$

TERMS AND NOTATION

electrolysis	subatomic particle	nucleus	neutron
gas discharge tube	electron	nuclear model of the atom	isotope
cathode glow	Millikan oil-drop experiment	electron volt, eV	mass number, A
cathode rays	radioactivity	atomic number, Z	discrete spectrum
Crookes tube	alpha rays	proton	spectral line
crossed-field experiment	beta rays	mass spectrometer	Balmer formula

EXERCISES AND PROBLEMS

Exercises

Section 37.3 Cathode Rays

Section 37.4 J. J. Thomson and the Discovery of the Electron

1. What was the significance of Thomson's experiment in which an off-center electrode was used to collect charge deflected by a magnetic field?

2. What is the evidence by which we know that an electron from an iron atom is identical to an electron from a copper atom?

3. The current in a Crookes tube is 10 nA. How many electrons strike the face of the glass tube each second?

4. An electron in a cathode-ray beam passes between 2.5-cm-long parallel-plate electrodes that are 5.0 mm apart. A 2.0 mT, 2.5-cm-wide magnetic field is perpendicular to the electric field between the plates. The electron passes through the electrodes without being deflected if the potential difference between the plates is 600 V.
   a. What is the electron's speed?
   b. If the potential difference between the plates is set to zero, what is the electron's radius of curvature in the magnetic field?

5. Electrons pass through the parallel electrodes shown in Figure EX37.5 with a speed of 5.0×10^6 m/s. What magnetic field strength and direction will allow the electrons to pass through without being deflected? Assume that the magnetic field is confined to the region between the electrodes.

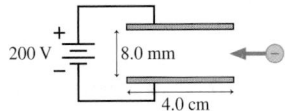

FIGURE EX37.5

Section 37.5 Millikan and the Fundamental Unit of Charge

6. A 0.80-μm-diameter oil droplet is observed between two parallel electrodes spaced 11 mm apart. The droplet hangs motionless if the upper electrode is 20 V more positive than the lower electrode. The density of the oil is 885 kg/m^3.
   a. What is the droplet's mass?
   b. What is the droplet's charge?
   c. Does the droplet have a surplus or a deficit of electrons? How many?

7. An oil droplet with 15 excess electrons is observed between two parallel electrodes spaced 12 mm apart. The droplet hangs motionless if the upper electrode is 25 V more positive than the lower electrode. The density of the oil is 860 kg/m^3. What is the radius of the droplet?

8. Suppose that in a hypothetical oil-drop experiment you measure the following values for the charges on the drops: 3.99×10^{-19} C, 6.65×10^{-19} C, 2.66×10^{-19} C, 10.64×10^{-19} C, and 9.31×10^{-19} C. What is the largest value of the fundamental unit of charge that is consistent with your measurements?

Section 37.6 Rutherford and the Discovery of the Nucleus

Section 37.7 Into the Nucleus

9. Express in eV (or keV or MeV if more appropriate):
   a. The kinetic energy of an electron moving with a speed of 5.0×10^6 m/s.
   b. The potential energy of an electron and a proton 0.10 nm apart.
   c. The kinetic energy of a proton that has accelerated from rest through a potential difference of 5000 V.

10. Express in eV (or keV or MeV if more appropriate):
    a. The kinetic energy of a Li^{++} ion that has accelerated from rest through a potential difference of 5000 V.
    b. The potential energy of two protons 10 fm apart.
    c. The kinetic energy, just before impact, of a 200 g ball dropped from a height of 1.0 m.

11. Determine:
    a. The speed of a 100 eV electron.
    b. The speed of a 5 MeV neutron.
    c. The specific type of particle that has 2.09 MeV of kinetic energy when moving with a speed of 1.0×10^7 m/s.

12. Determine:
    a. The speed of a 6 MeV proton.
    b. The speed of a 20 MeV helium atom.
    c. The specific type of particle that has 1.14 keV of kinetic energy when moving with a speed of 2.0×10^7 m/s.

13. a. Describe the experimental evidence by which we know that the nucleus is made up not just of protons.
    b. The neutron is not easy to isolate or control because it has no charge that would allow scientists to manipulate it. What evidence allowed scientists to determine that the mass of the neutron is almost the same as the mass of a proton?

14. How many electrons, protons, and neutrons are contained in the following atoms or ions: (a) ^{6}Li, (b) ^{13}C$^+$, and (c) ^{18}O^{++}?

15. How many electrons, protons, and neutrons are contained in the following atoms or ions: (a) ^{9}Be$^+$, (b) ^{12}C, and (c) ^{15}N^{+++}?

16. Write the symbol for an atom or ion with:
    a. three electrons, three protons, and five neutrons.
    b. five electrons, six protons, and eight neutrons.

17. Write the symbol for an atom or ion with:
    a. one electron, one proton, and one neutron.
    b. five electrons, seven protons, and seven neutrons.

18. Consider the gold isotope ^{197}Au.
    a. How many electrons, protons, and neutrons are in a neutral ^{197}Au atom?
    b. The gold nucleus has a diameter of 14.0 fm. What is the density of matter in a gold nucleus?
    c. The density of lead is 11,400 kg/m^3. How many times the density of lead is your answer to part b?

19. Consider the lead isotope ^{207}Pb.
    a. How many electrons, protons, and neutrons are in a neutral ^{207}Pb atom?
    b. The lead nucleus has a diameter of 14.2 fm. What are the electric potential and the electric field strength at the surface of a lead nucleus?

20. Explain how the observation of alpha particles scattered at very large angles led Rutherford to reject Thomson's model of the atom and to propose a nuclear model.

Section 37.8 The Emission and Absorption of Light

21. Figure 37.19b identified the wavelengths of four lines in the spectrum of hydrogen.
    a. Determine the Balmer formula n and m values for these wavelengths.
    b. Predict the wavelength of the fifth line in the spectrum.

22. Figure 37.19b identified the wavelengths of four lines in the spectrum of hydrogen.
    a. Determine the Balmer formula n and m values for these wavelengths.
    b. Figure 37.19b labels a feature called the *series limit,* although no spectral line is present at that point. Verify the wavelength of the series limit.

23. The wavelengths in the hydrogen spectrum with $m = 1$ form a series of spectral lines called the Lyman series. Calculate the wavelengths of the first four members of the Lyman series.

24. Two of the wavelengths emitted by a hydrogen atom are 102.6 nm and 1876 nm.
    a. What are the m and n values for each of these wavelengths?
    b. For each of these wavelengths, is the light infrared, visible, or ultraviolet?

Problems

25. What is the total energy, in MeV, of
    a. A proton traveling at 99% of the speed of light?
    b. An electron traveling at 99% of the speed of light?
    Hint: This problem uses relativity.

26. What is the velocity, as a fraction of c, of
    a. A proton with 500 GeV total energy?
    b. An electron with 2.0 GeV total energy?
    Hint: This problem uses relativity.

27. You learned in Chapter 36 that mass has an equivalent amount of energy. What are the energy equivalents in MeV of the rest masses of an electron and a proton?

28. The factor γ appears in many relativistic expressions. A value $\gamma = 1.01$ implies that relativity changes the Newtonian values by approximately 1% and that relativistic effects can no longer be ignored. At what kinetic energy, in MeV, is $\gamma = 1.01$ for (a) an electron, (b) a proton, and (c) an alpha particle?

29. The fission process n + ^{235}U → ^{236}U → ^{144}Ba + ^{89}Kr + 3n converts 0.185 u of mass into the kinetic energy of the fission products. What is the total kinetic energy in MeV?

30. An electron in a cathode-ray beam passes between 2.5-cm-long parallel-plate electrodes that are 5.0 mm apart. A 1.0 mT, 2.5-cm-wide magnetic field is perpendicular to the electric field between the plates. If the potential difference between the plates is 150 V, the electron passes through the electrodes without being deflected. If the potential difference across the plates is set to zero, through what angle is the electron deflected as it passes through the magnetic field?

31. The two 5.0-cm-long parallel electrodes in Figure P37.31 are spaced 1.0 cm apart. A proton enters the plates from one end, an equal distance from both electrodes. A potential difference $\Delta V = 500$ V across the electrodes deflects the proton so that it strikes the outer end of the lower electrode. What magnetic field strength and direction will allow the proton to pass through undeflected while the 500 V potential difference is applied? Assume that both the electric and magnetic fields are confined to the space between the electrodes.

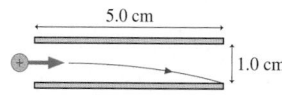

FIGURE P37.31   Trajectory at $\Delta V = 500$ V

32. An unknown charged particle passes without deflection through crossed electric and magnetic fields of strengths 187,500 V/m and 0.125 T, respectively. The particle passes out of the electric field, but the magnetic field continues, and the particle makes a semicircle of diameter 25.05 cm. What is the particle's charge-to-mass ratio? Can you identify the particle?

33. In one of Thomson's experiments he placed a thin metal foil in the electron beam and measured its temperature rise. Consider a cathode-ray tube in which electrons are accelerated through a 2000 V potential difference, then strike a 10 mg copper foil.
    a. How many electrons strike the foil in 10 s if the foil temperature rises 6.0° C?
    b. What is the current of the electron beam?

34. A lithium atom has three electrons. As you will discover in Chapter 41, two of these electrons form an "inner core," but the third—the valence electron—orbits at much larger radius. From the valence electron's perspective, it is orbiting a spherical ball of charge having net charge $+1e$ (i.e., the three protons in the nucleus and the two inner-core electrons). The energy required to ionize a lithium atom is 5.14 eV. According to Rutherford's nuclear model of the atom, what are the orbital radius and speed of the valence electron?
    Hint: Consider the energy needed to remove the electron *and* the force needed to give the electron a circular orbit.

35. The diameter of an atom is 1.2×10^{-10} m and the diameter of its nucleus is 1.0×10^{-14} m. What percent of the atom's volume is occupied by mass and what percent is empty space?

36. Balmer discovered the famous formula that bears his name by inspection and trial-and-error. See if you can discover the formula for each of the following series of wavelengths. Each formula involved an integer n, but, as in the Balmer formula, n may not start with 1.
    a. 125.00, 31.25, 13.89, 7.81, and 5.00 nm.
    b. 375, 900, 1575, 2400, 3375, and 4500 nm.

37. The diameter of an aluminum atom is approximately 1.2×10^{-10} m. The diameter of the nucleus of an aluminum atom is approximately 8×10^{-15} m. The density of solid aluminum is 2700 kg/m³.
    a. What is the average density of an aluminum atom?
    b. Your answer to part a was similar to but larger than the density of solid aluminum. This suggests that the atoms in solid aluminum have spaces between them rather than being tightly packed together. What is the average volume per atom in solid aluminum? If this volume is a sphere, what is the radius? What can you conclude about the average spacing between atoms compared to the size of the atoms?
    Hint: The volume *per* atom is not the same as the volume *of* an atom.
    c. What is the density of the aluminum nucleus? Compare this to the density of ordinary matter.

38. The charge-to-mass ratio of a nucleus, in units of e/u, is $q/m = Z/A$. For example, a hydrogen nucleus has $q/m = 1/1 = 1$.
    a. Make a graph of charge-to-mass ratio versus proton number Z for nuclei with $Z = 5, 10, 15, 20, \ldots, 90$. For A, use the average atomic mass shown on the periodic table of elements. Show each of these 18 nuclei as a dot, but don't connect the dots together as a curve.
    b. Describe any trend that you notice in your graph.
    c. What's happening in the nuclei that is responsible for this trend?

39. If the nucleus is a few fm in diameter, the distance between the centers of two protons must be ≈2 fm.
    a. Calculate the electric force between two protons that are 2.0 fm apart.
    b. Calculate the gravitational force between two protons that are 2.0 fm apart. Could gravity be the force that holds the nucleus together?
    c. Your answers to parts a and b imply that there must be some other force that binds the nucleus together and prevents the protons from pushing each other out. What characteristics of this force can you deduce from the discussion of the atom and the nucleus in this chapter?

40. In a head-on collision, the closest approach of a 6.24 MeV alpha particle to the center of a nucleus is 6.00 fm. The nucleus is in an atom of what element?

41. Through what potential difference would you need to accelerate an alpha particle, starting from rest, so that it will just reach the surface of a 15-fm-diameter ^{238}U nucleus?

42. The oxygen nucleus ^{16}O has a radius of 3.0 fm.
    a. With what speed must a proton be fired toward an oxygen nucleus to have a turning point 1.0 fm from the surface?
    b. What is the proton's kinetic energy in MeV?

43. To initiate a nuclear reaction, an experimental nuclear physicist wants to shoot a proton *into* a ^{12}C nucleus. The proton must impact the nucleus with a kinetic energy of 3.0 MeV. The nuclear radius is 2.75 fm.

a. With what speed must the proton be fired toward the target?
b. Through what potential difference must the proton be accelerated from rest to acquire this speed?

44. The cesium isotope ^{137}Cs, with $Z = 55$, is radioactive and decays by beta decay. A beta particle is observed in the laboratory with a kinetic energy of 300 keV. The nucleus of a ^{137}Cs atom has a diameter of 12.4 fm. With what kinetic energy was the beta particle ejected from the ^{137}Cs nucleus?

Challenge Problems

45. An alpha particle approaches a ^{197}Au nucleus with a speed of 1.50×10^7 m/s. As Figure CP37.45 shows, the alpha particle is scattered at a 49° angle at the slower speed of 1.49×10^7 m/s. In what direction does the ^{197}Au nucleus recoil, and with what speed?

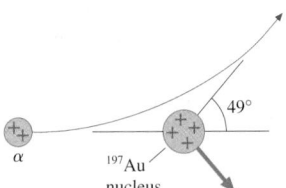

FIGURE CP37.45

46. Physicists first attempted to understand the hydrogen atom by applying the laws of classical physics. Consider an electron of mass m and charge $-e$ in a circular orbit of radius r around a proton of charge $+e$.
    a. Use Newtonian physics to show that the total energy of the atom is $E = -e^2/8\pi\epsilon_0 r$.
    b. Show that the potential energy is -2 times the electron's kinetic energy. This result is called the *virial theorem*.
    c. The minimum energy needed to ionize a hydrogen atom (i.e., to remove the electron) is found experimentally to be 13.6 eV. From this information, what are the electron's speed and the radius of its orbit?

47. Consider an oil droplet of mass m and charge q. We want to determine the charge on the droplet in a Millikan-type experiment. We will do this in several steps. Assume, for simplicity, that the charge is positive and that the electric field between the plates points upward.
    a. An electric field is established by applying a potential difference to the plates. It is found that a field of strength E_0 will cause the droplet to be suspended motionless. Write an expression for the droplet's charge in terms of the suspending field E_0 and the droplet's weight mg.
    b. The field E_0 is easily determined by knowing the plate spacing and measuring the potential difference applied to them. The larger problem is to determine the mass of a microscopic droplet. Consider a mass m falling through viscous medium in which there is a retarding or drag force. For very small particles, the retarding force is given by $F_{\text{drag}} = -bv$ where b is a constant and v the droplet's velocity. The sign recognizes that the drag force vector points upward when the droplet is falling (negative v). A falling droplet quickly reaches a constant speed, called the *terminal speed*. Write an expression for the terminal speed v_{term} in terms of m, g, and b.

c. A spherical object of radius r moving slowly through the air is known to experience a retarding force $F_{drag} = -6\pi\eta rv$ where η is the *viscosity* of the air. Use this and your answer to part b to show that a spherical droplet of density ρ falling with a terminal velocity v_{term} has a radius

$$r = \sqrt{\frac{9\eta v_{term}}{2\rho g}}$$

d. Oil has a density 860 kg/m³. An oil droplet is suspended between two plates 1.0 cm apart by adjusting the potential difference between them to 1177 V. When the voltage is removed, the droplet falls and quickly reaches constant speed. It is timed with a stopwatch, and falls 3.00 mm in 7.33 s. The viscosity of air is 1.83×10^{-5} kg/m s. What is the droplet's charge?

e. How many units of the fundamental electric charge does this droplet possess?

STOP TO THINK ANSWERS

Stop to Think 37.1: b. This observation says that all electrons are the same.

Stop to Think 37.2: b. From the right-hand rule with $\vec{v}$ to the right and $\vec{B}$ out of the page.

Stop to Think 37.3: 4. Neutral carbon would have six electrons. C^{++} is missing two.

Stop to Think 37.4: 6 protons and 8 neutrons. The number of protons is the atomic number, which is 6. That leaves $14 - 6 = 8$ neutrons.

Stop to Think 37.5: a is emission, b is absorption. All wavelengths in the absorption spectrum are seen in the emission spectrum, but not all wavelengths in the emission spectrum are seen in the absorption spectrum.

38 Quantization

A scanning tunneling microscope image of a "quantum corral" made from 60 iron atoms.

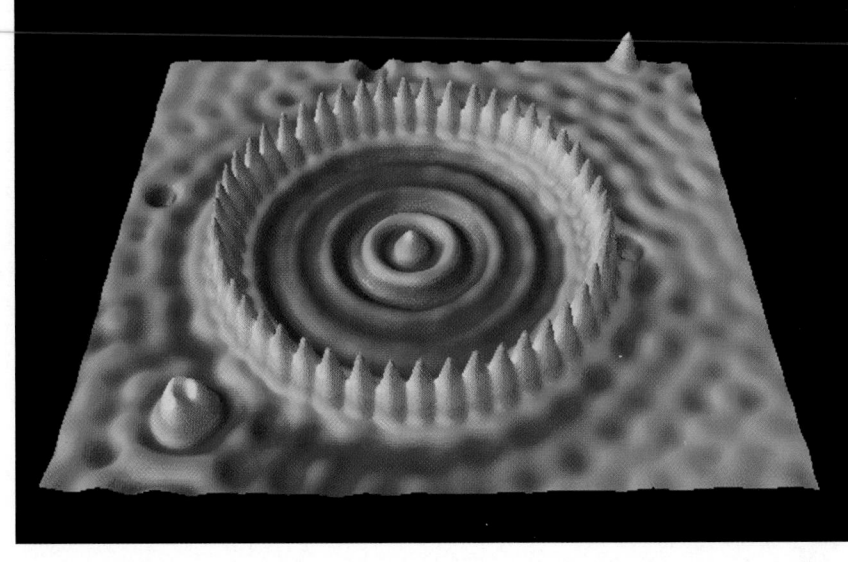

▶ **Looking Ahead**

The goal of Chapter 38 is to understand the quantization of energy for light and matter. In this chapter you will learn to:

- Understand the photoelectric effect in terms of Einstein's light quanta.
- Use the photon concept.
- Understand how de Broglie's matter waves lead to the quantization of energy.
- Use Bohr's model of quantization in atoms.
- Calculate energies and wavelengths for hydrogen and hydrogen-like ions.

◀ **Looking Back**

Many of the ideas in this chapter were introduced in Chapter 24, which is an essential prerequisite for this chapter. Please review:

- Sections 22.2, 22.3, and 22.6 Interference and interferometers.
- Sections 24.1–24.5 Photons, matter waves, and quantization.
- Section 37.6 Electron volts and Rutherford's nuclear model of the atom.

The picture shown here, called a "quantum corral," was made with a scanning tunneling microscope, a device we'll study in Chapter 40. The image shows the electron density in the vicinity of a circle of 60 iron atoms that have been carefully placed on a plane of carbon. But it's not the circle of electrons gathered around the iron atoms that is most interesting. Notice the circular ripple-like rings in the center of the corral. What you're seeing is an *electron standing wave*, rather like the standing wave on the head of a vibrating drum.

Recall from Chapter 24 what Einstein, de Broglie, and others found: that the classical either-or distinction between particles and waves, as useful as it is for macroscopic systems, does not exist in the microscopic world of electrons and atoms. Instead, light and matter exhibit characteristics of *both* particles *and* waves. This new *wave-particle duality,* as it is called, defies our commonsense picture of how things ought to behave. Nonetheless, the experimental evidence for wave-particle duality is now overwhelming and cannot be doubted. Wave-like electrons and particle-like photons of light are no longer just scientific curiosities. Modern engineering devices, such as *quantum-well lasers,* make explicit use of wave-particle duality.

This chapter will explore two critical ideas: Einstein's introduction of a particle-like nature of light and Bohr's development of a quantum atom. We will begin to think about and describe matter and light in terms of a quantum model rather than classical models. In addition, the ideas in this chapter are the final steps we need before introducing quantum mechanics in Chapter 39.

38.1 The Photoelectric Effect

In 1886, Heinrich Hertz was the first to demonstrate that electromagnetic waves can be artificially generated. By verifying the predictions of Maxwell's electromagnetic theory, Hertz cemented the last blocks of classical physics into place.

Yet in one of those ever-present ironies of history, Hertz happened, quite by chance, to discover the very phenomenon that would launch the quantum revolution. He noticed, in the course of his investigations, that a negatively charged electroscope could be discharged by shining ultraviolet light on it.

Hertz's observation caught the attention of J. J. Thomson. Figure 38.1 illustrates Thomson's interpretation of the observation: He inferred that the ultraviolet light was somehow causing the electrode to emit negative charges, thus restoring itself to electric neutrality. In 1899, using techniques similar to those with which he discovered the electron, Thomson showed that the emitted charges had exactly the same charge-to-mass ratio as electrons, and, presumably, were electrons. The emission of electrons from a substance due to light striking its surface came to be called the **photoelectric effect.** The emitted electrons are often called *photoelectrons* to indicate their origin, but they are identical in every respect to all other electrons.

Although this discovery might seem to be a minor footnote in the history of science, it soon became a, or maybe *the,* pivotal event that opened the door to new ideas. We will look at the photoelectric effect in a fair bit of detail. Our goals are to understand how classical physics was unable to explain the details of such a simple experiment and to recognize the startling new concept introduced by Einstein.

Characteristics of the Photoelectric Effect

It was not the discovery itself that dealt the fatal blow to classical physics, but the specific characteristics of the photoelectric effect found around 1900 by one of Hertz's students, Phillip Lenard. Lenard built a glass tube, shown in Figure 38.2, with two facing electrodes and a window. After removing the air from the tube, so that electrons could move freely from one electrode to the other, he allowed light to shine on the cathode.

Lenard found a steady counterclockwise current (clockwise flow of electrons) through the ammeter whenever ultraviolet light was shining on the cathode. There are no junctions in this circuit, so the current must be the same all the way around the loop. The current in the space between the cathode and the anode consists of electrons moving freely through space (i.e., not inside a wire) at the *same rate* (same number of electrons per second) as the current in the wire. There is no current if the electrodes are in the dark, so electrons don't spontaneously leap off the cathode. Instead, the light causes electrons to be ejected from the cathode at a steady rate.

Lenard used a battery to establish an adjustable potential difference ΔV between the two electrodes. He then studied how the current I varied as the potential difference and the light's wavelength and intensity were changed. Lenard found the photoelectric effect to have the following properties.

1. The current I is directly proportional to the light intensity. If the light intensity is doubled, the current also doubles.
2. The current appears without delay when the light is applied. To Lenard, this meant within the ≈ 0.1 s with which his equipment could respond. Later experiments showed that the current begins less than 1 ns after light hits the cathode!
3. Photoelectrons are emitted *only* if the light frequency f exceeds a **threshold frequency** f_0. This is shown in the graph of Figure 38.3.
4. The value of the threshold frequency f_0 depends on the type of metal from which the cathode is made.
5. If the potential difference ΔV is positive (anode positive with respect to the cathode), the current does not change as ΔV is increased. If ΔV is made negative (anode negative with respect to the cathode), by reversing the

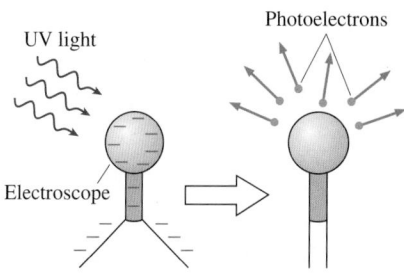

Ultraviolet light discharges a negatively charged electroscope by causing it to emit electrons.

FIGURE 38.1 Ultraviolet light discharges a negatively charged electroscope.

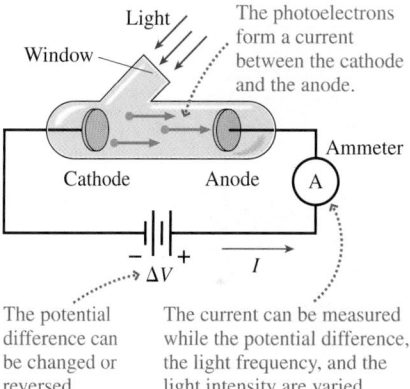

The photoelectrons form a current between the cathode and the anode.

The potential difference can be changed or reversed.

The current can be measured while the potential difference, the light frequency, and the light intensity are varied.

FIGURE 38.2 Lenard's experimental device to study the photoelectric effect.

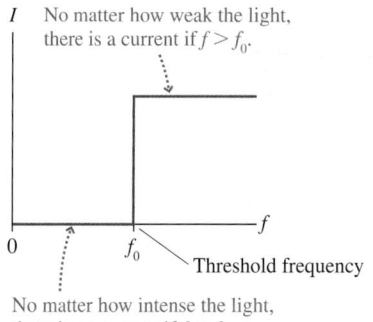

No matter how weak the light, there is a current if $f > f_0$.

No matter how intense the light, there is no current if $f < f_0$.

FIGURE 38.3 The photoelectric current as a function of the light frequency f.

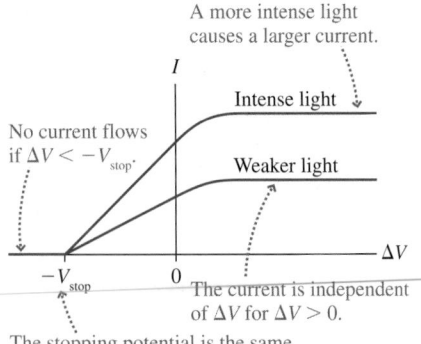

No current flows if $\Delta V < -V_{stop}$.

A more intense light causes a larger current.

Intense light

Weaker light

The current is independent of ΔV for $\Delta V > 0$.

The stopping potential is the same for intense light and weak light.

FIGURE 38.4 The photoelectric current as a function of the battery potential difference ΔV.

The *minimum* energy to remove a drop of water from the pool is mgh.

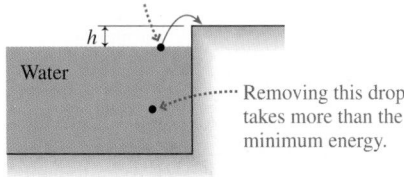

h

Water

Removing this drop takes more than the minimum energy.

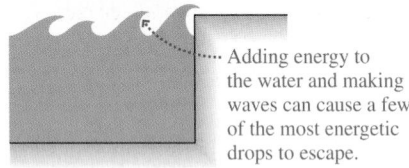

Adding energy to the water and making waves can cause a few of the most energetic drops to escape.

FIGURE 38.5 A swimming pool analogy of electrons in a metal.

TABLE 38.1 The work function for some of the elements

Element	E_0 (eV)
Potassium	2.30
Sodium	2.75
Aluminum	4.28
Tungsten	4.55
Copper	4.65
Iron	4.70
Gold	5.10

battery, the current decreases until, at some voltage $\Delta V = -V_{stop}$ the current reaches zero. The value of V_{stop} is called the **stopping potential.** This behavior is shown in Figure 38.4.

6. The value of V_{stop} is the same for both weak light and intense light. A more intense light causes a larger current, as Figure 38.4 shows, but in both cases the current ceases when $\Delta V = -V_{stop}$.

NOTE ▶ We're defining V_{stop} to be a *positive* number. The potential difference that stops the electrons is $\Delta V = -V_{stop}$, with an explicit minus sign. ◀

Classical Interpretation of the Photoelectric Effect

The mere existence of the photoelectric effect is not, as is sometimes assumed, a difficulty for classical physics. You learned in Chapter 28 that electrons are the charge carriers in a metal and move around freely inside like a sea of negatively charged particles. The electrons are bound inside the metal and do not spontaneously spill out of an electrode at room temperature. But a piece of metal heated to a sufficiently high temperature *does* emit electrons in a process called **thermal emission.** The electron gun in a television or computer display terminal starts with the thermal emission of electrons from a hot tungsten filament.

A useful analogy, shown in Figure 38.5, is the water in a swimming pool. Water molecules do not spontaneously leap out of the pool if the water is calm. To remove a water molecule, you must do *work* on it to lift it upward, against the force of gravity, to the edge of the pool. A minimum energy is needed to extract a water molecule, namely the energy needed to lift a molecule that is right at the surface. Removing a water molecule that is deeper requires more than the minimum energy. People playing in the pool add energy to the water, causing waves. If sufficient energy is added, a small fraction of the water molecules may gain enough energy to splash over the edge and leave the pool.

Similarly, a *minimum* energy is needed to free an electron from a metal. To extract an electron, you would need to exert a force on it and pull it (i.e., do *work* on it) until its speed is large enough to escape. The minimum energy E_0 needed to free an electron is called the **work function** of the metal. Some electrons, like the deeper water molecules, may require more energy than E_0 to escape, but all will require *at least* E_0. The values of the work function E_0 differ among metals according to the densities and crystal structures of the metals. Table 38.1 provides a short list. Notice that work functions are given in electron volts.

Heating a metal, like splashing in the pool, increases the thermal energy of the electrons. At a sufficiently high temperature, the kinetic energy of a small percentage of the electrons may exceed the work function. These electrons can "make it out of the pool" and leave the metal. In practice, the thermal emission of electrons requires $T > 1500°$ C, and there are only a few elements, such as tungsten, for which thermal emission can become significant before the metal melts!

Heating a metal increases the temperature of not only the electrons but also the much more massive crystal lattice of positive ions. But suppose we could raise the temperature of the electrons alone and not the crystal lattice. One possible way to do this is to shine a light wave on the surface. Because electromagnetic waves are absorbed by the conduction electrons, not by the positive ions, the light wave heats only the electrons. Eventually the electrons' energy is transferred to the crystal lattice, via collisions, but if the light is sufficiently intense, the *electron temperature* may be significantly higher than the temperature of the metal. In 1900, it was plausible to think that an intense light source could cause the thermal emission of electrons without melting the metal.

The Stopping Potential

Photoelectrons leave the cathode with kinetic energy. An electron with energy E_{elec} inside the metal loses energy ΔE as it escapes, so it emerges as a photoelectron with kinetic energy $K = E_{elec} - \Delta E$. The work function energy E_0 is the *minimum* energy needed to remove an electron, so the *maximum* possible kinetic energy of a photoelectron is

$$K_{max} = E_{elec} - E_0 \qquad (38.1)$$

The photoelectrons, after leaving the cathode, move out in all directions. Some electrons reach the anode, creating a measurable current, but many do not. However, as Figure 38.6 shows,

- A positive anode attracts *all* of the photoelectrons to the anode. Once all electrons reach the anode, a further increase in ΔV does not cause any further increase in the current I. That is why the graph lines become horizontal on the right side of Figure 38.4.
- A negative anode repels the electrons. However, a photoelectron leaving the cathode with sufficient kinetic energy can still reach the anode, just as a ball hits the ceiling if you toss it upward with sufficient kinetic energy. A slightly negative anode voltage turns back only the slowest electrons. The current steadily decreases as the anode voltage becomes increasingly negative until, at the stopping potential, *all* electrons are turned back and the current ceases. This was the behavior observed on the left side of Figure 38.4.

We can use conservation of energy to analyze the photoelectrons. Let the cathode be the point of zero potential energy, as shown in Figure 38.7. An electron emitted from the cathode with kinetic energy K_i has initial total energy

$$E_i = K_i + U_i = K_i + 0 = K_i$$

When the electron reaches the anode, which is at potential ΔV relative to the cathode, it has potential energy $U = q\Delta V = -e\Delta V$ and final total energy

$$E_f = K_f + U_f = K_f - e\Delta V$$

From conservation of energy, $E_f = E_i$, the electron's final kinetic energy is

$$K_f = K_i + e\Delta V \qquad (38.2)$$

The electron speeds up ($K_f > K_i$) if ΔV is positive. The electron slows down if ΔV is negative, but it still reaches the anode ($K_f > 0$) if K_i is large enough.

An electron with initial kinetic energy K_i will be stopped just as it reaches the anode if the potential difference is $\Delta V = -K_i/e$. The potential difference that will turn back the very fastest electrons, those with $K = K_{max}$, and thus stop the current is

$$\Delta V_{\text{stop fastest electrons}} = -\frac{K_{max}}{e}$$

By definition, the potential difference that causes the electron current to cease is $\Delta V = -V_{stop}$, where V_{stop} is the stopping potential. Thus the stopping potential seen in Figure 38.4 is

$$V_{stop} = \frac{K_{max}}{e} \qquad (38.3)$$

The stopping potential tells us the maximum kinetic energy of the photoelectrons.

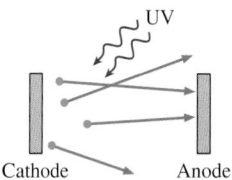

$\Delta V = 0$: The photoelectrons leave the cathode in all directions. Only a few reach the anode.

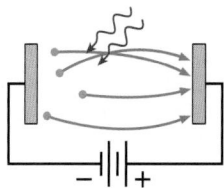

$\Delta V > 0$: Biasing the anode positive creates an electric field that pushes all the photoelectrons to the anode.

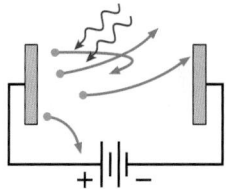

$\Delta V < 0$: Biasing the anode negative repels the electrons. Only the very fastest make it to the anode.

FIGURE 38.6 A positive anode attracts the photoelectrons. A negative anode repels them.

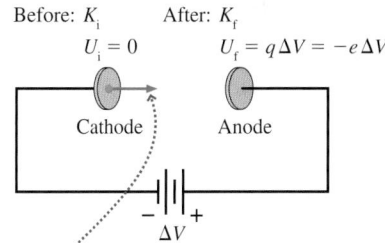

Energy is transformed from kinetic to potential as an electron moves from cathode to anode.

FIGURE 38.7 Energy is conserved.

EXAMPLE 38.1 **The classical photoelectric effect**
A photoelectric-effect experiment is performed with an aluminum cathode. An electron inside the cathode has a speed of 1.5×10^6 m/s. If the potential difference between the anode and cathode is -2.00 V, what is the highest possible speed with which this electron could reach the anode?

MODEL Energy is conserved.

SOLVE If this electron succeeds in escaping as a photoelectron, its maximum possible kinetic energy is $K_{max} = E_{elec} - E_0$, where $E_0 = 4.28$ eV is the work function of aluminum. If the electron escapes with the maximum possible kinetic energy, its kinetic energy at the anode will be given by Equation 38.2 with $\Delta V = -2.00$ V.

The electron's initial energy is

$$E_{elec} = \frac{1}{2}mv^2 = \frac{1}{2}(9.11 \times 10^{-31} \text{ kg})(1.5 \times 10^6 \text{ m/s})^2$$

$$= 1.025 \times 10^{-18} \text{ J} = 6.41 \text{ eV}$$

Its maximum possible kinetic energy as it leaves the cathode is

$$K_i = K_{max} = E_{elec} - E_0 = 2.13 \text{ eV}$$

Thus the kinetic energy at the anode is

$$K_f = K_i + e\Delta V = 2.13 \text{ eV} - (e)(2.00 \text{ V}) = 0.13 \text{ eV}$$

Notice that the electron loses 2.00 eV of *energy* as it moves through the *potential* difference of -2.00 V, so we can compute the final kinetic energy in eV without having to convert to joules. However, we must convert K_f to joules to find the final speed:

$$K_f = \frac{1}{2}mv_f^2 = 0.13 \text{ eV} = 2.1 \times 10^{-20} \text{ J}$$

$$v_f = \sqrt{\frac{2K_f}{m}} = 2.1 \times 10^5 \text{ m/s}$$

Limits of the Classical Interpretation

A classical analysis based on the thermal emission of electrons from a metal has provided a possible explanation of observations 1 and 5 above. But nothing in this explanation suggests that there should be a threshold frequency. If a weak intensity at a frequency just slightly above f_0 can generate a current, then certainly a strong intensity at a frequency just slightly below f_0 should be able to do so. There is no reason that a very slight change in frequency should matter. But Lenard found there to be a sharp threshold at f_0.

And what about his observation that the current starts instantly? If the photoelectrons are due to thermal emission, it should take some length of time for the light to raise the electron temperature sufficiently high for some to escape. In fact, fairly straightforward calculations show that, for a light of modest intensity, it should take several minutes before charge starts flowing! The experimental evidence was in sharp disagreement. If $f > f_0$, the current starts instantly for both weak light and intense light.

And last, more intense light would be expected to heat the electrons to a higher temperature. Doing so should increase the maximum kinetic energy of the photoelectrons and thus should increase the stopping potential V_{stop}. But as Lenard found, the stopping potential is the same for strong light as it is for weak light.

Although the mere presence of photoelectrons did not seem surprising, classical physics was unable to explain the observed behavior of the photoelectrons. The threshold frequency and the instant current seemed particularly anomalous.

38.2 Einstein's Explanation

Albert Einstein was a little-known young man of 26 in 1905. A photograph from the time, shown in Figure 38.8, bears little resemblance to the familiar picture of a white-haired older Einstein. He had recently graduated from the Polytechnic Institute in Zurich, Switzerland, with the Swiss equivalent of a Ph.D. in physics. Although his mathematical brilliance was recognized, his overall academic record was mediocre. Rather than pursue an academic career, Einstein took a job with the Swiss Patent Office in Bern. This was a fortuitous choice because it provided him with plenty of spare time to think about physics in his own unique way.

In 1905, within the span of a single year, Einstein published three papers on three different topics, all of which would revolutionize physics. One was his initial paper on the theory of relativity, the subject with which Einstein is most associated in the public mind. Interestingly, this paper received less attention at the time than the other two. A second paper explained a phenomenon called *Brownian motion.* In 1827, the Scottish botanist Robert Brown had used a microscope to examine small pollen grains suspended in water. He observed that the pollen grains jiggled about rather than floating at rest. Einstein, using the techniques of statistical mechanics, provided a convincing explanation that the jiggling resulted from the continual random collisions of water molecules with the pollen grains. His analysis provided one of the most definitive pieces of evidence for the reality of atoms and molecules.

But it is Einstein's third paper of 1905, on the nature of light, in which we are most interested. In it he offered an exceedingly simple but amazingly bold idea to explain Lenard's photoelectric-effect data. A few years earlier, in 1900, the German physicist Max Planck had been trying to understand the details of the rainbow-like spectrum of light emitted by a glowing, incandescent object. This problem didn't seem to yield to a classical physics analysis, but Planck found that he could calculate the spectrum perfectly if he made an unusual assumption. The atoms in a solid vibrate back and forth around their equilibrium positions with frequency f. You learned in Chapter 14 that the energy of a simple harmonic oscillator depends on its amplitude and can have *any* possible value. But to predict the spectrum correctly, Planck had to assume that the oscillating atoms are *not* free to have any possible energy. Instead, the energy of an atom vibrating with frequency f has to be one of the specific energies $E = 0, hf, 2hf, 3hf, \ldots$, where h is a constant. That is, the vibration energies are *quantized.*

Planck was able to determine the value of the constant h by comparing his calculations of the spectrum to experimental measurements. The constant that he introduced into physics is now called **Planck's constant.** Its contemporary value is

$$h = 6.63 \times 10^{-34} \, \text{J s} = 4.14 \times 10^{-15} \, \text{eV s}$$

The first value, with SI units, is the proper one for most calculations, but you will find the second to be useful when energies are expressed in eV.

Planck considered this to be a trick. He was skeptical that the atoms *really* had quantized energy, and he felt sure that further analysis would soon reveal how the trick worked. Einstein was the first to take Planck's idea seriously and to suggest that the quantization is real. Einstein went even further and suggested that **electromagnetic radiation itself is quantized!** That is, light is not really a continuous wave but, instead, arrives in small packets or bundles of energy. Einstein called each packet of energy a **light quantum,** and he postulated that the energy of one light quantum is directly proportional to the frequency of the light. That is, each quantum of light has energy

$$E = hf \tag{38.4}$$

where h is Planck's constant and f is the frequency of the light.

The idea of light quanta is subtle, so let's look at an analogy with raindrops. Although we often think of water as a continuous fluid, such as water in a beaker, rain consists of water that falls in discrete packets called raindrops. Raindrops are analogous to quanta of light. A downpour has a torrent of raindrops, but in a light shower the drops are few. The difference between "intense" rain and "weak" rain is the *rate* at which the drops arrive. An intense rain makes a continuous noise on the roof, so you are not aware of the individual drops, but the individual drops become apparent during a light rain.

FIGURE 38.8 A young Einstein.

For most light sources, the individual quanta are no more discernible than the individual raindrops in a downpour.

Similarly, a great number of light quanta arrive each second when the light is intense, but very weak light consists of only a few quanta per second. And just as raindrops come in different sizes, with larger-mass drops having larger kinetic energy, higher-frequency light quanta have a larger amount of energy. Although this analogy is not perfect, it does provide a useful mental picture of light quanta arriving at a surface.

EXAMPLE 38.2 The energy of a light quantum

What is the energy of one quantum of light having a wavelength of 500 nm?

SOLVE Light with a wavelength of 500 nm has frequency

$$f = \frac{v}{\lambda} = \frac{c}{\lambda} = \frac{3.00 \times 10^8 \text{ m/s}}{500 \times 10^{-9} \text{ m}} = 6.00 \times 10^{14} \text{ Hz}$$

One light quantum has energy

$$E = hf = 3.98 \times 10^{-19} \text{ J} = 2.49 \text{ eV}$$

ASSESS Because 500 nm is a typical wavelength for visible light (it would be perceived as green light), you can see that the electron volt is an energy unit of more appropriate size than the joule.

Einstein's Postulates

Einstein framed three postulates about light quanta and their interaction with matter:

1. Light of frequency f consists of discrete quanta, each of energy $E = hf$. Each photon travels at the speed of light c.
2. Light quanta are emitted or absorbed on an all-or-nothing basis. A substance can emit 1 or 2 or 3 quanta, but not 1.5. Similarly, an electron in a metal cannot absorb half a quantum but, instead, only an integer number.
3. A light quantum, when absorbed by a metal, delivers its entire energy to *one* electron.

NOTE ▶ These three postulates—that light comes in chunks, that the chunks cannot be divided, and that the energy of one chunk is delivered to one electron—are crucial for understanding the new ideas that will lead to quantum physics. They are completely at odds with the concepts of classical physics, where energy can be continuously divided and shared, so they deserve careful thought. ◀

Let's look at how Einstein's postulates apply to the photoelectric effect. If Einstein is correct, the light shining on the metal is a torrent of light quanta, each of energy hf. Each quantum is absorbed by *one* electron, giving that electron an energy $E_{\text{elec}} = hf$. This leads us to several interesting conclusions:

1. An electron that has just absorbed a quantum of light energy has $E_{\text{elec}} = hf$. (The electron's thermal energy at room temperature is so much less than hf that we can neglect it.) Figure 38.9 shows than this electron can escape from the metal, becoming a photoelectron, if

$$E_{\text{elec}} = hf \geq E_0 \tag{38.5}$$

In other words, there is a *threshold frequency*

$$f_0 = \frac{E_0}{h} \tag{38.6}$$

for the ejection of photoelectrons. If f is less than f_0, even by just a small amount, none of the electrons will have sufficient energy to escape no matter how intense the light. But even very weak light with $f \geq f_0$ will give a few electrons sufficient energy to escape **because each light quantum delivers all of its energy to one electron.** This threshold behavior is exactly what Lenard observed.

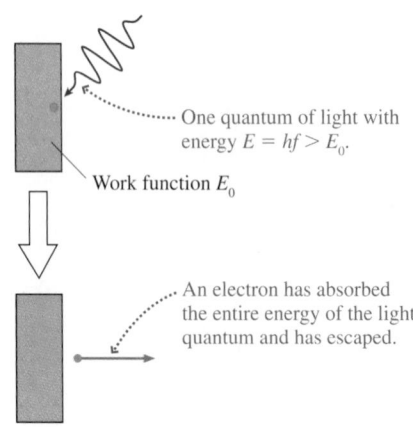

One quantum of light with energy $E = hf > E_0$.

Work function E_0

An electron has absorbed the entire energy of the light quantum and has escaped.

FIGURE 38.9 The creation of a photoelectron.

NOTE ▶ The threshold frequency is directly proportional to the work function. Metals with large work functions, such as iron, copper, and gold, exhibit the photoelectric effect only when illuminated by high-frequency ultraviolet light. Photoemission occurs with lower-frequency visible light for metals with smaller values of E_0, such as sodium and potassium. ◀

2. A more intense light delivers a larger number of light quanta to the surface. These quanta eject a larger number of photoelectrons and cause a larger current, exactly as observed.

3. There is a distribution of kinetic energies, because different photoelectrons require different amounts of energy to escape, but the *maximum* kinetic energy is

$$K_{\max} = E_{\text{elec}} - E_0 = hf - E_0 \qquad (38.7)$$

As we noted in Equation 38.3, the stopping potential V_{stop} is a measure of $K_{\max}$. Einstein's theory predicts that the stopping potential is related to the light frequency by

$$V_{\text{stop}} = \frac{K_{\max}}{e} = \frac{hf - E_0}{e} \qquad (38.8)$$

NOTE ▶ The stopping potential does *not* depend on the intensity of the light. Both weak light and intense light will have the same stopping potential, as Lenard had observed but which could not previously be explained. ◀

4. If each light quantum transfers its energy hf to just one electron, that electron *immediately* has enough energy to escape. The current should begin instantly, with no delay, exactly as Lenard had observed.

Using the swimming pool analogy again, Figure 38.10 shows a pebble being thrown into the pool. The pebble increases the energy of the water, but the increase is shared among all the molecules in the pool. The increase in the water's energy is barely enough to make ripples, not nearly enough to splash water out of the pool. But suppose *all* the pebble's energy could go to *one drop* of water that didn't have to share it. That one drop of water would easily have enough energy to leap out of the pool. Einstein's hypothesis that a light quantum transfers all its energy to one electron is equivalent to the pebble transferring all its energy to one drop of water.

A Prediction

Not only do Einstein's hypotheses explain all of Lenard's observations, they make a new prediction. According to Equation 38.8, the stopping potential should be a linearly increasing function of the light's frequency f. We can rewrite Equation 38.8 in terms of the threshold frequency $f_0 = E_0/h$ as

$$V_{\text{stop}} = \frac{h}{e}(f - f_0) \qquad (38.9)$$

A graph of the stopping potential V_{stop} versus the light frequency f should start from zero at $f = f_0$, then rise linearly with a slope of h/e. In fact, the slope of the graph provides a way to measure Planck's constant h.

Lenard had not measured the stopping potential for different frequencies, so Einstein offered this as an untested prediction of his postulates. Robert Millikan, who was well known for his oil-drop experiment to measure e, took up the challenge. Some of Millikan's data for a cesium cathode are shown in Figure 38.11. As you can see, Einstein's prediction of a linear relationship between f and V_{stop} was fully confirmed.

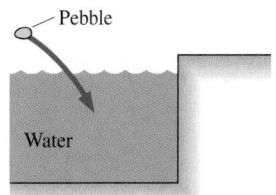

Classically, the energy of the pebble is shared by all the water molecules. One pebble causes only very small waves.

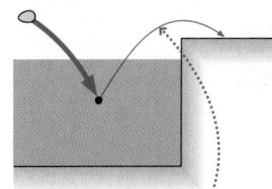

If the pebble could give *all* its energy to one drop, that drop could easily splash out of the pool.

FIGURE 38.10 A pebble transfers energy to the water.

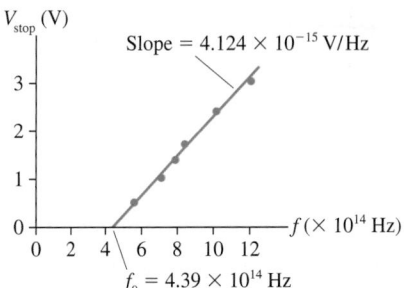

FIGURE 38.11 A graph of Millikan's data for the stopping potential as the light frequency is varied.

Millikan measured the slope of his graph and multiplied it by the value of e (which he had measured a few years earlier in the oil-drop experiment) to find h. His value agreed with the value that Planck had determined in 1900 from an entirely different experiment. Light quanta, whether physicists liked the idea or not, were real.

EXAMPLE 38.3 The photoelectric threshold frequency

What are the threshold frequencies and wavelengths for photo emission from sodium and from aluminum?

SOLVE Table 38.1 gives the sodium work function as $E_0 = 2.75$ eV. Aluminum has $E_0 = 4.28$ eV. We can use Equation 38.6, with h in units of eV s, to calculate

$$f_0 = \frac{E_0}{h} = \begin{cases} 6.64 \times 10^{14} \text{ Hz} & \text{sodium} \\ 10.34 \times 10^{14} \text{ Hz} & \text{aluminum} \end{cases}$$

These frequencies are converted to wavelengths with $\lambda = c/f$, giving

$$\lambda = \begin{cases} 452 \text{ nm} & \text{sodium} \\ 290 \text{ nm} & \text{aluminum} \end{cases}$$

ASSESS The photoelectric effect can be observed with sodium for $\lambda < 452$ nm. This includes blue and violet visible light but not red, orange, yellow, or green. Aluminum, with a larger work function, needs ultraviolet wavelengths $\lambda < 290$ nm.

EXAMPLE 38.4 Maximum photoelectron speed

What is the maximum photoelectron speed if sodium is illuminated with light of 300 nm?

SOLVE The light frequency is $f = c/\lambda = 1.00 \times 10^{15}$ Hz, so each light quantum has energy $hf = 4.14$ eV. The maximum kinetic energy of a photoelectron is

$$K_{max} = hf - E_0 = 4.14 \text{ eV} - 2.75 \text{ eV} = 1.39 \text{ eV}$$
$$= 2.22 \times 10^{-19} \text{ J}$$

Because $K = \frac{1}{2}mv^2$, where m is the electron's mass, not the mass of the sodium atom, the maximum speed of a photoelectron leaving the cathode is

$$v_{max} = \sqrt{\frac{2K_{max}}{m}} = 6.99 \times 10^5 \text{ m/s}$$

Note that we had to convert K_{max} to SI units of J before calculating a speed in m/s.

STOP TO THINK 38.1   The work function of metal A is 3.0 eV. Metals B and C have work functions of 4.0 eV and 5.0 eV, respectively. Ultraviolet light shines on all three metals, creating photoelectrons. Rank in order, from largest to smallest, the stopping voltages for A, B, and C.

38.3 Photons

Einstein was awarded the Nobel prize in 1921 not for his theory of relativity, as many would suppose, but for his explanation of the photoelectric effect. Although Planck had made the first suggestion, it was Einstein who showed convincingly that energy is quantized and that light, even though it exhibits interference, comes in some kind of particle-like packets of energy. These fundamental units of light energy were later given the name **photons.**

But just what are photons? Although particle-like, they clearly do not mesh with the classical idea of a particle. A classical particle, when faced with Young's double-slit apparatus, would go through one hole or the other. If light consisted of classical particles, we would see two bright spots on the screen. Instead, we see interference fringes behind a double slit. We even observed, in Chapter 24, that the interference pattern can be built up photon by photon if the light intensity is reduced to the point where only one photon at a time is traversing the apparatus. This behavior seems to indicate that a photon must, in some sense, go through *both* slits and interfere with itself! Photons seem to be both wave-like *and* particle-like at the same time.

Photons are sometimes visualized as **wave packets.** The electromagnetic wave shown Figure 38.12 has a wavelength and a frequency, yet it is also discrete and fairly localized. But this cannot be exactly what a photon is because a wave packet would take a finite amount of time to be emitted or absorbed. This is contrary to much evidence that the entire photon is emitted or absorbed in a single instant; there is no point in time at which the photon is "half absorbed." The wave packet idea, although useful, is still too classical to represent a photon.

The bottom line is that there simply is no "true" mental representation of a photon. Analogies such as raindrops or wave packets can be useful, but none are perfectly accurate. We can detect photons, measure the properties of photons, and put photons to practical use, but the ultimate nature of the photon remains a mystery. To paraphrase Gertrude Stein, "A photon is a photon is a photon."

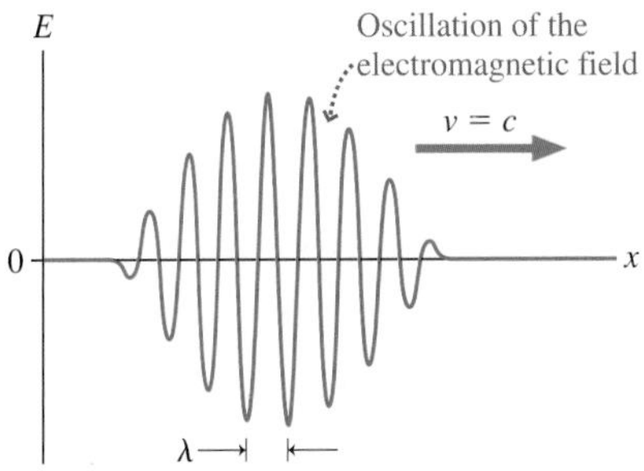

FIGURE 38.12 A wave packet has wave-like and particle-like properties.

The Photon Rate

Light, in the raindrop analogy, consists of a stream of photons. For monochromatic light of frequency f, N photons have a total energy $E_{\text{light}} = Nhf$. We are usually more interested in the *power* of the light, or the rate (in joules per second, or watts) at which the light energy is delivered. The power is

$$P = \frac{dE_{\text{light}}}{dt} = \frac{dN}{dt}hf = Rhf \qquad (38.10)$$

where $R = dN/dt$ is the *rate* at which photons arrive or, equivalently, the number of photons per second.

EXAMPLE 38.5 The photon rate in a laser beam

The 1.0 mW light beam of a helium-neon laser ($\lambda = 633$ nm) shines on a screen. How many photons strike the screen each second?

SOLVE The light-beam power, or energy delivered per second, is $P = 1.0$ mW $= 0.0010$ J/s. This is a realistic value. The frequency of the light is $f = c/\lambda = 4.74 \times 10^{14}$ Hz. The number of photons striking the screen per second, which is the *rate* of arrival of photons, is

$$R = \frac{P}{hf} = 3.2 \times 10^{15} \text{ photons per second}$$

ASSESS That is a lot of photons per second. No wonder that we are not aware of individual photons!

Photodetectors

Modern photodetectors are descendants of the photoelectric effect. These range from simple "electric eyes" to the detector array in a video camera. Most detectors use what is called a *photodiode* in which the photoelectrons are emitted internally in a semiconductor. Even so, they still have a threshold frequency, a stopping potential, and other attributes of the photoelectric effect.

Very low light levels can be detected photon by photon with a device called a *photomultiplier tube*, or PMT. Figure 38.13a on the next page shows that a PMT consists of a cathode, an anode, and a number of intermediate electrodes sealed inside an evacuated glass tube. The cathode is coated with a low-work-function material, allowing it to respond to most visible wavelengths of light. The cathode is at a fairly high negative voltage and the anode, at the other end, is at essentially zero volts. Steadily descending potentials are applied to the intermediate electrodes.

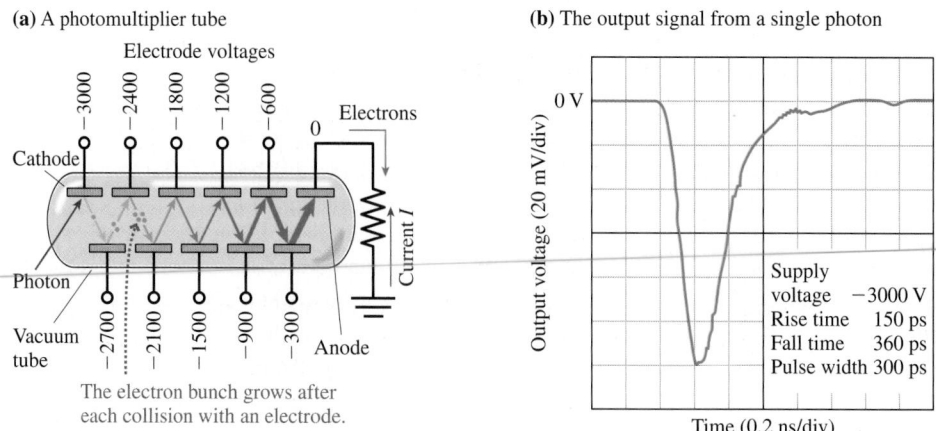

(a) A photomultiplier tube

Electrode voltages

−3000 −2400 −1800 −1200 −600

0   Electrons

Cathode

Photon

Vacuum tube

−2700 −2100 −1500 −900 −300   Anode

Current I

The electron bunch grows after each collision with an electrode.

(b) The output signal from a single photon

0 V

Output voltage (20 mV/div)

Supply voltage   −3000 V
Rise time   150 ps
Fall time   360 ps
Pulse width 300 ps

Time (0.2 ns/div)

FIGURE 38.13 A photomultiplier tube can detect individual photons.

A photon of light ejects a photoelectron from the cathode. The electric field between the cathode and the first intermediate electrode accelerates that electron through a potential difference of about 300 V, and it then strikes this electrode at high speed. When a fast electron collides with a metal surface, it can kick out two or three other electrons called *secondary electrons*. The secondary electrons of the first electrode are accelerated to the second electrode, where they kick out more electrons. These are accelerated to the third electrode, where they kick out yet more electrons, and so on. There is a chain-reaction *multiplication* of electrons— 1, 2, 4, 8, 16, . . . —as they move from the cathode toward the anode. For a typical PMT, a single photon at the cathode causes an electron bunch with 10^6 or 10^7 electrons to arrive at the anode.

The electrons are collected by the anode and flow through a resistor. Because these are negative charge carriers, we would say that a current pulse I travels upward through the resistor. This creates a *negative* voltage across the resistor, $\Delta V = IR$, for the length of time that the current lasts. Figure 38.12b, an actual measurement, shows a pulse generated by a single photon. The horizontal scale is 0.2 ns/division and the vertical scale is 20 millivolts (mV)/division. You can see that the width of the pulse is ≈0.3 ns and its height (measured downward from the baseline) is ≈120 mV = 0.12 V. This is not a large voltage, even after the multiplication, but it is a voltage easily detected with modern electronics.

NOTE ▶ The 0.3 ns pulse duration is *not* an indication of the duration of a photon. The photon absorption is instantaneous, but as the electron bunch grows in size, the electron-electron repulsion causes the bunch to spread out some. The observed pulse width is an artifact of the PMT, not a characteristic of the photon. ◀

STOP TO THINK 38.2   The intensity of a beam of light is increased but the light's frequency is unchanged. Which one (or perhaps more than one) of the following is true?

a. The photons travel faster.
b. Each photon has more energy.
c. The photons are larger.
d. There are more photons per second.

38.4 Matter Waves and Energy Quantization

Prince Louis-Victor de Broglie was a French graduate student in 1924. It had been 19 years since Einstein had shaken the world of physics by blurring the distinction between a particle and a wave. As de Broglie thought about these issues, it seemed that nature should have some kind of symmetry. If light waves could have a particle-like nature, why shouldn't material particles have some kind of wave-like nature? In other words, could **matter waves** exist?

With no experimental evidence to go on, de Broglie reasoned by analogy with Einstein's equation $E = hf$ for the photon and with some of the ideas of his theory of relativity. The details need not concern us, but they led de Broglie to postulate that *if* a material particle of momentum $p = mv$ has a wave-like nature, its wavelength must be given by

$$\lambda = \frac{h}{p} = \frac{h}{mv} \tag{38.11}$$

where h is Planck's constant. This wavelength is called the **de Broglie wavelength.**

EXAMPLE 38.6 The de Broglie wavelength of an electron

What is the de Broglie wavelength of a 1.0 eV electron?

SOLVE An electron with $1.0\,\text{eV} = 1.6 \times 10^{-19}\,\text{J}$ of kinetic energy has speed

$$v = \sqrt{\frac{2K}{m}} = 5.9 \times 10^5\,\text{m/s}$$

Although fast by macroscopic standards, this is a slow electron because it gains this speed by accelerating through a potential difference of a mere 1 V. Its de Broglie wavelength is

$$\lambda = \frac{h}{mv} = 1.2 \times 10^{-9}\,\text{m} = 1.2\,\text{nm}$$

ASSESS The electron's wavelength is small, but it is larger than the wavelengths of x rays and larger than the approximately 10^{-10} m spacing of atoms in a crystal.

What would it mean for matter—an electron or a proton or a baseball—to have a wavelength? Would it obey the principle of superposition? Would it exhibit interference and diffraction? These are questions we examined in Chapter 24, where we found that, indeed, matter *does* exhibit interference. For example, Figure 38.14 shows the intensity pattern recorded after 50 keV electrons passed through two slits separated by 1.0 μm. The pattern is clearly a double-slit interference pattern, and the spacing of the fringes is exactly as predicted for a wavelength given by de Broglie's formula. Because the electron beam was weak, with one electron at a time passing through the apparatus, it would appear that each electron somehow went through both slits, then recombined to interfere with itself!

Electrons are fundamental subatomic particles. Perhaps subatomic particles have wave-like aspects, but what about entire atoms, aggregates of many fundamental particles? Amazing as it seems, research during the 1980s demonstrated that whole atoms, and even molecules, can produce interference patterns.

Figure 38.15 on the next page shows an *atom interferometer.* You learned in Chapter 22 that an interferometer, such as the Michelson interferometer, works by dividing a wave front into two waves, sending the two waves along separate paths, then recombining them. For light waves, the wave division is accomplished by sending light through the *periodic* slits in a diffraction grating. In an atom interferometer, the atom's matter wave is divided by sending atoms through the *periodic* intensity of a standing light wave.

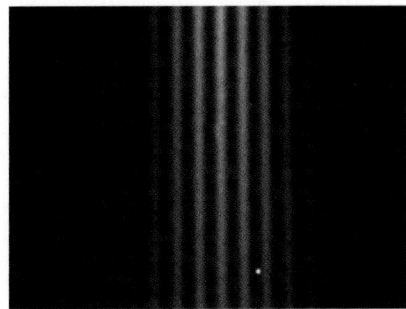

FIGURE 38.14 A double-slit interference pattern created with electrons.

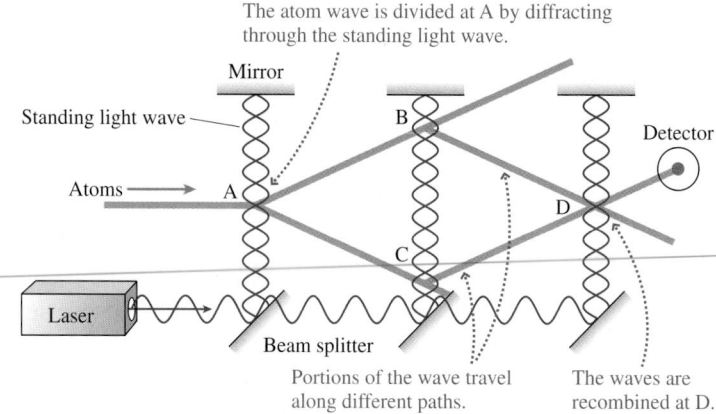

The atom wave is divided at A by diffracting through the standing light wave.

Mirror

Standing light wave

B

Detector

Atoms

A

D

C

Laser

Beam splitter

Portions of the wave travel along different paths.

The waves are recombined at D.

FIGURE 38.15 An atom interferometer.

You can see in the figure that a laser creates three parallel *standing waves* of light, each with nodes spaced a distance $\lambda/2$ apart. The wavelength is chosen so that the light waves exert small forces on an atom in the laser beam. Because the intensity along a standing wave alternates between maximum at the antinodes and zero intensity at the nodes, an atom crossing the laser beam experiences a *periodic* force field. A particle-like atom would be deflected by this periodic force, but a wave is *diffracted*. After being diffracted by the first standing wave at A, an atom is, in some sense, traveling toward both point B *and* point C.

The second standing wave diffracts the atom waves again at points B and C, directing them toward D where, with a third diffraction, they are recombined after having traveled along different paths. Depending on the phases of the waves as they recombine, the detector sometimes records atoms (constructive interference) but at other times does not (destructive interference). Altering one of the paths, such as by applying an electric field in the region around B but not around C, shifts the phases of the atom waves and causes the detector to record interference fringes.

The atom interferometer is fascinating because it completely inverts everything we previously learned about interference and diffraction. The scientists who studied the wave nature of light during the 19th century aimed light (a wave) at a diffraction grating (a periodic structure of matter) and found that it diffracted. Now we aim atoms (matter) at a standing wave (a periodic structure of light) and find that the atoms diffract. The roles of light and matter have been completely reversed!

Quantization of Energy

De Broglie considered a matter wave to be a traveling wave. But suppose that a "particle" of matter is *confined* to a small region of space and cannot travel? How do the wave-like properties manifest themselves?

This is the problem of "a particle in a box" that we looked at in Chapter 24. We will briefly summarize that discussion. Figure 38.16 shows a particle of mass m moving in one dimension as it bounces back and forth with speed v between the ends of a box of length L. A wave, if it reflects back and forth between two fixed points, sets up a standing wave. You learned in Chapter 21 that a standing wave of length L *must* have a wavelength given by

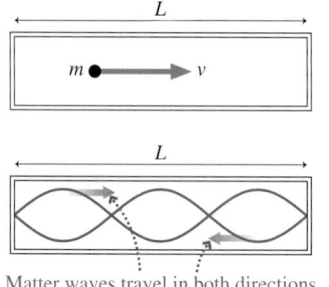

L

m v

L

Matter waves travel in both directions.

FIGURE 38.16 A particle in a box creates a standing de Broglie wave as it reflects back and forth.

$$\lambda_n = \frac{2L}{n} \qquad n = 1, 2, 3, 4, \ldots \qquad (38.12)$$

If the confined particle has wave-like properties, it should satisfy both Equation 38.12 *and* the de Broglie relationship $\lambda = h/mv$. That is, a particle in a box should obey the relationship

$$\lambda = \frac{h}{mv} = \frac{2L}{n}$$

This can be true only if the particle's speed is

$$v_n = n\left(\frac{h}{2Lm}\right) \qquad n = 1, 2, 3, \ldots \qquad (38.13)$$

In other words, the particle cannot bounce back and forth with just *any* speed. Rather, it can have *only* those specific speeds v_n, given by Equation 38.13, for which the de Broglie wavelength creates a standing wave in the box.

Thus the particle's energy, which is purely kinetic energy, is

$$E_n = \frac{1}{2}mv_n^2 = n^2\frac{h^2}{8mL^2} \qquad n = 1, 2, 3, \ldots \qquad (38.14)$$

De Broglie's hypothesis about the wave-like properties of matter leads us to the remarkable conclusion that **the energy of a confined particle is quantized.** The energy of the particle in the box can be $1(h^2/8mL^2)$, or $4(h^2/8mL^2)$, or $9(h^2/8mL^2)$, but it *cannot* have an energy between these values.

The possible values of the particle's energy are called **energy levels,** and the integer n that characterizes the energy levels is called the **quantum number.** The quantum number can be found by counting the antinodes, just as you learned to do for standing waves on a string. The standing wave shown in Figure 38.16 is $n = 3$, thus its energy is E_3.

We can rewrite Equation 38.14 in the useful form

$$E_n = n^2 E_1 \qquad (38.15)$$

where

$$E_1 = \frac{h^2}{8mL^2} \qquad (38.16)$$

is the **fundamental quantum of energy** for a particle in a box. It is analogous to the fundamental frequency f_1 of a standing wave on a string.

EXAMPLE 38.7 The energy levels of an oil droplet
What is the fundamental quantum of energy for one of Millikan's 1.0-μm-diameter oil droplets confined in a box of length 10 μm? The density of the oil is 900 kg/m^3.

SOLVE The mass of a droplet is $m = \rho V$, where the volume is $\frac{4}{3}\pi r^3$. A quick calculation shows that a 1.0-μm-diameter droplet has mass $m = 4.7 \times 10^{-16}$ kg. The confinement length is $L = 1.0 \times 10^{-5}$ m. From Equation 38.16, the fundamental quantum of energy is

$$E_1 = \frac{h^2}{8mL^2} = \frac{(6.63 \times 10^{-34}\,\text{J s})^2}{8(4.7 \times 10^{-16}\,\text{kg})(1.0 \times 10^{-5}\,\text{m})^2}$$

$$= 1.2 \times 10^{-42}\,\text{J} = 7.3 \times 10^{-24}\,\text{eV}$$

ASSESS This is such an incredibly small amount of energy that there is no hope of distinguishing between energies of E_1 or $4E_1$ or $9E_1$. For any macroscopic particle, even one this tiny, the allowed energies will *seem* to be perfectly continuous. We will not observe the quantization.

EXAMPLE 38.8 The energy levels of an electron
What are the first three allowed energies for an electron confined in a one-dimensional box of length 0.10 nm, about the size of an atom?

SOLVE We can use Equation 38.16, with $m_{elec} = 9.11 \times 10^{-31}$ kg and $L = 1.0 \times 10^{-10}$ m to find that the fundamental quantum of energy is $E_1 = 6.0 \times 10^{-18}$ J = 38 eV. Thus the first three allowed energies of an electron in a 0.10 nm box are

$$E_1 = 38 \text{ eV}$$
$$E_2 = 4E_1 = 152 \text{ eV}$$
$$E_3 = 9E_1 = 342 \text{ eV}$$

We see that confining a wave-like particle creates a standing de Broglie wave, and we know that a standing wave has only certain discrete wavelengths. Thus we find that a confined particle can have only certain discrete energies. In other words, **the confinement of a particle leads directly to the quantization of its energy.** The particle in a box, although not a realistic model of an atom, is a simple example to illustrate these ideas. An electron confined in a real atom will need to be a much more complex three-dimensional standing wave. But, just like the simple particle in a box, it will have quantized energies. Furthermore, we expect a typical energy difference between adjacent energy levels will be a few electron volts.

Now, this is an intriguing result. We found that visible and ultraviolet photons of light have energies of a few electron volts. We also know that atoms emit *discrete* wavelengths of visible and ultraviolet light, with photon energies of a few electron volts. Now we see that an electron confined into an atomic-size box has energy levels spaced a few electron volts apart. Might there be a connection between these phenomena? We will explore this topic in the next section.

STOP TO THINK 38.3 What is the quantum number of this particle confined in a box?

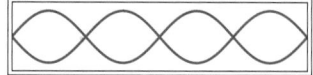

38.5 Bohr's Model of Atomic Quantization

18.1

Thomson's electron and Rutherford's nucleus made it clear that the atom has a *structure* of some sort. The challenge at the beginning of the 20th century was to deduce, from experimental evidence, the correct structure. The difficulty of this task cannot be exaggerated. The evidence about atoms, such as observations of atomic spectra, was very indirect, and experiments were carried out with only the simplest of measuring devices. Most observations were made by eye, and all calculations were carried out by hand. Using observations as a guide, physicists were attempting to construct a *model* of the atom that could successfully explain the various experiments.

Rutherford's nuclear model was the most successful of various proposals, but Rutherford's model failed to explain why atoms are stable or why their spectra are discrete. A missing piece of the puzzle, although not recognized as such for a few years, was Einstein's 1905 introduction of light quanta. If light comes in discrete packets of energy, which we now call photons, and if atoms emit and absorb light, what does that imply about the structure of the atoms?

This was the question posed by Niels Bohr. Bohr, shown as young man in Figure 38.17, was born, educated, and spent most of his life in Denmark. He later established an institute in Copenhagen that, for many decades, was the leading center for the development of quantum physics. Although few discoveries bear Bohr's name, he was the intellectual driving force behind the development of quantum mechanics and the mentor of many of the young physicists who reshaped physics in the 1920s and 1930s.

After receiving his doctoral degree in physics in 1911, Bohr went to England to work in Rutherford's laboratory. Rutherford had just, within the previous year, completed his development of the nuclear model of the atom. Rutherford's model certainly contained a kernel of truth, but Bohr wanted to understand how a solar-system-like atom could be stable and not radiate away all its energy. He soon recognized that Einstein's light quanta had profound implications about the structure of atoms. In 1913, Bohr proposed a radically new model of the atom in which he added quantization to Rutherford's nuclear atom.

The basic assumptions of the **Bohr model of the atom** are as follows:

1. An atom consists of negative electrons orbiting a very small positive nucleus, as in the Rutherford model.
2. Atoms can exist only in certain **stationary states.** Each stationary state corresponds to a particular set of electron orbits around the nucleus. These states are distinct and can be numbered $n = 1, 2, 3, 4, \ldots$ where n is the *quantum number.*
3. Each stationary state has a discrete, well-defined energy E_n. That is, atomic energies are *quantized.* The stationary states of an atom are numbered in order of increasing energy: $E_1 < E_2 < E_3 < E_4 < \cdots$
4. The lowest energy state of the atom, with energy E_1, is *stable* and can persist indefinitely. It is called the **ground state** of the atom. Other stationary states with energies $E_2, E_3, E_4, \ldots$ are called **excited states** of the atom.
5. An atom can "jump" from one stationary state to another by emitting or absorbing a photon of frequency

$$f_{photon} = \frac{\Delta E_{atom}}{h} \qquad (38.17)$$

where h is Planck's constant and $\Delta E_{atom} = |E_f - E_i|$. E_i and E_f are the energies of the initial and final states. Such a jump is called a **transition** or, sometimes, a **quantum jump.** Figure 38.18a is a schematic view of the emission and absorption of photons in an atom with stationary states.

6. An atom can move from a lower energy state to a higher energy state by absorbing energy $\Delta E_{atom} = E_f - E_i$ in an inelastic collision with an electron or another atom. This process, called **collisional excitation,** is shown in Figure 38.18b.
7. Atoms will seek the lowest energy state. An atom in an excited state, if left alone, will jump to lower and lower energy states until it reaches the ground state.

Bohr's model builds upon Rutherford's model, but it adds two new ideas that are derived from Einstein's ideas of quanta. The first, expressed in assumption 2, is that only certain electron orbits are "allowed" or can exist. The second, expressed in assumption 5, is that **the atom can jump from one state to another by emitting or absorbing a photon of just the right frequency to conserve energy.**

According to Einstein, a photon of frequency f has energy $E_{photon} = hf$. If an atom jumps from an initial state with energy E_i to a final state with *lower* energy E_f, energy will be conserved if the atom emits a photon with $E_{photon} = \Delta E_{atom}$. This photon must have exactly the frequency given by Equation 38.17 if it is to carry away exactly the right amount of energy. Similarly, an atom can jump to a higher energy state, for which additional energy is needed, by absorbing a photon

FIGURE 38.17 Niels Bohr.

(a) Emission and absorption of light

Excited-state electron   Allowed orbits

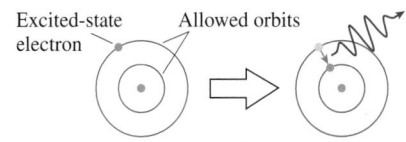

The electron jumps to a lower-energy stationary state and emits a photon.

Approaching photon

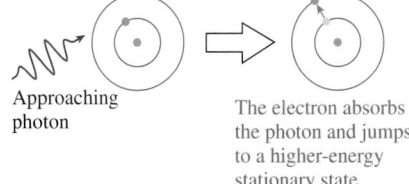

The electron absorbs the photon and jumps to a higher-energy stationary state.

(b) Collisional excitation

Particle loses energy

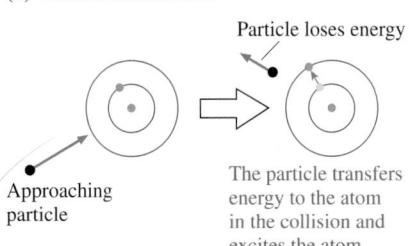

Approaching particle

The particle transfers energy to the atom in the collision and excites the atom.

FIGURE 38.18 An atom can change stationary states by emitting or absorbing a photon or by undergoing a collision.

of frequency $f_{photon} = \Delta E_{atom}/h$. The total energy of the atom-plus-light system is conserved.

NOTE ▶ When an atom is excited to a higher energy level by absorbing a photon, the photon vanishes. Thus energy conservation requires $E_{photon} = \Delta E_{atom}$. When an atom is excited to a higher energy level in a collision with a particle, such as an electron or another atom, the particle still exists after the collision and still has energy. Thus energy conservation requires the less stringent condition $E_{particle} \geq \Delta E_{atom}$. ◀

The implications of Bohr's model are profound. In particular:

1. **Matter is stable.** Once an atom is in its ground state, there are no states of any lower energy to which it can jump. It can remain in the ground state forever.

2. **Atoms emit and absorb a *discrete spectrum*.** Only those photons whose frequencies match the energy *intervals* between the stationary states can be emitted or absorbed. Photons of other frequencies cannot be emitted or absorbed without violating energy conservation.

3. **Emission spectra can be produced by collisions.** In a gas discharge tube, the current-carrying electrons moving through the tube occasionally collide with the atoms. A collision transfers energy to an atom and can kick it to an excited state. Once the atom is in an excited state, it can emit photons of light—a discrete emission spectrum—as it jumps back down to lower-energy states.

4. **Absorption wavelengths are a subset of the wavelengths in the emission spectrum.** Recall that all the lines seen in an absorption spectrum are also seen in emission, but many emission lines are *not* seen in absorption. According to Bohr's model, most atoms, most of the time, are in their lowest energy state, the $n = 1$ ground state. Thus the absorption spectrum consists of *only* those transitions such as $1 \rightarrow 2$, $1 \rightarrow 3$, . . . in which the atom jumps from $n = 1$ to a higher value of n by absorbing a photon. Transitions such as $2 \rightarrow 3$ are *not* observed because there are essentially no atoms in $n = 2$ at any instant of time. On the other hand, atoms that have been excited to the $n = 3$ state by collisions can emit photons corresponding to transitions $3 \rightarrow 1$ *and* $3 \rightarrow 2$. Thus the wavelength corresponding to $\Delta E_{atom} = E_3 - E_1$ is seen in both emission and absorption, but transitions with $\Delta E_{atom} = E_3 - E_2$ occur in emission only.

5. **Each element in the periodic table has a unique spectrum.** The energies of the stationary states are just the energies of the orbiting electrons. The atom has no other form of energy. Different elements, with different numbers of electrons, will have different stable orbits and thus different stationary states. States with different energies will emit and absorb photons of different wavelengths.

EXAMPLE 38.9 The wavelength of an emitted photon

An atom has stationary states with energies $E_j = 4.0\text{ eV}$ and $E_k = 6.0\text{ eV}$. What is the wavelength of a photon emitted in a quantum jump from state k to state j?

MODEL To conserve energy, the emitted photon must have exactly the energy lost by the atom in the quantum jump.

SOLVE The atom can jump from the higher energy state k to the lower energy state j by emitting a photon. The atom's change in energy is $\Delta E_{atom} = -2.0\text{ eV}$, so the photon energy must be $E_{photon} = 2.0\text{ eV}$.

The photon frequency is

$$f = \frac{E_{photon}}{h} = \frac{2.0\text{ eV}}{4.14 \times 10^{-15}\text{ eV s}} = 4.83 \times 10^{14}\text{ Hz}$$

The wavelength of this photon is

$$\lambda = \frac{c}{f} = 621\text{ nm}$$

ASSESS 621 nm is a visible-light wavelength.

Energy-Level Diagrams

An **energy-level diagram,** such as the one shown in Figure 38.19, is a useful pictorial representation of the stationary-state energies. An energy-level diagram is less a graph than it is a picture. The vertical axis represents energy, but the horizontal axis is not a scale. Think of this as a picture of a ladder in which the energies are the rungs of the ladder. The lowest rung, with energy E_1 is the ground state. Higher rungs are labeled by their quantum numbers, $n = 2, 3, 4, \ldots$

Energy-level diagrams are especially useful for showing transitions, or quantum jumps, in which a photon of light is emitted or absorbed. As examples, Figure 38.19 shows upward transitions in which a photon is absorbed by a ground-state atom ($n = 1$) and downward transitions in which a photon is emitted from an $n = 4$ excited state.

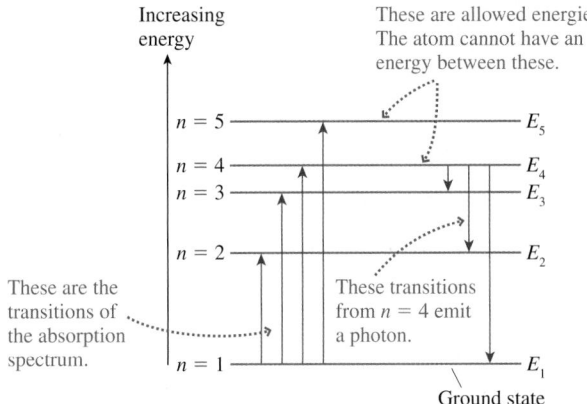

FIGURE 38.19 An energy-level diagram.

EXAMPLE 38.10 **Emission and absorption**

An atom has stationary states $E_1 = 0.0$ eV, $E_2 = 3.0$ eV, and $E_3 = 5.0$ eV. What wavelengths are observed in the absorption spectrum and in the emission spectrum of this atom?

MODEL Photons are emitted when an atom undergoes a quantum jump from a higher energy level to a lower energy level. Photons are absorbed in a quantum jump from a lower energy level to a higher energy level. But most of the atoms are in the $n = 1$ ground state, so the only quantum jumps seen in the absorption spectrum start from the $n = 1$ state.

VISUALIZE Figure 38.20 shows an energy-level diagram for the atom.

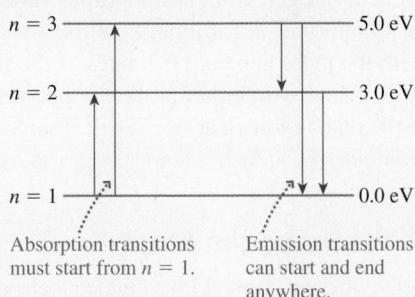

FIGURE 38.20 The atom's energy-level diagram.

SOLVE This atom will absorb photons on the $1 \rightarrow 2$ and $1 \rightarrow 3$ transitions, with $\Delta E_{1\rightarrow 2} = 3.0$ eV and $\Delta E_{1\rightarrow 3} = 5.0$ eV. From $f = \Delta E_{atom}/h$ and $\lambda = c/f$, we find that the wavelengths in the absorption spectrum are

$$1 \rightarrow 2 \quad f = 3.0 \text{ eV}/h = 7.25 \times 10^{14} \text{ Hz}$$
$$\lambda = 414 \text{ nm (blue)}$$
$$1 \rightarrow 3 \quad f = 5.0 \text{ eV}/h = 1.21 \times 10^{15} \text{ Hz}$$
$$\lambda = 248 \text{ nm (ultraviolet)}$$

The emission spectrum will also have the 414 nm and 248 nm wavelengths due to the $2 \rightarrow 1$ and $3 \rightarrow 1$ quantum jumps from excited states 2 and 3 to the ground state. In addition, the emission spectrum will contain the $3 \rightarrow 2$ quantum jump with $\Delta E_{3\rightarrow 2} = -2.0$ eV that is *not* seen in absorption because there are too few atoms in the $n = 2$ state to absorb. We found in Example 38.9 that a 2.0 eV transition corresponds to a wavelength of 621 nm. Thus the emission wavelengths are

$$2 \rightarrow 1 \quad \lambda = 414 \text{ nm (blue)}$$
$$3 \rightarrow 1 \quad \lambda = 248 \text{ nm (ultraviolet)}$$
$$3 \rightarrow 2 \quad \lambda = 621 \text{ nm (orange)}$$

STOP TO THINK 38.4   A photon with a wavelength of 414 nm has energy $E_{\text{photon}} = 3.0$ eV. Do you expect to see a spectral line with $\lambda = 414$ nm in the emission spectrum of the atom represented by this energy-level diagram? If so, what transition or transitions will emit it? Do you expect to see a spectral line with $\lambda = 414$ nm in the absorption spectrum? If so, what transition or transitions will absorb it?

$n = 3$	6.0 eV
$n = 2$	5.0 eV
$n = 1$	2.0 eV
$n = 0$	0.0 eV

38.6 The Bohr Hydrogen Atom

Bohr's hypothesis was a bold new idea, yet there was still one enormous stumbling block: What *are* the stationary states of an atom? Everything in Bohr's model hinges on the existence of these stationary states, of there being only certain electron orbits that are allowed. But nothing in classical physics provides any basis for such orbits. And Bohr's model describes only the *consequences* of having stationary states, not how to find them. If such states really exist, we will have to go beyond classical physics to find them.

To address this problem, Bohr did an explicit analysis of the hydrogen atom. The hydrogen atom, with only a single electron, was known to be the simplest atom. Furthermore, as we discussed in Chapters 24 and 37, Balmer had discovered a fairly simple formula that characterized the wavelengths in the hydrogen emission spectrum. Anyone with a successful model of an atom was going to have to *derive* Balmer's formula for the hydrogen atom.

Bohr's paper followed a rather circuitous line of reasoning. That is not surprising, because he had little to go on at the time. But our goal is a clear explanation of the ideas, not a historical study of Bohr's methods, so we are going to follow a different analysis using de Broglie's matter waves. De Broglie did not propose matter waves until 1924, 11 years after Bohr's paper, but with the clarity of hindsight we can see that treating the electron as a wave provides a more straightforward analysis of the hydrogen atom. Although our route will be different from Bohr's, we will arrive at the same point, and, in addition, we will be in a much better position to understand the work that came after Bohr.

> NOTE ▶ Bohr's analysis of the hydrogen atom is sometimes called the *Bohr atom* or the *Bohr model*. It's important not to confuse this analysis, which applies only to hydrogen, with the more general postulates of the *Bohr model of the atom*. Those postulates, which we looked at in the previous section, apply to any atom. To make the distinction clear, we'll call Bohr's analysis of hydrogen the *Bohr hydrogen atom*. ◀

The Stationary States of the Hydrogen Atom

Figure 38.21 shows a Rutherford hydrogen atom, with a single electron orbiting a nucleus that consists of a single proton. We will assume a circular orbit of radius r and speed v. We will also assume, to keep the analysis manageable, that the pro-

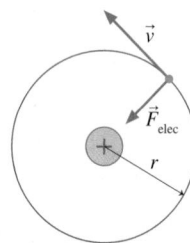

FIGURE 38.21 A Rutherford hydrogen atom. The size of the nucleus is greatly exaggerated.

ton remains stationary while the electron revolves around it. This is a reasonable assumption because the proton is roughly 1800 times as massive as the electron. With these assumptions, the atom's energy is the kinetic energy of the electron plus the potential energy of the electron-proton interaction. This is

$$E = K + U = \frac{1}{2}mv^2 + \frac{1}{4\pi\epsilon_0}\frac{q_{elec}q_{proton}}{r} = \frac{1}{2}mv^2 - \frac{e^2}{4\pi\epsilon_0 r} \quad (38.18)$$

where we used $q_{elec} = -e$ and $q_{proton} = +e$.

NOTE ▶ m is the mass of the electron, *not* the mass of the entire atom. ◀

Now, the electron, as we are coming to understand it, has both particle-like and wave-like properties. First, let us treat the electron as a charged particle. The proton exerts a Coulomb electric force on the electron,

$$\vec{F}_{elec} = \left(\frac{1}{4\pi\epsilon_0}\frac{e^2}{r^2}, \text{toward center}\right) \quad (38.19)$$

This force gives the electron an acceleration $\vec{a}_{elec} = \vec{F}_{elec}/m$ that also points to the center. This is a centripetal acceleration, causing the particle to move in its circular orbit. The centripetal acceleration of a particle moving in a circle of radius r at speed v *must* be v^2/r, thus

$$a_{elec} = \frac{F_{elec}}{m} = \frac{e^2}{4\pi\epsilon_0 mr^2} = \frac{v^2}{r} \quad (38.20)$$

Rearranging, we find

$$v^2 = \frac{e^2}{4\pi\epsilon_0 mr} \quad (38.21)$$

Equation 38.21 is a *constraint* on the motion. The speed v and radius r must obey Equation 38.21 if the electron is to move in a circular orbit. This constraint is not unique to atoms. We earlier found a similar relationship between v and r for orbiting satellites.

Now let's treat the electron as a de Broglie wave. In Section 38.4 we found that a particle confined to a one-dimensional box sets up a standing wave as it reflects back and forth. A standing wave, you will recall, consists of two traveling waves moving in opposite directions. When the round-trip distance in the box is equal to an integer number of wavelengths ($2L = n\lambda$), the two oppositely traveling waves interfere constructively to set up the standing wave.

Suppose that, instead of traveling back and forth along a line, our wave-like particle travels around the circumference of a circle. The particle will set up a standing wave, just like the particle in the box, if there are waves traveling in both directions and if the round-trip distance is an integer number of wavelengths. This is the idea we want to carry over from the particle in a box. As an example, Figure 38.22 shows a standing wave around a circle with $n = 10$ wavelengths.

The mathematical condition for a circular standing wave is found by replacing the round-trip distance $2L$ in a box with the round-trip distance $2\pi r$ on a circle. Thus a circular standing wave will occur when

$$2\pi r = n\lambda \qquad n = 1, 2, 3, \ldots \quad (38.22)$$

But the de Broglie wavelength for a particle *has* to be $\lambda = h/p = h/mv$. Thus the standing-wave condition for a de Broglie wave is

$$2\pi r = n\frac{h}{mv}$$

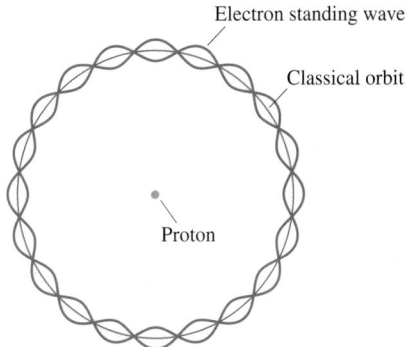

FIGURE 38.22 An $n = 10$ electron standing wave around the orbit's circumference.

This condition is true only if the electron's speed is

$$v_n = \frac{nh}{2\pi mr} \qquad n = 1, 2, 3, \ldots \qquad (38.23)$$

The quantity $h/2\pi$ occurs so often in quantum physics that it is customary to give it a special name. We define the quantity $\hbar$, pronounced "h bar," as

$$\hbar \equiv \frac{h}{2\pi} = 1.055 \times 10^{-34} \text{ J s} = 6.58 \times 10^{-16} \text{ eV s}$$

With this definition, we can write Equation 38.23 as

$$v_n = \frac{n\hbar}{mr} \qquad n = 1, 2, 3, \ldots \qquad (38.24)$$

This, like Equation 38.21, is another relationship between v and r. This is the constraint that arises from treating the electron as a wave.

Now if the electron can act as both a particle *and* a wave, then both the Equation 38.21 *and* Equation 38.24 constraints have to be obeyed. That is, v^2 as given by the Equation 38.21 particle constraint has to equal v^2 of the Equation 38.24 wave constraint. Equating these gives

$$v^2 = \frac{e^2}{4\pi\epsilon_0 mr} = \frac{n^2\hbar^2}{m^2 r^2}$$

We can solve this equation to find that the radius r is

$$r_n = n^2 \frac{4\pi\epsilon_0 \hbar^2}{me^2} \qquad n = 1, 2, 3, \ldots \qquad (38.25)$$

where we have added a subscript n to the radius r to indicate that it depends on the integer n.

The right-hand side of Equation 38.25, except for the n^2, is just a collection of constants. Let's group them all together and define the **Bohr radius** a_B as

$$a_B = \text{Bohr radius} \equiv \frac{4\pi\epsilon_0 \hbar^2}{me^2} = 5.29 \times 10^{-11} \text{ m} = 0.0529 \text{ nm}$$

With this definition, Equation 38.25 for the radius of the electron's orbit becomes

$$r_n = n^2 a_B \qquad n = 1, 2, 3, \ldots \qquad (38.26)$$

The first few allowed values of r_n are

$$r_n = \begin{cases} 0.053 \text{ nm} & n = 1 \\ 0.212 \text{ nm} & n = 2 \\ 0.476 \text{ nm} & n = 3 \\ \vdots & \vdots \end{cases}$$

We have discovered stationary states! That is, **a hydrogen atom can exist *only* if the radius of the electron's orbit is one of the values given by Equation 38.26.** Intermediate values of the radius, such as $r = 0.100$ nm, cannot exist because the electron cannot set up a standing wave around the circumference. The possible orbits are *quantized,* with only certain orbits allowed.

The key step leading to Equation 38.26 was the requirement that the electron have wave-like properties in addition to particle-like properties. This requirement leads to quantized orbits, or what Bohr called stationary states. The integer n is thus the *quantum number* that numbers the various stationary states.

Hydrogen Atom Energy Levels

Now we can make progress quickly. Knowing the possible radii, we can return to Equation 38.23 and find the possible electron speeds to be

$$v_n = \frac{n\hbar}{mr_n} = \frac{1}{n}\frac{\hbar}{ma_B} = \frac{v_1}{n} \qquad n = 1, 2, 3, \ldots \qquad (38.27)$$

where $v_1 = \hbar/ma_B = 2.19 \times 10^6$ m/s is the electron's speed in the $n = 1$ orbit. The speed decreases as n increases.

Finally, we can determine the energies of the stationary states. From Equation 38.18 for the energy, with Equations 38.26 and 38.27 for r and v, we have

$$E_n = \frac{1}{2}mv_n^2 - \frac{e^2}{4\pi\epsilon_0 r_n} = \frac{1}{2}m\left(\frac{\hbar^2}{m^2a_B^2n^2}\right) - \frac{e^2}{4\pi\epsilon_0 n^2 a_B} \qquad (38.28)$$

As a homework problem, you can show that this rather messy expression simplifies to

$$E_n = -\frac{1}{n^2}\left(\frac{1}{4\pi\epsilon_0}\frac{e^2}{2a_B}\right) \qquad (38.29)$$

Let's define

$$E_1 \equiv \frac{1}{4\pi\epsilon_0}\frac{e^2}{2a_B} = 13.60 \text{ eV}$$

We can then write the energy levels of the stationary states of the hydrogen atom as

$$E_n = -\frac{E_1}{n^2} = -\frac{13.60 \text{ eV}}{n^2} \qquad n = 1, 2, 3, \ldots \qquad (38.30)$$

This has been a lot of math, so we need to see where we are and what we have learned. Table 38.2 shows values of r_n, v_n, and E_n evaluated for quantum number values $n = 1$ to 5. We do indeed seem to have discovered stationary states of the hydrogen atom. Each state, characterized by its quantum number n, has a unique radius, speed, and energy. These are displayed graphically in Figure 38.23, in which the orbits are drawn to scale. Notice how the atom's diameter increases very rapidly as n increases. At the same time, the electron's speed decreases.

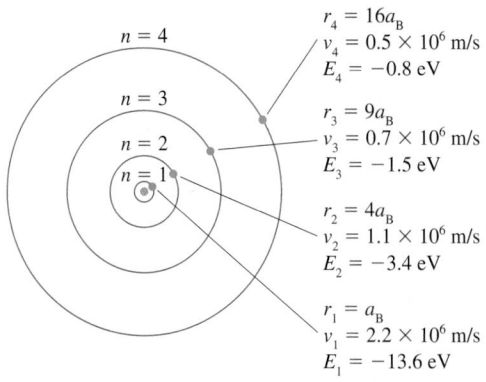

FIGURE 38.23 The first four stationary states, or allowed orbits, of the Bohr hydrogen atom drawn to scale.

TABLE 38.2 Radii, speeds, and energies for the first five states of the Bohr hydrogen atom

n	r_n (nm)	v_n (m/s)	E_n (eV)
1	0.053	2.19×10^6	-13.60
2	0.212	1.09×10^6	-3.40
3	0.476	0.73×10^6	-1.51
4	0.846	0.55×10^6	-0.85
5	1.322	0.44×10^6	-0.54

EXAMPLE 38.11 **Stationary states of the hydrogen atom**
Can an electron in a hydrogen atom have a speed of 3.60×10^5 m/s? If so, what are its energy and the radius of its orbit? What about a speed of 3.65×10^5 m/s?

SOLVE To be in a stationary state, the electron must have speed

$$v_n = \frac{v_1}{n} = \frac{2.19 \times 10^6 \text{ m/s}}{n}$$

where n is an integer. A speed of 3.60×10^5 m/s would require quantum number

$$n = \frac{2.19 \times 10^6 \text{ m/s}}{3.60 \times 10^5 \text{ m/s}} = 6.08$$

This is not an integer, so the electron can *not* have this speed. But if $v = 3.65 \times 10^5$ m/s, then

$$n = \frac{2.19 \times 10^6 \text{ m/s}}{3.65 \times 10^5 \text{ m/s}} = 6$$

This is the speed of an electron in the $n = 6$ excited state. An electron in this state has energy

$$E_6 = -\frac{13.60 \text{ eV}}{6^2} = -0.38 \text{ eV}$$

and the radius of its orbit is

$$r_6 = 6^2(5.29 \times 10^{-11} \text{ nm}) = 1.90 \times 10^{-9} \text{ m} = 1.90 \text{ nm}$$

Binding Energy and Ionization Energy

It is important to understand why the energies of the stationary states are negative. Because the potential energy of two charged particles is $U = q_1q_2/4\pi\epsilon_0 r$, the zero of potential energy occurs at $r = \infty$ where the particles are infinitely far apart. The state of zero total energy corresponds to having the electron at rest ($K = 0$) and infinitely far from the proton ($U = 0$). This situation, which is the case of two "free particles," occurs in the limit $n \to \infty$, for which $r_n \to \infty$ and $v_n \to 0$.

An electron and a proton bound into an atom have *less* energy than two free particles. We know this because we would have to do work (i.e., add energy) to pull the electron and proton apart. If the bound atom has less energy than two free particles, and if the total energy of two free particles is zero, then it must be the case that the atom has a *negative* amount of energy.

Thus $|E_n|$ is the **binding energy** of the electron in stationary state n. In the ground state, where $E_1 = -13.60$ eV, we would have to add 13.60 eV to the electron to free it from the proton and reach the zero energy state of two free particles. We can say that the electron in the ground state is "bound by 13.60 eV." An electron in an $n = 3$ orbit, where it is farther from the proton and moving more slowly, is bound by only 1.51 eV. That is the amount of energy you would have to supply to remove the electron from an $n = 3$ orbit.

Removing the electron entirely leaves behind a positive ion, H^+ in the case of a hydrogen atom. (The fact that H^+ happens to be a proton does not alter the fact that it is also an atomic ion.) Because nearly all atoms are in their ground state, the binding energy $|E_1|$ of the ground state is called the **ionization energy** of an atom. Bohr's analysis predicts that the ionization energy of hydrogen is 13.60 eV. Figure 38.24 illustrates the ideas of binding energy and ionization energy.

We can test this prediction by shooting a beam of electrons at hydrogen atoms. A projectile electron can knock out an atomic electron if its kinetic energy K is greater than the atom's ionization energy, leaving an ion behind. But a projectile electron will be unable to cause ionization if its kinetic energy is less than the atom's ionization energy. This is a fairly straightforward experiment to carry out, and the evidence shows that the ionization energy of hydrogen is, indeed, 13.60 eV.

Quantization of Angular Momentum

The angular momentum of a particle in circular motion, whether it is a planet or an electron, is

$$L = mvr$$

You will recall that angular momentum is conserved in orbital motion because a force directed toward a central point exerts no torque on the particle. Bohr used

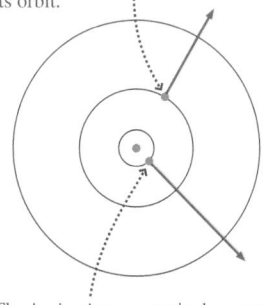

The *binding energy* is the energy needed to remove an electron from its orbit.

The *ionization energy* is the energy needed to create an ion by removing a ground-state electron.

FIGURE 38.24 Binding energy and ionization energy.

conservation of energy explicitly in his analysis of the hydrogen atom, but what role does conservation of angular momentum play?

The condition that a de Broglie wave for the electron set up a standing wave around the circumference was given, in Equation 38.22, as

$$2\pi r = n\lambda = n\frac{h}{mv}$$

We can rewrite this equation as

$$mvr = n\frac{h}{2\pi} = n\hbar \qquad (38.31)$$

But mvr is the angular momentum L for a particle in a circular orbit. It appears that the angular momentum of an orbiting electron cannot have just any value. Instead, it must satisfy

$$L = n\hbar \qquad n = 1, 2, 3, \ldots \qquad (38.32)$$

Thus angular momentum is quantized! The atom's angular momentum must be an integer multiple of Planck's constant $\hbar$.

The quantization of angular momentum is a direct consequence of this wave-like nature of the electron. We will find that the quantization of angular momentum plays a major role in the behavior of more complex atoms, leading to the idea of electron shells that you likely have studied in chemistry.

STOP TO THINK 38.5   What is the quantum number of this hydrogen atom?

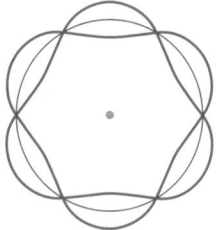

38.7 The Hydrogen Spectrum

Our analysis of the hydrogen atom has revealed stationary states, but how do we know whether the results make any sense? The most important experimental evidence that we have about the hydrogen atom is its spectrum, so the primary test of the Bohr hydrogen atom is whether it correctly predicts the spectrum.

The Hydrogen Energy-Level Diagram

Figure 38.25 is an energy-level diagram for the hydrogen atom. As we noted earlier, the energies are like the rungs of a ladder. The lowest rung is the ground state, with $E_1 = -13.6$ eV. The top rung, with $E = 0$ eV, corresponds to a hydrogen ion in the limit $n \to \infty$. This top rung is called the **ionization limit.** In principle there are an infinite number of rungs, but only the lowest few are shown. The higher values of n are all crowded together just below the ionization limit at $n = \infty$.

The figure shows a $1 \to 4$ transition in which a photon is absorbed and a $4 \to 2$ transition in which a photon is emitted. For two quantum states m and n, where $n > m$ and E_n is the higher-energy state, an atom can *emit* a photon in an $n \to m$ transition or *absorb* a photon in an $m \to n$ transition.

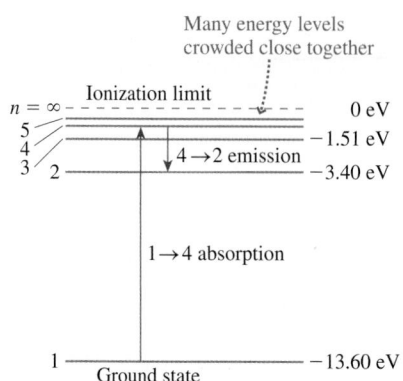

FIGURE 38.25 The energy-level diagram of the hydrogen atom.

The Emission Spectrum

According to the fifth assumption of Bohr's model of atomic quantization, the frequency of the photon emitted in an $n \rightarrow m$ transition is

$$f = \frac{\Delta E_{\text{atom}}}{h} = \frac{E_n - E_m}{h} \tag{38.33}$$

We can use Equation 38.29 for the energies E_n and E_m, to predict that the emitted photon has frequency

$$
\begin{aligned}
f &= \frac{1}{h}\left\{ \left[-\frac{1}{n^2}\left(\frac{1}{4\pi\epsilon_0}\frac{e^2}{2a_B} \right) \right] - \left[-\frac{1}{m^2}\left(\frac{1}{4\pi\epsilon_0}\frac{e^2}{2a_B} \right) \right] \right\} \\
&= \frac{1}{4\pi\epsilon_0}\frac{e^2}{2ha_B}\left(\frac{1}{m^2} - \frac{1}{n^2} \right)
\end{aligned}
\tag{38.34}
$$

The frequency is a positive number because $m < n$ and thus $1/m^2 > 1/n^2$.

We are more interested in wavelength than frequency, because wavelengths are the quantity measured by experiment. The wavelength of the photon emitted in an $n \rightarrow m$ quantum jump is

$$\lambda_{n \rightarrow m} = \frac{c}{f} = \frac{8\pi\epsilon_0 hca_B/e^2}{\left(\dfrac{1}{m^2} - \dfrac{1}{n^2} \right)} \tag{38.35}$$

This looks rather gruesome, but notice that the numerator is simply a collection of various constants. The value of the numerator, which we can call λ_0, is

$$\lambda_0 = \frac{8\pi\epsilon_0 hca_B}{e^2} = 9.112 \times 10^{-8} \text{ m} = 91.12 \text{ nm}$$

With this definition, our prediction for the wavelengths in the hydrogen emission spectrum is

$$\lambda_{n \rightarrow m} = \frac{\lambda_0}{\left(\dfrac{1}{m^2} - \dfrac{1}{n^2} \right)} \qquad m = 1, 2, 3, \ldots \quad n = m+1, m+2, \ldots \tag{38.36}$$

This should look familiar. It is the Balmer formula from Chapter 37! However, there is one *slight* difference: Bohr's analysis of the hydrogen atom has predicted $\lambda_0 = 91.12$ nm whereas Balmer found, from experiment, that $\lambda_0 = 91.18$ nm. Could Bohr have come this close but then fail to predict the Balmer formula correctly?

The problem, it turns out, is in our assumption that the proton remains at rest while the electron orbits it. In fact, *both* particles rotate about their common center of mass, rather like a dumbbell with a big end and a small end. The center of mass is very close to the proton, which is far more massive than the electron, but the proton is not entirely motionless. The good news is that a more advanced analysis can account for the proton's motion. It changes the energies of the stationary states ever so slightly—about 1 part in 2000—but that is precisely what is needed to give a revised value:

$$\lambda_0 = 91.18 \text{ nm when corrected for the nuclear motion}$$

It works! Unlike all previous atomic models, **the Bohr hydrogen atom correctly predicts the discrete spectrum of the hydrogen atom.** Figure 38.26 shows the *Balmer series* and the *Lyman series* transitions on an energy-level

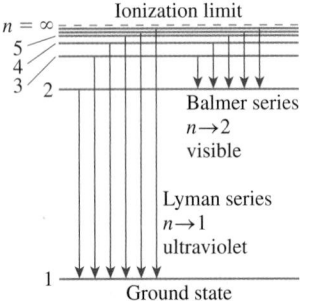

FIGURE 38.26 Transitions producing the Lyman series and the Balmer series of lines in the hydrogen spectrum.

diagram. Only the Balmer series, consisting of transitions ending on the $m = 2$ state, gives visible wavelengths, and this is the series that Balmer initially analyzed. The Lyman series, ending on the $m = 1$ ground state, is in the ultraviolet region of the spectrum and was not measured until later. These series, as well as others in the infrared, are observed in a discharge tube where collisions with electrons excite the atoms upward from the ground state to state n. They then decay downward by emitting photons. Only the Lyman series is observed in the absorption spectrum because, as noted previously, essentially all the atoms in a quiescent gas are in the ground state.

EXAMPLE 38.12 Hydrogen absorption

Whenever astronomers look at distant galaxies, they find that the light has been strongly absorbed at the wavelength of the $1 \rightarrow 2$ transition in the Lyman series of hydrogen. This absorption tells us that interstellar space is filled with vast clouds of hydrogen left over from the Big Bang. What is the wavelength of the $1 \rightarrow 2$ absorption in hydrogen?

SOLVE Equation 38.36 predicts the *absorption* spectrum of hydrogen if we let $m = 1$. The absorption seen by astronomers is from the ground state of hydrogen ($m = 1$) to its first excited state ($n = 2$). The wavelength is

$$\lambda_{1 \rightarrow 2} = \frac{91.18 \text{ nm}}{\left(\dfrac{1}{1^2} - \dfrac{1}{2^2} \right)} = 121.6 \text{ nm}$$

This wavelength is far into the ultraviolet. Ground-based astronomy cannot observe this region of the spectrum because the wavelengths are strongly absorbed by the atmosphere, but with space-based telescopes, first widely used in the 1970s, astronomers see 121.6 nm absorption in nearly every direction they look.

Hydrogen-Like Ions

An ion with a *single* electron orbiting Z protons in the nucleus is called a **hydrogen-like ion.** Z is the atomic number and describes the number of protons in the nucleus. He^+, with one electron circling a $Z = 2$ nucleus, and Li^{++}, with one electron and a $Z = 3$ nucleus, are hydrogen-like ions. So is U^{+91}, with one lonely electron orbiting a $Z = 92$ uranium nucleus.

Any hydrogen-like ion is simply a variation on the Bohr hydrogen atom. The only difference between a hydrogen-like ion and neutral hydrogen is that the potential energy $-e^2/4\pi\epsilon_0 r$ becomes, instead, $-Ze^2/4\pi\epsilon_0 r$. Hydrogen itself is the $Z = 1$ case. If we repeat the analysis of the previous sections with this one change, we find:

$$r_n = \frac{n^2 a_B}{Z}$$

$$v_n = Z\frac{v_1}{n}$$

$$E_n = -\frac{13.60 Z^2 \text{ eV}}{n^2} \quad (38.37)$$

$$\lambda_0 = \frac{91.18 \text{ nm}}{Z^2}$$

As the nuclear charge increases, the electron moves in to a smaller-diameter, higher-speed orbit. Its ionization energy $|E_1|$ increases significantly, and its spectrum shifts to shorter wavelengths. Table 38.3 compares the ground-state atomic diameter $2r_1$, the ionization energy $|E_1|$, and the first wavelength $3 \rightarrow 2$ in the Balmer series for hydrogen and the first two hydrogen-like ions.

TABLE 38.3 Comparison of hydrogen-like ions with $Z = 1, 2$, and 3

Ion	Diameter $2r_1$	Ionization energy $|E_1|$	Wavelength of $3 \rightarrow 2$
H ($Z = 1$)	0.106 nm	13.6 eV	656 nm
He$^+$ ($Z = 2$)	0.053 nm	54.4 eV	164 nm
Li^{++} ($Z = 3$)	0.035 nm	125.1 eV	73 nm

Success and Failure

Bohr's analysis of the hydrogen atom seemed to be a resounding success. By introducing stationary states, together with Einstein's ideas about light quanta, Bohr was able to provide the first solid understanding of discrete spectra and, in particular, to predict the Balmer formula for the wavelengths in the hydrogen spectrum. And the Bohr hydrogen atom, unlike Rutherford's model, was stable. There was clearly some validity to the idea of stationary states.

But Bohr was completely unsuccessful at explaining the spectra of any other atom. His method did not work even for helium, the second element in the periodic table with a mere two electrons. Something inherent in Bohr's assumptions seemed to work correctly for a single electron but not in situations with two or more electrons.

It is important to make a distinction between the Bohr model of atomic quantization, described in Section 38.5, and the Bohr hydrogen atom. The Bohr model assumes that stationary states exist, but it does not say how to find them. We found the stationary states of a hydrogen atom by requiring that an integer number of de Broglie waves fit around the circumference of the orbit, setting up standing waves. The difficulty with more complex atoms is not the Bohr model but the method of finding the stationary states. Bohr's model of the atomic quantization remains valid, and we will continue to use it, but the procedure of fitting standing waves to a circle is just too simple to find the stationary states of complex atoms. We need to find a better procedure.

Einstein, de Broglie, and Bohr carried physics into uncharted waters. Their successes made it clear that the microscopic realm of light and atoms is governed by quantization, discreteness, and a blurring of the distinction between particles and waves. Although Bohr was clearly on the right track, his inability to extend the Bohr hydrogen atom to more complex atoms made it equally clear that the complete and correct theory remained to be discovered. Bohr's theory was what we now call "semiclassical," a hybrid of classical Newtonian mechanics with the new ideas of quanta. Still missing was a complete theory of motion and dynamics in a quantized universe—a *quantum* mechanics.

SUMMARY

The goal of Chapter 38 has been to understand the quantization of energy for light and matter.

GENERAL PRINCIPLES

Light has particle-like properties

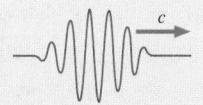

- The energy of a light wave comes in discrete packets called light quanta or photons.
- For light of frequency f, the energy of each photon is $E = hf$, where h is **Planck's constant.**
- For a light wave that delivers power P, photons arrive at rate R such that $P = Rhf$.
- Photons are "particle-like" but are not classical particles.

Matter has wave-like properties

- The **de Broglie wavelength** of a "particle" of mass m is $\lambda = h/mv$.
- The wave-like nature of matter is seen in the interference patterns of electrons, neutrons, and entire atoms.
- When a particle is confined, it sets up a de Broglie standing wave. The fact that standing waves have only certain allowed wavelengths leads to the conclusion that a confined particle has only certain allowed energies. That is, energy is quantized.

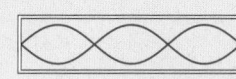

IMPORTANT CONCEPTS

Einstein's Model of Light

- Light consists of quanta of energy $E = hf$.
- Quanta are emitted and absorbed on an all-or-nothing basis.
- When a light quantum is absorbed, it delivers all its energy to *one* electron.

Bohr's Model of the Atom

- An atom can exist in only certain stationary states. The allowed energies are quantized. State n has energy E_n.
- An atom can jump from one stationary state to another by emitting or absorbing a photon with $E_{photon} = hf = \Delta E_{atom}$.
- Atoms can be excited in inelastic collisions.
- Atoms seek the $n = 1$ **ground state.** Most atoms, most of the time, are in the ground state.

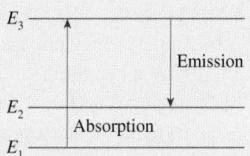

APPLICATIONS

Photoelectric effect

Light can eject electrons from a metal only if $f \geq f_0 = E_0/h$, where E_0 is the metal's **work function.**

The **stopping potential** that stops even the fastest electrons is

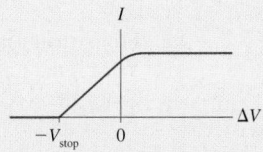

$$V_{stop} = \frac{h}{e}(f - f_0)$$

Particle in a box

A particle confined to a one-dimensional box of length L sets up de Broglie standing waves. The allowed energies are

$$E_n = \tfrac{1}{2}mv_n^2 = n^2 \frac{h^2}{8mL^2} \qquad n = 1, 2, 3, \ldots$$

The Bohr hydrogen atom

The stationary states are found by requiring an integer number of de Broglie wavelengths to fit around the circumference of the electron's orbit: $2\pi r = n\lambda$.

This leads to energy quantization with

$$r_n = n^2 a_B \qquad v_n = \frac{v_1}{n} \qquad E_n = -\frac{13.60\,\text{eV}}{n^2}$$

where $a_B = 0.0529$ nm is the **Bohr radius.** These energies successfully predict the Balmer formula for the hydrogen spectrum. Angular momentum is also quantized, with $L = n\hbar$.

TERMS AND NOTATION

photoelectric effect	de Broglie wavelength	quantum jump
threshold frequency, f_0	quantized	collisional excitation
stopping potential, V_{stop}	energy level	energy-level diagram
thermal emission	quantum number, n	Bohr radius, a_B
work function, E_0	fundamental quantum of energy, E_1	binding energy
Planck's constant, h or $\hbar$	Bohr model of the atom	ionization energy, $|E_1|$
light quantum	stationary state	ionization limit
photon	ground state	hydrogen-like ion
wave packet	excited state	
matter wave	transition	

EXERCISES AND PROBLEMS

Exercises

Section 38.1 The Photoelectric Effect

1. a. Explain why the graphs of Figure 38.4 are horizontal for $\Delta V > 0$.
   b. Explain why photoelectrons are ejected from the cathode with a range of kinetic energies, rather than all electrons having the same kinetic energy.
   c. Explain the reasoning by which we claim that the stopping potential V_{stop} measures the maximum kinetic energy of the electrons.

2. How would the graph of Figure 38.3 look *if* classical physics provided the correct description of the photoelectric effect? Draw the graph and explain your reasoning. Assume that the light intensity remains constant as its frequency and wavelength are varied.

3. How would the graphs of Figure 38.4 look *if* classical physics provided the correct description of the photoelectric effect? Draw the graph and explain your reasoning. Include curves for both weak light and intense light.

4. Figure Ex38.4 is the current-versus-potential-difference graph for a photoelectric-effect experiment with an unknown metal. *If* classical physics provided the correct description of the photoelectric effect, how would the graph look if:
   a. The light was replaced by an equally intense light with a shorter wavelength? Draw it.
   b. The metal was replaced by a different metal with a smaller work function? Draw it.

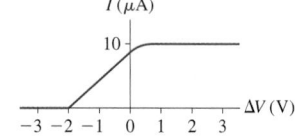

FIGURE EX38.4

5. How many photoelectrons are ejected per second in the experiment represented by the graph of Figure Ex38.4?

Section 38.2 Einstein's Explanation

6. Which metals in Table 38.1 exhibit the photoelectric effect for (a) light with $\lambda = 400$ nm and (b) light with $\lambda = 250$ nm?

7. Photoelectrons are observed when a metal is illuminated by light with a wavelength less than 388 nm. What is the metal's work function?

8. Electrons in a photoelectric-effect experiment emerge from a copper surface with a maximum kinetic energy of 1.10 eV. What is the wavelength of the light?

9. You need to design a photodetector that can respond to the entire range of visible light. What is the maximum possible work function of the cathode?

10. Use Millikan's photoelectric-effect data in Figure 38.11 to determine:
    a. The work function, in eV, of cesium.
    b. An experimental value of Planck's constant.

11. A photoelectric-effect experiment finds a stopping potential of 1.93 V when light of 200 nm is used to illuminate the cathode.
    a. From what metal is the cathode made?
    b. What is the stopping potential if the intensity of the light is doubled?

Section 38.3 Photons

12. a. Determine the energy, in eV, of a photon with a 700 nm wavelength.
    b. Determine the wavelength of a 5.0 keV x-ray photon.

13. What is the wavelength, in nm, of a photon with energy (a) 0.30 eV, (b) 3.0 eV, and (c) 30 eV? For each, is this wavelength visible, ultraviolet, or infrared light?

14. What is the energy, in eV, of (a) a 100 MHz radio-frequency photon, (b) a visible-light photon with a wavelength of 500 nm, and (c) an x-ray photon with a wavelength of 0.10 nm?

15. For what wavelength of light does a 100 mW laser deliver 2.5×10^{17} photons per s?

Section 38.4 Matter Waves and Energy Quantization

16. At what speed is an electron's de Broglie wavelength (a) 1.0 pm, (b) 1.0 nm, (c) 1.0 μm, and (d) 1.0 mm?

17. Through what potential difference must an electron be accelerated from rest to have a wavelength of 500 nm?

18. The diameter of the nucleus is about 10 fm. What is the kinetic energy, in MeV, of a proton with a de Broglie wavelength of 10 fm?

19. What is the length of a one-dimensional box in which an electron in the $n = 1$ state has the same energy as a photon with a wavelength of 600 nm?

20. The diameter of the nucleus is about 10 fm. A simple model of the nucleus is that protons and neutrons are confined within a one-dimensional box of length 10 fm. What are the first three energy levels, in MeV, for a proton in such a box?

21. An electron confined in a one-dimensional box is observed, at different times, to have energies of 12 eV, 27 eV, and 48 eV. What is the length of the box?

Section 38.5 Bohr's Model of Atomic Quantization

22. Figure Ex38.22 is an energy-level diagram for a simple atom. What wavelengths appear in the atom's (a) emission spectrum and (b) absorption spectrum?

$$n = 3 \underline{\hspace{3cm}} E_3 = 4.0 \text{ eV}$$

$$n = 2 \underline{\hspace{3cm}} E_2 = 1.5 \text{ eV}$$

FIGURE EX38.22 $\quad n = 1 \underline{\hspace{3cm}} E_1 = 0.0 \text{ eV}$

23. An electron with 2.0 eV of kinetic energy collides with the atom shown in Figure Ex38.22.
    a. Is the electron able to kick the atom to an excited state? Why or why not?
    b. If your answer to part a was yes, what is the electron's kinetic energy after the collision?

24. The allowed energies of a simple atom are 0.0 eV, 4.0 eV, and 6.0 eV.
    a. Draw the atom's energy-level diagram. Label each level with the energy and the quantum number.
    b. What wavelengths appear in the atom's emission spectrum?
    c. What wavelengths appear in the atom's absorption spectrum?

25. The allowed energies of a simple atom are 0.0 eV, 4.0 eV, and 6.0 eV. An electron traveling with a speed of 1.3×10^6 m/s collides with the atom. Can the electron excite the atom to the $n = 2$ stationary state? The $n = 3$ stationary state? Explain.

Section 38.6 The Bohr Hydrogen Atom

26. Show, by actual calculation, that the Bohr radius is 0.0529 nm and that the ground-state energy of hydrogen is -13.60 eV.

27. a. What quantum number of the hydrogen atom comes closest to giving a 500-nm-diameter electron orbit?
    b. What are the electron's speed and energy in this state?

28. a. Calculate the de Broglie wavelength of the electron in the $n = 1$, 2, and 3 states of the hydrogen atom. Use the information in Table 38.2.

b. Show numerically that the circumference of the orbit for each of these stationary states is exactly equal to n de Broglie wavelengths.
    c. Sketch the de Broglie standing wave for the $n = 3$ orbit.

29. How much energy does it take to ionize a hydrogen atom that is in its first excited state?

30. Show, by calculation, that the first three states of the hydrogen atom have angular momenta $\hbar$, $2\hbar$, and $3\hbar$, respectively.

31. Show that Planck's constant $\hbar$ has units of angular momentum.

Section 38.7 The Hydrogen Spectrum

32. Determine the wavelengths of all the possible photons that can be emitted from the $n = 4$ state of a hydrogen atom.

33. What is the wavelength of the series limit (i.e., the shortest possible wavelength) of the Lyman series in hydrogen?

34. Is a spectral line with wavelength 656.5 nm seen in the absorption spectrum of hydrogen atoms? Why or why not?

35. a. Find the radius of the electron's orbit, the electron's speed, and the energy of the atom for the first three stationary states of He^+.
    b. Show, by calculation, that the angular momentum in each state is equal to $n\hbar$.

Problems

36. An AM radio station broadcasts with a power of 10 kW at a frequency of 1000 kHz.
    a. How many photons does the antenna emit each second?
    b. Should the broadcast be treated as an electromagnetic wave or discrete photons? Explain.

37. A red laser with a wavelength of 650 nm and a blue laser with a wavelength of 450 nm emit laser beams with the same light power. How do their rates of photon emission compare? Answer this by computing R_{red}/R_{blue}.

38. A 100 W lightbulb emits about 5 W of visible light. (The other 95 W are emitted as infrared radiation or lost as heat to the surroundings.) The average wavelength of the visible light is about 600 nm, so make the simplifying assumption that all the light has this wavelength.
    a. What is the frequency of the emitted light?
    b. How many visible-light photons does the bulb emit per second?
    c. Should your answers to parts a and b be the same? Explain.

39. A ruby laser emits an intense pulse of light that lasts a mere 10 ns. The light has a wavelength of 690 nm, and each pulse has an energy of 500 mJ.
    a. How many photons are emitted in each pulse?
    b. What is the *rate* of photon emission, in photons per second, during the 10 ns that the laser is "on"?

40. In practice it is easier to measure the wavelength of light than to measure its frequency. Draw a graph of the stopping potential of a metal as a function of the wavelength of the incident light. Consider the full range of wavelengths from zero to infinity. Be sure to identify the threshold wavelength λ_0.

41. Potassium and gold cathodes are used in a photoelectric-effect experiment. For each cathode, find:
    a. The threshold frequency.
    b. The threshold wavelength.
    c. The maximum photoelectron ejection speed if the light has a wavelength of 220 nm.
    d. The stopping potential if the wavelength is 220 nm.

42. The maximum kinetic energy of photoelectrons is 2.8 eV. When the wavelength of the light is increased by 50%, the maximum energy decreases to 1.1 eV. What are (a) the work function of the cathode and (b) the initial wavelength?

43. In a photoelectric-effect experiment, the stopping potential at a wavelength of 400 nm is 25.7% of the stopping potential at a wavelength of 300 nm. Of what metal is the cathode made?

44. The graph in Figure P38.44 was measured in a photoelectric-effect experiment.
    a. What is the work function (in eV) of the cathode?
    b. What experimental value of Planck's constant is obtained from these data?

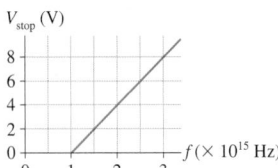

FIGURE P38.44

45. Figure P38.45 shows the stopping potential versus the light frequency for a metal cathode used in a photoelectric-effect experiment. Suppose this cathode is now illuminated with 10 μW of 300 nm light and that the efficiency of converting photons to photoelectrons is 10%. Draw a graph showing current I versus potential difference ΔV for potential difference values from -3 V to $+3$ V. Include a numerical scale on both axes.

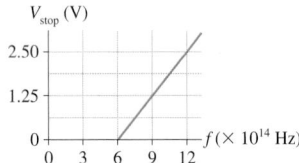

FIGURE P38.45

46. In a photoelectric-effect experiment, the stopping potential was measured for several different wavelengths of incident light. The data are shown below. Analyze these data to determine:
    a. The metal used for the cathode.
    b. An experimental value for Planck's constant. Your value should be found using *all* the data.

λ (nm)	V_{stop} (volts)
500	0.19
450	0.48
400	0.83
350	1.28
300	1.89
250	2.74

47. The relationship between momentum and energy from Einstein's theory of relativity is $E^2 - (pc)^2 = E_0^2$, where, in this context, $E_0 = mc^2$ is the rest energy rather than the work function.
    a. A photon is a massless particle. What is a photon's momentum p in terms of its energy E?
    b. Einstein also claimed that the energy of a photon is related to its frequency by $E = hf$. Use this and your result from part a to write an expression for the wavelength of a photon in terms of its momentum p.
    c. Your result for part b is for a "particle-like wave." Suppose you thought this expression should also apply to a "wave-like particle." What is your expression for λ if you replace p with the classical-mechanics expression for the momentum of a particle of mass m? Is this a familiar-looking expression?

48. The electron interference pattern of Figure 38.14 was made by shooting electrons with 50 keV of kinetic energy through two slits spaced 1.0 μm apart. The fringes were recorded on a detector 1.0 m behind the slits.
    a. What was the speed of the electrons? (The speed is large enough to justify using relativity, but for simplicity do this as a nonrelativistic calculation.)
    b. Figure 38.14 is greatly magnified. What was the actual spacing on the detector between adjacent bright fringes?

49. The neutron interference pattern of Figure P38.49 was made by shooting neutrons with a speed of 200 m/s through two slits spaced 0.10 mm apart.
    a. What was the energy, in eV, of the neutrons?
    b. What was the de Broglie wavelength of the neutrons?
    c. The pattern was recorded by using a neutron detector to measure the neutron intensity at different positions. Notice the 100 μm scale on the figure. By making appropriate measurements directly *on the figure,* determine how far the detector was behind the slits.

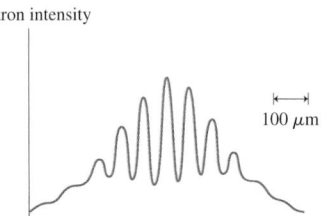

FIGURE P38.49

50. The electron beam in a cathode-ray tube is accelerated through a potential difference of 250 V. The electrons then pass through a small circular hole and are viewed on the screen. You observe that the central bright spot on the screen is the base of a cone with its apex at the hole. The outer edge of the cone makes a 0.50° angle with the original direction of the electron beam. What is the diameter of the hole?

51. An electron confined in a one-dimensional box emits a 200 nm photon in a quantum jump from $n = 2$ to $n = 1$. What is the length of the box?

52. A proton confined in a one-dimensional box emits a 2.0 MeV gamma-ray photon in a quantum jump from $n = 2$ to $n = 1$. What is the length of the box?

53. Imagine that the horizontal box of Figure 38.16 is instead oriented vertically. Also imagine the box to be on a neutron star where the gravitational field is so strong that the particle in the

box slows significantly, nearly stopping, before it hits the top of the box. Make a *qualitative* sketch of the $n = 3$ de Broglie standing wave of a particle in this box.

Hint: The nodes are *not* uniformly spaced.

54. The absorption spectrum of an atom consists of the wavelengths 200 nm, 300 nm, and 500 nm.
    a. Draw the atom's energy-level diagram.
    b. What wavelengths are seen in the atom's emission spectrum?

55. The first three energy levels of the fictitious element X are shown in Figure P38.55.
    a. What is the ionization energy of element X?
    b. What wavelengths are observed in the absorption spectrum of element X? Express your answers in nm.
    c. State whether each of your wavelengths in part b corresponds to ultraviolet, visible, or infrared light.
    d. An electron with a speed of 1.4×10^6 m/s collides with an atom of element X. Shortly afterward, the atom emits a 1240 nm photon. What was the electron's speed after the collision? Assume that, because the atom is so much more massive than the electron, the recoil of the atom is negligible.

    Hint: The energy of the photon is *not* the energy transferred to the atom in the collision.

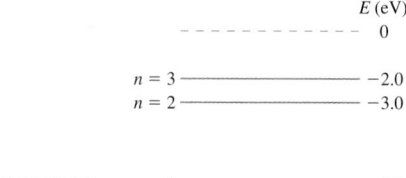

FIGURE P38.55

56. Starting from Equation 38.28, derive Equation 38.29.
57. What is the energy of a hydrogen atom with a 5.18 nm diameter?
58. Calculate *all* the wavelengths of *visible* light in the emission spectrum of the hydrogen atom.

    Hint: There are infinitely many wavelengths in the spectrum, so you'll need to develop a strategy for this problem rather than using trial and error.

59. A hydrogen atom in the ground state absorbs a 12.75 eV photon. Immediately after the absorption, the atom undergoes a quantum jump to the next-lowest energy level. What is the wavelength of the photon emitted in this quantum jump?
60. a. What wavelength photon does a hydrogen atom emit in a $200 \rightarrow 199$ transition?
    b. What is the *difference* in the wavelengths absorbed in a $2 \rightarrow 199$ transition and a $2 \rightarrow 200$ transition?
61. a. Calculate the orbital radius and the speed of an electron in both the $n = 99$ and the $n = 100$ state of hydrogen.
    b. Determine the orbital frequency of the electron in each of these states.
    c. Calculate the frequency of a photon emitted in a $100 \rightarrow 99$ transition.
    d. Compare the photon frequency of part c to the *average* of your two orbital frequencies from part b. By what percent do they differ?
62. Draw an energy-level diagram, similar to Figure 38.25, for the He^+ ion. On your diagram:
    a. Show the first five energy levels. Label each with the values of n and E_n.

b. Show the ionization limit.
c. Show all possible emission transitions from the $n = 4$ energy level.
d. Calculate the wavelengths (in nm) for each of the transitions in part c and show them alongside the appropriate arrow.

63. What are the wavelengths of the transitions $3 \rightarrow 2$, $4 \rightarrow 2$, and $5 \rightarrow 2$ in the hydrogen-like ion O^{+7}? In what spectral range do these lie?

64. Two hydrogen atoms collide head on. The collision brings both atoms to a halt. Immediately after the collision, both atoms emit a 121.6 nm photon. What was the speed of each atom just before the collision?

65. Ultraviolet light with a wavelength of 70 nm shines on a gas of hydrogen atoms in their ground states. Some of the atoms are ionized by the light. What is the kinetic energy of the electrons that are freed in this process?

66. A beam of electrons is incident upon a gas of hydrogen atoms.
    a. What minimum speed must the electrons have to cause the emission of 656 nm light from the $3 \rightarrow 2$ transition of hydrogen?
    b. Through what potential difference must the electrons be accelerated to have this speed?

Challenge Problems

67. The photomultiplier tube (PMT) of Figure 38.13 consists of a cathode, which the photon strikes; an anode, where the electrons are collected; and a number of intermediate electrodes called *dynodes*. The tube shown in the figure has nine dynodes, but consider a PMT with N dynodes. The cathode, when struck by a photon, ejects a single photoelectron. That electron is accelerated to the first dynode, where it causes (on average) the ejection of ϵ secondary electrons. The quantity ϵ is called the *secondary emission coefficient*. Each of these electrons ejects, on average, ϵ electrons from the second dynode, each of which in turn ejects ϵ electrons from the third dynode, and so on until a large pulse of electrons is collected by the anode.
    a. Write an expression, in terms of ϵ and N, for the average number of electrons arriving at the anode due to a single photon striking the cathode. This is called the *gain* of the PMT.
    b. The graph in Figure 38.13b shows the voltage pulse generated when the electron current flowed through a 50 Ω resistor. The baseline of the pulse is zero volts, and the voltage scale is 20 mV per division. What is the maximum *current* of this pulse?
    c. Because $I = dQ/dt$, the amount of charge delivered by a pulse of current is $Q = \int I \, dt$. This can be interpreted geometrically as the area under the I-versus-t curve. The area of a pulse is reasonably well approximated as its height multiplied by its width measured at half of its maximum height. Estimate the number of electrons in the current pulse shown in Figure 38.13b.
    d. The PMT that produced this pulse had 14 dynodes. By comparing your answers to parts a and c, determine the secondary emission coefficient for this PMT.

68. In the atom interferometer experiment of Figure 38.15, laser-cooling techniques were used to cool a dilute vapor of sodium atoms to a temperature of 0.001 K = 1 mK. The ultracold atoms passed through a series of collimating apertures to form the *atomic beam* you see entering the figure from the left. The standing light waves were created from a laser beam with a wavelength of 590 nm.
    a. What is the rms speed v_{rms} of a sodium atom ($A = 23$) in a gas at temperature 1 mK?
    b. By treating the laser beam as if it were a diffraction grating, calculate the first-order diffraction angle of a sodium atom traveling with the rms speed of part a.
    c. How far apart are points B and C if the second standing wave is 10 cm from the first?
    d. Because interference is observed between the two paths, each individual atom is apparently present at both point B *and* point C. Describe, in your own words, what this experiment tells you about the nature of matter.

69. Consider a hydrogen atom in stationary state n.
    a. Show that the orbital period of the electron is $T = n^3 T_1$, and find a numerical value for T_1.
    b. On average, an atom stays in the $n = 2$ state for 1.6 ns before undergoing a quantum jump to the $n = 1$ state. On average, how many revolutions does the electron make before the quantum jump?

70. Consider an electron undergoing cyclotron motion in a magnetic field. According to Bohr, the electron's angular momentum must be quantized in units of $\hbar$.
    a. Show that allowed radii for the electron's orbit are given by $r_n = (n\hbar/eB)^{1/2}$, where $n = 1, 2, 3, \ldots$
    b. Compute the first four allowed radii in a 1.0 T magnetic field.
    c. Find an expression for the allowed energy levels E_n in terms of $\hbar$ and the cyclotron frequency f_{cyc}.

71. The *muon* is a subatomic particle with the same charge as an electron but with a mass that is 207 times greater: $m_\mu = 207 m_e$. Physicists think of muons as "heavy electrons." However, the muon is not a stable particle; it decays with a half-life of 1.5 μs into an electron plus two neutrinos. Muons from cosmic rays are sometimes "captured" by the nuclei of the atoms in a solid. A captured muon orbits this nucleus, like an electron, until it decays. Because the muon is often captured into an excited orbit ($n > 1$), its presence can be detected by observing the photons emitted in transitions such as $2 \rightarrow 1$ and $3 \rightarrow 1$.
    Consider a muon captured by a carbon nucleus ($Z = 6$). Because of its large mass, the muon orbits well *inside* the electron cloud and is not affected by the electrons. Thus the muon "sees" the full nuclear charge Ze and acts like the electron in a hydrogen-like ion.
    a. What are the orbital radius and speed of a muon in the $n = 1$ ground state? Note that the mass of a muon differs from the mass of an electron.
    b. What is the wavelength of the $2 \rightarrow 1$ muon transition?
    c. Is the photon emitted in the $2 \rightarrow 1$ transition infrared, visible, ultraviolet, or x ray?
    d. How many orbits will the muon complete during 1.5 μs? Is this a sufficiently large number that the Bohr model "makes sense," even though the muon is not stable?

STOP TO THINK ANSWERS

Stop to Think 38.1: $V_A > V_B > V_C$. For a given wavelength of light, electrons are ejected faster from metals with smaller work functions because it takes less energy to remove an electron. Faster electrons need a larger negative voltage to stop them.

Stop to Think 38.2: d. Photons always travel at c, and a photon's energy depends only on the light's frequency, not its intensity.

Stop to Think 38.3: $n = 4$. There are four antinodes.

Stop to Think 38.4: Not in absorption. In emission from the $n = 2$ to $n = 1$ transition. The photon energy has to match the energy *difference* between two energy levels. Absorption is from the ground state, at $E_1 = 0$ eV. There's no energy level at 3 eV to which the atom could jump.

Stop to Think 38.5: $n = 3$. Each antinode is half a wavelength, so this standing wave has three full wavelengths in one circumference.

39 Wave Functions and Uncertainty

The surface of graphite, as imaged with atomic resolution by a scanning tunneling microscope. The hexagonal ridges show the most probable locations of the electrons.

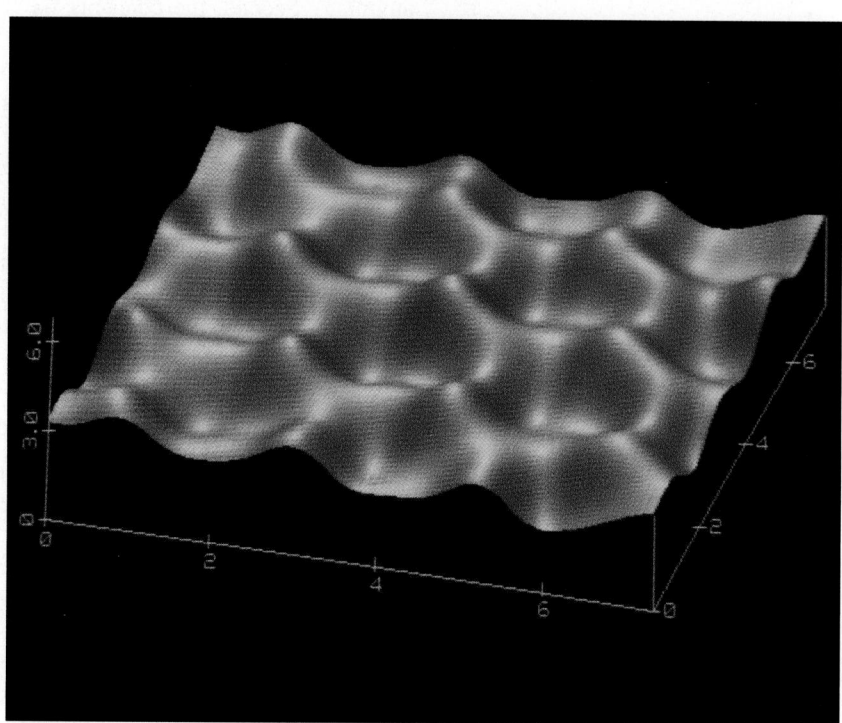

▶ **Looking Ahead**

The goal of Chapter 39 is to introduce the wave-function description of matter and learn how it is interpreted. In this chapter you will learn to:

- Connect the particle and wave descriptions of matter.
- Use basic ideas about probability.
- Use the wave function to calculate the probabilities of detecting particles.
- Recognize the limitations on knowledge imposed by the Heisenberg uncertainty principle.

◀ **Looking Back**

The ideas developed in this chapter are highly dependent on understanding the double-slit interference experiment for both light and matter. Please review:

- Sections 21.8 and 22.2 Interference, beats, and the double-slit experiment.
- Sections 24.3–24.4 Photons and matter waves.
- Section 38.4 The de Broglie wavelength and wave-particle duality.

You learned in the last two chapters that classical mechanics and electromagnetism were unable to explain the new phenomena associated with light, electrons, and atoms. Scientific theories that had triumphed during the 18th and 19th centuries stumbled over the smallest specks of matter. Many scientists refused to accept these limitations, thinking that it was only a question of time until someone discovered how to apply classical ideas to atoms. Their hopes were to go unfulfilled.

At the same time that classical physics was reaching its limits, the new ideas put forward by Einstein, Bohr, and de Broglie began pointing the way toward a new theory of light and matter. **Quantum mechanics,** as the theory came to be called, did not reach its completed form until the mid-1920s, but it has since proven to be the most successful physical theory ever devised.

This chapter and the next will introduce the essential ideas of quantum mechanics in one dimension. Although the full theory is beyond the scope of this textbook, we can delve far enough into quantum mechanics to learn how it solves the problems of atomic and nuclear structure. Our goal in this chapter is to introduce the concept of the *wave function*. The wave function, which reconciles the wave-like and particle-like aspects of matter, characterizes microscopic particles in terms of the *probability* of finding them at various points in space. This scanning tunneling microscope image of graphite shows that the electrons are most likely found along the ring-like structure created by the carbon-carbon bonds.

Interference fringes in an optical double-slit interference experiment.

39.1 Waves, Particles, and the Double-Slit Experiment

You may feel surprise at how slowly we have been building up to quantum mechanics. Why not just write it down and start using it? There are two reasons. First, quantum mechanics explains microscopic phenomena that we cannot directly sense or experience. It was important to begin by learning how light and atoms behave. Otherwise, how would you know if quantum mechanics explains anything? Second, the concepts we'll need in quantum mechanics are rather abstract. Before launching into the mathematics, we need to establish a connection between theory and experiment.

We will make the connection by returning to the double-slit interference experiment, an experiment that goes right to the heart of wave-particle duality. The significance of the double-slit experiment arises from the fact that both light and matter exhibit the same interference pattern. Regardless of whether photons, electrons, or neutrons pass through the slits, their arrival at a detector is a particle-like event. That is, they make a collection of discrete dots on a detector. Yet our understanding of how interference "works" is based on the properties of waves. Our goal is to find the connection between the wave description and the particle description of interference.

A Wave Analysis of Interference

The interference of light can be analyzed from either a wave perspective or a photon perspective. Let's start with a wave analysis. Figure 39.1 shows light waves passing through a double slit with slit separation d. You should recall that the lines in a wave-front diagram represent wave crests, spaced one wavelength apart. The bright fringes of constructive interference occur where two crests or two troughs overlap. The graphs and pictures below the detection screen (notice that they're aligned vertically) show the outcome of the experiment.

You studied interference and the double-slit experiment in Chapters 21 and 22. The two waves traveling from the slits to the viewing screen are traveling waves with displacements

$$D_1 = a\sin(kr_1 - \omega t)$$
$$D_2 = a\sin(kr_2 - \omega t)$$

where a is the amplitude of each wave, $k = 2\pi/\lambda$ is the wave number, and r_1 and r_2 are the distances from the two slits. The "displacement" of a light wave is not a physical displacement, as in a water wave, but a change in the electromagnetic field.

According to the principle of superposition, these two waves add together where they meet at a point on the screen to give a wave with net displacement $D = D_1 + D_2$. Previously (see Equations 21.24 and 22.12) we found that the amplitude of their superposition is

$$A(x) = 2a\cos\left(\frac{\pi d x}{\lambda L}\right) \tag{39.1}$$

where x is the horizontal coordinate on the screen, measured from $x = 0$ in the center.

The function $A(x)$, the top graph in Figure 39.1, is called the *amplitude function*. It describes the amplitude A of the light wave as a function of the position x on the viewing screen. The amplitude function has maxima where two crests from individual waves overlap and add constructively to make a larger wave with amplitude $2a$. $A(x)$ is zero at points where the two individual waves are out of phase and interfere destructively.

Approaching wave fronts

Double slit

d

λ λ

L

Screen

Crests overlap

$A(x)$

0 x

Wave amplitude along the screen

I

0 x

Interference fringes

Photon arrival positions

FIGURE 39.1 The double-slit experiment with light.

If you carry out a double-slit experiment in the lab, what you observe on the screen is the light's *intensity,* not its amplitude. A wave's intensity I is proportional to the *square* of the amplitude. That is, $I \propto A^2$, where the symbol $\propto$ means "is proportional to." Using Equation 39.1 for the amplitude at each point, the intensity $I(x)$ as a function of position x on the screen is

$$I(x) = C\cos^2\left(\frac{\pi dx}{\lambda L}\right) \quad (39.2)$$

where C is a proportionality constant.

The lower graph in Figure 39.1 shows the intensity as a function of position along the screen. This graph shows the alternating bright and dark interference fringes that you see in the laboratory. In other words, the intensity of the wave is the *experimental reality* that you observe and measure.

Probability

Before discussing photons, we need to introduce some ideas about probability. Imagine throwing darts at a dart board while blindfolded. Figure 39.2 shows how the board might look after your first 100 throws. From this information, can you predict where your 101st throw is going to land?

No. The position of any individual dart is *unpredictable.* No matter how hard you try to reproduce the previous throw, a second dart will not land at the same place as the first. Yet there is clearly an overall *pattern* to the where the darts strike the board. Even blindfolded, you had a general sense of where the center of the board was, so each dart was *more likely* to land near the center than at the edge.

Although we can't predict where any individual dart will land, we can use the information in Figure 39.2 to determine the *probability* that your next throw will land in region A or region B or region C. Because 45 out of 100 throws landed in region A, we could say that the *odds* of hitting region A are 45 out of 100, or 45%.

Now, 100 throws isn't all that many. If you throw another 100 darts, perhaps only 43 will land in region A. Then maybe 48 of the next 100 throws. Imagine that the total number of throws N_{tot} becomes extremely large. Then the **probability** that any particular throw lands in region A is defined to be

$$P_A = \lim_{N_{tot}\to\infty} \frac{N_A}{N_{tot}} \quad (39.3)$$

In other words, the probability that the outcome will be A is the fraction of outcomes that are A in an enormously large number of trials. Similarly, $P_B = N_B/N_{tot}$ and $P_C = N_C/N_{tot}$ as $N_{tot} \to \infty$. We can give probabilities either as a decimal fraction or a percentage. In this example, $P_A \approx 45\%$, $P_B \approx 35\%$, and $P_C \approx 20\%$. We've used $\approx$ rather than $=$ because 100 throws isn't enough to determine the probabilities with great precision.

What is the probability that a dart lands in either region A *or* region B? The number of darts landing in either A *or* B is $N_{A\,or\,B} = N_A + N_B$, so we can use the definition of probability to learn that

$$
\begin{aligned}
P_{A\,or\,B} &= \lim_{N_{tot}\to\infty} \frac{N_{A\,or\,B}}{N_{tot}} = \lim_{N_{tot}\to\infty} \frac{N_A + N_B}{N_{tot}} \\
&= \lim_{N_{tot}\to\infty} \frac{N_A}{N_{tot}} + \lim_{N_{tot}\to\infty} \frac{N_B}{N_{tot}} = P_A + P_B
\end{aligned}
\quad (39.4)
$$

That is, **the probability that the outcome will be A *or* B is the sum of P_A and P_B.** This important conclusion is a general property of probabilities.

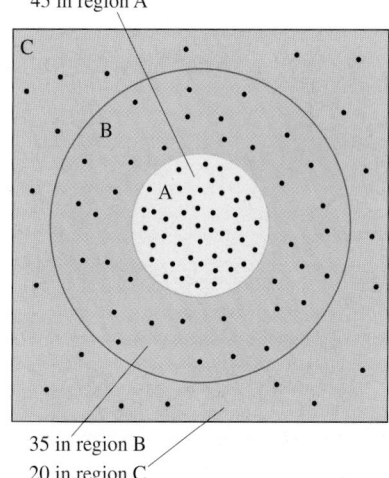

45 in region A

35 in region B
20 in region C

FIGURE 39.2 One hundred throws at a dart board.

Each dart lands *somewhere on* the board. Consequently, the probability that a dart lands in A *or* B *or* C must be 100%. And, in fact,

$$P_{\text{somewhere}} = P_{\text{A or B or C}} = P_A + P_B + P_C = 0.45 + 0.35 + 0.20 = 1.00$$

Thus another important property of probabilities is that **the sum of the probabilities of all possible outcomes must equal 1.**

Suppose exhaustive trials have established that the probability of a dart landing in region A is P_A. If you throw N darts, how many do you *expect* to land in A? This value, called the **expected value,** is

$$N_{\text{A expected}} = NP_A \tag{39.5}$$

The expected value is your best possible prediction of the outcome of an experiment.

If $P_A = 0.45$, your *best prediction* is that 27 of 60 throws (45% of 60) will land in A. Of course, predicting 27 and actually getting 27 aren't the same thing. You would predict 30 heads in 60 flips of a coin, but you wouldn't be surprised if the actual number were 28 or 31. Similarly, the number of darts landing in region A might be 24 or 29 instead of 27. In general, the agreement between actual values and expected values improves as you throw more darts.

STOP TO THINK 39.1   Suppose you roll a die 30 times. What is the expected numbers of 1's *and* 6's?

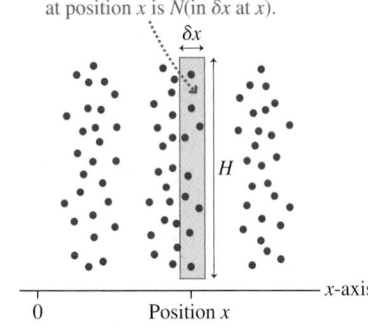

(a)   The number of photons in this narrow strip when it is at position x is N(in δx at x).

δx

H

x-axis

0   Position x

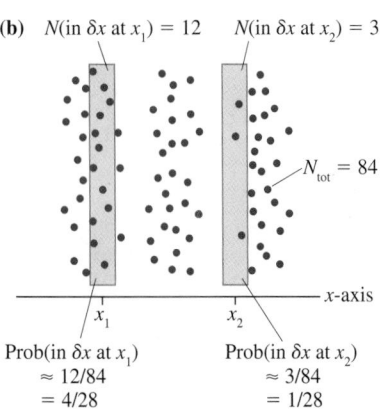

(b) N(in δx at x_1) = 12 N(in δx at x_2) = 3

$N_{\text{tot}} = 84$

x-axis

x_1 x_2

Prob(in δx at x_1)   Prob(in δx at x_2)
$\approx 12/84$ $\approx 3/84$
$= 4/28$ $= 1/28$

FIGURE 39.3 A strip of width δx at position x.

A Photon Analysis of Interference

Now let's look at the double-slit results from a photon perspective. We know, from experimental evidence, that the interference pattern is built up photon by photon. The bottom portion of Figure 39.1 shows the pattern made on a detector after the arrival of the first few dozen photons. It is clearly a double-slit interference pattern, but it's made, rather like a newspaper photograph, by piling up dots in some places but not others.

The arrival position of any particular photon is *unpredictable*. That is, nothing about how the experiment is set up or conducted allows us to predict exactly where the dot of an individual photon will appear on the detector. Yet there is clearly an overall pattern. There are some positions at which a photon is *more likely* to be detected, other positions at which it is *less likely* to be found.

If we record the arrival positions of many thousands of photons, we will be able to determine the *probability* that a photon will be detected at any given location. If 50 out of 50,000 photons land in one small area of the screen, then each photon has a probability of $50/50,000 = 0.001 = 0.1\%$ of being detected there. The probability will be zero at the interference minima because no photons at all arrive at those points. Similarly, the probability will be a maximum at the interference maxima. The probability will have some in-between value on the sides of the interference fringes.

Figure 39.3a shows a narrow strip, with width δx and height H. (We will assume that δx is very small in comparison with the fringe spacing, so the light's intensity over δx is very nearly constant.) Think of this strip as a very narrow detector that can detect and count the photons landing on it. Suppose we place the narrow strip at position x. We'll use the notation N(in δx at x) to indicate the number of photons that hit the detector at this position. The value of N(in δx at x) varies from point to point. N(in δx at x) is large if x happens to be near the center of a bright fringe; N(in δx at x) is small if x is in a dark fringe.

Suppose N_{tot} photons are fired at the slits. The *probability* that any one photon ends up in the strip at position x is

$$\text{Prob(in } \delta x \text{ at } x) = \lim_{N_{tot} \to \infty} \frac{N(\text{in } \delta x \text{ at } x)}{N_{tot}} \qquad (39.6)$$

As Figure 39.3b shows, Equation 39.6 is an empirical method for determining the probabilities of the photons hitting a particular spot on the detector.

Alternatively, suppose we can calculate the probabilities from a theory. In that case, the *expected value* for the number of photons landing in the narrow strip when it is at position x is

$$N(\text{in } \delta x \text{ at } x) = N \times \text{Prob(in } \delta x \text{ at } x) \qquad (39.7)$$

We cannot predict what any individual photon will do, but we can predict the fraction of the photons that should land in this little region of space. Prob(in δx at x) is the probability that it will happen.

39.2 Connecting the Wave and Photon Views

The wave model of light describes the interference pattern in terms of the wave's intensity $I(x)$, a continuous-valued function. The photon model describes the interference pattern in terms of the probability Prob(in δx at x) of detecting a photon. These two models are very different, yet Figure 39.1 shows a clear correlation between the *intensity of the wave* and the *probability of detecting photons*. That is, photons are more likely to be detected at those points where the wave intensity is high, less likely to be detected at those points where the wave intensity is low.

The intensity of a wave is $I = P/A$, the ratio of light power P (joules per second) to the area A on which the light falls. The narrow strip in Figure 39.3a has area $A = H \delta x$. If the light intensity at position x is $I(x)$, the amount of light energy falling onto this narrow strip during each second is

$$E(\text{in } \delta x \text{ at } x) = I(x)A = I(x)H\delta x \qquad (39.8)$$

The notation E(in δx at x) refers to the energy landing on this narrow strip if you place it at position x.

From the photon perspective, energy E is due to the arrival of N photons, each of which has energy hf. The number of photons that arrive in the strip each second is

$$N(\text{in } \delta x \text{ at } x) = \frac{E(\text{in } \delta x \text{ at } x)}{hf} = \frac{H}{hf}I(x)\,\delta x \qquad (39.9)$$

We can then use the Equation 39.6 definition of probability to write the *probability* that a photon lands in the narrow strip δx at position x as

$$\text{Prob(in } \delta x \text{ at } x) = \frac{N(\text{in } \delta x \text{ at } x)}{N_{tot}} = \frac{H}{hfN_{tot}}I(x)\,\delta x \qquad (39.10)$$

Equation 39.10 is the link between the wave model and the photon model.

As a final step, recall that the light intensity $I(x)$ is proportional to $|A(x)|^2$, the square of the amplitude function. Consequently,

$$\text{Prob(in } \delta x \text{ at } x) \propto |A(x)|^2 \delta x \qquad (39.11)$$

where the various constants in Equation 39.10 have all been incorporated into the unspecified proportionality constant of Equation 39.11.

In other words, **the probability of detecting a photon at a particular point is directly proportional to the square of the light-wave amplitude function at that point.** If the wave amplitude at point A is twice that at point B, then a photon is four times as likely to land in a narrow strip at A as it is to land in an equal-width strip at B.

NOTE ▶ Equation 39.11 is the connection between the particle perspective and the wave perspective. It relates the probability of observing a particle-like event—the arrival of a photon—to the amplitude of a continuous, classical wave. This connection will become the basis of how we interpret the results of quantum-physics calculations. ◀

Probability Density

We need one last definition. Recall that the mass of a wire or string of a length L can be expressed in terms of the linear mass density μ as $m = \mu L$. Similarly, the charge along a length L of wire can be expressed in terms of the linear charge density λ as $Q = \lambda L$. If the length had been very short—in which case we might have denoted it as δx, and if the density varied from point to point—we could have written

$$\text{mass(in length } \delta x \text{ at } x) = \mu(x)\,\delta x$$

$$\text{charge(in length } \delta x \text{ at } x) = \lambda(x)\,\delta x$$

where $\mu(x)$ and $\lambda(x)$ are the linear densities at position x. Writing the mass and charge this way separates the role of the density from the role of the small length δx.

Equation 39.11 looks similar. Using the mass and charge densities as analogies, as shown in Figure 39.4, let us define the **probability density** $P(x)$ such that

$$\text{Prob(in } \delta x \text{ at } x) = P(x)\,\delta x \tag{39.12}$$

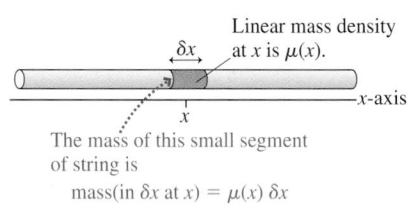

Linear mass density at x is $\mu(x)$.

The mass of this small segment of string is
$$\text{mass(in } \delta x \text{ at } x) = \mu(x)\,\delta x$$

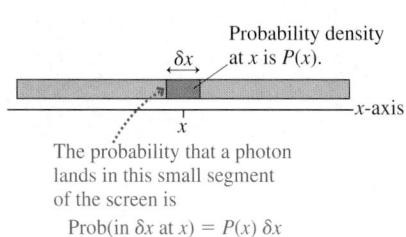

Probability density at x is $P(x)$.

The probability that a photon lands in this small segment of the screen is
$$\text{Prob(in } \delta x \text{ at } x) = P(x)\,\delta x$$

FIGURE 39.4 The probability density is analogous to the linear mass density.

Probability density has SI units of m^{-1}. Thus the probability density multiplied by a length, as in Equation 39.12, yields a dimensionless probability.

NOTE ▶ $P(x)$ itself is *not* a probability, just as the linear mass density λ is not, by itself, a mass. You must multiply the probability density by a length, as shown in Equation 39.12, to find an actual probability. ◀

By comparing Equation 39.12 to Equation 39.11, you can see that the photon probability density is directly proportional to the square of the light-wave amplitude:

$$P(x) \propto |A(x)|^2 \tag{39.13}$$

The probability density, unlike the probability itself, is independent of the width δx and depends only on the position x.

Although we were inspired by the double-slit experiment, nothing in our analysis actually depends on the double-slit geometry. Consequently, Equation 39.13 is quite general. It says that for *any* experiment in which we detect photons, **the probability density for detecting a photon is directly proportional to the square of the amplitude function of the corresponding electromagnetic wave.** We now have an explicit connection between the wave-like and the particle-like properties of the light.

EXAMPLE 39.1 **Calculating the probability density**

In an experiment, 6000 out of 600,000 photons are detected in a 1.0-mm-wide strip located at position $x = 50$ cm. What is the probability density at $x = 50$ cm?

SOLVE The probability that a photon arrives at this particular strip is

$$\text{Prob(in 1.0 mm at } x = 50 \text{ cm)} = \frac{6000}{600,000} = 0.010$$

Thus the probability density $P(x) = \text{Prob(in } \delta x \text{ at } x)/\delta x$ at this position is

$$P(50 \text{ cm}) = \frac{\text{Prob(in 1.0 mm at } x = 50 \text{ cm})}{0.0010 \text{ m}} = \frac{0.010}{0.0010 \text{ m}}$$

$$= 1.0 \times 10^{-5} \text{ m}^{-1}$$

STOP TO THINK 39.2 The figure shows the detection of photons in an optical experiment. Rank in order, from largest to smallest, the square of the amplitude function of the electromagnetic wave at positions A, B, C, and D.

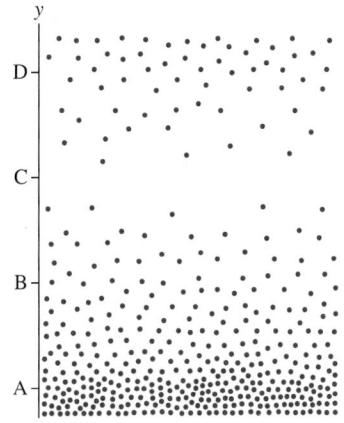

39.3 The Wave Function

Now let's look at the interference of matter. Electrons passing through a double-slit apparatus create the same interference patterns as photons. The pattern is built up electron by electron, but there is no way to predict where any particular electron will be detected. Even so, we can establish the *probability* of an electron landing in a narrow strip of width δx by measuring the positions of many individual electrons. The probability, as you might guess, turns out to be exactly the same as the arrival probability of a photon with the same wavelength.

For light, we were able to relate the photon probability density $P(x)$ to the amplitude of an electromagnetic wave. But there is no wave for electrons that is analogous to electromagnetic waves for light. So how do we find the probability density for electrons? We have reached the point where we must make an inspired leap beyond classical physics. Let us *assume* that there is some kind of continuous, wave-like function for matter that plays a role analogous to the electromagnetic amplitude function $A(x)$ for light. We will call this function the **wave function** $\psi(x)$, where ψ is a lowercase Greek psi. The wave function is a function of position, which is why we write it as $\psi(x)$.

To connect the wave function to the real world of experimental measurements, we will interpret $\psi(x)$ in terms of the *probability* of detecting a particle at position x. If a matter particle, such as an electron, is described by the wave function $\psi(x)$, then the probability Prob(in δx at x) of finding the particle within a narrow region of width δx at position x is

$$\text{Prob(in } \delta x \text{ at } x) = |\psi(x)|^2 \delta x = P(x) \delta x \qquad (39.14)$$

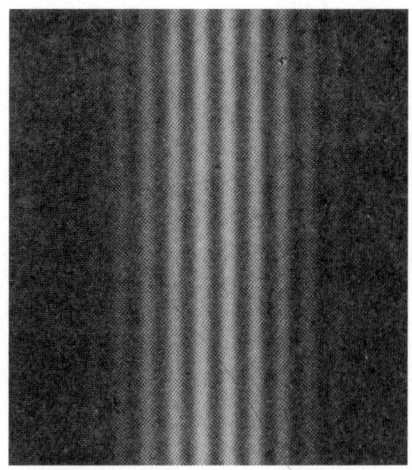

Electrons create interference fringes after passing through a double slit.

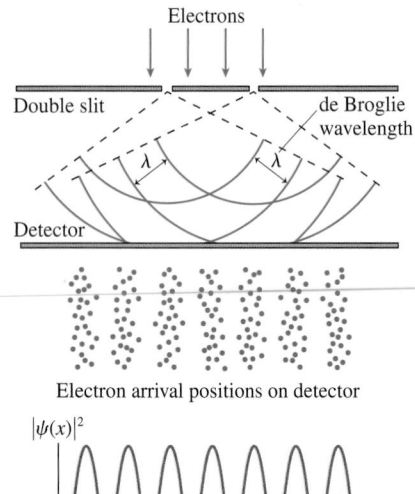

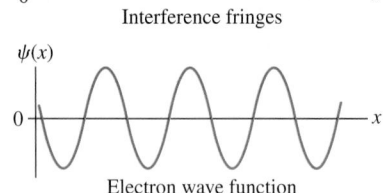

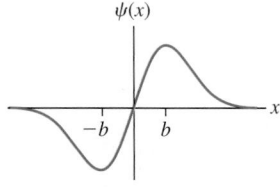

FIGURE 39.5 The double-slit experiment with electrons.

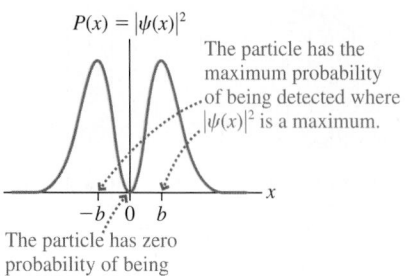

(a) Wave function

(b) Probability density

The particle has the maximum probability of being detected where $|\psi(x)|^2$ is a maximum.

The particle has zero probability of being detected where $|\psi(x)|^2 = 0$.

FIGURE 39.6 The square of the wave function is the probability density for detecting the electron at various values of the position x.

That is, the probability density $P(x)$ for finding the particle is

$$P(x) = |\psi(x)|^2 \tag{39.15}$$

With Equations 39.14 and 39.15, we are *defining* the wave function $\psi(x)$ to play the same role for material particles that the amplitude function $A(x)$ does for photons. The only difference is that $P(x) = |\psi(x)|^2$ is for particles, whereas Equation 39.13 for photons is $P(x) \propto |A(x)|^2$. The difference is due to the fact that the electromagnetic field amplitude $A(x)$ had previously been defined through the laws of electricity and magnetism. $|A(x)|^2$ is *proportional* to the probability density for finding a photon, but it is not directly *the* probability density. In contrast, we do not have any preexisting definition for the wave function $\psi(x)$. Thus we are free to define $\psi(x)$ so that $|\psi(x)|^2$ is *exactly* the probability density. That is why we used $=$ rather than $\propto$ in Equation 39.15.

Figure 39.5 shows the double-slit experiment with electrons. This time we will work backward. From the observed distribution of electrons, which represents the probabilities of their landing in any particular location, we can deduce that $|\psi(x)|^2$ has alternating maxima and zeros. The oscillatory wave function $\psi(x)$ is the square root *at each point* of $|\psi(x)|^2$. Notice the very close analogy with the amplitude function $A(x)$ in Figure 39.1.

NOTE ▶ $|\psi(x)|^2$ is uniquely determined by the data, but the wave function $\psi(x)$ is *not* unique. The alternative wave function $\psi'(x) = -\psi(x)$—an upside-down version of the graph in Figure 39.5—would be equally acceptable. ◀

Figure 39.6a is a different example of a wave function. After squaring it *at each point,* as shown in Figure 39.6b, we see that this wave function represents a particle most likely to be detected very near $x = -b$ or $x = +b$. These are the points where $|\psi(x)|^2$ is a maximum. There is zero likelihood of finding the particle right in the center. The particle is more likely to be detected at some positions than at others, but we cannot predict its exact location. The wave function, from which we can calculate probabilities, is all we know about the particle.

NOTE ▶ One of the difficulties in learning to use the concept of a wave function is coming to grips with the fact that there is no "thing" that is waving. There is no disturbance associated with a physical medium. The wave function $\psi(x)$ is simply a *wave-like function* (i.e., it oscillates between positive and negative values) that can be used to make probabilistic predictions about atomic particles. ◀

A Little Science Methodology

Equation 39.14 defines the wave function $\psi(x)$ for a particle in terms of the probability of finding the particle at different positions x. But our interests go beyond merely characterizing experimental data. We would like to develop a new *theory* of matter. But just what is a theory? Although this is not a book on scientific methodology, we can loosely say that a physical theory needs two basic ingredients:

1. A *descriptor.* This is a mathematical quantity used to describe our knowledge of a physical object.
2. One or more *laws* that govern the behavior of the descriptor.

For example, Newtonian mechanics is a theory of motion. The primary descriptor in Newtonian mechanics is a particle's *position* $x(t)$ as a function of time. This describes our knowledge of the particle at all times. The position is governed by *Newton's laws.* These laws, especially the second law, are mathematical statements of how the descriptor changes in response to forces. If we predict $x(t)$ for a known set of forces, we feel confident that an experiment carried out at time t will find the particle right where predicted.

Newton's theory of motion *assumes* that a particle's position is well-defined at every instant of time. The difficulty facing physicists early in the 20th century was the astounding discovery that **the position of an atomic-size particle is *not* well-defined.** An electron in a double-slit experiment must, in some sense, go through *both* slits to produce an electron interference pattern. It simply does not have a well-defined position as it interacts with the slits. But if the position function $x(t)$ is not a valid descriptor for matter at the atomic level, what is?

We will assert that the wave function $\psi(x)$ is the *descriptor* of a particle in quantum mechanics. In other words, the wave function tells us everything we can know about the particle. The wave function $\psi(x)$ plays the same leading role in quantum mechanics that the position function $x(t)$ plays in classical mechanics.

Whether this hypothesis has any merit will not be known until we see if it leads to predictions that can be verified. And before we can do that, we need to learn the "law of psi." What new law of physics determines the wave function $\psi(x)$ in a given situation? We will answer this question in the next chapter.

It may seem to you, as we go along, that we are simply "making up" ideas. Indeed, that is at least partially true. The inventors of entirely new theories use their existing knowledge as a guide, but ultimately they have to make an inspired guess as to what a new theory should look like. Newton and Einstein both made such leaps, and the inventors of quantum mechanics had to make such a leap. We can attempt to make the new ideas *plausible*, but ultimately a new theory is simply a bold new assertion that must be tested against experimental reality. The wave-function theory of quantum mechanics passed the only test that really matters in science—it works!

STOP TO THINK 39.3　This is the wave function of a neutron. At what value of x is the neutron most likely to be found?

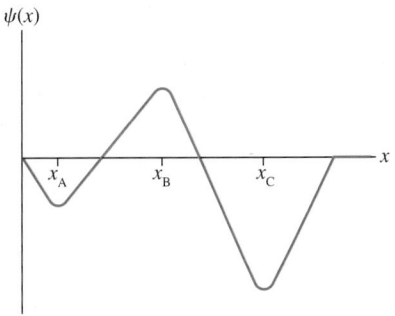

39.4 Normalization

In our discussion of probability we noted that the dart has to hit the wall *somewhere*. The mathematical statement of this idea is the requirement that $P_A + P_B + P_C = 1$. That is, the probabilities of all the mutually exclusive outcomes *must* add up to 1.

Similarly, a photon or electron has to land *somewhere* on the detector after passing through an experimental apparatus. Consequently, the probability that it will be detected at *some* position is 100%. To make use of this requirement, consider an experiment in which an electron is detected on the x-axis. As Figure 39.7 shows, we can divide the region between position x_L and x_R into N adjacent narrow strips of width δx.

The probability that any particular electron lands in the narrow strip i at position x_i is

$$\text{Prob(in } \delta x \text{ at } x_i) = P(x_i)\,\delta x$$

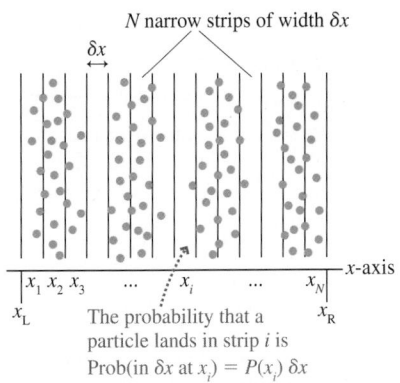

FIGURE 39.7 Dividing the entire detector into many small strips of width δx.

where $P(x_i) = |\psi(x_i)|^2$ is the probability density at x_i. The probability that the electron lands in the strip at x_1 or x_2 or x_3 or ... is the sum

$$\text{Prob(between } x_L \text{ and } x_R) = \text{Prob(in } \delta x \text{ at } x_1)$$
$$+ \text{Prob(in } \delta x \text{ at } x_2) + \cdots$$
$$= \sum_{i=1}^{N} P(x_i)\,\delta x = \sum_{i=1}^{N} |\psi(x_i)|^2 \delta x \qquad (39.16)$$

That is, **the probability that the electron lands *somewhere* between x_L and x_R is the sum of the probabilities of landing in each narrow strip.**

If we let the strips become narrower and narrower, then $\delta x \to dx$ and the sum becomes an integral. Thus the probability of finding the particles in the range $x_L \le x \le x_R$ is

$$\text{Prob(in range } x_L \le x \le x_R) = \int_{x_L}^{x_R} P(x)\,dx = \int_{x_L}^{x_R} |\psi(x)|^2 dx \qquad (39.17)$$

As Figure 39.8a shows, we can interpret Prob(in range $x_L \le x \le x_R$) as the area under the probability density curve between x_L and x_R.

NOTE ▶ The integral of Equation 39.17 is needed when the probability density changes over the range x_L to x_R. For sufficiently narrow intervals, over which $P(x)$ remains essentially constant, the expression Prob(in δx at x) = $P(x)\,\delta x$ is still valid and is easier to use. ◀

Now let the detector become infinitely wide, so that the probability that the electron will arrive *somewhere* on the detector becomes 100%. The statement that the electron has to land *somewhere* on the x-axis is expressed mathematically as

$$\int_{-\infty}^{\infty} P(x)\,dx = \int_{-\infty}^{\infty} |\psi(x)|^2 dx = 1 \qquad (39.18)$$

Equation 39.18 is called the **normalization condition.** Any wave function $\psi(x)$ must satisfy this condition; otherwise we would not be able to interpret $|\psi(x)|^2$ as a probability density. As Figure 39.8b shows, Equation 39.18 tells us that the total area under the probability density curve must be 1.

NOTE ▶ The normalization condition integrates the *square* of the wave function. We don't have any information about what the integral of $\psi(x)$ might be. ◀

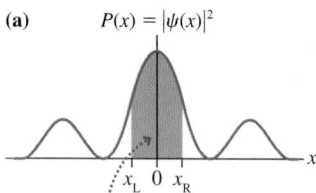

(a) $P(x) = |\psi(x)|^2$

The area under the curve between x_L and x_R is the probability of finding the particle between x_L and x_R.

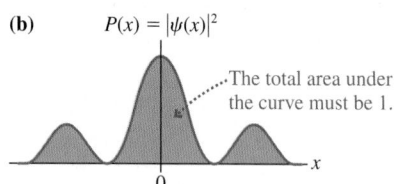

(b) $P(x) = |\psi(x)|^2$

The total area under the curve must be 1.

FIGURE 39.8 The area under the probability density curve is a probability.

EXAMPLE 39.2 Normalizing and interpreting a wave function

Figure 39.9 shows the wave function of a particle confined within the region between $x = 0$ nm and $x = L = 1.0$ nm. The wave function is zero outside this region.

a. Determine the value of the constant c.
b. Draw a graph of the probability density $P(x)$.
c. Draw a dot picture showing where the first 40 or 50 particles might be found.
d. Calculate the probability of finding the particle in a region of width $\delta x = 0.01$ nm at positions $x_1 = 0.05$ nm, $x_2 = 0.50$ nm, and $x_3 = 0.95$ nm.

MODEL The probability of finding the particle is determined by the probability density $P(x)$.

VISUALIZE The wave function is shown in Figure 39.9.

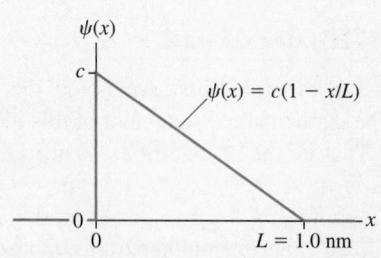

FIGURE 39.9 The wave function of Example 39.2.

SOLVE

a. The wave function is $\psi(x) = c(1 - x/L)$. This is a function that decreases linearly from $\psi = c$ at $x = 0$ to $\psi = 0$ at $x = L$. The constant c is the height of this wave function.

The particle *has* to be in the region $0 \le x \le L$ with probability 1, and only one value of c will make it so. We can determine c by using Equation 39.18, the normalization condition. Because the wave function is zero outside the interval from 0 to L, the integration limits are 0 to L. Thus

$$1 = \int_0^L |\psi(x)|^2 dx = c^2 \int_0^L \left(1 - \frac{x}{L}\right)^2 dx$$

$$= c^2 \int_0^L \left(1 - \frac{2x}{L} + \frac{x^2}{L^2}\right) dx$$

$$= c^2 \left[x - \frac{x^2}{L} + \frac{x^3}{3L^2}\right]_0^L = \frac{1}{3}c^2 L$$

The solution for c is

$$c = \sqrt{\frac{3}{L}} = \sqrt{\frac{3}{1 \text{ nm}}} = 1.732 \text{ nm}^{-1/2}$$

Note the unusual units for c. Although these are not SI units, we can correctly compute probabilities as long as δx has units of nm. A multiplicative constant such as c is often called a *normalization constant*.

b. The wave function is

$$\psi(x) = (1.732 \text{ nm}^{-1/2})\left(1 - \frac{x}{1.0 \text{ nm}}\right)$$

Thus the probability density is

$$P(x) = |\psi(x)|^2 = (3.0 \text{ nm}^{-1})\left(1 - \frac{x}{1.0 \text{ nm}}\right)^2$$

This probability density is graphed in Figure 39.10a.

c. Particles are most likely to be detected at the left edge of the interval, where the probability density $P(x)$ is maximum. The probability steadily decreases across the interval, becoming zero at $x = 1.0$ nm. Figure 39.10b shows how a group of particles described by this wave function might appear on a detection screen.

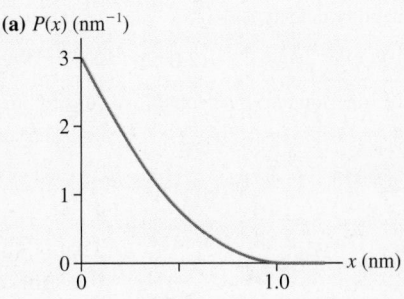

(b)

FIGURE 39.10 The probability density $P(x)$ and the detected positions of particles described by this probability density.

d. $P(x)$ is essentially constant over the small interval $\delta x = 0.01$ nm, so we can use

$$\text{Prob(in } \delta x \text{ at } x) = P(x)\delta x = |\psi(x)|^2 \delta x$$

for the probability of finding the particle in a region of width δx at the position x. We need to evaluate $|\psi(x)|^2$ at the three positions $x_1 = 0.05$ nm, $x_2 = 0.50$ nm, and $x_3 = 0.95$ nm. Doing so gives

$$\text{Prob(in 0.01 nm at } x_1 = 0.05 \text{ nm)} = c^2(1 - x_1/L)^2 \delta x$$
$$= 0.0270 = 2.70\%$$

$$\text{Prob(in 0.01 nm at } x_2 = 0.50 \text{ nm)} = c^2(1 - x_2/L)^2 \delta x$$
$$= 0.0075 = 0.75\%$$

$$\text{Prob(in 0.01 nm at } x_3 = 0.95 \text{ nm)} = c^2(1 - x_3/L)^2 \delta x$$
$$= 0.00008 = 0.008\%$$

EXAMPLE 39.3 The probability of finding a particle
A particle is described by the wave function

$$\psi(x) = \begin{cases} 0 & x < 0 \\ ce^{-x/L} & x \ge 0 \end{cases}$$

where $L = 1$ nm.

a. Determine the value of the constant c.
b. Draw graphs of $\psi(x)$ and the probability density $P(x)$.
c. Calculate the probability of finding the particle in the region $x \ge 1$ nm.

MODEL The probability of finding the particle is determined by the probability density $P(x)$.

SOLVE

a. The wave function is an exponential $\psi(x) = ce^{-x/L}$ that extends from $x = 0$ to $x = +\infty$. Equation 39.18, the normalization condition, is

$$1 = \int_{-\infty}^{\infty} |\psi(x)|^2 dx = c^2 \int_0^{\infty} e^{-2x/L} dx = -\frac{c^2 L}{2}e^{-2x/L}\Big|_0^{\infty} = \frac{c^2}{2L}$$

We can solve this for the normalization constant c:

$$c = \sqrt{\frac{2}{L}} = \sqrt{\frac{2}{1 \text{ nm}}} = 1.414 \text{ nm}^{-1/2}$$

b. The probability density is

$$P(x) = |\psi(x)|^2 = (2.0 \text{ nm}^{-1})e^{-2x/(1.0 \text{ nm})}$$

The wave function and the probability density are graphed in Figure 39.11.

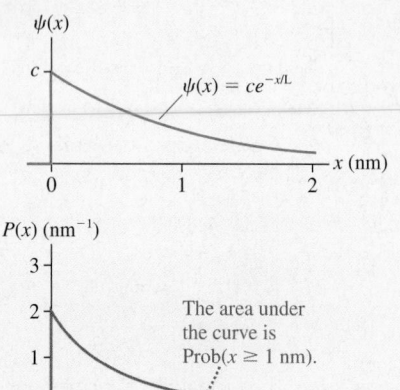

FIGURE 39.11 The wave function and probability density of Example 39.3.

c. The probability of finding the particle in the region $x \geq 1$ nm is the shaded area under the probability density curve in Figure 39.11. We must use Equation 39.17 and integrate to find a numerical value. The probability is

$$\text{Prob}(x \geq 1 \text{ nm}) = \int_{1 \text{ nm}}^{\infty} |\psi(x)|^2 dx$$

$$= (2.0 \text{ nm}^{-1})\int_{1 \text{ nm}}^{\infty} e^{-2x/(1.0 \text{ nm})} dx$$

$$= (2.0 \text{ nm}^{-1})\left(-\frac{1.0 \text{ nm}}{2}\right)e^{-2x/(1.0 \text{ nm})}\Big|_{1 \text{ nm}}^{\infty}$$

$$= e^{-2} = 0.135 = 13.5\%$$

ASSESS There is a 13.5% chance of finding the particle beyond 1 nm and thus an 86.5% chance of finding it within the interval $0 \leq x \leq 1$ nm. Unlike classical physics, we cannot make an exact prediction of the particle's position.

STOP TO THINK 39.4   The value of the constant a is

a. $a = 2.0 \text{ mm}^{-1}$.
b. $a = 1.0 \text{ mm}^{-1}$.
c. $a = 0.5 \text{ mm}^{-1}$.
d. $a = 2.0 \text{ mm}^{-1/2}$.
e. $a = 1.0 \text{ mm}^{-1/2}$.
f. $a = 0.5 \text{ mm}^{-1/2}$.

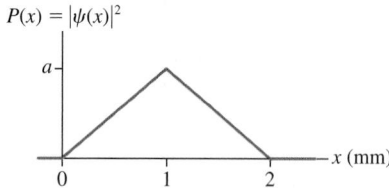

39.5 Wave Packets

The experimental evidence is overwhelming that light has particle-like characteristics and that matter has wave-like characteristics. These observations are completely at odds with the classical models of particles and waves that we used to study the physics of baseballs and sound waves. In particular, the classical ideas of particles and waves are mutually exclusive. An object can be one or the other, but not both. The classical models fail to describe the wave-particle duality seen at the atomic level. An alternative model with both particle and wave characteristics is a *wave packet*.

Consider the wave shown in Figure 39.12. Unlike the sinusoidal waves we have considered previously, which stretch through time and space, this wave is bunched up, or localized. The localization is a particle-like characteristic. The oscillations are wave-like. Such a localized wave is called a **wave packet.**

A wave packet travels through space with constant speed v, just like a photon in a light wave or an electron in a force-free region. A wave packet has a wavelength, hence it will undergo interference and diffraction. But because it is also localized, a wave packet has the possibility of making a "dot" when it strikes a detector. We can visualize a light wave as consisting of a very large number of these wave packets moving along together. Similarly, we can think of a beam of electrons as a series of wave packets spread out along a line.

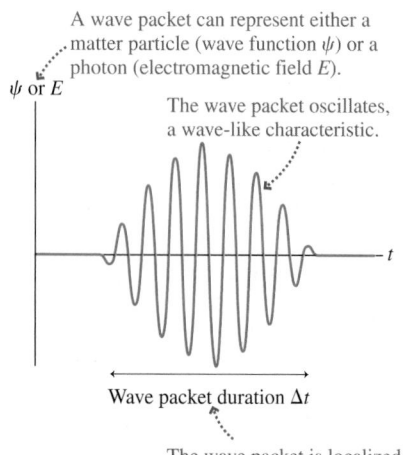

FIGURE 39.12 History graph of a wave packet with duration Δt.

Wave packets are not a perfect model of photons or electrons (we need the full treatment of quantum physics to get a more accurate description), but they do provide a useful way of thinking about photons and electrons in many circumstances.

You might have noticed that the wave packet in Figure 39.12 looks very much like one cycle of a beat pattern. You will recall that beats occur if we superimpose two waves of frequencies f_1 and f_2 where the two frequencies are very similar: $f_1 \approx f_2$. Figure 39.13, which is copied from Chapter 21 where we studied beats, shows that the loud, soft, loud, soft, . . . pattern of beats is a series of wave packets.

In Chapter 21, the beat frequency (number of pulses per second) was found to be

$$f_{\text{beat}} = f_1 - f_2 = \Delta f \qquad (39.19)$$

where Δf is the *range* of frequencies that are superimposed to form the wave packet. Figure 39.13 defines Δt as the duration of each beat or each wave packet. This interval of time is equivalent to the *period* T_{beat} of the beat. Because period and frequency are inverses of each other, the duration Δt is

$$\Delta t = T_{\text{beat}} = \frac{1}{f_{\text{beat}}} = \frac{1}{\Delta f}$$

We can rewrite this as

$$\Delta f \Delta t = 1 \qquad (39.20)$$

Equation 39.20 is nothing new; we are simply writing what we already knew in a different form. Equation 39.20 is a combination of three things: the relationship $f = 1/T$ between period and frequency, writing T_{beat} as Δt, and the specific knowledge that the beat frequency f_{beat} is the difference Δf of the two frequencies contributing to the wave packet. As the frequency separation gets smaller, the duration of each beat gets longer.

When we superimpose two frequencies to create beats, the wave packet repeats over and over. A more advanced treatment of waves, called Fourier analysis, reveals that a single, *nonrepeating* wave packet can be created through superposition of *many* waves of very similar frequency. Figure 39.14 illustrates this idea. At one instant of time, all the waves interfere constructively to produce the maximum amplitude of the wave packet. At other times, the individual waves get out of phase and their superposition tends toward zero.

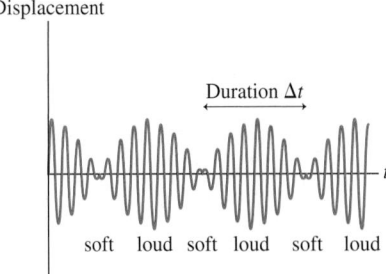

FIGURE 39.13 Beats are a series of wave packets.

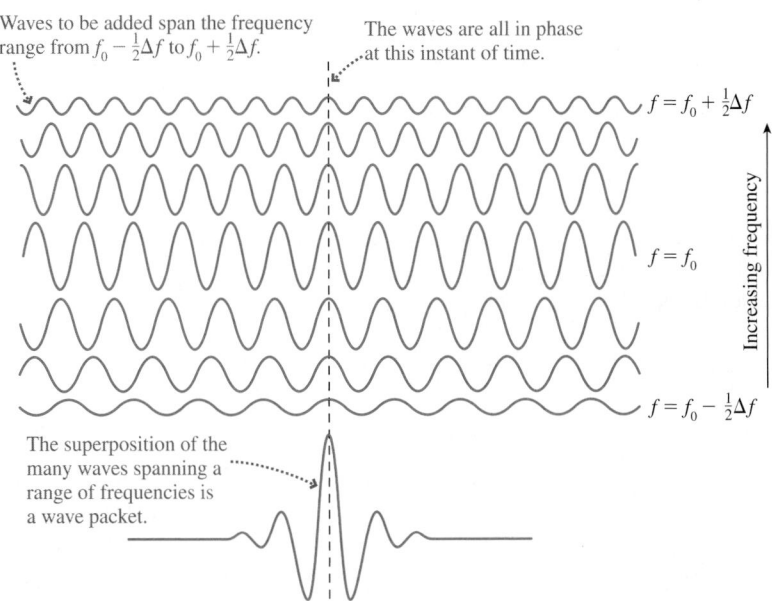

FIGURE 39.14 A single wave packet is the superposition of many component waves of similar wavelength and frequency.

Suppose a single nonrepeating wave packet of duration Δt is created by the superposition of *many* waves that span a range of frequencies Δf. We'll not prove it, but Fourier analysis shows that for *any* wave packet

$$\Delta f \Delta t \approx 1 \qquad\qquad (39.21)$$

The relationship between Δf and Δt for a general wave packet is not as precise as Equation 39.20 for beats. There are two reasons for this:

1. Wave packets come in a variety of shapes. The exact relationship between Δf and Δt depends somewhat on the shape of the wave packet.
2. We have not given a precise definition of Δt and Δf for a general wave packet. The quantity Δt is "about how long the wave packet lasts," while Δf is "about the range of frequencies needing to be superimposed to produce this wave packet." For our purposes, we will not need to be any more precise than this.

Equation 39.21 is a purely classical result that applies to waves of any kind. It tells you the range of frequencies you need to superimpose to construct a wave packet of duration Δt. Alternatively, Equation 39.21 tells you that a wave packet created as a superposition of various frequencies cannot be arbitrarily short but *must* last for a time interval $\Delta t \approx 1/\Delta f$.

EXAMPLE 39.4 Creating radio-frequency pulses

A short-wave radio station broadcasts at a frequency of 10.000 MHz. What is the range of frequencies of the waves that must be superimposed to broadcast a radio-wave pulse lasting 0.800 μs?

MODEL A pulse of radio waves is an electromagnetic wave packet, hence it must satisfy the relationship $\Delta f \Delta t \approx 1$.

VISUALIZE Figure 39.15 shows the pulse.

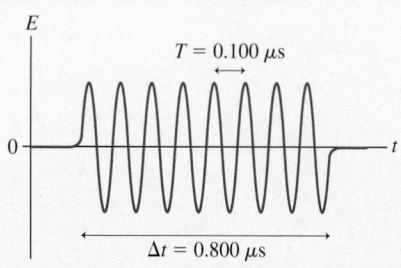

FIGURE 39.15 A pulse of radio waves.

SOLVE The period of a 10.000 MHz oscillation is 0.100 μs. A pulse 0.800 μs in duration is 8 oscillations of the wave. Although the station broadcasts at a nominal frequency of 10.000 MHz, this pulse is not a pure 10.000 MHz oscillation. Instead, the pulse has been created by the superposition of many waves whose frequencies span

$$\Delta f \approx \frac{1}{\Delta t} = \frac{1}{0.800 \times 10^{-6}\,\text{s}} = 1.250 \times 10^{6}\,\text{Hz} = 1.250\,\text{MHz}$$

This range of frequencies will be centered at the 10.000 MHz broadcast frequency, so the waves that must be superimposed to create this pulse span the frequency range

$$9.375\,\text{MHz} \le f \le 10.625\,\text{MHz}$$

Bandwidth

Short-duration pulses, like the one in Example 39.4, are used to transmit digital information. Digital signals are sent over a phone line by brief tone pulses, over satellite links by brief radio pulses like the one in the example, and through optical fibers by brief laser-light pulses. Regardless of the type of wave and the medium through which it travels, any wave pulse must obey the fundamental relationship $\Delta f \Delta t \approx 1$.

Sending data at a higher rate (i.e., more pulses per second) requires that the pulse duration Δt be shorter. But a shorter-duration pulse must be created by the superposition of a *larger* range of frequencies. Thus the medium through which a shorter-duration pulse travels must be physically able to transmit the full range of frequencies.

The range of frequencies that can be transmitted through a medium is called the **bandwidth** Δf_B of the medium. The shortest possible pulse that can be transmitted through a medium is

$$\Delta t_{min} \approx \frac{1}{\Delta f_B} \qquad (39.22)$$

A pulse shorter than this would require a larger range of frequencies than the medium can support.

The concept of bandwidth is extremely important in digital communications. A higher bandwidth transmits shorter pulses and allows a higher data rate. A standard telephone line does not have a very high bandwidth, and that is why a modem is limited to sending data at the rate of roughly 50,000 pulses per second. A 0.80 μs pulse can't be sent over a phone line simply because the phone line won't transmit the range of frequencies that would be needed.

An optical fiber is a high-bandwidth medium. A fiber has a bandwidth $\Delta f_B > 1$ GHz and thus can transmit laser-light pulses with duration $\Delta t < 1$ ns. More than 10^9 pulses per second can be sent along an optical fiber, which is why optical-fiber networks form the backbone of the Internet.

Uncertainty

There is another way of thinking about the time-frequency relationship $\Delta f \Delta t \approx 1$. Suppose you want to determine *when* a wave packet arrives at a specific point in space, such as at a detector of some sort. At what instant of time can you say that the wave packet is detected? When the front edge arrives? When the maximum amplitude arrives? When the back edge arrives? Because a wave packet is spread out in time, there is not a unique and well-defined time t at which the packet arrives. All we can say is that it arrives within some interval of time Δt. We are *uncertain* about the exact arrival time.

Similarly, suppose you would like to know the oscillation frequency of a wave packet. There is no precise value for f because the wave packet is constructed from many waves within a range of frequencies Δf. All we can say is that the frequency is within this range. We are *uncertain* about the exact frequency.

The time-frequency relationship $\Delta f \Delta t \approx 1$ tells us that the uncertainty in our knowledge about the arrival time of the wave packet is related to our uncertainty about the packet's frequency. The more precisely and accurately we know one quantity, the less precisely we will be able to know the other.

Figure 39.16 shows two different wave packets. The wave packet of Figure 39.16a is very narrow and thus very localized in time. As it travels, our knowledge of when it will arrive at a specified point is fairly precise. But a very wide range of frequencies Δf is required to create a wave packet with a very small Δt. The price we pay for being fairly certain about the time is a very large uncertainty Δf about the frequency of this wave packet.

Figure 39.16b shows the opposite situation: The wave packet oscillates many times and the frequency of these oscillations is pretty clear. Our knowledge of the frequency is good, with minimal uncertainty Δf. But such a wave packet is so spread out that there is a very large uncertainty Δt as to its time of arrival.

In practice, $\Delta f \Delta t \approx 1$ is really a lower limit. Technical limitations may cause the uncertainties in our knowledge of f and t to be even larger than this relationship implies. Consequently, a better statement about our knowledge of a wave packet is

$$\Delta f \Delta t \geq 1 \qquad (39.23)$$

The fact that waves are spread out makes it meaningless to specify an exact frequency and an exact arrival time simultaneously. This is an inherent feature of waviness that applies to all waves.

(a)

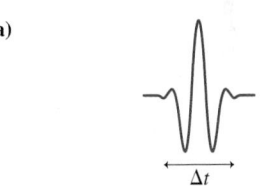

This wave packet has a large frequency uncertainty Δf.

(b)

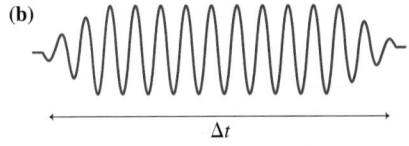

This wave packet has a small frequency uncertainty Δf.

FIGURE 39.16 Two wave packets with different Δt.

STOP TO THINK 39.5 What minimum bandwidth must a medium have to transmit a 100-ns-long pulse?

a. 1 MHz   b. 10 MHz   c. 100 MHz   d. 1000 MHz

39.6 The Heisenberg Uncertainty Principle

17.6, 17.7

If matter has wave-like aspects and a de Broglie wavelength, then the expression $\Delta f \Delta t \geq 1$ must somehow apply to matter. How? And what are the implications?

Consider a particle with velocity v_x as it travels along the x-axis with de Broglie wavelength $\lambda = h/p_x$. Figure 39.12 showed a *history graph* (ψ versus t) of a wave packet that might represent the particle as it passes a point on the x-axis. It will be more useful to have a *snapshot graph* (ψ versus x) of the wave packet traveling along the x-axis.

The time interval Δt is the duration of the wave packet as the particle passes a point in space. During this interval, the packet moves forward

$$\Delta x = v_x \Delta t = \frac{p_x}{m} \Delta t \qquad (39.24)$$

where $p_x = mv_x$ is the x-component of the particle's momentum. The quantity Δx, shown in Figure 39.17, is the length or spatial extent of the wave packet. Conversely, we can write the wave packet's duration in terms of its length as

$$\Delta t = \frac{m}{p_x} \Delta x \qquad (39.25)$$

You will recall that any wave with sinusoidal oscillations must satisfy the wave condition $\lambda f = v$. For a material particle, where λ is the de Broglie wavelength, the frequency f is

$$f = \frac{v}{\lambda} = \frac{(p_x/m)}{(h/p_x)} = \frac{p_x^2}{hm}$$

A small range of frequencies Δf is related to a small range of momenta Δp_x by

$$\Delta f = \frac{2p_x \Delta p_x}{hm} \qquad (39.26)$$

where we have assumed that $\Delta f \ll f$ and $\Delta p_x \ll p_x$ (a reasonable assumption) and thus treated the small ranges Δf and Δp_x as if they were differentials df and dp_x.

Multiplying together these expressions for Δt and Δf, we find that

$$\Delta f \Delta t = \frac{2p_x \Delta p_x}{hm} \frac{m \Delta x}{p_x} = \frac{2}{h} \Delta x \Delta p_x \qquad (39.27)$$

Because $\Delta f \Delta t \geq 1$ for any wave, one last rearrangement of Equation 39.27 shows that a matter wave must obey the condition

$$\Delta x \Delta p_x \geq \frac{h}{2} \qquad \text{(Heisenberg uncertainty principle)} \qquad (39.28)$$

This statement about the relationship between the position and momentum of a particle was proposed by Werner Heisenberg, creator of one of the first successful quantum theories. Physicists often just call it the **uncertainty principle.**

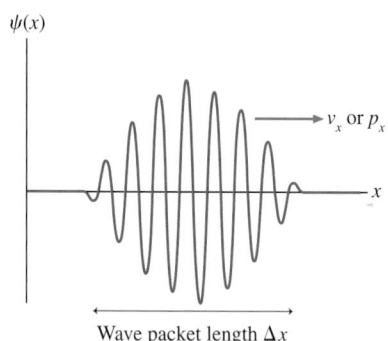

$\psi(x)$

v_x or p_x

Wave packet length Δx

FIGURE 39.17 A snapshot graph of a wave packet.

NOTE ▶ In statements of the uncertainty principle, the right side is sometimes $h/2$, as we have it, but other times it is just h or contains various factors of π. The specific number is not especially important because it depends on exactly how Δx and Δp are defined. The important idea is that the product of Δx and Δp_x for a particle cannot be significantly less than Planck's constant h. A similar relationship for $\Delta y \Delta p_y$ applies along the y-axis. ◀

What Does It Mean?

Heisenberg's uncertainty principle is a statement about our *knowledge* of the properties of a particle. If we want to know *where* a particle is located, we measure its position x. That measurement is not absolutely perfect, but has some uncertainty Δx. Likewise, if we want to know *how fast* the particle is going we need to measure its velocity v_x or, equivalently, its momentum p_x. This measurement also has some uncertainty Δp_x.

Uncertainties are associated with all experimental measurements, but better procedures and techniques can reduce those uncertainties. Newtonian physics places no limits on how small the uncertainties can be. A Newtonian particle at any instant of time has an exact position x and an exact momentum p_x, and with sufficient care we can measure both x and p_x with such precision that the product $\Delta x \Delta p_x \to 0$. There are no inherent limits to our knowledge about a classical, or Newtonian, particle.

Heisenberg, however, made the bold and original statement that our knowledge has real limitations. No matter how clever you are, and no matter how good your experiment, you *cannot* measure both x and p_x simultaneously with arbitrarily good precision. Any measurements you make are limited by the condition that $\Delta x \Delta p_x \geq h/2$. **Our knowledge about a particle is *inherently* uncertain.**

Why? Because of the wave-like nature of matter. The "particle" is spread out in space, so there simply is not a precise value of its position x. Similarly, the de Broglie relationship between momentum and wavelength implies that we cannot know the momentum of a wave packet any more exactly than we can know its wavelength or frequency. Our belief that position and momentum have precise values is tied to our classical concept of a particle. As we revise our ideas of what atomic particles are like, we will also have to revise our old ideas about position and momentum.

EXAMPLE 39.5 The uncertainty of a dust particle
A 1.0-μm-diameter dust particle ($m \approx 10^{-15}$ kg) is confined within a 10-μm-long box. Can we know with certainty if the particle is at rest? If not, within what range is its velocity likely to be found?

MODEL All matter is subject to the Heisenberg uncertainty principle.

SOLVE If we know *for sure* that the particle is at rest, then $p_x = 0$ with no uncertainty. That is, $\Delta p_x = 0$. But then, according to the uncertainty principle, the uncertainty in our knowledge of the particle's position would have to be $\Delta x \to \infty$. In other words, we would have no knowledge at all about the particle's position—it could be anywhere! But that is not the case. We know the particle is *somewhere* in the box, so the uncertainty in our knowledge of its position is at most $\Delta x = L = 10\ \mu$m. With a finite Δx, the uncertainty Δp_x *cannot* be zero. We cannot know with certainty if the particle is at rest inside the box. No matter how hard we try to bring the particle to rest, the

uncertainty in our knowledge of the particle's momentum will be $\Delta p_x \approx h/(2\Delta x) = h/2L$. We've assumed the most accurate measurements possible so that the $\geq$ in Heisenberg's uncertainty principle becomes $\approx$. Consequently, the range of possible velocities is

$$\Delta v_x = \frac{\Delta p_x}{m} \approx \frac{h}{2mL} \approx 3.0 \times 10^{-14}\ \text{m/s}$$

This range of possible velocities will be centered on $v_x = 0$ m/s if we have done our best to have the particle be at rest. Thus all we can know with certainty is that the particle's velocity is somewhere within the interval -1.5×10^{-14} m/s $\leq v \leq 1.5 \times 10^{-14}$ m/s.

ASSESS For practical purposes you might consider this to be a satisfactory definition of "at rest." After all, a particle moving with a speed of 1.5×10^{-14} m/s would need 6×10^{10} s to move a mere 1 mm. That is about 2000 years! Nonetheless, we can't know if the particle is "really" at rest.

EXAMPLE 39.6 The uncertainty of an electron

What range of velocities might an electron have if confined to a 0.10-nm-wide region, about the size of an atom?

MODEL Electrons are subject to the Heisenberg uncertainty principle.

SOLVE The analysis is the same as in Example 39.5. If we know that the electron's position is located within an interval $\Delta x \approx 0.1$ nm, then the best we can know is that its velocity is within the range

$$\Delta v_x = \frac{\Delta p_x}{m} \approx \frac{h}{2mL} \approx 4 \times 10^6 \text{ m/s}$$

Because the *average* velocity is zero, the best we can say is that the electron's velocity is somewhere in the interval -2×10^6 m/s $\leq v \leq 2 \times 10^6$ m/s. It is simply not possible to know the electron's velocity any more precisely than this.

ASSESS Unlike the situation in Example 39.5, where Δv was so small as to be of no practical consequence, our uncertainty about the electron's velocity is enormous—about 1% of the speed of light!

Once again, we see that even the smallest of macroscopic objects behaves very much like a classical Newtonian particle. Perhaps a 1-μm-diameter particle is slightly fuzzy and has a slightly uncertain velocity, but it is far beyond the measuring capabilities of even the very best instruments to detect this wave-like behavior. In contrast, the effects of the uncertainty principle at the atomic scale are stupendous. We are unable to determine the velocity of an electron in an atom-size container to any better accuracy than about 1% of the speed of light.

STOP TO THINK 39.6 Which of these particles, A or B, can you locate more precisely?

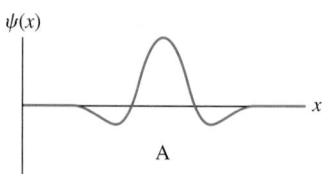

A

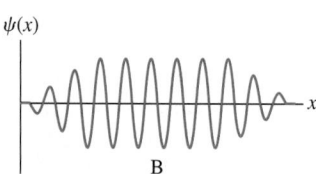

B

SUMMARY

The goal of Chapter 39 has been to introduce the wave-function description of matter and learn how it is interpreted.

GENERAL PRINCIPLES

Wave Functions and the Probability Density

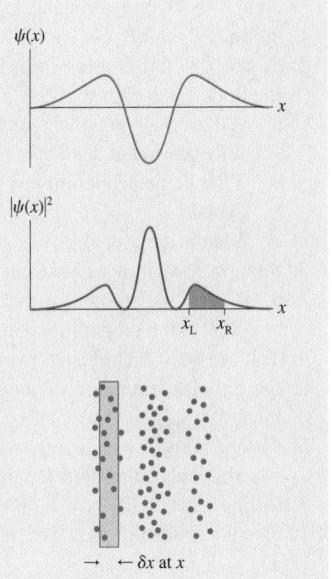

We cannot predict the exact trajectory of an atomic-level particle such as an electron. The best we can do is to predict the **probability** that a particle will be found in some region of space. The probability is determined by the particle's wave function $\psi(x)$.

- $\psi(x)$ is a continuous, wave-like (i.e., oscillatory) function.

- The probability that a particle will be found in the narrow interval δx at position x is
  $\text{Prob}(\text{in } \delta x \text{ at } x) = |\psi(x)|^2 \, \delta x$.

- $|\psi(x)|^2$ is the probability density $P(x)$.

- For the probability interpretation of $\psi(x)$ to make sense, the wave function must satisfy the normalization condition

$$\int_{-\infty}^{\infty} P(x)\,dx = \int_{-\infty}^{\infty} |\psi(x)|^2 \, dx = 1$$

That is, it is certain that the particle is *somewhere* on the x-axis.

- For an extended interval

$$\text{Prob}(x_{\text{L}} \le x \le x_{\text{R}}) = \int_{x_{\text{L}}}^{x_{\text{R}}} |\psi(x)|^2 \, dx = \text{area under the curve}$$

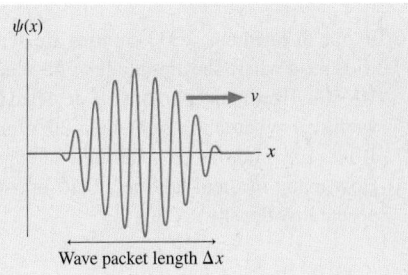

Heisenberg Uncertainty Principle

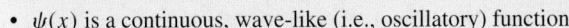

A particle with wave-like characteristics does not have a precise value of position x or a precise value of momentum p_x. Both are uncertain. The position uncertainty Δx and momentum uncertainty Δp_x are related by $\Delta x \, \Delta p_x \ge h/2$. The more you try to pin down the value of one, the less precisely the other can be known.

IMPORTANT CONCEPTS

The probability that a particle is found in region A is

$$P_{\text{A}} = \lim_{N_{\text{tot}} \to \infty} \frac{N_{\text{A}}}{N_{\text{tot}}}$$

If the probability is known, the expected number of A outcomes in N trials is $N_{\text{A}} = N P_{\text{A}}$.

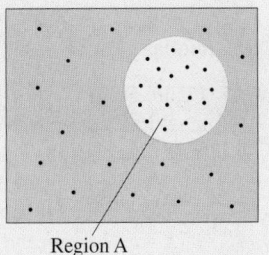

Region A

A wave packet of duration Δt can be created by the superposition of many waves spanning the frequency range Δf. These are related by

$$\Delta f \, \Delta t \approx 1$$

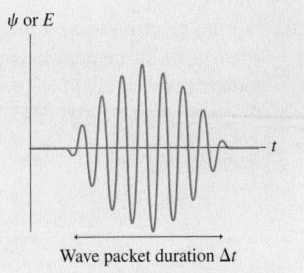

Wave packet duration Δt

TERMS AND NOTATION

quantum mechanics	wave function, $\psi(x)$	bandwidth, Δf_{B}
probability	normalization condition	uncertainty principle
probability density, $P(x)$	wave packet	

EXERCISES AND PROBLEMS

Exercises

Section 39.1 Waves, Particles, and the Double-Slit Experiment

1. An experiment has four possible outcomes, labeled A to D. The probability of A is $P_A = 40\%$ and of B is $P_B = 30\%$. Outcome C is twice as probable as outcome D. What are the probabilities P_C and P_D?

2. Suppose you toss three coins into the air and let them fall on the floor. Each coin shows either a head or a tail.
   a. Make a table in which you list all the possible outcomes of this experiment. Call the coins A, B, and C.
   b. What is the probability of getting two heads and one tail? Explain.
   c. What is the probability of getting *at least* two heads?

3. Suppose you draw a card from a regular deck of 52 cards.
   a. What is the probability that you draw an ace?
   b. What is the probability that you draw a spade?

4. You are dealt 1 card each from 1000 decks of cards. What is the expected number of picture cards (jacks, queens, and kings)?

5. Make a table in which you list all possible outcomes of rolling two dice. Call the dice A and B. What is the probability of rolling (a) a 7, (b) any double, and (c) a 6 or an 8? You can give the probabilities as fractions, such as 3/36.

Section 39.2 Connecting the Wave and Photon Views

6. In one experiment, 2000 photons are detected in a 0.10-mm-wide strip where the amplitude of the electromagnetic wave is 10 V/m. How many photons are detected in a nearby 0.10-mm-wide strip where the amplitude is 30 V/m?

7. 1.0×10^{10} photons pass through an experimental apparatus. How many of them land in a 0.10-mm-wide strip where the probability density is $20\ \text{m}^{-1}$?

Section 39.3 The Wave Function

8. What are the units of ψ? Explain.

9. What is the difference between the probability and the probability density?

10. For the electron wave function shown in Figure Ex39.10, at what position or positions is the electron most likely to be found? Least likely to be found? Explain.

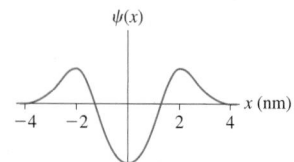

FIGURE EX39.10

11. Figure Ex39.11 shows the probability density for an electron that has passed through an experimental apparatus. If 1.0×10^6 electrons are used, what is the expected number that will land in a 0.010-mm-wide strip at (a) $x = 0.000$ mm and (b) $x = 2.000$ mm?

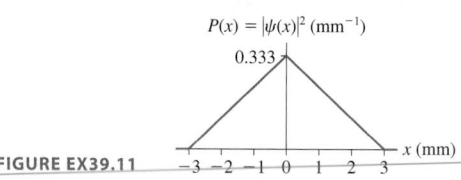

FIGURE EX39.11

12. In an interference experiment with electrons, you find the most intense fringe is at $x = 7.0$ cm. There are slightly weaker fringes at $x = 6.0$ and 8.0 cm, still weaker fringes at $x = 4.0$ and 10.0 cm, and two very weak fringes at $x = 1.0$ and 13.0 cm. No electrons are detected at $x < 0$ cm or $x > 14$ cm.
   a. Sketch a graph of $|\psi(x)|^2$ for these electrons.
   b. Sketch a possible graph of $\psi(x)$.
   c. Are there other possible graphs for $\psi(x)$? If so, draw one.

13. Figure Ex39.13 shows the probability density for an electron that has passed through an experimental apparatus. What is the probability that the electron will land in a 0.010-mm-wide strip at (a) $x = 0.000$ mm, (b) $x = 0.500$ mm, (c) $x = 1.000$ mm, and (d) $x = 2.000$ mm?

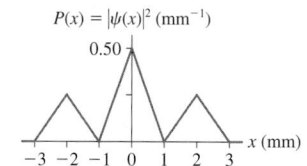

FIGURE EX39.13

Section 39.4 Normalization

14. Figure Ex39.14 is a graph of $|\psi(x)|^2$ for an electron.
   a. What is the value of a?
   b. Draw a graph of the wave function $\psi(x)$. (There is more than one acceptable answer.)
   c. What is the probability that the electron is located between $x = 1.0$ nm and $x = 2.0$ nm?

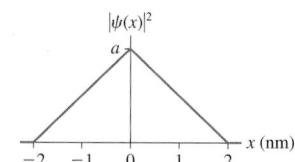

FIGURE EX39.14

15. Figure Ex39.15 is a graph of $|\psi(x)|^2$ for a neutron.
   a. What is the value of a?
   b. Draw a graph of the wave function $\psi(x)$. (There is more than one acceptable answer.)
   c. What is the probability that the neutron is located between $x = -2.0$ mm and $x = 2.0$ mm?

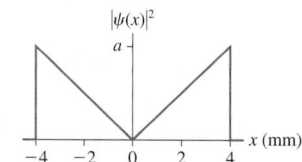

FIGURE EX39.15

16. Figure Ex39.16 shows the wave function of an electron.
    a. What is the value of c?
    b. Draw a graph of $|\psi(x)|^2$.
    c. What is the probability that the electron is located between $x = -1.0$ nm and $x = 1.0$ nm?

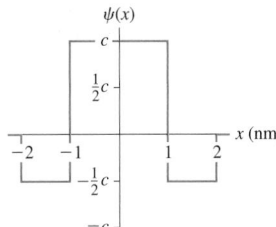

FIGURE EX39.16

17. Figure Ex39.17 shows the wave function of a neutron.
    a. What is the value of c?
    b. Draw a graph of $|\psi(x)|^2$.
    c. What is the probability that the neutron is located between $x = -1.0$ mm and $x = 1.0$ mm?

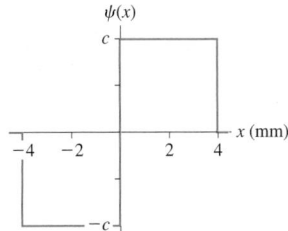

FIGURE EX39.17

Section 39.5 Wave Packets

18. A radar antenna broadcasts electromagnetic waves with a period of 0.100 ns. What range of frequencies would need to be superimposed to create a 1.0-ns-long radar pulse?

19. A radio-frequency amplifier is designed to amplify signals in the frequency range 80 MHz to 120 MHz. What is the smallest duration radio-frequency pulse that can be amplified without distortion?

20. What minimum bandwidth is needed to transmit a pulse that consists of 100 cycles of a 1.00 MHz oscillation?

21. A 1.5-μm-wavelength laser pulse is transmitted through a 2.0-GHz-bandwidth optical fiber. How many oscillations are in the shortest-duration laser pulse that can travel through the fiber?

Section 39.6 The Heisenberg Uncertainty Principle

22. Andrea thinks she's sitting at rest in her 5.0-m-long dorm room as she does her physics homework. Can Andrea be sure she's at rest? If not, within what range is her velocity likely to be?

23. What is the smallest box in which you can confine an electron if you want to know for certain that the electron's speed is no more than 10 m/s?

24. A thin solid barrier in the xy-plane has a 10-μm-diameter circular hole. An electron traveling in the z-direction with $v_x = 0$ m/s passes through the hole. Afterward, is v_x still zero? If not, within what range is v_x likely to be?

25. A proton is confined within an atomic nucleus of diameter 4.0 fm. Estimate the smallest range of speeds you might find for a proton in the nucleus.

Problems

26. A 1.0-mm-diameter sphere bounces back and forth between two walls at $x = 0$ mm and $x = 100$ mm. The collisions are perfectly elastic, and the sphere repeats this motion over and over with no loss of speed. At a random instant of time, what is the probability that the center of the sphere is
    a. At exactly $x = 50.0$ mm?
    b. Between $x = 49.0$ mm and $x = 51.0$ mm?
    c. At $x \geq 75$ mm?

27. Sound waves of 498 Hz and 502 Hz are superimposed at a temperature where the speed of sound in air is 340 m/s. What is the length Δx of one wave packet?

28. Ultrasound pulses of with a frequency of 1.000 MHz are transmitted into water, where the speed of sound is 1500 m/s. The spatial length of each pulse is 12 mm.
    a. How many complete cycles are contained in one pulse?
    b. What range of frequencies must be superimposed to create each pulse?

29. Figure P39.29 shows a *pulse train*. The period of the pulse train is $T = 2\Delta t$, where Δt is the duration of each pulse. What is the maximum pulse-transmission rate (pulses per second) through an electronics system with a 200 kHz bandwidth? (This is the bandwidth allotted to each FM radio station.)

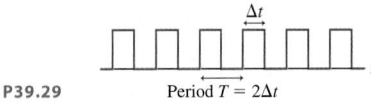

FIGURE P39.29   Period $T = 2\Delta t$

30. Consider a single-slit diffraction experiment using electrons. (Single-slit diffraction was described in Section 22.4.) Using Figure 39.5 as a model, draw
    a. A dot picture showing the arrival positions of the first 40 or 50 electrons.
    b. A graph of $|\psi(x)|^2$ for the electrons on the detection screen.
    c. A graph of $\psi(x)$ for the electrons. Keep in mind that ψ, as a wave-like function, oscillates between positive and negative.

31. An experiment finds electrons to be uniformly distributed over the interval 0 cm $\leq x \leq$ 2 cm, with no electrons falling outside this interval.
    a. Draw a graph of $|\psi(x)|^2$ for these electrons.
    b. What is the probability that an electron will land within the interval 0.79 to 0.81 cm?
    c. If 10^6 electrons are detected, how many will be detected in the interval 0.79 to 0.81 cm?
    d. What is the probability density at $x = 0.80$ cm?

32. In an experiment with 10,000 electrons, which land symmetrically on both sides of $x = 0$, 5000 are detected in the range -1.0 cm $\leq x \leq +1.0$ cm, 7500 are detected in the range -2.0 cm $\leq x \leq +2.0$ cm, and all 10,000 are detected in the range -3.0 cm $\leq x \leq +3.0$ cm. Draw a graph of a probability density that is consistent with these data. (There may be more than one acceptable answer.)

33. Figure P39.33 shows $|\psi(x)|^2$ for the electrons in an experiment.
    a. Is the electron wave function normalized? Explain.
    b. Draw a graph of $\psi(x)$ over this same interval. Provide a numerical scale on both axes. (There may be more than one acceptable answer.)
    c. What is the probability that an electron will be detected in a 0.0010-cm-wide region at $x = 0.00$ cm? At $x = 0.50$ cm? At $x = 0.999$ cm?
    d. If 10^4 electrons are detected, how many are expected to land in the interval -0.30 cm $\leq x \leq 0.30$ cm?

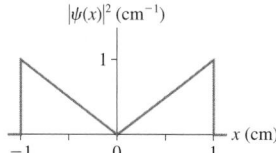

FIGURE P39.33

34. Figure P39.34 shows the wave function of a particle confined between $x = 0$ nm and $x = 1$ nm. The wave function is zero outside this region.
    a. Determine the value of the constant c, as defined in the figure.
    b. Draw a graph of the probability density $P(x) = |\psi(x)|^2$.
    c. Draw a dot picture showing where the first 40 or 50 particles might be found.
    d. Calculate the probability of finding the particle in the interval 0.0 nm $\leq x \leq 0.3$ nm.

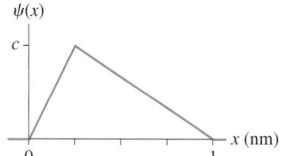

FIGURE P39.34

35. Figure P39.35 shows the wave function of a particle confined between $x = -4$ mm and $x = 4$ mm. The wave function is zero outside this region.
    a. Determine the value of the constant c, as defined in the figure.
    b. Draw a graph of the probability density $P(x) = |\psi(x)|^2$.
    c. Draw a dot picture showing where the first 40 or 50 particles might be found.
    d. Calculate the probability of finding the particle in the interval -2.0 mm $\leq x \leq 2.0$ mm.

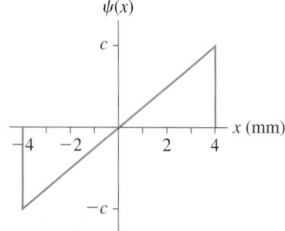

FIGURE P39.35

36. Figure P39.36 shows the probability density for finding a particle at position x.
    a. Determine the value of the constant a, as defined in the figure.
    b. At what value of x are you most likely to find the particle? Explain.
    c. Within what range of positions centered on your answer to part b are you 75% certain of finding the particle?
    d. Interpret your answer to part c by drawing the probability density graph and shading the appropriate region.

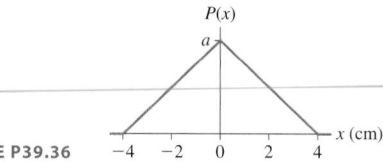

FIGURE P39.36

37. An electron that is confined to $x \geq 0$ nm has the normalized wave function

$$\psi(x) = \begin{cases} 0 & x < 0 \text{ nm} \\ (1.414 \text{ nm}^{-1/2})e^{-x/(1.0 \text{ nm})} & x \geq 0 \text{ nm} \end{cases}$$

where x is in nm.
    a. What is the probability of finding the electron in a 0.010-nm-wide region at $x = 1.0$ nm?
    b. What is the probability of finding the electron in the interval 0.50 nm $\leq x \leq 1.50$ nm?

38. A particle is described by the wave function

$$\psi(x) = \begin{cases} ce^{x/L} & x \leq 0 \text{ mm} \\ ce^{-x/L} & x \geq 0 \text{ mm} \end{cases}$$

where $L = 2.0$ mm.
    a. Sketch graphs of both the wave function and the probability density as functions of x.
    b. Determine the normalization constant c.
    c. Calculate the probability of finding the particle within 1.0 mm of the origin.
    d. Interpret your answer to part b by shading the region representing this probability on the appropriate graph in part a.

39. Consider the electron wave function

$$\psi(x) = \begin{cases} c\sqrt{1 - x^2} & |x| \leq 1 \text{ cm} \\ 0 & |x| \geq 1 \text{ cm} \end{cases}$$

where x is in cm.
    a. Determine the normalization constant c.
    b. Draw a graph of $\psi(x)$ over the interval -2 cm $\leq x \leq 2$ cm. Provide numerical scales on both axes.
    c. Draw a graph of $|\psi(x)|^2$ over the interval -2 cm $\leq x \leq 2$ cm. Provide numerical scales.
    d. If 10^4 electrons are detected, how many will be in the interval 0.00 cm $\leq x \leq 0.50$ cm?

40. Consider the electron wave function

$$\psi(x) = \begin{cases} c\sin\left(\dfrac{2\pi x}{L}\right) & 0 \leq x \leq L \\ 0 & x < 0 \text{ or } x > L \end{cases}$$

    a. Determine the normalization constant c. Your answer will be in terms of L.
    b. Draw a graph of $\psi(x)$ over the interval $-L \leq x \leq 2L$.
    c. Draw a graph of $|\psi(x)|^2$ over the interval $-L \leq x \leq 2L$.
    d. What is the probability that an electron is in the interval $0 \leq x \leq L/3$?

41. The probability density for finding a particle at position x is

$$P(x) = \begin{cases} \dfrac{a}{(1-x)} & -1 \text{ mm} \le x < 0 \text{ mm} \\ b(1-x) & 0 \text{ mm} \le x \le 1 \text{ mm} \end{cases}$$

and zero elsewhere.

   a. You will learn in Chapter 40 that the wave function must be a *continuous* function. Assuming that to be the case, what can you conclude about the relationship between a and b?
   b. Draw a graph of the probability density over the interval $-2 \text{ mm} \le x \le 2 \text{ mm}$.
   c. Determine values for a and b.
   d. What is the probability that the particle will be found to the left of the origin?

42. A pulse of light is created by the superposition of many waves that span the frequency range $f_0 - \frac{1}{2}\Delta f \le f \le f_0 + \frac{1}{2}\Delta f$, where $f_0 = c/\lambda$ is called the *center frequency* of the pulse. Laser technology can generate a pulse of light that has a wavelength of 600 nm and lasts a mere 6.0 fs (1 fs = 1 femtosecond $= 10^{-15}$ s).

   a. What is the center frequency of this pulse of light?
   b. How many cycles, or oscillations, of the light wave are completed during the 6.0 fs pulse?
   c. What range of frequencies must be superimposed to create this pulse?
   d. What is the spatial length of the laser pulse as it travels through space?
   e. Draw a snapshot graph of this wave packet.

43. A small speck of dust with mass 1.0×10^{-13} g has fallen into the hole shown in Figure P39.43 and appears to be at rest. According to the uncertainty principle, could this particle have enough energy to get out of the hole? If not, what is the deepest hole of this width from which it would have a good chance to escape?

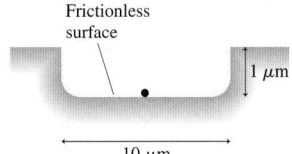

Frictionless surface

1 μm

FIGURE P39.43

10 μm

44. Physicists use laser beams to create an *atom trap* in which atoms are confined within a spherical region of space with a diameter of about 1 mm. The scientists have been able to cool the atoms in an atom trap to a temperature of approximately 1 nK, which is extremely close to absolute zero, but it would be interesting to know if this temperature is close to any limit set by quantum physics. We can explore this issue with a one-dimensional model of a sodium atom in a 1-mm-long box.

   a. Estimate the *smallest* range of speeds you might find for a sodium atom in this box.
   b. Even if we do our best to bring a group of sodium atoms to rest, individual atoms will have speeds within the range you found in part a. Because there's a distribution of speeds, suppose we estimate that the root-mean-square speed v_{rms} of the atoms in the trap is half the value you found in part a. Use this v_{rms} to estimate the temperature of the atoms when they've been cooled to the limit set by the uncertainty principle.

45. You learned in Chapter 37 that, except for hydrogen, the mass of a nucleus with atomic number Z is larger than the mass of the Z protons. The additional mass was ultimately discovered to be due to neutrons, but prior to the discovery of the neutron it was suggested that a nucleus with mass number A might contain A protons and $(A - Z)$ electrons. Such a nucleus would have the mass of A protons, but its net charge would be only Ze.

   a. We know that the diameter of a nucleus is approximately 10 fm. Model the nucleus as a one-dimensional box and find the minimum range of speeds that an electron would have in such a box.
   b. What does your answer imply about the possibility that the nucleus contains electrons? Explain.

46. a. Starting with the expression $\Delta f \Delta t \approx 1$ for a wave packet, find an expression for the product $\Delta E \Delta t$ for a photon.
   b. Interpret your expression. What does it tell you?
   c. The Bohr model of atomic quantization says that an atom in an excited state can jump to a lower-energy state by emitting a photon. The Bohr model says nothing about how long this process takes. You'll learn in Chapter 41 that the time any particular atom spends in the excited state before emitting a photon is unpredictable, but the *average lifetime* Δt of many atoms can be determined. You can think of Δt as being the uncertainty in your knowledge of how long the atom spends in the excited state. A typical value is $\Delta t \approx 10$ ns. Consider an atom that emits a photon with a 500 nm wavelength as it jumps down from an excited state. What is the uncertainty in the energy of the photon? Give your answer in eV.
   d. What is the *fractional uncertainty* $\Delta E/E$ in the photon's energy?

Challenge Problems

47. Figure CP39.47 shows 1.0-μm-diameter dust particles ($m = 1.0 \times 10^{-15}$ kg) in a vacuum chamber. The dust particles are released from rest above a 1.0-μm-diameter hole, fall through the hole (there's just barely room for the particles to go through), and land on a detector at distance d below.

   a. If the particles were purely classical, they would all land in the same 1.0-μm-diameter circle. But quantum effects don't allow this. If $d = 1.0$ m, by how much does the diameter of the circle in which most dust particles land exceed 1.0 μm? Is this increase in diameter likely to be detectable?
   b. Quantum effects would be noticeable if the detection-circle diameter increased by 10% to 1.1 μm. At what distance d would the detector need to be placed to observe this increase in the diameter?

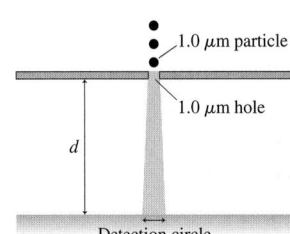

1.0 μm particle

1.0 μm hole

d

FIGURE CP39.47

Detection circle

48. The wave function of a particle is

$$\psi(x) = \sqrt{\frac{b}{\pi(x^2 + b^2)}}$$

    where b is a positive constant. Find the probability that the particle is located in the interval $-b \leq x \leq b$.

49. The wave function of a particle is

$$\psi(x) = \begin{cases} \dfrac{b}{(1 + x^2)} & -1 \text{ mm} \leq x < 0 \text{ mm} \\ c(1 + x)^2 & 0 \text{ mm} \leq x \leq 1 \text{ mm} \end{cases}$$

    and zero elsewhere.

    a. You will learn in Chapter 40 that the wave function must be a *continuous* function. Assuming that to be the case, what can you conclude about the relationship between b and c?
    b. Draw graphs of the wave function and the probability density over the interval -2 mm $\leq x \leq 2$ mm.
    c. What is the probability that the particle will be found to the right of the origin?

50. Consider the electron wave function

$$\psi(x) = \begin{cases} cx & |x| \leq 1 \text{ nm} \\ \dfrac{c}{x} & |x| \geq 1 \text{ nm} \end{cases}$$

    where x is in nm.

    a. Determine the normalization constant c.
    b. Draw a graph of $\psi(x)$ over the interval -5 nm $\leq x \leq 5$ cm. Provide numerical scales on both axes.
    c. Draw a graph of $|\psi(x)|^2$ over the interval -5 nm $\leq x \leq 5$ nm. Provide numerical scales.
    d. If 10^6 electrons are detected, how many will be in the interval -1.0 nm $\leq x \leq 1.0$ nm?

STOP TO THINK ANSWERS

Stop to Think 39.1: 10. The probability of a 1 is $P_1 = \frac{1}{6}$. Similarly, $P_6 = \frac{1}{6}$. The probability of a 1 *or* 6 is $P_{1 \text{ or } 6} = \frac{1}{6} + \frac{1}{6} = \frac{1}{3}$. Thus the expected number is $30(\frac{1}{3}) = 10$.

Stop to Think 39.2: A > B = D > C. $|A(x)|^2$ is proportional to the density of dots.

Stop to Think 39.3: x_C. The probability is largest at the point where the *square* of $\psi(x)$ is largest.

Stop to Think 39.4: b. The area $\frac{1}{2}a(2 \text{ mm})$ must equal 1.

Stop to Think 39.5: b. $\Delta t = 1.0 \times 10^{-7}$ s. The bandwidth is $\Delta f_B = 1/\Delta t = 1.0 \times 10^7$ Hz = 10 MHz.

Stop to Think 39.6: A. Wave packet A has a smaller spatial extent Δx. The wavelength isn't relevant.

40 One-Dimensional Quantum Mechanics

An example of atomic engineering. Thirty-five xenon atoms have been manipulated into position with the probe tip of a scanning tunneling microscope.

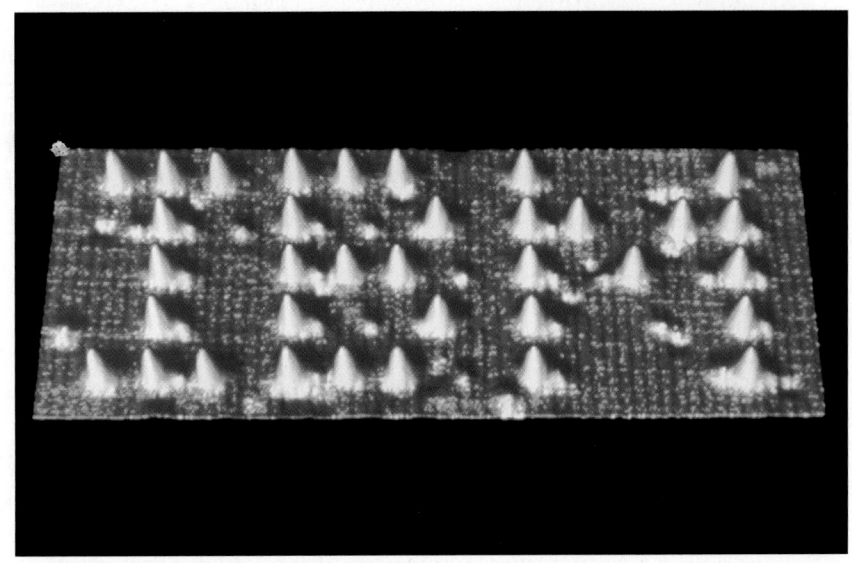

▶ **Looking Ahead**

The goal of Chapter 40 is to understand and apply the essential ideas of quantum mechanics. In this chapter you will learn to:

- Use a strategy for finding and interpreting wave functions.
- Draw wave functions with appropriate shapes.
- Use potential-energy functions to make quantum-mechanical models.
- Understand and use several important quantum-mechanical models.
- Calculate the probability of quantum-mechanical tunneling.

◀ **Looking Back**

Quantum mechanics will be developed around two fundamental ideas: energy diagrams and wave functions. A review of energy diagrams in Chapter 10 is especially important. Please review:

- Section 10.7 Energy diagrams.
- Sections 38.4–38.5 Matter waves and the Bohr model of quantization.
- Sections 39.3–39.4 Wave functions and normalization.

Quantum mechanics is not just for physicists any more. It is now an essential tool in the design of semiconductor devices such as diode lasers. Whole new classes of devices, called *quantum-well devices,* have been designed and built to exploit the quantized energy levels. We will look at some examples in this chapter.

Also at the cutting edge of engineering science is the design and manufacture of *nanostructures*—small machines or other devices only a few hundred nanometers in size. Many scientists and engineers envision a day in the near future when nanostructures will be constructed literally atom by atom. Quantum effects will be important in devices this small. This photograph, of a curious symbolic structure that scientists at IBM's research laboratory built by moving xenon atoms around on a metal surface, shows an early example of "atomic engineering."

Our goal for this chapter is to introduce the essential ideas of quantum mechanics. Although the real world is three-dimensional, we will limit our study of quantum mechanics to one dimension. This will allow us to focus on the fundamental concepts of quantum physics without becoming overwhelmed by mathematical complications. We will discuss some of the aspects of finding and using wave functions, then look at several applications of quantum mechanics. We'll conclude this chapter with a look at a phenomenon called *quantum-mechanical tunneling,* one of the more startling aspects of quantum physics.

40.1 Schrödinger's Equation: The Law of Psi

In the fall of 1925, just before Christmas, the Austrian physicist Erwin Schrödinger gathered together a few books and headed off to a villa in the Swiss Alps. He had recently learned of de Broglie's 1924 suggestion that matter has wave-like properties, and he wanted some time free from distractions to think about it. Before the trip was done, Schrödinger had discovered the law of quantum mechanics.

1277

Erwin Schrödinger.

Schrödinger's goal was to predict the outcome of atomic experiments, a goal that had eluded classical physics. The mathematical equation that he developed is now called the **Schrödinger equation.** It is the law of quantum mechanics in the same way that Newton's laws are the laws of classical mechanics. It would make sense to call it Schrödinger's law, but by tradition it is called simply the Schrödinger equation.

You learned in Chapter 39 that a matter particle is characterized in quantum physics by its wave function $\psi(x)$. If you know a particle's wave function, you can predict the probability of detecting it in some region of space. That's all well and good, but Chapter 39 didn't provide any method for determining wave functions. The Schrödinger equation is the missing piece of the puzzle. It is an equation for finding a particle's wave function $\psi(x)$ along the x-axis.

Consider an atomic particle with mass m and mechanical energy E whose interactions with the environment can be characterized by a one-dimensional potential-energy function $U(x)$. The Schrödinger equation for the particle's wave function is

$$\frac{d^2\psi}{dx^2} = -\frac{2m}{\hbar^2}[E - U(x)]\psi(x) \qquad \text{(the Schrödinger equation)} \qquad (40.1)$$

This is a differential equation whose solution is the wave function $\psi(x)$ that we seek. Our first goal is to learn what this equation means and how it is used.

Justifying the Schrödinger Equation

The Schrödinger equation can be neither derived nor proved. It is not an outgrowth of any previous theory. Its success depended on its ability to explain the various phenomena that had refused to yield to a classical-physics analysis and to make new predictions that were subsequently verified.

Although the Schrödinger equation cannot be derived, the reasoning behind it can at least be made *plausible.* De Broglie had postulated a wave-like nature for matter in which a particle of mass m, velocity v, and momentum $p = mv$ has a wavelength

$$\lambda = \frac{h}{p} = \frac{h}{mv} \qquad (40.2)$$

Schrödinger's goal was to find a *wave equation* for which the solution would be a wave function having the de Broglie wavelength.

An oscillatory wave-like function with wavelength λ is

$$\psi(x) = \psi_0 \sin\left(\frac{2\pi x}{\lambda}\right) \qquad (40.3)$$

where ψ_0 is the amplitude of the wave function. Suppose we take a second derivative of $\psi(x)$:

$$\frac{d\psi}{dx} = \frac{2\pi}{\lambda}\psi_0 \cos\left(\frac{2\pi x}{\lambda}\right)$$

$$\frac{d^2\psi}{dx^2} = \frac{d}{dx}\frac{d\psi}{dx} = -\frac{(2\pi)^2}{\lambda^2}\psi_0 \sin\left(\frac{2\pi x}{\lambda}\right)$$

We can use the definition of $\psi(x)$, from Equation 40.3, to write the second derivative as

$$\frac{d^2\psi}{dx^2} = -\frac{(2\pi)^2}{\lambda^2}\psi(x) \qquad (40.4)$$

Equation 40.4 relates the wavelength λ to a combination of the wave function $\psi(x)$ and its second derivative.

> NOTE ▶ These manipulations are not specific to quantum mechanics. Equation 40.4, which is well known for classical waves, applies equally well to sound waves or waves on a string. ◀

Schrödinger's insight was to identify λ with the de Broglie wavelength of a particle. We can write the de Broglie wavelength in terms of the particle's kinetic energy K as

$$\lambda = \frac{h}{mv} = \frac{h}{\sqrt{2m(\frac{1}{2}mv^2)}} = \frac{h}{\sqrt{2mK}} \qquad (40.5)$$

Notice that **the de Broglie wavelength increases as the particle's kinetic energy decreases.** This observation will play a key role as we develop an understanding of wave functions.

If we square this expression for λ and substitute it into Equation 40.4, we find

$$\frac{d^2\psi}{dx^2} = -\frac{(2\pi)^2 2mK}{h^2}\psi(x) = -\frac{2m}{\hbar^2}K\psi(x) \qquad (40.6)$$

where $\hbar = h/2\pi$. Equation 40.6 is a differential equation for the function $\psi(x)$. The solution to this differential equation is the sinusoidal wave function of Equation 40.3, where λ is the de Broglie wavelength for a particle with kinetic energy K.

Our derivation of Equation 40.6 assumed that the particle's kinetic energy K is constant. The energy diagram of Figure 40.1a reminds you that a particle's kinetic energy remains constant as it moves along the x-axis only if its potential energy U is constant. In this case, the de Broglie wavelength is the same at all positions.

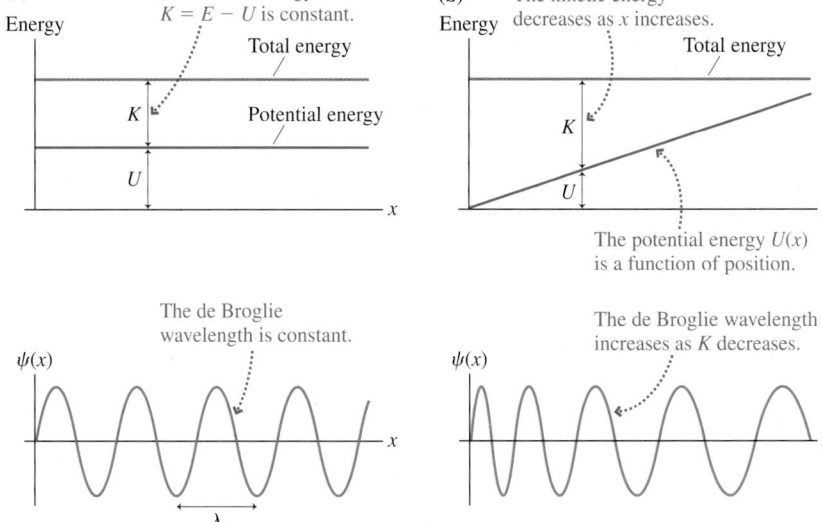

FIGURE 40.1 The de Broglie wavelength changes as a particle's kinetic energy changes.

In contrast, Figure 40.1b shows the energy diagram for a particle whose potential energy changes with x and whose kinetic energy is *not* constant. This particle speeds up and slows down as it moves along the x-axis, transforming potential energy to kinetic energy or vice versa. Consequently, its de Broglie wavelength changes with position.

Suppose a particle's potential energy—gravitational or electric or any other kind of potential energy—is described by the function $U(x)$ or $U(y)$. That is, the potential energy is a *function of position* along the axis of motion. For example, the gravitational potential energy near the earth's surface is the function $U(y) = mgy$.

If E is the particle's total mechanical energy, its kinetic energy at position x is

$$K = E - U(x) \qquad (40.7)$$

If we use this expression for K in Equation 40.6, that equation becomes

$$\frac{d^2\psi}{dx^2} = -\frac{2m}{\hbar^2}[E - U(x)]\psi(x)$$

This is Equation 40.1, the Schrödinger equation for the particle's wave function $\psi(x)$.

NOTE ▶ This has not been a derivation of the Schrödinger equation. We've made a *plausibility argument,* based on de Broglie's hypothesis about matter waves, but only experimental evidence will show if this equation has merit. ◀

STOP TO THINK 40.1   Three de Broglie waves are shown for particles of equal mass. Rank in order, from largest to smallest, the speeds of particles a, b, and c.

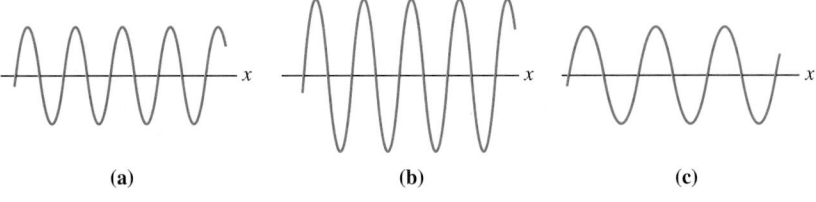

(a)   (b)   (c)

Quantum-Mechanical Models

Long ago, in your study of Newtonian mechanics, you learned of the importance of *models.* To understand the motion of an object, we made many simplifying assumptions: that the object could be represented by a particle, that friction could be described in a simple way, that air resistance could be neglected, and so on. In other words, we constructed a model that gave a good, though not perfect, description of reality. Models allowed us to understand the primary features of an object's motion without getting lost in the details.

The same holds true in quantum mechanics. The exact description of a microscopic atom or a solid is extremely complicated. Our only hope for using quantum mechanics effectively is to make a number of simplifying assumptions—that is, to make a **quantum-mechanical model** of the situation. Much of this chapter will be about building and using quantum mechanical models.

The test of a model's success is its agreement with experimental measurement. Laboratory experiments cannot measure $\psi(x)$, and they rarely make direct measurements of probabilities. Thus it will be important to tie our models to measurable quantities such as wavelengths, charges, currents, times, and temperatures.

There's one very important difference between models in classical mechanics and quantum mechanics. Classical models are described in terms of *forces,* and Newton's laws are a connection between force and motion. The Schrödinger equation for the wave function is written in terms of *energies.* Consequently, quantum-mechanical modeling involves finding a potential-energy function $U(x)$ that describes a particle's interactions with its environment

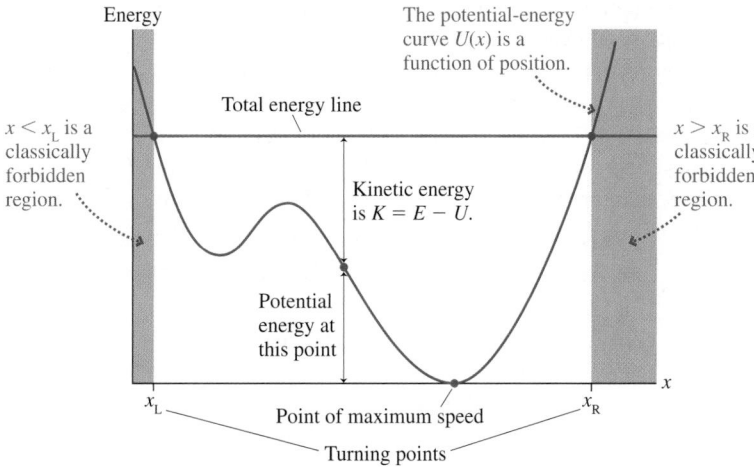

FIGURE 40.2 Interpreting an energy diagram.

Figure 40.2 reminds you how to interpret an energy diagram. We will use energy diagrams extensively in this and the remaining chapters to portray quantum-mechanical models. A review of Section 10.7, where energy diagrams were introduced, is highly recommended.

 20.1

The primary limitation to our quantum-mechanical models is the fact that the real world is three-dimensional. Unfortunately, solving the Schrödinger equation in three dimensions raises the mathematical level beyond what is appropriate for this text. Although the one-dimensional models that we will use are oversimplifications, there are many situations for which a one-dimensional model is a reasonably good approximation. Our major goal is to learn how quantum-mechanical concepts are used, and one-dimensional quantum mechanics will be sufficient for this purpose.

40.2 Solving the Schrödinger Equation

The Schrödinger equation is a second-order differential equation, meaning that it is a differential equation for $\psi(x)$ involving its second derivative. However, this textbook does not assume that you know how to solve differential equations. As we did with Newton's laws, we will restrict ourselves to situations where the mathematical skills are those you have been developing in calculus.

The solution to an algebraic equation is simply a number. For example, $x = 3$ is the solution to the equation $2x = 6$. In contrast, the solution to a differential equation is a *function*. You saw this idea in the previous section, where Equation 40.6 was constructed so that the function $\psi(x) = \psi_0 \sin(2\pi x/\lambda)$ would be a solution.

The Schrödinger equation can't be solved until the potential-energy function $U(x)$ has been specified. Different potential-energy functions result in different wave functions, just as different forces lead to different trajectories in classical mechanics. Once $U(x)$ has been specified, the solution of the differential equation is a *function $\psi(x)$* that is defined at all values of x. We will usually display the solution as a graph of $\psi(x)$ versus x.

Restrictions and Boundary Conditions

Not all functions $\psi(x)$ make *acceptable* solutions to the Schrödinger equation. That is, there may be functions that satisfy the Schrödinger equation but that are not physically meaningful. We have previously encountered restrictions in our solutions of algebraic equations. We insist, for physical reasons, that masses be positive rather than negative numbers, that positions be real rather than imaginary numbers, and so on. Mathematical solutions not meeting these restrictions are rejected as being unphysical.

Because we want to interpret $|\psi(x)|^2$ as a probability density, we have to insist that the function $\psi(x)$ be one for which this interpretation is possible. The conditions or restrictions on acceptable solutions are called the **boundary conditions.** You will see, in later examples, how the boundary conditions help us to choose the correct solution for $\psi(x)$. The primary conditions that the wave function must obey are

1. $\psi(x)$ is a continuous function.
2. $\psi(x) = 0$ if x is in a region where it is physically impossible for the particle to be.
3. $\psi(x) \rightarrow 0$ as $x \rightarrow +\infty$ and $x \rightarrow -\infty$.
4. $\psi(x)$ is a normalized function.

The last is not, strictly speaking, a boundary condition but is an auxiliary condition we require for the wave function to have a useful interpretation. Boundary condition 3 is needed to enable the normalization integral $\int |\psi(x)|^2 dx$ to converge.

Once boundary conditions have been established, there are three general approaches to solving the Schrödinger equation: Use general techniques for solving second-order differential equations, use a numerical technique to solve the equation on a computer, or guess.

More advanced courses make extensive use of the first and second approaches. However, we are not assuming a knowledge of differential equations, so you will not be asked to use these methods. The third, although it sounds almost like cheating, is widely used in simple situations where we can use physical arguments to infer the functional form of the wave function. The upcoming examples will illustrate this third approach.

A quadratic algebraic equation has two different solutions. Similarly, a second-order differential equation has two independent solutions $\psi_1(x)$ and $\psi_2(x)$. By "independent solutions" we mean that $\psi_2(x)$ is not just a constant multiple of $\psi_1(x)$, such as $3\psi_1(x)$, but that $\psi_1(x)$ and $\psi_2(x)$ are totally different functions.

Suppose that $\psi_1(x)$ and $\psi_2(x)$ are known to be two independent solutions of the Schrödinger equation. A theorem you will learn in differential equations states that a *general solution* of the equation can be written

$$\psi(x) = A\psi_1(x) + B\psi_2(x) \tag{40.8}$$

where A and B are constants whose values are determined by the boundary conditions. In other words, the general solution is a *superposition,* one of the primary characteristics of waves. Equation 40.8 is a powerful statement, although one that will make more sense after you see it applied in upcoming examples. The main point is that **if we can find two solutions $\psi_1(x)$ and $\psi_2(x)$ by guessing, then Equation 40.8 is *the* general solution to the Schrödinger equation.**

Quantization

We've asserted that the Schrödinger equation is the law of quantum mechanics, but thus far we've not said anything about quantization. Although the particle's total energy E appears in the Schrödinger equation, it is treated in the equation as an unspecified constant. However, it will turn out that there are *no* acceptable solutions for most values of E. That is, there are no functions $\psi(x)$ that satisfy both the Schrödinger equation *and* the boundary conditions. Acceptable solutions exist only for *discrete* values of E. The energies for which solutions exist are the quantized energies of the system. Thus, as you'll see, the Schrödinger equation has quantization as a built-in feature.

Problem Solving in Quantum Mechanics

Our problem-solving strategy for classical mechanics focused on identifying and using forces. In quantum mechanics we're interested in *energy* rather than forces. The critical step in solving a problem in quantum mechanics is to determine the particle's potential-energy function $U(x)$. Identifying the interactions that cause a potential energy is the *physics* of the problem. Once the potential-energy function is known, it is "just mathematics" to solve for the wave function.

 PROBLEM-SOLVING STRATEGY 40.1 Quantum-mechanics problems

MODEL Determine a potential-energy function that describes the particle's interactions. Make simplifying assumptions.

VISUALIZE The potential-energy curve is the pictorial representation.

- Draw the potential-energy curve.
- Identify known information.
- Establish the boundary conditions that the wave function must satisfy.

SOLVE The Schrödinger equation is the mathematical representation.

- Utilize the boundary conditions.
- Normalize the wave functions.
- Draw graphs of $\psi(x)$ and $|\psi(x)|^2$.
- Determine the allowed energy levels.
- Calculate probabilities, wavelengths, or other specific quantities.

ASSESS Check that your result has the correct units, is reasonable, and answers the question.

The solutions to the Schrödinger equation are the stationary states of the system. Bohr had postulated the existence of stationary states, but he didn't know how to find them. Now we have a strategy for finding them.

Bohr's idea of transitions, or quantum jumps, between stationary states remains very important in Schrödinger's quantum mechanics. The system can jump from one stationary state, characterized by wave function $\psi_i(x)$ and energy E_i, to another state, characterized by $\psi_f(x)$ and E_f, by emitting or absorbing a photon of frequency

$$f = \frac{\Delta E}{h} = \frac{|E_f - E_i|}{h}$$

Thus the solutions to the Schrödinger equation will allow us to predict the emission and absorption spectra of a quantum system. These predictions will test the validity of Schrödinger's theory.

40.3 A Particle in a Rigid Box: Energies and Wave Functions

Figure 40.3 shows a particle of mass m confined in a rigid, one-dimensional box of length L. The walls of the box are assumed to be perfectly rigid, and the particle undergoes perfectly elastic reflections from the ends. This situation is known as a "particle in a box."

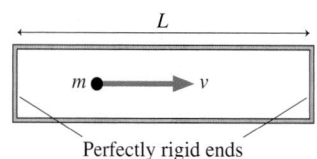

FIGURE 40.3 A particle in a rigid box of length L.

20.2   Act|v
ONLINE
Phys|cs

A classical particle bounces back and forth between the walls of the box. There are no restrictions on the speed or kinetic energy of a classical particle. In contrast, a wave-like particle characterized by a de Broglie wavelength sets up a standing wave as it reflects back and forth. In Chapters 24 and 38, we found that a standing de Broglie wave automatically leads to energy quantization. That is, only certain discrete energies are allowed. However, our hypothesis of a de Broglie standing wave was just a guess, with no real justification, because we had no theory as to how a wave-like particle ought to behave.

We will now revisit this problem from the new perspective of quantum mechanics. The basic questions we want to answer in this, and any quantum-mechanics problem, are

- What are the allowed energies of the particle?
- What is the wave function associated with each energy?
- In which part of the box is the particle most likely to be found?

We can answer these questions by following the steps of the problem-solving strategy.

Model: Identify a Potential-Energy Function

By a *rigid box* we mean a box whose walls are so sturdy that they can confine a particle no matter how fast the particle moves. Furthermore, the walls are so stiff that they do not flex or give as the particle bounces. No real container has these attributes, so the rigid box is a *model* of a situation in which a particle is extremely well confined. Our first task is to characterize the rigid box in terms of a potential-energy function.

Let's establish a coordinate axis with the boundaries of the box at $x = 0$ and $x = L$. The rigid box has three important characteristics:

1. The particle can move freely between 0 and L at constant speed and thus with constant kinetic energy.
2. No matter how much kinetic energy the particle has, its turning points are at $x = 0$ and $x = L$.
3. The regions $x < 0$ and $x > L$ are forbidden. The particle cannot leave the box.

A potential-energy function that describes the particle in this situation is

$$U_{\text{rigid box}}(x) = \begin{cases} 0 & 0 \leq x \leq L \\ \infty & x < 0 \quad \text{or} \quad x > L \end{cases} \tag{40.9}$$

Inside the box, the particle has only kinetic energy. The infinitely high potential-energy barriers prevent the particle from ever having $x < 0$ or $x > L$ no matter how much kinetic energy it may have. It is this potential energy for which we want to solve the Schrödinger equation.

Visualize: Establish Boundary Conditions

Figure 40.4 is the energy diagram of a particle in the rigid box. You can see that $U = 0$ and $E = K$ inside the box. The upward arrows labeled ∞ indicate that the potential energy becomes infinitely large at the walls of the box ($x = 0$ and $x = L$).

NOTE ▶ Figure 40.4 is not a picture of the box. It is a graphical representation of the particle's kinetic and potential energy. ◀

Next, we need to establish the boundary conditions that the solution must satisfy. Because it is physically impossible for the particle to be outside the box, we require

$$\psi(x) = 0 \quad \text{for } x < 0 \quad \text{or} \quad x > L \tag{40.10}$$

That is, there is zero probability of finding the particle outside the box.

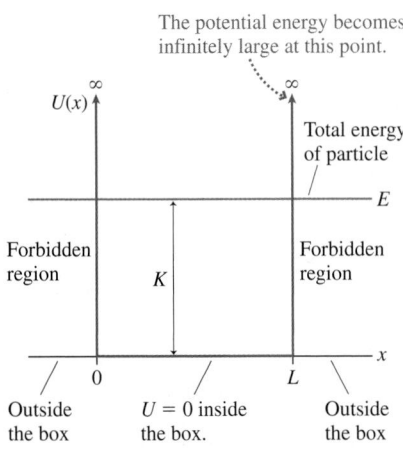

FIGURE 40.4 The energy diagram of a particle in a rigid box of length L.

Furthermore, the wave function must be a *continuous* function. That is, there can be no break in the wave function at any point. Because the solution is zero everywhere outside the box, continuity requires that the wave function inside the box obey the two conditions

$$\psi(\text{at } x = 0) = 0 \quad \text{and} \quad \psi(\text{at } x = L) = 0 \qquad (40.11)$$

In other words, as Figure 40.5 shows, the oscillating wave function inside the box must go to zero at the boundaries to be continuous with the wave function outside the box. This requirement of the wave function is equivalent to saying that a standing wave on a string must have a node at the ends because it is fixed at those points and cannot move.

Solve I: Find the Wave Functions

At all points *inside* the box the potential energy is $U(x) = 0$. Thus the Schrödinger equation inside the box is

$$\frac{d^2\psi}{dx^2} = -\frac{2mE}{\hbar^2}\psi(x) \qquad (40.12)$$

There are two aspects to solving this equation:

1. For what values of E does Equation 40.12 have physically meaningful solutions?
2. What are the solutions $\psi(x)$ for those values of E?

To begin, let's simplify the notation by defining $\beta^2 = 2mE/\hbar^2$. Equation 40.12 is then

$$\frac{d^2\psi}{dx^2} = -\beta^2\psi(x) \qquad (40.13)$$

We're going to solve this differential equation by guessing! Can you think of any functions whose second derivative is a *negative* constant times the function itself? Two such functions are

$$\psi_1(x) = \sin\beta x \quad \text{and} \quad \psi_2(x) = \cos\beta x \qquad (40.14)$$

Both are solutions to Equation 40.13 because

$$\frac{d^2\psi_1}{dx^2} = \frac{d^2}{dx^2}(\sin\beta x) = -\beta^2\sin\beta x = -\beta^2\psi_1(x)$$

$$\frac{d^2\psi_2}{dx^2} = \frac{d^2}{dx^2}(\cos\beta x) = -\beta^2\cos\beta x = -\beta^2\psi_2(x)$$

Furthermore, these are *independent* solutions because $\psi_2(x)$ is not a multiple or a rearrangement of $\psi_1(x)$. Consequently, according to Equation 40.8, the general solution to the Schrödinger equation for the particle in a rigid box is

$$\psi(x) = A\sin\beta x + B\cos\beta x \qquad (40.15)$$

where

$$\beta = \frac{\sqrt{2mE}}{\hbar} \qquad (40.16)$$

The constants A and B must be determined by using the boundary conditions of Equation 40.11. First, the wave function must go to zero at $x = 0$. That is,

$$\psi(\text{at } x = 0) = A \cdot 0 + B \cdot 1 = 0 \qquad (40.17)$$

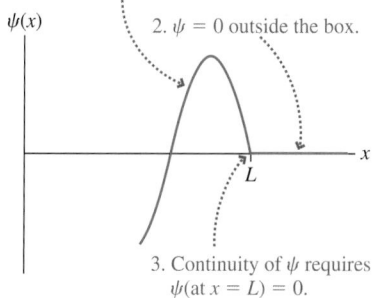

1. Inside the box, ψ is oscillating in some way still to be determined.

2. $\psi = 0$ outside the box.

3. Continuity of ψ requires $\psi(\text{at } x = L) = 0$.

FIGURE 40.5 Applying boundary conditions to the wave function of a particle in a box.

This boundary condition can be satisfied only if $B = 0$. The $\cos\beta x$ term may satisfy the differential equation in a mathematical sense, but it is not a physically meaningful solution for this problem because it does not satisfy the boundary conditions. Thus the physically meaningful solution is

$$\psi(x) = A\sin\beta x$$

The wave function must also go to zero at $x = L$. That is,

$$\psi(\text{at } x = L) = A\sin\beta L = 0 \tag{40.18}$$

This condition could be satisfied by $A = 0$, but then we wouldn't have a wave function at all! Fortunately, that isn't necessary. This boundary condition is also satisfied if

$$\beta L = n\pi \quad \text{or} \quad \beta = \frac{n\pi}{L} \qquad n = 1, 2, 3, \ldots \tag{40.19}$$

Notice that n starts with 1, not 0. The value $n = 0$ would give $\beta = 0$ and make $\psi = 0$ at all points, a physically meaningless solution.

Thus the solutions to the Schrödinger equation for a particle in a rigid box are

$$\psi_n(x) = A\sin\beta_n x = A\sin\left(\frac{n\pi x}{L}\right) \qquad n = 1, 2, 3, \ldots \tag{40.20}$$

We've found a whole *family* of solutions, each corresponding to a different value of the integer n. These wave functions represent the stationary states of the particle in the box. The constant A remains to be determined.

Solve II: Find the Allowed Energies

Equation 40.16 defined β. Equation 40.19 then placed restrictions on the possible values of β:

$$\beta_n = \frac{\sqrt{2mE_n}}{\hbar} = \frac{n\pi}{L} \qquad n = 1, 2, 3, \ldots \tag{40.21}$$

where the value of β and the energy associated with the integer n have been labeled β_n and E_n. We can solve for E_n by squaring both sides:

$$E_n = n^2\frac{\pi^2\hbar^2}{2mL^2} = n^2\frac{h^2}{8mL^2} \qquad n = 1, 2, 3, \ldots \tag{40.22}$$

where, in the last step, we used the definition $\hbar = h/2\pi$. For a particle in a box, **these energies are the only values of E for which there are physically meaningful solutions to the Schrödinger equation.**

We have found that the particle's energy is quantized! It is useful to write the energies of the stationary states as

$$E_n = n^2 E_1 \tag{40.23}$$

where E_n is the energy of the stationary state with *quantum number n*. The smallest possible energy $E_1 = h^2/8mL^2$ is the energy of the $n = 1$ *ground state*. These allowed energies are shown in the *energy-level diagram* of Figure 40.6, where you can see that the allowed energies increase with the square of the quantum number. Recall, from Chapter 38, that an energy-level diagram is not a graph (the horizontal axis doesn't represent anything) but a "ladder" of allowed energies.

Equation 40.22 is identical to the energies we found in Chapter 38 by requiring the de Broglie wave of a particle in a box to form a standing wave. Only now we have a theory that tells not only the energies but also the wave functions.

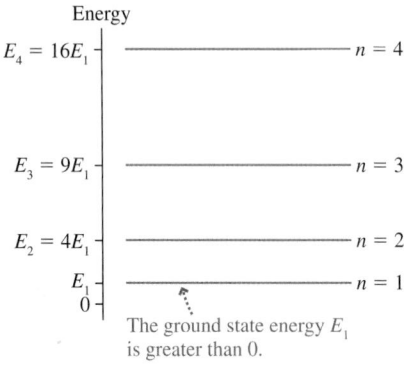

FIGURE 40.6 The energy-level diagram for a particle in a box.

EXAMPLE 40.1 An electron in a box

An electron is confined to a rigid box. What is the length of the box if the energy difference between the first and second states is 3.0 eV?

MODEL Model the electron as a particle in a rigid one-dimensional box of length L.

SOLVE The first two quantum states, with $n = 1$ and $n = 2$, have energies E_1 and $E_2 = 4E_1$. Thus the energy difference between the states is

$$\Delta E = 3E_1 = \frac{3h^2}{8mL^2} = 3.0 \text{ eV} = 4.8 \times 10^{-19} \text{ J}$$

The length of the box for which $\Delta E = 3.0$ eV is

$$L = \sqrt{\frac{3h^2}{8m\,\Delta E}} = 6.14 \times 10^{-10} \text{ m} = 0.614 \text{ nm}$$

ASSESS The expression for E_1 is in SI units, so energies must be in J, not eV.

Solve III: Normalize the Wave Functions

We can determine the constant A by requiring the wave functions to be normalized. The normalization condition, which we found in Chapter 39, is

$$\int_{-\infty}^{\infty} |\psi(x)|^2 dx = 1$$

This is the mathematical statement that the particle must be *somewhere* on the x-axis. The integration limits extend to $\pm\infty$, but here we need to integrate only from 0 to L because the wave function is zero outside the box. Thus normalization requires

$$\int_0^L |\psi_n(x)|^2 dx = A_n^2 \int_0^L \sin^2\left(\frac{n\pi x}{L}\right) dx = 1 \tag{40.24}$$

or

$$A_n = \left[\int_0^L \sin^2\left(\frac{n\pi x}{L}\right) dx\right]^{-1/2} \tag{40.25}$$

We placed a subscript n on A_n because it is possible that the normalization constant is different for each wave function in the family. This is a standard integral. We will leave it as a homework problem for you to show that its value, for any n, is

$$A_n = \sqrt{\frac{2}{L}} \qquad n = 1, 2, 3, \dots \tag{40.26}$$

We now have a complete solution to the problem. The normalized wave function for the particle in quantum state n is

$$\psi_n(x) = \begin{cases} \sqrt{\dfrac{2}{L}} \sin\left(\dfrac{n\pi x}{L}\right) & 0 \le x \le L \\ 0 & x < 0 \text{ and } x > L \end{cases} \tag{40.27}$$

40.4 A Particle in a Rigid Box: Interpreting the Solution

Our solution to the quantum-mechanical problem of a particle in a box tells us that

1. The particle must have energy $E_n = n^2 E_1$ where $n = 1, 2, 3, \dots$ is the quantum number and where $E_1 = h^2/8mL^2$ is the energy of the $n = 1$ ground state.

2. The wave function for a particle in quantum state n is

$$\psi_n(x) = \begin{cases} \sqrt{\dfrac{2}{L}} \sin\!\left(\dfrac{n\pi x}{L}\right) & 0 \le x \le L \\ 0 & x < 0 \text{ and } x > L \end{cases}$$

These are the stationary states of the system.

3. The probability density for finding the particle at position x inside the box is

$$P_n(x) = |\psi_n(x)|^2 = \frac{2}{L}\sin^2\!\left(\frac{n\pi x}{L}\right) \tag{40.28}$$

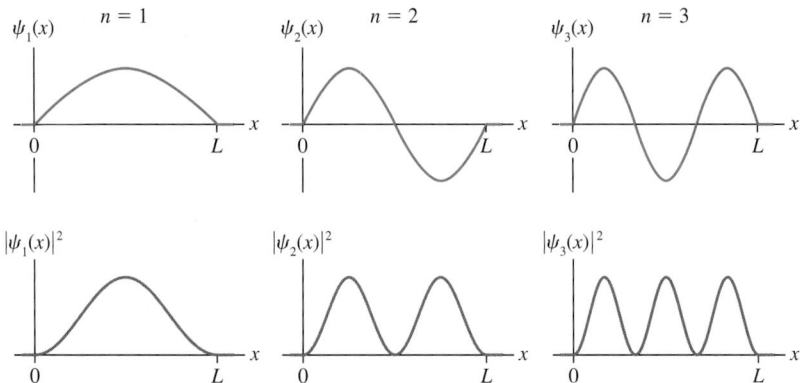

FIGURE 40.7 Wave functions and probability densities for a particle in a rigid box of length L.

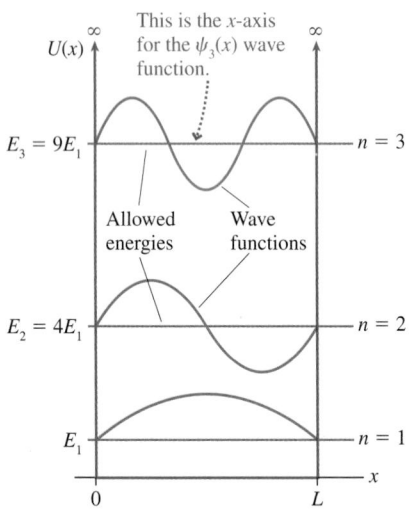

FIGURE 40.8 An alternative way to show the potential-energy diagram, the energies, and the wave functions.

A graphical presentation will make these results more meaningful. Figure 40.7 shows the wave functions $\psi(x)$ and the probability densities $P(x) = |\psi(x)|^2$ for quantum states $n = 1$ to 3. Notice that the wave functions go to zero at the boundaries and thus are continuous with $\psi = 0$ outside the box.

The wave functions $\psi(x)$ for a particle in a rigid box are analogous to standing waves on a string that is tied at both ends. You can see that $\psi_n(x)$ has $n - 1$ nodes (zeros), excluding the ends, and n antinodes (maxima and minima). This is a general result for any wave function, not just for a particle in a rigid box.

Figure 40.8 shows another way in which energies and wave functions are shown graphically in quantum mechanics. First, the graph shows the potential-energy function $U(x)$ of the particle. Second, the allowed energies are shown as horizontal lines (total energy lines) across the potential-energy graph. These are labeled with the quantum number n and the energy E_n. Third—and this is a bit tricky—the wave function for each n is drawn *as if* the energy line were the x-axis. That is, the graph of $\psi_n(x)$ is drawn on top of the E_n energy line. This allows energies and wave functions to be displayed simultaneously, but it does *not* imply that ψ_2 is in any sense "above" ψ_1. Both oscillate sinusoidally about zero, as Figure 40.7 shows.

EXAMPLE 40.2 Energy levels and quantum jumps
A semiconductor device known as a *quantum-well device* is designed to "trap" electrons in a 1.0-nm-wide region. Treat this as a one-dimensional problem:

a. What are the energies of the first three quantum states?
b. What wavelengths of light can these electrons absorb?

MODEL Model an electron in a quantum-well device as a particle confined in a rigid box of length $L = 1.0$ nm.

VISUALIZE Figure 40.9 shows the first three energy levels and the transitions by which an electron in the ground state can absorb a photon.

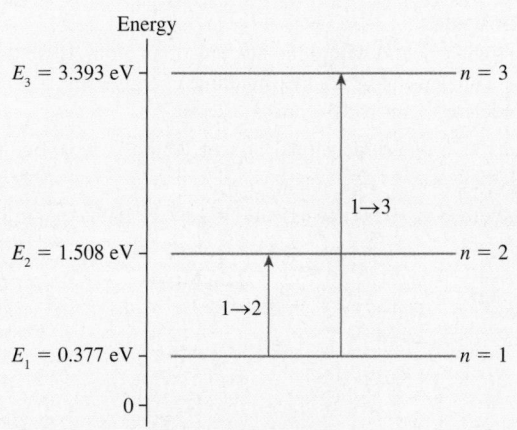

FIGURE 40.9 Energy levels and quantum jumps for an electron in a quantum-well device.

SOLVE

a. The particle's mass is $m = m_e = 9.11 \times 10^{-31}$ kg. The allowed energies, in both J and eV, are

$$E_1 = \frac{h^2}{8mL^2} = 6.03 \times 10^{-20} \text{ J} = 0.377 \text{ eV}$$

$$E_2 = 4E_1 = 1.508 \text{ eV}$$

$$E_3 = 9E_1 = 3.393 \text{ eV}$$

b. An electron spends most of its time in the $n = 1$ ground state. According to Bohr's model of stationary states, the electron can absorb a photon of light and undergo a transition, or quantum jump, to $n = 2$ or $n = 3$ if the light has frequency $f = \Delta E/h$. The wavelengths, given by $\lambda = c/f = hc/\Delta E$, are

$$\lambda_{1 \to 2} = \frac{hc}{E_2 - E_1} = 1098 \text{ nm}$$

$$\lambda_{1 \to 3} = \frac{hc}{E_3 - E_1} = 411 \text{ nm}$$

ASSESS In practice, various complications usually make the $1 \to 3$ transition unobservable. But quantum-well devices do indeed exhibit strong absorption and emission at the $\lambda_{1 \to 2}$ wavelength. In this example, which is typical of quantum-well devices, the wavelength is in the near-infrared portion of the spectrum. Devices such as these are used to construct the semiconductor lasers used in CD players and laser printers.

NOTE ▶ The wavelengths of light emitted or absorbed by a quantum system are determined by the *difference* between two allowed energies. Quantum jumps involve two stationary states. ◀

Zero-Point Motion

The lowest energy state in Example 40.2, the $n = 1$ ground state, has $E_1 = 0.38$ eV. There is no stationary state having $E = 0$. Unlike a classical particle, **a quantum particle in a box cannot be at rest!** No matter how much its energy is reduced, such as by cooling it toward absolute zero, it cannot have energy less than E_1.

The particle motion associated with energy E_1, called the **zero-point motion,** is a consequence of Heisenberg's uncertainty principle. Because the particle is somewhere in the box, its position uncertainty is $\Delta x = L$. If the particle were at rest in the box, we would know that its velocity and momentum are exactly zero with *no* uncertainty: $\Delta p_x = 0$. But then $\Delta x \Delta p_x = 0$ would violate the Heisenberg uncertainty principle. One of the conclusions that follows from the uncertainty principle is that **a confined particle cannot be at rest.**

Although the particle's position and velocity are uncertain, the particle's energy in each state can be calculated with a high degree of precision. This distinction between a precise energy and an uncertain position and velocity seems rather strange, but it is just our old friend the standing wave. In order to *have* a stationary state at all, the de Broglie waves have to form standing waves. Only for very precise frequencies, and thus precise energies, can the standing-wave pattern appear.

EXAMPLE 40.3 Nuclear energies

Protons and neutrons are tightly bound within the nucleus of an atom. If we use a one-dimensional model of a nucleus, what are the first three energy levels of a neutron in a 10-fm-diameter nucleus ($1 \text{ fm} = 10^{-15} \text{ m}$)?

MODEL Model the nucleus as a one-dimensional box of length $L = 10 \text{ fm}$. The neutron is confined within the box.

SOLVE The energy levels, with $L = 10 \text{ fm}$ and $m = m_n = 1.67 \times 10^{-27} \text{ kg}$, are

$$E_1 = \frac{h^2}{8mL^2} = 3.29 \times 10^{-13} \text{ J} = 2.06 \text{ MeV}$$

$$E_2 = 4E_1 = 8.24 \text{ MeV}$$

$$E_3 = 9E_1 = 18.54 \text{ MeV}$$

ASSESS An electron confined in an atom-size space has energies of a few eV. A neutron confined in a nucleus-size space has energies of a few *million* eV.

EXAMPLE 40.4 The probabilities of locating the particle

A particle in a rigid box of length L is in its ground state.

a. Where is the particle most likely to be found?
b. What are the probabilities of finding the particle in an interval of width $0.01L$ at $x = 0, 0.25L$, and $0.50L$?
c. What is the probability of finding the particle in the center half of the box?

MODEL The wave functions for a particle in a rigid box have been determined.

VISUALIZE Figure 40.10 shows the probability density $P_1(x) = |\psi_1(x)|^2$ in the ground state.

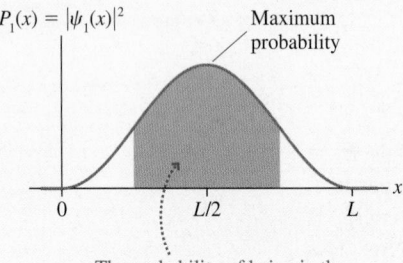

FIGURE 40.10 Probability density for a particle in the ground state.

SOLVE

a. The particle is most likely to be found at the point where the probability density $P(x)$ is a maximum. You can see from Figure 40.10 that the point of maximum probability for $n = 1$ is $x = L/2$.

b. For a *small* width δx, the probability of finding the particle in δx at position x is

$$\text{Prob(in } \delta x \text{ at } x) = P_1(x)\delta x = |\psi_1(x)|^2\delta x = \frac{2}{L}\sin^2\left(\frac{\pi x}{L}\right)\delta x$$

The interval $\delta x = 0.01L$ is sufficiently small for this to be valid. The probabilities of finding the particle are

$$\text{Prob(in } 0.01L \text{ at } x = 0.00L) = 0.000 = 0.0\%$$

$$\text{Prob(in } 0.01L \text{ at } x = 0.25L) = 0.010 = 1.0\%$$

$$\text{Prob(in } 0.01L \text{ at } x = 0.50L) = 0.020 = 2.0\%$$

c. The center half of the box stretches from $x = L/4$ to $x = 3L/4$. The probability that the particle is in this interval is the area under the probability-density curve:

$$\text{Prob}\left(\text{in interval } \frac{1}{4}L \text{ to } \frac{3}{4}L\right) = \int_{L/4}^{3L/4} P_1(x)\,dx$$

$$= \frac{2}{L}\int_{L/4}^{3L/4}\sin^2\left(\frac{\pi x}{L}\right)dx$$

$$= \left[\frac{x}{L} - \frac{1}{\pi}\sin\left(\frac{\pi x}{L}\right)\cos\left(\frac{\pi x}{L}\right)\right]_{L/4}^{3L/4}$$

$$= \frac{1}{2} + \frac{1}{\pi} = 0.818$$

ASSESS If a particle in a box is in the $n = 1$ ground state, there is an 81.8% chance of finding it in the center half of the box. The probability is greater than 50% because, as you can see in Figure 40.10, the probability density $P_1(x)$ is larger near the center of the box than near the boundaries.

This has been a lengthy presentation of the particle-in-a-box problem. However, it was important that we explore the method of solution completely. Future examples will now go more quickly because many of the issues discussed here will not need to be repeated.

STOP TO THINK 40.2   A particle in a rigid box in the $n = 2$ stationary state is most likely to be found

a. In the center of the box.
b. One-third of the way from either end.
c. One-quarter of the way from either end.
d. It is equally likely to be found at any point in the box.

40.5 The Correspondence Principle

Suppose we confine an electron in a microscopic box, then allow the box to get bigger and bigger. What started out as a quantum-mechanical situation should, when the box becomes macroscopic in size, eventually look like a classical-physics situation. Similarly, a classical situation such as two charged particles revolving about each other should begin to exhibit quantum behavior as the size becomes smaller and smaller.

These examples suggest that there should be some in-between size, or energy, for which the quantum-mechanical solution corresponds in some way to the solution of classical mechanics. Niels Bohr put forward the idea that the *average* behavior of a quantum system should begin to look like the classical solution in the limit that the quantum number becomes very large—that is, as $n \to \infty$. Because the radius of the Bohr hydrogen atom is $r = n^2 a_B$, the atom becomes a macroscopic object as n becomes very large. Bohr's idea, that the quantum world should blend smoothly into the classical world for high quantum numbers, is today known as the **correspondence principle.**

Our quantum knowledge of a particle in a box is given by its probability density

$$P_{\text{quant}}(x) = |\psi_n(x)|^2 = \frac{2}{L} \sin^2\left(\frac{n\pi x}{L}\right) \qquad (40.29)$$

To what classical quantity can the probability density be compared as $n \to \infty$?

Interestingly, we can also define a classical probability density $P_{\text{class}}(x)$. A classical particle follows a well-defined trajectory, but suppose we observe the particle at random times. For example, suppose the box containing a classical particle has a viewing window. The window is normally closed, but at random times, selected by a random-number generator, the window opens for a brief interval of time δt and you can measure the particle's position. When the window opens, what is the probability that the particle will be in a narrow interval δx at position x?

The probability of finding a classical particle within a small interval δx is equal to the *fraction of its time* that it spends passing through δx. That is, you're more likely to find the particle in those intervals δx where it spends lots of time, less likely to find it in a δx where it spends very little time.

If the particle oscillates between two turning points with period T, the time it spends moving from one turning point to the other is $\frac{1}{2}T$. As it moves between the turning points, it passes once through the interval δx at position x, taking time δt to do so. Consequently, the probability of finding the particle within this interval is

$$\text{Prob}_{\text{class}}(\text{in } \delta x \text{ at } x) = \text{fraction of time spent in } \delta x = \frac{\delta t}{\frac{1}{2}T} \qquad (40.30)$$

The amount of time needed to pass through δx is $\delta t = \delta x/v(x)$, where $v(x)$ is the particle's velocity at position x. Thus the probability of finding the particle in the interval δx at position x is

$$\text{Prob}_{\text{class}}(\text{in }\delta x\text{ at }x) = \frac{\delta x/v(x)}{\frac{1}{2}T} = \frac{2}{Tv(x)}\delta x \qquad (40.31)$$

You learned in Chapter 39 that the probability is related to the probability density by

$$\text{Prob}_{\text{class}}(\text{in }\delta x\text{ at }x) = P_{\text{class}}(x)\,\delta x$$

Thus the classical probability density for finding a particle at position x is

$$P_{\text{class}}(x) = \frac{2}{Tv(x)} \qquad (40.32)$$

where the velocity $v(x)$ is expressed as a function of x. **Classically, a particle is more likely to be found where it is moving slowly, less likely to be found where it is moving quickly.**

NOTE ▶ Our derivation of Equation 40.32 made no assumptions about the particle's motion other than the requirement that it be periodic. This is the classical probability density for any oscillatory motion. ◀

Figure 40.11a shows a motion diagram of a classical particle in a rigid box of length L. The particle's speed is a *constant* $v(x) = v_0$ as it bounces back and forth between the walls. The particle travels distance $2L$ during one round trip, so the period is $T = 2L/v_0$. Consequently, the classical probability density for a particle in a box is

$$P_{\text{class}}(x) = \frac{2}{(2L/v_0)v_0} = \frac{1}{L} \qquad (40.33)$$

$P_{\text{class}}(x)$ is independent of x, telling us that the particle is equally likely to be found *anywhere* in the box.

In contrast, Figure 40.11b shows a particle with nonuniform speed. A mass on a spring slows down near the turning points, so it spends more time near the ends of the box than in the middle. Consequently the classical probability density for this particle is a maximum at the edges and a minimum at the center. We'll look at this classical probability density again later in the chapter.

(a) Uniform speed

Particle in an empty box

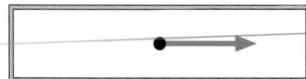

Motion diagram

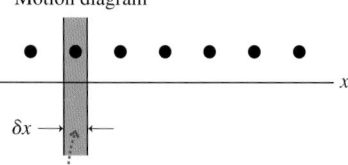

The probability of finding the particle in δx is the fraction of time the particle spends in δx.

(b) Nonuniform speed

Particle on a spring

Motion diagram

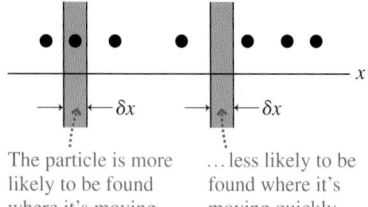

The particle is more likely to be found where it's moving slowly, …

…less likely to be found where it's moving quickly.

FIGURE 40.11 The classical probability density is indicated by the density of dots in a motion diagram.

EXAMPLE 40.5 The classical probability of locating the particle

A classical particle is in a rigid 10-cm-long box. What is the probability that, at a random instant of time, the particle is in a 1.0-mm-wide interval at the center of the box?

SOLVE The particle's probability density is

$$P_{\text{class}}(x) = \frac{1}{L} = \frac{1}{10\text{ cm}} = 0.10\text{ cm}^{-1}$$

The probability that the particle is in an interval of width $\delta x = 1.0\text{ mm} = 0.10\text{ cm}$ is

$$\text{Prob(in }\delta x\text{ at }x = 5\text{ cm}) = P(x)\delta x = (0.10\text{ cm}^{-1})(0.10\text{ cm})$$
$$= 0.010 = 1.0\%$$

ASSESS The classical probability is 1.0% because 1.0 mm is 1% of the 10 cm length.

Figure 40.12 shows the quantum and the classical probability densities for the $n = 1$ and the $n = 20$ quantum states of a particle in a rigid box. Notice that

■ The quantum probability density oscillates between a minimum of 0 and a maximum of $2/L$, so it oscillates around the classical probability density $1/L$.

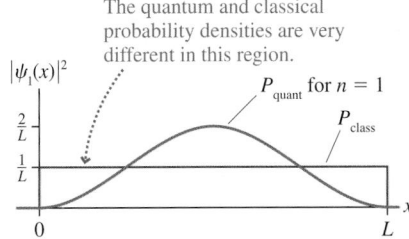

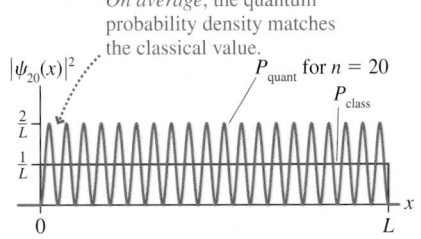

FIGURE 40.12 The quantum and classical probability densities for a particle in a box.

- For $n = 1$, the quantum and classical probability densities are quite different. The ground state of the quantum system will be very nonclassical.
- For $n = 20$, you can see that *on average* the quantum particle's behavior looks very much like that of the classical particle.

As n gets even bigger, and the number of oscillations increases, the probability of finding the particle in an interval δx will be the same for both the quantum and the classical particle as long as δx is large enough to include several oscillations of the wave function. As Bohr predicted, the quantum mechanical solution "corresponds" to the classical solution in the limit $n \rightarrow \infty$.

40.6 Finite Potential Wells

Figure 40.4, the potential-energy diagram for a particle in a rigid box, is an example of a **potential well,** so named because the graph of the potential-energy "hole" looks like a well from which you might draw water. The rigid box was an *infinite* potential well. There was no chance that a particle inside could escape the infinitely high walls.

No box is infinitely strong. A more realistic model of a confined particle is the *finite* potential well shown in Figure 40.13a. A particle with total energy $E < U_0$ is confined within the well, bouncing back and forth between turning points at $x = 0$ and $x = L$. The regions $x < 0$ and $x > L$ are **classically forbidden regions** for a particle with $E < U_0$. However, the particle will escape the well if it somehow manages to acquire energy $E > U_0$.

Recall that the zero of energy is arbitrary. Figure 40.13a defined $U = 0$ as the potential energy inside the well. Figure 40.13b has repositioned the zero of energy at the level of the "energy plateau" on both sides of the well. **Figures 40.13a and 40.13b are the same potential well.** Both have width L and depth U_0, and both have the same wave functions and the same allowed energies (relative to our choice of $E = 0$). Which we use is a matter of convenience for the situation we are modeling.

We've made no mention of the *force* that is responsible for this potential well. An electron confined within a semiconductor by an electric force has a potential energy that can be modeled as a finite potential well. So does a proton confined within the nucleus by the nuclear force. The Schrödinger equation depends on the *shape* of the potential-energy function, not the cause. Hence *any* situation in which a force confines a particle to a well-defined region can be modeled as a finite potential well.

Although it is possible to solve the Schrödinger equation exactly for the finite potential well, the result is cumbersome and not especially illuminating. Instead, we'll present the results of numerical calculations. The derivation of the wave functions and energy levels is not as important as understanding and interpreting the results.

(a) $U = 0$ inside the well.

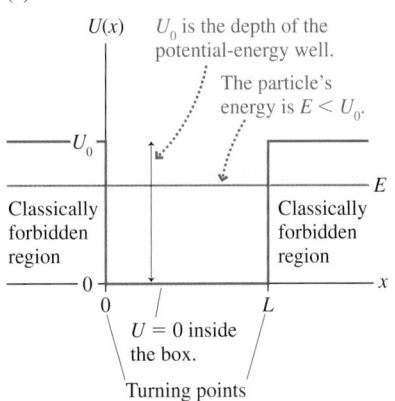

(b) $U = 0$ outside the well.

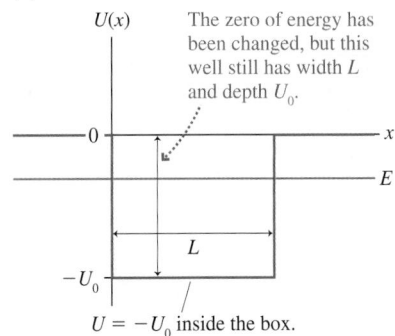

FIGURE 40.13 A finite potential well of width L and depth U_0.

As a first example, consider an electron in a 2.0-nm-wide potential well of depth $U_0 = 1.0$ eV. These are reasonable parameters for an electron in a semiconductor device. Figure 40.14a is a graphical presentation of the allowed energies and wave functions. For comparison, Figure 40.14b shows the first three energy levels and wave functions for a rigid box ($U_0 \to \infty$) with the same 2.0 nm width.

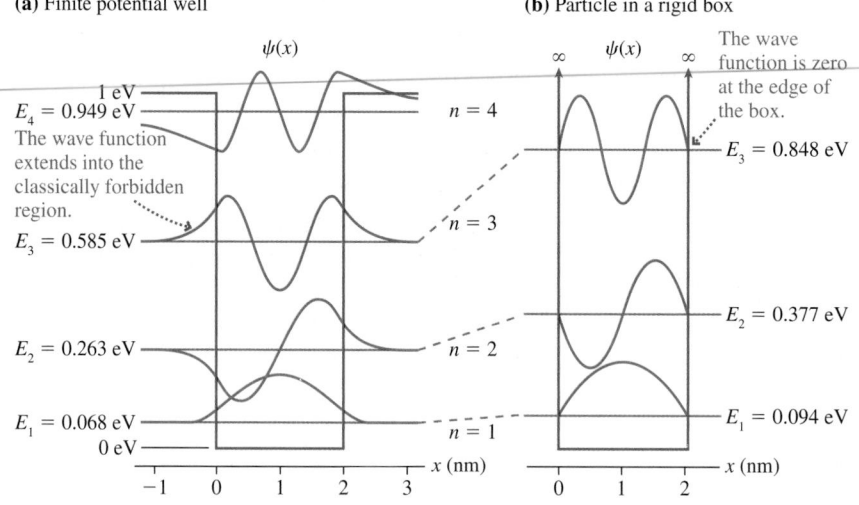

(a) Finite potential well

(b) Particle in a rigid box

The wave function is zero at the edge of the box.

$E_4 = 0.949$ eV

The wave function extends into the classically forbidden region.

$E_3 = 0.585$ eV

$E_2 = 0.263$ eV

$E_1 = 0.068$ eV

$E_3 = 0.848$ eV

$E_2 = 0.377$ eV

$E_1 = 0.094$ eV

FIGURE 40.14 Energy levels and wave functions for a finite potential well. For comparison, the energies and wave functions are shown for a rigid box of equal width.

20.3

The quantum-mechanical solution for a particle in a finite potential well has several important properties:

- The particle's energy is quantized. A particle in the potential well *must* be in one of the stationary states with quantum numbers $n = 1, 2, 3, \ldots$
- There are only a finite number of **bound states**—four in this example, although the number will be different in other examples. These wave functions represent electrons confined to, or bound in, the potential well. There are no stationary states with $E > U_0$ because such a particle would not remain in the well.
- The wave functions are qualitatively similar to those of a particle in a rigid box, but the energies are somewhat lower. This is because the wave functions are slightly more spread out. A slightly larger de Broglie wavelength corresponds to a lower velocity and thus a lower energy.
- Most interesting, perhaps, is that the wave functions of Figure 40.14a extend into the classically forbidden regions. It is as though a tennis ball penetrated partly *through* the racket's strings before bouncing back, but without breaking the strings.

EXAMPLE 40.6 Absorption spectrum of an electron

What wavelengths of light are absorbed by a semiconductor device in which electrons are confined in a 2.0-nm-wide region with a potential-energy depth of 1.0 eV?

MODEL The electron is in the finite potential well whose energies and wave functions were shown in Figure 40.14a.

SOLVE Photons of light can be absorbed if a photon's energy $E_{\text{photon}} = hf$ exactly matches the energy difference ΔE between two energy levels of the system. Because most electrons will be in the $n = 1$ ground state, the absorption transitions will be $1 \to 2$, $1 \to 3$, and $1 \to 4$.

The absorption wavelengths $\lambda = c/f$ are

$$\lambda_{n \to m} = \frac{hc}{\Delta E} = \frac{hc}{|E_n - E_m|}$$

For this example, we find

$$\Delta E_{1-2} = 0.195 \text{ eV} \qquad \lambda_{1 \to 2} = 6.37 \text{ } \mu\text{m}$$
$$\Delta E_{1-3} = 0.517 \text{ eV} \qquad \lambda_{1 \to 3} = 2.40 \text{ } \mu\text{m}$$
$$\Delta E_{1-4} = 0.881 \text{ eV} \qquad \lambda_{1 \to 4} = 1.41 \text{ } \mu\text{m}$$

ASSESS These transitions are all in the infrared portion of the spectrum.

STOP TO THINK 40.3   This is a wave function for a particle in a finite quantum well. What is the particle's quantum number?

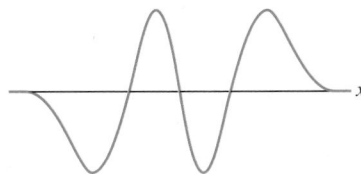

The Classically Forbidden Region

The extension of a particle's wave functions into the classically forbidden region is an important difference between classical and quantum physics. Let's take a closer look at the wave function in the region $x \geq L$. The potential energy in the classically forbidden region is U_0, thus the Schrödinger equation for $x \geq L$ is

$$\frac{d^2\psi}{dx^2} = -\frac{2m}{\hbar^2}(E - U_0)\psi(x)$$

We're assuming a confined particle, with E less than U_0, so $E - U_0$ is negative. It will be useful to reverse the order of these and write

$$\frac{d^2\psi}{dx^2} = \frac{2m}{\hbar^2}(U_0 - E)\psi(x) = \frac{1}{\eta^2}\psi(x) \tag{40.34}$$

where

$$\eta^2 = \frac{\hbar^2}{2m(U_0 - E)} \tag{40.35}$$

is a *positive* constant. As a homework problem, you can show that the units of η are m.

The Schrödinger equation of Equation 40.34 is one we can solve by guessing. We simply need to think of two functions whose second derivatives are a positive constant times the functions themselves. Two such functions, as you can quickly confirm, are $e^{x/\eta}$ and $e^{-x/\eta}$. Thus, according to Equation 40.8, the general solution of the Schrödinger equation for $x \geq L$ is

$$\psi(x) = Ae^{x/\eta} + Be^{-x/\eta} \quad \text{for } x \geq L \tag{40.36}$$

One requirement of the wave function is that $\psi \to 0$ as $x \to \infty$. The function $e^{x/\eta}$ diverges as $x \to \infty$, so the only way to satisfy this requirement is to set $A = 0$. This leaves

$$\psi(x) = Be^{-x/\eta} \quad \text{for } x \geq L \tag{40.37}$$

This is an exponentially decaying function. Notice that all the wave functions in Figure 40.14a look like exponential decays for $x > L$.

The wave function must also be continuous. Suppose the oscillating wave function within the potential well ($x \leq L$) has the value ψ_{edge} when it reaches the classical boundary at $x = L$. To be continuous, the wave function of Equation 40.37 has to match this value at $x = L$. That is,

$$\psi\,(\text{at } x = L) = Be^{-L/\eta} = \psi_{edge} \tag{40.38}$$

This boundary condition at $x = L$ is sufficient to determine that the constant B is

$$B = \psi_{edge}e^{L/\eta} \tag{40.39}$$

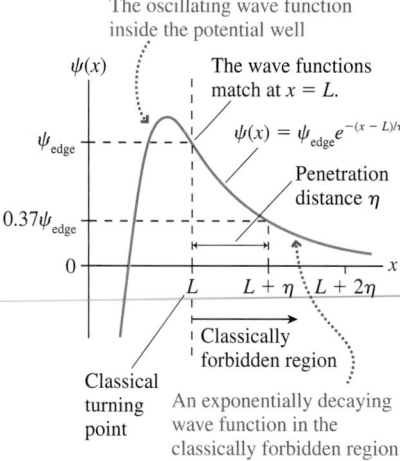

The oscillating wave function inside the potential well

The wave functions match at $x = L$.

$\psi(x) = \psi_{edge}e^{-(x-L)/\eta}$

Penetration distance η

Classically forbidden region

Classical turning point

An exponentially decaying wave function in the classically forbidden region

FIGURE 40.15 The wave function in the classically forbidden region.

If we use the Equation 40.39 result for B in Equation 40.37, we find that the wave function in the classically forbidden region of a finite potential well is

$$\psi(x) = \psi_{edge}e^{-(x-L)/\eta} \quad \text{for } x \geq L \qquad (40.40)$$

In other words, **the wave function oscillates until it reaches the classical turning point at $x = L$, then it decays exponentially within the classically forbidden region.**

Figure 40.15 shows the wave function in the classically forbidden region. You can see that the wave function at $x = L + \eta$ has decreased to

$$\psi(\text{at } x = L + \eta) = e^{-1}\psi_{edge} = 0.37\psi_{edge}$$

Although an exponential decay does not have a sharp ending point, the parameter η measures "about how far" the wave function extends past the classical turning point before the probability of finding the particle has decreased nearly to zero. This distance is called the **penetration distance:**

$$\text{penetration distance } \eta = \frac{\hbar}{\sqrt{2m(U_0 - E)}} \qquad (40.41)$$

A classical particle reverses direction at the $x = L$ turning point. But atomic particles are not classical. Because of wave-particle duality, an atomic particle is "fuzzy" and without a well-defined edge. Thus an atomic particle can spread a distance of roughly η into the classically forbidden region.

The penetration distance is unimaginably small for any macroscopic mass, but it can be significant for atomic particles. Notice that the penetration distance depends inversely on the quantity $U_0 - E$, which is the distance of the energy level below the top of the potential well. You can see in Figure 40.14a that η is much larger for the $n = 4$ state, near the top of the potential well, than for the $n = 1$ state.

NOTE ▶ In making use of Equation 40.41 you *must* use SI units of J s for $\hbar$ and J for the energies. The penetration distance η is then in m. ◀

EXAMPLE 40.7 Penetration distance of an electron
An electron is confined in a 2.0-nm-wide region with a potential-energy depth of 1.00 eV. What are the penetration distances into the classically forbidden region for an electron in the $n = 1$ and $n = 4$ states?

MODEL The electron is in the finite potential well whose energies and wave functions were shown in Figure 40.14a.

SOLVE The ground state has $U_0 - E_1 = 1.000 \text{ eV} - 0.068 \text{ eV} = 0.932 \text{ eV}$. Similarly, $U_0 - E_4 = 0.051 \text{ eV}$ in the $n = 4$ state. We can use Equation 40.41 to calculate

$$\eta = \frac{\hbar}{\sqrt{2m(U_0 - E)}} = \begin{cases} 0.20 \text{ nm} & n = 1 \\ 0.86 \text{ nm} & n = 4 \end{cases}$$

ASSESS These values are consistent with the penetration distances that you can estimate visually in Figure 40.14a.

Quantum-Well Devices

In Part VI we developed a model of electrical conductivity in which the valence electrons of a metal form a loosely bound "sea of electrons." The typical speed of an electron is the rms speed

$$v_{rms} = \sqrt{\frac{3k_B T}{m}}$$

where k_B is Boltzmann's constant. Hence at room temperature, where $v_{rms} \approx 1 \times 10^5$ m/s, the de Broglie wavelength of a typical conduction electron is

$$\lambda \approx \frac{h}{mv_{rms}} \approx 6 \text{ nm}$$

There is a range of wavelengths, because the electrons have a range of speeds, but this is a typical value.

You've now seen many times that wave effects are significant only when the sizes of physical structures are comparable to or smaller than the wavelength. This is why the interference and diffraction of light are hard to observe and why the wave-like nature of matter becomes important only on microscopic scales. Because the de Broglie wavelength of conduction electrons is only a few nm, quantum effects are insignificant in electronic devices whose features are larger than about 100 nm. The electrons in macroscopic devices can be treated as classical particles, which is how we analyzed electric current in Chapter 28.

However, devices smaller than about 100 nm do exhibit quantum effects. Some semiconductor devices, such as the semiconductor lasers used in fiber-optic communications, now incorporate features only a few nm in size. Quantum effects play an important role in these devices.

Figure 40.16a shows the construction of a *semiconductor diode laser.* Although the operating principles of diodes are beyond the scope of this textbook, we can note that a current travels through this device from left to right. In the center is a very thin layer of the semiconductor gallium arsenide (GaAs). It is surrounded on either side by layers of gallium aluminum arsenide (GaAlAs), and these in turn are embedded within the larger structure of the diode. The electrons within the central GaAs layer begin to emit laser light when the current through the diode exceeds some *threshold current.*

You can learn in a solid-state physics or materials engineering course that the electric potential energy of an electron is slightly lower in GaAs than in GaAlAs. This makes the GaAs layer a potential well for electrons, with higher-potential-energy GaAlAs "walls" on either side. As a result, the electrons become trapped within the thin GaAs layer. Such a device is called a **quantum-well laser.**

As an example, Figure 40.16b shows a quantum-well device with a 1.0-nm-thick GaAs layer in which the electron's potential energy is 0.30 eV less than in the surrounding GaAlAs layers. A numerical solution of the Schrödinger equation finds that this potential well has only a *single* quantum state, $n = 1$ with $E_1 = 0.125$ eV. Every electron trapped in this quantum well has the *same* energy—a very nonclassical result! The fact that the electron energies are so well defined, in contrast to the range of electron energies in bulk material, is what makes this a useful device. You can also see from the probability density $|\psi|^2$ that the electrons are more likely to be found in the center of the layer than at the edges. This concentration of electrons makes it easier for the device to begin laser action.

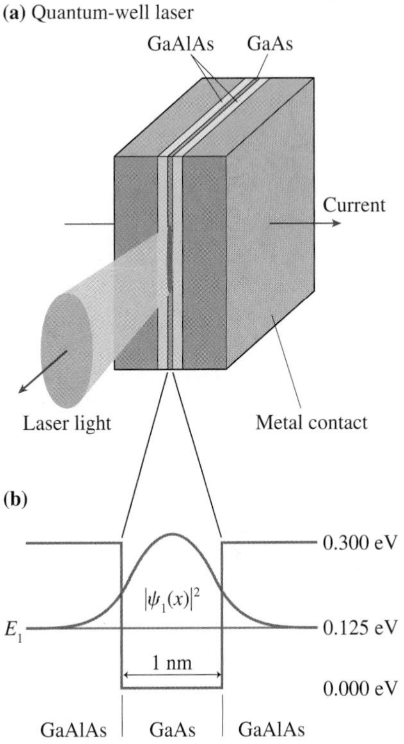

(a) Quantum-well laser

GaAlAs   GaAs

Current

Laser light

Metal contact

(b)

0.300 eV

$|\psi_1(x)|^2$

E_1

0.125 eV

1 nm

0.000 eV

GaAlAs | GaAs | GaAlAs

FIGURE 40.16 A semiconductor diode laser with a single quantum well.

Nuclear Physics

The nucleus of an atom consists of an incredibly dense assembly of protons and neutrons. The positively charged protons exert extremely strong electric repulsive forces on each other, so you might wonder how the nucleus keeps from exploding. During the 1930s, physicists found that protons and neutrons also exert an *attractive* force on each other. This force, one of the fundamental forces of nature, is called the *strong force.* It is the force that holds the nucleus together.

The primary characteristic of the strong force, other than its strength, is that it is a *short range* force. The attractive strong force between two *nucleons* (a nucleon is either a proton or a neutron; the strong force does not distinguish between them) rapidly decreases to zero if they are separated by more than about 2 fm. This is in sharp contrast to the long-range nature of the electric force.

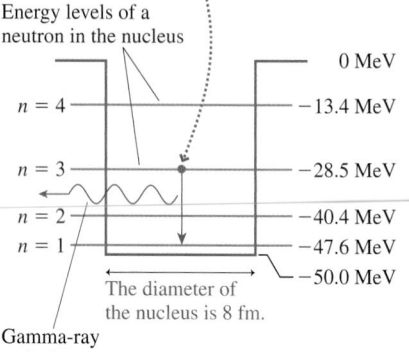

A radioactive decay has left the neutron in the $n = 3$ excited state. The neutron jumps to the $n = 1$ ground state, emitting a gamma-ray photon.

Energy levels of a neutron in the nucleus

	0 MeV
$n = 4$	−13.4 MeV
$n = 3$	−28.5 MeV
$n = 2$	−40.4 MeV
$n = 1$	−47.6 MeV
	−50.0 MeV

The diameter of the nucleus is 8 fm.

Gamma-ray emission

FIGURE 40.17 There are four allowed energy levels for a neutron in this nuclear potential well.

A reasonable model of the nucleus is to think of the protons and neutrons as particles in a nuclear potential well that is created by the strong force. The diameter of the potential well is equal to the diameter of the nucleus (this varies with atomic mass), and nuclear physics experiments have found that the depth of the potential well is ≈50 MeV.

The real potential well is three-dimensional, but let's make a simplified model of the nucleus as a one-dimensional potential well. Figure 40.17 shows the potential energy of a neutron along an *x*-axis passing through the center of the nucleus. Notice that the zero of energy has been chosen such that a "free" neutron, one outside the nucleus, has $E = 0$. Thus the potential energy inside the nucleus is −50 MeV. The 8 fm diameter shown is appropriate for a nucleus having atomic mass number $A \approx 40$, such as argon or potassium. Lighter nuclei will be a little smaller, heavier nuclei somewhat larger. (The potential-energy diagram for a proton is similar, but is complicated a bit by the electric potential energy.)

A numerical solution of the Schrödinger equation finds the four stationary states shown in Figure 40.17. The wave functions have been omitted, but they look essentially identical to the wave functions in Figure 40.14a. The major point to note is that the allowed energies differ by several *million* electron volts! These are enormous energies compared to those of an electron in an atom or a semiconductor. But recall that the energies of a particle in a rigid box, $E_n = n^2 h^2 / 8mL^2$, are proportional to $1/L^2$. Our previous examples, with nanometer-size boxes, found energies in the eV range. When the box size is reduced to femtometers, the energies jump up into the MeV range.

It often happens that the nuclear decay of a radioactive atom leaves a neutron in an excited state. For example, Figure 40.17 shows a neutron that has been left in the $n = 3$ state by a previous radioactive decay. This neutron can now undergo a quantum jump to the $n = 1$ ground state by emitting a photon with energy

$$E_{\text{photon}} = E_3 - E_1 = 19.1 \text{ MeV}$$

and wavelength

$$\lambda_{\text{photon}} = \frac{c}{f} = \frac{hc}{E_{\text{photon}}} = 6.50 \times 10^{-5} \text{ nm}$$

This photon is ≈10^7 times more energetic, and its wavelength ≈10^7 times smaller, than the photons of visible light! These extremely high-energy photons are called **gamma rays.** Gamma-ray emission is, indeed, one of the primary processes in the decay of radioactive elements.

Our one-dimensional model cannot be expected to give accurate results for the energy levels or gamma-ray energies of any specific nucleus. Nonetheless, this model does provide a reasonable understanding of the energy-level structure in nuclei and correctly predicts that nuclei can emit photons having energies of several million electron volts. This model, when extended to three dimensions, becomes the basis of the *shell model* of the nucleus in which the protons and neutrons are grouped in various shells analogous to the electron shells around an atom that you remember from chemistry. You can learn more about nuclear physics and the shell model in Chapter 42.

40.7 Wave-Function Shapes

Bound-state wave functions are standing de Broglie waves. In addition to boundary conditions, two other factors govern the shapes of wave functions:

1. The de Broglie wavelength is inversely dependent on the particle's speed. Consequently, the node spacing is smaller (shorter wavelength) where the kinetic energy is larger, and it is larger (longer wavelength) where the kinetic energy is smaller.

2. A classical particle is more likely to be found where it is moving more slowly. In quantum mechanics, the probability of finding the particle increases as the wave-function amplitude increases. Consequently, the wave-function amplitude is larger where the kinetic energy is smaller, and it is smaller where the kinetic energy is larger.

We can use this information to draw reasonably accurate wave functions for the different allowed energies in a potential-energy well.

TACTICS BOX 40.1 Drawing wave functions

❶ Draw a graph of the potential energy $U(x)$. Show the allowed energy E as a horizontal line. Locate the classical turning points.

❷ Draw the wave function as a continuous, oscillatory function between the turning points. The wave function for quantum state n has n anti-nodes and $n - 1$ nodes (excluding the ends).

❸ Make the wavelength longer (larger node spacing) and the amplitude higher in regions where the kinetic energy is smaller. Make the wavelength shorter and the amplitude lower in regions where the kinetic energy is larger.

❹ Bring the wave function to zero at the edge of an infinitely high potential-energy "wall."

❺ Let the wave function decay exponentially inside a classically forbidden region where $E < U$. The penetration distance η increases as E gets closer to the top of the potential-energy well.

EXAMPLE 40.8 Sketching wave functions

Figure 40.18a shows a potential-energy well and the allowed energies for the $n = 1$ and $n = 4$ quantum states. Sketch the $n = 1$ and $n = 4$ wave functions.

VISUALIZE The steps of Tactics Box 40.1 have been followed to sketch the wave functions shown in Figure 40.18b.

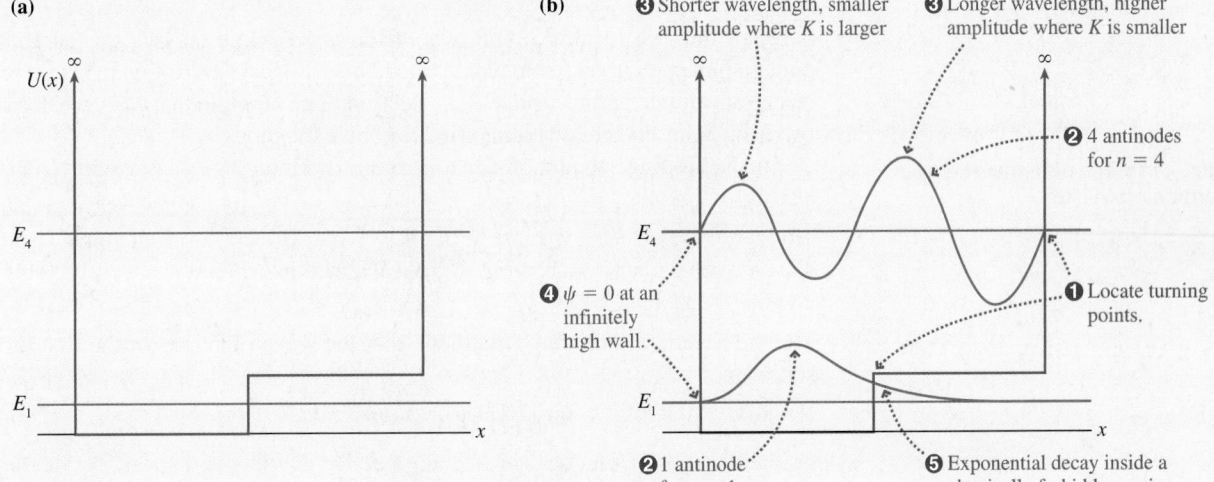

FIGURE 40.18 The $n = 1$ and $n = 4$ wave functions in a potential-energy well.

STOP TO THINK 40.4 For which potential energy is this an appropriate $n = 4$ wave function?

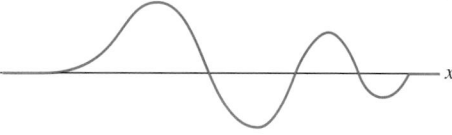

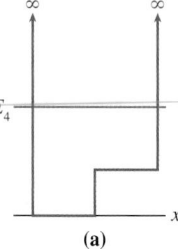

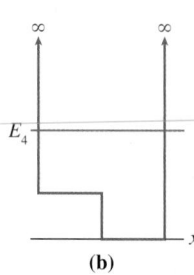

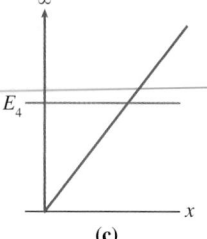

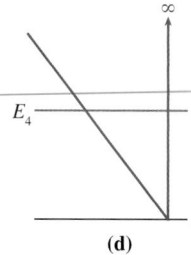

(a)   (b)   (c)   (d)

40.8 The Quantum Harmonic Oscillator

Simple harmonic motion is exceptionally important in classical physics, where it serves as a prototype for more complex oscillations. As you might expect, a microscopic oscillator—the **quantum harmonic oscillator**—is equally important as a model of oscillations at the atomic level.

The defining characteristic of simple harmonic motion is a linear restoring force: $F = -kx$, where k is the spring constant. The corresponding potential-energy function, as you learned in Chapter 10, is

$$U(x) = \frac{1}{2}kx^2 \tag{40.42}$$

where we'll assume that the equilibrium position is $x_e = 0$. The potential energy of a harmonic oscillator is shown in Figure 40.19. It is a potential-energy well with curved sides.

A classical particle of mass m oscillates with angular frequency

$$\omega = \sqrt{\frac{k}{m}} \tag{40.43}$$

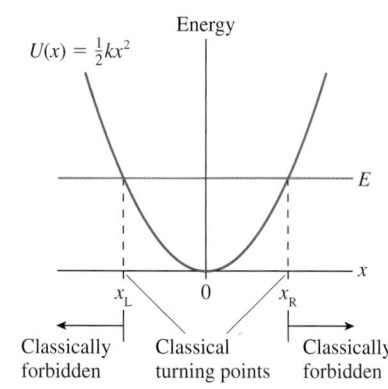

FIGURE 40.19 The potential energy of a harmonic oscillator.

between the two turning points where the energy line crosses the parabolic potential-energy curve. As you've learned, this classical description fails if m represents an atomic particle, such as an electron or an atom. In that case, we need to solve the Schrödinger equation to find the wave functions.

The Schrödinger equation for a quantum harmonic oscillator with $U(x) = \frac{1}{2}kx^2$ is

$$\frac{d^2\psi}{dx^2} = -\frac{2m}{\hbar^2}\left(E - \frac{1}{2}kx^2\right)\psi(x) \tag{40.44}$$

We will assert, without deriving them, that the wave functions of the first three states are

$$\psi_1(x) = A_1 e^{-x^2/2b^2}$$

$$\psi_2(x) = A_2 \frac{x}{b} e^{-x^2/2b^2}$$

$$\psi_3(x) = A_3\left(1 - \frac{2x^2}{b^2}\right)e^{-x^2/2b^2} \tag{40.45}$$

where b is

$$b = \sqrt{\frac{\hbar}{m\omega}} \qquad (40.46)$$

The constant b has dimensions of length. We will leave it as a homework problem for you to show that b is the classical turning point of an oscillator in the $n = 1$ ground state. The constants A_1, A_2, and A_3 are normalization constants. For example, A_1 can be found by requiring

$$\int_{-\infty}^{\infty} |\psi_1(x)|^2 dx = A_1^2 \int_{-\infty}^{\infty} e^{-x^2/b^2} dx = 1 \qquad (40.47)$$

The completion of this calculation will also be left as a homework problem.

As expected, stationary states of a quantum harmonic oscillator exist only for certain discrete energy levels, the quantum states of the oscillator. The allowed energies are given by the very simple equation

$$E_n = \left(n - \frac{1}{2}\right)\hbar\omega \qquad n = 1, 2, 3, \ldots \qquad (40.48)$$

where ω is the classical angular frequency of Equation 40.43 and n is the quantum number.

NOTE ▶ The ground-state energy of the quantum harmonic oscillator is $E_1 = \frac{1}{2}\hbar\omega$. An atomic mass on a spring can *not* be brought to rest. This is a consequence of the uncertainty principle. ◀

Figure 40.20 shows the first three energy levels and wave functions of a quantum harmonic oscillator. Notice that the energy levels are equally spaced by $\Delta E = \hbar\omega$. This result differs from the particle in a box, where the energy levels get increasingly farther apart. Also notice that the wave functions, like those of the finite potential well, extend beyond the turning points into the classically forbidden region.

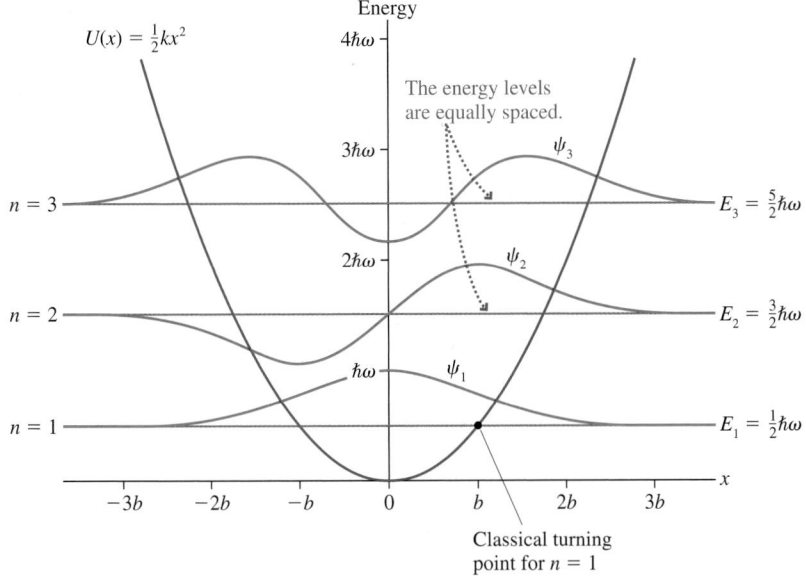

FIGURE 40.20 The first three energy levels and wave functions of a quantum harmonic oscillator.

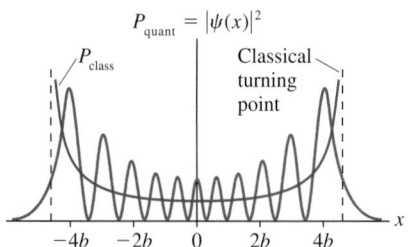

FIGURE 40.21 The quantum and classical probability densities for the $n = 11$ state of a quantum harmonic oscillator.

Figure 40.21 shows the probability density $|\psi(x)|^2$ for the $n = 11$ state of a quantum harmonic oscillator. Notice how the node spacing and the amplitude both increase as the particle moves away from the equilibrium position at $x = 0$. This is consistent with rule 3 of Tactics Box 40.1. The particle slows down as it moves away from the origin, causing its de Broglie wavelength *and* the probability of finding it to increase.

Section 40.5 introduced the classical probability density $P_{class}(x)$ and noted that a classical particle is most likely to be found where it is moving the slowest. Figure 40.21 shows $P_{class}(x)$ for a classical particle with the same total energy as the $n = 11$ quantum state. You can see that *on average* the quantum probability density $|\psi(x)|^2$ mimics the classical probability density. This is just what the correspondence principle leads us to expect.

EXAMPLE 40.9 Light emission by an oscillating electron
An electron in a harmonic-oscillator potential well emits light of wavelength 600 nm as it jumps from one level to the next. What is the spring constant of the restoring force?

MODEL The electron is a quantum harmonic oscillator.

SOLVE A photon is emitted as the electron undergoes the quantum jump $n \rightarrow n - 1$. We can use Equation 40.48 for the energy levels to find that the electron loses energy

$$\Delta E = E_n - E_{n-1} = \left(n - \frac{1}{2}\right)\hbar\omega_e - \left(n - 1 - \frac{1}{2}\right)\hbar\omega_e = \hbar\omega_e$$

$\Delta E = \hbar\omega_e$ for *all* transitions, independent of n, because the energy levels of the quantum harmonic oscillator are equally spaced. We need to distinguish the oscillations of the electron from the oscillations of the light wave, hence the subscript e on ω_e.

The emitted photon has energy $E_{photon} = hf_{ph} = \Delta E$. Thus

$$\hbar\omega_e = \frac{h}{2\pi}\omega_e = hf_{ph} = \frac{hc}{\lambda}$$

The wavelength of the light is $\lambda = 600$ nm, hence the classical angular frequency of the oscillating electron is

$$\omega_e = 2\pi\frac{c}{\lambda} = 3.14 \times 10^{15} \text{ rad/s}$$

The electron's angular frequency is related to the spring constant of the restoring force by

$$\omega_e = \sqrt{\frac{k}{m}}$$

Thus $k = m\omega_e^2 = 9.0$ N/m.

Molecular Vibrations

We've made many uses of the idea that atoms are held together by spring-like molecular bonds. We've always assumed that the bonds could be modeled as classical springs. The classical model is acceptable for some purposes, but it fails to explain some important features of molecular vibrations. Not surprisingly, the quantum harmonic oscillator is a better model of a molecular bond.

Figure 40.22 is the potential energy of two atoms connected by a molecular bond. Nearby atoms attract each other through a polarization force, much as a charged rod picks up small pieces of paper. If the atoms get too close, a *repulsive* force between the negative electrons pushes them away. The equilibrium separation at which the attractive and repulsive forces are balanced is r_0, and two classical atoms would be at rest at this separation. But quantum particles, even in their lowest energy state, have $E > 0$. Consequently, the molecule *vibrates* as the two atoms oscillate back and forth along the bond.

U_{dissoc} is the energy at which the molecule will *dissociate* and the two atoms will fly apart. Dissociation can occur at very high temperatures or after the molecule has absorbed a high-energy (ultraviolet) photon, but under typical conditions a molecule has energy $E \ll U_{dissoc}$. In other words, the molecule is in an energy level near the bottom of the potential well.

You can see that the lower portion of the potential well is very nearly a parabola. Consequently, we can model a molecular bond as a quantum harmonic

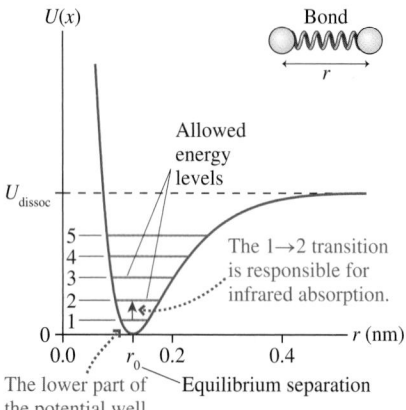

FIGURE 40.22 The potential energy of a molecular bond and a few of the allowed energies.

oscillator. The energy associated with the molecular vibration is quantized and can have *only* the values

$$E_{\text{vib}} \approx \left(n - \frac{1}{2}\right)\hbar\omega \qquad n = 1, 2, 3, \ldots \qquad (40.49)$$

where ω is the angular frequency with which the atoms would vibrate if the bond were a classical spring. The molecular potential-energy curve is not exactly that of a harmonic oscillator, hence the $\approx$ sign; but the model is very good for low values of the quantum number n. The energy levels calculated by Equation 40.49 are called the oscillatory function **vibrational energy levels** of the molecule. The first few vibrational energy levels are shown in Figure 40.22.

At room temperature, most molecules are in the $n = 1$ vibrational ground state. Their vibrational motion can be excited by absorbing photons of frequency $f = \Delta E/h$. This frequency is usually in the infrared region of the spectrum, and these *vibrational transitions* give each molecule a unique and distinctive infrared absorption spectrum.

As an example, Figure 40.23 shows the infrared absorption spectrum of acetone. The vertical axis is the percentage of the light intensity passing all the way through the sample. The sample is essentially transparent at most wavelengths (transmission $\approx$ 100%), but there are two prominent absorption features. The transmission drops to $\approx$75% at $\lambda = 3.3\ \mu$m and to a mere 7% at $\lambda = 5.8\ \mu$m. The 3.3 μm absorption is due to the $n = 1$ to $n = 2$ transition in the vibration of a $C - CH_3$ carbon-methyl bond. The 5.8 μm absorption is the $1 \rightarrow 2$ transition of a vibrating $C = O$ carbon-oxygen double bond.

Absorption spectra such as this are known for thousands of molecules, and chemists routinely use absorption spectroscopy to identify the chemicals in a sample. A specific bond has the same absorption wavelength regardless of the larger molecule in which it is embedded, thus the presence of that absorption wavelength is a "signature" that the bond is present within a molecule.

FIGURE 40.23 The absorption spectrum of acetone.

STOP TO THINK 40.5 Which probability density represents a quantum harmonic oscillator with $E = \frac{5}{2}\hbar\omega$?

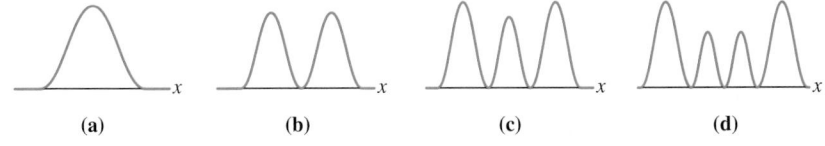

(a) (b) (c) (d)

40.9 More Quantum Models

In this section we'll look at two more examples of quantum-mechanical models.

A Particle in a Capacitor

Many semiconductor devices are designed to confine electrons within a layer only a few nanometers thick. If a potential difference is applied across the layer, the electrons act very much as if they are trapped within a microscopic capacitor.

Figure 40.24a on the next page shows two capacitor plates separated by distance L. The left plate is positive, so the electric field points to the right with strength $E = \Delta V_0/L$. Because of its negative charge, an electron launched from the left plate is slowed by a *retarding* force. The electron makes it across to the right plate if it starts with sufficient kinetic energy; otherwise it reaches a turning point and then is pushed back to the positive plate.

(a)

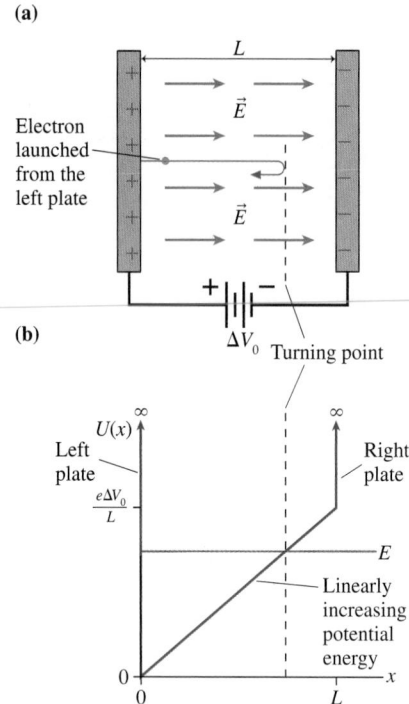

(b)

FIGURE 40.24 An electron in a capacitor.

This classical analysis is a valid model of a macroscopic capacitor. But if L becomes sufficiently small, comparable to the de Broglie wavelength of an electron, then the wave-like properties of the electron cannot be ignored. We need a quantum-mechanical model.

Let's establish a coordinate system with $x = 0$ at the left plate and $x = L$ at the right plate. Define the electric potential to be zero at the positive plate. The potential *decreases* in the direction of the field, so the potential inside the capacitor (see Section 29.5) is

$$V(x) = -Ex = -\frac{\Delta V_0}{L}x$$

The electron, with charge $q = -e$, has potential energy

$$U(x) = qV(x) = +\frac{e\Delta V_0}{L}x \qquad 0 < x < L \qquad (40.50)$$

This potential energy increases linearly for $0 < x < L$. If we assume that the capacitor plates act like the walls of a rigid box, then $U(x) \to \infty$ at $x = 0$ and $x = L$.

Figure 40.24b shows the electron's potential-energy function. It is the particle-in-a-rigid-box potential with a sloping "floor" due to the electric field. The figure also shows the total energy line E of an electron in the capacitor. The energy is purely kinetic at $x = 0$, where $K = E$, but it is converted to potential energy as the electron moves to the right. The right turning point occurs where the energy line E crosses the potential-energy curve $U(x)$. If the electron is a classical particle, it must reverse position at this point.

NOTE ▶ This is also the potential energy for a microscopic bouncing ball that is trapped between a floor at $y = 0$ and a ceiling at $y = L$. ◀

It is physically impossible for an electron to be outside the capacitor, so the wave function must be zero for $x < 0$ and $x > L$. The continuity of ψ requires the same boundary conditions as for a particle in a rigid box: $\psi = 0$ at $x = 0$ and at $x = L$. The wave functions inside the capacitor are too complicated to find by guessing, so we have solved the Schrödinger equation numerically and will present the results graphically.

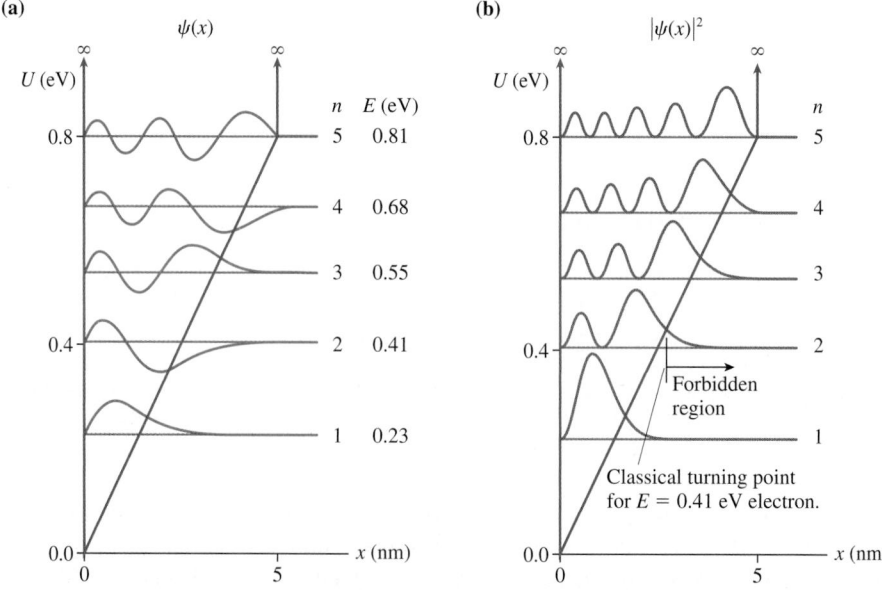

FIGURE 40.25 Energy levels, wave functions, and probability densities for an electron in a 5.0-nm-wide capacitor with a 0.80 V potential difference.

Figure 40.25 shows the wave functions and probability densities for the first five quantum states of an electron confined in a 5.0-nm-thick layer that has a 0.80 V potential difference across it. Each allowed energy is represented as a horizontal line, with the numerical values shown on the right. They range from $E_1 = 0.23$ eV up to $E_5 = 0.81$ eV. An electron *must* have one of the allowed energies shown in the figure. An electron cannot have $E = 0.30$ eV in this capacitor because no de Broglie wave with that energy can match the necessary boundary conditions.

NOTE ▶ Remember that each wave function is graphed as if its energy line is the *x*-axis. ◀

We can make several observations about the Schrödinger equation solutions:

1. The energies E_n become more closely spaced as n increases. This behavior is in contrast to the particle in a box, for which E_n became more widely spaced.
2. The spacing between the nodes of a wave function is not constant but increases toward the right. This is because an electron on the right side of the capacitor has less kinetic energy and thus a slower speed and a larger de Broglie wavelength.
3. The height of the probability density $|\psi|^2$ increases toward the right. That is, we are more likely to find the electron on the right side of the capacitor than on the left. But this also makes sense if, classically, the electron is moving more slowly when on the right side and thus spending more time there than on the left side.
4. The electron penetrates *beyond* the classical turning point into the classically forbidden region.

EXAMPLE 40.10 **The emission spectrum of an electron in a capacitor**

What are the frequencies of photons emitted by electrons in the $n = 4$ state of Figure 40.25?

SOLVE Photon emission occurs as the electrons make $4 \rightarrow 3$, $4 \rightarrow 2$, and $4 \rightarrow 1$ quantum jumps. In each case, the photon frequency is $f = \Delta E/h$ and the wavelength is

$$\lambda = \frac{c}{f} = \frac{hc}{\Delta E}$$

The energies of the quantum jumps, which can be read from Figure 40.25a, are $\Delta E_{4 \rightarrow 3} = 0.13$ eV, $\Delta E_{4 \rightarrow 2} = 0.27$ eV, and $\Delta E_{4 \rightarrow 1} = 0.45$ eV. Thus

$$\lambda_{4 \rightarrow 3} = 9500 \text{ nm} = 9.5 \text{ } \mu\text{m}$$

$$\lambda_{4 \rightarrow 2} = 4600 \text{ nm} = 4.6 \text{ } \mu\text{m}$$

$$\lambda_{4 \rightarrow 1} = 2800 \text{ nm} = 2.8 \text{ } \mu\text{m}$$

ASSESS The $n = 4$ electrons in this device emit three distinct infrared wavelengths.

The Covalent Bond

You probably recall from chemistry that a **covalent molecular bond,** such as the bond between the two atoms in molecules such as H_2 and O_2, is a bond in which the electrons are shared between the atoms. The basic idea of covalent bonding can be understood with a one-dimensional quantum-mechanical model.

The simplest molecule, the hydrogen molecular ion H_2^+, consists of two protons and one electron. Although it seems surprising that such a system could be stable, the two protons form a molecular bond with one electron. This is the simplest covalent bond.

How can we model the H_2^+ ion? To begin, Figure 40.26a on the next page shows a one-dimensional model of a hydrogen *atom* in which the electron's Coulomb potential energy, with its $1/r$ dependence, has been approximated by a finite potential well of width 0.10 nm ($\approx 2a_B$) and depth 24.2 eV. You learned in Chapter 38 that an electron in the ground state of the Bohr hydrogen atom orbits

(a) Simple one-dimensional model of an electron in a hydrogen atom

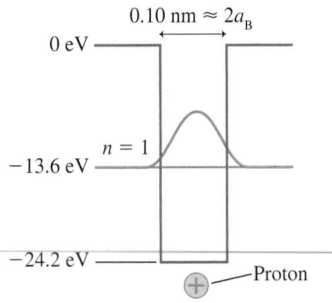

(b) An H_2^+ molecule modeled as an electron with two protons separated by 0.12 nm

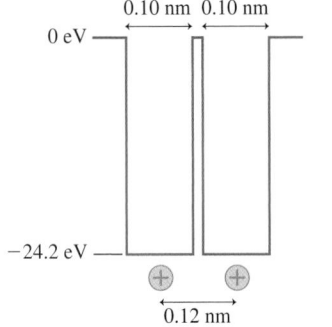

FIGURE 40.26 A molecule can be modeled as two closely spaced potential wells, one representing each atom.

the proton with radius $r_1 = a_B$ (the Bohr radius) and energy $E_1 = -13.6$ eV. A numerical solution of the Schrödinger equation finds that the ground-state energy of this finite potential well is $E_1 = -13.6$ eV. Clearly this model of a hydrogen atom is oversimplified, but it does have the correct size and the correct ground-state energy.

We can model H_2^+ by bringing two of these potential wells close together. The molecular bond length of H_2^+ is known to be ≈ 0.12 nm, so Figure 40.26b shows potential wells with 0.12 nm between their centers. This is a model of H_2^+, not a complete H_2 molecule, because this is the potential energy of a single electron. (Modeling H_2 is more complex because we would need to consider the repulsion between the two electrons.)

Figure 40.27 shows the allowed energies and wave functions for an electron with this potential energy. The $n = 1$ wave function has a high probability of being found within the classically forbidden region *between* the two protons. In other words, an electron in this quantum state really is "shared" by the protons and spends most of its time between them.

In contrast, an electron in the $n = 2$ energy level has zero probability of being found between the two protons because the $n = 2$ wave function has a node at the center. The probability density shows that an $n = 2$ electron is "owned" by one proton or the other rather than being shared.

To learn the consequences of these wave functions we need to calculate the total energy of the molecule: $E_{mol} = E_{p-p} + E_{elec}$. The $n = 1$ and $n = 2$ energies shown in Figure 40.27 are the energies E_{elec} of the electron. At the same time, the protons repel each other and have electric potential energy E_{p-p}. It's not hard to calculate that $E_{p-p} = 12.0$ eV for two protons separated by 0.12 nm. Thus

$$E_{mol} = E_{p-p} + E_{elec} = \begin{cases} 12.0\text{ eV} - 17.5\text{ eV} = -5.5\text{ eV} & n = 1 \\ 12.0\text{ eV} - 9.0\text{ eV} = +3.0\text{ eV} & n = 2 \end{cases}$$

The $n = 1$ molecular energy is less than zero, showing that this is a *bound state*. The $n = 1$ wave function is called a **bonding molecular orbital.** Although

(a) Bonding orbital

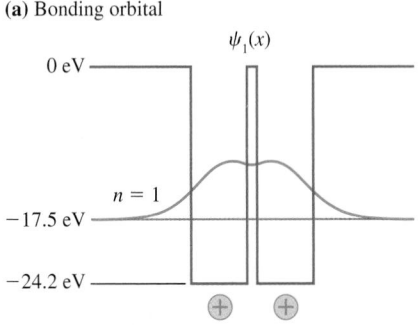

(b) Antibonding orbital

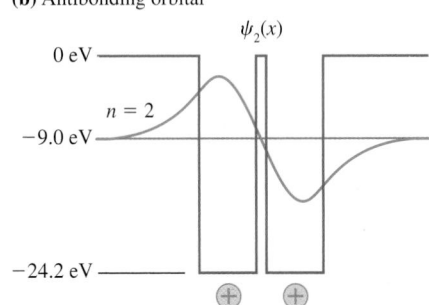

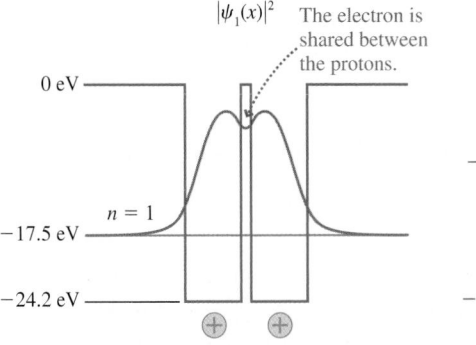

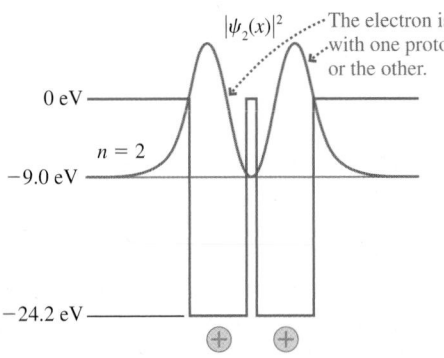

FIGURE 40.27 The wave functions and probability densities of the electron in H_2^+.

the protons repel each other, the shared electron provides sufficient "glue" to hold the system together. The $n = 2$ molecular energy is positive, so this is *not* a bound state. The system would be more stable as a hydrogen atom and a distant proton. The $n = 2$ wave function is called an **antibonding molecular orbital.**

Both E_{elec} and E_{p-p} depend on the separation between the protons, which we assumed to be 0.12 nm in this calculation. If we were to calculate and graph E_{mol} for many different values of the proton separation, the graph would look like the molecular-bond energy curve shown in Figure 40.22. In other words, a molecular bond has an equilibrium length where the bond energy is a minimum *because* of the interplay between E_{p-p} and E_{elec}.

Although real molecular wave functions are more complex than this one-dimensional model, the $n = 1$ wave function captures the essential idea of a covalent bond. Notice that a "classical" molecule cannot have a covalent bond because the electron would not be able to exist in the classically forbidden region. Covalent bonds can be understood only within the context of quantum mechanics. In fact, the explanation of molecular bonds was one of the earliest successes of quantum mechanics.

40.10 Quantum-Mechanical Tunneling

Figure 40.28a shows a ball rolling toward a hill. A ball with sufficient kinetic energy can go over the top of the hill, slowing down as it ascends and speeding up as it rolls down the other side. A ball with insufficient energy rolls partway up the hill, then reverses direction and rolls back down.

Activ Physics 20.4

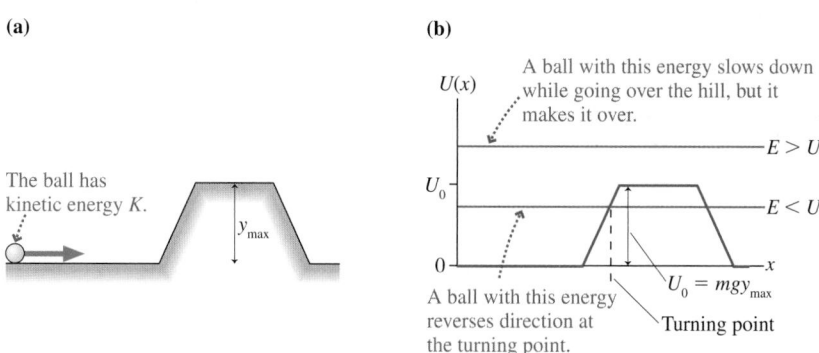

FIGURE 40.28 A hill is an energy barrier to a rolling ball.

We can think of the hill as an "energy barrier" of height $U_0 = mgy_{max}$. As Figure 40.28b shows, a ball incident from the left with energy $E > U_0$ can go over the barrier (i.e., roll over the hill), but a ball with $E < U_0$ will reflect from the energy barrier at the turning point. According to the laws of classical physics, a ball that is incident on the energy barrier from the left with $E < U_0$ will never be found on the right side of the barrier.

NOTE ▶ Figure 40.28b is not a "picture" of the energy barrier. And when we say that a ball with energy $E > U_0$ can go "over" the barrier, we don't mean that the ball is thrown from a higher elevation in order to go over the top of the hill. The ball rolls *on the ground* the entire time, as Figure 40.28a shows, and Figure 40.28b describes the kinetic and potential energy of the ball as it rolls. A higher total energy line means a larger initial kinetic energy, not a higher elevation. ◄

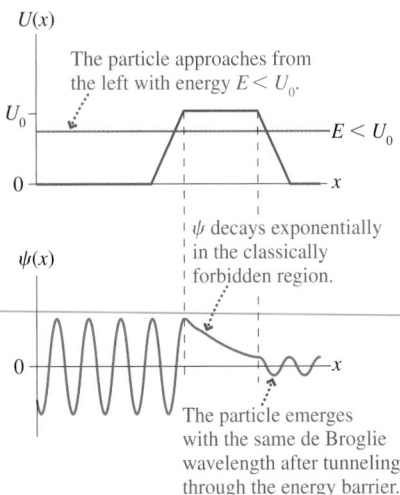

The particle approaches from the left with energy $E < U_0$.

ψ decays exponentially in the classically forbidden region.

The particle emerges with the same de Broglie wavelength after tunneling through the energy barrier.

FIGURE 40.29 A quantum particle can penetrate through the energy barrier.

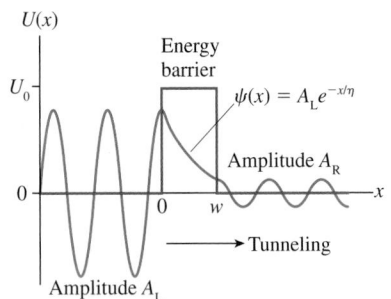

FIGURE 40.30 Tunneling through an idealized energy barrier.

Figure 40.29 shows the situation from the perspective of quantum mechanics. As you've learned, quantum particles can penetrate with an exponentially decreasing wave function into the classically forbidden region of an energy barrier. Suppose that the barrier is very narrow. Although the wave function decreases within the barrier, starting at the classical turning point, it hasn't vanished when it reaches the other side. In other words, there is some probability that a quantum particle will pass *through* the barrier and emerge on the other side!

It is very much as if the ball of Figure 40.28a gets to the turning point and then, instead of reversing direction and rolling back down, tunnels its way *through* the hill and emerges on the other side. Although this feat is strictly forbidden in classical mechanics, it is apparently acceptable behavior for quantum particles. The process is called **quantum-mechanical tunneling.**

The process of tunneling through a potential-energy barrier is one of the strangest and most unexpected predictions of quantum mechanics. Yet it does happen, and you will see that it even has many practical applications.

NOTE ▶ The word "tunneling" is used as a metaphor. If a classical particle really did tunnel, it would expend energy doing so and emerge on the other side with less energy. Quantum-mechanical tunneling requires no expenditure of energy. The total energy line is at the same height on both sides of the barrier. A particle that tunnels through a barrier emerges with *no* loss of energy. That is why the de Broglie wavelength is the same on both sides of the potential barrier in Figure 40.29. ◀

To simplify our analysis of tunneling, Figure 40.30 shows an idealized energy barrier of height U_0 and width w. We've superimposed the wave function on top of the energy diagram so that you can see how it aligns with the potential energy. The wave function to the left of the barrier is a sinusoidal oscillation with amplitude A_L. The wave function *within* the barrier is the decaying exponential we found in Equation 40.40:

$$\psi_{in}(0 \le x \le w) = \psi_{edge}e^{-x/\eta} = A_L e^{-x/\eta} \tag{40.51}$$

where we've assumed $\psi_{edge} = A_L$. The penetration distance η was given in Equation 40.41 as

$$\eta = \frac{\hbar}{\sqrt{2m(U_0 - E)}}$$

NOTE ▶ You *must* use SI units when calculating values of η. Energies must be in J and $\hbar$ in J s. The penetration distance η has units of m. ◀

The wave function decreases exponentially within the barrier, but before it can decay to zero it emerges again on the right side ($x > w$) as an oscillation with amplitude

$$A_R = \psi_{in}(\text{at } x = w) = A_L e^{-w/\eta} \tag{40.52}$$

The probability that the particle is to the left of the barrier is proportional to $|A_L|^2$ and the probability of finding it to the right of the barrier is proportional to $|A_R|^2$. Thus the probability that a particle striking the barrier from the left will emerge on the right is

$$P_{tunnel} = \frac{|A_R|^2}{|A_L|^2} = (e^{-w/\eta})^2 = e^{-2w/\eta} \tag{40.53}$$

This is the probability that a particle will tunnel through the energy barrier.

Now, our analysis, we have to say, has not been terribly rigorous. For example, we assumed that the oscillatory wave functions on the left and the right were exactly at a maximum where they reached the barrier at $x = 0$ and $x = w$. There is no reason this has to be the case. We have taken other liberties, which experts will spot, but—fortunately—it really makes no difference. Our result, Equation 40.53, turns out to be perfectly adequate for most applications of tunneling.

Because the tunneling probability is an exponential function, it is *very* sensitive to the values of w and η. The tunneling probability can be substantially reduced by even a small increase in the thickness of the barrier. The parameter η, which measures how far the particle can penetrate into the barrier, depends both on the particle's mass and on $U_0 - E$. A particle with E only slightly less than U_0 will have a larger value of η and thus a larger tunneling probability than will an identical particle with less energy.

EXAMPLE 40.11 Electron tunneling

a. Find the probability that an electron will tunnel through a 1.0-nm-wide energy barrier if the electron's energy is 0.10 eV less than the height of the barrier.

b. Find the tunneling probability if the barrier in part a is widened to 3.0 nm.

c. Find the tunneling probability if the electron in part a is replaced by a proton with the same energy.

SOLVE

a. An electron with energy 0.10 eV less than the height of the barrier has $U_0 - E = 0.10$ eV $= 1.60 \times 10^{-20}$ J. Thus its penetration distance is

$$\eta = \frac{\hbar}{\sqrt{2m(U_0 - E)}}$$

$$= \frac{1.05 \times 10^{-34} \text{ J s}}{\sqrt{2(9.11 \times 10^{-31} \text{ kg})(1.60 \times 10^{-20} \text{ J})}}$$

$$= 6.18 \times 10^{-10} \text{ m} = 0.618 \text{ nm}$$

The probability that this electron will tunnel through a barrier of width $w = 1.0$ nm is

$$P_{\text{tunnel}} = e^{-2w/\eta} = e^{-2(1.0 \text{ nm})/(0.618 \text{ nm})} = 0.039 = 3.9\%$$

b. Changing the width to $w = 3.0$ nm has no effect on η. The new tunneling probability is

$$P_{\text{tunnel}} = e^{-2w/\eta} = e^{-2(3.0 \text{ nm})/(0.618 \text{ nm})} = 6.0 \times 10^{-5}$$

$$= 0.006\%$$

Increasing the width by a factor of 3 decreases the tunneling probability by a factor of 660!

c. A proton is more massive than an electron. Thus a proton with $U_0 - E = 0.10$ eV has $\eta = 0.014$ nm. Its probability of tunneling through a 1.0-nm-wide barrier is

$$P_{\text{tunnel}} = e^{-2w/\eta} = e^{-2(1.0 \text{ nm})/(0.014 \text{ nm})} \approx 1 \times 10^{-64}$$

For practical purposes, the probability that a proton will tunnel through this barrier is zero.

ASSESS If the probability of a proton tunneling through a mere 1 nm is only 10^{-64}, you can see that a macroscopic object will "never" tunnel through a macroscopic distance!

Quantum-mechanical tunneling seems so obscure that it is hard to imagine practical applications. Surprisingly, there are many. We will look at two: the scanning tunneling microscope and resonant tunneling diodes.

The Scanning Tunneling Microscope

Diffraction limits the view of an optical microscope to objects no smaller than about a wavelength of light—roughly 500 nm. This is more than 1000 times the size of an atom, so there is no hope of resolving atoms or molecules via optical microscopy. Electron microscopes are similarly limited by the de Broglie wavelength of the electrons. Their resolution is much better than an optical microscope, but still not quite at the level of resolving individual atoms.

This situation changed dramatically in 1981 with the invention of the **scanning tunneling microscope,** or STM as it is usually called. The STM allowed scientists, for the first time, to "see" surfaces literally atom by atom. Figure 40.31 on the next page shows two pictures taken with an STM. In one you can see individual atoms of carbon on the surface of graphite. The other shows a

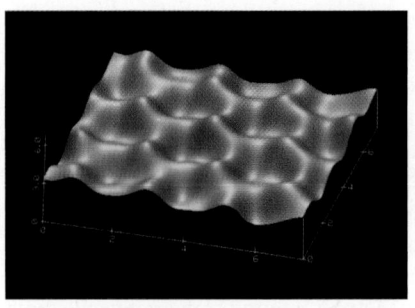

Individual atoms of carbon on   The surface of silicon
the surface of graphite

FIGURE 40.31 Two pictures made with a scanning tunneling microscope.

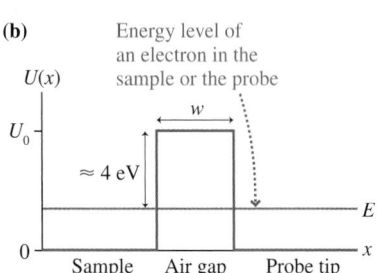

FIGURE 40.32 A scanning tunneling
microscope.

somewhat less magnified surface of silicon. These pictures, and many others you have likely seen (but may not have known where they came from) are stupendous, but how are they made?

Figure 40.32a shows how the scanning tunneling microscope works. A conducting probe with a *very* sharp tip, just a few atoms wide, is brought to within a few tenths of a nanometer of a surface. Preparing the tips and controlling the spacing are both difficult technical challenges, but scientists have learned how to do both. Once positioned, the probe can be mechanically scanned back and forth across the surface.

When we analyzed the photoelectric effect, you learned that electrons are bound inside metals by an amount of energy called the *work function* E_0. A typical work function is 4 or 5 eV. This is the energy that must be supplied—by a photon or otherwise—to lift an electron out of the metal. In other words, the electron's energy in the metal is E_0 less than its energy outside the metal.

This fact is the basis for the potential-energy diagram of Figure 40.32b. The small air gap between the sample and the probe tip is a potential-energy barrier. The energy of an electron in the metal of the sample or the probe tip is less than the energy of an electron in the air by ≈ 4 eV, the work function. The absorption of a photon with $E_{photon} > 4$ eV would lift the electron *over* the barrier, from the sample to the probe. This is just the photoelectric effect. Alternatively, electrons can tunnel *through* the barrier if it is sufficiently narrow. This creates a *tunneling current* from the sample into the probe.

In operation, the tunneling current is recorded as the probe tip scans across the surface. You saw above that the tunneling current is extremely sensitive to the barrier thickness. As the tip scans over the position of an atom, the gap decreases by ≈ 0.1 nm and the current increases. The gap is larger when the tip is between atoms, so the current drops. Today's STMs can sense changes in the gap of as little as 0.001 nm, or about 1% of an atomic diameter! The images you see, such as those in Figure 40.31, are computer-generated from the current measurements at each position.

The STM has revolutionized the science and engineering of microscopic objects. They are now used to study everything from how surfaces corrode and oxidize, a topic of great practical importance, to how biological molecules are structured. Another example of quantum mechanics working for you!

The Resonant Tunneling Diode

The semiconductor diode laser that we examined in Section 40.6 had a narrow GaAs layer surrounded by wide layers of GaAlAs. Because an electron's potential energy is ≈ 0.3 eV less in GaAs than in GaAlAs, this structure provides a quantum well in which electrons are confined in a single energy level.

Suppose we manufacture a device in which a thin layer of GaAs is surrounded by still thinner layers of GaAlAs, only a few nanometers thick. Figure 40.33a is the potential-energy diagram of an electron in such a device. Because the GaAlAs layers are very thin, an electron inside the quantum well can tunnel through to the outside.

Conversely, an electron coming from the outside and impinging on the GaAlAs barrier might tunnel *into* the quantum well. However, tunneling into the well from the outside is hindered by a serious energy mismatch. An electron inside the quantum well *must* have one of the allowed energies. Typically there is a single allowed quantum state with $E_1 \approx 0.15$ eV. Electrons on the outside have thermal energy

$$E_{th} \approx \frac{3}{2}k_B T = 6.0 \times 10^{-21}\,\text{J} = 0.040\,\text{eV}$$

at room temperature. Tunneling may be a strange phenomenon, but energy does still have to be conserved. An electron approaching the barrier with $E \approx 0.04$ eV cannot tunnel inside unless there is a quantum state with this allowed energy.

Figure 40.33b shows the effect of placing a potential difference ΔV across the three layers of the device, with the left side more positive. Because electrons are negative, ΔV *lowers* the potential energy on the left side. As the potential difference is increased, it will reach a value ΔV_{res} at which the energy level inside the quantum well matches the energy of an electron approaching from the right. We then have a *resonance,* much as when an external driving frequency matches the natural frequency of an oscillator.

Once the energies match, electrons approaching from the right can easily tunnel into the quantum well. They then tunnel through the opposite barrier and emerge on the left with kinetic energy $K \approx e\Delta V$. In other words, there is a current through the device when the potential difference is ΔV_{res}. This device is called a **resonant tunneling diode.**

Applying too much voltage destroys the resonance. As Figure 40.33c shows, a large ΔV drops the energy level in the quantum well too low, so again electrons from the right side have no matching energy level into which they can tunnel. Charge flows through a resonant tunneling diode only for a small range of voltages near ΔV_{res}.

Figure 40.34 is an experimental current-voltage graph for a device having a 4 nm GaAs quantum well surrounded by 10-nm-wide GaAlAs barriers. There is a small range of voltages around 0.25 volts for which the current shoots up by a factor of 10. This is ΔV_{res}, and the current is due to electrons tunneling through the diode. The current then drops back to near zero by the time $\Delta V = 0.40$ V. (The current increase for $\Delta V > 0.7$ V is "normal" diode behavior. A resonant tunneling diode would not be operated with voltages that large.)

The ability to drastically change current with just a small change in voltage makes tunneling diodes very useful in the digital circuits of high-speed computers. These diodes can also be used as very-high-speed oscillators, creating oscillating voltages with frequencies as high as 500 GHz.

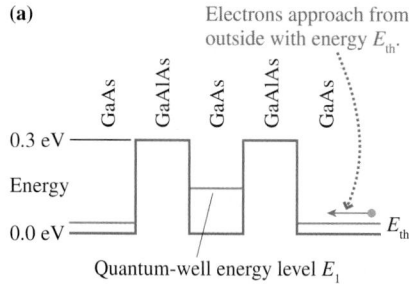

(a)

Electrons approach from outside with energy E_{th}.

Quantum-well energy level E_1

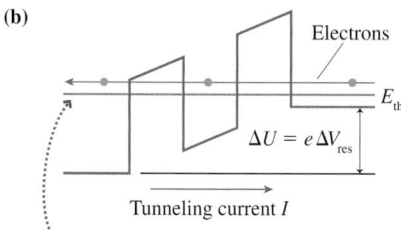

(b)

Electrons

$\Delta U = e\,\Delta V_{res}$

Tunneling current I

The quantum-well energy matches the electron energy, allowing the electrons to tunnel through.

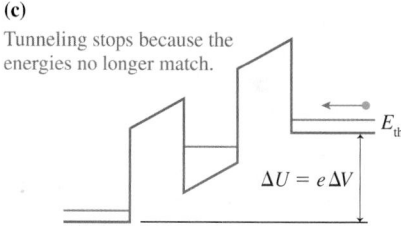

(c)

Tunneling stops because the energies no longer match.

$\Delta U = e\,\Delta V$

FIGURE 40.33 Electron potential energy in a resonant tunneling diode.

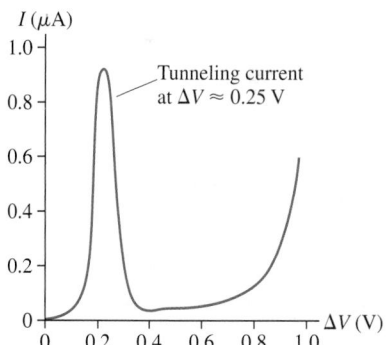

FIGURE 40.34 Experimental measurement of the current-voltage characteristics of a resonant tunneling diode.

STOP TO THINK 40.6   A particle with energy E approaches an energy barrier with height $U_0 > E$. If U_0 is slowly decreased, the probability that the particle reflects from the barrier

   a. Increases.
   b. Decreases.
   c. Does not change.

SUMMARY

The goal of Chapter 40 has been to understand and apply the essential ideas of quantum mechanics.

GENERAL PRINCIPLES

The Schrödinger Equation (The "law of psi")

$$\frac{d^2\psi}{dx^2} = -\frac{2m}{\hbar^2}[E - U(x)]\psi(x)$$

This equation determines the wave function $\psi(x)$ and, through $\psi(x)$, the probabilities of finding a particle of mass m with potential energy $U(x)$.

Boundary conditions

- $\psi(x)$ is a continuous function.
- $\psi(x) \rightarrow 0$ as $x \rightarrow \pm\infty$.
- $\psi(x) = 0$ in a region where it is physically impossible for the particle to be.
- $\psi(x)$ is normalized.

Shapes of wave functions

- The wave function oscillates between the classical turning points.
- State n has n antinodes.
- Node spacing and amplitude increase as kinetic energy K decreases.
- $\psi(x)$ decays exponentially in a classically forbidden region.

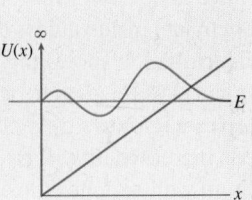

Quantum-mechanical models are characterized by the particle's potential-energy function $U(x)$.

- Wave-function solutions exist only for certain values of E. Thus energy is quantized.
- Photons are emitted or absorbed in quantum jumps.

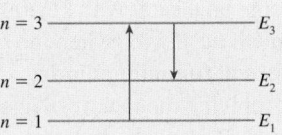

IMPORTANT CONCEPTS

Quantum-mechanical tunneling

A wave function can penetrate into a classically forbidden region with

$$\psi(x) = \psi_{\text{edge}}e^{-(x-L)/\eta}$$

where the **penetration distance** is

$$\eta = \frac{\hbar}{\sqrt{2m(U_0 - E)}}$$

The probability of tunneling through a barrier of width w is

$$P_{\text{tunnel}} = e^{-2w/\eta}$$

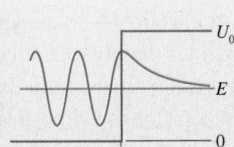

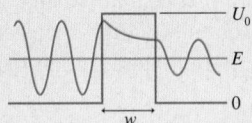

The correspondence principle says that the quantum world blends smoothly into the classical world for high quantum numbers. This is seen by comparing $|\psi(x)|^2$ to the classical probability density

$$P_{\text{class}} = \frac{2}{Tv(x)}$$

P_{class} expresses the idea that a classical particle is more likely to be found where it is moving slowly.

APPLICATIONS

Particle in a rigid box $E_n = n^2\dfrac{h^2}{8mL^2}$ $n = 1, 2, 3, \ldots$

Quantum harmonic oscillator $E_n = (n - \frac{1}{2})\hbar\omega$ $n = 1, 2, 3, \ldots$

Other applications were studied through numerical solution of the Schrödinger equation.

TERMS AND NOTATION

Schrödinger equation	bound state	bonding molecular orbital
quantum-mechanical model	penetration distance, η	antibonding molecular orbital
boundary conditions	quantum-well laser	quantum-mechanical tunneling
zero-point motion	gamma rays	scanning tunneling microscope (STM)
correspondence principle	quantum harmonic oscillator	resonant tunneling diode
potential well	vibrational energy levels	
classically forbidden regions	covalent molecular bond	

EXERCISES AND PROBLEMS

Exercises

Section 40.3 A Particle in a Rigid Box: Energies and Wave Functions

Section 40.4 A Particle in a Rigid Box: Interpreting the Solution

1. An electron in a rigid box absorbs light. The longest wavelength in the absorption spectrum is 600 nm. How long is the box?

2. The electrons in a rigid box emit photons of wavelength 1484 nm during the $3 \rightarrow 2$ transition.
   a. What kind of photons are they—infrared, visible, or ultraviolet?
   b. How long is the box in which the electrons are confined?

3. Figure Ex40.3 shows the wave function of an electron in a rigid box. The electron energy is 6.0 eV. How long is the box?

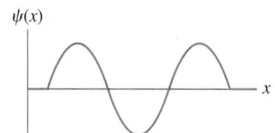

$\psi(x)$

FIGURE EX40.3

4. Figure Ex40.4 shows the wave function of an electron in a rigid box. The electron energy is 12.0 eV. What is the energy of the electron's ground state?

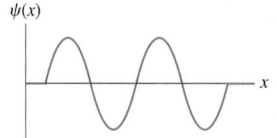

$\psi(x)$

FIGURE EX40.4

Section 40.6 Finite Potential Wells

5. Show that the penetration distance η has units of m.

6. a. Sketch graphs of the probability density $|\psi(x)|^2$ for the four states in the finite potential well of Figure 40.14a. Stack them vertically, similar to the Figure 40.14a graphs of $\psi(x)$.

b. What is the probability that a particle in the $n = 2$ state of the finite potential well will be found at the center of the well? Explain.

c. Is your answer to part b consistent with what you know about waves? Explain.

7. For a particle in a finite potential well of depth U_0, what is the ratio of the probability Prob(in δx at $x = L + \eta$) to the probability Prob(in δx at $x = L$)?

8. A finite potential well has depth $U_0 = 2.00$ eV. What is the penetration distance for an electron with energy (a) 0.50 eV, (b) 1.00 eV, and (c) 1.50 eV?

9. An electron in a finite potential well has a 1.0 nm penetration distance into the classically forbidden region. How far below U_0 is the electron's energy?

10. A helium atom is in a finite potential well. The atom's energy is 1.0 eV below U_0. What is the atom's penetration distance into the classically forbidden region?

Section 40.7 Wave-Function Shapes

11. Sketch the $n = 6$ wave function for the potential energy shown in Figure Ex40.11.

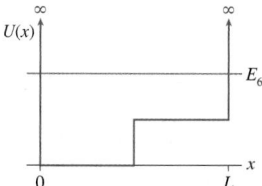

$U(x)$ ∞ ∞ E_6

FIGURE EX40.11 0 L x

12. Sketch the $n = 8$ wave function for the potential energy shown in Figure Ex40.12.

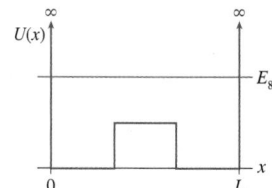

$U(x)$ ∞ ∞ E_8

FIGURE EX40.12 0 L x

13. The graph in Figure Ex40.13 shows the potential-energy function $U(x)$ of a particle. Solution of the Schrödinger equation finds that the $n = 3$ level has $E_3 = 0.5$ eV and that the $n = 6$ level has $E_6 = 2.0$ eV.
    a. Redraw this figure and add to it the energy lines for the $n = 3$ and $n = 6$ states.
    b. Sketch the $n = 3$ and $n = 6$ wave functions. Show them as oscillating about the appropriate energy line.

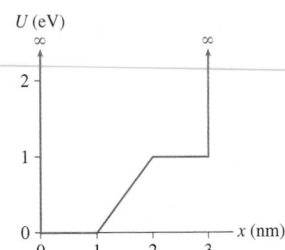

FIGURE EX40.13

14. Sketch the $n = 1$ and $n = 7$ wave functions for the potential energy shown in Figure Ex40.14.

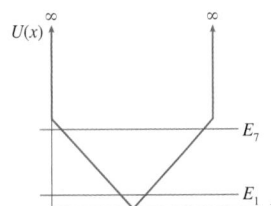

FIGURE EX40.14

Section 40.8 The Quantum Harmonic Oscillator

15. An electron in a harmonic potential well emits a photon with a wavelength of 300 nm as it undergoes a $3 \rightarrow 2$ quantum jump. What wavelength photon is emitted in a $3 \rightarrow 1$ quantum jump?
16. Consider a quantum harmonic oscillator.
    a. What happens to the spacing between the nodes of the wave function as $|x|$ increases? Why?
    b. What happens to the heights of the antinodes of the wave function as $|x|$ increases? Why?
    c. Sketch a reasonably accurate graph of the $n = 8$ wave function of a quantum harmonic oscillator.
17. An electron is confined in a harmonic potential well that has a spring constant of 2.0 N/m.
    a. What are the first three energy levels of the electron?
    b. What wavelength photon is emitted if the electron undergoes a $3 \rightarrow 1$ quantum jump?
18. An electron is confined in a harmonic potential well that has a spring constant of 12.0 N/m. What is the longest wavelength of light that the electron can absorb?
19. Two adjacent energy levels of an electron in a harmonic potential well are known to be 2.0 eV and 2.8 eV. What is the spring constant of the potential well?

Section 40.10 Quantum-Mechanical Tunneling

20. What is the probability that an electron will tunnel through a 0.45 nm gap from a metal to a STM probe if the work function is 4.0 eV?

21. A proton's energy is 1.0 MeV below the top of 10-fm-wide energy barrier. What is the probability that the proton will tunnel through the barrier?

Problems

22. A 2.0-μm-diameter water droplet is moving with a speed of 1.0 μm/s in a 20-μm-long box.
    a. Estimate the particle's quantum number.
    b. Use the correspondence principle to determine whether quantum mechanics is needed to understand the particle's motion or if it is "safe" to use classical physics.
23. Suppose that $\psi_1(x)$ and $\psi_2(x)$ are both solutions to the Schrödinger equation for the same potential energy $U(x)$. Prove that the superposition $\psi(x) = A\psi_1(x) + B\psi_2(x)$ is also a solution to the Schrödinger equation.
24. Figure 40.26a modeled a hydrogen atom as a finite potential well with rectangular edges. A more realistic model of a hydrogen atom, although still a one-dimensional model, would be the electron + proton electrostatic potential energy in one dimension:

$$U(x) = -\frac{e^2}{4\pi\epsilon_0|x|}$$

    a. Draw a graph of $U(x)$ versus x. Center your graph at $x = 0$.
    b. Despite the divergence at $x = 0$, the Schrödinger equation can be solved to find energy levels and wave functions for the electron in this potential. Draw a horizontal line across your graph of part a about one-third of the way from the bottom to the top. Label this line E_2, then, on this line, sketch a plausible graph of the $n = 2$ wave function.
    c. Redraw your graph of part a and add a horizontal line about two-thirds of the way from the bottom to the top. Label this line E_3, then, on this line, sketch a plausible graph of the $n = 3$ wave function.
25. a. Derive an expression for $\lambda_{2 \rightarrow 1}$, the wavelength of light emitted by a particle in a rigid box during a quantum jump from $n = 2$ to $n = 1$.
    b. In what length rigid box will an electron undergoing a $2 \rightarrow 1$ transition emit light with a wavelength of 694 nm? This is the wavelength of a ruby laser.
26. Model an atom as an electron in a rigid box of length 0.10 nm, roughly twice the Bohr radius.
    a. What are the four lowest energy levels of the electron?
    b. Calculate all the wavelengths that would be seen in the emission spectrum of this atom due to quantum jumps between these four energy levels. Give each wavelength a label $\lambda_{n \rightarrow m}$ to indicate the transition.
    c. Are these wavelengths in the infrared, visible, or ultraviolet portion of the spectrum?
    d. The stationary states of the Bohr hydrogen atom have negative energies. The stationary states of this model of the atom have positive energies. Is this a physically significant difference? Explain.
    e. Compare this model of an atom to the Bohr hydrogen atom. In what ways are the two models similar? Other than the signs of the energy levels, in what ways are they different?

27. Show that the normalization constant A_n for the wave functions of a particle in a rigid box has the value given in Equation 40.26.

28. A particle confined in a rigid one-dimensional box of length 10 fm has an energy level $E_n = 32.9$ MeV and an adjacent energy level $E_{n+1} = 51.4$ MeV.
    a. Determine the values of n and $n + 1$.
    b. Draw an energy-level diagram showing all energy levels from 1 through $n + 1$. Label each level and write the energy beside it.
    c. Sketch the $n + 1$ wave function on the $n + 1$ energy level.
    d. What is the wavelength of a photon emitted in the $n + 1 \rightarrow n$ transition? Compare this to a typical visible-light wavelength.
    e. What is the mass of the particle? Can you identify it?

29. Consider a particle in a rigid box of length L. For each of the states $n = 1$, $n = 2$, and $n = 3$:
    a. Sketch graphs of $|\psi(x)|^2$. Label the points $x = 0$ and $x = L$.
    b. Where, in terms of L, are the positions at which the particle is *most* likely to be found?
    c. Where, in terms of L, are the positions at which the particle is *least* likely to be found?
    d. Determine, by examining your $|\psi(x)|^2$ graphs, if the probability of finding the particle in the left one-third of the box is less than, equal to, or greater than $\frac{1}{3}$. Explain your reasoning.
    e. *Calculate* the probability that the particle will be found in the left one-third of the box.

30. For the quantum-well laser of Figure 40.16, *estimate* the probability that an electron will be found within one of the GaAlAs layers rather than in the GaAs layer. Explain your reasoning.

31. In a nuclear physics experiment, a proton is fired toward a $Z = 13$ nucleus with the diameter and neutron energy levels shown in Figure 40.17. The nucleus, which was initially in its ground state, subsequently emits a gamma ray with wavelength 1.73×10^{-4} nm. What was the *minimum* initial speed of the proton?
    Hint: Don't neglect the proton-nucleus collision.

32. Use the data from Figure 40.23 to calculate the first three vibrational energy levels of a $C{=}O$ carbon-oxygen double bond.

33. Verify that the $n = 1$ wave function $\psi_1(x)$ of the quantum harmonic oscillator really is a solution of the Schrödinger equation. That is, show that the right and left sides of the Schrödinger equation are equal if you use the $\psi_1(x)$ wave function.

34. Show that the constant b used in the quantum-harmonic-oscillator wave functions (a) has units of length and (b) is the classical turning point of an oscillator in the $n = 1$ ground state.

35. a. Determine the normalization constant A_1 for the $n = 1$ ground-state wave function of the quantum harmonic oscillator. Your answer will be in terms of b.
    b. Write an expression for the probability that a quantum harmonic oscillator in its $n = 1$ ground state will be found in the classically forbidden region.
    c. (Optional) Use a numerical integration program to evaluate your probability expression of part b.
    Hint: It helps to simplify the integral by making a change of variables to $u = x/b$.

36. a. Derive an expression for the classical probability density $P_{\text{class}}(x)$ for a simple harmonic oscillator with amplitude A.
    b. Graph your expression between $x = -A$ and $x = +A$.
    c. Interpret your graph. Why is it shaped as it is?

37. a. Derive an expression for the classical probability density $P_{\text{class}}(y)$ for a ball that bounces between the ground and height h. The collisions with the ground are perfectly elastic.
    b. Graph your expression between $y = 0$ and $y = h$.
    c. Interpret your graph. Why is it shaped as it is?

38. Even the smoothest mirror finishes are "rough" when viewed at a scale of 100 nm. When two very smooth metals are placed in contact with each other, the actual distance between the surfaces varies from 0 nm at a few points of real contact to ≈ 100 nm. The average distance between the surfaces is ≈ 50 nm. The work function of aluminum is 4.3 eV. What is the probability that an electron will tunnel between two pieces of aluminum that are 50 nm apart?

39. An electron approaches a 1.0-nm-wide potential-energy barrier of height 5.0 eV. What energy electron has a tunneling probability of (a) 10%, (b) 1.0%, and (c) 0.10%?

Challenge Problems

40. A typical electron in a piece of metallic sodium has energy $-E_0$ compared to a free electron, where E_0 is the 2.7 eV work function of sodium.
    a. At what distance *beyond* the surface of the metal is the electron's probability density 10% of its value *at* the surface?
    b. How does this distance compare to the size of an atom?

41. Consider a particle in a rigid box of length L with walls at $x = -L/2$ and $x = +L/2$.
    a. What is the wave function $\psi(x)$ for $x < -L/2$ and $x > L/2$? Explain.
    b. Write the Schrödinger equation in the region $-L/2 \leq x \leq L/2$ for a particle with energy E.
    c. Write down a general solution to the Schrödinger equation that is valid in the region $-L/2 \leq x \leq L/2$.
    d. What are the boundary conditions this wave function must satisfy?
    e. Apply the boundary conditions to determine the allowed energy levels. Note that there are two different ways to satisfy the boundary conditions, each giving a different set of wave functions and energy levels.
    f. Compare your results to the rigid box that was analyzed in this chapter. In what ways are the results the same and in what ways are they different? Are any differences physically meaningful?

42. A particle of mass m has the wave function $\psi(x) = A x \exp(-x^2/a^2)$ when it is in an allowed energy level with $E = 0$.
    a. Draw a graph of $\psi(x)$ versus x.
    b. At what value or values of x is the particle most likely to be found?
    c. Find and graph the potential-energy function $U(x)$.

43. In most metals, the atomic ions form a regular arrangement called a *crystal lattice*. The conduction electrons in the sea of electrons move through this lattice. Figure CP40.43 is a one-dimensional model of a crystal lattice. The ions have mass m, charge e, and an equilibrium separation b.

    a. Suppose the middle charge is displaced a very small distance ($x \ll b$) from its equilibrium position while the outer charges remain fixed. Show that the net electric force on the middle charge is given approximately by

    $$F = -\frac{e^2}{b^3 \pi \epsilon_0} x$$

    In other words, the charge experiences a linear restoring force.

    b. Suppose this crystal consists of aluminum ions with an equilibrium spacing of 0.30 nm. What are the energies of the four lowest vibrational states of these ions?

    c. What wavelength photons are emitted during quantum jumps between *adjacent* energy levels? Is this wavelength in the infrared, visible, or ultraviolet portion of the spectrum?

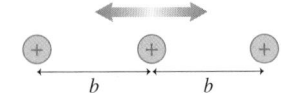

FIGURE CP40.43

44. a. What is the probability that an electron will tunnel through a 0.50 nm air gap from a metal to a STM probe if the work function is 4.0 eV?

    b. The probe passes over an atom that is 0.050 nm "tall." By what factor does the tunneling current increase?

    c. If a 10% current change is reliably detectable, what is the smallest height change the STM can detect?

45. Tennis balls traveling at greater than 100 mph routinely bounce off tennis rackets. At some sufficiently high speed, however, the ball will break through the strings and keep going. The racket is a potential-energy barrier whose height is the energy of the slowest string-breaking ball. Suppose that a 100 g tennis ball traveling at 200 mph is just sufficient to break the 2.0-mm-thick strings. Estimate the probability that a 120 mph ball will tunnel through the racket without breaking the strings. Give your answer as a power of 10 rather than a power of e.

STOP TO THINK ANSWERS

Stop to Think 40.1: $v_a = v_b > v_c$. The de Broglie wavelength is $\lambda = h/mv$, so slower particles have longer wavelengths. The wave amplitude is not relevant.

Stop to Think 40.2: c. The $n = 2$ state has a node in the middle of the box. The antinodes are centered in the left and right halves of the box.

Stop to Think 40.3: $n = 4$. There are four antinodes, three nodes (excluding the ends).

Stop to Think 40.4: d. The wave function reaches zero abruptly on the right, indicating an infinitely high potential-energy wall. The exponential decay on the left shows that the left wall of the poten-

tial energy is *not* infinitely high. The node spacing and the amplitude increase steadily in going from right to left, indicating a *steadily* decreasing kinetic energy and thus a steadily increasing potential energy.

Stop to Think 40.5: c. $E = (n - \frac{1}{2})\hbar\omega$, so $\frac{5}{2}\hbar\omega$ is the energy of the $n = 3$ state. An $n = 3$ state has 3 antinodes.

Stop to Think 40.6: b. The probability of tunneling through the barrier increases as the difference between E and U_0 decreases. If the tunneling probability increases, the reflection probability must decrease.

41 Atomic Physics

Lasers are one of the most important applications of the quantum-mechanical properties of atoms and light.

▶ Looking Ahead

The goal of Chapter 41 is to understand the structure and properties of atoms. In this chapter you will learn to:

- Use a quantum-mechanical model of the hydrogen atom.
- Understand the idea of electron spin.
- Apply Schrödinger's quantum theory to multielectron atoms.
- Interpret atomic spectra.
- Understand how lasers work.

◀ Looking Back

The material in this chapter depends on an understanding of the Bohr model of atomic quantization and one-dimensional quantum mechanics. Please review:

- Sections 38.5–38.7 Bohr's model of quantization and the hydrogen atom.
- Sections 39.3–39.4 Interpreting and using wave functions.
- Sections 40.1–40.2 The basic ideas of quantum mechanics.

The problem of discovering the structure of atoms is one that we have continued to revisit. The first model of an atom we looked at, Rutherford's solar-system model, was purely classical. This model incorporated Rutherford's discovery of a very small nucleus, but otherwise it had almost no agreement with the experimental evidence about atoms. It could not explain their discrete spectra, nor could it explain why atoms are stable!

The Bohr model of the hydrogen atom was a big step forward. The concept of stationary states provided a means of understanding both the stability of atoms and the quantum jumps that lead to discrete spectra. And Bohr's ability to derive the Balmer formula for the hydrogen spectrum indicated that he was on the right track. Yet, as we have seen, the Bohr model was not successful for any atom other than hydrogen.

Now it's Schrödinger's turn. Is Schrödinger's theory of quantum mechanics better at explaining atomic structure than other models? The answer, as you can probably anticipate, is a decisive yes. This chapter is an overview of how quantum mechanics finally provides us with an understanding of atomic structure and atomic properties.

41.1 The Hydrogen Atom: Angular Momentum and Energy

Let's begin with a quantum-mechanical model of the hydrogen atom. Figure 41.1 shows an electron at distance r from a proton. The proton is much more massive than the electron, so we will assume that the proton remains at rest at the origin.

As you learned in Chapter 40, the problem-solving procedure in quantum mechanics consists of two basic steps:

1. Specify a potential-energy function.
2. Solve the Schrödinger equation to find the wave functions, allowed energy levels, and other quantum properties.

The first step is easy. The proton and electron are charged particles with $q = \pm e$, so the potential energy of a hydrogen atom as a function of the electron distance r is

$$U(r) = -\frac{1}{4\pi\epsilon_0}\frac{e^2}{r} \tag{41.1}$$

The difficulty arises with the second step. The Schrödinger equation of Chapter 40 was for one-dimensional problems. Atoms are three-dimensional, and the three-dimensional Schrödinger equation turns out to be a partial differential equation whose solution is outside the scope of this textbook. Consequently, we'll present results that can be derived only in a more advanced course in quantum mechanics. The good news is that you have learned enough quantum mechanics to interpret and use the results.

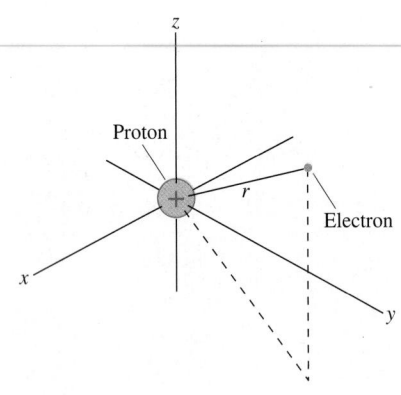

FIGURE 41.1 The electron in a hydrogen atom is distance r from the proton.

Stationary States of Hydrogen

In one dimension, energy quantization appeared as a consequence of *boundary conditions* on the wave function. That is, there were solutions to the Schrödinger equation that satisfied the boundary conditions only for certain discrete energies, characterized by the quantum number n. In three dimensions, the wave function must satisfy *three* different boundary conditions. Consequently, solutions to the three-dimensional Schrödinger equation have *three* quantum numbers and *three* quantized parameters.

Solutions to the Schrödinger equation for the hydrogen atom potential energy exist only if three conditions are satisfied:

1. The atom's energy must be one of the values

$$E_n = -\frac{1}{n^2}\left(\frac{1}{4\pi\epsilon_0}\frac{e^2}{2a_B}\right) = -\frac{13.60\ \text{eV}}{n^2} \qquad n = 1, 2, 3, \ldots \tag{41.2}$$

where $a_B = 4\pi\epsilon_0\hbar^2/me^2 = 0.0529$ nm is the Bohr radius. The integer n is called the **principal quantum number.** These energies are the same as those in the Bohr hydrogen atom.

2. The angular momentum L of the electron's orbit must be one of the values

$$L = \sqrt{l(l+1)}\hbar \qquad l = 0, 1, 2, 3, \ldots, n-1 \tag{41.3}$$

The integer l is called the **orbital quantum number.**

3. The z-component of the angular momentum L_z must be one of the values

$$L_z = m\hbar \qquad m = -l, -l+1, \ldots, 0, \ldots, l-1, l \qquad (41.4)$$

The integer m is called the **magnetic quantum number.**

In other words, each stationary state of the hydrogen atom is identified by a triplet of quantum numbers (n, l, m). Each quantum number is associated with a physical property of the atom.

NOTE ▶ The energy of the stationary state depends only on the principal quantum number n, not on l or m. ◀

EXAMPLE 41.1 Listing quantum numbers

List all possible states of a hydrogen atom that have energy $E = -3.40$ eV.

SOLVE Energy depends only on the principal quantum number n. States with $E = -3.40$ eV have

$$n = \sqrt{\frac{-13.60 \text{ eV}}{-3.40 \text{ eV}}} = 2$$

An atom with principal quantum number $n = 2$ could have either $l = 0$ or $l = 1$, but $l \geq 2$ is ruled out. If $l = 0$, the only

possible value for the magnetic quantum number m is $m = 0$. If $l = 1$, then the atom could have $m = -1$, $m = 0$, or $m = +1$. Thus the possible quantum numbers are

n	l	m
2	0	0
2	1	1
2	1	0
2	1	-1

These four states all have the same energy.

Hydrogen turns out to be unique. For all other elements, the allowed energies depend on both *n and l* (but not *m*). Consequently, it is useful to label the stationary states by their values of n and l. The lowercase letters shown in Table 41.1 are customarily used to represent the various values of quantum number l. These symbols come from spectroscopic notation used in prequantum-mechanics days, when some spectral lines were classified as **s**harp, others as **p**rincipal, and so on.

Using these symbols, the ground state of the hydrogen atom, with $n = 1$ and $l = 0$, is called the $1s$ state. The $3d$ state has $n = 3$, $l = 2$. In Example 41.1, we found one $2s$ state (with $l = 0$) and three $2p$ states (with $l = 1$), all with the same energy.

TABLE 41.1 Symbols used to represent quantum number l

l	Symbol
0	s
1	p
2	d
3	f

Angular Momentum Is Quantized

If the hydrogen atom were classical, the electron's orbit, like that of a planet in the solar system, would be an ellipse. Furthermore, the orbit need not lie in the xy-plane. Figure 41.2 shows a possible classical orbit tilted at angle θ below the xy-plane.

We introduced the angular momentum vector $\vec{L}$ in Chapter 13. It will be useful to call $\vec{L}$ the *orbital* angular momentum in order to distinguish it later from *spin* angular momentum. Figure 41.2 reminds you that the vector $\vec{L}$ is perpendicular to the plane of the electron's orbit. The angular momentum vector has a z-component $L_z = L\cos\theta$ along the z-axis.

Classically, L and L_z can have any values. Not so in quantum mechanics. Quantum conditions 2 and 3 tell us that **the electron's orbital angular momentum is quantized.** The magnitude of the orbital angular momentum must be one of the discrete values

$$L = \sqrt{l(l+1)}\,\hbar = 0, \sqrt{2}\hbar, \sqrt{6}\hbar, \sqrt{12}\hbar, \ldots$$

where l is an integer. Simultaneously, the z-component L_z must have one of the values $L_z = m\hbar$, where m is an integer between $-l$ and l. No other values of L or L_z allow the wave function to satisfy the boundary conditions.

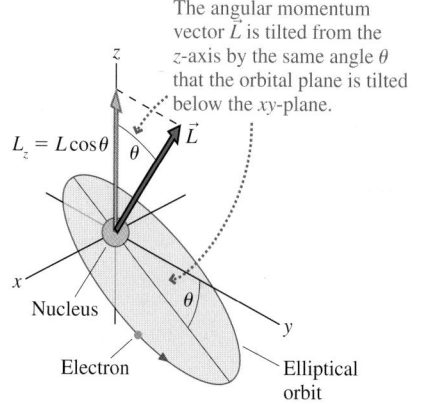

The angular momentum vector $\vec{L}$ is tilted from the z-axis by the same angle θ that the orbital plane is tilted below the xy-plane.

$L_z = L\cos\theta$

FIGURE 41.2 The angular momentum of an elliptical orbit.

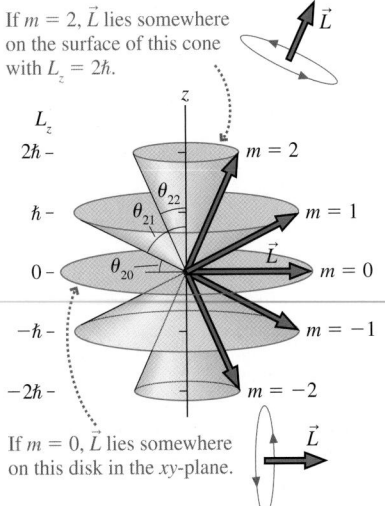

If $m = 2$, $\vec{L}$ lies somewhere on the surface of this cone with $L_z = 2\hbar$.

If $m = 0$, $\vec{L}$ lies somewhere on this disk in the xy-plane.

FIGURE 41.3 Five possible orientations of the angular momentum vector for $l = 2$. The angular momentum vectors all have length $L = \sqrt{6}\hbar = 2.45\hbar$.

The quantization of angular momentum places restrictions on the shape and orientation of the electron's orbit. To see this, consider a hydrogen atom with orbital quantum number $l = 2$. In this state, the *magnitude* of the electron's angular momentum must be $L = \sqrt{6}\hbar = 2.45\hbar$. Furthermore, the angular momentum vector must point in a *direction* such that $L_z = m\hbar$, where m is one of only five integers in the range $-2 \leq m \leq 2$.

The combination of these two requirements allows $\vec{L}$ to point only in certain directions in space, as shown in Figure 41.3. This is a rather unusual figure that requires a little thought to understand. Suppose $m = 0$ and thus $L_z = 0$. With no z-component, the angular momentum vector $\vec{L}$ must lie somewhere in the xy-plane. Furthermore, because the length of $\vec{L}$ is constrained to be $2.45\hbar$, the tip of $\vec{L}$ must lie somewhere on the circle labeled $m = 0$. These values of $\vec{L}$ correspond to classical orbits tipped into a vertical plane.

Similarly, $m = 2$ requires $\vec{L}$ to lie along the cone whose height is $2\hbar$ and whose side has length $2.45\hbar$. These values of $\vec{L}$ correspond to classical orbits tilted slightly out of the xy-plane. Notice that $\vec{L}$ **cannot point directly along the z-axis.** The maximum possible value of L_z, when $m = l$, is $(L_z)_{max} = l\hbar$. But $l < \sqrt{l(l+1)}$, so $(L_z)_{max} < L$. The angular momentum vector *must* have either an x- or a y-component (or both). In other words, the corresponding classical orbit cannot lie in the xy-plane.

An angular momentum vector $\vec{L}$ tilted at angle θ from the z-axis corresponds to an orbit tilted at angle θ out of the xy-plane. The quantization of angular momentum restricts the orbital planes to only a few discrete angles. For quantum state (n, l, m), the angle of the angular momentum vector is

$$\theta_{lm} = \cos^{-1}\left(\frac{L_z}{L}\right) = \cos^{-1}\left(\frac{m\hbar}{\sqrt{l(l+1)}\hbar}\right) = \cos^{-1}\left(\frac{m}{\sqrt{l(l+1)}}\right) \qquad (41.5)$$

Angles θ_{22}, θ_{21}, and θ_{20} are labeled in Figure 41.3. Orbital planes at other angles are not allowed because they don't satisfy the quantization conditions for angular momentum.

EXAMPLE 41.2 The angle of the angular momentum vector

What is the angle between $\vec{L}$ and the z-axis for a hydrogen atom in the stationary state $(n, l, m) = (4, 2, 1)$?

SOLVE The angle θ_{21} is labeled in Figure 41.3. The state $(4, 2, 1)$ has $l = 2$ and $m = 1$, thus

$$\theta_{21} = \cos^{-1}\left(\frac{1}{\sqrt{6}}\right) = 65.9°$$

ASSESS This quantum state corresponds to a classical orbit tilted $65.9°$ away from the xy-plane.

NOTE ▶ The ground state of hydrogen, with $l = 0$, has *no* angular momentum. A classical particle cannot orbit unless it has angular momentum, but apparently a quantum particle does not have this requirement. We will examine this issue in the next section. ◀

Energy Levels of the Hydrogen Atom

The energy of the hydrogen atom is quantized. Only those energies given by Equation 41.2 allow the wave function to satisfy the boundary conditions. The allowed energies of hydrogen depend only on the principal quantum number n, but for other atoms the energies will depend on both n and l. In anticipation of using both quantum numbers, Figure 41.4 is an *energy-level diagram* for the hydrogen atom in which the rows are labeled by n and the columns by l. The left column contains all of the $l = 0$ s states, the next column is the $l = 1$ p states, and so on.

Because the quantum condition of Equation 41.3 requires $n > l$, the s states begin with $n = 1$, the p states begin with $n = 2$, and the d states with $n = 3$. That is, the lowest-energy d state is $3d$ because states with $n = 1$ or $n = 2$ cannot have $l = 2$. For hydrogen, where the energy levels do not depend on l, the energy-level diagram shows that the $3s$, $3p$, and $3d$ states have equal energy. Figure 41.4 shows only the first few energy levels for each value of l, but there really are an infinite number of levels, as $n \to \infty$, crowding together beneath $E = 0$. The dotted line at $E = 0$ is the atom's *ionization limit,* the energy of a hydrogen atom in which the electron has been moved infinitely far away to form an H^+ ion.

The lowest energy state, the $1s$ state with $E_1 = -13.60$ eV, is the *ground state* of hydrogen. The value $|E_1| = 13.60$ eV is the **ionization energy,** the *minimum* energy that would be needed to form a hydrogen ion by removing the electron from the ground state. All of the states with $n > 1$ are *excited states.*

Quantum number l	0	1	2	3
Symbol	s	p	d	f

n	$E = 0$ eV			
		Ionization limit		
4	-0.85 eV	$4s$ $4p$	$4d$	$4f$
3	-1.51 eV	$3s$ $3p$	$3d$	
2	-3.40 eV	$2s$ $2p$		
1	-13.60 eV	$1s$ Ground state		

FIGURE 41.4 Energy-level diagram for the hydrogen atom.

> **STOP TO THINK 41.1** What are the quantum numbers n and l for a hydrogen atom with $E = -(13.60/9)$ eV and $L = \sqrt{2}\hbar$?

41.2 The Hydrogen Atom: Wave Functions and Probabilities

You learned in Chapter 40 that the probability of finding a particle in a small interval of width δx at the position x is given by

$$\text{Prob(in } \delta x \text{ at } x) = |\psi(x)|^2 \delta x = P(x)\,\delta x$$

where

$$P(x) = |\psi(x)|^2$$

is the probability density. This interpretation of $|\psi(x)|^2$ as a probability density lies at the heart of quantum mechanics. However, $P(x)$ was for a one-dimensional wave function. Because we're now looking at a three-dimensional atom, we need to consider the probability of finding a particle in a small *volume* of space δV at the position described by the three coordinates (x, y, z). This probability is

$$\text{Prob(in } \delta V \text{ at } x, y, z) = |\psi(x, y, z)|^2 \delta V \qquad (41.6)$$

We can still interpret $|\psi(x, y, z)|^2$ as a probability density.

In one-dimensional quantum mechanics we could simply graph $P(x)$ versus x. Portraying the probability density of a three-dimensional wave function is more of a challenge. One way to do so, shown in Figure 41.5 on the next page, is to use denser shading to indicate regions of larger probability density. That is, the amplitude of ψ is larger, and the electron is more likely to be found in regions where the shading is darker. These figures show the probability densities of the $1s$, $2s$, and $2p$ states of hydrogen. As you can see, the probability density in three dimensions creates what is often called an **electron cloud** around the nucleus.

These figures contain a lot of information. For example, notice how the p electrons have directional properties. These directional properties allow p electrons to "reach out" toward nearby atoms, forming molecular bonds. The quantum mechanics of bonding goes beyond what we can study in this text, but the electron-cloud pictures of the p electrons begin to suggest how bonds could form.

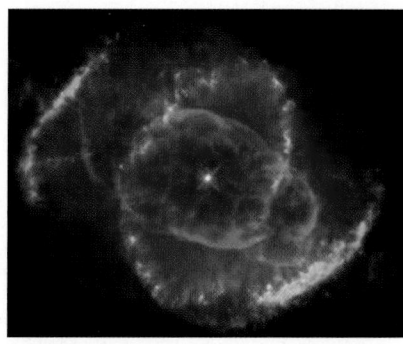

The red color of this nebula is due to the emission of light from hydrogen atoms. The atoms are excited by intense ultraviolet light from the star in the center. They then emit red light ($\lambda = 656$ nm) in a $3 \to 2$ transition, part of the Balmer series of spectral lines emitted by hydrogen.

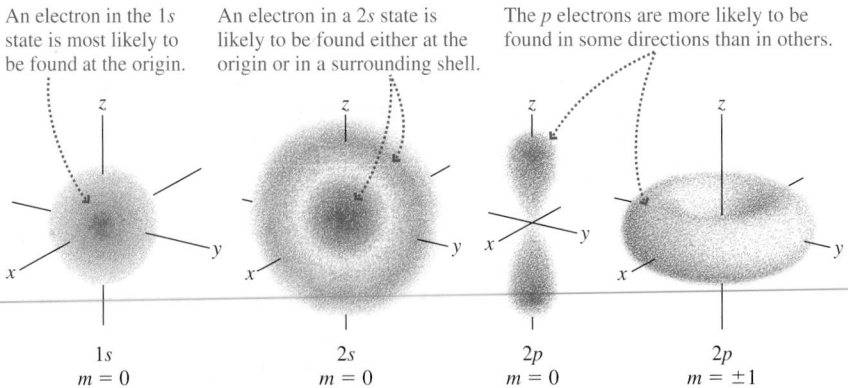

An electron in the 1s state is most likely to be found at the origin.

An electron in a 2s state is likely to be found either at the origin or in a surrounding shell.

The p electrons are more likely to be found in some directions than in others.

1s
$m = 0$

2s
$m = 0$

2p
$m = 0$

2p
$m = \pm 1$

FIGURE 41.5 The probability densities of the electron in the 1s, 2s, and 2p states of hydrogen.

Radial Wave Functions

Figures such as Figure 41.5 are useful for "seeing" the electron clouds, but these figures are hard to use. Often, we would simply like to know the probability of finding the electron at a certain *distance* from the nucleus. That is, what is the probability that the electron is to be found within the small range of distances δr at the distance r?

It turns out that the solutions to the three-dimensional Schrödinger equation can be written in a form that focuses on the electron's radial distance r from the proton. The portion of the wave function that depends only on r is called the **radial wave function.** These functions, which depend on the quantum numbers n and l, are designated $R_{nl}(r)$. The first three radial wave functions are

$$R_{1s}(r) = \frac{1}{\sqrt{\pi a_B^3}} e^{-r/a_B}$$

$$R_{2s}(r) = \frac{1}{\sqrt{8\pi a_B^3}} \left(1 - \frac{r}{2a_B}\right) e^{-r/2a_B} \qquad (41.7)$$

$$R_{2p}(r) = \frac{1}{\sqrt{24\pi a_B^3}} \left(\frac{r}{2a_B}\right) e^{-r/2a_B}$$

where a_B is the Bohr radius.

The radial wave functions may seem mysterious, because we haven't shown where they come from, but they are essentially the same as the one-dimensional wave functions $\psi(x)$ you learned to work with in Chapter 40. In fact, these radial wave functions are similar to the one-dimensional wave functions of the simple harmonic oscillator. One important difference, however, is that r ranges from 0 to ∞. For one-dimensional wave functions, x ranged from $-\infty$ to ∞.

Figure 41.6 shows the radial wave functions for the 1s and 2s states. Notice that the radial wave function is nonzero at $r = 0$, the position of the nucleus. This is surprising, but it is consistent with our observation in Figure 41.5 that the 1s and 2s electrons have a strong probability of being found at the origin.

We can gain some understanding of the s-state wave functions by considering the angular momentum. A classical particle, for which $L = mvr$, can have $L = 0$ only if the radius of its orbit shrinks to zero. This is impossible for a classical particle, but zero angular momentum *is* achievable for quantum particles because the uncertainty principle prevents a quantum particle from being localized at a single

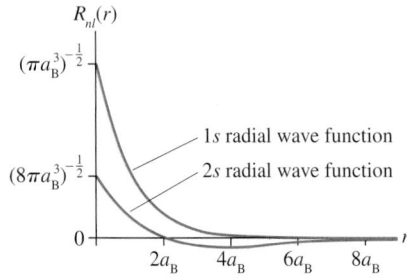

FIGURE 41.6 The 1s and 2s radial wave functions of hydrogen.

point. The *s*-state wave functions of Figure 41.6, with their maximum values at $r = 0$, are the quantum analogs of a classical particle orbiting with $r = 0$.

Our purpose for introducing the radial wave functions was to determine the probability of finding the electron a certain *distance* from the nucleus. Figure 41.7 shows a shell of radius r and thickness δr centered on the nucleus. The probability of finding the electron at distance r from the nucleus is equivalent to the probability that the electron is located somewhere within this shell. The volume of a thin shell is its surface area multiplied by its thickness δr. The surface area of a sphere is $4\pi r^2$, so the volume of this thin shell is

$$\delta V = 4\pi r^2 \delta r \tag{41.8}$$

We will assert, without proof, that the probability of finding the electron within this shell is

$$\text{Prob(in } \delta r \text{ at } r) = |R_{nl}(r)|^2 \delta V = 4\pi r^2 |R_{nl}(r)|^2 \delta r = P_r(r)\,\delta r \tag{41.9}$$

where

$$P_r(r) = 4\pi r^2 |R_{nl}(r)|^2 \tag{41.10}$$

is called the **radial probability density** for the state *nl*.

The radial probability density tells us the relative likelihood of finding the electron at distance r from the nucleus. The volume factor $4\pi r^2$ reflects the fact that more space is available in a shell of larger r, and this additional space increases the probability of finding the electron at that distance.

The probability of finding the electron between r_{min} and r_{max} is

$$\text{Prob}(r_{\text{min}} \le r \le r_{\text{max}}) = \int_{r_{\text{min}}}^{r_{\text{max}}} P_r(r)\,dr = 4\pi \int_{r_{\text{min}}}^{r_{\text{max}}} r^2 |R_{nl}(r)|^2\,dr \tag{41.11}$$

The electron must be *somewhere* between $r = 0$ and $r = \infty$, so the integral of $P_r(r)$ between 0 and ∞ must equal 1. This normalization condition was used to determine the constants in front of the radial wave functions of Equation 41.7.

Figure 41.8 shows the radial probability densities for the $n = 1$, 2, and 3 states of the hydrogen atom, all drawn to the same scale so that you can compare them to each other. The horizontal scale is in units of the Bohr radius a_{B}.

You can see that the $1s$, $2p$, and $3d$ states, with maxima at a_{B}, $4a_{\text{B}}$ and $9a_{\text{B}}$, respectively, are following the pattern $r_{\text{peak}} = n^2 a_{\text{B}}$. These are exactly the radii of the orbits in the Bohr hydrogen atom. There we simply bent a one-dimensional de Broglie wave into a circle of that radius. Now we have a three-dimensional wave function for which the electron is *most likely* to be this distance from the nucleus, although it *could* be found at other values of r. The physical situation is very different in quantum mechanics, but it is good to see that various aspects of the Bohr atom can be reproduced.

But why is it the $3d$ state that agrees with the Bohr atom rather than $3s$ or $3p$? All states with the same value of n form a collection of "orbits" having the same energy. As Figure 41.9 shows, the state with $l = n - 1$ has the largest angular momentum of the group. Consequently, the maximum-l state corresponds to a circular classical orbit and matches the circular orbits of the Bohr atom. Notice that the radial probability densities for the $2p$ and $3d$ states have a single peak, corresponding to a classical orbit at a constant distance.

States with smaller l correspond to elliptical orbits. You can see in Figure 41.8 that the radial probability density of a $3s$ electron has a peak in close to the nucleus. The $3s$ electron also has a good chance of being found *farther* from the nucleus than a $3d$ electron, suggesting an orbit that alternately swings in near the nucleus, then moves out past the circular orbit with the same energy. This distinction

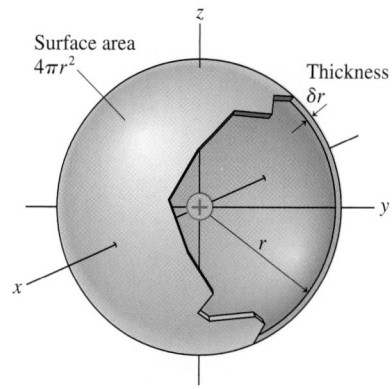

FIGURE 41.7 The radial probability density gives the probability of finding the electron in a spherical shell of thickness δr at radius r.

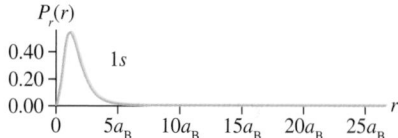
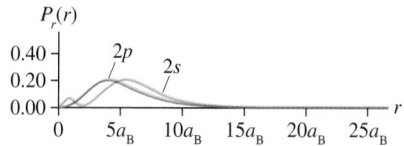
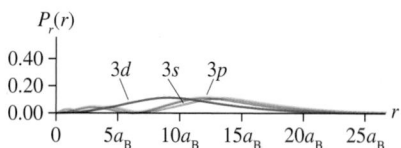

FIGURE 41.8 The radial probability densities for $n = 1$, 2, and 3.

The circular orbit has the largest angular momentum. The electron stays at a constant distance from the nucleus.

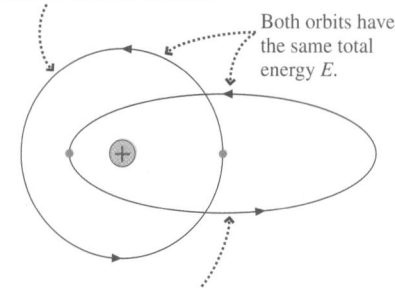

Both orbits have the same total energy E.

The elliptical orbit has a smaller angular momentum. Compared to the circular orbit, the electron gets both closer to and farther from the nucleus.

FIGURE 41.9 More circular orbits have larger angular momenta.

between circular and elliptical orbits will be important when we discuss the energy levels in multielectron atoms.

> **NOTE** ▶ In quantum mechanics, nothing is really orbiting. However, the probability densities for the electron to be, or not to be, at any given distance from the nucleus mimic certain aspects of classical orbits and provide us with a useful analogy. ◀

You can see in Figure 41.8 that the most likely distance from the nucleus of an $n = 1$ electron is approximately a_B. The distance of an $n = 2$ electron is most likely to be between about $3a_B$ and $7a_B$. An $n = 3$ electron is most likely to be found between about $8a_B$ and $15a_B$. In other words, the radial probability densities give the clear impression that each value of n has a fairly well defined range of radii where the electron is most likely to be found. This is the basis of the **shell model** of the atom that is used in chemistry.

However, there's one significant puzzle. In Figure 41.5, the fuzzy sphere representing the $1s$ ground state is densest at the center, where the electron is most likely to be found. This maximum density at $r = 0$ agrees with the $1s$ radial wave function of Figure 41.6, which is a maximum at $r = 0$, but it seems to be in sharp disagreement with the $1s$ graph of Figure 41.8, which is *zero* at the nucleus and peaks at $r = a_B$.

Resolving this puzzle requires distinguishing between the probability density $|\psi(x, y, z)|^2$ and the *radial* probability density $P_r(r)$. The $1s$ wave function, and thus the $1s$ probability density, really does peak at the nucleus. But $|\psi(x, y, z)|^2$ is the probability of being in a small volume δV, such as a small box with sides δx, δy, and δz, whereas $P_r(r)$ is the probability of being in a spherical shell of thickness δr. Compared to $r = 0$, the probability density $|\psi(x, y, z)|^2$ is smaller at any *one* point having $r = a_B$. But the volume of *all* points with $r \approx a_B$ (i.e., the volume of the spherical shell at $r = a_B$) is so large that the radial probability density P_r peaks at this distance.

To use a mass analogy, consider a fuzzy ball that is densest at the center. Even though the density away from the center has decreased, a spherical shell of modest radius r can have *more total mass* than a small-radius spherical shell of the same thickness simply because it has so much more volume.

EXAMPLE 41.3 Maximum probability
Show that an electron in the $2p$ state is most likely to be found at $r = 4a_B$.

SOLVE We can use the $2p$ radial wave function from Equation 41.7 to write the radial probability density

$$P_r(r) = 4\pi r^2 |R_{2p}(r)|^2 = 4\pi r^2 \left[\frac{1}{\sqrt{24\pi a_B^3}} \left(\frac{r}{2a_B} \right) e^{-r/2a_B} \right]^2$$

$$= Cr^4 e^{-r/a_B}$$

where $C = (24a_B)^{-5}$ is a constant. This expression for $P_r(r)$ was graphed in Figure 41.8.

The most probable value of r occurs at the point where the derivative of $P_r(r)$ is zero:

$$\frac{dP_r}{dr} = C(4r^3)(e^{-r/a_B}) + C(r^4) \left(\frac{-1}{a_B} e^{-r/a_B} \right)$$

$$= Cr^3 \left(4 - \frac{r}{a_B} \right) e^{-r/a_B} = 0$$

This expression is zero only if $r = 4a_B$, so $P_r(r)$ is maximum at $r = 4a_B$. An electron in the $2p$ state is most likely to be found at this distance from the nucleus.

STOP TO THINK 41.2 How many maxima will there be in a graph of the radial probability density for the $4s$ state of hydrogen?

41.3 The Electron's Spin

Recall, from Chapter 32, that an electron orbiting a nucleus is a microscopic *magnetic moment* $\vec{\mu}$. Figure 41.10 reminds you that a magnetic moment, like a compass needle, has a north and south pole. Consequently, a magnetic moment in an external magnetic field experiences forces and torques. In the early 1920s, the German physicists Otto Stern and Walter Gerlach developed a technique to measure the magnetic moments of atoms. Their apparatus, shown in Figure 41.11, prepares an *atomic beam* by evaporating atoms out of a hole in an "oven." These atoms, traveling in a vacuum, pass through a *nonuniform* magnetic field. The field is stronger toward the top of the magnet, weaker toward the bottom.

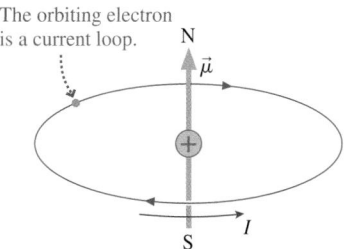

The orbiting electron is a current loop.

A current loop generates a magnetic moment with a north and south magnetic pole.

FIGURE 41.10 An orbiting electron generates a magnetic moment.

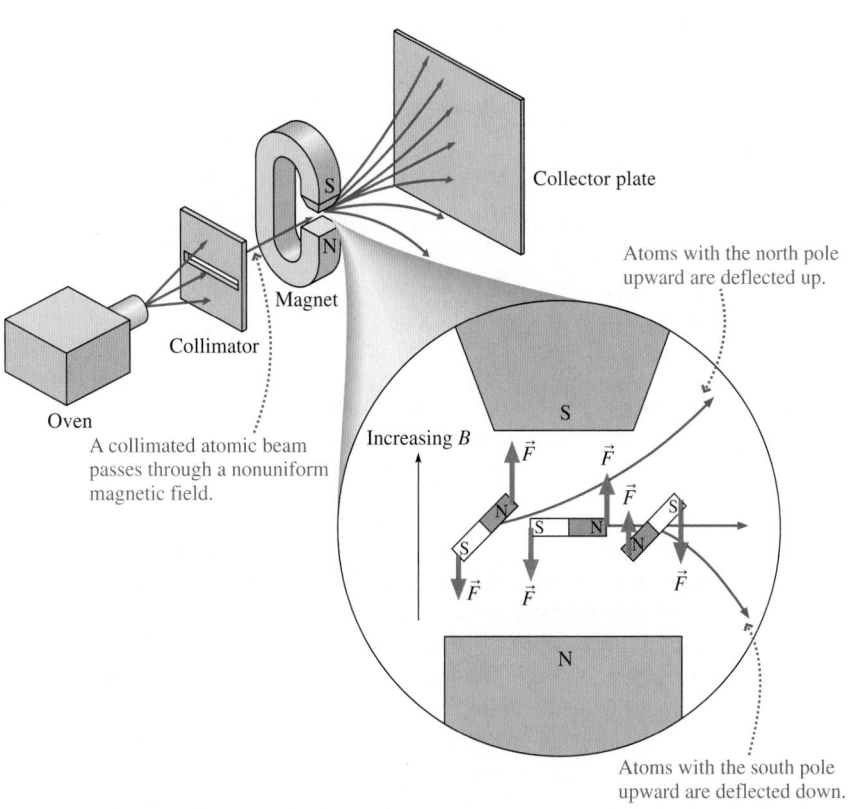

Collector plate

Magnet

Collimator

Oven

A collimated atomic beam passes through a nonuniform magnetic field.

Atoms with the north pole upward are deflected up.

Increasing B

$\vec{F}$

Atoms with the south pole upward are deflected down.

FIGURE 41.11 The Stern-Gerlach experiment.

A nonuniform field exerts forces of *different* strengths on the north and south poles of a magnetic moment, causing a *net force*. An atom whose magnetic moment vector $\vec{\mu}$ is tilted upward ($\mu_z > 0$) has an upward force on its north pole that is larger than the downward force on its south pole. As the figure shows, this atom is deflected upward as it passes through the magnet. A downward-tilted magnetic moment ($\mu_z < 0$) experiences a net downward force and is deflected downward. A magnetic moment perpendicular to the field ($\mu_z = 0$) feels no net force and passes through the magnet without deflection. In other words, an atom's deflection as it passes through the magnet is proportional to μ_z, the z-component of its magnetic moment.

It's not hard to show, although we will omit the proof, that an atom's magnetic moment is proportional to the electron's orbital angular momentum: $\vec{\mu} \propto \vec{L}$. Because the deflection of an atom depends on μ_z, measuring the deflections in a nonuniform field provides information about the L_z values of the atoms in the atomic beam. The measurements are made by allowing the atoms to stick on a collector plate at the end of the apparatus. After the experiment has been run for several hours, the collector plate is removed and examined to learn how the atoms are being deflected.

With the magnet turned off, the atoms pass through without deflection and land along a narrow line at the center, as shown in Figure 41.12a. If the orbiting electrons are classical particles, they should have a continuous range of angular momenta. Turning on the magnet should produce a continuous range of vertical deflections, and the distribution of atoms collected on the plate should look like Figure 41.12b. But if angular momentum is *quantized,* as Bohr had suggested several years earlier, the atoms should be deflected to discrete positions on the collector plate.

(a) Collector plate

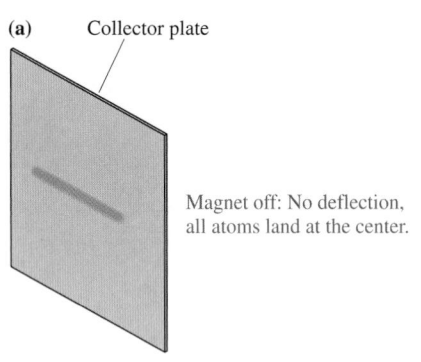

Magnet off: No deflection, all atoms land at the center.

(b)

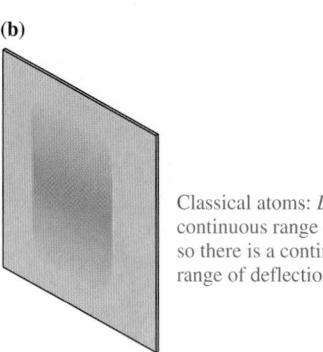

Classical atoms: L_z has a continuous range of values, so there is a continuous range of deflections.

(c)

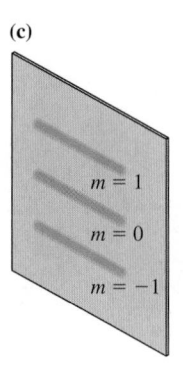

$m = 1$

$m = 0$

$m = -1$

Quantum atoms with $l = 1$: There are three values of L_z, hence three groups of atoms.

FIGURE 41.12 Distribution of the atoms on the collector plate.

Center of plate

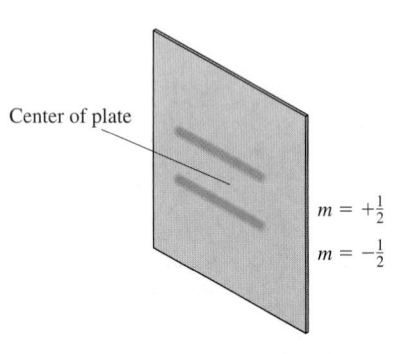

$m = +\frac{1}{2}$

$m = -\frac{1}{2}$

FIGURE 41.13 The outcome of the Stern-Gerlach experiment for hydrogen atoms.

For example, an atom with $l = 1$ has three distinct values of L_z corresponding to quantum numbers $m = -1, 0$, and 1. This leads to a prediction of the three distinct groups of atoms shown in Figure 41.12c. There should always be an *odd* number of groups because there are $2l + 1$ values of L_z.

In 1927, with Schrödinger's quantum theory brand new, the Stern-Gerlach technique was used to measure the magnetic moment of hydrogen atoms. The ground state of hydrogen is 1s, with $l = 0$, so the atoms should have *no* magnetic moment and there should be *no* deflection at all. Instead, the experiment produced the two-peaked distribution shown in Figure 41.13.

Because the hydrogen atoms were deflected, they *must* have a magnetic moment. But where does it come from if $L = 0$? Even stranger was the deflection into two groupings, rather than an odd number. The deflection is proportional to L_z, and $L_z = m\hbar$ where m ranges in integer steps from $-l$ to $+l$. The experimental results would make sense only if $l = \frac{1}{2}$, allowing m to take on the two possible values $-\frac{1}{2}$ or $+\frac{1}{2}$. But according to Schrödinger's theory, the quantum numbers l and m must be integers.

An explanation for these observations was soon suggested, then confirmed: The electron has an *inherent* magnetic moment. After all, the electron has an inherent gravitational character, its mass m_e, and an inherent electric character, its charge $q_e = -e$. These are simply part of what an electron is. Thus it is plausible that an electron should also have an inherent magnetic character described by a built-in magnetic moment $\vec{\mu}_e$. A classical electron, if thought of as a little ball of charge, could spin on its axis as it orbits the nucleus. A spinning ball of charge would have a magnetic moment associated with its angular momentum. This inherent magnetic moment of the electron is what caused the unexpected deflection in the Stern-Gerlach experiment.

If the electron has an inherent magnetic moment, it must have an inherent angular momentum. This angular momentum is called the electron's **spin,** which is designated $\vec{S}$. The outcome of the Stern-Gerlach experiment tells us that the z-component of this spin angular momentum is

$$S_z = m_s \hbar \quad \text{where } m_s = +\frac{1}{2} \quad \text{or} \quad -\frac{1}{2} \qquad (41.12)$$

The quantity m_s is called the **spin quantum number.**

The z-component of the spin angular momentum vector is determined by the electron's orientation. The $m_s = +\frac{1}{2}$ state, with $S_z = +\frac{1}{2}\hbar$, is called the **spin-up** state and the $m_s = -\frac{1}{2}$ state is called the **spin-down** state. It is convenient to picture a little angular momentum vector that can be drawn $\uparrow$ for an $m_s = +\frac{1}{2}$ state and $\downarrow$ for an $m_s = -\frac{1}{2}$ state. We will use this notation in the next section. Because the electron must be either spin-up or spin-down, a hydrogen atom in the Stern-Gerlach experiment will be deflected either up or down. This causes the two groups of atoms seen in Figure 41.13. No atoms have $S_z = 0$, so there are no undeflected atoms in the center.

NOTE ▶ The atom has spin angular momentum *in addition* to any orbital angular momentum that the electrons may have. Only in s states, for which $L = 0$, can we see the effects of "pure spin." ◀

The equation for the spin angular momentum S is analogous to Equation 41.3 for L:

$$S = \sqrt{s(s+1)}\hbar = \frac{\sqrt{3}}{2}\hbar \qquad (41.13)$$

where s is a quantum number with the single value $s = \frac{1}{2}$. S is the *inherent* angular momentum of the electron. Because of the single value of s, physicists usually say that the electron has "spin one-half." Figure 41.14, which should be compared to Figure 41.3, shows that the terms "spin up" and "spin down" refer to S_z not the full spin angular momentum. As was the case with $\vec{L}$, it's not possible for $\vec{S}$ to point along the z-axis.

NOTE ▶ The term "spin" must be used with caution. Although a classical charged particle could generate a magnetic moment by spinning, the electron most assuredly is *not* a classical particle. It is not spinning in any literal sense. It simply has an inherent magnetic moment, just as it has an inherent mass and charge, and that magnetic moment makes it look *as if* the electron is spinning. It is a convenient figure of speech, not a factual statement. **The electron has a spin, but it is *not* a spinning electron!** ◀

The electron's spin has significant implications for atomic structure. The solutions to the Schrödinger equation could be described by the three quantum numbers n, l, and m, but the Stern-Gerlach experiment implies that this is not a complete description of an atom. Knowing that a ground-state atom has quantum numbers $n = 1$, $l = 0$, and $m = 0$ is not sufficient to predict whether the atom will be deflected up or down in a nonuniform magnetic field. We need to add the spin quantum number m_s to make our description complete. (Strictly speaking, we also need to add the quantum number s, but it provides no additional information because its value never changes.) So we really need *four* quantum numbers (n, l, m, m_s) to characterize the stationary states of the atom. The spin orientation does not affect the atom's energy, so a ground-state electron in hydrogen could be in either the $(1, 0, 0, +\frac{1}{2})$ spin-up state or the $(1, 0, 0, -\frac{1}{2})$ spin-down state.

The fact that s has the single value $s = \frac{1}{2}$ has other interesting implications. The correspondence principle tells us that a quantum particle begins to "act classical" in the limit of large quantum numbers. But s cannot become large! **The electron's spin is an intrinsic quantum property of the electron that has *no* classical counterpart.**

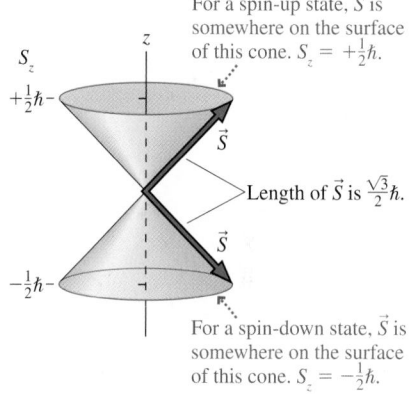

For a spin-up state, $\vec{S}$ is somewhere on the surface of this cone. $S_z = +\frac{1}{2}\hbar$.

Length of $\vec{S}$ is $\frac{\sqrt{3}}{2}\hbar$.

For a spin-down state, $\vec{S}$ is somewhere on the surface of this cone. $S_z = -\frac{1}{2}\hbar$.

FIGURE 41.14 The spin angular momentum has two possible orientations.

Can the spin angular momentum vector lie in the *xy*-plane? Why or why not?

41.4 Multielectron Atoms

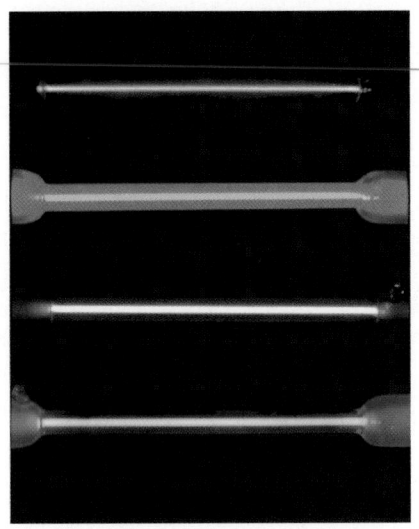

The distinctive color of these gas discharge tubes is due to the unique energy-level structure of each element in the periodic table. The discharges seen here are in hydrogen, neon, helium, and mercury.

The Schrödinger-equation solution for the hydrogen atom matches the experimental evidence, but so did the Bohr hydrogen atom. The real test of Schrödinger's theory is how well it works for multielectron atoms.

A neutral multielectron atom consists of Z electrons surrounding a nucleus with Z protons and charge $+Ze$. Z, the *atomic number*, is the order in which elements are listed in the periodic table. Hydrogen is $Z = 1$, helium $Z = 2$, lithium $Z = 3$, and so on.

The potential-energy function of a multielectron atom consists of Z electrons interacting with the nucleus *and* of the Z electrons interacting *with each other*. The electron-electron interaction makes the atomic-structure problem more difficult than the solar-system problem, and it proved to be the downfall of the simple Bohr model. The planets in the solar system do exert attractive gravitational forces on each other, but their masses are so much less than that of the sun that these planet-planet forces are insignificant for all but the most precise calculations. Not so in an atom. The electron charge is the same as the proton charge, so the electron-electron repulsion is just as important to atomic structure as is the electron-nucleus attraction.

The potential energy due to electron-electron interactions fluctuates rapidly in value as the electrons move and the distances between them change. Rather than treat this interaction in detail, we can reasonably consider each electron to be moving in an *average* potential due to all the other electrons. That is, electron i has potential energy

$$U(r_i) = -\frac{Ze^2}{4\pi\epsilon_0 r_i} + U_{\text{elec}}(r_i) \tag{41.14}$$

where the first term is the electron's interaction with the Z protons in the nucleus and U_{elec} is the average potential energy due to all the other electrons. Because each electron is treated independently of the other electrons, this approach is called the **independent particle approximation,** or IPA. This approximation allows the Schrödinger equation for the atom to be broken into Z separate equations, one for each electron.

A major consequence of the IPA is that **each electron can be described by a wave function having the same four quantum numbers n, l, m, and m_s used to describe the single electron of hydrogen.** Because m and m_s do not affect the energy, we can still refer to electrons by their n and l quantum numbers, using the same labeling scheme that we used for hydrogen.

A major difference, however, is that the energy of an electron in a multielectron atom depends on both n *and* l. Whereas the $2s$ and $2p$ states in hydrogen had the same energy, their energies are different in a multielectron atom. The difference arises from the electron-electron interactions that do not exist in a single-electron hydrogen atom.

Figure 41.15 shows an energy-level diagram for the electrons in a multielectron atom. For comparison, the hydrogen-atom energies are shown on the right edge of the figure. The comparison is quite interesting. States in a multielectron atom that have small values of l are significantly lower in energy than the corresponding state in hydrogen. For each n, the energy increases as l increases until the maximum-l state has an energy very nearly that of the same n in hydrogen. Can we understand this pattern?

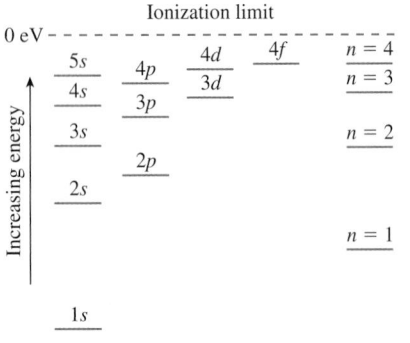

FIGURE 41.15 An energy-level diagram for electrons in a multielectron atom.

Indeed we can. Recall that states of lower *l* correspond to elliptical classical orbits and the highest-*l* state corresponds to a circular orbit. Except for the smallest values of *n*, an electron in a circular orbit spends most of its time *outside* the electron cloud of the remaining electrons. This is illustrated in Figure 41.16. The outer electron is orbiting a ball of charge consisting of *Z* protons and $(Z - 1)$ electrons. This ball of charge has *net* charge $q_{net} = +e$, so the outer electron "thinks" it is orbiting a proton. An electron in a maximum-*l* state is nearly indistinguishable from an electron in the hydrogen atom, thus its energy is very nearly that of hydrogen.

The low-*l* states correspond to elliptical orbits. A low-*l* electron penetrates in very close to the nucleus, which is no longer shielded by the other electrons. Its interaction with the *Z* protons in the nucleus is much stronger than the interaction it would have with the single proton in a hydrogen nucleus. This strong interaction *lowers* its energy in comparison to the same state in hydrogen.

As we noted earlier, a quantum electron does not really orbit. Even so, the probability density of a 3*s* electron has in-close peaks that are missing in the probability density of a 3*d* electron, as you should confirm by looking back at Figure 41.8. Thus a low-*l* electron really does have a likelihood of being at small *r*, where its interaction with the *Z* protons is strong, whereas a high-*l* electron remains farther from the nucleus.

The Pauli Exclusion Principle

By definition, the ground state of a quantum system is the state of lowest energy. What is the ground state of an atom having *Z* electrons and *Z* protons? Because the 1*s* state is the lowest energy state in the independent particle approximation, it seems that the ground state should be one in which all *Z* electrons are in the 1*s* state. However, this hypothesis is not consistent with the experimental evidence.

In 1925, the young Austrian physicist Wolfgang Pauli hypothesized that no two electrons in a quantum system can be in the same quantum state. That is, **no two electrons can have exactly the same set of quantum numbers** (n, l, n, m_s). If one electron is present in a state, it *excludes* all others. This statement is called the **Pauli exclusion principle.**

Wolfgang Pauli was a prodigy who burst onto the physics scene in 1921 when, at the age of 21, he wrote a masterful article on Einstein's theory of relativity. He made many contributions to theoretical physics in the 1920s and the 1930s, but he is most well known for his hypothesis that no two electrons can share the same quantum state. This turns out to be an extremely profound statement about the nature of matter.

The exclusion principle is not applicable to hydrogen, where there is a only a single electron. But in helium, with $Z = 2$ electrons, we must make sure that the two electrons are in different quantum states. This is not difficult. For a 1*s* state, with $l = 0$, the only possible value of the magnetic quantum number is $m = 0$. But there are *two* possible values of m_s, namely $+\frac{1}{2}$ and $-\frac{1}{2}$. If a first electron is in the spin-up 1*s* state $(1, 0, 0, +\frac{1}{2})$, a second 1*s* electron can still be added as long as it is in the spin-down state $(1, 0, 0, -\frac{1}{2})$. This is shown schematically in Figure 41.17a, where the dots represent electrons on the rungs of the "energy ladder" and the arrows represent spin up or spin down.

The Pauli exclusion principle does not prevent both electrons of helium from being in the 1*s* state as long as they have opposite values of m_s, so we predict this to be the ground state. A list of an atom's occupied energy levels is called its **electron configuration.** The electron configuration of the helium ground state is written $1s^2$, where the superscript 2 indicates two electrons in the 1*s* energy level. An excited state of the helium atom might be the electron configuration $1s2s$. This state is shown in Figure 41.17b. Here, because the two electrons have different values of *n*, there is no restriction on their values of m_s.

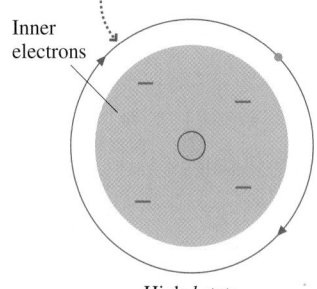

A high-*l* electron corresponds to a circular orbit. It stays outside the core of inner electrons and sees a net charge of +*e*, so it behaves like an electron in a hydrogen atom.

Inner electrons

High-*l* state

A low-*l* electron corresponds to an elliptical orbit. It penetrates into the core and interacts strongly with the nucleus. The electron-nucleus force is attractive, so this interaction lowers the electron's energy.

Low-*l* state

FIGURE 41.16 High-*l* and low-*l* orbitals in a multielectron atom.

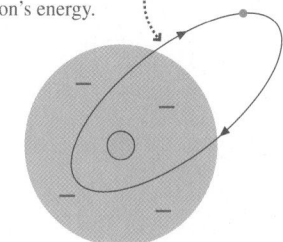

(a) He ground state

The horizontal lines are the allowed energies. 2*s*

Each circle represents an electron in that energy level. 1*s*

(b) He excited state

2*s*

The arrow indicates whether the electron's spin is up ($m_s = +\frac{1}{2}$) or down ($m_s = -\frac{1}{2}$).

1*s*

FIGURE 41.17 The ground state and first excited state of helium.

(a) Li ground state

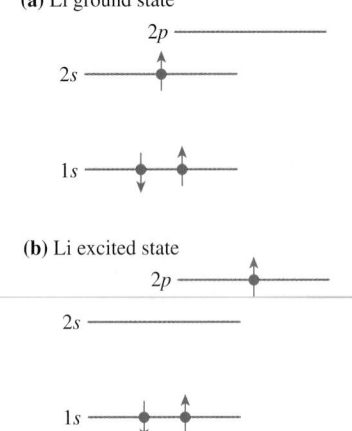

(b) Li excited state

FIGURE 41.18 The ground state and first excited state of lithium.

(a) Li ground state
$1s^2 2s$

(b) Li excited state
$1s^2 2p$

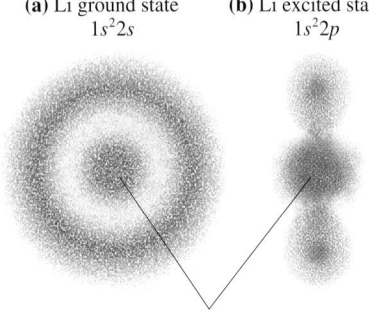

Inner core of two $1s$ electrons

FIGURE 41.19 Electron clouds for the lithium electron configurations $1s^2 2s$ and $1s^2 2p$.

The states $\left(1, 0, 0, +\frac{1}{2}\right)$ and $\left(1, 0, 0, -\frac{1}{2}\right)$ are the only two states with $n = 1$. The ground state of helium has one electron in each of these states, so all the possible $n = 1$ states are filled. Consequently, the electron configuration $1s^2$ is called a **closed shell.** Because the two electron magnetic moments point in opposite directions, we can predict that helium has *no* net magnetic moment and will be undeflected in a Stern-Gerlach apparatus. This prediction is confirmed by experiment.

The next element, lithium, has $Z = 3$ electrons. The first two electrons can go into $1s$ states, with opposite values of m_s, but what about the third electron? The $1s^2$ shell is closed, and there are no additional quantum states having $n = 1$. The only option for the third electron is the next energy state, $n = 2$. The $2s$ and $2p$ states had equal energies in the hydrogen atom, but they do *not* in a multielectron atom. As Figure 41.15 showed, a lower-l state has lower energy than a higher-l state with the same n. The $2s$ state of lithium is lower in energy than $2p$, so lithium's third ground-state electron will be $2s$. This requires $l = 0$ and $m = 0$ for the third electron, but the value of m_s is not relevant because there is only a single electron in $2s$. Figure 41.18a shows the electron configuration with the $2s$ electron being spin-up, but it could equally well be spin-down. The electron configuration for the lithium ground state is written $1s^2 2s$. This indicates two $1s$ electrons and a single $2s$ electron.

Figure 41.19a shows the probability density of electrons in the $1s^2 2s$ ground state of lithium. You can see the $2s$ electron shell surrounding the inner $1s^2$ core. For comparison, Figure 41.19b shows the *first excited state* of lithium, in which the $2s$ electron has been excited to the $2p$ energy level. This forms the $1s^2 2p$ configuration, also shown in Figure 41.18b.

The Schrödinger equation accurately predicts the energies of the $1s^2 2s$ and the $1s^2 2p$ configurations of lithium, but the Schrödinger equation does not tell us which states the electrons actually occupy. The electron spin and the Pauli exclusion principle were the final pieces of the puzzle. Once these were added to Schrödinger's theory, the initial phase of quantum mechanics was complete. Physicists finally had a successful theory for understanding the structure of atoms.

41.5 The Periodic Table of the Elements

The 19th century was a time when chemists were discovering new elements and studying their chemical properties. The century opened with the atomic model still not completely validated, with no clear distinction between atoms and molecules, and with no one having any idea how many elements there might be. But chemistry developed quickly, and by mid-century it was clear that there were dozens of elements, but not hundreds.

Several chemists in the 1860s began to point out the regular recurrence of chemical properties. For example, there are obvious similarities among the alkali metals lithium, sodium, potassium, and cesium. But attempts at organization were hampered by the fact that many elements had yet to be discovered.

The Russian chemist Dmitri Mendeléev was the first to propose, in 1867, a *periodic* arrangement of the elements. He did so by explicitly pointing out "gaps" where, according to his hypothesis, undiscovered elements should exist. He could then predict the expected properties of the missing elements. The subsequent discovery of these elements verified Mendeléev's organizational scheme, which came to be known as the *periodic table of the elements.*

Figure 41.20 shows a modern periodic table. A larger version is printed in Appendix B. The significance of the periodic table to a physicist is the implication that there is a basic regularity or periodicity to the *structure* of atoms. Any successful theory of the atom needs to explain *why* the periodic table looks the way it does.

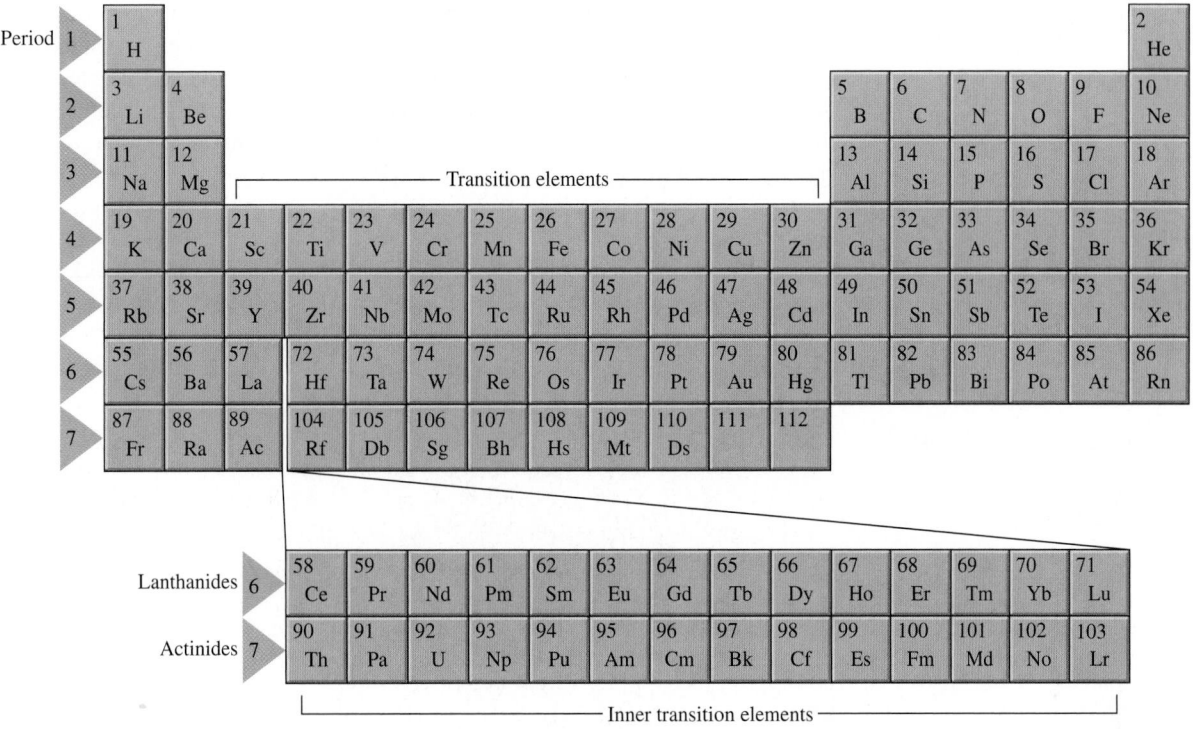

FIGURE 41.20 The modern periodic table of the elements, showing the atomic number Z of each.

The First Two Rows

Quantum mechanics successfully explains the structure of the periodic table. We need three basic ideas to see how this works:

1. The energy levels of an atom are found by solving the Schrödinger equation for multielectron atoms. Figure 41.15, a very important figure for understanding the periodic table, showed that the energy depends on quantum numbers n and l.
2. For each value l of the orbital quantum number, there are $2l + 1$ possible values of the magnetic quantum number m and, for each of these, two possible values of the spin quantum number m_s. Consequently, each energy *level* in Figure 41.15 is actually $2(2l + 1)$ different *states*. Each of these states has the same energy.
3. The ground state of the atom is the lowest-energy electron configuration that is consistent with the Pauli exclusion principle.

We used these ideas in the last section to look at the elements helium $(Z = 2)$ and lithium $(Z = 3)$. Four-electron beryllium $(Z = 4)$ comes next. The first two electrons go into $1s$ states, forming a closed shell, and the third goes into $2s$. There is room in the $2s$ level for a second electron as long as its spin is opposite that of the first $2s$ electron. Thus the third and fourth electrons occupy states $(2, 0, 0, +\frac{1}{2})$ and $(2, 0, 0, -\frac{1}{2})$. These are the only two possible $2s$ states. All of the states with the same values and n and l are called a **subshell,** so the fourth electron closes the $2s$ subshell. (The outer two electrons are called a subshell, rather than a shell, because they complete only the $2s$ possibilities. There are still spaces for $2p$ electrons.) The ground state of beryllium, shown in Figure 41.21, is $1s^2 2s^2$.

These principles can continue to be applied as we work our way through the elements. There are $2l + 1$ values of m associated with each value of l, and each of these can have $m_s = \pm\frac{1}{2}$. This gives, altogether, $2(2l + 1)$ distinct quantum states in each nl subshell. Table 41.2 lists the number of states in each subshell.

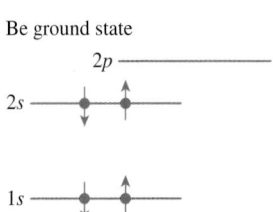

FIGURE 41.21 The ground state of beryllium $(Z = 4)$.

TABLE 41.2 Number of states in each subshell of an atom

Subshell	l	Number of states
s	0	2
p	1	6
d	2	10
f	3	14

Boron ($1s^2 2s^2 2p$) opens the $2p$ subshell. The remaining possible $2p$ states are filled as we continue across the second row of the periodic table. These elements are shown in Figure 41.22. With neon ($1s^2 2s^2 2p^6$), which has six $2p$ electrons, the $n = 2$ shell is complete, and we have another closed shell. The second row of the periodic table is eight elements wide because of the two $2s$ electrons *plus* the six $2p$ electrons needed to fill the $n = 2$ shell.

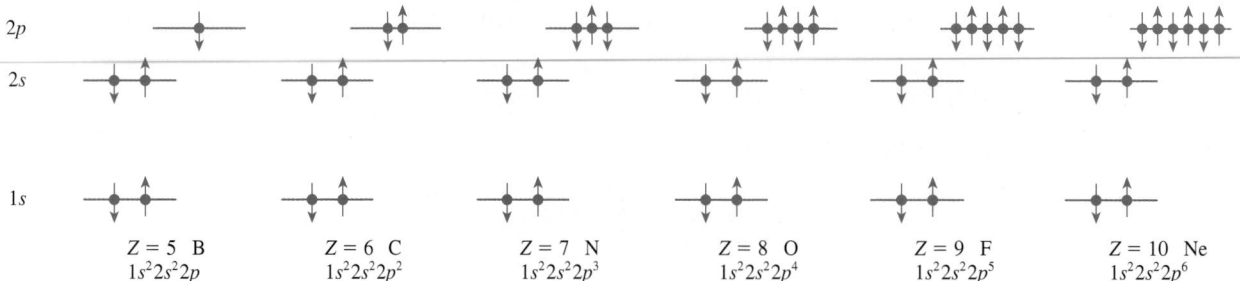

FIGURE 41.22 Filling the $2p$ subshell in the elements boron ($Z = 5$) through neon ($Z = 10$).

Elements with $Z > 10$

The third row of the periodic table is similar to the second. The two $3s$ states are filled in sodium and magnesium. The two columns on the left of the periodic table (groups I and II) represent the two electrons that can go into an s subshell. Then the six $3p$ states are filled, one by one, in aluminum through argon. The six columns on the right (groups III–VIII) represent the six electrons of the p subshell. Argon ($Z = 18$, $1s^2 2s^2 2p^6 3s^2 3p^6$) is another inert gas, although this seems perhaps surprising in that the $3d$ subshell is still open.

The fourth row is where the periodic table begins to get complicated. You might expect the closure of the $3p$ subshell in argon to be followed, starting with potassium ($Z = 19$), by filling the $3d$ subshell. But if you look back at Figure 41.15, where the energies of the different nl states are shown, you will see that the $3d$ state is slightly *higher* in energy than the $4s$ state. Because the ground state is the *lowest energy state* consistent with the Pauli exclusion principle, potassium finds it more favorable to fill a $4s$ state than to fill a $3d$ state. Thus the ground state configuration of potassium is $1s^2 2s^2 2p^6 3s^2 3p^6 4s$ rather than the expected $1s^2 2s^2 2p^6 3s^2 3p^6 3d$.

At this point, we begin to see a competition between increasing n and decreasing l. The highly elliptical characteristic of the $4s$ state brings part of its orbit in so close to the nucleus that its energy is less than that of the more circular $3d$ state. The $4p$ state, though, reverts to the "expected" pattern. We find that

$$E_{4s} < E_{3d} < E_{4p}$$

so the states across the fourth row are filled in the order $4s$, then $3d$, then finally $4p$.

Because there had been no previous d states, the $3d$ subshell "splits open" the periodic table to form the 10-element-wide group of *transition elements*. Most commonly occurring metals are transition elements, and their metallic properties are determined by their partially filled d subshell. The $3d$ subshell closes with zinc, at $Z = 30$, then the next six elements fill the $4p$ subshell up to krypton, at $Z = 36$.

Things get even more complex starting in the sixth row, but the ideas are familiar. The $l = 3$ subshell (f electrons) becomes a possibility with $n = 4$, but it turns out that the 5s, 5p, and 6s states are all lower in energy than 4f. Not until barium ($Z = 56$) fills the 6s subshell (and lanthanum ($Z = 57$) adds a 5d electron) is it energetically favorable to add a 4f electron. Immediately after lanthanum you have to switch down to the *lanthanides* at the bottom of the table. The lanthanides fill in the 4f states.

The 4f subshell is complete with $Z = 71$ lutetium. Then $Z = 72$ hafnium through $Z = 80$ mercury complete the transition-element 5d subshell, followed by the 6p subshell in the six elements thallium through radon at the end of the sixth row. Radon, the last inert gas, has $Z = 86$ electrons and the ground-state configuration

$$\text{radon } (Z = 86): 1s^2 2s^2 2p^6 3s^2 3p^6 4s^2 3d^{10} 4p^6 5s^2 4d^{10} 5p^6 6s^2 4f^{14} 5d^{10} 6p^6$$

This is frightening to behold, but we can now understand it!

EXAMPLE 41.4 The ground state of arsenic
Predict the ground-state electron configuration of arsenic.

SOLVE The periodic table shows that arsenic (As) has $Z = 33$, so we must identify the states of 33 electrons. Arsenic is in the fourth row, following the first group of transition elements. Argon ($Z = 18$) filled the 3p subshell, then calcium ($Z = 20$) filled the 4s subshell. The next 10 elements, through zinc ($Z = 30$) filled the 3d subshell. The 4p subshell starts filling with gallium ($Z = 31$), and arsenic is the third element in this group, so it will have three 4p electrons. Thus the ground-state configuration of arsenic is

$$1s^2 2s^2 2p^6 3s^2 3p^6 4s^2 3d^{10} 4p^3$$

The entire periodic table is well explained by quantum mechanics. Figure 41.23 summarizes the results, showing the subshells as they are filled. It is especially important to note the significance of the electron's spin. Although the introduction of the electron's spin and magnetic moment may have seemed obscure and unnecessary, we now find that the spin quantum number m_s is absolutely essential for understanding the periodic table.

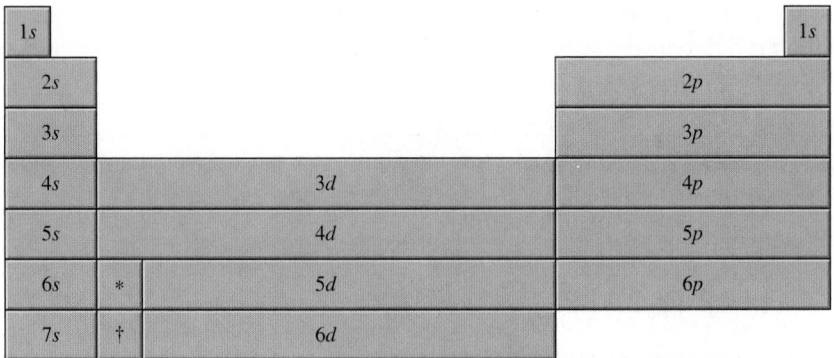

FIGURE 41.23 Summary of the order in which subshells are filled in the periodic table.

Ionization Energies

Ionization energy is the energy needed to remove a ground-state electron from an atom and leave a positive ion behind. The ionization energy of hydrogen is 13.60 eV, because the ground-state energy is $E_1 = -13.60$ eV. Figure 41.24 shows the ionization energies of the first 60 elements in the periodic table.

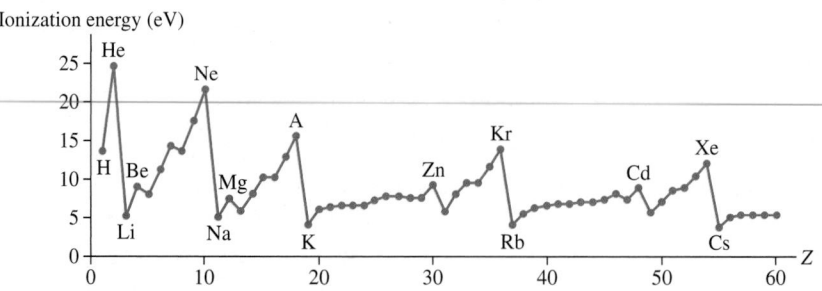

FIGURE 41.24 Ionization energies of the elements up to $Z = 60$.

The ionization energy is different for each element, but there's a clear pattern to the values. Ionization energies are ≈ 5 eV for the alkali metals, on the left edge of the periodic table, then increase steadily to ≥ 15 eV for the inert gases before plunging back to ≈ 5 eV. Can the quantum theory of atoms explain this recurring pattern in the ionization energies?

Indeed it can. The inert-gas elements (helium, neon, argon, . . .) in the right column of the periodic table have *closed shells*. A closed shell is a very stable structure, and that is why these elements are chemically nonreactive (i.e., inert). It takes a large amount of energy to pull an electron out of a stable closed shell, thus the inert gases have the largest ionization energies.

The alkali metals, in the left column of the periodic table, have a single *s*-electron outside a closed shell. This electron is easily disrupted, which is why these elements are highly reactive and have the lowest ionization energies. Between the edges of the periodic table are elements such as beryllium ($1s^2 2s^2$) with a closed $2s$ subshell. You can see in Figure 41.24 that the closed subshell gives beryllium a larger ionization energy than its neighbors lithium ($1s^2 2s$) or boron ($1s^2 2s^2 2p$). However, a closed subshell is not nearly as tightly bound as a closed shell, so the ionization energy of beryllium is much less than that of helium or neon.

All in all, you can see that the basic idea of shells and subshells, which follows from the Schrödinger-equation energy levels and the Pauli principle, provides a good understanding of the recurring features in the ionization energies.

STOP TO THINK 41.4 Is the electron configuration $1s^2 2s^2 2p^4 3s$ a ground-state configuration or an excited-state configuration?

a. Ground-state
b. Excited-state
c. It's not possible to tell without knowing which element it is.

41.6 Excited States and Spectra

18.2

The periodic table organizes information about the *ground states* of the elements. These states are chemically most important because most atoms spend most of the time in their ground states. All the chemical ideas of valence, bonding, reactivity, and so on are consequences of these ground-state atomic structures. But the periodic table does not tell us anything about the excited states of atoms. It is the excited states that hold the key to understanding atomic spectra, and that is the topic to which we turn next.

Sodium ($Z = 11$) is a multielectron atom that we will use as a prototypical atom. The ground-state electron configuration of sodium is $1s^2 2s^2 2p^6 3s$. The first 10 electrons completely fill the $n = 1$ and $n = 2$ shells, creating a *neon core*, while the $3s$ electron is a valence electron. It is customary to represent this configuration as $[Ne]3s$ or, more simply, as just $3s$.

The excited states of sodium are produced by raising the valence electron to a higher energy level. The electrons in the neon core are unchanged. Thus the excited states can be labeled $[Ne]nl$ or, more simply, as just nl. Figure 41.25 is an energy-level diagram showing the ground state and some of the excited states of sodium. Notice that the $1s$, $2s$, and $2p$ states of the neon core are not shown on the diagram. These states are filled and unchanging, so only the states available to the valence electron are shown.

Figure 41.25 has a new feature: The zero of energy has been shifted to the ground state. As we have discovered many times, the zero of energy can be located where it is most convenient. When we solved the Schrödinger equation, it was most convenient to let zero energy represent the energy of an electron infinitely far away. But for analyzing spectra it is more convenient to let the ground state have $E = 0$. With this choice, the excited-state energies tell us how far each state is above the ground state. The ionization limit now occurs at the value of the atom's ionization energy, which is 5.14 eV for sodium.

The first energy level above $3s$ is $3p$, so the *first excited state* of sodium is $1s^2 2s^2 2p^6 3p$, written as $[Ne]3p$ or, more simply, $3p$. The valence electron is excited while the core electrons are unchanged. This state is followed, in order of increasing energy, by $[Ne]4s$, $[Ne]3d$, and $[Ne]4p$. Notice that the order of excited states is exactly the same order ($3p$-$4s$-$3d$-$4p$) that explained the fourth row of the periodic table.

Other atoms with a single valence electron have energy-level diagrams similar to that of sodium. Things get more complicated when there is more than one valence electron, so we'll defer those details to more advanced courses. The point to remember is that quantum mechanics provides the correct framework for classifying and understanding the many interactions that take place within an atom. You can *utilize* the information shown on an energy-level diagram without having to understand precisely *why* each level is where it is.

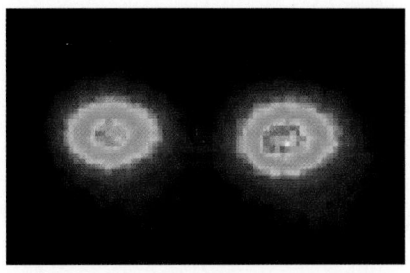

The dots of light are being emitted by two beryllium ions held in a device called an ion trap. Each ion, which is excited by an invisible ultraviolet laser, emits about 10^6 visible-light photons per second.

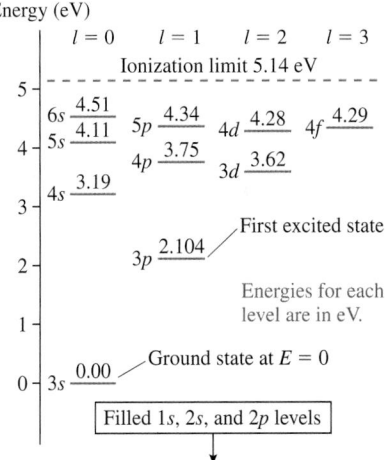

FIGURE 41.25 The $[Ne]3s$ ground state of the sodium atom and some of the excited states.

Excitation by Absorption

Left to itself, an atom will be in its lowest-energy ground state. How does an atom get into an excited state? The process of getting it there is called **excitation,** and there are two basic mechanisms: absorption and collision. We'll begin by looking at excitation by absorption.

One of the postulates of the basic Bohr model is that an atom can jump from one stationary state, of energy E_1, to a higher-energy state E_2 by absorbing a photon of frequency

$$f = \frac{\Delta E_{atom}}{h} = \frac{E_2 - E_1}{h} \tag{41.15}$$

Because we are interested in spectra, it is more useful to write Equation 41.15 in terms of the wavelength:

$$\lambda = \frac{c}{f} = \frac{hc}{\Delta E_{atom}} = \frac{1240 \text{ eV nm}}{\Delta E(\text{in eV})} \tag{41.16}$$

The final expression, which uses the value $hc = 1240$ eV nm, gives the wavelength in nanometers *if* ΔE_{atom} is in electron volts.

Bohr's idea of quantum jumps remains an integral part of our interpretation of the results of quantum mechanics. By absorbing a photon, an atom jumps from its ground state to one of its excited states. However, a careful analysis of how the electrons in an atom interact with a light wave shows that not every conceivable transition can occur. The **allowed transitions** must satisfy one or more **selection rules.**

The only selection rule that will concern us says that a transition (either absorption or emission) from a state in which the valence electron has orbital quantum number l_1 to another with orbital quantum number l_2 is allowed only if

$$\Delta l = |l_2 - l_1| = 1 \quad \text{(selection rule for emission and absorption)} \qquad (41.17)$$

For example, this selection rule says that an atom in an *s* state ($l = 0$) can absorb a photon and be excited to a *p* state ($l = 1$) but *not* to another *s*-state or to a *d* state. An atom in a *p* state ($l = 1$) can emit a photon by dropping to a lower-energy *s* state *or* to a lower-energy *d* state but not to another *p* state.

EXAMPLE 41.5 Absorption in hydrogen

What is the longest wavelength in the absorption spectrum of hydrogen? What is the transition?

SOLVE The longest wavelength corresponds to the smallest energy change ΔE_{atom}. Because the atom starts from the $1s$ ground state, the smallest energy change occurs for absorption to the first $n = 2$ excited state. The energy change is

$$\Delta E_{\text{atom}} = E_2 - E_1 = \frac{-13.6 \text{ eV}}{2^2} - \frac{-13.6 \text{ eV}}{1^2} = 10.2 \text{ eV}$$

The wavelength of this transition is

$$\lambda = \frac{1240 \text{ eV nm}}{10.2 \text{ eV}} = 122 \text{ nm}$$

This is an ultraviolet wavelength. Because of the selection rule, the transition is $1s \rightarrow 2p$, not $1s \rightarrow 2s$.

EXAMPLE 41.6 Absorption in sodium

What is the longest wavelength in the absorption spectrum of sodium? What is the transition?

SOLVE The sodium ground state is $[\text{Ne}]3s$. The lowest excited state is the $3p$ state. $3s \rightarrow 3p$ is an allowed transition ($\Delta l = 1$), so this will be the longest wavelength. You can see from the data in Figure 41.25 that $\Delta E_{\text{atom}} = 2.104$ eV for this transition.

The corresponding wavelength is

$$\lambda = \frac{1240 \text{ eV nm}}{2.104 \text{ eV}} = 589 \text{ nm}$$

ASSESS This wavelength (yellow color) is a prominent feature in the spectrum of sodium. Because the ground state has $l = 0$, absorption *must* be to a *p* state. The *s* states and *d* states of sodium cannot be excited by absorption.

Collisional Excitation

An electron traveling with a speed of 1.0×10^6 m/s has a kinetic energy of 2.85 eV. If this electron collides with a ground-state sodium atom, a portion of its energy can be used to excite the atom to its $3p$ state. This process is called **collisional excitation** of the atom.

Collisional excitation differs from excitation by absorption in one very fundamental way. In absorption, the photon disappears. Consequently, *all* of the photon's energy must be transferred to the atom. Conservation of energy requires $E_{\text{photon}} = \Delta E_{\text{atom}}$. In contrast, the electron is still present after collisional excitation and can carry away some kinetic energy. That is, the electron does *not* have to transfer its entire energy to the atom. If the electron has an incident kinetic energy of 2.85 eV, it could transfer 2.10 eV to the sodium atom, thereby exciting it to the 3p state, and still depart the collision with a speed of 5.1×10^5 m/s and an energy of 0.75 eV.

To excite the atom, the incident energy of the electron (or any other matter particle) merely has to *exceed* ΔE_{atom}. That is $E_{particle} \geq \Delta E_{atom}$. There's a threshold energy for exciting the atom, but no upper limit. It is all a matter of energy conservation. Figure 41.26 shows the idea graphically.

Collisional excitation by electrons is the predominant method of excitation in electrical discharges such as fluorescent lights, street lights, and neon signs. A gas is placed in a tube at reduced pressure (≈ 1 mm of Hg), then a fairly high voltage (≈ 1000 V) between electrodes at the ends of the tube causes the gas to ionize, creating a current in which both ions and electrons are charge carriers. The mean free path of electrons between collisions is large enough for the electrons to gain several eV of kinetic energy as they accelerate in the electric field. This energy is then transferred to the gas atoms upon collision. The process does not work at atmospheric pressure because the mean free path between collisions is too short for the electrons to gain enough kinetic energy to excite the atoms.

NOTE ▶ In contrast to photon absorption, there are no selection rules for collisional excitation. Any state can be excited if the colliding particle has sufficient energy. ◀

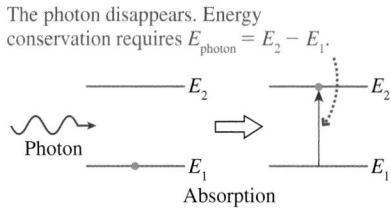

The photon disappears. Energy conservation requires $E_{photon} = E_2 - E_1$.

Absorption

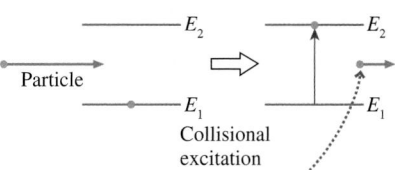

Collisional excitation

The particle carries away energy. Energy conservation requires $E_{particle} \geq E_2 - E_1$.

FIGURE 41.26 Excitation by photon absorption and electron collision.

EXAMPLE 41.7 Excitation of hydrogen
Can an electron traveling at 2.0×10^6 m/s cause a hydrogen atom to emit the prominent red spectral line ($\lambda = 656$ nm) in the Balmer series?

MODEL The electron must have sufficient energy to excite the upper state of the transition.

SOLVE The electron's energy is $E_{elec} = \frac{1}{2}mv^2 = 11.4$ eV. This is significantly larger than the 1.89 eV energy of a photon with wavelength 656 nm, but don't confuse the energy of the photon with the energy of the excitation. The red spectral line in the

Balmer series is emitted by an $n = 3$ to $n = 2$ quantum jump with $\Delta E_{atom} = 1.89$ eV. But to cause this emission, the electron must excite an atom from its *ground state*, with $n = 1$, up to the $n = 3$ level. The necessary excitation energy is

$$\Delta E_{atom} = E_3 - E_1 = (-1.51 \text{ eV}) - (-13.60 \text{ eV})$$
$$= 12.09 \text{ eV}$$

The electron does *not* have sufficient energy to excite the atom to the state from which the emission would occur.

Emission Spectra

The absorption of light is an important process, but it is the emission of light that really gets our attention. The overwhelming bulk of sensory information that we perceive comes to us in the form of light. The recognition and appreciation of light and color has formed the basis of aesthetics and art since the days of prehistory. With the small exception of cosmic rays, all of our knowledge about the cosmos comes to us in the form of light and other electromagnetic waves emitted in various processes.

The discovery of discrete emission spectra helped bring down classical physics, and it was the understanding of discrete emission spectra that provided the first major triumph of quantum mechanics. Emission spectra are more than just scientific curiosities. Many of today's artificial light sources, from fluorescent lights to lasers, are applications of emission spectra.

Understanding emission hinges upon the three ideas shown in Figure 41.27. Once we have determined the energy levels of an atom, by solving the Schrödinger equation, we can immediately predict its emission spectrum. Conversely, we can use the measured emission spectrum to determine an atom's energy levels.

As an example, Figure 41.28a on the next page shows some of the transitions and wavelengths observed in the emission spectrum of sodium. This diagram makes the point that each wavelength represents a quantum jump between two well-defined energy levels. Notice that the selection rule $\Delta l = 1$ is being obeyed

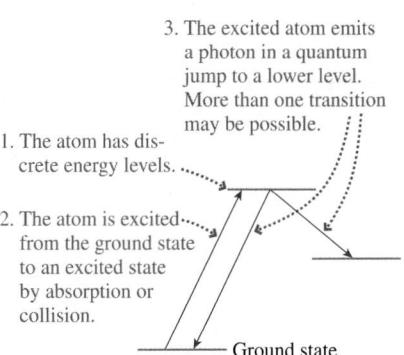

3. The excited atom emits a photon in a quantum jump to a lower level. More than one transition may be possible.

1. The atom has discrete energy levels.

2. The atom is excited from the ground state to an excited state by absorption or collision.

Ground state

FIGURE 41.27 Generation of an emission spectrum.

(a)

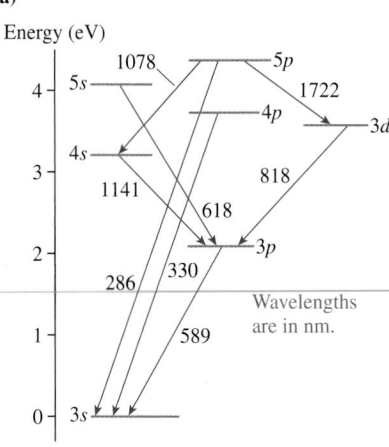

(b)

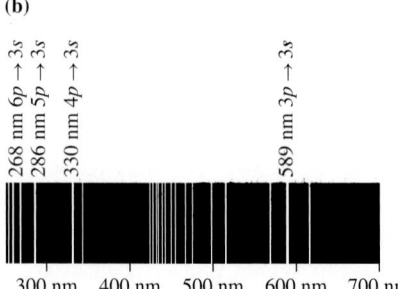

FIGURE 41.28 The emission spectrum of sodium.

The colors in a stained-glass window are due to the selective absorption of light.

in the sodium spectrum. The 5p levels can undergo quantum jumps to 3s, 4s, or 3d but *not* to 3p or 4p.

Figure 41.28b shows the emission spectrum of sodium as it would be recorded in a spectrometer. (Many of the lines seen in this spectrum start from higher excited states that are not seen in the rather limited energy-level diagram of Figure 41.28a.) By comparing the spectrum to the energy-level diagram, you can recognize the spectral lines at 589 nm, 330 nm, 286 nm, and 268 nm form a *series* of lines due to all the possible np → 3s transitions. They are the dominant features in the sodium spectrum.

The most obvious visual feature of sodium emission is its bright yellow color, produced by the emission wavelength of 589 nm. This is the basis of the *flame test* used in chemistry to test for sodium: A sample is held in a Bunsen burner, and a bright yellow glow indicates the presence of sodium. The 589 nm emission is also prominent in the pinkish-yellow glow of the common sodium-vapor street lights. These operate by creating an electrical discharge in sodium vapor. Most sodium-vapor lights use high-pressure lamps to increase their light output. The high pressure, however, causes the formation of Na_2 molecules, and these molecules emit the pinkish portion of the light.

Some cities close to astronomical observatories use low-pressure sodium lights, and these emit the distinctively yellow 589 nm light of sodium. The glow of city lights is a severe problem for astronomers, but the very specific 589 nm emission from sodium is easily removed with a *sodium filter*. The light from the telescope is passed through a container of sodium vapor, and the sodium atoms *absorb* just the unwanted 589 nm photons without disturbing any other wavelengths! However, this cute trick does not work for the other wavelengths emitted by high-pressure sodium lamps or light from other sources.

Color in Solids

It is worth concluding this section with a few remarks about color in solids. Whether it be the intense multihued colors of a stained glass window, the bright colors of flowers or paint, or the deep luminescent red of a ruby, most of the colors we perceive in our lives come from solids rather than free atoms. The basic principles are the same, but the details are different for solids.

An excited atom in a gas has little choice but to give up its energy by emitting a photon. Its only other option, which is rare for gas atoms, is to collide with another atom and transfer its energy into the kinetic energy of recoil. But the atoms in a solid are in intimate contact with each other at all times. Although an excited atom in a solid has the option of emitting a photon, it is often more likely that the energy will be converted, via interactions with neighboring atoms, to the thermal energy of the solid. A process in which an atom is de-excited without radiating is called a **nonradiative transition.**

This is what happens in pigments, such as those in paints, plants, and dyes. Pigments are molecules that absorb certain wavelengths of light but not other wavelengths. The energy-level structure of a molecule is complex, so the absorption consists of "bands" of wavelengths rather than discrete spectral lines. But instead of re-radiating the energy by photon emission, as a free atom would, the pigment molecules undergo nonradiative transitions and convert the energy into increased thermal energy. That is why darker objects get hotter in the sun than lighter objects.

When light falls on an object, it can be either absorbed or reflected. If *all* wavelengths are reflected, the object is perceived as white. Any wavelengths absorbed by the pigments are removed from the reflected light. A pigment with blue-absorbing properties converts the energy of blue-wavelength photons into thermal energy, but photons of other wavelengths are reflected without change. A blue-absorbing pigment reflects the red and yellow wavelengths, causing the object to be perceived as the color orange!

The mechanism for creating colors in colored glass and plastic is the selective absorption of some wavelengths, followed by nonradiative transitions. Blue glass contains molecules that absorb all wavelengths *except* blue. Nonradiative transitions convert the absorbed energy to the thermal energy of the glass. Only blue wavelengths pass through and are seen. The glass acts as a *filter* that passes some wavelengths but filters out others.

Some solids, though, are a little different. The color of many minerals and crystals is due to so-called *impurity atoms* embedded in them. For example, the gemstone ruby is a very simple and common crystal of aluminum oxide, called corundum, that happens to have chromium atoms present at the concentration of about one part in a thousand. Pure corundum is transparent, so all of a ruby's color comes from these chromium impurity atoms.

Figure 41.29 shows what happens when ruby is illuminated by white light. The chromium atoms have a group of excited states that absorb all wavelengths shorter than about 600 nm—that is, everything except orange and red. Unlike the pigments in red glass, which convert all the absorbed energy into thermal energy, the chromium atoms dissipate only a small amount of heat as they undergo a nonradiative transition to another excited state. From there they emit a photon with $\lambda = hc/(E_2 - E_1) \approx 690$ nm as they jump back to the ground state.

The net effect is that short-wavelength photons, rather than being completely absorbed, are *re-radiated* as longer-wavelength photons. This is why rubies sparkle and have such intense color, whereas red glass is a dull red color. The color of other minerals and gems is due to different impurity atoms, but the principle is the same.

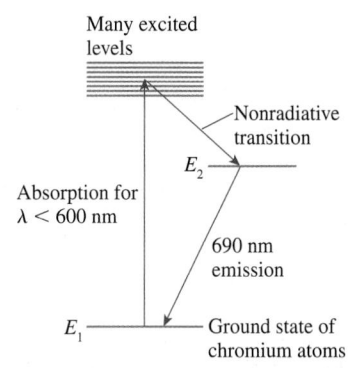

FIGURE 41.29 Absorption and emission in a crystal of ruby.

STOP TO THINK 41.5 In this hypothetical atom, what is the photon energy E_{photon} of the longest-wavelength photons emitted by atoms in the 5*p* state?

a. 1.0 eV
b. 2.0 eV
c. 3.0 eV
d. 4.0 eV

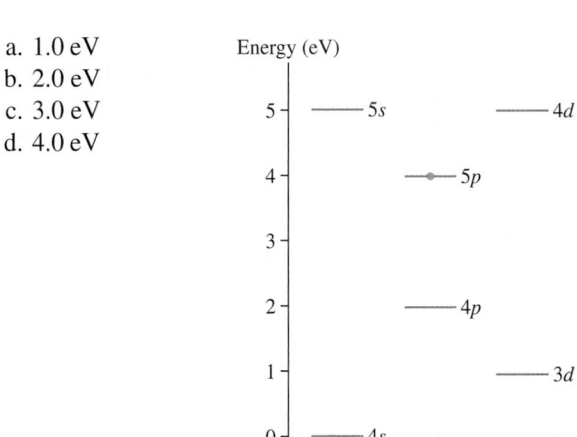

41.7 Lifetimes of Excited States

Excitation of an atom, by either absorption or collision, leaves it in an excited state. From there it jumps back to a lower energy level by emitting a photon. How long does this process take? There are actually two questions here. First, how long does an atom remain in an excited state before undergoing a quantum jump to a lower state? Second, how long does the transition last as the quantum jump is occurring?

Our best understanding of the quantum physics of atoms is that quantum jumps are instantaneous. The absorption or emission of a photon is an all-or-nothing event, so there is not a moment when a photon is "half emitted." The prediction that quantum jumps are instantaneous has troubled many physicists, but careful experimental tests have never revealed any evidence that the jump itself takes a measurable amount of time.

The time spent in the excited state, waiting to make a quantum jump, is another story. Figure 41.30 shows experimental data for the length of time that doubly charged xenon ions Xe^{++} spend in a certain excited state. In this experiment, a pulse of electrons was used to excite the atoms to the excited state. The number of excited-state atoms was then monitored by detecting the photons emitted—one-by-one!—as the excited atoms jumped back to the ground state. The number of photons emitted at time t is directly proportional to the number of excited-state atoms present at time t. As the figure shows, the number of atoms in the excited state decreases *exponentially* with time, and virtually all have decayed within 25 ms of their creation.

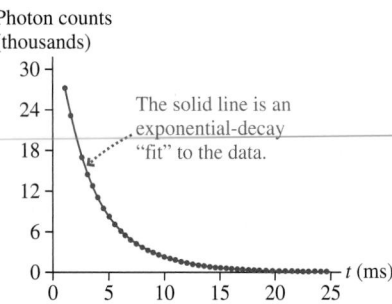

Photon counts (thousands)

The solid line is an exponential-decay "fit" to the data.

FIGURE 41.30 Experimental data for the photon emission rate from an excited state in Xe^{++}.

Figure 41.30 has two important implications. First, atoms spend time in the excited state before undergoing a quantum jump back to a lower state. Second, the length of time spent in the excited state is not a constant value but varies from atom to atom. If every excited xenon ion lived for 5 ms in the excited state, then we would detect *no* photons for 5 ms, a big burst right at 5 ms as they all decay, then no photons after that. Instead, the data tell us that there is a *range* of times spent in the excited state. Some undergo a quantum jump and emit a photon after 1 ms, others after 5 ms or 10 ms, and a few wait as long as 20 or 25 ms.

Consider an experiment in which N_0 excited atoms are created at time $t = 0$. As the solid curve in Figure 41.30 shows, the number of excited atoms remaining at time t is very well described by the exponential function

$$N_{\text{exc}} = N_0 e^{-t/\tau} \tag{41.18}$$

where τ is the point in time at which $e^{-1} = 0.368 = 36.8\%$ of the original atoms are left. Thus 63.2% of the atoms, nearly two-thirds, have emitted a photon and jumped to the lower state by time $t = \tau$. The interval of time τ is called the **lifetime** of the excited state. From Figure 41.30 we can deduce that the lifetime of this state in Xe^{++} is ≈ 4 ms because that is the point in time at which the curve has decayed to 36.8% of its initial value.

This lifetime in Xe^{++} is abnormally long, which is why the state was being studied. More typical excited-state lifetimes are a few nanoseconds. Table 41.3 gives some measured values of excited-state lifetimes. Whatever the value of τ, the number of excited-state atoms decreases exponentially. Can we understand why this is?

TABLE 41.3 Some excited-state lifetimes

Atom	State	Lifetime (ns)
Hydrogen	$2p$	1.6
Sodium	$3p$	17
Neon	$3p$	20
Potassium	$4p$	26

The Decay Equation

Quantum mechanics is about probabilities. We cannot say exactly where the electron is located, but we can use quantum mechanics to calculate the *probability* that the electron is located in a small interval Δx at position x. Similarly, we cannot say exactly when an excited electron will undergo a quantum jump and emit a photon. However, we can use quantum mechanics to find the *probability* that the electron will undergo a quantum jump during a small time interval Δt at time t.

Let us assume that the probability of an excited atom emitting a photon during time interval Δt is *independent* of how long the atom has been waiting in the excited state. For example, a newly excited atom may have a 10% probability of emitting a photon within the 1 ns interval from 0 ns to 1 ns. If it survives until $t = 7$ ns, our assumption is that it still has a 10% probability for emitting a photon during the 1 ns interval from 7 ns to 8 ns.

This assumption, which can be justified with a detailed analysis, is similar to flipping coins. The probability of a head on your first flip is 50%. If you flip seven heads in a row, the probability of a head on your eighth flip is still 50%. It is *unlikely* that you will flip seven heads in a row, but doing so does not influence the eighth flip. Likewise, it may be *unlikely* for an excited atom to live for 7 ns, but doing so does not affect its probability of emitting a photon during the next 1 ns.

If Δt is small, the probability of photon emission during time interval Δt is directly proportional to Δt. That is, if the emission probability in 1 ns is 1%, it will be 2% in 2 ns and 0.5% in 0.5 ns. (This logic fails if Δt gets too big. If the probability is 70% in 20 ns, we can *not* say that the probability would be 140% in 40 ns because a probability >1 is meaningless.) We will be interested in the limit $\Delta t \rightarrow dt$, so the concept is valid and we can write

$$\text{Prob(emission in } \Delta t \text{ at time } t) = r\Delta t \qquad (41.19)$$

where r is called the **decay rate** because the number of excited atoms decays with time. It is a probability *per second,* with units of s^{-1}, and thus is a rate. For example, if an atom has a 5% probability of emitting a photon during a 2 ns interval, its decay rate is

$$r = \frac{P}{\Delta t} = \frac{0.05}{2 \text{ ns}} = 0.025 \text{ ns}^{-1} = 2.5 \times 10^7 \text{ s}^{-1}$$

NOTE ▶ Equation 41.19 is directly analogous to Prob(found in Δx at x) $= P\Delta x$, where P, which had units of m^{-1}, was the probability density. ◀

Figure 41.31 shows N_{exc} atoms in an excited state. During a small time interval Δt, the number of these atoms that we expect to undergo a quantum jump and emit a photon is N_{exc} multiplied by the probability of decay. That is,

number of photons in Δt at time $t = N_{\text{exc}} \times$ Prob(emission in Δt at t)

$$= rN_{\text{exc}}\Delta t \qquad (41.20)$$

Now the *change* in N_{exc} is the *negative* of Equation 41.20. For example, suppose 1000 excited atoms are present at time t and that each has a 5% probability of emitting a photon in the next 1 ns. On average, the number of photons emitted during the next 1 ns will be $1000 \times 0.05 = 50$. Consequently, the number of excited atoms changes by $\Delta N_{\text{exc}} = -50$, with the minus sign indicating a decrease.

Thus the *change* in the number of atoms in the excited state is

$$\Delta N_{\text{exc}}(\text{in } \Delta t \text{ at } t) = -N_{\text{exc}} \times \text{Prob(decay in } \Delta t \text{ at } t) = -rN_{\text{exc}}\Delta t \qquad (41.21)$$

Now let $\Delta t \rightarrow dt$. Then $\Delta N_{\text{exc}} \rightarrow dN_{\text{exc}}$ and Equation 41.21 becomes

$$\frac{dN_{\text{exc}}}{dt} = -rN_{\text{exc}} \qquad (41.22)$$

Equation 41.22 is a *rate equation* because it describes the *rate* at which the excited-state population changes. If r is large, the population will decay at a rapid rate and will have a short lifetime. Conversely, a small value of r implies that the population decays slowly and will live a long time.

The rate equation is a differential equation, but we've solved similar equations before. First, rewrite Equation 41.22 as

$$\frac{dN_{\text{exc}}}{N_{\text{exc}}} = -r\,dt$$

Then integrate both sides from $t = 0$, when the initial excited-state population is N_0, to an arbitrary time t when the population is N_{exc}. That is,

$$\int_{N_0}^{N_{\text{exc}}} \frac{dN_{\text{exc}}}{N_{\text{exc}}} = -r\int_0^t dt \qquad (41.23)$$

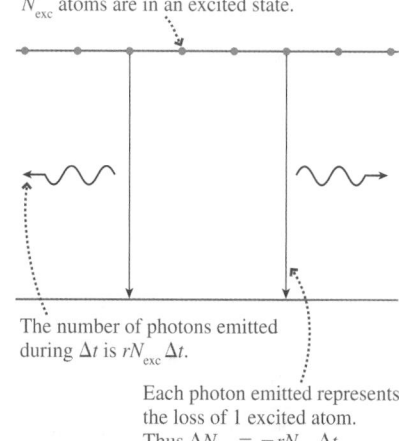

N_{exc} atoms are in an excited state.

The number of photons emitted during Δt is $rN_{\text{exc}}\Delta t$.

Each photon emitted represents the loss of 1 excited atom. Thus $\Delta N_{\text{exc}} = -rN_{\text{exc}}\Delta t$.

FIGURE 41.31 The number of atoms that emit photons during Δt is directly proportional to the number of excited atoms.

Both are well-known integrals, giving

$$\ln N_{\text{exc}} \Big|_{N_0}^{N_{\text{exc}}} = \ln N_{\text{exc}} - \ln N_0 = \ln\left(\frac{N_{\text{exc}}}{N_0}\right) = -rt$$

We can solve for the number of excited atoms at time t by taking the exponential of both sides, then multiplying by N_0. Doing so gives

$$N_{\text{exc}} = N_0 e^{-rt} \qquad (41.24)$$

Notice that $N_{\text{exc}} = N_0$ at $t = 0$, as expected. Equation 41.24, the *decay equation*, shows that the excited-state population decays exponentially with time, as we saw in the experimental data of Figure 41.30.

It will be more convenient to write Equation 41.24 as

$$N_{\text{exc}} = N_0 e^{-t/\tau} \qquad (41.25)$$

where

$$\tau = \frac{1}{r} = \text{the } \textit{lifetime} \text{ of the excited state} \qquad (41.26)$$

This is the definition of the lifetime we used in Equation 41.18 to describe the experimental results. Now we see explicitly that the lifetime is the inverse of the decay rate r.

EXAMPLE 41.8 The lifetime of an excited state in mercury

The mercury atom has two valence electrons. One is always in the $6s$ state, the other is in a state with quantum numbers n and l. One of the excited states in mercury is the state designated $6s6p$. The decay rate of this state is $7.7 \times 10^8 \text{ s}^{-1}$.

a. What is the lifetime of this state?
b. If 1.0×10^{10} mercury atoms are created in the $6s6p$ state at $t = 0$, how many photons will be emitted during the first 1.0 ns?

SOLVE

a. The lifetime is

$$\tau = \frac{1}{r} = \frac{1}{7.7 \times 10^8 \text{ s}^{-1}} = 1.3 \times 10^{-9} \text{ s} = 1.3 \text{ ns}$$

b. If there are $N_0 = 10^{10}$ excited atoms at $t = 0$, the number still remaining at $t = 1.0$ ns is

$$N_{\text{exc}} = N_0 e^{-t/\tau} = (1.0 \times 10^{10}) e^{-(1.0 \text{ ns})/(1.3 \text{ ns})} = 4.63 \times 10^9$$

This result implies that 5.37×10^9 atoms undergo quantum jumps during the first 1.0 ns. Each of these atoms emits one photon, so the number of photons emitted during the first 1.0 ns is 5.37×10^9.

The decay rates R for excited states can be calculated in quantum mechanics and compared to experimentally measured lifetimes of excited states. The agreement is very good, thus providing another validation of the quantum-mechanical description of atoms.

STOP TO THINK 41.6 An equal number of excited A atoms and excited B atoms are created at $t = 0$. The decay rate of B atoms is twice that of A atoms: $r_B = 2r_A$. At $t = \tau_A$ (i.e., after one lifetime of A atoms has elapsed), the ratio N_B/N_A of the number of excited B atoms to the number of excited A atoms is

a. > 2. b. 2. c. 1. d. $\frac{1}{2}$. e. $< \frac{1}{2}$.

41.8 Stimulated Emission and Lasers

We have seen that an atom can jump from a lower-energy level E_1 to a higher-energy level E_2 by absorbing a photon. Figure 41.32a illustrates the basic absorption process, with a photon of frequency $f = \Delta E_{atom}/h$ disappearing as the atom jumps from level 1 to level 2. Once in level 2, as shown in Figure 41.32b, the atom can emit a photon of the same frequency as it jumps back to level 1. Because this transition occurs spontaneously, without the introduction of outside energy, it is called **spontaneous emission.**

In 1917, four years after Bohr's proposal of stationary states in atoms but still prior to de Broglie and Schrödinger, Einstein was puzzled by how quantized atoms reach thermodynamic equilibrium in the presence of electromagnetic radiation. Einstein found that the process of absorption and spontaneous emission were not sufficient to allow a collection of atoms to reach thermodynamic equilibrium. To resolve this difficulty, Einstein proposed a third mechanism for the interaction of atoms with light.

The left half of Figure 41.32c shows a photon with frequency $f = \Delta E_{atom}/h$ approaching an *excited* atom. If a photon can induce the $1 \rightarrow 2$ transition of absorption, then Einstein proposed that it should also be able to induce a $2 \rightarrow 1$ transition. In a sense, this transition is a *reverse absorption*. But to undergo a reverse absorption, the atom must *emit* a photon of frequency $f = \Delta E_{atom}/h$. The end result, as seen in the right half of Figure 41.32c, is an atom in level 1 plus *two* photons! Because the first photon induced the atom to emit the second photon, this process is called **stimulated emission.**

Stimulated emission occurs only if the first photon's frequency exactly matches the $E_2 - E_1$ energy difference of the atom. This is precisely the same condition that absorption has to satisfy. More interesting, the emitted photon is *identical* to the incident photon. This means that as the two photons leave the atom they have exactly the same frequency and wavelength, are traveling in exactly the same direction, and are exactly in phase with each other. In other words, **stimulated emission produces a second photon that is an exact clone of the first.**

Stimulated emission is of no importance in most practical situations. Atoms typically spend only a few nanoseconds in an excited state before undergoing spontaneous emission, so the atom would need to be in an extremely intense light wave for stimulated emission to occur prior to spontaneous emission. Ordinary light sources are not nearly intense enough for stimulated emission to be more than a minor effect, hence it was many years before Einstein's prediction was confirmed. No one had doubted Einstein, because he had clearly demonstrated that stimulated emission was necessary to make the energy equations balance, but it seemed no more important than would pennies to a millionaire balancing her checkbook. At least, that is, until 1960, when a revolutionary invention appeared that made explicit use of stimulated emission: the laser.

Lasers

The word **laser** is an acronym for the phrase **l**ight **a**mplification by the **s**timulated **e**mission of **r**adiation. Lasers were an outgrowth of research with microwaves during the 1950s, which culminated in the invention of the *maser*, short for **m**icrowave **a**mplification by the **s**timulated **e**mission of **r**adiation. The first laser, a ruby laser, was demonstrated in 1960, and several other kinds of lasers appeared within a few months. The driving force behind much of the research was the American physicist Charles Townes. Townes was awarded the Nobel prize in 1964 for the invention of the maser and his theoretical work leading to the laser.

Today, lasers do everything from being the light source in fiber-optic communications to measuring the distance to the moon and from playing your CD to performing delicate eye surgery. But what is a laser? Basically it is a device that

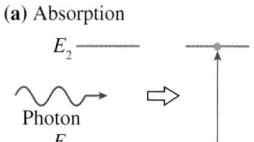

(a) Absorption

(b) Spontaneous emission

(c) Stimulated emission

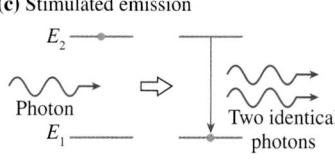

FIGURE 41.32 Three types of radiative transitions.

 18.3

Charles Townes.

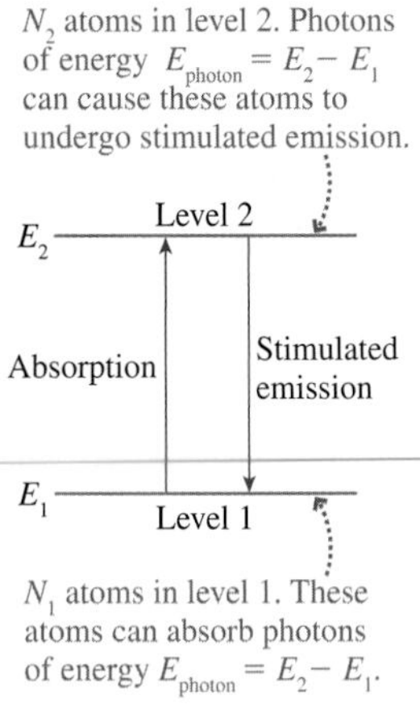

N_2 atoms in level 2. Photons of energy $E_{photon} = E_2 - E_1$ can cause these atoms to undergo stimulated emission.

E_2 ——————— Level 2

Absorption | Stimulated emission

E_1 ——————— Level 1

N_1 atoms in level 1. These atoms can absorb photons of energy $E_{photon} = E_2 - E_1$.

FIGURE 41.33 Energy levels 1 and 2, with populations N_1 and N_2.

produces a beam of highly *coherent* and essentially monochromatic (single-color) light as a result of stimulated emission. **Coherent** light is light in which all the electromagnetic waves have the same phase, direction, and amplitude. It is the coherence of a laser beam that allows it to be very tightly focused or to be rapidly modulated for communications.

Let's take a brief look at how a laser works. Figure 41.33 represents a system of atoms that have a lower energy level E_1 and a higher energy level E_2. Suppose that there are N_1 atoms in level 1 and N_2 atoms in level 2. Left to themselves, all the atoms would soon end up in level 1 because of the spontaneous emission $2 \rightarrow 1$. To prevent this, we can imagine that some type of excitation mechanism, perhaps an electrical discharge, is continuing to produce new excited atoms in level 2.

Let a photon of frequency $f = (E_2 - E_1)/h$ be incident on this group of atoms. Because it has the correct frequency, it could be absorbed by one of the atoms in level 1. Another possibility is that it could cause stimulated emission from one of the level 2 atoms. Ordinarily $N_2 \ll N_1$, so absorption events far outnumber stimulated emission events. Even if a few photons were generated by stimulated emission, they would quickly be absorbed by the vastly larger group of atoms in level 1.

But what if we could somehow arrange to place *every* atom in level 2, making $N_1 = 0$. Then the incident photon, upon encountering its first atom, will cause stimulated emission. Where there was initially one photon of frequency f, now there are two. These will strike two additional excited-state atoms, again causing stimulated emission. Then there will be four photons. As Figure 41.34 shows, there will be a *chain reaction* of stimulated emission until all N_2 atoms emit a photon of frequency f.

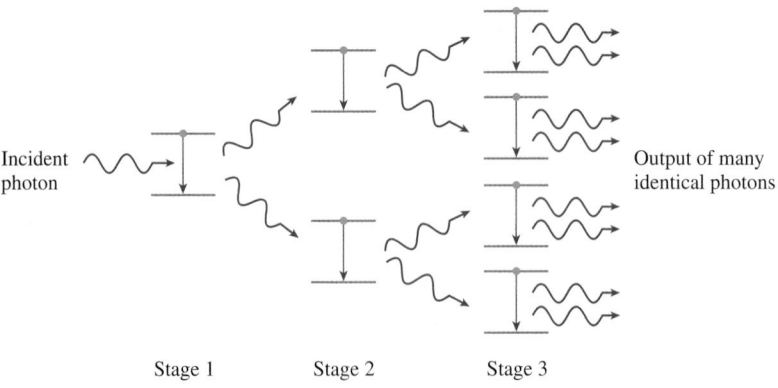

Incident photon

Output of many identical photons

Stage 1   Stage 2   Stage 3

FIGURE 41.34 Stimulated emission creates a chain reaction of photon production in a population of excited atoms.

In stimulated emission, each emitted photon is *identical* to the incident photon. The chain reaction of Figure 41.34 will lead not just to N_2 photons of frequency f, but to N_2 *identical* photons, all traveling together in the same direction with the same phase. If N_2 is a large number, as would be the case in any practical device, the one initial photon will have been *amplified* into a gigantic, coherent pulse of light! A collection of excited-state atoms is referred to as an *optical amplifier*.

The stimulated emission is sustained by placing the *lasing medium*—the sample of atoms that emits the light—in an **optical cavity** consisting of two facing mirrors. As Figure 41.35 shows, the photons interact repeatedly with the atoms in the medium as they bounce back and forth. This repeated interaction is necessary for the light intensity to build up to a high level. One of the mirrors will be partially transmitting so that some of the light emerges as the *laser beam*.

Laser medium | Counterpropagating light waves

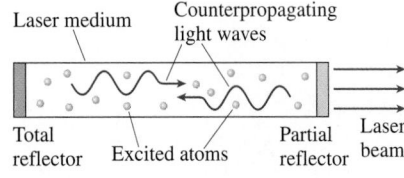

Total reflector | Excited atoms | Partial reflector | Laser beam

FIGURE 41.35 Lasing takes place in an optical cavity.

Although the chain reaction of Figure 41.34 illustrates the idea most clearly, it is not necessary for every atom to be in level 2 for amplification to occur. All that is needed is to have $N_2 > N_1$ so that stimulated emission exceeds absorption. Such a situation is called a **population inversion.** The process of obtaining a population inversion is called **pumping,** and we will look at two specific examples. Pumping is the technically difficult part of designing and building a laser because normal excitation mechanisms do not create population inversions. In fact, lasers would likely have been discovered accidentally long before 1960 if population inversions were easy to create.

The Ruby Laser

The first laser to be developed was a ruby laser. Figure 41.36a shows the energy-level structure of the chromium atoms that gives ruby its optical properties. Normally, the number of atoms in the ground-state level E_1 far exceeds the number of excited-state atoms with energy E_2. That is, $N_2 \ll N_1$. Under these circumstances 690 nm light is absorbed rather than amplified. But suppose that we could *rapidly* excite more than half the chromium atoms to level E_2. Then we would have a population inversion ($N_2 > N_1$) between levels E_1 and E_2.

This can be accomplished by *optically pumping* the ruby with a very intense pulse of white light from a *flashlamp*. A flashlamp is like a camera flash, only vastly more intense. In the basic arrangement of Figure 41.36b, a helical flashlamp is coiled around a ruby rod that has mirrors bonded to its end faces. The lamp is fired by discharging a high-voltage capacitor through it, creating a very intense light pulse that lasts just a few microseconds. This intense light excites nearly all the chromium atoms from the ground state to the upper energy levels. From there, they quickly ($\approx 10^{-8}$ s) decay nonradiatively to level 2. With $N_2 > N_1$, a population inversion has been created.

Once a photon initiates the laser pulse, the light intensity builds quickly into a brief but incredibly intense burst of light. A typical output pulse lasts 10 ns and has an energy of 1 J. This gives a *peak power* of

$$P = \frac{\Delta E}{\Delta t} = \frac{1 \text{ J}}{10^{-8} \text{ s}} = 10^8 \text{ W} = 100 \text{ MW}$$

One hundred megawatts of light power! That is more than the electrical power consumed by a small city. The difference, of course, is that a city consumes that power continuously but the laser pulse lasts a mere 10 ns. The laser cannot fire again until the capacitor is recharged and the laser rod cooled. A typical firing rate is a few pulses per second, so the laser is "on" only a few billionths of a second out of each second.

Ruby lasers have been replaced by other pulsed lasers that, for various practical reasons, are easier to operate. However, they all operate with the same basic idea of rapid optical pumping to upper states, rapid nonradiative decay to level 2 where the population inversion is formed, then rapid build-up of an intense optical pulse.

The Helium-Neon Laser

The familiar red laser used in lecture demonstrations, laboratories, and supermarket checkout scanners is the helium-neon laser, often called a HeNe laser. Its output is a *continuous,* rather than pulsed, wavelength of 632.8 nm. The medium of a HeNe laser is a mixture of $\approx 90\%$ helium and $\approx 10\%$ neon gases. As Figure 41.37a on the next page shows, the gases are sealed in a glass tube, then an electrical discharge is established along the bore of the tube. Two mirrors are bonded to the ends of the discharge tube, one a total reflector and the other having $\approx 2\%$ transmission so that the laser beam can be extracted.

(a)

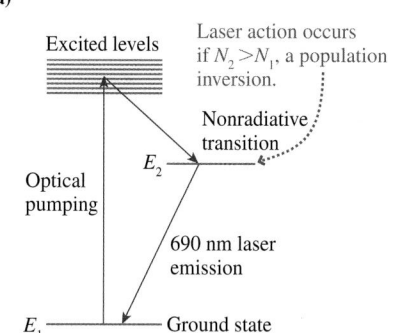

(b)

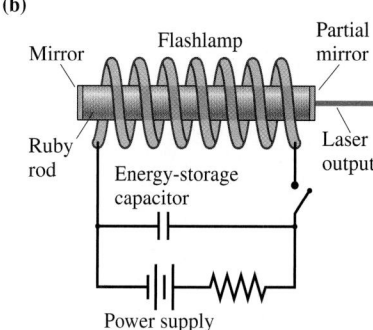

FIGURE 41.36 A flashlamp-pumped ruby laser.

(a)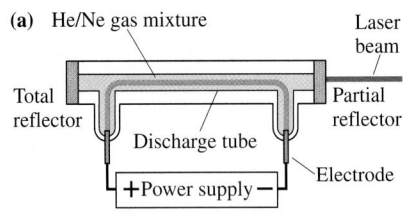

(b)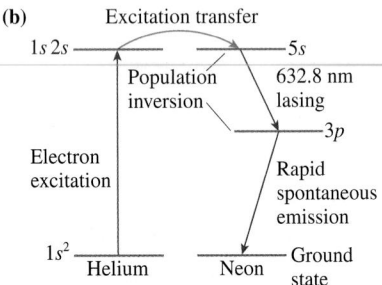

FIGURE 41.37 A HeNe laser.

The atoms that lase in a HeNe are the neon atoms, but the pumping method involves the helium atoms. The electrons in the discharge easily excite the $1s2s$ excited state of helium. This state has a very small spontaneous decay rate (i.e., a very long lifetime), so it is possible to build up a fairly large population (but not an inversion) of excited helium atoms in this state. The energy of the $1s2s$ state is 20.6 eV.

Interestingly, an excited state of neon, the $5s$ state, also has an energy of 20.6 eV. If a $1s2s$ excited helium atom collides with a ground state neon atom, as frequently happens, the excitation energy can be transferred from one atom to the other! Written as a chemical reaction, the process is

$$\text{He}^* + \text{Ne} \rightarrow \text{He} + \text{Ne}^*$$

where the asterisk indicates the atom is in an excited state. This process, called **excitation transfer,** is very efficient for the $5s$ state because the process is *resonant*—a perfect energy match. Thus the two-step process of collisional excitation of helium, followed by excitation transfer between helium and neon, pumps the neon atoms into the excited $5s$ state. This is shown in Figure 41.37b.

The $5s$ energy level in neon is ≈ 1.95 eV above the $3p$ state. The $3p$ state is very nearly empty of population, both because it is not efficiently populated in the discharge and because it undergoes very rapid spontaneous emission to the $3s$ states. Thus the large number of atoms pumped into the $5s$ state creates a population inversion with respect to the lower $3p$ state. These are the necessary conditions for laser action.

Because the lower level of the laser transition is normally empty of population, or very nearly so, placing only a small fraction of the neon atoms in the $5s$ state creates a population inversion. Thus a fairly modest pumping action is sufficient to create the inversion and start the laser. Furthermore, a HeNe laser can maintain a *continuous* inversion and thus sustain continuous lasing. The electrical discharge continuously creates $5s$ excited atoms in the upper level, and the rapid spontaneous decay of the $3p$ atoms from the lower level keeps its population low enough to sustain the inversion.

A typical helium-neon laser has a power output of 1 mW = 10^{-3} J/s at 632.8 nm in a 1-mm-diameter laser beam. As you can show in a homework problem, this output corresponds to the emission of 3.2×10^{15} photons per second. Other continuous lasers operate by similar principles, but can produce much more power. The argon laser, which is widely used in scientific research, can produce up to 20 W of power at green and blue wavelengths. The carbon dioxide laser produces output power in excess of 1000 W at the infrared wavelength of 10.6 μm. It is used in industrial applications for cutting and welding.

EXAMPLE 41.9 An ultraviolet laser
An ultraviolet laser generates a 10 MW, 5.0-ns-long light pulse at a wavelength of 355 nm. How many photons are in each pulse?

SOLVE The energy of each light pulse is the power multiplied by the duration:

$$E_{\text{pulse}} = P\Delta t = (1.0 \times 10^7 \text{ W})(5.0 \times 10^{-9} \text{ s}) = 0.050 \text{ J}$$

Each photon in the pulse has energy

$$E_{\text{photon}} = hf = \frac{hc}{\lambda} = 3.50 \text{ eV} = 5.60 \times 10^{-19} \text{ J}$$

Because $E_{\text{pulse}} = NE_{\text{photon}}$, the number of photons is

$$N = \frac{E_{\text{pulse}}}{E_{\text{photon}}} = 8.9 \times 10^{16} \text{ photons}$$

SUMMARY

The goal of Chapter 41 has been to understand the structure and properties of atoms.

IMPORTANT CONCEPTS

Hydrogen Atom

The three-dimensional Schrödinger equation has stationary-state solutions for the hydrogen atom potential energy only if three conditions are satisfied:

- Energy $E_n = -13.60 \text{ eV}/n^2 \qquad n = 1, 2, 3, \ldots$
- Angular momentum $L = \sqrt{l(l+1)}\,\hbar \qquad l = 0, 1, 2, 3, \ldots, n-1$
- z-component of angular momentum
  $L_z = m\hbar \qquad m = -l, -l+1, \ldots, 0, \ldots, l-1, l$

Each state is characterized by **quantum numbers** (n, l, m), but the energy depends only on n.

The probability of finding the electron within a small distance interval δr at distance r is

$$\text{Prob(in } \delta r \text{ at r)} = P_r(r)\,\delta r$$

where $P_r(r) = 4\pi r^2 |R_{nl}(r)|^2$ is the **radial probability density.**

Graphs of $P_r(r)$ suggest that the electrons are arranged in shells.

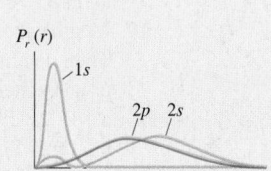

Multielectron Atoms

The potential energy is electron-nucleus plus electron-electron. In the **independent particle approximation,** each electron is described by the same quantum numbers (n, l, m, m_s) used for the hydrogen atom. The energy of a state depends on n and l. For each n, energy increases as l increases.

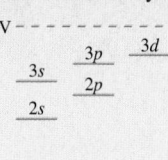

- High-l states correspond to circular orbits. These stay outside the core.
- Low-l states correspond to elliptical orbits. These penetrate the core to interact more strongly with the nucleus. This interaction lowers their energy.

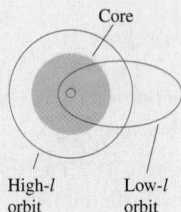

Electron spin

The electron has an inherent angular momentum $\vec{S}$ and magnetic moment $\vec{\mu}$ *as if* it were spinning. The spin angular momentum has a fixed magnitude $S = \sqrt{s(s+1)}\,\hbar$, where $s = \frac{1}{2}$. The z-component is $S_z = m_s\hbar$, where $m_s = \pm\frac{1}{2}$. These two states are called **spin-up** and **spin-down.** Each atomic state is fully characterized by the four quantum numbers (n, l, m, m_s).

The Pauli exclusion principle says that no more than one electron can occupy each quantum state. The periodic table of the elements is based on the fact that the ground state is the lowest-energy electron configuration compatible with the Pauli principle.

APPLICATIONS

Atomic spectra are generated by excitation followed by a photon-emitting quantum jump.

- Excitation by absorption or collision.
- Quantum-jump selection rule $\Delta l = \pm 1$.

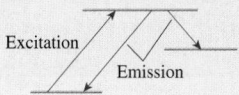

Lifetimes of excited states

The excited-state population decreases exponentially as

$$N_{\text{exc}} = N_0 e^{-t/\tau}$$

where $\tau = 1/r$ is the **lifetime** and r is the **decay rate.** It's not possible to predict when a particular atom will decay, but the *probability* is

$$\text{Prob(in } \delta t \text{ at t)} = r\,\delta t.$$

Stimulated emission of an excited state can be caused by a photon with $E_{\text{photon}} = E_2 - E_1$. Laser action can occur if $N_2 > N_1$, a condition called a **population inversion.**

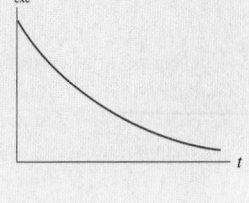

TERMS AND NOTATION

principal quantum number, n	spin-down	nonradiative transition
orbital quantum number, l	independent particle	lifetime, τ
magnetic quantum number, m	approximation (IPA)	decay rate, r
ionization energy	Pauli exclusion principle	spontaneous emission
electron cloud	electron configuration	stimulated emission
radial wave function, $R_{nl}(r)$	closed shell	laser
radial probability density, $P_r(r)$	subshell	coherent
shell model	excitation	optical cavity
spin	allowed transition	population inversion
spin quantum number, m_s	selection rule	pumping
spin-up	collisional excitation	excitation transfer

EXERCISES AND PROBLEMS

Exercises

Section 41.1 The Hydrogen Atom: Angular Momentum and Energy

Section 41.2 The Hydrogen Atom: Wave Functions and Probabilities

1. What is the angular momentum of a hydrogen atom in (a) a $4p$ state and (b) a $5f$ state? Give your answers as a multiple of $\hbar$.

2. List the quantum numbers, excluding spin, of (a) all possible $3p$ states and (b) all possible $3d$ states.

3. A hydrogen atom has orbital angular momentum 3.65×10^{-34} J s.
   a. What letter (s, p, d, or f) describes the electron?
   b. What is the atom's minimum possible energy? Explain.

4. What is the maximum possible angular momentum L (as a multiple of $\hbar$) of a hydrogen atom with energy -0.544 eV?

5. What are E and L (as a multiple of $\hbar$) of a hydrogen atom in the $6f$ state?

6. Explain the difference between l and L.

Section 41.3 The Electron's Spin

7. How many lines of atoms would you expect to see on the collector plate of a Stern-Gerlach apparatus if the experiment is done with (a) lithium and (b) beryllium? Explain.

8. When all quantum numbers are considered, how many different quantum states are there for a hydrogen atom with $n = 1$? With $n = 2$? With $n = 3$? List the quantum numbers of each state.

9. Explain the difference between s and S.

Section 41.4 Multielectron Atoms

Section 41.5 The Periodic Table of the Elements

10. Predict the ground-state electron configurations of Mg, Sr, and Ba.

11. Predict the ground-state electron configurations of Si, Ge, and Pb.

12. Identify the element for each of these electron configurations. Then determine whether this configuration is the ground state or an excited state.
    a. $1s^2 2s^2 2p^5$
    b. $1s^2 2s^2 2p^6 3s^2 3p^6 3d^{10} 4s^2 4p$

13. Identify the element for each of these electron configurations. Then determine whether this configuration is the ground state or an excited state.
    a. $1s^2 2s^2 2p^4 3d$
    b. $1s^2 2s^2 2p^6 3s^2 3p^6 3d^8 4s^2$

14. Figure 41.24 shows that the ionization energy of cadmium ($Z = 48$) is larger than that of its neighbors. Why is this?

Section 41.6 Excited States and Spectra

15. What is the electron configuration of the second excited state of lithium?

16. Show that $hc = 1240$ eV nm.

17. A neon discharge emits a bright reddish-orange spectrum. But a glass tube filled with neon is completely transparent. Why doesn't the neon in the tube absorb orange and red wavelengths?

18. The two spectra shown in Figure Ex41.18 belong to the same element, a fictional Element X. Explain why they are different.

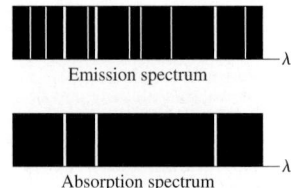

Emission spectrum

FIGURE EX41.18   Absorption spectrum

19. a. Is a $4p \rightarrow 4s$ transition allowed in sodium? If so, what is its wavelength? If not, why not?
    b. Is a $3d \rightarrow 4s$ transition allowed in sodium? If so, what is its wavelength? If not, why not?

Section 41.7 Lifetimes of Excited States

20. An atom in an excited state has a 1.0% chance of emitting a photon in 0.10 ns. What is the lifetime of the excited state?
21. An excited state of an atom has a 25 ns lifetime. What is the probability that an excited atom will emit a photon during a 0.50 ns interval of time?
22. 1.0×10^6 sodium atoms are excited to the $3p$ state at $t = 0$ s. How many of these atoms remain in the $3p$ state at (a) $t = 10$ ns, (b) $t = 30$ ns, and (c) $t = 100$ ns?
23. 1.0×10^6 atoms are excited to an upper energy level at $t = 0$ s. At the end of 20 ns, 90% of these atoms have undergone a quantum jump to the ground state.
    a. How many photons have been emitted?
    b. What is the lifetime of the excited state?

Section 41.8 Stimulated Emission and Lasers

24. A 1.0 mW helium neon laser emits a visible laser beam with a wavelength of 633 nm. How many photons are emitted per second?
25. A 1000 W carbon dioxide laser emits an infrared laser beam with a wavelength of 10.6 μm. How many photons are emitted per second?

Problems

26. a. Draw a diagram similar to Figure 41.3 to show all the possible orientations of the angular momentum vector $\vec{L}$ for the case $l = 3$. Label each $\vec{L}$ with the appropriate value of m.
    b. What is the minimum angle between $\vec{L}$ and the z-axis?
27. There exist subatomic particles whose spin is characterized by $s = 1$, rather than the $s = \frac{1}{2}$ of electrons. These particles are said to have a spin of one.
    a. What is the magnitude of the spin angular momentum S for a particle with a spin of one?
    b. What are the possible values of the spin quantum number?
    c. Draw a vector diagram similar to Figure 41.14 to show the possible orientations of $\vec{S}$.
28. For a hydrogen atom with $l = 2$, what are the (a) minimum and (b) maximum values of the quantity $(L_x^2 + L_y^2)^{1/2}$?
29. A hydrogen atom in its fourth excited state emits a photon with a wavelength of 1282 nm. What is the atom's maximum possible orbital angular momentum after the emission?
30. Calculate (a) the radial wave function and (b) the radial probability density at $r = \frac{1}{2}a_B$ for an electron in the $1s$ state of hydrogen. Give your answers in terms of a_B.
31. For an electron in the $1s$ state of hydrogen, what is the probability of being in a spherical shell of thickness $0.01a_B$ at distance (a) $\frac{1}{2}a_B$, (b) a_B, and (c) $2a_B$ from the proton?
32. The hydrogen atom $1s$ wave function is a maximum at $r = 0$. But the $1s$ radial probability density, shown in Figure 41.8, peaks at $r = a_B$ and is zero at $r = 0$. Explain this paradox.
33. Prove that the normalization constant of the $1s$ radial wave function of the hydrogen atom is $(\pi a_B^3)^{-1/2}$, as given in Equation 41.7.
    Hint: A useful definite integral is

$$\int_0^\infty x^n e^{-\alpha x}\, dx = \frac{n!}{\alpha^{n+1}}$$

34. Prove that the normalization constant of the $2p$ radial wave function of the hydrogen atom is $(24\pi a_B^3)^{-1/2}$, as shown in Equation 41.7.
    Hint: See the hint in Problem 33.
35. Prove that the radial probability density peaks at $r = a_B$ for the $1s$ state of hydrogen.
36. a. Calculate and graph the hydrogen radial wave function $R_{2p}(r)$ over the interval $0 \le r \le 8a_B$.
    b. Determine the value of r (in terms of a_B) for which $R_{2p}(r)$ is a maximum.
    c. Example 41.3 and Figure 41.8 showed that the radial probability density for the $2p$ state is a maximum at $r = 4a_B$. Explain why this differs from your answer to part b.
37. In general, an atom can have both orbital angular momentum and spin angular momentum. The *total* angular momentum is defined to be $\vec{J} = \vec{L} + \vec{S}$. The total angular momentum is quantized in the same way as $\vec{L}$ and $\vec{S}$. That is, $J = \sqrt{j(j+1)}\hbar$, where j is the total angular momentum quantum number. The z-component of $\vec{J}$ is $J_z = L_z + S_z = m_j\hbar$, where m_j goes in integer steps from $-j$ to $+j$. Consider a hydrogen atom in a p state, with $l = 1$.
    a. L_z has three possible values and S_z has two. List all possible combinations of L_z and S_z. For each, compute J_z and determine the quantum number m_j. Put your results in a table.
    b. The number of values of J_z that you found in part a is too many to go with a single value of j. But you should be able to divide the values of J_z into two groups that correspond to two values of j. What are the allowed values of j? Explain. In a classical atom, there would be no restrictions on how the two angular momenta $\vec{L}$ and $\vec{S}$ can combine. Quantum mechanics is different. You've now shown that there are only two allowed ways to add these two angular momenta.
38. In a multielectron atom, the lowest-l state for each n ($1s$, $2s$, $3s$, etc.) is significantly lower in energy than the hydrogen state having the same n. But the highest-l state for each n ($2p$, $3d$, $4f$, etc.) is very nearly equal in energy to the hydrogen state with the same n. Explain.
39. Draw a series of pictures, similar to Figure 41.22, for the ground states of K, Ti, Fe, Ge, and Br.
40. Draw a series of pictures, similar to Figure 41.22, for the ground states of Ca, Sc, Co, Zn, and Kr.
41. a. What downward transitions are possible for a sodium atom in the $6s$ state? (See Figure 41.25.)
    b. What are the wavelengths of the photons emitted in each of these transitions?
42. The ionization energy of an atom is known to be 5.5 eV. The emission spectrum of this atom contains only the four wavelengths 310.0 nm, 354.3 nm, 826.7 nm, and 1240.0 nm. Draw an energy-level diagram with the fewest possible energy levels that agrees with these experimental data. Label each level with an appropriate l quantum number.
    Hint: Don't forget about the Δl selection rule.
43. Suppose you put five electrons into a 0.50-nm-wide one-dimensional rigid box (i.e., an infinite potential well).
    a. Use an energy-level diagram to show the electron configuration of the ground state.
    b. What is the ground-state energy?
44. The $5d \rightarrow 3p$ transition in the emission spectrum of sodium has a wavelength of 499 nm. What is the energy of the $5d$ state?

45. A sodium atom emits a photon with wavelength 818 nm shortly after being struck by an electron. What minimum speed did the electron have before the collision?

46. Figure P41.46 shows a few energy levels of the mercury atom.
    a. Make a table showing all the allowed transitions in the emission spectrum. For each transition, indicate the photon wavelength, in nm.
    b. What minimum speed must an electron have to excite the 492-nm-wavelength blue emission line in the Hg spectrum?

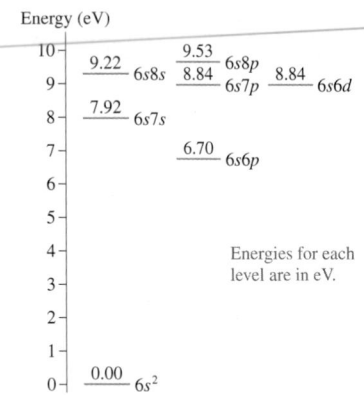

Energy (eV)

FIGURE P41.46

47. Figure P41.47 shows the first few energy levels of the lithium atom. Make a table showing all the allowed transitions in the emission spectrum. For each transition, indicate
    a. The wavelength, in nm.
    b. Whether the transition is in the infrared, the visible, or the ultraviolet spectral region.
    c. Whether or not the transition would be observed in the lithium absorption spectrum.

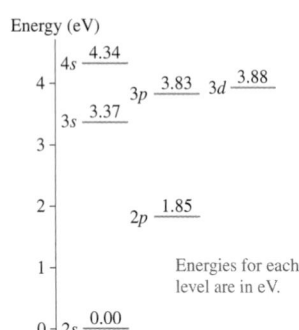

Energy (eV)

FIGURE P41.47

48. Three electrons are in a one-dimensional rigid box (i.e., an infinite potential well) of length 0.5 nm. Two are in the $n = 1$ state and one is in the $n = 6$ state. The selection rule for the rigid box allows only those transitions for which Δn is odd.
    a. Draw an energy-level diagram. On it, show the filled levels and show all transitions that could emit a photon.
    b. What are all the possible wavelengths that could be emitted by this system?

49. a. What is the decay rate for the $2p$ state of hydrogen?
    b. During what interval of time will 10% of a sample of $2p$ hydrogen atoms decay?

50. A hydrogen atom is in the $2p$ state. How much time must elapse for there to be a 1% chance that this atom will undergo a quantum jump to the ground state?

51. An atom in an excited state has a 1.0% chance of emitting a photon in 0.20 ns. How long will it take for 25% of a sample of excited atoms to decay?

52. a. Find an expression in terms of τ for the *half-life* $t_{1/2}$ of a sample of excited atoms. The half-life is the time at which half of the excited atoms have undergone a quantum jump and emitted a photon.
    b. What is the half-life of the $3p$ state of sodium?

53. An electrical discharge in a neon-filled tube maintains a *steady* population of 1.0×10^9 atoms in an excited state with $\tau = 20$ ns. How many photons are emitted per second from atoms in this state?

54. A ruby laser emits a 100 MW, 10-ns-long pulse of light with a wavelength of 690 nm. How many chromium atoms undergo stimulated emission to generate this pulse?

55. A laser emits 1.0×10^{19} photons per second from an excited state with energy $E_2 = 1.17$ eV. The lower energy level is $E_1 = 0$ eV.
    a. What is the wavelength of this laser?
    b. What is the power output of this laser?

Challenge Problems

56. Two excited energy levels are separated by the very small energy difference ΔE. As atoms in these levels undergo quantum jumps to the ground state, the photons they emit have nearly identical wavelengths λ.
    a. Show that the wavelengths differ by

$$\Delta\lambda = \frac{\lambda^2}{hc}\Delta E$$

    b. In the Lyman series of hydrogen, what is the wavelength difference between photons emitted in the $n = 20$ to $n = 1$ transition and photons emitted in the $n = 21$ to $n = 1$ transition?

57. What is the probability of finding a $1s$ hydrogen electron at distance $r > a_B$ from the proton?

58. What is the probability of finding a $1s$ hydrogen electron at distance $r < \frac{1}{2}a_B$ from the proton?

59. Prove that the most probable distance from the proton of an electron in the $2s$ state of hydrogen is $5.236a_B$.

60. Find the distance, in terms of a_B, between the two peaks in the radial probability density of the $2s$ state of hydrogen.
    Hint: This problem requires a numerical solution.

61. Suppose you have a machine that gives you pieces of candy when you push a button. Eighty percent of the time, pushing the button gets you two pieces of candy. Twenty percent of the time, pushing the button yields 10 pieces. The *average* number of pieces per push is $N_{avg} = 2 \times 0.80 + 10 \times 0.20 = 3.6$. That is, 10 pushes should get you, on average, 36 pieces. Mathematically, the average value when the probabilities differ is $N_{avg} = \Sigma(N_i \times$ Probability of $i)$. We can do the same thing in quantum mechanics, with the difference that the sum becomes an integral. If you measured the distance of the electron from the proton in many hydrogen atoms, you would get many values, as indicated by the radial probability density. But the *average* value of r would be

$$r_{avg} = \int_0^\infty rP_r(r)\,dr$$

Calculate the average value of r in terms of a_B for the electron in the $1s$ and the $2p$ states of hydrogen.

62. The 1997 Nobel prize in physics went to Steven Chu, Claude Cohen-Tannoudji, and William Phillips for their development of techniques to slow, stop, and "trap" atoms with laser light. To see how this works, consider a beam of rubidium atoms (mass 1.4×10^{-25} kg) traveling at 500 m/s after being evaporated out of an oven. A laser beam with a wavelength of 780 nm is directed against the atoms. This is the wavelength of the $5s \rightarrow 5p$ transition in rubidium, with $5s$ being the ground state, so the photons in the laser beam are easily absorbed by the atoms. After an average time of 15 ns, an excited atom spontaneously emits a 780-nm-wavelength photon and returns to the ground state.

   a. The energy-momentum-mass relationship of Einstein's theory of relativity is $E^2 = p^2 c^2 + m^2 c^4$. A photon is massless, so the momentum of a photon is $p = E_{photon}/c$. Assume that the atoms are traveling in the positive x-direction and the laser beam in the negative x-direction. What is the initial momentum of an atom leaving the oven? What is the momentum of a photon of light?

   b. The total momentum of the atom and the photon must be conserved in the absorption processes. As a consequence, how many photons must be absorbed to bring the atom to a halt?

   NOTE ▶ Momentum is also conserved in the emission processes. However, spontaneously emitted photons are emitted in random directions. Averaged over many absorption/emission cycles, the net recoil of the atom due to emission is zero and can be ignored. ◀

   c. Assume that the laser beam is so intense that a ground-state atom absorbs a photon instantly. How much time is required to stop the atoms?

   d. Use Newton's second law in the form $F = \Delta p/\Delta t$ to calculate the force exerted on the atoms by the photons. From this, calculate the atoms' acceleration as they slow.

   e. Over what distance is the beam of atoms brought to a halt?

STOP TO THINK ANSWERS

Stop to Think 41.1: $n = 3, l = 1$, or a $3p$ state.

Stop to Think 41.2: 4. You can see in Figure 41.8 that the ns state has n maxima.

Stop to Think 41.3: No. $m_s = \pm\frac{1}{2}$, so the z-component S_z cannot be zero.

Stop to Think 41.4: b. The atom would have less energy if the $3s$ electron were in a $2p$ state.

Stop to Think 41.5: c. Emission is a quantum jump to a lower-energy state. The $5p \rightarrow 4p$ transition is not allowed because $\Delta l = 0$ violates the selection rule. The lowest-energy allowed transition is $5p \rightarrow 3d$, with $E_{photon} = \Delta E_{atom} = 3.0$ eV.

Stop to Think 41.6: e. Because $r_B = 2r_A$, the ratio is $e^{-2}/e^{-1} = e^{-1} < \frac{1}{2}$.

42 Nuclear Physics

The technique known as *carbon dating* uses the radioactive decay of the naturally occurring carbon isotope ^{14}C to determine the age of fossils and archeological artifacts.

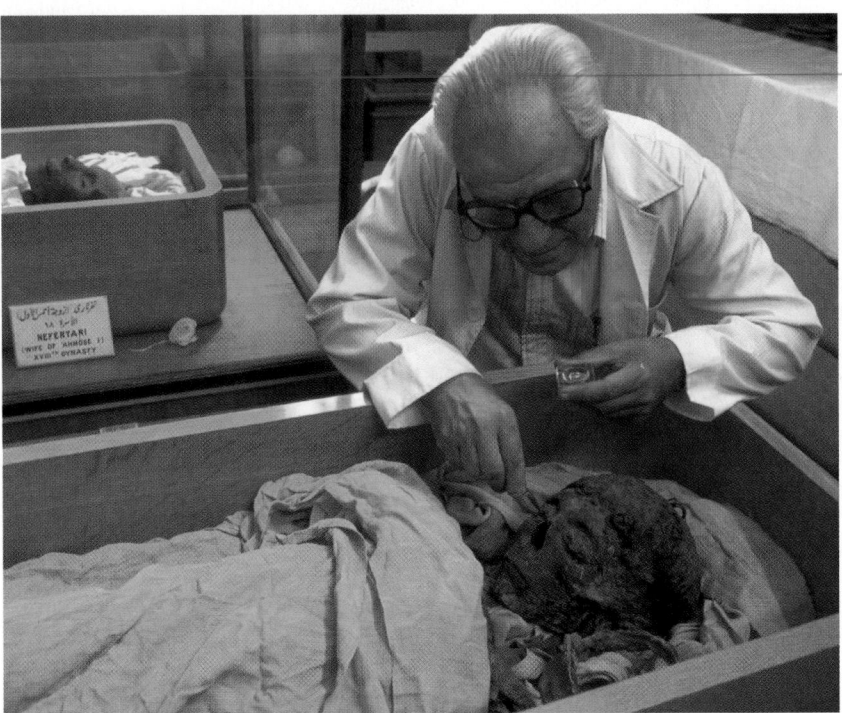

▶ **Looking Ahead**

The goal of Chapter 42 is to understand the physics of the nucleus and some of the applications of nuclear physics. In this chapter you will learn to:

- Interpret the basic structure of the nucleus.
- Understand how the strong force holds the nucleus together.
- Understand why some nuclei are unstable and undergo radioactive decay.
- Calculate the half-lives of radioactive decay.
- Apply nuclear physics to biology and medicine.

◀ **Looking Back**

The material in this chapter depends on basic atomic structure and on the quantized energy levels in potential-energy wells. Please review:

- Sections 37.6–37.7 Rutherford's model of the nucleus.
- Section 40.6 Finite potential-energy wells.

The nucleus of the atom is extremely remote from our everyday experience. Thus it comes as something of a surprise to notice the extent to which nuclear physics has become part of our modern technology and contemporary vocabulary: nuclear power and nuclear weapons, nuclear medicine and nuclear waste, nuclear fission and nuclear fusion.

Rutherford's discovery of the atomic nucleus marked the beginning of nuclear physics. Other physicists were soon designing experiments to probe within the nucleus and learn the properties of nuclear matter. In this final chapter, we'll explore the physics of the nucleus and look at some of the applications of nuclear physics.

One interesting application, shown in the photo, is the use of naturally occurring radioactivity to date archeological samples and geological formations. We'll also examine how exposure to nuclear radiation is measured and look at some of the uses of nuclear physics in medicine.

42.1 Nuclear Structure

The 1890s was a decade of mysterious rays. Cathode rays were being studied in several laboratories, and, in 1895, Röntgen discovered x rays. In 1896, after hearing of Röntgen's discovery, the French scientist A. H. Becquerel wondered if mineral crystals that fluoresce after exposure to sunlight were emitting x rays. He put a piece of film in an opaque envelope, then placed a crystal on top and left it in the sun. To his delight, the film in the envelope was exposed.

Becquerel thought he had discovered x rays coming from crystals, but his joy was short lived. He soon found that the film could be exposed equally well simply by being stored in a closed drawer with the crystals. Further investigation showed that the crystal, which happened to be a mineral containing uranium, was spontaneously emitting some new kind of ray. Rather than finding x rays, as he had hoped, Becquerel had discovered what became known as *radioactivity.*

Ernest Rutherford soon took up the investigation and found not one but three distinct kinds of rays emitted from crystals containing uranium. Not knowing what they were, he named them for their ability to penetrate matter and ionize air. The first, which caused the most ionization and penetrated the least, he called *alpha rays*. The second, with intermediate penetration and ionization, were *beta rays,* and the third, with the least ionization but the largest penetration, became *gamma rays.*

Within a few years, Rutherford was able to show that alpha rays are helium nuclei emitted from the crystal at very high velocities. These became the projectiles that he used in 1909 to probe the structure of the atom. The outcome of that experiment, as you learned in Chapter 37, was Rutherford's discovery that atoms have a very small, dense nucleus at the center.

Rutherford's discovery of the nucleus may have settled the question of atomic structure, but it raised many new issues for scientific research. Foremost among them were

- What is nuclear matter? What are its properties?
- What holds the nucleus together? Why doesn't the repulsive electrostatic force blow it apart?
- What is the connection between the nucleus and radioactivity?

These were the beginnings of **nuclear physics,** the study of the properties of the atomic nucleus.

Nucleons

The nucleus is a tiny speck in the center of a vastly larger atom. As Figure 42.1 shows, the nuclear diameter of roughly 10^{-14} m is only about 1/10,000 the diameter of the atom. What we call *matter* is overwhelming empty space!

You learned in Chapter 37 that the nucleus is composed of two types of particles: *protons* and *neutrons.* Together, these are referred to as **nucleons.** The role of the neutrons, which have nothing to do with keeping electrons in orbit, is an important issue that we'll address in this chapter. Table 42.1 summarizes the basic properties of protons and neutrons.

As you can see, protons and neutrons are virtually identical other than the fact that the proton has one unit of the fundamental charge e whereas the neutron is electrically neutral. The neutron is slightly more massive than the proton, but the difference is very small, only about 0.1%. Notice that the proton and neutron, like the electron, have an *inherent angular momentum* and magnetic moment with spin quantum number $s = \frac{1}{2}$. As a consequence, protons and neutrons obey the Pauli exclusion principle.

The number of protons Z is the element's **atomic number.** In fact, an element is identified by the number of protons in the nucleus, not by the number of orbiting electrons. Electrons are easily added and removed, forming negative and positive ions, but doing so doesn't change the element. The **mass number** A is defined to be $A = Z + N$, where N is the **neutron number.** The mass number is the total number of nucleons in a nucleus.

NOTE ▶ The mass number, which is dimensionless, is *not* the same thing as the atomic mass m. We'll look at actual atomic masses below. ◀

This picture of an atom would need to be 10 m in diameter if it were drawn to the same scale as the dot representing the nucleus.

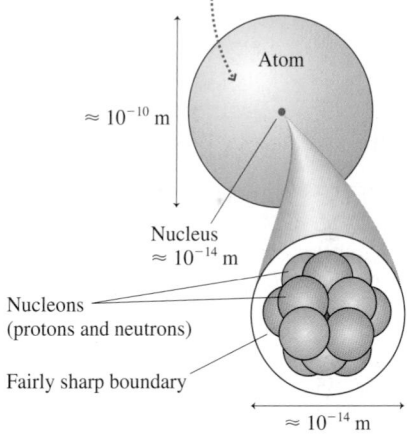

FIGURE 42.1 The nucleus is a tiny speck within an atom.

TABLE 42.1 Protons and neutrons

	Proton	Neutron
Number	Z	N
Charge q	$+e$	0
Spin s	$\frac{1}{2}$	$\frac{1}{2}$
Mass, in u	1.00728	1.00866

Isotopes and Isobars

It was discovered early in the 20th century that not all atoms of the same element (same Z) have the same mass. There are a *range* of neutron numbers that happily form a nucleus with Z protons, creating a series of nuclei having the same Z-value (i.e., they are all the same chemical element) but different A-values. Each A-value in a series of nuclei with the same Z-value is called an **isotope.**

Chemical behavior is determined by the orbiting electrons. All isotopes of one element have the same number of orbiting electrons (if the atoms are electrically neutral) and thus have the same chemical properties, but different isotopes of the same element can have quite different nuclear properties. In addition, macroscopic behavior that depends on mass, such as the diffusion of a gas, can slightly favor one isotope over another.

The notation used to label isotopes is ^{A}Z, where the mass number A is given as a *leading* superscript. The proton number Z is not specified by an actual number but, equivalently, by the chemical symbol for that element. Hence ordinary carbon, which has six protons and six neutrons in the nucleus, is written ^{12}C and pronounced "carbon twelve." The radioactive form of carbon used in carbon dating is ^{14}C. It has six protons, making it carbon, and eight neutrons.

More than 3000 isotopes are known. The majority of these are **radioactive,** meaning that the nucleus is not stable but, after some period of time, will either fragment or emit some kind of subatomic particle in an effort to reach a more stable state. Many of these radioactive isotopes are created by nuclear reactions in the laboratory and have only a fleeting existence. Only 266 isotopes are **stable** (i.e., nonradioactive) and occur in nature. We'll begin to look at the issue of nuclear stability in the next section.

The *naturally occurring* nuclei include the 266 stable isotopes and a handful of radioactive isotopes with such long half-lives, measured in billions of years, that they also occur naturally. The most well-known example of a naturally occurring radioactive isotope is the uranium isotope ^{238}U. For each element, the fraction of naturally occurring nuclei represented by one particular isotope is called the **natural abundance** of that isotope.

Although there are many radioactive isotopes of the element iodine, iodine occurs *naturally* only as ^{127}I. Consequently, we say that the natural abundance of ^{127}I is 100%. Most elements have multiple naturally occurring isotopes. The natural abundance of ^{14}N is 99.6%, meaning that 996 out of every 1000 naturally occurring nitrogen atoms are the isotope ^{14}N. The remaining 0.4% of naturally occurring nitrogen is the isotope ^{15}N, with one extra neutron.

A series of nuclei having the same A-value (the same mass number) but different values of Z and N are called **isobars.** For example, the three nuclei ^{14}C, ^{14}N, and ^{14}O are isobars with $A = 14$. Only ^{14}N is stable; the other two are radioactive.

Atomic Mass

You learned in Chapter 16 that atomic masses are specified in terms of the *atomic mass unit* u, defined such that the atomic mass of the isotope ^{12}C is exactly 12 u. The conversion to SI units is

$$1\ u = 1.6605 \times 10^{-27}\ kg$$

Alternatively, we can use Einstein's $E_0 = mc^2$ to express masses in terms of their energy equivalent. The energy equivalent of 1 u of mass is

$$E_0 = (1.6605 \times 10^{-27}\ kg)(2.9979 \times 10^8\ m/s)^2$$
$$= 1.4924 \times 10^{-10}\ J = 931.49\ MeV \tag{42.1}$$

Thus the atomic mass unit can be written

$$1\ u = 931.49\ MeV/c^2$$

It may seem unusual, but the units MeV/c^2 are units of mass.

NOTE ▶ We're using more significant figures than usual. Many nuclear calculations look for the small difference between two masses that are almost the same. Those two masses must be calculated or specified to four or five significant figures if their difference is to be meaningful. ◀

Table 42.2 shows the atomic masses of the electron, the nucleons, and three important light elements. Appendix C contains a more complete list. Notice that the mass of a hydrogen atom is the sum of the masses of a proton and an electron. But a quick calculation shows that the mass of a helium atom (2 protons, 2 neutrons, and 2 electrons) is 0.03038 u *less* than the sum of the masses of its constituents. The difference is due to the *binding energy* of the nucleus, a topic we'll look at in Section 42.2.

The isotope ^2H is a hydrogen atom in which the nucleus is not simply a proton but a proton and a neutron. Although the isotope is a form of hydrogen, it is called **deuterium.** The natural abundance of deuterium is 0.015%, or about 1 out of every 6700 hydrogen atoms. Water made with deuterium (sometimes written D_2O rather than H_2O) is called *heavy water.*

NOTE ▶ Don't let the name *deuterium* cause you to think this is a different element. Deuterium is an isotope of hydrogen. Chemically, it behaves just like ordinary hydrogen. ◀

The *chemical* atomic mass shown on the periodic table of the elements is the *weighted average* of the atomic masses of all naturally occurring isotopes. For example, chlorine has two stable isotopes: ^{35}Cl, with $m = 34.97$ u, is 75.8% abundant and ^{37}Cl, at 36.97 u, is 24.2% abundant. The average, weighted by abundance, is $0.758 \times 34.97 + 0.242 \times 36.97 = 35.45$. This is the value shown on the periodic table and is the correct value for most chemical calculations, but it is not the mass of any particular isotope of chlorine.

NOTE ▶ The atomic masses of the proton and the neutron are both ≈ 1 u. Consequently, the value of the mass number A is *approximately* the atomic mass in u. The approximation $m \approx A$ u is sufficient in many contexts, such as when we're calculating the masses of atoms in the kinetic theory of gases, but in nuclear physics calculations, we almost always need the more accurate mass values that you find in Table 42.2 or Appendix C. ◀

TABLE 42.2 Some atomic masses

Particle	Symbol	Mass (u)	Mass (MeV/c^2)
Electron	e	0.00055	0.51
Proton	p	1.00728	938.28
Neutron	n	1.00866	939.57
Hydrogen	^1H	1.00783	938.79
Deuterium	^2H	2.01410	1876.12
Helium	^4He	4.00260	3728.40

Nuclear Size and Density

Unlike the atom's electron cloud, which is quite diffuse, the nucleus has a fairly sharp boundary. Experimentally, the radius of a nucleus with mass number A is found to be

$$r = r_0 A^{1/3} \qquad (42.2)$$

where $r_0 = 1.2$ fm. Recall that 1 fm = 1 femtometer = 10^{-15} m.

As Figure 42.2 shows, the radius is proportional to $A^{1/3}$, but the volume of the nucleus (proportional to r^3) is directly proportional to A, the number of nucleons. A nucleus with twice as many nucleons will occupy twice as much volume. This finding has several implications:

- Nucleons are incompressible. Adding more nucleons doesn't squeeze the inner nucleons into a smaller volume.
- The nucleons are tightly packed, looking much like the drawing in Figure 42.1.
- Nuclear matter has a constant density.

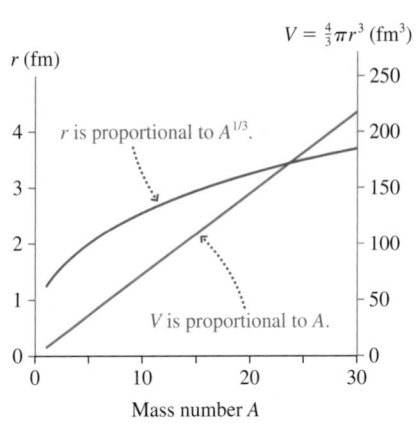

FIGURE 42.2 The nuclear radius and volume as a function of A.

In fact, we can use Equation 42.2 to estimate the density of nuclear matter. Consider a nucleus with mass number A. Its mass, within 1%, is A atomic mass units. Thus

$$\rho_{nuc} \approx \frac{A \text{ u}}{\frac{4}{3}\pi r^3} = \frac{A \text{ u}}{\frac{4}{3}\pi r_0^3 A} = \frac{1 \text{ u}}{\frac{4}{3}\pi r_0^3} = \frac{1.66 \times 10^{-27} \text{ kg}}{\frac{4}{3}\pi(1.2 \times 10^{-15} \text{ m})^3}$$

$$= 2.3 \times 10^{17} \text{ kg/m}^3 \qquad (42.3)$$

The fact that A cancels means that **all nuclei have this density.** It is a staggeringly large density, roughly 10^{14} times larger than the density of familiar liquids and solids. One early objection to Rutherford's model of a nuclear atom was that matter simply can't have a density this high. Although we have no direct experience with such matter, nuclear matter really is this dense.

Figure 42.3 shows the density profiles of three nuclei. The constant density right to the edge is analogous to that of a drop of incompressible liquid, and, indeed, one successful model of many nuclear properties is called the **liquid-drop model.** Notice that the range of nuclear radii, from small helium to large uranium, is not quite a factor of 4. The fact that ^{56}Fe is a fairly typical atom in the middle of the periodic table is the basis for our earlier assertion that the nuclear diameter is roughly 10^{-14} m, or 10 fm.

Imagine the nucleus is a drop of liquid. Its density is the same up to the edge of the drop.

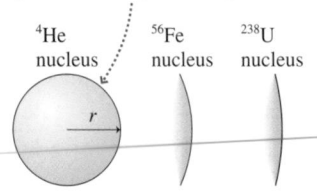

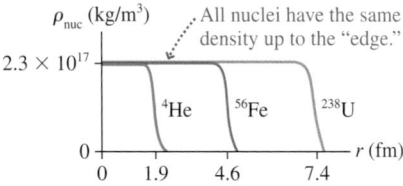

FIGURE 42.3 Density profiles of three nuclei.

STOP TO THINK 42.1 Three electrons orbit a neutral ^{6}Li atom. How many electrons orbit a neutral ^{7}Li atom?

42.2 Nuclear Stability

We've already noted that less than 10% of the known nuclei are stable (i.e., not radioactive). Because nuclei are characterized by two independent numbers, N and Z, it is useful to show the known nuclei on a plot of neutron number N versus proton number Z. Figure 42.4 shows such a plot. Stable nuclei are represented by blue diamonds and unstable, radioactive nuclei by red dots.

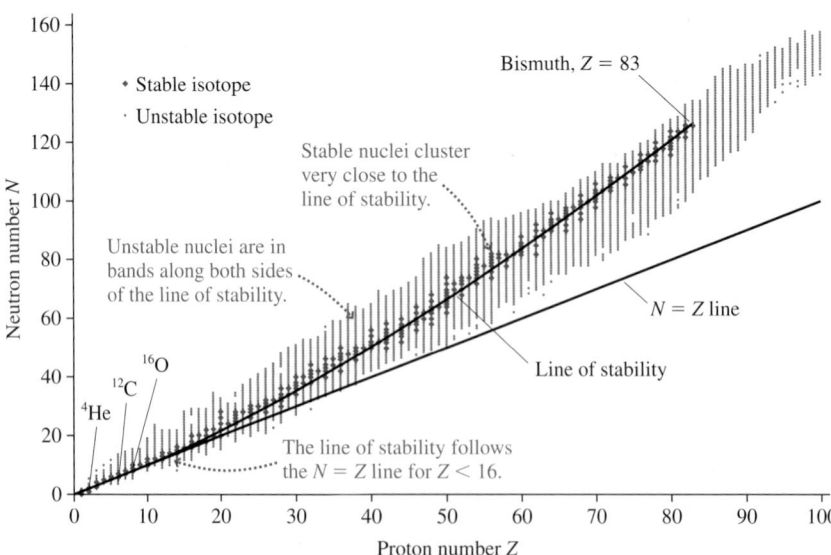

FIGURE 42.4 Stable and unstable nuclei shown on a plot of neutron number N versus proton number Z.

We can make several observations from this graph:

- The stable nuclei cluster very close to the curve called the **line of stability.**
- There are no stable nuclei with $Z > 83$ (bismuth).
- Unstable nuclei are in bands along both sides of the line of stability.
- The lightest elements, with $Z < 16$, are stable when $N \approx Z$. The familiar elements ^{4}He, ^{12}C, and ^{16}O all have equal numbers of protons and neutrons.
- As Z increases, the number of neutrons needed for stability grows increasingly larger than the number of protons. The N/Z ratio is ≈ 1.2 at $Z = 40$ but has grown to ≈ 1.5 at $Z = 80$.

These observations—especially $N \approx Z$ for small Z but $N > Z$ for large Z—cry out for an explanation. The quantum-mechanical model of the nucleus that we'll develop in Section 42.4 will provide the explanation we seek.

STOP TO THINK 42.2   The isobars corresponding to one specific value of A are found on the plot of Figure 42.4 along

a. A vertical line.

b. A horizontal line.

c. A diagonal line that goes up and to the right.

d. A diagonal line that goes up and to the left.

Binding Energy

A nucleus is a *bound system.* That is, you would need to supply energy to disperse the nucleons by breaking the nuclear bonds between them. Figure 42.5 shows this idea schematically.

You learned a similar idea in atomic physics. The energy levels of the hydrogen atom are negative numbers because the bound system has less energy than a free proton and electron. The energy you must supply to an atom to remove an electron is called the *ionization energy.*

In much the same way, the energy you would need to supply to a nucleus to disassemble it into individual protons and neutrons is called the **binding energy.** Whereas ionization energies of atoms are only a few eV, the binding energies of nuclei are tens or hundreds of MeV, energies large enough that their mass equivalent is not negligible.

Consider a nucleus with mass m_{nuc}. It is found experimentally that m_{nuc} is *less* than the total mass $Zm_p + Nm_n$ of the Z protons and N neutrons that form the nucleus, where m_p and m_n are the masses of the proton and neutron. That is, the energy equivalent $m_{nuc}c^2$ of the nucleus is less than the energy equivalent $(Zm_p + Nm_n)c^2$ of the individual nucleons. The binding energy B of the nucleus (not the entire atom) is defined as

$$B = (Zm_p + Nm_n - m_{nuc})c^2 \tag{42.4}$$

This is the energy you would need to supply to disassemble the nucleus into its pieces.

The practical difficulty is that laboratory scientists use mass spectroscopy to measure *atomic* masses, not nuclear masses. The atomic mass m_{atom} is m_{nuc} plus the mass Zm_e of Z orbiting electrons. (Strictly speaking, we should allow for the binding energy of the electrons, but these binding energies are roughly 10^6 smaller than the nuclear binding energies and can be neglected in all but the most precise measurements and calculations.)

Fortunately, we can switch from the nuclear mass to the atomic mass by the simple trick of both adding and subtracting Z electron masses. Begin by writing Equation 42.4 in the equivalent form

$$B = (Zm_p + Zm_e + Nm_n - m_{nuc} - Zm_e)c^2 \tag{42.5}$$

The binding energy is the energy that would be needed to disassemble a nucleus into individual nucleons.

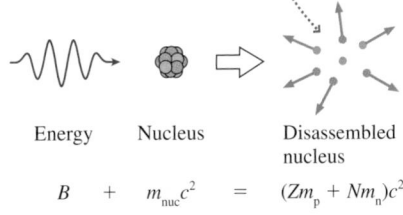

$$B + m_{nuc}c^2 = (Zm_p + Nm_n)c^2$$

FIGURE 42.5 The nuclear binding energy.

Actv
Physics   19.2

Now $m_{nuc} + Zm_e = m_{atom}$, the atomic mass, and $Zm_p + Zm_e = Z(m_p + m_e) = Zm_H$, where m_H is the mass of a hydrogen *atom*. Finally, use the conversion factor $1 \text{ u} = 931.49 \text{ MeV}/c^2$ to write $c^2 = 931.49 \text{ MeV/u}$. The binding energy is then

$$B = (Zm_H + Nm_n - m_{atom}) \times (931.49 \text{ MeV/u})$$
(binding energy)

(42.6)

where all the three masses are in atomic mass units.

EXAMPLE 42.1 **The binding energy of iron**

What is the binding energy of the ^{56}Fe nucleus?

SOLVE The isotope ^{56}Fe has $Z = 26$ and $N = 30$. The atomic mass of ^{56}Fe, found in Appendix C, is 55.9349 u. Thus the mass difference between the ^{56}Fe nucleus and its constituents is

$B = 26(1.0078 \text{ u}) + 30(1.0087 \text{ u}) - 55.9349 \text{ u} = 0.528 \text{ u}$

where, from Table 42.2, 1.0078 u is the mass of the hydrogen atom. Thus the binding energy of ^{56}Fe is

$B = (0.538 \text{ u}) \times (931.49 \text{ MeV/u}) = 492 \text{ MeV}$

ASSESS The binding energy is extremely large, the energy equivalent of more than half the mass of a proton or a neutron.

The nuclear binding energy increases as A increases simply because there are more nuclear bonds. A more useful measure for comparing one nucleus to another is the quantity B/A, called the *binding energy per nucleon*. Iron, with $B = 492 \text{ MeV}$ and $A = 56$, has 8.79 MeV per nucleon. This is the amount of energy, on average, you would need to supply in order to remove *one* nucleon from the nucleus. Nuclei with larger values of B/A are more tightly held together than nuclei with smaller values of B/A.

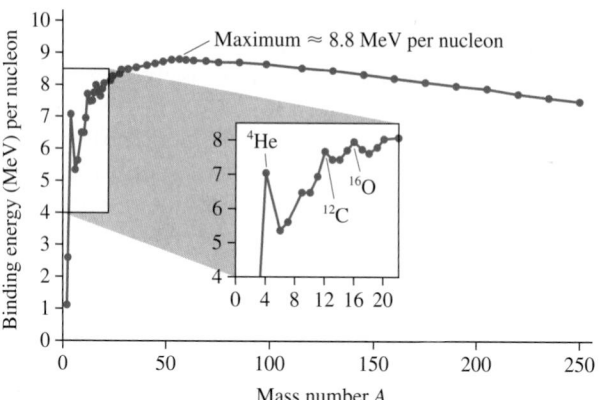

FIGURE 42.6 The curve of binding energy.

Figure 42.6 is a graph of the binding energy per nucleon versus mass number A. The dotted line connecting the points is often called the **curve of binding energy.** This curve has several important features:

- There are peaks in the binding energy curve at $A = 4$, 12, and 16. The one at $A = 4$, corresponding to ^{4}He, is especially pronounced. As you'll see, these peaks, which represent nuclei more tightly bound than their neighbors, are due to *closed shells* in much the same way that the graph of atomic ionization energies (see Figure 41.24) peaked for closed electron shells.
- The binding energy per nucleon is *roughly* constant at $\approx 8 \text{ MeV}$ per nucleon for $A > 20$. This suggests that, as a nucleus grows, there comes a point where the nuclear bonds are *saturated*. Each nucleon interacts only with its nearest neighbors, the ones it's actually touching. This, in turn, implies that the nuclear force is a *short-range* force.

■ The curve has a broad maximum at $A \approx 60$. This will be important for our understanding of radioactivity. In principle, heavier nuclei could become *more* stable (more binding energy per nucleon) by breaking into smaller pieces. Lighter nuclei could become *more* stable by fusing together into larger nuclei. There may not always be a mechanism for such nuclear transformations to take place, but *if* there is a mechanism, it is energetically favorable for it to occur.

42.3 The Strong Force

Rutherford's discovery of the atomic nucleus was not immediately accepted by all scientists. Their primary objection was that the protons would blow themselves apart at tremendously high speeds due to the extremely large electrostatic forces between them at a separation of a few femtometers. No known force could hold the nucleus together.

It soon became clear that a previously unknown force of nature operates within the nucleus to hold the nucleons together. This new force had to be stronger than the repulsive electrostatic force, hence it was named the **strong force.** It is also called the *nuclear force.*

The strong force has several important properties:

1. It is an *attractive* force between any two nucleons.
2. It does not act on electrons.
3. It is a *short-range* force, acting only over nuclear distances.
4. Over the range where it acts, it is *stronger* than the electrostatic force that tries to push two protons apart.

The fact that the strong force is short-range, in contrast to the long-range $1/r^2$ electric, magnetic, and gravitational forces, is apparent from the fact that we see no evidence for nuclear forces outside the nucleus.

Figure 42.7 summarizes the three interactions that take place within the nucleus. Whether the strong force between two protons is the same strength as the force between two neutrons or between a proton and a neutron is an important question that can be answered experimentally. The primary means of investigating the strong force is to accelerate a proton to very high velocity, using a cyclotron or some other particle accelerator, then to study how the proton is scattered by various target materials.

The conclusion of many decades of research is that the strong force between two nucleons is independent of whether they are protons or neutrons. Charge is the basis for electromagnetic interactions, but it is of no relevance to the strong force. Protons and neutrons are identical as far as nuclear forces are concerned.

Potential Energy

Unfortunately, there's no simple formula to calculate the strong force or the potential energy of two nucleons interacting via the strong force. Figure 42.8 is an experimentally determined potential-energy diagram for two interacting nucleons, with r the distance between their centers. The potential-energy minimum at $r \approx 1$ fm is a point of stable equilibrium.

Recall that the force is the negative of the slope of a potential-energy diagram. The steeply rising potential for $r < 1$ fm represents a strongly repulsive force. That is, the nucleon "cores" strongly repel each other if they get too close together. The force is attractive for $r > 1$ fm, where the slope is positive, and it is strongest where the slope is steepest, at $r \approx 1.5$ fm. Notice that the strong force quickly decreases for $r > 1.5$ fm and is zero for $r > 3$ fm. That is, the strong force represented by this potential energy is effective only over a very short range of distances.

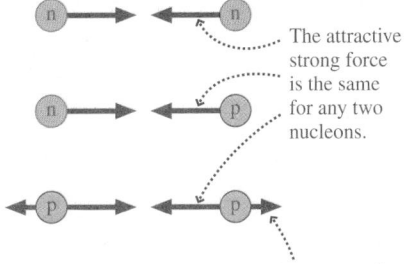

The attractive strong force is the same for any two nucleons.

Two protons also experience a smaller electrostatic repulsive force.

FIGURE 42.7 The strong force is the same between any two nucleons.

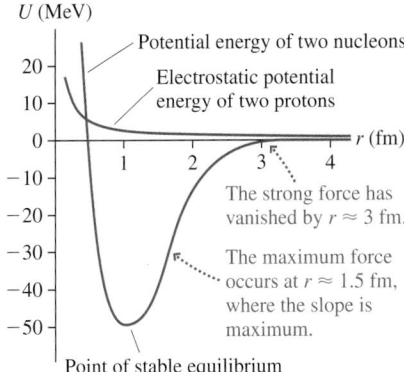

The strong force has vanished by $r \approx 3$ fm.

The maximum force occurs at $r \approx 1.5$ fm, where the slope is maximum.

Point of stable equilibrium

FIGURE 42.8 The potential-energy diagram for two nucleons interacting via the strong force.

Figure 42.8 also shows the electrostatic potential energy of two protons for comparison. Notice how small the electrostatic energy is in comparison to the potential energy of the strong force. At $r \approx 1.5$ fm, where the strong force is maximum, the aptly named attractive strong force is ≈ 100 times larger than the repulsive electrostatic force.

A question asked earlier was why the nucleus has neutrons at all. The answer is related to the short range of the strong force. Protons throughout the nucleus exert repulsive electrostatic forces on each other, but, because of the short range of the strong force, a proton feels an attractive force only from the very few other protons with which it is in close contact. Even though the strong force at its maximum is much larger than the electrostatic force, there wouldn't be enough attractive nuclear bonds for an all-proton nucleus to be stable. Because neutrons participate in the strong force but exert no repulsive forces, **the neutrons provide the extra "glue" that holds the nucleus together.** In small nuclei, where most nucleons are in contact, one neutron per proton is sufficient for stability. Hence small nuclei have $N \approx Z$. But as the nucleus grows, the repulsive force increasefaster than the binding energy. More neutrons are needed for stability, causing heavy nuclei to have $N > Z$.

42.4 The Shell Model

Figure 42.8 is the potential energy of *two* interacting nucleons. To solve Schrödinger's equation for the nucleus, we would need to know the total potential energy of *all* interacting nucleon pairs within the nucleus, including both the strong force and the electrostatic force. This is far too complex to be a tractable problem.

We faced a similar situation with multielectron atoms. Calculating an atom's exact potential energy, including electron-electron repulsion, is exceedingly complicated. To simplify the problem, we made a *model* of the atom in which each electron moves independently with an *average* potential energy due to the nucleus and all other electrons. That model, although not perfect, led to the correct prediction of electron shells and explained the periodic table of the elements.

The **shell model** of the nucleus, using multielectron atoms as an analogy, was proposed in 1949 by Marie Goeppert-Mayer, one of the first prominent women in physics. The shell model considers each nucleon to move independently with an *average* potential energy due to the strong force of all the other nucleons. For the protons, we also have to include the electrostatic potential energy due to the other protons.

Figure 42.9 shows the average potential energy of a neutron and a proton. You can see that, to a good approximation, a nucleon appears to be a particle in a finite potential well, a quantum-mechanics problem you learned how to deal with in Chapter 40.

Marie Goeppert-Mayer shows her Nobel Prize.

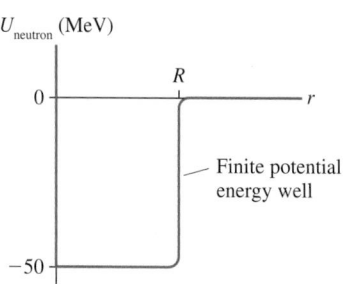

The average neutron potential energy is due to the strong force.

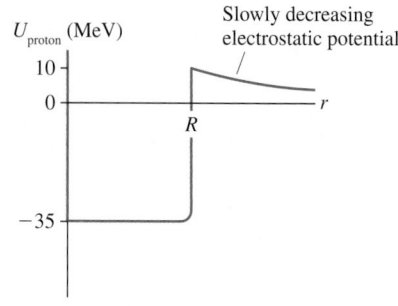

The average proton potential energy is due to the strong force and the electric force. This potential-well depth is for $Z \approx 30$.

FIGURE 42.9 The average potential energy of a neutron and a proton.

Several observations are worthwhile:

1. The depth of the neutron's potential-energy well is ≈ 50 MeV for all nuclei. The radius of the potential-energy well is the nuclear radius $R = r_0 A^{1/3}$.
2. For protons, the positive electrostatic potential energy "lifts" the potential-energy well. The lift varies from essentially none for very light elements to a significant fraction of the well depth for very heavy elements. The potential-energy well shown in the figure would be appropriate for a nucleus with $Z \approx 30$.
3. Outside the nucleus, where the strong force has vanished, a proton's potential energy is $U = (Z - 1)e^2/4\pi\epsilon_0 r$ due to its electrostatic interaction with the $(Z - 1)$ other protons within the nucleus. This positive potential energy decreases slowly with increasing distance.

The task of quantum mechanics is to solve for the energy levels and wave functions of the nucleons in these potential-energy wells. Once the energy levels are found, we build up the nuclear state, just as we did with atoms, by placing all the nucleons in the lowest energy levels consistent with the Pauli principle. The Pauli principle affects nucleons, just as it did electrons, because they are spin-$\frac{1}{2}$ particles. Each energy level can hold only a certain number of spin-up particles and spin-down particles, depending on the quantum numbers. Additional nucleons have to go into higher energy levels.

Low-Z Nuclei

As an example, we'll consider the energy levels of low-Z nuclei ($Z < 8$). Because these nuclei have so few protons, we can use a reasonable approximation that neglects the electrostatic potential energy due to proton-proton repulsion and considers only the much larger nuclear potential energy. In that case, the proton and neutron potential-energy wells and energy levels are the same.

Figure 42.10 shows the three lowest energy levels and the maximum number of nucleons that the Pauli principle allows in each. Energy values vary from nucleus to nucleus, but the spacing between these levels is several MeV. It's customary to draw the proton and neutron potential-energy diagrams and energy levels back to back. Notice that the radial axis for the proton potential-energy well points to the right, while the radial axis for the neutron potential-energy well points to the left.

Let's apply this model to the $A = 12$ isobar. Recall that an isobar is a series of nuclei with the same number of neutrons and protons. Figure 42.11 shows the energy-level diagrams of ^{12}B, ^{12}C, and ^{12}N. Look first at ^{12}C, a nucleus with six protons and six neutrons. You can see that exactly six protons are allowed in the $n = 1$ and $n = 2$ energy levels. Likewise for the six neutrons. Thus ^{12}C has a closed $n = 2$ proton shell and a closed $n = 2$ neutron shell.

NOTE ▶ Protons and neutrons are different particles, so the Pauli principle is not violated if a proton and a neutron have the same quantum numbers. ◀

The neutron radial distance is measured to the left.

The proton potential energy is nearly identical to the neutron potential energy when Z is small.

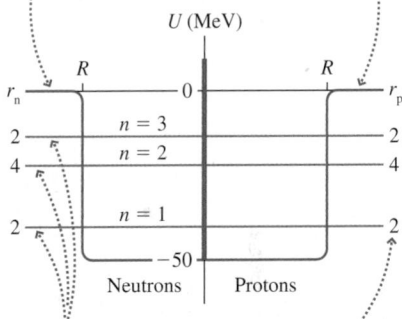

These are the first three allowed energy levels. They are spaced several MeV apart.

These are the maximum number of nucleons allowed by the Pauli principle.

FIGURE 42.10 The three lowest energy levels of a low-Z nucleus. The neutron energy levels are on the left, the proton energy levels on the right.

A ^{12}B nucleus could lower its energy if a neutron could turn into a proton.

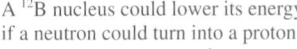

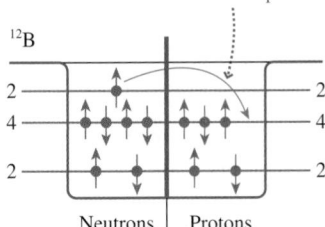

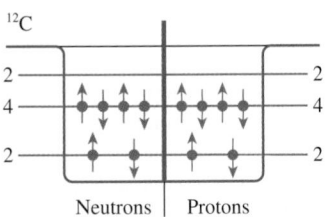

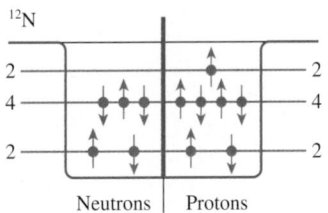

FIGURE 42.11 The $A = 12$ isobar has to place 12 nucleons in the lowest available energy levels.

^{12}N has seven protons and five neutrons. The sixth proton fills the $n = 2$ proton shell, so the seventh proton has to go into the $n = 3$ energy level. The $n = 2$ neutron shell has one vacancy because there are only five neutrons. ^{12}B is just the opposite, with the seventh neutron in the $n = 3$ energy level. You can see from the diagrams that the ^{12}B and ^{12}N nuclei have significantly more energy—by several MeV—than ^{12}C.

In atoms, electrons in higher energy levels decay to lower energy levels by emitting a photon as the electron undergoes a quantum jump. That can't happen here because the higher-energy nucleon in ^{12}B is a neutron whereas the vacant lower energy level is that of a proton. But an analogous process could occur *if* a neutron could somehow turn into a proton. And that's exactly what happens! We'll explore the details in Section 42.6, but both ^{12}B and ^{12}N decay into ^{12}C in the process known as *beta decay*.

^{12}C is just one of three low-Z nuclei in which both the proton and neutron shells are full. The other two are ^{4}He (filling both $n = 1$ shells with $Z = 2$, $N = 2$) and ^{16}O (filling both $n = 3$ shells with $Z = 8$, $N = 8$). If the analogy with closed electron shells is valid, these nuclei should be more tightly bound than nuclei with neighboring values of A. And indeed, we've already noted that the curve of binding energy (Figure 42.6) has peaks at $A = 4$, 12, and 16. The shell model of the nucleus satisfactorily explains these peaks. Unfortunately, the shell model quickly becomes much more complex as we go beyond $n = 3$. Heavier nuclei do have closed shells, but there's no evidence for them in the curve of binding energy.

High-Z Nuclei

We can use the shell model to give a qualitative explanation for one more observation, although the details are beyond the scope of this text. Figure 42.12 shows the neutron and proton potential-energy wells of a high-Z nucleus. In a nucleus with many protons, the electrostatic potential energy lifts the proton potential-energy well higher than the neutron potential-energy well. Protons and neutrons now have a different set of energy levels.

As a nucleus is "built," by the adding of protons and neutrons, the proton energy well and the neutron energy well must fill to just about the same height. If there were neutrons in energy levels above vacant proton levels, the nucleus would lower its energy by using beta decay to change the neutron into a proton. Similarly, beta decay would change a proton into a neutron if there were a vacant neutron energy level beneath a filled proton level. **The net result of beta decay is to keep the filled levels on both sides at just about the same height.**

Because the neutron potential-energy well starts at a lower energy, *more neutron states* are available than proton states. Consequently, a high-Z nucleus will have more neutrons than protons. This conclusion is consistent with our observation in Figure 42.2 that $N > Z$ for heavy nuclei.

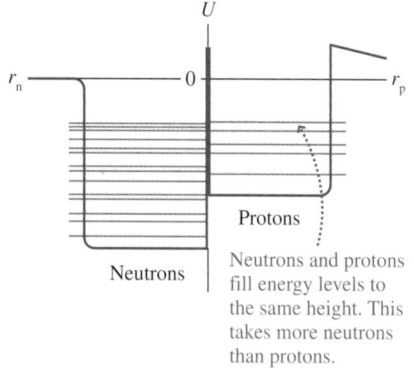

FIGURE 42.12 The proton energy levels are displaced upward in a high-Z nucleus.

42.5 Radiation and Radioactivity

Becquerel's 1896 discovery of "rays" from crystals of uranium prompted a burst of activity. Becquerel was soon joined in France by Marie Curie and Pierre Curie. They focused on isolating the element or elements responsible for the radiation, and, in the process, discovered the element radium.

In England, J. J. Thompson and, especially, his student and protégé Ernest Rutherford worked to identify the unknown rays. Using combinations of electric and magnetic fields, much as Thompson had done in his investigations of cathode rays, they found three distinct types of radiation. Figure 42.13 shows the basic experimental procedure, and Table 42.3 summarizes the results.

Marie Curie.

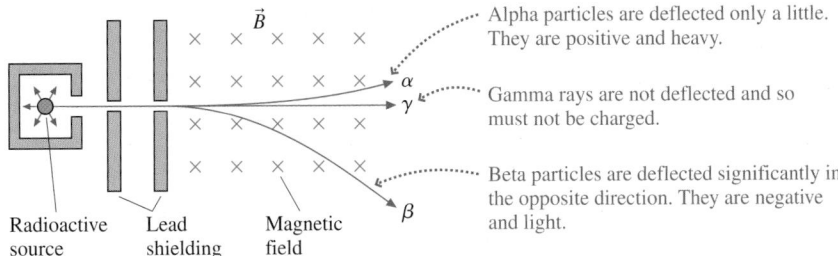

Alpha particles are deflected only a little. They are positive and heavy.

Gamma rays are not deflected and so must not be charged.

Beta particles are deflected significantly in the opposite direction. They are negative and light.

FIGURE 42.13 Identifying radiation by its deflection in a magnetic field.

TABLE 42.3 Three types of radiation

Radiation	Identification	Charge	Stopped by
Alpha, α	^{4}He nucleus	$+2e$	Sheet of paper
Beta, β	Electron	$-e$	Few mm of aluminum
Gamma, γ	High-energy photon	0	Many cm of lead

Within a few years, as Rutherford and others deduced the basic structure of the atom, it became clear that these emissions of radiation were coming from the atomic nucleus. We now define *radioactivity* or *radioactive decay* to be the spontaneous emission of particles or high-energy photons from unstable nuclei as they decay from higher-energy to lower-energy states. Radioactivity has nothing to do with the orbiting valence electrons.

NOTE ▶ The term "radiation" merely means something that is *radiated outward,* similar to the word "radial." Electromagnetic waves are often called *electromagnetic radiation.* Infrared waves from a hot object are referred to as "thermal radiation." Thus it was no surprise that these new "rays" were also called radiation. Unfortunately, the general public has come to associate the word "radiation" with *nuclear radiation,* something to be feared. It is important, when you use the term, to be sure you're not conveying a wrong impression to a listener or a reader. ◀

Ionizing Radiation

Electromagnetic waves, from microwaves through ultraviolet radiation, are *absorbed by* matter. The absorbed energy increases an object's thermal energy and its temperature, which is why objects sitting in the sun get warm.

In contrast to visible-light photon energies of a few eV, the energies of the alpha and beta particles and the gamma-ray photons of nuclear decay are typically in the range 0.1–10 MeV, a factor of roughly 10^6 larger. These energies are much larger than the ionization energies of atoms and molecules. Rather than simply being absorbed and increasing an object's thermal energy, nuclear radiation *ionizes* matter and *breaks* molecular bonds. Nuclear radiation and x rays, which behave much the same in matter, are called **ionizing radiation.**

An alpha or beta particle traveling through matter creates a trail of ionization, as shown in Figure 42.14a. Because the ionization energy of an atom is ≈ 10 eV, a particle with 1 MeV of kinetic energy can ionize $\approx 100,000$ atoms or molecules before finally stopping. The low-mass electrons are kicked sideways, but the much more massive positive ions barely move and form the trail. This behavior is the basis for the *cloud chamber* and the *hydrogen bubble chamber,* where microscopic water droplets or hydrogen gas bubbles coalesce around the positive ions to make the trail visible. Figure 42.14b is a picture of the ionization trails of high-energy particles in a bubble chamber. The curvature of the trajectories is due to a magnetic field.

(a)

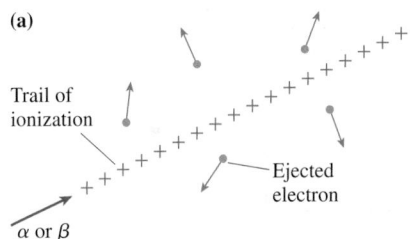

Trail of ionization

Ejected electron

α or β

(b)

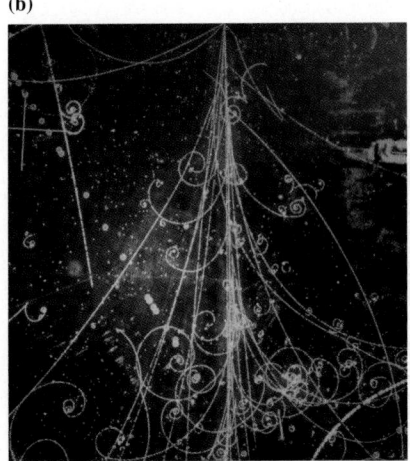

FIGURE 42.14 Alpha and beta particles create a trail of ionization as they pass through matter. This is the basis for the hydrogen bubble chamber.

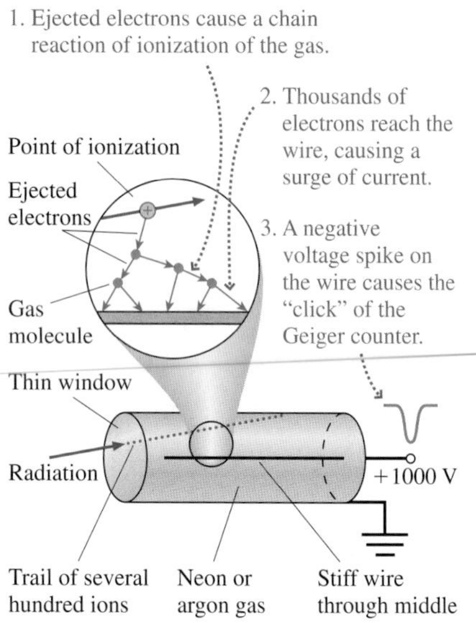

1. Ejected electrons cause a chain reaction of ionization of the gas.

2. Thousands of electrons reach the wire, causing a surge of current.

Point of ionization

Ejected electrons

3. A negative voltage spike on the wire causes the "click" of the Geiger counter.

Gas molecule

Thin window

Radiation

+1000 V

Trail of several hundred ions

Neon or argon gas

Stiff wire through middle

FIGURE 42.15 A Geiger counter.

Ionization is also the basis for the **Geiger counter,** one of the most well-known detectors of nuclear radiation. Figure 42.15 shows how a Geiger counter works. The important thing to remember is that a Geiger counter detects only *ionizing radiation.*

Ionizing radiation damages materials. Ions drive chemical reactions that wouldn't otherwise occur. Broken molecular bonds alter the workings of molecular machinery, especially in large biological molecules. It is through these mechanisms—ionization and bond breaking—that nuclear radiation can cause mutations or tumors. We'll look at the biological issues in Section 42.7.

NOTE ▶ Ionizing radiation causes structural damage to materials, but **irradiated objects do not become radioactive.** Ionization drives chemical processes involving the electrons. An object could become radioactive only if its nuclei were somehow changed, and that does not happen. ◀

STOP TO THINK 42.3 A very bright spotlight shines on a Geiger counter. Does it click?

Nuclear Decay and Half-Lives

Rutherford was the first to find that the number of radioactive atoms in a sample decreases exponentially with time. This is the expected time dependence if the decay is a *random process.* But to say that a process is random doesn't mean there are no patterns. Tossing a coin is a random process because you can't predict what one coin will do. Even so, if you tossed 1000 coins into the air, you'd certainly find very nearly 500 heads and 500 tails. Nuclear decay is similar.

Let r be the probability that one particular nucleus will decay in the next 1 s by emitting an alpha or beta particle or a gamma-ray photon. For example, $r = 0.010 \text{ s}^{-1}$ means that a nucleus has a 1% chance of decay in the next second. Notice that r, which is called the **decay rate,** has units s^{-1}, making it a *rate.*

The probability that a nucleus decays during the small interval of time Δt is

$$\text{Prob(in } \Delta t) = r\Delta t \qquad (42.7)$$

For example, a nucleus with $r = 0.010 \text{ s}^{-1}$ has a 0.1% chance of decay (Prob = 0.001) during a 0.1 s interval. If there are N independent nuclei, the number of nuclei expected to decay during Δt is

$$\text{number of decays} = N \times \text{Probability of decay} = rN\Delta t \qquad (42.8)$$

This is like saying you expect 500 heads when tossing 1000 coins, each coin with a 50% probability of landing heads up.

Each decay *decreases* the number of radioactive nuclei in the sample, hence the change in the number of radioactive nuclei during Δt is

$$\Delta N = -rN\Delta t \qquad (42.9)$$

The negative sign shows that N, the number of nuclei, decreases due to the decays. Finally, if we let $\Delta t \rightarrow dt$, Equation 42.9 becomes

$$\frac{dN}{dt} = -rN \qquad (42.10)$$

The *rate of change* in the number of radioactive nuclei depends both on the decay rate (a larger probability of decay per second means more decays per second) and on the number of radioactive nuclei present (more nuclei means that more are available to decay). And dN/dt is negative because N is decreasing.

Equation 42.10 is the same equation we solved in Chapter 31, with different symbols, for the voltage decay in an *RC* circuit. First, separate the variables onto opposite sides of the equation:

$$\frac{dN}{N} = -r\,dt \tag{42.11}$$

We need to integrate this equation, starting from $N = N_0$ nuclei at $t = 0$. Thus

$$\int_{N_0}^{N} \frac{dN}{N} = -r\int_0^t dt \tag{42.12}$$

By carrying out the integrations we find

$$\ln N - \ln N_0 = \ln\left(\frac{N}{N_0}\right) = -rt \tag{42.13}$$

We can now solve for N by taking the exponential of both sides and multiplying by N_0. The result is

$$N = N_0 e^{-rt} \tag{42.14}$$

Equation 42.13 predicts that the number of radioactive nuclei will decrease exponentially, a prediction that has been borne out in countless experiments during the last hundred years.

It is useful to define the **time constant** τ as

$$\tau = \frac{1}{r}$$

With this definition, Equation 42.13 becomes

$$N = N_0 e^{-t/\tau} \tag{42.15}$$

Figure 42.16 shows the decrease of N with time. The number of radioactive nuclei decreases from N_0 at $t = 0$ to $e^{-1}N_0 = 0.368N_0$ at time $t = \tau$. In practical terms, the number decreases by roughly two-thirds during one time constant.

NOTE ▶ An important aspect of exponential decay is that you can choose any instant you wish to be $t = 0$. The number of radioactive nuclei present at that instant is N_0. If at one instant you have 10,000 radioactive nuclei whose time constant is $\tau = 10$ min, you'll have roughly 3680 nuclei 10 min later. The fact that you may have had more than 10,000 nuclei earlier isn't relevant. ◀

Equation 42.14 is useful in the theoretical sense that we can relate τ directly to the probability of decay. But in practice, it's much easier to measure the time at which half of a sample has decayed than the time at which 36.8% has decayed. Let's define the **half-life** $t_{1/2}$ as the time interval in which half of a sample of radioactive atoms decays. The half-life is shown in Figure 42.16.

The half-life is easily related to the time constant τ because we know, by definition, that $N = \frac{1}{2}N_0$ at $t = t_{1/2}$. Thus, according to Equation 42.15

$$\frac{N_0}{2} = N_0 e^{-t_{1/2}/\tau} \tag{42.16}$$

The N_0 cancels, and we can then take the natural logarithm of both sides to find

$$\ln\left(\frac{1}{2}\right) = -\ln 2 = -\frac{t_{1/2}}{\tau} \tag{42.17}$$

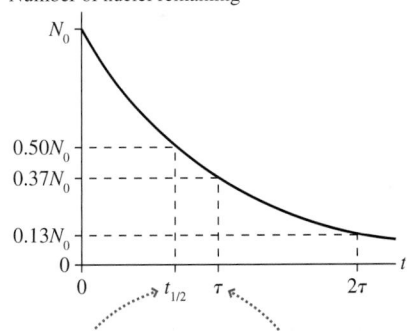

Number of nuclei remaining

The half-life is the time in which half the nuclei decay.

The time constant is the time at which the number of nuclei is e^{-1}, or 37%, of the initial number.

FIGURE 42.16 The number of radioactive atoms decreases exponentially with time.

With one final rearrangement we have

$$t_{1/2} = \tau \ln 2 = 0.693\tau \qquad (42.18)$$

We'll leave it as a homework problem for you to show that Equation 42.15 can be written in terms of the half-life as

$$N = N_0 \left(\frac{1}{2}\right)^{t/t_{1/2}} \qquad (42.19)$$

Thus $N = N_0/2$ at $t = t_{1/2}$, $N = N_0/4$ at $t = 2t_{1/2}$, $N = N_0/8$ at $t = 3t_{1/2}$, and so on. **No matter how many nuclei there are, the number decays by half during the next half-life.**

NOTE ▶ Half the nuclei decay during one half-life, but don't fall into the trap of thinking that all will have decayed after two half-lives. ◀

Figure 42.17 shows the half-life graphically. This figure also conveys two other important ideas:

1. Nuclei don't vanish when they decay. The decayed nuclei have merely become some other kind of nuclei.
2. The decay process is random. We can predict that half the nuclei will decay in one half-life, but we can't predict which ones.

Each radioactive isotope, such as ^{14}C, has its own half-life. That half-life doesn't change with time as a sample decays. If you've flipped a coin 10 times and, against all odds, seen 10 heads, you may feel that a tail is overdue. Nonetheless, the probability that the next flip will be a head is still 50%. After 10 half-lives have gone by, $(1/2)^{10} = 1/1024$ of a radioactive sample is still there. There was nothing special or distinctive about these nuclei, and, despite their longevity, each remaining nucleus has exactly a 50% chance of decay during the next half-life.

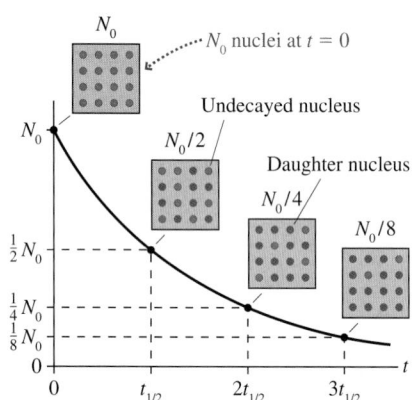

FIGURE 42.17 Half the nuclei decay during each half-life.

EXAMPLE 42.2 The decay of iodine
The iodine isotope ^{131}I, which has an eight-day half-life, is used in nuclear medicine. A sample of ^{131}I containing 2.00×10^{12} atoms is created in a nuclear reactor.

a. How many ^{131}I atoms remain 36 hours later when the sample is delivered to a hospital?
b. Although the sample is constantly getting weaker, it remains usable as long as there are at least 5.0×10^{11} ^{131}I atoms. What is the maximum delay before the sample is no longer usable?

MODEL The number of ^{131}I atoms decays exponentially.

SOLVE

a. The half-life is $t_{1/2} = 8$ days $= 192$ hr. After 36 hr have elapsed,

$$N = (2.00 \times 10^{12})\left(\frac{1}{2}\right)^{36/192} = 1.76 \times 10^{12} \text{ nuclei}$$

b. The time after creation at which 5.0×10^{11} ^{131}I atoms remain is given by

$$5.0 \times 10^{11} = 0.50 \times 10^{12} = (2.0 \times 10^{12})\left(\frac{1}{2}\right)^{t/8 \text{ days}}$$

To solve for t, first write this as

$$\frac{0.50}{2.00} = 0.25 = \left(\frac{1}{2}\right)^{t/8 \text{ days}}$$

Now take the logarithm of both sides. Either natural logarithms or base-10 logarithms can be used, but we'll use natural logarithms:

$$\ln(0.25) = -1.39 = \frac{t}{t_{1/2}}\ln(0.5) = -0.693\frac{t}{t_{1/2}}$$

Solving for t gives

$$t = 2.00t_{1/2} = 16 \text{ days}$$

ASSESS The weakest usable sample is one-quarter of the initial sample. You saw in Figure 42.17 that a radioactive sample decays to one-quarter of its initial number in 2 half-lives.

Activity

The **activity** R of a radioactive sample is the number of decays per second, or the decay rate. The decay rate is simply the absolute value of dN/dt, or

$$R = \left| \frac{dN}{dt} \right| = rN = rN_0 e^{-t/\tau} = R_0 e^{-t/\tau} = R_0 \left(\frac{1}{2} \right)^{t/t_{1/2}} \qquad (42.20)$$

where $R_0 = rN_0$ is the activity at $t = 0$. The activity of a sample decreases exponentially along with the number of remaining nuclei.

The SI unit of activity is the **becquerel,** defined as

$$1 \text{ becquerel} = 1 \text{ Bq} \equiv 1 \text{ decay/s or } 1 \text{ s}^{-1}$$

An older unit of activity, but one that continues in widespread use, is the **curie.** The curie was originally defined as the activity of 1 g of radium. Today, the conversion factor is

$$1 \text{ curie} = 1 \text{ Ci} \equiv 3.7 \times 10^{10} \text{ Bq}$$

1 Ci is a substantial amount of radiation. The radioactive samples used in laboratory experiments are typically $\approx 1 \ \mu\text{Ci}$, or, equivalently, $\approx 40,000 \text{ Bq}$. These samples can be handled with only minor precautions. Larger sources of activity require lead shielding and special precautions to prevent exposure to high levels of radiation.

EXAMPLE 42.3 A laboratory source

The isotope ^{137}Cs is a standard laboratory source of gamma rays. The half-life of ^{137}Cs is 30 years.

a. How many ^{137}Cs atoms are in a 5.0 μCi source?
b. What is the activity of the source 10.0 years later?

MODEL The number of ^{137}Cs atoms decays exponentially.

SOLVE

a. The number of atoms can be found from $N_0 = R_0/r$. The activity in SI units is

$$R = 5.0 \times 10^{-6} \text{ Ci} \times \frac{3.7 \times 10^{10} \text{ Bq}}{1 \text{ Ci}} = 1.85 \times 10^5 \text{ Bq}$$

To find the decay rate, first convert the half-life to seconds:

$$t_{1/2} = 30 \text{ years} \times \frac{3.15 \times 10^7 \text{ s}}{1 \text{ year}} = 9.45 \times 10^8 \text{ s}$$

Then

$$r = \frac{1}{\tau} = \frac{\ln 2}{t_{1/2}} = 7.33 \times 10^{-10} \text{ s}^{-1}$$

Thus the number of ^{137}Cs atoms is

$$N_0 = \frac{R_0}{r} = \frac{1.85 \times 10^5 \text{ Bq}}{7.33 \times 10^{-10} \text{ s}^{-1}} = 2.52 \times 10^{14} \text{ atoms}$$

b. The activity decreases exponentially, just like the number of nuclei. After 10 years,

$$R = R_0 \left(\frac{1}{2} \right)^{t/t_{1/2}} = (5.0 \ \mu\text{Ci}) \left(\frac{1}{2} \right)^{10/30} = 4.0 \ \mu\text{Ci}$$

ASSESS Although N_0 is a very large number, it is a very small fraction ($\approx 10^{-10}$) of a mole. The sample is about 60 ng (nanograms) of ^{137}Cs.

Radioactive Dating

Many geological and archeological samples can be dated by measuring the decays of naturally occurring radioactive isotopes. Because we have no way to know N_0, the initial number of radioactive nuclei, radioactive dating depends on the use of ratios.

The most well-known dating technique is carbon dating. The carbon isotope ^{14}C has a half-life of 5730 years, so any ^{14}C present when the earth formed 4.5 billion years ago would long since have decayed away. Nonetheless, ^{14}C is present in atmospheric carbon dioxide because high-energy cosmic rays collide with gas molecules high in the atmosphere. These cosmic rays are energetic enough to create ^{14}C nuclei from nuclear reactions with nitrogen and oxygen nuclei. The creation and decay of ^{14}C have reached a steady state in which the $^{14}\text{C}/^{12}\text{C}$ ratio is

1.3×10^{-12}. That is, atmospheric carbon dioxide has ^{14}C at the concentration of 1.3 parts per trillion. As small as this is, it's easily measured by modern chemical techniques.

All living organisms constantly exchange carbon dioxide with the atmosphere, so the $^{14}C/^{12}C$ ratio in living organisms is also 1.3×10^{-12}. As soon as an organism dies, the ^{14}C in its tissue begins to decay and no new ^{14}C is added. Objects are dated by comparing the measured $^{14}C/^{12}C$ ratio to the 1.3×10^{-12} value of living material.

Carbon dating is used to date skeletons, wood, paper, fur, food material, and anything else made of organic matter. It is quite accurate for ages to about 15,000 years. Beyond that, the difficulty of measuring such a small ratio and some uncertainties about the cosmic ray flux in the past combine to decrease the accuracy. Even so, items are dated to about 50,000 years with a fair degree of reliability.

Other isotopes with longer half-lives are used to date geological samples. Potassium-argon dating, using ^{40}K with a half-life of 1.25 billion years, is especially useful for dating rocks of volcanic origin.

EXAMPLE 42.4 Carbon dating

Archeologists excavating an ancient hunters' camp have recovered a 5.0 g piece of charcoal from a fireplace. Measurements on the sample find that the ^{14}C activity is 0.35 Bq. What is the approximate age of the camp?

MODEL Charcoal, from burning wood, is almost pure carbon. The number of ^{14}C atoms in the wood has decayed exponentially since the branch fell off a tree. Because wood rots, it is reasonable to assume that there was no significant delay from when the branch fell off the tree and when the hunters burned it.

SOLVE The $^{14}C/^{12}C$ ratio was 1.3×10^{-12} when the branch fell from the tree. We first need to determine the present ratio, then use the known ^{14}C half-life $t_{1/2} = 5730$ years to calculate the time needed to reach the present ratio. The number of ordinary ^{12}C nuclei in the sample is

$$N(^{12}C) = \left(\frac{5.0 \text{ g}}{12 \text{ g/mol}}\right) 6.02 \times 10^{23} \text{ atoms/mol}$$

$$= 2.5 \times 10^{23} \text{ nuclei}$$

The number of ^{14}C nuclei can be found from the activity to be $N(^{14}C) = R/r$, but we need to determine the ^{14}C decay rate r. After converting the half-life to seconds, $t_{1/2} = 5730$ years = 1.807×10^{11} s, we can compute

$$r = \frac{1}{\tau} = \frac{1}{t_{1/2}/\ln 2} = 3.84 \times 10^{-12} \text{ s}^{-1}$$

Thus

$$N(^{14}C) = \frac{R}{r} = \frac{0.35 \text{ Bq}}{3.84 \times 10^{-12} \text{ s}^{-1}} = 9.1 \times 10^{10} \text{ nuclei}$$

and the present $^{14}C/^{12}C$ ratio is $N(^{14}C)/N(^{12}C) = 0.36 \times 10^{-12}$. Because this ratio has been decaying with a half-life of 5730 years, the time needed to reach the present ratio is found from

$$0.36 \times 10^{-12} = 1.3 \times 10^{-12}\left(\frac{1}{2}\right)^{t/t_{1/2}}$$

To solve for t, first write this as

$$\frac{0.36}{1.3} = 0.277 = \left(\frac{1}{2}\right)^{t/t_{1/2}}$$

Now take the logarithm of both sides:

$$\ln(0.277) = -1.28 = \frac{t}{t_{1/2}}\ln(0.5) = -0.693\frac{t}{t_{1/2}}$$

Thus the age of the hunters' camp is

$$t = 1.85 t_{1/2} = 10,600 \text{ years}$$

ASSESS This is a realistic example of how radioactive dating is done.

STOP TO THINK 42.4 A sample starts with 1000 radioactive atoms. How many half-lives have elapsed when 750 atoms have decayed?

    a. 0.25
    b. 1.5
    c. 2.0
    d. 2.5

42.6 Nuclear Decay Mechanisms

This section will look in more detail at the mechanisms of the three types of radioactive decay.

Activ
Physics ONLINE 19.4

Alpha Decay

An alpha particle, symbolized as α, is a ^{4}He nucleus, a strongly bound system of two protons and two neutrons. An unstable nucleus that ejects an alpha particle will lose two protons and two neutrons, so we can write the decay as

$$^A X_Z \rightarrow {}^{A-4}Y_{Z-2} + \alpha + \text{energy} \qquad (42.21)$$

Figure 42.18 shows the alpha-decay process. The original nucleus X is called the **parent nucleus** and the decay-product nucleus Y is the **daughter nucleus.** This reaction can occur only when the mass of the parent nucleus is greater than the mass of the daughter nucleus plus the mass of an alpha particle. This requirement is met for heavy, high-Z nuclei well above the maximum on the Figure 42.6 curve of binding energy. It is energetically favorable for these nuclei to eject an alpha particle because the daughter nucleus is more tightly bound than the parent nucleus.

Although the mass requirement is based on the nuclear masses, we can express it—as we did the binding energy equation—in terms of atomic masses. The energy released in an alpha decay, essentially all of which goes into the alpha particle's kinetic energy, is

$$\Delta E \approx K_\alpha = (m_X - m_Y - m_{He})c^2 \qquad (42.22)$$

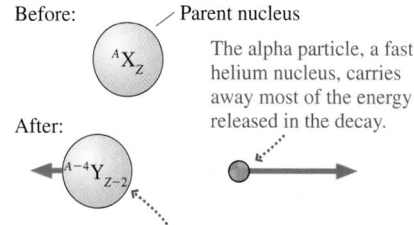

The alpha particle, a fast helium nucleus, carries away most of the energy released in the decay.

The daughter nucleus has two fewer protons and four fewer nucleons. It has a small recoil.

FIGURE 42.18 Alpha decay.

EXAMPLE 42.5 Alpha decay of uranium

The uranium isotope ^{238}U undergoes alpha decay to ^{234}Th. The atomic masses are 238.0508 u for ^{238}U and 234.0436 u for ^{234}Th. What is the kinetic energy, in MeV, of the alpha particle?

MODEL Essentially all of the energy release ΔE goes into the alpha particle's kinetic energy.

SOLVE The atomic mass of helium, from Table 42.2, is 4.0026 u. Thus

$$K_\alpha = (238.0508 \text{ u} - 234.0436 \text{ u} - 4.0026 \text{ u})c^2$$

$$= \left(0.0046 \text{ u} \times \frac{931.5 \text{ MeV}/c^2}{1 \text{ u}}\right)c^2 = 4.3 \text{ MeV}$$

ASSESS This is a typical alpha-particle energy. Notice how the c^2 canceled from the calculation so that we never had to evaluate c^2.

Alpha decay is a purely quantum-mechanical effect. Figure 42.19 shows the potential energy of an alpha particle, where the ^{4}He nucleus of an alpha particle is so tightly bound that we can think of it as existing "prepackaged" inside the parent nucleus. Both the depth of the energy well and the height of the Coulomb barrier are twice that of a proton because the charge of an α particle is $2e$.

Because of the high Coulomb barrier (alpha decay occurs only in high-Z nuclei), there may be one or more allowed energy levels with $E > 0$. Energy levels with $E < 0$ are completely bound, but an alpha particle in an energy level with $E > 0$ can *tunnel* through the Coulomb barrier and escape. That is exactly how alpha decay occurs.

Energy must be conserved, so the kinetic energy of the escaping α particle is the height of the energy level above $E = 0$. That is, potential energy is transformed into kinetic energy as the particle escapes. Notice that the width of the barrier decreases as E increases. The tunneling probability depends very sensitively on the barrier width, as you learned in conjunction with the scanning tunneling microscope. Thus an alpha particle in a higher energy level should have a

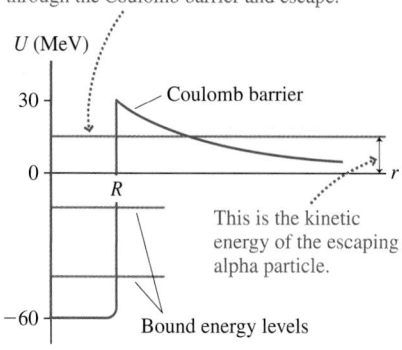

An alpha particle in this energy level can tunnel through the Coulomb barrier and escape.

This is the kinetic energy of the escaping alpha particle.

Bound energy levels

FIGURE 42.19 The potential-energy diagram of an alpha particle in the parent nucleus.

shorter half-life and escape with *more kinetic energy*. The full analysis is beyond the scope of this text, but this prediction is in excellent agreement with measured energies and half-lives.

Beta Decay

Beta decay was initially associated with the emission of an electron e⁻, the beta particle. It was later discovered that some nuclei can undergo beta decay by emitting a positron e⁺, the antiparticle of the electron, although this decay mode is not as common. A positron is identical to an electron except that it has a positive charge. To be precise, the emission of an electron is called *beta-minus decay* and the emission of a positron is *beta-plus decay*.

A typical example of beta decay occurs in the carbon isotope ^{14}C, which undergoes the beta-decay process ^{14}C $\rightarrow$ ^{14}N + e⁻. Carbon has $Z = 6$ and nitrogen has $Z = 7$. Because Z increases by 1 but A doesn't change, it appears that a neutron within the nucleus has changed itself into a proton and electron. That is, the basic beta-minus decay process appears to be

$$n \rightarrow p^+ + e^- \tag{42.23}$$

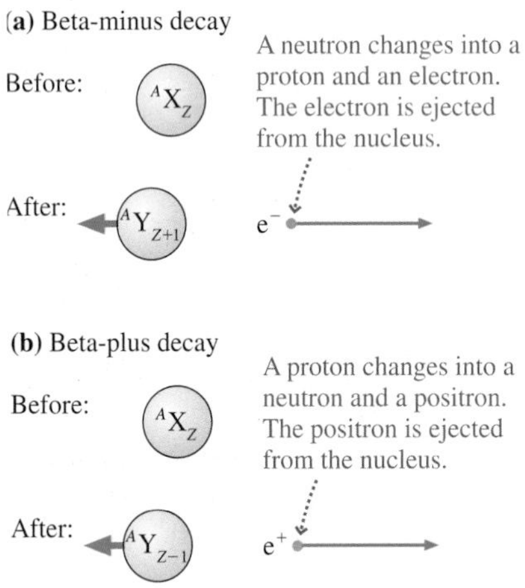

The electron is ejected from the nucleus but the proton is not. Thus the decay process, shown in Figure 42.20a, is

$$^A X_Z \rightarrow {}^A Y_{Z+1} + e^- + \text{energy} \qquad \text{(beta-minus decay)} \tag{42.24}$$

Indeed, a free neutron turns out to *not* be a stable particle. It decays with a half-life of approximately 10 min into a proton and an electron. This decay is energetically allowed because $m_n > m_p + m_e$. Furthermore, it conserves charge.

Whether a neutron *within* a nucleus can decay depends not only on the masses of the neutron and proton but also on the masses of the parent and daughter nuclei, because energy has to be conserved for the entire nuclear system. **Beta decay occurs only if $m_X > m_Y$.** ^{14}C can undergo beta decay to ^{14}N because $m(^{14}\text{C}) > m(^{14}\text{N})$. But $m(^{12}\text{C}) < m(^{12}\text{N})$, so ^{12}C is stable and its neutrons cannot decay.

Beta-plus decay is the conversion of a proton into a neutron and a positron:

$$p^+ \rightarrow n + e^+ \tag{42.25}$$

The full decay process, shown in Figure 42.20b, is

$$^A X_Z \rightarrow {}^A Y_{Z-1} + e^+ + \text{energy} \qquad \text{(beta-plus decay)} \tag{42.26}$$

Beta-plus decay does *not* happen for a free proton because $m_p < m_n$. It *can* happen within a nucleus as long as energy is conserved for the entire nuclear system.

In our earlier discussion of Figure 42.11 we noted that the ^{12}B and ^{12}N nuclei could reach a lower energy state if a proton could change into a neutron and vice versa. Now we see that such a change can occur if the energy conditions are favorable. And, indeed, ^{12}B undergoes beta-minus decay to ^{12}C while ^{12}N undergoes beta-plus decay to ^{12}C.

In general, beta decay is a process used by nuclei with too many neutrons or too many protons in order to move closer to the line of stability in Figure 42.4.

> **NOTE** ▶ The electron emitted in beta decay has nothing to do with the atom's valence electrons. The beta particle is created in the nucleus and ejected directly from the nucleus when a neutron is transformed into a proton and an electron. ◀

A third form of beta decay occurs in some nuclei that have too many protons but not enough mass to undergo beta-plus decay. In this case, a proton changes

(a) Beta-minus decay

Before: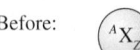

A neutron changes into a proton and an electron. The electron is ejected from the nucleus.

After: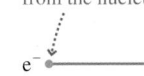

(b) Beta-plus decay

Before:

A proton changes into a neutron and a positron. The positron is ejected from the nucleus.

After: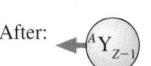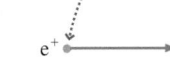

FIGURE 42.20 Beta decay.

into a neutron by "capturing" an electron from the innermost shell of orbiting electrons (an $n = 1$ electron). The process is

$$p^+ + \text{orbital } e^- \rightarrow n \qquad (42.27)$$

This form of beta decay is called **electron capture,** abbreviated EC. The net result, $^AX_Z \rightarrow {}^AY_{Z-1}$, is the same as beta-plus decay but without the emission of a positron. Electron capture is the only nuclear decay mechanism that involves the orbital electrons.

The Weak Interaction

We've presented beta decay as if it were perfectly normal for one kind of matter to change spontaneously into a completely different kind of matter. For example, it would be energetically favorable for a large truck to spontaneously turn into a Cadillac and a VW Beetle, ejecting the Beetle at high speed. But it doesn't happen.

Once you stop to think of it, the process $n \rightarrow p^+ + e^-$ seems ludicrous, not because it violates mass-energy conservation but because we have no idea *how* a neutron could turn into a proton. Alpha decay may be a strange process because tunneling in general goes against our commonsense notions, but it is a perfectly ordinary quantum-mechanical process. Now we're suggesting that one of the basic building blocks of matter can somehow morph into a different basic building block.

To make matters more confusing, measurements in the 1930s found that beta decay didn't seem to conserve either energy or momentum. Faced with these difficulties, the Italian physicist Enrico Fermi made two bold suggestions:

1. A previously unknown fundamental force of nature is responsible for beta decay. This force, which has come to be known as the **weak interaction,** has the ability to turn a neutron into a proton and vice versa.
2. The beta-decay process emits a particle that, at that time, had not been detected. This new particle has to be electrically neutral, in order to conserve charge, and it has to be much smaller than an electron. Fermi called it the **neutrino,** meaning "little neutral one." Energy and momentum really are conserved, but the neutrino carries away some of the energy and momentum of the decaying nucleus. Thus experiments that detect only the electron seem to violate conservation laws.

The neutrino is represented by the symbol ν, a lowercase Greek nu. The beta-decay processes that Fermi proposed are

$$
\begin{aligned}
n &\rightarrow p^+ + e^- + \bar{\nu} \\
p^+ &\rightarrow n + e^+ + \nu
\end{aligned}
\qquad (42.28)
$$

The symbol $\bar{\nu}$ is an *antineutrino,* although the reason why one is a neutrino and the other an antineutrino need not concern us here. Figure 42.21 shows that the electron and antineutrino (or positron and neutrino) *share* the energy released in the decay.

The neutrino interacts with matter so weakly that a neutrino can pass straight through the earth with only a very slight chance of a collision. Thousands of neutrinos created by nuclear fusion reactions in the core of the sun are passing through your body every second. Neutrino interactions are so rare that the first laboratory detection did not occur until 1956, over 20 years after Fermi's proposal.

It was initially thought that the neutrino had not only zero charge but zero mass. However, experiments within the last few years have shown that the neutrino mass, although very tiny, is not zero. The best current evidence suggests a mass about one-millionth the mass of an electron. Experiments now underway will attempt to determine a more accurate value. The result will have far more importance than simply understanding beta decay. Neutrinos are the most numerous of all particles in the universe, so it may turn out that the neutrino mass has cosmological significance for the evolution of the universe.

The Super Kamiokande neutrino detector in Japan looks for the neutrinos emitted from nuclear fusion reactions in the core of the sun.

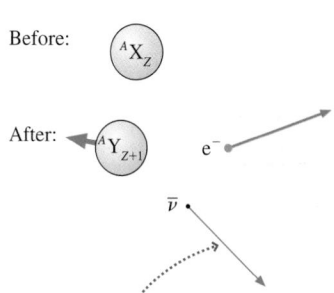

If only the electron and the daughter nucleus are measured, energy and momentum appear not to be conserved. The "missing" energy and momentum are carried away by the undetected antineutrino.

FIGURE 42.21 A more accurate picture of beta decay includes neutrinos.

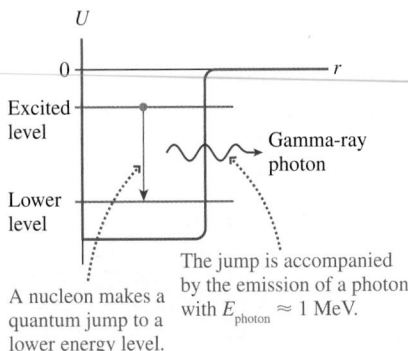

FIGURE 42.22 Gamma decay.

A nucleon makes a quantum jump to a lower energy level.

The jump is accompanied by the emission of a photon with $E_{photon} \approx 1$ MeV.

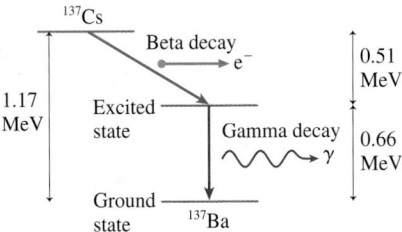

FIGURE 42.23 The decay of ^{137}Cs involves both beta and gamma decay.

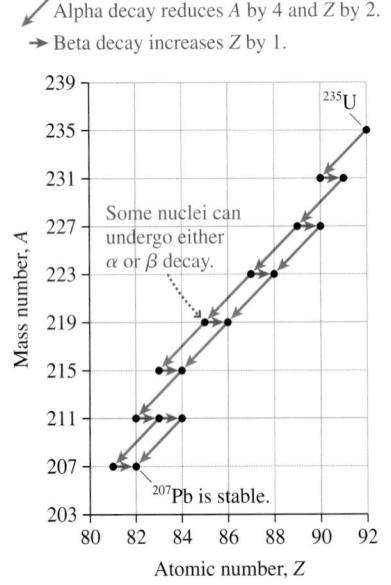

FIGURE 42.24 The nuclear decay series of ^{235}U.

EXAMPLE 42.6 Beta decay of ^{14}C

How much energy is released in the beta-minus decay of ^{14}C?

MODEL The decay is ^{14}C $\rightarrow$ ^{14}N $+ e^- + \bar{\nu}$.

SOLVE In Appendix C we find $m(^{14}C) = 14.003\,242$ u and $m(^{14}N) = 14.003\,074$ u. The mass difference is a mere $0.000\,168$ u, but this is the mass that is converted into the kinetic energy of the escaping particles. The energy released is

$$E = (\Delta m)c^2 = (0.000168 \text{ u}) \times (931.5 \text{ MeV/u})$$

$$= 0.156 \text{ MeV}$$

ASSESS This energy is shared between the electron and the antineutrino.

Gamma Decay

Gamma decay is the easiest form of nuclear decay to understand. You learned that an atomic system can emit a photon with $E_{photon} = \Delta E_{atom}$ when an electron undergoes a quantum jump from an excited energy level to a lower energy level. Nuclei are no different. A proton or a neutron in an excited nuclear state, such as the one shown in Figure 42.22, can undergo a quantum jump to a lower-energy state by emitting a high-energy photon. This is the gamma-decay process.

The spacing between atomic energy levels is only a few eV. Nuclear energy levels, by contrast, are typically 1 MeV apart. Hence gamma-ray photons have $E_{gamma} \approx 1$ MeV. Photons with this much energy have tremendous penetrating power and deposit an extremely large amount of energy at the point where they are finally absorbed.

Nuclei left to themselves are usually in their ground states and thus cannot emit gamma-ray photons. However, alpha and beta decay often leave the daughter nucleus in an excited nuclear state, so gamma emission is usually found to accompany alpha and beta emission.

The cesium isotope ^{137}Cs is a good example. We noted earlier that ^{137}Cs is used as a laboratory source of gamma rays. Actually, ^{137}Cs undergoes beta-minus decay to ^{137}Ba. Figure 42.23 shows the full process. A ^{137}Cs nucleus undergoes beta-minus decay by emitting an electron and an antineutrino, which share between them a total energy of 0.51 MeV. The half-life for this process is 30 years. This leaves the daughter ^{137}Ba nucleus in an excited state 0.66 MeV above the ground state. The excited Ba nucleus then decays within a few seconds to the ground state by emitting a 0.66 MeV gamma-ray photon. Thus a ^{137}Cs sample *is* a source of gamma-ray photons, but the photons are actually emitted by barium nuclei rather than cesium nuclei.

Decay Series

A radioactive nucleus decays into a daughter nucleus. In many cases, the daughter nucleus is also radioactive and decays to produce its own daughter nucleus. The process continues until reaching a daughter nucleus that is stable. The sequence of isotopes, starting with the original unstable isotope and ending with the stable isotope, is called a **decay series.**

Decay series are especially important for very heavy nuclei. As an example, Figure 42.24 shows the decay series of ^{235}U, an isotope of uranium with a 700-million-year half-life. This is a very long time, but it is only about 15% the age of the earth and most (but not all) of the ^{235}U nuclei present when the earth was formed have now decayed. There are many unstable nuclei along the way, but all ^{235}U nuclei eventually end as the ^{207}Pb isotope of lead, a stable nucleus.

Notice that some nuclei can decay by either alpha *or* beta decay. Thus there are a variety of paths that a decay can follow, but they all end at the same point.

STOP TO THINK 42.5 The cobalt isotope ^{60}Co ($Z = 27$) decays to the nickel isotope ^{60}Ni ($Z = 28$). The decay process is

a. Alpha decay.   b. Beta-minus decay.   c. Beta-plus decay.
d. Electron capture.   e. Gamma decay.

42.7 Biological Applications of Nuclear Physics

Nuclear physics has brought both peril and promise to society. Radioactivity can cause tumors. At the same time, radiation can be used to cure some cancers. This section is a brief survey of medical and biological applications of nuclear physics.

Radiation Dose

Nuclear radiation, which is ionizing radiation, disrupts a cell's machinery by altering and damaging the biological molecules. The consequences of this disruption vary from genetic mutations to uncontrolled cell multiplication (i.e., tumors) to cell death.

Beta and gamma radiation can penetrate the entire body and damage internal organs. Alpha radiation has less penetrating ability, but it deposits all its energy in a very small, localized volume. Internal organs are usually safe from alpha radiation, but the skin is very susceptible, as are the lungs if radioactive dust is inhaled.

Biological effects of radiation depend upon the **dose,** the amount of radiation received. Two factors enter into determining the dose. The first is the physical factor of energy absorbed by the body. The second is the biological factor of how tissue reacts to different forms of radiation.

The **rad,** an acronym for **r**adiation **a**bsorbed **d**ose, measures the energy deposited in an irradiated material. It is defined as

$$1 \text{ rad} \equiv 0.010 \text{ J/kg of absorbed energy}$$

The number of rads depends only on the energy absorbed, not at all on the type of radiation or on what the absorbing material is.

Biologists and biophysicists have found that a 1 rad dose of gamma rays and a 1 rad dose of alpha particles have different biological consequences. To account for such differences, the **relative biological effectiveness,** RBE, is defined as the biological effect of a given dose relative to the biological effect of an equal dose of x rays.

Table 42.4 shows the relative biological effectiveness of different forms of radiation. Larger values correspond to larger biological effects. Alpha and beta radiation have a range of values because the biological effect varies with the energy of the particle. Alpha radiation has the largest RBE because the energy is deposited in a smaller volume.

Combining these two measures, the **biologically equivalent dose** is the product of the energy dose in rads with the relative biological effectiveness. The biologically equivalent dose is measured in **rem,** an acronym for **r**öntgen **e**quivalent in **m**an. (This term is based on historical usage. The *röntgen* is a unit for measuring the ionization ability of x rays, but we need not be concerned with its definition.) To be precise,

$$\text{biologically equivalent dose in rem} \equiv \text{dose in rad} \times \text{RBE}$$

One rem of radiation produces the same biological damage regardless of the type of radiation.

TABLE 42.4 Relative biological effectiveness of radiation

Radiation type	RBE
X rays	1
Gamma rays	1
Beta particles	1–2
Alpha particles	10–20

EXAMPLE 42.7 **Radiation exposure**

A 75 kg laboratory technician working with the radioactive isotope ^{137}Cs receives an accidental 100 mrem (millirem) exposure. ^{137}Cs emits 0.66 MeV gamma-ray photons. How many gamma-ray photons are absorbed in the technician's body?

MODEL The radiation dose is a combination of deposited energy and biological effectiveness. The RBE for gamma rays is 1. Gamma rays are penetrating, so this is a whole-body exposure.

SOLVE The dose in rads is the dose in rems divided by the RBE. In this case, because the RBE = 1, the dose is 100 mrad = 0.10 rad = 0.0010 J/kg. This is a whole-body exposure, so the total energy deposited in the technician's body is 0.075 J. The energy of each absorbed photon is 0.66 MeV, but this value must be converted into joules. The number of photons in 0.075 J is

$$N = \frac{0.075 \text{ J}}{(6.6 \times 10^5 \text{ eV/photon})(1.60 \times 10^{-19} \text{ J/eV})}$$
$$= 7.1 \times 10^{11} \text{ photons}$$

ASSESS The energy deposited, 0.075 J, is very small. Radiation does its damage not by thermal effects, which would require substantially more energy, but by ionization.

TABLE 42.5 Radiation exposure

Radiation source	Typical exposure (mrem/year)
Natural background	300
Mammogram x ray	80
Chest x ray	30
Dental x ray	3

Table 42.5 shows some basic information about radiation exposure. We are all exposed to a continual *natural background* of radiation from cosmic rays and from naturally occurring radioactive atoms (uranium and atoms in the uranium decay series) in the ground, the atmosphere, and even the food we eat. This background averages about 300 mrem per year, although there are wide regional variations depending on the soil type and the elevation. (Higher elevations have a larger exposure to cosmic rays.)

Medical x rays vary significantly. The average person in the United States receives approximately 60 mrem per year from all medical sources. All other sources, such as fallout from atmospheric nuclear tests many decades ago, nuclear power plants, and industrial uses of radioactivity, amount to <10 mrem per year.

The question inevitably arises, "What is a safe dose?" This remains a controversial topic and the subject of ongoing research. The effects of large doses of radiation are easily observed. The effects of small doses are hard to distinguish from other natural and environmental causes. Thus there's no simple or clear definition of a safe dose. A prudent policy is to avoid unnecessary exposure to radiation but not to worry over exposures less than the natural background. It's worth noting that the μCi radioactive sources used in laboratory experiments provide exposures *much* less than the natural background, even if used on a regular basis.

Medical Uses of Radiation

Radiation can be put to good use killing cancer cells. This area of medicine is called *radiation therapy*. Gamma rays are the most common form of radiation, often from the isotope ^{60}Co. As Figure 42.25 shows, the gamma rays are directed along many different lines, all of which intersect the tumor. The goal is to provide a lethal dose to the cancer cells without overexposing nearby tissue. The patient and the radiation source are rotated around each other under careful computer control to deliver the proper dose.

Other tumors are treated by surgically implanting radioactive "seeds" within or next to the tumor. Alpha particles, which are very damaging locally but don't penetrate far, can be used in this fashion.

Radioactive isotopes are also used as *tracers* in diagnostic procedures. This technique is based on the fact that all isotopes of an element have identical chemical behavior. As an example, a radioactive isotope of iodine is used in the diagnosis of certain thyroid conditions. Iodine is an essential element in the body, and it concentrates in the thyroid gland. A doctor who suspects a malfunctioning thyroid gland gives the patient a small dose of sodium iodide in which some of the normal ^{127}I atoms have been replaced with ^{131}I. (Sodium iodide, which is harmless, dissolves in water and can simply be drunk.) The ^{131}I isotope, with a half-life of eight days, undergoes beta decay, and subsequently emits a gamma-ray photon.

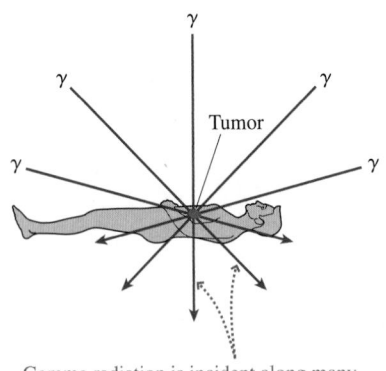

Gamma radiation is incident along many lines, all of which intersect the tumor.

FIGURE 42.25 Radiation therapy is designed to deliver a lethal dose to the tumor without damaging nearby tissue.

The radioactive iodine concentrates within the thyroid gland within a few hours. The doctor then monitors the gamma-ray photon emissions over the next few days to see how the iodine is being processed within the thyroid and how quickly it is eliminated from the body.

Other important radioactive tracers include the chromium isotope ^{51}Cr, which is taken up by red blood cells and can be used to monitor blood flow, and the xenon isotope ^{133}Xe, which is inhaled to reveal lung functioning. Radioactive tracers are *noninvasive,* meaning that the doctor can monitor the inside of the body without surgery.

Magnetic Resonance Imaging

The proton, like the electron, has an inherent angular momentum (spin) and an inherent magnetic moment. You can think of the proton as being like a little compass needle that can be in one of two positions, the positions we call spin up and spin down.

A compass needle aligns itself with an external magnetic field. This is the needle's lowest-energy position. Turning a compass needle by hand is like rolling a ball uphill; you're giving it energy, but, like the ball rolling downhill, it will realign itself with the lowest-energy position when you remove your finger. There is, however, an *unstable equilibrium* position in which the needle is anti-aligned with the field. The slightest jostle will cause it to flip around, but the needle will be steady in its upside-down configuration if you can balance it perfectly.

A proton in a magnetic field behaves similarly, but with a major difference: Because the proton's energy is quantized, the proton cannot assume an intermediate position. It's either aligned with the magnetic field (the spin-up orientation) or anti-aligned (spin-down). Figure 42.26a shows these two quantum states. Turning on a magnetic field lowers the energy of a spin-up proton and increases the energy of an anti-aligned, spin-down proton. In other words, the magnetic field creates an *energy difference* between these states.

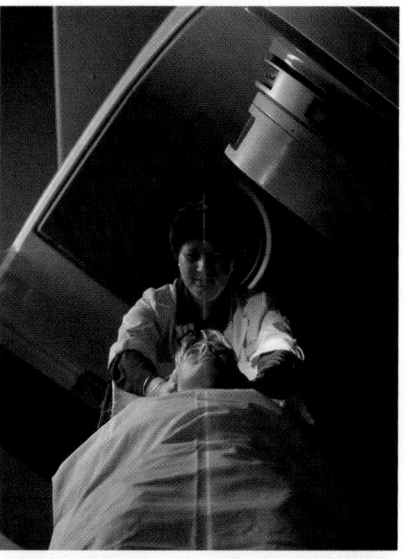

Radiation therapy is a beneficial use of nuclear physics.

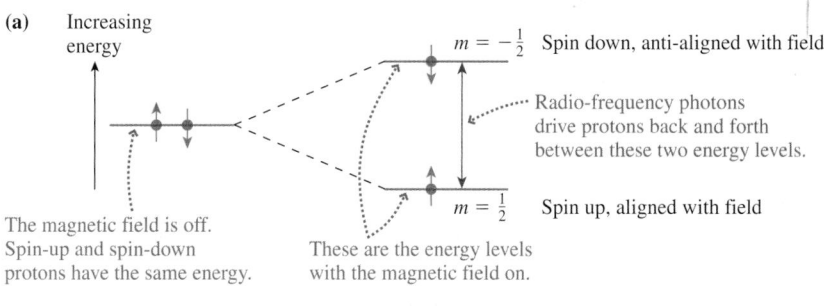

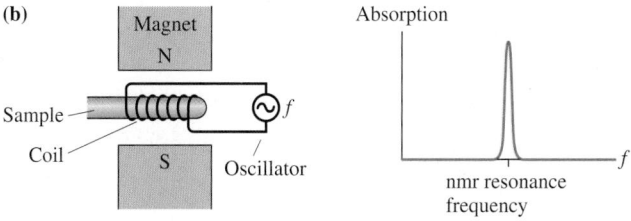

FIGURE 42.26 Nuclear magnetic resonance is possible because spin-up and spin-down protons have slightly different energies in a magnetic field.

The energy difference is very tiny, only about 10^{-7} eV. Nonetheless, photons whose energy matches the energy difference cause the protons to move back and forth between these two energy levels as the photons are absorbed and emitted. In effect, the photons are causing the proton's spin to flip back and forth rapidly. The

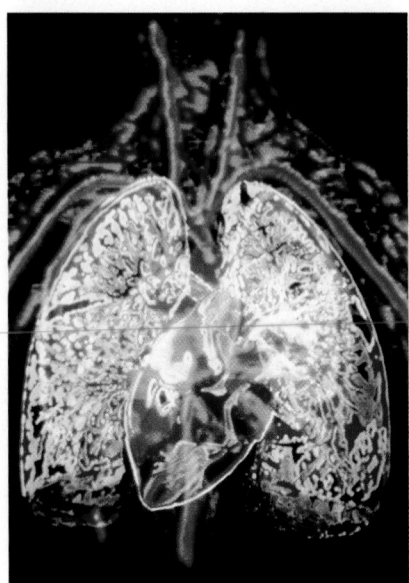

FIGURE 42.27 Magnetic resonance imaging shows internal organs in exquisite detail.

photon frequency, which depends on the magnetic field strength, is typically about 100 MHz, similar to FM radio frequencies.

Figure 42.26b shows how this behavior is put to use. A sample containing protons is placed in a magnetic field. A coil is wrapped around the sample, and a variable-frequency AC source drives a current through this coil. The protons absorb power from the coil when its frequency is just right to flip the spin back and forth; otherwise, no power is absorbed. A *resonance* is seen by scanning the coil through a small range of frequencies.

This technique of observing the spin flip of nuclei (the technique works for nuclei other than hydrogen) in a magnetic field is called **nuclear magnetic resonance,** or *nmr*. It has many applications in physics, chemistry, and materials science. Its medical use exploits the fact that tissue is mostly water, and two out of the three nuclei in a water molecule are protons. Thus the human body is basically a sample of protons, with the proton density varying as tissue density varies.

The medical procedure known as **magnetic resonance imaging,** or MRI, places the patient in a spatially varying magnetic field. The variations in the field cause the proton absorption frequency to vary from point to point. From the known shape of the field and measurements of the frequencies that are absorbed, and how strongly, sophisticated computer software can transform the raw data into detailed images such as the one shown in Figure 42.27.

As an interesting footnote, the technique was still being called *nuclear magnetic resonance* when it was first introduced into medicine. Unfortunately, doctors soon found that many patients were afraid of it because of the word "nuclear." Hence the alternative term "magnetic resonance imaging" was coined. It is true that the public perception of nuclear technology is not always positive, but equally true that nuclear physics has made many important and beneficial contributions to society.

SUMMARY

The goal of Chapter 42 has been to understand the physics of the nucleus and some of the applications of nuclear physics.

GENERAL PRINCIPLES

The Nucleus

The nucleus is a small, dense, positive core at the center of an atom.

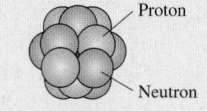

Z protons: charge $+e$, spin $\frac{1}{2}$

N neutrons: charge 0, spin $\frac{1}{2}$

The mass number is $A = Z + N$

The nuclear radius is $r = r_0 A^{1/3}$, where $r_0 = 1.2$ fm. Typical radii are a few fm.

Nuclear forces

Attractive strong force	**Repulsive electric force**
• Acts between any two nucleons	• Acts between two protons
• Is short range, <3 fm	• Is long range
• Is felt between nearest neighbors	• Is felt across the nucleus

Nuclear Stability

Most nuclei are not stable. Unstable nuclei undergo radioactive decay. Stable nuclei cluster along the **line of stability** in a plot of the isotopes.

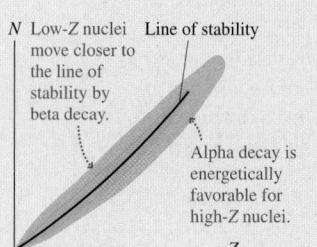

Three mechanisms by which unstable nuclei decay:

Decay	Particle	Mechanism	Energy	Penetration
α	^{4}He nucleus	tunneling	few MeV	low
β	e^-	$n \to p^+ + e^-$	≈ 1 MeV	medium
	e^+	$p^+ \to n + e^+$	≈ 1 MeV	medium
γ	photon	quantum jump	≈ 1 MeV	high

IMPORTANT CONCEPTS

Shell model

Each nucleon moves with an average potential energy due to all other nucleons.

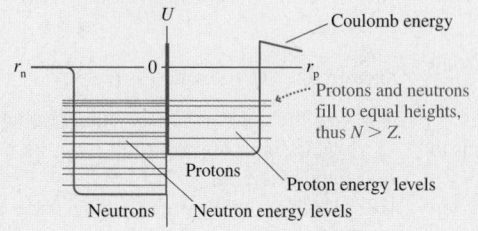

Curve of binding energy

The average binding energy per nucleon has a broad maximum at $A \approx 60$.

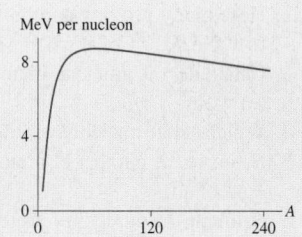

APPLICATIONS

Radioactive decay

The number of undecayed nuclei decreases exponentially with time t.

$$N = N_0 \exp(-t/\tau)$$
$$= N_0(1/2)^{t/t_{1/2}}$$

The **time constant** τ is $1/r$, where r is the **decay rate.** The **half-life**

$$t_{1/2} = \tau \ln 2 = 0.693\tau$$

is the time in which half of any sample decays.

Measuring radiation

The **activity** $R = rN$ of a radioactive sample, measured in becquerels or curies, is the number of decays per second.

The radiation **dose** is measured in rad, where

$$1 \text{ rad} \equiv 0.010 \text{ J/kg of absorbed energy}$$

The **relative biological effectiveness** RBE is the biological effect of a dose relative to the biological effects of x rays. The **biologically equivalent dose** is measured in rem, where rem = rad × RBE. One rem of radiation produces the same biological effect regardless of the type of radiation.

TERMS AND NOTATION

nuclear physics	curve of binding energy	parent nucleus
nucleon	strong force	daughter nucleus
atomic number, Z	shell model	electron capture
mass number, A	alpha decay	weak interaction
neutron number, N	beta decay	neutrino
isotope	gamma decay	decay series
radioactive	ionizing radiation	dose
stable	Geiger counter	rad
natural abundance	decay rate, r	relative biological effectiveness (RBE)
isobar	time constant, τ	biologically equivalent dose
deuterium	half-life, $t_{1/2}$	rem
liquid-drop model	activity, R	nuclear magnetic resonance
line of stability	becquerel, Bq	magnetic resonance imaging (MRI)
binding energy, B	curie, Ci	

EXERCISES AND PROBLEMS

See Appendix C for data on atomic masses, isotopic abundance, radioactive decay modes, and half-lives.

Exercises

Section 42.1 Nuclear Structure

1. How many protons and how many neutrons are in (a) ^{3}H, (b) ^{40}Ar, (c) ^{40}Ca, and (d) ^{239}Pu?

2. How many protons and how many neutrons are in (a) ^{3}He, (b) ^{20}Ne, (c) ^{60}Co, and (d) ^{226}Ra?

3. Calculate the nuclear diameters of (a) ^{4}He, (b) ^{40}Ar, and (c) ^{220}Rn.

4. Which stable nuclei have a diameter of 8.84 fm?

5. Estimate the number of protons and the number of neutrons in 1 m^3 of air.

6. Estimate the number of protons and the number of neutrons in your body.

7. What would be the mass of a 1.0-cm-diameter marble if it had nuclear density?

Section 42.2 Nuclear Stability

8. Use data in Appendix C to make your own chart of stable and unstable nuclei, similar to Figure 42.4, for all nuclei with $Z \leq 8$. Use a blue or black dot to represent stable isotopes, a red dot to represent isotopes that undergo beta-minus decay, and a green dot to represent isotopes that undergo beta-plus decay or electron-capture decay.

9. a. What is the smallest value of A for which there are two stable nuclei? What are they?
   b. For which values of A less than this are there *no* stable nuclei?

10. Calculate (in MeV) the total binding energy and the binding energy per nucleon for ^{3}H and for ^{3}He.

11. Calculate (in MeV) the total binding energy and the binding energy per nucleon for ^{40}Ar and for ^{40}Ca.

12. Calculate (in MeV) the binding energy per nucleon for ^{3}He and ^{4}He. Which is more tightly bound?

13. Calculate (in MeV) the binding energy per nucleon for ^{12}C and ^{13}C. Which is more tightly bound?

14. Calculate (in MeV) the binding energy per nucleon for (a) ^{14}N, (b) ^{56}Fe, and (c) ^{207}Pb.

15. Calculate the chemical atomic mass of neon.

16. Calculate the chemical atomic mass of magnesium.

Section 42.3 The Strong Force

17. Use the potential-energy diagram in Figure 42.8 to estimate the strength of the strong force between two nucleons separated by 1.5 fm.

18. Use the potential-energy diagram in Figure 42.8 to sketch an approximate graph of the strong force between two nucleons versus the distance r between their centers.

19. What is the ratio of the gravitational potential energy to the nuclear potential energy for two neutrons separated by 1.0 fm?

Section 42.4 The Shell Model

20. a. Draw energy-level diagrams, similar to Figure 42.11, for all $A = 10$ nuclei listed in Appendix C. Show all the occupied neutron and proton levels.
    b. Which of these nuclei are stable? What is the decay mode of any that are radioactive?

21. a. Draw energy-level diagrams, similar to Figure 42.11, for all $A = 14$ nuclei listed in Appendix C. Show all the occupied neutron and proton levels.
    b. Which of these nuclei are stable? What is the decay mode of any that are radioactive?

Section 42.5 Radiation and Radioactivity

22. The barium isotope ^{133}Ba has a half-life of 10.5 years. A sample begins with 1.0×10^{10} ^{133}Ba atoms. How many are left after (a) 2 years, (b) 20 years, and (c) 200 years?

23. The cadmium isotope ^{109}Cd has a half-life of 462 days. A sample begins with 1.0×10^{12} ^{109}Cd atoms. How many are left after (a) 50 days, (b) 500 days, and (c) 5000 days?

24. The radioactive hydrogen isotope ^{3}H is called *tritium*.
    a. What are the decay mode and the daughter nucleus of tritium?
    b. What are the time constant and the decay rate of tritium?

25. How many half-lives must elapse until (a) 90% and (b) 99% of a radioactive sample of atoms has decayed?

26. What is the age in years of a bone in which the ^{14}C/^{12}C ratio is measured to be 1.65×10^{-13}?

27. What is the half-life in days of a radioactive sample with 5.0×10^{15} atoms and an activity of 5.0×10^{8} Bq?

Section 42.6 Nuclear Decay Mechanisms

28. Identify the unknown isotope X in the following decays.
    a. ^{234}U $\rightarrow$ X $+ \alpha$.
    b. ^{32}P $\rightarrow$ X $+ e^- + \bar{\nu}$
    c. X $\rightarrow$ ^{30}Si $+ e^+ + \nu$
    d. ^{24}Na $\rightarrow$ ^{24}Mg $+ e^- + \bar{\nu} \rightarrow$ X $+ \gamma$

29. Identify the unknown isotope X in the following decays.
    a. X $\rightarrow$ ^{224}Ra $+ \alpha$
    b. X $\rightarrow$ ^{207}Pb $+ e^- + \bar{\nu}$
    c. ^{7}Be $+ e^- \rightarrow$ X $+ \nu$
    d. X $\rightarrow$ ^{60}Ni $+ \gamma$

30. What is the energy (in MeV) released in the alpha decay of ^{239}Pu?

31. What is the energy (in MeV) released in the alpha decay of ^{228}Th?

32. What is the total energy (in MeV) released in the beta-minus decay of ^{3}H?
    Hint: The daughter $^A Y_{Z-1}$ is a positive ion. Tabulated masses are for neutral atom.

33. What is the total energy (in MeV) released in the beta-minus decay of ^{19}O? See the hint for Problem 32.

34. What is the total energy (in MeV) released in the beta decay of a neutron?

Section 42.7 Biological Applications of Nuclear Physics

35. A 50 kg laboratory worker is exposed to 20 mJ of beta radiation with RBE = 1.5. What is the dose in mrem?

36. How many rad of gamma-ray photons cause the same biological damage as 30 rad of alpha radiation with an RBE of 15?

37. 150 rad of gamma radiation are directed into a 150 g tumor during radiation therapy. How much energy does the tumor absorb?

38. The doctors planning a radiation therapy treatment have determined that a 100 g tumor needs to receive 0.20 J of gamma radiation. What is the dose in rads?

Problems

39. a. What initial speed must an alpha particle have to just touch the surface of a ^{197}Au gold nucleus before being turned back?

    b. What is the initial energy (in MeV) of the alpha particle?
    Hint: The alpha particle is not a point particle.

40. Particle accelerators fire protons at target nuclei for investigators to study the nuclear reactions that occur. In one experiment, the proton needs to have 20 MeV of kinetic energy as it impacts a ^{207}Pb nucleus. With what initial kinetic energy (in MeV) must the proton be fired toward the lead target?
    Hint: The proton is not a point particle.

41. Stars are powered by nuclear reactions that fuse hydrogen into helium. The fate of many stars, once most of the hydrogen is used up, is to collapse, under gravitational pull, into a *neutron star*. The force of gravity becomes so large that protons and electrons are fused into neutrons in the reaction $p^+ + e^- \rightarrow n + \nu$. The entire star is then a tightly packed ball of neutrons with the density of nuclear matter.
    a. Suppose the sun collapses into a neutron star. What will its radius be? Give your answer in km.
    b. The sun's rotation period is now 27 days. What will its rotation period be after it collapses?
    Rapidly rotating neutron stars emit pulses of radio waves at the rotation frequency and are known as *pulsars*.

42. The chemical atomic mass of hydrogen, with the two stable isotopes ^{1}H and ^{2}H (deuterium), is 1.00798. Use this value to determine the natural abundance of these two isotopes.

43. You learned in Chapter 41 that the binding energy of the electron in a hydrogen atom is 13.6 eV.
    a. By how much does the mass decrease when a hydrogen atom is formed from a proton and an electron? Give your answer both in atomic mass units and as a percentage of the mass of the hydrogen atom.
    b. By how much does the mass decrease when a helium nucleus is formed from two protons and two neutrons? Give your answer both in atomic mass units and as a percentage of the mass of the helium nucleus.
    c. Compare your answers to parts a and b. Why do you hear it said that mass is "lost" in nuclear reactions but not in chemical reactions?

44. Use the graph of binding energy to estimate the total energy released if a nucleus with mass number 240 fissions into two nuclei with mass number 120.

45. Use the graph of binding energy to estimate the total energy released if three ^{4}He nuclei fuse together to form a ^{12}C nucleus.

46. Could a ^{56}Fe nucleus fission into two ^{28}Al nuclei? Your answer, which should include some calculations, should be based on the curve of binding energy.

47. a. What are the isotopic symbols of all $A = 17$ isobars?
    b. Which of these are stable nuclei?
    c. For those that are not stable, identify both the decay mode and the daughter nucleus.

48. a. What are the isotopic symbols of all $A = 19$ isobars?
    b. Which of these are stable nuclei?
    c. For those that are not stable, identify both the decay mode and the daughter nucleus.

49. Derive Equation 42.19 from Equation 42.15.

50. What energy (in MeV) alpha particle has a de Broglie wavelength equal to the diameter of a ^{238}U nucleus?

51. What is the activity in Bq and in Ci of a 2.0 mg sample of ^{3}H?

52. The activity of an ^{39}Ar sample is 1.25×10^9 Bq. What is the mass of the sample?

53. The activity of a sample of the cesium isotope ^{137}Cs is 2.0×10^8 Bq. Many years later, after the sample has fully decayed, how many beta particles will have been emitted?

54. A 115 mCi radioactive tracer is made in a nuclear reactor. When it is delivered to a hospital 16 hours later its activity is 95 mCi. The lowest usable level of activity is 10 mCi.
    a. What is the tracer's half-life?
    b. For how long after delivery is the sample usable?

55. The radium isotope ^{223}Ra, an alpha emitter, has a half-life of 11.43 days. You happen to have a 1.0 g cube of ^{223}Ra, so you decide to use it to boil water for tea. You fill a well-insulated container with 100 mL of water at 18° C and drop in the cube of radium.
    a. How long will it take the water to boil?
    b. Will the water have been altered in any way by this method of boiling? If so, how?

56. A sample of 1.0×10^{10} atoms that decay by alpha emission has a half-life of 100 min. How many alpha particles are emitted between $t = 50$ min and $t = 200$ min?

57. A sample contains radioactive atoms of two types, A and B. Initially there are five times as many A atoms as there are B atoms. Two hours later, the numbers of the two atoms are equal. The half-life of A is 0.50 hours. What is the half-life of B?

58. Radioactive isotopes often occur together in mixtures. Suppose a 100 g sample contains ^{131}Ba, with a half-life of 12 days, and ^{47}Ca, with a half-life of 4.5 days. If there are initially twice as many calcium atoms as there are barium atoms, what will be the ratio of calcium atoms to barium atoms 2.5 weeks later?

59. The technique known as potassium-argon dating is used to date old lava flows. The potassium isotope ^{40}K has a 1.28 billion year half-life and is naturally present at very low levels. ^{40}K decays by beta emission into ^{40}Ar. Argon is a gas, and there is no argon in flowing lava because the gas escapes. Once the lava solidifies, any argon produced in the decay of ^{40}K is trapped inside and cannot escape. A geologist brings you a piece of solidified lava in which you find the ^{40}Ar/^{40}K ratio to be 0.12. What is the age of the rock?

60. The half-life of the uranium isotope ^{235}U is 700 million years. The earth is approximately 4.5 billion years old. How much more ^{235}U was there when the earth formed than there is today? Give your answer as the then-to-now ratio.

61. A 75 kg patient swallows a 30 μCi beta emitter that is to be used as a tracer. The isotope's half-life is 5.0 days. The average energy of the beta particles is 0.35 MeV with an RBE of 1.5. Ninety percent of the beta particles are absorbed within the patient's body and 10% escape. What total dose (in mrem) does the patient receive?

62. What dose in rads of gamma radiation must be absorbed by a block of ice at 0°C to transform the entire block to liquid water at 0°C?

63. A chest x ray uses 10 keV photons with an RBE of 0.85. A 60 kg person receives a 30 mrem dose from one x ray that exposes 25% of the patient's body. How many x ray photons are absorbed in the patient's body?

64. The rate at which a radioactive tracer is lost from a patient's body is the rate at which the isotope decays *plus* the rate at which the element is excreted from the body. Medical experiments have shown that stable isotopes of a particular element are excreted with a 6.0 day half-life. A radioactive isotope of

the same element has a half-life of 9.0 days. What is the effective half-life of the isotope in a patient's body?

65. The plutonium isotope ^{239}Pu has a half-life of 24,000 years and decays by the emission of a 5.2 MeV alpha particle. Plutonium is not especially dangerous if handled because the activity is low and the alpha radiation doesn't penetrate the skin. However, there are serious health concerns if even the tiniest speck of plutonium is inhaled and lodges deep in the lungs. This could happen following any kind of fire or explosion that disperses plutonium as dust. Let's determine the level of danger.
    a. Soot particles are roughly 1 μm in diameter, and it is known that these particles can go deep into the lungs. How many atoms are in a 1.0-μm-diameter particle of ^{239}Pu? The density of plutonium is 19,800 kg/m^3.
    b. What is the activity, in Bq, of a 1.0-μm-diameter particle?
    c. The activity of the particle is very small, but the penetrating power of alpha particles is also very small. The alpha particles are all stopped, and each deposits its energy in a 50-μm-diameter sphere around the particle. What is the dose, in rem/year, to this small sphere of tissue in the lungs? Use an average RBE of 15 and assume that the tissue density is that of water.
    d. Is this exposure likely to be significant? How does it compare to the natural background of radiation exposure?

Challenge Problems

66. The uranium isotope ^{238}U is naturally present at low levels in many soils. One of the nuclei in the decay series of ^{238}U is the radon isotope ^{222}Rn with $t_{1/2} = 3.82$ days. Radon is a gas, and it tends to seep from soil into basements. The Environmental Protection Agency recommends that homeowners take steps to remove radon, by pumping in fresh air, if the radon activity exceeds 4 pCi per liter of air.
    a. How many ^{222}Rn atoms are there in 1 m^3 of air if the activity is 4 pCi/L?
    b. The range of alpha particles in air is $\approx$3 cm. Suppose we model a person as a 180-cm-tall, 25-cm-diameter cylinder with a mass of 65 kg. Only decays within 3 cm of the cylinder can cause exposure, and only $\approx$50% of the decays direct the alpha particle toward the person. Determine the dose in mrem per year for a person who spends the entire year in a room where the activity is 4 pCi/L. Assume an average RBE of 15.
    c. Does the EPA recommendation seem appropriate? Why or why not?

67. Estimate the stopping distance in air of a 5.0 MeV alpha particle. Assume that the particle loses an average 30 eV per collision.

68. Beta-plus decay is $^{A}X_Z \rightarrow {}^{A}Y_{Z-1} + e^+ + \nu$.
    a. Determine the mass threshold for beta-plus decay. That is, what is the minimum atomic mass m_X for which this decay is energetically possible? Your answer will be in terms of the atomic mass m_Y and the electron mass m_e.
    Hint: Start with the nuclear masses, then add an equal number of electrons to both sides of the reaction to get atomic masses.
    b. Can ^{13}N undergo beta-plus decay into ^{13}C? If so, how much energy is released in the decay?

69. All the very heavy atoms found in the earth were created long ago by nuclear fusion reactions in a supernova, an exploding star. The debris spewed out by the supernova later coalesced into the gases from which the sun and the planets of our solar system were formed. Nuclear physics suggests that the uranium isotopes ^{235}U and ^{238}U should have been created in roughly equal numbers. Today, 99.28% of uranium is ^{238}U and only 0.72% is ^{235}U. How long ago did the supernova occur?

70. It might seem strange that in beta decay the positive proton, which is repelled by the positive nucleus, remains in the nucleus while the negative electron, which is attracted to the nucleus, is ejected. To understand beta decay, let's analyze the decay of a free neutron that is at rest in the laboratory. We'll ignore the antineutrino and consider the decay n → p$^+$ + e$^-$. The analysis requires the use of relativistic energy and momentum, from Chapter 36.
   a. What is the total kinetic energy, in MeV, of the proton and electron?
   b. Write the equation that expresses the conservation of relativistic energy for this decay. Your equation will be in terms of the three masses m_n, m_p, and m_e and the relativistic factors γ_p and γ_e. Then rearrange your equation to get the mass loss $\Delta m = m_n - m_p - m_e$ on one side.
   c. Write the equation that expresses the conservation of relativistic momentum for this decay. Let v represent speed, rather than velocity, then write any minus signs explicitly.
   d. You have two simultaneous equations in the two unknowns v_p and v_e. To help in solving these, first prove that $\gamma v = (\gamma^2 - 1)^{1/2}c$.
   e. Solve for v_p and v_e. (It's easiest to solve for γ_p and γ_e, then find v from γ.) First get an algebraic expression for each, in terms of the masses. Then evaluate each, giving v as a fraction of c.
   f. Calculate the kinetic energy in MeV of the proton and the electron. Verify that their sum matches your answer to part a.
   g. Now explain why the electron is ejected in beta decay while the proton remains in the nucleus.

71. Alpha decay occurs when an alpha particle tunnels through the Coulomb barrier. Figure CP42.71 shows a simple one-dimensional model of the potential-energy well of an alpha particle in a nucleus with $A \approx 235$. The 15 fm width of this one-dimensional potential-energy well is the *diameter* of the nucleus. Further, to keep the model simple, the Coulomb barrier has been modeled as a 20-fm-wide, 30-MeV-high rectangular potential-energy barrier. The goal of this problem is to calculate the half-life of an alpha particle in the energy level $E = 5.0$ MeV.
   a. What is the kinetic energy of the alpha particle while inside the nucleus? What is its kinetic energy after it escapes from the nucleus?
   b. Consider the alpha particle within the nucleus to be a point particle bouncing back and forth with the kinetic energy you stated in part a. What is the particle's *collision rate*, the number of times per second it collides with the wall of the potential-energy barrier?
   c. What is the tunneling probability P_{tunnel}?
   d. P_{tunnel} is the probability that on any one collision with the wall the alpha particle tunnels through instead of reflecting. The probability of *not* tunneling is $1 - P_{tunnel}$. Hence the probability that the alpha particle is still inside the nucleus after N collisions is $(1 - P_{tunnel})^N \approx 1 - NP_{tunnel}$, where we've used the binomial approximation because $P_{tunnel} \ll 1$. The half-life is the *time* at which half the nuclei have not yet decayed. Use this information to determine (in years) the half-life of this nucleus.

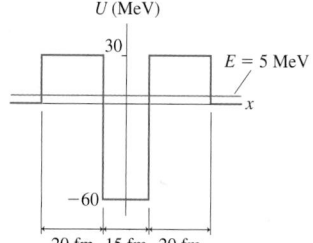

FIGURE CP42.71

STOP TO THINK ANSWERS

Stop to Think 42.1: 3. Different isotopes of an element have different numbers of neutrons but the same number of protons. The number of electrons in a neutral atom matches the number of protons.

Stop to Think 42.2: d. To keep A constant, increasing N by 1 requires decreasing Z by 1.

Stop to Think 42.3: No. A Geiger counter responds only to ionizing radiation. Visible light is not ionizing radiation.

Stop to Think 42.4: c. One-quarter of the atoms are left. This is one-half of one-half, or $(1/2)^2$.

Stop to Think 42.5: b. An increase of Z with no change in A occurs when a neutron changes to a proton and an electron, ejecting the electron.

Relativity and Quantum Physics

Niels Bohr was right on target with his remark, "Anyone who is not shocked by quantum theory has not understood it." Quantum mechanics *is* shocking. The predictability of Newtonian physics has been replaced by a mysterious world in which physical entities that by all rights should be waves sometimes act like particles. Electrons and neutrons somehow produce wave-like interference with themselves. These discoveries stood common sense on its head.

According to quantum mechanics, the wave function and its associated probabilities are *all we can know* about an atomic particle. This idea is so unsettling that many great scientists were reluctant to accept it. Einstein famously said, "God does not play dice with the universe." But Einstein was wrong. As strange as it seems, this is the way that nature really is.

As we conclude our journey into physics, the knowledge structure for Part VII summarizes the important ideas of relativity and quantum physics. Whether you're shocked or not, these are the scientific theories behind the emerging technologies of the 21st century.

KNOWLEDGE STRUCTURE VII Relativity and Quantum Physics

ESSENTIAL CONCEPTS	Reference frame, event, atom, photon, quantization, wave function, probability density
BASIC GOALS	What are the properties and characteristics of space and time?
	How do we know about light and atoms?
	How are atomic and nuclear phenomena explained by energy levels, wave functions, and photons?

GENERAL PRINCIPLES	**Principle of relativity**	All the laws of physics are the same in all inertial reference frames.
	Schrödinger's equation	$\dfrac{d^2\psi}{dx^2} = -\dfrac{2m}{\hbar^2}[E - U(x)]\psi(x)$
	Pauli exclusion principle	No more than one electron can occupy the same quantum state.
	Uncertainty principle	$\Delta x \Delta p \geq h/2$

RELATIVITY It follows from the principle of relativity that:

- The speed of light c is the same in all inertial reference frames. No particle or causal influence can travel faster than c.

- Length contraction: The length of an object in a reference frame in which the object moves with speed v is

$$L = \sqrt{1 - \beta^2}\,\ell \leq \ell$$

where ℓ is the proper length and $\beta = v/c$.

- Time dilation: The proper time interval $\Delta\tau$ between two events is measured in a reference frame in which the two events occur at the same position. The time interval Δt in a frame moving with relative speed v is

$$\Delta t = \Delta\tau/\sqrt{1 - \beta^2} \geq \Delta\tau$$

- $E = mc^2$ is the energy equivalent of mass. Mass can be transformed into energy and energy into mass.

QUANTUM PHYSICS Quantum systems are described by a wave function $\psi(x)$.

- The probability that a particle will be found in the narrow interval δx at position x is Prob(in δx at x) = $P(x)\,\delta x$. The probability density is $P(x) = |\psi(x)|^2$.

- The wave function must be normalized

$$\int_{-\infty}^{\infty} |\psi(x)|^2\,dx = 1$$

- The wave function can penetrate into a classically forbidden region with penetration distance

$$\eta = \hbar/\sqrt{2m(U_0 - E)}$$

- A particle can tunnel through an energy barrier of height U_0 and width w with probability $P_{\text{tunnel}} = e^{-2w/\eta}$

Properties of light

- A photon of light of frequency f has energy $E_{\text{photon}} = hf$.
- Photons are emitted and absorbed on an all-or-nothing basis.

Properties of atoms

- Quantized energy levels, found by solving the Schrödinger equation, depend on quantum numbers n and l.

- An atom can jump from one state to another by emitting or absorbing a photon of energy $E_{\text{photon}} = \Delta E_{\text{atom}}$

- The ground-state electron configuration is the lowest-energy configuration consistent with the Pauli principle.

Properties of nuclei

- The nucleus is held together by the strong force, an attractive short-range force between any two nucleons.

- Nuclei are stable only if the proton and neutron number fall along the line of stability.

- Unstable nuclei decay by alpha, beta, or gamma decay. The number of nuclei decreases exponentially with time.

Quantum Computers

All the systems we studied in Part VII were in a single, well-defined quantum state. For example, a hydrogen atom was in the $1s$ state or, perhaps, the $2p$ state. But there's another possibility. Some quantum systems can exist in a *superposition* of two or more quantum states.

We hinted at the possibility of superposition when we re-examined the double-slit interference experiment in the light of quantum physics. We noted that a photon or electron must, in some sense, go through both slits and then interfere with itself to produce the dot-by-dot buildup of an interference pattern on the screen. Suppose we say that an electron that has passed through the top slit in the figure is in quantum state ψ_a. An electron that has passed through the bottom slit is in state ψ_b.

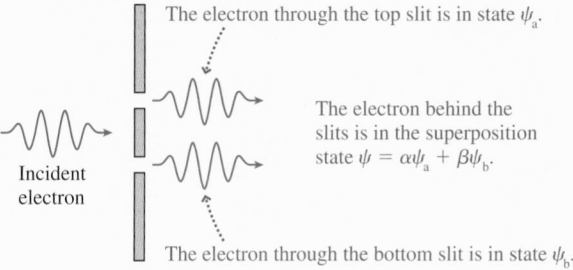

The electron through the top slit is in state ψ_a.

Incident electron

The electron behind the slits is in the superposition state $\psi = \alpha\psi_a + \beta\psi_b$.

The electron through the bottom slit is in state ψ_b.

FIGURE PSVII.1 The electron emerging from the double slit is in a superposition state.

To say that the electron goes through both slits is to say that the electron emerges from the double slit in the *superposition state* $\psi = \alpha\psi_a + \beta\psi_b$, where the coefficients α and β must satisfy $\alpha^2 + \beta^2 = 1$. (Notice that this is like finding the magnitude of a vector from its components.) If we were to detect the electron, α^2 and β^2 are the probabilities that we would find it to be in state ψ_a or state ψ_b, respectively. But until we detect it, the electron exists in the superposition of *both* state ψ_a and ψ_b. It is this superposition that allows the electron to interfere with itself to produce the interference pattern.

But what does this have to do with computers? As you know, everything a modern digital computer does, from surfing the Internet to crunching numbers, is accomplished by manipulating binary strings of 0s and 1s. The *concept* of computing with binary bits goes back to Charles Babbage in the mid-19th century, but it wasn't until the mid-20th century that scientists and engineers developed the technology that gives this concept a physical representation.

A binary bit is always a 1 or a 0; there's no in-between state. These are represented in a modern microprocessor

by small capacitors that are either charged or uncharged. Suppose we wanted to represent information not with capacitors but with a quantum system that has two states. We could say that the system represents a 0 when it is in state ψ_a and a 1 when it is in ψ_b. Such a quantum system is an ordinary binary bit as long as the system is in one state or the other.

But the quantum system, unlike a classical bit, has the possibility of being in a superposition state. Using 0 and 1, rather than ψ_a and ψ_b, we could say that the system can be the state $\psi = \alpha \cdot 0 + \beta \cdot 1$. This basic unit of quantum computing is called a *qubit*. It may seem at first that we could do the same thing with a classical system by allowing the capacitor charge to vary, but a partially charged capacitor is still a single, well-defined state. In contrast, the qubit—like the electron that goes through both slits— is simultaneously in both state 0 *and* state 1.

To illustrate the possibilities, suppose you have three classical bits and three qubits. The three bits can represent eight different numbers (000 to 111), but only one at a time. The three qubits represent all eight numbers *simultaneously*. To perform a mathematical operation, you must do it eight times on the three bits to learn all the possible outcomes. But you would learn all eight outcomes simultaneously from *one* operation on the three qubits. In general, computing with n qubits provides a theoretical improvement of 2^n over computing with n bits.

We say "theoretical" because quantum computing is still mostly in the concept stage, much as digital computers were 150 years ago. What kind of quantum systems can actually be placed in an appropriate superposition state? How do you manipulate qubits? How do you read information in and out? What kinds of computations would be improved by quantum computing?

These are all questions that are being actively researched today. Quantum computing is in its infancy, and the technology for making a real quantum computer is largely unknown. Just as Charles Babbage couldn't possibly have imagined today's computers, the uses of tomorrow's quantum computers are still unforeseen. But, quite possibly, there are uses that some of you may help to invent.

FIGURE PSVII.2 This string of beryllium ions held in an ion trap is being studied as a possible quantum computer. The quantum states of the ions are manipulated with laser beams.

Mathematics Review

Algebra

Using exponents:

$$a^{-x} = \frac{1}{a^x} \qquad a^x a^y = a^{(x+y)} \qquad \frac{a^x}{a^y} = a^{(x-y)} \qquad (a^x)^y = a^{xy}$$

$$a^0 = 1 \qquad a^1 = a \qquad a^{1/n} = \sqrt[n]{a}$$

Fractions:

$$\left(\frac{a}{b}\right)\left(\frac{c}{d}\right) = \frac{ac}{bd} \qquad \frac{a/b}{c/d} = \frac{ad}{bc} \qquad \frac{1}{1/a} = a$$

Logarithms:

If $a = e^x$, then $\ln(a) = x$ $\qquad \ln(e^x) = x \qquad\qquad e^{\ln(x)} = x$

$$\ln(ab) = \ln(a) + \ln(b) \qquad \ln\left(\frac{a}{b}\right) = \ln(a) - \ln(b) \qquad \ln(a^n) = n\ln(a)$$

The expression $\ln(a + b)$ cannot be simplified.

Linear equations: The graph of the equation $y = ax + b$ is a straight line. a is the slope of the graph. b is the y-intercept.

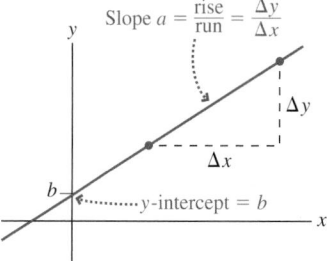

Proportionality: To say that y is proportional to x, written $y \propto x$, means that $y = ax$, where a is a constant. Proportionality is a special case of linearity. A graph of a proportional relationship is a straight line that passes through the origin. If $y \propto x$, then

$$\frac{y_1}{y_2} = \frac{x_1}{x_2}$$

Quadratic equation: The quadratic equation $ax^2 + bx + c = 0$ has the two solutions $x = \dfrac{-b \pm \sqrt{b^2 - 4ac}}{2a}$.

Geometry and Trigonometry

Area and volume:

Rectangle
$A = ab$

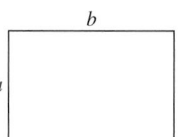

Rectangular box
$V = abc$

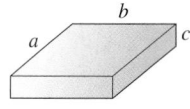

Triangle
$A = \frac{1}{2}ab$

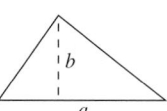

Right circular cylinder
$V = \pi r^2 l$

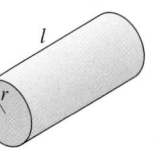

Circle
$C = 2\pi r$
$A = \pi r^2$

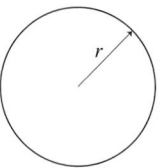

Sphere
$A = 4\pi r^2$
$V = \frac{4}{3}\pi r^3$

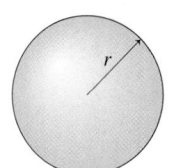

Arc length and angle: The angle θ in radians is defined as $\theta = s/r$.

The arc length that spans angle θ is $s = r\theta$.

2π rad $= 360°$

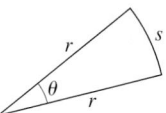

Right triangle: Pythagorean theorem $\quad c = \sqrt{a^2 + b^2}$ or $a^2 + b^2 = c^2$

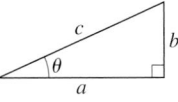

$$\sin\theta = \frac{b}{c} = \frac{\text{far side}}{\text{hypotenuse}} \qquad\qquad \theta = \sin^{-1}\left(\frac{b}{c}\right)$$

$$\cos\theta = \frac{a}{c} = \frac{\text{adjacent side}}{\text{hypotenuse}} \qquad\qquad \theta = \cos^{-1}\left(\frac{a}{c}\right)$$

$$\tan\theta = \frac{b}{a} = \frac{\text{far side}}{\text{adjacent side}} \qquad\qquad \theta = \tan^{-1}\left(\frac{b}{a}\right)$$

General triangle: $\alpha + \beta + \gamma = 180° = \pi$ rad

Law of cosines $c^2 = a^2 + b^2 - 2ab\cos\gamma$

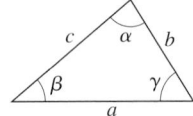

Identities:

$$\tan\alpha = \frac{\sin\alpha}{\cos\alpha} \qquad\qquad\qquad \sin^2\alpha + \cos^2\alpha = 1$$

$$\sin(-\alpha) = -\sin\alpha \qquad\qquad \cos(-\alpha) = \cos\alpha$$

$$\sin(\alpha \pm \beta) = \sin\alpha\cos\beta \pm \cos\alpha\sin\beta \qquad \cos(\alpha \pm \beta) = \cos\alpha\cos\beta \mp \sin\alpha\sin\beta$$

$$\sin(2\alpha) = 2\sin\alpha\cos\alpha \qquad\qquad \cos(2\alpha) = \cos^2\alpha - \sin^2\alpha$$

$$\sin(\alpha \pm \pi/2) = \pm\cos\alpha \qquad\qquad \cos(\alpha \pm \pi/2) = \mp\sin\alpha$$

$$\sin(\alpha \pm \pi) = -\sin\alpha \qquad\qquad \cos(\alpha \pm \pi) = -\cos\alpha$$

Expansions and Approximations

Binomial expansion: $(1 + x)^n = 1 + nx + \dfrac{n(n-1)}{2}x^2 + \ldots$

Binomial approximation: $(1 + x)^n \approx 1 + nx \quad$ if $\quad x \ll 1$

Trigonometric expansions: $\sin\alpha = \alpha - \dfrac{\alpha^3}{3!} + \dfrac{\alpha^5}{5!} - \dfrac{\alpha^7}{7!} + \ldots$ for α in rad

$\cos\alpha = 1 - \dfrac{\alpha^2}{2!} + \dfrac{\alpha^4}{4!} - \dfrac{\alpha^6}{6!} + \ldots$ for α in rad

Small-angle approximation: If $\alpha \ll 1$ rad, then $\sin\alpha \approx \tan\alpha \approx \alpha$ and $\cos\alpha \approx 1$.

The small-angle approximation is excellent for $\alpha < 5°$ (≈ 0.1 rad) and generally acceptable up to $\alpha \approx 10°$.

Periodic Table of Elements

Legend: 27 Co 58.9 — Atomic number / Symbol / Atomic mass

Transition elements · **Inner transition elements**

Period																	
1	1 H 1.0																2 He 4.0
2	3 Li 6.9	4 Be 9.0									5 B 10.8	6 C 12.0	7 N 14.0	8 O 16.0	9 F 19.0	10 Ne 20.2
3	11 Na 23.0	12 Mg 24.3									13 Al 27.0	14 Si 28.1	15 P 31.0	16 S 32.1	17 Cl 35.5	18 Ar 39.9

Period 4
19 K 39.1 | 20 Ca 40.1 | 21 Sc 45.0 | 22 Ti 47.9 | 23 V 50.9 | 24 Cr 52.0 | 25 Mn 54.9 | 26 Fe 55.8 | 27 Co 58.9 | 28 Ni 58.7 | 29 Cu 63.5 | 30 Zn 65.4 | 31 Ga 69.7 | 32 Ge 72.6 | 33 As 74.9 | 34 Se 79.0 | 35 Br 79.9 | 36 Kr 83.8

Period 5
37 Rb 85.5 | 38 Sr 87.6 | 39 Y 88.9 | 40 Zr 91.2 | 41 Nb 92.9 | 42 Mo 95.9 | 43 Tc 96.9 | 44 Ru 101.1 | 45 Rh 102.9 | 46 Pd 106.4 | 47 Ag 107.9 | 48 Cd 112.4 | 49 In 114.8 | 50 Sn 118.7 | 51 Sb 121.8 | 52 Te 127.6 | 53 I 126.9 | 54 Xe 131.3

Period 6
55 Cs 132.9 | 56 Ba 137.3 | 57 La 138.9 | 72 Hf 178.5 | 73 Ta 180.9 | 74 W 183.9 | 75 Re 186.2 | 76 Os 190.2 | 77 Ir 192.2 | 78 Pt 195.1 | 79 Au 197.0 | 80 Hg 200.6 | 81 Tl 204.4 | 82 Pb 207.2 | 83 Bi 209.0 | 84 Po 209.0 | 85 At 210.0 | 86 Rn 222.0

Period 7
87 Fr 223.0 | 88 Ra 226.0 | 89 Ac 227.0 | 104 Rf 261 | 105 Db 262 | 106 Sg 263 | 107 Bh 264 | 108 Hs 269 | 109 Mt 268 | 110 Ds 271 | 111 272 | 112 285

Lanthanides (6)
58 Ce 140.1 | 59 Pr 140.9 | 60 Nd 144.2 | 61 Pm 144.9 | 62 Sm 150.4 | 63 Eu 152.0 | 64 Gd 157.3 | 65 Tb 158.9 | 66 Dy 162.5 | 67 Ho 164.9 | 68 Er 167.3 | 69 Tm 168.9 | 70 Yb 173.0 | 71 Lu 175.0

Actinides (7)
90 Th 232.0 | 91 Pa 231.0 | 92 U 238.0 | 93 Np 237.0 | 94 Pu 239.1 | 95 Am 241.1 | 96 Cm 244.1 | 97 Bk 249.1 | 98 Cf 252.1 | 99 Es 257.1 | 100 Fm 257.1 | 101 Md 258.1 | 102 No 259.1 | 103 Lr 262.1

Atomic and Nuclear Data

Atomic Number (Z)	Element	Symbol	Mass Number (A)	Atomic Mass (u)	Percent Abundance	Decay Mode	Half-Life $t_{1/2}$
0	(Neutron)	n	1	1.008 665		β^-	10.4 min
1	Hydrogen	H	1	1.007 825	99.985	stable	
	Deuterium	D	2	2.014 102	0.015	stable	
	Tritium	T	3	3.016 049		β^-	12.33 yr
2	Helium	He	3	3.016 029	0.000 1	stable	
			4	4.002 602	99.999 9	stable	
			6	6.018 886		β^-	0.81 s
3	Lithium	Li	6	6.015 121	7.50	stable	
			7	7.016 003	92.50	stable	
			8	8.022 486		β^-	0.84 s
4	Beryllium	Be	7	7.016 928		EC	53.3 days
			9	9.012 174	100	stable	
			10	10.013 534		β^-	1.5×10^6 yr
5	Boron	B	10	10.012 936	19.90	stable	
			11	11.009 305	80.10	stable	
			12	12.014 352		β^-	0.020 2 s
6	Carbon	C	10	10.016 854		β^+	19.3 s
			11	11.011 433		β^+	20.4 min
			12	12.000 000	98.90	stable	
			13	13.003 355	1.10	stable	
			14	14.003 242		β^-	5 730 yr
			15	15.010 599		β^-	2.45 s
7	Nitrogen	N	12	12.018 613		β^+	0.011 0 s
			13	13.005 738		β^+	9.96 min
			14	14.003 074	99.63	stable	
			15	15.000 108	0.37	stable	
			16	16.006 100		β^-	7.13 s
			17	17.008 450		β^-	4.17 s
8	Oxygen	O	14	14.008 595		EC	70.6 s
			15	15.003 065		β^+	122 s
			16	15.994 915	99.76	stable	
			17	16.999 132	0.04	stable	
			18	17.999 160	0.20	stable	
			19	19.003 577		β^-	26.9 s
9	Fluorine	F	17	17.002 094		EC	64.5 s
			18	18.000 937		β^+	109.8 min
			19	18.998 404	100	stable	
			20	19.999 982		β^-	11.0 s
10	Neon	Ne	19	19.001 880		β^+	17.2 s
			20	19.992 435	90.48	stable	
			21	20.993 841	0.27	stable	
			22	21.991 383	9.25	stable	

Atomic Number (Z)	Element	Symbol	Mass Number (A)	Atomic Mass (u)	Percent Abundance	Decay Mode	Half-Life $t_{1/2}$
11	Sodium	Na	22	21.994 434		β^+	2.61 yr
			23	22.989 770	100	stable	
			24	23.990 961		β^-	14.96 hr
12	Magnesium	Mg	24	23.985 042	78.99	stable	
			25	24.985 838	10.00	stable	
			26	25.982 594	11.01	stable	
13	Aluminum	Al	27	26.981 538	100	stable	
			28	27.981 910		β^-	2.24 min
14	Silicon	Si	28	27.976 927	92.23	stable	
			29	28.976 495	4.67	stable	
			30	29.973 770	3.10	stable	
			31	30.975 362		β^-	2.62 hr
15	Phosphorus	P	30	29.978 307		β^+	2.50 min
			31	30.973 762	100	stable	
			32	31.973 908		β^-	14.26 days
16	Sulfur	S	32	31.972 071	95.02	stable	
			33	32.971 459	0.75	stable	
			34	33.967 867	4.21	stable	
			35	34.969 033		β^-	87.5 days
17	Chlorine	Cl	35	34.968 853	75.77	stable	
			36	35.968 307		β^-	3.0×10^5 yr
			37	36.965 903	24.23	stable	
18	Argon	Ar	36	35.967 547	0.34	stable	
			38	37.962 732	0.06	stable	
			39	38.964 314		β^-	269 yr
			40	39.962 384	99.60	stable	
			42	41.963 049		β^-	33 yr
19	Potassium	K	39	38.963 708	93.26	stable	
			40	39.964 000	0.01	β^+	1.28×10^9 yr
			41	40.961 827	6.73	stable	
20	Calcium	Ca	40	39.962 591	96.94	stable	
			42	41.958 618	0.64	stable	
			43	42.958 767	0.13	stable	
			44	43.955 481	2.08	stable	
			47	46.954 547		β^-	4.5 days
			48	47.952 534	0.18	stable	
24	Chromium	Cr	50	49.946 047	4.34	stable	
			51	50.944 772		EC	27 days
			52	51.940 511	83.79	stable	
			53	52.940 652	9.50	stable	
			54	53.938 883	2.36	stable	
25	Manganese	Mn	55	54.938 048	100	stable	
26	Iron	Fe	54	54.939 613	5.9	stable	
			55	54.938 297		EC	2.7 yr
			56	55.934 940	91.72	stable	

Atomic Number (Z)	Element	Symbol	Mass Number (A)	Atomic Mass (u)	Percent Abundance	Decay Mode	Half-Life $t_{1/2}$
			57	56.935 396	2.1	stable	
			58	57.933 278	0.28	stable	
27	Cobalt	Co	59	58.933 198	100	stable	
			60	59.933 820		β^-	5.27 yr
28	Nickel	Ni	58	57.935 346	68.08	stable	
			60	59.930 789	26.22	stable	
			61	60.931 058	1.14	stable	
			62	61.928 346	3.63	stable	
			64	63.927 967	0.92	stable	
29	Copper	Cu	63	62.929 599	69.17	stable	
			65	64.927 791	30.83	stable	
30	Zinc	Zn	64	63.929 144	48.6	stable	
			66	65.926 035	27.9	stable	
			67	66.927 129	4.1	stable	
			68	67.924 845	18.8	stable	
47	Silver	Ag	107	106.905 091	51.84	stable	
			109	108.904 754	48.16	stable	
48	Cadmium	Cd	106	105.906 457	1.25	stable	
			109	108.904 984		EC	462 days
			110	109.903 004	12.49	stable	
			111	110.904 182	12.80	stable	
			112	111.902 760	24.13	stable	
			113	112.904 401	12.22	stable	
			114	113.903 359	28.73	stable	
			116	115.904 755	7.49	stable	
53	Iodine	I	127	126.904 474	100	stable	
			129	128.904 984		β^-	1.6×10^7 yr
			131	130.906 124		β^-	8 days
54	Xenon	Xe	128	127.903 531	1.9	stable	
			129	128.904 779	26.4	stable	
			130	129.903 509	4.1	stable	
			131	130.905 069	21.2	stable	
			132	131.904 141	26.9	stable	
			133	132.905 906		β^-	5.4 days
			134	133.905 394	10.4	stable	
			136	135.907 215	8.9	stable	
55	Cesium	Cs	133	132.905 436	100	stable	
			137	136.907 078		β^-	30 yr
56	Barium	Ba	131	130.906 931		EC	12 days
			133	132.905 990		EC	10.5 yr
			134	133.904 492	2.42	stable	
			135	134.905 671	6.59	stable	
			136	135.904 559	7.85	stable	
			137	136.905 816	11.23	stable	
			138	137.905 236	71.70	stable	

Atomic Number (Z)	Element	Symbol	Mass Number (A)	Atomic Mass (u)	Percent Abundance	Decay Mode	Half-Life $t_{1/2}$
79	Gold	Au	197	196.966 543	100	stable	
80	Mercury	Hg	198	197.966 743	9.97	stable	
			199	198.968 253	16.87	stable	
			200	199.968 299	23.10	stable	
			201	200.970 276	13.10	stable	
			202	201.970 617	29.86	stable	
			204	203.973 466	6.87	stable	
81	Thallium	Tl	203	202.972 320	29.524	stable	
			205	204.974 400	70.476	stable	
			207	206.977 403		β^-	4.77 min
82	Lead	Pb	204	203.973 020	1.4	stable	
			205	204.974 457		EC	1.5×10^7 yr
			206	205.974 440	24.1	stable	
			207	206.975 871	22.1	stable	
			208	207.976 627	52.4	stable	
			210	209.984 163		α, β^-	22.3 yr
			211	210.988 734		β^-	36.1 min
83	Bismuth	Bi	208	207.979 717		EC	3.7×10^5 yr
			209	208.980 374	100	stable	
			211	210.987 254		α	2.14 min
			215	215.001 836		β^-	7.4 min
84	Polonium	Po	209	208.982 405		α	102 yr
			210	209.982 848		α	138.38 days
			215	214.999 418		α	0.001 8 s
			218	218.008 965		α, β^-	3.10 min
85	Astatine	At	218	218.008 685		α, β^-	1.6 s
			219	219.011 294		α, β^-	0.9 min
86	Radon	Rn	219	219.009 477		α	3.96 s
			220	220.011 369		α	55.6 s
			222	222.017 571		α, β^-	3.823 days
87	Francium	Fr	223	223.019 733		α, β^-	22 min
88	Radium	Ra	223	223.018 499		α	11.43 days
			224	224.020 187		α	3.66 days
			226	226.025 402		α	1 600 yr
			228	228.031 064		β^-	5.75 yr
89	Actinium	Ac	227	227.027 749		α, β^-	21.77 yr
			228	228.031 015		β^-	6.15 hr
90	Thorium	Th	227	227.027 701		α	18.72 days
			228	228.028 716		α	1.913 yr
			229	229.031 757		α	7 300 yr
			230	230.033 127		α	75.000 yr
			231	231.036 299		α, β^-	25.52 hr
			232	232.038 051	100	α	1.40×10^{10} yr
			234	234.043 593		β^-	24.1 days

Atomic Number (Z)	Element	Symbol	Mass Number (A)	Atomic Mass (u)	Percent Abundance	Decay Mode	Half-Life $t_{1/2}$
91	Protactinium	Pa	231	231.035 880		α	32.760 yr
			234	234.043 300		β^-	6.7 hr
92	Uranium	U	233	233.039 630		α	1.59×10^5 yr
			234	234.040 946		α	2.45×10^5 yr
			235	235.043 924	0.72	α	7.04×10^8 yr
			236	236.045 562		α	2.34×10^7 yr
			238	238.050 784	99.28	α	4.47×10^9 yr
93	Neptunium	Np	236	236.046 560		EC	1.15×10^5 yr
			237	237.048 168		α	2.14×10^6 yr
94	Plutonium	Pu	238	238.049 555		α	87.7 yr
			239	239.052 157		α	2.412×10^4 yr
			240	240.053 808		α	6 560 yr
			242	242.058 737		α	3.73×10^6 yr
			244	244.064 200		α	8.1×10^7 yr

Answers

Answers to Odd-Numbered Exercises and Problems

Solutions to questions posed in the Part Overview captions can be found at the end of this answer list.

Chapter 1

1.
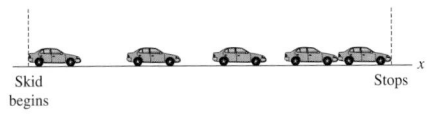

7.
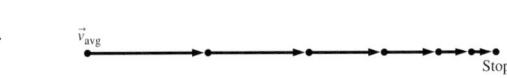

9. a.
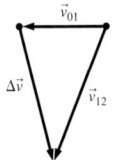
b. Greater

11.
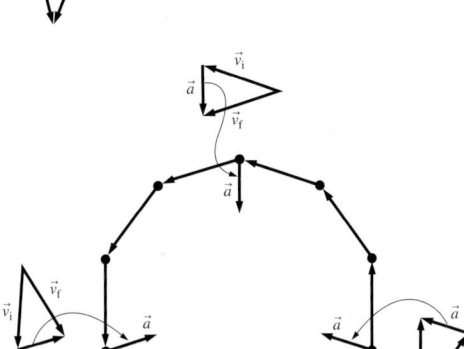

13.

15.
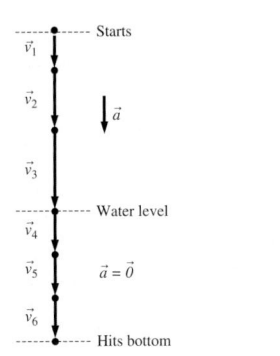

17.
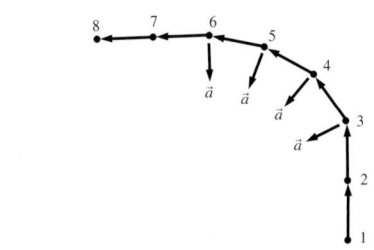

19.
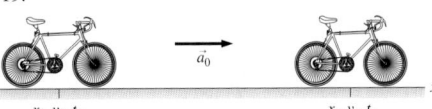

21. a. 9.12×10^{-6} s b. 3.42×10^3 m c. 440 m/s d. 22.2 m/s
23. a. 3.60×10^3 s b. 8.64×10^4 s c. 3.16×10^7 s d. 9.75 m/s^2
25. 6.40×10^3 m^2 and 8.25×10^3 m^2
27. a. 3 b. 3 c. 3 d. 2
29. a. 846 b. 7.9 c. 5.77 d. 13.1
35.
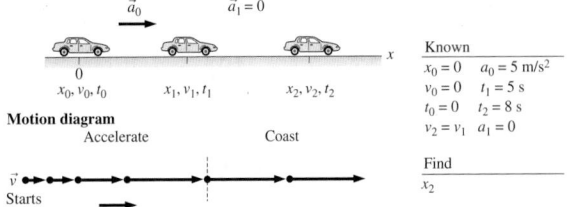

37.
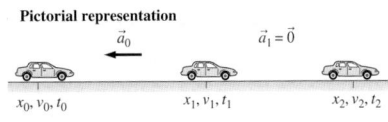
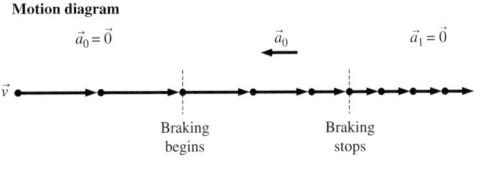

39.

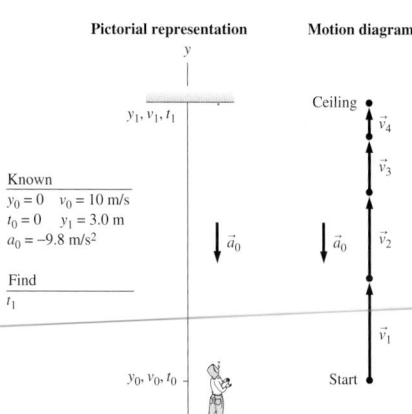

Pictorial representation

Known
$y_0 = 0$ $v_0 = 10$ m/s
$t_0 = 0$ $y_1 = 3.0$ m
$a_0 = -9.8$ m/s^2

Find
t_1

Motion diagram

41.

Known
$\theta = 20°$
$x_0 = 0$ $v_0 = 10$ m/s
$t_0 = 0$ $a < 0$
$v_1 = 0$

Find
$h = x_1 \sin \theta$

Pictorial representation

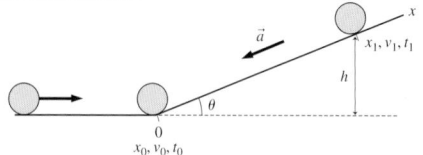

Motion diagram

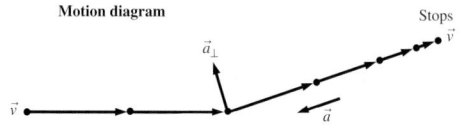

43.

Pictorial representation

Known
$x_{P0} = 0$ $x_{B0} = 15$ m
$t_0 = 0$ $v_{B0} = 4$ m/s
$a_P = a_B = 0$
$x_{P1} = x_{B1} = 20$ m
$v_{B1} = 4$ m/s

Find
v_{P0}

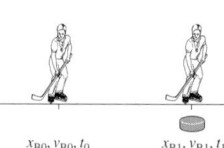
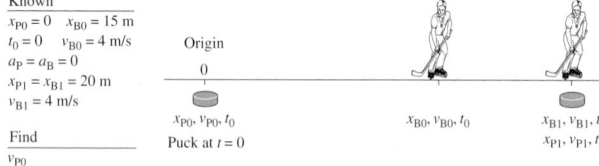

Motion diagram

51. a.

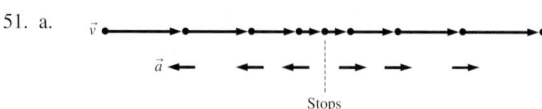

53. a.

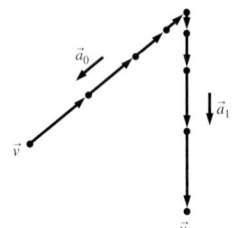

55. a.

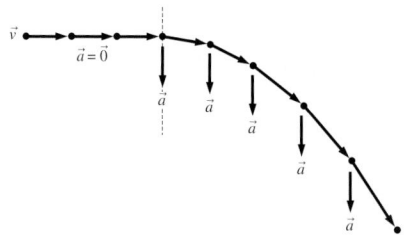

57. a. Neither is zero. b. Velocity is zero, acceleration is not zero.

Chapter 2

1. b.

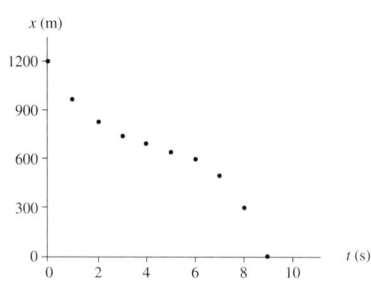

5. 450 m
7. a. Beth b. 20 min
9. a. 48 mph b. 50 mph
11. a. 26 m, 28 m, 26 m b. At $t = 3$ s
13. a.

b. 1 m/s^2
15. a. 2.68 m/s^2 b. 27.3% c. 134 m or 440 ft
17. -2.8 m/s^2
19. a. 78.4 m b. -39.2 m/s
21. 3.2 s
23. 134 m
25. a. 15 m b. 23 m/s c. 24 m/s^2
27. 16 m/s

29. a.

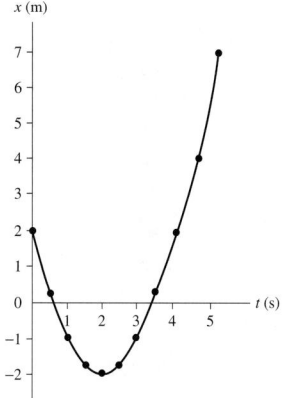

c. −2 m/s d. −2 m e. 2 m

f.

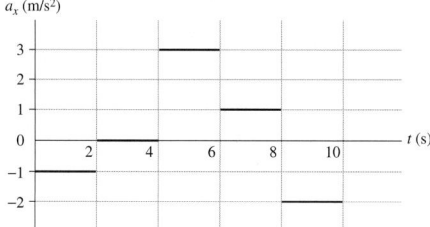

Turn around at t = 2.0 s

31. $v_A = -10$ m/s, $v_B = -20$ m/s $v_C = 75$ m/s
33. 0, 5, 20, 30, and 30 m/s
35. a.

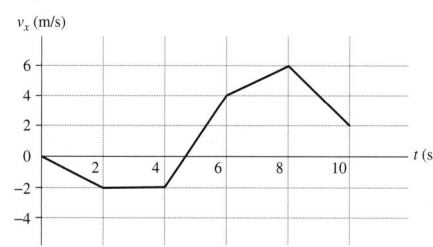

b. Displaced upward by 2.0 m/s
37. a. 0 s and 3 s b. 12 m and −18 m/s²; −15 m and 18 m/s²
39. 2.0 m/s³
41.

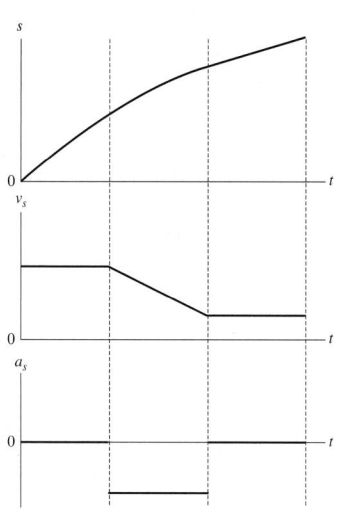

43.

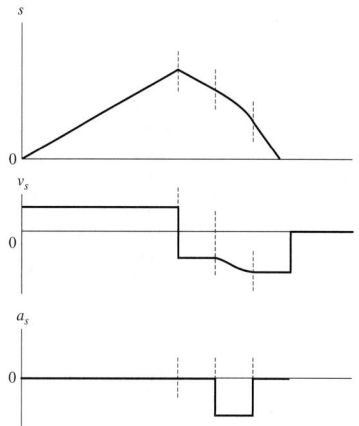

45.

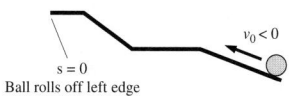

$v_0 < 0$

s = 0
Ball rolls off left edge

47. a. 179 mph b. Yes c. 35 s d. No
49. Yes
51. a. 100 m

b.

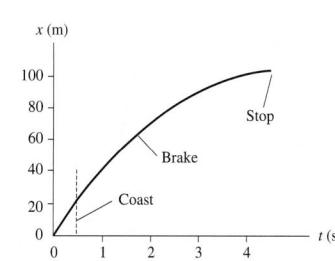

53. a. 54.8 km b. 228 s

c.

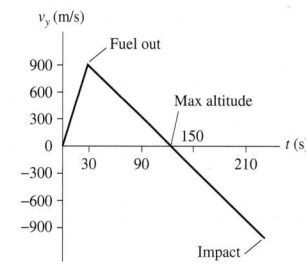

55. 19.7 m
57. 216 m
59. 9.9 m/s
61. a. 2.32 m/s b. 5.00 m/s c. 0%
63. Yes
65. a. 214 km/hr b. 16%
67. 14 m/s
69. a. 900 m b. 60 m/s
71. No
73. 5.5 m/s²
75. c. 17.2 m/s
77. c. $x_1 = 250$ m, $x_2 = 750$ m
79. 70 m/s
81. a. 10.0 s b. 3.83 m/s² c. 6.4%
83. −4500 m/s²

Chapter 3

1. a. Yes b. No
3. a. If $\vec{B}$ is in the same direction as $\vec{A}$ and $A > B$. b. If $\vec{B}$ is opposite to $\vec{A}$.
7. 11.9 m/s
9. a. $(70.7, -70.7)$ m b. $(282, 103)$ m/s c. $(0, -5.0)$ m/s^2
   d. $(-40, 30)$ N
11. $\vec{C}: (-3.04, 0.815)$ m; $\vec{D}: (12.8, -22.2)$
13. a. 7.21, 56.3° below +x-axis b. 94.3 m, 58.0° above the +x-axis
    c. 44.7 m/s, 63.4° above the $-x$-axis
    d. 6.3 m/s^2, 18.4° right of the $-y$-axis
15.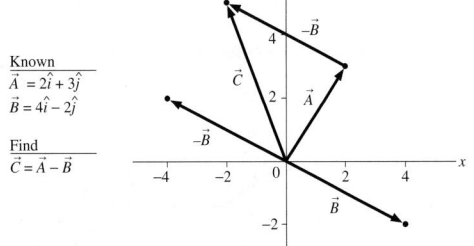

17. a. $8\hat{i} + 7\hat{j}$
    b.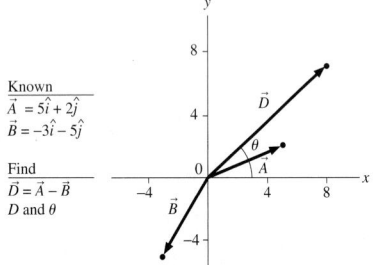

    c. 10.6, 41.2° above the +x-axis
19. a. $17\hat{i} + 22\hat{j}$
    b.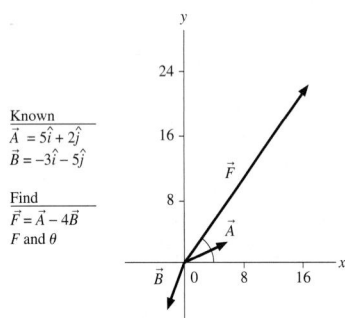

    c. 27.8, 52.3° above the +x-axis
21. Coordinate system 1: $\vec{A} = -4\hat{j}$ m, $\vec{B} = (-4.33\hat{i} + 2.50\hat{j})$ m;
    Coordinate system 2: $\vec{A} = (-2.00\hat{j} - 3.46\hat{j})$ m, $\vec{B} = (-2.50\hat{i} + 4.33\hat{j})$ m
23. a.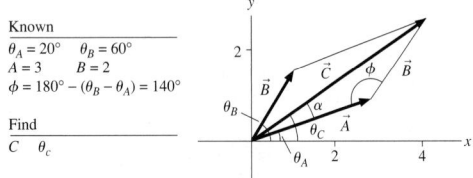

    b. 4.71, 35.8° above the +x-axis c. 4.71, 35.8° above the +x-axis
25. a. $-6\hat{i} + 2\hat{j}$ b. 6.32, 18.4° above the $-x$-axis
27. $(4.90\hat{i} + 2.83\hat{j})$ m

29. a. 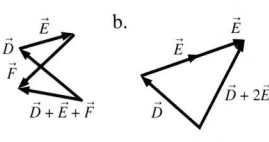 b.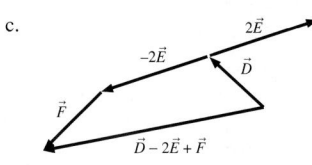

    c.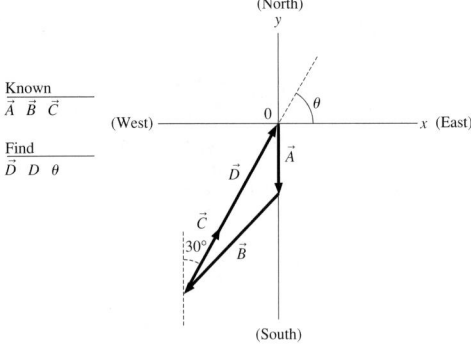

31. $0.707\hat{i} + 0.707\hat{j}$
33. a. 100 m lower
    b. (500 m, east) + (5000 m, north) − (100 m, vertical)
35. a.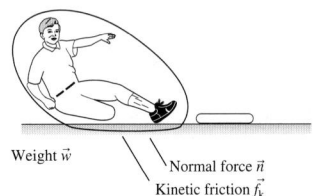

    b. 360 m, 59.4° north of east
37. 7.5 m
39. 86.6 m/s
41. 385 paces, 24.6° west of north
43. -15.0 m/s
45. a. -3.4 m/s b. -9.4 m/s
47. 4.36 units, 83.4° below the $-x$-axis
49. 7.29 N, 79.2° below the $-x$-axis

Chapter 4

3.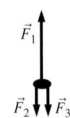

   Weight $\vec{w}$ Normal force $\vec{n}$
   Kinetic friction $\vec{f}_k$

5. $m_1 = 0.08$ kg; $m_3 = 0.50$ kg
9. 0.25 kg
11. a. ≈ 0.05 N b. ≈ 100 N
13.

   F_1
   F_2 F_3

19. **Force identification**   **Free-body diagram**

   Normal force $\vec{n}$ Weight $\vec{w}$

21.

Force identification   **Free-body diagram**

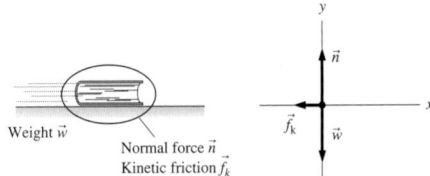

Weight $\vec{w}$

Normal force $\vec{n}$
Kinetic friction $\vec{f}_k$

23.

Motion diagrams

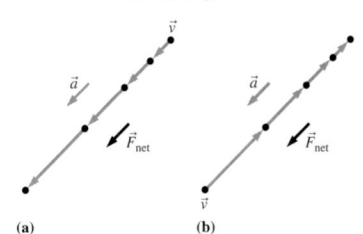

(a)   (b)

25.

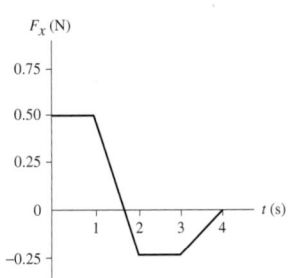

27.

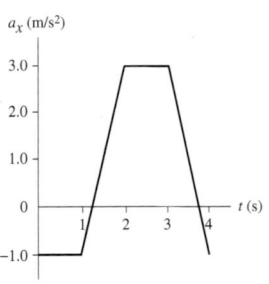

29. a. 16 m/s² b. 4 m/s² c. 8 m/s² d. 32 m/s²

31.

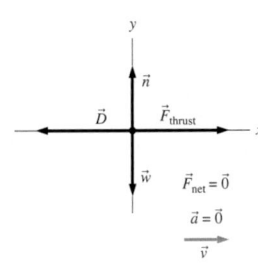

33.

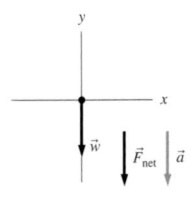

35.

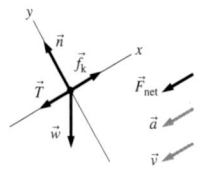

37.

Tension $\vec{T}$

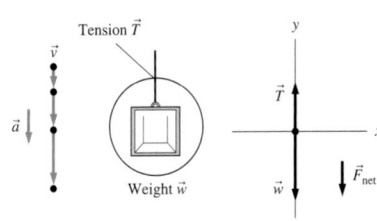

Weight $\vec{w}$

39.

Thrust $\vec{F}_{thrust}$ Drag $\vec{D}$

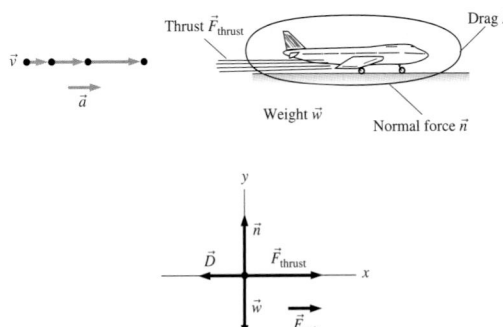

Weight $\vec{w}$ Normal force $\vec{n}$

41.

Wind $\vec{F}_w$

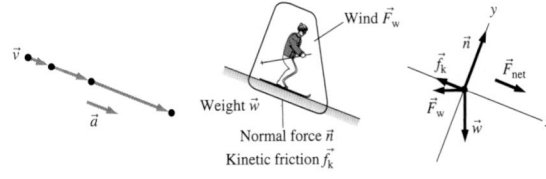

Weight $\vec{w}$

Normal force $\vec{n}$
Kinetic friction $\vec{f}_k$

43.

Drag $\vec{D}$

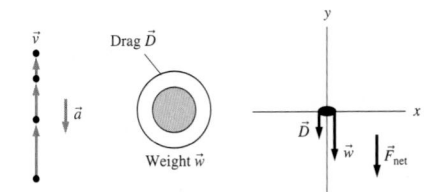

Weight $\vec{w}$

45.

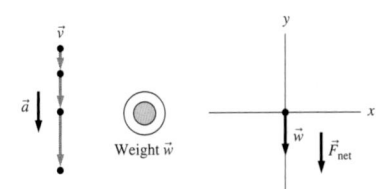

Weight $\vec{w}$

47.

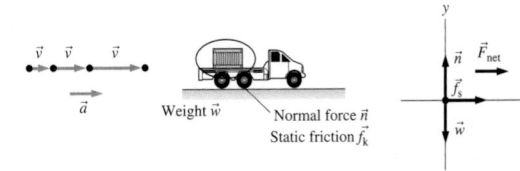

Weight $\vec{w}$ Normal force $\vec{n}$
Static friction $\vec{f}_k$

49. a.

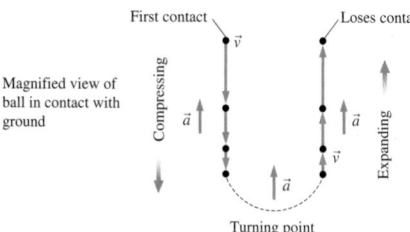

Magnified view of ball in contact with ground

First contact — Loses contact

Compressing — Expanding

Turning point

b.

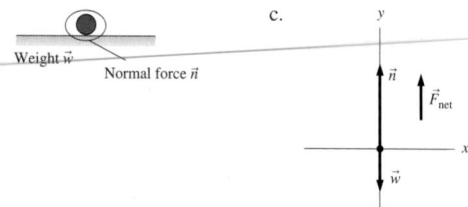

Weight $\vec{w}$ Normal force $\vec{n}$

c.

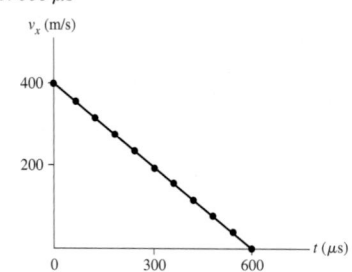

Chapter 5

1. $T_1 = 86.7$ N, $T_2 = 50.0$ N
3. 147 N
5. a. $a_x = 1.0$ m/s^2, $a_y = 0$ m/s^2 b. $a_x = 1.0$ m/s^2, $a_y = 0$ m/s^2
7. 8 N, 0 N, −12 N
9. a. 0 N b. 0 N c. 250 N
11. 307 N
13. a. 590 N b. 740 N c. 590 N
15. 0.25
17. 136 m
19. 2550 m
21. 192 m/s
23. ≈ 3 m/s^2
25. 4.0 m/s
27. Left first, then right.
29. a. 0.0036 N b. 0.0104 N
31. a. 784 N b. 1050 N
33. a. 58.8 N b. 67.8° c. 79.0 N
35. a. 6670 N b. 600 μs
   c.

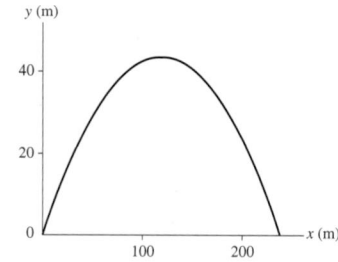

37. a. 3.96 N b. 2.32 N
39. a. 15.7 N b. 2.87 m/s c. 4.36 m/s
41. 0.165
43. 0.68 m
45. a. 3.79 m b. 6.97 m/s
47. 0.12
49. 14.3
51. 23.1 N
53. 51 m/s
55. b. 12.3 m/s^2
57. Defective cable
59. 13.0 m/s^2

67. $T = 144$ N
69. $\theta = 11.3°$
71. Green
73. b. 134 s and 402 s c. No

Chapter 6

1. a. $(2\hat{i} - 2\hat{j})$ m/s^2 b. $(22\hat{i} - 16\hat{j})$ m, $(9\hat{i} - 9\hat{j})$ m/s, 12.7 m/s
3. a.

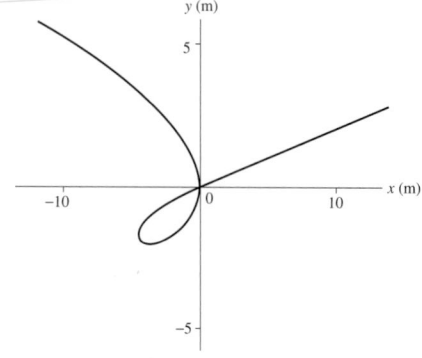

b. $\vec{0}$ m, 2.0 m/s at $t = 0$ s; $\vec{0}$ m, 8.3 m/s at $t = 4$ s
c. $-90°$ at $t = 0$ s; 14° at $t = 4$ s
5. 38.8 m
9. a. At $t = 6$ s, $x = 240$ m, $y = 3.6$ m, $v_x = 40$ m/s, $v_y = -28.8$ m/s, $v = 49.3$ m/s
b.

11. a. 0.0639 s b. 782 m/s
13. 2.0 km/hr
15. 0.40 m
17. a. 39.1 mi b. 19.5 mph
19. a. 55.6 hr b. 0.0917° c. Yes
21. 6.56×10^{12} m/s^2
23. a. $v_0^2 \sin^2\theta / 2g$ b. $h = 14.4$ m, 28.8 m, 43.2 m; $d = 99.8$ m, 115.2 m, 99.8 m
25. a. Launch point 80.8 m higher b. 34.4 m
   c. 49.8 m/s, 72.5° below horizontal
27. a. 276 m b. 12.75 s
29. Clears by 1.01 m
31. No
33. 34.3°
35. 678 m
37. a. 239 m b. 42.9 m
39. 106 m/s
41. 4.48 m/s
43. 105 m
45. Crocodile food.
47. 2.96 m
49. b. $\theta = 11.5°$
51. b. $x_1 = -29.2$ m

53. a. On the opposite bank 150 m east of where she started.

b.

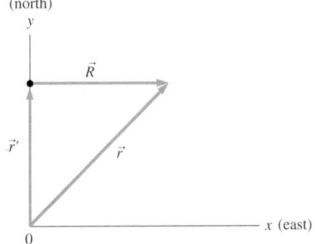

55. a. 44.4° above the $-x'$-axis

b.

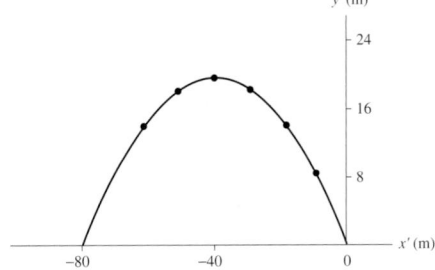

57. 10.1°

59. a. 30° toward the rear of the car b. 17.3 m/s

61. a. 7.18° south of east b. 2.48 hr

63. 3.0×10^8 m/s

65. 40.6° below horizontal

67. 4.78 m/s

69. a. Rotate the spacecraft 153.4° counterclockwise so that the exhaust is 26.6° below the positive x-axis. Fire with a thrust of 103,300 N for 433 s.

b.

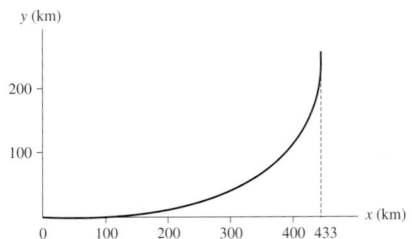

Chapter 7

1. b.

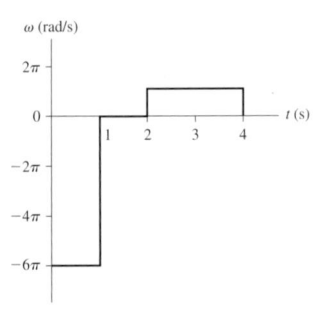

3. a. 1.5π rad/s b. 1.33 s

5. a. 3.0×10^4 m/s b. 2.0×10^{-7} rad/s c. 6.0×10^{-3} m/s^2

7. 5.65 m/s and 106 m/s^2

9. 34.3 m/s

11. a. 3.93 m/s b. 6.18 N

13. 7.27°

15. 2.0×10^{20} N

17. 1.58 m/s^2

19. 12.1 m/s

21. 19.8 m/s

23. $a_r = 2.72$ m/s^2; $a_t = 1.27$ m/s^2

25. a. -2.618 m/s^2 b. 31.25 rev

27. 49.5°

29. a. 0.967 m/s^2 b. 14.3g

31. 2.5 N higher at the north pole.

33. 172 N

35. 34.5 m/s

37. No

39. a. 5.00 N b. 30.2 rpm

41. 24.4 rpm

43. a. -9.80 m/s^2 b. -12.92 m/s^2 c. -6.68 m/s^2

45. a. 4.9 N b. 2.9 N c. 32.5 N

47. a. 319 N and 1397 N b. 5.68 s

49. 29.9 rpm

51. 2.63 m right of the point where the string was cut.

53. a. 1.90 m/s^2 at 20.6° from the r-axis b. 23.5 s

55. 3.75 rev

57. b. $\omega = 20$ rad/s

59. b. $\omega_f = 0.40$ rad/s

61. 14.19 N and 8.31 N

63. c. 94.5 rpm

Chapter 8

1.

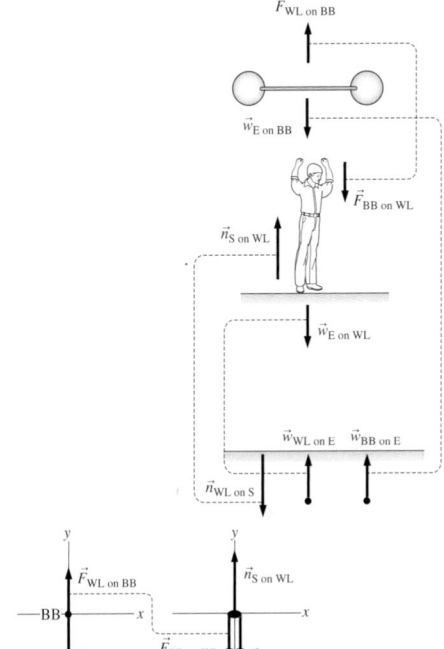

3.

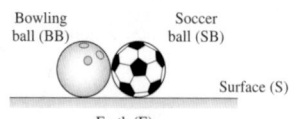

(i)

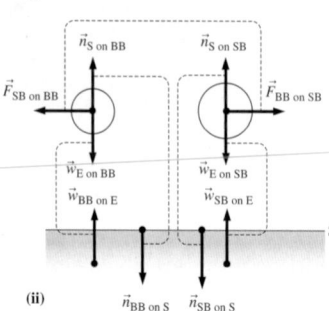

(ii)

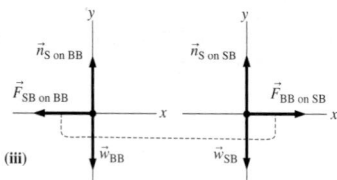

(iii)

5. a.

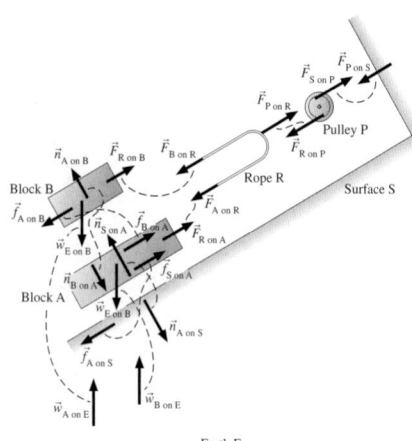

Earth E

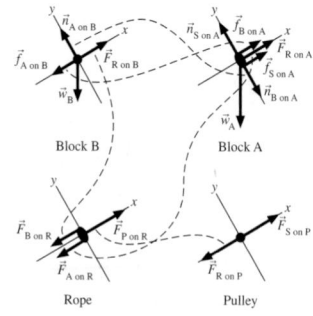

7. a. 784 N b. 1580 N
9. $F_{2\ on\ 3} = 6\ N; F_{2\ on\ 1} = 10\ N$

11. 588 N
13. a. 20 N b. 21 N
15. 66.6 N at 36°
17. 60 N
19.

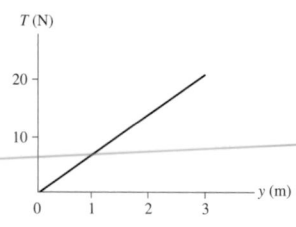

21. No
23. 99.0 m
25. 1.48 s
27. a. 3.92 N b. 2.16 m/s²
29. 154.7 N
31. 200 kg
33. 98.9 kg
35. 2.29 m/s²
37. a. 19.6 N b. Down c. 20.6 N
39. a. 3770 N b. 28.2 m/s
41. 3.27 m/s²
43. 3590 N
47. a. 1.0 m/s b. 90 N
49. b. 8.99 N

Chapter 9

1. a. 1.5×10^4 kg m/s b. 8.0 kg m/s
3. 1500 N
5. 5.0 N s
7. a. 1.5 m/s to the right b. 0.5 m/s to the right
9. 0.50 s
11. 0.20 s
13. 1.43 m/s
15. 0.20 m/s
17. 3.6 m/s
19. 2.0 mph
21. 1.7 m/s 45° north of east
23. 2.89×10^{34} kg m²/s
25. a. (1.083, 0.625) kg m/s when thrown, (1.083, 0) kg m/s at the top
27. a. 6.4 m/s b. 360 N
29. a. 0.432 N s upward c. 40 to 80 N is reasonable estimate
31. a. 0.588 N s b. −0.588 N s
33. a. 6.7×10^{-8} m/s b. 2×10^{-10}%
35. 13.3 s
37. 1.73 m/s at 54.7° south of east
39. 7.57 cm in the direction Brutus was running
41. 402 m
43. a. 286 μs, 26,200 N b. 0.0214 m/s
45. 27.8 m/s
47. 5 s⁻¹
49. 1.46×10^7 m/s in the forward direction
51. 14 u
53. b. and c. 1.40×10^{-22} kg m/s in the direction of the electron
55. 0.850 m/s, 72.5° below the x-axis
57. 1.97×10^3 m/s
59. 4.5 rpm

61. a. 2.80 m/s b. 2.22 m/s at a radius of 25.2 cm
63. c. $(v_{ix})_2 = 6.0$ m/s
65. c. $(v_{fx})_1 = -12$ m/s
67. 5.65 m/s
69. 90.3 m/s
71. 8

Chapter 10

1. The bullet
3. 112 km/hr
5. a. 6.75×10^5 J b. 45.9 m c. No
7. a. 12.9 m/s b. 14.0 m/s
9. 7.67 m/s
11. a. 1.403 m/s b. 30°
13. a. Yes b. 14.1 m/s
15. a. 49 N b. 1450 N/m c. 3.4 cm
17. 98 N/m
19. 10 J
21. 2.00 m/s
23. 3.00 m/s
25. 0.857 m/s and 2.86 m/s
27. a. −5.0 m/s and 5.0 m/s b. Both 2.5 m/s
29. a. Right b. 20.0 m/s at $x = 2.0$ m c. 1.0 and 6.0 m
31. 63.2 m/s
33. Yes
35. a. No b. 17.3 m/s
37. a. Left b. Yes c. 200 N/m d. 19.0 m/s
39. $v_0/\sqrt{2}$
41. 25.8 cm
43. 51.0 cm
45. 19.6 N/m
47. a. 14.8 m/s b. Go hungry.
49. a. 21,600 N/m b. 18.6 m/s
51. a. 3.33 m/s b. 11.8 cm c. 0.833 m/s and 6.45 cm
53. a. $\frac{3}{2}R$ b. 15 m
55. $2.5R$
57. 7.94 m/s
59. 100 g ball 0.80 m/s to the left; 400 g ball 2.2 m/s to the right
61. a. Vibrates about an equilibrium position on one side of the H_3 plane or the other.
    b. Oscillates from one side of the H_3 plane to the other side.
65. c. $k = 35.6$ N/m
67. $v_f = 2.65$ m/s
69. a. 1.46 m b. 19.6 cm
71. b. 453 m/s
73. 100 g ball to 79.3°, 200 g ball to 14.7°

Chapter 11

1. a. 15.3 b. −4.0 c. 0
3. a. −30 b. 0
5. a. 12.0 J b. −6.0 J
7. 0 J
9. 12,500 J by the weight, −7920 J by $\vec{T}_1$, −4580 J by $\vec{T}_2$
11. 4.0 J, −4.0 J, 4.0 J, 0 J, −3.0 J
13. 7.35 m/s, 9.17 m/s, 9.70 m/s
15. 8.0 N
17. −30 N at 1 m, 20 N at 3 m

19. a.

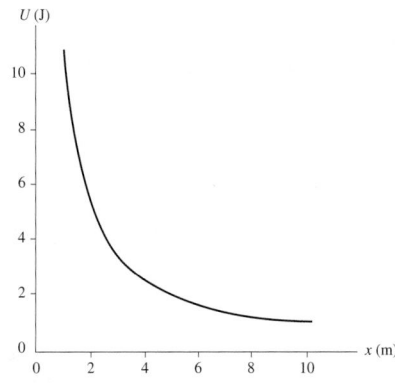

b. 2.5 N, 0.40 N, and 0.156 N
21. 1360 m/s
23. a. Potential energy is transformed to kinetic and thermal energy.
    b. 548 J
25.

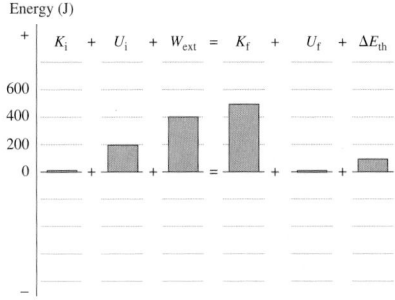

27. 6.26 m/s
29. a. 176.4 J b. 58.8 J
31. Night light
33. a. 102 N b. 416 W, 832 W, and 1248 W
35. a.

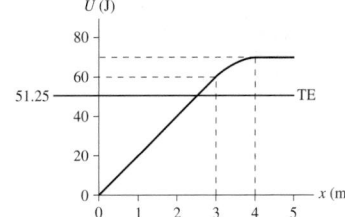

b. 51.25 J d. 2.56 m
37.

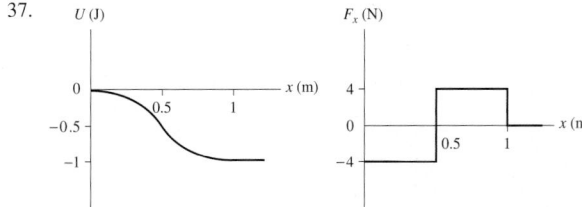

39. a. −98,000 J b. 108,000 J c. 10,000 J d. 4.47 m/s
41. a. and b. 3.97 m/s
43. 2.37 m/s
45. 0.037
47. a. 1.70 m/s b. No
49. a. 571 J b. −196 J c. −38.5 J d. 0 J
51. a. 2.16 m/s b. 0.0058
53. a. 9.90 m/s b. 9.39 m/s c. 93.9 cm d. 10
55. a. $\sqrt{2gh}$ b. $h - \mu_k L$

57. a. N/m³
    b.

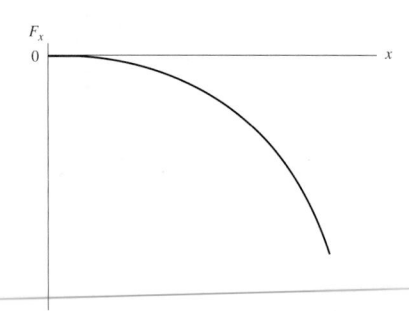

   c. $\frac{1}{4}qx^4$ d. 10 m/s
61. 233 W
63. a. −245 J b. 255,000 kg
65. ≈15 m/s
67. a. 6.53 m/s² b. 16.7 m/s c. 2.56 s
69. c. $v_1 = 6.34$ m/s
71. c. $P = 32.4$ kW
73. 6.68 m
75. a.

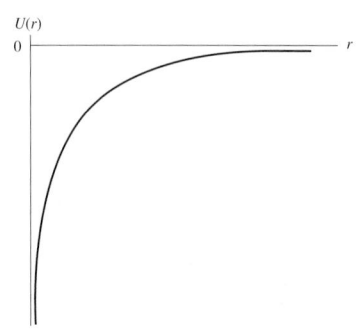

   b. Infinity c. 689,000 m/s and 172,000 m/s

Chapter 12

1. a. 3.53×10^{22} N b. 1.99×10^{20} N c. 0.56%
3. 6.00×10^{-4}
5. 1.60×10^{-7}
7. a. 8.97 N
9. a. 1.62 m/s² b. 25.9 m/s²
11. 2430 m
13. a. 3.0×10^{24} kg b. 0.889 m/s²
15. 418 km
17. 60.2 km/s
19. a. 1.80×10^7 m b. 9410 m/s
21. a. 7680 m/s b. 92.4 min
23. 4.2 hr
25. 2.01×10^{30} kg
27. 6.72×10^8 J
29. 46 kg and 104 kg
31. 1.19×10^{-3} rad or 0.0679°
33. (11.7 cm, 0 cm)
35. a. (4.72×10^{-7} N, 45° ccw from −y-axis)
   b. (4.56×10^{-7} N, 7.6° cw from y-axis)
37. a. -10.0×10^{-8} J b. -9.65×10^{-8} J
39. a. 1.38×10^7 m b. 4450 m/s
41. a. 2.82×10^6 m b. 3670 m/s
43. 32,600 m/s
45. a. $\sqrt{16GM/7R}$ b. $\sqrt{4GM/3R}$

47. a. 3.46×10^8
    b.

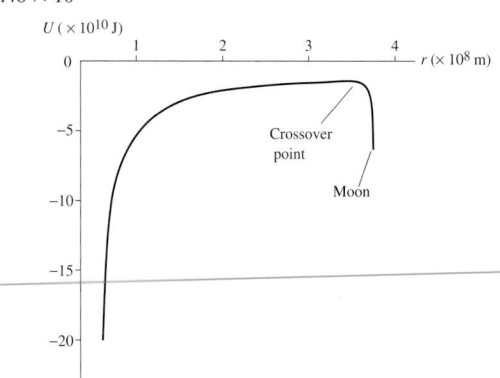

   c. -2.24×10^{10} J d. 9.60×10^9 J e. 11.0 km/s
49. 3.0×10^4 m/s
51. 1.405 hr
53. a. 6.95 m/s b. 12.3 m/s
55. 8.67×10^7 m
57. a. $y = (q/p)x + (\log C)/p$ b. linear c. q/p e. 1.996×10^{30} kg
59. 317 m (8.26×10^{-5}%) and 2.9 s (1.25×10^{-4}%)
61. 9.33×10^{10} m
63. 3.71 km/s
65. 4.49 km/s
67. Yes
69. c. 1.00×10^8 m
71. $v_{f1} = 596$ m/s, $v_{f2} = 298$ m/s
73. a. 2.05×10^8 yr d. 9.4×10^{10}
75. Crash.

Chapter 13

1.

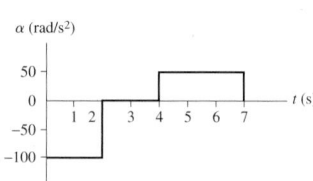

3. a.

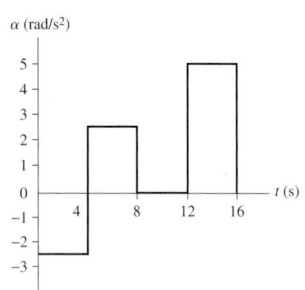

5. 13.2 m/s
7. a. −100.5 rad/s² b. 50.0
9. 36.3 cm/s
11. −0.20 N m
13. 175.5 N
15. 12,500 N m
17. a. (0.0571 m, 0.0571 m) b. 0.0080 kg m²
19. a. (0.060 m, 0.040 m) b. 0.0020 kg m² c. 0.00128 kg m²
21. 0.75 rad/s
23. 0.0471 N m
25. 11.76 N m

27. 1.40 m
29. 15.8 J
31. 1.75 J
33. 0.375 J
35. a. (20.78, out of page) b. (24, into page)
37. a. $\hat{j}$ b. $\hat{j}$
39. a. $\hat{i} + 3\hat{j} + 11\hat{k}$
    b.

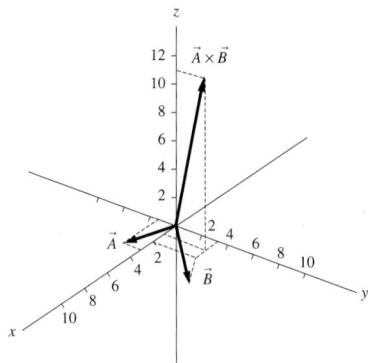

41. $-50\hat{k}$ N m
43. a. 8.97 s b. 0.448 kg m^2/s
45. 1.20$\hat{k}$ kg m^2/s or (1.20 kg m^2/s, out of page)
47. $-0.0251\hat{i}$ kg m^2/s or (0.0251 kg m^2/s, into page)
49. 28.3 m/s
51. a. 0.010 kg m^2 b. 0.030 kg m^2
53. a. $\frac{1}{2}M(R^2 + r^2)$ c. 1.37 m/s
55. $\frac{1}{6}ML^2$
57. Yes
59. 1.00 m
61. a. (20 cm, 80 cm) b. 0.48 kg m^2 c. 1.0 N m d. 56.4°
63. a. 24.4 yr b. 4080 m/s and 12,250 m/s
65. a. 177 s b. 5.55 × 10^5 J c. 139 kW d. 1300 N m
67. 1.11 s
69. 1.57 N
71. 4.25 m
73. a. $\sqrt{2g/R}$ b. $\sqrt{8gR}$
75. 20τ/13MR^2
77. a. 42.9 cm b. No
79. 50 rpm
81. a. No b. 2000 m/s c. 4000 m/s
83. a. 3v_0/2d b. No
85. 393 m/s
87. a. 68,700 m b. 4.32 × 10^6 m/s

Chapter 14

1. 2.27 ms
3. a. 3.3 s b. 0.303 Hz c. 1.904 rad/s d. 0.25 m e. 0.476 m/s
5. a. 10 cm b. 0.50 Hz c. π/3 rad or 60°
7.

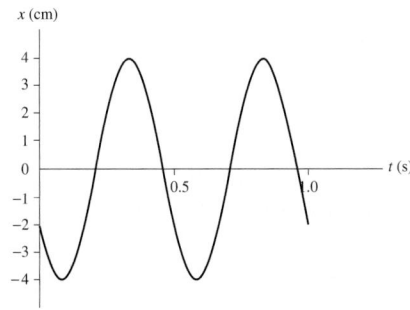

9. $x(t) = (4.0 \text{ cm}) \cos[(8.0\pi \text{ rad/s})t - \pi/2]$
11. a. $-2\pi/3$ rad or $-120°$ b. $-2\pi/3$ rad, 0 rad, $2\pi/3$ rad, $4\pi/3$ rad
13. 5.48 N/m
15. a. 0.50 s b. 4π rad/s c. 5.54 cm d. 0.445 rad e. 69.6 cm/s
    f. 875 cm/s^2 g. 0.484 J h. 3.81 cm
17. a. 10.0 cm b. 34.6 cm/s
19. a. 0.169 kg b. 0.565 m/s
21. c. 12° d. 10° e. 0° to 10°
23. 35.7 cm
25. 0.330 m
27. 5.0 s
29. 21
31.

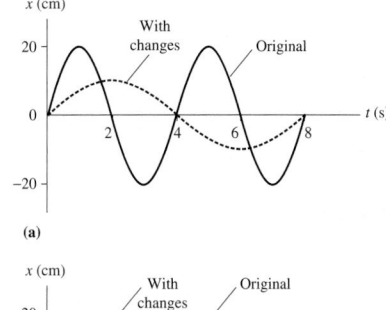

(a)

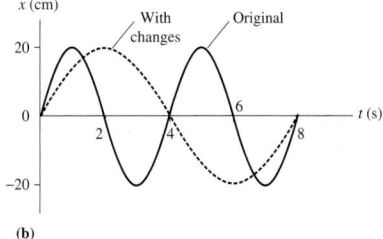

(b)

33. a. $-\pi/3$ rad or $-60°$ b. 6.80 cm/s b. 7.85 cm/s
35. a. 0.25 Hz, 3.0 s b. 6.0 s, 1.5 s c. 2.25
37. 1.405 s,

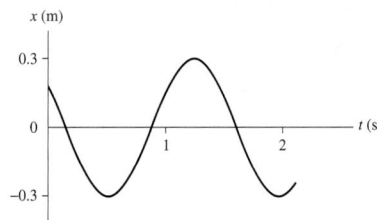

39. 0.0955 s
43. a. 6.40 cm b. 160 cm/s^2 c. -6.40 cm d. 0.283 m/s
45. 1.02 m/s
47. a. 3.18 Hz b. 0.0707 m c. 5.0 J
49. a. 1.125 Hz b. 23.5 cm c. -4.09 cm
51. a. 47,400 N/m b. 1.80 Hz
53. a. 0.314 s b. It would be unchanged.
55. 0.716
57. 0.669 s
59. a. 200.9 s b. 200.4 s c. Yes d. 9.77 m/s^2
61. 0.110 m at 1.72 s
63. a. 502 m/s b. No
65. $f = (1/2\pi) \sqrt{2T/mL}$
67. $T = 2\pi \sqrt{R/g}$
69. $g_X = 5.86$ m/s^2
71. a. 6.03 cm b. 6.32 s
73. 7.3°
77. 1.83 Hz
79. 2.23 cm

Chapter 15

1. 1200 kg/m³
3. 1.44×10^5 kg
5. 1097 atm
7. 2440 kg
9. 3153 m
11. 88,000 Pa
13. 55.2 cm
15. Ethyl alcohol
17. 45.8 kg
19. 1.87 N
21. 3.18 m/s
23. $1.27v_0$
25. 2.0 kg
27. 1.0 mm
29. 0.2%
31. a. 5830 N b. 5990 N
33. a. 0.377 N b. 20.4 m/s
35. 5.27×10^{18} kg
37. a. 10.85 m b. 10.21 m
39. a. 0.483 m b. 2.34 cm
41. 3.7 mm
43. a. $\frac{1}{2}\rho gwd^2$ b. 1.76×10^9 N
45. a. 8080 m b. 1.05 kg/m³, 82%
47. 667 kg/m³
49. 74.7 N
51. 43.9 N
53. 8.38 cm
55. $(\rho - \rho_1)/(\rho_2 - \rho_1)$
57. 5.22 cm
59. 14.1 cm
61. a. p_{atmos} b. 4.61 m
63. a. Lower b. 835 Pa c. 75,100 N
65. a. 144 m/s and 5.78 m/s b. 4.54×10^{-4} m³/s
67. a. 3.34 L/min b. 1.06 mm/min
69. 1.23 mm
71. 1.30 L
73. 3.6 g
75. e. 18.9 s

Chapter 16

1. 22.6 m³
3. 8.33 cm
5. 4.82×10^{23} atoms
7. a. 6.02×10^{28} atoms/m³ b. 3.28×10^{28} atoms/m³
9. 2.17 cm
11. Lowest: $-88°C = 185$ K; highest $58°C = 331$ K
13. a. 171°Z b. 671°C = 944 K
15. a. 32.02°F, 608 Pa b. $-68.8°F$, 5.06×10^5 Pa
17. Freezing point lower, boiling point higher.
19. a. 0.0497 m³ b. 1.33 atm
21. 18.8 atm
23. a. 55.1 mol b. 1.234 m³
25. a. 5.41×10^{23} atoms b. 3.59 g c. 2.30×10^{26} m⁻³ d. 1.52 kg/m³
27. a. 0.732 atm b. 0.523 atm
29. a. 9520 kPa
    b.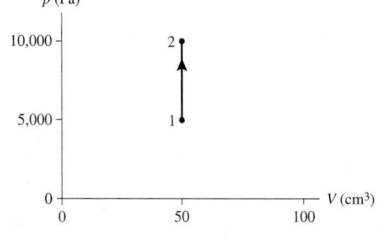

31. a. 12.02 atm
    b.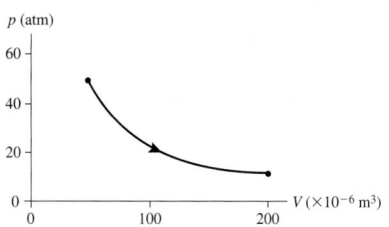

33. a. Isothermal b. Both are 914 K c. 300 cm³
35. 0.228 nm
37. a. 7.03×10^{-21} J b. 2060 m/s
39. 1.1×10^{15} m⁻³
41. a. 1.32×10^{-13} b. 1.24×10^{11} molecules
43. 174°C
45. 92.8 cm³
47. 34.7 psi
49. 174.3°C
51. a. 3.05×10^{21} b. 2.02 mg
53. No
55.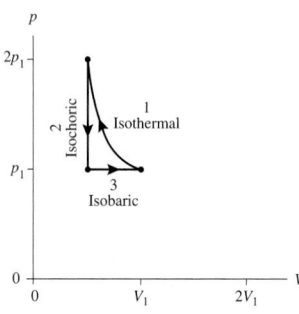

57. a. 884 kPa b. 323°C, $-49.5°C$, 397.5°C
59. a. Both 366 K = 93°C b. Isothermal c. 1098 K = 825°C
61. a. 0.509 atm b. $-112°C$
    c.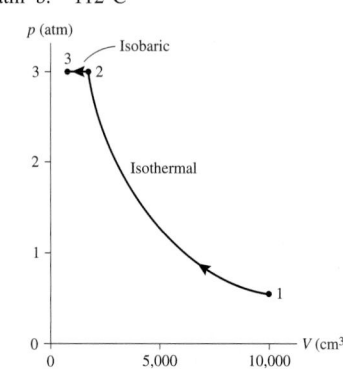

63. a. 4.0 atm, $-73°C$
    b.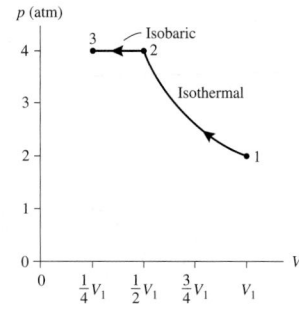

65. b.

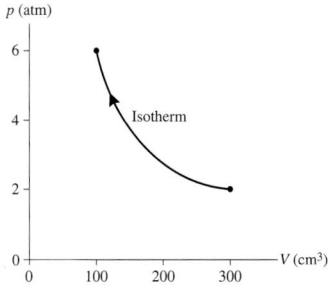

c. 6 atm

67. b.

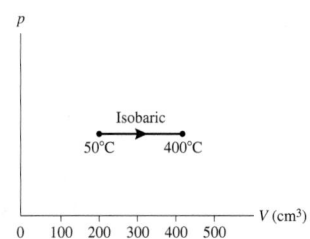

c. 417 cm³

69. a. 23.5 cm b. 7.8 cm
71. a. 2.73 m b. 10.96 atm
73. 1.02 cm

Chapter 17

1. 490 J
3. 40 J
5. 200 cm³
7.

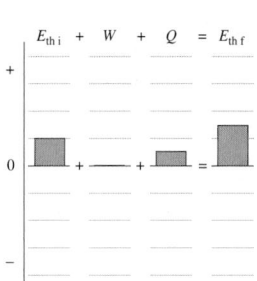

9.

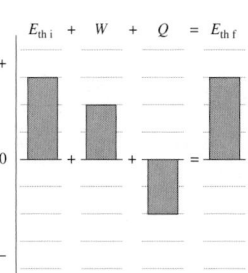

11.

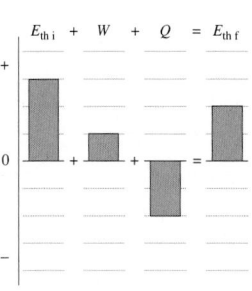

13. 700 J from the system
15. 12,500 J
17. 6864 J
19. 6.79×10^4 J
21. 27.5°C
23. 73.5°C
25. Iron
27. a. 91.2 J b. 140°C
29. a. 1.14 atm b. 48.5°C
31. a. 26.4 b. 7.07
33. 8.73 hr
35. 994 cm³
37. −56.4°C
39. Aluminum
41. 87.3 min
43. a. 83.3 J/kg K b. 2.0×10^5 J/kg
45. 0.0605 kg = 60.5 g
47. 5450 J
49. a. 245 J b. 944 m/s
51. 1660 J
53. a. 253°C b. 32.6 cm
55. 7750 J
57. a. 4290 cm³, 606°C b. 3050 J c. 1.0 atm d. 2180 J
e.

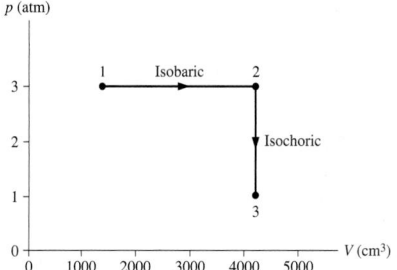

59. A: −1010 J, B: 1419 J
61. a. 1135 J b. −811 J c. 0 J
63. a. A: 2.46×10^{-3} m³, 300 K; B: 1.80×10^{-3} m³, 220 K
b.

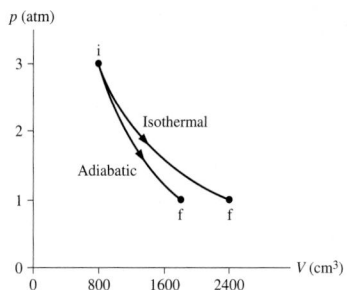

65. a. −50.7 J b. −15.2 J c. 35.5 J
67. 1100 K, 23.9 cm³
69. a. 39.3 b. 171
71. a. 0.5 atm b. −1074 J c. 1074 J d. 0 J
e.

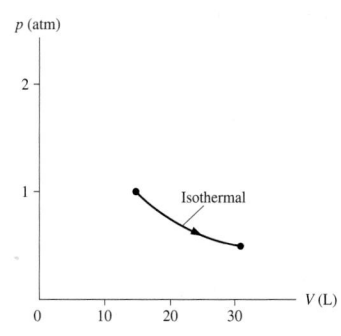

73. a. 5460 K b. 0 J c. 5.39×10^4 J d. 20
   e.

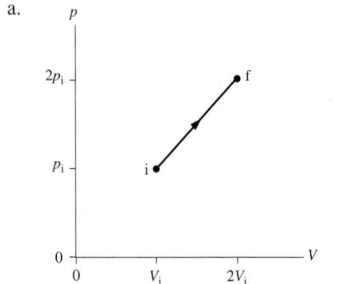

75. b. 0.0156 mol
77. b. 0.472 kg
79. a. At 1: 3.0 atm, 946°C, 1000 cm³. At 2: 1.0 atm, 946°C, 3000 cm³.
   At 3: 0.48 atm, 310°C, 3000 cm³. b. −334 J, 0 J, 239 J
   c. 334 J, −239 J, 0 J
81. 14.5 atm

Chapter 18

1. 2.69×10^{25} m⁻³
3. a. 3.30×10^{12} m⁻³ b. 1.71×10^6 m
5. 61
7. a. $0\hat{\imath} + 0\hat{\jmath}$ b. 59.2 m/s c. 61.6 m/s
9. a. 9.16×10^4 Pa b. 332 K
11. 1.91×10^{24} s⁻¹
13. 2820 m/s, 891 m/s
15. 283 m/s
17. a. 68.3 K b. 1090 K
19. 7.22×10^{12} K
21. a. 3400 J b. 3400 J c. 3400 J
23. a. 4×10^{-16} J b. 7×10^5 m/s
25. 3.65×10^7 J
27. a. 0.0800°C b. 0.0481°C c. 0.0400°C
29. a. 62.4 J b. 104 J c. 104 J d. 145 J
31. a. Gas B b. A: 5200 J, B: 7800 J
33. 8.48
35. a. Helium b. 1367 m/s c. 1.86×10^{-6} m
37. 9.6×10^{-5} m/s
39. a. $\lambda = 1/[\sqrt{2}\,\pi(N/V)r^2]$ b. 1.82×10^{-6} Pa or 1.80×10^{-11} atm
41. a. 1.273×10^{25} m⁻³ b. 449 m/s c. 259 m/s d. 1.296×10^{25} s⁻¹
   e. 56,800 Pa f. 56,700 Pa
43. a. Helium: 30.4 J; Argon: 121.6 J b. Helium: 47.3 J; Argon: 104.7 J
   c. 16.9 J is transferred from the argon to the helium. d. 580 K
   e. Helium: 3.11 atm; Argon: 3.45 atm
45. 482 K
51. a. $R = 8.31$ J/molK b. $2R = 16.6$ J/molK
53. a. 4 b. 1 c. 16
55. 1/2
57. a. 2.03×10^6 J b. 4.83×10^{-6} c. 0.00132 K
59. a.   b. $9p_i V_i$

61. c. 436 K, 850 J is transferred from the oxygen to the helium.

Chapter 19

1. a. 10 J, 110 J b. 0.0833
3. a. 0.273 b. 15 kJ
5. a. 250 J b. 150 J
7. a. 200 J b. 250 J
9. 96,000
11.

	ΔE_{th}	W_s	Q
A	+	0	+
B	0	+	+
C	−	+	0
D	−	−	−

13. 40.5 J
15. a. 0.0952 b. 285 J
17. 24.7
19. a. b only b. a only
21. 7°C
23. a. 40% b. 215°C
25. 233 K
27. a. 6.32 b. 32 W c. 232 J/s
29. a. 60 J b. −23°C
31. 8.44×10^5 J
33. 5.34×10^4 J
39. 8.25%
41. 47°C and −33°C
43. 2/3
45. a. 2.5 kW b. $270 and $45
47. a. 48 m b. 32.1%
49. 37%
51.

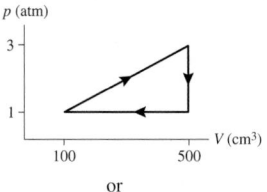

or

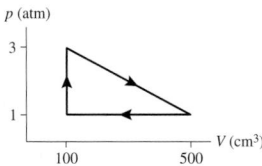

53. a.

	W_s (J)	Q (J)	ΔE_{th} (J)
1 → 2	3.04	16.97	13.93
2 → 3	0	−10.13	−10.13
3 → 1	−1.52	−5.32	−3.80
Net	1.52	1.52	0

   b. 8.96% c. 12.7 W

55. a.

	W_s (J)	Q (J)	ΔE_{th} (J)
1 → 2	0	282.2	282.2
2 → 3	207.2	0	−207.2
3 → 1	−50.0	−125.0	−75.0
Net	157.2	157.2	0

   b. 52.2%

57. a. 5.743×10^4 Pa, 4000×10^{-6} m³, 229.7 K
    b.

	ΔE_{th} (J)	W_s (J)	Q (J)
1 → 2	425.7	−425.7	0
2 → 3	0	554.5	554.5
3 → 1	−425.7	0	−425.7
Net	0	128.8	128.8

    c. 23.2%
59. a. Point 1: 1.013×10^5 Pa, 1.0×10^{-3} m³, 406 K;
    Point 2: 5.06×10^5 Pa, 1.0×10^{-3} m³, 2030 K;
    Point 3: 1.013×10^5 Pa, 5.0×10^{-3} m³, 2030 K b. 28.8% c. 80%
61. a. 1620 K, 2407 K, 6479 K
    b.

	ΔE_{th} (J)	W_s (J)	Q (J)
1 → 2	327	−327	0
2 → 3	1692	677	2369
3 → 1	−2019	0	−2019
Net	0	350	350

    c. 14.8%
63. 345.6 J, 24.0%
65. b. $T_H = 1092°C$
67. b. $Q_H = 100$ J, $Q_C = 80$ J
69. a.

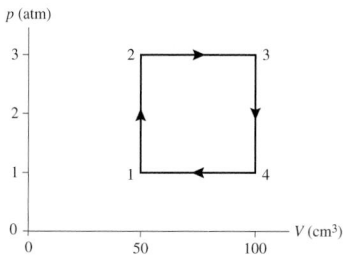

    b. 10.13 J c. 12.9%
71. c.

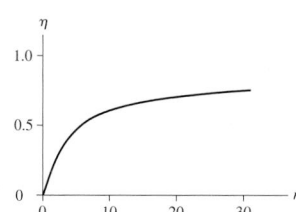

Chapter 20

1.

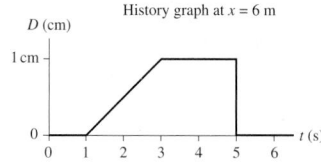

3.

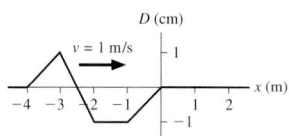

5.

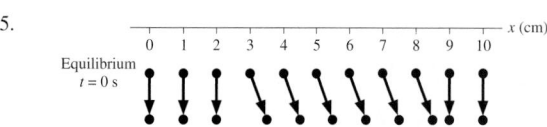

7. 283 m/s
9. 140 m/s
11. a. 4.19 m b. 47.7 Hz
13. a. 11.5 Hz b. 1.14 m c. 13.1 m/s
15. 4.0 cm, 12 m, 2.0 Hz
17. 40 cm
19. 34 Hz, 68 Hz
21. 0.076 s
23. 793 m
25. a. 1715 Hz b. 1.50 GHz c. 987 nm
27. a. 10 GHz b. 0.167 ms
29. a. 1.50×10^{-11} s b. 3.38 mm
31. 459 nm
33. 6.05×10^5 J
35. a. 1.11×10^{-3} W/m² b. 1.11×10^{-7} J
37. 38.1 m/s
39. a. 432 Hz b. 429 Hz
41. a.

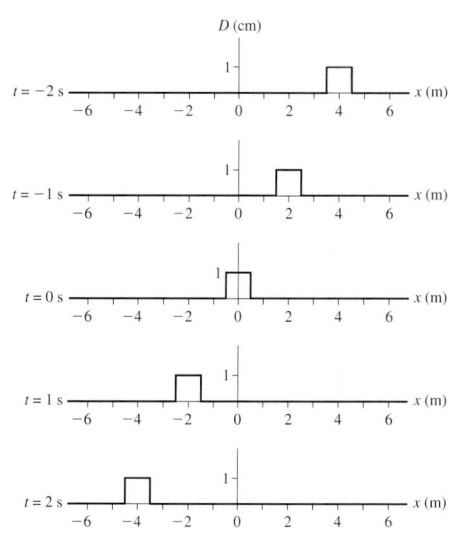

    b. 2 m/s c. 2 m/s
43. a. 0.80 m b. $-\frac{1}{2}\pi$ rad
    c. $D(x, t) = (2.0 \text{ mm})\sin(2.5\pi x - 10\pi t - \frac{1}{2}\pi)$
45. 25 g
47. 2.34 m, 1.66 m
49.

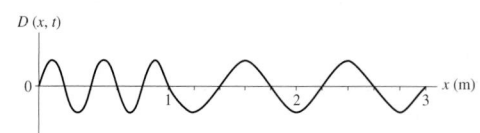

51. 1, 4.31, 4.31
53. 0.07°C
55. a. $-x$ direction b. 12.0 m/s, 5.0 Hz, 2.62 rad/m c. −1.50 cm
57. $D(y, t) = (5.0 \text{ cm})\sin[(4\pi \text{ rad/m})y + (16\pi \text{ rad/s})t]$
61. $\pi/2$ rad = 90°

63. a.

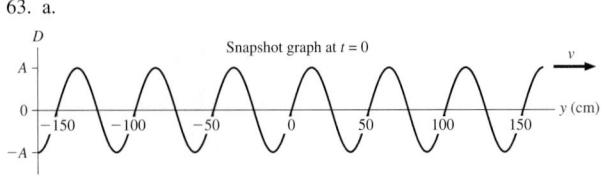

b. 0 rad c. $D(y, t) = A \sin[(12.57 \text{ m}^{-1})y - (4310 \text{ s}^{-1})t]$
d. and e.

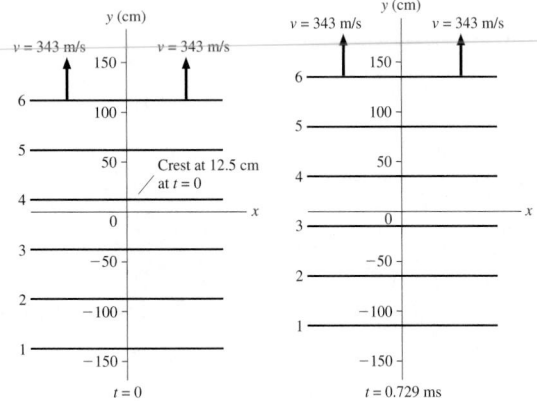

f. $-\frac{1}{2}\pi$ rad and $-\frac{3}{2}\pi$ rad

65.

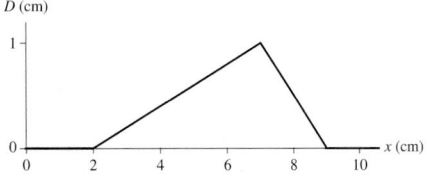

67. 15.9 Hz, 2.0 cm
69. a. 0.040 W/m² b. 637,000 W/m²
71. a. 250 μW/m² b. 15.8 km
73. 85.8 m/s, away from you
77. 796 nm, infrared
79. $200 million
81. 8
83. b. 5.17 × 10⁻¹¹ s

Chapter 21

1.

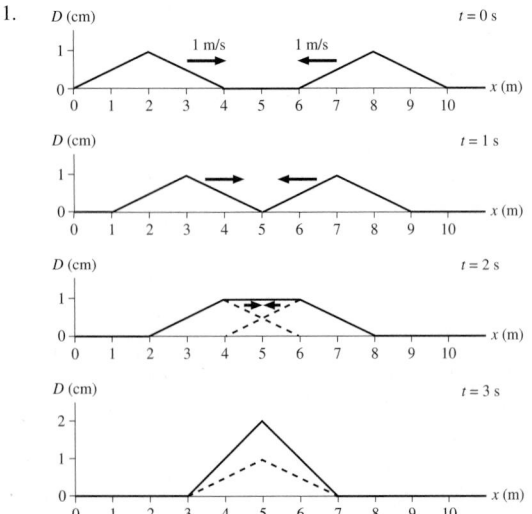

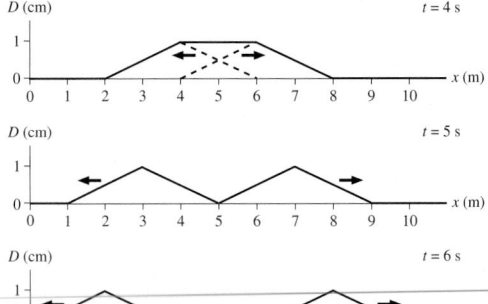

5. a. $t = 4$ s
b.

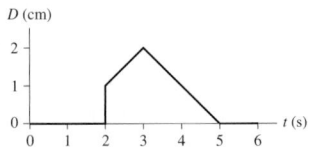

7. 60 Hz
9. a. 6 b. $2f_0$
11. a. 12 Hz, 24 m/s
b.

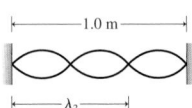

13. a. 700 Hz b. 56.4 N
15. 400 m/s
17. 10.5 m
19. 4.8 cm
21. a. 0.25 m b. 0.25 m
23. 1.0 m, 3.0 m, 5.0 m
25. 200 nm
27. a. Out of phase b.

	r_1	r_2	Δr	C/D
P	2λ	3λ	λ	D
Q	3λ	1.5λ	1.5λ	C
R	2.5λ	3λ	0.5λ	C

29. Perfect destructive
31. 527 Hz
33.

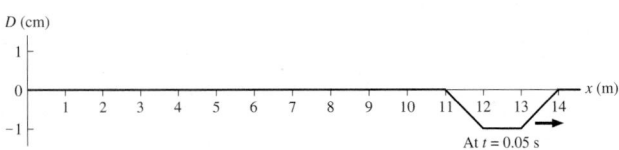

35. 0.62 cm, 1.18 cm, 1.62 cm, 1.90 cm, 2.00 cm
37. 1.41 cm
39. 1.23 m
41. 28.4 cm
43. 8.19 m/s²
45. 18 cm
47. 13.0 cm
49. 328 m/s
51. 26.1 cm, 55.6 cm, 85.2 cm
53. 450 N
55. 1605 Hz
59. 7.89 cm

61. a. 850 Hz b. $-\frac{1}{2}\pi$ rad
63. 345 nm
65. 7.15 cm
67. 20
69. a. 170 Hz b. 510 Hz and 850 Hz
71. 150 MHz
73. a. a b. 1.0 m c. 9
75. a. 5 beats/s b. 4.6 mm
77. 7.0 m/s
79. b. 2.0%
81. 8.00 m/s^2
83. c. 2.09 cm/s d. 2.2 mm
85. a. $\frac{1}{4}\lambda$ b. $\frac{1}{2}\pi$ rad   c. $\frac{1}{4}T$ d. 75 m, 250 ns

Chapter 22

1. a.

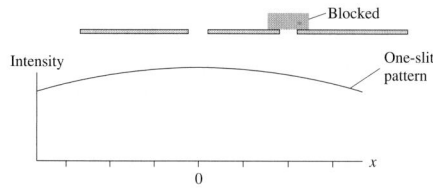

3. 0.020 rad = 1.15°
5. 500 nm
7. 0.40 mm
9. 530
11. a. 1.258 m b. 7
13. 14.5 cm
15. 4.0 mm
17. 611 nm
19. 7.56 m
21. 0.25 mm
23. 0.01525 rad = 0.874°
25. 30,467
27. 19
29. a. Single slit b. 0.15 mm
31. 0.286°
33. 1.33 mm
35. 500 nm
37. 500 nm
39. 667.8 nm
41. 396 nm
43. 533 to 700 nm
45. 500 nm
47. 0.118 mm
49. a. 2 b. 1.15 c. 1
51. 0.10 mm
53. 0.122 mm
55. a. 550 nm b. 0.40 mm
57. a. No b. 0.0295° c. 0.30 cm d. 103 cm
59. a. 3.0 mm b. $\frac{1}{4}$ c. $\frac{1}{2}\pi$ rad   d. 0.75 mm toward the slit
61. 14.2 μm closer to the beam splitter
63. a. 376 nm b. 1319 c. 1319
65. 1.5525
67. 12.0 μm
69. b. 0.022°, 0.058°
71. b. $-11.5°$, $-53.1°$
73. c. 1.3 m

Chapter 23

1. a. 3.33 ns b. 0.75 m, 0.67 m, 0.51 m
3. 8.0 cm
5. 668 m
7. 9.0 cm
9. 42°
11. 433 cm
13. 65.0°
15. 1.37
17. 76.7°
19. 3.18 cm
21. 1.52
23. b. 1.1°
25. 1580 nm
27. Inverted image 15 cm behind the lens
29. Upright image 6 cm in front of the lens
31. 68 cm
33. -203 cm
35. 1.54 cm
37. 54.6 km
39. b. Relative to the intersection of the two mirrors, 3 images are at coordinates $(+1$ m, -2 m$)$, $(-1$ m, $+2$ m$)$, and $(+1$ m, $+2$ m$)$
    c.

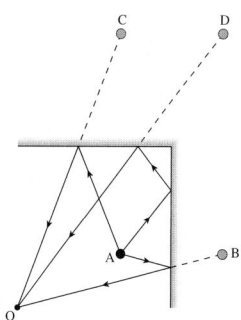

41. 10.0 m
43. 41.8°
45. 82.8°
47. a. Bottom of tank coming up b. 60.0 cm
49. 4.73 m
51. a. Deep b. 17.5 m
53. 1.552
55. a. 17.9° b. 27.9° to the left of the normal
57. 3.0 cm
59. b. 40 cm, 2 cm
61. b. -60 cm, 8.0 cm
63. b. -8.6 cm, 1.14 cm
65. 44.4 cm, 67 cm
67. c. ≈ 3.6 cm
69. 15.7 cm
71. b. 20 cm in front of second lens, 2.0 cm tall
73. 11.5 km
75. a. 2 μm b. 165 MB
77. b. 1.574
79. b. 40 cm, 156.5 cm
81. a. -200 cm

Chapter 24

1. 410.3 nm, 389.0 nm, 379.9 nm
3. $n = 8$
5. 63.8°
7. 4

9. 1.99×10^{-16} J

11. 1.2×10^5 J

13. a. 3.6×10^6 m/s b. 2.0×10^3 m/s

15. a. 1.1×10^{-34} m b. 1.7×10^{-23} m/s

17. 0.20 nm

19. a. 121.6 nm, 102.6 nm, 97.3 nm, 95.0 nm b. 91.18 nm c. 31.4 cm

21. a. 2.0×10^{-12} m b. 2.51×10^5

23. a. 3.14×10^{-19} J b. 3.19×10^{15}

25. 18.7°, 50.8°, and 71.6°

27. b. 2.4 nm and 1.2 nm

29. a. 0.818 μm b. 1.09×10^3 m/s

31. 170 m/s

33. a. 1.23×10^{-19} J, 4.92×10^{-19} J, 1.11×10^{-18} J
    b. 3.69×10^{-19} J c. 539 nm

35. 1.35 nm

37. a. $(h/2mL)n$ b. 1.819×10^6 m/s, 3.64×10^6 m/s, 5.46×10^6 m/s

39. a. 72.5°, 53.1°, and 25.8° b. 64.9° and 31.9°
    c. 19.9° and 76.9°, matching the peaks in Figure 24.7c

41. b. 7.28×10^4 m/s, 1.46×10^5 m/s, 2.18×10^5 m/s, 2.91×10^5 m/s

Chapter 25

1. a. Electrons removed from glass b. 3.13×10^{10}

3. 3.04×10^{-11}

7. Right negatively charged, left positively charged

11. a. 9.0×10^9 N b. 9.0×10^9 m/s²

13. -10 nC

15. $\vec{F}_{\text{B on A}} = 4.50 \times 10^{-3} \, \hat{\jmath}$ N, $\vec{F}_{\text{A on B}} = -4.50 \times 10^{-3} \, \hat{\jmath}$ N

17. 30 N/kg

19. a. (9.83 N/kg, toward earth) b. (2.70×10^{-3} N/kg, toward earth)

21. 0.111 nC

23. -8.0 nC

25. (3.27×10^6 N/C, downward)

27. a. $3.6 \times 10^4 \, \hat{\imath}$ N/C, $(-1.27 \times 10^4 \, \hat{\imath} + 1.27 \times 10^4 \, \hat{\jmath})$ N/C,
    $(-1.27 \times 10^4 \, \hat{\imath} - 1.27 \times 10^4 \, \hat{\jmath})$ N/C
    b.

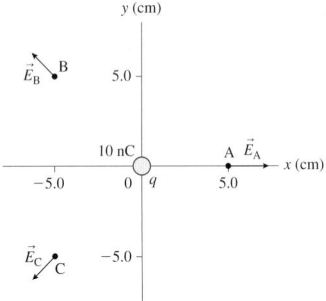

29. 1.36×10^5 C, -1.36×10^5 C

31. a. Electrons removed from sphere and added to rod b. 2.5×10^{10}

33. -160 nC and 0 nC

35. a. 498 N b. 2.98×10^{29} m/s²

37. a. 0.45 N b. 1.0×10^{-6} C, 5.0×10^{-7} C c. 4.5 m/s²

39. 1.80×10^{-4} N to the right

41. 4.74×10^{-3} N, 71.6° above $-x$-axis

43. 1.74×10^{-4} N, 51.75° below $+x$-axis

45. $-1.02 \times 10^{-3} \, \hat{\imath}$ N

47. $(1.02 \times 10^{-5} \, \hat{\imath} + 2.16 \times 10^{-5} \, \hat{\jmath})$ N

49. 0.68 nC

51. $-2KQqa/(y^2 + a^2)^{3/2}$

53. a.

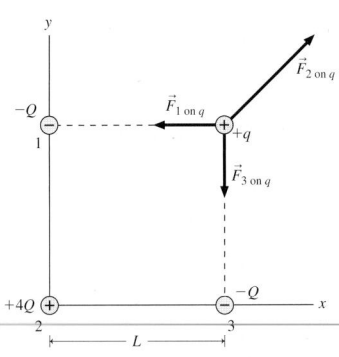

b. $(2 - \sqrt{2})KQq/L^2$

55. a. 243 N b. Yes. Any difference must therefore be smaller than 1 part in 10^9.

57. 3.2×10^{15}

59. 4.42°

61. $1.0 \times 10^5 \, \hat{\jmath}$ N/C, $(2.88 \times 10^4 \, \hat{\imath} + 2.16 \times 10^4 \, \hat{\jmath})$ N/C,
    $5.63 \times 10^4 \, \hat{\imath}$ N/C

63. $(4.02 \times 10^4 \, \hat{\imath} + 8.05 \times 10^4 \, \hat{\jmath})$ N/C, $4.5 \times 10^5 \, \hat{\imath}$ N/C,
    $(4.02 \times 10^4 \, \hat{\imath} - 8.05 \times 10^4 \, \hat{\jmath})$ N/C

65. a. $(-1$ cm, 2 cm) b. (3 cm, 3 cm) c. (4 cm, -2 cm)

67. a. $(3.20 \, \hat{\imath} + 6.40 \, \hat{\jmath}) \times 10^{-17}$ N b. $(-3.20 \, \hat{\imath} - 6.40 \, \hat{\jmath}) \times 10^{-17}$ N
    c. 4.28×10^{10} m/s² d. 7.85×10^{13} m/s²

69. 14.3°

71. b. 22.4 nC

73. b. 5.13 nC

75. 4.06 g

Chapter 26

1. (2550 N/C, 0° above horizontal)

3. (3975 N/C, 9.3° above horizontal)

5. a. 18.0 N/C b. 36.0 N/C

7. 2.28×10^5 N/C, 1.67×10^5 N/C, 2.28×10^5 N/C

9. (8.78×10^{-4} N, toward rod)

11. -0.056 nC

13. a. 0 N/C b. 4110 N/C

15. a. 0 N/C b. 1.49×10^5 N/C

17. 1.39×10^{-3} nC

19. 1.41×10^5 N/C

21. 1.86 cm

23. 6.13×10^5 N/C, down

25. 5.93×10^5 N/C

27. 0.185 m

29. (9.0×10^{-13} N, direction opposite $\vec{p}$)

31. $(132,600 \, \hat{\imath} - 12,130 \, \hat{\jmath})$ N/C; (133,200 N/C, 5.23° below the
    $+x$-axis)

33. $(675 \, \hat{\imath} - 78,400 \, \hat{\jmath})$ N/C; (78,400 N/C, 89.5° below the $+x$-axis)

35. a. $\vec{E}_1 = [q/(4\pi\epsilon_0)5\sqrt{5}a^2](-3 \, \hat{\imath} + 2 \, \hat{\jmath})$; $\vec{E}_2 = [7q/(4\pi\epsilon_0)9a^2] \, \hat{\imath}$;
    $\vec{E}_3 = [17q/(4\pi\epsilon_0)9a^2] \, \hat{\imath}$; $\vec{E}_4 = [q/(4\pi\epsilon_0)5\sqrt{5}a^2](-3 \, \hat{\imath} - 2 \, \hat{\jmath})$
    b.

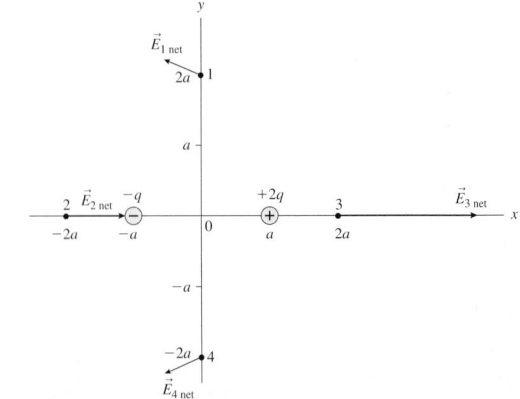

39. 1.08×10^5 N/C

41. a. $\dfrac{8\lambda d}{4\pi\epsilon_0(4y^2 + d^2)}$

b.

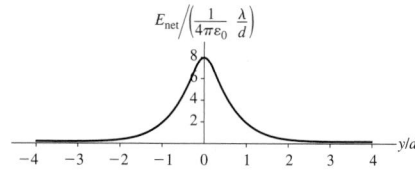

43. -2.29 nC/m

45. $\dfrac{Q}{4\pi\epsilon_0}\dfrac{1}{x\sqrt{x^2 + L^2}}\,\hat{i} - \dfrac{Q}{4\pi\epsilon_0 Lx}\left(1 - \dfrac{x}{\sqrt{x^2 + L^2}}\right)\hat{j}$

47. b. $(1/4\pi\epsilon_0)(2Q/3\sqrt{3}R^2)$

49. c. $(1/4\pi\epsilon_0)(2Q/\pi R^2)(\hat{i} + \hat{j})$

51. 0.9995 cm

53. 1.19×10^7 m/s

55. a. Positive b. 37,500 N/C c. 2.5 mm

57. a. 8.84×10^5 N/C b. ± 0.188 nC

59. -9.89×10^{-12} C

61. 18.6 nm

63. a. mC^2/N or C^2s^2/kg b. $((1/4\pi\epsilon_0)^2(2\alpha q^2/r^5)$, toward ion)

65. b. 1.0 mm

67. b. $z = R/\sqrt{3}$

69. 4.16×10^{-4} N

71. a. $(\eta/\pi\epsilon_0)\tan^{-1}(L/2z)\,\hat{k}$

c.

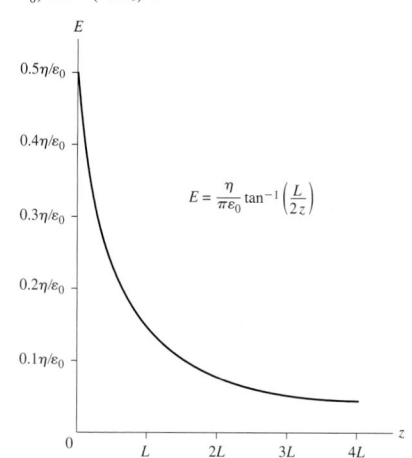

73. c. 2.0×10^{12} Hz

Chapter 27

1.

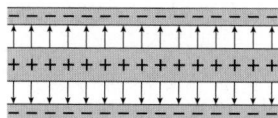 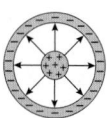

3.

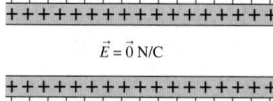

5. No charge

7. 5 N/C, pointing in

9. -1.0 N m^2/C

11. a. 6.0×10^{-2} N m^2/C b. 0 N m^2/C

13. 1.26 N m^2/C

15. a. 0 b. $2\pi R^2 E$

17. -1.0 N m^2/C

19. 5.31 nC

21.

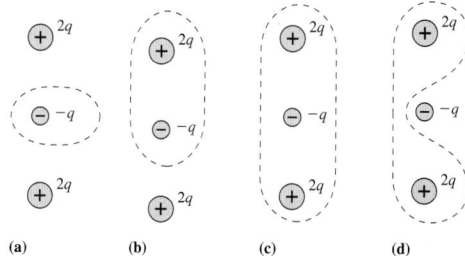

23. 113 N m^2/C

25. 2.66×10^{-5} C/m^2

27. a. -0.390 N m^2/C, 0.225 N m^2/C, 0.390 N m^2/C, -0.225 N m^2/C
b. 0 N m^2/C

29. a. -3.46 N m^2/C b. 1.15 N m^2/C

31. $-2Q/\epsilon_0$

33. a. 2000 N/C b. 251 N m^2/C c. 2.22 nC

35. a. -100 nC b. $+50$ nC

37. a. 2.39×10^{-6} C/m^3 b. 1.25 nC, 10.0 nC, 80.0 nC
c. 4500 N/C, 9000 N/C, 18,000 N/C

39. a. -1.068×10^{-8} C b. $+1.068 \times 10^{-8}$ C c. 4.82×10^{-8} C

41.

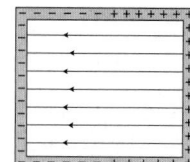

43. a. $(q/4\pi\epsilon_0 r^2)\,\hat{r}$ b. $-(q/4\pi\epsilon_0 r^2)\,\hat{r}$

45. a. $(-Qr/4\pi\epsilon_0 a^3)\,\hat{r}$, $-(Q/4\pi\epsilon_0 r^2)\,\hat{r}$, $\vec{0}$, $(Q/4\pi\epsilon_0 r^2)\,\hat{r}$
b.

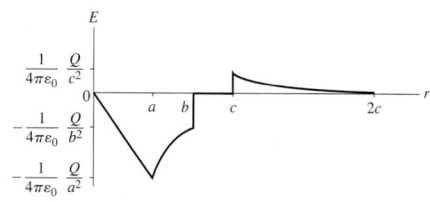

47. a. $\rho_0 z/\epsilon_0$ b. $\rho_0 z_0/\epsilon_0$
c.

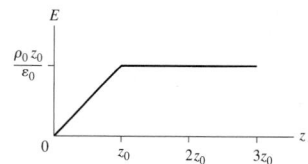

49. a. $3Q/2A\epsilon_0$, 0, $Q/2A\epsilon_0$, 0, $3Q/2A\epsilon_0$ b. $\frac{3}{2}Q/A$, $-\frac{1}{2}Q/A$, $\frac{1}{2}Q/A$, $\frac{3}{2}Q/A$
51. a. $(\lambda/2\pi\epsilon_0 r)\,\hat{r}$ b. $(\lambda r/2\pi\epsilon_0 R^2)\,\hat{r}$
53. b. 0 N/C c. 4.64×10^{13} N/C
55. a.

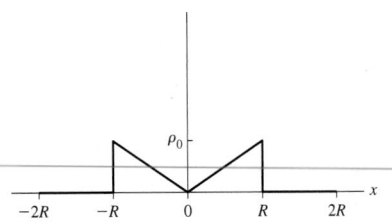

  c. $(\lambda r^2/2\pi\epsilon_0 R^3)\,\hat{r}$

Chapter 28

1. 7.5×10^{-5} m/s
3. 2.62×10^5 s
5. Aluminum
7. a. 4.63×10^{21} b. 4.32×10^{-12} m
9. 0.31 N/C
11. 9.4×10^{18}
13. 3.2 mA
15. a. 6.25×10^5 A/m^2 b. 6.51×10^{-5} m/s
17. a. 1.73×10^7 A/m^2 b. 5.31×10^{18} s^{-1}
19. 0.141 mA
21. 2.08×10^{-14} s, 4.19×10^{-15} s
23. 1.68 A
25. 5.01×10^{-8} Ωm
27. a. 1.64×10^{-3} N/C b. 1.10×10^{-5} m/s
31. a. Doubled b. Unchanged c. Unchanged d. Doubled
33. a. 3.12×10^{14} b. 398 A/m^2 c. 9.11×10^5 N/C d. 0.227 W
35. 22.6 mA
37. a. 120 C b. 0.449 mm
39. 1/4
41. 10.4 A
43. a. $I/4\pi\sigma r^2$ b. 3.32×10^{-4} N/C, 5.31×10^{-5} N/C
45. a.

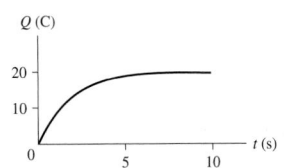

  b. $(10\text{ A})e^{-t/(2.0\text{ s})}$ c. 10.0 A
  d.

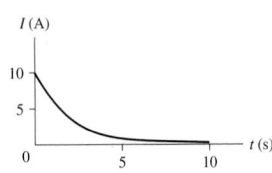

47. 2.43 A
49. 5.6×10^{-6} m/s
51. 1/2
53. 1.01×10^{23}

55. a. 1.15×10^5 m/s b. 1.5 nm
57. a. $(\epsilon_0 I/A)(1/\sigma_2 - 1/\sigma_1)$ b. 3.68×10^{-18} C

Chapter 29

1. 1.38×10^5 m/s
3. 7.07×10^4 m/s
5. 2.82×10^{-6} J
7. -2.24×10^{-19} J
9. 1.61×10^8 N/C
11. 1.87×10^7 m/s
13. -2.09×10^4 V
15. a. Lower b. -0.712 V
19. a. 200 V b. 3.54×10^{-10} C
21. a. Right plate b. 1.0×10^5 V/m c. 2.40×10^{-17} J
23. a. 1800 V, 1800 V, 900 V b. -2.88×10^{-16} J, -2.88×10^{-16} J, -1.44×10^{-16} J c. 0 V, -900 V
25. 4.17×10^{-10} C
27. $+1410$ V
29. a. 3140 V b. 5.02×10^{-16} J
31. $x = 3$ cm and 6 cm
33. a. q_a is positive, q_b is negative with the same magnitude
  b.

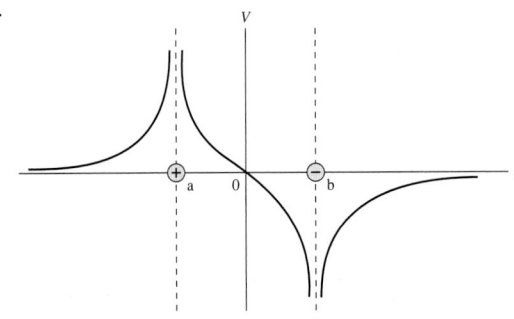

35. 0 V
37. 1.44×10^{-3} N
39. a. $x = \pm\infty$ b. $x = \pm\infty$ and 0
  c.

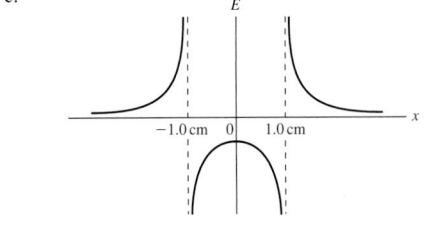

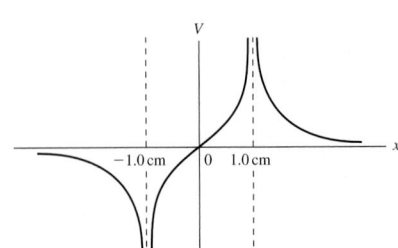

41. a. 0.720 J b. 14.4 N c. 21.9 m/s and 10.95 m/s
43. 25.3×10^{-6} J
45. a. 1000 V b. 1.39×10^{-9} C c. 7.0×10^6 m/s

Answers to Odd-Numbered Exercises and Problems

47. a.

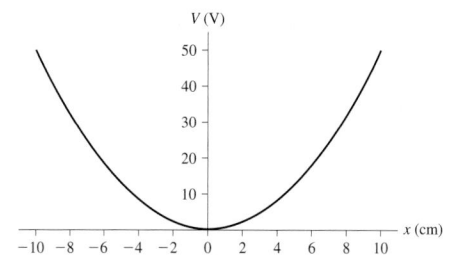

   b. SHM c. 3.20×10^{-7} J d. 2.53 cm/s
49. 1.17×10^6 m/s
51. Disk
53. 3.99×10^7 m/s
55. 4.07×10^7 m/s
57. a. 2.09×10^{-10} C, 3000 V/m, 15 V b. 2.09×10^{-10} C, 3000 V/m, 30 V c. 2.09×10^{-10} C, 750 V/m, 3.75 V
59. a. V_0/R b. 100,000 V/m
61. b. 8.33 μC c. 0 V/m, 3.33×10^6 V/m
63. 2126 V, point b higher
65. a. $4q/(4\pi\epsilon_0 s) + 16qx^2/(4\pi\epsilon_0 s^3)$ b. SHM
67. $(2qs^2)/(4\pi\epsilon_0 y^3)$
69. $(Q/4\pi\epsilon_0 L)\ln[(x + L/2)/(x - L/2)]$
71. $Q/4\pi\epsilon_0 R$
73. b. q_1 and q_2 are 10 nC and 30 nC
75. b. $Q = 0.35$ nC
77.

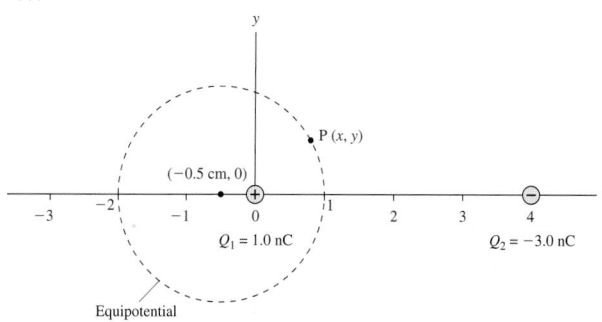

79. $v_A = 0.0548$ m/s, $v_B = 0.110$ m/s
83. c. 2.30×10^{-13} J

Chapter 30

1. -200 V
3. -1000 V/m
5.

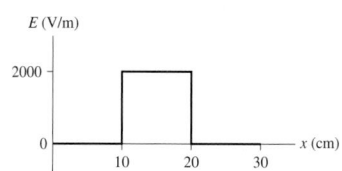

7. 10,000 V/m to the left
9. -20 V
11. 1.5×10^{-6} J
13. 12 V
15. a. 0.087 Ω b. 3.5 Ω
17. 3.0 V
19. 2.29 mA
21. 4.75 cm

23. 24.0 V
25. 32 μF
27. 200 pF
29. 1414 V
31. 1/2
33. a. 1.11×10^{-7} J b. 0.708 J/m³
35. a.

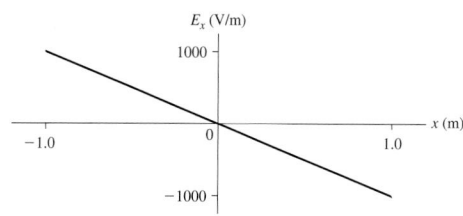

   b. $+25$ V
37. $V_0 - (\lambda/2\pi\epsilon_0)\ln(r/R)$
39. a.

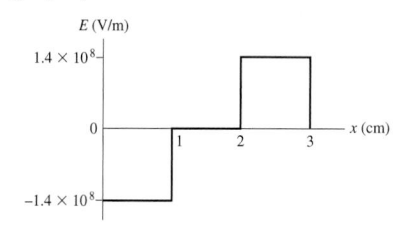

   b.

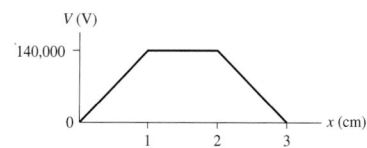

41. $(Q/2\pi\epsilon_0 R^2)[1 - z/(R^2 + z^2)^{1/2}]$
43. Point 1: 3750 V/m, downward; point 2; 7500 V/m, upward
45. 1000 V/m, 53.1° above the $-x$-axis
47. 2 nC and 4 nC
49. 1.1 nC
51. 9.1 A
53. 1800 C
55. a. $\pm 3.19 \times 10^{-11}$ C, 9 V b. $\pm 3.19 \times 10^{-11}$ C, 18 V
57. 5.90 cm and 6.10 cm
59. 150 μF, in series
61. 37 μF
63. 60 μC on each; 5.0 V, 15.0 V, 10.0 V
65. 45 μC, 9 V; 21.6 μC, 5.4 V; 21.6 μC, 3.6 V
67. 1.67 pF
69. 1.33×10^{-12} F = 1.33 pF
71. 20 μF
73. 2.4 J
75. 0.177 J/m³
77. 179 km × 179 km; not feasible
79. b. $L = 4.86$ m
81. b. $C = 2 \mu$F
83. $-\rho R^2/4\epsilon_0$
85. a.

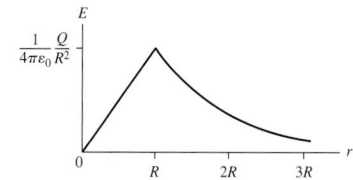

b. $(Q/4\pi\epsilon_0 R)[3/2 - r^2/2R^2]$ c. 3/2

d.

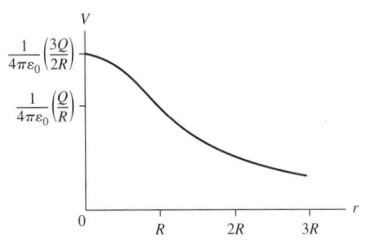

Chapter 31

1. 5.5 Ω
3. 2.0 A
5. 0.64 mm
7.

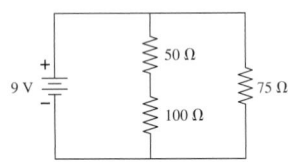

9. 5 A, toward the junction
11. a. 0.50 A, left to right

b.

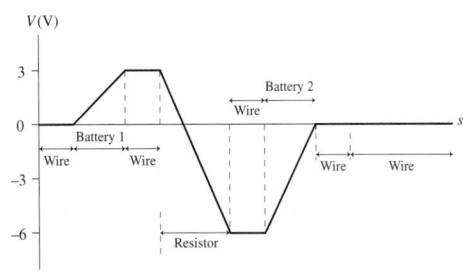

13. 9.6 Ω, 12.5 A
15. 23.6 μm
17. a. 11.6 A b. 10.4 Ω
19. 78.4 mΩ
21. 25 Ω
23. 93.4 W
25. 3.23%
27. R/4
29. 12.0 Ω
31. 24 Ω
33. 183.3 Ω
35. 13 V, 9 V, 0 V, −2 V
37. 2 ms
39. 6.93 ms
41. 869 Ω
43. A > D = E > B = C
45. Increase
47. 8.4×10^{-8} Ω
49. 7 Ω
51. 60 V, 10 Ω
53. 9.00 V, 0.50 Ω
55. 2.0 V for each
57. 1.0 A, 2.0 A, 15.0 V
59. 3.0 A
61. a. $14.40 b. 34.7 months
63. a. 8 A, 8 V b. 9.14 A, 0 V
65. a. 0.505 Ω b. 0.50 Ω

67.

R (Ω)	I (A)	ΔV (V)
6	2.0	12.0
15	0.8	12.0
6	1.2	7.2
4	1.2	4.8

69.

R (Ω)	I (A)	ΔV (V)
4	2	8
6	$\frac{4}{3}$	8
8	1	8
24	$\frac{1}{3}$	8
24	$\frac{2}{3}$	16

71. a. 10 A b. 80 W c. 60 V
73. 36.4 Ω
75. 0.69 ms
77. a. 80 μC b. 0.23 ms
81. b. 5140 Ω

Chapter 32

1. Out of the page
3. (2.0 mT, into the page), (4.0 mT, into the page)
5. a. $1.60 \times 10^{-15} \hat{k}$ T b. 0 T c. 0 T
7. $1.13 \times 10^{-15} \hat{k}$ T
9. 6.25×10^6 m/s in the +z-direction
11. 4.0 cm, 0.4 mm, 20 μm to 2 μm, 0.20 μm
13. a. 20 A b. 1.60×10^{-3} m
15. $2.0 \times 10^{-4} \hat{i}$ T, $4.0 \times 10^{-4} \hat{i}$ T, and $2.0 \times 10^{-4} \hat{i}$ T
17. a. 0.025 A m² b. 5.0 μT
19. 8.75×10^{-5} m²
21. 0.0707 T m
23. 1.0 A
25. 2390 A
27. a. Into the page b. No deflection
29. a. In the plane of the paper, 45° cw from straight up
    b. In the plane of the paper, 45° ccw from straight down
31. a. $-8.0 \times 10^{-13} \hat{k}$ N b. $5.66 \times 10^{-13} (-\hat{j} - \hat{k})$ N
33. 1.61×10^{-3} T
35. 2.9×10^{28} m⁻³
37. 0.025 N, to the right
39. 3.0 Ω
41. 7.5×10^{-4} N m
43. a. 1.26×10^{-11} N m b. Rotated 90°
45. a. 1.0 μT b. 2.0% c. 2.0 μT d. 2.0 μT; twice field in (a)
47. $(5.2 \times 10^{-5}$ T, out of the page); 0 T
49.

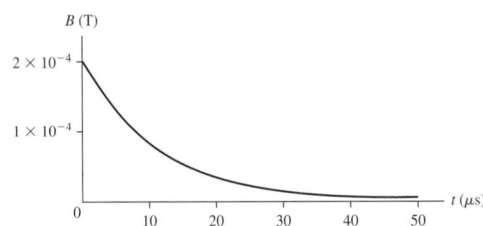

51. 750 A
53. a. 1.13×10^{10} A b. 0.014 A/m²
    c. The current density in the earth is much less than the current density in the wire.

55. #18 gauge; 4.06 A
57. a. $(1.25)^{-3/2}\mu_0 NI/R$ b. 1.80×10^{-4} T
59. $\mu_0 I/4R$
61. 0; $(\mu_0 I/2\pi r)[(r^2 - R_1^2)/(R_2^2 - R_1^2)]$; $\mu_0 I/2\pi r$
63. 2.9×10^{-3} T
65. a. $(2.4 \times 10^{10}$ m/s^2, down) b. $(2.2 \times 10^{11}$ m/s^2, up)
67. a. 4.6×10^{-13} J b. 2850
69. 2.10 T
71. 2.0 A
73. (0.00864 T, down)
75. 12.5 T
77. 0.036 mm
81. a. $\mu_0 IL/4\pi d\sqrt{(L/2)^2 + d^2}$ b. $\sqrt{2}\mu_0 I/\pi R$ c. 90.0%
83. a. $\mu_0 Ir/2\pi R_1^2$, $\mu_0 I/2\pi r$, 0
    b.

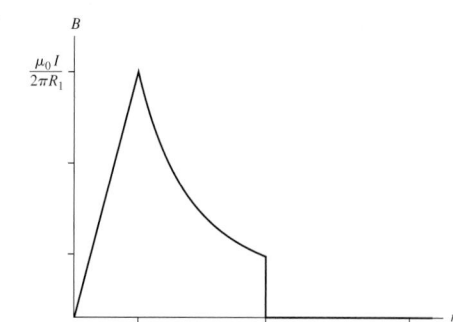

Chapter 33

1. (0.10 T, out of the page)
3. a. 4.0 m/s b. 2.24 T
5. 6.28×10^{-5} Wb in both cases
7. a. Right to left b. No
9. cw
11. 3.14 V
13. Increasing at 2.34 T/s
15. 100 V, increase
17. 9.47×10^{-5} J
19. 0.253 μH
21. 900 Ω
23. 3.54×10^{-4} Wb
25. a.

t (s)	B (T)	$\mathcal{E}$ (V)	I (A)
0.0	0.00	0.16	1.6
0.5	1.50	0.08	0.8
1.0	2.00	0.00	0.0
1.5	1.50	-0.08	-0.8
2.0	2.00	-0.16	-1.6

   b.

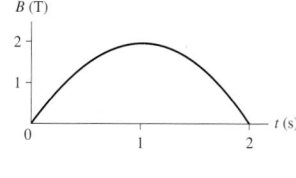

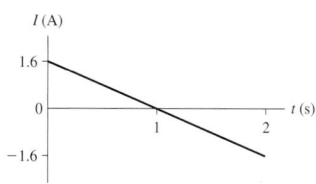

27. a.

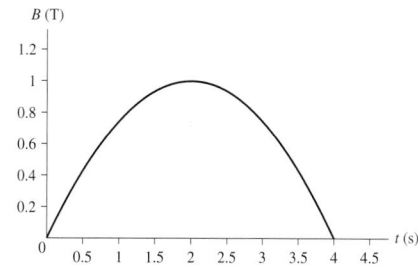

   b. $0.0628(1 - \frac{1}{2}t)$ A   c. 31.4 mA, 0.0 A, -31.4 mA
29. 41.7 mV
31. 43.9 μA
33. 0.853 mA
35. i

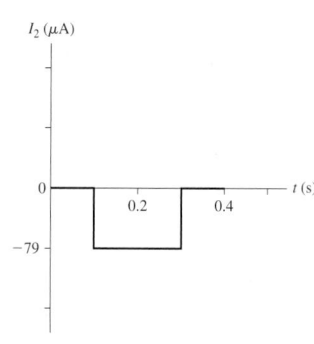

37. $(7.44$ mA$)\cos(2\pi ft)$
39. a. 12,500 b. 2.0 A
41. 39.5 nA
43. a. 0.20 A b. 4.0×10^{-3} N c. 11°C
45. a. $(4.93 \times 10^{-3})f\sin(2\pi ft)$ A b. 405 Hz; not feasible
47. a. $(vlB\cos\theta)/R$ b. $(mgR\tan\theta)/(l^2 B^2 \cos\theta)$
49. a. $(mgR)/(l^2 B^2)$ b. 0.98 m/s
51. 12 V
53. 500
55. a. 0.0637 J/m^3 b. 0.628 J/m^3
57. a. 1.0×10^{16} J b. 0.25%
59. 3.0 s
61.

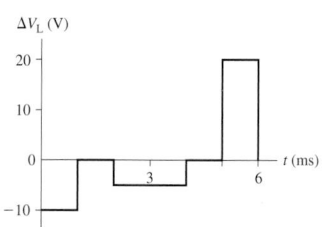

63.

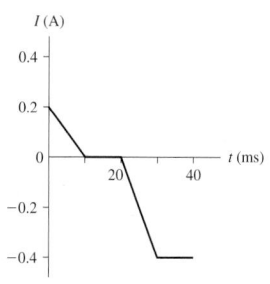

65. a. $(LI_0/\tau)e^{-t/\tau}$ b. 1.0 V, 0.37 V, 0.13 V, 0.05 V
c.

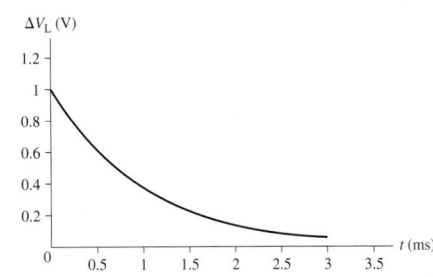

67. a. 0.628 ms b. 25 V
69. $0.707Q_0$
71. a. 76 mA b. 0.50 ms
73. a. 1.0 A b. 2.0 A
75. a. $\Delta V_{bat}/R$ b. $I_0(1 - e^{-t/(L/R)})$
c.

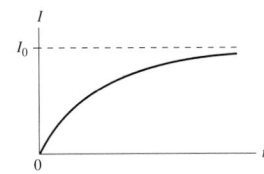

77. $(\mu_0 vI/2\pi)\ln[(d + l)/d]$
79. a. $v_0\exp[-(l^2B^2/mR)t]$
b.

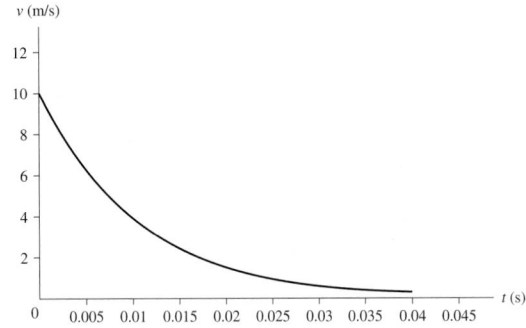

81. a. $(\mu_0/2\pi)\ln(r_2/r_1)$ b. 0.36 μH/m

Chapter 34

1. (2.0 T, into the face)
3. $(-3.2\,\hat{i} - 4.8\,\hat{j}) \times 10^{-14}$ N
5. $(1.73 \times 10^6$ V/m, left)
7. a. $(2.0 \times 10^6$ m/s, 45° from the y-axis)
   b. $(1.47 \times 10^6$ m/s, 16.2° from the y'-axis)
9. $-1.0 \times 10^6\,\hat{j}$ V/m, $1.11 \times 10^{-5}\,\hat{k}$ T
11. 16.3°
13. a. 0 V/m b. 0.040 V/m
17. 1.0×10^6 V/s
19. 22.1 μA
21. 6.0×10^5 V/m
23. a. 1.00×10^{-8} m b. 3.0×10^{16} Hz c. 6.67×10^{-8} T
27. 1.2×10^{-10} W/m^2
29. a. 2.21×10^{-6} W/m^2 b. 0.041 V/m
31. 3.3×10^{-6} N
33. 60°
35. 131 W/m^2
37. a. (0.10 T, into the page) b. 0 V/m, (0.10 T, into the page)
39. 1.0×10^7 m/s parallel to the current
41. $(R^2/2r)(dB/dt)$

43. b. 1.48×10^{-13} A
45. a. $(2.83 \times 10^3t^2)$ V m
b.

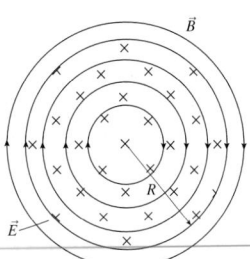

c. $1.11 \times 10^{-9}rt$ T; 4.44×10^{-12} T
d. $1.00 \times 10^{-14}t/r$ T; 5.0×10^{-12} T
47. a. 3.85×10^{26} W b. 589 W/m^2
49. Yes
51. 73.5 W
53. 0.408 m/s
55. $I_0/8$
57. b. 6.67×10^{-6} J/m^3
59. a. IR/L; $\mu_0 I/2\pi r$ b. $(I^2R/2\pi rL$, radially inward)

Chapter 35

1. a. 175 rad/s b. -8.66 V
3.

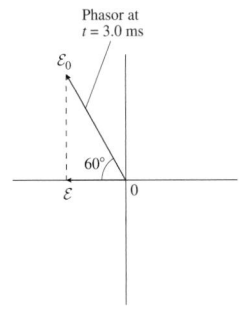

5. a. 50 mA b. 50 mA
7. a. 1.88 mA b. 1.88 A
9. a. 79.6 kHz b. 0 V
11. 125 Ω
13. 6.02 V, 7.99 V
15. 1.59 μF
17. a. 0.796 A b. 0.796 mA
19. a. 3.18×10^4 Hz b. 0 V
21. 80 Ω
23. 1.27 μF
25. a. 200 kHz b. 141 kHz
27. a. $-27.2°$ b. $+26.3°$
29. 9.6 Ω
31. 43.5 Ω
33. 395 W
35. a. $1/\sqrt{3}RC$ b. $(\sqrt{3}/2)\mathcal{E}_0$ c. 3630 rad/s
37. a. 9.95 V, 9.57 V, 7.05 V, 3.15 V, 0.990 V
b.

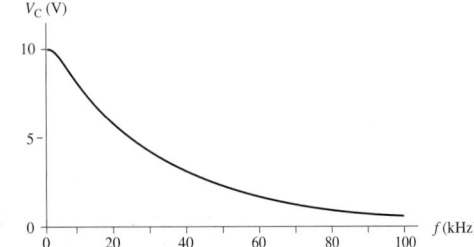

41. a. 10.0 Hz b. 4.47 V, 3.45 V, 2.24 V
43. a. 25.1 mA b. 6.67 V
45. 44.1 Hz
47. a. $\mathcal{E}_0/\sqrt{R^2 + \omega^2 L^2}$, $\mathcal{E}_0 R/\sqrt{R^2 + \omega^2 L^2}$, $\mathcal{E}_0 \omega L/\sqrt{R^2 + \omega^2 L^2}$
   c. Low d. R/L
49. a. 1.62 A b. $-17.7°$ c. 137 W
51. a. 3.16×10^4 rad/s $= 5.03 \times 10^3$ Hz b. 10.0 V, 31.6 V
53. a. 69.53 Ω, 0.072 A, $-44.0°$ b. 50.0 Ω, 0.100 A, 0°
   c. 62.42 Ω, 0.080 A, 36.8°
55. a. 3.60 V b. 3.47 V c. -3.60 V
59. a. 11.6 pF b. 1.5×10^{-3} Ω
61. 40 W: 14.4 W; 60 W: 9.6 W; 100 W: 100 W
63. a. 0.833 b. 100 V c. 12.5 Ω d. 320 μF
65.

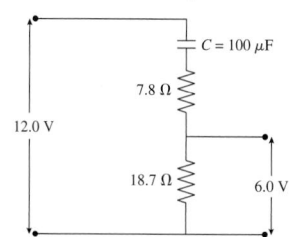

67. b. 10.0 V, 11.55 V
69. c. $\sqrt{1/LC}$
   d.

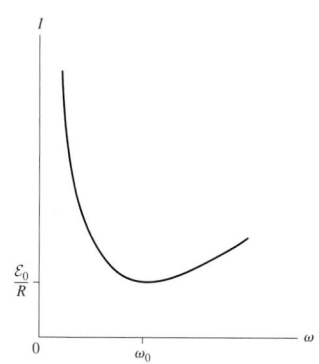

Chapter 36

1. 5 m, 1 s; -5 m, 5 s
3. 345 m/s, 15 m/s
5. a. 15 m/s b. 5 m/s c. 11.2 m/s
7. 3.00×10^8 m/s
9. 167 ns
11. 2 μs
13. Bolt 2 first, by 20 μs
15. Simultaneously
17. 0.866c
19. a. 0.8c b. 16 y
21. 4.8 ns
23. Yes
25. 0.78 m
27. 9.5×10^4 m/s
29. (8.25×10^{10} m, 325 s)
31. 0.36c
33. 0.71c
35. 0.8c
37. 0.71c
39. a. 1.8×10^{16} J b. 9.0×10^9
41. 0.943c
43. 50 g ball: 1.33 m/s to the right; 100 g ball: 3.33 m/s to the right

45. 1st ball: 4.0 m/s to the right; 2nd ball: 2.0 m/s to the left
47. 11.2 hr
49. a. No b. 67.1 y
51. a. 0.9965c b. 59.8 ly
53. 4600 kg/m^3
55. a. 8.50 ly, 17 y b. 7.36 ly, 14.7 y
57. 0.96c
59. 0.9997c
61. a. 0.98c b. 8.49×10^{-11} J
63. b. Lengths perpendicular to the motion are not affected.
65. a. $u_y' = u_y/\gamma(1 - u_x v/c^2)$ b. 0.877c
67. 3.87mc
69. 0.786c
71. a. 7.56×10^{16} J b. 0.84 kg
73. 7.5×10^{-13} J
75. 1.06×10^{-12} m
77. 22 m
79. 0.845c
81. Yes

Chapter 37

3. 6.25×10^{10}
5. (5.0×10^{-3} T, out of page)
7. 0.521 μm
9. a. 71.2 eV b. -14.4 eV   c. 5.0 keV
11. a. 5.93×10^6 m/s b. 3.10×10^7 m/s c. Alpha particle
15. a. 3, 4, and 5 b. 6, 6, and 6 c. 4, 7, and 8
17. a. ^{2}H b. ^{14}N^{++}
19. a. 82 protons, 82 electrons, 125 neutrons
   b. 1.66×10^7 V, 2.34×10^{21} V/m
21. a. 2 and 3; 2 and 4; 2 and 5; 2 and 6 b. 397.1 nm
23. 121.6 nm, 102.6 nm, 97.3 nm, 95.0 nm
25. a. 6.66 GeV b. 3.63 MeV
27. 0.512 MeV and 939 MeV
29. 173 MeV
31. (0.0457 T, into page)
33. a. 7.2×10^{13} b. 1.16 μA
35. 0.000000000058% occupied, 99.999999999942% empty
37. a. 50,000 kg/m^3 b. 3.2×10^{-10} m   c. 1.7×10^{17} kg/m^3
39. a. 57.6 N b. 4.65×10^{-35} N
   c. Very strong, very short range, independent of charge
41. 1.77×10^7 V
43. a. 3.43×10^7 m/s b. 6.14×10^6 V
45. 2.52×10^5 m/s, 65° below the $+x$-axis
47. a. mg/E_0 b. mg/b d. 2.40×10^{-18} C e. 15

Chapter 38

3.

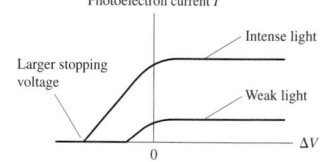

5.
   6.25×10^{13}
7. 3.20 eV
9. 1.78 eV
11. a. Aluminum b. 1.93 V
13. a. 4140 nm; infrared b. 414 nm; visible c. 41.4 nm; ultraviolet
15. 497 nm
17. 6.0×10^{-6} V
19. 0.427 nm

21. 0.354 nm
23. a. Yes b. 0.5 eV
25. Yes to $n = 2$, no to $n = 3$
27. a. 69 b. 3.2×10^4 m/s, -0.0029 eV
29. 3.40 eV
33. 91.18 nm
35. a.

n	r_n (nm)	v_n (m/s)	E_n (eV)
1	0.026	4.38×10^6	-54.4
2	0.106	2.19×10^6	-13.6
3	0.238	1.46×10^6	-6.0

37. 1.44
39. a. 1.74×10^{18} b. 1.74×10^{26} photons/s
41. Potassium: a. 5.56×10^{14} Hz b. 540 nm c. 1.08×10^6 m/s d. 3.35 V;
    Gold: a. 1.23×10^{15} Hz b. 244 nm c. 4.4×10^5 m/s d. 0.55 V
43. Sodium
45.

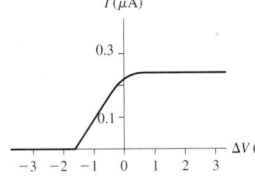

47. a. $p = E/c$ b. $\lambda = h/p$ c. $\lambda = h/mv$
49. a. 2.09×10^{-4} eV b. 1.985 nm c. 3.5 m
51. 0.427 nm
53.

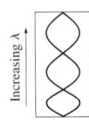

55. a. 6.5 eV b. 355 nm, 276 nm c. Both ultraviolet d. 6.16×10^5 m/s
57. -0.278 eV
59. 1876 nm
61. a. $n = 99$: 518 nm, 2.21×10^4 m/s; $n = 100$: 529 nm, 2.19×10^4 m/s
    b. 6.79×10^9 Hz, 6.59×10^9 Hz c. 6.68×10^9 Hz d. 0.15%
63. 10.28 nm, 7.62 nm, 6.80 nm; ultraviolet
65. 4.16 eV
67. a. ϵ^N b. 2.4 mA c. 4.5×10^6 d. 3.0
69. a. 1.518×10^{-16} s b. 1.32×10^6
71. a. 4.26×10^{-5} nm, 1.31×10^7 m/s b. 0.0164 nm c. X ray
    d. 7.3×10^{13} orbits

Chapter 39

1. 20%, 10%
3. a. 7.7% b. 25%
5. a. 1/6 b. 1/6 c. 5/18
7. 2.0×10^7
11. a. 3333 b. 1111
13. a. 5.0×10^{-3} b. 2.5×10^{-3} c. 0 d. 2.5×10^{-3}
15. a. 0.25 mm^{-1} b. 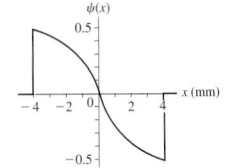 c. 0.25

17. a. 0.354 mm$^{-1/2}$ b. 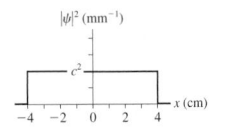 c. 0.25

19. 25 ns
21. 100,000
23. 18 μm
25. 0 m/s $\leq v \leq 2.5 \times 10^7$ m/s
27. 85 m
29. 1.0×10^5 pulses/s
31. a.

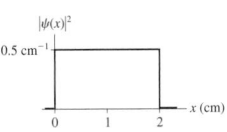

    b. 1.0% c. 10^4 d. 0.50 cm^{-1}
33. a. Yes b.

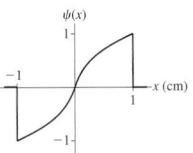

    c. 0.000, 0.0050, 0.0010 d. 900
35. a. $\sqrt{3/8}$ mm$^{-1/2}$
    b.

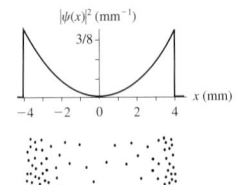

    c.

    d. 0.125
37. a. 0.27% b. 31.8%
39. a. 0.866 cm$^{-1/2}$
    b.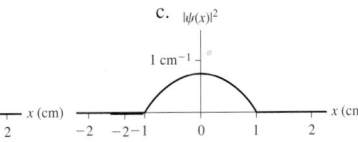

    d. 3440
41. a. $a = b$ b.

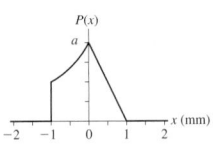

    c. Both 0.838 d. 58.1%
43. No; 1.4×10^{-27} m
45. a. $0 \leq v \leq 1.8 \times 10^{10}$ m/s b. Not possible
47. a. 1.5×10^{-13} m; no b. 4.4×10^{11} m
49. a. $b = c$
    b.

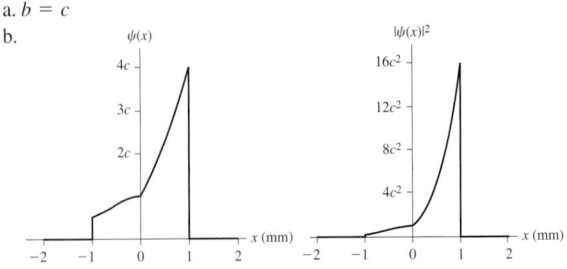

    c. 91%

Chapter 40

1. 0.739 nm
3. 0.752 nm
7. 0.135
9. 0.038 eV
11.

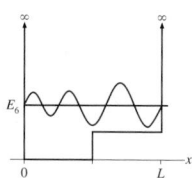

13. a.

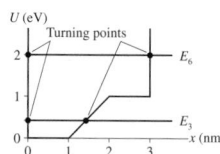

b.

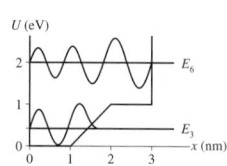

15. 150 nm
17. a. 0.49 eV, 1.46 eV, 2.43 eV b. 640 nm
19. 1.35 N/m
21. 1.22%
25. a. $\lambda = 8mcL^2/3h$ b. 0.795 nm
29. a.

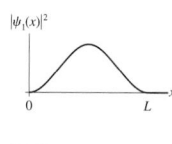

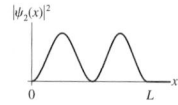

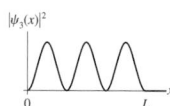

n	b. Most	c. Least	d. Probability	e. Probability
1	$\frac{1}{2}L$	0 and L	$<\frac{1}{3}$	0.195
2	$\frac{1}{4}L, \frac{3}{4}L$	$0, \frac{1}{2}L, L$	$<\frac{1}{3}$	0.402
3	$\frac{1}{6}L, \frac{3}{6}L, \frac{5}{6}L$	$0, \frac{1}{3}L, \frac{2}{3}L, L$	$=\frac{1}{3}$	0.333

31. 4.77×10^7 m/s
35. a. $(\pi b^2)^{-1/4}$ b. $2(\pi b^2)^{-1/2} \int_b^\infty \exp(-x^2/b^2) \, dx$ c. 15.7%

37. a. $\dfrac{1}{2h\sqrt{1-(y/h)}}$ b.

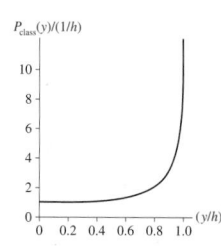

39. a. 4.95 eV b. 4.80 eV c. 4.55 eV
43. b. 0.0091 eV, 0.0272 eV, 0.0453 eV, 0.0634 eV c. 69 μm; infrared
45. $10^{1.17\times10^{-32}}$ or $10^{-117000000000000000000000000000000}$

Chapter 41

1. a. $\sqrt{2}\hbar$ b. $\sqrt{12}\hbar$
3. a. f b. -0.85 eV
5. -0.378 eV; $\sqrt{12}\hbar$
7. a. 2 b. 1
11. Si: $1s^2 2s^2 2p^6 3s^2 3p^2$; Ge: $1s^2 2s^2 2p^6 3s^2 3p^6 4s^2 3d^{10} 4p^2$;
    Pb: $1s^2 2s^2 2p^6 3s^2 3p^6 4s^2 3d^{10} 4p^6 5s^2 4d^{10} 5p^6 6s^2 4f^{14} 5d^{10} 6p^2$.
13. a. Fluorine; excited state b. Nickel; ground state
15. $1s^2 3s^1$
19. a. Yes; 2.21 μm b. No; $\Delta l \neq 1$
21. 0.020
23. a. 9.0×10^5 b. 8.7 ns
25. 5.3×10^{22}
27. a. 1.48×10^{-34} Js b. $-1, 0, 1$
    c.

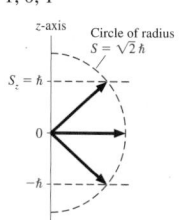

29. $\sqrt{6}\hbar$
31. a. 3.68×10^{-3} b. 5.41×10^{-3} c. 2.93×10^{-3}
37. a.

L_z	S_z	J_z	m_j
$\hbar$	$\frac{1}{2}\hbar$	$\frac{3}{2}\hbar$	$\frac{3}{2}$
$\hbar$	$-\frac{1}{2}\hbar$	$\frac{1}{2}\hbar$	$\frac{1}{2}$
0	$\frac{1}{2}\hbar$	$\frac{1}{2}\hbar$	$\frac{1}{2}$
0	$-\frac{1}{2}\hbar$	$-\frac{1}{2}\hbar$	$-\frac{1}{2}$
$-\hbar$	$\frac{1}{2}\hbar$	$-\frac{1}{2}\hbar$	$-\frac{1}{2}$
$-\hbar$	$-\frac{1}{2}\hbar$	$-\frac{3}{2}\hbar$	$-\frac{3}{2}$

b. $\frac{1}{2}, \frac{3}{2}$

39.

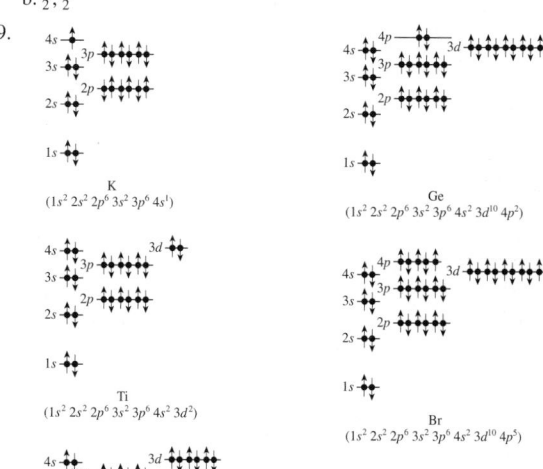

41. a. $6s \rightarrow 5p$, $6s \rightarrow 4p$, and $6s \rightarrow 3p$ b. 7290 nm; 1630 nm; 515 nm
43. a.

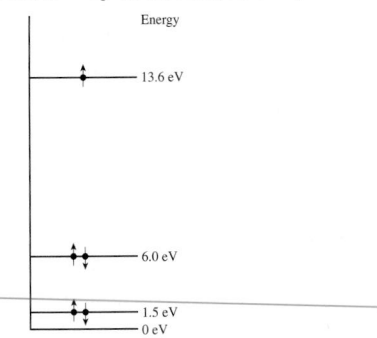

b. 21.7 eV
45. 1.13×10^6 m/s
47.

Transition	a. Wavelength	b. Type	c. Absorption
$2p \rightarrow 2s$	670 nm	VIS	Yes
$3s \rightarrow 2p$	816 nm	IR	No
$3p \rightarrow 2s$	324 nm	UV	Yes
$3p \rightarrow 3s$	2696 nm	IR	No
$3d \rightarrow 2p$	611 nm	VIS	No
$3d \rightarrow 3p$	25 μm	IR	No
$4s \rightarrow 2p$	498 nm	VIS	No
$4s \rightarrow 3p$	2430 nm	IR	No

49. a. 6.25×10^8 s^{-1} b. 0.17 ns
51. 5.72 ns
53. 5.0×10^{16} s^{-1}
55. a. 1.06 μm b. 1.87 W
57. 0.677
61. $1.5a_B$; $5a_B$

Chapter 42

1. a. 1 proton; 2 neutrons b. 18 protons; 22 neutrons
   c. 20 protons; 20 neutrons d. 94 protons; 145 neutrons
3. a. 3.8 fm b. 8.2 fm c. 14.5 fm
5. 3.6×10^{26} protons; 3.6×10^{26} neutrons
7. 1.2×10^{11} kg
9. a. ^{36}S and ^{36}Ar b. 5, 8
11. ^{40}Ar: 344 MeV, 8.59 MeV/nucleon; ^{40}Ca: 342 MeV, 8.55 MeV/nucleon
13. ^{12}C: 7.68 MeV/nucleon; ^{13}C: 7.47 MeV/nucleon; ^{12}C

15. 20.179 u
17. 8000 N
19. 2.3×10^{-38}
21. a.

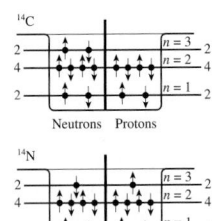

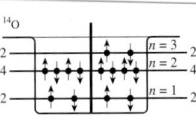

b. ^{14}N stable; ^{14}C beta-minus decay; ^{14}O beta-plus decay
23. a. 9.3×10^{11} b. 4.7×10^{11} c. 5.5×10^8
25. a. 3.32 b. 6.64
27. 80.2 days
29. a. ^{228}Th b. ^{207}Tl c. ^{7}Li d. ^{60}Ni
31. 5.52 MeV
33. 4.82 MeV
35. 60 mrem
37. 0.225 J
39. a. 3.50×10^7 m/s b. 25.6 MeV
41. a. 12.7 km b. 780 μs
43. a. 1.46×10^{-8} u; 1.45×10^{-6}% b. 0.0304 u; 0.76%
45. 6.0 MeV
47. a. ^{17}N, ^{17}O, ^{17}F b. ^{17}O
    c. ^{17}N $\rightarrow$ ^{17}O by beta-minus; ^{17}F $\rightarrow$ ^{17}O by EC
51. 7.16×10^{11} Bq or 19.4 Ci
53. 2.73×10^{17}
55. a. 18.9 s b. No
57. 1.19 hr
59. 210 million years
61. 69.7 mrem
63. 3.31×10^{12}
65. a. 2.61×10^{10} b. 0.0239 Bq c. 1.436×10^7 rem/year
    d. Yes; $\approx$50 million times the natural background.
67. 15 cm
69. $\approx$6 billion years
71. a. 65.0 MeV; 5.0 MeV b. 3.7×10^{21} s^{-1} c. 6.6×10^{-39}
    d. 650 million years

Part Overview Solutions

PART I Overview

If the acceleration of a Podracer can reach 50 m/s², what are the maximum tensions in the two large cables? To find out, what quantities do you need to estimate?

MODEL The cables from the two engines pull the passenger car, or Pod, forward. Treat the Pod as a particle. The size of the Pod suggests a mass of about 2000 kg. Because planets that have atmospheres and support life are likely to be similar in size to the earth, we'll estimate that the acceleration due to gravity is 10 m/s². The cables are curved and are angle θ above horizontal where they attach to the Pod. We'll neglect air resistance.

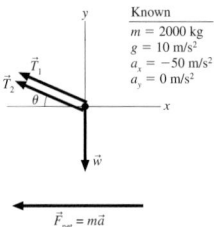

Known
$m = 2000$ kg
$g = 10$ m/s²
$a_x = -50$ m/s²
$a_y = 0$ m/s²

VISUALIZE The figure shows a free-body diagram of the Pod. Tensions $\vec{T}_1$ and $\vec{T}_2$ are due to the two cables. We'll assume equal magnitudes, so $T_1 = T_2 = T$. We want to find the tension when the acceleration reaches its maximum value $a_x = -50$ m/s², where the negative sign shows that the Podracer is accelerating to the left.

SOLVE Newton's second law for the Pod is

$$\sum F_x = -2T\cos\theta = ma_x$$
$$\sum F_y = 2T\sin\theta - mg = ma_y = 0$$

Thus $2T\sin\theta = mg$ and $2T\cos\theta = -ma_x$. Dividing the first of these equations by the second gives $\tan\theta = (-g/a_x)$, from which we find that

$$\theta = \tan^{-1}\left(-\frac{g}{a_x}\right) = \tan^{-1}\left|\frac{10 \text{ m/s}^2}{50 \text{ m/s}^2}\right| = 11.3°$$

We can see from the picture of the Podracer that this is a reasonable value for the angle. We can now use the angle to find the tension:

$$T = \frac{mg}{2\sin\theta} = \frac{(2000 \text{ kg})(10 \text{ m/s}^2)}{2\sin(11.3°)} = 51,000 \text{ N}$$

ASSESS The value of the tension would be slightly larger if we included air resistance, but 51,000 N is a reasonable estimate.

PART II Overview

The exploding Death Star releases 10^{33} J of energy. If the expanding shock wave exerts an impulse of 2.5×10^6 N s on the *Millennium Falcon*, by how much does the starship's velocity change? To find out, what property of the starship do you need to estimate?

MODEL The impulse-momentum theorem relates an object's change in momentum or change in velocity to the impulse exerted on it. Assume that the mass of the *Millennium Falcon* is 250,000 kg.

SOLVE The impulse momentum theorem is

$$\Delta p_x = m\Delta v_x = J_x$$

Thus the shock wave, when it hits, causes the starship's velocity to suddenly change by

$$\Delta v_x = \frac{J_x}{m} = \frac{2.5 \times 10^6 \text{ N s}}{2.5 \times 10^5 \text{ kg}} = 10 \text{ m/s}$$

ASSESS The starship experiences a sudden change of about 20 mph. This change will cause a noticeable jolt, but it is probably not sufficient to damage the starship.

PART III Overview

At what altitude above the surface of Alderaan does the Death Star orbit with a period of 10 hours? To find out, what properties of the planet do you need to estimate?

MODEL Assume that the Death Star is in a circular orbit around Alderaan. We need to know the mass and radius of Alderaan. Because planets that have atmospheres and support life are likely to be similar in size to the earth, we'll estimate that $M = 6.0 \times 10^{24}$ kg and $R = 6.1 \times 10^6$ m.

SOLVE The period T of a satellite in a circular orbit of radius r is given by

$$T^2 = \left(\frac{4\pi^2}{GM}\right)r^3$$

The radius of the orbit must be

$$r = \left[\left(\frac{GM}{4\pi^2}\right)T^2\right]^{1/3}$$
$$= \left[\frac{(6.67 \times 10^{-11} \text{ N m}^2/\text{kg}^2)(6.0 \times 10^{24} \text{ kg})(36,000 \text{ s})^2}{4\pi^2}\right]^{1/3}$$
$$= 23.6 \times 10^6 \text{ m}$$

Thus the Death Star's height above the surface is

$$h = r - R = 17.2 \times 10^6 \text{ m} = 17,200 \text{ km}$$

ASSESS The height is about 10,000 mi. This is much higher than the space shuttle orbits with a period of about 90 min, but less than the altitude of a geosynchronous satellite that orbits with a period of 24 hours. Thus the height is reasonable for a 10-hour period.

PART IV Overview

R2-D2's internal systems have been on standby while inside the ice cave on the planet Hoth. How much energy must R2-D2's fuel cells provide to raise the temperature of his internal systems from the standby temperature of $-20°C$ to a working temperature of $30°C$? To find out, what properties of R2-D2 do you need to estimate?

MODEL Assume that R2–D2 is well insulated so that all the energy goes into raising his temperature, and none of the energy escapes into the ice cave. Although R2-D2 is made of many parts, they are mostly aluminum (structural materials) and silicon (semiconductor electronics). Thus we'll estimate that the *average* specific heat is 800 J/kg K. R2-D2 isn't terribly large, so we'll estimate the mass of his internal systems to be 40 kg.

SOLVE The heat energy needed to change the temperature by ΔT is

$$Q = mc\Delta T = (40 \text{ kg})(800 \text{ J/kg K})(30°C - (-20°C))$$

$$= 1.6 \text{ MJ}$$

ASSESS The conversion of electrical energy from the fuel cells into heat energy is nearly 100% efficient. Thus the fuel cells must supply 1.6 MJ of energy.

PART V Overview

If violent storms drive waves across the ocean of planet Kamino at a speed of 75 m/s, what is the wavelength of an ocean wave whose period is 10 s?

MODEL Assume the waves are sinusoidal with period $T = 10$ s.

SOLVE The fundamental relationship of sinusoidal waves is $v = \lambda f$. The frequency is related to the period by $f = 1/T$. Consequently, the wavelength is

$$\lambda = vT = (75 \text{ m/s})(10 \text{ s}) = 750 \text{ m}$$

ASSESS 750 is much larger than the wavelength of ocean waves on earth because the 75 m/s ≈ 150 mph wave speed on Kamino is much larger than the wave speed on earth.

PART VI Overview

Anakin's and Obi-Wan's speeder discharges the power coupler on the planet Coruscant by flying through it. If the discharge current through the speeder is 5000 A, what voltage is developed across the speeder? To find out, what properties of the speeder do you need to estimate?

MODEL The speeder's shell is built from carbon-nanotube fibers, an advanced material that is both lightweight and extremely rigid. Carbon-nanotube fibers are poor conductors, with resistivity 5.0×10^{-6} Ω m. To estimate the resistance of the speeder's shell, model it as a 5.0-m-long tube with a 2.0 m × 5.0 m rectangular cross section. The shell's thickness is 2 mm.

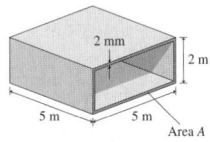

VISUALIZE The figure shows the simplified geometry of the speeder's shell.

SOLVE The cross-section area of the material is the perimeter of the shell multiplied by the thickness, or

$$A = (5.0 \text{ m} + 2.0 \text{ m} + 5.0 \text{ m} + 2.0 \text{ m}) \times (0.002 \text{ m})$$

$$= 0.028 \text{ m}^2$$

With this information, we can calculate that the speeder's resistance is

$$R = \frac{\rho L}{A} = \frac{(5.0 \times 10^{-6} \text{ Ω m})(5.0 \text{ m})}{0.028 \text{ m}^2} = 9.0 \times 10^{-4} \text{ Ω}$$

According to Ohm's law, the potential difference produced across the speeder by a 5000 A current though the speeder is

$$\Delta V = IR = (5000 \text{ A})(9.0 \times 10^{-4} \text{ Ω}) = 4.5 \text{ V}$$

ASSESS This potential difference is easily measured, but it is not sufficient to disrupt the speeder's electronic systems.

PART VII Overview

If the fusion reactor in the core of the Death Star generates 10^{15} W of power, how much mass does the Death Star lose each year as it converts mass into energy?

MODEL The Death Star is powered by controlled thermonuclear fusion, the same nuclear reaction that powers the sun. Mass is converted into energy as hydrogen nuclei are fused into helium nuclei at extremely high temperatures.

SOLVE There are 3.15×10^{7} s in 1 yr, so the energy produced each year is

$$E = (1.0 \times 10^{15} \text{ J/s})(3.15 \times 10^{7} \text{ s}) = 3.15 \times 10^{22} \text{ J}$$

According to Einstein's famous equation $E = (\Delta m)c^2$, the mass lost (i.e., converted to energy) each year is

$$\Delta m = \frac{E}{c^2} = \frac{3.15 \times 10^{22} \text{ J}}{(3.0 \times 10^{8} \text{ m/s})^2} = 350{,}000 \text{ kg}$$

ASSESS 350,000 kg is about the mass of the water in a medium-size swimming pool. It is a completely insignificant fraction of the total mass of the Death Star.

Credits

All Part Overview images are courtesy of Lucasfilm Ltd. Addison Wesley would like to give special thanks to Lucy Wilson, Christopher Holm, and the staff of Lucasfilm Ltd. for granting us permission to use these images and for their help in selecting them.

INTRODUCTION

Page **xxvi**: Courtesy of International Business Machines Corporation. Unauthorized use not permitted.

TITLE PAGE

Page **iii**: Rainbow/PictureQuest.

PART I

Part I Overview image: *Star Wars: Episode I – The Phantom Menace* © 1999 Lucasfilm Ltd. & ™. All rights reserved. Used under authorization. Unauthorized duplication is a violation of applicable law. Page **2**: Herman Eisenbeiss/Photo Researchers.

CHAPTER 1

Page **3**: Al Bello/Getty Images. Page **4** UL: David Woods/Corbis. Page **4** UR: Joseph Sohm/Corbis. Page **4** LL: Richard Megna/ Fundamental Photos. Page **4** LR: Fredrick M. Brown/Getty Images. Page **11**: Kevin Muggleton/Corbis. Page **13**: Wm. Sallaz/Corbis. Page **19**: Tony Freeman/PhotoEdit. Page **23**: Steve Smith/Getty Images. Page **26** T: U.S. Department of Commerce. Page **26** B: Bureau Int. des Poids et Mesures.

CHAPTER 2

Page **35**: Patrik Giardino/Corbis. Page **39**: Spencer Grant/PhotoEdit. Page **43** T: Tim Wright/Corbis. Page **43** B: Phil Boorman/Getty Images. Page **60**: James Sugar/Stockphoto.com. Page **63**: Scott Markewitz/Getty Images.

CHAPTER 3

Page **78**: Michael Yamashita/Corbis. Page **80**: Paul Chesley/Getty Images.

CHAPTER 4

Page **97**: George Lepp/Getty Images/Stone. Page **98** (a): Tony Freeman/ PhotoEdit. Page **98** (b): Brian Drake/Index Stock. Page **98** (c): Duomo/ Corbis. Page **98** (d): Dorling Kindersley Media Library. Page **98** (e): Chuck Savage/Corbis. Page **98** (f): Jeff Coolidge Photography. Page **101**: Jeff J. Daly/Fundamental Photographs. Page **112**: David Woods/Corbis.

CHAPTER 5

Page **122**: Joe McBride/Getty Images. Page **123**: PhotoDisc. Page **129**: Jonathan Nourak/PhotoEdit. Page **132**: Roger Ressmeyer/Corbis. Page **137** T: Corbis Digital Stock. Page **137** B: Jeff Coolidge Photography. Page **138**: Patrick Behar/Agence Vandystadt/Photo Researchers, Inc.

CHAPTER 6

Page **151**: Gerard Planchenault/Getty Images. Page **159**: Tony Freeman/ PhotoEdit. Page **161**: Richard Megna/Fundamental Photographs.

CHAPTER 7

Page **177**: Robin Smith/Getty Images. Page **185**: Robert Laberge/Getty Images. Page **192**: Sightseeing Archive/Getty Images. Page **194**: Tony Freeman/PhotoEdit.

CHAPTER 8

Page **207**: Chris Cole/Getty Images. Page **209**: Chuck Savage/Corbis. Page **212**: Pete Saloutos/Corbis.

PART II

Part II Overview image: *Star Wars: Episode VI – Return of the Jedi* © 1983 and 1997 Lucasfilm Ltd. & ™. All rights reserved. Used under authorization. Unauthorized duplication is a violation of applicable law. Page **238**: M.C. Escher Heirs/Cordon Art, Baarn, Holland.

CHAPTER 9

Page **239**: Russ Kinne/Comstock. Page **254**: Roger Ressmeyer/Corbis. Page **257**: Richard Megna/Fundamental Photos. Page **259**: Frederick M. Brown/Getty Images.

CHAPTER 10

Page **268**: Wally McNamee/Corbis. Page **270**: Roger Ressmeyer/Corbis. Page **276**: Lester Lefkowitz/Corbis. Page **280**: Gary Buss/Getty Images. Page **282**: Paul Harris/Getty Images. Page **287**: Dorling Kindersley Media Library.

CHAPTER 11

Page **304**: Shaun Botterill/Getty Images. Page **308**: Bettmann/Corbis. Page **323**: Al Behrman/AP Wide World. Page **329**: AFP/Corbis.

PART III

Part III Overview image: *Star Wars: Episode IV – A New Hope* © 1977 and 1997 Lucasfilm Ltd. & ™. All rights reserved. Unauthorized duplication is a violation of applicable law. Page **342**: Courtesy Sandia National Laboratories/SUMMiT™ Technologies.

CHAPTER 12

Page **343**: PhotoDisc/Getty Images. Page **344**: Library of Congress. Page **346**: American Institute of Physics/Emilio Segre Visual Archives/ Regents of the University of California. Page **349**: Corbis Digital Stock. Page **358**: NASA.

CHAPTER 13

Page **369**: Glyn Kirk/Getty Images. Page **377**: Hart, G. K. & Vikki/Getty Images: Page **390**: Alain Choisnet/Getty Images. Page **393**: Richard Megna/Fundamental Photos.

CHAPTER 14

Page **413**: Michael Neveux/Corbis. Page **425**: Courtesy of Professor Thomas D. Rossing, Northern Illinois University. Page **430**: Richard Megna/Fundamental Photos. Page **432**: AP Photo/Toby Talbot. Page **436**: Martin Bough/Fundamental Photographs.

CHAPTER 15

Page **444:** Stu Forster/Getty Images. Page **450:** Richard Megna/ Fundamental Photos. Page **453:** Joseph Sinnot/Fundamental Photographs. Page **462:** Diane Hirsch/Fundamental Photos. Page **463:** Andy Sacks/ Getty Images. Page **464:** Don Farrall/Getty Images.

PART IV

Part IV Overview image: *Star Wars: Episode V – The Empire Strikes Back* © 1980 and 1997 Lucasfilm Ltd. & ™. All rights reserved. Used under authorization. Unauthorized duplication is a violation of applicable law.

CHAPTER 16

Page **485:** Luis Veiga/Getty Images. Page **486:** Kevin Schafer/Corbis. Page **489:** Richard Megna/Fundamental Photos. Page **491 T:** David Young-Wolff/PhotoEdit. Page **491 B:** David Young-Wolff/PhotoEdit.

CHAPTER 17

Page **512:** W.A. Sharman/Milepost 92/Corbis. Page **514:** Robert & Linda Mostyn/Eye Ubiquitous/Corbis. Page **520:** Kevin Fleming/Corbis. Page **529:** Roger Ressmeyer/Corbis.

CHAPTER 18

Page **547:** Pete Turner/Getty Images.

CHAPTER 19

Page **573:** Mark Wagner/Getty Images. Page **574:** Spencer Grant/ PhotoEdit. Page **575:** W. Cody/Corbis. Page **577:** Larry Chiger/ SuperStock. **581:** Carrier Corporation. Page **586:** Malcolm Fife/Getty Images. Page **589:** Paul Silverman/Fundamental Photos.

PART IV SUMMARY

Page **607:** Kristian Hilsen/Getty Images.

PART V

Part V Overview image: *Star Wars: Episode II – Attack of the Clones* © 2002 Lucasfilm Ltd. & ™. All rights reserved. Used under authorization. Unauthorized duplication is a violation of applicable law.

CHAPTER 20

Page **611:** Reuters NewMedia Inc./Corbis. Page **612 B:** Doug Wilson/ Corbis. Page **612 T:** Hoard Dratch/The Image Works. Page **614:** Uri Haber-Schaim. Page **629:** V.C.L./Getty Images. Page **631:** David Parker/ Photo Researchers, Inc. Page **633:** Doug Sokell/Visuals Unlimited. Page **638:** Space Telescope Science Institute.

CHAPTER 21

Page **646:** Rosco Permacolor™ filter: Rosco Laboratories, Inc. Page **648:** Richard Megna/Fundamental Photographs. Page **652:** Education Development Center, Newton, MA. Page **654:** Tom Pantages. Page **658:** Richard Gross/Corbis. Page **659:** Uri Haber-Schaim. Page **665:** Peter Aprahamian/Photo Researchers.

CHAPTER 22

Page **684:** Chris Collins/Corbis. Page **685 T:** Richard Megna/ Fundamental Photographs. Page **685 B:** Todd Gipstein/Getty Images. Page **686:** Springer-Verlag GmbH & Co KG. Page **687:** M. Cagnet et al., Atlas of Optical Phenomena Springer-Verlag,1962. Page **692:** M. Cagnet et al., Atlas of Optical Phenomena Springer-Verlag,1962. Page **693:** Courtesy Holographix LLC. Page **699 T, B:** Ken Kay/Fundamental Photographs. Page **700:** Springer-Verlag GmbH & Co KG. Page **704:** CENCO Physics/Fundamental Photographs. Page **706:** Dr. Rod Nave, Georgia State University.

CHAPTER 23

Page **714:** Charles O'Rear/Corbis. Page **718:** Sylvester Allred/ Fundamental Photographs. Page **721:** Richard Megna/Fundamental Photographs. Page **725:** Richard Megna/Fundamental Photographs. Page **726:** Francisco Cruz/SuperStock. Page **730:** Tony Freeman/ PhotoEdit. Page **731:** Benjamin Rondel/Corbis. Page **732 L, R:** Richard Megna/Fundamental Photographs. Page **736:** Photodisc Blue/Getty Images. Page **747:** Benjamin Rondel/Corbis. Page **748 L, M, R:** Springer-Verlag, GmbH & Co KG.

CHAPTER 24

Page **757:** IBM/Phototake NYC. Page **758:** Courtesy of Ocean Optics, Inc. Page **759:** Gerard Herzberg/Atomic Spectra and Atomic Structure, Prentice-Hall,1937. Page **762:** General Electric Corporate Research & Development Center. Page **763:** General Electric. Page **763 L, M, R:** Eugene Hecht/Optics 2e, p.11, Addison Wesley. Page **764 T, M, B:** E.R. Huggins, Physics I/W.A. Benjamin, 1968, Reading, MA, Addison Wesley Longman, reprinted with permission. Page **769 L, M:** Film Studio/Education Development Center, Newton, MA. Page **769 R:** Tipler and Llewellyn/Modern Physics 3e, 1987, p. 207, New York, NY, Worth Publishers, courtesy of C.G. Shull. Page **769:** Dr. Claus Jonsson. Page **770:** P. Merli and G. Missiroli, *American Journal of Physics*, 44, p. 306, 1976. Page **774:** P. Merli and G. Missiroli, *American Journal of Physics*, 44, p. 306, 1976.

PART VI

Part VI Overview image: *Star Wars: Episode II – Attack of the Clones* © 2002 Lucasfilm Ltd. & ™. All rights reserved. Used under authorization. Unauthorized duplication is a violation of applicable law. Page **782:** UHB Trust/Getty Images.

CHAPTER 25

Page **783:** Gandee Vasan/Getty Images. Page **784:** Tony Freeman/ PhotoEdit. Page **794:** Courtesy Xerox Corporation. Page **802 T, B:** Richard Megna/Fundamental Photographs.

CHAPTER 26

Page **817:** Rachel Epstein/PhotoEdit. Page **832 U:** Jody Dole/Getty Images. Page **832 L:** Hannu-Pekka Hedman, University of Turku.

CHAPTER 27

Page **849:** Paul A. Souders/Corbis.

CHAPTER 28

Page **879:** Visuals Unlimited. Page **895:** IBM Research/Peter Arnold, Inc.

CHAPTER 29

Page **900:** Corbis Digital Stock. Page **920:** Christopher Johnson, University of Utah.

CHAPTER 30

Page **932:** Martyn F. Chillmaid/Science Library Photo. Page **939:** Paul Silverman/Fundamental Photos. Page **942:** Tom Pantages. Page **946:** Tom Pantages. Page **953:** Adam Hart-Davis/Photo Researchers, Inc.

CHAPTER 31

Page **961:** Courtesy Intel Corporation. Page **963:** Tom Ridley/Dorling Kindersley Media Library. Page **972:** Maya Barnes/The Image Works. Page **983:** Francisco Cruz/SuperStock.

CHAPTER 32

Page **996:** Photodisc Green/Getty Images. Page **1000:** Richard Megna/Fundamental Photographs. Page **1016:** Charles Thatcher/Getty Images. Page **1019:** Richard Megna/Fundamental Photographs. Page **1020:** Courtesy Dr. L.A. Frank, University of Iowa. Page **1021:** CERN Geneva. Page **1022:** Ernest Orlando Lawrence Berkeley National Laboratory.

CHAPTER 33

Page **1041:** Photodisc Green/Getty Images. Page **1042:** Hulton Archive/Getty Images. Page **1063:** Lester Lefkowitz/Corbis. Page **1064:** Jonathan Nourok/PhotoEdit.

CHAPTER 34

Page **1084:** Lawrence Manning/Corbis. Page **1113** L, M, R: Richard Megna/Fundamental Photographs.

CHAPTER 35

Page **1121:** Inga Spence/Visuals Unlimited. Page **1138:** Courtesy Edwards, Inc.

PART VII

Part VII Overview image: *Star Wars: Episode VI – Return of the Jedi* © 1983 and 1997 Lucasfilm Ltd. & ™. All rights reserved. Used under authorization. Unauthorized duplication is a violation of applicable law. Page **1150:** IBM Research, Almaden Research Center.

CHAPTER 36

Page **1151:** John Y. Fowler. Page **1152:** Topham/The Image Works. Page **1172:** Stanford Linear Accelerator Center. Page **1186:** Science Photo Library/Photo Researchers. Page **1187:** Wellcome Dept. of Cognitive Neurology/Science Photo Library/Photo Researchers.

CHAPTER 37

Page **1195:** Richard Megna/Fundamental Photos. Page **1197:** DaimlerChrysler. Page **1201** T: Science Photo Library/Photo Researchers. Page **1201** B: Science Museum/Science and Society Picture Library. Page **1212:** Gerard Herzberg/Atomic Spectra and Atomic Structure, Prentice-Hall, 1937.

CHAPTER 38

Page **1220:** Courtesy of International Business Machines Corporation. Unauthorized use not permitted. Page **1225** B: DPA/HM/The Image Works. Page **1225** T: Bettman/Corbis. Page **1231:** Dr. Claus Jonsson. Page **1235:** Bettman/Corbis.

CHAPTER 39

Page **1253:** Digital Instruments Inc. Page **1254:** Dr. Claus Jonsson.

CHAPTER 40

Page **1277:** IBM Corporate Archives. Page **1278:** Bettman/Corbis. Page **1310** L: Digital Instruments Inc. Page **1310** R: Prelim Ed., from G. Binnign and H. Rohres, *Surface Science*, 144, p. 321, 1984.

CHAPTER 41

Page **1317:** Ray Nelson/Phototake. Page **1321:** Tom Pantages. Page **1328:** Prelim Ed., from G. Binnign and H. Rohres, *Surface Science*, 144, p. 321, 1984. Page **1335:** Courtesy National Institute of Standards and Technology. Page **1338:** Archivo Iconografico, S.A./Corbis. Page **1343:** Meggers Gallery/American Institute of Physics/Science Photo Library/Photo Researchers.

CHAPTER 42

Page **1352:** Landmann Patrick/Corbis SYGMA. Page **1360:** Bettman/Corbis. Page **1362:** Hulton/Getty Images. Page **1363:** Kevin Fleming/Corbis. Page **1371:** ICRR Institute for Cosmic Ray Research. Page **1375:** Lonnie Duka/Index Stock. Page **1376:** Howard Sochurek/Corbis.

Index

For users of the five-volume edition: pages 1–481 are in Volume 1; pages 482–607 are in Volume 2; pages 608–779 are in Volume 3; pages 780–1194 are in Volume 4; and pages 1148–1383 are in Volume 5.
Pages 1195–1383 are not in the Standard Edition of this textbook.

Astronomical Data

Planetary body	Mean distance from sun (m)	Period (years)	Mass (kg)	Mean radius (m)
Sun	—	—	1.99×10^{30}	6.96×10^8
Moon	3.84×10^8*	27.3 days	7.36×10^{22}	1.74×10^6
Mercury	5.79×10^{10}	0.241	3.18×10^{23}	2.43×10^6
Venus	1.08×10^{11}	0.615	4.88×10^{24}	6.06×10^6
Earth	1.50×10^{11}	1.00	5.98×10^{24}	6.37×10^6
Mars	2.28×10^{11}	1.88	6.42×10^{23}	3.37×10^6
Jupiter	7.78×10^{11}	11.9	1.90×10^{27}	6.99×10^7
Saturn	1.43×10^{12}	29.5	5.68×10^{26}	5.85×10^7
Uranus	2.87×10^{12}	84.0	8.68×10^{25}	2.33×10^7
Neptune	4.50×10^{12}	165	1.03×10^{26}	2.21×10^7

*Distance from earth

Typical Coefficients of Friction

Material	Static μ_s	Kinetic μ_k	Rolling μ_r
rubber on concrete	1.00	0.80	0.02
steel on steel (dry)	0.80	0.60	0.002
steel on steel (lubricated)	0.10	0.05	
wood on wood	0.50	0.20	
wood on snow	0.12	0.06	
ice on ice	0.10	0.03	

Melting/Boiling Temperatures and Heats of Transformation

Substance	T_m (°C)	L_f (J/kg)	T_b (°C)	L_v (J/kg)
water	0	3.33×10^5	100	22.6×10^5
nitrogen (N_2)	-210	0.26×10^5	-196	1.99×10^5
ethyl alcohol	-114	1.09×10^5	78	8.79×10^5
mercury	-39	0.11×10^5	357	2.96×10^5
lead	328	0.25×10^5	1750	8.58×10^5

Properties of Materials

Substance	ρ (kg/m³)	c (J/kg K)
air at STP*	1.2	
ethyl alcohol	790	2400
gasoline	680	
glycerin	1260	
mercury	13,600	140
oil (typical)	900	
seawater	1030	
water	1000	4190
aluminum	2700	900
copper	8920	385
gold	19,300	129
ice	920	2090
iron	7870	449
lead	11,300	128
silicon	2330	703

*Standard temperature (0°C) and pressure (1 atm)

Molar Specific Heats of Gases

Gas	C_P (J/mol K)	C_V (J/mol K)
Monatomic Gases		
He	20.8	12.5
Ne	20.8	12.5
Ar	20.8	12.5
Diatomic Gases		
H_2	28.7	20.4
N_2	29.1	20.8
O_2	29.2	20.9

Indices of Refraction

Material	Index of refraction
vacuum	1 exactly
air	1.0003
water	1.33
glass	1.50
diamond	2.42